DUBBEL

Handbook of Mechanical Engineering

DUBBEL

Handbook of
MECHANICAL
ENGINEERING

Edited by W. Beitz and K.-H. Küttner
English Edition edited by B.J. Davies
Translation by M.J. Shields

With 1258 Figures

Springer-Verlag
London Berlin Heidelberg New York
Paris Tokyo Hong Kong
Barcelona Budapest

Wolfgang Beitz, Professor Dr.-Ing.
Technische Universität Berlin,
Institut für Maschinenkonstruktion, 10623 Berlin, Germany

Karl-Heinz Küttner, Professor Dipl.-Ing.
Formerly at Technische Fachhochschule Berlin
Address for correspondence: Müllerstrasse 120,
13449 Berlin, Germany

Chairman, UK Advisory Board

B.J. Davies, Professor
7 Queens Crescent, Putnoe, Bedford MK41 9BN, UK

Translator

M.J. Shields, FIInfSc, MITI
Literary and Technical Language Services,
Unit 10, Centenary Business Centre,
Attleborough Fields Industrial Estate,
Nuneaton, Warwickshire CV11 6RY, UK

ISBN 3-540-19868-7 Springer-Verlag Berlin Heidelberg New York
ISBN 0-387-19868-7 Springer-Verlag New York Berlin Heidelberg

British Library Cataloguing in Publication Data
Dubbel: Handbook of Mechanical Engineering
    I. Beitz, Wolfgang II. Küttner,
    Karl-Heinz III. Shields, Michael J.
    621
ISBN 3-540-19868-7

Library of Congress Cataloging-in-Publication Data
Dubbel, Heinrich, 1873-1947.
    [Taschenbuch für den Maschinenbau. English]
    Handbook of mechanical engineering / Dubbel ; [edited by] W. Beitz and K.-H. Küttner.
      p. cm.
    Includes bibliographical references and index.
    ISBN 3-540-19868-7 (Berlin : acid-free paper). –
    ISBN 0-387-19868-7 (New York : acid-free paper)
    1. Mechanical engineering–Handbooks, manuals, etc. I. Beitz, Wolfgang.
    II. Küttner, Karl-Heinz. III. Title.
TJ151.D813 1994   94-16420
621–dc20   CIP

Typeset by Photo-graphics, Honiton, Devon
Printed and bound by The Bath Press, Bath
69/3830-543210 Printed on acid-free paper

UK Advisory Board

Chairman

Professor B. J. Davies, University of Manchester Institute of Science and Technology

Members

Dr. J. N. Ashton, University of Manchester Institute of Science and Technology

Dr. N. C. Baines, Imperial College of Science, Technology and Medicine, London

Professor C. B. Besant, Imperial College of Science, Technology and Medicine, London

Dr. B. Lengyel, Imperial College of Science, Technology and Medicine, London

D. A. Robb, Imperial College of Science, Technology and Medicine, London

Dr. C. Ruiz, University of Oxford

Professor J. E. E. Sharpe, Lancaster University

Dr. D. A. Yates, University of Manchester Institute of Science and Technology

Contributors

B. Behr, Rheinisch-Westfälische Technische Hochschule Aachen

Professor W. Beitz, Technische Universität Berlin

Professor A. Burr, Fachhochschule Heilbronn

E. Dannenmann, Universität Stuttgart

Professor L. Dorn, Technische Universität Berlin

Dr. K.A. Ebert†, Hattersheim

Professor K. Ehrlenspiel, Technische Universität München

Professor D. Föller, Batelle-Institut e.V., Frankfurt a.M.

Professor H. Gelbe, Technische Universität Berlin

Professor K.-H. Habig, Bundesanstalt für Materialforschung und-prüfung (BAM), Berlin

Professor G. Harsch, Fachhoschschule Heilbronn

Dr. K. Herfurth, Verein Deutscher Gießereifachleute VDG, Düsseldorf

Dr. H. Kerle, Technische Universität Braunschweig

Professor L. Kiesewetter, Technische Universität Cottbus

Professor K.H. Kloos, Technische Hochschule Darmstadt

Professor K.-H. Küttner, Technische Fachhochschule Berlin

J. Ladwig, Universität Stuttgart

G. Mauer, Rheinisch-Westfälische Technische Hochschule Aachen

Professor H. Mertens, Technische Universität Berlin

Professor H.W. Müller, Technische Hochschule Darmstadt

Professor R. Nordmann, Universität Kaiserslautern

Professor G. Pahl, Technische Hochschule Darmstadt

Professor H. Peeken, Rheinisch-Westfälische Technische Hochschule Aachen

Professor G. Pritschow, Universität Stuttgart

W. Reuter, Rheinisch-Westfälische Technische Hochschule Aachen

Professor R. Röper, Universität Dortmund

Professor J. Ruge, Technische Universität Braunschweig

Professor G. Rumpel, Technische Fachhochschule Berlin

Professor G. Seliger, Technische Universität Berlin

Professor K. Siegert, Universität Stuttgart

Professor H.D. Sondershausen, Technische Fachhochschule Berlin

Professor G. Spur, Technische Universität Berlin

Professor K. Stephan, Universität Stuttgart

Professor H.K. Tönshoff, Universität Hannover

Professor H.-J.Warnecke, Universität Stuttgart

Professor M. Weck, Rheinisch-Westfälische Technische Hochschule Aachen

T. Werle, Universität Stuttgart

Professor H. Winter, Technische Universität München

H. Wösle, Technische Universität Braunschweig

Preface to the English Edition

It has been an education and a pleasure to assist in the preparation of this first English version of the widely used "DUBBEL: Taschenbuch für den Maschinenbau", which has been a standard mechanical engineering reference book in German-speaking countries since 1914.

All the chapters of primary interest to English-speaking mechanical engineers have been translated. I trust that this "Pocket Book" will be a ready and authoritative source of the best current practice in mechanical engineering. It is up to date, having been revised regularly, with the last revision appearing in 1990.

It provides an easily accessible theoretical and practical treatment of a wide range of mechanical engineering topics with comprehensive explanatory diagrams, tables, formulae and worked examples.

Much care has been given to ensuring a correct and easily understood translation of the German text. For completeness, it was felt necessary to retain many German references and also DIN Standards. Where possible, ISO equivalents have been given. It is unlikely that this complex exercise is entirely error free but I believe that faith has been kept with the original text.

B. John Davies
Emeritus Professor, Department of Mechanical Engineering, UMIST
May 1994

Introduction

Since 1914 the Dubbel Handbook of Mechanical Engineering has been the standard reference text used by generations of students and practising engineers in the German-speaking countries. The book covers all fundamental Mechanical Engineering subjects.

Contributions are written by leading experts in their fields. This handbook is not primarily intended for specialists in particular areas, but for students and practitioners, who, within the framework of their responsibilities, also need to know about the basics outside their own special area.

The handbook deliberately focuses on fundamentals and on the solutions of problems, but it also covers a wide range of applications. Charts and tables with general material values and specific parameters are included. As a German handbook, it relies more on the German Industrial Standards (DIN) and focuses on the components of German manufacturers. This should not be a problem in this English–international edition owing to the exemplary character of these applications and examples; and with the increasing referencing of EN- and ISO/IEC-standards, the national DIN standard becomes less significant.

In parallel with the complete German edition, the selected subjects in this edition combine the fundamentals of theoretical sciences, materials and engineering design with important mechanical engineering applications.

I would like to thank all those involved in the production of this handbook for their enthusiastic co-operation, since this has made an important standard mechanical engineering text available to an international readership.

W. Beitz
Technische Universität Berlin
November 1993

Contents

A Mechanics

1 Statics of Rigid Bodies **A1**

1.1 Introduction A1
1.2 Combination and Resolution of Concurrent Forces A2
1.3 Combination and Resolution of Non-Concurrent Forces A4
1.4 Conditions of Equilibrium A5
1.5 Types of Support; the 'Free Body' A7
1.6 Support Reactions A7
1.7 Systems of Rigid Bodies A10
1.8 Pin-Jointed Frames A10
1.9 Cables and Chains A12
1.10 Centre of Gravity A13
1.11 Friction A15

2 Kinematics **A19**

2.1 Motion of a Particle A19
2.2 Motion of a Rigid Body A22

3 Dynamics **A27**

3.1 Basic Concepts of Energy, Work, Power, Efficiency A27
3.2 Particle Dynamics, Straight-Line Motion of Rigid Bodies A28
3.3 Dynamics of Systems of Particles A30
3.4 Dynamics of a Rigid Body A33
3.5 Dynamics of Relative Motion A39
3.6 Impact A39

4 Mechanical Vibrations **A40**

4.1 One-Degree-of-Freedom Systems A40
4.2 Multi-Degree-of-Freedom Systems (Coupled Vibrations) A44
4.3 Non-linear Vibrations A48

5 Hydrostatics **A49**

6 Hydrodynamics and Aerodynamics (Dynamics of Fluids) **A51**

6.1 One-Dimensional Flow of Ideal Fluids A51
6.2 One-Dimensional Flow of Viscous Newtonian Fluids A52
6.3 One-Dimensional Flow of Non-Newtonian Fluids A59
6.4 Forces Due to the Flow of Incompressible Fluids A60
6.5 Multi-Dimensional Flow of Inviscid Fluids A60
6.6 Multi-Dimensional Flow of Viscous Fluids A63

7   Similarity Mechanics **A69**

7.1 Introduction A69
7.2 Similarity Laws A69

8   References **A72**

B   Strength of Materials

1   General Fundamentals **B1**

1.1 Stress and Strain B1
1.2 Strength and Properties of Materials B4
1.3 Failure Criteria, Equivalent Stresses B6

2   Stresses in Bars and Beams **B7**

2.1 Tension and Compression B7
2.2 Transverse Shear Stresses B7
2.3 Contact Stresses and Bearing Pressures B8
2.4 Bending B8
2.5 Torsion B27
2.6 Combined Stresses B31
2.7 Statically Indeterminate Systems B32

3   Theory of Elasticity **B36**

3.1 General B36
3.2 Axisymmetric Stresses B36
3.3 Plane Stresses B37

4   Hertzian Contact Stresses (Formulae of Hertz) **B38**

4.1 Spheres B38
4.2 Cylinders B38
4.3 Arbitrarily Curved Surfaces B38

5   Plates and Shells **B39**

5.1 Plates B39
5.2 Discs, Plates Under In-Plane Loads B41
5.3 Shells B41

6   Centrifugal Stresses in Rotating Components **B43**

6.1 Rotating Bars B43
6.2 Rotating Thin Rings B43
6.3 Rotating Discs B43

7   Stability Problems **B45**

7.1 Buckling of Bars B45
7.2 Lateral Buckling of Beams B48
7.3 Buckling of Plates and Shells B48

8   Finite-Element and Boundary-Element Methods **B50**

8.1 Finite Elements B50
8.2 Boundary Elements B53

9   Theory of Plasticity **B55**

9.1   Introduction to Theory of Plasticity B55
9.2   Uses B56

10   Appendix B: Diagrams and Tables **B59**

11   References **B76**

C   Thermodynamics

1   Scope of Thermodynamics. Definitions **C1**

1.1   Systems, Boundaries of Systems, Surroundings C1
1.2   Description of the State of a System. Thermodynamic
      Processes C1

2   Temperatures. Equilibria **C2**

2.1   Adiabatic and Diathermal Walls C2
2.2   Zeroth Law and Empirical Temperature C2
2.3   Temperature Scales C2

3   First Law **C3**

3.1   General Formulation C3
3.2   The Various Forms of Energy C4
3.3   Application to Closed Systems C4
3.4   Application to Open Systems C5

4   Second Law **C6**

4.1   The Principle of Irreversibility C6
4.2   General Formulation C6
4.3   Special Formulations C7

5   Exergy and Anergy **C7**

5.1   Exergy of a Closed System C7
5.2   Anergy C8
5.3   Exergy of an Open System C8
5.4   Exergy and Heat C8
5.5   Exergy Losses C8

6   Thermodynamics of Substances **C9**

6.1   Thermal State Variables of Gases and Vapours C9
6.2   Caloric Properties of Gases and Vapours C10
6.3   Solids . . . : C12
6.4   Mixing Temperature, Measurement of Specific Heat
      Capacities C13

7   Changes of State of Gases and Vapours **C14**

7.1   Changes of State of Gases and Vapours at Rest C14
7.2   Changes of State of Gases and Vapours in Motion C15

8   Thermodynamic Processes **C16**

8.1   Combustion Processes C16

8.2 Internal Combustion Engines C18
8.3 Cyclic Processes C19
8.4 Cooling and Heating C22

9   Ideal Gas Mixtures **C23**

9.1 Dalton's Law. Thermal and Caloric Properties of State C23
9.2 Mixtures of Gas and Vapour C24
9.3 Humid Air C24

10   Heat Transfer **C26**

10.1 Steady-State Heat Conduction C26
10.2 Heat Transfer and Heat Transmission C27
10.3 Instationary Heat Transmission C28
10.4 Heat Transfer by Convection C30
10.5 Radiative Heat Transfer C33

11   Tables **C34**

12   References **C54**

D   Materials Technology

**1   Fundamental Properties of Materials and Structural
      Parts** **D1**

1.1 Load and Stress Conditions D1
1.2 Causes of Failure D2
1.3 Materials Design Values D5
1.4 Effect of Materials Structure, Manufacturing Process and
     Environment Conditions on Strength and Ductility Behaviour   D8
1.5 Strength Properties and Constructional Design D10
1.6 Loadbearing Capability of Structural Components D13

2   Materials Testing **D17**

2.1 Fundamentals D17
2.2 Test Methods D18

3   Properties and Application of Materials **D26**

3.1 Iron Base Materials D26
3.2 Non-Ferrous Metals D43
3.3 Non-Metallic Materials D49
3.4 Materials Selection D53

4   Plastics **D54**

4.1 Introduction D54
4.2 Structure and Properties D54
4.3 Properties D54
4.4 Important Thermoplastics D55
4.5 Fluorinated Plastics D57
4.6 Thermosets D57
4.7 Plastic Foams (Cellular Plastics) D58
4.8 Elastomers D58
4.9 Testing of Plastics D59

4.10 Processing of Plastics D62
4.11 Design and Tolerances of Formed Parts D66
4.12 Finishing D67

5   Tribology **D67**

5.1   Friction D67
5.2   Friction States of Oil-Lubricated Sliding Pairs D67
5.3   Elastohydrodynamic Lubrication D68
5.4   Wear D70
5.5   Systems Analysis of Friction and Wear Processes D71
5.6   Lubricants D72

6   Appendix D: Diagrams and Tables **D76**

7   References **D121**

E   Fundamentals of Engineering Design

1   Fundamentals of Technical Systems **E1**

1.1   Energy, Material and Signal Transformation E1
1.2   Functional Interrelationship E1
1.3   Working Interrelationships E2
1.4   Constructional Interrelationship E4
1.5   System Interrelationship E4
1.6   General Objectives and Constraints E4

2   Fundamentals of a Systematic Approach **E4**

2.1   General Working Method E4
2.2   General Problem-Solving E4
2.3   Abstracting to Identify Functions E4
2.4   Search for Solution Principles E5
2.5   Evaluation of Solutions E6

3   The Design Process **E10**

3.1   Defining Requirements E10
3.2   Conceptual Design E11
3.3   Embodiment Design E12
3.4   Detail Design E12
3.5   Types of Engineering Design E12

4   Fundamentals of Embodiment Design **E13**

4.1   Basic Rule of Embodiment Design E13
4.2   Principles of Embodiment Design E14
4.3   Guidelines for Embodiment Design E16

**5   Fundamentals of the Development of Series and
       Modular Design** **E20**

5.1   Similarity Laws E20
5.2   Decimal-Geometric Series of Preferred Numbers (Renard
       Series) E20
5.3   Geometrically Similar Series E21
5.4   Semi-similar Series E21

5.5   Use of Exponential Equations E21
5.6   Modular System E22

**6   Fundamentals of Standardisation and Engineering
         Drawing** **E23**

6.1   Standardisation E23
6.2   Basic Standards E24
6.3   Engineering Drawings and Parts Lists E25
6.4   Item Numbering Systems E30

7   References **E31**

F   Mechanical Machine Components

1   Connections **F1**

1.1   Welding F1
1.2   Soldering and Brazing F18
1.3   Adhesive Bonding F21
1.4   Connections with Force Transmission by Friction F23
1.5   Positive Connections F28
1.6   Bolted Connections F34
1.7   Selecting Types of Connection F47

2   Elastic Connections (Springs) **F50**

2.1   Uses, Characteristics, Properties F50
2.2   Metal Springs F51
2.3   Rubber Springs and Anti-vibration Mountings F59
2.4   Fibre Composite Springs F62
2.5   Gas Springs F63

3   Couplings, Clutches and Brakes **F64**

3.1   Survey, Functions F64
3.2   Permanent Torsionally Stiff Couplings F65
3.3   Permanent Elastic Couplings F66
3.4   Clutches F69
3.5   Automatic Clutches F74

4   Rolling Bearings **F75**

4.1   Fundamentals F75
4.2   Types of Rolling Bearings F77
4.3   Load Capacity, Fatigue Life, Service Life F79
4.4   Lubrication of Rolling Bearings F84
4.5   Friction and Heating F87
4.6   Design of Rolling Bearing Assemblies F87

5   Plain Bearings **F89**

5.1   Fundamentals of Plain Bearing Design F89
5.2   Calculation of Plain Journal Bearings Under Steady Radial
      Load F89
5.3   Calculation of Plain Journal Bearings Under Variable Radial
      Load F92
5.4   Turbulent Film Flow F92

5.5 Calculation of Plain Thrust Bearings F92
5.6 Form Design of Plain Bearings F96
5.7 Lobed and Multi-pad Plain Bearings F98
5.8 Bearing Seals F98
5.9 Dry Bearings F99
5.10 Bearing with Hydrostatic Jacking Systems F99
5.11 Hydrostatic Bearings F99

6   Belt and Chain Drives **F101**

6.1 Types, Uses F101
6.2 Flat Belt Drives F101
6.3 V-Belts F107
6.4 Synchronous Belts F109
6.5 Chain Drives F109

7   Friction Drives **F110**

7.1 Mode of Operation, Definitions F110
7.2 Types, Examples F111
7.3 Principles of Calculation F112
7.4 Hints on Use and Operation F115

8   Gearing **F116**

8.1 Spur and Helical Gears – Gear Tooth Geometry F117
8.2 Tooth Errors and Tolerances, Backlash F124
8.3 Lubrication and Cooling F125
8.4 Materials and Heat Treatment – Gear Manufacture F126
8.5 Load Capacity of Spur and Helical Gears F128
8.6 Bevel Gears F137
8.7 Crossed Helical Gears F139
8.8 Worm Gears F139
8.9 Epicyclic Gear Arrangements F143
8.10 Design of Geared Transmissions F153

9   Kinematics **F157**

9.1 Systematics of Mechanisms F157
9.2 Analysis of Mechanisms F161
9.3 Synthesis of Mechanisms F166
9.4 Special Mechanisms F168

10   Crank Mechanisms **F168**

10.1 Kinematics F168
10.2 Dynamics F170
10.3 Components of Crank Mechanism F172

11   Appendix F: Diagrams and Tables **F176**

12   References **F194**

G   Hydraulic and Pneumatic Power Transmission

**1   Fundamentals of Fluid Power Transmission
      Systems** **G1**

1.1 The Flow Process G1
1.2 Hydraulic Fluids G2
1.3 Systematology G2

2   Components of Hydrostatic Transmissions **G4**

2.1 Pumps G4
2.2 Hydraulic Motors G8
2.3 Valves G9
2.4 Hydraulic Equipment G12

3   Structure and Function of Hydraulic Transmissions   **G12**

3.1 Hydraulic Circuits G12
3.2 Operation of Hydraulic Transmissions G12
3.3 Control G13

**4   Configuration and Design of Hydraulic
      Transmissions** **G15**

4.1 Hydraulic Circuit Arrangements G15
4.2 Design of Hydraulic Circuits G15

5   Pneumatic Installations **G16**

5.1 Pneumatic Components G16
5.2 Circuits G17

6   Water Hydraulic Systems **G17**

7   Appendix G: Diagrams and Tables **G18**

8   References **G20**

H Components of Thermal Apparatus

1   Fundamentals **H1**

1.1 Heat Exchanger Characteristics H1
1.2 Thermodynamic and Fluid Dynamic Design H1
1.3 Heat Exchanger Flow Arrangements and Operating
     Characteristics H4
1.4 Efficiency, Exergy Losses H4

2   Apparatus and Piping Components **H5**

2.1 Basis for Design Calculations H5
2.2 Cylindrical Shells and Tubes Under Internal Pressure H6
2.3 Cylindrical Shells Under External Pressure H6
2.4 Flat End Closures and Tube Plates H7
2.5 Domed End Closures H7
2.6 Cutouts H8
2.7 Flange Joints H8
2.8 Piping H10
2.9 Shutoff and Control Valves H14
2.10 Seals H18

3   Types of Heat Exchanger **H21**

3.1 Tube-Bundle (Shell-and-Tube) Heat Exchangers H21
3.2 Other Types H22

4   Condensers and Reflux Coolers **H23**

4.1 Principles of Condensation H23
4.2 Surface Condensers H24
4.3 Injection (Direct-Contact) Condensers H25
4.4 Air-Cooled Condensers H26
4.5 Auxiliary Equipment H26
4.6 Indirect Air Cooling and Cooling Towers H27

5   Appendix H: Diagrams and Tables **H29**

6   References **H31**

J   **Machine Dynamics**

**1   Crank Operation, Forces and Moments of Inertia,
      Flywheel Calculations** **J1**

1.1 Graph of Torque Fluctuations in Multi-Cylinder Reciprocating
    Machines J1
1.2 Forces and Moments of Inertia J4

2   Vibrations **J10**

2.1 The Problem of Vibrations in Machines J10
2.2 Some Fundamental Concepts J11
2.3 Basic Problems in Machine Dynamics J14
2.4 Representation of Vibrations in the Time and Frequency
    Domains J16
2.5 Origin of Machine Vibrations, Excitation Forces $F(t)$ J18
2.6 Mechanical Equivalent Systems, Equations of Motion J21
2.7 Application Examples for Machine Vibrations J25

3   Acoustics in Mechanical Engineering **J29**

3.1 Basic Concepts J29
3.2 The Generation of Machine Noise J31
3.3 Methods for Reducing Machine Noise J33

4   References **J36**

K   **Manufacturing Processes**

1   Survey of Manufacturing Processes **K1**

1.1 Definition and Criteria K1
1.2 Classification K1

2   Primary Shaping **K2**

2.1 General K2
2.2 Shaping of Metals by Casting K3
2.3 Forming of Plastics K15

2.4   Forming of Metals and Ceramics by Powder Metallurgy K17
2.5   Other Methods of Primary Shaping K19

3   Metal Forming **K19**

3.1   Classification and Introduction K19
3.2   Fundamentals of Metal Forming K20
3.3   Theoretical Models K23
3.4   Stresses and Forces in Selected Metal Forming Processes   . . .   K25
3.5   Technology K28

4   Cutting **K35**

4.1   General K35
4.2   Machining with Geometrically Well-defined Tool Edges K35
4.3   Machining with Geometrically Non-defined Tool Edges K50
4.4   Chipless Machining K57
4.5   Shearing and Blanking K61

5   Special Technologies **K67**

5.1   Thread Production K67
5.2   Gear Cutting K70
5.3   Manufacturing in Precision Engineering and Microtechnology   K81
5.4   Surface Coating K90

6   Assembly **K91**

6.1   Definitions K92
6.2   Tasks of Assembly K93
6.3   Realisation of Assembly K93

7   Production and Works Management **K96**

7.1   Job Planning K96
7.2   Manufacturing Systems K100
7.3   Quality Engineering K103
7.4   Operational Costing K105
7.5   Basic Ergonomics K107

8   Appendix K: Diagrams and Tables **K109**

9   References **K118**

L   Manufacturing Systems

1   Machine Tool Components **L1**

1.1   Fundamentals L1
1.2   Drives L4
1.3   Frames L21
1.4   Linear and Rotary Guides and Bearings L25

2   Control Systems **L34**

2.1   Fundamentals of Control L34
2.2   Means of Control L37
2.3   Programmable Logic Controller (PLC) L41

2.4   Numerical Control (NC) L42
2.5   Equipment for Position Measurement at NC Machines L48

3   Shearing and Blanking Machines **L52**

3.1   Shearing Machines L52
3.2   Blanking Machines L53
3.3   Nibbling Machines L53
3.4   Beam Cutting Machines L54

4   Presses and Hammers for Metal Forging **L54**

4.1   Characteristics of Presses and Hammers L54
4.2   Mechanical Presses L55
4.3   Hydraulic Presses L60
4.4   Hammers and Screw Presses L61
4.5   Safety L65

5   Metal-Cutting Machine Tools **L66**

5.1   Lathes L66
5.2   Drilling and Boring Machines L73
5.3   Milling Machines L79
5.4   Horizontal Boring and Milling Machines L83
5.5   Machining Centres L83
5.6   Planing, Shaping and Slotting Machines L85
5.7   Broaching Machines L86
5.8   Sawing and Filing Machines L87
5.9   Grinding Machines L89
5.10 Honing Machines L92
5.11 Lapping Machines L94
5.12 Multi-machine Systems L97

6   Welding and Soldering (Brazing) Machines **L97**

6.1   Arc Welding Machines L97
6.2   Resistance Welding Machines L99
6.3   Soldering and Brazing Equipment L100

7   Industrial Robots **L100**

7.1   Systematics of Handling Systems L100
7.2   Components of Robots L101
7.3   Kinematic and Dynamic Models L102
7.4   Characteristics, Accuracy L102
7.5   Industrial Robot Control Systems L102
7.6   Programming L104
7.7   Main Applications and Selection of Robots L107

8   References **L108**

Index 1

Mechanics

G. Rumpel and H. D. Sondershausen, Berlin

1 Statics of Rigid Bodies

1.1 Introduction

Statics is the study of the equilibrium of solid bodies or of systems of solid bodies. Equilibrium prevails if a body is at rest or is in uniform motion in a straight line. Rigid bodies as understood in statics are bodies of which the deformations are so small that the points at which force is applied undergo negligible displacement.

Forces. These are vectors of varying direction and displaceable in their lines of action, which cause changes in the motions or shapes of bodies. The determinant factors of forces are magnitude, direction and location (**Fig. 1a**).

$$\boldsymbol{F} = \boldsymbol{F}_x + \boldsymbol{F}_y + \boldsymbol{F}_z = F_x \boldsymbol{e}_x + F_y \boldsymbol{e}_y + F_z \boldsymbol{e}_z$$

$$= (F \cos \alpha)\boldsymbol{e}_x + (F \cos \beta)\boldsymbol{e}_y + (F \cos \gamma)\boldsymbol{e}_z, \quad (1)$$

where

$$F = |\boldsymbol{F}| = \sqrt{F_x^2 + F_y^2 + F_z^2}. \quad (2)$$

For the cosines of direction, $\cos \alpha = F_x/F$, $\cos \beta = F_y/F$, $\cos \gamma = F_z/F$, and $\cos^2 \alpha + \cos^2 \beta + \cos^2 \gamma = 1$.

There are applied forces and reaction forces, as well as external and internal forces. External forces are all the forces that exert an effect on a body capable of free motion from the outside (see A1.5) (loads and supporting forces). Internal forces are all the cutting forces and binding forces occurring inside a system.

Moments or Couples. These consist of two equal and opposite forces with parallel lines of action (**Fig. 1b**) or of a vector which is perpendicular to the plane of effect. In this situation, $\boldsymbol{r}, \boldsymbol{F}, \boldsymbol{M}$ form a right helix (a right-handed system). Couples can be displaced at random in their planes of effect and perpendicular to it; in other words, the moment vector is a free vector, determined by the vector product

$$\boldsymbol{M} = \boldsymbol{r} \times \boldsymbol{F} = \boldsymbol{M}_x + \boldsymbol{M}_y + \boldsymbol{M}_z = M_x \boldsymbol{e}_x + M_y \boldsymbol{e}_y + M_z \boldsymbol{e}_z$$

$$= (M \cos \alpha^*)\boldsymbol{e}_x + (M \cos \beta^*)\boldsymbol{e}_y + (M \cos \gamma^*)\boldsymbol{e}_z. \quad (3)$$

$$\boldsymbol{M} = |\boldsymbol{M}| = |\boldsymbol{r}| \cdot |\boldsymbol{F}| \cdot \sin \varphi = Fb = \sqrt{M_x^2 + M_y^2 + M_z^2}. \quad (4)$$

M signifies the magnitude or amount of the moment, and provides a graphic representation of the area of the parallelogram formed by $\boldsymbol{r}$ and $\boldsymbol{F}$. In this situation, b is the moment arm perpendicular to $\boldsymbol{F}$. For the direction cosines (**Fig. 1c**), the following applies:

$$\cos \alpha^* = M_x/M, \quad \cos \beta^* = M_y/M, \quad \cos \gamma^* = M_z/M.$$

Moment of a Force Relative to a Point (Moment of Displacement). The effect of an individual force with a random point of contact in relation to a point O becomes clear if a null vector is added; i.e. two mutually opposed forces of equal value, $\boldsymbol{F}$ and $-\boldsymbol{F}$, at the point O (**Fig. 2a**). A single force $\boldsymbol{F}$ is derived at the point O, as well as a pair of forces or moment $\boldsymbol{M}$ (displacement moment), the vector of which is perpendicular to the plane formed by

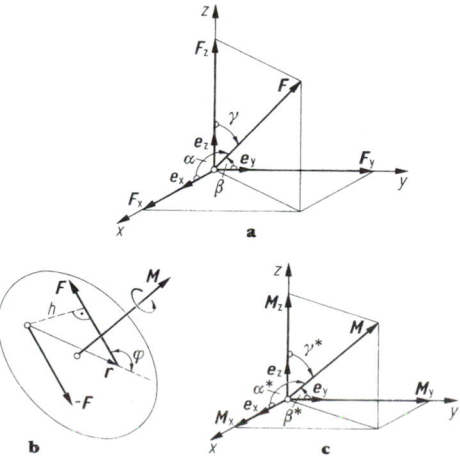

Figure 1. Representation of vectors: **a** force; **b** couple; **c** moment.

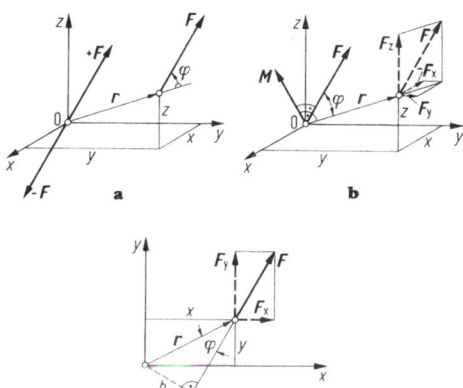

Figure 2. Force and moment: **a** and **b** force displacement; **c** plane moment.

r and F. If r and F are given in components x, y, z or F_x, F_y, F_z (**Fig. 2b**), then the following applies:

$$M = r \times F = \begin{vmatrix} e_x & e_y & e_z \\ x & y & z \\ F_x & F_y & F_z \end{vmatrix}$$

$$= (F_z y - F_y z)e_x + (F_x z - F_z x)e_y + (F_y x - F_x y)e_z$$

$$= M_x e_x + M_y e_y + M_z e_z. \tag{5}$$

The following applies to the components, the value of the moment vector, and the direction cosines:

$$M_x = F_z y - F_y z, \quad M_y = F_x z - F_z x,$$

$$M_z = F_y x - F_x y;$$

$$M = |M| = |r| \cdot F \cdot \sin \varphi = Fb = \sqrt{M_x^2 + M_y^2 + M_z^2};$$

$$\cos \alpha^* = M_x/M, \quad \cos \beta^* = M_y/M,$$

$$\cos \gamma^* = M_z/M.$$

If the force vector is located in the x, y plane, i.e. if z and F_z are equal to zero, then it follows that (**Fig. 2c**):

$$M = M_z = (F_y x - F_x y)e_z;$$

$$M = |M| = M_z = F_y x - F_x y = Fr \sin \varphi = Fb.$$

Projection of a moment vector onto a given axis (direction): if ψ is the angle between the vector and the axis, and e_1 is the unit vector of the axis, then from the scalar product we derive: $M_1 = Me_1 = M \cos \psi$.

1.2 Combination and Resolution of Concurrent Forces

1.2.1 Systems of Coplanar Forces

Combination of Forces to One Resultant Force. Forces are added together geometrically (vectorially), for two forces with the parallelogram or triangle of forces (**Fig. 3**), and for several forces with the polygon of forces (**Fig. 4**; force scale 1 cm $\hat{=}$ κ N). The calculated solution is

$$F_R = \sum_{i=1}^{n} F_i = \sum_{i=1}^{n} F_{ix} e_x + \sum_{i=1}^{n} F_{iy} e_y$$

$$= F_{Rx} e_x + F_{Ry} e_y, \tag{6}$$

where $F_{ix} = F_i \cos \alpha_i$, $F_{iy} = F_i \sin \alpha_i$. Size and direction of the resultant force are given by

$$F_R = \sqrt{F_{Rx}^2 + F_{Ry}^2}, \qquad \tan \alpha_R = F_{Ry}/F_{Rx}. \tag{7}$$

Resolution of a Force in the Plane. This is unique only in two directions, and the solution is multi-valued in three and more directions (statically indeterminate). For a graphical solution, see **Fig. 5a, b**.

Calculated Solution. (**Fig. 5c**) $F = F_1 + F_2$, or, in components:

$$F \cos \alpha = F_1 \cos \alpha_1 + F_2 \cos \alpha_2,$$

$$F \sin \alpha = F_1 \sin \alpha_1 + F_2 \sin \alpha_2;$$

i.e. $F_2 = (F \sin \alpha - F_1 \sin \alpha_1)/\sin \alpha_2$ and therefore

$$F \cos \alpha = F_1 \cos \alpha_1 + \cos \alpha_2 (F \sin \alpha - F_1 \sin \alpha_1)/\sin \alpha_2.$$

$$F \cos \alpha \sin \alpha_2 - F \sin \alpha \cos \alpha_2$$

$$= F_1 \cos \alpha_1 \sin \alpha_2 - F_1 \sin \alpha_1 \cos \alpha_2.$$

In other words, $F_1 = F \sin (\alpha_2 - \alpha)/\sin (\alpha_2 - \alpha_1)$ and accordingly $F_2 = F \sin (\alpha_1 - \alpha)/\sin (\alpha_2 - \alpha_1)$. These formulae correspond to the sine law for a scalene triangle (**Fig. 5c**).

1.2.2 Forces in Space (Three-Dimensional Force System)

Combination of Forces to Form One Resultant Force. Forces are added geometrically (vectorially), as the polygon of force is plotted in space (**Fig. 6a**). For this purpose, the projection is made on two planes, i.e. the task is resolved into horizontal and vertical projections (**Fig. 6a**). In the x, y-plane (horizontal plane), vector addition gives the projection F_R' of the resultant force F_R in the y,z-plane (vertical projection) F_R'' (**Fig. 6b**). The projection of the F_R'' in the z-direction produces the true component F_{Rz} of the resultant force. F_{Rz} is combined with F_R' in the extended plane of both components to form the resultant force F_R (**Fig. 6c**).

Calculated Solution. This is

$$F_R = \sum_{i=1}^{n} F_i = \sum_{i=1}^{n} F_{ix} e_x + \sum_{i=1}^{n} F_{iy} e_y + \sum_{i=1}^{n} F_{iz} e_z$$

$$= F_{Rx} e_x + F_{Ry} e_y + F_{Rz} e_z, \tag{8}$$

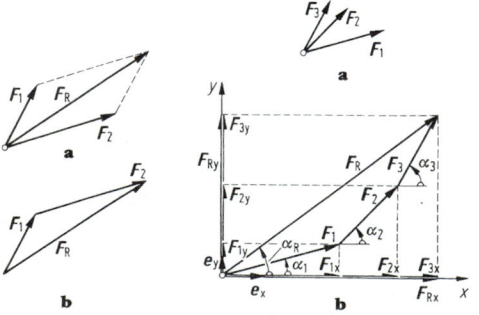

Figure 3. Combination of two plane forces: **a** parallelogram of forces, **b** triangle of forces.

Figure 4. Combination of several plane forces: **a** location plan, **b** polygon of forces.

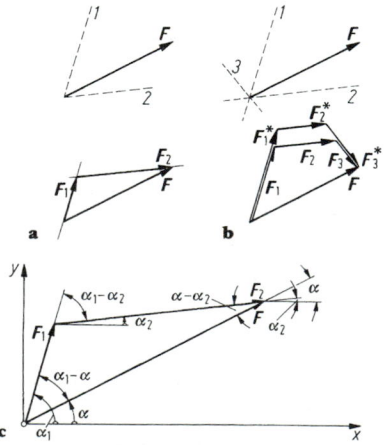

Figure 5. Resolution of a plane force: **a** in two directions, (single value); **b** in three directions (multiple value); **c** calculated.

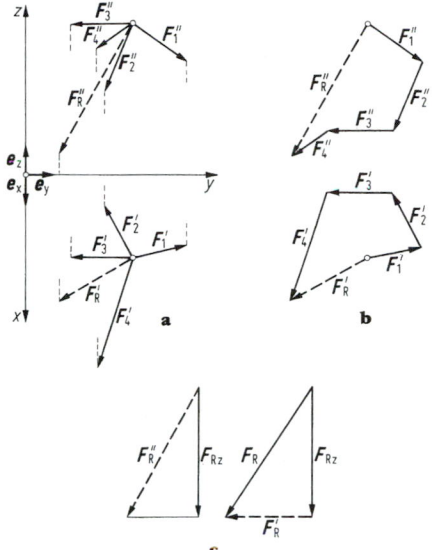

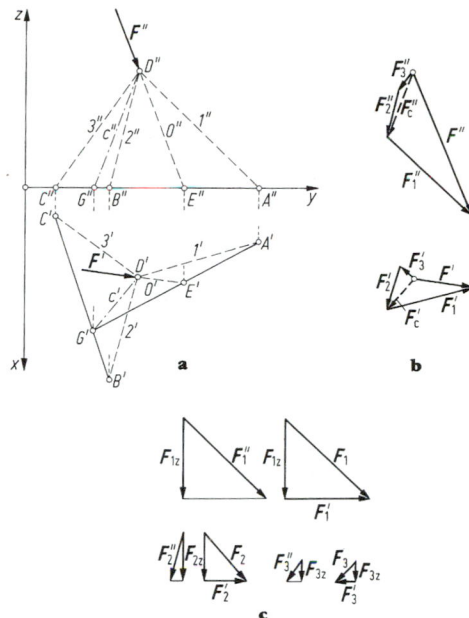

Figure 6. Combination of forces in space: **a** location plan, **b** polygon of forces, **c** overall resultant force.

where $F_{ix} = F_i \cos \alpha_i$, $F_{iy} = F_i \cos \beta_i$, $F_{iz} = F_i \cos \gamma_i$. The size and direction of the resultant force are

$$F = \sqrt{F_{Rx}^2 + F_{Ry}^2 + F_{Rz}^2};$$

$$\cos \alpha_R = F_{Rx}/F, \quad \cos \beta_R = F_{Ry}/F, \quad \cos \gamma_R = F_{Rz}/F. \quad (9)$$

The Resolution of a Force in a plane is uniquely possible only in three directions; in four and more directions the solution is multi-valued (statically indeterminate). In the graphical solution (**Fig. 7a, b**), the point E at which force $\boldsymbol{F}$ (line of action 0) intersects the abscissa is initially determined in the vertical and horizontal projection, and therefore the plane $A'D'E'$ extended from 0 and 1 is determined in the horizontal projection. It meets the line of intersection C' of plane $B'C'D'$ formed by 2 and 3 in the horizontal projection at the projected intersection point of line G'. The projection of G' into the vertical projection provides G'' and therefore c''. Force $\boldsymbol{F}$ is now resolved in the horizontal and vertical projections in directions 1 and c, which are located in two different planes, namely ABD and BCD. $\boldsymbol{F}_c$ is known as Culmann's auxiliary force, which is subsequently resolved in directions 2 and 3 in the horizontal and vertical projections. The final value of $\boldsymbol{F}_1$ is obtained by combining the component $\boldsymbol{F}_{1z}$, obtained from the horizontal projection, with the force $\boldsymbol{F}_1'$ obtained from the horizontal projection in the plane extended from both these forces. The same applies correspondingly to $\boldsymbol{F}_2$ and $\boldsymbol{F}_3$ (**Fig. 7c**).

Calculated Solution. This is

$$\boldsymbol{F}_1 + \boldsymbol{F}_2 + \boldsymbol{F}_3 = \boldsymbol{F};$$
$$F_{1x} + F_{2x} + F_{3x} = F_x,$$
$$F_{1y} + F_{2y} + F_{3y} = F_y,$$
$$F_{1z} + F_{2z} + F_{3z} = F_z.$$

As in **Fig. 8**, the following applies to the direction cosines of the three directions given:

$$\cos \alpha_i = x_i / \sqrt{x_i^2 + y_i^2 + z_i^2},$$

$$\cos \beta_i = y_i / \sqrt{x_i^2 + y_i^2 + z_i^2},$$

Figure 7. Resolution of a force in space: **a** location plan, **b** polygon of forces, **c** final forces.

$$\cos \gamma_i = z_i / \sqrt{x_i^2 + y_i^2 + z_i^2}.$$

It follows that

$$F_1 \cos \alpha_1 + F_2 \cos \alpha_2 + F_3 \cos \alpha_3 = F \cos \alpha,$$
$$F_1 \cos \beta_1 + F_2 \cos \beta_2 + F_3 \cos \beta_3 = F \cos \beta,$$
$$F_1 \cos \gamma_1 + F_2 \cos \gamma_2 + F_3 \cos \gamma_3 = F \cos \gamma.$$

These three linear equations for the three unknown forces F_1, F_2 and F_3 have a unique solution only if their system determinant does not equal zero, i.e. if the three direction vectors are not in one plane. According to **Fig. 8**, $F_1 \boldsymbol{e}_1 + F_2 \boldsymbol{e}_2 + F_3 \boldsymbol{e}_3 = \boldsymbol{F}$, and, after multiplication by $\boldsymbol{e}_2 \times \boldsymbol{e}_3$,

$$F_1 \boldsymbol{e}_1 (\boldsymbol{e}_2 \times \boldsymbol{e}_3) + F_2 \boldsymbol{e}_2 (\boldsymbol{e}_2 \times \boldsymbol{e}_3) + F_3 \boldsymbol{e}_3 (\boldsymbol{e}_2 \times \boldsymbol{e}_3)$$
$$= \boldsymbol{F}(\boldsymbol{e}_2 \times \boldsymbol{e}_3).$$

Because the vector $(\boldsymbol{e}_2 \times \boldsymbol{e}_3)$ is perpendicular to both $\boldsymbol{e}_2$ as well as to $\boldsymbol{e}_3$, the scale products are zero, and it follows that

$$F_1 \boldsymbol{e}_1 (\boldsymbol{e}_2 \times \boldsymbol{e}_3) = \boldsymbol{F}(\boldsymbol{e}_2 \times \boldsymbol{e}_3)$$

and

$$F_1 = \boldsymbol{F}\boldsymbol{e}_2 \boldsymbol{e}_3 / (\boldsymbol{e}_1 \boldsymbol{e}_2 \boldsymbol{e}_3),$$
$$F_2 = \boldsymbol{e}_1 \boldsymbol{F}\boldsymbol{e}_3 / (\boldsymbol{e}_1 \boldsymbol{e}_2 \boldsymbol{e}_3), \quad F_3 = \boldsymbol{e}_1 \boldsymbol{e}_2 \boldsymbol{F} / (\boldsymbol{e}_1 \boldsymbol{e}_2 \boldsymbol{e}_3). \quad (10)$$

$\boldsymbol{F}\boldsymbol{e}_2 \boldsymbol{e}_3$, $\boldsymbol{e}_1 \boldsymbol{e}_2 \boldsymbol{e}_3$ etc. are parallelipipedal products, i.e. scalar quantities, whose values determine the spatial volume of the parallelipiped formed by the three vectors. The solution is single-value if the parallelipiped product is $\boldsymbol{e}_1 \boldsymbol{e}_2 \boldsymbol{e}_3 \neq 0$, i.e. the three vectors may not occur in one plane.

With $\boldsymbol{e}_i = \cos \alpha_i \boldsymbol{e}_x + \cos \beta_i \boldsymbol{e}_y + \cos \gamma_i \boldsymbol{e}_z$ it follows that

$$F_1 = \begin{vmatrix} F \cos \alpha & \cos \alpha_2 & \cos \alpha_3 \\ F \cos \beta & \cos \beta_2 & \cos \beta_3 \\ F \cos \gamma & \cos \gamma_2 & \cos \gamma_3 \end{vmatrix} : \begin{vmatrix} \cos \alpha_1 & \cos \alpha_2 & \cos \alpha_3 \\ \cos \beta_1 & \cos \beta_2 & \cos \beta_3 \\ \cos \gamma_1 & \cos \gamma_2 & \cos \gamma_3 \end{vmatrix},$$
$$(11)$$

corresponding to F_2 and F_3.

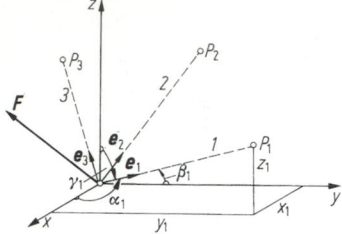

Figure 8. Calculated resolution of a force in space.

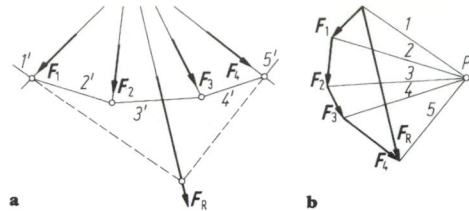

Figure 9. Composition of several plane forces: **a** location plan (funicular polygon), **b** polygon of forces (solid angle).

1.3 Combination and Resolution of Non-Concurrent Forces

1.3.1 Coplanar Forces

Combination of Several Forces to Form One Resultant Force

Graphical Process with Polygon of Forces and Funicular Polygon. The forces are added geometrically to the resultant force in the polygon of forces (**Fig. 9**), a random pole P is selected, and the radius vectors 1 to n are drawn. The parallels to these are transferred to the location plan (**Fig. 9a**) as rays of the funicular polygon $1'$ to n', such that the forces of a triangle of forces of the polar solid angle intersects it in the ground plan at a point (point–triangle rule). The point of intersection of the first and last rays of the funicular polygon provides the contact point of the resultant force, the value and direction of which are derived from the polygon of forces.

Calculation Procedure. By reference to the zero point, the plane group of forces provides a resultant force and a resultant (displacement) moment (**Fig. 10a**):

$$F_R = \sum_{i=1}^{n} F_i, \quad M_R = \sum_{i=1}^{n} M_i, \quad \text{or} \quad F_{Rx} = \sum_{i=1}^{n} F_{ix},$$

$$F_{Ry} = \sum_{i=1}^{n} F_{iy}, \quad M_R = \sum_{i=1}^{n} (F_{iy} x_i - F_{ix} y_i) = \sum_{i=1}^{n} F_i b_i.$$

For a random point, the effect of the group of forces is the same as that of their resultant force. If the resultant force is displaced parallel from the zero point so far that M_R becomes zero, it follows for its position that $M_R = F_R b_R$ etc. (**Fig. 10b**)

$$b_R = M_R/F_R \quad \text{or} \quad x_R = M_R/F_{Ry} \quad \text{or}$$

$$y_R = -M_R/F_{Rx}.$$

Resolution of a Force. The resolution of force in a plane is possible as a single value, in three given directions that do not intersect at a point, and of which a maximum of two may be parallel. A force is graphically resolved with the aid of Culmann's auxiliary vectors (**Fig. 11a, b**). In addition to this, the force F is made to intersect one of the three lines of application, and the other two lines of application are made to intersect each other. The line joining the points of intersection A and B is the so-called Culmann's auxiliary vector c. After resolution of the force F in the polygon of forces in the directions 3 and c, F_3 and F_c are derived. The force F_c is then again resolved in directions 1 and 2, giving F_1 and F_2.

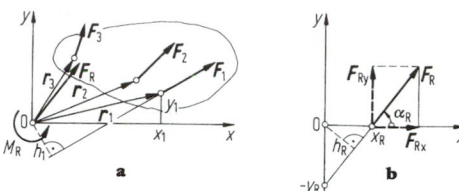

Figure 10. Resultant of plane forces.

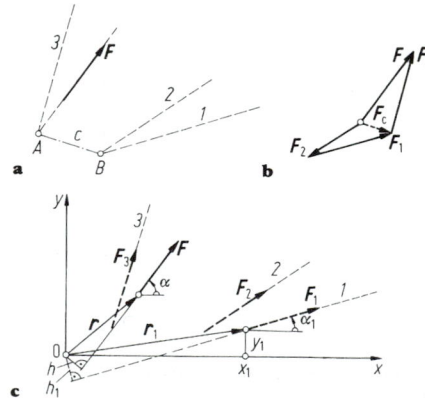

Figure 11. Resolution of a plane force: **a** location plan with Culmann lines c, **b** polygon of forces, **c** calculated solution.

The calculated solution follows from the condition that the application of force and moment of the individual forces F_i and the force F must be equal in relation to the zero point (**Fig. 11c**):

$$\sum_{i=1}^{n} F_i = F, \quad \sum_{i=1}^{n} (r_i \times F_i) = r \times F,$$

$$F_1 \cos \alpha_1 + F_2 \cos \alpha_2 + F_3 \cos \alpha_3 = F \cos \alpha,$$

$$F_1 \sin \alpha_1 + F_2 \sin \alpha_2 + F_3 \sin \alpha_3 = F \sin \alpha;$$

$$F_1 (x_1 \sin \alpha_1 - y_1 \cos \alpha_1) + F_2(x_2 \sin \alpha_2 - y_2 \cos \alpha_2)$$

$$+ F_3(x_3 \sin \alpha_3 - y_3 \cos \alpha_3) = F(x \sin \alpha - y \cos \alpha);$$

or, instead of the last equation, $F_1 b_1 + F_2 b_2 + F_3 b_3 = Fb$, where anticlockwise moments are positive. There are three equations for the three unknowns F_1, F_2, F_3. Their denominator determinants may not be zero; i.e. the con-

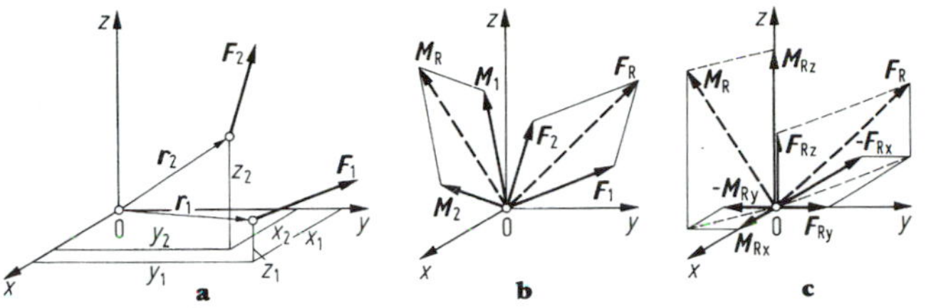

Figure 12. Reduction of forces in space: **a** location plan, **b** force and moment resultant, **c** force and moment components.

ditions specified for the graphical solution regarding the position of the lines of action must be fulfilled, if it is intended that the solution should be unique.

1.3.2 Forces in Space

Force Combination (Reduction). A group of forces in space, $\mathbf{F}_i = (F_{ix}; F_{iy}; F_{iz})$, the contact points of which are given by the radius vectors $\mathbf{r}_i = (x_i; y_i; z_i)$, can be combined (reduced) in relation to a random point to give a resultant force $\mathbf{F}_R$ and a resultant moment $\mathbf{M}_R$. The complex graphical solution is acquired in the projection planes [1]. The calculated solution (**Fig. 12**), in relation to the zero point, is

$$\mathbf{F}_R = \sum_{i=1}^{n} \mathbf{F}_i,$$

$$\mathbf{M}_R = \sum_{i=1}^{n} (\mathbf{r}_i \times \mathbf{F}_i) = \sum_{i=1}^{n} \begin{vmatrix} \mathbf{e}_x & \mathbf{e}_y & \mathbf{e}_z \\ x_i & y_i & z_i \\ F_{ix} & F_{iy} & F_{iz} \end{vmatrix} \quad \text{or}$$

$$F_{Rx} = \sum_{i=1}^{n} F_{ix}, \quad F_{Ry} = \sum_{i=1}^{n} F_{iy}, \quad F_{Rz} = \sum_{i=1}^{n} F_{iz};$$

$$M_{Rx} = \sum_{i=1}^{n} (F_{iz} y_i - F_{iy} z_i), \quad M_{Ry} = \sum_{i=1}^{n} (F_{ix} z_i - F_{iz} x_i).$$

$$M_{Rz} = \sum_{i=1}^{n} (F_{iy} x_i - F_{ix} y_i).$$

Parallelogram of Forces. A further simplification of the reduced system of forces is possible if there is an axis that has a specific position in which the force vector and the moment vector are parallel (**Fig. 13**). This axis is called the central axis. It is derived by resolving $\mathbf{M}_R$ in the plane E formed by $\mathbf{M}_R$ and $\mathbf{F}_R$ into the components $M_F = M_R \cos \varphi$ (parallel to $\mathbf{F}_R$) and $M_S = M_R \sin \varphi$ (perpendicular to $\mathbf{F}_R$). Here, φ follows from the scalar product $\mathbf{M}_R \cdot \mathbf{F}_R = M_R F_R \cos \varphi$, i.e. $\cos \varphi = \mathbf{M}_R \cdot \mathbf{F}_R$ $(M_R F_R)$. M_S is then made equal to zero by displacing $\mathbf{F}_R$ perpendicular to the plane E by the amount $a = M_S/F_R$. The vector that pertains to this is $\mathbf{a} = (\mathbf{F}_R \times \mathbf{M}_R)/F_R^2$, since its value is $|\mathbf{a}| = a = F_R M_R \sin \varphi/F_R^2 = M_S/F_R$. The

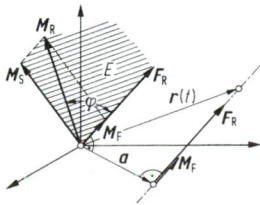

Figure 13. Parallelogram of forces.

vector equation for the central axis, in the direction of which $\mathbf{F}_R$ and $\mathbf{M}_F$ take effect, then reads, with t as parameter:

$$\mathbf{r}(t) = \mathbf{a} + \mathbf{F}_R \cdot t$$

Force Resolution in Space. A force can be resolved in space as a single value in six given directions. If the directions are given by their direction cosines, and if the forces are designated $\mathbf{F}_1 \dots \mathbf{F}_6$, then

$$\sum_{i=1}^{6} F_i \cos \alpha_i = F \cos \alpha, \quad \sum_{i=1}^{6} F_i \cos \beta_i = F \cos \beta,$$

$$\sum_{i=1}^{6} F_i \cos \gamma_i = F \cos \gamma;$$

$$\sum_{i=1}^{6} F_i(y_i \cos \gamma_i - z_i \cos \beta_i) = F(y \cos \gamma - z \cos \beta),$$

$$\sum_{i=1}^{6} F_i(z_i \cos \alpha_i - x_i \cos \gamma_i) = F(z \cos \alpha - x \cos \gamma),$$

$$\sum_{i=1}^{6} F_i(x_i \cos \beta_i - y_i \cos \alpha_i) = F(x \cos \beta - y \cos \alpha).$$

From these six linear equations a unique solution may be derived, if the denominator determinant is not equal to zero.

1.4 Conditions of Equilibrium

A body is in equilibrium if it is at rest or in uniform motion. Since all acceleration values are zero, it follows from the basic principles of dynamics that no resultant force and no resultant moment are exerted on the body.

1.4.1 System of Forces in Space

The conditions of equilibrium are

$$\mathbf{F}_R = \Sigma \mathbf{F}_i = 0 \quad \text{and} \quad \mathbf{M}_R = \Sigma \mathbf{M}_i = 0; \qquad (12)$$

or, in components

$$\begin{array}{lll} \Sigma F_{ix} = 0, & \Sigma F_{iy} = 0, & \Sigma F_{iz} = 0; \\ \Sigma M_{ix} = 0, & \Sigma M_{iy} = 0, & \Sigma M_{iz} = 0. \end{array} \qquad (13)$$

Each of the three conditions of equilibrium for the forces can be replaced by another for the moments about another random axis, which may not pass through the origin 0.

Six unknown values (forces or moments) can be calculated from the six conditions of equilibrium. If more than six unknowns exist, the problem is described as statically indeterminate. Its solution is possible only by invoking deformation values (see B2.7). If forces with *common*

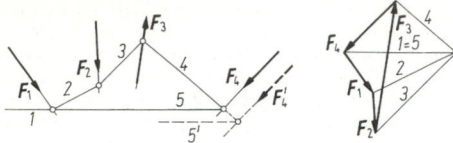

Figure 14. Graphical conditions of equilibrium.

points of contact are present, then the moment conditions of Eq. (13) are fulfilled with respect to the point of intersection (and therefore also for all other points, since M_R is a free vector). Only the equilibrium conditions of forces of Eq. (13) then apply, from which three unknown forces can be determined. For the graphical solution, in this case, the polygon of force in space must be closed due to $F_R = \Sigma F_i = 0$ (performed in horizontal and vertical projection in accordance with **Fig. 7**).

1.4.2 System of Coplanar Forces

The equation system (13) is reduced to three conditions of equilibrium:

$$\Sigma F_{ix} = 0, \quad \Sigma F_{iy} = 0, \quad M_{iz} = 0. \tag{14}$$

Both force equilibrium conditions can be replaced by two further moment conditions. The three points of reference for the moment equations must not lie on a straight line. From the three conditions of equilibrium of the plane, three unknown values (forces or moments) can be determined. If more unknowns are present, the problem is statically indeterminate.

The graphical solution for equilibrium in the plane follows from the principle that polygons of force and funicular polygons must close (**Fig. 14**). If the polygon of force closes, but the funicular polygon does not, then no equilibrium prevails; a couple of forces remains (see **Fig. 14**, force F_4 and couple of forces, consisting of funicular polygons 1 and 5′). Special cases: Two forces are in equilibrium if they have the same lines of application, are of equal magnitude, and are in opposing directions. Three forces must intersect at a point, and the polygon of force must close. In the case of four forces, the polygon of force must close, and the resultant force of each pair must lie on the same line of action, be of equal magnitude, and be in opposing directions (**Fig. 15**).

Forces with a Common Point of Contact in the Plane. For these, the moment condition in Eq. (14) is

fulfilled in an identical manner, and there remain only the two force conditions

$$\Sigma F_{ix} = 0, \quad \Sigma F_{iy} = 0. \tag{15}$$

The graphical solution follows from the vector equation $F_R = 0$, i.e. the polygon of force must be closed (**Fig. 16**).

1.4.3 Principle of Virtual Work

The principle replaces the equilibrium conditions, and states: If a rigid body is subjected to a minor (virtual) displacement that is compatible with its geometrical constraints, and if the body is in equilibrium (**Fig. 17**), then the total virtual work of all the external forces and moments affected – identified by superscript (e) – are equal to zero:

$$\delta W^{(e)} = \Sigma F_i^{(e)} \delta r_i + \Sigma M_i^{(e)} \delta \varphi_i = 0; \tag{16}$$

or, in components,

$$\delta W^{(e)} = \Sigma(F_{ix}^{(e)} \delta x_i + F_{iy}^{(e)} \delta y_i + F_{iz}^{(e)} \delta z_i)$$
$$+ \Sigma(M_{ix}^{(e)} \delta \varphi_{ix} + M_{iy}^{(e)} \delta \varphi_{iy} + M_{iz}^{(e)} \delta \varphi_{iz}) = 0;$$

here $r_i = (x_i; y_i; z_i)$ radius vectors to the points of contact of the forces; $\delta r_i = (\delta x_i; \delta y_i; \delta z_i)$ variations (vector differentials expressed mathematically) of the radius vectors, which result from the formation of the first derivative; $\delta \varphi_i$ = angle of rotation differentials of the torsional motions φ_i.

In natural coordinates, the principle takes the form

$$\delta W^{(e)} = \Sigma F_{is}^{(e)} \delta s_i + \Sigma M_{i\varphi}^{(e)} \delta \varphi_i = 0, \tag{17}$$

where $F_{is}^{(e)}$ are the components of force in the direction of the displacement, and $M_{i\varphi}^{(e)}$ are the components of the moments taking effect about the axis of rotation. The principle is used, among other things, in statics to examine the equilibrium of displaceable systems and to calculate the effect of migrating loads on intersection and contact forces (lines of application).

Example *Drafting Machine* (**Fig. 18**). A counterweight F_0 and its lever arm l are to be determined in such a way that the drafting machine will be in equilibrium from its own weight F_G in any position.

The system has two different degrees of freedom:

$$r_G = (-c \sin \varphi + b \sin \psi; b \cos \psi - c \cos \varphi),$$
$$r_Q = (l \sin \varphi - a \sin \psi; -a \cos \psi + l \cos \varphi),$$
$$\delta r_G = (-c \cos \varphi \, \delta \varphi + b \cos \psi \, \delta \psi; -b \sin \psi \, \delta \psi + c \sin \varphi \, \delta \varphi),$$
$$\delta r_Q = (l \cos \varphi \, \delta \varphi - a \cos \psi \, \delta \psi; a \sin \psi \, \delta \psi - l \sin \varphi \, \delta \varphi).$$

With $F_G = (0; -F_G)$ and $F_Q = (0; -F_Q)$, it follows that

$$\delta W^{(e)} = \Sigma F_i^{(e)} \delta r_i = -F_G (-b \sin \psi \, \delta \psi + c \sin \varphi \, \delta \varphi)$$
$$-F_Q (a \sin \psi \, \delta \psi - l \sin \varphi \, \delta \varphi)$$
$$= \sin \psi \, \delta \psi \, (F_G b - F_Q a) + \sin \varphi \, \delta \varphi \, (-F_G c + F_Q l).$$

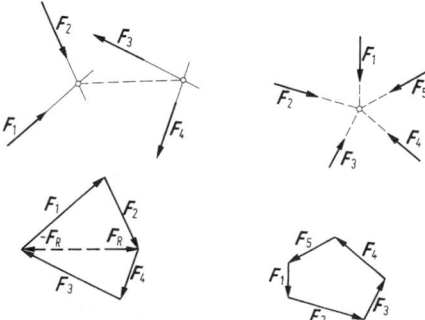

Figure 15. Equilibrium of four plane forces.   **Figure 16.** Equilibrium of forces with common points of contact.

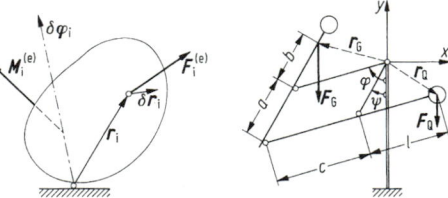

Figure 17. Principle of virtual displacements.   **Figure 18.** Drafting machine.

From $\delta W^{(e)} = 0$, due to the random nature of φ and ψ

$$F_G b - F_Q a = 0 \quad \text{and} \quad -F_G c + F_Q l = 0$$

and therefore

$$F_Q = F_G b/a \quad \text{and} \quad l = c\, F_G/F_Q = ca/b.$$

In addition,

$$\delta^2 W^{(e)} = \cos\psi\,\delta\psi^2\,(F_G b - F_Q a) + \cos\varphi\,\delta\varphi^2\,(-F_G c + F_Q l).$$

It follows from this, with the solution values calculated, that $\delta^2 W^{(e)} = 0$, i.e. neutral equilibrium prevails (see A1.4.4).

1.4.4 Types of Equilibrium

A distinction is drawn between stable, unstable, and neutral equilibria (see **Fig. 19**). Stable equilibrium prevails when a body reverts to its initial position under a displacement that is compatible with its geometrical constraints; unstable equilibrium when the body seeks to leave its initial position; neutral equilibrium prevails when any adjacent position is a new position of equilibrium. If, in accordance with A1.4.3, the minor displacement is perceived as virtual, then, in accordance with the principle of virtual work, $\delta W^{(e)} = 0$ applies at the equilibrium position. If the body is moved as in **Fig. 19a**, from a position *1* to a position *2*, through the position of equilibrium *0*, then the work is $\delta W^{(e)} = F_s\,\delta s > 0$, in the range *1* to *0*; i.e., positive, and in the range *0* to *2*, $\delta W^{(e)} < 0$, i.e. negative. It follows from the function $\delta W^{(e)} = f(s)$ that the increase of $\delta W^{(e)}$ is negative, i.e. $\delta^2 W^{(e)} < 0$, if equilibrium is stable. The general application for the equilibrium is: stable $\delta^2 W^{(e)} < 0$, unstable $\delta^2 W^{(e)} > 0$, neutral $\delta^2 W^{(e)} = 0$.

If the problems involved are such that only weight forces are acting, then with the potential $U = F_G z$ or $\delta U = F_G \delta z$ the following applies:

$$\delta W^{(e)} = \mathbf{F}^{(e)}\delta\mathbf{r} = (0,0, -F_G)\,(\delta x;\, \delta y;\, \delta z)$$

$$= -F_G\delta z = -\,\delta U$$

and $\delta^2 W^{(e)} = -\delta^2 U$, i.e. in stable equilibrium $\delta^2 U > 0$, and therefore the potential energy U is a minimum, while in unstable equilibrium $\delta^2 U < 0$, and the potential energy is a maximum.

1.4.5 Stability

Bodies having supports that can only absorb compressive forces are in danger of overturning. This is prevented if the sum of the righting moments around the potential tilting edges A or B (**Fig. 20**) is greater than the sum of the tilting moments, i.e. if the resultant of the system of forces within the tilting edges intersects the standing surface. Stability is the ratio of the total of all righting moments to the total of all tilting moments by reference to a tilting edge: $S = \Sigma M_S / \Sigma M_K$. For $S \geq 1$, stability and equilibrium prevail.

1.5 Types of Support; the 'Free Body'

Bodies are supported by bearings. The supporting forces function as reactive forces to the forces imposed on the body from outside. Depending on the mode of construction of the bearings, a maximum of three forces and a maximum of three moments can be transferred in space. The reactive forces and moments become external forces owing to the so-called 'freeing' of a body. A body is freed by being separated from its surroundings due to a closed section through all the bearings, and by all the bearing forces being applied as external forces (**Fig. 21**, freeing principle). Equal and opposite forces of action and reaction (Newton's third law) then take effect on the bearings. A distinction is drawn between a bearing with a single value and one with up to six values, depending on the bearing's structural nature and the number of the reaction values (**Fig. 22**).

1.6 Support Reactions

1.6.1 Plane Problems

In the plane, a body has three degrees of freedom in respect of its possibilities of motion (displacement in the *x*- and *y*-axes, rotation about the *z*-axis). It therefore requires a bearing with a total of three values to provide stable and statically-determinant retention. This may consist of a firm clamping effect or of one fixed and one free bearing, or of three free (slide) bearings (in the latter case, the three lines of action of the reactive forces may not intersect at a point). If the bearing is *n*-valued ($n > 3$), then the system is $(n - 3)$ times statically indeterminate in its bearings. If the bearings are less than three-valued, the system is statically underdeterminated, i.e. unstable and movable. The calculation of the contact reactions is carried out by freeing the body and determining the conditions of equilibrium.

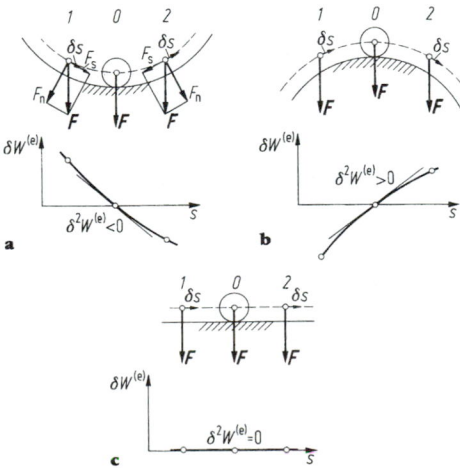

Figure 19. Types of equilibrium: **a** stable, **b** unstable, **c** neutral.

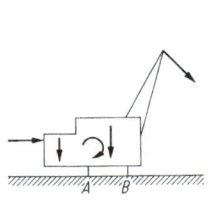

Figure 20. Stability.

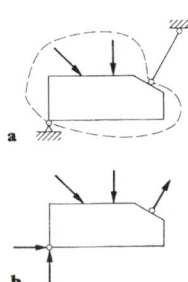

Figure 21. Freeing principle: **a** supported body with closed line of intersection, **b** free body.

Type	Symbol	Reaction values in plane	in space	Value factor	
Movable bearings:					
Radial bearing		F_{Ay}	F_{Ay}, F_{Az}	1	2
Slide bearing		F_{Ay}	F_{Az}	1	1
Roller bearing		F_{Ay}	$(F_{Ay}), F_{Az}$	1	1 (2)
Vibrating rod, cable		F_A	F_A	1	1
Fixed bearings:					
Thrust and axial bearings		F_{Ax}, F_{Ay}	F_{Ax}, F_{Ay}, F_{Az}	2	3
Fixed joint		F_{Ax}, F_{Ay}	F_{Ax}, F_{Ay}, F_{Az}	2	3
Fixed clamping		F_{Ax}, F_{Ay}, M_E	M_{Ex}, M_{Ey}, M_{Ez} F_{Ax}, F_{Ay}, F_{Az}	3	6

Figure 22. Types of bearings.

Example *Shaft* (**Fig. 23a**). The contact forces in A and B are sought in terms of the forces F_1 and F_2.

Calculated Solution. The following applies to the freed shaft (**Fig. 23b**):

$$\Sigma M_{i/a} = 0 = -F_1 a + F_B l - F_2 (l + c), \quad \text{i.e.}$$

$$F_B = [F_1 a + F_2 (l + c)]/l;$$

$$\Sigma M_{iB} = 0 = -F_{Ay} l + F_1 b - F_2 c, \quad \text{i.e.}$$

$$F_{Ay} = (F_1 b - F_2 c)/l;$$

$$\Sigma F_{ix} = 0 = F_{Ax}.$$

The condition of equilibrium $F_{iy} = 0$ must likewise be fulfilled, and can be used as a control equation.

$$\Sigma F_{iy} = F_{Ay} - F_1 + F_B - F_2$$
$$= (F_1 b - F_2 c)/l - F_1 + [F_1 a + F_2 (l + c)]/l - F_2$$
$$= F_1 (a + b - l)/l + F_2 (-c + l + c - l)/l = 0.$$

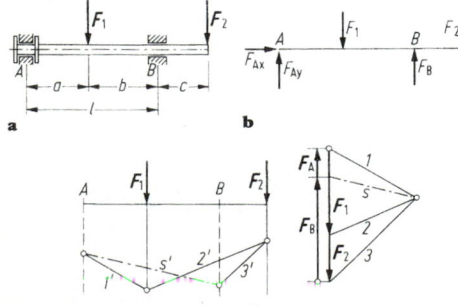

a

b

c

Figure 23. Shaft: **a** system, **b** freeing, **c** graphical solution.

Graphical Solution (**Fig. 23c**). The polygon of force and the polar solid angle, together with the funicular polygon pertaining to them, are shown by the forces F_1 and F_2. The points of intersection of the two outer radius vectors $1'$ and $3'$ with the known lines of action of both contact forces (since in this case $F_{Ax} = 0$) provides the closing side s', which 'closes' the radius vector. Parallel transfer to the polar solid angle provides the two contact forces F_A and F_B while maintaining the point-triangle rule (see A1.3.1). Both the funicular polygon and polygon of forces are then closed; i.e. equilibrium exists between the forces F_1, F_2, F_B, F_A.

Example *Angled Carrier Element* (**Fig. 24a**). The angled carrier element is loaded by the two point forces F_1 and F_2 and the constant distributed load q. The applied force in the fixed bearing A and the force in the pendulum rod at B are to be determined.

Calculated Solution. With the resultant of the distributed load $F_q = qc$ (**Fig. 24b**):

$$\Sigma M_{iA} = 0 = -F_1 \sin \alpha_1 \, a - qc(a + b + c/2)$$
$$- F_2 e + F_s \cos \alpha_s d + F_s \sin \alpha_s b,$$

and from this

$$F_s = [F_1 \sin \alpha_1 \, a + qc(a + b + c/2) + F_2 e]/(l \cos \alpha_s + b \sin \alpha_s).$$

From

$$\Sigma F_{ix} = 0 = F_{Ax} + F_1 \cos \alpha_1 + F_2 - F_s \sin \alpha_s \quad \text{and}$$
$$\Sigma F_{iy} = 0 = F_{Ay} - F_1 \sin \alpha_1 - qc + F_s \cos \alpha_s$$

it follows that

$$F_{Ax} = -F_1 \cos \alpha_1 - F_2 + F_s \sin \alpha_s \quad \text{and}$$
$$F_{Ay} = F_1 \sin \alpha_1 + qc - F_s \cos \alpha_s,$$

in which the calculated value for F_s is to be used.

Graphical Solution (**Fig. 24c**). After drawing the polygon of force from the given forces F_1, F_2, and F_q and the polar solid angle, the funicular polygon pertaining to them is drawn, for which the first radius vector $1'$ must be laid through the fixed bearing A; this is due to the fact that it is the only known point on the line of action of F_A. The point of intersection of the last vector radius $4'$ with the (known) line of action of F_s provides the closing side s', which closes the funicular polygon. The transfer of this to the polar solid angle, maintaining the point-triangle rule (see A1.3.1), provides first F_s and then, by closing the polygon of force, the force F_A.

Example *Wagon on an Inclined Plane* (**Fig. 25a**). The wagon, loaded by the force of gravity F_G, and the trailer tractive force F_Z, is held in equilibrium on the inclined plane by a cable

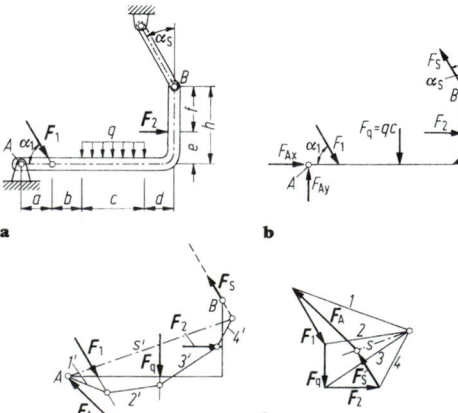

a

b

c

Figure 24. Angled bearing element: **a** system, **b** freeing, **c** graphical solution.

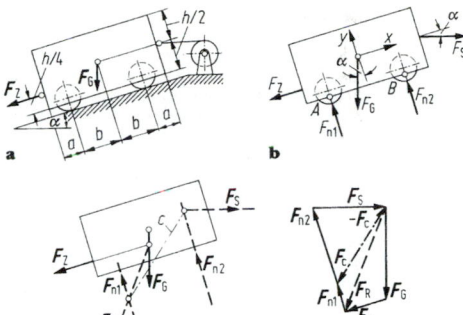

Figure 25. Wagon on an inclined plane: **a** system, **b** freeing, **c** graphical solution.

winch. The tensile force in the retaining cable and the supporting force on the wheels are to be calculated, neglecting friction forces.

Calculated Solution. On the freed wagon (**Fig. 25b**), equilibrium conditions prevail:

$$\Sigma F_{ix} = 0 = -F_Z - F_G \sin \alpha + F_S \cos \alpha,$$

$$F_S = F_G \tan \alpha + F_Z/\cos \alpha;$$

$$\Sigma M_{iA} = 0 = F_Z b/4 + F_G(b/2) \sin \alpha - F_G b \cos \alpha + 2F_{n2} b$$
$$- F_S(b/2) \cos \alpha - F_S(a + 2b) \sin \alpha;$$

$$\Sigma M_{iB} = 0 = F_Z b/4 - 2F_{n1}b + F_G(b/2) \sin \alpha + F_G b \cos \alpha$$
$$- F_S(b/2) \cos \alpha - F_S a \sin \alpha.$$

from which follows

$$F_{n2} = - F_Z b/(8b) - F_G[(b/2) \sin \alpha - b \cos \alpha]/(2b)$$
$$+ F_S[(b/2) \cos \alpha + (a + 2b) \sin \alpha]/(2b) \quad \text{and}$$

$$F_{n1} = F_Z b/(8b) + F_G[(b/2) \sin \alpha + b \cos \alpha]/(2b)$$
$$- F_S[(b/2) \cos \alpha + a \sin \alpha]/(2b),$$

where the calculated value of F_S is to be used. The condition $\Sigma F_{iy} = 0 = F_{n1} + F_{n2} - F_G \cos \alpha - F_S \sin \alpha$ can then be used as a check equation.

Graphical Solution (**Fig. 25c**). The imposed forces, F_G and F_Z, are combined to form the resultant imposed force F_R, the position of which is given by the point of intersection of the lines of action of F_G and F_Z. The equilibrium between the four forces F_R, F_S, F_{n1}, F_{n2} requires that the resultant of every two forces (e.g. F_R and F_{n1} or F_{n2} and F_S) must be opposed forces (see A1.4.2). The point of intersection of the lines of action of F_R and F_{n1} or of F_{n2} and F_S produces the Culmann's auxiliary vector c, on which the two resultants must lie as opposing forces. F_R can be combined in the polygon of forces with F_{n1} to F_c, and the opposed force $-F_c$ is then resolved into F_{n2} and F_S.

Example *Carrier Element Under Variable Load: Line of Action* (**Fig. 26a**). The contact force F_A for a force F in a random load

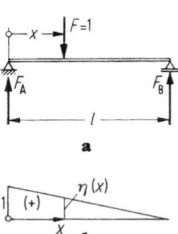

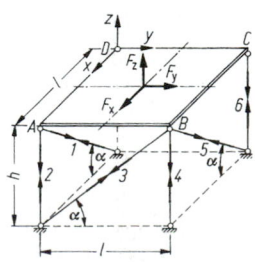

Figure 26. Load-bearing element with variable load: **a** system, **b** line of action.

Figure 27. Plate supported in space by six rods.

setting x derives from $\Sigma M_{iB} = 0 = -F_A l + F(l - x)$ as $F_A = F(l - x)/l = F\eta(x)$. For $F = 1$, it follows that $F_A = \eta(x) = (l - x)/l$. This function is a straight line (**Fig. 26b**), the ordinates of which represent the effect of the varying load $F = 1$ on the contact force F_A. If, for example, a force $F_1 = 300$ N acts at the point $x_1 = 3l/4$, then $F_A = F_1 \eta(x_1) = 300$ N $(1/4) = 75$ N. For several single forces F_i at the points x_i, it follows that $F_A = \Sigma F_i \eta(x_i)$. With a distributed load $q(x)$ in the range $a \leq x \leq b$, $F_A = \int_{x=a}^{b} q(x)\eta(x)$ dx. It follows that, for a constant distributed load q_0 on the whole carrier element length,

$$F_A = q_0 \int_{x=0}^{l} \eta(x) \, dx = (q_0/l) \int_{x=0}^{l} (l - x) \, dx$$

$$= (q_0/l) [lx - x^2/2]_{x=0}^{l} = q_0 l/2.$$

The maximum ordinate of the line of action provides the most unsatisfactory position for the supporting force.

1.6.2 Body in Space

In space, a body has six degrees of freedom (three displacements and three rotations). For stable positioning it therefore requires a six-valued bearing. If the bearing is n-valued ($n > 6$), then the system bearing is ($n - 6$) times statically indeterminate. If $n < 6$, it is statically underdeterminate, in other words movable and unstable.

Example *A Plate Supported in Space by Six Rods* (**Fig. 27**). The axial forces F_1 to F_6 are to be determined by calculation. The conditions of equilibrium are presented in the form of force or moment equations, as far as possible in such a way that only one unknown is contained.

$$\Sigma F_{iy} = 0 \quad \text{gives} \quad F_4 = F_y/\cos \alpha;$$

$$\Sigma M_{iBz} = 0 \quad \text{gives} \quad F_1 = (F_x - F_y)/(2 \cos \alpha);$$

$$\Sigma F_{ix} = 0 \quad \text{gives} \quad F_5 = (F_x + F_y)/(2 \cos \alpha);$$

$$\Sigma M_{iBx} = 0 \quad \text{gives} \quad F_2 = F_z/2 - [(F_x - F_y) \tan \alpha]/2;$$

$$\Sigma M_{iAy} = 0 \quad \text{gives} \quad F_6 = F_z/2.$$

$$\Sigma M_{L\overline{AC}} = 0 = F_3 l \sin 45° + F_5 l \sin \alpha \sin 45° + F_1 l \sin \alpha \sin 45°,$$

gives

$$F_3 = - [(F_x + 3F_y) \tan \alpha]/2.$$

Check

$$F_{iz} = -F_1 \sin \alpha - F_2 - F_3 \sin \alpha - F_4$$
$$- F_5 \sin \alpha - F_6 + F_z = \ldots = 0.$$

Example *A Shaft with Helical Teeth* (**Fig. 28**). The contact forces of the shaft are to be calculated. The shaft can rotate about the x-axis, i.e. $\Sigma M_{ix} = 0$ does not apply. The remaining five conditions of equilibrium are:

$$\Sigma F_{ix} = 0 \quad \text{gives} \quad F_{Ax} = F_{1x} - F_{2x};$$

$$\Sigma M_{iBz} = 0 \quad \text{gives} \quad F_{Ay} = -(F_{1x}r_1 + F_{1y}b + F_{2x}r_2 + F_{2y}c)/l;$$

$$\Sigma M_{iBy} = 0 \quad \text{gives} \quad F_{Az} = (F_{1z}b + F_{2z}c)/l;$$

$$\Sigma M_{iAz} = 0 \quad \text{gives} \quad F_{By} = [F_{1x}r_1 - F_{1y}a + F_{2x}r_2 + F_{2y}(l + c)]/l;$$

$$\Sigma M_{iAy} = 0 \quad \text{gives} \quad F_{Bz} = [F_{1z}a + F_{2z}(l + c)]/l.$$

The conditions $\Sigma F_{iy} = 0$ and $\Sigma F_{iz} = 0$ can be used as a check.

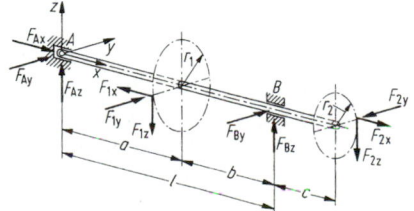

Figure 28. Shaft with inclined teeth.

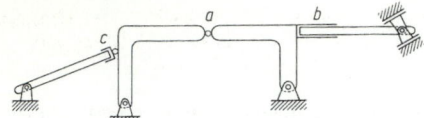

Figure 29. System of rigid bodies.

1.7 Systems of Rigid Bodies

These consist of several bodies connected by elements, such as (*a*) joints or (*b*) guides, or (*c*) guides with jointed connections (**Fig. 29**). A joint transfers forces in two directions, but transfers no moment; a guide transfers one force transverse to the guide and a moment, but transfers no parallel force; a jointed guide transfers one force transverse to the guide, but transfers no parallel force and no moment. Thus, we speak of two-value or single-value connecting elements. If i is the sum of the values of the supports, and j the sum of the values of the connecting elements, then in a system consisting of k bodies with $3k$ conditions of plane equilibrium, the condition $i + j = 3k$ is fulfilled if it is intended that a stable system should be statically determinate.

If $i + j > 3k$, then the system is statically indeterminate; i.e. if $i + j = 3k + n$, it is n times statically indeterminate. If $i + j < 3k$, the system is statically underdetermined, and is in any event unstable. For the stable system in **Fig. 29**, $i + j = 7 + 5 = 12$ and $3k = 3 \cdot 4 = 12$, i.e. the system is statically determinate. In statically determinate systems the support reactions and reactions in the connecting elements are ascertained, inasmuch as the conditions of equilibrium for the freed individual bodies are fulfilled.

Example *Three-Jointed Frame or Three-Jointed Arch* (**Fig. 30a**)

Calculated Solution. After freeing the two individual bodies (**Fig. 30b**), conditions of equilibrium for body I are:

$$\Sigma F_{ix} = 0 \quad \text{gives} \quad F_{Ax} = F_{Cx} - F_{1x}; \quad (18a)$$

$$\Sigma F_{iy} = 0 \quad \text{gives} \quad F_{Ay} = F_{1y} + F_2 - F_{Cy}; \quad (18b)$$

$$\Sigma M_{iA} = 0 = F_{Cx}H + F_{Cy}a - F_{1x}y_1 - F_{1y}x_1 - F_2 x_2; \quad (18c)$$

and those for body II are:

$$\Sigma F_{ix} = 0 \quad \text{gives} \quad F_{Bx} = F_{Cx} - F_{5x}; \quad (18d)$$

$$\Sigma F_{iy} = 0 \quad \text{gives} \quad F_{By} = F_{Cy} + F_{3y}; \quad (18e)$$

$$\Sigma M_{iB} = 0 = -F_{Cx}b + F_{Cy}b$$
$$+ F_{5x} [y_3 - (H - b)] + F_{3y} (l - x_3). \quad (18f)$$

From Eqs (18c) and (18f) are derived the pin forces F_{Cx} and F_{Cy}, used in Eqs (18a), (18b), (18d) and (18e), and then the support forces F_{Ax}, F_{Ay}, F_{Bx}, F_{By}. $\Sigma M_{iC} = 0$ is used as a check for the whole system.

Graphical Solution (**Fig. 30c**). The resultants F_{R1} and F_{R2} of the imposed forces are formed for each body, and their effect on each

other are examined. F_{R1} must be in equilibrium with the forces F_{A1} and F_{B1} in the bearings A and B. The line of action of F_{B1} must in this instance pass through the points B and C, since moment equilibrium must prevail for random points in the body II, initially still to be regarded as load-free. However, the line of action of F_{A1} must also pass through the point of intersection D of this line of action with F_{R1}, if it is intended that equilibrium should prevail between the three forces F_{R1}, F_{A1} and F_{B1} (see A1.4.2). The values of F_{A1} and F_{B1} are derived from the polygon of forces which is now ascertained. From a similar construction for F_{R2} (where the polygon of forces F_{R2} usefully applies to F_{R1}), the forces F_{A2} and F_{B2} then follow. The vectorial addition of F_{A1} and F_{A2} gives F_A, and that of F_{B1} and F_{B2} gives F_B. Finally, the equilibrium condition in the polygon of forces, $F_1 + F_2 + F_3 + F_B + F_A = 0$, is fulfilled as required.

1.8 Pin-Jointed Frames

1.8.1 Plane Frames

Pin-jointed frames consist of rods connected by freely rotating joints at nodal points. The joints are assumed to be free of friction; i.e. only forces in the direction of the rods are transferred. The friction torque that actually exists at the nodal points, and the deflection-resistant connections, lead to secondary stresses, which as a rule are negligible. The external forces act on the nodal points, or, in accordance with the lever principle on the rod, are distributed over these points.

If a pin-jointed frame has n nodes and s rods, and if it is externally statically determinate, bearing on three support forces, then $2n = s + 3$, $s = 2n - 3$ applies to a statically determinate and stable pin-jointed frame (**Fig. 31a**), since two conditions of equilibrium exist for each node; i.e., of the $2n - 3$ conditions of equilibrium, s unknown axial forces can be calculated. A pin-jointed frame with $2 < 2n - 3$ rods is statically underdeterminate and kinematically unstable (**Fig. 31b**), and a pin-jointed frame with $s > 2n - 3$ is internally statically indeterminate (**Fig. 31c**). The following laws apply to the formation of statically determinate and stable pin-jointed frames:

Starting from a stable basic triangle, new nodal points are connected one after another by two rods, **Figs 31a, 32a**.
A new pin-jointed frame is formed from two statically determinate pin-jointed frames by means of three connecting rods, the lines of action of which have no common point of intersection (**Fig. 32b**). Two rods may in this situation be replaced by a node common to both frames (**Fig. 32b**, right).
Any frame formed in accordance with these rules can be transformed into another statically determinate and stable frame by transposing rods, provided that the transposed rod is fitted between two points that are capable of moving by reason of their distance from one another (**Fig. 32c**).

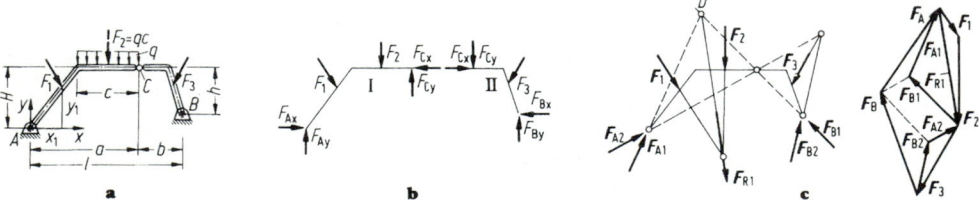

Figure 30. Three-jointed frame. **a** system, **b** freeing, **c** graphical solution.

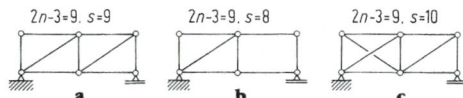

Figure 31. Framework: **a** statically determinate, **b** statically underdetermined, **c** statically indeterminate.

New stable pin-frame systems can be formed from several stable frames in accordance with the rules of rigid body systems as per A1.7 (**Fig. 32d**).

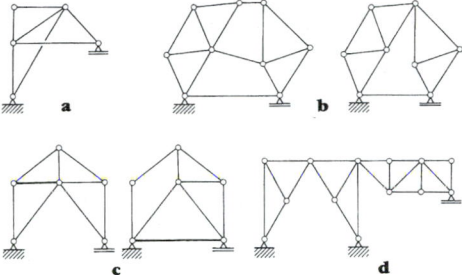

Figure 32. Framework: **a** to **d** laws of formation 1 to 4.

Determining Axial Forces

Node Intersection Procedure. In general, the axial forces s and the three supporting forces for a statically determinate pin-jointed frame are derived by setting up equilibrium conditions. $\Sigma F_{ix} = 0$ and $\Sigma F_{iy} = 0$ at all the n nodes freed by circular cuts. $2n$ linear equations are derived. If the denominator determinant of the equation system does not equal zero, the frame is stable; if it equals zero, the frame is unstable (displaceable) [1]. It frequently occurs (e.g. after the support forces have been ascertained beforehand from the conditions of equilibrium of the entire system) that an initial node exists with only two unknown axial forces, and further nodes connect to the initial node with only two unknown axial forces in each case, so that they can be calculated one after another from the equilibrium conditions without an equation system needing to be solved. In the graphical solution, this leads to what is known as the *Cremona force diagram*. If there are only two unknowns progressing from node to node, they are graphically determined from the closed polygon of forces. The arrangement of the polygons of force next to one another produces the Cremona force diagram, where all the nodes follow one another in the same direction of rotation. Axial forces that impose on one node in this situation provide a closed polygon of force in the force diagram (see example).

Ritter's Method of Sections. An analytical procedure in which an entire pin-jointed frame is freed by cutting three rods, and, after application of the three conditions of equilibrium for this part, the three unknown axial forces are calculated (see example).

Henneberg's Rod Transposition Process. Complex-structured frames can be referred back to simple structures by the rod transposition process. The axial force in the equivalent member as a consequence of external load and the force of the equivalent member must together total zero; from this is derived the force in the equivalent member. The method is also well suited for establishing the stability of a frame, since in the event of instability the force in the equivalent member approaches infinity.

Influence Lines Resulting From Variable Loads

The calculation of an axial force F_{Si} as a function of x resulting from a variable load $F = 1$ gives the application function $\eta(x)$; the graphical representation of this is called the influence line. The evaluation of several concentrated loads F_i provides the axial force $F_{Si} = \Sigma F_i \eta(x_i)$ (see example).

Example *Frame Supports* (**Fig. 33a**). Given: $F_1 = 5$ kN, $F_2 = 10$ kN, $F_3 = 20$ kN, $a = 2$ m, $b = 3$ m, $h = 2$ m, $\alpha = 45°$, $\beta = 33.69°$.
Required: axial forces.

Node Intersection Process. The unknown axial forces F_{Si} are applied as positive tensile forces (**Fig. 33b**). For nodes E the following apply:

$$\Sigma F_{iy} = 0 \quad \text{gives} \quad F_{S2} = -F_2/\sin \alpha = -14.14 \text{ kN}$$
$$\text{(pressure)};$$

$$\Sigma F_{ix} = 0 \quad \text{gives} \quad F_{S1} = F_1 - F_{S2} \cos \alpha = +15.00 \text{ kN}$$
$$\text{(tensile force)}.$$

For nodes C, the following applies:

$$\Sigma F_{ix} = 0 \quad \text{gives} \quad F_{S4} = F_{S1} = +15.00 \text{ kN}$$
$$\text{(tensile force)};$$

$$\Sigma F_{iy} = 0 \quad \text{gives} \quad F_{S5} = -F_5 = -20.00 \text{ kN}$$
$$\text{(pressure)}.$$

For nodes D, the following applies:

$$\Sigma F_{iy} = 0 \quad \text{gives} \quad F_{S5} = -(F_{S2} \sin \alpha + F_{S4})/\sin \beta$$
$$= +54.08 \text{ kN} \quad \text{(tensile force)};$$

$$\Sigma F_{ix} = 0 \quad \text{gives} \quad F_{S6} = F_{S2} \cos \alpha - F_{S5} \cos \beta$$
$$= -55.00 \text{ kN} \quad \text{(pressure)}.$$

For nodes B, the following applies:

$$\Sigma F_{iy} = 0 \quad \text{gives} \quad F_{S7} = 0;$$
$$\Sigma F_{ix} = 0 \quad \text{gives} \quad F_B = -F_{S6} = 55.00 \text{ kN}.$$

For nodes A, the following applies:

$$\Sigma F_{ix} = 0 \quad \text{gives} \quad F_{Ax} = F_{S4} + F_{S5} \cos \beta = 60.00 \text{ kN};$$
$$\Sigma F_{iy} = 0 \quad \text{gives} \quad F_{Ay} = F_{S5} \sin \beta + F_{S7} = 30.00 \text{ kN}.$$

The support forces also derive from the conditions of equilibrium in the (non-intersected) total system.

Cremona Force Diagram (**Fig. 33c**). Moving in a clockwise direction.

Ritter's Method of Sections. The axial forces F_{S4}, F_{S5} and F_{S6} are determined by Ritter's method of sections (**Fig. 33d**).

$$\Sigma M_{iD} = 0 \quad \text{gives} \quad F_{S4} = (F_2 a + F_1 b)/b = +15.00 \text{ kN};$$
$$\Sigma M_{iA} = 0 \quad \text{gives} \quad F_{S6} = -[F_2(a + b) + F_3 b]/b = -55.00 \text{ kN};$$
$$\Sigma F_{iy} = 0 \quad \text{gives} \quad F_{S5} = (F_2 + F_3)/\sin \beta = +54.08 \text{ kN}.$$

Influence Lines for Axial Force F_{S6}. The influence of a vertical variable load F_y on the axial force F_{S6} is examined (in any random position x on the top chord) (**Fig. 33e**). From

$$\Sigma M_{iA} = 0 = F_y (a + b - x) + F_{S6} b$$

it follows, with $F_y = 1$

$$\eta(x) = -1 \cdot (a + b - x)/b = -5/2 + x/(2 \text{ m})$$

in other words, a straight line (**Fig. 33f**). In view of the fact that F_1 has no influence on F_{S6} (see $\Sigma M_{iA} = 0$), the evaluation for the given loads provides

$F_{S6} = F_2\eta \ (x = 0) + F_3\eta \ (x = a)$

$= 10 \text{ kN}(-5/2) + 20 \text{ kN}(-3/2) = -55 \text{ kN}.$

1.8.2 Space Frames

In view of the fact that three conditions of equilibrium pertain per node in space, and six bearing forces are required for the stable, statically determined bearing of the pin-frame as a whole, the numeric criterion $3n = s + 6$ or $s = 3n - 6$ applies. Otherwise, methods apply to the plane frames which are analogous to the axial force calculation, etc. [2].

1.9 Cables and Chains

Cables and chains are regarded as flexible and bending; i.e. they can transfer only tensile forces. If the elongation of the individual elements is disregarded (first-order theory), then the following apply for the plane problem, as a result of a vertical uniformly distributed load, from the conditions of equilibrium at the cable element (**Fig. 34a**):

With a Given Load $q(s)$. $\Sigma F_{ix} = 0$, i.e. $dF_H = 0$, $\Sigma F_{iy} = 0$, i.e. $dF_y = q(s)$ d, i.e. $dF_v = q(s)$ ds, in other words F_H = const. and $dF_v/ds = qs$. According to **Fig. 34a**, there further applies

$$\tan \varphi = y' = F_V/F_H, \text{ i.e. } F_V = F_H y' \text{ or }$$
$$F'_V = dF_V/dx = F_H y''.$$

With $ds = \sqrt{1 + y'^2}dx$ this becomes

$$dF_V/ds = (dF_V/dx)(dx/ds) = F_H y''/\sqrt{1 + y'^2} = q(s).$$

It follows that

$$y'' = [q(s)/F_H] \sqrt{1 + y'^2}. \qquad (19)$$

At a Given Load $q(x)$. According to **Fig. 34a**, $q(s) \, ds = q(x) \, dx$, i.e.

$$q(s) = q(x) \, dx/ds = q(x) \cos \varphi = q(x)/\sqrt{1 + y'^2}$$

and therefore, according to Eq. (19),

$$y'' = q(x)/F_H. \qquad (20)$$

The solutions to these differential equations produce the catenary curve $y(x)$. The two integration constants that occur in the process and the unknown (constant) horizontal tensile force F_H follow from the boundary con-

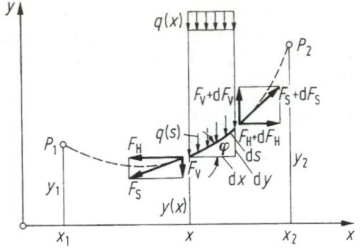

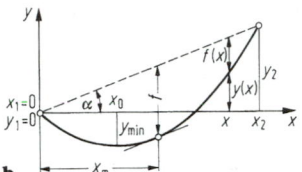

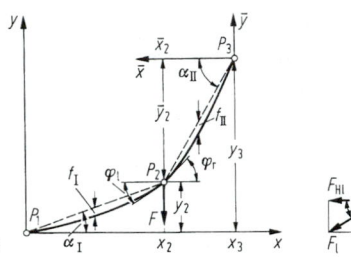

Figure 34. Cable: **a** element, **b** cable under its own weight, **c** cable under point load.

ditions $y(x = x_1) = y_1$ and $y(x = x_2) = y_2$, as well as from the given cable length $L = \int ds = \int \sqrt{1 + y'^2} \, dx$.

1.9.1 The Catenary

For a cable with a constant cross-section, with $q(s) = $ const. $= q$ from Eq. (19) with $a = F_H/q$ after separation of the variables and integration, it follows that $\text{arcsinh } y' = (x - x_0)/a$ or $y = \sinh[(x - x_0)/a]$, and thus for the catenary

$$y(x) = y_0 + a \cosh[(x - x_0)/a]. \qquad (21)$$

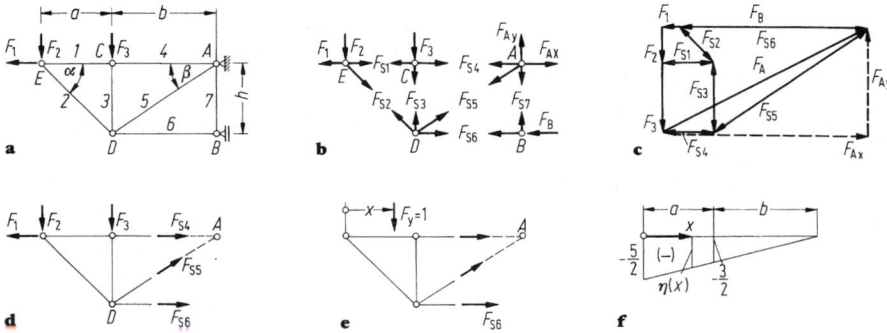

Figure 33. Framework bearings: **a** system, **b** node intersections, **c** Cremona diagram, **d** Ritter's section, **e** variable load, **f** line of action.

The extreme value of $x(y)$ follows from $y' = 0$ at the point $x = x_0$ to $y_{min} = y_0 + a$. The unknown constants x_0, y_0 and $a = F_H/q$ are derived from the following three conditions (**Fig. 34b**):

$$y(x_1 = 0) = 0 = y_0 + a \cosh(x_0/a),$$

$$y(x = x_2) = y_2 = y_0 + a \cosh[(x_2 - x_0)/a],$$

$$L = \int_{x=0}^{x_2} \sqrt{1 + \sinh^2[(x - x_0)/a]} \, dx$$

$$= a \sinh[(x_2 - x_0)/a] + a \sinh(x_0/a).$$

From this are derived

$$y_0 = -a \cosh(x_0/a), \quad x_0 = x_2/2 - a \operatorname{arctanh}(y_2/L) \text{ and}$$

$$\sinh(x_2/2a) = \sqrt{L^2 - y_2^2}/(2a).$$

From the latter (transcendental) equation, a can be calculated, and subsequently x_0 and y_0. The maximum sag f against the chord follows at the point $x_m = x_0 + a \operatorname{arcsinh}(y_2/x_2)$ for $f = y_2 x_m/x_2 - y(x_m)$.

For the forces, the following applies:

$$F_H = aq = \text{const.} \quad F_V(x) = F_H y'(x),$$

$$F_s(x) = \sqrt{F_H^2 + F_V^2(x)}. \tag{22}$$

The maximum cable force occurs at the point at which y' is a maximum, i.e. in one of the fixing points.

Example *Catenary.* Fixing point P_1 (0; 0) and P_2 (300 m; -50 m). Cable length $L = 340$ m, loading $q(x) = 30$ N/m. From the transcendental equation, by iteration, we derive $a = 179.2$ m and therefore $x_0 = 176.5$ m and $y_0 = -273.4$ m, from which the catenary is determined according to Eq. (21). The maximum sag against the chord occurs at the point $x_m = 146.8$ m and has the value $f = 67.3$ m. The horizontal tensile force amounts to $F_H = aq = 5.375$ kN = const. The greatest cable force occurs at the point P_1:

$$F_V(x = 0) = F_H [y'(x = 0)] = 6.192 \text{ kN} \quad \text{and therefore}$$

$$F_{s,max} = F_s(x = 0) = 8.20 \text{ kN}.$$

1.9.2 Cable with Uniform Load over the Span

This includes not only cables with attached uniform loads over the span $q(x) = $ const., but also cables with plane sags under dead weight, since at $q(s) = q_0 = $ const., because of $q(s) \sqrt{1 + y'^2} = q_0/\cos \varphi = q(x)$ with $\cos \varphi \approx \cos \alpha = $ const., $q(x) = $ const. $= q$ also pertains. Double integration of Eq. (20) provides $y(x) = (q/F_H)x^2/2 + C_1 x + C_2$; boundary conditions with given sag f in the centre; $y(x_1 = 0) = 0$, $y(x = x_2) = y_2$, $y(x = x_2/2) = y_2/2 - f$.

From this $C_2 = 0$, $C_1 = (y_2 - 4f)/x_2$, $F_H = qx_2^2/(8f)$ and therefore

$$y(x) = (y_2/x_2)x - (4f/x_2^2)(x_2 x - x^2) = (y_2/x_2) x - f(x),$$

where $f(x)$ is the sag against the chord (**Fig. 34b**). In addition, $F_V(x) = F_H y'(x)$ and $F_s(x) = \sqrt{F_H^2 + F_V^2(x)}$; $F_{s,max}$ applies at the point of maximum slope.

The length L of the cable follows from $L = \int_{x=0}^{x_2} \sqrt{1 + y'^2} \, dx$ with $a = F_H/q$ to

$$L = (a/2) [(C_1 + x_2/a) \sqrt{1 + (C_1 + x_2/a)}$$

$$+ \ln(C_1 + x_2/a + \sqrt{1 + (C_1 + x_2/a)^2})$$

$$- C_1 \sqrt{1 + C_1^2} - \ln(C_1 + \sqrt{1 + C_1^2})].$$

For cables with plane sags, with the chord length $l = \sqrt{x_2^2 + y_2^2}$, the approximation formula applies:

$$L \approx l[1 + 8x_2^2 f^2/(3l^4)]. \tag{23}$$

Example *Cable with plane sag.* The example from A1.9.1 is calculated as approximately a plane sagging cable. Given: P_1 (0; 0), P_2 (300 m; -50 m), $f = 67.3$ m, $q_0 = 30$ N/m. From $\tan \alpha = -50/300$ it follows that $\alpha = -9.46°$ and $\cos \alpha = 0.9864$ so that $q \approx q_0/\cos \alpha = 30.41$ N/m. It follows that $C_1 = -1.064$ and $F_H = 5.083$ kN. The cable line is accordingly

$$y(x) = -0.1667 \cdot x - 0.003 \text{ m}^{-1} (300 \text{ m} \cdot x - x^2)$$

$$= -1.064 \cdot x + 0.003 \text{ m}^{-1} \cdot x^2.$$

At the point $x = 0$, $y'_{max} = |y'(0)| = 1.064$, i.e. $F_{V,max} = F_H Y'_{max} = 5.408$ kN and therefore $F_{s,max} = 7.42$ kN.

The approximation formula Eq. (23) for the cable length then provides, with $l = 304.1$ m, the value $L \approx 342.7$ m. The results show that the approximation solution does not deviate substantially from the exact values (see A1.9.1), although the 'plane' sag applies here only to a minor degree.

1.9.3 Cable with Point Load

A cable with only plane sags with respect to the chord is considered (**Fig. 34c**, left). If x_2, y_2, x_3, y_3 are given, then with $F_{HI} = F_{HII} = F_H$ the following relationships apply:

$$q_I = q_0/\cos \alpha_I, \qquad q_{II} = q_0/\cos \alpha_{II},$$

$$f_I = q_I x_2^2/(8F_{II}), \qquad f_{II} = q_{II} \bar{x}_2^2/(8F_H),$$

$$y(x) = (y_2/x_2)x - (q_I/2F_H)(x_2 x - x^2),$$

$$\bar{y}(\bar{x}) = (\bar{y}_2/\bar{x}_2)\bar{x} - (q_{II}/2F_{II})(\bar{x}_2 \bar{x} - \bar{x}^2),$$

$$y'(x) = (y_2/x_2) - (q_I/2F_H)(x_2 - 2x),$$

$$\bar{y}'(\bar{x}) = (\bar{y}_2/\bar{x}_2) - (q_{II}/2F_H)(\bar{x}_2 - 2\bar{x}).$$

From the condition of equilibrium $\Sigma F_{iv} = 0 = F_{VI} + F - F_{Vr}$ at the node P_2 (**Fig. 34c**, right) it follows that $F_V = F_H \cdot |y'|$, taking into consideration that y' is negative, and therefore $|y'| = -y$.

$$F_H y_2/x_2 + q_I x_2/2 + F + F_H \bar{y}_2/\bar{x}_2 + q_{II} \bar{x}_2/2 = 0, \text{ i.e.}$$

$$F_H = [-q_I x_2 - q_{II} \bar{x}_2 - 2F]/[2(y_2/x_2 + \bar{y}_2/\bar{x}_2)].$$

This enables f_I and f_{II} to be calculated as shown, $F_V(x)$ and $F_s(x)$ to be calculated by Eq. (22), and L_I and L_{II} to be calculated by Eq. (23).

1.10 Centre of Gravity

The mass elements of a body of mass m are affected by the forces of gravity $d\mathbf{F}_0 = dm\mathbf{g}$, all of which are parallel to one another. The point of contact of their resultant

$$\mathbf{F}_G = \int d\mathbf{F}_G \text{ is termed the centre of gravity (Fig. 35a). Its}$$

position is determined by the condition that the moment of the resultant must be equal to that of the individual forces, i.e.

$$\mathbf{r}_s \times \mathbf{F}_G = \int \mathbf{r} \times d\mathbf{F}_G \quad \text{or, with } d\mathbf{F}_G = dF_G \mathbf{e},$$

$$(\mathbf{r}_s \mathbf{F}_G - \int \mathbf{r} \, dF_G) \times \mathbf{e} = 0, \quad \text{i.e.}$$

$$\mathbf{r}_s = (\int \mathbf{r} \, dF_G)/F_G \quad \text{or, in components,}$$

$$x_s = (1/F_G) \int x \, dF_G, \quad y_s = (1/F_G) \int y \, dF_G,$$

$$z_s = (1/F_G) \int z \, dF_G. \tag{24}$$

By analogy, with a constant acceleration of gravity g for the centre of gravity, a constant density ρ for the centre of volume, and for the centroids of areas and centres of lines in vector form, the following applies:

$$\boldsymbol{r}_s = (1/m) \int \boldsymbol{r}\, dm; \quad \boldsymbol{r}_s = (1/V) \int \boldsymbol{r}\, dV;$$

$$(25)$$

$$\boldsymbol{r}_s = (1/A) \int \boldsymbol{r}\, dA \quad \text{and} \quad \boldsymbol{r}_s = (1/s) \int \boldsymbol{r}\, ds.$$

If the bodies consist of an infinite number of parts with known centres of gravity, then the following applies in the components, for example for the centroid of area:

$$x_s = (1/A) \sum x_i A_i;$$

$$y_s = (1/A) \sum y_i A_i; \quad (26)$$

$$z_s = (1/A) \sum z_i A_i.$$

The values $\int x\, dA$ or $\sum x_i A_i$ etc. are designated as static moments. If they are zero, it also follows that $x_s = 0$ etc., i.e. the static moment in respect of an axis through the centre of gravity (centroid axis or axis of gravity) always equals zero. All axes of symmetry fulfil this condition; i.e. they are always axes of gravity.

Centres of gravity determined by integration, of homogeneous bodies and of surfaces and lines, are shown in **Tables 1** to **3**.

Example *Centre of Gravity of a Cross-Section Through a Bearing Element.* The centre of gravity for the combined bearing element cross-section is to be determined (**Fig. 35b**). The centre of gravity lies on the axis of symmetry. y_s is determined by tabular means, for which the hole is regarded as a 'negative surface'.

Table 1. Centres of gravity of homogeneous bodies

Prism, cylinder (right or oblique)	Angled circular cylinder	Cones, pyramids, (right or oblique)	Truncated pyramid or cone
$z_S = h/2$	$x_S = r^2 \tan\alpha/(4h)$ $z_S = h/2 + r^2 \tan^2\alpha/(8h)$	$z_S = h/4$	$z_S = \dfrac{h}{4}\cdot\dfrac{A_1 + 2\sqrt{A_1 A_2} + 3A_2}{A_1 + \sqrt{A_1 A_2} + A_2}$ or $z_S = \dfrac{h}{4}\cdot\dfrac{r_1^2 + 2r_1 r_2 + 3r_2^2}{r_1^2 + r_1 r_2 + r_2^2}$
Wedge	Truncated wedge	Ungula of a cylinder	Segment of a sphere
$z_S = \dfrac{h}{2}\cdot\dfrac{a_1 + a_2}{2a_1 + a_2}$	$z_S = \dfrac{h}{2}\cdot\dfrac{a_1 b_1 + a_1 b_2 + a_2 b_1 + 3a_2 b_2}{2a_1 b_1 + a_1 b_2 + a_2 b_1 + 2a_2 b_2}$	$x_S = 3\pi\, r/16$ $z_S = 3\pi h/32$	$z_S = \dfrac{3}{4}\cdot\dfrac{(2r-h)^2}{(3r-h)}$
Hemisphere	Sector of a sphere	Paraboloid of rotation	Ellipsoid
$z_S = 3r/8$ Hollow hemisphere $z_S = \dfrac{3}{8}\cdot\dfrac{r_a^4 - r_i^4}{r_a^3 - r_i^3}$	$z_S = 3r(1+\cos\alpha)/8$ $= 3(2r - h)/8$	$z_S = h/3$	$z_S = 3h/8$

Table 2. Centres of gravity of surfaces

Plane surfaces			
Triangle	Parallelogram	Trapezoid	Segment of a circle
$y_S = h/3$	$y_S = h/2$	$y_S = \dfrac{h}{3} \cdot \dfrac{a+2b}{a+b}$	$y_S = 2r \sin\alpha/(3\alpha)$ $= 2rl/(3b)$ Semicircle area: $y_S = 4r/(3\pi)$
Sector of a circle	Sector of an annulus	Parabolic surfaces	Sector of a parabola
$y_S = \dfrac{2}{3} \cdot \dfrac{r \sin^3\alpha}{\alpha - \sin\alpha \cos\alpha}$ Halbkreisfläche: $y_S = 4r/(3\pi)$	$y_S = \dfrac{2}{3} \cdot \dfrac{(r_a^3 - r_i^3)\sin\alpha}{(r_a^2 - r_i^2)\alpha}$	$x_{S1} = 3a/8$ $y_{S1} = 2h/5$ $x_{S2} = 3a/4$ $y_{S2} = 3h/10$	$y_S = 2h/5$
Segment of an ellipse	Three-dimensional surfaces		
	Zone or segment of sphere	Surface of pyramid and cone	Surface of truncated circular cone
$y_S = \dfrac{2}{3} \cdot \dfrac{b \sin^3\alpha}{\alpha - \sin\alpha \cos\alpha}$	$z_S = (r/2)(\cos\alpha_1 + \cos\alpha_2) = h_0 + h/2$ or $z_S = (r/2)(1 + \cos\alpha_2) = (h_0 + r)/2$	$z_S = h/3$	$z_S = \dfrac{h}{3} \cdot \dfrac{r_1 + 2r_2}{r_1 + r_2}$

Surface	A_i cm²	y_i cm	$y_i A_i$ cm³
1. U300	58.8	38.30	2252.0
2. 2L 100 × 14	2 × 26.2	37.02	1939.8
3. ⌀400 × 20	80.0	20.00	1600.0
4. 2L 150 × 100 × 14	2 × 33.2	4.97	330.0
5. Hole dia. 25	− 12.0	7.50	−90.0
	Σ 245.6		Σ 6031.8

$y_S = 6031.8 \text{ cm}^3/245.6 \text{ cm}^2 = 24.56 \text{ cm}$
⌀ cross section of rectangle

1.11 Friction

1.11.1 Static and Sliding Friction

Static Friction (Friction at Rest). If a body remains under the influence of a resultant force F, which holds it against a supporting base, then static friction pertains (**Fig. 36**). The distribution of the surface pressure between the body and the supporting base is in most cases unknown, and is substituted by the reactive force F_n. For reasons of equilibrium, $F_n = F_s = F \cos\alpha$ and $F_r = F_t = F \sin\alpha$, i.e. $F_r = F_n \tan\alpha$. The body remains at rest until the reactive force F_r reaches the limiting value $F_{r0} = F_n \tan\rho_0 = F_n\mu_0$, i.e. for as long as F, considered in

Table 3. Centres of gravity of lines

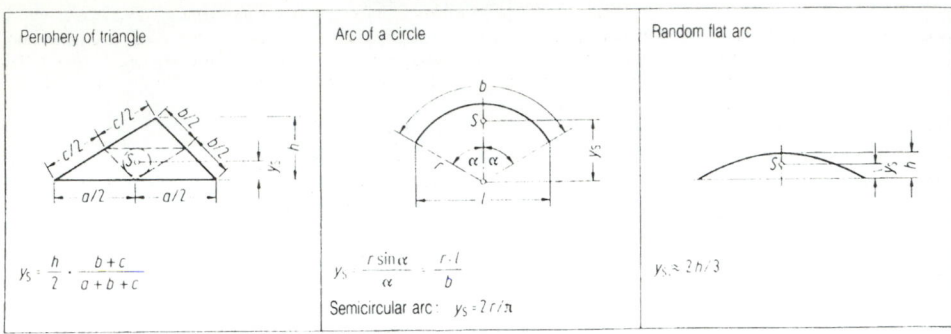

Periphery of triangle	Arc of a circle	Random flat arc
$y_S = \dfrac{h}{2} \cdot \dfrac{b+c}{a+b+c}$	$y_S = \dfrac{r \sin \alpha}{\alpha} = \dfrac{r \cdot l}{b}$ Semicircular arc: $y_S = 2r/\pi$	$y_S \approx 2h/3$

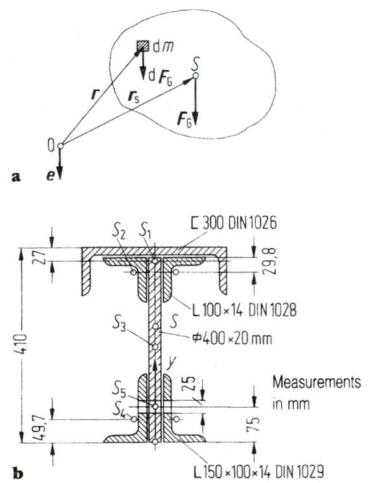

Figure 35. Centre of gravity: **a** of a body, **b** of a bearing element cross-section. ⊕ cross section of rectangle.

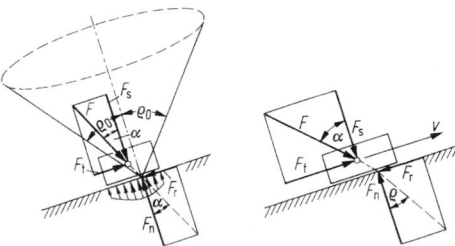

Figure 36. Static friction.   **Figure 37.** Sliding friction.

spatial terms, lies within what is known as the cone of friction with the angular aperture $2\rho_0$. The inequality

$$F_r \le F_n \tan \rho_0 = F_n \mu_0 \qquad (27)$$

applies to the reactive force F_r.

The coefficient of static friction μ_0 depends on the materials in contact, on the composition of their surfaces, on the presence of a lubricant film, on the temperature and the humidity, on the contact pressure, and on the

value of the normal pressure force; μ_0 accordingly fluctuates between specific limits and may, if necessary, be determined by experiment [3]. For reference values for μ_0 see **Table 4**.

Sliding Friction (Friction of Motion). If the static friction is overcome, and the body is set in motion, then Coulomb's law of friction applies to the friction force (**Fig. 37**):

$$F_r/F_n = \text{const.} = \tan \rho = \mu \quad \text{or} \quad F_r = \mu F_n. \qquad (28)$$

The sliding friction force is an imposed force that is directed against the speed or displacement vector. The sliding friction coefficient μ (or sliding friction angle ρ) depends not only on the influences described under A1.11.1 but particularly on the lubrication conditions (dry friction, semi-fluid friction, fluid friction; see D5.1), as well as, in part, on the velocity [4, 5]. For reference values for μ, see **Table 4**.

Example *Stability of a Ladder* (**Fig. 38**). How far may a person (gravitational force F_Q) climb a ladder (length l, angle of inclination β, weight negligible), without the ladder slipping, if the static frictional angle between ladder and wall and ground is ρ_0? How great would the forces then be at the upper and lower points of contact? The cone of friction is drawn in at the head and foot points. As long as the force F_Q lies within the shaded area, it can be resolved in an infinite number of ways so that the forces F_o and F_u at the head and foot points lie within the cone of friction, and equilibrium prevails (i.e. the problem is statically indeterminate, and the force are thus indeterminable). The limit value is attained when F_Q passes through point A (**Fig. 38**). The results follow from the illustration (owing to the right angle between the upper surface line and the lower one).

$$a = l \cos(\beta + \rho_0), \quad x = a \cos \rho_0 = l \cos \rho_0 \cos(\beta + \rho_0).$$

From the polygon of force, we read off:

$$F_o = F_Q \sin \rho_0, \quad F_{no} = F_Q \sin \rho_0 \cos \rho_0, \quad F_{ro} = F_Q \sin^2 \rho_0,$$
$$F_u = F_Q \cos \rho_0, \quad F_{nu} = F_Q \cos^2 \rho_0, \quad F_{ru} = F_Q \sin \rho_0 \cos \rho_0.$$

1.11.2 Applications of Static and Sliding Friction

Friction on a Wedge. The force F is being sought, which is required for the raising or lowering of a load at constant speed. The solution is derived most simply from the sine law at the polygon of forces; e.g. for raising the load according to **Fig. 39**,

$$\frac{F_2}{F_Q} = \frac{\sin(90° + \rho_3)}{\sin[90° - (\alpha + \rho_2 + \rho_3)]}, \quad \frac{F}{F_2} = \frac{\sin(\alpha + \rho_1 + \rho_2)}{\sin(90° - \rho_1)};$$

Table 4. Static and sliding friction coefficients

Substance pairs	Coefficient of static friction μ_o		Coefficient of sliding friction μ	
	Dry	Lubricated	Dry	Lubricated
Iron–iron			1.0	
Copper–copper			0.60 to 1.0	
Steel–steel	0.45 to 0.80	0.10	0.40 to 0.70	0.10
Chrome–chrome			0.41	
Nickel–nickel			0.39 to 0.70	
Aluminium–aluminium			0.94 to 1.35	
St.37–St.37, polished			0.15	
Steel–grey cast iron	0.18 to 0.24	0.10	0.17 to 0.24	0.02 to 0.21
Steel–white metal			0.21	
Steel–lead			0.50	
Steel–tin			0.60	
Steel–copper			0.23 to 0.29	
Brake lining–steel			0.50 to 0.60	0.20 to 0.50
Cup leather metal	0.60	0.20	0.20 to 0.25	0.12
Steel–polytetra-fluorethylene (PTFE)			0.04 to 0.22	
Steel–polyamide			0.32 to 0.45	0.10
Wood–metal	0.50 to 0.65	0.10	0.32 to 0.45	0.10
Wood–wood	0.40 to 0.65	0.16 to 0.20	0.20 to 0.40	0.04 to 0.16
Steel–ice	0.027		0.014	

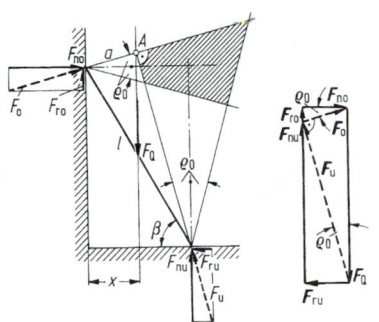

Figure 38. Ladder with cones of friction.

from which

$$F = F_Q \frac{\tan(\alpha + \rho_2) + \tan \rho_1}{1 - \tan(\alpha + \rho_2) \tan \rho_3}.$$ Accordingly,

$$F = F_Q \frac{\tan(\alpha - \rho_2) - \tan \rho_1}{1 + \tan(\alpha - \rho_2) \tan \rho_3}$$ (29)

for the lowering of the load. If $F \leq 0$, self-locking occurs; then:

$$\tan(\alpha = \rho_2) \leq \tan \rho_1 \quad \text{or} \quad \alpha \leq \rho_1 + \rho_2.$$

The wedge must then be taken out or pushed out from the other side. The efficiency of the wedge system when

raising the load is $\eta = F_0/F$; in this case, $F_0 = F_Q \cdot \tan \alpha$ is the force required without friction. For $\rho_1 = \rho_2 = \rho_3 = \rho$ there applies $F = F_Q \tan(\alpha \pm 2\rho)$; self-locking for $\alpha \leq 2\rho$, efficiency $\eta = \tan \alpha/\tan(\alpha + 2\rho)$. In the case of self-locking, $\eta = \tan 2\rho/\tan 4\rho = 0.5 - 0.5 \tan^2 2\rho < 0.5$.

Screws

Rectangular Thread (Square-Threaded Screw) (**Fig. 40a**). The torque M for uniform raising or lowering of the load is required.

$$\Sigma F_{iz} = 0 = \int dF \cos(\alpha + \rho) - F_Q, \quad F = F_Q/\cos(\alpha + \rho),$$

$$\Sigma M_{iz} = 0 = M - \int dF \sin(\alpha + \rho)r_m,$$

$$M = F_Q r_m \tan(\alpha + \rho).$$

Efficiency for raising $\eta = M_0/M = \tan \alpha/\tan(\alpha + \rho)$; M_0 is the required moment without friction. During lowering, $-\rho$ occurs instead of ρ; $M = F_Q r_m \tan(\alpha - \rho)$. Self-locking for $M \leq 0$, i.e. $\tan(\alpha - \rho) \leq 0$ and therefore $\alpha \leq \rho$. A negative moment is then required for lowering the load. For $\alpha = \rho$, it follows that

$$\eta = \tan \rho/\tan 2\rho = 0.5 - 0.5 \tan^2 \rho < 0.5.$$

Trapezoidal and Triangular Threads (**Fig. 40b**). The same equations apply as for the rectangular thread, if instead of $\mu = \tan \rho$ the friction coefficient $\mu' = \tan \rho' = \mu/\cos(\beta/2)$, is used, i.e. instead of ρ the angle of friction $\rho' = \arctan[\mu/\cos(\beta/2)]$. Proof is according to **Fig. 40b**, since instead of dF_n the force $dF_n'' = dF_n/\cos(\beta/2)$ obtains, and instead of $dF_r = \mu dF_n$ the force $dF_r' = \mu dF_n'' = [\mu/\cos(\beta/2)]dF_n = \mu' dF_n$. In this case, β is the flank angle of the thread. *Note*: For screws used for securing purposes, self-locking is required, i.e. $\alpha \leq \rho_o'$.

Rope Friction (**Fig. 41**). Sliding friction occurs when there is relative motion between rope and pulley (belt brake, bollard with running rope), static friction when a state of rest prevails between rope and pulley (belt drive, belt brake as retaining brake, bollard with rope or cable at rest). Accordingly, μ or μ_0 is taken as the coefficient of friction. Equilibrium in the normal and tangential direction at the rope element (**Fig. 41b**) gives $dF_n = F_s d\varphi$, $dF_s = dF_r$; with $dF_r = \mu dF_n$ it follows that $dF_s = \mu F_s d\varphi$. After integration over the angle of belt contact α, there follows Euler's rope friction formula

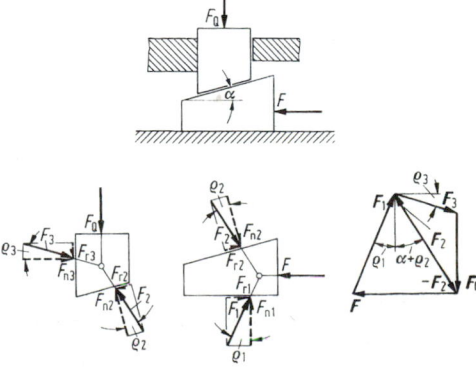

Figure 39. Friction on wedge.

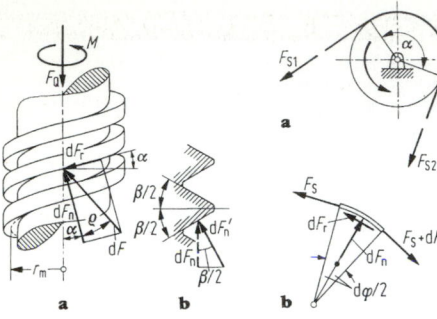

Figure 40. Friction on **a** flat-threaded and **b** sharp-threaded screw.

Figure 41. Cable friction: **a** forces, **b** element.

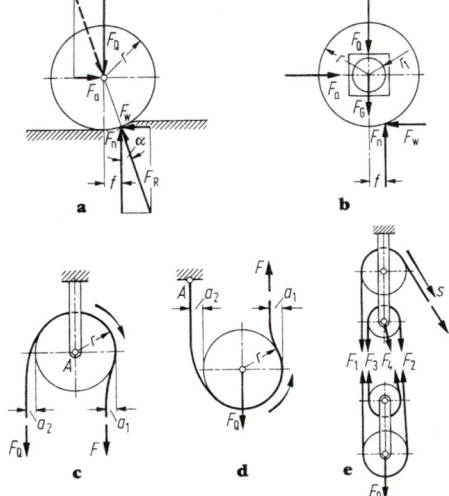

Figure 42. Resistances: **a** rolling resistance, **b** tractive resistance, **c** fixed and **d** loose rope pulley, **e** pulley block.

$$F_{S2} = F_{S1} \cdot \exp \mu\alpha. \tag{30}$$

The friction force is derived from $F_r = F_{S2} - F_{S1}$ and the friction moment from $M_r = F_r r$.

In cases in which the velocity of the rope cannot be disregarded (e.g. on belt drives), centrifugal forces $q_F = m^* v^2/r$ occur on the rope (m^* = mass per length unit of the rope). In such cases, F_S is replaced by $F_S - m^* v^2$.

1.11.3 Rolling Resistance

If a cylindrical body or a body of similar shape rolls on a supporting surface (**Fig. 42a**), the deformation of the supporting surface and of the body produces a resultant that is directed at an oblique angle, the horizontal component of which is the resisting force F_w. If the motion is uniform, the driving force F_a must retain the body in equilibrium. With $F_n = F_Q$ and $f \ll r$, i.e. $\tan \alpha \approx \sin \alpha = f/r$, it follows that

$$F_w = F_Q f/r = F_Q \mu_r$$

and $M_w = F_w r = \mu_r F_Q r = F_Q f$ as what is known as the moment of rolling friction, where $\mu_r = f/r$ is the coefficient of rolling friction. The lever arm f of the rolling friction is determined empirically. For steel wheels on rails, $f \approx 0.05$ cm, and for rolling bearings $f \approx 0.0005$ to 0.001 cm.

The sum of the rolling resistance and the bearing friction resistance is designated the *tractive resistance* (**Fig. 42b**),

$$F_{w.\text{tot}} = (F_Q + F_G) f/r + F_Q \mu_z r_1/r$$

(where F_G is the weight of the wheel and u_z the journal friction coefficient).

1.11.4 Resistance at Pulleys

As a result of the bending rigidity of the rope, a 'raising' effect occurs at the nip point by a_2 (see **Fig. 42c**), and an 'osculation' effect at the run-off point by a_1. If simultaneous consideration is taken of the bearing friction, at uniform velocity it follows for the

Fixed Pulley (**Fig. 42c**). During lifting

$$\Sigma M_A = 0 = F(r - a_1) - F_Q (r + a_2) - (F + F_Q) r_z,$$

$$F = F_Q (r + a_2 + r_z)/(r - a_1 - r_z) = F_Q/\eta.$$

η is the efficiency of the fixed pulley during lifting ($\eta \approx 0.95$). When lowering, η is replaced by $1/\eta$ (r_z = radius of the journal friction).

Free Pulley (**Fig. 42d**). During lifting

$$\Sigma M_A = 0 = F(2r + a_2 - a_1) - F_Q (r + a_2 + r_z), \text{ i.e.}$$

$$F = (F_Q/2) (r + a_2 + r_z)/(r + a_2/2 - a_1/2)$$
$$= (F_Q/2)/\eta.$$

η = effective work/work done on a system = $(F_Q s/2)/(Fs)$. By way of approximation, it is likewise taken that $\eta \approx 0.95$. During lowering η is replaced by $1/\eta$.

Pulley Tackle (**Fig. 42e**). With the results for the fixed and free pulley, $F_1 = \eta F$, $F_2 = \eta^2 F_1 = \eta^2 F$, etc. Equilibrium for the freed lower block leads to

$$\Sigma F_y = 0 = F_1 + F_2 + F_3 + F_4 - F_Q, \text{ i.e.}$$

$$F(\eta + \eta^2 + \eta^3 + \eta^4) = F_Q.$$

With

$$1 + \eta + \eta^2 + \eta^3 = (1 - \eta^4)/(1 - \eta) \quad \text{it follows that}$$

$$F = F_Q/[\eta(1 - \eta^4)/(1 - \eta)].$$

With n load-bearing rope lines, the force and the overall degree of efficiency for lifting become

$$F = F_Q/[\eta(1 - \eta^n)/(1 - \eta)] \quad \text{and}$$

$$\eta_{\text{tot}} = W_n/W_z = (F_Q s/n)/(Fs) = \eta (1 - \eta^n)/[(1 - \eta)n].$$

For lowering, η is again to be substituted by $1/\eta$.

Standards and Guidelines. DIN 1305: Mass, Weight, Weight Force, Acceleration of Fall, Terms. – DIN 1311: Oscillation Study. – DIN 1342: Viscosity of Newtonian Fluids. – DIN 5492: Formula Symbols for Fluid Mechanics. – DIN 5497: Mechanics; Rigid Bodies; Formula symbols.

2 Kinematics

Kinematics is the study of the geometrical and analytical description of the states of motion of points and bodies. It does not take account of forces and moments as causes of the motion.

2.1 Motion of a Particle

2.1.1 Introduction

Trajectory. A particle moves as a function of time in space along a trajectory. The spatial coordinates of the particle are determined by the spatial vector (**Fig. 1a**):

$$\boldsymbol{r}(t) = x(t)\boldsymbol{e}_x + y(t)\boldsymbol{e}_y + z(t)\boldsymbol{e}_z = (x(t); y(t); z(t)). \tag{1}$$

A particle has three degrees of freedom in space, two in guided motion along a surface and one degree of freedom along a line.

Velocity. The velocity vector is acquired by differentiating the spatial vector with respect to time:

$$\boldsymbol{v}(t) = \mathrm{d}\boldsymbol{r}/\mathrm{d}t = \dot{\boldsymbol{r}}(t) = \dot{x}(t)\boldsymbol{e}_x + \dot{y}(t)\boldsymbol{e}_y + \dot{z}(t)\boldsymbol{e}_z \tag{2}$$

$$= (\dot{x}(t); \dot{y}(t); \dot{z}(t)) = (v_x; v_y; v_z).$$

The velocity vector is always tangential to the trajectory, since in natural coordinates t, n, b (accompanying trihedral, where t is the tangent direction in what is termed the osculating plane, n is the normal direction in the osculating plane, and b is the binormal direction vertical to t and n; see **Fig. 1a**),

$$\boldsymbol{v}(t) = \frac{\mathrm{d}\boldsymbol{r}(t)}{\mathrm{d}t} = \frac{\mathrm{d}\boldsymbol{r}}{\mathrm{d}s}\frac{\mathrm{d}s}{\mathrm{d}t} = \boldsymbol{e}_t v \tag{3}$$

applies ($\boldsymbol{e}_t$ is the tangent unit vector). The sum of the velocity is

$$|\boldsymbol{v}| = v = \mathrm{d}s/\mathrm{d}t = \dot{s} = \sqrt{v_x^2 + v_y^2 + v_z^2} = \sqrt{\dot{x}^2 + \dot{y}^2 + \dot{z}^2}. \tag{4}$$

Acceleration. The acceleration vector is ascertained by differentiating the velocity vector with respect to time:

$$\boldsymbol{a}(t) = \frac{\mathrm{d}\boldsymbol{v}}{\mathrm{d}t} = \frac{\mathrm{d}^2\boldsymbol{r}}{\mathrm{d}t^2} = \ddot{\boldsymbol{r}}(t) = \ddot{x}(t)\boldsymbol{e}_x + \ddot{y}(t)\boldsymbol{e}_y + \ddot{z}(t)\boldsymbol{e}_z \tag{5}$$

$$= (\ddot{x}(t); \ddot{y}(t); \ddot{z}(t)) = (a_x; a_y; a_z)$$

or, in natural coordinates,

$$\boldsymbol{a}(t) = \frac{\mathrm{d}}{\mathrm{d}t}(v\boldsymbol{e}_t) = \frac{\mathrm{d}v}{\mathrm{d}t}\boldsymbol{e}_t + v \cdot \frac{\mathrm{d}\boldsymbol{e}_t}{\mathrm{d}t}.$$

With $\dfrac{\mathrm{d}\boldsymbol{e}_t}{\mathrm{d}t} = \dfrac{\mathrm{d}\boldsymbol{e}_t}{\mathrm{d}s}\dfrac{\mathrm{d}s}{\mathrm{d}t} = \dfrac{\mathrm{d}\varphi\,\boldsymbol{e}_n}{\mathrm{d}s}v = \dfrac{1}{R}\boldsymbol{e}_n v$ (see **Fig. 1b**) follows,

$$\boldsymbol{a}(t) = \dot{v}\boldsymbol{e}_t + (v^2/R)\boldsymbol{e}_n = \boldsymbol{a}_t + \boldsymbol{a}_n, \tag{6}$$

i.e. the acceleration vector always lies in the osculating plane (**Fig. 1a**). Its components in the tangential and normal direction are called tangential and normal acceleration

$$a_t = \mathrm{d}v/\mathrm{d}t = \dot{v}(t) = \ddot{s}(t) \tag{7}$$

and

$$a_n = v^2/R, \tag{8}$$

where R is the radius of curvature of the trajectory. The normal acceleration is always directed to the mid-point of the curvature; in other words, it is always a centripetal acceleration. Applicable to the value of the (resultant) acceleration vector is

$$a = |\boldsymbol{a}| = \sqrt{a_x^2 + a_y^2 + a_z^2} = \sqrt{a_t^2 + a_n^2}. \tag{9}$$

Uniform Motion. This pertains if $v(t) = \dot{s}(t) = v_o =$ const. By integration, it follows that

$$s(t) = \int \dot{s}(t)\,\mathrm{d}t = v_0 t + C_1$$

or, with the initial condition $s(t = t_1) = s_1$, from this $C_1 = s_1 - v_0 t_1$ and therefore

$$s(t) = v_0(t - t_1) + s_1.$$

Graphical representations of $v(t)$ and $s(t)$ provide the velocity–time diagram and the distance–time diagram (**Fig. 2**). From $s(t)$, by differentiation, $v(t)$ follows inversely.

Uniformly Accelerated (and Retarded) Motion (Fig. 3a). This pertains when

$$a_t(t) = \dot{v}(t) = \ddot{s}(t) = a_{t0} = \text{const., i.e.}$$

$$v(t) = a_{t0}t + C_1 \quad \text{and} \quad s(t) = a_{t0}t^2/2 + C_1 t + C_2.$$

From this, with the initial conditions $v(t = t_1) = v_1$ and $s(t = t_1) = s_1$, the constants follow:

$$C_1 = v_1 - a_{t0}t_1 \quad \text{and} \quad C_2 = s_1 - v_1 t_1 + a_{t0}t_1^2/2$$

and therefore

$$a_t(t) = a_{t0} = \text{const.}, \quad v(t) = a_{t0}(t - t_1) + v_1,$$

$$s(t) = a_{t0}(t - t_1)^2/2 + v_1(t - t_1) + s_1.$$

After elimination of $(t - t_1)$ the relationships derived are

$$t - t_1 = (v - v_1)/a_{t0}, \quad a_{t0} = (v^2 - v_1^2)/[2(s - s_1)],$$

$$v = \sqrt{v_1^2 + 2a_{t0}(s - s_1)}, \quad s = (v^2 - v_1^2)/(2a_{t0}) + s_1.$$

For the special case of $t_1 = 0$, $v_1 = 0$, $s_1 = 0$ it follows that

$$v(t) = a_{t0}t, \quad s(t) = a_{t0}t^2/2, \quad t = v/a_{t0},$$

$$a_{t0} = v^2/(2s), \quad v = \sqrt{2a_{t0}s}, \quad s = v^2/(2a_{t0}).$$

The mean velocity is derived by

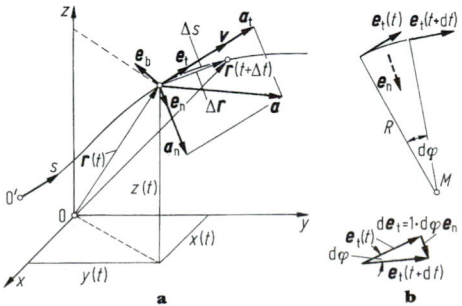

a

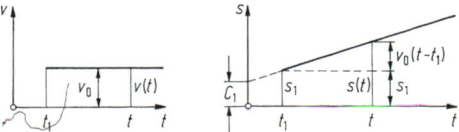

b

Figure 1. Particle motion: **a** trajectory, velocity and acceleration vector, **b** differentiation of tangent unit vector.

Figure 2. Uniform motion, diagrams of motion.

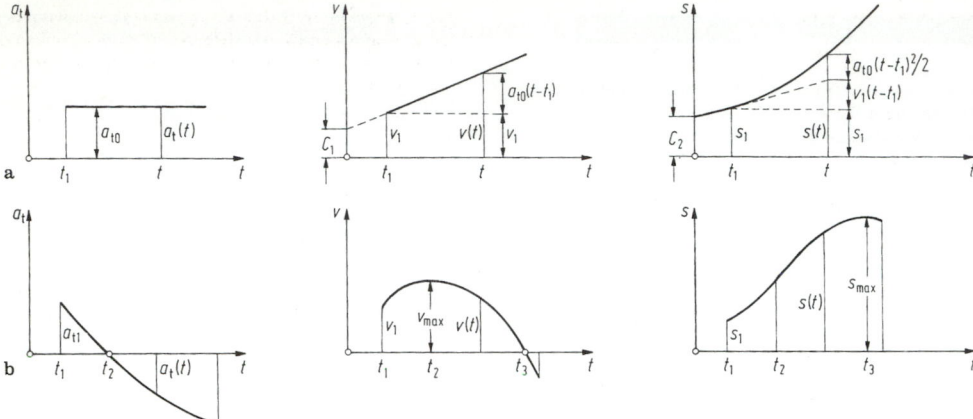

Figure 3. Accelerated motion: **a** uniform, **b** non-uniform.

$$v_m = \int_{t_1}^{t_2} v(t)\, dt/(t_2 - t_1)$$

$$= (s_2 - s_1)/(t_2 - t_1) = (v_1 + v_2)/2.$$

In all the equations, a_t can be positive or negative: Positive a_t signifies acceleration during the movement of a particle in the positive s-direction, but retardation in motion in the negative s-direction; negative a_t signifies retardation in motion in the positive s-direction, but acceleration in motion in the negative s-direction. If $s(t)$ is given, $v(t)$ and $a_t(t)$ are obtained by differentiation.

Non-uniformly Accelerated (and Retarded) Motion. This pertains when $a_t(t) = f_1(t)$ (**Fig. 3b**). Integration leads to

$$v(t) = \int a_t(t)\, dt = \int f_1(t)\, dt = f_2(t) + C_1 \quad \text{and}$$

$$s(t) = \int v(t)\, dt = \int [f_2(t) + C_1]\, dt = f_3(t) + C_1 t + C_2.$$

The constants are determined from the initial conditions $v(t = t_1) = v_1$ and $s(t = t_1) = s_1$ or equivalent conditions. From $\dot{v}(t) = a_t(t)$ it follows that, in cases in which $v(t)$ assumes an extreme value (where $v = 0$), in the a_t, t-diagram the function $a_t(t)$ passes through zero. Similarly, it follows from $\dot{s}(t) = v(t)$ that $s(t)$ has an extreme value when $v(t)$ in the v, t-diagram passes through zero. The mean velocity is given by $v_m = (s_2 - s_1)/(t_2 - t_1)$. In accordance with the graphic significance of the integral as the surface area, with a given $a_t(t)$ the values $v(t)$ and $s(t)$ are determined with the methods of graphical or numerical integration.

2.1.2 Plane Motion

Trajectory (Path), Velocity, Acceleration. The formulae of A2.1.1 apply, reduced to the two components x and y (**Fig. 4a**):

$$r(t) = x(t)e_x + y(t)e_y = (x(t); y(t)),$$

$$v(t) = \dot{x}(t)e_x + \dot{y}(t)e_y = (\dot{x}(t); \dot{y}(t)) = (v_x; v_y),$$

$$a(t) = \ddot{x}(t)e_x + \ddot{y}(t)e_y = (\ddot{x}(t); \ddot{y}(t)) = (a_x; a_y)$$

or, in natural coordinates t and n:

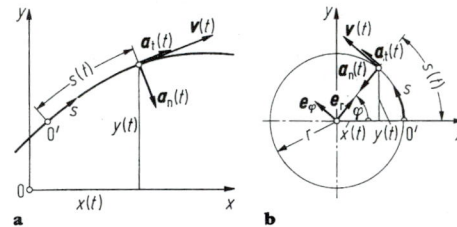

Figure 4. Motion in a plane: **a** general, **b** circle.

$$a(t) = \dot{v}(t)e_t + (v^2/R)e_n = (\dot{v}(t); v^2/R) = (a_t; a_n).$$

If the trajectory is given with $y(x)$ and the position of the particle with $s(t)$, then a connection is derived between t and x across the arc length $s(x) = \int \sqrt{1 + y'^2}\, dx$ from $s(x) = s(t)$. From this, $t(x)$ or $x(t)$ are explicitly calculable only in simple cases (see next example).

Example *Motion on a Trajectory $y(x)$* (**Fig. 4b**). The motion of a particle on the orbit $y(x) = \sqrt{r^2 - x^2}$ is examined in accordance with the distance-time law $s(t) = At^2$. According to Eqs (4), (7) and (8), we derive

$$v(t) = \dot{s}(t) = 2At, \quad a_t(t) = \dot{s}(t) = \ddot{s}(t) = 2A \quad \text{and}$$
$$a_n(t) = v^2/R = 4A^2t^2/r$$

and therefore $a(t) = \sqrt{a_t^2 + a_n^2} = 2A\sqrt{1 + 4A^2t^4/r^2}$. For the orbit, using $y' = -x/\sqrt{r^2 - x^2}$ the length of the arc is obtained as

$$s(x) = \int_x^r \sqrt{1 + y'^2}\, dx = \int_x^r \sqrt{r^2/(r^2 - x^2)}\, dx = r \arccos(x/r),$$

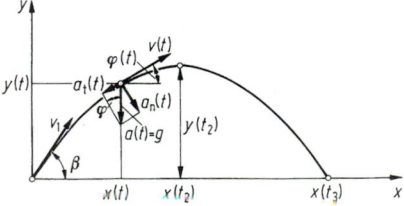

Figure 5. Offset trajectory, trajectory.

from which it follows that

$$s(x) = s(t) = At^2$$

$$t(x) = \sqrt{r \arccos(x/r)/A} \quad \text{or} \quad x(t) = r \cos(At^2/r).$$

Accordingly,

$$s(x) = r \arccos(x/r), \quad v(x) = 2\sqrt{Ar \arccos(x/r)}, \quad a_t(x) = 2A,$$

$$a_n(x) = 4A \arccos(x/r), \quad a(x) = 2A\sqrt{1 + 4[\arccos(x/r)]^2}.$$

Parametric solution:

$$x(t) = r \cos(At^2/r), \quad y(t) = \sqrt{r^2 - x^2} = r \sin(At^2/r),$$

$$v_x(t) = \dot{x}(t) = -2At \sin(At^2/r), \quad v_y(t) = \dot{y}(t) = 2At \cos(At^2/r),$$

and accordingly

$$v(t) = \sqrt{v_x^2 + v_y^2} = 2At\sqrt{\sin^2(At^2/r) + \cos^2(At^2/r)} = 2At,$$

$$a_x(t) = \dot{v}_x(t) = \ddot{x}(t) = -2A[\sin(At^2/r) + (2t^2 A/r)\cos(At^2/r)],$$

$$a_y(t) = \dot{v}_y(t) = \ddot{y}(t) = 2A[\cos(At^2/r) - (2t^2 A/r)\sin(At^2/r)].$$

from which it follows that

$$a(t) = \sqrt{a_x^2 + a_y^2} = 2A\sqrt{1 + (2t^2 A/r)^2}.$$

Example *The Oblique Trajectory* (**Fig. 5**). Non-uniform acceleration. Launch velocity is v_1 at angle β. Disregarding air resistance, gravity is the only force at work. Accordingly, $a_x(t) = 0$ and $a_y(t) = -g = \text{const.}$ Integration provides

$$v_x(t) = C_1, \quad x(t) = C_1 t + C_2$$

and

$$v_y(t) = -gt + C_3, \quad y(t) = -gt^2/2 + C_3 t + C_4.$$

Initial conditions

$$x(0) = 0, \quad y(0) = 0, \quad v_x(0) = v_1 \cos\beta, \quad v_y(0) = v_1 \sin\beta$$

give $C_2 = 0$, $C_4 = 0$, $C_1 = v_1 \cos\beta$, $C_3 = v_1 \sin\beta$ and therefore

$$x(t) = v_1 t \cos\beta, \quad y(t) = v_1 t \sin\beta - gt^2/2$$

(parametric representation).
Elimination of t produces trajectory $y = f(x)$:

$$y(x) = x \tan\beta - x^2 g/(2v_1^2 \cos^2\beta) \cdot \text{(trajectory parabola)}.$$

Velocity

$$v_x(t) = \dot{x}(t) = v_1 \cos\beta, \quad v_y(t) = \dot{y}(t) = v_1 \sin\beta - gt,$$

$$v(t) = \sqrt{(v_1 \cos\beta)^2 + (v_1 \sin\beta - gt)^2}.$$

Acceleration

$$a_x(t) = \ddot{x}(t) = 0, \quad a_y(t) = \ddot{y}(t) = -g,$$

$$a(t) = \sqrt{0 + g^2} = g = \text{const.}$$

From v_y/v_x the slope of the trajectory is obtained, and from this the natural components of acceleration (see **Fig. 5**):

$$a_n(t) = g \cos\varphi(t) \quad \text{and} \quad a_t(t) = -g \sin\varphi(t) \neq \text{const.!}$$

Time to maximum height and value of maximum height are obtained from

$$t_2 = v_1 \sin\beta/g, \quad y(t_2) = v_1^2 \sin^2\beta/(2g),$$

duration and range of trajectory from

$$t_3 = 2v_1 \sin\beta/g = 2t_2, \quad x(t_3) = t_1^2 \sin 2\beta/g.$$

Because $\sin(180° - 2\beta) = \sin 2\beta$, the same range is derived for launch angle β and $(90° - \beta)$. Maximum range at a given v_1 is achieved with the jettison angle $\beta = 45°$.

Plane Motion in Polar Coordinates. The track and location of a particle are determined by $r(t)$ and $\varphi(t)$. With the accompanying vectors $\boldsymbol{e}_r$ and $\boldsymbol{e}_\varphi$ (**Fig. 6a**), it follows that

$$\boldsymbol{r}(t) = r(t)\boldsymbol{e}_r. \tag{10}$$

By deriving the velocity vector, it follows from this that

$$\boldsymbol{v}(t) = \dot{\boldsymbol{r}}(t) = \dot{r}(t)\boldsymbol{e}_r + r(t)\dot{\boldsymbol{e}}_r = \dot{r}\boldsymbol{e}_r + \dot{\varphi}r\boldsymbol{e}_\varphi = \boldsymbol{v}_r + \boldsymbol{v}_\varphi, \tag{11}$$

since, according to **Fig. 6c**, $\dot{\boldsymbol{e}}_r = d\boldsymbol{e}_r/dt = 1 \cdot d\varphi \cdot \boldsymbol{e}_\varphi/dt = \dot{\varphi}\boldsymbol{e}_\varphi$. Here, $\dot{\varphi} = d\varphi/dt$ is the rotational speed of the radius vector r, also known as the angular velocity ω. Deriving the velocity vector gives the acceleration (**Fig. 6b**):

$$\boldsymbol{a}(t) = \dot{\boldsymbol{v}}(t) = \ddot{\boldsymbol{r}}(t) = \ddot{r}\boldsymbol{e}_r + \dot{r}\dot{\boldsymbol{e}}_r + \dot{\varphi}r\dot{\boldsymbol{e}}_\varphi + (\dot{\varphi}\dot{r} + \ddot{\varphi}r)\boldsymbol{e}_\varphi$$

$$= (\ddot{r} - \dot{\varphi}^2 r)\boldsymbol{e}_r + (\ddot{\varphi}r + 2\dot{r}\dot{\varphi})\boldsymbol{e}_\varphi = \boldsymbol{a}_r + \boldsymbol{a}_\varphi \tag{12}$$

with $\dot{\boldsymbol{e}}_\varphi = d\boldsymbol{e}_\varphi/dt = -1 \cdot d\varphi \cdot \boldsymbol{e}_r/dt = -\dot{\varphi}\boldsymbol{e}_r$, as in **Fig. 6c**. Here, $\ddot{\varphi} = \dot{\omega}$ is the change in the angular velocity of the radius vector r with time, referred to as the angular acceleration α.

Representing in Cartesian coordinates (**Fig. 6a, b**):

$$\boldsymbol{r}(t) = r \cos\varphi\boldsymbol{e}_x + r \sin\varphi\boldsymbol{e}_y = x(t)\boldsymbol{e}_x + y(t)\boldsymbol{e}_y, \tag{13}$$

$$\boldsymbol{v}(t) = \dot{\boldsymbol{r}}(t) = (\dot{r} \cos\varphi - r\dot{\varphi} \sin\varphi)\boldsymbol{e}_x$$

$$+ (\dot{r} \sin\varphi + r\dot{\varphi} \cos\varphi)\boldsymbol{e}_y = v_x\boldsymbol{e}_x + v_y\boldsymbol{e}_y, \tag{14}$$

$$\boldsymbol{a}(t) = \dot{\boldsymbol{v}}(t) = (\ddot{r} \cos\varphi - 2\dot{r}\dot{\varphi} \sin\varphi - r\dot{\varphi}^2 \cos\varphi$$

$$- r\ddot{\varphi} \sin\varphi)\boldsymbol{e}_x + (\ddot{r} \sin\varphi + 2\dot{r}\dot{\varphi} \cos\varphi - r\dot{\varphi}^2 \sin\varphi$$

$$+ r\ddot{\varphi} \cos\varphi)\boldsymbol{e}_y$$

$$= a_x\boldsymbol{e}_x + a_y\boldsymbol{e}_y. \tag{15}$$

Relationship between components in r, φ- and x, y- direction (**Fig. 6b**):

$$v_r = v_x \cos\varphi + v_y \sin\varphi, \quad v_\varphi = -v_x \sin\varphi + v_y \cos\varphi,$$

$$v_x = v_r \cos\varphi - v_\varphi \sin\varphi, \quad v_y = v_r \sin\varphi + v_\varphi \cos\varphi.$$

Analogous equations apply to the acceleration $\boldsymbol{a}$.
Resultant velocity and acceleration:

$$v = \sqrt{v_r^2 + v_\varphi^2} = \sqrt{v_x^2 + v_y^2}, \quad a = \sqrt{a_r^2 + a_\varphi^2} = \sqrt{a_x^2 + a_y^2}.$$

The acceleration vector $\boldsymbol{a}$ can also be reduced to the natural components a_t and a_n, since the direction t is given as vertical to this by the velocity vector and the direction n (**Fig. 6b**).

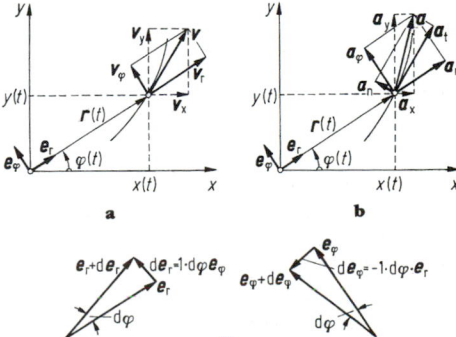

Figure 6. Polar coordinates: **a** velocities, **b** accelerations, **c** differentiation of unit vectors.

Circular Motion in a Plane (Fig. 4b). From the representation in polar coordinates, it follows, with $r =$ const., i.e. with $\dot{r} = \ddot{r} = 0$, and because now the e_φ- and e_r-direction coincide with the e_t and the negative e_n direction,

$$v(t) = \dot{\varphi} r e_t = \omega r e_t \quad \text{and}$$

$$a(t) = -\dot{\varphi}^2 r e_r + r\ddot{\varphi} e_\varphi = \omega^2 r e_n + r\alpha e_t. \qquad (16)$$

$$v = \omega r, \qquad (17)$$

$$a_t = \ddot{\varphi} r = \dot{\omega} r = \alpha r, \qquad (18)$$

$$a_n = \dot{\varphi}^2 r = \omega^2 r, \qquad (19)$$

$$a = |a| = \sqrt{a_t^2 + a_n^2} = r\sqrt{\alpha^2 + \omega^4}. \qquad (20)$$

2.1.3 Motion in Space

The equations of A2.1.1 apply. As an application, *motion in a cylindrical helix* is considered (**Fig. 7a**; see also the example in A3.2.4). Solution in cylindrical coordinates: $r_0(t)$, $\varphi(t)$, $z(t)$.

With $r_0(t) = r_0 =$ const., a random function $\varphi(t)$, and $z(t) = \varphi(t)\, h/2\pi$, $r(t) = r_0 e_r + z(t) e_z$. It follows from this, by analogy with Eqs [11] and [12] with $\dot{r}_0 = 0$, $\ddot{r}_0 = 0$,

$$v(t) = v_r + v_\varphi + v_z = \dot{\varphi} r_0 e_\varphi + \dot{z} e_z = \dot{\varphi} r_0 e_\varphi + (\dot{\varphi} h/2\pi) e_z$$

or

$$a(t) = a_r + a_\varphi + a_z = -\dot{\varphi}^2 r_0 e_r + \ddot{\varphi} r_0 e_\varphi + \ddot{z} e_z$$
$$= -\dot{\varphi}^2 r_0 e_r + \ddot{\varphi} r_0 e_\varphi + (\ddot{\varphi} h/2\pi) e_z.$$

For the values of velocity, path and acceleration, it can be derived, with the pitch angle

$$\beta = \arctan[h/(2\pi r_0)]$$

$$v(t) = |v| = \sqrt{v_r^2 + v_\varphi^2 + v_z^2} = r_0 \dot{\varphi} \sqrt{1 + h^2/(2\pi r_0)^2}$$

$$= r_0 \dot{\varphi}/\cos\beta; \quad s(t) = r_0 \varphi/\cos\beta,$$

$$a(t) = |a| = \sqrt{a_r^2 + a_\varphi^2 + a_z^2}$$

$$= r_0 \sqrt{\dot{\varphi}^4 + \ddot{\varphi}^2 [1 + h^2/(2\pi r_0)^2]}$$

$$= r_0 \sqrt{\dot{\varphi}^4 + (\ddot{\varphi}/\cos\beta)^2}.$$

Natural Components of Acceleration. For the components normal to the slope of the helix (**Fig. 7b**), the following applies:

$$-a_\varphi \sin\beta + a_z \cos\beta = -\ddot{\varphi} r_0 \sin\beta + (\ddot{\varphi} h/2\pi) \cos\beta$$
$$= -\ddot{\varphi} r_0 \sin\beta + \ddot{\varphi} r_0 \tan\beta \cos\beta$$
$$= 0.$$

Similarly, the unit binormals e_b are located in this direction, in which according to A2.1.1 there is no acceleration. That is, $e_n = -e_t$, and therefore $a_n = a_r = r_0\, \dot{\varphi}^2$.

In addition (see **Fig. 7b**),

$$a_t = a_\varphi \cos\beta + a_z \sin\beta = \ddot{\varphi} r_0 \cos\beta + \ddot{\varphi} r_0 \tan\beta \sin\beta$$
$$= r_0 \ddot{\varphi}/\cos\beta = r_0 \ddot{\varphi} \sqrt{1 + h^2/(2\pi r_0)^2}.$$

Solution in Cartesian coordinates:

$$r(t) = x(t) e_x + y(t) e_y + z(t) e_z$$
$$= r_0 \cos\varphi e_x + r_0 \sin\varphi e_y + (\varphi h/2\pi) e_z.$$

By analogy with Eqs (14) and (15), it follows that

$$v(t) = v_x e_x + v_y e_y + v_z e_z$$
$$= -r_0 \dot{\varphi} \sin\varphi e_x + r_0 \dot{\varphi} \cos\varphi e_y + (\dot{\varphi} h/2\pi) e_z,$$

$$a(t) = a_x e_x + a_y e_y + a_z e_z$$
$$= -(r_0 \dot{\varphi}^2 \cos\varphi + r_0 \ddot{\varphi} \sin\varphi) e_x$$
$$+ (r_0 \ddot{\varphi} \cos\varphi$$
$$- r_0 \dot{\varphi}^2 \sin\varphi) e_y + (\ddot{\varphi} h/2\pi) e_z,$$

from which, in turn, it follows that

$$v = |v| = \sqrt{v_x^2 + v_y^2 + v_z^2} = r_0 \dot{\varphi} \sqrt{1 + h^2/(2\pi r_0)^2} \quad \text{and}$$

$$a = |a| = \sqrt{a_x^2 + a_y^2 + a_z^2}$$
$$= r_0 \sqrt{\dot{\varphi}^4 + \ddot{\varphi}^2 [1 + h^2/(2\pi r_0)^2]}.$$

2.2 Motion of a Rigid Body

2.2.1 Rigid Body Translation

All particles describe congruent paths (**Fig. 8a**); i.e. the body does not rotate. The laws and equations of the movement of particles as under A2.1 also apply to translation, since the movement of *one* body particle is sufficient for the description.

2.2.2 Rigid-Body Rotation

The term rotation is understood to mean the rotation of a rigid body about an axis that is fixed in space (**Fig. 8b**).

Vectorial Representation. If the angular acceleration is allocated to the vector $\omega = \omega e$, i.e. if the plane OPO' rotates with ω, then the particle P, and therefore all particles, describe circular trajectories or orbits. The vector of the circumferential velocity v is derived from the vector product

$$v = \dot{r}_P = \omega e \times r_P \quad \text{with} \quad |v| = v = \omega r_P \sin\beta = \omega r;$$
$$(21)$$

v is a vector perpendicular to e and r_p in the sense of a right helix. With $r_p = r_0 + r$, it follows that

$$v = \omega e \times (r_0 + r) = \omega e \times r_0 + \omega e \times r.$$

Since e and r_0 are parallel to each other, $e \times r_0 = 0$, i.e. $|v| = v = \omega r \sin 90° = \omega r$. Accordingly

$$v = \omega r e_t. \qquad (22)$$

In Cartesian coordinates

$$v = \omega e \times r_P = \omega \times r_P = \begin{vmatrix} e_x & e_y & e_z \\ \omega_x & \omega_y & \omega_z \\ x & y & z \end{vmatrix}$$

$$= (\omega_y z - \omega_z y) e_x + (\omega_z x - \omega_x z) e_y + (\omega_x y - \omega_y x) e_z$$

$$= v_x e_x + v_y e_y + v_z e_z. \qquad (23)$$

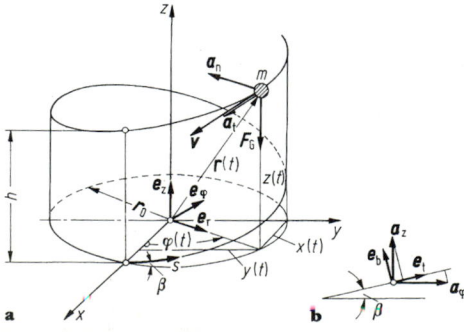

Figure 7. a and b. Mass point on a helix.

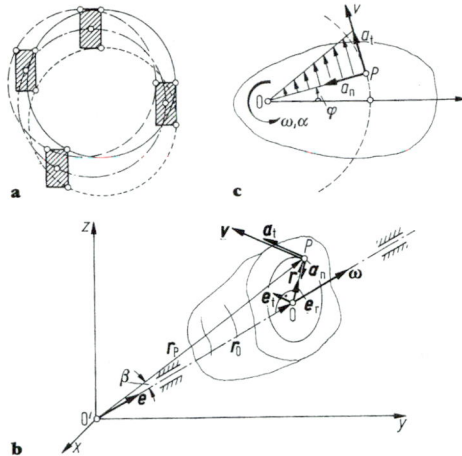

Figure 8. Motion of rigid bodies: **a** translation, **b** rotation in space, **c** rotation in the plane.

Acceleration of particle P:

$$\boldsymbol{a} = \dot{\boldsymbol{v}} = \ddot{\boldsymbol{r}}_P = (\omega\boldsymbol{e} \times \dot{\boldsymbol{r}}_P) + (\dot{\omega}\boldsymbol{e} \times \boldsymbol{r}_P)$$

$$= (\omega\boldsymbol{e} \times \boldsymbol{v}) + (\dot{\omega}\boldsymbol{e} \times \boldsymbol{r}_P). \tag{24a}$$

With $\dot{\omega} = \alpha$ (angular acceleration), in natural coordinates

$$\boldsymbol{a} = -\omega v \boldsymbol{e}_r + \alpha r_P \sin\beta\boldsymbol{e}_t = -\omega^2 r\boldsymbol{e}_r + \alpha r\boldsymbol{e}_t$$

$$= -a_n\boldsymbol{e}_r + a_t\boldsymbol{e}_t. \tag{24b}$$

In Cartesian coordinates, the following is derived from Eq. (23) by differentiation

$$\boldsymbol{a} = [-(\omega_y^2 + \omega_z^2)x + (\omega_x\omega_y - \alpha_z)y + (\omega_x\omega_z + \alpha_y)z]\boldsymbol{e}_x$$
$$+ [(\omega_x\omega_y + \alpha_z)x - (\omega_x^2 + \omega_z^2)y + (\omega_y\omega_z - \alpha_x)z]\boldsymbol{e}_y$$
$$+ [(\omega_x\omega_z - \alpha_y)x + (\omega_y\omega_z + \alpha_x)y - (\omega_x^2 + \omega_y^2)z]\boldsymbol{e}_z. \tag{25a}$$

Or, with single rotation about the z-axis,

$$\boldsymbol{a} = (-\omega_z^2 x - \alpha_z y)\boldsymbol{e}_x + (\alpha_z x - \omega_z^2 y)\boldsymbol{e}_y. \tag{25b}$$

Since, in rotation, all points describe orbits in planes perpendicular to the axis of rotation, it suffices to have the plane representation, as follows.

Plane Representation (Fig. 8c). In this situation, the axis of rotation perpendicular to the drawing plane passes through the point 0. It follows that

$$s(t) = r\varphi(t); \ v(t) = r\dot{\varphi}(t) = r\omega(t);$$

$$a_t(t) = r\ddot{\varphi}(t) = r\dot{\omega}(t) = r\alpha(t); \tag{26}$$

$$a_n(t) = r\dot{\varphi}^2(t) = r\omega^2(t).$$

i.e. all values increase linearly with r, so that, to describe the rotational movement of a rigid body, the angle of rotation $\varphi(t)$, the angular velocity $\omega(t) = \dot{\varphi}(t)$, and the angular acceleration $\alpha(t) = \dot{\omega}(t) = \ddot{\varphi}(t)$ are sufficient. In application, the number of revolutions n is frequently taken for calculation; then $\omega = 2\pi n$ and $v = 2\pi rn$. For the orbital period with $\omega =$ const., $T = 2\pi/\omega$. For uniform and non-uniform rotation, the laws of particle motion and the pertinent diagrams according to A2.1.1 apply, if in these cases a_t is replaced by α, v by ω, and s by φ.

2.2.3 General Rigid-Body Motion

Motion in Space. A body has six degrees of freedom in space: three of translation (displacement in the x-, y- and z-directions) and three of rotation (rotation about the x-, y- and z-axes). In general the motion of each point of the body is accordingly composed of translation and rotation (compound motion). For translation, it is sufficient to know the trajectory of one single fixed-body point, e.g. of the centre of gravity (see A2.2.1), to provide an adequate description; i.e. knowing the position vector $\boldsymbol{r}_0(t)$. For rotation, the description of the rotational motion through the angular velocity vector $\boldsymbol{\omega}$ about the fixed-body point is sufficient (see A2.2.2); i.e. $\boldsymbol{\omega}$ is a free vector. (**Fig. 9a**) applies

$$\boldsymbol{r}_P(t) = \boldsymbol{r}_0(t) + \boldsymbol{r}_1(t) \tag{27}$$

$$\boldsymbol{v}(t) = \dot{\boldsymbol{r}}_P(t) = \dot{\boldsymbol{r}}_0 + \dot{\boldsymbol{r}}_1 = \dot{\boldsymbol{r}}_0 + \omega(t)\boldsymbol{e} \times \boldsymbol{r}_1$$

$$= \boldsymbol{v}_0(t) + \omega r\boldsymbol{e}_\varphi = \boldsymbol{v}_0(t) + \boldsymbol{v}_1(t). \tag{28}$$

Here, $\boldsymbol{v}_0$ is the translational component and $\boldsymbol{v}_1$ the rotational component (Euler's velocity formula). From Eq. (28), it follows after multiplication by dt that

$$d\boldsymbol{r}_P = d\boldsymbol{r}_0 + d\varphi \times \boldsymbol{r}_1 = d\boldsymbol{r}_0 + r\,d\varphi\boldsymbol{e}_\varphi. \tag{29}$$

This equation (Euler's equation) states that a very small change in location of a point can be composed of a trans-

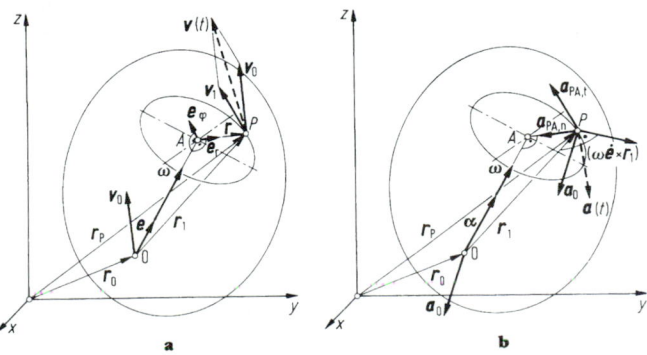

Figure 9. Motion in space: **a** velocities, **b** accelerations.

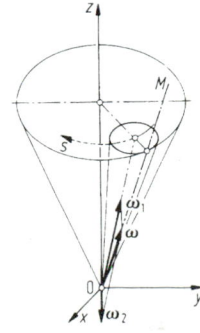

Figure 10. Spherical motion.

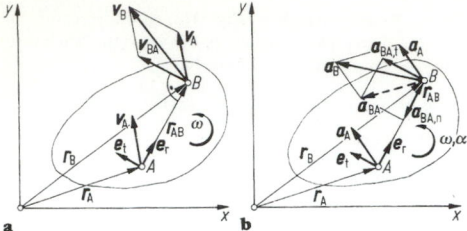

Figure 11. General plane motion: **a** velocities, **b** accelerations.

lation $d\boldsymbol{r}_0$ and of a rotation with the value $ds = r\,d\varphi$ (derived from rotation about the axis). For the acceleration of the point P of the body, it follows from Eq. (28) that

$$
\begin{aligned}
\boldsymbol{a}(t) &= \dot{\boldsymbol{v}}(t) = \ddot{\boldsymbol{r}}_P(t) \\
&= \ddot{\boldsymbol{r}}_0(t) + \omega(t)\boldsymbol{e} \times \dot{\boldsymbol{r}}_1 + (\dot{\omega}\boldsymbol{e} + \omega\dot{\boldsymbol{e}}) \times \boldsymbol{r}_1 \\
&= \boldsymbol{a}_0(t) + \omega\boldsymbol{e} \times (\omega\boldsymbol{e} \times \boldsymbol{r}_1) + \dot{\omega}\boldsymbol{e} \times \boldsymbol{r}_1 \\
&\quad + \omega\dot{\boldsymbol{e}} \times \boldsymbol{r}_1 \\
&= \boldsymbol{a}_0(t) + \omega\boldsymbol{e} \times \omega r\boldsymbol{e}_\varphi + \dot{\omega} r\boldsymbol{e}_\varphi + \omega\dot{\boldsymbol{e}} \times \boldsymbol{r}_1 \\
&= \boldsymbol{a}_0 - \omega^2 r\boldsymbol{e}_r + \alpha r\boldsymbol{e}_\varphi + \omega\dot{\boldsymbol{e}} \times \boldsymbol{r}_1 \\
&= \boldsymbol{a}_0 + \boldsymbol{a}_{PA,n} + \boldsymbol{a}_{PA,t} + (\omega\dot{\boldsymbol{e}} \times \boldsymbol{r}_1),
\end{aligned}
\tag{30}
$$

i.e. the overall acceleration is composed of the translational component $\boldsymbol{a}_0$, the normal acceleration component $\boldsymbol{a}_{PA,n}$ for rotation about O, the tangential acceleration component $\boldsymbol{a}_{PA,t}$ for rotation about O, and the component arising from the change in direction of the rotational axis (**Fig. 9b**).

Rotation About a Point (Spherical Motion). In this case, the body has only three degrees of freedom of rotation, i.e. in Eqs (27) to (30) $\boldsymbol{r}_0$, $\boldsymbol{v}_0$ and $\boldsymbol{a}_0$ are eliminated if the point O in **Fig. 9** is selected as the point of reference. The angular velocity vector is now a sliding vector, i.e. translatable only in its lines of application. The momentary axis of rotation (instantaneous axis $\overline{OM}$) describes the space cone (axode or fixed cone of instantaneous axes) during rotation of the body with reference to a coordinate system fixed in space and with reference to a coordinate system fixed in the body, describing a rolling axode that rolls on the space cone. For the angular velocity in relation to the instantaneous axis, $\omega = \omega_1 + \omega_2$ (**Fig. 10**).

Plane Motion. A body has three degrees of freedom in plane motion: two of translation (displacement in the x- and y-direction) and one of rotation (rotation about the z-axis vertical to the page). As with motion in space, ran-

dom plane motion is obtained by superposition of translation and rotation. In view of the facts that, in plane motion, the vector $\boldsymbol{e}$ is always vertical to the page, and that its direction does not change, it follows, from Eqs (27) to (30) with $\dot{\boldsymbol{e}} = 0$ and the designations according to **Fig. 11**, that

$$
\boldsymbol{r}_B(t) = \boldsymbol{r}_A(t) + \boldsymbol{r}_{AB}(t),
\tag{31}
$$

$$
\boldsymbol{v}_B = \dot{\boldsymbol{r}}_B = \dot{\boldsymbol{r}}_A + \omega\boldsymbol{e}_z \times \boldsymbol{r}_{AB}
$$

$$
= \boldsymbol{v}_A + \omega r_{AB}\boldsymbol{e}_t = \boldsymbol{v}_A + \boldsymbol{v}_{BA},
\tag{32}
$$

$$
\boldsymbol{a}_B = \ddot{\boldsymbol{r}}_B = \boldsymbol{a}_A - \omega^2 r_{AB}\boldsymbol{e}_r + \alpha r_{AB}\boldsymbol{e}_t
$$

$$
= \boldsymbol{a}_A + \boldsymbol{a}_{BA,n} + \boldsymbol{a}_{BA,t}.
\tag{33}
$$

Equations (32) and (33) are Euler's velocity equation and Euler's acceleration equation. According to these, the velocity of the points of a disc moving evenly is derived according to Eq. (32), if the velocity of a point A and the angular velocity ω of the disc are known, and acceleration according to Eq. (33) is derived if the acceleration of a point A and the angular velocity and angular acceleration α of the disc are known. The vectors $\boldsymbol{v}_B$ and $\boldsymbol{a}_B$ are often determined graphically, since the calculated solution is complicated.

Example *Crank Mechanism (**Fig. 12**).* The piston A of the crank mechanism ($l = 500$ mm, $r = 100$ mm) has, in the position sketched ($\varphi = 35°$) velocity $v_A = 1.2$ m/s and acceleration $a_A = 20$ m/s^2. To be determined for this position are: Velocity and acceleration vectors of crankpin B, angular velocities and accelerations of crank K and connecting rod S, and velocity and acceleration vectors of any random point C of the connecting rod. Velocities (**Fig. 12a**): From the vectors of Eq. (32), $\boldsymbol{v}_A$ is known from the magnitude and direction, and $\boldsymbol{v}_B$ and $\boldsymbol{v}_{BA}$ of the direction from ($\boldsymbol{v}_B \perp r$, $\boldsymbol{v}_{BA} \perp l$). From the polygon of velocities it follows that $v_B = 1.4$ m/s, $v_{BA} = 1.2$ m/s, and from this $\omega_K = v_B/r = 14$ s^{-1}, $\omega_S = v_{BA}/l = 2.4$ s^{-1}. The velocity of the point C is then derived according to Eq. (32) as $\boldsymbol{v}_C = \boldsymbol{v}_A + \boldsymbol{v}_{CA}$, where $v_{CA} = \omega_S \cdot \overline{AC} = v_{BA} \cdot \overline{AC}/l$, and is geometrically acquired from the set of radii. Accelerations (**Fig. 12b**): Euler's acceleration equation (Eq. 33) assumes the form $\boldsymbol{a}_{B,n} + \boldsymbol{a}_{B,t} = \boldsymbol{a}_A + \boldsymbol{a}_{BA,n} + \boldsymbol{a}_{BA,t}$, since B moves on a trajectory. From this, $\boldsymbol{a}_{B,n}$ are known from the magnitude ($a_{B,n} = r\omega_K^2 = 19.6$ m/s^2) and direction (in the direction of r), $\boldsymbol{a}_{B,t}$ from the direction ($\perp r$), $\boldsymbol{a}_A$ from the magnitude and direction ($a_A = 20$ m/s^2, given), $\boldsymbol{a}_{BA,n}$ from the magnitude ($a_{BA,n} = l\omega_S^2 = 2.88$ m/s^2) and direction (in the direction of l), and $\boldsymbol{a}_{BA,t}$ from the direction ($\perp l$). From this polygon of acceleration is obtained $a_{B,t} = 5.3$ m/s^2, $a_{BA,t} = 6.5$ m/s^2 and therefore $\alpha_K = a_{B,t}/r = 53$ s^{-2}, $\alpha_S = a_{BA,t}/l = 13$ s^{-2}. The acceleration of the point C is $\boldsymbol{a}_C = \boldsymbol{a}_A + \boldsymbol{a}_{CA,n} + \boldsymbol{a}_{CA,t}$, where $a_{CA,n} = \omega_S^2 \cdot \overline{AC}$ and $a_{CA,t} = \alpha_S \cdot \overline{AC}$ in each case increases linearly with $\overline{AC}$, so that $\boldsymbol{a}_{CA} = \boldsymbol{a}_{CA,n} + \boldsymbol{a}_{CA,t}$ also increases linearly with $\overline{AC}$ and must be parallel to the vector $\boldsymbol{a}_{BA}$. From the set of radii $\boldsymbol{a}_{CA}$ is derived, and geometrical composition with $\boldsymbol{a}_A$ gives $\boldsymbol{a}_C$.

Instantaneous Centre of Rotation. There is always a point round which plane motion can be instantaneously regarded as a pure rotation (instantaneous centre of rotation or polygon of velocities), i.e. a point that is tem-

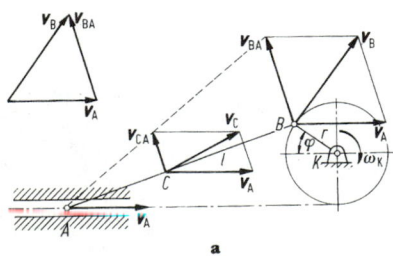

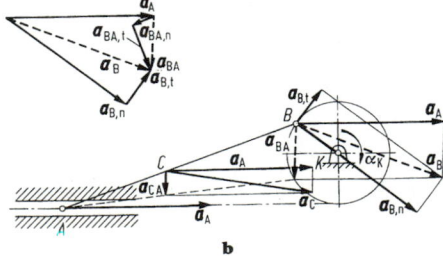

Figure 12. Crank drive: **a** velocities, **b** accelerations.

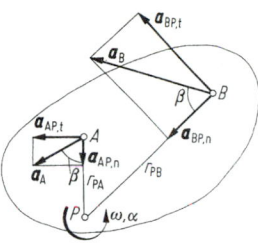

porarily at rest. This is obtained as the point of intersection of the normal units of two velocity directions (**Fig. 13a**). If, in addition to the two velocity directions, the value of a velocity is given (e.g. v_A), then the instantaneous angular velocity $\omega = v_A/r_{MA}$, and further

$$v_B = \omega r_{MB} = v_A r_{MB}/r_{MA} \quad \text{and} \quad v_C = \omega r_{MC} = v_A r_{MC}/r_{MA},$$

etc. By graphical means the value of the accelerations is derived by the method of 'rotated' velocities; i.e., v_A is rotated by 90° in the direction r_{MA} and describes the parallels to the line $\overline{AB}$. The sections separated from the radii r_{MB} and r_{MC}, $\overline{BB}'$ and $\overline{CC}'$, provide the values of the velocities v_B and v_C (radii set).

As an application, the velocities of the example of the crank mechanism are examined: With the given directions of v_A and v_B, from **Fig. 13b** is obtained the instantaneous centre of rotation M to $r_{MA} = 495$ mm, and therefore $\omega_S = v_A/r_{MA} = (1.2 \text{ m/s})/0.495 \text{ m} = 2.42 \text{ s}^{-1}$, and with $r_{MB} = 580$ mm then $v_B^2 = \omega_S r_{MB} = 1.40$ m/s. The graphical formulation by means of the rotated velocities provides the same results.

The instantaneous centre of rotation describes the herpolhode during motion in relation to a coordinate system fixed in space (fixed curve of instantaneous centres, polhode) and in relation to a fixed-body coordinate system of the polhode (roulette, herpolhode). During motion, the polhode rolls on the herpolhode. **Figure 13c** shows a slipping bar. In the fixed-space coordinate system, the equation for the herpolhode (R) is $x^2 + y^2 = l^2$ and in the fixed-body, ζ-, η system that of the polhode is $\zeta^2 + \eta^2 = (l/2)^2$; i.e., the two polhodes are circles.

Instantaneous Centre of Acceleration. It is the point P which momentarily has no acceleration. For other points A and B there then applies (**Fig. 14**) $a_A = a_{AP,t} + a_{AP,n}$, with $a_{AP,t} = \alpha r_{PA}$ and $a_{AP,n} = \omega^2 r_{PA}$, and $a_{AP,t}/a_{AP,n} = \alpha/\omega^2 = \tan \beta$, and further $a_B = a_{BP,t} + a_{BP,n}$ with $a_{BP,t} = \alpha r_{PB}$ and $a_{BP,n} = \omega^2 r_{PB}$ and also $a_{BP,t}/a_{BP,n} = \alpha/\omega^2 = \tan \beta$. The instantaneous centre of acceleration, then, is the point of intersection of two radii, which are located beneath the angle β to two given acceleration vectors.

Relative Motion. If a particle P moves at a relative velocity v_r or relative acceleration a_r on a given path relative to a body, the motion in space of which is determined by

Figure 14. Acceleration pole.

the translation of the fixed-body point O and the rotation about this particle (see motion in space, **Fig. 9**), then the problem is distinguished from that of the motion of a body inasmuch as now the vector $r_1(t)$ not only changes its direction as a result of the rotation of the vehicle, but also its direction and value as a result of relative motion. In accordance with the representation for the motion of a body in space according to Eqs (27) to (30), the following apply here (**Fig. 15a**):

$$r_P(t) = r_0(t) + r_1(t), \tag{34}$$

$$v(t) = \dot{r}_P(t) = \dot{r}_0(t) + \dot{r}_1(t),$$

$$= \dot{r}_0(t) + \omega(t)e \times r_1 + d_r r_1/dt = v_F + v_r. \tag{35}$$

Here, $d_r r_1/dt = v_r$ is the relative velocity of the particle in respect of the vehicle and $\dot{r}_0 + \omega e \times r_1 = v_F$ is the driving velocity or the speed of the vehicle. Equation (35) contains the rule: the derivative $\dot{r}_1$ of a vector in the fixed-body system with respect to time contains the component $\omega e \times r_1$ of the rotation of the system, and the so-called relative derivative in the system itself. Accordingly, the following is derived for the acceleration (**Fig. 15b**):

$$a(t) = \dot{v}(t) = \dot{v}_F + \dot{v}_r = \ddot{r}_0 + \frac{d}{dt}(\omega e \times r_1) + \frac{d}{dt}v_r$$

$$= \ddot{r}_0 + [(\dot{\omega}e + \omega\dot{e}) \times r_1] + \omega e \times \dot{r}_1 + \dot{v}_r.$$

With r_1 from Eq. (35) and $\dot{v}_r = \omega e \times v_r + d_r v_r/dt = \omega e \times v_r + d_r^2 r_1/dt^2 = \omega e \times v_r + a_r$, it follows that

$$a(t) = \ddot{r}_0 + [(\dot{\omega}e + \omega\dot{e}) \times r_1] + \omega e \times (\omega e \times r_1)$$

$$+ d_r^2 r_1/dt^2 + 2\omega e \times v_r = a_F + a_r + a_C. \tag{36}$$

The first three elements of this equation concur with those of the spatial motion of the rigid body according to Eq. (30), and therefore represent the driving acceleration or vehicle acceleration a_F. The fourth element is the relative acceleration a_r, and the last element is known as Coriolis acceleration a_C, which is additionally derived as a result of relative motion. This becomes zero when $\omega = 0$ (i.e. when the vehicle carries out a pure translation), or

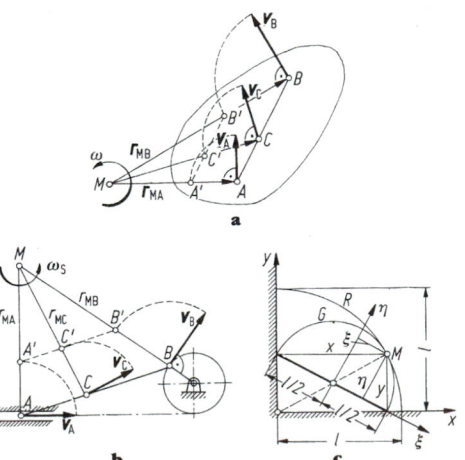

Figure 13. Instantaneous centre of rotation: **a** "rotated" velocities, **b** crank drive, **c** polar curves.

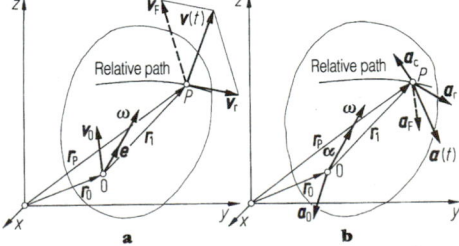

Figure 15. Relative motion: **a** velocities, **b** accelerations.

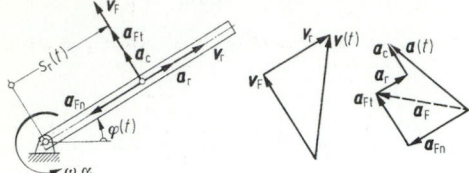

Figure 16. Motion in a rotating tube.

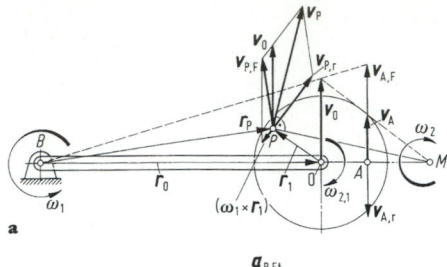

a

e and v_r are parallel to each other (relative velocity in the direction of the momentary axis of rotation) or if $v_r = 0$. The value of this is $a_C = 2\omega v_r \sin \beta$, where β is the angle between ω and v_r, and it stands perpendicular to the vectors e and v_r in the sense of a right helix. During *plane motion* (motion of a particle on a plane disc), the vectors e and v_r are perpendicular to each other; i.e. $\sin \beta = 1$, and therefore $a_C = 2\omega v_r$. Moreover, here too,

$$v = v_F + v_r \quad \text{and} \quad a = a_F + a_r + a_C, \qquad (37)$$

and therefore all the vectors are in the plane of the disc.

Example *Motion in a Rotating Tube* (**Fig. 16**). In a tube that is rotating according to the (randomly) given $\varphi(t)$ law, a mass point moves outwards relatively accordingly to the distance-time law $s_r(t)$, likewise given. For any desired point in time t, absolute velocity and acceleration of the mass point are to be determined. From $s_r(t)$ is derived, for relative velocity and acceleration, $v_r(t) = \dot{s}_r$ and $a_r(t) = \ddot{s}_r$, while the driving motion is described by $v_F(t) = s_r(t)$ $\omega(t)$ and $a_{Ft}(t) = s_r(t)\alpha(t)$, $a_{Fn}(t) = s_r(t)\omega^2(t)$, with $\omega(t) = \varphi$ and $\alpha(t) = \ddot{\varphi}$. The Coriolis acceleration is then $a_C = 2 \omega(t) v_r(t)$ with the direction vertical v_r. Absolute velocity and acceleration are obtained according to Eq. (37) by geometric composition (**Fig. 16**).

Example *Epicyclic Gears* (**Fig. 17**). The crank, rotating with angular velocity ω_1, guides the planetary wheel, which rotates at $\omega_{2,1}$ with respect to the crank, on the fixed sun wheel. According to Eq. (37), $v_P = v_F + v_r$, with the value $v_P = \omega_1 (l + r) + \omega_{2,1} r$ and accordingly $v_{P'} = \omega_1 (l - r) - \omega_{2,1} r$. Because the sun wheel is fixed, $v_{P'} = 0$, from which it follows that

$$\omega_{2,1} = \omega_1 (l - r) r \quad \text{and} \quad v_P = \omega_1 (l + r) + \omega_1 (l - r) = 2 \omega_1 l.$$

The motion of the planetary wheel can be regarded as an extension motion with $\omega_2 = \omega_1 + \omega_{2,1} = \omega_1 l/r$ around its instantaneous centre of rotation P' (point of contact of planetary wheel and sun wheel), from which likewise it follows that $v_P = \omega_2 2r = 2 \omega_1 l$. From this, it is derived in general that the resultant of two angular velocities ω_1 and ω_2 is found about parallel axes at a distance L, and with two forces (lever law), namely with $\omega_{\text{res}} = \omega_1 + \omega_2$ at a distance $l_1 = L \omega_2/(\omega_1 + \omega_2)$ from the axis of ω_1.

Example *Rotation of Two Discs About Parallel Axes* (**Fig. 18**). A bar rotating about the fixed bearing B has angular velocity ω_1 and angular acceleration α_1. A disc is mounted at its point O, which rotates in the same moment in respect of the point at $\omega_{2,1} > \omega_1$ and $\alpha_{2,1}$. The momentary velocity and acceleration vectors of a random point P are sought.

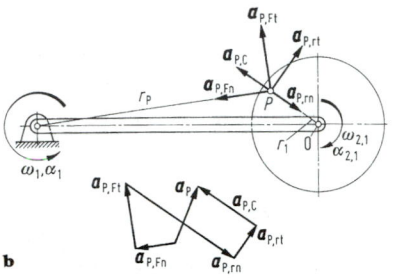

b

Figure 18. Rotation of two discs: **a** velocities, **b** accelerations.

For point A, according to Eq. (37),

$$v_A = v_{A,F} + v_{A,r} \quad \text{with} \quad v_{A,r} = \omega_{2,1} \cdot \overline{OA} \quad \text{and}$$

$$v_{A,F} = \omega_1 \cdot \overline{BA} = \omega_1 \cdot \overline{BC} + \omega_1 \cdot \overline{OA} = v_0 + \omega_1 \cdot \overline{OA},$$

so that $v_A = v_{A,F} - v_{A,r} = v_0 - (\omega_{2,1} - \omega_1) \cdot \overline{OA}$. With $\omega_{2,1} - \omega_1 = \omega_2$ and $v_0/\omega_2 = l_2 = \overline{OM}$, it follows that

$$v_A = \omega_2 (\overline{OM} - \overline{OA}) = \omega_2 \overline{MA},$$

i.e. a pure rotational velocity about the instantaneous centre of rotation M (**Fig. 18a**). Since $v_0 = r_0\omega$, and therefore $l_2 = r_0\omega_1/(\omega_{2,1} - \omega_1)$ apply, this is a confirmation of the equation regarding the combination of the angular velocities for parallel axes, where, in the case of mutually opposed rotations for ω_{res}, the difference between the two angular velocities applies, and their axis lies outside the two given axes. If both angular velocities are opposed in equal value, $\omega_{\text{res}} = 0$, and the disc performs a pure translation (in this case, with v_0). For the random point P, according to Eq. (37), there applies $v_P = v_{P,F} + v_{P,r}$, whereby, according to Eq. (35)

$$v_{P,F} = \dot{r}_0 + \omega_1 \times r_1 = v_0 + \omega_1 \times r, \quad \text{and also}$$

$$v_{P,F} = \omega \times (r_0 + r_1) = \omega \times r_P \quad \text{and} \quad v_P.r = d_r r/dt = \omega_{2,1} \times r_1.$$

This result is also derived from the pure rotation about M to $|v_P| = \omega_2 \cdot \overline{MP}$ with $v_P \perp \overline{MP}$ (**Fig. 18a**). The acceleration of point P follows from Eq. (37) or (36) $a_P = a_{P,F} + a_{P,r} + a_{P,C}$. Here, $a_{P,F} = a_{P,Fn} + a_{P,Ft}$ with $a_{P,Fn} = \omega_1^2 r_P$ and $a_{P,Ft} = \alpha_1 r_P$, $a_{P,r} = a_{P,rn} + a_{P,rt}$ with $a_{P,rn} = \omega_{2,1}^2 r_1$ and $a_{P,rn} = \alpha_{2,1} r_1$ as well as $a_{P,C} = 2\omega \times v_{P,r}$ with

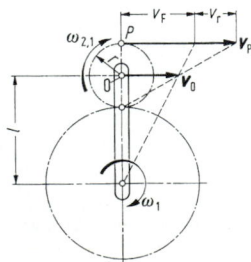

Figure 17. Epicyclic gears.

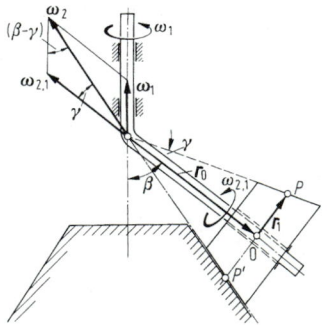

Figure 19. Bevel gear.

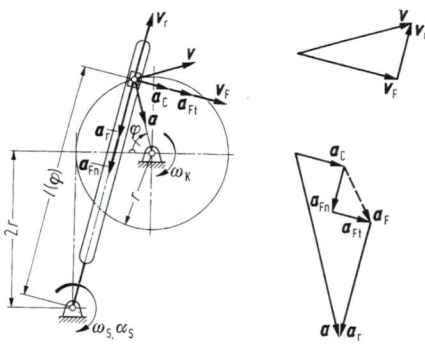

Figure 20. Rotating slider crank.

the value $a_{P,C} = 2\omega_1 v_{P,r} = 2\omega_1 \omega_{2,1} r_1$. The geometrical composition then provides $\boldsymbol{a}_P$ (**Fig. 18b**).

Example *Rotation About Two Intersecting Axes* (**Fig. 19**). An inclined axis rotates with $\boldsymbol{\omega}_1$ and guides a bevel gear that rotates with $\boldsymbol{\omega}_{2,1}$ relative to this axis, and rolls on a fixed sphere. According to Eq. (35), then,

$$\boldsymbol{v}_P = \boldsymbol{v}_F + \boldsymbol{v}_r = (\boldsymbol{r}_0 + \boldsymbol{\omega}_1 \times \boldsymbol{r}_1) + \boldsymbol{\omega}_{2,1} \times \boldsymbol{r}_1$$

$$= (\boldsymbol{\omega}_1 \times \boldsymbol{r}_0 + \boldsymbol{\omega}_1 \times \boldsymbol{r}_1) + \boldsymbol{\omega}_{2,1} \times \boldsymbol{r}_1 \quad \text{with the value}$$

$$v_P = \omega_1 r_0 \sin\beta + \omega_1 r_1 \sin(90° - \beta) + \omega_{2,1} r_1$$

$$= \omega_1 r_0 \sin\beta + \omega_1 r_1 \cos\beta + \omega_{2,1} r_1$$

and accordingly

$$v_{P'} \ \omega_1 r_0 \sin\beta - \omega_1 r_1 \cos\beta - \omega_{2,1} r_1.$$

From $v_{P'} = 0$ there follows, with $\cot\gamma = r_0/r_1$ the relationship between the angular velocities (constrained motion)

$$\omega_{2,1} = \omega_1 (\cot\gamma \sin\beta - \cos\beta) = \omega_1 \sin(\beta - \gamma)/\sin\gamma.$$

This means that the angular velocities ω_1 and $\omega_{2,1}$ may be combined to form a resultant ω_2 in accordance with $\omega_2 = \omega_1 + \omega_{2,1}$ (**Fig. 19**), since the sine law provides the result given for the polygon of vectors. The motion of the bevel gear can therefore be described as pure rotation with ω_2 about the line of contact, as an instantaneous axis. Two angular velocities, ω_1 and ω_2 about two mutually intersecting axes in general produce a resultant $\omega_{rls} = \omega_1 + \omega_2$.

Example *Rotating Slide Crank* (**Fig. 20**). The crank ($r = 150$ mm) rotates at $\omega_K = 4$ s^{-1} = const. For the position $\varphi = 75°$, angular velocity ω_K and acceleration α_s of the slide is to be determined. The sliding block P performs a relative motion with respect to the slide. Its absolute motion is given by the crank movement: $v = \omega_K r = 0.6$ m/s, $a = a_n = \omega^2{}_K r = 2.40$ m/s, and since ω_K = const., then $= \alpha_K = 0$ and $a_t = \alpha_K r = 0$. Since the relative motion is a straight line, the relative velocity $\boldsymbol{v}_r$ and acceleration $\boldsymbol{a}_r$ have the direction of the relative path, in other words, that of the slide.

According to Eq. (37), $\boldsymbol{v} = \boldsymbol{v}_F + \boldsymbol{v}_r$ follows with the known vector $\boldsymbol{v}$ and the known directions of $\boldsymbol{v}_F$ ($\perp$ slide) and $\boldsymbol{v}_r$ ($\parallel$ slide) from the polygon of velocities (**Fig. 20**) $v_r = 0.29$ m/s and $v_F = 0.52$ m/s. With $l(\varphi = 15°) \approx 460$ mm the angular velocity of the slide is $\omega_s = v_F/l = 1 \cdot 13$ s^{-1} and therefore $a_{Fn} = l \omega_s^2 = 0.59$ m/s^2 (direction of $\parallel$ circuit). The Coriolis acceleration $a_C = 2\omega_s v_r = 0.66$ m/s^2 is perpendicular to the slide, so that, with a known vector $\boldsymbol{a}$ and the known directions of $\boldsymbol{a}_{Ft}$ ($\perp$ slide) and $\boldsymbol{a}_r$ ($\parallel$ slide) according to Eq. (37), $\boldsymbol{a} = \boldsymbol{a}_{Fn} + \boldsymbol{a}_{Ft} + \boldsymbol{a}_r + \boldsymbol{a}_C$, from the polygon of acceleration (**Fig. 20**) $a_r = 1.45$ m/s^2 and $a_{Ft} = 0.50$ m/s^2 is obtained, from which it then follows that $\alpha_s = a_{Ft}/l = 1.09$ s^{-2}.

3 Dynamics

Dynamics examines the motion of mass particles, mass particle systems, bodies and body systems, in terms of the forces and moments taking effect on them, and taking into consideration the laws of kinematics.

3.1 Basic Concepts of Energy, Work, Power, Efficiency

Work. The work differential is defined as a scalar product of the force vector and the vector of the distance element (**Fig. 1a**) $dW = \boldsymbol{F}\,d\boldsymbol{r} = F\,ds\cos\beta = F_t\,ds$. According to this, only the tangential components of a force perform work. The overall work is derived with

$$dW = F_x dx + F_y dy + F_z dz \quad \text{as}$$

$$W = \int_{s_1}^{s_2} \boldsymbol{F}(s)d\boldsymbol{r} = \int_{s_1}^{s_2} F_t(s)ds = \int_{(P_1)}^{(P_2)} (F_x dx + F_y dy + F_z dz).$$

(1)

This is equal to the content of the tangential force–distance diagram (**Fig. 1b**). For $F = F_0 = $ const., it follows that $W = F_0(s_2 - s_1)$. If forces have a potential, i.e. if

$$\boldsymbol{F} = -\operatorname{grad} U = -\frac{\partial U}{\partial x}\boldsymbol{e}_x - \frac{\partial U}{\partial y}\boldsymbol{e}_y - \frac{\partial U}{\partial z}\boldsymbol{e}_z,$$

then it follows that

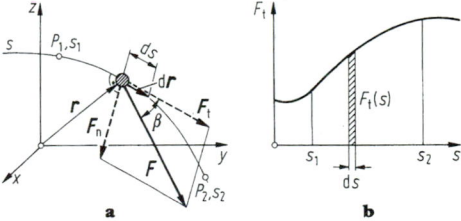

Figure 1. a Work of a force. **b** Tangential force-path diagram.

$$W = -\int_{(P_1)}^{(P_2)} \left(\frac{\partial U}{\partial x}dx + \frac{\partial U}{\partial y}dy + \frac{\partial U}{\partial z}dz\right)$$

$$= -\int_{(P_1)}^{(P_2)} dU = U_1 - U_2.$$

(2)

Work is therefore independent of the integration distance and equal to the difference between the potentials between the initial point P_1 and the final point P_2. Forces with potential are forces of gravity and spring forces (actual shape-changing forces).

Special work (Fig. 2a-d)

(a) Force of Gravity. Potential (potential energy) $U = F_G z$,

$$\text{work} \quad W_G = U_1 - U_2 = F_G(z_1 - z_2). \tag{3}$$

(b) Spring Force. Potential (potential spring energy) $U = cs^2/2$, spring force $\mathbf{F}_c = -\text{grad } U = -\dfrac{\delta U}{\delta s} \mathbf{e} = -cs\mathbf{e}$

or $|\mathbf{F}_c| = F = cs$ (c = spring rate).

$$\text{Work} \quad W_c = \int_{s_1}^{s_2} cs \, ds = c(s_2^2 - s_1^2)/2. \tag{4}$$

(c) Frictional Force. No potential, since frictional work is lost in the form of heat.

$$\text{Work} \quad W_r = \int_{s_1}^{s_2} \mathbf{F}_r(s) \, d\mathbf{r} = \int_{s_1}^{s_2} F_r(s) \cos 180° \, ds$$

$$= -\int_{s_1}^{s_2} F_r(s) \, ds. \tag{5}$$

For $F_r = \text{const.} = F_{r0}$, $W_r = -F_{r0}(s_2 - s_1)$.

(d) Torque

$$\text{Work} \quad W_M = \int_{\varphi_1}^{\varphi_2} \mathbf{M}(\varphi) \, d\varphi = \int_{\varphi_1}^{\varphi_2} M(\varphi) \cos \gamma \, d\varphi$$

$$= \int_{\varphi_1}^{\varphi_2} M_t(\varphi) \, d\varphi, \tag{6}$$

i.e. only the moment components $\mathbf{M}_t$ parallel to the axis of rotation perform work. For $\mathbf{M} = \text{const.} = M_0$,

$$W_M = M_0 \cos \gamma(\varphi_2 - \varphi_1) = M_{t0}(\varphi_2 - \varphi_1).$$

Total Work. If forces and moments are at work on a body, then

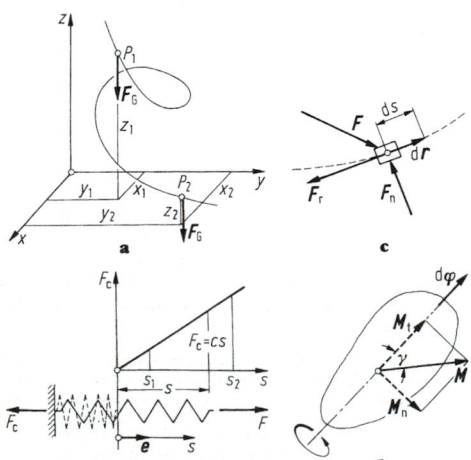

Figure 2. Work: **a** gravity, **b** spring force, **c** frictional force, **d** torque.

$$W = \int_{s_1}^{s_2} (\Sigma \, \mathbf{F}_i \, d\mathbf{r}_i) + \sum_{\varphi_1}^{\varphi_2} (\Sigma \, \mathbf{M}_i \, d\varphi_i)$$

$$= \int_{s_1}^{s_2} (\Sigma \, F_i \cos \beta_i \, ds_i) + \int_{\varphi_1}^{\varphi_2} (\Sigma \, M_i \cos \gamma_i \, d\varphi_i)$$

$$= \int_{s_1}^{s_2} (\Sigma \, F_{ti} \, ds_i) + \int_{\varphi_1}^{\varphi_2} (\Sigma \, M_{ti} \, d\varphi_i) \tag{7}$$

or for $F_i = \text{const.} = F_{i0}$ and $M_i = \text{const.} = M_{i0}$, work $W =$

$$\Sigma \, [F_{i0} \, (s_{i2} - s_{i1})] + \Sigma \, [M_{i0} \, (\varphi_{i2} - \varphi_{i1})].$$

Power is work per time unit

$$P(t) = dW/dt = \Sigma \, \mathbf{F}_i \mathbf{v}_i + \Sigma \, \mathbf{M}_i \omega_i = \Sigma \, F_{ti} v_i + \Sigma \, M_{ti} \omega_i$$

$$= \Sigma \, (F_{xi} v_{xi} + F_{yi} v_{yi} + F_{zi} v_{zi})$$

$$+ \Sigma \, (M_{xi} \omega_{xi} + M_{yi} \omega_{yi} + M_{zi} \omega_{zi}). \tag{8}$$

That is to say, for *one* force $P = F_t v$ and for *one* moment $P = M\omega$. Integration over time gives the work

$$W = \int_{t_1}^{t_2} dW = \int_{t_1}^{t_2} P(t) \, dt = P_m \, (t_2 - t_1).$$

Mean power:

$$P_m = \int_{t_1}^{t_2} P(t) \, dt/(t_2 - t_1) = W/(t_2 - t_1). \tag{9}$$

Efficiency. This is the ratio of useful work to work done, in which the latter consists of useful work and lost work:

$$\eta_m = W_n/W_z = W_n/(W_n + W_v), \tag{10}$$

where η_m = mean efficiency (work changes with the time). Instantaneous efficiency

$$\eta = \frac{dW_n}{dW_z} = \frac{dW_n}{dt} \bigg/ \frac{dW_z}{dt} = P_n/P_z = P_n/(P_n + P_v). \tag{11}$$

If there are several elements involved with the process, then:

$$\eta = \eta_1 \eta_2 \eta_3 \cdots .$$

3.2 Particle Dynamics, Straight-Line Motion of Rigid Bodies

3.2.1 Newton's Law of Motion

If several external forces are at work on a free particle (mass element, body moving in a straight line), then the resultant force $\mathbf{F}_R$ is equal to the time rate of change of the momentum vector $\mathbf{p} = m\mathbf{v}$, or, if the mass m is constant, to the product of the mass m and the acceleration vector $\mathbf{a}$ (**Fig. 3a**):

$$\mathbf{F}_{Res}^{(a)} = \mathbf{F}_R^{(a)} = \Sigma \, \mathbf{F}_i = \frac{d}{dt} (m\mathbf{v}), \tag{12}$$

$$\mathbf{F}_R^{(a)} = \Sigma \, \mathbf{F}_i = m\mathbf{a} = m \, d\mathbf{v}/dt. \tag{13}$$

The components in natural or Cartesian coordinates (**Fig. 3a, c**) are

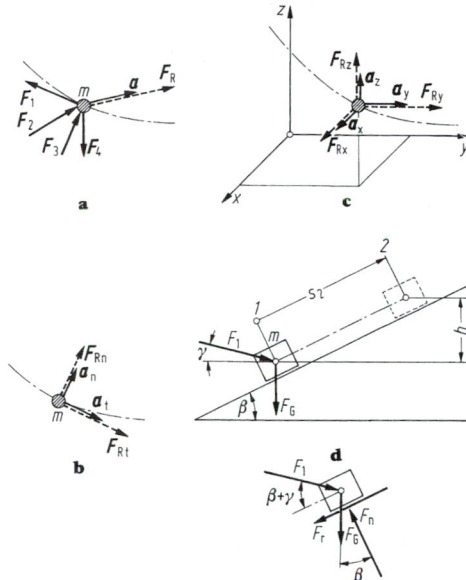

Figure 3. Dynamic basic law: **a** vectorial, **b** in natural coordinates, **c** in Cartesian coordinates, **d** mass point on inclined plane.

$$F_{Rt}^{(a)} = \Sigma F_{it} = ma_t, \quad F_{Rn}^{(a)} = \Sigma F_{in} = ma_n \quad \text{or}$$
$$F_{Rx}^{(a)} = \Sigma F_{ix} = ma_x, \quad F_{Ry}^{(a)} = \Sigma F_{iy} = ma_y, \qquad (14)$$
$$F_{Rz}^{(a)} = \Sigma F_{iz} = ma_z.$$

In solving the tasks by means of Newton's law, the particle or the body moving in a straight line must be free; i.e. all imposed forces and all reactive forces are to be applied as external forces.

Example *Particle on Inclined Plane* (**Fig. 3d**). The mass $m = 2.5$ kg is moved from the position of rest 1 by the force $F_1 = 50$ N ($\gamma = 15°$) onto the inclined plane ($\beta = 25°$) (coefficient of sliding friction $\mu = 0.3$). To be determined are the acceleration, time and velocity upon attaining position 2 ($s_2 = 4$ m). Since the motion is in a straight line, a_n must equal 0. According to Eq. (14) there applies $F_{Rn}^{(a)} = \Sigma F_{in} = 0$, i.e.

$$F_n = m g \cos\beta + F_1 \sin(\beta + \gamma) = 54.37 \text{ N} \quad \text{and}$$
$$ma_t = F_{Rt}^{(a)} = \Sigma F_{it} = F_1 \cos(\beta + \gamma) - F_G \sin\beta - F_r,$$

from which, with $F_r = \mu F_n$, it then follows that $ma_t = 11.63$ N and $a_t = 4.65$ m/s².

With the laws of uniform acceleration from rest (see A2.1.1), it is possible to derive

$$t_2 = \sqrt{2s_2/a_t} = 1.31 \text{ s} \quad \text{and} \quad v_2 = \sqrt{2a_t s_2} = 6.10 \text{ m/s}.$$

3.2.2 Energy Equation

From Eq. (13) it follows, after multiplication by $d\boldsymbol{r}$ and integration of the energy equation, that

$$W_{1,2} = \int_{(r_1)}^{(r_2)} \boldsymbol{F}_R \, d\boldsymbol{r} = \int_{(r_1)}^{(r_2)} m \frac{d\boldsymbol{v}}{dt} \, d\boldsymbol{r} = \int_{v_1}^{v_2} m\boldsymbol{v} \, d\boldsymbol{v}$$

$$= \frac{m}{2} v_2^2 - \frac{m}{2} v_1^2 = E_2 - E_1, \qquad (15)$$

i.e. the work is equal to the difference of the dynamic

energies. If all the forces involved with the process have a potential, then the process then runs without energy loss, and $W_{1,2} = U_1 - U_2$ applies (see A3.1) and from Eq. (15) there follows the energy equation

$$U_1 + E_1 = U_2 + E_2 = \text{const.} \qquad (16)$$

Example *Particle on an Inclined Plane* (**Fig. 3d**). For the example in A3.2.1 the velocity v_2 is to be determined in accordance with the work equation.

With $v_1 = 0$, i.e. $E_1 = 0$,

$$mv_2^2/2 = W_{1,2} = F_1 \cos(\beta + \gamma)s_2 - F_r s_2 - F_G b = 46.51 \text{ Nm}.$$

Accordingly,

$$v_2 = \sqrt{2 \cdot 46.51 \text{ Nm}/2.5 \text{ kg}} = 6.10 \text{ m/s}.$$

3.2.3 Momentum Equation

From Eq. (13), it follows, after multiplication by dt and integration for constant mass m, that

$$P_{1,2} = \int_{t_1}^{t_2} \boldsymbol{F}_R \, dt = \int_{v_1}^{v_2} m \, d\boldsymbol{v} = m\boldsymbol{v}_2 - m\boldsymbol{v}_1 = \boldsymbol{p}_2 - \boldsymbol{p}_1. \qquad (17)$$

The time integral of the force, known as impetus, is therefore equal to the difference in momentum.

3.2.4 D'Alembert's Principle

From Newton's law, it follows for the particle $\boldsymbol{F}_R - m\boldsymbol{a} = 0$, i.e. external forces and forces of inertia (negative mass acceleration, d'Alembert's ancillary force) form a 'state of equilibrium'. In the event of an imposed motion, the resultant $\boldsymbol{F}_R$ is composed of the imposed forces $\boldsymbol{F}_e$, the reactive forces $\boldsymbol{F}_z$, and the forces of friction $\boldsymbol{F}_r$:

$$\boldsymbol{F}_e + \boldsymbol{F}_z + \boldsymbol{F}_r - m\boldsymbol{a} = 0. \qquad (18)$$

If the principle of virtual work (see A1.4.3) is applied to this 'system of equilibrium' then it follows (**Fig. 4**) that $\delta W = (\boldsymbol{F}_e + \boldsymbol{F}_z + \boldsymbol{F}_r - m\boldsymbol{a}) \, \delta\boldsymbol{r} = 0$. Here $\delta\boldsymbol{r}$ is a displacement tangential to the path, geometrically compatible to the direction of guidance. Since the guide forces $\boldsymbol{F}_z$ are normally on the path, and therefore do not perform any work, then

$$\delta W = (\boldsymbol{F}_e + \boldsymbol{F}_r - m\boldsymbol{a}) \, \delta\boldsymbol{r} = 0; \qquad (19)$$

or, in Cartesian coordinates,

$$\delta W = (F_{ex} + F_{rx} - ma_x) \, \delta x + (F_{ey} + F_{ry} - ma_y) \, \delta y$$
$$+ (F_{ez} + F_{rz} - ma_z) \, \delta z = 0; \qquad (20)$$

or, in natural coordinates,

$$\delta W = (F_{et} - F_r - ma_t) \, \delta s = 0. \qquad (21)$$

(For the corresponding expression in cylindrical coordinates etc.: see following example.) The equations (19) to (21) represent the d'Alembert principle in the Lagrange representation. The principle is particularly well suited for tasks without friction, since it saves the calculation of the reactive forces.

Example *Particle on Helical Curve* (see A2, **Fig. 7**). The mass m moves, free of friction as a consequence of its gravitational force, below a cylindrical helical curve, which is described by cylindrical coordinates

$$r_0(t) = r_0 = \text{const.} \quad \varphi(t) \quad \text{and} \quad z(t) = (b/2\pi)\varphi(t)$$

(see A2.1.3). From

$$\boldsymbol{r}(t) = r_0\boldsymbol{e}_r + 0 \cdot \boldsymbol{e}_\varphi + z(t)\boldsymbol{e}_z \quad \text{it follows that} \quad \delta\boldsymbol{r} = r_0\delta\varphi\boldsymbol{e}_\varphi + \delta z\boldsymbol{e}_z.$$

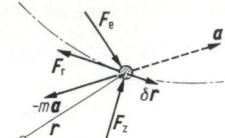

Figure 4. The d'Alembert principle.

With $F_e = F_G = -mge_z$ and $a(t) = -\dot\varphi^2 r_0 e_r + \ddot\varphi r_0 e_\varphi + \ddot\varphi(h/2\pi)e_z$ according to A2.1.3, and according to Eq. (19),

$$\delta W = (F_e - ma)\,\delta r = -mg\,\delta z - mr_0^2\ddot\varphi\,\delta\varphi - m\ddot\varphi(h/2\pi)\,\delta z = 0$$

and with $\delta z = (h/2\pi)\delta\varphi$

$$m\,\delta\varphi\,[gh/2\pi + r_0^2\ddot\varphi + h^2/(2\pi)^2 \cdot \ddot\varphi] = 0,$$

from which it follows that

$$\ddot\varphi = -\frac{gh/(2\pi r_0^2)}{1 + h^2/(2\pi r_0)^2} = \text{const.} = -A$$

Integration gives $\dot\varphi(t) = At + C_1$ and $\varphi(t) = At^2/2 + C_1 t + C_2$, where the integration constants are determined from initial conditions. The equations in A2.1.3 then provide, with $\beta = \arctan [h/(2\pi r_0)]$, the laws of motion of the particle:

$$s(t) = r_0 (-At^2/2 + C_1t + C_2)/\cos \beta,$$

$$v(t) = r_0 (-At + C_1)/\cos \beta,$$

$$a_n(t) = r_0 (-At + C_1)^2, \quad a_t(t) = -r_0 A/\cos \beta = \text{const.},$$

i.e. a uniformly accelerated (reverse) motion.

3.2.5 Angular Momentum Equation

After vectorial multiplication with a radius vector r, it follows from Eq. (13) that $r \times F_R = M_R = r \times ma$. Because $v \times mv = 0$,

$$M_R = \frac{d}{dt}(r \times mv) = \frac{dD}{dt} \tag{22}$$

Angular Momentum Equation. The lateral change of momentum $D = r \times mv$ (also known as angular momentum or moment of impulse) is equal to the resulting moment.

Now $r \times mv = m(r \times dr/dt)$ and $r \times dr = 2dA$ is a vector, the sum of which is equal to double the area of the surface covered by the vector r (**Fig. 5**). Accordingly, Eq. (22) takes the form of

$$M_R = \frac{d}{dt}\left(2m\frac{dA}{dt}\right) = 2m\frac{d^2A}{dt^2}. \tag{23}$$

Surface Equation. The resultant moment is equal to the product of twice the mass and the derivation of the surface velocity dA/dt. If F_R is a central force, i.e. always directed in the direction of r, then $M_R = r \times F_R = 0$, and

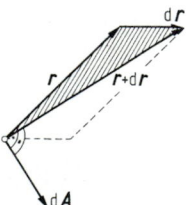

Figure 5. Law of moment of momentum (law of areas).

therefore, according to Eq. (23), $dA/dt = \text{const.}$, i.e. the surface velocity is constant, and the radius vector covers the same surfaces in the same time (Kepler's second law).
From Eq. (22) it follows that

$$\int_{t_1}^{t_2} M_R\,dt = \int_{t_1}^{t_2} d(r \times mv) = \int_{t_1}^{t_2} dD = D_2 - D_1. \tag{24}$$

Angular Momentum Conservation Law. The time integral over the moment is equal to the difference of the angles of momentum. If $M_R = 0$, then $D_1 = D_2 = \text{const.}$

3.3 Dynamics of Systems of Particles

A system of particles is a structure of n particles (**Fig. 6a**) held together by internal forces (e.g. general gravitation, spring forces, longitudinal forces). Newton's third law of action and reaction applies to the inner forces, i.e. $F_{ik}^{(i)} = F_{ki}^{(i)}$.

3.3.1 Motion of the Centroid

Newton's law for free particles and the summation across the entire structure provides

$$\sum_{i=1}^{n} F_{Ri}^{(a)} + \sum_{i,k=1}^{n} F_{ik}^{(i)} = \sum_{i=1}^{n} m_i\,a_i. \tag{25}$$

Since $\Sigma F_{ik}^{(i)} = 0$ for the internal forces, and, according to B1 Eq. (25) $\ddot r_s m = \Sigma m_i \ddot r_i$, it follows that

$$\sum_{i=1}^{n} F_{Ri}^{(a)} = ma_s \tag{26}$$

Centre of Mass Theorem. The mass mid-point (centre of gravity) of a system of particles moves as if the total mass were united in it, and all external forces were to take effect on it.

3.3.2 Energy Equation

From Eq. (25) it follows, after multiplication by dr_i (differential of small displacement vector of the ith mass point), and after integration between two points in time 1 and 2, that

$$\sum \int_{(1)}^{(2)} F_{Ri}^{(a)}\,dr_i + \sum \int_{(1)}^{(2)} F_{ik}^{(i)}\,dr_i = \sum \int_{(1)}^{(2)} m_i v_i\,dv_i$$

$$W_{1,2}^{(a)} + W_{1,2}^{(i)} = \Sigma (m_i/2)(v_{i2}^2 - v_{i1}^2) \tag{27}$$

Work Law. The work of the external and internal forces in the system of particles (where the reactive forces are again zero) is equal to the difference between the dynamic energies. The internal forces provide no work if there are rigid connections between the particles. If all the forces involved have a potential, then the energy equation (Eq. 16) applies.

Example *Systems of Particles on Inclined Planes* (**Fig. 6b**). The two masses, connected by an inextensible cable, are drawn out of the position of rest by the force F along the inclined planes. The value being sought is the velocity after covering a distance s_1. After freeing, the normal pressure forces (reactive forces) occur at $F_{n2} = F_{G2}\cos\beta_2$ and $F_{n1} = F_{G1}\cos\beta - F\sin\beta_1$, as a precondition for the mass not rising, there must apply $F \le F_{G1}\cot\beta$. Accordingly, the friction forces are $F_{r2} = \mu_2 F_{n2}$ and $F_{r1} = \mu_1 F_{n1}$. The work equation (Eq. 27) provides

$$F \cos \beta_1 s_1 + F_{G_1} b_1 - F_{r1} s_1 - F_s s_1 + F_s s_2 - F_{G_2} b_2 - F_{r2} s_2$$

$$= m_1 v_1^2/2 + m_2 v_2^2/2,$$

and with $s_2 = s_1$, $v_2 = v_1$ (inextensible chain) and with $b = s_1 \sin \beta_1$ and $b_2 = s_2 \sin \beta_2$ it then follows that

$$v_1^2 = 2s_1 \, [F \cos \beta_1 + F_{G_1} \sin \beta_1 - \mu_1 \, (F_{G_1} \cos \beta_1 - F \sin \beta_1)$$

$$- F_{G_2} \sin \beta_2 - \mu_2 F_{G_2} \cos \beta_2]/(m_1 + m_2).$$

3.3.3 Momentum Equation

From Eq. (25) it follows, after multiplication by dt and integration, that:

$$\sum \int_{t_1}^{t_2} \boldsymbol{F}_{Ri}^{(a)} \, dt + \sum \int_{t_1}^{t_2} \boldsymbol{F}_{ik}^{(i)} \, dt = \sum \int_{t_1}^{t_2} m_i \, \frac{d\boldsymbol{v}_i}{dt} \, dt$$

$$= \sum m_i \, (\boldsymbol{v}_{i2} - \boldsymbol{v}_{i1}) = \boldsymbol{p}_2 - \boldsymbol{p}_1.$$

Since $\sum \int_{t_1}^{t_2} \boldsymbol{F}_{ik}^{(i)} \, dt = 0$ and according to A1 Eq. (25) $m\boldsymbol{v}_s$

$= \sum m_i \boldsymbol{v}_i$, it follows that

$$\boldsymbol{p}_2 - \boldsymbol{p}_1 = \sum \int_{t_1}^{t_2} \boldsymbol{F}_{Ri}^{(a)} \, dt - \sum m_i(\boldsymbol{v}_{i2} - \boldsymbol{v}_{i1}) = m(\boldsymbol{v}_{s2} - \boldsymbol{v}_{s1})$$
(28)

Momentum Law. The time integral across the external forces of the system is equal to the difference of all momenta or equal to the difference of the centre of gravity momenta. If there are no external forces present, then it follows from Eq. (28) that

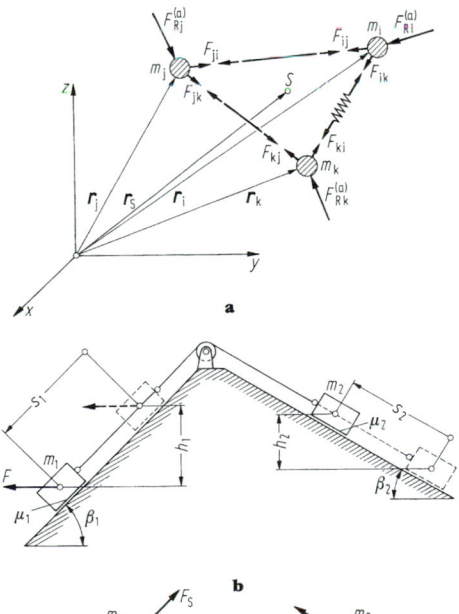

a

b

Figure 6. System of particles: **a** general, **b** two masses.

$$\sum m_i \boldsymbol{v}_{i1} = \sum m_i \boldsymbol{v}_{i2} = \text{const.} \quad \text{or}$$

$$m\boldsymbol{v}_{s1} = m\boldsymbol{v}_{s2} = \text{const.}, \tag{29}$$

i.e. the total momentum is retained.

Example *Systems of Particles and Momentum Law* (**Fig. 7**). A spring (spring rate c), which is prestressed by the value s_1, thrusts the masses m_1 and m_2 apart. Their velocities are to be determined. Disregarding friction forces during the relaxation process of the spring, there are no external forces in the direction of movement, so that with $v_{11} = 0$ and $v_{21} = 0$, it follows from Eq. (29) that $m_2 v_{12} - m_2 v_{22} = 0$, i.e. $m_1 v_{12} = m_2 v_{22}$. Accordingly, the energy equation, Eq. (16), provides $cs_1^2/2 = +m_1 v_{12}^2/2 + m_2 v_{22}^2/2$, and then

$$v_{12} = \sqrt{cs_1^2/(m_1 + m_1^2/m_2)} \quad \text{and} \quad v_{22} = \sqrt{cs_1^2/(m_2 + m_2^2/m_1)}.$$

3.3.4 D'Alembert's Principle, Constrained Motion

From Eq. (25) it follows that $\sum \boldsymbol{F}_{Ri}^{(a)} + (- \sum m_i \boldsymbol{a}_i) = - \sum \boldsymbol{F}_{ik}^{(i)}$. Because $\sum \boldsymbol{F}_{ik}^{(i)} = 0$, the lost forces, that is, the entirety of the external forces, plus the forces of inertia (negative mass accelerations), on the system of particles are in equilibrium:

$$\sum \boldsymbol{F}_{Ri}^{(a)} + (- \sum m_i \boldsymbol{a}_i) = 0. \tag{30}$$

The principle is well suited in this formulation, especially for calculating intersecting loads of dynamically stressed systems, on which the intersecting loads are introduced as external forces. In the event of imposed motions, the resultant is composed of the external forces on the individual systems of particles from the imposed forces $\boldsymbol{F}_i^{(e)}$, the reactive forces $\boldsymbol{F}_i^{(z)}$ and the frictional forces $\boldsymbol{F}_i^{(r)}$. For rigid systems, it follows from Eq. (30), using the equilibrium principle of virtual work (see A1.4.3), in which every particle is allocated a displacement $\delta \boldsymbol{r}_i$ compatible with the geometrical connections, that

$$\sum [\boldsymbol{F}_{Ri}^{(e)} + \boldsymbol{F}_{Ri}^{(z)} + \boldsymbol{F}_{Ri}^{(r)} + (-m_i \boldsymbol{a}_i)] \, \delta \boldsymbol{r}_i = 0.$$

Since the reactive forces do not exert any work during displacements, the d'Alembert principle follows in the Lagrange formulation:

$$\sum [\boldsymbol{F}_{Ri}^{(e)} + \boldsymbol{F}_{Ri}^{(r)} + (-m_i \boldsymbol{a}_i)] \, \delta \boldsymbol{r}_i = 0. \tag{31}$$

In Cartesian or natural coordinates, Eq. (31) reads according to Eqs (20) and (21) for the mass particle. This principle is especially well suited for calculating the state of acceleration of imposed motion without friction, since it saves calculating the reactive forces.

Example *Physical Pendulum* (**Fig. 8**). The oscillation differential equation is set out for the pendulum consisting of two particles m_1 and m_2 on 'massless' rods (given r_1, r_2, b and therefore $\beta = \arcsin (b/r_2)$). If frictional forces are absent, the d'Alembert principle in the Lagrange formulation in natural coordinates, analogously to Eq. (21), assumes the form

$$\delta W = \sum (F_{ti}^{(e)} - m_i a_{ti}) \, \delta s_i = 0$$

so that

$$\delta W = (-F_{G_1} \sin \varphi - m_1 a_{t1}) \, \delta s_1$$

$$+ (-F_{G_2} \sin(\beta + \varphi) - m_2 a_{t2}) \, \delta s_2 = 0.$$

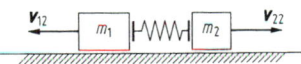

Figure 7. The principle of linear momentum and energy conservation law.

With $\delta s_1 = r_1 \, \delta\varphi$, $\delta s_2 = r_2 \, \delta\varphi$ and $a_{t1} = r_1\ddot\varphi$, $a_{t2} = r_2\ddot\varphi$ is derived $[m_1 (gr_1 \sin\varphi + r_1^2\ddot\varphi) + m_2(gr_2 \sin(\beta + \varphi) + r_2^2\ddot\varphi)] \, \delta\varphi = 0$, from which the non-linear differential equation of this pendulum oscillation follows:

$$\ddot\varphi(m_1r_1^2 + m_2r_2^2) + m_1gr_1 \sin\varphi + m_2gr_2 \sin(\varphi + \beta) = 0.$$

For minor deviations φ is taken, since $\sin\varphi \approx \varphi$ and $\sin(\varphi + \beta) \approx \varphi \cos\beta + \sin\beta$, this assumes the form

$$\ddot\varphi(m_1r_1^2 + m_2r_2^2) + \varphi(m_1gr_1 + m_2gr_2 \cos\beta) = -m_2gr_2 \sin\beta,$$

the solution of which is described in A4.

3.3.5 Angular Momentum Equation

From Newton's law $\mathbf{F}_{Ri}^{(a)} + \mathbf{F}_{ik}^{(i)} = m_i\mathbf{a}_i$ it follows, after vectorial multiplication with a radius vector $\mathbf{r}_i$ and summation across the entire system of particles, that

$$\Sigma \, (\mathbf{r}_i \times \mathbf{F}_{Ri}^{(a)}) + \Sigma(\mathbf{r}_i \times \mathbf{F}_{ik}^{(i)}) = \Sigma \, (\mathbf{r}_i \times m_i\mathbf{a}_i).$$

It follows from this, by analogy with the derivation from Eq. (22), that

$$\mathbf{M}_R^{(a)} = \Sigma \, (\mathbf{r}_i \times \mathbf{F}_{Ri}^{(a)}) = \frac{d}{dt} \Sigma \, (\mathbf{r}_i \times m_i\mathbf{v}_i) = \frac{d\mathbf{D}}{dt}. \tag{32}$$

Law of Angular Momentum or Moment of Linear Momentum. The temporal change in the moment of linear momentum (angular momentum) $\mathbf{D} = \Sigma \, (\mathbf{r}_i \times m_i\mathbf{v}_i)$ is equal to the resultant moment of the external forces on the particle system.

Equation (32) applies in respect of a point fixed in space or with regard to the randomly moving centre of gravity. From this follows after integration across the time, the angular momentum law, analogously to Eq. (24).

3.3.6 Lagrange's Equations

These provide the equations of motion for the system, by means of differentiation processes relating to the dynamic energy. A system with n mass points may well have $3n$ degrees of freedom, but owing to mechanical bonding there are frequently relationships between certain coordinates, causing the number of degrees of freedom to be reduced to m (in extreme cases, down to $m = 1$). If holonomic systems are involved, in which the relationships between the coordinates can be represented in finite form and not in differential form, then Lagrange's equations apply (second kind):

$$\frac{d}{dt}\left(\frac{\partial E}{\partial \dot{q}_k}\right) - \frac{\partial E}{\partial q_k} = Q_k \quad (k = 1,2,\dots,m). \tag{33}$$

Here, E is the total dynamic energy of the system, q_k are the generalised coordinates of the m degrees of freedom, and Q_k are the generalised forces. If q_k is a length, then the Q_k pertaining to it is a force; if q_k is an angle, then the Q_k pertaining to it is a moment.

The Lagrange force Q_k is obtained from

$$Q_k \, \delta q_k = \Sigma \, \mathbf{F}_i^{(a)} \, \delta\mathbf{s}_i \quad \text{or} \quad Q_k = (\Sigma \, \mathbf{F}_i^{(a)} \, \delta\mathbf{s}_i)/\delta q_k, \tag{34}$$

where $\delta\mathbf{s}_i$ are translations of the system as the result of one single variation of the coordinates, q_k ($\delta q_i = 0$; $i \neq k$). If the forces involved have a potential, then the following applies:

$$Q_k = -\frac{\partial U}{\partial q_k} \quad \text{and} \quad \frac{\partial U}{\partial \dot{q}_k} = 0.$$

It follows from this, from Eq. (33), that

$$\frac{d}{dt}\left(\frac{\partial E}{\partial \dot{q}_k}\right) - \frac{\partial E}{\partial q_k} = -\frac{\partial U}{\partial q_k} \quad \text{or}$$

$$\frac{d}{dt}\left(\frac{\partial L}{\partial \dot{q}_k}\right) - \frac{\partial L}{\partial q_k} = 0, \tag{35}$$

where $L = E - U = L(q_1\dots q_m; \dot{q}_1\dots \dot{q}_m)$ is the Lagrange function.

Example *Oscillator with One Degree of Freedom* (**Fig. 9**). The oscillation is being sought for minor deflections φ, i.e. for $x = l_1\varphi$ and $y = l_2\varphi$ and disregarding the masses of the rod and spring. To this there applies

$$E = m_1\dot{x}^2/2 + m_2\dot{y}^2/2 = m_1l_1^2\dot\varphi^2/2 + m_2l_2^2\dot\varphi^2/2, \quad \text{i.e.}$$

$$\frac{\partial E}{\partial \varphi} = 0 \quad \text{and}$$

$$\frac{\partial E}{\partial \dot\varphi} = (m_1l_1^2 + m_2l_2^2)\dot\varphi, \text{ i.e. } \frac{d}{dt}\left(\frac{\partial E}{\partial \dot\varphi}\right) = (m_1l_1^2 + m_2l_2^2)\ddot\varphi.$$

In addition,

$$U = m_1g(l_1 + l_2) + m_2gl_2(1 - \cos\varphi) + c(l_2\varphi)^2/2, \quad \text{i.e.}$$

$$\frac{\partial U}{\partial \varphi} = m_2gl_2\sin\varphi + cl_2^2\varphi \text{ with } \sin\varphi \approx \varphi \text{ becomes}$$

$$\frac{\partial U}{\partial \varphi} = (m_2gl_2 + cl_2^2)\varphi.$$

It then follows from Eq. (35), with $q_k = \varphi$, that

$$\ddot\varphi(m_1l_1^2 + m_2l_2^2) + \varphi(m_2gl_2 + cl_2^2) = 0 \quad \text{(for solution see A4).}$$

3.3.7 Hamilton's Principle

While Lagrange's equations represent a differential principle, Hamilton's principle is an integral principle (from which the Lagrange equations can also be derived). This reads

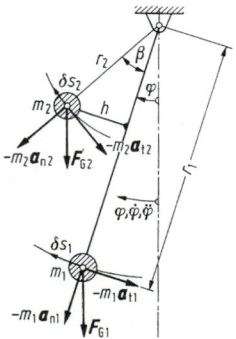

Figure 8. Physical pendulum.

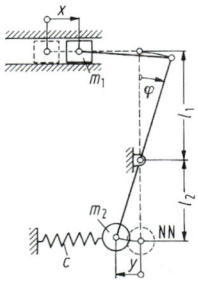

Figure 9. Oscillator.

$$\int_{t_1}^{t_2} (\delta W^{(e)} + \delta E)\, dt = 0.$$

If the imposed forces have a potential, then $\delta W^{(e)} = -\delta U$ is therefore a total differential, so that from this

$$\int_{t_1}^{t_2} (\delta E - \delta U)\, dt = \delta \int_{t_1}^{t_2} (E - U)\, dt = \delta \int_{t_1}^{t_2} L\, dt = 0,$$

i.e. the variation of the time integral from the Lagrange function becomes zero; the time integral takes on an extreme value.

3.3.8 Systems with Variable Mass

Fundamental Equation of Rocket Drive. As a result of the mass flow $\dot{\mu}(t)$ ejected with the relative velocity $\boldsymbol{v}_r(t)$ (relative motion), the rocket mass $m(t)$ is variable. From the dynamic law, Eq. (12), it then follows that

$$\boldsymbol{F}_R^{(a)} = \frac{d}{dt}\,[m(t)\boldsymbol{v}(t)] = \dot{m}(t)\boldsymbol{v}(t) + m(t)\dot{\boldsymbol{v}}(t).$$

Now $\dot{m}(t)\boldsymbol{v}(t) = -\dot{\mu}(t)\boldsymbol{v}(t)$ (the mass decreases) and therefore $\boldsymbol{F}_R^{(a)} = m(t)\boldsymbol{a}(t) - \dot{\mu}(t)$ or $m(t)\boldsymbol{a}(t) = \boldsymbol{F}_R^{(a)} + \dot{\mu}(t)\boldsymbol{v}_r(t)$. If no other forces ($\boldsymbol{F}_R^{(a)} = 0$) take effect, then

$$m(t)\boldsymbol{a}(t) = \dot{\mu}(t)\boldsymbol{v}_r(t) = \boldsymbol{F}_s(t), \qquad (36)$$

i.e. $\boldsymbol{a}$ is parallel to $\boldsymbol{v}_r$ and $\boldsymbol{F}_s(t)$ is the thrust of the rocket.

If, in addition, $\dot{\mu} = \dot{\mu}_0 = $ const. and $\boldsymbol{v}_r = \boldsymbol{v}_{r0} = $ const. and $\boldsymbol{v}_r$ is parallel to $\boldsymbol{v}$, then the path becomes a straight line. Then $m(t)a_t(t) = \dot{\mu}_0 v_{r0} = F_{s0}$. The mass lost up to the time t is $\mu(t) = \dot{\mu}_0 t$ and therefore $m(t) = m_0 - \dot{\mu}_0 t$. With $a_t = dv/dt$, it then follows that

$$\frac{dv}{dt} = \frac{\dot{\mu}_0 v_{r0}}{m_0 - \dot{\mu}_0 t} = \frac{\dot{\mu}_0 v_{r0}}{m_0[1 - (\dot{\mu}_0/m_0)t]}.$$

Integration with the initial conditions $v(t=0) = 0$ and $s(t=0) = 0$ provides

$$v(t) = -v_{r0} \ln\left(1 - \frac{\dot{\mu}_0}{m_0}\, t\right) \quad \text{and}$$

$$s(t) = \frac{m_0 v_{r0}}{\dot{\mu}_0}\left[\left(1 - \frac{\dot{\mu}_0}{m_0}\, t\right) \ln\left(1 - \frac{\dot{\mu}_0}{m_0}\, t\right) + \frac{\dot{\mu}_0}{m_0}\, t\right].$$

3.4 Dynamics of a Rigid Body

A rigid body is a continuous particle system with an infinite number of mass elements rigidly bound to one another. The dynamic principles are described in A2.2. A rigid body can be a translation or a rotation, or describe a general motion in a plane or in space.

3.4.1 Rigid Body Rotation About a Fixed Axis

In accordance with Eq. (26), in this case the centre-of-mass law applies at integration across the entire body:

$$\boldsymbol{F}_R^{(a)} = \boldsymbol{F}_R^{(e)} + \boldsymbol{F}_R^{(z)} = \Sigma\, \boldsymbol{F}_i^{(a)} = m\boldsymbol{a}_s; \qquad (37)$$

or, in components (with extension about the z-axis, **Fig. 10a**),

$$
\left.
\begin{aligned}
F_{Rx}^{(e)} + F_{Rx}^{(z)} &= \Sigma\, F_{ix}^{(e)} + F_{Ax} + F_{Bx} = ma_{Sx}, \\
F_{Ry}^{(e)} + F_{Ry}^{(z)} &= \Sigma\, F_{iy}^{(e)} + F_{Ay} + F_{By} = ma_{Sy}, \\
F_{Rz}^{(e)} + F_{Rz}^{(z)} &= \Sigma\, F_{iz}^{(e)} + F_{Az} = 0
\end{aligned}
\right\} \qquad (38a\text{-}c)
$$

with $a_{Sx} = -\omega_z^2 x_S - \alpha_z y_S$ and $a_{Sy} = \alpha_z x_S - \omega_z^2 y_S$ (see A2, Eq. 25b). These equations apply both for a fixed-body system that is fixed in space and to a system rotating in common, with zero point on the axis of rotation. In addition, the law of angular momentum applies analogously to the particle system

$$\boldsymbol{M}_R^{(a)} = \boldsymbol{M}_R^{(e)} + \boldsymbol{M}_R^{(z)} = \frac{d}{dt}\int (\boldsymbol{r} \times \boldsymbol{v})\, dm = \frac{d\boldsymbol{D}}{dt}. \quad (39)$$

In accordance with A2, Eq. (23), in Cartesian coordinates (with rotation about the z-axis, i.e. with $\omega_x = \omega_y = 0$), there applies

$$v_x = (\omega_y z - \omega_z y) = -\omega_z y, \quad v_y = (\omega_z x - \omega_x z) = \omega_z x,$$
$$v_z = (\omega_x y - \omega_y x) = 0. \qquad (40)$$

Accordingly, from Eq. (39),

$$\boldsymbol{M}_R^{(e)} + \boldsymbol{M}_R^{(z)} = \frac{d}{dt}\int \begin{vmatrix} \boldsymbol{e}_x & \boldsymbol{e}_y & \boldsymbol{e}_z \\ x & y & z \\ v_x & v_y & 0 \end{vmatrix} dm$$

$$= \frac{d}{dt}\left[\int -\omega_z xz\, dm\, \boldsymbol{e}_x\right.$$

$$\left. + \int -\omega_z yz\, dm\, \boldsymbol{e}_y + \int \omega_z(x^2 + y^2)\, dm\, \boldsymbol{e}_z\right]$$

$$= \frac{d}{dt}\,[-\omega_z J_{xz}\boldsymbol{e}_x - \omega_z J_{yz}\boldsymbol{e}_y + \omega_z J_z \boldsymbol{e}_z]; \quad (41)$$

$J_{xz} = \displaystyle\int xz\, dm$, $J_{yz} = \displaystyle\int yz$ deviation or centrifugal moments, $J_z = \displaystyle\int (x^2 + y^2)\, dm = \displaystyle\int r_z^2\, dm$ axial mass moment of inertia. In components

$$
\left.
\begin{aligned}
M_{Rx}^{(e)} + M_{Rx}^{(z)} &= \Sigma\, M_{ix}^{(e)} + F_{Ay}l_1 - F_{By}l_2 \\
&= -d(\omega_z J_{xz})/dt = -J_{xz}\alpha_z + \omega_z^2 J_{yz}, \\
M_{Ry}^{(e)} + M_{Ry}^{(z)} &= \Sigma\, M_{iy}^{(e)} + F_{Bx}l_2 - F_{Ax}l_1 \\
&= -d(\omega_z J_{yz})/dt = -J_{yz}\alpha_z - \omega_z^2 J_{xz}, \\
M_{Rz}^{(e)} &= \Sigma\, M_{iz}^{(e)} = d(\omega_z J_z)/dt = J_z \alpha_z.
\end{aligned}
\right\} \quad (42a\text{-}c)
$$

These equations apply both to an x-, y-, z-coordinate system that is fixed in space and to one rotating in common, with zero point on the axis of rotation. In the first instance, J_{xz} and J_{yz} are temporally changeable, and in the

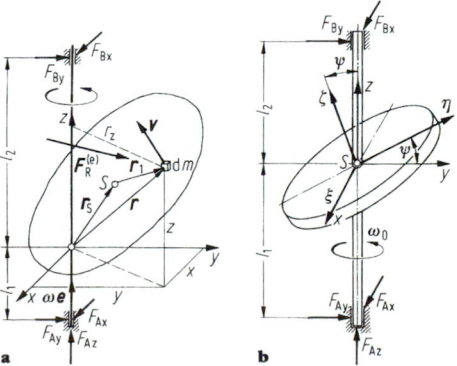

Figure 10. Dynamic bearing pressures: **a** general, **b** shaft with off-set disc.

second instance constant. Equations (38a-c) and (42a, b) provide the five unknown bearing reactions, where α_z and ω_z follow from Eq. (42c). The purely static bearing reactions are derived from the imposed forces $F_i^{(e)}$ and moments $M_{ix}^{(e)}$ and $M_{iy}^{(e)}$ while the dynamic bearing reactions can be calculated from $\boldsymbol{F}_i^{(e)} = 0$, $M_{ix}^{(e)} = M_{iy}^{(e)} = 0$

$$F_{Ax}^{(k)} + F_{Bx}^{(k)} = ma_{Sx}, \quad F_{Ay}^{(k)} + F_{By}^{(k)} = ma_{Sy}, \quad F_{Az}^{(k)} = 0, \tag{43}$$

$$F_{Ay}^{(k)}l_1 - F_{By}^{(k)}l_2 = -J_{xz}\alpha_z + \omega_z^2 J_{yz},$$
$$F_{Bx}^{(k)}l_2 - F_{Ax}^{(k)}l_1 = -J_{yz}\alpha_z - \omega_z^2 J_{xz} \tag{44}$$

According to these equations, the reactions disappear when $\boldsymbol{a}_S = 0$; i.e. when the axis of rotation passes through the centre of gravity and if it is a principal axis of inertia; that is, if the centrifugal moments J_{xz} and J_{yz} become zero. The axis of rotation is then known as a free axis. Equations (38a-c) and (42a, b) translate into the known conditions of equilibrium for these, while the *dynamic fundamental law for rotational motion* according to Eq. (42c) reads

$$M_R^{(e)} = \Sigma M_i^{(e)} = J\alpha, \tag{45}$$

$J = \int r^2 dm$, where r is the distance perpendicular to the axis of rotation.

Work and Moment-of-Momentum Laws. From Eq. (45) it follows that

$$W_{1,2} = \int_{\varphi_1}^{\varphi_2} M_R^{(e)} d\varphi = \int_{\varphi_1}^{\varphi_2} J \frac{d\omega}{dt} d\varphi$$

$$= J \int_{\omega_1}^{\omega_2} \omega \, d\omega = \frac{J}{2}(\omega_2^2 - \omega_1^2), \tag{46}$$

$$D_2 - D_1 = \int_{t_1}^{t_2} M_R^{(e)} dt = \int_{t_1}^{t_2} J \frac{d\omega}{dt} dt$$

$$= J \int_{\omega_1}^{\omega_2} d\omega = J(\omega_2 - \omega_1). \tag{47}$$

Example *A Shaft with a Disc at a Skewed Angle* (**Fig. 10b**). A fully cylindrical disc (radius $= r$, density b, mass m) is keyed onto a shaft rotating with $\omega_z = $ const. $= \omega_0$. The bearing forces are to be determined. As the only imposed force, the weight $F_G = mg$ acting through the centre exerts no moments, so Eqs (38a-c) and (42a, b) with $a_{Sx} = a_{Sy} = 0$ and (because $\omega = $ const.) $\alpha_z = 0$ become

$$F_{Ax} + F_{Bx} = 0,$$

$$F_{Ay} + F_{By} = 0, \quad -F_G + F_{Az} = 0,$$

$$F_{Ay}l_1 - F_{By}l_2 = \omega_0^2 J_{yz}, \quad F_{Bx}l_2 - F_{Ax}l_1 = -\omega_0^2 J_{xz}.$$

With the angles of direction of the x-axis with reference to the main axes, ξ, η, ζ (see A3.4.2), $\alpha_1 = 0$, $\beta_1 = 90°$, $\gamma_1 = 90°$, those with the y-axis $\alpha_2 = 90°$, $\beta_2 = \psi$, $\gamma_2 = 90° + \psi$ and those with the z-axis $\alpha_3 = 90°$, $\beta_3 = 90° - \psi$, $\gamma_3 = \psi$, according to Eq. (52) there is obtained

$$J_{yz} = -J_1 \cos \alpha_2 \cos \alpha_3 - J_2 \cos \beta_2 \cos \beta_3 - J_3 \cos \gamma_2 \cos \gamma_3$$

$$= -J_2 \cos \psi \sin \psi + J_3 \sin \psi \cos \psi,$$

and accordingly $J_{xz} = 0$. According to Table 1, $J_2 = J_\eta = m(3r^2 + b^2)/12$, $J_4 = J_\zeta = mr^2/2$ and therefore $J_{yz} = [m(3r^2 - b^2)/24] \sin 2\psi$, so the bearing forces are derived:

$$F_{Ax} = F_{Bx} = 0, \quad F_{Az} = F_G,$$

$$F_{Ay} = -F_{By} = \{\omega_0^2 m(3r^2 - b^2)/[24(l_1 + l_2)]\} \sin 2\psi.$$

3.4.2 Moment of Inertia (Fig. 11)

Axial moments of inertia:

$$\left.\begin{aligned} J_x &= \int (y^2 + z^2) \, dm = \int r_x^2 \, dm, \\ J_y &= \int (x^2 + z^2) \, dm = \int r_y^2 \, dm, \\ J_z &= \int (x^2 + y^2) \, dm = \int r_z^2 \, dm. \end{aligned}\right\} \tag{48}$$

Polar moment of inertia and deviation or centrifugal moments:

$$J_P = \int r^2 dm = \int (x^2+y^2+z^2) dm = (J_x+J_y+J_z)/2;$$

$$J_{xy} = \int xy \, dm, \quad J_{xz} = \int xz \, dm, \quad J_{yz} = \int yz \, dm. \tag{49}$$

The moments of inertia can be combined with $J_x = J_{xx}$, $J_y = J_{yy}$ and $J_z = J_{zz}$ to form the moment of inertia tensor, a symmetrical tensor of the second rank. In matrix form,

$$\boldsymbol{J} = \begin{pmatrix} J_{xx} & -J_{xy} & -J_{xz} \\ -J_{yx} & J_{yy} & -J_{yz} \\ -J_{zx} & -J_{zy} & J_{zz} \end{pmatrix}.$$

Main Axes. If $J_{\xi\eta} = J_{\zeta\zeta} = J_{\eta\zeta} = 0$, then the main axes of inertia pertain. The main axial moments of inertia that pertain, J_1, J_2, J_3, behave in such a manner that one is the absolute maximum and another the absolute minimum of all the moments of inertia of the body. If a body has a plane of symmetry, then every axis vertical to it is a main axis. In general, the main moments of inertia are obtained as external values of Eq. (50), with the constraining condition $b = \cos^2 \alpha + \cos^2 \beta + \cos^2 \gamma - 1 = 0$. With the abbreviations $\cos \alpha = \lambda$, $\cos \beta = \mu$, $\cos \gamma = v$ there follow with

$$J = J_x\lambda^2 + J_y\mu^2 + J_z v^2 - 2J_{xy}\lambda\mu - 2J_{yz}\mu v - 2J_{xz}\lambda v \text{ and}$$

$f = J = cb$ from $\partial f/\partial \lambda = 0$ etc., three homogeneous linear equations for λ, μ, v, which have a non-trivial solution only if their coefficient determinants are zero. From this is derived the cubic equation for c with the solutions $c_1 = J_1$, $c_2 = J_2$ and $c_3 = J_3$.

Ellipsoid of Inertia. If the values $1/\sqrt{J_x}$, $1/\sqrt{J_y}$, $1/\sqrt{J_z}$ are carried in the direction of the axes x, y, z, then the end points lie on the ellipsoid of inertia with the main axes $1/\sqrt{J_1}$ etc. and the equation $J_1\xi^2 + J_2\eta^2 + J_3\zeta^2 = 1$. If the coordinate initial point in this instance lies in the centre of gravity, we may speak of a central ellipsoid, and the main axes pertaining to it are then free axes.

Moments of Inertia in Relation to Rotated Axes. For an axis $\bar{x}$ inclined with respect to x, y, z at angles α, β, γ, $\boldsymbol{e}_{\bar{x}} = (\cos \alpha, \cos \beta, \cos \gamma)$ where $J_{\bar{x}} = \boldsymbol{e}_{\bar{x}} \boldsymbol{J} \boldsymbol{e}_{\bar{x}}^T$ and $J_{xy} = J_{yx}$, etc.

$$J_{\bar{x}} = -J_x \cos^2 \alpha + J_y \cos^2 \beta + J_z \cos^2 \gamma$$
$$- 2J_{xy} \cos \alpha \cos \beta - 2J_{yz} \cos \beta \cos \gamma$$
$$- 2J_{xz} \cos \alpha \cos \gamma. \tag{50}$$

If, by contrast, α_1, β_1, γ_1 are the direction angles of the

Table 1. Moments of inertia of homogeneous bodies

Circular cylinder	Hollow cylinder	Sphere	Circular cone
$m = \varrho \pi r^2 h$	$m = \varrho \pi (r_0^2 - r_i^2) h$	$m = \varrho \frac{4}{3} \pi r^3$	$m = \varrho \pi r^2 h / 3$
$J_x = \frac{m r^2}{2}$ $J_y = J_z = \frac{m(3r^2 + h^2)}{12}$	$J_x = \frac{m(r_0^2 + r_i^2)}{2}$ $J_y = J_z = \frac{m(r_0^2 + r_i^2 + h^2/3)}{4}$	$J_x = J_y = J_z = \frac{2}{5} m r^2$	$J_x = \frac{3}{10} m r^2$ $J_y = J_z = \frac{3m(4r^2 + h^2)}{80}$
Zylinderschale Wanddicke $\delta \ll r$:		Kugelschale Wanddicke $\delta \ll r$:	Cylindrical shell wall thickness $\delta \ll r$:
$m = \varrho 2 \pi r h \delta$		$m = \varrho 4 \pi r^2 \delta$	$m = \varrho \pi r s \delta$
$J_x = m r^2$ $J_y = J_z = \frac{m(6r^2 + h^2)}{12}$		$J_x = J_y = J_z = \frac{2}{3} m r^2$	$J_x = \frac{m r^2}{2}$

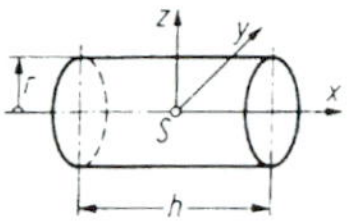

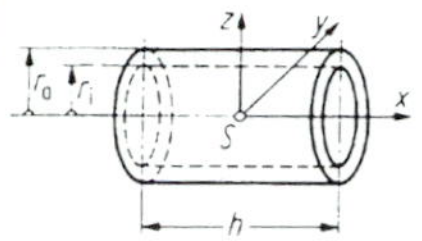

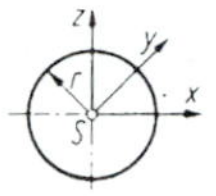

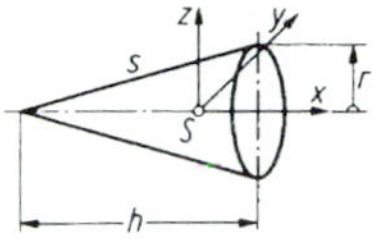

Cuboid	Thin bar	Hollow sphere	Truncated circular cone
$m = \varrho a b c$	$m = \varrho A l$	$m = \varrho \frac{4}{3} \pi (r_0^3 - r_i^3)$	$m = \varrho \frac{1}{3} \pi h (r_2^2 + r_2 r_1 + r_1^2)$
$J_x = \frac{m(b^2 + c^2)}{12}$ $J_y = \frac{m(a^2 + c^2)}{12}$	$J_y = J_z = \frac{m l^2}{12}$	$J_x = J_y = J_z = \frac{2}{5} m \frac{r_0^5 - r_i^5}{r_0^3 - r_i^3}$	$J_x = \frac{3}{10} m \frac{r_2^5 - r_1^5}{r_2^3 - r_1^3}$
$J_z = \frac{m(c^2 + b^2)}{12}$			

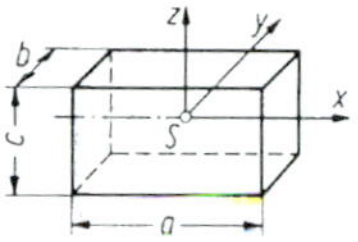

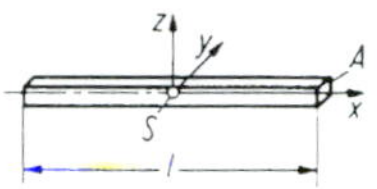

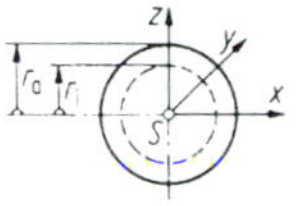

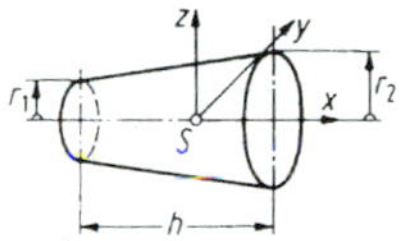

Square-based pyramid	Circular torus	Hemisphere	Arbitrary solid of rotation
$m = \varrho a b h / 3$	$m = \varrho 2 \pi^2 r^2 R$	$m = \varrho \frac{2}{3} \pi r^3$	$m = \varrho \pi \int_{x_1}^{x_2} f^2(x) dx$
$J_x = \frac{m(a^2 + b^2)}{20}$ $J_y = \frac{m(b^2 + \frac{3}{4} h^2)}{20}$	$J_x = J_y = \frac{m(4R^2 + 5r^2)}{8}$	$J_x = J_y = \frac{83}{320} m r^2$ $J_z = \frac{2}{5} m r^2$	$J_x = \frac{1}{2} \varrho \pi \int_{x_1}^{x_2} f^4(x) dx$
$J_z = \frac{m(a^2 + \frac{3}{4} h^2)}{20}$	$J_z = \frac{m(4R^2 + 3r^2)}{4}$		

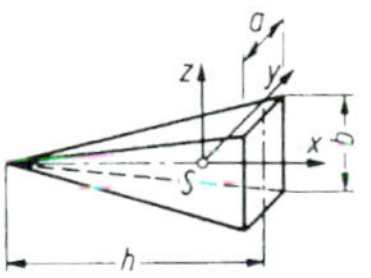

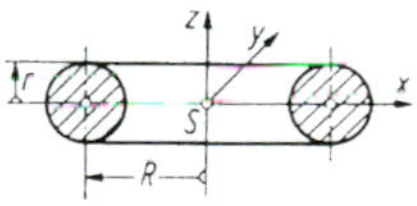

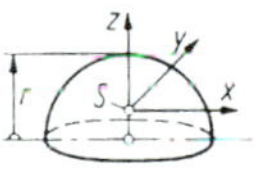

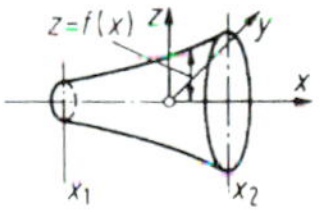

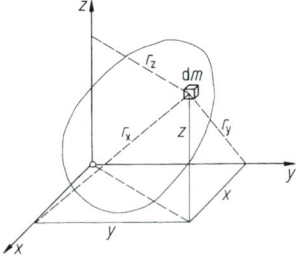

Figure 11. Mass moment of inertia.

x-axis against the main axes ζ, η, ζ, then the following applies for the axial moment of inertia:

$$J_x = J_1 \cos^2 \alpha_1 + J_2 \cos^2 \beta_1 + J_3 \cos^2 \gamma_1; \quad (51)$$

and analogously for J_y, J_z with direction angles $\alpha_2, \beta_2, \gamma_2$ or $\alpha_3, \beta_3, \gamma_3$ of the *y*- or *z*-axis against the main axes. The deviation moments which pertain are (and analogously for J_{xz} and J_{yz})

$$J_{xy} = -J_1 \cos \alpha_1 \cos \alpha_2 - J_2 \cos \beta_1 \cos \beta_2$$
$$- J_3 \cos \gamma_1 \cos \gamma_2. \quad (52)$$

Steiner's Principles. For parallel axes, the following applies:

$$J_x = J_{\bar{x}} + (y_S^2 + z_S^2)m, \quad J_y = J_{\bar{y}} + (z_S^2 + x_S^2)m,$$

$$J_z = J_{\bar{z}} + (x_S^2 + y_S^2)m, \quad J_{xy} = J_{\bar{x}\bar{y}} + x_S y_S m, \qquad (53)$$

$$J_{xz} = J_{\bar{x}\bar{z}} + x_S z_S m, \quad J_{yz} = J_{\bar{y}\bar{z}} + y_S z_S m;$$

$\bar{x}, \bar{y}, \bar{z}$ are parallel axes to x, y, z through the centre of gravity.

Radius of Inertia. If the total mass is concentrated at distance i from the axis of rotation (with J and m given), then $J = i^2 m$ or $i = \sqrt{J/m}$.

Reduced Mass. If the mass m_{red} is thought of as being a random distance d from the axis of rotation (with J given), then $J = d^2 m_{red}$ or $m_{red} = J/d^2$ applies.

Calculating the Mass Moments of Inertia. For individual bodies, by means of triple integrals,

$$J_x = \int r_x^2 \, dm = \iiint \rho(y^2 + z^2) \, dx \, dy \, dz.$$

Depending on the shape of the body, both cylindrical and spherical coordinates are used. For example, for the solid circular cylinder (**Table 1**), use is made of

$$J_x = \int_{r=0}^{r_a} \int_{\varphi=0}^{2\pi} \int_{z=-b/2}^{+b/2} \rho r^2 (r \, d\varphi \, dr \, dz)$$

$$= \rho(r_a^4/4)2\pi b = mr_a^2/2.$$

For combined bodies, using Steiner's principles, there applies $J_x = \Sigma[J_{xi} + (y_{Si}^2 + z_{Si}^2)m_i]$, etc.

3.4.3 General Plane Motion of a Rigid Body

Plane motion signifies $z = $ const. or $v_z = \omega_x = \omega_y = 0$ and $\alpha_z = \alpha_x = \alpha_y = 0$. As with the particle system, the *centre-of-mass law* and *the law of moment of momentum* apply (lever law)

$$F_R^{(a)} = \Sigma F_i^{(a)} = ma_S, \qquad (54)$$

$$M_R^{(a)} = \Sigma M_i^{(a)} = \frac{d}{dt}\int (r \times v)dm$$

$$= \frac{d}{dt}\int \begin{vmatrix} e_x & e_y & e_z \\ x & y & z \\ \dot{x} & \dot{y} & 0 \end{vmatrix} dm = \frac{dD}{dt}. \qquad (55)$$

(The law of lever applies in relation to a point fixed in space or a centre of gravity moving at random.)

In Cartesian coordinates

$$F_{Rx}^{(a)} = \Sigma F_{ix}^{(a)} = ma_{Sx}, \quad F_{Ry}^{(a)} = \Sigma F_{iy}^{(a)} = ma_{Sy},$$

$$F_{Rz}^{(a)} = \Sigma F_{iz}^{(a)} = 0,$$

$$M_{Rx}^{(a)} = -\frac{d}{dt}\int z\dot{y}\,dm = -\frac{d^2}{dt^2}\int zy\,dm = -\frac{d^2 J_{yz}}{dt^2},$$

$$M_{Ry}^{(a)} = \frac{d}{dt}\int z\dot{x}\,dm = \frac{d^2}{dt^2}\int zx\,dm = \frac{d^2 J_{xz}}{dt^2}, \qquad (56)$$

$$M_{Rz}^{(a)} = \frac{d}{dt}\int (x\dot{y} - \dot{x}y)\,dm$$

or, with Eq. (40) and $\omega_z = \omega$,

$$M_{Rz}^{(a)} = \frac{d}{dt}\int \omega(x^2 + y^2)\,dm = \frac{d}{dt}\int \omega r_z^2\,dm = \frac{d}{dt}(\omega J_z).$$

$M_{Rx}^{(a)}$ and $M_{Ry}^{(a)}$ are the external moments required to force the movement in the plane, if z is not a principal axis of inertia. If z is a principal axis of inertia ($J_{yz} = J_{xz} = 0$), then it follows that $M_{Rx}^{(a)} = 0$, $M_{Ry}^{(a)} = 0$, $M_{Rz}^{(a)} = \frac{d}{dt}(\omega J_z)$, or, in relation to the fixed-body centre of gravity with J_S = const.,

$$M_{RS}^{(a)} = \Sigma M_{iS}^{(a)} = J_S \alpha. \qquad (57)$$

Work Law

$$W_{1,2} = \int F_R^{(a)}\,dr + \int M_{RS}^{(a)}\,d\varphi = \left(\frac{m}{2}v_{S2}^2 + \frac{J_S}{2}\omega_2^2\right)$$

$$-\left(\frac{m}{2}v_{S1}^2 + \frac{J_S}{2}\omega_1^2\right) = E_2 - E_1 \qquad (58)$$

If the external forces and moments have a potential, then the *energy law* $U_1 + E_1 = U_2 + E_2 = $ const. applies.

Centre-of-Mass and Moment-of-Momentum Law

$$p_2 - p_1 = \int_{t_1}^{t_2} F_R^{(a)}\,dt = m(v_{S2} - v_{S1}) \qquad (59)$$

$$D_2 - D_1 = \int_{t_1}^{t_2} M_{RS}^{(a)}dt = J_S(\omega_2 - \omega_1) \qquad (60)$$

D'Alembert's Principle. The lost forces, i.e. the total of imposed forces and forces of inertia, maintain the equilibrium of the whole body. With the equilibrium principle of virtual displacements, there then applies, in the Lagrange formulation

$$(F_R^{(e)} - ma_S)\,\delta r_s + (M_{RS}^{(e)} - J_S\alpha)\,\delta\varphi = 0. \qquad (61)$$

Example *Rolling Motion on an Inclined Plane* (**Fig. 12**). A cylindrical body (r, m, J_S) is to be rolled from its position of rest by the force F up an inclined plane (angle of inclination β), without sliding. The acceleration of its centre of gravity is to be calculated, as well as the time and velocity upon reaching position 2 after covering distance s_2. Because the centre of gravity moves in a straight line, its acceleration vector falls in the direction of motion. The centre-of-mass law, Eq. (54), and the law of moment of momentum, Eq. (57), provide (**Fig. 12a**) $ma_S = F \cos\beta - F_G \sin\beta - F_r$ and $J_S\alpha = F_r r$, from which, with $\alpha = a_S/r$ owing to the pure rolling motion, there follows

$$a_S = (F \cos\beta - F_G \sin\beta)/(m + J_S/r^2).$$

With the laws of uniformly accelerated motion from a position of rest (see A2.1.1) is derived $v_{S2} = \sqrt{2a_S s_2}$ and $t_2 = v_{S2}/a_S$. The work law, Eq. (58)

$$(F \cos\beta - F_G \sin\beta)s_2 = mv_{S2}^2/2 + J_S\omega_2^2/2$$

in turn provides, with $\omega_2 = v_{S2}/r$,

$$v_{S2} = \sqrt{2(F \cos\beta - F_G \sin\beta)s_2/(m + J_S/r^2)}.$$

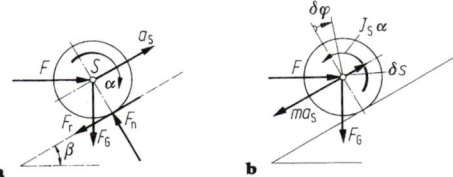

Figure 12. a and **b** Rolling motion on inclined plane.

Centre-of-mass and moment-of-momentum law, Eqs (59), (60),

$$(F \cos \beta - F_G \sin \beta - F_r)t_2 = m v_{s2} \quad \text{and} \quad F_r r t_2 = J_s \omega_2$$

likewise give

$$t_2 = v_{s2}(m + J_s/r^2)/(F \cos \beta - F_G \sin \beta) = v_{s2}/a_s.$$

The d'Alembert principle in the Lagrange formulation according to Eq. (61) leads to (**Fig. 12b**)

$$(F \cos \beta - F_G \sin \beta - m a_s) \, \delta s + (0 - J_s \alpha) \, \delta \varphi = 0;$$

with $\alpha = a_s/r$, $\delta \varphi = \delta s/r$ it follows that

$$\delta s[F \cos \beta - F_G \sin \beta - m a_s - J_s a_s/r^2] = 0, \quad \text{i.e., again}$$

$$a_s = (F \cos \beta - F_G \sin \beta)/(m + J_s/r^2).$$

Plane Rigid-Body Systems. The motion can be calculated in several ways:

- Freeing each individual body and applying the centre-of-mass law, Eq. (54), and the lever law, Eq. (57), if z is a principal axis of inertia;
- Application of the d'Alembert principle, Eq. (61), to the system consisting of n bodies:

$$\Sigma \, (\boldsymbol{F}_{Ri}^{(e)} - m_i \boldsymbol{a}_{is}) \, \delta \boldsymbol{r}_{is} + \Sigma \, (\boldsymbol{M}_{Ri}^{(e)} - J_{is} \boldsymbol{\alpha}_i) \, \delta \varphi_i = 0; \tag{62}$$

- Application of the Lagrange equations of motion, Eqs (33) to (35).

Example *Accelerations of a Rigid-Body System (**Fig. 13**).* The system moves in the directions indicated, with the friction force F_{r1} taking effect on the guiding of m_1 and the roller describing a rolling motion. The d'Alembert principle in the Lagrange formulation, Eq. (62), provides

$$(F_{G1} - F_{r1} - m_1 a_1) \, \delta z - J_2 \alpha_2 \, \delta \varphi - (F_{G3} \sin \beta + m_3 a_{3s}) \, \delta s$$

$$- J_{3s} \alpha_3 \, \delta \psi = 0.$$

With

$$\delta z = r_a \delta \varphi, \quad \delta s = r_i \delta \varphi \quad \text{and} \quad \delta \psi = \delta s/r_3 = \delta \varphi r_i/r_3 \quad \text{or}$$

$$a_1 = \ddot{z} = r_a \ddot{\varphi} = r_a \alpha_2, \quad a_{3s} = \ddot{s} = r_i \ddot{\varphi} = r_i \alpha_2 \quad \text{and}$$

$$\alpha_3 = \ddot{\psi} = \ddot{s}/r_3 = \alpha_2 r_i/r_3$$

We find

$$\delta \varphi[(F_{G1} - F_{r1})r_a - m_1 \, r_a^2 \alpha_2 - J_2 \alpha_2 - F_{G3} r_i \sin \beta$$

$$- m_3 r_i^2 \alpha_2 - J_{3s}(r_i/r_3)^2 \alpha_2] = 0.$$

The angle acceleration of the rope pulley is therefore

$$\alpha_2 = [(F_{G1} - F_{r1})r_a - F_{G3} r_i \sin \beta]/[m_1 r_a^2 + J_2 + m_3 r_i^2 + J_{3s}(r_i/r_3)^2],$$

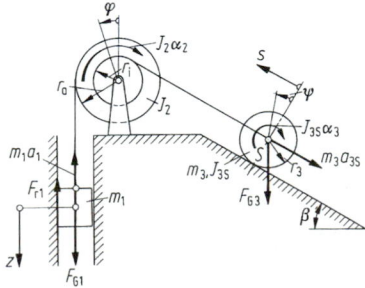

Figure 13. Rigid-body system.

with also

$$a_1 = r_a \alpha_2, \quad a_{3s} = r_i \alpha_2 \quad \text{and} \quad \alpha_3 = \alpha_2 r_i/r_3 \quad \text{being determined.}$$

3.4.4 General Motion in Space

Equations of Motion. These are provided by the centre-of-mass law and the law of moment of momentum (the lever law):

$$\boldsymbol{F}_R^{(a)} = \Sigma \, \boldsymbol{F}_i^{(a)} = m \boldsymbol{a}_s \tag{63}$$

$$\boldsymbol{M}_R^{(a)} = \Sigma \, \boldsymbol{M}_i^{(a)} = \frac{d\boldsymbol{D}}{dt} = \frac{d}{dt} \int (\boldsymbol{r} \times \boldsymbol{v}) \, dm \tag{64}$$

(for explanations, see Eqs (26) and (32)).

The lever law applies with reference to a particle fixed in space or the randomly moving centre of gravity. In Cartesian coordinates with v according to A2, Eq. (23),

$$\boldsymbol{M}_R^{(a)} = \frac{d}{dt} \int \begin{vmatrix} \boldsymbol{e}_x & \boldsymbol{e}_y & \boldsymbol{e}_z \\ x & y & z \\ v_x & v_y & v_z \end{vmatrix} dm$$

$$= \frac{d}{dt} \left[(\omega_x J_x - \omega_y J_{xy} - \omega_z J_{xz}) \boldsymbol{e}_x \right.$$

$$+ (\omega_y J_y - \omega_x J_{xy} - \omega_z J_{yz}) \boldsymbol{e}_y$$

$$\left. + (\omega_z J_z - \omega_x J_{xz} - \omega_y J_{yz}) \boldsymbol{e}_z \right]. \tag{65}$$

This equation relates to an $\boldsymbol{x}, \boldsymbol{y}, \boldsymbol{z}$ coordinate system fixed in space (**Fig. 14**), the coordinate initial point of which may also lie at the centre of gravity; i.e. the values J_x, J_{xy}, etc. are time-dependent, since the position of the body changes.

If, according to Euler, a fixed-body coordinate system in common motion, ξ, η, ζ, is introduced (for the sake of simplicity, in the direction of the principal axes of inertia of the body), and the angular velocity vector in this coordinate system is reduced to its components $\boldsymbol{\omega} = \omega_1 \boldsymbol{e}_1 + \omega_2 \boldsymbol{e}_2 + \omega_3 \boldsymbol{e}_3$, then Eq. (65) takes the form

$$\boldsymbol{M}_R^{(a)} = \frac{d}{dt} [\omega_1 J_1 \boldsymbol{e}_1 + \omega_2 J_2 \boldsymbol{e}_2 + \omega_3 J_3 \boldsymbol{e}_3], \tag{66}$$

where J_1, J_2, J_3 are now constant, and $\omega_1 J_1$ etc. are the components of the angular momentum vector $\boldsymbol{D}$ in the moving coordinate system. With the rule for the derivation of a vector in the moved coordinate system (see A2, Eq. (35)), $d\boldsymbol{D}/dt = d_r \boldsymbol{D}/dt + \boldsymbol{\omega} \times D$, where $d_r \boldsymbol{D}/dt$ is the derivation of the vector $\boldsymbol{D}$ relative to the coordinate system moving in common. From Eq. (66) it follows, in components, that

$$\begin{rcases} M_{R\xi}^{(a)} = [\dot{\omega}_1 J_1 + \omega_2 \omega_3 (J_3 - J_2)], \\ M_{R\eta}^{(a)} = [\dot{\omega}_2 J_2 + \omega_1 \omega_3 (J_1 - J_3)], \\ M_{R\zeta}^{(a)} = [\dot{\omega}_3 J_3 + \omega_1 \omega_2 (J_2 - J_1)]. \end{rcases} \tag{67}$$

These are the *Euler's equations of motion* of a body in space in relation to the main axes, with a particle fixed in space or with the randomly moved centre of gravity as the source. From the three linked differential equations, however, only the angular velocities $\omega_1(t), \omega_2(t), \omega_3(t)$ are derived, relating to the coordinate system moving in common, but not to the position of the body in respect of the spatially fixed directions x, y, z. For this, the introduction of the Euler's angles φ, ψ, ϑ is required [1]. The position of the centre of gravity of a body freely moving in space can be calculated from the centre-of-mass law, Eq. (63), in the same way as for a particle (see A3.2).

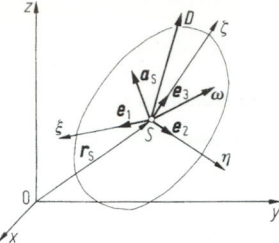

Figure 14. General motion in space.

Angular Momentum Law.

$$\int_{t_1}^{t_2} \boldsymbol{M}_{R}^{(a)} \, dt = \int_{t_1}^{t_2} d\boldsymbol{D} = \boldsymbol{D}_2 - \boldsymbol{D}_1.$$

For $\boldsymbol{M}_{r}^{(a)} = 0$, $\boldsymbol{D}_2 = \boldsymbol{D}_1$, i.e. without the effect of external forces the angular momentum vector retains its direction in space.

Energy Conservation Law. If the forces taking effect have a potential, then the following applies:

$$U_1 + E_1 = U_2 + E_2 = \text{const.}$$

Dynamic energy

$$E = m v_s^2/2 + (J_1 \omega_1^2 + J_2 \omega_2^2 + J_3 \omega_3^2)/2.$$

Gyroscopic Motion (Fig. 15). This is understood to mean the rotation of a rigid body about a fixed point. Euler's equations of motion, Eqs (67), apply.

Zero-Force Gyroscopic Motion. If all the moments of the external forces are zero, i.e. bearing on the centre of gravity (**Fig. 15a**), and if no forces or moments are otherwise taking effect, then the motion is free of forces; the angular momentum vector retains its direction and value in space. In this instance, the possible forms of motion of the gyroscope are derived from

$$J_1 \dot{\omega}_1 = (J_2 - J_3) \omega_2 \omega_3, \quad J_2 \dot{\omega}_2 = (J_3 - J_1) \omega_1 \omega_3,$$

$$J_3 \dot{\omega}_3 = (J_1 - J_2) \omega_1 \omega_2;$$

that is, either

$$\omega_1 = \text{const.}, \quad \omega_2 = \omega_3 = 0 \quad \text{or}$$

$$\omega_2 = \text{const.}, \quad \omega_1 = \omega_3 = 0 \quad \text{or}$$

$$\omega_3 = \text{const.}, \quad \omega_1 = \omega_2 = 0,$$

i.e., in each case rotation about a principal axis of inertia (motion is stable, if the rotation is about the axis of the largest or smallest moment of inertia).

For the *symmetrical gyro*, there follow, with $J_1 = J_2$, the equations (see [2, 3])

$$\omega_3 = \text{const.}, \quad \ddot{\omega}_1 + \lambda^2 \omega_1 = 0 \quad \text{and} \quad \ddot{\omega}_2 + \lambda^2 \omega_2 = 0,$$

with the solutions

$$\omega_1 = c \sin(\lambda t - \alpha) \quad \text{and} \quad \omega_2 = c \cos(\lambda t - \alpha),$$

where $\lambda = (J_3/J_1 - 1)\omega_3$.

With $\omega_1^2 + \omega_2^2 = c^2 = \text{const.}$, it follows that the angular velocity vector $\boldsymbol{\omega} = \omega_1 \boldsymbol{e}_\xi + \omega_2 \boldsymbol{e}_\eta + \omega_3 \boldsymbol{e}_\zeta$ (the momentary axis of rotation) describes a circular cone in the fixed-body system, the moving axode, which rolls on the herpolhode cone, the axis of which is the fixed angular momentum vector (**Fig. 15a**). The axis of symmetry in this instance describes the precession cone (constant precession).

Centroid Gyro. In this instance, consideration is given in particular to the rapidly rotating symmetrical gyro, under its own weight (**Fig. 15b**). In the case of the rapidly rotating gyro, $\boldsymbol{D} \approx \omega_3 J_3 \boldsymbol{e}_\zeta$, i.e. the angular inertia vector and the axis of symmetry approximately coincide. It follows from the law of moment of momentum that $d\boldsymbol{D} = \boldsymbol{M}_{R}^{(a)} \, dt = (\boldsymbol{r} \times \boldsymbol{F}_G) \, dt$, i.e. the gyro endeavours to set its axis of symmetry parallel to and in the same direction of rotation as the moment that is taking effect on it (Poinsot's law). According to **Fig. 15b**, $M = F_G r \sin \vartheta$, $dD = D \sin \vartheta \cdot d\varphi$. From $dD = M \, dt$, it follows that $\omega_p = d\varphi/dt = F_G r / D \approx F_G r/(J_3 \omega_3)$. ω_p is the angular velocity of precession of the gyro. Because of ω_p, the angular momentum vector does not lie precisely on the axis of symmetry, and therefore the nutation still overrides the precession [2, 3].

Guided Gyro. This is a rotating body, as a rule symmetrical in rotation, on which forces of transportation compel a change in the angular momentum vector, causing the emergence of the moment of gyroscopic effect and the bearing forces associated with it, which are in part considerable (pug mill, slewing motion of wheel sets, propeller shafts, etc.). For a vehicle on a curve, the gyroscopic force of the wheels provides an additional tilting moment. Conversely, guided gyros are used as stabilising elements for ships, monorails, etc. With the gyrocompass, arranged to float horizontally, the axis of angular momentum is forced into a north–south direction by the rotation of the earth.

For the bodies of rotation represented in **Fig. 15c**, and guided with ω_F, there applies

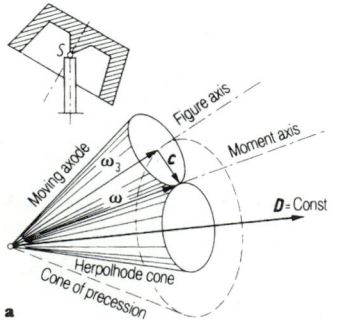

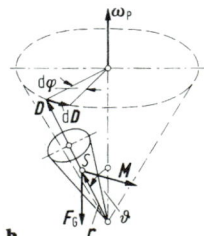

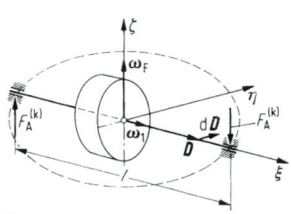

Figure 15. Gyro: **a** free of forces; **b** heavier; **c** guided.

$$M^{(a)} = \frac{d\boldsymbol{D}}{dt} = \omega_F \times \boldsymbol{D} = \begin{vmatrix} \boldsymbol{e}_\xi & \boldsymbol{e}_\eta & \boldsymbol{e}_\zeta \\ 0 & 0 & \omega_F \\ \omega_1 J_1 & 0 & \omega_F J_3 \end{vmatrix} = \omega_F \omega_1 J_1 \boldsymbol{e}_\eta$$

or $M^{(a)} = F_A^{(k)} l = \omega_F \omega_1 J_1$, i.e. $F_A^{(k)} = \omega_F \omega_1 J_1 / l$. The moment of the gyro effect produces the opposing bearing pressure to $F_A^{(k)}$ in the bearings.

3.5 Dynamics of Relative Motion

In the case of guided relative motion, A2, Eq. (36) applies to the acceleration, and therefore for Newton's law

$$\boldsymbol{F}_R^{(a)} = m\boldsymbol{a}_F + m\boldsymbol{a}_r + m\boldsymbol{a}_C. \tag{68}$$

For an observer located in the vehicle, only the relative acceleration can be perceived:

$$m\boldsymbol{a}_r = \boldsymbol{F}_R^{(a)} - m\boldsymbol{a}_F - m\boldsymbol{a}_C = \boldsymbol{F}_R^{(a)} + \boldsymbol{F}_F + \boldsymbol{F}_C, \tag{69}$$

i.e. the force of transport and Coriolis force are to be added to the external forces.

Example *Motion in a Rotating Tube* **(Fig. 16).** In a tube that rotates about a vertical axis with $\alpha_F(t)$ and $\omega_F(t)$, the mass m with the relative velocity $v_r(t)$ is drawn inwards by means of a frictionless thread. For a random position $r(t)$, the thread force and the normal force between mass and tube are to be determined. With $\boldsymbol{a}_F = \boldsymbol{a}_{Fn} + \boldsymbol{a}_{Ft}$ ($a_{Fn} = r\omega_F^2$, $a_{Ft} = r\alpha_F$) and $a_C = 2\omega_F v_r$, the following are derived from the free mass, according to Eq. (68):

$$F_s = m(a_r + a_{Fn}) = m(a_r + r\omega_F^2) \quad \text{and}$$

$$F_n = m(a_C - a_{Ft}) = m(2\omega_F v_r - r\alpha_F).$$

3.6 Impact

When an impact occurs between two bodies, relatively large forces exert an effect in a short period of time, in respect of which other forces such as weight and friction are negligible. The perpendiculars to the contact surfaces are referred to as the shock normal. If this passes through the centres of gravity of both bodies, the impact is referred to as centric, and otherwise as eccentric. If the velocities are in the direction of the shock normal, then it is a straight impact; otherwise it is oblique. There are only a few results available regarding the forces transferred in the contact surface during the impact, and about the collision time [4, 5]. The impact process is subdivided into the compression period K, during which impact force increases, until both bodies have reached the common velocity u, and the resolution period R, during which impact force is reduced and the bodies achieve their different final velocities c_1 and c_2 **(Fig. 17).** The momenta of the impact or impulses of force in the compression period and in the restitution period are:

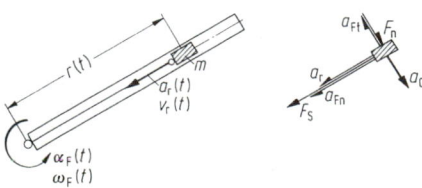

Figure 16. Relative motion.

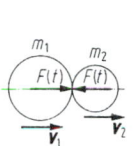

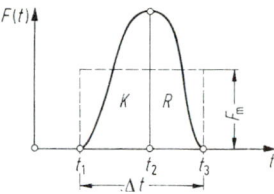

Figure 17. Course of force during impact.

$$p_K = \int_{t_1}^{t_2} F_K(t)\, dt, \quad p_R = \int_{t_2}^{t_3} F_R(t)\, dt. \tag{70}$$

p_K and p_R are set in relationship to one another by means of Newton's impact theorem

$$p_R = kp_K, \tag{71}$$

where $k = 1$ is the impact coefficient. Fully elastic impact: $k = 1$; partially elastic impact: $k = 0$. Mean force of impact $F_m = (p_K + p_R)/\Delta t$.

3.6.1 Normal Impact

With v_1 and v_2 as the velocities of both bodies prior to the impact **(Fig. 17)**, u and c_1 or c_2 as explained follow from Eqs (70) and (71):

$$u = (m_1 v_1 + m_2 v_2)/(m_1 + m_2),$$

$$c_1 = [m_1 v_1 + m_2 v_2 - km_2(v_1 - v_2)]/(m_1 + m_2),$$

$$c_2 = [m_1 v_1 + m_2 v_2 + km_1(v_1 - v_2)]/(m_1 + m_2),$$

$$k = p_R/p_K = (c_2 - c_1)/(v_1 - v_2).$$

Energy Loss on Impact.

$$\Delta E = \frac{m_1 m_2}{2(m_1 + m_2)}(v_1 - v_2)^2(1 - k^2).$$

Special Cases.

$m_1 = m_2, k = 1$:

$$u = (v_1 + v_2)/2, c_1 = v_2, c_2 = v_1;$$

$m_1 = m_2, k = 0$:

$$u = c_1 = c_2 = (v_1 + v_2)/2;$$

$m_2 \to \infty, v_2 = 0, k = 1$:

$$u = 0, c_1 = -v_1, c_2 = 0;$$

$m_2 \to \infty, v_2 = 0, k = 0$:

$$u = 0, c_1 = 0, c_2 = 0.$$

Determination of Impact Coefficient. In free fall against an infinitely great mass m_2, $k = (c_2 - c_1)/(v_1 - v_2) = \sqrt{h_2/h_1}$ applies where h_1 is the height of the fall prior to the impact, h_2 is the height of lift after the impact. k is dependent on the velocity of arrival, at $v \approx 2.8$ m/s, for ivory $k = 8/9$, for steel $k = 5/9$, for glass $k = 15/16$, and for wood $k = 1/2$.

Impact Force and Collision Time. For the purely elastic impact of two spheres with radii r_1 and r_2, Hertz [4] derived max $F = k_1 v^{6/5}$, where v is the relative velocity and $k_1 = [1/25 \cdot m_1 m_2/(m_1 + m_2)]^{3/5} c_1^{2/5}$ with

$$c_1 = (16/3)\left[\sqrt{1/r_1 + 1/r_2}(\vartheta_1 + \vartheta_2)\right];$$

$$\vartheta = (2/G)(1 - v),$$

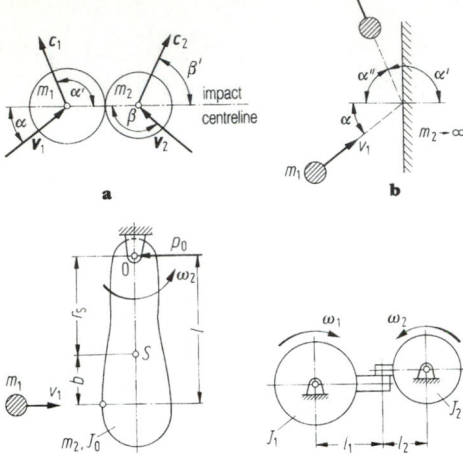

Figure 18. Impact: **a** offset central impact, **b** law of reflection, **c** eccentric impact, **d** torsional impact.

G is the shear modulus, v the transverse strain. In addition, for the collision time

$$T = k_2/\sqrt[5]{v} \text{ with } k_2 = 2.943 \left(\frac{5}{4c_1} \frac{m_1 m_2}{m_1 + m_2} \right)^{2/5}.$$

3.6.2 Oblique Impact

With the designations as per **Fig. 18a**, the following equations apply:

$$v_1 \sin \alpha = c_1 \sin \alpha', \quad v_2 \sin \beta = c_2 \sin \beta',$$

$$c_1 \cos \alpha' = \eta_1 \cos \alpha$$

$$\qquad - [(v_1 \cos \alpha - v_2 \cos \beta)(1 + k)/(1 + m_1/m_2)],$$

$$c_2 \cos \beta' = v_2 \cos \beta$$

$$\qquad - [(v_2 \cos \beta - v_1 \cos \alpha)(1 + k)/(1 + m_2/m_1)],$$

from which are obtained α', β', c_1 and c_2.

Example *Impact of a Sphere Against a Wall* (**Fig. 18b**). With $v_2 = c_2 = 0$ and $m_2 = -\infty$, from the equations above are derived

$$c_1 \cos \alpha' = -kv_1 \cos \alpha, \quad -\tan \alpha' = \tan \alpha'' = (\tan \alpha)/k \text{ and}$$

$$c_1 = -kv_1 \cos \alpha/\cos \alpha' = -v_1 \cos \alpha \sqrt{k^2 + \tan^2 \alpha}.$$

For $k = 1$, $\alpha' = \pi - \alpha$ or $\alpha'' = \alpha$ and $c_1 = v_1$, i.e. the angle of incidence is equal to the angle of emergence (law of reflection) with constant velocities.

3.6.3 Eccentric Impact

If a mass m_1 impacts against a suspended body capable of swinging (**Fig. 18c**) with the moment of inertia J_0 about the centre of rotation 0, then all the formulae for the straight central impact apply if, in this case, m_2 is replaced by the reduced mass $m_{2\text{red}} = J_0/l^2$. In addition, the dynamic relationships $v_2 = \omega_2 l$ etc. also apply. For the impulse of force on the point of suspension, there applies (if $\omega_2 = 0$)

$$p_0 = (1 + k)m_1 v_1 (J_0 - m_2/r_s)/(J_0 + m_1 l^2).$$

This impulse becomes zero or

$$l = l_r = J_0/(m_2 r_s) \quad \text{or} \quad r_s = r_{sr} = J_s/(m_2 b).$$

l_r or r_{sr} provides the position of the impact mid-point, which remains free of forces at impact, or around which a freely impacted body rotates (instantaneous centre of rotation). l_r is at the same time the reduced pendulum length when replaced by a mathematical thread pendulum.

3.6.4 Rotary Impact

For two rotating bodies that impact with one another (**Fig. 18d**), $m_1 = J_1/l_1^2$, $m_2 = J_2/l_2^2$, $v_1 = \omega_1 l_1$, $v_2 = \omega_2 l_2$ etc. are taken, and the problem thereby reverts to the straight central impact. The formulae in A3.6.1 then apply.

4 Mechanical Vibrations

4.1 One-Degree-of-Freedom Systems

Examples of these are the spring–mass system, the physical pendulum, or a rigid-body system reduced to one degree of freedom by constraints (**Fig. 1**). Initially, only linear systems are examined; in these cases, the differential equations themselves as well as the coefficients are linear. A precondition for this is the linear spring characteristic $F_c = cs$ (**Fig. 2b**).

4.1.1 Free, Undamped Vibrations

Spring-mass System (Fig. 1a). From the dynamic law there follows, with the displacement $\hat{s}$ from the zero position and the spring rate $\hat{c}$, the differential equation

$$F_G - c\hat{s} = m\ddot{\hat{s}} \quad \text{or} \quad \ddot{\hat{s}} + \omega_1^2 \hat{s} = g \quad \text{with} \quad \omega_1^2 = c/m.$$

This is derived from the energy equation $U + E = \text{const.}$ or from

$$\frac{d}{dt}(U + E) = \frac{d}{dt}\left[mg(b - \hat{s}) + \frac{c}{2}\hat{s}^2 + \frac{m}{2}\dot{\hat{s}}^2 \right] = 0, \quad \text{i.e.}$$

$$-mg\hat{s} + c\hat{s}\dot{\hat{s}} + m\dot{\hat{s}}\ddot{\hat{s}} = 0, \quad \text{i.e.}$$

$$\ddot{\hat{s}} + (c/m)\hat{s} = g. \qquad (1)$$

The solution is $\hat{s}(t) = C_1 \cos \omega_1 t + C_2 \sin \omega_1 t + mg/c$. The particular solution mg/c corresponds to the static displacement $\hat{s}_{st} = F_G/c$; the vibration therefore takes place about the static position of rest:

$$s(t) = \hat{s}(t) - \hat{s}_{st}(t) = C_1 \cos \omega_1 t + C_2 \sin \omega_1 t$$

$$= A \sin(\omega_1 t + \beta). \qquad (2)$$

In this situation, the amplitude of the vibration is $A = \sqrt{C_1^2 + C_2^2}$ and the phase displacement $\beta = \arctan (C_1/C_2)$. C_1 and C_2 and A and β are to be determined from the initial conditions; e.g. $s(t = 0) = s_1$ and $\dot{s}(t = 0) = 0$ gives

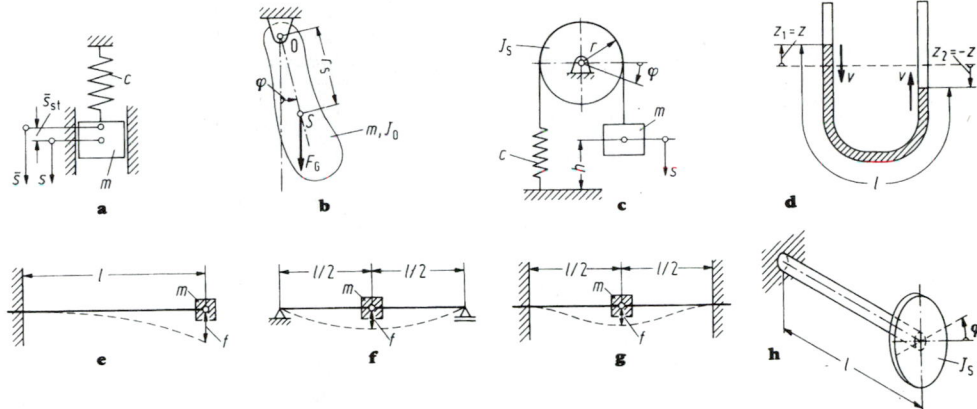

Figure 1. Vibrating elements with one degree of freedom: **a** spring–mass system; **b** physical pendulum; **c** rigid-body system; **d** oscillating water column; **e** beams with concentrated mass, secured on one side; **f** mounted on jointed bearings, and **g** secured on both sides; **h** torsional vibrat.

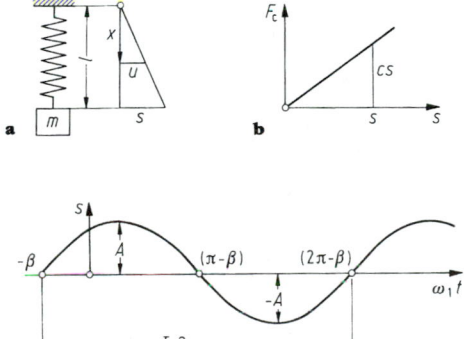

Figure 2. Harmonic vibration: **a** vibrator; **b** spring characteristic, **c** path-time function.

$C_2 = 0$ and $C_1 = s_1$ or $A = s_1$ and $\beta = \pi/2$. The vibration is a harmonic movement with the fundamental or angular frequency (number of vibrations in 2π seconds) $\omega_1\sqrt{c/m}$ or the Hertzian frequency $\nu_1 = \omega_1/2\pi$ and the period of vibration $T = 1/\nu_1 2\pi\omega$ (**Fig. 2c**).

Maximum values: Velocity $v = A\omega_1$, acceleration $a = A\omega_1^2$, spring force $F_c = cA$.

For the angular frequency, there applies, with the static displacement $\bar{s}_{st} = F_G/c$, i.e. $c = mg/\bar{s}_{st}$ and also $\omega_1 = \sqrt{g/\bar{s}_{st}}$.

Determining the Spring Rate. Every elastic system represents a spring. The spring rate is $c = F/f$, if f is the displacement of the mass as a consequence of the force F. For springs according to **Fig. 1e–g**, $c = F/(Fl^3/3EI_y) = 3EI_y/l^3$, $c = 48EI_y/l^3$ and $c = 192EI_y/l^3$.

Connecting Springs. Connecting in parallel (**Fig. 3a, b**):

$$c = c_1 + c_2 + c_3 + \ldots = \Sigma\, c_i. \qquad (3)$$

Connecting in series or in sequence (**Fig. 3c**):

$$1/c = 1/c_1 + 1/c_2 + \ldots = \Sigma\, 1/c_i. \qquad (4)$$

Considering the Spring Mass. On the assumption that the displacements are equal to those for static dis-

placement, i.e. $u(x) = (s/l)x$ (**Fig. 2a**), it follows, with $dm = (m_F/l)\,dx$, by equating the dynamic energies, that

$$(1/2)\int \dot{u}^2\, dm = (1/2)\dot{s}^2 \int_{x=0}^{l} (x^2/l^3)m_F\, dx$$

$$= (\dot{s}^2/2)(m_F/3) = km_F\dot{s}^2/2;$$

in other words, $k = 1/3$, i.e. a third of the spring mass is to be added to the vibrating mass m. For springs according to **Fig. 1e, f**, $k = 33/140$ and $k = 17/35$.

Pendulum Oscillation. For the physical pendulum (**Fig. 1b**), the law of dynamics provides the rotational motion in respect of the zero point

$$J_0\ddot{\varphi} = -F_G r_s \sin\varphi \quad \text{or} \quad \ddot{\varphi} + (mgr_s/J_0)\sin\varphi = 0.$$

For minor displacements $\sin\varphi \approx \varphi$, i.e. $\ddot{\varphi} + \omega_1^2\varphi = 0$ with $\omega_1^2 = g/l_r$ and $l_r = J_0/(mr_s)$ ($l_r =$ reduced pendulum length). For the mathematical thread pendulum, with the mass m at the end, $r_s = l$, $J_0 = ml^2$ and $\omega_1^2 = g/l$.

Rotary Oscillation. For the pulley according to **Fig. 1h**, A3 Eq. (45) gives $J_S\ddot{\varphi} = -M_t = -(GI_t/l)\varphi$ or $\ddot{\varphi} + \omega_1^2\varphi = 0$, with $\omega_1 = \sqrt{GI_t/(lJ_S)}$. Here, I_t is the torsion surface moment of the torsion rod. The rotary inertia of the torsion spring is taken into consideration by adding $J_F/3$ to J_S of the pulley.

Rigid Body System (e.g. **Fig. 1c**).

$$E+U=m\dot{s}^2/2+J_S\dot{\varphi}^2/2+cs^2/2+mg(h-s)=\text{const.},$$

$$d(E+U)/dt=m\ddot{s}\dot{s}+J_S\dot{\varphi}\ddot{\varphi}+cs\dot{s}-mg\dot{s}=0.$$

From this, with $\varphi = s/r$, $\dot{\varphi} = \dot{s}/r$ and $\ddot{\varphi} = \ddot{s}/r$, is derived

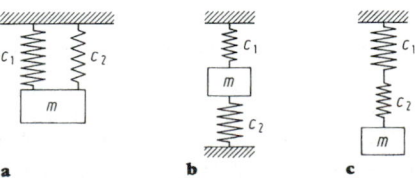

Figure 3. Springs: **a** and **b** parallel connection, **c** connected in series.

$$\ddot{s} + \omega_1^2 s = mg/(m + J_s/r^2),$$

where $\omega_1^2 = c/(m + J_s/r^2)$. Further solution as for the spring–mass system.

4.1.2 Free, Damped Vibration

Damping Due to Constant Friction Force (Coulomb's law of friction). For the spring–mass system,

$$\ddot{s} + \omega_1^2 s = \mp F_r/m$$

(minus for forwards and plus for return). The solution for the first backwards movement with the initial conditions $s(t_0 = 0) = s_0$, $\dot{s}(t_0 = 0) = 0$ is $s(t) = (s_0 - F_r/c)\cos \omega_1 t + F_r/c$. First reversal for $\omega_1 t_1 = \pi$ at the point $s_1 = -(s_0 - 2F_r/c)$, and accordingly there follow $s_2 = +(s_0 - 4F_r/c)$ and $|s_n| = s_0 - n \cdot 2F_r/c$. The oscillation is retained for as long as $c|s_n| \geq F_r$ i.e. for $n \leq (cs_0 - F_r)/(2F_r)$. The oscillation amplitude undergoes linear decrease with time, i.e. $A_n - A_{n-1} = 2F_r/c = $ const.; the amplitudes form an arithmetical series.

Damping Proportional to Velocity. In oscillation dampers (gas or liquid dampers), a friction force $F_r = kv = k\dot{s}$ occurs. For the spring–mass system, there applies (**Fig. 4a**):

$$\ddot{s} + (k/m)\dot{s} + (c/m)s = 0 \quad \text{or} \quad \ddot{s} + 2\delta\dot{s} + \omega_1^2 s = 0, \tag{5}$$

$k = $ damping constants, $\delta = k/(2m)$.

Solution for *weak damping*, i.e. for $\lambda^2 = \omega_1^2 - \delta^2 > 0$: $s(t) = Ae^{-\delta t} \sin(\lambda t + \beta)$, i.e. an oscillation with reducing amplitude according to $e^{-\delta t}$ and the angular frequency $\lambda = \sqrt{\omega_1^2 - \delta^2}$ (**Fig. 4b**). The normal frequency becomes smaller as the damping increases, and the period of oscillation $T = 2\pi/\lambda$ becomes correspondingly greater.

Zero setting of $s(t)$ with $t = (n\pi - \beta)/\lambda$.
Extreme values at $t_n = [\arctan(\lambda/\delta) + n\pi - \beta]/\lambda$.
Points of contact at $t'_n = [(2n + 1)\pi/2 - \beta]/\lambda$,

$$t'_n - t_n = \text{const} = [\arctan(\delta/\lambda)]/\lambda.$$

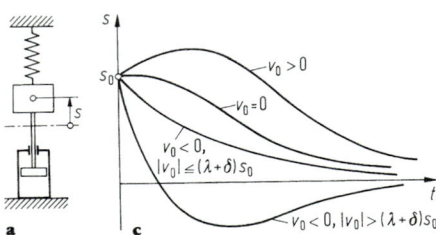

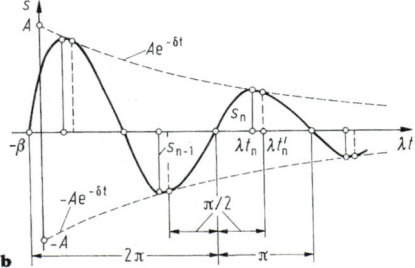

Figure 4. Damped free vibration: **a** vibrator, **b** weak and **c** strong damping.

Ratio of amplitudes

$$|s_{n-1}|/|s_n| = \text{const.} = e^{\delta\pi/\lambda} = e^{\delta T/2} = q.$$

Logarithmic decrement $\vartheta = \ln q = \delta T/2$ provides $\delta = 2\vartheta/T$ or $k = 2m\delta$ from the measurement of the period of oscillation.

With *strong damping*, i.e. $\lambda^2 = \delta^2 - \omega_1^2 \geq 0$ an aperiodic motion occurs with the solutions

$$s(t) = e^{-\delta t}(C_1 e^{\lambda t} + C_2 e^{-\lambda t}) \quad \text{for} \quad \lambda^2 > 0 \quad \text{and}$$

$$s(t) = e^{-\delta t}(C_1 + C_2 t) \quad \text{for} \quad \lambda^2 = 0.$$

Depending on the individual initial conditions (s_0, v_0), differing sequences of motion occur (**Fig. 4c**).

4.1.3 Forced, Undamped Vibrations

Forced vibrations are caused by kinematic independent or outside excitation (e.g. movement of the point of suspension), or dynamic outside excitation (unbalance forces at the mass).

With kinematic excitation (e.g. as per **Fig. 5a**), there applies

$$m\ddot{s} + c(s - r\sin\omega t) = 0, \quad \text{i.e.} \quad \ddot{s} + \omega_1^2 s = \omega_1^2 r\sin\omega t, \tag{6}$$

and with dynamic excitation (e.g. as per **Fig. 5b**)

$$(m + 2m_1)\ddot{s} + cs = 2m_1 e\omega^2 \sin\omega t, \quad \text{i.e.}$$

$$\ddot{s} + \omega_1^2 s = \omega^2 R\sin\omega t, \tag{7}$$

with $\omega_1^2 = c/(m + 2m_1)$, $R = 2m_1 e/(m + 2m_1)$. The two equations differ only in the factor on the right-hand side.

For random periodic excitation $f(t)$ the following applies:

$$\ddot{s} + \omega_1^2 s = f(t), \tag{8}$$

where $f(t)$ can be represented by a Fourier series (harmonic development):

$$f(t) = \Sigma(a_j \cos j\omega t + b_j \sin j\omega t), \quad \omega = 2\pi/T, \tag{9}$$

with the Fourier coefficients $a_j = (2/T) \int_0^T f(t) \cos j\omega t \, dt$,

$b_j = (2/T) \int_0^T f(t) \sin j\omega t \, dt$. If $s_j(t)$ is a solution for the differential equation $\ddot{s}_j + \omega_1^2 s_j = a_j \cos j\omega t + b_j \sin j\omega t$, then the total solution is $s(t) = \Sigma s_j(t)$.

Examination of the basic case $\ddot{s} + \omega_1^2 s = b \sin \omega t$ shows that the solution is composed of one homogeneous and one particular portion

$$s(t) = s_h(t) + s_p(t) = A\sin(\omega_1 t + \beta)$$
$$+ [b/(\omega_1^2 - \omega^2)]\sin\omega t.$$

For the initial conditions $s(t = 0) = 0$ and $\dot{s}(t = 0) = 0$ is derived

$$s(t) = [b/(\omega_1^2 - \omega^2)][\sin\omega t - (\omega/\omega_1)\sin\omega_1 t],$$

i.e. the overlap of the harmonic fundamental oscillation with the harmonic excitation oscillation. For $\omega \approx \omega_1$ the curve of $s(t)$ represents a beat (**Fig. 5c**). This solution fails in the resonance case $\omega = \omega_1$. It then reads

$$s(t) = A\sin(\omega t + \beta) - (b/\omega)t\cos\omega t$$

or for $s(t = 0)$ and $\dot{s}(t = 0) = 0$

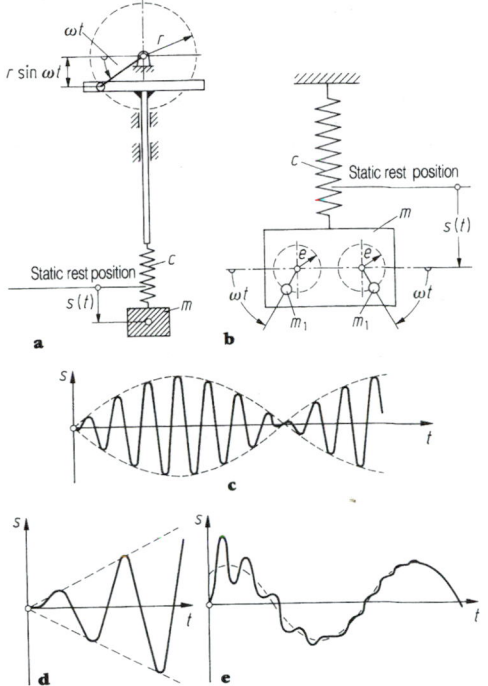

a **b**

c

d **e**

Figure 5. Forced vibration: **a** kinematic and **b** dynamic excitation; **c** beating; **d** resonance behaviour; **e** transient process.

$$s(t) = (b/\omega^2)(\sin \omega t - \omega t \cos \omega t);$$

i.e. in the resonance case, the deflections become infinite with time (**Fig. 5d**). If the excitation function applies in accordance with Eq. (9), then resonance will also occur for $\omega_1 = 2\omega_1, 3\omega \dots$.

4.1.4 Forced Damped Vibrations

With velocity-proportional damping and harmonic excitation (see A4.1.3), there applies:

$$\ddot{s} + 2\delta\dot{s} + \omega_1^2 s = b \sin \omega t \quad \text{or}$$

$$s(t) = Ae^{-\delta t} \sin (\lambda t + \beta) + C \sin (\omega t - \psi). \quad (10)$$

The first section, the damped fundamental oscillation, decays with time (fundamental oscillation process), and thereafter the forced oscillation has the same frequency as the excitation (**Fig. 5e**). Factor C and the phase displacement ψ in the second section (excited oscillation or particular solution) are derived by inclusion in the differential equation and coefficient comparison with

$$C = b \Big/ \sqrt{(\omega_1^2 - \omega^2)^2 + 4\delta^2\omega^2} \quad \text{and}$$

$$\psi = \arctan[2\delta\omega/(\omega_1^2 - \omega^2)]. \quad (11)$$

With $b = \omega_1^2 r$ with kinematic excitation and $b = \omega^2 R$ with dynamic excitation, the amplification factors are derived (**Fig. 6a, b**):

$$V_k = 1 \Big/ \sqrt{(1 - \omega^2/\omega_1^2)^2 + (2\delta\omega/\omega_1^2)^2} \quad \text{and}$$

$$V_d = V_k(\omega/\omega_1)^2.$$

From $dV_k/d\omega = 0$ it follows for the points of resonance ω^* with kinematic excitation $\omega^*/\omega_1 = \sqrt{1 - 2\delta^2/\omega_1^2}$ or with dynamic excitation $\omega^*/\omega_1 = 1/\sqrt{1 - 2\delta^2/\omega_1^2}$. The points of resonance therefore lie in the subcritical range with kinematic excitation, and in the supercritical range with dynamic excitation (**Fig. 6a, b**). The resonance amplitude is $C^* = (b/2\delta)/\sqrt{\omega_1^2 - \delta^2}$. For the phase angle ψ according to Eq. (11), **Fig. 6c** applies to both types of excitation. For $\omega < \omega_1$, $\psi < \pi/2$, and for $\omega > \omega_1$, $\psi \sqrt{} \pi/2$. Without friction ($\delta = 0$), excitation and deflection are in phase for $\omega < \omega_1$, while for $\omega > \omega_1$ they are mutually opposed.

4.1.5 Critical Speed of Shafts

Critical speed and (Hertzian) fundamental bending frequency are identical if the gyro effect due to the disc not being located in the centre of the supporting shaft (**Fig. 7a**) and the springing capacity of the bearings are ignored [1, 2]. For the fundamental bending frequency, $\omega_1 = \sqrt{c/m_1}$ (disregarding the mass of the shaft), with $c = 3EI_y l/(a^2b^2)$ (see A4.1.1 and B2, **Table 5a**). If e is the eccentricity of the disc and w_1 is the elastic defor-

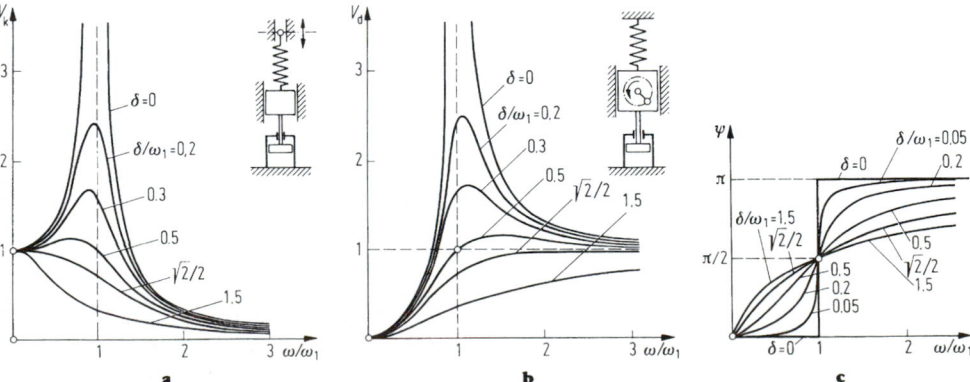

a **b** **c**

Figure 6. Damped forced vibration: **a** amplification factor with kinematic and **b** dynamic excitation, **c** phase angle.

mation due to the centrifugal forces, then it follows, from the equilibrium between the elastic restoring and centrifugal forces, that

$$cw_1 = m_1\omega^2(e + w_1), \quad w_1 = e\frac{(\omega/\omega_1)^2}{1 - (\omega/\omega_1)^2}. \quad (12)$$

For $\omega = \omega_1$, it follows that $w_1 \to \infty$, in other words resonance (**Fig. 7b**). By contrast, the value $w_1 = -e$ occurs for $\omega/\omega_1 \to \infty$, i.e. the shaft centres itself above ω_1, and the centre of gravity for $\omega \to \infty$ is located precisely on the connecting line of the bearings. For $e = 0$, it follows from Eq. (12) that $w_1(c - m_1\omega^2) = 0$, i.e. $w_1 \neq 0$ for $\omega = \sqrt{c/m_1} = \omega_1$, in other words critical speed $n = \omega/(2\pi) = \omega_1/(2\pi) = v_1$.

For other types of bearings, a corresponding c is to be applied (see A4.1.1). Damping is as a rule very low for rotating shafts, and has hardly any influence on critical speed.

4.2 Multi-Degree-of-Freedom Systems (Coupled Vibrations)

Figure 8a–e represents a single mass system with three degrees of freedom in the plane and several two-mass systems with two degrees of freedom, which are connected or elastically coupled etc. A system with n degrees of freedom has n natural frequencies. The derivation of n coupled differential equations is performed in cases of several degrees of freedom using the Lagrange equations (see A3.3.6).

4.2.1 Free Multi-Degree-of-Freedom Vibrations

For a non-damped system as per **Fig. 8b**

$$m_1\ddot{s}_1 = -c_1 s_1 + c_2(s_2 - s_1), \quad m_2\ddot{s}_2 = -c_2(s_2 - s_1) \quad \text{or}$$

$$m_1\ddot{s}_1 + (c_1 + c_2)s_1 - c_2 s_2 = 0, \quad m_2\ddot{s}_2 + c_2 s_2 - c_2 s_1 = 0; \quad (13)$$

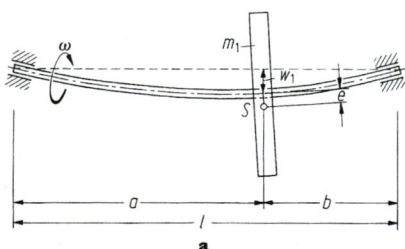

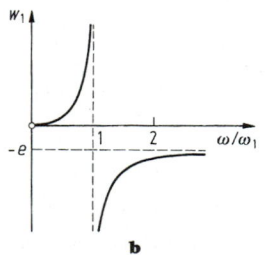

Figure 7. Critical speed: **a** single engaged shaft, **b** resonance curves.

s_1, s_2 are displacements from the static position of rest. The solution condition (see A4.1.1)

$$s_1 = A \sin(\omega t + \beta) \quad \text{and} \quad s_2 = B \sin(\omega t + \beta) \quad (14)$$

provides, with $c = c_1 + c_2$

$$A(m_1\omega^2 - c) + Bc_2 = 0 \quad \text{and} \quad (15a)$$

$$Ac_2 + B(m_2\omega^2 - c_2) = 0. \quad (15b)$$

This linear homogeneous equation system for A and B only has solutions that differ from zero when the denominator determinants disappear, i.e.

$$m_1 m_2\omega^4 - (m_1 c_2 + m_2 c)\omega^2 + (cc_2 - c_2^2) = 0.$$

The two solutions ω_1 and ω_2 of this characteristic equation are the natural frequencies of the system. In view of the fact that the differential equations are linear, the superposition law applies and the whole solution reads

$$s_1 = A_1 \sin(\omega_1 t + \beta_1) + A_2 \sin(\omega_2 t + \beta_2), \quad (16a)$$

$$s_2 = B_1 \sin(\omega_1 t + \beta_1) + B_2 \sin(\omega_2 t + \beta_2). \quad (16b)$$

According to Eq. (15a), $A_1/B_1 = c_2/(c - m_1\omega_1^2) = 1/\kappa_1$ or $A_2/B_2 = c_2/(c - m_1\omega_2^2) = 1/\kappa_2$ and therefore from Eq. (16b)

$$s_2 = \kappa_1 A_1 \sin(\omega_1 t + \beta_1) + \kappa_2 A_2 \sin(\omega_2 t + \beta_2). \quad (16c)$$

Equations (16a, c) contain four constants, $A_1, A_2, \beta_1, \beta_2$, for matching to the four initial conditions. The oscillation process is periodic only if ω_1 and ω_2 are in a rational ratio to each other. If $\omega_1 \approx \omega_2$, beat frequencies occur.

With more than two degrees of freedom, a statement according to Eq. (14) is made for each. From the coefficient determinants set equal to zero, a characteristic equation is derived for n degrees of freedom, from which n natural frequencies follow.

For the damped oscillation, the differential equations at two degrees of freedom for the system read as per **Fig. 8b**:

$$m_1\ddot{s}_1 + k_1\dot{s}_1 + (c_1 + c_2)s_1 - c_2 s_2 = 0,$$

$$m_2\ddot{s}_2 + k_2\dot{s}_2 + c_2 s_2 - c_2 s_1 = 0.$$

With the statement $s_1 = \bar{A}e^{\kappa t}$ and $s_2 = \bar{B}e^{\kappa t}$ an equation is again derived of the fourth degree with complex roots in conjugated pairs $\kappa_1 = -\rho_1 + i\omega_1$ etc. and therefore the final solution

$$s_1(t) = e^{-\rho_1 t} A_1 \sin(\omega_1 t + \beta_1) + e^{-\rho_2 t} A_2 \sin(\omega_2 t + \beta_2),$$

$$s_2(t) = e^{-t_1 t} B_1 \sin(\omega_1 t + \beta_1) + e^{-\rho_2 t} B_2 \sin(\omega_2 t + \beta_2).$$

Between A_1 and B_1 and A_2 and B_2 there is again a linear relationship analogous to the undamped oscillation.

4.2.2 Forced Multi-Degree-of-Freedom Vibrations

For an *undamped system* according to **Fig. 8b**, with kinematic or dynamic excitation $b_1 \sin \omega t$ of the mass m_1,

$$m_1\ddot{s}_1 + (c_1 + c_2)s_1 - c_2 s_2 = b_1 \sin \omega t,$$

$$m_2\ddot{s}_2 + c_2 s_2 - c_2 s_1 = 0. \quad (17)$$

In view of the fact that the homogeneous part of the solution decays as a result of the weak damping that is always present during the transient process, it is sufficient to consider the particular solution. For this, it follows, with the statement

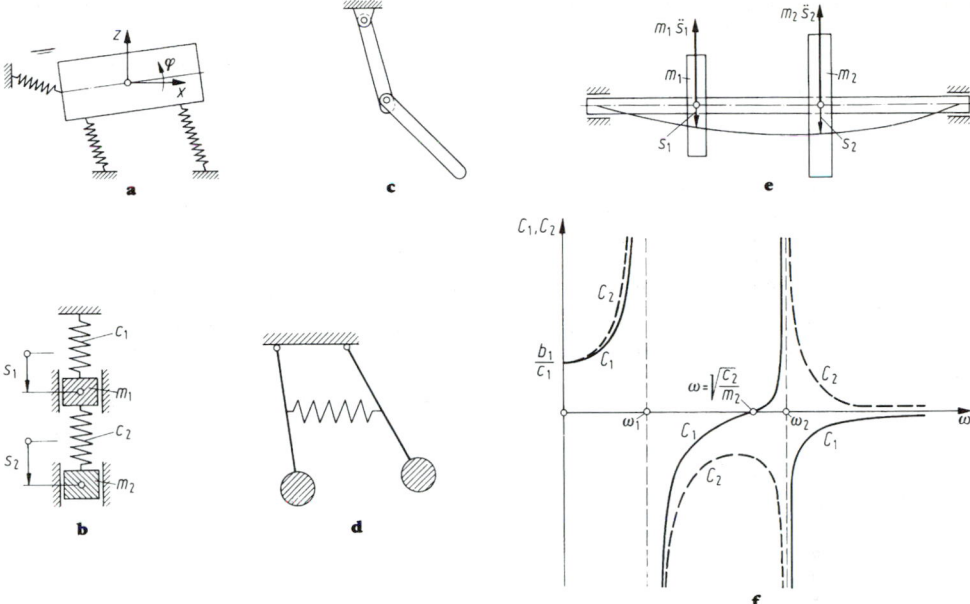

Figure 8. Oscillation of coupled circuits: **a** three and **b–e** up to two degrees of freedom, **f** resonance curves with two degrees of freedom.

$$s_1 = C_1 \sin(\omega t - \psi_1), \quad s_2 = C_2 \sin(\omega t - \psi_2) \quad (18)$$

by insertion in Eq. (17) and coefficient comparison $\psi_1 = 0$, $\psi_2 = 0$, and with $c_1 + c_2 = c$, that

$$C_1(m_1\omega^2 - c) + C_2 c_2 = -b_1,$$

$$C_1 c_2 + C_2(m_2\omega^2 - c_2) = 0. \quad (19)$$

From this, $C_1 = Z_1/N$ and $C_2 = Z_2/N$, where the denominator determinants $N = m_1 m_2 \omega^4 - (m_1 c_2 + m_2 c)\omega^2 + (cc_2 - c_2^2)$ coincide with those in the characteristic equation in A4.2.1. Resonance occurs if $N = 0$, i.e. for natural frequencies ω_1 and ω_2 of the free oscillator. The numerator determinants are $z_1 = b_1(c_2 - m_2\omega^2)$, $Z_2 = b_1 c_2$. For kinematic excitation $(b_1 = \omega_1^2 r)$, in **Fig. 8f** the amplitudes C_1 and C_2 are represented as function of ω. For $\omega = \sqrt{c_2/m_2}$, $C_1 = 0$ and C_2 is relatively small, i.e. the mass m_1 is at rest (mass m_2 has the effect of an attenuator). With n masses, resonance occurs at the n natural frequencies. In this instance, the deflections need not always increase to infinity, and some may even remain finite (apparent resonance [1]).

For *damped forced vibration*, Eq. (17) for example takes the form

$$m_1\ddot{s}_1 + k\dot{s}_1 + cs_1 - c_2 s_2 = b_1 \sin \omega t,$$

$$m_2\ddot{s}_2 + k_2\dot{s}_2 + c_2 s_2 - c_2 s_1 = 0 \quad (20)$$

$(c = c_1 + c_2)$. Without the transient process, i.e. the homogeneous solution part, and with the forced (particular) part of the solution according to Eq. (18), after insertion in Eq. (20) and after comparison of coefficient, the values for the amplitudes C_1, C_2 and the phase angles ψ_1 and ψ_2 follow. Resonance pertains if $C_1 - C_2 =$ extr., i.e. ω_1 and ω_2 follow from $d(C_1 - C_2)/dt = 0$.

In a system of n masses, the expenditure of effort on calculation is very great. Accordingly, in cases of weak damping, a satisfactory result can be achieved by the

determination of the natural frequencies for the undamped system.

4.2.3 Natural Frequency of Undamped Systems

Flexural Vibrations and Critical Speeds of Shafts with Multiple Engagement. The Hertz frequencies of the flexural vibrations and critical speeds (without gyroscopic effect) are identical. With $s_i = w_i \sin \omega t$ it follows, taking into account the forces of inertia $-m_i\ddot{s}_i = m_i\omega^2 w_i \sin \omega t$ for the flexural vibration (**Fig. 8c**), that

$$s_1 = -\alpha_{11} m_1\ddot{s}_1 - \alpha_{12} m_2\ddot{s}_2,$$

$$s_2 = -\alpha_{21} m_1\ddot{s}_1 - \alpha_{22} m_2\ddot{s}_2 \quad (21)$$

or

$$w_1 = \alpha_{11} m_1\omega^2 w_1 + \alpha_{12} m_2\omega^2 w_2,$$

$$w_2 = \alpha_{21} m_1\omega^2 w_1 + \alpha_{22} m_2\omega^2 w_2. \quad (22)$$

Equation (22) also occurs for the circulating shaft with the centrifugal forces $m_i\omega^2 w_i$. The α_{ik} are influence coefficients; they are equal to the sagging under load w_i resulting from a force $F_k = 1$. Calculation is conveniently carried out to by the principle of the virtual displacement for elastic bodies from or according to Mohr's process or by other methods (tabular values, integration, etc.: see B2.4.8). $\alpha_{ik} = \alpha_{ki}$ applies (Maxwell's law). From Eq. (22) there follows

$$w_1(\alpha_{11} m_{11} - 1/\omega^2) + w_2 \alpha_{12} m_2 = 0,$$

$$w_1 \alpha_{21} m_1 + w_2(\alpha_{22} m_2 - 1/\omega^2) = 0. \quad (23)$$

These only have non-trivial solutions if the determinants are zero, i.e. (with $1/\omega^2 = \Omega$) where

$$\Omega^2 - (m_1\alpha_{11} + m_2\alpha_{22})\Omega + (\alpha_{11}\alpha_{22} - \alpha_{12}\alpha_{21})m_1 m_2 = 0.$$

From these there follow two solutions, $\Omega_{1,2}$ or $\omega_{1,2}$, for the natural frequencies. For the amplitude ratios there is derived from Eq. (23) $w_2/w_1 = (1/\omega^2 - \alpha_{11}m_1)/(\alpha_{12}m_2)$. For the shaft occupied n times, n natural frequencies are obtained analogously from an equation of the n-th degree.

Approximate Values with the Rayleigh Quotient. From $U_{max} = E_{max} = \omega^2 \bar{E}_{max}$ there follows the Rayleigh quotient

$$R = \omega^2 = U_{max}/\bar{E}_{max}. \tag{24}$$

$$U_{max} = (1/2) \int M_b^2(x) \, dx/(EI_y),$$

$$\bar{E}_{max} = (1/2) \int w^2(x) \, dm + (1/2) \, \Sigma \, m_i w_i^2.$$

$w(x)$ and $M_b(x) = EI_y w''(x)$ are bending line and bending moment line during vibration. For the true bending line (eigenfunction), R becomes a minimum. For an equational function which satisfies the peripheral conditions (e.g. bending line and bending moment line as a result of net weight), good approximate values are derived for R_1 and ω_1 (first natural frequency). The approximate value is always greater than the true value. By means of a Ritz formula of several functions $w(x) = \Sigma \, c_k v_k(x)$ there follow from

$$I = U_{max} - \omega^2 \bar{E}_{max} = (1/2) \int [EI_y w''^2(x)$$

$$- \, \omega^2 w^2(x)\rho A] \, dx - (1/2)\omega^2 \, \Sigma \, m_i w_i^2 = \text{extr.},$$

i.e. $\partial I/\partial c_j = 0$ $(j = 1,2,\dots,n)$, n homogeneous linear equations and, by zero-setting of the determinants, an equation of the n-th degree for the n natural frequencies as an approximate value. It is also possible to estimate the proper function for each higher proper value, determine it directly from Eq. (24), and possibly improve it step by step [1–3].

Torsional Vibration of a Multiple-Occupied Shaft. Processes similar to the flexural vibration are available (see J2.1).

4.2.4 Vibration of Continuous Systems

A continuum subjected to a mass has an infinite number of natural frequencies. Partial differential equations are obtained from the laws of dynamics as equations of motion. Satisfying the peripheral conditions provides transcendent proper number equations. For approximate solutions, the Rayleigh quotient and the Ritz process are taken as the basis (see A4.2.3).

Flexural Vibrations of Rods. The differential equation reads

$$\rho A \frac{\partial^2 w}{\partial t^2} = -p(x,t) - \frac{\partial^2}{\partial x^2}\left[EI_y \frac{\partial^2 w}{\partial x^2}\right]$$

or for free vibration and constant cross-section,

$$\partial^2 w/\partial t^2 = -c^2 \partial^4 w/\partial x^4, \quad c^2 = EI_y/(\rho A). \tag{25}$$

The product statement by Bernoulli

$$w(x,t) = X(x)T(t),$$

used in Eq. (25) provides

$$X\ddot{T} = -c^2 X^{(4)}T \quad \text{or} \quad \ddot{T}/T = -c^2 X^{(4)}/X = -\omega^2,$$

i.e. $\ddot{T} + \omega^2 T = 0$ and $X^{(4)} - (\omega^2/c^2)X = 0$. With $\lambda^4 = (\omega^2/c^2)l^4$ the solution becomes:

$w(x,t) = A \sin(\omega t + \beta)[C_1 \cos(\lambda x/l) + C_2 \sin(\lambda x/l)$
$$+ C_3 \cosh(\lambda x/l) + C_4 \sinh(\lambda x/l)]. \tag{26}$$

For the rod as in **Fig. 9a**, the peripheral conditions are $X(0) = 0$, $X'(0) = 0$, $X''(l) = 0$, $X'''(l) = 0$. Accordingly, from Eq. (26) the proper value equation $\cosh \lambda \cos \lambda = -1$ follows, with the eigenvalues $\lambda_1 = 1.875$; $\lambda_2 = 4.694$; $\lambda_3 = 7.855$ etc. For the rods according to **Fig. 9b-d**, the first three eigenvalues are derived as $\lambda_1 = \pi$; 3.927; 4.730; $\lambda_2 = 2\pi$; 7.069; 7.853; $\lambda_3 = 3\pi$; 10.210; 10.996.

For rods with additional individual masses, the solution of Eq. (26) is to be applied for each section. After the fulfilment of the transition conditions etc., the frequency equation is obtained. In view of the great effort involved, the approximation is used with the Rayleigh quotient and the Ritz process (see A4.2.3).

Extensional Vibration of Rods. The differential equation reads

$$\rho A \frac{\partial^2 u}{\partial t^2} = \frac{\partial}{\partial x}\left[EA \frac{\partial u}{\partial x}\right] \quad \text{or, for } A = \text{const.,}$$

$$\partial^2 u/\partial t^2 = c^2 \partial^2 u/\partial x^2, \quad c^2 = E/\rho, \tag{27}$$

with the solution

$$u(x,t) = A\sin(\omega t + \beta)[C_1 \cos(\omega x/c) + C_2 \sin(\omega x/c)]. \tag{28}$$

After fulfilment of the boundary conditions, the following fundamental frequencies are obtained:

Rod secured at one end, and free at the other:
$$\omega_k = (k - 1/2)\pi c/l \qquad (k = 1,2,\dots);$$
Rod secured at both ends:
$$\omega_k = k\pi c/l \qquad (k = 1,2,\dots);$$
Rod free at both ends:
$$\omega_k = k\pi c/l \qquad (k = 1,2,\dots).$$

In the case of a rod additionally occupied by individual masses, the observations made for flexural vibration apply accordingly. The Rayleigh quotient is

$$R = \omega^2 = U_{max}/\bar{E}_{max} \quad \text{with}$$

$$U_{max} = (1/2) \int E \, Af'^2(x) \, dx, \quad \bar{E} = (1/2) \int \rho A f^2(x) \, dx,$$

when $f(x)$ is a comparative function fulfilling the boundary conditions (see also A4.2.3).

Torsional Vibrations of Rods. In this case, there applies

$$J \frac{\partial^2 \varphi}{\partial t^2} = \frac{\partial}{\partial x}\left[GI_t \frac{\partial \varphi}{\partial x}\right]$$

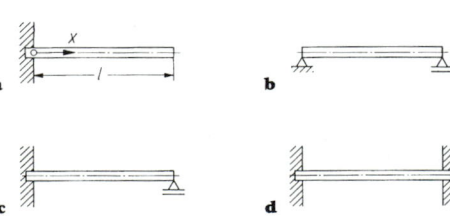

Figure 9. Flexural vibration of rods: **a** clamped on one side, **b** mounted on jointed bearings, **c** mounted on jointed bearings and clamped, **d** clamped on both sides.

or, for I_t = const.,

$$\partial^2\varphi/\partial t^2 = c^2\partial^2\varphi/\partial x^2, \quad c^2 = GI_t/J. \tag{29}$$

Solution and proper values as for extensional vibration. For rods that are additionally occupied by rotational moments of inertia, the appropriate remarks apply as for flexural vibrations. The Rayleigh quotient is $R = \omega^2 = U_{max}/\bar{E}_{max}$, with

$$U_{max} = (1/2)\int GI_t f'^2(x)\,dx, \quad \bar{E} = (1/2)\int (J/l)f^2(x)\,dx.$$

Vibrations of Strings (tautly-stretched strings). In this case there applies

$$\partial^2 w/\partial t^2 = c^2\partial^2 w/\partial x^2, \quad c^2 = S/\mu \tag{30}$$

(S = tension, μ = mass per unit length). For solution of Eq. (30), see Eq. (28). Frequencies $\omega_k = k\pi c/l$ ($k = 1,2,\ldots$), where l = string length. Rayleigh quotient $R = \omega^2 = U_{max}/\bar{E}_{max}$, with $U_{max} = (1/2)S\int f'^2(x)\,dx$, $\bar{E}_{max} = (1/2)\mu\int f^2(x)\,dx$. $f(x)$ is a comparative function that satisfies the peripheral conditions (see also A4.2.3).

Vibrations of Membranes. For *rectangular membranes* there applies

$$S(\partial^2 w/\partial x^2 + \partial^2 w/\partial y^2) = \mu\partial^2 w/\partial t^2 \tag{31}$$

(S = tension per unit length, μ = mass per unit area), with the solution

$$w(x,y,t) = A\sin(\omega t + \beta)[C_1\cos\lambda x + C_2\sin\lambda x]$$
$$\cdot [D_1\cos\kappa y + D_2\sin\kappa y]. \tag{32}$$

With a and b as lateral lengths, there apply for the eigenvalues $\lambda_j j\pi/a$, $\kappa_k = k\pi/b$ ($j,k = 1,2,\ldots$) the following eigenfrequencies:

$$\omega_{jk} = \pi\sqrt{(S/\mu)[j^2/a^2 + k^2/b^2]} \quad (j,k = 1, 2, \ldots).$$

Rayleigh quotient: $R = \omega^2 = U_{max}/\bar{E}_{max}$ with

$$U_{max} = (S/2)\int\int\left[\left(\frac{\partial f}{\partial x}\right)^2 + \left(\frac{\partial f}{\partial y}\right)^2\right]dx\,dy,$$

$$\bar{E}_{max} = (\mu/2)\int\int f^2(x,y)\,dx\,dy.$$

$f(x,y)$ is a comparative function that fulfils the peripheral conditions (see also A4.2.3).

For *circular membranes* there applies, in polar coordinates with $c^2 = S/\mu$,

$$\frac{\partial^2 w}{\partial t^2} = c^2\left(\frac{\partial^2 w}{\partial r^2} + \frac{1}{r}\frac{\partial w}{\partial r} + \frac{1}{r^2}\frac{\partial^2 w}{\partial\varphi^2}\right) \tag{33}$$

with the solution

$$w(r,\varphi,t) = A\sin(\omega t + \beta)(C\cos n\varphi + D\sin n\varphi)\cdot J_n(\omega r/c)$$
$$(n = 0, 1, 2, \ldots). \tag{34}$$

$J_n(\omega r/c)$ are Bessel functions of the first kind [4]. (For rotationally symmetrical vibrations, $n = 0$.) Values $\omega_{nj} = (c/a)x_{nj}$ (a radius of the membrane, x_{nj} zero setting of the Bessel functions): $x_{01} = 2.405$; $x_{02} = 5.520$; $x_{11} = 3.832$; $x_{12} = 7.016$; $x_{21} = 5.135$ etc. Rayleigh quotient: $R = \omega^2 = U_{max}/\bar{E}_{max}$. For rotationally symmetrical vibrations,

$$U_{max} = (S/2)\int\left(\frac{df}{dr}\right)^2 2\pi r\,dr \quad \text{and}$$

$$\bar{E}_{max} = (\mu/2)\int f^2(r)2\pi r\,dr.$$

Flexural Vibrations of Plates. The differential equation reads, with $N = Eb^3/[12(1 - v^2)]$, for the *square plate*

$$\frac{\partial^2 w}{\partial t^2} = -\frac{N}{\rho b}\Delta\Delta w = -\frac{N}{\rho b}\left(\frac{\partial^4 w}{\partial x^4} + 2\frac{\partial^4 w}{\partial x^2\partial y^2} + \frac{\partial^4 w}{\partial y^4}\right). \tag{35}$$

With a and b as the lateral lengths, there applies, for the pivot-bearing mounted plate:

$$w(x,y,t) = A\sin(\omega t + \beta)\sin(j\pi x/a)\sin(k\pi y/b). \tag{36}$$

Proper values: $\omega_{jk} = (j^2/a^2 + k^2/b^2)\pi^2\sqrt{N/(\rho b)}$ (j, $k = 1, 2, \ldots$). Rayleigh quotient: $R = \omega^2 = U_{max}/\bar{E}_{max}$ with

$$U_{max} = (N/2)\int\int\left[\left(\frac{\partial^2 f}{\partial x^2} + \frac{\partial^2 f}{\partial y^2}\right)^2\right.$$
$$\left. - 2(1 - v)\left(\frac{\partial^2 f}{\partial x^2}\frac{\partial^2 f}{\partial y^2} - \left(\frac{\partial^2 f}{\partial x\partial y}\right)^2\right)\right]dx\,dy \quad \text{and}$$

$$\bar{E}_{max} = (\rho b/2)\int\int f^2(x,y)\,dx\,dy.$$

$f(x, y)$ is a comparative function which satisfies the peripheral conditions (see A4.2.3).

For the *circular plate*, with rotationally symmetrical vibration, $w = w(r,t) = f(r)\sin(\omega t + \beta)$ and therefore, according to Eq. (35),

$$(\omega^2\rho b/N)f(r) = \lambda^4 f(r) = \Delta\Delta f(r), \quad \text{i.e.}$$

$$\Delta\Delta f - \lambda^4 f = 0 \quad \text{or} \quad (\Delta + \lambda^2)(\Delta - \lambda^2)[f] = 0.$$

From this there follow the differential equations

$$\Delta f + \lambda^2 f = 0 \quad \text{and} \quad \Delta f - \lambda^2 f = 0 \tag{37}$$

or $\quad d^2 f/dr^2 + (1/r)df/dr + \lambda^2 f = 0$
and $\quad d^2 f/dr^2 + (1/r)df/dr - \lambda^2 f = 0$.

Superposed solutions of the Bessel differential equations (Eqs 37) are

$$f(r) = C_1 J_0(\lambda r) + C_2 N_0(\lambda r) + C_3 I_0(\lambda r) + C_4 K_0(\lambda r) \tag{38}$$

(N_0 = Neumann function; I_0, K_0 = modified Bessel functions [8]). For the pivot-bearing mounted plate with radius a, there follows from Eq. (38) the eigenvalue equation

$$J_0(\lambda a)\left[I_0(\lambda a) - \frac{I_1(\lambda a)}{\lambda a}\right] + I_0(\lambda a)\left[J_0(\lambda a) - \frac{J_1(\lambda a)}{\lambda a}\right] = 0 \tag{39}$$

with the solutions

$$\lambda_1 a = 2.108; \lambda_2 a = 5.42; \lambda_3 a = 8.59.$$

From this, $\omega = \lambda^2\sqrt{N/(\rho b)}$.

For the stressed circular plate, there follows from Eq. (38) the eigenvalue equation $J_0(\lambda a)I_1(\lambda a) + I_0(\lambda a)J_1(\lambda a) = 0$ with the solutions $\lambda_1 a = 3.190$; $\lambda_2 a = 6.306$; $\lambda_3 a = 9.425$. From this, $\omega = \lambda^2\sqrt{N/(\rho b)}$.

Rayleigh quotient: $R = \omega^2 = U_{max}/\bar{E}_{max}$. For rotationally symmetrical vibration,

$$U_{max} = (N/2) \int \left[\left(\frac{d^2f}{dr^2} + \frac{1}{r}\frac{df}{dr} \right)^2 \right.$$

$$\left. -2(1-v)\frac{1}{r}\frac{df}{dr}\frac{d^2f}{dr^2} \right] 2\pi r\, dr \quad \text{and}$$

$$\bar{E}_{max} = (\rho b/2) \int f^2(r) 2\pi r\, dr.$$

4.3 Non-linear Vibrations

Vibration problems of this kind lead to non-linear differential equations. Non-linear vibrations arise, for example, owing to non-linear spring characteristics or displacement forces (physical pendulum with large deflection values), or owing to displacement forces dependent not only on deflection but also on time (e.g. pendulum with displaced suspension point).

4.3.1 Systems with Non-linear Spring Characteristics

Here, $m\ddot{s} = F(s)$ (**Fig. 10a**), by way of approximation:

$$F(s) = -cs(1 + \varepsilon s^2)$$

($\varepsilon > 0$ superlinear, $\varepsilon < 0$ sublinear characteristic).

Freely Undamped Vibration. The differential equation reads

$$\ddot{s} + \omega_1^2 s(1 + \varepsilon s^2) = 0 \quad \text{or} \quad \ddot{s} + \omega_1^2 s + \omega_1^2 \varepsilon s^3 = 0. \tag{40}$$

Multiplication by $\dot{s}$ provides $\dot{s}\ddot{s} + \omega_1^2 s\dot{s} + \omega_1^2 \varepsilon s^3 \dot{s} = 0$ and from this integration with the initial conditions $s(t=0) = s_0$, $\dot{s}(t=0) = v_0$ and separation of the variables also gives

$$\dot{s}^2 + \omega_1^2(s^2 + \varepsilon s^4/2) = v_0^2 + \omega_1^2(s_0^2 + \varepsilon s_0^4/2) = C^2, \tag{41}$$

$$t(s) = \int_{s_0}^{s} ds / \sqrt{C^2 - \omega_1^2 s^2 - \omega_1^2 \varepsilon s^4/2}. \tag{42}$$

After conversion [5, 6], the integral provides an ellipti-

cal integral of the first kind [7]. Duration and frequency of vibration are dependent on the maximum deflection. For small deflections, by iterative approximation [1] for the frequency, $\omega = \sqrt{\omega_1^2(1 + 0.75\varepsilon A^2)}$, where A is the amplitude of the vibration deflection.

The physical pendulum with the reduced pendulum length $l = J_0/(mr_s)$ (see A3.6.3) can be reduced to a mathematical pendulum with $\ddot{\varphi} + (g/l)\sin\varphi = 0$. The solution again leads to an elliptical integral of the first kind, with the vibration duration $T = \sqrt{l/g}\, F(\pi/2,\kappa)$ for the moving pendulum ($\kappa^2 = \omega_1^2 l/(4g) < 1$). For smaller deflections, the following approximate solution [1] is derived:

$$T = 2\pi \sqrt{l/g}(1 + A^2/16).$$

Forced Vibrations. The differential equation reads

$$\ddot{s} + 2\delta\dot{s} + \omega_1^2(1 + \varepsilon s^2)s = a_0 \cos(\omega t + \beta) \tag{43}$$

for velocity-proportional damping and periodic excitation force. With $s = A\cos\omega t$, it follows from Eq. (43), after comparison of coefficients that

$$[(\omega_1^2 - \omega^2 + 0.75\omega_1^2\varepsilon A^2)^2 + 4\delta^2\omega^2]A^2 = a_0^2. \tag{44}$$

Figure 10b shows amplitudes as a function of the exciter frequency ω (resonance curves) for $\varepsilon > 0$ and $\varepsilon < 0$. In certain ranges, there are solutions with several interpretations. The central dotted line is not stable, and will not persist. Depending on whether ω becomes larger or smaller, a sudden increase in amplitude occurs at points P, Q, R, S (tilting) [5].

4.3.2 Vibration of Systems with Periodically Varying Parameters (Parametrically Excited Vibrations)

In this case, the restoring force is dependent not only on the deflection, but also on a variable coefficient $c = c(t)$ (e.g. pendulum with moving suspension, locomotive rod vibration [1]). For undamped vibration, $m\ddot{s} + [c - f(t)]s = 0$ or $\ddot{s} + [\lambda + \gamma\Phi(t)]s = 0$. This equation is termed

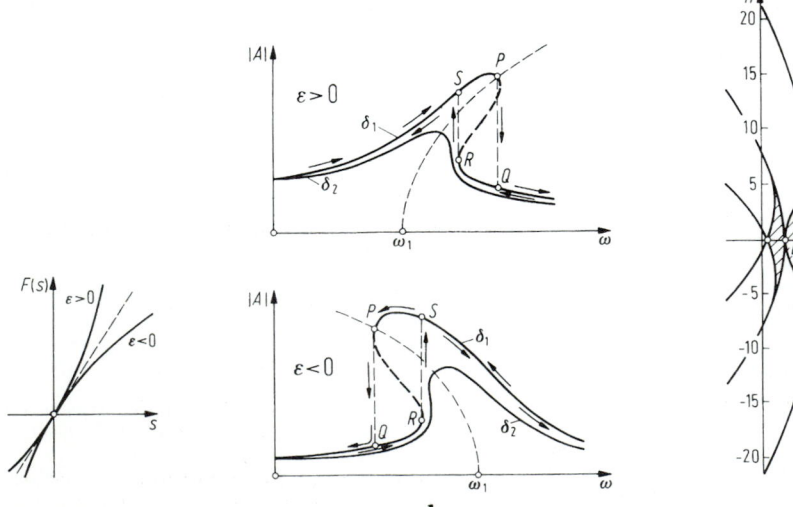

Figure 10. Non-linear vibrations: **a** spring characteristic, **b** resonance diagrams, **c** strutt map (shaded solution ranges are stable)

Hill's differential equation, when $\Phi(t)$ is periodic [8]. A special form of this equation is Mathieu's differential equation [1, 5, 8]

$$\ddot{s} + (\lambda - 2h \cos 2t)s = 0. \tag{45}$$

(This applies, for example, to pendulum vibration with a periodically varying suspension point, or for flexural vibrations of a rod under a pulsing axial load.) For solutions with Mathieu's functions etc., see [8]. $s(t)$ shows, as a function of λ and h, areas of stable and unstable behaviour; i.e. whether the deflections are becoming smaller or greater. Stable and unstable areas were determined by Strutt, and are represented in Strutt's map, named after its inventor (**Fig. 10c**).

5 Hydrostatics

Liquids and gases differ from each other essentially in their low and high degrees of compressibility, respectively. They have many properties in common, and are together designated as fluids. They are easily displaced and take on any external form without substantial resistance; in most cases, they can be regarded as a homogeneous continuum.

Pressure. $p = dF/dA$ is independent of direction in liquids at rest, i.e. a scalar position function, since from Newton's law of shearing stress

$$\tau_{xy} = \eta(\partial v_x/\partial y + \partial v_y/\partial x)$$

for $v_x = v_y = 0$, $\tau_{xy} = 0$, and accordingly $\tau_{xz} = \tau_{yz} = 0$ results. It therefore follows from the conditions of equilibrium $p_x = p_y = p_z = p(x, y, z)$. On the plane faces, because $\tau = 0$, p is perpendicular to the surface.

Density. $\rho = dm/dV$. Liquids are only slightly compressible; thus $dV/V = dp/E$ or $\rho = \rho_0/(1 - \Delta p/E)$. Modulus of elasticity E at 0°C: for water $2.1 \cdot 10^5$ N/cm², for benzene $1.2 \cdot 10^5$ N/cm², for mercury $2.9 \cdot 10^6$ N/cm² (by contrast, for steel $2.1 \cdot 10^7$ N/cm²). For most problems, liquids can be regarded as non-compressible. Gases are compressible; i.e. the densities change in accordance with $\rho = p/(RT)$ (see C6.1.1).

Capillary and Surface Tension. Liquids rise or fall in capillaries as a result of the molecular forces between the liquid and the wall or between the liquid and the air. Molecular forces produce surface tension σ (e.g. at 20°C for water against air, 0.073 N/m; for alcohol against air, 0.025 N/m; and for mercury against air, 0.47 N/m). The capillary rise amounts to $h = 4\sigma/(d\rho\, g)$ (d = capillary diameter). With non-wetting liquids (e.g. mercury), the level in the capillary falls.

Pressure Distribution in the Liquid. Owing to the equilibrium for an element (**Fig. 1a**), there applies

$$p\, dA + \rho g\, dA\, dz - (p + dp)dA = 0,$$

i.e. $\quad dp/dz = \rho g$

or, after integration

$$p = p(x,y,z) = \rho g z + C.$$

With $p(z = 0) = p_0$,

$$p = p(z) = p_0 + \rho g z, \tag{1}$$

i.e. the pressure depends in a linear relationship on the depth z, and is independent of x and y. For $\rho g = 0$, i.e. without taking the weight into consideration, it follows from Eq. (1) that $p(x, y, z) = p_0$, i.e. the compressive pressure p_0 acts uniformly on all points (Pascal's law).

Pressure on Plane Walls. For a container with an overpressure $p_{ü}$ (**Fig. 1b**), the equivalent level height is first calculated, $h_{ü} = p_{ü}/(\rho g)$. From this, the coordinates z and η are counted ($z = \eta \sin \beta$). The resultant pressure force

$$F = \int \rho g z\, dA = \rho g A_S \tag{2}$$

is imposed at the centre of pressure M. The location of the centre of pressure is given by

$$e_\varsigma = I_x/(A\eta_s), \quad e_x = I_{x\varsigma}/(A\eta_s), \tag{3}$$

where I_x is the axial surface moment of the second degree, and $\bar{x}$ and $\bar{y}$ are the axes through the centre of gravity of the surface. For symmetrical surfaces, $I_{x\varsigma} = 0$. For cases according to **Fig. 1c**, there applies, with $\beta = 90°$,

Wall:

$$I_x = bh^3/12, \quad F = \rho g b h^2/2, \quad e_\varsigma = h/6;$$

Rectangular flap:

$$I_x = bh^3/12, \quad F = \rho g b h z_s, \quad e_\varsigma = h^2/(12 z_s);$$

Circular flap:

$$I_x = \pi d^4/64, \quad F = \rho g z_s \pi d^2/4, \quad e_\varsigma = d^2/(16 z_s).$$

Example *Container with Outlet Flap-valve.* Given $P_{ü} = 0.5$ bar; $H = 2$ m, $\beta = 60°$. To be calculated are the size and position of the resultant pressure force on a circular flap of diameter $d = 500$ mm.

With

$$h_{ü} = p_{ü}/(\rho g) = (0.5 \cdot 10^5 \text{ N/m}^2)/(1000 \text{ kg/m}^3 \cdot 9.81 \text{ m/s}^2)$$

$$= 5.097 \text{ m}$$

$z_s = H + h_{ü} = 7.097$ m, according to Eq. (2) $F = \rho g(\pi d^2/4)z_s = 13.67$ kN and according to Eq. (3) $e_\varsigma = (\pi d^4/64)/[(\pi d^2/4)z_s/\sin \beta]$ $= 1.9$ mm.

Pressure on Curved Walls (Fig. 2a). The force components are

$$F_x = \rho g \int z\, dA_x = \rho g z_{sx} A_x,$$

$$F_y = \rho g \int z\, dA_y = \rho g z_{sy} A_y, \tag{4}$$

$$F_z = \rho g \int z\, dA_z = \rho g \int dV = \rho g V.$$

Here, A_x and A_y are the projection faces of the curved surface on the y, z- or x, z-plane. F_z is the weight imposed in the centre of volume. With random surfaces, the three forces do not pass through one point. In the case of

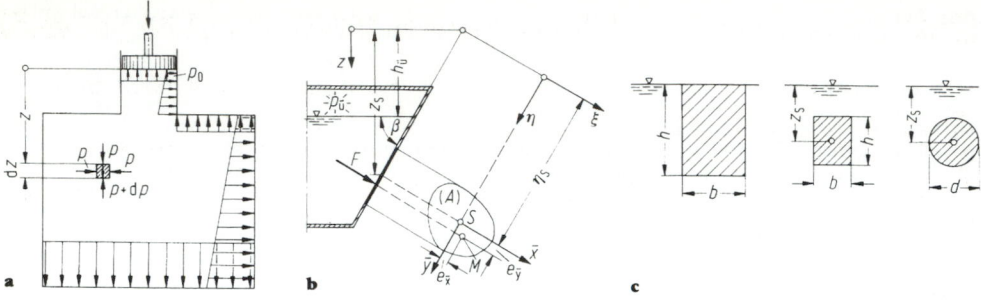

Figure 1. Hydrostatic pressure: **a** distribution, **b** on inclined and **c** on vertical walls.

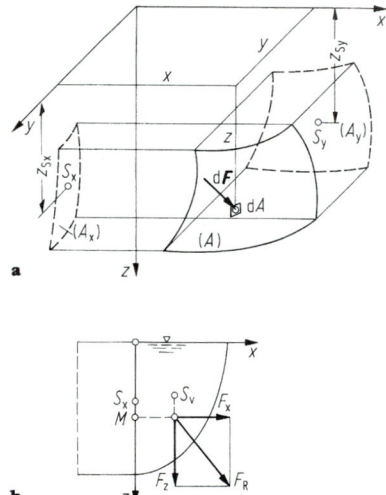

Figure 2. Pressure on curved walls: **a** general; **b** cylindrical and spherical surfaces.

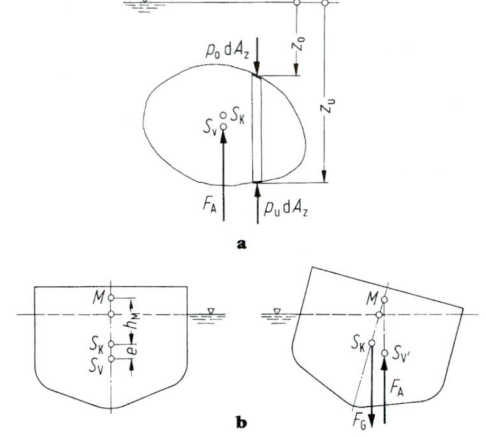

Figure 3. a Lift. **b** Floating stability.

spherical or cylindrical surfaces, the projection on the y, z plane is sufficient. F_x and F_z then lie in one plane and have the resultant $F_R = \sqrt{F_x^2 + F_z^2}$ (**Fig. 2b**). According to Eq. (4), the horizontal pressure force on a curved surface in a random direction is as great as on a projecting surface standing perpendicular to the direction of force. The point of contact of the pressure forces is derived according to Eq. (3) to e_x and e_y, when x and y are the axes through the centre of gravity of the individual projection surface. In the case of spherical and cylindrical surfaces, the resultant F_R always passes through the centre of curvature.

Lift (Fig. 3a). For bodies wholly (or partially) immersed, the force $d\mathbf{F} = p_0\, dA_x \mathbf{e} + p_0\, dA_y \mathbf{e}_y + p_0\, dA_z \mathbf{e}_z$ takes effect on a surface element facing upwards. Because the components dF_x and dF_y maintain the equilibrium on the closed body, i.e. $F_x = 0$ and $F_y = 0$, there remains only one force in the z-direction:

$$F_A = F_z = \int dF_z = \int (p_u - p_o)\, dA_z$$

$$= \int \rho\, g(z_u - z_o)\, dA_z = \rho g V. \qquad (5)$$

This lift force is equal to the weight of the displaced liquid. It is imposed at the centre of volume of the displaced liquid (and not at the centre of gravity of the body; in homogeneous bodies, both centres coincide).

Stability of Floating Bodies (Fig. 3b). An immersed body floats when $F_G = F_A$. It floats in a stable manner when the metacentre M lies above the centre of gravity of the body S_K, and is unstable when both coincide. For the metacentric height there applies

$$b_M = (I_x/V) - e.$$

Here, I_x is the surface moment of the second order of the flotation area (waterline cross-section) about the longitudinal axis, V is the displaced volume, and e is the distance between the centre of gravity of the body and the centre of volume. In the case of bodies in suspension (submarines), $I_x = 0$ and $b_M = -e$. If e is negative, i.e. if the centre of gravity of the body lies beneath the centre of volume, then it follows that $b_M > 0$, and the body in suspension floats stably.

6 Hydrodynamics and Aerodynamics (Dynamics of Fluids)

The object of fluid dynamics is to examine the values of velocity, pressure and density of a fluid as a function of the space coordinates x, y, z, or, in the case of one-dimensional problems (e.g. pipeline flows), as a function of arc length s. In many flow processes, compression is negligible, even with gaseous fluids (e.g. if bodies have air flowing around them at normal temperature and at less than 0.5 times the velocity of sound). In such cases the laws of incompressible media apply (for flows with changes in volume, see C7.2).

Ideal and Non-ideal Liquid. An ideal liquid is incompressible and frictionless, i.e. no transverse stress occurs ($\tau_{xy} = 0$). The pressure on an element is equally great in all directions (see A5). With non-ideal or viscous liquids, transverse stresses depend on the velocity gradients, and the pressures p_x, p_y, p_z differ from one another. If the transverse stresses depend on the velocity gradient in a linear relationship perpendicular to the direction of the flow (**Fig. 1**), $\tau = \eta \, (dv/dz)$ applies, and a Newtonian liquid is present (e.g. water, air, and oil). Here, η is the absolute or dynamic viscosity. Non-Newtonian liquids with non-linear characteristics are, for example, suspensions, pastes, and thixotropic liquids.

Stationary and Non-stationary Flow. With stationary flows, the values of velocity v, pressure p, and density ρ depend only on the space coordinates, i.e. $v = v(x, y, z)$ etc. With non-stationary flows, the flow also changes at one location with the time, i.e. $v = v(x, y, z, t)$ etc.

Streamline, Stream Tube, Stream Filament. The streamline is the line that touches each point of the velocity vectors at a specific moment in time (**Fig. 2**); i.e. $v_x : v_y : v_z = dx : dy : dz$. In the case of stationary flows, the streamline is a spatially fixed space curve; in addition, it is identical to the path curve of the individual section. With non-stationary flows, the streamlines change their locations in space with time; they are not identical to the path curves of the sections. A bundle of streamlines which is twisted about by a closed curve is called a stream tube (**Fig. 2**). Parts of the stream tube with cross-section dA, across which p and A are to be regarded as constant, form a stream filament. In cases of tube flows of ideal liquids, p and v are approximately constant across the entire cross-section A, i.e. the entire tube content forms a stream filament.

6.1 One-Dimensional Flow of Ideal Fluids

Euler's Equation for the Stream Filament. For an element dm along the length of the streamline shown in **Fig. 3a**, Euler's equation of motion (in the tangential direction) becomes

$$a_t = \frac{dv}{dt} = \frac{\partial v}{\partial t} + \frac{\partial v}{\partial s}\frac{ds}{dt} = -g\frac{\partial z}{\partial s} - \frac{1}{\rho}\frac{\partial p}{\partial s}$$

or, with $\dfrac{ds}{dt} = v$,

$$\frac{\partial}{\partial s}\left(\frac{v^2}{2} + \frac{p}{\rho} + gz\right) + \frac{\partial v}{\partial t} = 0. \qquad (1)$$

In the case of stationary flow, $\partial v/\partial t = 0$. For the normal direction,

$$a_n = \frac{v^2}{r} = -\frac{1}{\rho}\frac{\partial p}{\partial n} - g\frac{\partial z}{\partial n} \quad \text{or} \quad \frac{\partial p}{\partial n} = -\rho\frac{v^2}{r^2} - \rho g\frac{\partial z}{\partial n};$$

or, ignoring the inherent weight, $\partial p/\partial n = -\rho v^2/r$. Accordingly, the pressure increases from the concave to the convex side of the stream filament.

Bernoulli's Equation for the Stream Filament. From Eq. (1) along the stream filament, it follows, for the non-stationary flow

$$\rho v^2/2 + p + \rho gz + \rho \int \frac{\partial v}{\partial t}\,ds = \text{const.}, \qquad (2a)$$

or

$$\rho v_1^2/2 + p_1 + \rho gz_1 = \rho v_2^2/2 + p_2 + \rho gz_2 + \rho \int_{s_1}^{s_2} \frac{\partial v}{\partial t}\,ds. \qquad (2b)$$

For the stationary case ($\partial v/\partial t = 0$) there applies

$$\rho v_1^2/2 + p_1 + \rho gz_1 = \rho v_2^2/2 + p_2 + \rho gz_2 = \text{const.} \qquad (3)$$

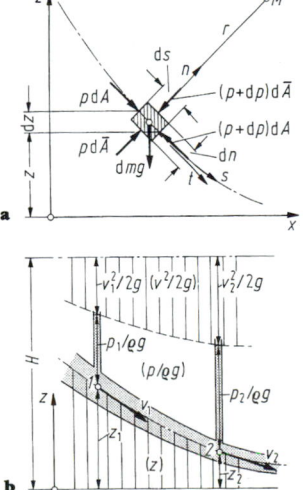

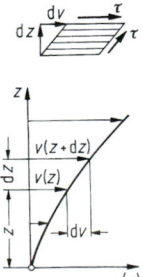

Figure 1. Transverse stress in a liquid.

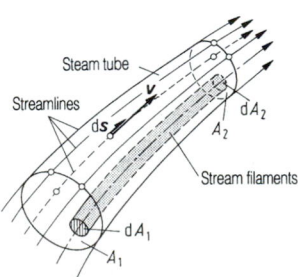

Figure 2. Flow tubes and flow filaments.

Figure 3. Flow filament. **a** Element. **b** Bernoulli's peaks.

Accordingly, the total energy, consisting of dynamic, pressure, and potential energy, is maintained for the mass unit along the length of the stream filament. From Eq. (3), there is derived, after division by ρg,

$$v_1^2/(2G) + p_1/(\rho g) + z_1$$
$$= v_2^2/(2g) + p_2/(\rho g) + z_2 = \text{const.} = H, \quad (4)$$

i.e. the entire energy level H, consisting of velocity, pressure and elevation, remains constant (Bernoulli's equation; **Fig. 3b**).

Continuity Equation. For a stream filament, the mass flowing through each cross-section per time unit (mass flow) must be constant:

$$d\dot{m} = \rho v \, dA = \rho_1 v_1 \, dA_1 = \rho_2 v_2 \, dA_2 = \text{const.} \quad (5)$$

In the case of incompressible media ($\rho = \text{const.}$), the volume flow must be constant:

$$d\dot{V} = v \, dA = v_1 \, dA_1 = v_2 \, dA_2 = \text{const.} \quad (6)$$

In the case of stream tubes with a constant mean velocity v across the cross-section A, it follows from Eqs (5) and (6) that

$$\dot{m} = \rho v A = \text{const.} \quad \text{or} \quad \dot{V} = v A = \text{const.}$$

6.1.1 Application of Bernoulli's Equation for Steady Flow Problems

Dynamic Pressure. In cases in which a flow encounters a fixed obstacle, dynamic pressure occurs (**Fig. 4a**). Bernoulli's equation, Eq. (3), has, without the height element, the form

$$\rho v_1^2/2 + p_1 = \rho v_2^2/2 + p_2. \quad (7)$$

From this it follows when $v_2 = 0$, $p_2 = p_1 + \rho v_1^2/2$. At a critical point, the pressure is composed of the static pressure $p_{st} = p_1$ and the (dynamic) stagnation pressure $p_{dyn} = \rho v_1^2/2$.

Example *Dynamic Pressure of Wind Against a Wall.* With wind velocity $v = 100$ km/h $= 27.8$ m/s with $\rho_{air} = 1.2$ kg/m³ the dynamic pressure is $p_{dyn} = \rho v^2/2 = 464$ N/m².

Pitot Tube. The Pitot tube is well suited for measuring the flow velocity in open channels (**Fig. 4b**). For point *1*, according to A5 Eq. (1), $p_1 = p_L + \rho g z_1$. For the stream-line *1-2* there applies $p_1 + \rho v_1^2/2 = p_2$, i.e. $p_2 = p_L + \rho g z_1 + \rho v_1^2/2$. The hydrostatic pressure in the Pitot tube is $p_2 = p_L + \rho g(z_1 + h)$ and therefore $\rho v_1^2/2 = \rho g h$ or $v_1 = \sqrt{2gh}$. The head h is a flow velocity. For measuring the air velocity, the arrangement shown in **Fig. 4c** is well suited.

If ρ_M is the density of the pressure gauge liquid, then for point *2* the following applies: $p_{dyn} = \rho v_1^2/2 = \rho_M g h$, i.e.

$$v_1 = \sqrt{2(\rho_M/\rho)gh}.$$

Venturi Tube. This is used to measure the flow velocity in pipelines (**Fig. 5**). Bernoulli's equation (7) between positions *1* and *2* reads $\rho v_1^2/2 + p_1 = \rho v_2^2/2 + p_2$ and the continuity equation $v_1 A_1 = v_2 A_2$. From this, we obtain

$$\Delta p = p_2 - p_1 = (\rho v_1^2/2)[(A_1/A_2)^2 - 1]$$

or, with $\Delta p = (\rho_M - \rho)gh$,

$$v_1 = \sqrt{2gh(\rho_M/\rho - 1)/[(A_1/A_2)^2 - 1]}.$$

In reality, pressure loss still has to be taken into account between points *1* and *2* because of friction (see A6.2ff.).

6.1.2 Application of Bernoulli's Equation for Unsteady Flow Problems

Investigated here is the exit flow from a container with falling surface level, disregarding friction (**Fig. 6**).

Solution. From Eqs (2) and (6),

$$v_1 = \sqrt{2g\left(z - \frac{1}{g}\int_{s_1}^{s_2} \frac{\partial v}{\partial t}\,ds\right) \Big/ [(A_1/A_2)^2 - 1]}.$$

With $v_1 = -dz/dt$, $A_1/A_2 = \alpha$, and, disregarding the integral (small in comparison with z), it follows from Eq. (2b) $v_1 = -dz/dt = \sqrt{2gz/(\alpha^2 - 1)}$ and from this, after integration, $t = -\sqrt{2(\alpha^2 - 1)z/g} + C$. For $z(t = 0) = H$, $C = \sqrt{2(\alpha^2 - 1)H/g}$ and therefore

$$t = \left(1 - \sqrt{z/H}\right)\sqrt{2(\alpha^2 - 1)H/g} \quad \text{or}$$

$$z = H\left\{1 - t\sqrt{g/[2H(\alpha^2 - 10]}\right\}^2.$$

From this, there follows for $z = 0$ the outflow time

$$T = \sqrt{2(\alpha^2 - 1)H/g},$$

the velocity

$$v_1 = -\,dz/dt =$$
$$\left\{1 - t\sqrt{g/[2H(\alpha^2 - 1)]}\right\}\sqrt{2gH/(\alpha^2 - 1)}$$

and the outflow velocity $v_2 = v_1 A_1/A_2$. The velocities decrease linearly with the time.

6.2 One-Dimensional Flow of Viscous Newtonian Fluids

In *laminar flows*, the particles move in parallel paths (layers), while in *turbulent flows* the main flow is over-

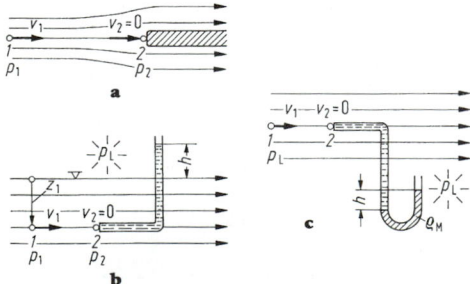

Figure 4. Dynamic pressure; **a** stagnation point, **b** Pitot tube for liquids and **c** gases.

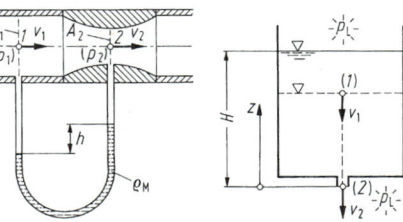

Figure 5. Venturi tube.   **Figure 6.** Unstable outflow.

laid by additional velocity components in the x-, y- and z-directions (rotational or vortex flows). Transition from laminar to turbulent flow occurs when the Reynolds number $Re = vd/\nu$ reaches the critical value (e.g. $Re_k = 2320$ for tubes with circular cross-sections).

For *laminar flow*, Newton's law applies to the transverse stress between the particles:

$$t = \eta(dv/dz) \qquad (8)$$

(**Fig. 1**). Here, η is the *dynamic viscosity*. This is temperature-dependent, and for gases also pressure-dependent (which is negligible, however, provided that no substantial changes in density occur).

For *turbulent flow*, according to Prandtl and von Kármán [1, 11, 12] by approximation the transverse stress law, $\tau = \eta\, dv/dz + \rho l^2(dv/dz)^2$ applies. Here l is the free path length of a particle.

As a result of the transverse stresses, pressure losses (energy losses) occur along the length of the flow filament.

Kinematic Viscosity. This is $\nu = \eta/\rho$. For water at 20 °C, $\eta = 10^{-3}$ N s/m² and $\nu = 10^{-6}$ m²/s (for further values, see **Appendix C10, Table 2**).

Bernoulli's Equation with Loss Element. If there is no energy input or extraction between two points *1* and *2* (e.g. by pump or turbine), then Bernoulli's equation becomes

$$\rho v_1^2/2 + p_1 + \rho g z_1$$
$$= \rho v_2^2/2 + p_2 + \rho g z_2 + \Delta p_v + \rho \int_{s_1}^{s_2} \frac{\partial v}{\partial t}\, ds. \qquad (9)$$

For the stationary situation, $\partial v/\partial t = 0$, and the last element does not apply. In this case Δp_v is the pressure loss between the points *1* and *2* as a result of pipe friction, internal resistance, etc. If Eq. (9) is divided by ρg, then

$$v_1^2/(2g) + p_1/(\rho g) + z_1 = v_2^2/(2g) + p_2/(\rho g) + z_2 + b_v. \qquad (10)$$

Here the individual elements indicate energy levels, and $b_v = \Delta p_v/(\rho g)$ indicates the loss level.

Pressure Loss and Head Loss (Fig. 16). Let the tube diameter be constant between two points *1* and *2*. Then

$$\Delta p_v = (\lambda l/d)\rho v^2/2 + \Sigma\, \zeta\rho v^2/2 \quad \text{or} \qquad (11a)$$

$$b_v = (\lambda l/d)v^2/(2g) + \Sigma\, \zeta v^2/(2g), \qquad (11b)$$

where λ is the tube friction coefficient and ζ is the drag coefficient of installed items. For *compressible fluids*, which expand as a result of the pressure drop from *1* to *2*, it follows from the continuity equation, Eq. (5), and from the statement $dp = -(\lambda/d)\, dx\, \rho v^2/2$ for the isothermic case, $p_1/\rho_1 = p/\rho = \text{const.}$, $p_1^2 - p_2^2 = \lambda v_1^2\rho_1 p_1 l/d$, for the pressure loss due to tube friction,

$$\Delta p_v = p_1 - p_2 = p_1\left[1 - \sqrt{1 - \lambda v_1^2\rho_1 l/(p_1 d)}\,\right]. \quad (12)$$

In the case of small pressure losses, the expansion is negligible, and Eq. (11a) can also be used for compressible fluids. The error which arises in this case is

$$f \approx 0.5 \cdot \Delta p_v/p_1 \; [6].$$

6.2.1 Steady Laminar Flow in Pipes of Circular Cross-section

In accordance with **Fig. 7a**, it follows from $\Sigma F_{ix} = 0 =$ $(p_1 - p_2)\pi r^2 - \tau \cdot 2\pi r l$ with $\tau = -\eta\, dv/dr$ and the

adhesion condition $v(r = d/2) = 0$ after integration $v(r) = \Delta p_v(d^2/4 - r^2)/(4 - r^2)$. The velocity distribution is therefore parabolic (Stokes's law). For the transverse stresses, $\tau(r) = -\eta dV/dr = \Delta p_v r/(2l)$; i.e. they increase in a linear progression outwards. For the volume flow there applies

$$\dot{V} = \int_{r=0}^{d/2} v(r)2\pi r\, dr = \Delta p_v \pi d^4/(128\eta l)$$

(Hagen–Poiseuille formula), and therefore for the mean velocity and the pressure loss $v_m = v = \dot{V}/A = \Delta p_v d^2/(32\eta l)$ and $\Delta p_v = v_m\, 32\eta l/d^2$. The pressure loss and therefore also the transverse stresses accordingly increase linearly with velocity. With the Reynolds number $Re = vd/V$ are derived $\Delta p_v = (64/Re)(l/d)(\rho v^2/2)$ and $b_v = (64/Re)(l/d)(v^2/2g)$. Accordingly, from Eqs (11a, b), the tube friction coefficient $\lambda = 64/Re$, i.e. with laminar flow, regardless of the roughness of the tube wall.

6.2.2 Steady Turbulent Flow in Pipes of Circular Cross-section

When $Re > 2320$, transition to turbulent flow takes place. The tube friction coefficient λ depends on the tube or pipe roughness k (wall protrusions in mm, see **Table 1**) and on Re. The velocity profile is essentially flatter (**Fig. 7b**) than with laminar flow. In the peripheral range it consists of a laminar boundary layer of thickness $\delta = 34.2d/(0.5Re)^{0.875}$ (Prandtl). The velocity distribution likewise depends on Re and k; according to Nikuradse, it can be represented by $v(r) = v_{max}(1 - 2r/d)^n$ (e.g. $n = 1/7$ for $Re = 10^5$). Exponent n increases with tube roughness. The ratio $v/v_{max} = 2/[(1 + n) \cdot (2 + n)]$ is on average about 0.84. The friction forces, i.e. pressure loss or loss value, increase with the square of velocity in turbulent flow.

Determining the Tube Friction Coefficient

Hydraulically Smooth Tubes and Pipes. These occur when the barrier width is greater than the wall protrusion, i.e. for $\delta/k \geq 1$ or $Re < 65d/k$.

Blasius formula (applicable to $2320 < Re < 10^5$):

$$\lambda = 0.3164\big/\sqrt[4]{Re}\,.$$

Nikuradse's formula (applicable to $10^5 < Re < 10^8$):

$$\lambda = 0.0032 + 0.221/Re^{0.237}.$$

The formula of Prandtl and von Kármán (applicable to

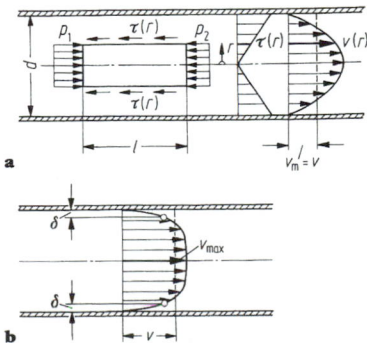

a

b

Figure 7. Pipe flow: **a** laminar, **b** turbulent.

the entire turbulent range, but complicated owing to its implicit form):

$$\lambda = 1/[2 \lg(Re\sqrt{\lambda}/2.51)]^2.$$

Instead of this, the approximative formula $\lambda = 0.309/[\lg(Re/7)]^2$ can be used.

Hydraulically Rough Tubes or Pipes. These exist when the wall protrusions are greater than the boundary layer thickness, i.e. for $\delta/k < 1$ or $Re > 1300d/k$. The tube fric-

Table 1. Reference values for wall roughness [2]

Material and type of pipe	Condition of pipe	k in mm
New drawn and pressed pipes of Cu, mild steel, bronze, Al, other light metals, glass, plastics	Technically smooth	0.001 to 0.0015
New rubber pressure hose	Technically smooth	approx. 0.0016
Cast-iron pipes	New, commercially conventional	0.25 to 0.5
	Rusted	1.0 to 1.5
	Scaled	1.5 to 5.0
New seamless steel pipe, rolled or drawn	With rolling skin	0.02 to 0.06
	Pickled	0.03 to 0.04
	For thin pipes	Up to 0.1
New lap-welded steel pipes	With rolling skin	0.04 to 0.1
New steel pipes with coating	Metal spray coating	0.08 to 0.09
	Dip-galvanised	0.07 to 0.1
	Conventional zinc coating	0.1 to 0.16
	Bituminised	Approx. 0.05
	Cemented	Approx. 0.18
	Galvanised	Approx. 0.008
Used steel pipe	Uniform rust scarring	Approx. 0.15
	Light scaling	0.15 to 0.4
	Medium scaling	Approx. 1.5
	Heavy scaling	2.0 to 4.0
Asbestos cement pipes	New, conventional	0.03 to 0.1
Concrete pipe, new	Conventional smooth finish	0.3 to 0.8
	Conventional medium smooth	1.0 to 2.0
	Conventional rough	2.0 to 3.0
Concrete pipes after several years use with water		0.2 to 0.3
Wood cladding, rough		1.0 to 2.5
Unworked stone		8 to 15
Mean value for pipe lengths without joints		0.2
Mean value for pipes with joints		2.0

tion coefficient λ is only dependent on the relative roughness d/k, and Nikuradse's formula applies:

$$\lambda = 1/[2 \lg(3.71d/k)]^2$$

for the range located above the limit curve (**Fig. 8**). The limit curve can be determined by means of $\lambda = [(200d/k)/Re]^2$.

Pipes and Tubes in the Transitional Range. In these $65d/k < Re < 1300d/k$, i.e. in the area below the limit curve in **Fig. 8**. The tube friction coefficient λ is dependent on Re and d/k. As a good approximation

$$\lambda = 1 \left/ \left[2 \lg \left(\frac{2.51}{Re\sqrt{\lambda}} + \frac{0.27}{d/k} \right) \right]^2 \right.$$

(Colebrook's formula). This relates to tubes with technical roughness. For tubes with adhesively attached sand grains, Nikuradse measured the curves entered as hatched lines in **Fig. 8**.

Colebrook–Nikuradse Diagram. The formulae given are represented in graphical form in **Fig. 8**, so that λ is read off as a function of Re and d/k, and, if required, can be recalculated or improved (for further refinements, see [1, 3]). If λ is known, the pressure loss or the head loss is calculated in accordance with Eqs (11) or (12), and subsequently the pipe section to be examined is calculated by the Bernoulli's equation with the loss element according to Eqs (9) or (10).

Example Compressed air at $V = 400$ m³/h is passed through a steel pipe (used, $k = 0.15$ mm) with a diameter $d = 150$ mm and length $l = 400$ m. Pressure and density in the boiler: $p_1 = 6$ bar, $\rho_1 = 6.75$ kg/m³. To be determined is the pressure loss at the end of the pipe.

With the delivery velocity

$$v = \dot{V}/A = \dot{V}/(\pi d^2/4) = 6.29 \text{ m/s} \quad \text{and}$$

$$v = \eta/\rho = (2 \cdot 10^{-5} \text{ Ns/m}^2)/(6.75 \text{ kg/m}^3) = 2.963 \cdot 10^{-6} \text{ m}^2/\text{s}$$

$Re = vd/v = 318\,427$. With $d/k = 150/0.15 = 1000$, we derive from **Fig. 8** or from Colebrook's formula $\lambda = 0.0205$. From Eq. (12) it follows, for the pressure loss at the end of the pipe

$$\Delta p_v = p_1 \left[1 - \sqrt{1 - \lambda v^2 \rho_1 l/(p_1 d)} \right] = 0.261 \text{ bar}.$$

If expansion as a result of pressure decrease is disregarded, Eq. (11a) gives

$$\Delta p_v = (\lambda l/d)\rho v^2/2 = 25\,550 \text{ N/m}^2 = 0.256 \text{ bar},$$

i.e. an error $f = (0.261 - 0.256)/0.261 = 1.92\%$, which also concurs well with the estimation formula $f = 0.5 \cdot \Delta p_v/p_1 = 2.13\%$. The change in density of the compressed air accordingly exerts hardly any effect.

6.2.3 Flow in Pipes of Non-circular Cross-section

After the introduction of the hydraulic diameter $d_h = 4A/U$ (A = cross-sectional area, U = wetted circumference), calculation is made as in A6.2.1 and A6.2.2. However, in laminar flow, $\lambda = \varphi \cdot 64/Re$ is to be inserted [5]. For annular and square cross-sections,

Annulus	d_a/d_i	1	5	10	20	50	100
	φ	1.50	1.45	1.40	1.35	1.28	1.25

Square	b/b	0	0.1	0.3	0.5	0.8	1.0
	φ	1.50	1.34	1.10	0.97	0.90	0.88

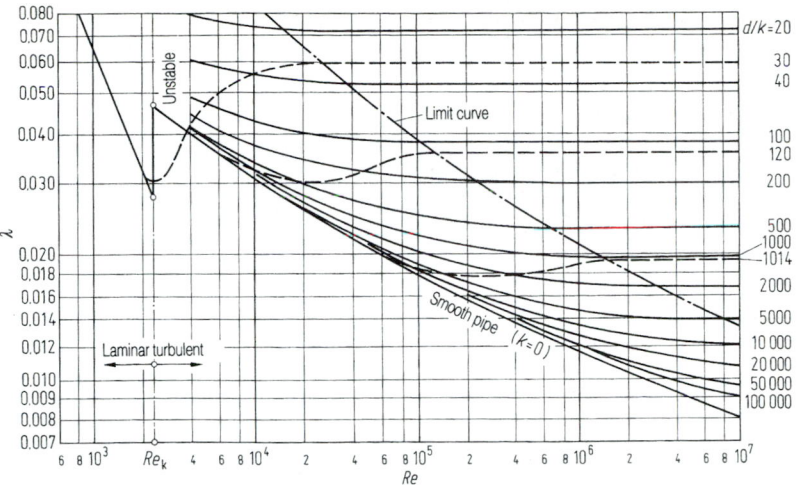

Figure 8. Pipe friction coefficient after Colebrook and (dotted) after Nikuradse.

6.2.4 Loss Factors for Pipe Fittings and Bends

In addition to the wall friction losses of the pipe elements, the following also applies for the pressure loss and the lost head

$$\Delta p_V = \zeta \rho v^2/2 \quad \text{or} \quad h_V = \zeta v^2/(2g).$$

Drag Coefficient ζ for Bends (Fig. 9) [5].

(a) Annular bends:

$\varphi = 90°$ for $\varphi \neq 90°$: $\zeta = k\zeta_{90°}$

R/d	1	2	4	6	10	φ	30°	60°	120°	150°	180°
Smooth $\zeta_{90°}$	0.21	0.14	0.11	0.09	0.11	k	0.4	0.7	1.25	1.5	1.7
Rough	0.51	0.30	0.23	0.18	0.20						

(b) Segment bends

φ	30°	45°	60°	90°
Number of seams	2	3	3	3
ζ	0.10	0.15	0.20	0.25

(c) Grey cast-iron 90° bends

NW	50	100	200	300	400	500
ζ	1.3	1.5	1.8	2.1	2.2	2.2

(d) Folded pipe bend: $\zeta = 0.4$.
(e) Bend with baffle plates: $\zeta = 0.15$ to 0.20 [1].
(f) Double bend: $\zeta = 2\zeta_{90°}$.
(g) Space bends: $\zeta = 3\zeta_{90°}$.
(h) Swan-neck bends: $\zeta = 4\zeta_{90°}$.
(i) Bends of square cross-sections: for $b/b < 1$, $\zeta = \zeta_0 b/b$; for $b/b > 1$, $\zeta = \zeta_0 \sqrt{b/b}$. ζ_0 as for bends of circular cross-sections, if the value $d_h = 22bb(b + b)$ is substituted for d.

Elbow Pieces [5]

With circular cross-sections

(δ = bend angle):

δ	22.5°	30°	45°	60°	90°
Smooth ζ	0.07	0.11	0.24	0.47	1.13
Rough	0.11	0.17	0.32	0.68	1.27

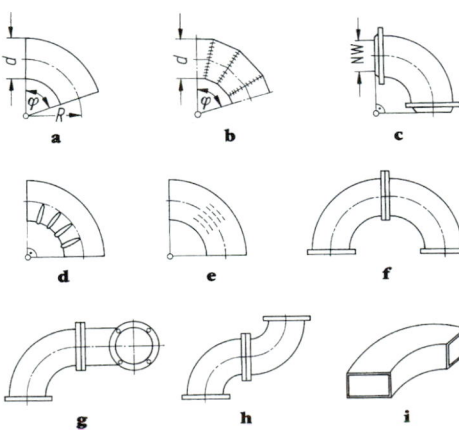

Figure 9. a–i Pipe bends.

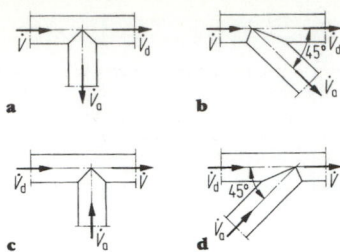

Figure 10. a–d Pipe branches and unions.

With square cross-sections:

δ		30°	45°	60°	75°	90°
ζ		0.15	0.52	1.08	1.48	1.60

Pipe Branches and Unions [6]

$\dot{V}$ = total flow, $\dot{V}_a$ = flow out or in, ζ_d = resistance in main pipe, ζ_a = resistance in branch pipe. Minus sign indicates pressure gain.

	Separation				Combination			
	Fig. 10a		Fig. 10b		Fig. 10c		Fig. 10d	
$\dot{V}_a/\dot{V}$	ζ_a	ζ_d	ζ_a	ζ_d	ζ_a	ζ_d	ζ_a	ζ_d
0	0.95	0.04	0.90	0.04	−1.2	0.04	−0.92	0.04
0.2	0.88	−0.08	0.68	−0.06	−0.4	0.17	−0.38	0.17
0.4	0.89	−0.05	0.50	−0.04	0.08	0.30	0.00	0.19
0.6	0.95	0.07	0.38	0.07	0.47	0.41	0.22	0.09
0.8	1.10	0.21	0.35	0.20	0.72	0.51	0.37	−0.17
1.0	1.28	0.35	0.48	0.33	0.91	0.60	0.37	−0.54

Expansion Bellows (Fig. 11) [5]

(a) Corrugated tube compensator: $\zeta = 0.20$ per corrugation (can be rendered almost zero by installing a guide).

(b) U-bends:

a/d	0	2	5	10
ζ		0.33	0.21	0.21	0.21

(c) Compensation tube bends: smooth pipe bends, $\zeta = 0.7$; folded pipe bends, $\zeta = 1.4$.

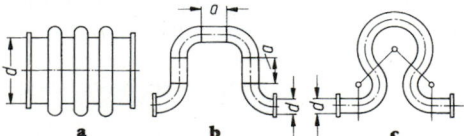

Figure 11. a–c Expansion bellows.

Pipe Inlets (Fig. 12a–e)

(a) Sharp-edged: $\zeta = 0.5$; broken, $\zeta = 0.25$.
(b) and (c) Sharp-edged: $\zeta = 3.0$; broken, $\zeta = 0.6$ to 10.
(d) Depending on wall roughness: $\zeta = 0.01$ to 0.05.

$(d/d_c)^2$	1	1.25	2	5	10
ζ	0.5	1.17	5.45	54	245

Change in Cross-section From A_1 to A_2 (Fig. 13)

(a) Discontinuous expansion: The loss coefficient can be derived from Bernoulli's equation (see A6.4) and the momentum theorem: $\zeta = (A_2/A_1 - 1)^2$. (See **Fig. 13a**.)
(b) Continuous expansion (diffuser): The loss coefficient for pipes of average roughness can be derived from the diagram **Fig. 13b**.
(c) Discontinuous narrowing: From Bernoulli's equation and the momentum theorem, it follows that $\zeta = (A_2/A_0 - 1)^2$. In view of the fact that the constricted cross-section A_0 is unknown, ζ is derived from the diagram **Fig. 13c** for the ratio A_2/A_1 in the case of sharp-edged connections [5].
(d) Continuous narrowing (nozzle): The energy losses from friction are small. On average, $\zeta = 0.05$.

Shutoff and Control Elements

(a) Slide valves, open, without guide pipes: $\zeta = 0.2$–0.3; with guide pipe $\zeta \approx 0.1$. For slide valves with various opening conditions, see [5].
(b) Valves: The drag coefficients fluctuate depending on the type of valve construction between $\zeta = 0.6$ (free-flow valve) and $\zeta = 4.8$ (DIN valve). The details in the literature vary (1, 2, 4–6). With partially open valves, the drag coefficients are greater.
(c) Non-return flap valves, flap valves, cocks: The drag coefficients of non-return flap valves measure, according to [5], $\zeta = 0.8$ at NW 200 and $\zeta = 1.4$ at NW 50. With flap valves, values occur from $\zeta = 0.5$ in the almost open condition ($\varphi = 10°$) and from $\zeta = 4.0$ at $\varphi = 30°$. In the case of cocks, $\zeta = 0.3$ ($\varphi = 10°$) and $\zeta = 5.5$ ($\varphi = 30°$) [5].

Throttle Devices. These are used to measure the velocity and volumes of flows, and are standardised as standard diaphragms, standard nozzles, and standard venturi nozzles (DIN 1952). For drag coefficients see [2].

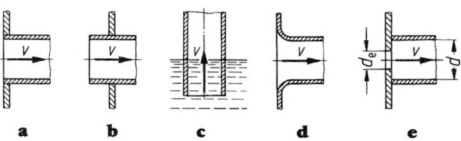

Figure 12. a–e Pipe inlets.

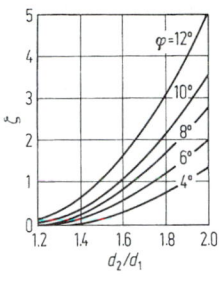

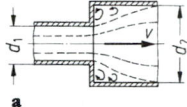

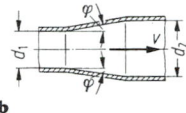

a **b**

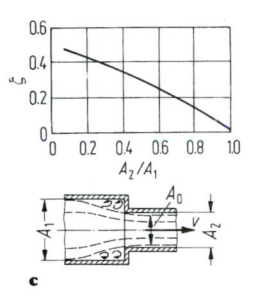

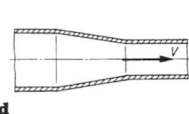

c **d**

Figure 13. a-d Changes in cross-section.

Round Bar Grids, Screens and Flow Straighteners [5]

(a) Round bar grids as per **Fig. 14a**:

$$\zeta = \frac{0.8 s/t}{(1 - s/t)^2}$$

(b) Screens according to **Fig. 14b**:

s	2	2	2.5	3.1	mm
t	20	25	25	25	mm
ζ	0.34	0.27	0.32	0.39	

(c) Flow straighteners: For conventional commercial flow straighteners with foot valves at the start of a pipeline ζ = 4 to 5.

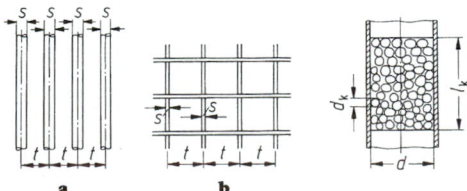

a **b**

Figure 14. a Round bar cascade, **b** Screen. **Figure 15.** Solid-body filling.

Solid Body Filling [5]

For flow through the filling in accordance with **Fig. 15**, $\zeta = \lambda_F l_k/d_k$. Up to $Re_k = v d_k/\nu = 10$ (v = mean velocity in empty pipe), laminar flow pertains, and it is $F = 2000/Re_k$. For $Re_k > 10$ (turbulent flow), λ_F still depends only on d/d_k:

d/d_k	25	17	8	3.5
λ_F	50	40	30	15

Example *Pipeline with Special Resistance Values* (**Fig. 16**). It is intended that $\dot V = 8$ litres/second of water should be passed through a pipeline. The value to be determined is the pressure p_0 in the pressure container. Given: $b_1 = 7$ m, $b_2 = 5$ m, $l_1 = 35$ m, $l_2 = 25$ m, $l_3 = 13$ m, $l_4 = 25$ m, $d_1 = d_6 = 80$ mm, $d_2 = 60$ mm, wall roughness $k = 0.04$ mm (new, longitudinally welded steel pipe). Resistance coefficients: pipe intake $\zeta_1 = 0.5$; nozzle $\zeta_2 = 0.05$; elbow ($\delta = 22.5°$) $\lambda_3 = 0.11$; diffusor $\zeta_5 = 0.3$, kinematic viscosity at 20°C: $\nu = 10^{-6}$ m²/s. Air pressure: $p_L = 1$ bar.

From the continuity equation (6), it follows for the flow velocities $v_1 = v_6 = \dot V/A_1 = \dot V/(\pi d_1^2/4) = 1.59$ m/s and $v_2 = \dot V/(\pi d_2^2/4) = 2.83$ m/s. With the Reynolds number $Re_1 = v_1 d_1/\nu = 127\,200$, $Re_2 = v_2 d_2/\nu = 169\,800$ and the relative roughness values $d_1/k = 2000$, $d_2/k = 1500$ there follow from the formula or from Colebrook's diagram (**Fig. 8**) the pipe friction coefficients $\lambda_1 = 0.0197$ and $\lambda_2 = 0.0200$. According to Eq. (11b), this provides the head loss values

$h_{V1} = \zeta_1 v_1^2/2g = 0.06$ m;

$h_{V2} = h_{V1} + (\lambda_1 l_1/d_1)v_1^2/2g + \zeta_2 v_2^2/2g$

$\qquad = (0.06 + 1.11 + 0.02)$ m $= 1.19$ m;

$h_{V3} = h_{V3} + (\lambda_2 l_2/d_2)v_2^2/2g + \zeta_4 v_2^2/2g$

$\qquad = (1.19 + 3.40 + 0.04)$ m $= 4.63$ m;

$h_{V4} = h_{V3} + (\lambda_2 l_3/d_2)v_2^2/2g + \zeta_4 v_2^2/2g$

$\qquad = (4.63 + 1.77 + 0.04)$ m $= 6.44$ m;

$h_{V5} = h_{V4} + (\lambda_2 l_4/d_2)v_2^2/2g = (6.44 + 3.40)$ m $= 9.84$ m;

$h_{V6} = h_{V5} + \zeta_5 v_6^2/2g = (9.84 + 0.04)$ m $= 9.88$ m.

Bernoulli's equation (10) between points 0 and 6 then provides, with $v_0 \approx 0$ (because $A_6 \gg A_0$)

$p_0/(\rho g) + b_1 = v_6^2(2g + p_L/(\rho g) + b_2 + h_{V6}$, i.e.

$p_0 = p_L + \rho v_6^2/2 + \rho g(b_2 + h_{V6} - b_1)$

$\qquad = p_L + 1264$ N/m² $+ 77\,303$ N/m² $= 1.786$ bar.

With the velocity heads $v_1^2(2g) = v_6^2/(2g) = 0.13$ m, $v_2^2/(2g) = 0.41$ m and pressure heads $p_0/(\rho g) = 18.21$ m, $p_L/(\rho g) = 10.19$ m. the Bernoulli heads can be plotted (**Fig. 16**).

6.2.5 Steady Flow from Vessels

From Bernoulli's equation (10), between the points 1 and 2 (**Fig. 17**) there follows, with Eq. (11b) for the outflow velocity $v = \sqrt{[2gb + 2(p_1 - p_2)/\rho]/(1 + \zeta)}$. With vessels, this is usually written

$$v = \varphi\sqrt{2gb + 2(p_1 - p_2)/\rho}, \tag{13}$$

where $\varphi = \sqrt{1/(1 + \zeta)}$ is the velocity coefficient. For volume flow V, the flow constrictions must still be taken

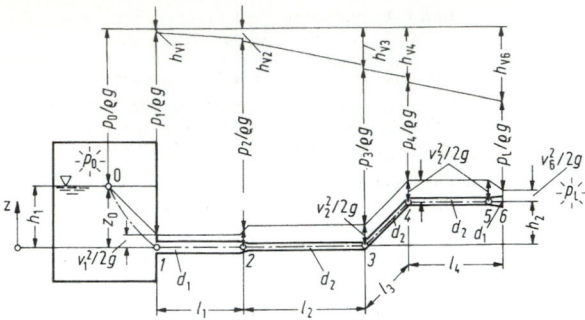

Figure 16. Pipeline.

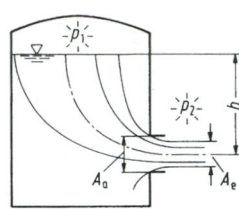

Figure 17. Outflow from container.

into account. With the contraction coefficient $\alpha = A_c/A_a$ is derived:

$$\dot{V} = \alpha\varphi A \sqrt{2gb + 2(p_1 - p_2)/\rho}$$

$$= \mu A_a \sqrt{2gb + 2(p_1 - p_2)/\rho}. \qquad (14)$$

$\mu = \alpha\varphi$ is the outflow coefficient. The following values apply to φ, α and μ (**Fig. 18**):

(a) Sharp-edged orifice:

$\varphi = 0.97;$ $\alpha = 0.61$ to $0.64;$ $\mu = 0.59$ to $0.62.$

(b) Rounded orifice

$\varphi = 0.97$ to $0.99;$ $\alpha = 1;$ $\mu = 0.97$ to $0.99.$

(c) Cylindrical extension pipe, $l/d = 2$ to 3:

$\varphi = 0.82;$ $\alpha = 1;$ $\mu = 0.82.$

(d) Conical extension pipe: $\varphi = 0.95$ to $0.97.$

$(d_2/d_1)^2$	0.1	0.2	0.4	0.6	0.8	1.0
α	0.83	0.84	0.87	0.90	0.94	1.0

Equations (13) and (14) do not apply to outflow cross-sections in which v is constantly over the cross-section. For large openings, with a flow filament at depth z (without overpressure), $v = \sqrt{2gz}$, and the volume flow is $\dot{V} = \mu \int_{z_1}^{z_2} b(z)\sqrt{2gz}\, dz$, e.g. for a square aperture $\dot{V} = 2\mu b \sqrt{2g}\,(z_2^{3/2} - z_1^{3/2})/3$. The outflow coefficient is at $\mu = 0.60$ for sharp-edged apertures, and at $\mu = 0.75$ for rounded apertures.

6.2.6 Steady Flow in Open Channels

In the case of steady flows, level and bottom gradients are parallel. From Bernoulli's equation (Eq. 10) it follows that

$$z_1 - z_2 = b_v \quad \text{or} \quad (z_1 - z_2)/l = \sin \alpha = (\lambda/d_h)v^2/(2g). \qquad (15)$$

If d_h is here the hydraulic diameter according to A6.2.3, then the formulae of pipe flows according to A6.2.1 to A6.2.4 apply. v is the mean velocity, i.e. $\dot{V} = vA$ or $v = \dot{V}/A$ applies. If $\dot{V}$ or v are known, then the required gradient follows from Eq. (15), or, if the gradient is known, the flow velocity v is derived (for reference values for k, see **Table 1**).

6.2.7 Non-steady Flow of Viscous Newtonian Fluids

The equations that apply in this case are given by the Bernoulli's equation, in the form of Eq. (9), taking into consideration Eq. (11a) and the continuity equation in the form of Eqs (5) and (6).

6.2.8 The Free Jet

If a jet flows with a constant velocity profile from an aperture into a surrounding fluid at rest, of the same nature (**Fig. 19**), then at the peripheries particles from the surroundings will be dragged in owing to the friction. That is to say, the volume flow will increase with the length of the jet, while the velocity will drop. In this situation, a broadening of the jet occurs. The pressure in the interior of the jet is equal to the ambient pressure, i.e. the pulse is constant in each jet cross-section:

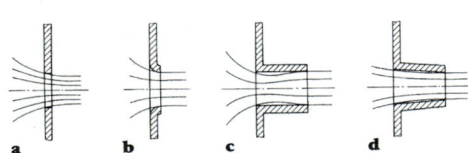

Figure 18. a–d Forms of orifice.

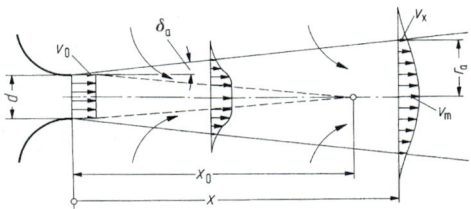

Figure 19. Fine jet.

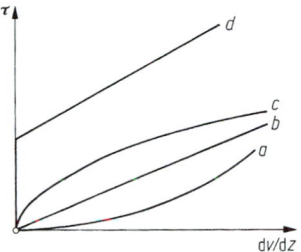

Figure 20. Flow curves: **a** dilatant, **b** Newtonian, and **c** structurally viscous fluid; **d** Bingham medium.

$$I = \int_{-\infty}^{+\infty} \rho v^2 \, dA = \text{const.}$$

The conical core of the jet, in which $v = \text{const.}$, dissipates along the path lengths x_0. The velocity profiles are accordingly in affinity with one another. Results for the round jet [1]: Core length $x_0 = d/m$, with $m = 0.1$ for laminar jets, and $m = 0.3$ for completely turbulent jets ($0.1 < m < 0.3$). Mean velocity $v_m = v_0 x_0 / x$. Energy absorption $E = 0.667 E_0 x_0 / x$ (E_0 dynamic energy at outlet). Jet dispersion

$$r_a = m \sqrt{0.5 \ln 2} \cdot x = 0.5887 \, mx,$$

with, at the expansion periphery, $v_x = 0.5 v_m$. Jet expansion angle

$$\delta_a = \arctan[0.707m \sqrt{\ln(v_x/v_m)}],$$

i.e. for $v_x/v_m = 0.5$ and $m = 0.3$, there derived $\delta_a = 10°$. The volume flow is $\dot{V} = 2m\dot{V}_0 x/d$ [1, 3].

6.3 One-Dimensional Flow of Non-Newtonian Fluids

In the case of non-Newtonian fluids, according to Eq. (8), there is *no* linear relationship between the transverse stress τ and the shear velocity [9]. For these rheological substances, a distinction is made between the following flow laws (**Fig. 20**):

Dilatant Liquids. The viscosity increases with the shear velocity $\dot{\gamma}$ (e.g. coating paints, glass frits etc.). $\dot{\gamma} = dv/dz = k\tau^m$, $m < 1$ (formula from Ostwald de Waele [7]). k is the fluidity factor and m the flow coefficient. Dilatant liquids can also be determined using the

Prandtl–Eyring formula: $\gamma = dv/dz = c \sinh(\tau/a)$, where c and a are substance-dependent constants.

Structurally Viscous Liquids. The viscosity increases with the growing shear velocity (e.g. silicones, spinning solutions, consistent grease). The laws given apply, but with $m > 1$ and with corresponding constants c and a.

Bingham Medium. The material does not start to flow until the yield point τ_F is reached. Below τ_F, it behaves as an elastic body, and above it as a Newtonian fluid (e.g. toothpaste, waste water sludge, granular suspensions) $\dot{\gamma} = dv/dz = k(\tau - \tau_F)$ (Bingham's law).

Elastoviscous Substances (Maxwell medium). These have both the properties of liquids as well as those of elastic bodies (e.g. dough, polyethylene resins). The transverse stress is time-dependent; that is to say, it is still available when $\dot{\gamma}$ is already zero. $\dot{\gamma} = dv/dz = (\tau/\eta) + (1/G)(dr/dt)$ (Maxwell's law).

Thixotropic and Rheopexic Liquids. In this case too, the transverse stress values are time-dependent; in addition, the flow behaviour changes with the mechanical stress. With thixotropic liquids, the flow capacity increases with time (e.g. during stirring or spreading), while with rheopexic liquids it decreases with mechanical stress (e.g. plaster paste). Flow laws have not so far been derived.

Calculation of Pipe Flows

For *dilatant and structurally viscous liquids*, the pressure drop can be calculated in accordance with Eq. (11a) after Metzner [7], in the same way as for Newtonian liquids with the generalised Reynolds number:

$$Re^* = v^{(2m-1)/m} d^{1/m} \rho/\eta^*:$$

$$\eta^* = 8^{(1-m)/m} (1/k^m)[(3+m)/4]^{1/m}.$$

In the laminar range ($Re^* < 2300$) $\lambda = 64/Re^*$ applies, and in the turbulent range ($Re^* > 3000$)

$$\lambda = 0.0056 + 0.5/(Re^*)^{0.32}.$$

For *Bingham media*, the pressure drop is derived from Eq. (11a), with the pipe friction coefficient [7]

$$\lambda = \frac{64}{Re} + \frac{32}{3} \frac{He}{Re^2} - \frac{4096}{3} \frac{1}{\lambda^3} \left(\frac{He}{Re^2}\right)^4,$$

where the influence of the yield point is expressed by the Hedström coefficient He:

$$He = \tau_F \rho d^2/\eta^2 = \tau_F d^2/(\rho v^2).$$

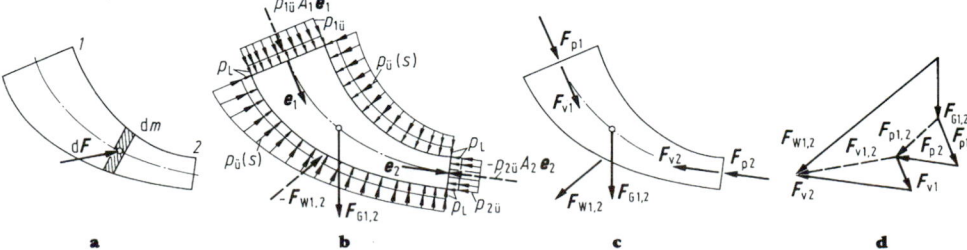

Figure 21. a–d Force effect on a flowing fluid.

6.4 Forces Due to the Flow of Incompressible Fluids

6.4.1 Equation of Momentum

From Newton's basic law, there follows, for the mass element $dm = \rho A\, ds$ of the flow pipes in **Fig. 21a**,

$$d\boldsymbol{F} = \frac{d}{dt}(dm\boldsymbol{v}) = \frac{d(dm)}{dt}\,\boldsymbol{v} + dm\,\frac{d\boldsymbol{v}}{dt}.$$

For incompressible fluids, $d(dm)/dt = 0$, and with $v = v(s, t)$ there applies, for the non-steady flow,

$$d\boldsymbol{F} = dm\left(\frac{\partial \boldsymbol{v}}{\partial t} + \frac{\partial \boldsymbol{v}}{\partial s}\frac{ds}{dt}\right)$$

or for steady flow with $\partial \boldsymbol{v}/\partial t = 0$

$$d\boldsymbol{F} = dm\,\frac{\partial \boldsymbol{v}}{\partial s}\,v = \rho A v\,d\boldsymbol{v} = \rho \dot{V}\,d\boldsymbol{v}.$$

For the total control area between *1* and *2*, it follows, after integration, that

$$\boldsymbol{F}_{1,2} = \rho \dot{V}(\boldsymbol{v}_2 - \boldsymbol{v}_1). \tag{16}$$

Here, $\boldsymbol{F}_{1,2}$ is the force on the fluid enclosed in the control area. It is composed of the portions according to **Fig. 21b**, where the resultant of the air pressure is zero. With $-\boldsymbol{F}_{w1,2}$ as the resultant of the overpressure $p_{\ddot{u}}(s)$, $\boldsymbol{F}_{1,2} = -\boldsymbol{F}_{w1,2} + \boldsymbol{F}_{G1,2} + p_{1\ddot{u}}A_1\boldsymbol{e}_1 - p_{2\ddot{u}}A_2\boldsymbol{e}_2$. From this it follows, for the force imposed by the fluid on the 'wall', with Eq. (16), that

$$\boldsymbol{F}_{w1,2} = \boldsymbol{F}_{G1,2} + (p_{1\ddot{u}}A_1\boldsymbol{e}_1 - p_{2\ddot{u}}A_2\boldsymbol{e}_2)$$

$$+ (\rho\dot{V}v_1\boldsymbol{e}_1 - \rho\dot{V}v_2\boldsymbol{e}_2)$$

$$= \boldsymbol{F}_{G1,2} + (\boldsymbol{F}_{p1} + \boldsymbol{F}_{p2}) + (\boldsymbol{F}_{v1} + \boldsymbol{F}_{v2}) \tag{17}$$

$$= \boldsymbol{F}_{G1,2} + \boldsymbol{F}_{p1,2} + \boldsymbol{F}_{v1,2}.$$

The wall force is composed of the weight fraction $\boldsymbol{F}_{G1,2}$, the pressure fraction $\boldsymbol{F}_{p1,2}$ and the velocity fraction $\boldsymbol{F}_{v1,2}$ (**Fig. 21c** and **d**).

6.4.2 Application (Fig. 22)

(a) *Jet Impact Force Against Walls*. Disregarding the inherent weight, and taking account of the fact that inside the jet the pressure everywhere is equal to the air pressure (i.e. $p_{\ddot{u}} = 0$, see A6.2.8), it follows from Eq. (17), for the x-direction and for the control area *1-2-3*

$$F_{wx} = (\rho\dot{V}v_1\boldsymbol{e}_1 - \rho\dot{V}_2v_2\boldsymbol{e}_2 - \rho\dot{V}_3v_3\boldsymbol{e}_3)\boldsymbol{e}_x = \rho\dot{V}v_1\cos\beta.$$

For the y-direction, it follows from Eq. (17) that

$$F_{wy} = 0 = (\rho\dot{V}v_1\boldsymbol{e}_1 - \rho\dot{V}_2v_2\boldsymbol{e}_2 - \rho\dot{V}_3v_3\boldsymbol{e}_3)\boldsymbol{e}_y,$$

i.e. $\dot{V}v_1\sin\beta - \dot{V}_2v_2 + \dot{V}_3v_3 = 0$. With $v_1 = v_2 = v_3$ from Bernoulli's equation, and $\dot{V} = \dot{V}_2 + \dot{V}_3$ from the continuity equation, there is derived

$$\dot{V}_2/\dot{V}_3 = (1 + \sin\beta)/(1 - \sin\beta).$$

For $\beta = 0$ (impact against vertical wall) there applies

$$F_{wx} = \rho\dot{V}v_1 = \rho A_1v_1^2 \quad \text{and} \quad \dot{V}_2/\dot{V}_3 = 1.$$

If the vertical wall moves in the x-direction at the velocity u, then

$$F_{wx} = \rho\dot{V}(v_1 - u) = \rho A_1v_1(v_1 - u).$$

For the curved plate, $F_{wx} = \rho\dot{V}v_1(1 + \cos\beta)$ can be derived accordingly. If the curved plate moves at velocity u (free jet turbine), then there applies

$$F_{wx} = \rho\dot{V}(v_1 - u)(1 + \cos\beta).$$

(b) *Force on Pipe Bends*. From Eq. (17) it follows, disregarding the inherent weight and with $A_1 = A_2 = A$ or $v_1 = v_2 = v$ or $p_{1\ddot{u}} = p_{2\ddot{u}} = p_{\ddot{u}}$, that

$$\boldsymbol{F}_{w1,2} = (p_{\ddot{u}}A + \rho\dot{V}v)\boldsymbol{e}_1 - (p_{\ddot{u}}A + \rho\dot{V}v)\boldsymbol{e}_2 \quad \text{and}$$

$$|\boldsymbol{F}_{w1,2}| = F_{w1,2} = F_x = 2(p_{\ddot{u}}A + \rho\dot{V}v)\cos(\beta/2).$$

Tensile forces in the flange bolt connections function as reaction forces.

(c) *Force on Nozzle*. With $p_{2\ddot{u}} = 0$ and $v_2 = v_1A_1/A_2 = v_1\alpha$ and $p_{1\ddot{u}} = \rho(v_2^2 - v_1^2)/2$ it follows from Eq. (17)

$$\boldsymbol{F}_{w1,2} = (\rho/2)v_1^2A_1(\alpha - 1)^2\boldsymbol{e}_x.$$

Tensile forces in the flange bolt connection function as reaction forces.

(d) *Forces with Sudden Broadening of Pipes*. According to Carnot, the wall force is determined by the fact that the pressure p across the cross-section *1* is set constantly equal to p_1 (as in narrower cross-sections): $\boldsymbol{F}_w = -p_1(A_2 - A_1)\boldsymbol{e}_x$. The following then applies for the control area *1-2*, according to Eq. (17):

$$\boldsymbol{F}_{w1,2} = -p_1(A_2 - A_1)\boldsymbol{e}_x$$

$$= (p_1A_1 + \rho v_1^2 A_1 - p_2A_2 - \rho v_2^2A_2)\boldsymbol{e}_x.$$

With $v_1 = v_2A_2/A_1 = v_2\alpha$, it follows that $p_1 = \rho v_2^2\alpha + p_2 + \rho v_2^2$. From Eq. (9) is derived, for the steady case with $z_1 = z_2$ and $\Delta p_v = \zeta\rho v^2/2$ for the loss coefficient $\zeta = (\alpha - 1)^2$ (Borda–Carnot equation).

(e) *Rocket Transverse Thrust*. With the relative velocities $v_{r1} = 0$ and $v_{r2} = v_r$, it follows from Eq. (17) for the transverse thrust

$$\boldsymbol{F}_w = \rho\dot{V}(0 - \boldsymbol{v}_{r2}) = -\rho\dot{V}v_r\boldsymbol{e}_x = -\rho A_2v_r^2\boldsymbol{e}_x.$$

(f) *Propeller Transverse Thrust*. When a propeller or screw rotates, the fluid is sucked in and accelerated. The flow tubes are selected such that $v_1A_1 = v_sA_3 = v_5A_5$. v_1 is the velocity of the vehicle and therefore the inlet velocity of the fluid. From the principle of linear momentum (17), it follows, with the shear force

$$F_s = \rho\dot{V}(v_5 - v_1) = \rho A_3v_3(v_5 - v_1).$$

From Bernoulli's equation for the ranges *1-2* and *4-5* there follows, with $p_1 = p_5$ (free jet), the pressure differential $p_4 - p_2 = \rho(v_5^2 - v_1^2)/2$ and therefore $F_s = \rho A_3(v_5^2 - v_1^2)/2$. Equating the expressions for F_s leads to $v_3 = (v_1 + v_5)/2$ and therefore to $F_s = c_s\rho v_1^2A_3/2$, where $c_s = (v_5/v_1)^2 - 1$ is the degree of thrust loading. If the output applied is $P_z = F_sv_3$ and the actual output is $P_n = F_sv_1$, then the theoretical degree of efficiency of the propeller $\eta = P_n/P_z = v_1/v_3$. In addition, with $k = 2P_z/(\rho v_1^3 A_3)$, the equation $k = 4(1 - \eta)\eta^3$ applies, as does $\eta = 2/(1 + \sqrt{1 + c_s})$. From this are derived, with P_z and v_1 given, the values k, η, F_s etc.

6.5 Multi-Dimensional Flow of Inviscid Fluids

6.5.1 Fundamentals

Euler's equations of motion. These follow from Newton's basic law in the x-direction (analogously for the y- and z-directions), with the mass force $\boldsymbol{F} = (X; Y; Z)$ related to the element

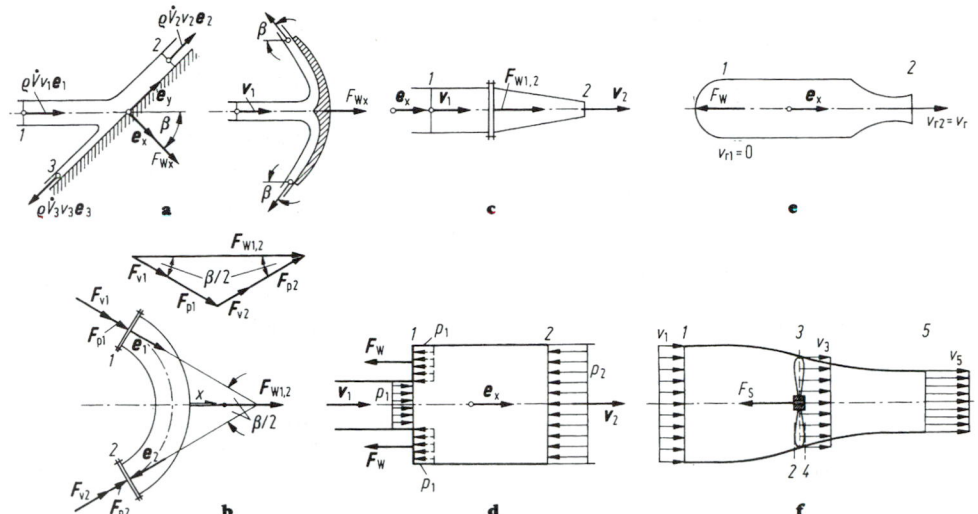

Figure 22. a-f Applications for effect of force.

$$\frac{dv_x}{dt} = \frac{\partial v_x}{\partial t} + v_x \frac{\partial v_x}{\partial x} + v_y \frac{\partial v_x}{\partial y} + v_z \frac{\partial v_x}{\partial z} = X - \frac{1}{\rho}\frac{\partial p}{\partial x}.$$
(18)

The change in velocity $\partial v_x/\partial t$ with the time at a fixed position is referred to as local, while that with $(v_x\partial v_x/dx + v_y\partial v_x/\partial y + v_z\partial v_x/\partial z)$ at a specific time with a change in position is referred to as convective. Vectorially there applies

$$\frac{dv}{dt} = \frac{\partial v}{\partial t} + (v\nabla)v = F - \frac{1}{\rho}\,\text{grad}\,p,$$
(19)

where, with the del or nabla operator ∇ and curl $v = \nabla \times v$, $(v\nabla)v = \text{grad}\,v^2/2 - v \times \text{curl}\,v$. Here, $(1/2)$ curl $v = w$ is the angular velocity with which individual fluid particles rotate. If a stream is irrotational, i.e. curl $v = 0$, then a potential flow exists. Lines that are touched by curl v are called vortex lines, and several of these lines form the vortex tubes.

Flow Circulation. This is the line integral across the scalar product $v\,dr$ along the length of a closed curve:

$$\Gamma = \oint_{(O)} v\,dr = \oint_{(O)} (v_x\,dx + v_y\,dy + v_z\,dz).$$

Using Stokes's law, this equation can also be written

$$\Gamma = \oint_{(C)} v\,dr = \iint_{(A)} \text{rot}\,v\,dA,$$
(20)

where A is a surface stretched across C. With potential flows, curl $v = 0$, i.e. $\Gamma = 0$.

Helmholtz Equation for Vorticity. If Eq. (20) is applied to curves surrounding vortex tubes, there follows

$$\Gamma_1 = \oint_{(C_1)} v\,dr = \Gamma_2 = \oint_{(C_2)} v\,dr = \text{const.}$$

First Helmholtz Law. The circulation has the same value for every curve surrounding vortex tubes, i.e. vortex tubes can neither begin nor end inside a fluid range (in other

words, they either form closed tubes - known as ring vortices - or pass on to the end of the fluid range). For $F = -\,\text{grad}\,U$ and barotropic fluid $\rho = \rho(p)$, it follows from Eqs (19) and (20)

$$\frac{d\Gamma}{dt} = \oint \frac{dv}{dt}\,dr = \iint \text{curl}\,\frac{dv}{dt}\,dA = 0.$$

Second Helmholtz Law. The circulation has a temporally unchangeable value when the mass forces have a potential and the fluid is barotropic (i.e. for example, potential flows always remain potential flows).

Continuity Equation. The mass flowing into an element $dx\,dy\,dz$ must be equal to the local change in density plus the outflowing mass:

$$\frac{\partial \rho}{\partial t} + \frac{\partial(\rho v_x)}{\partial x} + \frac{\partial(\rho v_y)}{\partial y} + \frac{\partial(\rho v_z)}{\partial z} = 0$$

or, in vectorial form,

$$\frac{\partial \rho}{\partial t} + \nabla(\rho v) = \frac{\partial \rho}{\partial t} + \text{div}(\rho v) = 0.$$

For incompressible fluids ($\rho = \text{const.}$) it follows that

$$\frac{\partial v_x}{\partial x} + \frac{\partial v_y}{\partial y} + \frac{\partial v_z}{\partial z} = \text{div}\,v = 0.$$
(21)

Equations (19) and (20) form four coupled partially differential equations for calculating the four unknowns v_x, v_y, v_z and p of a flow. Solutions can in general only be obtained for potential flows, i.e. if curl $v = 0$.

6.5.2 Potential Flows

Euler's equations can be integrated when the vector v has a velocity potential $\Phi(x, y, z)$, i.e. when

$$v = \text{grad}\,\Phi = \frac{\partial \Phi}{\partial x}e_x + \frac{\partial \Phi}{\partial y}e_y + \frac{\partial \Phi}{\partial z}e_z$$

and F likewise has a potential, i.e.

$$F = -\text{grad}\,U = -\frac{\partial U}{\partial x}e_x - \frac{\partial U}{\partial y}e_y - \frac{\partial U}{\partial z}e_z.$$

It therefore follows for the potential flow curl $\boldsymbol{v}$ = curl grad $\Phi = \nabla \times \nabla\Phi = 0$ and from Eq. (19) after integration

$$\text{grad}\left[\frac{\partial\Phi}{\partial t} + \frac{v^2}{2} + \frac{p}{\rho} + U\right] = 0 \quad \text{and}$$

$$\frac{\partial\Phi}{\partial t} + \frac{v^2}{2} + \frac{p}{\rho} + U = C(t),$$

or, for the steady flow,

$$v^2/2 + p/\rho + U = C = \text{const.} \tag{22}$$

This is the generalised Bernoulli equation for the potential flow, which has the same constant C for the entire field of flow.

From the continuity equation (Eq. 21) there follows

$$\text{div } \boldsymbol{v} = \text{div grad } \Phi$$

$$= \nabla\nabla\Phi = \Delta\Phi = \frac{\partial^2\Phi}{\partial x^2} + \frac{\partial^2\Phi}{\partial y^2} + \frac{\partial^2\Phi}{\partial z^2} = 0 \tag{23}$$

(Laplace potential equation). Equations (22) and (23) are used to calculate p and v. The latter has an infinite number of solutions; accordingly, known solutions are examined and interpreted as flows. For example, $\Phi(x, y, z) = C/r = C/\sqrt{x^2 + y^2 + z^2}$ is a solution. From this is derived $v_x = \partial\Phi/\partial x = -Cx/\sqrt{r^3}$, $v_y = \partial\Phi/\partial y = -Cy/\sqrt{r^3}$ and $v_z = \partial\Phi/\partial z = -Cz/\sqrt{r^3}$, as well as $v = \sqrt{v_x^2 + v_y^2 + v_z^2} = C/r$. This is a flow directed radially around the centre point, i.e. a sink (or a source, if C is replaced by $-C$).

Plane Potential Flow. In this case, all the analytical (complex) functions form solutions, because

$$w = f(z) = f(x + iy) = \Phi(x,y) + i\Psi(x,y) \tag{24}$$

suffice as analytical functions of the Cauchy–Riemann differential equations

$$\partial\Phi/\partial x = \partial\Psi/\partial y \quad \text{and} \quad \partial\Phi/\partial y = -\partial\Psi/\partial x \tag{25}$$

and therefore also the potential equations

$$\frac{\partial^2\Phi}{\partial x^2} + \frac{\partial^2\Phi}{\partial y^2} = 0 \quad \text{and} \quad \frac{\partial^2\Psi}{\partial x^2} + \frac{\partial^2\Psi}{\partial y^2} = 0. \tag{26}$$

$\Phi(x, y)$ = const. are the potential lines to which the velocity vector is perpendicular and $\Psi(x, y)$ = const. are the flow lines tangential to the velocity vector; i.e. both groups of curves are perpendicular to each other. From Eqs (24) and (25) it follows that

$$f'(z) = \frac{dw}{dz} = \frac{\partial\Phi}{\partial x} + i\frac{\partial\Psi}{\partial x} = v_x - iv_y = \bar{\boldsymbol{v}}, \quad \text{i.e.} \tag{27a}$$

$$\boldsymbol{v} = \bar{f}'(\bar{z}) = \partial\Phi/\partial x - i\partial\Psi/\partial x = v_x + iv_y. \tag{27b}$$

The horizontal line above the letters indicates the conjugated complex value. $w = f(z)$ is termed the complex velocity potential. If s and n coordinates are tangential and vertical to the potential line Φ (**Fig. 23**), the volume flow is

$$\dot{V} = \int_{(1)}^{(2)} v_n \, ds = \int_{(1)}^{(2)} \frac{\partial\Phi}{\partial n} \, ds = \int_{(1)}^{(2)} \frac{\partial\Psi}{\partial s} \, ds = \Psi_2 - \Psi_1;$$

i.e. it is equal to the difference of the flow line values. The velocity is inversely proportional to the interval between the flow lines. **Figure 24** shows some examples of complex velocity potentials:

(a) *Parallel Flow.* From the velocity potential $w = v_0 z = v_0 x + iv_0 y = \Phi + i\Psi$ the potential lines follow

to $\Phi = v_0 x$ = const., i.e. x = const.; the potential lines are, then, straight lines parallel to the y-axis. The flow lines are straight lines parallel to the x-axis, because $\Psi = v_0 y$ = const., i.e. y = const. Moreover, $v_x = \partial\Phi/\partial x = v_0$ and $v_y = \partial\Phi/\partial y = 0$ also apply.

(b) *Vortex Line Flow (Potential Vortex).* C may be real. $w = iC \log z = -C \arctan(y/z) + i(C/2) \ln(x^2 + y^2) = \Phi + i\Psi$, or $\Phi = -C \arctan(y/z)$ = const. gives $y = cx$; the potential lines are therefore straight. $\Psi = (1/2)C \ln(x^2 + y^2)$ = const. provides $x^2 + y^2 = c$; the flow lines are therefore circles.

$$f'(z) = \frac{iC}{z} = \frac{iC}{x + iy} = \frac{iC(x - iy)}{x^2 + y^2}$$

$$= C\frac{y}{x^2 + y^2} + iC\frac{x}{x^2 + y^2}$$

$$= \frac{Cy}{r^2} + i\frac{Cx}{r^2} = v_x - iv_y,$$

i.e. v_x is positive in the first quadrant and v_y is negative. The flow is therefore circulating clockwise.

$$v = |\boldsymbol{v}| = \sqrt{v_x^2 + v_y^2} = \sqrt{C^2(x^2 + y^2)/r^4} = C/r.$$

Despite the potential present, a circulation exists:

$$\Gamma = \oint \boldsymbol{v} \, d\boldsymbol{r} = \oint v \, ds \cos\beta = -(C/r)2\pi r = -2\pi C.$$

(c) *Dipole Current Flow.*

$$w = \frac{\mu}{z} = \frac{\mu x}{x^2 + y^2} + i\frac{-\mu y}{x^2 + y^2} = \Phi + i\Psi.$$

$\Phi = \mu x/(x^2 + y^2)$ = const. gives $x^2 + y^2 = cx$ or $(x - c/2)^2 + y^2 = (c/2)^2$; the potential lines are therefore circles with centres on the x-axis. $\Psi = -\mu y/(x^2 + y^2)$ = const. provides $x^2 + y^2 = cy$, or $x^2 + (y - c/2)^2 = (c/2)^2$; the flow lines are therefore circles with centres on the y-axis. All the circles pass through the zero point. The sum of the velocities $v = |w'(z)| = \mu/z^2 = \mu/(x^2 + y^2) = \mu/r^2$ decreases outwards with $1/r^2$.

(d) *Parallel Outflow from a Circular Cylinder.* When the parallel flow and dipole current flow overlap, there derives for the cylinder with radius a, $w = f(z) = v_0(z + a^2/z)$. For $z \to \pm\infty$ is derived the parallel flow. There further applies

$$\Phi + i\Psi = \left(v_0 x + \frac{v_0 a^2 x}{x^2 + y^2}\right) + i\left(v_0 y - \frac{v_0 a^2 y}{x^2 + y^2}\right).$$

For $\Psi = 0$, $v_0 y(1 - a^2/(x^2 + y^2)) = 0$, i.e. $y = 0$ (x-axis) and $x^2 + y^2 = a^2$ (boundary of cylinder) form a flow line. The velocity of the flow follows from $f'(z) = v_0(1 - a^2/z^2) = v_x - iv_y$ to

$$v = |f'(z)| = |v_0(1 - a^2/z^2)|.$$

For $z = \pm a$, $v = 0$ (critical points), and for $z = \pm ia$, $v = 2v_0$ (points of intersection); the velocity is therefore symmetrical to the vertical axis. From Eq. (22) a symmetrical pressure distribution then also follows from (22), i.e. the force taking effect on the body when flowed around by an inviscid fluid equals zero (d'Alembert's hydrostatic paradox). Flow forces pertain only owing to the friction of the fluids.

(e) *Unsymmetrical Flow Round a Circular Cylinder.* If the flow round the body according to (d) overlaps the potential vortex according to (b), then we obtain:

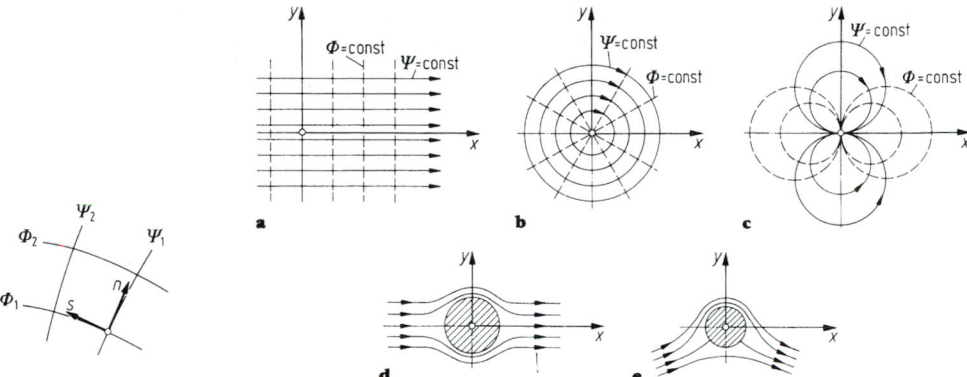

Figure 23. Potential and flow lines.

Figure 24. a-e Potential flows.

$$w = f(z) = v_0(z + a^2/z) + iC \log z,$$

$$\Psi = v_0 y \left(1 - \frac{a^2}{x^2 + y^2}\right) + \frac{C}{2} \ln (x^2 + y^2),$$

$$\Phi = v_0 x \left(1 + \frac{a^2}{x^2 + y^2}\right) - C \arctan (y/x).$$

The flow function Ψ is symmetrical about the x-axis, but not about the y-axis; i.e. by integration of the pressure along the length of the circumference a force is derived in the y-direction. This 'lift force' can be calculated to

$$F_A = \rho v_0 \Gamma = \rho v_0 2\pi C$$

(Kutta-Joukowski law); it is dependent only on the free-stream velocity and the circulation, but not on the contour of the cylinder.

Equiangular Representation of the Circle. With the method of equiangular representation it is possible to represent the circle on any desired number of other, easily connected contours, and vice versa, thereby determining the flow around these contours since the random flow around the circle is known [3].

6.6 Multi-Dimensional Flow of Viscous Fluids

6.6.1 Navier-Stokes Equations

In the case of flow in space of Newtonian fluids, for additional stresses arising due to friction, as the generalisation of Newton's transverse stress law, the following equations apply (with the additional viscosity constant η^* [3]):

$$\sigma_x = 2\eta \frac{\partial v_x}{\partial x} + \eta^* \text{ div } \boldsymbol{v}, \quad \sigma_y = 2\eta \frac{\partial v_y}{\partial y} + \eta^* \text{ div } \boldsymbol{v},$$

$$\sigma_z = 2\eta \frac{\partial v_z}{\partial z} + \eta^* \text{ div } \boldsymbol{v}, \quad (28a)$$

$$\tau_{xy} = \eta \left(\frac{\partial v_x}{\partial y} + \frac{\partial v_y}{\partial x}\right), \quad \tau_{xz} = \eta \left(\frac{\partial v_x}{\partial z} + \frac{\partial v_z}{\partial x}\right),$$

$$\tau_{yz} = \eta \left(\frac{\partial v_y}{\partial z} + \frac{\partial v_z}{\partial y}\right). \quad (28b)$$

Newton's basic law for a fluid element reads, for the x-direction,

$$\frac{dv_x}{dt} = \frac{\partial v_x}{\partial t} + \frac{\partial v_x}{\partial x} v_x + \frac{\partial v_x}{\partial y} v_y + \frac{\partial v_x}{\partial z} v_z$$

$$= X - \frac{1}{\rho} \frac{\partial p}{\partial x} + \frac{1}{\rho} \left(\frac{\partial \sigma_x}{\partial x} + \frac{\partial \tau_{xy}}{\partial y} + \frac{\partial \tau_{xz}}{\partial z}\right). \quad (29)$$

From Eqs (28) and (29) there follow for the incompressible fluids (div $\boldsymbol{v} = 0$) the equations of motion by Navier-Stokes (analogous equations apply for the y- and z-directions):

$$\frac{dv_x}{dt} = X - \frac{1}{\rho} \frac{\partial p}{\partial x} + \frac{\eta}{\rho} \left(\frac{\partial^2 v_x}{\partial x^2} + \frac{\partial^2 v_x}{\partial y^2} + \frac{\partial^2 v_x}{\partial z^2}\right)$$

$$= X - \frac{1}{\rho} \frac{\partial p}{\partial x} + \frac{\eta}{\rho} \Delta v_x, \quad (30)$$

or, in vectorial form,

$$\frac{d\boldsymbol{v}}{dt} = \frac{\partial \boldsymbol{v}}{\partial t} + (\boldsymbol{v}\nabla)\boldsymbol{v} = \boldsymbol{F} - \frac{1}{\rho} \text{ grad } p + \frac{\eta}{\rho} \Delta \boldsymbol{v}. \quad (31)$$

Here, p is the mean pressure, since from div $\boldsymbol{v} = 0$ there follows $\sigma_x + \sigma_y + \sigma_z = 0$, i.e. the sum of the additional stresses σ_x, σ_y, σ_z to the mean pressure is zero. Equations (28) to (31) apply for laminar flow; for turbulent flow, the turbulence force is to be introduced as a further element [3]. Solutions to the Navier-Stokes equations exist only for a few special cases (see A6.6.2), for small Reynolds coefficients. At large Reynolds numbers, in other words low viscosities, many problems are solved by boundary layer theory, the origin of which is attributable to Prandtl. In this case, the flowing viscous fluid, always subjected to the non-slip condition at the body, is regarded as frictionally viscous only in a thin boundary layer, but otherwise regarded as inviscid.

6.6.2 Some Solutions at Low Reynolds Number (Laminar Flow (Fig. 25) [10]

(a) *Couette Flow.* An external cylinder rotates uniformly about a core at rest, driven by an external torque M. The Navier-Stokes equation (31), in polar coordinates in this instance, in the r- and φ-direction (with $v_r = 0$, $v_\varphi = v(r)$, $p = p(r)$ for reasons of symmetry and $\boldsymbol{F} = 0$), assumes the

form $-\frac{v^2}{r} = -\frac{1}{\rho} \frac{\partial p}{\partial r}$ and

$$\frac{\eta}{\rho}\left(\frac{d^2v}{dr^2} + \frac{1}{r}\frac{dv}{dr} - \frac{v}{r^2}\right) = \frac{\eta}{\rho}\frac{d}{dr}\left[\frac{1}{r}\frac{d}{dr}(rv)\right] = 0.$$

From this is derived by integration $v = C_1 r/2 + C_2/r$. The constants C_1 and C_2 are obtained from $v(r_i) = 0$ and $v(r_a) = \omega r_a$ as $C_2 = -C_1 r_i^2/2$ and $C_1 = 2\omega r_a^2/(r_a^2 - r_i^2)$; and

therefore $v = \dfrac{\omega r_a^2}{r_a^2 - r_i^2}\left(r - \dfrac{r_i^2}{r}\right)$.

Equation (28) applies analogously in polar coordinates for the transverse stresses:

$$\tau = \eta\left(\frac{1}{r}\frac{\partial v_r}{\partial \varphi} + \frac{\partial v_\varphi}{\partial r} - \frac{v_\varphi}{r}\right)$$

$$= \eta\left(\frac{dv}{dr} - \frac{v}{r}\right) = \frac{2\eta\omega r_a^2}{r_a^2 - r_i^2}\frac{r_i^2}{r^2},$$

$$\tau(r = r_a) = 2\eta\omega r_i^2/(r_a^2 - r_i^2).$$

For the external moment required on the cylinder, $M = \tau \cdot 2\pi r_a l r_a$, it follows that $M = 4\pi\eta\omega l r_a^2 r_i^2/(r_a^2 - r_i^2)$. By measuring M, the viscosity η can be determined from this (Couette viscosimeter).

(b) *Lubricant Friction.* If a slightly curved (or flat) plate moves parallel to another, with a small intermediate space, then a flow pressure is derived, which prevents the two surfaces from coming in contact and prevents their friction.

With $v_y \approx 0$, $\partial v_y/\partial y \approx 0$, $v_x = v$ it follows from the continuity equation, Eq. (21), that $\partial v/\partial x = -\partial v_y/\partial y = 0$, i.e. $\partial^2 v/\partial x^2 = 0$. Because $\partial v_x/\partial t = 0$, there derives from Eqs (29) and (30) $\partial p/\partial x = \eta \partial^2 v/\partial y^2$ with the solution $v(y) = \dfrac{1}{\eta}\dfrac{\partial p}{\partial x}\dfrac{y^2}{2} + C_1 y + C_2$. With C_1 and C_2 from the condition that the fluid adheres to the plate, there is derived

$$v(y) = \frac{1}{\eta}\frac{\partial p}{\partial x}\frac{y}{2}(y - b) + v_0\left(1 - \frac{y}{b}\right).$$

From $\dot{V} = 1\displaystyle\int v\,dy = $ const. there follows $\dfrac{\partial p}{\partial x} = \dfrac{6\eta}{b^3}v_0(b - b_0)$, with $\partial p/\partial x = 0$ for $b = b_0$. For the transverse stress at $y = 0$ there applies $\tau = \eta v_0(3b_0 - 4b)/b^2$.

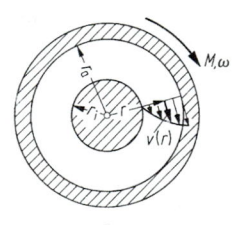

a

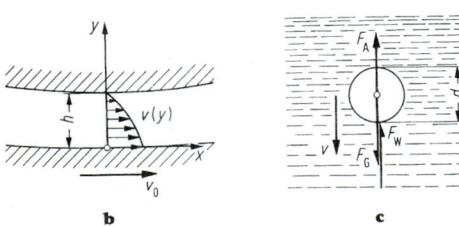

b   c

Figure 25. a–c Flows of viscous fluids.

(c) *Stokes Resistance Formula for the Sphere.* At low Reynolds numbers ($Re \leq 1$), flow takes place around a sphere. The resistance force is derived from Stokes to

$$F_W = 3\pi\eta \, dv_0. \tag{32}$$

This formula was improved by Oseen, taking account of the acceleration fractions, to

$$F_W = 3\pi\eta \, dv_0[1 + (3/8)Re].$$

Example   Viscosity determination.   If a sphere falls at $v = $ const. through a viscous fluid, then there applies $F_G - F_W - F_A = 0$, i.e.

$$\rho_K \, g\pi d^3/6 - 3\pi\eta \, dv - \rho_F g\pi d^3/6 = 0,$$

and from this $\eta = gd^2(\rho_K - \rho_F)/(18v)$.

6.6.3 Boundary Layer Theory

If a substance of low viscosity (air, water) flows around a body, then, because of the adherence of the fluid to the surface of the body, a boundary layer is formed, of thickness $\delta(x)$, in which a pronounced velocity drop pertains, and therefore large shear stresses. Outside this layer the velocity drop is small, with the result that, with small η, the transverse stresses are negligible; i.e. the fluid can be regarded as inviscid. As a rule, the initial range of the boundary layer is laminar and then passes into turbulent flow at the transition point, with increased transverse stresses. For approximation, the transition point is located at the point of minimum pressure of the outer flow [8]. From the Navier–Stokes equation (Eq. 31) it follows, for the plane case, with stable flow and without mass forces, with the continuity equation (Eq. 21) and the simplifications $v_y \ll v_x$; $\partial v_y/\partial x \ll \partial v_y/\partial y$; $\partial v_x/\partial x \ll \partial v_x/\partial y$; $\partial p/\partial y \approx 0$, that

$$\rho v_x \frac{\partial v_x}{\partial x} = -\frac{dp}{dx} + \eta\frac{\partial^2 v_x}{\partial y^2}. \tag{33}$$

With a slightly curved profile (**Fig. 26**), it follows for the wall $y = 0$ with $v_x = 0$ (adherence) from Eq. (33)

$$\frac{dp}{dx} = \eta\left(\frac{\partial^2 v_x}{\partial y^2}\right)_{y=0}. \tag{34}$$

If $dp/dx < 0$ (initial range, **Fig. 26**), then it follows from Eq. (34) $\partial^2 v_x/\partial y^2 < 0$; the velocity profile is therefore convex. For $dp/dx = 0$ $\partial^2 v_x/\partial y^2 = 0$; the velocity profile therefore has no curvature. For $dp/dx > 0$, $\partial^2 v_x/\partial y^2 > 0$; the profile is therefore curved concave, and a point is reached at which $\partial v_x/\partial y = 0$. v_x is then negative; i.e. a reverse flow sets in, which transforms into a concentrated vortex. Because of the vortex, an underpressure occurs behind the body, which, together with the transverse stresses along the length of the boundary layer, provides the overall flow resistance of the body [3, 8, 10].

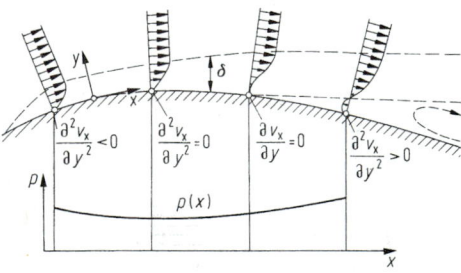

Figure 26. Boundary layer.

6.6.4 Drag of Solid Bodies

The resistance that occurs along the length of the boundary layer arising from the transverse stresses becomes frictional resistance, and the resistance derived from the underpressure caused behind the body by the separation of the flow and the vortex formation is referred to as pressure resistance or pressure drag. Both together give the total resistance. While the frictional resistance can be further calculated with the help of the boundary layer theory, the pressure resistance, which is difficult to calculate theoretically, is essentially determined empirically. Depending on the shape of the body, either the frictional resistance or the pressure resistance predominates. For the bodies in **Fig. 27**, the percentage ratio between them amounts to (a) $100:0$, (b) $90:10$, (c) $10:90$ and (d) $0:100$.

Frictional Resistance. With the very slim and streamlined bodies, the boundary layer surrounds the entire body; i.e. there is no vortex and no pressure resistance, but only frictional resistance.

$$F_r = c_r\,(\rho v_0^2/2)A_0,$$

where A_0 is the surface of the body around which the flow is taking place. Similar dependencies apply to the friction coefficient c_r as for pipes with a flow through them. The results for the thin plate of length l (**Fig. 27a**), around which the flow takes place, are taken as the basis. Transition from laminar to turbulent flow occurs at $Re = 5 \cdot 10^5$. Here, $Re = v_0 l/v$.

The transition point from laminar to turbulent flow on the plate is, then, at $x_u = v\,Re_k/v_0$. The thickness of the laminar boundary layer is $\delta = 5 \cdot \sqrt{vx/v_0}$, and that of the turbulent boundary layer $\delta = 0.37\sqrt[5]{vx^4/v_0}$. Friction coefficients $c_r = 1.327/\sqrt{Re}$ for laminar flow, $c_r = 0.074/\sqrt[5]{Re}$ for turbulent flow – smooth plate, and $c_r = 0.418/(2 + \lg(l/k))^{2.53}$ for turbulent flow – rough plate ($k = 0.001$ mm for polished surface, $k = 0.05$ mm for cast surface). For $k = 100l/Re$, the plate is to be regarded as hydraulically smooth. For a diagram see [3].

Pressure Drag (Form Drag). This is derived by integration via the pressure components in the flow direction in front of and behind the body. It can be compiled to

$$F_d = c_d(\rho v_0^2/2)A_p$$

(A_p = projection surface of the body, also known as the shadow surface). c_d can be determined by measuring the pressure distribution. As a rule, however, the measurements lead directly to the total resistance value.

Total Resistance. This is composed of the friction drag and form drag:

$$F_w = c_w(\rho v_0^2/2)A_p. \tag{35}$$

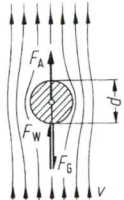

Figure 28. Suspension state.

For bodies with rapid jet separation (practically pure resistance), c_w depends only on the shape of the body, and for all other bodies it depends on the Reynolds number. For some bodies, drag coefficients c_w can be derived from **Table 2**.

Wind Pressure on Building Structures. The determinant wind velocities and coefficients c_w are to be taken from DIN 1055, page 4.

Air Resistance (Aerodynamic Drag) of Motor Vehicles. The resistance is calculated from Eq. (35), with the resistance coefficients c_w to be taken from tables.

Suspension Velocity of Particles. If a falling particle is blown upwards by air directed from below with a velocity v, then a suspension effect occurs (**Fig. 28**), when $F_G = F_A + F_w$, i.e. $\rho_K Vg = \rho Vg + c_w(\rho_F v^2)A_p$ and from this

$$v = \sqrt{4d(\rho_K - \rho_F)g/(3c_w\rho_F)}$$

Friction Resistance on Rotating Discs. If a thin rotating disc with an angular velocity ω is in motion in a fluid, then a boundary layer forms, the particles of which adhere to the surface of the disc. The frictional forces that occur on both sides create a torque that takes effect in opposition to the motion (**Fig. 29**):

$$M = 2\int r\,dF_r = 2\int rc_F \frac{\rho v^2}{2}\,dA = \int_0^{d/2} rc_F\rho\omega^2 r^2\,2\pi r\,dr$$

$$= \frac{4\pi c_F}{5}\left(\frac{\rho\omega^2}{2}\right)\left(\frac{d}{2}\right)^5 = c_M\frac{\rho\omega^2}{2}\left(\frac{d}{2}\right)^5.$$

For the torque coefficient c_M the following apply, as a function of the Reynolds number $Re = \omega d^2/(2v)$, according to [1]:

For Extended Liquids at Rest.
for $Re < 5 \cdot 10^5$ (laminar flow) $c_M = 5.2/\sqrt{Re}$;
for $Re > 5 \cdot 10^5$ (turbulent flow) $c_M = 0.168/\sqrt[5]{Re}$.

For Fluids in Housings (here, s is the distance between the disc and the wall of the housing).

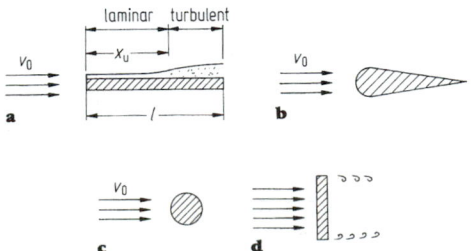

Figure 27. a–d Types of drag.

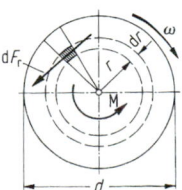

Figure 29. Disc friction.

Table 2. Drag coefficients c_w of bodies in a flow

Sphere	Rotational ellipsoid	Circular cylinder	Profile
$10^3 < Re < 2 \cdot 10^5 : c_w = 0.47$ $Re = 4 \cdot 10^5 : \quad 0.09$ $Re = 10^6 : \quad 0.13$	$\frac{a}{b} = \frac{1}{0.75}$ $Re < 5 \cdot 10^5 : c_w = 0.6$ $Re > 5 \cdot 10^5 : c_w = 0.21$ $\frac{a}{b} = \frac{1}{1.80}$ $Re > 10^5 : c_w = 0.05 ... 0.1$	$Re < 9 \cdot 10^4 : l/d = 1 : c_w = 0.63$ $\quad 2 \quad 0.68$ $\quad 5 \quad 0.74$ $\quad 10 \quad 0.82$ $\quad 40 \quad 0.99$ $\quad \infty \quad 1.20$ $Re > 5 \cdot 10^5 : \quad \infty : \quad 0.35$	$Re > 10^5 : t/d = 2 : c_w = 0.2$ $\quad 3 \quad 0.1$ $\quad 5 \quad 0.06$ $\quad 10 \quad 0.083$ $\quad 20 \quad 0.094$
Hemisphere without base: $c_w = 0.34$ with base: $\quad 0.40$	**Hemisphere** without base: $c_w = 1.33$ with base: $\quad 1.17$	**Cone (without base)** $\alpha = 30° : c_w = 0.34$ $\quad 60° \quad 0.51$	**Cone (slender)** $c_w = 0.58$
Circular cylinder $l/d = 1 : c_w = 0.91$ $\quad 2 \quad 0.85$ $\quad 4 \quad 0.87$ $\quad 7 \quad 0.99$	**Prism** $l/a = 2.5 : c_w = 0.81$	**Prism** $\alpha = 90° : l/a = 5 : c_w = 1.56$ $\quad \infty \quad 2.03$ $\alpha = 45° \quad 5 \quad 0.92$ $\quad \infty \quad 1.54$	**I – profile bar** $c_w = 2.04$ $c_w = 0.86$
Circular plate $c_w = 1.1$	**Circular annulus** $\frac{d}{D} = 0.5 : c_w = 1.22$	**Two circular plates behind one another** $\frac{l}{d} = 1 : c_w = 0.93$ $\quad 1.5 \quad 0.78$ $\quad 2 \quad 1.04$ $\quad 3 \quad 1.52$	**Rectangular plate** $\frac{a}{b} = 1 : c_w = 1.10$ $\quad 2 \quad 1.15$ $\quad 4 \quad 1.19$ $\quad 10 \quad 1.29$ $\quad 18 \quad 1.40$ $\quad \infty \quad 2.01$

for $Re \quad < 3 \cdot 10^4 \qquad c_M = 2\pi d/(sRe);$

$3 \cdot 10^4 < Re < 6 \cdot 10^5 \qquad c_M = 3.78/\sqrt{Re};$

$Re \quad > 6 \cdot 10^5 \qquad c_M = 0.0714/\sqrt[5]{Re}.$

6.6.5 Aerofoils and Blades

An aerofoil subjected to a flow under an angle of incidence α with v_0 experiences a lifting force F_A vertical to the direction of the imposed flow, and a resistance force F_W parallel to the direction of flow (**Fig. 30a, b**):

$$F_a = c_A \, (\rho v_0^2/2)A, \quad F_W = c_w \, (\rho v_0^2/2)A. \quad (36a, b)$$

Here, c_a is the lift coefficient and A is the wing surface projecting vertically to the chord l.

The objective is to achieve the most favourable possible fineness ratio $\varepsilon = c_w/c_a$. The forces axial and tangential to the chord follow from the resultant $F_R = \sqrt{F_A^2 + F_W^2}$ and $\beta = \arctan(F_W/F_A)$ (**Fig. 30c**):

$$F_n = F_R \cos(\beta - \alpha), \quad F_t = F_R \sin(\beta - \alpha).$$

The position of the point of imposition of the resultant on the chord (pressure point D) is determined by the distance s from the initial point of the chord or from the moment coefficient c_m: $F_n s = F_n' l = c_m \, (\rho v_0^2/2)Al$ (F_n' is an imaginary force taking effect on the rear edge). With $F_n \approx F_A' = c_a(\rho v_0^2/2)A$ is derived $s = (c_m/c_a)l$.

Lift. The sole element determinant for the lift, according to the Kutta-Joukowski equation (see A6.5.2), is the circulation Γ:

$$F_A = \rho v_0 \Gamma = \rho v_0 2\pi C = c_a(\rho v_0^2/2)A. \quad (37)$$

The constant C is determined in such a manner that the flow on the rear edge flows off smoothly (Kutta flow condition; there is no flow round the rear edge). As a result of the circulation, the flow on the upper side (suction side) is more rapid, and slower on the underside (pressure side), i.e. according to Bernoulli's formula $\rho v^2/2 + p = $ const., the pressure is less on the top and greater below. Underpressure Δp_1 and overpressure Δp_2 are applied in **Fig. 30d** along the length of the profile circumference. The lift can be determined by means of the circulation according to Eq. (37), or by integration via the pressure Δp, with the same result. The calculation using the circulation can be made for an aerofoil or infinite length in two ways: either by equiangular representation of the profile on a circle, since the potential flow with circulation is known for this (see A6.5.2), or by using the singularity method (approximation process), in which case the profile around which the flow is moving is approximated by a series of vortices, sources, sinks and dipoles [3].

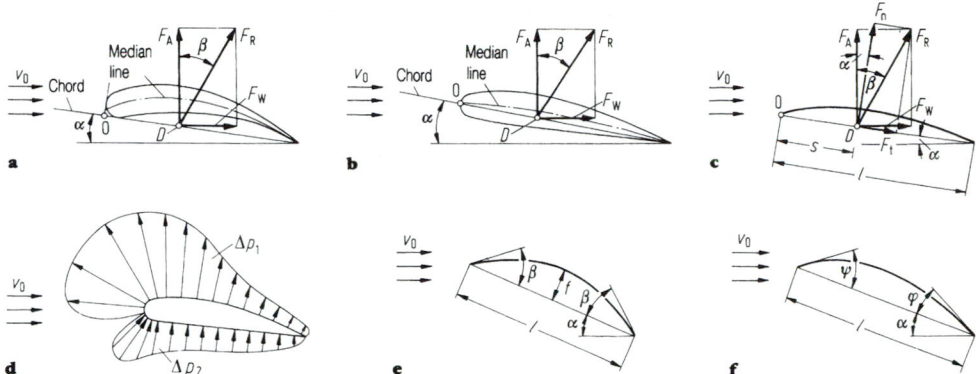

Figure 30. Aerofoils: **a** curved profile, **b** teardrop profile, **c** force resolution, **d** pressure distribution, **e** and **f** thin-walled profiles.

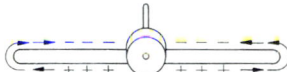

Figure 31. Transverse flow on aerofoil.

By using these methods, the lift coefficient $c_a = 2\pi \sin (\alpha + \beta/2) \approx 2\pi(\alpha + 2f/l)$ is derived for an arc profile of curvature f (**Fig. 30e**), and for a randomly curved profile with the final angles ψ and φ (**Fig. 30f**), $c_a = 2\pi \sin (\alpha + \psi/8 + 3\varphi/8)$. The result for the arc profile can be used as a good approximation for all profiles, if the angle of incidence is not too great. The lift accordingly increases in linear progression with the angle of incidence and the relative curvature f/l. For $\alpha_0 = -2f/l$, the lift is zero.

For aerofoils of finite width, the pressure difference between the underside and top side forces a flow to the ends of the wing, since at that point the pressure difference must be zero (**Fig. 31**), i.e. a flow in space exists, which is no longer capable of being acquired with the methods of plane potential theory. In this case, the lift and therefore the circulation constantly decrease towards zero from the centre to the ends, in a manner closely approximating an ellipse. At the ends of the wing, a circulation exists permanently, which floats in the form of a free vortex, and which incurs 'induced drag' because of the energy it consumes.

Drag Force. The total resistance according to Eq. (36b) is composed of the frictional and pressure drag (see A6.6.4), and the induced drag resulting from the vortex formation at the ends of the wing: $F_W = F_{Wo} + F_{Wi}$,

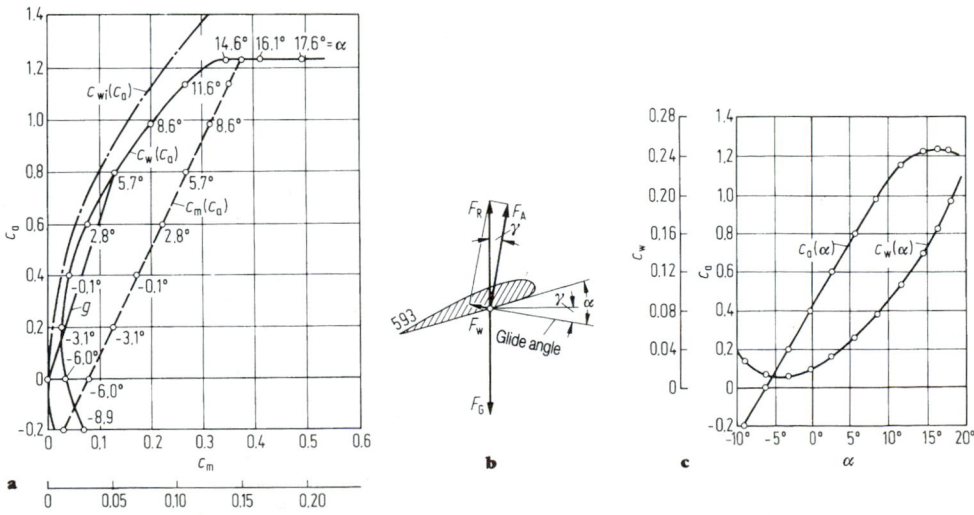

Figure 32. Aerofoil theory: **a** polar curve, **b** glide angle, **c** lift and drag coefficients.

$c_w = c_{wo} + c_{wi}$. According to Prandtl, there applies for the coefficient of the induced drag in elliptical lift distribution

$$c_{wi} = \lambda c_a^2/\pi, \tag{38}$$

where $\lambda = A/b^2$ is known as the aspect ratio and b is the wingspan. The induced drag accordingly increases with the square of the lift, or linearly with the aspect ratio. The profile drag coefficient c_{wo} is independent of λ and changes only slightly with c_a or α.

Polar Curve. The values c_a, c_w and c_m, calculated or measured, are entered on the polar curve, in **Fig. 32a** for example, for the Göttingen-type profile 593 with $\lambda = 1 : 5$. Here, the coefficients c_w and c_m form the abscissa and the coefficient c_a forms the ordinates. The angles of incidence α pertaining to the individual values are likewise entered. The parabola of the induced drag according to Eq. (38) is represented as a broken line. The straight line g to a point on the c_w-curve has the gradient $\tan \gamma = c_w/c_a = \varepsilon$. The angle γ can be represented as the limiting angle of friction of an aircraft without drive (**Fig. 32b**). **Figure 32c** shows the values c_a and c_w for the same profile as a function of the angle of incidence α. Up to about 13°, lift increases linearly with angle of incidence, reaches the peak at about 15°, and then falls off. The cause for this drop is the break in the flow on the upper side of the profile, which is to be equated with a reduction of the angle of incidence. The drag coefficient c_w is minimal for the angle of incidence $\alpha = -4°$; it increases quadratically in both directions.

General Results. If a geometrical comparison is made of similar profiles, then there applies for c_a, c_w and α

$$c_{a2} = c_{a1} = c_a, \quad c_{w2} = c_{w1} + (c_a^2/\pi)(A_2/b_2^2 - A_1/b_1^2),$$

$$\alpha_2 = \alpha_1 + (c_a/\pi)(A_2/b_2^2 - A_1/b_1^2). \tag{39}$$

Both lift and profile drag increase at more or less the same rate with increasing profile thickness. With the same thickness, the lift becomes greater as the curvature increases. Below an $Re = vl/v \neq$ value of 60 000 to 80 000 (subcritical range), profiles are considerably less favourable than blades. The lift decays to a maximum value of c_a of 0.3 to 0.4, depending on the thickness of the profile, while the drag increases sharply. In the supercritical range, the lift with Re with moderately curved profiles becomes greater, and smaller with strongly curved profiles. Flaps on the rear end and front wings increase the lift substantially, as do air suction or the emission of gas

jets at the end of the wing. At larger Re numbers, laminar friction resistance is substantially smaller than turbulent resistance. With suitable shaping, the transition point is set as far as possible towards the end of the profile (laminar wing, e.g. the thickest part of the profile is displaced towards the rear and the boundary layer is suctioned off). This can reduce the c_w-value by 50% and more.

6.6.6 Blade Rows (Cascades)

In the cascade (**Fig. 33a-c**), the friction losses play a decisive part. If the blade distribution is too narrow, the surface friction becomes too great, and if the distribution is too broad, separation losses occur. In both cases, the degree of efficiency will be impaired. The most favourable blade distribution is determined according to Zweifel's results [1]. Cascades without friction losses are calculated below:

(a) *Cascades at Rest with an Infinite Number of Blades.* From the continuity equation there follows $v_m = v_1 \cos \alpha_1 = v_2 \cos \alpha_2 = $ const., and from the centre-of-mass theorem and Bernoulli's equation it follows that

$$F_y = bt\rho v_m(v_{1u} - v_{2u}), \quad F_x = bt\rho(v_{1u}^2 - v_{2u}^2)/2, (40)$$

where b is cascade depth vertical to the drawing plane. In addition, there applies

$$\tan \alpha_\infty = F_x/F_y = \left(\frac{v_{1u} + v_{2u}}{2}\right) \Big/ v_m, \quad F_A = \sqrt{F_x^2 + F_y^2}. \tag{41}$$

(b) *Cascade in Motion with an Infinite Number of Blades.* If the cascade moves at velocity u, Eqs (40) and (41) apply, if the absolute velocities v at that point are replaced by the relative velocities w. The force F_y provides the product

$$P = F_y u = bt\rho w_m u(w_{1u} - w_{2u}).$$

(c) *Cascade with a Finite Number of Blades.* Displacement from α_1 to α_2 is possible only if the ends of the blades are angled, or shaped in such a way that $\alpha_1 < \alpha_1'$ and $\alpha_2 > \alpha_2'$. Equations (40) and (41) apply to the balanced flow, i.e. for the substitute cascade width a'. The force F_A which takes effect on a blade is vertical to α_∞ and can be calculated according to profile theory from

$$F_A = c_A (\rho v_\infty^2/2)bl \quad \text{and} \quad v_\infty = \sqrt{v_m^2 + [(v_{1u} + v_{2u})/2]^2}$$

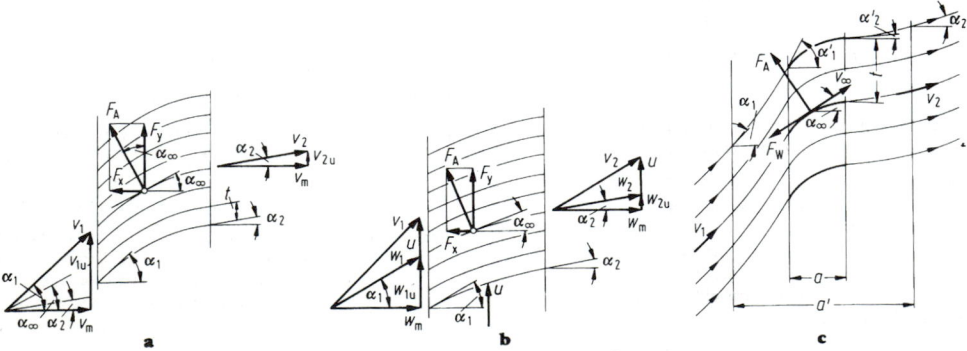

Figure 33. a-c Blade cascade.

The corresponding calculation applies to the drag force $F_W = c_W(\rho v_\infty^2/2)bl$. For the cascade in motion, which absorbs work (turbine) or puts out work (pump), there applies, with $\Delta p = (p_2 + \rho v_2^2/2) - (p_1 + \rho v_1^2/2)$, $c_a = 2t\,\Delta p/(uw_\infty\rho l)$. For the optimum blade distribution, the research by Zweifel [1] is determinant: with $F_A = \psi_A\,(\rho w_2^2/2)l$ and $\psi_A = (2\,\sin^2\alpha_2/\sin\,\alpha_\infty)(\cot\,\alpha_2 - \cot\,\alpha_1)t/l$, the most favourable blade distribution is derived,

and an optimum degree of efficiency for $0.9 < A < 1.0$. For F_y there applies accordingly

$$F_y = \psi_T\,(\rho w_2^2/2)a \quad \text{with}$$

$$\psi_T = 2\,\sin^2\,\alpha_2(\cot\,\alpha_2 - \cot\,\alpha_1)t/a.$$

For optimum blade distribution, there applies $0.9 < \psi_T < 1.0$.

7 Similarity Mechanics

7.1 Introduction

The task of similarity mechanics is to establish laws by which the experimental results gained from the model (as a rule reduced in size) can be transferred to the real arrangement (main arrangement). Model experiments are necessary if an exact mathematical–physical solution to a technical problem is not possible, or when it is appropriate to confirm theoretical principles and working hypotheses by experiment. The model laws of similarity mechanics accordingly form the basis for a wide range of experimentation in statics, strength-of-materials research, vibration mechanics, fluid mechanics, shipbuilding and marine engineering, aeronautical engineering, hydraulics and turbine construction, heating engineering problems, and so on.

Physical Similarity [1]. A prerequisite is the geometrically similar design of the model, i.e. equiangular (of equal shape) to the real object (angles have no identity, and therefore the scale factor is always equal to 1). Complete mechanical similarity exists when all the values involved in the physical process, such as distances, times, forces, stresses, velocities, pressures, work etc., are scaled in a similar manner in accordance with the laws of physics. This is however in general not possible, since only the SI base units m, kg, s and K or their scale factors are available for transfer, supplemented by substance parameters such as density ρ, modulus of elasticity E, etc. It follows from this that only a limited number of physical fundamental equations are transferable in a similar manner; i.e., as a rule, only incomplete similarity can in reality be achieved.

Scale Factors. For the basic values of length l, time t, force F, and temperature T, geometrical, temporal, dynamic or thermal similarity between the actual design (H) and the model (M) applies if

$$l_M/l_H = l_V, \quad t_M/t_H = t_V, \quad F_M/F_H = F_V \quad \text{or}$$

$$T_M/T_H = T_V$$

is maintained for all the points of the system (l_V, t_V, F_V and T_V are ratio coefficients, known as scale factors).

Units. If a physical value $B = F^{n_1}\,l^{n_2}\,t^{n_3}\,T^{n_4}$ has units $N^{n_1}\,m^{n_2}\,s^{n_3}\,K^{n_4}$, then the transfer scale $B_V = B_M/B_H$ follows directly from the units to $B_V = F_V^{n_1}\,l_V^{n_2}\,t_V^{n_3}\,T_V^{n_4}$. For example, the transfer law for the mechanical work W is derived directly from the unity N m to $W_M/W_H = F_V l_V$ instead of the more unwieldy form $W_M/W_H = (F_M l_M)/(F_H l_H) = F_V l_V$.

Similarity Parameters. The limiting quantities that play a determinant part in a process, and which are given

units, can be compiled in the form of power products to create similarity parameters, which have no units (e.g. Froude number, Reynolds number). This reduces the number of variables, and every determinant equation that determines a process or every differential equation can be transformed into a function of similarity parameters without units. In this the following applies, according to [1]: The ratio between two values of random type can be replaced by the ratio between any other random values, provided that the new values lead to the same unities as the first ones.

Extended Similarity. Because of the large number of limiting quantities, strict similarity can frequently not be attained. For this reason, a restriction is imposed (on grounds of economy also) only to the similarity of the values that predominate in a process, and free availability is provided for the remainder.

7.2 Similarity Laws

7.2.1 Static Similarity

Scale Factor for Gravitational Forces. For weights $F_M = \rho_M V_M g_M$ on the model and $F_H = \rho_H V_H g_H$ on the main design (V = volume, g = acceleration of gravity) the transition law applies:

$$F_M/F_H = \rho_M V_M g_M/(\rho_H V_H g_H), \quad \text{i.e.}$$

$$F_{V1} = (\rho_M/\rho_H)l_V^3 \qquad (1)$$

(since $g_M = g_H$ on earth). If ρ_M, ρ_H and l_V are freely selected, this equation accordingly determines the force scale.

Example It is intended to manufacture a model in a scale $l_V = l_M/l_H = 1 : 10$, made of aluminium ($\rho_M = 2700$ kg/m³) from the real design of a steel structure ($\rho_H = 7850$ kg/m³), the model reproducing the inherent gravitational forces in a mechanically similar manner. In what proportion are the inherent gravitational forces, and in what proportion will other imposed forces have to be? In what proportion will the stresses and (Hooke's) shape changes be transferred ($E_H = 210$ kN/mm², $E_M = 70$ kN/mm²)? According to Eq. (1), $F_{V1} = (2.70/7.85)/10^3 = 1/2907 = F_M/F_H$, i.e. the forces on the model are 2907 times smaller. For the stresses, there follows $\sigma_M/\sigma_H = F_V/l_V^2 = 100/2907 = 1/29 = \sigma_V$. For the shape changes, there is derived from $\Delta l = l\sigma/E$ the ratio

$$\Delta l_M/\Delta l_H = \Delta l_V = l_V\sigma_V E_H/E_M = (1/10)\,(1/29)\,210/70 = 1/96.7.$$

Scale Factor for Equal Expansion (for so-called elastic forces). If it is intended that the elastic (Hooke's) expansion on the model and on the main design are to be equal, there applies for the forces, from the condition

$$\varepsilon_M = F_M/(E_M A_M) = \varepsilon_H = F_H/(E_H A_H),$$

$$F_M/F_H = E_M A_M/(E_H A_H), \quad \text{i.e.} \quad F_{V_2} = (E_M/E_H)\,l_V^2. \quad (2)$$

Hooke's Model Law. Two bodies are mechanically similar in respect of elastic expansion if the Hooke's coefficients Ho coincide:

$$Ho = F_M/(E_M\, l_M^2) = F_H/(E_H\, l_H^2). \tag{3}$$

Example A true-to-scale model is being made of a steel strut, in a ratio of $l_V = 1:8$, in aluminium ($E_H = 210\ \text{kN/mm}^2$, $E_M = 70\ \text{kN/mm}^2$), and a critical force of 1.2 kN is to be measured on the model. How great is the critical force F_K on the real design, and in what ratio do the stresses and the deformations have to one another?

$$F_V = (70/210)/64 = 1/192;\quad F_K = 192\cdot 1.2\ \text{kN} = 230.4\ \text{kN};$$

$$\sigma_V = \sigma_M/\sigma_H = F_V/l_V^2 = 1/3.0;\ \Delta l_M/\Delta l_H = l_V\,\sigma_V\,E_H/E_M = 1/8.0.$$

Simultaneous Consideration of Gravitational and Elastic Forces. If it is intended that gravitational forces and elastic expansions should be simultaneously transferred in a mechanically similar manner, then the force scales according to Eqs (1) and (2) must be equal. From $F_{V1} = F_{V2}$ it follows

$$(\rho_M/\rho_H)\, l_V^3 = (E_M/E_H)\, l_V^2, \quad \text{i.e.}$$

$$l_V = (E_M/E_H)\,(\rho_H/\rho_M). \tag{4}$$

The longitudinal scale is no longer freely selectable; it now depends only on the material parameters.

Example For the first example in A7.2.1, the scale factor is being sought for mechanical similarity of gravitational forces and expansions. $l_V = (70/210)(7850/2700) = 1:1.03$, i.e. simultaneous consideration of gravitational forces and expansions is possible only on the real design. Accordingly, a restriction is imposed on the extended similarity, inasmuch as the similarity for the elastic forces is fulfilled for the scale 1 : 10. It then follows according to Eq. (2) that $F_V = (70/210)/100 = 1/300 = F_M/F_H$, while the gravitational forces, as in the first example, are transferred in the ratio of 1/2907. The difference between the gravitational forces $((1/300)-(1/2907))\cdot F_{GH}$ can be applied to the model as an additional external load.

7.2.2 Dynamic Similarity

Newton–Bertrand Similarity Law. Accelerated movement processes satisfy Newton's basic law, $\boldsymbol{F} = \boldsymbol{ma}$. It follows from this, for the force scale with mechanical similarity of the forces of inertia on the model and on the main design, with $a_V = l_V/t_V^2$, that

$$F_M/F_H = \rho_M V_M a_M/(\rho_H V_H a_H), \quad \text{i.e.}$$

$$F_{V3} = (\rho_M/\rho_H)\,(l_V^4/t_V^2). \tag{5}$$

With the force of inertia taking effect alone, and the free selection of ρ_M, ρ_H, l_V and t_V, Eq. (5) established the force scale. It follows from this that

$$F_M/[\rho_M(l_M/t_M)^2 l_M^2] = F_H/[\rho_H(l_H/t_H)^2 l_H^2]$$

and with $l_M/t_M = v_M$ and $l_H/t_H = v_H$ that

$$Ne = F_M/(\rho_M v_M^2 l_M^2) = F_M/(\rho_H v_H^2 l_H^2). \tag{6}$$

Newton's Similarity Law. Two processes are similar in respect of the forces of inertia if the Newtonian similarity parameters Ne coincide.

Example It is intended to create a model in wood ($\rho_M = 600\ \text{kg/m}^3$) in a scale of 1 : 20 of a wagon made of steel ($\rho_H = 7850\ \text{kg/m}^3$, $v_H = 1\ \text{m}^3$, $F_H = 10\ \text{kN}$), moving on a horizontal path. What forces must be imposed on the model if the time scale is to be $t_V = t_M/t_H = 1 : 100$? In what ratio will velocities and accelerations be translated?

$$F_{V3} = (600/7850)\,(100^2/20^4) = 1/209.3;$$

$$F_M = F_H F_{V_3} = 47.8\ \text{N};$$

$$v_M/v_H = l_V/t_V = 100/20 = 5;$$

$$a_M/a_H = l_V/t_V^2 = 100^2/20 = 500.$$

Cauchy's Similarity Law. If forces of inertia and elastic forces are involved in a determinant manner in a movement process, it follows from $F_{V3} = F_{V2}$, according to Eqs (5) and (2), that

$$t_V = l_V\sqrt{(E_H/E_M)\,(\rho_M/\rho_H)}; \tag{7}$$

i.e., only the longitudinal scale (or the time scale) is still freely selectable. With $t_V = t_M/t_H$ and $l_V = l_M/l_H$ it follows from this that

$$v_M/v_H = \sqrt{(E_M/E_H)\,(\rho_H/\rho_M)}\quad \text{or}$$

$$Ca = v_M/\sqrt{E_M/\rho_M} = v_H/\sqrt{E_H/\rho_H} \tag{8}$$

Cauchy's Similarity Law. Two processes, which are predominantly under the influence of forces of inertia and elastic forces, are mechanically similar if their Cauchy similarity parameters Ca coincide.

Froude's Similarity Law. If forces of inertia and gravitational forces are predominantly involved in a movement process, then it follows from $F_{V1} = F_{V3}$, according to Eqs (1) and (5), that

$$t_V = \sqrt{l_V}; \tag{9}$$

i.e. only the longitudinal scale (or the time scale) is still freely selectable. It follows from this that $t_M^2/t_H^2 = l_M/l_H$ or $l_M^2/(l_M\,t_M^2) = l_H^2/(l_H t_H^2)$ and therefore

$$Fr = v_M^2/(l_M g_M) = v_H^2/(l_H g_H). \tag{10}$$

Froude's Model Law. Two processes are mechanically similar, in respect of forces of inertia and gravitational forces, if the Froude numbers Fr coincide.

Example It is intended to make a model in wood ($\rho_M = 600\ \text{kg/m}^3$) in a scale of 1 : 4 of a physical pendulum made of steel ($\rho_H = 7850\ \text{kg/m}^3$). How great is the transfer scale t_V, and how do forces, stresses, frequencies, velocities and accelerations behave in relation to each other?

$$t_V = \sqrt{1/4} = 1/2;$$

$$F_V = F_M/F_H = (600/7850)/64 = 1/837;$$

$$\sigma_M/\sigma_H = F_V/l_V^2 = 1/52;$$

$$\omega_M/\omega_H = t_H/t_M = 1/t_V = 2.0;$$

$$v_M/v_H = l_V/t_V = 2/4 = 1/2;$$

$$a_M/a_H = l_V/t_V^2 = 4/4 = 1.0.$$

Reynolds Similarity Law. If the forces of inertia and frictional forces of Newtonian fluids are predominantly involved in a movement process, then there follows, for the latter, with $F = \eta(dv/dz)A$ according to Eq. (8), the force scale

$$\frac{F_M}{F_H} = \frac{\eta_M}{\eta_H}\cdot\frac{dv_M/dz_M}{dv_H/dz_H}\cdot\frac{A_M}{A_H}, \quad \text{i.e.}\quad F_{V4} = \frac{\eta_M}{\eta_H}\cdot\frac{l_V^2}{t_V} \tag{11}$$

and therefore, from $F_{V4} = F_{V3}$ according to Eqs (11) and (5),

$$t_V = (\rho_M/\rho_H)\,(\eta_H/\eta_M)\, l_V^2 = (\nu_H/\nu_M)\, l_V^2 \tag{12}$$

(η absolute, $\nu = \eta/\rho$ kinematic viscosity). Only the longitudinal scale is still freely selectable and within the framework of the substance parameter media available ν_M. From Eq. (12) there follows

$$t_M/t_H = (\nu_H/\nu_M)\ l_M^2/l_H^2, \quad \text{i.e.}$$

$$Re = \upsilon_M\ l_M/\nu_M = \upsilon_H\ l_H/\nu_H. \qquad (13)$$

Reynolds Similarity Law. Two flows of viscous Newtonian fluids are mechanically similar, under the predominant influence of the forces of inertia and frictional forces, if the Reynolds numbers Re coincide.

Example The flow resistance of an installed component in an oil pipe is to be determined by experiments on a model in a scale of 1 to 10, by measuring the pressure drop, for which water is provided as the model medium. How do the flow velocities and the forces or the pressure drop behave? ($\nu_M = 10^{-6}\ \text{m}^2/\text{s}$; $\nu_H = 1.1 \cdot 10^{-4}\ \text{m}^2/\text{s}$; $\eta_M = 10^{-3}\ \text{N s/m}^2$; $\eta_H = 10^{-1}\ \text{N s/m}^2$)

$$l_v = l_M/l_H = 1/10;$$

$$\upsilon_v = \upsilon_M/\upsilon_H = (\nu_M/\nu_H)/l_v$$

$$= (10^{-6}/1.1 \cdot 10^{-4})/(1/10) = 1/11$$

$$F_v = F_M/F_H = (\eta_M/\eta_H)l_v^2/t_v = (\eta_M/\eta_H)\upsilon_v l_v$$

$$= (10^{-3}/10^{-1})\ (1/11)(1/10) = 1/11\,000;$$

$$\Delta p_M/\Delta p_H = (F_M/F_H)/l_v^2 = 100/11\,000 = 1/110.$$

Weber's Similarity Law. If, in addition to the forces of inertia, surface tension σ is also predominantly involved in a process, i.e. the surface forces $F_\sigma = \sigma l$ (where σ is to be considered as a material constant), then there follows, as the transfer scale for the surface forces,

$$F_{\sigma M}/F_{\sigma H} = \sigma_M l_M/(\sigma_H l_H), \quad \text{i.e.} \quad F_{VS} = (\sigma_M/\sigma_H)l_v,\,(14)$$

and therefore, from $F_{VS} = F_{V3}$, according to Eqs (14) and (15),

$$(\rho_M/\sigma_M)l_M^3/t_M^2 = (\rho_H/\sigma_H)\ l_H^3/t_H^2 \quad \text{or}$$

$$We = \rho_M\upsilon_M^2 l_M/\sigma_M = \rho_H\upsilon_H^2 l_H/\sigma_H. \qquad (15)$$

Weber's Similarity Law. Processes under the predominant influence of forces of inertia and surface forces are mechanically similar if the Weber numbers We coincide.

Further Similarity Laws for Flow Problems

Euler's Similarity Parameters. In the case of flow problems in which friction can be disregarded, i.e. in which pressure forces and forces of inertia predominate (e.g. in the measurement of dynamic pressure Δp), mechanical similarity prevails if the Euler's similarity parameters Eu are equal:

$$Eu = \Delta p_M/(\rho_M\upsilon_M^2) = \Delta p_H/(\rho_H\upsilon_H^2). \qquad (16)$$

Mach's Similarity Parameter. With gaseous fluids, the flow velocities of which lie close to the speed of sound c, mechanical similarity prevails if the Mach numbers Ma are equal:

$$Ma = \upsilon_M/c_M = \upsilon_H/c_H. \qquad (17)$$

7.2.3 Thermal Similarity

Fourier's Similarity Law. The Fourier differential equation applies to the unstable thermal conductivity process:

$$\frac{\partial T}{\partial t} = b\left(\frac{\partial^2 T}{\partial x^2} + \frac{\partial^2 T}{\partial y^2} + \frac{\partial^2 T}{\partial z^2}\right), \qquad (18)$$

where $b = \lambda/(c\rho)$ temperature conductivity, $\lambda =$ thermal

conductivity, $c =$ specific thermal capacity, $\rho =$ density. According to the rule regarding identities, it follows that

$$T_V/t_V = (b_M/b_H)\ (T_V/l_V^2) \quad \text{or} \quad t_V = (b_H/b_M)\ l_V^2 \ (19)$$

and from this

$$Fo = t_M b_M/l_M^2 = t_H b_H/l_H^2. \qquad (20)$$

Fourier's Similarity Law. Two thermal conductivity processes are similar if the Fourier numbers Fo coincide (see C10.4).

Example For a model in a scale of 1 : 10, there follows, in the same material ($b_M = b_H$) : $t_M = (l_M/l_H)^2 t_H = (1/100)t_H$, i.e. the temperature distribution in the model is attained in 1/100 of the time taken in the main design.

Péclet's Similarity Law. If it is intended that two flow processes should match thermally in respect of thermal conductivity, the Péclet numbers Pe must be equal:

$$Pe = \upsilon_M l_M/b_M = \upsilon_H l_H/b_H. \qquad (21)$$

Prandtl's Similarity Law. If it is intended that two flow processes should match in respect of thermal conductivity and thermal convection, the Reynolds and the Péclet similarity parameters must coincide. From this is derived an equality of the Prandtl numbers Pr:

$$Pr = Pe/Re = \nu_M/b_M = \nu_H/b_H. \qquad (22)$$

Nusselt's Similarity Law. Similarity prevails for heat transfer between two substances if the Nusselt numbers coincide:

$$Nu = \alpha_M l_M/\lambda_M = \alpha_H l_H/\lambda_H, \qquad (23)$$

where $\alpha =$ heat transfer coefficient and $\lambda =$ thermal conductivity capacity.

7.2.4 Dimensional Analysis and Π-Theorem

If the limiting quantities with attributed units of a process are known, then power products in the form of similarity parameters without units can be formed from them. The similarity parameters required to represent a problem form a complete set. Every physically correct equation between quantities can be represented as a function of the similarity parameters of a complete set (Buckingham's Π-theorem).

For example, Bernoulli's equation for the friction-free flow $\rho \upsilon^2/2 + p + \rho gz = \text{const.}$ or $1/2 + p/(\rho \upsilon^2) + gz/\upsilon^2 = \text{const.}$ can also be written as $1/2 + Eu + 1/Fr = \text{const.}$, i.e. Euler and Froud numbers form a complete set for the friction-free and temperature-dependent flow. The five limiting quantities ρ, υ, p, g, z can therefore be replaced by two non-dimensional parameters without units, which are sufficient for the complete description of the problem.

One method for determining the complete set of similarity parameters of a problem, even in cases in which the physical fundamental equations are not known, is the analysis of the units based upon Buckingham's theorem [2] as a basis. This states: If the relationship $f(x_1, x_2, \ldots, x_n) = 0$ applies to n limiting quantities with attributed units, then this can always be written in the form $f^*(\Pi_1, \Pi_2, \ldots, \Pi_m) = 0$, where Π_i are the m similarity parameters without units, and $m = n - q$. Here, q is the number of basic units involved. For m, kg, s, $q = 3$ for mechanical problems, and for m, kg, s, K, $q \doteq 4$ with thermal problems. With a product statement

$$\Pi = x_1^a\ x_2^b\ x_3^c\ x_4^d \ldots, \qquad (24)$$

and after inserting the units for x_i, the total of the

exponents of the basic units m, kg, s and K are in each case zero, since the right-hand side must also be non-dimensional because of the left-hand side. For example, the values ρ, v, z, g, p are involved in the flow mentioned above. There then applies

$$\Pi = (\text{kg/m}^3)^a \ (\text{m/s})^b \ (\text{m})^c \ (\text{m/s}^2)^d \ (\text{kg/m s}^2)^e. \ (25)$$

For the exponents of kg, m, s it then follows that

$$a + e = 0,$$
$$-3a + b + c + d - e = 0, \qquad (26)$$
$$-b - 2d - 2e = 0.$$

Two exponents can be freely chosen. For example, if p

and g are to be command variables, d and e are freely selectable. It then follows from Eq. (26) that $a = -e$, $b = -2d - 2e$ and $c = d$, and therefore

$$\Pi = \rho^a v^b z^c g^d p^e = \rho^{-e} \ v^{-2d-2e} \ z^d g^d p^e = (zg/v^2)^d \ (p/\rho v^2)^e$$

or with $d = 1$ and $e = 1$

$$\Pi = (1/Fr)Eu, \quad \text{i.e.} \quad \Pi_1 = Fr, \quad \Pi_2 = Eu. \quad (27)$$

The problem, then, of frictionless flow can be described with $m = n - q = 5 - 3 = 2$ similarity parameters, namely with the Froud and Euler numbers. A functional connection in the form of Bernoulli's equation cannot of course be derived with this process (for further details, see [1-5]).

8 References

A1 Statics of Rigid Bodies. [1] Föppl A. Vorlesungen über technische Mechanik, vol I, 13th edn.; vol II, 9th edn. Oldenbourg, Munich Berlin, 1943 and 1942. - [2] Schlink W. Technische Statik, 3rd edn. Springer, Berlin, 1946. - [3] Drescher H. Die Mechanik der Reibung zwischen festen Körpern. VDI-Z 1959; 101: 697-707. - [4] Krause H., Poll G. Mechanik der Festkörperreibung. VDI, Düsseldorf, 1980. - [5] Kragelski, Dobyčin, Kombalov: Grundlagen der Berechnung von Reibung und Verschleiss. Hanser, Munich, 1983.

A3 Dynamics. [1] Sommerfeld A. Mechanik, vol I, 3rd edn. Akad Verlagsges Geest u Portig, Leipzig, 1947. - [2] Klein I., Sommerfeld A. Theorie des Kreisels (4 vols). Teubner, Leipzig, 1897-1910. - [3] Grammel R. Der Kreisel (2 vols), 2nd edn. Springer, Berlin, 1950. - [4] Hertz H. Über die Berührung fester elastischer Körper. J f reine u angew Math. 1881; 92. - [5] Berger F. Das Gesetz des Kraftverlaufs beim Stoss. Vieweg, Brunswick, 1924.

A4 Mechanical Vibrations. [1] Söchting F. Berechnung mechanischer Schwingungen. Springer, Vienna, 1951. - [2] Biezeno, Grammel. Technische Dynamik, vol II, 2nd edn. Springer, Berlin, 1953. - [3] Collatz L. Eigenwertaufgaben. Akad Verlagsges Geest u Portig, Leipzig, 1963. - [4] Hayashi K. Tafeln für die Differenzenrechnung sowie für die Hyperbel-, Besselschen, elliptischen und anderen Funktionen. Springer, Berlin, 1933. - [5] Magnus K. Schwingungen, 2nd edn. Teubner, Stuttgart, 1969. - [6] Klotter K. Technische Schwingungslehre, vol 1, pt B, 3rd edn. Springer, Berlin, 1980. - [7] Jahnke, Emde, Lösch. Tafeln höherer Funktionen. Stuttgart, 1966. - [8] Rothe,

Szabó. Höhere Mathematik, pt VI, 2nd edn. Teubner, Stuttgart, 1958.

A6 Hydrodynamics and Aerodynamics. [1] Eck B. Technische Strömungslehre, 7th edn. Springer, Berlin, 1966. - [2] Kalide W. Einführung in die technische Strömungslehre, 5th edn. Hanser, Munich, 1980. - [3] Truckenbrodt E. Strömungsmechanik. Springer, Berlin, 1968. - [4] Jogwich A. Strömungslehre. Girardet, Essen, 1974. - [5] Bohl W. Technische Strömungslehre. Vogel, Würzburg, 1971. - [6] Herning F. Stoffströme in Rohrleitungen, 4th edn. VDI, Düsseldorf, 1966. - [7] Ullrich H. Mechanische Verfahrenstechnik. Springer, Berlin, 1967. - [8] Schlichting H. Grenzschicht-Theorie, 5th edn. Braun, Karlsruhe, 1965. - [9] Brauer H. Grundlagen der Einphasen- und Mehrphasenströmungen. Sauerlander, Aarau Frankfurt-on-Main, 1971. - [10] Szabó I. Höhere Technische Mechanik, 5th edn. Springer, Berlin, 1972. - [11] Sigloch H. Technische Fluidmechanik. Schrödel, Hanover, 1980. - [12] Prandtl, Oswatitsch, Wieghardt. Führer durch die Strömungslehre, 8th edn. Vieweg, Brunswick, 1984.

A7 Similarity Mechanics. [1] Weber M. Das allgemeine Ähnlichkeitsprinzip in der Physik und sein Zusammenhang mit der Dimensionslehre und der Modellwissenschaft. Jahrb Schiffbautecht Ges, 1930; 274-388. - [2] Katanek S, Gröger R, Bode C. Ähnlichkeitstheorie. VEB Deutscher Verlag f Grundstoffindustrie, Leipzig, 1967. - [3] Feucht W. Einführung in die Modelltechnik. Handbuch der Spannungs- und Dehnungsmessung (Fink, Rohrbach). VDI, Düsseldorf, 1958. - [4] Zierep J. Ähnlichkeitsgesetze und Modellregeln der Strömungslehre. Braun, Karlsruhe, 1972. - [5] Görtler H. Dimensionsanalyse. Springer, Berlin, 1975.

B Strength of Materials

G. Rumpel and H. D. Sondershausen, Berlin

1 General Fundamentals

The purpose of studying the strength of materials is to determine the stresses and strains in a structural member and prove that they are borne sufficiently safely to prevent failure. Failure may consist of excessive deformation or expansion, a fracture may occur, or the structural member may become unstable (e.g. due to buckling or bulging). The material characteristics that are relevant in this case depend upon the state of stress (one-, two- or three-dimensional), the types of stress (tensile, compression or shearing stresses), the load state (static or dynamic), the working temperature and the size and surface condition of the structural member.

1.1 Stress and Strain

1.1.1 Stresses

The external forces and moments on a body (as well as the inertia forces) are balanced by corresponding reaction forces inside the body. If the mass of the body is assumed to be distributed homogeneously, the internal reaction forces are distributed evenly within it.

Fundamental planes of section dA may pass through every point on a body in an infinite number of directions identified by the normal vector $\boldsymbol{n}$ (**Fig. 1a**). The stress vector $\boldsymbol{s} = dF/dA$ can be broken down into a normal stress $\sigma = dF_n/dA$ and a tangential or shear stress $\tau = dF_t/dA$. Cartesian coordinates (**Fig. 1b**) give one normal stress $\sigma_z = dF_n/dA$ and two shear stresses $\tau_{zx} = dF_{tx}/dA$ and $\tau_{zy} = dF_{ty}/dA$. Three planes or a cubic element (**Fig. 1c**) with three stress vectors and the stress tensor are required to describe the complete state of stress at one point.

$$\boldsymbol{s}_x = \sigma_x \boldsymbol{e}_x + \tau_{xy} \boldsymbol{e}_y + \tau_{xz} \boldsymbol{e}_z ,$$

$$\boldsymbol{s}_y = \tau_{yx} \boldsymbol{e}_x + \sigma_y \boldsymbol{e}_y + \tau_{yz} \boldsymbol{e}_z , \quad \boldsymbol{S} = \begin{pmatrix} \sigma_x & \tau_{xy} & \tau_{xz} \\ \tau_{yx} & \sigma_y & \tau_{yz} \\ \tau_{zx} & \tau_{zy} & \sigma_z \end{pmatrix} . \quad (1)$$

$$\boldsymbol{s}_z = \tau_{zx} \boldsymbol{e}_x + \tau_{zy} \boldsymbol{e}_y + \sigma_z \boldsymbol{e}_z ;$$

From the conditions of moment equilibrium around the coordinate axes for the element shown in **Fig. 1c**, $\tau_{xy} = \tau_{yx}$, $\tau_{xz} = \tau_{zx}$, $\tau_{yz} = \tau_{zy}$ (principle of equality of the assigned shear stresses), i.e. to fully describe the state of stress at one point, three normal stresses and three shear stresses are required.

One-dimensional State of Stress. This applies if a normal stress is applied to a cubic element (**Fig. 2a**), e.g. $\sigma_x = dF/dA$, $\sigma_y = \sigma_z = 0$, $\tau_{xy} = \tau_{xz} = \tau_{yz} = 0$. For an area element at angle φ the relevant stresses σ and τ are produced as $\sigma = (\sigma_x/2) \cdot (1 + \cos 2\varphi)$ and $\tau = -(\sigma_x/2) \sin 2\varphi$ from the equilibrium conditions in directions n and t. This

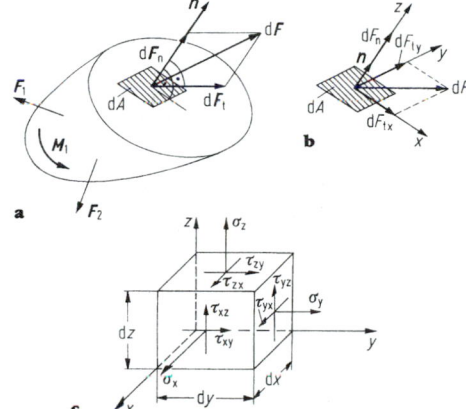

Figure 1. Stresses: **a, b** definition; **c** tensor.

gives the equation for Mohr's circle, $(\sigma - \sigma_x/2)^2 + \tau^2 = (\sigma_x/2)^2$ (**Fig. 2b**). Taking $2\varphi = 90°$ or $\varphi = 45°$, the maximum shear stress $\tau = -\sigma_x/2$, the corresponding normal stress is also $\sigma = \sigma_x/2$. The maximum and minimum normal stress ($\sigma = \sigma_x$ and $\sigma_2 = 0$) and the maximum shear stress ($\tau_1 = -\sigma_x/2$) are called the principal normal and shear stress. Lines that are at all points tangential to the principal normal and principal shear stresses are called principal normal stress and principal shear stress trajectories (**Fig. 2c, d**).

Two-dimensional (Plane) State of Stress. If stresses occur only in one plane (e.g. the x, y plane), then a plane state of stress applies (**Fig. 3a**). For stresses σ

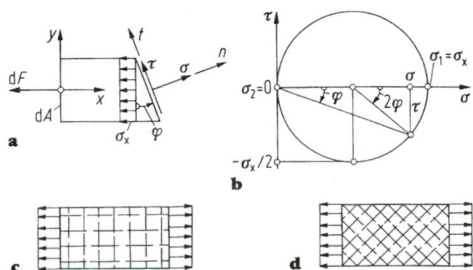

Figure 2. One-dimensional state of stress: **a** stresses on the element; **b** Mohr's circle; **c, d** principal normal and principal shear stress trajectories.

and τ lying in the plane of section inclined at angle φ, the equilibrium conditions in directions n and t, taking $\tau_{xy} = \tau_{yx}$, give

$$
\left.
\begin{aligned}
\sigma &= \sigma_x \cos^2\varphi + \sigma_y \sin^2\varphi + 2\tau_{xy}\sin\varphi\cos\varphi \\
&= \tfrac{1}{2}(\sigma_x + \sigma_y) + \tfrac{1}{2}(\sigma_x - \sigma_y)\cos 2\varphi + \tau_{xy}\sin 2\varphi , \\
\tau &= (\sigma_y - \sigma_x)\sin\varphi\cos\varphi + \tau_{xy}(\cos^2\varphi - \sin^2\varphi) \\
&= -\tfrac{1}{2}(\sigma_x - \sigma_y)\sin 2\varphi + \tau_{xy}\cos 2\varphi .
\end{aligned}
\right\}
\tag{2}
$$

After squaring and adding the equation for Mohr's circle (**Fig. 3b**) we obtain the following with radius r:

$$
\left.
\begin{aligned}
&\left(\sigma - \frac{\sigma_x + \sigma_y}{2}\right)^2 + \tau^2 = \left(\frac{\sigma_x - \sigma_y}{2}\right)^2 + \tau_{xy}^2 , \\
&r = \sqrt{\left(\frac{\sigma_x - \sigma_y}{2}\right)^2 + \tau_{xy}^2} .
\end{aligned}
\right\}
\tag{3}
$$

The centre of the circle is at point $(\sigma_x + \sigma_y)/2$. Taking $\tau = 0$, the principal normal stresses are obtained from Eqs (2) at angles φ_{01} and $\varphi_{02} = \varphi_{01} + 90°$ as

$$
\tan 2\varphi_0 = 2\tau_{xy}/(\sigma_x - \sigma_y) \tag{4}
$$

and become

$$
\sigma_{1,2} = (\sigma_x + \sigma_y)/2 \pm \sqrt{[(\sigma_x - \sigma_y)/2]^2 + \tau_{xy}^2} , \tag{5}
$$

In accordance with Eqs (2), maximum shear stresses are given by $d\tau/d\varphi = 0$ at angles φ_{11} and $\varphi_{12} = \varphi_{11} + 90°$, which are given by

$$
\tan 2\varphi_1 = (\sigma_y - \sigma_x)/(2\tau_{xy}) , \tag{6}
$$

where $\varphi_{11} = \varphi_{01} + 45°$ and $\varphi_{12} = \varphi_{02} + 45°$ (**Fig. 3c**). The size of these principal shear stresses corresponds to the radius of the Mohr's circle, i.e.

$$
\tau_{1,2} = \mp \sqrt{[(\sigma_x - \sigma_y)/2]^2 + \tau_{xy}^2} . \tag{7}
$$

The corresponding normal stresses are equal for both angles, i.e. $\sigma_M = (\sigma_x + \sigma_y)/2$.

The direction of the principal normal stress trajectories which follow from Eq. (4),

$$
\tan 2\varphi_0 = \frac{2\tan\varphi_0}{1 - \tan^2\varphi_0} = \frac{2y'}{1 - y'^2} = \frac{2\tau_{xy}}{\sigma_x - \sigma_y} , \quad \text{is}
$$

$$
y'_{1,2} = \frac{\sigma_y - \sigma_x}{2\tau_{xy}} \pm \sqrt{\left(\frac{\sigma_y - \sigma_x}{2\tau_{xy}}\right)^2 + 1} ,
$$

and the direction of the principal shear stress trajectories rotated through an angle of 45° which follow from Eqs (6),

$$
\tan 2\varphi_1 = \frac{2\tan\varphi_1}{1 - \tan^2\varphi_1} = \frac{2y'}{1 - y'^2} = \frac{\sigma_y - \sigma_x}{2\tau_{xy}} , \quad \text{is}
$$

$$
y'_{3,4} = \frac{2\tau_{xy}}{\sigma_x - \sigma_y} \pm \sqrt{\left(\frac{2\tau_{xy}}{\sigma_x - \sigma_y}\right)^2 + 1} .
$$

Three-dimensional (Spatial) State of Stress. If stresses occur in three perpendicular planes, a spatial state of stress applies (**Fig. 1c**). It is determined from the six stress components $\sigma_x, \sigma_y, \sigma_z, \tau_{xy} = \tau_{yx}, \tau_{xz} = \tau_{zx}$ and $\tau_{yz} = \tau_{zy}$. For any tetrahedral plane of section, the position of which is specified by the normal vector

$$
\boldsymbol{n} = \cos\alpha\,\boldsymbol{e}_x + \cos\beta\,\boldsymbol{e}_y + \cos\gamma\,\boldsymbol{e}_z = n_x\boldsymbol{e}_x + n_y\boldsymbol{e}_y + n_z\boldsymbol{e}_z
$$

(**Fig. 4**), the stress vector: $\boldsymbol{s} = s_x\boldsymbol{e}_x + s_y\boldsymbol{e}_y + s_z\boldsymbol{e}_z$ and its components are produced from the equilibrium conditions in directions x, y and z as

$$
\left.
\begin{aligned}
s_x &= n_x\sigma_x + n_y\tau_{yx} + n_z\tau_{zx} , \\
s_y &= n_x\tau_{xy} + n_y\sigma_y + n_z\tau_{zy} , \quad s = \sqrt{s_x^2 + s_y^2 + s_z^2} \\
s_z &= n_x\tau_{xz} + n_y\tau_{yz} + n_z\sigma_z ;
\end{aligned}
\right\}
\tag{8}
$$

The normal stress perpendicular to the tetrahedral plane of section is

$$
\sigma = \boldsymbol{s}\boldsymbol{n} = s_x n_x + s_y n_y + s_z n_z = n_x^2\sigma_x + n_y^2\sigma_y + n_z^2\sigma_z
$$
$$
+ 2(n_x n_y\tau_{xy} + n_x n_z\tau_{xz} + n_y n_z\tau_{yz}) .
$$

The resulting shear stress (**Fig. 4**), $\tau = \sqrt{s^2 - \sigma^2}$.

The principal normal stresses occur in the three planes perpendicular to each other where τ becomes zero. The stress tensor then takes the form

$$
\boldsymbol{S} = \begin{pmatrix} \sigma_1 & 0 & 0 \\ 0 & \sigma_2 & 0 \\ 0 & 0 & \sigma_3 \end{pmatrix} ,
$$

and the following applies for the stress vectors $\boldsymbol{s}_i = \boldsymbol{n}_i\sigma_i (i = 1, 2, 3)$, i.e.

$$
s_{ix} = n_{ix}\sigma_i , \quad s_{iy} = n_{iy}\sigma_i , \quad s_{iz} = n_{iz}\sigma_i , \tag{9}
$$

If Eqs (8) and (9) are equated, this produces

$$
\left.
\begin{aligned}
(\sigma_x - \sigma_i)n_{ix} + \tau_{yx}n_{iy} + \tau_{zx}n_{iz} &= 0 , \\
\tau_{xy}n_{ix} + (\sigma_y - \sigma_i)n_{iy} + \tau_{zy}n_{iz} &= 0 , \\
\tau_{xz}n_{ix} + \tau_{yz}n_{iy} + (\sigma_z - \sigma_i)n_{iz} &= 0 .
\end{aligned}
\right\}
\tag{10}
$$

The solution to this linear homogeneous system of equations for components n_{ix}, n_{iy} and n_{iz} of the principal normal vectors is significant only if the coefficient determinant becomes zero. This gives a cubic equation for σ_i in the form

$$
\sigma_i^3 - J_1\sigma_i^2 + J_2\sigma_i - J_3 = 0 , \tag{11}
$$

where

$$
J_1 = \sigma_x + \sigma_y + \sigma_z ,
$$
$$
J_2 = \sigma_x\sigma_y + \sigma_x\sigma_z + \sigma_y\sigma_z - \tau_{xy}^2 - \tau_{xz}^2 - \tau_{yz}^2 ,
$$
$$
J_3 = \sigma_x\sigma_y\sigma_z - \sigma_x\tau_{yz}^2 - \sigma_y\tau_{zx}^2 - \sigma_z\tau_{xy}^2 + 2\tau_{xy}\tau_{yz}\tau_{zx} .
$$

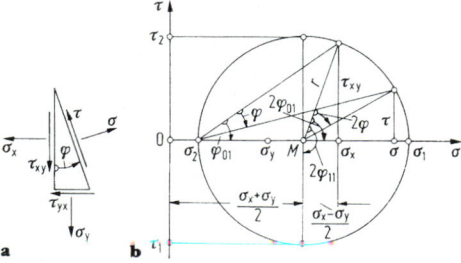

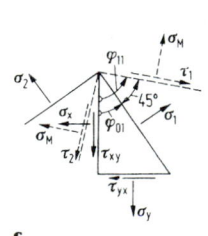

Figure 3. Plane state of stress: **a** stresses on the element; **b** Mohr's circle; **c** principal stresses.

Figure 4. Three-dimensional state of stress.

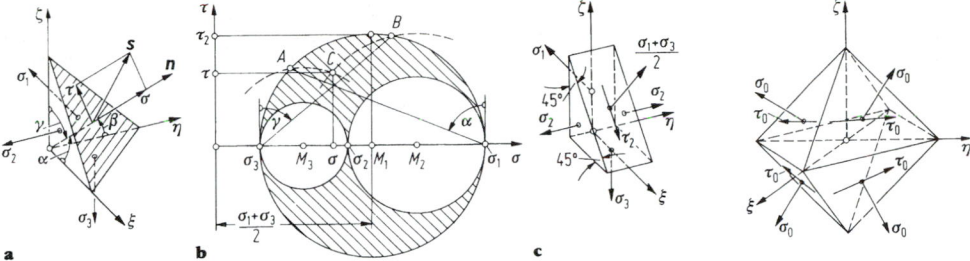

Figure 5. Three-dimensional state of stress: **a** principal stress axis; **b** Mohr's circle; **c** principal shear stress.

Figure 6. Octahedral stresses.

J_1, J_2, J_3 are invariants of the stress tensor since they take the same value for all reference systems, i.e. for the principal directions $J_1 = \sigma_1 + \sigma_2 + \sigma_3$, $J_2 = \sigma_1\sigma_2 + \sigma_1\sigma_3 + \sigma_2\sigma_3$, $J_3 = \sigma_1\sigma_2\sigma_3$. If from Eq. (11) σ_i ($i = 1, 2, 3$) are determined then three linear equations for components n_{ix}, n_{iy}, n_{iz} of a principal normal direction follow from Eq. (10) after inserting σ_i ($i = 1, 2, 3$). Since two of the three equations are linearly dependent upon each other, the relation $n_{ix}^2 + n_{iy}^2 + n_{iz}^2 = 1$, which always applies, must also be used.

If the principal normal vectors $\boldsymbol{n}_i$ ($i = 1, 2, 3$) are determined from this, then the value and direction of the principal normal stresses will be known. For the principal stress system ξ, η, ζ (directions $i = 1, 2, 3$; **Fig. 5a**), where $\sigma_3 = 0$, a plane state of stress is obtained with the principal stresses σ_1 and σ_2 and the equation for the Mohr's circle as shown in Eq. (3):

$$\left(\sigma - \frac{\sigma_1 + \sigma_2}{2}\right) + \tau^2 = \left(\frac{\sigma_1 - \sigma_2}{2}\right)^2.$$

Corresponding circles are obtained for $\sigma_2 = 0$ and $\sigma_1 = 0$ (**Fig. 5b**).

Components σ and τ of the stress vector $\boldsymbol{s}$ for any area element described by $\boldsymbol{n} = (\cos\alpha; \cos\beta; \cos\gamma)$ (**Fig. 5a**) are given by the Mohr's circle (**Fig. 5b**) when angle α is extended from σ_1, and angle γ is extended from σ_3 and circles that are concentric to the adjacent circles are drawn through intersection points A and B on the main circle. Intersection point C determines the relevant values of σ and τ [1-5].

The stresses for any normal angle always lie within the shaded area in **Fig. 5b**. The largest principal shear stress is $\tau_2 = (\sigma_1 - \sigma_3)/2$. It lies in the ξ, ζ plane in an area element, the perpendicular of which is at 45° to the ξ- and ζ-axes (**Fig. 5c**). Correspondingly, $\tau_1 = (\sigma_2 - \sigma_3)/2$ and $\tau_3 = (\sigma_1 - \sigma_2)/2$. The planes of the principal shear stresses are not perpendicular to each other, but form the sides of a regular dodecahedron [4].

The *octahedral* shear and normal stresses are of great significance when assessing complex spatial states of stress. They belong to the eight cutting planes, the perpendiculars of which form equal angles with the three principal axes and form a regular octahedron (**Fig. 6**). Their value is [4]

$$\sigma_0 = (\sigma_1 + \sigma_2 + \sigma_3)/3 = (\sigma_x + \sigma_y + \sigma_z)/3\ ,$$

$$\tau_0 = \frac{1}{3}\sqrt{(\sigma_1 - \sigma_2)^2 + (\sigma_2 - \sigma_3)^2 + (\sigma_1 - \sigma_3)^2}\ ,$$

$$= \frac{1}{3}\sqrt{(\sigma_x - \sigma_y)^2 + (\sigma_y - \sigma_z)^2 + (\sigma_z - \sigma_x)^2 + 6(\tau_{xy}^2 + \tau_{yz}^2 + \tau_{xz}^2)}.$$

1.1.2 Strains

Every body undergoes strains under the action of external forces and moments. The corner point P of a cubic element with sides of length dx, dy, dz (only the x, y plane is shown in **Fig. 7**) undergoes a displacement $\boldsymbol{f} = u\boldsymbol{e}_x + v\boldsymbol{e}_y + w\boldsymbol{e}_z$ with components u, v, w. At the same time the element is stretched, i.e. the side lengths increase (or decrease) to dx', dy' and dz' and it becomes a parallelepiped with shear angles γ_1, γ_2 etc. In the case of small strains (**Fig. 7**) the following apply for *elongations* ε and *shear* γ:

$$\varepsilon_x = \frac{dx' - dx}{dx} = \frac{\dfrac{\partial u}{\partial x}dx}{dx} = \frac{\partial u}{\partial x},\quad \varepsilon_y = \frac{\partial v}{\partial y},\quad \varepsilon_z = \frac{\partial w}{\partial z}\ ;(12)$$

$$\gamma_{xy} = \gamma_1 + \gamma_2 = \frac{\dfrac{\partial v}{\partial x}dx}{dx + \dfrac{\partial u}{\partial x}dx} + \frac{\dfrac{\partial u}{\partial y}dy}{dy + \dfrac{\partial v}{\partial y}dy} = \frac{\partial v}{\partial x} + \frac{\partial u}{\partial y}\ ,$$

$$\gamma_{xz} = \frac{\partial w}{\partial x} + \frac{\partial u}{\partial z},\quad \gamma_{yz} = \frac{\partial w}{\partial y} + \frac{\partial v}{\partial z}\ . \qquad (13)$$

Taking

$$\varepsilon_{xy} = \left(\frac{\partial v}{\partial x} + \frac{\partial u}{\partial y}\right)\bigg/2,\quad \varepsilon_{xz} = \left(\frac{\partial w}{\partial x} + \frac{\partial u}{\partial z}\right)\bigg/2\ ,$$

$$\varepsilon_{yz} = \left(\frac{\partial w}{\partial y} + \frac{\partial v}{\partial z}\right)\bigg/2$$

the strain condition can be described with the strain tensor

$$V = \begin{pmatrix} \varepsilon & \varepsilon_{xy} & \varepsilon_{xz} \\ \varepsilon_{yx} & \varepsilon_y & \varepsilon_{yz} \\ \varepsilon_{zx} & \varepsilon_{zy} & \varepsilon_z \end{pmatrix},$$

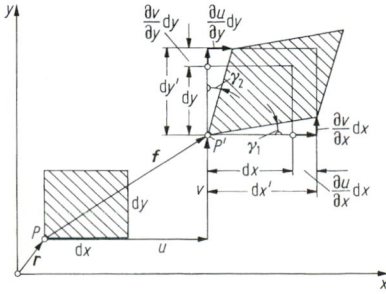

Figure 7. Distortion.

for which similar characteristics and methods of calculation apply as for the stress tensor, Eqs (8). For the principal elongations ε_1, ε_2, ε_3, the characteristic cubic equation

$$\varepsilon_i^3 - J_4\varepsilon_i^2 + J_5\varepsilon_i - J_6 = 0 , \tag{14}$$

where $J_4 = \varepsilon_x + \varepsilon_y + \varepsilon_z$, $J_5 = \varepsilon_x\varepsilon_y + \varepsilon_y\varepsilon_z + \varepsilon_z\varepsilon_x - \varepsilon_{xy}^2 - \varepsilon_{yz}^2 - \varepsilon_{zx}^2$ and $J_6 = \varepsilon_x\varepsilon_y\varepsilon_z - \varepsilon_x\varepsilon_{yz}^2 - \varepsilon_y\varepsilon_{zx}^2 - \varepsilon_z\varepsilon_{xy}^2 + 2\varepsilon_{xy}\varepsilon_{yz}\varepsilon_{zx}$ are again invariants is produced by

$$(\varepsilon_x - \varepsilon_i)n_{ix} + \qquad \varepsilon_{xy}n_{iy} + \qquad \varepsilon_{xz}n_{iz} = 0 ,$$

$$\varepsilon_{xy}n_{ix} + (\varepsilon_y - \varepsilon_i)n_{iy} + \qquad \varepsilon_{yz}n_{iz} = 0 , \tag{15}$$

$$\varepsilon_{xz}n_{ix} + \qquad \varepsilon_{yz}n_{iy} + (\varepsilon_z - \varepsilon_i)n_{iz} = 0$$

by setting the coefficient determinants at zero. If ε_i has been calculated from Eq. (14), then Eqs (15) (two of which are linearly dependent) with $n_{ix}^2 + n_{iy}^2 + n_{iz}^2 = 1$ produce components n_{ix}, n_{iy}, n_{iz} $(i = 1, 2, 3)$ for the three principal strain directions, i.e. the directions for which elongation occurs but no shear and for which the strain tensor assumes the form

$$\boldsymbol{V} = \begin{pmatrix} \varepsilon_1 & 0 & 0 \\ 0 & \varepsilon_2 & 0 \\ 0 & 0 & \varepsilon_3 \end{pmatrix}.$$

The invariants are

$$J_4 = \varepsilon_1 + \varepsilon_2 + \varepsilon_3, \quad J_5 = \varepsilon_1\varepsilon_2 + \varepsilon_2\varepsilon_3 + \varepsilon_1\varepsilon_3, \quad J_6 = \varepsilon_1\varepsilon_2\varepsilon_3 .$$

In spatial and planar cases, (Mohr's) strain circles can be developed for elongation and shear as a function of angles α, β, γ, as they can with stresses. For homogeneous isotropic material, which is always assumed to be the case in the following, the principal directions of stress and strain coincide, i.e. stress and strain tensors are coaxial.

Volume Strain. This refers to

$$\varepsilon = \frac{dV' - dV}{dV} = \frac{dx'\,dy'\,dz'}{dx\,dy\,dz} 1$$

$$= \frac{(1 + \varepsilon_x)dx(1 + \varepsilon_y)\,dy(1 + \varepsilon_z)dz}{dx\,dy\,dz} - 1$$

$$= \varepsilon_x + \varepsilon_y + \varepsilon_z + \varepsilon_x\varepsilon_y + \varepsilon_x\varepsilon_z + \varepsilon_y\varepsilon_z + \varepsilon_x\varepsilon_y\varepsilon_z ,$$

or, if the small values of higher degree are disregarded,

$$\varepsilon = \varepsilon_x = \varepsilon_y + \varepsilon_z . \tag{16}$$

1.1.3 Strain Energy

On an element of volume element $dx\,dy\,dz$ with strains $\varepsilon_x = \dfrac{\partial u}{\partial x}$ etc., stress σ_x, for example, performs work

$$dW = \int \sigma_x dy\, dz\, d\!\left(\frac{\partial u}{\partial x}dx\right) = \int_0^{\varepsilon_x} \sigma_x d\varepsilon_x dV .$$

As a consequence of all normal and shear stresses, after integration we obtain strain energy

$$W = \int_{(V)}\left[\int_0^{\varepsilon_x} \sigma_x\,d\varepsilon_x + \int_0^{\varepsilon_y} \sigma_y\,d\varepsilon_y + \int_0^{\varepsilon_z} \sigma_z\,d\varepsilon_z + \int_0^{\gamma_{xy}} \tau_{xy}\,d\gamma_{xy}\right.$$

$$\left. + \int_0^{\gamma_{xz}} \tau_{xz}\,d\gamma_{xz} + \int_0^{\gamma_{yz}} \tau_{yz}\,d\gamma_{yz}\right]dV , \tag{17}$$

For principal axes *1, 2, 3*,

$$W = \int_{(V)}\left[\int_0^{\varepsilon_1} \sigma_1 d\varepsilon_1 + \int_0^{\varepsilon_2} \sigma_2 d\varepsilon_2 + \int_0^{\varepsilon_3} \sigma_3 d\varepsilon_3\right]dV . \tag{18}$$

In the case of materials that obey Hooke's law, i.e. where stresses σ or τ are proportional to elongation ε or shear γ,

$$W = (1/2) \int_{(V)} (\sigma_x\varepsilon_x + \sigma_y\varepsilon_y + \sigma_z\varepsilon_z$$

$$+ \tau_{xy}\gamma_{xy} + \tau_{xz}\gamma_{xz} + \tau_{yz}\gamma_{yz})\,dV , \tag{19}$$

or

$$W = (1/2) \int_{(V)} (\sigma_1\varepsilon_1 + \sigma_2\varepsilon_2 + \sigma_3\varepsilon_3)\,dV . \tag{20}$$

1.2 Strength and Properties of Materials

See D2.2 for explanations regarding material properties such as limit of proportionality, apparent limit of elasticity or limit of elasticity and breaking point which can be taken from the stress–strain curve of a material.

Hooke's Law. For normal stresses, in the case of a bar which is subjected to tensile stress along one axis, the law

$$\sigma = E\varepsilon \tag{21}$$

applies in the proportionality range of the stress–strain curve (**Fig. 8a**)

Here $\sigma = F/A_0$ is the stress, $\varepsilon = \Delta l/l_0$ the elongation (Δl is the change in the length of the rod) and E the modulus of elasticity. When the rod is extended its diameter is reduced by $\Delta d = d - d_0$. Thus $\varepsilon_q = \Delta d/d_0$ is the transverse extension. The relation $\varepsilon_q = -\nu\varepsilon$ between the longitudinal and transverse extension where ν is the transverse extension or Poisson ratio applies in accordance with DIN 1304 ($\nu_{\text{steel}} = 0.30$). In recent literature the reciprocal value $m = 1/\nu$ is used as Poisson's ratio.

The equivalent Hooke's law applies for shear stresses (**Fig. 8b**):

$$\tau = G\gamma , \tag{22}$$

where $\gamma = du/dy$ is the shear and G the modulus of elasticity in shear. $G = E/[2(1 + \nu)]$.

For values for E, G and ν (see Appendix D3) and for expanded Hooke's laws for any states of stress, see B3.

Safety and Permissible Stress in the Case of Static Loading. If a structure fails as a result of excess-

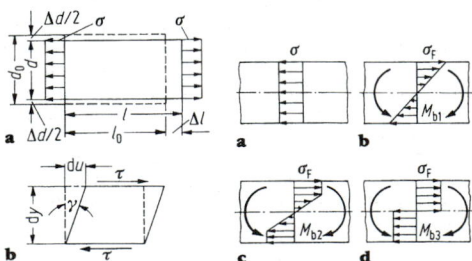

Figure 8. Hooke's law: **a** for elongation, **b** for shear.

Figure 9. Distribution of stresses: **a** uniform; **b** non-uniform; **c** partially plastic; **d** fully plastic.

ive strain (materials with an apparent limit of elasticity), fracture (brittle material) or becoming unstable (due to bending, tipping, buckling) and failure occurs with stress $\sigma = K$ (where K is a material constant), then the safety or the permissible stress is given by

$$S = \frac{K}{\sigma_{exist}}, \qquad \sigma_{perm} = \frac{K}{S}. \qquad (23)$$

Uniform Distribution of Stress. If stresses are distributed uniformly over the cross-section (**Fig. 9a**), then $K = R_e$ should be used for tough materials and $K = R_m$ or σ_{dB} for brittle materials. To protect against deformation it is assumed that $S_F = 1.2$ to 2.0, to protect against fracture that $S_B = 2.0$ to 4.0 and to protect against instability that $S_K = 1.5$ to 4.0.

Non-uniform Distribution of Stress. For *brittle materials* where stresses are distributed non-uniformly over the cross-section (**Fig. 9b**), $K = \sigma_{bB}$ (bending strength) should be used in the case of bending shown in Eqs (23) ($\sigma_{bB} \approx 1.6$ to $2.0R_m$). For torsional strain $\tau_{perm} = K/S$ applies where $K = 1.0$ to $1.1R_m$. For combined strains K is to be calculated from the formulae for equivalent stresses (see B1.3).

For *tough materials*, $K = R_e$ can be used in the case of bending shown in Eqs (23) $K = R_e$; in an initial approximation the strains already seem excessive if the fibres under the greatest stress begin to yield. However, since all the other fibres are still within the elastic range, the external fibre is prevented from significant yielding by the supporting effect of the internal fibres, i.e. there are no excessive deformations yet. As a result the yield stresses are allowed to continue to spread over the cross-section until the edge fibre has attained a permanent elongation of 0.2% (**Fig. 9c**; method for determining the permanent elongation limit [6–10]).

No really excessive strains occur until the yield stresses have spread over the whole cross-section (**Fig. 9d**). For example, the bending moment shown in **Fig. 9b** for a rectangular cross-section which can still be taken as elastic is $M_{b1} = \sigma_F bb^2/6$, while the moment of inertia in the fully plastic state in **Fig. 9b** is $M_{b3} = \sigma_F bb^2/4$, i.e. $M_{b3} = 1.5 \cdot M_{b1}$. In reality, the moment that can be applied before fracture occurs is even greater because $n_{f,pl} = M_{b3}/M_{b1}$ is called the fully plastic support parameter and forms the basis of the limit design in steel construction.

In accordance with the method for determining the *permanent elongation limit*, the value $K = K^*_{0.2}$ can be used in Eqs (23). Here the permanent elongation limit $K^*_{0.2}$ is an imaginary equivalent stress in accordance with the theory of elasticity, which (e.g. in the case of bending) produces the same moment of inertia as the actual stresses involved in a permanent 0.2% elongation of the edge fibres. Here it is assumed that the cross-sections remain plane even in the plastic range. For a rectangular cross-section, e.g. for an ideal elasto-plastic stress–strain curve as shown in **Fig. 10a**, taking $\sigma_F = 210$ N/mm², i.e. $\varepsilon_{el} = 210/210\,000 = 0.1\%$, for $\varepsilon_{pl} = 0.2\%$, it follows that the total elongation is $\varepsilon = \varepsilon_{el} + \varepsilon_{pl} = 0.3\%$. Thus the elongation of the fibres is below height $b/6$ in the elastic range and above height $b/6$ in the plastic range (**Fig. 10b**), which produces the stress distribution shown in **Fig. 10c**. The moment of inertia is

$$M^*_{b,el} = K^*_{0.2}bb^2/6;$$

$$M_{b,pl} = M_{b2}$$

$$= \sigma_F \frac{bb}{3}\frac{2}{3}b + \sigma_F \frac{bb}{12}\cdot\frac{2}{9}b$$

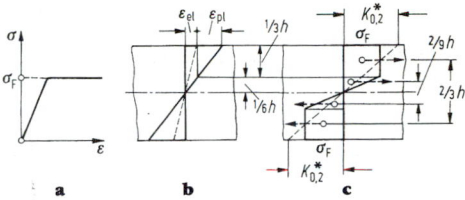

Figure 10. Permanent elongation limit: **a** idealised stress–strain diagram; **b** strains; **c** stresses.

$$= \sigma_F \frac{13}{9}\frac{bb^2}{6} = 1.44 \cdot \sigma_F \frac{bb^2}{6}.$$

$K^*_{0.2} = 1.44 \cdot \sigma_F$ follows from $M_{b,pl} = M^*_{b,el}$. The permanent elongation limit stress $K^*_{0.2}$ is dependent upon the height of the yielding point and the shape of the stress–strain curve. The permanent elongation limit ratio $\delta_{0.2} = K^*_{0.2}/\sigma_F$ or $\delta_{0.2} = K^*_{0.2}/R_{p0.2}$, also known as support figure $n_{0.2}$ [5], is however not greatly dependent upon the value for the strain and yield limit and only dependent upon the shape of the stress–strain curve. **Table 1** gives the support figures $\delta_{0.2}$ for various cross-sections and for two typical stress–strain curves (in accordance with [9]). For the failure value K in Eq. (23), $K = K^*_{0.2} = \delta_{0.2}\sigma_F = \delta_{0.2}R_{p0.2}$ then applies.

Safety and Permissible Strain for Dynamic Loading. See D1.5 and 1.6.

Table 1. Ratios for permanent elongation limit $\delta_{0.2}$

Structural member	Shape of cross-section	$\delta_{0.2}$	
		$\sigma_F = 300$ N/mm²	$R_{p0.2} = 500$ N/mm²
Straight bars with bending	▨	1.40	1.30
	◉	1.55	1.40
	◆	1.75	1.55
	I	1.15	1.10
Cylindrical hollow bars with torsion	r_i/r_0		
	0	1.30	1.20
	0.4	1.25	1.17
	0.8	1.10	1.07
Rotating disc with hole	r_i/r_0		
	0.2	2.00	1.70
	0.4	1.46	1.60
	0.6	1.26	1.35
	0.8	1.10	1.15
Hollow cylinder under internal pressure	r_i/r_0		
	1.5	1.45	1.35
	2.0	1.80	1.55
	2.5	1.95	1.65
	3.0	2.05	1.75
Perforated flat bar under tensile stress/pressure	b/d		
	1.0	2.05	1.80
	2.0	2.25	2.00
	4.0	2.55	2.20
	9.0	2.70	2.35

1.3 Failure Criteria, Equivalent Stresses

In the case of multi-dimensional states of stress it is necessary to go back to a single-dimensional equivalent stress σ_v, since there are generally no material constants for multi-dimensional states. The following failure criteria take into account the type of cause of failure as a result of the different types of material behaviour.

1.3.1 Maximum Principal Stress Criterion

This is to be applied if a cleavage fracture occurs perpendicular to the principal tensile stress, i.e. in the case of brittle materials (e.g. grey cast-iron castings or weld seams), or if the state of stress limits the possibility of deformation of the material (e.g. in the case of triaxial tensile stress or impact loading). For the three-dimensional (spatial) state of stress $\sigma_v = \sigma_1$ applies (determination of σ_1 in accordance with B1.1.1) and for the two-dimensional (plane) state of stress (see B1.1.1),

$$\sigma_v = \sigma_1 = 0.5[\sigma_x + \sigma_y + \sqrt{(\sigma_x - \sigma_y)^2 + 4\tau^2}] .$$

1.3.2 Maximum Shear Stress (Tresca) Criterion

If a shear fracture leads to failure (e.g. in the case of static tensile and pressure loading of ductile materials and pressure loading of brittle materials), then according to Mohr the principal shear stresses can be seen as determining factors. The equivalent stress σ_v is then

$$\sigma_v = 2\tau_{max} = \sigma_3 - \sigma_1$$

for the three-dimensional (spatial) state of stress (where $\sigma_1 > \sigma_2 > \sigma_3$ (see **Fig. 5b**); determination of σ_1 and σ_3 is in accordance with B1.1.1).

For the two-dimensional (plane) state of stress,

$$\sigma_v = 2\tau_{max} = \sqrt{(\sigma_x - \sigma_y)^2 + 4\tau^2}$$

applies.

1.3.3 Maximum Shear Strain Energy Criterion

The maximum shear strain energy criterion, also known as von Mises's criterion, compares the amounts of work required to produce deformation (not volumetric changes) due to shearing at the onset of yielding in the case of multi-dimensional and single-dimensional states of stress, and produces the equivalent stress σ_v. It applies to plastic materials that fail with the occurrence of plastic deformation as well as when they are subjected to oscillating loads with failure resulting from fatigue fracture. For the three-dimensional (spatial) state of stress,

$$\sigma_v = (1/\sqrt{2})\sqrt{(\sigma_1 - \sigma_2)^2 + (\sigma_2 - \sigma_3)^2 + (\sigma_3 - \sigma_1)^2}$$
$$= \sqrt{\sigma_x^2 + \sigma_y^2 + \sigma_z^2 - (\sigma_x\sigma_y + \sigma_y\sigma_z + \sigma_x\sigma_z) + 3(\tau_{xy}^2 + \tau_{yz}^2 + \tau_{xz}^2)}$$

(determination of $\sigma_1, \sigma_2, \sigma_3$ in accordance with B1.1.1) and for the two-dimensional (plane) state of stress

$$\sigma_v = \sqrt{\sigma_1^2 + \sigma_2^2 - \sigma_1\sigma_2} = \sqrt{\sigma_x^2 + \sigma_y^2 - \sigma_x\sigma_y + 3\tau^2} .$$

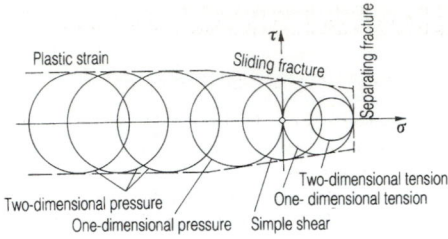

Figure 11. Mohr's limit strength.

It should be mentioned that the criterion can also be derived by equating the octahedral shear stresses (see B1.1.1).

1.3.4 Mohr's Criterion

According to Mohr this is based on various measured limit stress states. The envelope of the relevant Mohr's circle is then the limit strength curve $\tau = f(\sigma)$ and represents a comprehensive material constant. Since for the most part there are insufficient material constants (especially for spatial states of stress), the envelope is replaced by three straight lines (**Fig. 11**).

1.3.5 Bach's Correction Factor

Since σ and τ are often subject to different loading cases (see D1.1) (e.g. σ for case III and τ for case I), τ is converted to the loading case of σ. To do this τ is replaced by $\alpha_0 \tau$. The correction factor is $\alpha_0 = \sigma_{limit}/(\varphi\tau_{limit})$. If $\sigma = 0$ is used, factor φ is produced for the relevant failure criterion, i.e.

$\sigma_v = \tau$	produces $\varphi = 1$	for the normal stress criterion
$\sigma_v = 2\tau$	produces $\varphi = 2$	for the shear stress criterion
$\sigma_v = \sqrt{3}\tau$	produces $\varphi = 1.73$	for the maximum shear strain energy criterion.

For the important loading case of simultaneous bending and torsion of a bar the following applies for the correction factor approximated from the limit stresses for steel:

for alternating bending, static torsion $\alpha_0 \approx 0.7$,

for alternating bending, alternating torsion $\alpha_0 = 1.0$,

for static bending, alternating torsion $\alpha_0 \approx 1.5$,

while the equivalent stresses take the form

$$\sigma_v = 0.5[\sigma_b + \sqrt{\sigma_b^2 + 4(\alpha_0\tau_t)^2}] \quad \text{(normal stress criterion)}$$

$$\sigma_v = \sqrt{\sigma_b^2 + 4(\alpha_0\tau_t)^2} \quad \text{(shear stress criterion)}$$

$$\sigma_v = \sqrt{\sigma_b^2 + 3(\alpha_0\tau_t)^2} \quad \text{(maximum shear stress}$$

energy criterion)   (24)

2 Stresses in Bars and Beams

2.1 Tension and Compression

2.1.1 Uniform Bars Under Constant Axial Load

In the area of constant axial or normal load $F_N = F$, $\sigma = F_N/A$; $\varepsilon = du/dx = \Delta l/l = \sigma/E$; $u(x) = (\sigma/E)x$; $u(l) = \Delta l = \varepsilon l = (\sigma/E)l$ apply for stress, strain and shear (**Fig. 1a**). Hooke's law is always assumed to apply here and in the following. In accordance with B1.1.3, the strain energy is

$$W = (1/2) \int \sigma\varepsilon \, dV = \sigma^2 Al/(2E) = F_N^2 l/(2EA) \,.$$

These equations apply to tensile and compressive loads. In the case of compressive loads it is also necessary to demonstrate that no buckling occurs (see B7).

2.1.2 Bars with Variable Axial Loads

Variable axial load F_N occurs, e.g., as a result of own weight (density ρ) (**Fig. 1a**). For cross-section A = const. it follows that

$$F_N(x) = \rho g V = \rho g A(l - x), \quad \sigma(x) = \rho g(l - x) \,,$$

$$u(x) = \int du = \int \varepsilon(x) dx = \int \left(\frac{1}{E}\right) \rho g(l - x) dx$$

$$= \left(\frac{\rho g}{E}\right)(lx - x^2/2) + C \,;$$

$C = 0$ from $u(x = 0) = 0$, i.e. $\Delta l = u(l) = \rho g l^2/(2E)$; strain energy

$$W = \frac{1}{2} \int \sigma\varepsilon \, dV = \frac{1}{2} \int_{x=0}^{l} \frac{\sigma^2}{E} A \, dx = \frac{F_G^2 l}{6EA} \,.$$

2.1.3 Bars of Variable Cross-Section

If the axial load $F_N = F$ is constant (**Fig. 1b**),

$$\sigma(x) = F/A(x), \quad u(x) = \int \varepsilon(x) dx = \int \frac{F}{EA(x)} dx \,;$$

$$W = \frac{1}{2} \int \sigma\varepsilon \, dV = \frac{1}{2} \int_{x=0}^{l} \frac{F^2}{EA(x)} dx \,.$$

2.1.4 Notched Bars

Here the principal designs apply, first of all for stability of shape and notch effect (see D1.5.1). Nominal stress $\sigma_n = F/A_n$, maximum stress $\sigma_{max} = \alpha_k \sigma_n$ (for values of α_k see VDI 2226, Figs 7 to 12). In the case of dynamic loading the effective stress is $\sigma_{max,eff} = \beta_k \sigma_n$ (for values β_k or calculated with relevant stress cases see D1.5.2).

2.1.5 Bars with Variation of Temperature

Hooke's law takes the form $\varepsilon(x) = \sigma(x)/E + \alpha_t \Delta t$. This yields $u(k) = \int \varepsilon(x) dx$, or for σ = const.: $u(l) = \Delta l =$

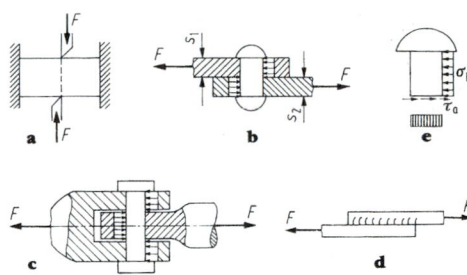

Figure 2a-e. Shear stresses.

$(\sigma/E + \alpha_t \Delta t)l$; α_t coefficient of thermal expansion (steel $1.2 \cdot 10^{-5}$, cast iron $1.05 \cdot 10^{-5}$, aluminium $2.4 \cdot 10^{-5}$, copper $1.65 \cdot 10^{-5}$). If axial expansion is impeded (e.g. if the bar is fixed between two rigid walls or fixed by the substructure of a railway girder of infinite length), then $u(l) = 0$ yields the relevant stress. If σ = const. along the length of the bar, then it follows from $\Delta l = 0$ that the thermal stress $\sigma = E\alpha_t \Delta t$. For St 37 for example, where $\sigma_F = 240$ N/mm^2, $E = 2.1 \cdot 10^5$ N/mm^2 and $\alpha_t = 1.2 \cdot 10^{-5}$ K^{-1}, the yielding point is reached at $\Delta t = \sigma_F/(E\alpha_t) = 95.2$ K.

2.2 Transverse Shear Stresses

Shear stress occurs as a result of two equal, slightly offset loads in bolts, pins, screws, rivets, weld seams, etc. (**Figs 2a-d**). Here, in the case of press fits for rivet, pin and other joints, the bending moments occurring in the rivet, pin, etc. are negligible, because the surrounding material prevents the connecting elements from bending. This creates a spatial state of stress that is difficult to calculate. Bolts or screws that are loosely fitted require additional evidence of bending. Shear is established by assuming that the shear stresses (which are also present when tough materials reach the fully plastic state; **Fig. 2e**) are uniformly distributed:

$$\tau_a = F/(nmA)$$

where $n = 1, 2, 3 \ldots$ indicates a single-, double- or multishear joint, and $m = 1, 2, 3 \ldots$ the number of rivets, screws, etc. In machine construction the permissible shear stress is $\tau_{a,perm} = \sigma_S/\sqrt{3}S$ for tough materials, where $S \approx 1.5$ for static loading and $S \approx 2.0$ for dynamic and variable loading.

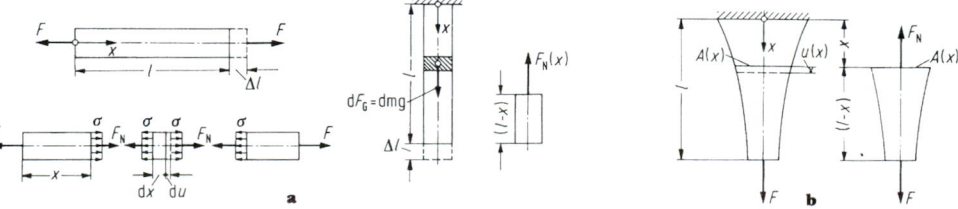

Figure 1. a Bar with a constant cross-section; **b** variable cross-section.

2.3 Contact Stresses and Bearing Pressures

Two parts that are pressed together with their surfaces in contact with each other are subject to contact stresses (point contact: see B4).

2.3.1 Plane Surfaces

Stress distribution depends upon the stiffness of the bodies in contact with each other. An approximation is made with the mean value (**Fig. 3a**)

$$\sigma_p = F_n/A \quad \text{or} \quad \sigma_p = F_n/A_{proj} .$$

A_{proj} is the area projected onto the line perpendicular for the direction of the force. Thus for the wedge shown in **Fig. 3a** $\sigma_{p1} = F_1/A_1 = F_1/(A/\sin \alpha)$ applies and because $F_1/F_n = \sin \beta/\sin(\alpha + \beta)$ then

$$\sigma_{p1} = F_n\sin \alpha \sin \beta/[A\sin(\alpha + \beta)] = F_n/[A(\cot \alpha + \cot \beta)]$$

$$= F_n/(A_{1proj} + A_{2proj}) = F_n/A_{proj} ,$$

$$\sigma_{p2} = F_2/A_2 = F_n/A_{proj}$$

also apply accordingly.

The permissible contact stress is greatly dependent upon the loading case (static, dynamic, alternating). The strength of the weaker structural member is a determining factor. Reference values for $\sigma_{p.perm}$: for tough materials $\sigma_{p.perm} \approx \sigma_{dF}/1.2$ with static loads and $\sigma_{p.perm} \approx \sigma_{dF}/2.0$ with dynamic loads; for brittle materials $\sigma_{p.perm} \approx \sigma_{dB}/2.0$ with static loads and $\sigma_{p.perm} \approx \sigma_{dB}/3.0$ with dynamic loads. Incidentally, $\sigma_{p.perm}$ is dependent upon operating conditions such as running speed and temperature (see F6.2).

2.3.2 Curved Surfaces

Journals. In the calculation, a stress which varies across the area is replaced by the average stress on the projected area (**Fig. 3b**):

$$\sigma_p = F/A_{proj} = F/(dl) .$$

In each case $\sigma_{p.perm}$ depends on the operating conditions (e.g. 2 to 30 N/mm² for large diesel or small gasoline engines; see F10.3.1)

Bolts, Pins, Rivets, Screws. Contact stress is also known as bearing pressure in the case of rivets and

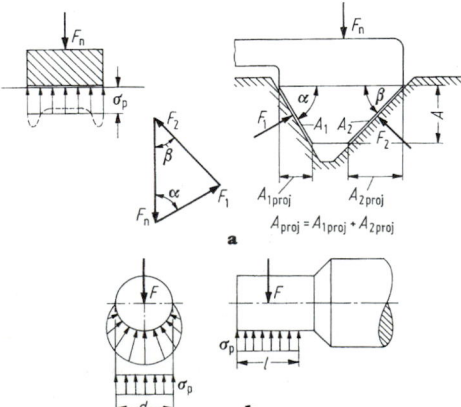

Figure 3. Contact stresses: **a** plane surfaces; **b** journals.

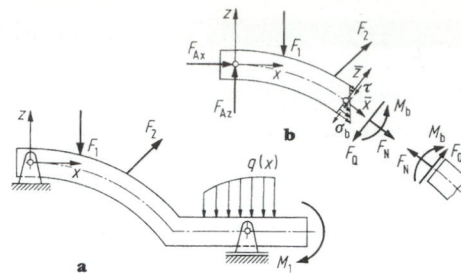

Figure 4a and **b**. Sectional loads.

screws. The following (**Fig. 2b, c, e**) applies again with reference to the projection area:

$$\sigma_p = \sigma_1 = F/A = F/(ds) ;$$

here F is the proportion of force on the transmission area A and s is the thickness of the material. In machine construction $\sigma_{p.perm}$ is as for plane surfaces.

2.4 Bending

2.4.1 Forces and Moments: Axial Force, Shear Force, Bending Moment

Bar-shaped bodies such as girders or beams with a straight, curved or angled axis, which are balanced by supporting reactions (see A1.6), transfer external loads (single loads, line loads, single moments) to the supports by means of internal axial and shear stresses (in **Fig. 4a, b** for the plane case). The resultants of these stresses produce the three forces and moments M_b, F_Q, F_N, in the plane, i.e. a bending moment of which the moment vector acts in the direction of $\hat{y}$, a shear force perpendicular and an axial or longitudinal force tangential to the beam axis. Shear forces and bending moments will be positive if the directions of their vectors at the left edge of the section are opposite the positive coordinate directions $\hat{y}$ and $\hat{z}$; axial force (and torsional moment) if their vectors are directed in the positive coordinate direction $\hat{x}$. According to Newton's law, positive sectional loads should be calculated at the right-hand edge of the section opposite those at the left-hand edge of the section (**Fig. 4b**).

The three sectional loads in the plane are calculated from the three conditions of equilibrium at the freed part of the beam:

$$\Sigma F_{ix} = 0, \quad \Sigma F_{iz} = 0, \quad \Sigma M_i = 0 . \quad (1)$$

As a rule, $\Sigma M_i = 0$ is formed with regard to the intersection, so that indeterminates F_Q and F_N do not appear in this equation. There are six equilibrium conditions for six sectional loads in space (see B2.4.4). To enable simple calculations to be carried out, the system has to be statically determinate (see A1.7).

2.4.2 Forces and Moments in Straight Beams

First of all supporting reactions F_{Ax}, F_{Az} and F_B on the whole of the beam are determined from the three conditions of equilibrium, and then the sectional loads are calculated from Eqs (1) (see **Table 1** for common cases).

Table 1. Bending moments and shear stress curves for standard examples

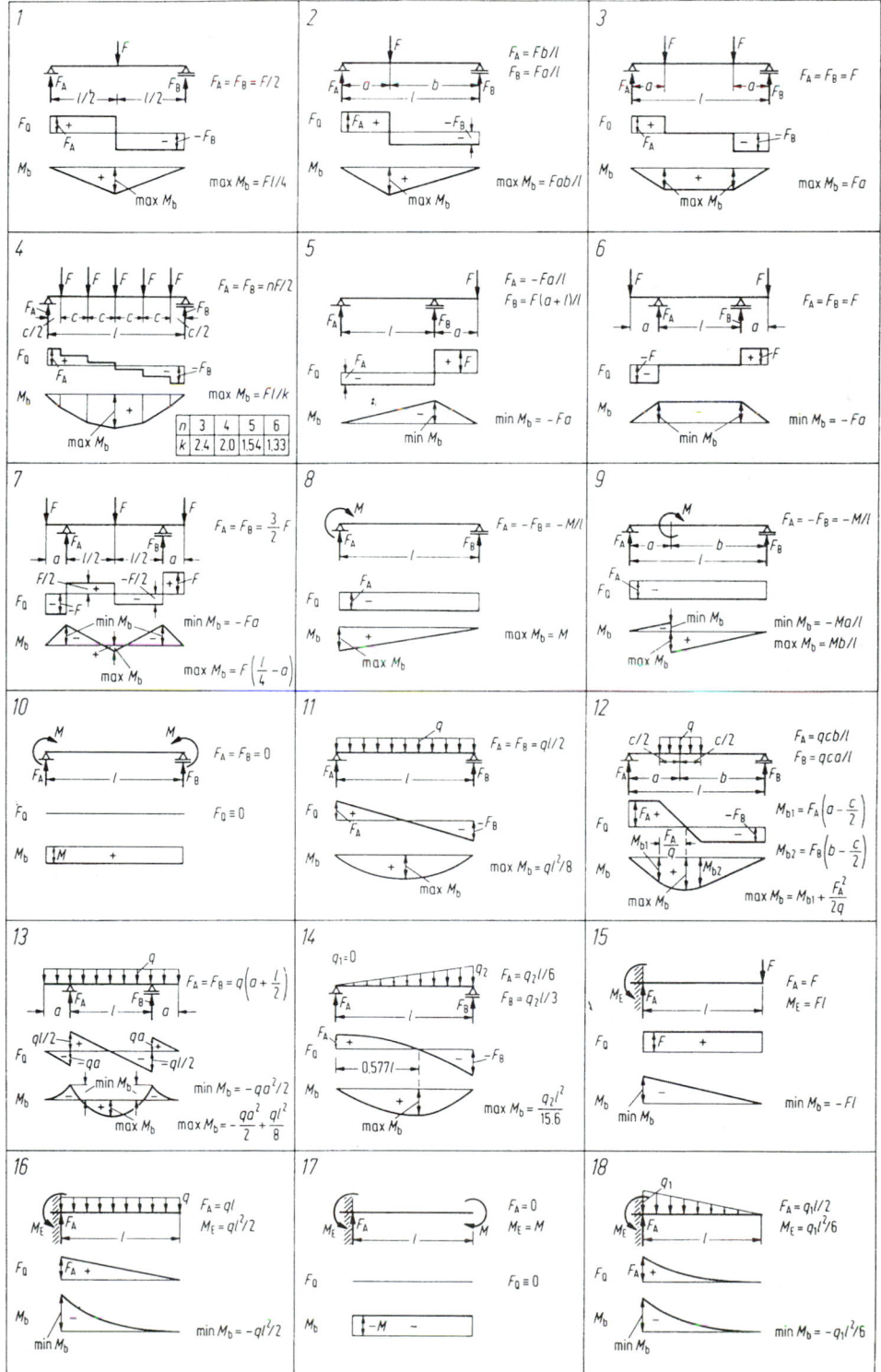

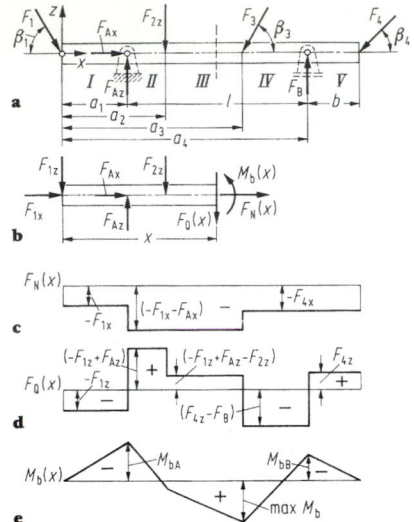

a

b

c

d

e

Figure 5a-e. Beams with single loads and sectional loads.

Beams with Single Loads. (**Figs 5a-e**) In order to calculate the sectional loads, it is necessary to divide the beam into sections since transverse forces, longitudinal forces and bending moments are not constant at points of application. The corresponding sectional loads are obtained from a cross-section in each section; e.g. for section *III* as shown in **Fig. 5b**,

$$\Sigma F_{ix} = 0, \quad F_N(x) = -F_{Ax} - F_{1x};$$

$$\Sigma F_{iz} = 0, \quad F_Q(x) = F_{Az} - F_{1z} - F_{2z};$$

$$\Sigma M_i = 0, \quad M_b(x) = F_{Az}(x - a_1) - F_{1z}x - F_{2z}(x - a_2).$$

Axial and transverse forces are constant in sections and the bending moments are linear functions of *x*, i.e. straight lines. The graphical representation of the sectional loads gives the axial stress, the transverse stress and the bending moment curve (**Figs 5c-e**). The first derivative of $M_b(x)$ gives

$$dM_b/dx = M'_b(x) = F_{Az} - F_{1z} - F_{2z} = F_Q. \quad (2)$$

This result is generally valid, i.e. the first derivative of the bending moment is equal to the transverse force. Since a function where its first derivative is zero has extremities, the extreme values of the bending moments are located at the zero position on the shear stress curve. Since the bending moment curve for single loads has straight sections, it is sufficient to calculate values $M_b(x = a_1)$, $M_b(x = a_2)$, etc. and connect these by means of straight lines.

Example *Calculate the shear stress and moment curve for a chain wheel shaft* (**Fig. 6a**) $\Sigma M_{iB} = 0$ gives $F_{Az} = 17\,250$ N first of all, and $\Sigma M_{iA} = 0$ gives the bearing force $F_B = 27\,750$ N. From $\Sigma F_{iz} =$

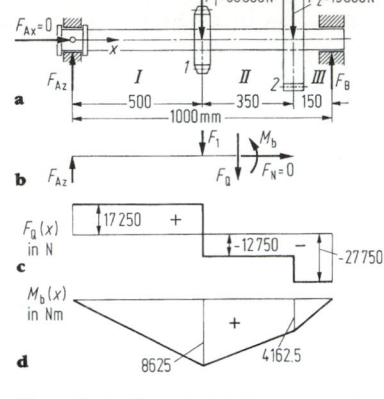

a

b

c

d

Figure 6a-d. Chain wheel shaft. Sectional loads.

$0 = F_{Az} - F_1 - F_0$ a section in range *II* (**Fig. 6b**) produces the transverse force $F_Q = -12\,750$ N. See **Fig. 6c** for shear stress curve $F_Q(x)$ ("stepped curve"). By taking a section at points *I* and *2*, bending moments are obtained from $\Sigma M_{i1} = 0 = -F_{Az} \cdot 0.5$ m $+ M_{b1}$ as $M_{b1} = 8625$ Nm and from $\Sigma M_{i2} = 0 = -F_{Az} \cdot 0.85$ m $+ F_1 \cdot 0.35$ m $+ M_{b2}$ as $M_{b2} = 4162.5$ Nm. The straight lines connecting these values to each other and to the zero position at the supports give the bending moment curve $M_b(x)$ (**Fig. 6d**).

Beams with Single Moments. For a beam that is loaded with a single moment *M* (see B2, Table 1, case no. 9), the bending moment curve has a steady slope because $M'_b(x) = F_Q(x) = \text{const}$. However, at the point of application of moment *M* it jumps by the amount of that moment.

Beams with Line Loads. (**Fig. 7**) As with beams under single loads – apart from single-span beams with continuous line loads – these need to be divided into sections. If a cross-section is taken in each section then for example for section *II* (**Fig. 7a**)

$$\Sigma F_{iz} = 0 = -\int_0^x q(\xi)d\xi + F_{Az} - F_{Q11}(x)$$

$$F_{Q11}(x) = F_{Az} - f(x) \quad (3)$$

and then $M'_b(x) = F_Q(x)$ gives

$$M_{b11}(x) = \int F_{Q11}(x)dx = F_{Az}x - \int f(x)dx + C. \quad (4)$$

Constant *C* is obtained from $M_{b11}(x = a) = M_{bA}$, where M_{bA} is obtained from calculating section *I*. The bending moment is equal to the area of shear force plus the initial value M_{bA}. From Eq. (3) differentiation followed by integration gives

$$dF_Q/dx = F'_Q(x) = M''_b(x) = -q(x),$$

$$F_Q(x) = M'_b(x) = -\int q(x)dx = f(x) + C_1,$$

$$M_b(x) = \int F_Q(x)dx = g(x) + C_1x + C_2. \quad (5)$$

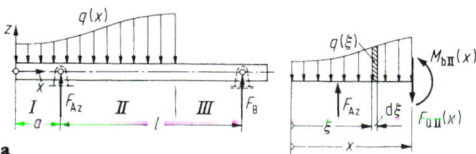

a

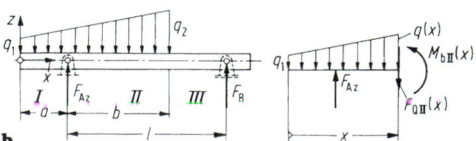

b

Figure 7. Beams with line loads: **a** arbitrary; **b** linear.

Transverse force $F_Q(x)$ and bending moment $M_b(x)$ can be calculated from Eq. (5) instead of Eqs (3) and (4). Constants C_1 and C_2 are obtained from

$$F_{QII}(x = a) = F_{QI}(x = a) + F_{Az} \quad \text{and}$$

$$M_{bII}(x = a) = M_{bI}(x = a) ,$$

where $F_{QI}(x = a)$ and $M_{bI}(x = a)$ are known from the calculations for section I. If the line loads are constant or linearly rising straight lines (**Fig. 7b**), then for section II for example,

$$q(x) = q_1 + \frac{q_2 - q_1}{(a + b)}x,$$

$$F_{QII}(x) = F_{Az} - q_1 x - \frac{q_2 - q_1}{(a + b)}\frac{x^2}{2},$$

$$M_{bII}(x) = F_{Az}(x - a) - q_1\frac{x^2}{2} - \frac{q_2 - q_1}{(a + b)}\frac{x^3}{6}.$$

In the case of linearly rising or constant line loads, the bending moment curves are parabolas of the third or second order (see Table 1).

Straight Beams Under Arbitrary Load (Fig. 8). First of all, from the three conditions of equilibrium $\Sigma F_{ix} = 0$, $\Sigma M_{iB} = 0$, $\Sigma M_{iA} = 0$, the bearing forces are calculated as $F_{Ax} = -3.5$ kN, $F_{Az} = 3.68$ kN and $F_B = 2.17$ kN. The shear forces are calculated as appropriate directly to the left and right of a section boundary and the bending moments are calculated for the section boundary itself. For example, for a cross-section through section II at the left-hand boundary limit, i.e. to the right of support A,

$$\Sigma F_{ix} = 0, \ F_{NAr} = -F_{Ax} = +3.5 \text{ kN} ;$$

$$\Sigma F_{iz} = 0, \ F_{QAr} = F_{Az} - F_1 = 1.68 \text{ kN} ;$$

$$\Sigma M_i = 0, \ M_{bA} = -F_1 a = -2.0 \text{ kNm} .$$

Accordingly, for the right-hand boundary limit of section II,

$$\Sigma F_{ix} = 0 = F_{Ax} + F_{NII} \qquad F_{NII} = +3.5 \text{ kN} ;$$

$$\Sigma F_{iz} = 0 = -F_1 + F_{Az} - qb - F_{QII} \qquad F_{QII} = -1.32 \text{ kN} ;$$

$$\Sigma M_i = 0 = F_1(a + b) - F_{Az}b + qb^2/2 + M_{bI}$$

$$M_{bI} = -1.64 \text{ kN m} .$$

In area II, $F_Q(x) = 0$ gives the zero position of shear force as $x_0 = 2.12$ m with moment $M_b(x_0) = -1.06$ kN m. By calculating the sectional loads for the other areas, the sectional loading curves shown in **Fig. 8** are obtained.

Graphic Determination of Bending Moments. After replacing the line loads (**Figs 9a, b**) by single loads $F_E = qc$, a polar angle is produced (force scale 1 cm ≈ κ kN, length scale 1 cm ≈ λ cm). Transferring the polar lines to the layout plan (line polygon) and drawing the final curve as shown in A1.6.1 on the layout plan and the pole plan produces bearing forces F_A and F_B. For intersection x, forces F_A and F_1 should be replaced by forces F_S and F_{S2} from the pole plan, which can be broken down into horizontal and vertical components (**Fig. 9b**). The bending moment is then $M_b(x) = F_H m(x)$, where $F_H =$ const. in accordance with the pole plan; the values $m(x)$ therefore represent the moment area. If F_H and $m(x)$ are read off in M

$$M_b(x) = \kappa \lambda F_H m(x) \quad \text{in N m} \tag{6}$$

2.4.3 Forces and Moments in Plane Curved Beams

Plane Angled Beams. The suspended structure of a crane trolley as shown in **Fig. 10a** is taken as an example. The system is divided into individual straight sections (because it is symmetrical, it is only necessary to observe one half), and in the revolving system of coordinates $\bar{x}$, $\bar{y}$, $\bar{z}$ the sectional loads are calculated from $\Sigma F_{i\bar{x}} = 0$, $\Sigma F_{i\bar{z}} = 0$ and $\Sigma M_i = 0$. Therefore for area II, for example (**Fig. 10b**), $F_N = F/2$ is obtained from $\Sigma F_{i\bar{x}} = 0 = F_N - F/2$, $F_Q = 0$ from $\Sigma F_{i\bar{z}} = 0 = -F_Q$, and $M_b = Fa/2$ from $\Sigma M_i = 0 = -(F/2)a + M_b$. See **Fig. 10c** for a complete set of results.

Plane Curved Beams. In the case of a slotted annular beam (piston ring) under constant radial load q (**Fig. 11a**), a section at angle φ in the revolving coordinate system $\bar{x}$, $\bar{y}$, $\bar{z}$ as shown in **Fig. 11b** gives

$$\Sigma F_{i\bar{x}} = 0 = \int_0^\varphi qr \sin(\varphi - \psi)d\psi + F_N(\varphi) ,$$

$$F_N(\varphi) = -qr(1 - \cos\varphi) ;$$

$$\Sigma F_{i\bar{z}} = 0 = -\int_0^\varphi qr \cos(\varphi - \psi)d\psi - F_Q(\varphi) ,$$

$$F_Q(\varphi) = -qr \sin\varphi ;$$

$$\Sigma M_i = 0 = \int_0^\varphi qr^2 \sin(\varphi - \psi)d\psi + M_b(\varphi) ,$$

$$M_b(\varphi) = -qr^2(1 - \cos\varphi) .$$

See **Fig. 11c** for a graphical representation of the sectional loads.

2.4.4 Forces and Moments in Beams in Space

If a beam is statically determinate there are six conditions of equilibrium. These produce the six forces and moments F_N, $F_{Q\bar{y}}$, $F_{Q\bar{z}}$, $M_{b\bar{y}}$, $M_{b\bar{z}}$, M_t. For cantilever beams in space

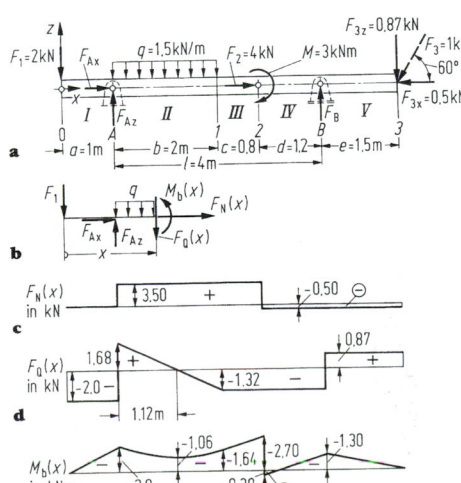

Figure 8a–e. Beams with arbitrary loads, sectional loads.

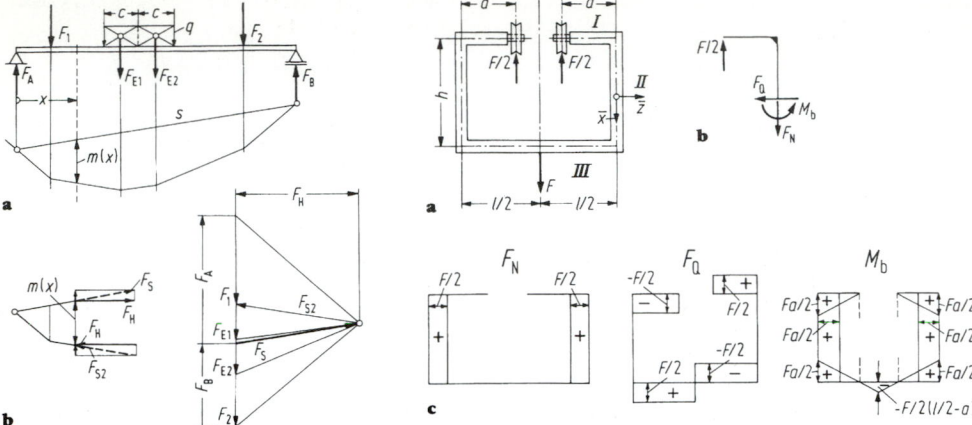

Figure 9a and **b**. Bending moment line, graphical determination.

Figure 10a-c. Suspended structure, sectional loads.

Figure 11a-c. Piston ring, sectional loads.

in accordance with **Fig. 12a, b** the following applies; e.g. for area *III*,

$\Sigma F_{ix} = 0$ $F_N = 0$; $\Sigma F_{iy} = 0$ $F_{Qy} = 0$;

$\Sigma F_{iz} = 0$ $F_{Qz} = F$; $\Sigma M_{ix} = 0$ $M_t = -Fa$;

$\Sigma M_{iy} = 0$ $M_{by} = -F(l - x)$; $\Sigma M_{iz} = 0$ $M_{bz} = 0$.

The forces and moments are similar to those shown in **Fig. 10c**.

2.4.5 Bending Stresses in Straight Beams

Simple Bending. This refers to the effect of all loads parallel to a cross-sectional axis that is simultaneously the principal axis - see Eq. (17). If it is the *z*-axis, then only bending moments M_{by} are produced by the load in direction *z* (**Fig. 13a**). If it is assumed that the loading plane passes through the shear centre *M* (see B2.4.6) Hooke's law $\sigma = E\varepsilon$ applies and the cross-sections remain plane, i.e. any bulging of cross-sections caused by shear stresses is negligible (Bernoulli's criteria), and it follows that

$$\sigma = E\varepsilon = mz \qquad (7)$$

and thus from the conditions of equilibrium that

$$\Sigma F_{ix} = 0 = \int \sigma \, dA = \int mz \, dA, \quad \int z \, dA = 0 ,$$

i.e. the stress zero line passes through the centre of mass and

$$\Sigma M_{iz} = 0 = \int \sigma y \, dA = \int myz \, dA, \quad \int yz \, dA = I_{yz} = 0 ,$$

i.e. the biaxial moment of area I_{yz} must be zero and *y* and *z* must be principal axes. In addition,

$$M_{by} = M_b = -\int \sigma z \, dA = -\int mz^2 \, dA$$

$$= -m \int z^2 \, dA = -mI_y ;$$

I_y is the axial moment of area of the second order. If $m = -M_b/I_y$, Eq. (7) gives

$$\sigma = -(M_b/I_y)z . \qquad (8)$$

Thus the bending stresses increase linearly with the dis-

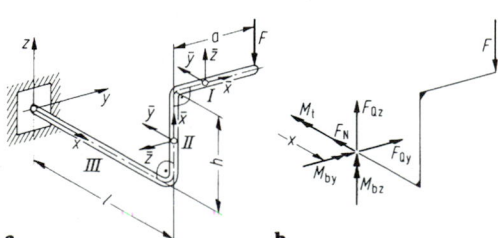

Figure 12a and **b**. Forces and moments in space.

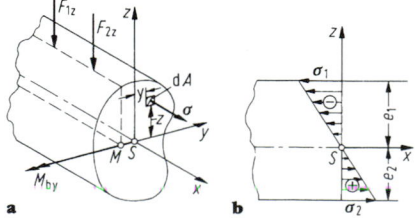

Figure 13a and **b**. Bending stresses.

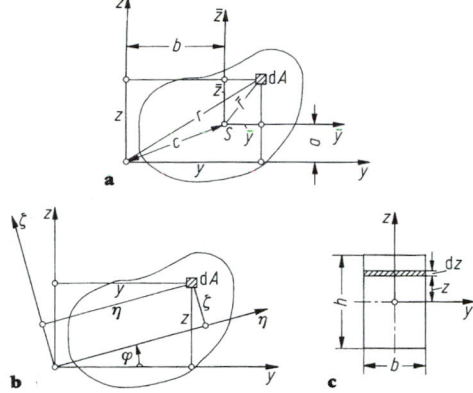

Figure 14. Area moments for **a** parallel axes, **b** rotated axes, **c** rectangular cross-section.

tance from the zero line. Taking $z = e_1$ and $z = -e_2$, the extreme stresses (**Fig. 13**) are

$$\sigma_1 = -M_b/W_{y1} \quad \text{and} \quad \sigma_2 = +M_b/W_{y2} \qquad (9)$$

$$W_{y1} = W_{b1} = I_y/e_1 \quad \text{and} \quad W_{y2} = W_{b2} = I_y/e_2 \qquad (10)$$

are the (axial) section moduli against bending (see **Table 2**). The absolute maximum bending stress for $W_{y\,min}$ is

$$\sigma_{max} = |M_b|/W_{y\,min} \qquad (11)$$

In the case of cross-sections symmetrical to the y-axis, $e_1 = e_2$ and $W_{y1} = W_{y2} = W_y$.

Second Moments of Area. In general beam bending theory the following second moments of area are required (**Fig. 14a**):

$$I_y = \int z^2 \, dA, \ I_z = \int y^2 \, dA; \ I_{yz} = \int yz \, dA \ ;$$

$$I_p = \int r^2 \, dA = \int (y^2 + z^2) \, dA = I_y + I_z . \qquad (12)$$

The axial area moments I_y, I_z and the polar area moment I_p are always positive; the biaxial area moment (centrifugal moment) I_{yz} may be positive, negative or zero.

Radii of Inertia.

$$i_y = \sqrt{I_y/A}, \quad i_z = \sqrt{I_z/A}, \quad i_p = \sqrt{I_p/A} . \qquad (13)$$

Steiner's Principles. For mutually parallel axis systems y, z and $\bar{y}$, $\bar{z}$ (**Fig. 14a**),

$$I_y = \int z^2 \, dA = \int (\bar{z} + a)^2 \, dA$$

$$= \int \bar{z}^2 \, dA + 2a \int \bar{z} \, dA + a^2 \int dA = I_{\bar{z}} + 2aS_{\bar{y}} + a^2A. \quad (14)$$

If the $\bar{y}$- and $\bar{z}$-axes pass through the centre of gravity, the static moment $S_{\bar{y}}$ (and likewise $S_{\bar{z}}$) becomes zero, and Steiner's principles,

$$I_y = I_{\bar{y}} + a^2A, \qquad I_z = I_{\bar{z}} + b^2A ,$$

$$I_{yz} = I_{\bar{y}\bar{z}} + ab\,A, \quad I_p = I_{\bar{p}} + c^2A , \qquad (15)$$

follow (same for the other area moments). For $a = b = c = 0$ axes y and z pass through the centre of gravity and the axial and polar second moments of area become minimum. These equations are used to calculate the area moments of combined cross-sections with known individual area moments.

Rotating the Coordinate System. For a rotated coordinate system η, ζ (**Fig. 14b**),

$$\left.\begin{aligned}
&\eta = y \cos \varphi + z \sin \varphi, \quad \zeta = z \cos \varphi - y \sin \varphi , \\
&I_\eta = \int \zeta^2 \, dA = (I_y + I_z)/2 \\
&\qquad + [(I_y - I_z)/2] \cos 2\varphi - I_{yz} \sin 2\varphi , \\
&I_\zeta = \int \eta^2 \, dA = (I_y + I_z)/2 \\
&\qquad - [(I_y - I_z)/2] \cos 2\varphi + I_{yz} \sin 2\varphi , \\
&I_{\eta\zeta} = \int \eta\zeta \, dA = [(I_y - I_z)/2] \sin 2\varphi + I_{yz} \cos 2\varphi .
\end{aligned}\right\} \quad (16)$$

These equations can be represented graphically in the form of Mohr's circle [1]. This also produces the invariant relations $I_\eta + I = I_y + I_z$, $I_\eta I_\zeta - I_{\eta\zeta}^2 = I_y I_z - I_{yz}^2$, which are not dependent upon φ.

Principal Axes and Principal Second Moments of Area. Axes for which the biaxial moment $I_{\eta\zeta}$ becomes zero are known as principal axes 1 and 2. Their positioning angle φ_0 is obtained from

$$\tan 2\varphi_0 = 2I_{yz}/(I_z - I_y) , \qquad (17)$$

taking $I_{\eta\zeta} = 0$ in accordance with Eqs (16). The relevant principal area moments I_1 and I_2 follow, with φ_0 from Eqs (16) or directly from

$$I_{1,2} = (1/2)[I_y + I_z + \sqrt{(I_y - I_z)^2 + 4I_{yz}^2}] . \qquad (18)$$

I_1 and I_2 are the maximum and minimum second moments of area of a cross-section. Each axis of symmetry of a cross-

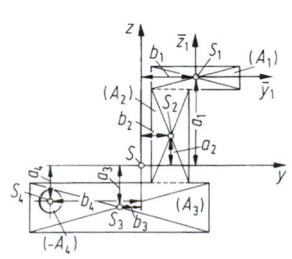

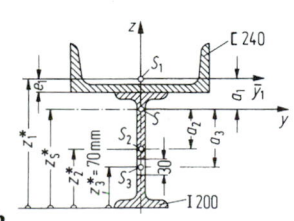

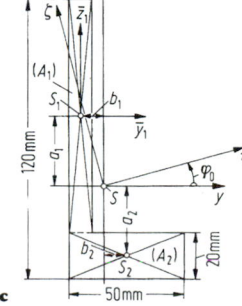

Figure 15a-c. Combined cross-sections.

section and all axes perpendicular to it are always principal axes. If a principal axis system is rotated through angle β then in accordance with Eqs (16)

$$
\left.\begin{aligned}
I_\eta &= (I_1 + I_2)/2 + [(I_1 - I_2)/2]\cos 2\beta \,, \\
I_\zeta &= (I_1 + I_2)/2 - [(I_1 - I_2)/2]\cos 2\beta \,, \\
I_{\eta\zeta} &= [(I_1 - I_2)/2]\sin 2\beta \,.
\end{aligned}\right\} \quad (19)
$$

If $I_1 = I_2$ for a cross-section, then it follows from Eqs (19) that $I_{\eta\zeta} = 0$ regardless of β, i.e. all axes through the reference point are principal axes where $I_\eta = I_\zeta = I_1 = I_2 =$ const. The change of I_η and I_ζ in accordance with Eqs (19) can be represented graphically by the ellipse of inertia [1].

Calculation of Area Moments. For simple areas the boundaries of which can be expressed mathematically, calculation is carried out by means of integration. For example,

$$
I_y = \int_{z=-b/2}^{+b/2} bz^2\, dz = [bz^3/3]_{-b/2}^{+b/2} = bb^3/12
$$

applies for the rectangular cross-section shown in **Fig. 14c**. **Table 2** contains the second moments of area of significant cross-sections (see **Appendix B2 Tables 1-17**).

For combined cross-sections (**Fig. 15**) with Steiner's principles in accordance with Eqs (15),

$$
I_y = \Sigma(I_{\bar{y}i} + a_i^2 A_i) \,, \quad I_z = \Sigma(I_{\bar{z}i} + b_i^2 A_i) \,,
$$
$$
I_{yz} = \Sigma(I_{\bar{y}\bar{z},i} + a_i b_i A_i) \,. \quad (20)
$$

Cavities in areas (e.g. surface A_4 in **Fig. 15a**) are to be taken into account by means of negative I and negative A.

Example 1 *Calculate the height of the centre of mass z_s^* and the second moment of area I_y for the cross-section shown in **Fig. 15b** consisting of profiles U240 and I200 (with bole $d = 30$ mm).* Areas $A_1 = 4230$ mm^2 and $A_2 = 3340$ mm^2 and dimension $e_1 = 22.3$ mm are taken from profile tables. This gives

$$
z_s^* = (\Sigma z_i^* A_i)/A = (4230 \cdot 222.3 + 3340 \cdot 100
$$
$$
-7.5 \cdot 30 \cdot 70)\ \text{mm}^3/7345\ \text{mm}^2 = 171.4\ \text{mm}
$$

for the height of the centre of mass in accordance with A1.10. Thus the distances a_i are

$$
a_1 = (222.3 - 171.4)\ \text{mm} = 50.9\ \text{mm} \,,
$$
$$
a_2 = (100 - 171.4)\ \text{mm} = -71.4\ \text{mm} \,,
$$
$$
a_3 = (70 - 171.4)\ \text{mm} = -101.4\ \text{mm} \,.
$$

According to the profile tables,

$$
I_{y1} = 248 \cdot 10^4\ \text{mm}^4 \quad \text{and} \quad I_{y2} = 2140 \cdot 10^4\ \text{mm}^4 \,,
$$

where it follows from Eqs (20) that

$$
I_y = [248 \cdot 10^4 + 50.9^2 \cdot 4230 + 2140 \cdot 10^4 + 71.4^2 \cdot 3340
$$
$$
-7.5 \cdot 30^3/12 - 101.4^2 \cdot (7.5 \cdot 30)]\ \text{mm}^4 = 4954 \cdot 10^4\ \text{mm}^4 \,.
$$

Example 2 *For the angular cross-section shown in **Fig. 15c**, calculate I_y, I_z, I_{yz}, I_1, I_2, φ_0, i_1, i_2.* In accordance with Eq. (20), taking $I_y = bb^3/12$ according to Table 2, for the rectangular cross-section, division into two areas $A_1 = 10 \cdot 100$ mm^2 = 1000 mm^2 and $A_2 = 50 \cdot 20$ mm^2 = 1000 mm^2 where $a_1 = 30$ mm, $b_1 = 10$ mm, $a_2 = 30$ mm, $b_2 = 10$ mm, gives

$$
I_y = (10 \cdot 100^3/12 + 30^2 \cdot 1000 + 50 \cdot 20^3/12 + 30^2 \cdot 1000)\ \text{mm}^4
$$
$$
= 266.7 \cdot 10^4\ \text{mm}^4 \,,
$$
$$
I_z = (100 \cdot 10^3/12 + 10^2 \cdot 1000 + 20 \cdot 50^3/12 + 10^2 \cdot 1000)\ \text{mm}^4
$$
$$
= 41.7 \cdot 10^4\ \text{mm}^4 \,.
$$

For the individual rectangles, $I_{\bar{y}\bar{z}} = 0$, since $\bar{y}$ and $\bar{z}$ are their principal axes. Thus in accordance with Eq. (20),

$$
I_{yz} = \Sigma a_i b_i A_i = [30 \cdot (-10) \cdot 1000 + (-30) \cdot 10 \cdot 1000]\ \text{mm}^4
$$
$$
= -60 \cdot 10^4\ \text{mm}^4 \,.
$$

The principal area moments in accordance with Eq. (18) are

$$
I_{1,2} = 0.5 \cdot [(266.7 + 41.7) \cdot 10^4
$$
$$
\pm\sqrt{(266.7 - 41.7)^2 \cdot 10^8 + 4 \cdot 60^2 \cdot 10^8}\ \text{mm}^4
$$
$$
= (154.2 \cdot 10^4 \pm 127.5 \cdot 10^4)\ \text{mm}^4 \,;
$$
$$
I_1 = 281.7 \cdot 10^4\ \text{mm}^4; \ I_2 = 26.7 \cdot 10^4\ \text{mm}^4 \,.
$$

The angle of the principal axes in accordance with Eq. (17) is

$$
\varphi_0 = 0.5 \cdot \arctan\frac{-2 \cdot 60 \cdot 10^4\ \text{mm}^4}{(41.7 - 266.7) \cdot 10^4\ \text{mm}^4} = 14.04° \,.
$$

The radii of inertia in accordance with Eq. (13) are

$$
i_1 = \sqrt{281.7 \cdot 10^4/2000}\ \text{mm} = 37.5\ \text{mm} \,;
$$
$$
i_2 = \sqrt{26.7 \cdot 10^4/2000}\ \text{mm} = 11.6\ \text{mm} \,.
$$

Oblique Bending. If the loading plane is not parallel to one of the principal axes, or loads act in the direction of both principal axes (**Fig. 16a, b**), this is known as oblique bending. Loading of each loading plane produces bending moments, the assigned vectors of which are normal to the loading plane in respect of a right-hand screw. They will be positive if they are at the left-hand edge of the section opposite the positive direction of the coordinates (**Fig. 16c, d**). In the case of non-symmetrical cross-sections it is necessary to determine the bending moment vectors in the direction of the principal axes η, ζ.

If M_{by} and M_{bz} are known, then (**Fig. 17**)

$$
M_{b\eta} = M_{by}\cos \varphi_0 + M_{bz}\sin \varphi_0 \,,
$$
$$
M_{b\zeta} = -M_{by}\sin \varphi_0 + M_{bz}\cos \varphi_0 \,. \quad (21)
$$

Assuming that Hooke's law is linear with $\sigma = E\varepsilon$ and that the cross-sections remain plane, the formulation for a linear distribution $\sigma = a\eta + b\zeta$ applies for the stresses and thus for the bending moments

$$
M_{b\eta} = -\int \sigma\zeta\, dA = -\int(a\eta\zeta + b\zeta^2)\, dA = -bI_\eta \,,
$$
$$
M_{b\zeta} = +\int \sigma\eta\, dA = +\int(a\eta^2 + b\eta\zeta)\, dA = aI_\zeta \,,
$$

and therefore for the stresses

$$
\sigma = -(M_{b\eta}/I_\eta)\zeta + (M_{b\zeta}/I_\zeta)\eta \,. \quad (22)
$$

For the stress zero line (neutral fibres) and its rise, $\sigma = 0$ produces

$$
\zeta = (M_{b\zeta}/M_{b\eta})(I_\eta/I_\zeta)\eta \quad \text{or}
$$
$$
\tan \alpha = (M_{b\zeta}/M_{b\eta})(I_\eta/I_\zeta) \,. \quad (23)
$$

Maximum stress is produced at each point P that is furthest from the zero line (**Fig. 18a**: y and z are identical to the principal axes η and ζ).

Double Bending. This applies for the special case of circular cross-sections. Since in a circle every axis is a principal axis, $M_{b,res} = \sqrt{M_{by}^2 + M_{bz}^2}$ always falls in the direction of a principal axis (**Fig. 18b**). Thus for the stresses and their zero line

$$
\sigma = -(M_{b,res}/I_\eta)\zeta \,, \quad \tan \alpha = M_{bz}/M_{by} \,. \quad (24)
$$

For $\zeta = \pm R$ the extreme bending stresses are

$$
\sigma_{extr} = \mp M_{b,res}/W_\eta \quad \text{with} \quad W_\eta = I_\eta/R \,. \quad (25)
$$

Example *For the axle of the rope pulley shown in **Fig. 19**, calculate M_{by}, M_{bz}, $M_{b,res}$, α and σ_{extr}, where $F = 7500$ N, $l = 300$ mm and $d = 50$ mm.* The moments obtained are shown in **Table 1** as $M_{by} = M_{bz} = Fl/4 = 562.5$ N m. Thus $M_{b,res} = \sqrt{562.5^2 + 562.5^2}$ N m $= 795.4$ N m, $\alpha = \arctan(562.5/562.5) = 45°$; and where

Table 2. Axial second moments area and resistance moments

$I_y = \dfrac{bh^3}{12}$ $I_z = \dfrac{hb^3}{12}$ $W_y = \dfrac{bh^2}{6}$ $W_z = \dfrac{hb^2}{6}$	$I_y = I_z = \dfrac{a^4}{12}$ $W_y = W_z = \dfrac{a^3}{6}$ $I_y = I_z = \dfrac{a^4}{12}$ $W_y = W_z = \dfrac{\sqrt{2}}{12}\, a^3 = 0.118\, a^3$
$I_y = I_z = \dfrac{5\sqrt{3}}{16} R^4 = 0.5413\, R^4$ $W_y = \dfrac{5}{8} R^3 = 0.625\, R^3$ $W_z = \dfrac{5\sqrt{3}}{16} R^3 = 0.5413\, R^3$	$I_y = I_z = (1+2\sqrt{2})\,\dfrac{R^4}{6} = 0.638\, R^4$ $W_y = W_z = 0.6906\, R^3$ $I_y = I_z = (1+2\sqrt{2})\,\dfrac{R^4}{6} = 0.638\, R^4$ $W_y = W_z = 0.638\, R^3$
$I_y = \dfrac{bh^3}{36}$ $I_z = \dfrac{hb^3}{48}$ $W_y = \dfrac{bh^2}{24}$ for $e = \dfrac{2}{3} h$ $W_z = \dfrac{hb^2}{24}$	$I_y = \dfrac{h^3}{36}\cdot\dfrac{b_1^2 + 4b_1 b_2 + b_2^2}{b_1 + b_2}$ $W_y = \dfrac{h^2}{12}\cdot\dfrac{b_1^2 + 4b_1 b_2 + b_2^2}{2b_1 + b_2}$ for $e = \dfrac{h}{3}\cdot\dfrac{2b_1 + b_2}{b_1 + b_2}$
$I_y = I_z = \dfrac{\pi d^4}{64}$ $W_y = W_z = \dfrac{\pi d^3}{32}$	$I_y = I_z = \dfrac{\pi (D^4 - d^4)}{64}$ $W_y = W_z = \dfrac{\pi (D^4 - d^4)}{32D}$ with thin walls $\left(\dfrac{s}{d_m}\right)^2 \ll 1$: $I_y = I_z = \dfrac{\pi d_m^3 s}{8}$ $\quad W_y = W_z = \dfrac{\pi d_m^2 s}{4}$
$I_y = \dfrac{\pi a^3 b}{4}$ $I_z = \dfrac{\pi b^3 a}{4}$ $W_y = \dfrac{\pi a^2 b}{4}$ $W_z = \dfrac{\pi b^2 a}{4}$	$I_y = \dfrac{\pi}{4}(a_1^3 b_1 - a_2^3 b_2)$ $W_y = \dfrac{\pi(a_1^3 b_1 - a_2^3 b_2)}{4a_1}$ with thin walls $I_y = \dfrac{\pi a^2(a+3b)s}{4}$, $W_y = \dfrac{\pi a(a+3b)s}{4}$
$I_y = \left(\dfrac{\pi}{8} - \dfrac{8}{9\pi}\right) r^4 = 0.1098\, r^4$ $W_y = I_y/e = 0.1908\, r^3$ for $e = \left(1 - \dfrac{4}{3\pi}\right) r = 0.5756\, r$	$I_y = 0.1098(R^4 - r^4) - 0.283 R^2 r^2 \dfrac{R-r}{R+r}$ $W_{y1,2} = I_y/e_{1,2}$ for $e_1 = \dfrac{4}{3\pi}\cdot\dfrac{R^2 + Rr + r^2}{R+r}$ or $e_2 = R - e_1$

	$I_y = \dfrac{B(H^3 - h^3) + b(h^3 - h_1^3)}{12}$ $W_y = \dfrac{B(H^3 - h^3) + b(h^3 - h_1^3)}{6H}$
	$I_y = \dfrac{BH^3 + bh^3}{12}$ $W_y = \dfrac{BH^3 + bh^3}{6H}$ taking $B = B_1 + B_2$ $b = b_1 + b_2$
	$I_y = \dfrac{BH^3 - bh^3}{12}$ $W_y = \dfrac{BH^3 - bh^3}{6H}$ taking $b = b_1 + b_2$
	$I_y = \dfrac{BH^3 + bh^3}{3} - (BH + bh) e_1^2$ taking $B = B_1 + B_2$, $b = b_1 + b_2$ $W_{y1,2} = I_y/e_{1,2}$ for $e_1 = \dfrac{1}{2}\cdot\dfrac{BH^2 + bh^2}{BH + bh}$ or $e_2 = H - e_1$

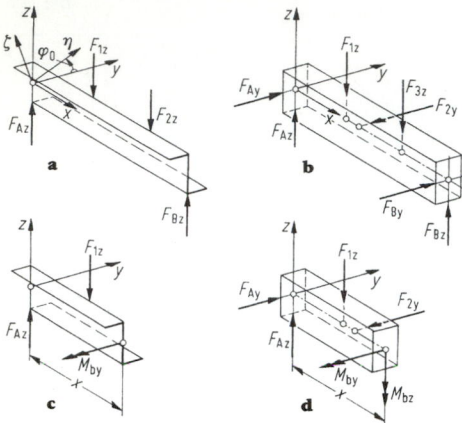

Figure 16a-d. Oblique bending.

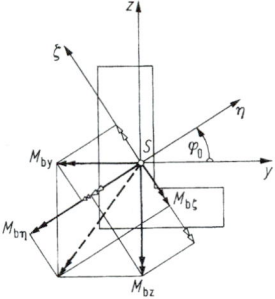

Figure 17. Moment vectors in the directions of the principal axes.

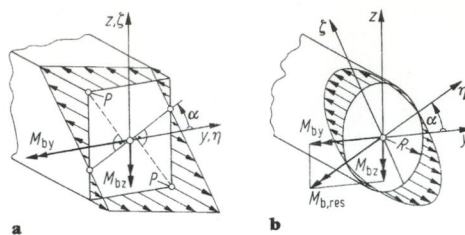

Figure 18. Stresses for **a** oblique bending, **b** double bending.

$$W_\eta = \pi d^3/32 = 12\,272 \text{ mm}^3, \quad \sigma_{\text{extr}} = (795\,400/12\,272) \text{ N/mm}^2 = 64.8 \text{ N/mm}^2.$$

Beams Under Equal Bending Loads. With a view to minimising weight, beams are given a shape whereby the permissible bending load σ_{perm} is applied to the boundary fibres at every point. For example, for beams supported at both ends, as shown in **Fig. 20**, with constant height b_0, variable width $b(x)$ and length $l' = 2l$ (or for a canti-lever beam of length l),

$$\sigma = M_b(x)/W_y(x) = Fx/[b(x)b_0^2/6] = \sigma_{\text{perm}} = \text{const},$$

$$b(x) = 6Fx/(b_0^2\sigma_{\text{perm}}), \quad b_0 = b(l) = 6Fl/(b_0^2\sigma_{\text{perm}}) \;.$$

Thus the width increases linearly with x. For further cases see **Table 3**. The boundary that is shown in the form of a shaded straight line through the tangents is often selec-ted as an approximate shape.

2.4.6 Shear Stress Distribution in Straight Beams, Shear Centre

Shear Stresses. Shear stresses occur in every cross-section of a beam that bends under transverse load. The resultant of these is the shear force F_Q (**Fig. 21**). The shear stresses at the edge are tangential to the boundary

Table 3. Beams under equal bending loads

	Loading case	Cross-sections	Progression of the cross-section, deflection f of the point of application of force		Loading case	Cross-sections	Progression of the cross-section, deflection f of the point of application of force
1a			$b(x) = b_0 = \text{const}$ $h(x) = h_0\sqrt{x/l}$ Squared parabola $h_0 = \sqrt{\dfrac{6Fl}{b_0\sigma_{\text{perm}}}}$	3			$d(x) = d_0\sqrt[3]{x/l}$ Cubic parabola $d_0 = \sqrt[3]{\dfrac{32Fl}{\pi\sigma}}_{\text{perm}}$ $f = \dfrac{192\,F}{5\pi d_0 E}\left(\dfrac{l}{d_0}\right)^3$
1b			$f = \dfrac{8F}{b_0 E}\left(\dfrac{l}{h_0}\right)^3$	4	Cases _1_ to _3_ also apply to beams hinged at both ends of length $l' = 2\,l$ under a single load in the centre $F' = 2F$ (see also Fig.20)		
2			$h(x) = h_0 = \text{const}$ $b(x) = b_0 x/l$ Straight line $b_0 = \dfrac{6Fl}{h_0^2\sigma_{\text{perm}}}$ $f = \dfrac{6F}{b_0 E}\left(\dfrac{l}{h_0}\right)^3$	5			$b(x) = b_0 = \text{const}$ $h(x) = h_0\sqrt{x/a_1}$ $h(\bar{x}) = h_0\sqrt{\bar{x}/a_2}$ (squared parabola) $h_0 = \sqrt{\dfrac{6Fa_1 a_2}{b_0 l\sigma}}_{\text{perm}}$

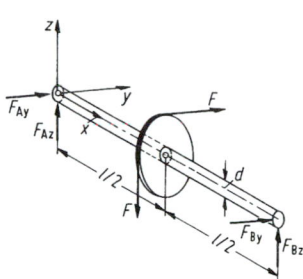

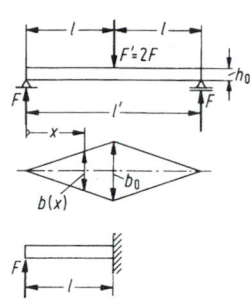

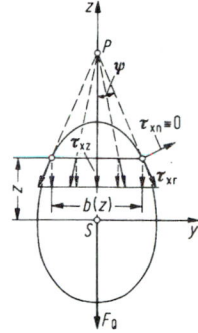

Figure 19. Shaft with double bending.   **Figure 20.** Beams under equal bending loads.   **Figure 21.** Shear stresses for bending under transverse load.

since, because $\tau_{xn} = \tau_{nx}$ (law of assigned shear stresses), for a surface that is free from shear load $\tau_{nx} = \tau_{xn} = 0$ applies. Assuming that all shear stresses at height z pass through the same point P and components τ_{xz} are constant across width $b(z)$ (**Fig. 21**), from the conditions of equilibrium for a beam element of length dx, because $\tau_{zx} = \tau_{xz}$ (**Fig. 22**), it follows that

$$\Sigma F_{ix} = 0 = \tau_{xz}b(z)dx + \int_{z}^{e_1}(\partial\sigma/\partial x)\,dx\,dA\ ,$$

and where $\sigma = -(M_b/I_y)\zeta$ in accordance with Eq. (8) and $dM_b/dx = F_Q$ in accordance with Eq. (2), if I_y is constant,

$$\tau_{xz} = \frac{F_Q}{I_yb(z)}\int_{\zeta=z}^{e_1}\zeta\,dA = \frac{F_Q S_y(z)}{I_y b(z)}\quad\text{where}$$

$$S_y(z) = \int_{z}^{e_1}\zeta\,dA = \int_{z}^{e_1}\zeta b(\zeta)\,d\zeta\ . \tag{26}$$

Here S_y is the first moment of area of the part of the cross-section lying above height z in relation to the y-axis. The maximum shear stress at the edge (**Fig. 21**) is then $\tau_{xr} = \tau_{xz}/\cos\psi$ in each case. In reality, though, as a result of the transverse strain etc., shear stresses τ_{x2} across width b are not constant [1, 2]. Shear stress distributions are determined below for various cross-sections.

Rectangular Cross-Section. (**Fig. 23a**)

$$S_y(z) = \int_{z}^{b/2}\zeta b\,d\zeta = \frac{b}{2}\left(\frac{b^2}{4} - z^2\right) = \frac{bb^2}{8}\left[1 - \left(\frac{z}{b/2}\right)^2\right];$$

$$\tau_{xz} = \frac{3}{2}\frac{F_Q}{bb}\left[1 - \left(\frac{z}{b/2}\right)^2\right],$$

$$\max\tau = \tau_{xz}(z=0) = \frac{3}{2}\frac{F_Q}{bb}\ ,$$

$$\tau_{xz}(z = \pm b/2) = 0\ .$$

The shear stresses are distributed parabolically across the height and the maximum shear stress is $\max\tau = 1.5\cdot F_Q/A = 1.5\cdot\tau_m$, i.e. 50% greater than for uniform distribution. A more accurate theory gives an increase in shear stresses at the boundary and a decrease in the centre. The maximum boundary shear stress for $z = 0$ is obtained from

$$\max\tau_{xz}(z=0) = f\frac{3}{2}\frac{F_Q}{A}, \text{ with } f \text{ in accordance with}$$

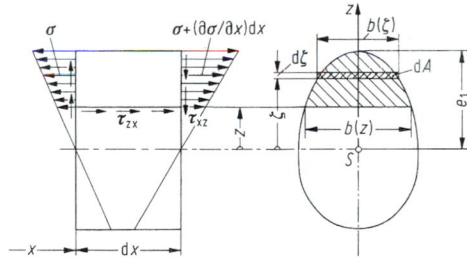

Figure 22. Stresses on the beam element.

b/b	0.5	1	2	4
f	1.03	1.13	1.40	1.99

Circular Cross-Section. (**Fig. 23b**) With $S_y(z) = \int_{z}^{r}\zeta b(\zeta)\,d\zeta$, it follows that

$$b(\zeta) = 2r\cos\varphi,\quad \zeta = r\sin\varphi,\quad d\zeta = r\cos\varphi\,d\varphi$$

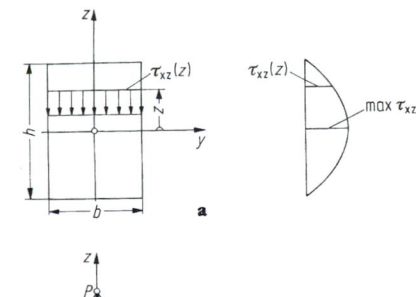

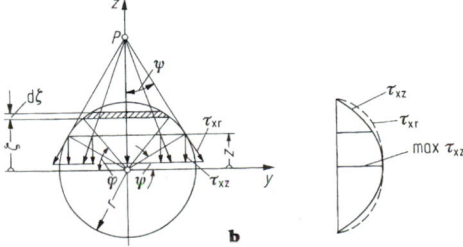

Figure 23. Shear stress distribution for **a** rectangular cross-section, **b** circular cross-section.

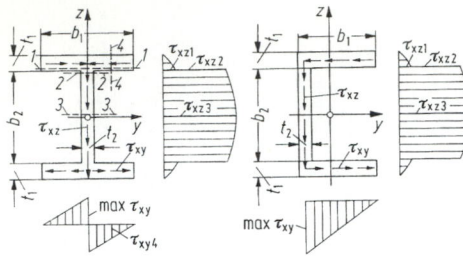

Figure 24. Shear stresses in thin-walled profiles.

$$S_y(z) = \int_\psi^{\pi/2} 2r^3 \sin \varphi \cos^2 \varphi \, d\varphi = \left[-\frac{2}{3} r^3 \cos^3 \varphi \right]_\psi^{\pi/2}$$

$$= \frac{2}{3} r^3 \cos^3 \psi \,,$$

$$\tau_{xz} = \frac{F_Q}{(\pi r^4/4) \cdot 2r \cos \psi} \cdot \frac{2}{3} r^3 \cos^3 \psi = \frac{4 F_Q \cos^2 \psi}{3 \pi r^2}$$

$$= \frac{4}{3} \frac{F_Q}{\pi r^2} \left[1 - \left(\frac{z}{r} \right)^2 \right] \,,$$

$$\tau_{xr} = \tau_{xz}/\cos \varphi = \frac{4 F_Q}{3 \pi r^2} \cos \psi = \frac{4 F_Q}{3 \pi r^2} \sqrt{1 - \left(\frac{z}{r} \right)^2} \,.$$

τ_{xz} is distributed in the form of a parabola across the height and τ_{xr} in the form of an ellipse along the boundary (**Fig. 23b**). For $z = 0$ we obtain

$$\max \tau_{xz} = \frac{4}{3} \frac{F_Q}{\pi r^2} = \frac{4}{3} \frac{F_Q}{A} = \frac{4}{3} \tau_m \,.$$

Annular Cross-section. With internal and external radius r_i and r_a,

$$\max \tau_{xz} = \tau_{xz}(z = 0) = k \frac{F_Q}{A} \,, \quad \text{where}$$

$$k = \left(\frac{4}{3} \right) \frac{r_i^2 + r_i r_a + r_a^2}{r_i^2 + r_a^2} \,.$$

For thin-walled cross-sections, with $r_i \approx r_a \approx r$, the value becomes $k = 2.0$.

I-Cross-section, C-Cross-section and Similar Thin-Walled Profiles (**Fig. 24**). Where $A_1 = b_1 t_1$, $A_2 = b_2 t_2$ and $A = 2A_1 + A_2$,

$$I_y = 2b_1 t_1^3/12 + 2A_1(b_2/2 + t_1/2)^2 + t_2 b_2^3/12 \,.$$

$$S_{y1} = A_1(b_2 + t_1)/2, \quad \tau_{xz1} = F_Q S_{y1}/(I_y b_1) \,;$$

$$S_{y2} = A_1(b_2 + t_1)/2 = S_{y1} \,,$$

$$\tau_{xz2} = F_Q S_{y1}/(I_y t_2) = \tau_{xz1}(b_1/t_2) \,;$$

$$S_{y3} = S_{y1} + A_2 b_2/8, \quad \tau_{xz3} = F_Q S_{y3}/(I_y t_1) = \max \tau_{xz} \,.$$

Distribution of Shear Stresses τ_{xz} (**Fig. 24**). While τ_{xz} is very small in the flanges, τ_{xy} attains considerable magnitudes at these points. For section *4-4*,

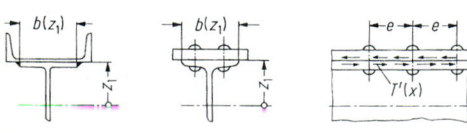

Figure 25. Combined profiles.

$$S_{y4} = (b_1/2 - y)t_1(b_2 + t_1)/2, \quad \tau_{xy4} = F_Q S_{y4}/(I_y t_1) \,.$$

τ_{xy} reaches its maximum for $y = 0$:

$$\max S_{y4} = b_1 t_1(b_2 + t_1)/4 = A_1(b_2 + t_1)/4 = S_{y1}/2 \,,$$

$$\max \tau_{xy} = F_Q S_{y1}/(2I_y t_1) = \tau_{xz2}(t_2/t_1)/2 \approx \tau_{xz2}/2 \,.$$

Accordingly, for the C-profile, $\max \tau_{xy} = \tau_{xz2}(t_2/t_1) \approx \tau_{xz2}$, if $t_2 \approx t_1$. In practice, it is usually enough to indicate the maximum shear stresses in the web using the approximation formula $\max \tau_{xz} = F_Q/A_{web}$.

Shear Stresses in Fasteners on Combined Beams. If profiles are reinforced by means of boom plates or other profiles, then they need to be joined together by means of weld seams or rivets and screws (**Fig. 25**). For shear flow $T'(x)$ per unit length, in accordance with Eqs (26):

$$T'(x) = \tau(x)b(z_1) = F_Q S_y(z_1)/I_y \,.$$

Here $S_y(z_1)$ is the first moment of area of the part of the cross-section above the parting plane in relation to the axis through the centre of mass of the total cross-section and I_y the axial second moment of area of the total cross-section.

At the weld seams of thickness a and in rivets or screws with pitch e and shear area A, the shear stresses are

$$\tau_a = T'/(2a) \quad \text{or} \quad \tau_a = T'e/(2A) \,. \tag{27}$$

Shear Centre. For torsion-free bending under transverse force it is assumed that the load plane passes through the point of application of the resultants of the shear stress, i.e. through the shear centre M (e.g. for loading in the direction of the principal axis z through the point at distance y_M as shown in **Fig. 26**).

Calculation of Coordinates y_M and z_M of the Shear Centre. Because the moment of the shear flow forces must be equal to that of the transverse force F_{Qz} around the centre of mass, the following applies:

$$F_{Qz} y_M = \int_0^l T'(s)b(s) \, ds$$

$$= \int_0^l [T'(s)z \cos \varphi \, ds + T'(s)y \sin \varphi \, ds] \,,$$

$$T'(s) = F_{Qz} S_y(s)/I_y, \quad S_y(s) = \int_0^s z \, dA = \int_0^s zt \, ds \,,$$

$$y_M = \frac{1}{I_y} \int_0^l S_y(s)b(s) \, ds = \frac{1}{I_y} \int_0^l S_y(s)(y \sin \varphi + z \cos \varphi) \, ds \,,$$

where $S_y(s)$ is the first moment of area of the part of the cross-section above the point of intersection s. Accordingly, for a force in the direction of the principal axis y,

$$z_M = -\frac{1}{I_z} \int_0^l S_z(s)b(s) \, ds$$

$$= -\frac{1}{I_z} \int_0^l S_z(s)(y \sin \varphi + z \cos \varphi) \, ds \,,$$

$$S_z(s) = \int_0^s y \, dA = \int_0^s y t \, ds \,.$$

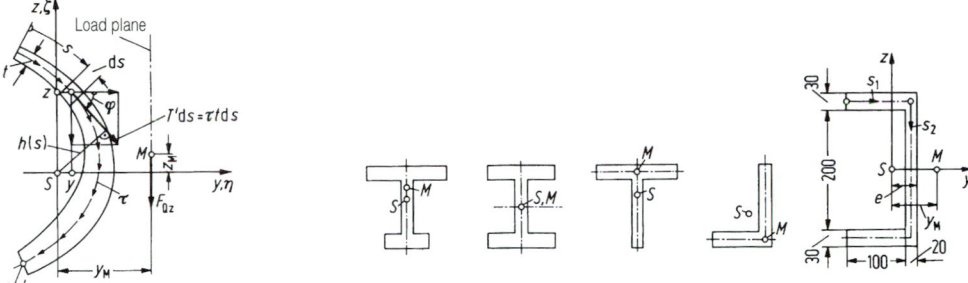

Figure 26. Shear centre.   **Figure 27.** Shear centre of thin-walled cross-sections.

If a cross-section has one axis of symmetry, then the shear centre will lie on this axis. If it has two axes of symmetry, then the shear centre will coincide with the point of symmetry, i.e. at the centre of gravity. In the case of cross-sections made up of two rectangles it will be at the point where the centrelines of the rectangles intersect (**Fig. 27**).

Example I-*profile* (**Fig. 27**). Position of the centre of gravity gives $e = 4.214$ cm and thus $I_y = 10\,909$ cm⁴. For the upper flange, $S_y(s_1) = 3$ cm $\cdot$ 11.5 cm $\cdot s_1 = 34.5$ cm² $\cdot s_1$; $S_y(s_1 = 11$ cm$) = 379.5$ cm³; for the web up to the centre

$$S_y(s_2) = 379.5 \text{ cm}^3 + 2 \text{ cm} \cdot s_2 (11.5 \text{ cm} - s_2/2) = 379.5 \text{ cm}^3$$

$$+ 23 \text{ cm}^2 \cdot s_2 - 1 \text{ cm} \cdot s_2^2; \ S_y(s_2 = 11.5 \text{ cm}) = 511.75 \text{ cm}^3 \ .$$

The cross-section is symmetrical to the y-axis, i.e. the same values are obtained for the lower half. Thus

$$y_M = \frac{2}{I_y} \left[\int_0^{11 \text{ cm}} 34.5 \text{ cm}^2 \cdot s_1 \cdot 11.5 \text{ cm} \cdot ds_1 + \right.$$

$$\left. \int_0^{11.5 \text{ cm}} (379.5 \text{ cm}^3 + 23 \text{ cm}^2 \cdot s_2 - 1 \text{ cm} \cdot s_2^2) \cdot 3.214 \text{ cm} \cdot ds^2 \right]$$

$$= \frac{2 \cdot 41\,289 \text{ cm}^5}{10\,909 \text{ cm}^4} = 7.57 \text{ cm} \ .$$

2.4.7 Bending Stresses in Highly Curved Beams

While the formulae for bending stresses of straight bars (Eqs (8) to (11)) apply to slightly curved bars, the various lengths of the external and internal fibres have to be taken into account for highly curved bars, i.e. for $R \gg d$. This produces a hyperbolic stress distribution for σ; the external stresses are smaller and internal stresses greater than with a linear stress distribution.

Under the effect of an axial force F_N and bending moment M_b, assuming that the cross-sections remain plane the following applies (**Fig. 28**):

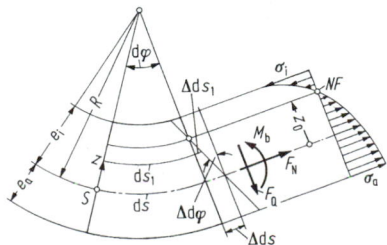

Figure 28. Bending of a highly curved beam.

$$\varepsilon(z) = \frac{\Delta ds_1}{ds_1} = \frac{\Delta ds - z \Delta d\varphi}{(R - z)\, d\varphi} = \varepsilon_0 + \left(\varepsilon_0 - \frac{\Delta d\varphi}{d\varphi} \right) \frac{z}{R - z} \ .$$

Here $\varepsilon_0 = \Delta ds/ds = \Delta ds/(R\, d\varphi)$ is the elongation in the axis of the centre of gravity. The following also applies:

$$\sigma(z) = E\varepsilon(z) = E \left[\varepsilon_0 + \left(\varepsilon_0 - \frac{\Delta d\varphi}{d\varphi} \right) \frac{z}{R - z} \right], \quad (28)$$

i.e. elongation and bending stresses are distributed in accordance with a hyperbolic law (**Fig. 28**). ε_0 and $\Delta d\varphi/d\varphi$ follow from

$$F_N = \int \sigma(z)\, dA = \varepsilon_0 EA + E \left(\varepsilon_0 - \frac{\Delta d\varphi}{d\varphi} \right) \int \frac{z}{R - z}\, dA \ , \quad (29)$$

$$-M_b = \int \sigma(z) z\, dA = E \left(\varepsilon_0 - \frac{\Delta d\varphi}{d\varphi} \right) \int \frac{z^2}{R - z}\, dA \ . \quad (30)$$

Where $\displaystyle\int \frac{z}{R - z}\, dA = \kappa A$ and

$$\int \frac{z^2}{R - z}\, dA = \int \left(\frac{R_z}{R - z} - z \right) dA = R \int \frac{z}{R - z}\, dA = R\kappa A \ ,$$

it follows from Eqs (30) and (29) that

$$\varepsilon_0 - \frac{\Delta d\varphi}{d\varphi} = -\frac{M_b}{ER\kappa A} \quad \text{and}$$

$$\varepsilon_0 = \frac{F_N}{EA} - \left(\varepsilon_0 - \frac{\Delta d\varphi}{d\varphi} \right) \kappa = \frac{F_N}{EA} + \frac{M_b}{ERA}$$

and thus from Eq. (28) that

$$\sigma(z) = \frac{F_N}{A} + \frac{M_b}{RA} \left(1 - \frac{1}{\kappa} \frac{z}{R - z} \right). \quad (31)$$

This gives stresses in the boundary fibres for $z = e_i$ and $z = -e_a$. $\sigma(z) = 0$ gives the stress zero line as

$$z_0 = \frac{F_N R + M_b}{\dfrac{M_b}{\kappa R} + \dfrac{F_N R + M_b}{R}} = \frac{\kappa R}{\kappa + \dfrac{1}{1 + F_N R/M_b}} \ .$$

For $M_b = -F_N R$, $z = 0$, i.e. the neutral fibre lies in the axis of the centre of gravity, if the single load $F = F_N$ acts on the centre of curvature. For pure bending ($F_N = 0$), $z_0 = \kappa R/(1 + \kappa) < R$ and for pure axial load ($M_b = 0$), $z_0 = R$, i.e. the zero line lies at the centre of curvature. Form factor κ for various cross-sections:

Rectangle. Where $\psi = e/R = b/(2R)$ the following applies:

$$\kappa = -1 + \frac{1}{2\psi} \ln \frac{1 + \psi}{1 - \psi} \approx \frac{\psi^2}{3} + \frac{\psi^4}{5} + \frac{\psi^6}{7} .$$

Circle, Ellipse. Where $\psi = e/R$ (e being the semi-axis in the plane of curvature) the following applies:

$$\kappa \approx \psi^2/4 + \psi^4/8 + 5\psi^6/64 .$$

Triangle (Equilateral). Where $\psi = e_i/R = h/(3R)$ the following applies:

$$\kappa = -1 + \frac{2}{3\psi} \left[\left(0.67 + \frac{0.33}{\psi} \right) \ln \frac{1 + 2\psi}{1 - \psi} - 1 \right] .$$

The maximum stress from the bending moment always occurs on the inside of the curved bar. Comparison with nominal stress $\sigma_n = M_b/W_{yi}$ for linear stress distribution gives

$$\sigma_i = \max \sigma_b = \alpha_{ki}\sigma_n . \tag{32}$$

Form factor $\alpha_{ki} = \sigma_i/\sigma_n$ is dependent upon the shape of the cross-section and curvature (**Table 4**):

Table 4. Form factors α_{ki}

$\psi = e_i/R$	0.1	0.2	0.3	0.4	0.5	0.6	0.7	0.8	0.9
Circle, ellipse	1.05	1.17	1.29	1.43	1.61	1.89	2.28	3.0	5.0
Rectangular	1.07	1.14	1.25	1.37	1.53	1.74	2.26	2.59	3.94
Isosceles triangle	-	-	-	1.43	1.64	1.95	2.24	2.88	4.5

Since form factors depend only slightly upon the shape of the cross-section, these values may also be used in the same way for other cross-sectional shapes.

2.4.8 Deflection of Beams

Deflection Line of Straight Beams. Assuming that cross-sections remain plane (disregarding shear stress), as shown in **Fig. 29**, it follows that

$$\varepsilon = \frac{ds_1 - ds}{ds} = \frac{(\rho - z)\,d\alpha - \rho\,d\alpha}{\rho\,d\alpha} = -\frac{z}{\rho}$$

and as a result with Hooke's law $\varepsilon = \sigma/E$ and Eqs (8)

$$k = \frac{1}{\rho} = \frac{M_b(x)}{EI_y(x)} , \tag{33}$$

i.e. the curvature is proportional to bending moment $M_b(x)$ and, in reverse, proportional to bending stiffness $EI_y(x)$. It follows from Eq. (33), with the formula of curvature of a curve

$$k = d\alpha/ds = \pm w''(x)/(1 + w'^2(x))^{3/2}$$

that the differential equation of the deflection line of the beam axis (Euler's elastica) is

$$\frac{w''(x)}{(1 + w'^2(x))^{3/2}} = -\frac{M_b(x)}{EI_y(x)} .$$

Small deflections, i.e. $w'^2(x) \ll 1$, give the linearized differential equation used in the study of beam bending,

$$w''(x) = -M_b(x)/(EI_y(x)) . \tag{34}$$

For the special case of constant cross-section the second moments of area $I_y(x) = I_0$, integration produces

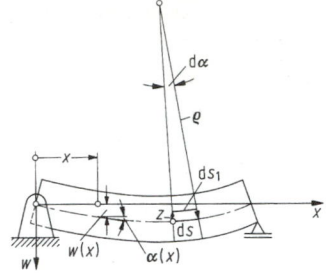

Figure 29. Deflection of a straight beam.

$$w'(x) \approx \alpha(x) = -\frac{1}{EI_0} \int M_b(x)\,dx$$

$$= -\frac{1}{EI_0} f(x) + C_1 , \tag{35a}$$

$$w(x) = \int \left[-\frac{1}{EI_0} f(x) + C_1 \right] dx$$

$$= -\frac{1}{EI_0} g(x) + C_1 x + C_2 . \tag{35b}$$

Constants C_1 and C_2 are determined from the boundary conditions (**Fig. 30a, b**): for beams hinged at both ends $w(x = 0) = 0$ and $w(x = 1) = 0$ and for cantilever beams $w(x = 0) = 0$ and $w'(x = 0) = 0$ (or $w(x = 1) = 0$ and $w'(x = 1) = 0$ if they are fastened at the right-hand end). The common cases (**Table 5**) were calculated using this method.

Expanded Differential Equation. In accordance with Eqs (2) and (5), $dM_b/dx = F_Q(x)$ and $dF_Q/dx = -q(x)$. From Eq. (34) it follows therefore that

$$\frac{d}{dx}[EI_y(x)w''(x)] = -\frac{dM_b}{dx} = -F_Q(x) ,$$

$$\frac{d^2}{dx^2}[EI_y(x)w''(x)] = -\frac{d^2M_b}{dx^2} = -\frac{dF_Q}{dx} = q(x) .$$

For $I_y = I_0 = \text{const.}$,

$$EI_0 w''''(x) = q(x) . \tag{36}$$

Quadruple integration then gives

$$\left.\begin{array}{l}
EI_0 w'''(x) = -F_Q(x) = \int q(x)\,dx = f_1(x) + C_1 , \\[4pt]
EI_0 w''(x) = -M_b(x) = -\int F_Q(x)\,dx \\[4pt]
\qquad = f_2(x) + C_1 x + C_2 , \\[4pt]
EI_0 w'(x) \approx EI_0 \alpha(x) = -\int M_b(x)\,dx \\[4pt]
\qquad = f_3(x) + C_1 x^2/2 + C_2 x + C_3 , \\[4pt]
EI_0 w(x) = f_4(x) + C_1 x^3/6 + C_2 x^2/2 + C_3 x + C_4 .
\end{array}\right\} \tag{37}$$

C_1 to C_4 are determined from the boundary conditions shown in **Fig. 30a, b**. If a moment M or force F acts on the free end of the beam shown in **Fig. 30b**, then the relevant boundary condition will be

$$EI_0 w''(x = l) = \pm M \quad \text{or} \quad EI_0 w'''(x = l) = \pm F .$$

Table 5a. Bending lines of statically determinate beams with constant cross-section

Loading case	Equation for bending line	Deflection	Angle of inclination
1	$0 \le x \le l/2$: $w(x) = \dfrac{Fl^3}{48EI_y}\left[3\dfrac{x}{l} - 4\left(\dfrac{x}{l}\right)^3\right]$	$f_m = \dfrac{Fl^3}{48EI_y}$	$\alpha_A = \alpha_B = \dfrac{Fl^2}{16EI_y}$
2	$0 \le x \le a$: $w_1(x) = \dfrac{Fab^2}{6EI_yl}\left[\left(1+\dfrac{l}{b}\right)\dfrac{x}{l} - \dfrac{x^3}{abl}\right]$ $a \le x \le l$: $w_2(x) = \dfrac{Fa^2b}{6EI_yl}\left[\left(1+\dfrac{l}{a}\right)\dfrac{l-x}{l} - \dfrac{(l-x)^3}{abl}\right]$	$f = \dfrac{Fa^2b^2}{3EI_yl}$ $a>b: f_m = \dfrac{Fb\sqrt{(l^2-b^2)^3}}{9\sqrt{3}\,EI_yl}$ in $x_m = \sqrt{(l^2-b^2)/3}$ $a<b: f_m = \dfrac{Fa\sqrt{(l^2-a^2)^3}}{9\sqrt{3}\,EI_yl}$ in $x_m = l - \sqrt{(l^2-a^2)/3}$	$\alpha_A = \dfrac{Fab(l+b)}{6EI_yl}$ $\alpha_B = \dfrac{Fab(l+a)}{6EI_yl}$
3a	$w(x) = \dfrac{Ml^2}{6EI_y}\left[2\dfrac{x}{l} - 3\left(\dfrac{x}{l}\right)^2 + \left(\dfrac{x}{l}\right)^3\right]$	$f = \dfrac{Ml^2}{16EI_y}$ at $x = \dfrac{l}{2}$ $f_m = \dfrac{Ml^2}{9\sqrt{3}EI_y}$ at $x_m = l - \dfrac{l}{\sqrt{3}}$	$\alpha_A = \dfrac{Ml}{3EI_y}$ $\alpha_B = \dfrac{Ml}{6EI_y}$
3b	$w(x) = \dfrac{Ml^2}{6EI_y}\left[\dfrac{x}{l} - \left(\dfrac{x}{l}\right)^3\right]$	$f = \dfrac{Ml^2}{16EI_y}$ at $x = \dfrac{l}{2}$ $f_m = \dfrac{Ml^2}{9\sqrt{3}EI_y}$ at $x_m = \dfrac{l}{\sqrt{3}}$	$\alpha_A = \dfrac{Ml}{6EI_y}$ $\alpha_B = \dfrac{Ml}{3EI_y}$
4	$w(x) = \dfrac{ql^4}{24EI_y}\left[\dfrac{x}{l} - 2\left(\dfrac{x}{l}\right)^3 + \left(\dfrac{x}{l}\right)^4\right]$	$f_m = \dfrac{5}{384}\dfrac{ql^4}{EI_y}$	$\alpha_A = \alpha_B = \dfrac{ql^3}{24EI_y}$
5	$w(x) = \dfrac{q_2 l^4}{360EI_y}\left[7\dfrac{x}{l} - 10\left(\dfrac{x}{l}\right)^3 + 3\left(\dfrac{x}{l}\right)^5\right]$	$f_m = \dfrac{q_2 l^4}{153.3EI_y}$ at $x_m = 0.519\,l$	$\alpha_A = \dfrac{7}{360}\dfrac{q_2 l^3}{EI_y}$ $\alpha_B = \dfrac{8}{360}\dfrac{q_2 l^3}{EI_y}$
6	$w(x) = \dfrac{Fl^3}{6EI_y}\left[2 - 3\dfrac{x}{l} + \left(\dfrac{x}{l}\right)^3\right]$	$f = \dfrac{Fl^3}{3EI_y}$	$\alpha = \dfrac{Fl^2}{2EI_y}$
7	$w(x) = \dfrac{Ml^2}{2EI_y}\left[1 - 2\dfrac{x}{l} + \left(\dfrac{x}{l}\right)^2\right]$	$f = \dfrac{Ml^2}{2EI_y}$	$\alpha = \dfrac{Ml}{EI_y}$

(continued)

Table 5a. Continued

Loading case	Equation for bending line	Deflection	Angle of inclination
8	$w(x) = \dfrac{ql^4}{24EI_y}\left[3 - 4\dfrac{x}{l} + \left(\dfrac{x}{l}\right)^4\right]$	$f = \dfrac{ql^4}{8EI_y}$	$\alpha = \dfrac{ql^3}{6EI_y}$
9	$w(x) = \dfrac{q_2 l^4}{120EI_y}\left[4 - 5\dfrac{x}{l} + \left(\dfrac{x}{l}\right)^5\right]$	$f = \dfrac{q_2 l^4}{30EI_y}$	$\alpha = \dfrac{q_2 l^3}{24EI_y}$
10	$w(x) = \dfrac{q_1 l^4}{120EI_y}\left[11 - 15\dfrac{x}{l} + 5\left(\dfrac{x}{l}\right)^4 - \left(\dfrac{x}{l}\right)^5\right]$	$f = \dfrac{11}{120}\dfrac{q_1 l^4}{EI_y}$	$\alpha = \dfrac{q_1 l^3}{8EI_y}$
11	$0 \leq x \leq l:$ $w(x) = -\dfrac{Fal^2}{6EI_y}\left[\dfrac{x}{l} - \left(\dfrac{x}{l}\right)^3\right]$ $0 \leq \bar{x} \leq a:$ $w(\bar{x}) = \dfrac{Fa^3}{6EI_y}\left[2\dfrac{l}{a}\dfrac{\bar{x}}{a} + 3\left(\dfrac{\bar{x}}{a}\right)^2 - \left(\dfrac{\bar{x}}{a}\right)^3\right]$	$f = \dfrac{Fa^2(l+a)}{3EI_y}$ $f_m = \dfrac{Fal^2}{9\sqrt{3}\,EI_y}$ at $x_m = \dfrac{l}{\sqrt{3}}$	$\alpha = \dfrac{Fa(2l+3a)}{6EI_y}$ $\alpha_A = \dfrac{Fal}{6EI_y}$ $\alpha_B = \dfrac{Fal}{3EI_y}$
12	$0 \leq x \leq l:$ $w(x) = -\dfrac{qa^2 l^2}{12EI_y}\left[\dfrac{x}{l} - \left(\dfrac{x}{l}\right)^3\right]$ $0 \leq \bar{x} \leq a:$ $w(\bar{x}) = \dfrac{qa^4}{24EI_y}\left[4\dfrac{l}{a}\dfrac{\bar{x}}{a} + 6\left(\dfrac{\bar{x}}{a}\right)^2 - 4\left(\dfrac{\bar{x}}{a}\right)^3 + \left(\dfrac{\bar{x}}{a}\right)^4\right]$	$f = \dfrac{qa^3(4l+3a)}{24EI_y}$ $f_m = \dfrac{qa^2 l^2}{18\sqrt{3}\,EI_y}$ at $x_m = \dfrac{l}{\sqrt{3}}$	$\alpha = \dfrac{qa^2(l+a)}{6EI_y}$ $\alpha_A = \dfrac{qa^2 l}{12EI_y}$ $\alpha_B = \dfrac{qa^2 l}{6EI_y}$

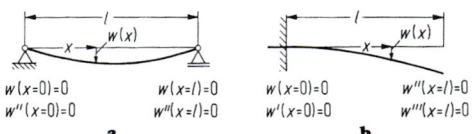

$w(x=0)=0$ $w(x=l)=0$ $w(x=0)=0$ $w'''(x=l)=0$
$w''(x=0)=0$ $w''(x=l)=0$ $w'(x=0)=0$ $w'''(x=l)=0$

      a   **b**

Figure 30a and b. Boundary conditions.

Superposition Method. Suitable superposition of the results recorded in **Table 5** gives the deformations for beams with several single loads, moments and line loads from $w = \Sigma w_i = w_1 + w_2 + w_3 + \ldots$ and $\alpha = \Sigma \alpha_i = \alpha_1 + \alpha_2 + \alpha_3 + \ldots$, where index i corresponds to one of the cases recorded in **Table 5** in each case.

Example *Beams with a cantilever arm* (**Fig. 31**). Taking $I_1 = 30$ cm⁴, $I_2 = 12$ cm⁴, $E = 2.1 \cdot 10^5$ N/mm², $l = 600$ mm, $a = 300$ mm and $F = 2$ kN, find the deflection of the cantilever arm. See **Fig. 31b**; in accordance with **Table 5a**, case 3b. Deflection f_2 resulting from the bending of the cantilever arm (**Fig. 31c**) follows from **Table 5a**, case 6, as $f_2 = Fa^3/(3EI_2)$. Thus $f = f_1 + f_2 = Fa^2 l/(3EI_1) + Fa^3/(3EI_2) = (0.057 + 0.071)$ cm = 0.128 cm.

Mohr's Method. This method has proved particularly effective for beams with varying cross-section or arbitrary complex loading. It is based on the analogy found when comparing Eqs (5) and (34):

$$M_b''(x) = -q(x)$$

$$EI_0 w''(x) = -M_b(x)I_0/I_y(x) = -q^*(x)\,,$$

$$M_b'(x) = F_Q(x) = -\int q(x)\,dx$$

$$EI_0 w'(x) = F_Q^*(x) = -\int q^*(x)\,dx\,,$$

$$M_b(x) = \int F_Q(x) = -\int\!\int q(x)\,dx$$

$$EI_0 w(x) = M_b^*(x) = -\int\!\int q^*(x)\,dx\,.$$

Thus if a beam is loaded with the imaginary load $q^*(x) = M_b(x)I_0/I_y(x)$ then the relevant imaginary transverse load $F_Q^*(x)$ is equal to the EI_0 simple angle of inclination $\alpha \approx \tan\alpha = w'$ and the relevant imaginary bending moment $M_b^*(x)$ is equal to the EI_0-multiple deflection. In any case,

Table 5b. Bending moments and bending lines of statically indeterminate beams of constant cross-section

Loading case	Supporting forces bending moments	Equation for bending line	Deflection	Angle of inclination
1	$F_A = \dfrac{5}{16} F, \quad F_B = \dfrac{11}{16} F$ $M_B = -\dfrac{3}{16} Fl$ $M_F = \dfrac{5}{32} Fl$	$0 \le x \le l/2:$ $w(x) = \dfrac{Fl^3}{96EI_y}\left[3\dfrac{x}{l} - 5\left(\dfrac{x}{l}\right)^3\right]$ $0 \le \bar{x} \le l/2:$ $w(\bar{x}) = \dfrac{Fl^3}{96EI_y}\left[9\left(\dfrac{\bar{x}}{l}\right)^2 - 11\left(\dfrac{\bar{x}}{l}\right)^3\right]$	$f = \dfrac{7}{768}\dfrac{Fl^3}{EI_y}$ $f_m = \dfrac{Fl^3}{48\sqrt{5}\,EI_y}$ At $x_m = \dfrac{l}{\sqrt{5}} = 0.447l$	$\alpha_A = \dfrac{Fl^2}{32EI_y}$
2	$F_A = F\left(\dfrac{b}{l}\right)^2\left(1+\dfrac{a}{2l}\right)$ $F_B = F\left(\dfrac{a}{l}\right)^2\left(1+\dfrac{b}{2l}+\dfrac{3}{2}\dfrac{b}{a}\right)$ $M_B = -F\dfrac{ab}{l}\left(1-\dfrac{b}{2l}\right)$ $M_F = F\dfrac{ab^2}{l^2}\left(1+\dfrac{a}{2l}\right)$	$0 \le x \le a:$ $w(x) = \dfrac{Flb^2}{4EI_y}\left[\dfrac{a}{l}\dfrac{x}{l}\right.$ $\left.-\dfrac{2}{3}\left(1+\dfrac{a}{2l}\right)\left(\dfrac{x}{l}\right)^3\right]$ $0 \le \bar{x} \le b:$ $w(\bar{x}) = \dfrac{Fl^2 a}{4EI_y}\left[\left(1-\dfrac{a^2}{l^2}\right)\left(\dfrac{\bar{x}}{l}\right)^2\right.$ $\left.-\left(1-\dfrac{a^2}{3l^2}\right)\left(\dfrac{\bar{x}}{l}\right)^3\right]$	$f = \dfrac{Fa^2b^3}{4EI_y l^2}\left(1+\dfrac{a}{3l}\right)$ For $a \le 0.414l:\ f_m = w(\bar{x}_m)$ At $\bar{x}_m = \dfrac{b(1+l/a)}{1+3b/2a+b/2l}$ For $a \ge 0.414l:\ f_m = w(x_m)$ At $x_m = l\sqrt{\dfrac{a/2l}{1+a/2l}}$	$\alpha_A = \dfrac{Fab^2}{4EI_y l}$
3	$F_A = \dfrac{3}{8} ql, \quad F_B = \dfrac{5}{8} ql$ $M_B = -\dfrac{1}{8} ql^2$ $M_F = \dfrac{9}{128} ql^2$ At $x_0 = \dfrac{3}{8} l$	$w(x) = \dfrac{ql^4}{48EI_y}\left[\dfrac{x}{l} - 3\left(\dfrac{x}{l}\right)^3\right.$ $\left.+ 2\left(\dfrac{x}{l}\right)^4\right]$	$f_m = \dfrac{ql^4}{185EI_y}$ At $x_m = 0.4215l$	$\alpha_A = \dfrac{ql^3}{48EI_y}$
4	$F_A = \dfrac{1}{10} q_2 l, \quad F_B = \dfrac{4}{10} q_2 l$ $M_B = -\dfrac{1}{15} q_2 l^2$ $M_F = 0.0298 q_2 l^2$ At $x_0 = \dfrac{l}{\sqrt{5}} = 0.447l$	$w(x) = \dfrac{q_2 l^4}{120EI_y}\left[\dfrac{x}{l} - 2\left(\dfrac{x}{l}\right)^3\right.$ $\left.+ \left(\dfrac{x}{l}\right)^5\right]$	$f_m = \dfrac{q_2 l^4}{419EI_y}$ At $x_m = \dfrac{l}{\sqrt{5}} = 0.447l$	$\alpha_A = \dfrac{q_2 l^3}{120EI_y}$
5	$F_A = \dfrac{11}{40} q_1 l, \quad F_B = \dfrac{9}{40} q_1 l$ $M_B = -\dfrac{7}{120} q_1 l^2$ $M_F = 0.0423 q_1 l^2$ At $x_0 = 0.329l$	$w(x) = \dfrac{q_1 l^4}{240EI_y}\left[3\dfrac{x}{l} - 11\left(\dfrac{x}{l}\right)^3\right.$ $\left.+ 10\left(\dfrac{x}{l}\right)^4 - 2\left(\dfrac{x}{l}\right)^5\right]$	$f_m = \dfrac{q_1 l^4}{328EI_y}$ At $x_m = 0.4025l$	$\alpha_A = \dfrac{q_1 l^3}{80EI_y}$

(continued)

Table 5b. Continued

Loading case	Supporting forces bending moments	Equation for bending line	Deflection	Angle of inclination
6	$F_A = F_B = \frac{1}{2}F$ $M_A = M_B = -\frac{1}{8}Fl$ $M_F = \frac{1}{8}Fl$	$0 \le x \le l/2:$ $w(x) = \frac{Fl^3}{48EI_y}\left[3\left(\frac{x}{l}\right)^2 - 4\left(\frac{x}{l}\right)^3\right]$	$f_m = \frac{Fl^3}{192EI_y}$	—
7	$F_A = F\left(\frac{b}{l}\right)^2\left(1+2\frac{a}{l}\right)$ $F_B = F\left(\frac{a}{l}\right)^2\left(1+2\frac{b}{l}\right)$ $M_A = -Fa\left(\frac{b}{l}\right)^2$ $M_B = -Fb\left(\frac{a}{l}\right)^2$ $M_F = 2Fl\left(\frac{a}{l}\right)^2\left(\frac{b}{l}\right)^2$	$0 \le x \le a:$ $w(x) = \frac{Flb^2}{6EI_y}\left[3\frac{a}{l}\left(\frac{x}{l}\right)^2 - \left(1+\frac{2a}{l}\right)\left(\frac{x}{l}\right)^3\right]$ $0 \le \bar{x} \le b:$ $w(\bar{x}) = \frac{Fla^2}{6EI_y}\left[3\frac{b}{l}\left(\frac{\bar{x}}{l}\right)^2 - \left(1+\frac{2b}{l}\right)\left(\frac{\bar{x}}{l}\right)^3\right]$	$f = \frac{Fa^3b^3}{3EI_yl^3}$ $a>b:\ f_m = \frac{2}{3}\frac{Fa^3b^2}{EI_yl^2}\left(\frac{1}{1+2a/l}\right)^2$ At $x_m = l\frac{1}{1+l/2a}$ $a<b:\ f_m = \frac{2}{3}\frac{Fa^2b^3}{EI_yl^2}\left(\frac{1}{1+2b/l}\right)^2$ At $x_m = l\frac{1}{1+l/2b}$	—
8	$F_A = F_B = \frac{1}{2}ql$ $M_A = M_B = -\frac{1}{12}ql^2$ $M_F = \frac{1}{24}ql^2$	$w(x) = \frac{ql^4}{24EI_y}\left[\left(\frac{x}{l}\right)^2 - 2\left(\frac{x}{l}\right)^3 + \left(\frac{x}{l}\right)^4\right]$	$f = \frac{ql^4}{384EI_y}$	—
9	$F_A = \frac{3}{20}q_2l$ $F_B = \frac{7}{20}q_2l$ $M_A = -\frac{1}{30}q_2l^2$ $M_B = -\frac{1}{20}q_2l^2$ $M_F = 0.0214\,q_2l^2$ At $x_0 = l\sqrt{\frac{3}{10}} = 0.548l$	$w(x) = \frac{q_2l^4}{120EI_y}\left[2\left(\frac{x}{l}\right)^2 - 3\left(\frac{x}{l}\right)^3 + \left(\frac{x}{l}\right)^5\right]$	$f_m = \frac{q_2l^4}{764EI_y}$ At $x_m = 0.525l$	—
10	$F_A = 0,\quad F_B = F$ $M_A = \frac{1}{2}Fl$ $M_B = -\frac{1}{2}Fl$	$w(\bar{x}) = \frac{Fl^3}{12EI_y}\left[3\left(\frac{\bar{x}}{l}\right)^2 - 2\left(\frac{\bar{x}}{l}\right)^3\right]$	$f = \frac{Fl^3}{12EI_y}$	—

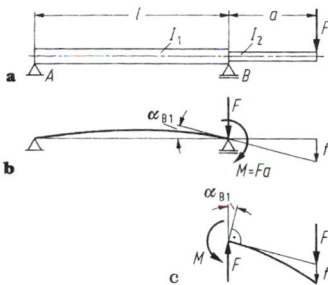

Figure 31a-c. Superposition method.

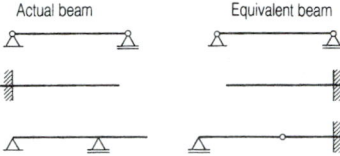

Figure 32. Equivalent beam for Mohr's method.

the imaginary load $q^*(x)$ should be applied to an equivalent beam, because the boundary conditions for the deflections and angle on the actual beam must correspond with those for the imaginary bending moments and transverse loads on the equivalent beam, e.g. $(w = 0) \approx (M_b^* = 0)$ and $w' \neq 0) \approx (F_Q^* \neq 0)$. The resultant equivalent beam is shown in **Fig. 32**.

Example *Find the deflections at points 1 and 2 and the angle of inclination at the supports for the beam* (**Fig. 33a**). $F_1 = 10$ kN, $F_2 = 20$ kN, $I_0 = 1000$ cm^4, $I_1 = 500$ cm^4, $E = 2.1 \cdot 10^5$ N/mm^2 are specified. $F_A = 0$ and $F_B = -10$ kN are obtained from the conditions of equilibrium $\Sigma M_{iB} = 0$ and $\Sigma M_{iA} = 0$. This gives the momentum line shown in **Fig. 33b** with the extreme moment $M_{b2} = F_B \cdot 0.75$ m $= -7.5$ kN m. The distortion function $I_0/I_y(x)$ follows the course shown in **Fig. 33c**, where $q^*(x) = M_b(x)I_0/I_y(x)$ follows as shown in **Fig. 33d**. For this load function we calculate the imaginary support forces as $F_A^* = -3.52$ kN m^2 and $F_B^* = -4.92$ kN m^2 and the imaginary bending moments at points 1 and 2 as

$$M_{b1}^* = F_A^* \cdot 0.5 \text{ m} = -1.76 \text{ kN m}^3 \quad \text{and}$$

$$M_{b2}^* = F_A^* \cdot 1.25 \text{ m} + 0.5 \cdot 15 \text{ kN m} \cdot 0.75 \text{ m} \cdot 0.75 \text{ m}/3$$

$$= -2.99 \text{ kN m}^3.$$

According to Mohr's method the support angles obtained are

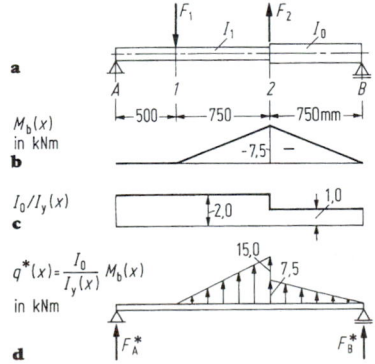

Figure 33a-d. Mohr's method in the form of calculation.

$$\alpha_A = F_{QA}^*/(EI_0) = F_A^*/(EI_0) = -0.00168 \approx -0.097° \quad \text{and}$$

$$\alpha_B = F_{QB}^*/(EI_0) = -F_B^*/(EI_0) = +0.00234 \approx 0.134°$$

and the deflections are $w_2 = M_{b1}^*/(EI_0) = -0.084$ cm and $w_2 = M^*_{b2}/(EI_0) = -0.142$ cm.

The graphic version of Mohr's method (**Fig. 34**) is based on the graphic determination of the bending moments as shown in B2.4.2 (**Fig. 34b**). They are distorted $[I_0/I_y(x)]$ times in the mathematical calculation (**Fig. 34c**). For this imaginary loading, after converting the areas into single forces A_i the imaginary bending moments are again determined in the graph, i.e. the bending line (**Fig. 34d**). With the length scale 1 cm $\approx \lambda$ cm and force scale 1 cm $\approx \kappa$ kN and A_i area scale 1 cm $\approx \varphi$ cm^2 (**Figs 34a-d**). By expanding Eq. (6) the actual deflection follows from

$$w(x) = \kappa\varphi\lambda^5 F_H A_H m^*(x)/(EI_0) \quad \text{in cm}, \quad (38)$$

where F_H, A_H and $m^*(x)$ are expressed in cm and EI_0 in kN cm^2.

Example *The deflections for the beam* (**Fig. 31a**) *were determined graphically.* See **Fig. 34a-d**. By using the factors in Eq. (38) we obtain at point 2 the value $w_2 = 0.127$ cm^2, which corresponds very closely to the result obtained by means of calculation. We obtain $w_1 = 0.21$ mm for point 1, i.e. $x = l/2$.

Deflection for Oblique Bending. If $M_{b\eta}(x)$ and $M_{b\zeta}(x)$ are the bending moments about the principal axes η and ζ (see B2.4.5), then deflections $v(x)$ and $w(x)$ in direction η and ζ are obtained using one of the specified methods. The resultant dislocation follows from $f(x) = \sqrt{v^2 + w^2}$ and represents a spatial curve. At each point, $f(x)$ is perpendicular to the corresponding neutral fibre [1].

Effect of Shear Strains on the Bending Line. As a result of the transverse loads F_Q, the shear stresses that vary across the height of a beam τ are given in accordance with Eq. (26). With regard to shears, it follows from Hooke's law (see B1, Eq. (22)) and **Fig. 35a** that $\gamma =$

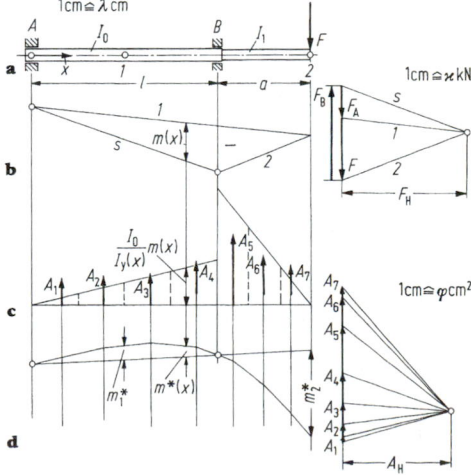

Figure 34a-d. Mohr's method, in the form of a graph.

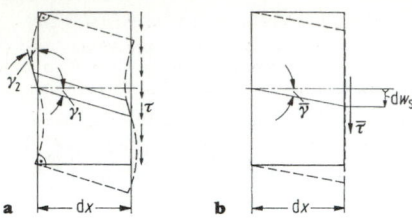

Figure 35a and b. Shear deformation.

$\gamma_1 + \gamma_2 = \tau/G$. They also vary across the height, i.e. the cross-sections bulge. A standard shear stress $\bar{\tau} = \alpha F_Q/A$, for which factor α follows from the fact that the strain energy is the same for the actual and standard state of stress, serves as an approximation:

$$\frac{1}{2}F_Q \, dw_s = \frac{1}{2G}\int \tau^t \, dV, \quad \text{thus}$$

$$\frac{1}{2}F_Q\bar{\gamma}\, dx = \frac{1}{2G}\int \left(\frac{F_Q S_y}{I_y b}\right)^2 dA \, dx, \quad \text{i.e.}$$

$$\frac{1}{2}F_Q\frac{\bar{\tau}}{G} = \frac{1}{2}\frac{F_Q^2}{AG}\alpha = \frac{F_Q^2}{2G}\int\left(\frac{S_y}{I_y b}\right)^2 dA \, ,$$

and thus $\alpha = A\int\left(\dfrac{S_y}{I_y b}\right)^2 dA$.

For a rectangular cross-section $\alpha = 1, 2$; for a circular cross-section $\alpha = 10/9 \approx 1.1$. For the value of shear deformation the following then applies (**Fig. 35b**):

$$dw_s/dx = \bar{\gamma} = \bar{\tau}/G = \alpha F_Q/(GA) \quad \text{and}$$

$$w_s(x) = \frac{\alpha}{GA}\int F_Q(x)\, dx = \frac{\alpha}{GA}M_b(x) + C \, .$$

For example, for a cantilever fixed at the right-hand end with a single load at the (left-hand) free end, $M_b(x) = -F_x$ and thus $w_s(x) = -(\alpha/GA)Fx + C$. From $w_s(x = l) = 0$ it follows that $C = (\alpha/GA)Fl$ and thus $w_s(x) = (\alpha/GA) \cdot F(l-x)$, or $w_s(x=0) = (\alpha/GA)Fl$. The corresponding value resulting from bending is $w(x=0) = Fl^3/(3EI_y)$. For a rectangular cross-section $w_s/w = (0.3 \cdot E/G)(b/l)^2$. Now $0.3 \cdot E/G \approx 1$, thus $w_s/w \approx (b/l)^2$.

For $b/l = 1/5$, $w_s \approx 0.04 \cdot w$, i.e. the shear deformations for low beams are negligible compared to the bending deformations.

Deflection of Slightly Curved Beams. According to the result for straight beams, see Eq. (33), the change in the curve will be (**Fig. 36a**)

$$\frac{1}{\rho} - \frac{1}{R} = -\frac{M_b}{EI_y} \, .$$

This produces the differential equation

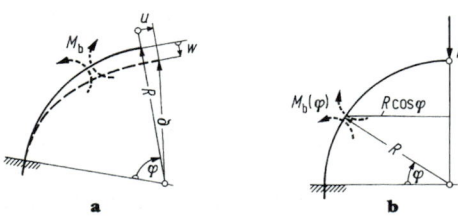

Figure 36a and b. Deflection of a slightly curved beam.

$$\frac{d^2w}{d\varphi^2} + w = \frac{R^2}{EI_y}M_b(\varphi)$$

for the radial displacement w of an originally circular beam [3, 4].

The tangential displacement u follows as

$$u(\varphi) = \int w(\varphi)d\varphi \, .$$

Example *Calculate the displacements of the point of application of force for the quadrant beam* (**Fig. 36b**). *If* $M_b(\varphi) = -FR\cos\varphi$, *the differential equation obtained is*

$$w''(\varphi) + w(\varphi) = -(FR^3/EI_y)\cos\varphi \, ,$$

with the solution

$$w(\varphi) = C_1\sin\varphi + C_2\cos\varphi - (FR^3/2EI_y)\varphi\sin\varphi \, .$$

The boundary conditions $w(0) = 0$ and $w'(0) = 0$ give $C_1 = C_2 = 0$ and thus $w(\varphi) = (FR^3/2EI_y)\varphi\sin\varphi$ with $w(\pi/2) = \pi FR^3/(4EI_y)$. Thus for $u(0) = 0$

$$u(\varphi) = (FR^3/2EI_y)\int\varphi\sin\varphi \, d\varphi = (FR^3/2EI_y)(\sin\varphi - \varphi\cos\varphi)$$

and $u(\pi/2) = FR^3/(2EI_y)$.

2.4.9 Bending Strain Energy, Energy Methods for Deflection Analysis

Strain Energy.

$$W_b = \frac{1}{2}\int M_b d\varphi = \frac{1}{2}\int\frac{M_b^2}{EI_y} \, ds \, . \tag{39}$$

Castigliano's Theorem. The following applies for systems made of materials that obey Hooke's law (**Fig. 37a**):

$$w_F = \frac{\partial W}{\partial F}, \quad \alpha_M = \frac{\partial W}{\partial M} \, . \tag{40}$$

Differentiation of strain energy with a single load produces displacement in the direction of the single load, differentiation with a moment produces the angle of rotation at the point of application. (If displacements are sought at points or in directions that are not subject to a single load, then an auxiliary force $\bar{F}$ is applied and then set at zero again after the calculation has been carried out; the same applies to angles of rotation and moments.)

Example *Calculate the horizontal displacement u of the point of application of force for the quadrant beam shown in* **Fig. 36b**. *With auxiliary force $\bar{F}$ in the horizontal direction* (**Fig. 37b**), $M_b(\varphi) = -FR\cos\varphi - \bar{F}R(1 - \sin\varphi)$ *applies for the bending moment and for the strain energy and displacement*

$$W = \frac{1}{2EI_y}\int_0^{\pi/2}[-FR\cos\varphi - \bar{F}R(1 - \sin\varphi)]^2R \, d\varphi \, ,$$

$$u = \frac{\partial W}{\partial\bar{F}} = -\frac{1}{EI_y}\int_0^{\pi/2}[-FR\cos\varphi - \bar{F}R(1 - \sin\varphi)](1 - \sin\varphi)R^2 \, d\varphi \, ;$$

or, with $\bar{F} = 0$,

$$u = +\frac{1}{EI_y}\int_0^{\pi/2}FR\cos\varphi(1 - \sin\varphi)R^2 \, d\varphi$$

$$= \frac{FR^3}{EI_y}\left[\sin\varphi - \frac{1}{2}\sin^2\varphi\right]_0^{\pi/2} = \frac{FR^3}{2EI_y} \, .$$

Principle of Virtual Work. If an elastic system is subjected to an arbitrary (virtual) displacement, i.e. consistent with the geometric conditions, then in the case of equilibrium the total external and internal virtual work is equal to zero:

$$\delta W^{(a)} + \delta W^{(i)} = 0 \, .$$

If we select only one virtual auxiliary force $\bar{F} = 1$ as

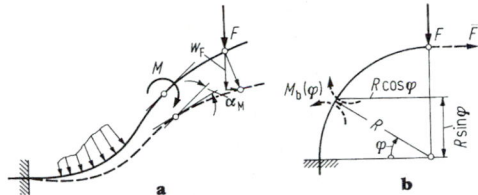

Figure 37. Castigliano's theorem: **a** general; **b** quadrant beam.

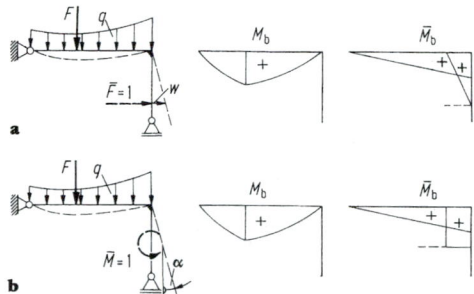

Figure 38a and **b**. Principle of virtual work.

external force and the actual displacements as the external force and displacement respectively (principle of virtual forces) (**Fig. 38a**), then

$$\delta W^{(a)} = -\delta W^{(i)} \quad \text{gives}$$

$$\bar{F}w = 1 \cdot w = \int \bar{M}_b \, d\varphi = \int \frac{\bar{M}_b M_b}{EI_y} \, ds . \qquad (41)$$

This gives displacement w in the direction of the auxiliary force $\bar{F} = 1$. In this case $\bar{M}_b$ are the bending moments resulting from this auxiliary force and M_b the bending moments resulting from the actual load. If a virtual auxiliary moment $\bar{M} = 1$ is selected as the external load and again the actual displacements as the displacement, then the following applies (**Fig. 38b**):

$$\bar{M}\alpha = 1 \cdot \alpha = \int \bar{M}_b \, d\varphi = \int \frac{\bar{M}_b M_b}{EI_y} \, ds . \qquad (42)$$

This gives the angle of rotation at the point of appli-

cation of the auxiliary moment. For beams where $EI_y = $ constant the integrals in Eqs (41) and (42) should only be formed for the product $\bar{M}_b M_b$ and are given for the most important basic cases in **Table 6**.

Example Cantilever beams with line load (**Fig. 39**). Find the deflection and angle of inclination at the free end. The following applies for the deflection shown in **Table 6**, column 8, line b, where $i = ql^2/2$ and $k = 1$

$$1 \cdot f = \int_0^l \bar{M}_b M_b \frac{dx}{EI_y} = \frac{1}{EI_y} \cdot \frac{1}{4} lik = \frac{ql^4}{8EI_y}$$

and for the angle of inclination shown in line a, where $i = ql^2/2$ and $k = 1$

$$1 \cdot \alpha = \int_0^l \bar{M}_b M_b \frac{dx}{EI_y} = \frac{1}{EI_y} \cdot \frac{1}{3} lik = \frac{ql^3}{6EI_y}$$

(cf. **Table 5a**, line 8).

2.5 Torsion

2.5.1 Bars of Circular Cross-section and Constant Diameter

No warping occurs if bars of circular cross-section are subjected to torsion, i.e. the cross-sections remain plane. In addition the radii of the circular cross-sections remain straight, i.e. the cross-sections rotate as a rigid whole. Linear generators on the surface become helixes which can, however, be seen as straight lines because of the small deformations (**Fig. 40a**).

Where $\gamma l = \varphi r$ and with Hooke's law $\gamma = \tau/G$

$$\tau = (G\varphi/l)r , \qquad (43)$$

is produced, i.e. the shear stresses τ increase linearly with radius r (**Fig. 40a**). The moment of all shear stresses about the centre of the circle must be equal to the torque

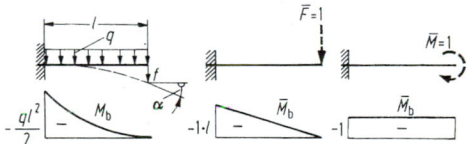

Figure 39. Deformations of a cantilever beam.

Table 6. Values for $\int \bar{M}M \, ds$

$\bar{M}$		1 $i\ \boxed{}\ i$	2 $i\ \diagdown\ i$	3 $i\ \diagup\ i$	4 $i_1\ \boxed{}\ i_2$	5 Squared parabola $i\ \smallsmile\ i$	6 Squared parabola $i\ \diagdown\!\smallfrown\ i$	7 Squared parabola $i\ \smallfrown\!\diagup\ i$	8 Squared parabola $i\ \smallfrown\ i$	9 Squared parabola $i\ \smallfrown\ i$
a	$k\ \boxed{}\ k$	lik	$\frac{1}{2}lik$	$\frac{1}{2}lik$	$\frac{1}{2}l(i_1+i_2)k$	$\frac{2}{3}lik$	$\frac{2}{3}lik$	$\frac{2}{3}lik$	$\frac{1}{3}lik$	$\frac{1}{3}lik$
b	$\diagdown\ k$	$\frac{1}{2}lik$	$\frac{1}{3}lik$	$\frac{1}{6}lik$	$\frac{1}{6}l(i_1+2i_2)k$	$\frac{1}{3}lik$	$\frac{5}{12}lik$	$\frac{1}{4}lik$	$\frac{1}{4}lik$	$\frac{1}{12}lik$
c	$k_1\ \diagup\ k_2$	$\frac{1}{2}li(k_1+k_2)$	$\frac{1}{6}li(k_1+2k_2)$	$\frac{1}{6}li(2k_1+k_2)$	$\frac{1}{6}l[i_1(2k_1+k_2)\ +i_2(k_1+2k_2)]$	$\frac{1}{3}li(k_1+k_2)$	$\frac{1}{12}li(3k_1+5k_2)$	$\frac{1}{12}li(5k_1+3k_2)$	$\frac{1}{12}li(k_1+3k_2)$	$\frac{1}{12}li(3k_1+k_2)$
d	$\mid\!\alpha l\!\mid\!\!-\beta l\!\mid\ k$	$\frac{1}{2}lik$	$\frac{1}{6}l(1+\alpha)ik$	$\frac{1}{6}l(1+\beta)ik$	$\frac{1}{6}lk[(1+\beta)i_1\ +(1+\alpha)i_2]$	$\frac{1}{3}l(1+\alpha\beta)ik$	$\frac{1}{12}l(5-\beta-\beta^2)ik$	$\frac{1}{12}l(5-\alpha-\alpha^2)ik$	$\frac{1}{12}l(1+\alpha+\alpha^2)ik$	$\frac{1}{12}l(1+\beta+\beta^2)ik$

$$M_t = \int_0^{d/2} \tau r \, dA = (G\varphi/l) \int_0^{d/2} r^2 \, dA = (G\varphi/l)I_p, \quad (44)$$

$$I_p = \int_0^{d/2} r^2 \, dA = \int_0^{d/2} r^2 2\pi r \, dr = \pi d^4/32. \quad (45)$$

I_p is the polar second area moment of the circular cross-section. For the torsion shear stresses and with the polar section modulus $W_p = I_p/d(2) = \pi d/16$ of the circular cross-section, Eqs (44) and (43) give

$$\tau(r) = (M_t/I_p)r \quad \text{and}$$

$$\tau_{max} = (M_t/I_p)(d/2) = M_t/W_p. \quad (46)$$

For the total angle of twist per unit length Eq. (44) gives

$$\varphi = \frac{M_t l}{GI_p} \quad \text{and} \quad \vartheta = \frac{\varphi}{l} = \frac{M_t}{GI_p}. \quad (47)$$

The strain energy is

$$W = \frac{1}{2}M_t\varphi = \frac{1}{2}\frac{M_t^2 l}{GI_p}. \quad (48)$$

If the torques $m_d(x)$ acting on the bar are continuously distributed then $M_t(x) = \int m_d(x)\,dx$

$$\vartheta(x) = \frac{d\varphi}{dx} = \frac{M_t(x)}{GI_p}, \quad \varphi(x) = \frac{1}{GI_p}\int M_t(x)\,dx,$$

$$W = \frac{1}{2}\int M_t(x)\,d\varphi = \frac{1}{2GI_p}\int M_t^2(x)\,dx.$$

The equations also apply for circular hollow cross-sections, where $I_p = \pi(d_a^4 - d_i^4)/32$ and $W_p = I_p/(d_a/2)$ (see **Table 7**).

Example *For the shaft shown in **Fig. 41a** where $G = 81\,kN/mm^2$, $\tau_{perm} = 12\,N/mm^2$ and speed $n = 1000$ r.p.m. find (a) the initial and final torques, (b) the curve for the moments of torsion, (c) the diameters required for each section, (d) the twist and angle of rotation for each section and the total angle of rotation.*
(a) The initial torque M_{d1} with the transferred power, $P_1 = 4.4$ kW is produced from $P = M_{d}\omega$ where $\omega = 2\pi n = 2\pi \cdot 16.67\ 1/s = 104.7\ 1/s$, as $M_{d1} = P_1/\omega = (4400\,N\,m\,s)/(104.7\ 1/s) = 42.0\,N\,m$. The torques produced are $M_{d2} = (1470\,W)/(104.7\ 1/s) = 14.0\,N\,m$ and $M_{d3} = (2930\,W)/(104.7\ 1/s) = 28.0\,N\,m$.
(b) Therefore the torsion moments will be $M_{t1.2} = M_{d1} = 42.0\,N\,m$ and $M_{t2.3} = M_{Md1} - M_{d2} = M_{d3} = 28.0\,N\,m$ (**Fig. 41b**).
(c) The diameters are given by $W_{p.req} = \pi d^3/16 = M_t/\tau_{perm}$ as $d_1 = \sqrt[3]{16M_{t1.2}/(\pi\tau_{perm})} = 26.1$ mm (having selected 27 mm) and $d_2 = 22.8$ mm (having selected 23 mm).
(d) Twist $\vartheta_{1.2} = M_{t1.2}/(GI_{p1}) = M_{t1.2}/(G\pi d_1^4/32) = 0.99 \cdot 10^{-5}$ 1/mm, angle of torsion $\varphi_{1.2} = \vartheta_{1.2}l_{1.2} = 0.004\,95 \approx 0.284°$, accordingly, $\vartheta_{2.3} = 1.26 \cdot 10^{-5}$ 1/mm, $\varphi_{2.3} = 1.26 \cdot 10^{-5} \cdot 250 = 0.003\,15 \approx 180°$. The

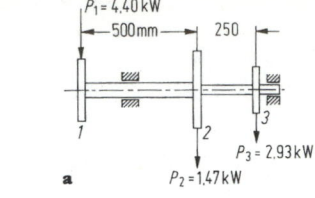

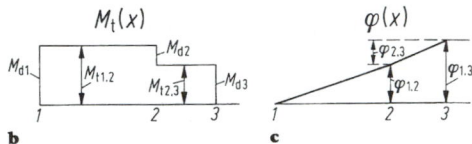

a

$M_t(x)$ $\varphi(x)$

b   **c**

Figure 41. Torsion of a shaft.

total angle of rotation (**Fig. 41c**) is then $\varphi_{1.3} = \varphi_{1.2} + \varphi_{2.3} = 0.284° + 0.180° = 0.464°$.

2.5.2 Bars of Circular Cross-section and Variable Diameter

If $I_p(x) = \pi d^4(x)/32$, the following applies as an approximation for total angle of twist per unit length

$$\vartheta(x) = \frac{M_t(x)}{GI_p(x)}, \quad \varphi(x) = \int\frac{M_t(x)}{GI_p(x)}dx.$$

The stresses are again calculated from $\tau(r) = M_t/I_p$ or $\tau_{max} = M_t/W_p$. In offset shafts, peak stresses (stress concentrations) occur and these are taken into account with the form factor α_k in accordance with $\tau = \alpha_k M_t/W_p$ (cf. B2.1.4).

2.5.3 Thin-walled Tubes (Bredt–Batho Theory)

Assuming that shear stress τ over wall thickness t is constant, equilibrium on the element in the x-direction gives

$$-\tau t\,dx + \tau t\,dx + \frac{\partial}{\partial s}(\tau t\,dx)\,ds = 0, \quad \text{thus} \quad \tau t = T = \text{const., i.e.}$$

the shear flow is constant along the circumference (**Fig. 40b**). The relationship between shear stress and torsion moment is given by $M_t = \oint \tau t h\,ds = \tau t\oint h\,ds = \tau t \cdot 2A_m$ and produces $\tau = M_t/(2A_m t)$ (Bredt's first equation), where A_m is the area of the tube enclosed by the centreline.

The following applies for the angle of torsion:

$$\varphi = \frac{M_t l}{GI_t} \quad \text{with} \quad I_t = \frac{4A_m^2}{\oint\dfrac{ds}{t(s)}}, \quad \text{where}$$

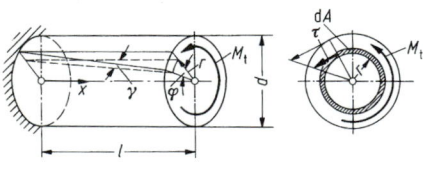

a

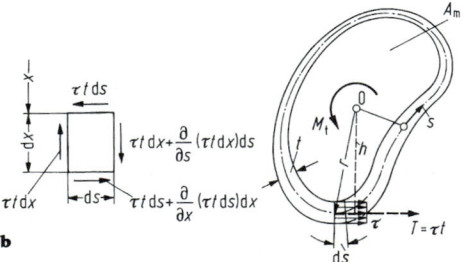

b

Figure 40. Torsion of a bar with **a** circular cross-section, **b** thin-walled hollow cross-section.

Table 7. Torsional moments of area I_t and section moduli W_t

	Cross-section	I_t	W_t	Notes
1		$\dfrac{\pi d^4}{32} = I_p$	$\dfrac{\pi d^3}{16} = W_p$	τ_{max} on the circumference
2		$\dfrac{\pi (d_a^4 - d_i^4)}{32} = I_p$ For thin walls, i.e. $\pi d_m^3 t/4$	$\dfrac{\pi (d_a^4 - d_i^4)}{16 \, d_a} = W_p$ $\left(\dfrac{t}{d_m}\right)^2 \ll 1$: $\pi d_m^2 t/2$	τ_{max} on the circumference
3		$\dfrac{\pi d^4}{32} = I_p$	$\dfrac{W_p}{\lambda} = \dfrac{\pi d^3}{16\lambda}$ $\lambda = \dfrac{2-\xi}{1-2\xi^2+(16/3\pi)\xi^3}$ For law ξ: $\lambda \approx 2$	τ_{max} at the base of the notch (at P) $\xi = \dfrac{\varrho}{d/2}$
4		$\dfrac{\pi a^3 b^3}{a^2+b^2} = \dfrac{\pi n^3 b^4}{n^2+1}$	$\dfrac{\pi ab^2}{2} = \dfrac{\pi n b^3}{2}$	Prerequisite : $a/b = n \geqslant 1$ τ_{max} At P_1 in P_2: $\tau_2 = \tau_{max}/n$
5		$\dfrac{\pi n^3 (b_1^4 - b_2^4)}{n^2+1}$	$\dfrac{\pi n (b_1^4 - b_2^4)}{2 b_1}$	Prerequisite: $a_1/b_1 = a_2/b_2 = n \geqslant 1$ τ_{max} At P_1 in P_2: $\tau_2 = \tau_{max}/n$
6		$\dfrac{b^4}{46.19} \approx \dfrac{h^4}{26}$	$\dfrac{b^3}{20} \approx \dfrac{h^3}{13}$	τ_{max} at the mid-point of the side (P_1) at the corner (P_2): $\tau_2 = 0$
7		$0.133 b^2 A = 0.115 b^4$	$0.217 bA = 0.188 b^3$	τ_{max} at the mid-point of the side (P)
8		$0.130 b^2 A = 0.108 b^4$	$0.223 bA = 0.185 b^3$	τ_{max} at the mid-point of the side (P)
9		$0.141 b^4$	$0.208 b^3$	τ_{max} at the mid-point of the side (P_1) at the corner (P_2): $\tau_2 = 0$

Table 7. Continued

Cross-section	I_t	W_t	Notes

Row 10

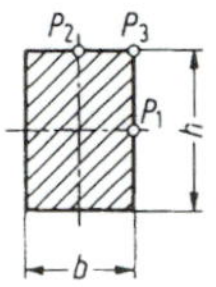

I_t: $c_1 h b^3 = c_1 n b^4$
W_t: $c_2 h b^2 = c_2 n b^3$

Notes: Prerequisite: $h/b = n \geq 1$
τ_{max} at P_1
At P_2: $\tau_2 = c_3 \tau_{max}$ At P_3: $\tau_3 = 0$

$n = h/b$	1	1,5	2	3	4	6	8	10	∞
c_1	0.141	0.196	0.229	0.263	0.281	0.298	0.307	0.312	0.333
c_2	0.208	0.231	0.246	0.267	0.282	0.299	0.307	0.312	0.333
c_3	1.000	0.858	0.796	0.753	0.745	0.743	0.743	0.743	0.743

Row 11 — Thin-walled profiles

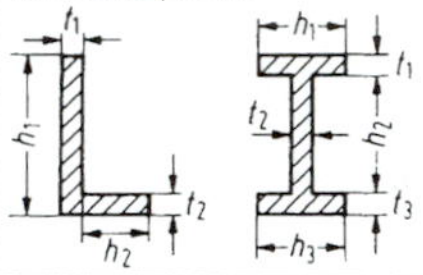

I_t: $\dfrac{\eta}{3} \Sigma h_i t_i^3$
W_t: I_t / t_{max}

Notes: Prerequisite: $h_i / t_i \gg 1$
τ_{max} in the centre of the long side of the rectangle
mit t_{max}

Profile	L	C	⊥	I	I PB	+
η	0.99	1.12	1.12	1.31	1.29	1.17

Row 12 — Thin-walled hollow cross-sections

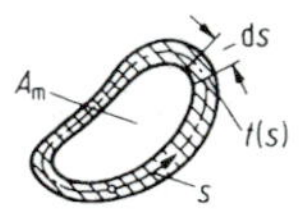

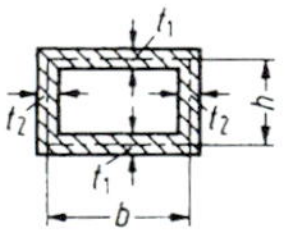

I_t: $\dfrac{4 A_m^2}{\oint ds / t(s)}$
W_t: $2 A_m t_{min}$

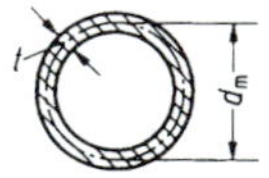

For constant wall thickness:
I_t: $4 A_m^2 t / U$
W_t: $2 A_m t$

Notes: A_m = Area enclosed by the centre line
U = Circumference of the centre line
τ_{max} at a point where
$\tau(s) \cdot t(s) = M_t / 2 A_m = \text{const.}$ applies

Row 12a

I_t: $\dfrac{4(bh)^2}{2(b/t_1 + h/t_2)}$
W_t: $2 b h t_{min}$

Notes: τ_{max}, where $t = t_{min}$

Row 12b

I_t: $\pi d_m^3 t / 4$
W_t: $\pi d_m^2 t / 2$

I_t is the torsional moment of area (Bredt's second equation). When subjected to torsion the cross-section does not remain plane; bulging occurs in the x-direction (longitudinal direction). The Bredt's formulae apply only to unrestrained warping where the axis of rotation corresponds to the shear centre (see B2.4.6). If warp is constrained, additional axial stresses σ occur, causing altered shear stresses and angles of rotation (cf. B2.5.5).

2.5.4 Bars of Arbitrary Cross-section

In principle, the cross-section warps in this case when the bar is subject to torsion. If warping is not constrained, Saint-Venant's theory applies [4]. The solution to the problem can be traced back to a warp function $\psi(y, z)$ or a stress function $\Psi(y, z)$, where $\psi(y, z)$ must satisfy the potential equation $\Delta \psi = 0$ and $\Psi(y,z)$ must satisfy Poisson's equation $\Delta \Psi = 1$. Exact solutions can be obtained only for a few cross-sections (e.g. ellipse, triangle, rectangle). The following applies for angle of torsion and maximum shear stress:

$$\varphi = \frac{M_t l}{G I_t}, \quad \tau_{max} = \frac{M_t}{W_t}, \tag{49}$$

where I_t is the torsional moment of area. Thus

$$I_t = \int \left(y^2 + z^2 = y \frac{\partial \psi}{\partial z} - z \frac{\partial \psi}{\partial y} \right) dA = -4 \int \Psi(y, z) \, dA,$$

i.e. I_t is proportional to the volume of the stress warp over the cross-section. W_t is the torsional section modulus. The following applies:

$$W_t = I_t \left/ \left[2 \left(\frac{\partial \Psi}{\partial n} \right)_{max} \right] \right.,$$

where $(\partial \Psi / \partial n)_{max}$ is the largest gradient of the stress warp. The corresponding shear stress is then vertical to the plane that cuts through the stress warp (**Fig. 42a**). See **Table 7** for results for I_t and W_t.

Estimation of the position of the maximum shear stresses and experimental determination of the shear stresses allow the following analogies:

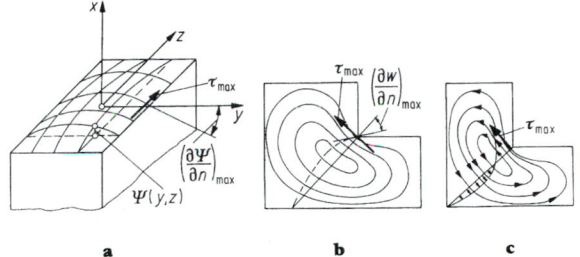

Figure 42. Arbitrary cross-section: **a** torsion function; **b** soap film analogy; **c** flow analogy.

Prandtl's Soap Film Analogy. Since the differential equations are equivalent for the stress function and for a soap film subject to overpressure, and the boundary conditions with $\Psi = 0$ or $w = 0$ also correspond, the slope is equivalent to a film of soap stretched over a cross-section and the density of the contour lines is equivalent to the value of the shear stresses the assigned direction of which is perpendicular to the slope (**Fig. 42b**).

Flow Analogy. Owing to the analogies of the different equations, the course of the flow lines for a potential flow of constant circulation, in a vessel of the same cross-section as that of the twisted bar, corresponds to the direction of the resulting shear stress. Here the density of the flow lines is therefore a measure for the value of the shear stresses (**Fig. 42c**).

2.5.5 Torsion with Warping Constraints

If, in the case of the bars discussed in B2.5.3 and B2.5.4, warping in any cross-section (e.g. due to fastening) is constrained, normal stresses σ_x occur in the longitudinal direction causing associated additional shear stresses τ_{xy} and τ_{xz}. The angle of rotation is smaller than for torsion where warping is not constrained. The problem can be solved for thin-walled open or singly and multiply closed cross-sections [5]. Note that, for example, the cross-sections shown in **Fig. 43**, i.e. all polygons of constant wall thickness formed by tangents from a circle and all star-shaped cross-sections, do not warp, and so remain plane, so that no torsion with warping constraint occurs. For solid cross-sections, approximate solutions can be found only in a few cases [4]; however, the effect of the warp constraint is negligible in most cases.

2.6 Combined Stresses

2.6.1 Bending and Axial Load

Figure 44a shows an angled beam the vertical section of which is subjected to axial (normal) loads and bending moments, as shown by the course of the sectional loads in **Fig. 44b–d**. For bending around a principal axis of the cross-section for normal stress or for the extreme stresses in the edge fibres, the following applies (**Fig. 44a**):

$$\sigma = \sigma_N + \sigma_M = F_N/A - M_b z/I_y \quad \text{or}$$

$$\sigma_{1,2} = F_N/A \mp M_b/W_{y1,2}. \tag{50}$$

The position of the zero line follows from this equation, where $\sigma = 0$, as $z_0 = F_N I_y/(M_b A)$.

In the case of oblique bending, i.e. loading in both planes of the principal axes, the following applies from Eq. (22) for stress and zero line:

$$\left. \begin{aligned} \sigma &= \frac{F_N}{A} - \frac{M_{by}}{I_y}z + \frac{M_{bz}}{I_z}y \\[2mm] y &= \frac{M_{by}I_z}{M_{bz}I_y}z - \frac{F_N I_z}{M_{bz}A}. \end{aligned} \right\} \tag{51}$$

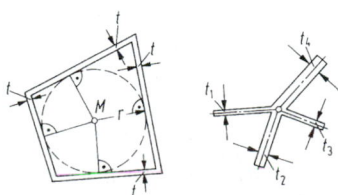

Figure 43. Cross-sections without warp.

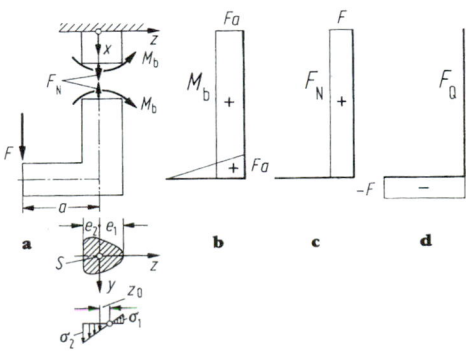

Figure 44a–d. Bending and axial load.

The extreme stresses occur at the furthest points perpendicular to the zero line with coordinates (y_1, z_1) and (y_2, z_2). The simplest method of determination is by graph and calculation.

Core of a Cross-section. If the stresses in a cross-section are to have the same sign, i.e. zero at the edge in the limit case, then force F (**Fig. 44a**), in the case of simple bending with axial load and $M_b = Fa$ in accordance with Eq. (50), must be applied at distance $a_{1,2} \leq I_y/(A \cdot e_{1,2}) = W_y/A$. For oblique bending with axial load, it must lie within the core (**Fig. 45**). For a description of the core see [6].

2.6.2 Bending and Shear

Bending and shear generally occur simultaneously in most cross-sections of beams, shafts, axles, etc. (plane state of stress). Since the axial bending stresses σ are extreme at the edge although the shear stresses τ are zero (**Fig. 46a**), the equivalent stress σ_V has to be determined at various heights using one of the formulae given in B1.3. σ and τ are obtained from Eqs (8) and (26). For example, for an I-section, calculate σ_V at the upper edge, at the transition between flange and web and in the centre: according to the maximum shear stress energy criterion (cf. B1.3.2), then $\sigma_V = \sigma_{edge}$ or $\sigma_V = \sqrt{\sigma_0^2 + 3\tau_0^2}$ or $\sigma_V = 1.73\tau_{centre}$ and it must be max $\sigma_V \leq \sigma_{perm}$. However, accurate determination of σ_V is unnecessary in most cases and axial and shear stresses are determined separately and compared with σ_{perm} and τ_{perm}. For long beams ($l \geq 4h$ to $5h$) only the normal stresses are determining factors, while for short beams ($l \leq h$) it is only the shear stresses.

2.6.3 Bending and Torsion

If axial bending stresses σ and torsional shear stresses τ are applied at the same time (**Fig. 46b**), there is a plane state of stress. The extreme values of σ and τ occur in the edge fibres. They are calculated using Eqs (9) and (46) or (49). The equivalent stress σ_V is determined by applying one of the criteria as specified in B1.3.

Example *The shaft shown in* **Fig. 41a** *or a corresponding example has to transmit a maximum bending moment* $M_b = 75$ mm *in area 1 to 2. Calculate* σ_V. *Taking* $\sigma = M_b/W_y$ *and* $\tau = M_t/W_p$ *as well as* $W_y = \pi d^3/32$ *and* $W_p = 2W_y = \pi d^3$, *using the maximum shear strain energy criterion, it follows from B1 Eq. (24) that for* σ_V

$$\sigma_V = \sqrt{M_b^2 + 0.75\alpha_0^2 M_t^2}/W_y = M_V/W_y. \qquad (52)$$

If the load is alternating for bending and dynamic for torsion, $\alpha_0 \approx 0.85$. For $d = 27$ mm, $W_y = \pi d^3/32 = 1932$ mm³ and $\sigma_V = \sqrt{75\,000^2 + 0.75 \cdot 0.85^2 \cdot 42\,000^2}$ N mm/1932 mm³ = 42 N/mm².

2.6.4 Axial Load and Torsion

This load, which occurs e.g. on expansion screws and spindles owing to σ and τ, corresponds to a plane state

$$d_K = \frac{d_a}{4}\left[1 + \left(\frac{d_i}{d_a}\right)^2\right]$$

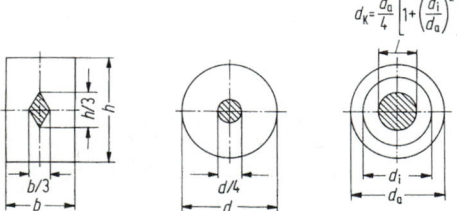

Figure 45. Core of the cross-section.

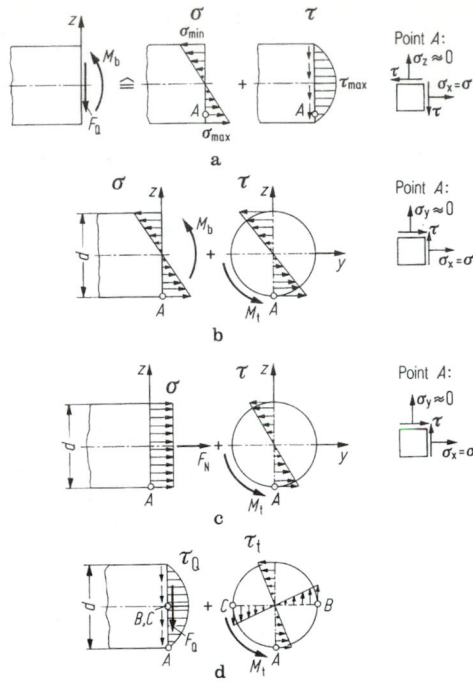

Figure 46. Combined stresses: **a** bending and shear; **b** bending and torsion; **c** axial force and torsion; **d** shear and torsion.

of stress (**Fig. 46c**). The extreme stresses occur in the fibres at the edge and this is where the equivalent stress σ_V is calculated using one of the criteria specified in B1.3.

2.6.5 Shear and Torsion

This stress, which occurs e.g. on short journals (**Fig. 46d**), yields a resultant maximum shear stress only where τ_Q complies with Eq. (26) and τ_t with Eq. (46) or (49);

at point A $\qquad \tau_{res} = \tau_t$,

at point $\qquad$ B $\tau_{res} = \tau_Q - \tau_t$,

at point $\qquad$ C $\tau_{res} = \tau_Q + \tau_t$.

Conversion to σ_V, e.g. in accordance with the maximum shear strain energy criterion, gives $\sigma_V = 1.73 \cdot \alpha_0\tau_{res}$.

2.6.6 Combined Bending, Axial Load, Shear and Torsion

In this case, for points A, B and C as shown in **Fig. 46d**, $\sigma_A = \sigma_N + \sigma_M$, $\tau_A = \tau_t$; $\sigma_B = \sigma_N$, $\tau_B = \tau_Q - \tau_t$; $\sigma_C = \sigma_N$, $\tau_C = \tau_Q + \tau_t$. Here σ_A, τ_A, etc. create a plane state of stress in each case and are to be combined in the same way as equivalent stress σ_V in B1.3.

2.7 Statically Indeterminate Systems

A distinction is made between externally and internally statically indeterminate systems where a system may be externally and internally statically indeterminate at the same time. Systems that are supported by more than three supporting reactions in a plane or more than six supporting reactions in space are externally statically indetermi-

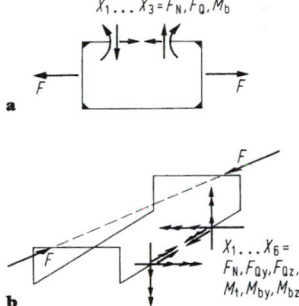

$X_1 \ldots X_3 = F_N, F_Q, M_b$

a

$X_1 \ldots X_6 =$
$F_N, F_{Qy}, F_{Qz},$
M_t, M_{by}, M_{bz}

b

Figure 47. Closed frame: **a** plane; **b** spatial.

nate. An *n*-times supported system is $m = (n - 3)$ times externally statically determinate in a plane and $m = (n - 6)$ times externally statically indeterminate in space. A closed frame is 3 times internally statically determinate as a plane system (**Fig. 47a**) and 6 times internally statically indeterminate as a spatial system (**Fig. 47b**).

The principal method for calculating statically indeterminate systems is the *force method*. The system can be traced back to a basic statically indeterminate system by removing supporting reactions (forces or moments) or taking sections e.g. as shown in **Fig. 48** (for every indeterminate system there are several possible basic systems from which one should be selected). The quantities that are taken out are known as static indeterminates X_1, X_2 ... X_m. The solution is based on the following superposition method:

1. Calculation of the deformation differentials δ_{10}, δ_{20}, δ_{30} ... between both edges of the section in the basic system in the direction of X_1, X_2, X_3 ... by means of the external load (0). (The deformations will be positive in the direction of the statically indeterminate quantities.)
2. Calculation of the deformation differentials δ_{ik} (*i*, *k* = 1, 2, 3 ...) in the basic system, where *i* denotes the direction of X_1, X_2, X_3 ... and *k* = 1, 2, 3 ... denotes the loading $X_1 = 1$, $X_2 = 1$, $X_3 = 1$ ... .
3. The deformation differentials in the actual system must be zero, i.e. for three unknowns for example

$$X_1\delta_{11} + X_2\delta_{12} + X_3\delta_{13} + \delta_{10} = 0 ,$$
$$X_1\delta_{21} + X_2\delta_{22} + X_3\delta_{23} + \delta_{20} = 0 , \qquad (53)$$
$$X_1\delta_{31} + X_2\delta_{32} + X_3\delta_{33} + \delta_{30} = 0 .$$

The three unknowns X_1, X_2, X_3 (the indeterminates X_1, ..., X_m for *m*-times indeterminate systems) are calculated from these linear systems of equations.

4. After superimposing the external loads and static indeterminates in the basic system the final supporting

reactions, bending moments etc. are calculated. It is also to be noted that $\delta_{ik} = \delta_{ki}$ always applies if $i \neq k$ (Maxwell's theory) whereby the number of δ_{ik} to be calculated is considerably reduced.

Deformation characteristics are calculated using one of the methods specified in B2.4.8 and B2.4.9. In simple, clear-cut cases the results shown in **Table 5a** are used and in complex unclear cases the methods specified in B2.4.9. The advantage of the latter is that it also automatically gives the correct sign for the δ_{ik} elements.

Example *Calculation of both static indeterminates on a constrained cantilever (**Fig. 49a**).* A cantilever (**Fig. 49b**) is selected for the basic statically determinate system. Deformation characteristics δ_{ik} should be calculated by means of two methods, i.e. by consulting **Table 5a** and generally by applying the principle of virtual work as specified in B2.4.9. In accordance with **Table 5a** (**Fig. 49c-e**),

$$\delta_{10} = f_{10} = -ql^4/(8EI_y), \quad \delta_{20} = \alpha_{20} = -ql^3/(6EI_y) ,$$
$$\delta_{11} = f_{11} = l^3/(3EI_y), \quad \delta_{21} = \alpha_{21} = l^2/(2EI_y) = \delta_{12} ,$$
$$\delta_{22} = \alpha_{22} = l/(EI_y) .$$

The principle of virtual forces as shown in Eqs (41) and (42) and **Table 6** gives

$$\delta_{10} = \int M_1 M_0 \, dx/(EI_y) = l ik/(4EI_y) = -ql^4/(8EI_y) ,$$
$$\delta_{20} = \int M_2 M_0 \, dx/(EI_y) = l ik/(3EI_y) = -ql^3/(6EI_y) ,$$
$$\delta_{11} = \int M_1 M_1 \, dx/(EI_y) = l ik/(3EI_y) = l^3/(3EI_y) ,$$
$$\delta_{21} = \delta_{12} = \int M_1 M_2 \, dx/(EI_y) = l ik/(2EI_y) = l^2/(2EI_y) ,$$
$$\delta_{22} = \int M_2 M_2 dx/(EI_y) = l ik/(EI_y) = l/(EI_y) .$$

Both methods thus produce the same deformations. From the two linear equations shown in Eqs (53),

$$X_1 = (-\delta_{10}\delta_{22} + \delta_{20}\delta_{12})/(\delta_{11}\delta_{22} - \delta_{12}^2) = ql/2 ,$$
$$X_2 = (-\delta_{11}\delta_{20} + \delta_{21}\delta_{10})/(\delta_{11}\delta_{22} - \delta_{12}^2) = -ql^2/12 .$$

Then as a result of external load and X_1 and X_2 the final supporting reactions in the basic system are calculated as

$$F_A = ql - X_1 = ql/2 = F_B,$$
$$M_{EA} = -ql^2/2 + X_1 l + X_2 = -ql^2/12 = M_{EB}$$

and the maximum field moment as

$$M_F = M_b(l/2) = ql^2/24$$

The results for simple statically indeterminate beams are given in **Table 5b**. The results shown below apply to several significant statically indeterminate systems:

Continuous Beams (Fig. 50). With the support moments as static indeterminates X_i, the equation

$$X_{i-1}\beta_{i,i-1} + X_i(\alpha_{i,i-1} + \alpha_{i,i+1}) + X_{i+1}\beta_{i,i+1} + (\alpha_{i0,1} + \alpha_{i0,r}) = 0 \qquad (54)$$

applies for point *i*, since the angular differential on the actual beam must be zero and because $X_i = 1$ at the point of application of the moment, the angular differential $\delta_{ii} = \alpha_{i,i-1} + \alpha_{i,i+1}$ and because $X_{i-1} = 1$ at point *i* angle $\beta_{i,i-1}$ occurs etc.

For each internal support of the *m*-times indeterminate

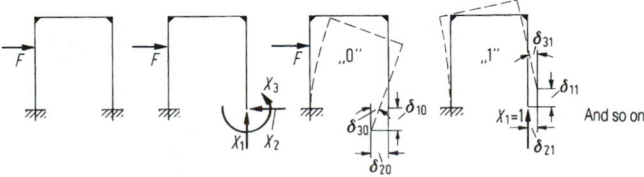

Figure 48. Force quantity method.

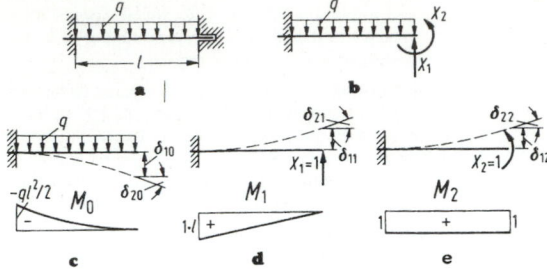

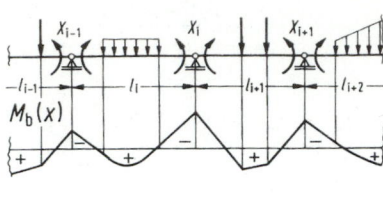

Figure 49a–e. Beams constrained at both ends.   **Figure 50.** Continuous beams.

beam a three-moment equation is produced. m indeterminates $X_1 \ldots X_m$ can then be calculated from the linear system of equations. Angles α and β are positive in the direction of X_i and can be taken from **Table 5a** or calculated using one of the methods specified in B2.4.8 and B2.4.9.

Closed Rectangular Frame. (a) For support moments M_s and field moments M_f, where $k = l_1 l_2/(l_2 l_1)$:

$$M_s = -\frac{Fl_2}{8(k+1)}, \quad M_f = \frac{Fl_2}{4} + M_s = \frac{Fl_2(2k+1)}{8(k+1)}$$

applies for single loads (**Fig. 51a**).
(b) For line loads (**Fig. 51b**)

$$M_s = -\frac{q_1 l_1^2 k + q_2 l_2^2}{12(k+1)}, \quad M_{f1} = \frac{q_1 l_1^2}{8} + M_s,$$

$$M_{f2} = \frac{q_2 l_2^2}{8} + M_s.$$

Double-jointed Frame (Fig. 52a). Where $k = bl_2/(l l_1)$

$$F_{Az} = F_{Bz} = F/2, \quad F_{Ax} = F_{Bx} = \frac{3Fl}{8b(2k+3)};$$

$$M_s = -\frac{3Fl}{8(2k+3)}, \quad M_f = \frac{Fl(4k+3)}{8(2k+3)}.$$

Constrained Frame (Fig. 52b). Where $k = bl_2/(l l_1)$

$$F_{Az} = F_{Bz} = F/2, \quad F_{Ax} = F_{Bx} = \frac{3Fl}{8b(k+2)};$$

$$M_A = M_B = \frac{Fl}{8(k+2)}, \quad M_s = \frac{Fl}{4(k+2)},$$

$$M_f = \frac{Fl(k+1)}{4(k+2)}.$$

Double-jointed Frame with Supports at Different Heights (Fig. 52c). Where $k = bl_2/(l l_1)$,

$$F_{Az} = \frac{F(8k+11)}{16(k+1)}, \quad F_{Bz} = \frac{F(8k+5)}{16(k+1)},$$

$$F_{Ax} = F_{Bx} = \frac{3Fl}{16b(k+1)};$$

$$M_x = -\frac{3Fl}{16(k+1)}, \quad M_f = \frac{Fl(8k+5)}{32(k+1)}.$$

Projecting Frame (Fig. 53). Where $j = aEI_1/(lGI_t)$,

$$F_{Az} = F_{Bz} = F/2, \quad M_{Ab} = M_{Bb} = Fa/2,$$

$$M_{At} = M_{Bt} = \frac{Fl}{8(1+2j)};$$

$$M_s = -\frac{Fl}{8(1+2j)}, \quad M_f = \frac{Fl(1+4j)}{8(1+2j)}.$$

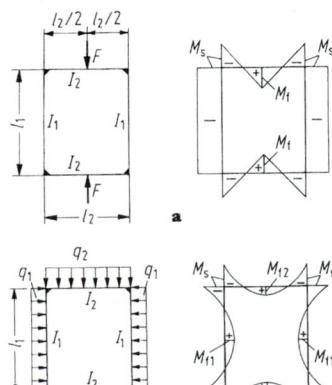

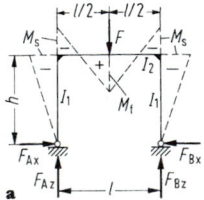

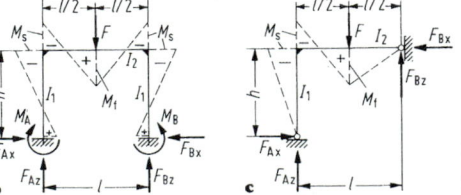

Figure 51. Closed rectangular frames under **a** single loads, **b** line loads.

Figure 52a–c. Frame under single load.

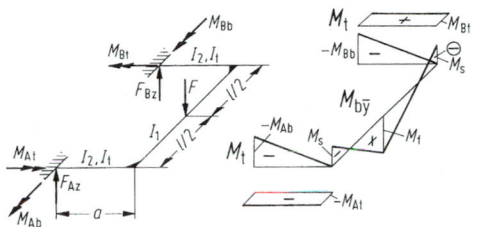

Figure 53. Frame beam.

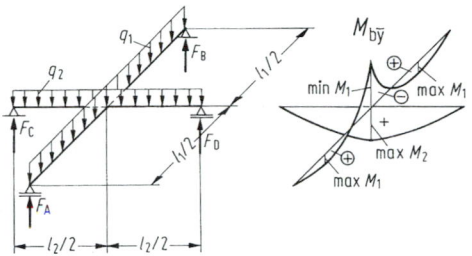

Figure 54. Grid.

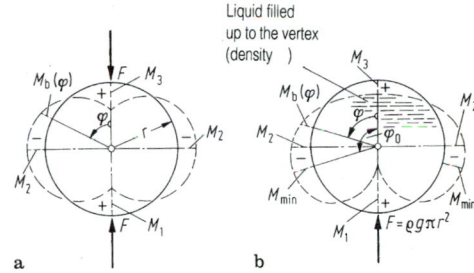

Figure 55. Annular beams **a** under single loads, **b** filled with liquid.

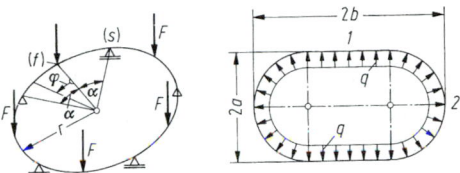

Figure 56. Annular beam.   **Figure 57.** Frame with semi-circular curves.

Grid (Fig. 54). Where $k = l_1^3 I_2/(l_2^3 I_1)$, for the support load X_1 between both beams (with positive downward direction for beam 2 and upward for beam 1) and for the final support loads and bending moments,

$$X_1 = \frac{5(q_1 l_1 k - q_2 l_2)}{8(k+1)}$$

$$F_A = F_B = \frac{q_1 l_1 - X_1}{2}, \quad F_C = F_D = \frac{q_2 l_2 + X_1}{2},$$

$$\min M_1 = -X_1 l_1/4 + q_1 l_1^2/8,$$

$$\max M_1 = F_A^2/(2q_1), \quad \max M_2 = X_1 l_2/4 + q_2 l_2^2/8.$$

Annular Frames. (a) Under single loads (**Fig. 55a**):

$$M_b(\varphi) = Fr\left(\frac{1}{\pi} - \frac{1}{2}\sin \varphi\right) \quad (0° \le \varphi \le 180°),$$

$$M_1 = M_3 = 0.318 \cdot Fr, \quad M_2 = -0.182 \cdot Fr.$$

(b) Filled with liquid up to the vertex (**Fig. 55b**), where $F = \rho g \pi r^2$:

$$M_b(\varphi) = (2 - \cos \varphi - 2\varphi \sin \varphi)Fr/(4\pi)$$

$$(0° \le \varphi \le 180°),$$

$$M_1 = 0.75 \cdot Fr/\pi,$$

$$M_2 = -0.285 \cdot Fr/\pi,$$

$$M_3 = 0.25 \cdot Fr/\pi,$$

$$M_{min} = -0.321 \cdot Fr/\pi$$

with $\varphi_0 = 105.2°$.

Annular Beams. (a) With n supports equally spaced:

$$F_A = F, \quad M_f = -M_s = (Fr/2) \cdot \tan(\alpha/2),$$

$$M_{tmax} = M_t(\varphi = \alpha/2) = (Fr/2)\left[\frac{1}{\cos(\alpha/2)} - 1\right];$$

for support loads, bending moments (field and support) and torsional moments where $\alpha = \pi/n$ for single loads (**Fig. 56**).
(b) For constant line load q along the circumference:

$$F_A = q \cdot 2\pi r/n, \quad M_f = qr^2(\alpha/\sin \alpha - 1),$$

$$M_s = -qr^2(1 - \alpha/\tan \alpha), \quad M_t(\varphi) = qr^2\left(\varphi - \frac{\alpha}{\sin \alpha}\sin \varphi\right),$$

extreme value for

$$\varphi_0 = \arccos\left(\frac{\sin \alpha}{\alpha}\right).$$

Frame with Semi-circular Curves (Fig. 57). For constant internal pressure:

$$M_1 = \frac{qa^2}{2} - \frac{qa^3}{2b + (\pi - 2)a}\left[\frac{1}{3}\left(\frac{b}{a} - 1\right)^3 + \frac{\pi}{2}\left(\frac{b}{a} - 1\right)^2\right.$$

$$\left. + 3\left(\frac{b}{a} - 1\right) + \frac{\pi}{2}\right],$$

$$M_2 = \frac{qb^2}{2} - \frac{qa^3}{2b + (\pi - 2)a}\left[\frac{1}{3}\left(\frac{b}{a} - 1\right)^3 + \frac{\pi}{2}\left(\frac{b}{a} - 1\right)^2\right.$$

$$\left. + 3\left(\frac{b}{a} - 1\right) + \frac{\pi}{2}\right];$$

e.g. for $a/b = 0.5$: $M_1 = -0.76 \cdot qa^2$, $M_2 = +0.74 \cdot qa^2$.

3 Theory of Elasticity

3.1 General

The task of the theory of elasticity is to determine the state of stress and deformation of a body taking into account the given boundary conditions, i.e. to calculate the values $\sigma_x, \sigma_y, \sigma_z, \tau_{xy}, \tau_{xz}, \tau_{yz}, \varepsilon_x, \varepsilon_y, \varepsilon_z, \gamma_{xy}, \gamma_{xz}, \gamma_{yz}, u, v, w$. Initially there are equations B1 Eq. (12) and B1 Eq. (13) for these 15 unknowns. In addition there are three equilibrium conditions (**Fig. 1**) with the volume forces X, Y, Z.

$$\left.\begin{array}{l} \dfrac{\partial \sigma_x}{\partial x} + \dfrac{\partial \tau_{yx}}{\partial y} + \dfrac{\tau_{zx}}{\partial z} + X = 0 , \\[2mm] \dfrac{\partial \tau_{xy}}{\partial x} + \dfrac{\partial \sigma_y}{\partial y} + \dfrac{\partial \tau_{zy}}{\partial z} + Y = 0 , \\[2mm] \dfrac{\partial \tau_{xz}}{\partial x} + \dfrac{\partial \tau_{yz}}{\partial y} + \dfrac{\partial \sigma_z}{\partial z} + Z = 0 , \end{array}\right\} \quad (1)$$

and for isotropic bodies the six generalized Hooke's laws

$$\left.\begin{array}{l} \varepsilon_x = [\sigma_x - \nu(\sigma_y + \sigma_z)]/E , \;\; \varepsilon_y = [\sigma_y - \nu(\sigma_x + \sigma_z)]/E , \\[2mm] \varepsilon_z = [\sigma_z - \nu(\sigma_x + \sigma_y)]/E , \\[2mm] \gamma_{xy} = \tau_{xy}/G , \quad \gamma_{xz} = \tau_{xz}/G , \;\; \gamma_{yz} = \tau_{yz}/G . \end{array}\right\} \quad (2)$$

Thus there are 15 equations for 15 unknowns. If *all* stresses are eliminated from these, three partial differential equations are obtained for the unknown strains:

$$\left.\begin{array}{l} G\left(\Delta u + \dfrac{1}{1 - 2\nu}\dfrac{\partial \varepsilon}{\partial x}\right) + X = 0 , \\[3mm] G\left(\Delta v + \dfrac{1}{1 - 2\nu}\dfrac{\partial \varepsilon}{\partial y}\right) + Y = 0 , \\[3mm] G\left(\Delta w + \dfrac{1}{1 - 2\nu}\dfrac{\partial \varepsilon}{\partial z}\right) + Z = 0 , \end{array}\right\} \quad (3)$$

where $\Delta u = \partial^2 u/\partial x^2 + \partial^2 u/\partial y^2 + \partial^2 u/\partial z^2$ etc. and $\varepsilon = \varepsilon_x + \varepsilon_y + \varepsilon_z = \partial u/\partial x + \partial v/\partial y + \partial w/\partial z$.

Navier's equations (3) can be used to solve problems where strains are specified as boundary conditions. If *all* strains and their derivatives are eliminated from the 15 equations quoted, this leaves 6 equations for the unknown stresses:

$$\Delta \sigma_x + \frac{1}{1 + \nu}\frac{\partial^2 \sigma}{\partial x^2} + 2\frac{\partial X}{\partial x} + \frac{\nu}{1 - \nu}\left(\frac{\partial X}{\partial x} + \frac{\partial Y}{\partial y} + \frac{\partial Z}{\partial z}\right) = 0 \quad (4a)$$

(as appropriate for directions y and z) and

$$\Delta \tau_{xy} + \frac{1}{1 + \nu}\frac{\partial^2 \sigma}{\partial x \partial y} + \frac{\partial X}{\partial y} + \frac{\partial Y}{\partial x} = 0 \quad (4b)$$

(as appropriate for directions y and z).

In this case $\sigma = \sigma_x + \sigma_y + \sigma_z$. Beltram's equations (4) can be used to solve problems where stresses are specified as boundary conditions. In the case of combined boundary conditions both systems of equations shall be used. In the main there are solutions to differential equations (3) and (4) for axisymmetrical and plane problems.

3.2 Axisymmetric Stresses

If we assume that the z-axis is symmetrical, then only stresses $\sigma_r, \sigma_t, \sigma_z, \tau_{rz} = \tau_{zr} = \tau$ occur (**Fig. 2**). The equilibrium conditions in direction r and z are

$$\left.\begin{array}{l} \dfrac{\partial}{\partial r}(r\sigma_r) + \dfrac{\partial}{\partial z}(r\tau) - \sigma_t + rR = 0 , \\[3mm] \dfrac{\partial}{\partial r}(r\tau) + \dfrac{\partial}{\partial z}(r\sigma_z) + rZ = 0 . \end{array}\right\} \quad (5)$$

Hooke's laws take the form

$$\left.\begin{array}{l} \varepsilon_r = \partial u/\partial r = [\sigma_r - \nu(\sigma_t + \sigma_z)]/E , \\[2mm] \varepsilon_t = u/r = [\sigma_t - \nu(\sigma_r + \sigma_z)]/E , \\[2mm] \varepsilon_z = \partial w/\partial z = [\sigma_z - \nu(\sigma_r + \sigma_t)]/E , \\[2mm] \gamma_{rz} = \partial u/\partial z + \partial w/\partial r = \tau/G = 2(1 + \nu)\tau/E. \end{array}\right\} \quad (6)$$

If they are broken down in terms of stresses

$$\left.\begin{array}{l} \sigma_r = 2G\left(\dfrac{\partial u}{\partial r} + \dfrac{\nu}{1 - 2\nu}\varepsilon\right), \;\; \sigma_t = 2G\left(\dfrac{u}{r} + \dfrac{\nu}{1 - 2\nu}\varepsilon\right), \\[3mm] \sigma_z = 2G\left(\dfrac{\partial w}{\partial z} + \dfrac{\nu}{1 - 2\nu}\varepsilon\right), \;\; \tau = G\left(\dfrac{\partial u}{\partial z} + \dfrac{\partial w}{\partial r}\right), \end{array}\right\} \quad (7)$$

where

$$\varepsilon = \varepsilon_r + \varepsilon_t + \varepsilon_z = \frac{\partial u}{\partial r} + \frac{u}{r} + \frac{\partial w}{\partial z} . \quad (8)$$

If Love's stress function Φ is introduced, then it must satisfy the bipotential equation

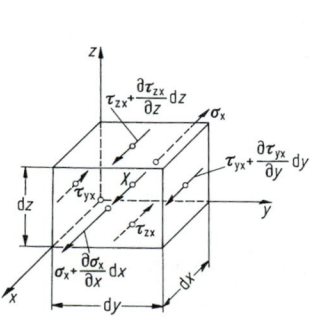

Figure 1. Element equilibrium.

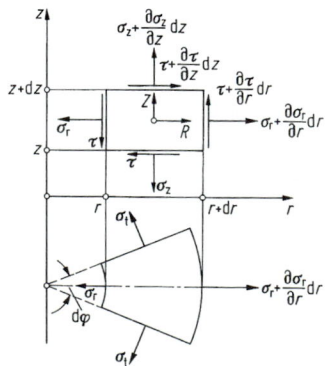

Figure 2. Axisymmetric state of stress.

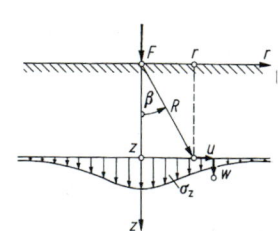

Figure 3. Single load on half-space.

$$\left(\frac{\partial^2}{\partial z^2} + \frac{\partial^2}{\partial r^2} + \frac{1}{r}\frac{\partial}{\partial r}\right)\left(\frac{\partial^2 \Phi}{\partial z^2} + \frac{\partial^2 \Phi}{\partial r^2} + \frac{1}{r}\frac{\partial \Phi}{\partial r}\right) = \Delta\Delta\Phi = 0 \ .(9)$$

Solutions to the bipotential equation include, for example, $\Phi = r^2$, $\ln r$, $r^2 \ln r$, z, z^2 and $\sqrt{r^2 + z^2}$ as well as linear combinations of these [1, 3]. The stresses and strains are then produced from

$$u = -\frac{1}{1-2\nu}\frac{\partial^2 \Phi}{\partial r \partial z},$$

$$w = \frac{2(1-\nu)}{1-2\nu}\Delta\Phi - \frac{1}{1-2\nu}\frac{\partial^2 \Phi}{\partial z^2},$$

$$\sigma_r = \frac{2G\nu}{1-2\nu}\frac{\partial}{\partial z}\left(\Delta\Phi - \frac{1}{\nu}\frac{\partial^2 \Phi}{\partial r^2}\right),$$

$$\sigma_z = \frac{2(2-\nu)G}{1-2\nu}\frac{\partial}{\partial z}\left(\Delta\Phi - \frac{1}{2-\nu}\frac{\partial^2 \Phi}{\partial z^2}\right), \qquad (10)$$

$$\sigma_t = \frac{2G\nu}{1-2\nu}\frac{\partial}{\partial z}\left(\Delta\Phi - \frac{1}{\nu}\frac{1}{r}\frac{\partial \Phi}{\partial r}\right),$$

$$\tau = \frac{2(1-\nu)G}{1-2\nu}\frac{\partial}{\partial r}\left(\Delta\Phi - \frac{1}{1-\nu}\frac{\partial^2 \Phi}{\partial z^2}\right).$$

Example *Single force on half-space (Boussinesq's formulae)* (**Fig. 3**). The boundary conditions are

$$\sigma_z(z = 0, r \neq 0) = 0, \ \tau(z = 0, r \neq 0) = 0 \ .$$

With the formulation $\Phi = C_1 R + C_2 z \ln(z + R)$, where $R = \sqrt{r^2 + z^2}$, it follows from Eq. (10) that

$$\sigma_z = -2G\left[\left(C_1 - \frac{2\nu}{1-2\nu}C_2\right)\frac{z}{R^3} + \frac{3}{1-2\nu}(C_1 + C_2)\frac{z^3}{R^5}\right] \ \text{and}$$

$$\tau = -2G\left[\left(C_1 - \frac{2\nu}{1-2\nu}C_2\right)\frac{r}{R^3} + \frac{3}{1-2\nu}(C_1 + C_2)\frac{rz^2}{R^5}\right].$$

While the first boundary condition is automatically satisfied,

$$C_2 = \frac{1-2\nu}{2\nu}C_1 \ \text{follows from the second and thus}$$

$$\sigma_z = -C_1\frac{3G}{\nu(1-2\nu)}\frac{z^3}{R^5}.$$

$C_1 = F\nu(1 - 2\nu)/(2\pi G)$ follows from

$$F = -\int_{r=0}^{\infty}\sigma_z 2\pi r \, dr$$

and thus from Eq. (10),

$$u = \frac{F}{4\pi G}\left[\frac{rz}{R^3} - (1 - 2\nu)\frac{r}{R(z+R)}\right],$$

$$w = \frac{F}{4\pi G}\left[2(1-\nu)\frac{1}{R} + \frac{z^2}{R^3}\right],$$

$$\sigma_z = -\frac{3F}{2\pi}\frac{z^3}{R^5}, \quad \sigma_r = \frac{F}{2\pi}\left[(1-2\nu)\frac{1}{R(z+R)} - 3\frac{zr^2}{R^5}\right], \qquad (11)$$

$$\sigma_t = \frac{F}{2\pi}(1-2\nu)\left[\frac{z}{R^3} - \frac{1}{R(z+R)}\right], \quad \tau = -\frac{3F}{2\pi}\frac{rz^2}{R^5}.$$

Since $\sigma_z/\tau = z/r$, σ_z and τ can be combined to form the stress vector, which is always in direction R. In accordance with $s_R = \sqrt{\sigma_z^2 + \tau^2} = 3Fz^2/(2\pi R^4)$, $\sigma_r = 0$ zero points are produced from $\sin^2 \beta \cos \beta (1 + \cos \beta) = (1 - 2\nu)/3$ for σ_r where $\nu = 0.3$ at $\beta_1 = 15.4°$ and $\beta_2 = 83°$. Between the circular cones determined between $2\beta_1 = 30.8°$ and $2\beta_2 = 166°$, σ_r is negative (compressive stress); outside, it is positive (tensile stress). $\sigma_t = 0$ gives $\cos^2 \beta + \cos \beta = 1$, i.e. $\beta = 52°$. For $\beta < 52°$, σ_t is positive (tensile stress); for $\beta > 52°$ it is negative (compressive stress).

3.3 Plane Stresses

Plane stresses occur where $\sigma_z = 0$, $Z = 0$, $\tau_{xz} = \tau_{yz} = 0$, i.e. if stresses occur only in the x, y plane. The equilibrium conditions for constant volume forces are

$$\frac{\partial\sigma_x}{\partial x} + \frac{\partial\tau_{yx}}{\partial y} + X_0 = 0, \quad \frac{\partial\sigma_y}{\partial y} + \frac{\partial\tau_{xy}}{\partial x} + Y_0 = 0 \ . \quad (12)$$

Hooke's laws take the form

$$\varepsilon_x = (\sigma_x - \nu\sigma_y)/E, \quad \varepsilon_y = (\sigma_y - \nu\sigma_x)/E,$$

$$\gamma_{xy} = \tau_{xy}/G, \qquad (13)$$

and for deformations the following apply:

$$\frac{\partial u}{\partial x} = \varepsilon_x, \quad \frac{\partial v}{\partial y} = \varepsilon_y, \quad \frac{\partial u}{\partial y} + \frac{\partial v}{\partial x} = \gamma_{xy} \ . \quad (14)$$

These are eight equations for eight unknowns. Equation (14) gives the compatibility condition

$$\frac{\partial^2 \varepsilon_x}{\partial y^2} + \frac{\partial^2 \varepsilon_y}{\partial x^2} = \frac{\partial^2 \gamma_{xy}}{\partial x \partial y}, \qquad (15)$$

and by inserting Eq. (15) in Eq. (13) the following is produced:

$$\frac{1}{E}\left(\frac{\partial^2\sigma_x}{\partial y^2} - \nu\frac{\partial^2\sigma_y}{\partial y^2} + \frac{\partial^2\sigma_y}{\partial x^2} - \nu\frac{\partial^2\sigma_x}{\partial x^2}\right) = \frac{1}{G}\frac{\partial^2\tau_{xy}}{\partial x \partial y}. \quad (16)$$

If the equilibrium conditions (12) are now satisfied by introducing Airy's stress function $F = F(x, y)$ such that

$$\sigma_x = \frac{\partial^2 F}{\partial y^2}, \quad \sigma_y = \frac{\partial^2 F}{\partial x^2}, \quad \tau_{xy} = \frac{\partial^2 F}{\partial x \partial y} - X_0 y - Y_0 x \ , (17)$$

then it follows from Eq. (16) for $F(x, y)$ that

$$\frac{\partial^4 F}{\partial x^4} + 2\frac{\partial^4 F}{\partial x^2 \partial y^2} + \frac{\partial^4 F}{\partial y^4} = \Delta\Delta F = 0 \ , \qquad (18)$$

i.e. Airy's stress function must satisfy the bipotential equation. There are infinite solutions to the bipotential equation, e.g. $F = x$, x^2, x^3, y, y^2, y^3, xy, x^2y, x^3y, xy^2, xy^3, $\cos \lambda x \cdot \cosh y$, $x \cos \lambda x \cdot \cosh \lambda y$ etc., in addition to biharmonic polynomials [2] and the real and imaginary components of analytical functions $f(z) = f(x \pm iy)$ etc. [1]. Suitable linear combinations of these solutions can be added to satisfy the specified boundary conditions and thus to solve the plane problem.

Example *Half-plane under single load.* Polar coordinates are used to solve this (**Fig. 4a**). Then

$$\Delta\Delta F = \left(\frac{\partial^2}{\partial r^2} + \frac{1}{r}\frac{\partial}{\partial r} + \frac{1}{r^2}\frac{\partial^2}{\partial \varphi^2}\right)\left(\frac{\partial^2 F}{\partial r^2} + \frac{1}{r}\frac{\partial F}{\partial r} + \frac{1}{r^2}\frac{\partial^2 F}{\partial \varphi^2}\right) = 0$$

applies for Airy's stress function and

$$\sigma_r = \frac{1}{r}\frac{\partial F}{\partial r} + \frac{1}{r^2}\frac{\partial^2 F}{\partial \varphi^2}, \quad \sigma_t = \frac{\partial^2 F}{\partial r^2}, \quad \tau_{rt} = -\frac{\partial}{\partial r}\left(\frac{1}{r}\frac{\partial F}{\partial \varphi}\right).$$

for the stresses (where $X = Y = 0$).

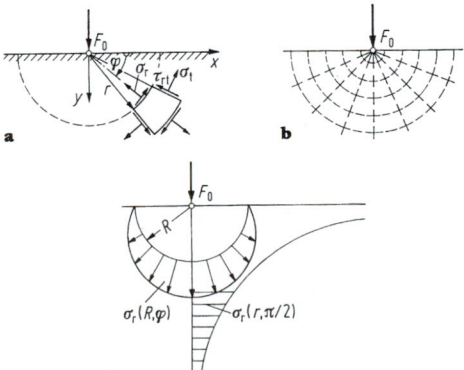

Figure 4a–c. Half-plane under single load.

The boundary conditions are

$$\sigma_t(r, \varphi = 0) = 0, \quad \sigma_t(r, \varphi = \pi) = 0,$$

$$\tau_{rt}(r, \varphi = 0) = 0, \quad \tau_{rt}(r, \varphi = \pi) = 0.$$

With the formulation $F(r, \varphi) = Cr\varphi \cos \varphi$,

$$\Delta\Delta F = 0, \quad \sigma_r = -C(2/r)\sin \varphi, \quad \sigma_t = 0, \quad \tau_{rt} = 0.$$

The solution fulfils the boundary conditions. With plate thickness b, constant C follows from the equilibrium condition

$$\Sigma F_{iy} = 0 = \int_0^\pi \sigma_r \sin\varphi \cdot br\, d\varphi + F_0 = 0 \text{ as } C = F_0/(\pi b).$$

Since $\tau_{rt} = 0$, σ_r and σ_t are principal normal stresses, i.e. the corresponding trajectories are straight lines through zero or the concentric circles about zero (**Fig. 4b**). The principal shear stress trajectories lie at $45°$ (see B1.1.1). The gradient of the stresses σ_r is $\sigma_r = -2F_0/(\pi bR) \cdot \sin\varphi$ for $r = R = $ const. and $\sigma_r = -[2F_0/(\pi b)]/r$ for $\varphi = \pi/2$ (**Fig. 4c**).

4 Hertzian Contact Stresses (Formulae of Hertz)

If two bodies come into contact with each other at a point or along a line, then according to Hertzian theory deformations and stresses occur due to the influence of compressive forces [1, 2]. Boussinesq's formulae, B3 Eq. (11), are a starting point for Hertzian solutions. It is assumed that the material is homogeneous and isotropic, that Hooke's law applies and that the only influence on the surface of contact is from normal stresses. Moreover, the deformation, i.e. magnitude w_0 of approach (also called flattening), of both bodies (**Fig. 1a**) must be small in relation to the dimensions of the bodies. If the bodies that come into contact are of different material, $E = 2E_1E_2/(E_1 + E_2)$ applies. $\nu = 0.3$ is applied uniformly for the lateral extension factor.

4.1 Spheres

Sphere-Sphere Contact (Fig. 1b). Taking $1/r = 1/r_1 + 1/r_2$,

$$\max \sigma_z = \sigma_0 = -\frac{1}{\pi}\sqrt[3]{\frac{1.5 \cdot FE^2}{r^2(1 - \nu^2)^2}} \quad \text{and}$$

$$w_0 = \sqrt[3]{\frac{2.25 \cdot (1 - \nu^2)^2 F^2}{E^2 r}}.$$

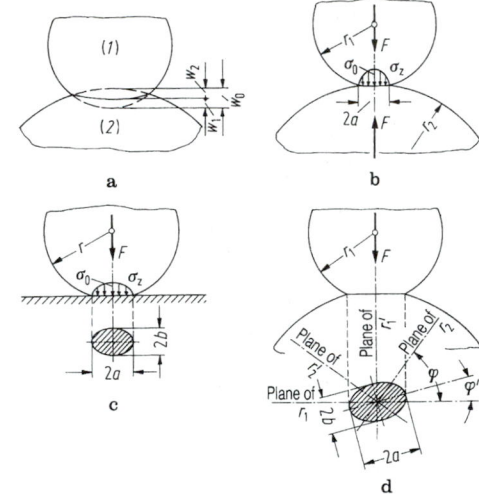

Figure 1a–d. Hertz formulas.

apply. The compressive stress is distributed over the compression surface in the form of a hemisphere. The projection of the compression surface is a circle of radius $a = \sqrt[3]{1.5 \cdot (1 - \nu^2)Fr/E}$. The stresses σ_r and σ_t at the centre volume element of the compression surface are $\sigma_r = \sigma_t = \sigma_0(1 + 2\nu)/2 = 0.8 \cdot \sigma_0$ in the centre and $\sigma_r = -\sigma_t = 0.133 \cdot \sigma_0$ at the edge. If the larger sphere (as a hollow sphere) envelopes the smaller one, r_2 will be negative.

Sphere-Plane Contact. Taking $r_2 \to \infty$, i.e. $r = r_1$, these results apply in the same way. The gradient of the stress in direction z [3] produces the maximum shear stress for $z = 0.47a$ as $\max \tau = 0.31 \cdot \sigma_0$ and the corresponding values $\sigma_z = 0.8 \cdot \sigma_0$, $\sigma_r = \sigma_t = 0.18 \cdot \sigma_0$. As Föppl [3] has shown, flow lines develop from the point of max τ. However, proof of max $\sigma_z = \sigma_0$ is usually taken to be sufficient.

4.2 Cylinders

Cylinder-Cylinder Contact(Fig. 1b). The projection of the pressure surface is a rectangle of width $2a$ and cylinder length l. The compressive stresses are distributed over width $2a$ in the form of a hemisphere. Taking $1/r = 1/r_1 + 1/r_2$,

$$\max \sigma_z = \sigma_0 = -\sqrt{\frac{FE}{2\pi rl(1 - \nu^2)}}, \quad a = \sqrt{\frac{8Fr(1 - \nu^2)}{\pi El}}$$

applies. It is assumed that $q = F/l$ is evenly distributed over the length as a linear load. Hertz did not determine flattening since the limited length of the cylinder makes it more difficult to solve the problem. Stresses σ_x and σ_y on an element of the pressure surface (x in the longitudinal direction and y in the transverse direction) are $\sigma_x = 2\nu\sigma_z = 0.6 \cdot \sigma_0$, $\sigma_y = \sigma_z = \sigma_0$ in the centre of the cylinder. The gradient of the stress in direction z [3] produces the maximum shear stress at depth $z = 0.78 \cdot a$ as $\max \tau = 0.30 \cdot \sigma_0$. At the centre volume element of the surface of contact,

$$\max \tau = 0.5(\sigma_1 - \sigma_3) = 0.5(\sigma_0 - 0.6 \cdot \sigma_0) = 0.2\,\sigma_0$$

in the centre of the cylinder and $\max \tau = 0.5 \cdot \sigma_0$ at the end of the cylinder. In this case $\max \tau$ is diagonal to the surface in surface elements, since in accordance with assumptions in the surface elements themselves and thus in accordance with the theory regarding allocated shear stresses τ also $= 0$ in surface elements perpendicular to it, i.e. the surface stresses are principal stresses.

Cylinder-Plane Contact. Taking $r_2 \to \infty$, the corresponding results apply.

4.3 Arbitrarily Curved Surfaces

Contact Between an Arbitrarily Curved Surface and a Plane(Fig. 1c). If the principal bending radii are

Table 1.

ϑ	90°	80°	70°	60°	50°	40°	30°	20°	10°	0°
ζ	1	1.128	1.284	1.486	1.754	2.136	2.731	3.778	6.612	∞
η	1	0.893	0.802	0.717	0.641	0.567	0.493	0.408	0.319	0
ψ	1	1.12	1.25	1.39	1.55	1.74	1.98	2.30	2.80	∞

at contact point r and r', then an ellipse is formed with semi-axes a and b in the direction of the principal bending planes. The compressive stresses are distributed in the form of an ellipsoid. The following apply:

$$\max \sigma_z = \sigma_0 = 1.5 \cdot F/(\pi ab) ,$$

$$a = \sqrt[3]{3\xi^3(1 - \nu^2)F/[E(1/r + 1/r')]} ,$$

$$b = \sqrt[3]{3\eta^3(1 - \nu^2)F/[E(1/r + 1/r')]} ,$$

$$w_0 = 1.5 \cdot \psi(1 - \nu^2)F/Ea.$$

The values ξ, η, ψ depend upon the auxiliary angle

$$\vartheta = \arccos[(1/r' - 1/r)/(1/r' + 1/r)] ; \text{ see } \textbf{Table 1.}$$

Contact Between Arbitrarily Curved Surfaces (Fig. 1d). Given: principal bending radii r_1 and r_1', r_2 and r_2' and angle φ between the planes of r_1 and r_2 [4].

Refer back to the case above, assuming $r_1 > r_1'$ and $r_2 > r_2'$, by introducing

$$1/r' + 1/r = 1/r_1' + 1/r_1 + 1/r_2' + 1/r_2 \tag{1}$$

and

$$\frac{1}{r'} - \frac{1}{r} =$$

$$\sqrt{\left(\frac{1}{r_1'} - \frac{1}{r_1}\right)^2 + \left(\frac{1}{r_2'} - \frac{1}{r_2}\right)^2 + 2\left(\frac{1}{r_1'} - \frac{1}{r_1}\right)\left(\frac{1}{r_2'} - \frac{1}{r_2}\right)\cos 2\,\varphi} \,. \tag{2}$$

The projection of the pressure surface is again an ellipse with semi-axes a and b. The axis a lies between the planes r_1 and r_2. The angle φ' is obtained from

$$(1/r' + 1/r)\sin 2\varphi' = (1/r_1' - 1/r_1)\sin 2\varphi .$$

If a larger body (hollow profile) envelopes the smaller one, then corresponding radii will be negative. The value specified in Eq. (2) may not exceed the value specified in Eq. (1).

5 Plates and Shells

5.1 Plates

Assuming that the plate thickness b is small for the surface dimensioning and the deflection w is also small, the bipotential equation

$$\Delta\Delta w = \frac{\partial^4 w}{\partial x^4} + 2\frac{\partial^4 w}{\partial x^2 \partial y^2} + \frac{\partial^4 w}{\partial y^4}$$
$$= \frac{p(x, y)}{N} \tag{1}$$

is produced with surface loading $p(x, y)$ and plate rigidity $N = Eb^3/[12(1 - \nu^2)]$ for deflections $w(x, y)$. Bending moments M_x and M_y and torsional moment M_{xy} follow from

$$\left. \begin{array}{l} M_x = -N(\partial^2 w/\partial x^2 + \nu\partial^2 w/\partial y^2) , \\[4pt] M_y = -N(\partial^2 w/\partial y^2 + \nu\partial^2 w/\partial x^2) , \\[4pt] M_{xy} = -(1 - \nu)N\partial^2 w/(\partial x\, \partial y) . \end{array} \right\} \tag{2}$$

The extreme stresses on the upper and lower side of the plate are produced from

$$\sigma_x = M_x/W , \qquad \sigma_y = M_y/W , \qquad \tau = M_{xy}/W , \tag{3}$$

where the section modulus $W = b^2/6$. In the case of axisymmetrically loaded circular plates $w = w(r)$ and Eq. (1) changes into the usual Euler's differential equation

$$w''''(r) + \frac{2}{r}w'''(r) - \frac{1}{r^2}w''(r) + \frac{1}{r^3}w'(r) = \frac{p(r)}{N} . \tag{4}$$

In addition, the following apply:

$$M_r = -N\left(w'' + \frac{\nu}{r}w'\right), \quad M_t = -N\left(\nu w'' + \frac{1}{r}w'\right), \tag{5}$$

$$\sigma_r = M_r/W, \ \sigma_t = M_t/W \ \text{ where } W = b^2/6 . \tag{6}$$

Torsional moments do not occur, owing to the axisymmetry. In the following the most significant results are given for various types of plates (lateral extension factor $\nu = 0.3$).

5.1.1 Rectangular Plates

Uniformly Loaded Plates (Fig. 1)

Circular Hinged Edge [1–3]. The maximum stresses and deformations occur in the centre of the plate:

$$\sigma_x = c_1 pb^2/b^2, \quad \sigma_y = c_2 pb^2/b^2, \quad f = c_3 pb^4/Eb^3 . \tag{7}$$

Rising single forces $F = c_4 pb^2$ which are to be anchored occur at the corners (c_i = correction values; see **Table 1**).

Circular Constrained Edge. In addition to the stresses and deformations in the centre of the plate in accordance with Eq. (7), maximum bending stresses occur in the centre of the long edge (for values of c_i see **Table 1**):

$$\sigma_y = c_5 pb^2/b^2; \quad \text{correspondingly, } \sigma_x = 0.3\sigma_y .$$

No rising bearing forces occur in the form of single forces at the corners. Detailed representation of all sectional loads and bearing reactions is given in [4, 7].

Uniformly Loaded, Infinite Plates on Single Supports (Fig. 2). For stresses and deformations the following are produced with the support force $F = 4a^2 p$ and $2b \geq b$.

Table 1.

a/b	\multicolumn{4}{c}{Hinged plate}				\multicolumn{4}{c}{Circular constrained plate}			
	c_1	c_2	c_3	c_4	c_1	c_2	c_3	c_5
1.0	1.15	1.15	0.71	0.26	0.53	0.53	0.225	1.24
1.5	1.20	1.95	1.35	0.34	0.48	0.88	0.394	1.82
2.0	1.11	2.44	1.77	0.37	0.31	0.94	0.431	1.92
3.0	0.97	2.85	2.14	0.37	-	-	-	-
4.0	0.92	2.96	2.24	0.38	-	-	-	-
∞	0.90	3.00	2.28	0.38	0.30	1.00	0.455	2.00

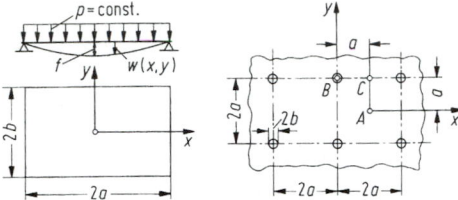

Figure 1. Rectangular plate.

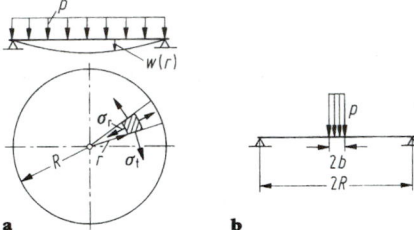

Figure 3. Circular plate with **a** area load, **b** single load.

$\sigma_{xA} = \sigma_{yA} = 0.861 \cdot pa^2/b^2$,

$\sigma_{xB} = \sigma_{yB} = -0.62 \cdot F[\ln(a/b) - 0.12]/b^2$,

$f_A = 0.092 \cdot pa^4/N, \quad f_C = 0.069 \cdot pa^4/N$.

$\sigma_r = \sigma_t = 1.95(b/R)^2[0.77 - 0.135(b/R)^2 - \ln(b/R)]pR^2/b^2$,

$f = 0.682(b/R)^2[2.54 - (b/R)^2(1.52 - \ln(b/R))]pR^4/(Eb^3)$.

Constrained Edge. In the centre,

$\sigma_r = \sigma_t = 1.95(b/R)^2[0.25(b/R)^2 - \ln(b/R)]pR^2/b^2$,

5.1.2 Circular Plates

Uniformly Loaded Plate

Hinged Edge (**Fig. 3a**). The maximum stresses and deformations occur in the centre of the plate:

$\sigma_r = \sigma_t = 1.24 \cdot pR^2/b^2, \quad f = 0.696 \cdot pR^4/(Eb^3)$.

Constrained Edge. In the centre,

$\sigma_r = \sigma_t = 0.488 \cdot pR^2/b^2, \quad f = 0.171 \cdot pR^4/(Eb^3)$;

at the edge,

$\sigma r = 0.75 \cdot pR^2/b^2, \quad \sigma_t = \nu\sigma_r = 0.225 \cdot pR^2/b^2$.

Plate Under Single Load (Fig. 3b)

For a force $F = \pi b^2 p$ in the centre, which is uniformly distributed on a circular surface of radius b, the following applies:

Hinged Edge. Maximum stresses and deformations occur in the centre

$f = 0.682(b/R)^2[1 - (b/R)^2 \cdot (0.75 - \ln(b/R))]pR^4/(Eb^3)$;

at the edge,

$\sigma_r = -0.75(b/R)^2[2 - (b/R)^2]pR^2/b^2, \quad \sigma_t = \nu\sigma_r$.

Further detailed results for circular and annular plates under various loads are given in [5].

5.1.3 Elliptical Plates

Under Uniform Load p

Semi-axes: $a > b$ (a in direction x, b in direction y).

Hinged Edge. Maximum bending stress in the centre

$\sigma_y \approx (3.24 - 2b/a)pb^2/b^2$.

Constrained Edge. Taking $c_1 = 8/[3 + 2(b/a)^2 + 3(b/a)^4]$,

$\sigma_x = 3c_1 pb^2[(b/a)^2 + 0.3]/(8b^2)$,

$\sigma_y = 3c_1 pb^2[1 + 0.3(b/a)^2]/(8b^2)$,

$f = 0.171 \cdot c_1 pb^4/(Eb^3)$

applies in the centre.

At the end of the minor axis,

$\min \sigma = \sigma_y = -0.75 \cdot c_1 pb^2/b^2, \quad \sigma_x = \nu\sigma_y$.

At the end of the major axis,

$\sigma_x = -0.75 \cdot c_1 pb^4/(a^2b^2), \quad \sigma_y = \nu\sigma_x$.

5.1.4 Triangular Plate

Under Uniform Load p

Circular Hinged. (**Fig. 4**)

$\sigma_x = \sigma_y = 0.145 \cdot pa^2/b^2, \quad f = 0.00103 \cdot pa^4/N$

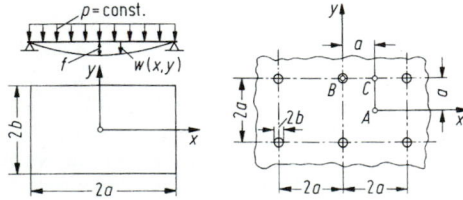

Figure 2. Plate on single supports.

applies for the centre of inertia of the plate S. Where $x = 0.129a$ and $y = 0$, maximum stress occurs which is $\sigma_y = 0.155 \cdot pa^2/b^2$.

5.1.5 Thermal Stresses in Plates

With a temperature difference of Δt between the top and bottom no stresses occur in the case of plates with completely free edges and in the case of plates hinged on all sides stresses occur in accordance with the plate theory [6].

In the case of completely constrained plates,

$$\sigma_x = \sigma_y = \alpha_t \Delta t E/[2(1 - v)] = \sigma_r = \sigma_t .$$

5.2 Discs, Plates Under In-Plane Loads

These are plane load-bearing structures that are loaded in their plane. To determine the stresses theoretically using the Airy's stress function see B3.3. In the following the stresses are given for a few cases of technical significance. Let the thickness of the disc be b.

5.2.1 Circular Discs

Radial Uniform Line Load q (Fig. 5)

$$\sigma_r = \sigma_t = -q/b , \quad \tau_{rt} = 0 .$$

Uniform Heating Δt. Only radial displacements $u(r) = \alpha_t \Delta tr$, but no stresses occur in a disc with a displaced edge. If the edge is not displaced ($u = 0$) the following applies:

$$\sigma_r = \sigma_t = -E\alpha_t \Delta t/(1 - v) , \quad \tau_{rt} = 0 .$$

5.2.2 Annular Discs

Internal and External Radial Line Load (Fig. 6a)

$$\sigma_r = -\frac{q_i r_i^2}{b(r_a^2 - r_i^2)}\left(\frac{r_a^2}{r^2} - 1\right) - \frac{q_a r_a^2}{b(r_a^2 - r_i^2)}\left(1 - \frac{r_i^2}{r^2}\right) ,$$

$$\sigma_t = +\frac{q_i r_i^2}{b(r_a^2 - r_i^2)}\left(\frac{r_a^2}{r^2} + 1\right) - \frac{q_a r_a^2}{b(r_a^2 - r_i^2)}\left(1 + \frac{r_i^2}{r^2}\right) ,$$

$$\tau_{rt} = 0 .$$

Uniform Heating Δt. Only radial displacements $u(r) = \alpha_t \Delta tr$, but no stresses occur in a disc with a displaced edge. If the edge is not displaced ($u = 0$) the following applies:

$$\sigma_r = -E\alpha_t \Delta t \frac{r_a^2}{(1 - v)r_a^2 + (1 + v)r_i^2}\left(1 - \frac{r_i^2}{r^2}\right) ,$$

$$\sigma_t = -E\alpha_t \Delta t \frac{r_a^2}{(1 - v)r_a^2 + (1 + v)r_i^2}\left(1 + \frac{r_i^2}{r^2}\right) , \quad \tau_{rt} = 0 .$$

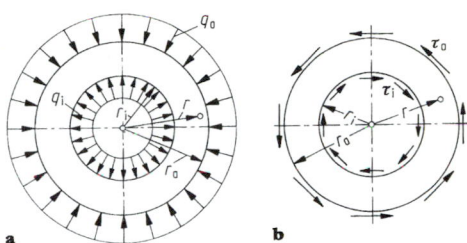

Figure 6a and b. Annular disc.

Annular Shear Load (Fig. 6b). If τ_i and $\tau_a = \tau_i r_i^2/r_a^2$ are the effective shear stresses, then the following applies:

$$\tau_{rt} = \tau_i r_i^2/r^2 , \quad \sigma_r = \sigma_t = 0 .$$

5.2.3 Infinite Plate with a Hole (Fig. 7)

As a result of internal pressure $p = q/b$ the following stresses arise

$$\sigma_r = -pr_i^2/r^2 , \quad \sigma_t = +pr_i^2/r^2 , \quad \tau_{rt} = 0 .$$

5.2.4 Wedge-shaped Plate Under Point Load (Fig. 8)

The following apply for the stresses:

$$\sigma_r = -\frac{2F_1\cos \varphi}{rb(2\beta + \sin 2\beta)} + \frac{2F_2\sin \varphi}{rb(2\beta - \sin 2\beta)} ,$$

$$\sigma_t = 0 , \quad \tau_{rt} = 0 .$$

5.3 Shells

Here we are referring to three-dimensionally curved components under loads from normal stresses σ_x and σ_y and shear stresses τ_{xy} (and in the case of shells of rotation from σ_φ and σ_ϑ and $\tau_{\varphi\vartheta}$), which all lie in the shell surface. This is called the membrane stress condition since membranes (soap bubbles, air balloons, thin metal sheets etc.), i.e. shells weak in bending, can only support loads in this way (**Fig. 9a, b**). Thin-walled metal constructions as a rule fulfil the membrane stress condition in wide areas. In the case of certain shell forms, at fault points (e.g. transition from wall to floor) and in all thick-walled shells, bending moments and transverse forces do occur, i.e. normal bending and transverse shear stresses (as in the case

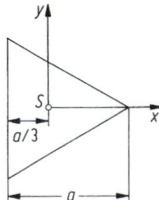

Figure 4. Triangular plate.

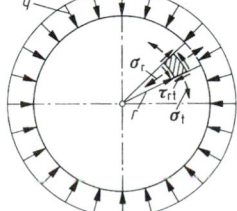

Figure 5. Circular disc.

Figure 7. Plate with hole

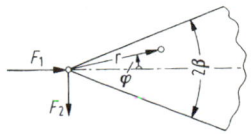

Figure 8. Wedge-shaped plate.

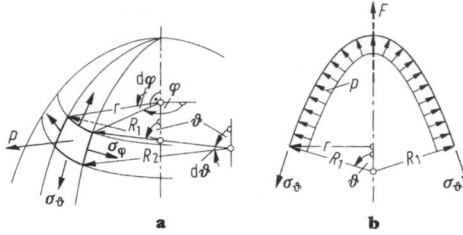

Figure 9a and **b.** Membrane stress condition.

Table 2.

a/b	0.5	0.6	0.7	0.9	0.8	1.0
c_1	3.7	2.3	1.4	0.7	0.3	0
c_2	5.1	2.9	1.7	0.8	0.3	0

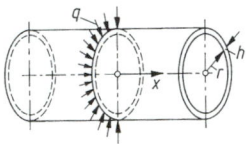

Figure 12. Constrained hollow cylinder.

of plates) which have to be taken into account. This then involves rigid shells and the bending stress condition. However, this usually falls away rapidly with the removal of the fault point.

5.3.1 Shells Under Internal Pressure, Membrane Stress Theory

The equilibrium conditions on the element (**Fig. 9a**) in the direction of the normal and at the shell section (**Fig. 9b**) produce the following in the vertical direction

$$\sigma_\varphi/R_1 + \sigma_\vartheta/R_2 = p/b, \quad \sigma_\vartheta = F/(2\pi R_1 b \sin^2 \vartheta) .$$

Here σ_ϑ is the stress in the meridianal direction, σ_φ the stress in the direction of the circle of latitude and b the shell thickness. F is the resulting external force in the vertical direction, i.e.

$$F = \int_{\vartheta=0}^{\vartheta} p(\vartheta) R_2(\vartheta) \cdot 2\pi R_1(\vartheta) \cdot \sin \vartheta \cos \vartheta \, d\vartheta .$$

In the case of constant internal pressure F is equal to the force on the projection surface i.e. $F = p\pi r^2 = p\pi(R_1 \sin \vartheta)^2$.

Circular Cylindrical Shell Under Constant Internal Pressure

$$\sigma_\varphi = pr/b = pd/(2b), \quad \sigma_\vartheta = \sigma_x = 0 .$$

Cylindrical Shells with Hemispherical Bases Under Constant Internal Pressure (Fig. 10). In the cylinder,

$$\sigma_\varphi = pr/b = pd/(2b), \quad \sigma_x = pr/(2b) = pd/(4b);$$

in the circular shell,

$$\sigma_\varphi = \sigma_\vartheta = pr/(2b) = pd/(4b) .$$

5.3.2 Bending Rigid Shells

Elliptical Hollow Cylinders Under Internal Pressure (Fig. 11). If the bending stresses are superposed upon membrane stresses, this gives, for points A and B (**Table 2**):

$$\sigma_A = pa/b + c_1 pa^2/b^2, \quad \sigma_B = pb/b + c_2 pa^2/b^2 .$$

Bound Hollow Cylinders (Fig. 12). Owing to the shear load q, the following peripheral stresses arise:

$$\sigma_\varphi(x) = -\frac{qr}{\sqrt{2}Lb} e^{-x/L} \sin\left(\frac{x}{L} + \frac{\pi}{4}\right), \quad \sigma_\varphi(x=0) = -\frac{qr}{2LH},$$

where $L = \sqrt[4]{\dfrac{r^2 b^2}{3(1-\nu^2)}}$ and the following bending stresses occur in the direction x:

$$\sigma_x(x) = \frac{3qL}{\sqrt{2}b^2} e^{-x/L} \cos\left(\frac{x}{L} + \frac{\pi}{4}\right),$$

$$\sigma_x(x=0) = \max \sigma_x = 1.5qL/b^2 .$$

Tube Bends Under Internal Pressure (Fig. 13). In the longitudinal direction of the bend, the stresses $\sigma_x = pr/(2b) = pd/(4b)$ arise, i.e. the same stresses as in the case of sealed straight tubes. In the peripheral direction the following applies:

$$\sigma_\varphi = \frac{pd}{2b} \cdot \frac{R/d + 0.25 \sin \varphi}{R/d + 0.5 \sin \varphi} .$$

For the upper side and lower side of the bend ($\varphi = 0$ or 180°), $\sigma_\varphi(0) = pd/(2b)$, i.e. a stress in the case of circular cylindrical tubes. For the outside and inside of the bends,

$$\sigma_\varphi(90°) = \frac{pd}{2b} \cdot \frac{R/d + 0.25}{R/d + 0.50} \quad \text{and}$$

$$\sigma_\varphi(-90°) = \frac{pd}{2b} \cdot \frac{R/d - 0.25}{R/d - 0.50},$$

i.e. $\sigma_\varphi(90°)$ is less and $\sigma_\varphi(-90°)$ greater than $\sigma_\varphi(0)$.

Curved Bends Under Constant Internal Pressure (Fig. 14). For the stresses in the spherical

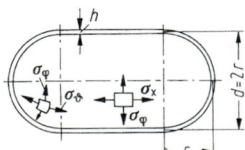

Figure 10. Sealed cylindrical shells.

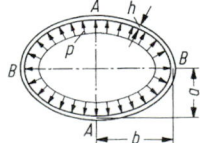

Figure 11. Elliptical hollow cylinders.

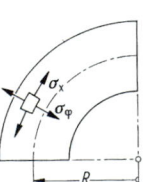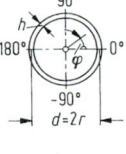

Figure 13. Tube bend.

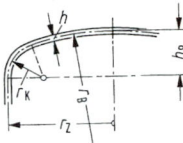

Figure 14. Curved bend.

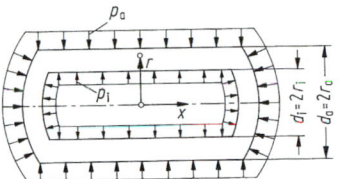

Figure 15. Thick-walled circular cylinder.

Table 3.

b_b/r_z	0.2	0.4	0.6	0.8	1.0
c_1	6.7	3.8	2.0	1.3	1.0

$$\sigma_x = p_i \frac{r_i^2}{r_a^2 - r_i^2} - p_a \frac{r_a^2}{r_a^2 - r_i^2} \, ,$$

$$\sigma_\varphi = p_i \frac{r_i^2}{r_a^2 - r_i^2} \left(\frac{r_a^2}{r^2} + 1 \right) - p_a \frac{r_a^2}{r_a^2 - r_i^2} \left(1 + \frac{r_i^2}{r^2} \right) ,$$

$$\sigma_r = -p_i \frac{r_i^2}{r_a^2 - r_i^2} \left(\frac{r_a^2}{r^2} - 1 \right) - p_a \frac{r_a^2}{r_a^2 - r_i^2} \left(1 - \frac{r_i^2}{r^2} \right) .$$

In the case of only internal or external pressure the maximum stress occurs on the inside as $\sigma_\varphi(r = r_i)$. The bending stress of the cylinder on the base is not taken into account in this case.

curve, $\sigma_\varphi = \sigma_\vartheta = pr_B/(2b)$ applies (as in the case of the spherical shell), while

$$\sigma_\vartheta = c_1 pr_z/(2b) = c_1 pd_z/(4b)$$

applies for the (maximum) meridianal stresses (**Table 3**).

Thick-Walled Circular Cylinders Under Internal and External Pressure (Fig. 15). There is a three-dimensional state of stress with stresses (in the central area of the cylinder):

Thick-Walled Hollow Spheres Under Internal and External Pressure. There is a three-dimensional stress condition with the stresses

$$\sigma_\varphi = \sigma_\vartheta = p_i \frac{r_i^3}{r_a^3 - r_i^3} \left(1 + \frac{r_a^3}{2r^3} \right) - p_a \frac{r_a^3}{r_a^3 - r_i^3} \left(1 + \frac{r_i^3}{2r^3} \right) ,$$

$$\sigma_r = -p_i \frac{r_i^3}{r_a^3 - r_i^3} \left(\frac{r_a^3}{r^3} - 1 \right) - p_a \frac{r_a^3}{r_a^3 - r_i^3} \left(1 - \frac{r_i^3}{r^3} \right) .$$

The maximum stress is produced by $\sigma_\varphi(r = r_i)$.

6 Centrifugal Stresses in Rotating Components

Stresses and strains with angular velocity ω in rotating components can be determined in accordance with the rules of statics and the theory of strength if, in the sense of d'Alembert's principle, the centrifugal forces (inertial force, negative mass accelerations) $\omega^2 r \, dm = \omega^2 r \rho \, dA \, dr$ (ρ = density) are calculated as external forces on the elements of mass. In the following, results are given only for stresses (in the case of discs for the transverse extension factor $\nu = 0.3$) and for radial displacements.

6.1 Rotating Bars

(Fig. 1) With bar cross-section A and modulus of elasticity E,

$$\sigma_r(r) = \rho\omega^2(l^2 - r^2)/2 + m_1\omega^2 l_1/A \, ,$$

$$\max \sigma_r = \sigma_r(r = 0) = \rho\omega^2 l^2/2 + m_1\omega^2 l_1/A \, ,$$

$$u(r) = \rho\omega^2(3l^2 r - r^3)/(6E) + m_1\omega^2 l_1 r/(AE) \, ,$$

$$u(r = l) = \rho\omega^2 l^3/(3E) + m_1\omega^2 l_1 l/(AE) \, .$$

6.2 Rotating Thin Rings

(Fig. 2)

$$\sigma_t = \rho\omega^2 R^2 \, , \quad u = \rho\omega^2 R^3/E \, .$$

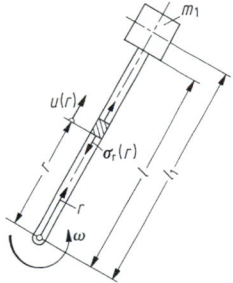

Figure 1. Rotating bar.

6.3 Rotating Discs

6.3.1 Discs of Uniform Thickness (Fig. 3)

$$\sigma_r(r) = 0.4125\rho\omega^2 R^2(1 - r^2/R^2) \, ,$$

$$\max \sigma_r = \sigma_r(r = 0) = 0.4125\rho\omega^2 R^2 \, ,$$

$$\sigma_t(r) = 0.4125\rho\omega^2 R^2(1 - 0.576r^2/R^2) \, ,$$

$$\max \sigma_t = \sigma_t(r = 0) = 0.4125\rho\omega^2 R^2 \, ,$$

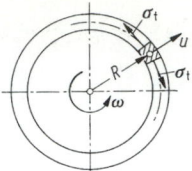

Figure 2. Rotating ring.

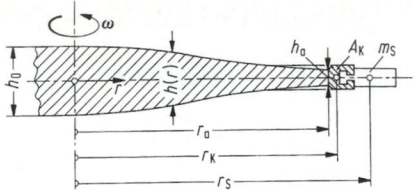

Figure 5. Disc of constant strength.

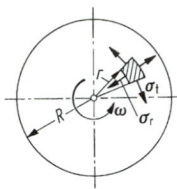

Figure 3. Rotating full disc.　　**Figure 4.** Rotating annular disc.

$$u(r) = r[\sigma_t(r) - \nu\sigma_r(r)]/E ,$$

$$u(r = R) = \rho\omega^2 R^3(1 - \nu)/(4E) .$$

6.3.2 Annular Discs of Constant Thickness (Fig. 4)

For $\sigma_i = \sigma_a = 0$

$$\sigma_r(r) = 0.4125\rho\omega^2 r_a^2(1 + r_i^2/r_a^2 - r_i^2/r^2 - r^2/r_a^2) ,$$

$$\sigma_r(r = r_i) = \sigma_r(r = r_a) = 0 ,$$

$$\sigma_t(r) = 0.4125\rho\omega^2 r_a^2(1 + r_i^2/r_a^2 + r_i^2/r^2 - 0.576r^2/r_a^2) ,$$

$$\max \sigma_t = \sigma_t(r = r_i) = 0.825\rho\omega^2 r_a^2(1 + 0.212r_i^2/r_a^2) .$$

For $r_i \rightarrow 0$, i.e. if the hole is very small, max $\sigma_t = 0.825\rho\omega^2 R^2$ is twice as great as in the case of the full disc!

$$u(r) = r[\sigma_t(r) - \nu\sigma_r(r)]/E ,$$

$$u_i = u(r = r_i) = \rho\omega^2 r_i[2c_1 r_a^2 + (c_1 - c_2)r_i^2]/E ,$$

$$u_a = u(r = r_a) = \rho\omega^2 r_a[2c_1 r_i^2 + (c_1 - c_2)r_a^2]/E ,$$

where $c_1 = (3 + \nu)/8$ and $c_2 = (1 + 3\nu)/8$.
For any value of σ_i and σ_a,

$$\sigma_r(r) = A_1 + A_2/r^2 - c_1\rho\omega^2 r^2 ,$$

$$\sigma_t(r) = A_1 - A_2/r^2 - c_2\rho\omega^2 r^2 ,$$

where

$$A_1 = (\sigma_a r_a^2 - \sigma_i r_i^2)/(r_a^2 - r_i^2) + c_1\rho\omega^2(r_a^2 + r_i^2) ,$$

$$A_2 = -(\sigma_a - \sigma_i)r_a^2 r_i^2/(r_a^2 - r_i^2) - c_1\rho\omega^2 r_a^2 r_i^2 ;$$

Displacements $u(r)$ and c_1 and c_2 are as above.
In the case of discs with rim and hub, σ_i and σ_a are statically indeterminate quantities, which can be determined from the conditions of the same displacement at points $r = r_i$ and $r = r_a$ [1].

6.3.3 Discs of Constant Strength (Fig. 5)

For the case where $\sigma_r = \sigma_t = \sigma$ is constant throughout, the differential equations for rotating discs [1] give the disc thickness $h(r) = h_0 e^{-\rho(\omega r)^2/(2\sigma)}$ (de Laval's disc of constant strength with no centre hole). h_0 is the disc thickness where $r = 0$. The profile curve has a turning point for $r = \sqrt{\sigma/(\rho\omega^2)}$. The radial displacement is $u(r) = (1 - \nu)\sigma r/E$, $u(r = r_a) = (1 - \nu)\sigma r_a/E$. The disc thickness $h(r = r_a) = h_a$ is obtained from the effect of the vanes (total mass m_s) and the rim (cross-section A_K) to which the vanes are attached as [1]:

$$h_a = \frac{1}{r_a}\left\{\left(\frac{m_s r_s}{2\pi} + \rho r_K^2 A_K\right)\frac{\omega^2}{\sigma} - A_K\left[\nu + (1 - \nu)\frac{r_a}{r_K}\right]\right\}$$

and thus $h_0 = h_a e^{\rho(\omega r_a)^2/(2\sigma)}$.

6.3.4 Discs with Varying Thickness

Solutions can be found in [1] for discs with hyperbolic or conical profiles. This also gives approximate methods for arbitrary profiles.

6.3.5 Rotating Thick-Walled Cylinder

In addition to stresses σ_r and σ_t in the radial and tangential direction, restricted transverse extension causes additional stresses σ_x in the longitudinal direction (three-dimensional state of stress):

$$\sigma_r(r) = \rho\omega^2 r_a^2 \frac{3 - 2\nu}{8(1 - \nu)}\left(1 + \frac{r_i^2}{r_a^2} - \frac{r_i^2}{r^2} - \frac{r^2}{r_a^2}\right) ,$$

$$\sigma_t(r) = \rho\omega^2 r_a^2 \frac{3 - 2\nu}{8(1 - \nu)}\left(1 + \frac{r_i^2}{r_a^2} + \frac{r_i^2}{r^2} - \frac{(1 + 2\nu)r^2}{(3 - 2\nu)r_a^2}\right) ,$$

$$\sigma_x(r) = \rho\omega^2 r_a^2 \frac{2\nu}{8(1 - \nu)}\left(1 + \frac{r_i^2}{r_a^2} - 2\frac{r^2}{r_a^2}\right) .$$

7 Stability Problems

7.1 Buckling of Bars

When subjected to compressive stress, when the critical stress or load is reached, slender bars or systems of bars change from the position of equilibrium where they are not bent outwards (unstable) to an adjacent position where they are bent (stable). If the bar shifts in the direction of an axis of symmetry it is flexural buckling, otherwise it is torsional–flexural buckling (see B7.1.7).

7.1.1 Elastic (Euler) Buckling

If we consider the deformed position of equilibrium of the bar in **Fig. 1**, in the case of small deflections, the differential equation for buckling around the principal axis of the cross-section y (where I_y is the smaller second area moment) is

$$EI_y w''(x) = -M_b(x) = -Fw(x) \quad \text{or}$$

$$w''(x) + \alpha^2 w(x) = 0 \quad \text{where} \quad \alpha = \sqrt{F/(EI_y)} \quad (1)$$

and the solution

$$w(x) = C_1 \sin \alpha x + C_2 \cos \alpha x . \quad (2)$$

The boundary conditions $w(x = 0) = 0$ and $w(x = 1) = 0$ give $C_2 = 0$ and $\sin \alpha l = 0$ (eigenvalue equation) with the eigenvalues $\alpha_K = n\pi/l; n = 1, 2, 3, \ldots$.
Thus in accordance with Eqs (1) and (2)

$$F_K = \alpha_K^2 EI_y = n^2 \pi^2 EI_y/l^2, \quad w(x) = C_1 \sin(n\pi x/l) . (3)$$

The smallest (Euler) buckling load for $n = 1$ is $F_K = \pi^2 EI_y/l^2$. Corresponding eigenvalues are produced for other types of support, but all can be restored to the form $\alpha_K = n\pi/l_K$ with the reduced or effective buckling length l_K (**Fig. 2**). The following then generally applies for Euler's buckling load:

$$F_K = \pi^2 EI_y/l_K^2 . \quad (4)$$

With radius of gyration $i_y = \sqrt{I_y/A}$ and slenderness $\lambda = l_K/i_y$, the buckling stress is

$$\sigma_K = F_K/A = \pi^2 E/\lambda^2 . \quad (5)$$

The function $\sigma_K(\lambda)$ represents Euler's hyperbola (line *1* in **Fig. 3**).
These equations only apply in the linear, elastic material range, therefore provided that

$$\sigma_K = \pi^2 E/\lambda^2 \leq \sigma_P \quad \text{or} \quad \lambda \geq \sqrt{\pi^2 E/\sigma_P} .$$

The transition from the elastic to the inelastic (plastic) range takes place at the limit slenderness:

$$\lambda_0 = \sqrt{\pi^2 E/\sigma_P} . \quad (6)$$

For example, for St 37 with

$$R_e = 240 \text{ N/mm}^2, \quad \sigma_P \approx 0.8 R_e = 192 \text{ N/mm}^2$$

and $E = 2.1 \cdot 10^5 \text{ N/mm}^2$ the limit slenderness $\lambda_0 = 104$. See **Table 1** for other limit slenderness values.

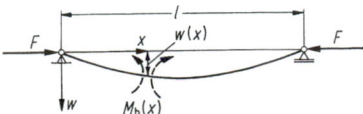

Figure 1. Buckling of a bar.

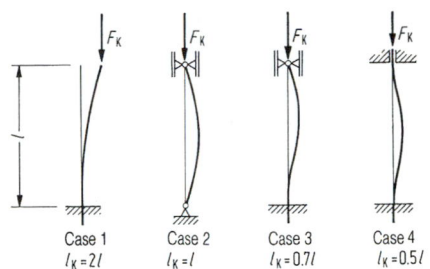

Figure 2. Euler's four cases of buckling.

Case 1 $l_K = 2l$
Case 2 $l_K = l$
Case 3 $l_K = 0.7l$
Case 4 $l_K = 0.5l$

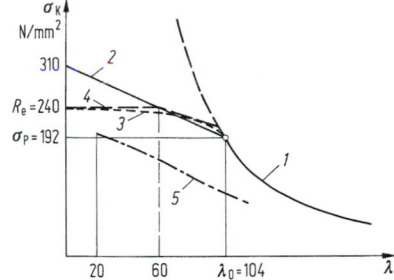

Figure 3. Buckling stress diagram for St 37: *1* Euler's hyperbola; *2* Tetmajer's line; *3* Engesser-von Kármán curve; *4* von Kármán's lines; *5* Jäger's load curve.

Buckling Safety

$$S_K = F_K/F_{exist} \quad \text{or} \quad S_K = \sigma_K/\sigma_{exist} . \quad (7)$$

In general mechanical engineering, $S_K \approx 5$ to 10 in the elastic range and $S_K \approx 3$ to 8 in the inelastic range.

Bending Outward with Buckling. The solution of the linear differential equation (1) gives the shape of the bending line, Eq. (3), but not the extent of the deflection (bending arrow). Inserting the actual expression for the curvature in Eq. (1) instead of w'' gives a non-linear differential equation. An approximate solution of this gives the value [1]

$$f = \sqrt{8(Fl^2 - \pi^2 EI_y)/(\pi^2 F)} ,$$

as the bending arrow, i.e. $f(F = F_K) = 0$ and $f(F = 1.01 \cdot F_K) \approx 0.091$; exceeding the buckling load by 1% gives as much as 9% of the length of the bar as the deflection.

Table 1. Values of a and b according to Tetmajer

Material	E N/mm²	λ_0	a N/mm²	b N/mm²
St 37	$2.1 \cdot 10^5$	104	310	1.14
St 50, St 60	$2.1 \cdot 10^5$	89	335	0.62
5% Ni-steel	$2.1 \cdot 10^5$	86	470	2.30
Grey cast iron	$1.0 \cdot 10^5$	80	$\sigma_K = 776 - 12\lambda + 0.053\lambda^2$	
Pine	$1.0 \cdot 10^4$	100	29.3	0.194

7.1.2 Inelastic Buckling (Tetmajer's Method)

According to Engesser and von Kármán's theory, the effect of the shape (curvature) of the stress-strain line in this range is taken into account by introducing the buckling modulus $T_K < E$:

$$\sigma_K = \pi^2 T_K/\lambda^2, \quad T_K = 4TE/(\sqrt{T} + \sqrt{E})^2 \tag{8}$$

$T = T(\sigma) = d\sigma/d\varepsilon$ is the tangent modulus and corresponds to the rise of the stress-strain line. T_K applies for rectangular cross-sections but can also be used with a low degree of error for other cross-sections. T should be determined for different σ from the stress-strain line and thus $T_K(\sigma)$ and $\lambda(\sigma_K) = \sqrt{\pi^2 T_K/\sigma_K}$ are calculated in accordance with Eq. (8).

The inverse function $\sigma_K(\lambda)$ is then buckling stress line 3, Engesser-von Kármán's curve, in **Fig. 3**. Von Kármán replaced the line by two tangential lines, the horizontals of which pass through the yield point (line 4 in **Fig. 3**).

Shanley [2] has shown that initial deflections for the value $\sigma_K = \pi^2 T/\lambda^2$ (Engesser's first formula) are already possible if the load is increased further. This value thus represents the bottom limit of the buckling stresses in the inelastic range and the value from Eq. (8) represents the top limit.

Practical Calculation According to Tetmajer. On the basis of experiments Tetmajer recorded the buckling stresses by a line that is still used today in mechanical engineering (line 2 in **Fig. 3**):

$$\sigma_K = a - b\lambda . \tag{9}$$

The values a, b for different materials should be taken from **Table 1**.

Example *Dimensioning a connecting rod. Determine the necessary diameter of a connecting rod made of St 37 of length l =* 2000 mm, (a) *for compressive force F =* 96 kN *with buckling safety* $S_K = 8$, (b) *for F =* 300 kN *with* $S_K = 5$. If the connecting rod is hinged at both ends Euler's second case applies, i.e. $l_K = l =$ 2000 mm. If elastic buckling is assumed in case (a) it follows from Eqs (4) and (7) that

$$\text{erf } I_y = FS_K l_K^2/(\pi^2 E)$$

$$= 96 \cdot 10^3 \, N \cdot 8 \cdot 2000^2 \, mm^2/(\pi^2 \cdot 2.1 \cdot 10^5 \, N/mm^2)$$

$$= 148.2 \cdot 10^4 \, mm^4$$

and where $I_y = \pi d^4/64$ then $d = \sqrt[4]{64 \cdot 148.2 \cdot 10^4 \, mm^4/\pi} =$ 74 mm.
Where $i_y = \sqrt{I_y/A} = d/4 = 18.5$ mm the slenderness is

$$\lambda = l_K/i_y = 2000 \, mm/18.5 \, mm = 108 > 104 = \lambda_0 \,,$$

so that the assumption of elastic buckling has been confirmed. In case (b) under this assumption,

$$\text{erf } I_y = FS_K l_K^2/(\pi^2 E) = 289.5 \cdot 10^4 \, mm^4 \quad \text{and} \quad \text{erf } d = 88 \, mm \,,$$

thus $\lambda = l_K/i_y = 91 < \lambda_0$, i.e. buckling in the inelastic range. According to Tetmajer, Eq. (9), for this slenderness in accordance with Table 1,

$$\sigma_K = (310 - 1.14 \cdot 91) \, N/mm^2 = 206 \, N/mm^2$$

and where

$$\sigma_{exist} = F/A = 300 \cdot 10^3 \, N/(\pi \cdot 88^2/4) \, mm^2 = 49.3 \, N/mm^2$$

the buckling safety $S_K = \sigma_K/\sigma_{exist} = 206/49.3 = 4.2 < 5$. For $d =$ 95 mm $\lambda = l_K/i_y = 84$ and $\sigma_K = a - b\lambda = 214 \, N/mm^2$, and where $\sigma_{exist} = F/(\pi d^2/4) = 42.3 \, N/mm^2$ then $S_K = \sigma_K/\sigma_{exist} = 5.06 \approx 5$.

7.1.3 The Omega Method

This method is stipulated by the authorities for the construction of cranes, high buildings and bridges in accordance with DIN 4114. For the elastic range it is based on Euler's hyperbola and a buckling safety $S_K = 2.5$, and for the inelastic range it is based on a bearing stress (line 5 in **Fig. 3**) that was determined by Jäger on the basis of eccentricity $u = i_y/20 + l_K/500$ of the point of application of the force in the case of an ideally elasto-plastic stress-strain line, and safety $S_K = 1.5$.

With $\sigma_{K,perm} = \sigma_K/S_K$ and $F/A \leq \sigma_{K,perm}$

$$(F/A) \cdot (\sigma_{d,perm}/\sigma_{K,perm}) \leq \sigma_{d,perm}, \quad \text{i.e.}$$

$$\omega F/A \leq \sigma_{d,perm} \tag{10}$$

The value $\omega(\lambda) = \sigma_{d,perm}/\sigma_{K,perm}$ is taken from **Table 2**. A compression bar should therefore be indicated for ω times the loads if the permissible compression stress $\sigma_{d,perm} = 140 \, N/mm^2$ is maintained for St 37 and $\sigma_{d,perm} =$ 210 N/mm² for St 52 (DIN 18 800).

Example *Dimensioning a support. For a support made of St 52 of length l =* 6.00 m, *which is constrained at both ends, determine the dimensions for axial compression force F =* 150 kN *as an IPB profile.* Reduced buckling length (support case 4) is $l_K = 0.5 \cdot l =$ 3.00 m. If the profile IPB 100 is selected, with $i_{min} = i_y = 2.53$ cm, slenderness $\lambda = l_K/i_y = 300/2.53 = 119$ and according to Table 2, $\omega = 3.59$. Therefore in accordance with Eq. (10)

$$\omega F/A = 3.59 \cdot 150 \cdot 10^3 \, N/2600 \, mm^2$$

$$= 207 \, N/mm^2 < 210 \, N/mm^2 = \sigma_{d,perm} \,.$$

7.1.4 Approximate Methods for Estimating Critical Loads

Energy Method. Since in the case of outward buckling the bar assumes a stable adjacent position of equilibrium, the external energy must be equal to the strain energy (**Fig. 4a**). B2 Eq. (39) and B2 Eq. (34) give

$$W^{(a)} = F_K v = W = \frac{1}{2}\int_0^l M_b^2 \frac{dx}{EI_y} = \frac{1}{2}\int_0^l EI_y w''^2 \, dx \quad \text{and}$$

$$v = \int_0^l (ds - dx) = \int_0^l (\sqrt{1 + w'^2} - 1) \, dx \approx \frac{1}{2}\int_0^l w'^2 \, dx \,.$$

Thus Rayleigh's quotient

$$F_K = \frac{2W}{2v} = \frac{\displaystyle\int_0^l EI_y(x)w''^2(x) \, dx}{\displaystyle\int_0^l w'^2(x) \, dx} . \tag{11}$$

With the exact bending line $w(x)$, this equation gives the exact buckling force for the elastic range. In bars with

Table 2. Buckling factors ω in accordance with DIN 4114 and DIN 1052

λ	20	40	60	80	100	120	140	160	180	200	220	250
St 37	1.04	1.14	1.30	1.55	1.90	2.43	3.31	4.32	5.47	6.75	8.17	10.55
St 52	1.06	1.19	1.41	1.79	2.53	3.65	4.96	6.48	8.21	10.13	12.26	15.83
Wood	1.08	1.26	1.62	2.20	3.00	4.32	5.88	7.68	9.72	12.00	14.52	18.75

varying cross-section, comparison with the buckling force $F_K = \pi^2 EI_{y0}/l_K^2$ from the corresponding case according to Euler of a bar with constant cross-section gives the equivalent area moment

$$I_{y0} = F_K l_K^2/(\pi^2 E) \ .$$

This then also approximately applies for proof of buckling in the inelastic range.

In reality the exact bending line (characteristic function) is unknown. Therefore according to Ritz a comparison function $w(x)$ that satisfies the boundary conditions is inserted in Eq. (11). An approximate value is produced for F_K that is always greater than the exact buckling load, since for the exact characteristic function the strain energy is at a minimum and always somewhat too high for the comparison function. The bending lines for the corresponding beam at any load can be considered as examples of comparison functions.

For other and improved approximation methods see [1–5].

Example *Comparative calculation of the buckling load for a bar of constant cross-section and support in accordance with Euler's case 2 using the energy method.* The bending line under single load in accordance with B2, **Table 5a**, case 1, is selected as the comparison function: $w(x) = c_1(3l^2x - 4x^3)$ for $0 \le x \le l/2$. Taking $w'(x) = c_1(3l^2 - 12x^2)$ and $w''(x) = -24c_1x$, after integration in accordance with Eq. (11), $2W = c_1^2 \cdot 48EI_y l^3$, $2v = c_1^2 l^5 \cdot 4.8$ and thus $F_K = 10.0EI_y/l^2$. This result is 1.3% greater than the exact result $\pi^2 EI_y/l^2$.

7.1.5 Columns with Variable Cross-section or Axial Load

This can be calculated in accordance with B7.1.4. DIN 4114 Part 2, Table 4, gives the equivalent area moments I_m for I-sections and Table 5 gives the equivalent buckling lengths for linearly and parabolically varying longitudinal force. For other cases, see [4].

7.1.6 Buckling of Rings, Frames and Systems of Bars

Closed annular beams under external load $q =$ const. (**Fig. 4b**). [4] applies for buckling in the loading plane if the load is always perpendicular to the axis of the bar, $q_K = 3EI_y/R^3$, and if the load maintains its original direction, $q_K = 4EI_y/R^3$. Outward buckling perpendicular to the plane of the beam is produced for

$$q_K = 9EI_zGI_t/[R^3(4GI_t + EI_z)] \ .$$

Closed Frames (**Fig. 4c**). For outward buckling in the plane of the frame, $F_K = \alpha^2 EI_1$ is produced for the critical load from the eigenvalue equation [4] for α:

$$\frac{\alpha l_1}{\tan(\alpha l_1)} - \frac{l_1(\alpha^2 l_2^2 I_1^2 - 36I_2^2)}{12I_2 I_1 I_2} = 0 \ .$$

For further results, including those for systems of bars, see [2, 4].

7.1.7 Torsional Buckling

In addition to pure flexural bending in a bar loaded with a longitudinal force (and torsion moment), a three-dimensional curved and twisted position of equilibrium, torsional buckling, can occur. Torsional buckling alone (without outward buckling) is also possible, owing to longitudinal force.

Bars with Circular Cross-section (Shafts)

Suitable differential equations for the problem are given in [3]. Torsional buckling due to the torsion moment occurs for $M_{tK1} = 2\pi EI_y/l$.

It is significant only for very slender shafts and wires. If longitudinal force F and torsion moment M_t are applied together, then for a bar hinged at both ends

$$F_K = \frac{\pi^2 EI_y}{l^2}\left(1 - \frac{M_t^2}{M_{tK1}^2}\right), \quad M_{tK} = M_{tK1}\sqrt{1 - \frac{Fl^2}{\pi^2 EI_y}} \ .$$

Bars with Arbitrary Cross-section Under Longitudinal Load

Double Symmetrical Cross-section. Shear centre and centre of gravity coincide and the following three differential equations apply:

$$EI_y w'''' + Fw'' = 0, \quad EI_z v'''' + Fv'' = 0 \ , \\ EC_M \varphi'''' + (Fi_p^2 - GI_t)\varphi'' = 0 \tag{12}$$

The first two give the known Euler buckling loads; the third means that pure torsional buckling (without flexure) is possible and for a bearing which is hinged at both ends, from $\varphi(x) = C \sin(\pi x/l)$, i.e. where $\varphi = 0$, at the ends, it gives the buckling load

$$F_{Kt} = (GI_t + \pi^2 EC_M/l^2)/i_p^2 \ . \tag{13}$$

C_M is the deflection resistance due to constrained warping [2], e.g. for an IPB cross-section $C_M = I_y b^2/4$, b being the distance between the centre of the flanges. For full cross-sections, $C_M \approx 0$. F_{Kt} can be significant only for small buckling lengths l. For standard I-profiles I_z, i.e. buckling in the y-direction, is always significant and not torsional buckling.

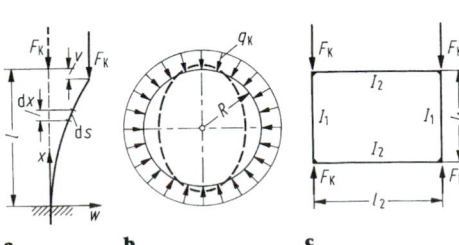

Figure 4.

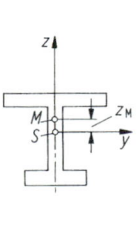

Figure 5.

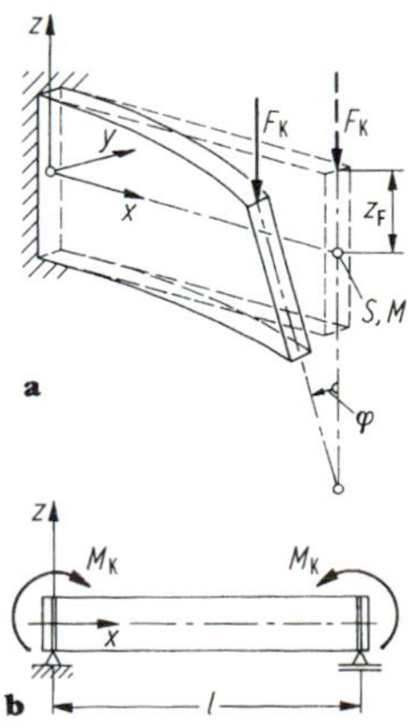

Figure 6.

Single Symmetric Cross-section (**Fig. 5**). If z is the axis of symmetry, the second and third of Eqs (12) apply together [2, 5], i.e. torsional–flexural buckling is possible. For buckling around the y-axis (in direction z) the standard Euler buckling load $F_{Ky} = \pi^2 EI_y / l^2$ applies. The other two critical loads are obtained for forked support at the ends from

$$\frac{1}{F_K} = \frac{1}{2}\left[\frac{1}{F_{Kz}} + \frac{1}{F_{Kt}}\right.$$
$$\left. \pm \sqrt{\left(\frac{1}{F_{Kz}} - \frac{1}{F_{Kt}}\right)^2 + \frac{4}{F_{Kz}F_{Kt}}\left(\frac{z_M}{i_M}\right)^2}\right],$$

where F_{Kt} is in accordance with Eq. (13), $F_{Kz} = \pi^2 EI_z / l^2$, i_M is the polar radius of gyration in respect of the shear centre and z_M is the distance of the shear centre from the centre of gravity.

7.2 Lateral Buckling of Beams

When the critical load is reached due to bending and twisting, short beams assume a marked adjacent position of equilibrium (**Fig. 6a**). The corresponding differential equation for double symmetrical cross-sections is

$$EC_M \varphi'''' - GI_t \varphi'' - (M_y^2/EI_z - M_y'' z_F)\varphi = 0 , \quad (14)$$

where φ is the angle of torsion, z_F the height of the point at which the load is applied above the shear centre (centre of gravity in this case) and C_M the deflection resistance. The non-linear differential equation cannot generally be solved by itself. For approximate solutions see [1, 4, 5]. For full cross-sections, $C_M \approx 0$.

7.2.1 Beams of Rectangular Cross-section

(a) **Forked support** and *application of two equal moments M_K at the ends* (**Fig. 6b**). In this case Eq. (14) becomes $\varphi''(x) + \dfrac{M_K^2}{EI_z GI_t}\varphi(x) = 0$. The solution which satisfies the boundary conditions $\varphi(x) = C \sin(\pi x / l)$ gives the following for the critical buckling moment:

$$M_K = (\pi/l)\sqrt{EI_z GI_t} = (\pi/l)K .$$

If deformations of the basic condition [4] are taken into account more accurately then $K = \sqrt{EI_z GI_t(I_y - I_z)/I_y}$.

(b) **Forked bearing** and *single load F_K in the centre of the beam* (load applied at height z_F):

$$F_K = \frac{16.93}{l^2}K\left(1 - z_F \cdot \frac{3.48}{l}\sqrt{\frac{EI_z}{GI_t}}\right).$$

(c) **Cantilever** with *single load F_K at the end* (load applied at height z_F) as shown in **Fig. 6a**:

$$F_K = \frac{4.013}{l^2}K\left(1 - \frac{z_F}{l}\sqrt{\frac{EI_z}{GI_t}}\right).$$

7.2.2 I-Beams

The deflection resistance $C_M \approx I_z h^2/4$ should be taken into account. With the abbreviation $\chi = \dfrac{EI_z}{GI_t}\left(\dfrac{h}{2l}\right)^2$ the following similarly applies for the cases given in B7.2.1 (h being the distance between flange centres):

(a) $M_K = (\pi/l)K\beta_1, \qquad \beta_1 = \sqrt{1 + \pi^2\chi}$.

(b) When the load is applied at the height of the centre of gravity ($z_F = 0$)

$$F_K = (16.93/l^2)K\beta_1, \quad \beta_1 = \sqrt{1 + 10.2\chi} ;$$

when the load is applied to the top or bottom flange,

$$F_K = (16.93/l^2)K\beta_1\left(\sqrt{1 + 3.24\chi/\beta_1^2} \mp 1.80\sqrt{\chi/\beta_1^2}\right).$$

(c) When the load is applied at the height of the centre of gravity ($z_F = 0$)

$$F_K = (4.013/l^2)K\beta_1, \quad \beta_1 = \left(\frac{1 + 1.61\sqrt{\chi}}{1 + 0.32\sqrt{\chi}}\right)^2 .$$

7.3 Buckling of Plates and Shells

When the critical load is reached, plates and shells change to an adjacent (outwardly bulging) stable position of equilibrium.

7.3.1 Buckling of Plates

Rectangular Plates (**Fig. 7a–c**). With plate thickness h and plate stiffness $N = Eh^3/[12(1 - \nu^2)]$, assuming that Hooke's law applies, the differential equation for the problem is

$$N\Delta\Delta w + h\left(\sigma_x \frac{\partial^2 w}{\partial x^2} + \sigma_y \frac{\partial^2 w}{\partial y^2} + \tau \frac{\partial^2 w}{\partial x\partial y}\right) = 0 . \quad (15)$$

(a) *Plate hinged on all sides under longitudinal stresses σ_x*. With the product formulation satisfying the boundary conditions

$$w(x,y) = c_{mn}\sin(m\pi x/a) \sin(n\pi y/b) ,$$

insertion in differential Eq. (15) gives

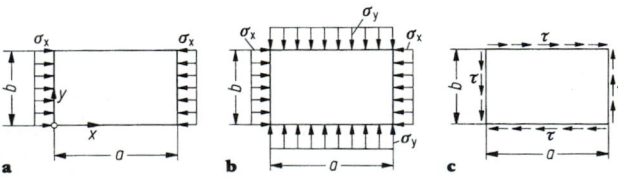

Figure 7.

$$\pi^2 N \left(\frac{m^2}{a^2} + \frac{n^2}{b^2} \right)^2 = b\sigma_x \frac{m^2}{a^2} \quad \text{or}$$

$$\sigma_x = \frac{\pi^2 N}{b^2 b} \left(m \frac{b}{a} + \frac{n^2}{m} \frac{a}{b} \right)^2.$$

This gives the (minimum) critical buckling stresses:

For $a < b$, $m = n = 1$: $\sigma_{xK} = \frac{\pi^2 N}{b^2 b} \left(\frac{b}{a} + \frac{a}{b} \right)^2$.

For $a = b$, $m = n = 1$: $\sigma_{xK} = \frac{4\pi^2 N}{b^2 b}$.

For $a > b$: If the side ratio a/b is a whole number, the plate is divided into individual squares by nodal lines and $\sigma_{xK} = 4\pi^2 N/(b^2 b)$ again applies. This value is also used for side ratios, which are not whole numbers as the actual values are only slightly above this.

(b) *Plates hinged on all sides under longitudinal stresses σ_x and σ_y.* With the formulation as in (a),

$$\sigma_x = \frac{\pi^2 N}{b^2 b} \frac{(m^2 b^2/a^2 + n^2)^2}{m^2 b^2/a^2 + n^2 \sigma_y/\sigma_x}.$$

When the aspect ratios b/a and stress ratio σ_y/σ_x are given, the (whole-number) values m and n should be selected so that σ_x is at minimum σ_{xK}.

For the special case of equal pressure on all sides $\sigma_x = \sigma_y = \sigma$,

$$\sigma = \frac{\pi^2 N}{b^2 b} \left(m^2 \frac{b^2}{a^2} + n^2 \right)$$

with the minimum for $m = n = 1$

$$\sigma_K = \frac{\pi^2 N}{b^2 b} \left(\frac{b^2}{a^2} + 1 \right).$$

(c) *Plates hinged on all sides under shear stresses.* There is no exact solution. With a five-part Ritz formulation, using the energy method, i.e. from $\Pi = W - W^{(a)} = \min$, the following approximate formulae are obtained (see [4, 6]):

For $a \le b$: $\tau_K = \frac{\pi^2 N}{b^2 b} \left(4.00 + 5.34 \frac{b^2}{a^2} \right)$,

For $a \ge b$: $\tau_K = \frac{\pi^2 N}{b^2 b} \left(5.34 + 4.00 \frac{b^2}{a^2} \right)$.

(d) *Infinitely long hinged strips of plate under single loads* **(Fig. 8)**.

$$F_K = \frac{8b}{\pi} \frac{\pi^2 N}{b^2} = \frac{8\pi N}{b}.$$

For further results for rectangular plates, see [4].

Circular Plate. (Fig. 9a-c)

(a) *Full circular plate with constant radial pressure σ.* This problem is relatively easy to solve [1]. In accordance with B5.2.1 $\sigma_r = \sigma_t = \sigma$ and $\tau_{rt} = 0$ apply for the state of stress of the disc. Differential equation (15) therefore takes the form

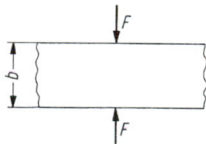

Figure 8.

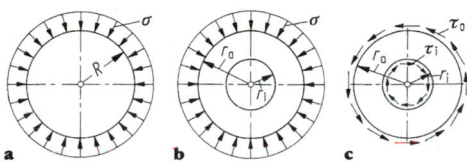

Figure 9.

$$N\Delta\Delta w + b\sigma\Delta w = 0 \quad \text{or} \quad \Delta(\Delta + \alpha^2)w = 0, \quad \alpha^2 = b\sigma/N.$$

It is satisfied if

$$(\Delta + \alpha^2)w = 0 \quad \text{and} \quad \Delta w = 0$$

or because $\Delta = d^2/dr^2 + (1/r)d/dr$, if

$$\frac{d^2 w}{dr^2} + \frac{1}{r}\frac{dw}{dr} + \alpha^2 w = 0 \quad \text{and} \quad \frac{d^2 w}{dr^2} + \frac{1}{r}\frac{dw}{dr} = 0.$$

The solution to this equation is

$$w(r) = C_1 J_0(\alpha r) + C_2 N_0(\alpha r) + C_3 + C_4 \ln r$$

where J_0 and N_0 are respectively Bessel's and Neumann's zero function. Satisfying the boundary conditions $w(R) = 0$ and $M_r(R) = 0$ (for the hinged plate) or $w(R) = 0$ and $w'(R) = 0$ (for the constrained plate) as well as the additional conditions $w'(0) = 0$ and finally $w(0)$ lead to the eigenvalue equations

$$\alpha R J_0(\alpha R) - (1 - \nu)J_1(\alpha R) = 0 \quad \text{(hinged plate)}$$

and

$$J_1(\alpha R) = 0 \quad \text{(constrained plate)}$$

This gives the buckling stresses

$$\sigma_K = 4.20 N/(R^2 b) \quad \text{(hinged plate, } \nu = 0.3) \quad \text{and}$$

$$\sigma_K = 14.67 N/(R^2 b) \quad \text{(constrained plate)}.$$

(b) *Circular plate with constant radial pressure.* The mathematical solution is more complicated than in (a) (see [3]). If the internal edge is free

$$\sigma_K = c_1 N/(r_a^2 b) \quad \text{(hinged plate)} \quad \text{and}$$

$$\sigma_K = c_2 N/(r_a^2 b) \quad \text{(constrained plate)} \quad \textbf{(Table 3)}$$

Table 3. Correction values c_1 and c_2 for $\nu = 0.3$

$r_i/r_a =$	0	0.2	0.4	0.6	0.8
c_1	4.2	3.6	2.7	1.5	2.0
c_2	14.7	13.4	18.1	≈ 40	–

(c) *Circular plate with shear stresses.* If τ_a and $\tau_i = \tau_a r_a^2/r_i^2$ are the shear stresses applying, for constrained edges

$$\tau_{aK} = c_3 N/(r_a^2 b).$$

For $\nu = 0.3$ and $r_i/r_a = 0.1; 0.2; 0.3; 0.4$, $c_3 \approx 17.8; 37.0; 61.0; 109.0$.

For further results for circular and annular plates see [4].

7.3.2 Buckling of Shells

Spherical Shells Under Constant External Pressure p. The complicated differential equations are given in [7] and [8] for example. The smallest critical buckling pressure (in accordance with this theory as a branch problem) is given as

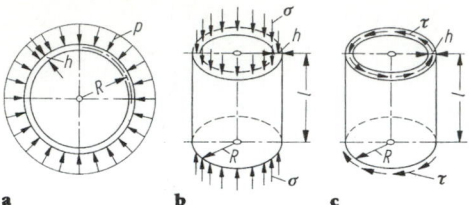

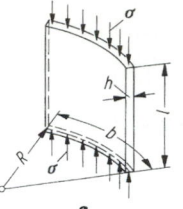

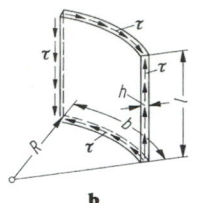

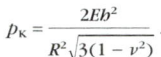

Figure 10.

Figure 11.

$$p_K = \frac{2Eb^2}{R^2\sqrt{3(1-\nu^2)}} .$$

However, shells can puncture, i.e. assume stable positions of equilibrium in the case of infinitely large deformations. In that case, in accordance with [9]

$$p_K = 0.365Eb^2/R^2 ,$$

i.e. this buckling load is only around one-third that of the branching problem!

Cylindrical Shells (Fig. 10a–c)

(a) *Under Constant External Radial Pressure p.* For infinitely long shells,

$$p_K = 0.25Eb^3/[R^3(1-\nu^2)] .$$

For results for short shells see [4].

(b) *Under Longitudinal Axial Stress σ.* For the derivative of the exact differential equations, see [8] and [9]. The following applies approximately for the smallest critical longitudinal load [9]:

$$\sigma_K = Eb/[R\sqrt{3(1-\nu^2)}] ,$$

if a sufficient number of bending shafts can be inserted in the longitudinal direction. This is the case if $l \geq 1.73\sqrt{bR}$ (for materials with $\nu = 0.3$). In the case of short lengths the shell can be recorded as at the circumference of supported strips of shell (see below for solutions). Furthermore, in the case of cylindrical shells the problem of puncturing should also be taken into account as this leads to small buckling stresses. In accordance with [9], the following approximate formula applies in this case:

$$\sigma_K = \frac{0.605 + 0.000\,369R/b}{1 + 0.006\,22R/b} \cdot \frac{Eb}{R} .$$

Collapse of the shell as a whole, i.e. as a longer bar, occurs for $\sigma_K = \pi^2ER^2/(2l^2)$.

(c) *Under Torsional Shear Stresses τ.* In accordance with [9], $\tau_K = 0.747 \dfrac{Eb^2}{l^2}\left(\dfrac{l}{\sqrt{Rb}}\right)^{3/2}$ applies for buckling stress. This factor should be further multiplied by the factor 0.7 to take account of pre-buckling.

Cylindrical Shell Strips (Fig. 11a, b)

(a) *Under Longitudinal Stress σ if the Longitudinal Edges are Hinged*

For $b/\sqrt{Rb} \leq 3.456$: $\sigma_K = \dfrac{\pi^2Eb^2}{3(1-\nu^2)b^2} + \dfrac{Eb^2}{4\pi^2R^2}$;

for $b/\sqrt{Rb} \geq 3.456$: $\sigma_K = \dfrac{2E}{\sqrt{12(1-\nu^2)}}\dfrac{b}{R}$.

(b) *Under Shear Stress τ if the Longitudinal Edges are Hinged.* The critical shear stresses are given by

$$\tau_K = 4.82\left(\frac{b}{b}\right)^2 E\sqrt[4]{1 + 0.0146\frac{b^4}{R^2b^2}} .$$

7.3.3 Inelastic (Plastic) Buckling

The formulae in B7.3.1 and B7.3.2 give the buckling stresses assuming that the material has elastic behaviour. They can also be used as a basis for the inelastic range if they are reduced in the same proportion as given for buckling stress of bars from Euler's curve and the Engesser-von Kármán curve (approximately Tetmajer's line). See DIN 4114 Part 1, Table 7, in this respect for St 37 and St 52.

---

8 Finite-Element and Boundary-Element Methods

8.1 Finite Elements

The finite-element method (FEM) is an efficient method for numerical solution of problems relating to stress and strain of all types (including problems relating to stability) in the elastic and plastic range. It is based on the solution of high-order linear systems of equations by means of powerful computers. For clarity the systems of equations are constructed as appropriate using the matrix method. The system to be calculated, called the structure, is divided into consistent elements connected by nodes

(**Fig. 1**). In the *displacement method* the node displacements are introduced as unknowns and in the *flexibility matrix method* it is the stresses. The rigidity matrix (generalised spring parameter), for which the system of equations for the unknown displacements follows from the equilibrium conditions for all nodes [1–4], is produced for each element as a result of the unit displacements of its nodes taking into account the definitive material law (e.g. Hooke's law).

The most important basic elements (**Fig. 2a–f**) of all structures are the compression–tension rod element (a),

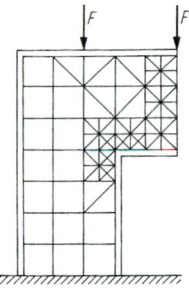

Figure 1. Machine supports and finite elements.

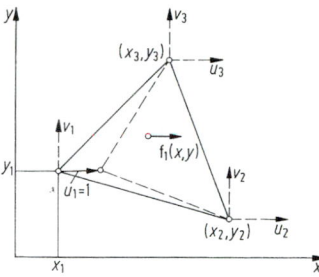

Figure 3. Plane triangular element with displacement condition $u_1 = 1$.

the beam element (b), the triangular and rectangular disc element (c), the triangular and rectangular plate element (d), the three-dimensional curved shell element (e) and the tetrahedral volume element (f). In the following the plane state of stress and especially division into triangular elements (**Fig. 2c**) are observed:

Displacements. In the first approximation these are linear for the element boundaries and the interior of the element. For unit displacement $u_1 = 1$ the displacement function is then (**Fig. 3**)

$$f_1(x,y) = \frac{1}{2A}[x(y_3 - y_2) + y(x_2 - x_3) + x_3 y_2 - x_2 y_3] ,$$ (1)

A is the area of the element. The same function arises for $v_1 = 1$. Corresponding functions $f_2(x, y)$ and $f_3(x,y)$ follow for $u_2 = 1$ and $v_2 = 1$ and $u_3 = 1$ and $v_3 = 1$:

$$f_2(x,y) = \frac{1}{2A}[x(y_1 - y_3) + y(x_3 - x_1) + x_1 y_3 - x_3 y_1] ,$$

$$f_3(x,y) = \frac{1}{2A}[x(y_2 - y_1) + y(x_1 - x_2) + x_2 y_1 - x_1 y_2] .$$

$$\left. \begin{array}{l} u(x,y) = f_1(x,y)u_1 + f_2(x,y)u_2 + f_3(x,y)u_3 , \\ v(x,y) = f_1 v_1 + f_2 v_2 + f_3 v_3 \end{array} \right\} .$$ (2)

then applies for the total displacement inside the element (and on the boundary) as a result of the unit displacements.

u and v make up displacement vector $\boldsymbol{v}$. In matrix form,

$$\begin{pmatrix} u \\ v \end{pmatrix} = \begin{pmatrix} f_1 & f_2 & f_3 & 0 & 0 & 0 \\ 0 & 0 & 0 & f_1 & f_2 & f_3 \end{pmatrix} \begin{pmatrix} u_1 \\ u_2 \\ u_3 \\ v_1 \\ v_2 \\ v_3 \end{pmatrix} ;$$ (3)

or, in abbreviated form,

$$\boldsymbol{v}(x,y) = \boldsymbol{f} \boldsymbol{v}_k \quad (k = 1, 2, 3) .$$ (4)

Elongations and Shears

$$\varepsilon_x = \frac{\partial u}{\partial x} = \frac{1}{2A}[(y_3 - y_2)u_1 + (y_1 - y_3)u_2 + (y_2 - y_1)u_3]$$

$$= g_1 u_1 + g_2 u_2 + g_3 u_3 ,$$

$$\varepsilon_y = \frac{\partial v}{\partial y} = \frac{1}{2A}[(x_2 - x_3)v_1 + (x_3 - x_1)v_2 + (x_1 - x_2)v_3]$$

$$= g_4 v_1 + g_5 v_2 + g_6 v_3 ,$$

$$\gamma_{xy} = \frac{\partial u}{\partial y} + \frac{\partial v}{\partial x} = g_4 u_1 + g_5 u_2 + g_6 u_3 + g_1 v_1 + g_2 v_2 + g_3 v_3$$

follow from Eq. (2) for the constant elongations and

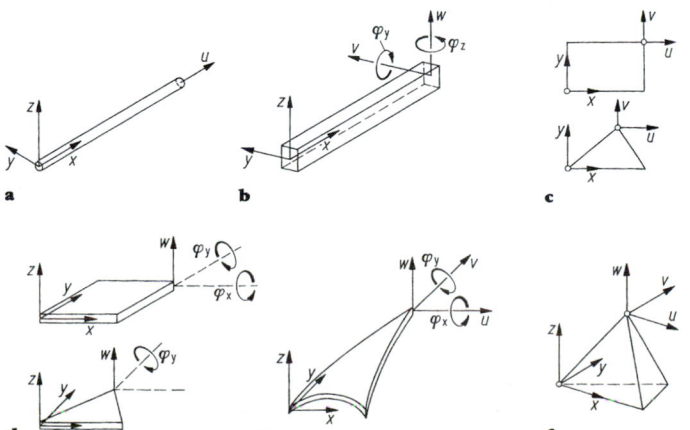

Figure 2a–f. Types of finite elements.

shearing of the element $\varepsilon_x, \varepsilon_y, \gamma_{xy}$ (see B1 Eqs (12, 13)); or, in matrix form,

$$
\begin{pmatrix} \varepsilon_x \\ \varepsilon_y \\ \gamma_{xy} \end{pmatrix} = \frac{1}{2A} \begin{pmatrix} g_1 & g_2 & g_3 & 0 & 0 & 0 \\ 0 & 0 & 0 & g_4 & g_5 & g_6 \\ g_4 & g_5 & g_6 & g_1 & g_2 & g_3 \end{pmatrix} \begin{pmatrix} u_1 \\ u_2 \\ u_3 \\ v_1 \\ v_2 \\ v_3 \end{pmatrix},
$$

in abbreviated form

$$\varepsilon = gv_k . \tag{5}$$

Stresses

Using a material law (stress–strain relationship), e.g. Hooke's law (see B3 Eq. (13)),

$$\sigma = E\varepsilon = Egv_k \tag{6}$$

applies in matrix form and with Eq. (5). With Poisson's ratio v

$$
E = \frac{E}{1 - v^2} \begin{pmatrix} 1 & v & 0 \\ v & 1 & 0 \\ 0 & 0 & (1-v)/2 \end{pmatrix}. \tag{7}
$$

Nodal forces are produced as a function of displacements v_k via the principle of equilibrium of virtual work (see B2.4.9) in matrix [1 to 7]

$$F\delta v_k^T = \iint_{(A)} \sigma \delta \varepsilon^T h \, dx \, dy . \tag{8}$$

$F = F_k = \{F_{kx}, F_{ky}\}$ is the vector of nodal forces of an element, T is the transposed matrix and h the thickness of the element.

$$F\delta v_k^T = \iint_{(A)} Egv_k g^T \, \delta v_k^T h \, dx \, dy$$

follows with Eqs (5) and (6) and, since v_k and δv_k are independent of x and y and likewise E, g and g^T are constant with regard to the element, this gives

$$F = Egg^T hAv_k = kv_k . \tag{9}$$

A is the area of the element. The rigidity matrix of the element is found with k. The elements are then combined to form the total structure by creating an equilibrium at each node. They are combined either using the direct method of superposing the element rigidity matrices for one node or mathematically by transformation via a Boolean matrix [5].

$$F^{(a)} = Kv , \tag{10}$$

a matrix equation for n nodal points with $2n$ displacements where K is the system rigidity matrix, follows, with $F^{(a)}$ as the vector of the external forces. Taking into account m displacement boundary conditions, Eq. (10) represents a system of $2n - m$ equations for the displacements of the nodes. When these have been calculated the associated stresses at the nodal points follow from Eq. (7). Many computers have systems of programs for carrying out the complex calculations. See [3, 4, 7] for some introductory examples; see [5, 6] for theoretical developments of the FEM.

Applications. Without exception the following examples were calculated assuming a linear elastic material law (Hooke's law) on a 386 PC.

1 Beam Elements: Frames with Semi-circular Curves (B2.7 **Fig. 57**). Given: $a = 600$ mm, $b = 1200$ mm, line load $q = 18$ kN/m, cross-sectional measurements $B = 100$ mm, $H = 40$ mm. 32 beam elements connected by 32 nodes were selected as the structure (**Fig. 4**). The computer program produced the nodal displacements from 136 equations. The extreme values were as follows: nodes *1* and *17* in direction y 17.6 mm to the outside, nodes *9* and *25* in direction x 7.25 mm to the inside. The stresses calculated from the displacements by the program have the following extreme values: at external nodes *1* and *17*, $\sigma_a = +186.1$ N/mm² and internally $\sigma_i = -180.8$ N/mm², at external nodes *9* and *25* $\sigma_a = -175.7$ N/mm² and internally $\sigma_i = +186.4$ N/mm². In comparison, according to the formulae in B2.7 the following are produced: the moments $M_1 = -0.76qa^2 = -4.925$ kN m, $M_2 = +0.74qa^2 = +4.795$ kN m and with the longitudinal forces $F_{N1} = qa = 10.8$ kN, $F_{N2} = qb = 21.6$ kN as with $A = BH = 4000$ mm², $W_b = BH^2/6 = 26667$ mm³ the stresses at nodes *1* and *17*: $\sigma_{a,i} = F_{N1}/A \mp M_1/W_b = (2.7 \pm 184.7)$ N/mm² $= +187.4$ N/mm² and -182.0 N/mm² and at nodes *9* and *25*: $\sigma_{a,i} = F_{N2} \mp M_2/W_b = (5.4 \mp 179.8)$ N/mm² $= -174.4$ N/mm² and $+185.2$ N/mm².

Thus the maximum deviation in the results produced using the FEM from the results produced by using the formulae (node 9, external) is only 0.75%.

2 Shell Elements: Plate Rod with Hole Under Uni-axial Tensile Load (**Fig. 5a**). Given $l = 480$ mm, $b = 120$ mm, $d = 60$ mm, thickness of disc $h = 10$ mm, tensile load $\sigma = 80$ N/mm². The structure was constructed using 336 disc elements connected by 384 nodes (**Fig. 5b**). From a total of 758 equations the computer program produced the displacements of all node points where the maximum values (at the nodes of the free boundary) equal $u = 0.218$ mm. The stresses on all elements calculated from the displacements have their maximum values at node points *1* and *209* with $\sigma_x = 328.4$ N/mm², while the stress at nodes *92* and *300* is $\sigma_x = 84.9$ N/mm².

According to the FEM, the nominal stress $\sigma_n = \sigma \cdot b/(b - d) = 160$ N/mm² gives the stress concentration $\alpha_k = \sigma_x/\sigma_n = 328.4/160 = 2.05$, while the value $\alpha_k = 2.15$ is produced in the original stress concentration factor diagram according to Wellinger-Dietmann [8] for $a/b = 600/1200 = 0.5$. The elongation of the rod according to Hooke's law is $\Delta l = l \cdot \sigma/E = 480$ mm $\cdot 80$ N/mm²/$(2.1 \cdot 10^5$ N/mm²$) = 0.183$ mm, where the difference from the FEM result represents the influence of the hole. If an approximate calculation is made along the length of the hole taking the nominal cross-section, then $u = (l - d) \cdot \sigma/E + d \cdot \sigma_n/E = 0.16$ mm $+ 0.046$ mm $=$

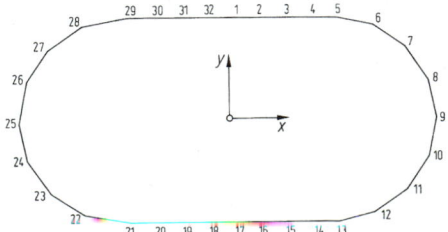

Figure 4. Frames in accordance with B2.7 **Figure 57**, with 32 nodes.

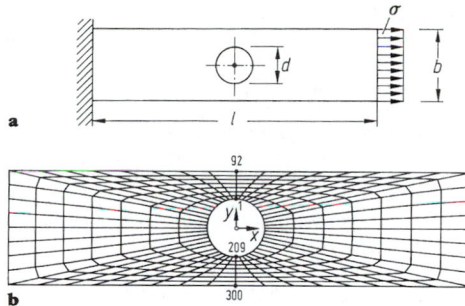

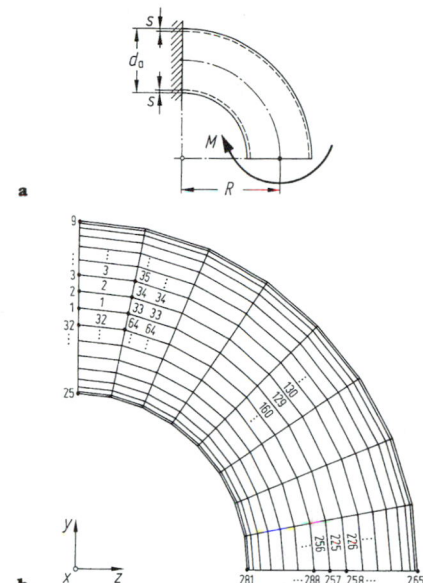

Figure 5. Plate rod with hole: **a** loading diagram; **b** structure made up of 336 disc elements and 384 nodes.

0.206 mm is produced. This approximation only produces a deviation of 5.5% from the result produced using the FEM, which is obviously more accurate.

3 Plate Elements: Restrained Cover Plate with Feed Opening (Annular Plate) (**Fig. 6a**). Given: $d_1 = 2400$ mm, $d_2 = 600$ mm, $b = 10$ mm, surface load $p = 5$ kN/m². After dividing the structure into 216 plate elements with 240 nodes (**Fig. 6b**) the computer program gives the displacement (deflections) of all node points from 1296 equations and from this the stresses on all elements. Then the maximum deflection at the free internal boundary (node *1*) is given as $f = 8.02$ mm and the maximum peripheral stress as $\sigma_t = 40.7$ N/mm² and at the restraint (node *10*) the maximum radial stress $\sigma_r = 54.2$ N/mm². The plate theory (see B5 [5]) gives the same value 8.02 mm for the deflection of the internal boundary and for the stresses at the free boundary $\sigma_t = 40.9$ N/mm² and $\sigma_r = 51.1$ N/mm² at the restrained boundary, so that, for the latter, the deviation of the result obtained using the FEM from that obtained using the plate theory is 6.1%, which is due to the element being divided too coarsely.

4 Shell Elements: Tube Bends Under Bending Load (**Fig. 7a**). Given: Steel tube DN 500 with $d_a =$

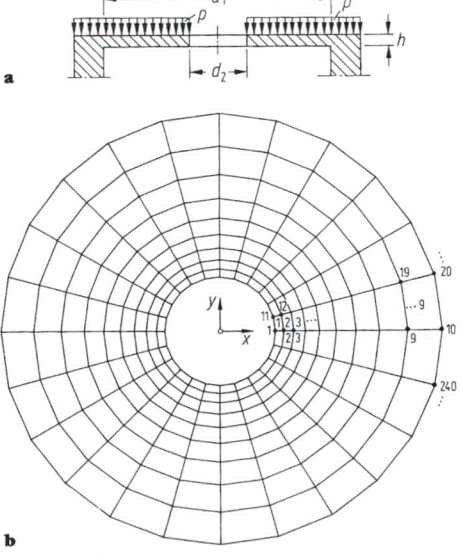

Figure 6. Annular plate: **a** construction and loading; **b** structure made up of 216 plate elements and 240 nodes.

Figure 7. Tube bends: **a** arrangement and loading; **b** structure with 256 shell elements and 288 nodes.

508 mm, wall thickness $s = 6.3$ mm, $R = 762$ mm, $M = 25$ kN m. The structure was divided into $8 \cdot 32 = 256$ shell elements connected at $9 \cdot 32 = 288$ node points (**Fig. 7b**). The computer program calculated the displacements of all nodes in directions x, y and z of all nodes from 1536 equations, and from this the stresses on all elements. Then at the free end, for example, the resulting displacement of point P on the tube axis is given as $f = 4.91$ mm, and at element *129* the maximum occurring reduced stress as $\sigma_V = 149.4$ N/mm². In contrast, the rod statics (cross-sections maintain their circular shape) for the displacement of point P produces the value $f_{St} = 0.254$ mm, from which with the Kármán factor $K = Rs/(1.65 r_m^2) = 762$ mm $\cdot$ 6.3 mm/$(1.65 \cdot 250 \cdot 85^2$ mm²$) = 0.0462$ follows taking into account the cross-sectional deformation [9] and finally $f = f_{St}/K = 5.50$ mm.

The result produced by the FEM deviates from this value by 10.7% and this means that Kármán's theory still applies to curved tube bends of this type. With regard to stress the nominal stress (from rod statics) is given by $\sigma_V = i_{V,el} \cdot \sigma_0$, where $\sigma_0 = M_b/W_b$ where $W_b = \pi d_m^2 s/4 = 1245.4 \cdot 10^3$ mm³, that is $\sigma_0 = 25 \cdot 10^6$ N mm/$(1245.4 \cdot 10^3$ mm³$) = 20.07$ N/mm² and the stress intensification factor in the elastic range as $i_{V,el} = 149.4/20.07 = 7.44$. For tube bends certain plastic deformations are generally permitted, which leads to an increase in load-bearing capacity (support figure *n*; see B9) and is taken into account by $i_{V,pl} = i_{V,el}/n$, i.e. by a reduced stress intensification factor. For example, in accordance with ASME Code ANSI B31.1, for $i_{V,pl} = 0.9/\sqrt[3]{b^2}$ where $b = 4Rs/d_m^2 = 0.07629$, therefore $i_{V,pl} = 5.00$, so that in this case the support figure $n = i_{V,el}/i_{V,pl} = 7.44/5.00 = 1.49$ which is approximately the same as the value $n = 1.43$ for a rectangular cross-section (see B9).

8.2 Boundary Elements

The boundary element method (BEM) is an integral equation method which originates from the fact that it is possible to trace back the solution for a differential equation to an integral equation via Green's function and the loading function. Green's function (influence coefficient) is a function that satisfies the boundary condition and the differential equation as a result of a single load $F = 1$.

Beam. The differential equation for deflections is $w''''(x) = -q(x)/EI_y$ for the known case of beam bending (cf. B2.4.8). In the case of a hinged beam with boundary conditions $w(x = 0) = w''(x = 0) = w(x = 1) = w''(x = 1) = 0$ (**Fig. 8a**) the solution for deflections applies in the form of an integral equation:

$$w(x) = \int_0^l G_0(x,\xi)q^*(\xi)\,d\xi = \int_0^l \eta_0(x,\xi)q^*(\xi)\,d\xi \,, \quad (11)$$

taking $q^*(x) = q(x)/EI_y$, where $G(x, \xi)$ is Green's function (influence coefficient) for deflection at point x due to a moving load $F = 1$ at point ξ (**Fig. 8b**). In recent literature y is used for the running variable, instead of the Greek letter ξ, and this is the case below. As the differential equation $w''''(x) = 0$ applies for $F = 1$, quadruple integration gives a parabola of the third order for the Green function, which must also satisfy the boundary conditions. We are already familiar with a function of this kind from B2, **Table 5a**, case 2, if we take $a = x$, $b = (1 - x)$ and $x = y$, as well as $F = 1$. It reads

$$G_0(x,y) = \eta_0(x,y) =$$

$$\frac{1}{6EI_y}\begin{cases} x(l-x)(2l-x)y - (l-x)y^3 & \text{for } 0 \le y \le x \,, \\ x(l^2-x^2)(l-y) + x(l-y)^3 & \text{for } x \le y \le l \,. \end{cases} \quad (12)$$

If we insert the influence coefficient (12) in Eq. (11), we obtain the elastic curve $w(x)$ for each loading function $q(x)$. By single differentiation with the point coordinate x, Green's function (12) also gives the line of influence for bending angle $\eta_\alpha(x,y) = \partial \eta_0/\partial x$, by double differentiation with x it gives the line of influence for the bending moment $\eta_M(x,y) = EI_y\partial^2\eta_0/\partial x^2$ and by triple differentiation with x it gives the line of influence for the transverse forces $\eta_Q(x,y) = -EI_y\partial^3\eta_0/\partial x^3$. Conversely, for a fixed place of loading $y = x$, by derivation with running variables y from Eq. (12), the first derivative gives the line of the angle of inclination $\alpha(y, x)$, the second derivative gives the line of the bending moment $M_b(y) = -EI_y\partial^2\eta_0/\partial y^2$ and the third derivative with y gives the shear stress curve $F_Q(y)$.

Summary. If Green's function is known for a problem involving a differential equation, i.e. a solution that satisfies the boundary conditions as a result of a moving load $F = 1$, which also satisfies the differential equation, according to Eq. (11), the problem can be solved for any loading function.

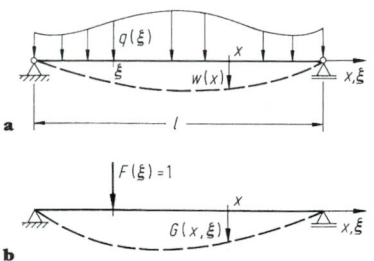

Figure 8. Single-span girders: **a** with line load; **b** with moving load.

Discs, Plates and Dishes. For these, it is only in extremely rare cases that Green's functions are known; the solution is known e.g. for a plate under a single load at any point (y_1, y_2) for each location (x_1, x_2) that satisfies the boundary conditions. However, so-called basic or fundamental solutions for $w(x_1, x_2, y_1, y_2)$ due to a single load $F = 1$ in (y_1, y_2) are always known for discs, plates and dishes [11] and these may be considered as a solution for an infinitely extended disc, plate or dish. In this case, as a solution to the actual boundary value problem, the boundary element method is used as follows. Imagine, e.g., the actual plate cut out of the infinite range Ω, first apply the actual load $q(y_1, y_2)$ and then the single load $\hat{F}(x_1, x_2) = 1$ and all boundary section sizes and boundary deformations (**Fig. 9a, b**) and use *Betti's law:* for two equilibrium conditions of a system (F, M) and $(\hat{F}, \hat{M})$ with the relevant deformations (w, α) and $(\hat{w}, \hat{\alpha})$, for the work:

$$\sum \hat{F}w + \sum \hat{M}\alpha = \sum F\hat{w} + \sum M\hat{\alpha} \,, \quad W_{1.2} = W_{2.1} \,.$$

If Betti's law is applied to the plates shown in **Fig. 9a, b** it follows that

$$W_{1.2} = 1 \cdot w(x_1, x_2) + \int_\Gamma (\hat{V}_n w + \hat{M}_n\alpha_n)\,ds + \sum \hat{F}_e w_e =$$

$$W_{2.1} = \int_\Omega p\hat{w}\,d\Omega + \int_\Gamma (V_n\hat{w} + M_n\hat{\alpha}_n)\,ds + \sum F_e\hat{w}_e \,; \quad (13a)$$

and so, for the deflection sought (influence coefficient), this gives:

$$w(x_1, x_2) = \int_\Omega p\hat{w}\,d\Omega + \int_\Gamma (V_n\hat{w} + M_n\hat{\alpha}_n)\,ds + \sum F_e\hat{w}_e$$

$$- \int_\Gamma (\hat{V}_n w + \hat{M}_n\alpha_N)\,ds - \sum \hat{F}_e w_e \quad (13b)$$

or

$$w(x_1, x_2) = \int_\Omega p\hat{w}\,d\Omega + W_{\text{edge } 2.1} - W_{\text{edge } 1.2} \,. \quad (13c)$$

The integral above Ω is an area integral and the integrals above Γ are boundary integrals; n is the direction of the perpendiculars at the edge and V_n and M_n are Kirchhoff boundary transverse forces (equivalent shear loads) and the bending moment in a boundary area perpendicular to n.

Infinitely Expanded Plate. Since the area solution due to $\hat{F} = 1$ at point (x_1, x_2) for deflection $w(x_1, x_2, y_1, y_2)$ is known and, according to [11, 12] reads as follows (so-called basic or fundamental solution):

$$\hat{w}_0(r) = \hat{g}_0(r) = \frac{1}{8\pi N} \cdot r^2 \ln r \,, \quad (14)$$

where $r = \sqrt{(y_1 - x_1)^2 + (y_2 - x_2)^2}$ is the distance of the loading point (x_1, x_2), e.g. from a boundary point (y_1, y_2), and $N = Eb^3/12(1 - v^2)$ is the so-called rigidity of the plate (see B5.1); then, by corresponding differentiations, all angles of inclination, bending moments and shear loads, i.e. all boundary characteristics marked with a "circumflex" in Eq. 13, are also known, such as $\hat{w}_0, \hat{\alpha}_{0n}, \hat{M}_{0n}$ and $\hat{V}_{0n}$.

Actual Plate. Here, 2 out of the 4 boundary functions w, α_n, M_n, V_n are known, while 2 are given by the boundary conditions of the plate, e.g. in the case of a plate which is hinged on all sides the values α_n and V_n are unknown, while $w = 0$ and $M_n = 0$ along the boundary are specified. The unknown functions α_n and V_n are now determined

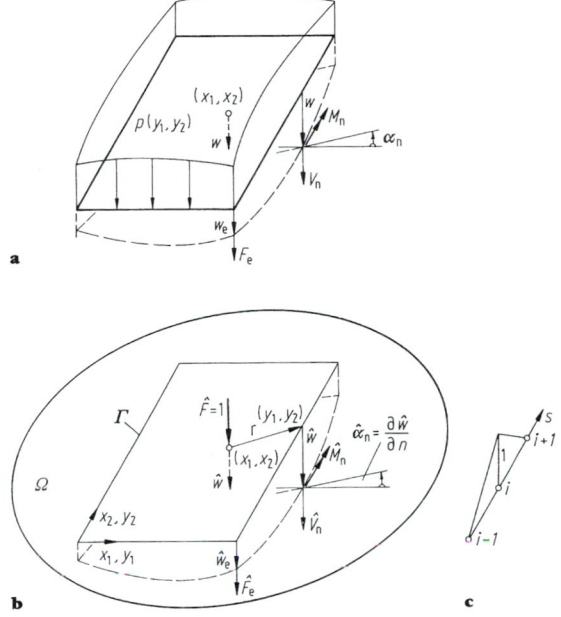

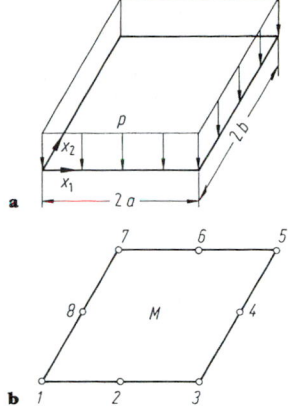

Figure 10. Steel plate hinged on all sides: **a** with constant area load; **b** boundary elements with 8 nodes.

Figure 9. Rectangular plate: **a** under area load; **b** under auxiliary load $F = 1$, **c** boundary elements with "^" function.

numerically using the boundary element method for m discrete nodes connected by m boundary elements, where the single load $F_i = 1$ is applied at each individual node, i.e. m times, and *Betti's law* is written in accordance with Eq. (13b) m times, and thus m linear equations for the $2m$ unknowns α_{ni} and V_{ni} are obtained ($i = 1 \ldots m$).

A further m equations are obtained by applying a boundary moment $\hat{M} = 1$ at each node that has the basic solution:

$$\hat{g}_1(r) = \frac{\partial}{\partial r} \, \hat{g}_0(r) = \frac{1}{8\pi N} \, r \, (1 + 2 \ln r)\frac{\partial r}{\partial n}. \quad (15)$$

Thus the boundary characteristics $\hat{w}_1$, $\hat{\alpha}_{1n}$, $\hat{M}_{1n}$, $\hat{V}_{1n}$ are known in turn in *Betti's law* m times.

In order to be able to perform a numerical integration on the boundary, the unknowns α_{ni} and V_{ni} are linked with elementary functions, $\alpha_{ni}(s) = \alpha_{ni}\varphi(s)$ and $V_{ni}(s) = V_{ni}\psi(s)$ for which linear "^ functions" as shown in **Fig. 9c** are usually sufficient (for plates with free elementary boundaries Hermitian polynomials are required for w_i,

cf. [12, 13, 14]). If all integrations are carried out we have $2m$ equations for the $2m$ unknowns.

The solution obtained (with additional considerations for the corner forces) is inserted in Eq. (13b) to give the deflections $w(x_1, x_2)$ for any points (x_1, x_2) and by differentiation the angle of inclination and sectional loads. See [12, 13, 14] for details of the procedure.

Example Determine the deflection and bending moments or bending stresses in the centre of the plate for a hinged square steel plate, 10 mm thick ($E = 2.1 \cdot 10^8 \, \text{kN/m}^2$) and with constant area load $p = 10 \, \text{kN/m}^2$ and sides $2a = 2b = 1.0$ m, using the boundary element method (**Fig. 10a**).

Solution. The boundaries are divided into $m = 8$ boundary elements with $m = 8$ nodes and the calculation is carried out using a boundary element method program. The result obtained for plate centre M (**Fig. 10b**) is deflection $w = 2.19$ mm and bending moments $m_{x1} = m_{x2} = 0.48$ kN m/m and from this the bending stresses $\sigma = 28.8 \, \text{N/mm}^2$. The formulae given in B5.1.1 are used as a comparison: $w = f = c_s pb^4/Eb^3$ and $\sigma = c_1 pb^2/b^2$, which, with the coefficients $c_s = 0.71$ and $c_1 = 1.15$ given in B5 **Table 1**, give the values $w = 2.11$ mm and $\sigma = 28.8 \, \text{N/mm}^2$, i.e. the result obtained using the boundary element method deviates from the values in the table by 3.8% for w and by 0% for σ, which is a very good result if the boundary is roughly divided into sections.

9 Theory of Plasticity

9.1 Introduction to Theory of Plasticity

If when stress is applied to a material the limit of elasticity is exceeded and permanent elongations ε_b remain after the stress is removed (**Fig. 1**), this is strain in the plastic (non-elastic) region. When the stress is reapplied the behaviour of the material is elastic and the stress–strain

curve is the line $\overline{AP_1}$ parallel to Hooke's straight line $\overline{OP}$, i.e. as a result of the cold strain the yield point is raised. Further loading up to stress σ_{p2} raises the yield point to this value. This is connected with embrittlement of the material and thus a reduction in ductility before fracture.

If a test rod is then subjected to compressive stress, the yield point will be considerably reduced in the compression range, i.e. the stress–strain curve bends very early on, and if the stress is reapplied then a hysteresis loop is formed (**Fig. 2**). Its area represents the strain energy lost during a cycle. If the cycle is repeated several times, this energy is used up each time. These types of dynamic processes often cause premature fracture of the structural member (Bauschinger effect) and are related to fatigue strength.

The theory of plasticity mainly concerns behaviour under static load. Distinctions are made between the following:

Ideal Elasto-plastic Material (Unalloyed Structural Steels) (curve *1* in **Fig. 1**). Here

$$\sigma = E\varepsilon \quad \text{for} \quad -\varepsilon_F \leq \varepsilon \leq \varepsilon_F,$$

$$\sigma = \sigma_F \quad \text{for} \quad \varepsilon \geq \varepsilon_F.$$

Elastic Hardening Material (Hardened and Tempered Steels) (curve *2* in **Fig. 1**) where

$$\sigma = E\varepsilon \quad \text{for} \quad -\varepsilon_F \leq \varepsilon \leq \varepsilon_F,$$

$$\sigma = A|\varepsilon|^k \quad \text{for} \quad \varepsilon \geq \varepsilon_F$$

or by approximation if curve *2* is replaced by straight line *3* with the hardening modulus $E_2 = \tan \alpha_2$

$$\sigma = \sigma_F + E_2(\varepsilon - \varepsilon_F).$$

For other material laws see [2, 3]; for plastics [4]. When the load is removed from the material, the linear (Hooke's) law always applies:

$$\sigma = E(\varepsilon - \varepsilon_b) = \sigma_{P1} - E(\varepsilon_{P1} - \varepsilon).$$

Creep. Above the recrystallisation temperature at which hardening is increased due to cold forming (for steel at $T_K \geq 400\,°C$), deformation which increases with time under constant load creep occurs. Thus creep strength $R_{m/t/T}$ and creep strain limit under tensile load $R_{P1/t/T}$ which lead to fracture or elongation of 1% after $t = 100\,000$ h at temperature T are to be determined (see D1.6.4).

Relaxation. If the elongation of steel at high temperatures ($T \geq 400$ K) is kept constant, then any forced stresses will be removed with time (by creep).

Deformation Technology. This involves the processes of forming (rolling, pressing, forging). In this case plastic deformations are so great that the theory [3] does not take elastic deformations into account.

Theory of Viscoelasticity. This concerns the elasto-plastic behaviour of plastics with particular emphasis on the time-dependency of deformations and stresses (creep and relaxation). Based on the material laws of Maxwell and Kelvin [4].

9.2 Uses

9.2.1 Bending of Rectangular Beams

Taking a material which is ideally plastic (the results for hardening material differ only slightly in the plastic range of initial elongation), assuming that the cross-sections remain plane even in the plastic range (Bernouilli's criterion) with height h and width b of the beam as shown in **Fig. 3a**,

$$M_{bF} = 2 \int_0^{h/2} \sigma(z) z b \, \mathrm{d}z \quad \text{where} \quad \sigma(z) = \sigma_F z / a$$

for $0 \leq z \leq a$ and $\sigma(z) = \sigma_F$ for $a \leq z \leq h/2$, i.e.

$$M_{bF} = 2 \int_0^a \sigma_F(z^2/a) b \, \mathrm{d}z + 2 \int_a^{h/2} \sigma_F z b \, \mathrm{d}z$$

$$= 2\sigma_F b a^2/3 + \sigma_F b [(h/2)^2 - a^2]$$

$$= \sigma_F(b h^2/6)(3/2 - 2a^2/h^2)$$

$$= \sigma_F W_b [1.5 - (2a^2/h^2)] = M_{bE} n_{pl}.$$

Here M_{bE} is the moment of inertia of the rectangular cross-section upon leaving the range of elasticity and n_{pl} the support figure that specifies the ratio in which the moment of inertia increases as a function of the plastic range of extension. For $a = 0$ (fully plastic cross-section), $n_{pl} = 1.5$, i.e. the load capacity is 50% greater than upon leaving the elastic range. For elongation

$$\varepsilon(z) = (\varepsilon_F/a) z = (\sigma_F z)/(Ea), \quad \varepsilon_{max} = \sigma_F h/(2Ea) ;$$

i.e. for $a = 0$ (fully plastic cross-section), ε_{max} is infinite and very large deformations are required to exhaust the load capacity completely (a "plastic hinge" is formed at the point of the maximum moment). Therefore, in practice, elongation ε_p is restricted to 0.2%. For St 37, where $\sigma_F = 240$ N/mm² and $E = 2.1 \cdot 10^5$ N/mm², $\varepsilon_F = \sigma_F/E = 0.114\%$, thus $\varepsilon_{max} = \varepsilon_p + \varepsilon_F = 0.314\%$ and thus $a = \sigma_F h/(2\varepsilon_{max} E) = 0.182h$. This gives the support figure $n_{pl} = 1.5 - 2(a/h)^2 = 1.43$. For this case, that is, for $\varepsilon_p = 0.2\%$, $n_{pl}\sigma_F = K_{0.2}^*$, which is equal to the permanent elongation limit specified in B1.2. See [1, 2] for results for various other cross-sections and basic types of stress.

Residual Stress. If the moment M_{bF} acting on the cross-section is removed, then this is equivalent to the application of an opposite moment $-M_{bF}$ (**Fig. 3b**). Since the material follows Hooke's straight lines $\overline{AP_1}$ when the load is removed, stresses $\sigma_c(z) = -M_{bF} z / I_y$ are produced with linear distribution and maximum value $\sigma_{c.\,max} = -M_{bF}/W_b$. Superposition with stresses $\sigma(z)$ as shown in **Fig. 3** produces the residual stresses $\sigma_r(z) = \sigma(z) - \sigma_c(z)$, as shown in **Fig. 3c**, which remain after each elongation above the yield limit and subsequent removal of load in non-uniform states of stress.

9.2.2 Three-dimensional and Plane Stresses

Flow Conditions

Tresca. According to Tresca, for ideally elasto-plastic material, the following applies:

$$[(\sigma_1 - \sigma_2)^2 - \sigma_F^2][(\sigma_2 - \sigma_3)^2 - \sigma_F^2][(\sigma_3 - \sigma_1)^2 - \sigma_F^2] = 0.$$

According to this, the onset of flow begins when the maximum principal stress differential reaches the value σ_F. If σ_1 and σ_3 are the maximum and minimum principal

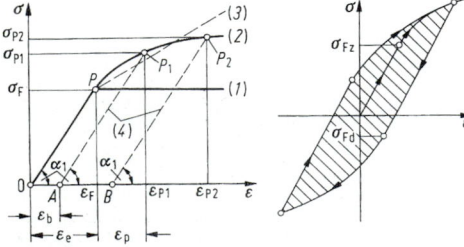

Figure 1. Stress–strain curves in the plastic range.

Figure 2. Hysteresis loop upon loading in the plastic range.

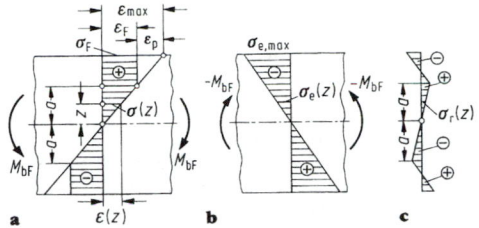

Figure 3. Bending stresses in the plastic range: **a** partially plastic cross-section; **b** stress superposition when load is removed, **c** residual stresses after load has been removed.

stress, then $\sigma_1 - \sigma_3 = 2\tau_{max} = \sigma_F$ follows. If $\sigma_V = \sigma_F$ is assumed to be a one-dimensional equivalent stress, then Tresca's law is identical to the maximum shear stress criterion (see B1.3.2).

Von Mises. According to von Mises,

$$(\sigma_1 - \sigma_2)^2 + (\sigma_2 - \sigma_3)^2 + (\sigma_3 - \sigma_1)^2 = 2\sigma_F^2.$$

According to this, the onset of flow begins where

$$\sigma_V = (1/\sqrt{2})\sqrt{(\sigma_1 - \sigma_2)^2 + (\sigma_2 - \sigma_3)^2 + (\sigma_3 - \sigma_1)^2} = \sigma_F.$$

This law is identical to the maximum shear strain energy criterion (see B1.3.3).

Stress-Strain Laws

Prandtl-Reuss's Law. This takes the infinite (differential) form

$$d\mathbf{V}_D = d\mathbf{V}_{D.e} + d\mathbf{V}_{D.p} = (d\mathbf{S}_D + \mathbf{S}_D d\lambda)/(2G) ;$$

or, after introducing the distortion rates,

$$\dot{\mathbf{V}}_D = (\dot{\mathbf{S}}_D + \mathbf{S}_D \cdot \lambda)/(2G) .$$

Here $\mathbf{V}_D$ is the "deviator" of the distortion tensor $\mathbf{V}$ (see B1.1.2), i.e. $\mathbf{V}_D = \mathbf{V} - e \cdot \mathbf{I}$ applies where $e = (\varepsilon_x + \varepsilon_y + \varepsilon_z)/3$ and represents the unit spherical tensor. The distortion deviator represents the maximum shear strain for constant volume. $\mathbf{S}_D$ is the deviator of the stress tensor [5]. G is the shear modulus and $d\lambda$ or $\dot{\lambda}$ is a scalar proportionality factor, which by balancing the maximum shear strain energies of the three-dimensional and one-dimensional equivalent state is given as $d\lambda = \dfrac{3}{2} \dfrac{d\sigma_V}{T_p(\sigma_V)\sigma_V}$, where $T_p = d\sigma_V/d\varepsilon_{Vp}$ is the plastic tangent modulus (rise of the σ_V-ε_{Vp} line).

Hencky's Law. This has the finite form

$$\mathbf{V}_D = \mathbf{V}_{D.e} + \mathbf{V}_{D.p} = \left(\frac{1}{2G} + \frac{1}{2G_p}\right)\mathbf{S}_D .$$

Here G_p is the variable modulus of plasticity which by applying the law to the one-dimensional equivalent state is obtained from $\varepsilon_{Vp} = \dfrac{1}{2G_p} \cdot \dfrac{\sigma_V}{3}$ as $G_p(\varepsilon_{Vp}) = \dfrac{1}{3} \dfrac{\sigma_V}{\varepsilon_{Vp}}$, i.e. from the corresponding stress-strain curve.

Closed, Thick-walled Pipe Under Internal Pressure. Investigate the state of stress in the pipe at the onset of plasticisation at the internal fibres (i.e. pipe still just within the elastic range), when plasticisation has reached the centre of the wall and when the wall has become plasticised.

Fully Elastic State. By Eqs (5) with $\tau_{rz} = \tau_{zr} = \tau = 0$ and $R = 0$ gives the condition of equilibrium

$$\frac{d}{dr}(r\sigma_r) - \sigma_t = r\frac{d\sigma_r}{dr} + \sigma_r - \sigma_t = 0 . \quad (1)$$

This gives the stresses as

$$\sigma_r = -p \cdot \frac{r_i^2}{r_a^2 - r_i^2}\left(\frac{r_a^2}{r^2} - 1\right),$$

$$\sigma_t = p \cdot \frac{r_i^2}{r_a^2 - r_i^2}\left(\frac{r_a^2}{r^2} + 1\right), \quad \sigma_z = p \cdot r_i^2/(r_a^2 - r_i^2) . \quad \left.\right\} \quad (2)$$

Partially Plastic State. For ideally elasto-plastic material, von Mises's flow condition with

$$\sigma_1 = \sigma_r, \quad \sigma_2 = \sigma_t, \quad \sigma_3 = \sigma_z = 0.5(\sigma_r + \sigma_t)$$

gives the flow condition

$$\sigma_t - \sigma_r = 2\sigma_F/\sqrt{3} . \quad (3)$$

For a cylinder that is plasticised up to radius r_p, the stress formulae for the elastic range $(r \geq r_p)$ are as specified in Eq. (2):

$$\sigma_r = -\frac{\sigma_F}{\sqrt{3}} \frac{r_p^2}{r_a^2}\left(\frac{r_a^2}{r^2} - 1\right),$$

$$\sigma_t = \frac{\sigma_F}{\sqrt{3}} \frac{r_p^2}{r_a^2}\left(\frac{r_a^2}{r^2} + 1\right), \quad \sigma_z = \frac{\sigma_F}{\sqrt{3}} \frac{r_p^2}{r_a^2} . \quad (4)$$

For the plastic range $(r \leq r_p)$ Eq. (1) and Eq. (3) give the condition of equilibrium

$$r\frac{d\sigma_r}{dr} - \frac{2\sigma_F}{\sqrt{3}} = 0 \quad (5)$$

and thus the stresses

$$\sigma_r = -\frac{\sigma_F}{\sqrt{3}}\left(1 - \frac{r_p^2}{r_a^2} + 2\ln\frac{r_p}{r}\right), \quad (6a)$$

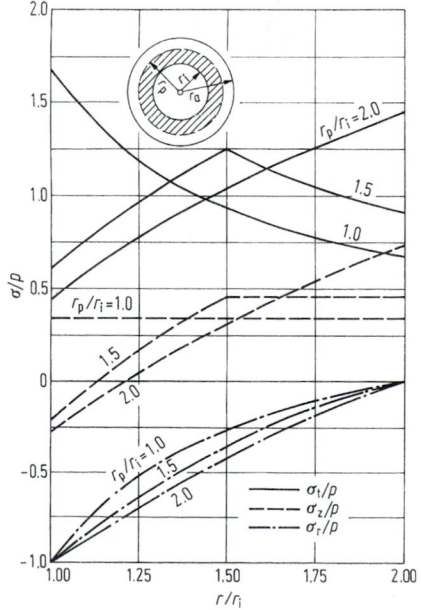

Figure 4. Stresses in the pipe where $r_a/r_i = 2.0$.

$$\sigma_t = \frac{\sigma_F}{\sqrt{3}} \left(1 + \frac{r_p^2}{r_a^2} - 2 \ln \frac{r_p}{r} \right), \quad \sigma_z = \frac{\sigma_F}{\sqrt{3}} \left(\frac{r_p^2}{r_a^2} - 2 \ln \frac{r_p}{r} \right). \quad (6b)$$

For the internal pressure where $\sigma_r(r_i) = -p$, Eqs (6a) and (6b) give

$$p = \frac{\sigma_F}{\sqrt{3}} \left(1 - \frac{r_p^2}{r_a^2} + 2 \ln \frac{r_p}{r_i} \right). \quad (7)$$

The radius of plasticisation r_p can be determined from this as a function of the internal pressure and vice versa. At the onset of plasticisation at the inside edge of the cylinder, i.e. for $r_p = r_i$, Eq. (7) gives the corresponding internal pressure as

$$p_1 = \frac{\sigma_F}{\sqrt{3}} \left(1 - \frac{r_i^2}{r_a^2} \right).$$

For full plasticisation, where $r_p = r_a$, the internal pressure is given as

$$p_2 = \frac{2\sigma_F}{\sqrt{3}} \ln \frac{r_a}{r_i}.$$

Thus for a pipe where $r_a/r_i = 2$ the increase in load capacity between the elastic and the fully plastic state is

$$p_2/p_1 = 2 \ln 2/0.75 = 1.85 .$$

Figure 4 shows the progression of stresses for a pipe where $r_a/r_i = 2.0$ and the state of stress is still just elastic (i.e. $r_p = r_i$, $p = p_1 = 0.43\sigma_F$), when it is semi-plasticised ($r_p = 1.5r_i$, $p = 0.72\sigma_F$) and when it is fully plasticised ($r_p = r_a$, $p = p_2 = 0.80\sigma_F$). It can be seen that for σ_t and σ_z there is significant redistribution of stress between the elastic and plastic state but only slight redistribution for σ_r.

10 Appendix B: Diagrams and Tables

Table 1. Hot-rolled I-profiles, narrow I-profiles, I-series in accordance with DIN 1025 Part 1 (extract)

I second moment of area
W section modulus
i radius of inertia
S_x area moment of the first order for half cross-section
$s_x = I_x/S_x$ distance between the centre of compression and tension

Abbreviation I	Dimensions for b mm	s mm	l mm	r_1 mm	r_2 mm	Cross-section A cm²	Weight G kg/m	x-x I_x cm⁴	W_x cm³	i_x cm	y-y I_y cm⁴	W_y cm³	i_y cm	S_x cm³	s_x cm
80	42	3.9	5.9	3.9	2.3	7.57	5.94	77.8	19.5	3.20	6.29	3.00	0.91	11.4	6.84
100	50	4.5	6.8	4.5	2.7	10.6	8.34	171	34.2	4.01	12.2	4.88	1.07	19.9	8.57
120	58	5.1	7.7	5.1	3.1	14.2	11.1	328	54.7	4.81	21.5	7.41	1.23	31.8	10.3
140	66	5.7	8.6	5.7	3.4	18.2	14.3	573	81.9	5.61	35.2	10.7	1.40	47.7	12.0
160	74	6.3	9.5	6.3	3.8	22.8	17.9	935	117	6.40	54.7	14.8	1.55	68.0	13.7
180	82	6.9	10.4	6.9	4.1	27.9	21.9	1450	161	7.20	81.3	19.8	1.71	93.4	15.5
200	90	7.5	11.3	7.5	4.5	33.4	26.2	2140	214	8.00	117	26.0	1.87	125	17.2
220	98	8.1	12.2	8.1	4.9	39.5	31.1	3060	278	8.80	162	33.1	2.02	162	18.9
240	106	8.7	13.1	8.7	5.2	46.1	36.2	4250	354	9.59	221	41.7	2.20	206	20.6
260	113	9.4	14.1	9.4	5.6	53.3	41.9	5740	442	10.4	288	51.0	2.32	257	22.3
280	119	10.1	15.2	10.1	6.1	61.0	47.9	7590	542	11.1	364	61.2	2.45	316	24.0
300	125	10.8	16.2	10.8	6.5	69.0	54.2	9800	653	11.9	451	72.2	2.56	381	25.7
320	131	11.5	17.3	11.5	6.9	77.7	61.0	12510	782	12.7	555	84.7	2.67	457	27.4
340	137	12.2	18.3	12.2	7.3	86.7	68.0	15700	923	13.5	674	98.4	2.80	540	29.1
360	143	13.0	19.5	13.0	7.8	97.0	76.1	19610	1090	14.2	818	114	2.90	638	30.7
380	149	13.7	20.5	13.7	8.2	107	84.0	24010	1260	15.0	975	131	3.02	741	32.4
400	155	14.4	21.6	14.4	8.6	118	92.4	29210	1460	15.7	1160	149	3.13	857	34.1
425	163	15.3	23.0	15.3	9.2	132	104	36970	1740	16.7	1440	176	3.30	1020	36.2
450	170	16.2	24.3	16.2	9.7	147	115	45850	2040	17.7	1730	203	3.43	1200	38.3
475	178	17.1	25.6	17.1	10.3	163	128	56480	2380	18.6	2090	235	3.60	1400	40.4
500	185	18.0	27.0	18.0	10.8	179	141	68740	2750	19.6	2480	268	3.72	1620	42.4
550	200	19.0	30.0	19.0	11.9	212	166	99180	3610	21.6	3490	349	4.02	2120	46.8
600	215	21.6	32.4	21.6	13.0	254	199	139000	4630	23.4	4670	434	4.30	2730	50.9

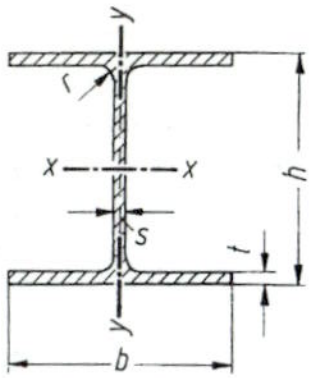

Table 2. Hot-rolled I-profiles, wide I-profiles, lightweight design, IPBl-series in accordance with DIN 1025 Part 3 (extract)

I — second moment of area
W — section modulus
i — radius of inertia
S_x — first moment of area for half cross-section
$s_x = I_x/S_x$ — distance between the centre of compression and tension

Abbreviation IPBl	Dimensions for					Cross-section A	Weight G	For the bending axis						S_x	s_x
								x–x			y–y				
	b	b	s	t	r			I_x	W_x	i_x	I_y	W_y	t_t		
	mm	mm	mm	mm	mm	cm²	kg/m	cm⁴	cm³	cm	cm⁴	cm³	cm	cm³	cm
100	96	100	5	8	12	21.2	16.7	349	72.8	4.06	134	26.8	2.51	41.5	8.41
120	114	120	5	8	12	25.3	19.9	606	106	4.89	231	38.5	3.02	59.7	10.1
140	133	140	5.5	8.5	12	31.4	24.7	1030	155	5.73	389	55.6	3.52	86.7	11.9
160	152	160	6	9	15	38.8	30.4	1670	220	6.57	616	76.9	3.98	123	13.6
180	171	180	6	9.5	15	45.3	35.5	2510	294	7.45	925	103	4.52	162	15.5
200	190	200	6.5	10	18	53.8	42.3	3690	389	8.28	1340	134	4.98	215	17.2
220	210	220	7	11	18	64.3	50.5	5410	515	9.17	1950	178	5.51	284	19.0
240	230	240	7.5	12	21	76.8	60.3	7760	675	10.1	2770	231	6.00	372	20.9
260	250	260	7.5	12.5	24	86.8	68.2	10450	836	11.0	3670	282	6.50	460	22.7
280	270	280	8	13	24	97.3	76.4	13670	1010	11.9	4760	340	7.00	556	24.6
300	290	300	8.5	14	27	112	88.3	18260	1260	12.7	6310	421	7.49	692	26.4
320	310	300	9	15.5	27	124	97.6	22930	1480	13.6	6990	466	7.49	814	28.2
340	330	300	9.5	16.5	27	133	105	27690	1680	14.4	7440	496	7.46	925	29.9
360	350	300	10	17.5	27	143	112	33090	1890	15.2	7890	526	7.43	1040	31.7
400	390	300	11	19	27	159	125	45070	2310	16.8	8560	571	7.34	1280	35.2
450	440	300	11.5	21	27	178	140	63720	2900	18.9	9470	631	7.29	1610	39.6
500	490	300	12	23	27	198	155	86970	3550	21.0	10370	691	7.24	1970	44.1
550	540	300	12.5	24	27	212	166	111900	4150	23.0	10820	721	7.15	2310	48.4
600	590	300	13	25	27	226	178	141200	4790	25.0	11270	751	7.05	2680	52.8
650	640	300	13.5	26	27	242	190	175200	5470	26.9	11720	782	6.97	3070	57.1
700	690	300	14.5	27	27	260	204	215300	6240	28.8	12180	812	6.84	3520	61.2
800	790	300	15	28	30	286	224	303400	7680	32.6	12640	843	6.65	4350	69.8
900	890	300	16	30	30	320	252	422100	9480	36.3	13550	903	6.50	5410	78.1
1000	990	300	16.5	31	30	347	272	553800	11190	40.0	14000	934	6.35	6410	86.4

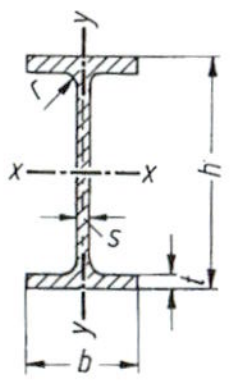

Table 3. Hot-rolled I-profiles, medium width I-profiles, IPB-series in accordance with DIN 1025 Part 5 (extract)

I second moment of area
W section modulus
i radius of inertia
S_x first moment of area for half cross-section
$s_x = I_x/S_x$ distance between the centre of compression and tension

| Abbreviation IPE | Dimensions for | | | | | Cross-section A | Weight G | For the bending axis | | | | | | S_x | s_x |
| | b | b | s | t | r | | | x–x | | | y–y | | | | |
	mm	mm	mm	mm	mm	cm²	kg/m	I_x cm⁴	W_x cm³	i_x cm	I_y cm⁴	W_y cm³	i_y cm	cm³	cm
80	80	46	3.8	5.2	5	7.64	6.0	80.1	20.0	3.24	8.49	3.69	1.05	11.6	6.90
100	100	55	4.1	5.7	7	10.3	8.1	171	34.2	4.07	15.9	5.79	1.24	19.7	8.68
120	120	64	4.4	6.3	7	13.2	10.4	318	53.0	4.90	27.7	8.65	1.45	30.4	10.5
140	140	73	4.7	6.9	7	16.4	12.9	541	77.3	5.74	44.9	12.3	1.65	44.2	12.3
160	160	82	5.0	7.4	9	20.1	15.8	869	109	6.58	68.3	16.7	1.84	61.9	14.0
180	180	91	5.3	8.0	9	23.9	18.8	1320	146	7.42	101	22.2	2.05	83.2	15.8
200	200	100	5.6	8.5	12	28.5	22.4	1940	194	8.26	142	28.5	2.24	110	17.6
220	220	110	5.9	9.2	12	33.4	26.2	2770	252	9.11	205	37.3	2.48	143	19.4
240	240	120	6.2	9.8	15	39.1	30.7	3890	324	9.97	284	47.3	2.69	183	21.2
270	270	135	6.6	10.2	15	45.9	36.1	5790	429	11.2	420	62.2	3.02	242	23.9
300	300	150	7.1	10.7	15	53.8	42.2	8360	557	12.5	604	80.5	3.35	314	26.6
330	330	160	7.5	11.5	18	62.6	49.1	11770	713	13.7	788	98.5	3.55	402	29.3
360	360	170	8.0	12.7	18	72.7	57.1	16270	904	15.0	1040	123	3.79	510	31.9
400	400	180	8.6	13.5	21	84.5	66.3	23130	1160	16.5	1320	146	3.95	654	35.4
450	450	190	9.4	14.6	21	98.8	77.6	33740	1500	18.5	1680	176	4.12	851	39.7
500	500	200	10.2	16.0	21	116	90.7	48200	1930	20.4	2140	214	4.31	1100	43.9
550	550	210	11.1	17.2	24	134	106	67120	2440	22.3	2670	254	4.45	1390	48.2
600	600	220	12.0	19.0	24	156	122	92080	3070	24.3	3390	308	4.66	1760	52.4

Table 4. Hot-rolled I-profiles, wide I-profiles, IPB-series in accordance with DIN 1025 Part 2 (extract)

I second moment of area
W section modulus
i radius of inertia
S_x first moment of area for half cross-section
$s_x = I_x/S_x$ distance between the centre of compression and tension

Abbreviation IPB	Dimensions for					Cross-section A cm²	Weight G kg/m	For the bending axis						S_x cm³	s_x cm
								x–x			y–y				
	b mm	b mm	s mm	t mm	r_1 mm			I_x cm⁴	W_x cm³	i_x cm	I_y cm⁴	W_y cm³	i_y cm		
100	100	100	6	10	12	26.0	20.4	450	89.9	4.16	167	33.5	2.53	52.1	8.63
120	120	120	6.5	11	12	34.0	26.7	864	144	5.04	318	52.9	3.06	82.6	10.5
140	140	140	7	12	12	43.0	33.7	1510	216	5.93	550	78.5	3.58	123	12.3
160	160	160	8	13	15	54.3	42.6	2490	311	6.78	889	111	4.05	177	14.1
180	180	180	8.5	14	15	65.3	51.2	3830	426	7.66	1360	151	4.57	241	15.9
200	200	200	9	15	18	78.1	61.3	5700	570	8.54	2000	200	5.07	321	17.7
220	220	220	9.5	16	18	91.0	71.5	8090	736	9.43	2840	258	5.59	414	19.6
240	240	240	10	17	21	106	83.2	11260	938	10.3	3920	327	6.08	527	21.4
260	260	260	10	17.5	24	118	93.0	14920	1150	11.2	5130	395	6.58	641	23.3
280	280	280	10.5	18	24	131	103	19270	1380	12.1	6590	471	7.09	767	25.1
300	300	300	11	19	27	149	117	25170	1680	13.0	8560	571	7.58	934	26.9
320	320	300	11.5	20.5	27	161	127	30820	1930	13.8	9240	616	7.57	1070	28.7
340	340	300	12	21.5	27	171	134	36660	2160	14.6	9690	646	7.53	1200	30.4
360	360	300	12.5	22.5	27	181	142	43190	2400	15.5	10140	676	7.49	1340	32.2
400	400	300	13.5	24	27	198	155	57680	2880	17.1	10820	721	7.40	1620	35.7
450	450	300	14	26	27	218	171	79890	3550	19.1	11720	781	7.33	1990	40.1
500	500	300	14.5	28	27	239	187	107200	4290	21.2	12620	842	7.27	2410	44.5
550	550	300	15	29	27	254	199	136700	4970	23.2	13080	872	7.17	2800	48.9
600	600	300	15.5	30	27	270	212	171000	5700	25.2	13530	902	7.08	3210	53.2
650	650	300	16	31	27	286	225	210600	6480	27.1	13980	932	6.99	3660	57.5
700	700	300	17	32	27	306	241	256900	7340	29.0	14440	963	6.87	4160	61.7
800	800	300	17.5	33	30	334	262	359100	8980	32.8	14900	994	6.68	5110	70.2
900	900	300	18.5	35	30	371	291	494100	10980	36.5	15820	1050	6.53	6290	78.5
1000	1000	300	19	36	30	400	314	644700	12890	40.1	16280	1090	6.38	7450	86.8

Table 5. Cold-drawn round steel: Tolerances in accordance with the ISO tolerance zone h8 in DIN 670 (extract)
Standard designs: $d < 45$ mm cold-drawn (K) and then ground, $d \geq 45 \leq 150$ mm bright turned (SH) and then ground

Nominal diameter d range mm	Tolerance for d	Preferred values for d^a
≥ 1≤ 3	0 to 0.014	1–1.5–2–2.5–3
> 3≤ 6	0 to 0.018	3.5–4–4.5–5–5.5–6
> 6≤ 10	0 to 0.022	6.5–7–7.5–8–8.5–9–9.5–10
> 10≤ 18	0 to 0.027	11–12–13–14–15–16–17–18
> 18≤ 30	0 to 0.033	19–20–21–22–23–24–25–26–27–28–29–30
> 30≤ 50	0 to 0.039	32–34–35–36–38–40–42–45–48–50
> 50≤ 80	0 to 0.046	52–55–58–60–63–65–70–75–80
> 80≤120	0 to 0.054	85–90–100–110–120
>120≤150	0 to 0.063	125–130–140–150

[a] Other nominal diameters can also be supplied upon agreement.

Table 6. Cold drawn round steel: Tolerances in accordance with the ISO tolerance zone h 11 in DIN 668 (extract)
Standard designs: $d < 45$ mm cold drawn (K) and then ground, $d \geq 45 \leq 200$ mm bright turned (SH)

Nominal diameter range	Tolerance for d	Preferred values for d^a
≥ 1≤ 3	0 to 0.060	1–1.5–2–2.5–3
> 3≤ 6	0 to 0.075	3.5–4–4.5–5–5.5–6
> 6≤ 10	0 to 0.090	6.5–7–7.5–8–8.5–9–9.5–10
> 10≤ 18	0 to 0.110	11–12–13–14–15–16–17–18
> 18≤ 30	0 to 0.130	19–20–21–22–23–24–25–26–27–28–29–30
> 30≤ 50	0 to 0.160	32–34–35–36–38–40–42–45–48–50
> 50≤ 80	0 to 0.190	52–55–58–60–63–65–70–75–80
> 80≤120	0 to 0.220	85–90–100–110–120
>120≤180	0 to 0.250	125–130–140–150–160–180
>180≤200	0 to 0.290	200

[a] Other nominal diameters can also be supplied upon agreement.

Table 7. Hot-rolled I-profiles, reinforced design, IPBv-series in accordance with DIN 1025 Part 4 (extract)

l second moment of area
W section modulus
i radius of inertia
S_x first moment of area for half cross-section
$s_x = I_x/S_x$ distance between the centre of compression and tension

Abbreviation IPBv	Dimensions for					Cross-section A	Weight G	For the bending axis						S_x	s_x
								x–x			y–y				
	b	b	s	l	r			I_x	W_x	i_x	I_y	W_y	i_y		
	mm	mm	mm	mm	mm	cm²	kg/m	cm⁴	cm³	cm	cm⁴	cm³	cm	cm³	cm
100	120	106	12	20	12	53.2	41.8	1140	190	4.63	399	75.3	2.74	118	9.69
120	140	126	12.5	21	12	66.4	52.1	2020	288	5.51	703	112	3.25	175	11.5
140	160	146	13	22	12	80.6	63.2	3290	411	6.39	1140	157	3.77	247	13.3
160	180	166	14	23	15	97.1	76.2	5100	566	7.25	1760	212	4.26	337	15.1
180	200	186	14.5	24	15	113	88.9	7480	748	8.13	2580	277	4.77	442	16.9
200	220	206	15	25	18	131	103	10640	967	9.00	3650	354	5.27	568	18.7
220	240	226	15.5	26	18	149	117	14600	1220	9.89	5010	444	5.79	710	20.6
240	270	248	18	32	21	200	157	24290	1800	11.0	8150	657	6.39	1060	22.9
260	290	268	18	32.5	24	220	172	31310	2160	11.9	10450	780	6.90	1260	24.8
280	310	288	18.5	33	24	240	189	39550	2550	12.8	13160	914	7.40	1480	26.7
300	340	310	21	39	27	303	238	59200	3480	14.0	19400	1250	8.00	2040	29.0
320/305	320	305	16	29	27	225	177	40950	2560	13.5	13740	901	7.81	1460	28.0
320	359	309	21	40	27	312	245	68130	3800	14.8	19710	1280	7.95	2220	30.7
340	377	309	21	40	27	316	248	76370	4050	15.6	19710	1280	7.90	2360	32.4
360	395	308	21	40	27	319	250	84870	4300	16.3	19520	1270	7.83	2490	34.0
400	432	307	21	40	27	326	256	104100	4820	17.9	19330	1260	7.70	2790	37.4
450	478	307	21	40	27	335	263	131500	5500	19.8	19340	1260	7.59	3170	41.5
500	524	306	21	40	27	344	270	161900	6180	21.7	19150	1250	7.46	3550	45.7
550	572	306	21	40	27	354	278	198000	6920	23.6	19160	1250	7.35	3970	49.9
600	620	305	21	40	27	364	285	237400	7660	25.6	18970	1240	7.22	4390	54.1
650	668	305	21	40	27	374	293	281700	8430	27.5	18980	1240	7.13	4830	58.3
700	716	304	21	40	27	383	301	329300	9200	29.3	18800	1240	7.01	5270	62.5
800	814	303	21	40	30	404	317	442600	10870	33.1	18630	1230	6.79	6240	70.9
900	910	302	21	40	30	424	333	570400	12540	36.7	18450	1220	6.60	7220	79.0
1000	1008	302	21	40	30	444	349	722300	14330	40.3	18460	1220	6.45	8280	87.2

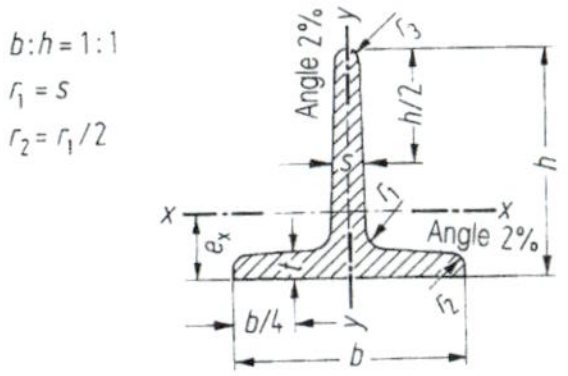

Table 8. Hot-rolled, round-edged deep-webbed T-sections in accordance with DIN 1024 (extract)

b:h = 1:1
r₁ = s
r₂ = r₁/2

I second moment of area
W section modulus
i radius of inertia

Abbreviation	Dimensions for				Cross-section	Weight	e_x	For the bending axis					
T					A	G		x–x			y–y		
	b mm	$s=t$ $=r_1$ mm	r_3 mm		cm^2	kg/m	cm	I_x cm^4	W_x cm^3	i_x cm	I_y cm^4	W_y cm^3	i_y cm
20	20	3	1		1.12	0.88	0.58	0.38	0.27	0.58	0.20	0.20	0.42
25	25	3.5	1		1.64	1.29	0.73	0.87	0.49	0.73	0.43	0.34	0.51
30	30	4	1		2.26	1.77	0.85	1.72	0.80	0.87	0.87	0.58	0.62
35	35	4.5	1		2.97	2.33	0.99	3.10	1.23	1.04	1.57	0.90	0.73
40	40	5	1		3.77	2.96	1.12	5.28	1.84	1.18	2.58	1.29	0.83
45	45	5.5	1.5		4.67	3.67	1.26	8.13	2.51	1.32	4.01	1.78	0.93
50	50	6	1.5		5.66	4.44	1.39	12.1	3.36	1.46	6.06	2.42	1.03
60	60	7	2		7.94	6.23	1.66	23.8	5.48	1.73	12.2	4.07	1.24
70	70	8	2		10.6	8.32	1.94	44.5	8.79	2.05	22.1	6.32	1.44
80	80	9	2		13.6	10.7	2.22	73.7	12.8	2.33	37.0	9.25	1.65
90	90	10	2.5		17.1	13.4	2.48	119	18.2	2.64	58.5	13.0	1.85
100	100	11	3		20.9	16.4	2.74	179	24.6	2.92	88.3	17.7	2.05
120	120	13	3		29.6	23.2	3.28	366	42.0	3.51	178	29.7	2.45
140	140	15	4		39.9	31.3	3.80	660	64.7	4.07	330	47.2	2.88

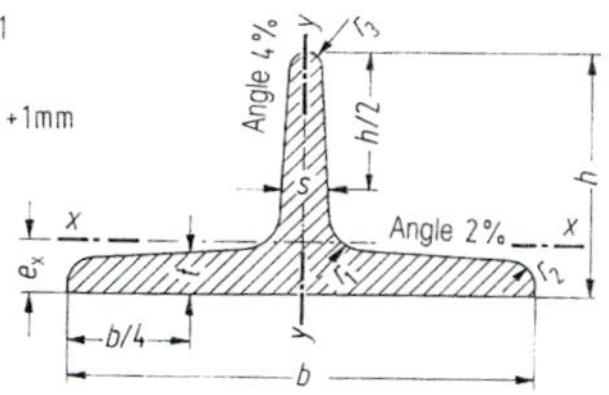

Table 9. Hot-rolled, round-edged, broad-flanged T-sections in accordance with DIN 1024 (extract)

$b:h = 2:1$
$h = b/2$
$t = 0{,}15h + 1\,mm$
$r_1 = s$
$r_2 = r_1/2$

I second moment of area
W section modulus
i radius of inertia

Abbreviation TB	Dimensions for				Cross-section A cm²	Weight G kg/m	e_x cm	For the bending axis					
								x–x			y–y		
	b mm	b mm	$s=t$ $=r_1$ mm	r_3 mm				I_x cm⁴	W_x cm³	i_x cm	I_y cm⁴	W_y cm³	i_y cm
30	30	60	5.5	1.5	4.64	3.64	0.67	2.58	1.11	0.75	8.62	2.87	1.36
35	35	70	6	1.5	5.94	4.66	0.77	4.49	1.65	0.87	15.1	4.31	1.59
40	40	80	7	2	7.91	6.21	0.88	7.81	2.50	0.99	28.5	7.13	1.90
50	50	100	8.5	2	12.0	9.42	1.09	18.7	4.78	1.25	67.7	13.5	2.38
60	60	120	10	2.5	17.0	13.4	1.30	38.0	8.09	1.49	137	22.8	2.84

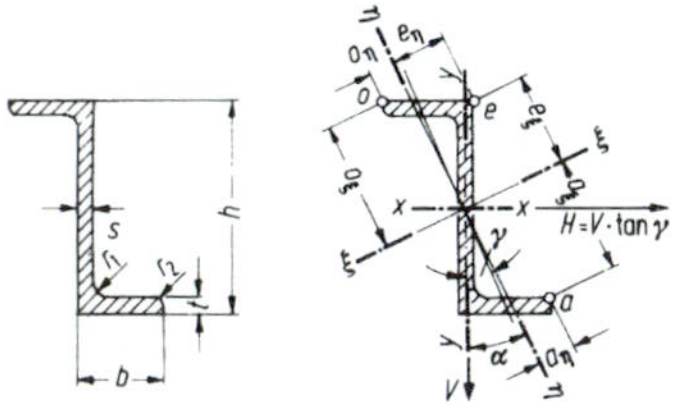

Table 10. Hot-rolled, round-edged Z-sections in accordance with DIN 1027 (extract)

I = second moment of area
W = section modulus
i = radius of inertia

Abbreviation	Dimensions for						Cross-section	Weight	Position of axis	Distance between axes ξ-ξ and η-η					
	b mm	b mm	s mm	t mm	r_1 mm	r_2 mm	A cm²	G kg/m	η-η tan α	o_ξ cm	o_η cm	e_ξ cm	e_η cm	a_ξ cm	a_η cm
30	30	38	4	4.5	4.5	2.5	4.32	3.39	1.655	3.86	0.58	0.61	1.39	3.54	0.87
40	40	40	4.5	5	5	2.5	5.43	4.26	1.181	4.17	0.91	1.12	1.67	3.82	1.19
50	50	43	5	5.5	5.5	3	6.77	5.31	0.939	4.60	1.24	1.65	1.89	4.21	1.49
60	60	45	5	6	6	3	7.91	6.21	0.779	4.98	1.51	2.21	2.04	4.56	1.76
80	80	50	6	7	7	3.5	11.1	8.71	0.558	5.83	2.02	3.30	2.29	5.35	2.25
100	100	55	6.5	8	8	4	14.5	11.4	0.492	6.77	2.43	4.34	2.50	6.24	2.65
120	120	60	7	9	9	4.5	18.2	14.3	0.433	7.75	2.80	5.37	2.70	7.16	3.02
140	140	65	8	10	10	5	22.9	18.0	0.385	8.72	3.18	6.39	2.89	8.08	3.39
160	160	70	8.5	11	11	5.5	27.5	21.6	0.357	9.74	3.51	7.39	3.09	9.04	3.72
180	180	75	9.5	12	12	6	33.3	26.1	0.329	10.7	3.86	8.40	3.27	9.99	4.08
200	200	80	10	13	13	6.5	38.7	30.4	0.313	11.8	4.17	9.39	3.47	11.0	4.39

Abbreviation	Static values for the bending axis												Centrifugal moment	For vertical loading V and with		Free outward bending on side W
	x-x			y-y			ξ-ξ			η-η				Sides prevented from bending outwards by H		
	I_x cm⁴	W_x cm³	i_x cm	I_y cm⁴	W_y cm³	i_y cm	I_ξ cm⁴	W_ξ cm³	i_ξ cm	I_η cm⁴	W_η cm³	i_η cm	I_{xy} cm⁴	W_x cm³	$\frac{H}{V}=\tan\gamma$	cm³
30	5.96	3.97	1.17	13.7	3.80	1.78	18.1	4.69	2.04	1.54	1.11	0.60	7.35	3.97	1.227	1.26
40	13.5	6.75	1.58	17.6	4.66	1.80	28.0	6.72	2.27	3.05	1.83	0.75	12.2	6.75	0.913	2.26
50	26.3	10.5	1.97	23.8	5.88	1.88	44.9	9.76	2.57	5.23	2.76	0.88	19.6	10.5	0.752	3.64
60	44.7	14.9	2.38	30.1	7.09	1.95	67.2	13.5	2.81	7.60	3.73	0.98	28.8	14.9	0.647	5.24
80	109	27.3	3.13	47.4	10.1	2.07	142	24.4	3.58	14.7	6.44	1.15	55.6	27.3	0.509	10.1
100	222	44.4	3.91	72.5	14.0	2.24	270	39.8	4.31	24.6	9.26	1.30	97.2	44.4	0.438	16.8
120	402	67.0	4.70	106	18.8	2.42	470	60.6	5.08	37.7	12.5	1.44	158	67.0	0.392	25.6
140	676	96.6	5.43	148	24.3	2.54	768	88.0	5.79	56.4	16.6	1.57	239	96.6	0.353	38.0
160	1060	132	6.20	204	31.0	2.72	1180	121	6.57	79.5	21.4	1.70	349	132	0.330	52.9
180	1600	178	6.92	270	38.4	2.84	1760	164	7.26	110	27.0	1.82	490	178	0.307	72.4
200	2300	230	7.71	357	47.6	3.04	2510	213	8.06	147	33.4	1.95	674	230	0.293	94.1

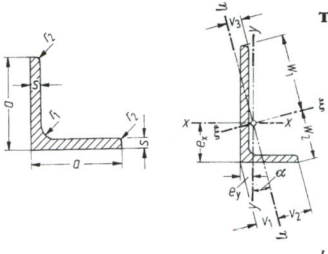

Table 11. Hot-rolled, round-edged unequal angle in accordance with DIN 1029 (extract)

I second moment of area
W section modulus
i radius of inertia

Table contains only standardised preferred values.

Abbreviation L	Dimensions for					Cross-section cm²	Weight kg/m	Lateral area m²/m	Distance between axes				
	a mm	b mm	s mm	r_1 mm	r_2 mm				e_x cm	e_y cm	w_1 cm	w_2 cm	v_1 cm
30 × 20 × 3	30	20	3	3.5	2	1.42	1.11	0.097	0.99	0.50	2.04	1.51	0.86
30 × 20 × 4			4			1.85	1.45		1.03	0.54	2.02	1.52	0.91
40 × 20 × 3	40	20	3	3.5	2	1.72	1.35	0.117	1.43	0.44	2.61	1.77	0.79
40 × 20 × 4			4			2.25	1.77		1.47	0.48	2.57	1.80	0.83
45 × 30 × 4	45	30	4	4.5	2	2.87	2.25	0.146	1.48	0.74	3.07	2.26	1.27
45 × 30 × 5			5			3.53	2.77		1.52	0.78	3.05	2.27	1.32
50 × 30 × 4	50	30	4	4.5	2	3.07	2.41	0.156	1.68	0.70	3.36	2.35	1.24
50 × 30 × 5			5			3.78	2.96		1.73	0.74	3.33	2.38	1.28
50 × 40 × 5	50	40	5	4	2	4.27	3.35	0.177	1.56	1.07	3.49	2.88	1.73
60 × 30 × 5	60	30	5	6	3	4.29	3.37	0.175	2.15	0.68	3.90	2.67	1.20
60 × 40 × 5			5			4.79	3.76		1.96	0.97	4.08	3.01	1.68
60 × 40 × 6	60	40	6	6	3	5.68	4.46	0.195	2.00	1.01	4.06	3.02	1.72
65 × 50 × 5	65	50	5	6	3	5.54	4.35	0.224	1.99	1.25	4.52	3.61	2.08
70 × 50 × 6	70	50	6	6	3	6.88	5.40	0.235	2.24	1.25	4.82	3.68	2.20
75 × 50 × 7	75	50	7			8.30	6.51	0.244	2.48	1.25	5.10	3.77	2.13
75 × 55 × 5			5			6.30	4.95		2.31	1.33	5.19	4.00	2.27
75 × 55 × 7	75	55	7	7	3.5	8.66	6.80	0.254	2.40	1.41	5.16	4.02	2.37
80 × 40 × 6	80	40	6	7	3.5	6.89	5.41	0.234	2.85	0.88	5.21	3.53	1.55
80 × 40 × 8			8			9.01	7.07		2.94	0.95	5.15	3.57	1.65
80 × 60 × 7	80	60	7	8	4	9.38	7.36	0.274	2.51	1.52	5.55	4.42	2.70
80 × 65 × 8	80	65	8	8	4	11.0	8.66	0.283	2.47	1.73	5.59	4.65	2.79
90 × 60 × 6	90	60	6	7	3.5	8.69	6.82	0.294	2.89	1.41	6.14	4.50	2.46
90 × 60 × 8			8			11.4	8.96		2.97	1.49	6.11	4.54	2.56
100 × 50 × 6			6			8.73	6.85		3.49	1.04	6.50	4.39	1.91
100 × 50 × 8	100	50	8	9	4.5	11.5	8.99	0.292	3.59	1.13	6.48	4.44	2.00
100 × 50 × 10			10			14.1	11.1		3.67	1.20	6.43	4.49	2.08
100 × 65 × 7			7		5	11.2	8.77		3.23	1.51	6.83	4.91	2.66
100 × 65 × 9	100	65	9	10	5	14.2	11.1	0.321	3.32	1.59	6.78	4.94	2.76
100 × 75 × 9	100	75	9	10		15.1	11.8	0.341	3.15	1.91	6.91	5.45	3.22
120 × 80 × 8			8		5.5	15.5	12.2		3.83	1.87	8.23	5.99	3.27
120 × 80 × 10	120	80	10	11		19.1	15.0	0.391	3.92	1.95	8.18	6.03	3.37
120 × 80 × 12			12			22.7	17.8		4.00	2.03	8.14	6.06	3.46
130 × 65 × 8			8		5.5	15.1	11.9		4.56	1.37	8.50	5.71	2.49
130 × 65 × 10	130	65	10	11		18.6	14.6	0.381	4.65	1.45	8.43	5.76	2.58

v_2 cm	v_A cm	Position of axis η-η tan α	x-x I_x cm⁴	W_x cm³	i_x cm	y-y I_y cm⁴	W_y cm³	i_y cm	ξ-ξ I_ξ cm⁴	i_ξ cm	η-η I_η cm⁴	i_η cm
1.04	0.56	0.431	1.25	0.62	0.94	0.44	0.29	0.56	1.43	1.00	0.25	0.42
1.03	0.58	0.423	1.59	0.81	0.93	0.55	0.38	0.55	1.81	0.99	0.33	0.42
1.19	0.46	0.259	2.79	1.08	1.27	0.47	0.30	0.52	2.96	1.31	0.30	0.42
1.18	0.50	0.252	3.59	1.42	1.26	0.60	0.39	0.52	3.79	1.30	0.39	0.42
1.58	0.83	0.436	5.78	1.91	1.42	2.05	0.91	0.85	6.65	1.52	1.18	0.64
1.58	0.85	0.430	6.99	2.35	1.41	2.47	1.11	0.84	8.02	1.51	1.44	0.64
1.67	0.78	0.356	7.71	2.33	1.59	2.09	0.91	0.82	8.53	1.67	1.27	0.64
1.66	0.80	0.353	9.41	2.88	1.58	2.54	1.12	0.82	10.4	1.66	1.56	0.64
1.84	1.27	0.625	10.4	3.02	1.56	5.89	2.01	1.18	13.3	1.76	3.02	0.84
1.77	0.72	0.256	15.6	4.04	1.90	2.60	1.12	0.78	16.5	1.96	1.69	0.63
2.09	1.10	0.437	17.2	4.25	1.89	6.11	2.02	1.13	19.8	2.03	3.50	0.86
2.08	1.12	0.433	20.1	5.03	1.88	7.12	2.38	1.12	23.1	2.02	4.12	0.85
2.38	1.50	0.583	23.1	5.11	2.04	11.9	3.18	1.47	28.8	2.28	6.21	1.06
2.52	1.42	0.497	33.5	7.04	2.21	14.3	3.81	1.44	39.9	2.41	7.94	1.07
2.63	1.38	0.433	46.4	9.24	2.36	16.5	4.39	1.41	53.3	2.53	9.56	1.07
2.71	1.58	0.530	35.5	6.84	2.37	16.2	3.89	1.60	43.1	2.61	8.68	1.17
2.70	1.62	0.525	47.9	9.39	2.35	21.8	5.52	1.59	57.9	2.59	11.8	1.17
2.42	0.89	0.259	44.9	8.73	2.55	7.59	2.44	1.05	47.6	2.63	4.90	0.84
2.38	1.04	0.253	57.6	11.4	2.53	9.68	3.18	1.04	60.9	2.60	6.41	0.84
2.92	1.68	0.546	59.0	10.7	2.51	28.4	6.34	1.74	72.0	2.77	15.4	1.28
2.94	2.05	0.645	68.1	12.3	2.49	40.1	8.41	1.91	88.0	2.82	20.3	1.36
3.16	1.60	0.442	71.7	11.7	2.87	25.8	5.61	1.72	82.8	3.09	14.6	1.30
3.15	1.69	0.437	92.5	15.4	2.85	33.0	7.31	1.70	107	3.06	19.0	1.29
2.98	1.15	0.263	89.7	13.8	3.20	15.3	3.86	1.32	95.2	3.30	9.78	1.06
2.95	1.18	0.258	116	18.0	3.18	19.5	5.04	1.31	123	3.28	12.6	1.05
2.91	1.22	0.252	141	22.2	3.16	23.4	6.17	1.29	149	3.25	15.5	1.04
3.48	1.73	0.419	113	16.6	3.17	37.6	7.54	1.84	128	3.39	21.6	1.39
3.46	1.78	0.415	141	21.0	3.15	46.7	9.52	1.82	160	3.36	27.2	1.39
3.63	2.22	0.549	148	21.5	3.13	71.0	12.7	2.17	181	3.47	37.8	1.59
4.20	2.16	0.441	226	27.6	3.82	80.8	13.2	2.29	261	4.10	45.8	1.72
4.19	2.19	0.438	276	34.1	3.80	98.1	16.2	2.27	318	4.07	56.1	1.71
4.18	2.25	0.433	323	40.4	3.77	114	19.1	2.25	371	4.04	66.1	1.71
3.86	1.47	0.263	263	31.1	4.17	44.8	8.72	1.72	280	4.31	28.6	1.38
3.82	1.54	0.259	321	38.4	4.15	54.2	10.7	1.71	340	4.27	35.0	1.37

Table 12. Hot-rolled, round-edged equal-angle in accordance with DIN 1028 (extract)

I second moment of area
W section modulus
i radius of inertia

Table contains only standardised preferred values.

Abbreviation	Dimensions for				Cross-section	Weight	Lateral area	Distance between axes				Static values for the bending axis							
												$x-x = y-y$			$\xi-\xi$		$\eta-\eta$		
L	a mm	s mm	r_1 mm	r_2 mm	cm²	kg/m	m²/m	e cm	w cm	v_1 cm	v_2 cm	I_x cm⁴	W_x cm³	i_x cm	I_ξ cm⁴	i_ξ cm	I_η cm⁴	W_η cm³	i_η cm
20 × 3	20	3	3.5	2	1.12	0.88	0.077	0.60	1.41	0.85	0.70	0.39	0.28	0.59	0.62	0.74	0.15	0.18	0.37
25 × 3	25	3	3.5	2	1.42	1.12	0.097	0.73	1.77	1.03	0.87	0.79	0.45	0.75	1.27	0.95	0.31	0.30	0.47
30 × 3	30	3	5	2.5	1.74	1.36	0.116	0.84	2.12	1.18	1.04	1.41	0.65	0.90	2.24	1.14	0.57	0.48	0.57
35 × 4	35	4	5	2.5	2.67	2.1	0.136	1.00	2.47	1.41	1.24	2.96	1.18	1.05	4.68	1.33	1.24	0.88	0.68
40 × 4	40	4	6	3	3.08	2.42	0.155	1.12	2.83	1.58	1.40	4.48	1.55	1.21	7.09	1.52	1.86	1.18	0.78
45 × 5	45	5	7	3.5	4.3	3.38	0.174	1.28	3.18	1.81	1.58	7.83	2.43	1.35	12.4	1.70	3.25	1.80	0.87
50 × 5	50	5	7	3.5	4.8	3.77	0.194	1.40	3.54	1.98	1.76	11.0	3.05	1.51	17.4	1.90	4.59	2.32	0.98
60 × 6	60	6	8	4	6.91	5.42	0.233	1.69	4.24	2.39	2.11	22.8	5.29	1.82	36.1	2.29	9.43	3.95	1.17
70 × 7	70	7	9	4.5	9.4	7.38	0.272	1.97	4.95	2.79	2.47	42.4	8.43	2.12	67.1	2.67	17.6	6.31	1.37
80 × 8	80	8	10	5	12.3	9.66	0.311	2.26	5.66	3.20	2.82	72.3	12.6	2.42	115	3.06	29.6	9.25	1.55
90 × 9	90	9	11	5.5	15.5	12.2	0.351	2.54	6.36	3.59	3.18	116	18.0	2.74	184	3.45	47.8	13.3	1.76
100 × 10	100	10	12	6	19.2	15.1	0.390	2.82	7.07	3.99	3.54	177	24.7	3.04	280	3.82	73.3	18.4	1.95
110 × 10	110	10	12	6	21.2	16.6	0.430	3.07	7.78	4.34	3.89	239	30.1	3.36	379	4.23	98.6	22.7	2.16
120 × 12	120	12	13	6.5	27.5	21.6	0.469	3.40	8.49	4.80	4.26	368	42.7	3.65	584	4.60	152	31.6	2.35
150 × 15	150	15	16	8	43	33.8	0.586	4.25	10.6	6.01	5.33	898	83.5	4.57	1430	5.76	370	61.6	2.93
180 × 18	180	18	18	9	61.9	48.6	0.705	5.10	12.7	7.22	6.41	1870	145	5.49	2970	6.93	757	105	3.49
200 × 20	200	20	18	9	76.4	59.9	0.785	5.68	14.1	8.04	7.15	2850	199	6.11	4540	7.72	1160	144	3.89

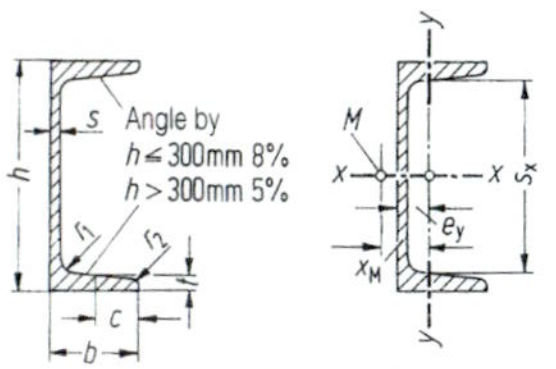

Table 13. Hot-rolled, round-edged U-section in accordance with DIN 1026 (extract)

I area moment of the second order
W section modulus
i radius of inertia
S_x first moment of area for half cross-section
$s_x = I_x/S_x$ distance between centre of compression and tension
x_M distance of shear centre M from axis y-y

Angle by
$h \leq 300$mm 8%
$h > 300$mm 5%

$c = b/2$ by $h \leq 300$
$c = (b-s)/2$ by $h > 300$

| Abbreviation U | Dimensions for | | | | | | Cross-section A | Weight G | For the bending axis | | | | | | S_x | s_x | Distance of axis y-y e_y | x_M |
| | | | | | | | | | x-x | | | y-y | | | | | | |
	b mm	b mm	s mm	t mm	r_1 mm	r_2 mm	cm²	kg/m	I_x cm⁴	W_x cm³	i_x cm	I_y cm⁴	W_y cm³	i_y cm	cm³	cm	cm	cm
30 × 15	30	15	4	4.5	4.5	2	2.21	1.74	2.53	1.69	1.07	0.38	0.39	0.42	—	—	0.52	0.74
30	30	33	5	7	7	3.5	5.44	4.27	6.39	4.26	1.08	5.33	2.68	0.99	—	—	1.31	2.22
40 × 20	40	20	5	5.5	5	2.5	3.66	2.87	7.58	3.79	1.44	1.14	0.86	0.56	—	—	0.67	1.01
40	40	35	5	7	7	3.5	6.21	4.87	14.1	7.05	1.50	6.68	3.08	1.04	—	—	1.33	2.32
50 × 25	50	25	5	6	6	3	4.92	3.86	16.8	6.73	1.85	2.49	1.48	0.71	—	—	0.81	1.34
50	50	38	5	7	7	3.5	7.12	5.59	26.4	10.6	1.92	9.12	3.75	1.13	—	—	1.37	2.47
60	60	30	6	6	6	3	6.46	5.07	31.6	10.5	2.21	4.51	2.16	0.84	—	—	0.91	1.50
65	65	42	5.5	7.5	7.5	4	9.03	7.09	57.5	17.7	2.52	14.1	5.07	1.25	—	—	1.42	2.60
80	80	45	6	8	8	4	11.0	8.64	106	26.5	3.10	19.4	6.36	1.33	15.9	6.65	1.45	2.67
100	100	50	6	8.5	8.5	4.5	13.5	10.6	206	41.2	3.91	29.3	8.49	1.47	24.5	8.42	1.55	2.93
120	120	55	7	9	9	4.5	17.0	13.4	364	60.7	4.62	43.2	11.1	1.59	36.3	10.0	1.60	3.03
140	140	60	7	10	10	5	20.4	16.0	605	86.4	5.45	62.7	14.8	1.75	51.4	11.8	1.75	3.37
160	160	65	7.5	10.5	10.5	5.5	24.0	18.8	925	116	6.21	85.3	18.3	1.89	68.8	13.3	1.84	3.56
180	180	70	8	11	11	5.5	28.0	22.0	1350	150	6.95	114	22.4	2.02	89.6	15.1	1.92	3.75
200	200	75	8.5	11.5	11.5	6	32.2	25.3	1910	191	7.70	148	27.0	2.14	114	16.8	2.01	3.94
220	220	80	9	12.5	12.5	6.5	37.4	29.4	2690	245	8.48	197	33.6	2.30	146	18.5	2.14	4.20
240	240	85	9.5	13	13	6.5	42.3	33.2	3600	300	9.22	248	39.6	2.42	179	20.1	2.23	4.39
260	260	90	10	14	14	7	48.3	37.9	4820	371	9.99	317	47.7	2.56	221	21.8	2.36	4.66
280	280	95	10	15	15	7.5	53.3	41.8	6280	448	10.9	399	57.2	2.74	266	23.6	2.53	5.02
300	300	100	10	16	16	8	58.8	46.2	8030	535	11.7	495	67.8	2.90	316	25.4	2.70	5.41
320	320	100	14	17.5	17.5	8.75	75.8	59.5	10870	679	12.1	597	80.6	2.81	413	26.3	2.60	4.82
350	350	100	14	16	16	8	77.3	60.6	12840	734	12.9	570	75.0	2.72	459	28.6	2.40	4.45
380	380	102	13.5	16	16	8	80.4	63.1	15760	829	14.0	615	78.7	2.77	507	31.1	2.38	4.58
400	400	110	14	18	18	9	91.5	71.8	20350	1020	14.9	846	102	3.04	618	32.9	2.65	5.11

Table 14. Hot-rolled flat steel for general use in accordance with DIN 1017 (extract)

Weight in kg/m

Width b mm	Tolerance	Thickness s in mm											
		5	6	8	10	12	15	20	25	30	40	50	60
		±0.5							±1.0			±1.5	
												±1.5	
20	±0.75	0.785	0.942	1.26	1.57	(1.88)	(2.36)	–	–	–	–	–	–
25		0.981	1.18	1.57	1.96	(2.36)	(2.94)	–	–	–	–	–	–
30		1.18	1.41	1.88	2.36	2.83	3.53	4.71	–	–	–	–	–
35		1.37	1.65	2.20	2.75	(3.30)	(4.12)	(5.50)	(6.87)	–	–	–	–
40	±1.0	1.57	1.88	2.51	3.14	3.77	4.71	6.28	(7.85)	(9.42)	–	–	–
45		(1.77)	(2.12)	(2.83)	(3.53)	(4.24)	(5.30)	(7.07)	(8.83)	(10.6)	–	–	–
50		1.96	2.36	3.14	3.93	4.71	5.89	7.85	9.81	11.8	(15.7)	–	–
60		2.36	2.83	3.77	4.71	5.65	7.07	9.42	(11.8)	14.1	(18.8)	–	–
70		2.75	3.30	4.40	5.50	6.59	8.24	11.0	(13.7)	(16.5)	(22.0)	–	–
80	±1.5	3.14	3.77	5.02	6.28	7.54	9.42	12.6	(15.7)	(18.8)	(25.1)	(31.4)	(37.7)
90		(3.53)	(4.24)	(5.65)	(7.07)	(8.48)	(10.6)	(14.1)	(17.7)	(21.2)	(28.3)	(35.3)	–
100		3.93	4.71	6.28	7.85	9.42	11.8	15.7	(19.6)	23.6	(31.4)	(39.3)	(47.1)
110	±2.0	–	–	(6.91)	(8.64)	(10.4)	–	(17.3)	(23.6)	(25.9)	–	–	–
120		–	(5.65)	7.54	9.42	11.3	14.1	18.8	(23.6)	(28.3)	(37.7)	(47.1)	(56.5)
130	±2.5	–	–	(8.16)	(10.2)	(12.2)	(16.5)	(22.0)	–	–	–	–	–
140		–	–	(8.79)	(11.0)	(13.2)	14.1	23.6	–	–	–	–	(66.0)
150		–	(7.06)	9.42	11.8	14.1	17.7	23.6	(29.4)	(35.3)	(47.1)	(58.9)	(70.7)

Values in brackets should be avoided wherever possible.

Table 15. Hot-rolled flat steel for general use in accordance with DIN 1013 (extract)

Diameter d in mm			Cross-section cm²	Weight kg/m	Lateral area cm²/m
Series A[a]	Series B	Standard tolerance			
8		±0.4	0.503	0.395	251
10			0.785	0.617	314
12			1.13	0.888	377
	13		1.33	1.04	408
14			1.54	1.21	440
	15		1.77	1.39	471
16		±0.5	2.01	1.58	503
	17		2.27	1.78	534
18			2.54	2.00	565
	19		2.84	2.23	597
20			3.14	2.47	628
	21		3.46	2.72	660
22			3.80	2.98	691
	23		4.15	3.26	723
24			4.52	3.55	754
25			4.91	3.85	785
	26	±0.6	5.31	4.17	817
27			5.73	4.49	848
28			6.16	4.83	880
30			7.07	5.55	942
31			7.55	5.92	974
32			8.04	6.31	1010
	34		9.08	7.13	1070
35			9.62	7.55	1100
	36	±0.8	10.2	7.99	1130
37			10.8	8.44	1160
38			11.3	8.90	1190
40			12.6	9.86	1260
42			13.9	10.9	1320
44			15.2	11.9	1380
45			15.9	12.5	1410
	47		17.3	13.6	1480
	48		18.1	14.2	1510
50			19.6	15.4	1570
52		±1	21.2	16.7	1630
	53		22.1	17.3	1670
55			23.8	18.7	1730
60			28.3	22.2	1880
	63		31.2	24.5	1980
65			33.2	26.0	2040
70			38.5	30.2	2200
75			44.2	34.7	2360
80			50.3	39.5	2510
	85	±1.3	56.7	44.5	2670
90			63.6	49.9	2830
	95		70.9	55.6	2980
100			78.5	61.7	3140
110		±1.5	95.0	74.6	3460
120			113	88.8	3770
	130	±2	133	104	4080
140			154	121	4400
150			177	139	4710
160			201	158	5030
	170	±2.5	227	178	5340
180			254	200	5650
	190		284	223	5970
200			314	247	6280

[a] Series A contains preferred diameters.

Table 16. Seamless steel pipes in accordance with DIN 2448 (extract). Bold type shows the preferred wall thicknesses for pipes in series 1

Masses for unit length in kg/m for wall thicknesses in mm.

Ext. diameter – Series 1	Series 2	Series 3	Normal wall thickness mm	1.6	1.8	2	2.3	2.6	2.9	3.2	3.6	4	4.5	5	5.4	5.6	6.3	7.1	8	8.8	10	11	12.5	14.2	16	17.5	20	22.2	25	28
10.2			1.6	0.339	0.373	0.404	0.448	0.487																					
13.5			2		0.519	**0.567**	0.635	0.699	0.758	0.813	0.879																		
	16		2		0.630	0.691	0.777	0.859	0.937	1.01	1.10	1.18																		
17.2			2		0.684	**0.750**	0.845	0.936	1.02	1.10	1.21	1.30	1.41																
		19	2			0.838	0.947	1.05	1.15	1.25	1.37	1.48	1.61	1.73															
		20	2			0.888	1.00	1.12	1.22	1.33	1.46	1.58	1.72	1.85															
21.3			2			**0.952**	1.08	1.20	1.32	**1.43**	1.57	**1.71**	1.86	2.01	2.12														
25			2			1.13	1.29	1.44	1.58	1.72	1.90	2.07	2.28	2.47	2.61	2.68	2.91												
	25.4		2			1.15	1.31	1.46	1.61	1.75	1.94	2.11	2.32	2.52	2.66	2.73	2.97												
26.9			2			**1.23**	1.40	1.56	1.72	**1.87**	2.07	**2.26**	2.49	2.70	2.86	2.94	3.20	3.47											
		30	2.3				1.57	1.76	1.94	2.11	2.34	2.56	2.83	3.08	3.28	3.37	3.68	4.01	4.34											
	31.8		2.3				1.67	1.87	2.07	2.26	2.50	2.74	3.03	3.30	3.52	3.62	3.96	4.32	4.70											
33.7			2.3				**1.78**	1.99	2.20	**2.41**	2.67	2.93	**3.24**	3.54	3.77	3.88	4.26	4.66	5.07	5.40										
	38		2.6					2.27	2.51	2.75	3.05	3.35	3.72	4.07	4.34	4.47	4.93	5.41	5.92	6.34	6.91									
42.4			2.6					**2.55**	2.82	3.09	**3.44**	3.79	4.21	**4.61**	4.93	5.08	5.61	6.18	6.79	7.29	7.99	8.52								
	44.5		2.6					2.69	2.98	3.26	3.63	4.00	4.44	4.87	5.21	5.37	5.94	6.55	7.20	7.75	8.51	9.09	9.86							
48.3			2.6					**2.93**	3.25	3.56	**3.97**	4.37	4.86	**5.34**	5.71	5.90	6.53	7.21	7.95	8.57	9.45	10.1	11.0							
		51	2.6					3.10	3.44	3.77	4.21	4.64	5.16	5.67	6.07	6.27	6.94	7.69	8.48	9.16	10.1	10.9	11.9							
	54		2.9						3.65	4.01	4.47	4.93	5.49	6.04	6.47	6.68	7.41	8.21	9.08	9.81	10.9	11.7	12.8							
		57	2.9						3.87	4.25	4.74	5.23	5.83	6.41	6.87	7.10	7.88	8.74	9.67	10.5	11.6	12.5	13.7	15.0						
60.3			2.9						**4.11**	4.51	5.03	**5.55**	6.19	6.82	7.31	**7.55**	8.39	9.32	10.3	11.2	12.4	13.4	14.7	16.1	17.5					
		63.5	2.9						4.33	4.76	5.32	5.87	6.55	7.21	7.74	8.00	8.89	9.88	11.0	11.9	13.2	14.2	15.7	17.3	18.7					
	70		2.9						4.80	5.27	5.90	6.51	7.27	8.01	8.60	8.89	9.90	11.0	12.2	13.3	14.8	16.0	17.7	19.5	21.3	22.7				
		73	2.9						5.01	5.51	6.16	6.81	7.60	8.38	9.00	9.31	10.4	11.5	12.8	13.9	15.5	16.8	18.7	20.6	22.5	24.0				
76.1			2.9						**5.24**	5.75	6.44	7.11	7.95	**8.77**	9.42	9.74	10.8	**12.1**	13.4	14.6	16.3	17.7	19.6	21.7	23.7	25.3	27.7			
		82.5	3.2							6.26	7.00	7.74	8.66	9.56	10.3	10.6	11.8	13.2	14.7	16.0	17.9	19.4	21.6	23.9	26.2	28.1	30.8	33.0		
88.9			3.2							**6.76**	7.57	8.38	9.37	10.3	11.1	**11.5**	12.8	14.3	**16.0**	17.4	19.5	21.1	23.6	26.2	28.8	30.8	34.0	36.5	39.4	
	101.6		3.6								8.70	9.63	10.8	11.9	12.8	13.3	14.8	16.5	18.5	20.1	22.6	24.6	27.5	30.6	33.8	36.3	40.2	43.5	47.2	50.8
		108	3.6								9.27	10.3	11.5	12.7	13.7	14.1	15.8	17.7	19.7	21.5	24.2	26.3	29.4	32.8	36.3	39.1	43.4	47.0	51.2	55.2
114.3			3.6								**9.83**	10.9	12.2	13.5	14.5	15.0	**16.8**	18.8	21.0	**22.9**	25.7	28.0	31.4	35.1	38.8	41.8	46.5	50.4	55.1	59.6
		127	4									12.1	13.6	15.0	16.2	16.8	18.8	21.0	23.5	25.7	28.9	31.5	35.3	39.5	43.8	47.3	52.8	57.4	62.9	68.4

Table 17. Welded pipes in accordance with DIN 2458 (extract). Bold type shows the preferred wall thicknesses for pipes in series 1

External pipe diameter in mm — Series 1	Series 2	Series 3	Normal wall thickness mm	Masses for unit length in kg/m for wall thicknesses in mm — 1.4	1.6	1.8	2	2.3	2.6	2.9	3.2	3.6	4	4.5	5	5.4	5.6	6.3	7.1	8	8.8	10	11
10.2			1.6	0.304	**0.339**	0.373	0.404	0.448	0.487														
13.5			1.6	0.418	**0.490**	0.519	0.567	0.635	0.699	0.758	0.813	0.879											
16			1.6	0.504	**0.568**	0.630	0.691	0.777	0.859	0.937	1.01	1.10											
17.2			1.6	0.546	**0.616**	0.684	0.750	0.845	0.936	1.02	1.10	1.21	1.30										
19			1.8	0.608	0.687	**0.764**	0.838	0.947	1.05	1.15	1.25	1.37	1.48										
20			1.8	0.642	0.726	**0.808**	0.888	1.00	1.12	1.22	1.33	1.46	1.58										
21.3			1.8	0.687	0.777	**0.866**	0.952	1.08	1.20	1.32	1.43	1.57	1.71	1.86									
25			1.8	0.815	0.923	**1.03**	1.13	1.29	1.44	1.58	1.72	1.90	2.07	2.28	2.47								
		25.4	1.8	0.829	0.939	1.05	1.15	1.31	1.46	1.61	1.75	1.94	2.11	2.32	2.52								
26.9			1.8	0.880	0.998	**1.11**	1.23	1.40	1.56	1.72	1.87	2.07	2.26	2.49	2.70								
		30	2	0.987	1.12	1.25	1.38	1.57	1.76	1.94	2.11	2.34	2.56	2.83	3.08	3.28	3.37	3.68					
31.8			2	1.05	1.19	1.33	**1.47**	1.67	1.87	2.07	2.26	2.50	2.74	3.03	3.30	3.52	3.62	3.96	4.32					
33.7			2	1.12	1.27	1.42	**1.56**	1.78	1.99	2.20	2.41	2.67	2.93	3.24	3.54	3.77	3.88	4.26	4.66	5.07				
38			2	1.26	1.44	1.61	**1.78**	2.02	2.27	2.51	2.75	3.05	3.35	3.72	4.07	4.34	4.47	4.93	5.41	5.92	6.34			
42.4			2.3	1.42	1.61	1.80	1.99	**2.27**	2.55	2.82	3.09	3.44	3.79	4.21	4.61	4.93	5.08	5.61	6.18	6.79	7.29			
		44.5	2.3	1.49	1.69	1.90	2.10	2.39	2.69	2.98	3.26	3.63	4.00	4.44	4.87	5.21	5.37	5.94	6.55	7.20	7.55			
48.3			2.3	1.62	1.84	2.06	2.28	**2.61**	2.93	3.25	3.56	3.97	4.37	4.86	5.34	5.71	5.90	6.53	7.21	7.95	8.57			
51			2.3	1.71	1.95	2.18	2.42	**2.76**	3.10	3.44	3.77	4.21	4.64	5.16	5.67	6.07	6.27	6.94	7.69	8.48	9.16			
		54	2.3	1.82	2.07	2.32	2.56	2.93	3.30	3.65	4.01	4.47	4.93	5.49	6.04	6.47	6.68	7.41	8.21	9.08	9.81	10.9		
57			2.3	1.92	2.19	2.45	2.71	**3.10**	3.49	3.87	4.25	4.74	5.23	5.83	6.41	6.87	7.10	7.88	8.74	9.67	10.5	11.6		
60.3			2.3	2.03	2.32	2.60	2.88	**3.29**	3.70	4.11	4.51	5.03	5.55	6.19	6.82	7.31	7.55	8.39	9.32	10.3	11.2	12.4		
63.5			2.3		2.44	2.74	3.03	**3.47**	3.90	4.33	4.76	5.32	5.87	6.55	7.21	7.74	8.00	8.89	9.88	10.9	11.9	13.2		
70			2.6		2.70	3.03	3.35	3.84	**4.32**	4.80	5.27	5.90	6.51	7.27	8.01	8.60	8.89	9.90	11.0	12.2	13.3	14.8		
		73	2.6		2.82	3.16	3.50	4.01	4.51	5.01	5.51	6.16	6.81	7.60	8.38	9.00	9.31	10.4	11.5	12.8	13.9	15.5		
76.1			2.6		2.94	3.30	3.65	4.19	**4.71**	5.24	5.75	6.44	7.11	7.95	8.77	9.42	9.74	10.8	12.1	13.4	14.6	16.3		
		82.5	2.6		3.19	3.58	3.97	4.55	5.12	5.69	6.26	7.00	7.74	8.66	9.56	10.3	10.6	11.8	13.2	14.7	16.0	17.9		
88.9			2.6		3.44	3.87	4.29	4.91	**5.53**	6.15	6.76	7.57	8.38	9.37	10.3	11.1	11.5	12.8	14.3	16.0	17.4	19.5		
101.6			2.9				4.91	5.63	6.35	**7.06**	7.77	8.70	9.63	10.8	11.9	12.8	13.3	14.8	16.5	18.5	20.1	22.6		
		108	2.9				5.23	6.00	6.76	7.52	8.27	9.27	10.3	11.5	12.7	13.7	14.1	15.8	17.7	19.7	21.5	24.2		26.3
114.3			3.2				5.54	6.35	7.16	7.97	**8.77**	9.83	10.9	12.2	13.5	14.5	15.0	16.8	18.8	21.0	22.9	25.7		28.0
127			3.2				6.17	7.07	7.98	8.88	**9.77**	11.0	12.1	13.6	15.0	16.2	16.8	18.8	21.0	23.5	25.7	28.9		31.5

11 References

B1 General Fundamentals. [1] Leipholz H. Einführung in die Elastizitätstheorie. Braun, Karlsruhe, 1968 – [2] Biezeno C, Grammel R. Technische Dynamik, 2nd edn. Springer, Berlin, 1971. – [3] Müller W. Theorie der elastischen Verformung. Akad. Verlagsgesell. Geest u. Portig, Leipzig, 1959. – [4] Neuber H. Technische Mechanik, pt II. Springer, Berlin, 1971. – [5] Betten J. Elastizitäts- und Plastizitätstheorie, 2nd edn. Vieweg, Brunswick, 1986. – [6] Siebel E. Neue Wege der Festigkeitsrechnung. VDI-Z. 1948; 90: 135–9. – [7] Siebel E, Rühl K. Formdehngrenzen für die Festigkeitsberechnung. Die Technik 1948; 3: 218–23. – [8] Siebel E, Schwaigerer S. Das Rechnen mit Formdehngrenzen. VDI-Z 1948; 90: 335–41. – [9] Schwaigerer S. Werkstoffkennwert und Sicherheit bei der Festigkeitsberechnung. Konstruktion 1951; 3: 233–39. – [10] Wellinger K, Dietmann H. Festigkeitsberechnung, 3rd edn. Kröner, Stuttgart, 1976.

B2 Stresses in Bars and Beams. [1] Szabó I. Einführung in die Technische Mechanik, 8th edn. Springer, Berlin, 1975. – [2] Weber C. Biegung und Schub in geraden Balken. Z. angew. Math. u. Mech. 1924; 4: 334–48. – [3] Schultz-Grunow F. Einführung in die Festigkeitslehre. Werner, Düsseldorf, 1949. – [4] Szabó I. Höhere Technische Mechanik, 5th edn. Springer, Berlin, 1977. – [5] Neuber H. Technische Mechanik, pt II. Springer, Berlin, 1971. – [6] Leipholz H. Festigkeitslehre für den Konstrukteur. Springer, Berlin, 1969. – [7] Roark, Young. Formulas for Stress and Strain, 5th edn. McGraw-Hill, Singapore, 1986.

B3 Theory of Elasticity. [1] Szabó I. Höhere Technische Mechanik, 5th edn. Springer, Berlin, 1977. – [2] Girkmann K. Flächentragwerke, 3rd edn. Springer, Vienna, 1954. – [3] Timoshenko S, Goodier J N. Theory of Elasticity, 3rd edn. McGraw-Hill, Singapore, 1982.

B4 Hertzian Contact Stresses (Formulae of Hertz). [1] Hertz H. Über die Berührung fester elastischer Körper. Ges. Werke, vol. I. Barth, Leipzig, 1895. – [2] Szabó I. Höhere Technische Mechanik, 5th edn. Springer, Berlin, 1977. – [3] Föppl L. Der Spannungszustand und die Anstrengung der Werkstoffe bei der Berührung zweier Körper. Forsch. Ing.-Wes. 1936; 7: 209–21. – [4] Timoshenko S, Goodier J N. Theory of elasticity, 3rd edn. McGraw-Hill, Singapore, 1982.

B5 Plates and Shells. [1] Girkmann K. Flächentragwerke, 3rd edn. Springer, Vienna, 1954. – [2] Nádai A. Die elastischen Platten. Springer, Berlin, 1925 (reprinted 1968). – [3] Wolmir A S. Biegsame Platten und Schalen. VEB Verlag f. Bauwesen, Berlin, 1962. – [4] Czerny F. Tafeln für vierseitig und dreiseitig gelagerte Rechteckplatten. Betonkal. 1984. vol. I. Ernst, Berlin, 1984. – [5] Beyer K. Die Statik im Stahlbetonbau. Springer, Berlin, 1948. – [6] Worch G. Elastische Platten. Betonkal 1960, vol. II. Ernst, Berlin, 1960. – [7] Timoshenko S, Woinowsky-Krieger S. Theory of plates and shells, 2nd edn. McGraw-Hill, Kogakusha, 1983.

B6 Centrifugal Stresses in Rotating Components. [1] Biezeno C, Grammel R. Technische Dynamik, 2nd edn. Springer, Berlin, 1971.

B7 Stability Problems. [1] Szabó I. Höhere Technische Mechanik, 5th edn. Springer, Berlin, 1977. – [2] Kollbrunner C F, Meister M. Knicken, Biegedrillknicken, Kippen, 2nd edn. Springer, Berlin, 1961. – [3] Biezeno C, Grammel R. Technische Dynamik, 2nd edn. Springer, Berlin, 1971. – [4] Pflüger A. Stabilitätsprobleme der Elastostatik. Springer, Berlin, 1950. – [5] Bürgermeister G, Steup H. Stabilitätstheorie. Akademie-Verlag, Berlin, 1959. – [6] Timoshenko S. Theory of elastic stability. McGraw-Hill, New York, 1936. – [7] Wolmir A S. Biegsame Platten und Schalen. VEB Verlag f. Bauwesen, Berlin, 1962. – [8] Flügge W. Statik und Dynamik der Schalen, 2nd edn. Berlin, 1957. – [9] Schapitz E. Festigkeitslehre für den Leichtbau, 2nd edn. VDI-Verlag, Düsseldorf, 1963.

B8 Finite-Element and Boundary Element Methods. [1] Zienkiewicz O C. Methoden der finiten Elemente. Hanser, Munich, 1975. – [2] Gallagher R H. Finite-Element-Analysis. Springer, Berlin, 1976. – [3] Schwarz H R. Methode der finiten Elemente. Teubner, Stuttgart, 1980. – [4] Link M. Finite Elemente in der Statik und Dynamik. Teubner, Stuttgart, 1984. – [5] Argyris J, Mlejnek H.-P. Die Methode der finiten Elemente, vols 1–III. Vieweg, Brunswick, 1986–8. – [6] Bathe K.-J. Finite-Element-Methoden. Springer, Berlin, 1986. – [7] Oldenburg W. Die Finite-Elemente-Methode auf dem PC. Vieweg, Brunswick, 1989. – [8] Wellinger K, Dietmann H. Festigkeitsberechnung, Grundlagen und technische Anwendung. 3 vols. Kröner, Stuttgart, 1976. – [9] Hampel H. Rohrleitungsstatik, Grundlagen, Gebrauchsformeln, Beispiele. Springer, Berlin, 1972. – [10] Collatz L. Numerische Behandlung von Differentialgleichungen, 2nd edn. Springer, Berlin, 1955. – [11] Girkmann K. Flächentragwerke, 3rd edn. Springer, Vienna, 1954. – [12] Hartmann F. Methode der Randelemente. Springer, Berlin, 1987. – [13] Brebbia C A, Telles J C F, Wrobel L C. Boundary Element Techniques. Springer, Berlin, 1987. – [14] Zotemantei R. Berechnung von Platten nach der Methode der Randelemente, Dissertation 1985: University of Dortmund.

B9 Theory of Plasticity. [1] Wellinger K, Dietmann H. Festigkeitsberechnung. Grundlagen und technische Anwendung, 3rd edn. Kröner, Stuttgart, 1976. – [2] Reckling K A. Plastizitätstheorie und ihre Anwendung auf Festigkeitsprobleme. Springer, Berlin, 1967. – [3] Lippmann H, Mahrenholtz O. Plastomechanik der Umformung metallischer Werkstoffe. Springer, Berlin, 1967. – [4] Schreyer G. Konstruieren mit Kunststoffen. Hanser, Munich, 1972. – [5] Szabó I. Höhere Technische Mechanik, 5th edn, revised reprint. Springer, Berlin, 1977. – [6] Ismar H, Mahrenholtz O. Technische Platomechanik. Vieweg, Brunswick, 1979. – [7] Kreissig R, Drey K-D, Naumann J. Methoden der Plastizität. Hanser, Munich, 1980. – [8] Lippmann H. Mechanik des Plastischen Fliessens. Springer, Berlin, 1980.

C Thermodynamics

K. Stephan, Stuttgart

1 Scope of Thermodynamics. Definitions

The general science of thermodynamics is a subdivision of physics. It is concerned with the various forms in which energy arises and with conversion between such forms. It provides the general laws governing energy conversion.

1.1 Systems, Boundaries of Systems, Surroundings

A thermodynamic system, or *system* for short, means any material configuration whose thermodynamic characteristics are the subject of investigation. Typical systems are quantities of gas, liquid or its vapour, mixtures of liquids, or a crystal. The system is separated from its surroundings, the "environment", by the *system boundary*. A system boundary may move during the process under investigation, for example if a quantity of gas expands, and it may also be permeable to energy and matter. A system is *closed* if the system boundary is impermeable and *open* if it is permeable. While the mass of a closed system is constant, the mass of an open system is variable if the mass flowing into the system during a specific period is different from that flowing out of the system. If input and output masses are the same, then the mass of the open system will remain constant. Examples of closed systems are solid bodies or mass elements in mechanics, and fluid flows (gases or liquids) in ducts. The term *closed* is applied to a system which is isolated from all influence from all environmental influences, such that neither energy nor matter can be exchanged with its surroundings.

The distinction between closed and open systems corresponds to the distinction between the *Lagrange* and the *Euler* datum systems in fluid mechanics. In the Lagrange datum system, which corresponds to the closed system, we investigate the motion of a fluid by dividing it into small elements of constant mass and deriving their equation of motion. In the Euler datum system, which corresponds to the open system, concepts are related to a constant volume element and fluid flow through the volume element is investigated. Both definitions are equivalent, and the decision whether to treat a system as closed or open is often regarded as merely a question of expediency.

1.2 Description of the State of a System. Thermodynamic Processes

A system is characterised by specific physical variables that can be measured, for example pressure, temperature, density, electrical conductivity, refractive index, etc. The *state* of a system is defined by the fact that all these physical variables, the *state variables*, acquire constant values. The transfer of a system from one state into another is called a *change of state*.

Example A balloon is filled with gas. Let the thermodynamic system be the gas. Its volume, as revealed by measurement, is determined by pressure and temperature. Accordingly, the system's state variables are volume, pressure and temperature, and the state of the system (gas) is characterised by three fixed values: volume, pressure and temperature. The transition to a different set of values, for example if a certain quantity of gas escapes, is referred to as a *change of state*.

The mathematical relationship between state variables is referred to as the *equation of state* or *state function*.

Example The volume of gas in a balloon has been demonstrated as a function of pressure and temperature. The mathematical relationship between these state variables is an example of such an equation of state.

State variables are divided into three categories:

1. *Intensive* state variables are independent of the size of the system and thus retain their values on division of the system into subsystems.

Example If a space filled with gas of constant temperature is divided into smaller spaces, the temperature remains unchanged. It is an intensive state variable.

2. State variables which are proportionate to the quantity of the system are referred to as *extensive* state variables.

Example Volume, energy or quantity as such.

3. If an extensive state variable X is divided by the quantity of the system, then a *specific* state variable x is obtained.

Example Let an extensive state variable be the volume of a gas; the specific state variable is then the *specific volume* $v = V/m$, if m is the mass of the gas. The SI unit of specific volume is m^3/kg.

Changes of state arise from interactions with the surroundings of the system, for example because energy is subtracted from or added to the system via its boundaries. For the purposes of defining a state variable, it is enough to state the chronological curve of state variables. Description of a process also requires information concerning the magnitude and type of interactions with the surroundings. A *process* is thus understood to refer to the changes of state caused by specific external influences.

2 Temperatures. Equilibria

2.1 Adiabatic and Diathermal Walls

When a system is brought into contact with its surroundings – an environment whose state we postulate as unchanging – then the state variables of the system vary in time and eventually take on new constants. The system is then said to be in equilibrium with its surroundings. The rate at which the system achieves a state of equilibrium depends on the nature of contact with its surroundings. If the system and the environment are separated from each other by only a thin metal panel then equilibrium will be reached rapidly, but if they are separated by thick polystyrene foam panels then equilibrium will be set up very slowly. A partition that purely impedes any exchange of matter, and any mechanical, magnetic or electrical interaction, is referred to as *diathermal*. A diathermal wall is "thermally" conductive. The ultimate state to be reached will be *thermal equilibrium*.

Systems whose partitions permit only exchange of work with the surroundings are called *adiabatic*. Thermally, they are completely separated from their surroundings.

2.2 Zeroth Law and Empirical Temperature

If there is thermal equilibrium between systems A and C and between systems B and C, then, empirically, systems A and B are still in thermal equilibrium if they are brought into contact with each other via a diathermal panel. This empirical law is described as the *zeroth main law of thermodynamics*. It runs as follows:

Two systems in thermal equilibrium with a third are also in thermal equilibrium with each other.

In order to discover whether two systems A and B are in thermal equilibrium, they are consecutively brought into contact with a system C whose mass is postulated as low in comparison with that of systems A and B, such that changes of state in systems A and B are negligible during the process of reaching equilibrium. If C is first brought into contact with A, then certain state variables of C will vary, for example its electrical resistance. These state variables will remain unchanged in subsequent contact between B and C, if preceded by thermal equilibrium between A and B. System C enables the position of thermal equilibrium between A and B to be ascertained. After equilibration, the state variables of C can be allocated with constants as desired. These constants are referred to as *empirical temperatures*, and the measurement instrument itself is a *thermometer*.

2.3 Temperature Scales

In order to devise empirical temperature scales, the gas thermometer is used (**Fig. 1**), with which pressure p is measured as exerted by gas volume V.

Gas volume V is held constant by change in height Δz of the mercury column. Pressure p exerted by the mercury column and the environment is measured and the product pV is formed. Measurements with various pressures which are adequately low will, by extrapolation, produce a limit value:

$$\lim_{p \to 0} pV = A.$$

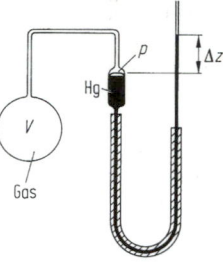

Figure 1. Gas thermometer with gas volume V in the bulb up to the mercury column.

To this value (obtained from measurements) there is allocated an empirical temperature in accordance with a linear formula:

$$T = \text{const} \cdot A. \tag{1}$$

After establishing the "const" constants, all we then need to ascertain is the value of A from measurements enabling calculation of empirical temperature T from Eq. (1). The "fixed point" required for determination of the empirical temperature scale was fixed by the 10th General Conference of Weights and Measures (CGPM) at Paris in 1954 as the triple point of water at temperature $T_{tr} = 273.16$ K. At the triple point of water, its vapour, liquid water and ice are in equilibrium with each other at a defined pressure of 0.006 112 bar. The temperature scale introduced by this means is called the *Kelvin scale*. It is identical with the *thermodynamic temperature scale*. The following formula applies:

$$T = T_{tr} A/A_{tr}, \tag{1a}$$

if A_{tr} is the value for variable A measured using a gas thermometer at the triple point of water. It is ascertained by measurements that temperature $T_{tr} = 273.26$ K at the triple point of water is approximately 0.01 K higher than temperature $T = 273.15$ K at the freezing point. The scale counted from freezing point $T = 273.15$ K is designated the Celsius scale, whose temperatures t are set out in °C. For temperature T in K,

$$T = t + 273.15 \ °C \tag{2}$$

At a pressure of 0.101 325 MPa of water, the following precise temperatures apply at the *freezing point*:

$$T_0 = (273.15 \pm 0.0002) \ \text{K}$$

And the precise temperature at the *boiling point*:

$$T_1 = (373.1464 \pm 0.0036) \ \text{K}.$$

In the English-speaking world, the *Fahrenheit scale* is still commonly employed, with a temperature of 32°F at the freezing point of water and of 212°F at its boiling point (for a pressure of 0.101 325 MPa). For conversion of a temperature t_F stated in °F into the Celsius temperature t in °C, the following formula is applied:

$$t = \frac{5}{9} (t_F - 32). \tag{3}$$

The scale counted from absolute zero in °F is designated the *Rankine scale* (°R). The following formula applies:

$$T_R = \frac{9}{5} T, \qquad (4)$$

T_R in °R, T in K. The freezing point of water is 491.67°R.

2.3.1 The International Practical Temperature Scale

Since the precise measurement of temperatures using gas thermometers is a difficult and time-consuming process, the international practical temperature scale has been made law. It has been laid down by the International Committee for Weights and Measures such that temperature on the scale approximates as closely as possible to the thermodynamic temperature of specific materials. The international practical temperature scale is laid down by the melting and freezing points of these substances, which have been ascertained as precisely as possible using gas thermometers in various countries' national scientific institutes. By means of resistance thermometers, thermocouples and radiation measurement instruments, interpolative calculations are made between these fixed points, and specific regulations are issued for the relationships between the indirectly measured variables and temperature.

The main international temperature scale provisions, which apply equally in all countries, run as follows:

1. In the 1948 international temperature scale, temperatures are designated by "°C" or "°C (Int. 1948)" and represented by the letter symbol t.
2. The scale is based first on a number of fixed and continuously reproducible equilibrium temperatures (fixed points), to which there are allocated specific numerical values, and second on precisely established formulae which embody the relationships between temperature and readouts of measurement instruments calibrated at these fixed points.
3. Fixed points and the numerical values allocated to them are compiled in tables (see **Appendix C2, Table 1**). With the exception of triple points and one fixed point in equilibrium hydrogen (17.042 K), the allocated temperatures correspond to equilibrium states at the pressure of normal physical atmosphere, i.e. as per the definition for 0.101 325 MPa.
4. Between fixed temperatures, interpolation is performed by means of formulae that are also laid down by international conventions. These formulae enable numerical values for international practical temperature to be assigned to the indications of "standard" instruments.

In order to facilitate temperature measurement, a series of further thermometric fixed points of substances that are sufficiently easy to produce in pure form has been appended as precisely as possible to the statutory temperature scale. The most important fixed points are set out in **Appendix C2, Table 2**. As a standard instrument, the platinum resistance thermometer is used between the triple point of 13.81 K of hydrogen in equilibrium and the solidification point of antimony at 903.89 K ($= 630.74$ °C). The standard instrument employed between the solidification points of antimony and gold (1337.58 K ($= 1064.43$ °C)) is a platinum-rhodium (10% rhodium)/platinum thermocouple. Above the solidification point of gold, the international practical temperature is defined by *Planck's radiation law*:

$$\frac{J_t}{J_{Au}} = \frac{\exp\left[\dfrac{c_2}{\lambda(t_{Au} + T_0)}\right] - 1}{\exp\left[\dfrac{c_2}{\lambda(t + T_0)}\right] - 1;} \qquad (5)$$

here J_t and J_{Au} relate to radiation energies, emitted by a black body at wavelength λ per surface, time and wavelength interval at temperature t and at the gold point of t_{Au}; c_2 is the value of constant c_2 as a value established as $0.014\ 388$ m K (metre kelvin); $T_0 = 273.15$ K is the numerical value for the temperature of the melting point of ice; and λ is the numerical value of one wavelength of the visible spectrum range in m.

For practical temperature measurement see [1].

---

3 First Law

3.1 General Formulation

The first law is an *empirical law*. It cannot be proven and is applicable only because all conclusions drawn from it are in line with experience. It is generally formulated as follows:

> Each system possesses an extensive state variable of energy. It is constant in a closed system.

If a positive value is agreed for all energy introduced to the system, and a negative value for all energy subtracted from it, then the first law can also be formulated as follows:

> In a closed system, the sum of all energy variations is equal to zero.

3.2 The Various Forms of Energy

In order to enable mathematical formulation of the first law, the distinction must be drawn between the various forms of energy, which must be defined.

3.2.1 Work

In thermodynamics, the concept of *work* is taken from mechanics and is defined as follows:

> If a force is applied to a system, then the work applied to the system equals the product of force and displacement of the point of application of force.

Work applied along a path z between points 1 and 2 of force F is as follows:

$$W_{12} = \int_1^2 \boldsymbol{F} \cdot d\boldsymbol{z}. \qquad (1)$$

Mechanical work W_{m12} is interpreted as the work of forces accelerating a closed system of mass m from velocity w_1 to w_2, in raising it in a gravitational field against acceleration g from height z_1 to z_2.

$$W_{m12} = m\left(\frac{w_2^2}{2} - \frac{w_1^2}{2}\right) + mg\,(z_2 - z_1). \qquad (2)$$

Volumetric work is the work that must be applied in

order to change the volume of a system. In a system of volume V possessing variable pressure p, an element dA of the surface travels by distance dz. The applied work is as follows:

$$dW = -p \int dA \cdot dz = -p \, dV, \qquad (3)$$

and the following formula applies:

$$W_{12} = - \int_1^2 p \, dV. \qquad (4)$$

The minus sign is introduced because an applied unit of work is agreed to be positive and causes a reduction in volume. Equation (4) applies only because pressure p within the system is (at each moment in the process of change of state) an unequivocal function of volume and is equal to the pressure exerted by the environment. A small overpressure or underpressure in the environment will then bring about either a reduction or an increase in the volume of the system. Such changes of state, where an arbitrarily small "overweight" is adequate to allow it to move in one or the other direction, are referred to as *reversible*. Equation (4) is consequently the volumetric work for a reversible change of state. In actual processes, in order to overcome the friction within the system, a finite overpressure in the environment is required. Such changes of state are referred to as *irreversible*. The applied work is greater by the dissipated proportion $(W_{\text{diss}})_{12}$. The volumetric work in an irreversible change of state is

$$W_{12} = - \int_1^2 p \, dV + (W_{\text{diss}})_{12}. \qquad (5)$$

The continuously positive *work of dissipation* increases the energy of the system and brings about a different state curve $p(V)$ from that in the reversible case. The prerequisite for calculation of the integral in Eq. (5) is that t is an unequivocal function of V. Equation 5, for example, is no longer applicable in a system range through which a sound wave is passing.

Generally, work can be derived as a product of a generalised load F_k and a generalised displacement dX_k. For actual processes, we have to add dissipated work:

$$dW = \Sigma \, F_k \, dX_k + dW_{\text{diss}}. \qquad (6)$$

It is found that in irreversible processes, $W_{\text{diss}} > 0$, more work has to be applied, or less work is obtained than in reversible processes, $W_{\text{diss}} = 0$.

Table 1 sets out various forms of work.

Technical work relates to the work applied to a flow of material by a machine, compressor, turbine, jet engine etc. If a mass m along a path dz through a machine experiences a pressure increase dp, then technical work is

$$dW_t = mv \, dp + dW_{\text{diss}}.$$

If, furthermore, kinetic and potential energy are modified, then further mechanical work is applied. The technical work applied along path *1-2* is

$$W_{t12} = \int_1^2 V \, dp + (W_{\text{diss}})_{12} + W_{m12} \qquad (7)$$

with W_{m12} as in Eq. (2).

3.2.2 Internal Energy

Apart from kinetic and potential energy, all systems also possess *stored* energy internally in the form of the translational, rotational and vibrational energy of elementary particles. This energy is referred to as the *internal energy* U of the system. It is an extensive state variable. Overall energy E of a system of mass M consists of internal energy, kinetic energy E_{kin} and potential energy E_{pot}:

$$E = U + E_{\text{kin}} + E_{\text{pot}}. \qquad (8)$$

3.2.3 Heat

Application of work changes a system's internal energy. However, the internal energy of a system can be changed without applying work if the system is brought into contact with its environment. The energy exchanged between a system and its environment without the application of work is designated *heat*. For this purpose we set out Q_{12} if the system is transferred by heat from state *1* to state *2*. It is the convention that an *applied* heat is *positive* and a *derived* heat is *negative*.

3.3 Application to Closed Systems

Heat Q_{12} and work W_{12} applied to a system during a change of state from *1* to *2* bring about the effect of variation of energy E of a system by

$$E_2 - E_1 = Q_{12} + W_{12}. \qquad (9)$$

W_{12} covers all work applied to a system. If no mechanical work is applied, then only the internal energy level is changed, and according to Eq. (8), $E = U$. If it is further assumed that only volume work is applied to the system, then Eq. (9) applies as follows:

$$U_2 - U_1 = Q_{12} - \int_1^2 p \, dV + (W_{\text{diss})12}. \qquad (10)$$

3.4 Application to Open Systems

3.4.1 Steady State Processes

In technology, work is usually applied by a material flow that flows constantly through a machine. If applied work is constant in time, then the process is described as a *steady state flow process*. **Figure 1** illustrates a typical example: a material flow of a fluid (gas or liquid) at pressure p_1 and temperature T_1 flows at velocity w_1 into system σ. In a machine, work is applied and is obtained from the shaft as technical work W_{t12}. The fluid flows through a heat exchanger in which heat Q_{12} is exchanged with the environment, then leaving the system σ at a pressure p_2, temperature T_2 and velocity w_2. Tracing the path of a constant mass Δm through system σ, an observer moving with the flow would perceive mass m as a closed system. The first law of thermodynamics applies in this instance; see Eq. (9) for closed systems. Work applied at Δm is composed of $\Delta m p_1 v_1$, displacing Δm from the environment across the system boundary (**Fig. 2**), technical work W_{t12} and work $-\Delta m p_2 v_2$ to bring Δm through the system boundary back into the environment. Accordingly, work applied to the closed system is

$$W_{12} = W_{t12} + m(p_1 v_1 - p_2 v_2). \qquad (11)$$

Table 1. Various forms of work. SI units are shown in brackets [ ]

Type of work	Generalised force	Generalised displacement	Applied work
Linear elastic displacement	Force F [N]	Displacement dz [N]	$dW = F\,dz = \sigma\,d\varepsilon V$ [N m]
Rotation of a rigid body	Torque moment M_d [N m]	Angle of rotation $d\alpha$ [-]	$dW = M_d\,d\alpha$ [N m]
Volume work	Pressure p [N/m^2]	Volume dV [m^3]	$dW = -p\,dV$ [N m]
Surface enlargement	Surface stress σ' [N/m]	Surface A [m^2]	$dW = \sigma'\,dA$ [N m]
Electrical work	Voltage U_e [V]	Load Q_e [C]	$dW = U_e\,dQ_e$ [W s] in a linear conductor of resistance R $dW = U_e I\,dt$ $= RI^2\,dt$ $= (U^2/R)\,dt$ [W s]
Magnetic work in vacuum	Magnetic field strength H_0 [A/m]	Magnetic induction $dB = d\mu_0 H_0$ [V s/m^2]	$dW_v = \mu_0 dH_0$ [W s/m^3]
Magnetisation	Magnetic field strength H [A/m]	Magnetic induction $dB = d(\mu_0 H + M)$ [V s/m^2]	$dW_v = H \cdot dB$ [W s/m^3]
Electrical polarisation	Electrical field strength E [V/m]	Dielectrical displacement $dD = d(\varepsilon_0 E + P)$ [A s/m^2]	$dW = E \cdot dD$ [W s/m^3]

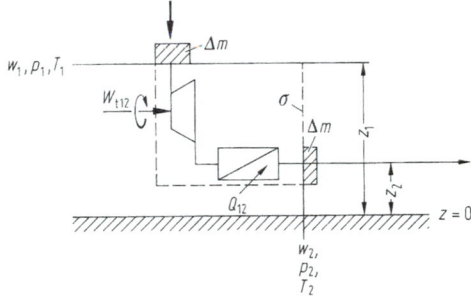

w_1, p_1, T_1

Figure 1. Work on an open system.

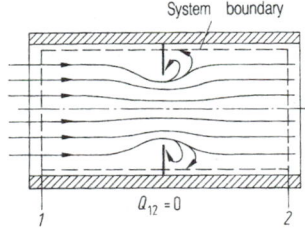

Figure 2. Adiabatic throttling.

Displacement work is the term applied to

$$\Delta m(p_1 v_1 - p_2 v_2).$$

It represents the amount by which technical work W_{t12} differs from the work on a closed system. Accordingly, the first law of thermodynamics, Eq. (9), runs as follows:

$$E_2 - E_1 = Q_{12} + W_{t12} + \Delta m(p_1 v_1 - p_2 v_2), \quad (12)$$

with E as in Eq. (8). The state variable of *enthalpy H* is defined as follows:

$$H = U + pV \quad \text{and} \quad h = u + pv, \quad (13)$$

and Eq. (12) can then be set out as follows:

$$\left(h_2 + \frac{w_2^2}{2} + gz_2\right)\Delta m - \left(h_1 + \frac{w_1^2}{2} + gz_1\right)\Delta m = Q_{12} + W_{t12}. \quad (14)$$

In this form, the first law of thermodynamics is used for steady state flow processes of open systems. Changes in kinetic and potential energy are frequently negligible. Accordingly, the following applies:

$$H_2 - H_1 = Q_{12} + W_{t12}. \quad (15)$$

Special cases are as follows:

1. *Adiabatic* changes of state, such as those arising as approximations in compressors, turbines and engines:

$$H_2 - H_1 = W_{t12}. \quad (16)$$

2. *Throttling* of a flow in an adiabatic duct by installed obstacles (**Fig. 2**). These bring about a pressure drop. Upstream and downstream of the throttling point, the following formula applies:

$$H_1 = H_2. \quad (17)$$

Enthalpy remains constant during throttling. It should be noted that the change in kinetic and potential energy has been ignored.

3.4.2 Unsteady State Processes

If the matter Δm_1 introduced into the system according to **Fig. 2** during a given period is different from matter Δm_2 derived during the same period, then matter is stored within the system, which will result in a chronological variation of its internal energy and possibly of its kinetic and potential energy. The energy stored within the system will vary by $E_2 - E_1$ during a change of state *1-2*, such that instead of Eq. (14), the following form of the first law applies:

$$\left(h_2 + \frac{w_2^2}{2} + gz_2\right)\Delta m_2 - \left(h_1 + \frac{w_1^2}{2} + gz_1\right)\Delta m_1 \quad (18)$$

$$+ E_2 - E_1 = Q_{12} + W_{t12}.$$

If fluid states *1* are chronologically variable on input and fluid states *2* are chronologically variable on output, then it is logical to change to a differential mode of formulation:

$$\left(h_2 + \frac{w_2^2}{2} + gz_2\right)dm_2 - \left(h_1 + \frac{w_1^2}{2} + gz_1\right)dm_1$$

$$+ dE = dQ + dW_t. \quad (19)$$

In the process of investigating the filling or emptying of tanks, variations in kinetic and potential energy can usually be ignored, and often, furthermore, no technical work is applied, such that Eq. (19) becomes abbreviated to

$$h_2 \, dm_2 - h_1 \, dm_1 + dU = dQ \qquad (20)$$

with the (chronologically variable) internal energy $U = um$ of the matter closed in the tank. In accordance with the convention, dm_1 is the quantity of matter introduced into the system and dm_2 is the quantity of matter derived from it. Thus if matter only is introduced then $dm_2 = 0$, while if matter only is derived then $dm_1 = 0$.

4 Second Law

4.1 The Principle of Irreversibility

If a system is brought into contact with its environment, then exchange processes take place, and after a sufficiently long period a fresh state of equilibrium will be set up. For example, let a system be brought into contact with an environment of different temperature. In the final state, the system and the environment will each possess the same temperature. Thermal equilibrium will have been established. Until equilibrium is set up, non-equilibrium states will be passed through in a continuous sequence. The same end-state of equilibrium could also be achieved if the system were brought temporarily into contact with its environment then insulated from it, the experimenter waiting until a state of equilibrium was set up in the system, then repeatedly bringing the system into temporary contact with its environment, and so forth. In the process, a continuous sequence of equilibrium states would be experienced.

A process consisting of a *continuous* sequence of equilibrium states is referred to as *reversible*. Since it is stipulated that the system is in equilibrium at each moment, an arbitrarily small "load", for example an overpressure or an overtemperature in the surrounding system, would set off, depending on its polarity, a process both in one or the other direction and, for example, bring about a decrease in volume or an increase in temperature in the system. Reversible processes are idealised boundary cases of actual processes and do not arise in nature. All *natural* processes are *irreversible*, because a finite "force" is required in order to set off a process, for example a finite force in order to displace a body in the context of friction, or a finite difference in temperature in order to contribute heat to it. The finite force means that they must take place in a given direction. This empirically observed fact results in the following formulations of the *second law*:

All natural processes are irreversible.
All processes entailing friction are irreversible.
Heat can never *spontaneously* transfer from a body of lower temperature to a body of higher temperature.

In this context, "spontaneously" denotes that the process cannot take place without changes in nature itself. In addition to the above-mentioned formulations, there are many others that apply to other special processes.

4.2 General Formulation

Mathematical formulation of the second law in conjunction with the concept of *entropy* becomes a further state variable of a system. The value of introducing a state variable of this type can be explained in the light of the example of thermal exchange between a system and its environ-

ment. The first law requires that a system must exchange work and heat with its environment. Application of work brings about a variation in internal energy, for example by variation in the volume of the system at the expense of the volume of the environment. Accordingly, $U = U(V, \ldots)$. Volume is an *exchange* variable: it is an extensive state variable which is "exchanged" between system and environment. Furthermore, application of heat between a system and its environment can be conceived such that an extensive state variable is exchanged between the system and the environment. This is purely to postulate the existence of such a state variable, whose introduction is justifiable by the fact that all statements made with this variable are in accordance with empirical observation. The new extensive state variable is referred to as *entropy* and designated S. Accordingly, $U = U(V, S, \ldots)$. If volume work only is applied and heat is introduced, then $U = U(V, S)$. By differentiation we then obtain *Gibbs' fundamental equation*:

$$dU = T dS - p \, dV \qquad (1)$$

with thermodynamic temperature

$$T = (\partial U/\partial S)_V \qquad (2)$$

and pressure

$$p = - (\partial U/\partial V)_S. \qquad (3)$$

An equation equivalent to Eq. (1) is obtained if U is eliminated and replaced by enthalpy $H + U + pV$:

$$dH = T \, dS + V \, dp. \qquad (4)$$

It can be illustrated that thermodynamic temperature is identical to the temperature measured using a gas thermometer (see C2.3). Study of the characteristics of entropy reveals that in a closed system it can only increase, attaining a maximum value in the boundary case of equilibrium. In a non-closed system, entropy also varies due to exchange of heat with the environment. Entropy also enters the system together with introduced matter. Chronological increase in entropy S_i within the system is termed *entropy generation*, and entropy exchanged with the environment per unit of time is termed *entropy flux*. Entropy flux is positive if heat is introduced into the system, but negative if heat is taken out of the system and zero in the case of adiabatic systems. Accordingly, the second law can be formulated as follows:

There exists a state variable S, the entropy of a system, whose chronological change $\dot{S}$ is composed of entropy flux $\dot{S}_a$ and entropy generation $\dot{S}_i$.

$$\dot{S} = \dot{S}_a + \dot{S}_i. \qquad (5)$$

The following applies to entropy generation:

$$S_i = 0 \text{ for reversible processes;}$$
$$S_i > 0 \text{ for irreversible processes;} \qquad (6)$$
$$S_i < 0 \text{ impossible.}$$

4.3 Special Formulations

4.3.1 Adiabatic Systems

For adiabatic systems, $S_a = 0$ and accordingly $S = S_i$. Accordingly, the following applies:

1. In adiabatic systems, entropy can never decrease; it can only increase in the context of irreversible processes or remain constant in the case of reversible processes.
2. If an adiabatic system consists of α subsystems, then the following formula applies for the total of changes in entropy $\Delta S^{(\alpha)}$ of the subsystems:

$$\sum_\alpha \Delta S^{(\alpha)} \geq 0. \qquad (7)$$

3. In an adiabatic system, in accordance with Eq. (1) where $dS = dS_i$:

$$dU = T dS_i - p dV.$$

4. On the other hand, there follows from the first law as C3 Eq. (10):

$$dU = dW_{diss} - p dV$$

and hence:

$$dW_{diss} = T dS_i = d\Psi \qquad (8)$$

or

$$(W_{diss})_{12} = T(S_i)_{12} = \Psi_{12}.$$

Ψ_{12} is the term which is applied to the *energy dissipated* during a change of state *1-2*. The following principle applies: dissipated energy is always positive.

This statement applies not only for adiabatic systems but also in quite general terms, because entropy generation by definition is a component of entropy variation, which occurs if the system is adiabatically isolated, i.e. if $S_a = 0$ is postulated.

4.3.2 Systems with Heat Addition

For systems with heat addition, Eq. (1) can be set out as follows:

$$dU = T dS_a + T dS_i - p dV = T dS_a + dW_{diss} - p dV.$$

From comparison with the first law, C3 Eq. (10), we obtain

$$dQ = T dS_a. \qquad (9)$$

Accordingly, heat is energy that flows by entropy through the system boundary, while work is transferred without an exchange of entropy.

If the constantly positive term $T dS_i$ is added to the right-hand side of Eq. (9), then *Clausius's inequality* follows:

$$dQ \leq T dS \quad \text{or} \quad S \geq \int_1^2 \frac{dQ}{T} \qquad (10)$$

In irreversible processes, entropy variation is greater than the integral over all values of dQ/T, and it is only in the case of reversible processes that the equality sign applies.

◼5◼ Exergy and Anergy

The first law requires that energy remains constant in a closed system. Since all non-closed systems can be converted into closed systems by involvement of the environment, it is always possible to configure a system in which energy remains constant during a thermodynamic process. For this reason, a loss of energy is not possible. In a thermodynamic process, energy is merely converted. The question of how much of the system's stored energy is converted will depend on the state of the environment. If the environment is in a state of equilibrium with the system, then no energy is converted, and the greater is the deviation from equilibrium then the larger is the portion of the system's energy that can be converted.

Many thermodynamic processes occur in the earth's atmosphere, which represents the environment of most thermodynamic systems. The earth's atmosphere can be regarded, in comparison with the very large number of smaller thermodynamic systems, as an infinitely large system whose intensive state variables of pressure, temperature and composition do not vary during a process, if we ignore daily and seasonal fluctuations in intensive state variables.

In many technical processes, work is obtained by bringing a system from a given initial state into equilibrium with its environment. The maximum work is then obtained if all changes of state are reversible. The maximum work

obtainable on setting up equilibrium with the environment is termed *exergy* and is symbolised by W_{ex}.

5.1 Exergy of a Closed System

In order to calculate the exergy of a closed system, we begin by conceiving of it as being brought to ambient temperature by reversibly adiabatic means. The work expended in this process is composed of the maximum work W_{ex}, energy which can then be exploited and work $p_u(V_2 - V_1)$, which has to be expended in order to overcome the pressure of the environment; $W_{12} = W_{ex} - p_u(V_2 - V_1)$. Next, heat is reversibly exchanged with the environment at constant temperature T_u: $Q_{12} = T_u (S_2 - S_1)$. In accordance with the first law for closed systems, C3 Eq. (10) is as follows if we ignore mechanical work:

$$U_2 - U_1 = W_{ex} - p_u (V_2 - V_1) + T_u(S_2 - S_1).$$

In state *2*, the system is in equilibrium with its environment, designated by subscript u. The exergy of the closed system is accordingly

$$- W_{ex} = U_1 - U_u - T_u(S_1 - S_u) + p_u (V_1 - V_u). \qquad (1)$$

If the system has rigid walls, then $V_1 = V_u$ and the last term is dispensed with.

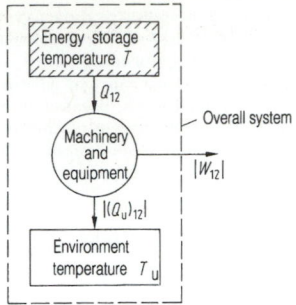

Figure 1. For conversion of heat into work.

If the system in its initial state is already in equilibrium with the environment, state $1 = $ state u, then in accordance with Eq. (1) no work can be obtained. The following law therefore applies:

The internal energy of the environment cannot be converted into exergy.

Accordingly, the powerful energy sources in the atmosphere around us cannot be exploited for propulsion of vehicles.

5.2 Anergy

The proportion of a form of energy that is not converted into exergy is referred to as *anergy*, symbolised by B. From Eq. (1) we obtain anergy B_U of an internal energy:

$$B_U = U_u + T_u (S_1 - S_u) - p_u (V_1 - V_u),$$

and the following applies:

$$U_1 = (- E_{ex}) + B_U. \qquad (2)$$

All energy is made up of exergy and anergy.

5.3 Exergy of an Open System

The maximum technical work or the exergy of a flow of matter is obtained by bringing the flow of matter into equilibrium with its environment reversibly by applying work and by adding or subtracting heat. There follows from the first law for steady state processes of open systems, C3 Eq. (12), ignoring the variation in kinetic and potential energy:

$$- W_{ex} = H_1 - H_u - T_u (S_1 - S_u). \qquad (3)$$

Only a reduced proportion of enthalpy H_1, a proportion which is reduced by $H_u + T_u (S_1 - S_u)$ is converted into technical work. If heat from the environment is added to a flow of matter, then $T_u (S_1 - S_u)$ is negative and exergy is greater than the change in enthalpy by the proportion of such added heat.

5.4 Exergy and Heat

Let heat Q_{12} from an energy store of temperature T be added to a machine and converted into work W_{12} (**Fig. 1**). Heat $(Q_u)_{12}$ which is not convertible into work

is subtracted and absorbed into the environment. The maximum work can be obtained if all changes of state occur reversibly. The same applies to the exergy of heat. All changes of state are reversible if

$$\int_1^2 \frac{dQ}{T} + \int_1^2 \frac{dQ_u}{T_u} = 0$$

where $dQ + dQ_u + dW_{ex} = 0$ in accordance with the first law. Accordingly, there is obtained the exergy of the levels of heat applied to machinery and apparatus:

$$- W_{ex} = \int_1^2 \left(1 - \frac{T_u}{T}\right) dQ; \qquad (4)$$

or, set out differentially,

$$- dW_{ex} = \left(1 - \frac{T_u}{T}\right) dQ. \qquad (5)$$

In a reversible process, only the component of added heat dQ which has been multiplied by the *"Carnot" factor* $1 - (T_u/T)$ can be converted into work. Component $dQ_u - T_u (dQ/T)$ is the anergy of heat which is once again given off into the environment and cannot be captured as work.

It is also observed that heat which is available at ambient temperature cannot be converted into exergy.

5.5 Exergy Losses

Dissipated energy is not fully lost; it increases entropy and thus, due to $U (S,V)$, also increases a system's internal energy. Dissipated energy can also be added in a reversible substitution process as heat from outside, such that it produces the same increase in entropy as in the irreversible process. Since added heat, Eq. (5), can be converted into work, it is also possible to obtain the proportion

$$- dW_{ex} = \left(1 - \frac{T_u}{T}\right) d\Psi \qquad (6)$$

of dissipated energy $d\Psi$ as work (exergy). The remaining component $T_u d\Psi/T$ of added dissipation energy must be reabsorbed into the environment as heat and is not convertible into work. This is designated *exergy loss*. It is equivalent to the anergy of dissipation energy and is expressed by

$$(W_v)_{12} = \int_1^2 \frac{T_u}{T} d\Psi = \int_1^2 T_u dS_i. \qquad (7)$$

The following formula applies to an adiabatic process, as $dS_i = dS$:

$$(W_v)_{12} = \int_1^2 T_u dS = T_u (S_2 - S_1). \qquad (8)$$

For exergy as opposed to energy, there is no law of conservation. The amounts of exergy added to a system are equivalent to those derived from it and exergy losses. Losses due to irreversibilities produce thermodynamic repercussions, which become increasingly unfavourable the lower is the temperature T at which a process occurs; see Eq. (7).

6 Thermodynamics of Substances

For application of the generally relevant laws of thermodynamics to any given substance, and for calculation of exergy and anergy levels, numerical values must be ascertained for state variables U, H, S, p, V and T. Of these, variables U, H and S are defined as *caloric* state variables and p, V and T are designated *thermal* state variables. The relationships between them are material-related and can generally be ascertained only by measurements.

6.1 Thermal State Variables of Gases and Vapours

A thermal equation of state for pure substances is of the following form:

$$F(p, v, T) = 0 \tag{1}$$

or $p = p(v, T)$, $v = v(p, T)$ and $T = T(p, v)$. For technical calculations, preference is given to equations of state of the following form: $v = v(p, T)$, since pressure and temperature are usually postulated as independent variables.

6.1.1 Ideal Gases

The thermal equation of state of ideal gases is particularly simple:

$$pV = mRT \quad \text{or} \quad pv = RT, \tag{2}$$

where p = absolute pressure, V = volume, v = specific volume, R = individual gas constants, and t = thermodynamic temperature. Gases behave approximately ideally only if their pressure is adequately low, $p \rightarrow 0$.

6.1.2 Gas Constant and the Law of Avogadro

The unit of material quantity, the mole, is represented by the unit symbol mol.

The number of particles (molecules, atoms, elementary particles) of a substance is referred to as 1 mol, if this substance consists of as many identical particles as are contained in exactly 12 g of pure atomic carbon of the nuclide ^{12}C.

The number of equal particles contained in a mol is termed *Avogadro's constant*. It is a universal physical constant and its numerical value is as follows:

$$N_A = (6.022\ 131\ 8 \pm 0.000\ 007\ 6) \cdot 10^{26}/\text{kmol}.$$

The mass of a mol, i.e. of N_A identical particles, is a substance-specific variable and is termed the *molecular weight M* (for values see **Appendix C6, Table 1**):

$$M = m/n \tag{3}$$

(where M is measured in the SI unit kg/kmol, m = the mass in kg and n = the molecule count in kmol).

Avogadro (1831) produced the following well-known hypothesis:

> Ideal gases at the same pressure and temperature contain the same number of molecules in the same volume.

It follows, after introduction of molecular weight into the thermal equation of state for ideal gas, Eq. (2), that $pV/nT = MR$ is a value which is fixed for all gases.

$$MR = R. \tag{4}$$

The universal gas constant is symbolised by R. It is a physical constant. The following formula applies:

$$R = 8.314\ 41 \pm 0.000\ 26\ \text{kJ/kmol K}.$$

In conjunction, we have the thermal equation of state for an ideal gas:

$$pV = nRT. \tag{5}$$

Example In a steel flask with a volume $V_1 = 200$ l, there is hydrogen at $p_1 = 120$ bar and $t_1 = 20\ °C$. How much space will the hydrogen occupy at $c_2 = 1$ bar and $t_2 = 0\ °C$, if we ignore the slight deviations exhibited by hydrogen from the characteristics of the ideal gas?

In accordance with Eq. (5), $p_1V_1 = nRT_1$; $p_2V_2 = nRT_2$ and accordingly:

$$V_2 = \frac{p_1T_2}{p_2T_1}V_1 = \frac{120\ \text{bar} \cdot 273.15\ \text{K}}{1\ \text{bar} \cdot 283.15\ \text{K}}\ 0.2\ \text{m}^3 = 23.15\ \text{m}^3.$$

6.1.3 Real Gases

The equation of state for the ideal gas applies to actual gases and vapours only as a boundary law for infinitely low pressures. The deviation in characteristics of gaseous water from the equation of state for ideal gases is illustrated by **Fig. 1**, where pv/RT over t for various pressures is illustrated. The real gas factor of $Z = pv/RT$ is equal to 1 for ideal gases but deviates from that for real gases. In the atmosphere between 0 and 200 °C and for hydrogen from − 15 to 200 °C, deviations in Z amount to approximately 1% from unity at pressures of 20 bar. At atmospheric pressures, almost all gases exhibit negligible deviations from the law for the ideal gas. Various types of state equations have proved valuable for describing the properties of real gases. One of these is that real gas factor Z is represented in the form of a series, with correction terms successively added to the value of 1 for the ideal gas:

$$Z = 1 + \frac{B(T)}{v} + \frac{C(T)}{v^2} + \frac{D(T)}{v^3}. \tag{6}$$

Here B is termed the second, C the third and D the fourth virial coefficient. Tabular works [2, 3] set out a list of secondary virial coefficients for many gases. The virial equation with two or three virial coefficients is applicable only in the range of moderate pressures. The equation of state of Benedict–Webb–Rubin [4] represents a balanced compromise between computation workload and achievable accuracy in describing the properties of dense gases. It runs as follows:

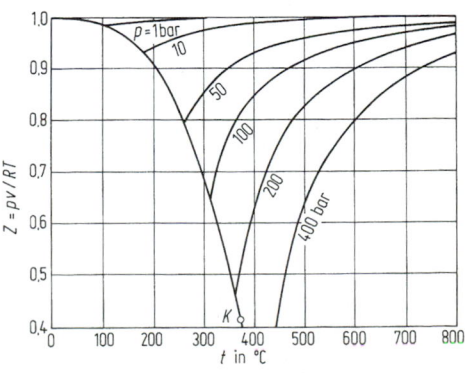

Figure 1. Real gas factor of water vapour.

$$Z = 1 + \frac{B(T)}{v} + \frac{C(T)}{v^2} + \frac{a\alpha}{v^5\,RT} + \frac{c}{v^3\,RT^3}\left(1 + \frac{\gamma}{v^2}\right)\exp\left(-\frac{\gamma}{v^2}\right); \tag{7}$$

where

$$B(T) = B_0 - \frac{A_0}{RT} - \frac{C_0}{RT^3} \quad \text{and} \quad C(T) = b\left(\frac{a}{RT}\right).$$

The equation contains the eight constants of A_0, B_0, C_0, a, b, c, α, γ, which are tabulated for many substances [4]. Highly accurate equations of state are required for the ingredients of water [5], air [6] and coolants [7] that are used in power stations and in refrigeration plants. The equations for these substances are more laborious, contain more constants and can be analysed only by data-processing systems.

6.1.4 Vapours

Vapours are gases approaching their liquefaction point. A vapour is said to be *saturated* if an arbitrarily low temperature decrease would cause it to liquefy, and is referred to as *superheated* if a finite temperature drop would produce the same effect. If heat is applied to a liquid at a constant pressure, then – once a specific temperature is reached – vapour of the same temperature starts to be formed. Vapour and liquid are in equilibrium. This state is referred to as *saturation*; it is characterised by the related values of *saturation temperature* and *saturation pressure*, the relationship of which is represented by the *vapour-pressure curve* (**Fig. 2**). It starts at a substance's *triple point* and ends at its critical point K. This is referred to as the state point of $p_k T_k$, above which vapour and liquid are no longer separated by a markedly noticeable boundary but continuously intersect (see **Appendix C6, Table 1**). The critical point, in common with the triple point at which vapour, liquid and the solid phase of a substance are in equilibrium, is a characteristic point for all substances. The vapour pressure of many substances can be represented from the triple point to the boiling point at atmospheric pressure by *Antoine's equation*,

$$\ln p = A - B/(C + T), \tag{8}$$

in which variables A, B and C are substance-related constants (see **Appendix C6, Table 2**).

If superheated vapour is condensed at a constant temperature by a reduction in volume, then pressure increases over a virtually hyperbolic curve similar to that for an ideal gas, for example see *isothermals* for 300 °C in **Fig. 3**.

Condensation starts as soon as the saturation pressure is attained, and volume decreases without an increase in pressure until all vapour is liquefied. The segment of curve shown in **Fig. 3** is a characteristic graphic illustration of an equation of state for many substances. If we combine the specific volumes of liquid at saturation temperatures prior to evaporation and the volume of saturated vapour v' and v'', we obtain two curves a and b, referred to as the *left and right boundary curves*, which intersect at critical point K. If x is *vapour content*, defined as the mass of saturated vapour m'' in relation to the total mass of saturated vapour m'' and boiling liquid m', and if v' is the specific volume of boiling liquid and v'' the specific volume of saturated vapour, then the following formula applies for saturated vapour:

$$v = xv'' + (1 - x)\,v' \tag{9}$$

(**Fig. 3** illustrates lines $x = $ const).

Example   In a boiler of 2 m³ capacity, there are 1000 kg of water and steam at 121 bar and at saturation temperature. What is the specific volume of steam? The specific volume of steam can be determined by consulting the vapour table (**Appendix C6, Table 5**), the specific volume of steam $v'' = 0.014\,13$ m³/kg by interpolation at 121 bar and the specific volume of liquid is similarly obtained as $v' = 0.001\,531$ m³/kg. Mean specific volume $v = V/m$ is $v = 2$ m³/1000 kg $= 0.002$ m³/kg. There follows from Eq. (9), $x = (v - v')/(v'' - v') = (0.002 - 0.001\,531)/(0.014\,13 - 0.001\,531) = 0.037\,25 = m''/m$, i.e. $m'' = 1000 - 0.037\,25$ kg $= 37.25$ kg, $m' = 1000 - 37.25$ kg $= 962.75$ kg.

The equation of state can also be represented as a surface in space with coordinates p, v, t as shown in **Fig. 4**. Projection of the boundary curve into the p, T plane produces the vapour pressure curve, and the illustration in **Fig. 3** provides projection of the surface into the p, v plane.

6.2 Caloric Properties of Gases and Vapours

6.2.1 Ideal Gases

The internal energy of ideal gases is dependent only upon temperature, $u = u\,(T)$, and consequently enthalpy

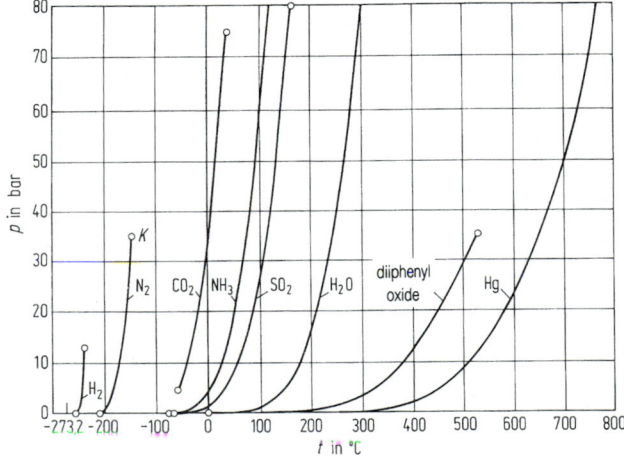

Figure 2. Vapour pressure curves of some substances.

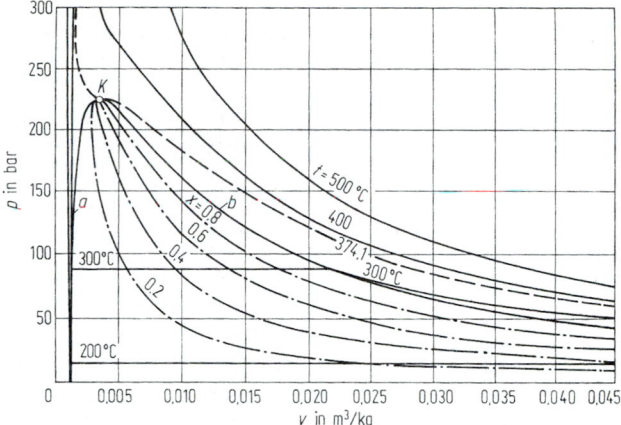

Figure 3. p, v curve of water.

$h = u + pv = u + RT$ is purely a temperature function $h = h(T)$. Derivations of u and h according to temperature are called *specific heat capacities*, and increase hand-in-hand with temperature (see **Appendix C6, Table 3** containing values for atmosphere). The following formulae apply:

$$du/dT = c_v \tag{10}$$

and

$$dh/dT = c_p. \tag{11}$$

Derivation of $h - u = RT$ produces

$$c_p - c_v = R. \tag{12}$$

The difference in molecular heat capacities or molecular heat levels $C_p = Mc_p$, $C_v = Mc_v$ is equal to the universal gas constant:

$$C_p - C_v = \mathbf{R}.$$

Relationship $\kappa = c_p/c_v$ plays a part in reversible adiabatic changes of state and is consequently termed the *adiabatic exponent*. For monoatomic gases, adequate accuracy is obtained by $\kappa = 1.66$, for diatomic gases $\kappa = 1.40$ is adequate and so is $\kappa = 1.30$ for triatomic gases. The mean specific heat capacity is the integral mean value defined by

$$\left[c_p\right]_{t_1}^{t_2} = \frac{1}{t_2 - t_1} \int_{t_1}^{t_2} c_p dt; \quad \left[c_v\right]_{t_1}^{t_2} = \frac{1}{t_2 - t_1} \int_{t_1}^{t_2} c_v \, dt. \tag{13}$$

Equations (10) and (11) give rise to the following equations for *variations* in *internal energy and enthalpy*:

$$u_2 - u_1 = \left[c_v\right]_{t_1}^{t_2} (t_2 - t_1) = \left[c_v\right]_0^{t_2} t_2 - \left[c_v\right]_0^{t_1} t_1. \tag{14}$$

and

$$h_2 - h_1 = \left[c_p\right]_{t_1}^{t_2} (t_2 - t_1) = \left[c_p\right]_0^{t_2} t_2 - \left[c_p\right]_0^{t_1} t_1. \tag{15}$$

The numerical values of $[c_v]_0^t$ and $[c_p]_0^t$ are obtained from the mean molecular heat levels stated in **Appendix C6**,

Table 4. Entropy is obtained from C4 Eq. (1), taking account of Eq. (10):

$$ds = \frac{du + p \, dv}{T} = c_v \frac{dT}{T} + R \frac{dv}{v}$$

by integration with $c_v = \text{const}$, arriving at

$$s_2 - s_1 = c_v \ln \frac{T_2}{T_1} + R \ln \frac{v_2}{v_1}. \tag{16}$$

An equivalent expression is obtained by integration of C4 Eq. (4), where $c_p = \text{const}$:

$$s_2 - s_1 = c_p \ln \frac{T_2}{T_1} - R \ln \frac{p_2}{p_1}. \tag{17}$$

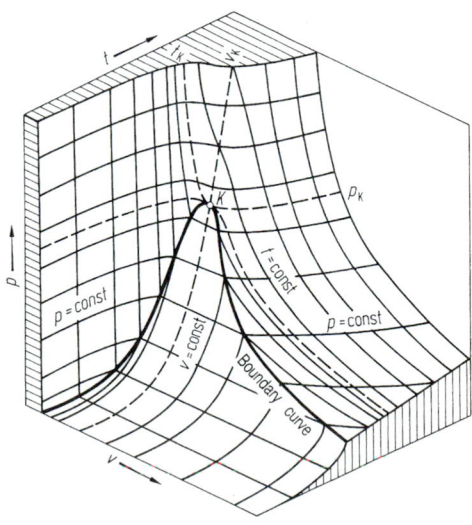

Figure 4. State surface of water in perspective view.

6.2.2 Real Gases and Vapours

The caloric properties of real gases and vapours are generally ascertained from measurements, but can also be derived, up to an initial value, from the thermal equation of state. They are set out as follows in tables or curves $u = u(v, T)$, $h = h(p, T)$, $s = s(p, T)$, $c_v = c_v(v, T)$, $c_p = c_p(p, T)$. Recourse is frequently had to state equations that require a computer.

For vapours, the following rule applies:

Enthalpy h'' of the saturated vapour differs from enthalpy h' of liquid in the saturation state where $p, T = $ const by enthalpy of evaporation.

$$r = h'' = h', \tag{18}$$

which decreases as temperature increases and is zero at the critical point where $h'' = h'$. The enthalpy of saturated vapour is

$$h = (1 - x)h' + xh'' = h' + xr. \tag{19}$$

Accordingly, internal energy is

$$u = (1 - x)\,u' + xu'' = u' + x\,(u'' - u'), \tag{20}$$

and the entropy is

$$s = (1 - x)s' + xs'' = s' + xr/T, \tag{21}$$

since the enthalpy of evaporation and the entropy of evaporation $s'' - s'$ are interconnected by

$$r = T(s'' - s'). \tag{22}$$

In accordance with *Clausius-Clapeyron*, the enthalpy of evaporation is connected with the slope dp/dT of the vapour pressure curve $p(T)$ (if T is a boiling temperature at pressure p) by:

$$r = T\,(v'' - v')\,\frac{dp}{dT'}. \tag{23}$$

This equation can be used in order to take two of the three variables r, $v'' - v'$ and dp/dT for calculation of the third.

If state variables are not commonly accessible to calculation, or if a sufficiently powerful computer is not available, then vapour tables are used for practical calculations, and these tables provide the results of theoretical and experimental investigations for state variables. For the main materials employed in technology, the vapour tables set out in **Appendix C6, Tables 5 to 11** can be consulted. In order to obtain guideline values and to illustrate changes of state, it is advantageous to refer to curves (**Fig. 5**). In practice, Mollier curves are most commonly used; these have enthalpy as one of the coordinates (**Fig. 6**).

A vapour's specific heat capacity $c_p = (\partial h/\partial T)_p$ depends not only on temperature but also, to a substantial degree, upon pressure, and $c_v = (\partial u/\partial T)_v$ depends on specific volume as well as temperature. As value c_p of superheated steam approaches the limit curve, it increases sharply with decrease in temperature and is virtually infinite at the critical point. For vapours, $c_p - c_v$ is no longer the constant variable that it represents in the case of ideal gases.

6.3 Solids

6.3.1 Thermal Expansion

In the equation of state $V = v\,(p, T)$ of a solid, the influence of pressure on volume is usually negligible, as in the case of liquids. Almost all solids expand like liquids with temperature increase, and conversely contract with temperature decrease; the exception is water, which has its greatest density at 4 °C and which also expands at temperatures both above and below 4 °C. If the equation of state is developed into a Taylor series based on temperature and is truncated after the linear term, we obtain volumetric expansion in accordance with the cubic volumetric expansion coefficient γ_v (SI unit : 1/K):

$$V = V_0[1 + \gamma_v(t - t_0]].$$

Accordingly, superficial expansion is

$$A = A_0[1 + \gamma_A(t - t_0)],$$

and longitudinal expansion is

$$l = l_0[1 + \gamma_L(t - t_0)].$$

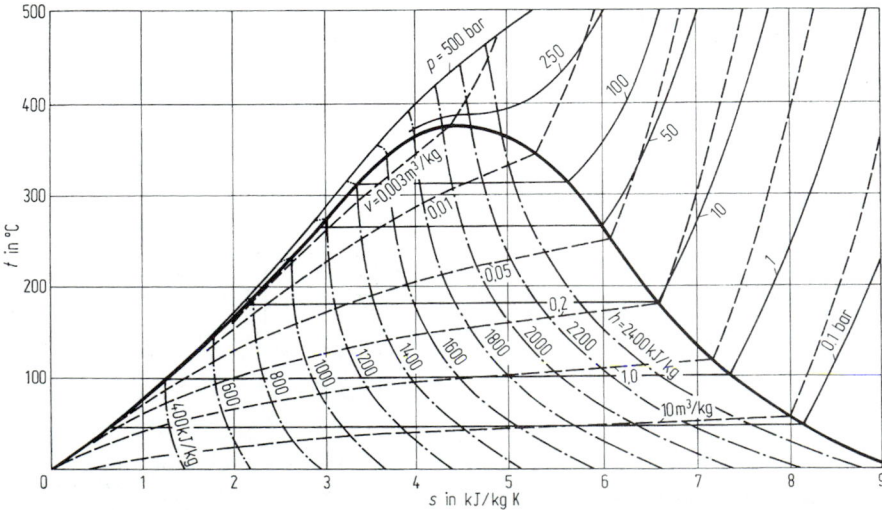

Figure 5. t, s curve of water with curves $p = $ const. (continuous), $v = $ const. (dotted) and curves of equal enthalpy (dot dash).

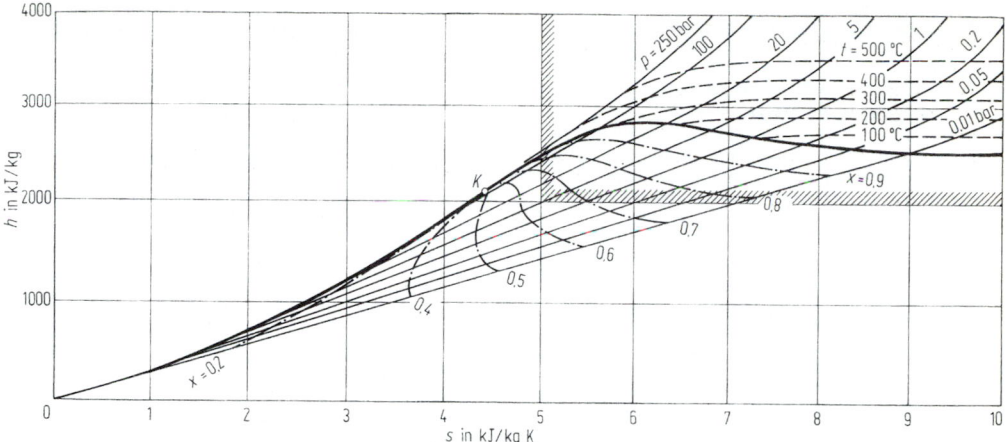

Figure 6. h, s curve of water. The range for steam technology is within the hatched border.

The following equations apply: $\gamma_A = (2/3)\ \gamma_v$ and $\gamma_L = (1/3)\ \gamma_v$. **Appendix C6, Table 12** provides values for γ_L.

6.3.2 Melting and Sublimation Curve

At every pressure in a liquid, there is a temperature, within certain limits, at which it is in equilibrium with its solid phase. This relationship $p(T)$ is established by the *melting and sublimation curve* (**Fig. 7**), while the *sublimation curve* reproduces the equilibrium between the gaseous and solid phases. **Figure 7** also illustrates the *vapour pressure curve*. All three curves intersect at the triple point at which a substance's solid, liquid and gaseous phases are in equilibrium with each other. The triple point of water is by definition 273.16 K, and pressure at the triple point is 0.006 112 bar.

6.3.3 Caloric Properties of State

When a liquid solidifies, the enthalpy of melting Δh_E (E = rigidification) is given off (**Appendix C6, Table 13**). In the process, the liquid experiences a reduction in entropy $\Delta s_E = \Delta h_E/T_E$, where T_E is the melting or freezing

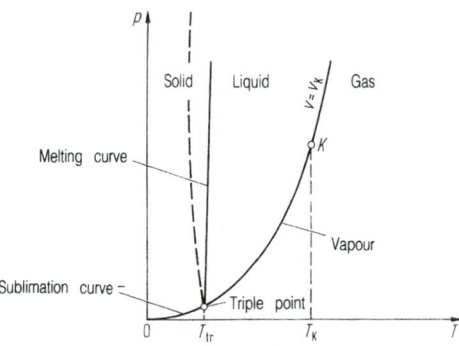

Figure 7. p, T curve with the three limit curves for the phases (the slope of the melting curve for water is negative, shown by the broken line).

temperature. According to the *Dulong-Petit rule*, molecular heat capacity at temperatures above ambient is approximately 25.9 kJ/(kmol K) divided by the number of atoms in the molecule. As absolute zero is approached this rule of thumb no longer applies. In that range, the molecular heat capacity for a constant volume of all solids is as follows:

$$\bar{C} = a(T/\Theta)^3, \quad \text{where} \quad T/\Theta < 0.1$$

where $a = 472.5$ J/(mol K) and where Θ is the Debye temperature (**Appendix C6, Table 14**).

6.4 Mixing Temperature, Measurement of Specific Heat Capacities

If several substances of differing weights m_i, temperatures t_i and specific heat capacities c_{pi} ($i = 1, 2 \dots$) are mixed at constant pressure and without external addition of heat, there will eventually arise a mixing temperature t_m. The following formula applies:

$$t_m = \frac{\Sigma m_i c_{pi} t_i}{\Sigma m_i c_{pi}},$$

where c_{pi} are the mean specific heat capacities between 0 °C and t_i °C. By measuring t_m it is possible to calculate an unknown specific heat capacity if all other values are known.

Example $m_a = 0.2$ kg aluminium at $t_a = 100$ °C is placed in a fully insulated calorimeter which is filled with 0.8 kg water at 15 °C, $c_p = 4.186$ kJ/kg K   and   which   is   made   of   0.25 kg of silver $c_{ps} = 0.234$ kJ/kg K. After compensation, a mixing temperature of 19.24 °C is measured. What is the value for the specific heat capacity of aluminium? The following formula applies:

$$t_m = [(mc_p + m_s c_{ps})t + m_a c_{pa} t_a]/[mc_p + m_s c_{ps} + m_a c_{pa}],$$

with the following resolution:

$$c_{pa} = [(mc_p + m_s c_{ps})(t - t_m)]/[m_a(t_m - t_a)],$$

$$c_{pa} = \frac{(0.8\text{kg} \cdot 4.186\text{kJ/kgK} + 0.25\text{kg} \cdot 0.234\text{kJ/kgK})(15°\text{C} - 19.24°\text{C})}{0.2\text{ kg }(19.24°\text{C} - 100°\text{C})}$$

$$= 0.894 \text{ kJ/kg K}.$$

7 Changes of State of Gases and Vapours

7.1 Changes of State of Gases and Vapours at Rest

Let the thermodynamic system be of mass Δm, which is not in motion as a complete whole. The following changes of state, as idealised boundary cases, are differentiated from actual changes of state.

Changes of state at *constant volume* or *isochorous changes of state*. In this context the gas volume remains unchanged; for example if a gas volume is contained in a tank with rigid walls. No work is applied. The added heat performs the function of changing internal energy.

Changes of state at *constant pressure* or *isobaric state changes*. In order to maintain a constant pressure, a gas must expand adequately in volume on addition of heat. Specific heat in the case of reversible changes of state brings about an increase in enthalpy.

Changes of state at *constant temperature* or *isothermic changes of state*. In order for temperature to remain constant during expansion of a gas, heat must be added, and heat must be taken away in compression (apart from a few exceptions). In the case of the ideal gas, $U(T) =$ const., so the added heat equals the work done, in accordance with the first law $(dQ + dW = 0)$. The isothermal of the ideal gas $(pV = mRT = $ const.) is set out in the p, V curve as a hyperbola.

Adiabatic Changes of State are characterised by isolation of the system from its environment in terms of heat density. Such changes are approximately demonstrated in compressors and in expansion engines, because the compression and expansion of gases occur so rapidly that very little heat is exchanged with the environment during a change of state. In accordance with the second law (see C4.3.1), all entropic variation is brought about by irreversibilities within the system, $\dot{S} = \dot{S}_i$.

A reversible adiabatic event occurs at constant entropy $\dot{S} = 0$. Such a change of state is called *isentropic*. A reversible adiabatic event is therefore simultaneously *isentropic*. But an isentropic event does not need to be adiabatic (because the result of $\dot{S} = \dot{S}_a + \dot{S}_i = 0$ is not $\dot{S}_a = 0$ either).

Figure 1 illustrates the various changes of state in the p, V and T, S curves, and the main relationships for changes of state in ideal gases.

Polytropic Changes of State. While complete heat interchange is the prerequisite for isothermal change of state, in the case of adiabatic changes of state all heat exchange with the environment is inhibited. In reality, neither situation can be fully attained. Accordingly, there is introduced a polytropic change of state by the following equation:

$$pv^n = \text{const.} , \qquad (1)$$

where in practical cases n is mostly between 1 and κ. Isochores, isobars, isothermals and reversible adiabatics are special cases of polytropes with the following exponents (**Fig. 2**): isochore, $n = \infty$; isobars, $n = 0$; isothermals, $n = 1$; *reversible adiabatic*, $n = \kappa$. The following equation also applies:

$$v_2/v_1 = (p_1 p_2)^{1/n} = (T_1/T_2)^{1/(n-1)} , \qquad (2)$$

$$W_{12} = mR(T_1 - T_1) / (n - 1)$$

$$= (p_2 V_2 - p_1 V_1) / (n - 1)$$

$$= p_1 V_1 [(p_2/p_1)^{(n-1)/n} - 1] / (n - 1)$$

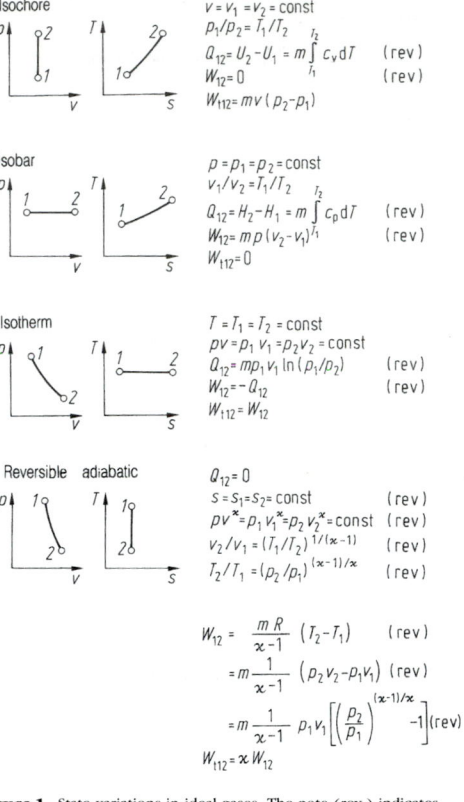

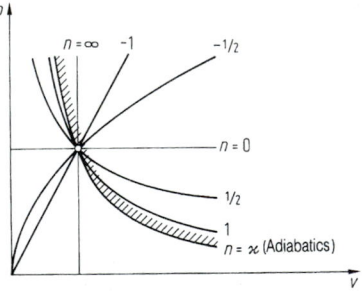

Isochore

$$V = V_1 = V_2 = \text{const}$$
$$p_1/p_2 = T_1/T_2$$
$$Q_{12} = U_2 - U_1 = m \int_{T_1}^{T_2} c_v \, dT \quad (\text{rev})$$
$$W_{12} = 0 \quad (\text{rev})$$
$$W_{t12} = mv(p_2 - p_1)$$

Isobar

$$p = p_1 = p_2 = \text{const}$$
$$v_1/v_2 = T_1/T_2$$
$$Q_{12} = H_2 - H_1 = m \int_{T_1}^{T_2} c_p \, dT \quad (\text{rev})$$
$$W_{12} = mp(v_2 - v_1)_{T_1} \quad (\text{rev})$$
$$W_{t12} = 0$$

Isotherm

$$T = T_1 = T_2 = \text{const}$$
$$pv = p_1 v_1 = p_2 v_2 = \text{const}$$
$$Q_{12} = mp_1 v_1 \ln(p_1/p_2) \quad (\text{rev})$$
$$W_{12} = -Q_{12} \quad (\text{rev})$$
$$W_{t12} = W_{12}$$

Reversible adiabatic

$$Q_{12} = 0$$
$$S = S_1 = S_2 = \text{const} \quad (\text{rev})$$
$$pv^\kappa = p_1 v_1^\kappa = p_2 v_2^\kappa = \text{const} \quad (\text{rev})$$
$$v_2/v_1 = (T_1/T_2)^{1/(\kappa-1)} \quad (\text{rev})$$
$$T_2/T_1 = (p_2/p_1)^{(\kappa-1)/\kappa} \quad (\text{rev})$$

$$W_{12} = \frac{mR}{\kappa - 1}(T_2 - T_1) \quad (\text{rev})$$

$$= m \frac{1}{\kappa - 1}(p_2 v_2 - p_1 v_1) \quad (\text{rev})$$

$$= m \frac{1}{\kappa - 1} p_1 v_1 \left[\left(\frac{p_2}{p_1}\right)^{(\kappa-1)/\kappa} - 1 \right] (\text{rev})$$

$$W_{t12} = \kappa W_{12}$$

Figure 1. State variations in ideal gases. The note (rev.) indicates that the state variation should be reversible.

Figure 2. Polytropes with various exponents.

and

$$W_{t12} = nW_{12} . \qquad (3)$$

The exchanged heat is

$$Q_{12} = mc_v (n - \kappa)(T_2 - T_1) / (n - 1) . \qquad (4)$$

Example A compressed-air system is to supply 1000 m³ of compressed air per hour at 15 bar (note: 1 m³ = 1 standard cubic metre is the volume of gas converted to 0 °C and 1.013 25 bar), with an inlet pressure of $p_1 = 1$ bar and a temperature of $t_1 = 20\,°C$. For atmosphere, $\kappa = 1.4$. What power is required for compression to take place polytropically at $n = 1.3$? What heat flux must be absorbed under such circumstances?

According to the objective, the volumetric flow of air intake will be 1000 m³ at 0 °C and 1.013 5 bar.

$$\dot{V}_1 = \frac{p_0 T_1}{p_1 T_0}\,\dot{V}_0 = \frac{1.013\,25 \cdot 293.15}{1 \cdot 273.15}\,1000\,\frac{m^3}{h} = 1087.44\,\frac{m^3}{h}.$$

In the case of a polytropic change of state, in accordance with Eq. (3):

$$P_{12} = \dot{m}W_{t12} = \frac{np_1\dot{V}_1}{n-1}\left[\left(\frac{p_2}{p_1}\right)^{\frac{n-1}{n}} - 1\right]$$

$$= \frac{1.3 \cdot 10^5\,\frac{N}{m^2}\,1087.44\,\frac{m^3}{h}}{1.3 - 1}\left[15^{\frac{1.3-1}{1.3}} - 1\right] = 113.6\,kW.$$

In accordance with Eqs (4) and (3),

$$\frac{\dot{Q}_{12}}{W_{t12}} = \frac{\dot{Q}_{12}}{P_{12}} = c_v\,\frac{n-\kappa}{nR}$$

or, since

$$R = c_p - c_v \quad \text{and} \quad \kappa = c_p/c_v,$$

then

$$\frac{\dot{Q}_{12}}{P_{12}} = \frac{1}{n}\frac{n-\kappa}{\kappa-1}.$$

Accordingly,

$$\dot{Q} = \frac{1}{1.3} \cdot \frac{1.3 - 1.4}{1.4 - 1}\,113.6\,kW = -21.85\,kW.$$

7.2 Changes of State of Gases and Vapours in Motion

In order to characterise the flow of a fluid mass Δm, we need to know not only thermodynamic state variables but also the magnitude and direction of velocity at each point of the field. In this context we restrict ourselves to steady-state flows in ducts of constant, expanding or contracting cross-section.

In addition to the first and second laws, the law of conservation of mass also applies:

$$\dot{M} = Aw\rho = \text{const.} \tag{5}$$

In a flow which does not give up work to the environment, $W_{t12} = 0$, the first law as C3 Eq. (14) is converted into

$$\Delta m(h_2 - h_1) + \Delta m\left(\frac{w_2^2}{2} - \frac{w_1^2}{2}\right) + \Delta mg(z_2 - z_1) = Q_{12}, \tag{6}$$

irrespective of whether the flow events are reversible or irreversible. Omitting the usually negligible feed work, then the following equation applies for an adiabatic flow:

$$h_2 - h_1 + \frac{w_2^2}{2} - \frac{w_1^2}{2} = 0. \tag{7}$$

An increase in kinetic energy equates to a decrease in fluid enthalpy. In accordance with C4 Eq. (4), enthalpy variation is caused by a pressure change, $dh = v\,dp$ in the case of reversibly adiabatic flow.

7.2.1 Flow of Ideal Gases

Application of Eq. (7) to an ideal gas which escapes from a tank (**Fig. 3**), where the gas is in a constant state of

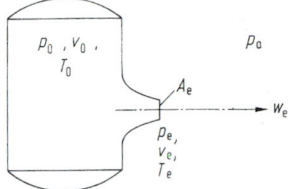

Figure 3. Outlet flow from the pressure tank.

p_0, v_0, T_0 and where $w_0 = 0$, gives rise to the following values due to $h_e = h_0 = c_p\,(T_e - T_0)$ and $w_0 = 0$:

$$\frac{w_e^2}{2} = c_p\,(T_0 - T_e) = c_p\,T_0\left(1 - \frac{T_e}{T_0}\right).$$

In the case of a reversibly adiabatic change of state, $T_e/T_0 = (p_e/p_0)^{(\kappa-1)/\kappa}$, Eq. (2), and the following formulae apply: $T_0 = p_0v_0/R$, C6 Eq. (2), and $c_p/R = \kappa/(\kappa-1)$, C6 Eq. (12). Accordingly the output speed is

$$w_e = \sqrt{2\,\frac{\kappa}{\kappa-1}p_0v_0\left[1 - \left(\frac{p_e}{p_0}\right)^{(\kappa-1)/\kappa}\right]} \tag{8}$$

Quantitative output $M = A_ew_e\,v_e$ is consequently, taking account of $p_0v_0 = p_ev_e$,

$$\dot{M} = A\Psi\,\sqrt{2p_0/v_0}, \tag{9}$$

with the output function of

$$\Psi = \sqrt{\frac{\kappa}{\kappa-1}}\,\sqrt{\left(\frac{p}{p_0}\right)^{2/\kappa} - \left(\frac{p}{p_0}\right)^{(\kappa+1)/\kappa}} \tag{10}$$

It is a function of adiabatic exponent κ and of compression ratio p/p_0 (**Fig. 4**) and exhibits a maximum value Ψ_{max}, which is obtained from $d\Psi/d(p/p_0) = 0$. The maximum value is at a specific compression ratio which is called *Laval's compression ratio*:

$$\frac{p_s}{p_0} = \left(\frac{2}{\kappa+1}\right)^{\kappa/(\kappa-1)} \tag{11}$$

The following formula applies in this compression ratio:

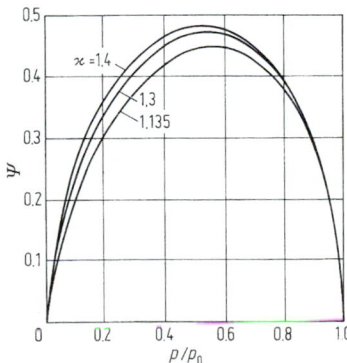

Figure 4. Outlet flow function.

$$\Psi_{max} = \left(\frac{2}{\kappa+1}\right)^{\kappa/(\kappa-1)} \sqrt{\frac{\kappa}{\kappa+1}} . \qquad (12)$$

For compression ratio p_s/p_0, there is according to Eq. (8) a velocity of $w_c = w_s$, where $p_c\,p_0 = p_s/p_0$. The following formula applies:

$$w_s = \sqrt{2\,\frac{\kappa}{\kappa+1}\,p_0 v_0} = \sqrt{\kappa p_s v_s} = \sqrt{\kappa R T_s}. \qquad (13)$$

This equals the *velocity of sound* in state p_s, v_s.

In general, the velocity of sound is the velocity at which pressure and density fluctuations propagate and – in the case of reversible adiabatic changes of state – is expressed by

$$w_s = \sqrt{(\partial P/\partial \rho)_s},$$

from which the following formula is derived for ideal gases:

$$w_s = \sqrt{\kappa R T}.$$

The velocity of sound is a state variable.

Example A steam boiler produces 10 t of saturated steam at $p_0 = 15$ bar hourly. Steam can be regarded as an ideal gas ($\kappa = 1.3$); what is the minimum open aperture section of the safety valve?

The safety valve must be capable of disposing of the full steam supply. Since $\dot{m}$ is constant at all sections during output, therefore $A\Psi = $ const., in accordance with Eq. (9). Since flow is restricted, i.e. A is decreased, Ψ increases. The maximum attainable value is Ψ_{max}. In this context the back pressure is equal to or less than Laval's pressure. In this instance the back pressure from the atmosphere of $p_0 = 1$ bar is less than Laval's pressure, which is calculated as 2.7 bar in accordance with Eq. (11). Accordingly, the section required from Eq. (9) is obtained if we substitute $\Psi = \Psi_{max} = 0.472$ into Eq. (12).

According to Eq. (9), we obtain $A = 12.93$ cm^2, where

$$\dot{m} = 10 \cdot 10^3 \frac{1}{3600} \text{ kg/s}$$

and

$$v_0 = v'' = 0.1317 \text{ m}^3/\text{kg}$$

(in accordance with **Appendix C6, Table 5** where $p_0 = 15$ bar) from Eq. (9) where $A = 12.93$ cm^2. Due to the jet restriction whose magnitude is a function of the form of valve, an additional supplement must be applied.

7.2.2 Jet Flow and Diffusor Flow

In accordance with Fig. 4, a specific value for output function κ will relate, at a prescribed adiabatic exponent Ψ, to a given compression ratio p/p_0. Since mass flow $\dot{m}$ is constant at every section, Eq. (9) also gives $A\Psi = $ const. For each compression ratio, therefore, a specific section

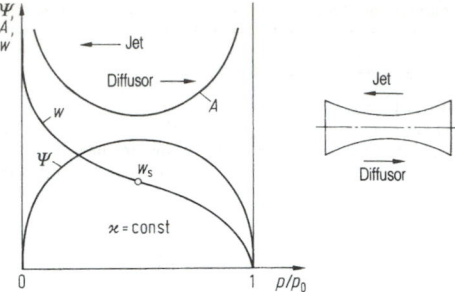

Figure 5. Jet flow and diffusor flow.

A can be allocated; see **Fig. 5**. The distinction must be drawn between the two cases:

a: Pressure Decreases in the Direction of Flow. Curves Ψ, A and w pass from right to left in **Fig. 5**. Section A initially decreases then begins to increase again. Velocity increases from subsonic to supersonic, while flow kinetic energy increases. Such a system is termed a "jet". For a subsonic jet, the section continuously decreases but with a supersonic jet it continuously increases.

In a jet which contracts in the direction of flow, it is not possible for pressure to drop below Laval's pressure at the outlet section, even if the pressure in the external space is made arbitrarily low. This is the logical consequence of $A\Psi = $ const. Since A decreases in the direction of flow, there can be only increases in Ψ. The maximum level it can reach is Ψ_{max}, which corresponds to Laval's compression ratio.

If pressure at the outlet section of a jet becomes lower than the pressure value prevailing at the outlet cross-section, then the steam will expand after leaving the jet. If back pressure is increased over and above the correct level, then the pressure increase will move upstream if gas escapes at subsonic velocity. If gas output is at the velocity of sound or – in an expanded jet – at supersonic velocity, then there is a compression shock wave at the aperture of the jet, in which there will occur a stepwise pressure transition to ambient level.

b: Pressure Increases in the Direction of Flow. Curves Ψ, A and w will run from left to right as in **Fig. 4**. Cross-section also initially decreases, then increases once again. Velocity drops from supersonic to subsonic. Kinetic energy decreases and pressure increases. Such a system is referred to as a *diffusor*. In a diffusor, which operates only in the subsonic range, the section continuously increases, but in the supersonic range it continuously decreases.

8 Thermodynamic Processes

8.1 Combustion Processes

Heat in technical processes is usually obtained from combustion, in most cases. Combustion is the chemical reaction of a substance, often carbon, hydrogen and hydrocarbons, with oxygen, which occurs highly exothermically, i.e. involving the release of heat. Fuels

may be solid, liquid or gaseous and the oxygen carrier is usually atmospheric air. In order to induce combustion, fuel must initially be brought to ignition temperature, which is a function of fuel type. The main components of all technically important fuels are carbon (C) and hydrogen (H), and fuels will also in many cases contain oxygen (O) and, with the exception of methane, a certain quantity of sulphur (S) which on combustion gives rise to the unwanted sulphur dioxide (SO_2).

8.1.1 Equations of Reactions

Elements H, C and S which are present in fuels are burnt in full combustion to form CO_2, H_2O and SO_2. Reaction equations determine the required amount of oxygen and the physical quantity in the form of flue gas. The following formulae apply:

$$C \quad + \quad O_2 \qquad = CO_2$$

$$1 \text{ kmol C} + \quad 1 \text{ kmol } O_2 = 1 \text{ kmol } CO_2$$

$$12 \text{ kg C} + 32 \text{ kg } O_2 \quad = 44 \text{ kg } CO_2$$

From the above, we determine the *minimum oxygen requirement* that is needed for complete combustion:

$$O_{min} = (1/12) \text{ kmol/kg C}$$

or

$$O_{min} = 1 \text{ kmol/kmol C,}$$

the *minimum air requirement*:

$$l_{min} = (O_{min}/0.21) \text{ kmol air/kg C}$$

or

$$L_{min} = (O_{min}/0.21) \text{ kmol air/kmol C}$$

and the quantity of CO_2 in the flue gas is calculated as (1/12) kmol/kg C. The following equations of reaction are applicable:

$$H_2 + \qquad \tfrac{1}{2}O_2 = H_2O$$

$$1 \text{ kmol } H_2 + \tfrac{1}{2} \text{ kmol } O_2 = 1 \text{ kmol } H_2O$$

$$2 \text{ kgH}_2 + \quad 16 \text{ kg } O_2 = 18 \text{ kg } H_2O$$

$$S + \qquad O_2 = SO_2$$

$$1 \text{ kmol S} + 1 \text{ kmol } O_2 = 1 \text{ kmol } SO_2$$

$$32 \text{ kg S} + \quad 32 \text{ kg } O_2 = 64 \text{ kg } SO_2.$$

With carbon, hydrogen, sulphur and oxygen contents in kg per kg of fuel, designated respectively by c, h, s and o, the minimum oxygen requirement is calculated in accordance with the above calculation:

$$O_{min} = \left(\frac{c}{12} + \frac{h}{4} + \frac{s}{32} - \frac{o}{32} \right) \text{ kmol/kg.} \qquad (1)$$

The following abbreviated form is employed:

$$O_{min} = \frac{1}{12} c\sigma \text{ kmol/kg} \qquad (2)$$

where σ is a characteristic of fuel (O_2 requirement in kmol related to kmol C in fuel) (values for σ in **Appendix C8, Tables 3 and 4**). The actual air requirement (related to 1 kg of fuel) is:

$$l = \lambda l_{min} = (\lambda O_{min}/0.21) \text{ kmol air/kg} \qquad (3)$$

where λ is the air surplus. In flue gases there arise not only products of combustion CO_2, H_2O, and SO_2 but also water content $w/18$ (SI unit: kmol per kg of fuel) and the intake combustion air l minus the consumed oxygen quantity O_{min}. Flue gas quantity amounts to:

$$n_R = l + \frac{1}{12} \left(3h + \frac{3}{8} + \frac{2}{3} w \right) \text{ kmol/kg.} \qquad (4)$$

Example In a furnace, 500 kg of coal of composition $c = 0.78$, $h = 0.05$, $o = 0.08$, $s = 0.01$, $w = 0.02$ and an ash content of $a = 0.06$, with surplus air $\lambda = 1.4$, is fully consumed hourly. How much air must be fed into the furnace, how much flue gas is produced and what is its composition?

Minimum oxygen requirement according to Eq. (1) is $O_{min} = 0.78/12 + 0.05/4 + 0.01/32 - 0.08/32$ kmol/kg = 0.0753 kmol/kg. Minimum air requirement is $l_{min} = O_{min}/0.21 = 0.3586$ kmol/kg. The air intake $l = \lambda l_{min} = 1.4 \cdot 0.3586 = 0.502$ kmol/kg, i.e. 0.502 kmol/kg · 500 kg/h = 251 kmol/h. This produces, given molecular weight $M = 28.953$ kg/kmol of air, an air requirement of $0.502 \cdot 28.953$ kg/kg = 14.54 kg/kg, i.e. 14.54 kg/kg; 500 kg/h = 7270 kg/h. Flue gas quantity according to Eq. (4) is $n_R = 0.0502 + \frac{1}{12}(3 \cdot 0.05 + \frac{3}{8} \cdot 0.08 + \frac{2}{3} \cdot 0.02)$ kmol/kg = 0.518 kmol/kg, i.e. 0.518 kmol/kg, 500 kg/h = 259 kmol/h with 0.065 kmol CO_2/kg, 0.0225 kmol H_2O/kg, 0.0003 kmol SO_2/kg, 0.0445 kmol N_2/kg and 0.0301 kmol O_2/kg.

8.1.2 Net Calorific Value and Gross Calorific Value

Net calorific value is the heat released on combustion if gases of combustion have been cooled to the temperature at which fuel and air are taken in. Water is contained in flue gases in gaseous form. If water vapour is condensed, the term given to the released heat is the *gross calorific value*. DIN 51 900 lays down net and gross calorific values for combustion at atmospheric pressure if the involved substances before and after combustion are at a temperature of 25 °C. The net and gross calorific values (see **Appendix C8, Tables 1 to 4**) are independent of surplus air and are merely a characteristic of the fuel. Gross calorific value Δh_o is greater by evaporation enthalpy r of the water contained in the flue gas then net calorific value Δh_u:

$$\Delta h_o = \Delta h_u + (8.937h + w)r.$$

Since – in most cases – water leaves technical furnaces in the form of steam, it is usually possible to make use only of the net calorific value. Experience shows that the net calorific value of heating fuels can be effectively reproduced [8] by the following equation of numerical values:

$$\Delta h_u = 54.04 - 13.29\rho - 29.31s \text{ MJ/kg}, \qquad (5)$$

where ρ is the density of heating fuel in kg/dm³ at 15 °C and s is sulphur content in kg/kg.

Example What is the net calorific value of a light heating fuel of density $\rho = 0.86$ kg/dm³, whose sulphur content $s = 0.8$ weight per cent? In accordance with Eq. (5), $\Delta h_u = 54.04 - 13.29 \cdot 0.86 - 29.31 \cdot 0.8 \cdot 10^{-2} = 42.38$ MJ/kg.

8.1.3 Combustion Temperature

The theoretical combustion temperature is the temperature of flue gas in fully isobaric/adiabatic combustion, if no dissociation occurs. Heat released in combustion is used to increase internal energy and thus the temperature of the gases and to apply work of displacement. Theoretical combustion temperature is calculated from the condition of equal enthalpy before and after combustion:

$$\Delta h_u(0 \text{ °C}) + \left[c_B \right]_0^{t_B} \cdot t_B + l \left[\bar{C}_{pL} \right]_0^{t_L} \cdot t_L = n_R \left[\bar{C}_{pR} \right]_0^t \cdot t. \qquad (6)$$

where t_B = fuel temperature, t_L = air temperature and t = theoretical combustion temperature, $[c]_0$ is the mean specific heat capacity of fuel, $[\bar{C}_{pL}]_0$ is the mean molar heat capacity of air and $[\bar{C}_{pR}]_0$ that of flue gas. The mean molar heat capacity is made up of the mean molar heat capacities of individual constituents:

$$n_R \left[\bar{C}_{pR} \right]_0^t = \frac{c}{12} \left[\bar{C}_{pCO_2} \right]_0^t + \left(\frac{h}{2} + \frac{w}{18} \right) \left[\bar{C}_{pH_2O} \right]_0^t \qquad (7)$$
$$+ 0.21 (\lambda - 1)l_{min} \left[\bar{C}_{pO_2} \right]_0^t + 0.79\lambda l_{min} \left[\bar{C}_{pN_2} \right]_0^t.$$

Theoretical combustion temperature has to be calculated iteratively from Eqs (6) and (7).

The actual combustion temperature is lower, even given full combustion of the fuel, than the theoretical temperature, because of heat given off to the environment, mainly by radiation, the disintegration of molecules which starts at temperatures over 1500 °C and the noticeable dissociation over and above 2000 °C. The heat of dissociation is released again when temperature drops below dissociation level.

8.2 Internal Combustion Engines

In an internal combustion engine, gasified fuel is the working material. It does not pass through a self-contained process but is fed as an exhaust gas to the environment, after it has performed work in a turbine or in a piston engine. Internal combustion engines include open gas turbine plants and petrol and diesel combustion engines.

In an *open gas turbine system*, air drawn in is raised to a high pressure in a compressor, preheated and heated in a combustion chamber by combustion of the injected fuel. Combustion gases are released in a turbine at operating power, they release part of their residual heat in a heat exchanger to the air preheating system and then they emerge into the open atmosphere. In a generator, the effective work is converted into electrical energy.

Overall energy efficiency $\eta = -P/(\dot{m}_B \, \Delta h_u)$ is used to characterise energy conversion.

P is the useful performance of the system, $\dot{m}_B$ is the bulk flow of the fuel taken in. The exergetic overall efficiency level $\zeta = -P/(\dot{m}_B \, (w_{ex})_B)$ states what component of the flow of exergy taken in with fuel is converted into useful work.

Generally, $(w_{ex})_B$ is only slightly greater than the net calorific value, such that η and ζ differ only slightly from each other. For large engines (diesels), the overall efficiency is approximately 42%, for steam-driven power stations up to 40%, for vehicle engines approximately 25% and for open gas turbine plants 20 to 30%. Gasoline and diesel engines are the most important examples of internal combustion engines.

In the *spark ignition engine*, the cylinder is at the end of the inlet stroke in stage *1* (**Fig. 1**) and is charged with the combustible mixture at ambient temperature and atmospheric pressure. The mixture is compressed along adiabatics *1 2* from initial volume $V_k + V_h$ to compression volume V_k. V_h is the swept volume. At TDC *2*, electrical ignition usually initiates combustion, causing pressure to increase from point *2* to point *3*. In **Fig. 1** the simplificatory assumption is made that the gas remains unchanged and that heat $Q_{23} = Q$ released in combustion is brought in from outside. On return of the piston, the gas expands along adiabatic *3 4 4'' 4'*. The exhaust stroke starting at *4* is replaced by subtraction of a heat value $|Q_0|$ at constant volume, where pressure decreases from point *4* to point *1*. At point *1*, gases of combustion must be replaced by a new mixture, for which purpose a double stroke is required (which is not illustrated).

The heat taken in is

$$Q = Q_{23} = mc_v(T_3 - T_2). \tag{8}$$

The heat taken out is

$$|Q_0| = |Q_{41}| = mc_v(T_4 - T_1) . \tag{9}$$

The work applied is

$$|W_t| = Q - |Q_0| \tag{10}$$

and thermal efficiency is

$$\eta = \frac{|W_t|}{Q} = 1 - \frac{T_4 - T_1}{T_3 - T_2} = 1 - \frac{T_1}{T_2} = 1 - \left(\frac{p_1}{p_2}\right)^{(\kappa - 1)/\kappa}. \tag{11}$$

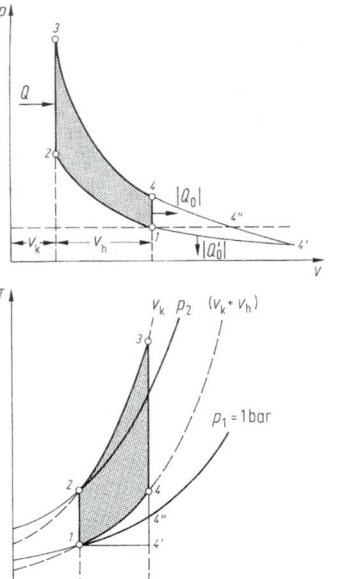

Figure 1. Theoretical process of spark-ignition engine in p, V and T, S curves.

Accordingly, thermal efficiency depends only on compression ratio p_2/p_1 (apart from adiabatic exponent κ), and not on the magnitude of heat taken in. The further the compression process is promoted, the better the exploitation of heat. The pressure of compression is limited by the fuel-air mix ignition temperatures. This restriction does not apply in the case of the *diesel engine*, where the air of combustion is heated by high compression above the ignition temperature of the fuel, which is injected into the heated air. The simplified process of the diesel engine is illustrated in **Fig. 2**. It consists of adiabatic compression *1 2* of the air of combustion, isobaric compression *2 3'* after injection of fuel into the hot, compressed air of combustion, adiabatic release *3' 4* and exhaust *4 1*, which is replaced by an isochore with introduction of heat $|Q_0|$ in **Fig. 2**. The heat taken in is

$$Q'_{23} = Q = mc_p(T_{3'} - T_2) , \tag{12}$$

and the exhaust heat which is regarded as being carried along isochore *4 1* is

$$|Q_{41}| = |Q_0| = mc_v (T_4 - T_1) , \tag{13}$$

the applied work is

$$|W_t| = Q - |Q_0|$$

and thermal efficiency is

$$\eta = \frac{|W_t|}{Q} = 1 - \frac{1}{\kappa} \frac{T_4 - T_1}{T_{3'} - T_2} = 1 - \frac{1}{\kappa} \frac{T_4}{T_{3'}} \frac{T_3}{T_2} \frac{T_1}{T_2} - 1}{\frac{T_{3'}}{T_2} - 1}. \tag{14}$$

For thermal efficiency, we obtain, via compression ratio $\varepsilon = V_1/V_2 = (V_k + V_h)/V_k$ and injection ratio $\varphi = (V_k + V_e)/V_k$:

$$\eta = 1 - \frac{1}{\kappa \varepsilon^{\kappa - 1}} \frac{\varphi^\kappa - 1}{\varphi - 1}. \tag{15}$$

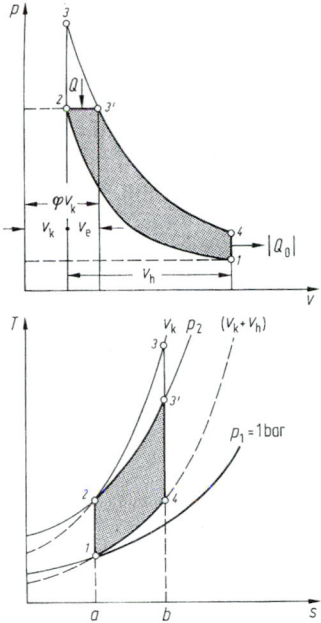

Figure 2. Theoretical diesel engine process in the p, V and T, S curves.

The thermal efficiency of the diesel process is dependent only on compression ratio ε (apart from adiabatic exponent κ) and on injection ratio φ, which increases as loading increases.

8.3 Cyclic Processes

A process which brings the system back to its initial state is called a *cyclic process*. After it has been completed, all state variables of the system, such as pressure, volume, internal energy and enthalpy, resume their original state. According to the first law, C3 Eq. (9), the following formula applies after passing through processes $E_2 = E_1$, and therefore

$$\Sigma Q_{ik} + \Sigma W_{ik} = 0 . \qquad (16)$$

The total applied work is $-W = -\Sigma W_{ik} = \Sigma Q_{ik}$: work done equals excess heat taken in relative to heat given off. According to the second law, heat taken in cannot be fully converted into work.

If heat taken in is greater than heat given off, then the process functions in the same way as a *power station* or *thermal engine* whose function is to produce work. If heat given off is greater than heat taken in, work must be added. In this type of process, heat can be taken out of a substance at lower temperatures, and then at a higher temperature, e.g. at ambient temperature, more heat may be removed along with the input work. A process of this type operates as a *refrigeration process*. In a *heat pump process*, heat is taken from the environment and given off together with the input work at a higher temperature.

8.3.1 Carnot Cycle

In the course of historic development, though not in general practice, the cycle process introduced by Carnot has

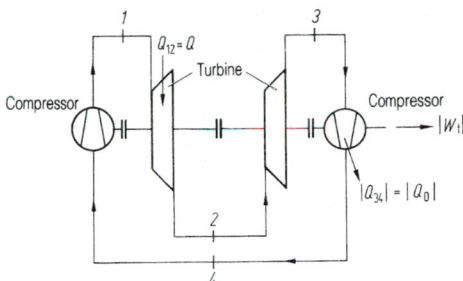

Figure 3. Flowchart of a heat engine operating according to the Carnot cycle.

played a decisive role (**Figs 3 and 4**). It consists of the following changes of state (clockwise process):

1-2: Isothermal expansion temperature T with input of heat Q.

2-3: Adiabatic expansion from pressure p_2 to pressure p_3.

3-4: Isothermal compression at temperature T_0 with input of $|Q_0|$.

4-1: Adiabatic compression from pressure p_4 to pressure p_1.

Where changes of state are reversible, the input heat is

$$Q = mRT \ln V_2/V_1 = T (S_2 - S_t) . \qquad (17)$$

and the heat given off is

$$|Q_0| = mRT_0 \ln V_3/V_4 = T_0 (S_2 - S_1) . \qquad (18)$$

Applied technical work is $-W_t = Q - |Q_0|$ and thermal efficiency is as follows:

$$\eta = |W_t|/Q = 1 - (T_0/T) . \qquad (19)$$

In reverse sequence *4-3-2-1* of changes of state, with input of technical work W_t, heat Q_0 is taken from a body at low temperature t_0 and heat Q given off at high tem-

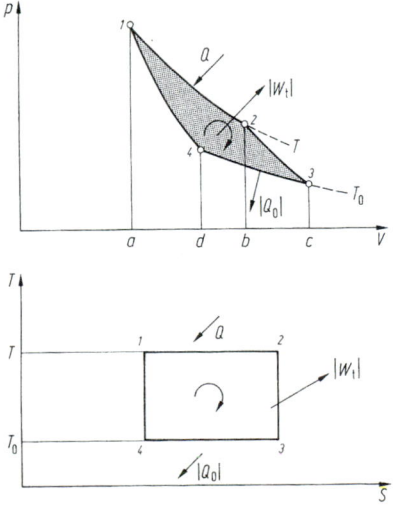

Figure 4. Carnot cycle of heat engine in the p, V and T, S curves.

perature T. An anticlockwise Carnot cycle of this type enables heat Q_0 to be extracted from an item for cooling at low temperature t_0, i.e. it can function as a refrigerator, and heat $|Q| = W_t + Q_0$ can be given off once again at higher temperature T. If the purpose of the process is to release heat $|Q|$ at the higher temperature T for heating purposes, then the process operates as a heat pump. Carnot cycles have achieved no practical significance, because their output is very low in relation to their structural volume.

8.3.2 Thermal Power Plants

In thermal power plants the process material is subjected to the energy of the combustion gases in the form of heat. The process material undergoes a cyclic process.

The *Ackeret-Keller process* consists of the following changes of state as shown in **Fig. 5**.

1-2 Isothermal compression at temperature T_0 from pressure p_0 to pressure p.

2-3 Isobaric heat input at pressure p.

3-4 Isothermal expansion at temperature T from pressure p to pressure p_0.

4-1 Isobaric heat input at pressure p_0.

This process is based on a proposal made by the Swedish engineer J. Ericson (1803–99) and is therefore also referred to as the *Ericson process*. However, it was not until 1941 that it was used by Ackeret and Keller as a comparative process for gas turbine plants.

The heat required for isobaric heating *2-3* of the compressed process material is provided by isobaric cooling *4-1* of the released process material, $Q_{23} = |Q_{41}|$. The thermal efficiency figure is the same as that for the Carnot process, since

$$-W_t = Q_{34} - |Q_{21}| \qquad (20)$$

and

$$\eta = 1 - \frac{|Q_{21}|}{Q_{34}} = 1 - \frac{T_0}{T}. \qquad (21)$$

However, technical implementation of the process is difficult, because isothermic compression and release are virtually impossible to realise. Since these processes can only

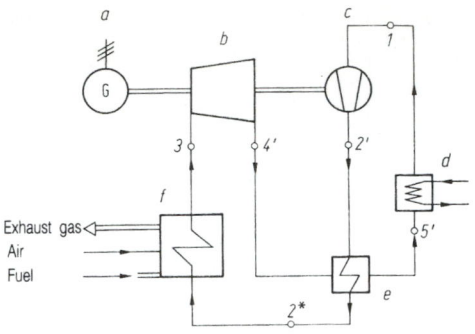

Figure 6. Gas turbine process with enclosed circuit: *a* generator; *b* turbine; *c* compressor; *d* cooler; *e* heat exchanger; *f* gas heater.

be approximated by multi-stage adiabatic compression with intermediate cooling, the prime value of the Ackeret-Keller process is as a control for gas turbine processes with multi-stage compression and release.

In a *closed gas turbine plant* (**Fig. 6**), gas is compressed in the compressor, heated in the heat exchanger and gas heater to a high temperature, then released in a turbine with application of work and cooled in the heat exchanger and the downstream cooler to the initial temperature, whereupon the gas is once again taken in by the compressor. The process materials include air and also other gases such as helium or nitrogen.

The closed gas turbine system is readily controllable, and pollution of the turbine blades can be avoided by using suitable gases. The high energy costs arising in comparison with open systems are disadvantageous, because a cooler is required and high-quality steels are required for the heater. **Figure 7** illustrates the process in the p, v and T, s curves.

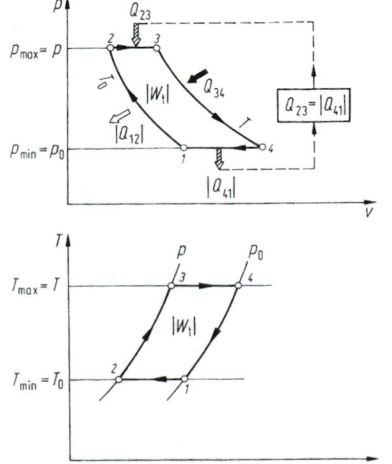

Figure 5. Ackeret-Keller process in the p, v and T, s curves.

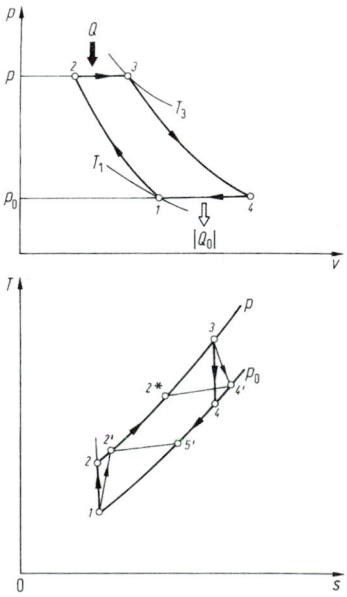

Figure 7. Gas turbine process in the p, v and T, s curves. The p, v curve illustrates only the reversible process (Joule process) *1, 2, 3, 4*.

The reversible cyclic process consisting of two isobaric and two isentropic components is referred to as the *Joule process* (state points *1, 2, 3, 4*). The input heat is

$$\dot{Q} = \dot{m}c_p\,(T_3 - T_2)\,. \tag{22}$$

The output heat is

$$|\dot{Q}_0| = \dot{m}c_p(T_4 - T_1)\,. \tag{23}$$

The applied work is

$$-P = -\dot{m}w_t = \dot{Q} - |\dot{Q}_0| = \dot{m}c_p(T_3 - T_2)\left(1 - \frac{T_4 - T_1}{T_3 - T_2}\right) \tag{24}$$

and the thermal efficiency

$$\eta = \frac{|P|}{\dot{Q}} = \left(1 - \frac{T_4 - T_1}{T_3 - T_2}\right). \tag{25}$$

Owing to the isentropic equation

$$\left(\frac{p_0}{p}\right)^{(\kappa-1)/\kappa} = \frac{T_1}{T_2} = \frac{T_4}{T_3} \quad \text{is} \tag{26}$$

$$\frac{T_4 - T_1}{T_3 - T_2} = \frac{T_1}{T_2} = \left(\frac{p_0}{p}\right)^{(\kappa-1)/\kappa}$$

and thermal efficiency

$$\eta = \frac{|P|}{\dot{Q}} = 1 - \left(\frac{p_0}{p}\right)^{(\kappa-1)/\kappa} \tag{27}$$

is dependent only on compression ratio p/p_0 or temperature ratio T_2/T_1 of the compression process. Compression power increases more rapidly with compression ratio than turbine power, so that the obtained useful power as Eq. (24), taking account of Eq. (26),

$$-P = \dot{m}c_p\,T_1\left(\frac{T_3}{T_1} - \left[\frac{p}{p_0}\right]^{(\kappa-1)/\kappa}\right)\left(1 - \left[\frac{p_0}{p}\right]^{(\kappa-1)/\kappa}\right), \tag{28}$$

reaches a maximum level at a given compression ratio for prescribed values of maximum temperature T_3 and of minimum temperature T_1. This optimum compression ratio is obtained by differentation from Eq. (28):

$$\left(\frac{p}{p_0}\right)^{(\kappa-1)/\kappa}_{\text{opt}} = \sqrt{(T_3/T_1)}\,, \tag{29}$$

which from Eq. (26) is synonymous with $T_4 = T_2$. Taking account of efficiency η_T for the turbine, η_V of the compressor and of mechanical efficiency η_m for energy transfer between turbine and compressor, the following optimum compression ratio is obtained:

$$\left(\frac{p}{p_0}\right)^{(\kappa-1)/\kappa}_{\text{opt}} = \sqrt{\eta_m\eta_T\eta_V(T_3/T_1)}\,. \tag{30}$$

More than half of the turbine output of a gas turbine system is required to drive the compressor. Consequently, the total installed power is four to six times the useful output.

Steam power stations are driven by a process material, usually water, which is evaporated and recondensed during the process. It is used for generation of most of the electrical energy in our electrical grids. In the simplest form of the process (**Fig. 8**), in boiler *a* the process material is heated at a high pressure isobarically to boiling point, evaporated and then superheated in superheater *b*. Steam is then adiabatically released in turbine *c* with appli-

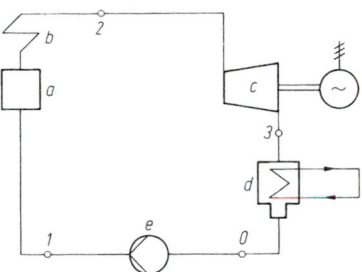

Figure 8. Steam power station: *a* boiler; *b* superheater; *c* turbine; *d* condenser; *e* feed-water pump.

cation of work and precipitated in condenser *d* with heat being given off.

The liquid is brought by feed pump *e* to boiler pressure and fed back into the boiler. The reversible cyclic process *01'23'0* (**Fig. 9**), consisting of two isobars and two isentropes, is called the *Clausius-Rankine process*. The actual cycle process follows the changes of state *01230* in **Fig. 9**.

Intake of heat in the steam generator is

$$Q_{12} = m(h_2 - h_1)\,. \tag{31}$$

The output of the adiabatic turbine is

$$|P_{23}| = |\dot{m}w_{t23}| = \dot{m}(h_2 - h_3) = \dot{m}\eta_T(h_2 - h_3')\tag{32}$$

with isentropic turbine efficiency η_T. The heat flux carried away in the condenser is

$$-Q_{30} = m\,(h_3 - h_0)\,. \tag{33}$$

The effective output of the cyclic process is

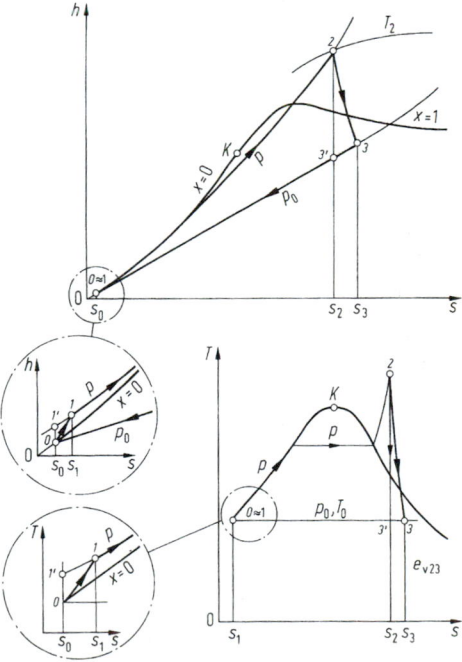

Figure 9. State variation of water in cyclic process of simple steam power station in the T, s and h, s curves.

$$-P = -\dot{m}w_t = -P_{23} - P_{01} , \qquad (34)$$

with pump output

$$P_{01} = \dot{m}\,(b_1 - b_0) = \dot{m}\,\frac{1}{\eta_V}(b_1 - b_0) , \qquad (35)$$

where η_V is the feed-pump efficiency. Effective output differs only slightly from turbine output. Thermal efficiency is

$$\eta = -\frac{\dot{m}w_t}{\dot{Q}_{12}} = \frac{(b_2 - b_3) - (b_1 - b_0)}{b_2 - b_1} . \qquad (36)$$

Values of $\eta \approx 0.42$ for thermal efficiency are achieved with a back pressure of $p_0 = 0.05$ bar, a fresh steam pressure of 150 bar and steam temperature of 500 °C.

8.4 Cooling and Heating

8.4.1 Compression Refrigeration Plant

In a refrigeration plant gases or vapours are used as the process material in the same way as in a thermal power station, but here they are referred to as *coolants*. The purpose of a refrigeration plant is to extract heat from an item so as to cool it. For this purpose work must be applied, and will be given off to the environment in the form of heat together with the heat taken from the item for cooling. Compressor-type refrigerators are normally used to generate temperatures down to – 100 °C approximately.

The flowchart of a *compression refrigerator* is illustrated in **Fig. 10**. Compressor a, which is usually configured as a piston engine for low outputs or as a turbo compressor for high outputs, takes in vapour from evaporator b at pressure p_0 and at the corresponding saturation temperature T_0 and compresses it along the adiabatics *1 2* (**Fig. 11**) to pressure p. The vapour is then precipitated in the condenser c at pressure p. The liquid coolant is released at throttle valve d and then returns to the evaporator where it is subjected to heat. The refrigerator extracts from the item for cooling an amount of heat Q_0, which is transferred to the evaporator b. In condenser c it gives off heat $|Q| = Q_0 + W_t$ to the environment.

Since water freezes at 0 °C and water vapour has an inconveniently high specific volume, other fluids are selected for the coolant, such as ammonia (NH_3), carbon dioxide (CO_2), monochloromethane (CH_3Cl), monofluorotrichloromethane ($CFCl_3$), difluorodichloromethane (CF_2Cl_2) and difluoromonochloromethane

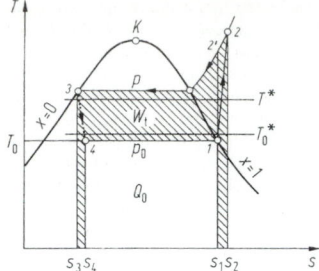

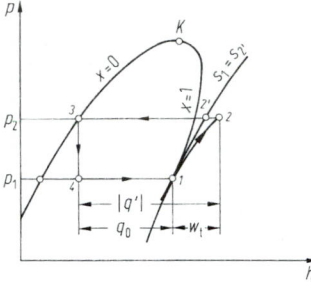

Figure 11. Circuit process of a vapour refrigerator in the T, s and Mollier p, b curves.

(CHF_2Cl). **Appendix C6, Tables 7 to 11**, contains vapour tables for coolants. Refrigerating performance, with m as the bulk flow of the circulating coolant, is as follows:

$$\dot{Q}_0 = \dot{m}q_0 = \dot{m}(b_1 - b_4) = \dot{m}(b_0'' - b') , \qquad (37)$$

because $b_4 = b_3 = b'$. The compressor power is

$$P = \dot{m}w_t = \dot{m}(b_2 - b_1) = \dot{m}\,\frac{1}{\eta_V}(b_2 - b_0'') , \qquad (38)$$

where η_V is its isentropic efficiency. The heat flux given off by the condenser is

$$|\dot{Q}| = \dot{m}|q| = \dot{m}(b_2 - b_3) = \dot{m}(b_2 - b') . \qquad (39)$$

The output index of a refrigerator is defined as the relationship of refrigeration power Q_0 to power draw P of the compressor:

$$\varepsilon_{KM} = \frac{\dot{Q}_0}{P} = \frac{q_0}{w_t} = \eta_V\,\frac{b_0'' - b'}{b_2 - b_0''} . \qquad (40)$$

It is dependent only upon the two pressures p and p_0 (apart from the isentropic compression efficiency).

8.4.2 Compression Heat Pump

This operates by a similar process to the compressor-type refrigeration plant illustrated in **Figs 10 and 11**. Its purpose is to apply heat to a body. Accordingly, heat (anergy) Q_0 is taken from the environment and fed together with the applied work W_t (exergy) as heat to the body for heating $|Q| = Q + W_t$. The output index of a heat pump is defined as the ratio of heat output $|Q|$ given off by the heat pump to power draw P of the compressor:

$$\varepsilon_{WP} = \frac{|\dot{Q}|}{P} = \frac{|q|}{w_t} = \eta_V\,\frac{b_2 - b'}{b_2 - b_0''} . \qquad (41)$$

As illustrated by the T, s curve (**Fig. 11**), surface w_t

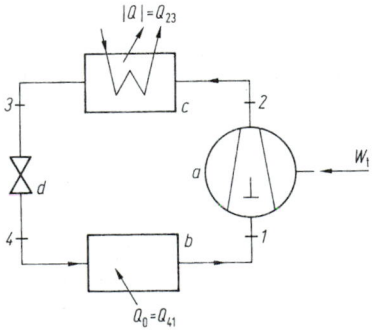

Figure 10. Circuit diagram of a vapour refrigerator. Explanatory notes in main text.

decreases at higher ambient temperatures T_0 and at lower heating temperatures T^*. Less power is required for the compressor. The output index increases. For economical operation of heat pumps in heating residential rooms, heater temperature must be kept low, for example by means of floor heating, where t^* is approximately $\leq 29\,°C$. Also, heat pumps are uneconomical where external temperatures are too low. If output index ε_{WP} drops to levels below about 2.3, then no primary energy is saved by comparison with conventional heating, because the efficiency of conversion of primary energy P_{Pr} in a power station into electrical energy P for driving the heat pump $\eta_{el} = P/P_{Pr}$ is approximately 0.35. Therefore the heating index $\zeta = |Q|/P_{Pr}$ is 0.8, which approximately equals the efficiency of a conventional heating system. Present-day electrically powered heat pumps seldom attain annual heat indexes of 2.3, unless the heat pump is switched off at excessively low external temperatures of below about $3\,°C$, with a transfer to conventional heating. Powered heat pumps with exploitation of heat given off, in common with sorption heat pumps, exploit primary energy better than electrically powered heat pumps.

8.4.3 Combined Power and Heat Generation (Co-generation)

Simultaneous generation of heating power and electrical energy in thermal power stations is termed *combined power and heat generation*. In this process, the heat given off by the power station, which is always plentiful, is used for space heating rather than going to waste. Since the heat used in this way is predominantly made up of more than 90% anergy, less primary energy is converted into heat for heating – as primary energy predominantly consists of exergy – than in the case of conventional heating. Low-pressure steam is taken from the steam turbine, because it contains not only anergy but so much exergy that the heating energy and the exergy losses in heat distri-

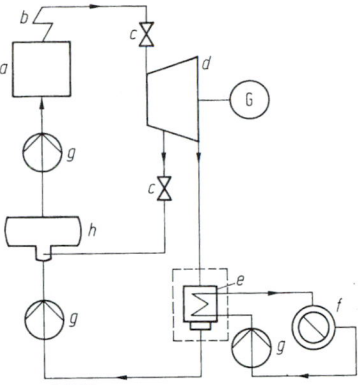

Figure 12. Flowchart of combined power and heat generation in bleed/backpressure mode: *a* steam generator; *b* superheater; *c* throttle; *d* turbine; G generator; *e* condenser (heat generator); *f* heat consumer; *g* pump; *h* storage system.

bution, usually via remote heating grid, can be covered. Although there is a work penalty by comparison with a dedicated power station – arising from steam bleed-off – nevertheless the primary energy conversion for simultaneous generation of work and heat is less than in the case of separate recovery of work in the power station and heat in a conventional heating system. **Figure 12** illustrates a simplified circuit. Depending on the type of circuit, heat indexes $\zeta = |Q|/P_{Pr}$ of up to 2.2 can be attained [9], where P_{Pr} is the component of primary energy that is derived purely for heating. Heating indexes are markedly above those of most heat pump heating systems.

9 **Ideal Gas Mixtures**

9.1 Dalton's Law. Thermal and Caloric Properties of State

A mixture of ideal gases that do not interact chemically will normally behave as an ideal gas. The following thermal equation of state is applicable:

$$pV = nRT. \tag{1}$$

Each individual gas, referred to as a *component*, is distributed over the whole of volume V as though other gases were not present. Consequently the following formula applies for each component i:

$$p_i V = n_i RT, \tag{2}$$

where p_i is the pressure exerted by each individual gas, which is designated a partial pressure. Calculating a total of all individual gases, there follows $\Sigma p_i V = \Sigma n_i RT$ or $V \Sigma p_i = RT \Sigma n_i$.

It is illustrated by comparison with Eq. (1) that

$$p = \Sigma p_i \tag{3}$$

is applicable: total pressure p of the gas mix is equal to the sum of the individual gases' partial pressures, if such

gases occupy volume V of the mix at temperature T (*Dalton's law*). The thermal state equation, Eq. (1), of an ideal gas mix, can also be set out as follows:

$$pV = mRT, \tag{4}$$

together with gas constant R of the mix:

$$R = \Sigma R_i m_i / m . \tag{5}$$

Caloric state variables of a mix at pressure p and temperature t are obtained by incrementation of the caloric state variable at the same values p, T of the individual gases in accordance with their quantitative proportions. The following formulae apply:

$$c_v = \frac{1}{m} \Sigma m_i c_{vi}, \quad c_p = \frac{1}{m} \Sigma m_i c_{pi},$$
$$u = \frac{1}{m} \Sigma m_i u_i, \quad h = \frac{1}{m} \Sigma m_i h_i . \tag{6}$$

An exception is represented by entropy, since in the mixing of individual gases in state p, T to form a mix of the same state there occurs an increase in entropy. If n_i are the molecular quantities of individual gases and n those of the mix, then the following formula applies:

$$s = \frac{1}{m} \sum m_i s_i - \sum m_i R_i \ln \frac{n_i}{n} . \qquad (7)$$

The following formulae apply: $n_i = m_i/M_i$ and $n = \Sigma n_i$, together with mass m_i and molecular weight M_i of individual gases. Mixes of real gases and liquids deviate from the above formulae, particularly at higher pressures.

9.2 Mixtures of Gas and Vapour

Mixtures of gases and readily condensing vapours frequently occur in physics and in technology. Atmospheric air consists mainly of dry air (i.e. oxygen plus nitrogen) and water vapour. Desiccation and conditioning processes are determined by application of steam-air mix laws, together with the formation of fuel-vapour-air mixes in the combustion engine.

9.3 Humid Air

Dry air consists of 78.04 mol% nitrogen, 21.00 mol% oxygen, 0.93 mol% argon and 0.03 mol% carbon dioxide. Atmospheric air can be regarded as a dual mix consisting of dry air and water which can be present in vapour form or in vapour and liquid form or in vapour and solid form. The mix is also termed *humid* air. *Dry* air is regarded as a single substance. Since the total pressure is almost always close to atmospheric pressure during changes of state, humid air, consisting of dry air and water vapour, can be regarded as a mix of ideal gases. The following formula then applies for dry air and for water vapour:

$$p_L V = m_L R_L T \quad \text{and} \quad p_W V = m_W R_W t . \qquad (8)$$

Where $p = p_L + p_W$, it is possible to obtain from the above equations the quantity of water vapour which is mixed in with 1 kg of dry air.

$$x = \frac{m_w}{m_L} = \frac{R_L p_W}{R_W (p - p_W)} . \qquad (9)$$

Variable $x = m_w/m_L$ is termed *water content*. It may be between 0 (dry air) and ∞ (pure water). Where humid air of temperature T is saturated with water vapour, then the partial pressure of water vapour becomes the same as saturation pressure $p_W = p_{WS}$ at temperature T and water content becomes

$$x_S = \frac{R_L p_{WS}}{R_W (p - p_{WS})} . \qquad (10)$$

Example Let water content x_S of saturated humid air be calculated at a temperature of 20 °C and a total pressure of 1000 mbar. The following formulae apply: $R_L = 0.2872$ kJ/kg K, $R_w = 0.4615$ kJ/kg K. Vapour pressure p_{WS} (20 °C) = 23.37 mbar is obtained from the water vapour table in **Appendix C6, Table 5**.

Accordingly, the following is obtained:

$$x_S = \frac{0.2872 \cdot 23.37}{0.4615 (1000 - 23.37)} \cdot 10^3 \frac{g}{kg} = 14.887 \text{ g/kg} .$$

Other x_S values are set out in **Appendix C9, Table 1**.

Humidity Level, Relative Humidity. Humidity level $\psi = x/x_S$ is defined as a relative index for vapour content. In meteorology, on the other hand, calculations are usually based on relative humidity $\varphi = p_w(t)/p_{ws}(t)$. Both values deviate from each other only slightly in the vicinity of saturation, because the following formulae apply:

$$\frac{x}{x_S} = \frac{p_W}{p_{WS}} \frac{(p - p_{WS})}{(p - p_W)}$$

or

$$\psi = \varphi \frac{(p - p_{WS})}{(p - p_W)} .$$

At saturation, $\psi = \varphi = 1$. If pressure is increased or if the temperature is decreased for saturated humid air, then the excess water vapour will condense. The condensed vapour will precipitate as fog or precipitation (rain); at temperatures below 0 °C, ice crystals (snow) will be formed. Relative atmospheric humidity can be calculated by direct-indication instruments (e.g. hair hygrometers) or using an aspiration psychrometer of the Assmann type.

Enthalpy of Humid Air. Since the quantity of air involved in changes of state of humid air will remain the same, and vary only in terms of the admixed water quantity arising due to condensation or evaporation, all state variables are related to 1 kg of dry air. This will then contain $x = m_w/m_L$ kg of water. The following formula applies for enthalpy h_{1+x} of the mix of 1 kg dry air and x kg vapour:

$$h_{1+x} = c_{pL}t + x (c_{pD}t + r) . \qquad (11)$$

$c_{pL} = 1.005$ kJ kg K is the isobaric specific heat capacity of air, $c_{pD} = 1.852$ kJ kg K that of water vapour and $r = 2501.6$ kJ/kg the enthalpy of evaporation of water at 0°C. In the relevant temperature range of $- 60$ to $+ 100$°C, constant values of c_p can be assumed. At saturation, we obtain $x = x_S$ and $h_{1+x} = (h_{1+x})_S$. If the proportion of water $x - x_S$ is contained in the mix in the form of fog or as a ground covering, then the following applies:

$$h_{1+x} = (h_{1+x})_S + (x - x_S) c_w t . \qquad (12)$$

If the component of water $x - x_S$ precipitates in the form of snow or ice, then the following formula applies:

$$h_{1+x} = (h_{1+x})_S + (x - x_S) (\Delta h_S - c_e t) . \qquad (13)$$

$c_w = 4.19$ kJ/kg K is the specific heat capacity of water, $c_e = 2.04$ kJ/kg K that of ice and $\Delta h_S = 333.5$ kJ/kg the melting enthalpy of ice. **Appendix C9, Table 1** sets out the saturation pressures, vapour contents and enthalpy figures for saturated humid air at temperatures between $- 20$ and $+100$°C for a total pressure of 1000 mbar.

9.3.1 Mollier Curve of Humid Air

For graphic representation of changes of state in humid air, Mollier proposed an h_{1+x}, x curve, **Fig. 1a**. Here, enthalpy h_{1+x} of $(1 + x)$ kg of humid air is plotted against water content in an oblique coordinate system. Axis $h = 0$, corresponding to humid air at 0°C, is set out obliquely to the bottom right, such that the 0°C isothermal for humid unsaturated air runs horizontally. **Figure 1b** illustrates the construction of isothermals in accordance with Eqs (11) and (12). Lines $x = $ const. are perpendicular, lines $h = $ const. are lines parallel to axis $h_{1+x} = 0$. **Figure 1a** plots boundary curve $\varphi = 1$ for total pressure 1000 mbar. It separates the field of non-saturated mixes (top) from the *fog range* (below), in which humidity is contained in the mix partly as vapour, partly as liquid (fog, precipitation) or solid (freezing fog, snow). Isothermals in the unsaturated range according to Eq. (11) are lines rising slightly to the right, which bend downwards at the boundary curve and run virtually parallel to the lines of constant enthalpy in the fog range, according to Eq. (12). For a point in the fog range of temperature t and water content x, we observe the vapour component

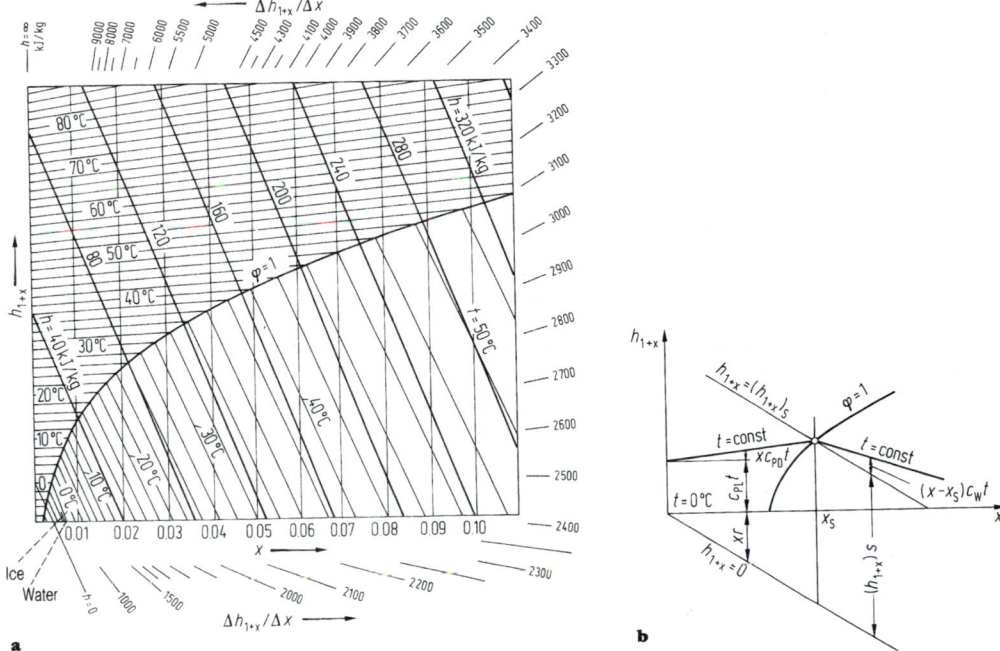

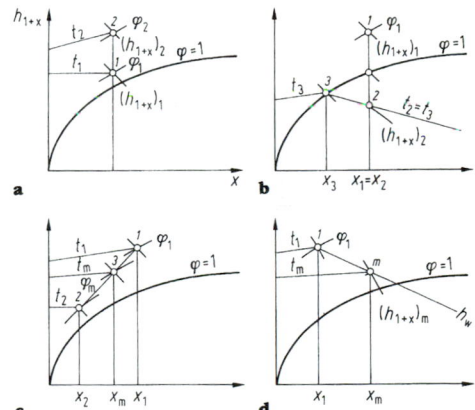

Figure 1. h_{1+x}, x curve of humid air according to Mollier.

by tracing isothermal t to its intersection with the boundary curve. Component x_s observed at the point of intersection is contained as vapour and accordingly component $x - x_s$ is contained as liquid in the mix. The oblique, radiating line segments $\Delta h_{1+x}/\Delta x$ establish, together with the zero point, the plane in which we move from a given point on a curve, if water or water vapour are added to the mix, such water having enthalpy in kJ/kg equal to the figures at the boundary line. In order to establish the direction of change of state, a line must be drawn through the state point of humid air parallel to the line which is established by the zero point ($h = 0$, $x = 0$) and the boundary line.

9.3.2 Changes of State of Humid Air

Heating or Cooling. If a given mix is heated up, there is upward movement on a perpendicular line (*1-2* in **Fig. 2a**). If it is cooled, then there is downward movement on a perpendicular line (*2-1*), whereby the perpendicular distance between two state points measured on the enthalpy scale is the exchanged heat related to 1 kg of dry air:

$$Q_{12} = m_L(c_{pL} + x_{pD}x)(t_2 - t_1), \tag{14}$$

where $c_{pL} = 1.005$ kJ/kg K and $c_{pD} = 1.852$ kJ/kg K. When humid air cools below the dew point of water (*1-2* in **Fig. 2b**) precipitation occurs. The output heat is:

$$Q_{12} = m_L[(h_{1+x})_2 - (h_{1+x})_1], \tag{15}$$

where $(h_{1+x})_1$ is stated by Eq. (11) and $(h_{1+x})_2$ is stated by Eq. (12). The following quantity of water precipitates:

$$m_w = m_L(x_1 - x_s). \tag{16}$$

Example 1000 kg of humid air at $t_1 = 30°C$, $\varphi_1 = 0.6$ and $p = 1000$ mbar are cooled to 15°C. How much condensate is produced? Water content x_1 is obtained from Eq. (9) where

Figure 2. State variations in humid air: **a** heating and cooling; **b** cooling below the dew point; **c** mix; **d** addition of water or water vapour.

$p_w = \varphi_1 p_{ws}$. In accordance with **Appendix C6, Table 1**, p_{ws} (30°C) = 42.41 mbar. Hence

$$x_1 = \frac{R_L(\varphi_1 p_{ws})}{R_w(p - \phi_1 p_{ws})} = \frac{0.2872 \cdot 0.6 \cdot 42.41}{0.4615(1000 - 0.6 \cdot 42.41)}$$

$$= 16.25 \cdot 10^{-3} \text{ kg/kg} = 16.25 \text{ g/kg}.$$

The 1000 kg of humid air consists of $1000/(1 + x_1) = 1000/1.01625$ kg $= 984.01$ kg of dry air and $1000 - 984.01 = 15.99$ kg water vapour. Water content at point 3 $x_s = x_s$ follows, from **Appendix C9, Table 1**, where $t_s = 15°C$, as $x_s = 10.78$ (41) g/kg. Correspondingly, the following values are obtained:

$$m_w = 984.01 \cdot (16.25 - 10.78) \cdot 10^{-3} \text{ kg} = 5.38 \text{ kg}.$$

Mix of Two Quantities of Air. Where two quantities of air in states *1* and *2* (**Fig. 1c**) are mixed, and where it is ensured that no heat is exchanged with the environment, then state *m* after mix is on connecting lines *1-2*. Point *m* is obtained by dividing lines *1-2* in proportion to the quantities of dry air $m_{L2} m_{L1}$. The following formula applies:

$$x_m = (m_{L1}x_1 + m_{L2}x_2) / (m_{L1} + m_{L2}) . \qquad (17)$$

Mixing of saturated quantities of air at different temperatures always causes fog accompanied by separation of water quantity $x_m - x_s$, where x_s is the level of saturation on the fog isothermal through the mixing point.

Example   1000 kg of humid air at $t_1 = 30°C$ and $\varphi_1 = 0.6$ are mixed with 1500 kg of saturated humid air of $t_2 = 10°C$ at 1000 mbar. What is the temperature level after mixing? As already calculated in the previous example, $x_c = 16.25$ g/kg. From **Appendix C9, Table 1**, we obtain the following water content for $t_2 = 10°C$: $x_{2s} = 7.7283$ g/kg. Dry air quantities are $m_{L1} = 1000/(1 + x_1)$ kg = $1000/(1 + 16.25 \cdot 10^{-3})$ kg = 984.01 kg and $m_{L2} = 1500 (1 + x_{2s})$ kg = $1500/(1 + 7.7283 \cdot 10^{-3})$ kg = 1488.57 kg. Accordingly, the following formula is obtained:

$$x_m = (984.01 \cdot 16.25 + 1488.5 \cdot 7.7283)/(984.01 + 1488.5) \text{ g/kg}$$
$$= 11.12 \text{ g/kg} .$$

Enthalpy is calculated in accordance with Eq. (11). The following formulae apply:

$$(h_{1+x})_1 = (1.005 \cdot 30 + 16.25 \cdot 10^{-3}$$
$$\cdot (1.852 \cdot 30 + 2501.6)) \text{ kJ/kg} = 71.70 \text{ kJ/kg},$$
$$(h_{1+x})_2 = (1.005 \cdot 10 + 7.7283 \cdot 10^{-3}$$
$$\cdot (1.852 \cdot 10 + 2501.6) \text{ kJ/kg} = 29.53 \text{ kJ/kg} .$$

The enthalpy of the mix is

$$(h_{1+x})_m = (m_{L1}(h_{1+x})_1 + m_{L2}(h_{1+x})_2)/(m_{L1} + m_{L2})$$
$$= (984.01 \cdot 71.70 + 1488.5 \cdot 29.53)/$$
$$(984.01 + 1488.5) \text{ kJ/kg} = 46.31 \text{ kJ/kg} .$$

On the other hand, in accordance with Eq. (11):

$$(h_{1+x})_m = [1.005t_m + 11.12 \cdot 10^{-3} (1.852t_m$$
$$+ 2501.6)] \text{ kJ/kg} .$$

From this there follows $t_m = 18°C$.

Addition of Water or Water Vapour. If air is mixed with m_w kg water or water vapour, then water content after mix amounts to

$$x_m = (m_{L1}x + m_w)/m_{L1} .$$

The enthalpy is

$$(h_{1+x})_m = (m_{L1}(h_{1+x})_1 + m_wh_w)/m_{L1} . \qquad (18)$$

In the Mollier curve for humid air (**Fig. 2d**), the final state after mixing is indicated on the line through initial state *1* of humid air, which runs parallel to the line passing through the coordinate origin with rise h_w, where $h_w = \Delta h_{1+x}/\Delta x$ is stated by the line sections of the boundary scale.

Limit Cooling Temperature. If unsaturated humid air in state t_1, x_1 extends over the surface of water or ice, then water evaporates or sublimates and is absorbed by the atmosphere, and consequently its water content increases. In the process, the temperature of the water or of the ice will decrease, eventually reaching a steady-state final value which is called the limit cooling temperature. Limit cooling temperature t_g is ascertained by means of the Mollier curve, by looking for the fog isothermal t_g whose extension passes through state point *1*.

10 Heat Transfer

Where temperature differences exist between various bodies which are not insulated one from another or within various areas of a body, then heat will flow from the *higher* temperature point to the *lower* temperature point until the temperatures of the various parts have equalised. This process is called *heat transfer*. Distinctions must be drawn between three cases of heat transfer:

Heat transfer by *conduction* in static or non-moving liquid and gaseous bodies. In this process, kinetic energy is transferred from each elementary particle to its neighbour.

Heat transfer by *convection* by liquid or gaseous bodies in motion.

Heat transfer by *radiation*, which is performed without material carriers, by electromagnetic waves.

In technology, all three types of heat transfer often act in conjunction.

10.1 Steady-State Heat Conduction

Steady-State Heat Conduction Through a Plane Wall. Where both surfaces of a plane wall of thickness δ

are maintained at different temperatures T_1 and T_2, then the following heat will flow through surface A in period t in accordance with *Fourier's law*:

$$Q = \lambda A \frac{T_1 - T_2}{\delta} t .$$

Here, λ is a physical value (SI unit: W/km) which is called *heat conductivity* (see **Appendix C10, Table 1**). $Q/t = \dot{Q}$ is termed *heat flux* (SI unit: W) and the term given to $Q/(tA) = q$ (SI unit: W/m²) is *heat flux density*. The following formulae apply:

$$\dot{Q} = \lambda A \frac{T_1 - T_2}{\delta} \quad \text{and} \quad \dot{q} = \lambda \frac{T_1 - T_2}{\delta} . \qquad (1)$$

By analogy with an electrical cable, a current I flows only if a voltage U is connected, in order to overcome resistance R ($I = U/R$), and in this context a heat flux $\dot{Q}$ will flow only if a temperature difference $\Delta T = T_1 - T_2$ is present:

$$\dot{Q} = \frac{\lambda A}{\delta} \Delta T .$$

By analogy with Ohm's law, the term $R_w = \delta/(\lambda A)$ is a *thermal resistance* (SI unit: K/W).

Fourier's Law. If instead of considering a wall of infinite thickness δ, we consider a disc of thickness dx cut out from it perpendicular to heat flux, then we obtain Fourier's law in the following form:

$$\dot{Q} = -\lambda A \frac{\mathrm{d}T}{\mathrm{d}x} \quad \text{and} \quad \dot{q} = -\lambda \frac{\mathrm{d}T}{\mathrm{d}x}, \tag{2}$$

where the minus sign denotes that heat if flowing in the direction of decreasing temperature. Here, $\dot{Q}$ is heat flux in the direction of the x-axis, and a corresponding statement applies for $\dot{q}$. Heat flux in the direction of three coordinates x, y, z is a vector:

$$\dot{\boldsymbol{q}} = -\lambda \left(\frac{\partial T}{\partial x} \boldsymbol{e}_x + \frac{\partial T}{\partial y} \boldsymbol{e}_y + \frac{\partial T}{\partial z} \boldsymbol{e}_z \right) \tag{3}$$

with unit vectors $\boldsymbol{e}_x$, $\boldsymbol{e}_y$, $\boldsymbol{e}_z$. Equation (3) is also the general form of Fourier's law. In this form, it applies for isotropic bodies, i.e. bodies whose heat conductivity is equally great in the direction of the three coordinate axes.

Steady-State Heat Conduction Through a Pipe Wall. In accordance with Fourier's law, a heat flux $\dot{Q} = -\lambda 2\pi r l (\mathrm{d}T/\mathrm{d}r)$ is transferred through a cylindrical surface of radius r and length l. In the case of steady-state heat conduction, heat flux is the same for all radii, $\dot{Q} = $ const., such that variables T and r can be separated and integrated with temperature T from the inner surface where $r = r_i$ of the cylinder at temperature T_i to a given point r of temperature T. The temperature curve obtained in a tubular shell of thickness $r - r_i$ is

$$T_i - T = \frac{\dot{Q}}{\lambda 2\pi l} \ln \frac{r}{r_i}.$$

At temperature T_a of the external surface of radius r_a we obtain heat flux in a pipe of thickness $r_a - r_i$ and of length l:

$$\dot{Q} = \lambda 2\pi l \frac{T_i - T_a}{\ln r_a / r_i}. \tag{4}$$

In order to obtain formal consistency with Eq. (1), we can also set out the following:

$$\dot{Q} = \lambda A_m \frac{T_i - T_a}{\delta}, \tag{5}$$

where $\delta = r_a - r_i$ and

$$A_m = \frac{A_a - A_i}{\ln A_a / A_i},$$

if $A_a = 2\pi r_a l$ is the external surface of the pipe and $A_i = 2 r_i l$ its internal surface. A_m is the logarithmic average between the external and the internal surfaces of the pipe.

The "heat conductivity resistance" of the pipe $R_W = \delta/(\lambda A_m)$ (SI unit: K/W) must be overcome by a temperature difference in order for a heat flux to occur.

10.2 Heat Transfer and Heat Transmission

Where heat is transferred to a panel by a fluid and the heat propagates and is transferred to a second fluid on the other side of the wall, the process is termed *heat transfer*. In this context, two *heat transfers* and a *heat conduction* process take place consecutively. Temperature decreases sharply in a layer immediately adjacent to the wall (**Fig. 1**), while temperatures at a small distance from the wall differ only slightly. As a simplification it can be assumed that a thin static layer of fluid of film thickness

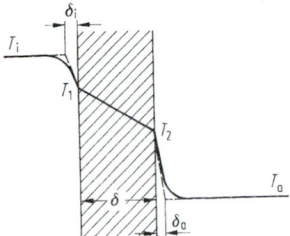

Figure 1. Heat transfer through a plane wall.

δ_i or δ_a adheres to the wall while the fluid also cancels out temperature differences. In the thin fluid film, heat is transferred by conduction, and, from the Fourier equation, we obtain, for the heat flux transferred to the left-hand side of the wall:

$$\dot{Q} = \lambda A \frac{T_i - T_1}{\delta_i}.$$

where λ is the heat conductivity of the fluid. Film thickness δ_i is a function of several variables, such as the velocity of the fluid along the wall, the form of the wall and its surface characteristics. It has been found constructive to use the ratio $\lambda/\delta_i = \alpha$ instead of film thickness δ_i. We arrive at Newton's principle for heat transfer:

$$\dot{Q} = \alpha A (T_f - T_0), \tag{6}$$

where, in general, $T_f = $ fluid temperature and $T_0 = $ surface temperature. Variable α is termed the *heat transfer coefficient* (SI unit: W/m^2 K). **Table 1** sets out the orders of magnitude of heat transfer coefficients.

In conjunction with Ohm's law $I = (1/R) U$, we term $1/\alpha A = R_W$ the *thermal resistance* (SI unit: K/W). This must be overcome by temperature difference $\Delta T = T_f - T_0$ in order to enable flow of heat flux $\dot{Q}$.

Figure 1 sets out the three consecutive individual resistances which the heat flux has to overcome. These are added together to form the total resistance.

Heat Transfer Through Plane Walls. The heat flux passing through a plane wall (**Fig. 1**) is

$$\dot{Q} = kA (T_i - T_a), \tag{7}$$

together with total heat resistance $1/(kA)$, which is additively made up of the following individual resistances:

$$\frac{1}{kA} = \frac{1}{\alpha_i A} + \frac{\delta}{\lambda A} + \frac{1}{\alpha_a A}. \tag{8}$$

Table 1. Heat transfer coefficients α in W/m^2 K

	α
Free convection in:	
Gases	3 to 20
Water	100 to 600
Boiling water	1 000 to 20 000
Forced convection in:	
Gases	10 to 100
Liquids	50 to 500
Water	500 to 10 000
Condensing steam	1 000 to 100 000

The variable k defined by Eq. (7) is called the *heat transfer coefficient* (SI unit: $W/m^2 K$). If the wall consists of several homogeneous layers (**Fig. 2**) of thickness δ_1, δ_2, ... and thermal conductivities λ_1, λ_2, ..., then Eq. (7) will also apply but the total thermal resistance will in this case be:

$$\frac{1}{kA} = \frac{1}{\alpha_i A} + \sum \frac{\delta_j}{\lambda_j} A + \frac{1}{\alpha_a A}. \tag{9}$$

Example The wall of a refrigerator consists of an internal concrete layer which is 5 cm thick ($\lambda = 1$ W/K m), a cork block insulation layer which is 10 cm thick ($\lambda = 0.04$ W/K m) and an external brick wall which is 50 cm thick ($\lambda = 0.75$ W/K m). The heat transfer coefficient on the inside is $\alpha_i = 7$ W/m^2 K, and the coefficient on the outside is $\alpha_a = 20$ W/m^2 K. How much heat flows through 1 m^2 of wall at an internal temperature of -5 °C and an external temperature of 25 °C? From Eq. (9), thermal transmission resistance is as follows:

$$\frac{1}{kA} = \left(\frac{1}{7 \cdot 1} + \frac{0.05}{1 \cdot 1} + \frac{0.1}{0.04 \cdot 1} + \frac{0.5}{0.75 \cdot 1} + \frac{1}{20 \cdot 1}\right) \frac{m^2\,K}{W}$$

$$= 3.41\ m^2\ K/W\ .$$

Heat flux is

$$\dot{Q} = \frac{1}{3.41}(-5-25)W, \quad |\dot{Q}| = 8.8\ W\ .$$

Heat Transmission Through Pipes. Equation (7) also applies for heat transmission through a pipe. Heat resistance is additively composed of individual resistances $1/kA = 1/\alpha_i A_i + \delta/\lambda A_m + 1/\alpha_a A_a$.

It is common to relate heat transmission coefficient k to the external surface of the pipe which is usually easy to ascertain: $A = A_a$, such that the total heat resistance is conveyed by:

$$\frac{1}{kA_a} = \frac{1}{\alpha_i A_i} + \frac{\delta}{\lambda A_m} + \frac{1}{\alpha_a A_a} \tag{10}$$

where $A_m = (A_a - A_i)/\ln(A_a/A_i)$.

If the pipe consists of several homogeneous individual pipes of thickness δ_1, δ_2, ... and heat conductivities λ_1, λ_2, then Eq. (7) is again applicable but now the total resistance to heat is:

$$\frac{1}{kA_a} = \frac{1}{\alpha_i A_i} + \sum \frac{\delta_j}{\lambda_j A_{mj}} + \frac{1}{\alpha_a A_a}, \tag{11}$$

where the sum of all individual pipes has to be added and A_{mj} is the mean logarithmic surface area of an individual pipe,

$$A_{mj} = (A_{aj} - A_{ij})/\ln A_{aj}/A_{ij}.$$

10.3 Instationary Heat Transmission

In the case of instationary heat transmission, temperatures vary with time. In a plane wall with predetermined surface temperatures, the temperature curve is no longer linear, because heat flowing into a disc differs from heat flowing out. The difference between an incoming and an outgoing heat flux is latent as internal energy in the disc and will increase or decrease its temperature as a function of time. For plane walls with a heat flux in the direction of the x-axis, Fourier's heat transmission equation applies:

$$\frac{\partial T}{\partial t} = a \frac{\partial^2 T}{\partial x^2}. \tag{12}$$

In multi-dimensional heat transmission, the following applies:

$$\frac{\partial T}{\partial t} = a \left(\frac{\partial^2 T}{\partial x^2} + \frac{\partial^2 T}{\partial y^2} + \frac{\partial^2 T}{\partial z^2}\right). \tag{13}$$

In this form, both equations assume the prerequisite of constant heat conductivity λ. Variable $a = \lambda/(\rho c)$ is *temperature conductivity* (SI unit: m^2/s; numerical values in **Appendix C10, Table 2**).

To solve Fourier's heat conductivity equation, it is logical, as in the case of other heat transmission problems, to use non-dimensional quantities so as to reduce the number of variables. In order to illustrate the fundamental principle, we refer to Eq. (12). Stipulating $\Theta = (T - T_c)/(T_0 - T_c)$, where T_c is a characteristic constant temperature and T_0 is a datum temperature, then T_c, for example, can signify, in the case of cooling of a plate from an initially constant temperature T_0 in a cold environment: ambient temperature $T_c = T_u$. All lengths are related to a characteristic length X, i.e. half the plate thickness. It is, furthermore, logical to introduce a dimensionless period by $Fo = at/X^2$, which is called the *Fourier index*.

Solutions to the heat conductivity equation are then of the following form:

$$\Theta = f(x/X, Fo).$$

In many problems, the heat arriving at the surface of a body by transmission is removed by convection into the ambient fluid of temperature T_u. In this case the energy budget at the surface (index w = wall) is as follows:

$$-\lambda \left(\frac{\partial T}{\partial x}\right)_w = \alpha(T_w - T_u) \quad \text{or} \quad \frac{1}{\Theta_w}\left(\frac{\partial \Theta}{\partial \zeta}\right)_w = -\frac{\alpha X}{\lambda},$$

where $\Theta = (T - T_u)/(T_0 - T_u)$ and $\Theta_w = (T_w - T_u)/(T_0 - T_u)$. The solution is also a function of dimensionless variable $\alpha X/\lambda$: we refer to $\alpha X/\lambda$ as the *Biot index, Bi*, in which λ is the heat conductivity of the body and α is the heat transmission coefficient to the ambient fluid. Solutions are of the following form:

$$\Theta = f(x/X, Fo, Bi). \tag{14}$$

10.3.1 The Semi-infinite Body

Temperature changes normally occur in a *boundary zone*, which is thin in relation to the size of the body. Such a body is referred to as *semi-infinite*. If a semi-infinite plane wall (**Fig. 3**) of constant initial temperature T_0 is considered, let wall temperature at time $t = 0$ reduce to T_0 and subsequently remain constant. For various periods, t_1, t_2, ... we obtain temperature profiles. These are stated as follows:

$$\frac{T - T_u}{T_0 - T_u} = f\left(\frac{x}{2\sqrt{at}}\right), \tag{15}$$

with the Gaussian error function $f(x/(2\sqrt{at}))$ (**Fig. 4**). The transferred heat flux density is obtained by differentiation of $\dot{q} = -\lambda(\partial T/\partial x)_{x=0}$ to produce

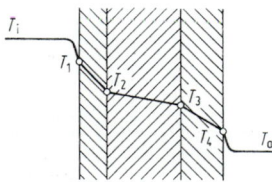

Figure 2. Heat transmission through a plane, multi-layer wall.

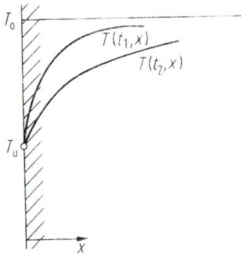

Figure 3. Semi-infinite body.

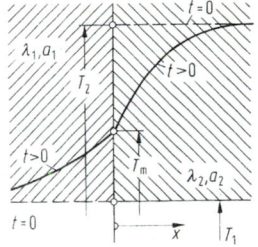

Figure 5. Contact temperature T_m between two semi-infinite bodies.

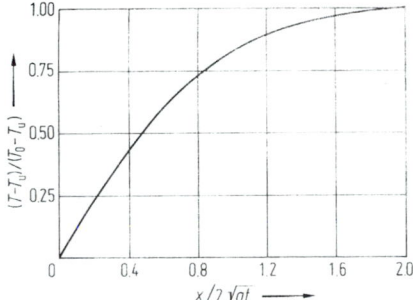

Figure 4. Temperature curve in a semi-infinite body.

$$q = -\frac{b}{\sqrt{\pi t}}(T_u - T_0), \qquad (16)$$

with *heat penetration coefficient* $b = \sqrt{\lambda \rho c}$ (SI unit: W s$^{1/2}$/m^2 K) (**Table 2**), which is an index for the magnitude of heat flux to have penetrated the body over a given period, if surface temperature is suddenly increased by a certain value $T_u - T_0$ relative to initial temperature T_0.

Example On a sudden change in the weather, temperature at the earth's surface drops from $+5$ to -5 °C. What level will temperature decrease to at 1 m depth after 20 days? Temperature conductivity of the earth's surface is $a = 6.94 \cdot 10^{-7}$ m^2/s. In accordance with Eq. (15):

$$\frac{T-(-5)}{5-(-5)} = f\left(\frac{1}{2(6.94 \cdot 10^{-7} \cdot 20 \cdot 24 \cdot 3600)^{1/2}}\right) = f(0.456).$$

In **Fig. 4** we see $f(0.456) = 0.48$. Accordingly $T = -0.2$ °C.

Finite Heat Transfer at the Surface. If heat is transferred by convection to the environment, such that it is the case that at the surface of the body according to **Fig. 3**: $\dot{q} = -\lambda\,(\partial T/\partial x) = \alpha\,(T_w - T_u)$, where T_u is the ambient temperature and $T_w = T(x=0)$ is the chronologically variable wall temperature, then Eq. (15) no longer applies, but the following formula is applied:

$$q = \frac{b}{\sqrt{\pi t}}(T_u - T_0)\,\Phi\,(z), \qquad (17)$$

Table 2. Heat penetration coefficient $b = \sqrt{\lambda \rho c}$ in W s$^{1/2}$/m^2 K

Copper	36 000	Sand	1 200
Iron	15 000	Wood	400
Concrete	1 600	Foams	40
Water	1 400	Gases	6

where

$$\Phi(z) = 1 - \frac{1}{2z^2} + \frac{1 \cdot 3}{2^2 z^4} - \ldots + (-1)^{n-1}\frac{1 \cdot 3 \ldots (2n-3)}{2^{n-1} z^{2n-2}}$$

and where $z = \alpha\sqrt{at/\lambda}$.

10.3.2 Two Semi-infinite Bodies in Thermal Contact

Two semi-infinite bodies of differing but initially constant temperatures T_1 and T_2 with thermal characteristics λ_1, a_1 and λ_2, a_2, are suddenly brought into contact at time $t = 0$ (see **Fig. 5**). After a very short period, there arises a temperature T_m at both sides of the contact surface, which remains constant. The following formula is applicable:

$$\frac{T_m - T_1}{T_2 - T_1} \frac{b_2}{b_1 + b_2}.$$

Contact temperature T_m will approximate more closely to the temperature of the body having the greater heat penetration coefficient b. It is possible to ascertain one of the values b by measurement of T_m, given knowledge of the other.

10.3.3 Temperature Equalisation in Simple Bodies

A simple body – plate, cylinder or sphere – is, at time $t = 0$, at a consistent temperature T_0 and is subsequently cooled or heated at $t > 0$ by transfer of heat to a fluid surrounding it at temperature T_u according to boundary condition $-\lambda\,(\partial T/\partial n)_w = \alpha(T_w - T_u)$ (let n be the coordinate normal to the surface of the body).

Plane Sheet. The conditions in **Fig. 6** apply, where a temperature profile is also set out.

The temperature profile is described by an infinite

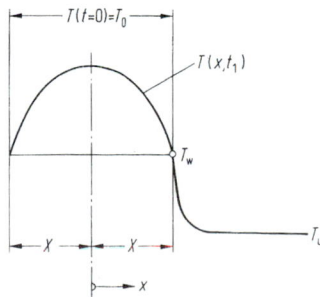

Figure 6. Cooling of a plane sheet.

Table 3. Constants C and α in Eq. (18)

Bi	∞	10	5	2	1	0.5	0.2	0.1	0.01
C	1.2732	1.2620	1.2402	1.1784	1.1191	1.0701	1.0311	1.0161	1.0017
δ	1.5708	1.4289	1.3138	1.0769	0.8603	0.6533	0.4328	0.3111	0.0998

Table 4. Constants C and α in Eq. (19)

Bi	∞	10	5	2	1	0.5	0.2	0.1	0.01
C	1.6020	1.5678	1.5029	1.3386	1.2068	1.1141	1.0482	1.0245	1.0025
δ	2.4048	2.1795	1.9898	1.5994	1.2558	0.9408	0.6170	0.4417	0.1412

series, but can be approximated, where $at/X^2 \geq 0.24$ ($a = \lambda/(\rho c)$ is temperature conductivity) with an error in temperature of $< 1\%$, by

$$\frac{T - T_u}{T_0 - T_u} = C \exp\left(-\delta^2 \frac{at}{X^2}\right) \cos\left(\delta \frac{x}{X}\right). \tag{18}$$

As in **Table 3**, constants C and δ are a function of Biot index $Bi = \alpha X/\lambda$.

Wall temperature T_w is obtained from Eq. (18) by substituting $x = X$. Heat flux follows from $\dot{Q} = -\lambda A(\partial T/\partial x)_{x=X}$.

Cylinder. Instead of the local coordinate x in **Fig. 6**, we employ radial coordinate r. Cylinder radius is R. Temperature profile is once again described by an infinite series, which can be approximated with less than 1% error where $at/R^2 \geq 0.21$ by the following:

$$\frac{T - T_u}{T_0 - T_u} = C \exp\left(-\delta \frac{at}{R^2}\right) I_0\left(\delta \frac{r}{R}\right). \tag{19}$$

I_0 is a 0th-order Bessel function whose values are found in tabular reference works, e.g. [10]. In accordance with **Table 4**, constants C and α are a function of the Biot index.

Wall temperature can be obtained from Eq. (19), by substituting $r = R$, and heat flux from $\dot{Q} = -\lambda A(\partial T/\partial r)_{r=R}$. Accordingly, the derivation of the Bessel function is as follows: $I_0' = -I_1$. The first-order Bessel function I_1 is also set out in tables [10].

Sphere. The cooling or heating of a sphere of radius R is also defined by an infinite series. It can be approximated with less than 2% error, where $at/R^2 \geq 0.18$, by the following formula:

$$\frac{T - T_u}{T_0 - T_u} = C \exp\left(-\delta \frac{at}{R^2}\right) \frac{\sin(\delta r/R)}{\delta r/R}. \tag{20}$$

Constants C and δ are a function of the Biot index, in accordance with **Table 5**.

10.4 Heat Transfer by Convection

With heat transfer involving fluids in motion, energy is mainly transported by convection, via (molecular) heat conduction. Each volumetric element of the fluid has internal energy which is transmitted by the flow, and in this case of heat transfer it is transferred to a solid body by convection in the form of heat.

Non-dimensional Characteristics. Similarity mechanics (see A7) illustrates the processes of convective transfer. The number of influencing variables can be greatly reduced, and heat transfer laws can be generally and consistently formulated for geometrically similar bodies and for the widest range of substances. The following non-dimensional coefficients are relevant:

Nusselt number	$Nu = \alpha l/\lambda$
Reynolds number	$Re = wl/v$
Prandtl number	$Pr = v/a$
Péclet number	$Pe = wl/a = Re\, Pr$
Grashof number	$Gr = l^3 g\beta\, \Delta T/v^2$
Stanton number	$St = \alpha/(\rho w c_p) = Nu/(Re\, Pr)$
Geometrical characteristics	l_n/l; $n = 1, 2, \ldots$

λ = the thermal conductivity of fluid, L = a characteristic dimension of the flow space $l_1, l_2, \ldots$, v = the kinematic viscosity of the fluid; ρ = the fluid density, $a = \lambda/(\rho c_p)$ = the thermal conductivity, c_p = the specific heat capacity of the fluid at a constant pressure, g = the gravitational acceleration, $\Delta T = T_w - T_f$ = the difference between the wall temperatures; T_f = the mean temperature of the fluid flowing along the wall, β = the coefficient of thermal expansion for wall temperature, where $\beta = 1/T_w$ for ideal gases. The Prandtl number is a physical value (see **Appendix C10, Table 2**). Heat transfer in the case of forced convection is described by equations of the following form:

Table 5. Constants C and α in Eq. (20)

Bi	∞	10	5	2	1	0.5	0.2	0.1	0.01
C	2.0000	1.9249	1.7870	1.4793	1.2732	1.1441	1.0592	1.0298	1.0030
δ	3.1416	2.8363	2.5704	2.0288	1.5708	1.1656	0.7593	0.5423	0.1730

$$Nu = f_1 \, (Re, \, Pr, \, l_n/l) \tag{21}$$

and in the case of free convection by

$$Nu = f_2 \, (Gr, \, Pr, \, l_n/l) \,. \tag{22}$$

The target heat transfer coefficient is obtained from the Nusselt number as $\alpha = Nu\lambda/l$. Functions f_1 and f_2 cannot often be measured theoretically, but must usually be measured empirically. They are dependent on the form of the heating and cooling surfaces (plane or curved, smooth, rough or ribbed), flow direction (parallel or normal to the surface) and also, albeit to a small extent, on the direction of the heat flux (heating or cooling of the flowing fluid).

10.4.1 Heat Transfer Without Change of Phase

Flat Plate in Laminar Flow with Longitudinal Inflow. The following formula (after Pohlhausen) applies for the mean Nusselt number of a plate of length l:

$$Nu = 0.664 \, Re^{1/2} Pr^{1/3} \,, \tag{23}$$

where $Nu = \alpha l/\lambda$, $Re = wl/\nu < 10^5$ and $0.6 \le Pr \le 2000$. Physical values are substituted at mean fluid temperature $T_m = (T_w + T_\infty)/2$. T_w is the wall temperature, as T_∞ is the temperature at a point remote from the wall.

Plane Sheet in Turbulent Flow with Longitudinal Inflow. The boundary layer becomes turbulent above approx $Re = 5 \cdot 10^5$. The mean Nusselt number of a plate of length l is

$$Nu = \frac{0.037 \, Re^{0.8} \, Pr}{1 + 2.443 \, Re^{-0.1} \, (Pr^{2/3} - 1)}, \tag{24}$$

where $Nu = \alpha l/\lambda$, $Re = wl/\nu$, $5 \cdot 10^5 < Re < 10^7$ and $0.6 \le Pr \le 2000$. Physical values must be configured at a mean fluid temperature $T_m = (T_w + T_\infty)/2$. T_w is the wall temperature, and, again, T_∞ is the temperature remote from the wall.

Heat Transfer in the Context of Flow Through Pipes (General). Below a Reynolds number of $Re = 2300$ ($Re = wd/\nu$, where w = mean section velocity; d = pipe diameter), flow is always laminar, and above $Re = 10^4$ it is turbulent. In the range of $2300 < Re < 10^4$, laminarity or turbulence of flow is dictated by the type of inflow and the form of the pipe inlet. Mean heat transfer coefficient α over pipe length l is defined by $q = \alpha \, \Delta\vartheta$, with mean logarithmic temperature difference:

$$\Delta\vartheta = \frac{(T_w - T_E) - (T_w - T_A)}{\ln \dfrac{T_w - T_E}{T_w - T_A}}, \tag{25}$$

where T_w = the wall temperature, T_E = the inlet temperature and T_A = the outlet temperature.

Heat Transfer in the Context of Laminar Flow Through Pipes. Flow is termed *hydrodynamic* if the velocity profile does not vary with the flow path. In laminar flow of a high-viscosity fluid, a Poiseuille parabola will be set up as a velocity profile even after a short flow distance. The mean Nusselt number at constant wall temperature can be calculated precisely by an infinite series (Graetz solution), but it converges unsatisfactorily. Stephan provides the following approximate solution for hydrodynamically configured laminar flow:

$$Nu_0 = \frac{3.657}{\tanh (2.264X^{1/3} + 1.7X^{2/3})} + \frac{0.0499}{X} \tanh X \,, \tag{26}$$

where $Nu_0 = \alpha_0 d/\lambda$, $X = L/(dRePr)$, $Re = wd/\nu$, $Pr = \nu/a$. The equation is applicable to laminar flow $Re \le 2300$

throughout the range of $0 \le X \le \infty$, and the greatest deviation from precise values of the Nusselt number is 1%. Physical values are to be substituted at mean fluid temperature $T_m = (T_w + T_B)/2$ where $T_B = (T_E + T_A)/2$.

If a fluid enters a pipe at an approximately constant velocity, then the velocity profile varies with the flow path until it transfers into the Poiseuille parabola after a flow distance of $1/(dRe) = 5.7 \cdot 10^{-2}$. For this case of a *laminar flow without hydrodynamic configuration*, Stephan provides the following formula in the range of $0.1 \le Pr \le \infty$:

$$\frac{Nu}{Nu_0} = \frac{1}{\tanh (2.43 \, Pr^{1/6} X^{1/6}} \,, \tag{27}$$

where $Nu = \alpha d/\lambda$, and the other factors are as defined above. Where $1 \le Pr \le \infty$, there is less than 5% error, but it may amount to 10% where $0.1 \le Pr < 1$. Physical values are substituted for mean fluid temperature $T_m = (T_w + T_B)/2$, where $T_B = (T_E + T_A)/2$.

Heat Transfer in the Context of Turbulent Flow Through Pipes. For a hydrodynamically configured flow $1/d \ge 60$, the *Dittus-Boelter-Kraussold equation* applies in the range of $10^4 \le Re \le 10^5$ and $0.5 < Pr < 100$:

$$Nu = 0.02 \, Re^{0.8} \, Pr^{1/3} \,. \tag{28}$$

Physical values are substituted for mean temperature $T_m = (T_w + T_B)2$ where $T_B = (T_E + T_A)/2$. For the non-hydrodynamically configured and for the configured flow in the range of $10^4 \le Re \le 10^7$ and $0.5 \le Pr \le 1000$, *Petukhov's equation*, as modified by Gnielinski, applies:

$$Nu = \frac{(Re - 1000)Pr\zeta/8}{1 + 12.7 \, \sqrt{\zeta/8}(Pr^{2/3} - 1)} \left[1 + \left(\frac{d}{l}\right)^{2/3} \right] \tag{29}$$

with drag coefficient $\zeta = (0.79 \ln Re - 1.64)^{-2}$. $Nu = \alpha d/\lambda$, $Re = wd/\nu$. The physical values are to be configured for mean temperature $T_m = (T_w + T_B)/2$ where $T_B = (T_E + T_S)/2$.

At *pipe bends*, heat transfer coefficients are greater, under otherwise equal conditions, than in straight pipes of the same aperture. In accordance with Hausen's formula for turbulent flow, the following formula applies for a pipe bend of curvature diameter D:

$$\alpha = \alpha_{\text{even}} [1 + (21/Re^{0.14})(d/D)] \,. \tag{30}$$

Heat Transfer at an Individual Pipe with Transverse Inflow. For an individual pipe with transverse inflow, the following mean heat-transfer coefficient is obtained from *Gnielinski's equation*:

$$Nu = 0.3 + (Nu_l^2 + Nu_t^2)^{1/2} \tag{31}$$

with Nusselt number Nu_l for laminar plane flow in accordance with Eq. (23) and Nu_t for turbulent plane flow in accordance with Eq. (24). $Nu = \alpha l/\lambda$, $1 < Re = wl/\nu < 10^7$ and $0.6 < Pr < 1000$. The wetted length $l = d\pi/2$ is substituted as length l. Physical values are configured for mean temperature $T_m = (T_w + T_\infty)/2$, T_w is wall temperature, T_∞ is temperature remote from the wall. The equation is applicable to technical applications where 6 to 10% inflow turbulence level is anticipated.

Heat Transfer to an Array of Pipes with Transverse Inflow. For an individual array of pipes with transverse inflow (**Fig. 7**), Eq. (31) again applies. However, the Reynolds number is to be configured with velocity w_c in the narrowest section. Now, $Re = w_c d/\nu$ where $w_c = w/\varphi$, where w is inflow velocity and $\varphi = 1 - \pi/4a$ is the packing factor.

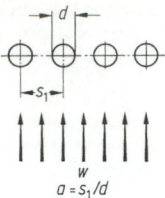

Figure 7. Array of pipes with transverse inflow.

Heat Transfer to a Pipe Bundle. In a straight square layout, the centrelines of all pipes are in line in the direction of flow, and in an offset arrangement, the centrelines of a row of pipes are displaced in relation to the previous row, see **Fig. 8**. Heat transfer is also dependent on the transverse and longitudinal spacing of pipes $a = s_1/d$ and $b = s_2/d$. In order to calculate heat transfer coefficients, the Nusselt number is initially calculated on a transverse inflow individual pipe in accordance with Eq. (31), in which the Reynolds number with velocity in the narrowest cross-section to be configured is $Re = w_c l/v$ where $w_c = 2/\varphi$, and where w is inflow velocity and φ is the packing factor $\varphi = 1 - \pi/(4a)$, where $b > 1$ and $\varphi = 1 - \pi/(4ab)$ and where $b < 1$. The Nusselt number Nu must be multiplied by a layout factor f_A. Hence, the Nusselt number $Nu_B = \alpha_B l/\lambda$ (where $l = d\pi/2$) of the bundle is then obtained:

$$Nu_B = f_A Nu . \tag{32}$$

In a straight square layout,

$$f_A = 1 + 0.7(b/a - 0.3)/(\psi^{1/2}(b/a + 0.7)^2) \tag{33}$$

and in a staggered layout,

$$f_A = 1 + 2/(3b) . \tag{34}$$

Heat flux density is $\dot{q} = \alpha \Delta \vartheta$ with $\Delta \vartheta$ according to Eq. (25). Equations (33) and (34) apply to pipe bundles with 10 or more rows of pipes. In exchangers with fewer pipe arrays, the heat transfer coefficient also has to be multiplied by a factor $[1 + (n - 1)]/n$, where n signifies the number of rows of pipes.

Free Convection. Free convection occurs because of density differences, which are generally caused by temperature differences, and less frequently by pressure differences. The heat transfer coefficient on a perpendicular wall is calculated from the *Churchill and Chu equation* as:

$$Nu^{1/2} = 0.825 + 0.387 \, Ra^{1/6} / (1 + (0.492/Pr)^{9/16})^{8/27} , \tag{35}$$

in which the mean Nusselt number $Nu = \alpha l/\lambda$ is con-

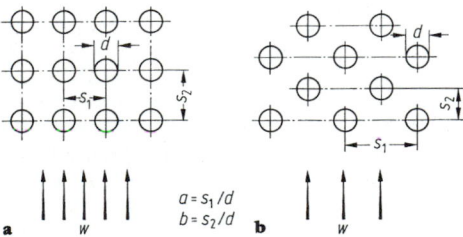

Figure 8. Arrangement of pipes in pipe bundles: **a** straight square pipe layout; **b** offset pipe layout.

figured by planar height l and the Rayleigh number is defined by $Ra = Gr \, Pr$ with the Grashof number

$$Gr = \frac{gl^3}{v^2} \frac{\rho_\infty - \rho_w}{\rho_w}$$

and the Prandtl number $Pr = v/a$. (g = gravitational acceleration, l = planar height, v = kinematic viscosity, ρ_∞ = density of fluid outside the temperature boundary layer, ρ_w = density of the fluid at the wall, a = temperature conductivity.)

If free convection is caused only by temperature differences, then the Grashof number can be set out as follows:

$$Gr = \frac{gl^3}{v^2} \beta (T_w - T_\infty) ,$$

β is the coefficient of thermal expansion. For ideal gases, $\beta = 1/T_w$.

Equation (35) applies in the range $0 < Pr < \infty$ and $0 < Ra < 10^{12}$. Physical values must be configured for mean temperature $T_m = (T_w + T_\infty)/2$. A similar equation is provided by *Churchill and Chu* for free convection about a *horizontal cylinder*:

$$Nu^{1/2} = 0.60 + 0.387 \, Ra^{1/6} / 1 + (0.559/Pr)^{9/16})^{8/27} . \tag{36}$$

Definitions according to Eq. (35) apply, characteristic length is $l = d\pi/2$ and the range of validity is $0 < Pr < \infty$ and $10^5 < Ra < 10^{12}$. For *horizontal rectangular panels*, the following formulae apply:

$$Nu = 0.70 Ra^{1/4} \quad \text{where} \quad Ra < 4 \cdot 10^7 \tag{37}$$

and

$$Nu = 0.155 Ra^{1/3} \quad \text{where} \quad Ra \geq 4 \cdot 10^7 , \tag{38}$$

where $Nu = \alpha l/\lambda$, if l is the shortest rectangle side.

10.4.2 Heat Transfer in Condensation and in Boiling

If the temperature of a wall surface is less than the saturation temperature of adjacent vapour, then vapour will precipitate on the wall surface. Depending on the wetting characteristics, condensate can form either as drops or as a closed film of liquid. In the case of *dropwise condensation*, heat-transfer coefficients are usually greater than for *filmwise condensation*. However, it can only be sustained under special precautions such as the use of anti-wetting agents and therefore seldom occurs.

Filmwise Condensation. Where condensate flows as a laminar film over a *perpendicular wall* of height l, then mean heat transfer coefficient α is provided by

$$\alpha = 0.943 \left(\frac{\rho g r \lambda^3}{v(T_S - T_w)} \frac{1}{l} \right)^{1/4} . \tag{39}$$

The following formula applies for condensation on *horizontal individual pipes* of external diameter d:

$$\alpha = 0.728 \left(\frac{\rho g r \lambda^3}{v(T_S - T_w)} \frac{1}{d} \right)^{1/4} , \tag{40}$$

where ρ = the density of the liquid, g = gravitational acceleration, r = the enthalpy of evaporation, λ = the heat conductivity of the liquid, v = the kinematic viscosity of liquid, T_S = the saturation temperature and T_w = the wall temperature.

A prerequisite for the equations is that the vapour should exert no noticeable transverse load on the condensate film.

At Reynolds numbers $Re_\delta = \bar{w}\delta/v$ ($\bar{w}$ = the mean velocity of the condensate, δ = the film thickness, ν = the kinematic viscosity) of between 75 and 1200, the transition to turbulent flow in the condensate film occurs gradually. In the transition range, the following formula applies:

$$\alpha = 0.22\lambda/(\nu^2/g)^{1/3} , \qquad (41)$$

while in the case of turbulent film flow $Re_\delta > 1200$, the following equation applies in accordance with *Grigull*:

$$\alpha = 0.003 \left(\frac{\lambda^3 g\,(T_s - T_w)}{\rho\nu^3 r}l\right)^{1/2} . \qquad (42)$$

Equations (41) and (42) also apply to vertical pipes and panels, but not to horizontal pipes.

Evaporation. If a liquid is heated in a container, then evaporation occurs after boiling temperature T_s is exceeded. At low superheating temperatures of $T_w - T_s$ on the wall, the liquid evaporates only at its free surface (*non-effervescent boiling*). In this context, heat is transported by convection from the heating surface to the surface of the liquid. At higher superheating temperatures $T_w - T_s$, steam bubbles are formed at the heating surface (*effervescent boiling*) and rise. They increase the motion of the liquid and hence heat transfer. As the superheating temperature increases, the bubbles increasingly converge to form a film of steam, which causes heat transfer to reduce again (*transitional boiling*), and at adequately high superheating temperatures it increases once again (*film boiling*). **Figure 9** illustrates the various heat transfer ranges. The heat transfer coefficient α is defined by

$$\alpha = \dot{q}/(T_w - T_s) ,$$

where $\dot{q}$ = the heat flux density in W/m².

Technical Evaporators operate in the range of non-effervescent boiling or – more frequently – in the range of effervescent boiling. In the range of non-effervescent boiling, the laws for heat transfer in free convection, Eqs (35) and (36) apply. In the range of effervescent boiling, the following formulae apply:

$$\alpha = \dot{c}q^n F(p) \quad \text{where } 0.5 < n < 0.8.$$

For water, the following formula (Fritz) applies to boiling pressures between 0.5 and 20 bar:

$$\alpha = 1.95\dot{q}^{0.72}p^{0.24} , \qquad (43)$$

with α stated in W/m² K, q in W/m² and p in bar.

For any given liquid, the following formula for ambient pressure by *Stephan and Preusser* applies for effervescent evaporation in the region of ambient pressure:

$$Nu = 0.0871 \left(\frac{\dot{q}d}{\lambda'T_s}\right)^{0.674} \left(\frac{\rho''}{\rho'}\right)^{0.156} \left(\frac{rd^2}{a'^2}\right)^{0.371}$$
$$\cdot \left(\frac{a'^2\rho'}{\sigma d}\right)^{0.350} (Pr')^{-0.162} \qquad (44)$$

where $Nu = \alpha d/\lambda'$, d = the separation diameter of the steam bubbles $= 0.851\beta_0[2\sigma/g\,(\rho' - \rho'')]^{1/2}$ with boundary angle $\beta_0 = 45°$ for water, 1° for low-boiling and 35° for other liquids, λ' = the heat conductivity of the liquid, $\dot{q}$ = the heat flux density, T_s = the boiling temperature, ρ'' = the steam density, ρ' = the liquid density, r = the evaporation enthalpy, a' = the temperature conductivity of the liquid, σ = the surface tension and Pr' = the Prandtl number of the liquid. (Variables marked "$'$" relate to the boiling liquid; those marked "$''$" refer to saturated steam.) The above equations do not apply to forced-flow boiling.

10.5 Radiative Heat Transfer

Heat can be transferred not only by direct contact but also by radiation. Thermal radiation consists of a spectrum of electromagnetic waves in the waveband between 0.76 and 360 µm and is distinguished from visible light by its greater wavelength.

If a heat flux $\dot{Q}$ reaches a body by radiation, a fraction $r\dot{Q}$ is reflected, a second component $a\dot{Q}$ is absorbed and a component $d\dot{Q}$ is transmitted, where $r + d + a = 1$. A body which reflects all radiation ($r = 1$, $d = a = 0$) is termed an *ideal mirror*, a body which absorbs all incoming radiation ($a = 1$, $r = 0$) is called a *black body*. A body is termed *diathermic* ($d = 1$, $r = a = 0$) if it allows all radiation to pass through. Examples of this are gases such as O_2, N_2 etc.

10.5.1 The Stefan–Boltzmann Law

All bodies emit *radiation* in proportion to their temperature. The maximum possible level of radiation is emitted by a black body. This can be experimentally approximated by a surface which has been blackened, e.g. treated with soot, or by a hollow body whose walls are of the same temperature throughout, and where a small aperture is applied to allow radiation to emerge. The total radiation emitted by a black body per unit surface area is stated as follows:

$$E_s = \sigma T^4 . \qquad (45)$$

E_s is referred to as the *emission* (W/m²) of the black radiator, while $\sigma = 5.67 \cdot 10^{-8}$ W/m² K⁴ is the *radiation coefficient*.

Where E_n is emission in the normal plane, E_φ the radiation in the plane φ against the normal plane, then *Lambert's cosine* law applies for black bodies: $E_\varphi = E_n \cos \varphi$.

The radiation of actual bodies frequently deviates from these laws.

10.5.2 Kirchhoff's Law

Real bodies emit less radiation than black bodies. The energy they emit is:

$$E = \varepsilon E_s = \varepsilon , \qquad (46)$$

where $\varepsilon \leq 1$ is the emission number which is generally dependent on temperature (see **Appendix C10, Table 3**). Many technical surfaces can be regarded as *grey radi-*

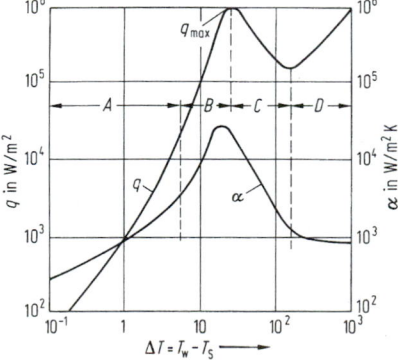

Figure 9. Ranges of boiling for water at 1 bar. *A* free convection (non-effervescent boiling), *B* effervescent boiling, *C* transitional boiling, *D* film boiling.

ators within limited temperature ranges (with the exception of blank metal surfaces). Their radiated energy is distributed in the same way over wavelength as in the case of a black radiator, except that for them it is reduced by a factor of $\varepsilon < 1$. Strictly speaking, the following formula applies to grey radiators $\varepsilon = \varepsilon(T)$, but within narrow temperature ranges ε can be taken as constant. If energy E emitted by a surface unit of a radiator at temperature T reaches a body of temperature T' and surface dA, then it will absorb energy

$$d\ddot{Q} = aE\, dA \,. \tag{47}$$

The *absorption number* defined by this equation is dependent on temperature T of the radiator and temperature T' of the body receiving radiation. For black bodies, $A = 1$, because they absorb all incoming radiation and for non-black surfaces, $a < 1$. For grey radiators, $a = \varepsilon$. From Kirchhoff's law, all surfaces in thermal equilibrium with their environment (i.e. where surface temperature does not vary with time) have an *emission index* equal to their absorption number, $\varepsilon = a$.

10.5.3 Heat Exchange by Radiation

A heat flux

$$\dot{Q}_{12} = \sigma A(T_1^4 - T_2^4) \tag{48}$$

is exchanged between two black surfaces that are very large in relation to their spacing, being of size A and temperatures T_1 and T_2, by radiation. Grey radiators whose emission numbers are ε_1 and ε_2 exchange a heat flux

$$\dot{Q}_{12} = C_{12}A(T_1^4 - T_2^4) \tag{49}$$

of *radiation exchange number*

$$C_{12} = \sigma \left/ \left(\frac{1}{\varepsilon_1} + \frac{1}{\varepsilon_2} - 1 \right) \right. . \tag{50}$$

A heat flux in accordance with Eq. (49) will also flow between an internal pipe of external surface A_1 and a surrounding pipe of internal surface A_2, and both grey radiators possess emission numbers of ε_1 and ε_2, but now the following equation applies:

$$C_{12} = \sigma \left/ \left(\frac{1}{\varepsilon_1} + \frac{A_t}{A_2}\left(\frac{1}{\varepsilon_2} - 1\right) \right) \right. . \tag{51}$$

Where $A_1 \ll A_2$, for example in the case of a pipe in a wide space, $C_{12} = \sigma\varepsilon_1$.

Between two surfaces arbitrarily arranged in space and having temperatures T_1, T_2 and emission numbers of ε_1, ε_2, the following heat flux is obtained, ignoring the components of reflective radiation:

$$Q_{12} = e_{12}A_1\varepsilon_1\varepsilon_2\sigma\,(T_1^4 - T_2^4)$$

in which e_{12} is the irradiation number dependent on the geometrical arrangement of surfaces. Values are set out in [11].

10.5.4 Radiation of Gas

Most gases are permeable to thermal radiation, and neither emit nor absorb it. An exception is constituted by certain gases such as carbon dioxide, carbon monoxide, hydrocarbons, water vapour, sulphur dioxide, ammonia, hydrochloric acid and alcohols. They emit and absorb radiation only in certain wavebands. The emission and absorption numbers of these gases depend not only upon temperature but also upon the geometrical configuration of the gas body.

11 Tables

Appendix C2, Table 1. Fixed reference points on the international temperature scale of 1968 (IPTS-68)

State of equilibrium	Assigned values of international practical temperature	
	T_{68} (K)	t_{68} (°C)
Equilibrium between the solid, liquid and vapour form phases of hydrogen in equilibrium (triple point of hydrogen in equilibrium)	13.81	−259.34
Equilibrium between the liquid and vapour form phases of hydrogen in equilibrium at 0.333 306 bar pressure	17.042	−256.108
Equilibrium between the liquid and vapour form phases of hydrogen in equilibrium (boiling point of hydrogen in equilibrium)	20.28	−252.87
Equilibrium between the liquid and vapour form phases of neon (boiling point of neon)	27.102	−246.048
Equilibrium between the solid, liquid and vapour form phases of oxygen (triple point of oxygen)	54.361	−218.789
Equilibrium between the liquid and vapour form phases of oxygen (boiling point of oxygen)	90.188	−182.962
Equilibrium between the solid and vapour form phases of water (triple point of water)[a]	273.16	0.01
Equilibrium between the liquid and vapour form phases of water (boiling point of water)[a,b]	373.15	100
Equilibrium between the solid and liquid phases of zinc (setting point of zinc)	692.73	419.58
Equilibrium between the solid and liquid phases of silver (setting point of silver)	1235.08	961.93
Equilibrium between the solid and liquid phases of gold (setting point of gold)	1337.58	1064.43

[a] The water used must have the isotopic composition of seawater.
[b] The state of equilibrium between the solid and liquid phases of tin (setting point of tin) has been assigned the value of $t_{68} = 231.9681$°C. This state of equilibrium can be employed instead of the boiling point of water.

Appendix C2, Table 2. Thermometric fixed reference points for pressure 0.101 325 0 MPa. E = setting point, Sd = boiling point, Tr = triple point

		°C
Normal hydrogen	Tr	-259.194
Normal hydrogen	Sd	-252.753
Nitrogen	Sd	-195.802
Carbon dioxide	Sd	-78.476
Mercury	E	-38.862
Water (air-saturated)	E	0
Diphenyl ether	Tr	26.87
Benzoic acid	Tr	122.37
Indium	E	156.634
Bismuth	E	271.442
Cadmium	E	321.108
Lead	E	327.502
Mercury	Sd	356.66
Sulphur	Sd	444.674
Antimony	E	630.74
Copper	E	1084.5
Nickel	E	1455
Palladium	E	1554
Platinum	E	1772
Rhodium	E	1963
Iridium	E	2447
Tungsten	E	3387

Appendix C6, Table 1. Critical data for certain substances, arranged according to critical temperatures[a]

	Formula	M (kg/kmol)	p_k (bar)	T_k (K)	v_k (dm³/kg)
Mercury	Hg	200.59	1490	1765	0.213
Aniline	C_6H_7N	93.1283	53.1	698.7	2.941
Water	H_2O	18.0153	220.64	647.14	3.106
Benzol	C_6H_6	78.1136	48.98	562.1	3.311
Ethyl alcohol	C_2H_5OH	46.0690	61.37	513.9	3.623
Diethyl ether	$C_4H_{10}O$	74.1228	36.42	466.7	3.774
Monochloroethane	C_2H_5Cl	64.5147	52.7	460.4	2.994
Sulphur dioxide	SO_2	64.0588	78.84	430.7	1.901
Monochloromethane	CH_3Cl	50.4878	66.79	416.3	2.755
Ammonia	NH_3	17.0305	113.5	405.5	4.255
Hydrogen chloride	HCl	36.4609	83.1	324.7	2.222
Nitrous oxide	N_2O	44.0128	72.4	309.6	2.212
Acetylene	C_2H_2	26.0379	61.39	308.3	4.329
Ethane	C_2H_6	30.0696	48.72	305.3	4.926
Carbon dioxide	CO_2	44.0098	73.84	304.2	2.156
Ethylene	C_2H_4	28.0528	50.39	282.3	4.651
Methane	CH_4	16.0428	45.95	190.6	6.173
Nitrogen monoxide	NO	30.0061	65	180	1.901
Oxygen	O_2	31.999	50.43	154.6	2.294
Argon	Ar	39.948	48.65	150.7	1.873
Carbon monoxide	CO	28.0104	34.98	132.9	3.322
Air	–	28.953	37.66	132.5	3.195
Nitrogen	N_2	28.0134	33.9	126.2	3.195
Hydrogen	H_2	2.0159	12.97	33.2	32.26
Helium-4	He	4.0026	2.27	5.19	14.29

[a] Compiled in accordance with:

Rathmann D, Bauer J, Thompson PA. A table of miscellaneous thermodynamic properties for various substances with emphasis on the critical properties. Bericht no. 6, 1978, Max-Planck-Institut für Strömungsforschung, Göttingen.

Atomic weight of elements. Pure Appl Chem 1983; 55: 1102–18.

Ambrose D. Vapour-liquid critical properties. National Physical Laboratory, Teddington, 1980.

Appendix C6, Table 2. Antoine equation. Constants for certain substances[a]

$\lg p = A - \dfrac{B}{C + t}$, where p is in hPa, t in °C.

Substance	A	B	C
Methane	6.820 51	405.42	267.777
Ethane	6.959 42	663.70	256.470
Propane	6.928 88	803.81	246.99
Butane	6.933 86	935.86	238.73
2-Methylpropane	7.035 38	946.35	246.68
Pentane	7.001 22	1 075.78	233.205
Isopentane	6.958 05	1 040.73	235.445
Neopentane	6.729 17	883.42	227.780
Hexane	6.995 14	1 168.72	224.210
Heptane	7.018 75	1 264.37	216.636
Octane	7.034 30	1 349.82	209.385
Cyclopentane	7.011 66	1 124.162	231.361
Methyl cyclopentane	6.987 73	1 186.059	226.042
Cyclohexane	6.966 20	1 201.531	222.647
Methyl cyclohexane	6.947 90	1 270.763	221.416
Ethylene	6.872 46	585.00	255.00
Propylene	6.944 50	785.00	247.00
But-1-ene	6.967 80	926.10	240.00
cis-But-2-ene	6.994 16	960.100	237.000
trans-But-2-ene	6.994 42	960.80	240.00
Isobutylene	6.966 24	923.200	240.000
Pent-1-ene	6.971 40	1 044.895	233.516
Hex-1-ene	6.990 63	1 152.971	225.849
Propadiene	5.838 6	458.06	196.07
Buta-1,3-diene	6.974 89	930.546	238.854
Methylbuta-1,3-diene	7.010 54	1 071.578	233.513
Benzene	7.030 55	1 211.033	220.790
Toluene	7.079 54	1 344.800	219.482
Ethyl benzene	7.082 09	1 424.255	213.206
m-Xylol	7.133 98	1 462.266	215.105
p-Xylol	7.115 42	1 453.430	215.307
Isopropyl benzene	7.061 56	1 460.793	207.777
Water (90 to 100 °C)	8.073 299 1	1 656.390	226.86

[a] From: Wilhoit RC, Zwolinski BJ. Handbook of vapor pressures and heats of vaporization of hydrocarbons and related compounds. Publication 101, Thermodynamics Research Centre, Department of Chemistry, Texas A&M University, 1971 (American Petroleum Institute Research Project 44).

Appendix C6, Table 3. Specific heat capacity of air at various pressures, calculated by Baehr and Schwier's state equation [6]

$p =$		1	25	50	100	150	200	300	bar
$t = \ \ \ 0$ °C	$c_p =$	1.0065	1.0579	1.1116	1.2156	1.3022	1.3612	1.4087	kJ/(kg K)
$t = \ \ 50$ °C	$c_p =$	1.0080	1.0395	1.0720	1.1335	1.1866	1.2288	1.2816	kJ/(kg K)
$t = 100$ °C	$c_p =$	1.0117	1.0330	1.0549	1.0959	1.1316	1.1614	1.2045	kJ/(kg K)

Appendix C6, Table 4. Mean molecular heat $[\bar{C}_p]_0^t$ of ideal gases in kJ/(kmol K) between 0 °C and t °C. Mean molecular heat capacity $[\bar{C}_v]_0^t$ is obtained by reducing the numbers on the table by 8.3143 kJ/(kmol K). The numbers are divided by the molecular weight values stated on the last line in order to convert to 1 kg

t (°C)	$[\bar{C}_p]_0^t$ (kJ/kmol K)							
	H_2	N_2	O_2	CO	H_2O	CO_2	Air	NH_3
0	28.620 2	29.089 9	29.264 2	29.106 3	33.470 8	35.917 6	29.082 5	34.99
100	28.942 7	29.115 1	29.526 6	29.159 5	33.712 1	38.169 9	29.154 7	36.37
200	29.071 7	29.199 2	29.923 2	29.288 2	34.083 1	40.127 5	29.303 3	38.13
300	29.136 2	29.350 4	30.387 1	29.498 2	34.538 8	41.829 9	29.520 7	40.02
400	29.188 6	29.563 2	30.866 9	29.769 7	35.048 5	43.329 9	29.791 4	41.98
500	29.247 0	29.820 9	31.324 4	30.080 5	35.588 8	44.658 4	30.092 7	44.04
600	29.317 6	30.106 6	31.749 9	30.408 0	36.154 4	45.846 2	30.406 5	46.09
700	29.408 3	30.400 6	32.140 1	30.735 6	36.741 5	46.906 3	30.720 3	48.01
800	29.517 1	30.694 7	32.492 0	31.051 9	37.341 3	47.860 9	31.026 5	49.85
900	29.646 1	30.980 4	32.815 1	31.357 1	37.948 2	48.723 1	31.320 5	51.53
1 000	29.789 2	31.254 8	33.109 4	31.645 4	38.557 0	49.501 7	31.599 9	53.08
1 100	29.948 5	31.518 1	33.378 1	31.919 8	39.162 1	50.205 5	31.863 8	54.50
1 200	30.115 8	31.767 3	33.624 5	32.171 7	39.758 3	50.852 2	32.112 3	55.84
1 300	30.289 1	31.999 8	33.854 8	32.409 7	40.341 8	51.437 3	32.345 8	57.06
1 400	30.470 5	32.218 2	34.072 3	32.630 8	40.912 7	51.978 3	32.565 1	58.14
1 500	30.654 0	32.425 5	34.277 1	32.838 0	41.467 5	52.471 0	32.771 3	59.19
1 600	30.839 4	32.618 7	34.469 0	33.031 2	42.004 2	52.928 5	32.965 3	60.20
1 700	31.024 8	32.797 9	34.651 3	33.210 3	42.522 9	53.350 8	33.148 2	61.12
1 800	31.210 3	32.968 8	34.830 5	33.381 1	43.025 4	53.742 3	33.320 9	61.95
1 900	31.393 7	33.128 4	35.000 0	33.537 9	43.508 1	54.103 0	33.484 3	62.75
2 000	31.575 1	33.279 7	35.166 4	33.689 0	43.974 5	54.441 8	33.639 2	63.46
M (kg/kmol)	2.015 88	28.013 40	31.999	28.010 40	18.015 28	44.009 80	28.953	17.030 52

Appendix C6, Table 5. Steam table. Saturation state (temperature table)

t (°C)	p (bar)	v' (dm³/kg)	v'' (m³/kg)	b' (kJ/kg)	b'' (kJ/kg)	r (kJ/kg)	s' (kJ/kg K)	s'' (kJ/kg K)
0	0.006 108	1.000 2	206.3	-0.04	2 501.6	2 501.6	-0.000 2	9.157 7
2	0.007 055	1.000 1	179.9	8.39	2 505.2	2 496.8	0.030 6	9.104 7
4	0.008 129	1.000 0	157.3	16.80	2 508.9	2 492.1	0.061 1	9.052 6
6	0.009 345	1.000 0	137.8	25.21	2 512.6	2 487.4	0.091 3	9.001 5
8	0.010 720	1.000 1	121.0	33.60	2 516.2	2 482.6	0.121 3	8.951 3
10	0.012 270	1.000 3	106.4	41.99	2 519.9	2 477.9	0.151 0	8.902 0
12	0.014 014	1.000 4	93.84	50.38	2 523.6	2 473.2	0.180 5	8.853 6
14	0.015 973	1.000 7	82.90	58.75	2 527.2	2 468.5	0.209 8	8.806 0
16	0.018 168	1.001 0	73.38	67.13	2 530.9	2 463.8	0.238 8	8.759 3
18	0.020 062	1.001 3	65.09	75.50	2 534.5	2 459.0	0.267 7	8.713 5
20	0.023 37	1.001 7	57.84	83.86	2 538.2	2 454.3	0.296 3	8.668 4
22	0.026 42	1.002 2	51.49	92.23	2 541.8	2 449.6	0.324 7	8.624 1
24	0.029 82	1.002 6	45.93	100.59	2 545.5	2 444.9	0.353 0	8.580 6
26	0.033 60	1.003 2	41.03	108.95	2 549.1	2 440.2	0.381 0	8.537 9
28	0.037 78	1.003 7	36.73	117.31	2 552.7	2 435.4	0.408 8	8.495 9
30	0.042 41	1.004 3	32.93	125.66	2 556.4	2 430.7	0.436 5	8.454 6
32	0.047 53	1.004 9	29.57	134.02	2 560.0	2 425.9	0.464 0	8.414 0
34	0.053 18	1.005 6	26.60	142.38	2 563.6	2 421.2	0.491 3	8.374 0
36	0.059 40	1.006 3	23.97	150.74	2 567.2	2 416.4	0.518 4	8.334 8
38	0.066 24	1.007 0	21.63	159.09	2 570.8	2 411.7	0.545 3	8.296 2
40	0.073 75	1.007 8	19.55	167.45	2 574.4	2 406.9	0.572 1	8.258 3
42	0.081 98	1.008 6	17.69	175.81	2 577.9	2 402.1	0.598 7	8.220 9

t (°C)	p (bar)	v' (dm³/kg)	v'' (m³/kg)	b' (kJ/kg)	b'' (kJ/kg)	r (kJ/kg)	s' (kJ/kg K)	s'' (kJ/kg K)
100	1.013 3	1.043 7	1.673	419.06	2 676.0	2 256.9	1.306 9	7.355 4
105	1.208 0	1.047 7	1.419	440.17	2 683.7	2 243.6	1.363 6	7.296 2
110	1.432 7	1.051 9	1.210	461.32	2 691.3	2 230.0	1.418 5	7.238 8
115	1.690 6	1.056 2	1.036	482.50	2 698.7	2 216.2	1.473 3	7.183 2
120	1.985 4	1.060 6	0.891 5	503.72	2 706.0	2 202.2	1.527 6	7.129 3
125	2.321 0	1.065 2	0.770 2	524.99	2 713.0	2 188.0	1.581 3	7.076 9
130	2.701 3	1.070 0	0.668 1	546.31	2 719.9	2 173.6	1.634 4	7.026 1
135	3.131	1.075 0	0.581 8	567.68	2 726.6	2 158.9	1.686 9	6.976 6
140	3.614	1.080 1	0.508 5	589.10	2 733.1	2 144.0	1.739 0	6.928 4
145	4.155	1.085 3	0.446 0	610.60	2 739.3	2 128.7	1.790 6	6.881 5
150	4.760	1.090 8	0.392 4	632.15	2 745.4	2 113.2	1.841 6	6.835 8
155	5.433	1.096 4	0.346 4	653.78	2 751.2	2 097.4	1.892 3	6.791 1
160	6.181	1.102 2	0.306 8	675.47	2 756.7	2 081.3	1.942 5	6.747 5
165	7.008	1.108 2	0.272 4	697.25	2 762.0	2 064.8	1.992 3	6.704 8
170	7.920	1.114 5	0.242 6	719.12	2 767.1	2 047.9	2.041 6	6.663 0
175	8.924	1.120 9	0.216 5	741.07	2 771.8	2 030.7	2.090 6	6.622 1
180	10.027	1.127 5	0.193 8	763.12	2 776.3	2 013.1	2.139 3	6.581 9
185	11.233	1.134 4	0.173 9	785.26	2 780.4	1 995.2	2.187 6	6.542 4
190	12.551	1.141 5	0.156 3	807.52	2 784.3	1 976.7	2.235 6	6.503 6
195	13.987	1.148 9	0.140 8	829.88	2 787.8	1 957.9	2.283 3	6.465 4
200	15.549	1.156 5	0.127 2	852.37	2 790.9	1 938.6	2.330 7	6.427 8
205	17.243	1.164 4	0.115 0	874.99	2 793.8	1 918.8	2.377 8	6.390 6

T								
44	0.091 00	1.009 4	16.04	184.17	2 581.5	2 397.3	0.625 2	8.184 2
46	0.100 86	1.010 3	14.56	192.53	2 585.1	2 392.5	0.651 6	8.148 1
48	0.111 62	1.011 2	13.23	200.89	2 588.6	2 387.7	0.677 6	8.112 5
50	0.123 35	1.012 1	12.05	209.26	2 592.2	2 382.9	0.703 5	8.077 6
52	0.136 13	1.013 1	10.98	217.62	2 595.7	2 378.1	0.729 3	8.043 2
54	0.150 02	1.014 0	10.02	225.98	2 599.2	2 373.2	0.755 0	8.009 3
56	0.165 11	1.015 0	9.159	234.35	2 602.7	2 368.4	0.780 4	7.975 9
58	0.181 47	1.016 1	8.381	242.72	2 606.2	2 363.5	0.805 8	7.943 1
60	0.199 20	1.017 1	7.679	251.09	2 609.7	2 358.6	0.831 0	7.910 8
62	0.218 4	1.018 2	7.044	259.46	2 613.2	2 353.7	0.856 0	7.879 0
64	0.239 1	1.019 3	6.469	267.84	2 616.6	2 348.8	0.880 9	7.847 7
66	0.261 5	1.020 5	5.948	276.21	2 620.1	2 343.9	0.905 7	7.816 8
68	0.285 6	1.021 7	5.476	284.59	2 623.5	2 338.9	0.930 3	7.786 4
70	0.311 6	1.022 8	5.046	292.97	2 626.9	2 334.0	0.954 8	7.756 5
72	0.339 6	1.024 1	4.656	301.35	2 630.3	2 329.0	0.979 2	7.727 0
74	0.369 6	1.025 3	4.300	309.74	2 633.7	2 324.0	1.003 4	7.697 9
76	0.401 9	1.026 6	3.976	318.13	2 637.1	2 318.9	1.027 5	7.669 3
78	0.436 5	1.027 9	3.680	326.52	2 640.4	2 313.9	1.051 4	7.641 0
80	0.473 6	1.029 2	3.409	334.92	2 643.8	2 308.8	1.075 3	7.613 2
82	0.513 3	1.030 5	3.162	343.31	2 647.1	2 303.8	1.099 0	7.585 8
84	0.555 7	1.031 9	2.935	351.71	2 650.4	2 298.7	1.122 5	7.558 8
86	0.601 1	1.033 3	2.727	360.12	2 653.6	2 293.5	1.146 0	7.532 1
88	0.649 5	1.034 7	2.536	368.53	2 656.9	2 288.4	1.169 3	7.505 8
90	0.701 1	1.036 1	2.361	376.94	2 660.1	2 283.2	1.192 5	7.479 9
92	0.756 1	1.037 6	2.200	385.36	2 663.4	2 278.0	1.215 6	7.454 3
94	0.814 6	1.039 1	2.052	393.78	2 666.6	2 272.8	1.238 6	7.429 1
96	0.876 9	1.040 6	1.915	402.20	2 669.7	2 267.5	1.261 5	7.404 2
98	0.943 0	1.042 1	1.789	410.63	2 672.9	2 262.2	1.284 2	7.379 6

T								
210	19.077	1.172 6	0.104 2	897.74	2 796.2	1 898.5	2.424 7	6.353 9
215	21.060	1.181 1	0.094 6	920.63	2 798.3	1 877.6	2.471 3	6.317 6
220	23.198	1.190 0	0.086 0	943.67	2 799.9	1 856.2	2.517 8	6.281 7
225	25.501	1.199 2	0.078 4	966.89	2 801.2	1 834.3	2.564 1	6.246 1
230	27.976	1.208 7	0.071 5	990.26	2 802.0	1 811.7	2.610 2	6.210 7
235	30.632	1.218 7	0.065 3	1 013.8	2 802.3	1 788.5	2.656 2	6.175 6
240	33.478	1.229 1	0.059 7	1 037.6	2 802.2	1 764.6	2.702 0	6.140 6
245	36.523	1.239 9	0.054 6	1 061.6	2 801.6	1 740.0	2.747 8	6.105 7
250	39.776	1.251 3	0.050 0	1 085.8	2 800.4	1 714.6	2.793 5	6.070 8
255	43.246	1.263 2	0.045 9	1 110.2	2 798.7	1 688.5	2.839 2	6.035 9
260	46.943	1.275 6	0.042 1	1 134.9	2 796.4	1 661.5	2.884 8	6.001 0
265	50.877	1.288 7	0.038 7	1 159.9	2 793.5	1 633.6	2.930 6	5.965 8
270	55.058	1.302 5	0.035 6	1 185.2	2 789.9	1 604.6	2.976 3	5.930 4
275	59.496	1.317 0	0.032 7	1 210.9	2 785.5	1 574.7	3.022 3	5.894 7
280	64.202	1.332 4	0.030 1	1 236.8	2 780.4	1 543.6	3.068 3	5.858 6
285	69.186	1.348 7	0.027 7	1 263.2	2 774.5	1 511.3	3.114 6	5.822 0
290	74.461	1.365 9	0.025 5	1 290.0	2 767.6	1 477.6	3.161 1	5.784 8
295	80.037	1.384 4	0.023 5	1 317.3	2 759.8	1 442.6	3.207 9	5.746 9
300	85.927	1.404 1	0.021 7	1 345.0	2 751.0	1 406.0	3.255 2	5.708 1
310	98.700	1.448 0	0.018 3	1 402.4	2 730.0	1 327.6	3.351 2	5.627 8
320	112.89	1.499 5	0.015 5	1 462.6	2 703.7	1 241.1	3.450 0	5.542 3
330	128.63	1.561 5	0.013 0	1 526.5	2 670.2	1 143.6	3.552 8	5.449 0
340	146.05	1.638 7	0.010 8	1 595.5	2 626.2	1 030.7	3.661 6	5.342 7
350	165.35	1.741 1	0.008 8	1 671.9	2 567.7	895.7	3.780 0	5.217 7
360	186.75	1.895 9	0.006 9	1 764.2	2 485.4	721.3	3.921 0	5.060 0
370	210.54	2.213 6	0.005 0	1 890.2	2 342.8	452.6	4.110 8	4.814 4
374.15	221.20	3.17	0.003 2	2 107.4	2 107.4	0.0	4.442 9	4.442 9

Appendix C6, Table 6. State variables of water and superheated steam[a]

$p \rightarrow$	1 bar t_s = 99.63 °C			5 bar t_s = 151.84 °C			10 bar t_s = 179.88 °C			15 bar t_s = 198.29 °C			25 bar t_s = 223.94 °C		
	v''	b''	s''	v''	b''	s''	v''	b''	s''	v''	b''	s''	v''	b''	s''
	1.694	2 675.4	7.359 8	0.374 7	2 747.5	6.819 2	0.194 3	2 776.2	6.582 8	0.131 7	2 789.9	6.440 6	0.079 9	2 800.9	6.253 6
t (°C)	v (dm³/kg)	b (kJ/kg)	s (kJ/kg K)	v (dm³/kg)	b (kJ/kg)	s (kJ/kg K)	v (dm³/kg)	b (kJ/kg)	s (kJ/kg K)	v (dm³/kg)	b (kJ/kg)	s (kJ/kg K)	v (dm³/kg)	b (kJ/kg)	s (kJ/kg K)
0	1.000 2	0.1	-0.000 1	1.000 0	0.5	-0.000 1	0.999 7	1.0	-0.000 1	0.999 5	1.5	0.000 0	0.999 0	2.5	0.000 0
20	1.001 7	84.0	0.296 3	1.000 15	84.3	0.296 2	1.001 3	84.8	0.296 1	1.001 0	85.3	0.296 0	1.000 6	86.2	0.295 8
40	1.007 8	167.5	0.572 1	1.007 6	167.9	0.571 9	1.007 4	168.3	0.571 7	1.007 1	168.8	0.571 5	1.006 7	169.7	0.571 1
60	1.017 1	251.2	0.830 9	1.016 9	251.5	0.830 7	1.016 7	251.9	0.830 5	1.016 5	252.3	0.830 2	1.016 0	253.2	0.829 7
100	1.696	2 676.2	7.361 8	1.043 5	419.4	1.306 6	1.043 2	419.7	1.306 2	1.043 0	420.1	1.306 2	1.042 5	420.9	1.305 0
120	1.793	2 716.5	7.467 0	1.060 5	503.9	1.527 3	1.060 2	504.3	1.526 9	1.059 9	504.6	1.526 4	1.059 3	505.3	1.525 5
150	1.936	2 776.3	7.613 7	1.090 8	632.2	1.841 6	1.090 4	632.5	1.841 0	1.090 1	632.8	1.840 5	1.089 4	633.4	1.839 4
200	2.172	2 875.4	7.834 9	0.425 0	2 855.1	7.059 2	0.205 9	2 826.8	6.692 2	0.132 4	2 794.7	6.450 8	1.155 5	852.8	2.329 2
250	2.406	2 974.5	8.034 2	0.474 4	2 961.1	7.272 1	0.232 7	2 943.0	6.925 9	0.152 0	2 923.5	6.709 9	0.087 0	2 879.5	6.407 7
300	2.639	3 074.5	8.216 6	0.522 6	3 064.8	7.461 4	0.258 0	3 052.1	7.125 1	0.169 7	3 038.9	6.920 7	0.098 9	3 010.4	6.647 0
350	2.871	3 175.6	8.385 8	0.570 1	3 168.1	7.634 3	0.282 4	3 158.5	7.303 1	0.186 5	3 148.7	7.104 4	0.109 8	3 128.2	6.844 2
400	3.102	3 278.2	8.544 2	0.617 2	3 272.1	7.794 8	0.306 5	3 264.4	7.466 5	0.202 9	3 256.6	7.270 9	0.120 0	3 240.7	7.017 8
450	3.334	3 382.4	8.693 4	0.664 0	3 377.2	7.945 4	0.330 3	3 370.8	7.619 0	0.219 1	3 364.3	7.425 3	0.130 0	3 351.3	7.176 3
500	3.565	3 488.1	8.834 8	0.710 8	3 483.8	8.087 9	0.354 0	3 478.3	7.762 7	0.235 0	3 472.8	7.570 3	0.139 9	3 461.7	7.324 0
550	3.797	3 595.6	8.969 5	0.757 4	3 591.8	8.223 3	0.377 5	3 587.1	7.899 1	0.250 9	3 582.4	7.707 7	0.149 6	3 572.9	7.463 3
600	4.028	3 704.8	9.098 2	0.803 9	3 701.5	8.352 6	0.401 0	3 697.4	8.029 2	0.266 7	3 693.3	7.838 5	0.159 2	3 685.1	7.595 6
650	4.259	3 815.7	9.221 7	0.850 4	3 812.8	8.476 6	0.424 4	3 809.3	8.153 7	0.282 4	3 805.7	7.963 6	0.168 8	3 798.6	7.722 0
700	4.490	3 928.2	9.340 5	0.896 8	3 925.8	8.595 7	0.447 7	3 922.7	8.273 4	0.298 0	3 919.6	8.083 8	0.178 3	3 913.4	7.843 1
750	4.721	4 042.5	9.454 9	0.943 2	4 040.3	8.710 5	0.471 0	4 037.6	8.388 5	0.313 6	4 034.9	8.199 3	0.187 7	4 029.5	7.959 5
800	4.952	4 158.3	9.565 4	0.989 6	4 156.4	8.821 3	0.494 3	4 154.1	8.499 7	0.329 2	4 151.7	8.310 8	0.197 1	4 147.0	8.071 6

[a] The horizontal lines in the columns for pressure which are below critical pressure p_k = 221.20 bar separate the liquid (top) from the vapour form state (below); v above the lines in dm³/kg, below the lines in m³/kg, but for pressures above the critical pressure: dm³/kg, v in m³/kg. Taken from: Schmidt E. Properties of water and steam in SI units, 3rd edn, ed. Grigull U. Springer, Berlin, 1982.

(continued)

Appendix C6, Table 6. Continued

$p \rightarrow$	50 bar t_s = 263.91 °C			100 bar t_s = 310.96 °C			150 bar t_s = 342.13 °C			200 bar t_s = 365.70 °C			220 bar t_s = 373.69 °C		
	v'' 0.03943	b'' 2794.2	s'' 5.9735	v'' 0.01804	b'' 2727.7	s'' 5.6198	v'' 0.01034	b'' 2615.0	s'' 5.3178	v'' 0.00588	b'' 2418.4	s'' 4.941	v'' 0.00373	b'' 2195.6	s'' 4.5799
t °C	v (dm³/kg)	b (kJ/kg)	s (kJ/kg K)	v (dm³/kg)	b (kJ/kg)	s (kJ/kg K)	v (dm³/kg)	b (kJ/kg)	s (kJ/kg K)	v (dm³/kg)	b (kJ/kg)	s (kJ/kg K)	v (dm³/kg)	b (kJ/kg)	s (kJ/kg K)
0	0.9977	5.1	0.0002	0.9953	10.1	0.0005	0.9928	15.1	0.0007	0.9904	20.1	0.0008	0.9895	22.1	0.0009
20	0.9995	88.6	0.2952	0.9972	93.2	0.2942	0.9950	97.9	0.2931	0.9929	102.5	0.2919	0.9920	104.4	0.2914
40	1.0056	171.9	0.5702	1.0034	176.3	0.5682	1.0013	180.7	0.5663	0.9992	185.1	0.5643	0.9983	186.8	0.5635
60	1.0149	255.3	0.8283	1.0127	259.4	0.8257	1.0105	263.6	0.8230	1.0083	267.8	0.8204	1.0075	269.5	0.8194
100	1.0412	422.7	1.3030	1.0386	426.5	1.2992	1.0361	430.3	1.2954	1.0337	434.0	1.2916	1.0327	435.6	1.2902
120	1.0579	507.1	1.5233	1.0551	510.6	1.5188	1.0523	514.2	1.5144	1.0497	517.7	1.5101	1.0486	519.2	1.5084
150	1.0877	635.0	1.8366	1.0843	638.1	1.8312	1.0811	641.3	1.8259	1.0779	644.5	1.8207	1.0767	645.7	1.8186
200	1.1530	853.8	2.3253	1.1480	855.9	2.3176	1.1433	858.1	2.3102	1.1387	860.4	2.3030	1.1369	861.4	2.3001
250	1.2494	1085.8	2.7910	1.2406	1085.8	2.7792	1.2324	1086.2	2.7692	1.2247	1086.7	2.7574	1.2218	1087.0	2.7532
300	0.04530	2925.5	6.2105	1.3979	1343.4	3.2488	1.3779	1338.2	3.2277	1.3606	1334.3	3.2088	1.3543	1332.9	3.2018
350	0.05194	3071.2	6.4545	0.02242	2925.8	5.9489	0.01146	2694.8	5.4467	1.6664	1647.1	3.7310	1.6362	1637.0	3.7096
400	0.05779	3198.3	6.6508	0.02641	3099.9	6.2182	0.01566	2979.1	5.8876	0.00995	2820.5	5.5585	0.00825	2738.8	5.4102
450	0.06325	3317.5	6.8217	0.02974	3243.6	6.4243	0.01845	3159.7	6.1468	0.01271	3064.3	5.9089	0.01111	3022.3	5.8179
500	0.06849	3433.7	6.9770	0.03276	3374.6	6.5994	0.02080	3310.6	6.3487	0.01477	3241.1	6.1456	0.01312	3211.7	6.0716
550	0.07360	3549.0	7.1215	0.03560	3499.8	6.7564	0.02291	3448.3	6.5213	0.01655	3394.1	6.3374	0.01481	3371.6	6.2721
600	0.07862	3664.5	7.2578	0.03832	3622.7	6.9013	0.02488	3579.8	6.6764	0.01816	3535.5	6.5043	0.01633	3517.4	6.4441
650	0.08356	3780.7	7.3872	0.04096	3744.7	7.0373	0.02677	3708.3	6.8195	0.01967	3671.1	6.6554	0.01774	3656.1	6.5986
700	0.08845	3897.9	7.5108	0.04355	3866.8	7.1660	0.02859	3835.4	6.9536	0.02111	3803.8	6.7953	0.01907	3791.4	6.7410
750	0.09329	4016.1	7.6292	0.04608	3989.1	7.2886	0.03036	3962.1	7.0806	0.02250	3935.0	6.9267	0.02036	3924.1	6.8743
800	0.09809	4135.3	7.7431	0.04858	4112.0	7.4058	0.03209	4088.6	7.2013	0.02385	4065.3	7.0511	0.02160	4055.9	7.0001

ᵃ The horizontal lines in the columns for pressure which are below critical pressure p_k = 221.20 bar separate the liquid (top) from the vapour form state (below); v above the lines in dm³/kg, below the lines in m³/kg. but for pressures above the critical pressure: dm³/kg, v in m³/kg. Taken from: Schmidt E. Properties of water and steam in SI units, 3rd edn, ed. Grigull U. Springer, Berlin, 1982.

(continued)

Appendix C6, Table 6. Continued

p →	230 bar			250 bar			300 bar			400 bar			500 bar		
t (°C)	v (dm³/kg)	b (kJ/kg)	s (kJ/kg K)	v (dm³/kg)	b (kJ/kg)	s (kJ/kg K)	v (dm³/kg)	b (kJ/kg)	s (kJ/kg K)	v (dm³/kg)	b (kJ/kg)	s (kJ/kg K)	v (dm³/kg)	b (kJ/kg)	s (kJ/kg K)
0	0.9890	23.1	0.0009	0.9881	25.1	0.0009	0.9857	30.0	0.0008	0.9811	39.7	0.0004	0.9768	49.3	−0.0002
20	0.9916	105.3	0.2912	0.9907	107.1	0.2907	0.9886	111.7	0.2895	0.9845	120.8	0.2870	0.9804	129.9	0.2843
40	0.9979	187.8	0.5631	0.9971	189.4	0.5623	0.9951	193.8	0.5604	0.9910	202.5	0.5565	0.9872	211.2	0.5525
60	1.0070	270.3	0.8189	1.0062	272.0	0.8178	1.0041	276.1	0.8153	1.0001	284.5	0.8102	0.9961	292.8	0.8052
100	1.0322	436.3	1.2894	1.0313	437.8	1.2879	1.0289	441.6	1.2143	1.0244	449.2	1.2771	1.0200	456.8	1.2701
120	1.0481	519.9	1.5076	1.0470	521.3	1.5059	1.0445	524.9	1.5017	1.0395	532.1	1.4935	1.0347	539.4	1.4856
150	1.0760	646.4	1.8176	1.0748	647.7	1.8155	1.0718	650.9	1.8105	1.0660	657.4	1.8007	1.0605	664.1	1.7912
200	1.1360	861.8	2.2987	1.1343	862.8	2.2960	1.1301	865.2	2.2891	1.1220	870.2	2.2759	1.1144	875.4	2.2632
250	1.2204	1087.2	2.7512	1.2175	1087.5	2.7472	1.2107	1088.4	2.7374	1.1981	1090.8	2.7188	1.1866	1193.6	2.7015
300	1.3512	1332.3	3.1983	1.3453	1331.1	3.1916	1.3316	1328.7	3.1756	1.3077	1325.4	3.1469	1.2874	1323.7	3.1213
320	1.4304	1441.4	3.3854	1.4214	1438.9	3.3764	1.4012	1433.6	3.3556	1.3677	1425.9	3.3193	1.3406	1421.0	3.2882
340	1.5431	1563.4	3.5876	1.5273	1558.3	3.5743	1.4939	1547.7	3.5447	1.4434	1532.9	3.4965	1.4055	1523.0	3.4573
360	1.7575	1714.1	3.8296	1.6981	1701.0	3.8036	1.6285	1678.0	3.7541	1.5425	1650.5	3.6855	1.4865	1633.9	3.6355
380	4.7472	2362.5	4.8303	2.2402	1941.0	4.1757	1.8737	1837.7	4.0021	1.6818	1776.4	3.8814	1.5889	1746.8	3.8110
400	7.476	2692.3	5.3294	6.014	2582.0	5.1455	2.8306	2161.8	4.4896	1.9091	1934.1	4.1190	1.7291	1877.7	4.0083
420	8.872	2843.0	5.5502	7.580	2774.1	5.4271	4.9216	2558.0	5.0706	2.3709	2145.7	4.4285	1.9378	2026.6	4.2262
440	9.944	2953.2	5.7070	8.696	2901.7	5.6087	6.227	2754.0	5.3495	3.1997	2399.4	4.7893	2.2689	2199.7	4.4723
460	10.851	3044.0	5.8327	9.609	3002.3	5.7479	7.189	2887.7	5.5349	4.137	2617.2	5.0907	2.7470	2387.2	4.7316
480	11.659	3123.8	5.9402	10.407	3088.5	5.8640	7.985	2993.9	5.6779	4.941	2779.8	5.3097	3.3082	2565.9	4.9709
500	12.399	3196.7	6.0357	11.128	3165.9	5.9655	8.681	3085.0	5.7972	5.616	2906.8	5.4762	3.882	2723.0	5.1782
550	14.053	3360.2	6.2407	12.721	3337.0	6.1801	10.166	3277.4	6.0386	6.982	3151.6	5.7835	5.113	3021.1	5.5255
600	15.530	3508.3	6.4154	14.126	3489.9	6.3604	11.436	3443.0	6.2340	8.088	3346.4	6.0135	6.111	3248.3	5.8207
650	16.896	3648.6	6.5717	15.416	3633.4	6.5203	12.582	3595.0	6.4033	9.053	3517.0	6.2035	6.960	3438.9	6.0351
700	18.188	3784.7	6.7153	16.630	3771.9	6.6664	13.647	3739.7	6.5560	9.930	3674.8	6.3701	7.720	3610.2	6.2138
750	19.427	3918.6	6.8495	17.789	3907.7	6.8025	14.654	3880.3	6.6970	10.748	3825.5	6.5210	8.420	3770.9	6.3749
800	20.623	4051.2	6.9761	18.906	4041.9	6.9306	15.619	4018.5	6.8288	11.521	3971.7	6.6606	9.076	3925.3	6.5222

Appendix C6, Table 7. State variables of ammonia, NH₃, at saturation[a]

Temperature t	Pressure p	Specific volume of liquid v'	Specific volume of vapour v''	Density of liquid ρ'	Density of vapour ρ''	Enthalpy of liquid b'	Enthalpy of vapour b''	Enthalpy of evaporation r = b'' − b'	Entropy of liquid s'	Entropy of vapour s''	r/T = s'' − s'
(°C)	(bar)	(dm³/kg)	(dm³/kg)	(kg/m³)	(kg/m³)	(kJ/kg)	(kJ/kg)	(kJ/kg)	(kJ/kg K)	(kJ/kg K)	(kJ/kg K)
−50	0.41	1.424	2626	702.1	0.3808	136.2	1552	1416	4.787	11.13	6.346
−45	0.55	1.436	2005	696.2	0.4987	158.3	1561	1402	4.885	11.05	6.146
−40	0.72	1.449	1552	690.1	0.6445	180.5	1569	1388	4.981	10.93	5.954
−35	0.93	1.462	1215	684.0	0.8228	202.8	1577	1374	5.076	10.84	5.768
−30	1.19	1.475	962.9	677.8	1.039	225.2	1584	1359	5.168	10.76	5.589
−25	1.52	1.489	770.9	671.5	1.297	247.7	1592	1344	5.260	10.68	5.415
−20	1.90	1.504	623.3	665.1	1.604	270.3	1599	1328	5.350	10.60	5.247
−15	2.36	1.518	508.5	658.6	1.967	293.1	1605	1312	5.438	10.52	5.084
−10	2.91	1.534	418.3	652.0	2.391	315.9	1612	1296	5.526	10.45	4.924
−5	3.55	1.550	346.7	645.3	2.885	338.9	1618	1279	5.612	10.38	4.770
0	4.29	1.566	289.4	638.6	3.456	362.0	1624	1262	5.697	10.32	4.619
5	5.16	1.583	243.2	631.7	4.113	385.2	1629	1244	5.781	10.25	4.471
10	6.15	1.601	205.6	624.6	4.865	408.5	1634	1225	5.863	10.19	4.327
15	7.28	1.619	174.7	617.5	5.723	432.0	1638	1206	5.945	10.13	4.186
20	8.57	1.639	149.3	610.2	6.697	455.7	1642	1186	6.025	10.07	4.047
25	10.03	1.659	128.2	602.8	7.801	479.5	1645	1166	6.105	10.02	3.910
30	11.67	1.680	110.5	595.2	9.046	503.6	1648	1145	6.184	9.960	3.775
35	13.50	1.702	95.70	587.4	10.45	527.9	1650	1122	6.263	9.905	3.643
40	15.55	1.726	83.15	579.5	12.03	552.4	1652	1099	6.341	9.852	3.511
45	17.82	1.750	72.48	571.3	13.80	577.2	1653	1076	6.418	9.799	3.381
50	20.33	1.777	63.37	562.9	15.78	602.4	1653	1051	6.495	9.746	3.251

[a] After: Ahrendts J, Baehr HD. Die thermodynamischen Eigenschaften von Ammoniak (Thermodynamic characteristics of ammonia). VDI research volume 596. Düsseldorf, 1979. The zero point for internal energy u' is at the triple point (−77.6 °C; 0.0603 bar). Entropy values are absolute.

Appendix C6, Table 8. State variables of carbon dioxide, CO_2, at saturation[a]

Temperature t (°C)	Pressure p (bar)	Specific volume of liquid v' (dm³/kg)	of vapour v'' (dm³/kg)	Density of liquid ρ' (kg/m³)	of vapour ρ'' (kg/m³)	Enthalpy of liquid b' (kJ/kg)	of vapour b'' (kJ/kg)	Enthalpy of evaporation $r = b'' - b'$ (kJ/kg)	Entropy of liquid s' (kJ/kg K)	of vapour s'' (kJ/kg K)	$r/T = s'' - s'$ (kJ/kg K)
−50	6.84	0.865 3	55.68	1156	17.96	−201.4	139.0	340.5	2.692	4.218	1.526
−45	8.34	0.879 8	45.94	1137	21.77	−191.0	140.5	331.4	2.738	4.191	1.453
−40	10.07	0.895 3	38.19	1117	26.19	−180.6	141.7	322.2	2.782	4.164	1.382
−35	12.05	0.911 7	31.96	1097	31.29	−170.2	142.6	312.7	2.826	4.139	1.313
−30	14.30	0.929 3	26.90	1076	37.18	−159.7	143.1	302.9	2.868	4.114	1.246
−25	16.85	0.948 4	22.75	1054	43.96	−149.2	143.4	292.6	2.910	4.089	1.179
−20	19.72	0.969 0	19.31	1032	51.77	−138.6	143.2	281.9	2.951	4.065	1.113
−15	22.93	0.991 6	16.45	1008	60.78	−127.8	142.7	270.5	2.992	4.040	1.048
−10	26.51	1.017	14.05	983.6	71.20	−116.7	141.5	258.3	3.033	4.015	0.981 5
−5	30.47	1.045	12.00	957.2	83.31	−105.4	139.8	245.2	3.075	3.989	0.914 5
0	34.86	1.077	10.26	928.8	97.49	−93.58	137.4	231.0	3.117	3.962	0.845 7
5	39.70	1.114	8.749	897.8	114.3	−81.24	134.1	215.4	3.159	3.934	0.774 3
10	45.01	1.158	7.430	863.3	134.6	−68.16	129.7	197.8	3.204	3.903	0.698 7
15	50.85	1.214	6.258	823.8	159.8	−54.00	123.6	177.6	3.251	3.867	0.616 4
20	57.25	1.289	5.187	776.1	192.8	−38.14	115.1	153.2	3.303	3.826	0.522 7
25	64.28	1.404	4.152	712.3	240.8	−18.92	102.0	120.9	3.365	3.770	0.405 4
30	72.06	1.700	2.892	588.2	345.7	12.99	72.60	59.60	3.467	3.663	0.196 6
31.06	73.84	2.156	2.156	463.7	463.7	42.12	42.12	0	3.561	3.561	0

[a] After: (1) Bender E. Equation of state exactly representing the phase behavior of pure substances. Proc. 5th Symp. Thermophys. Properties. ASME, New York, 1970, pp 227–35; (2) Sievers U, Schulz S. Korrelation thermodynamisbhen Eigenschaften der idealen Gase Ar, CO, H₂, N₂, O₂, CO₂, H₂O, CH₄, and C₂H₄, Chem. Ing. Tech. 1981; 53: 459–61.

Appendix C6, Table 9. State variables of monofluorotrichloromethane, $CFCl_3$ (R11) at saturation[a]

Temperature t (°C)	Pressure p (bar)	Specific volume of liquid v' (dm³/kg)	Specific volume of vapour v'' (dm³/kg)	Enthalpy of liquid h' (kJ/kg)	Enthalpy of vapour h'' (kJ/kg)	Enthalpy of evaporation $r = h'' - h'$ (kJ/kg)	Entropy of liquid s' (kJ/kg K)	Entropy of vapour s'' (kJ/kg K)
−30	0.0917	0.6250	1594.6	175.72	375.26	199.54	0.9059	1.7267
−25	0.1206	0.6292	1235.2	179.63	377.80	198.17	0.9219	1.7205
−20	0.1568	0.6335	967.9	183.60	380.36	196.76	0.9377	1.7150
−15	0.2014	0.6379	766.7	187.62	382.92	195.30	0.9534	1.7100
−10	0.2560	0.6424	613.5	191.70	385.49	193.79	0.9690	1.7055
−5	0.3221	0.6470	495.6	195.82	388.06	192.24	0.9845	1.7015
0	0.4014	0.6517	403.9	200.00	390.63	190.63	1.0000	1.6979
5	0.4958	0.6565	332.0	204.23	393.20	188.97	1.0153	1.6947
10	0.6071	0.6615	275.0	208.53	395.77	187.24	1.0306	1.6919
15	0.7376	0.6666	229.4	212.87	398.33	185.46	1.0457	1.6894
20	0.8892	0.6718	192.7	217.26	400.88	183.62	1.0608	1.6872
25	1.0644	0.6772	163.0	221.71	403.43	181.72	1.0758	1.6854
30	1.2655	0.6828	138.6	226.20	405.96	179.76	1.0907	1.6837
35	1.4950	0.6885	118.6	230.73	408.47	177.74	1.1055	1.6823
40	1.7553	0.6944	102.05	235.32	410.97	175.65	1.1202	1.6812
45	2.049	0.7005	88.22	239.95	413.45	173.50	1.1348	1.6802
50	2.379	0.7067	76.63	244.62	415.91	171.29	1.1493	1.6794

[a] After: Kältemaschinenregeln, 7th edn. Müller, Karlsruhe, 1981.

Appendix C6, Table 10. State variables of difluorodichloromethane, CF₂Cl₂ (R12) at saturation[a]

Temperature t	Pressure p	Specific volume of liquid v'	of vapour v''	Enthalpy of liquid b'	of vapour b''	Enthalpy of evaporation r = b'' − b'	Entropy of liquid s'	of vapour s''
(°C)	(bar)	(dm³/kg)	(dm³/kg)	(kJ/kg)	(kJ/kg)	(kJ/kg)	(kJ/kg K)	(kJ/kg K)
−70	0.1227	0.6248	1128.7	137.73	319.79	182.06	0.7379	1.6341
−65	0.1681	0.6301	842.50	142.04	322.16	180.12	0.7588	1.6242
−60	0.2263	0.6355	639.13	146.36	324.53	178.17	0.7793	1.6153
−55	0.2999	0.6410	492.11	150.71	326.92	176.21	0.7995	1.6073
−50	0.3916	0.6467	384.11	155.06	329.30	174.24	0.8192	1.6001
−45	0.5045	0.6526	303.59	159.45	331.69	172.24	0.8386	1.5936
−40	0.6420	0.6587	242.72	163.85	334.07	170.22	0.8576	1.5878
−35	0.8074	0.6650	196.11	168.27	336.44	168.17	0.8763	1.5826
−30	1.0045	0.6716	160.01	172.72	338.80	166.08	0.8948	1.5779
−25	1.2374	0.6783	131.72	177.20	341.15	163.95	0.9120	1.5737
−20	1.5101	0.6853	109.34	181.70	343.48	161.78	0.9308	1.5699
−15	1.8270	0.6926	91.45	186.23	345.78	159.55	0.9485	1.5666
−10	2.1927	0.7002	77.02	190.78	348.06	157.28	0.9658	1.5636
−5	2.6117	0.7081	65.29	195.38	350.32	154.94	0.9830	1.5609
0	3.0889	0.7163	55.678	200.00	352.54	152.54	1.0000	1.5585
5	3.6294	0.7249	47.736	204.66	354.72	150.06	1.0167	1.5563
10	4.2383	0.7338	41.131	209.35	356.86	147.51	1.0333	1.5543
15	4.9208	0.7433	35.601	214.09	358.96	144.87	1.0497	1.5525
20	5.6824	0.7532	30.942	218.88	361.01	142.13	1.0660	1.5509
25	6.528	0.7637	26.934	223.72	363.00	139.28	1.0822	1.5494
30	7.465	0.7747	23.629	228.62	364.94	136.32	1.0982	1.5479
35	8.498	0.7864	20.745	233.59	366.81	133.22	1.1142	1.5466
40	9.633	0.7989	18.261	238.62	368.60	129.98	1.1301	1.5452
45	10.877	0.8122	16.110	243.75	370.31	126.56	1.1460	1.5439
50	12.236	0.8265	14.259	248.96	371.92	122.96	1.1619	1.5425

[a] After: Kältemaschinenregeln, 7th edn. Müller, Karlsruhe, 1981.

Appendix C6, Table 11. State variables of difluoromonochloromethane, CHF_2Cl (R22) at saturation[a]

Temperature t (°C)	Pressure p (bar)	Specific volume — of liquid v' (dm³/kg)	of vapour v'' (dm³/kg)	Enthalpy — of liquid b' (kJ/kg)	of vapour b'' (kJ/kg)	Enthalpy of evaporation $r = b'' - b'$ (kJ/kg)	Entropy — of liquid s' (kJ/kg K)	of vapour s'' (kJ/kg K)
−80	0.1052	0.6594	1757.8	113.62	367.85	254.23	0.6303	1.9466
−75	0.1487	0.6649	1273.9	118.27	370.41	252.14	0.6541	1.9266
−70	0.2061	0.6706	940.1	123.02	372.97	249.95	0.6777	1.9081
−65	0.2808	0.6765	705.3	127.87	375.52	247.65	0.7013	1.8911
−60	0.3762	0.6825	537.2	132.84	378.07	245.23	0.7248	1.8754
−55	0.4966	0.6888	415.0	137.92	380.59	242.67	0.7483	1.8608
−50	0.6463	0.6954	324.8	143.11	383.09	239.98	0.7718	1.8473
−45	0.8301	0.7021	257.2	148.40	385.55	237.15	0.7952	1.8347
−40	1.0533	0.7092	205.9	153.81	387.97	234.16	0.8185	1.8229
−35	1.3213	0.7165	166.5	159.31	390.34	231.03	0.8418	1.8120
−30	1.6402	0.7241	135.9	164.90	392.65	227.75	0.8649	1.8016
−25	2.0160	0.7320	111.96	170.57	394.89	224.32	0.8879	1.7919
−20	2.4550	0.7403	92.93	176.34	397.07	220.73	0.9108	1.7828
−15	2.9640	0.7490	77.69	182.16	399.16	217.00	0.9334	1.7741
−10	3.5498	0.7581	65.40	188.06	401.18	213.12	0.9558	1.7658
−5	4.2193	0.7676	55.39	194.00	403.10	209.10	0.9780	1.7579
0	4.9797	0.7776	47.18	200.00	404.93	204.93	1.0000	1.7503
5	5.8385	0.7882	40.398	206.03	406.65	200.62	1.0216	1.7429
10	6.803	0.7994	34.754	212.11	408.27	196.16	1.0430	1.7358
15	7.881	0.8112	30.026	218.21	409.77	191.56	1.0640	1.7289
20	9.081	0.8238	26.041	224.34	411.14	186.80	1.0848	1.7221
25	10.411	0.8373	22.661	230.50	412.38	181.88	1.1053	1.7153
30	11.879	0.8517	19.779	236.69	413.48	176.79	1.1254	1.7087
35	13.495	0.8673	17.305	242.93	414.42	171.49	1.1454	1.7020
40	15.268	0.8841	15.171	249.22	415.19	165.97	1.1651	1.6952
45	17.208	0.9024	13.320	255.57	415.76	160.19	1.1847	1.6882
50	19.326	0.9226	11.704	262.03	416.11	154.08	1.2042	1.6811
55	21.635	0.9449	10.286	268.62	416.20	147.58	1.2238	1.6736
60	24.145	0.9700	9.033	275.41	415.99	140.58	1.2436	1.6656

[a] After: Kältemaschinenregeln, 7th edn. Müller, Karlsruhe, 1981.

Appendix C6, Table 12. Thermal coefficient of longitudinal expansion γ for some solid bodies between 0 °C and t in 10^{-5} 1/K, in relation to $l_0 = 1$ m per °C

Substance	0 to -190	0 to 100	0 to 200	0 to 300	0 to 400	0 to 500	0 to 600	0 to 700	0 to 800	0 to 900	0 to 1000
Aluminium	-3.43	2.38	4.90	7.65	10.60	13.70	17.00				
Lead	-5.08	2.90	5.93	9.33							
Al-Cu-Mg [0.95 Al; 0.04 Cu + Mg, Mn, St, Fe]		2.35	4.90	7.80	10.70	13.65					
Iron/nickel alloys [0.64 Fe; 0.36 Ni]		0.15	0.75	1.60	3.10	4.70	6.50	8.5	10.5	12.55	
Iron/nickel alloys [0.77 Fe; 0.23 Ni]	-1.13			2.80	4.00	5.25	6.50	7.80	9.25	10.50	11.85
Glass: Jena 16III		0.81	1.67	2.60	3.59	4.63					
Glass: Jena 1565III		0.345	0.72								
Gold	-2.48	1.42	2.92	4.44	6.01	7.62	9.35	11.15	13.00	14.90	
Grey cast iron	-1.59	1.04	2.21	3.49	4.90	6.44	8.09	9.87	11.76		
Constantan [0.60 Cu; 0.40 Ni]	-2.26	1.52	3.12	4.81	6.57	8.41					
Copper	-2.65	1.65	3.38	5.15	7.07	9.04	11.09				
Sintered magnesia			2.45	3.60	4.90	6.30	7.75	9.30	10.80	12.35	13.90
Magnesium	-4.01	2.60	5.41	8.36	11.53	14.88					
Manganese bronze [0.85 Cu; 0.09 Mn; 0.06 Sn]	-2.84	1.75	3.58	5.50	7.51	9.61	11.90	14.3	16.80		
Manganin [0.84 Cu; 0.12 Mn; 0.04 Ni]		1.75	3.65	5.60	7.55	9.70					
Brass [0.62 Cu; 0.38 Zn]	-3.11	1.84	3.85	6.03	8.39						
Molybdenum	-0.79	0.52	1.07	1.64	2.24						
Nickel	-1.89	1.30	2.75	4.30	5.95	7.60	9.27	11.05	12.89	14.80	16.80
Palladium	-1.93	1.19	2.42	3.70	5.02	6.38	7.79	9.24	10.74	12.27	13.86
Platinum	-1.51	0.90	1.83	2.78	3.76	4.77	5.80	6.86	7.94	9.05	10.19
Platinum-iridium alloy [0.80 Pt; 0.20 Ir]	-1.43	0.83	1.70	2.59	3.51	4.45	5.43	6.43	7.47	8.53	9.62
Quartz glass	+0.03	0.05	0.12	0.19	0.25	0.31	0.36	0.40	0.45	0.50	0.54
Silver	-3.22	1.95	4.00	6.08	8.23	10.43	12.70	15.15	17.65		
Fused corundum			1.30	2.00	2.75	3.60	4.45	5.30	6.25	7.15	8.15
Soft steel	-1.67	1.20	2.51	3.92	5.44	7.06	8.79	10.63			
Hard steel	-1.64	1.17	2.45	3.83	5.31	6.91	8.60	10.40			
Zinc	-1.85	1.65									
Tin	-4.24	2.67									
Tungsten	-0.73	0.45	0.90	1.40	1.90	2.25	2.70	3.15	3.60	4.05	4.60

Appendix C6, Table 13. Caloric values: density ρ, specific thermal capacity c_p for 0 to 100 °C, fusion temperature t_E, enthalpy of fusion Δb_E, boiling temperature t_s and enthalpy of evaporation r

	ρ (kg/dm³)	c_p (kJ/kg K)	t_E (°C)	Δb_E (kJ/kg)	t_s (°C)	r (kJ/kg)
Solids (metals and sulphur) at 1.0132 bar						
Aluminium	2.70	0.921	660	355.9	2 270	11 723
Antimony	6.69	0.209	630.5	167.5	1 635	1 256
Lead	11.34	0.130	327.3	23.9	1 730	921
Chromium	7.19	0.506	1 890	293.1	2 642	6 155
Iron (pure)	7.87	0.465	1 530	272.1	2 500	6 364
Gold	19.32	0.130	1 063	67.0	2 700	1 758
Iridium	22.42	0.134	2 454	117.2	2 454	3 894
Copper	8.96	0.385	1 083	209.3	2 330	4 647
Magnesium	1.74	1.034	650	209.3	1 100	5 652
Manganese	7.3	0.507	1 250	251.2	2 100	4 187
Molybdenum	10.2	0.271	2 625	—	3 560	7 118
Nickel	8.90	0.444	1 455	293.1	3 000	6 197
Platinum	21.45	0.134	1 773	113.0	3 804	2 512
Mercury	13.55	0.138	−38.9	11.7	357	301
Silver	10.45	0.234	960.8	104.7	1 950	2 177
Titanium	4.54	0.471	1 800	—	3 000	—
Bismuth	9.80	0.126	271	54.4	1 560	837
Tungsten	19.3	0.134	3 380	251.2	6 000	1 815
Zinc	7.14	0.385	419.4	112.2	907	1 800
Tin	7.28	0.226	231.9	58.6	2 300	2 596
Sulphur (rhombic)	2.07	0.720	112.8	39.4	444.6	293
Liquids at 1.0132 bar						
Ethyl alcohol	0.79	2.470	−114.5	104.7	78.3	841.6
Ethyl ether	0.71	2.328	−116.3	100.5	34.5	360.1
Acetone	0.79	2.160	−94.3	96.3	56.1	523.4
Benzol	0.88	1.738	5.5	127.3	80.1	395.7
Glycerine[a]	1.26	2.428	18.0	200.5	290.0	854.1
Salt solution of water (saturated)	1.19	3.266	−18.0	—	108.0	—
Seawater (3.5% salt content)	1.03	—	−2.0	—	100.5	—
Methyl alcohol	0.79	2.470	−98.0	100.5	64.5	1 101.1
n-Heptane	0.68	2.219	−90.6	141.5	98.4	318.2
n-Hexane	0.66	1.884	−95.3	146.5	68.7	330.8
Turpentine oil	0.87	1.800	−10.0	116.0	160.0	293.1
Water	1.00	4.183	0.0	333.5	100.0	2 257.1
Gases at 1.0132 and 0 °C	kg/m³					
Ammonia	0.771	2.060	−77.7	332.0	−33.4	1371
Argon	1.784	0.523	−189.4	29.3	−185.9	163
Ethylene	1.261	1.465	−169.5	104.3	−103.9	523
Helium	0.178	5.234	—	37.7	−268.9	21
Carbon dioxide	1.977	0.825	−56.6	180.9	−78.5[b]	574
Carbon monoxide	1.250	1.051	−205.1	30.1	−191.5	216
Air	1.293	1.001	—	—	−194.0	197
Methane	0.717	2.177	−182.5	58.6	−161.5	548
Oxygen	1.429	0.913	−218.8	13.8	−183.0	214
Sulphur dioxide	2.926	0.632	−75.5	115.6	−10.2	390
Nitrogen	1.250	1.043	−210.0	25.5	−195.8	198
Hydrogen	0.09	14.235	−259.2	58.2	−252.8	454

[a] Setting point at 0 °C. The melting and freezing points do not always coincide with each other.
[b] CO_2 does not boil but sublimates at 1.0132 bar.

Appendix C6, Table 14. Debye temperatures for certain substances

Metal	Θ/K	Metal	Θ/K
Pb	88	Al	398
Hg	97	Fe	453
Cd	168	Other substances	
Na	172	KBr	177
Ag	215	KCl	230
Ca	226	NaCl	281
Zn	235	C	1860

Appendix C8, Table 1. Net calorific values of the simplest fuels at 25 °C and 1.013 25 bar

	C	CO	H_2 (Gross calorific value)	H_2 (Net calorific value)	S
Calorific value (kJ)					
Per kmol	393 510	282 989	285 840	241 840	296 900
Per kg	32 762	10 103	141 800	119 972	9 260

Appendix C8, Table 2. Composition and net calorific value of solid fuels

Fuel	Ash (wt %)	Water (wt %)	Composition of ash-free dry substance (wt %)					Gross calorific value	Net calorific value
			C	H	S	O	N	(MJ/kg in operating state)	
Air-dry wood	< 0.5	10 to 20	50	6	0.0	43.9	0.1	15.91 to 18.0	14.65 to 16.75
Air-dry peat	< 15	15 to 35	50 to 60	4.5 to 6	0.3 to 2.5	30 to 40	1 to 4	13.82 to 16.33	11.72 to 15.07
Raw lignite	2 to 8	50 to 60	65 to 75	5 to 8	0.5 to 4	15 to 26	0.5 to 2	10.47 to 12.98	8.37 to 11.30
Lignite brick	3 to 10	12 to 18						20.93 to 21.35	19.68 to 20.10
Hard coal	3 to 12	0 to 10	80 to 90	4 to 9	0.7 to 1.4	4 to 12	0.6 to 2	29.31 to 35.17	27.31 to 34.12
Anthracite	2 to 6	0 to 5	90 to 94	3 to 4	0.7 to 1	0.5 to 4	1 to 1.5	33.49 to 34.75	32.66 to 33.91
By-product cokes	8 to 10	1 to 7	97	0.4 to 0.7	0.6 to 1	0.5 to 1	1 to 1.5	28.05 to 30.56	27.84 to 30.35

Appendix C8, Table 3. Combustion of liquid fuels

Fuel		Mol weight (kg/kmol)	Content (wt %)		Characteristic σ	Gross calorific value (kJ/kg)	Net calorific value (kJ/kg)
			C	H			
Ethanol C_2H_5OH		46.069	52	13	1.50	29 730	26 960
Alcohol C_2H_5OH	95%	—	—	—	1.50	28 220	25 290
	90%	—	—	—	1.50	26 750	23 860
	85%	—	—	—	1.50	25 250	22 360
Benzene (pure) C_6H_6		78.113	92.2	7.8	1.25	41 870	40 150
Toluol (pure) C_7H_8		92.146	91.2	8.8	1.285	42 750	40 820
Xylol (pure) C_8H_{10}		106.167	90.5	9.5	1.313	43 000	40 780
Standard commercial benzene I (90 benzene)[a]		—	92.1	7.9	1.26	41 870	40 190
Standard commercial benzene II (50 benzene)[b]		—	91.6	8.4	1.30	42 290	40 400
Naphthalene (pure) $C_{10}H_8$ (Melting temperature)		128.19	93.7	6.3	1.20	40 360	38 940
Tetralin (undiluted) $C_{10}H_{12}$		132.21	90.8	9.2	1.30	42 870	40 820
Pentane C_5H_{12}		72.150	83.2	16.8	1.60	49 190	45 430
Hexane C_6H_{14}		86.177	83.6	16.4	1.584	48 360	44 670
Heptane C_7H_{16}		100.103	83.9	16.1	1.571	47 980	44 380
Octane C_8H_{18}		114.230	84.1	15.9	1.562	48 150	44 590
Benzene (mean values)		—	85	15	1.53	46 050	42 700

[a] 0.84 benzene, 0.13 toluol, 0.03 xylol (fractions by weight).
[b] 0.43 benzene, 0.46 toluol, 0.11 xylol (fractions by weight).

Appendix C8, Table 4. Combustion of some simple gases at 25 °C and 1.013 25 bar

Type of gas	Mol wt[a] (kg/kmol)	Density (kg/m³)	Characteristic σ	Gross calorific value[a] (MJ/kg)	Net calorific value[a] (MJ/kg)
Hydrogen H_2	2.0158	0.082	∞	141.80	119.97
Carbon dioxide CO_2	28.0104	1.14	0.50	10.10	10.10
Methane CH_4	16.043	0.656	2.00	55.50	50.01
Ethane C_2H_6	30.069	1.24	1.75	51.88	47.49
Propane C_3H_8	44.09	1.80	1.67	50.35	46.35
Butane C_4H_{10}	58.123	2.37	1.625	49.55	45.72
Ethylene C_2H_4	28.054	1.15	1.50	50.28	47.15
Propylene C_3H_6	42.086	1.72	1.50	48.92	45.78
Butylene C_4H_8	56.107	2.90	1.50	48.43	45.29
Acetylene C_2H_2	26.038	1.07	1.25	49.91	48.22

[a] As per DIN 51 850: Gross and net calorific values of gaseous fuels, April 1980.

Appendix C9, Table 1. Partial p_{ws}, vapour content x_s and enthalpy b_{1+xs} of saturated humid air at temperature t, relative to 1 kg dry air at a total pressure of 1000 mbar (at 0 °C over ice)

t (°C)	p_{ws} (mbar)	x_s (g/kg)	b_{1+x_s} (kJ/kg)	t (°C)	p_{ws} (mbar)	x_s (g/kg)	b_{1+x_s} (kJ/kg)
−20	1.029	0.640 82	−18.520 66	41	77.77	52.462	176.427
−19	1.133	0.705 66	−17.354 5	42	81.98	55.556	185.510
−18	1.247	0.776 76	−16.172 7	43	86.39	58.827	195.061
−17	1.369	0.852 85	−14.978 3	44	91.00	62.281	205.097
−16	1.504	0.937 08	−13.763 5	45	95.82	65.929	215.647
−15	1.651	1.028 82	−12.529 8	46	100.86	69.786	226.751
−14	1.809	1.127 46	−11.278 9	47	106.12	73.857	238.424
−13	1.981	1.234 87	−10.005 5	48	111.62	78.166	250.728
−12	2.169	1.352 32	−8.707 0	49	117.36	82.720	263.684
−11	2.373	1.479 81	−7.383 2	50	123.35	87.537	277.338
−10	2.594	1.617 99	−6.032 4	51	129.61	92.641	291.755
−9	2.833	1.767 49	−4.652 9	52	136.13	98.035	306.945
−8	3.094	1.930 83	−3.238 4	53	142.93	103.749	322.987
−7	3.376	2.107 41	−1.790 4	54	150.02	109.804	339.936
−6	3.581	2.298 50	−0.305 6	55	157.41	116.223	357.856
−5	4.010	2.504 77	1.217 7	56	165.11	123.033	376.820
−4	4.368	2.729 37	2.787 5	57	173.13	130.26	396.894
−3	4.754	2.971 71	4.402 5	58	181.47	137.926	418.141
−2	5.172	3.234 36	6.069 0	59	190.16	146.082	440.695
−1	5.621	3.516 74	7.785 9	60	199.20	154.754	464.628
0	6.108	3.823 3	9.564 3	61	208.6	163.98	490.042
1	6.566	4.111 8	11.298 6	62	218.4	173.83	517.122
2	7.055	4.420 2	13.083 9	63	228.6	184.36	546.020
3	7.575	4.748 5	14.920 2	64	239.1	195.49	576.528
4	8.129	5.098 7	16.812 6	65	250.1	207.48	609.333
5	8.718	5.471 4	18.762 9	66	261.5	220.29	644.333
6	9.345	5.868 6	20.776 1	67	273.3	233.97	681.666
7	10.012	6.291 7	22.855 8	68	285.6	248.71	721.834
8	10.720	6.741 4	25.004 1	69	298.4	264.59	765.054
9	11.472	7.219 8	27.226 3	70	311.6	281.60	811.307
10	12.270	7.728 3	29.526 2	71	325.3	299.95	861.150
11	13.116	8.268 2	31.907 1	72	339.6	319.91	915.304
12	14.014	8.842 4	34.376 6	73	354.3	341.36	973.461
13	14.965	9.451 5	36.936 4	74	369.6	364.74	1 036.790
14	15.973	10.098 5	39.594 2	75	385.5	390.28	1 105.909
15	17.039	10.784 1	42.352 0	76	401.9	418.04	1 180.988
16	18.168	11.511 9	45.219 2	77	418.9	448.47	1 263.231
17	19.362	12.283 4	48.199 8	78	436.5	481.91	1 353.550
18	20.62	13.098	51.292 5	79	454.7	518.76	1 453.023
19	21.96	13.968	54.528 8	80	473.6	559.72	1 563.523
20	23.37	14.887	57.892 7	81	493.1	605.18	1 686.107
21	24.85	15.853	61.379 4	82	513.3	656.12	1 823.400
22	26.42	16.882	65.029 8	83	534.2	713.48	1 977.929
23	28.08	17.974	68.844 37	84	555.7	778.11	2 151.988
24	29.82	19.122	72.805 52	85	578.0	852.10	2 351.175
25	31.66	20.340	76.949 2	86	601.1	937.47	2 580.917
26	33.60	21.630	81.281 1	87	624.9	1 036.43	2 847.112
27	35.64	22.992	85.801 4	88	649.5	1 152.84	3 160.269
28	37.78	24.426	90.510 7	89	674.9	1 291.52	3 533.190
29	40.04	25.948	95.450 1	90	701.1	1 459.26	3 984.164
30	42.41	27.552	100.604 8	91	728.1	1 665.94	4 539.734
31	44.91	29.253	106.013 7	92	756.1	1 928.61	5 245.675
32	47.53	31.045	111.622 0	93	784.9	2 270.14	6 163.447
33	50.29	32.943	117.588 5	94	814.6	2 733.46	7 408.356
34	53.18	34.942	123.781 1	95	845.3	3 399.37	9 197.424
35	56.22	37.059	130.283 9	96	876.9	4 431.70	11 970.74
36	59.40	39.288	137.082 2	97	909.4	6 244.61	16 840.81
37	62.74	41.645	144.217 8	98	943.0	10 292.37	27 713.906
38	66.24	44.133	151.699 0	99	977.6	27 151.38	72 999.538
39	69.91	46.762	159.552	100	1 013.3	—	—
40	73.75	49.535	167.786				

Appendix C10, Table 1. Heat conductivity values λ (W/km)

Solid bodies at 20 °C

Silver	458
Copper, pure	393
Copper, standard commercial	350 to 370
Gold, pure	314
Aluminium (99.5%)	221
Magnesium	171
Brass	80 to 120
Platinum, pure	71
Nickel	58.5
Iron	67
Grey cast iron	42 to 63
0.2% C steel	50
0.6% C steel	46
Constantan: 55% Cu, 45% Ni	40
V2A, 18% Cr, 8% Ni	21
Monel metal: 67% Ni, 28% Cu, 5% Fe + Mn + Si + C	25
Manganin	22.5
Graphite, increasing with density and purity	12 to 175
Natural hard coal	0.25 to 0.28
Various stones	1 to 5
Quartz glass	1.4 to 1.9
Concrete, steel-reinforced concrete	0.3 to 1.5
Fireproof stones	0.5 to 1.7
Glass (2 500 °C)	0.81
Ice at 0 °C	2.2
Moist soil	2.33
Dry soil	0.53
Dry quartz sand	0.3
Dry brickwork	0.25 to 0.55
Moist brickwork	0.4 to 1.6

Insulating materials at 20 °C

Alfol	0.33
Asbestos	0.08
Asbestos sheet	0.12 to 0.16
Glass wool	0.04
Cork boards (150)[a]	0.05
Fossil meal brick, burnt	0.08 to 0.13
Slag wool, mineral wool mat (120)[a]	0.035
Slag wool, filled (250)[a]	0.045
Synthetic resin – foamed products (15)[a]	0.035
Silk (100)[a]	0.055
Peat turfs, air-dry	0.04 to 0.009
Wool	0.04

Liquids

Water[b] at 1 bar and	0 °C	0.562
	20 °C	0.5996
	50 °C	0.6405
	80 °C	0.6668
Saturation state:	99.63 °C	0.6773
Carbon dioxide	0 °C	0.109
	20 °C	0.086
Lubricating oils		0.12 to 0.18

Gases at 1 bar and at temperature ϑ in °C

Hydrogen	$\lambda = 0.171 \ (1 + 0.003\ 49\ \vartheta)$	$-100\ °C \le \vartheta \le 1\ 000\ °C$
Air	$\lambda = 0.0245 \ (1 + 0.002\ 25\ \vartheta)$	$0\ °C \le \vartheta \le 1\ 000\ °C$
Carbon dioxide	$\lambda = 0.014\ 64 \ (1 + 0.005\ \vartheta)$	$0\ °C \le \vartheta \le 1\ 000\ °C$

[a] In brackets: density in kg/m³.
[b] According to Schmidt E. Properties of water and steam in SI units, 3rd edn. ed. Grigull U. Springer, Berlin, 1982.

Appendix C10, Table 2. Physical values of liquids, gases and solids

	ϑ (°C)	ρ (kg/m³)	c_p (J/kg K)	λ (W/K m)	$a \cdot 10^6$ (m²/s)	$\eta \cdot 10^6$ (Pa s)	Pr
Liquids and gases at a pressure of 1 bar							
Mercury	20	13 600	139	8 000	4.2	1 550	0.027
Sodium	100	927	1 390	8 600	67	710	0.011 4
Lead	400	10 600	147	15 100	9.7	2 100	0.02
Water	0	999.8	4 217	0.562	0.133	1 791.8	13.44
	5	1 000	4 202	0.572	0.136	519.6	11.16
	20	998.3	4 183	0.5996	0.144	1 002.6	6.99
	99.3	958.4	4 215	0.6773	0.168	283.3	1.76
Thermal oil S	20	887	1 000	0.133	0.083 3	426	576
	80	835	2 100	0.128	0.073	26.7	43.9
	150	822	2 160	0.126	0.071	18.08	31
Air	−20	1.376 5	1 006	0.023 01	16.6	16.15	0.71
	0	1.275 4	1 006	0.024 54	17.1	19.1	0.7
	20	1.188 1	1 007	0.026 03	21.8	17.98	0.7
	100	0.932 9	1 012	0.031 81	33.7	21.6	0.69
	200	0.725 6	1 026	0.038 91	51.6	25.7	0.68
	300	0.607 2	1 046	0.045 91	72.3	29.2	0.67
	400	0.517 0	1 069	0.052 57	95.1	32.55	0.66
Water vapour	100	0.589 5	2 032	0.024 78	20.7	12.28	1.01
	300	0.379	2 011	0.043 49	57.1	20.29	0.938
	500	0.684 6	1 158	0.053 36	67.29	34.13	0.741

(Continued)

Appendix C10, Table 2. Physical values of liquids, gases and solids (continued)

	ϑ (°C)	ρ (kg/m³)	c_p (J/kg K)	λ (W/K m)	$a \cdot 10^6$ (m²/s)	$\eta \cdot 10^6$ (Pa s)	Pr
Solids							
99.99% Aluminium	20	2 700	945	238	93.4		
Annealed V2A steel	20	8 000	477	15	3.93		
Lead	20	11 340	131	35.3	23.8		
Chromium	20	6 900	457	69.1	21.9		
Gold (pure)	20	19 290	128	295	119		
UO₂	600	11 000	313	4.18	1.21		
UO₂	1 000	10 960	326	3.05	0.854		
UO₂	1 400	10 900	339	2.3	0.622		
Gravel concrete	20	2 200	879	1.28	0.612		
Roughcast	20	1 690	800	0.79	0.58		
Pine, radial	20	410	2 700	0.14	0.13		
Cork boards	30	190	1 880	0.041	0.11		
Glass wool	0	200	660	0.037	0.28		
Earth	20	2 040	1 840	0.59	0.16		
Quartz	20	2 300	780	1.4	0.78		
Marble	20	2 600	810	2.8	1.35		
Fireclay	20	1 850	840	0.85	0.52		
Wool	20	100	1 720	0.036	0.21		
Hard coal	20	1 350	1 260	0.26	0.16		
Snow (solid)	0	560	2 100	0.46	0.39		
Ice	0	917	2 040	2.25	1.2		
Sugar	0	1 600	1 250	0.58	0.29		
Graphite	20	2 250	610	155	1.14		

Appendix C10, Table 3. Emission index ε at temperature t

Material	Surface	t (°C)	ε	Material	Surface	t (°C)	ε
Roofing fabric	—	21	0.91	Brass	polished	19	0.05
Oak wood	planed	21	0.89	Brass	polished	300	0.031
Enamel paint	zinc white	24	0.91	Brass	matt	56 to 338	0.22
Glass	smooth	22	0.94	Nickel	polished	230	0.071
Lime mortar	matt white	21 to 83	0.93	Nickel	polished	380	0.087
Marble	light grey, polished	22	0.93	Silver	polished	230	0.021
Porcelain	glazed	22	0.92	Steel	polished	—	0.29
Carbon black	smooth	—	0.93	Zinc	galvanised iron sheet	28	0.23
Fireclay	glazed	1000	0.75	Zinc	polished	230	0.045
Spirit varnish	black gloss	25	0.82	Zinc	blank galvanised sheet	24	0.057 to 0.087
Brick	red, rough	22	0.93 to 0.95				
Water	vertical radiation	—	0.96	Oxidised metals			
Oil	thick coat	—	0.82				
Oil paint	—	—	0.78	Iron	partial red surface rust	20	0.61
Aluminium	rough	26	0.071 to 0.087	Iron	rusted through	20	0.69
Aluminium	polished	230	0.038	Iron	smooth or rough		
Lead	polished	130	0.057		with casting skin	23	0.81
Grey cast iron	machined	22	0.44	Copper	black	25	0.78
Grey cast iron	molten	1330	0.28	Copper	oxidised	600	0.56 to 0.7
Gold	polished	630	0.035	Nickel	oxidised	330	0.40
Copper	polished	23	0.049	Nickel	oxidised	1 330	0.74
Copper	rolled	—	0.16	Steel	matt oxidised	26 to 356	0.96

12 References

[1] Knoblauch O, Hencky K. Anleitung zu genauen technischen Temperaturmessungen, 2nd edn. Munich and Berlin, 1926, and VDI-Temperaturmessregeln. Temperaturmessungen bei Abnahmeversuchen und in der Betriebsüberwachung DIN 1953, 3rd edn. Berlin, 1953, re-issued in July 1964 as VDE/VDI Richtlinie 3511, Technische Temperaturmessungen. – [2] Landolt–Bornstein. Zahlenwerte und Funktionen aus Physik, Chemie, Astronomie, Geophysik und Technik, 6th edn, vol. II, pt 1. Springer, Berlin, 1971, pp. 245-97. – [3] Dymond JR, Smith EB. The virial coefficients of pure gases and mixtures. Clarendon, Oxford, 1980. – [4] Reid RC, Prausnitz JM, Poling BE. The properties of gases and liquids, 4th edn. McGraw-Hill, New York, 1986. – [5] Schmidt E. Properties of water and steam in SI units, 3rd edn, ed. Grigull U. Springer, Berlin, 1982. – [6] Baehr HD, Schwier K. Die thermodynamischen Eigenschaften der Luft. Springer, Berlin, 1961. – [7] Polt H. Thermochemical properties of refrigerants. Springer, Berlin, 1988. – [8] Brandt F. Brennstoffe und Verbrennungsrechnung. Vulkan, Essen, 1981. – [9] Baehr HD. Zur Thermodynamik des Heizens. Brennst. Wärme Kraft 32 (1980). Pt I, pp 9-15; pt II, pp 47-57). – [10] Bronstein IN, Semendjajew KA. Taschenbuch der Mathematik. Deutsch, Frankfurt-on-Main. – [11] VDI-Wärmeatlas, 5th edn. VDI-Verlag, Düsseldorf, 1988.

Materials Technology

A. Burr, Schwäbisch Gmünd; K.-H. Habig, Berlin; G. Harsch, Beilstein; K. H. Kloos, Darmstadt

The properties of components are greatly influenced by choice of material. Optimisation of properties can be achieved only if the properties of the final component are related to material properties as affected by production, changes (deliberate or otherwise) in material properties due to manufacturing processes at the semi-finished and finished stages (initial forming, re-forming, cutting, jointing, coating, material properties alteration), the effect of design configuration (external stress systems), and to inherent stresses induced by manufacture and loading (internal stress systems).

Apart from these functional factors in the choice of material, shortages of energy and raw materials may also be decisive criteria; for example, increase in product life due to improved corrosion and wear protection, reusability of materials and components for mass-produced items, and use of energy-saving production and manufacturing processes. Questions of environmental protection and safety at work are also gaining in importance.

1 Fundamental Properties of Materials and Structural Parts

K. H. Kloos, Darmstadt

Function-related choice of material is based on a comprehensive calculated or experimental stress analysis (see B2 to 9), and a comparison with suitable material parameters, which are frequently determined under ideal conditions. The working stresses that occur in practice include mechanical, mechanical-thermal, mechanical-chemical and tribological stresses, which can occur either individually or in combination.

In cases of component damage, therefore, the combined effect of several damage mechanisms may be involved.

1.1 Load and Stress Conditions

Mechanical and mechanical-thermal working stresses can be characterised by different load conditions, where the chronological sequence of all possible types of load at room temperature and at high and low temperatures is described.

1.1.1 Fundamental Load Conditions

The chronological sequence of external normal forces, transverse forces and moments is described as the fundamental load condition, which causes internal stresses, such as tensile, compressive, bending and torsional stresses in the sections under consideration (**Fig. 1**). Provided that linear–elastic stress–strain behaviour can be assumed, the load–time curves shown in **Fig. 1** correspond to the stress–time and strain–time curves (see B1).

Stress conditions can occur both at room temperature

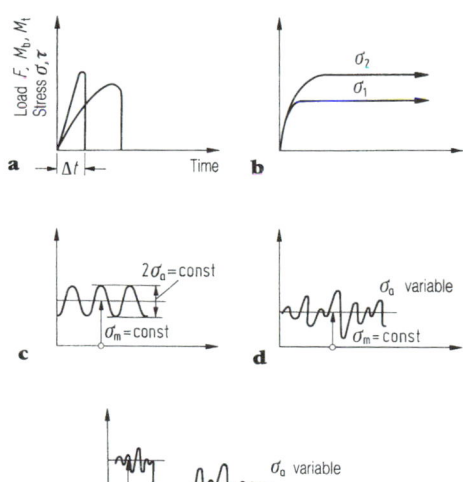

Figure 1. Curves of stress-time functions resulting from the various fundamental load cases: **a** short-term or impact load, **b** steady long-term load, **c** periodic alternating load with constant pre-load and alternating load, **d** aperiodic alternating load with constant pre-load and variable alternating load, **e** aperiodic alternating load with variable pre-load and alternating load.

and particularly at higher temperatures, in which the linear–elastic relationship between stresses and strains no longer applies. Stress– and strain-time functions as in **Fig. 2** are obtained for steady or alternating stresses. These load conditions apply for constant temperatures.

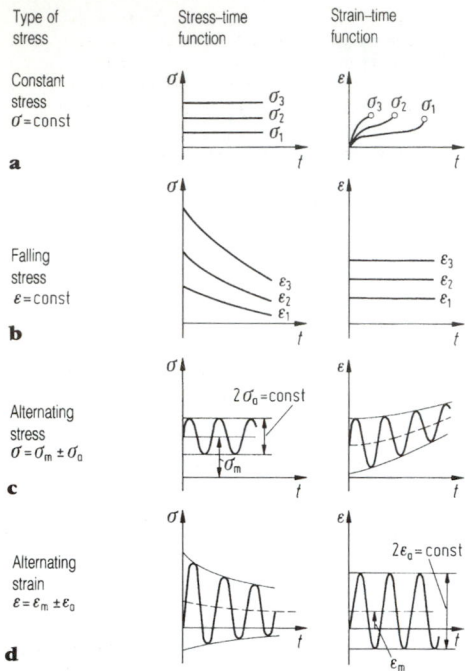

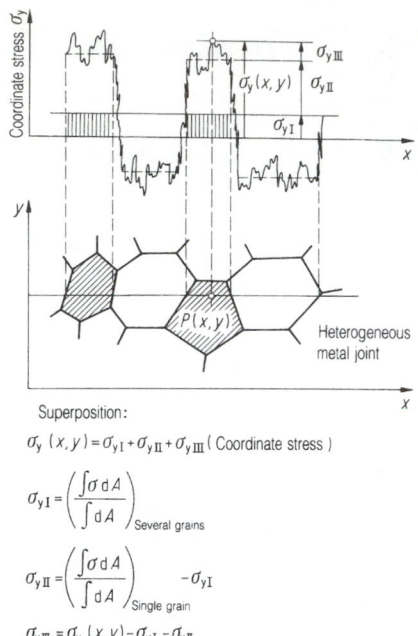

Figure 3. Superposition of first-, second- and third-order internal stresses in a heterogeneous metal joint.

Figure 2. Fundamental load conditions for mechanical–thermal stress: **a** non-deforming steady stress, **b** deforming steady stress, **c** non-deforming alternating stress, **d** deforming alternating stress.

1.1.2 Load Conditions at Surface Under Load by Force

The fundamental load conditions in D1.1.1 above refer to types of loading, where there is no direct introduction of force over the area of the stress cross-sections considered. In many applications, pairs of surfaces are subjected to combined pressure–thrust loading, depending on whether the stresses on the loaded areas are static or sliding [1] (Hertz pressure: see B4; Coulomb friction: see A1.11; combined pressure: see F1.4.2).

1.1.3 Loading States by Residual Stress

In many practical applications, internal stresses are superposed on the fundamental load conditions by internal forces and moments, which can cause a number of changes of properties, particularly near the surface. Internal stresses are fundamentally static multiaxial stresses, which frequently have the same directional character as the main load stresses.

Figure 3 shows a schematic representation of the definition of internal stresses carried out in [2] and a division into internal stresses of the first, second and third order. Depending on the area of the material element, the resultant internal stress peaks can be regarded as a superposition on internal stresses of the first or higher order. This representation makes it clear that, in a heterogeneous metal joint, the excess stress can be several times the first-order internal stress state. As the load stresses in all cases of fundamental loading are related only to macroscopic areas, only first-order internal stresses can have load stresses superposed in the process of calculation.

The number of internal stress states can be traced back to causes that are either metallurgical or that arise from

manufacture [3, 4] or loading. For example in the diffusion of metal atoms into the fundamental lattice and in conversion processes with a changed specific volume within the diffusion depth or the depth effect of the conversion, internal stresses with a particular source function (e.g. nitriding, hardening, surface hardening) are created.

1.2 Causes of Failure

The many possible causes of failure in practice can be traced back to mechanical and also to complex causes. Failure of components caused by mechanical overstressing arises from excessive deformation, from fracture processes and from instability, e.g. buckling and bulging. Depending on stress type and material condition, the various types of fracture can be divided into *cleavage fracture* or *gradual fracture* for continuous stress application, or *fatigue fracture* for alternating stresses. Gradual or deformation fracture is preceded by plastic flow, so that the stress condition causing the fracture is not the same as the stress condition at yield.

1.2.1 Failure Due to Mechanical Stress Conditions

Continuous Stress

Owing to their crystalline structure, the technically important metals and metal alloys have pronounced elasticity with a linear elastic stress–strain behaviour up to the yield point (see **D2 Fig. 3**). While elastic deformation is based on reversible lattice expansions, compressions and distortions at yield, an irreversible slip of whole lattice areas occurs along preferred planes, which coincides with

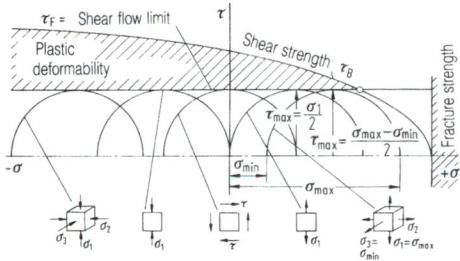

Figure 4. Effect of state of stress on the shear flow limit τ_F and the curve of shear strength τ_B.

the direction of maximum shear stresses in homogeneous isotropic materials. Owing to the existence of one-dimensional lattice defects (displacements), the onset of yield occurs at considerably lower shear stresses than would be expected from calculations for an ideal lattice. For corresponding multiplication of the yield processes at the atomic level, a macroscopic yield formation occurs in the direction of the greatest shear stress (see B1). For a triaxial stress state, without taking the mean main normal stress into account, the flow equation

$$\sigma_1 - \sigma_3 = 2\tau_{max} = R_e \approx \sigma_F$$

applies (R_e is the designation for yield or elastic limit according to ISO).

After exceeding the yield limit, ductile materials can deform up to fracture, depending on the stress state; with multi-axial compressive stress conditions there is a greater possibility of deformation than in tensile stress conditions. At the limit, multi-axial tensile stresses can cause cleavage fracture without deformation.

Figure 4 shows the effect of uni- and multi-axial tensile and compressive stress conditions on the curve of shear yield limit τ_F, shear strength τ_B, and the corresponding Mohr stress circles for the initiation of flow [5]. The distance between τ_F and τ_B gives a direct comparison for plastic deformability. Above the point of intersection of both parameters, there is a danger of cleavage fracture in the region of multi-axial tensile stresses. For brittle material states, the point of intersection between τ_F and τ_B can be displaced into the pressure region.

In contrast to macroscopic consideration of homogeneous multi-axial stress conditions, in fracture mechanics critical stress conditions are assumed at the region of the crack tip, which have a marked gradient effect [6, 7]. For continuous loading, the stress condition at the crack tip may lead to unstable propagation of the crack to cleavage fracture; **Fig. 5** shows the stress condition

assumed in the linear elastic fracture mechanism of a finite disc with a crack of length $2a$ assumed to be symmetrical about the single axis of stress and the curve of the coordinate stress σ_y before the crack tip. At the crack tip itself, a biaxial stress condition occurs, where the stress σ_y declines quickly, depending on the width of the disc. In polar coordinates, the stress coordinates are:

$$\sigma_y = \sigma_1 \frac{\sqrt{\pi a}}{\sqrt{2\pi r}} \cdot f(\varphi), \quad \sigma_x = \sigma_1 \frac{\sqrt{\pi a}}{\sqrt{2\pi r}} \cdot f(\varphi),$$

$$\tau_{xy} = \sigma_1 \frac{\sqrt{\pi a}}{\sqrt{2\pi r}} \cdot f(\varphi).$$

For given disc stress values σ_1 and half the crack length a, the numerator of the expressions represents a parameter for stress concentration in the region of the crack tip. For Mode I of crack propagation (at right angles to the greatest normal stress), the stress intensity factor K is defined as $K_1 = \sigma_1 \sqrt{\pi a}$. If K_1 reaches a critical value then unstable crack propagation may be expected. The following fracture condition applies here:

$$K_1 - K_{1c} \text{ (fracture criterion)}.$$

The critical value of the stress concentration factor is a material characteristic for a given testpiece geometry, which is described as the ductility at fracture. For known values of K_{1c}, with a known crack length, the fracture load can be calculated, while for a known external load, the critical crack length which leads to unstable crack propagation can be estimated. For an example see **Fig. 13**.

Alternating Stress

Alternating stress continues to be one of the most frequent causes of damage since, on the one hand, the actual stresses are not known, and, on the other, no fixed theory can be established due to the many material parameters involved. Progressive damage to samples subject to alternative stress and fatigue fracture is shown in **Fig. 6** [8]. For alternating stresses below the static yield limit, microyielding occurs over the fatigue strength range in smooth and notched samples, which preferentially leads to submicroscopic cracking near the surface. After the phase in which the micro-cracks consolidate, an actual crack is finally formed, running at right angles to the maximum normal stress and causing a considerable excess stress at the crack tip. The fracture load criterion (cycles to failure) N_B can be subdivided into a crack load criterion N_A (actual crack length 0.1 to 1 mm) and a crack load propagation factor ΔN_F, so that $N_B = N_A + \Delta N_F$. The ratio of the crack and fracture criteria N_A/N_B gives the percentage time of the crack initiation phase, from which the

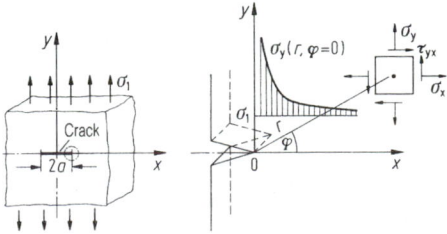

Figure 5. Stress state near the crack tip for a uniaxially loaded finite disc.

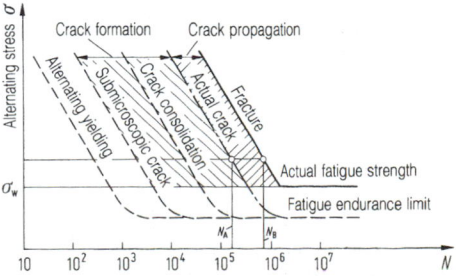

Figure 6. Schematic diagram showing progress of alternating stress damage.

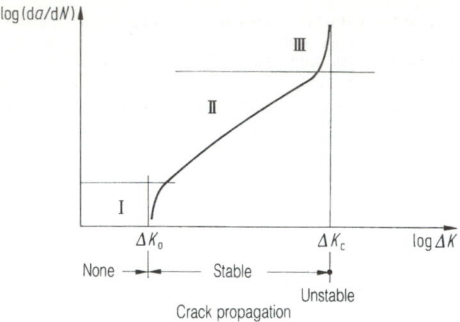

Figure 7. Crack growth in relation to cyclic stress intensity factor.

proportion of the crack propagation phase can be derived. For smooth samples, N_A/N_B is about 90 to 95%, but this can drop considerably for notched samples, so that the crack propagation phase is of greater importance.

Crack propagation under alternating stress is limited to tensile stress amplitudes and occurs at a lower growth rate (subcritical crack growth). Only when the crack has reached a critical length can fracture be caused by unstable crack propagation.

The duration of the subcritical crack propagation phase can also be estimated from the stress intensity factor K_1, if the change in stress intensity over crack propagation is taken into account. Crack propagation per load cycle N can be estimated by:

$$\frac{da}{dN} = C_0 \, \Delta K^n \text{ (Paris equation)}.$$

Figure 7 shows a schematic representation in a log–log scale of the three different stages of crack growth. In region I, subcritical crack growth begins only after the fatigue limit K_0 is exceeded. Below this value, a fatigue crack does not spread any further. In region II, there is stable crack growth and this obeys the Paris equation. The constants C and n depend on the material. The rate of increase n has values of $n = 2$ to 3 for steels. In more developed concepts, the effect of variable stress amplitudes on progress of the cracks is also taken into account in addition to the effect of the mean stresses or internal stresses [9]. In region III, crack progress is accelerated and becomes a fracture when $\Delta K = \Delta K_c$.

1.2.2 Strength Theories

Strength theories are intended to facilitate comparison between multi-axial component stress and strength values of a material, usually determined under uniaxial stress conditions [10, 11] (see B1.3). **Figure 8** shows the funda-

mentals of a strength calculation. The most important failures under mechanical stresses are:

Material parameter

Yield point	R_e, $R_{p0.2}$ (σ_s, $\sigma_{0.2}$)
Cleavage fracture	R_m (σ_B)
Fatigue fracture	σ_w

As soon as the resultant stress reaches the strength limit of the material, component failure occurs. Unlike the failure condition in which σ_v = material parameter K, the strength condition requires $\sigma_v \leq \sigma_{perm} = K/S$. By stating a safety factor $S > 1$, the permissible stress is maintained as a defined proportion of ultimate stress.

For *multi-axial alternating stresses* [12, 13], the failure criterion is the fatigue or alternating stress fracture, which is normally initiated at the surface (nominal stress concept). The triaxial stress condition composed of steady stresses and alternating stresses is

$$\sigma_{1, 2, 3} = \sigma_{m1, 2, 3} \pm \sigma_{a1, 2, 3}.$$

Similarly, the resultant stress can be separated into static and dynamic components: $\sigma_v = \sigma_{vm} \pm \sigma_{va}$.

As fatigue cracks always start at the surface in materials largely free of internal stresses, the strength theories for alternating stresses can be limited to biaxial tensile stress conditions. In the case where $\sigma_{m1, 2, 3} = 0$, the strength condition for alternating stress is $\sigma_{va} = k = \sigma_w$ (σ_w is the tensile–compressive fatigue strength).

Where steady stresses are concerned, it should be noted that, for $\sigma_{m1, 2, 3} \neq 0$, fatigue strength properties depend on their sign and magnitude.

Many tests have shown that, depending on material state, the calculation of the resultant of alternating stresses can be carried out using static stress theory, i.e. normal stress theory for brittle materials, shear stress and deformation energy theory for ductile materials.

Tests have also been carried out and equations developed for phase-displaced superposition of multi-axial alternating stresses with fixed variable main stress lines of action [13–15].

1.2.3 Modes of Failure Under Complex Stress Conditions

This type of failure is marked by time-dependent changes in strength and ductility influences such as high temperature, solid, liquid or gaseous corrosion media, or by wear

Multi-axial
component stress

Example: shaft stressed in bending and torsion

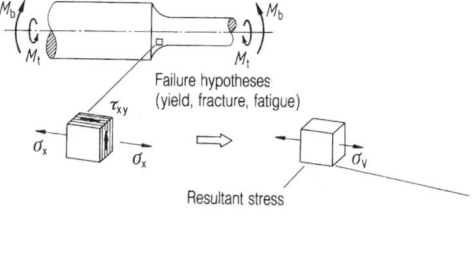

Failure hypotheses
(yield, fracture, fatigue)

Resultant stress

Uniaxial material
characteristic K

Yield point, tensile strength, fatigue strength

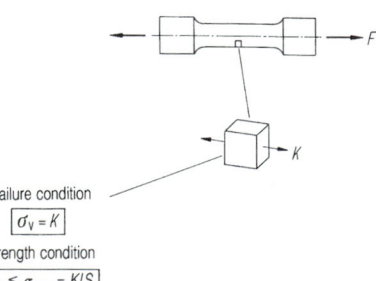

Failure condition

$$\boxed{\sigma_v = K}$$

Strength condition

$$\boxed{\sigma_v \leq \sigma_{perm} = K/S}$$

Figure 8. Fundamentals for strength calculation.

processes. The surface of the materials is therefore attached by chemical and/or electrochemical corrosion, or the component pair properties are altered by adhesive-abrasive wear processes. Quantitative stress analysis can only be carried out in terms of mechanical and thermal stresses if time- and temperature-dependent strength and ductility properties are taken into account.

Explanation of failure by corrosive media and tribological stresses must remain limited to phenomenological data here.

Corrosion Processes. These are based on phase-boundary surface reactions between metals and solid, liquid or gaseous corrosive media. In contrast to chemical corrosion by non-conductors (dry gases, organic materials), many damage mechanisms are triggered particularly by electrochemical corrosion in liquid media with ion conduction, which can lead both to even and uneven material removal and selective change of material properties [16, 17]. Local stress conditions in particular are uncontrollably affected.

Figure 9 shows the most important types of electrochemical corrosion damage, where, in particular, the types of damage connected with mechanical stresses (stress corrosion cracking, alternating stress corrosion cracking) cause considerable problems in practice.

Wear Processes. In contrast to the types of failure outlined above, tribological stresses differ in that both friction surface must be considered, as well as the intermediate medium. Unlike material or component properties, the various wear mechanisms of surfaces stressed by sliding friction can be described as system properties [18, 19] (see F5.1.2).

1.3 Materials Design Values

Materials design values are available for static and for alternating stresses, determined for various fundamental loads (see **Figs 1 and 2**). If no time-dependent changes of material properties have to be considered (e.g. at room temperature or high/low temperatures) then static material design values from short-term tests can be used [11].

Apart from the alternating-stress strength values from stress-controlled experiments, materials design values from strain-controlled vibration tests at room temperature and higher temperatures are gaining importance in the determination of crack characteristics [20].

For certain design problems, ductility properties are required in addition to strength properties. A quantitative evaluation of brittle fracture strength in the particular stress conditions of thick-walled components is made possible by fracture ductility values [21].

1.3.1 Static Stress Conditions

Important materials design values for the calculation of stresses and deformations in the linear-elastic range are represented by the elastic constants, i.e. the modulus of elasticity (Young modulus) and Poisson's ratio ($m = 1/\mu$). The modulus of elasticity, which by definition can be used as a direct comparison parameter for component stiffness, shows a dependence on the material and temperature, according to **Appendix D1, Table 1**, which must be observed for compound structures of various materials and for stresses at high temperature. For some anisotropic alloys, the directional nature of the modulus of elasticity must also be taken into account.

For strength calculations at room temperature and above, materials design values related to flow failures and fractures are needed, depending on the type of stress. **Appendix D1, Table 2** shows a survey of the usual materials strength values under different fundamental loads.

With bending loads, which differ from uniaxial homogeneous tensile loads, there is a 20 to 30% increase in yield limit, depending on the sample thickness, given the same plastic edge strain. This effect, known as the support effect, leads to an apparent increase in the bending yield limit $\sigma_{bo.2}$, without an actual increase of the local edge zone yield limit at maximum stress [10] (see σ_{bf} and $R_{po.2}$ values in **Appendix D1, Table 3**).

The torsion limit (biaxial stress) can be estimated from the tensile yield point R_e, using deformation energy theory:

$$\sigma_v = R_e = \sqrt{3\sigma_1^2} = \sigma_1 \sqrt{3} = \tau_F \sqrt{3}; \quad \tau_F = 0.577 R_e.$$

For mechanical–thermal stresses, short-term values must be used below the regeneration temperature T_K, but, at temperatures above T_K, time-dependent strain limits and creep strength values at constant temperature must be used. See **Fig. 10**.

1.3.2 Dynamic Stress Conditions

Dynamic stress conditions differ from static materials design values at room temperature; they are independent of time, but there is a clear dependence on the number of alternating cycles. For non-positive stresses, materials design values are related to the number of stress cycles; for positive stresses they are related to numbers of strain cycles to cracking or fracture.

Non-positive Dynamic Stress Conditions

Steel alloys frequently have a clear fatigue strength range above a certain limiting number of load cycles (2 to

Type of attack	Marks	Schematic
Even	Corrosion (a) developing hydrogen (b) using oxygen	Me
Uneven	Gap corrosion	Me — Me
	Contact corrosion	Me1 Me2 / Me2 baser than Me1
	Selective corrosion	Heterogeneous structure
	Pitting	Me
	Intercrystalline corrosion	Crack / Grain boundary attack
Uneven, tied to mechanical loads	Stress crack corrosion	Crack / F (constant)
	Alternating stress crack corrosion	Crack / F (alternating)

Figure 9. Appearance of electrochemical corrosion.

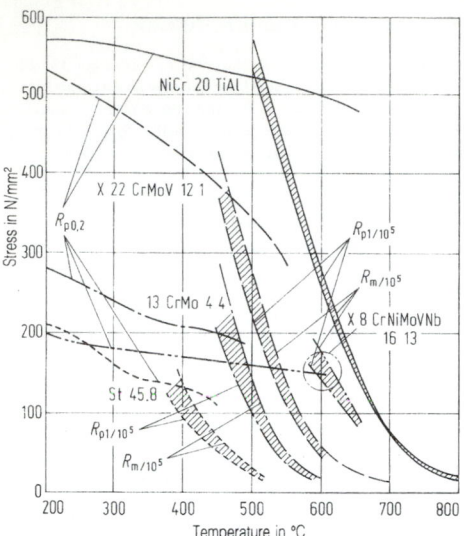

Figure 10. Materials design values of different steels and a nickel-based alloy at high temperature. Comparison between short-term and long-term values.

$20 \cdot 10^6$). For certain applications (soft steels, welded structural parts) the number of stress cycles in the fatigue strength range can be limited to $2 \cdot 10^6$ cycles. Depending on the given mean stress, the following materials design values may be distinguished:

Fatigue strength $\sigma_W \left(\sigma_m = 0, S = \dfrac{\sigma_U}{\sigma_o} = -1 \right).$

Threshold strength $\sigma_{Thr} \ (\sigma_m = \sigma_A, S = 0).$

Fatigue endurance limit $\sigma_D = \sigma_m \pm \sigma_A.$

Depending on stress type, metallic materials have a different sensitivity to mean stress [22], which can be read off directly from the fatigue strength diagrams (**Fig. 11**). In both diagrams, the permissible excess stress σ_0 is constrained by the flow limit $R_{p0.2}$ ($\sigma_{0.2}$).

Owing to the support effect, the bending fatigue strength in the diametrical range of 7 to 15 mm is about 10 to 20% higher than the tensile-compressive fatigue strength which is almost independent of size. From deformation energy theory, the alternating torsion strength values would be about 58% of the tensile-compressive alternative values. The results of tests on small samples gave τ_W / τ_{zdW} values ranging from 0.54 to 0.62.

Appendix D1, Table 2 gives, from the example of some heat-treatable steels, each representing a certain alloy group, a table of static dynamic strength values for tensile, bending and torsional stress. From the comparison of the values it can be seen that ratios of static and dynamic materials design values, which permit conversion to heat-treatable steels with different strength properties, can be given if freedom of the dynamic samples from internal stresses can be guaranteed, e.g. numerical equation, $\sigma_{bw} = 0.383 R_m + 94 \text{ N/mm}^2$ [24]. For further fatigue strength values see **Appendix D1, Figs 1 to 8**.

Positive Dynamic Stress Conditions

In the creep strength range between 10 and 10^4 cycles, and particularly at high temperatures, the linear-elastic stress–strain relationship no longer applies for cyclic loading, so that in elastic–plastic dynamic deformation, closed stress–strain diagrams called hysteresis loops are obtained. Such stresses occur in components designed for relatively few alternating cycles in the creep strength range, or for alternating stresses at high temperature.

Under positive strain-controlled loads, materials may gain or lose strength, the consequence of which is an increase or decrease of the alternating stress amplitude σ_a [11]. Depending on material condition and temperature, this change in alternating stress amplitude is completed after about 10 to 20% of the number of cycles to failure, so that until macro-cracks form in samples subject to alternating strain, almost stable hysteresis loops occur.

Figure 12 shows the change of the elastic–plastic strain part of a material with reduction in strength depending on the number of alternating cycles. The spontaneous reduction of alternating stress amplitude during the tensile phase can be attributed to the formation of macro-cracks. The point of intersection between the actual curve of alternating stress amplitude and a stress value of the stabilised curve reduced by 5% is defined as the number of cycles to failure N_A.

The crack curves determined from the strain-controlled loads represent important materials criteria for the design of components at room temperature and higher temperatures in the low cycle fatigue range [25].

In recent design concepts for single-stage, multi-stage, and randomly stressed components, which differ from the nominal stress concept (see D1.2.2), the notch concept or "local concept" is used to predict fatigue life [20]. With this concept, the plastic alternating strains in the critical cross-section for failure causing the damage are used to estimate fatigue life. The endurance component in the stable crack propagation phase can be estimated using fracture mechanics equations as in **Fig. 7**.

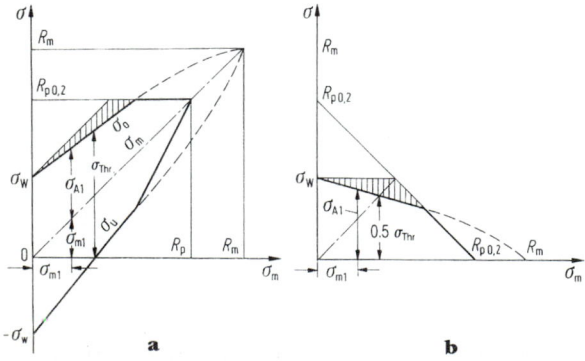

Figure 11. Creep strength diagrams according to Smith (**a**) and Haigh (**b**) and diagram of mean stress sensitivity (shaded area).

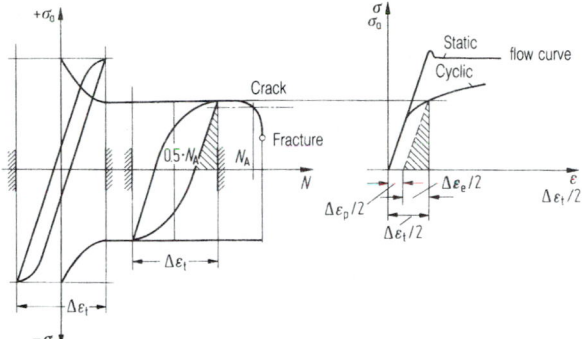

Figure 12. Elastic–plastic alternating strain and cyclic flow curve of a material with reduction in strength.

1.3.3 Characteristics of Ductility and Fracture Toughness

Apart from the materials design values for static and alternating stresses, the plastic deformation properties of a material from the start of flow to fracture are important. The tensile test values (fracture strain, "necking") are not sufficient. Under the influence of actual stress conditions (number of axes, stress rate, size effect), various ambient conditions (low temperatures, corrosive media) and material parameters (purity, grain size, dispersion), there can be considerable loss of ductility, which favours cleavage fractures with little deformation.

No universal test method is yet known by which all external stress conditions can be simulated to evaluate brittle fracture problems. The simulation of defined rates of deformation poses particular difficulties. The notch impact bending test represents a compromise solution which, with further development (instrumentation [26], variable notch sharpness to cracking [27]), gives better information on the quantitative evaluation of various parameters. The position of the transition temperature from the ductile to the brittle state has proved to be a suitable parameter for comparison (see **D2, Fig. 7**).

The critical stress intensity factors at the crack tip (K_c values) determined by various test processes (see D2.2.6) [28, 29] in linear-elastic fracture mechanics also give design values suitable for assessing failure conditions for unstable propagation of a crack of given geometry and position relative to a uniaxial stress field. The preconditions of almost elastic stress distribution at the crack tip are only present with very strong materials, or with medium strength materials and large component dimensions. An even strain condition can be produced at the crack tip by large dimensions which also prevent plastic flow.

For a series of materials, K_{Ic} values (fundamental case 1 of crack propagation) are known, depending on the state of the material and the test temperature [30]. From the fundamental relationship of fracture mechanics,

$$K_1 = \sigma_1 \, Y \, \sqrt{\pi a} \quad (Y \text{ is the geometry factor});$$

any one of the three variable parameters K (stress intensity factor), σ (uniaxial normal stress outside the crack tip) and a (crack length) can be found if the other two are known. The main technical applications of fracture mechanics then follow [31]: choice of material from higher K_{Ic} values in relation to temperature, calculation of permissible stress σ_{perm} for known K_{Ic}, and experimentally determined crack length (by non-destructive test methods), calculation of critical crack length for known stress condition and K_{Ic} value, and calculation of crack progress with alternating stress.

Figure 13 shows a plot of ductility at fracture against test temperature for several heat-treatable steels [30]. Owing to the ductility of these steels, very large samples are required to produce an even strain condition (up to CT5: $125 \times 300 \times 215$ mm). From the comparison of the K_{Ic} values at room temperature, the improvements achieved with high-alloy heat-treatable steels for large turbine and generator shafts become clear. Assuming an elliptical internal fault with an axis ratio $a/2c = 0.1$ and uniaxial normal stress $\sigma = 700$ N/mm^2, critical crack lengths were calculated, which also show the improvement in fracture toughness by the increase of the values of critical crack length.

The assumptions of linear-elastic fracture mechanics apply only to brittle materials. For small partially plastic deformations at the crack tip, the fundamental equations

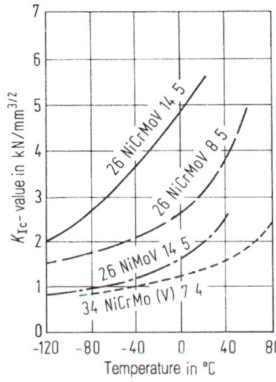

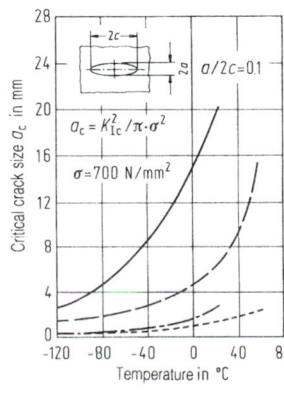

Figure 13. Fracture toughness (K_{Ic} values) and calculated critical crack lengths of steel for turbine and generator shafts against temperature [30].

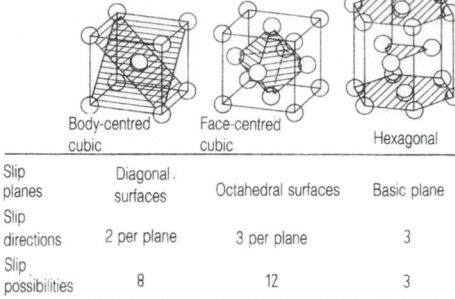

	Body-centred cubic	Face-centred cubic	Hexagonal
Slip planes	Diagonal surfaces	Octahedral surfaces	Basic plane
Slip directions	2 per plane	3 per plane	3
Slip possibilities	8	12	3

Figure 14. Effect of lattice type on the slip possibilities and formability of pure metals.

of fracture mechanics can be retained if the size of the plastic zone r_{pl} is taken into account in stating half the crack length a. The stress intensity factors then become $K = \sigma_1 \sqrt{\pi a_{eff}} = \sigma_1 \sqrt{\pi (a + r_{pl})}$.

If the conditions of small plastic zones at the crack tip are no longer fulfilled, various calculation processes of flow fracture mechanics are used which are particularly suitable for ductile material states. At high temperatures, with high stress rates, the creep crack growth can be calculated from the laws of creep fracture mechanics.

1.4 Effect of Materials Structure, Manufacturing Process and Environment Conditions on Strength and Ductility Behaviour

The choice of materials for components subject to continuous or alternating stresses requires, depending on design configuration (external notch effect), a given strength/ductility ratio, which itself depends on many material, manufacturing and design parameters.

The ductile properties of pure metals depend on the number of sliding systems (direction of sliding, side

planes) of their crystal lattices, where, as can be seen in **Fig. 14**, cubic lattices (e.g. γ-Fe, α-Fe), in contrast to hexagonal lattices (e.g. Ti, Zn), have considerably greater sliding possibilities and therefore better ductility. Homogeneous structures (interstitial or substitution solid solutions) also have better ductility than heterogeneous structures.

The strength properties of metallic materials primarily depend on the micro-structural ability of an alloy to resist displacement motion (onset of flow). Increases in strength by resisting displacement can be achieved by the mechanisms listed below (see **Fig. 15**). These fundamental mechanisms may also be classified by the magnitude of the resistance which prevents displacement at the onset of plastic flow.

While material and structure of the whole cross-section is decisive in terms of static strength properties, for fatigue strength the state of the surface and of the area near the edge is of primary significance.

1.4.1 Metallurgical Effects

In the individual steelmaking processes (LD process, electric arc, electroslag), different proportions of oxide, sulphide and silicate inclusions remain in the material, and their size, shape and distribution have an adverse effect on strength and ductility properties [33]. Depending on their melting or softening point, in hot forming the non-metallic inclusions may change their original solid form and take up a directional character, depending on the degree of deformation (see K3).

The micro-geometric configuration of the inclusions and their position relative to external stress direction have an internal notch effect with different excess stresses. The peak stress value depends not only on the geometry of the inclusion and its position relative to the load stress system, but also on the yield point of the material.

Apart from the excess stresses due to loading, internal stress effects may be superposed that can be ascribed, e.g., to different thermal coefficients of expansion of the inclusions compared with the fundamental material [34].

In Stahl-Eisen Prüfblatt (Steel–Iron Test Sheet) 1570–71, based on the guide series developed by H. Diergarten, the different types of inclusions are graded according to their type and size, so that the inclusions of the next high-

Magnitude of resistance	Strength-increasing mechanisms	Prevention of displacement movement		Stress component dependent on
0	Formation of solid solutions; impurities	(a) substitution (b) dispersion (c) vacancies		Concentration of impurities
1	Cold-forming displacement	Slip prevention by intersecting slip planes		Total displacement density
2	Grain boundaries slip system (coarse or fine grain) stacking fault	Prevention of even displacement	Grain boundary	Grain diameter ($\approx 1/\sqrt{d}$)
3	Separation hardening, dispersion hardening	(a) Bypass mechanism (b) Shearing of particles		Size, distance and mechanical properties of displacement obstacles

Figure 15. Fundamental strength-increasing mechanisms in metallic materials.

est guide number correspond to double the area content. The quality of a batch of steel can therefore be immediately evaluated by stating the danger factors.

By using the electroslag process or by vacuum casting, the fatigue strength properties of heat-treatable steels can be improved by 30 to 40% in the electric furnace, compared with conventional smelting [35]. The negative effects of non-metallic inclusions can be reduced by alloying. For example, by the addition of calcium and cerium, sulphide inclusions are distributed more finely and become spherical, thus reducing the internal notch effect.

Both strength and fatigue properties of welded joints in air and in corrosive media are subject to metallurgical effects such as segregation zones, dispersion processes in the transition area between the seam and the base material, and by dendritic solidification of the weld. In mechanical–chemical stress conditions, it must be ensured that strain energy, welding sequence, and shape of the melting zone avoid grain boundary separation and therefore preclude intercrystalline corrosion.

1.4.2 Production-Technological Effects

Cold Forming. The increase in dislocation density produced by cold forming causes cold strengthening, which is frequently combined with an increase in fatigue strength. The increase in fatigue strength depends upon whether total or partial cold forming was carried out, and whether the direction of cold forming coincides with the direction of component stress. Using the example of heat-treatable steels, it was shown that the increase in tensile-compressive fatigue strength is proportional to the increase in tensile strength, regardless of whether the higher tensile strength was produced by heat treatment or by cold forming. By contrast, fatigue strength cannot be increased by cold working if it is based on the same tensile strength [36]. For partial cold forming of the area near the edge (by shot-blasting, surface pressure), above an edge hardness value of about 350 HV 10, there is always an increase in bending fatigue strength compared with heat-treated samples owing to pressure deformation, as shown in **Fig. 16** [37]. This favourable effect is even more strongly apparent in notched components. Under optimised rolling conditions, notched samples can reach a higher fatigue strength than unnotched (smooth) samples [38].

Heat Treatment. The static strength and ductility and the fatigue strength of steels can be affected within wide limits by tempering. While high static strength values may be obtained via the depth of the tempering process up to through-tempering, for the fatigue strength of components with non-homogeneous stress distribution, the properties of the edge area play a decisive part.

In martensitic hardening of components made of C steel with different cross-sections but the same quenching medium, the increasing diameter there is a reduced edge hardness and a lower depth of hardening, which can be ascribed to different cooling rates, depending on sample size. The different area/volume ratio of the samples is also responsible for a different internal stress configuration (thermal and transformational internal stresses). The alloying elements Mn, Cr, Cr + Mo, Cr + Ni + Mo, Cr + V increase hardenability over that of C steels in that order, and therefore guarantee higher fatigue strength increases for larger dimensions.

In contrast to conventional tempering, conversion in the bainite phase (intermediate stage tempering) can produce better ductility and fatigue strength properties [39].

1.4.3 Surface Effects

The mechanical properties for steady and alternating stresses can be adversely affected by component surface properties. Surface properties include the total effect of micro-roughness, edge strength, and edge internal stresses, and individual weightings have previously only been examined in individual cases [40].

Force-free Surfaces. Surface properties only play a subsidiary part under continuous stress, as the depth effect of surfaces produced by cutting or cold forming is small compared with the total cross-section. For vibration stresses, the properties of the area near the edge are of great importance, as the introductory crack phase depends mainly on surface properties, while the crack propagation is affected by stress mechanics factors at the crack tip.

Figure 17 shows the effect of manufacturing surface roughness on surface factor C_O of steels for tensile, compressive and bending stresses, depending on tensile strength. The changes in the fatigue-strength properties may be due to the micro-notch effect increasing with greater depth of roughness [41].

Different mechanical or thermo-chemical surface-strengthening processes (e.g. shotblasting, nitriding) change edge internal stress conditions as well as increasing edge strength. If compressive internal stresses occur, then for superposition of load stresses, the mean stress is displaced towards lower values.

Figure 18 shows the combined effect of increase in edge strength and distribution of compressive internal stresses in a modified Smith creep-strength diagram, where the outer diagram applies to the strengthened edge zone. In contrast to non-strengthened materials sites, the

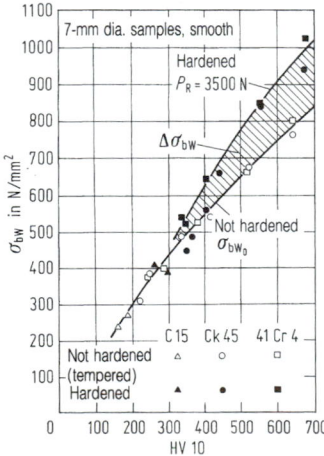

Figure 16. Fatigue bending strength of tempered and tempered-rolled samples in relation to edge hardness [37].

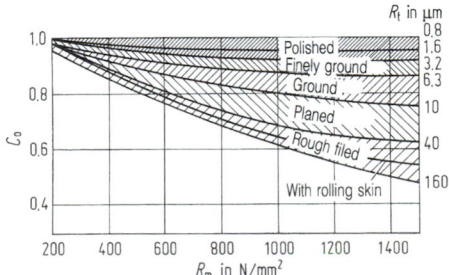

Figure 17. Surface effect factor C_0 for bending and tension-compression.

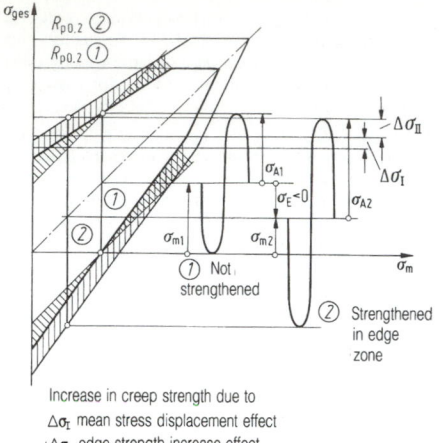

Increase in creep strength due to
$\Delta\sigma_I$ mean stress displacement effect
$\Delta\sigma_{II}$ edge strength increase effect

Figure 18. Joint effect of changes in edge strength and first-order internal stresses on creep strength of smooth samples.

increase of fatigue strength can be ascribed to the effect of mean stress displacement, owing to first-order compressive internal stresses and the increase of edge strength effect.

On the other hand, there are a number of surface effects, which can reduce fatigue strength properties (e.g. cracks in hard chrome coatings [42] or edge decarburisation [43]).

Force-Bearing Surfaces. The force transmission across a pair of surfaces is considerably distorted by microroughness, so that a considerable loss of rolled strength may be expected, particularly for hardened surfaces. It should be noted that the Hertz surface pressure for rolled bodies with and without slip rises considerably in the micro-region. While for tempered surfaces, owing to partial plastic deformation of roughness peaks, a progressive approximation of the true surface pressures to the apparent surface pressure values can be achieved, for hardened surfaces the high micro-surface pressure values are largely resisted by elastic deformation of roughness peaks, which can reduce the pitting limit considerably [44].

1.4.4 Environmental Effects

Materials design values depend greatly on ambient temperature, ambient medium, and radiation loading. The effect of temperature can be seen primarily in changes in the sliding mechanisms of the lattice structures of homogeneous and heterogeneous alloys, and has an effect on the total cross-section of samples and components. In contrast, boundary reactions are caused on surfaces under the influence of corrosive media, leading to macroscopic and microscopic wear, damage to passive layers, or partial embrittlement by hydrogen diffusion. Such damage mechanisms favour the formation of cracks for superposed static or alternating stresses and therefore reduce strength and ductility values.

Effect of Temperature. Between room and elevated temperatures, there is a fundamental tendency to reduce static and dynamic strength values of metals, with a simultaneous increase in ductility values. At high temperatures, it must be remembered that, apart from creep strength, fatigue strength values fall with time owing to time- and temperature-dependent structural changes. The fatigue

endurance limit ceases to exist at high temperature. Because of the marked dependence of experimental results on frequency and hence on time, the stress amplitude σ_a is often plotted not against cycles to fracture N_B, but against the time to fracture $t_B = N_B/f$ ($f =$ frequency) [45, 46].

With decreasing temperatures, strength and fatigue properties of metals generally rise with a simultaneous loss of ductility down to low-temperature embrittlement. Metals with face-centred cubic lattice structure, nickel alloy steels and fine-grained structural steels have a transition temperature from ductile to brittle fracture displaced towards low temperatures.

Effect of Corrosive Media. Damage processes with continuous and alternating stresses under the simultaneous influence of corrosive media show differences specific to the material and media. While stress-crack corrosion only occurs with certain alloys and media, fatigue-crack corrosion causes progressive reduction of the permissible amplitude of vibration in all metals, in comparison with the values in air or vacuum. A fatigue strength limit for corrosion fatigue has not so far been experimentally confirmed. The start of damage both in the active and in the passive state of material could be ascribed to alternating slip processes, as plastically deformed surface areas act as local anodes under the influence of an electrolyte and are therefore subject to an increased rate of material removal [47]. Micro-notches are therefore formed, which finally lead to the formation of a microcrack.

Because of the many parameters involved, the assessment of the sensitivity of a material to intercrystalline stress crack corrosion is very difficult, particularly as this type of attack is produced in the important group of stainless austenitic steels, especially by halogenide and hydroxyl ions [48]. The formation of micro-cracks is preceded by an incubation phase depending on stress and electrolyte composition, but this can be appreciably shortened by the existence of internal tensile stresses due to manufacture. Stress crack corrosion may be caused solely by internal tensile stresses.

In alkaline, nitrate and carbonate solutions, unalloyed and low alloy steels are subject to intercrystalline stress-crack corrosion in certain potential areas.

A special form of stress-crack corrosion is represented by hydrogen embrittlement, which can occur due to electrochemical action in various manufacturing processes (e.g. galvanising) and in electrochemical corrosion. The precondition for this type of damage is the availability of monatomic hydrogen. The hydrogen is dissolved in the iron lattice as an interstitial solid solution [49].

The Effect of High-Energy Radiation. In the irradiation of metal materials with neutrons, ions or electrons, there are many interactions with the lattice atoms, which can lead to a change of mechanical, physical and chemical material properties. Possible radiation damage is of particular importance for the choice of material in reactor construction, depending on the working temperature and neutron flux, which can be divided into irradiation hardening due to slip blockage, creep induced by irradiation at high temperatures, high-temperature embrittlement, and barriers induced by irradiation due to pore formation [50]. Control of pore formation, which is based on the agglomeration of gaps, plays a decisive part for the design of fuel elements for fast breeder reactors and helium-cooled high-temperature reactors.

1.5 Strength Properties and Constructional Design

Component strength properties under static and alternating stresses can only be estimated within certain limitations by setting up strength and failure conditions. This is due to the fact that failure theories apply to homogeneous multi-axial stress conditions and do not take account of a gradial effect for notch stress conditions. The importance of this can be seen in that essential design elements of machine, apparatus, and steel construction (e.g. changes of cross-section, transverse holes, shrunk-on seats, screwed connections, welded joints) typically have multi-axial notch stress conditions.

Also, strength calculation does not take into account the fact that component size and the size of the sample from which uni-axial design values were determined may not be the same. Finally, in setting up a strength condition for fatigue fracture, there is no guarantee that the surface properties of the component are the same as those of the sample bars used to determine uniaxial fatigue values.

1.5.1 Constructional Design and Static Strength Properties

Effect of Notches. In contrast to the uni-axial homogeneous stress distribution present in bars subject to tensile stress, the strength behaviour of components affected by multi-axial notch stress conditions shows marked stress peaks of the surface, depending on the design. Taking linear-elastic behaviour into account as in **Fig. 19**, the stress peaks occurring in the notch bottom for tension, bending or torsion can be defined by the stress form factor α_k (e.g. $\alpha_{k\,ten} = \sigma_{1\,max} / \sigma_{1n}$).

For the same notch geometry, difference α_k values in the inequality $\alpha_{k\,ten} > \alpha_{k\,ben} > \alpha_{k\,tor}$ are obtained, depending on the type of stress.

From calculations (e.g. by the finite-element method) and from many experimental investigations, values of stress form factor α_k are known for different notch cases in design practice. Using the equation given in **Appendix D1, Table 4**, and the associated factors and exponents, determined by the finite-element method, stress form factors can be calculated for flat and circular bars notched and stepped on both sides for various stress cases [51].

If, for a ductile material under continuous stress, a notched bar is only loaded to the flow limit R_e/α_k, the material would be under-utilised. Loads on ductile materials, therefore, can be considerably higher than that which would initiate flow at the notch bottom, where, without appreciable increase in edge yield stress, the plastic zone reaches a greater depth, until the limit of load-bearing capacity is obtained in the fully plastic state. This

applies for ideal elastic-plastic materials without cold hardening (**Fig. 20**).

From the notched bar yield curve [52], it can be seen that strength increases with plastic edge strain and depth effect in the plastic zone. The ratio of load increase after yield F_{pl} to yield load limit F_F, also known as the section number n_{pl}, is a suitable indication of increased strength [11]: $n_{pl} = F_{pl}/F_F > 1$.

For brittle conditions, these considerations do not apply. In this case, there is no flow condition, but a fracture condition $R_{mk} = \sigma_{1\,n} = \sigma_{1\,max}/\alpha_k$.

The relative notch tensile strength $\gamma_k = R_{mk}/R_m$ as a function of α_k proves to be a suitable criterion for judging ductile or brittle behaviour under notch stress conditions. With increasing stress form factors, ductile materials show relative notch tensile strength values $R_{mk}/R_m > 1$, while brittle conditions give values of $R_{mk}/R_m < 1$. For ideal brittle material conditions, there is the limit

$$R_{mk}/R_m = (\sigma_{1\,max}/\alpha_k)/\sigma_{1\,max} = 1/\alpha_k \leq 1.$$

Size Effect. To transfer design values obtained on samples to actual components, the size effect must be taken into account. Assuming elasto-mechanical similarity, it has been shown on geometrically similarly notched test bars that yield point and yield curve are virtually independent of geometric size effect for small plastic deformations [53]. However, in notch tensile tests in the diameter range of 6 to 180 mm, it has also been shown that notch specimens of C60 steel ($\alpha_k = 3.85 = $ constant) below

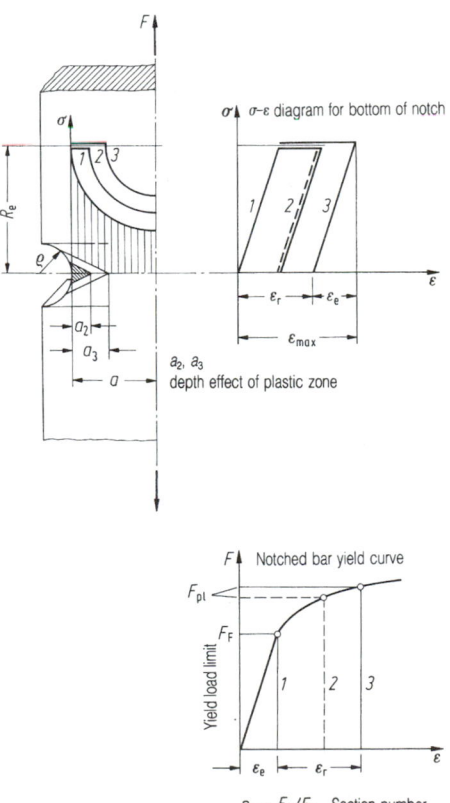

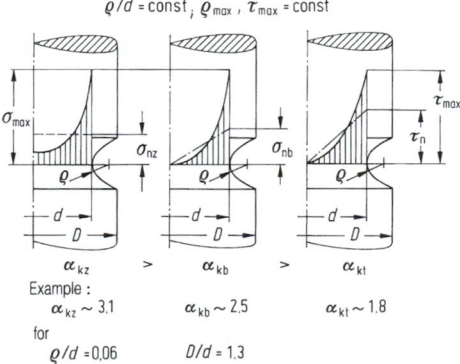

Figure 19. Stress – definition for tensile, bending and torsional stress.

Figure 20. Support effect in notched bars with partial plastic deformation.

80 mm external diameter have a notch tensile strength ratio < 1. This indicates a clear size effect, and therefore even with quasi-static stress there may be a transition from ductile to brittle component behaviour at certain critical diameters.

1.5.2 Design of Structures and Dynamic Strength Properties

Effect of Notches. Assuming linear-elastic behaviour in the fatigue strength range, it can be expected that for notched bars and therefore for notched components, the alternating stress amplitude at the notch bottom is increased by α_k times the rated stress and therefore that fatigue strength σ_{Dk} of notched samples or components can be reduced to the lowest value of rated stress $\sigma_{Dk} = \sigma_D/\alpha_k$ given by elasticity theory. Many investigations have shown that the reduction in fatigue strength of notched samples is less than would be expected from minimum values of elasticity theory [54]. Depending on the sharpness of the notch and the size of the notch bottom diameter, considerably higher fatigue strength values are obtained due to the support effect. This means that in the diameter region < 100 mm, fatigue strength properties of notched samples depend not only on the stress form factor α_k, but also on other factors. Rather than the stress form factor α_k, fatigue strength properties can be summarised by the fatigue notch factor β_k:

$$\sigma_{Dk} = \sigma_D/\beta_k; \quad \beta_k = \sigma_D/\sigma_{Dk}; \quad \beta_k \leq \alpha_k.$$

The fatigue notch factor β_k can not only be determined experimentally, but can also be calculated by various processes [55].

The notch sensitivity factor defined by A. Thum, $\eta_k = (\beta_k - 1)/(\alpha_k - 1)$, can be estimated as a first approximation by the following equation, according to H. Neuber [56]:

$$\eta_k = \frac{1}{1 + \sqrt{\dfrac{a}{r}}}$$

where r = notch radius, a = material constant, $a = f(R_m)$.

Figure 21 shows the shape of $\eta_k = f(r)$ for heat-treatable steels of different strength properties. The hardness values converted from tensile strength values (HB $\approx$ $R_m/3.5$) allow values of η_k to be calculated for tensile-compressive, bending and torsional stresses in notched bars with varying support effect. Similar relationships are known for high-strength aluminium alloys.

Size Effect. In order to be able to apply fatigue strength data from single-stage tests on smooth and notched samples to components stressed in a single stage, all

important size effect parameters must be known. They can be divided into the following individual mechanisms [54]: *technological* size effect, *stress mechanics (geometric)* size effect, *statistical* size effect [57] and *surface technology* size effect.

Figure 22 shows, as examples of the stress mechanics size effect, the fatigue strength properties of iron-graphite materials compared with a heat-treatable steel, related to sample size [54]. As statistical test evaluations show, the support effect due to a drop in stress declines very rapidly in iron-graphite materials, but the tensile-compressive alternating strength becomes largely independent of diameter above about 20 mm.

Only a few experimental results for estimating the size effect mechanism which permit a decoupling of the individual mechanisms are available in the literature. **Figure 23** shows an evaluation by R. Hänchen to determine the size effect factor C_D, which takes into account both the stress-mechanics and the technological size effect of heat-treatable steels for bending or torsional stress [58].

A final interpretation of the surface technology size effect is not yet possible. By the use of mechanical or thermo-chemical surface hardening processes (e.g. surface rolling, nitriding, case hardening), considerable increases in fatigue strength can be achieved in notched components [59].

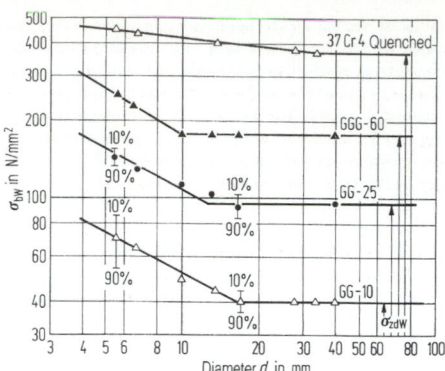

Figure 22. Alternating bending strength values of iron-graphite materials related to sample size.

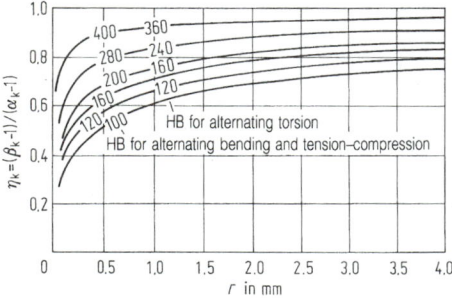

Figure 21. Notch sensitivity η_k related to notch radius r.

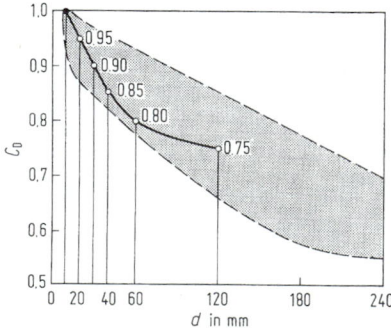

Figure 23. Size effect factor C_D in comparison with scatter area of test results.

Figure 24 shows a comparison of the fatigue strength properties for bending load of heat-treatable steel 30 CrNiMo 8 in the annealed state ($R_m = 900$ N/mm² for all diameters) and in the plasma-nitrided state (24 hours at 530 °C; mean nitrided depth 0.3 mm) relative to sample size in the range of 80 to 100 mm diameter. For small sample diameters in particular, great increases in fatigue strength are obtained by plasma nitriding.

1.6 Loadbearing Capability of Structural Components

Estimation in strength calculations is done without taking into account internal stress conditions. Strength calculations for components under vibratory stress assume that the surface properties of sample bars are identified with those of components as regards strength and internal edge stress.

The choice of safety factor for yield fracture or instability is important in dimensioning components. Account must also be taken of scatter in material properties, inaccurate calculation of comparative stress, and unsafe assumptions about loadings.

1.6.1 Static Load

With this type of load and under multi-axial homogeneous stress, strength calculation is carried out for the most highly stressed cross-section. With safety factor S, calculated reference stress σ_V must be less than the uni-axial material factor K:

$$\sigma_v \leq \sigma_{perm} = K/S.$$

For components with only a small degree of inhomogeneity of multi-axial stresses (e.g. a pipe with internal

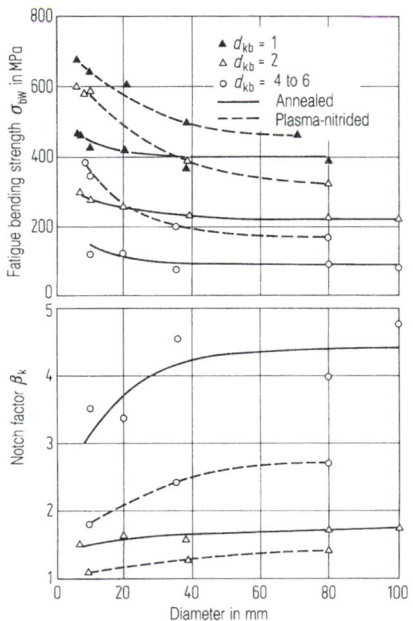

Figure 24. Alternating bending strength properties and β_k values of annealed and plasma-nitrided samples of 30 CrNiMo 8 ($R_m = 900$ N/mm²) of different diameters. Solid lines: annealed. Broken lines: plasma-nitrided.

pressure), a reference stress mean value $\overline{\sigma}_v$ calculated from the mean load stress values can be assumed.

For ductile components with marked inhomogeneity of multi-axial stress condition (e.g. notch effect on cross-sectional transitions), the flow limit can be exceeded at the most highly stressed location, so that the loadbearing limit is only reached on approaching the fully plastic state. The materials factor K for the yield condition has the section number n_{pl}: $K = n_{pl} R_e$ (see D1.5.1).

With S_F as the yield safety factor, the yield condition is $\sigma_v \leq \sigma_{perm} = (n_{pl} R_e)/S_F$.

If, on notched components, permissible stress is related to rated stresses, then

$$\sigma_{perm} = (n_{pl} R_e)/(\alpha_k S_F).$$

With sufficient deformation capability of the material, the ratio n_{pl}/α_k can be set > 1, i.e. the reduction in loadbearing capability due to the notch effect is compensated by the support effect [11]. The section number n_{pl} is either calculated or determined experimentally from component yield curves [52]. On the assumption of a permissible plastic strain of 0.2% at the root of the notch of a component, the section number can be calculated from the following numerical equation [55]:

$$n_{pl} = n_{0.2} \approx 1 + 0.75 \, (\alpha_k - 1) + \sqrt[4]{\frac{300}{R_e}},$$

where α_k is a non-dimensional factor and R_e is expressed in N/mm². Possible instability following buckling represents a special case of static component loading (see B7).

The component loadbearing capability under steady load can also be affected by multi-axial internal stress conditions. Depending on the depth effect of the source of the internal stress, multi-axial internal tensile stresses raise the component yield point, where the increasing partial plastic deformation of the internal stress condition is again reduced. At the limit, three axial hydrostatic tensile internal stresses can cause a danger of cleavage fracture, which can be estimated as follows, using the normal stress theory:

$$\sigma_{1 \, max} = \sigma_{1 \, load} + \sigma_{1 \, int}.$$

The superposition of load stress and internal stress presupposes that the tri-axial internal stress condition of the main axis system is known in magnitude and direction.

1.6.2 Loadbearing Capability Under Single-Stage Dynamic Load

The component capability under single-stage dynamic load known as the *fatigue limit* or *material service life* can be calculated from the properties of notched samples taking the most important factors (notch effect, size effect, surface effect) into account. The component properties under dynamic stress are affected by material, manufacturing technique and design factors, where by using mechanical and thermochemical edge-hardening processes, the most effective increase in fatigue limit can be achieved.

The dynamic strength σ_D of unnotched components, the surface of which has neither partial hardening nor softening (e.g. edge decarburisation), can be estimated from known dynamic strength properties of the material, taking the surface and size effect into account [24], as follows:

$$\sigma_D = \sigma_w C_D C_O,$$

where σ_w = dynamic strength, C_D = size effect factor, and C_O = surface factor.

The effect of surface quality on the fatigue limit of unnotched components can be assessed by the surface factor as in **Fig. 17**.

The dynamic strength of notched components σ_{Dk} with single-stage stressing can be calculated as follows:

$$\sigma_{Dk} = (\sigma_w\, C_D\, C_O)/\beta_{kO},$$

where β_{kO} is the corrected notch factor taking certain surface quality values into account.

The corrected notch effect factor is

$$\beta_{kO} = 1 + (\beta_k - 1)\, C_O.$$

In contrast to these simplified nominal stress equations, dynamic strength calculation can be improved by further factors (e.g. cross-sectional form factor for non-circular cross-sections or anisotropy coefficient) [60].

The dynamic strength properties of machines and plant from the whole range of machine apparatus and steel construction are greatly affected by the fatigue properties of individual design elements [61]. in the first place, screwed connections and welded joints [62] must be mentioned, the dynamic strength values of which have only been evaluated statistically in recent years.

Example. *Fatigue Bending Strength of a Notched Shaft.* A shaft of heat-treatable steel 37 Cr 4 (R_m = 830 N/mm^2) under bending has a circumferential notch of radius ρ = 3.6 mm. The outside diameter of the shaft is D = 50 mm, the diameter at the base of the notch D = 40 mm. The surface is finely ground ($R_r \le 5\,\mu$m). What is the alternating bending strength σ_{bwk} of the notched shaft? From the geometric data, there is a stress form factor of α_{kb} = 2.0. The bending fatigue strength of the shaft material is

$$\sigma_{bw} = 0.5 R_m = 415\ \text{N/mm}^2;$$
$$\sigma_{bwk} = (\sigma_w C_D C_O)/\beta_{kO}.$$

Size effect factor $C_D = 0.85$ (**Fig. 23**).
Surface factor $C_O = 0.9$ (**Fig. 17**).
$\beta_{kO} = 1 + (\beta_k - 1)\, C_O$; β_k is determined from the notch sensitivity (**Fig. 21**):

$$\eta_k = \frac{\beta_k - 1}{\alpha_k - 1} = 0.87$$

$\beta_k = 1.87$. From this it follows that $\beta_{kO} = 1.78$.
Result: $\sigma_{bwk} = (415 \times 0.85 \times 0.9)/1.78 = 178\ \text{N/mm}^2$.

1.6.3 Loadbearing Capability of Randomly Loaded Structural Components

In working conditions, components are usually subject to uncontrolled loads with statistically distributed vibration amplitudes and constant or variable mean loads, so that the component fatigue strength properties obtained from single-stage experiments have only limited value as design rules. In many applications of machine and steel construction, and particularly in light structures, vibratory stresses must be permitted whose stress deflection is more than twice the permanent fatigue strength, so that damage may occur due to alternating slip processes in the endurance strength range.

For the quantitative assessment of component damage in the endurance strength range (*cumulative damage*), measurement processes are required that refer irregular loading to a sequence of alternating load cycles of a certain magnitude and frequency. Using various measurement processes (e.g. the peak-value process, the endurance process, and the class passage process as in DIN 45 667), frequency distributions and the frequency of working loads or nominal stresses can be compiled. However, the information content of real stress–time curves is lost in such a collective picture.

Figure 25 shows three different stress–time curves

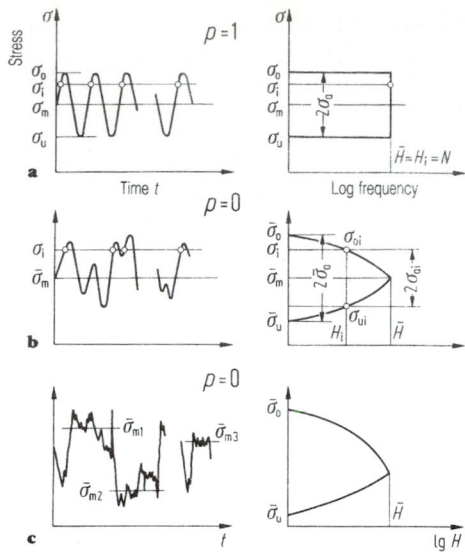

Figure 25. Effect of different stress–time functions on the group stress: **a** constant amplitude and mean stress, **b** variable amplitude and constant mean stress, **c** variable amplitude and variable mean stress.

and the associated collective stresses that were determined by the class passage process [63]. For a clear identification of a group of stresses, the total frequency H, the group form according to a given statistical distribution law, the upper and lower maximum stress values $\overline{\sigma}_o\,\overline{\sigma}_u$ or the maximum stress amplitude $\overline{\sigma}_a$, and the associated mean stress $\overline{\sigma}_m$ are required.

Starting from the stationary random process with normal distribution (**Fig. 26**), for stress–time functions, the mixed group above the normal distribution can be approximated by the normal group in a certain load range. The group coefficient p represents the ratio of minimum to maximum amplitude in the group and lies within the limits $0 \le p \le 1$, according to **Fig. 26**. The life prediction of components under random load–time functions can be made by using calculation processes and by programme random testing.

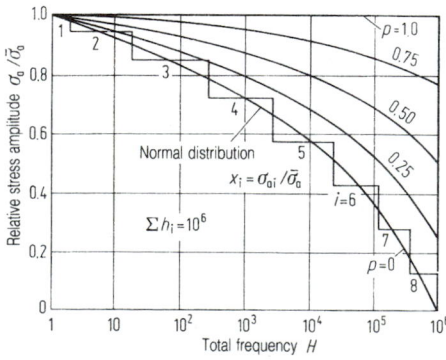

Figure 26. p-group value and possible distribution for block programme experiments.

Calculated Life Prediction

This can be done for known group loads and experimentally determined component Wöhler curves in the fatigue and endurance strength range using a suitable damage accumulation theory. The theory postulated by Palmgren and Miner assumes a linear growth of damage with the number N_i of alternating cycles, where a partial damage of $1/N_i$ occurs for each load cycle if N_i is the number of breaking load cycles for a stress deflection σ_{ai}. If the group load is replaced by a multi-stage load according to **Fig. 27**, the individual partial damage n_i/N_i for m load stages are summed as follows:

$$S = \frac{n_1}{N_1} + \frac{n_2}{N_2} + \frac{n_3}{N_3} + \ldots = \sum_{i=1}^{m} \frac{n_i}{N_i} = 1.$$

A fatigue fracture occurs if the damage sum $S = 1$.

The group load can be divided into a number of sequences, whose total damage per stage and sequence is $S_i = b_i/N_i$, where b_i is the number of alternating cycles (partial damage) per load stage of a sequence.

The total damage at fracture is obtained from Z = number of sequences as

$$S = \sum \frac{n_i}{N_i} = Z \sum \frac{b_i}{n_i} \quad (\text{for } Z = 1, \, n_i = b_i).$$

The frequency maximum of the total damage calculated according to Miner is at $S = 1$ [64]. However, considerable deviations ($S \gtrless 1$) are possible. Total damage values of $S = 0.3$ are frequently recommended for the predimensioning of components.

To reduce the difference between the experimentally determined life and the calculated life, various improvements were proposed, such as, for example, extending the fatigue strength line to $\sigma_a = 0$ with half the angle (see **Fig. 27** [65]).

Experimental Determination of Service Life

This is done with programmed experiments and random tests. A programme in which the group amplitude is divided into eight stages and sequences with $0.5 \cdot 10^6$ load cycles each has gained great importance [66]. In order to achieve a practical mix of high- and low-stress amplitudes, the stress values of each sequence are applied first in ascending and then in descending order. The results of a programmed experiment can be shown like the Wöhler curve as a life line in the fatigue strength range.

Vibration tests which try to simulate actual stress–time functions are called random tests. They are divided into working load tests (simulations), randomised programmed experiments, and random process tests [67]. In experimental comparisons between programmed and random tests, it was shown that random tests with realistic stresses give a shorter component life than different block programme tests [68].

1.6.4 Loadbearing Capability of Structural Components Under Creep Conditions

Unlike room temperature design, strength calculations for components working above the crystal regeneration temperature T_k of the heat-resisting material require design values depending on time and temperature. For calculating permitted stresses, the following equations are obtained with working temperature T, loading time t, deformation safety factor S_F and fracture safety factor S_B:

$$T < T_k: \quad \sigma_{perm} = R_{p\,0.2/T} / S_F,$$
$$T > T_k: \quad \sigma_{perm} = R_{p\,0.2/t/T} / S_F,$$
$$\sigma_{perm} = R_{m/t/T} / S_B.$$

Taking into account the 100 000-hour creep-strength values at working temperatures above T_k, fracture safety factors down to the lowest value of $S_B = 1.5$ can be selected.

Smooth samples are usually used for strength calculations of components at high working temperatures. This is permissible only if the creep strength of notched samples ($R_{mk/t/T}$) after long loading periods is greater than the creep strength of smooth samples. With sufficiently ductile materials, component creep strength is raised compared with smooth samples due to design notch effects and the prevention of deformation caused by these due to tri-axial tensile stress conditions. This can be established by the creep notch strength ratio γ_k of notched samples or components:

$$\gamma_k = R_{mk/t/T}/R_{m/t/T} \geq 1,$$

where $R_{mk/t/T}$ is the creep strength of notched samples or components.

The loadbearing capability of structural components calculated for given operating periods at high temperature based on long-term strength values (e.g. 100 000-hour creep strength) only applies for constant assumed load and temperature. The life of components operated at high temperature is changed by variable operating conditions (e.g. excess and low temperature).

Figure 28 shows the permitted stress $\sigma_{perm} = R_{m/t/T}/S_B$ for 13 CrMo 44 relative to the duration of stressing at different working temperatures. If 10 000-hour and 100 000-hour creep-strength values are available, straight lines can be drawn approximately through these points on a log–log scale. Care must be taken in extrapolation, however, as the lines are frequently slightly curved downwards [69].

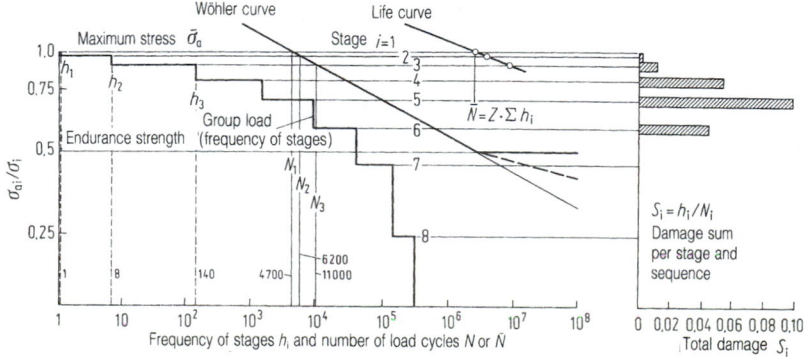

Figure 27. Calculation of damage sum according to Palmgren and Miner (eight-stage test).

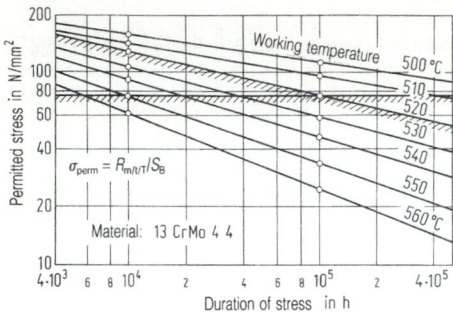

Figure 28. Determination of stress safety of components operating at high temperature.

The rule of linear damage accumulation for heat resistant steels applies only to a limited extent for calculating service life at variable working temperatures [70].

A similar life calculation is possible for variable stress at constant or variable working temperature.

1.6.5 Factors of Safety (for Guidance)

The magnitude of the safety factor $S > 1$ guarantees that permissible stress has a defined margin over the failure limit stress at yield, fracture or instability. By stating safety factors, all the uncertainties involved in the strength calculation should be compensated.

The safety factors represent minimum values (S_{min}), which must be increased further depending on uncertainties in the calculation process, load assumptions, and the scatter of materials design values. According to VDI Guideline 2226 [55], the total safety factor is

$$S = S_{min} \cdot S_1 \cdot S_2 \cdots S_n,$$

where S_{min} = the minimum safety factor for the basic stresses (yield, fracture, instability) and S_1 to S_n are the uncertainties. **Table 1** shows the safety factors for static load, depending on working temperature and type of failure [11]. In choosing these, account must be taken of the increased danger of fracture, the lower the plastic deformation reserve of the material used. Conversely, with

Table 1. Safety factors for static stress [11]

Temperature	Safety factors for		
	Yield S_F	Fracture S_B	Instability S_k
$T < T_k$	1.2 to 2	2 to 4	3 to 5
$T > T_k$	1 to 1.5	1.5 to 2	3 to 5

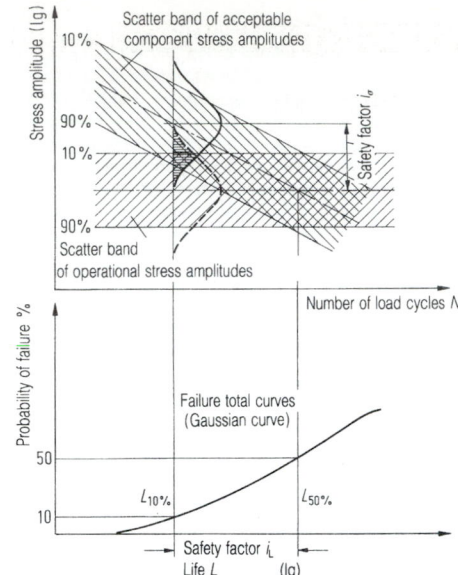

Figure 29. Safety and probability of failure of components under alternating stress.

increasing plastic deformation capability, stress peaks in notches due to local yielding have less effect.

Components that are exposed to high-temperature fatigue require a temperature safety factor in addition to the stress safety factor. This is the difference between working temperature and that temperature at which the creep strength $R_{m/10^5}$ is reached for the same stress [11]:

$$S_T = T_{fracture/10^5} - T_{working}.$$

For alternating stresses, statistical safety factors are increasingly being used, where taking the scatter into account, acceptable operational stress amplitudes and distributions can be given for a clear probability of failure, depending on the number of alternating load cycles.

Figure 29 shows the scatter band of acceptable stress amplitudes of a component with the probability of survival $P_{ü} = 90$, 50 and 10%, and the scatter band of operational stress amplitudes [71]. Depending on the number of load cycles, the probability of failure can be shown for the failure total curve and can be calculated for each life value from the mean and standard deviations of the two scatter distributions. As can be seen from the curve of the 50% values of acceptable stress amplitudes and the working stresses, the stress amplitude scatter bands are displaced relative to one another, so that the probabilities of failure rise with increasing life. According to this process, statistically certain safety factors can be given for components under alternating stress, which must be allocated to the mean of the two scatter distributions (**Fig. 29**).

2 Materials Testing

K. H. Kloos, Darmstadt

Materials testing supplies values which characterise both test bar and component properties under mechanical, thermal or chemical stress conditions. Test processes are also available to assess machining properties, which can frequently only be specified by several values. These materials testing processes are also used for the objective assessment of material properties in incoming inspections, product and manufacturing monitoring, and to establish causes of damage.

2.1 Fundamentals

Mechanical, technical and chemical test processes simulate characteristic working stresses, often assuming idealised stress conditions. The transferability of design values obtained from small samples frequently presents considerable difficulties in determining realistic dimensions.

Test processes are divided into destructive and nondestructive processes. For special safety requirements (e.g. the aviation industry), random sample testing is no longer permitted and must be replaced by 100% testing, which is partly integrated in the manufacturing process. Various components from reactor technology are not only monitored during manufacture and during final acceptance, but can be inspected during stressing by well-established non-destructive test procedures.

2.1.1 Sampling

Normally smelted steels exhibit a distinctive anisotropy in their ductile properties, so that the position of the samples in the component must be stated in terms of length, width and thickness. In large components, due to solidification conditions, there may be great differences between the strength properties of the core and the edge. **Figure 1**

shows an example of a shaft for a large generator, with different solidification areas in the block and the possibility of taking radial and axial samples to ensure quality. Differences in the mechanical properties can occur particularly in the segregation areas of carbon, phosphorus and sulphur. In components stressed on several axes, samples should be taken in the direction of the greatest normal stress. Special requirements for sampling are made in ensuring the quality of cast components. The mechanical properties of cast samples can agree with the material properties of the casting only if the cooling conditions are the same in both cases. This applies particularly for iron-graphite materials, the mechanical properties of which greatly depend on the shape and distribution of the graphite.

Standards. DIN 1605: Mechanical testing of metals, general and acceptance – DIN 50 108: Testing of cast iron with lamellar graphite, sampling for tensile and bending tests.

2.1.2 Evaluation of Tests

When determining material properties, the spread of results caused by differences in chemical composition of the samples and by manufacturing and test technique effects is also important. Design values determined statistically are therefore often needed when determining safety factors in strength calculation.

Evaluation Process for Static Materials Design Values

Most static materials design factors are determined from mean values. Apart from the arithmetic mean, it is frequently necessary to state the lower limit (minimum value) below which no sample falls. When using the minimum value in component strength calculation a 100% survival rate may be assumed, while the use of the arithmetic mean gives a survival probability of only 50%.

The variance or standard deviation and the coefficient of variation derived from it have proved to be useful for specifying the spread of materials design factors.

For economic reasons, it is only rarely possible to determine the materials design factors from a fundamental whole. If the above parameters are known for a sample, then by stating the confidence range (confidence limit), conclusions can be drawn from the sample which apply to the fundamental whole. A certain probability may be stated, so that the true mean value of the fundamental whole lies within a precisely defined interval around the mean value of the sample.

Evaluation Process for Dynamic Strength Values

Owing to the large number of dynamic parameters, all definitive fatigue strength values should be coupled with a statement of a given *probability of survival or fracture*, for which a large number of samples are required.

With only a few samples per load horizon and a small number of sample figures per Wöhler curve, an improvement of the evaluation process is possible where, based on the observed distribution diagram of the test values, appropriate distribution laws can be formulated with sufficient accuracy. The best-known distribution laws are the Gaussian normal distribution, the Gumbel extreme value distribution (the Weibull distribution is a special case of

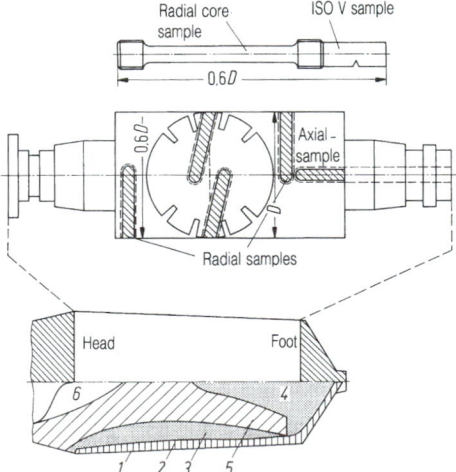

Figure 1. Sampling for large forged parts and effects due to solidification in the lattice formation and segregation zones: *1* edge zone (fine crystallite), *2* stem crystals, *3* contact dendrites, *4* coarse spherical crystals, *5* segregation zone, *6* solidification with little segregation.

this) and the arcsin $\sqrt{p}$ transformation. To obtain an acceptable probability level, at least ten samples are required per load horizon. On the assumption of a normal distribution, two evaluation processes are at present being used to determine fatigue strength [1-4].

Staircase Method. A large number of samples (17 to 35) are tested one after the other at several load stages, where the level of stress depends on whether the previously tested sample broke or reached the cycle number limit [2]. In the case of a fracture, the load is reduced by one stage. The size of the step should be about 4 to 5% of the stress amplitude if small scatter is anticipated, and about 8 to 12% of the stress deflection with large scatter. The optimum number of steps appears to be four.

Table 1, from VDI Guideline 2227 [5], shows calculations and evaluations to determine the dynamic strength values by the staircase method.

Limitation Method. Here a sample is again stressed at the level of the expected fatigue strength. If the sample breaks, the load is reduced until the first sample survives. If the series starts with an unbroken sample, the load is increased until the first fracture occurs. At least eight samples are then tested at the level of the first survivor or fracture. The second load horizon σ_{a2} can be calculated for the number of fractures r, and the total number of samples n:

$$\sigma_{a2} = \sigma_{a1} + \left(1 - \frac{r}{n}\right) \cdot 0.1\sigma_{a1} \quad \text{for } \frac{r}{n} \leq 0.5,$$

$$\sigma_{a2} = \sigma_{a1} \frac{r}{n} \cdot 0.1\sigma_{a1} \quad \text{for } \frac{r}{n} \geq 0.5.$$

The same number of samples as on the first load horizon is tested on the second one. The probability of fracture

$$P_B = \frac{3r - 1}{3n + 1} \times 100\% \quad \text{or} \quad P_B = \frac{r}{n + 1} \times 100\%$$

is calculated for both load horizons and is entered in a probability network (e.g. normal distribution or extreme value distribution) on the selected load horizon. The straight line drawn through the two points allows load horizons for probability of fracture for 10, 50 and 90% (**Fig. 2**) to be determined.

Standards. DIN 1319 Sheets 1-3: Basic terms of measurement technique. - DIN 53 589 Sheet 1: Statistical evaluation on samples with examples of elastomer and plastic tests. - DIN 55 302 Sheets 1 and 2: Frequency distribution, mean value and scatter. - DIN 50 100: Fatigue test.

2.2 Test Methods

Within the group of mechanical test processes, strength and ductility tests, and the test procedures of linear-elastic fracture mechanics for determining ductility at fracture, occupy a central position. Most strength tests can be divided as follows for different fundamental loads: static short-term test procedures: tensile test, compression test, bending test, torsion test; static long-term test procedures: creep test, relaxation test; dynamic short-term procedures: notch impact bending test (Izod test), impact tearing test; dynamic long-term test procedures: long-term vibration test, single-stage, multi-stage and reproduction test.

2.2.1 Tensile Test

Purpose. Used to determine mechanical properties with homogeneous uni-axial tensile stresses.

Table 1. Calculations for staircase method [5]

No. of stress stages	Stress amplitude, $\pm\sigma_a$ (N/mm²)	After 2×10^6 cycles, 37 successively tested samples		Number at each stage of		Frequency of event[a] (no fracture)		
				Broken samples	Unbroken samples			
i	χ_i	To fracture ○ Not to fracture ●		r	l	f_i	if_i	i^2f_i
3	± 130.5		Failed specimen	2				
2	± 127.0			8	2	2	4	8
1	± 123.5		Unbroken specimen	9	7	7	7	7
0	±120.0				9	9		
3	3		Totals	19	18	$F = 18$	$A = 11$	$B = 15$

Mean $m = \chi_0^b + d \left(\dfrac{A^c}{F} \pm \dfrac{1}{2}\right) = 123.9$ N/mm².

Standard deviation $s = 1.62d \left(\dfrac{FB - A^2}{F^2} + 0.029\right) = 2.77$ N/mm².

Check whether formula is valid for standard deviation s:

$\dfrac{FB - A^2}{F^2}$ must be > 0.3: $\dfrac{FB - A^2}{F^2} = 0.46$.

Probability of survival $P_{\ddot{u}} = 90\% \approx m - s \cdot 1.28 = 120.4$ N/mm².

$P_{\ddot{u}} = 10\% \approx m + s \cdot 1.28 = 127.4$ N/mm².

Preconditions for staircase process:
1. Number of samples sufficiently large, if possible ≥ 40.
2. Step $d \geq 2s$.
3. Frequency distribution of dynamic strength is a normal distribution.
[a] Enter the less frequent event here, i.e. "not broken".
[b] χ_0 = lowest stage of less frequent event, i.e. "not broken".
[c] + less frequent event "not broken", in example therefore +; − less frequent event "broken".

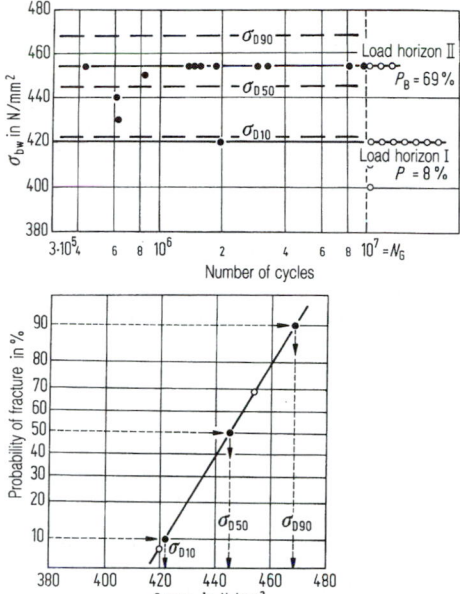

Figure 2. Example of determining probability of fracture values for alternating bending stress by the limitation method.

Sample Geometry. Values are determined using samples of circular, square or rectangular cross-section. In order to be able to compare strains at fracture, certain measurement length ratios must be maintained. The measured length is usually selected to be 5 or 10 times the rod diameter. For bars of different cross-sectional form, the diameter of a circle of the same area is used to determine the measured length:

$$S_0 = \frac{\pi d_0^2}{4} \rightarrow d_0 = 2\sqrt{\frac{S_0}{\pi}}$$

(S_0 = initial cross-section).

In general, proportional bars are used, where the measured length $L_0 = 5d_0$ (short proportional bar: $L_0 = 5.65\sqrt{S_0}$) or $L_0 = 10d_0$ (long proportional bar: $L_0 = 11.3\sqrt{S_0}$). When testing brittle materials it must be ensured that clamps do not introduce bending.

Values

Strength. When passing steadily from the elastic to the plastic range, the 0.2% strain limit $R_{p\,0.2}$ (0.2 limit) is determined, where the 0.01% strain limit is called the technical elastic limit. For a sudden transition, an elastic limit R_e which can be divided into a lower and upper elastic limit is determined (**Fig. 3**).

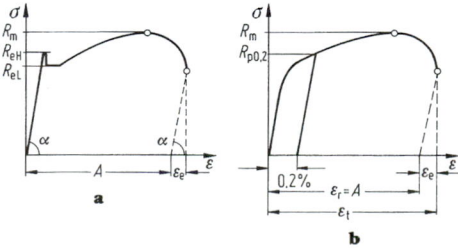

Figure 3. Strength and deformation values in tensile test: **a** with characteristic elastic limit, **b** with strain limit.

The tensile strength $R_m = \dfrac{F_{max}}{S_0}$ is the stress which is obtained from the maximum force relative to the initial cross-section S_0. The usual symbols for strain limits, elastic limits and tensile strength are $\sigma_{0.2}$, σ_s and σ_B.

Deformation. Fracture strain A is the change of length relative to the initial measured length L_0 after the sample has fractured:

$$A = \frac{L_u - L_0}{L_0} \times 100\%.$$

The strain at fracture consists of uniform strain plus elongation due to necking. It depends on the material and the length of the reference section L_0. As the necking elongation for a measured length $L_0 = 5d_0$, compared with the even strain, is more important, the A_5 values are greater than the A_{10} values. The necking at fracture Z is obtained from the difference between initial and fracture area, relative to the initial area:

$$Z = \frac{S_0 - S_u}{S_0} \times 100\%.$$

Elastic Modulus. By Hooke's law, the elastic modulus in the elastic range of the stress–strain diagram can be determined as follows:

$$E = \sigma / \varepsilon_e = (F/S_0)/(\Delta L/L_0).$$

For materials with non-linear stress–strain curves (e.g. iron–graphite materials), the tangential modulus can be given as the slope of the σ–ε curve at the point $\sigma = 0$:

$$E_0 = \left| \frac{d\sigma}{d\varepsilon} \right|.$$

Special Test Procedures

Hot Tensile Test. This is used to determine mechanical material properties at high temperature. The hot elastic limit, hot tensile strength, extension at fracture and necking at fracture are determined. The hot elastic limit and hot tensile strength depend on the test time as well as temperature. To obtain reproduceability of the material design values, it is necessary to keep within limits for the rate of increase of stress and strain.

Impact Test. This is used to determine the liability to brittle fracture of smooth or notched tensile samples at impact speeds between 5 and 15 m/s, in exceptional cases up to 100 m/s (high-speed forming). To determine impact strength, necking of the sample at fracture is determined. Determining the impact strength or impact elastic limit presupposes dynamic force and deformation measurements.

Standards. DIN 50 145: Tensile test; terms, symbols. – DIN 50 125: Tensile tests. DIN 50 114; Tensile tests on sheets. – DIN 50 140: Tensile test: on tubes and strip for tubes. – DIN 50 120: Tensile test on welded butt joints (steel). – DIN 50 123: Non-ferrous metals. – DIN 50 127: Notch tensile, pipe notch tensile sample of welded butt and fillet joints. – DIN 40 148: Tensile samples for non-ferrous metal castings. – DIN 50 109: Tensile test on cast iron with lamellar graphite. – DIN 50 149: Tensile test on tempered casting. – DIN 52 188: Testing timber, tensile test. DIN 53 455: Testing plastics, tensile test. DIN 53 504: Testing elastomers, tensile test. – DIN 51 211: Tensile test machines.

2.2.2 Compression Test

Purpose. This is used to determine mechanical material properties under homogeneous, uni-axial compression stress, and is used on metals, minerals, concrete and other building materials. The compression test can also be used to determine the yield curve of ductile materials.

Sample Geometry. The test is carried out on circular or prismatic bodies between two parallel plates. In normal cases, the length of the sample is equal to its thickness. When fine strain is measured, a greater sample length is required, but no more than 2.5 to 3 times the thickness of the sample (danger of folding).

Materials Design Values

Brittle Materials. The compression strength is the maximum load relative to the initial cross-section at which fracture occurs: $\sigma_{dB} = F_B/S_0$.

For geometrically similar samples, the compression strength is comparable. For the same test diameter, compression strength decreases with sample height, owing to the different support effect of the "pressure cones" (**Fig. 4**).

Ductile Materials. The start of plastic flow is characterised by the compression yield limit σ_{dF}, which is analogous to the tensile yield point. Owing to surface friction there is a bulge at the centre of the samples. Total fracture of the sample does not occur, but some cracks are formed as a result of transverse tensile stresses.

Special Test Procedures. To determine the yield stress k_f (former designation: deformation strength), a cylinder compression test is used. To ensure uni-axial deformation under compression, friction must be kept low. The k_f values make it possible to calculate the ideal forces and work required in hot and cold forming processes.

Standards. DIN 50 106: Pressure test on metal materials. – DIN 1048: Regulations for concrete tests for works consisting of concrete and reinforced concrete. – DIN 52 105: Testing natural rock: compression test. – DIN 52 185: Testing timber: compression test. – DIN 53 454: Testing plastics: compression test. – DIN 51 223: Compression test machines.

2.2.3 Bending Test

Purpose. This is used to determine mechanical properties of steel, cast materials, wood, concrete, and building materials under inhomogeneous uni-axial bending stress. For ductile materials, it is used to determine bending yield point and maximum bending angle, and for brittle materials it is used to determine bending strength.

Sample Geometry. The test is carried out on either samples or components. The sample is supported at both ends and is loaded by a single force in the centre.

Design Values

Brittle Materials. The bending strength σ_{bB} can be calculated from the maximum bending moment $M_{b\,max}$ and the section modulus of the sample. It is determined as a design value preferably for tool steel, fast-cutting steel, hard metals and oxide ceramic materials. The bending strength of iron–graphite materials with non-linear stress–strain curves is calculated by the same equation, where the bending strength exceeds the tensile strength, depending on sample cross-section.

Ductile Materials. The start of plastic flow is determined by the bending yield point σ_{bF} (**Fig. 5**).

Special Test Procedures. Notch impact bending test: see D2.2.5. Technical tests: see D2.2.9.

Standards. DIN 50 110: Testing cast iron bending test – DIN 1048: Regulations for concrete tests for works consisting of concrete and reinforced concrete. – DIN 52 186: Testing timber; bending test. – DIN 53 452: Testing plastics; bending test. – DIN 51 227: Bending test machines. – DIN 51 230: Dynstat equipment to determine the bending strength and impact strength on small samples.

2.2.4 Hardness Test Methods

Purpose. Within limits, hardness tests can be described as non-destructive test procedures. Depending on the procedure used, the hardness values represent a direct means of comparison for material abrasive wear resistance. In individual procedures, there are some approximate relationships between hardness and tensile strength. Further, macro- and micro-hardness test methods are suitable for evaluating ductility for small volumes.

Methods. Static hardness test methods can be described as penetration processes, in which the penetration resistance of defined bodies (spheres, pyramids, cones) is determined at a material surface. Depending on the test method, penetration resistance is either determined as the ratio of test force to surface area of the impression (Brinell hardness, Vickers hardness) or as the remaining depth of penetration of the penetrating body (Rockwell hardness).

Design Values

Brinell Hardness Test. Brinell hardness is calculated from the ratio of test force to surface area of the remaining spherical impression. The numerical equation is

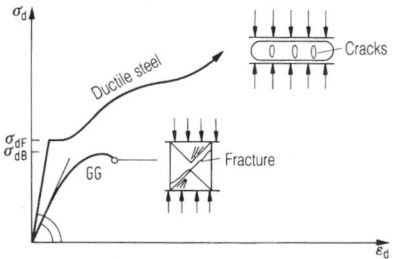

Figure 4. Stress–strain diagram of ductile steel and iron–graphite material in compression.

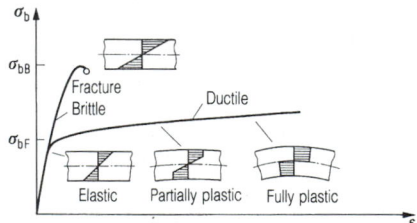

Figure 5. Stress–strain diagram of a brittle and a ductile steel in bending.

$$HB = \frac{0.102 \times 2F}{A} = \frac{0.102 \times 2F}{\pi D \left(D - \sqrt{D^2 - d^2}\right)}$$

where F = the test force in N, D = the sphere diameter in mm and d = the indentation diameter in mm.

The results obtained with different sphere diameters (10, 5, 2, 2.5 mm) are only comparable if the loads are the square of the sphere diameter. Numerical equation with load X: $F = SD^2$.

Depending on material hardness, loading is in steps between $X = 30$ and $X = 0.5$. For different loads, different hardness values are obtained, as the indentations are not geometrically similar.

The symbols used in Brinell hardness are hardness HB, ball diameter D, test force F in N multiplied by 0.102 and the period of action in seconds, e.g. 120 HB 5/250/30 (non-dimensional).

The relationship between Brinell hardness and tensile strength of steel is R_m in N/mm^2 = 3.5 × HB 30.

Vickers Hardness Test. The Vickers hardness is calculated from ratios of test force and surface area of the remaining pyramidal indentation. The numerical equation of test force F in newtons with diagonal length of indentation d in mm is

$$HV = 0.102F/A = 0.190F/d^2.$$

The usual loads are 98 and 294 N. Owing to the geometrical similarity of the indentations, the Vickers process is independent of load above 100 N.

The symbols used in Vickers hardness are hardness value HV, test force F in N multiplied by 0.102, and the time the test force acts, e.g. 640 HV 30/10.

The use of test loads between 2 and 50 N (small load range) makes hardness measurement of thin layers possible. With test loads below 2 N, hardness measurement of individual components of the structure (micro-hardness test) is possible.

Rockwell Hardness Test. In this process, the penetrating body (cone or ball) is pressed into the sample in two load stages and the remaining depth of penetration e is measured. The Rockwell hardness is obtained from the difference between a fixed value A and the depth of penetration e. The dimension for e is 0.002 mm (DIN 50 103). If a conical body is used: HRC = 100 − e.
For a spherical body: HRB = 130 − e.

There is no direct conversion of Rockwell hardness into Vickers or Brinell hardness. The individual hardness values from all three test procedures can be given in hardness comparison tables.

Special Test Procedures

Dynamic hardness test procedures (drop hardness test, recoil hardness test).
Hardness test at high temperature (hot hardness test).

Standards. DIN 50 351: Brinell hardness test. – DIN 50 133: Vickers hardness test. – DIN 50 103: Rockwell hardness test. – DIN 50 150: Conversion tables for Vickers hardness, Brinell hardness, Rockwell hardness and tensile strength. – DIN 51 200: Hardness test, guidelines for configuration of holding devices. – DIN 50 132: Brinell hardness test at temperatures up to 400 °C. – DIN 51 224: Hardness test equipment with depth of penetration measuring device. – DIN 51 225: Hardness test with optical indentation measuring device.

2.2.5 Notched-Bar Impact Bending Test

Purpose. This is used to assess the ductility of metallic materials under special test conditions. Owing to high stress rate and multi-axial tensile stress conditions, the transition from ductile fracture to brittle fracture can be determined at certain temperatures, where the position of the transition temperature acts as a comparison for ductility. To determine the transition temperature, various criteria have proved useful, e.g. a certain value of notch impact work or a defined proportion of honeycomb or cleavage fracture areas.

Sample Geometry. Design values are obtained from samples of mainly square cross-section, with notches of defined geometry on the tension side. **Figure 6** shows an example of the frequently used DVM sample and the ISO acute notch sample, together with dimensions and arrangement of the bending sample in the mechanism. The similarity law does not apply; therefore it is essential to state sample geometry in all notched-bar impact tests.

Materials Design Values. In the notched-bar impact bending test, the impact work $A_v = G\,(h_1 - h_2)$ absorbed by the counterbearing to break the sample is given in newton metres (N m) or joules (J). The shape of sample is also added when stating the notch impact work, e.g. A_v (DVM) = 80 J.

The impact work absorbed relative to test cross-section A is given as the notch impact ductility a_k: $a_k = A_v/A$.

If no test temperature is stated, the notch impact ductility properties refer to 20 °C. To be able to determine the transition temperature T_0 notch impact tests at different temperatures are required (**Fig. 7**).

When comparing steels with different transition temperatures, the material with the highest transition temperature is most liable to brittle fracture. To determine the liability of ageing of steels, samples are compressed 10% in the transverse direction and are aged at 250 °C for $\frac{1}{2}$ hour (artificial ageing).

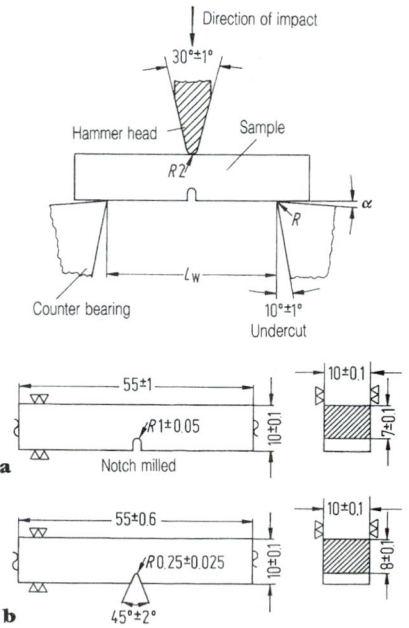

Figure 6. Dimensions and arrangement of bending samples in notched-bar impact bending test: **a** DVM sample, **b** ISO acute notch sample (ISO V-sample).

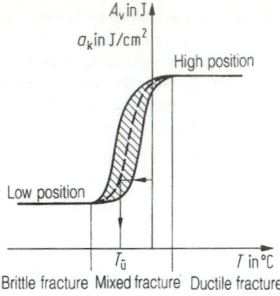

Brittle fracture Mixed fracture Ductile fracture

Direction of displacement of curve	Parameter
	Material
←— —→	Metallurgical manufacture
←— —→	Heat treatment
—→	Cold forming, ageing
	Test condition
—→	Increasing speed of impact
—→	Sample size
—→	Notch sharpness
—→	Notch depth

Figure 7. Effect of impact work used in the notch impact bending test.

Standards. DIN 50 115: Notched-bar impact bending test. – DIN 50 122: Notched-bar impact bending test of zinc and zinc alloys. – DIN 50 122: Notched-bar impact bending test on welded butt joints. – DIN 53 453: Notched-bar impact bending test on plastics.

2.2.6 Fracture Mechanics Testing

Purpose. Fracture mechanics [6–9] presupposes the existence of cracks in materials, and tries to develop criteria on how a crack with certain geometrical dimensions propagates, becomes unstable, and finally leads to complete failure under the effect of an external stress field. Fracture mechanics supplies critical stress intensity factors for the three basic cases of crack propagation shown in **Fig. 8**, for faults of defined size, where unstable crack growth occurs without increase of load. Under these conditions, the K_{Ic} values for fracture mode I show the stresses that lead directly to fracture, or enable a critical fault size that leads to unstable crack growth to be calculated.

Sample Geometry. The requirement that for ductile conditions at the crack tip under tensile stress, no plastic deformation will occur, can only be fulfilled by a suf-

ficiently large sample size. In this way, an even strain condition is set at the crack tip. These requirements can be most nearly obtained by geometric minimum sizes of the samples. In the equations, K_{Ic} is the crack ductility in $N/mm^{3/2}$ and $R_{p\,0.2}$ in N/mm^2.

Sample thickness:

$$B \geq 2.5 \left(\frac{K_{Ic}}{R_{p\,0.2}} \right)^2,$$

Crack length:

$$a \geq 2.5 \left(\frac{K_{Ic}}{R_{p\,0.2}} \right)^2.$$

As shown in **Fig. 9**, three-point bending samples and square or round compact tensile samples (CT samples) are frequently used for fracture mechanics tests. Starting from a machined macro-notch on the tension side, a fatigue crack with defined shape and length is produced in tensile threshold experiments. The test forces required for this purpose are established, in order to cause only slight plastic dynamic deformation at the crack tip.

Figure 10 shows the dimensions for the three-point bending sample and **Fig. 11** those for the CT sample, from which the geometric function $f(a/W)$ can be calculated. As the valid value of K_{Ic} has to fulfil a number of validity criteria, the preliminary K_{Ic} value calculated from a test is usually called K_Q.

Design Values. The K_{Ic} value corresponds to the critical stress intensity factor of the basic case I for unstable crack growth. The triggering point for unstable crack propagation can be determined by crack border displacement, by measurement of sample bending, by potential measurement of crack length, or by ultrasonic testing. The force F_Q required for unstable crack growth is frequently determined from a valid crack propagation diagram (see **Fig. 9**). The force F_Q and therefore the stress is obtained from the point of intersection of the 95% straight line with the crack propagation curve. After fracture, the crack length a is determined exactly and, with a known calibration function Y for the shape of sample used, crack ductility in $N/mm^{3/2}$ is obtained from the equation $K_{Ic} = Y\sigma\sqrt{\pi a_{\text{eff}}}$, where σ is in N/mm^2 and a_{eff} in mm.

Special Test Procedure. This concerns the determination of K_{Ic} values in media causing stress crack corrosion. Stress intensity factors can be determined with fracture mechanics samples, where, under the influence of an electrolyte, the crack no longer propagates.

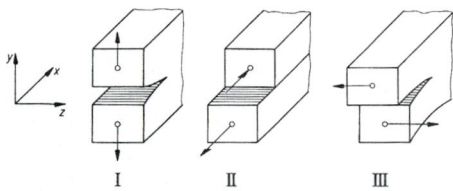

I   II   III

Figure 8. Basic cases of crack propagation.

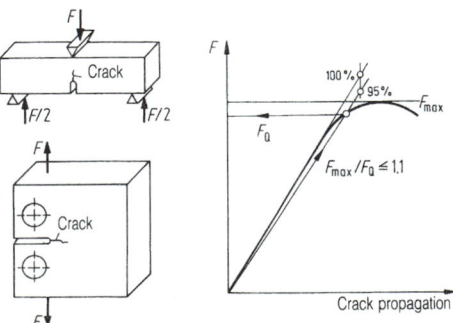

Figure 9. Types of samples for fracture mechanics tests (three-point bending sample, CT sample) and load–crack propagation diagram.

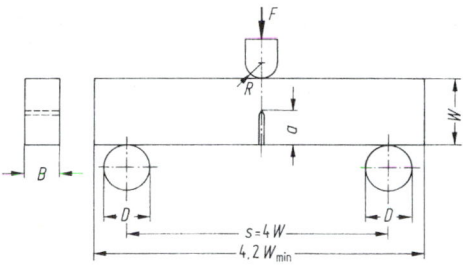

Figure 10. Standard bending sample.

$$\frac{W}{2} < D < W; \; R \geq \frac{W}{8}; \; B = \frac{W}{2}.$$

$$K_Q = \frac{P_Q \cdot s}{BW^{3/2}} \cdot f\left(\frac{a}{W}\right);$$

$$f\left(\frac{a}{W}\right) =$$

$$\frac{3\left(\frac{a}{W}\right)^{1/2}\left\{1.99 - \frac{a}{W}\left(1 - \frac{a}{W}\right)\left[2.15 - 3.93\frac{a}{W} + 2.7\left(\frac{a}{W}\right)^2\right]\right\}}{2 \cdot \left(1 + 2\frac{a}{W}\right) \cdot \left(1 - \frac{a}{W}\right)^{3/2}}$$

In the range $0 \leq \dfrac{a}{W} \leq 1$, K_Q is obtained for $s = 4W$ with an accuracy of $\pm 0.5\%$.

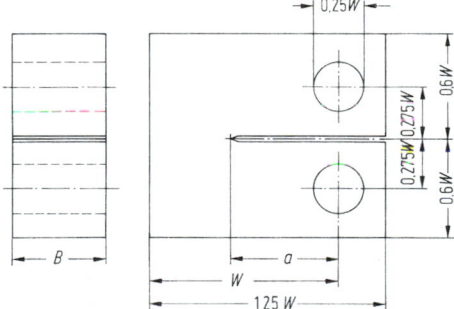

Figure 11. Standard compact tensile sample.

$$B = \frac{W}{2};$$

$$K_Q = \frac{P_Q}{BW^{1/2}} f\left(\frac{a}{W}\right);$$

$$f\left(\frac{a}{W}\right) =$$

$$\frac{\left(2 + \frac{a}{W}\right)\left\{0.886 + 4.64\frac{a}{W} - 13.32\left(\frac{a}{W}\right)^2 + 14.72\left(\frac{a}{W}\right)^3 - 5.6\left(\frac{a}{W}\right)^4\right\}}{\left(1 - \frac{a}{W}\right)^{3/2}}$$

In the range $0.2 \leq \dfrac{a}{W} \leq 1$, K_Q is obtained with an accuracy of $\pm 0.5\%$.

2.2.7 Chemical and Physical Analysis Methods

Purpose. To identify metallic materials, composition is determined qualitatively or quantitatively by chemical and physical methods of analysis. In the analysis of metal and non-metallic alloys and accompanying elements, pro-

cedures which determine the gas content are gaining in importance. Apart from determining the composition of the basic material, the identification of surface coatings is necessary for assessing corrosion and wear processes that occur owing to the interaction of the atmosphere, corrosive media, or lubricants.

Taking Samples. The size of sample for chemical analysis must be taken in such a quantity that the elements will be present in their average concentrations. Depending on the weighed amount or the analytically measured sample volume, the procedures may be described as macro-, semimicro and micro-analytical. **Figure 12** shows a comparison of methods of analysis and the smallest detectable quantities or ranges.

Analysis Processes

Wet Chemical Process. This involves analysis by titration. The desired material is determined by a reaction of a solution with a given reagent. The concentration of the element can be calculated from the amount of reagent solution used for complete conversion. The end of conversion is usually recognised visually as a change of colour or by apparatus, e.g. by a change of conductivity.

Spectral Analysis. In emission spectral analysis, composition is determined by the characteristic wavelengths of the elements in the optical spectrum, and their intensities. Discharges, arcs or lasers are used for excitation. For detection, the light beam is split into its components via gratings or prisms. The absorption of characteristic spectrum lines can also be used for analysis.

X-ray Fluorescence Analysis. X-ray fluorescence analysis works for wavelengths in which X-rays are emitted by secondary excitation when hard X-rays hit a sample. Spectra are analysed by refraction in suitable single crystals (wavelength dispersion).

Electron Beam Micro-analysis. Thin surface layers emit X-ray spectra when highly accelerated electrons impinge upon them. By focusing the electrons into a fine beam, a very small range ($\sim 1 \, \mu m^3$) can be detected. Elements with atomic numbers from 4 (Be) upwards can be identified.

The application of electron beam micro-analysis permits point, line or grid surface analysis and correlation of the results with metallurgical findings.

2.2.8 Metallographic Investigation Methods

Purpose. The purpose of metallographic investigations is to make visible and to describe the macroscopic and

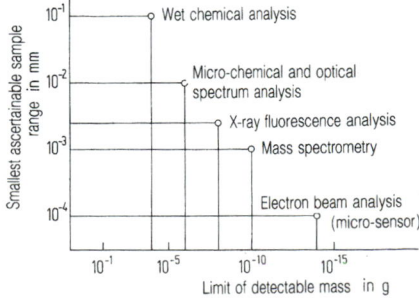

Figure 12. Smallest possible sample range for chemical and physical analysis processes.

microscopic structures of a specimen, and to use these data to interpret the properties of the material in the broadest sense. After the investigation, predictions can often be made about the behaviour of an alloy under certain load conditions or in specific treatment processes. Metallography is a method of investigating metals that is of especial importance in selecting structures suitable for particular uses or processes, in checking and identifying treatment flaws and in explaining material failures.

Removing and Preparing Specimens

Investigation of Macrostructures. The surfaces of specimens for fractographic assessment and cross-sections can be examined macroscopically (up to 50-fold enlargement) without special preparation.

Investigation of Microstructures. Specimens are removed by cutting, grinding or spark erosion, the surfaces to be investigated being made as flat as possible. Heating of the samples is absolutely to be avoided. For better manipulation, ground samples are often clamped in frames or embedded in a suitable material. A mirror surface is obtained by grinding and final polishing (mechanical, electrolytic, chemical).

Transmission Electron Micoscopy. Apart from foils permeable to radiation, produced from metals by various methods, surface imprints are investigated; these are obtained by the lacquer replica technique, vacuum metallising, the oxidation process and the extraction replica technique.

Investigation Methods

Macrostructure. This method is used to detect cracks and pores, and to test the quality of welds and cold-formed products; macro-fractography is used to identify various types of fracture.

Microstructure. The structure of the metal is revealed by chemical etching, electrolytic potentiostatic etching and vacuum etching (etching by ion bombardment), either the grain boundaries (grain boundary etching) or the individual crystallites (grain surface etching) being made visible. Through the differential removal of material, contrasts are produced which reflect oblique light differently. Microstructure investigation gives indications of composition, method of production (e.g. cast structure, forged structure) and heat and surface treatment, and allows the degree of local forming in cold-formed semi-products and components to be determined. Quantitative investigations of structures enable grain sizes and non-metallic inclusions and impurities to be classified (e.g. by standard series) and degrees of porosity in fine castings to be determined.

Electron Microscopy. Transmission, reflection and emission electron microscopy are distinguished by the different kind of interaction between the electron beams and the specimen which is used to generate the image. Whereas in the first two procedures the electron radiation strikes the object from outside, in emission electron microscopy the radiation is formed within the object itself. The lower limit of resolution at present is approximately 50^{-1} nm. The main fields of application for electron microscopy are in detecting dislocation structures, submicroscopic precipitations and phase boundaries. By applying the extraction replica technique inclusions can be isolated, their crystal structures being identified by electron diffraction.

Scanning Electron Microscopy. When surfaces are scanned with focused electron beams, secondary elec-

trons are released and diverted to a scintillation counter. The impulses measured yield a topographic image of the surface. Apart from its use in investigating the morphology of working surfaces and their alteration by corrosion or wear processes, scanning electron microscopy is employed in the fractographic analysis of break surfaces (e.g. in determining the contributions of honeycomb and cleavage fissures to failure).

Special Investigation Methods

Thermoanalysis. Through fluctuations in the temperature–time gradient when heating or cooling metal samples, melting and solidification processes as well as alterations in the solid state (lattice changes) can be detected.

Attenuation Measurement. This involves ascertaining the attenuation function of mechanical oscillations.

Dilatometer Measurement. This method correlates the linear extension of metals to changes in the solid state (e.g. by determining the hardening temperature).

2.2.9 Production Technological Tests

Purpose. Tests are referred to as technological when the behaviour of materials or components is observed without the measurement of forces under load conditions like those to be encountered predominantly in subsequent processing or use. Determining the capacity of materials and semi-products for cold or hot forming is an especially important application.

Test Procedures. *Folding test*: DIN 1605 Sheet 4; *reverse bending test* on wires: DIN 51 211; *torsion test* on wires: DIN 51 212; *cupping test* on sheet and strip: DIN 50 101 (width ≥ 90 mm), DIN 50 102 (width from 30 to 90 mm); *internal pressure test* on hollow bodies of any shape up to a particular internal pressure: DIN 50 104; *expansion test* on tubes: DIN 50 135; *flattening test* on tubes: DIN 50 136; *bending test* on fusion-welded butt welds: DIN 50 121.

Special Investigation Methods

Machinability Tests. Machinability refers to the capacity of a material to be worked by cutting tools (Stahl-Eisen-Prüfblatt 1160-52).

Wear Resistance Test. Wear resistance refers to the capacity of a tool with a given cutting form to withstand the demands imposed in cutting a material under prescribed conditions over a certain time. The following test procedures are used:

Tool-life test at elevated temperatures (Stahl-Eisen-Prüfblatt 1161-52)
Wear endurance test   (Stahl-Eisen-Prüfblatt 1162-52)
Groove wear test   (Stahl-Eisen-Prüfblatt 1164-52)

2.2.10 Non-destructive Testing

Purpose. In non-destructive testing the usefulness of the sample is not impaired; in the narrower sense these tests comprise the X-ray and γ-ray test, ultrasonic testing, magnetic powder testing and electrical and magnetic investigations. As distinct from sampling tests, the non-destructive gross testing of various components achieves a higher degree of certainty. Non-destructive tests apply both to partial areas of the test specimen (e.g. the surface) and its entire cross-section. To identify flaws (e.g. cracks, blowholes, slag inclusions) and segregation zones various physical properties of the material are exploited (e.g. X-ray absorption, reflection of ultrasound waves, sound emission, magnetic properties).

Types of Procedure

X-Ray and γ-Ray Test (DIN 5410 and 5411). This is based on the absorption and dispersion of X-rays passing through the material. By means of a luminescent screen, photographic plate or counting tube, points of varying radiation intensity that occur at faults can be detected. Differences of brightness at material faults can only be registered above a particular value. As radiation sources X-ray tubes with acceleration voltages up to 400 kV, Betatron devices (electron centrifuges) or γ-rays produced by radioactive decay processes are used. The advantages of the latter method of testing lie in the small size of the radiation emitter, good accessibility and independence from an electric current supply. The penetrative capacity of X-rays and γ-rays increases with rising radiation energy. By using a Betatron, wall thicknesses of up to 500 mm can be tested. The chief application of X-ray and γ-ray testing is in checking welds for flaws. To increase the radiation effect, X-ray films are placed between reinforcing sheets, which can heighten the photographic contrast. To quantify fault sizes, wire lattices divided into sixteen stages, with wire gauges between 0.1 and 3.2 mm, are used.

Ultrasonic Testing (DIN 5411). In solid bodies ultrasonic waves in the frequency band between 100 kHz and 25 MHz spread in a straight line and almost without diminution, and are reflected at the body–air interface and at flaws (e.g. cracks, blowholes, inclusions). In the transmission technique the test specimen is placed between the sound emitter and receiver. The sound waves passing through the workpiece are converted back into electrical impulses by the receiver (piezo effect) and transmitted to the display instrument. It is not possible to determine the depth of the fault by this method. When using the impulse-echo technique the sound head is used as both transmitter and receiver, short sound impulses being sent into the workpiece and, after total or partial reflection, converted back into a receiver impulse by the same sound head. The transmission impulse, back-wall echo and possible fault echoes can be registered with a cathode ray oscillograph, which allows the depth of the fault to be ascertained. The use of angular testing heads with acoustic irradiation angles between 35 and 80° makes the technique especially suited to weld seams, since the probe-to-specimen contact can be outside the rough weld surface to permit the locating of weld flaws.

Sound Emission Analysis. This is based on the reception and analysis of sound impulses generated by high-frequency oscillations of the workpiece and converted into electrical signals by piezo-electric receivers. Such sound emissions can be triggered by plastic deformation, or crack formation and progression. Sound emission analysis techniques are used especially for approving welded pressure containers. Both from the amplitude profile and from the frequency spectrum, important data on plastic deformations can be acquired from macroscopic and microscopic voltage disturbances. Through an arrangement of several receivers, the position of the sound emission can be located from differences in the duration of the sound impulses.

Magnetic Fracture Testing (DIN 4113). Fractures, slag lines and pores at or near the surface of ferromagnetic materials can be detected by magnetic fields through the retention of scattered iron powder at fracture surfaces. The indication limit of this method is at a fracture width

of 10^{-3} to 10^{-4} mm. With strong magnetisation an outer zone up to a depth of approximately 8 mm can be tested. While surface fractures running transversely to the specimen axis can be detected by a pole magnetisation, longitudinal fractures can be detected by a current flux resulting from the induced circular magnetic field. At cross-section transitions a false error signal can be triggered by overmagnetisation. A discontinuity in the lines of force can also be caused by a sudden change in ferromagnetic properties (e.g. a transition from ferritic to austenitic areas of weld joints).

2.2.11 Longtime Tests

Purpose. All long-duration tests under mechanical, mechanical–thermal and mechanical–chemical stresses in which the stress period or the number of load–extension cycles is of primary importance for the properties of the material or component, are called longtime tests. Longtime tests are always needed when a change in the damage mechanism occurs in short-duration tests and a correlation between short-duration and long-duration stresses is possible. This applies especially to changes in material properties dependent on time, temperature or loading.

Investigation Procedures

Stress Rupture Test. This is used to determine the properties of the material or component under a static traction load in a temperature range between room temperature and 1100 °C, and can be continued at constant temperature up to a certain deformation (creep strain limit) or up to the fracture of the specimen (longtime rupture strength): DIN 50 118 and 51 226.

The creep strain limit $R_{p\,0.2/t/T}$ at a certain test temperature T is the test load which, after a certain load duration t, leads to a prescribed plastic total extension A_p.

The longtime rupture strength $R_{m/t/T}$ at a certain test temperature T is the test load which, after a certain load duration t, leads to the rupture of the specimen.

The test results are evaluated either by a time–extension diagram or by a time–rupture diagram on a log–log scale.

To shorten the test times extrapolation techniques can be used. Extrapolation is frequently performed by extending the isothermal time–rupture curve or the creep strain limit curve in the time–rupture diagram (ISO/TC 17 WG 10 ETP-SG).

Relaxation Test. In this test a specimen or component (e.g. a screw fixing) in a close-fitting clamp is subjected at constant temperature to an initial deformation and the time-dependent reduction in load is measured while the deformation is kept constant.

Longtime Pulsation Test. This test is used to establish characteristic mechanical values for a material or component under pulsating or changing traction, bending or torsional loads: DIN 50 100 and DIN 50 113 [10]. Flat or indented specimens or components produced in the same way are exposed to oscillating loads of varying amplitudes, so that either fractures occur in the time–rupture range or specimens remain unfractured in the fatigue endurance range from 2 to $50 \cdot 10^6$ load cycles. By plotting the alternating stress amplitude against the logarithm of the number of load cycles the Wöhler curve is obtained. To establish a statistical foundation for this in the fatigue endurance range at least 10 to 15 samples per stress horizon are needed. From several Wöhler curves derived from test rods or components at various mean stress values, complete fatigue endurance diagrams can be obtained (see **D1 Fig. 11**).

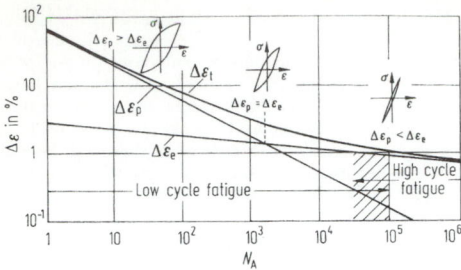

Figure 13. Incipient fracture curve with pulsating extension load at low and high numbers of load cycles (schematically from Coffin).

Oscillating Extension Test. Both in the fatigue endurance range at room temperature and under pulsating loads at elevated temperatures, the linear relation between loads and elastic changes of form no longer applies. Together with load-controlled tests, cyclic extension tests (**Fig. 13**), using either constant total extensions or constant amplitudes of oscillating plastic extension, are gaining importance in determining the number of load cycles N_A at the point of incipient fracture.

The endurable oscillation amplitude for a total extension $\Delta\varepsilon_t$ under traction–compression loading or pulsating traction loading can be broken down into an elastic component $\Delta\varepsilon_c$ and a fixed component $\Delta\varepsilon_p$. Within the fatigue limit of a material, the plastic deformation components recede, and only the elastic component remains (load change test). In the fatigue range between 10^2 and 10^4 extension cycles, however, the plastic deformation component has a determining influence on the total extension component. The characteristic incipient fracture curves for metallic materials can be shown as approximately straight lines on the double logarithmic scale.

---

3 Properties and Application of Materials

K. H. Kloos, Darmstadt

3.1 Iron Base Materials

Metal alloys usable for components and tools are referred to as iron base materials when the mean proportion of iron by weight exceeds that of any other constituent. They are divided into two groups: steels and cast-iron materials. The two groups are distinguished primarily by their carbon content and in some cases show very different properties. While steels represent iron base materials that are generally suited to hot forming, cast iron materials are shaped by primary moulds (see K2). Except in a few Cr-rich steels the C-content of steels is below approximately 2%, while the C-content of cast iron materials is above 2%. Whereas in steels the carbon is dissolved in the iron lattice or is present in a chemically bonded form, it appears in cast iron partly as graphite. Cast steel, also shaped by moulding, is included in the steel group.

3.1.1 The Iron-Carbon Constitutional Diagram

In the stable iron–carbon system carbon appears as graphite in a hexagonal lattice structure. This equilibrium phase is only achieved by very extended heating. In the usual heat treatment of steels, carbon is present in a chemically bonded form as iron carbide Fe_3C (cementite). For technical applications, therefore, the metastable system iron-cementite is as a rule considered instead of the iron–carbon system, even though partial graphite formation occurs in the cast iron range (C > approximately 2%), so that the real state of the material is between that of the stable and the metastable system.

At temperatures above the liquidus line ACD of the metastable system (**Fig. 1**) an iron–carbon solution exists in the molten state. This solution does not solidify like a pure metal at a certain temperature, but within a temperature range lying between the liquidus line ACD and the solidus line AECF. As temperature falls the proportion of crystals precipitated in the melt increases, until the melt solidifies completely as the solidus line is reached. Fixed solidification points only occur at the contact points of the liquidus and solidus lines (A and C). The melting point of pure iron (C = 0%) lies at point A

(1536 °C); the lowest melting point of the iron–carbon system is reached at point C at 1147 °C, where C = 4.3%. The structure which comes into being as solidification occurs at this point is a eutectic alloy system known as ledeburite. In the hypereutectic range (C > 4.3%) pure iron-carbide crystals Fe_3C (primary cementite) are precipitated from the melt; in the hypoeutectic range (C < 4.3%) γ solid solution (austenite: face-centred cubic iron crystals with high solvent power for carbon) is precipitated. Ledeburite consists of an ordered conglomerate of both phases.

The phase field IESG contains a structure consisting solely of austenite. With a C content of 0.86%, below the transformation value at point S (723 °C), the austenite is converted into the eutectoid pearlite, consisting of a fine conglomeration of ferrite (α solid solution) and cementite.

When C > 0.86% (hypereutectoid steels) secondary cementite is precipitated along the line SE; when C < 0.86% (hypoeutectoid steels) ferrite is precipitated along the line GOS. The solvent power of ferrite for carbon is very limited (0.02% at 723 °C, about 10^{-5}% at room temperature), as the narrow area GPQ indicates. The line GOSE is called the upper transformation line, and the transformation temperatures that can be read off along it are called A_3 points. If the temperature falls below the lower transformation line PSK (A_1 point) the remaining γ solid solution of the dual-phase areas below the lines GOS and SE decomposes into pearlite, so that hypoeutectoid steel at room temperature, after slow cooling, consists of ferrite and pearlite, while hypereutectoid steel consists of pearlite and secondary cementite. Above the A_2 point (769 °C) steel loses its magnetic properties. The transformation points A_1, A_2 and A_3 can be moved to higher or lower temperature values according to the rate of temperature change during heating or cooling. In heating the symbol A_c is used instead of A, and in cooling the symbol A_r.

3.1.2 Steelmaking

Steel Smelting Processes

Two main processes are now used in steelmaking throughout the world, as shown in **Fig. 2**:

1. Production of pig iron from ore in a blast furnace and processing of pig iron into crude steel in an oxygen converter (where possible on line without solidification by use of a torpedo car).
2. Melting down of sorted steel scrap to crude steel in an electric arc furnace.

In both cases the crude steel is submitted in a second

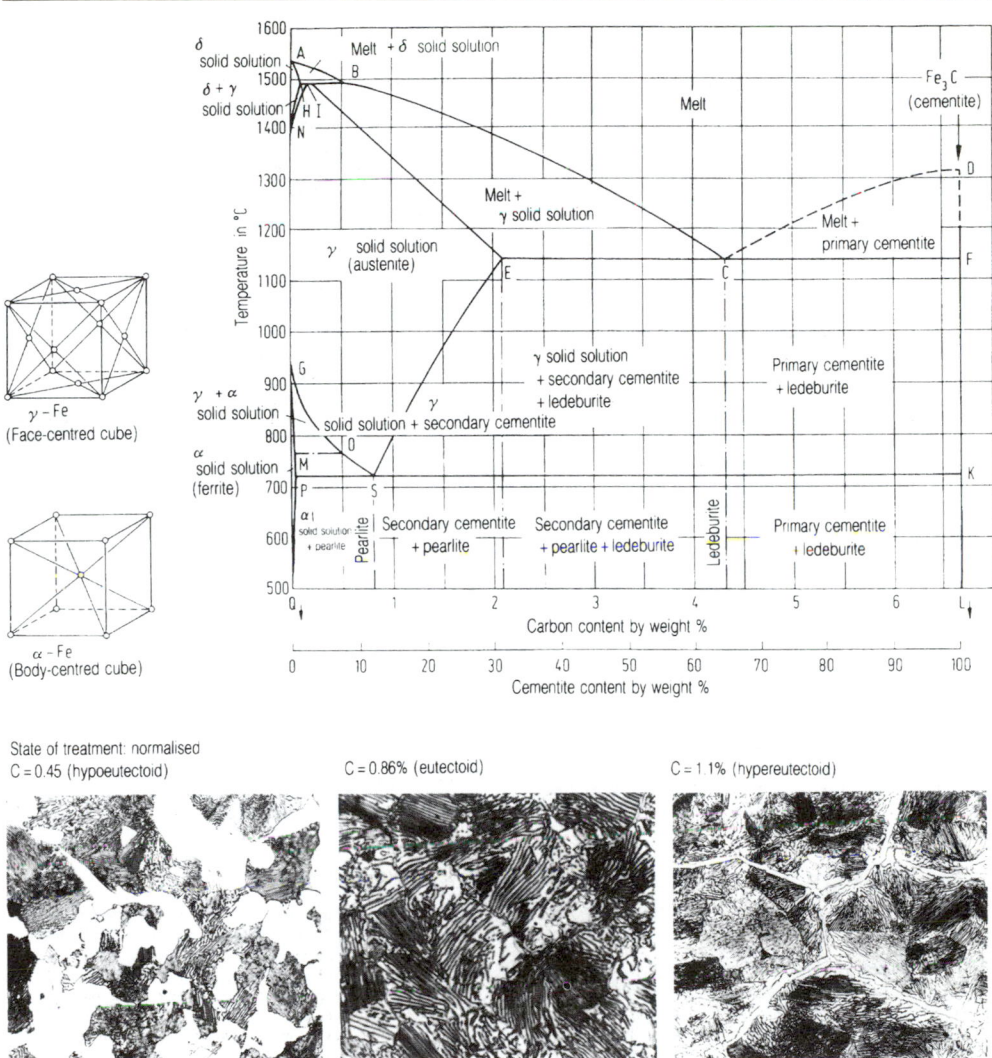

Figure 1. Metastable phase diagram iron–carbon.

stage to secondary metallurgical processes of pan or vacuum treatment to achieve the desired alloy composition, the required degree of purity and the optimal pouring temperature.

By separating the basic melting process from the time-consuming metallurgical reactions it has been possible to use increasingly large and powerful units (oxygen converters and "ultra-high-power"-arc (UHP-arc) furnaces) more economically.

This technology has more or less entirely superseded traditional primary processes such as the following:

Thomas process (air converter; Thomas steels have a pronounced liability to ageing because of high N content).
Siemens–Martin process (primary hearth process with crude iron ore or pig iron and scrap).
LD converter process (oxygen is blown on to the melt; about 20% scrap content is possible).

In the Federal German Republic in 1987, 82.5% of crude steel was produced by blast furnace and 17.5% by arc furnace.

With the *combined blow technique*, in which oxygen is blown simultaneously from the top and through the bottom of the furnace, primary outputs of up to about 400 t/h can be achieved. Further developments of converter technology aim to vary the charge up to 100% scrap, carbon being blown with the oxygen into the furnace and the CO produced by the primary reaction in the upper part of the converter being after-burned to generate heat.

In *high-power arc furnaces* approximately 100 t/h of scrap can be melted. Performance is improved, e.g. by computer control of the process and by additional blowing of oxygen, combustible materials and gas through the floor (improved mixing).

The important measures taken in secondary metallurgy are preventing slag from flowing with the melt, mixing and homogenising in the flushed pan, deoxidising,

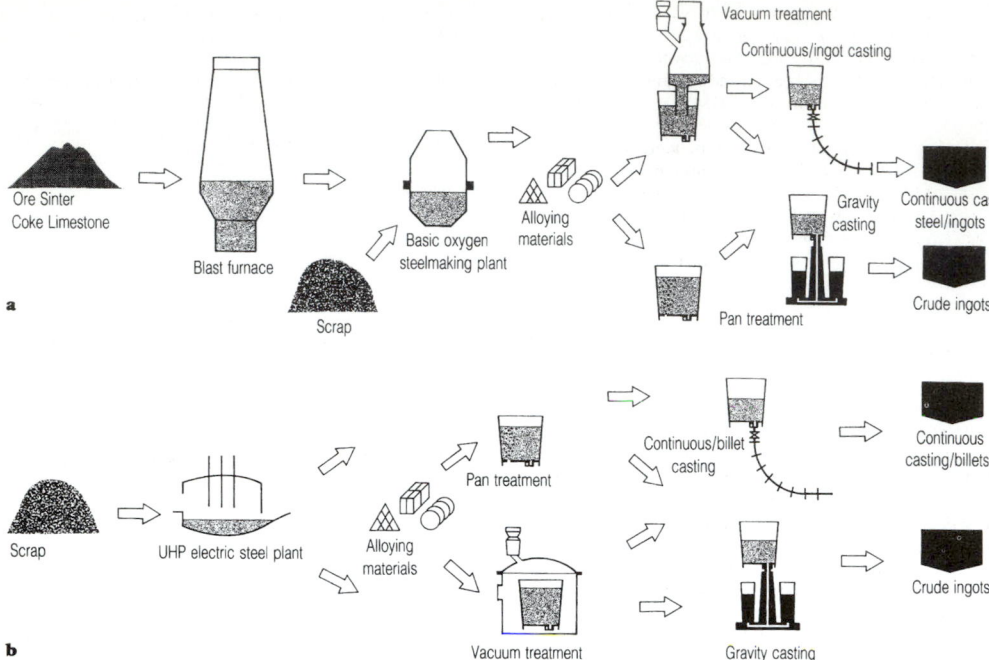

Figure 2. The most important steelmaking processes. Metallurgical basis: **a** pig-iron–oxygen blast furnace, **b** high-powered arc furnace for scrap.

alloying and microalloying in the ppm range in the pan, heating in pan furnaces, vacuum treatment and shielding of the molten stream.

Special Techniques

To improve the properties of steel (especially purity), vacuum and remelting techniques are increasingly used.

Vacuum-Pouring. This technique prevents air access to the molten steel between the pouring pan and the mould. The steel is melted and poured under vacuum conditions.

Electric Slag Remelting. A conventionally produced steel ingot is melted as a self-consuming electrode in a slag bath. In this remelting the resulting steel drops react intensively with the slag.

Core Zone Remelting. For the production of slugs as fault-free as possible for large forged parts the core zone of an ingot formed in a block mould is removed through holes and the hollow ingot is remelted by the electric slag remelting process.

Casting of Steels

Casting can be done in two different ways (primary mould technique):

1. Casting into preformed blanks (ingot or continuous casting). In 1987 about 89% of steel was produced by continuous casting. In general, ingot casting is now used only for producing large forging slugs.
2. Casting into finished parts (see K2).

Plastic Forming

Hot, semi-cold and cold processes of metal forming are distinguished. The heating limit between cold and hot for-

ming is determined by the recrystallisation temperature and is about half the absolute melting temperature (see K3.2).

Current Tendencies

These include shortening of the process chain, or approximating the billet cross-section to the final dimensions of the semiproduct, and using the pouring machine to adapt the flow to variable cross-sections (e.g. in producing thin plate that can be further processed in a cold rolling mill).

Powder Metallurgy

Apart from smelting, metallic materials can also be produced by powder metallurgy. In this process metal powders are the starting point. Depending on the degree of purity, particularly if pre-alloyed powders are used, the powdering process is very protracted. Numerous combinations of materials are possible by mechanical mixing (mechanical alloying), which permits the product to be closely adapted to the needs of the component being produced.

Component production from metal powders involves the following steps:

Powder production (reduction of raw material by compressed air and deflector plates, water atomisation, melting open by electron radiation and rotation pulverising).

Mixing and compressing the powder to slugs.

Sintering (e.g. single-sinter process, double-sinter process, powder forging).

Aftertreatment, e.g. hot isostatic pressing (HIP treatment) to eliminate microporosity.

Up to certain limiting dimensions, fully finished components can be produced in this way.

The use of powder metallurgy allows raw materials to be economically used and, with very pure powders, enables components to be produced with approximately isotropic properties of strength and toughness. In particular, sinter-forged components have good mechanical properties because of their high density. Powder metallurgy is also increasingly used in producing tool steels.

Sintered steels are also gaining importance as substitutes for forged and cast materials. The rigidity of sintered steels is largely determined by their density or porosity. Rigidity properties similar to those of compact materials can be obtained by subsequent mechanical compacting of the surface layer by processes such as shot-blasting or rolling (see "Sintered Materials" in D3.1.4, p. D41).

3.1.3 Heat Treatment

The purpose of heat treatment is to impart to a material the desired properties for use or further treatment. The material is subjected to certain temperature–time sequences and in some cases to additional thermomechanical or thermochemical treatments. For many steels the temperature–dependent appearance of α and γ solid solutions (ferrite and austenite) (**Fig. 1**) with variable solvent power for carbon is the basis of the capacity of their properties to be modified within wide limits.

The kinetics of the transformation of austenite into other phases result from the isothermal time-temperature transformation diagram (or TTT-curve). Using the example of Ck 45 steel, **Fig. 3** shows the beginning and end of transformation after rapid cooling of the austenite to a certain temperature and subsequent maintenance of an isothermal state. Above the M_s line the transformation begins with a time lag which is at a minimum at about 550 °C. This is because on the one hand the tendency of austenite to undergo transformation increases as it is cooled, while on the other the reduced diffusion velocity hinders the ability of atoms to change places to form a new crystal lattice. Whereas ferrite-pearlite transformation takes place at temperatures above this peak, in the area below it the structure bainite is produced, a substance consisting of ferrite needle crystals with included carbides. With rapid undercooling to temperatures below the M_s line there is, without time lag, a diffusionless shearing of the austenite lattice into the martensite lattice, the proportion of martensite formed rising as temperature falls. The configuration of the transformation lines in the time-temperature transformation diagram is determined

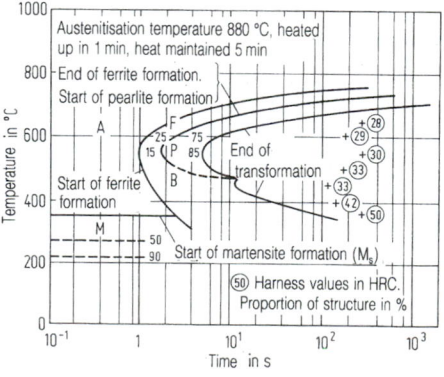

Figure 3. Isothermal time-temperature transformation diagram for Ck 45 steel: A austenite, F ferrite, P pearlite, B bairate, M martensite.

by the level of the austenitisation temperature and the chemical composition of the steel.

The processes discussed in relation to isothermal transformation follow a similar pattern if cooling continues below the austenitisation temperature, as happens in many heat treatment techniques. With slow cooling, in the case of Ck 45 steel, a ferritic–pearlitic structure is produced, as can be seen from the iron–carbon diagram. As the rate of cooling increases the proportions of bainite and martensite in the structure rise, until above a critical upper cooling rate only martensite is formed.

Hardening

Martensite formation brings about a considerable increase in the hardness of steel. Therefore heat treatment that results in martensite formation in greater or lesser areas of the cross-section of a workpiece after austenitisation is called *hardening*, and the temperature from which the workpiece is cooled is called the *hardening temperature*. For hypoeutectoid steels the hardening temperature lies above the line GOS in the FeC diagram in the zone of pure γ solid solutoin, but for hypereutectoid steels it lies above the line SK in the zone of γ solid solution and secondary cementite. Dissolving of the naturally hard secondary cementite is not necessary provided it is finely divided and not present in lattice form. The high hardness of martensite results from the low solvent power of the α lattice of iron for carbon atoms as compared with the γ lattice. With rapid cooling the C atoms released at the hardening temperature cannot diffuse out of the γ solid solution as it undergoes transformation and, in remaining forcibly detached, they cause a warping of the martensite crystal as it forms, which is expressed in extreme hardness. The warping increases with the number of forcibly free C atoms; this is why the hardening capability of a steel rises with C content. However, significant hardening only occurs if the C content is at least 0.3%.

To achieve a cooling rate high enough for martensite formation in the interior of the workpiece as well, heat must be conducted away as quickly as possible. This is achieved by quenching means such as oil, water, iced water or salt solutions, but above a certain cross-section through-hardening is not possible.

In alloyed steels, as compared to unalloyed steels, the critical cooling rate is reduced because of the obstruction of carbon diffusion by the atoms of the alloying elements intercalated in the solid solution. For this reason, with alloy steels, larger cross-sections can be hardened or milder quenching fluids can be used, e.g. air instead of oil or oil instead of water. High temperature differences between the core and periphery of a workpiece give rise to high internal heat stresses which, together with the internal transformational stresses resulting from the increase in volume with martensite formation, can cause distortion and heat cracking. The danger of distortion and cracking during quenching can be reduced, e.g. by step hardening, in which the temperature within the workpiece is first equalised at a temperature just above the M_s value, before martensite formation begins as the piece is cooled to room temperature.

The most important alloying elements for heightening the through-hardening properties of steels are Mn, Cr, Mo and Ni in proportions of about 1 to 3%. The behaviour of a material undergoing core hardening can be ascertained by the end quench test (DIN 50 191).

Annealing and Tempering

The martensite structure produced during hardening is very brittle. For this reason a workpiece is usually *annealed* after hardening, i.e. heated to temperatures between room temperature and Ac_1. In the lower temperature range for annealing (up to about 300 °C) the extreme warping of the martensite is moderated by the diffusion of carbon atoms; brittleness is reduced without hardness being significantly affected. ε-carbide, more carbon-rich in comparison with cementite, is precipitated; the remaining austenite in the hardened structure decomposes.

At annealing temperatures above 300 °C, toughness (rupture elongation, necking, notch impact strength) increases sharply, while strength and hardness diminish (**Fig. 4**). These changes result from the decomposition of the martensite into ferrite and the formation of finely divided cementite from the ε-carbide formed at lower temperature. In the range of annealing temperatures between 450 °C and Ac_1 a fine-grained structure of good toughness and high strength is produced, as is desired for structural parts. The process of hardening and annealing within this temperature range is called *tempering*. Tempering strength, like through-hardening capability, depends on the chemical composition of the steel and the cross-section of the workpiece.

At annealing temperatures between about 450 and 600 °C alloy steels, particularly with Mo, W and V as alloying elements, show a sharp rise in hardness and strength as a result of secondary hardening. In this process finely divided deposits (usually special carbides or intermetallic phases) with a slip-blocking effect are formed from the supersaturated solid solution produced after austenitisation (solution heat treatment) and rapid cooling. This process is used with tool steels, heat-resistant and precipitation-hardened steels to increase their strength.

Heat Treatment

Heat treatment refers to a treatment of a workpiece at a certain temperature for a certain duration with subsequent cooling, to obtain specific material properties.

Normalising. This is done at a temperature little above Ac_3 (with hypereutectoid steels above Ac_1), with subsequent cooling in a still atmosphere. This treatment is used to eliminate the coarse-grained structure to be found in cast steel parts and sometimes in the vicinity of weld seams (Widmannstätten structure). The effect of a previous treatment or cold forming process is also removed by normalisation. If the austenitisation temperature chosen is too high there is an increase in γ solid solution, which leads to a coarse-grained structure even after transformation (fine-grain steels are less prone to grain coarsening). Similarly, too slow cooling gives rise to a coarse ferrite grain.

Coarse-Grain Annealing. When matching soft steels, a coarse-grained structure, which produces a short-brittle shearing chip, may be desired. This structure is obtained by heat treating far above Ac_3. The coarse γ solid solution produced by grain growth is converted during slow cooling into a similarly coarse-grained ferrite–pearlite structure.

Homogenising. This process is used to eliminate segregation zones in ingots and billets and within crystals (crystal segregation). The heat treatment is carried out just below the solidus temperature, the material being held at this temperature for a long period to equalise concentration through diffusion. If no hot forming is undertaken after homogenisation the piece must be normalised to eliminate the coarse grain.

Soft Annealing. To improve the formability of C steels they are soft- (or dead)-annealed at temperatures in the range of Ac_1. At these temperatures the cementite lamellae present in the barred pearlite take on a spherical form (*spheroidising treatment*). Then the material is slowly cooled to minimise internal stresses. The spheroidising of the cementite lamellae and, in hypereutectic steels, of the cementite network, is facilitated by repeatedly raising the temperature above Ac_1. The spheroidal form of the cementite can also be achieved by austenitisation and controlled cooling.

Stress Relief Heat Treatment. Internal stresses, which are superimposed on working stresses, can arise in workpieces through uneven heating or cooling, through structural changes or cold forming. To remove or reduce these internal stresses, e.g. after straightening or welding, or to reduce internal stresses in cast parts, stress relief treatment is carried out. The heating temperature is usually below 650 °C, but in the case of tempered steels it is below the annealing temperature in order not to reduce the tempered strength of the workpiece. During treatment the internal stresses in the piece are reduced by plastic deformation to the value determined by the high-temperature yield point.

Recrystallisation Annealing. The extent of a cold forming process is limited by the increase in the hardness and the reduction in the ductility of a material with the degree of forming. Recrystallisation annealing after cold forming allows a re-formation of the structure at temperatures above the recrystallisation temperature, producing mechanical properties like those that existed before forming, so that any number of forming processes can alternate with recrystallisation treatment. The danger of a coarse grain being formed in the recrystallised structure is present with low deformations, particularly in steels with a low C content (< 0.2%), if high temperatures are used for a long duration. The recrystallisation temperature of steels falls with the degree of forming, as the deforming energy stored in the lattice favours new grain formation. Recrystallisation annealing is used with cold-rolled strip and thin sheet, cold-drawn wire and deep-drawn parts. To protect against scaling the treatment is done in airtight containers (bright annealing).

Solution Heat Treatment. This process is used to dissolve segregated constituents in solid solutions. To achieve a homogeneous structure austenitic and ferritic steels that do not undergo γ-α transformation are solution treated at about 950 to 1150 °C and then quenched, to avoid the formation of intermetallic phases that would cause brittleness in slow cooling.

For steels which are subjected to precipitation harden-

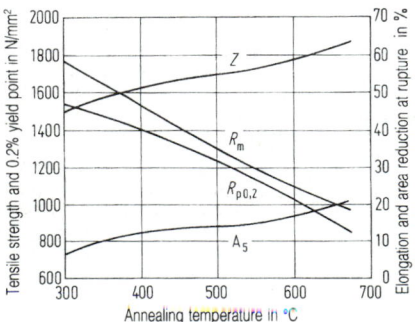

Figure 4. Tempering diagram for the material 42 CrMo 4.

ing as well as martensite hardening during transformation (alloyed tool steels, heat-resistant and martensite hardening steels) solution treatment is carried out at the same time as austenitisation, giving rise after quenching to a supersaturated solution which is separated through the formation of precipitates during hardening.

The quantity of dissolved constituents rises with the solution temperature and the duration of treatment. This increases the capacity of the structure to segregate constituents during hardening, so that the strength attained by the material is also increased.

Boundary Layer Hardening

For many workpieces which need a hard, wear-resistant surface, it is sufficient for hardening to be limited to the boundary layer. There are various forms of boundary layer hardening: *flame-hardening*, *induction hardening* and *laser surface hardening*.

Flame Hardening. In this process a workpiece surface is heated to the austenitisation temperature and then quenched with water (water spray) before the heat has penetrated the interior of the piece. Thus martensite hardening occurs only in the austenitised boundary layer. The depth of the hardened layer is determined by the flame temperature, the duration of heating and the heat conductivity of the steel.

Induction Heating. In this process the boundary layer is heated by induced current in a high-frequency coil and, after the austenitisation temperature has been reached, quenched in a water spray or bath. As a result of the skin-effect the depth of the heated boundary layer is reduced as the frequency increases, so that hardening depths of only a few tenths of a millimetre can be attained. Tempered steels with 0.35 to 0.55% C can be used. With low C contents the hardening is insufficient, and with high C contents there is an increased risk of deformation or hardening cracks, particularly as higher austenitisation temperatures have to be used than with normal hardening. After boundary layer hardening the piece is generally subjected to stress relief treatment at 150 to 180 °C.

Laser Surface Hardening. By continuous radiation from a CO_2 laser, individual working areas of a component can be subjected to localised boundary layer hardening. Laser hardening is one of the group of short-duration hardening processes. The hardening is produced by self-quenching and can be limited to thin boundary layers. With a correct selection of the radiation parameters fatigue strength as well as surface hardness can be increased [1]. As with induction hardening, tempered steels with 0.35 to 0.55% C or tool steels can be used in this process.

Thermochemical Treatments

Thermochemical treatments are heat treatments in which the chemical composition of a workpiece is deliberately changed by diffusing several elements into or out of the material. The objective is usually to impart certain properties, such as scale or corrosion resistance, or increased wear resistance, to the boundary layer of a workpiece. As the workpieces are exposed to a high temperature for a long period, changes to the core properties should be taken into account. As compared with galvanic surface treatments, the advantage of the diffusion process lies in the even thickness of the surface layer treated, regardless of corners, grooves or drill holes.

Case Hardening. A high degree of boundary layer hardness in parts made of steels with C contents of approxi-

mately 0.1 to 0.25% can be attained by hardening after the thermochemical treatments of carburisation or carbonitriding. In carburisation the boundary layer of the workpiece is enriched in carbon by being heated to 850 to 950 °C (above the GOS line) in carbon-releasing substances. Powder, gas, salt-bath and paste carburisation are distinguished according to the means of carburisation used. The C content of the boundary layer after carburisation should be not more than about 0.8 to 0.9% to avoid excessive cementite formation, which might impair the properties of the boundary layer. After carburising the boundary layer of a piece is hardenable. Because of its higher C content the structure of the surface has a lower transformation temperature than that of the core. If the hardening temperature is adjusted to the C content of the boundary layer, the core is not fully transformed, so that steels which are liable to grain growth retain a coarse-grained structure in their core as a result of the long carburising time (*single hardening*). A fine core grain can be obtained by *double hardening*. In this the piece is first cooled from a temperature corresponding to the C content of the core, causing the core to recrystallise; then the boundary layer is hardened. This produces a high degree of surface hardness combined with extreme toughness of the core. However, the repeated heating and cooling increases the danger of distortion. This can be countered by quenching in a hot bath.

The hardening of the carburised boundary layer can also be effected directly by the carburising temperature (*direct hardening*), the workpiece being in some cases previously cooled to a hardening temperature corresponding to the C content of the boundary layer. Up to now this process has been preferred for mass-produced components of secondary importance, or for steels with a low tendency to grain growth (fine-grain steels).

Case-hardening steels with a higher proportion of alloying ingredients, such as the material 20 NiCrMo 63, have been specially developed for direct hardening, to achieve better strength and toughness.

In *carbo-nitriding* the boundary layer of a workpiece is enriched simultaneously with carbon and nitrogen. This treatment is carried out e.g. in special cyanide baths at 800 to 830 °C. After carbo-nitriding the material is usually quenched, to further heighten the hardness attained by nitride formation through additional martensite conversion.

After case-hardening the material is treated for stress relief at temperatures of 150 to 250 °C.

Nitriding. This involves a saturation of the boundary layer of a workpiece with nitrogen through diffusion, to enhance hardness, wear resistance, fatigue strength or corrosion resistance. Compared with carburisation, nitriding in the presence of special nitride-forming elements can achieve higher boundary hardness; however, because of the lower diffusion depth the drop in hardness towards the interior of the material is steeper. After nitriding the boundary layer consists of an outer nitride layer (white layer) and then a layer of nitrogen-enriched solid solution and segregated nitrides (diffusion layer). Gas nitriding in an ammonia stream at 500 to 550 °C is distinguished from salt-bath nitriding in cyanide baths at 520 to 580 °C.

Gas nitriding requires long nitriding times (e.g. 100 h for a nitriding depth of about 0.6 mm). Nitriding times can be shortened by additional measures such as added oxygen or ionisation of the nitrogen by brush discharge (*plasma nitriding*). A further reduction of nitriding times can be achieved by *salt bath nitriding*, though the cyanide baths always also cause a carburisation of the boundary layer; however, at the low bath temperatures used here this effect is small. The low bath temperatures and the slow cooling (no quenching) lead to very low distortion of the workpiece (measuring instruments).

In *nitro-carburisation* the treating medium contains carbon-releasing constituents as well as nitrogen. The process can be carried out in powder, salt bath, gas or plasma. The gas nitrocarburisation processes, referred to by the blanket term shorttime gas nitriding, need considerably shorter treatment times as compared with the other gas nitriding processes. These times are of the same order as for salt bath nitiriding, at temperatures from 570 to 590 °C.

Alloying elements with an especially high affinity to nitrogen, such as chromium, molybdenum, aluminium, titanium or vanadium, yield particularly hard boundary layers with high resistance to wear by sliding friction (nitride steels). In the case of tempered steels with low retention of hardness care must be taken to ensure that longtime nitriding does not reduce material strength at the core. Hardness retention can be improved by alloying elements such as chrome and molybdenum, so that high core strength can be combined with high boundary layer hardness in low-alloy CrMo steels.

Aluminisation. This generally refers to the production of Al coatings. Of the diffusion processes, calorising and alitising have proved particularly useful.

In *calorisation* the workpieces (usually small parts) are heated in Al powder with certain additives in a rotating reaction drum at 450 °C. They are then heated for a short period outside the drum to 700 to 800 °C to increase diffusion. A brittle, tightly bonded Fe-Al alloy layer (Al > 10%) is formed below a hard layer of Al_2O_3, which has good resistance to scaling.

A less brittle protective layer with better formability and the same scaling resistance is produced by *alitising*. In this process the article is heated in a powder of Fe-Al alloy at 800 to 1200 °C.

Both processes are also applicable to metallic materials other than steel, e.g. calorising with copper and brass and alitising with nickel alloys for gas turbine blades.

Siliconising. A brittle but very scale-resistant surface is obtained for low-carbon steel by treatment with hot $SiCl_4$ vapour. The Si content of the layer can be as much as 20%.

Sherardisation. This process is carried out in a similar way to calorisation. After morduranting or sand blasting, the workpieces are heated at 370 to 400 °C in zinc dust with certain additives. Apart from increased corrosion protection, a good keying surface for painting is obtained.

Boronising. By boronising hard and wear-resistant boundary layers are produced. The process can be carried out in powder (950 to 1050 °C), gas and salt baths.

Chromising. The process is carried out at about 1000 to 1200 °C with chrome-releasing substances in the gas or molten phase. The boundary layer of the workpiece is enriched to as much as 35% Cr. It thus becomes scale-resistant up to temperatures of over 800 °C. Because of the corrosion resistance of the layer this process can obviate the need for non-corroding material in the workpiece itself.

Special Heat Treatments

Isothermal Transformation in the Bainite Stage. In this process a workpiece is rapidly cooled after austenitising to a temperature at which, if this temperature is maintained, bainite conversion takes place. The temperature appropriate to a particlar material can be ascertained from the isothermal TTT-curve. The best strength and toughness properties result from conversion in the lower temperature band of the bainite stage. Apart from good mechanical properties the process offers economic advantages over tempering, since a second heating stage is not required. In particular, small parts made from structural steel are treated by this process.

Patenting. This refers to a heat treatment of wire and strip in which, after austenitisation, the workpiece is rapidly cooled to a temperature above M_s, to obtain a favourable structure for subsequent cold forming. In wire production the material is usually cooled in a hot bath at temperatures that give rise to close-barred pearlite, since this structure is especially suited to tension loads.

Martensite Hardening. In carbon-poor Fe-Ni alloys with more than 6 to 7% nickel the conversion of the γ solid solution takes place even under slow cooling from the austenite range (820 to 850 °C); it no longer involves diffusion into ferrite, but diffusionless shearing into nickel martensite, a metastable solid solution supersaturated with nickel (instead of carbon). In the subsequent hot-hardening process below the re-austenitising temperature (450 to 500 °C), alloying elements such as Ti, Nb, Al and especially Mo bring about a considerable increase in strength coupled with good toughness through the segregation of finely divided intermetallic phases and the production of slip-resistant ordering phases.

Thermomechanical Treatments

Thermomechanical treatments are a combination of forming processes with heat treatments, to obtain certain properties in the material.

Austenitic Form Hardening. In this process a steel is subjected to forming after cooling from the austenitising temperature and before or during austenite transformation. In this way strength can be increased while toughness is improved by a finer bainite and martensite structure.

Temperature-Controlled Hot Forming. By manipulating the temperature in the last stages of a hot forming process involving a sufficient change of form and subsequent cooling, a similar structure to that produced by normalising is aimed at.

Hot-Cold Compacting. An increase in strength is achieved by forming at a high temperature below the recrystallisation threshold using reduced forming force as compared with room temperature. This process is especially suited to austenitic materials.

3.1.4 Steels

Designation of Steels

To create a clear means of classifying the multiplicity of steel varieties now known, a system of abbreviated names is needed: DIN 17 006 and EU-Norm 27-74. The most recent survey of abbreviated names and material numbers for iron base materials is contained in the 7th edition of DIN Standards List 3 (1983).

Unalloyed steels not suited to heat treatment (*ordinary steels*) are classified in terms of strength. For example, St 37 is a steel with a minimum tensile strength of 360 N/mm². Modern weldable structural steels are classified by their lower yield point, referred to by the letter E, e.g. St E 36. Letters can also be used to denote the method of production: E: electric steel, R: killed steel, U: unkilled steel.

Example. R St 37-2.
Unalloyed steels intended for heat treatment (*high-grade or special steels*) are usually case-hardened or tempered. They are named according to chemical composition. The centuple of the mean carbon content in % is placed after the symbol C for carbon.

Example. C 10 (C content 0.1 wt%).
A k after the letter C indicates a low phosphorus and sulphur content, e.g. Ck 10.
Unalloyed tool steels are denoted by the letter C and the centuple

of their mean carbon content in %. This is followed by the symbol for their merit number or quality factor.

Example. C 100 W 1, tool steel, merit no. 1 with a mean carbon content of 1 wt%.

Low-Alloy Steels (alloying content < 5 wt%). Here the chemical composition is indicated. The letter C for carbon is dropped. The classification begins with the centuple of the mean carbon content in %. This is followed by the chemical symbols for the alloying elements, in descending order of content. At the end are figures denoting the mean alloying contents that, to obtain sufficient differentiation, are multiplied by the multiplier

    4 for the elements Cr, Co, Mn, Ni, Si, W.
   10 for the elements Al, Be, Cu, Mo, Nb, Pb, Ta, Ti, V, Zr.
  100 for the elements P, S, N, Ce.
1000 for B.

Examples. 15 Cr 3 is a low-alloy steel with a mean carbon content of 0.15 wt% and a mean chromium content of 0.75 wt%.

100 V 1 is a low-alloy tool steel with a mean carbon content of 1 wt% and a mean vanadium content of 0.1 wt%.

High-Alloy Steels (alloying content > 5 wt%). These are named, like low-alloy steels, by their chemical composition. Multipliers are not used, except in the case of the carbon content, which is multiplied by a factor of 100. High-alloy steels are denoted by the prefix X.

Example. X 5 CrNiMo 17 12 2 is a high-alloy steel with 0.05 wt% carbon, 17 wt% chromium, 12 wt% nickel and 2 wt% molybdenum.
   To denote additional qualities, letters indicating the melting and deoxidising conditions and the state of treatment of a steel have been allocated.

Material Numbers

DIN 17 007 contains a systematic list of material numbers; EU Norm 20 deviates from DIN 17 007. DIN List of Standards 3 indicates the future numbers.

Alloying Elements

In addition to its mechanical properties, these elements influence the behaviour of steel in heat treatment (shifting the transformation temperatures, changing the critical cooling rate, widening or narrowing the γ range).

Effects of the Most Important Alloying Elements (in Alphabetical Order)

Aluminium (Al) is regarded as the strongest means of deoxidation and denitriding. In small quantities it assists fine grain formation. It is used in nitride steels, as it forms extremely hard nitrides with nitrogen. It increases resistance to scaling.

Boron (B) improves through-hardening in structural steels and increases core strength in case-hardened steels.

Carbon (C) is the most important and influential alloying element in steel. With increasing C content the strength and hardening properties of steel rise. Rupture elongation, malleability and ease of working with cutting tools are reduced with higher C contents.

Chromium (Cr) increases strength (by about 80 to 100 N/mm² per 1% Cr) and only slightly reduces elasticity, improves heat and scaling resistance and though-hardening properties. It is a powerful carbide former, so increasing the hardness of tool and bearing steels. With chromium content > 12% steels become rustproof.

Cobalt (Co) does not form carbides. It inhibits grain growth at high temperatures and improves tempering properties and heat resistance. It is an alloying element in high-speed and hot-working steels, and in high temperature and very high temperature resistant materials.

Copper (Cu) increases the strength of steel but reduces its rupture elongation: in low quantities (0.2 to 0.5%) it improves the rust resistance of steel under atmospheric influences.

Lead (Pb), through its fine, suspension-like distribution, forms short chips and clean machined surfaces in free-cutting steel.

Manganese (Mn) increases the strength of steel and only slightly reduces rupture elongation. Mn has a beneficial effect on malleability and welding properties. In combination with carbon, Mn improves wear resistance. With Mn contents of up to 3% the tensile strength of steels is increased by about 100 N/mm² per 1% Mn. With contents of 3 to 8% the increase is less, and above 8% tensile strength declines again. Mn widens the γ range and improves through-hardening.

Molybdenum (Mo) increases tensile strength and especially high temperature strength, and is beneficial to welding properties. Mo is a strong carbide-forming agent and is used in high-speed and hot-working steels, in austenitic steels, in case-hardening and tempering steels and in high-temperature resistant steels. Mo reduces the tendency for annealing to cause brittleness.

Nickel (Ni) increases strength with only slight loss of ductility. Ni allows good through-hardening. Cr–Ni steels are rust and scale resistant and have good high-temperature strength. Weldability is not impaired by Ni. It improves impact strength when notched, especially at low temperatures.

Silicon (Si) increases the scaling resistance and the tensile strength and raises the yield point of steel (by about 100 N/mm² per 1% Si). Toughness is only slightly influenced.

Sulphur (S) makes steel brittle and red short. In machining steels sulphur is deliberately added in quantities up to 0.3% to allow the formation of short chips.

Titanium (Ti), *tantalum* (Ta) and *niobium* (Nb) are strong carbide-forming agents and are used predominantly in austenitic steels to stabilise the material against intercrystalline corrosion.

Tungsten (W) is a strong carbide-forming agent. W increases strength, hardness and edge-holding properties, and produces good high-temperature hardness; hence its use in high-speed and hot-working steels.

Vanadium (V) improves high-temperature strength and suppresses sensitivity to overheating. It increases the edge-holding properties of high-speed steels. V is a strong carbide-forming agent. It increases tensile strength and raises the yield point.

Rolled and Forged Steels

The classification of steels can be done from various standpoints, e.g. according to the *smelting procedure* (oxygen steels, electric steels, electroslag remelted steels, etc.), their *uses* (structural and tool steels), their *alloying content* (unalloyed, low-alloy and high-alloy steels), their *chemical and physical properties* (non-corroding, high-temperature and scale resistant, non-magnetic and soft magnetic, wear-resistant steels) and according to their *merit group* (ordinary, high-quality and special steels).

Rolling and forging steels are divided into *structural steels*, *tool steels* and *steels for special components*.

General Structural Steels. These are unalloyed and low-alloy steels with 0.15 to 0.5% C, < 0.045% P, 0.003 to 0.3% S and 0.006 to 0.014% N, which are normally used in construction, bridge building, water engineering, container construction, and vehicle and machine construction. They may be used in the hot-formed state, after normalising or tempering, or they may be cold-formed after soft annealing, owing to their tensile strength and high yield point: DIN 17 100, **Appendix D3, Table 1.** They are generally supplied in the hot-formed, i.e. hot-rolled or hot-forged state. They should be neither brittle when cold nor red short. To ensure sufficient insensibility to brittle fracture, impact values are guaranteed. The general suitability of these steels for the various welding procedures is not guaranteed, as the behaviour of a steel during and after welding depends not only on the material but also on its dimensions and shape and on the conditions under which the component is made and used.

Structural Steels for Certain Kinds of Products. Unalloyed structural steels in mill finish and in specified forms are standardised with regard to their use. As regards their mechanical and technical properties and their chemical composition they deviate more or less widely from the steels in DIN 17 100.

Standards. DIN 1614: Hot-rolled strip and sheet made of soft unalloyed steels. The standard applies to continuously hot-rolled flat products (strip and sheet) up to 15 mm thick made of soft unalloyed steels; it contains varieties of steel characterised by their chemical composition, that are suited to further processing by cold rolling into strip or sheet to DIN 1623 or into cold-rolled strip to DIN 1624, and varieties characterised by their mechanical properties, that are intended for immediate cold forming or for producing welded tubes not of prescribed strength. The varieties are St 22, St 23, St 24, St W 22, St W 23 and St W 24.

DIN 1623 Part 1: Cold-rolled strip and sheet of soft unalloyed steels. The standard applies to cold-rolled flat product under 3 mm thick made of soft unalloyed steels intended for forming and surface upgrading but not for case-hardening and quenching or for tempering. Steels in this group are: St 12, St 13 and St 14.

DIN 1623 Part 2: Thin sheet of unalloyed steels, thin sheet of general structural steels. The standard applies to sheet under 3 mm thick made of general structural steels to DIN 17 100, used essentially in climatic temperature conditions in mill finish because of its tensile strength and yield point.

DIN 1624: Cold-rolled strip of soft unalloyed steels. The standard applies to cold-rolled strip of soft unalloyed steels intended for cold forming and surface upgrading but not for quench hardening or tempering. Case hardening is possible, though suitability for this is only guaranteed for cold-rolled strip of steels conforming to DIN 17 210, which do not come under this standard. Steels from St 0 to St 4 are listed, in the following qualities: basic, folding, drawing, deep-drawing and special deep-drawing.

DIN 1652: Bright unalloyed steel. Under this standard, in addition to general structural steels conforming to DIN 17 100, case-hardening steels to DIN 17 210 and tempering steels to DIN 17 200 for bright steel are listed, with indications of the production process, chemical composition, mill finish and strength. Bright steel refers to steel bar which, as compared to the hot-formed state, has a

smooth, bright surface through descaling and drawing or through shelling, polishing and dressing.

Weatherproof Structural Steels. Adhesive rust layers can be produced by adding 0.65% Cr, 0.4% Cu, ≤ 0.4% Ni and increased phosphorus content. Coatings of this kind prevent further rusting of these steels, which are listed in Stahl-Eisen-Werkstoffblatt 087 (WT St 37-2, WT St 37-3 and Wt St 52-3).

However, the high quantities of Cu added impair the scrap quality (danger of red shortness) for recycling, and the reddish brown surface is often not desired. For this reason steels with higher Cr content are recommended instead, e.g. X 2 Cr 11 (material no. 1.4003) or X 7 Cr 14 (material no. 1.4001), which, depending on strength requirements, are used especially to combat the effects of acid rain in bridge construction, and in concrete structures threatened by corrosion.

High-Strength Weldable Structural Steels (Fine-Grained Structural Steels). The highest yield point of weldable structural steels conforming to DIN 17 110 is that of St 52-3 at 355 N/mm². Although higher yield points can be attained by increasing the carbon content, the weldability and toughness of the material are impaired with C contents above 0.22%. For this reason weldable structural steels with a high yield point are produced by reducing grain size. Apart from aluminium, the alloying elements niobium, titanium and vanadium have a grain-reducing effect and even in small quantities cause the precipitation of carbides and nitrides which produce a further rise in the yield point. The carbon content can thus be kept low (*low-pearlite fine-grained structural steels*). Temperature-controlled rolling (thermomechanical treatment) further reduces grain size, while strength and yield point are further raised by controlled cooling. By adding small quantities of Cr, Ni, Mo and Cu, hardness-increasing processes (partly through precipitation) are made possible. Water quenching followed by tempering after rolling (the so-called tempcore process) can contribute to a further increase in strength.

Example. TStE 690 VA, material no. 1.89820 with a yield point of R_{eH} = 690 N/mm²

Some modern weldable fine-grained steels are listed in Stahl-Eisen-Werkstoffblatt 083, e.g. BStE 550 TM with a yield point of R_{eH} = 550 N/mm², which has been given the dual-phase structure of ferrite and 5 to 35% martensite by thermomechanical treatment.

Pearlitic Forging Steels. By adding niobium, titanium or vanadium forged parts can be used without further heat treatment after controlled cooling from the forging temperature (BY). An important area of application is vehicle manufacture. The steels are converted to a pearlitic structure with a defined grain size, are suited to machining and sufficiently tough. Pearlitic forging steels are listed in Stahl-Eisen-Werkstoffblatt 101.

Example.   38 MnSiVS 6 BY.

Non-Ageing Steels. These are steels that even when stored for a long period suffer only a slight loss of toughness as compared with their initial state. A sign of ageing resistance is that impact strength when notched does not fall more than 5 or 10% below certain values after artificial ageing through cold forming, with subsequent ageing of ½ h at 250 °C.

Non-ageing steels are standardised in DIN 17 135, in which impact values for temperatures down to about −50 °C and yield point values for temperatures up to about + 400 °C are guaranteed with respect to areas of use.

The steels are listed as A St 35, A St 41, A St 45 and A St 52.

Maraging Steels. These are nickel steels with extremely low contents of C, Si and Mn. Like fine-grained structural steels, therefore, they have good welding properties.

Example. X 2 NiCoMo 18 8 5 with 18% nickel, 8% cobalt, 5% molybdenum, 0.4% titanium and 0.1% aluminium.

In their solution-treated mill finish these steels have a structure of practically carbon-free nickel martensite with relatively low strength (tensile strength about 1000 N/mm^2).

By hot ageing at not more than 500 °C, these steels can be hardened to tensile strengths of about 2200 N/mm^2 with sufficient toughness, through precipitating intermetallic compounds such as Ni$_3$(Ti, Al) and Fe$_2$Mo from the martensite.

Tempering Steels. These are given the properties needed for their uses by a suitable heat treatment. It can take the form of tempering (hardening and annealing the whole component), which transforms the structure into the martensite or bainite stage. With unalloyed structural steels, proportions of pearlite and ferrite can also arise as a mixed structure. On the other hand, certain properties can be obtained by only or additionally subjecting the component to boundary layer hardening (flame or induction hardening). By tempering, certain properties of strength (tensile strength, 0.2% yield point) and toughness (rupture elongation, area reduction when breaking and impact strength when notched) can be achieved in a continuous progression (see **Fig. 4**).

In DIN 17 200 the mechanical properties of tempering steels up to a diameter of 100 mm for unalloyed steels and some alloy steels, and up to a diameter of 250 mm for other alloy steels, are standardised (**Appendix D3, Table 2**).

As defined by this standard, tempering steels are structural steels that are suited to hardening on grounds of their chemical composition, particularly their carbon content, and which have high toughness for a given tensile strength in the tempered state.

The standard distinguishes between different unalloyed high-grade steels and between unalloyed and alloyed special steels. Special steels are distinguished from high-grade steels as follows:

Minimum values in notched-bar impact work in the tempered state (for unalloyed steels, only those with medium carbon percentage by mass, C < 0.50%).
Limiting values for hardenability in the end quench test (for unalloyed steels, only those with medium carbon percentage by mass, C > 0.30%).
More even response to heat treatment.
Limited content of oxidic inclusions.
Lower admissible phosphorus and sulphur content.

In the group of special steels two series of steel varieties are listed:

Special steel series for which only the maximum percentage of sulphur by mass of 0.035% is stated.
Special steel series for which a controlled percentage of sulphur by mass from 0.02 to 0.04%, for better machinability, is stated.

DIN 17 200 contains, among other data, information on the limiting values of Rockwell-C hardness when tested for hardenability in the end quench test, and on the guaranteed mechanical properties of steels in the tempered and normalised state (unalloyed steels) in relation to product thickness.

In the steels listed in **Appendix D3, Table 2**, hardenability rises in the order of listing. For example, the same tensile strength of 900 to 1100 N/mm^2 is achieved by 30 CrNiMo 8 steel with a diameter of 200 mm as by 34 Cr 4 steel with a diameter of only 15 mm.

Steels for Boundary Layer Hardening. These, too, are tempering steels that have been specially developed to produce a hard, wear-resistant surface while retaining a tough core.

In boundary layer hardening, flame hardening is distinguished from induction hardening.

Steels for boundary layer hardening are standardised in DIN 17 212. They are generally taken from Stahl-Eisen-Werkstoffblatt 830 and largely resemble the varieties conforming to DIN 17 200. They are distinguished from them only by certain necessary additional guarantees of suitability for flame and induction hardening, and especially by narrower limits for carbon content and lower maximum values for phosphorus content.

Nitriding Steels. These steels too are in principle tempering steels alloyed with Cr, Mo and Ni. Nitriding steels are standardised in DIN 17 211; see **Appendix D3, Table 3**. As defined by the norm, they are martensitic-bainitic steels which, because of the nitride-forming agents such as aluminium and vanadium – and to a limited extent Cr – contained in them, are especially suitable for nitriding (apart from steels with elements in the following concentrations: ≤ 0.4% Si, 0.40 to 1.10% Mn, < 0.025% P, < 0.030% S).

The most important nitriding techniques are described in D3.1.3.

Case Hardening Steels. These are structural steels with relatively low carbon content which are carburised on the surface, and sometimes nitrogenised (carbonitrided) at the same time, and then hardened. After hardening the steels have a high degree of hardness in the boundary layer, good wear-resistance and, above all, very tough core material.

The steels developed for case-hardening are high-grade and special steels (**Appendix D3, Table 4**).

In the group of special steels some varieties with a certain minimum sulphur content (0.020% S by weight) are standardised. Their average content of sulphide inclusions is higher than for other steels, which guarantees better machining properties.

Molybdenum–chromium steels, in particular, have been developed for direct hardening. The alloying elements Mn, Cr, Mo and Ni heighten core strength with large cross-sections and improve boundary hardenability (except for elements in the concentrations: ≤ 0.04% Si, 0.30 to 1.30% Mn, < 0.035 to 0.045% P, < 0.035 to 0.045% S). Case hardening steels are supplied in the following states of finish: hot-formed (untreated (U)); treated to a certain strength (TS), treated for shearability (C), soft-annealed (A) or treated for good workability (GW) with a ferrite–pearlite structure for machining.

Free-Cutting Steels. These steels are characterised by good machinability and good chip formation, these properties being obtained by a higher sulphur content, sometimes with other additives, e.g. lead.

The steels listed in DIN 1651 are divided into steels not intended for heat treatment (except for elements in the concentrations: ≤ 0.05% Si, 0.90 to 1.50% Mn, < 0.1% P), free-cutting case hardening steels (except for elements in the code designation: 0.10 to 0.30% Si, 0.70 to 1.10% Mn, < 0.06% P) and free-cutting tempering steels (except for elements in the code designation: 0.01 to 0.30% Si, 0.70 to 1.10% Mn, < 0.06% P).

They are generally intended for mass production on automatic machines with high cutting speeds. Free-cutting steels are supplied in untreated (hot-formed), normalised, shelled or cold-drawn condition. The mechanical properties are dimension-dependent, especially in the cold-drawn state.

Other free-cutting steels are micro-alloyed with Ca to obtain a spheroidal sulphidic form instead of the usual elongated form. This substantially improves the lateral toughness of the steel. If the S content is to be kept as low as possible, Te is also added in the parts-per-million range. Manganese telluride is formed in the steel, guaranteeing good machining properties, but not representing an inhomogeneous phase like a normal sulphide. Like lead, however, tellurium is toxic, which is especially noticeable through its high vapour pressure in the molten steel.

Stainless Steels. According to DIN 17 440 (ISO 9328-1-1991/9328-5-1991) stainless steels are distinguished by special resistance to chemically corrosive substances; in general they have a chromium content of at least 12% by weight. The stainless steels standardised in DIN 17 440 are subdivided into ferritic and martensitic, and ferritic–austenitic and austenitic steels. They are suited to both cold and hot forming. Except for the steels X 30 Cr 13, X 38 Cr 13, X 46 Cr 13 and X 45 CrMoV 15, stainless steels are suitable for welding. However, with carbon-rich steels welding is only possible if special precautionary measures are taken. Their resistance to intercrystalline corrosion in mill finish is an important property of stainless steels. With low-carbon ferritic chromium steels this property is guaranteed as generally supplied. In the welded state this often no longer applies. For this reason, some ferritic and martensitic chromium steels should be annealed or retempered after normal welding. After high-speed welding (e.g. laser beam welding) this is not necessary. Statistical information on chemical composition is not given in DIN 17 440, as the resistance values obtained under laboratory conditions do not always apply to behaviour in use.

A special kind of stainless steel is the dual-phase steel X2 CrNiMoN 22 5 (material no. 1.4462) with about 50% austenite and about 50% soft martensite or nitrogen pearlite, in which the two phases fulfil different functions: the austenite guarantees corrosion protection, e.g. seawater resistance in this case; the soft martensite guarantees component strength.

Appendix D3, Table 5 gives a survey of some stainless steels conforming to DIN 17 440.

Other Standards. Stahl-Eisen-Werkstoffblatt 400: Stainless rolling and forging steels. It contains the steels which are not included in DIN 17 440 because of their limited uses. – DIN 17 445: Stainless cast steel. – SEW 410: Stainless cast steel. This contains varieties of cast stainless steel primarily used for special purposes. – DIN 1694: Austenitic cast iron. – DIN 17 224: Spring wire and spring strip made of stainless steels. – Stahl-Eisen-Einsatzliste 430: Stainless steels and alloys for medical purposes. – SEE 432: Stainless steels for use in building construction.

High-Temperature and Very High-Temperature Steels (Alloys). These have good mechanical properties, such as high time yield limits and longtime rupture strength or high relaxation resistance, under long-duration loading at high temperatures. High-temperature steels show these properties up to an operating temperature of 540 °C, and very high-temperature steels up to about 800 °C. They are used predominantly in power generation and the chemical industry.

High-temperature steels include unalloyed steels with an upper operating temperature in longtime use of about 400 °C, and low-alloy steels with maximum operating temperatures of about 540 °C; in terms of alloy type these belong mainly to the Mo, CrMo, MoV and CrMoV group of steels. Their high heat resistance is produced by the atoms of the alloying elements intercalated in the α solid solution (solid-solution hardening), and by precipitation, above all, of Mo and V carbides.

Very high-temperature steels include the group of 12% Cr steels (operating temperature up to about 600 °C), which contain alloying additives of Mo, V and in some cases N and Nb to cause precipitations. More heat-resistant than the 12% Cr steels (up to about 650 °C) are the very high-temeprature austenitic steels with about 16 to 18% Cr and 10 to 13% Ni, since the austenitic lattice has a higher resistance to form changes at high temperatures. To increase strength Mo, V, W, Nb, Ti and B are admixed in some cases.

For longtime loadings at temperatures of 700 °C and above (e.g. in gas turbines or chemical plants), Ni-base and Co-base alloys with high Cr contents (usually > 12%) to guarantee scale and corrosion resistance are predominantly used, as well as high-alloy Fe alloys. Some of these materials, called super-alloys, do not belong to the group of iron-base materials, and are therefore included under this heading.

The nickel-base alloys contain, above all, Co, Mo and W, apart from Cr, to increase the strength of the solid solution, and usually Ti and Al to form creep-resisting deposits in the form of intermetallic phases ($Ni_3(TiAl) = \gamma'$-phase). In addition, small quantities of rare earths such as cerium, hafnium, zircon and yttrium are added in many cases because of their corrosion-resisting influence.

The Co-base alloys (cheaper to produce as they do not need vacuum pouring) lack an effective hardening mechanism comparable to that of the Ni-base alloys. For this reason their longtime rupture strength is generally lower. Additions of Cr, Ni, W, Mo, V and Nb cause complex carbides and intermetallic phases to be formed.

The temperature limits above which components are assigned the time-dependent strength values established in the stress–rupture test (e.g. time yield limit, longtime rupture strength), instead of short-time values (e.g. high-temperature yield point), depend on the chemical composition of the material, the prescribed loading time and the chosen safety correction values.

Applications of High-Temperature and Very High-Temperature Steels (Alloys)

Heavy Boiler Plate. The steels listed in **Appendix D3, Table 6** are suitable in plate form for boiler installations, pressure vessels, large pressurised pipelines and similar components. Apart from unalloyed steels (HI to HIV), low-alloy steels, some of which are also employed as tube steels, are used for higher temperatures. As the steels must have good fusion welding properties they have a low C content and increased Mn content.

Standards. DIN 17 155 (ISO 9328-1-1991/9328-2-1991): Heavy boiler plate.

Steels for Seamless Tubes and Collectors (up to about 600 °C). For seamless tubes including headers for steam boilers, pipelines, pressure vessels and apparatus where temperatures rise to about 500 °C with high pressures, unalloyed and alloy steels are used, some of which are also employed for boiler plate (**Appendix D3, Table 6**). Depending on scaling conditions, the limiting temperature indicated can shift to lower values (unalloyed steels) or higher values (12% Cr steels). If hot processing produces a flawless structure of sufficient uniformity then no heat treatment may be needed, except in the case of CrMo steels, which have to be annealed, and the steels 14 MoV 6 3 and X 20 CrMoV 12 1, which must always be tempered.

Standards. DIN 17 175 (ISO/DIS 9329-2-1992): Seamless tubes and collectors made of high-temperature steels.

Materials for Nuts, Bolts and Forgings. Nuts, bolts and similar threaded and shaped parts for use at temperatures above about 300 °C are as a rule produced in the materials shown in **Appendix D3, Table 6**. To minimise loss of pre-stress as a result of creep in threaded joints, the materials listed have high relaxation resistance. For very high temperatures (≥ 700 °C) over long durations the nickel-base alloy NiCr 20 Ti Al should be specified, though it is also increasingly used for lower-temperature applications (smaller threaded cross-sections, narrower flanges).

Forgings for boilers, pipelines and pressure vessels are made predominantly of steels to DIN 17 243, some of which correspond to those in DIN 17 155 and 17 175.

For larger free-formed forgings for turbines (shafts, rotors) at temperatures up to about 540 °C, CrMo(Ni)V steels are predominantly used (examples 21 CrMoV 57, 30 CrMoNiV 5 11, X 22 CrMoV 12 1), in which the nickel additive improves through-tempering where large cross-sections are used.

Standards. DIN 17 240: High-temperature and very high-temperature materials for nuts and bolts. – DIN 17 243: Forgings and rolled or forged steel bar of high-temperature weldable steels. – SEW 555: Steels for large forgings used in turbines and generators.

High-Temperature Steels of Various Forms. In the temperature range above about 550 °C high alloy martensitic steels with about 12 Cr and austenitic CrNi steels are used; they are distinguished by good performance under long-time loading. They are supplied as bar of various cross-sections, sheet, strip, tube or forgings. Some important steels are indicated in **Appendix D3, Table 6**.

The steels X 20 CrMoV 12 1 and X 8 CrNiMoVNb 16 13 are also known as pressurised-hydrogen-resistant steels (SEW 590); the steel X 22 CrMoV 12 1 is listed as a screw steel in DIN 17 420.

Standards. DIN 17 459: Seamless round tubing in high-temperature austenitic steels. – DIN 17 460: High-temperature steels. Technical supply conditions for sheet, cold- and hot-rolled strip, bar and forgings. – SEW 670: High-temperature steels. – SEL 675: Seamless tubes of high-temperature steels.

High-Temperature Fe, Ni and Co Alloys. Various high-alloy materials with an Fe, Ni and Co base are used in the temperature range above about 650 °C in aircraft, industrial turbines, rockets and high-temperature chemical installations. As resistance to corrosive influences as well as high-temperature strength are demanded of these alloys, the boundary between them and heat-resistant materials is often difficult to draw clearly.

The alloys mentioned in **Appendix D3, Table 7** are supplied as wrought alloys in the form of sheet, bar, tube and/or forgings; some are also processed into castings. Hot forming and mechanical processing are difficult with alloys with a high proportion of intermetallic phases. In some cases the only possible forming technique is casting.

Heat-Resisting Steels. The chief requirement for heat-resisting steels is not especially good high-temperature strength but sufficient resistance to hot gas corrosion in the temperature range above 550 °C. The highest temperature at which a heat-resisting steel can be used depends on operational conditions. Recommended temperatures for air and hydrogen atmospheres are contained in **Appendix D3, Table 8**. These values do not apply to molten metals and salts.

The scaling limit temperature for heat-resisting steels is defined as the temperature at which the material loss in clean air is 0.5 mg cm^{-2} h^{-1}.

The scale resistance of heat-resisting steels is based on the formation of dense, adhesive surface layers of oxides of the alloying elements Cr, Si and Al. The protective effect starts when the Cr content is 3 to 5%, but Cr contents up to 30% can be alloyed. The protective effect of these layers is limited by the corrosive low-melting-point eutectics and by carburising. Carburising produces chromium carbides, so that protection is reduced by chromium loss in the wall mass and by the formation of internal tensions (which rupture the coating). To increase heat resistance the alloying element Ni is added in addition to Cr, in quantities that give rise to a stable austenitic structure (Cr + Ni = 25 to 35%); however, ferritic–austenitic steels are also used, as well as ferritic steels of the types Cr, Al, Si (**Appendix D3, Table 8**).

Ferritic steels with a Cr content of over 12% can become brittle at temperatures about 475 °C; for this reason they should not be kept at this temperature for long periods during heat treatment or in use. At temperatures above 800 °C rapid growth of the ferrite grain can set in, producing a decline of toughness at lower temperatures. Finally, in all steels with a Cr content above about 20%, precipitation of σ phases (Fe–Cr bonding) can occur at 600 to 850 °C, which also causes brittleness.

Unlike austenitic steels and Ni alloys, heat-resisting ferritic steels are resistant to reducing sulphurous gases. Heat-resisting steels are used in chemical plants and industrial furnaces, e.g. for pipes in ethylene plants and conveyor parts for continuous furnaces.

In this context the heat-conducting alloys should be mentioned; their chemical composition has an Ni–Cr, Ni–Cr–Fe or an Fe–Cr–Al base (examples: NiCr 80 20, NiCr 60 15, CrNi 25 20, CrAl 25 5).

Standards. SEW 470: Heat-resisting rolled and forged parts. – Euronorm 95: Heat-resisting rolled and forged parts. – DIN 17 470: Heat-conducting alloys.

Pressurised-Hydrogen-Resisting Steels. In chemical plants such as oil refineries, hydrogenation plants and synthetics containers, materials are exposed at the same time to high temperatures and high partial pressures of hydrogen. Hydrogen diffuses into the steel, decarbonising it by forming carbon–hydrogen compounds such as methane (CH_4). This causes dissolution of the grain boundary carbides, fissure at the grain boundaries and brittleness in the material. By alloying the steel with elements with very stable carbides at operating temperature, the susceptibility to pressurised hydrogen can be sharply reduced. The most important alloying element of these steels is chromium; they also contain Mo and in some cases V to increase heat resistance and reduce annealing brittleness (examples: 25 CrMo 4, 24 CrMo 10, 17 CrMoV 10, X 20 CrMoV 12 1). Finely divided chromium-rich special carbides are formed during tempering. Owing to their high alloying content, austenitic very high-temperature steels have good resistance to pressurised hydrogen, coupled to good strength at temperatures above 550 °C (example: X 8 CrNiMoVNb 16 13).

Standards. SEW 590: Pressurised-hydrogen-resistant steels. – DIN 17 176: Seamless round tubing in pressurised-hydrogen-resistant steels.

Low-Temperature Steels. As temperature falls the resistance of steels to changes of form increases, i.e. yield point and tensile strength are raised, while their forming properties (rupture elongation, area reduction when breaking) are impaired. This increases the liability to brittle fracture through stress concentrations (notch effect, internal stresses), even without forming, as temperature falls. This behaviour is more pronounced with ferritic than with austenitic steels. Ferritic steels show a steep drop in the curve of their temperature-dependent toughness values, so that their use is limited to temperatures above this part of the curve. By contrast, austenitic steels show only a gradual drop in toughness, which allows their use even at very low temperatures. The characteristic value for toughness is normally established by notched-bar impact work.

For low-temperature steels according to DIN 17 820 in mill finish a minimum value for notched-bar impact work of 27 J (established in ISO tests with transverse or tangential V-notch) is guaranteed at a temperature of – 60 °C or below. Low-alloy, low-temperature steels can be used at minimum temperatures of −100 °C. Low-temperature steels are used in apparatus for the chemical industry, in containers and in low-temperature technology. They are predominantly alloyed with nickel in proportions between 1.5 and 5%. The structure of these steels is ferritic–pearlitic. With Fe, nickel forms a solid solution that has high toughness at low temperatures. Ni is not involved in carbide formation. With higher demands for toughness, high-alloy steels such as X 9 Ni 9 are used, and for extreme demands austenitic steels. Their notched-bar impact work is satisfactory at temperatures as low as 200 °C. With low-temperature steels attention should be paid to good weldability, if possible without heat treatment.

A number of typical low-temperature steels are listed in **Appendix D3, Table 9**. Steels to DIN 17 135 and DIN 17 440 can also be used at low temperatures.

Standards. DIN 17 820 (ISO 9328-1-1991/9328-3-1991): Low-temperature steels. – DIN 17 440 (ISO 9328-1-1991/9328-5-1991): Stainless steels.

Tool Steels. Owing to their diverse uses and varying specific operating conditions, tool steels have chemical compositions and properties that range between very wide limits. Tool steels have no typical alloying agents or properties that characterise them specifically as tool steels. Even their hardness varies within wide limits; what matters is their difference in hardness to the material to be worked. Because of this diversity the definition of tool steels rests on their use in tools (DIN 17 350).

An ordering scheme and a classification for the multiplicity of tool steels is derived from their various application groups (DIN 8580):

Steels for cutting (including punching, machining).
Steels for hot forming (forging and pressing).
Steels for cold forming (stamping and extruding).
Steels for hot forming (diecasting, chill casting and glass moulding).
Steels for cold forming (plastic moulding).
Steels for hand tools.

Alloys and heat treatment for the various uses are chosen so that the steels do not fracture under the relevant loadings (toughness, ductility), are not permanently

distorted (sufficiently high yielding point; hardness) and their surfaces remain unimpaired for as long as possible (high wear and corrosion resistance). The desirable properties are not always attainable with traditional forgeable or cast tool steels. For this reason steels produced by powder metallurgy (PM steels) are increasingly used as tool steels; they remain ductile with very high alloying content. Additional surface coating techniques are used to heighten their wear resistance.

Steels for Cutting/Punching. These are predominantly ledeburitic chromium steels with 1.5 to 2% C, 12% Cr and admixtures of Mo, W and V, containing up to 20% carbides. In addition, for sheet above 5 mm thick, tougher varieties with 0.6 to 1% C and alloying contents of 3 to 6% (chromium, manganese, molybdenum, tungsten, vanadium) are used.

For cutting and punching, the steels used for machining purposes (high-speed steels) are suitable. In powder-metallurgical steels wear-resistant alloys with 4 to 10% V are used for cutting and stamping (**Appendix D3, Table 10a**).

Steels for Machining. These should have good hardness, high retention of hardness and good high-temperature hardness (to approx. 600 °C, red heat hardness). The most frequently used alloy has 0.9% C, 4% Cr, 6% W, 5% Mo and 2% V; W and Mo are exchangeable in proportions of one part molybdenum to two parts tungsten. Additions of up to 5% vanadium increase wear resistance, while additions of cobalt (usually 5 to 8%) improve hardness, high-temperature hardness and hardness retention. PM steels with higher alloy contents are also used for machining purposes, especially because of their good workability but also because of their high resistance to fracture (**Appendix D3, Table 10b**).

Steels for Hot Forming (Forging and Pressing). Such steels are exposed to stress by temperature change and heat wear, e.g. by scaling. For hammer forging, low alloy Ni–Cr–V–Mo steels with 0.5 to 0.6% C, are used. For pressing with long contact times, hot-forming steels with 0.3 to 0.4% C, 5 to 7% Cr, 1 to 3% Mo and up to 2% V are used. Austenitic high-temperature steels are also suitable as press tools where working temperature rises above 500 °C (**Appendix D3, Tables 10c, d**).

Steels for Cold Forming (Stamping, Massive and Continuous Forming). These must possess high crush strength and therefore have a minimum C content of 0.5%. In most cases the C content is about 1%. When adapted to the required crushing pressures and working temperatures, ledeburitic Cr steels (see "Steels for Cutting/Punching" above) and high-speed steels (see "Steels for Machining") are used as stresses become more acute (**Appendix D3, Table 10c**).

Steels for Hot Forming (Diecasting, Glass Moulding). These are exposed to special stress by temperature change. For diecasting tools and chill forming moulds, hot-working steels with 0.3 to 0.4% C, 5% Cr, 1 to 3% Mo and up to 1% V have proved appropriate. Tool behaviour under these stresses depends largely on the degree of homogeneity (the amount of liquation) of the steel used. Glass pressing moulds with C contents of 0.2% are, like stainless steels, alloyed with 12.5 to 17% Cr (**Appendix D3, Tables 10f, g**).

Steels for Cold Forming (Plastic Moulds). These are used for plastics manufacture at low temperatures. For simple plastics moulds, pre-tempered steels with a hardness of approx. 300 HB are predominantly used, to avoid form changes during hardening. Most frequently a material with 0.4% C, 1.5% Cr, 1.5% Mn and molybdenum additive is used. To reduce deformation steels with good hardening properties, having a Ni content of 3 to 4%, are always used for plastics moulds.

If the plastics contain abrasive additives like glass fibre, carbon fibre or rock flour, very hard, wear-resisting steel alloys, of the kind used for cutting or machining steels, are needed for the forming tools. Powder-metallurgical varieties with very high V contents are also used in injection machines. For plastic materials with tool corroding properties, alloy variants with 12 to 20% chromium are needed to give the tools sufficient corrosion resistance (**Appendix D4, Table 10h**).

Steels for Hand Tools. Hammers, pincers, axes, etc. are made predominantly of unalloyed tool steels with 0.45 to 0.60% C. Materials for saws have a somewhat higher carbon content of 0.75 to 0.85%. The highest C content is found in files, at 1.2 to 1.4%. Steels for spanners and screwdrivers have a carbon content of 0.3% and are alloyed with 0.3 to 0.8% Cr and vanadium additive.

Steels for Nuts and Bolts

Owing to their diverse uses, nuts and bolts have to fulfil a number of demands, which must be considered in selecting the material.

The technical standards for mechanical joining elements therefore set minimum requirements for materials, depending on the range of demands the product has to meet. For single demands particular varieties of material are specified. The materials that can be used for nuts and bolts, which must be predominantly cold-formable, are listed in the various materials standards:

The minimum requirements for the chemical composition of materials for screws in the property classes 3.6 to 12.9 specified in DIN ISO 898 Part 1 can be met by steels conforming to: DIN 1651: Machining steels; DIN 1654: Cold-upsetting steel and cold extruding steels; DIN 17 100: General structural steels; DIN 17 111: Low-carbon unalloyed steels for nuts, bolts and rivets; DIN 17 200: Tempering steels; DIN 17 210: Case-hardening steels.
Appendix D3, Table 11 shows the materials normally used in practice for bolts, taking account of property class, dimensions and production process. For nuts of lower property classes unalloyed steels are used, as for bolts, as well as machining steels. From property class 10 onwards nuts are made of tempering steels.
Rust- and acid-resistant steels for nuts and bolts according to DIN 267 Part 11 are standardised in DIN 1654 Part 5 and DIN 17 440.
In DIN 267 Part 12 (*self-tapping screws for sheet*) and DIN 7500 (*thread-cutting screws*) case-hardening steels to DIN 17 210 and tempering steels to DIN 17 200 are referred to.
Materials for *low- and high-temperature fasteners* are prescribed in DIN 267 Part 13; their characteristics are described more fully in DIN 17 240 and DIN 17 280 (ISO 9328-1-1991/9328-3-1991).
Materials for *mechanical fastenings of non-ferrous metals* are specified in DIN 267 Part 18. Their chemical composition and the required material properties are stated in the standards for light and heavy metals which they also list.

Spring Steels

In conformity with DIN 17 221 – hot rolled steels for heat-treatable springs – spring steels in the tempered state are distinguished by their high limit of elasticity and are thus especially suited for resilient components of all kinds. The properties desired in springs are obtained by higher carbon content and alloying ingredients such as Si, Mn, Cr, Mo and V, as well as by heat treatment, i.e. hardening in oil or water followed by annealing. Spring steels are defined by various DIN standards in terms of product form, state of treatment and use. For a selection of spring steels see **Appendix D3, Table 12**.

Standards. DIN 17 221: Hot-rolled steels for heat-treatable springs. Their technical standards are in DIN 17 222: Cold-rolled steel strip for springs. – DIN 17 223 Part 1: Patented drawn spring wire of unalloyed steel. The varieties of wire A, B, C and D listed in DIN 17 223 Part 1 are characterised by their tensile strength, which is guaranteed in relation to dimensions. – DIN E 17 223 Part 2: Oil-tempered spring steel wire of unalloyed and alloyed steel. – DIN 17 224: Spring wire and spring strip of stainless steel. The standard includes three varieties of steel, X 12 CrNi 17 7, X 7 CrNiAl 17 7 and X 5 CrNiMo 17 12 2. In general, material X 12 CrNi 17 7 is used. The steel X 5 CrNiMo 17 12 2 is used to meet a requirement for increased corrosion resistance, while the hardenable austenitic–ferritic steel X 7 CrNiAl 17 7 is specified for highly stressed springs at temperatures between 20 and 350 °C.

All standards for spring steels contain special requirements regarding surface condition and boundary decarbonisation, as the durability of springs under changing loads depends largely on a notch-free, hard surface.

Steels for Rolling Bearings. The balls, rollers, needles, races and discs of rolling bearings are generally exposed to high transient local tension and compression loads and wear stresses. The steels used must therefore be free of flaws such as blowholes, bubbles and segregation, and must be generally free of microscopic non-metallic inclusions; they must have good hot- or cold-forming properties and good machinability, must guarantee a high degree of hardness and must be dimensionally stable under long storage conditions.

According to Stahl-Eisen-Werkstoffblatt 350, suitable steels are those such as 100 Cr 6, 100 CrMn 6 and 100 CrMo 7, with surface hardnesses of HRC 56 to 66 after tempering.

For special purposes beyond this the tempering steel 41 Cr 4 (spring rollers) and the high-alloy steels X 46 Cr 13 and X 89 CrMoV 18 1 (balls, rollers, needles and tracks for stainless bearings) are standardised, their maximum hardness being 58 HRC. Steels for rolling bearings are also standardised in Euronorm EU 94.

A special application is steel X 75 WCrV 18 4 1, used for high-temperature bearings in aircraft turbines.

Valve Materials

The valves of internal-combustion engines, especially exhaust valves, are subjected to high mechanical loads at high temperatures, as well as the corrosive effects of, above all, Pb, S and V, and combustible residues in the hot exhaust gases. Valve materials must therefore be resistant to heat, temperature change and longtime vibration, impact, wear and corrosion stresses; they must also be suited to hot forming. High heat conductivity and low thermal expansion are also desirable, so that temperature differences and the resulting heat tension are kept to a minimum.

Today the three materials X 45 CrSi 9 3, X 60 CrMnMoVNbN 21 10 and NiCr 20 TiAl are used for valves in internal-combustion engines.

Standards. SEW 490-52 (obsolete): Valve steels. – Euronorm 90-71: Steels for the exhaust valves of internal-combustion engines.

Steels for Pipes and Tubes

The material used for steel tubing is determined by the production process and the use of the finished part.

DIN 1626 (ISO 9330-1-1990): Welded steel tubes of unalloyed and low alloy steels. – DIN 1629 (ISO 9329-1-1989): Seamless tubes of unalloyed steel. – DIN 2391: Seamless precision steel tubes, cold drawn or cold rolled. – DIN 2393: Welded precision steel tubes of special dimensional accuracy. – DIN 2394: Welded precision steel tubes, cold-drawn or cold-rolled once. – DIN 2440: Steel tubes; medium-weight threaded tubes. – DIN 2441: Steel tubes; heavy threaded tubes. – DIN 2442: Threaded tubes of prescribed merit; nominal pressure 1 to 100. – DIN 2462: Seamless tubes of stainless steel. – DIN 2463: Welded tubes of austenitic stainless steel. – DIN 2464: Seamless precision tubes of stainless steel. – DIN 2465: Welded precision tubes of austenitic stainless steel. – DIN 17 175 (ISO/DIS 9329-2-1992): Seamless tubes of high-temperature steels. – DIN TAB 15: Standards for steel pipe-lines.

Steels for Electrical Machines

In these materials magnetic properties play a decisive part. For magnetic steel sheet and strip to DIN 46 400 demands for minimum re-magnetisation losses and high magnetic induction are made. Although the chemical composition of these steels is not prescribed, the required properties are generally achieved by alloying with silicon. The sheet and strip is delivered heat-treated and may not be cold-formed by hammering, bending or twisting, as this impairs its magnetic properties.

Appendix D3, Table 13 contains a selection of cold and hot rolled magnetic sheets to DIN 46 400 and their magnetic and technical properties.

Other Standards. DIN 17 405: Soft magnetic materials for DC relays. – DIN 17 410: Materials for permanent magnets. – DIN 41 301: Magnetic sheets; magnetic materials for translators.

Apart from materials with good magnetic properties, non-magnetisable materials are also needed in electric machine construction. These are steels with an austenitic structure. According to Stahl-Eisen-Werkstoffblatt 390 their magnetic permeability, which is calculated from induction B (previously in gausses, now in teslas) with a field of 100 oersteds (1 Oe = 79.58 A/m), has a maximum value of 1.08 G/Oe = $1.08 \cdot 10^{-4}$ T/Oe.

Examples. X 120 Mn 13 (1.3802), X 40 MnCr 18 (1.3817), X 8 CrMnN 18 8 (1.3965) and cast steel G-X 120 Mn 13 (1.3802).

Steels for Aircraft and Spacecraft

These steels are subject to special national and international specifications. They are taken from the groups of structural steels and non-rusting steels and are listed under their own material numbers, e.g. 15 CrMoV 6 9 (material no. LW 1.7734), derived from 14 CrMoV 6 9 (material no. 1.7735). Such steels are frequently electroslag-remelted or electron-beam-remelted steels of extremely good purity and low inhomogeneity through segregation.

Steels for Nuclear Energy Plants

Here too special specifications are introduced. For example, the steel producer is obliged to keep production data available for at least ten years so that complete damage investigation is possible.

Cast Steel

Cast steel refers to Fe–C alloys with a C content up to about 2%, which are cast into components in sand moulds, or less often in permanent moulds of graphite or metal. The smelting and alloying of cast steel resembles that of rolling and forging steel, which is cast in chill moulds and further processed by hot forming. This can give rise to considerable differences in the mechanical properties of components, especially toughness, both laterally and longitudinally in relation to the direction of forming. In cast steel, strength is largely independent of direction. To avoid gas bubbles cast steel is always killed. As the material solidifies a coarse, inhomogeneous structure of low toughness is formed. By normalising or tempering a structure and properties resembling those of forging steels are obtained. After welding or mechanical working, cast steel parts are frequently subjected to stress-relief heat treatment.

Compared with cast iron, the pourability of cast steel is inferior and its tendency to cavitation greater, owing to the higher melting temperature and greater shrinkage (about 2%); yet cast steel shows greater strength and in some cases higher toughness. As compared with corresponding rolled and formed material, cast steel often has a higher C content, to improve its pourability by lowering the liquidus temperature. Steel casting has cost advantages for many components, owing to the simple forming process. It is also used with alloys that present difficulties for hot or cold forming (e.g. permanent-magnet casting, straight manganese steel casting).

The general data on rolling and forging steels also apply to the corresponding cast steel varieties.

Cast Steel for General Uses. *Unalloyed* or *low alloy* cast steel for general purposes accounts for by far the greater part (75%) of cast steel production. Depending on C content, its strength ranges from 370 to 690 N/mm², while toughness is also high (**Appendix D3, Table 14**). It has good welding properties, particularly when C content is low. The classification of varieties is based on mechanical properties at room temperature. It has a broad range of uses for highly-stressed components. Normalising is the form of heat treatment predominantly used.

Standards. DIN 1681 (ISO/DIS 8062-1991): Cast steel for general uses.

Tempering Cast Steel. If high strength and a high limit of elasticity, good toughness and good through-hardenability are required in a cast steel component, tempering cast steel is used.

Standards. SEW 510: Tempering cast steel.

High-Temperature Cast Steel is used for housings, valves and flanges in steam and gas turbine installations and for components in high temperature chemical plant. In chemical plants heat resisting or pressurised hydrogen resisting cast steel can be superior to high-temperature cast steel (**Appendix D3, Table 15**).

Standards. DIN 17 425: High-temperature ferritic cast steel.

Very High-Temperature Investment Casting Alloys. Gas turbine materials used in blades at temperatures up to about 1050 °C can be found in **Appendix D3, Table 7**.

Heat-Resisting Cast Steel (Appendix D3, Table 15). Like heat-resisting rolling and forging steel, this material is used in industrial furnaces, in the cement industry, in ore processing, smelting and casting technology and in the chemical industry.

Cast steel for oil and gas installations must have good resistance to pressurised hydrogen, to carburisation and to aggressive media (oleic acids, lyes, sulphur compounds). Some high-temperature ferritic cast steels to DIN 17 245 and heat-resisting cast steels to SEW 471 are also suited to these applications. Spun-cast pipes made of the often-used steel G-X 40 CrNiSi 25 20 for reformer furnaces and ethylene plants deserve special mention.

To meet the highest stresses Ni base alloys are used. The borderline between these materials and very high temperature steels and alloys is fluid.

Standards. DIN 17 465: Heat-resisting cast steel. – SEW 471: Heat-resisting cast steel. – SEW 595: Cast steel for oil and natural gas installations.

Low-Temperature Cast Steel must show sufficiently high toughness at low temperatures (below about −10 °C). At the minimum operating temperature for a steel, a lower limiting value in notched-bar impact work of 27 J (ISO-V-test) should not be exceeded (examples: GS-Ck 24, GS-10 Ni 14, G-X 6 CrNi 18 10).

Standards. SEW 685: Low-temperature cast steel.

Non-Rusting Cast Steel. For the rotors of water turbines, valves and armatures, and for acid-resisting parts in the chemical industry, non-rusting cast steel usually having a Cr content of more than 12% is used. Pearlitic–martensitic cast steel with 13 to 17% Cr and 0.1 to 0.25% C is distinguished from the frequently used austenitic CrNi cast steel, which has greater toughness (**Appendix D3, Table 14**).

Standards. DIN 17 445, SEW 410: Non-rusting cast steel.

Wear-Resisting Cast Steel. This is used for components in size reduction machines, wear-resisting parts in construction machines and conveyors, and tools for cold working (wood and plastics processing) and hot working (rolling, drawing dies).

The groups of cast austenitic manganese steel (C: 1.2 to 1.5%, Mn: 12 to 17%), tempered hardened cast steel (C: approx. 0.6%, Cr; 2 to 3%) and martensitic–carbidic cast steel (C: 1.0 to 2.0%, Cr: 12 to 25%: for hot-working additives of W and V), the first group being the most important.

Cast Steel for Electric Technology. Here the most important material is non-magnetisable cast steel with an austenitic structure stabilised by Mn or Ni, sometimes with strength-increasing and corrosion-inhibiting alloying additives such as Cr, Mo and V.

Examples. G-X 120 Mn 12, G-X 45 MnNiCrV 8 8 5, G-X 10 CrNiNb 16 13.

For permanent-magnet casting iron alloys to DIN 17 410 with an Al content of 6 to 13%, Ni from 13 to 28%, Co from 0 to 34% and Cu from 2 to 5% are suitable. Some alloys contain Ti additives (examples of abbreviations: AlNi 120, AlNiCo 400).

These alloys cannot be hot-formed and are therefore sandcast. They are finished by grinding.

Sintered Materials

Moulded parts of sintered iron and sintered steel are produced from iron powder which is usually consolidated in compression moulds under pressures of 400 to 700 MPa and then sintered in electric furnaces at 1100 to 1300 °C (see K2.3).

Iron materials produced from powders are divided into three groups:

Sintered iron with low carbon content.

Sintered steel, which has higher strength because of its content of carbon and alloying elements and low porosity, and can have its mechanical properties improved by heat treatment.

Iron composite materials which like cast iron contain free carbon.

Sintered steel is produced under higher pressures to densities of up to 7.3 g/cm³ and tensile strengths up to 1000 N/mm². Sintered steel with carbon content can be tempered. Sintered steels with the alloying ingredients Cu, Cr, Cu–Ni and Cr–Ni have gained practical importance above all in the mass production of components. Copper admixtures of about 2% compensate for the contraction of iron particles during sintering, making the parts stronger and more dimensionally stable. Corrosion-resistant sintered steels are alloyed with Cr or Cr–Ni. A special kind of alloy is formed by filling the pores with a metal with a low melting point (*infiltration alloys*), e.g. copper or copper alloys, brass or manganese. In this way oil- and gas-tight components can be produced which are at the same time very strong and very tough. Sintered iron with intercalated graphite particles (graphite proportion up to 20 wt%) is used as a sliding friction material (brake linings, clutches, sliding bearings, contact makers).

In **Appendix D3, Table 16** a number of important sintered materials are listed, with their chemical compositions and mechanical properties.

3.1.5 Cast Iron

The term cast iron refers to all iron–carbon alloys with more than 2% C. However, the maximum C content is seldom above 4.5%. Smelting normally takes place in a cupola or an electric furnace using pig iron, steel scrap, coke and ferrous alloys.

When cooled rapidly cast iron solidifies according to the metastable FeC system, i.e. carbon is combined with iron in the form of carbides. Because of the bright appearance of a fracture surface it is also called *white cast iron*. It is very hard, brittle and only workable by grinding.

As the rate of cooling is slowed down, carbon is precipitated to an increasing degree in elemental form as graphite. The fracture surface is now dark, the material being called *grey cast iron*.

Apart from the rate of cooling (dependent on wall thickness), graphite precipitation is influenced by the C, Si and Mn content. As can be seen from the Maurer diagram (**Fig. 5**), a higher C and Si content favours graphite

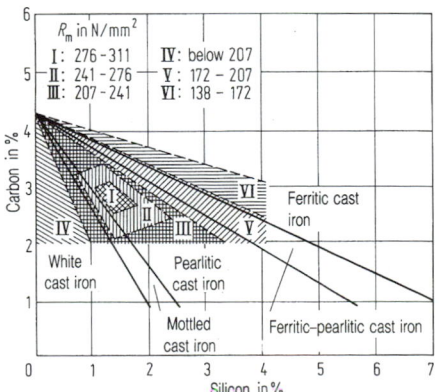

Figure 5. Maurer diagram of the structure of cast iron, showing strength ranges as defined by Coyle.

formation. A higher Mn content promotes the precipitation of Fe_3C at the expense of the graphite component.

Cast iron is a very inexpensive structural material, but its toughness and formability are in general far lower than those of steel.

The various varieties of cast iron (**Appendix D3, Table 17**) are classified in terms of the tensile strength of specimens taken from separately cast rods of different unfinished diameters. The numerical values (e.g. GG-35) show tensile strength in kp/mm^2 (1 kp mm^2 = 9.81 N/mm^2).

Lamellar-Graphite Cast Iron (Grey Cast Iron)

This is the most commonly used form of cast iron. Its graphite component is largely arranged in laminated form. Because of their low mechanical strength the graphite platelets play no part in force transmission but act as hollow spaces, which reduce the effective cross-section and produce stress concentrations at their edges through the notch effect. The formability and impact toughness of cast iron are therefore very low. Its strength increases as the graphite becomes more finely divided. Coarse graphite platelets are found in ferritic cast iron that has solidified slowly, whereas fine platelets are found in high-grade pearlitic cast iron with very finely divided pearlite after more rapid cooling.

Because of the close link between rate of cooling and strength, greater strength can be expected with thinner walls, and vice versa.

No exact recommended analysis is given for the chemical composition of varieties of cast iron. The proportions of Si, P, S and Mn should be arranged so that the desired properties for a cast iron part are achieved.

The mechanical and physical properties of cast iron are largely determined by the form of the graphite and the basic structure. Because of its special structure, grey cast iron has a far lower modulus of elasticity than steel. For ferritic grey cast iron the modulus is approx. 90 000 N/mm^2 and for high-grade pearlitic grey cast iron approx. 150 000 N/mm^2. It decreases as tension increases, i.e. there is no linear relation between tension and extension. Its compression strength is about four times its tensile strength and its bending strength about twice its tensile strength.

Because of the strong internal notch effect of the graphite platelets, the influence of external notching on its mechanical properties is relatively slight.

Grey cast iron has a high buffering capacity, good sliding properties and especially good emergency running properties. It is therefore used, e.g. for engine beds, for bushes and for internal-combustion engine cylinder heads.

Internally pressurised parts must be tested for pressure tightness, because linked graphite platelets can sometimes cause porosity.

The properties of cast iron can be adjusted to certain uses by heat treatment (e.g. hardening, tempering) and by alloying admixtures. For example, Cr, Ni, Mo and Cu increase the strength of low-alloy cast iron.

Volume changes of cast iron through graphite formation as iron carbides decay at temperatures above 350 °C (growth of the cast iron) can be prevented by a low Si content, by alloying with Cr and Ni and by the formation of fine graphite.

Standards. DIN 1691: Cast iron with laminated graphite (grey cast iron).

Spheroidal-Graphite Cast Iron (SG Iron)

The formation of graphite in spheroidal (nodulised) form produces a significant increase in strength and toughness,

as compared with cast iron with laminated graphite (**Appendix D3, Table 17**). The spheroidal graphite configuration is achieved by the admixture of small quantities of magnesium (up to 0.5% by weight) in combination with cerium and calcium. Magnesium is introduced either by adding pre-alloys or by insertion with a nitrogen lance. The properties of cast iron with spheroidal graphite lie between those of cast iron with laminated graphite and those of steel. The modulus of elasticity is about 175 000 N/mm^2. Buffering capacity is reduced as compared with cast iron with laminated graphite, but machinability is good. The properties of this variety of cast iron can be improved by heat treatment to a greater extent than with grey cast iron. To achieve the highest impact toughness, heat treatment producing a ferritic basic structure is normally used.

Spheroidal-graphite cast iron is used for parts with higher vibration stresses, e.g. rollers, crankshafts and housings. The properties of the basic structure can be modified by alloying elements in the same way as for grey cast iron.

Standards. DIN 1693: Cast iron with spheroidal graphite.

Malleable Cast Iron

Parts with complex shapes that have to be tough, workable and have good impact strength are made of malleable cast iron. The starting point is a cast iron in which the carbon and silicon content are so arranged that the casting is graphite-free on solidifying, the entire carbon content being bonded to the iron carbide cementite. If the part is then heat-treated, the cementite decomposes without residue. A further heat treatment produces a casting which is tempered within certain limits.

Two kinds of malleable cast iron are distinguished (**Appendix D3, Table 17**):

White malleable iron, which is decarbonised during heat treatment.

Black malleable iron which is not decarbonised during heat treatment.

White malleable iron is produced by heating for 50 to 80 hours at about 1050 °C in a decarbonising atmosphere (CO, CO_2, H_2, H_2O). In this process C is removed from the casting, so that after cooling a purely ferritic structure is left behind in the boundary layer of the part. With large cross-sections the core contains proportions of graphite (temper carbon). Malleable iron with small cross-sections is weldable, though the S and Si contents must be low.

Black malleable iron is produced by heating in a neutral atmosphere, first for about 30 h at 950 °C. In this process the cementite of the ledeburite decomposes into austenite and graphite (temper carbon), which is precipitated in flaky clusters. In a second heat treatment the austenite is converted during slow cooling from 800 to 700 °C into ferrite and temper carbon.

After cooling the structure consists uniformly of a ferritic–pearlitic basic structure with temper carbon; the pearlite content can be increased by faster cooling. This increases toughness and wear resistance.

The higher strength and toughness of malleable iron as compared with grey cast iron result from the precipitation of the graphite in clusters. As with spheroidal graphite, this produces a considerably lower internal notch effect, as compared with laminated graphite.

Standards. DIN 1692: Malleable iron.

White Iron

Cast iron that solidifies white is called *white iron* or *chill cast iron*. Full white iron, in which the whole cross-section sets white, is distinguished from clear chill iron, in which only the boundary layer (partly aided by chill plates) remains free of graphite, separated from the core by a mottled zone (mottled cast iron) containing areas of grey and white cast iron side by side. The core of a clear chill casting consists of grey iron.

The depth of hardness, or the thickness of the white layer, depends on the rate of cooling and the chemical composition (Mn, Cr, Si) of the casting. With white iron, especially the clear chill variety, casting is more difficult than with grey iron, because of the varying shrinkage and the high internal tensions. White iron is very susceptible to impact fracture, but has high wear resistance. It is therefore used for parts exposed to high wear conditions, such as rollers, camshafts and deep-draw tools.

Special Cast Irons

Austenitic Cast Iron. Owing to its high alloying content (especially Ni and Cr), this material has an austenitic basic structure in which carbon is precipitated mainly as graphite. Depending on the form of the graphite, austenitic cast iron with *laminated graphite* is distinguished from austenitic cast iron with *spheroidal graphite* (**Appendix D3, Table 17**). Austenitic cast iron fulfils numerous requirements, e.g. for corrosion heat and wear resistance, non-magnetic behaviour and, in the case of material with spheroidal graphite, low-temperature toughness. The material is used for pump parts, exhaust ducts, oven parts and suchlike.

Standards. DIN 1694: Austenitic cast iron.

Silicon Iron. The material contains up to 18% of Si. This favours graphite formation, so that with the C content of only about 0.8% common with this material, graphite is still formed. This variety is resistant to hot concentrated nitric and sulphuric acids.

Aluminium Iron. With an aluminium content of about 7% this material has good resistance to scaling.

Chromium Iron. With a Cr content of up to 35% this material is resistant to scaling and acids; it can also contain Ni, Cu and Al.

3.2 Non-Ferrous Metals

(Physical properties of non-ferrous metals and their alloys: see **Appendix D3, Table 39** and **Figs 1 to 3**).

3.2.1 Copper and Copper Alloys

Because of its excellent conductivity to electricity and heat, its malleability and its resistance to air humidity, hot water and some acids, copper is the second most important metal after iron. The low strength of pure copper can be considerably increased by cold forming. The mechanical properties of copper show no deterioration at low temperatures. Impurities and admixtures reduce its electric conductivity.

Crude copper obtained by treatment in the air furnace and converter, like the precipitated copper obtained by hydrometallurgy, has a purity of about 99%. Both varieties are refined pyrometallurgically into refined copper A, B, C, D and F (99.0 to 99.9% pure). Refined copper as the anode is converted by electrolysis into cathode or electrolytic copper.

In ordering semiproducts (strip and sheet) of copper and copper alloys different characteristics can be specified. DIN 17 670 offers the following possibilities:

1. F-number order. Characteristics: tensile strength, 0.2% yield point and rupture elongation.
2. H-number order. Characteristic: hardness.
3. K-number order. Characteristic: grain size.

Best Selected Copper

Molten copper can absorb considerable quantities of oxygen, which after solidification remains almost entirely in the metal in the form of copper oxide inclusions (Cu_2O). This makes the copper sensitive when heated in a reducing atmosphere (welding, soldering). The hydrogen diffuses into the metal and reduces the copper oxide. The water vapour thereby formed is under high pressure and disrupts the structure. If contact with reducing gases is unavoidable, oxygen-free varieties of copper should be used (denoted by the prefixed letter S, e.g. SF-Cu).

Copper is easily soldered. All welding techniques are possible. Techniques using protective gases (TIG, MIG) are particularly suited to copper.

Materials and properties: **Appendix D3, Tables 18** and **19**.

Standards. DIN 1708: Copper; cathodes and mould formats. – DIN 1718: Copper alloys; terms. – DIN 1787: Copper; semiproduct. – DIN 17 655: Copper; casting materials, unalloyed and low alloy; castings. – DIN 17 666: Low-alloy wrought copper alloys; composition. – DIN 17 677: Copper and wrought copper wire; properties. – DIN 40 500: Copper for electrical technology.

Dimensional Standards. DIN 1754: Seamless drawn copper tubing. – DIN 1756: Round bar of copper and copper alloys. – DIN 1757: Copper and wrought copper wire, drawn. – DIN 1759: Rectangular bar of copper and wrought copper, drawn, with sharp edges. – DIN 1761: Square bar of copper and wrought copper, drawn, with sharp edges. – DIN 1763: Hexagonal bar of copper and wrought copper, drawn, with sharp edges; dimensions. – DIN 46 415: Copper and wrought copper strip, cold-rolled, with rounded (rolled) edges.

Copper-Zinc Alloys (Brass)

Brass is the most commonly used copper alloy, with up to 45% zinc and up to 3% lead (to improve machinability). It is characterised by ductility and corrosion resistance. Alloys with less than 33% zinc are frequently called *tombac* (red brass).

The abbreviated names of copper alloys contain the most important elements in wt% (if no figure is given the alloy proportion is generally < 1%). The remainder is the Cu proportion; e.g. CuZn 37: 37%Zn, ≈ 63%Cu.

Three groups of structures that have a dominant influence on the properties of brass are distinguished:

α Brass with Zn content < 39%.
(α + β) Brass with a Zn content of 39 to 46%.
β Brass with 46 to 50% Zn.

α brass has good cold-forming properties, but is more difficult to hot form and has poor machinability. β brass is harder to cold form but has good hot-forming and machining properties. The most important alloys for technical purposes are CuZn 37 (α brass), CuZn 40 and CuZn (α + β brass). Alloys with a pure β structure (Zn > 45%) have few technical uses. Copper-zinc alloys are not hardenable by precipitation. High hardness and strength values can only be obtained by cold forming.

Guidelines for selection and use are in **Appendix D3, Table 20**.

A shrinkage of 1.5% (brass) and 2% (tombac) must be allowed for when casting. Melting loss is approx. 10%.

Processing. Deep drawing, chasing, bending, pressing, stamping, machining, casting.

Heat Treatment. Soft annealing at 450 to 600 °C, stress relief treatment at 200 to 275 °C, annealing to a certain hardness at 300 to 450 °C with holding times of approx. 3 h.

Welding and Soldering. Brass is well suited to soft and hard soldering. With gas and fusion welding, excess oxygen should be avoided. Arc welding causes heavy zinc evaporation, so that zinc-free electrodes should be used. For welding in protective gas only TIG welding (especially suited to thin sheet) is appropriate. Electric resistance welding requires highly controllable machines of adequate power. Spot-welding machines need a special control. Alloys with less than 20% zinc cause difficulties.

Corrosion. Especially with β brass, local "dezincing" can occur under certain corrosive conditions, red copper being isolated in the form of a plug. Small admixtures of arsenic and phosphorus reduce this phenomenon.

In conjunction with internal tension stresses and/or tension loadings concurrent with the effects of certain aggressive substances (mercury, mercury salt, ammonia), intercrystalline or transcrystalline fracture can occur without forming. Low-copper alloys are most susceptible to such damage. This "tension corrosion" can be largely obviated by careful stress-relief of the finished component.

Mechanical Strength Properties. The usual values for important copper-zinc alloys can be found in **Appendix D3, Table 20**.

Casting. Copper-zinc alloys can be sand-cast (dry or wet), chill-cast, continuously cast, spin-cast and diecast.

Copper-Zinc Alloys with Further Alloying Elements (Special Brass)

An admixture of nickel increases strength, hardness, density, corrosion resistance and grain fineness, as compared with pure copper-zinc alloys. Aluminium has a similar effect to nickel, but also increases scale resistance. Manganese and tin increase heat and seawater resistance. Silicon heightens elasticity and wear resistance (springs, sliding bearings). However, resistance to deformation is also sharply increased. Lead admixtures improve machinability. Iron produces a finer grain and improves sliding properties (in corrosive conditions Fe < 0.5%). Phosphorus and/or arsenic prevent dezincing. Alloys such as CuZn 35 Ni have high resistance to seawater. Brasses free of aluminium and silicon are used for hard soldering. Aluminium-free special brasses can be fusion-welded. With an aluminium content up to 2.3% satisfactory results can be obtained from welding, using protective gas and high-frequency current.

The mechanical strength values of some special brass alloys, and information on properties and uses, are to be found in **Appendix D3, Table 21**.

Casting Brass and Special Casting Brass

These alloys have high corrosion resistance, and though their strength and hardness are somewhat less than those of wrought copper alloys they are surprisingly tough for cast materials (**Appendix D3, Table 22**). In the abbreviated names G means cast, K chill-cast, D diecast and Z spin-cast.

Bronze

Copper alloys with more than 60% Cu are called bronzes, although zinc need not be the main alloying element.

Depending on the main admixtures, copper-zinc alloys (tin bronze CuSn ...), copper-aluminium alloys (aluminium bronze CuAl ...), copper-lead-zinc casting alloys (casting zinc-lead bronze CuPb ... Sn) and special bronzes are distinguished.

Copper-Tin Alloys (Tin Bronze)

These materials combine good hardness and ductility with very good corrosion resistance. For wrought alloys a tin content up to 9% and for casting tin bronze a tin content up to 20% is appropriate. Tin bronzes cannot be precipitation-hardened. Strengthening is only possible by cold forming. Most copper-tin alloys are shaped by casting. Because of its excellent sliding and wearing properties, highly stressed sliding bearings and worm wheels are made of this material.

Processing. Tin bronzes have good cold forming properties but are not suited to hot forming (especially Sn < 10%). Machining is possible.

Heat Treatment. Homogenisation annealing is done at 700 °C for 3 h, soft annealing at 600 °C for 3 h.

Welding and Soldering. Copper-tin alloys are weldable only under certain conditions. Gas welding is possible with a neutral flame using an additional rod of special brass. They are generally well suited to hard and soft soldering.

Casting. Copper-tin alloys can be sandcast, chill-cast, continuously cast or spin-cast. Shrinkage is 0.75 to 1.5%. Ingotism can be largely avoided by slow cooling.

Corrosion. Copper-tin alloys have good resistance to corrosion and cavitation. They are also seawater-resistant.

Data on Mechanical Strength and Uses. See **Appendix D3, Table 23**.

Copper-Aluminium Alloys (Aluminium Bronze)

When used in wrought or cast material these alloys with up to 11% aluminium are distinguished by high resistance to heat, scale and corrosion, as they form an adhesive Al$_2$O$_3$ layer when oxidised. Mechanical vibration is well damped. Multi-alloy bronzes containing nickel can be precipitation-hardened and can attain tensile strength values of 1000 MN/m^2, with a yield point of about 700 MN/m^2. While hot forming by forging, pressing or drawing does not normally present problems, cold forming is difficult, especially with multi-alloy bronzes. It can be helped by work annealing at approx. 650 °C.

Machining is also difficult. Soldering and welding are impeded by the aluminium oxide layer. Using suitable fluxes or electrode coatings aluminium bronzes can be welded autogenously and electrically. Weldability declines as Al content rises. The material is usually chill- or spin-cast at temperatures about 1150 to 1200 °C.

A survey of mechanical properties and a guide to uses are in **Appendix D3, Table 24**.

Copper-Lead-Tin Casting Alloys (Lead Bronze)

These alloys contain at least 60% copper. The main alloying additive is lead in quantities up to 35%. Tin, nickel or zinc are also admixed. Because of the differences of specific gravity between the alloying elements there is a tendency to gravity segregation. As lead is insoluble in copper, the spheroidal lead inclusions give good lubrication and emergency running properties. Because of their low strength, pure Cu-Pb alloys are only used for lining steel supports. Thin friction layers are especially resistant to impact loads. With an admixture of tin these alloys are used for bushes, thrust rings etc. (**Appendix D3, Table 25**.)

Special Bronzes. Copper–nickel alloys (nickel bronzes) with up to 44% Ni have high heat resistance, good cavitation and erosion resistance and good resistance to seawater (used for condenser pipes in ships, chemical plants). Alloys with 30 to 40% Ni and 3% Mn are used for producing electric resistance wire. The alloys CuNi 10 Fe, CuNi 20 Fe and CuNi 30 Fe have good welding properties.

Copper–manganese alloys (manganese bronzes) with up to 15% Mn are used as resistance materials in the electrical industry. With a composition of 45 to 60% Cu, 25 to 30% Mn and 25% Sn they are strongly ferromagnetic. Copper–beryllium alloys (beryllium bronzes) with up to 2% Be are precipitation-hardenable (750 °C/W/350 °C), and in this state attain tensile strength values of 1400 N/mm². Castings can also be hardened. Beryllium bronzes are used for springs with good conductivity, sparkless tools, electrodes of spot-welding machines and for surgical instruments. Special bronzes are processed by rolling, pressing, drawing or casting. Soft soldering is possible after hardening, while hard soldering and welding are possible before heat treatment.

Standards. DIN 1705: Copper–tin and copper–tin–zinc casting alloys (casting tin-bronze and red bronze); castings. – DIN 1709: Copper–zinc casting alloys (casting brass and special casting brass); castings. – DIN 1714: Copper–aluminium casting alloys (casting aluminium bronze). castings. – DIN 1716: Copper–lead–tin casting alloys (casting tin-lead bronze); castings. – DIN 17 660: Wrought copper alloys; copper–zinc alloys (brass, special brass); composition. – DIN 17 662: Wrought copper alloys; copper–tin alloys (tin bronze); composition. – DIN 17 663: Wrought copper alloys; copper–nickel– zinc alloys (nickel silver); composition. – DIN 17 664: Wrought copper alloys; copper–nickel alloys; composition. – DIN 17 665: Wrought copper alloys, copper–aluminium alloys (aluminium bronze); composition. – DIN 17 666: Low-alloy wrought copper alloys; composition.

3.2.2 Aluminium and Aluminium Alloys

To produce 1 t of aluminium, 2 t of oxide of aluminium or 5 t of bauxite and 20 000 kW h of electricity are needed. Despite this high energy requirement, aluminium is the most-used material after steel. Its advantages are low weight (only about one-third of the weight of steel for certain alloys of almost equal tensile strength), very good conductivity of heat and electricity, good resistance to corrosion by atmospheric conditions and weak alkaline and acid solutions through forming a natural oxide skin, and good formability (rolling, drawing, pressing, extruding, cold forming).

Aluminium's relatively low modulus of elasticity of 70 000 N mm² (one-third that of steel) causes three times as much elastic deformation as that of steel under the same load (**Appendix D3, Table 1**). For this reason designers aim at high moments of inertia with small cross-sections (tubes or rectangular hollow sections). Reinforcing ribs and beads are also used to raise the moment of inertia. Extrusion techniques allow highly complex profiles with wall thicknesses as low as 1.5 mm to be produced.

With high-grade aluminium and soft Al alloys difficulties arising during machining can be obviated by admixtures of Pb. Increased Si content causes faster tool wear. The fast permissible cutting speeds allow short working times. To join Al parts all the normal techniques are applicable. Fusion welding is done predominantly by inert-gas-shielded TIG and MIG techniques. Bonded and clamped joints are gaining in importance. Contact with copper and its alloys involves a high risk of corrosion because of the electronegative potential of aluminium; this can be averted by insulating inserts. Sandwich construction in steel and aluminium (e.g. in vehicle construction) often allows optimised weight, price and appearance.

High-Grade Aluminium. The primary aluminium produced in the smelting works is sold in the form of pig, shot or grains. "High-grade aluminium" is unalloyed aluminium with a purity between 98 and 99.9%. "Highest-grade aluminium" is obtained from aluminium pig or recycling aluminium by a special refining process and is 99.99% pure for pig and at least 99.98% pure for semi-product. As purity increases strength diminishes, while chemical stability rises.

Cold forming can increase strength by more than 100% (**Appendix D3, Table 26**). Work annealing at 300 to 460 °C reduces the strength attained by cold forming in favour of better workability.

The chief uses of the material are in construction, container and appliance production, the chemical and food industries, packaging (foils) and electrical engineering (bars, cables).

Aluminium Forging Alloys

The most important alloying elements in aluminium forging alloys are Cu, Mg, Mn, Zn and Ni.

The main alloying ingredients are in some cases indicated in % in the abbreviated name, but are sometimes not indicated.

Alloys containing copper are more prone to corrosion than those without copper. A number of Al alloys can be strengthened by precipitation hardening without their ductility or formability being seriously impaired. The most important representatives of this group are: AlCuMg (hh and ch – see below), AlCuNi (hh and ch), AlZnMg (hh and ch) and AlMgSi (hh and ch). After solution annealing in salt baths or in a furnace at 500 ± 10 °C the material is quenched and aged at room temperature (cold hardening: ch) or at annealing temperatures of 100 to 200 °C (hot hardening: hh). This gives the alloys better strength properties as a result of precipitation processes. However, these properties can be lost by subsequent heating to 100 to 200 °C (e.g. with a welding or soldering torch) so that a repeat of the hardening may be needed. The highest strength (up to 650 MN/m²) is attained by AlZnMgCu (**Appendix D3, Table 27**).

Al alloys have low heat resistance, the best in this respect being those of the class AlCuNi (tensile strength 200 N/mm² at 300 °C).

Fatigue strength values for aluminium forging alloys are given in **Appendix D3, Table 28**.

To protect them against corrosion, Al alloys containing Cu are coated with layers of pure aluminium, AlMg or AlMgSi by hot rolling.

Soft soldering is possible after the destruction of the oxide layer, but is not common. Hard soldering (except with alloys with 2% Mg) is done at about 540 to 570 °C, using aluminium solder (L AlSi 13; L Al 80). Flux residues should be removed by rinsing. Resistance welds have about 80% of the strength of the original material. Non-hardenable alloys can be fusion-welded and hardenable alloys can be fusion-welded before hardening. With forging alloys, the long setting period should be respected, owing to the danger of chill cracking. Alloys with more than 7% Mg are difficult to weld. Gas fusion welding is possible if fluxes are used (residues to be removed by rinsing). The material can also be TIG- or MIG-welded, and in exceptional cases it can be electron-beam-welded.

Aluminium Casting Alloys

Of these about 28% are at present processed by sand casting, 50% by chill casting, 17% by die casting and 5% by spin and composite casting. The most important alloying element is Si, which in a proportion of 12% forms a eutectic with Al that has very good casting properties (G-AlSi 12). By modifying with small admixtures of sodium this alloy is given a particularly fine grain. Admixtures of Si produce good mechanical strength. By adding Mg heat and corrosion resistance are improved, although casting and melting properties are produced. The AlCuTi casting alloys attain the highest strength through hardening, while at the same time having good extension properties (parts subject to impact and vibration stresses in aircraft and vehicle construction). A prerequisite for this is a dense, fault-free structure (**Appendix D3, Table 29**).

The unstandardised piston alloys AlSi 12 CuMgNi, AlSi 18 CuMgNi and AlSi 25 CuMgNi, with a low thermal coefficient of expansion, for use at high operating temperatures, are produced by casting or forging.

The parts are characterised by high dimensional accuracy, surface quality and uniform properties. Because of the numerous small oxide inclusions, the fatigue strength of pressure-cast parts is lower and welding can be problematic.

Sintered Aluminium Materials

A material with especially good heat resistance is produced from pure aluminium powder with an oxide content of 6 to 15% by compression, sintering and hot forming. Tensile strength at 500 °C is 90 to 100 N/mm^2, with a breaking elongation of 1.5 to 4%. This alloy, frequently used in the nuclear industry under the name SAP, can be hot- and cold-formed and pressure-welded.

Powders of Al alloys with admixtures of Cu, Si and Mg attain, after pressing and sintering and in some cases hardening, tensile strength values up to 410 N/mm^2 with rupture elongation of 18%. These materials are used primarily for connecting rods, pistons and gearbox parts.

Standards. DIN 1712 Part 1: Aluminium: pig. – DIN 1712 T3: Aluminium; semiproduct. – DIN 1725 Part 1: Aluminium alloys; wrought alloys. – DIN 1725 Part 2: Aluminium alloys; cast alloys, sand castings, chill castings, die-castings, precision castings. – DIN 1746 Part 1: Tubing of aluminium and wrought aluminium alloys; strength properties. – DIN 1749 Part 1: Drop-forged parts of aluminium and wrought aluminium alloys; properties. – DIN 17 606 Part 1: Open-die forgings of wrought aluminium alloys; strength properties. – DIN 1745 Part 1: Strip and sheet of aluminium and wrought aluminium alloys with thicknesses over 0.35 mm; properties. – DIN 1745 Part 2: Strip and sheet of aluminium and wrought aluminium alloys with thicknesses of 0.021 to 0. 350 mm; properties. – DIN 1747 Part 1: Bar of aluminium and wrought aluminium alloys. – DIN 1748 Part 1: Bar extrusion profiles for aluminium and wrought aluminium alloys; properties.

3.2.3 Magnesium Alloys

Pure magnesium is used to a limited extent for conductor rails. Of the alloying additives, manganese improves its weldability and corrosion resistance, zinc improves formability, while aluminium enhances strength and hardness. Small admixtures of cerium produce a finer grain and improve heat resistance. If Al content rises above 6% strength is reduced, although this effect can be avoided by homogenising heat treatment (ho) if Al content is not more than 11%.

As compared with Al alloys, Mg alloys have low strength values at room and elevated temperatures (**Appendix D3, Table 30**). While the difference in fatigue strength is slight (**Appendix D3, Table 28**), notch sensitivity is higher. An unscored surface and the avoidance of notches are therefore essential in using Mg alloys. Their low modules of elasticity makes Mg alloys less sensitive to impact stresses and gives them better sound-damping properties (gear housings).

All magnesium alloys have excellent machinability, but care should be taken to ensure that coarse chips are produced. Fine chips and dust are liable to catch light and explode (extinguish with cast-iron chips or sand, never with water). No coolants containing water should be used during machining.

The high tendency of molten magnesium to oxidise demands special measures when casting and welding. TIG welding has proved useful, but gas fusion welding using welding rods of the same composition as the workpiece and fluxes is also possible. Alloys of the varieties MgMn and MgAl 6 can be well joined by electric resistance welding, using machines with special converters. Soldering is not possible. Mg alloys are usually formed by extrusion, die pressing, forging, rolling or drawing above 210 °C. The hexagonal lattice structure makes cold forming difficult, and in view of the risk of stress corrosion cracking this technique should be avoided where possible (relaxation treatment can be carried out at 280 to 300 °C).

The extreme electronegative potential of Mg and its alloys makes corrosion protection against humidity and climate effects necessary; this is usually done by spraying with or dipping in corrosion protective liquids (e.g. baths containing chromium). Magnesium castings that are to be exposed to aggressive atmospheres for only short periods should be coated in non-porous paint. Special care should be taken to avoid corrosion through contact with other materials. Steel screws should be zinc- or cadmium-plated.

Standards. DIN 1729 Part 1: Magnesium alloys; wrought alloys. – DIN 1729 Part 2: Magnesium alloys; casting alloys, sand casting, chill casting, die casting. – DIN 17 800: Primary magnesium. – DIN 9715: Semiproducts of wrought magnesium alloys; properties.

3.2.4 Titanium Alloys

The strength properties of Ti alloys (**Appendix D3, Table 31**) are comparable with those of high-tempered steels. The corresponding values for Ti alloys fall only slightly at temperatures up to 300 °C. In practice, temperatures up to 500 °C are of interest. Some alloys can be heat-treated. Hot forming is carried out by forging, pressing, drawing or rolling at 700 to 1000 °C. Pure titanium can be readily cold-formed, but this is possible only to a limited extent with Ti alloys (which need to be soft-annealed at 500 to 600 °C). Soft soldering is possible if the surface has been plated with silver, copper or zinc under inert gas (argon). Hard soldering is performed in a vacuum or an inert gas with appropriate fluxes. Welding is best done with MIG or TIG machines (or by electron beam welding). Because of the formation of brittle intermetallic compounds, bonding with other metals is problematic. Spot welding is possible without inert gas. It is helpful to employ low cutting speeds with a high feed rate (hard metal tool) on machining, because of the poor heat conduction and tendency to scuffing.

Ti and its alloys have high corrosion resistance, particularly against nitric acid, nitro-hydrochloric acid, chloride solutions, organic acids and brine.

Standards. DIN 17 850: Titanium; composition. – DIN 17 851: Titanium alloys; semi-finished products; composition. – DIN 17 860: Strips and sheet titanium and its alloys. – DIN 17 862: Bars of titanium and titanium wrought alloys; technical delivery conditions. – DIN 17 863: Wires made of titanium. – DIN 17 864: Forgings of titanium and titanium wrought alloys.

3.2.5 Nickel and Its Alloys

Pure nickel is supplied in purity levels of 98.5 to 99.98%. Small additions of Fe, Cu and Si barely exert any influence other than in respect of electrical characteristics. Mn increases the tensile strength and yield point without a penalty in terms of toughness. Nickel can be hardened by the addition of up to 3% beryllium. For soft pure nickel, tensile strength is 400 to 500 N/mm^2, the 0.2% proof stress is 120 to 200 N/mm^2 and ductile yield is 35 to 50% (in the cold drop-forged hardened state: $R_m = 750$ to 850 N/mm^2; $R_{p\,0.2} = 700$ to 800 N/mm^2 and $A_5 = 2$ to 4%).

There is barely any decrease in strength up to 500 °C; it is only from 800 °C upwards that the surface exhibits greater tendency to scale. In the low-temperature range, nickel retains its toughness; it is ferromagnetic up to 360 °C.

Ni can be alloyed with Cu in any proportions and can be brazed and welded. Alloys with 10% to 45% Ni can be used for precision electrical resistances up to 500 °C. Their temperature coefficient of electrical resistance is particularly low (DIN 17 471). The excellent temperature strength and corrosion resistance of 63% Ni alloys with approximately 30% Cu and the remainder of Fe and Mn (Monel metal) mean that they are suitable for the manufacture of chemical apparatus, pickling vessels, steam turbine blades and valves (DIN 17 743 and DIN 17 730).

Ni-Cr alloys are distinguished by high corrosion resistance (not applicable to sulphur-based gases) and high heat resistance (up to 1200 °C). Fe, Mo and Mn are commonly alloyed together, and Mn increases resistance against S (DIN 17 742 and DIN 17 744).

For NiCr alloys with high temperature strength, see Section D3.1.4, subsection on "High-Temperature and Very High-Temperature Steels (Alloys)" (pp. D36 ff.).

NiFe alloys are used for special applications: at 25% Ni a steel becomes non-magnetic, and at 30% Ni the temperature coefficient of the modulus of elasticity disappears (balance springs in clocks); at 36% Ni (Invar steel) the coefficient of thermal expansion reaches a minimum (metrology equipment), at 45 to 55% Ni it reaches the same level as for glass (sealing wires for incandescent lamps, DIN 17 745), and at 78% Ni an alloy with the maximum permeability is obtained. Ni is also important for soft magnetic alloys (14 to 17% Fe, DIN 17 745) and for permanent magnet alloys with a high coercive force (Al-Ni and Al-Ni-Co alloys).

Standards. DIN 1701: Refined nickel. – DIN 17 740: Nickel as a semi-finished product; composition. – DIN 17 741: Low-alloy nickel wrought alloys; composition. – 17 742: Nickel wrought alloys with chromium; composition. – DIN 17 743: Nickel wrought alloys with copper; composition. – DIN 17 744: Nickel wrought alloys containing molybdenum and chromium; composition. – DIN 17 745: Wrought alloys of nickel and iron; composition. – DIN 17 471: Alloys for resistances. – DIN 17 730: Nickel and nickel/copper casting alloys. – DIN 17 750 to DIN 17 754: Characteristics of alloys in accordance with DIN 17 740 to DIN 17 744 in various forms of products.

3.2.6 Zinc and its Alloys

Zinc is readily amenable to hot and cold forming (sheets, wires). Under the influence of atmospheric air, layers form which cover the surface and protect it from further attack, except in the presence of a highly acidic atmosphere. In the rolled state, the tensile strength of zinc is approximately 200 N/mm^2 with a ductile yield of approximately 20%, but it exhibits a tendency to creep even at ambient temperature (less so in the transverse plane). Zinc can be readily soldered using tin and cadmium. Weld joints can be performed by all methods except arc welding. Approximately 30% of zinc production is for sheets (roof coverings, roof gutters, drain pipes, etching panels, drying components), and approximately 40% for hot dipping of steel.

Zn diecastings, mostly consisting of alloys of Zn with Al and Cu (*fine zinc casting alloys*, **Appendix D3, Table 32**) are of high dimensional precision, but are more susceptible to corrosion than pure zinc.

Main alloy elements are stated as percentages in their abbreviated form, and the other information is the proportion of zinc.

Standards. DIN 1706: Zinc. – DIN 1743 Part 1: Fine zinc casting alloys; block metals.

3.2.7 Lead

Pure lead (*white lead*) at purity grades of 99.94 to 99.99% is frequently used in the chemical industry because of its good corrosion resistance (especially against sulphuric acid). It is relatively amenable to forming, welding, brazing (pipes, sheets, foils, wires) and casting (bearing materials, accumulator plates). Owing to its position in the periodic system, Pb is a very effective protection against X-rays and γ-rays.

Lead–antimony alloys (*hard lead*) are used for the manufacture of cable sleeves, tubes and claddings and for hot dipping in lead. Printers' metals contain not only antimony (up to 19%) but also tin (up to 31%). Lead diecastings are of high dimensional precision.

On lead and its alloys see **Appendix D3, Table 33**.

The proportion of lead is stated in its abbreviation in %; the presence of other alloy components is mentioned, but not their percentages.

Standards. DIN 1719: Lead; composition. – DIN 1741: Lead diecasting alloys: diecastings. – DIN 17 640 Part 1: Lead alloys; alloys for general application. – DIN 17 640 Part 2: Lead alloys for cable sleeves. – DIN 17 640 Part 3: Lead alloys; alloys for accumulators.

3.2.8 Tin

Tin is employed in purity grades of 98 to 99.90%, because of its effective anti-corrosive action in the manufacture of metal coatings (hot dipping in tin, galvanisation with tin) on Cu and steel (white sheet) and for the manufacture of solder rods. Tin foil (Stanniol) has now been substantially replaced by aluminium foil.

Sn diecastings are of particularly high dimensional accuracy. Components of pure tin may disintegrate at temperatures around zero (tin plague).

On tin and its alloys see **Appendix D3, Table 34**.

The component of tin is stated by its abbreviation and a percentage; other alloy components are mentioned without stating percentage.

Standards. DIN 1703: Lead and tin alloys for plain bearings. – DIN 1704: Tin. – DIN 1742: Tin diecasting alloys;

diecastings. – DIN 17 810: Equipment manufactured from tin; composition of tin alloys.

3.2.9 Coatings on Metals

These are used for anti-corrosion purposes, protection against wear, for purposes of greater surface hardness or cosmetic appearance, improvement of slip characteristics or application of material to wear points.

Metal Coatings. Coatings are applied galvanically, by hot dipping, metal spray, plating or diffusion.

Galvanic Coatings. These are produced by electrolysis in suitable baths (acids or aqueous solutions) of the relevant metal salts. The thickness of the coating depends on current density and exposure time (coating thicknesses normally up to 10 μm). Owing to changes in current density at edges and recesses, the coating thickness is not completely consistent. A prerequisite for effective adherence of the coating is a grease-free and oxide-free surface (degreasing, pickling) so as to achieve effective protection of the substrate metal by means of a dense, non-porous coating. Components are galvanically plated with tin, copper, zinc, cadmium, nickel or chromium.

Not only pure metals but also alloys (e.g. brass) may be deposited. Today, nickel plating is increasingly employed by non-electrical means. The placing of the substrate and coating materials in the electrochemical series, which allocates metals according to their electrode potential relative to hydrogen, is important with regard to corrosion protection. Electronegative metals are called base, electropositive metals noble. In the presence of an electrolyte, the baser of the two metals will always be attacked, unless the original potential is altered by surface passivation (for example, in the case of Al, on the more noble side). The following voltage series (in volts) applies with regard to the potentials of the main metals against hydrogen:

Mg −2.40	Cr −0.51	Ni −0.25	Cu +0.35
Al −1.69	Fe −0.44	Sn −0.16	H = ±0	Ag +0.81
Zn −0.76	Cd −0.40	Pb −0.13		Au +1.38

In the case of cosmetic chromium plating, it is common practice firstly to copper plate, then to nickel plate and to apply a coating of chrome less than 1 μm thick. Hard chrome layers (in baths with higher current density and higher tempertaure) produce a very high wear resistance with Vickers hardness levels of 800 to 1000 HV. In thicker hard chrome coating, there are formed inherent tensile stresses which, in the event of cracking, can cause impairment of mechanical characteristics, especially vibration strength.

Hot-Dip Coatings. By dipping in liquid-metal baths (hot dipping of tin, zinc, lead and aluminium), the corresponding alloy coatings are formed (with the exception of lead coating) by diffusion processes taking place between the metal atoms of the liquid coating metal and the atoms of the substrate metal. When the components are taken out of the bath, they possess a pure covering of the coating metal.

Hot-dip platings impart a greater coating thickness and thus longer corrosion protection than galvanic coatings (hot-dipping coating thickness 25 to 100 μm; or, for aluminium, 25 to 50 μm). One advantage of hot-dip coatings is that the solution also reaches cavities and inaccessible places. Components should never contain fully enclosed cavities (because of the risk of an explosion).

Currently, Zn and Al coatings are applied to wide strip sheets in a continuously operating process (the Sendzimir process). Al coatings give steel sheet good heat and scaling resistance together with mechanical characteristics which are better than pure Al. Not only Zn but also Al coatings can be transferred by diffusion annealing in Fe-Zn or Fe-Al alloy coatings (*galvanealing process, calorisation*).

Metal Spray Coatings. These are applied to particularly large components or those of which only certain parts have to be treated. The metal is melted in wire or powder form by a fuel gas mix or by an arc, and sprayed onto the component for treatment in the form of fine droplets propelled by compressed air. Adhesion to the surface is purely mechanical, so therefore the surface must previously be roughened by sandblasting to a medium surface roughness. The process is suitable for metals with a melting point of up to 1600 °C. To compensate for the porosity of spray coatings, they are soaked in solutions of synthetic resins or densified by rolling or pressing. The main fields of application are corrosion protection and the repair of wear points.

Plating. Today, this is usually performed by the roll-weld-plating method. Either (1) the substrate and plating materials are coated in thin bonding layes, heated and rolled out, with the bonding layer removed by pickling, or (2) the substrate is wound with the plating material, then heated and rolled out at high rolling pressure. It is common to plate Al alloys with pure aluminium, or steel with stainless steel, copper, nickel, monel metal or aluminium. Chemical industry tanks are also clad using weld plating techniques.

Diffusion Coatings. These are produced by annealing of workpieces in powder of the coating metal (e.g. Zn, Cr, Al, W, Mn, Mo, Si) in an oxygen-free atmosphere, or by addition of chlorides at temperatures below the melting point (400 °C for zinc coatings in the case of "Sherardisation", 1000 °C for aluminium in the context of "Alitisation", 1200 °C for chromium in the case of "chromising").

Gas Phase Deposition of Thin Coatings (CVD/PVD Coatings)

In order to improve the protection of components against wear and/or corrosion, the CVD (chemical vapour deposition) or PVD (physical vapour deposition) processes can be used to deposit metals, carbides, nitrides, borides and oxides from the gas phase onto the surfaces of tools and components.

The CVD process is based on deposition of solids by chemical gas phase reactions in the range of temperatures between 800 and 1100 °C [2]. The process of greatest technical value is the deposition of TiC and TiN coatings as wear protection layers. The high deposition temperatures entailed in CVD make it ideal for coating of cutter materials, especially hard metals, with ledeburitic chromium steels (e.g. X 210 CrW 12) being predominantly used for cold-machining steels. By contrast, deposition temperatures of less than 500 °C can be employed for plasma-assisted vacuum coating technologies in the PVD field, thus making it possible, for example, for rapid-machining steels or annealing steels to be employed as the substrate [2].

Non-Metallic Coatings

Oxidisation. Oxide coatings on a metal surface, which are in effect the result of a corrosion process, can represent corrosion protection in the form of passive coatings, provided the coatings are adequately dense and that they reform if damaged (formation of oxide coatings on Al, Cu alloys containing Al, stainless steel). Steel can also be given an oxide coating with some temporary protection by heating and dipping in oil (black annealing) or by oxidising pickle (browning). The very thin natural coating of oxide (0.01 μm) on Al can be thickened by chemical oxidation (MBV process) to 1 to 2 μm (providing a well-keyed surface for painting). In the case of anodic oxidisation (e.g. in sulphuric acid), components are connected to the positive pole of a DC power source. The eloxal coating formed by this process can be coloured as desired, owing to its porosity, and is electrically non-conductive. The pores are closed by redensification in hot water. Hard-wearing hard eloxal coatings have a Vickers hardness of approximately 500 HV in coating thicknesses up to 50 μm.

Phosphating and Chromating. By dipping steel or aluminium components in hot solutions of phosphoric acid and heavy metal phosphates (atramentisation, bonding), protective coatings up to 15 μm thick are formed, and their effectiveness can be enhanced by subsequent oil treatment. It is also possible to exploit the absorption power of the coating as a keying surface for paints. Thin coatings of manganese phosphate prevent scuffing of sliding components (gears, cylinder sleeves). In the case of dipping in, or irrigation with, potassium dichromate, temporary corrosion protection is achieved at a coating thickness of less than 1 μm.

Enamelling. This process is restricted to steel and grey castings. The basic enamel solution consisting of silicates and fluorides is applied by dipping, pouring or spraying, and stoved at approx. 900 °C. The covering enamel is sprinkled in powder form onto the heated components and fused smooth. The glassy coating is resistant to many chemicals, temperature variations and impact.

Painting. This is used not only for corrosion protection but also for decorative purposes. Paint consists of bonding agents (linseed oil, nitro-cellulose, synthetic resin, chloride rubber), pigment (e.g. white lead, red lead, iron oxide, mica, zinc white, chromium compounds, graphite, Al powder), the solvent (e.g. turpentine, benzine, benzol, alcohol) and, where applicable additives to achieve certain characteristics. After careful cleaning of the surface (sandblasting, brushing, pickling, degreasing), the paint is applied in single or multiple primer and top coats by brushing, rolling, spraying or stoving.

Chloride rubber paints are useful for aggressive atmospheres, and if there is additional mechanical loading it is best to use single- or double-component epoxy or polyurethane-based paints. For steel, the double system (hot-dipping plus painting) imparts significant advantages, because it avoids rusting of the substrate if there are any cracks in the paintwork. In standard production, there are many applications for electrostatic spraying – which saves paint – and infra-red drying.

Hot-Dip Solutions. Hot-dip solutions, which consist of cellulose derivatives, form a dense skin on the surface of the metal after the components have been dipped, thus providing protection for storage and transport.

3.3 Non-Metallic Materials

3.3.1 Ceramics

In addition to the conventional silicate ceramic materials such as porcelain, stoneware and glass ceramics, two new groups of ceramic materials have been redeveloped:

Oxide ceramics.
Non-oxide ceramics.

These have been put to numerous technical applications in machine building and plant building, in electronics and in electrical engineering, and for tools in the main categories of production equipment for primary forming, reforming and separation. The advantages of high-performance ceramics include:

High temperature hardness and compression strength.
High resistance to abrasion and erosion.
High resistance to corrosion and hot gas corrosion.
High resistance to creep at high temperatures.
Low specific weight.

They are counterbalanced by some disadvantages, such as:

High cold embrittlement under uni- and multi-axial loadings.
Relatively high spread of material characteristics.
Additional expense of powder manufacture and reprocessing operations.
Complications of bonding to metal materials.

The manufacturing process for oxidic and non-oxidic ceramics is similar to powder metallurgy, where the end characteristics of components are largely dictated by powder characteristics (purity), "green densification" in forming, and finally the sintering process. Sintering is a solid diffusion process in which the ceramic material is condensed and recrystallised by dissolution and dispersion, enabling it to achieve its strength characteristics. The range of scatter of strength characteristics can be substantially reduced by hot isostatic pressing (HIP) after sintering.

Oxide Ceramic Materials Aluminium oxide, Al_2O_3, is the technically most important member of this group of materials. Densely sintered aluminium oxide possesses excellent strength and hardness and resistance to high temperature and corrosion. Zirconium oxide ZrO_2 possesses high flexural strength at high temperatures and is frequently also used with success for purposes of prevention of wear in cold-forming tools (wire extrusion nozzles). Because of its low thermal conductivity, ZrO_2 is used for heat insulation in engine manufacture.

Appendix D3, Table 35 sets out important mechanical and physical characteristics of oxide ceramic materials.

Non-Oxidic Ceramic Materials. These include carbides, nitrides, borides and silicides, which are also designated hard materials. The profile of characteristics for this material category is distinguished by a high E modulus, high temperature strength and hardness, with good thermal conductivity and resistance to corrosion. In addition to compact materials, non-oxidic ceramics are employed particularly for the manufacture of composite coating materials for cutting tools, forming tools and components with high tribological loading. As the priority, the CVD and PVD coating methods are applied for deposition of coatings (mechanical and physical characteristics are given in **Appendix D3, Table 35**).

Table 1 sets out an overall survey of the main potential applications for high-performance ceramics in mechanical engineering.

Bricks. These are formed from coarse or fine clay or clay mixes with or without additives and are kilned. The clay must contain no lime enclosures because these could re-slake after kilning and explode owing to volumetric expansion. After preliminary air drying, bricks are kilned in tunnel ovens at 900 to 1300 °C. By kilning until

Table 1. Applications of high-performance ceramics in mechanical engineering

Fields of application	Components	Materials
Engine building	Heat insulation in combustion chambers	Aluminium oxide
	Valve crowns	Aluminium titanate
	Turbocharging rotors	Silicon carbide
	Gas turbines	Silicon nitride
	Sparkplug electrodes	Zircon oxide
Process technology	Extrusion nozzles: wires	Zircon oxide
Manufacturing	Cutting tools	Aluminium oxide
	Sandblast nozzles	Boron carbide
	Thread guide components	Zircon oxide
High-temperature technology	Burners	Silicon nitride
	Welding nozzles	Silicon carbide
	Crucibles	Boron nitride

sintering is achieved, high-strength clinkers are obtained. Solid bricks are of Mz 4 to Mz 28 (compression strength category 4 to 28: mean compression strength 5 to 35 N/mm², solid clinkers KMz 36 to KMz 60 (60 N/mm²). They are used in high buildings and as ducting clinkers in underground construction (municipal water systems). Bricks with vertical holes provide improved heat insulation, and longitudinally holed bricks are made with through-holes parallel to the mating surface. Furthermore, lightweight vertical-hole bricks are manufactured in a maximum bulk density of 1.0 kg/dm³. Roofing tiles (crown tiles, interlocking tiles) must meet certain standards for loadbearing capacity, impermeability to water and frost resistance.

Fireproof Bricks. For bricking of raised ovens, Siemens Martin ovens, fusion ovens, annealing ovens, rotating duct ovens, distillation ovens, Rost ovens, furnaces for steam power stations and waste combustion plants, etc., bricks with a very high fusion point (greater than 1500 °C) due to their composition (e.g. silicic acid and alumina) are needed.

Types. These consist of dead-burned fireclay ($\approx$60% SiO_2, $\approx$40% Al_2O_3), silica ($\approx$95% SiO_2, $\approx$2% Al_2O_3), sillimanite ($\approx$90% Al_2O_3), magnesite ($\approx$88% MgO, $\approx$5% SiO_2), carborundum (45 to 80% SiC, 10 to 25% SiO_2), carbon ($\approx$90% C). Fireproof bricks must also have a high compression firing strength (temperature at which the brick begins to soften under load) and good fluctuation resistance. Finally, bricks in fusion ovens must not be corroded by acid or basic slags, depending on the management of the melt operation.

Fused quartz (quartz glass is transparent, quartz is opaque) is a highly fireproof material also possessing the characteristic of very high acid resistance. It possesses the lowest coefficient of expansion of any material, so that it can resist sudden temperature changes.

Stoneware. This is kilned from good-quality silicic acid to which there are added flux agents such as feldspar, quartzspar or pegmatite for the manufacture of high-quality components. Brown and white stoneware products possess the same physical characteristics. Stoneware is supplied as a building material in the form of clinker tiles, clinker plates and acid-resistant bricks. For the chemicals industry, hollow stoneware items are manufactured for the production of acid-resistant apparatus and machine components (reciprocating and centrifugal pumps, fans, stirrers, mixing machines). The strength and toughness characteristics of normal stoneware are as follows: tensile strength 6.5 to 13 N/mm², compression strength 320 to 580 N/mm², flexural strength 23 to 40 N/mm², impact strength 1.3 to 1.9 N mm/cm².

Standards. DIN 1081ff.: Fireproof building materials, fireproof bricks. – DIN 51 061, 51 064, 51 067, 51 068: Test methods for fireproof building materials.

3.3.2 Concrete

Concrete (DIN 1045) is an artificial stone, which is produced from a mix of cement, concrete additive and water by hardening of the cement lime (cement–water mix).

Depending on the composition and processing of concrete, it achieves excellent strength characteristics, resists weathering and frost and can be used in a wide range of applications without restrictions as to form and size.

Types of Cement. The main bonding agent for concrete is Portland cement (PC). It is manufactured by specialised fine grinding of Portland cement clinkers with the addition of calcium sulphite.

Ferrous Portland Cement is manufactured by the specialised fine grinding together of Portland cement clinker in proportions of at least 65% by weight and refined sand in maximum proportions of 35% by weight (granulated blast furnace slag), with the addition of calcium sulphate.

Blast Furnace Cement is manufactured by specialised fine-grinding of 50 to 64% by weight of Portland cement clinker and the corresponding secondary proportion of 85 to 36% by weight of refined sand (granulated blast furnace slag), with the addition of calcium sulphate.

Trass cement is manufactured by specialised fine grinding of 60 to 80% by weight of Portland cement clinker together with the corresponding secondary component of 40 to 20% by weight of trass, with the addition of calcium sulphate. Trass is a natural pozzualonic material; it must comply with DIN 51 043. Percentages for trass always relate to the total weight of Portland cement clinker and trass, in the case of ferrous Portland and blast furnace cement, percentages always relate to the total weight of Portland cement clinker and refined sand. DIN 1164 introduced a total of four cement strength categories: Z25, 35, 45 and 55.

There is further subdivision of Z35 and Z45 to Groups L (= slower initial hardening) and F (= higher initial strength) and a strength limit for Z24, Z35 and Z45 as above.

Strength categories for standard cements are given in **Appendix D3, Table 36**.

Steel-Reinforced Concrete. Since the tensile strength of concrete is very low in relation to its compressive strength, steel inserts are used to strengthen the areas that have to transfer tensile loads. These include shear stresses arising from transverse loadings, because these cause tensile stresses at 45° to the plane of application. The important characteristics of steel in relation to this composite construction are its high modulus of elasticity, its high yield point and its thermal expansion, which is approximately the same as concrete. Steel adheres well to concrete, and adhesion can be increased by profiling its surface. The steel is protected against corrosion by an adequately dense enclosure of concrete. Concrete bar steel is manufactured by the following processes in accordance with DIN 488:

Hot-rolled without subsequent treatment (previously RUS).

Hot-rolled and heat-treated by the heat of rolling (previously RTS).

Cold forming, by twisting or stretch forming of the hot-rolled original material (previously RK).

The only further specifications in DIN 488 T1 are the three steel types which are suitable for welding: BSt 420 S (III S), BSt 500 S (IV S) and BSt 500 M (IV M) (minimum apparent limit of elasticity 420 or 500 N/mm², tensile strength > 500 or 550 N/mm²). Types III S and IV S are supplied as corrugated concrete steel, type IV M as welded wire mesh with individually corrugated bars. "Reinforcement wires" BSt 500 G (smooth) and BSt 500 P (profiled) have been incorporated for certain supply conditions and application conditions.

Smooth concrete steel BSt 220/340 GU (IG) is omitted, as is IR, which is no longer manufactured. This is replaced for the future by building steel St 37-2 to DIN 17 100, suitable for welding.

Standards. DIN 488 Part 1: Concrete steel: types, characteristics, marking. – DIN 488 Part 2: Concrete steel, concrete bar steel; sizes and weights. – DIN 488 Part 3 (ISO 10065-1990): Concrete steel; concrete bar steel; tests. – DIN 488 Part 4 (ISO/DIS 10544-1991): Concrete steel; concrete steel mesh and reinforcement wire; construction, dimensions and weights. – DIN 488 Part 5 (ISO/DIS 10287-1991): Concrete steel; concrete steel mesh and reinforcement wire; tests. – DIN 488 Part 6 (ISO/DIS 10144-1990): Concrete steel; control (quality control). – DIN 488 Part 7: Concrete steel; certification of suitability for welding of concrete steel; performance and assessment of tests.

Prestressed Concrete. One development of steel-reinforced concrete is prestressed concrete, in which steel inlays are prestressed and therefore generate compressive loads in the concrete in the unloaded building state. This system can be refined to the point where all tensile loads in the concrete are taken up by the concrete's inherent weight and by the design loads. This is the only way to make full use of the high limits of elasticity of high-quality steels and the high compression strength values of high-quality types of concrete.

The distinctions are drawn between prestressing with immediate bonding, prestressing without bonding and prestressing with subsequent bonding.

Standards. DIN 1045: Concrete and steel-reinforced concrete construction; sizing and specification.

Lightweight Concrete. The distinctions are drawn, according to bulk density, between the following:

Heavy concrete, with more than 2800 kg/m³ bulk density.
Standard concrete, with more than 2000 kg/m³ and not more than 2800 kg/m³ bulk density.
Lightweight concrete, with a maximum bulk density of 2000 kg/m³.

A solid, dense concrete with 2200 to 2400 kg/m³ bulk weight is a relatively good conductor of heat, and is consequently rather unsuitable for heat insulation (domestic buildings). Low-specific-weight additives, such as sawdust, wood wool and chaff will increase heat insulation. Air in particular is a poor conductor of heat, and therefore pores and cavities improve heat insulation capabilities in proportion to the extent to which they decrease specific weight. Pumice, boiler slag, brick chips, pumice slag (expanded blast furnace slag) and sintered pumice are lightweight additives whose inherent porosity exerts a favourable action in this respect. If grains of equal size (standardised-grain-size concrete) are used, then additional heat-insulation cavities are produced. In general there is a conflict between the requirement for strength in concrete on the one hand and good heat insulation on the other.

Lightweight types of concrete, with a dry bulk density of between 300 and 500 kg/m³, possess no significant loadbearing capacity and are useful exclusively for heat insulation. Lightweight concrete types with densities greater than 500 kg/m³ can be used as loadbearing components.

Steel-reinforced lightweight concrete is lightweight concrete with reinforcement. Lightweight concrete is categorised in accordance with DIN 1045 in strength categories LB 10 to LB 55. In the case of lightweight concrete, strength category LB 25 of concrete group BII is also allocated to DIN 1045.

Standards. DIN 4226 page 2: Additives for concrete, additive with porous structure (lightweight additive). – DIN 42 232: Lightweight concrete walls with high bulk porosity structure; specification and sizing.

Concrete Blocks and Slabs. Concrete is also treated with the various types of additive to produce finished components, large-size full and hollow building bricks, wall panels, paving slabs, kerbstones, roof blocks, concrete blocks, oven bricks etc.

Standards. DIN 398: Oven bricks (full bricks, cavity bricks, hollow bricks). – DIN 18 153: Hollow bricks and T-section hollow bricks made of concrete with enclosed structure. – DIN 4165: Aerated concrete blocks. – DIN 18 151: Hollow lightweight concrete blocks. – DIN 18 148: Hollow lightweight concrete wall panels. – DIN 18 149: Lightweight concrete holed bricks. – DIN 18 150: Lightweight concrete formed components. – DIN 485: Lightweight concrete decking slabs. – DIN 483: Kerbstones. – DIN 1115: Concrete roof blocks. – DIN 18 500: Concrete blocks. – DIN 18 501: Concrete paving stones.

3.3.3 Glass

Technical Glass

Glass types which are manufactured from glass sands (e.g. quartz sand SiO_2), flux (e.g. potassium oxide NaO_2) and stabilisers (e.g. carbonates of alkaline earths) are categorised, according to their chemical composition, into lime-soda glass, lead glass and boron silica glass. Glasses that are fused at temperatures of between 1300 and 1500 °C are usually cooled slowly, owing to the risk of internal stresses. They are worked at approximately 1000 °C by blowing, pressing, drawing and rolling. Mass-produced components are manufactured in large runs on glass-blowing machines. Glass does not have a crystalline structure and there is no fixed melting point at which it changes to the viscous state. Stress-free glass can be machined, drilled, milled and planed by hardened metals and diamonds. Compression strength is 400 to 1300 N/mm², and tensile strength is only 30 to 90 N/mm². The softening temperature of most glasses is approximately 500 °C, while that of quartz glasses is more than 1200 °C. Transparency is 85 to 90%, heat conductivity is 0.7 to 1.0 W/m K, thermal expansion is 80 to 100·10⁻⁷ m/m K. Glass is vulnerable to impact and sudden temperature changes, is resistant to acids apart from hydrofluoric acid, but is less resistant to alkaline solutions. It possesses good dielectric characteristics (insulators).

Glass Products. These include flat and concave products, wired glass, glass blocks, insulating glass and ornamental glass products. Very thin threads of liquid glass become glass fibres, glass wool and glass cloth (heat and noise insulation, glass-fibre-reinforced plastics).

Special Glasses. By quenching a hot glass plate in its definitive form (e.g. by means of air), inherent compressive stresses are set up in the peripheral zone, such that flexural strength is approximately 3 to 8 times that of standard glasses. This also provides a substantial increase in the capacity to withstand temperature changes and its impact strength. On fracture, prestressed glass will disintegrate into minute fragments without sharp edges. In addition to this single-ply safety glass, it is possible to obtain glass that is bonded to both sides of a transparent plastic sheet – to which the fragments will adhere in the event of fracture (bonded glass).

Molten quartz is manufactured at more than 1700 °C and can be used up to approximately 1200 °C. Quartz is far more permeable to ultra-violet light than any glasses, is chemically resistant apart from attack by alkaline solutions, and possesses excellent temperature-change resistance characteristics (protective tubes of pyrometers, chemicals barrels).

Standards. DIN 1259: Glass. – DIN 1249: Flat glass in building. – DIN 52 290: Corrosion-resistant glazing. – DIN 52 292: Ascertainment of flexural strength *in conjunction with*: – DIN 52 303: Ascertainment of flexural strength. – DIN 52 337: Pendulum strike test. – DIN 52 338: Ball drop test for bonded glass. – DIN 52 349: Fracture structure of glass for buildings. – DIN 1286: Laminated insulation glass. – DIN 52 293: Testing of gas permeability of gas-filled laminated insulation glass. – DIN 52 294: Ascertainment of loading of desiccants into laminated insulation glass. – DIN 52 344: Climate-change test on laminated insulation glass. – DIN 52 345: Ascertainment of dew point temperature on laminated insulation glass. – DIN 58 925 Part 1: Optical glass; terminology, classification. – DIN 58 925 Part 2: Optical glass; terms for optical characteristics.

3.3.4 Wood

Wood, because of its low bulk weight combined with relatively high strength and the fact that it is easy to work, is used in building, shipbuilding, vehicle manufacture and textiles. It is used as a raw material for the production of chemical pulp and paper.

Structure and Strength. Wood consists mainly of cellulose, lignin, resins and, in some cases, tanning agents. It has a fibrous structure and consists of cells which are arranged radially about the trunk centreline. Annual variations in growth rate produce soft, light cells in the first part of the year, and darker, harder cells in the summer and autumn (annual rings). The soft sap wood, which conducts the sap, surrounds the solid, dead heartwood within the trunk. The strength characteristics of wood are very much dependent upon fibre grain, and will generally deteriorate with increasing moisture content. The highest tensile strength values are achieved on loading along the grain direction, while the lowest tensile strength is across the grain (**Appendix D3, Table 37**). Owing to fibre kinking, compressive strength is no more than approximately one half of the tensile strength. For the same reason, the flexural strength of wood is lower than its tensile strength. Thrust and shear loads across the grain naturally indicate the highest strength values. The possibilities for irregularity in the various woods mean that the permitted loads for wood in buildings have to be lower (higher safety coefficient) (**Appendix D3, Table 38**). Slow-

growing, dense woods (hardwoods) such as oak, beech, ash, hickory and pockwood possess high hardness and strength values. Poplar, lime and spruce are examples of softwoods. Resinous woods such as pine, larch and pitch pine have good weathering capabilities. Wood is jointed by mortising, scarfing, dovetailing, pinning, nailing or screwing.

Influence of Humidity. Wood for basic carpentry work can have approximately 20 to 25% moisture content, joinery wood can have approximately 15%, wood for furniture approximately 12%, wood for inlay work approximately 8%, and wood for check plates approximately 6%. Wood expands and contracts in the various directions as it absorbs and gives out moisture (axial : radial : tangential = 1 : 10 : 20). Tensile strength parallel to the grain decreases by 2 to 3% (and compression strength by 4 to 6%) per 1% absorption of water, with a converse increase in strength as water is lost. Wood of approximately 40% moisture has approximately two-thirds the tensile strength and approximately one-half the flexural strength of air-dried wood of 10% humidity.

Wood can be stored for very long periods at a consistent level of dryness or continuously under water. Humid air and dry–humid alternation will cause wood to rot.

Protection of Wood. Wood protection agents whose ingredients include fungicides and insecticides are used to protect wood against harmful fungi and insects. They are designated according to the following attributes:

P   Effective against fungi (rot prevention).
Iv   Effective as prevention against insects.
Ib   Effective for combating insects.
S   Suitable for painting, spraying and dipping of wood in buildings.
St Suitable for painting and dipping of wood for building and for spraying in static systems.
W   Also suitable for wood that is exposed to weathering but is not in contact with the ground and is not constantly in contact with water.
E   Also suitable for wood that is exposed to extreme loading (in contact with ground and in constant contact with water).
K_1 Treated wood that will not cause corrosion holing in chromium–nickel steels.

Minimum quantities in accordance with the protective categories for wood in buildings (DIN 68 800) are required in order to achieve effective action.

The distinction is drawn between water-soluble protective agents (which contain salt) whose ingredients are based on silicon fluorides, chromates, borates and copper salts (and combinations of them) and carbolinea (mineral tar oil) with oil-based wood protection agents based on organic materials in a solvent. Commercially available oil-based protection agents include the xyligen (furmecyclox) fungicide and lindane (γ-hexachlorocyclohexane) and, increasingly, synthetic pyrethroids for use as insecticides.

Pentachlorophenol, which had previously seen widespread use as a fungicide has now been banned in Germany (and in other countries).

Certain wood protection agents containing salts may exert a corrosive effect on metal or on glass.

Certain phosphate-, carbonate- and silicate-based products that are applied by painting, spraying or dipping, reduce the flammability of wood.

Standards. DIN 52 180 Part 1: Testing of wood; taking samples, basic principles. – DIN 52 175: Protection of wood; terms, basic principles. – DIN 68 800 Part 1: Protection of wood in high buildings; general information. – DIN 68 800 Part 2: Protection of wood in high buildings; preventive structural measures. – DIN 68 800 Part 3: Protection of wood in high buildings; preventive chemical treatment of solid wood. – DIN 68 800 Part 4: Protection of wood in high buildings; measures to combat attack by fungi and insects.

Wood-Based Materials

Wood is divided into thin layers of veneer by sawing, cutting or rotary cutting, and is then reglued.

Where wood strips are glued along their grain, the result is plywood with good strength properties along the grain. Where the strips are rotated through 90° (plywood) or 45° (star plywood), the result is consistent strength properties irrespective of the loading plane and very low distortion. Sheets of blockboard consist of glued wood strips (usually coniferous), veneered on either side with covering strips whose grain is perpendicular to that of the core.

In *building board*, the distinction is drawn between chipboard, particle board and flat pressed board.

Chipboard is made by pressing together mostly small particles of wood and/or other wood-type fibre materials with bonding agents. *Particle board* consists only of wood chips and bonding agents.

Flat pressed board is chipboard whose chips are normally arranged parallel to the plane of the batten. It is made in one ply, in several plies or with a continuous transition in its structure. The distinction is drawn between boards of ground or non-ground surface.

The distinction is drawn between the following standard types, depending on the type of gluing and which wood protection agents are used:

V20: Gluing will be resistant to use in rooms where atmospheric humidity is generally lower (non-weather-resistant gluing).
V100: Gluing will be resistant to high atmospheric humidity (gluing of limited weather resistance).
V100G: Gluing will be resistant to high atmospheric humidity (gluing of limited weather resistance). Treated with a wood protection agent to repel fungi that attack wood.

Where chipboard is to be used in building, especially with regard to the sizing and specification of wooden panel-built houses, special attention should be paid to the guidelines for using chipboards, in the interests of avoiding unacceptable formaldehyde concentrations in ambient air. Only chipboards of emission category E1, which are also clearly marked with this categorisation, may be used (emission category E1 means ≤ 10 mg HCHO/100 g atro board).

Standards. DIN 68 705: Plywood. – DIN 68 763: Flat pressed boards. – DIN 68 754 Part 1: Wood fibre boards. – DIN 68 754 Part 2: Panelled extruded particle boards. – DIN 68 750: Porous and hardwood fibre boards. – DIN 1052 Part 1: Wood buildings; sizing and specification. – DIN 1052 Part 2: Wood buildings; mechanical joints. – DIN 1052 Part 3: Wood buildings; panel-built wood houses; sizing and specification. – DIN 4076 Part 1: Specialist terms and abbreviations used for woods; wood types. – DIN 68 620: Assessment of adhesives for jointing of wood and wood-related materials.

3.4 Materials Selection

The purpose of *functionally appropriate* material selection is to guarantee component compatibility for a given

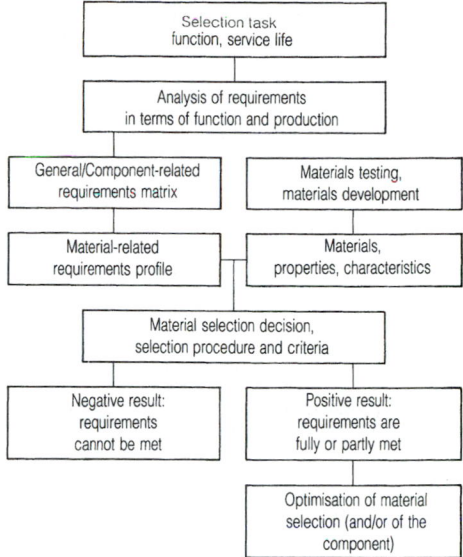

Figure 6. Basic system for material selection [5].

lifetime, with an adequate safety margin. Such selection is always guided by component loading and the envisaged design configuration, since material production and processing methods may have an effect on applications [3]. Where specific surface loadings have to be taken into account, adequate corrosion and/or wear protection must be guaranteed, in addition to component compatibility.

Above all, materials selection is subject to optimisation of properties within cost constraints. In the case of mass-produced components, a goal is to achieve material reusability so as to reduce demands on raw materials.

Owing to the wide range of influencing variables, materials are often selected empirically, since material properties provide only a limited indication as to the properties of components and component combinations. To the extent that component properties can be authoritatively defined in terms of material properties, materials can be selected with computer assistance on the basis of stored properties [4].

3.4.1 Fundamental System of Materials Selection

As shown in **Fig. 6**, the decision process consists of a series of partial stages. It is based primarily on a comparison between the component-related requirements and the profile of requirements for the material, together with typical properties of certain materials [5]. Component-related requirements such as loadbearing capacity as a measure of fracture strength, or rigidity as resistance to deformation, are in most cases covered by "strength certificates". For *static loads*, material selection is predominantly guided by the yield point or the 0.2% proof stress.

In the case of *oscillating loading* with a constant or random load–time curve, cycle-dependent variations in loadbearing capacity under the predominant influence of the surface layer and of the design configuration (notch effect, scale effect) must be considered. Material properties can provide only a limited basis for derivation of these influences.

Objective material selection can be attained by

technical–financial property weightings using analytical methods [6]. The MWC (mean weighted characteristics) method and the cost-related utility value method are both useful in this respect [7]. In many cases, the process of specific material selection will be carried out iteratively during preparation of design solutions, in the interests of improvement or of achieving optimum decisions (see E2.5). Where components are subject not only to mechanical or mechanical–thermal base-material loads but also surface loads due to corrosion and/or wear, the material selection process raises additional requirements in relation to adequate corrosion and/or wear prevention. If it is not possible to meet these requirements by means of a compromise between the properties of the base material

and those of its surface, then composite materials can be produced by diffusion methods, claddings or other coatings, in order to permit optimum matching to corrosive and/or tribological surface loads [8].

Material selection for component pairings under tribological loads must refer not only to an analysis of the surface complex loading condition but also the structure of the tribological system [9–11]. Low-wear component surfaces can be produced by surface stabilisation methods (thermal, thermo-chemical diffusion methods, see D3.1.3) or by coating methods (galvanisation, CVD process, PVD process, flame spraying, plasma spray coatings, see D3.2.9).

4 Plastics

A. **Burr**, Schwäbisch Gmünd, and G. **Harsch**, Beilstein

4.1 Introduction

Plastics are organic, high-molecular materials which are manufactured by predominantly synthetic means. They are produced as *polymers* (and are hence also called *polymer materials*) from *monomers* by *polymerisation*, *polycondensation* or *polyaddition*. Monomers are substances which contain carbon C, hydrogen H, oxygen O and nitrogen N, chlorine Cl, sulphur S and fluorine F. Their properties vary according to the type of resultant polymer: linear polymers are *thermoplastics*; cross-linked polymers are *thermosets*; and polymers with a more or less coarse-meshed cross-lining are *elastic synthetics*, which are also referred to as *elastomers*.

The scope for variation in plastics manufacture gives rise to a wide range of possible variety: *plastics* are tailored materials. The properties of *homopolymerisates* are influenced by chain length (level of polymerisation). Further variations are possible by *copolymerisation*, production of *polymer blends or alloys*. The wealth of options in manufacture means that some plastics embody totally new properties, thus enabling certain technical problems to be overcome for the first time: quick-fit joints, film hinges, sliding components, structural expanded foam components, lubrication-free bearings and integral manufacture of very complicated formations.

Standardisation and Designation of Plastics. DIN 7728 sets out abbreviations for plastics according to chemical composition, and gives data for reinforced plastics. DIN 16 780 covers the designation of polymer alloys. ISO 472 and ISO 1043 describe a new classification system for plastics, as used for DIN thermoplastic moulding material standards. These standard designations include nomenclature and identification blocks together with standard numbers and characteristic data blocks containing information on chemical structure with abbreviations, qualitative features (processing capabilities, additives), (coded) quantitative properties (viscosity number, density, modulus of elasticity, strength figures, etc.) and the type and form of fillers and reinforcing materials.

Moulding materials are non-formed raw products which are processed by technical methods to produce basic materials (semi-finished products, formings). The standard designations for thermosets are simpler and only contain information concerning the resin base and the filler and reinforcing material, though some thermosetting moulding materials are categorised.

4.2 Structure and Properties

Thermoplastics generally consist of chained molecules with up to 10^6 atoms at a length of approximately 10^{-6} to 10^{-3} mm. The strength of thermoplastics is temperature-dependent, owing to the lack of principal valency bonds between the chained molecules, and is influenced by "mechanical" twists in the chained molecules and the subsidiary valency forces between them. In the case of *amorphous* thermoplastics, the chained molecules lie in a swab-like form; strength properties are *isotropic*, i.e. equal in all directions. Processes such as extrusion, injection moulding or mechanical drawing can change the alignment of macromolecules, resulting in *anisotropy*. In the case of *semi-crystalline* thermoplastics, there are local arrangements of macromolecules; within these ordered, "crystalline" areas, plastics are very rigid and in the amorphous "joints" they are flexible. The properties of thermoplastics depend on the chemical structure of chains, on chain length, crystalline proportions and the type of subsidiary valency forces (dipole bonds, hydrogen bridging bonds, dispersion forces, etc.).

In the case of *elastomers*, the number of cross-linking points is of decisive importance to their elasticity properties: weakly elastic if there are few cross-linking points and harder as the number of cross-linking points increases. Re-forming and welding are therefore impossible. Because of their different structure, however, *thermoplastic elastomers* (TPEs) can be re-formed and welded.

The completeness of cross-linking of *thermosetting plastics* means that there is no facility for slip, so that this group of plastics can only be mechanically machined after moulding.

Figure 1 illustrates the ranges of conditions of plastics and their processing capabilities.

4.3 Properties

The molecular structure of plastics, as opposed to metals, which have an atomic structure, gives them different properties: *relatively low strength* (without reinforcements), *low modulus of elasticity* (low rigidity), *time dependence* of mechanical properties (load relief –

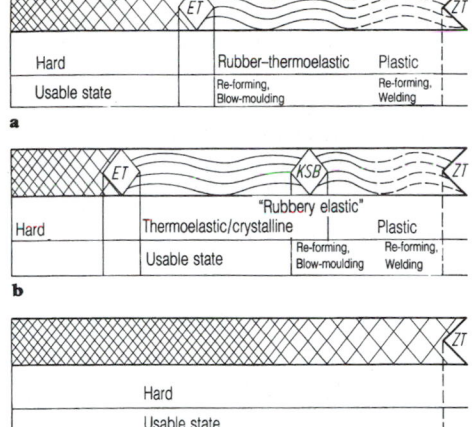

Figure 1. Diagram of ranges of states for plastics: **a** amorphous thermoplastics; **b** semi-crystalline thermoplastics; **c** thermosets: *ET* range of setting and softening temperatures, *KSB* crystalline melt range, *ZT* disintegration temperature range.

creep), high *temperature dependence*, especially in the case of thermoplastics, high thermal expansion and low *thermal conductivity*. Their favourable properties are good electrical *insulation properties*, good permanence, some degree of physiological safety and some excellent *frictional properties*, even without lubrication.

The properties of plastics can be varied in many ways, e.g. by *processing conditions*, *softening* (externally in the case of PVC-P, internally in the case of PVC-HI), production of *mixes* (polymer alloys, blends, e.g. ABS + PC or PBT + PC), *copolymerisation* (SAN, ASA, ABS), reinforcing materials (glass, carbon, aramide fibres), *fillers* (wood or stone dust, glass spheres, talcum) or other *auxiliary materials* (slip agents, stabilisers, dyes, pigments, fire inhibitors, propellants).

Appendix D4, Table 1 provides guideline values for the properties of important plastics groups.

4.4 Important Thermoplastics

Polyamide (PA) to DIN 16 773 (ISO 1874) and VDI guideline 2479 (Aculon, Bergamide, Durethane, Rilsan, Grilamide, Grilon, Maranyl, Minlon, Nylon, Technyl, Stanyl, Ultramide, Vestamide and Zytel). Semi-crystalline PA46, PA6, PA66, PA610, PA11, PA12 and amorphous PA6-3-T are normally used. They have a milky colour, and a strong tendency to absorb water with consequent variation in properties; as water content increases, toughness increases and strength decreases. Polyamides can be drawn. Hygroscopy decreases from PA6 to PA12. Electrical insulation properties depend on moisture content. Range of operating temperatures: −40 °C to 80 to 120 °C. Resistant to many solvents, fuels and oils. Not resistant to acids and alkaline solutions. In most cases it is necessary to condition polyamide components.

Typical *mouldings* are structural parts in applications where requirements call for strength, viscosity and low-friction properties, e.g. as in the case of sliding components, gears, idler rollers; and for casings, fan pulleys, bearing bushes, transport chains, dowel pins, guides; tow-ropes and mountaineering ropes; technical toy modules.

Typical *semi-finished products* include panels, pipes, sections, bars and films.

Polyacetal Resins (POMS) to DIN 16 781 (ISO 9988) (Delrin, Hostaform, Kematal and Ultraform). White, semi-crystalline plastics with practically no water absorption. Favourable rigidity and strength combined with adequate toughness and good elastic properties. Very favourable friction and wear properties. Good electrical insulation properties. Operating temperature range: −40 to 100 °C. Very good resistance to chemicals.

Typical *mouldings* are structural parts in applications where requirements call for high dimensional precision, strength, rigidity and good elastic and friction properties, e.g. as sliding bearings, bearing bushes, control discs, quick-fit and spring components, casings, pump components, hinges, fittings handles.

Typical *semi-finished products* are panels, sections, bars and pipes.

Linear Polyesters (PET PBT) (Polyalkylene Terephthalates) to DIN 16 779 (Arnite, Crastin, Grilpet, Hostadur, Pocan, Techster, Ultradur, Valox, Vandar and Vestodur). Semi-crystalline thermoplastics of various degrees of crystallinity (partly amorphous PET, milky-white PBT). Favourable mechanical properties, even at low temperatures and at temperatures up to 110 °C. Favourable long-term properties and low erosion with good slip properties. Very low absorption of moisture. No thermal expansion. Very good electrical insulation properties. Not resistant to hot water or steam, acetone or halogenic solvents, or strong alkaline or acid solutions.

Typical *mouldings* are structural parts of high dimensional accuracy with good running and friction properties in mechanical engineering, light engineering, for domestic applications and in office machines.

Typical *semi-finished products* are panels, sections and pipes; films for audio tapes, condensers, adhesive tapes, insulating films; backing foils and drawn packing strips.

Polycarbonate (PC) to DIN 7744 (ISO 7391) (Macrolon, Lexane, Sinvet and Xanter). Amorphous, transparent thermoplastics with high strength and good toughness. Very good electrical insulation properties. Operating temperatures: −100 to +130 °C. Resistant to greases and oils; not resistant to benzene and alkaline solutions. Tendency to crack when exposed to certain solvents.

Typical *mouldings* are primarily in electrical applications, including lamp covers; fuse boxes, coil units, plug-in connectors, tube sockets. Casings for light engineering and optical equipment; crockery, safety helmets and shields; safety glasses, helmet visors; triangular signs. Typical *semi-finished products* include pipes, sections, bars, panels and films.

Modified Polyphenyl Ether (PPE) (Noryl, Vestoren, Luranyl and Vestoblend), amorphous thermoplastics, usually modified with PS or PA, beige in colour. Very low water absorption. High strength and rigidity and good impact strength. Low tendency to creep and good temperature loading capacity up to 120 °C. Very good electrical insulation properties, virtually independent of frequency. Not resistant to aromatic, polar and chlorinated hydrocarbons.

Typical *mouldings* are casings in electronics and electrical engineering, with high thermal loading; extruded connectors, precision components for office machines and light engineering components.

Typical *semi-finished* products are sections, pipe, bars and panels.

Polyacrylate (PMMA) to DIN 7745 (ISO 8257) (Diakon, Perspex, Plexiglas, Resartglas and Paraglas). Transparent amorphous thermoplastic with very good optical properties ("organic glass"). Hard and brittle, with high strength. Good electrical insulation properties. Operating temperatures: up to +70 °C. Good resistance to light, ageing and weathering; not resistant to concentrated acids, halogenated hydrocarbons, benzene or alcohol. Easy to bond. Low-molecular types can be thermoplastically worked; high-molecular types can only be supplied as semi-finished products.

Typical *mouldings* are primarily suitable for optical applications, e.g. spectacles, magnifying glasses, lenses, prisms, reflectors; glazing items, inspection glasses, rows of windows. Domestic apparatus; writing and drawing implements. Roof glazing items, advertising and informational notices; bathtubs, sanitation components; theoretical/training models. Typical *semi-finished* products include blocks, panels, sections, pipes and optical fibres.

Polystyrene (PS) to DIN 7741 (ISO 1622) (Edistir, Laqcrene, Lastirol, Polystyrene, Scopyrol and Vestyron). Amorphous, transparent thermoplastics. Rigid, hard and very brittle. Very good electrical insulation properties; high capacity for electrostatic charge. Poor thermal loading capacity. Tendency for stress cracking even in air. Low resistance to organic solvents.

Typical *mouldings* include transparent packing, domestic apparatus, drawer inserts, filing boxes, projection slide frames, film and photographic spools, coil units, electrical engineering components, disposable crockery and cutlery.

Styrene Butadiene (SB) to DIN 16 771 (ISO 2897) (Polystyrol, K-resin, Saxerol, Styrolux and Vestyron). Amorphous but not usually very transparent thermoplastics (with the exception of Styrolux). Improved impact strength. Good electrical insulation properties, but a generally high capacity for electrostatic charge. Range of operating temperatures: up to +75 °C.

Typical *mouldings* have high impact loading capacity, such as toilet components, stacking boxes, projection slide frames, shoe horns, heels and casing components.

Typical *semi-finished products* are films for heat forming.

Styrene Acrylonitrile Copolymer (SAN) to DIN 16 775 (ISO 4894) (Kostil, Luran, Lustran and Vestyron). Amorphous, transparent thermoplastics with high surface gloss. Good mechanical strength properties, higher impact strength than PS, highest modulus of elasticity of all styrene polymers. Good electrical insulation properties. Operating temperatures up to +95 °C; good resistance to changes in temperature.

Typical *mouldings* possess high rigidity and dimensional stability, combined in some cases with transparency, e.g. scale discs, inspection glasses, casing components, packing items and warning triangles.

Acrilonitrile Butadiene Styrene Polymers (ABS) to DIN 16 772 (ISO 2580) (Cycolac, Lustran, Novodur, Ronfalin, Terluran, Urtal). Amorphous and in most cases non-transparent thermoplastics in the form of polymer blends or copolymers. Good mechanical strength properties and favourable impact strength. Good electrical insulation properties combined with very low capacity for electrostatic charge. Operating temperatures from −45 to +110 °C. To produce polymer alloys, ABS is mixed with PC (Bayblend T) or PVC (Ronfalloy) with special properties.

Typical *mouldings* are, in particular, casings of all kinds in domestic, TV and video manufacturing, office machines, furniture components of all kinds; cases, heels, safety helmets; sanitary installation components; toy modules.

Typical *semi-finished* products are panels, primarily for heat forming, and also for technical mouldings.

ASA (Luran S). This is an amorphous thermoplastic which is similar to ABS but possesses much greater weathering capability, and is therefore used particularly for outdoor applications.

Cellulose Derivatives (CA, CP and CAB) to DIN 7742 (Bergacell, Cellidor and Saxetat). Amorphous, transparent thermoplastics which are produced by esterification of cellulose with acids; usually compounded with plasticisers; some increase in hygroscopy. Good mechanical properties combined with high toughness. Operating temperatures up to +100 °C. Good chemical resistance.

Typical *mouldings* possess the required high viscosity, and include inserts for metal components, e.g. tool handles, hammerheads, writing and drawing implements; spectacle frames, brush handles, toys.

Typical *semi-finished products* include blocks, sections and panels.

Polysulphone (PSU/PES) (Udel, Radel, Ultrason, Victrex and PES). Amorphous thermoplastics with slight inherent colouring. Good strength and rigidity; up to 180 °C, low tendency to creep. Hygroscopy similar to PA. Good electrical insulation characteristics.

Typical *mouldings* are used for mechanical, thermal and electrical loadings.

Polyphenyl Sulphide (PPS) (Craston, Fortron, Ryton and Tedur). Semi-crystalline thermoplastics with high glass content. High strength and rigidity combined with low toughness; low tendency to creep and good friction characteristics. Operating temperatures up to 240 °C. Very high resistance to chemicals.

Typical *mouldings* are used for high mechanical, thermal, electrical and chemical loadings, e.g. in light engineering and electronics, such as plug-in connectors, carbon brush holders, casings, sockets, seals, condenser films, flexible conductor tracks; sleeves for semiconductor components; stove handles.

Polyimide (PI) (Kinel, Ultem, Torlon, Vespel and Kapton). Duroplastic, cross-linked or linear amorphous depending on structure. High strength and rigidity and low toughness; very good long-term properties. Favourable abrasion and wear properties. Very high electrical insulation action. Very low thermal expansion. Wide range of operating temperatures, −240 to +260 °C in the case of PI. Very good resistance to chemicals and to high-energy radiation.

Typical *mouldings* are used for high mechanical, thermal and electrical loadings and lubrication-free slip friction, e.g. in spacecraft, data processing, nuclear plant and high-vacuum technology. Insulation films with high insulating action.

Polyolefins

Polyethylene (PE) to DIN 16 776 (ISO 1872) (Baylon, Eltex, Moplen, Hostalen, Lacqtene, Lotrene, Lupolen, Natene and Vestolen). Various properties, depending on structure; linear PE-HD (high-density PE) with high strength in the form of branched PE-LD (low-density PE). Semi-crystalline thermoplastics. Low strength combined with high toughness (PE-LD). Good electrical insulation action. Very high chemical resistance. Range of operating temperatures −50 to 80 °C (PE-LD up to 100 °C). Ultra-high molecular PE (PE-UHMW) with very good mechanical and slip properties, can only be mechanically machined; special types can be worked thermally.

Typical *mouldings* are handles, seals, seal plugs, fittings, bottles, tanks, heating oil tanks, sewage containers; bottle boxes, cable sleeves, slip linings on skis. Typical *semi-finished* products are films, hoses, tubes and panels.

Polypropylene (PP) to DIN 16 774 (ISO 1873) (Bergaprop, Hostalen PP, Lacqtene P, Moplen, Novolen, Propathene, Stamylan P and Vestolen P). Semi-crystalline thermoplastics with more favourable mechanical and thermal characteristics than PE. Operating temperature range up to 110 °C.

Typical *mouldings* are transport boxes, tanks, coffers, mould components with film hinges, battery boxes, cable sleeves, heating ducts, pump casings and cables. Typical *semi-finished* products are films, monofilaments, bars, tubes, sections and panels.

Polyvinyl Chloride (PVC) to DIN 7746 (ISO 1060), DIN 7748 (ISO 1163) and DIN 7749 (ISO 2898) (Benvic, Welvic, Decelith, Hostalit, Lacqvyl, Solvic, Trovidur, Varlon, Vestolit, Vinidur, Vinnol and Viplast).

Unplasticised PVC (PVC-U) (or hard PVC). Amorphous, polar thermoplastics with good strength and rigidity. Operating temperatures only up to 60 °C approx. Low flammability. High dielectrical losses due to polarity, hence suitable for high-frequency welding. Good resistance to chemicals.

Typical *mouldings* include tanks in the photographic, chemical and galvanising industries; pipeline components, acid-resistant casings and apparatus items, acoustic panels, diffusion proof disposable bottles. Typical *semi-finished products* include sections, panels, films, blocks, bars, pipes and welding rods.

Plasticised PVC (PVC-P) (soft PVC). Amorphous, polar thermoplastic with varying degrees of flexibility, depending on plasticiser content. Low thermal loading capacity. Less resistance to chemicals than PVC-U. Not generally suitable for purposes of foodstuffs, owing to plasticiser.

Typical *mouldings* include dolls, inflatable animals, cable sleeves, floor coatings, bags, rain-shoes and rainwear, protective gloves and book inserts. Typical *semi-finished products* include films, hoses, sections, seals, floor coverings and sealing strips.

4.5 Fluorinated Plastics

Polytetrafluoroethylene (PTFE) (Fluon, Hostaflon and Teflon). Semi-crystalline thermoelastic (not meltable, but softenable). Laborious manufacturing process: by press sintering from powders to produce the semi-finished product, after which it can only be machined mechanically. Low strength, flexible, high creep ("cold flow"). Highly anti-adhesive, low coefficients of static and dynamic friction, hence no "stick-slip". Very good electrical insulation properties. Wide temperature operating range from −200 to +270 °C. Maximum chemical resistance. Expensive to process.

Typical *semi-finished* products include panels, bars, pipes and hoses which are mechanically machined for use as mouldings for the highest thermal and chemical loadings, such as laboratory equipment, pump components, corrugated pipe compensators, piston rings, sliding bearings and insulators. Also used for non-stick coatings.

Other Fluorinated Thermoplastics: FEP, PFA, ETFE, WCTFE and PVDF. In the form of semi-crystalline thermoplastics, they do not possess quite the extreme properties of PTFE, but can be worked more economically by injection moulding.

Typical *mouldings* are the same as for PTFE, but with slightly restricted characteristics in some cases.

4.6 Thermosets

Thermosetting moulding materials are covered by a DIN 7702 quality symbol containing the company logo and the type of thermoset.

Thermosets are processed in the form of *moulding resins*, *mould materials* or *pre-pregs*.

Moulding resins are used for the manufacture of mouldings or are processed with glass, carbon or aramide fibres to produce resin-fibre composites (GRP, CRFP, RFRP).

Moulding materials, which are preliminary resin products that have been treated with filling and reinforcing agents, are processed by pressing or injection moulding to produce mouldings.

Bulk moulding compounds (BMCs) are processed as pourable *granulates* or mould materials of doughy consistency, by pressing or injection moulding, and *sheet moulding compounds* are usually processed in the form of flat *pre-pregs* to produce large mouldings.

Laminated sheets are manufactured by pressing of resin-soaked flat forms (paper, fabric, wood veneer etc.). These materials can be machined.

Phenolic Resins (PF) to DIN 16 916 (ISOs 8244, 8975), DIN 7702 and DIN 7708 (ISO 800) (Bakelite, Resinol and Supraplast). Cross-linked, polar thermosets with yellow colouration. Electrical characteristics are partially affected by water arising on polycondensation. Practically always employed in the filled state, consequently characteristics are very highly dependent on type and quantity of filling and reinforcing agents. Usually relatively brittle with high strength and rigidity. Operating temperatures up to 150 °C. Good chemical resistance; not permitted for purposes of foodstuffs.

Typical *mouldings* include casings, handles and electrical installation components, in some cases with pressed-in metal components. Typical *semi-finished products* include laminated panels, sections for machining, paint resins, adhesives, bonding agents and friction linings, and mould sand.

Amino Plastics (MF, UF) to DIN 7708 Part 3 (ISO 2112) (Bakelite, Melbrite, Resart, Supraplast, Skanopal, Hornit, Resopal and Supraplast). Cross-linked polar thermosets; practically colourless, so can also be dyed in bright colours. Practically always used filled, so properties are very largely dependent on type and amount of filler. Usually relatively brittle with high strength and rigidity. Operating temperature for MF up to 130 °C. Good electrical insulating characteristics. Good chemical resistance; MF type 152.7 authorised for purposes of foodstuffs.

Typical *mouldings* include brightly coloured casings, installation components, electrical insulating components, switches, plug-in sockets, handles and crockery. Decorative laminated panels (HPL) used in furniture and as frontage boards.

Unsaturated Polyester Resins (UP) to DIN 16 911, DIN 16 945 and DIN 16 946 (ISOs 3672, 3673) (resins: Apolit, Leguval, Palatal, Rutapal, Stratyl and Vestopal;

mould materials in the form of SMC or BMC: Bakelite, Keripol, Menzolit, Norsomix, Resipol and Supraplast). Cross-linked thermosets of reaction resins which are usually processed with reinforcing agents. In the case of laminates, specific reinforcement is possible. Characteristics are dependent on the structure of the polyester, the level of cross-linking, the type and amount of the reinforcing material and the processing methods involved. High strength values (approaching those of non-alloyed steels) but with a low modulus of elasticity. Favourable electrical insulating properties. Operating temperatures up to 100 °C, and in some cases up to 180 °C. Chemical resistance is good, even in outdoor applications; also authorised for purposes of foodstuffs, depending on the resin/hardener system.

Typical *mouldings* include laminates for large-surface-area construction components such as vehicle components, boat and yacht hulls, tanks, heating oil tanks, containers, fishing rods, sports equipment, chairs and traffic signs. Mouldings are pressings and injection mouldings for technical mouldings with high requirements for mechanical and thermal properties combined with good electrical properties, such as ignition distributors, coil units, plug-in connectors and switch components.

Epoxy Resins (EP) to DIN 16 912 (ISO 3673), DIN 16 913 (ISO 8606), DIN 16 945, DIN 16 946, (ISOs 3672, 3673), DIN 16 947 (resins: Araldite, Hostapox, Epikote, Eposir, Eposin, Grilonit, Leuktherm and Rutapox; moulding materials for SMC or BMC: Araldite press compound, Bakelite, Melopas and Supraplast). Cross-linked thermosets of reaction resins, which are normally processed with very high-quality reinforcing agents (carbon and aramid fibres). In the case of laminates, specific reinforcements are possible. Properties depend on structure of the epoxy resin, level of cross-linking, type and amount of reinforcing agent and processing method. Very high strength and rigidity values, especially in the case of carbon fibres (CRP); not vulnerable to impact. Optimum electrical insulating properties in a wide temperature range, including outdoor applications. Operating temperature ranges are dependent on processing; cold-cured systems up to 80 °C, hot-cured up to 130 °C, in some cases up to 200 °C. Good chemical resistance, even in outdoor applications.

Typical *mouldings* include laminates for high-strength and rigidity components in aircraft and spacecraft manufacture (control surfaces, wings, helicopter rotor blades), copying tools, foundry models, adhesives, mouldings (pressings and injection mouldings) for construction components with high dimensional precision, especially in electrical engineering, also for sleeves, precision components in light engineering and in mechanical engineering. Also used for high-performance sports equipment.

4.7 Plastic Foams (Cellular Plastics)

The properties of foamed plastics depend on the plastic used, *cell structure*, and *bulk density*. Foams with a compact external skin (structural or integral foams) exhibit favourable rigidity combined with low weight. Mechanical loadability and insulating capacities depend largely on porosity (bulk density). The minimum bulk densities for foams are 50% of the non-foamed material. In principle, all plastics are foamable, but *thermoplastic foams TSG* based on SB, ABS, PE, PP, PC, PPE and PVC, together with *reaction foams RSG* based on PUR, are of particular significance. Cell structure is achieved by inclusion of gases, release of contained propellants and release of propellants in the chemical reaction for the raw products.

Expanded Polystyrene (EPS) (Styropor and Hostapor). With bulk densities between 13 and 80 kg/m³, used in the form of sheets, blocks, foils and mouldings for heat and noise insulation, in packing and for buoyancy devices.

Thermoplastic Foam Moulding (TSG). Used as a structural foam, normally for large-surface-area mouldings in furniture manufacture, office machinery, television and data processing equipment, transport containers and sports equipment.

Hard RSG Foams Based on PUR with bulk densities between 200 and 800 kg/m³, they possess good mechanical rigidity combined with low weight. Applications in furniture manufacture for office machines and television equipment, window sections, chassis components and sports equipment.

Soft RSG Foams Based on PUR. Have very good shock-absorbing characteristics and are used, for example, as foam cushions, steering wheel covers, shock absorber systems and shoe soles.

4.8 Elastomers

Elastomers are polymer materials with high elasticity. The moduli of elasticity of such elastomers are between 1 and 500 N/mm². Owing to the coarse-meshed chemical cross-linking, there is no need for heat forming and welding after vulcanisation forming.

Thermoplastically Processable Elastomers (TPE) represent a special category of elastomers, which can be processed and worked by all thermoplastic working methods. The elastic properties of these materials are achieved by physical cross-linking.

Rubber. Rubbers are manufactured from natural or synthetic latex and have a wide range of additives. More or less coarse-meshed cross-linking is performed by *vulcanisation* with sulphur or other cross-linking agents at temperatures of more than 150 °C and under pressure.

The *caoutchouc* determines rubber quality in terms of mechanical characteristics and resistance to chemicals. *Vulcanising agents* are sulphur or substances which release sulphur (less than 3%), and peroxide in the case of special rubbers. Sulphur bridging enables cross-linking of linear rubber molecules. The quantity of vulcanisation agent determines the level of cross-linking and hence strength characteristics (hard rubber – soft rubber).

Active (reinforcing) filling agents are carbon black in the case of black rubber types, and for light rubber types: silicic acid, magnesium carbonate and kaolin. Filling agents improve the strength and abrasion resistance of the vulcanate. *Inactive filling agents* are chalk, diatomite and talcum; they reduce the cost of end-products and partially increase electrical insulation and hardness. *Plasticisers* are mineral oils, stearic acid and pitch; they improve processability. If larger quantities are used, there is an increase in elasticity under impact, and hardness and mechanical strength are reduced. *Activators* such as zinc oxide improve vulcanisation. *Accelerators* increase the speed of reaction with a reduced sulphur content; they also improve thermal loadability. *Anti-ageing agents* protect rubber materials from ageing due to heat, oxygen and ozone, and against sunlight. *Dyes* can be added to carbon-free rubber mixes.

Natural Rubber (NR) (and in some cases also polyisoprene IR as a "synthetic" natural rubber). These possess high dynamic strength and elasticity and good abrasion strength. Poor weathering capabilities and swelling in mineral oils, lubricating greases and gasoline. Operating temperatures −60 to +80 °C. *Applications* include HGV tyres, rubber springs, rubber bearings, membranes and windscreen wiper blades.

Styrene Butadiene Rubbers (SBR) (Buna). These have better abrasion strength and greater ageing resistance than NR, but they are of inferior elasticity and their processing characteristics are poorer. Swelling is similar to NR. Operating temperatures −50 to +100 °C. Typical *applications* include car tyres, bellows, hoses and conveyor belts.

Polychloroprene Rubbers (CR) (Baypren and Neoprene). These possess better weathering and ozone resistance than NR, but they are of lower elasticity and resistance to cold. Adequately resistant to lubricating oils and grease, but not to hot water and fuels. Operating temperatures −30 to +100 °C. Typical *applications* include building seals, sleeves, cable insulation, mining conveyor belts and bridge bearings.

Acrylonitrile Butadiene Rubbers (NBR) (Hycar and Perbunan). Also known as nitrile rubbers; particularly resistant to oils and aliphatic hydrocarbons, but not resistant to aromatic and chlorinated hydrocarbons or brake fluids. Good abrasion strength and good resistance to ageing. Elasticity and resistance to cold are inferior to NR. Operating temperatures −40 to +100 °C. *Applications* include shaft ring seals, other types of ring seal, membranes, seals and fuel pipes.

Acrylate Rubber (ACM). These have higher resistance to heat and chemicals than NR, but have an inferior performance in cold environments and are hard to work. Resistant to mineral oils and greases but not resistant to hot water, steam and aromatic solvents. Operating temperatures −25 to 150 °C. *Applications* include heatproof ring seals, shaft ring seals and seals in general.

Butyl Rubbers (IIR) (Butyl and Polysar). These possess very low gas permeability and good electrical insulation characteristics, hot vapour resistance, weathering and ageing capability, but low elasticity combined with high internal damping. Not resistant to mineral oil, greases and fuels. Operating temperatures −40 to +100 °C. *Applications* include inner tubes for tyres, roof coverings, hot-water hoses, and damping components.

Ethylene Propylene Rubbers (EPM, EPDM) (Buna AP and Vistalon). These possess good weathering and ozone resistance combined with good electrical insulation characteristics. EPDM is cross-linked by peroxide and is hard to work. Strength is similar to NR; very good resistance to hot alkaline washing solutions. Operating temperatures −50 to +120 °C. *Applications* include washing and dishwashing machine seals, car window seals and car radiator hoses.

Silicone Rubber (VOM) (Elastosil and Silopren). These possess excellent resistant to heat, cold, light and ozone, low gas permeability and very good electrical insulating characteristics, but low resistance to cracking. Resistant to greases and oils, physiologically safe, not resistant to fuels and water vapour. Anti-adhesive. Operating temperatures −100 to 200 °C. *Applications* include seals in car, aircraft and machine building, for stoves and drying cabinets, cable insulations, conveyor belts for hot substances, medical equipment and hoses.

Fluorine Rubber (FKM) (Viton, Fluorel and Tecnoflon). Have excellent temperature, oil and fuel resistance but low resistance to cold. Operating temperatures −25 to +200 °C, and in some cases up to 250 °C. *Applications* include seals of all kinds for high temperatures, combined with high levels of hardness.

Pressable and mouldable polyurethane elastomers (PUR) (Adiprene, Baytec, Elastpal, Urepan and Vulkollan). These possess high mechanical strength and very high wearing strength, combined with a very high modulus of elasticity by comparison with rubbers; high damping. Resistant to fuels, low-grade greases and oils; not resistant to hot water or steam; embrittlement due to UV radiation. Operating temperatures −25 to +80 °C. *Applications* include track rollers, seals, couplings, bearing components, V-belts, wear linings, cutting underlays and damping components for metal claddings.

Thermoplastically Processable Elastomers (TPE). As polyurethane PUR (Daltomold, Desmopan and Elastollan), polyether amides (Pebax), polyester elastomers (Arnitel, Hytrel, Pibiflex and Riteflex) and polyolefin-based elastomers EVA (Levaflex, Dutral, Evatane, Lupolen V, Hostalen LD-EVA, Nordel and Santoprene), they have the advantage that they can be thermoplastically worked. They are used for applications similar to those for other types of rubber, but their properties are very different according to structure and composition, particularly in the case of EVA, because of the facility for variation of vinyl acetate content.

Range of operating temperatures −60 to 120°C, depending on type. Typical *applications* include sprockets, coupling and damping components, roller linings, buffers, seals, cable sleeves, bellows, skiing shoes and shoe soles.

4.9 Testing of Plastics

The characteristics of plastic mouldings depend very largely on manufacturing conditions. Consequently it is not possible to transfer characteristics measured on separately manufactured testpieces directly to plastic mouldings. For that reason, distinctions are drawn in testing of plastics between testing of separate manufactured samples, testing of samples taken from mouldings and testing of the whole moulding.

4.9.1 Measurement of Characteristics Using Testpieces

Material characteristics of plastics are determined by the same methods as for metals (see D2), except that time and temperature have a greater effect, so that *long-term tests* at ambient temperature and at higher temperatures are more important than with metals. In the case of plastics, not only *processing conditions* (material and tool temperature, pressures) but also *environmental influences* (technical conditioning, humidity, ageing, migration of plasticiser), *configuration influences* (wall thickness distribution, position and type of sprue) and *additives* exert a considerable influence on characteristics.

Testpieces (e.g. multi-purpose testpieces as in ISO 3167) need to be produced and tested in compliance with standard guidelines, in order to obtain comparable test results (see also standards for plastics and the CAMPUS database of raw material manufacturers).

Stresses σ are obtained as load F in relation to initial section A_o. Whereas with metals strain ε is a permanent deformation (i.e. can be measured after load relief), in the case of plastics it is always *total elongation* (i.e. deformation under load) that is measured.

Specimens are manufactured separately by injection

moulding or pressing and are taken from semifinished products or mouldings. Specimens are usually flat.

Owing to the effects of temperature and climate, testing is performed at *standard climate* DIN 50 014 (ISO 291) 23/50, i.e. at 23 °C and 50% relative atmospheric humidity.

Mechanical Characteristics

Mechanical material characteristics are designated by *limiting stresses* or *limiting deformations*. These are normally static short-term and long-term tests or dynamic impact or fatigue tests. Because the DIN test standards will, in the foreseeable future, be replaced by DIN ISO- and DIN EN-standards, the values will predominantly be listed according to the ISO-standards. Values as in DIN are shown in brackets.

Tensile Testing. In the DIN 53 455 (ISO 527) tensile test, values are determined under uni-axial, quasi-static tensile load. The *stress–strain curve* is indicative.

Figure 2 illustrates some characteristic stress–strain curves with the specified *values* (stresses in N/mm², strain in %):

σ_y	(σ_{zS})	Yield stress.
σ_M	(σ_{zM})	Tensile strength.
σ_B	(σ_{zR})	Tensile stress on fracture (ultimate tensile strength).
σ_x	(σ_{zx})	Tensile loading at x% elongation.
ε_y	(ε_{zS})	Yield elongation.
ε_M	(ε_{zM})	Elongation at tensile strength.
ε_B	(ε_{zR})	Elongation on fracture (ultimate elongation).

(Note: Designations in brackets correspond to ISO/DP 527.)

It is known that in the case of *brittle* plastics $\sigma_M = \sigma_B$, while in the case of *deformable* plastics it is possible to have $\sigma_B = \sigma_M > \sigma_y$ or $\sigma_y = \sigma_M > \sigma_B$.

Compression Testing. In the DIN 53 454 (ISO 604) compression test, values are determined under uni-axial, quasi-static compression loading. Specimens are to be selected such that no kinking occurs.

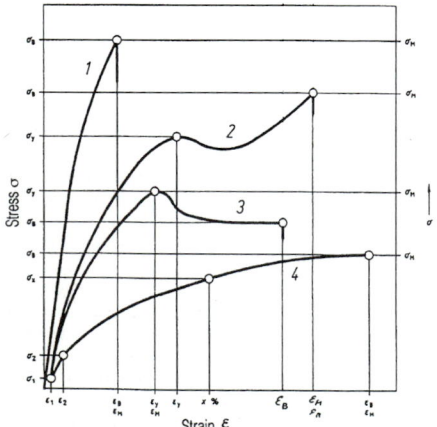

Figure 2. Tensile stress–strain curves. *1* Brittle plastics, e.g. PS, SAN, duroplastics ($\sigma_M = \sigma_B$). *2* Tough plastics, e.g. PC, ABS ($\sigma_M > \sigma_y$ or $\sigma = \sigma_y$). *3* Ductile plastics, e.g. PA, PE, PP ($\sigma_M = \sigma_y > \sigma_B$). *4* Plasticised plastics, e.g. PVC-P ($\sigma_M = \sigma_B$, σ_y not available).

Characteristics (strength values in MPa or N/mm², deformations in %):

σ_{cy}	(σ_{dQ})	Crush load.
σ_{cM}	(σ_{dM})	Compressive strength.
σ_{cB}	(σ_{dR})	Compressive load on fracture.
σ_x	(σ_{dx})	Compression at x% crush strain.
ε_{cy}	(ε_{dQ})	Crush strain.
ε_{cM}	(ε_{dM})	Crush strain at maximum load.
ε_{cB}	(ε_{dR})	Crush strain at fracture.

Bend Testing. In the DIN 53 453 (ISO 179) bend test, characteristics are ascertained by triple-point bend loading.

Characteristics (strength values in N/mm², deformation values in %):

σ_{fM}	(σ_{bM})	Bending strength.
σ_{fB}	(σ_{fR}) $(\sigma_{b3.5})$	Bending stress at fracture. 3.5% bending strength.
σ_{fc}		Bending strength at bending through s_c (see below).
ε_{fM}	(ε_{bM})	Boundary fibre elongation at maximum load.
ε_{fB}	(ε_{bR}) $(\varepsilon_{b3.5})$	Boundary fibre elongation at fracture. Boundary 3.5% fibre elongation.
s_c		Agreed degree of bending through equivalent 1.5 times thickness h.

Modulus of Elasticity E. This is ascertained by tensile or bend testing. However, since there is no clear Hooke's line for most plastics, with few exceptions in general, a *secant modulus* is calculated for elongations $\varepsilon_1 = 0.05$% and $\varepsilon_2 = 0.25$%, in accordance with DIN 53 457 (ISO 527).

The determination of the elasticity modulus E_t for tensile testing is given in ISO 527. The elasticity modulus E_c for the compression test is given in ISO 604 and E_f for the bend test in ISO 178. The elasticity modulus is determined as a secant modulus for strains $\varepsilon_1 = 0.05$% and $\varepsilon_2 = 0.25$%. According to **Fig. 2** the following formula applies for tensile testing:

$$E_t = (\sigma_2 - \sigma_1)/(\varepsilon_2 - \varepsilon_1).$$

Hardness of plastics is determined by the DIN ISO 2039 indentation test or, for plasticised plastics and elastomers, in accordance with Shore A or D DIN 53 505 (ISOs 868, 7619).

Characteristics. Indentation hardness H in N/mm² after 30 seconds' test time, Shore A or Shore B hardness after 3 seconds' test time.

Impact or Notch Bend Tests; Impact Tensile Test. In the DIN 53 453 , DIN 53 753, ISO 179, ISO 180 *impact or notch bend tests*, or in the DIN 53 448 (ISO 8256) *impact tensile test*, especially by testing at differing temperatures, the result is an indication of the toughness/brittleness properties and toughness/brittleness transition points. The notch form (U-notch, double V-notch, single V-notch) and the type of load (bilateral application in the case of Charpy tests, or unilateral loading in Izod tests) will exert a very strong influence on the properties. In the case of Charpy-impact tests as in ISO 179 there is a distinction between an impact on the narrow side (index "e", edgewise) and an impact on the broad side (index "f", flatwise). In addition, there are three forms of notches A, B or C with differing notch sharpnesses.

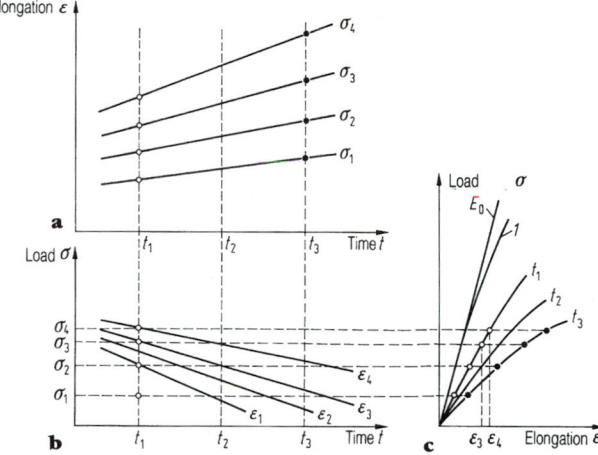

Figure 3. Test results from long-term tests: **a** creep curves $\varepsilon = f(t)$, parameter loading σ; **b** fatigue pattern $\sigma = f(t)$, parameter elongation ε; **c** isochronous load–elongation curves $\sigma = f(\varepsilon)$, parameter time t, 1 short-term test.

Properties (in kJ/m² or mJ/mm²):

a_{cU} Charpy impact toughness unnotched (ISO 179).

a_{cN} Charpy impact toughness notched (ISO 179) (N is equivalent to notch forms A, B or C).

a Load impact toughness unnotched (ISO 180).

a_k Izod impact toughness notched (ISO 180).

In cases where testpieces do not break, even when the sharpest notches are used, impact tensile tests are performed according to DIN 53448 (ISO 8256).

Fatigue Test. In this, characteristics for dynamic loading are obtained in conjunction with metals to DIN 50 100. Stress curves for various loading conditions (see D2.2) produce a Smith-type *fatigue test pattern*.

Since plastics generally exhibit no long-term fatigue endurance limit, the fatigue of 10^7 load cycles is normally used for fatigue strength. Secondly, the test frequency should not be more than 10 Hz, because of heating.

Characteristics (in N/mm²):

$\sigma_{W(10^7)}$ long-term load cycle strength for 10^7 load cycles.

$\sigma_{Sch\,(10^7)}$ long-term fluctuating stress strength for 10^7 load cycles.

In the DIN 53 444 (ISO 899/ISO 6622) *fatigue test*, which is a *retardation test, long-term expansion curves* $\varepsilon f(t)$ are recorded at constant loading. These produce *long-term pattern* $\sigma = f(t)$ and ultimately *isochronous load–elongation curves* $\sigma = f(\varepsilon)$. *Characteristics* are obtained (in N/mm²) from the isochronous stress–strain curve (**Fig. 3**):

$\sigma_{\varepsilon/t}$ long-term elongation stress (e.g. $\sigma_{2/1000}$ is the stress that produces total elongation of $\varepsilon = 2\%$ after 1000 hours).

σ_B fatigue strength.

$E_c(t, \sigma)$ creep modulus.

Creep moduli are a function of load, time and temperature. Currently, creep moduli are normally determined for stresses which cause elongations of $\varepsilon < 0.5\%$.

Electrical Characteristics

The following *characteristics* are ascertained at various electrical loads on testpieces:

E_D	breakdown voltage in kV/cm.	DIN/VDE 0303 Part 2 and IEC 243
R_{OG}	surface resistance in Ω.	DIN 53 482 and IEC 93
ρ_D	specific breakdown resistance in $\Omega\cdot$cm.	DIN 53 482
ε_r	dielectric constant.	VDE 0303 Part 4 and IEC 250
$\tan\delta$	dissipation loss factor.	VDE 0303 Part 4 and IEC 250
CTI	comparative tracking index.	DIN IEC 112
PTI	proof tracking index.	DIN IEC 112

Thermal Characteristics

Plastics, as organic materials, are greatly temperature-dependent. They also possess *low* thermal conductivity but high coefficients of linear expansion. The following *characteristics* are determined, although they do not provide any indication as to actual thermal loading capacity and are generally used only as comparative values:

$t_{Martens}$	Martens thermal dimensional stability (DIN 53 462).
HDT/A (B, C)	heat distortion temperature method A (B, C) (ISO 75).
VST/A (B)	Vicat softening temperature method A (B) (ISO 306).

Tables often state *operating temperature ranges*, but these are applicable for low loads only, in most cases. A further option for characterisation of plastic is to plot *temperature curves for moduli of elasticity in shear* from DIN 53 445 torsional vibration test.

Chemical Characteristics

The chemical resistance of plastics is dependent on their structure. *Thermosets*, because of chemical cross-linking, are extensively resistant to chemical attack. For *thermoplastics*, however, each individual material must be investigated for resistance to aggressive chemicals. Raw-materials manufacturers supply tables setting out the properties of plastics in combination with chemicals, and for various temperatures. A peculiarity of plastics is *stress cracking* under the simultaneous effect of inherent, assembly, or operating loads and chemical agencies. These can cause minor, but observable, cracks which develop into very obvious cracks and culminate in total failure.

DIN 53 449 describes stress crack investigations using *ball imprint or bend strip methods*; see also ISOs 4599, 4600 and 6252.

Processing Characteristics

The *MFR melt index* (g/10 min) and the *MVR volume flow index* (cm³/10 min) to DIN 53 735 (ISO 1133) are intended for assessment of the *flow characteristics* of plastics. *Viscosity number J* is also a processing variable for solutions of thermoplastics. Damage to plastics during processing will be reflected in variation of these characteristics.

In the case of thermosets, the DIN 53 465 "cup closing time" provides information as to flow characteristics.

The *contraction characteristics* of plastics are important in the design of plastic mouldings and the required tools. Contraction has repercussions on the dimensions and tolerances of mouldings.

Processing contraction VS is production-related and is determined according to DIN 16 901 (ISOs 2577, 3521); it is dependent on the plastic (amorphous, semi-crystalline, filled), on processing parameters (pressures, temperatures), and mould configuration. Recrystallisations in the case of semi-crystalline plastics, breakdown of internal stresses and recuring effects in the case of thermosets will culminate over a period of time in *recontraction NS*, which is mainly governed by materials, processing and environmental conditions. At higher tempertaures, recontraction may be accelerated, i.e. promoted.

DIN 53 466 *Powder density* (ISOs 60, 61, 171), DIN 53 466 *Apparent density* and DIN 53 492 (ISO 6186) *Pourability* are product tests for plastics raw materials, while DIN 53 713 *Humidity content* and DIN 53 715 *Volatility* are also relevant.

Other Tests

The *burn characteristics* of organic plastics are very important. There is a wide range of test methods, the most important of which are set out below. DIN VDE 0304 Part 4 and IEC 707 set out the *burn characteristics* of solid electrical insulating materials; these are test methods for determining *flammability* at various arrangements of the test bar and the combustion source (method BH, FH or FV). The *flammability tests* of UL regulation 94 are very important. In that system, plastics are allotted various categories, e.g. category 94 V-0 to 94 V-2 for vertical testpiece arrangement.

Various methods of *colour assessment* are important, for example for colour sampling and for purposes of enabling objective colour assessment using specified light sources.

The disintegration processes of plastics are investigated by *weathering tests* (DIN 53 384 (ISOs 4892, 878), DIN 53 386, DIN 53 387 (ISOs 4892, 879) and DIN 53 388 (ISO 877)) under weathering effects such as solar radiation, temperatures, precipitation and atmospheric oxygen. Such influences can severely affect the useful characteristics of plastic mouldings.

4.9.2 Testing of Plastic Parts

If it is possible to take specimens from plastic parts, then tests can be performed in accordance with the methods described in D4.9.1.

This is referred to as *testing the mould material in the moulding*, although there are generally limits to the comparaibility of the obtained test results with those obtained on standard testpieces. However, it is of greater value to test the part as a *complete moulding* (DIN 53 760).

Non-destructive test methods include: visual inspection, measuring moulding weight, dimensional tests and photoelastic investigations (on transparent mouldings only).

Destructive tests include: DIN 53 497 (ISO 8382) hot storage tests, DIN 53 449 (ISOs 4599, 4600, 6252) assessment of stress cracking characteristics and DIN 53 443 (ISO 6603) photomicroscopic structure investigations on transparent cuts or thin-ground sections in the case of semi-crystalline plastic, assessment of filling material orientations by reflected-light microscopy of sections, resistance tests, impact and drop tests.

Destructive tests are performed merely on a *random basis* and the results analysed by statistical rules.

Characteristics under operating conditions are ascertained by *operating tests* of integral mouldings and units. In order to compress time, individual test parameters can be deliberately accentuated, but it must be confirmed that the failure mode corresponds to that which would occur in practical conditions. The corresponding test methods must be defined in accordance with conditions.

Efforts are currently being made to monitor and control production (process monitoring) such that components do not need to be tested provided that the prescribed process parameters are adhered to (see D4.10).

4.10 Processing of Plastics

The low processing temperature (RT up to 400 °C) of plastics makes it possible to manufacture very complex components at low investment costs.

A further consideration is the saving in work stages in comparison with conventional materials, and the ease of modification by a wide range of filling materials. However, large production runs are needed in order for such mouldings to be economical, hence the need for automated manufacture of mass-produced components.

The question of reusability and of the recycling characteristics of polymers is also becoming increasingly important.

The pattern of characteristics depends not only on the characteristics of the individual plastics but also very substantially on processing conditions. For that reason, the optimisation, reproduction and constancy of process parameters are particularly important.

4.10.1 Primary Shaping

Primary shaping refers to the direct forming of finished and semifinished components from the raw material, which may be either a moulding material (granulate, powder, chips etc.) or a liquid preliminary product (see K2.3).

Injection Moulding. Injection moulding is a cycled production process in which components are made primarily from mould materials. Homogenised melts prepared in the cylinder are normally injected by the forward motion of a screw at a high pressure into the mould nest of a compartmentalised steel mould (see **K2, Fig. 25**).

Thermoplastics set in the mould nest by cooling, while thermosets and elastomers, on the other hand, stabilise by exothermal cross-linking reactions in the mould nest. Not only complex small components (springs, gears, etc.) but also large-surface-area mouldings (e.g. car bumpers) can be economically mass-produced in one stage, followed (or not as the case may be) by finishing work. The facility for integration of several functions in one mould (multifunctionality, e.g. snap-shut connectors and film hinges) is particularly important.

The mechanical characteristics and manufacturing pre-

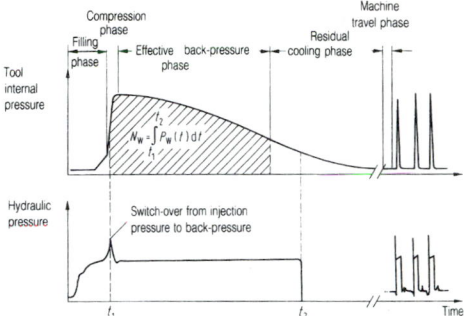

Figure 4. Synchronous plot of internal tool pressure (near the sprue) and hydraulic pressure. N_w: index for back pressure effect.

cision of injection-mouldings are dependent not only on the selected plastic and its batch consistency, but also on the configuration of the mould, the design and the quality of manufacture of the tool and the process.

The individual phases of injection moulding can be illustrated on the basis of the pressure curve in the vicinity of the sprue in the mould nest, synchronous with the hydraulic pressure curve (**Fig. 4**).

Thermosets are usually processed on the same type of injection moulding machines as thermoplastics. All that is changed is the plasticiser unit (liquid temperature control etc.). In order to maintain flow capability, major cross-linking of the mould material in the cylinder should be avoided. The relatively low viscosity of the melt in the injection moulding process for thermosets means that the mouldings have a lot of flash, which has to be removed afterwards.

Pressing and Injection Moulding

Pressing (see K2.3.6 and K2.3.7) is important for thermosets and elastomers and in the manufacture of laminates. In this process, the press material becomes plastic under the effects of pressure and heat; consequently the tool compartment fills up. Thermosetting powders are usually pre-formed and pre-heated by induction. The tablet is laid in the heated tool and pressure enables it to fill the tool compartment. Any resultant gases escape by a tool venting movement. After extensive cross-linking of the mould material, the stable, hot moulding can be taken out.

Whereas in *mould pressing* the mould material is placed directly in the tool compartment between the stamp and the die, in *injection pressing* the material is initially heated in a filling compartment. After plastic softening, the material is pressed through injection ducts into the compartments of the previously closed mould. Injection pressing is particularly suitable for multiple tools. In pressing of glass-fibre-reinforced injection resins, both components – the glass fibre reinforcement and the resin/hardener mix – are placed as pre-pregs (pre-impregnated glass fibre products) or separately in the press mould.

For large-surface-area components, e.g. chassis components in vehicle manufacture, polyester resin mesh ("OP-SMC pre-pregs") are used (SMC: sheet moulding compound). Large components are manufactured on a short-stroke vacuum press with a hydrostatically mounted clamping plate. These presses enable high positional accuracy of tool components to be achieved. Glass fibre mesh thermoplastics (GMT), for example, are used to manufacture sump guards or for hall seating systems with a grained surface. A commonly employed matrix is polypropylene

with approximately 30% proportion by weight of glass fibre. The advantage relative to SMC is its higher impact viscosity, even at low temperatures and with a medium modulus of elasticity.

Calendering. In the plastics and rubber processing industry, calendering refers to forming, at the processing temperature, of high-viscosity mixes in the gap between two or several rollers on a continuous conveyor (see **K2, Fig. 24**). Calendering is particularly important in the manufacture of films and sheets of hard and soft PVC (PVC-U, PVC-P). In rubber processing, roof lining foils, building insulation foils, floor coverings, sections, pulleys, transport belts and tyre casings are produced by calendering.

Extruding and Blowing. In extruding, a typically granulated or powder mould material is drawn from the filling hopper and plasticised under constant rotation of the screw (see **K2, Fig. 23**). The feed pressure which is built up forces the highly viscous material through a forming tool. Then calibration takes place before the extruded material becomes rigid. Tubes, sections, hoses, strips, panels, films and wire sleeves can be manufactured continuously by the extrusion method. The main components of an extrusion line are the plasticiser system (extruder), the die, the sizing tool, the cooling system, the take-up and the stacking system.

Extruded sections are frequently reprocessed in a second system that is combined with the extruder. This applies in particular to *blow forming*. In the case of *extrusion blow forming*, an extruded hose is fed out of a blowing tool and blown by means of an air pipe (**Fig. 5**). These extrusions exhibit a visible extrusion seam at their base. Bottles, canisters and fuel tanks are examples of products that can be manufactured by this method. Other frequently employed methods for manufacture of packing components and bottles are injection blowing and stretch blow forming.

Manufacture of Fibre-Reinforced Components. Glass fibre, carbon fibre and other synthetic fibres such as aramid and polyethylene are commonly embedded in a thermosetting matrix (polyester, epoxy or phenolic resin). In addition to rovings, flat semifinished products such as cloth, mesh and laminates are also used.

In *manual lamination* mesh or cloth is laid in a mould, which is typically made of wood. The fibre mesh is impregnated by brushing and the mesh is next compressed with a laminating roller. A smooth surface is achieved by application of a non-reinforced, filled pure resin coat (the gel coat). This method is suitable for the manufacture of large components and individual components.

Fibre injection, which is also regarded as an automated form of manual lamination, is suitable for small to medium-sized production runs. Resin, hardeners, accelerators and short fibres are injected into the mould by compressed air.

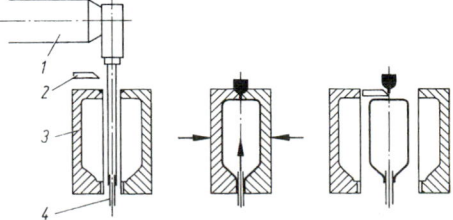

Figure 5. Extrusion blowing (diagram): *1* extruder, *2* separation knife, *3* tool, *4* air supply tube.

Using a rotating cutter, short fibres can be continuously generated from continuously input fibres. This applies exclusively to polyester resins. Typical components are bath tubs, swimming baths, tanks and roof components. Hollow bodies made of fibre-reinforced plastics are manufactured by an extensively automated *winding method*. The reinforcements fibres are wound around a core. In the impregnation bath, the rovings, which are bonded by the size are fanned out, impregnated with resin and thoroughly soaked in a "pressing section".

In order to manufacture components of maximum strength with the minimum of inherent weight, fibres must be located as precisely as possible in a subsequent main load axis and the core must be covered as consistently as possible. The roving is laid on the "geodetic" line (shortest connection between two points on a curved surface).

Foaming Process. Polymer materials can be foamed in the plastic or thermally softened state. The foaming process is performed by chemically separated gases, evaporating liquids or gas additives under pressure (chemical or physical propellants) (see K2.3.8).

The modulus of elasticity of foamed products decreases approximately in proportion to their solid content, but the rigidity of a workpiece increases as the cube of wall thickness. Components with a porous structure are several times more rigid than solid components of the same weight. "*Structural*" or "*integral foams*" have an inhomogeneous density distribution such that the foam core gradually transforms into a dense outer skin. **Figure 6** illustrates some fields of application for foams with various bulk weights.

In *thermoplastic foam casting* (TSG) a mould material is processed with small amounts of chemical propellant (e.g. ADC) by the injection moulding method. The gas-charged thermoplastic melt foams up in the incompletely filled mould nest. In the process, the outer skin is extensively compacted. This TSG method is used for production of imitation wood in the furniture industry. Other methods include TSE extrusion and TSB hollow body blowing. RIM, or reaction injection moulding, consists of

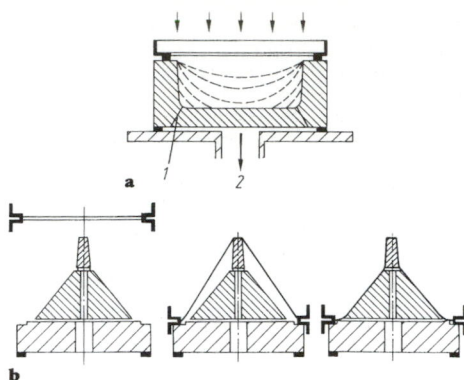

Figure 7. Vacuum forming. **a** Negative method (drawing into the form cavity): *1* suction channels, *2* vacuum. **b** Positive method (with vacuum and mechanical preliminary stretch).

the following stages: metering of the reaction partner, mixing, injection into the tool cavity, reaction in the cavity with formation of the foamed moulding, and removal of moulding. Di-isocyanates and polyhydroxyl compounds (polyalcohols) are the raw materials for polyurethane foaming materials (PUR). Reinforced PUR structure foamed products are manufactured by the RRIM process (reinforced reaction-injection moulding). SMC resin meshes and BMC moulding materials can be foamed up by micro-encapsulated physical propellants.

4.10.2 Forming and Joining

Hot Forming of Thermoplastics. For hot forming, the thermoplastic semifinished product (e.g. films, sheets) is rapidly and consistently heated up to the temperature of optimum thermoelasticity and formed by vacuum, compressed air or mechanical loads, then stabilised by cooling. Apart from the manual hot forming operations (bending, drawing), the operation is mostly performed with automated hot forming machines. Heating of the semifinished product which is stabilised in a clamping

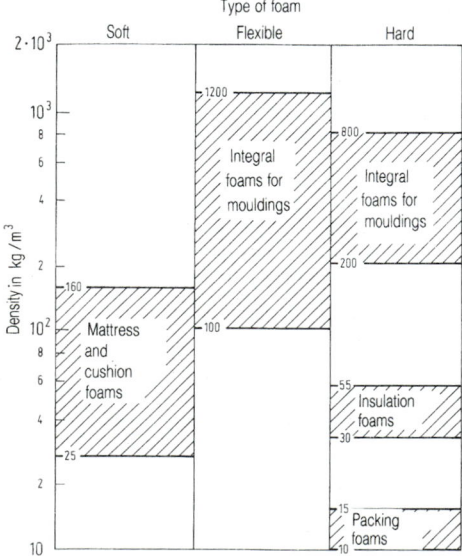

Figure 6. Application of foams with various bulk weights.

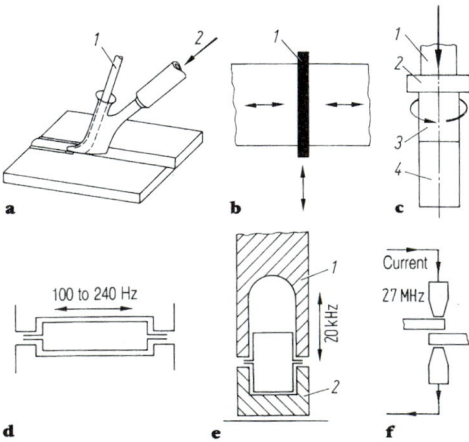

Figure 8. Welding methods for thermoplastics. **a** Hot gas welding: *1* additional rod, *2* hot gas. **b** Heating element welding: *1* heating element. **c** Friction welding: *1* pressure transmitter, *2* carrier, *3* rotating component, *4* static component. **d** Vibration welding. **e** Ultrasonic welding: *1* sonotrode, *2* anvil. **f** High-frequency welding.

frame is normally performed by infra-red surface heaters (ceramic or quartz radiators).

In *vacuum forming* the distinction is basically drawn between negative and positive processes (**Fig. 7**). In negative forming, the heated semifinished product is drawn into the concave moulding cavity, while in positive forming it is based onto a convex model (positive mould core). The side adjacent to the tool is smoother and more dimensionally accurate. The range of components manufactured by this means includes packing containers and extends to large mouldings such as swimming pools. For panels, large-surface-area components such as frontage items, sanitary cell mouldings, containers, and refrigerator casings, can be economically hot-formed. This method is also important for car components.

The *skin packing method* is used mostly for small and lightweight components. In this method, the product for packing is transported on heat-sealable cardboard to the heated film, which is then formed closely to the product by vacuum. In *blister packing*, the product for packing is laid in transparent pre-formed shells and connected with a cardboard counterpart by hot sealing. The ideal materials are amorphous thermoplastics PVC, PS, ABS, SB, SAN, PMMA, PC and the semi-crystalline materials PP and PE, as well as composite films.

Welding. Workpieces of identical or similar thermoplastic materials may be welded together by heating the materials to the temperature of viscous flow, pressing them together and allowing the joint to cool down under pressure (DIN 1910 Part 3, DIN 16 960 Part 1; ISO 857). A satisfactory joint is usually dependent on using materials of the same type, because it is necessary for the welding partners to be of equal viscosity.

Hot Gas Welding W. The base material and the additive material are converted by hot gas in the plastic state and welding under pressure (**Fig. 8a**). This method is applied in specimen production, individual component production and for large components. Equipment items made of PE, PP and PVC are often jointed by a V-, X- or fillet weld.

Heating Element Welding H. The jointing surfaces are heated by contact pressure on coated metallic heating elements. Next, the plasticised jointing surfaces are pressed together (**Fig. 8b**). This method is particularly suitable for polyolefins (PE, PP). Temperature-susceptible materials, such as PVC and POM, are less suitable for this method due to the longer heat-up time required in the context of the relatively high temperatures required.

Friction Welding (FR). With rotationally symmetrical components (up to approximately 100 mm diameter), one of the partners is rotated, and fusion is attained on the welding surfaces by relative motion under pressure. After a sudden stop, the welding surfaces are allowed to cool off while maintaining welding pressure (**Fig. 8c**).

Jointing parts clamped together in noise-insulated machines (up to approximately 500 mm in diameter, 60 to 80 cm² welding surface area) are subjected to a few degrees of angular or linear friction against each other in the process of *vibration welding* by means of electromagnetically operated oscillators with 100 or 240 Hz frequency (**Fig. 8d**). These welding techniques are used in fuel tanks, car bumpers and casings.

Ultrasonic Welding (US). A piezo-electric or magnetostrictive vibration transducer converts high frequency AC (20 000 Hz) into mechanical vibrations (**Fig. 8e**). The sonotrode matches the amplitude of the workpiece and induces vibration. The US method can be installed as a fully automated function in cycled production lines, and its short welding times make it particularly suitable for mass-produced items in the automotive, electrical and packing industries. Amorphous plastics up to approximately 350 mm in diameter, semi-crystalline plastics up to approximately 150 mm in diameter.

High-Frequency Welding HF. Polar plastics, such as PVC and CA, with high dielectric losses, can be heated up rapidly using a high-frequency electrical field. The usual welding frequency is 27 MHz (**Fig. 8f**). Main areas of application are flat form welds on soft PVC foils, envelopes, book inserts, confectionery packing, rainwear, automotive dashboards, car headlinings, automotive seat fittings and door linings.

Bonding. Bonding can be used to joint dissimilar materials (those of different types), e.g. glass to plastic,

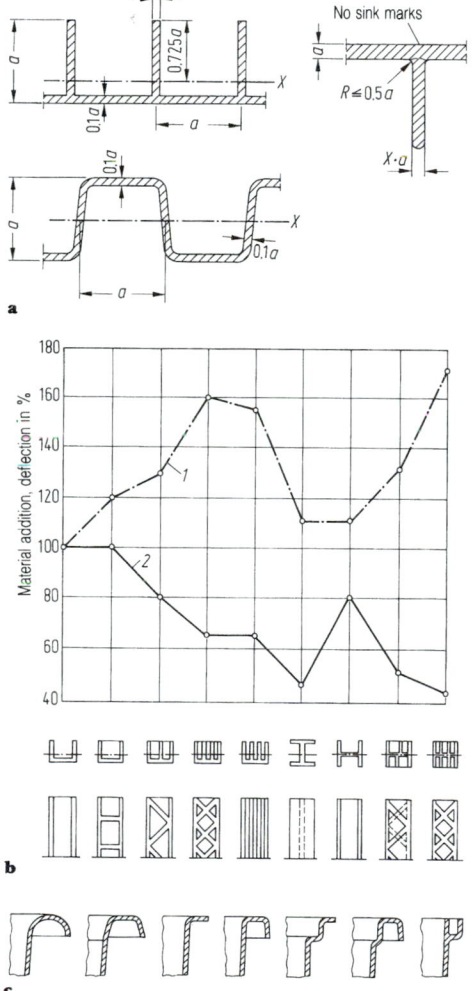

Figure 9. Stiffening of mouldings. **a** Rib and bead design, $X \approx 0.5$ for amorphous thermoplastics, $X \approx 0.35$ for PA non-reinforced, $X \approx 0.25$ for PA-GF30. **b** Deflection and material addition of various section forms: *1* material addition, *2* deflection. **c** Different edge configurations to increase inherent stiffness of large-surface-area mouldings.

Table 1. Allocation of plastic mould materials to tolerance groups (DIN 16 901)

Semi-crystalline thermoplastics	Amorphous thermoplastics	Thermosets	Tolerance categories for dimensions		
			General tolerances (without tolerances)	With directly stated dimensions (with statement of tolerances) Series 1	Series 2
PA-GF, POM-GF, PBT filled, PES, PPS filled	PS, SAN, SB, ABS, PVC hard, PMMA, PA amorphous, PC, PET amorphous, PPE modified	PF, UF/MF, EP and UP moulding material with inorganic filling	130	120	110
PP-GF (PP inorganically filled), POM (length < 150 mm), PA6, PA66, PA11, PA12, PET crystalline, PBT	CA, CAB, CAP, CP, PUR thermoplastic over 40 Shore D	PF, UF/MF with organic filling, UP resin meshes	140	130	120
PE, PP, POM (length > 150 mm), fluorinated thermoplastics such as FEP, ETFE	PUR thermoplastic 70 to 90 Shore A	150	140	130	

Table 2. Tolerances for dimensions of plastic formed components (DIN 16 901)

Tolerance category (Table 1)	Designating letter[a] tolerances and permitted deviations	Nominal dimension ranges															
		Above 1 to 3	3 6	6 10	10 15	15 22	22 30	30 40	40 53	53 70	70 90	90 120	120 160	160 200	200 250	250 315	315 400	400 500
110	A	0.1	0.12	0.14	0.16	0.18	0.2	0.22	0.26	0.3	0.34	0.4	0.48	0.58	0.7	0.86	1.06	1.3
	B	0.2	0.22	0.24	0.26	0.28	0.3	0.32	0.36	0.4	0.44	0.5	0.58	0.68	0.8	0.96	1.16	1.4
120	A	0.14	0.16	0.18	0.2	0.22	0.26	0.3	0.34	0.4	0.48	0.58	0.7	0.86	1.04	1.3	1.6	2.0
	B	0.34	0.36	0.38	0.4	0.42	0.46	0.5	0.54	0.6	0.68	0.78	0.9	1.06	1.24	1.5	1.8	2.2
130	A	0.18	0.2	0.22	0.26	0.3	0.34	0.4	0.48	0.56	0.68	0.82	1.0	1.3	1.6	2.0	2.4	3.0
	B	0.38	0.4	0.42	0.46	0.5	0.54	0.6	0.68	0.76	0.88	1.02	1.2	1.5	1.8	2.2	2.6	3.2
140	A	0.22	0.24	0.28	0.34	0.4	0.48	0.56	0.66	0.8	1.0	1.2	1.5	1.9	2.3	2.9	3.6	4.4
	B	0.42	0.44	0.48	0.54	0.6	0.68	0.76	0.86	1.0	1.2	1.4	1.7	2.1	2.5	3.1	3.8	4.8
150	A	0.3	0.34	0.4	0.48	0.56	0.66	0.78	0.94	1.16	1.42	1.74	2.2	2.8	3.4	4.2	5.4	6.6
	B	0.5	0.54	0.6	0.68	0.76	0.86	0.98	1.14	1.36	1.62	1.94	2.4	3.0	3.6	4.4	5.6	6.8
Light engineering	A	0.06	0.07	0.08	0.1	0.12	0.14	0.16	0.18	0.21	0.25	0.3	0.4	—	—	—	—	—
	B	0.12	0.14	0.16	0.2	0.22	0.24	0.26	0.28	0.31	0.35	0.4	0.5	—	—	—	—	—

[a] A for tool-related dimensions, B for non-tool-related dimensions.

ceramic to metal). In some cases this is the only available jointing process (see F1.3).

In bonding of plastic with metals, the parts must be configured correctly for joining, the surfaces must be pre-treated, the adhesives must be specifically selected and a suitable application technique identified.

Pre-treatment of the jointing surfaces is particularly important for plastics. The purpose of all pre-treatment is to activate the surface so that it is wettable and hence bondable. Various *mechanical* (grinding, sand blasting), *chemical* (degreasing, pickling) and *physical* (radiation, heat treatment) processes are recommended. Surfaces can be cleaned or degreased using solvents or flushing agents in a steam, dipping or US bath. The value of the "corona discharge" pre-treatment process in production has been confirmed for plastics. By this process, an air flow is blown between two electrodes (voltage 7 kV) and bombards the plastic surface as a stream of ionised molecules. Chemical stabilisation is achieved by coupling agents (silane coupling agent).

4.11 Design and Tolerances of Formed Parts

The design of formed parts as appropriate to their materials and production process is an essential prerequi-site for high-quality, reliably operating components (see VDI Guidelines 2001 and 2006).

Configuration Guidelines. Sink marks and bubbles (vacuoles) in the formed part will arise from *accumu-lations of material* on the component, which also lead to inconsistent cooling and the *tendency for distortion* (cause: differences in contraction). Adequate round-off radii must be selected (i.e. radii greater than the wall thickness). The preferred orientation of macro-molecules and fibrous additives and the position of joint lines, conflu-ence lines and air inclusions in the formed part are affec-ted by the position and geometry of the gate. An incor-rectly planned *tool temperature selection* can cause varying cool-off gradients in the component and consider-able distortion, owing to the resultant differences in con-traction.

Moulding distortion can often be minimised by using different *setting geometries* (**Fig. 9**).

Tolerances and permissible deviations for dimensions are set out in DIN 16 901 (ISOs 2577, 3521) for injection mouldings, injection pressings and pressings. It does not include form, positional and sectional deviations. For pur-poses of definition of tolerances, distinctions are drawn between tool-related dimensions (dimensions in one tool-half only) and non-tool-related dimensions (e.g. in the tool opening plane or movable slides). Tool related dimensions have closer tolerances.

DIN 16 901 sets out the various plastics according to their contraction characteristics in categories of tolerances (**Table 1**). The distinction is to be drawn between dimensions with *general tolerances* (dimensions without statements of tolerances) and dimensions with *directly stated* tolerances (dimensions with statement of tolerances). Series 1 applies to normal injection moulding, Series 2 to precision injection moulding. **Table 2** sets out the corresponding ranges of tolerance.

Tool tolerances, i.e. tolerances for manufacturing of tools, are set out in DIN 16 749; they are a maximum of one-third of the formed part tolerances. Allowances for *mould breakout, finishing* and *processing contraction* must also be made.

4.12 Finishing

In most cases components can be used after forming, without further machining. However, finishing work may be required for technical or cosmetic reasons.

Surface Treatments. For specific variation of surfaces or surface structure, or for specialist technical reasons, the following processes can further be applied: *painting, stamping, hot stamping, galvanising, steaming* and *dry-coating.*

Cutting. Plastics can be cut by methods that are familiar for metals, but particular tool geometries and different cut speeds must be used. In the case of thermosets and PTFE, cutting is the only facility for changing form after manufacture. With thermoplastics, springback effects and melt processes must be allowed for (see K4).

5 Tribology

K.-H. Habig, Berlin

Tribology is the "science and technology of interacting surfaces in relative motion and practices related thereto" [1]. It is also defined in DIN 50 323, Section 1. Tribology includes the subsidiary fields of *friction, wear,* and *lubrication.* It is directly affected by the materials of the involved bodies; hence its inclusion in this section.

5.1 Friction

Friction acts against relative motion between bodies in contact with each other (DIN 50 281; ISO 4378/2). It occurs in the form of friction force or friction energy. The relationship of friction force F_R to the effective normal load F_N is termed the coefficient of friction (see F4.5 and F5.2).

A distinction is drawn between different types of friction (**Fig. 1**) (see A1.11) as a function of the type of motion by the friction partners:

Sliding Friction. Friction of motion between bodies whose speed at the surface of contact differs as to magnitude and/or direction.

Rolling Friction. Theoretical friction of motion between bodies in point or linear contact and whose speeds are of equal magnitude and direction in the area of contact and in which at least one body rotates about a momentary axis within the contact area.

Sliding/Rolling Friction. Rolling friction that is combined with a component of sliding.

Drilling Friction. Friction of motion between two bodies in relative motion about an axis on the point of contact perpendicular to the surface.

Different types of friction also arise as a function of the contact condition of the friction partners:

Solid Friction. Friction entailing direct contact between the friction partners.

Boundary Friction. A special case of solid friction in which the surfaces of the friction partners are covered with adsorbed lubricant molecules.

Liquid Friction. Friction in a liquid film that continuously separates the friction partners and that can be generated hydrodynamically or hydrostatically.

Gas Friction. Friction in a gaseous film that separates the friction partners and that can be aerodynamically or aerostatically generated.

Mixed Friction. Friction in which the frictions for solids and liquid or gas are adjacent.

Table 1 sets out ranges of friction coefficients for various friction types and conditions. Generally, however, it should be noted that the friction coefficient is not a constant characteristic for a material or a material couple, but depends on loading conditions and the characteristics of all physical elements involved in the friction process. **Figure 2** sets out the influences of surface pressure, sliding speed and temperature in sliding friction between solids, on the basis of the example of friction by solids in the PTFE–steel sliding pair [2].

5.2 Friction States of Oil-Lubricated Sliding Pairs

The friction states of oil-lubricated sliding pairs with conformal contact can be designated by the Stribeck curve (see F5.2.1) (**Fig. 3**). This curve plots the friction coefficient via a parameter combination consisting mainly of oil viscosity, sliding speed and the normal load. If the total of the surface roughnesses of the sliding partners is less than the lubrication film thickness, then a condition of pure *liquid friction* or *hydrodynamic lubrication* is present.

F5.2.1 sets out the conditions for this friction state. If,

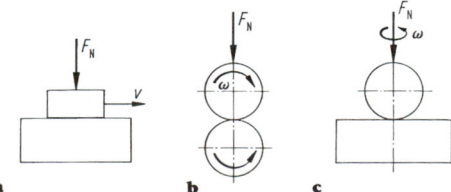

Figure 1. Types of motion between friction partners: **a** sliding friction, **b** rolling sliding friction, **c** drilling friction; F_N normal force, v sliding speed, ω angular velocity.

Table 1. Friction coefficients for different friction modes and conditions

Friction mode	Friction condition	Friction coefficient
Sliding friction	Friction by solids	0.1 to 1
	Boundary friction	0.1 to 0.2
	Mixed friction	0.01 to 0.1
	Liquid friction	0.001 to 0.01
	Gas friction	0.0001
Rolling friction	(Grease lubrication)	0.001 to 0.005

at a constant viscosity, the thickness of the lubrication film decreases with sliding speed or with the increase in normal load, to the extent that it reaches the total surface roughness of both sliding partners, then the external load is only partially supported by the lubrication film, and a different component is absorbed by the surface peaks of the sliding partner. This friction state is called *mixed friction*. If the lubricating film thickness decreases to zero with continuing decrease in the parameter combination, we enter the field of *boundary friction*.

5.3 Elastohydrodynamic Lubrication

The above statements refer to sliding friction in *conformal* contact. Two rollers or the flanks of gear teeth form a *contraformal* contact. In the contact areas, the

Table 2. Wear types and wear mechanisms as in DIN 50 320

System structure	Tribological loading (symbol)		Wear type	Active mechanisms (individual or in combination)			
				Adhesion	Abrasion	Surface disruption	Tribochemical reactions
Solid Intermediate material (complete film separation)	Sliding Rolling Rolling/sliding Collision Impact					×	×
Solid Solid (in the case of solid friction, boundary friction, mixed friction)	Sliding		Sliding wear	×	×	×	×
	Rolling Rolling/sliding		Rolling wear Rolling/sliding wear	×	×	×	×
	Collision Impact		Collision wear Impact wear	×	×	×	×
	Oscillation		Vibrational wear	×	×	×	×
Solid Solid and particle	Sliding		Striation wear			×	
	Sliding		Grain sliding wear			×	
	Rolling/sliding		Grain rolling wear			×	
Solid Liquid with particles	Flow		Flushing wear (erosion wear)		×	×	×
Solid Gas with particles	Flow		Jet-blast wear (erosion wear)		×	×	×
	Collision		Direct or oblique impact wear		×	×	×
Solid Liquid	Flow oscillation		Material cavitation, cavitational erosion			×	×
	Impact		Droplet strike			×	×

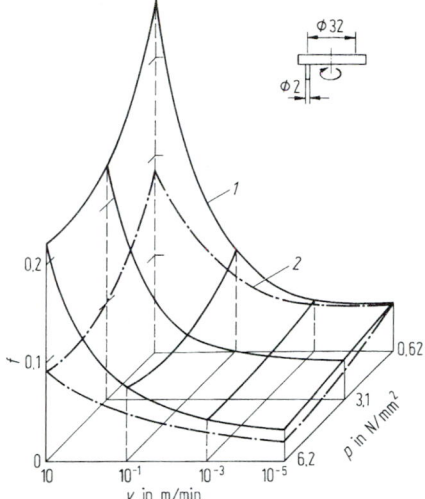

Figure 2. Friction coefficient f for PTFE-steel. p surface pressure, v sliding speed, steel: $R_z = 0.03\ \mu m$; ambient medium: synthetic air, $1\ T_a = 23\ °C,\ 2\ T_a = 70\ °C$.

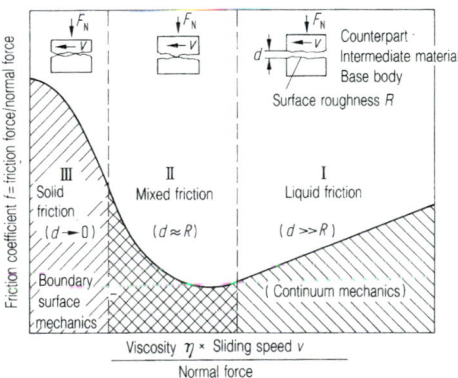

Figure 3. The Stribeck curve (schematic).

pressures encountered are generally much higher than in the case of conformal contact. In this type of contact, a separating lubricating film can be formed between the contact partners, whose thickness can be estimated by elastohydrodynamic theory [3, 4]. This theory combines the elastic theory for deformable bodies with hydrodynamics theory, whereby the increase in lubricant viscosity is taken into account as pressure increases.

Figure 4 contains a diagram of lubricating film thickness and pressure distribution in an elastohydrodynamic contact. The pressure peaks at the oil outlet side where the lubricating film is thinnest must be considered.

On the assumption of ideally smooth surfaces and isothermal conditions, lubricating film thickness can be estimated by the following formulae [5-7]:

$$H_{min} = 3.63U^{0.68} \cdot G^{0.49} \cdot W^{-0.073}\ (1 - e^{-0.68k}),$$

$$H_{min} = h_{min}/R_x \quad \text{(oil flow in x-axis)}.$$

Here:

$$U = \frac{\eta_0(u_1 + u_2)}{2ER_x}; \quad G = \alpha \cdot E; \quad W = F/(ER_x^2)$$

where h_{min} = the minimum lubricating film thickness.

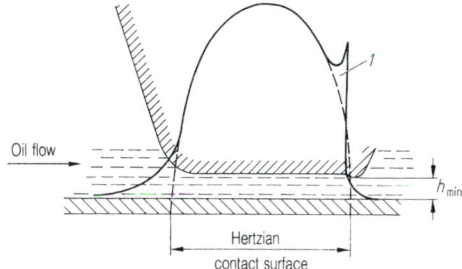

Figure 4. Pressure distribution in an elastohydrodynamic (EHD) contact; 1 Hertzian pressure distribution.

From **Fig. 5**:

$$R_x = R_{x1} \cdot R_{x2}/(R_{x1} + R_{x2});$$

$$R_y = R_{y1} \cdot R_{y2}/(R_{y1} + R_{y2});$$

η_0 = the oil viscosity at oil inlet; u_1, u_2 = the surface velocities of bodies 1 and 2;

$$\frac{1}{E} = \frac{1}{2}\left(\frac{1 - \nu_1^2}{E_1} + \frac{1 - \nu_2^2}{E_2}\right);$$

where ν_1, ν_2 = Poisson's ratios for bodies 1 and 2, E_1, E_2 = moduli of elasticity of bodies 1 and 2, α = the viscosity pressure coefficient and F = the normal load;

$$k = a/b = 1.04 \left[\frac{R_y}{R_x}\right]^{0.636}$$

Under real conditions, owing to the internal friction of the lubricant in the lubrication space, it may be assumed that there will be an increase in temperature which will cause a decrease in viscosity. In order to combine isothermal theory with non-isothermal theory, a thermal correction

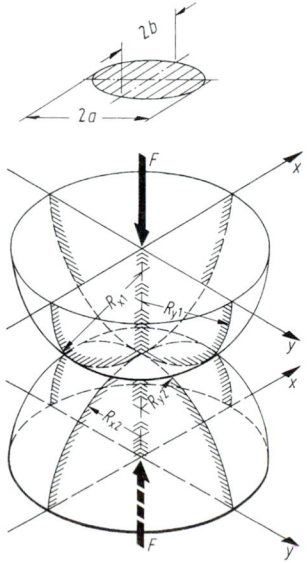

Figure 5. Contraformal contact between two bodies.

factor C_{th} is introduced, which is in turn determined from thermal load factor L_{th} [6]:

$$C_{th} = \frac{h_{min.th}}{h_{min.isoth}} = \frac{3.94}{3.94 + L_{th}^{0.62}}$$

where

$$L_{th} = \eta_{0M}\frac{\alpha^* \cdot u^2}{K}$$

In this context, $h_{min.th}$ = the theoretical minimum lubricating film thickness, taking account of temperature increase, $h_{min.isoth}$ = the theoretical minimum lubricating film thickness in the isothermal state, η_{0M} is the dynamic viscosity at a pressure of 1 bar and the material surface temperature T_M, α^* is the viscosity temperature coefficient of the lubricant, and K is the coefficient of thermal conductivity (for mineral oils: $K = 0.133$ W/m K).

The following formula is adopted for the viscosity–temperature function:

$$\eta_0 = \eta_{0M}e^{-\alpha^*}(T_0 - T_M).$$

Under real conditions, it should also be noted that the surfaces in contact are not theoretically smooth but have a measurable surface roughness [6]. For designation of such roughness, measurement variables F_{r1}/M_{r1}, R_K and $F_{r2}/(100 - M_{r2})$ are introduced, the meanings of which are given in **Fig. 6**. These variables provide the basis for derivation of a correction factor C_{RS} for the combination of a smooth surface with a rough surface, into which there are also included the average surface roughness $\bar{R}_z$:

$$C_{RS} = 0.8\left(\frac{\bar{R}_K}{\bar{R}_z}\right)^{0.61}\left(\frac{F_{r1}/M_{r1}}{F_{r2}/(100 - \bar{M}_{r2})}\right)^{0.25}$$

The condition for separation between the contact partners is then set out as follows:

$$\frac{h_{min.th}}{C_{RS}} > \bar{R}_z.$$

In this context, $\bar{R}_z$ is the arithmetical mean of three measurements taken at the circumference of the rough contact partner.

5.4 Wear

If the lubricating film thickness is inadequate for total separation of two sliding or rolling partners, then wear occurs. Tribological systems that have previously been operated without lubrication, such as dry bearings, friction brakes, transport systems for minerals etc., are subject to gradual wear.

DIN 50 320 defines wear as follows: "Wear is the progressive loss of material from the surface of a solid, caused by mechanical influences, i.e. contact and relative motion of a solid, liquid or gaseous counterpart."

Three instructions are given:

The loading of a solid by contact and relative motion of a solid, liquid or gaseous counterpart is also termed tribological loading.

Wear is exhibited in the occurrence of small detached particles (wear particles) and in changes in the substance and form of the tribologically loaded layer.

In technology, wear is normally undesirable, i.e. a factor that reduces value. In exceptional cases, as in the case of running-in processes, wear processes may nevertheless be technically desirable. Machining processes as value-enhancing technological processes do not count as wear with regard to the piece in the process of manufacture, although tribological processes occur in the boundary surface area between the tool and the workpiece, as in the case of wear.

DIN 50 320 also sets out the following basic, important terms with regard to wear:

Wear Types. Distinction between wear processes according to the type of tribological loading and the materials involved.

Wear Mechanisms. Physical and chemical processes occurring in the wear process.

Wear Phenomena. Wear-induced variations in the surface layer of a body and the type and form of the wear particles encountered.

Wear Measurement Variables. Wear measurement variables directly or indirectly designate the variation in the configuration or mass of a body due to wear (DIN 50 321). Wear is ultimately caused by the effect of wear mechanisms. Four wear mechanisms are regarded as particularly important [8] (see D1.2.3):

Adhesion. Formation and separation of atomic links (microwelds) between the friction partners.

Tribochemical Reaction. Chemical reaction of the base body and/or counterpart with the constituents of the lubricant or ambient medium due to friction-induced, chemical activation of the loaded surface areas.

Abrasion. Scratching and micro-machining of the friction partner by hard roughness peaks on the other or by hard particles in the intermediate material.

Surface Disruption. Cracking, crack propagation and separation of particles due to alternating loads in the surface areas of the friction partners.

Wear mechanisms can occur individually, consecutively or simultaneously. **Table 2** on p. **D**68 is a compilation of wear mechanisms for the various wear types.

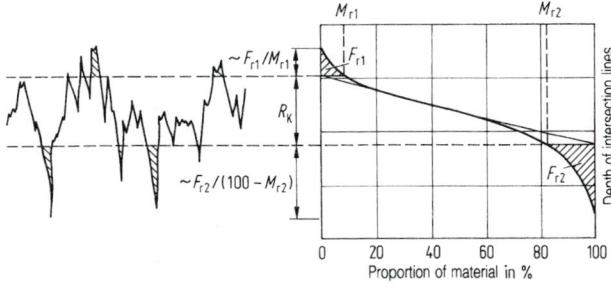

Figure 6. Surface roughness indices [6].

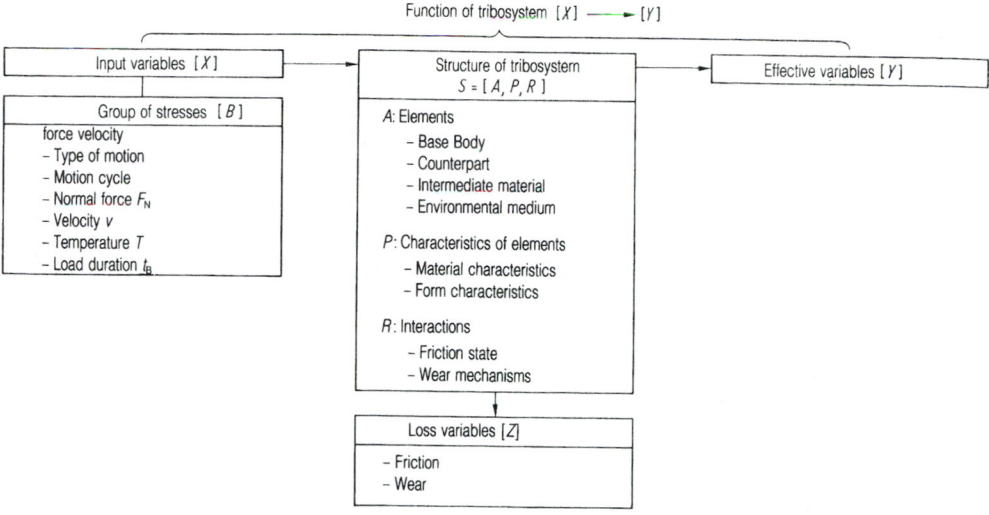

Figure 7. Diagram of a tribological system.

5.5 Systems Analysis of Friction and Wear Processes

Friction and wear are dependent on a wide range of influencing variables, which can best be categorised by the systems analysis method (**Fig. 7**) [9]. Accordingly, friction and wear are regarded as loss variables in a *tribosystem*, in which certain input variables, which are decisive for the *group of stresses*, are transformed by the *structure* of the tribosystem into effective variables. The *function* of the tribosystem is achieved by this transformation.

5.5.1 Function of Tribosystems

Tribosystems are used for implementation of various functions. For example, a bearing has to absorb loads and permit motion. Using friction brakes, on the other hand, movements have to be inhibited. Power trains transmit torque or change speeds; timing drives transmit information. The range of possible functions includes the recovery, transportation and processing of raw materials. Consequently, it is useful to state the function of tribosystems because this in itself provides certain indications of the type of components and the materials employed. If the function of a tribosystem, for example, consists of opening and closing an electrical circuit, then it will frequently be necessary to use switch contacts manufactured from special contact materials.

5.5.2 Operating Variables

The main variables of the group of stresses are set out in **Fig. 7**.

The *types of movement* can be distinguished from each other in the same way as for friction types: "sliding, rolling, rolling/sliding, drilling". However, further types of motion arise, such as "impact, collision or flow". The *motion cycle* can be continuous, intermittent, oscillatory or reversing. The *normal load* provides the basis for determining material strain, given knowledge of the dimensions of the components, the moduli of elasticity of the materials and the coefficients of friction. On the one hand, some importance accrues to *velocity* as the relative speed between the friction partners; on the other hand,

for *heat derivation*, it is necessary to know whether the friction partners are both in motion, or only one body. In addition to the *load duration* (or *load path*), we also have to take account of static periods in which the characteristics of the surface areas may vary, for example owing to corrosion.

5.5.3 Structure of Tribological Systems

Within the *structure* of tribological systems, the distinctions can generally be drawn between four components or materials which are designated "elements" (**Fig. 7**). The friction partners are present in any tribosystem, while the intermediate material or ambient medium may in some circumstances be absent. In order to reduce friction and wear, a lubricant will be employed as the intermediate material in many practical applications. The intermediate material, however, may consist of hard particles, e.g. ore to be ground in a ball mill.

It is often necessary to distinguish between *open* and *closed* tribosystems with regard to the prevention of wear.

For example, in the case of open systems, the surface of a tool is loaded by continuously fresh surface areas of the workpiece for machining. Its function primarily depends on wear of the tool which is one partner, while the load is generated by the other without its wear being of any significance.

In the case of enclosed systems, e.g. a cam tappet couple, on the other hand, the surface areas of both partners are repeatedly in contact. Functionality depends on wear of the cam and of the tappet.

Elements are characterised by their properties, the distinctions having to be drawn between material and form characteristics and between volume and surface characteristics.

Ultimately, friction and wear are dictated by interactions between elements which are characterised by the friction state (see D6.2) and the wear mechanism (see D6.4).

5.5.4 Tribological Characteristics

Tribological characteristics are used for quantitative and qualitative designation of friction and wear processes.

Friction is characterised by friction load F_R and the friction coefficients by f. Friction load F_R is a function of the variables of load collective B and system structure S. Consequently the following formula applies:

$$F_R = f(B, S).$$

A similar formula can be postulated for wear magnitude W:

$$W = f(B, S)$$

If wear magnitude is examined over the load period, two different curves are commonly obtained (**Fig. 8**). In the running-in phase, increased initial wear may occur, gradually decreasing and transferring to a prolonged permanent condition with a continual increase in the magnitude of wear (constant wear rate), before progressive increase culminates in failure (**Fig. 8a**).

If the prime wear mechanism is surface disruption, then a measurable wear will not occur until after a certain incubation period during which microstructural variations, cracking and crack propagation will take place before wear particles are separated (**Fig. 8b**).

Since wear is always a consequence of the effect of wear mechanisms, we should not only state the magnitude of wear or the wear rate but also set out the wear phenomena in the form of microscopic or electronmicroscopic plates, from which it is possible to identify the constellation of wear mechanisms. This is the only way to make the results of a wear examination usefully applicable to other and similar cases.

5.5.5 Checklist for Tribological Characteristics

It has been shown that friction and wear depend on a range of influencing variables. For reproducible implementation of friction and wear investigations in operation and at experimental level, it is useful to tabulate the main variables. **Table 3** may provide some assistance in this respect.

5.6 Lubricants

Lubricants are used to reduce friction and wear in tribological systems. They are employed in various aggregate states as lubricating oils, lubricating greases, or solid lubricants. Occasionally, water or liquid metals are used as lubricants, and operating conditions often permit forma-

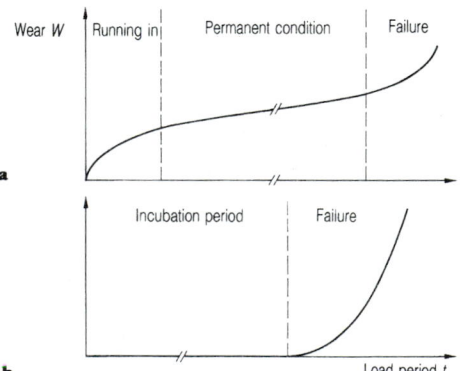

a

b   Load period t

Figure 8. Wear as a function of load duration.

Table 3. Checklist for classification of main friction and wear variables

Description of the tribosystem

Structure of the tribosystem	
Base body	
Designation	
Dimensions	
Material	Designation:
	Hardness:
	Structure:
Roughness	R_z (μm): R_a (μm):
Counterparts	
Designation	
Dimensions	
Roughness	Designation:
	Hardness:
	Structure:
Material	R_z (μm): R_a (μm)
Intermediate material	
Designation	
Aggregate state	□ solid □ liquid □ gaseous
Viscosity[a]	At ambient temperature
	At operating temperature
Ambient medium	
Designation	
Aggregate state	□ liquid □ gaseous
Relative humidity[b]	
Friction state:	□ solid friction
	□ boundary friction
	□ mixed friction
	□ gas friction
	□ liquid friction

Group of stresses	
– Type of motion	□ slip
	□ roll/slip
	□ impact
– Motion cycle	□ continuous
	□ intermittent
	□ reversing
	□ oscillating
– Normal force F_N (N)	
– Pressure p (N/mm²)	
– Velocity v (m/s)	
– Operating temperature T (°C)	
– Load duration t_B (h)	

Tribological characteristics
– Friction coefficient f
– Wear magnitude W_v (mm³)
– Wear rate $W_{1/t}$ (μ/h)
– Forms of wear phenomenon

Comments

[a] In liquid aggregate state.
[b] In gaseous ambient medium and at ambient temperature.

tion of a hydrodynamically generated film that separates the contact partners.

5.6.1 Lubricating Oils

Lubricating oils can be subdivided according to origin into:

Mineral oils, animal and vegetable oils, synthetic oils, others.

Mineral oils, which can be obtained from petroleum and

to some extent from coal, are the most important oils. They consist of paraffins, naphthenes or aromatics. *Animal* and *vegetable* oils such as castor oil, fish oil, olive oil, etc., are employed for special applications, e.g. in light engineering.

Synthetic oils are increasing in importance for high-temperature lubrication and for reducing friction. These involve in particular the following: polyether oils (polyalkyline glycols, perfluoropolyalkyl ether, polyphenyl ether), carbonic acid esters, ester oils, phosphoric acid esther, silicone oils and halogenated hydrocarbons.

Lubricating oils have to possess a range of physical and chemical properties in order to fulfil their complex functions [10, 11].

Properties of Lubricating Oils

Viscosity. Viscosity is of decisive importance in order to achieve a hydrodynamic or elastohydrodynamic lubrication state; it is an index for internal friction of lubricating oil.

In accordance with A6.2, the following formulae apply:

Dynamic viscosity:

$$\eta = \tau/(dv/dz) = \tau/D.$$

Kinematic viscosity:

$$\nu = \eta/\rho.$$

In this context, τ is the thrust load arising from shear of a laminar flow, $D = dv/dz$ shear or velocity gradient, and ρ is the density of the oil.

The unit of dynamic viscosity η is the pascal second, Pa·s (= 10 poises), and the unit of kinematic viscosity ν is the square metre per second (= 10^4 stokes). Viscosity is not a pure physical constant, but is generally dependent on various parameters such as speed and shear gradient D, time t, temperature T and pressure p. Where there is no dependency of viscosity on the shear gradient, then we refer to *Newtonian fluids* or *Newtonian lubricants.* These include pure mineral oils and synthetic oils of comparable molecular weight. Lubricating oils whose viscosity is dependent on the shear gradient are designated *non-Newtonian oils.* Where viscosity decreases as the shear gradient increases, this is a *structural viscous oil.* The admixture of additives to Newtonian base oils can cause intrinsic viscosity, e.g. the addition of polymers to engine or industrial oils, to improve the "viscosity index".

Where viscosity is dependent on time t, a distinction is to be drawn between the following:

Thixotropy. Reduction of viscosity owing to continuous shear stress, and its return after removal of stress.

Rheopexy. Increase in viscosity owing to continuous shear stress and its decrease after removal of stress.

The viscosity of lubricating oils decreases as temperature increases, so that temperature must be stated at each viscosity measurement. The dependence of viscosity upon temperature can be stated by various approximation formulae. Ubbelohde-Walther's transformation is often used for lubricating oils:

$$\lg \lg (\nu + C) = K - m \lg T.$$

Here, ν = kinematic viscosity, C = a constant (for mineral oils: 0.6 to 0.9), K = a constant, m = the slope of the curves of appropriately scaled viscosity temperature data

sheets and T = absolute temperature in K. The following formula is often used to describe the relationship of viscosity to pressure:

$$\eta_p = \eta_0 \cdot \exp (\alpha \cdot p),$$

where η_0 = viscosity at 1 bar, α = the "viscosity pressure coefficient" and p = the pressure. Viscosity accordingly increases very sharply (exponentially) with pressure; see **Appendix D5, Table 1**.

Density. This is required for conversion of dynamic viscosity into kinematic viscosity. DIN 51 757 (ISO 3675) states various methods for its determination. Density is temperature and pressure dependent (see A5).

Viscosity Index. This, in accordance with DIN ISO 2909, is a quantitative index for characterisation of the relationship of viscosity to temperature. It was introduced in 1928 with a scale ranging from 0 to 100, where oil with the greatest viscosity temperature dependence known at that time was given a viscosity index of VI = 0, and the oil with the lowest viscosity temperature dependence an index of 100. Improved refinery methods and the development of synthetic oils have now enabled the viscosity index of 100 to be markedly exceeded.

Shear Stability. The addition of oil-soluble polymers enables the viscosity of lubricating oils to be increased and their viscosity index to be improved. Shear processes allow the destruction of polymer molecules, causing a decrease in viscosity. In order to determine the irreversible decrease in viscosity due to shear, loads are applied in a gear test rig, in laboratory test rigs with high-pressure hydraulics, and in high-pressure diesel injection systems (to DIN 51 382 (ASTM D2603)).

Cloud and Pour Point. The viscosity of lubricating oils increases with decreasing temperature. The cloud point states the temperature at which the oil begins to cloud under test conditions laid down by DIN ISO 3015. The pour point represents the temperature at which the oil is still just capable of pouring (DIN ISO 3016).

Neutralisation Capacity. Lubricating oils may contain alkaline and acid constituents. Acid components in fresh oils may originate from refining or from lubricant additives. They can also be formed during operation by oxidation of the lubricating oil. Alkaline-action additives are added to engine oils in particular, in order to neutralise acid compounds that arise from combustion processes in the engine.

Neutralisation Number (NN). This is the number of mg of potassium hydroxide required to neutralise the acids present in 1 g of oil. For this purpose, a 0.1M KOH solution is slowly added to a solution of oil (titration) as set out in DIN 51 558, Section 1 (ISO 6618), until the *p*-naphthol benzoin indicator reaches the final point of titration, thus indicating neutralisation.

Total Base Number (TBN). This is the quantity of acid required in order to neutralise the basic components of the oil. It is stated as the equivalent quantity of potassium hydroxide that corresponds to the quantity of acid for 1 g oil. The TBN is ascertained by electrometric titration in accordance with DIN ISO 3771.

Flashpoint. The flashpoint is the lowest temperature at which under specified conditions an adequate quantity of vapours will be given off from the oil specimen under test for them to form a flammable mix with the air above the liquid level. If the flashpoint is over 79 °C, then it can

Table 4.   Compilation of important lubricant additives

Additive	Function	Material	Mode of operation
Viscosity index improver (VI improver)	To mitigate the loss in viscosity with temperature increase	Polymerised olefins and iso-olefins, polymethacrylate, polyalkyl styrols, etc.	Extension of convoluted molecules as temperature increases
Pour point depressant	To prevent solidification (failure to flow at low temperature)	Condensation products of chlorinated paraffin and naphthalene, polymethacrylate, etc.	Adsorption at the surfaces of paraffin crystals; preventing the growth of paraffin crystals
High-pressure additive (EP additives, anti-wear additives)	To prevent seizing or adhesive wear at high loads	Organic sulphur, phosphorus and chlorine compounds and their combinations, etc.	Formation of reaction coats on tribologically loaded surfaces
Friction reducer	To reduce coefficient of sliding friction	Fatty acids, fatty acid esters, fatty acid amides, fatty acid salts, etc.	Formation of adsorption and reaction coats on tribologically loaded surfaces
Corrosion inhibitor	To restrict metal corrosion	Fatty acids, nitrogen, phosphorus and sulphur compounds, etc.	Formation of protective coats which impede access of oxygen and water to the surface of the metal
Oxidation inhibitor	To reduce lubricating oil oxidation	Sulphur and phosphorus compounds, phenol derivatives, amines, etc.	Interruption of the radical chain mechanism of oxidation
Detergent	To prevent deposits on material surfaces	Metallo-organic compounds such as phenolates, sulphonates, phosphates, naphthenates, etc.	Prevention of coagulation of products of oxidation
Dispersant	To prevent cold sludge formation	Amides, imides of polybasic, organic acids	Peptisation of oil-insoluble products of oxidation
Demulsifying agent	To separate oil and water	Polar, boundary-layer active compounds	Increasing boundary surface-tension between oil and water
Emulsifying agent	Formation of emulsions (for cold lubricants)	Alkaline salts of carbonic acid, etc.	Reducing boundary surface-tension between water and oil
Anti-foaming agent	To prevent foaming	Silicon polymers, etc.	Destruction of oil films which surround air bubbles

be determined by application of Cleveland's method as standardised in DIN ISO 2592, where the oil is heated in an open crucible. Oils with lower flashpoints are investigated in the Abel–Pensky enclosed crucible (DIN 51 755 (ISO 2592), flashpoint 5 to 65 °C) or the Pensky–Martens crucible (DIN 51 758, flash point 65 to 165 °C). The flashpoint is not relevant to lubricating characteristics.

Thermal Capacity c_p and Thermal Conductivity. These values are important for purposes of calculating heat balance and transport. Both variables are temperature-dependent. See **Appendix D5, Figs 4 and 3**.

Air Entrainment. Lubricating oils can contain substantial quantities of air in solution. Solubility has a low dependency on temperature and a high dependency on pressure. The volume of air released can be determined by the Henry–Dalton law:

$$V_{air} = K \cdot V_{oil} \cdot p_2/p_1.$$

The Bunsen coefficient K for mineral oils is between 0.07 and 0.09; for silicon oils it is between 0.15 and 0.25. Lubricating oils in operation may contain not only dissolved air but also air in the form of a finely distributed second phase, for which the term "aero-emulsion" or "foaming" is used. Unlike dissolved air, aero-emulsions will cause deterioration of tribological characteristics, because viscosity and heat conductivity are reduced and oxidation processes and cavitation phenomena are increased. Oil transport may also be impeded.

There is a particularly disadvantageous effect from a stable surface foam that can arise owing to migration of the aero-emulsion to the surface. Air separation capability (aero-emulsion) can be determined by DIN 51 381 (ISO DIS 9120), while foaming characteristics can be ascertained by DIN 51 566 (ISO DIS 6247).

Water in Lubricating Oil. As a rule, lubricating oils should be water-free, because water accelerates oil ageing and the corrosion of materials, and also impedes formation of the lubricating film. Water content can be ascertained by DIN ISO 3733 or DIN 51 777 (ISO DIS 6296).

Solid Foreign Bodies in Lubricating Oil. Solid foreign bodies exert a negative effect in proportion to their hardness, size and quantity, because they can block up oil bores and filters and cause wear due to abrasion. Metal foreign particles often accelerate oil oxidation. The level of solid foreign bodies is generally determined by means of a DIN 51 365 (ISOs 4021, 4405) centrifuging process or by a membrane filter process.

Lubricant Additives. These are additives that improve the useful characteristics of lubricating oils. They can be divided into two groups according to function (**Table 4**): additives that improve the tribological characteristics of lubricants – such as the viscosity–temperature characteristics or the friction–wear characteristics under boundary or mixed friction conditions – and additives that influence other important characteristics, such as oxidation inhibitors, detergents, anti-foaming agents etc.

Additives can exert a mutually supportive effect, i.e.

synergy, or impede each other's function and operate antagonistically. Modern additives often exhibit several functions, and this reduces the risk of their impeding each other's action.

Classification of Lubricating Oils

Lubricating oils can be classified as follows according to application:

Machine lubricating oils (see F4.4.2).

Cylinder oils.

Turbine oils.

Engine oils.

Transmission oils (see F8.3).

Compressor oils.

Recirculating oils.

Hydraulic fluids (see G1.2).

Oils for processing metals, cooling lubricants (see K4.3.1).

Textile oils and textile machine oils.

DIN Taschenbuch nos. 20, 32, 57, 58, 192 and 228 contain extensive information on oils. The largest category of lubricating oils is represented by engine oils, which are classified according to viscosity. The classification was issued by the (US) Society of Automotive Engineers (SAE) in collaboration with the American Society for Testing and Materials (ASTM) and incorporated by DIN 51 511 (ISO DIS 10 369) (see **Appendix D5, Table 2**). Viscosity levels for −18 and +100 °C were laid down for SAE viscosity categories 5W to 20W, and the only values laid down for 20/50 oils are those for 100 °C. Categories 5W to 20W can be combined with categories 20 to 50 to include multigrade oils, which cover several viscosity categories thanks to their improved viscosity–temperature characteristics, and thus enable summer and winter operation.

5.6.2 Lubricating Greases

Lubricating greases are solid or semi-liquid products of a dispersion consisting of a thickening agent and a liquid lubricant.   They primarily fulfil the following functions in lubrication technology:

Provision of an adequate quantity of liquid lubricant by slow separation, in order to prevent friction and wear over wide temperature ranges and over long periods of time.
Sealing against water and foreign particles.

Most lubricating greases consist of a soap (alkali or alkaline earth soap) in 4 to 20% by weight, lubricating oil in 75 to 95% by weight and additives in 0 to 5% by weight.

Consistency Categories. Depending on their deformability (Walk penetration), lubricating greases are classified into various NLGI consistency categories (NLGI: National Lubrication Grease Institute), **Appendix D5, Table 3** to DIN 51 818 (ISO 2137).

Consistency is determined under DIN ISO 2137 by penetration of a standard cone into a grease specimen under defined test conditions, with measurement of penetration depth after a given period.

Flow Characteristics. The flow characteristics of lubricating greases cannot be adequately defined by consistency categories. Lubricating greases are substances which

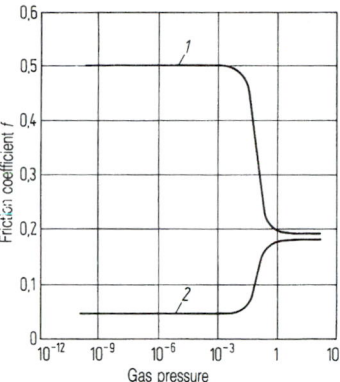

Figure 9. Friction coefficients for *1* graphite and *2* molybdenum disulphide [12].

have non-Newtonian flow characteristics which are dictated by temperature, shear gradient, shear time and historical data. The viscosity of lubricating greases generally decreases with an increase in shear gradient and shear time.

Applications. Lubricating greases are used in the temperature range of −70 to 350 °C approximately, for lubrication of machine components such as rolling and sliding bearings, slideways, transmissions, etc., and simultaneously provide a sealing function (see F4.4.3).

5.6.3 Solid Lubricants

Solid lubricants are present in the solid aggregate state. They are required for lubrication under extreme conditions, such as at very high or very low temperatures, in aggressive media, in vacuum, etc. Solid lubricants consist of the following groups of substances:

- Compounds with a *layer lattice structure*. These include: graphite, molybdenum disulphide, dichalcogenides, metal halogenides, graphite fluoride and hexagonal boron nitride.
- *Oxide* and *fluoride* compounds of transition and alkaline earth metals. These include: lead oxide, molybdenum oxide, tungsten oxide, zinc oxide, cadmium oxide, copper oxide, titanium dioxide etc., calcium fluoride, barium fluoride, strontium fluoride, lithium fluoride and sodium fluoride.
- *Soft metals*, such as lead, indium, silver, etc.
- *Polymers*, especially polytetrafluoroethylene (PTFE).

Particular importance accrues to solid lubricants which consist entirely or partially of graphite or molybdenum disulphide. Where graphite is employed, it is essential to note that it has low friction only when water molecules are dissolved in its lattice; these reduce the shear strength of the hexagonal basic surfaces. Consequently, graphite is unsuitable for use as a solid lubricant in a vacuum – see **Fig. 9**. Molybdenum disulpide, on the other hand, has a particularly low coefficient of friction in a vacuum, while in humid air it has a higher friction coefficient and – above all – is broken down at high temperatures [11].

When PTFE is used it should be noted that the coefficient of friction increases sharply with sliding speed (**Fig. 2**).

6 Appendix D: Diagrams and Tables

Appendix D1

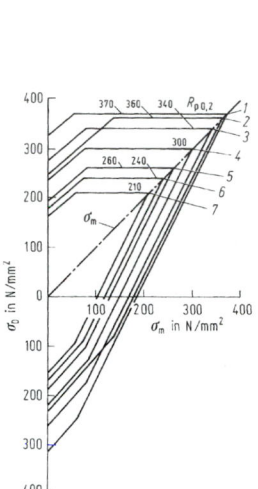

Figure 1. Fatigue strength pattern (Smith curve) for tensile-compressive loading [58]. General building steels to DIN 17 100: *1* St 70; *2* St 52 3; *3* St 60; *4* St 50; *5* St 42; *6* St 37; *7* St 34.

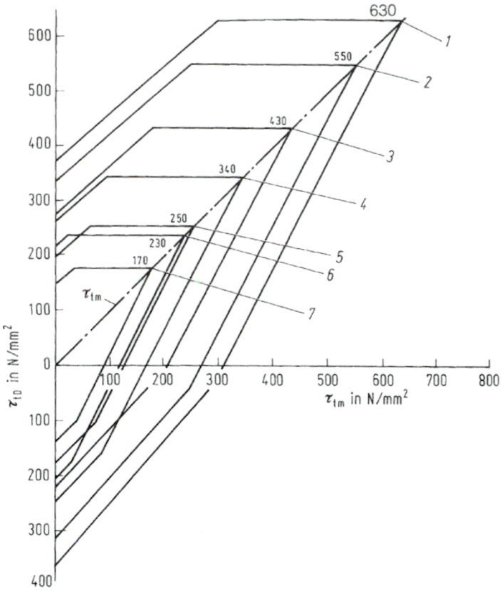

Figure 2. Fatigue strength pattern (Smith curve) for torsional loading [58]. *1* 42 CrMo 4; *2* 34 Cr 4; *3* 16 MnCr 5; *4* C 45, Ck 45; *5* C 22, Ck 22; *6* St 60; *7* St 37.

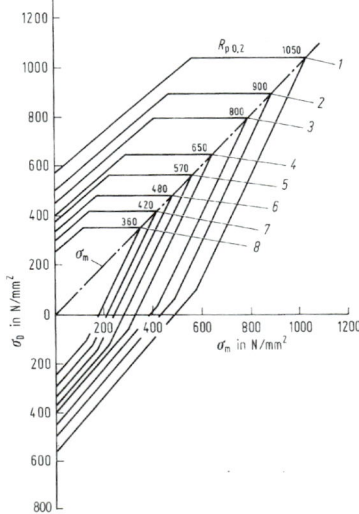

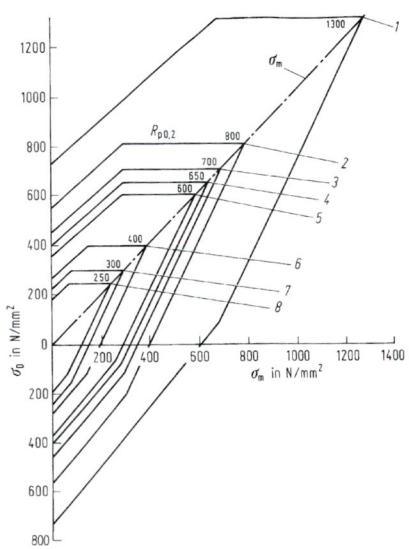

Figure 3. Fatigue strength pattern (Smith curve) for tensile-compressive loading [58]. Annealing steels to DIN 17 200: *1* 30 CrMoV 9, 30 CrNiMo 8; *2* 42 MnV 7, 42 CrMo 4; 42 CrV 6, 36 CrNiMo 4; 50 CrMo 4, 50 CrV 4; 34 CrNiMo 6; *3* 37 MnSi 5, 34 Cr 4; 36 Cr 6, 41 Cr 4; 34 CrMo 4; *4* 40 Mn 4, 27 MnCr 4, 30 Mn 5, 27 MnCrV 4, 40 Mn 4; *5* C 60, Ck 60; *6* C 45, Ck 45; *7* C 35, Ck 35; *8* C 22, Ck 22.

Figure 4. Fatigue strength pattern (Smith curve) for tensile-compressive loading [58]. Precipitation hardening steels to DIN 17 210: *1* 41 Cr 4; *2* 18 CrNi 8; *3* 20 MnCr 5; *4* 15 CrNi 6; *5* 16 MnCr 5; *6* 15 Cr 3; *7* C 15, Ck 15; *8* C 10, Ck 10.

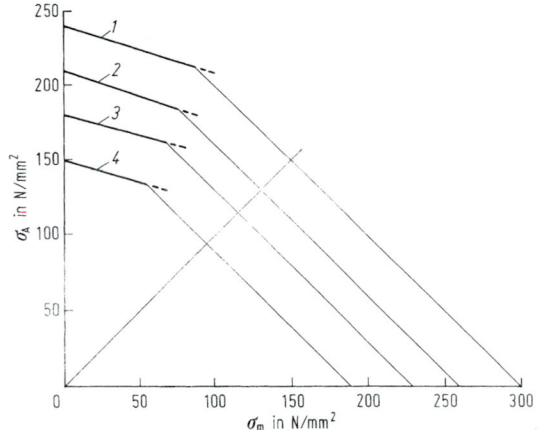

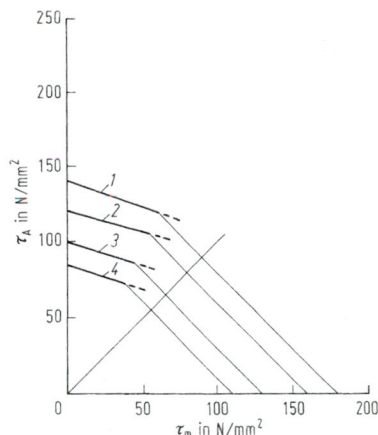

Figure 5. Fatigue strength pattern (Haigh curve) for tensile-compressive loading [58]. Cast steel to DIN 1681: *1* GS 60; *2* GS 52; *3* GS 45; *4* GS 38.

Figure 6. Fatigue strength pattern (Haigh curve) for torsional loading [58]. Cast steel to DIN 1681: *1* GS 60; *2* GS 52; *3* GS 45; *4* GS 38.

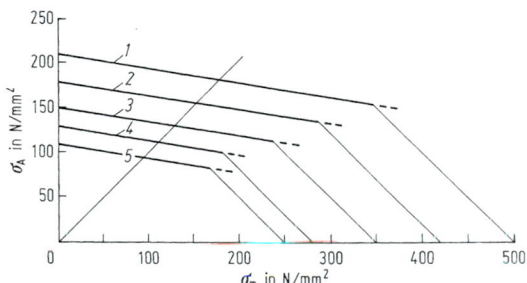

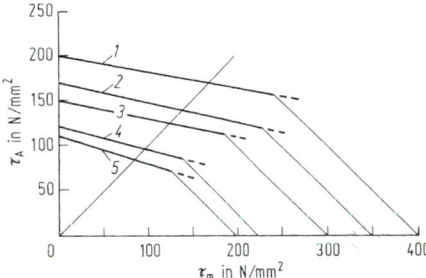

Figure 7. Fatigue strength pattern (Haigh curve) for tensile-compressive loading [58]. Cast iron with spheroidal graphite to DIN 1693: *1* GGG 70; *2* GGG 60; *3* GGG 50; *4* GGG 42; *5* GGG 38.

Figure 8. Fatigue strength pattern (Haigh curve) for torsional loading [58]. Cast iron with spheroidal graphite to DIN 1693: *1* GGG 70; *2* GGG 60; *3* GGG 50; *4* GGG 42; *5* GGG 38.

Appendix D1, Table 1. Statically determined modulus of elasticity and transverse contraction coefficient for various materials

Materials	Modulus of elasticity E (10^3 N/mm²)				Transverse contraction coefficient μ at 20 °C
	20 °C	200 °C	400 °C	600 °C	
Ferritic steels	211	196	177	127	approx. 0.3
Steels with approx. 12% Cr	216	200	179	127	approx. 0.3
Austenitic steels	196	186	174	157	approx. 0.3
Ni Cr 20 Ti Al	216	208	196	179	
Cast iron GG-20	90 to 115				0.25 to 0.26
GG-30	110 to 140				0.24 to 0.26
GG-40	125 to 155				0.24 to 0.26
GGG-38 to GGG-72	170 to 185	168 to 180	135 to 145		0.28 to 0.29
Aluminium alloys	60 to 80	54 to 72			approx. 0.33
Titanium alloys	112 to 130	99 to 113	88 to 93	77 to 80	0.32 to 0.38

Appendix D1, Table 2. Survey of material characteristics at ambient and elevated temperatures

Temperature T (°C)	Stress type	Material characteristic K				
		Yield			Fracture	
		Symbols	Description		Symbols	Description
Ambient temperature	Tensile[a]	$R_{p\,0.2}(\sigma_{0.2})$ $R_e(\sigma_s)$	0.2% Proof stress Yield point		$R_m(\sigma_b)$	Tensile strength
	Compression	σ_{dF}	Compressive yield point		σ_{dB}	Compressive strength
	Flexure	$\sigma_{bF}\sigma_{b\,0.2}$	Flexural yield point		σ_{bB}	Flexural strength
	Torsion	$\tau_F, \tau_{0.4}$	Torsional yield point		τ_B	Torsional strength
Elevated temperature	Tensile[a] $T < T_K$[b]	$R_{p\,0.2/T}(\sigma_{0.2/T})$	High-temperature yield point		$R_{m/T}(\sigma_{B/T})$	High-temperature strength
	Tensile[a] $T > T_K$[b]	$R_{p\,0.2/t/T}(\sigma_{0.2/t/T})$	Creep strain limit		$R_{m/t/T}(\sigma_{B/t/T})$	Creep strength

[a] The designations set out in these columns comply with the recommendations of the International Organisation for Standardisation (ISO) and the European standard published by the EGKS (European Coal and Steel Association). The previously designations figures are set out in brackets.
[b] T_K = re-crystallisation temperature.

Appendix D1, Table 3. Strength characteristics of annealing steels under static and oscillating load for the tensile, flexural and torsional stress types [23]

Material	Static material characteristics in N/mm²				Oscillatory strength characteristics in N/mm² (for 90% survival probability)					
	Tensile		Flexural	Torsional	Reciprocating strength characteristics			Dynamic strength characteristics		
	R_m	$R_{p\,0.2}$	σ_{bF}	τ_F	σ_{zdW}	σ_{bW}	τ_W	σ_{zSch}	σ_{bSch}	τ_{Sch}
Ck 35	650	420	540	270	250	310	180	400	470	270
Ck 45	750	480	620	310	300	370	210	480	550	310
40 Mn 4	900	650	720	410	350	430	260	560	660	410
34 Cr 4 34 CrMo 4	1 000	800	890	450	380	490	280	620	750	450
42 CrMo 4 36 CrNiMo 4	1 100	900	980	500	420	520	310	680	800	500
50 CrV 4	1 200	1 000	1 080	520	440	550	340	720	850	520

Appendix D1, Table 4. Equation for calculation of form numbers on symmetrical notch bars

	Flat bar				Round bar				
	Notched		Stepped		Notched			Stepped		

	z	b	z	b	z	b	t	z	b	t
A	0.10	0.08	0.55	0.40	0.10	0.12	0.40	0.44	0.40	0.40
B	0.7	2.2	1.1	3.8	1.6	4.0	15.0	2.0	6.0	25.0
C	0.13	0.20	0.20	0.20	0.11	0.10	0.10	0.30	0.80	0.20
k	1.00	0.66	0.80	0.66	0.55	0.45	0.35	0.60	0.40	0.45
l	2.00	2.25	2.20	2.25	2.50	2.66	2.75	2.20	2.75	2.25
m	1.25	1.33	1.33	1.33	1.50	1.20	1.50	1.60	1.50	2.00

z = tensile; b = flexural; t = torsion.

$$\alpha_k = 1 + \cfrac{1}{\sqrt{\cfrac{A}{\left(\frac{t}{\rho}\right)^k} + B\,\cfrac{\left[1 + \frac{a}{\rho}\right]^l}{\frac{a}{\rho}\sqrt{\frac{a}{\rho}}} + C\,\cfrac{\frac{a}{\rho}}{\left(\frac{a}{\rho}+\frac{t}{\rho}\right)\left(\frac{t}{\rho}\right)^m}}}$$

Appendix D3

Appendix D3, Table 1. Selection of constructional steels to DIN 17 100

Steel type		Tensile strength R_m for product thicknesses mm		Upper yield point R_{eH} for product thicknesses mm		Testpiece position	Elongation at fracture A_5 for product thicknesses mm		Notch impact work A_v[b] ISO V-notch testpieces (longitudinal) for product thicknesses mm			
Abbreviation	Material no.	< 3 N/mm²	≥ 3 ≤ 100	≤ 16 N/mm² min.	> 16 ≤ 40		≥ 3 ≤ 40 % min.	> 40 ≤ 63	Treatment state[a]	Test temperature °C	≥ 10 ≤ 16 J min.	> 16 ≤ 63
St 33	1.0035	310 to 340	290	185	175[c]	Longitudinal	18	—	U, N	—	—	—
						Transverse	16					
St 37-2	1.0037								U, N	+20	27	—
USt 37-2	1.0036							25	U, N	+20	27	—
RSt 37-2	1.0038	360 to 510	340 to 470	235	225	Longitudinal	26	23	U, N	+20	27	27
						Transverse	24					
St 37-3	1.0116								U	±0	27	27
									N	−20	27	27
St 44-2	1.0044								U, N	+20	27	27
		430 to 580	410 to 540	275	265	Longitudinal	22	21				
						Transverse	20	19				
St 44-3	1.0144								U	±0	27	27
									N	−20	27	27
St 52-3	1.0570	510 to 680	490 to 630	355	345	Longitudinal	22	21	U	±0	27	27
						Transverse	20	19	N	−20	27	27
St 50-2	1.0050	490 to 660	470 to 610	295	285	Longitudinal	20	19	U, N	—	—	—
						Transverse	18	17				
St 60-2	1.0060	590 to 770	570 to 710	335	325	Longitudinal	16	15	U, N	—	—	—
						Transverse	14	13				
St 70-2	1.0070	690 to 900	670 to 830	365	355	Longitudinal	11	10	U, N	—	—	—
						Transverse	10	9				

[a] U: formed, untreated; N: normalised.
[b] Means of 3 ISO V-notch testpieces.
[c] This value valid only for thicknesses up to 25 mm.

Appendix D3, Table 2. Guaranteed strength values of some annealing steels to DIN 17 200 in the annealed state

Steel type		Up to 16 mm diameter		Over 16 and up to 40 mm diameter		Over 40 and up to 100 mm diameter		Over 100 and up to 160 mm diameter		Over 160 and up to 250 mm diameter	
Abbreviation	Material number	$R_{p\,0.2\,min}$ N/mm²	R_m N/mm²	$R_{p\,0.2\,min}$ N/mm²	R_m N/mm²	$R_{p\,.2\,min.}$ N/mm²	R_m N/mm²	$R_{p\,0.2\,min.}$ N/mm²	R_m N/mm²	$R_{p\,0.2\,min.}$ N/mm²	R_m N/mm²
C 35[a]	1.0501	430	630/ 780	370	600/ 750	320	550/ 700	—	—	—	—
C 45[a]	1.0503	500	700/ 850	430	650/ 800	370	630/ 780	—	—	—	—
C 60[a]	1.0601	580	850/1 000	520	800/ 950	450	750/ 900	—	—	—	—
Ck 35[b]	1.1181	430	630/ 780	370	600/ 750	320	550/ 700	—	—	—	—
Ck 45[b]	1.1191	500	700/ 850	430	650/ 800	370	630/ 780	—	—	—	—
Ck 60[b]	1.1221	580	850/1 000	520	800/ 950	450	750/ 900	—	—	—	—
28 Mn 6[b]	1.1170	590	780/ 930	490	690/ 840	440	640/ 790	—	—	—	—
34 Cr 4[b]	1.7033	700	900/1 100	590	800/ 950	460	700/ 850	—	—	—	—
41 Cr 4[b]	1.7035	800	1 000/1 200	660	900/1 100	560	800/ 950	—	—	—	—
34 CrMo 4[b]	1.7220	800	1 000/1 200	650	900/1 100	550	800/ 950	500	750/ 900	450	700/ 850
42 CrMo 4[b]	1.7225	900	1 100/1 300	750	1 000/1 200	650	900/1 100	550	800/ 950	500	750/ 900
34 CrNiMo 6[b]	1.6582	1 000	1 200/1 400	900	1 100/1 300	800	1 000/1 200	700	900/1 100	600	800/ 950
30 CrNiMo 8[b]	1.6580	1 050	1 250/1 450	1 050	1 250/1 450	900	1 100/1 300	800	1 000/1 200	700	900/1 100
50 CrV 4[b]	1.8159	900	1 100/1 300	800	1 000/1 200	700	900/1 100	650	850/1 000	600	800/ 950

[a] Quality steel. [b] Stainless steel.

Appendix D3, Table 3. Nitriding steels to DIN 17 211

Steel type		Mechanical characteristics in annealed state		
Abbreviation	Material number	Diameter mm	$R_{p\,0.2\,min.}$ N/mm²	R_m N/mm²
31 CrMo 12	1.8515	≤ 100	800	1 000 to 1 200
		> 100 ≤ 250	700	900 to 1 100
31 CrMoV 9	1.8519	≤ 100	800	1 000 to 1 200
		> 100 ≤ 250	700	900 to 1 100
15 CrMoV 59	1.8521	≤ 100	750	900 to 1 100
		> 100 ≤ 250	700	850 to 1 050
34 CrAlMo 5	1.8507	≤ 70	600	800 to 1 000
34 CrAlNi 7	1.8550	≤ 100	650	850 to 1 050
		> 100 ≤ 250	600	800 to 1 000

Appendix D3, Table 4. Case-hardening steels to DIN 17 210

Steel type		Mechanical characteristics on blank-hardened cross-sections		
Abbreviation	Material number	Diameter mm	$R_{e\,min.}$ N/mm²	R_m N/mm²
C 10[a]	1.0301	11	390	640/ 780
		30	295	490/ 640
Ck 15[b]	1.1141	11	440	740/ 880
		30	355	590/ 780
17 Cr 3[b]	1.7016	11	510	780/1 030
		30	440	690/ 880
16 MnCr 5[b]	1.7131	11	635	880/1 180
		30	590	780/1 080
		63	440	640/ 930
20 MnCr 5	1.7147	11	735	1 080/1 360
		30	685	980/1 280
15 CrNi 6[b]	1.5919	11	685	960/1 270
		30	635	880/1 180
		63	540	780/1 080
17 CrNiMo 6[b]	1.6587	11	835	1 180/1 420
		30	785	1 080/1 320
		63	685	980/1 270
Cm 15[c]	1.1140	11	440	740/ 880
		30	355	590/ 780
20 MoCrS 4[c]	1.7323	11	635	880/1 180
		30	590	780/1 080

[a] Quality steel. [b] Stainless steel. [c] Stainless steel with guaranteed range of sulphur content.

Appendix D3, Table 5. Selection of stainless steels to DIN 17 440

Steel type		Heat treatment state	R_m N/mm²	Resistance to intercrystalline corrosion[b]	
Abbreviation	Material number			In the delivery state	In the welded state[a]
Ferritic and martensitic steels					
X 20 Cr 13	1.4021	Tempered	750/950	ng	ng
X 45 CrMoV 15	1.4116	Annealed	≤ 900	ng	ns
X 6 CrTi 17	1.4510	Annealed	450/600	g	g
X 12 CrMoS 17	1.4104	Tempered	640/840	ng	ng
Austenitic steels					
X 5 CrNi 18 10	1.4301	Quenched	500/700	g	g
X 6 CrNiTi 18 10	1.4541	Quenched	500/730	g	g
X 6 CrNiMoTi 17 12 2	1.4571	Quenched	500/730	g	g
X 2 CrNiMoN 17 12 2	1.4406	Quenched	580/800	g	g

[a] Without heat treatment.
[b] g = guaranteed; ng = not guaranteed; ns = not weldable.

Appendix D3, Table 8. Selection of heat-resistant steels to SEW 470[a]

Steel type	Abbreviation	Material number	Mechanical characteristics						Maximum application temperature in air °C
			$R_{p\,0.2\,min.}$ N/mm²	R_m N/mm²	$A_{5\,min.}$ %	$R_{m/10}^2$ N/mm²			
						600 °C	800 °C	900 °C	
Ferritic steels	X 10 CrAl 7	1.4713	220	420/620	20	20	2.3	1.0	800
	X 10 Cr 13	1.4724	250	450/650	15				850
	X 10 CrAlSi 24	1.4762	280	520/720	10				1 150
[b]	X 15 CrNiSi 25 4	1.4821	400	600/850	16	20	2.3	1.0	1 100
Austenitic steels	X 10 CrNiTi 18 10	1.4878	210	500/700	35	65	10		850
	X 15 CrNiSi 25 20	1.4841	230	550/750	30	80	7	3	1 150
	X 12 CrNiSi 35 16	1.4864	230	550/750	30	75	7	3	1 100

[a] In Euronorm 95: Heat-resistant steels, in identical or similar composition.
[b] Ferritic–austenitic steel.

Appendix D3, Table 9. Examples of cryogenic steels to DIN 17 280 and DIN 17 440

Steel type	Abbreviation	Material number	Mechanical characteristics					
			$R_{p\,0.2}$ N/mm²	R_m N/mm²	A_5 %	Minimum notch impact work (J)		
						−80 °C	−120 °C	−195 °C
Ni-alloyed	10 Ni 14	1.5637	355	470/640	20	35	27	27
	12 Ni 19	1.5680	390	510/710	19	40	35	
	X 8 Ni 9	1.5662	490	640/840	18	50		
Austenitic CrNi steel	X 5 CrNi 18 10	1.4301	195	500/700	45	a	a	a
	X 6 CrNiTi 18 10	1.4541	200	500/730	40	a	a	a
	X 6 CrNiNb 18 10	1.4550	205	510/740	40	a	a	a

[a] DIN 17 440 does not set out any values for minimum notch impact work for austenitic steels at low temperatures. However, it does indicate excellent low-temperature toughness properties down to less than −200 °C.

Appendix D3, Table 6. Selection of heat-resistant and highly heat-resistant steels

Application	Abbreviation	Material number	Mechanical characteristics $R_{p\,0.2\,min}$ N/mm²	R_m N/mm²	$A_{5\,min}$ %	$R_{m/10^5}$ N/mm²	400 °C	450 °C	500 °C	550 °C	600 °C	650 °C	700 °C	800 °C	900 °C
Boiler plates to DIN 17 155	HI(< 0.16% C, 0.8% Mn)	1.0345	225[a]	350/480	23		132	69	41						
	17 Mn 4[b,c]	1.0844	285[a]	440/580	20		179	85	93						
	15 Mo 3[b,c]	1.5415	275[a]	420/590	19			245							
	13 CrMo 4 4[b,c]	1.7335	295[a]	420/590	19			285	137	49					
Seamless tubes and manifolds to DIN 17 175	St 35.8 (< 0.17% C, 0.6% Mn)	1.0305	225[a]	360/480	25		132	69							
	10 CrMo 9 10[c]	1.7380	280[a]	450/600	20			221	125	68	34				
	14 MoV 63[c]	1.7715	320[a]	460/610	20				170	85					
	X 20 CrMoV 12 1[c]	1.4922	490[a]	690/840	17				235	128	59	23			
Bolts and nuts to DIN 17 240[f]	Ck 35 (0.65% Mn)	1.1181	280	500/650	22		138	69	34						
	24 CrMo 5	1.7258	440	600/750	18			226	36						
	21 CrMoV 57	1.7709	550	700/850	16			328	188	95					
	X 22 CrMoV 12 1[d]	1.4923	600	800/950	14			432	275	137	59				
	X 8 CrNiMoBNb 16 16[d,e]	1.4986	500	650/850	16						275	157	75		
	NiCr 20 TiAl (Nimonic 80 A)	2.4952	600	≥ 1 000	12					416	272	157			
Tubes, sheets, strips, bar steel, forgings to DIN 17 459 and DIN 17 460	X6 CrNi 18 11	1.4948	185	500/700	40				192	140	89	52	28	25	
	X 6 CrNiMo 17 13	1.4919	205	490/690	35					175	120	69	34		
	X 8 CrNiNb 16 13	1.4961	205	510/690	35						108	64	34		
	X 8 CrNiMoVNb 16 13	1.4988	255	540/740	30						172	98	46		
	X 8 NiCrAlTi 32 21 (Alloy 800)	1.4959	170	500/750											9.9

[a] Sheets or wall thickness > 16 ≤ 40 mm.
[b] Also in DIN 17 175: Seamless tubes and manifolds of heat-resistant steels.
[c] Also in DIN 17 243: Forgings and bar steel of heat-resistant steels suitable for welding.
[d] Also in SEW 670: Highly heat-resistant steels.
[e] Hot work-hardened and aged.
[f] Relaxation characteristics: residual load after 3·10⁴ h at an initial elongation of 0.2%.

Temperature (°C)	400	450	500	550	600
Material 21 CrMoV 5 7	250	154	56		
X 22 CrMoV 12 1	216	155	85	38	
NiCr 20 TiAl		342	310	237	149

Appendix D3, Table 7. Selection of highly heat-resistant Fe, Co and Ni alloys

Alloy type	Abbreviation[a]	Material number	C	Co	Cr	Fe	Ni	Mo	Ti	Al	W	V	Other	$R_{p0.2}$ N/mm²	R_m N/mm²	A_5 %	$R_{m/10}$[f] N/mm²
			% weight											Guideline values for mechanical characteristics[b]			
Fe basic alloys	X 5 NiCrTi 26 15 (A-286)	1.4980	0.05		15.0	Remainder	26.0	1.3	2.15	0.20		0.3		690	1 000	25	732 °C: 145
	X 40 CoCrNi 20 20 (S-590)[c]	1.4977	0.43	20	21	Remainder	20	4.0			4.0		Nb: 4.0				600 °C: 303[f] 700 °C: 156[f] 800 °C: 72[f]
Co basic alloy	S-816		0.40	Remainder	20.0	3.0	20.0	4.0			4.0		Nb: 4.0	485	970	35	815 °C: 145
Ni basic alloys	NiCr 20 TiAl[d] (Nimonic 80 A)	2.4952	0.06	1.0	20.5	<5.0	Remainder		2.05	1.4				735	1 180		700 °C: 333 800 °C: 118 900 °C: 26
	NiCo 20 Cr 15 MoAlTi (similar to Nimonic 115)	2.4634	0.15	20.0	15.0	<1.0		5.0	1.2	4.5				785	1 180	16	700 °C: 500 800 °C: 250 900 °C: 93
	NiCr 21 Fe 18 Mo (Hastalloy X)	2.4603	0.10	1.5	22.0	18.5	Remainder	9.0			0.6			360	790	43	760 °C: 128 871 °C: 34 982 °C: 16
	NiCr 23 Co 12 Mo[e] (Alloy 617)	2.4663	0.08	12.0	22.0	2.0	Remainder	9.0	0.35	1.0				300	680/ 950	30	800 °C: 65[f] 900 °C: 30[f] 1000 °C: 10[f]
	Inconel 718	2.4668	0.08	1.0	19.0	17.0	Remainder	3.1	0.9	0.5			Nb + Ta: 4.9	1 185	1 430	21	649 °C: 593 760 °C: 172
	Udimet 520		0.05	12.0	19.0		Remainder	6.0	3.0	2.0	1.0			860	1 310	21	649 °C: 385 760 °C: 345 871 °C: 150
Co basic alloy	X-40		0.50	Remainder	25.5	2.0	10.5				7.5			525	750	9	815 °C: 352
	FSX 414		0.25	Remainder	29.0	1.0	10				7.5			440	740	11	816 °C: 118
Ni basic alloys	MAR-M-247		0.1	10.0	8.2	1.5	Remainder	0.6	1.0	5.5	10.0		Ta: 3 Hf: 1.5 B, Zr	815	905	6	871 °C: 290 927 °C: 195 982 °C: 125
	IN-100		0.18	15.0	9.5	1.0	Remainder	3.0	4.8	5.5		0.95	B, Zr	850	1 010	9	760 °C: 517 871 °C: 255 982 °C: 103
	IN 713 C		0.12		12.5		Remainder	4.2	0.8	6.1			Nb: 2.0 Zr: 0.1	740	850	7.9	760 °C: 448 871 °C: 193 982 °C: 90
	IN 738 LC		0.05	8.5	16.0	0.5	Remainder	1.8	3.5	3.5	2.6		Ta: 1.8 Nb, Zr	950	1 100	5.5	760 °C: 475 871 °C: 215 982 °C: 83

[a] Some common commercial designations.
[b] Consult material manufacturers for further information.
[c] In accordance with report FVHT/FVV No. 2-87, Düsseldorf 1987, "Results of common long-term fatigue tests".
[d] Also in DIN 17 240: Heat-resistant and highly heat-resistant materials for bolts and nuts.
[e] In accordance with VdTÜV material data sheet 485 ($R_{p0.2}$ and A_5: minimum values).
[f] $R_{m/10}$.

Appendix D3, Table 10. Important tool steels and their most common application

Steel types		Hardness values		Common applications
Abbreviation	Material number	Soft-annealed HB max.	Working hardness HRC	
a. Steels for punching and cutting (separating)				
X 210 CrW 12	1.2436	255	58 to 64	Cutting tools ≤ 3 mm
X 155 CrVMo 12 1	1.2379	255	58 to 64	Cutting tools ≤ 6 mm
			58 to 60	Fine cutting tools > 12 mm
X 100 CrMoV 5 1	1.2363	240	58 to 64 ⎫	Cutting tools ≤ 12 mm
90 MnCrV 8	1.2842	229	55 to 60 ⎭	
60 WCrV 7	1.2550	229	55 to 60 ⎫	Cutting tools > 12 mm
45 WCrV 7	1.2542	225	48 to 50 ⎭	
X 220 CrVMo 13 4	1.2380	260	60 to 65	Billet shears as for material no. 1.2379, but with greater wear resistance
b. Steels for machining (separating)				
S 6-5-2	1.3343	300	64	Twist drills, saws
S 6-5-2-5	1.3243	300	65	Milling machines, thread taps
S 6-5-3	1.3344	300	65	Threat taps, countersinks, reamers
S 10-4-3-10	1.3207	300	66	Lathe cutters
S 12-1-4-5	1.3202	300	66	Form millers
80 CrV 2	1.2235	248	42 to 48	Wood saws
C 125 W	1.1663	213	63 to 66	Files
c. Steels for hot forming (forging)				
55 NiCrMoV 6	1.2713	248	32 to 40	Hammer dyes for steel, full dyes
56 NiCrMoV 7	1.2714	248	32 to 40	As for material no. 1.2713
145 V 33	1.2838	230	45	Hammer dyes for flat impressions
X 48 CrMoV 8 11	1.2360	250	45	High creep-resistance, wear-resistant press dyes
X 38 CrMoV 5 1	1.2343	229	42 to 46	Press tools
d. Steels for hot forming (extruding)				
X 40 CrMoV 51	1.2344	229	42 to 52	Extrusion press matrices, recipients, press stamps, press mandrels, press tools
X 45 CoCrW V 5 5 5	1.2678	250	44 to 48	Press matrices, press tools
X 20 CoCrW Mo 10 9	1.2888	320	50 to 55	Press matrices, press mandrels
NiCr 19 CoMo	2.4973		a	Matrices, mandrels for pressing of heavy metal
X 50 NiCrW V 13 13	1.2731		35 to 40	Matrices for pressing of heavy metal
e. Steels for cold forming (stamping, flow pressing and rolling)				
60 WCrV 7	1.2550	229	58 to 62	Stamping tools
75 NiCrMo 5 3 3	1.2773	240	60 to 64	Stamping tools
S 6-5-2	1.3343	300	62 to 65	Flow press stamps, pressing bushes, threaded rollers, cold rollers
X 155 CrVMo 12 1	1.2379	255	59 to 62	Drawing stamps, drawing rings, threaded rollers, work rollers in multi-roller apparatus
85 CrMo 7	1.2304	230	61 to 65	Cold rolling
f. Steels for hot forming (diecast forming)				
X 38 CrMoV 5 1	1.2343	229	42 to 46	Diecast forming
X 40 CrMoV 5 1	1.2344	229	42 to 46	
X 40 CrMoV 5 3	1.2367	229	40 to 46	Diecast forming for heavy metals
X 20 CoCrWMo 10 9	1.2888	320	50 to 55	Cores, mould inserts
X 3 NiCoMoTi 18 9 5	1.2709	330[b]	50 to 53	Cores, moulded components
g. Steels for hot forming (glass working)		HB		
X 21 Cr 13	1.2082	220	250 to 300	Glass forming for low-melt glasses, glass rollers
X 23 CrNi 17	1.4057	275	220 to 300	As X 21 Cr 13 but large production runs
X 16 CrNiSi 25 20	1.4841	223	ca. 200	Glass forms for high-melt glasses and high production runs, glass rollers
h. Steels for cold forming (plastic working)				
X 6 CrMo 4	1.2341	108	62	Hobbing moulds, multiple forms
21 MnCr 5	1.2162	212	62	Medium-sized moulds, case-hardened
X 19 NiCrMo 4	1.2764	255	62	Large moulds, case-hardened
X 45 NiCrMo 4	1.2767	262	48 to 56	Large moulds, readily hardenable
40 CrMnMo 7	1.2311	230	32	Large moulds up to 400 mm thick, pre-tempered
40 CrMnNiMo 8 6 4	1.2738	240	32	Very large moulds above 400 mm thick, pre-tempered
40 CrMnMoS 8 6	1.2312	230	32	Large moulds, pre-tempered, readily machinable
X 42 Cr 13	1.2083	225	52	Moulds for processing of aggressive plastics
X 36 CrMo 17	1.2316	285	27 to 31	Moulds for processing of aggressive plastics
X 20 Cr 13	1.2082	230	30	For corrosion-resistant moulds, relatively viscous

[a] Tensile strength in solution-annealed and hardened state 1300 N/mm².

[b] In solution-annealed state.

Appendix D3, Table 11. Suitable standard materials for bolts of strength categories in accordance with DIN ISO 898 T 1[a], production methods and dimensions

Strength category	Production method			Bolt dimensions	Heat treatment to		
	Cold forming	Hot forming	Machining		Cold forming	Hot forming	Machining
3.6 4.6	QSt 36-2 ≙ 1.0203 QSt 36-2 ≙ 1.0204 QSt 38-2 ≙ 1.0217 QSt 38-2 ≙ 1.0224	RSt 37-2 ≙ 1.0038 RSt 44-2 ≙ 1.0419	9 S 20 ≙ 1.0711	Up to M 39	Annealing	Annealing	None
4.8	QSt 36-2 ≙ 1.0203 QSt 38-2 ≙ 1.0204		9 S 20 ≙ 1.0711	Common up to M 16	None		None
5.6	Cq22 ≙ 1.1152	St 50-2 ≙ 1.0533		Up to M 39	Annealing	Annealing	
5.8	Cq 22 ≙ 1.1152 Cq 35 ≙ 1.1172		9 SMn 28 ≙ 1.0715 10 S 20 ≙ 1.0721	Up to M 39	None		None or tempering
6.8	Cq 35 ≙ 1.1172 35 B 2 ≙ 1.5511 Cq 45 ≙ 1.1192	C 45 ≙ 1.0503 46 Cr 2 ≙ 1.7006	10 S 20 ≙ 1.0721	Up to M 39	None or tempering	Tempering	None or tempering
8.8	22 B 2 ≙ 1.5508 28 B 2 ≙ 1.5510	22 B 2 ≙ 1.5508 28 B 2 ≙ 1.5510	Not common	Up to M 12			
	19 MnB 4 ≙ 1.5523 35 B 2 ≙ 1.5511 Cq 35 ≙ 1.1172 Cq 45 ≙ 1.1192	C 45 ≙ 1.0503 46 Cr 1 ≙ 1.7002		Up to M 22			
	34 Cr 4 ≙ 1.7033 37 Cr 4 ≙ 1.7034	46 Cr 2 ≙ 1.7006		From M 24 Up to M 39			
10.9	19 MnB 4 ≙ 1.5523 35 B 2 ≙ 1.5511 Cq 35 ≙ 1.1172	19 MnB 4 ≙ 1.5523 35 B 2 ≙ 1.5511 Cq 35 ≙ 1.1172		Up to M 8			
	34 Cr 4 ≙ 1.7033	41 Cr 4 ≙ 1.7035	Quite uncommon	From M 8 Up to M 18		Tempering	
	41 Cr 4 ≙ 1.7035 34 CrMo 4 ≙ 1.7220 42 CrMo 4 ≙ 1.7225	41 Cr 4 ≙ 1.7035 34 CrMo 4 ≙ 1.7220 42 CrMo 4 ≙ 1.7225		Up to M 39			
12.9	34 CrMo 4 ≙ 1.7220 37 Cr 4 ≙ 1.7034 41 Cr 4 ≙ 1.7035	34 CrMo 4 ≙ 1.7220 37 Cr 4 ≙ 1.7034 41 Cr 4 ≙ 1.7035		Up to M 18			
	42 CrMo 4 ≙ 1.7225	42 CrMo 4 ≙ 1.7225		Up to M 24			
	30 CrNiMo 8 ≙ 1.6580 34 CrNiMo 6 ≙ 1.6582	30 CrNiMo 8 ≙ 1.6580 34 CrNiMo 6 ≙ 1.6582		Up to M 39			

[a] DIN ISO 898 T1 applies only to bolts of nominal diameter up to M39. Bolts of larger diameter can be manufactured from materials envisaged for utilisation up to M 39, and their mechanical characteristics must comply with the requirements of ISO 898 T1.

Appendix D3, Table 12. Materials for springs

Material type	Abbreviation		Material number	Tensile strength R_m N/mm²	E/G modulus kN/mm²
Hot-rolled steels for temperable springs DIN 17 221	Special steels	38 SiCr 7	1.5023		$E \approx 206$
		54 SiCr 6	1.7102		$G \approx 80$
		60 SiCr 7	1.7108		
		55 Cr 3	1.7176		
		50 CrV 4	1.8159		
		51 CrMoV 4	1.7701		
Cold-rolled steel strip for DIN 17 222 springs				Cold rolled + hardened + aged	
	Quality steels	C 55	1.0535	1150 to 1650	E: 206
		C 60	1.0601	1180 to 1680	
		C 67	1.0603	1230 to 1770	
		C 75	1.0605	1320 to 1870	
		55 Si 7	1.0904	1300 to 1800	
	Special steels	Ck 55	1.1203	1150 to 1650	G: 78
		Ck 60	1.1221	1180 to 1680	(to DIN
		Ck 67	1.1231	1230 to 1770	17 222,
		Ck 75	1.1248	1320 to 1870	Section 8.5.6)
		Ck 85	1.1269	1400 to 1950	
		Ck 101	1.1274	1500 to 2100	
		71 Si 7	1.5029	1500 to 2200	
		67 SiCr 5	1.7103	1500 to 2200	
		50 CrV 4	1.8159	1400 to 2000	
Patented drawn spring wire of non-alloyed steels DIN 17 223, Section 1	A			1720 to 1970 at ⌀ 1 mm 1060 to 1230 at ⌀ 10 mm	E: 206
	B			1980 to 2220 at ⌀ 1 mm 1240 to 1400 at ⌀ 10 mm	G: 81.5 (to DIN
	C			1980 to 2200 at ⌀ 2 mm 1410 to 1570 at ⌀ 10 mm	17 223, Section
	D			2230 to 2470 at ⌀ 1 mm 1410 to 1570 at ⌀ 10 mm	1.5.4.3)
Oil-quenched spring steel wire of non-alloyed or alloy steel DIN 17 223, Section 2	VD			1850 to 2000 at ⌀ 0.5 mm	E: 206
	VD CrV			1910 to 2060 at ⌀ 0.5 mm	
	VD SiCr			2080 to 2230 at ⌀ 0.5 mm	
	FD			1900 to 2100 at ⌀ 0.5 mm	
	FD CrV			2000 to 2100 at ⌀ 0.5 mm	(to DIN
	FD SiCr			2100 to 2300 at ⌀ 0.5 mm	17 223)
Spring wire and spring strip of stainless steels DIN 17 224	X 12 CrNi 17 7		1.4310	1900 to 2150 up to ⌀ 1 mm 1250 to 1500 up to ⌀ 10 mm	E:[a] 185 (195) G: 70 (73)
	X 7 CrNiAl 17 7		1.4568	1800 to 2050 up to ⌀ 1 mm 1300 to 1550 up to ⌀ 6 mm	E: 195 (200) G: 73 (78)
	X 5 CrNiMo 17 12 2		1.4401	1500 to 1750 up to ⌀ 1 mm 1050 to 1300 up to ⌀ 8 mm	E: 180 (190) G: 68 (71)

[a] Values for roll-hardened state K (values for rolled and aged or hot-aged state K + A).

Specimen applications, ranges	Comments		
	Limiting sizes for core hardening		
	Flat products (thickness in mm)	Round section steel (diameter in mm)	
These spring steels are generally worked to produce tempered leaf, rotating bar, tapered, bolt and plate springs, circular springs and all other types of spring component	— 12 14 14 20 35	— $\varnothing$ 18 $\varnothing$ 22 $\varnothing$ 22 $\varnothing$ 30 $\varnothing$ 60	
Cold-rolled strip in thicknesses $\leq$ 5 mm and widths $\leq$ 600 mm, but also for other highly loaded components of all types	Strip thickness up to which tensile strength values are applicable[b] (mm) 2.0 2.0 2.5		Ideal hardening method: Oil quench (peripheral decarburisation to be avoided)
Cold-rolled steel strip for springs, excellent dimensional precision and good surface state, with the facility, in the cold-rolled + hardened + aged state (H + A) to attain high values of hardness, tensile strength and elastic limit	2.5 2.0		
	2.0 2.0 2.5 2.5 2.5 2.0 3.0 3.0 3.0		Springs for highest loadings should wherever possible have a polished surface
Tension, pressure, rotary and form springs with low static and infrequent dynamic loading	$d = (1.0 \text{ to } 10 \text{ mm})$		
Tension, pressure, rotary and form springs with medium static and low dynamic loading	$d = (0.3 \text{ to } 20 \text{ mm})$		
Tension, pressure, rotary and form springs with high static and low dynamic loading	$d = (2.0 \text{ to } 20 \text{ mm})$		
Tension and pressure springs with high static and medium dynamic loading and rotary and form springs with high static and high dynamic loading	$d = (0.07 \text{ to } 20 \text{ mm})$		
High dynamic torsion loading at ambient temperature Very high dynamic torsion loading up to 80 °C operating temperature Very high dynamic torsion loading up to 100 °C operating temperature	$d = (0.50 \text{ to } 10 \text{ mm})$		
G 79.5 Static loading Section 2.5.5.3)	$d = (0.50 \text{ to } 17 \text{ mm})$		
Drawn wire of spring hardness for pressure, tension, rotary (arm) and form springs Rolled strip of spring hardness for leaf, flat arm, plate springs and other form springs Form springs	Springs and sprung components of all types, which are exposed to the effects of corrosion by air, water vapour or other chemically-aggressive media Operating temperature should not exceed $\approx$ 250 °C for steels X 12 CrNi 17 7 and X 5 CrNiMo 17 12 2 and $\approx$ 350 °C for steel X 7 CrNi Al 17 7		

Appendix D3, Table 12. Continued.

Material type			Abbreviation	Material number	Tensile strength[a] R_m N/mm²	E/G modulus kN/mm²
Strip and sheet made of copper wrought alloys (choice of several hardnesses which are suitable for springs)	Work hardening	Brass	CuZn 37 F 61[c,f]	2.0321.34	Min. 610 for thicknesses from 0.2 to 2 mm	$E^{d,e}$ 110 in the aged state
		Bronze	CuSn 6 F63	2.1020.34	Min. 630 for thicknesses from 0.1 to 2 mm	$E^{d,e}$ 115 in the aged state
			Cu Sn 8 F66	2.1030.34	Min. 660 for thicknesses from 0.1 to 2 mm	$E^{d,e}$ 115 in the aged state
			CuSn 6 Zn 6 F 76	2.1080.34	Min. 760 for thicknesses from 0.1 to 2 mm	
		Argentan	CuNi 18 Zn 20 H 210	2.0740.34	Min. 680 for thicknesses from 0.1 to 2 mm	$E^{d,e}$ 140 in the aged state
				2.0742.34	Min. 700 for thicknesses from 0.1 to 2 mm	
			CuNi 18 Zn 27 H 220			
	Hardenable		CuBe 1.7 F 124	2.1245.76	Hardened 1240 to 1380 0.2 to 3 mm thick	E: 135
			CuBe 2 F 131	2.1247.76	Hardened 1310 to 1480, 0.2 to 3 mm thick	E: 130 to 135
			CuCo 2 Be F 85	2.1285.79	Hardened 850 to 1000, 0.2 to 3 mm thick	E: 138
Round-section spring wire made of copper wrought alloys, DIN 17 682	Work hardening	Brass	CuZn 36 F 70[f]	2.0335.39	750 to 930 at $\varnothing \geq 0.3$ to 0.8 mm 700 to 800 at $\varnothing$ 0.8 to 1.5 mm 650 to 770 at $\varnothing$ 1.5 to 3.0 mm According to agreement for diameter > 3.0 mm	E: $110^{d,e}$ G: $39^{d,e}$
		Bronze	CuSn 6 F 95[f]	2.1020.39	1050 to 1230 at $\varnothing$ 0.1 to 0.3 mm 1000 to 1180 at $\varnothing$ 0.3 to 0.8 mm 950 to 1100 at $\varnothing$ 0.8 to 1.5 mm 900 to 1020 at $\varnothing$ 1.5 to 3.0 mm According to agreement for diameter > 3.0 mm	E: $115^{d,e}$ G: $42^{d,e}$
		Argentan	CuNi 18 Zn 20 F 83[f]	2.0740.39	860 to 1040 at $\varnothing > 0.3$ to 0.8 mm 830 to 980 at $\varnothing$ 0.8 to 1.5 mm 800 to 920 at $\varnothing$ 1.5 to 3.0 mm According to agreement for diameter > 3.0 mm	E: $135^{d,e}$ G: $45^{d,e}$
	Hardenable		CuBe 2	2.1247	420 to 1550 according to material state	E: 120 to $135^{d,e}$ according to material state G: $47^{d,e}$
			CuCoBe	2.1285	250 to 1000 according to material state	E: 130 to $138^{d,e}$ according to material state G: $48^{d,e}$

[a] Note: The tensile strength ranges stated here provide purchasers with the basis for establishing an accurate tensile strength range generally of > 200 N/mm² for purposes of ordering. If it is necessary to adhere to tensile strength ranges less than 200 N/mm² in special cases, for example for supplies for springs which are to be bent, then this must be particularly agreed at the time of order. For a given tensile strength, the steel type should be selected primarily on the basis of thickness and the operating conditions of the springs.

Specimen applications, ranges	Comments

Leaf springs, sheet springs, strip springs; use other materials if any risk of stress crack corrosion

Strength characteristics of copper wrought alloys
Sheets and strips:	DIN 17 670, Section 1
Bars:	DIN 17 672, Section 1
Wires:	DIN 17 677, Section 1

All types of springs, especially for electrical industry: diaphragms, spring units, tubes

Composition:
Copper–zinc alloys (brass) (special brass): DIN 17 660
Copper–tin alloys (tin/bronze): DIN 17 662
Copper–nickel/zinc alloys (argentan): DIN 17 663

Leaf springs

All types of springs, diaphragms

All types of springs: especially where risk of stress crack corrosion.
Especially for current-carrying springs as DIN 43 801, Section 1

All types of springs: especially where risk of stress crack corrosion.
Especially for current-carrying springs as DIN 43 801, Section 1

Relay springs

All types of springs
Prestressing in helical tension springs is lost during hardening

---

[b] Note: At greater thicknesses, tensile strength values must be agreed at the time of order.
[c] Tempering CuZn 37 springs is recommended in order to reduce susceptibility to stress crack corrosion.
[d] Not applicable to acceptance.
[e] In general there is a ± 5% scatter in modulus of elasticity and modulus of slip.
[f] The F value in the abbreviation corresponds to one-tenth of the minimum tensile strength value for the diameter range of 0.8 to 1.5 mm.

Appendix D3, Table 13. Magnetic and technological characteristics of some cold- and hot-rolled magnetic steel sheets and strips to DIN 46 400

Steel type		Nominal thickness (mm)	Remagnetisation loss (W/kg) max. at		Magnetic induction T (tesla) min. at a field strength of A/m				Loss anisotropy % max.	Minimum stacking factor	Minimum bend number	Density kg/dm³
Abbreviation	Material number		P 1.0	P 1.5	2 500 (B 25)	5 000 (B 50)	10 000 (B 100)	(30 000)[a] (B 300)				
Cold-rolled												
V 110-35A	1.0899	0.35	1.1	2.7	1.49	1.60	1.71	(1.89)	± 14	0.95	2	7.60
V 135-50A	1.0897	0.50	1.35	3.3	1.49	1.60	1.71	(1.89)	± 14	0.97	3	7.60
V 230-50A	1.0893	0.50	2.3	5.3	1.54	1.64	1.75	(1.97)	± 12	0.97	10	7.70
Hot-rolled												
V 90-35 B	1.0883	0.35	0.9	2.3	1.47	1.59	1.70	(1.88)	± 8	0.93	2	7.55
V 110-50 B	1.0879	0.50	1.1	2.7	1.47	1.59	1.70	(1.88)	± 8	0.95	2	7.55
V 200-50 B	1.0874	0.50	2.0	4.8	1.49	1.61	1.72	(1.93)	± 6	0.95	10	7.65

[a] Satisfactory measurement of values for field strength 30 000 A/m (B 300) is possible only where special precautions are taken, owing to the heating up that occurs in the 25 cm Epstein frame.

Appendix D3, Table 14. Selection of various cast-steel materials

Type	Abbreviation		Material number	Mechanical characteristics		
				$R_{p\,0.2}$ N/mm²	R_m N/mm²	A_5 %
Cast steel for general application to DIN 1681	GS-38		1.0420	200	360	25
	GS-52		1.0552	260	520	18
	GS-60		1.0558	300	600	15
Tempering cast steel to SEW 510-62	GS-Ck 25		1.1155	275	490/635	20[a]
	GS-25 CrMo 4		1.7218	510	735/885	12[a]
	GS-42 CrMo 4	0.65% Mn	1.7225	665	885/1030	9[a]
	GS-40 NiCrMo 6 5 6		1.6748	685	885/1030	11[a]
Stainless cast steel to DIN 17 445	G-X 8 CrNi 13[b]		1.4008	440	590/780	15
	G-X 22 CrNi 17[b]	1.0% Mn	1.4059	590	780/980	4
	G-X 6 CrNi 18 9[c]		1.4308	175	440/640	20
	G-X 6 CrNiMo 18 10[c]	1.5% Mn	1.4408	185	440/640	20
	G-X 5 CrNiMoNb 18 10[c]		1.4581	185	440/640	20

[a] Strength category II, up to 30 mm wall thickness.
[b] Ferritic cast steel.
[c] Austenitic cast steel.

Appendix D3, Table 15. Selection of some heat-resistant cast-steel materials

Type	Abbreviation		Material number	$R_{p\,0.2}$ min. N/mm²	R_m N/mm² %	A_5	Mechanical characteristics $R_{m\,10}$ N/mm²					Maximum operating temperature in air °C
							400 °C	450 °C	500 °C	550 °C	600 °C	
Ferritic cast steel	GS-C25[a,c]		1.0619	245	440/590 22		160	83	40			500
	GS-22 Mo 4[a]		1.5419	245	440/590 22			196	92	29		530
	GS-17 CrMoV 5 11[a]	0.65% Mn	1.7706	440	590/780 15			275	171	96	28	540
	GS-18 CrMo 9 10[a]		1.7379	400	590/740 18			255	142	66	29	600
	G-X22 CrMoV 12 1[a,c]	0.6% Mn	1.4931	590	690/880 15			309	207	118	49	620
							600 °C	700 °C	800 °C	900 °C	1000 °C	
Ferritic cast steel	G-X 30 CrSi 6[b]	0.75% Mn	1.4710	No data in DIN 17 465			25[d]	10[d]	4[d]	1.5[d]		750
	G-X 40 CrSi 23[b]		1.4745	(expired 3.76)			25[d]	10[d]	4[d]	1.5[d]		1 050
Austenitic cast steel + Ni alloy	G-X 8 CrNiNb 19 10[c]		1.4827	175	440/640 20		100	35				800
	G-X 40 CrNiSi 25 20[b,c]	1.0% Mn	1.4848	220	440/640 8			50	25	11	4	1 100
	G-X 40 CrNiSi 35 25[b,c]		1.4857	220	440/640 8			55	29	13	4.3	1 150
	G-NiCr 50 Nb[c] 0.5% Mn		2.4813	270	540/740 8			50	25	10	3.5	1 050

[a] To DIN 17 245: Heat-resistant ferritic cast steel.
[b] To DIN 17 465: Heat-resistant cast steel.
[c] To SEW 595: Cast steel for petroleum and natural gas plant.

Appendix D3, Table 16. Chemical composition and mechanical characteristics of important sintering materials (taken from DIN V 30 910, Section 4)

Material	Abbreviation	Permitted ranges Density ρ g/cm³	Porosity $\frac{\Delta V}{V}\cdot 100$ %	Representative examples Density ρ g/cm³	C	Cu	Ni	Mo	Sn	Cr	Fe	Others	Tensile strength R_m N/mm²	Apparent limit of elasticity $R_{p\,0.1}$ N/mm²	Elongation at rupture A %	Hardness HB	Modulus of elasticity $E\cdot 10^3$ N/mm²
Sintering iron	C 00	6.4 to 6.8	15 ± 2.5	6.6	—	—	—	—	—	—	Remainder	< 0.5	130	60	4	40	100
	D 00	6.8 to 7.2	10 ± 2.5	6.9									190	90	10	50	130
	E 00	> 7.2	< 7.5	7.3									260	130	18	65	160
Sintering steel Cupriferous	C 10	6.4 to 6.8	15 ± 2.5	6.6	—	1.5	—	—	—	—	Remainder	< 0.5	230	160	3	55	100
	D 10	6.8 to 7.2	10 ± 2.5	6.9									300	210	6	85	130
	E 10	> 7.2	< 7.5	7.3									400	290	12	120	160
Containing Cu, Ni and Mo	C 30	6.4 to 6.8	15 ± 2.5	6.6	0.3	1.5	4.0	0.5	—	—	Remainder	< 0.5	390	310	2	105	100
	D 30	6.8 to 7.2	10 ± 2.5	6.9									510	370	3	130	130
	E 30	> 7.2	< 7.5	7.3									680	440	5	170	160
Stainless sintering steel AISI 316	C 40	6.4 to 6.8	15 ± 2.5	6.6	0.06	—	13	2.5	—	18	Remainder	< 0.5	330	250	1	110	100
	D 40	6.8 to 7.2	10 ± 2.5	6.9									400	320	2	135	130
Sintering bronze	C 50	7.2 to 7.7	15 ± 2.5	7.4	—	Remainder	—	—	10	—	—	< 0.5	150	90	4	40	50
	D 50	7.7 to 8.1	10 ± 2.5	7.9									220	120	6	55	70
Cupriferous sintering aluminium	D 73	2.45 to 2.55	10 ± 2.5	2.5	—	4.5	Mg 0.6	Si 0.7	—	—	Al remainder	< 0.5	160	130	1	50	50
	E 73	2.55 to 2.65	6 ± 1.5	2.6									210	150	2	60	60

Appendix D3, Table 17. Selection of various types of cast iron

Material	Abbreviation	Material number	$R_{p0.2}$ N/mm² min.	R_m N/mm² min.	A_3 % min.	Structure
Lamellar graphite cast iron to DIN 1691	GG-10	0.6010		100[a]		No data in DIN 1691
	GG-20	0.6020		200[a]		
	GG-35	0.6035		350[a]		
Spheroidal graphite cast iron to DIN 1693	GGG-40	0.7040	250[c]	390[c]	15[d]	Predominantly ferritic
	GGG-60	0.7060	360[c]	600[c]	2[d]	Pearlitic–ferritic
	GGG-70	0.7070	400[c]	700[c]	2[d]	Predominantly pearlitic
White malleable cast iron to DIN 1692	GTW-35-04	0.8035		350[b]	4	Greater ranges of fluctuation relative to GTW-40-05 are permitted
	GTW-45-07	0.8045	260[b]	450[b]	7	Core: (grainy) pearlite[e] + temper carbon
	GTW-40-05	0.8040	220[b]	400[b]	5	Core: (lamellar to grainy) pearlite + temper carbon
Black temper carbon to DIN 1692	GTS-35-10	0.8135	200[c]	350[c]	10	Ferrite + temper carbon
	GTS-45-06	0.8145	270[c]	450[c]	6	Pearlite + ferrite + temper carbon
	GTS-65-02	0.8165	430[c]	650[c]	2	Pearlite + temper carbon
	GTS-70-02	0.8170	530[c]	700[c]	2	Tempering structure + temper carbon
Austenitic lamellar graphite to DIN 1694	GGL-NiMn 13 7	0.6652		140/220	—	
	GGL-NiCr 20 3	0.6661		190/240	1 to 2	
	GGL-NiSiCr 30 5 5	0.6680		170/240	—	
Austenitic spheroidal graphite cast iron to DIN 1694	GGG-NiMn 13 7	0.7652	210	390	15[d]	
	GGG-NiCr 20 3	0.7661	210	390	7[d]	
	GGG-NiSiCr 30 5 5	0.7680	240	390	—	
	GGG-NiCr 35 3	0.7685	210	370	7[d]	

[a] Values relate to separately cast specimens with 30 mm blank casting diameter corresponding to 15 mm wall thickness. The anticipated tensile strength values for the casting are dependent on wall thickness.
Example GG-20:

Wall thickness (mm)	2.5 to 5	5 to 10	10 to 20	20 to 40	40 to 80	80 to 150
R_m (N/mm²)	230	205	180	155	130	115

[b] Applicable to specimen bar diameter of 12 mm.
Example of dependence of strength on specimen bar diameter in the case of GTW-45-07:

Specimen bar diameter (mm)	9	12	15
R_m (N/mm²)	400	450	480
$R_{p0.2}$ (N/mm²)	230	260	280

[c] Diameter of tensile testpiece: 12 or 15 mm. For castings with < 6 mm average wall thickness, it is possible to use tensile testpieces of a different cross-section appropriate to wall thickness.
[d] A_5.
[e] Preferably by atmospheric tempering.

Appendix D3, Table 18. Copper, semifinished products. Taken from DIN 1787

Abbreviation	Material number	Notes on characteristics and applications	Semifinished product types[c]				
			Tubes	Bars	Wires	Die forgings	Extrusion press sections
Oxidised copper							
E-Cu 58	2.0065	Oxidised (tough-pitch) copper with electrical conductivity in the soft state of at least 58.0 m/Ω·mm^2, but without requirements concerning welding and hard soldering capability			○		
E-Cu 57[d]	2.0060	Oxidised (tough-pitch) copper with electrical conductivity in the soft state of at least 57.0 m/Ω·mm^2, but without requirements for welding and hard soldering capability	○	○	○	×	○
Oxygen-free copper, not deoxidised							
OF-Cu	2.0040	High-purity copper, largely free of elements which evaporate in vacuum, with electrical conductivity in the soft state of at least 58.0 m/Ω·mm^2 Semifinished product with high requirements for hydrogen resistance[e], welding and hard soldering capability For vacuum technology, electronics	○	○	○		○
Oxygen-free copper, phosphorus-deoxidised							
SE-Cu[a]	2.0070[a]	Deoxidised copper with low residual phosphorus content[b] and high electrical conductivity High-electrical-conductivity semi-finished product with high requirements for formability, with good welding and hard soldering capability and hydrogen resistance For electronics, plating material	○	○	○		○
SW-Cu	2.0076	Deoxidised copper with limited, low residual phosphorus content Semifinished product without defined electrical conductivity (approx. 52.0 m/Ω·mm^2), but with good welding and hard soldering capability and hydrogen resistance[e] For apparatus engineering, building	×	×	×		×
SF-Cu	2.0090	Deoxidised copper with limited, high-residual phosphorus content Semifinished product without requirements as to electrical conductivity but with good welding and hard soldering capability and hydrogen resistance[e] For ducts, apparatus engineering, building	×	×	×	×	×

[a] SE–Cu is generally supplied with electrical conductivity in the soft state of ≥ 57.0 m/Ω·mm^2. By agreement it can also be supplied with an electrical conductivity of ≥ 58.0 m/Ω·mm^2 and with a lower phosphorus content.

[b] Phosphorus can be completely or partially replaced by other deoxidants.

[c] The standard commercial delivery forms are designated as follows:
× = Semifinished product for general application according to DIN 17 670 to DIN 17 674 and DIN 17 677.
○ = Copper semifinished product for electrical engineering to DIN 40 500.

[d] Previous designation E-Cu (see DIN 40 500, pages 1 to 3).

[e] Hydrogen resistance is tested as laid down in technical delivery conditions. If these test conditions do not comply with the requirements, then other test conditions must be agreed upon at the time of order.

Appendix D3, Table 19a. Copper sheets and strips to DIN 17 670

Material		Thickness mm	Tensile strength R_m N/mm²	0.2% elongation limit $R_{p\,0.2}$ N/mm²	Elongation at rupture		Vickers hardness		Brinel hardness	
Abbreviation	Number				A_5 % min.	A_{10} % min.	min.	max.	min.	max.
SW-Cu	· 2.0076									
SF-Cu	2.0090									
F.20	.10	over 5 to 15	200 to 250	max. 100	42	36	—	—	—	—
H40			—	—	—	—	—	—	40	60
F22	.10	from 0.2 to 5	220 to 260	max. 140	42	36	—	—	—	—
H40			—	—	—	—	40	70	40	65
F24	.26	from 0.2 to 15	240 to 300	min. 180	15	12	—	—	—	—
H70			—	—	—	—	70	95	65	90
F29	.30	from 0.2 to 10	290 to 360	min. 250	6	—	—	—	—	—
H90			—	—	—	—	90	110	85	105
F36	.32	from 0.2 to 2	min. 360	min. 320	—	—	—	—	—	—
H110			—	—	—	—	110	—	105	—

Appendix D3, Table 19b. Sheets and strips for electrical engineering. From DIN 40 500

Abbreviation	Strength state abbreviation	Dimensions in accordance with dimensional standards Thickness mm	Mechanical characteristics						Electrical characteristics	
			R_m N/mm²	$R_{p\,0.2}$ N/mm²	A_5 % min.	A_{10} % min.	HB10	E N/mm² Guideline	Specific resistance at 20 °C ρ $\Omega\cdot mm^2/m$ max.	Conductivity at 20 °C $\kappa = 1/\rho$ $m/\Omega\cdot mm^2$ min.
SE-Cu	F20	0.1 to 1 >1 to 5	200 to 250	max. 120	38 45	32 38	45 to 70		0.01754	57
E-Cu 57 SE-Cu CuAg 0.1 CuAg 0.1 P	F 25	0.1 to 1 >1 to 5	250 to 300	min. 200	17 20	14 16	70 to 90		0.01786	56
	F30	0.1 to 1 >1 to 5	300 to 360	min. 250	7 8	4 5	85 to 105	11.10⁴	0.01818 0.01786	55 56
	F37	0.1 to 1 >1 to 3	min. 360	min. 320	3 5	2 3	95 to 120		0.01818	55

Appendix D3, Table 20. Copper–zinc wrought alloys. Strength characteristics. From DIN 17 670 and DIN 17 671.

Z = machinability, U = formability, V = application.

Abbreviation	Material number	Thickness mm	R_m N/mm²	$R_{p\,0.2}$ N/mm²	A_5 % min.	A_{10} % min.	Notes on characteristics and application	Sheet	Strip	Tube	Bar	Wire	Forgings	Sections
CuZn 28														
p	2.0261.08	As agreed	Without prescribed strength values				Readily cold-formable by deep drawing, pressing, riveting, flanging; very readily solderable; easily plated onto steel. Instruments, sleeves of all types.	x	x	x	x			
F27	.10		270 to 350	≤ 160	50	45								
F35	.26	0.2 to 5	350 to 420	≥ 200	33	30								
F42	.30		420 to 520	≥ 340	15	12								
CuZn 33														
p	2.0280.08	As agreed	Without prescribed strength values				Very readily cold-formable, particularly suitable for flanging and cold-heading. Wire braiding, radiator cores, hollow rivets.	x	x	x	x	x		
F 28	.10	0.2 to 5	280 to 360	≤ 170	50	45								
F 36	.26	0.2 to 5	360 to 430	≥ 200	31	28								
F 43	.30	0.2 to 5	430 to 530	≥ 360	13	10								
F 53	.32	0.2 to 2	≥ 530	≥ 480	—	—								
CuZn 36														
p	2.0335.08	As agreed	Without prescribed strength values				Main alloy for cold-forming by deep drawing, pressing, cold-heading, rolling, threaded rollers, stamping and bending; readily soldered and welded. Metal and wooden bolts, pressure bolts, radiator cores, zips, leaf springs, hollow products, ball point pen cartridges.	x	x	x	x	x		
F 30	.10	0.2 to 5	300 to 370	≤ 180	48	43								
F 37	.26	0.2 to 5	370 to 440	≥ 200	28	24								
F 44	.30	0.2 to 5	440 to 540	≥ 370	12	8								
F 54	.32	0.2 to 2	540 to 610	≥ 490	—	—								
CuZn 36 Pb 1.5														
p	2.0331.08	As agreed	Without prescribed strength values				Z: Good U: Readily cold-formed V: Pressing, stamping, machining, punching		x	x	x	x		
F29	.10	0.3 to 5	290 to 370	≤ 200	50	44								
F37	.20	0.3 to 5	370 to 440	≥ 200	28	24								
F44	.30	0.3 to 5	440 to 540	≥ 370	12	8								
F54	.32	0.3 to 2	≥ 540	≥ 490	—	—								

(continued)

Appendix D3, Table 20. Continued

Abbreviation	Material number	Thickness mm	R_m N/mm²	$R_{p\,0.2}$ N/mm²	A_5 % min.	A_{10} % min.	Notes on characteristics and application	Sheet	Strip	Tube	Bar	Wire	Forgings	Sections
CuZn 37 Pb 0.5														
P	2.0332.08	As agreed	Without prescribed strength values											
F29	.10	0.03 to 5	290 to 370	≤ 200	50	44	Z: Adequate		×	×	×			
F37	.26	0.03 to 5	370 to 440	≥ 200	28	24	U: Very readily cold-formable							
F44	.30	0.03 to 5	440 to 540	≥ 370	12	8	V: Deep drawing, pressing							
F54	.32	0.03 to 2	≥ 540	≥ 490	—	—								
CuZn 36 Pb 3														
P	2.0375.08	As agreed	Without prescribed strength values											
F34	.10	up to 10	≥ 340	≤ 210	40	—	Z: Good			×	×			
F40	.26	up to 10	≥ 400	≥ 210	18	—	U: Readily cold formable							
F46	.30	up to 5	≥ 460	≥ 330	10	—	V: Alloy for all machining processes; suitable for automats							
CuZn 38 Pb 1.5														
P	2.0371.08	As agreed	Without prescribed strength values											
F34	.10	0.3 to 15	≥ 340	≤ 240	43	38	Z: Good		×	×	×	×	×	×
F41	.26	0.3 to 15	≥ 410	≥ 240	23	20	U: Readily hot-workable, readily cold-workable							
F47	.30	0.3 to 5	≥ 470	≥ 390	12	9	V: Bending, riveting, cold-heading. Alloy for all machining processes							
F54	.32	0.3 to 2	≥ 540	≥ 490	—	—								

(continued)

Appendix D3, Table 20. Continued

Abbreviation Material number	Thickness mm	R_m N/mm²	$R_{p\,0.2}$ N/mm²	A_5 % min.	A_{10} % min.	Notes on characteristics and application	Sheet	Strip	Tube	Bar	Wire	Forging	Section
Cu/Zn 39 Pb 0.5													
p 2.0372.08	As agreed	Without prescribed strength values					X	X		X	X	X	X
F34 .10	0.3 to 15	≥ 340	≤ 240	43	38	Z: Adequate							
F41 .26	0.3 to 15	≥ 410	≤ 240	23	20	U: Readily hot-formable, readily cold-formable							
F47 .30	0.3 to 5	≥ 470	≥ 390	12	9	V: Bending, riveting, cold-heading, flanging							
F54 .32	0.3 to 2	≥ 540	≥ 490	—	—								
Cu/Zn 39 Pb 3													
p 2.0401.08	As agreed	Without prescribed strength values							X	X	X	X	X
zh .20	up to 10	≥ 360	≤ 250	35	35	Z: Very good							
F36 .10	up to 10	≥ 430	≤ 250	15	15	U: Readily hot-formable, limited capability for cold forming by bending, riveting, flanging							
F43 .26	up to 5	≥ 500	≥ 370	10	10	V: Drilling and milling quality; readily punchable							
F50 .30													
Cu/Zn 40 Pb 2													
p 2.0402.08	As agreed	Without prescribed strength values					X	X	X	X	X	X	X
zh .20		≥ 360	≤ 270	0	35	Z: Very good							
F36 .10		≥ 430	≤ 270	20	17	U: Readily heat formable, of limited capacity for cold-forming							
F43 .26	0.3 to 5	≥ 490	≥ 420	9	6	V: Alloy for all machining processes; clock brass for wheels and discs							
F49 .30													
Cu/Zn 40													
p 2.0360.08	As agreed	Without prescribed strength values					X	X	X	X	X	X	
F34 .10	0.3 to 15	≥ 340	≤ 240	43	38	Readily hot and cold-formable (forging brass, coining metal); suitable for bending, riveting, cold-heading and flanging and (in the soft state) for stamping and also for deep drawing							
F41 .26	0.3 to 5	≥ 410	≤ 240	23	20								
F47 .30	0.3 to 5	≥ 470	≥ 390	12	9								

Appendix D3, Table 21. Copper–zinc alloys with other alloy components (special brass). From DIN 17670 and DIN 17671

Abbreviation Material number	Thickness mm	R_m N/mm² min.	$R_{p\,0.2}$ N/mm² min.	A_5 % min.	Approximate mean value	Notes on characteristics and applications	Sheet	Strip	Tube	Bar	Wire	Forgings	Sections
CuZn 20 Al 2 2.0460.08 p		Without prescribed strength values					×		×			×	
F33 .10	3 to 15	330	90	30	85	Tubes and tube sheets for condensers and heat exchangers							
F39 .26		390	240	25	190								
CuZn 31 Si 1 2.0490.08 p		Without prescribed strength values						×	×	×			
F44 .27	1 to 8	440	200	30	120	For sliding loads, including high loads. Bearing bushes, guides and other slip components							
F49 .31	1 to 8	490	290	15	160								
CuZn 35 Ni 2 2.0540.08 p		Without prescribed strength values							×	×		×	
F49 .27	3 to 12	490	290	18	130	Medium- to high-strength structural material. Apparatus engineering, shipbuilding							
F54 .31	3 to 8	540	390	14	150								
CuZn 40 Al1 2.0561.08 p		Without prescribed strength value							×	×		×	×
F39 .09	3 to 12	390	150	25	110	Medium strength and high toughness structural material. Good resistance to weathering. Suitable for sliding loads							
F44 .27	3 to 8	440	200	20	120								
F49 .31	3 to 8	490	260	15	140								
CuZn 40 Al 2 2.0550.08 p		Without prescribed strength values							×	×		×	×
F54 .27	4 to 12	540	230	15	150	Good resistance to weathering. For severe requirements in terms of sliding loads							
F59 .31	4 to 10	590	250	10	160								
CuZn 40 Mn 2 2.0572.08 p		Without prescribed strength values							×	×		×	×
F44 .27	3 to 12	440	180	20	125	Medium-strength constructional material; aluminium-free; solderable; weather-resistant. Apparatus engineering, architecture							
F49 .31	3 to 8	490	270	18	140								
CuZn 40 Mn 1 Pb 2.0580.08 p		Without prescribed strength values							×	×		×	×
F39 .09	3 to 12	390	150	22	110	Medium strength free-cutting alloy with good machinability. Roller-bearing cages							
F44 .27	3 to 8	440	180	18	125								
F49 .31	2 to 5	490	290	15	140								

Appendix D3, Table 22. Casting brass and special casting brass to DIN 1709

Abbreviation	Material number	Delivery form	Material characteristics in specimen bar				Density kg/dm³	Comments	Notes on application
			$R_{p0.2}$ N/mm² min.	R_m N/mm² min.	A_5 % min.	HB 10/1000 min.	$\approx$		
G-CuZn 15	2.0241.01	Sand casting	70	170	25	45	8.6	Structural material, good resistance to sea water, very good capacity for soft and hard soldering, electrical conductivity approx. 15 m/Ω·mm²	For components to be soldered, e.g. flanges and other components for shipbuilding, mechanical engineering, electrical engineering, fine mechanics, optics etc.
G-CuZn 33 Pb	2.0290.01	Sand casting	70	180	12	45	8.5	Structural material, corrosion resistant in process liquids up to approx. 90 °C, electrical conductivity approx. 10 to 14 m/Ω·mm²	Casing for gas and water fittings, construction and fitting items for mechanical engineering, electrical engineering, fine mechanics, optics etc.
GD-CuZn 37 Pb	2.0340.05	Die casting	120	280	4	75	8.5	Structural material, readily machined	Fittings and structural components of general types, sanitation and stacking fittings; die castings for mechanical engineering, electrical engineering, fine mechanics, optics etc.
GK-CuZn 37 Pb	2.0340.02	Chill casting	90	280	25	70			
GK-CuZn 38 Al	2.0591.02	Chill casting	130	380	20	75	8.5	Structural material, easy to cast, cryogenic, corrosion resistant to atmosphere, electrical conductivity approx. 12 m/Ω·mm²	For wound construction components of all types, predominantly in the electrical industry and in mechanical engineering

(continued)

Appendix D3, Table 22. Continued

Abbreviation	Material number	Delivery form	Material characteristics in specimen bar				Density kg/dm³	Comments	Notes on application
			$R_{p\,0.2}$ N/mm² min.	R_m N/mm² min.	A_5 % min.	HB 10/1000 min.			
G-CuZn 40 Fe GZ-CuZn 40 Fe	2.0590.01 2.0590.03	Sand casting Spin casting	130 150	300 325	15 15	75 85	≈ 8.6	Structural material, cryogenic, readily soft and hard soldered, electrical conductivity approx. 10 m/Ω·mm²	Fittings casings for high gas and water pressures, components in low-temperature technology
GK-CuZn 37 Al 1	2.0595.02	Chill casting	170	450	25	105	8.5	Structural material	Structural components for mechanical engineering, electrical engineering, fine mechanics etc.
G-CuZn 35 Al 1 GZ-CuZn 35 Al 1	2.0592.01 2.0592.03	Sand casting Spin casting	170 200	450 500	20 18	110 120	8.6	Structural material, moderate frictional characteristics	Compression nuts for roller systems and spindle presses, bottom and gland bushings, ships' propellers
G-CuZn 34 Al 2 GZ-CuZn 34 Al 2	2.0596.01 2.0596.03	Sand casting Spin casting	250 260	600 620	15 14	140 150	8.6	Structural material with high static strength and hardness	Statically-loaded structural components, valve and control components, seats, tapers
G-CuZn 25 Al 5 GZ-CuZn 25 Al 5	2.0598.01 2.0598.03	Sand casting Spin casting	450 480	750 750	8 5	180 190	8.2	Structural material with very high static loadbearing capacity	Structural components with very high static loading, e.g. bearings at very high load and low rotational speeds, highly loaded, slow-running worm gear crowns, internal components of high-pressure fittings
G-CuZn 15 Si 4 GD-CuZn 15 Si 4 GK-CuZn 15 Si 4	2.0492.01 2.0492.05 2.0492.02	Sand casting Die casting Chill casting	230 300 300	400 550 500	10 8 10	100 125 120	8.6	Structural material, good resistance to corrosion and seawater, very good casting capability	Highly-loaded, thin-walled wound structural components for mechanical engineering and shipbuilding, electrical industry, fine mechanics etc.

Appendix D3, Table 23. Copper-tin alloys (tin bronze) to DIN 17 662, DIN 17 670, cast tin bronze and red bronze to DIN 1705

Abbreviation		Material number	Thickness mm	R_m N/mm^2	$R_{p\,0.2}$ N/mm^2	A_5 % min.	A_{10} % min.	HB 2.5/62.5	Notes on characteristics and application
CuSn 4	F33	2.2016.10	0.1 to 5	330 to 380	≤ 190	50	45		Strips for metal hoses, tubes, live springs
CuSn 6	F35	2.1020.10	0.1 to 5	350 to 410	≤ 300	55	50		All types of springs, particularly for
	F41	.26	0.1 to 5	410 to 500	≥ 300	30	25		the electrical industry.
	F48	.30	0.1 to 5	480 to 580	≥ 450	20	15		Window and door seals, tubes and
	F55	.32	0.1 to 2	550 to 650	≥ 510	10	8		sleeves for spring units, hose tubes
	F63	.34	0.1 to 2	≥ 630	≥ 600	6	—		and spring tubes for pressure measurement instruments, diaphragms and filter wire, gong bars, damper bars, components for chemical industry
CuSn 8	F37	2.1030.10	0.1 to 5	370 to 450	≤ 300	60	55		Slip components, particularly for thin-
	F45	.26	0.1 to 5	450 to 540	≥ 300	33	28		walled sliding bearing bushes and
	F54	.30	0.1 to 5	540 to 630	≥ 470	25	20		guide rails.
	F59	.32	0.1 to 5	590 to 690	≥ 520	10	7		Beater knives; abrasion and corrosion
	F66	.34	0.1 to 2	≥ 660	≥ 600	6	—		resistances higher than CuSn 6
CuSn 6 Zn 6	F61 F76	2.1080.30 .34	0.1 to 2 0.1 to 2	610 to 690 ≥ 760	≥ 570 ≥ 690	15	12	225	All types of spring, diaphragms

		Delivery form						HB 10/1000	
G-CuSn 12		2.1052.01	Sand casting	260	140	12		80	Dome bricks, shaft nuts, worms and
GZ-CuSn 12		.03	Spin casting	280	150	5		95	helical gear wheels, highly-loaded
GC-CuSn 12		.04	Extrusion casting	280	150	8		90	actuation and gibs
G-CuSn 12 Ni		2.1060.01	Sand casting	280	160	14		90	As for material no. 2.1052, but for
GZ-CuSn 12 Ni		.03	Spin casting	300	180	8		100	higher strength, wear strength and standby characteristics. Resistant to corrosion and seawater; resistant to cavitation loading
G-CuSn 12 Pb		2.1061.01	Sand casting	260	140	10		80	Sliding bearings with high load peaks,
GZ-CuSn 12 Pb		.03	Spin casting	280	150	5		90	piston gudgeon bushes, shaft nuts.
GC-CuSn 12 Pb		.04	Extrusion casting	280	150	7		90	Good emergency running properties and wear resistance. Resistant to corrosion and seawater
G-CuSn 10		2.1050.01	Sand casting	270	130	18		70	Fittings, pump casings, guide wheels and impeller wheels. High elongation, resistant to corrosion and seawater
G-CuSn 10 Zn		2.1086.01	Sand casting	260	130	15		75	Sliding bearing shells, moderately
GZ-CuSn 10 Zn		.03	Spin casting	270	150	7		85	loaded dome bricks, stern tubes
GC-CuSn 10 Zn		.04	Extrusion casting	270	150	7		80	
G-CuSn 7 ZnPb		2.1090.01	Sand casting	240	120	15		65	Axle bearing shells, sliding bearings,
GZ-CuSn 7 ZnPb		.03	Spin casting	270	130	13		75	piston gudgeon bushes, friction rings, slide strips and guides. Medium hard
GC-CuSn 7 ZnPb		.04	Extrusion casting	270	130	16		70	sliding bearing material, resistant to seawater
G-CuSn 6 ZnNi		2.1093.01	Sand casting	270	140	15		75	Fittings, pump casing, pressure-proof castings. Good casting characteristics, resistant to seawater
G-CuSn 5 ZnPb		2.1096.01	Sand casting	240	90	18		60	Fittings for exposure to water and steam up to 225 °C, thin-walled intricate castings. Good casting characteristics, can be soft-soldered, in some instances can be hard-soldered, resistant to seawater
G-CuSn 2 ZnPb		2.1098.01	Sand casting	210	90	18		60	For thin-walled fittings for exposure to temperatures up to 225 °C. Good casting characteristics, resistant to corrosion from process fluids even at high temperatures

Appendix D3, Table 24. Copper–aluminium alloys (aluminium bronze) to DIN 17 665, DIN 17 672 and DIN 1714

Abbreviation	Strength	Material number	R_m N/mm² min.	$R_{p0.2}$ N/mm² min.	A_5 % min.	HB 2.5/62.5 Approximate mean value	Characteristics and application	
CuAl 8	P	2.0920.08	Without prescribed strength values				Chemicals industry, particularly resistant to sulphuric acid and acetic acid	
	F37	.10	370	120	35	90		
	F49	.30	490	270	15	130		
CuAl 8 Fe 3	P	2.0932.08	Without prescribed strength value				Condenser bottoms, sheets, cold-formable	High strength even at high temperatures; high long-term fatigue loading capacity, even under corrosion loading
	F47	.97	470	200	25	110		
	F59	.30	590	270	10	150		
CuAl 10 Fe 3 Mn 2	P	2.0936.08	Without prescribed strength values				Scale-resistant components, shafts, bolts	Good corrosion resistance in neutral and acid aqueous media and sea-water
	F59	.97	590	250	12	150		
	F69	.98	690	340	7	180		
CuAl 9 Mn 2	P	2.0960.08	Without prescribed strength values				Transmission and worm gears, valve seats	Good resistance to scaling, erosion and cavitation
	F49	.97	490	200	25	110		
	F59	.98	590	250	15	150		
CuAl 10 Ni 5 Fe 4	P	2.0966.08	Without prescribed strength values				Condenser bottoms, control components for hydraulics	
	F64	.97	640	270	15	180		
	F74	.98	740	390	10	195		
CuAl 11 Ni 6 Fe 5	P	2.0978.08	Without prescribed strength values				Maximum strength components, bearings, valves	
	F73	.97	730	440	5	210		
	F83	.98	830	590	—	240		

						HB 10/1000 mean value	
G-CuAl 10 Fe		2.0940.01	500	180	15	115	Levers, casings, mountings, pinions, taper gears, minimal temperature dependency between −200 and +200 °C
GK-CuAl 10 Fe		.02	550	200	25	115	
G-CuAl 9 Ni		2.0970.01	500	200	20	110	Fittings, controllable-pitch propellers, stern components, pickling baskets; very good welding capacity; resistant to seawater and non-oxidising acids
GK-CuAl 9 Ni		.02	600	250	20	120	
G-CuAl 10 Ni		2.0975.01	600	270	12	140	Highly loaded components, marine propellers, stern tubes, U-bends, impellers, pump casings, good resistance to reciprocating fatigue load
GK-CuAl 10 Ni		.02	600	300	12	150	
GZ-CuAl 10 Ni		.03	700	300	13	160	
GC-CuAl 10 Ni		.04	700	300	13	160	
G-CuAl 11 Ni		2.0980.01	680	320	5	170	As above, but for more stringent requirements for cavitation and/or wear resistance; turbine and pump impellers
GK-CuAl 11 Ni		.02	680	400	5	200	
GZ-CuAl 11 Ni		.03	750	400	5	185	
G-CuAl 8 Mn		2.0962.01	440	180	18	105	Corrosion-loaded components with low magnetisation capacity and low electrical conductivity
GK-CuAl 8 Mn		.02	450	200	30	105	

Appendix D3, Table 25. Copper–lead–tin alloy castings (tin–lead–bronze casting) to DIN 1716

Abbreviation	Material number	$R_{p\,0.2}$ N/mm² min.	R_m N/mm² min.	A_s % min.	HB 10/1000 min.	Characteristics and application
G-CuPb 5 Sn	2.1170.01	130	240	15	70	Corrosion and acid-resistant fittings (diluted hydrochloric and sulphuric acid, fatty acids)
G-CuPb 10 Sn	2.1176.01	80	180	8	65	Sliding bearings with high surface pressures,
GZ-CuPb 10 Sn	.03	110	220	8	70	composite bearing in combustion engines
GC-CuPb 10 Sn	.04	110	230	12	70	(P_{max} = 10 000 N/cm²)
G-CuPb 15 Sn	2.1182.01	90	180	8	60	Bearings with high surface pressures
GZ-CuPb 15 Sn	.03	110	220	7	65	(P_{max} = 5000 N/cm²), composite bearings for
GC-CuPb 15 Sn	.04	110	220	8	65	combustion engines (P_{max} = 7000 N/cm²)
G-CuPb 20 Sn	2.1188.01	90	160	6	50	Sliding bearings for high sliding speeds, resistant to sulphuric acid, composite bearings, fittings
G-CuPb 22 Sn	2.1166.09	—	—	—	30	Highly loaded composite bearings (crankshaft, conrod and camshaft bearings, (P_{max} = 7000 N/cm²)

Appendix D3, Table 26. Highest-grade aluminium and high-grade aluminium to DIN 1790

Abbreviation		Material number	Diameter of wire mm	Tensile strength R_m N/mm² min.	0.2-Limit $R_{p\,0.2}$ N/mm² min.	Elongation at rupture		Brinel hardness HB ≈	States
						A_{10} % min.	$A_{L=100}$ % min.		
Al 99.98R	W4	3.0385.10	to 18	40	—	25	20	15	soft
	F7	.26	to 15	70	—	8	6	20	drawn
	F11	.30	to 10	110	—	4	3	25	drawn
Al 99.9	W4	3.0305.10	to 18	40	—	25	20	15	soft
	F7	.26	to 15	70	—	8	6	20	drawn
	F11	.30	to 10	110	—	4	3	25	drawn
Al 99.8	W6	3.0285.10	to 18	55	≤50	23	16	18	soft
	F9	.26	to 15	90	60	7	5	25	drawn
	F12	.30	to 10	120	95	4	2	30	drawn
Al 99.5	W7	3.0255.10	to 18	60	≤ 55	22	18	20	soft
	F10	.26	to 15	100	70	6	4	30	drawn
	F14	.30	to 10	140	115	3	2	38	drawn
Al 99	W8	3.0205.10	to 18	75	≤ 70	18	14	22	soft
	F11	.26	to 15	110	80	4	2	32	drawn
	F15	.30	to 10	150	125	2	1	40	drawn

Appendix D3, Table 27. Aluminium wrought alloys to DIN 1745

Abbreviation	Material number	Thickness Sheet mm over	Sheet up to	Strip mm over	Strip up to	Tensile strength R_m N/mm² min.	max.	0.2-Limit $R_{p0.2}$ N/mm²	Elongation A_5 % min.	A_{10} % min.	Brinel hardness HB ≈		Characteristics
AlMn 1	3.05												
W9	15.10	0.35	10	0.35	3.0	90	140	35	24	21	28	s	Very good cold-forming characteristics; weldable; corrosion resistant
F12	.24	0.35	10	0.35	3.0	120	160	90	7	5	40	cr	
F14	.26	0.35	10	0.35	2.5	140	180	120	5	4	45	cr	
F17	.30	0.35	3.0	0.35	3.0	165	205	145	4	3	50	cr	
F19	.32	0.35	2.5	0.35	2.0	185	—	165	3	2	55	cr	
AlR Mg1	3.3319												
Al 99.9 Mg 1	3.3318												
Al 99.85 Mg 1	3.3317												
W10	.10	0.35	6.0	0.35	3.0	100	140	35 to 60	23	20	30	s	Not precipitation hardenable, very good capacity for cold working, weldable; seawater resistant
G12	.25	0.35	3.0	0.35	3.0	120	160	70	15	12	40	r	
G14	.27	0.35	2.5	0.35	2.5	140	180	100	10	8	45	r	
G16	.31	0.35	2.0	0.35	2.0	160	200	130	8	6	50	r	
F18	.32	0.35	2.0	0.35	2.0	180	—	160	4	3	55	cr	
AlMg 1.5	3.3316												
W13	.10	0.35	6.0	0.35	3.0	130	170	45	23	20	37	s	
G18	.27	0.35	3.0	0.35	3.0	175	215	130	10	8	55	r	
F20	.28	0.35	3.0	0.35	3.0	200	240	175	4	3	60	cr	
F23	.30	0.35	3.0	0.35	3.0	225	—	200	3	2	65	cr	
G23	.31	0.35	2.0	0.35	2.0	225	—	180	6	5	65	r	
AlMg 2.5	3.3523												
W17	.10	0.35	10	0.35	3.0	170	215	60	20	17	50	s	Not amenable to precipitation hardening, good cold-forming capacity, weldable; very good resistance to seawater
F21	.24	0.35	10	0.35	3.0	210	250	160	10	8	65	cr	
G21	.25	0.35	10	0.35	3.0	210	250	130	12	10	65	r	
F23	.26	0.35	10	0.35	3.0	230	270	180	5	4	73	cr	
G23	.27	0.35	10	0.35	3.0	230	270	150	10	8	73	r	
F25	.28	0.35	4.0	0.35	3.0	250	290	210	4	3	80	cr	
G25	.29	0.35	4.0	0.35	3.0	250	290	180	7	6	80	r	
F27	.30	0.35	3.0	0.35	3.0	270	—	240	3	2	85	cr	
G27	.31	0.35	3.0	0.35	3.0	270	—	210	6	5	85	r	
AlMg 3	3.3535												
W19	.10	0.35	6.0	0.35	3.0	190	230	80	20	17	50	s	
W19	.10	6.0	50	—	—	190	230	80	18	—	50	s	
F19	.07	25	50	—	—	190	—	80	12	—	50	hr	
F20	.07	10	25	—	—	200	—	120	10	—	60	hr	

Alloy	State	Material No.	c1	c2	c3	c4	Rm₁	Rm₂	Rp	A₁	A₂	HB	Form	Characteristics
AlMg 3	F21	3.3535.07	5.0	10	3.0	1 0	210	—	140	12	—	60	hr	Not amendable to precipitation hardening, good cold-forming capacity, weldable; very good resistance to seawater
	F22	.24	0.35	10	0.35	3.0	220	260	165	9	7	65	cr	
	G22	.25	0.35	10	0.35	3.0	220	260	130	14	12	65	r	
	F24	.26	0.35	10	0.35	3.0	240	280	190	5	4	73	cr	
	G24	.27	0.35	10	0.35	3.0	240	280	160	10	8	73	r	
	F27	.28	0.35	4.0	0.35	3.0	265	305	215	4	3	80	cr	
	G27	.29	0.35	4.0	0.35	3.0	265	305	190	7	6	80	r	
	F29	.30	0.35	3.0	0.35	3.0	290	—	250	3	2	85	cr	
AlMg 2 Mn 0.8	W19	3.3527.10	0.35	6.0	0.35	3.0	190	230	80	20	17	50	s	Good cold-forming characteristics, good weldability characteristics; corrosion resistant, good heat resistance
	W19	.10	6.0	50	—	—	190	230	80	18	—	50	s	
	F19	.07	25	50	—	—	190	230	80	12	—	50	hr	
	F20	.07	10	25	—	—	200	240	120	10	—	60	hr	
	F21	.24	6.0	10	0.35	3.0	210	250	140	12	7	60	hr	
	F22	.25	0.35	10	0.35	3.0	220	260	165	9	12	65	cr	
	G22	.26	0.35	10	0.35	3.0	220	260	130	14	4	65	r	
	F24	.27	0.35	10	0.35	3.0	240	280	190	5	8	73	cr	
	G24	.28	0.35	4.0	0.35	3.0	240	280	160	10	3	73	r	
	F27	.29	0.35	4.0	0.35	3.0	265	305	215	4	6	80	cr	
	G27	.30	0.35	4.0	0.35	3.0	265	305	190	7	2	80	r	
	F29		0.35	3.0	0.35	3.0	290	—	250	3		85	cr	
AlMg 4 Mn	W24	3.3545.10	1.0	6.0	—	—	240	310	100	18	—	65	s	Not amenable to precipitation hardening, corrosion resistance, good heat resistance
	W24	.10	6.0	50	—	—	240	310	95	17	—	60	s	
	F28	.24	4.0	6.0	—	—	275	330	200	7	—	80	cr	
	G28	.25	1.0	6.0	—	—	275	330	190	12	—	80	r	
	G30	.27	1.0	6.0	—	—	300	360	230	8	—	90	r	
AlMgSi 0.8	F20	3.2316.51	0.35	6.0	0.35	3.0	200	—	110ᵃ	16	14	60	cp	Good formability, polishing and anodic oxidation; good corrosion resistance
	F28	.71	0.35	6.0	0.35	3.0	275	—	200	12	10	85	hp	
AlMgSi 1	W	3.2315.10	0.35	10	0.35	3.0	—	—	≤ 85	18	15	35	s	
	F21	.51	0.35	3.0	0.35	3.0	205	—	110ᵃ	16	14	65	cp	
	F21	.51	3.0	20	—	—	205	—	110	14	12	65	cp	
	F28	.71	0.35	3.0	0.35	3.0	275	—	200	14	12	85	hp	
	F28	.71	3.0	60	—	—	275	—	200	12	—	85	hp	
	F32	.72	3.0	10	0.35	3.0	315	—	255	10	8	95	hp	
	F30	.72	0.35	20	—	—	295	—	245	9	—	95	hp	
	F30	.72	20	1 00	—	—	295	—	240	8	—	90	hp	
AlCuMg 1	W	3.1325.10	0.35	12	0.35	3.0	—	215	≤ 140	13	11	50	s	High-strength alloy with capacity for cold precipitation hardening; not very corrosion resistant
	F40	.51	0.35	3.0	0.35	3.0	395	—	265	13	11	100	cp	
	F39	.51	3.0	12	—	—	390	—	265	13	—	100	cp	
	F39	.51	12	60	—	—	385	—	245	12	—	95	cp	

Appendix D3, Table 27. Continued

Abbreviation		Material number	Thickness Sheet mm over	up to	Strip mm over	up to	Tensile strength R_m N/mm² min.	max.	0.2-Limit $R_{p0.2}$ N/mm²	Elongation at rupture A_5 % min.	A_{10} % min.	Brinel hardness HB ≈		Characteristics
AlCuMg 2	W	3.1355.10	0.35	12	0.35	3.0	–	220	≤ 140	13	11	55	s	
	F44	.51	0.35	3.0	0.35	3.0	440	–	290	13	11	110	cp	
AlCuSiMn	W	3.1255.10	6.0	12	–	–	–	220	≤ 140	13	–	55	s	
	F40	.51	1.5	25	–	–	400	–	250	12	–	105	cp	
	F40	.51	25	50	–	–	400	–	250	11	–	100	cp	
	F39	.51	50	100	–	–	390	–	250	8	–	100	hp	
	F46	.71	1.5	25	–	–	460	–	400	7	–	125	hp	
AlZn 4.5 Mg 1	W	3.4335.10	1.5	6.0	–	3.0	–	220	≤ 140	15	13	45	w	High-strength alloy for weld constructions, highest-strength hot precipitation hardenable alloy; not very corrosion-resistant
	F35	.71	0.35	15	0.35	–	350	–	275	10	8	105	hp	
	F34	.71	15	60	–	–	340	–	270	9	–	105	hp	
AlZnMgCu 0.5	F45	3.4345.71	6.0	25	–	–	450	–	370	8	–	125	hp	
	F45	.71	25	50	–	–	450	–	370	7	–	125	hp	
	F43	.71	50	100	–	–	430	–	350	5	–	110	hp	
	F41	.71	100	200	–	–	410	–	330	3	–	100	hp	
AlZnMgCu 1.5	F53	3.4365.71	6.0	12	–	–	530	–	450	8	–	140	hp	
	F53	.71	12	25	–	–	530	–	450	5	–	140	hp	
	F53	.71	25	50	–	–	530	–	450	3	–	140	hp	
	F50	.71	50	63	–	–	500	–	430	2	–	130	hp	
	F48	.71	63	75	–	–	480	–	410	2	–	130	hp	
	F48	.71	75	100	–	–	480	–	390	2	–	130	hp	
AlSi 12 CuMgNi[b]							370		340	1	–	125	hp	High wear resistance, good running characteristics, good heat resistance; alloys for pistons
AlSi 18 CuMgNi[b]							300		260	0.5	–	125	hp	
AlSi 25 CuMgNi[b]							210		200	0.1	–	125	hp	

[a] Maximum value 180 N/mm² with regard to forming work.
[b] Not standardised.

Appendix D3, Table 28. Fatigue strength of light metal alloys

hp = hot precipitation-hardened; cp = cold precipitation-hardened.

Material	Alloy		$R_{p\,0.2}$	R_m	Fatigue limit under alternating stress			Fatigue strength under fluctuating stresses	
					Tensile-compressive	Flexural	Torsional	Tensile-compressive	Flexural
			N/mm²	N/mm²	N/mm²	N/mm²	N/mm²	N/mm²	N/mm²
Aluminium	AlMg 5	F26	180	260	90	100	65	160	180
wrought alloys		F23	140	230	80	90	60	130	150
		F18	80	180	65	75	45	80	90
	welded				45			65	
	AlMgMn	F26	180	260	90	100	65	160	180
		F23	140	230	80	90	60	130	150
		F18	80	180	65	75	45	80	90
	welded				45			65	
	AlMg 5	F32	240	320	100	115	70	180	200
		F28	180	280	90	100	65	160	180
		F24	110	240	80	90	60	110	125
	welded				45			70	
	AlMgSi 1	F32	250	320	100	115	70	180	200
		F28	180	280	90	90	65	150	170
		F20	100	200	70	80	50	100	115
	welded				45			70	
	AlZnMg 1	hp	280	360	100	115	70	180	200
		cp	230	320	80	100	60	150	170
	welded				45			70	

Reciprocating load count > 10⁷; surface, rolling skin; wall thickness (diameter) ≤ 10 mm; weld seam; butt joint not machined.

Mg wrought alloys			< 200	$0.36 \cdot R_m$	$(0.3$ to $0.5) \cdot R_m$	$0.25 \cdot R_m$	$(0.5$ to $0.6) \cdot R_m$	$(0.5$ to $0.6) \cdot R_m$
			> 250	$0.30 \cdot R_m$		$0.14 \cdot R_m$	$(0.4$ to $0.55) \cdot R_m$	
Al casting alloys				—		$(0.15$ to $0.3) \cdot R_m$	—	—
Mg casting alloys				$(0.19$ to $0.34) \cdot R_m$		$(0.17$ to $0.26) \cdot R_m$	$0.25 \cdot R_m$	$(0.45$ to $0.55) \cdot R_m$

Appendix D3, Table 29. Aluminium casting alloys to DIN 1725 Sheet 2

pp = partially precipitation-hardened. cp = cold precipitation-hardened. hp = hot precipitation-hardened.

Abbreviation	$R_{p\,0.2}$ N/mm²	R_m N/mm²	A_5 %	HB 5/250 approx.	Fatigue strength under alternating bending stresses at $N = 50\cdot10^6$ N/mm²	Castability	Polishability	Anodic oxidation	Effects of weathering	Seawater	Machinability	Weldability
G-AlSi 12	70 to 100	160 to 210	5 to 10	45 to 60	55 to 65	×	–	–	×	×	×	×
GK-AlSi 12	80 to 110	180 to 240	6 to 12	50 to 60	70 to 80							
GD-AlSi 12	140 to 180	220 to 280	1 to 3	60 to 80	60 to 70							
G-AlSi 12 (Cu)	80 to 100	150 to 220	1 to 4	55 to 65	60 to 70	×	–	–	–	–	×	×
GK-AlSi 12 (Cu) hp	90 to 120	180 to 260	2 to 4	55 to 75	70 to 80							
GD-AlSi 12 (Cu)	140 to 200	220 to 300	1 to 3	60 to 80	70 to 80							
G-AlSi 10 Mg	80 to 110	170 to 220	2 to 6	50 to 60	65 to 75	×	×	–	×	×	×	×
G-AlSi 10 Mg hp	180 to 260	220 to 320	1 to 4	80 to 110	90 to 110							
GK-AlSi 10 Mg	90 to 120	180 to 240	2 to 6	60 to 80	80 to 100							
GK-AlSi 10 Mg hp	210 to 280	240 to 320	1 to 4	85 to 115	100 to 110							
GD-AlSi 10 Mg	140 to 200	220 to 300	1 to 3	70 to 90	70 to 90							
G-AlSi 8 Cu 3	100 to 150	160 to 200	1 to 3	65 to 90	50 to 70	×	×	–	–	–	×	×
GK-AlSi 8 Cu 3	110 to 160	170 to 220	1 to 3	70 to 100	60 to 80							
GD-AlSi 8 Cu 3	160 to 240	240 to 310	0.5 to 3	80 to 110	70 to 90							
G-AlSi 6 Cu 4	100 to 150	160 to 200	1 to 3	60 to 80	50 to 60	×	×	–	–	–	×	×
GK-AlSi 6 Cu 4	120 to 180	180 to 240	1 to 3	70 to 100	60 to 70							
GD-AlSi 6 Cu 4	150 to 220	220 to 300	0.5 to 3	70 to 100	70 to 90							
G-AlSi 5 Mg	100 to 130	140 to 180	1 to 3	55 to 70	60 to 65	×	×	–	×	×	×	×
G-AlSi 5 Mg cp	150 to 180	180 to 250	2 to 5	70 to 85	70 to 75							
G-AlSi 5 Mg hp	220 to 290	240 to 300	0.5 to 2	80 to 110	70 to 75							
GK-AlSi5 Mg	120 to 160	160 to 200	1.5 to 4	60 to 75	70 to 75							
GK-AlSi 5 Mg cp	160 to 190	210 to 270	2 to 8	70 to 90	80 to 85							
GK-AlSi 5 Mg hp	240 to 290	260 to 320	1 to 3	90 to 110	80 to 85							
G-AlMg 3	70 to 100	140 to 190	3 to 8	50 to 60	60 to 65	–	×	×	×	×	×	–
GK-AlMg 3	70 to 100	150 to 200	5 to 12	50 to 60	70 to 75							

Material															
G-AlMg3 Si		80 to 100	140 to 190	3 to 8	50 to 60	60 to 65	×	×	×	×	×	×	×	—
G-AlMg 3 Si	hp	120 to 160	200 to 280	2 to 8	65 to 90	75 to 80							×	
GK-AlMg 3 Si		80 to 100	150 to 200	4 to 10	50 to 65	70 to 80	×	×	×	×	×	×	×	×
GK-AlMg 3 Si	hp	120 to 180	220 to 300	3 to 10	65 to 90	80 to 90							×	×
G-AlMg 5		100 to 120	160 to 220	3 to 8	55 to 70	60 to 70	—	×	×	×	×	×	×	×
GK-AlMg 5		100 to 140	180 to 240	4 to 10	60 to 75	70 to 80							×	
G-AlMg 5 Si		110 to 130	160 to 200	2 to 4	60 to 75	60 to 65	×	×	×	×	×	×	×	×
GK-AlMg 5 Si		110 to 150	180 to 240	2 to 5	65 to 85	70 to 75							×	×
GD-AlMg 9		140 to 220	200 to 300	1 to 5	70 to 100	55 to 65	×	×	—	×	×	×	×	—
G-AlSi 9 Mg	hp	200 to 270	250 to 300	2 to 5	75 to 110	70	×	×	—	×	×	×	×	×
GK-AlSi 9 Mg	hp	200 to 280	260 to 340	4 to 7	80 to 115	80								
G-AlCu 4 Ti	pp	180 to 230	280 to 380	5 to 10	85 to 105	80 to 90	—	×	—	—	—	—	—	
G-AlCu 4 Ti	hp	200 to 260	300 to 380	3 to 8	95 to 110	80 to 90							×	
GK-AlCu 4 Ti	pp	180 to 230	320 to 400	8 to 18	90 to 105	90 to 100								
GK-AlCu 4 Ti	hp	220 to 270	330 to 400	7 to 12	95 to 110	90 to 100								
G-AlCu 4 TiMg	cp	220 to 280	300 to 400	5 to 15	90 to 115	80 to 90	—	×	—	—	—	—	×	—
G-AlCu 4 TiMg	hp	240 to 350	350 to 420	3 to 10	95 to 125	80 to 90								
GK-AlCu 4 TiMg	cp	220 to 300	320 to 420	8 to 18	95 to 115	90 to 100								
GK-AlCu 4 TiMg	hp	260 to 380	350 to 440	3 to 12	100 to 130	90 to 100								

Appendix D3, Table 30. Magnesium alloys to DIN 1729 and DIN 9715
ho = homogenised; hp = hot precipitation-hardened.

Abbreviation		$R_{p0.2}$	R_m	A_{10}	HB 5/250	Fatigue strength under alternating bending stresses at $N = 50 \cdot 10^6$	Characteristics and application
		N/mm^2 min.	N/mm^2 min.	% min.	approx.	N/mm^2	
MgMn 2	F20	145	200	1.5	40		Good weldability and formability
MgAl 3 Zn	F24	155	240	10	45		Weldable and formable
MgAl 6 Zn	F27	175	270	8	55		Of limited weldability
MgAl 8 Zn	F29	205	290	6	60		Maximum strength
G-MgAl 6		80 to 110	180 to 240	8 to 12	50 to 65	70 to 90	High elongation and impact strength, e.g. for car wheels
GD-MgAl 6		120 to 150	190 to 230	4 to 8	55 to 70	50 to 70	
GD-MgAl 6 Zn 1		130 to 160	200 to 240	3 to 6	55 to 70	50 to 70	Components with vibration loading, impact-loaded components, good frictional characteristics, weldable
G-MgAl 8 Zn 1		90 to 110	160 to 220	2 to 6	50 to 65	70 to 90	
G-MgAl 8 Zn 1	ho	90 to 120	240 to 280	8 to 12	50 to 65	80 to 100	
GK-MgAl 8 Zn 1		90 to 110	160 to 220	2 to 6	50 to 65	70 to 90	
GK-MgAl 8 Zn 1	ho	90 to 120	240 to 280	8 to 12	50 to 65	80 to 100	
GD-MgAl 8 Zn 1		140 to 160	200 to 240	1 to 3	60 to 85	50 to 70	
G-MgAl 9 Zn 1	ho	110 to 140	240 to 280	6 to 12	55 to 70	80 to 100	Maximum values for tensile strength and 0.2-limit, homogenised and hot precipitation-hardened for castings of high structural strength; good frictional characteristics, weldable
G-MgAl 9 Zn 1	hp	150 to 190	240 to 300	2 to 7	60 to 90	80 to 100	
GK-MgAl 9 Zn 1	ho	120 to 160	240 to 280	6 to 10	55 to 70	80 to 100	
GK-MgAl 9 Zn 1	hp	150 to 190	240 to 300	2 to 7	60 to 90	80 to 100	
GD-MgAl 9 Zn 1		150 to 170	200 to 250	0.5 to 3.0	65 to 85	50 to 70	

Appendix D3, Table 31. Titanium and its alloys to DIN 17 860

Abbreviation		Material number	State	$R_{p0.2}$	R_m	A_5 %	HB 30	Notch impact work (DVM) A_v	Bend radius r for thickness		Fatigue limit under alternating stress	
				N/mm^2 min.	N/mm^2 min.	min.	approx.	min.	$s \leq 2$ mm	$2 < s < 5$ mm	Tensile-compressive N/mm^2	Flexural N/mm^2
Ti 99.8		3.7025.10	annealed	180	290 to 410	30	120	60	1s	1.5s		
Ti 99.7		3.7035.10	annealed	250	390 to 540	22	150	35	1.5s	2s		
Ti 99.6		3.7055.10	annealed	320	460 to 590	18	170	25	2s	2.5s		
Ti 99.5		3.7065.10	annealed	390	540 to 740	16	200	20	2.5s	3s		520
TiAl 6 V 4	F89	3.7165.10	annealed	820	890	6			4.5s	5.5s	580	560
TiAl 5 Sn 2	F79	3.7115.10	annealed	760	790	6			4s	4.5s	450	540
TiAl 7 Mo 4	hp	—	hot precipitation-hardened	1000	1080	8	300					
TiAl 6 Zr 5	hp	—	hot precipitation-hardened	1140	1270	6						

Appendix D3, Table 32. Fine zinc casting alloys to DIN 1743

Abbreviation	$R_{p\,0.2}$	R_m	A_s	HB	Fatigue strength under alternating bending stresses at $N = 20 \cdot 10^6$
	N/mm²	N/mm²	%		N/mm²
GD-ZnAl 4	200 to 230	250 to 300	3 to 6	70 to 90	6 to 8
GD-ZnAl 4 Cu 1	220 to 250	280 to 350	2 to 5	85 to 105	7 to 10
G-ZnAl 4 Cu 3	170 to 200	220 to 260	0.5 to 2	90 to 100	—
GK-ZnAl 4 Cu 3	200 to 230	240 to 280	1 to 3	100 to 110	—
G-ZnAl 6 Cu 1	150 to 180	180 to 230	1 to 3	80 to 90	—
GK-ZnAl 6 Cu 1	170 to 200	220 to 260	1.5 to 3	80 to 90	—

Appendix D3, Table 33. Lead and its alloys to DIN 1719, DIN 1741 and DIN 17 641

Abbreviation	R_m N/mm²	A_s %	HB 2.5/31.25 approx.
Pb 99.99			4
Pb 99.90			5
Pb 98.5			7
R-Pb			8
GD-Pb 95 Sb	50	15	10
GD-Pb 87 Sb	60	10	14
GD-Pb 85 SbSn	70	8	18
GD-Pb 80 SbSn	74	8	18

Appendix D3, Table 34. Tin and tin alloys to DIN 1704 and DIN 1742

Abbreviation	R_m N/mm²	A_s %	HB 2.5/31.25
GD-Sn 80 Sb	115	2.5	30
GD-Sn 60 SbPb	90	1.7	28
GD-Sn 50 SbPb	80	1.9	26

Appendix D3, Table 35. Mechanical and physical characteristics of ceramics

Characteristic	Dimension	Temperature °C	Oxide ceramics			Non-oxide ceramics		
			Al_2O_3	Al_2TiO_5	ZrO_2	SSiC	SiSiC	SSN
Density	g/cm³	20	3.85	3.2	5.95	3.15	3.05	3.25
Flexural strength	N/mm²	20	350	40	950	410	380	750
(4-point)		1000	230	50	400	400	350	450
Modulus of elasticity	10³ N/mm²	20	370	18	200	410	350	280
Tensile strength	MN/m³/²	20	4.9	—	12	3.3	3.3	7.0
Thermal elongation	10⁻⁶/K	20 to 1000	8.0	1.0	10	4.7	4.5	3.2
Thermal conductivity	W/m K	20	28	2.0	2.5	110	140	35
		1000	15	1.5	1.8	45	50	17

Appendix D3, Table 36. Cements to DIN 1164

Strength class		Compressive strength (N/mm²) after				Identifying colour (basic colour of bag)	Colour of impression
		2 days min.	7 days min.	28 days			
				min.	max.		
250	a	—	10.0	25.0	45.0	Violet	Black
350	L[b]	—	17.5	35.0	55.0	Light brown	Black
	F[c]	10.0	—				Red
450	L[b]	10.0	—	45.0	65.0	Green	Black
	F[c]	20.0	—				Red
550		30.0	—	55.0	—	Red	Black

[a] Only for cement with low heat of hydration and/or high sulphate resistance.
[b] L = cement with slow initial hardening.
[c] F = cement with high initial strength.

Appendix D3, Table 37. Strength characteristics[a] of air-dried structural timber (average humidity content approx. 15%)

Type of wood	Position in relation to grain	Weight per unit volume kg/dm³	R_m N/mm²	σ_{dB} N/mm²	Bend strength σ_{bB} N/mm²	Shear strength τ_B N/mm²	Long-term bend strength σ_{bw} N/mm²
Oak	‖	0.4 to 0.7 to 0.95	50 to 90 to 180	40 to 50 to 60	70 to 90 to 100	5 to 10 to 15	—
	⊥		5	10		30	
Ash	‖	0.5 to 0.7 to 0.9	30 to 100 to 220	30 to 50 to 60	50 to 100 to 180	7	35
	⊥		7	10			
Hickory	‖	0.7 to 0.8 to 1	150	50	110 to 120	10	—
	⊥		10	10			
Walnut	‖	0.6 to 0.7 to 0.75	100	40 to 60 to 70	80 to 120 to 140	—	40
	⊥		4	10			
Elm	‖	0.5 to 0.7 to 0.85	60 to 80 to 210	30 to 40 to 60	50 to 70 to 160	7	—
	⊥		4	10		25	
Copper beech	‖	0.5 to 0.7 to 0.9	60 to 140 to 180	40 to 50 to 80	60 to 110 to 180	5 to 10 to 20	—
	⊥		7	10		35	
Common beech	‖	0.5 to 0.8 to 0.85	50 to 110 to 200	40 to 70 to 80	50 to 110 to 140	10	—
	⊥		6	10		30	
Pine	‖	0.3 to 0.5 to 0.9	40 to 100 to 190	30 to 50 to 80	40 to 90 to 200	5 to 10 to 15	25
	⊥		3	10		90	20
Pitch pine	‖	0.5 to 0.7 to 0.9	100	30 to 50 to 80	90	10	—
	⊥		3	7			
Spruce	‖	0.3 to 0.5 to 0.7	40 to 90 to 240	30 to 50 to 70	40 to 70 to 120	5 to 10	20
	⊥		3	5 to 10		25	
Fir	‖	0.3 to 0.45 to 0.7	50 to 80 to 120	30 to 40 to 50	40 to 60 to 100	5	—
	⊥		2	4		25	
Gaboon	‖	0.2 to 0.3 to 0.5	20 to 30 to 40	10 to 15 to 20	25	—	15

[a] The middle figures of the three figures set out represent the most common values.

Appendix D3, Table 38. Permitted loads for structural wood in load case H

Line	Type of load		Permitted loads (N/mm²) for					
			Coniferous woods (European) Quality category			Laminated wood (glued from European coniferous woods) as in Section 11.5.5		Oak and beech
			III	II	I	II	I	Medium quality
1	Bend		7	10	13	11	14	11
2	Tensile	permitted σ_B	0	8.5	10.5	8.5	10.5	10
3	Compressive	permitted σ_z II	6	8.5	11	8.5	11	10
4	Compressive	permitted σ_D II	2	2	2	2	2	3
		permitted σ_D	2.5[a]	2.5[a]	2.5[a]	2.5[a]	2.5[a]	4[a]
5	Shear	permitted τ	0.9	0.9	0.9	0.9	0.9	1.0
6	Shear from transverse load	permitted τ	0.9	0.9	0.9	1.2	1.2	1.0

[a] Where these values are applied, large indentations must be taken into account and possibly allowed for at structural level. These values must not be applied for connections with different fasteners.

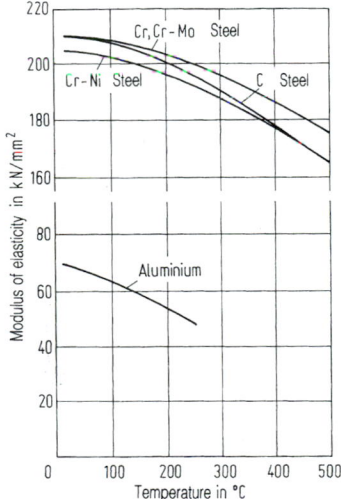

Figure 1. Temperature dependence of modulus of elasticity for aluminium and for steel.

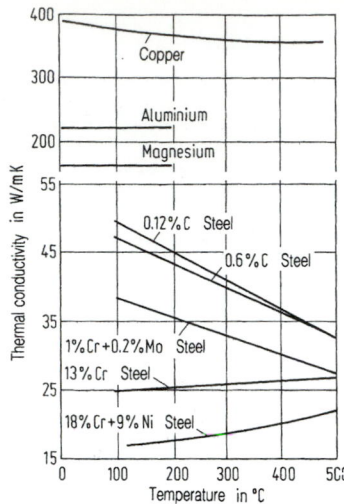

Figure 2. Temperature dependence of thermal conductivity of various steels and non-ferrous metals.

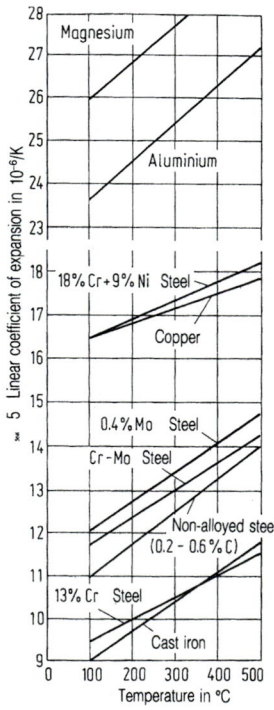

Figure 3. Temperature dependence of linear heat expansion coefficients of various non-ferrous metals and steels.

Appendix D3, Table 39. Physical characteristics of non-ferrous metals and their alloys

Material	Density g/cm³	Melting point and solidification range °C	Heat forming temperature °C	Index of linear contraction %	Modulus of elasticity E kN/mm²	Shear modulus G kN/mm²	Poisson's ratio μ	Linear thermal expansion coefficient 20 to 100 °C 10⁻⁶/K	Specific heat 20 to 100 °C J/g·K	Thermal conductivity at 20 °C J/cm·s·K	Specific resistance at 20 °C Ω·mm²/m
Aluminium	2.70	660	480 to 500	1	72.2	27.2	0.34	24	0.896	2.11	0.026
AlCuMg	2.8	530 to 645	380 to 460	1.2	71.5			22.8	0.92	1.59	0.050
AlMgSi	2.7	600 to 640	450 to 500	1.1	70.0			23.1	0.92	1.76	0.035
AlMg 5	2.6	580 to 630	380 to 420	1.2	69.5			23.5	0.92	1.17	0.060
G-AlSi 12	2.65	570 to 600	—	1.1	76.0			20.5	0.88	1.59	0.048
Lead	11.34	327	—	—	16.0	5.7	0.44	29.1	0.125	0.347	0.2
Copper	8.93	1 083	800 to 950	1.5	125	46.4	0.35	16.86	0.385	3.85	0.017
Copper-zinc alloy	8.3	895 to 1 025	700 to 850	2.0	104	40	0.37	19.2	0.39	1.17	0.07
Copper-tin alloy	8.8	910 to 1 040	600 to 900	2.0	116	43	0.35	17	0.37	0.71	0.11
Copper-beryllium alloy	8.9	950	600 to 900	2.0	120	45	0.38	17.5		0.84	0.07
Copper-aluminium alloy	7.73	1 030 to 1 080			123	47		17.9	0.45	0.71	0.114
Constantan 54 Cu, 45 Ni, 1 Mn	8.9	1 250	850 to 1 100	—	—	—	—	15.2		0.21	0.50
Magnesium	1.74	650	250 to 450	1.9	45.15	17.7	0.33	26.0	0.102	1.575	0.045
MgMn 2	1.8	645 to 650	280 to 320	1.4	45		0.3	26.0	0.105	1.42	0.06
MgAl 6 Zn	1.8	430 to 600	—	1.4	44		0.3	26.0	0.105	0.84	0.14
GD-MgAl 6 Zn 1	1.8	400 to 600			44		0.3	26.5	0.105	0.84	0.15
Nickel	8.86	1 453	870 to 1 150	2.0	197	75	0.31	13.3	0.444	0.92	0.069
67 Ni, 32 Cu, 1 Mn (Monel)	8.9	1 300 to 1 350		2.0	200			14	0.42	0.25	0.44
84 Ni, 9 Si, 4 Cu 1 Cr	7.8	1 100 to 1 120			205			11	0.45	0.21	0.11
Titanium	4.5	1 668	700 to 1 000		105.2	38.7	0.33	8.35	0.616	0.15	0.42
Titanium alloys	4.45 to 4.6	1 668			105						
Zinc	7.14	419.5	200 to 260		94	37.9	0.25	29	0.41	1.11	0.061
GD-ZnAl 4	6.6	380 to 386		1.3	130			27	0.42	1.13	0.06
GD-ZnAl 4 Cu	6.7	380 to 386		1.3	130			27	0.42	1.09	0.06
Tin	7.29	231.9			55	20.6	0.33	21.4	0.222	0.64	0.115

Appendix D4

Appendix D4, Table 1. Characteristics of the main groups of plastics (selection)

NB = non-breakable; dr = dry; h = humid; italics: characteristics for filled or reinforced plastics.

Plastic	Abbreviation DIN 7728	Density g/cm³	Strength characteristics N/mm²		Elongation values %		Modulus of elasticity N/mm²	ISO 180 impact strength kJ/m²
Polyamide	PA 6	1.12 to 1.14	60 to 90 dr	(σ_{zS})	6 to 12 dr	(ε_{zS})	1 500 to 3 200 dr	NB dr
			35 to 70h	(σ_{zS})	10 to 20 h	(ε_{zS})	600 to 1 600 b.	NB h
		to 1.4	*150 to 220 dr*	(σ_{zM})	*4 to 6*	(ε_{zM})	*10 000 to 18 000 dr*	*40 to 65 dr*
			120 to 170 b	(σ_{zM})			*5 000 to 10 000 b*	*60 to 80 h*
	PA 66	1.13 to 1.15	70 to 90 dr	(σ_{zS})	6 to 12 dr	(ε_{zS})	2 000 to 3 500 dr	NB dr
			55 to 75 h	(σ_{zS})	10 to 20 h	(ε_{zS})	1 200 to 2 100 h	NB h
		1.4	*180 to 230 dr*	(σ_{zM})	*2 to 5 dr*	(ε_{zM})	*9 000 to 17 000 dr*	*40 to 60 dr*
			130 to 180 b	(σ_{zM})			*6 000 to 10 000 b*	*60 to 80 b*
	PA 11	1.03 to 1.05	40 to 60	(σ_{zS})	9 to 22	(ε_{zS})	800 to 1 400	
		to 1.26	*60 to 150*	(σ_{zM})			*3 000 to 4 000*	
	PA 12	1.01 to 1.02	35 to 50	(σ_{zS})	8 to 26		1 200 to 1 600	15 to 60
		to 1.25	*50 to 120*	(σ_{zM})	*3 to 8*		*4 000 to 5 000*	
Amorphous polyamide	PA 6-3-T	1.04 to 1.12	70 to 110	(σ_{zS})	6 to 10		2 800 to 3 000	
		to 1.4	*149 to 160*	(σ_{zM})	*3*		*9 000 to 10 000*	
Polyacetal resin	POM	1.4 to 1.45	60 to 80	(σ_{zS})	8 to 15	(ε_{zS})	2 500 to 3 500	80 to 130
		to 1.6	*90 to 140*	(σ_{zM})	*2 to 6*	(ε_{zM})	*5 000 to 12 000*	*15 to 40*
Linear polyester	PET	1.31 to 1.37	50 to 75	(ε_{zS})	3 to 4	(ε_{zS})	2 500 to 3 200	
		1.5 to 1.8	*120 to 180*	(ε_{zM})	*2 to 3*	(ε_{zM})	*6 500 to 12 000*	
	PBT	1.29 to 1.3	50 to 60	(σ_{zS})	3 to 4	(ε_{zS})	2 600 to 2 900	*100 to 170*
		1.5 to 1.6	*110 to 160*	(σ_{zM})	*2 to 3*	(ε_{zM})	*6 500 to 11 000*	*25 to 60*
Polycarbonate	PC	1.2 to 1.23	55 to 70	(σ_{zS})	5 to 7	(ε_{zS})	2 000 to 2 500	40 to 60
		1.27 to 1.45	*70 to 150*	(σ_{zM})	*2 to 5*	(ε_{zM})	*3 500 to 9 000*	*35 to 45*
Polyphenylene ether	PPE	1.04 to 1.11	36 to 70	(σ_{zS})	3 to 8	(ε_{zS})	2 000 to 2 500	30 to 170
		to 1.38	*70 to 140*	(σ_{zR})	*1 to 3*	(ε_{zM})	*3 500 to 9 000*	*12 to 28*
Polyacrylate	PMMA	1.17 to 1.2	60 to 90	(σ_{zM})	2 to 10	(ε_{zM})	2 400 to 4 500	
Polystyrene	PS	1.05	45 to 65	(σ_{zM})	2 to 4	(ε_{zM})	3 000 to 3 600	5 to 15
Styrene butadiene	SB	1.04 to 1.05	15 to 50	(σ_{zS})	2 to 3	(ε_{zS})	1 500 to 3 000	15 to 90
Styrene acrylonitrile	SAN	1.08	70 to 80	(σ_{zM})	5	(ε_{zM})	3 600	
		1.2 to 1.4	*to 140*	(σ_{zM})	*3*	(ε_{zM})	*5 000 to 10 000*	
Acrylonitrile butadiene	ABS	1.06 to 1.08	30 to 55	(ε_{zS})	2 to 3	(ε_{zS})	1 500 to 2 900	
Styrene		*1.09 to 1.5*	*70*	(σ_{zM})	*1*	(ε_{zS})	*4 500 to 6 000*	
SAN with acrylic ester	ASA	1.07	45 to 60	(σ_{zM})	10 to 20	(ε_{zR})	2 500 to 2 800	
Cellulose ester	CA	1.22 to 1.35	30 to 65	(σ_{zS})	3 to 5	(ε_{zS})	2 000 to 3 600	
	CP	1.19 to 1.24	18 to 28	(σ_{zS})	3 to 5	(ε_{zS})	1 000 to 2 500	
	CAB	1.15 to 1.24	16 to 25	(σ_{zS})	3 to 5	(ε_{zS})	800 to 2 200	
Polysulphone	PSU	1.24	70 to 100	(σ_{zS})	5 to 6	(ε_{zS})	2 100 to 2 500	
	PES	1.38	85 to 95	(σ_{zS})	5 to 6	(ε_{zS})	2 500 to 3 100	
Polyphenylene sulphide	PPS	1.35	70 to 80	(σ_{zM})	3	(ε_{zM})	3 500	15 to 30
		to 2.06	*80 to 150*	(σ_{zM})	*1 to 2*	(ε_{zM})	*12 000 to 16 000*	
Polyimide	PI	1.4 to 1.5	70 to 100	(σ_{zM})			3 000 to 3 500	
		to 1.9	*100 to 200*	(σ_{zM})	*1 to 6*	(ε_{zR})	*6 000 to 30 000*	
Polyethylene	PE-HD	0.94 to 0.96	20 to 35	(σ_{zS})	12 to 20	(ε_{zS})	400 to 1 500	
	PE-LD	0.92 to 0.94	8 to 20	(σ_{zS})	8 to 14	(ε_{zR})	150 to 600	
Polypropylene	PP	0.9	18 to 38	(σ_{zS})	10 to 20	(ε_{zS})	650 to 1 400	50 to NB
		to 1.32	*40 to 75*	(σ_{zM})	*7 to 70*	(ε_{zR})	*2 500 to 6 000*	*25 to 50*
Polyvinyl chloride (hard)	PVC-U	1.32 to 1.45	50 to 80	(σ_{zM})	3 to 7	(ε_{zS})	2 900 to 3 600	5 to 100
(soft)	PVC-P	1.2 to 1.35	15 to 30	(σ_{zR})	50 to 300	(ε_{zR})	450 to 600	
Fluorinated plastic	PTFE	2.1 to 2.2	9 to 12	(σ_{zS})	250 to 500	(ε_{zR})	450 to 750	
Fluorinated thermoplastic	FEP	2.1 to 2.17	19 to 22	(σ_{zS})	250 to 350	(ε_{zR})	350 to 600	
	ETFE	1.7	27	(σ_{zS})	150 to 200	(ε_{zR})	800 to 1 400	
	PVDF	1.77	50	(σ_{zS})	20 to 25	(ε_{zR})	1 000 to 2 000	
Phenol formaldehyde	PF	*1.4 to 1.9*	*15 to 40*	(σ_{zM})	*to 1*	(ε_{zR})	*6 000 to 10 000*	
Amino plastic	UF/MF	*1.5 to 2.0*	*15 to 30*	(σ_{zM})	*to 1*	(ε_{zR})	*5 000 to 9 000*	

Notch impact strength ISO 180 kJ/m²	Impact strength DIN 53 453 kJ/m²	U-notch toughness DIN 53 453 kJ/m²	Creep stress $\sigma_{1/1000}$ N/mm²	Thermal conductivity J/(m·K)	Coefficient of linear expansion 10^{-5} 1/K	Moulding shrinkage %	Crystalline melting point C	Abbreviation DIN 7728
	NB dr	3 to 20 dr	6 dr	0.27 to 0.30	7 to 11 dr	0.8 to 2.0 dr	215 to 225	PA 6
	NB h	19 to NB h	4 h					
10 to 18 dr	35 to 60 dr	3 to 14 dr	40 to 50 dr	0.30 to 0.32	2 to 5 dr	0.2 to 1.0 dr		
16 to 25 h		12 to 20 h	30 to 40 h					
	NBr dr	2 to 14 dr	7 dr	0.27 to 0.28	6 to 10 dr	0.8 to 2.2 dr	250 to 265	PA 66
	NBr h	10 to 20 h	6 h					
7 to 12 dr	30 to 40 dy	10 to 14	50 to 60 dr	0.28 to 0.30	1 to 5 dy	0.2 to 0.8 dr		
12 to 20 h								
	NB	10 to NB	5	0.28	9 to 13	0.5 to 1.5	180 to 190	PA 11
5 to 28			12		2 to 4	0.4 to 1.0		
	NB	10 to NB	4 to 5	0.27	12 to 15	0.5 to 1.5	175	PA 12
	50 to 70	7 to 10			3 to 5			
	NB	10 to 20	12		6 to 8	0.4 to 0.7		PA 6-3-T
	28 to 32	6 to 8						
4 to 7	100 to NB	4 to 10	12 to 18	0.29 to 0.36	11 to 13	1.6 to 2.8	175 (homo-polym.)	POM
3 to 5	10 to 30	3 to 6		0.40	2 to 4	0.4 to 1.0	165 to 168 (co-polym.)	
	NB	3 to 4	26	0.24 to 0.29	7	1.3 to 2.0	255 to 258	PET
	25 to 35	7 to 10		0.33 to 0.34	2 to 3	0.3 to 0.8		
8	NB	3 to 5	12 to 15	0.21	3 to 7	1.3 to 2.0	220 to 225	PBT
6 to 13	20 to 45	6 to 11	55	0.23 to 0.26	3 to 4	0.3 to 0.8		
	NB	20 to 36	18	0.21 to 0.23	6 to 7	0.7 to 0.8		PC
10 to 16	30 to 70	6 to 15	40	0.23 to 0.25	2 to 5	0.2 to 0.5		
9 to 60	15 to NB	15 to 20	18	0.17 to 0.22	5 to 10	0.5 to 0.7		PPE
6 to 12	8 to 10	8 to 10	35	0.22 to 0.28	3 to 5	0.1 to 0.5		
	12 to 22	2 to 3	15 to 20	0.18 to 0.19	7 to 9	0.3 to 0.8		PMMA
2	5 to 25	2 to 3	18 to 20	0.15 to 0.17	7 to 8	0.4 to 0.7		PS
4 to 12	40 to 80 to NB	5 to 12	12	0.16 to 0.17	8 to 10	0.4 to 0.7		SB
	8 to 25	2 to 3	15 to 25	0.15 to 0.17	6 to 8	0.4 to 0.6		SAN
			60					
	50 to 80 to NB	8 to 25	9 to 15	0.15 to 0.17	8 to 11	0.4 to 0.8		ABS
	15	5 to 8	30 to 40		3 to 4	0.1 to 0.4		ASA
	NB	7 to 20	12	0.17	10 to 11	0.4 to 0.7		
	80 to NB	3 to 25	5 to 10	0.20 to 0.22	9 to 12	0.4 to 0.7		CA
	NB	3 to 25	5 to 10	0.20 to 0.22	12 to 15	0.4 to 0.7		CP
	NB	3 to 25	5 to 10	0.20 to 0.22	12 to 15	0.4 to 0.7		CAB
7	NB	2 to 4	18	0.26 to 0.28	5 to 6	0.7 to 0.8		PSU
9	NB	3 to 5	23	0.18	5 to 6	0.5 to 0.7		PES
					6		280 to 288	PPS
	4 to 15	3 to 7	20	0.25	4	0.2		
			30 (PEI)	0.22 (PEI)	5 to 6			PI
4 to 8			60 (PEI-GF)		2 to 3	0.1 to 0.5		
3 to 20	NB	3 to 20 to NB	2 to 5	to 0.51	13 to 20	2.0 to 5.0	125 to 140	PE-HD
10 to 70	NB	NB	1 to 3	0.29 to 0.40	18 to 24	1.5 to 3.0	105 to 115	PE-LD
7 to 15	NB	4 to 14	5 to 6	0.20 to 0.22	10 to 18		158 to 168	PP
2 to 7	15 to 50	4 to 8	6 to 20	0.25 to 0.51	6 to 10	1.0 to 2.5		
	20 to NB	4 to 8	20 to 25	0.14 to 0.17	7 to 8	0.5 to 1.0		PVC-U
	NB	6 to NB		0.12 to 0.15	18 to 21	1.0 to 3.0		PVC-P
	NB	13 to 16	1 to 2	0.25			327	PTFE
				0.20 to 0.23	12 to 16	3.0 to 4.0	285 to 295	FEP
				0.24	8 to 10		270	ETFE
		20		0.14 to 0.15	9	2.0 to 2.5	171	PVDF
	3 to 15	1 to 15		0.30 to 0.7	1 to 5	0.2 to 0.8		PF
	4 to 18	1 to 10		0.35 to 0.70	2 to 6	0.2 to 1.2		UF/MF

Appendix D4, Table 1 Continued

Plastic	Abbreviation DIN 7728	Density g/cm³	Strength characteristics N/mm²		Elongation values %	Modulus of elasticity N/mm²	Impact strength ISO 180 kJ/m²
Unsaturated polyester	UP	1.5 to 2.0	20 to 200 Laminates to 1000	(σ_{zM}) (σ_{zM})	up to 1 (ε_{zR})	3 000 to 19 000	
Epoxy resin	EP	1.5 to 1.9	60 to 200 Laminate to 1000	(σ_{zM}) (σ_{zM})	2 to 5 (ε_{zR})	5 000 to 20 000	
Steel	Fe	7.8	300 to 1 500	(R_{m})	2 to 30 (A_{s})	210 000	
Aluminium (alloys)	Al	2.7	50 to 500	(R_{m})	2 to 40 (A_{s})	70 000	
Copper (alloys)	Cu	8.9	200 to 1 200	(R_{m})	2 to 60 (A_{s})	100 000	

Appendix D5

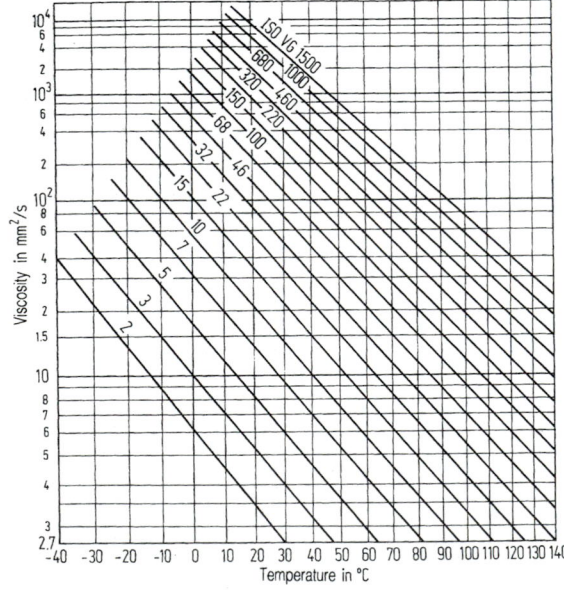

Figure 1. Viscosity temperature sheet (ISO VG series; plot with $VI = 100$).

10	15	2	3	5	7
100	150	22	32	46	68
1 000	1 500	220	320	460	680

Mid-point viscosity values in mm²/s at 40 °C with ± 10% tolerance.
Formula: $v_{n+1} =$ approx. $1.5 \cdot v_{n}$.

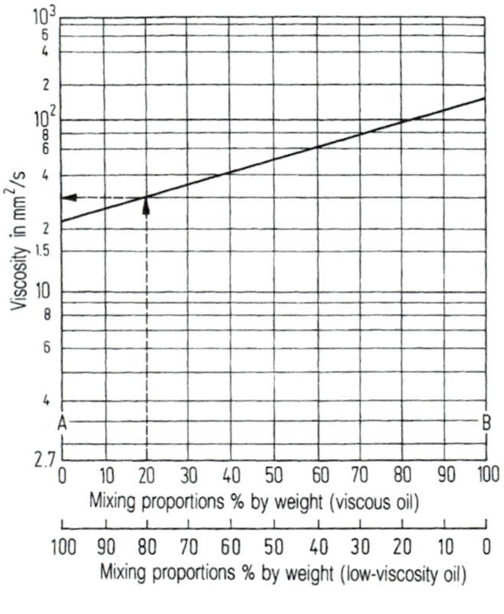

Figure 2. Mixing diagram for mineral oils. Note: It should only be used for plotting oils whose viscosity data relate to the *same temperature*.

Notch impact strength ISO 180 kJ/m²	Impact strength DIN 53 453 kJ/m²	U-notch toughness DIN 53 453 kJ/m²	Creep stress $\sigma_{1/1000}$ N/mm²	Thermal conductivity J/(m·K)	Coefficient of linear expansion 10^{-5} 1/K	Moulding shrinkage %	Crystalline melting point C	Abbreviation
	4 to 40 Laminate to 150	4 to 20 Laminate to 60	Laminate 50 to 150	0.50 to 0.70	2 to 10	0.3 to 0.8		UP
	15 to 200	5 to 20 Laminate to 60	Laminate 100 to 150	0.40 to 0.80	2 to 6	0.0 to 0.5		EP
				75	1.2			Fe
				230	2.35			Al
				390	1.65			Cu

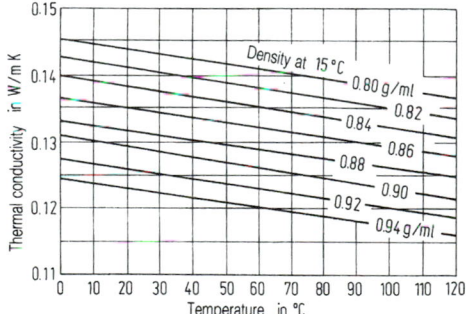

Figure 3. Liquid lubricants dependence of thermal conductivity on temperature.

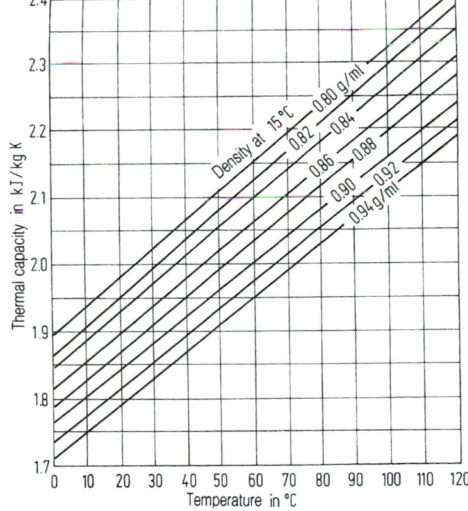

Figure 4. Liquid lubricants: dependence of thermal capacity on temperature.

Appendix D5, Table 1. Viscosity-pressure coefficients α of lubricants and increases in viscosity due to pressure [10]

Type of oil	$\alpha_{25\,°C} \cdot 10^5$ bar^{-1}	$\dfrac{\eta_{2000\,bar}}{\eta_{1\,bar}}$ at 25 °C	$\dfrac{\eta_{2000\,bar}}{\eta_{1\,bar}}$ at 80 °C
		ca.	ca.
Paraffin-based mineral oils	1.5 to 2.4	15 to 100	10 to 30
Naphthene-based mineral oils	2.5 to 3.5	150 to 800	40 to 70
Aromatic solvent extracts	4 to 8	1000 to 200 000	100 to 1000
Polyolefins	1.3 to 2.0	10 to 50	8 to 20
Ester oils (diester, blanched)	1.5 to 2.0	20 to 50	12 to 20
Polyether oils (aliphatic)	1.1 to 1.7	9 to 30	7 to 13
Silicone oils (aliphatic substitute)	1.2 to 1.4	9 to 16	7 to 9
Silicone oils (aromatic substitute)	2 to 2.7	300	—
Chlorinated paraffins (depending on degree of halogenisation)	0.7 to 5	5 to 20 000	

Appendix D5, Table 2. SAE viscosity categories for engine lubrication oils to DIN 51 511 (ISO DIS 10 369)

SAE viscosity category	Maximum apparent viscosity[a] (mPa·s) at temperature (°C)	Maximum pumping limit temperature[b] (°C)	Kinematic viscosity[c] at 100 °C (mm²/s)	
			min.	max.
0 W	3250 at −30	−35	3.8	—
5 W	3500 at −25	−30	3.8	—
10 W	3500 at −20	−25	4.1	—
15 W	3500 at −15	−20	5.6	—
20 W	4500 at −10	−15	5.6	—
25 W	6000 at − 5	−10	9.3	—
20	—	—	5.6	below 9.3
30	—	—	9.3	below 12.5
40	—	—	12.5	below 16.3
50	—	—	16.3	below 21.9

[a] Testing to DIN 51 377 (ASTM 2602).
[b] Testing to ASTM D 3829 and CEC L-32-T-82 (ISO DIS 9262).
[c] Testing to DIN 51 550 (ISO DS 3104) in conjunction with DIN 51 561 (ISO DS 3105) and DIN 51 562, Section 1 (ISO DS 3105).

Appendix D5, Table 3. Consistency classes of lubricating oils to DIN 51 818 (ISO 2137) and applications [11]

NLGI class	Penetration mm/10	Consistency	Sliding bearing	Roller bearing	Central lubrication systems	Transmission lubrication	Water pumps	Block greases
000	445 to 475	Almost liquid			×	×		
00	400 to 430	Semi-liquid			×	×		
0	355 to 385	Exceptionally soft			×	×		
1	310 to 340	Very soft			×	×		
2	265 to 295	Soft	×	×				
3	220 to 250	Medium consistency	×	×				
4	175 to 205	Fairly soft		×			×	
5	130 to 160	Solid					×	
6	85 to 115	Very solid and rigid						×

7 References

D1   Fundamental Properties of Materials and Structural Parts. [1] Kloos KH, Broszeit E. Verschleissschäden durch Oberflächenermüdung. VDI-Ber 1975; 243: 189–204. – [2] Macherauch E, Wohlfahrt H, Wolfstieg U. Zur zweckmässigen Definition von Eigenspannungen. Härterei-Techn Mitt 1973; 28: 200–11. – [3] Klein HD, Eigenspannungen und ihre Verminderung in metallischen Werkstücken durch spannende Bearbeitung. Diss., TH Hannover, 1969. – [4] Staudinger H. Metallurgische Vorgänge beim Schleifen von Stahl. Schweizer Archiv 1957; 23: 231–40. – [5] Siebel E. Werkstoffmechanik. VDI-Sonderheft Werkstoffe, I. VDI-Verlag, Düsseldorf, 1953, pp 27–33. – [6] Heckel K. Einführung in die technische Anwendung der Bruchmechanik. Hanser, Munich, 1970. – [7] Macherauch E. Bruchmechanik. Grundlagen des Festigkeits- und Bruchverhaltens. Stahleisen, Düsseldorf, 1974. – [8] Jacoby G. Schwingfestigkeit. Neuzeitliche Verfahren der Werkstoffprüfung. Stahleisen, Düsseldorf, 1973. – [9] Schütz W. Versuchsmethoden der Bruchmechanik. Der Maschinenschaden 1975; 48: 137–48. – [10] Rühl K. Tragfähigkeit metallischer Baukörper. Ernst & Sohn, Berlin, 1952. – [11] Wellinger K, Dietmann H. Festigkeitsberechnung. Grundlagen und technische Anwendung. Kröner, Stuttgart, 1976. – [12] Dietmann H. Werkstoffverhalten unter mehrachsiger schwingender Beanspruchung. Pt 1: Berechnungsmöglichkeiten. Z Werkstofftechnik 1973; 4: 255–63. – [13] Troost A, El-Magd E. Beurteilung der Schwingfestigkeit bei mehrachsiger Beanspruchung auf der Grundlage kritischer Schubspannungen. Metall 1976; 30: 37–41. – [14] Issler L. Festigkeitsverhalten metallischer Werkstoffe bei mehrachsiger phasenverschobener Schwingbeanspruchung. Diss., Univ. Stuttgart, 1973. Auszug in VDI-Ber 268: Werkstoff- und Bauteilverhalten unter Schwingbeanspruchung, pp 93–100. VDI-Verlag, Düsseldorf, 1976. – [15] Simbürger A. Festigkeitsverhalten zäher Werkstoffe bei einer mehrachsigen, phasenverschobenen Schwingbeanspruchung mit körperfesten und veränderlichen Hauptspannungsrichtungen. Diss., TH Darmstadt, 1975. – [16] Das Verhalten mechanisch beanspruchter Werkstoffe und Bauteile unter Korrosionseinwirkung. VDI-Ber 235. VDI-Verlag, Düsseldorf, 1975. – [17] Spähn H. Korrosionsgerechte Gestaltung. VDI-Ber 1977; 277: 37–45. – [18] Czichos H. The principles of systems analysis and their application to tribology. ASLE Trans 1974; 17. – [19] Kloos KH. Material selection and material pairing. Tribotechnical considerations. Wear 1975; 34: 95–107. – [20] Seeger T. Werkstoffmechanisches Konzept der Dauer- und Zeitfestigkeit. In: VDI-Ber 661. VDI-Verlag, Düsseldorf, 1988. – [21] Dahl W. Grundlagen und Anwendungsmöglichkeiten der Bruchmechanik bei der Sprödbruchprüfung. Z Metallkunde 1970; 61: 794–804. – [22] Schütz W. Schwingfestigkeit von Werkstoffen. VDI-Ber 1974; 214: 45–57. – [23] DDR-Standard TGL 19 340, Maschinenbauteile, Dauerschwingfestigkeit. Verlag für Standardisierung, Leipzig, 1983. – [24] Buch A. Einige Bemerkungen über das Einflussfaktorenverfahren zur Berechnung der Dauerfestigkeit von Maschinenbauteilen. Materialprüfung 1976; 18: 194–99. – [25] Kloos KH, Granacher J, Rieth P, Barth H. Hochtemperaturverhalten warmfester Stähle unter zeitlich veränderter Beanspruchung. VGB Kraftwerkstechnik 1984; 64: 1020–34. – [26] Schmidtmann E, Mall HP. Die Kennzeichnung der Sprödbruchneigung eines Stahles durch Auswertung von Kraft-Durchbiegung – Kurven aus Kerbschlagbiegeversuchen. Arch Eisenhüttenwes 1967; 38: 571–6. – [27] Degenkolbe

J, Müsgen B. Studium des Rissauslösungsverhaltens von Baustählen, Versuche mit Scharfkerbbiegeproben. Materialprüfung 1969; 11: 365–72. – [28] Dahl W. Prüfung der Sprödbruchunempfindlichkeit. Neuzeitliche Verfahren der Werkstoffprüfung. Stahleisen, Düsseldorf, 1973. – [29] Schinn R, Schieferstein U. Anforderungen und Abnahmekriterien für schwere Schmiedestücke des Turbogeneratorenbaues. VGB-Kraftwerkstechn 1973; 53: 182–95. – [30] Hochstein F. Beitrag zur Herstellung schwerer Schmiedestücke aus Stahl, metallurgisch bedingte Eigenschaften und neuere Prüfkriterien. Stahl Eisen 1975; 95: 777–84. – [31] Hagedorn KE. Messmethoden und technische Anwendungen der Bruchmechanik. Z Werkstofftechnik 1972; 3: 122–9. – [32] Riedel H. Fracture at high temperatures. Springer, Berlin, 1987. – [33] Buch A. Einfluss der Stahlreinheit und der Stahlhärte auf die Anisotropie der mechanischen Eigenschaften von Schmiedestücken. IfL-Mitt 1967; 6: 402–8. – [34] Schlicht H. Einfluss der Stahlherstellung auf das Ermüdungsverhalten und Bauteilen bei kräftefreier und kräftegebundener Oberfläche. VDI-Ber 268. VDI-Verlag, Düsseldorf, 1976. – [35] Randak A, Stanz A, Verderber W. Eigenschaften von nach Sonderschmelzverfahren hergestellten Werkzeug- und Wälzlagerstählen. Stahl Eisen 1972; 92: 891–3. – [36] Flemming G. Mechanische Eigenschaften von Stahl bei statischer und wechselnder Beanspruchung nach einer Massivformgebung. Diss., Darmstadt, 1972. – [37] Wiegand H, Strigens P. Die Steigerung der Dauerfestigkeit durch Oberflächenverfestigung in Abhängigkeit von Werkstoff und Vergütungszustand. Draht 1969; 20: 189–94 and 302–8. – [38] Kloos KH, Adelmann J. Schwingfestigkeitssteigerung durch Festwalzen. Matwiss Werkstofftechnik 1988; 19: 15–23. – [39] VDI-Richtlinie 2227: Festigkeit bei wiederholter Beanspruchung, Zeit- und Dauerfestigkeit metallischer Werkstoffe, insbesondere von Stählen. VDI-Handbuch Konstruktion. – [40] Syren B, Wohlfahrt H, Macherauch E. Der Einfluss von Bearbeitungseigenspannungen auf das Biegewechselfestigkeitsverhalten von Stahl Ck 45 im weichgeglühten Zustand. Arch Eisenhüttenwes 1975; 46: 735–9. – [41] Tauscher H. Berechnung der Dauerfestigkeit. Fachbuchverlag, Leipzig, 1960. – [42] Wiegand H, Fürstenberg U. Hartverchromung. Eigenschaften und Auswirkungen auf den Grundwerkstoff. Maschinenbau-Verlag, Frankfurt, 1968. – [43] Funke P, Heye W, Randak A, Sikora E. Einfluss unterschiedlicher Randentkohlungen auf die Dauerschwingfestigkeit von Federstählen. Stahl Eisen 1976; 96: 28–32. – [44] Niemann G, Rettig H. Steigerungsmöglichkeiten für die Zahnflankentragfähigkeit. Konstruktion 19 (1968) 262–7. – [45] Hempel M. Zug-Druck-Wechselfestigkeit ungekerbter und gekerbter Proben warmfester Werkstoffe im Temperaturbereich von 500 bis 700 °C. Arch Eisenhüttenwes 1972; 43: 479–88. – [46] Wiegand H, Jahr O. Langzeiteigenschaften einiger warmfester und hochwarmfester Werkstoffe. Z Werkstofftechnik 1976; 7: 177–81 and 212–19. – [47] Spähn H. Grundlagen und Erscheinungsformen der Schwingungsrisskorrosion. VDI-Ber 1975; 235: 103–15. – [48] Speckhardt H. Grundlagen und Erscheinungsformen der Spannungsrisskorrosion. Massnahmen zu ihrer Vermeidung. VDI-Ber 1975; 235 – [49] Paatsch W. Probleme der Wasserstoffversprödung unter besonderer Berücksichtigung galvanotechnischer Prozesse. VDI-Ber 1975; 235: 97–101. – [50] Böhm H. Bedeutung des Bestrahlungsverhaltens für die Auswahl und Entwicklung warmfester Legierungen im Reaktorbau. Arch Eisenhüttenwes 1974; 45: 821-30. – [51] Rainer G. Kerbwirkung an gekerbten und abgesetzten Flach- und Rundstäben. Diss., TH Darmstadt, 1978. – [52] Dietmann H. Berechnung der Fliesskurven von Bauelementen bei kleinen Verformungen. Habil., Univ. Stuttgart, 1969. – [53] Wellinger K, Pröger M. Der Grösseneffekt beim Kerbzugversuch mit Stahl. Materialprüfung 1968; 10: 401-6. – [54] Kloos KH. Einfluss des Oberflächenzustandes und der Probengrösse auf die Schwingfestigkeitseigenschaften. VDI-Ber 1976; 268: 63-76. – [55] VDI-Richtlinie 2226: Empfehlung für die Festigkeitsberechnung metallischer Bauteile. VDI-Verlag, Düsseldorf; 1965. – [56] Neuber H. Über die Berücksichtigung der Spannungskonzentration bei Festigkeitsberechnungen. Konstruktion 1968; 20: 245-51. – [57] Heckel H, Köhler G. Experimentelle Untersuchung des statistischen Grösseneinflusses im Dauerschwingversuch an ungekerbten Stahlproben. Z Werkstofftechnik 1975; 6: 52-4. – [58] Hänchen R, Decker KH. Neue Festigkeitsberechnungen für den Maschinenbau. Hanser, Munich, 1967. – [59] VDI-Bericht 354: Übertragbarkeit von Versuchs- und Prüfergebnissen auf Bauteile. VDI-Verlag, Düsseldorf, 1979. – [60] Schuster C, Wirtgen G. Aufbau und Anwendung des DDR-Standards. TGL 19 340 "Maschinenbauteile, Dauerschwingfestigkeit". IfL-Mitt. 1975; 14: 3-29. – [61] Nowak B, Saal H, Seeger T. Ein Vorschlag zur Schwingfestigkeitsbemessung von Bauteilen aus hochfesten Baustählen. Stahlbau 1975; 44: 257-68 and 306-13. – [62] Minner HH, Seeger T. Erhöhung der Schwingfestigkeiten von Schweissverbindungen aus hochfesten Feinkornbaustählen durch das WIG-Nachbehandlungsverfahren. Stahlbau 1977; 46: 257-63. [63] Gassner E. Betriebsfestigkeit, eine Bemessungsgrundlage für Konstruktionsteile mit statistisch wechselnden Betriebsbeanspruch- ungen. Konstruktion 1954; 6: 97-104. – [64] Schütz W, Zenner H. Schadensakkumulationshypothesen zur Lebensdauervorhersage bei schwingender Beanspruchung – ein kritischer Überblick. Z Werkstofftechnik 1973; 4: 25-33 and 97-102. – [65] Haibach E. Modifizierte lineare Schadensakkumulations-Hypothese zur Berücksichtigung des Dauerfestigkeitsabfalles mit fortschreitender Schädigung. LBF-TM no. 50/70. – [66] Gassner E. Betriebsfestigkeit. In: Lueger Lexikon der Technik, Band Fahrzeugtechnik. – [67] Jacoby, G. Schwingfestigkeit. Neuzeitliche Verfahren der Werkstoffprüfung, pp 80-107. Stahleisen, Düsseldorf, 1973. – [68] Jacoby G. Beitrag zum Vergleich der Aussagefähigkeit von Programm- und Randomversuchen. Z Flugwissenschaften 1970; 18: 253-8. – [69] Verein Deutscher Eisenhüttenleute (ed.): Ergebnisse deutscher Zeitstandversuche langer Dauer. Düsseldorf, 1969. – [70] Wiegand H, Granacher J, Sander M. Zeitstandbruchverhalten einiger warmfester Stähle unter rechteckzyklisch veränderter Spannung oder Temperatur. Arch Eisenhüttenwes 1975; 46: 533-9. [71] Haibach E. Beurteilung der Zuverlässigkeit schwingbeanspruchter Bauteile. Luftfahrttechnik-Raumfahrttechnik 1967; 13: 188-93.

D2 Materials Testing. [1] Sachs K. Angewandte Statistik. Springer, Berlin; 1974. – [2] Bühler HY, Schreiber W. Lösung einiger Aufgaben der Dauerschwingfestigkeit mit dem Treppenstufen-Verfahren. Arch Eisenhüttenwes 1957; 28: 153-6. – [3] Maennig W-W. Bemerkungen zur Beurteilung des Dauerschwingfestigkeitsverhaltens von Stahl und einige Untersuchungen zur Bestimmung des Dauerfestigkeitsbereiches. Materialprüfung 1970; 12: 124-31. – [4] Little RE, Jebe EH. Statistical design of fatigue experiments. Applied Science Publishers, London. – [5] VDI-Richtlinie 2227: Festigkeit bei wiederholter Beanspruchung, Zeit- und Dauerfestigkeit metallischer

Werkstoffe, insbesondere von Stählen. VDI-Verlag, Düsseldorf. 1974. – [6] Heckel K. Einführung in die Techn. Anwendung der Bruchmechanik. Hanser, Munich, 1970. – [7] Schütz W. Versuchsmethoden der Bruchmechanik. Der Maschinenschaden 1975; 48: 137-48. – [8] Schwalbe KH. Bruchmechanik metallischer Werkstoffe. Hanser, Munich, 1980. – [9] Macherauch E. Bruchmechanik, Grundlagen des Festigkeits- und Bruchverhaltens. Stahleisen, Düsseldorf, 1974. – [10] Kloos KH, Granacher J, Rieth P, Barth H. Hochtemperaturverhalten warmfester Stähle unter zeitlich veränderter Beanspruchung. VGB Kraftwerkstechnik 1984; 64: 1020-34.

D3 Properties and Application of Materials. [1] Winderlich B, Brenner B. Bestimmende Faktoren für die Biegewechselfestigkeit laserstrahlgehärteter Proben aus Stahl C 70 W 2. HTM 1989; 44: 166-73. – [2] Gabriel HM. Oberflächenschutzschichten aus der Gasphase (CVD-, PVD-Verfahren) – Stand und Entwicklungstendenzen. Z Werkstofftechnik 1983; 14: 70-1. – [3] Wiegand H. Betrachtungen zur Werkstoffauswahl und Werkstoffnutzung. Z Werkstofftechnik 1973; 4: 93-6. – [4] Grosch J. Einsatzmöglichkeiten der EDV für die Werkstoffauswahl. VDI-Ber 1974; 214: 115-21. – [5] Grosch J. Systematische Erfassung von Anforderungen an Werkstoffe für Apparate und Anlagen - Anforderungsgerechte Auswahl von Werkstoffen. Z Werkstofftechnik 1978; 9: 338-43. – [6] Spies HJ. Beitrag zu den Grundlagen und der Methodik der Werkstoffauswahl. IFL-Mitt 1977; 16; 107-13. – [7] Schott G. Kostenbezogener Gebrauchswertfaktor als Grundlage für eine technisch und ökonomisch begründete Werkstoffauswahl. Maschinenbautechnik 1975; 24: 482-6. – [8] Gräfen H, Gerischer K, Horn E-M. Die Bedeutung der Werkstoffauswahl für die Gebrauchstauglichkeit von Chemieapparaten - Auswahlkriterien und Prüfverfahren. Z Werkstofftechnik 1973; 4: 169-86. – [9] Czichos H. Die systemtechnischen Grundlagen der Tribologie und ihre Anwendung zur Bearbeitung von Reibungs- und Verschleissproblemen. Schmiertechnik Tribologie 1977; 24: 109. – [10] Kloos KH. Werkstoffauswahl und Oberflächenbehandlung unter tribotechnischen Gesichtspunkten. Z Werkstofftechnik 1979; 10: 456-66. – [11] DIN 50 320: Verschleiss; Begriffe, Systemanalyse von Verschleissvorgängen, Gliederung des Verschleissgebietes.

D5 Tribology. [1] Jost P. Lubrication (Tribology). Her Majesty's Stationery Office, London, 1966. – [2] Mittmann HU, Czichos H. Reibungsmessungen und Oberflächenuntersuchungen an Kunststoff-Metall-Gleitpaarungen. Materialprüfung 1975; 17: 366-72. – [3] Dowson D, Higginson GR. A new roller bearing lubrication formula. Engineering (London) 1961; 19: 158-9. – [4] Dowson D. Elastohydrodynamic lubrication - the fundamentals of roller and gear lubrication, 2nd edn, Pergamon Press, Oxford, 1977. – [5] Hamrock BJ, Dowson D. Isothermal elastohydrodynamic lubrication of point contacts. Part III, Fully flooded results. Trans ASME J Lubr Eng 1977; 99 Ser. F, 264-75. – [6] Schmidt H, Bodschwinna H, Schneider U. Mikro-EHD: Einfluss der Oberflächenrauheit auf die Schmierfilmbildung in realen EHD-Wälzkontakten. Pt I: Grundlagen. Antriebstechnik 1987; 26: H.11, 55-60. Pt III: Ergebnisse und rechnerische Auslegung eines realen EHD-Wälzkontaktes. Antriebstechnik 1987; 26: H.12, 55-60. – [7] Winer WO, Cheng HS. Film thickness, contact stress and surface temperatures. In: Peterson MB, Winer WO (eds): Wear Control Handbook. The American Society of Mechanical Engineers, New York, 1980. – [8] Habig K-H. Verschleiss und Härte von Werkstoffen. Hanser, Munich, 1980. – [9] Czichos H. Tribology - a systems approach to the science and technology of friction, lubrication and wear. Elsevier, Amsterdam; 1978. - [10] Klamann D. Schmierstoffe und verwandte Produkte - Herstellung, Eigenschaften, Anwendung. Verlag Chemie, Weinheim, 1982. - [11] Möller UJ, Boor U. Schmierstoffe im Betrieb. VDI-Verlag, Düsseldorf, 1987. - [12] Buckley DH. Surface effects in adhesion, friction, wear, and lubrication. Elsevier, Amsterdam, 1981.

Standards and Guidelines. DIN 50 281: Reibung in Lagerungen; Begriffe, Arten, Zustände, physikalische Grössen. - DIN 50 320: Verschleiss; Begriffe, Systemanalyse von Verschleissvorgängen, Gliederung des Verschleissgebietes. - DIN 50 321: Verschleiss-Messgrössen. - DIN 50 322: Kategorien der Verschleissrüfung. - DIN 50 323: Tribologie; Pt 1: Begriffe. - DIN 51 365: Prüfung von Schmierstoffen; Bestimmung der Gesamtverschmutzung von gebrauchten Motorenschmierölen; Zentrifugierverfahren. - DIN 51 381: Prüfung von Schmierölen, Reglerölen und Hydraulikflüssigkeiten; Bestimmung des Luftabscheidevermögens. - DIN 51558, Pt 1: Prüfung von Mineralölen; Bestimmung der Neutralisationszahl, Farbindikator-Titration. - DIN 51 566: Prüfung von Schmierölen; Bestimmung des Schaumverhaltens. - DIN 51 755: Prüfung von Mineralölen und anderen brennbaren Flüssigkeiten; Bestimmung des Flammpunktes im geschlossenen Tiegel nach Abel-Pensky. - DIN 51 758: Prüfung von Mineralölen und anderen brennbaren Flüssigkeiten; Bestimmung des Flammpunktes im geschlossenen Tiegal nach Pensky-Martens. - DIN 51 777, Pt 1: Prüfung von Mineralöl-Kohlenwasserstoffen und Lösemitteln; Bestimmung des Wassergehaltes nach Karl-Fischer; Direktes Verfahren. - DIN ISO 3733: Mineralölerzeugnisse und bituminöse Bindemittel; Bestimmung des Wassergehaltes, Destillationsverfahren. - DIN ISO 3771: Mineralölerzeugnisse; Gesamtbasenzahl; Bestimmung durch potentiometrische Perchlorsäure-Titration. - DIN-Taschenbuch 20: Mineralöle und Brennstoffe 1. Eigenschaften und Anforderungen. - DIN-Taschenbuch 32: Mineralöle und Brennstoffe 2. Prüfverfahren. - DIN-Taschenbuch 57: Mineralöle und Brennstoffe 3. Normen über Prüfverfahren. - DIN-Taschenbuch 58: Mineralöle und Brennstoffe 4. Prüfverfahren. - DIN-Taschenbuch 228: Mineralöle und Brennstoffe 5. Prüfverfahren. - DIN-Taschenbuch 192: Schmierstoffe. Normen über Eigenschaften, Anwendung, Prüfung. Beuth, Berlin, 1983. - DIN-Taschenbuch 203: Schmierstoffe; Prüfung.

English-Language equivalents for Standards (mostly not identical). DIN 50 281: see ISO 4378/2: Plain bearings; terms, definitions and classification. Part 2: Friction and wear. - DIN 50 320: no equivalent. - DIN 50 321: no equivalent. - DIN 50 323 Pt 1. no equivalent. - DIN 51 365: see ISO 4021: Hydraulic fluid power; particulate contamination analysis; extraction of fluid samples from lines of an operating system; ISO 4405: Hydraulic fluid power; fluid contamination; determination of particulate contamination by the gravimetric method. - DIN 51 377: see ASTM D 2602: Standard test method for apparent viscosity of engine oils at low temperature using the cold-cranking simulator. - DIN 51 381: see ISO DIS 9120: Petroleum-type steam turbine and other oils; determination of air release properties; Impinger method. - DIN 51 382: see ASTM D 2603: Test method for sonic shear stability of polymer-containing oils. - DIN 51 511: see ISO DIS 10369: Engine oil viscosity classification.- DIN 51 550: see ISO DIS 3104: Petroleum products; transparent and opaque liquids; determination of kinematic viscosity and calculation of dynamic viscosity and calculation of dynamic viscosity. - DIN 51 558: see ISO 6618: Petroleum products and lubricants; neutralization number; colour-indicator titration method. - DIN 51 561: see: ISO DIS 3105: Glass capillary kinematic viscometers; specification and operating instructions. - DIN 51 562: Part 1: see ISO DIS 3105: Glass capillary kinematic viscometers; specifications and operating instructions. - DIN 51 566: see ISO DIS 6247: Petroleum products; lubricating oils; determination of foaming characteristics. - DIN 51 755: see ISO 2592: Petroleum products; determination of flash and fire points, Cleveland open cup method. - DIN 51 757: see ISO 3675: Crude petroleum and liquid petroleum products; laboratory determination of density or relative density; hydrometer method. - DIN 51 758: see ISO 2719: Petroleum products and lubricants; determination of flash point; Pensky–Martens closed cup method. - DIN 51 777: see ISO 6296: Liquid petroleum products; determination of water; Karl Fischer method. - DIN 51 818: see ISO 2137: Petroleum products; lubricating grease; determination of cone penetration.

Other standards of interest. ASTM D 3829: Test method for predicting the borderline pumping temperature of engine oil. - CEC L-32-T-82: Predicting the borderline pumping temperature of engine oils using the Brookfield viscometer. See also ISO DIS 9262: Automotive fluid lubricants; low-temperature viscosity; Brookfield viscometer methods.

Fundamentals of Engineering Design

G. Pahl, Darmstadt

G. Pahl, Darmstadt

1 Fundamentals of Technical Systems

1.1 Energy, Material and Signal Transformation

Technical entities (systems, apparatus, machines, equipment, assemblies, individual components) are artificial and concrete *systems* which consist of a totality of organised elements, linked together by relationships caused by their characteristics. A *system* is characterised by the fact that it is separate from its surroundings, while the links with its surroundings – the input and output – are intersected by the *boundary* of the system. A system may be subdivided into *part systems*. For each particular purpose, such system subdivisions may be made more or less broadly, depending on viewpoint.

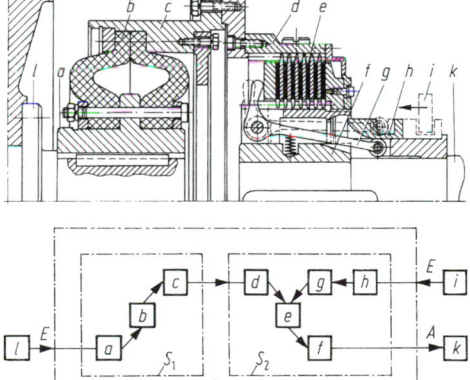

Figure 1. System "coupling": *a* to *h* system elements (examples), *i* to *l* connection elements, *S* total system, S_1 subsystem "elastic coupling", S_2 subsystem "switching coupling", *E* inputs, *A* outputs.

Thus, for example, in **Fig. 1** the system "coupling" forms a module within a machine, but it may be divided into the two independent subsystems, "elastic coupling" and "switched coupling". These subsystems may be further subdivided into system *elements*, in this case individual components. This subdivision depends upon design. However, it is also possible to analyse it by function: the "coupling" system could be functionally classified into subsystems "compensation" and "switching", with the latter further subdivided into the subsystems "converting switching force to normal force" and "transferring friction force", etc.

Technical systems represent a process by which energy, material and signals are routed and/or transformed (**Fig. 2**). It is therefore concerned with *energy, material* and/or *signal transformation*. In technical processes, the flow of energy, material or signals can be the most important factor, depending on the task or the nature of the solution, and it should then be considered the principal flow for practical purposes. A secondary flow often accompanies it, and it is frequently the case that all three are involved.

In each transformation operation, the *quantity* and *quality* of the factors involved should be noted so that the criteria for the precise definition of the task and the evaluation of any solution are unambiguous.

1.2 Functional Interrelationship

In a technical system with energy, material and signal transformation, there must be unambiguous, reproducible interrelationships both between main system and subsystem input and output, as well as between the subsystems themselves. They are always intentionally task-oriented (e.g. introducing torque, converting electrical energy into mechanical energy, blocking the flow of material, storing signals). Interrelationships of this type, which exist between system input and output to fulfil a task, are called

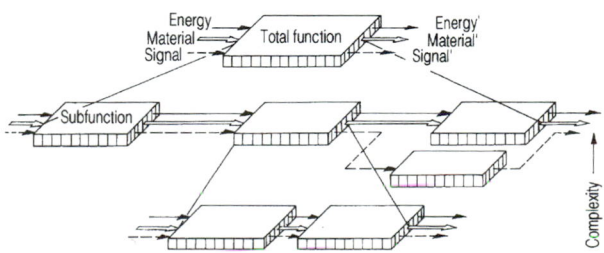

Figure 2. Function structure for energy, material and signal flow in terms of total and subfunctions.

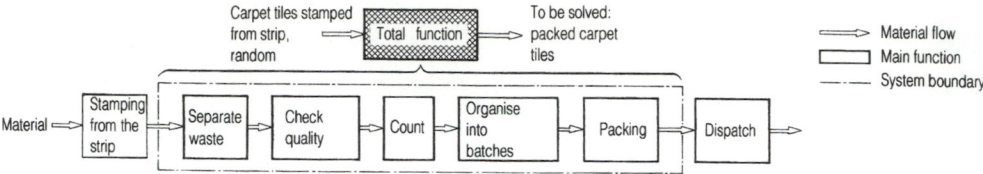

Figure 3. Function structure for processing carpet tiles.

its *function*. The function is an abstract formulation of the task without regard to the solution. If it refers to the task as a whole, it is known as the *overall function*. It may often be possible to divide this into recognisable *subfunctions* which correspond to the subtasks within the overall task (**Fig. 2**). The way in which the subfunctions are linked to the overall function usually leads automatically to a *function structure*. The basis for differing solutions may frequently be produced by varying the allocation. Subfunctions must be linked to the overall function in a way which is practical and compatible.

A distinction should be made between the primary and secondary function for practical purposes. *Primary functions* serve the overall function directly. *Secondary functions* contribute only indirectly to the overall function; they are of a supportive or supplementary character and are frequently dependent on the nature of the solution (**Figs 3** and **4**). The functions presuppose a physical action for their fulfilment, while the physical dimensions must correspond to each other from subfunction to subfunction; otherwise intermediate functions must be interpolated.

There are also logical interrelationships which determine or influence function structure. Thus certain subfunctions have to be fulfilled first before it is practical to insert others (e.g. in **Fig. 4** the subfunction "count" only makes sense after "check quality"). However, logical interrelationships are also necessary with reference to circuit logic. *Logical functions* serve this purpose, enabling statements to be made such as true/untrue, yes/no, on/off, fulfilled/unfulfilled in a two-valued logic. A distinction is made between AND-, OR- and NOT-operations and their combination to form complexes such as NOR- (OR

with NOT), NAND- (AND with NOT) and memory operations with the aid of flip-flops.

1.3 Working Interrelationships

1.3.1 Physical Effects

In general, subfunctions represent physical events caused by *physical effects*. The physical effect may also be described quantitatively by means of the laws of physics which interrelate the physical factors involved. If these effects are assigned to a subfunction in a concrete situation, the subfunction's *physical operative principle* is obtained (**Fig. 5**). A part function may be fulfilled by differing physical effects (see **E2, Table 1**).

1.3.2 Geometrical and Material Features

The place where a physical event occurs is called the *effective location*. The fulfilment of the function produced when the physical effects occur is caused by the *effective geometry* (arrangement of *effective areas* and selection of *effective motions*). The form of the effective area can be varied or fixed by the type, shape, position, size and number of these features. The requisite effective motion is determined in a similar way (see **E2, Table 2**).

In addition, there must be at least a concept in principle with regard to the type of *material* with which the effective geometry is to be realised. Only the joint action of the physical effect and the geometrical and material features (effective area, effective movement and material) enables the *effective principle* to be visible (**Fig. 5**).

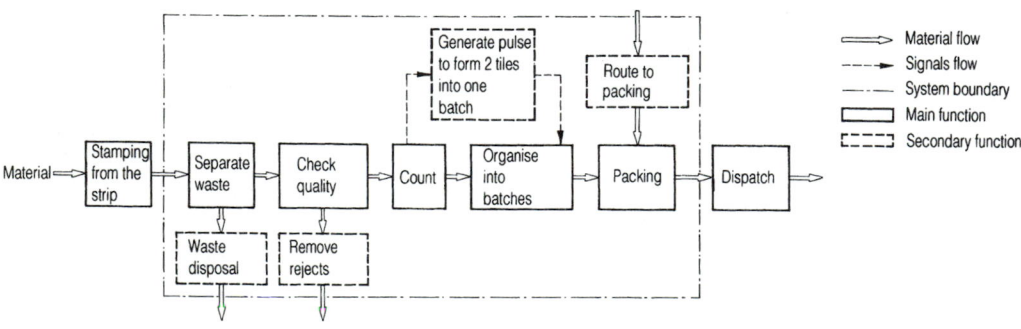

Figure 4. Function structure for processing carpet tiles as in Fig. 3, showing secondary functions.

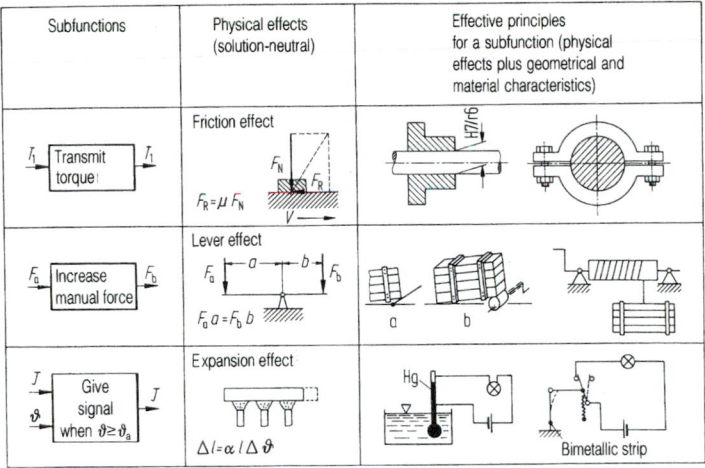

Figure 5. Fulfilment of subfunctions by effective principles arising from physical effects and geometrical and material characteristics.

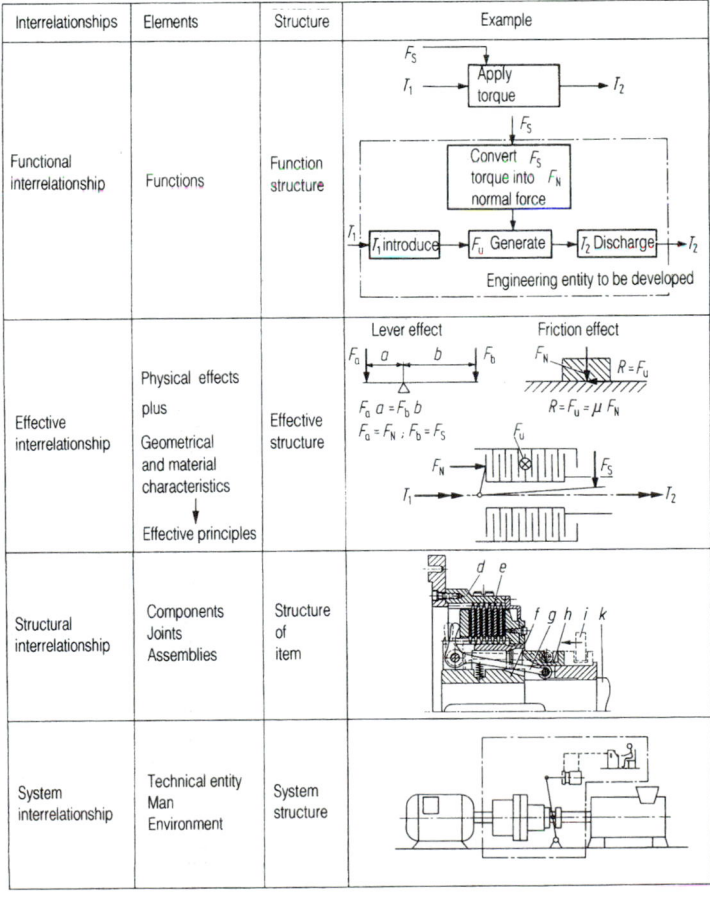

Figure 6. Interrelationships in engineering systems.

The combination of several effective principles leads to the *effective structure* in which the principle of the solution is recognisable.

1.4 Constructional Interrelationship

The effective interrelationship recognisable in the effective structure forms the basis for further concretisation, which leads to the *production structure*, which refers to requirements for manufacture, assembly etc. The components, assemblies and their interrelationship in the product are stipulated in the latter (**Fig. 6**).

1.5 System Interrelationship

Technical products do not occur in isolation; they are constituent parts of a higher-order system. Mostly, human beings work with the system or interact with it. In the latter case, they experience *reactions* which make them take further actions. Humans therefore support the deliberate *intended effects* of the technical system. However, *disruptive effects* also occur in the form of accidental input dimensions and *secondary effects* (**Fig. 7**). All these effects must be noted.

1.6 General Objectives and Constraints

The solution of technical tasks is determined by objectives and constraints. Despite the latter, the *general objective* always remains the fulfilment of the technical function,

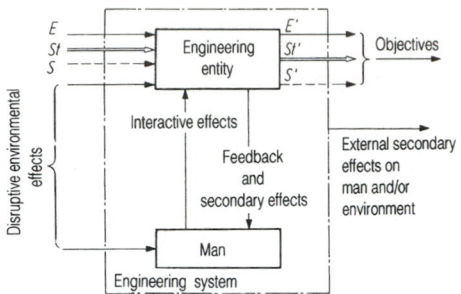

Figure 7. Interrelationships in engineering systems involving humans.

its economic realisation, and safety for both humans and their environment.

The *constraints* may arise from the task (conditions specific to the task), the state of the art, the economic and the general situation (general constraints).

The following features may be used to describe the objective and constraints briefly and comprehensively: function – effective principle – realisation – safety – ergonomics – manufacture – monitoring – assembly – transportation – use – maintenance – recycling – expenditure.

Standards and Guidelines: VDI-Richtlinie 2221: Methodik zum Entwickeln und Konstruieren technischer Systeme und Produkte. VDI-Verlag, Düsseldorf, 1986. – VDI-Richtlinie 2222: Konzipieren technischer Produkte. VDI-Verlag, Düsseldorf, 1977. – VDI-Richtlinie 2225: Technisch-wirtschaftliches Konstruieren. VDI-Verlag, Düsseldorf, 1977.

2 Fundamentals of a Systematic Approach

2.1 General Working Method

Essentially, problem-solving consists of analysis and synthesis. *Analysis* involves obtaining and breaking down information, as well as dissecting it and classifying and investigating the characteristics of individual elements and the interrelationships between them. This involves recognition, definition, structuring and ordering. Similarly, *synthesis* consists of processing information by forming links, connecting elements to produce totally new effects, bringing them together in an orderly summary. It is a process of searching and finding (creation) plus composition and combination.

In addition, the following preconditions must be fulfilled for a systematic approach: ensuring problem-solving *motivation*; *clarification* of limiting and initial conditions; removal of *prejudices*, discovering *variants*, taking *decisions*.

The search for solutions is assisted both by intuitive thought (characterised by imagination, mainly in the subconscious, almost impossible to influence or reconstruct) and discursive thought (conscious, step by step, communicative).

If the problems are complex and extensive, it is necessary to divide them into manageable subproblems. Complex problems are solved step by step, but it is quite poss-

ible or even desirable for subsidiary results to be found by intuitive means.

2.2 General Problem-Solving

During the working and decision stages, the problem-solving process normally moves from the *qualitative* to the *quantitative*. First, the process generally produces a *confrontation* with the problems and possible solutions which are not (yet) known.

The next stages in the problem-solving process consist of collecting *information* about the process, *definition* of the most significant problems, *evaluation* of the solutions in terms of process objectives and *decisions* on further action [1]. A suitable procedure for many areas of development and design (**Fig. 1**) is given in VDI-Richtlinie 2221 [2].

2.3 Abstracting to Identify Functions

In abstracting, individual and random factors are ignored, and generally valid and significant factors are clarified by analysing the list of requirements. This makes the core of a problem clearly visible. If this is formulated accurately,

Divisions of work   Results of work

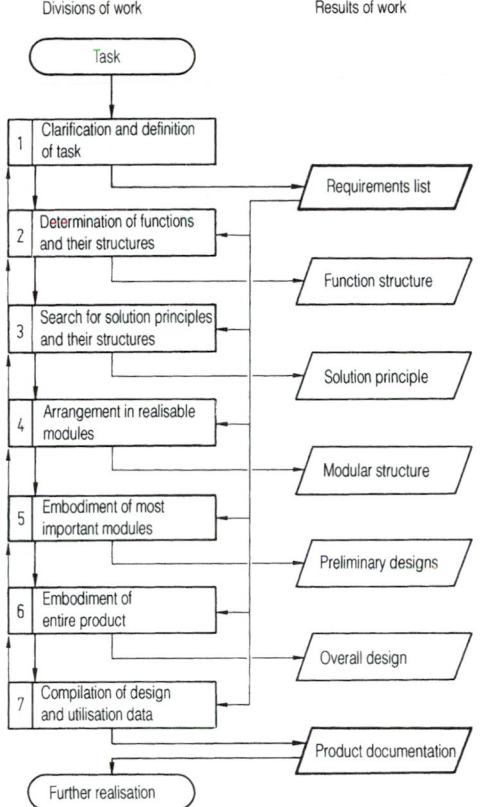

Figure 1. General procedure for development and design [2].

the overall function (see E1.2) and any significant constraints will also be clearly visible.

2.4 Search for Solution Principles

2.4.1 Generally Applicable Methods

Obtaining and processing information using analysis and synthesis are the major factors in seeking solutions. Conventional aids in this matter are to be found in *bibliographical and patent searches, analyses of natural and known technical systems, consideration of analogies, measurements, and model tests.*

Creative techniques employ the methods listed below, which can therefore be considered to be generally applicable [3]: *objective questioning, elimination and reconception, advance, withdrawal, subclassification (factorising), and systematisation.*

2.4.2 Intuitive Methods

These methods work mainly by association of ideas based on unbiased comments by colleagues, consideration of analogies and group-dynamic effects. They have been more or less formalised as *brainstorming* [4], the *gallery method* [5], *synthetics* [6], the *635 method* [7] and the *Delphi method* [8]. Brainstorming is the simplest and least expensive, while the gallery method is particularly helpful in problems involving construction.

2.4.3 Discursive Methods

These methods move stepwise towards a solution, without, however, excluding intuition. They begin with systematic investigation of the actual or potential physical factors involved, then derive *organisational viewpoints* from previously known functional, physical or constructional interrelationships; this can provide the stimulus for new or different solutions within a search plan (organisation plan).

Systematic Investigation of the Physical Phenomena. This leads (especially if several physical factors are involved) to different solutions, since the relationships between them (i.e. between dependent and independent variables) can be analysed successively while keeping the remaining influencing factors constant. For the equation $y = f(u, v, w)$, solution variations are sought for the equations $y_1 = f(u, \underline{v}, \underline{w})$, $y_2 = f(\underline{u}, v, \underline{w})$, and $y_3 = f(\underline{u}, \underline{v}, w)$, where the underlined factors are constant. The resulting relationships are then realised in concrete form by differing solutions methods, effective areas, or previously known components [9].

Systematic Search using Organisational Plans. A systematic, organised representation of information stimulates the search for further solutions. It allows recognition of significant features of the solution which may be a further stimulus for completion, and produces an outline of feasible possibilities and interconnections. Organisational plans can be used in many ways in the design process as search plans, compatibility matrices, or lists [10].

The two-dimensional plan normally used consists of columns and lines in which parameters are allocated from a *hierarchical viewpoint*. The solutions are entered in the intersecting areas of the plan (matrix). In the example shown in **Fig. 2**, the line hierarchy is the way in which the strip moves, and the column hierarchy is the motion of the supporting device, which can have the parameters at rest, translation, oscillation and rotation, including feasible combinations. **Tables 1 and 2** can help to select hierarchies and parameters.

If the column heading is "Subfunctions" and the line heading is "Problem-solving features", the intersecting areas become solutions for the individual subfunctions which can be combined in each case to produce the respective *overall function*. If m_1 solutions are available for the subfunction F_1, m_2 for the subfunction F_2 etc., then $N = m_1 m_2 \ldots m_n$ theoretically possible variations of the overall solution can be obtained for a complete combination (**Fig. 3**). Of course, not all the combinations are practical and compatible. Only those which seem promising are pursued further [11].

Systematic Search Using Lists. *Lists* can be used very advantageously in recurring problems and those which have a certain general validity [12]. These can be lists of suppliers, or can even be more or less complete collections of solutions. If the features of the solutions are allocated systematically to the constraints of the respective problem, an appropriate solution may be obtained directly, or alternatively further new stimuli may be produced [13].

Systematically constructed lists are particularly advantageous because they are not only more comprehensive; they also enable the characteristic features and nature of the solutions to be recognised as comparisons. Systems that are recognisable in this way also form excellent bases for an ongoing independent search for a solution. In addition to a large number of different lists, Roth [10] has provided a detailed explanation of the structure and use

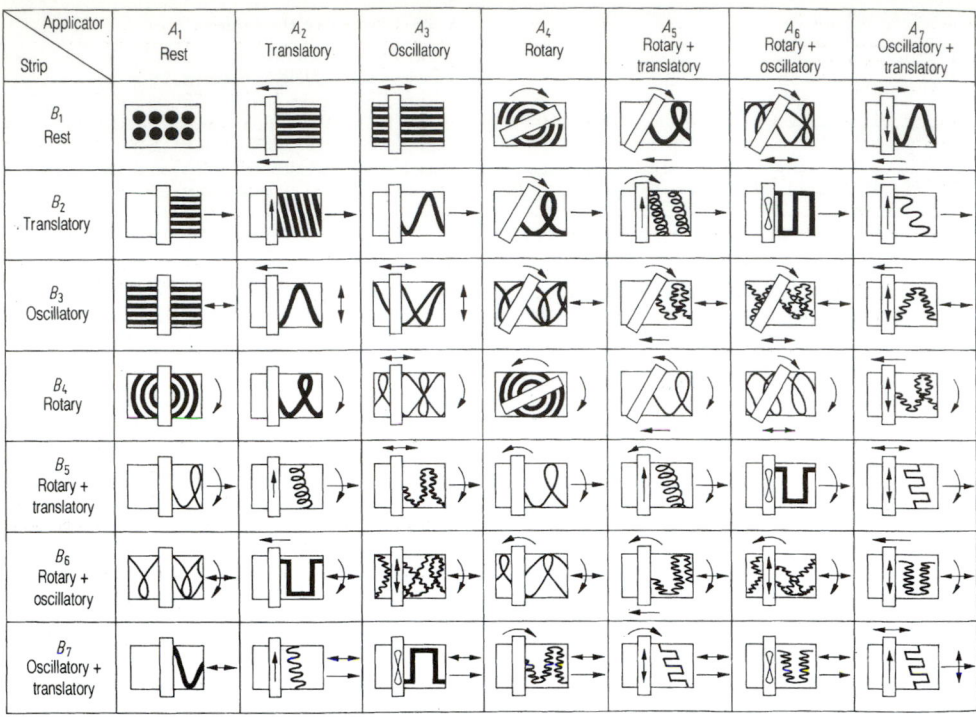

Figure 2. Possibilities for coating carpet strips by combinations of movement of the strip and the applicator (summary).

Table 1. Organisational hierarchy and features for variation on the physical search plane

Organisational hierarchy:

Types of energy, physical effects and types of phenomena

Characteristics	Examples
Mechanical	Gravitation, inertia, centrifugal force
Hydraulic	Hydrostatic, hydrodynamic
Pneumatic	Aerostatic, aerodynamic
Electrical	Electrostatic, electrodynamic, inductive, capacitative, piezo-electric, transformation, rectification
Magnetic	Ferromagnetic, electromagnetic
Optical	Reflection, refraction, diffraction, interference, polarisation, infrared, visible, ultraviolet
Thermal	Expansion, bimetal effect, heat storage, heat transference, conduction of heat, heat insulation
Chemical	Combustion, oxidation, reduction, solution, bonding, converting, electrolysis, exothermic, endothermic reaction
Nuclear	Radiation, isotopes, energy source
Biological	Fermentation, decay, disintegration

of such *lists*: as a rule it should consist of a *classifying section* (hierarchical viewpoints for subdivision from which scope and completeness are apparent), *main section* (content in the form of objects with explanatory for-

mulae and sketches) and the *access section* (characteristic features, which enable a reliable and simple selection to be made).

2.5 Evaluation of Solutions

2.5.1 Selection Procedure

A formalised selection procedure makes selection easier via *elimination* and *preference*, especially when there are a large number of proposals or combinations. In principle, a selection procedure of this kind should be undertaken after each work stage in which variations occur. Only those that are *compatible* with the task and/or with each other *fulfil the demands* of the list of requirements, and lead to *feasibility of realisation* in terms of effectiveness, magnitude, arrangement etc. and *an acceptable cost* will be pursued. If a large number of variations still remain, selection may be based on paths that offer *direct compliance with the safety regulations* or advantageous ergonomic preconditions, or that appear *easily realisable within the user's own area* with known methods, materials or work processes plus an advantageous patent situation [1].

2.5.2 Evaluation Procedure

An evaluation should determine the value of a solution with reference to targets that have been set in advance so that a more precise assessment can be provided for solutions that are to be pursued further in accordance with the user's own selection procedure. Technical and economic aspects are to be taken into consideration in the

Table 2. Organisational hierarchies and features for variation on the structural search plane

Organisational hierarchy:

Effective geometry, effective movement and material characteristics

Effective geometry (effective body, effective area)

Features	Examples
Type	Point, line, area, body
Shape	Curve, circle, ellipse, hyperbola, parabola, triangle, quadrilateral, rectangle, pentagon, hexagon, octagon, cylinder, cone, rhombus, cube, sphere, symmetrical, asymmetrical
Position	Axial, radial, vertical, horizontal; parallel, series
Size	Small, large, narrow, wide, high, low
Number	Undivided, divided, single, double, multiple

Effective movement

Features	Examples
Type	Rest, translatory, rotatory
Shape	Regular, irregular, oscillating; plane, solid
Direction	In x-, y-, z-direction and/or around x-, y-, z-axis
Amount	Speed
Number	One, several, combined movements

Material characteristics

Features	Examples
Status	Solid, liquid, gaseous
Behaviour	Rigid, elastic, plastic, viscous
Shape	Solid, granular, powder, dust

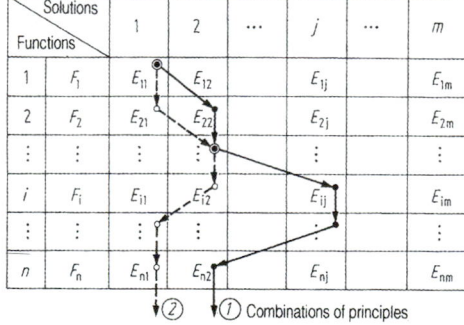

Figure 3. Combination of principles that fulfil the overall function by differing solution principles for the individual subfunctions.

latter. The methods used are value analysis [14] and techno-economic evaluation in accordance with VDI-Richtlinie 2225, which is based on Kesselring [15, 16] for the most part. General work stages for the assessment procedure are as follows.

Table 3. Guidelines showing principal features for evaluation

Principal feature	Examples
Function	Features of necessary secondary function carriers automatically generated by the solution principle selected or by variants
Effective principle	Feature of the principle(s) selected for simple and unambiguous fulfilment of function; adequate effect; low disturbance levels
Embodiment	Few components; little complexity; low space requirement; no special material and layout problems
Safety	Preference for direct safety technology (intrinsically safe); no additional protective measures; operational and environmental safety guaranteed
Ergonomics	Man–machine relationship satisfactory; no stress or impairment; good embodiment of shape
Manufacture	Few, standard manufacturing processes, no expensive devices; few, simple components
Monitoring	Few checks or tests; simply and reliably performed
Assembly	Easy, convenient and fast; no special tools
Transport	Normal transport facilities; no risks
Use	Simple operation; long life; low wear; easy and obvious operation
Maintenance	Low and easy maintenance and cleaning; easy inspection; problem-free repairs
Recycling	Good utilisability; problem-free disposal
Expenditure	No special operating or other secondary costs; no deadline risks

Recognition of Evaluation Criteria. A target concept usually includes several objectives. The evaluation criteria are directly derived from the latter. Because they will later be allocated to value concepts, they are formulated in a positive manner (e.g. "low noise" rather than "loud"). The minimum demands and requests in the requirements list are no longer taken into consideration (see E2.5.1) and general technical characteristics (**Table 3**) give indications for the evaluation criteria. The evaluation criteria must be independent of each other so that double evaluations are avoided.

Investigation of Significance for the Overall Value. If possible, only equivalents are to be evaluated. Unimportant evaluation criteria are eliminated. Differences in importance are to be taken into consideration by means of weighting factors. **Table 4** shows both possibilities.

Collating Characteristic Values. The allocation of value concepts is simplified if quantitative coefficients can be assigned to characteristic values, which, however, is not always possible. In this case qualitative verbal statements are to be formulated (**Table 4**).

Evaluation in Accordance with Value Concepts. The actual assessment takes place by awarding values (points). These arise from the characteristic values, which are determined by assigning value concepts (w_{ij} or wg_{ij}). Value analysis utilises a broader spectrum (0 = useless, 10 = ideal), while the VDI version uses a narrower one (0 to 4). There is a danger of a subjective influence in the allocation of the values. Therefore the

Table 4. Evaluation list with examples (summary)

Evaluation criteria			Characteristic dimensions		Variant V_1 (i.e. M_1)			Variant V_2 (i.e. M_v)		
Number		Weight		Unit	Characteristic e_{i1}	Value w_{i1}	Weighted value wg_{i1}	Characteristic e_{i2}	Value w_{i2}	Weighted value wg_{i2}
1	Low fuel consumption	0.3	Fuel consumption	$\frac{g}{kWh}$	240	8	2.4	300	5	1.5
2	Simple design	0.15	Weight per unit power	$\frac{kg}{kW}$	1.7	9	1.35	2.7	4	0.6
3	Simple manufacture	0.1	Simplicity of cast components	–	Complicated	2	0.2	Mean	5	0.5
4	Long life	0.2	Life	km driven	80000	4	0.8	150000	7	1.4
⋮	⋮	⋮	⋮	⋮	⋮	⋮	⋮	⋮	⋮	⋮
i		g_i			e_{i1}	w_{i1}	wg_{i1}	e_{i2}	w_{i2}	wg_{i2}
⋮	⋮	⋮	⋮	⋮	⋮	⋮	⋮	⋮	⋮	⋮
n		g_n			e_{n1}	w_{n1}	wg_{n1}	e_{n2}	w_{n2}	wg_{n2}
		$\sum_{i=1}^{n} g_i = 1$				Gw_1 W_1	Gwg_1 Wg_1		Gw_2 W_2	Gwg_2 Wg_2

awarding of points should be undertaken by a group of assessors, and should be done criterion by criterion for all variants (line by line) and never variant by variant.

Determining the Overall Value. The addition of the unweighted or weighted part values (w_3 or wg_3) produces the overall value.

Comparison of the Variants. This is most practically undertaken by determining the significance of the variants, in that the total value is set against the maximum possible overall value. In many cases it is advisable to determine a technical significance W_t and an economic significance W_w, especially if the manufacturing costs or prices are known for the item concerned. The technical significance W_t is determined by

$$W_j = \frac{\sum_{i=1}^{n} w_{ij}}{w_{max} n} \quad \text{(unweighted) or}$$

$$Wg_j = \frac{\sum_{i=1}^{n} g_i w_{ij}}{w_{max} \sum_{i=1}^{n} g_i} \quad \text{(weighted).}$$

Both can be classified within a significance diagram and checked for mutual balance [14, 15].

Estimating Uncertainties in Evaluation. Before a decision is taken, it is necessary to estimate the uncertainties involved in awarding values because of lack of information and differing individual approaches. If necessary, confidence limits, or a trend should be additionally indicated. Slight differences in significance found in the latter do not produce a fixed order of precedence.

Search for Weak Points. Values which are below average for individual evaluation criteria indicate weak points. In general, a variant with a lower overall value but evenly distributed individual values is more advantageous than one with higher overall values but a marked weak point which could prove unsatisfactory.

2.5.3 Estimating Production Costs

Production costs HK are composed of *material costs MK* (manufactured and bought-in materials) and *manufacturing costs FK* [17], so that $HK = MK + FK$. If applicable, special costs for manufacture are also added. In the case of differential job cost calculations, which are usual for technical products, the material costs *MK* consist of the costs for production materials *FM* (if applicable, plus bought-in materials) plus the *general material costs MGK*, which cover the costs of material management, while the *manufacturing costs FK* consist of the *labour costs FL* and the *general manufacturing costs FGK*: thus $MK = FM + MGK$ and $FK = FL + FGK$. Material costs and labour costs are variable costs (dependent on level of employment). The extra manufacturing costs in addition to labour costs are subdivided into stable (fixed) general costs (e.g. amortisation of equipment, rental of premises, salaries) and variable (proportional) general direct costs (e.g. energy costs, tooling costs, maintenance, wages for additional workers).

To increase accuracy, a *cost centre calculation* is frequently undertaken, which determines and includes a separate surcharge rate for each cost centre taken from the ratio of general costs to individual costs which apply. The manufacturing costs are then produced from the total costs of all the cost centres $FM_1 + MGK_1 + FL_1 + FGK_1 + FM_2 + MGK_2 + FL_2 + FGK_2 + \ldots = \Sigma FM_i (1 + g_{Mi}) + FL_i (1 + g_{Li})$. The labour costs consist of the total of the basic time, rest periods and distribution time, if applicable, plus equipping time, multiplied by a wages rate (wages group) in currency unit per unit of time.

One important factor in determining price is *prime cost SK*, which consists of the manufacturing costs *HK*, the development and design costs *EKK*, the general administration costs *VwGK*, and the general costs for sales *VtGK*: thus $SK = HK + EKK + VwGK + VtGK$. See VDI-Richt-

Table 5. Indices for unit times for various manufacturing operations with geometrical similarity [22]

Machine type	Process	Index		Confidence
		Calculated	Rounded	
Universal lathe	Internal and external turning	2	2	++
	Thread-cutting	≈1	1	+
	Cropping, grooving	≈1.5	1	+
	Chamfering	≈1	1	+
Vertical borer	Internal and external turning	2	2	++
Radial borer	Drilling, thread-cutting, counter-sinking	≈1	1	0
Milling and boring machines	Turning, drilling, milling	≈1	1	0
Slot miller	Grooves on adjusting springs, milling	≈1.2	1	+
Universal cylindrical grinder	External grinding	≈1.8	2	++
Circular saw	Profiling	≈2	2	0
Table cutter	Sheet metal cutting	1.5 to 1.8	2	+
Edging machine	Edging sheet metal	≈1.25	1	+
Press	Straightening	1.6 to 1.7	2	+
Chamferer	Chamfering sheet metal	1	1	++
Flame cutter	Flame-cutting sheet metal	1.25	1	++
MIG and manual welding	I-welds	2	2	++
	V, X-fillet welds, corner welds	2.5	2	++
Annealing		3	3	++
Sandblasting	(depending on calculation via weight or surface)	2 or 3	2 or 3	++
Assembly		1	1	++
Tack welding		1	1	++
Hand polishing		1	1	++
Painting		2	2	++

++ good accuracy.
+ poorer than ++.
0 high degree of scatter possible.

linie 2225 (Section 10.4) for instructions on the concrete determination of costs.

2.5.4 Costing

It is helpful to the designer to already be in a position to recognise cost trends at the stage where varying solutions are being considered. It is usually sufficient merely to consider the variable costs for this purpose. The following possibilities have been developed.

List of Relative Costs. Prices or costs are related to a comparison factor in this method. The information is therefore valid for a very much longer period than is the case for absolute costs. Relative costs lists are usual for materials, semi-finished parts and standard components. Basic principles for the formation of lists of relative costs have been compiled in DIN 32991. Relative material costs, for example, are listed in [16].

Cost Estimation by Material Cost Proportion. If, in a given field of application, the ratio m of material costs MK to manufacturing costs HK is known, and is approximately equal, it is possible to estimate the manufacturing costs in accordance with [16] if the material costs have been established. The equation $H = MK/m$ is then produced. However, this process fails where major changes have been made in the size of the item.

Cost Estimation Using Regression Analysis Calculations. Costs are determined as a function of characteristic factors (e.g. output, weight, diameter, axle height) by means of statistical evaluation of calculation documents. By means of regression analysis a relationship is sought that defines the regression equation using regression coefficients and exponents. Costs can then be calculated for a given degree of scatter. The expenditure for compilation may be considerable and compilation itself is not usually possible without the use of a computer. The regression equation should be constructed in such a way that factors that change when information is updated, such as hourly rates, take the form of independent factors or relative costs. The exponents and coefficients of the regression equation do not usually permit conclusions to be drawn regarding the relationships of costs to selected geometrical or technical characteristic dimensions, they are of a mathematically formal nature. See [18, 19] for further information on the procedure and examples of its application.

Cost Estimation Using Similarity Equations. If geometrically similar or partially similar components are present in a series (see E5), or are variants of previously known series, the stipulations of cost growth laws from similarity equations are of practical use. The progressive ratio of the costs φ_{HK} is represented by the ratio of the costs of the *subsequent design* HK_q (unknown costs) to those of the *original design* HK_0 (known costs) and is determined via a similarity analysis:

$$\varphi_{HK} = \frac{HK_q}{HK_0} = \frac{MK_q + \Sigma FK_q}{MK_0 + \Sigma FK_0}.$$

The ratio of the material costs and the individual manufacturing costs, e.g. for turning, boring, grinding, to the production costs is determined using basic design:

$$a_m = MK_0/HK_0; \qquad a_{F,k} = FK_{k,0}/HK_0$$

per k manufacturing operations.

If there are known cost growth laws for the individual proportions, the cost growth law for the whole becomes

$$\varphi_{HK} = a_m\varphi_{MK} + \sum_k a_{F,k}\varphi_{FK,k}.$$

In a general form with $a_i = 1$ and $a_i \geq 0$, the following may be produced as a function of a characteristic length:

$$\varphi_{HK} = \sum_i a_i\varphi_L^{x_i}; \quad \varphi_L = L_q/L_0$$

(see E5.1), $a_i = 1$ and $a_i \leq 0$.

The definition of the exponents x_i as a function of the corresponding dimensions (characteristic length) is simple for geometrically similar components. It is still possible to work with integer exponents:

$$\varphi_{HK} = a_3\varphi_L^3 + a_2\varphi_L^2 + a_1\varphi_L^1 + (a_0/\varphi_z), \quad \text{with } \varphi_z = z_q/z_0;$$

z = batch size.

In general, $\varphi_{MK} = \varphi_L^3$ is applicable to material costs. **Table 5** covers the manufacturing operations.

The proportions a_i are calculated in a plan (example in **Table 6**) from the basic design, using allocation to the individual integral exponent. The cost growth law for this example would then be

$$\varphi_{HK} = 0.49\ \varphi_L^3 + 0.26\ \varphi_L^2 + 0.20\ \varphi_L + 0.05.$$

A variant that is twice the size but geometrically similar with $\varphi_L = 2$ would then produce a cost increase with a progressive ratio $\varphi_{HK} = 5.41$.

In the case of partially similar variants, only the respective changing lengths are to be inserted with corresponding relevant exponents. The proportions that remain constant then go into the last element of the equation. Examples and application to modules plus determination

Table 6. Proportions a_i for cost growth law based on the standard schedule plan and principal basic design costs (example)

Operation		Costs rising by φ_L^3	Costs rising by φ_L^2	Costs rising by φ_L	Constant costs
Material		800			
Flame cutting	⎫			60	15
Chamfering	⎬ Joints			35	
Tacking	⎪			105	
Welding	⎭		500		
Annealing		80			
Sandblasting		40			
Marking	⎫			40	
Turning	⎬ Mechanical finishing			100	70
Raboma	⎭			30	15
1890 DM = H_0 =		Σ_3 (= 920)	$+\Sigma_2$ (= 500)	$+\Sigma_1$ (370)	$+\Sigma_0$ (100)
		Σ_3/H_0 (= 0.49)	$+\Sigma_2/H_0$ (= 0.26)	$+\Sigma_1/H_0$ (= 0.20)	$+\Sigma_0/H_0$ (= 0.05)

of cost structures are given in [20, 21]. For rules for cost reduction see [19, 22].

2.5.5 Value Analysis

Value analysis is a plan-based procedure for minimising costs under the influence of comprehensive aspects (DIN 69910 [23-25]). Calculated costs of the individual components are used to establish those costs that arise in fulfilling the overall function and the requisite subfunctions. Such "function costs" are a reliable basis for evaluating such variants as coverage and critical assessment is given equally to the aspects of sales (are all the functions absolutely essential?), design (selection of suitable function structures and solution concepts plus the subfunctions therefore required) and manufacture (embodiment of the individual components). This investigation produces important indications in the search for new solutions with worthwhile cost reductions. Value analysis utilises the same methods and aids for subsequent checking as for systematic design. The two are therefore compatible and supplement each other.

3 The Design Process

The general solution procedure put forward in E2.2 is applied to various concretising stages via individual methods (see E2.3 to E2.5) and observing the principles of embodiment design (see E4). It is divided into the *principal phases* clarifying the problem, conceptual design, embodiment and detailed design.

3.1 Defining Requirements

This phase involves collecting information on requirements placed on the solution, plus existing constraints and their significance. It leads to the compilation of a

requirements list. Performance specifications are another form of requirement. However, they usually only contain the customer's requirements and are not formulated in the designer's language.

3.1.1 The Requirements List

This comprises the objectives of and constraints concerning the problem to be solved, in the form of demands and preferences:

- *Demands* must be fulfilled in all circumstances (minimum demands are to be formulated and stated, e.g. $P > 20$ kW, $L \leq 400$ mm).

- *Preferences* (of varying significance) should be taken into consideration if possible, perhaps with the concession that a limited additional cost is permissible.

Without already stipulating a fixed solution, the demands and preferences are given data on *quantity* (number, quantity, batch size, etc.) and quality (permissible deviations, tropicalised, etc.). Only after this is sufficient information available. For practical purposes, the *source* of the demands or preferences which have arisen is also stated.

Additions and *amendments* to the requirements, which may arise in the course of development as better information on possible solutions becomes available or because of time-dependent displacement of the main points, must always be added to the requirements list.

3.1.2 Formulation of Requirements

A list of principal features (**Table 1**) is recommended as an aid to recognising requirements. It produces an associ-

Table 1. Guidelines showing main features for compiling requirements list

Principal features	Examples
Geometry	Size, height, breadth, length, diameter, number
Kinematics	Direction of motion, velocity, acceleration
Forces	Direction of force, size of force, frequency of force, weight, load
Energy	Power, efficiency, pressure, temperature, heating, cooling, connection energy
Material	Material flow, input and output characteristics, auxiliary materials
Signal	Input and output measured variables, signal form, display, operating and monitoring equipment
Safety	Direct safety engineering, protection systems, workplace and environmental safety
Ergonomics	Man–machine relationship: operation, operating height, operating method, lighting, embodiment
Manufacture	Maximum manufacturing size, preferred manufacturing processes, tolerances
Monitoring	Measuring and testing facilities, special regulations (TÜV, ASME, DIN, ISO, AD-specification)
Assembly	Special assembly regulations, assembly, installation, on-site assembly, foundations
Transportation	Lifting gear restrictions, track profile, size and weight-related transport methods, dispatch method and conditions
Application	Low noise, wear rate, use and sales area, location (e.g. sulphurous atmosphere, tropics)
Maintenance	Maintenance-free or specified service interval, inspection, exchange and repair, cleaning
Recycling	Reusing, reprocessing, final storage, disposal
Costs	Max. permissible manufacturing costs, tooling costs, investment and amortisation
Deadlines	End of development, network plan for intermediate stages, delivery time

ation in the mind of the compiler so that he or she transfers the terms stated there to the actual problem under consideration, and generates the questions to which answers are required. The necessary functions and the specific constraints are covered along with the energy, material and signal transformation (features – geometry, kinematics, forces, energy, material, signal). The other features take account of the remaining general and specific constraints. Listing the terms ensures that essential factors are not omitted.

3.2 Conceptual Design

Conceptual design (**Fig. 1**) is that part of the design process in which the basic solution is laid down, and, after the problem has been clarified by abstracting, function structures and appropriate solution principles are determined.

Abstracting to recognise the significant problems brings out the core of the task and frees the designer from fixed ideas and conventional solutions so that new and more practical methods of solution become apparent. The overall function (see E1.2) is then defined as rigorously as possible in the context of energy, material and signal transformation, using input and output factors in a manner which is neutral to the solution and is divided into recognisable subfunctions (function structure).

The search then follows for *effective principles* (see

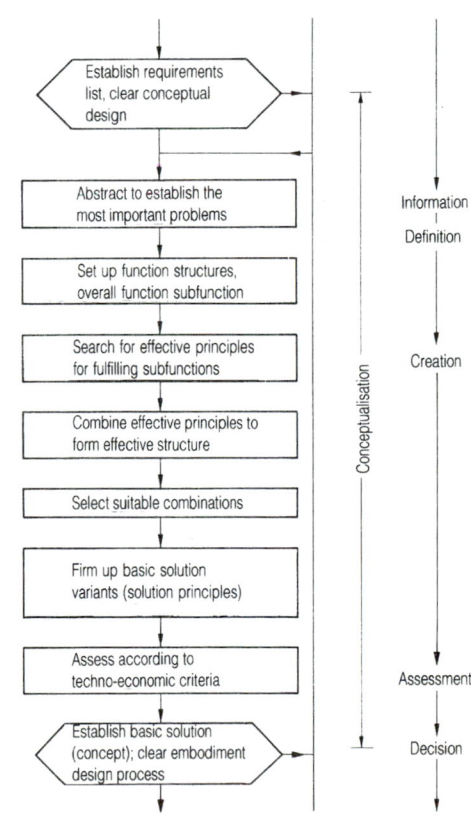

Figure 1. Stages in conceptual design.

E1.3 and E1.4) that fulfil individual subfunctions. These are then *combined* using the function structure so that they are compatible, fulfil the demands of the requirements list and remain within acceptable cost limits, after which a selection procedure is applied (see E2.5.1). The most appropriate combinations are then *concretised as primary solution variants*, so that they can be evaluated and assessed (see E2.5.2). Their most significant technical and economic characteristics must be apparent at this point.

3.3 Embodiment Design

Embodiment design is that part of the design process in which the techno-economic structure of the item is compiled unambiguously and completely, starting from the effective structure or solution principle.

In addition to *creative work stages*, embodiment design also requires a large number of *corrective stages* in which analysis and synthesis alternate. Here too it is necessary to move from the qualitative to the quantitative, i.e. *from general embodiment design to detailed embodiment design*. **Figure 2** shows work stages to be passed through more or less completely, depending on the complexity of the solution concept.

Embodiment design is characterised by a thinking and checking procedure which is effectively supported by following the *guidelines* (**Table 2**). The first feature should usually be considered before the next is processed or examined more intensively. This sequence has nothing to do with the significance of the features, but simply acts as a labour-saving procedure.

3.4 Detail Design

Detail design is that part of the design process which supplements embodiment design with final specifications on the arrangement, shape, dimensioning and surface quality of all individual components, by stipulating all the materials, checking manufacturing feasibilities and costs, and producing drawings, plus mandatory and other documentation for its construction and use [1].

The principal activity is the compilation of manufacturing documents, especially the individual component drawings, including group and general drawings plus parts lists. In addition to the latter, specifications for manufacture, assembly and use may also be necessary. A check on completeness and correctness plus the application of standards concludes this phase (**Fig. 3**).

As with the conceptual design and embodiment phases, work stages in the embodiment and detail design phase frequently overlap.

3.5 Types of Engineering Design

It is not always necessary to go through all the main phases for the entire technical system. New designs fre-

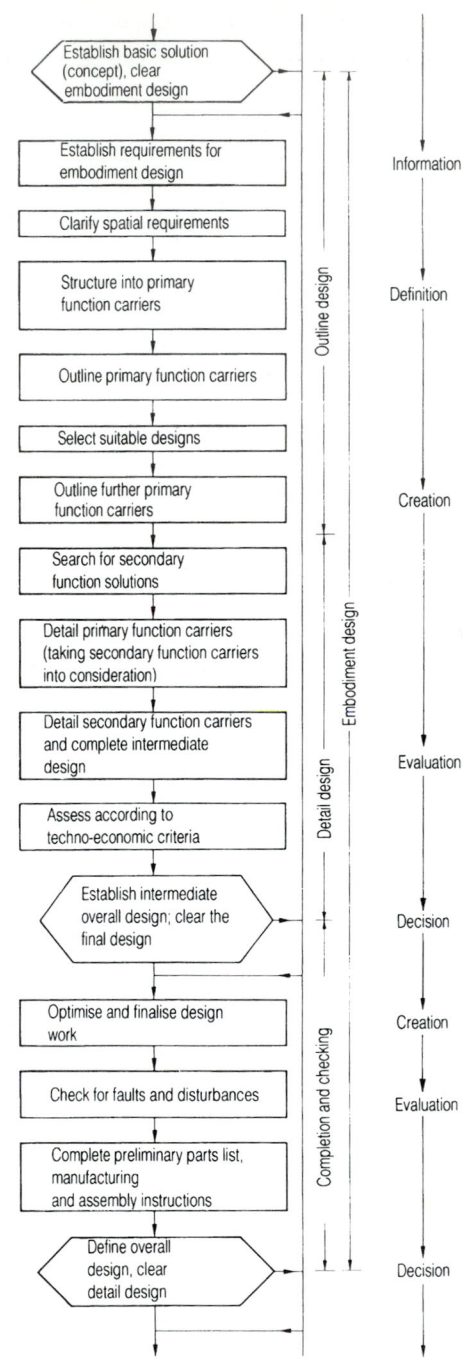

Figure 2. Stages in embodiment design. Primary function carriers: individual components and assemblies fulfilling a primary function; secondary function carriers: individual components and assemblies fulfilling a secondary function.

Table 2. Guidelines showing principal features for embodiment design

Principal features	Examples
Function	Is the intended function fulfilled? Which secondary functions are necessary?
Effective principle	Do the selected effective principles produce the desired effect? Which malfunctions are to be expected?
Design	Does the design material, in the shapes and dimensions selected, over the service life and stress conditions envisaged, guarantee acceptable durability, deformation, stability, freedom from resonance, expansion, corrosion and wear characteristics?
Safety	Are factors that affect component, function, operational and environmental safety taken into consideration?
Ergonomics	Are the man–machine relationships taken into account? Are stresses or impairment avoided? Was good design practice observed?
Manufacture	Are manufacturing aspects taken into consideration with techno-economic factors?
Monitoring	Are the necessary checks feasible and arranged for?
Assembly	Can all assembly processes in- and ex-works be undertaken simply and unambiguously?
Transportation	Are transport conditions and risks in- and ex-works checked and taken into consideration?
Use	Are events occurring during use, operation, or handling taken into consideration?
Maintenance	Are maintenance, inspection and repair procedures feasible and capable of monitoring?
Recycling	Has reuse or recycling been facilitated?
Costs	Are preset cost limits to be observed? Do additional operating or secondary costs arise?
Deadlines	Can the deadlines be met? Can a different design improve the deadline situation?

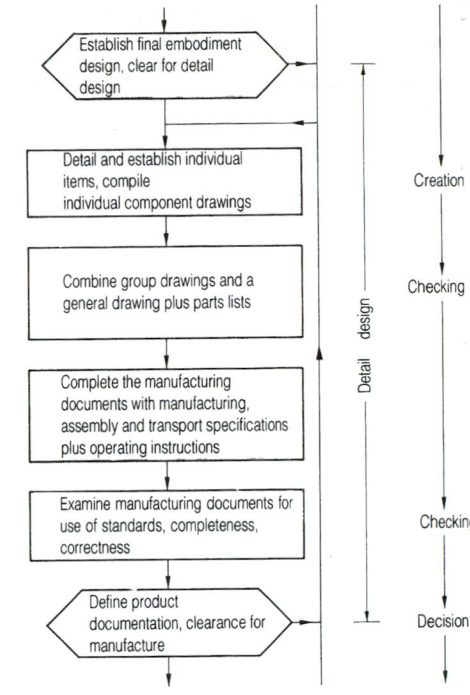

Figure 3. Stages in detail design.

quently involve only certain modules or system components. In other cases, an adjustment to different circumstances is sufficient, without having to alter the solution principle, or it is only necessary to vary dimensions or arrangements within a system that has been thought through in the past.

This produces three types of engineering design, the boundaries of which may be fluid:

Basic design. This consists of the compilation of a new solution principle for similar, different or new problems in a system (plant, apparatus, machine or module).

Redesign. By this is meant the adjustment of the embodiment (shape and material) of a known system (solution principle remaining the same) to an altered problem (including advancing of previous boundaries). Redesign of individual modules or components is frequently necessary.

Variant design. This means the variation of size and/or arrangement within the boundaries of systems that have been thought through in the past. Function, solution principle and embodiment are usually retained.

4 Fundamentals of Embodiment Design

4.1 Basic Rule of Embodiment Design

The basic rule, "unambiguous, simple, reliable", is an instruction for embodiment design and is derived from the general objective (see E1.5).

"Unambiguous" instructs the designer to predict action and behaviour, clearly and in a manner that is easily recognisable (fulfilment of the technical function).

"Simple" is an enjoinder to aim for embodiment by means of a few combined, clearly defined shapes and to keep down manufacturing costs (economic realisation).

"Reliable" tells the designer to include life, reliability, safety and environmental protection in the embodiment design process (safety for both humans and the environment).

If this basic rule is observed during embodiment design, a good realisation of objectives may be expected.

Linking guidelines (see **E3 Table 2**) with basic rules provides a stimulus for questions and ensures that significant factors are not missed so as to achieve good results.

4.2 Principles of Embodiment Design

The principles of embodiment design involve strategies which are not applicable generally.

4.2.1 Principle of the Division of Tasks

During embodiment design the practical selection and allocation of function carriers for the functions to be fulfilled leads to the following questions: which subfunctions can be fulfilled jointly using only a single function carrier, and which subfunctions must be fulfilled by a separate function carrier?

In general, the aim is to realise a large number of functions with only a few function carriers. However, function analyses and searches for weak points and defects can indicate whether restrictions or interactive hindrances or disruptions may occur. This is usually the case when *performance limits* are desired, or if the *behaviour* of the function carrier must remain *unambiguous* and uninfluenced. It is practical to divide the task in cases where the function is fulfilled by a function carrier allocated to this aspect.

The *principle of the division of tasks*, by which each function is allocated to a specific function carrier, produces better utilisation levels because of unambiguity in calculations (clarity), a higher performance because absolute limits can be reached (if these alone are decisive), unambiguous behaviour in operation (fulfilment of function, characteristics, length of life, etc.), and a better manufacturing and assembly cycle (simpler, parallel). The disadvantage lies in higher construction costs, which have to be matched by greater efficiency or safety.

Example   (**Fig. 1**) *Embodiment design for a helicopter rotor head.* Centrifugal force is transmitted to the central core section solely via element Z of the rotor blade which is compliant in torsion. The bending moment arising from aerodynamic loading is transferred only via component B to the roller bearing in the rotor head. In this way each component can be designed optimally according to its task.

Further examples are: separation of the radial and axial force absorption in fixed bearings; design of containers for process technology using an austenitic feed pipe to prevent corrosion, combined with a ferritic container wall to take the pressure; V-belts with internal tension cords bedded into the rubber, the surface of which has a high friction coefficient to transfer power.

4.2.2 Principle of Self-help

The objective of this principle is to make the system support itself so as to fulfil its function more effectively and to avoid damage under overload conditions.

The necessary *overall effect* of this principle arises from a *basic effect* and an *auxiliary effect* (e.g. **Fig. 2**). The same design may be *self-helping* or *self-impairing*, depending on arrangement. If the pressure in the container is greater than the external pressure, the left-hand arrangement is self-helping. If, on the other hand, it is less, the left-hand arrangement is self-impairing, and the right self-helping. A distinction is made between solutions of the following kinds:

Self-reinforcing Solutions. Under normal load, the design ensures that main or auxiliary factors provide a *reinforcing overall effect*.

Self-compensating Solutions. Under normal load, the design ensures that an auxiliary factor *counteracts* the original effect and therefore provides *compensation* to achieve a greater overall effect.

Self-protecting Solutions. Under overload conditions, the design provides for an *additional, force transmission path* for the overloaded main component. This leads to a redistribution of loading, which is within the load-bearing capacity of the affected components.

4.2.3 Principles of Force and Energy Transmission

Force transmission should include the transmission of bending moments and torques. It is accompanied by deformation.

Embodiment Design and the Flow of Force. The "flow of force" is a concept which is not physically valid but which is a visual concept for the transmission of forces. The forces and moments which are transmitted via the component cross-section are defined as the force flow. The following principal requirements are derived from this model for an embodiment incorporating the force flow concept:

The force flow must always be closed (action = reaction). Sharp deviations in the flow and sudden changes in its density due to sharp changes in cross-section should be avoided (notch effect).

Principle of Equal Strength. Aim for equal distribution of strength by suitable selection of materials and form, if economic reasons do not dictate otherwise (see **B2 Table 3** and D1.5).

Principle of Minimum Force Transmission Path. Forces and moments must be transmitted at mini-

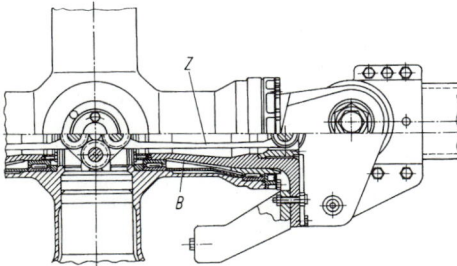

Figure 1. Helicopter rotor blade attachment on the principle of distribution of tasks (design: Messerschmitt-Bölkow).

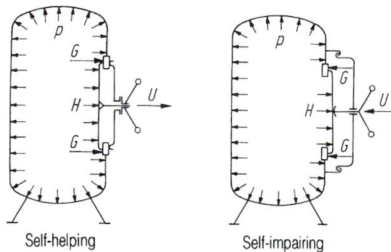

Figure 2. Arrangement of a manhole cover. *U* original effect, *H* auxiliary effect, *G* overall effect, *P* internal pressure.

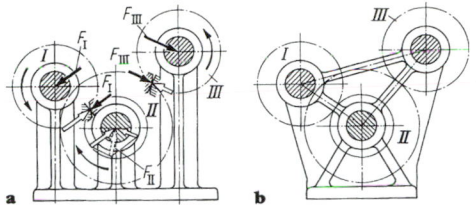

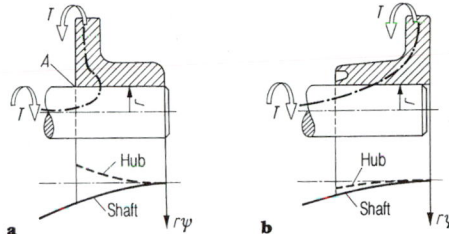

Figure 3. Bearing support for two-stage Leyer open gear system. **a** Bad: long force transmission paths, high bending proportions, poor casting design. **b** Good: bearing forces taken up directly, rigid support with mainly tensile and compressive stresses.

Figure 4. Shaft/hub connection: **a** with sharp deflection of force flow: in this case with torsional deformation between shaft and hub in the opposite direction at A (φ = torque angle); **b** with gradual deflection of force flow: in this case with torsional deformation in the same direction over the whole hub length (φ = torque angle).

mum material cost. *Low deformation* requires paths of minimum length and only tensile and compressive component stress where possible (e.g. **Fig. 3**). *High elastic deformation* requires a long force transmission path and, preferentially, bending or torsional stress (e.g. compression springs, pipeline with compensation elbows).

Principle of Harmonised Deformation. Components are to be designed so that an extensive adjustment takes place under loading with *equal deformation* and minimum *relative deformation*. The objective is to avoid or reduce excess stress and friction corrosion, and to obviate disruptions caused by deformation. Harmonisation may be achieved by adjustment of position, shape, or dimensions, or by selection of material (modulus of elasticity) (**Fig. 4**).

Principle of Force Compensation. In a fixed design, the main functional factors, such as applied loads, drive moments or peripheral forces, are frequently connected with auxiliary factors such as axial thrust, or tensile, mass

and flow forces. These auxiliary factors add to the loadings on force transmission zones, and can require a correspondingly more expensive design. The principle of force compensation recommends *compensating elements* for average forces and *symmetrical arrangements* for relatively large forces (**Fig. 5**).

4.2.4 Safety and Reliability Principles

According to DIN 31000, a distinction is made between direct, indirect and indicative safety principles. Basically, *direct* safety principles are those which ensure inherently and intrinsically that no danger can arise. *Indirect* safety involves the creation of protective systems and devices. *Indicative* safety principles, which only warn about danger or indicate the danger area, do not solve any safety problems. The principle of division of tasks (see E4.2.1)

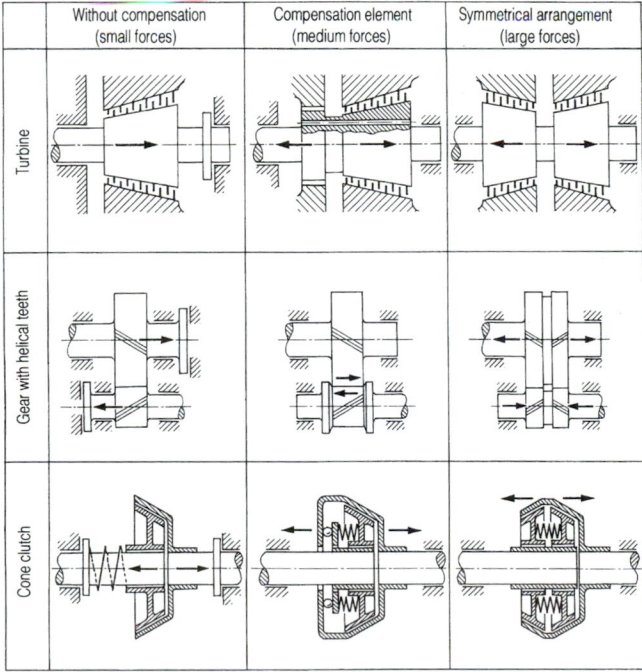

	Without compensation (small forces)	Compensation element (medium forces)	Symmetrical arrangement (large forces)
Turbine			
Gear with helical teeth			
Cone clutch			

Figure 5. Basic solutions for force compensation using the examples: a turbine, a gear, and a clutch.

and the basic rule of "unambiguity" (see E4.1) also contribute to the achievement of safe conditions.

Safe-Life Principle. This means that all components and their interrelationships must survive for the envisaged period of use under all probable or possible conditions without failure or breakdown.

Fail-Safe Principle. This allows breakdown and/or a fracture but avoids serious consequences. In this case it is necessary for

A function or capacity to be retained, even in limited form, to avoid a dangerous condition;
The restricted function to be taken over by the failed component or by another, and to continue to be exercised until the system or machine can safely be shut down;
The fault or failure to be recognisable;
The failure point to be such that its significance for overall safety can be evaluated.

Principle of Multiple or Redundant Arrangement. This means increased safety so long as the system element which fails does not in itself present a danger, and that the system elements in series or in parallel take over the defective function or at least part of it. In the case of *active redundancy* (**Fig. 6**) all system elements take an active part; in *passive redundancy* they are kept in reserve and their activation requires a switching process. The *principle of redundancy* is present if the function is the same but the principle of activation differs. However, the system elements themselves must be in accordance with one of the above principles.

Indirect Safety. Indirect safety includes *protective systems and devices* [1]. The latter safeguard danger points (e.g. cladding, covers, guards) in an ergonomic context (see E4.3.7). Under a dangerous condition, protective systems may shut down a system or machine automatically, cut off power or the material flow, or prevent startup.

The following requirements must be observed in the design of protective systems:

1. *Warning or signal.* Before a protective system initiates a change in operating conditions a warning must be given so that the operator or supervisor may remove the danger if possible, or at least initiate any necessary measures. If a protective system prevents startup then it should indicate the reason.
2. *Self-monitoring.* A protective system must monitor itself for constant availability; i.e. the system must be triggered not only by the danger situation against which protection is to be provided, but also by a fault in the protective system itself. Zero-signal current is the best guarantee of this requirement because power for activating the safety device is always stored in the system and a breakdown or fault releases it, switching off the machine or system during this process. The zero-signal current principle can be used not only in electrical systems, but also in systems using other types of power.
3. *Multiple independent protective systems with different principles.* If human lives are endangered or damage of a considerable magnitude is to be expected, a backup system must be provided, using different principles and independent of the primary system (primary and secondary protective circuit).
4. *Bistability.* Protective systems must be designed for a defined response. Triggering must take place immediately, without interim states.
5. *Restart barrier.* Systems must not restart automatically after a danger has been removed. They require startup to be triggered again.
6. *Ability to be tested.* It must be possible to test protective systems. The protective function must be retained during this process.

4.3 Guidelines for Embodiment Design

Guidelines for embodiment design arise from general constraints (see E1.5), from the special guideline given in **E3 Table 2**, and last but not least from the natural and probabilistic laws related to machine elements (see Section F).

4.3.1 Design for Strength

The propositions of strength theory (see B1 to 9), materials engineering (see D1), and the principles of force transmission (see E4.2.3) must be observed. Design should aim at optimum and uniform utilisation of construction and system components (principle of equal embodiment strength), provided that economic considerations do not dictate otherwise. Utilisation refers to the ratio of calculated to permissible stress.

4.3.2 Design for Controlled Deformation

Stresses are always accompanied to a greater or lesser degree by deformations (see E4.2.3). *Deformations* can also be restricted for functional reasons (e.g. limited sag of gear shafts, electric motors or turbines). Deformations must not lead to functional disruption during operation, because otherwise the flow of force or expansion is no longer guaranteed and can lead to overload or fracture. Stress deformations must be noted as well as (if applicable) values of transverse expansion (transverse contraction) and harmonised deformation (see E4.2.3).

4.3.3 Design for Stability and to Avoid Resonance

Stability refers to all the problems of rigidity and toppling, plus the risk of buckling and bending (see B7), but

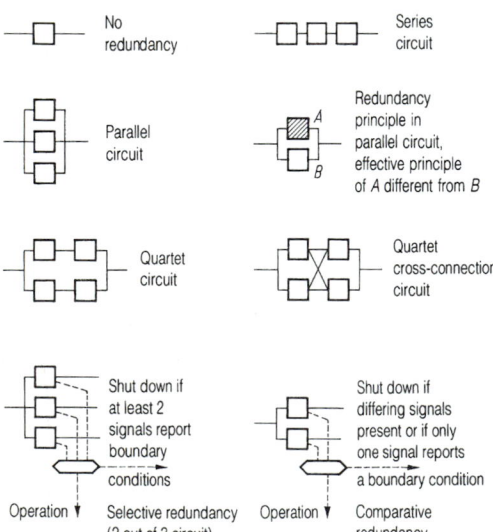

No redundancy

Series circuit

Parallel circuit

Redundancy principle in parallel circuit, effective principle of A different from B

Quartet circuit

Quartet cross-connection circuit

Shut down if at least 2 signals report boundary conditions

Operation ▼ Selective redundancy (2 out of 3 circuit)

Shut down if differing signals present or if only one signal reports a boundary condition

Operation ▼ Comparative redundancy

Figure 6. Redundant arrangements (circuits with system elements).

also includes the stable operation of a machine or system. Breakdowns should be avoided by stabilising behaviour, i.e. automatic return to the initial or normal position. Indifferent or unstable behaviour must not reinforce or amplify faults or cause them to become uncontrollable.

Resonances result in increased stresses which cannot be reliably estimated. They must therefore be avoided if the amplitudes cannot be properly damped (see A4). Consideration should be given not only to strength problems but also to accompanying phenomena such as noise and vibration amplitudes.

4.3.4 Design to Accommodate Thermal Expansion

Machines, apparatus and equipment only work correctly if the effect of *expansion* is taken into consideration.

Expansion of Components. The coefficient of expansion is a mean value over the temperature range in question; it is dependent on material and temperature (see Section C). Component expansion depends on the coefficient of linear expansion β, the length component l, and the average temperature change $\Delta\vartheta_m$.

Thermal expansion has implications for embodiment design. Each component must be located unambiguously and may be allowed only the degree of freedom required to fulfil its function correctly. A datum point is usually established, and guides are then arranged for the desired directions of movement. These guides must have only one degree of freedom; they should be aligned optically through the datum point so that they lie in the line of symmetry along which distortion takes place. Distortion may be caused by expansion or by load- and temperature-dependent stresses. As stress and temperature distribution also depend on the shape of the component, the line of symmetry of distortion is most likely to lie along the line of symmetry of the component and the temperature field imposed on it.

Relative Expansion Between Components. This arises from

$$\delta_{Rel} = \beta_1 l_1 \Delta\vartheta_{m1t} - \beta_2 l_2 \Delta\vartheta_{m2t}.$$

Steady Relative Expansion. If the average temperature difference is independent of time, design measures should concentrate on equalising temperature if the coefficients of linear expansion are the same and/or if the temperatures differ, by the selection of materials with different coefficients of expansion.

Unsteady Relative Expansion. If the temperature pattern changes with the time (e.g. in heating or cooling processes) the relative expansion may be much greater than in the steady condition, because the temperature difference may vary considerably in the individual components. In the frequently occurring situation where components are of the same length and have the same expansion coefficient, the following applies:

$$\delta_{Rel} = \beta l (\Delta\vartheta_{m1(t)} - \Delta\vartheta_{m2(t)}).$$

The form of the heating curve with respect to time is characterised by the coefficient of thermal conductivity. For example if the increase in temperature $\Delta\vartheta_m$ of a component is considered in terms of a sudden rise in temperature $\Delta\vartheta^*$ of the medium being heated, the pattern is as shown in **Fig. 7**, which follows the equation

$$\Delta\vartheta_m = \Delta\vartheta^* (1 - e^{-t/T}),$$

where t = time, T = thermal conductivity = $cm/(\alpha A)$,

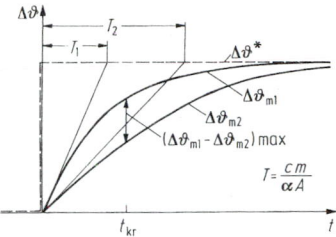

Figure 7. Temperature change with time for a temperature change $\Delta\vartheta^*$ in the heated medium for two components with different time constants.

c = specific heat of component material, $m = \rho V$ = mass of component, α = heat transfer coefficient of the heated surface, A = heated area of component. However, this is based on a general assumption that the surface and average component temperatures are the same, which is only approximately accurate in practice for relatively thin wall thicknesses and high heat transfer coefficients. If components 1 and 2 have differing thermal conductivities, different temperature patterns, which have a maximum difference at a certain critical time, are generated. If it is possible to equalise the thermal conductivities of the components, no relative expansion takes place. There are two design methods for this: harmonising the V/A (volume-to-heated-surface) ratios, or changing the heat transfer coefficient α using e.g. protective coverings or different incident flow velocities.

4.3.5 Design to Avoid Corrosion

Corrosion phenomena can only be reduced rather than avoided because the reason for the corrosion cannot be removed. The use of non-corroding materials is often uneconomic. Corrosion phenomena can be counteracted by an appropriate concept and more practical embodiment design. The measures used will depend on the nature of the corrosion phenomena (see D1.2.3 and [2, 3]).

Even Corrosion: Cause and Appearance. Such corrosion arises from moisture (in the form of a slightly alkaline or acid electrolyte) and the simultaneous presence of atmospheric or local oxygen, especially when at temperatures below the dew point. This results in mainly even surface corrosion (for steel e.g. approximately 0.1 mm per year in normal atmosphere). The remedy is to increase wall thickness and material; to employ process management to avoid corrosion or make it economically acceptable; to have small, smooth surfaces with a maximum volume/area ratio; to avoid moisture collection points; and to have zero temperature variation, good insulation and no heat transfer points.

Localised Corrosion. This is particularly dangerous because it generates a considerable notch effect and is often not easily foreseeable. Types of corrosion include crevice corrosion, contact corrosion, fatigue crack corrosion, or stress crack corrosion. For causes and remedies see D1 and [2, 3]. The following measures can help combat various types of corrosion:

1. *Crevice corrosion.* Smooth, crevice-free surfaces, including transition points; welds without permanent root gaps, butt welds or full penetration fillet welds; sealed crevice, moisture protection provided by sheaths or coatings; crevices enlarged to prevent con-

centration because of material flowing through or being exchanged.

2. *Contact corrosion.* Metal combinations with low potential difference and thus small contact corrosion current; electrolyte kept away from contact point so that the two metals are locally insulated; no electrolyte at all; if necessary, controlled corrosion by intentional removal of electrochemically less pure "wearing material", sacrificial anodes.

3. *Fatigue crack corrosion.* Low mechanical or thermal alternating loadings, avoid resonances; avoid notch stress points; prestress by shot-blasting, burnishing, nitriding etc. (longer life); keep corrosive media (electrolyte) away, protective surface coatings (rubberising, stoving, prestressed galvanic platings).

4. *Stress crack corrosion.* Select suitable materials, reduce or avoid tensile stress on an attack-prone surfaces; apply compressive stress (e.g. shrink bindings, prestressed multi-layer construction, shotblasting); relieve internal stresses by annealing; apply cathodic coatings; avoid or reduce the effect of corrosive agents by reducing concentration and temperature.

In general, embodiment design must be carried out so that the longest and most even life is possible for the components, even when they are subject to corrosion. If this requirement cannot be met economically by an appropriate selection of material and layout, zones and components particularly at risk from corrosion must be capable of being monitored and exchanged [4].

4.3.6 Design to Limit Wear

Wear means the undesirable release of particles due to mechanical or chemical effects (see D1.2.3). Like corrosion, wear is not always avoidable. From a design point of view, wear is regarded as the output of a *tribological system* arising from the interaction of functioning elements, their characteristics and environment, plus those of the intermediary material (lubricant). It therefore follows that *lubricant selection* alone cannot be sufficient to define the phenomenon but that design features also have a decisive effect on it. As a first step, the following measures can be taken:

An acceptable, clearly defined, and locally *even stress* (using elastically yielding or automatically adjusting elements etc.)

A *motion* of the contact surfaces to build up or support a lubricant film

Component *geometry* that remains constant when subject to temperature or other influences (e.g. crevice geometry, feed zone)

A *functional surface* (embodiment and roughness) that does not deteriorate during the wear process

An appropriate *material selection* to reduce adhesive or abrasive wear due to material combination

The following remedies may be appropriate for the basic mechanisms discussed in D1 and [4] (types of wear):

Adhesive wear. Select other materials and introduce different intermediate layers (e.g. solid lubricants).

Abrasive wear. Increase hardness of the softer component (e.g. nitriding, application of a layer of carbide [5, 6]).

Fatigue wear. Reduce and distribute localised stress.

Laminar wear. As this process usually arises in the so-called deep layer in the presence of wear processes that do not affect function (amount removed per unit of time or travel is small), it can be tolerated until the

thickness of the component is no longer sufficient (e.g. to meet the strength requirements).

Friction corrosion. This is a complex process (mechanical–chemical) and leads to hard oxides separating out which endanger function while the abraded point itself suffers from a notch effect which is damaging in many ways. The remedy is to avoid relative movements at joint positions by reinforcing the component, altering the introduction and withdrawal of load; and/or the incorporation of stress-relief grooves.

4.3.7 Design for Ergonomics and User Safety

Embodiment Design for User Safety. Personnel and their environment must be protected from harmful effects. DIN 31000 indicates basic requirements for the correct embodiment of technical products for user safety. DIN 31001, Parts 1, 2 and 10, gives instructions on protective devices. *Specifications* from the employers' liability insurance associations, trade boards and the technical monitoring associations (TÜVs) are to be followed for the appropriate branch of industry and products. However, *the law on the safety of equipment* binds the designer to responsible action. Domestic standards and other rulings or regulations with a safety engineering content are listed together in a general administrative regulation and in the lists attached to this law [7]. Lack of understanding or tiredness in the personnel involved must also be taken into consideration. **Table 1** gives *minimum requirements* for an embodiment appropriate to user safety covering mechanical entities.

Ergonomics. VDI-Richtlinie 2242 [8] gives information on the ergonomic design of products. In doing so it falls back on search lists for subjects and their effects and refers to the appropriate literature. Only a few references important to the designer can be given here as examples: physically appropriate operation and handling – see DIN 33400 to DIN 33402 plus [9, 10]; lighting in the workplace – see [11]; ventilation in the workplace – see DIN 33403; monitoring and control activities – see

Table 1. General minimum requirements for safety at work in the presence of machines

---

Avoid projecting or moving components in the contact area!

Protective devices are necessary for the following regardless of speed:

Gears, belts, chains and ropes.

All rotating components longer than 50 mm (even if they are completely smooth!).

All clutches.

Where there is a danger of projectiles.

Positions that can trap (carriage against stop; parts that pass each other or rotate against each other).

Components that fall or move downwards (tensioning weights, counterweights).

Components that are inserted or drawn in. The gap remaining between tools may not exceed 8 mm. In the case of rollers, take special note of the geometry, and if necessary provide protection strips or contacts to protect against the danger of being drawn into the machine.

Electrical systems should only be planned by an electrical engineer. Where there are *acoustic, chemical or radioactive dangers*, technical experts should be employed to compile auxiliary and protective measures.

DIN 3304, 33413, 33414 and [12]; reduction of noise – see [13, 14].

4.3.8 Design for Aesthetics

Recommendations to the designer on the aesthetics of technical products are drawn up in VDI-Richtlinie 2224 (with instructive pictorial examples). A systematic analysis is also to be found in [15] on shape, colour and graphics including the use of [20].

4.3.9 Design for Ease of Manufacture and Inspection

During embodiment and detailed design work, care must be taken that *component structure* lends itself to ease of manufacture and that *workpiece embodiment* is also aligned towards ease of manufacture and inspection, which is closely linked to *material selection* adjusted to manufacture.

Component Structure for Ease of Manufacture. This can be undertaken from the points of view of differential, integral and composite design.

Differential design refers to the division of a single component (fulfilling one or more functions) into several workpieces that are advantageous from the point of view of manufacturing technology.

Integral design refers to the combination of several individual components to form one workpiece. Typical examples are the use of casting instead of fabrication designs, extruded instead of jointed standard profiles, and forged-on rather than welded flanges.

Composite design refers to the permanent connection of several separately manufactured blanks to form a workpiece on which further work takes place (e.g. the connection of formed or unformed components). It also includes the simultaneous use of several jointing procedures to connect workpieces, and the combination of different materials for the optimum utilisation of their characteristics. Examples are the combination of steel castings with welded structures plus rubber–metal elements.

Embodiment of Workpieces for Ease of Manufacture. This influences shape, dimensions, surface quality, tolerances and joint fits, manufacturing processes, tools and quality control. The objective of workpiece embodiment is to reduce manufacturing *cost* and to improve workpiece *quality* during the various manufacturing processes and their individual process stages. Embodiment suggestions are: original embodiment – see K2.2.3; reshaping – see K3.5; joining – see F1; separating – see K4.

4.3.10 Design for Ease of Assembly

Since automatic assembly gained in importance, assembly linked *structure* has been decisive, as has embodiment of the *joint positions* and *joined components* [19].

The following operations can be recognised in assembly with varying completeness, sequence and frequency [16, 17, 18]: storage – handle workpiece (recognise, hold, move) – positioning – joining – arranging (adjusting) – securing – checking.

General Guidelines on Assembly. The aim should be uniform assembly methods, the minimum of simple and automatic assembly operations, and parallel assembly of modules.

Improvement of Individual Assembly Operations

Storage is facilitated by stacking workpieces with adequate bearing areas and contours for clear positioning orientation in the case of non-symmetrical components.

Workpiece Handling. Confusion with similar components must be excluded. Positive and reliable handling is particularly important for automated assembly. Aims should be minimum work path, observation of ergonomic and safety aspects, and simple workpiece handling.

Positioning. Aim for symmetry if no preferred position is required; if a preferred position is required, it should be indicated via the shape. Components to be joined should be self-aligning or, if this is not possible, adjustable connections should be provided.

Joining. Points which have to be released frequently (e.g. for changing worn parts) should be fitted with quick-release connections. Joints that are rarely or never released after initial assembly can use more costly methods. Simultaneous connecting and positioning is desirable. To facilitate economically justifiable tolerances, compensation must be provided for workpieces with a high level of spring rigidity by using sprung spacers or compensation pieces (design for tolerance). Insertion of a component into the joint surfaces is facilitated by good access for assembly tools, optical inspections, simple movements on joint surfaces; provision of insertion aids, avoidance of simultaneous joining operations and avoidance of double fits also helps.

Adjusting. Enable sensitive, reproducible adjustments to be made. Avoid feedback to other areas. Make result of adjustment measurable and monitorable.

Securing. The aim is the avoidance of any independent alteration; select self-securing connections or provide additional locking devices with the same substance or with positive locking, which can be fitted without a high expenditure.

Monitoring. A simple method of monitoring (measuring) the function-based requirements is to be provided during embodiment design. It must be possible to undertake monitoring and further adjustments without dismantling components that are already assembled.

4.3.11 Ensuring Operability and Maintainability

Embodiment design has to take account of the requirements of operation and maintenance, which is divided into *maintenance, inspection and repair*. In general, use or commissioning should be possible in a *reliable* and *simple* manner. Operating output in the form of signals, monitoring data and measured dimensions should be clearly visible. Operation should produce no serious burden on the environment. It should be possible to undertake maintenance simply and in a manner that can be monitored; inspections must enable critical states to be recognised, and repair should be feasible, if possible, without time-wasting assembly operations.

4.3.12 Designing for Ease of Recycling

The saving and recovery of raw materials is increasing in importance. VDI-Richtlinie 2243 [21] refers to recycling processes and gives design indications: economic dismantling, easy separation of materials, appropriate selection and marking of compatible materials.

5 Fundamentals of the Development of Series and Modular Design

A *series* refers to technical entities (machines, assemblies, individual components), which fulfil the same function with the same solution *in several size ranges* using so far as possible the same method of manufacture in a wide field of application. If other functions are to be fulfilled in addition to the size graduation, a modular system must be developed in addition to the series (see E5.6). Similarity laws apply automatically in the development of series, and decimal-geometric preferred numbers are advisable.

5.1 Similarity Laws

A purely geometrical increase in size is admissible only if similarity laws permit it. These laws are used as evaluation criteria just as in model technology (see A7.2), and it is clear that this can be transferred to series development. Mentally, the "model" can be equated with the original design, while the "basic design" and the "detailed design" of the model can equate with a component in the series as a "follow-up design".

Series design differs from model technology in its objective: an equally high level of utilisation for all series components using the same materials and the same technology. It follows that, if the function is fulfilled equally over a wide range of dimensions, the loading must also remain the same.

In engineering systems, the most frequently occurring forces are inertial (mass forces, acceleration forces, centrifugal forces) and elastic forces associated with the stress–strain interrelationship.

Constant stress can be achieved if all the velocities remain constant. If the scale factor for the length between follow-up and basic design is defined as $\varphi_L = L_1/L_0$, similar scale factors can be derived for all significant dimensions such as output and torque, subject to the condition that $\varphi_L = \varphi_t = \text{const.}$ and $\varphi_\rho = \varphi_E = \varphi_\sigma = \varphi_v = 1$; these are listed in **Table 1**.

It should be noted that material utilisation and safety are constant only if the effect of size on the limit values of the material can be disregarded within the graduation. It must be taken into consideration if necessary.

Table 1. Relationships for geometrical similarity and equal stress: relationship of basic dimensions to scale (Ca = Cauchy number)

With $Ca = \rho v^2/E = \text{const.}$ and with the same material, i.e. $\rho = E = \text{const.}$, therefore $v = \text{const.}$

Values then change in geometric similarity with the length scale φ_L:	
Rotational speed n, ω	
Transverse or torsional critical speeds n_{kr}, ω_{kr}	φ_L^{-1}
Strains ε, stresses σ, surface pressures p, caused by inertial and elastic forces, velocities v	φ_L^0
Spring rigidities c, elastic deformations Δl	φ_L^1
Gravitational factors:	
Strains ε, stresses σ, surface pressures p, forces F, loadings P	φ_L^2
Weights G, torques M_t, torsional rigidity c_t, section moduli W, W_t	φ_L^3
Geometric moments of inertia I, I_t	φ_L^4
Mass moments of inertia J	φ_L^5

Note: Material and safety are only constant if size effect is negligible at the material limit values.

5.2 Decimal-Geometric Series of Preferred Numbers (Renard Series)

5.2.1 Properties of Decimal-Geometric Series

The *decimal-geometric series* is produced by multiplying by a constant factor φ and is developed within a decade in each case. φ is the scale factor of the series and is defined by

$$\varphi = \sqrt[n]{a_n/a_0} = \sqrt[n]{10},$$

where n is the scale factor within a decade. Thus, for 10 stages, the series would have a scale factor $\varphi = \sqrt[10]{10} = 1.25$, and would be designated R 10. The number of terms in the series is $z = n + 1$; an extract from DIN 323 in which the main values of the basic series are stipulated is reproduced as **Table 2**.

Derived Series. In this case only every kth term of a basic series is used. To indicate this, the number k is placed as a denominator after the series designation with the first number of the series following in brackets, e.g.

R 20/4 (1.4 ...)   1.4   2.24   3.55   5.6   etc.,

R 10/3 (1   ...)   1   2   4   8   etc.

The scale factor, then, is always the ratio of two numbers which follow each other in sequence, or $\varphi = 10^{k/n}$.

5.2.2 Choice of Ratio

The ratio depends on anticipated market requirements for the individual structural sizes, on the market behaviour

Table 2. Principal values for preferred numbers (extract from DIN 323)

Basic series				Basic series			
R5	R10	R20	R40	R5	R10	R20	R40
1.00	1.00	1.00	1.00	4.00	4.00	4.00	4.00
			1.06				4.25
		1.12	1.12			4.50	4.50
			1.18				4.75
	1.25	1.25	1.25		5.00	5.00	5.00
			1.32				5.30
		1.40	1.40			5.60	5.60
			1.50				6.00
1.60	1.60	1.60	1.60	6.30	6.30	6.30	6.30
			1.70				6.70
		1.80	1.80			7.10	7.10
			1.90				7.50
	2.00	2.00	2.00		8.00	8.00	8.00
			2.12				8.50
		2.24	2.24			9.00	9.00
			2.36				9.50
2.50	2.50	2.50	2.50				
			2.65				
		2.80	2.80				
			3.00				
	3.15	3.15	3.15				
			3.35				
		3.55	3.55				
			3.75				

when models are taken out leaving associated gaps, on manufacturing costs and times for different ratios, and on the characteristics of the product in different scale sizes.

It will not always be practical to divide up the series size range using a constant scale factor. For technical and economic reasons, it is often more advantageous to break the range down into differing size intervals (i.e. by moving within and/or between coarser and finer preferred number series (R 5 to R 40)). As a rule, the ratio has to be finer the greater is the demand, and the more precisely defined is the technical characteristics which have to be observed.

5.2.3 Logarithmic Plotting of Preferred Numbers

Almost all technical relationships can be brought into the general form $y = cx^p$, the logarithmic form of which is $\lg y = \lg c + p \lg x$. Each preferred number NZ can be described by $NZ = 10^{m/n}$ or again by $\lg NZ = m/n$, where m designates the respective stage in the NZ series and n the scale factor of the NZ series within a decade.

$$m_y/n = m_c/n + p(m_x/n).$$

In a log–log diagram all dependent factors can be represented by straight lines, the slope of which corresponds to the exponent p of the technical relationship in each case (**Fig. 1**). The preferred numbers themselves are described on the coordinates instead of the logarithms [1].

The basic design is designated by index 0, the first follow-on design by index 1, and the kth by index k. If the nominal dimension x has been entered on the abscissa, the scale factor $\varphi_x = x_1/x_0$. In the presence of a geometrically similar series of sizes, it is, for practical purposes equal to the scale factor φ_L. All other magnitudes such as dimensions, torques, output levels and speeds are derived from the known exponent of its physical or technical laws (**Table 1**) if the basic design is known and can be entered as a straight line with an appropriate slope (e.g. weight $\varphi_G = \varphi_L^3$, i.e. with a 3 : 1 slope). Example: **Fig. 2**.

5.3 Geometrically Similar Series

Starting from a basic design it is necessary to determine generally whether only inertial and/or elastic forces are

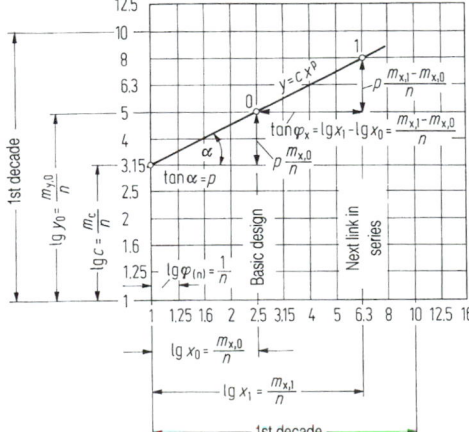

Figure 1. Technical relationships in the preferred number (NZ) diagram: n = basic scale ratio for the most detailed series of preferred numbers; each point on the grid is a preferred number from this series; each integer exponent again leads to a preferred number.

present. If this is so, and if circumferential speed is constant the similarity laws derived in **Table 1** can be used across the series. The exponent to be stated is that which defines the slope of the lines in the preferred-numbers diagram (see E5.2.3), and then the design data can be read off for the other nominal dimensions of the follow-on designs (**Fig. 2**).

The following, however, should be noted:

Fits and tolerances are not graded in the geometrically similar manner of the nominal dimensions, but follow the equation $i = 0.45 \cdot D^{1/3} + 0.001 \cdot D$; i.e. for the most part the scale factor for a tolerance i follows the pattern $\varphi^i = \varphi_i^{1/3}$.

Technological restrictions often lead to deviations; e.g. if the thickness of a casting cannot be less than a given value, or if a wall thickness cannot be through-tempered (size effect).

Higher-ranking standards are not always based consistently on preferred numbers. Components influenced by the latter have to be adjusted correspondingly.

Higher-ranking similarity laws or other requirements may force a major deviation from the geometrical similarity. In this case semi-similar series must be planned (see E5.4).

5.4 Semi-similar Series

Significant deviations from geometrical similarity (requiring a different series scale law, and producing what is called a "semi-similar" series) can be generated by the following considerations:

Higher-ranking similarity laws introduced by the effect of gravity, thermal processes, and/or other similarity considerations [2, 3].

Higher-ranking problem. Components which humans have to work with must correspond to the dimensions of the human body: they cannot change with the series elements. A higher-ranking problem can also arise because of technical difficulties if the initial or final products do not have any geometrically similar dimensions.

Higher-ranking economic requirements. Individual components and assemblies designed in fewer, larger size grades can produce longer series runs and thus remain economic in production. Semi-similar series are then obtained for adjacent components.

These examples show that it is not always possible to keep to geometrically similar series. Rather, while physical and other requirements are observed, it is necessary to derive scales which determine the sizes or other characteristic dimensions. In this case it is no longer possible to ensure equally high strength utilisation; instead, the dimension which gives a higher overall utilisation level is preferred for the series. Depending on physical elements, this dimension can even change between the size grades.

5.5 Use of Exponential Equations

Exponential equations are useful in conditions as given in E5.4 to describe the type of similarity laws which produce semi-similar series. If the preferred numbers diagram is used, only the exponent is significant in the growth law

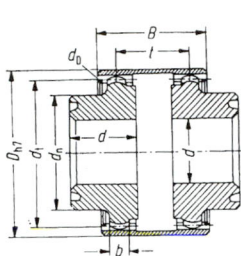

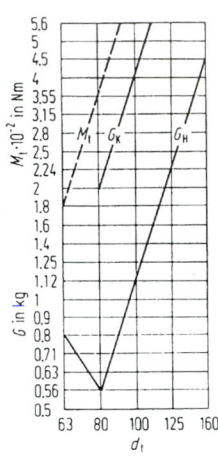

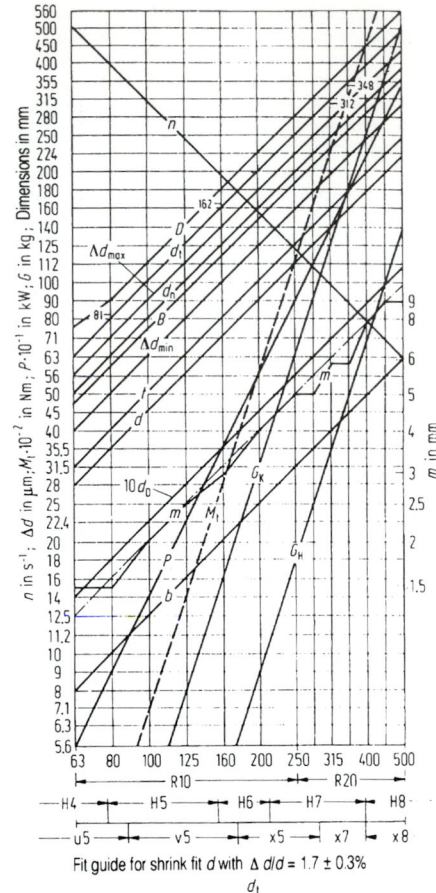

Fit guide for shrink fit d with $\Delta\ d/d = 1.7 \pm 0.3\%$

d_1

Figure 2. Specification sheet for toothed clutch series based on nominal diameter d_1; dimensions geometrically similar; exceptions: sleeve external diameter D for smallest size (for stiffness), modules not graded in accordance with preferred numbers and the requirement for even integral numbers of teeth (some reference diameters slightly adjusted); adjusted fits shown below the diagram.

for exponential function, provided that a fundamental design can be used as a basis.

The equation for the kth element of the series frequently takes the form

$$y_k = c_k\ x_k^{p_x}\ z_k^{p_z}.$$

The dependent variable y and the independent variables x and z can always be expressed with preferred numbers, starting from the fundamental design (index 0).

$$y_k = y_0\varphi_L^{y_e k}; \quad x_k = x_0\varphi_L^{x_e k}; \quad z_k = z_0\varphi_L^{z_e k}.$$

With $y_0 = c\ x_0^{p_x}\ z_0^{p_z}$ and $c_k = c$

$$y_0\varphi_L^{y_e k} = y_0\ \gamma_L^{(x_e k p_x\ +\ z_e k p_z)}.$$

Independently of k,

$$y_e = x_e p_x + z_e p_z.$$

is obtained by comparison of indices.

Here, y_e, x_e and z_e are the exponents which are to be established or determined for the grade, and p_x and p_z are the physical exponents of x and z.

The index y_e is to be determined as a function of x_e and z_e in each case. To do this, the physical dependent

factors are represented in an equation, existing special conditions are introduced, and only exponential equations are used for the calculations [4].

Example *Series of electric motors.* Motor power P is proportional to angular velocity ω, current density G, magnetic induction B, winding dimensions b, h, t (winding volume) and average distance $D/2$ of the conductor from the centre of the shaft. Let D be the nominal dimension of the series.

How does the power P grow? $P \sim \omega \cdot G \cdot B \cdot b \cdot h \cdot t \cdot D$. Expressed exponentially, $P_e = \omega_e + G_e + B_e + b_e + h_e + t_e + D_e$. Let ω, G and B be constant, where $\omega_e = G_e = B_e = 0$. Then b, h, t and D may increase in a geometrically similar manner, with $b_e = h_e = t_e = D_e$. The exponent of the motor power as a function of D is then $P_e = 4D_e$. Thus motor output increases as the fourth power of D if the size increase is geometrically similar.

How would the diameter of the journal bearing on the driven side have to change if the torsional loading remained constant? $\tau_t = M_t/W_t = M_t/(d_L^3\pi/16)$, $M_t \sim P$, $P \sim D^4$. Written as an exponent $\tau_{te} = 0 = 4D_e - 4D_{Le}$ or $d_{Le} = (4/3)D_e$. The diameter of the journal d_L increases by the exponent $4/3$ *vis-à-vis* the nominal dimension D ($\varphi_{d_L} = \varphi_D^{4/3}$).

5.6 Modular System

A *modular system* refers to machines and grouped or individual components which are produced as blocks, fre-

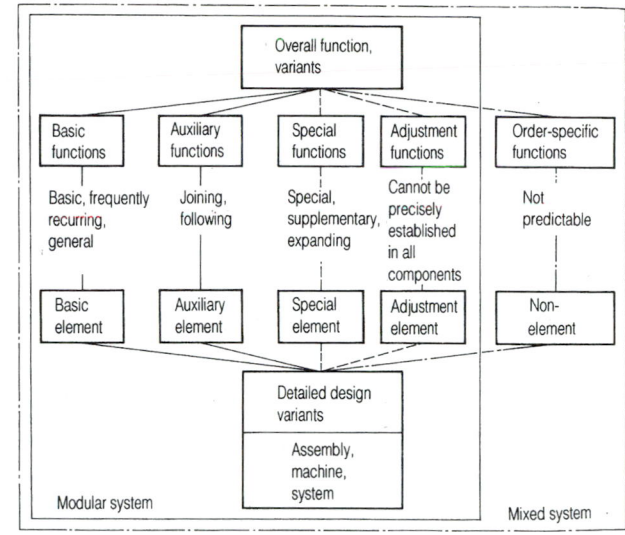

Figure 3. Types of function and elements in modular and mixed systems.

quently with differing solutions because of combinations, and which fulfil *different overall functions*. If there are several size ranges, these blocks often contain modules or modular series.

Modular systems are constructed from such blocks. It is possible to orient and define them in accordance with recurring function types, which – combined as subfunctions – fulfil differing overall functions (function variations). An arrangement for functions of this kind is proposed in **Fig. 3**. A similar order for the types of assembly (types of function carrier) is also produced. Depending on whether a block must be present in all function variants, it is referred to as an *essential* or *an optional block*.

Client-specific functions not foreseen in the modular

system are realised via "non-blocks", which must be developed as individual designs for a given task. This leads to a "mixed system" – a combination of blocks and non-blocks.

To limit the modules, *construction programmes* are defined with final predictable numbers of variants (closed modular system) and *type plans* with a wide range of possible combinations which are not planned and represented in full (open-module system).

Products from modular systems are not normally utilised to an equal degree in all zones. They are therefore often more difficult and more space-intensive than a specialised one-off item. Their economic viability will be found in the use of the system as a whole and not in the comparison of a combination with an individual design.

6 Fundamentals of Standardisation and Engineering Drawing

6.1 Standardisation

6.1.1 National and International Standards

According to DIN 820, standardisation is the planned unification of material and non-material items undertaken jointly by groups of interested parties for general use [1].

Sources of Standards. German standards, including the VDE (Association of German Electrical Engineers) regulations, come from the DIN (German Institute of Standardisation), European standards (EN-standards) are set by CEN (European Standards Committee) and CENELEC (European Electrotechnical Standards Committee). There are also recommendations of the IEC (International Electrotechnical Commission) and recommendations (more recently world standards) from the ISO (International Organisation for Standardisation), as well as VDI (Association of German Engineers) guidelines.

The standardisation process covers the content, range and level of standards (DIN 820).

The following areas are covered by standardisation as regards *content*: information, sorting, allocation to types, plans, dimensions, materials, quality, process, fitness for purpose, testing, delivery and safety.

A distinction is made as regards *range* between basic standards (standards of general, fundamental significance not restricted to a particular subject) and technnical standards (standards for a specific technical field).

The *level* of a standard is defined as regards breadth, depth and scope. It is usual for a standard to belong to several section groups. It can fully represent all relationships of breadth and depth, when it is a full standard, or it may be a part standard, which leaves details open, or an outline standard providing a rough sketch of the items under discussion (so that standardisation does not impede technical development).

Standards are to be found in the "DIN List of Technical Regulations", the most important of which are found in the "Introduction to DIN standards".

Alongside national and international standards there are further international *regulations and guidelines* (cf. DIN List):

VDE regulations of the Association of German Electrical Engineers, which are now considered equivalent to DIN standards

Regulations of the Technical Monitoring Associations (TÜV) e.g. AD-specifications (pressure vessels study group), which also have the character of a standard

VDI guideline of the VDI (Association of German Engineers).

6.1.2 Industrial Standards

Industrial standards are compiled to simplify and rationalise design and manufacture. For practical purposes, they should be written on the model of national and international standards (DIN 820).

Industrial standards may cover: lists of standards as a selection from national and international standards or *limitations* in accordance with company policy; catalogues, lists and information documents on *outside products*; catalogues or lists of company components; information sheets on technical/economic *optimisation* (e.g. on production equipment, production processes, cost comparisons); regulations or guidelines on the *calculation and design* of components, modules, machines and systems; information sheets on *storage and transportation equipment*; stipulations on *quality assurance* (e.g. production regulations, test instructions); regulations and guidelines for *drawings and parts lists*, for numbering techniques and electronic data processing.

6.1.3 Use of Standards

Standards are not absolutely *binding* in the legal sense. However, national and international standards are valid as recognised rules of engineering, the observation of which is advantageous, practical and essential in many cases.

Also, all working standards (adapted national, international and industrial standards) are considered binding within their area of validity, principally for economic reasons, so that compulsory use may be graduated.

The limits of application of a standard usually arise because it can only be valid and binding as long as it does not conflict with technical, economic, safety, ethical or even aesthetic requirements.

Recommendations and suggestions on the use of standards: primarily, basic DIN standards [2] are to be observed, as other standards are based on them. Disregarding basic standards results in consequences no longer being clear, especially in the long term. Depending on the technical field, appropriate standards or guidelines, especially *safety standards* (DIN 31 000/VDE 1000 [3, 4] should be sought in the list of standards and guidelines. *Preferred numbers and series of preferred numbers* for size ratios and type allocation are to be used where possible, especially for series and module development (see E5.2).

6.2 Basic Standards

Basic standards are of general, basic significance [2].

6.2.1 Engineering Surfaces

Basic Terms. A solid body is distinguished from its surroundings by its *actual surface*. The geometrically perfect body has an ideal surface, *the geometrical surface*, which is defined by the geometrical description, e.g. in a drawing or a computer model. It is almost impossible to achieve the geometric surface. During manufacture an actual surface is produced which deviates in shape (e.g. irregularities, ripples and roughnesses) from the ideal surface. It is possible to define these deviations insofar as they are capable of measurement.

Characteristic dimensions for roughness are established, starting from a *reference surface*, which usually has the same shape as the geometric surface and corresponds to the principal direction of the real surface with regard to its position in space. The *actual* or *geometric profile* and the *reference profile* are obtained by taking vertical sections. The reference profile is represented by the *reference line* to which the characteristic dimensions for roughness refer.

The following applies according to DIN 4762 ($\equiv$ ISO 4287/1). The *profile deviation y* is the distance of a point on the profile, determined in the direction of measurement, from the reference line (**Fig. 1**). *The reference line* is a line on the geometrical profile within a *reference section l* which intersects the actual profile in such a way that the sum of the squares of the profile deviations from this line is a minimum: the *central line m* of the least-square deviations for the profile is termed the "centreline".

Starting from the centreline, the following vertical dimensions are defined for roughness:

Maximum profile peak height R_p is the distance of the highest point of the profile from the centreline *m* within the reference section.

Maximum trough depth R_m is the distance of the lowest point of the profile from the centreline *m*.

Maximum profile height R_y is the distance between the line of the profile groups (upper contact line) and that of the profile troughs (lower contact line). $R_y = R_p + R_m$. R_y corresponds to the former characteristic dimension R_t and the maximum depth of roughness DIN 4768. Therefore the latter is no longer to be used.

Arithmetical mean roughness value R_a is the arithmetical mean value of the absolute values for the profile deviations within the reference section. As shown in **Fig.**

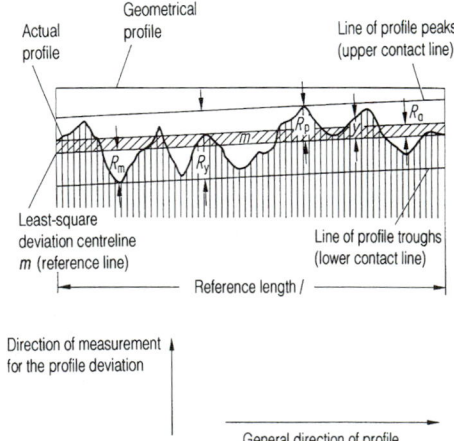

Figure 1. Position of geometrical and actual profiles and characteristic roughness values perpendicular to the centreline *m*.

2, it is to be entered in field *a* of the symbol for the nature of the surface.

Ten-point height R_z (ISO) is the mean value of the absolute values of the heights of the five highest peaks and the absolute values of the five highest troughs within the reference section.

Averaged depth of roughness R_z (DIN 4768) is the mean of the characteristic dimension for roughness for five reference sections within an evaluation length. A maximum permissible value for the ISO ten-point height is considered to have been observed if it is not exceeded by the DIN R_z value. The value for R_z is to be entered in field (*f*) of the symbol (cf. **Fig. 2**).

Profile support proportion t_p is the ratio of the supporting length of the profile to the reference section. It should be stated as a percentage. The supporting proportion by area is described in DIN 4765.

Establishing Roughness Depth. The permissible roughness depth of a surface depends on the function to be fulfilled (load-bearing proportion, subsidence, primer coat, top coat, etc.: cf. DIN 4764). On the other hand, only certain manufacturing processes can achieve low roughness depths, where manufacturing costs are to be taken into consideration (**Table 1** with R_z according to DIN).

Surface Roughness Data for Drawings. The surface symbol and the allocation of roughness depths are regulated by DIN ISO 1302. According to this standard, a distinction must be made between whether the surface is to have a machined finish or not. Moreover, the mean roughness value R_a or the roughness class N 1 to N 12 is to be stated if requirements are indicated with regard to roughness (**Fig. 2**). **Table 2** shows a comparison for orientation purposes between the previous stipulations in accordance with ISO 1302 and the former characteristic dimension R_t (max. roughness depth). Series 3 would normally be preferred for engineering purposes.

6.2.2 Tolerance Systems and Fits

Establishing Tolerances. The *nominal dimension* is marked on a drawing as an indication of size. It is not possible to manufacture a workpiece to this size with absolute precision. As a result, an *actual dimension* is

Symbol	Meaning
✓ ✓	This symbol alone is not significant. The surface may be produced by an additionally specified process.
✓ ✓	The surface must be produced by machining.
✓ ✓	The surface may not be produced by machining. (Permits either non-machined forming or retaining production by preceding manufacturing process (semi-finished part).)

Figure 2. Surface markings on drawings using symbols, roughness measurements, and additional information as DIN ISO 1302, or roughness class no. 1 to N 12: *a* average roughness R_a in μm; *b* manufacturing process, treatment, coating, etc.; *c* reference length, limiting wavelength; *d* groove direction (marking see DIN ISO 1302); *e* processing allowance; *f* other roughness measurements, e.g. R_z in μm.

established for the workpiece by measurement, which may lie within a *tolerance*, i.e. between a given *maximum* and *minimum dimension*, depending on application. Naturally, the tolerances of the measuring instruments must be taken into consideration.

The *tolerance* is the difference between the permissible maximum and minimum dimensions and it is determined by size and position. The *size* of a tolerance is determined by the *basic tolerances* which are graded first of all by nominal dimension ranges and secondly by qualities (abbreviated to IT) from IT 01 to IT 18 (ISO 286). The unit tolerance *i* is given by $i = 0.45 \cdot D^{1/3} + 0.001 \cdot D$ (*i* in μm, *D* in mm as the geometric mean of the nominal dimension range). From IT 5 onwards it is increased by a corresponding factor as the quality decreases (**Table 3**). The *position* of the tolerance range with regard to the nominal dimension (zero line) is determined by *allowances* (**Fig. 3** and **Table 4**); in the case of *internal dimensions* capital letters from A to H are used for positive allowances, J and K for mixed, and M onwards for negative allowances; for *external dimensions* lower-case letters from a to h are used for negative allowances, j for mixed, and m onwards for positive allowances.

The algebraic difference between maximum dimension and nominal dimension produces the *upper allowance* and the difference between the smallest dimension and the nominal dimension produces the *lower allowance*.

If dimensions are stated without a *tolerance being established, general tolerances* (free dimensional tolerances) are applicable. Decisions of this kind should be stated on the drawing. The tolerances and permitted deviations for cast blanks (DIN 1680) and forged-steel components (DIN 7526) and other standards should also be observed.

Fits. These arise from the relationship of the tolerances of pairs of components, and guarantee a specific function (e.g. sliding and guiding requirements, or friction contact in shrink joints) as well as interchangeability. **Table 4** shows the different types of fit. The allocation of tolerance ranges determines the type of fit and the amount of clearance of oversizing. A distinction is made here between the *fit systems*, which can be either hole-based or shaft-based.

Hole Basis. All internal dimensions have zero lower allowance, i.e. tolerance class H. Fit is determined by varying the tolerance on the external features (e.g. H7/f7, H7/g6, H7/h6, H7/k6, H7/s6). This basis is the preferred one for small quantities, with a limited range of tools and gauges for internal finishing.

Shaft Basis. All external dimensions have zero upper allowance, i.e. tolerance class h (e.g. G7/h6, F8/h6, E9/h9). Preferred for drawn semi-finished parts, unstepped shafts, or interchangeable sliding bearings.

Mixed fit systems can be useful. DIN 7157 recommends a limited *selection of fits* to save tools and gauges, summarised in **Table 4**. Other fits are to be found in DIN 7154 and DIN 7155.

6.3 Engineering Drawings and Parts Lists

6.3.1 Types of Engineering Drawing

DIN 199 distinguishes between technical drawings based on type of representation, type preparation, their contents and their purpose.

Table 1. Allocation of achievable surface roughness to manufacturing processes according to DIN 4766

Achievable average surface roughness R_z in μm (scale columns: 0.04, 0.06, 0.1, 0.16, 0.25, 0.4, 0.63, 1, 1.6, 2.5, 4, 6.3, 10, 16, 25, 40, 63, 100, 160, 250, 400, 630, 1000)

Main group	Designation
Basic forming	Sand casting
	Shell casting
	Die casting
Forming	Forging
	Burnishing
	Drawing
	Pressing
	Stamping
	Section rolling
Cutting	Cutting
	Vertical boring
	Horizontal boring
	Plunge cut boring
	Planing
	Jointing
	Scraping
	Drilling
	Open drilling
	Countersinking
	Reaming
	Circumferential milling
	Face milling
	Broaching
	Filing
	Circumferential vertical grinding
	Circumferential surface grinding
	Circumferential plunge-grinding
	Flat circumferential grinding
	Flat face grinding
	Polishing
	Roller burnishing
	Long-stroke honing
	Short-stroke honing
	Cylindrical lapping
	Flat lapping
	Lapping in
	Ultrasonic machining
	Polishing by lapping
	Shot-blasting
	Tumbling
	Flame cutting

Under *type of representation*, a distinction is made between sketches, scale drawings, dimensional drawings, plans and other graphic representations.

Nature of preparation distinguishes between original or master drawings (pencil or ink drawings) used for reproduction, and preliminary drawings which are often not to scale. It may be practical to build up drawings on a modular principle: general drawings are divided into part drawings so that new versions of the general drawings can be produced when required.

Content allows for a large number of possible distinctions. One aspect is the completeness of an item in a drawing. In this case a distinction is made between general, group, part, blank component, group component, model and plan drawings.

Moreover, collective drawings can rationalise the production of drawings; they can be built up as type drawings (for embodiment variations) with an inset or separate table of dimensions or as sets of drawings (combinations of related parts).

Having an appropriate *structure* for the set of drawings is useful when compiling manufacturing drawings. This corresponds to a breakdown of the product in terms of manufacture or assembly, and consists in principle of three sets of drawings: first, a general assembly drawing showing the construction of the product, from which further drawings may possibly be derived (e.g. for dispatch, erection and assembly plus approval); then several group drawings of varying levels of priority (complexity) which show the combination of individual components

Table 2. Comparison of symbols and roughness data in DIN ISO 1302 and previous standard DIN 3141

Surface symbol according to DIN 3141	Surface data R_a in µm (permitted roughness depth R_t in µm according to DIN ISO 1302)				
	Classification according to DIN 3141,				
	series 1	series 2	series 3	series 4	
Surface without symbol					Surfaces without specific requirements
		Smooth			Surfaces which only need improved evenness and appearances
					Surfaces which may not be produced by machining
		6.3/ or $\sqrt{R_t\,40}$			Clean, untreated surface with roughness requirements to be produced using any production process
▽	25/ (R_t =160)	12.5/ (R_t =100)	6.3/ (R_t =63)	3.2/ (R_t = 25)	Surfaces to be produced by machining which may not exceed the given max roughness value
▽▽	6.3/ (R_t =40)	3.2/ (R_t =25)	1.6/ (R_t =16)	0.8/ (R_t =10)	
▽▽▽	1.6/ (R_t =16)	0.8/ (R_t =6.3)	0.4/ (R_t =4)	0.2/ (R_t =2.5)	
▽▽▽▽		0.1/ (R_t = 1)	0.1/ (R_t = 1)	0.025/ (R_t =0.4)	

to form one manufacturing or assembly unit; and finally, component drawings, which can be further subdivided for various manufacturing stages (e.g. blank, model, preliminary processing, final finishing).

Drawings should be built up so that they can be reused for other applications. Repeat components and spare components should therefore be shown on separate drawings. The set of parts lists and the system of drawing numbers should also be constructed in the same way as the set of drawings (see E6.3.4 and E6.4).

6.3.2 Drawing Sizes and Formats, Lines and Lettering

Drawing formats are laid down in DIN 6771; they can be used as landscape or portrait (**Table 5**). The ratio of the size is $\sqrt{2} : 1$.

Line widths and letter heights should be suitable for the requirements of microfilming and also scale by $\sqrt{2}$. Series 1 is preferable for line widths (DIN 15) plus cursive and vertical standard type (DIN 6776).

Letter height refers to upper-case letters. Lower-case letters are produced at 10/14 letter height for type A and 7/10 for type B. Preferred letter heights are 2.5, 3.5, 5 and 7 mm. The line width for medium-sized type should be 1/10 of the letter height.

6.3.3 Drawing Conventions, Dimensioning

DIN 823 specifies the following *scales*:

Reductions:	1 : 2	1 : 5	1 : 10
	1 : 20	1 : 50	1 : 100
	1 : 200	1 : 500	1 : 1000
	1 : 2000	1 : 5000	1 : 10000
Enlargements:	50 : 1	20 : 1	10 : 1
	5 : 1	2 : 1	

Views and sections are normally arranged in *standard projection* (**Fig. 4**).

Items should be represented *in the position in which they appear* in general drawings and group drawings; in component drawings, they should preferably be in the *manufacturing position*. During this process, the minimum number of views (DIN 6, Part 1) or sections (DIN 6, Part 2) should be used, but they should also be sufficient to produce an unambiguous picture of the design.

Sections make drawings simpler (removing edges not in view) and should always be used for cylindrical hollow bodies (do not forget visible peripheral edges).

Folding: simple representations of cross-sections in the plane of the drawing reduce the number of views required.

Frequently occurring components are only drawn once. *Invisible edges* should only be shown on drawings if it avoids ambiguities and the need for additional views. Simplified representations are permissible so long as recognition of function, three-dimensional compatibility, and significant design elements are not impaired in any individual case.

Dimensioning must be clear and unambiguous. Rules are to be found in the standards [5].

6.3.4 Parts Lists

A parts list or a set of parts lists goes with each set of drawings, so that a product can be described completely. From left to right a parts list contains columns for item number, quantity, quantity unit, designation of the group or component (including standard components, bought-in components and auxiliary materials), reference number and/or abbreviation of standard for identification and remarks. Designation should be based on structural shape rather than function. A parts list usually consists of a text field and a list field, the formal structure of which is laid down in DIN 6771, Part 1 and Part 2.

Table 3. ISO basic tolerances in μm (DIN 7151: extract)

Nominal dimension range		IT 5	6	7	8	9	10	11	12	13	14	15	16
Factor		7i	10i	16i	25i	40i	64i	100i	160i	250i	400i	640i	1000i
from to	1 3	4	6	10	14	25	40	60	100	140	250	400	600
from to	3 6	5	8	12	18	30	48	75	120	180	300	480	750
from to	6 10	6	9	15	22	36	58	90	150	220	360	580	900
from to	10 18	8	11	18	27	43	70	110	180	270	430	700	1100
from to	18 30	9	13	21	33	52	84	130	210	330	520	840	1300
from to	30 50	11	16	25	39	62	100	160	250	390	620	1000	1600
from to	50 80	13	19	30	46	74	120	190	300	460	740	1200	1900
from to	80 120	15	22	35	54	87	140	220	350	540	870	1400	2200
from to	120 180	18	25	40	63	100	160	250	400	630	1000	1600	2500
from to	180 250	20	29	46	72	115	185	290	460	720	1150	1850	2900
from to	250 315	23	32	52	81	130	210	320	520	810	1300	2100	3200
from to	315 400	25	36	57	89	140	230	360	570	890	1400	2300	3600
from to	400 500	27	40	63	97	155	250	400	630	970	1550	2500	4000

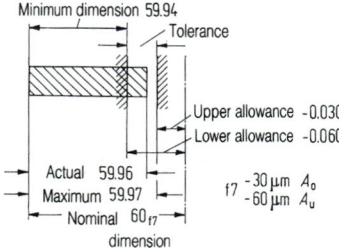

Figure 3. Allocation of nominal dimension, actual dimension and maximum dimension with corresponding upper and lower allowances (mm).

Quantity List. This only contains the individual components of the product (**Fig. 5a**) with their quantity data. Components which occur several times appear only once, but all the part numbers of the products are listed. Function- or manufacture-oriented groups cannot be recognised. This simplest type of parts list is sufficient for simple items which consist only of a few production stages (**Table 6**), with product breakdown as in **Fig. 5a**.

Structural Parts List. This reproduces product structure with all assemblies and components so that each group is immediately linked right through to the highest stage (organisation of product breakdown). The breakdown of groups and components normally corresponds to the manufacturing cycle (**Table 7**). The quantity data refer to the product described at the head of the parts list. Structural parts lists may be compiled for a product

Table 4. Examples of fits based on [6], with reference to fits recommended in DIN 7157
Fits marked * for uniform shafts, those marked () are only formed from series 2.

		Fit recommended in DIN 7157	Tolerance zone position and size for normal dimension 60 mm in μm / External dimen. / Internal dimension	Characteristic for assembly	Application
Press fit (oversize)	Tight press fit	H8/x8≤24mm / H8/u8>24mm		Needs pressure or temperature difference, high adhesion	Hubs of gears, rotors and fly wheels; flanges on shafts
	Medium press fit	H7/s6 / H7/r6		Needs pressure or temperature difference, medium adhesion	Clutch hubs, bearing bushes in housings, wheels or connecting rods Bronze collars on grey cast iron hubs
Transition fit (over- or undersize)	Interference fit	H7/n6		Needs pressure	Motor shaft armatures, toothed collars on wheels
	Wringing fit	H7/k6		Easily tapped with hammer	Pulleys, clutches, gears, flywheels fixed handwheels and permanent levers
	Close sliding fit	H7/j6		Hand pressure	Pulleys, gears, handwheels and easy-fit bearing bushes
Clearance fit (undersize)	Sliding fit	H7/h6 / H8/h9 / H11/h9 / (H11/h11)		Hand pressure with lubrication	Gears, tailstock sleeves, adjusting rings, loose bushes for piston bolts and pipelines Easy-fit components, spacer bushes, pinned or permanently fixed agricultural machinery components, h11 shafts of bright round steel DIN 668
	Tight running fit	G7/h6 * / H7/g6		Push-fit without noticeable clearance	Push-on gear wheels and clutches, connecting rod bearings, indicator pistons
	Running fit	H7/f7 / F8/h6 * / H8/f7 / F8/h9 *		Noticeable clearance	Machine-tool main bearings, crankshaft and connecting rod bearings, regulator mountings, shaft sleeves, clutch sleeves, guide blocks, cross-heads
	Light running fit	H8/e8 / E9/h9 *		Relatively large clearance	Long-shaft mountngs, agricultural machine bearings
	Broad running fit	H8/d9 / D10/h9 * / (H11/d9) / D10/h11 *		Large clearance	Multiple bearing shafts in machine tools and reciprocating engines, shafts h9 DIN 699 or Din 671 Hydraulic piston in cylinder, lever bolts, removable levers, bearings for rollers and guide system
	Loose clearance fit	C11/h9 * / C11/h11 * / (H11/c11) / (A11/h11) * / (H11/a11)		Loose fit	Pivot pins, locking pins, vehicle brake linkage bolts Spring and brake suspension, brake-shaft bearings, knuckle pins

Table 5. Drawing formats in mm according to DIN 823

Format series A	A0	A1	A2	A3	A4	A5	A6
Cut sheet	841 × 1189	594 × 841	420 × 594	297 × 420	210 × 297	148 × 210	105 × 148
Uncut sheet	880 × 1230	625 × 880	450 × 625	330 × 450	240 × 330	165 × 240	120 × 165

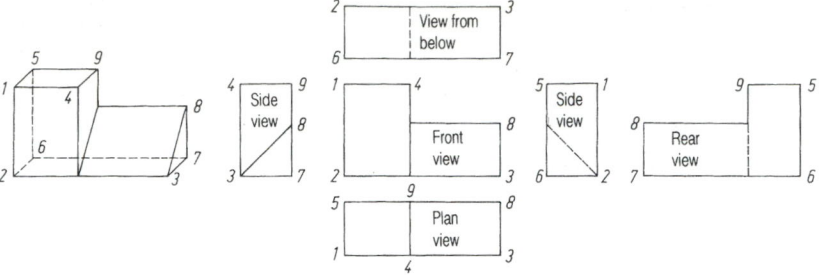

Figure 4. Views and sections in standard projection.

Table 6. Quantity summary parts list for product breakdown

Quantity			Designation E1	Quantity summary parts list
Item	Quantity	Quantity unit	Designation	Item number
1	1	ST	T1	
2	2	ST	T2	
3	2	ST	T3	
4	1	ST	T4	
5	2	ST	T5	
6	5	ST	T6	
7	4	KG	T7	
8	9	M	T8	

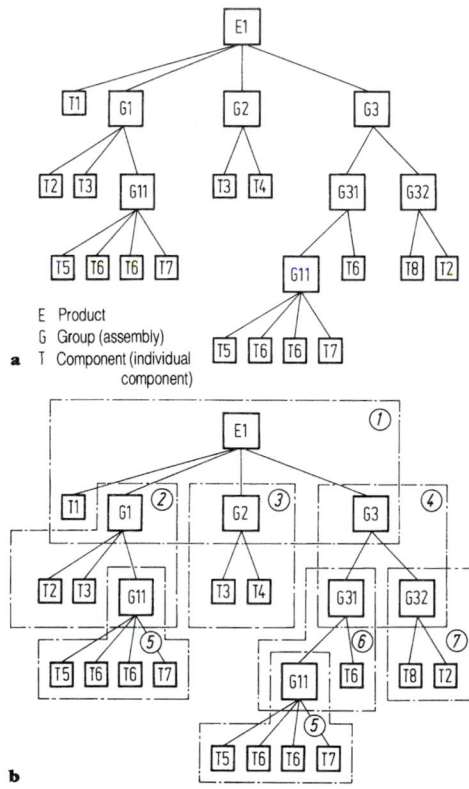

Figure 5. Product breakdown: **a** breakdown; **b** modular parts list.

as a whole or for individual parts of it, and their advantage is that the overall structure of a product or a group can be recognised. However, parts lists with a large number of item numbers are not clear, especially if a number of repeat groups recur at several different points. This also produces disadvantages for the amendment service.

Modular Parts List. This includes groups and components that belong together *without* reference to a definite product. The quantity data refer only to the assembly listed in the heading. Several modular parts lists of this kind must be combined, with other parts lists if necessary, to form a set of parts lists for a product, e.g. **Fig. 5b**. Parts list E1 consists of T1 and the parts lists G1, G2 and G3. These independent parts lists refer to others, e.g. G11, G31 and G32. Use of modular parts lists is recommended where there is a wide range of products, and assemblies are held in store and manufactured in relatively large numbers as repeat groups.

6.4 Item Numbering Systems

Item numbering systems cover the numbering and characteristics of items. It is useful to give the same number to a component drawing, the item in the relevant parts list, the work schedule concerned and the workpiece itself (manufactured component, spare component, stores component, bought-in component).

Item numbers must *identify* an item, and they may also *classify* it. An item numbering system can be constructed as a parallel number system or composite number system if it is to identify and classify.

Figure 6 shows the general structure of a number system with parallel coding. In parallel number systems, one or more classification numbers *independent* of the identification are allocated to an identification number (ID number). The advantage of this coding lies in the high degree of flexibility and the possibility of expansion as the two subsystems are independent of each other.

In a composite number system, the number (item number) as a whole consists of classifying and identifying (counting) number components which are *rigidly linked together* so that the counting number sections depend on the classifying number sections (**Fig. 7**). The disadvantage lies in the extreme rigidity where extension is concerned.

Table 7. Structural parts list for product breakdown

	Quantity			Designation E1	Structural parts list
Item	Quantity	Quantity unit	Grade	Designation	Item number
1	1	ST	.1		T1
2	1	ST	.1		G1
3	1	ST	..2		T2
4	1	ST	..2		T3
5	1	ST	..2		G11
6	1	ST	...3		T5
7	2	ST	...3		T6
8	2	KG	...3		T7
9	1	ST	.1		G2
10	1	ST	..2		T3
11	1	ST	..2		T4
12	1	ST	.1		G3
13	1	ST	..2		G31
14	1	ST	...3		G11
15	1	ST	4		T5
16	2	ST	4		T6
17	2	KG	4		T7
18	1	ST	...3		T6
19	1	ST	..2		G32
20	9	M	...3		T8
21	1	ST	...3		T2

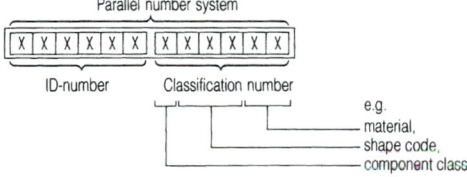

Figure 6. Basic structure of an item number for a parallel number system [7].

Classification of items and item characteristics – whether within the item number or by a separate classification system independent of the identification number – is important so that components can be used repeatedly and information on items can be retrieved. A graded classification system is normally used (broad and detailed classification).

If the features of one group are allocated to only one

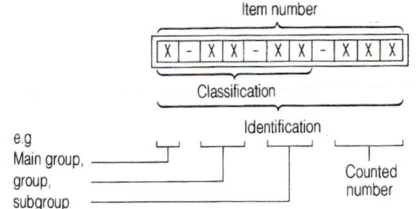

Figure 7. Basic structure of an item number for a composite number system [7].

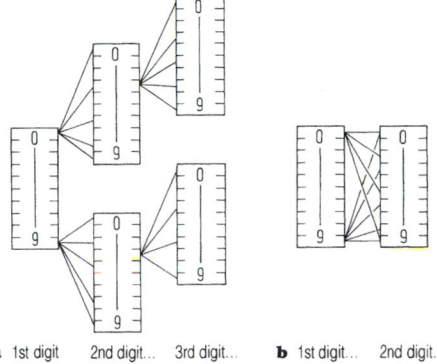

a 1st digit... 2nd digit... 3rd digit... **b** 1st digit... 2nd digit...

Figure 8. Possible links between characteristics in classification systems [8, 9].

feature of the previous group, the classification system must branch accordingly (**Fig. 8a**). If, in contrast, the features of a group can be allocated to every feature of the preceding group, a corresponding overlap in the arrangement is also possible (**Fig. 8b**). The advantages of the arrangement shown in **Fig. 8a** lie in the independent linking of the individual branches and the high storage capacity, while the advantages of the arrangement as **Fig. 8b**, in contrast, lie in the smaller memory requirement. In practice therefore both types of interrelationship are used in mixed systems.

Item features have commonly been introduced to describe components and groups, especially standard components; they are used to characterise specific features suitable for describing and distinguishing individual items within a group (DIN 4000). For principles and application, see [10].

7 References

E2 Fundamentals of Systematic Approach. [1] Pahl G, Beitz W. Konstruktionslehre, 2nd edn. Springer, Berlin, 1986. – [2] VDI-Richtlinie 2221: Methodik zum Entwickeln und Konstruieren technischer Systeme und Produkte. VDI-Verlag, Düsseldorf, 1986. – [3] Holliger H. Morphologie – Idee und Grundlage einer interdisziplinären Methodenlehre. Kommunikation 1, vol. 1. Schnelle, Quickborn, 1970. – [4] Osborn A F. Applied imagination – principles and procedures of creative thinking. Scribner, New York, 1957. – [5] Hellfritz H. Innovation via Galeriemethode. Eigenverlag, Königstein/Taunus, 1978. – [6] Gordon W J J. Synthetics, the development of creative capacity. Harper, New York, 1961. – [7] Rohrbach B. Kreativ nach Regeln – Methode 635, eine neue Technik zum Lösen von Problemen. Absatzwirtschaft 1969; 12: 73-5. – [8] Dalkey N D, Helmer O. An experimental application of the Delphi method to the use of experts. Management Sci. 1963; 9: 458-67. – [9] Rodenacker W G. Methodisches Konstruieren, 3rd edn. Konstruktionsbücher, vol. 27. Springer, Berlin, 1984. – [10] Roth K. Konstruieren mit Konstruktionskatalogen. Springer, Berlin, 1982. – [11] Zwicky F. Entdecken, Erfinden, Forschen im Morphologischen Weltbild. Droemer-Knaur, Munich, 1966, 1971. – [12] VDI-Richtlinie 2222 Bl. 2: Konstruktionsmethodik. Erstellung und Anwendung von Konstruktionskatalogen. VDI-Verlag, Düsseldorf, 1982. – [13] Kiper G. Katalog einfachster Getriebebauformen. Springer, Berlin, 1982. – [14] Zangemeister C. Nutzwertanalyse in der Systemtechnik. Wittemannsche Buchhandlung, Munich, 1970. – [15] Kesselring F. Bewertung von Konstruktionen, ein Mittel zur Steuerung von Konstruktionsarbeit. VDI-Verlag, Düsseldorf, 1951. – [16] VDI-Richtlinie 2225: Technisch-wirtschaftliches Konstruieren. VDI-Verlag, Düsseldorf, 1977. – [17] REFA vol. 3: Methodenlehre des Arbeitsstudiums, Kostenrechnung, Arbeitsgestaltung. Hanser, Munich, 1971. – [18] VDI-Berichte No. 457: Konstrukteure senken Herstellkosten – Methoden und Hilfsmittel. VDI-Verlag, Düsseldorf, 1982. – [19] VDI-Richtlinie 2235: Wirtschaftliche Entscheidungen beim Konstruieren, Methoden und Hilfen. VDI-Verlag, Düsseldorf, 1982. – [20] Pahl G, Rieg F. Kostenwachstumsgesetze für Baureihen. Hanser, Munich, 1984. – [21] Pahl G, Beelich K H. Kostenwachstumsgesetze nach Ähnlichkeitsbeziehungen für Schweissverbindungen. VDI-Berichte No. 457. VDI-Verlag, Düsseldorf, 1982. – [22] Ehrlenspiel K, Kiewert A, Lindemann U. Kostenfrüherkennung im Konstruktionsprozess. VDI-Berichte No. 347. VDI-Verlag, Düsseldorf, 1979. – [23] VDI-Richtlinien 2801 and 2802: Wertanalyse. VDI-Verlag, Düsseldorf, 1970, 1971. – [24] VDI: Wertanalyse. VDI-Taschenbuch T35. VDI-Verlag, Düsseldorf, 1972. – [25] VDI-Berichte No. 293: Wertanalyse 77. VDI-Verlag, Düsseldorf, 1977 (with extensive literature references). Standards: DIN 69 910: Wertanalyse; Begriffe, Methode (1973).

E3 The Design Process. [1] VDI-Richtlinie 2223: Begriffe und Bezeichnungen im Konstruktionsbereich. VDI-Verlag, Düsseldorf, 1969.

E4 Fundamentals of Body Design. [1] Peters U H, Meyna A. Handbuch der Sicherheitstechnik. Hanser, Munich, 1985. – [2] Spähn H, Fässler K. Zur konstruktiven Gestaltung korrosionsbeanspruchter Apparate in der chemischen Industrie. Konstruktion 1972; 24: 249–58, 321–25. – [3] Uhlig H H. Korrosion und Korrosionsschutz. Akademie-Verlag, Berlin, 1970. – [4] Rubo E. Der chemische Angriff von Werkstoffe aus der Sicht des Konstrukteurs. Der Maschinenschaden 1966: 65–74. – [5] Kloos K H. Werkstoffoberfläche und Verschleissverhalten in der Fertigung und konstruktiven Anwendung. VDI-Berichte No. 194. VDI-Verlag, Düsseldorf, 1973. – [6] Wahl W. Abrasive Verschleissschäden und ihre Verminderung. VDI-Berichte No. 243, "Methodik der Schadensuntersuchung". VDI-Verlag, Düsseldorf, 1975. – [7] Gerätesicherheitsgesetz (Gesetz über technische Arbeitsmittel): BGBl of 13.8.1979. Deutsches Informationszentrum für technische Regeln (DITR), Berlin. – [8] VDI-Richtlinie 2242. Konstruieren ergonomiegerechter Erzeugnisse. VDI-Verlag, Düsseldorf, 1986. – [9] Kroemer K H. Was man von Schaltern, Kurbeln und Pedalen wissen muss. Beuth, Berlin, 1967. – [10] Kroemer K H, Hettinger T. Körperkräfte im Bewegungsraum. RKW-Reihe Arbeitsphysiologie – Arbeitspsychologie. Beuth, Berlin, 1963. – [11] Bücker W. Künstliche Beleuchtung: ergonomisch und energiesparend. Campus, Frankfurt/Main, 1981. – [12] Schmidtke H. Überwachungs-, Kontroll- und Steuerungstätigkeiten. RKW-Reihe Arbeitsphysiologie – Arbeitspsychologie. Beuth, Berlin, 1966. – [13] VDI-Bericht No. 239: Beispiele

für lärmarme Maschinenkonstruktionen. VDI-Verlag, Düsseldorf, 1975. – [14] VDI-Richtlinie 3720: Lärmarm Konstruieren – Allgemeine Grundlagen. VDI-Verlag, Düsseldorf, 1975. – [15] VDI-Richtlinie 2224: Formgebung technischer Erzeugnisse für den Konstrukteur. VDI-Verlag, Düsseldorf, 1972. – [16] VDI-Richtlinie 3239: Sinnbilder für Zubringefunktionen. VDI-Verlag, Düsseldorf, 1966. – [17] Andresen U. Die Rationalisierung der Montage beginnt im Konstruktionsbüro. Konstruktion 1975; 27: 478–84. – [18] Andreasen M M, Kähler S, Lund T. Montagegerechtes Konstruieren. Springer, Berlin, 1985. – [19] Pahl G, Beitz W. Konstruktionslehre. Springer, Berlin, 1977, 2nd edn, 1986. – [20] Seger H. Industrie Designs. Expert Verlag, Grafenau, 1983. – [21] VDI-Richtlinie 2243 (Entwurf): Recyclingorientierte Gestaltung technischer Produkte. VDI-Verlag, Düsseldorf, 1984.

Standards: DIN 7521 to 7527: Schmiedestücke aus Stahl. – DIN 8580: Fertigungsverfahren; Einteilung. – DIN 8588: Fertigungsverfahren Zerteilen; Einordnung, Unterteilung, Begriffe. – DIN 8593: Fertigungsverfahren Fügen; Einordnung, Unterteilung, Begriffe. – DIN 9005: Gesenkschmiederstücke aus Magnesium-Knetlegierungen. – DIN 31 000: Sicherheitsgerechtes Gestalten technischer Erzeugnisse; Allgemeine Leitsätze. – DIN 31 001: Sicherheitsgerechtes Gestalten technischer Erzeugnisse. – DIN 31 051: Instandhaltung; Begriffe. – DIN 33 400: Gestalten von Arbeitssystemen nach arbeitswissenschaftlichen Erkenntnissen; Begriffe und allgemeine Leitsätze. – DIN 33 401: Stellteile; Begriffe Eignung, Gestaltungshinweise. – DIN 33 402: Körpermasse des Menschen; Begriffe. Messverfahren. – DIN 33 403: Klima am Arbeitsplatz und in der Arbeitsumgebung. – DIN 33 404: Gefahrensignale für Arbeitsstätten. – DIN 33 413: Ergonomische Gesichtspunkte für Anzeigeeinrichtungen. – DIN 33 414: Ergonomische Gestaltung von Warten.

E5 Fundamentals of Development of Series and Modular Design. [1] Berg S. Konstruieren in Grössenreihen mit Normzahlen. Konstruktion 1965; 17: 15–21. – [2] Gerhard E. Ähnlichkeitsgesetze beim Entwurf elektromechanischer Geräte. VDI-Z 1969; 111: 1013–19. – [3] Matz W. Die Anwendung des Ähnlichkeitsgesetzes in der Verfahrenstechnik. Springer, Berlin, 1954. – [4] Pahl G, Beitz W. Konstruktionslehre, 2nd edn, Springer, 1986.

Standards: DIN 323 pt 2: Normzahlen und Normzahlreihen: Einführung (1974).

E6 Fundamentals of Standardisation and Engineering Drawing. [1] DIN, Gesamtbearbeitung Krieg K G. Nationale und internationale Normung. Handbuch der Normung, vol. 1, 3rd edn. Beuth, Berlin, 1975. – [2] DIN-Taschenbuch 1: Mechanische Technik, Grundnormen. Beuth, Berlin, 1980. – [3] DNA: Normenverzeichnis mit sicherheitstechnischen Festlegungen. Beuth, Berlin, 1971. – [4] Gerätesicherheitsgesetz: Gesetz über technische Arbeitsmittel (from 24.06.1968 BGBl. 1717, revised from 13.08.1979, BG paper I: 1432 ff. Bezug durch Deutsches Informationszentrum für technische Regeln (DITR), Berlin. – [5] DIN-Taschenbuch 2: Zeichnungswesen, pt 1. DIN-Taschenbuch 148: Zeichnungswesen, pt 2. Beuth, Berlin, 1988. – [6] Reimpell J, Pautsch E, Stangenberg R. Die normgerechte technische Zeichnung für Konstruktion und Fertigung, vol. 1. VDI-Verlag, Düsseldorf, 1967. – [7] Bernhardt R. Nummerungstechnik. Vogel, Würzburg, 1975. – [8] Eversheim W, Wiendahl H P. Rationelle Auftragsabwicklung im Konstruktionsbereich. Girardet, Essen, 1971. – [9] VDI-Richtlinie 2215 (Entwurf): Datenverarbeitung in der Konstruktion. Organisatorische Voraussetzungen und allgemeine Hilfsmittel. VDI-Verlag, Düsseldorf, 1974. – [10] DIN: Sachmerkmale DIN 4000, Anwendung in der Praxis. Beuth, Berlin, 1979.

Standards: DIN 6: Darstellungen in Normalprojektion. – DIN 15: Linien in Zeichnungen. – DIN 6776: Normschriften für Zeichnungen. – DIN 199: Begriffe im Zeichnungs- und Stücklistenwesen. – DIN 820: Normungsarbeit. – DIN 823[1]: Zeichnungen; Blattgrössen, Massstäbe. – DIN 1680: Gussrohteile, Allgemeintoleranzen und Bearbeitungszugaben. – DIN ISO 1302: Angabe der Oberflächenbeschaffenheit in Zeichnungen. – DIN 4000 pt I: Sachmerkmal Leisten, Grundsätze. – DIN 4760: Gestaltabweichung; Begriffe, Ordnungssystem. – DIN 4761: Oberflächencharakter; Geometrische Oberflächentextur-Merkmale, Begriffe, Kurzzeichen. – DIN 4762[2]: Oberflächenrauheit; Begriffe. – DIN 4763: Stufung der Zahlenwerte für Rauheitsmessgrössen. – DIN 4764: Oberfläche an Teilen für Maschinenbau und Feinwerktechnik; Begriffe nach der Beanspruchung. – DIN 4765: Bestimmen des Flächentraganteils von Oberflächen; Begriffe. – DIN 4766: Herstellverfahren der Rauheit von Oberflächen. Erreichbare gemittelte Rauhtiefe R_z nach DIN 4768 pt 1 and Mikroflächentraganteil t_{ai}. – DIN 6771 pt 1: Schriftfelder für Zeichnungen, Pläne und Listen; pt 2: Vordrucke für technische Unterlagen; Stückliste; E DIN 6771 pt 6: Vordrucke für technische Unterlagen; Zeichnungen. – DIN 7150 pt 1: ISO-Toleranzen und ISO-Passungen für Längenmasse von 1 bis 500 mm; Einführung. – DIN 7151[3]: ISO-Grundtoleranzen für Längenmasse von 1 bis 500 mm Nennmass. DIN 7154: ISO-Passungen für Einheitsbohrung. – DIN 7157: Passungsauswahl; Toleranzfelder, Abmasse, Passtoleranzen. – DIN 7168: Allgemeintoleranzen (Freimasstoleranzen). – DIN 7526: Schmiedestücke aus Stahl; Toleranzen und zulässige Abweichungen für Gesenkschmiedestücke. – DIN 31000/VDE 1000: Allgemeine Leitsätze für das Sicherheitsgerechte Gestalten technischer Erzeugnisse. – DIN/VDE 31000 pt 2: Allgemeine Leitsätze für das sicherheitsgerechte Gestalten technischer Erzeugnisse Begriffe der Sicherheitstechnik, Grundbegriffe.

---

1 Now superseded by DIN 6771.
2 Identical with ISO 4287/1.
3 Now superseded by ISO 286.

Mechanical Machine Components

W. Beitz, H. Mertens and J. Ruge, Berlin; K.-A. Ebert, Hattersheim; K. Ehrlenspiel and H. Winter, Munich; H. Kerle and H. Wösle, Brunswick; K.-H. Küttner, Berlin; H. W. Müller, Darmstadt; H. Peeken, Aachen

1 Connections

1.1 Welding

J. Ruge, Munich, and **H. Wösle**, Brunswick
(Section 1.1.1 by **K.-A. Ebert**, Hattersheim)

In *joint welding* components are joined by means of welds at the welding point to become a welded part. A number of welded parts combine to make up a welded assembly and a number of welded assemblies combine to make up a welded structure. Welding has therefore become a manufacturing process which is a determining factor for design. *Build-up welding* enables worn areas of workpieces to be resurfaced, surfaces of less wear-resistant materials to be reinforced with layers of hard-wearing material (hard facing), base materials not resistant to corrosion to be "clad" with corrosion-resistant materials (cladding) or a reliable bond between materials of a dissimilar composition to be obtained using filler material (buttering). Besides metals, plastics can also be joined by welding.

1.1.1 Welding Processes

K.-A. Ebert, Hattersheim

Methods of Joining. When welding metal the metallic materials are joined:

By Heating the Points of Contact Until They Reach the Melting Range (fusion welding), in most cases adding material of a similar composition (filler metal) with a melting range identical or almost identical to the materials to be joined. A liquid zone is therefore present at the point of contact, which leaves cast structures after cooling.

By Heating the Points of Contact (to melting point, if required) *and Applying Pressure* (pressure welding). As there is no fused mass but in most cases considerable plastic deformation at the joint, the structure generally becomes fine-grained after cooling.

By Appling Pressure in the Cold Condition of the material (cold pressure welding). The joint can only be produced by considerable plastic deformation (beyond the crushing yield point) of the oxide free surfaces at the points of contact; the structure has been very thoroughly cold-worked.

By Heating the Weld Area in a Vacuum or Shield Gas employing little pressure with no plastic deformation at the joint (diffusion welding). The temperature at the joint must be high enough to diffuse the metal atoms.

Heat Sources. The following are used to produce the required welding temperature: gas flame (gas welding), electric arc (arc welding), Joule effect in the workpiece (resistance welding), induction (induction welding), Joule effect in the molten slag (electroslag welding), relative motion between the mating surfaces (friction welding and ultrasonic welding), energy from highly accelerated electrons (electron beam welding), light energy from extreme focusing or concentration (laser welding), exothermic chemical reaction (thermit welding), liquid heat transfer medium (casting welding) and furnace (forge welding).

Processes The *manual welding process*, whereby the heat source, the gas flame or the electric arc, is directed manually by the welder, still prevails in gas and arc welding. To increase the speed of welding the filler material can be directed to the weld from coils (wire electrode) (*partially mechanised process*) whereby it is possible to have a considerably greater current density than is the case with manual welding because the current is directed to the electrode in the immediate vicinity of the arc. In tank construction or hard-facing in particular the progress of the heat source along the weld can be effected by the travelling motion of the welding head or by the motion – travelling or rotating – of the workpiece – *fully mechanised welding process*. In mass production (industrial scale manufacture) welding is carried out in clamping and holding devices with automated – and if necessary, computer-controlled – means to complete the welding process – *automatic welding* – using welding robots, if necessary.

The processes most frequently encountered today have been arranged together with their characteristic features and principal applications in **Table 1**. A total of well over 200 welding processes have been included. Some of these are now only of historical importance, others failed to be adopted, many only differ from known processes by slight variations and a few have not gone beyond the stage of being special applications, with the result that it still cannot be said with certainty what importance they will attain.

Besides the characteristics of the heat sources and the degree of mechanisation previously mentioned, the processes vary in their methods of application. In many cases only certain welding positions are possible. Likewise, the form of joint and type of weld are to a greater or lesser degree dependent on the welding process. In the case of arc welding there are also differences in fusion penetration, by which is meant the fused depth of the joint flanks achieved by the arc.

The selection of the optimum welding process for fabrication is determined by a number of both technical and economic factors, with the result that no general rules can be laid down in this regard.

Table 1. List of the major welding processes

Welding process	Characteristic features	Principal application
Gas fusion (autogenous) welding	The low or balanced pressure torch heats the weld to melting point by burning a gas mixture, mainly an oxygen–acetylene mixture in the ratio 1 : 1 to 1 : 1.1. Any material lacking in the weld groove is supplemented by filler wire (gas welding filler rod).	Especially for butt and angle joints in all weld positions, mainly in steel sheet and tube as well as in copper. Normal wall thickness up to 5 mm, maximum approx. 15 mm. Wall thickness up to 3 mm for leftward welding, over 3 mm for rightward welding.
Open arc welding	The arc burns visibly in the atmosphere.	
Manual arc welding (consumable electrode)	The open arc is struck between the electrode, which acting as filler is consumed at the same time, and the workpiece. The welding current – 15 to 20 A per mm² of the core wire diameter of the electrode, at 10 to 45 V arc drop voltage – is supplied by specially constructed devices as direct current fed from motor-generators or rectifiers or as alternating current fed from transformers. The core wire of the electrodes is made mostly from materials of identical or similar chemical composition to the parts to be welded. The type of coating (e.g. oxide, rutile, basic or containing cellulose) has an effect on the welding action of the electrode and the properties of the finished weld. Besides the metallurgical effect of the components of the coatings (reaction between slag and weld deposit) these may also contribute to an increase in yield (heavy-duty electrode) or to the alloying of the weld deposit (alloy-coated electrodes).	For all types of welds and joints in all welding positions for all ferrous and non-ferrous metals with appropriate selection of electrodes and welding requirements (pre-heating, heat transfer when welding, cooling, post-weld treatment). Minimum wall thickness approx. 1 mm.
Metal arc welding with filler wire electrode	The arc is struck, with no additional shield gas supply, between the consumable electrode, fed from the coil, and the work. The electrode also acts as a filler. The tubular electrode (outside diameter 1.0 mm and over) contains mainly mineral components to deoxidise the fused mass as well as metal alloys to alloy the fused mass.	Mainly for single-run fillet welds (in multi-run welds there is a risk of pore formation) of non-alloy carbon steels and for hardfacing (wear-resistant deposits).
Carbon arc welding (non-consumable electrode)	The arc is struck between the carbon electrode and the work or between two carbon electrodes, so that it can be blown on to the work, having been magnetised and directed by a coil with the welding current flowing through it. The electrode holder is guided manually or by partially or fully mechanised means.	Preferred for corner or edge-formed welds as a fully mechanised welding method for mass-produced steel sheet. Hardly ever used now.
Submerged arc welding	The arc is struck between a coiled bare wire electrode and the work beneath a layer of special welding flux. The welding head is guided manually (semimechanised) or by fully mechanised means. The rate at which the wire is advanced can be controlled by the length of the arc; ignition is achieved underneath the layer of flux powder by the high-frequency voltage overlaying the welding arc voltage. In order to increase fusion efficiency, an arrangement of up to three welding heads is possible, the arcs of which are maintained in the same cavity.	Mainly in horizontal welding position for butt and fillet welds but also horizontal to and level with vertical wall with special equipment for holding the flux powder. Minimum plate thickness is approximately 2 mm, owing to its considerable fusion efficiency, but it is mostly used for thick plate and long joints.
Submerged arc strip welding	The arc is struck between a coiled strip electrode (up to 100 mm wide) and the surface of the work underneath a layer of a special composition of flux powder. The welding head is guided by machine. The rate at which the strip is advanced can be controlled by the length of the arc.	Refinement of submerged arc welding for the large-scale build-up welding of corrosion-inhibiting layers (cladding). Can only be used for greater thicknesses of workpiece, owing to the distortion caused by the welding heat.
One-side welding	As with submerged arc welding, the arc is struck between the wire electrode and the work in the welding groove under a layer of flux powder. In order to increase fusion efficiency up to three welding heads may be arranged behind each other. Iron alloy granules can also be introduced into the welding groove in front of the weld point. Owing to the large weld pool and the considerable localised heat supply a pool retaining method (high root gate or strong root position) is required.	Mainly in shipbuilding for welding long butt joints on one side only without turning the workpiece (sectional construction) for workpiece thickness up to approx. 40 mm in non-alloy and fine-grain steels up to StE 360.
Gas-shielded arc welding	The visible arc is struck inside an inert gas shield.	

Table 1. Continued

Welding process	Characteristic features	Principal application
Tungsten–inert-gas (TIG) arc welding	The arc is struck between the tungsten electrode (with added thorium) and the work in an envelope of inert gas. The filler material is added manually or by machine from coils. Argon is used almost exclusively as the shield gas in Germany, plus (rarely) argon–helium mixtures and pure helium. Direct-current arc welding, alternating-current used only for aluminium and its alloys. High-frequency unit to facilitate ignition.	For all butt and fillet welds and in all welding positions for almost all metallic materials, but mainly corrosion- and scale-resistant CrNi steels, aluminium and its alloys (without flux), copper and copper alloys (with flux) up to medium plate thicknesses.
(Tungsten) arc plasma welding	The arc plasma (in single or multi-atom gases – preferably argon, nitrogen or hydrogen – split into electrons and ions) melts parent and filler material.	
Plasma-beam welding	The arc is struck between a tungsten electrode and the inside wall of the nozzle (arc not transferred). The pressurised plasma (ionised shielding gas) beam melts the material (and the filler material supplied as wire or rod) at the weld point or heats the surface of the workpiece to bonding temperature and the filler material fed as a powder (predominantly hard alloys) to melting temperature.	Mainly for joint welding in high-alloy steels with low wall thicknesses (e.g. straight bead welding of pipes) and deposition (cladding) of alloys containing ingredients which are difficult to melt (carbides) where there is little fusion of the parent material.
Plasma arc welding	The arc is struck between a tungsten electrode and the work (transferred arc). Ignition is facilitated by an arc with a low current density (pilot arc) struck between a tungsten electrode and the inside of the nozzle. Filler material fed in the form of powder. Greater melting of the parent metal than in plasma-beam welding.	Mainly for deposition (cladding) of corrosion and wear inhibiting layers and materials of high-temperature stability onto low-resistance parent metals.
Metal–inert-gas (MIG) welding	The arc is struck in an inert gas shield between the coiled consumable electrode and the work. The electrode is also the filler material and must therefore match the material to be welded. Shielding gas is pure argon. As the current is fed to the electrode in the immediate vicinity of the arc high current density of 100 A/mm^2 is possible, with the resulting high burn-off rate. Electrode diameter mainly below 2.4 mm. Spray arc (high current density) for greater wall thicknesses and depositions in horizontal welding position, short arc (low current density and fine wire electrode) for smaller wall thicknesses, weld-sensitive materials and in all welding positions. For sensitive materials and in other special cases the arc can be broken in pulse mode by electronic control of the welding current (pulsed arc) to limit additional heating of the weld point.	For almost all types of seam and joint in all welding positions for all alloy steels, aluminium and its alloys, copper and copper alloys (with flux) with plate thickness over approx. 1 mm.
Metal-arc welding with non-inert gas mixture (MAGM)	Gas mixtures of argon, carbon dioxide (18%) and oxygen (up to 5%) are claimed to reduce the disadvantages of inert shielding gases (price, pore formation in some materials) and carbon dioxide (spatter, burn-off of alloying elements). Spray transfer, short and pulsed arc as in MIG welding.	For non-alloy, low-alloy and some high-alloy steels of all plate thicknesses. When welding high-alloy, corrosion-resistant steels, a reduction in corrosion resistance caused by chromium carbide formation as a function of the CO_2 content of the shielding gas must be borne in mind.
Metal-arc welding with carbon dioxide (CO_2) (MAGC)	*Carbon dioxide* is used as a substitute for expensive argon or helium but at high temperatures oxygen is separated from the gas which reacts with the material and filler material to be bonded (oxidation). Alloying elements (silicon, manganese) to be introduced must be added by the filler material (deposition) – also for deoxidising the weld deposit. *Carbon dioxide* or mixed gas with *flux-coated wire* or filler wire, a metal strip tacked onto a tube with flux powder incorporated as electrode and filler material is a further development of the metal-arc welding process with non-inert gas for a better metallurgical influence on the weld deposit.	Mainly for killed, non-alloy steels of all plate thicknesses using the spray transfer or short-arc technique (small thicknesses, constrained positions). Mainly for non-alloy steels in horizontal welding position and for deposition (wear-resistant layers).
Beam welding	A high-energy focused beam produces the heat required for the welding process or striking or penetrating the work.	
Electron beam welding	The kinetic energy of electrons accelerated to high velocity by a high voltage (up to 150 kV) heats the workpiece to melting temperature at the point of impact. By focusing the electron beam (electromagnetic lenses) to a focal spot diameter below 0.1 mm a considerable depth of penetration can be achieved with limited local heating. Because there are high energy losses in normal atmosphere (air ionisation), the welding process takes place in a high vacuum.	Mainly for weld-sensitive materials, motor industry and special jobs. High capital expenditure (equipment) is incurred in mass-production, accurate preparation of mating edges.

Table 1. Continued

Welding process	Characteristic features	Principal application
Laser beam welding	A laser beam produced in a solid-state or gas discharge laser, after being focused in a lens, heats the weld point to welding temperature on impact with the workpiece. A shielding gas is directed onto the weld point by a nozzle to protect the weld deposit. The laser must be machine-controlled, but at the same time this offers the possibility of programmed guidance of the welding beam.	Previously confined to special cases, owing to restricted energy level. Greater possibilities for use in separation of synthetic materials (including fabrics).
Resistance fusion welding	The fused mass is produced by electrical resistance.	
Electroslag welding	Molten slag of a similar composition to the flux powder used in submerged arc welding is heated by the current flowing through it. It melts the material to be welded and the filler material. Current is conducted to the resistive slag via the filler wire fed from drive rolls. The pool of molten slag is contained and formed by cooled copper shoes.	For butt joints in the vertical welding position moving upwards in non-alloy and low-alloy steels with workpiece thicknesses from 8 mm to approx. 1000 mm. Also suitable for buildup welding in both vertical and horizontal welding positions (cladding).
Resistance pressure welding	Electrical resistance in the weld area produces the heat required for welding when current is passed through it. The bond between the points to be joined is produced by pressing the parts together.	
Spot welding	The two workpieces, one on top of the other, are pressed together between two, usually domed, copper electrodes. The welding current, either alternating or high-capacity, low-voltage direct current, heats the parts to be joined to melting temperature or slightly below by means of contact resistance in a series of spots.	For joining sheets of non-alloy and alloy steel, light metals and other non-ferrous metals. With steel, sheet thickness is normally restricted to approx. 2×6 mm, with light metal to approx. 2×3 mm. Thicker sheets (up to 30 mm and 6 mm respectively) require very high electrical power.
Resistance butt welding (upset welding)	The clean butting faces, machined until plane-parallel, are pressed against each other. The welding current – high-capacity, low-voltage alternating current – heats a narrow strip of the workpieces to welding temperature, which is slightly below melting temperature, by contact resistance of the contact surface. The weld is completed under constant upset pressure, forming a beaded edge. Heating may also be inductive instead of by direct current passage.	Butt welds in round and flat sections of non-alloy steels up to 500 mm² cross-section.
Flash-butt welding	The rough butting faces are held in contact under such slight pressure, while the current is passing through, that there is constant flashing at the localised contact areas of the material. Any molten metal is expelled from the contact area. When the flashing area has reached sufficient depth, welding is completed by a sudden upsetting force and the current is cut off. A fin is produced at the weld area by molten material extruded from the gap at the joint.	Butt joints in sections and sheets of non-alloy and alloy steels, light metals and copper with cross-section up to 100 000 mm². It is also possible to joint dissimilar materials, e.g. high-speed steel and tool steel.
Projection welding	The two workpieces lying flat on top of each other, one of which is provided with pressed projections or studs (also annular for nuts), are pressed together by platen electrodes. The welding current – high-capacity, low-voltage alternating or direct current – heats the parts at the contact areas to welding temperature, just below melting temperature. Projections and studs are flattened by forging pressure.	For fastening small components, nuts, etc. to sheets. Used especially for steel in mass-production (pressings), where several weld points are situated close together and can be gripped simultaneously by the platen electrodes.
Seam welding	The current, usually high-capacity, low-voltage alternating current, is supplied to the overlapping or butting parts by disc-shaped electrodes, which at the same time transmit forging pressure, or sliding contacts. An unbroken seam is produced. Foil sheets are required to overlap one or both sides of the seam in butt joints (foil seam welding).	Mostly for joining non-alloy steel sheets, especially in tank construction; the sheet thickness in steel is limited to 2×3 mm, and in light metal to 2×2 mm.

Pressure welding with different energy supplies

Gas pressure welding	The workpieces to be joined are heated externally at the contact area by gas burners, e.g. with ring burners (closed gas pressure welding) or by burners introduced into the gap at the joint (open gas pressure welding) to temperatures above or below the melting point of the parts to be joined and united under pressure.	For butt joints in mainly round, non-alloy steel sections; also for small-diameter pipes. Flash-butt welding is preferable for high (static) stresses.

Table 1. Continued

Welding process	Characteristic features	Principal application
Arc pressure welding (e.g. *Cyc-Arc process, Nelson process*)	The usually round workpiece (stud) to be welded onto a flat surface is brought into contact with the surface with the welding current switched on, then retracted by removing the arc; after a preset arcing time the stud is suddenly pressed onto the surface with the current switched off.	Mainly for welding threaded bolts or stay rods onto flat surfaces.
Capacitor discharge stud welding	Welding heat is generated by capacitors, which discharge when the workpieces are brought into contact with them. The parts are joined in the molten metal and the contact pressure is maintained until the molten pool solidifies. As there is concentrated heat input, with little dissipation, it is also possible to weld parts that have very dissimilar melting temperatures.	Mainly for welding thin bolts and pegs onto thick sheets; also for butt welding wires.
Friction welding	The rotationally symmetrical parts are forced together in a high-speed lathe-type machine where one part is held firmly while the other part rotates. When sufficient heat has been generated rotational traint is released and the parts are joined together under pressure.	Mainly for joining small and medium-size rotationally symmetrical tube and solid sections in series production.
Ultrasonic welding	The sheets are subjected under pressure to mechanical vibrations in the ultrasonic range, which cause them to be joined. This method not only generates heat but also breaks up the surface films (oxides) that prevent bonding.	Mainly for joining materials that cannot be spot-welded. Up to now limited to special cases.

1.1.2 Weldability of Metals

In DIN 8528 the weldability of metallic materials is classed according to *suitability for welding* (joint can be produced by reason of the properties of the material), *possibility of welding* (technical feasibility) and *reliability of welding* (satisfactory operation of the component). If an appropriate welding process and a suitable finish are selected then almost all types of steel can be welded. (See also the related ISO 581.)

Suitability of Steel for Welding

Effects Related to Materials. These are categorised as follows:

Steelmaking Process. Low-carbon steels (non-alloy steels) and low-alloy steels are produced in a Bessemer converter, whereas special steels are mainly produced in a high-frequency induction or electric arc furnace (case-hardened steel).

Casting Process (Deoxidation). Segregation zones in the core of rimming steels should not be melted ("gated") during the welding process **(Fig. 1)**, as they contain concentrations of sulphur (hot short), phosphorus (cold cracking), nitrogen (ageing) and carbon (hardening). Segregation during the solidifying process is avoided by killing the molten metal (adding between 0.1 and 0.3% Si or double killing with silicon and aluminium).

Ageing (Age Hardening). The most important characteristic of the ageing of steel is a decrease in ductility as a result of weathering after cold-forming, i.e. the transition from ductile to brittle fracture (near room temperature in the notched bar test). Ageing increases the risk of brittle fracture should a conjunction of unfavourable circumstances arise.

Chemical Composition. In addition to sulphur, phosphorus and nitrogen there are some further elements whose importance for welding suitability must be emphasised:

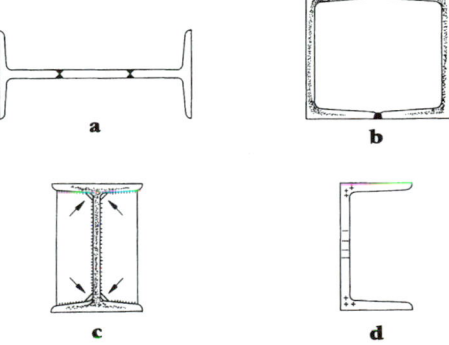

Figure 1. Joint positions in rolled steel sections: **a** section modulus enlarged by welded plate in I-beam, **b** welds in zones free from segregation in two U-sections, **c** web stiffeners with recesses in the corners of the rolled sections (rimming steel), **d** residual stresses in U-sections (+ tension, − compression).

C Content. Up to 0.25%, under normal welding conditions no significant hardness increase can be expected adjacent to the weld in non-alloy steels; it only occurs when the critical cooling rate is reduced: through an increase in carbon content alone (over 0.25%), or a combination of carbon and alloying elements including manganese, molybdenum, chromium and nickel. Such alloyed materials are readily weldable, e.g. Mn steels with up to 4% Mn when the C content is low.

Mn Content. Up to approximately 4%, manganese has a beneficial effect in non-alloy steels (increased toughness and notch ductility). It is therefore the principal element (up to approximately 1.5%) in higher strength fine-grained steels. Where the content exceeds 12% (austenitic manganese steel) special measures are required when welding (very rapid cooling rate) due to the formation of

ε-martensite. In austenitic CrNi steels manganese (up to approximately 6%) reduces the tendency to fracture.

Si Content. Above approximately 0.6% non-alloy steels have a tendency to form pores and fractures. In wire electrodes used for metal arc welding with non-inert gas shield (e.g. CO_2), however, approximately 1.1% is required to deoxidise the weld deposit.

Cu Content. Generally only present in the form of an impurity. A 0.5% content in weathering steels may combine with a high C content (over approximately 0.20%) to produce a risk of fracturing and embrittlement.

Cr Content. Only present as an impurity (below 0.2%) in non-alloy steels. Sharp reduction in critical cooling rate in high-temperature (air-hardened) steels (up to 5%); they can therefore only be welded after preheating (up to approximately 400 °C). Ferritic and martensitic Cr steels (9 to 30% Cr) can only be welded where possible to austenitic materials using preheating and post-weld treatment, due to sigma formation and coarse grain in and around the weld. In austenitic CrNi steels (16 to 25%) there is a risk of sigma phase embrittlement in the event of unfavourably high Cr content and unsuitable welding conditions.

Ni Content. Mainly in high-tensile fine-grained and tempering steels (up to about 2%). Accurate control of the welding conditions and use of hydrogen controlled electrodes are required owing to the improvement in quenchability (martensite). Low temperature Ni steels (usually 5 to 9%) are also tempering steels but with low C content (below 0.1%). They can be welded to austenitic materials or materials with a high nickel content. Ni is effective in forming austenite in CrNi steels and in general does not have a detrimental effect on weldability.

Mo Content. In higher tensile fine grained steels (up to 0.5%) and in high-temperature steels (up to 1%) it has no direct influence on weldability. In austenitic CrNi steels above about 3% there is a risk of embrittlement as sigma and Laves phases are promoted in unfavourable welding conditions.

Ti and Nb Content. In fine-grained steels (up to about 0.3%) it has no direct influence on weldability. Ti is added by alloying to austenitic CrNi steels to prevent disintegration of the grain (by bonding the carbon to special carbides). If the content is too high (above 1%) there is a risk of embrittlement in the ground mass.

Al Content. Present in fine-grained steels as deoxidation and denitration medium with simultaneous effect on fine grained structure. If the content is too high (above about 0.03%) a tendency to fracture through grain boundary separation in the weld deposit and the heat-affected area is fostered.

Cracking Due to the Intrinsic Properties of the Material. Weld joints which are subject to high stresses should react to overstressing by plastic deformation and not by fracturing without deformation (brittle fracture). The tendency to brittle fracture grows in proportion to falling temperature, rising stress rate, increasing multiaxial stress (e.g. notch effect of incipient cracks, poor design) and increasing plate thickness. The tendency to brittle fracture is further increased by those additions in the steel that favour or intensify the hardening or ageing process. The tendency towards brittle fracture increases starting from fine grain steel (Al-killed) by way of killed steel to rimmed steel (cf. DIN EN 10025). There is a danger of lamellar tearing in rolled products if they are stressed

through the thickness (production stress, e.g. brought about by residual welding stress or operating load). This is caused by linear sulphide inclusions.

Reliability of the Weld

In a structure this is conditional on structural design (magnetic flux, position of weld, thickness of workpiece, notch effect, stiffness) and the stress states (kind and extent of stresses, degree of multiaxiality, speed of stress applied, temperature, corrosion).

Basic Rules for Positioning Welds. Keep the number of welds to a minimum, do not position welds in areas of high stress concentration, avoid weld intersections, take magnetic flux into consideration when positioning welds, provide suitable weld position in rolled steel sections, e.g. avoid unkilled areas when welding the web plate of an I-beam (**Fig. 1a**), welding U-beams (**Fig. 1b**) and web stiffeners (**Fig. 1c**) and weld at section ends. Avoid welding in areas of residual tensile stress (**Fig. 1d**).

Component Thickness. In thin plate, a mainly biaxial residual stress state remains on the surface plane of the plate after welding (**Fig. 2a, b**), stress in the third direction rises as plate thickness increases. A triaxial tensile stress state means increased danger of brittle fracture as tensile stress in the third direction (plate thickness) prevents plastic deformation and therefore stress relief. Moreover, with thicker plate the danger of hardening adjacent to the weld (heat affected zone) increases as a function of welding process and welding conditions.

According to the standards for welded steel structures with predominantly static loadings (DIN 18801), Group 1 killed steels and Group 2 rimming steels (DIN EN 10025) must only be used for maximum thickness of 16 mm, otherwise permissible stresses must be halved. For cold-formed structural steels, welding in the forming range, including the range of the adjacent surfaces of width $5s$ is only permissible with a bending radius $R \geq 10s$ for any sheet thickness s, and with $R \geq 3.0s$ for sheet thickness $s \geq 24$ mm, with $R \geq 1.5s$ for thickness $s \geq 8$ mm, and with $R \geq 1.0s$ for thickness $s \leq 4$ mm.

In the case of non-alloy steels, from a plate thickness of about 25 mm onwards, preheating to between 100 and 400 °C is applied, according to material and thickness and/or stress relief heat treatment, e.g. at between 600 and 650 °C. With alloy steels the preheating and postweld treatment temperatures must be determined as a function of the alloying elements, the sections to be welded and the welding process (steel works' material lists).

According to the Technical Regulations for Boilers (TRD) and the Pressure Vessel Regulations (AD), nor-

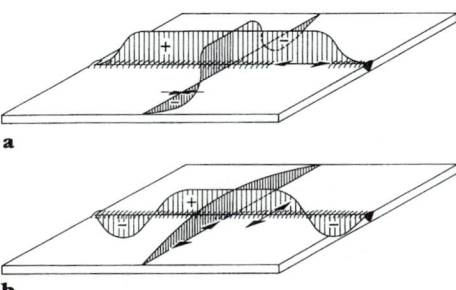

a

b

Figure 2. Residual welding stresses: **a** in the direction of the joint (longitudinal stresses), **b** across the joint (transverse stresses).

malising or tempering is required after welding when the required weld joint properties are only attainable in this way, and when, with cold forming, extreme fibre strain exceeds 5% ($R > 10s$, operating temperature $> - 10\,°C$), or 2% ($R > 25s$, operating temperature $> - 10\,°C$), or when the component is to be hot-formed at a working temperature outside the normalising temperature range before or after welding.

Stress-free annealing or tempering is carried out depending on material composition, wall thickness, and component shape (Pressure Vessel Code of Practice HP 7/1 and HP 7/2).

Weld Soundness as Conditioned by Fabrication. This is influenced by weld preparation (welding process, filler material, type of weld, form of joint, preheating, the execution of the work (directing the heat, applying the heat, welding sequence) and the post-weld treatment (heat treatment, machining, pickling).

With thick cross-sections *welding processes* are preferred in which a high degree of heat is supplied (except fine-grained steels, high-tensile quenched and drawn structural steels, fully austenitic steels, chromium steels). The *form of the joint* should be selected so that the amount of weld deposit required to ensure fusion of the flanks of the joint is kept as small as possible. *Multi-run welding* is preferable to *single-run welding* when welding larger cross-sections as the initial run is heat-treated (normalised) by subsequent runs. Like the single run, the final run has a cast structure.

Shrinkage of the weld means dimensional and geometrical changes in the welded part or residual stress caused by contraction of the weld deposit on cooling. This effect is intensified by the fact that the material was previously deformed when the weld zone was being heated owing to the restraining effect of the surrounding cold material. *Transverse shrinkage* is a function of the welding process, workpiece thickness and number of welding runs (**Fig. 3a**). *Angular shrinkage* occurs mainly in welds with asymmetrical forms of joint (**Fig. 3b**). Allowances must be made for changes to dimensions and angles by making both oversize. *Longitudinal shrinkage* in thinner workpieces, and in fillet welds in particular, leads to shortening (0.1 to 0.3 mm/m), buckling, bulging and slipping. The buckling effect is, however, used purposely and in a controlled manner in bridge and crane construction. If bonded parts are not allowed to follow the shrinkage particularly dangerous "reactive stresses" are set up, which makes it difficult if not impossible to achieve a crack-free backing run.

Descaling of structural parts before and after welding can be carried out either by means of external forces or through the shrinking effect of cooling components (flame descaling). Cold descaling is best avoided where possible, because there is a danger of cracking.

The *welding sequence,* i.e. the sequence in which welding operations are carried out within a joint and throughout the component, influences dimensional and geometrical change as well as residual stresses. Both of these may be kept within limits by setting out the individual steps in a welding schedule. In drums, for example, first the longitudinal seams and then the circumferential seams are welded; the welding sequence for longitudinal and transverse welds in plates is as shown in **Fig. 4a**. Sectional welding following the Pilger step-by-step procedure is recommended for longitudinal welds, **Fig. 4b**.

The degree of difficulty in welding grows in relation to the sequence of *welding positions* which range from flat (f) and horizontal (h) by way of vertically downward (d),

Cross-section of weld	Welding process and weld structure	Transverse shrinkage in mm
⌀6	Arc welding, coated electrode, 2 runs	1.0
12	Arc welding, coated electrode, 5 runs, root gouged, 2 backing runs	1.8
12	Rightwards gas welding	2.3
20 / 35	Arc welding, coated electrode, 20 runs, no back weld	3.2

a

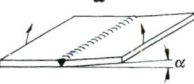

Cross-section of weld	Welding process and weld structure	Angular shrinkage α
12	Arc welding, coated electrode, 5 runs	3½°
12	Arc welding, coated electrode, 5 runs, root gouged, backing runs	0°
20	Arc welding, coated electrode, 8 broad runs	7°
20	Arc welding, coated electrode, 22 narrow passes	13°

b

Figure 3. Shrinkage in a butt weld after *Malisius*: **a** transverse shrinkage, **b** angular shrinkage.

vertically upward (u) and transversal (t) to overhead position (o): **Fig. 5**. Position (d) is only possible with certain electrodes (downward weld electrodes) under certain welding conditions (short arc with MIG or MAG welding).

If welding has to be carried out at temperatures below freezing point the *welding area* must be *heated* to at least + 10 °C and the workpiece must be preheated (50 to 100 °C); a windbreak must be provided when working at high altitude.

Filler Material. This should be selected so that the toughness values (yield point, tensile strength, elongation and notch ductility) of the welded joint at least attain the guaranteed (calculated) or standard values of the parent material. It is particularly important for the weld deposit to have sufficient malleability in cases where the parent material is little suited to welding or where for other reasons there is a danger of brittle fracture. In this case,

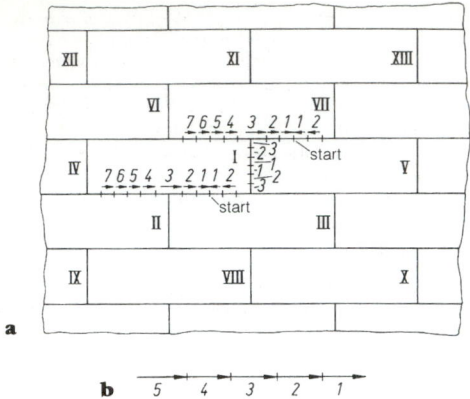

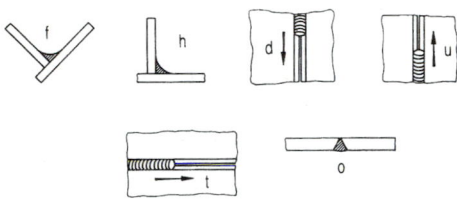

Figure 4. Welding procedure: **a** sequence of welding steps in longitudinal joints (1 to 7) and transverse joints I to XIII (welding steps 1 to 3) in a plate wall, **b** step-back welding.

Figure 5. Welding positions (see text).

electrodes with hydrogen-controlled basic coating and an increased Mn content (1.0 to 1.8%) or equivalent wire electrodes are preferred.

Standards. DIN 1913: Stabelektroden für das Verbindungsschweissen von Stahl, unlegiert und niedriglegiert. DIN 8554: Gasschweisstäbe für Verbindungsschweissen von Stählen, unlegiert und niedriglegiert. DIN 8555: Schweisszusatzwerkstoffe zum Auftragschweissen. DIN 8556: Schweisszusatzwerkstoffe für das Schweissen nichtrostender und hitzebeständiger Stähle. DIN 8557: Schweisszusätze und Schweisspulver für das Unterpulver-Schweissen. DIN 8559: Schweisszusatz für das Schutzgasschweissen. (See "Note on Standards" after references.) For welding suitability of individual steels, see D3.1.

Weldability of Cast Iron, Malleable Cast Iron and Non-ferrous Metals

Grey Cast Iron (GG-15 to GG-35) is welded mainly for patching and repair purposes. Gas welding is recommended for smaller wall thicknesses and, for thicker cross-sections, manual metal-arc welding using specially alloyed cast iron filler rods with a flux or using electrodes with or without a flux and preheating the workpiece to between 600 and 700 °C (welding with pre- and post-heating). Cold welding (manual metal-arc welding) is performed using nickel, nickel–copper (Monel) or nickel–iron stick electrodes with preheating from 100 to 200 °C. The weld deposit is readily machinable, and the heat-affected zone usually readily machinable (depending on welding conditions) but not if normal steel electrodes (type B) or special steel electrodes (increased C content) are used without post-weld heat treatment.

Blackheart Malleable Cast Iron (GTS) and Whiteheart Malleable Cast Iron (GTW) always lend themselves to soft soldering. Weldability must be specially agreed with the manufacturer. In the case of GTW-S38-12 with a wall thickness up to 8 mm, welding approval is always available for structural welding (without post-weld heat treatment). For subsidiary purposes GTS (temper carbon throughout the section) and GTW (decarburised rim zone) can also be welded using normal or low alloy filler metals whereby GTS produces welds that are hard and in danger of cracking owing to the additional carbon (molten temper carbon) dispersed in the weld deposit (preheat to between 200 and 250 °C).

Nodular Graphite Cast Iron (GGG) can be welded using special electrodes (Ni alloyed) with preheat (500 °C), post-weld heat treatment (900 to 950 °C) and tempering (700 to 750 °C). Action as with blackheart cast iron but without heat treatment.

Aluminium, when unalloyed, is weldable in almost all processes. Strain-hardening is eliminated in the heat-affected zone by crystal recovery and recrystallisation.

Precipitation-Hardenable Aluminium Alloys of the usual composition for precipitation hardening by thermal treatment and cold working can for the most part be welded by almost all processes. There is no precipitation hardening in the weld deposit and heat-affected zone or it has been eliminated by the heating effect. AlZnMg is welded in the hardened condition. An increase in toughness is then produced in the weld zone by spontaneous ageing. Welding processes which involve a narrow heat-affected zone are to be preferred to maintain toughness. Where the filler material is of a similar composition post-weld heat treatment can produce the same toughness characteristics as in the parent material.

Non-precipitation-Hardenable Aluminium Alloys can, as a rule, be readily welded by all processes. With magnesium as the alloying element difficulties may be experienced at levels over 5% Mg with the result that these alloys are not used for welded structures.

Copper presents no difficulties in the grades that are low in oxygen. High-oxygen copper content is however used in electrical engineering and this produces a froth in gas welding. Satisfactory results for both toughness and conductivity can be obtained using the inert-gas-shielded welding process and, where necessary, specially alloyed filler materials.

Copper Alloys such as CuZn (brass), CuSn (bronze) and CuSnZn (gunmetal) can be satisfactorily welded with experience. In the metal-arc welding process, however, zinc is vaporised out of the brass so that the weld becomes richer in copper; segregation may take place in many of the bronzes.

Nickel and Nickel Alloys are readily weldable (except nickel–iron alloys). The high gas pickup (oxygen, hydrogen) and the tendency to coarse graining require special measures to be taken when welding (low heat input, inert-gas shield) and choosing filler materials (deoxidising components). It is essential for the areas around the joints to be clean (free of grease). Metal arc welding processes are to be preferred.

Filler Materials. The principle of welding *similar* materials invariably applies and should be deviated from only in exceptional cases if justified or if it is not technically feasible to weld similar materials.

Standards. DIN 1732: Schweisszusatzwerkstoffe für Aluminium. DIN 1733: Schweisszusätze für Kupfer und Kupferlegierungen. DIN 1736: Schweisszusätze für Nickel und Nickellegierungen. DIN 8573: Schweisszusatzwerkstoffe zum Schweissen von Gusseisen; Part 1 Umhüllte Stabelektroden für das Lichtbogenhandschweissen an Gusseisen mit Lamellengraphit oder mit Kugelgraphit und an Temperguss; Part 2 Nicht umhüllte Stabelektroden und Schweissstäbe zum Schweissen von Gusseisen und Lamellengraphit oder mit Kugelgraphit. (See "Note on Standards" after references.)

1.1.3 Types of Weld and Joint

The type of weld is a consequence of the structural configuration of the parts to be joined. It is a factor to be taken into account when deciding the type of joint. Standards provide guidelines for the forms of joint as a function of the welding process with regard to workpiece thickness, included angle, root gap, root and flank height.

Standards. DIN 8551 Part 1: Schweissnahtvorbereitung, Fugenformen an Stahl, Gasschweissen, Lichtbogenschweissen und Schutzgasschweissen (see also the equivalent ISO 9692). DIN 8551 Part 4: Schweissnahtvorbereitung, Fugenformen an Stahl, Unter-Pulver-Schweissen. DIN 8552 Part 1: Schweissnahtvorbereitung, Fugenformen an Aluminium und Aluminiumlegierungen, Gasschweissen und Schutzgasschweissen. DIN 8552 Part 3: Schweissnahtvorbereitung, Fugenformen an Kupfer und Kupferlegierungen, Gasschmelzschweissen und Schutzgasschweissen. DIN 8553: Verbindungsschweissen plattierter Stähle, Richtlinien.

Joint Preparation. By mechanical cutting, in particular flame cutting. The suitability of steels for cutting is determined by their alloying constituents.

Carbon: up to 0.3% (up to 1.6% with preheating)
Silicon: up to 2.5% (upper limit 4%)
Manganese: up to 13% (maximum 18% Mn and 1.3% C)
Chromium: up to 1.5% (up to 3% with preheating to 600 °C)
Tungsten: up to 10% (with $C < 0.8\%$, $Ni < 0.2\%$ and $Cr < 5.0\%$)
Molybdenum: up to 0.8% (upper limit 2.5%)
Copper: up to 0.5% (reduced cutting speed with higher content)
Nickel: up to 7% (up to 35% with $C < 0.3\%$)

With modern nozzles cutting speeds of 550 mm/min can be achieved in non-alloy steels, e.g. with 20-mm plate thickness and quality of cut I according to DIN 2310. The torch must be directed by machine (crosshair on drawing), magnetic rollers (steel template), light beam (photocell following contours in drawing, even drawing reduced to 1 : 100) or digitally (computer-controlled). To produce an economical fabrication it is necessary to draw up a cutting plan showing the parts to be cut out on the metal sheet so as to avoid unnecessary waste. Materials which cannot be flame-cut (e.g. CrNi steels, copper, nickel, aluminium) can be cut by the *arc–plasma* process and the material which has only been melted in a narrow zone by high energy is expelled from the joint by the gas jet. Although, unlike flame cutting, subsequent machining of the joint faces is usually required the process does avoid the high costs involved in mechanical cutting. In non-alloy and low-alloy steels arc–plasma cutting can be used without post-treatment at up to four times the cutting speed of traditional flame cutting.

Back-gouging the root so that the root side can be welded may be achieved by chipping (air-operated hammers with forming cutters), grinding (manual grinders), planing, autogenous flame gouging (special torches similar to those used in flame cutting but with tangentially converging cutting path) or carbon arc flame gouging (material melted by carbon arc is expelled from the joint by compressed air). The applicability of these processes is dependent on the material (cf. limits of use in flame cutting), form of the seam (straight, curved), structural factors and accessibility. Thin steel plates can be cut very economically by laser.

Butt Joint

Square Butt Weld. The simplest type of weld; to accommodate greater loads it is necessary to give the joint a backing run after gouging.

Single-V Butt Joint (**Figs 3** and **6a**). To reduce shrinkage the included angle must be kept small ($\approx 60°$). Smallest included angle for clean back welding: $> 45°$. Even smaller included angles are possible in the partially and fully mechanised welding processes.

Double-V Butt Joint (X-Joint) (**Fig. 6c**). Used for thicker plate than the single-V butt joint, as only half the amount of weld deposit is needed for the same included angle. Angular shrinkage can be avoided to a great extent if runs are applied from each side alternately. The root should be back-gouged (depending on the welding process) before applying the backing run.

Other types of joint: double-flanged butt joint, steep flank joint, single-V butt joint with broad root face, U-groove (bell) joint, double-U-groove joint. Both of the last-named are limited to special cases owing to the usually high production costs.

Butt Joint in Workpieces of Unequal Thickness (**Fig. 6**). Where possible, position the cross-section symmetrically in the direction of the force (**Fig. 6c,f**); where differences in thickness are below $s_1 - s_2 = 10$ mm and there is static stress no aligning is required; otherwise bevel (**Fig. 6d**). Bevel where dynamic stress is above $s_1 - s_2 = 10$ mm (inclination 1 : 4 to 1 : 5) in order to obtain a favourable magnetic flux. At maximum stress in thicker plate machine off along a length $b \geq 2s_2$ (**Fig. 6g**).

Lap Joint (Fig. 7). The direction of the force lines in a fillet weld is more favourable in a concave fillet weld (**Fig. 7c**) than in a flush weld (**Fig. 7b**); the convex weld (**Fig. 7a**) is the least favourable. If dynamic stress

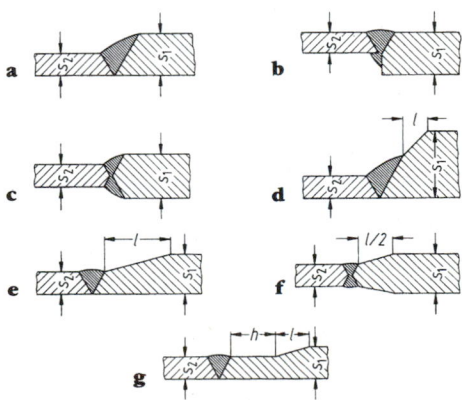

Figure 6. Typical forms of butt welds in unequal cross-sections: **a–d** for static stress, **e–g** for dynamic stress.

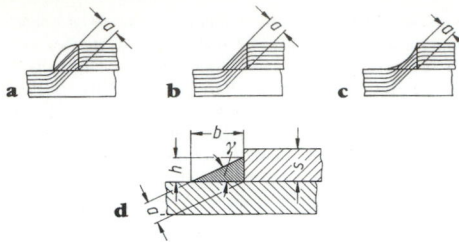

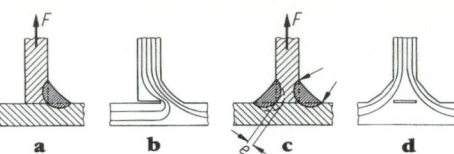

Figure 9. Fillet welds on T-joint: **a** single fillet weld, **b** cohesive pattern and magnetic flux, **c** double fillet weld, **d** cohesive pattern and magnetic flux.

Figure 7. Forms of weld and magnetic flux: **a** convex fillet weld, **b** flush fillet weld, **c** concave fillet weld, **d** asymmetrical fillet weld in normal shear.

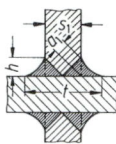

Figure 10. Double bevel weld with double fillet weld on double T-joint.

is present any force deflection is generally detrimental. The design throat thickness a is determined by the vertical leg length of the inscribed equilateral triangle. It should not be made stronger than the design requirements, with a maximum of $a = 0.7s$. Where fillets are in normal shear, steel construction requires a weld face width of at least $a = 0.5s$ and formation at $b : b = 1 : 1$ or flatter in case of static stress (**Fig. 7d**). Where there is dynamic stress (e.g. railway bridge construction) $\gamma \leq 25°$, and the weld face width $a = 0.5s$.

Parallel Joint (Fig. 8a). Where possible, fillet welds must be used as no edge preparation is required. In order to avoid fusion of the edges $q \geq 1.4a + 3$ mm is recommended as the size of the projecting end.

With regard to rolled steel sections, the weld face width $a = 0.7\,t$ is dependent on the thickness t of the thinnest part, **Fig. 8b**. Here also the welds should not be made thicker or longer than design requirements. In steel construction (DIN 18 800 Part 1) there is a minimum length of fillet weld $= 15a$ and a maximum length of fillet weld $= 100a$ for bar joints, whereas for other welds at right angles to the longitudinal axis of the bar and all-around welds the minimum length of each longitudinal weld is limited to $10a$. At least 2 mm or $\sqrt{\max s} - 0.5$ is prescribed for the weld face width. In addition, plug or slot welds are used in mechanical engineering (not bridge construction or steel framed structures) (**Fig. 8c, d**). $s \leq 15$ mm should be maintained as the thickness of the top plate, whereas $b \geq 2.5s$ (minimum 25 mm) and $l \geq 3b$ (tank construction) or $l \geq 2b$ (mechanical engineering) are recommended for the dimensions of the slot. The slot is not filled up with weld deposit owing to the high welding stresses which would be caused by this;

if there is a danger of corrosion the slot is filled up with, for example, mastic.

Tee Joint (Fig. 9). The simplest type of weld is the fillet weld which is particularly suitable for transmitting shear stress. The single fillet weld (**Fig. 9a, b**) is only to be used if a low magnitude of shear stress is to be transmitted. If there is a double fillet weld which has been produced using a process with deep weld penetration (e.g. fully mechanised gas-shielded metal arc welding or submerged arc welding) then half the penetration e (**Fig. 9c**) may be included in the strength calculations (DIN 18 800 Part 1). There is no gap in the joint with the resulting notch effect at the weld zone (**Fig. 9d**) if the section is joined as in **Fig. 10** by a double bevel weld with a covering fillet weld on both sides. This form of joint is used for maximum static and dynamic stress. It is $t = s_1 + 2b/3$ with an unequal fillet weld. Unwelded root gaps and undercut must be avoided or ground down, especially where there is dynamic stress.

Double T-Joint (Fig. 10). Weld types are as for the T-joint, but where there is tensile stress at the welded webs the transverse plate in between must be inspected for laminations (e.g. by ultrasonics) and its ability to withstand transverse tension must be guaranteed (DASt 014).

Oblique Joint (Fig. 11). Weld type as per T-joint. The quality of the weld is a function of the angle γ. Welding is often done without edge preparation if there are no large forces to be transmitted.

Fillet welds can be performed cleanly only if $b \leq 2$ mm where the face is at right angles and $\gamma \geq 60°$ where welding is done on both sides. Welds with smaller angles may be used in strength calculations as load bearing only if the solidity of the root point is guaranteed by the welding process used. A finish such as that shown in **Fig. 11b** must be avoided unless the face is machined.

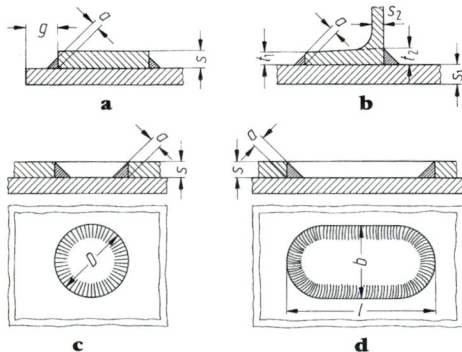

Figure 8. Welded joints in sheet metal: **a** parallel joint, **b** rolled section joined to a plate, **c** plug weld, **d** slot weld.

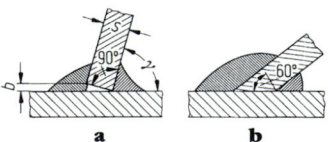

Figure 11. Fillet welds on oblique joint: **a** without edge preparation, **b** with poor edge preparation.

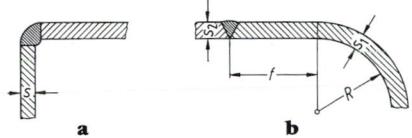

Figure 12. Structural corners: **a** corner joint, **b** corner construction using preformed parts, e.g. boiler bottoms.

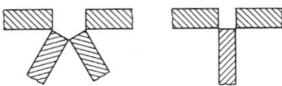

Figure 13. Multiple joint.

Corner Joint (Fig. 12a). In its execution the corner joint is a T-joint. It is generally true that welds should not be made in places where there are stress concentrations. In pressure vessels the weld is positioned away from the curve (**Fig. 12b**). The minimum distance between the weld and the curve should be $f \geq 5s_1$.

The information provided in the section on component thickness (see F1.1.2) must be taken into account when welding is done in cold-worked areas. If there are deviations from the dimensions given there, either the minimum distance f (**Fig. 12b**) must be adhered to or the cold-worked part must be normalised.

Multiple Joint (Fig. 13). Owing to the insecure bonding of the bottom plates (fusion penetration) when welding from one side, this type of joint is to be used only where there is an opportunity for careful fabrication or in cases where strength is of minor importance. If accessible from both sides the root must be gouged and given a backing run.

1.1.4 Graphical Symbols for Welds

For symbols and graphical representation see DIN 1912.

Types of Weld. These can be represented symbolically (**Fig. 14a, c**) or pictorially (**Fig. 14b, d**). Symbolical representation is to be preferred. The position of the symbol relative to the reference line denotes the location of the weld on the joint. Appendix F1, **Table 1**, shows basic and supplementary symbols as well as graphical representations of welds.

Welding Processes. Abbreviations and process reference numbers are as per ISO 4063: G – Gas welding 311, E – Manual metal arc welding 111, UP – Submerged arc welding 12, US – Firecracker welding 118, TIG – Tungsten–inert-gas welding 130, MAG – Metal arc welding with non-inert-gas shield 135. Supplement: m – manual welding, t – partially mechanised, v – fully mechanised, a – automatic welding.

Quality of the Welded Joint. The following quality classifications are established in DIN 8563 (Sicherung der Güte von Schweissarbeiten) based on production and testing requirements:

Butt welds: AS, BS, CS and DS.
Fillet welds: AK, BK and CK.

(See "Note on Standards" after references.)

The classifications to be selected must be determined by the designer in conjunction with production departments, quality control and if necessary supervisory boards and other bodies. They are dependent on the type of load (static, dynamic), environmental influences (chemical attacks, temperature) and additional requirements (e.g. integrity, safety requirements). They are to be guaranteed by: suitability of the material for welding with regard to process and intended use; workmanlike and supervised preparation; selecting the welding process according to material, workpiece thickness and stress on the welded joint; filler material compatible, tested and approved for use with the parent material; qualified welders supervised by welding overseers; proof of satisfactory execution of welding work (e.g. radiography); special requirements (e.g. vacuum tightness, all-round grinding of welds).

Welding Position. For a brief description see Fig. 5.

Examples **Figure 15a**: V-U-weld, single-V butt weld produced by metal arc welding with non-inert gas shield (135), single-U butt weld produced by UP welding (12), quality classification required BS, position flat w. **Figure 15b**: Intermittent fillet weld with throat thickness a, front measurement v, space e, length l and number n of individual welds, produced by manual metal arc welding (111), quality classification required CK, position horizontal b.

1.1.5 Strength Calculations for Welded Joints

Load-Bearing Capacity

In welded joints this is dependent on the *properties of the parent material*, the heat-affected weld junction and

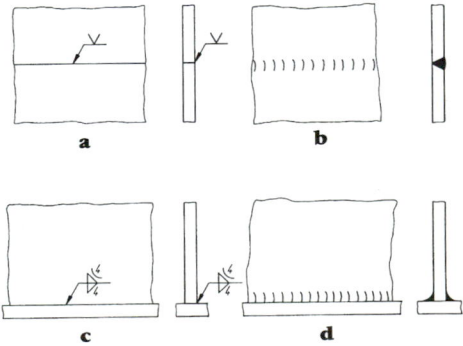

Figure 14. Forms of representation: **a** butt weld, symbolic representation; **b** butt weld, graphical representation; **c** double fillet weld, symbolic representation; **d** double fillet weld, graphical representation.

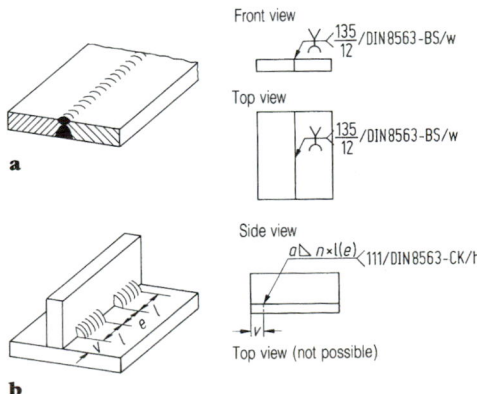

Figure 15. Graphical representation: **a** butt weld (single V-U-weld) with additional fabrication details, **b** intermittent fillet weld with end space and additional fabrication details.

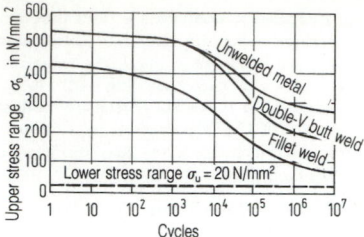

Figure 16. Stress–number curve. Parent material St 52. Stress running vertical to the joint.

deposit, the *type of stress* (tension, compression, shear, static or dynamic stress), the form of weld, position of the weld and machining of the weld, the *combined effect of the operating stresses* and the *residual welding stresses* (especially where stability is required and under certain conditions when dynamic stress is present) and the *quality of weld*. The highest standards in design and execution are required in cases of dynamic stress.

Static Load. If static load is present in a butt weld situated perpendicular to the contraction direction plastic deformation and cracking generally take place adjacent to the weld and where the load is parallel to the weld the parent material and the weld deposit are subject to an equivalent *amount* of deformation, resulting in tearing and fracturing in those types of structure which have low ductility (e.g. martensite in the heat-affected zone).

Dynamic Load. In this case failure due to vibration fatigue occurs most frequently in the transition zone of the parent material and weld, even in specimens which have been machined all round. The fatigue strengths of welded structural components are lower than the fatigue strength of the parent material and lower in unmachined welds than in machined welds. Fatigue strength diagrams are available for most materials, forms of welds and weld locations (see E1). Examples: **Figs 16 and 17**.

The following are not considered in the diagrams: **(a)** *Static initial loads caused by residual stresses,* which increase or decrease the mean stresses according to the preceding sign. Normally these residual stresses in service are, however, reduced in the course of fluctuating stress.

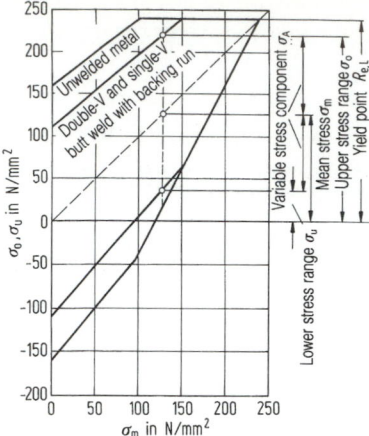

Figure 17. Fatigue-strength diagram. Parent material St 37. Stress running vertical to the joint.

(b) *The scale effect.* Intermittent overstressing has no effect provided certain limit values for the number of cycles and the stress (damage curve) are not exceeded.

Small inclusions *in* the weld (globular pores or slag) lower the fatigue strength of unmachined welded joints slightly or not at all. Cracks and surface defects such as undercut, end craters, unclean shoulder areas and ignition points caused by stray arcing adjacent to the weld may however be starting points for fatigue failure and therefore lower fatigue strength.

Figure 18 gives material limit values for *aluminium alloys* (Aluminium-Zentrale, Düsseldorf). In wall thicknesses greater than 10 mm, up to 70 mm a reduction (scale effect) to 0.8 is noted, but is constant thereafter (see D1.5.2). (See "Note on Standards" after references.)

Any stress arising must be set against the toughness of the material to determine load-bearing capacity or safe load, as the case may be.

Strength Calculations

These are carried out in the following sequence:

Determining the Applied Loads. When determining design loads, additional loads, impact coefficients and

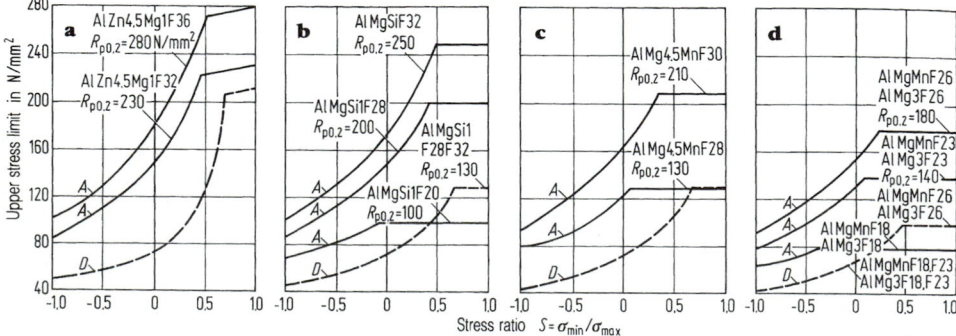

Figure 18. Limiting values for AlZn4.5Mg1, AlMgSi1, AlMg4.5Mn and AlMgMn/AlMg3, applicable to semi-finished tempered steel (DIN 1745–1749), surfaces untreated (e.g. with rolling skin), wall thickness (diameter) $s = 10$ mm. Compression-tension, bending and shear, number of cycles to failure $N = 10^7$. Continuous line A: unwelded material. Broken line D: butt-welded joint using MIG and TIG (with bead): **a** Broken line D applies to AlZn4.5Mg1F36 and F32. **c** Broken line D applies to AlMg4.5MnF28. When full cross-sections are bent the values of curve A for $s = 0$–50 mm must be multiplied by 1.2. In the case of shear stress the values of curve A must be multiplied by 0.65.

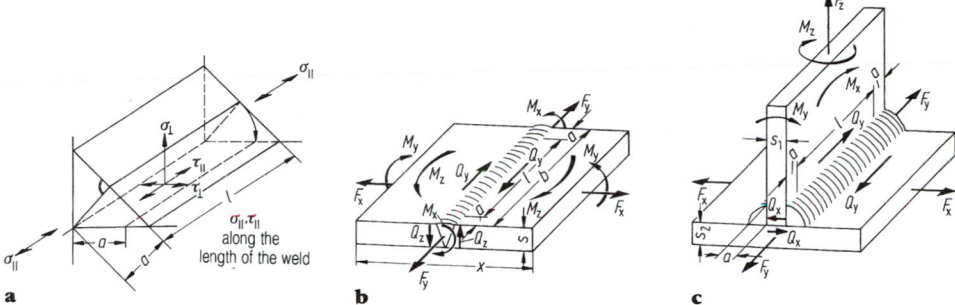

Figure 19. Stresses in welds, section sizes in welded joints and their symbols: **a** fillet weld, **b** butt joint (single V-weld), **c** T-joint.

safety margins for components covered by regulations enforced by law or by the customer (e.g. German railways), the specifications detailed therein must be taken into account (**Appendix F1, Table 2**). In all other cases these regulations may be used as reference points. Uncertainties in calculating strength are allowed for by establishing an appropriate safe stress or selecting suitable safety margins.

Calculating Nominal Stresses in Welds and Joint Cross-sections. Nominal stresses are calculated on the basis of loads as set out in the rules of the science of the strength of materials (see B1.3). Some of the equations to be used are laid down in regulations, **Appendix F1, Table 2. Figure 19a** shows the symbols for nominal stress and shear stress.

Often several kinds of stress occur in the component at the same time and these must be classified accordingly. Tension, compression and bending are the result of direct stresses which must be added arithmetically if they are in the same direction. If the directions are perpendicular to each other then a comparative value σ_v is established from the stresses, which is compared with the apparent limit of elasticity under static stress, and with the *fatigue limit* or *strength under load* under dynamic stress:

$$\sigma_v = \sqrt{\sigma_\perp^2 + \sigma_\parallel^2 - \sigma_\perp \sigma_\parallel + \alpha(\tau_\perp^2 + \tau_\parallel^2)} \le z_{\text{perm}}\,\sigma$$

with $\alpha = 1$ as in DIN 18800 Part 1 in the case of static load, $\alpha = 2$ as in DIN 15018 in the case of dynamic load. (See "Note on Standards" after references.)

In welded girders subjected to bending, the direct stress $\sigma_\parallel$ in the flange or web joint produced using double fillet welds or single bevel butt welds with broad root face plus fillet weld, has no effect under static load with the result that the comparative value becomes

$$\sigma_v = \sqrt{\sigma_\perp^2 + \tau_\perp^2 + \tau_\parallel^2}.$$

In the strength calculation the τ appropriate to $\sigma_{\max}$ and the σ appropriate to $\tau_{\max}$ must be selected. In addition, however, it must *always* be demonstrated separately that the shear stress τ alone does not exceed the critical shear value.

When calculating the strength of *butt welds* the thickness s of the thinner plate is always used as the throat thickness (**Fig. 6**). The length of the weld allows for end-craters by deducting a throat thickness, $l = b - 2a$. If end-pieces are used (**Fig. 20**), $l = b$. In *fillet welds* (**Fig. 19c**) the throat thickness a is equal to the vertical leg length of the inscribed equilateral triangle. The stress is calculated for the cross-section in the plane of the joint

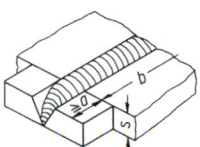

Figure 20. Butt joints with end projection for weld run-out.

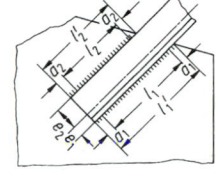

Figure 21. Weld dimensions when joining an angle section to a gusset (bar joint).

using side a. *Oblique-joint* fillet welds with smaller included angles than $\gamma = 60°$ may no longer be considered as load-bearing in strength calculations (except where the welding process ensures the root is secure). In *non-symmetrical bar joints* (**Fig. 21**) it is permissible for weld cross-sections to be equal ($a_1 l_1 = a_2 l_2$). Fillet welds in bar joints should be made no shorter than $15a$ and no longer than $100a$. In *cylindrical boiler shells, drums and collectors* the required plate thickness is calculated according to the formula for boilers (see H2.2), provided $D_a/D_i \le 1.2$ (AD) or ≤ 1.7 (TRD) is adhered to. In the case of *anchor bolts, stay tubes and stay rods* the shearing cross-section of the welded joints must be at least 125% of the bolt and anchor cross-section. The anchor bolts must be welded on both sides of the walls to be fixed.

Determining Safe Loads

Static Stress. **Figure 22** and **Table 2** apply to steel-framed structures. The roots of butt joints must be chipped and back-welded prior to final welding, or another way must be found to ensure perfect root penetration. **Table 3** applies to mechanical engineering and a limiting stress ratio of $S = +1$. Further values are given in DIN 15018.

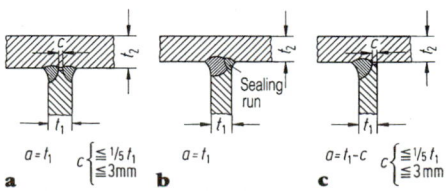

Figure 22. Forms of welds to illustrate Table 2. a = Throat thickness.

Table 2. Safe stresses under static load (DIN 18800 Part 1) for double-bevel butt weld with broad root face and double fillet weld

1	Type of weld	Weld quality	Type of stress	St 37 Load		St 52 Load	
				H N/mm²	HZ N/mm²	H N/mm²	HZ N/mm²
2	Parent material permissible in accordance with DIN 18800 Part 1	—	Tension	160	180	240	270
		—	Compression[a]	140	160	210	240
3, 4	Butt weld. Double-bevel butt weld with double fillet weld (root penetration) as shown in **Fig. 10**	All weld qualities	Compression and flexural compression	160	180	240	270
	Double-bevel butt weld with broad root face with double fillet weld as shown in **Fig. 22a**[b]	Freedom from cracks, weld and root defects proved	Tension and bending tension across weld direction	160	180	240	270
5	Single-bevel butt weld with fillet weld (back-welded sealing run) as shown in **Fig. 22b**	Weld quality not proved		135	150	170	190
6	Single-bevel butt-welded web with fillet weld as shown in **Fig. 22c**	All weld qualities	Compression and flexural compression Tension and flexural tension Comparative value	135	150	170	190
7	All welds		Shear	135	150	170	190

[a]If proof is required in DIN 4114.
[b]Line 4 is not applicable owing to the root gap.

DIN 18800 Part 1 (Table 2) applies to *light steel constructions* and *structural steel tubing*, as used in building construction, in respect of joining tube sections (round and rectangular tubes) to gussets or other structural members. DIN 18808 provides information concerning direct steel tube welding in frame structures. The minimum wall thickness is 1.5 mm for structural members having normal corrosion fatigue strength within enclosed spaces, and is 2 mm for members which are arc-welded and 3 mm for all other structural members.

Welded plate *road bridges* are not subjected to any pronounced dynamic stress. DIN 4101 (18809) therefore indicates the safe loads irrespective of the type of fatigue stress involved. Where road bridges are carrying a traffic load as specified in DIN 1072 and/or railway lines, an operating safety certificate must be provided showing the permissible values in accordance with DS 804 (see also the related ISO/TR 9492).

Boiler and *pipeline construction* are governed by the regulations and instruction sheets of the Vereinigung der Technischen Überwachungsvereine (Association of Technical Control Boards). The usual welding factor (quality assessment of a welded joint) $v = 0.8$ can be increased to $v = 1.0$ by additional testing of welders and work. In the construction of pressure vessels (AD) $v = 1.0$ is usual, but if the testing standards are lowered this is reduced to 0.85.

Dynamic Stress. Maximum safe tensile or compressive stresses, or shear stresses as the case may be, for *welded rail bridges* are laid down in DS 804 (*or* DIN EN 10025 (1991)) in respect of St 37 and St 52 steels (DIN EN 10025 equivalents: Fe 360 and Fe 510, respectively). They are determined according to the stress ratio $\kappa = \min \sigma/\max \sigma$ or $\kappa = \min \tau/\max \tau$, using the same stress ranges as in DV 952 (where the designation S is used instead of κ).

The safe loads for St 37 and St 52 for *welded rolling stock*, machines and equipment belonging to the German railways are also given in DV 952 according to the stress ratio S (**Fig. 23** and **Table 4**), but using values which differ from DS 804. The κ or S values denote the stress ranges: simple alternating stress $(= -1)$, alternating range (< 0), simple fluctuating stress $(= 0)$, zones for fluctuating tensile and compressive stress (> 0) and static tensile and compressive stress $(= 1)$.

The lines A to H of the safe loads correspond to various types of weld and joint (**Table 4**).

Both the DV 952 regulations and DIN 15018 cover the operating safety certificate used in mechanical engineering.

Table 3. Safe stresses under static load ($S = +1$) (DV 952)

		St 37 N/mm²	St 52 N/mm²
Principal stress	Parent material	160	240
Tension, compression, bending	Butt weld with backing run, radiographed	160	240
	Butt weld with backing run, radiographed at random	150	216
	Butt weld not radiographed	150	216
	Fillet weld	105	155
Shear	Butt weld	112	168
	Neck weld	98	152

DS 804 was drawn up for rail bridges on the basis of experience and is therefore only applicable to or pre-scribed for similar situations. Both DV 952 and DIN 15 018 do, however, cover mechanical engineering.

Dynamic Stress (Operating Safety). The operating cer-tificate of safety against fracture caused by frequently repeated stress amplitudes which alter with time may according to DIN 15 018 be used for stress cycles above $2 \cdot 10^4$. The maximum safe loads for normal and shear stresses are dependent on the set of stresses (**Fig. 24**), the number of stress cycles (**Table 5**), the notch effect (**Table 6**), the material and the stress ratio $\kappa = \min \sigma/\max \sigma$ or $\kappa = \min \tau/\max \tau$. The set of stresses obtained by measurement, calculated from the load cycle or known by other means, is compared with the idealised set of stresses in **Fig. 24** and allocated to a line (S_0 to S_3). The set of stresses and rate of cyclic stress N1 to N4 determine the stress group B1 to B6 (**Table 5**). The notch effect K0 to K4 present in the component or joint corresponds to the safe stresses for materials St 37 and St 52-3 at $\kappa = -1$ contained in **Table 7**. The safe stress at $-1 < \kappa \leq +1$ is obtained using the equations shown in **Table 8** based on the correlations in **Fig. 25**.

Comparison Between Nominal Stresses and Safe Stresses

Static Stress. $\sigma_{weld} \leq$ perm σ_{weld} (Tables 2 and 3), $\tau_{weld} \leq$ perm τ_{weld} or $S = R_e/\sigma$ for factor of safety for plastic deformation. In boiler and pipeline construction heat resistance must be taken into account where possible.

Dynamic Stress. $S = \sigma_A/\sigma_a$ or $= \tau_A/\tau_a$ (σ_A, τ_A fatigue-resistant, σ_a)τ_a existing variable stress component). DS 804, DV 952 and DIN 15 018 already specify safe stress-es. Some standard values from *Erker*:

1. Static stress: factor of safety for breaking point R_m of the material $S > 1.8$, normal $S = 2.5$.
2. Mainly static stress (up to 10 000 vibration cycles): factor of safety for yield point R_e of the material $S = 1.5$ to 2.0 (depending on notch acuity). Mean value $S = 1.7$.
3. Fatigue strength (up to 500 000 vibration cycles): fac-tor of safety for vibration fatigue failure $S = 1.0$ to 1.8, mean value $S = 1.3$ to 1.5.

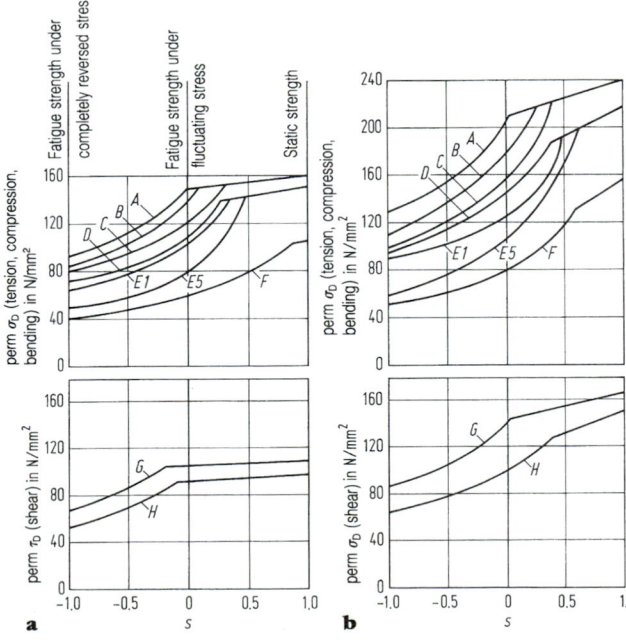

Figure 23. Safe stresses (DV 952): **a** St 37, **b** St 52. *Note:* DV 952 is contrary to current opi-nion and gives different fatigue strength values for St 37 and St 52 also when $s < 0.5$.

Table 4. Relating the forms of weld and joint to the lines in Fig. 23

Parent material	Butt joint (flange splice) Equal			Flange splice unequal	Transverse joint in web plate	Flange splice in box girder	Butt joint in pipe				Pipe connection using round material	Angled pipe connection
	Plate thickness				(principal stress)							

| Joint and weld type | Root back-welded | | Without Cross-sectional transition | With | Root back-welded | Strapped joint | With \| With \|With \| Without Backing ring | | | | Relief chamfer With Relief chamfer | With–out |

Machined Weld 1 — 2 Unmachined	Line A Line	Radiography B Full D At random E1 Without	E1	B Radiographed D At random	Shear G	D Radiographed At random D	Radiography D Full D At random E1 Without	E1	E1	E1	E5	F

	Plate structure	Butt welds in corner joints Sections with gusset plates Sections without gusset plates		with flush relief chamfer	T-joint (neck joints, transverse joints)		Double T-joint (welded both sides)		

| Joint and weld type | | | | | Double HV with double fillet weld | Double HY with double fillet weld | Double fillet weld |

Machined Weld 1 — 2 Unmachined	Line B Radiographed Line D Radiographed At random	D Radiographed E1 Not Dradiographed at random	F Not radiographed	E Not radiographed Shear:	Direction xx: C Direction yy: B H	E1 E5	F F	F F

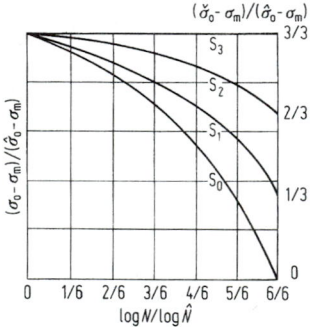

Figure 24. Idealised related sets of stresses. Here $\sigma_m = \frac{1}{2}$ (max σ + min σ) amplitude of constant mean stress, σ_0 = amplitude of the elastic limit which is reached or exceeded N times, $\hat{\sigma}$ = amplitude of the maximum elastic limit of the idealised set of stresses, $\check{\sigma}_0$ = amplitude of the minimum elastic limit of the idealised set of stresses, $\hat{N} = 10^6$ extent of the idealised set of stresses.

4. Endurance limit (over 500 000 vibration cycles): factor of safety for endurance failure $S = 1.5$ to 3.0, mean value $S = 1.5$ to 2.0, in special cases minimum value $S = 1.2$.

Combined Stress. In this case the values of the safe stresses must be taken into consideration when applying stress and carrying out calculations to compare the nominal stresses with the safe stresses (e.g. DIN 18 800 Part 1, DIN 15 018, DS 804, DV 952).

If it is desired to dispense with the calculation of equivalent tensile stress, individual fractional loads can be used as a basis for calculation and the factor of safety in the event of overstressing can be estimated. If, for example,

in bending and shear the safety margin for the first fractional load (bending)

$$S_I = \sigma^*/\sigma = 2.4$$

and the safety margin for the second fractional load (shear)

$$S_{II} = \tau^*/\tau = 1.6$$

(σ^*, τ^* reference quantities, e.g. toughness, yield point, etc.), then the factor of safety obtained from **Fig. 26** for the total stress is $S = 1.3$.

Example of Calculation from DIN 15 018 Components joined using standard quality butt weld across the direction of the force lines → notch value K1 as in Table 6, material St 37, $\sigma_{max} = 150$ N/mm² = $\hat{\sigma}_0$, $\sigma_{min} = -60$ N/mm² = $\check{\sigma}_u$ → $\kappa = 0.4$ and $\sigma_m = 45$ N/mm². Known from the course of the stress $\check{\sigma}_0 = 0.5\hat{\sigma}_0$ = 75 N/mm², $(\check{\sigma}_0 - \sigma_m)/(\hat{\sigma}_0 - \sigma_m) = 0.286$. The safe weld or parent material stresses at the weld joint are provided in tabular form for three different stress cycles:

Number of combined rated cyclic stresses	N_1 10^5	N_2 10^6	N_3 $5 \cdot 10^6$
$\dfrac{\log N}{\log \hat{N}} =$	$\dfrac{5}{6}$	$\dfrac{6}{6}$	$> \dfrac{6}{6}$
Set of stresses from Fig. 24	S_0	S_1	S_1
Stress group as in Table 5	B1	B4	B5
Perm $\sigma_{D(-1)}$ from Table 7 (N/mm²)	180	150	106.1
Perm $\sigma_{D(-1)}$ as in Table 8 with 1.315 perm $\sigma_{D(-1)} \leq 0.75\,R_{eff}$	180	180	139.5
The stress present is	smaller	smaller	greater

Table 5. Stress groups according to range of cyclic stress and set of stresses (DIN 15 018)

Range of cyclic stress	N1	N2	N3	N4
Total number of rated stress cycles N	Above $2 \cdot 10^4$ Below $2 \cdot 10^5$ Occasional irregular use with long periods of repose	Above $2 \cdot 10^5$ Below $6 \cdot 10^5$ Regular use with intermittent operation	Above $6 \cdot 10^5$ Below $2 \cdot 10^6$ Regular use in continuous operation	Above $2 \cdot 10^6$ Regular use in strenuous and continuous operation

Set of stresses	Stress group			
S_0 very light	B1	B2	B3	B4
S_1 light	B2	B3	B4	B5
S_2 medium	B3	B4	B5	B6
S_4 heavy	B4	B5	B6	B6

Pressure Welded Joints

Pressure and Flash Butt Welding. The cross-section to be used for strength calculations is the smallest cross-section adjacent to the weld. For standard values for safe stresses see **Table 9**.

Spot and Seam Welding. These joints are generally stressed to shearing point. A low fatigue strength is obtained owing to considerable notch effect. As the diameter of the spot weld is not known and is not readily determined even by non-destructive testing procedures, acceptable breaking loads are determined by experiment. Compare also Instruction Sheet DVS 1603 "Widerstandspunktschweissen von Stahl im Schienenfahrzeugbau", DVS 2902 Part 3 "Widerstandspunktschweissen von Stählen bis 3 mm Einzeldicke, Konstruktion und Berechnung", "Luftfahrt-Tauglichkeitsforderungen für das Widerstandspunkt- und Nahtschweissen LTF 3400-001" and DIN 18 801: Stahlhochbau.

1.1.6 Removal by Thermal Operations

Gas Cutting Processes

The heat required for removal is provided by oxidation and material is removed in the jet of oxygen.

Oxygen Cutting. The workpiece which has been locally heated to ignition temperature by an oxy-fuel gas flame burns in the cutting oxygen jet, the cutting slag (oxides and molten metal) being expelled from the joint by the O_2 jet. The requirements for cutting are that the metal must burn in the O_2 jet, and the ignition temperature must be below the melting temperature of the material. These requirements are met in non-alloy and low-alloy steels, titanium and molybdenum but not in aluminium, copper, grey cast iron and in high-alloy steels in general. Preheating is required where the carbon content > 0.3% owing to hardening. Accuracy of joint face (A, B) and surface quality (I, II) are dependent on the gas cutting machine, guiding mechanism, cutting speed and requirements (vertical, diagonal, straight, curved, manual, machine cutting with single or multiple torch arrangement). Machine blanks are cut round a plate template using a magnetic guide roller, and an electronic eye following a line drawing on a 1 : 1 scale or smaller, as well as on NC machines.

Metal Powder Flame Cutting. Metal powder is fed to the reaction point, producing additional heat and liquid slag. Suitable for cutting stainless steel, plated materials, cast iron and aluminium (no longer much used).

Metal Powder Fusion Cutting. Metal powder burns in the cutting oxygen stream, becoming metal oxide, and converts non-combustible, generally molten, mineral mass to slag (lava). Suitable for cutting all metallic, non-metallic and mineral materials (hardly used).

Mineral Powder Cutting (flame and fusion cutting). The kinetic energy of quartz particles added to the cutting oxygen assists in the expulsion of the molten slag. Hardly used.

Flame Gouging. Removal of material from workpiece surfaces by delivering a larger volume of cutting oxygen. In flame gouging, material is removed by specially shaped nozzles, forming a trough. The breadth and depth of the trough are determined by the size and angle of the cutting nozzle. The main range of application is the partial descaling of welds prior to final welding or in repairs to materials suitable for flame cutting.

Flame Scarfing. By removing material in layers it is used for cleaning steel ingots, billets and tube blanks prior to subsequent treatment. Manual descaling is used for removing localised defects and machine scarfing for treating large areas.

Oxygen Lancing (also Oxygen/Powder (OPL) and Oxygen Core (OCL) Lancing). This is a thermal hole boring process which is mainly used on mineral materials (concrete, reinforced concrete). OL works using a tube only and is increasingly being replaced by OPL which works using a tube plus iron or iron/aluminium powder. A tube packed with wires is used in OCL. The end of the tube which has been heated to white heat is in all three cases placed on the workpiece and burns when fed with oxygen. Metallic materials burn and minerals melt combining with metal oxide to form liquid slag. Oxygen lancing can be used with all metals, non-metals and mineral materials.

Table 6. Notch effects of forms of weld and configuration (notch values) (DIN 15 018, extract)

Form of weld configuration loading	Weld quality (as shown in **Appendix F1, Table 3**)	Notch value (notch effect)
	Special quality	K0 (weak)
	Standard quality	K1 (moderate)
	Standard quality	K0 (weak)
	Standard quality Sampling test	K1 (moderate)
a	Special quality^a $a \leq 1:4, b \leq 1:3$	K0 (weak)
	Standard quality^a $a \leq 1:4, b \leq 1:3$	K1 (moderate)
b	Standard quality^a $a \leq 1:3, b \leq 1:2$	K2 (medium)
	Standard quality^a $a \leq 1:2, b \leq 1:0$	K3 (strong)
	Special quality (double-bevel butt weld)	K1 (moderate)
	Special quality (double fillet weld)	K2 (medium)
	Standard quality (double fillet weld)	K3 (strong)
	Special quality (double-bevel butt weld)	K2 (medium)
	Standard quality (double-bevel butt weld)	K3 (strong)
	Standard quality (double fillet weld)	K4 (particularly strong)
	Special quality (double-bevel butt weld)	K2 (medium)
	Standard quality (double-bevel butt weld)	K3 (strong)
	Special quality (fillet weld)	K3 (strong)
	Standard quality (fillet weld)	K4 (particularly strong)

^{a}Sloping areas

Flame Descaling. This is used for removing (by burning or conversion) scale and coatings and for cleaning or pre-treating metal or mineral workpieces.

Electrical Discharge in Gas

Oxy-arc Cutting. The arc is struck between a tubular coated electrode and the workpiece. Oxygen is supplied through the bore of the electrode to the cut edge. The process is preferred for scrapping purposes.

Arc-Air Gouging. This is used for gouging out welds and cracks in metallic materials. Localised melting of the parent material by an arc struck between a copper-coated carbon electrode and the workpiece. Compressed air supplied parallel to the electrode assists in burning off some of the molten material and blows molten metal and slag out of the groove as it is created.

Arc-Plasma Cutting. A constricted arc brings about the dissociation of polyatomic and monoatomic gases for ionisation. The material melts and partly vaporises in the high-temperature and high kinetic energy plasma jet. A cut edge is produced if the workpiece or the torch is moved. The plasma gases are argon, hydrogen or mixtures of the two and depending on the material argon, nitrogen, hydrogen or mixtures of these can be used as cutting gases with non-alloy and low-alloy steels and compressed air. Electrically conductive materials are cut using a transferred arc and non-electrically conductive materials using a non-transferred arc. A good quality of cut can be achieved at high cutting speeds. The process can be used with all steels and non-ferrous metals.

Removal by Beam

A beam of energy is used (laser, electrons). High energy density of the YAG solid-state or CO_2 gas laser beam results in the fusion, vaporisation or sublimation (immediate change into the gaseous state) of the material. The cutting process is accompanied by inert gas for easily combustible materials and oxygen for metals, especially steel: *laser oxygen cutting*. Fusion of the workpiece and use of inert gas: *laser fusion cutting*. Immediate conversion of the material into the gaseous state: *laser sublimation cutting*. The advantages of laser cutting are minimal heat effect, narrow kerf, minimal distortion and high cutting speed. In addition to metal the following can also be cut: organic materials and plastics, wood, leather, rubber, paper, ceramics, quartz glass, porcelain, mica, stone and graphite.

The electron beam with increased power density in the focal spot (up to 10^8 W/cm^2, in welding 10^6 W/cm^2) causes an increased vaporisation rate in the material. If one electron beam pulse is sufficient to pierce the workpiece, we can speak of perforation. Pulsation cutting with the electron beam is referred to as boring. Bores with a diameter of between 0.1 and 1.2 mm and a maximum depth of 8 mm can be produced by means of perforation. Bores with a maximum diameter/depth ratio of 1 : 30 may be produced up to a thickness of 20 mm. Ten holes with a diameter of 0.5 mm and 8 mm deep, or 700 holes with a diameter of 0.2 mm and 0.5 mm deep, can be produced in one second. The process can be used for metals and some non-metals.

1.2 Soldering and Brazing

J. Ruge, Munich, and **H. Wösle**, Brunswick

1.2.1 Procedure

Soldering and brazing means the joining of heated metals while in their solid state by means of molten filler metals

Table 7. Safe stresses in N/mm² when verifying operating strength (DIN 15018, extract)

Grade of steel / Stress group	St 37					St 52-3				
Notch value	K0	K1	K2	K3	K4	K0	K1	K2	K3	K4
Stress group — Safe stresses perm $\sigma_{D(-1)}$ for $\kappa = -1$										
B1			180	180	(152.7)	270	270	270	(254)	(152.7)
B2	180	180		(180)	108			(252)	180	108
B3			(178.2)	127.3	76.4	(237.6)	(212.1)	178.2	127.3	76.4
B4	(168)	(150)	126	90	54	168	150	126	90	54
B5	118.8	106.1	89.1	63.6	38.2	118.8	106.1	89.1	63.6	38.2
B6	84	75	63	45	27	84	75	63	45	27

The progressive ratio between the stresses in two consecutive stress groups amounts to 1.4142 between notch values K0 and K4 for St 37 and St 52-3. This does not apply in respect of the transition to the values in brackets.

(solders). The workpieces must have at least reached *working temperature* at the soldering joint. This is always higher than the lower melting point (solidus temperature) of the solder and may be lower than the upper melting point (liquidus temperature). Bonding between the workpiece and the solder also takes place if the workpiece has not reached working temperature, providing the solder has a much higher temperature. This workpiece temperature is often described as the *bonding temperature* or wetting temperature. It is always lower than the working temperature and is only of technical significance in braze welding.

To allow the liquid solder to wet and flow the surfaces of the workpiece must be metallurgically pure. Thick oxide film is removed mechanically and thin oxide film, which is produced as a result of heating to soldering temperature, is dissolved by fluxes or reduced by fluxes or gases.

The *bond* is dependent on the reactions between the solder and the base material and on the working temperature. In the event of insufficient alloying between the base material and solder adjacent to the clean surface bond, in most cases there is diffusion of one or more of the components of the solder into the base material and vice-versa. When hard-soldering mild steel copper often diffuses along the grain boundaries, which results in brittleness of the solder. The strength of the soldered joint is dependent on joint clearance. Below the minimum joint clearance (≈ 0.02 mm) the strength drops sharply owing to increasing lack of bonding. Conversely, increased joint clearance also brings with it a reduction in strength. The upper limiting value for joint clearance of approximately 0.5 mm should not, therefore, be exceeded. The optimum range has proved to be between 0.05 and 0.2 mm. Tool marks from turning or planing, if they exceed 0.02 mm in depth, should be in the flow direction of the solder.

1.2.2 Soft Soldering

Soft soldering is carried out mainly on steel, copper and Cu alloys at a working temperature below 450 °C. The solders are mainly alloys of lead, tin, antimony, cadmium and zinc; for aluminium materials: alloys of aluminium, zinc, tin and cadmium, if necessary with additions of aluminium; DIN 1707: Weichlote (equivalents: ISO 3677, ISO/DIS 9453) and DIN 8512 Parts 1 to 5: Hartlote.

Heating the Soldering Joint. Heat is provided by a heated copper work coil, a torch, in a furnace, by electrical resistance or in a bath of molten flux. In heavy metals the following are used to remove oxide film: fluxes based on metal chlorides, including zinc chloride and/or ammonium chloride plus organic acids (citric, oleic, stearic and benzoic acids), as well as amines, diamines and urea, natural or modified natural resins with additions of halogenated or halogen-free activated fluxes. It should be noted that flux residues can have either a corrosive or a non-corrosive effect. Care must be taken to make a suitable selection and choose a suitable finish, DIN 8511 Part 1 and Sheet 2: Flussmittel zum Weichlöten von Schwermetallen (equivalent: prEN 1045).

Strength of the Soldered Joint. Strength continues to be lost the longer the joint is subject to load, as soft solders tend to creep under load (**Fig. 27**). Experimental values are given in **Table 10**. Strength also decreases as temperature rises (**Fig. 28**).

1.2.3 Hard Soldering and Brazing

At temperatures over 450 °C. Filler metals: **Table 11**.

Standards. DIN 8513: Hartlote für Schwermetalle (Part 1: Kupferlote, Part 2: Silberhaltige Hartlote mit weniger als 20 Masse-% Silber, Part 3: Silberhaltige Hartlote mit mindestens 20 Masse-% Silber). DIN 8513 Part 4: Hartlote für Aluminium-Werktoffe. (See also the related prEN 1044.)

Heating the Joint. Heat is provided mainly by flame, in a furnace with a protective atmosphere, or by passing an electric current through it. Suitable fluxes for removing metal oxides are boron compounds and complex fluorides at an effective temperature between 550 °C and 800 °C, chlorides and fluorides without boron compounds between 600 °C and 1000 °C, boron compounds between 750 °C and 1100 °C and boron compounds, phosphates

Table 8. Equations for safe upper stress ranges and shear stresses for structural members and welded joints (DIN 15018)

a Equations for safe upper stress ranges as a function of and perm $\sigma_{D(-1)}$.

Range of alternating stresses $-1 < \kappa < 0$	Tension	$\text{perm } \sigma_{Dz(\kappa)} = \dfrac{5}{3 - 2\kappa} \cdot \text{perm } \sigma_{D(-1)}$
	Compression	$\text{perm } \sigma_{Dd(\kappa)} = \dfrac{2}{1 - \kappa} \cdot \text{perm } \sigma_{D(-1)}$
Range of fluctuating stresses $0 < \kappa < +1$	Tension	$\text{perm } \sigma_{Dz(\kappa)} = \dfrac{\text{perm } \sigma_{Dz(0)}}{1 - \left(1 - \dfrac{\text{perm } \sigma_{Dz(0)}}{0.75\, R_m}\right) \cdot \kappa}$
	Compression	$\text{perm } \sigma_{Dd(\kappa)} = \dfrac{\text{perm } \sigma_{Dd(0)}}{1 - \left(1 - \dfrac{\text{perm } \sigma_{Dd(0)}}{0.90\, R_m}\right) \cdot \kappa}$

b Safe stresses perm $\tau_{D(\kappa)}$ for structural members and welded joints.

Structural members	$\text{perm } \tau_{D(\kappa)} =$	$\text{perm } \sigma_{Dz(\kappa)}$ after W 0
	$\dfrac{\text{perm } \sigma_{Dz(\kappa)}}{\sqrt{3}}$	
Welded joint	$\text{perm } \tau_{D(\kappa)} =$	$\text{perm } \sigma_{Dz(\kappa)}$ after K 0
	$\dfrac{\text{perm } \sigma_{Dz(\kappa)}}{\sqrt{2}}$	

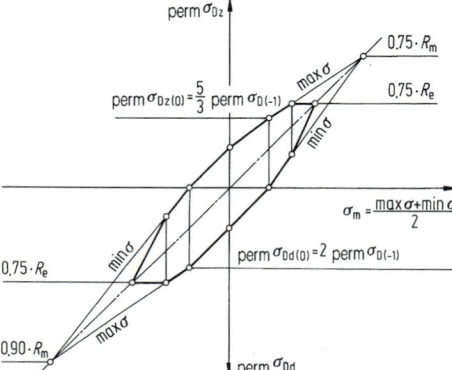

Figure 25. Interrelationships between perm $\sigma_{D(\kappa)}$ and perm $\sigma_{D(-1)}$.

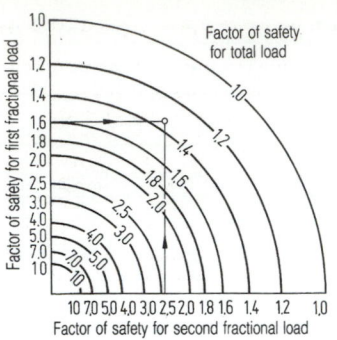

Figure 26. Safety circle to determine the overall factor of safety for combined stress after Thum and Erker.

Table 9. Standard values for safe stresses in upset and flash butt-welded joints

Type of stress	Weld machined	Weld unmachined	Observations
Static	0.9 to 1.0 perm σ	0.9 to 1.0 perm σ	Upset or flash butt welding
Dynamic	0.6 to 0.8 perm σ_a	0.6 to 0.8 perm σ_a	Upset butt welding
	0.8 to 0.9 perm σ_a	0.6 to 0.8 perm σ_a	Flash butt welding

perm σ_a = safe variable stress component of the parent material

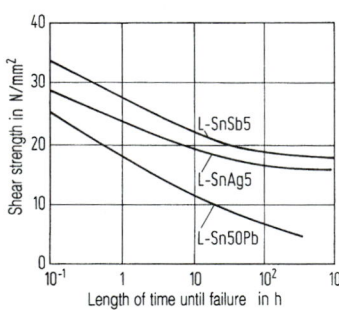

Figure 27. Shear strength of soft soldered joints after Zürn and Nesse.

and silicates above 1000 °C (DIN 8511 Part 1: Flussmittel zum Löten metallischer Werkstoffe; equivalent: prEN 1045).

Strength of the Joint. This very much depends on the base and filler materials, is slightly reduced depending on the filler material under extended periods of stress in relation to the short-time test and is strongly influenced by joint clearance, service temperature and number of cycles to failure when under cyclic loading. Reference value: endurance limit under reversed stresses 180 N/mm².

Table 10. Tensile and shear strengths of soft-soldered joints after Spengler

Symbols		Short-time loading (experimental values)		Permanent loading
New	Old	Tensile strength N/mm²	Shear strength N/mm²	Tensile strength N/mm²
	L Sn	10 to 20	5 to 15	0.5 to 1
L-Sn 60 Pb	L SnPb 38	35 to 45	25 to 35	—
L-Sn 50 Pb⎫	L SnPb 38-50 —	—	—	2 to 2.5
L-Sn 60 Pb⎭				
L-PbSn 40⎫	L PbSn 40-50 —	—	—	2 to 2.5
L-Sn 50 Pb⎭				
L-SnSb 5	L SnSb	40 to 50	30 to 40	4 to 6
L-SnAg 5	L SnAg 5	40 to 50	30 to 40	—
—	L CdZn	70 to 80	40 to 70	—
L-CdZnAg 5	L CdZnAg	80 to 90	70 to 80	—
	L SnCdZn	—	—	5 to 9

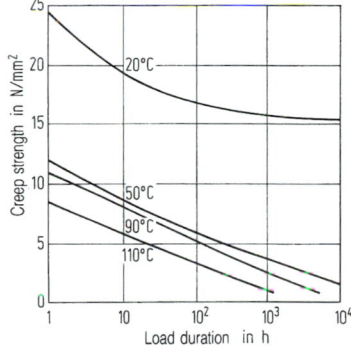

Figure 28. Creep strength of soft soldered joints after Haug.

1.3 Adhesive Bonding

J. Ruge, Munich, and **H. Wösle**, Brunswick

1.3.1 Uses and Procedures

Uses. Adhesive bonding makes it possible to join non-weldable materials without using rivets or screws. It is used to join metals to non-metals such as wood, plastic, rubber, glass and porcelain or in cases where welding the materials to be joined would cause adverse changes to their mechanical properties (e.g. precipitation-hardened duralumin). Thin workpieces in particular which can only be riveted or welded at great expense or not at all, may be joined together by adhesive bonding. Moreover, adhesive bonding of metals can offer engineering and economic advantages in mass production. Sandwich construction enables large-scale adhesive bonding of metals to take place in aircraft production, as in this way a high degree of rigidity can be combined with low mass.

Adhesiveness. In adhesive materials based on artificial resin this is mainly attributable to the adhesion between the material and the metal. Much less importance is attached to the mechanical bond produced by mechanical reinforcement. The following requirements must be met in order to produce satisfactory metal bonds: thorough and even wettability of the surfaces to be bonded by the

adhesive and internal stresses as low as possible after the adhesive has set, i.e. little tendency to shrink after setting. Residual stresses may result in the reduction of adhesive strength, especially where there is insufficient wettability at the same time. Further requirements are the lack of gas or air bubbles in the adhesive film and cleanliness of the components to be joined, i.e. freedom from dirt, grease and other impurities.

Surface Pretreatment of the components to be joined: degreasing and *mechanical pretreatment* by turning, planing, grinding or blasting with grease-free, fine-grain sand, corundum or wire shot for bonding iron, steel and non-ferrous metals.

Chemical Processes such as pickling with non-oxidising acid, etching with oxidising acid or *electrochemical treatment* produce greater adhesive strengths in aluminium and Al alloys, magnesium and Mg alloys, copper and Cu alloys than the mechanical processes.

1.3.2 Adhesive Materials

Epoxy resin adhesives are preferred over others for use in metal bonding. The adhesive may be cured either by a chemical reaction as in two-component adhesives or by a physical process such as evaporation of the solvent.

Polycondensation Adhesives are applied cold. After the components are joined they are subjected to a usually brief heating effect while under pressure to achieve a bond, while the chemical reaction of the condensation is taking place. Curing temperature and pressure must be determined by experiment in the absence of manufacturer's instructions. By adding hardeners condensation can also be effected without heat input.

Polymerisation Adhesives are solvent-free, reactive systems – heat input and the application of pressure have an accelerating effect and improve strength.

Polyaddition Adhesives cure without freeing cleavage products by means of an additive reaction. They can be hot or cold setting with no pressure. For a selection of adhesives with their conditions for use, see **Table 12**.

1.3.3 Strength of Bonded Joints

The strength of bonded joints is influenced by the mechanical properties of the materials to be joined and the adhesive, production requirements, geometric shape and type of stress.

Aluminium- and magnesium-based light metals are particularly well suited to adhesive bonding, non-ferrous heavy metals less so. Shear strength, i.e. the ratio of the breaking load to the bonding surface of a single-shear bonded joint, decreases as the yield point or elongation limit of the metal grows and the thickness of the adhesive film increases, **Fig. 29**.

The *strength of the adhesive* is a function of its structure and the conditions under which it is used (Table 12).

The *geometric shape* of the joint has a considerable influence on strength. The *single-shear joint* (**Fig. 30**) produces lower shear tensile strengths than the *double-shear joint* owing to additional bending and the associated tendency to peel off, whereas the *scarfed* joint, owing to the even distribution of shearing stresses in the bonded joint, obtains the highest values (**Fig. 31a**), which however reduce as the length of the overlap is increased. On the other hand the shear tensile strength of the single-shear bonded joint increases in proportion to plate thickness at a constant overlap length l_0 until it reaches a limiting value as the flexural rigidity of the plate is also increasing (**Fig. 31b**).

Shear tensile strength is a function of the overlap ratio $\ddot{u}$ = length of overlap l_0/plate thickness s. Increasing $\ddot{u}$ beyond an optimum

Table 11. Brazing solders (DIN 8513, selection)

Codes 8513 1734	Standardised in DIN 8513, Part	Working temperature (°C)	Steel	Special steel	Malleable cast iron	Cast iron	Copper	Copper alloys	Nickel	Nickel alloys	Precious metals	Hard metals	Tungsten and molybdenum materials
L-Ag 40 Cd	3	640	×		×		×	×	×	×			
L-Ag 50 Cd	3	640		×				×			×		
L-Ag 30 Cd	3	680	×		×		×	×	×	×			
L-Ag 49	3	690										×	×
L-CuP 8	1	710					×						
L-Ag 15 P	2	710					×	×					
L-Ag 2 P	2	710					×	×					
L-Ag 60	3	710									×		
L-Ag 20 Cd	3	750	×		×		×	×	×	×			
L-Ag 25	3	780	×		×		×	×	×	×			
L-Ag 72	3	780					×	×	×	×			
L-Ag 12	2	830	×		×		×	×	×	×			
L-Ag 83	3	830									×		
L-Ag 27	3	840										×	×
L-ZnCu 42	1	845						×					
L-Ag 5	2	860	×		×		×	×	×	×			
L-CuZn 39 Sn	1	900	×		×	×	×	×	×	×			
L-CuZn 40	1	900	×		×		×	×	×	×			
L-Ag 85	3	960	×						×	×			
L-CuSn 12	1	990							×				
L-CuSn 6	1	1040							×				
L-SCu	1	1100	×										
L-Cu	1	1100	×										

value ceases to be of any advantage, which can be attributed to the stress concentrations occurring at the ends of the overlap.

The formula for calculating dimensions is $\ddot{u} = l_0/s < 30$. Standard value: $\ddot{u} = 20$.

After ten weeks of weathering, shear tensile strength drops to about 70%; it is only dependent on temperature from about 80 °C, above which it falls away sharply. Bonded joints are also subject to dynamic stress. A pronounced fatigue strength as in metals is still absent up to 10^8 cycles (**Fig. 32**).

Table 12. Basic synthetic materials for the adhesive bonding of steel (Advisory Service for the Application of Steel, Guideline 382)

	Setting conditions	Strength	Ductility	Ageing stability	Heat resistant up to °C
1. Epoxy resin, 2 K	20 °C, no pressure	1/2	2	3	60 to 80
2. Epoxy resin, 1 K	120 °C, no pressure	1	2	2	200
3. Phenolic resin, 1 K	150 °C, 0.8 N/mm²	2	3	1	200
4. Polyurethane resin, 2 K	20 °C, no pressure	2/3	1	3	60 to 80
5. Copolymers, 2 K	20 °C, no pressure	2/3	2	2/3	60 to 80
6. Epoxy phenolic resin, 1 K	150 °C, 0.8 N/mm²	1	2/3	1/2	Up to 250
7. Epoxy nylon resin, 1 K	150 °C, 0.05 N/mm²	1	1	1/3	80
8. Polyimide resin, 1 K	180 °C, 0.5 N/mm²	2/3	3	1	Up to 400
9. Cyanate resin, 1 K	180 °C, no pressure	2/3	3	2	Up to 200
10. Experimental epoxy products	170 °C, no pressure	2/3	3	2	Up to 250

Quick-setting adhesives

11. Cyanoacrylate, 1 K	RT, no pressure	2	3	3	80
12. Diacrylic acid ester, 1 K	RT, no pressure	2	3	3	80 to 120

Naturally setting adhesives

13. PVC paste, 1 K	150 to 250 °C, no pressure	3/4	2	2	80 to 100
14. Hot-melt adhesives, 1 K	Above 100 °C, contact pressure	3/4	1	2	80 to 150

1 very high, 2 high, 3 medium, 4 low, 1 K one-component, 2 K two-component, RT room temperature

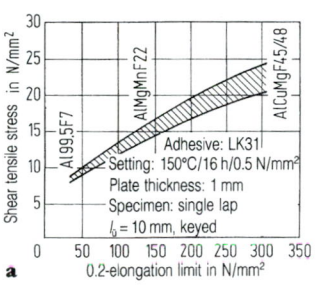

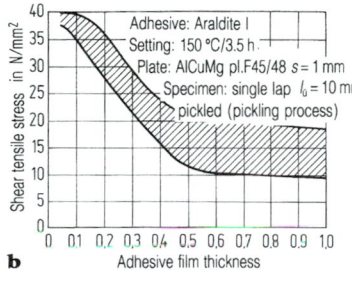

Figure 29. Shear strength of bonded joints: **a** as a function of the elongation limit in light metals, **b** as a function of the thickness of the adhesive film.

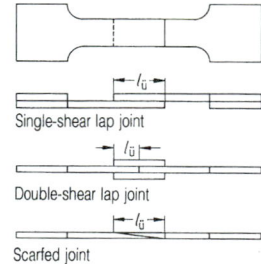

Single-shear lap joint

Double-shear lap joint

Scarfed joint

Figure 30. Forms of joint in bonded joints (test specimens).

1.4 Connections with Force Transmission by Friction

H. Mertens, Berlin

1.4.1 Types, Uses

Connections with force transmission by friction (1–39) are used first and foremost as shaft–hub assemblies with cylindrical or tapered working surfaces to transmit torque between shaft and hub with and without spacers (**Fig. 33**) or to introduce axial forces into axles or stub heads (e.g. **Fig. 34**). In addition to transmitting force – tight-fitting in operation and slipping when overstressed, with basic factors as described in A1.11 – the self-centring

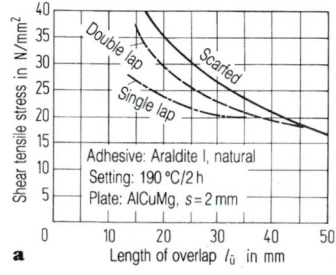

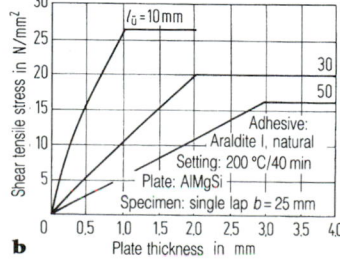

Figure 31. Shear tensile strength of bonded joints depending on **a** overlap length; **b** plate thickness and overlap length.

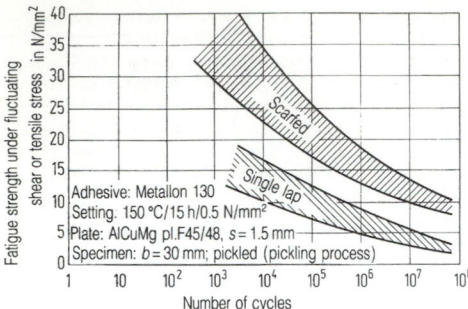

Figure 32. Fatigue strength of bonded joints under fluctuating shear tensile stresses.

facility, adjustability about the circumference, manufacturing and assembly costs, the required process tolerances, ease of disengaging or re-use as the case may be, are all factors to be taken into account when selecting these joints. Cylindrical interference fits are difficult to disengage as shown in **Fig. 33d**; interference fits with tapered working surfaces as shown in **Fig. 33e** are easier to disengage and joints with spacers are easier to engage and disengage. The clamp joint as shown in **Fig. 33b**, the joint with a flat or saddle key as shown in **Fig. 33c**, the annular spring tensioning device as shown in **Fig. 33i** and the joint with star wheels as shown in **Fig. 33j** and with shaft clamping sleeve as shown in **Fig. 33m** are not self-centring. Interference fits (longitudinal, transverse, tapered interference fits) must be manufactured to a high

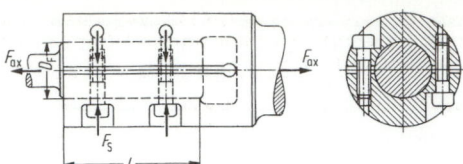

Figure 34. Axially (longitudinally) loaded cylindrical clamp joint ($z = 4$).

degree of accuracy, hydraulic hollow sheathed clamping sleeves less so [38]. For a selection of shaft–hub assemblies with design data see [10], manufacturers [39].

Connections with force transmission by friction with *smooth working surfaces* are frequently used today instead of riveted joints to transmit force between plates in steel and crane construction as *friction grip joints with high-tensile screws* (GV joints) [4]. Friction is also used to transmit frequently occurring working loads in rigid non-clutch type *shaft flange couplings* [22].

1.4.2 Interference Fits

Design Calculations. These are carried out as described in DIN 7190 for cylindrical interference fits to calculate the maximum safe turning moment M_t or the maximum safe axial force F_{ax} for two concentric rings having the same axial length l_F without in the first instance allowing for centrifugal forces; these calculations can also be used by approximation for clamping joints as shown in **Fig. 34**. The calculations should ensure that the minimum *contact pressure p* produced by the minimum effec-

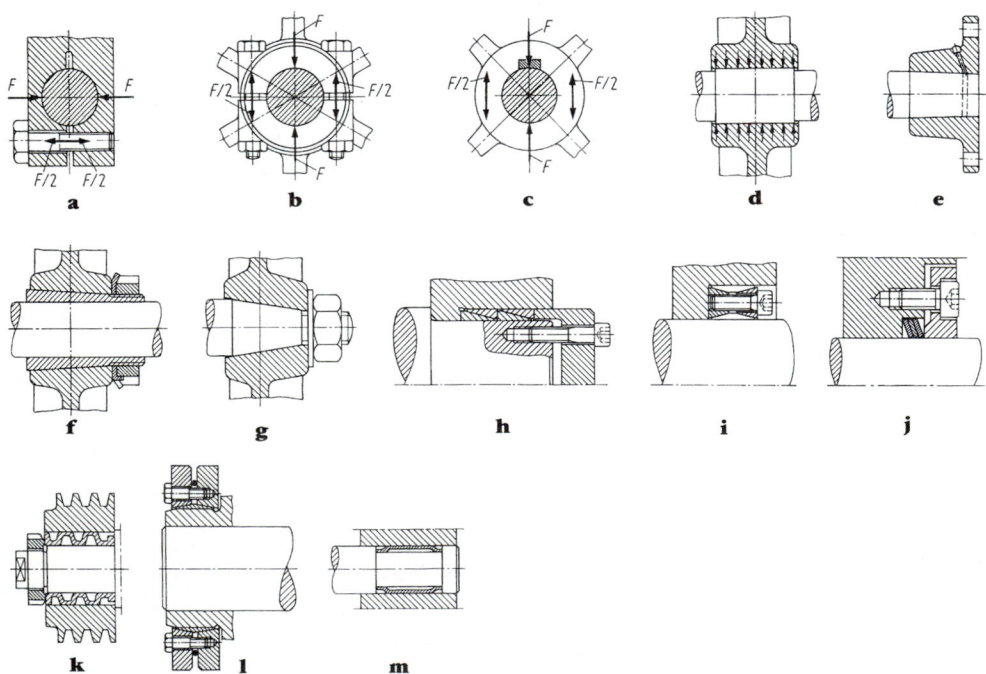

Figure 33. Friction-grip joints as in [54]: **a** clamp joint with split hub, **b** with divided hub, **c** with saddle key, **d** cylindrical interference fit, **e** oiled interference fit, **f** interference fit with tapered spring collet, **g** cone interference fit, **h** bracing joint with tapered straining rings (after Ringfeder), **i** bracing assembly (after Ringfeder), **j** star wheel (after Ringspann), **k** shaft clamping sleeve (after Spieth), **l** shrink fitted joint (after Stüwe), **m** shaft clamping sleeve (after Deutsche Star).

Table 13. Coefficients of friction for transverse interference fits in the longitudinal and circumferential directions during slippage (as in DIN 7190) for design calculations

Material combination	Lubrication, assembly	Coefficients of friction $\mu_{rb}\ \mu_{ru}$
Steel/steel combinations	Process A	0.12
	Process B	0.18
	Process C	0.14
	Process D	0.20
Steel/cast iron combinations	Process A	0.10
	Process B	0.16
Steel/MgAl combinations	Dry	0.10 to 0.15
Steel/CuZn combinations	Dry	0.17 to 0.25

Process A: Standard hydraulic fits joined using mineral oil
Process B: Hydraulic fits with degreased fit surfaces (joined using glycerine)
Process C: Standard shrink fit after heating outside component to 300 °C in electric furnace
Process D: Shrink fit with degreased fit surfaces after heating to 300 °C in electric furnace

tive *interference* $|P_w|$ between the shaft diameter and the hub bore brings about the required *adhesive force (friction force)* and the *contact pressure* $\hat{p}$ effected by the maximum interference $|P_w|$ does not cause the safe stresses or elongations for the member to be exceeded; contact pressure is expressed as $\check{p} \leq p \leq \hat{p}$.

The minimum contact pressure required to transmit M_t is $p_{min} = 2M_t S_r/(\pi D_F^2 l_F \mu_{ru})$, with theoretical factor of safety S_r for slippage, adhesion correction value μ_{ru} or μ_{rl} where there is circumferential or longitudinal slippage as in **Table 13**, joint diameter D_F after assembly (using nominal dimension for calculation), length of joint l_F.

For *interference fits under elastic stress* alone without allowing for end pressures, the relative effective interference is generally $\xi_w = |P_w|/D_F$ and at the same time $\xi_w = K p/E_A$ with the auxiliary size (index A or I for external and internal parts respectively):

$$K = \frac{E_A}{E_I}\left(\frac{1 + Q_I^2}{1 - Q_I^2} - \nu_I\right) + \frac{1 + Q_A^2}{1 - Q_A^2} + \nu_A.$$

E_A and E_I are moduli of elasticity, $Q_A = D_F/D_{aA}$ and $Q_I = D_{iI}/D_F$ are diameter ratios, and ν_A and ν_I are Poisson's ratios ($\nu \approx 0.3$ for St; $\nu \approx 0.25$ for GG20 to GG25).

As a result of the smoothing of points of roughness during assembly the effective *interference* $|P_w|$ is smaller than the actual fit $|P_i|$ measurable before assembly, which on the basis of deviations from the drawing with regard to the shaft diameter and the hub bore lies within the limits $|\check{P}|$ and $|\hat{P}|$; $|\check{P}| \leq |P_i| \leq |\hat{P}|$. If no experimental values are available longitudinal and transverse interference fits are represented by $|P_w| = |P_i| - 0.8(R_{zA} + R_{zl})$, R_{zA} or R_{zl} being the average height of irregularities on the mating surfaces. If allowance is made for the average deviation from the mean line R_a the average values of the ten-point height of irregularities R_z as determined in Appendix 1 of DIN 4768 Part 1 may be used for this. Owing to $\xi_w = |P_w|/D_F = Kp/E_A$ and $|\hat{P}_w| = |\hat{P}| - 0.8(R_{zA} + R_{zl})$ the effective interference $|P_w|$ and contact pressure $\check{p}$ are determined at a given fit p or the effective interference $|\hat{P}_w|$ or the fit $|P|$ at a given contact pressure $\hat{p}$ can be calculated once the principal deviations of the outer and inner ring have been established. Correspondingly $\xi_w = |\hat{P}_w|/D_F = K\hat{p}/E_A$ applies, so that the maximum contact pressure $\hat{p}$ at a given fit $|P|$ is known. For maximum heights of irregularities obtainable according to manufacturing process see DIN 4766 and **E6 Table 1**.

The maximum radial stress $\sigma_r = -\hat{p}$ occurs at the mating surfaces of the outer and inner component (**Fig. 35**), the maximum circumferential stress in the outer ring again amounts to $\sigma_{\varphi A} = (1 + Q_A^2)\hat{p}/(1 - Q_A^2)$, the maximum tangential stress in the inner ring amounts to $\sigma_{\varphi l} = -2\hat{p}/(1 - Q_l^2)$ for $Q_l > 0$ and is at the inner edge or $\sigma_{\varphi l} = -\hat{p}$ everywhere for a solid shaft with $Q_l = 0$. According to the shear stress theory (SST) the maximum equivalent tensile stresses are set up in the outer ring in relation to $\sigma_V = 2\hat{p}/(1 - Q_A^2)$, in the inner ring with $Q_l > 0$ in relation to $\sigma_V = 2\hat{p}/(1 - Q_l^2)$ or the solid shaft in relation to $\sigma_V = \hat{p}$. In DIN 7190 these equivalent tensile stresses are equated with toughness parameters ($2R_{eLA}/\sqrt{3}$) or ($2R_{eLl}/\sqrt{3}$) (modified SST), which are established with the lower elastic limits R_{eL} of the outer and inner component; e.g. $2\hat{p}/(1 - Q_A^2) \leq 2R_{eLA}/(\sqrt{3} \cdot S_{PA})$ with the theoretical factor of safety S_P for plastic elongation. There is a corresponding evaluation for inner ring or solid shaft, using a flow chart for dimensioning elasticity [10].

For ductile materials with an ultimate elongation $A \geq 10\%$ and a percentage reduction in DIN 7190 for calculating *interference fits under elastic/plastic stress* for solid shafts and $E_A = E_I = E$ as well as $\nu_A = \nu_I = \nu$. An internal plastic region is formed in the outer component, which is separated from an external residual elastic region by a cylinder surface with a plasticity diameter D_{PA} (**Fig. 35**). The relative plasticity diameter $\zeta = D_{PA}/D_F$ is determined by solving the transcendental equation $2\ln\zeta - (Q_A\zeta)^2 + 1 - \sqrt{3} \cdot p/R_{eLA} = 0$, where $1 \leq \zeta \leq 1/Q_A$

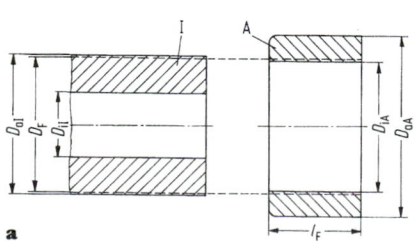

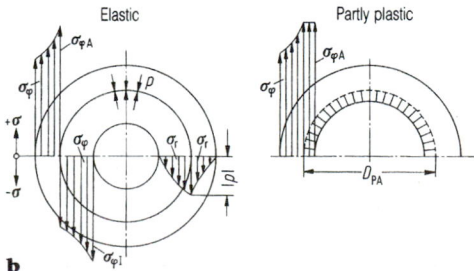

a   **b**

Figure 35. Stress distribution in elastic interference fits: **a** before assembly, **b** after assembly, σ circumferential and σ_r radial stresses and p contact pressure; hub elastic or partially plastic after joining.

must apply. The relative effective interference $\xi_w = |P_w|/D_F$ required for contact pressure p is produced in relation to $\xi_w = 2\zeta^2 R_{eLA}/(\sqrt{3} \cdot E)$. Finally the portion of the ring surface under plastic stress q_{PA} remains to be checked against the complete cross-section q_A of the outer component, with $q_{PA}/q_A = (\zeta^2 - 1)Q_A^2/(1 - Q_A^2) \leq 0.3$ for high-stress interference fits in machine construction. Checking whether a solid shaft remains under only elastic compression p, takes place in the same way as with interference fits under elastic stress. Checking for full plastic stress in the outer component with $p \leq 2R_{eLA}/(\sqrt{3} \cdot S_{PA})$ for $Q_A < 1/e = 0.368$ or $p \leq -2R_{eLA}(\ln Q_A/(\sqrt{3} \cdot S_{PA})$ for $Q_A > 0.368$ with theoretical factor of safety S_{PA} for full plastic stress. For flow diagrams see DIN 7190.

It is best for *assessment of the fatigue strength* of shaft–hub assemblies to take place when the nominal stress amplitudes and the associated mean stresses due to bending and torsion in the shaft are being calculated, taking into account any test results on similar shaft–hub assemblies. A suitable method of calculation, as shown with slight modifications in [10], is given in GDR Standard TGL 19 340. Fatigue notch factors β_{kb} for bending and β_{kt} for torsion have been introduced in order to document test results (see D1.5.2). An initial overview is provided in **Table 14**.

Similar fatigue notch factors must also be adopted for comparable tapered interference fits and commercially available frictionally engaged shaft–hub assemblies with spacers [23–39] (**Fig. 33h** to **m**). For a summary of fatigue notch factors see [10, 12].

Macrostructure. As a rule $l_F/D_F \leq 1.5$, if static torque stress is allowed for, as greater lengths hardly ever result in increased slipping moments. In the case of alternating or continuous bending moments, $l_F/D_F \geq 0.5$ with if possible solid inner components, in order to avoid axial excursion of the shaft from the hub due to microscopic slippage. In order to enable it to withstand high amplitudes of torque a solid shaft should if possible be mated with a hub which is not too thin-walled ($Q_A \leq 0.5$). The greatest possible gain in contact pressure p in opposition to the purely elastic extension is produced in the range $0.3 \leq Q_A \leq 0.4$. The optimum forms of interference fit for alternating or continuous bending moment are achieved by increasing the diameter of the shaft D_w to the diameter of the joint D_F as in $D_F/D_w \approx 1.1$ to 1.15 with transitional

radius r as in $r/D_F \approx 0.22$ to 0.18, and the appropriate limiting value must be selected for high-tensile shaft materials [11]. If the shaft cannot be stepped, suitable circular shaft indentations can be used with the hub projecting slightly. On no account, however, should grooves or indentations be provided inside the interference fit, e.g. for parallel keys. In the event that shaft and hub are made from materials with dissimilar elastic constants, the shaft should have the larger modulus of elasticity ($E_I > E_A$).

Note. Hydraulically assembled fits may only be subjected to stress after the oil film has decomposed (10 min to 2 h). Joining temperatures are a maximum of 350 °C for hubs made of structural steel with low fatigue strength, cast steel or nodular cast iron and a maximum of 200 °C for hubs made of high-temper structural steel or case-hardened steel (DIN 7190).

Macrostructure of Tapered Interference Fits. Construction is as shown in **Fig. 33g**. The amount of taper (relative to diameter as described in DIN 254) must in any case [21] be selected so as to permit self-locking, and therefore be less than or equal to $1 : 5$ in steel–steel combinations. As the outer component performs a helical restraining motion caused by torque when first subjected to load the effective coefficient of friction is practically eliminated in the axial direction. For this reason tapered interference fits, which must withstand greater torque, have to be axially braced as otherwise even a "self-locking" interference fit will loosen momentarily if the maximum permissible torque is exceeded. Feather or Woodruff keys, which are used to secure a position on the circumference in tapered interference fits, e.g. DIN 1448 and DIN 1449, prevent the helical restraining motion, so that the contact pressure cannot be used fully for transmitting torque: no feather or Woodruff keys should therefore be provided in tapered interference fits under heavy load. Assembly procedures combining torque stress and tightening have been proposed in order to avoid the restraining motion and variable circumferential positions during operation [21]. Rough calculation can be made assuming a cylindrical interference fit with average joint diameter D_{Fm} and axial joint length l_F. The minimum contact pressure required to transmit M_t is $p_{min} = 2M_t S_r/$ $[\pi D_{Fm}^2 (l_F/\cos \beta) \cdot \mu_{ru}]$ with the theoretical factor of safety S_r for slippage and the angle of taper $\alpha = 2\beta$. The necessary assembly force $F_e \geq p_{min}D_{Fm}\pi l_F (\tan \beta + \mu_{rt})$; the loosening force prior to stressing by torque M_t follows with negative μ_{rt}. The required restraining method is determined by the minimum required interference $|P_w|$ and the maximum permissible interference $|P_w|$, taking into account the angle of taper $\alpha = 2\beta$. For calculations with regard to the angle deviation between the inner and outer component see [10, 21].

Microstructure. Interference fits are often subjected to alternating or fluctuating torsion and/or continuous bending while in operation. Moments of cyclic stress may cause sliding motions (*slippage*) in the joint in alternating directions. As slippage increases the service life of some frictionally engaged mating components is sharply reduced [14]. In accordance with the *principle of harmonic deformation* relative displacements between shaft and hub can be reduced by a suitable force guide and hub design, e.g. under torsional strain as shown in **E4, Fig. 4**. Joint pressure and relative displacement can be accurately determined using finite-element calculations as shown in [11]. Interference fits with low notch effect and high strength are produced if the design ($D_F/D_w \geq 1.1$), manufacture (hub elastic/plastic) and thermal treatment

Table 14. Stress concentration factors for interference fits (as in GDR Standard TGL 19 340) with joint diameter $D_F = 40$ mm and transmission of moments [10]

Hub type	Fit	Stress concentration factor (D=40mm)	R_m in N/mm²								
			400	500	600	700	800	900	1000	1100	1200
	H8/u8	β_{kb}	1.8	2.0	2.1	2.3	2.5	2.7	2.8	2.8	2.9
		β_{kt}	1.2	1.3	1.4	1.5	1.6	1.7	1.8	1.8	1.9
$r/D \geq 0.06$	H8/u8 Hardened steel hub	β_{kb}	1.6	1.7	1.8	1.9	2.0	2.1	2.2	2.3	2.3
		β_{kt}	1.0	1.1	1.2	1.2	1.3	1.4	1.4	1.5	1.5
	H8/u8	β_{kb}	1.5	1.6	1.7	1.8	1.9	2.0	2.1	2.1	2.2
		β_{kt}	1.0	1.0	1.1	1.2	1.3	1.3	1.4	1.4	1.5
$r/D = 0.5$	H8/u8	β_{kb}	1.0	1.0	1.1	1.1	1.2	1.3	1.3	1.4	1.4
		β_{kt}	1.0	1.0	1.0	1.0	1.1	1.1	1.2	1.2	1.2

(inductive surface hardening, carburising or gas nitriding) are carefully selected or coordinated as shown in [10] on the basis of fatigue strength tests fully backed up by statistics.

The relatively low notch effect in elastically–plastically assembled transverse interference fits, as compared with those assembled elastically, confirms the action of frictional fatigue strength as described in [14, 15]. Thus the highest possible contact pressure should be selected in order to avoid relative displacements; this restricts the measures which can be taken to accommodate torsional rigidity, as shown in **E4, Fig. 4** when additional torsion effects are present. The optimum structure is thus a function of the ratio of the amplitude of the bending moment to be transmitted M_{ba} to the amplitude of the torsional moment M_{ta}. The interaction formula

$$\left[\frac{M_{ba}}{(M_{ba})_{acc}}\right]^2 + \left[\frac{M_{ta}}{(M_{ta})_{acc}}\right]^2 \leq \frac{1}{S_D}$$

can be used to make a judgement [11], if the *acceptable* amplitudes of the bending and torsional moments $(M_{ba})_{acc}$ and $(M_{ta})_{acc}$ in respect of the static portions of the moments are known from tests; S_D is the factor of safety for fatigue fracture. In practice a fractional load often dominates because of the quadratic relationship, with the result that design measures can then be directed towards the principal stress components.

If an interference fit is subjected to additional stress by centrifugal forces it may be necessary, owing to additional expansion of the hub in particular, to refine calculations in order to determine the contact pressure. For simplified computation see DIN 7190 or [20].

1.4.3 Clamp Joints

Easily removable clamp joints are created at their simplest, when adjacent parts are pressed together by means of screws. Such unifacial, level clamp joints are also used in a variety of ways to fasten slideways as shown in **L1, Fig. 53**. Clamp joints are employed in steel and crane construction as friction-grip joints using high-tensile screws.

As described in DIN 18 800, the screws used in clamp joints must be systematically prestressed in accordance with DIN 1000. In this way force can be transmitted by friction perpendicular to the axis of the screws in specially pretreated contact surfaces of the parts to be joined. When used with high-tensile dowel screws, force is transmitted at the same time by shear and pressure on the face of the hole (see F1.5). Friction-grip joints should be executed with a hole clearance $d \leq 2$ or 3 mm (clamp joints) and with a hole clearance $d \leq 0.3$ mm (clamp joints with dowel screws). In DIN 18 800, the permissible transferred force allowable Q_{GV} per friction surface perpendicular to the axis of the screw is set out for screw sizes M12 to M36 as proof of strength. A coefficient of friction $\mu = 0.5$ is used for calculation purposes with a factor of safety S_G for slippage (principal stresses) with the prescribed friction surface treatment (blasting with cast steel chips or flame descaling twice or sand blasting or surfacing with a coating of anti-slip material). For component cross-sections which are weakened because of holes, it may be assumed in general stress detection that 40% of allowable Q_{GV} of high-tensile screws in the given cross-section lie within the area deducted for holes, and have been locked by frictional engagement before the initiation of hole weakening. In addition it must be demonstrated that the solid cross-section bears the total force.

Clamp joints with *cylindrical* working surfaces as shown in **Fig. 34** or **Fig. 36 (Fig. 33a)** with a split

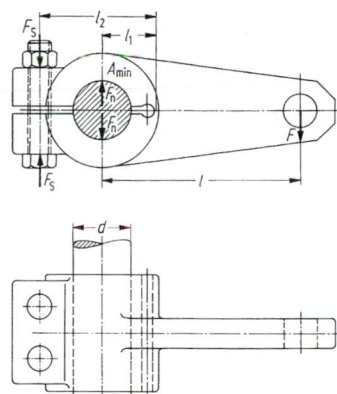

Figure 36. Clamp joint with split lever loaded by moments.

hub (lever) or **Fig. 33b** with a divided hub transmit torques M_t or axial forces F_{ax} in the same way as press fits (F1.4.2) if there is a transition fit rather than a clearance fit in the unclamped condition. If there is a clearance fit, however, a linear contact is present. If there is a divided hub the bolt forces and the contact forces set up must be in equilibrium, but if there is a split hub (lever), greater bolt forces must be selected owing to the statically indeterminate distribution of force into contact forces and hub deformation when there are uniform forces and moments to be transmitted, allowance for which may be made within the design calculation by increasing factors of safety S over and above the required factor of safety S_G $(S > S_G)$.

Design Calculation *For a clamp joint as shown in* **Fig. 34** *with z bolts and preload F_s for each bolt and linear contact.* Transmittable longitudinal force $F = 2\mu z F_s/S$. If the coefficient of friction μ can be reduced by combined oscillating motions or impacts, the coefficient of friction of motion μ_r may be selected. For reference values see **Table 13**. If it can be assumed that, instead of linear contact, a uniformly distributed unit pressure p is created in a fit with no clearance over the bore circumference πd and the grip length l, then the transmittable longitudinal force is $F = \pi \mu F_s/S$. The coefficient of friction μ may be increased by suitable surface treatment, by use of carborundum powder in the joint or non-metallic spacers adhesive-bonded or riveted on one side.

For a split lever as shown in **Fig. 36**. Transmittable torque $M_t = F_n \mu D_F/S$ with locally concentrated clamping forces F_n at two diametrical points on the pressed joint; for coefficient of friction μ and factor of safety S see the previous paragraph (clamp joint). As an initial approximation, $F_n = (l_2/l_1) \cdot F_s$ applies if the lever is sufficiently split. The cross-section A_{min} of the lever is subjected to flexural stress by a bending moment $M_b \approx F_s (l_2 - l_1)$. This flexural stress, as well as that at the end of the split and that in the locking screws, is reduced if the distance $(l_2 - l_1)$ is kept as short as possible and a transition fit rather than a clearance fit is present in the joint in its unclamped condition. Such clamp joints are used only to transmit torques which are slight and not subject to much fluctuation. They have the advantage that the position of the lever or hub can be easily altered longitudinally or circumferentially.

For clamp joints with eccentric application of force as

*shown in **Fig. 37.*** In order to calculate the self-locking limit it is assumed that the bending moment (kF) and the longitudinal force F are taken up by locally concentrated forces F_{res} in the centrelines of the friction cone at the limits of the hub at distance b. The requirement for secure clamping under static force F is $k \geq b/(2\mu_{ri})$, where $\mu_{ri} = 0.07$ for St/St $k \geq 7.0b$. Clamping can, however, take place as soon as $k \approx 2b$ when $\mu_{ri} = 0.16$ and application of the resulting normal combination of forces in the bore at distance $(2/3)b$. In order to calculate the unit pressure a linear unit pressure distribution as in **Fig. 39** is assumed: $p_{perm} = 50$ to 90 N/mm² for St/St combination and $p_{perm} = 32$ to 50 N/mm² for St/GG apply as standard value for permissible unit pressures.

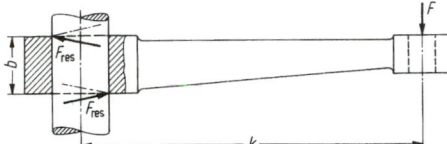

Figure 37. Longitudinally loaded clamp joint with force applied eccentrically.

1.5 Positive Connections

H. Mertens, Berlin

1.5.1 Types, Uses

The simplest fasteners in mechanical engineering are pins, bolts, screws, parallel keys, Woodruff keys and taper keys [41–72]. They are used for securing components together, flexible joints and bearings and for transmitting force. The joints are created by engaging the contours of parts of the fasteners. If the fasteners are integrated into components, positive connections are created which are more costly to manufacture but generally more accurate and capable of bearing greater stresses, such as keying and spline joints between shaft and hub toothing for connecting shafts to each other. It is generally possible to dismantle these joints using only a small expenditure of force in the preferred directions. Positive connections that have not been prestressed have a very low dynamic load capacity, owing to the unfavourable magnetic flux and relatively high stress concentrations. The static load capacity must on the other hand be considered much more favourable if suitable materials are selected, so that in practice combinations of friction-tight joints are created for frequently occurring working loads and positive connections for infrequent heavy loads, e.g. rigid shaft-flange joints using bolts, screws and pins. In positive connections riveted joints, the dismantling of which can be effected, for example, by reboring the rivet, can be treated as a special case.

1.5.2 Pinned and Taper-Pinned Joints

Pins for the positive connection of hubs, levers and adjusting rings to shafts or axles and for locking screwed components and as guide pins (fasteners subject to bending stress and restrained on one side to introduce force into, among other things, coil springs and traction ropes) are driven into bores using longitudinal press fits and interference [56]. Bores for straight pins are reamed to fit; bores for spring dowel sleeves (clamping sleeves) are produced using H12 and for grooved straight pins H11 is gen-

erally used. Taper pints in bores which have been reamed together before assembly provide the best fastening. The tolerance zones for the diameters of straight pins (DIN 7) are differentiated by the end shape of the pin (**Fig. 38**). For manufacturers' literature see [67].

Standards. DIN 1: Kegelstifte. DIN 7: Zylinderstifte. DIN 258: Kegelstifte, mit Gewindezapfen und konstanten Kegellängen. DIN 1469: Passkerbstifte mit Einfuhr-Ende. DIN 1471: Kegelkerbstifte. DIN 1472: Passkerbstifte. DIN 1473: Zylinderkerbstifte. DIN 1474: Steckkerbstifte. DIN 1475: Knebelkerbstifte. DIN 1476: Halbrundkerbnägel. DIN 1477: Senkkerbnägel. DIN 1481: Spannstifte (Spannhülsen), schwere Ausführung. DIN 6325: Zylinderstifte, gehärtet, Toleranzfeld m6. DIN 7343: Spiral-Spannstifte, Regelausführung. DIN 7344: Spiral-Spannstifte, schwere Ausführung. DIN 7346: Spannstifte (Spannhülsen), leichte Ausführung. DIN 7977: Kegelstifte, mit Gewindezapfen und konstanten Zapfenlängen. DIN 7978: Kegelstifte mit Innengewinde. DIN 7979: Zylinderstifte mit Innengewinde.

Guide pins as shown in **Fig. 39** are mainly subjected to bending stress in the clamping cross-section with bending moment $M_b = Fl$. If a linear unit pressure distribution between pin and bore (rigid pin) is assumed, a maximum pressure $p_{max} = p_d + p_b = F(4 + 6l/t)/(dt)$ is calculated in addition to the unit pressure due to interference. This is a more accurate model for calculation than a bedded-in girder with shear deformation [63]. Similar considerations permit the unit pressure p_{max} to be computed between a transverse pin and a shaft in a shaft–hub assembly under torsional moment M_t as shown in **Fig. 45a** at $p_{max} = 6M_t/(dD^2)$. For standard values for permissible unit pressures of pinned joints see **Table 15**, for those of stresses **Table 16**.

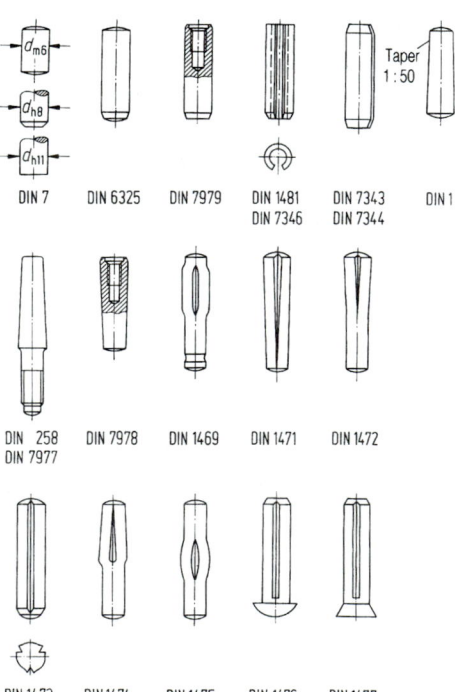

Figure 38. Standardised dowel pins (selection).

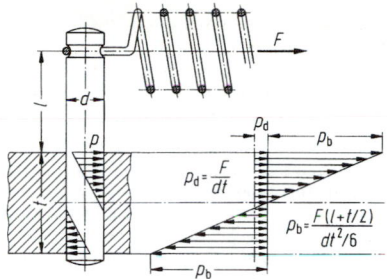

Figure 39. Socket joint under transverse load with linear distribution of unit pressure.

Table 15. Standard values for permissible unit pressures in clevis and pinned joints

p_{perm} in N/mm² for material combination	Interference fits[a]			Free fits[b]
	Static load	Dynamic load	Fluctuating load	
St 50 K/GG 9 S 20/GG	70	50	32	5
St 50 K/GS 9 S 20/GS	80	56	40	7
St 50 K/red brass, bronze	32	22	16	8
St hardened/red brass, bronze				10
St 50 K/St 37	90	63	45	
St 50 K/St 50	125	90	56	
St hardened/St 60	160	100	63	
St hardened/St 70	180	110	70	
St hardened/St hardened				16

[a]Bearing portion of grooved pin 70%.
[b]For knuckle joints.

1.5.3 Clevis Joints and Pivots

Standardised clevis and pivot pins as shown in **Fig. 40** with diameters (3, 4, 5, 6), 8, 10, 12, 14, 16, 18, 20, 24 up to 100, have many applications as axle and hinge pins using the single degree of freedom system (see **Fig. 41**).

Standards. DIN 1443: Bolzen ohne Kopf, Masse nach ISO. DIN 1444: Bolzen mit Kopf, Masse nach ISO. DIN 1445:

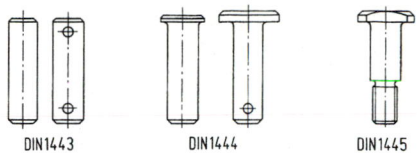

DIN 1443   DIN 1444   DIN 1445

Figure 40. Standardised clevis pins (selection).

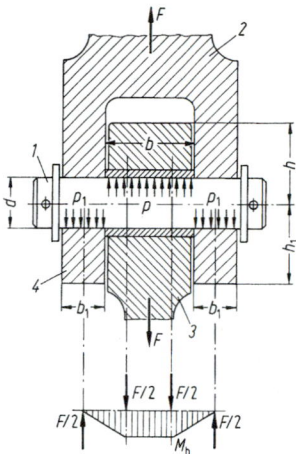

Figure 41. Clevis joint used as a knuckle joint (with simplified moment distribution as a basis for calculations): *1* clevis pin, *2* fork, *3* rod, *4* cover plate.

Bolzen mit Kopf und Gewindezapfen. Nicht mehr für Neukonstruktionen verwenden. DIN 1433: Bolzen ohne Kopf, Ausführung m. DIN 1434: Bolzen mit kleinem Kopf, Ausführung m. DIN 1435: Bolzen mit kleinem Kopf, Ausführung mg. DIN 1436: Bolzen mit grossem Kopf, Ausführung mg.

Design Calculation. For **Fig. 41**: clevis pins stressed under bending moment $M_b = (F/2)$ $(b_1/2 + b/4)$; unit pressure inside $p = F/(bd)$, outside $p = F/(2b_1d)$; shear stress in the clevis pin $\tau_s = 2F/(\pi d^2)$ is usually disregarded. Stress on rods or forks resulting from tensile stresses in the remaining cross-section of the rods or forks in transverse plane through the axis of the pin (width of rod head t, width of bracket t_1) and from shear stresses

Table 16. Standard values for permissible nominal bending and shear stresses for clevis and pinned joints

Pin or clevis material	$\sigma_{b\,perm}$ (N/mm²)			$\tau_{s\,perm}$ (N/mm²)		
	Static load	Dynamic load	Fluctuating load	Static load	Dynamic load	Fluctuating load
9 S20, 4.6	80	56	35	50	35	25
St 50 K, 6.8 9 SMnPb 28 K	110	80	50	70	50	35
St 60, 8.8 C 35, C 45	140	100	63	90	63	45
St 70	160	110	70	100	70	50

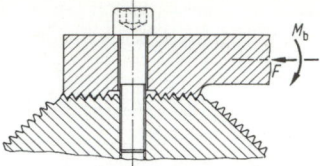

Figure 42. Positive locking joint with serrations.

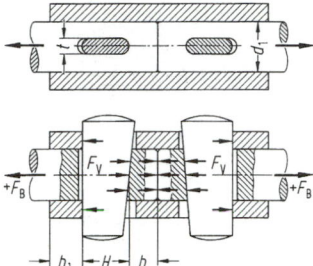

Figure 43. Cottered joint for connecting rods under tensile load.

in the ends of the rod head and bracket in the longitudinal plane in danger of shearing $b(b - d/2)$ or $2b_1(b_1 - d/2)$ on both sides of the pin. Standard values for dimensions: $b/d = 1.5$ to 1.7; $b_1/d = 0.4$ to 0.5; $b_1/d \approx b/d = 1.2$ to 1.5; $t_1/d \approx t/d = 2$ to 2.5. For standard values for permissible unit pressures see **Table 15** and for stresses **Table 16**. The microstructure of the clevis joint [66] – as well as the selection of fit between pin, bracket and fork – has a considerable influence on the assumed load distribution.

1.5.4 Cottered Joints

Positive connections require suitably applied preloads, at least where alternating loads are present, in order to be free from play. In general the cotter action is used for restraint with wedge angles in the area of self-locking (see A1.11). In **Fig. 42** a positive connection using *serrations* is prestressed by a holding-down bolt. Such joints can be used for example in addition to friction grip to make it easier to change tools on lathes, as the tool can be positioned free from play in the direction of the teeth at right angles to them. Radial tooth couplings work in a similar way using *serrations* (see **F3, Fig. 3**). Cottered joints as shown in **Fig. 43** are used to connect rods together, to connect rods to sleeves (e.g. crosshead) or rods to tie-bars; cottered joints as shown in **Fig. 44a** are used with

a stop collar on the rod or as shown in **Fig. 44b** with a tapered fit. They block any degrees of freedom a joint may have.

Design Calculations. To give a rough design for **Fig. 44a** the maximum unit pressures p between the cotter and the rod and the maximum tensile stresses σ_z in the remaining section of the rod are calculated on the basis of 1.5 times the longitudinal tensile force to be absorbed F_B, as in $pd_1t = 1.5F_B$ or $\sigma_z(d_1^2 \pi/4 - d_1t) = 1.5F_B$ owing to the statically indeterminate load distribution. If tensile and compressive forces $+ F_B$ are transmitted, then the unit pressure p in the annular contact surface between the rod and the sleeve follows on the basis of $p(d^2 - d_1^2) \pi/4 = 1.5F_B$. The outside diameter of the sleeve must be selected so that the maximum tensile stresses in the remaining sections of the sleeve and rod correspond to the respective limits of elasticity of the material. Calculation of the bending stress in the cotter is as for a pin joint as shown in **Fig. 41**; consideration of the shear stress is not usually required. As far as sleeves with collars are concerned, examination of the unit pressure between cotter and sleeve is usually unnecessary. Permissible unit pressures under working forces as in pinned joints, frequently $p_{perm} = 80$ to 100 N/mm^2 under alternating working forces F_B where preload $F_v > |F_B|$ and materials where $R_m = 500$ to 700 N/mm^2.

Should a preload F_v be obtained in joints as shown in **Fig. 44**, then the flat key with the included angle α at one side must be driven in with force $F_Q = F_v$ $\tan(\alpha + 2p)$. For angle of friction ρ and coefficient of friction μ, $\tan \rho = \mu$ applies (see A1.11). In the case of lubricated steel surfaces $\mu = 0.12$ can be assumed. The cotter automatically locks in place as soon as $2 > \alpha$. The force $F_{QL} = F_v \tan(2\rho - \alpha)$ is required to drive out (loosen) the cotter. For cotters which are seldom loosened or tightened: $\tan \alpha = 1:15$ to $1:25$; for permanent joints: $\tan \alpha = 1:100$. Taper cotters to take up the play where $\tan \alpha = 1:7$ require to be fixed by set screws along the length of the cotter to adjust and maintain the grip.

Design calculation for **Fig. 43** is similar. The sections with height b at the end of the rod must be examined for shear: standard value: $b \approx b_1 \approx 0.5$ to $0.6H$.

Microstructure of the cottered joints taking into account the deformation of the components to be fastened on the basis of the distortion diagrams (e.g. Fig. 62) known during the design of bolted joints with calculations of endurance limit.

1.5.5 Parallel Keys and Woodruff Keys

The parallel key assembly is the shaft–hub assembly most frequently used when uni-directional (dynamic) loading is present (**Fig. 45c**). Axial relative displacements between hub and shaft are possible if a suitable fit is selected (**Fig. 45d**); the parallel key (sliding key) is fixed in the keyway in the shaft using cheese head screws. The economical Woodruff key (**Fig. 45b**) is used for small torques, especially in machine tools and motor vehicles.

Standards. DIN 6880: Blanker Keilstahl, Masse, zulässige Abweichungen, Gewichte. DIN 6885 Bl. 1: Passfedern-Nuten, hohe Form. DIN 6885 Bl. 2: Passfedern-Nuten, hohe Form für Werkzeugmaschinen, Abmessungen und Anwendung. DIN 6885 Bl. 3: Passfedern - niedrige Form, Abmessungen und Anwendung. DIN 6888: Scheibenfedern, Abmessungen und Anwendung.

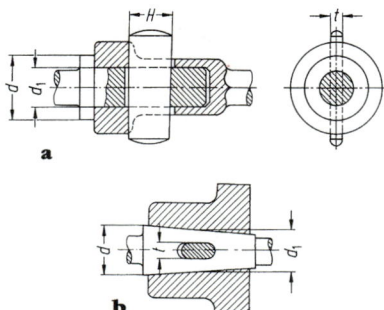

Figure 44. Flat key joint for connecting rods to sleeves for tensile or compressive loads: **a** rod with collar, **b** rod with taper end.

Design Calculations. For parallel keys as shown in **Fig. 45c**: unit pressure p between parallel key and hub:

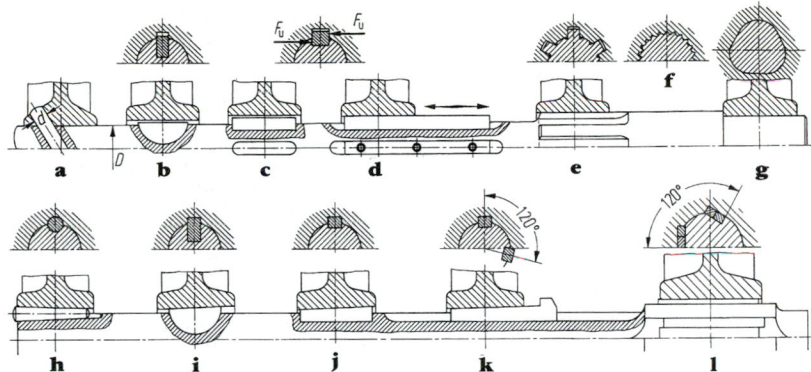

Figure 45. Positive locking joints as in [54]. **a** Angled pin. **b** Woodruff key. **c** Parallel key. **d** Sliding key. **e** Spline shaft. **f** Serration profile. **g** Spline profile. **h** Taper key. **i** Disc key. **j** Flat key. **k** Gib-head key. **l** Tangential key. **h** to **l** Prestressed positive locking.

$p = 2M_t/[D(b - t_1)l_{tr}]$, where M_t = the torsional moment, D = the diameter of the shaft, b = the height of the parallel key, t_1 = the depth of the keyway in the shaft and l_{tr} = the active length. Active length l_{tr} is dependent on the front shape of the parallel key (straight, round). On account of manufacturing tolerances and in order to avoid double fits only one parallel key is generally used. For infrequent heavy torques where behaviour of the material is ductile a second parallel key is occasionally also permitted and calculations are based on one and a half parallel keys being used. *Standard values* for permissible unit pressures as in [54]: for GG hub $p_{perm} \leq 50$ N/mm² for $l_{tr}/D = 1.6$ to 2.1; St hub $p_{perm} \leq 90$ N/mm² for $l_{tr}/D = 1.1$ to 1.4, where in special cases $p = 200$ N/mm² is also permissible for infrequent high extra loads.

Fatigue Strength of the shaft with stress concentration factors β_k as shown in the table in [55]. *Reference values*: shaft diameter $D = 34$ mm, shaft Ck35/St50; bending $\beta_{kb} = 2.4$ to 2.6, torsion $\beta_{kt} = 1.7$ to 1.8, whereby the nominal stresses are calculated using the outside diameter of the shaft. Stress concentration factors increase with the diameter!

Macrostructure. Fits for parallel keys with tolerance zone h9 as described in DIN 6885: *free fit* (keyway width H9 for shaft, D10 for hub; nominal diameter g6 for shaft, H7 for hub); *normal fit*, easy to assemble (keyway width N9 for shaft, JS9 for hub; nominal diameter h7 for shaft, H8 for hub); *close fit, still easy to withdraw*, for small alternating moments (keyway width P9 for shaft and hub; nominal diameter j6 for shaft, H7 for hub); *close fit, difficult to withdraw* (keyway width P9 for shaft and hub; nominal diameter for shaft k6 and hub H7). As is the case with frictionally engaged shaft–hub assemblies (see **E4, Fig. 4**), the unit pressure between the parallel key and the hub can be equalised by a favourable magnetic flux if torque initiation and decay are structurally decoupled in relatively thin hubs. In the case of thick-walled and standard hubs (where $D_i/D_a \leq 0.6$) the maximum unit pressure is hardly ever dependent on the place where the load decays on the hub side. As far as sliding keys are concerned the surfaces of both key and shaft must if possible be made harder than the surface of the hub in order to avoid wear.

Microstructure. More precise calculations of load distribution as in [59]. Measures to increase fatigue strength by providing keyways with larger root radii (not standardised). Assessment of fatigue strength using Neu-

ber's microsupporting effect theory [62] or fracture mechanics theory for short incipient cracks [60].

Design Calculations for Woodruff Keys (**Fig. 45b**). Similar to parallel key assembly but with increased weakening of the shaft. Matching Woodruff key to shaft diameter as shown in DIN 6888: for Woodruff keys used predominantly to fix the position of the hub in relation to the shaft, shaft diameters are provided that are larger than for Woodruff keys used only to transmit torque. If Woodruff keys are used in conjunction with tapered interference fits they must, in principle, be designed to meet the total torque (see also F1.4.2).

1.5.6 Splined Joints

Keyed or pinned joints are unsuitable for heavy reversing torque stresses or backlash, besides which they generally have an out-of-balance effect to a greater or lesser degree. Higher amplitudes of torque can be transmitted using splined joints (**Fig. 45a**) or serrations (**Fig. 45f**).

Standards. DIN ISO 14: Keilwellen-Verbindungen mit geraden Flanken und Innenzentrierung (previously published in DIN 5461, DIN 5462, DIN 5463). DIN 5466, Part 1: Tragfähigkeitsberechnung von Zahn- und Keilwellen-Verbindungen, Grundlagen. DIN 5471 to 5472: Werkzeugmaschinen; Keilwellen- und Keilnabenprofile mit 4 bzw. 6 Keilen, Innenzentrierung, Masse. DIN 5480: Zahnwellen-Verbindungen mit Evolventenflanken. DIN 5481: Kerbzahnnaben- und Kerbzahnwellen-Profile (Kerbverzahnungen).

Design Calculations. Unit pressure p between teeth and hub: $p = 2M_t/(D_m b_{tr} l_{tr} zk)$ where M_t = the torsional moment, D_m = the average diameter, b_{tr} = the depth of engagement, l_{tr} = the length of engagement, z = the number of teeth and k = the load factor. *Standard values* for permissible unit pressure as in [54] for percussive (non-percussive) service: for GG hub $p_{perm} \leq 40(60)$ N/mm²; for St hub $p_{perm} \leq 70(100)$ N/mm², where in isolated cases $p = 200$ N/mm² is also permissible for infrequent high extra loads. Load factor $k \approx 0.75$ for minor-diameter fit; $k \approx 0.75$ for side fit serrations; $k \approx 0.9$ for side fit splines is probably too favourable.

Fatigue Strength of the shaft with nominal stress τ_{twk} for fatigue strength under reversed torsional stresses of the spline in relation to the profile major diameter D as shown in the table in [55]. *Reference values*: $D \approx 34$ mm; 34CrNiMo6 hardened to $R_m \approx 1000$ to 1060 N/mm² gives

$\tau_{twk} \approx 76$ to 90 N/mm² and C35/St50 with $R_m \approx 610$ N/mm² gives $\tau_{twk} \approx 79$ to 84 N/mm² with a length of 60 mm and splined joints. In the case of splines made from material C35 $\tau_{twk} \approx 160$ N/mm² for $D = 34$ mm was measured and fatigue strength under reversed bending stresses of ≈ 69 N/mm² with a length $l = 30$ mm.

Microstructure. For an estimation of the carrying capacity of side fit splined joints with clearance and transition fits, see DIN 5466 Part 1, including an assessment of wearing behaviour. The hub should be checked for expansion – especially in the case of serrations.

1.5.7 Joints with Polygon Profile

Whereas in the case of splined joints the characteristic elements of the positive fit (keys, teeth) increase the stress concentration, this is to a very large extent reduced in polygon-type shaft connections (**Fig. 45g**). More precise calculations are available in the manufacturer's literature [68]. In practice it is mainly the standardised P3G and P4C designs as described in DIN 32 711 and DIN 32 712 that are used. Hubs using the P4C design can be displaced relative to the shaft under torque stress, which is impossible with P3G designs. As the hubs are subjected to very heavy loads by the stress concentration in the polygon surfaces, hardened steel hubs are often used; only the P3G design, which can be ground internally, can be considered for this.

Standards. DIN 32 711: Antriebselemente; Polygonprofile P3G. DIN 32 712: Antriebselemente; Polygonprofile P4C.

1.5.8 Prestressed Shaft-Hub Connections

Structural shapes as shown in Fig. 45b–l. In a similar way to the cottered joints described in F1.5.4, these combine the advantage of the positive fit with prestressing, but tend to produce eccentricity between shaft and hub; also as the saddle key lacks a keyway in the shaft, it relies solely on frictional engagement (F1.4).

Standards: DIN 268: Tangentkeile und Tangentkeilnuten, für stossartige Wechselbeanspruchungen. DIN 271: Tangentkeile und Tangentkeilnuten, für gleichbleibende Beanspruchung. DIN 6681: Hohlkeile, Abmessungen und Anwendung. DIN 6883: Flachkeile, Abmessungen und Anwendungen. DIN 6884: Nasenkeile, Abmessungen und Anwendungen. DIN 6886: Keile-Nuten, Abmessungen und

Anwendung. DIN 6887: Nassenkeile-Nuten, Abmessungen und Anwendung. DIN 6889: Nasenhohlkeile, Abmessungen und Anwendung.

Design Calculations. The torque which can be transmitted by friction-grip is a function of the ram pressure of the key and therefore uncertain, for example with saddle keys. For this reason positive prestressed connections are *only* checked for positive locking and allowance is made for the freedom from play for variable or alternating stresses by means of a permissible unit pressure gained from experience.

Reference Values. F1.5.4. With the exception of tangential keys restrained shaft–hub assemblies are suitable only for transmitting smaller torques and for axial positioning. They can be used only at relatively low circumferential speeds as on the one hand restraint on one side results in greater out-of-balance while on the other hand the centrifugal forces of the hub reduce the restraint. In the course of design calculations care must be taken with tangential keys that only one pair of keys takes up the torque and with divided hubs that the parting line bisects the 120° angle.

1.5.9 Axial Locking Devices

Locking devices on shafts or axles are used for retaining or guiding purposes, sometimes with considerable axial forces. Shaft collars, nuts and caps perform the same function. In **Fig. 46** are shown friction and positive locking devices. For large forces the use of *positive locking devices* is preferred.

Standards. DIN 94: Splinte. DIN 471: Sicherungsringe (Halteringe) für Wellen, Regelausführung und schwere Ausführung. DIN 472: Sicherungsringe (Halteringe) für Bohrungen, Regelausführung und schwere Ausführung. DIN 983: Sicherungsringe mit Lappen (Halteringe) für Wellen. DIN 984: Sicherungsringe mit Lappen (Halteringe) für Bohrungen. DIN 5417: Sprengringe für Wälzlager mit Ringnut. DIN 6799: Sicherungsscheiben (Haltescheiben) für Wellen. DIN 7993: Runddraht-Sprengringe und -Sprengringnuten für Wellen und Bohrungen. DIN 9045: Sprengringe. DIN 15 058: Achshalter (Hebezeuge und Fördermittel). DIN 82 242: Achshalter (Schiffbau). For *frictionally engaged locking devices:*

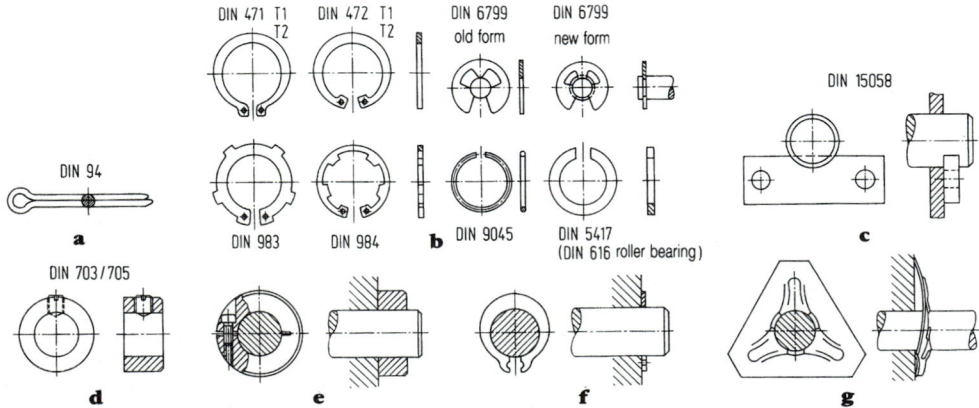

Figure 46. Axial locking devices: **a** cotter pins, **b** retaining rings, **c** axle stirrup, **d** setting rings, **e** clamping rings, **f** self-locking retaining rings, **g** self-locking triangular ring.

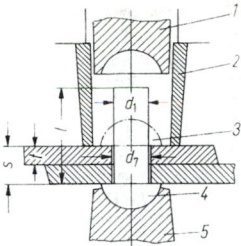

Figure 47. Driving a single shear solid rivet joint: *1* snap head die, *2* holding-down clamp for setting rivets in machine riveting, *3* snap head (round head as in DIN 124), *4* primary head, *5* dolly.

DIN 703: Blanke Stellringe, schwere Reihe. DIN 705: Blanke Stellringe, leichte Reihe.

Design Calculations. The loading capacity of the locking devices must either be in compliance with respective standards or the manufacturer's literature [69]. Retaining rings as described in DIN 471 require separate calculations for the load capacities of the groove and the retaining ring [70], as well as the control of the number of loosening turns as a function of the shaft diameter. The load capacities given in the standard do not contain factors of safety for yielding under static load or endurance failure under dynamic load; a minimum double factor of safety is present for fracture under static load. Numerical values are given for sharp corner abutment and chamfered or radiused abutment with regard to the axial load capacity of the retaining ring. Test results are available concerning the reduction in the endurance limit of shafts due to retaining rings being loaded by axial force [71].

1.5.10 Riveted Joints

Riveting consists of making a joint by reshaping a fastening to create a bearing joint which is generally non-detachable and, at least under heavy loads, is positive between the two parts to be joined [54]. Depending on the type of rivet and its accessibility, the rehsaping may be effected by axially upsetting (driving home) the shank of a *solid rivet* and clenching a *snap head* (**Fig. 47**), by flanging or expanding a collar on a *tubular rivet* as well as by upsetting a *retaining ring* round the *retaining ring stud* of a two-component rivet fastening (**Fig. 48** and **49**). For technical drawings for metal construction see DIN ISO 5261.

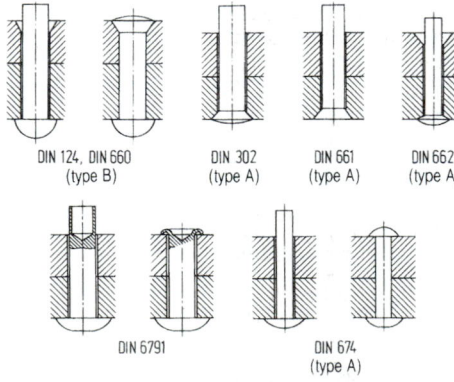

DIN 124, DIN 660   DIN 302   DIN 661   DIN 662
(type B)   (type A)   (type A)   (type A)

DIN 6791   DIN 674
             (type A)

Figure 48. Standardised forms of rivet (selection).

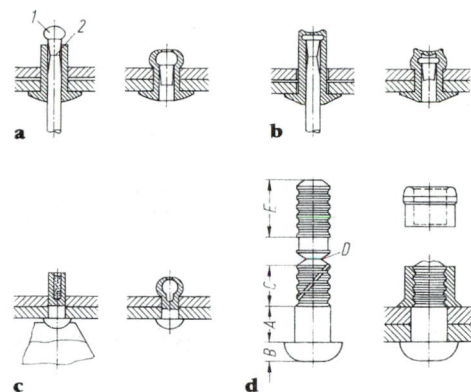

Figure 49. Blind rivet forms and retaining ring-pin joint: **a** DIN 7337 blind rivet, **b** POP cup blind rivet, **c** explosive rivet, **d** DIN 61155 tolerance rivet; *1* drift, *2* theoretical point of failure.

In the last fifty years the riveted joint has to a large extent been replaced by the welded joint as an impervious and load-transmitting joint in high-pressure boilers, tanks and pipes. In steel construction too its importance has declined in relation to welded joints and high-tensile DPH screwed joints (positive and/or friction-grip) [43, 45, 46]. The classic welding technique attracts relatively high labour costs and requires a high level of experience, especially when producing tight lap joints. In light metal construction heavily loaded components are joined individually by fusion welding or even adhesive bonding rather than by riveting, even though these processes have disadvantages. At the higher temperatures used in welding, microstructural changes, residual stresses and distortion may occur and in adhesive bonding careful attention must be paid to the temperature used and creep characteristics. Occasionally adhesive-bonded joints are provided with rivets to provide added protection against peeling. Rivets are still also used, for example where joining steel to aluminium makes welding impossible (for tight joints in plate-metal smokestacks or non-pressurised pipes) [53].

Where possible *solid rivets* are replaced by steel or aluminium *tubular rivets*, *blind rivets* and *collar-pin joints*. Blind rivets (as shown in **Fig. 49**) can be inserted and fastened from one side. The explosive rivets previously in common use are nowadays replaced by new systems such as compression rivets with enclosed drift, cup rivets (air and watertight due to the cup-shaped rivet shank) or modifications [72]. These riveting systems require suitable riveting tools, which are also supplied by the rivet manufacturers. A requirement of collar–pin joints as shown in **Fig. 49** is that the members to be joined should be accessible from both sides whereas the tool used is generally applied from one side. It engages the pin inserted in the prepared bore outside the collar in the corrugated tension component *E*, exerting tensile force on the pin, while at the same time exerting a compressive force on the conical stud on the collar. In this way when the tool is moved, first the members to be joined are pressed together using the tensile force permissible in the pin and then the collar is driven into the retaining grooves in component *C*. As soon as the deformation of the collar is completed the tension component of the pin breaks off at the predetermined breaking point *D*.

Design Calculations. The relative rules in force for estimating must be observed for design purposes. For boiler construction [51, 61], for steel structures [46], for cranes [45], for steel road bridges [43], for aluminium structures [44], for the aircraft industry [47, 48]. Riveted joints as shown in **Fig. 50** fail under static load if the shear strength of the rivet material or the bearing strength of the material of the member are exceeded, or if the deformation of the hole face becomes too great. In order to provide a simplified design, distances from the edge and between holes e, e' and a are given in the rules as a function of the diameter of the hole d_7 and/or the minimum thickness of the material to be joined t. For example, in DIN 18 800, Part 1, the following applies: $2d_7 \leq e \leq 3d_7$ or $6t$; $1.5d_7 \leq e' \leq 3d_7$ or $6t$; $3d_7 \leq a \leq 6d_7$ or $12t$ in regions under pressure for resistance to buckling; $3d_7 \leq a \leq 10d_7$ or $20t$ in regions under tension for tacking as well as in regions under pressure; further, the following combinations of unfinished rivet diameters d_1 and t – given as t [mm]/d_1 [mm] – are recommended: 4 to 5/10; 4 to 6/12; 6 to 8/16; 8 to 11/20; 10 to 14/22; 13 to 17/24; 16 to 21/27; 20 to 24/33. The same rules also apply to HV joints! In the case of bar joints a maximum of six screws or rivets may be arranged behind each other in the direction of the lines of force. The permissible values for shearing stress perm τ_a and bearing stress perm σ_1 depend on whether the rivets (as in DIN 124 and DIN 302) or set screws (DIN 7968) are loaded in single shear or double shear and whether load type H (principal load) or HZ (principal and subsidiary loads) is assumed. As an initial estimate for structural members made of St37 (rivets in USt36) perm $\tau_a = 84$ N/mm² and perm $\sigma_1 = 84$ N/mm² can be calculated for single-shear joints and perm $\tau_a = 113$ N/mm² and perm $\sigma_1 = 280$ N/mm² for multiple-shear joints; for members in St52 (rivets in RSt44-2) perm $\tau_a = 126$ N/mm² and perm $\sigma_1 = 315$ N/mm² apply to single-shear joints and perm $\tau_a = 168$ N/mm² and perm $\sigma_1 = 420$ N/mm² to multiple-shear joints in accordance with the data in DIN 15 018, Part 1, Table 12.

Information in Standards

Overview of standards for rivets in DIN 4000 Part 9: Tabular layout of characteristics, Table 3; selection: DIN 124: Halbrundniete. DIN 302: Senkniete. DIN 660: Halbrundniete. DIN 661: Senkniete. DIN 662: Linsenniete. DIN 674: Flachrundniete. DIN 675: Flachsenkniete. DIN 6791: Halbhohlniete mit Flachrundkopf. DIN 6792: Halbhohlniete mit Senkkopf. DIN 7337: Blindniete mit Sollbruchdorn. DIN 7338: Niete für Brems- und Kupplungsbeläge. DIN 7339: Hohlniete, einteilig. DIN 7340: Rohrniete. DIN 65 155: Passniete. DIN 65 156: Passniete.

Where steel rivets are used, when $d_1 > 10$ mm they generally have to be heated to incandescent heat before riveting. Smaller steel rivets up to about 10 mm in diameter, light metal, brass and copper rivets are driven cold.

If a riveted joint is to be made tight without additional means then solid rivets must be driven hot so that longitudinal tensile stresses σ_z remain in the shank when it contracts upon cooling. Longitudinal tensile stresses give rise to some frictional engagement between the plates where $\mu \approx 0.3$ to 0.5.

Butt joints and connections must be formed under pressure. The aim in butt joints must therefore be to cover the joint immediately by providing a double symmetrical cover plate, because in this way additional peeling stresses resulting from the bending stresses in the plate or bar are reduced.

In steel construction rivet or screw holes may only be drilled, punched or mechanically flame-cut to Quality II in compliance with DIN 2310 Part 3 or Quality I in compliance with DIN 2310 Part 4. In members over 16 mm thick subjected to tensile load the diameter of the punched hole must be increased by a minimum of 2 mm by reaming before assembly. This must be specified in the job sheet. Holes which belong together must match, and if there is misalignment of the holes a clear passage for rivets and screws must be drilled or reamed, but not drifted. Unless members are mainly subjected to static loading, holes must be de-burred and protruding edges of the hole broken off; the punching of holes is permissible only if the diameter holes has been enlarged by at least 2 mm by reaming before assembly. Rivets must be driven in such a way that the rivet holes are completely filled. The snap head must be fully hammered out; no harmful indentations must be made in the material during this operation. Once driven in, the rivets must be checked for a tight fit.

Rivet material and component material must be compatible in terms of resistance to corrosion. **Table 17** provides details of compatibility between rivet material and component material in accordance with [53]. It is often the case that improved protection against corrosion must be provided by a (sealing) coat of paint. There are special regulations which must be observed for aircraft (LN 9198) [47, 48] and high buildings (DIN 4113 Part 1).

1.6 Bolted Connections

H. Mertens, Berlin

1.6.1 Uses

A bolted or screwed connection [73–95] is a detachable assembly of two or more components fastened by one or more bolts or screws. The main types of joint are shown in **Fig. 51** [86]. The *fastener-type bolts and screws* used in these bolted or screwed connections must secure the static and dynamic working forces acting on the compo-

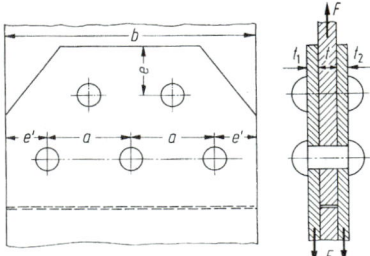

Figure 50. Example of a double strap butt joint (double shear).

Table 17. Matching rivet and component materials [53]

Rivet material	Material of the components to be joined
Al 99.5	Al 99.5 and higher percentage purities
Al 99	Al 99, AlMn
AlMg 3	AlMg 3, AlMg 5, AlMgMn, AlMg 4.5 Mn, AlMgSi 0.5, AlMgSi 0.8
AlMg 5	AlMg 5, AlMg 4.5 Mn, AlMgSi 1, AlZnMg 1
AlMgSi 1	AlMgSi 1, AlMg 5, AlZnMg 1
AlCuMg 0.5	AlCuMg 1 and AlCuMg 2
AlCuMg 1	AlCuMg 1, AlCuMg 2, AlZnMgCu 0.5, AlZnMgCu 1.5

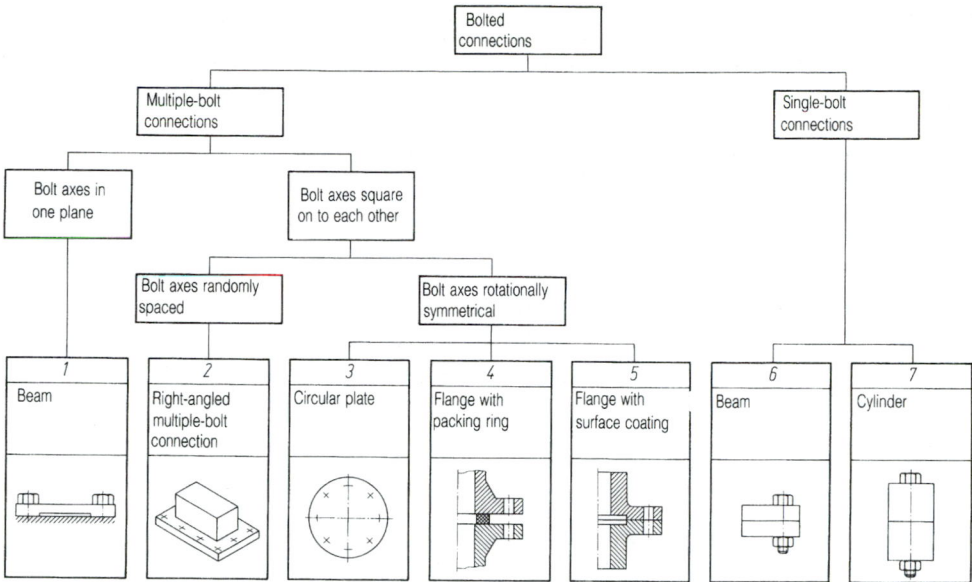

Figure 51. Classification of connection types [86].

nents without noticeable relative motion of the parts against each other unless this task is partly performed by positive locking devices as described in F1.5 or centring collars. If on the other hand defined relative motion is obtained, *translation-type bolts and screws* are suitable, because they convert rotary motions into longitudinal motions, as for example in machine tool spindles or vices.

1.6.2 Characteristics of Screw Motion

When tightening or loosening fastener-type bolts or screws or torquing translation-type bolts or screws a screw motion is made about and along a fixed axis, the screw axis. When one complete turn of the screw is made a (relative) axial displacement is created along the screw axis, which corresponds to the *flank lead* P_h in **Fig. 52**. The development of a helix lying on a cylinder where the radius $r_m = d_m/2$ results in a rising straight line with the *lead angle* β_m, where $\tan \beta_m = P_h/(\pi d_m)$. Generally, the lead angle β, where $\tan \beta = (r_m/r) \tan \beta_m$, is produced for the radius r and is larger for smaller radii than larger ones. The distance measured parallel to the axis between successive flanks acting in the same direction is referred to as *flank pitch P*. In single-start threads the lead P_h equals the pitch P. The formula $P_h = nP$ is used for n start threads.

1.6.3 Types of Thread

An overview of threads in general use or as used for the larger specialised areas is to be found in DIN 202. Terms and definitions for cylindrical threads are set out in DIN 2244 (German, English, French). The *thread profile* is the outline of a thread in axial section, whereas in general the *thread flanks* are the straight parts of the thread profile which are not parallel to the screw axis.

V-Thread for Fastener-Type Bolts

The *metric ISO thread* as described in DIN 13 Sheet 19 is an improved thread, unified throughout the world, which in practice is usually interchangeable with the earlier 'metric' thread. For the manufactured profile of the bolt and nut (design profile where the thread fit has no flank clearance) see **Fig. 53**. The major diameter d of the bolt thread is equal to the major diameter D of the nut thread; this is also referred to as the *nominal diameter*. The *minor diameter* d_3 is used to calculate the cross-section of the core $A_3 = \pi d_3^2/4$. The groove and ridge of the thread are the same width along the axis on the *pitch diameter* d_2 of the bolt or D_2 of the nut. The (average) lead angle is expressed as $\tan \beta = P/(\pi d_2) \cdot H$, where H is the height of the theoretical sharp V-profile with a *pitch*

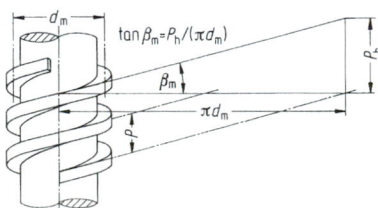

Figure 52. Bolt shank with double start flat thread: P_h lead, P pitch ($P_h = 2P$), β_m mean lead angle.

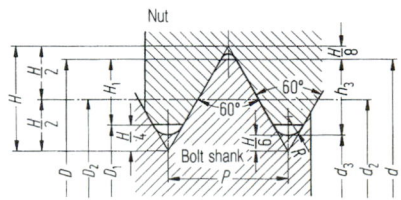

Figure 53. Metric ISO thread (DIN 13 Part 19). $D_1 = d - 2H_1$, $d_2 = D_2 = d - 0.649\,52P$, $d_3 = d - 1.226\,87P$, $H = 0.866\,03P$, $H_1 = 0.541\,27P$, $h_3 = 0.613\,43P$, $R = H/6 = 0.144\,34P$.

diameter $\alpha = 60°$. The flank overlap H_1 is also referred to as the *depth of thread engagement*. The fillet radius R in the thread root of the bolt, which is larger than the earlier 'metric' thread, coupled with the larger core cross-section A_3, results in an increase in fatigue strength and at the same time a reduction in the depth of thread. The fillet radius at the major diameter of the nut is not specified, because it necessarily emerges during manufacture and because the loads are not so great there. The stressed cross-section $A_s = \pi(d_2 + d_3)^2/16$ is required as a reference cross-section for strength calculations. In DIN 14 standards are given for metric ISO threads with diameters below 1 mm.

In **Appendix F1, Table 4** are listed nominal diameter d, lead P, core cross-section A_3 and stressed cross-section A_s for a selected series of (metric ISO) coarse- and fine-pitch threads as specified in DIN 13 Parts 12 and 28. Coarse-pitch threads, i.e. threads with a larger flank lead, are preferred to fine-pitch threads from the point of view of loading capacity.

The *Whitworth pipe thread* as specified in DIN 259 Parts 1 to 5 and DIN ISO 228, with cylindrical internal and external threads is still used in pipes and pipe joints. It is not self-sealing. The thread letter symbol (e.g. R1 1/2) in DIN 259 is no longer to be used in new designs in order to avoid confusion with similar letter symbols for tapered external threads. Instead of this the symbols (e.g. G1 1/2) in DIN ISO 228 must be used. With thread sizes up to 26 mm, tapered external threads as in DIN 158, e.g. for screwed sealing plugs and lubricating nipples, can be used for self-sealing joints [80]. See also DIN 2999 Parts 1 to 6 or ISO 7/I for Whitworth pipe threads for threaded pipes and fittings with cylindrical internal thread and tapered external thread or DIN 3858 for screwed pipe joints.

Flat Thread for Translation-Type Bolts

Acme and buttress threads produce less friction between nut and bolt than the V-thread. For nominal profiles of the nut and bolt of a 'metric Acme' thread, as specified in DIN 103 Part 1, with a clearance in the major and minor diameter but without flank clearance and using standardised designations, see **Fig. 54**. The Acme thread is side-fitting and should therefore be loaded only by axial forces (and torques); it locks if tilted. Multiple-start Acme threads have the same profile as single-start threads where the flank lead P_h = flank pitch P. **Appendix F1, Table 5** contains nominal sizes for Acme threads. The 'metric buttress' thread as described in DIN 513 Part 1 with an asymmetrical thread profile has bearing flanks with sectional flank angles (angles between the flank and a line perpendicular to the axis of the thread at the intersection)

of 3° and clearance in the minor diameter and between the non-bearing thread flanks.

Round Thread, Roller Threads

Round threads (in general see DIN 405 or for large depth of thread see DIN 20 400) are used for fastener-type or translation-type bolts where there exists a danger of dirty conditions. Screws with rollers between the threaded surfaces of the nut and the shank (*roller threads*) produce even lower frictional moments than buttress threads; the generatrix of the threaded surfaces usually consists of curved lines (e.g. arc of a circle or pointed arch) [93].

1.6.4 Types of Bolt and Nut

The international terms for bolts, nuts and accessories are set out in DIN ISO 1891. This standard is supplemented by DIN 918, which in an appendix provides an overview of standard types of nut and bolt. **Figure 55** shows basic and special forms of bolted joints.

Cap Bolts and Screws (Fig. 55a). The distinguishing features of these are the head form, shank form and point form. The *head form* is defined by the method of driving; examples: hexagonal head (DIN 931, DIN 933, DIN 960, DIN 961), hexagon socket head (DIN 912, DIN 6912, DIN 7991, DIN 7984), slotted head and crosshead (DIN 84, DIN 63, DIN 87, DIN 7987); also of the triangle, square, octagon and double hexagon type or with wing or hammer head. In DIN 74 standards are provided for countersinks, in Part 1 for countersunk bolts and screws, in Part 2 for cheese-head bolts and screws and in Part 3 for hexagonal head cap bolts and screws. Diameters for cylindrical countersinks are given in DIN 974.

The *point form* is determined, among other things, by the method of manufacture or type of assembly. Bolts and screws used for automatic assembly on production lines require points with a locating facility; tapping points are used to accommodate a certain amount of paint which may have got into the nut thread. For thread ends see DIN 78; for thread runouts and undercuts as well as for tapped holes (blind holes) see DIN 76.

The *shank form* is determined by the method of manufacture or additional requirements. In *reduced-shaft bolts* (anti-fatigue shaft bolts) with high elasticity the shank diameter is smaller than the minor diameter. In *set bolts or screws* (e.g. hexagon head, interference-body bolts or screws as in DIN 609) the shank diameter is manufactured so as to have a tight fit (e.g. k6) which provides a locking facility. In *full-diameter bolts or screws* the shank diameter is equal to the body diameter, whereas in *scant*

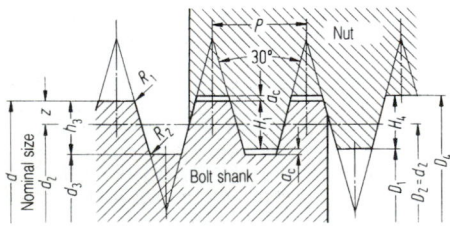

Figure 54. Metric ISO Acme thread (DIN 103 Parts 1 and 2):
$D_1 = d - 2H_1 = d - P$, $H_4 = H_1 + a_c = 0.5P + a_c$, $b_s = H_1 + a_c = 0.5P + a_c$, $z = 0.25P = H_1/2$, $D_4 = d + 2a_c$, $d_3 = d - 2b_s$, $d_2 = D_2 = d - 2z = d - 0.5P$, $R_1 = \max 0.5a_c$, $R_2 = \max a_c$, $a_c =$ clearance (index c of crest ≅ point).

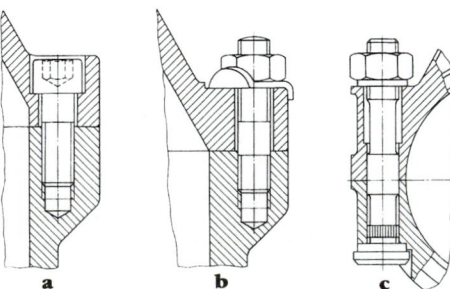

Figure 55. Basic and special forms of bolted and screwed connections: **a** DIN 912 socket head cap screw as a setting screw, **b** DIN 939 stud bolt in cast housing with tap washer, **c** through bolt in special structural shape for threaded joint in connecting rod cover.

shank bolts or screws it is approximately equal to the pitch (diameter of the base material for rolled thread).

Stud Bolts (Fig. 55b). These have a 2d long inserted end, as described in DIN 835, mainly for screwing into aluminium alloys, a 1.25d long inserted end, as described in DIN 939, for screwing into cast iron or a 1d long inserted end, as described in DIN 938, mainly used for screwing into steel.

Screw Bolts (DIN 2509). These are used, for example, to join components by means of nuts screwed onto both ends. A double-edged dog point at one end of the thread is said to make it possible to prevent the screw bolt from turning during assembly. *Screw bolts* and *through bolts* require *clearance holes* which are laid down in accordance with the relative design factors; details of standardised clearance holes are in DIN ISO 273 (fine, medium, coarse; e.g. $d_h = 10.5$ mm, $= 11$ mm, $= 12$ mm for M10).

Headless Screws. These have a continuous thread, a slot or hexagon socket at one end and a blunt start thread (DIN 427, DIN 551 or DIN 913), dog point (DIN 417 or DIN 915), retaining ring (DIN 438 or DIN 916) or point (DIN 553 or DIN 914) at the other end. They are also manufactured with thrust points as described in DIN 6332 and are suitable as components for clamp bolts with palm grip as in DIN 6335, star grip as in DIN 6336 and clamping lever as in DIN 99 (up to M20) or with thrust piece as in DIN 6311.

Special Forms of Bolt (Fig. 55c). These may have, for example, tight fit and corrugations to provide a rotary locking facility; see also DIN 4000 Part 2 (Sachmerkmal-Leisten für Schrauben und Muttern), DIN 6914 (Sechskantschrauben mit grossen Schlüsselweiten für HV-Verbindungen in Stahlkonstruktionen) and DIN 7999 (Sechskant-Passchrauben, hochfest, mit grossen Schlüsselweiten für Stahlkonstruktionen).

Nuts. In mechanical engineering hexagon nuts are the most frequently used; the former standard size of 0.8d still applies to steel nuts as described in DIN 934. For new designs it is recommended that hexagon nuts as described in DIN 970 (Series 1 coarse pitch thread) should be used with body diameters in the range of 5 to 39 mm and as described in DIN 971 Part 1 or 2 (Series 1 or 2 fine pitch thread) with a larger size of nut (Series 1 with m/d of 0.84 to 0.94 or Series 2 with m/d of 0.93 to 1.03). The width across flats s is somewhat reduced in places, its width across corners e being about 2d. If a smaller size of nut is required for special applications, then DIN 439 may possibly be used. *Cap nuts* (**Fig. 56a**) as in DIN 917 (low cap nut) and DIN 1587 (domed cap nut) provide protection against injury and are also used in conjunction with sealing washers to prevent liquids from seeping out or in. Special forms of nut developed for machine tool construction, such as grooved nuts (**Fig. 56b**) as described in DIN 1804, are used to provide axial location of hubs and rings on shafts or transmit axial force and must be tightened with a sickle spanner; this occasionally includes *capstan nuts* (**Fig. 56f**) as described in DIN 548 and DIN 1816. *Knurled nuts* as described in DIN 6303, *slotted round nuts* as in DIN 546 or *wingnuts* (**Fig. 56c**) as in DIN 315 can be considered for low stresses. In steel structures and coachwork square weld nuts as described in DIN 928 or *hexagon weld nuts* (**Fig. 56e**) as in DIN 929 which are fastened to the base material by spot welding. *Hexagon nuts with centring shoulder* (**Fig. 56g**) as in DIN 2510 Part 5 and *capped nuts* (**Fig.**

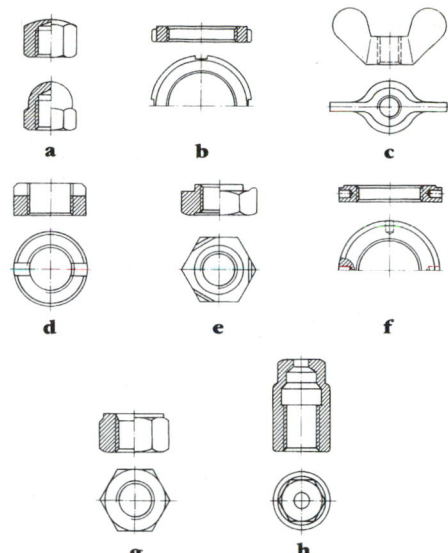

Figure 56. Standardised special forms of nut: **a** cap nut as in DIN 918 and DIN 1587, **b** groove nut as in DIN 1804 and DIN 981, fine thread (M6 to M200), **c** DIN 315 wingnut (M5 to M24), **d** DIN 546 slotted round nut (up to M16), **f** DIN 1816 capstan nut (fine thread M6 to M200), **g** hexagon nut with centring shoulder as in DIN 2510 Part 5, **h** capped nut for reduced shank bolted connections as in DIN 2510 Part 6.

56h) as in DIN 2510 Part 6 have been developed for threaded joints using reduced shanks. *Tension nuts* provide an even transition of the magnetic flux from tension in the bolt to compression in the bearing surface of the nut. Hexagon nuts with large widths across flats as described in DIN 6915 for M12 to M36 have been developed for (HV) joints subject to high preloads as used in steel construction.

Washers. These must be used under bolts and nuts if the base material has a tendency to set or would be over-stressed; examples of form in DIN 125. In the case of U-beams and I-beams square washers must be used to allow for the 8% or 14% slope, DIN 434 or DIN 435. DIN 7968 close-tolerance bolts generally require DIN 7968 washers. Hexagon head bolts with large widths across flats (DIN 6914, DIN 7999) may only be used with DIN 6916 washers.

1.6.5 Material Specification for Bolts and Nuts

In DIN ISO 898 Part 1 bolt materials are identified according to property class. The designation for each property class consists of two figures which are separated by a full-stop. *Example:* 5.6, 6.8, 8.8, 9.8, 10.9, 12.9, …. The *first* figure is equal to a hundredth of the nominal tensile strength R_m in N/mm²; the *second* figure indicates 10 times the ratio of the nominal yield stress R_{eL} or $R_{p0.2}$ to the nominal tensile strength R_m (ratio yield point). If the two figures are multiplied together the result is a tenth of the nominal yield stress in N/mm².

In DIN ISO 898 Part 2 nuts with specified proof load stresses are designated by a property class between 4 and 12. The designation number is equal to a hundredth of the minimum tensile strength of a bolt in N/mm², which can be loaded up to the minimum yield stress when assembled with the nut. *Example*: Bolt 8.8; nut 8, capable of being loaded up to the minimum yield stress. The previous property classes according to DIN 267 Part 1 do not provide a definite assurance that no stripping of the thread will take place in the course of tightening – especially during the tightening process determined by the yield stress. In general, nuts belonging to higher property classes are used instead of nuts belonging to the low property classes. This is advisable for nut-and-bolt connections where loads exceed the yield stress or the proof load stress.

DIN ISO 898 does not apply to headless screws or similar threaded fasteners, nor does it apply to special requirements such as weldability, corrosion resistance (see ISO 3506 and DIN 267 Part 11 respectively), heat resistance above + 300 °C and ductility at temperatures below − 50 °C (see DIN 267 Part 13) or retaining properties (see ISO 2320 or DIN 267 Part 15).

The required depth of tapped holes depends on the material used in the threaded part of the nut. Recommended depths for bottoming blind holes are given in **Appendix F1, Table 6**. Stud bolts with nuts are to be recommended for use in grey cast iron or light metal instead of cap screws [79].

1.6.6 Forces and Deformations in Joints due to Preload

Tightening Torque. If a symmetrical bolted connection as shown in **Fig. 57** is tightened by turning the nut a tensile load, known as *preload* F_v, is placed on the bolt and an equal compressive load between the plates. In this way the bolt is elongated by f_s and the plates are compressed by f_P. The plates are pressed together in the area of the *Rötscher cones* which extend from circles under the bolt head or the nut, each with an inscribed circle diameter s and a general bearing surface diameter for the bolt head and the nut (d_W and D_W respectively) below 45°.

When the nut is turned the *frictional moment in the thread* M_G increasing with F_v and *the frictional moment in the bearing surface of the nut* M_K must be overcome; tightening torque $M_A = M_G + M_K$. According to A1.11.2, $M_G = F_v(d_2/2) \tan (\beta_m + \rho')$, where d_2 is the pitch diam-

eter and β_m is the average lead angle, and the coefficient of friction in the thread $\mu' = \tan \rho' = \mu_G/\cos (\alpha/2)$, where α is the angle of thread and μ_G is the coefficient of friction in the thread. For triangular thread where $\alpha = 60°$, the formula $\mu' = 1.155\mu_G$. The amplitude of the moment is $M_K = F_v\mu_K D_{km}/2$ where μ_K is the coefficient of friction in the bearing surface of the nut and D_{km} is the effective diameter for the respective frictional moment. For coefficients of friction see [76], e.g. steel bolt, phosphorised and steel nut, plain, dry: $\mu_G = \mu_K = 0.12$ to 0.18; oiled: $\mu_G = \mu_K = 0.10$ to 0.16; MoS_2: 0.08 to 0.12. For triangular thread where $\alpha = 60°$ and P is the flank lead, thus $\tan \beta_m = P/(\pi d_2)$,

$$M_A \approx F_v[0.159P + \mu_G 0.577d_2 + D_{km}\mu_K/2] \quad (1)$$

follows in a simplified way because $\tan(\beta_m + \rho')\, \alpha \tan \beta_m + \tan \rho'$.

Example A hexagon head bolt M10 with metric ISO V-thread ($d_2 = 9.03$ mm, $P = 1.5$ mm) as in DIN 931, DIN 970 nut in Product Class A as in DIN ISO 4759 Part 1 ($d_w = 15.6$ mm), clearance hole as in DIN ISO 273 (average: $d_h = 11$ mm) without countersink is expressed, approximately, as: $D_{Km} = (d_w + d_h)/2 = 13.3$ mm. Where, for example, $\mu_G = \mu_K = 0.16$ it follows that $M_A = F_v$ (0.236 + 0.834 + 1.064) mm. The sum of the coefficients of friction therefore amounts to some 90% of the total tightening torque. In the case of lubricated bolts and usually also galvanised protective coatings, the element of friction is smaller, with the result that such bolts receive a higher preload F_v with the same tightening torque.

The frictional moment M_{GL} in the thread required to loosen it is $M_{GL} = F_v(d_2/2) \tan (\rho' - \beta_m)$. We talk about *self-locking* provided a moment $M_{GL} > 0$ is required for loosening. Self-locking stops as soon as $M_{GL} = 0$, i.e. $\beta_m = \rho'$, if the frictional moment M_K in the bearing surface of the nut or bolt head is ignored. The total moment M_L required for loosening is approximately equal to 0.7 to 0.9 times the tightening torque M_A in metric ISO V-threads so long as no vibrations reduce the effective coefficient of friction μ'.

Preload F_v and tightening torque M_A have an effect on *tensile and torsional stresses* in the bolt. The nominal tensile stress σ_z is calculated using either the stressed cross-section of the thread A_s or the cross-section of the reduced shank if smaller, the nominal torsional stress τ being similar to the corresponding section moduli. The Mises equivalent tensile stress σ_v then provides the effort on the material. If a 90% utilisation of the minimum yield stress of the bolt material is considered permissible, then *permissible assembly forces* F_{sp} and their associated *tightening torques* M_{sp} for specified coefficients of friction can be found in tables as in VDI [German Engineers' Association] Recommendation 2230 or determined using ready reckoners for bolts and screws obtainable from manufacturers. An extract from such tables is given in **Appendix F1, Table 7**.

Tightening Method. The tightening torque required is dependent on the tightening method. The ratio of the maximum preload resulting in practice when tightening to the minimum preload F_{Mmax}/F_{Mmin} is designated as the *tightening factor* α_A, the width across flats is $F_M = F_{Mmin} = F_{Mmin}(\alpha_A - 1)$. It is known from experience that the portion that accounts for the coefficients of friction alone lies within the limits 1.25 : 1 to 2 : 1. Standard values for dimensioning bolted connections may be indicated for α_A (values in brackets) on the basis of VDI Guideline 2230 [80]:

For *pulse-controlled tightening* an *impact screwdriver* (2.5 to 4) is used and for *torque-controlled tightening* a *dynamometric screwdriver* (1.7 to 2.5) in which the screwdriver is adjusted according to a tightening moment

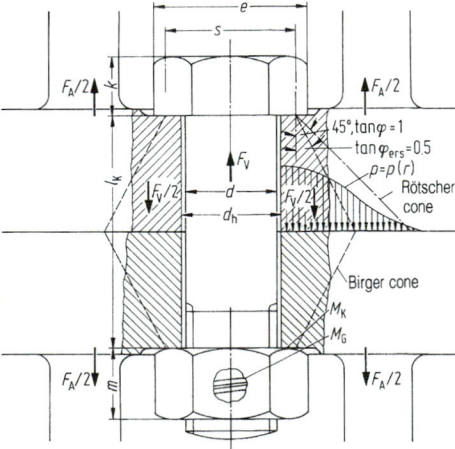

Figure 57. Tightening nut on through bolt to hold two plates (flange). (F_v is the preload in the bolted connection in the absence of external working load F_A.)

obtained by experiment. For torque-controlled tightening a *torque-setting spanner, dynamometric spanner* or *precision dynamometric screwdriver* is used with dynamic torque measurement: (1.6 to 1.8), if the theoretical tightening moment is determined by estimating the current coefficient of friction or (1.4 to 1.6) if the theoretical tightening moment is determined by measuring F_M at the threaded connection.

Hydraulic tightening is achieved by adjusting according to the length or pressure measured (1.2 to 1.6) where the preloading takes place by means of additional nuts on the extended thread and additional turns of the nut. *Measuring the elongation* of the calibrated bolt (1.2) follows. *Rotation angle controlled tightening* can be either motor-driven or manual (1.1 to 1.3), with an experimentally determined tightening torque and angle of rotation. Dispersion is to a great extent determined by the dispersion of the yield stress in the obstructed screwless state so that when dimensioning in accordance with F_{Mmin} the value $\alpha_A = 1$ can be formally set. *Yield stress controlled tightening*, motor-driven or manual (1.1 to 1.3, formally when dimensioning for F_{Mmin} again $\alpha_A = 1$). *Thermally controlled tightening* is used in turbine construction and is comparable with hydraulic tightening in respect of its advantages and disadvantages; the bolts for fastening the casing cover are provided with a central bore for heating and monitoring the temperature.

Assembly Force. Forces and deformations arising after tightening are dependent on the effective *assembly force* F_M. Assuming linear stiffness behaviour, individual graphical representations of the characteristic force-deformation lines can be combined in a straight-line diagram, the so-called *distortion triangle* (**Fig. 58**). Using the designations as indicated, the stiffness c_s of the bolts is expressed as $c_s = F_s/f_s$ and the resilience δ_s of the bolts is $\delta_s = 1/c_s$. The stiffness of the plates between the bearing surfaces of the bolt head and the nut is $c_P = F_P/f_P$ and the elastic resilience $\delta_P = 1/c_P$ if retained in the centre. After the nut is tightened the assembly forces in the bolt and plates are expressed as $F_{MS} = F_{MP} = F_M$; deformation expressed as $f_{SM} + F_{SP} = s_M$, where s_M is the axial displacement of the nut on the thread, provided that before being tightened the bolt head and nut are well set on all sides on the smooth plates or in suitable countersinks.

Resilience of Bolts. Bolts consist of a number of individual elements which can be readily substituted by imaginary cylinders of varying lengths l_i and cross-sections

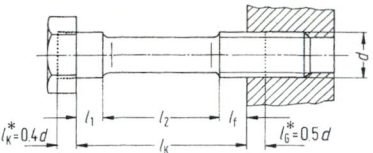

Figure 59. Dividing a bolt into separate cylindrical units in order to calculate their elastic resilience (VDI Guideline 2230) [76].

A_i (**Fig. 59**). It follows that the resilience of an individual cylindrical element is $\delta_i = l_i/(E_S A_i)$, where E_S is the modulus of elasticity of the bolt material. The total resilience of the bolt δ_S becomes $\delta_S = \delta_i$. The elastic resilience of the bolt head is given in VDI Guideline 2230 for standardised hexagon head and hexagon socket bolts as $\delta_K = 0.4d/(E_S A_N)$, where $A_N = \pi d^2/4$, the resilience of the thread root once screwed in is expressed as $\delta_G = 0.5d/(E_S A_3)$ where the root cross-section is $A_3 = \pi d_3^2/4$ and the resilience of the bolt and nut profiles is expressed as $\delta_M = 0.4d/(E_S A_N)$ for nuts as in DIN 934, the open portion of the thread is $\delta_f = l_f/(E_S A_3)$, where l_f is the length and A_3 is the root cross-section. **Figure 59** can therefore be expressed as $\delta_s = \delta_K + \delta_1 + \delta_2 + \delta_f + \delta_G + \delta_M$.

Resilience of Centrally Retained Plates. The resilience of plates δ_P which are centrally retained can be specified by approximation according to Birger [84] by determining the resilience of a double cone spreading out at an angle φ_{im} (where $\varphi_{im} = 0.5$) under the bolt head and nut with a bore d_h and evenly distributed compressive strain in the individual sections (Fig. 57). VDI Recommendation 2230 also provides approximation formulae for such plates; the stiffness c_P or resilience δ_P of the plates are calculated from the stiffness or resilience of an imaginary cylinder with a cross-section A_{im}: A_{im} as in **Fig. 60**;

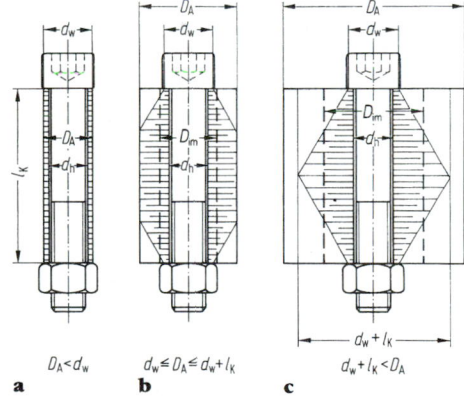

$D_A < d_w$ $d_w \leq D_A \leq d_w + l_K$ $d_w + l_K < D_A$

a   **b**   **c**

Figure 60. Imaginary pressure cylinders used for calculating the elastic resilience of restrained shells and plates.

$$\mathbf{a}\ A_{im} = \frac{\pi}{4}(D_A^2 - d_h^2);$$

$$\mathbf{b}\ A_{im} = \frac{\pi}{4}(d_w^2 - d_h^2) + \frac{\pi}{8}d_w(D_A - d_w)\left[\left(\sqrt[3]{\frac{l_K d_w}{D_A^2}} + 1\right)^2 - 1\right];$$

$$\mathbf{c}\ A_{im} = \frac{\pi}{4}(d_w^2 - d_h^2) + \frac{\pi}{8}d_w l_K\left[\left(\sqrt[3]{\frac{l_K d_w}{(l_K + d_w)^2}} + 1\right)^2 - 1\right]$$

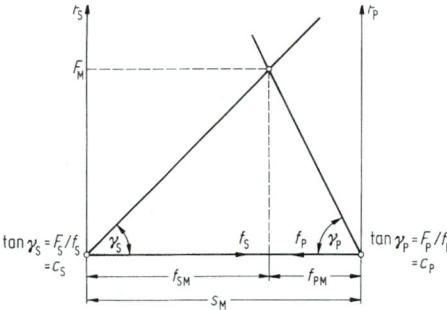

Figure 58. Deflection triangle as graphic representation of the force deflection conditions during tightening; F_s tensile force in the bolt $F_s = F_s(f_s)$, f_s elongation of the bolt, F_P compressive force in the plates, $F_P = F_P(f_P)$, f_P compression of the plates, F_M preload during assembly, s_M path of the nut on the thread.

$\delta_P = l_K/(A_{im}E_P)$, where E_P is the modulus of elasticity of the retained plates.

Dispersions During Tightening. The dispersions of assembly force F_M between F_{Mmin} and F_{Mmax} that occur during tightening can be clearly seen in the distortion diagram in **Fig. 61**. The maximum preload F_{Mmax} must remain smaller than the permissible bolt force, which according to VDI Guideline 2230 is equal to a 90% utilisation of yield stress for bolts up to M39 – limit $F_{Mmax} = F_{Sp}$ – in respect of tightening methods which are not controlled by yield stress or rotation angle.

Set. During tightening up to assembly preload F_M in the region F_{Mmin} to F_{Mmax}, the bearing surfaces under bolt head and nut and the junction lines between the plates are flattened. But even after this, set takes place in the junction lines with the continued flattening out of surface irregularities caused by working loads which vary with time. According to statistics from currently available test results the volume of set f_z is almost completely independent both of the number of junction lines and of the size of the irregularities on the mating surfaces. It clearly increases with the ratio of the length of thread engagement (l_K/d). Solid joints using bolts as in DIN 931 (in practice also DIN 933) are expressed as [76]

$$f_z \approx 3.29(l_K/d)^{0.34} \cdot 10^{-3} \text{ mm.} \qquad (2)$$

Owing to the setting of the joint by the amount f_z the assembly preload F_M is reduced, again by the amount F_z. The preload $F_V = F_{Mmin} - F_z$ is therefore all that remains of F_{Mmin} (**Fig. 61**). F_V must be greater than or equal to the required preload F_{Vreq}. The amount of set causes a reduction in the elongation of the bolt by $F_z\delta_S$ and the compression of the plates by $F_z\delta_P$; therefore $f_z = F_z\delta_S + F_z\delta_P$ and consequently $F_z = f_z/(\delta_S + \delta_P)$. In order to avoid increasing the setting unnecessarily when using high-strength, high-preload bolts no locking plates, washers or spring rings must be used under the bolt head or nut. The bearing surfaces under the bolt head and nut should always be properly finished and be at right angles to the axis of the bolt [76, 79].

1.6.7 Superposition of Preload and Working Loads

Central Retention and Loading. If an axial tensile load F_A acts centrally on a symmetrically shaped and (centrally) preloaded bolted connection (as in **Fig. 57**) under the bolt head and nut of the through bolt, the bolt is elongated by an amount f_{SA} and the compression of the plates is reduced by the equivalent amout f_{PA}; e.g. the bolt and plate are engaged parallel to the tensile load F_A, pro-

vided no misalignment of the bolted joint occurs in the junction line. The *additional bolt force* is expressed as $F_{SA} = c_S f_{SA}$ and $F_A = (c_S + c_P)f_{SA}$; the *clamping force* in the plates is reduced by $F_{PA} = F_A - F_{SA}$. The forces can usefully be included in the distortion diagram (**Fig. 62**). Further, $F_{SA} = (c_S/(c_S + c_P)) F_A \equiv \Phi_K F_A$, where Φ_K is the *force ratio* is the formula for the action of the external force F_A directly below the bolt head and nut. Where $\delta_S = 1/c_S$ and $\delta_P = 1/c_P$ it then follows that

$$\Phi_K = c_S/(c_S + c_P) = \delta_P/(\delta_S + \delta_P). \qquad (3)$$

The *residual clamping force* in the junction line F_{KR} after loading and setting is $F_{KR} = F_V - F_{PA} = F_V - (1 - \Phi_K)F_A$; it must be at least equal to the required clamping force: $F_{KR} \geq F_{Kreq}$. The result for the required preload is $F_{Vreq} = F_{Vreq} + F_{PA} \leq F_V$ and for the minimum assembly preload $F_{Mmin} = F_{Vreq} + F_z$, where F_z is the loss of preload due to setting. With α_A as the tightening factor the maximum assembly preload becomes

$$F_{Mmax} = \alpha_A F_{Mmin} = \alpha_A [F_{Kreq} + (1 - \Phi_K)F_A + F_z].(4)$$

If a 90% utilisation of yield stress is permissible after the tightening procedure, then F_{Mmax} may reach a maximum of F_{Sp} as described in **Appendix F1, Table 7** or VDI Guideline 2230. To ensure that the yield stress is not exceeded after the working load F_A is applied, F_{SA} may not be greater than about 13% of the maximum assembly preload F_{Mmax}, which requires the lowest values possible of the force ratio Φ_K.

Distribution of Force Over the Retained Members. In general the external axial force does not act directly below the bolt head and nut, even when acting centrally, but within the retained members. If it is assumed that the points where the force is acting are not at a distance l_K between the bearing surfaces of the bolt head and nut, but only at the distance nl_K (e.g. $n = 0.5$), then all the areas of the plate are no longer relaxed by the axial force F_A as the stiffness ratios of the loaded and relaxed areas of the bolted joint change. The interrelationships are shown in **Fig. 63**, where setting and preload dispersions are not taken into account. The additional bolt force F_{SA} is thus calculated using $F_{SA} = \Phi_n F_A$, Φ_n being the force ratio, for introducing the axial force F_A centrally into planes at the distance (nl_K):

$$\Phi_n = n\Phi_K = nc_S/(c_S + c_P) = n\delta_P/(\delta_S + \delta_P). \qquad (5)$$

Vibrating External Loads. In the event of vibrating external load both the maximum working load F_{Ao} and the

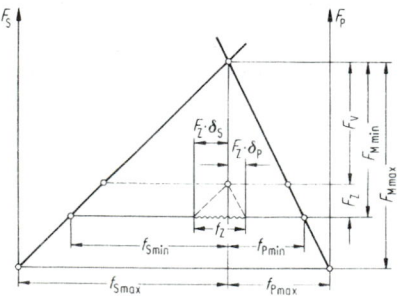

Figure 61. Deflection diagram used to determine the influence of yielding and preload distribution.

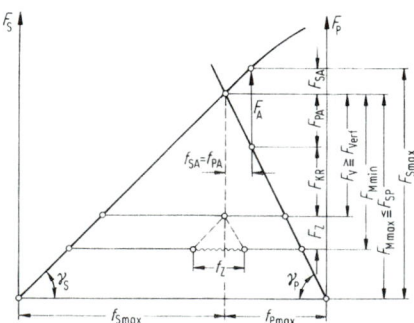

Figure 62. Deflection diagram used to determine additional bolt force F_{SA}, maximum bolt force F_{Smax} and residual clamping force F_{KR} where $\tan \gamma_s = c_s$ and $\tan \gamma_P = c_P$.

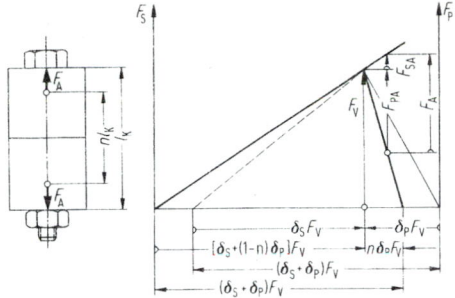

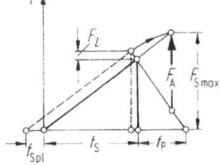

Figure 65. Deflection diagram showing loading of the bolt into the plastic region (influenced by the working load F_A).

Figure 63. Deflection diagram for working load F_A introduced inside the restrained members (yielding and preload distribution not taken into consideration)

minimum working load F_{Au} are included in the distortion diagram (**Fig. 64a**) using the algebraic sign and the vibrating load for the bolt is deduced from this. In the case of alternating working load the expression is $F_{Ao} = -F_{Au}$, so that the alternating portion of the load F_{SAo} of the additional bolt force is equal to F_{SAo}. With dynamic working load $F_{Ao} = F_A$ and $F_{Au} = 0$, where the alternating portion of the load is given by $F_{SAa} = \Phi_K F_A/2$ or $F_{SAa} = \Phi_n F_A/2$ (**Fig. 64b**). A centrally acting static compressive force F_A has been drawn into **Fig. 64c**.

Loading up to the Plastic Region. If a bolt is stressed by a centrally acting external tensile load F_A into the plastic region, a change in the preload triangle (represented by a broken line) then follows, as shown in **Fig. 65**. After the relaxation, the removal of the external load F_A, only the preload reduced by F_Z remains; F_Z is obtained with $F_Z = f_{Spl}/(\delta_S + \delta_P)$, where f_{Spl} is the portion of plastic deformation under the total bolt force F_{Smax} after F_A is applied. Similar considerations are necessary in the case of compressive forces and setting of the retained plates.

Eccentric Retention and Loading. The case dealt with above of a centrally retained and centrally loaded bolted connection can only rarely be put into practice with any degree of accuracy. If the axis of the bolt and the resultant of the external load F_A do not coincide with the centroid of the retained members, but lie parallel to it as in **Fig. 66**, this may have a considerable influence on the additional bolt force; in addition a bending moment is usually produced in the junction line of the bolted connection with the result that the eccentrically loaded bolted connection tends to lift (gape) in the junction line. It is desirable to prevent gaping in the bolted connection by means of suitable design; for design information for threaded connections see **Fig. 67** (cylinder connections) and **Fig. 68** (beam connections).

The results of various model analyses have been set out in **Table 18** to enable the forces and moments in eccen-

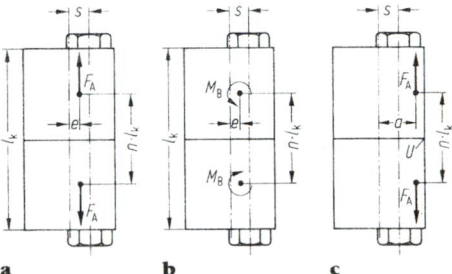

Figure 66. Prestressed and loaded prismatic bolted connections: **a** with tensile force F_A when $e = \Phi_K s$, **b** loaded solely by bending moment M_B, **c** with tensile force F_A at a distance a from the centroid of the prismatic beam with the bore.

trically loaded bolted connections to be calculated; this is conditional on there being no gaping in the junction line and the forces being introduced over the retained members at the distance $(nl_K)/2$ from the junction line of the bolted connection. Besides the (compression-tension) resilience of the bolt $\delta_S = 1/c_S$ and the (compression-tension) resilience of the plates $\delta_P = 1/c_P$ as in **Fig. 58**, the bending resilience of the bolt β_S and a bending resilience of the plates β_P are required. Prismatic bending members are given by the expression $\beta = l_K/(EI_B)$, where E = the modulus of elasticity, l_K = the length of engagement and I_B = the second moment of area of the bending member.

The distance e of the force F_A from the centroid of the retained members as in **Fig. 66a** has been expressed as $e = \Phi_K s$ and Φ_K in Eq. (3) in such a way that the additional bolt load F_{SA} becomes equal to the additional screwing load of a centrally retained and loaded bolted connection and the additional bending moment in the bolt M_{Sb} becomes equal to zero; a force of pressure present in the plates is reduced by the relaxation of the plates F_{PA} and a bending moment present in the junction line altered by M_{Pb}. A pure bending moment load M_B as in **Fig. 66b** produced an additional bolt load $F_{SA} = n\Phi_{mK}M_B/s$ with the force ratio

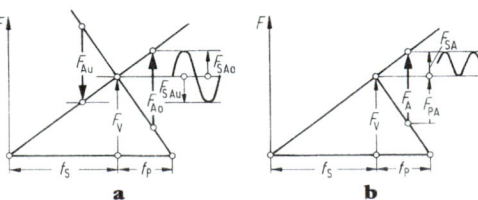

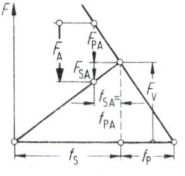

Figure 64. Deflection diagrams for external working loads F_A: **a** vibrating tensile-compressive load, **b** dynamic tensile load, **c** static compressive load.

Cylinder connections

Design recommendations	Unsuitable	Suitable
1 Preload: Preload as high as possible Higher property class - Accurate tightening procedure - Low coefficient of friction	Low preload	High preload (select tightening procedure with low tightening factor α_A)
2 Stiffness ratio: The resilience of the bolt should, where possible, be much greater than that of the plate (if possible, reduced-shaft bolt) $\delta_S \gg \delta_P$	Thin narrow cylinder (with given nominal diameter)	Cylinder diameter $G = d_w + h_{min}$
3 Eccentricity of the bolt: The eccentricity of the bolt location should be kept to a minimum (especially with concentric loading)	Great eccentricity s	Minimal eccentricity s
4 Eccentricity of application of force: Minimum eccentricity usually leads to lower additional bolt forces when $a > s$	Great eccentricity a	Minimal eccentricity a
5 Amount of load distribution: Force to be applied as far downwards as possible towards the junction line	Force applied in upper region	Force applied near the junction line
Standard values for	$n \approx 0.7$	$n \approx 0.3$

Figure 67. Recommendations for the design of cylinder joints taken from [76, 86], updated.

Beam connections

Design recommendations	Unsuitable	Suitable
1 Preload: Preload as high as possible Higher property class - Accurate tightening procedure - Low coefficient of friction	Low preload	High preload (select tightening procedure with low tightening factor α_A)
2 Width of beam (load centric) Where possible use the recommended beam width $b = d_w + h$	Very narrow connections	Beam width $b = d_w + h$
3 Height of beam: Greater beam heights produce lower additional bolt forces	Low beam heights	Great beam heights
4 Projecting end (load eccentric): A projecting end should always be provided so that the supporting effect can fully develop. Limited bearing surface at possible point of separation U	Minimum projecting end	Projecting end $\ddot{u} \approx h$
5 Connecting members: The additional bolt force becomes smaller when the members to be joined cause a parallel displacement in the beam, e.g. using symmetry.	Loose coupling	Fixed coupling e as small as possible

Figure 68. Recommendations for the design of beam joints taken from [76, 86], updated.

$$\Phi_{mK} = \frac{\beta_S \beta_P s^2 / (\beta_S + \beta_P)}{\delta_P + \delta_S + (\beta_S \beta_P s^2)/(\beta_S + \beta_P)}$$

$$\approx \frac{\beta_P s^2}{\delta_P + \delta_S + (\beta_P s^2)}, \qquad (6)$$

as usually $\beta_P \ll \beta_S$. For an eccentric bolt load F_A as in **Fig. 66c** the additional bolt load $F_{SA} = n\Phi_{cK}F_A/s$ then obtained through the superposition of loads as in **Fig. 66a, b** where $M_B = F_A(a - e)$ and

$$\Phi_{cK} \approx \frac{\delta_P + (\beta_P as)}{\delta_P + \delta_S + (\beta_P s^2)}. \qquad (7)$$

The bolt preload F_V (tension) produces in the junction line of the bolted connection an equally large force of pressure F_P, also a bolt bending moment $M_{Sb} = - F_V s \beta_P/(\beta_S + \beta_P) = - \Psi_K F_V s$ and in the junction line a bending moment $M_{Pb} = F_V s \beta_S/(\beta_S + \beta_P) \approx - F_V s$.

The expression for prismatic beams with an imaginary surface A_{im} and an imaginary second moment of area I_{Bim} is $\beta_P/\delta_P = A_{im}/I_{Bim}$. In VDI Guideline 2230 the strength calculations for a threaded joint on a connecting rod bearing cover and strength calculations for a threaded joint on a cylinder cover are dealt with as examples [76].

Separation Limit. In order to determine the limit load F_{Aab} or M_{Bab} below which no gaping occurs in the junction

line of the bolted connection, compressive stress is calculated from the minimum preload F_V and the working loads F_A and M_B in the junction line. To avoid gaping it is necessary that this compressive stress should not occur at any point on the junction line, e.g. at the point U in **Fig. 66c** with positive F_A, in the tension area.

Reverse Drawing of Flanges. In flange joints with thin flange plates these can be reverse drawn under external forces such as spring washers or crimped under external moments such as cap edges. For design information on multiple bolted connections using flanges see **Fig. 69**. If there are flexible seals between the flanges their resilience must be added to the resilience of the flange.

1.6.8 Static and Fatigue Strength of Bolted Connections

Working Loads. In order to design bolted connections the external loads that occur during operation must be known as precisely as possible. It is useful for design purposes to draw a distinction between rarely occurring heavy extra loads and frequently occurring working loads. The rare heavy extra loads are statistically assessed in terms of strength calculations, whereas for the frequently occurring working loads it is usually an assessment of fatigue strength that is aimed for, at least at the design stage. The ideal – not often practicable – is a bolted connection that fully compensates the connecting sections of

Table 18. Bolt and (additional) plate forces or (additional) moments resulting from external loading and preload

Forces and moments in bolt and junction line	Load as shown in			Bolt preload F_v
	Fig. 66a Bolt force when $e = \Phi_K s$	**Fig. 66b** Moment force only	**Fig. 66c** Eccentric load $(\beta_P \ll \beta_S)$	
F_{SA}	$n\,\Phi_K F_A$	$n\,\dfrac{\Phi_{mK}}{s}\cdot M_B$	$\approx n\,\Phi_{aK} F_A$	$F_v\;(=F_S)$
F_{PA}	$(1 - n\Phi_K)\,F_A$	$-\pi\,\dfrac{\Phi_{mK}}{s}\cdot M_B$	$\approx (1 - n\,\Phi_{cK})\,F_A$	$-F_v\;(=F_P)$
M_{Sb}	0	$n\cdot\dfrac{\delta_P + \delta_S}{\beta_B \cdot s^2}\,(\Phi_{mK}\,M_S)$	$\dfrac{n\,\beta_P(a - e)}{\beta_S + \beta_P + \dfrac{\beta_S\beta_P \cdot s^2}{\delta_S + \delta_P}}\cdot F_A$	$-\psi_K \cdot F_v \cdot s$
M_{Pb}	$(1 - n)\cdot F_A\,e$	$M_B - M_{Sb} - (F_S \cdot s)$	$\approx F_A\,a - F_S \cdot s$	$\approx -F_v \cdot s$

With load distributed to restrained members at distance $nl_k/2$ from the junction line: $\Phi_K = \delta_P/(\delta_P + \delta_S)$

where $\psi_K = \beta_P/(\beta_P + \beta_S)$

the members for any working or extra loads that may arise.

Bolted Connection. As a first step the loads acting on the bolted connection subjected to the greatest stresses must be derived from the external loads acting on the multiple bolted connection. A variety of computer programs are available for this step, e.g. [87]. In simple cases the working and extra loads acting on the bolted connection can also be determined without the use of computers, but this requires considerable experience when it comes to estimating the initial load amplitude nl_K. In order to design the bolted connection the external axial force F_A, the shearing forces F_x and F_y to be transmitted if necessary via the junction line and the bending moments M_x and M_y must be known both for the extra loads and the working loads (**Fig. 70**). Further, the required normal force F_N must be indicated and the associated tightening method with the tightening factor α_A must be defined in order to lay down a minimum residual clamping force, the required tight-fitting force and/or to absorb frictional forces at the mating edges $F_Q = \sqrt{F_x^2 + F_y^2} = \mu F_N$.

Predimensioning. The maximum bolt force F_{Smax} can be assumed for an initial rough calculation as $F_{Smax} \approx \alpha A$ $(F_{Kreq} + F_A)$. This must be lower than $F_{Sp}/0.9$ as in **Appendix F1, Table 7** or a table in VDI Guideline 2230 if a 90% yield stress duty is deemed to be permissible for tightening. Using these, the required bolt size d can be found for any desired property class, or the required property class of a bolt can be found for a bolt size defined for the first time in an initial design. It is on the basis of the structural design, which can if necessary be corrected at this point, that the length of engagement l_K must be determined, knowledge of which is required to calculate the bolt resilience δ_S and the plate resilience δ_P. If the unit pressure p under the bolt head and nut, roughly calculated at $F_{Smax} = 1.2F_{Sp}/0.9$ for tightening up to or beyond yield stress, makes it necessary to alter the length of engagement owing to an additional requirement for high-tensile washers, this must also be allowed for. The unit pressure

is calculated using the size of the bearing surface A_P according to the formula $p = F_{Smax}/A_P$ and may not be greater than the interfacial pressure p_G as in **Appendix F1, Table 8**. Eccentrically retained and eccentrically loaded bolted connections must therefore be checked to see whether gaping in the junction line can be prevented, at least in the area of the Birger cone, under the most unfavourable loads, as in **Fig. 57**, and whether the required minimum clamping force F_{Kreq} is ensured to allow for set and eccentricity. At all events, the aim must be that the frequently occurring working loads should be transmitted by friction at right angles to the axis of the bolt; additional positive locking devices (pins) can be used if required to transmit rarely occurring heavy extra loads [91].

Force Ratios. The compression–tension resilience values δ_S and δ_P must be determined in accordance with **Figs 59 and 60**, the bending resilience values β_S and β_P are estimated by adding the governing component resilience values $\beta_i = l_i/(EI_{Bi})$, where l_i represents component lengths, E the modulus of elasticity and I_{Bi} the geometrical moments of inertia. Similarly, δ_S is determined for a bolt as shown in **Fig. 59** as: $\beta_S \approx \beta_K + \beta_1 +_{\beta 2} + \beta_f + \beta_G + 8\delta_M/d^2$ where the geometrical moments of inertia are $I_{Bi} = \pi d_i^4/64$ and the resilience for nut displacement δ_M. The bending resilience of the imaginary beam must be calculated in a much less precise way. As an initial approximation $\beta_P = \delta_P A_{im}/I_{im}$ is used with A_{im} as in **Fig. 60** and $I_{im} = bb_B^3/12$ with estimated values for the width b and the height b_b of a rectangular beam subjected to bending; maximum values may be selected for b and b_B that do not exceed the diameter of the Birger cone (**Fig. 57**) in the plane of the junction line.

Bolt Loads. Table 18 is used to determine the bolt forces F_S and bolt bending moments M_{Sb} for the bolted connections as shown in **Fig. 66**. For standard values for the factor n see **Fig. 67**. In case of doubt, the less favourable value of n must be selected. The formulae assume plane-

Multiple-bolt connections

	Design recommendations	Unsuitable	Suitable
1	Preload: Preload as high as possible Higher property class Low coefficient of friction	Low preload	High preload (select tightening procedure with low tightening factor a_A)
2	Number of bolts z: Provide as large a number of bolts as possible limited by the external dimensions of the spanner	Small number of bolts or few large bolts	Large number of bolts used with rotationally symmetrical connection: $z = \dfrac{d_t \cdot \pi}{d_w + h}$ (rounded off)
3	Height of flange plate: Flange plate to be designed as thick as possible, standard value: plate thickness > eccentricity f	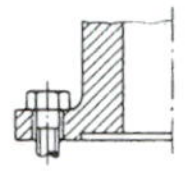 $h > f$	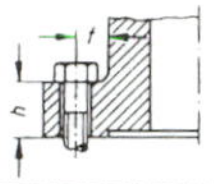
4	Eccentricity: To be minimised, if possible select hexagon socket head bolt but dimension transitional radius according to strength (e.g. fatigue strength)	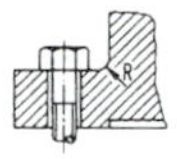	$f \rightarrow$ Minimum 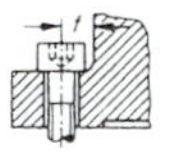
5	Projecting end of flange plate: Projecting end of flange plate $\ddot{u}$ to be equal to or greater than flange plate height h	$\ddot{u} < h$ 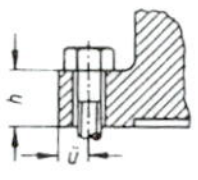	$\ddot{u} \approx h$ 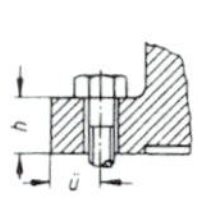
6	Bearing surface: A defined surface to be created in the junction line by means of an indentation. Maximum depth of indentation h_e to be 10 % of the flange plate height h	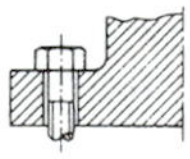	$l_4 \approx (d_w + h)/2$ 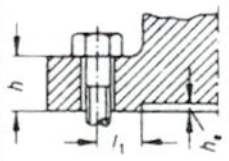
7	Stiffness of joint: Produce as much stiffness in the joint as possible. The ideal is the solid cross-section of the joint	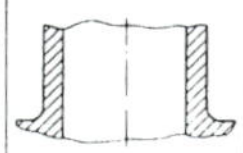	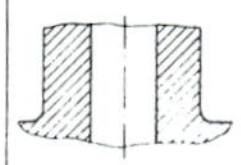

Figure 69. Recommendations for the design of multiple bolted connections taken from [76, 86], updated.

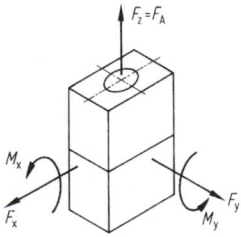

Figure 70. Possible loads on a bolted connection: F_A axial load, $F_0 = \sqrt{F_x^2 + F_y^2}$ transverse load, M_x, M_y bending moments.

parallel bearing surfaces for the bolt head and nut. The maximum assembly preload is specified in Eq. (4) as $\Phi_K \rightarrow n\Phi_{eK}$. For the frictional moment of the thread during tightening see F1.6.6.

Maximum Bolt Stress. At maximum assembly preload a full diameter bolt in the male thread is loaded by the nominal tensile stress $\sigma_{zM} = F_{Mmax}/A_S$ and a nominal torsional stress $\tau_{tM} = M_G/W_P$ (where W_P is the polar section modulus of the stressed cross-section A_S); the bending moment M_{Sb} produced by F_{Mmax} in an eccentrically retained bolted connection may usually be ignored. The additional force F_{SAo} resulting from axial working loads and moments is also acting and therefore, for example, in the case of eccentric working load F_{Ao} the additional tensile stress $\sigma_{SAo} = n\Phi_{eK}F_{Ao}/A_S$. The superposition of these stresses as stated in the Mises theory produces the reduced stress σ_{zred} which according to VDI Guideline 2230 may reach as much as $R_{P0.2}$ in the case of elastic tightening and in the case of tightening $R_{P0.2}$ at beyond yield stress may even be exceeded in a theoretically limited way. This rough-and-ready approach presupposes that even in the notched condition the material remains sufficiently free for the thread not to be stripped, and that any signs of set arising at working load are taken into account when determining the remaining length of engagement. This simple method of calculation is not suitable for determining component failure where large bolts are used [81, 89]. For reduced shank and anti-fatigue shank bolts it is the narrowest cross-section A_T that has to be taken into account rather than the stressed section A_S, and the same applies to the section modulus W_P.

Unit Pressure Under Bolt Head and Nut. It must be checked whether the permissible unit pressure p in the bearing surface of the bolt head and nut has been adhered to so that a maximum calculated bolt force $F_{Smax} = f_a(F_{Mmax} + \Phi F_{Ao})$ may be obtained, where $f_a = 1$ stands for elastic tightening and $f_a = 1.2$ for tightening at or beyond yield stress and Φ is the least favourable force ratio. For permissible unit pressures see **Appendix F1, Table 8**.

Unit Pressure in the Thread. For bolt-and-nut combinations with specified proof loads in accordance with DIN ISO 898 the unit pressure in the thread must not be checked when under continuous load. In translation-type bolts the unit pressure p in the thread determines what height of nut is required. The bearing face of one turn of thread is calculated from the dimensions given in **Fig. 52** and **Fig. 54**. Heights of nut where $h > 1.5d$ are not used and are no longer to be considered; $p_{perm} = 7.5$ N/mm^2 for non-alloy engineering steels and $p_{perm} = 15$ N/mm^2 for high-tensile steel can then be assumed as the permissible unit pressure, assuming a uniform pressure distribution for translation-type bolts using bronze nuts.

Stripping Strength of Bolt and Nut Threads. The load-carrying capacity of the threaded connection under continuous load is determined by the shear strength $\tau_B \approx 0.6\ R_m$ of the bolt or nut and by the corresponding effective shear areas A_{SG} which in turn depend on the expansion of the nut, the plastic warping of the thread and manufacturing tolerances. For example, for nuts with a ratio of width over flats/nominal diameter = 1.5 an additional reduction in the geometrical shear area by 25% can be assumed for nut expansion and plastic warping of the thread; the influence of friction when tightening the bolted connection can be taken into account by making a 10 to 15% allowance [76].

Table 19. Service life of the thread of bolts in property classes 8.8, 10.9 and 12.9 (reference values) with nominal thread side d up to 40 mm and forcing nuts, bolt force at 2% elongation limit $F_{0.2}$, preload F_v [76]

Service life	Coated thread		Rolled thread	
$\pm \sigma_A$ (N/mm)	$\sigma_{ASV} \approx 0.75 \left(\dfrac{180 \ [mm]}{d} + 52 \right)$		$\sigma_{ASG} \approx \left(2 - \dfrac{F_v}{F_{0.2}} \right) \sigma_{ASV}$	
Dependent on preload	No		Yes	
Scope	$0.2F_{0.2} < F_v < 0.8F_{0.2}$		$0.2F_{0.2} < F_v < 0.8F_{0.2}$	

Fatigue Loading. The limiting variable stress component σ_{za} in fatigue loading with 10^6 or more stress cycles is determined from the nominal stress amplitude $\sigma_{za} = n\Phi_{eK}(F_{Ao} - F_{Au})/(2A_s)$ and the associated nominal bending stress amplitude $\sigma_{ba} = [(M_{Sb})_o - (M_{Sb})_u]/(2W_s)$ (see **Fig. 64** and **Table 18**), core cross-section A_s and associated flexural section modulus $W_s = \pi d_s^3/32$. The *variable stress component* σ_{za} must remain smaller than the permissible value σ_{Apcrm}, which can be estimated using the required factor of safety S_D as in $\sigma_{Apcrm} = \sigma_A/S_D$ and **Table 19** for bolts in property classes 8.8, 10.9 and 12.9 as specified in VDI Guideline 2230. The factor 0.75 in the formula for σ_{ASV} allows for the fact that service life can vary by as much as 25% from experimental mean values [76].

Owing to the sharp roots in the V-thread and the introduction of force by means of a forcing nut, e.g. as in **Fig. 57**, service life σ_A is very short in comparison with the service life of a plain rod made from the same material. Introducing load by means of a forcing nut is thus very unsatisfactory as owing to deformation of the thread under load the tensile force in the shank is not uniformly distributed over the whole thread and owing to magnetic flux reversal a tensile force in the shank becomes a compressive force in the nut. It can be assumed that, using a standard forcing nut, up to 40% of the tensile force F_s in the first turn of thread is transmitted, unless the load is redistributed by means of plastic flow (set) when the nut is tightened.

Bolted connections with coated bolts up to $d = 40$ mm are shown to be relatively insensitive to mean stress owing to such plastic load redistribution (**Fig. 71**). The extended service life of rolled bolts decreases with increasing preload.

Thread Run-out and Bolt-Shaft Transition. If the fatigue strength of the shank thread is considerably increased by coating and rolling, stress raisers at other places on the bolts, such as the thread run-out as in DIN 76 Part 1, must be checked with regard to service life and structurally improved if necessary. The variable stress component which can be withstood at the centre in various transitions between thread and shank is shown in **Fig. 72** [78, 95]. The transitional radius for the head-shank transition is specified in the standards for bolts. The tolerances of the transitional radius as drawn into a design are still usually adequate for standard bolts with relatively flat heads, providing the thread is brought to maximum service life values by strain-hardening. By increasing the transitional radii to $0.08d$, particularly durable bolts can be designed but without a significant increase in the external diameter of the head.

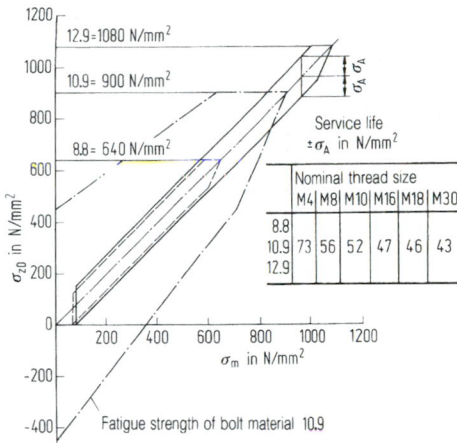

Figure 71. Limits on service life for coated bolts with cut thread and forcing nut.

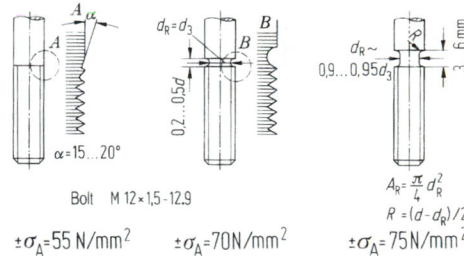

Figure 72. Influence of thread runout design on the alternating stress component σ_A to be withstood at the change in section from the thread to the shank [95].

Large Bolts [81, 89, 90]. Size requirements impose limits on the increase in fatigue strength achievable by rolling. Additional measures to increase the service life of large bolts are therefore employed when heavy dynamic fractional loads are present. In **Fig. 73** the first threads are relaxed on the bolt by displacing the magnetic flux reversal and in the tapped hole by the conical transitional radius to the shank using the mating thread. The expansion bolt is non-rigid and is retained at the bottom of the blind hole by a dog cap. The relatively non-rigid expansion bolt is hydraulically preloaded and relaxed by a shear pin

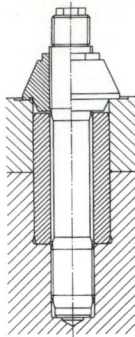

Figure 73. Structural devices to increase the service life of large bolts.

from rarely occurring transverse loads, whereas frequently occurring transverse working loads are transmitted by frictional engagement. Preloads produced by tightening must be specified in accordance with fracture mechanics calculations [81].

Bolted Connections with Special Requirements. These are treated to withstand higher or lower temperatures and/or corrosion (examples in [78–80]).

1.6.9 Thread Locking Devices

A structurally well-designed threaded connection which has been securely preloaded does not generally require any additional thread locking facility, especially when high-strength bolt or screw materials, adequate bolt or screw resilience δ_S, adequate length of engagement ($l_K \geq 6d$) and a minimum length of junction line are provided. Measures adopted to increase the length of engagement or resilience δ_S (**Fig. 74**) not only have the advantage that they lower the additional bolt force F_{SA} but also have the advantage of an increased factor of safety against loosening.

In many cases *relaxation* resulting from *set* or *creep* in the members being joined or automatic *loosening* as a consequence of relative motions between the contact surfaces may however mean that the required preload is not reached, so that suitable locking devices must be specified at the structural design stage. For example, creep may be observed when low-strength copper plates or lacquered steel plates are being retained, even at room temperature, while relative motions occur between the contact surfaces especially in thin retained components and under loads perpendicular to the axis of the bolt or screw when preload is inadequate. A distinction is made between *"anti-set locking devices"* to compensate for creep and set and *"anti-loosening locking devices"* which are capable of

blocking or preventing the "internal" loosening torque produced by relative motion; *"anti-loss locking devices"* may not prevent a partial loosening, but will certainly prevent a complete separation of the threaded connection.

Table 20 provides an overview of the operation and effectiveness of various locking devices [76, 78–80].

Tensioned Spring Devices are not, as a rule, capable of preventing the loosening process as a result of alternating transverse displacement. Their use as anti-set locking devices can be recommended for very short axially loaded bolts or screws belonging to the lower property classes (≤ 6.8). The spring action, however, must also be present at full preload and heaviest working load. The danger of crevice corrosion under certain atmospheric conditions must be borne in mind [76, 79, 83].

Positive Locking Devices can only absorb a limited loosening torque and should only therefore be used with bolts or screws in the lower strength range (≤ 6.8). As they generally retain only a low residual preload, they lock the joint against loss, especially after set (**Fig. 75**). In general, but particularly in machine tool construction, DIN 462 locking plates with internal boss are used for DIN 1804 slotted nuts (**Fig. 75d**).

Clamping Devices in "self-locking" nuts as specified in DIN 980/982/985/986/6924/6925 (example in **Fig.**

Table 20. Classification of locking devices by function and effectiveness as in [76, 78, 80]

Grouping according to function	Example	Effectiveness
Spring stop devices	Belleville spring washers Tension disc DIN 6769, 6908	Seating lock for axially loaded short bolts belonging to the lower property classes (≤ 6.8)
Positive locking devices	Castle nut DIN 935 Bolt with pinhole DIN 962 Wire locking Safety plate tongued outside DIN 432	Anti-loss locking for transversely stressed bolted connections belonging to the lower property classes (≤ 6.8)
Clamping devices	All-metal nuts with clamping component Self-tapping screws Nuts with plastic insert[a] Bolts with plastic coating on thread[a]	Anti-loss locking
Locking devices	Self-locking bolt with serrated bearing surface Self-locking nut with serrated bearing surface	Quick-release; except hardened surfaces (HRC > 40)
Adhesives	Microencapsulated bolts[a] Liquid adhesive[a]	Quick-release

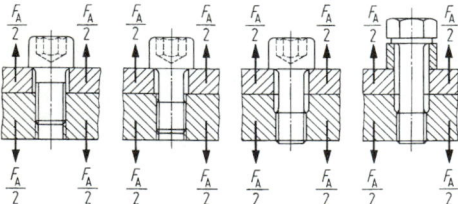

Figure 74. Structural devices showing gradually increasing service life and increasing safety from loosening for the bolted connection.

[a]Note temperature requirements.

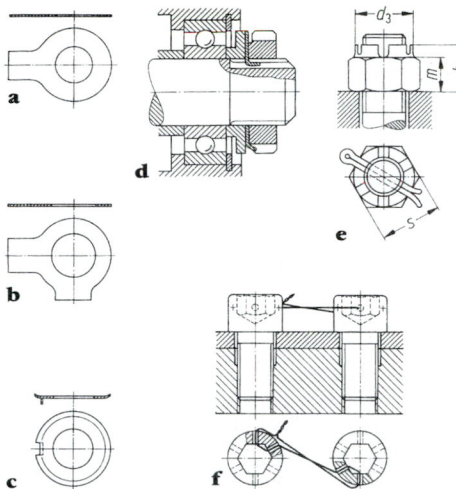

Figure 75. Positive locking devices: **a** DIN 93 tab washer, **b** DIN 463 safety plate with two flaps, **c** DIN 432 safety plate tongued outside, **d** DIN 462 safety plate tongued inside for DIN 1804 slotted round nut, **e** DIN 935 crown nut with DIN 94 split pin, **f** wire locking.

76a) provide a very tight fit and can at least be regarded as anti-loss locking devices. Jam nuts (with a flatter nut as the lower nut) as shown in **Fig. 76b** and DIN 7967 lock nuts as shown in **Fig. 76c** are unreliable as a protection against loosening.

Retaining Devices (ribs or teeth) in the bearing surface of the bolt or nut as shown in **Fig. 76d** and **Fig. 76e** are capable in most applications of blocking the internal loosening torque, thereby retaining preload to its full extent, as they dig into non-hardened surfaces; it must be borne in mind, however, that stress concentration may arise due to surface deformation [85].

Adhesive Coated Devices cause material retention in the thread, thereby preventing relative motions between shank and nut thread flanks, with the result that internal loosening torques do not come into play [79]. Adhesive coated locking devices are particularly suitable for hard-

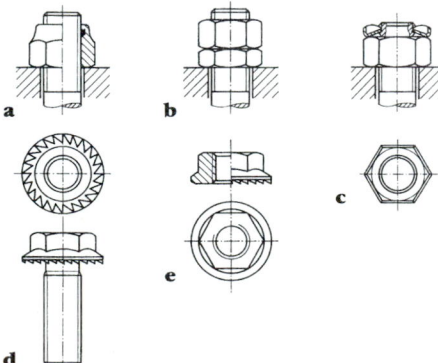

Figure 76. Friction-grip and bolt-locking devices: **a** DIN 982 self-locking nut, **b** jam nut, **c** DIN 7987 locknut, **d** self-locking bolt with serrated bearing surface, **e** self-locking nut with serrated bearing surface.

ened surfaces where retaining devices can no longer be used. Attention must be paid to the sometimes strongly interfering thread friction which is present during tightening, and the limit of application of about 90 °C. In heavy machine construction bolts and nuts are often secured against loosening by fillet welds at one or two sides of the hexagon.

1.7 Selecting Types of Connection

W. Beitz, Berlin

On looking at the variety of possible types of connection

Table 21. Requirements for joints, arranged according to constructional features

Features	Requirements
Geometry	Suitable for parts to be joined
	Small effective geometry
Kinematics	Simple assembly motions
Forces	Able to absorb high working loads
	Low resilience
	Adaptable resilience
	Low assembly forces
Energy	Suitable for energy used in assembly and
	dismantling
Material	Suitable for materials of parts to be joined
	Resistance to corrosion
	Insulating properties
Signal	Clear component location
Safety	High factor of safety against loosening
	High overload capacity
	Long service life
	Low risk of injury
Ergonomics/design	Flexible design
Fabrication	Simple component fabrication
	Capable of being standardised
Assembly	Suitable for assembly conditions
	Simple assembly tools (equipment)
	Adjustable (rough tolerances permitted)
	Ease of storage, access, handling
Control	Simple quality control
Transport	Simple packing
Use	Capable of being repeatedly loosened
	Capable of being sealed
	Temperature independence
Recycling/	Capable of being recycled
maintenance	Capable of being re-used
Costs	Low production costs
	Low operating costs

Table 22. Criteria for selecting connections

Main features	Individual features	Welding	Soldering and brazing	Adhesive bonding	Friction grip	Form grip	Bolting and screwing
Function	Multiplicity of loads	Very good (polydirectional loads possible)	Restricted (shear stress preferred)	Restricted (shear stress preferred)	Good (polydirectional loads possible in direction of friction grip)	Restricted (especially in joints that have not been prestressed)	Good (where mating parts are frictionally engaged)
	Centring facility	Not available (only possible through additional design features)	Not available (only possible through additional design features)	Not available (only possible through additional design features)	Limited (to direction at right angles to friction grip)	Good (especially in prestressed joints)	Limited (only possible through additional design features)
	Damping, stiffness	Hardly any additional damping. Good stiffness	Hardly any additional damping. Good stiffness	Hardly any additional damping. Good stiffness	Additional damping possible. Transitional stiffness satisfactory	Additional damping possible. Transitional stiffness satisfactory (preload)	Damping and transitional stiffness heavily dependent on design
	Additional functions	Few (sealing limited)	Sealing, electrical and thermal conductivity	Sealing, electrical insulation	None	Movement (limited to single directions)	Movement (in special thread types)
Design, layout	Multiplicity of designs	Very good (with regard to form), satisfactory (with regard to material)	Restricted (with regard to form), good (with regard to material)	Restricted (with regard to form), good (with regard to material)	Good (special forms of effective surface not usually required)	Restricted (as special locking devices are required)	Restricted (fastened to standard parts)
	Material utilisation	Good (with suitable design)	Good (low stress concentration, various component materials)	Good (low stress concentration, various component materials)	Good (with suitable design)	Poor (as there is often adverse stress distribution)	Poor (as there is adverse stress distribution)
	Load capacity (static)	Very good (as component material)	Good (through design of shear surfaces)	Good (through design of shear surfaces)	Limited (by friction coefficient and clamping force)	Limited (by adverse stress distribution)	Limited (by quality and number of bolts or screws)
	Load capacity (dynamic)	Restricted (form-related and metallurgical stress raisers)	Good (low stress concentration)	Good (low stress concentration)	Good (if designed with magnetic flux and deformation in mind)	Poor (high stress concentration due to form and magnetic flux)	Limited (stress concentration in thread, preload important)
	Space required	Small (as form of joint can be adapted to suit design features)	Large (as large mating surfaces are required)	Large (as large mating surfaces are required)	Medium (depending on the clamping force applied)	Medium (depending on positive locking devices)	Medium (depending on shape of bolts or screws)
Safety, ergonomics	Operating safety	Very good (in seamless welding)	Good (in seamless soldering and brazing)	Restricted (behaviour under long-term stress in the open air)	Good (if no reduction in clamping force)	Limited (by possibility of loosening and clearances)	Problematic (yielding, loosening)
	Design	Good – restricted (smooth surfaces, limited by standard profiles)	Good (smooth surfaces)	Good (smooth surfaces)	Restricted (depending on clamping force applied)	Restricted (depending on locking devices)	Restricted (depending on form and number of bolts or screws)
Manufacture, assembly, control	Degree of difficulty	Low (using suitable design and selecting correct material, risk of distortion)	High (due to gap widths and surface treatment)	High (in multi-component adhesives and hot-setting, gap widths)	Medium (narrow tolerances, simple forms of effective surface)	Medium (depending on locking devices)	Low (simple operation, standard parts)

	Property						
	Degree of automation	Good (advanced stage of development)	Restricted (equipment required by technology awkward)	Restricted (equipment required by technology awkward)	Restricted (as tools and assembly are expensive)	High (with appropriate quantity)	High (simple assembly)
	Detachability	Not possible (destruction required)	Possible under certain conditions (with soft soldering)	Possible under certain conditions (with some adhesives)	Restricted (only with easily disengaged clamping force)	Good (depending on preload)	Very good (easy to dismantle)
	Quality assurance	Good (small weld sizes, surface cracks easy to detect)	Problematic (poor soldering difficult to detect, no distortion)	Problematic (poor adhesion difficult to detect, no distortion)	Good (when clamp joints are separated, expensive (with interference fits)	Good (measurements easy to check)	Good (with suitable tightening process)
Use	Overload capacity	Problematic (only with plastic deformation)	Not possible (plastic deformation of mating parts not permissible)	Not possible (plastic deformation of mating parts not permissible)	Good (non-destructive slippage)	Not possible (plastic deformations endanger operations)	Possible (if it is possible to replace plastically deformed bolts or screws)
	Re-utilisation	Rare (only if mating parts are refinished)	Rare (only with refinishing in hard soldering)	Problematic (condition of surface, gap width)	Good (only if surface is not damaged during dismantling)	Good (depending on preload)	Good (if no plastic deformation)
	Temperature resistance	Very good (as component material)	Limited heat resistance (depending on the solder)	Limited heat resistance (depending on the adhesive)	Problematic (owing to change in clamping force)	Non-problematic (restricted operation if necessary)	Non-problematic (if taken into account in the layout)
	Corrosion stability	Problematic (gap corrosion can only be avoided by design)	Good (owing to seamless joint)	Problematic (owing to ageing tendency)	Good (owing to seamless contact between effective surfaces)	Problematic (if gaps are present)	Problematic (owing to gap corrosion)
Maintenance	Inspection, care	Simple (surface testing procedures possible)	Expensive (X-ray and ultrasonic testing)	Expensive (X-ray and ultrasonic testing)	Problematic (as the condition of the friction joint is not detectable)	Simple (as loose positive locking joint easy to detect)	Simple (loose bolts or screws easy to detect)
	Repair	Good (welding repairs possible)	Possible (resoldering especially with soft solder)	Rare (new adhesive must be applied)	Easy (only if clamping force can be re-applied)	Rare (in the case of damaged locking devices), easy (with preload)	Possible (by replacing the bolts or screws)
	Recycling	Good (compatible materials)	Restricted (unsuitable material combinations possible)	Restricted (unsuitable material combinations possible)	Restricted (dismantling expensive if material combination unsuitable)	Suitable (can be dismantled)	Suitable (can be dismantled)
Costs	Production costs	Low (especially with single-piece production and large parts)	High (equipment, accuracy of finish)	High (equipment, accuracy of finish)	Medium (simple effective surfaces, narrow tolerances, clamping devices)	High (narrow tolerances, adapting the locking devices)	Low (simple process, standard parts, simple assembly)
	Operating costs	None	None	None	None (except where clamping force has to be re-applied)	None (except for replacing or readjusting locking devices)	Only if tightening is required

and structural constraints, it is difficult to make general recommendations for selecting a specific connection to solve a practical problem. Deciding on a suitable and reliable connection depends too much on the loads that are acting and other operating conditions, on safety requirements, on component size, on quantity and the desired degree of automation, on user requirements and on design considerations. It is not the connection alone, but the whole area of design or even the subassembly or the product as a whole that must be considered if an optimum design is to be produced.

In order to assist in selection, feasible types of connection can be assessed in accordance with requirements or in accordance with assessment criteria derived from them which are relevant to the specific application. A list of general requirements of connections, arranged according to construction features (see **E2, Table 3**), is given in **Table 21**.

A further aid to selection is provided in **Table 22** in which, also arranged according to characteristics, the main properties of the types of connection discussed in F1.4 to F1.6 have been listed in terms of quality. In this table the characteristics used in **Table 21** such as geometry, kinematics, forces, energy, material and signal have been included in the characteristics function, design and layout. The remaining characteristics agree. The properties formulated in **Table 22** for each connection, compared with the respective individual characteristics, can only show tendencies owing to the extent and complexity of the subject matter. For this reason the user must

determine the precise design constraints for each duty and if necessary make a different assessment and come to a different decision. This is determined in particular by the optimal objectives that apply in each case. In any event the assessment criteria may however be used to provide a stimulus for establishing some selection considerations.

The following general guidelines for use may be formulated:

Positive Locking Connections are preferably used for

Frequent and easy loosening
Unidirectional arrangement of members
Absorbing relative motions
Joining members made of dissimilar materials

Friction-Grip Connections are preferably used for

Simple and economic joining of members made of dissimilar materials
Absorbing overstressing due to slippage
Positioning the members in relation to each other
Making possible considerable freedom of design for members

Connections Using Retention of Self Substance are preferably used for

Absorbing multiaxial, as well as dynamic, loads
Economic joining of individual components and small lots with ease of repair
Sealing the mating surfaces
Use with standardised members and profiles

2 Elastic Connections (Springs)

H. Mertens, Berlin

2.1 Uses, Characteristics, Properties

2.1.1 Uses

A spring is a machine element capable of absorbing energy at a relatively large deflection and then storing some or all of it as strain energy. If the spring is unloaded some or all of the stored energy is released. A spring can therefore be described according to its energy-storing and energy-absorbing characteristics (according to *storage and damping capacity*). From these the following uses can be determined:

Maintaining an almost constant pressure when there are small changes in deflection caused by motion, permanent set and wear, e.g. contact springs, spring washers used for locking bolts, compression springs in slip clutches.

Preventing high pressures when there are small relative displacements between components caused by thermal expansion, permanent set or other load independent deformations, e.g. bellows expansion joints in pipelines and electricity conduits, equalisation of expansion joints in plate construction, cover plates or diaphragms in couplings and clutches.

Load compensation or spatially uniform distribution of forces, e.g. for spring systems in vehicles, for spring mattresses.

Guiding machine components without play, e.g. using parallel leaf springs or flexible rubber couplings.

Storing energy, e.g. clock springs or clockwork motors for toys.

Returning a component to its home position after an excursion, e.g. valve springs, pull-back springs in hydraulic valves and meters and check valves.

Measuring forces and moments in measuring and control equipment when there is a reproducible, sufficiently linear interrelationship between force and deformation, e.g. spring balances.

Influencing the vibration characteristics of drive trains, especially eliminating or damping vibrations excited in steady or unsteady operation, but also inversely to produce sympathetic vibrations, e.g. in oscillating conveyors or fatigue-testing machines (see A4).

Vibration isolation, vibration damping, detuning; active and passive isolation of machinery and equipment (see J2.3).

Alleviating shocks by trapping shock energy on long deflections, e.g. vehicle air spring dampers, cushioning springs, insulating the foundations of hammers against shock.

Springs can be classified irrespective of their intended use according to the spring material: *metal springs, rubber springs, fibre composite springs, and gas springs*. The damping capacity of the material in metal springs is relatively low, but is technically efficient in rubber or composite springs. The elastic characteristics of metals can only be utilised with specific shape (*elasticity of shape*); even rubber is relatively stiff and practically incompressible. *Elasticity of volume* can only be utilised in gas springs.

2.1.2 Load-Deformation Diagrams, Spring Rate (Stiffness), Deformation Rate (Flexibility)

Load-Deformation Diagram. This shows the extent to which the elastic force F (or the elastic torque M_t) acting on the spring is dependent on the range of spring s (or angle of twist φ), the difference in excursion between the points of application of force (**Fig. 1**). The increase in the characteristic dF/ds is referred to as *stiffness* c or, as in DIN 2089, *spring rate R*. Provided the spring material complies with Hooke's law and the springs are free of friction, linear spring characteristics may arise for small ranges of spring. Thus

$$c = dF/ds = F/s = F_{max}/s_{max}$$

or

$$c_t = dM_t/d\varphi = M_t/\varphi = M_{t\,max}/\varphi_{max}. \qquad (1)$$

The reciprocal value of the spring rate is referred to as *deformation rate* δ,

$$\delta = 1/c = ds/dF \quad \text{or} \quad \delta_t = 1/c_t = d\varphi/dM_t. \qquad (2)$$

2.1.3 Energy Storage, Energy Storage Efficiency Factor, Damping Capacity, Damping Factor

The area below the characteristic line (**Fig. 1**) is a measure of the resilience or energy capacity of a spring (see A3.2),

$$W = \int_0^{s_{max}} F\, ds \quad \text{or} \quad W_t = \int_0^{\varphi_{max}} M_t\, d\varphi. \qquad (3)$$

For springs with linear characteristic between $s = 0$ and $s = s_{max}$,

$$W = F_{max}\, s_{max}/2 = cs_{max}^2/2 = F_{max}^2/(2c)$$

or

$$W_t = M_{t\,max}\, \varphi_{max}/2 = c_t\, \varphi_{max}^2/2 = M_{t\,max}^2/(2c_t). \qquad (4)$$

With Hooke's law $\sigma = E\varepsilon = E(s/l)$ the following applies for the resilience of a material where tensile or compressive stress is evenly distributed over spring cross-section A and spring length l and the volume is $V = Al$:

$$W = \int_0^{s_{max}} F\, ds = \int_0^{s_{max}} (F/A)\,(Al)\,d(s/l) = V\,\sigma_{max}^2/(2E)$$

or

$$W_t = V\, \tau_{max}^2/(2G). \qquad (5)$$

in the case of shear stress. In the case of unevenly distributed stress,

$$W = \eta_A\, V\, \sigma_{max}^2/(2E)$$

or

$$W_t = \eta_A\, V\tau_{max}^2/(2G), \qquad (6)$$

with volume energy storage efficiency factor η_A, which is a function of the shape of the spring and the type of stress and provides a useful comparison between different types of spring in terms of material utilisation.

In the case of cyclic deformation, e.g. dynamic range of spring as shown in **Fig. 2a** or alternating range of spring as shown in **Fig. 2b**, the area encircled by the characteristic line is a measure of the energy W_D dissipated during one stress cycle. In order to identify the resulting damping capacity for linearly elastic-viscous spring materials the *damping factor* is used; this shows, in the case of variable deformation as in **Fig. 2b**, the ratio of the area encircled by the characteristic line and proportional to W to the triangular area using the deformation amplitude $\hat{s}$ as the datum line and the associated elastic force amplitude F_C as the height; the triangular area is a measure of the elastic deformation energy W_{pot} stored in the reverse position:

$$\psi = W_D/W_{pot}. \qquad (7)$$

Amplification of non-linear characteristics takes place in cyclic deformation [15]. In order to distinguish non-linear elastic characteristics, especially under distributed stress, enlarged spring damping simulation models [88] are required.

2.2 Metal Springs

Metal springs [1–77] are usually manufactured from high-strength elastic materials (see D3.1.4 and **Appendix D3, Table 12**). All standards governing spring steels contain requirements concerning surface finish as the fatigue and endurance strength of springs depends to a large extent on their having a notch-free surface. These requirements must be applied to finished and mounted springs, which means that cracks and abrasion marks are to be avoided when mounting or operating them, and quality assurance is essential. Useful life can also be greatly reduced by the effects of corrosion. Organic or inorganic coatings may be applied to protect against corrosion [73]. It should be borne in mind that there is a danger of hydrogen embrittlement when using electroplated protective coatings [70]. Various chromium–nickel steels or non-fer-

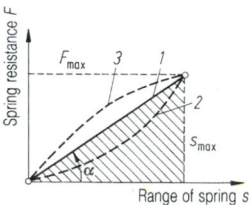

Figure 1. Spring characteristics under continuous load: *1* linear spring characteristic, *2* progressive spring characteristic, *3* decreasing spring characteristic; energy storage W for characteristic *1*, shaded area.

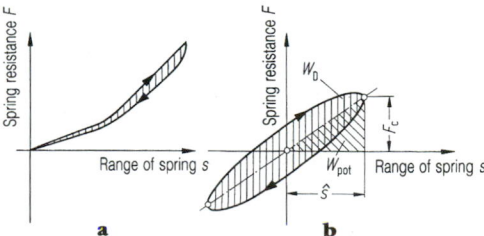

Figure 2. Spring characteristics under cyclic load: **a** characteristic curve for leaf springs layered in two stages under dynamic load, **b** hysteresis loop in the form of an ellipse for an elastic-viscous spring material under alternating stress with damping force proportional to speed.

rous metals can be used to suit the type of corrosion stress. The DIN standards listed in **Table 1** should be taken into consideration when calculating and designing springs. In engineering drawings springs are represented in accordance with DIN ISO 2162.

2.2.1 Axially Loaded Straight Bars and Ring Springs

Tension Bars, Compression Bars

Application. Owing to the high spring rate these are used only in high-frequency testing machines and vibration

Table 1. Design and calculation of steel springs (survey)

Stress type	Spring shape (DIN ISO 2162) Type of load	Design Load distribution (tolerances)	Semifinished standards Material standards	Calculation (static or dynamic load)
Tension, compression	Ring spring, under compressive force	Rings with conical effective surface, stressed alternately by tension and compression	(Work standard, Ringfeder GmbH, Krefeld)	Design calculation; see F2.2.1
Bending stress	Single leaf spring, loaded by transverse force (rectangular, triangular and trapezoidal type)	In springs subjected to dynamic high stress as shown in **Fig. 4**	DIN 1544 (Cold-rolled strip) DIN 17 221 (Hot-rolled) DIN 17 222 (Cold-rolled)	Design calculation; see **Table 2**
	Laminated leaf springs, loaded by transverse force	DIN 1573 (Shims, keys) DIN 2094 (Road vehicles) DIN 4621 (Hooks) DIN 5542 (Ends) DIN 5543 (Suspension) DIN 5544 (Rail vehicles) DIN 1147 (Farm machinery)	DIN 1570 (Hot-rolled, ribbed) DIN 4620 (Hot-rolled)	DIN 5544 (Load deformation curve of springs)
	Cylindrical helical spring (leg spring) with circular and rectangular cross-section	DIN 2088 (Design notes, fixing requirements)	DIN 17 223 Sheets 1, 2 DIN 17 224 (Rustproof) DIN 2076, DIN 2077	DIN 2088 (Equations, examples, nomograms, safe stresses)
	Spiral springs, torsionally loaded	DIN 8255 Part 1 (Rollers) DIN 8287 (For clocks)	DIN 17 222 (Cold-rolled) DIN 1544 (Cold-rolled strip) (DIN 43 801 Part 1)	DIN 43 801 Part 1 (see also F2.2.3)
	Belleville springs, loaded by compressive force (single springs, spring sets, spring columns)	DIN 2093 (Type, play) (DIN 6796 Conical spring washers)		DIN 2092 (Equations, characteristic curves, combinations, examples, literature)
Torsional stress	Torsion bar spring, with circular cross-section, torsionally loaded	DIN 2091 (Torsion bar heads) DIN 5481 (Groove toothing) SAE J 498	DIN 17 221 DIN 2077	DIN 2091 (Equations, equivalent length, pre-setting, endurance and fatigue strength diagram, relaxaton, see also F2.2.5)
	Cylindrical helical extension spring with circular cross-section	DIN 2097 (Loops) DIN 2099 Sheet 2 (Order form)	DIN 17 223 Sheet 1, 2 DIN 17 224 (Rustproof) DIN 17 225 (High-temperature) DIN 17 221 (Hot-formed) DIN 2076, DIN 2077	DIN 2089 Sheet 2 (Equations, coefficients, examples, nomograms)
	Cylindrical helical compression springs with circular cross-section	DIN 2099 Sheet 1 (Order form) DIN 2098 Sheet 1, 2	DIN 2095 DIN 2096 DIN 2098 Sheet 1	DIN 2089 Part 1 (Equations, characteristic curves, buckling, transverse springiness, endurance and fatigue strength diagram)
	Cylindrical helical compression springs with rectangular cross-section	DIN 2090 (For testing machines also cut from the solid)		DIN 2090 (Equations, coefficients: edgewise and flatwise wound)
	Conical helical compression springs	(Circular or rectangular cross-section)		Approximate equations, see [20, 42]

generators and as individual elements in bolted connections (see F1.6).

Principles. For a bar with length *l*, cross-section *A* and modulus of elasticity *E*, spring rate is represented by $c = EA/l$. The energy storage efficiency factor of the elastic volume is $\eta_A = 1$, if stress concentration due to clamping is avoided by means of appropriate transitions: stepped bars.

Ring Springs

Application. Owing to the high level of dissipated energy, used as a volute spring as well as an overload release and a damping device in press manufacture [7, 18].

Structural Shape (**Fig. 3a**). Axially loaded rings with conical work surfaces: (internal cross-section of ring A_i relative to external cross-section of ring A_a) ≈ 0.8; (external diameter of outer ring d_a relative to width of ring *b*) ≈ 5 to 6.

Principles. In order to prevent self-locking in fine-machined rings with angle of friction $\rho \approx 7°$, an angle of slope $\alpha \approx 12°$ is selected; an angle of slope $\alpha \approx 14°$ is selected for larger, unmachined, forged rings where $\rho \approx 9°$. Load applied $F\uparrow$ and load relieved $F\downarrow$, as with leading screws (see A1.11.2), are represented by:

$$F\uparrow = F_c \tan(\alpha + \rho)/\tan\alpha \approx (1.5\ \text{to}\ 1.6)F_c, \quad (8)$$

$$F\downarrow = F_c \tan(\alpha - \rho)/\tan\alpha, \quad (9)$$

where spring force is F_c without allowing for friction as shown in **Fig. 3b**. Energy absorption $W\uparrow$ when load applied is represented by $W\uparrow = (F\uparrow)s/2$, energy release $W\downarrow$ when load relieved is $W\downarrow = (F\downarrow)s/2$, dissipated energy $W_D = W\uparrow - W\downarrow \approx 3/4 W\uparrow$.

The tensile stress σ_z in the outer ring and the compressive strength σ_d in the inner ring are, for reasons of equilibrium, expressed as $\sigma_z A_a = \sigma_d A_i$. The unit pressure *p* in the friction surface becomes $p = \sigma_z A_a/(ld_m)$, where *l* is the overlap length of a pair of cones. Tangential force F_t in the outer ring $F_t = \sigma_z A_a$ limits the maximum carrying force F_{max}, as

$$F\uparrow = F_t \pi \tan(\alpha + \rho) = \sigma_z A_a \pi \tan(\alpha + \rho). \quad (10)$$

The pressing rate *s* of a ring spring support containing *n* rings in total, including two half-hooks, becomes

$$s = 0.5n(\sigma_z d_{ma} + \sigma_a d_{mi})/(E \tan\alpha). \quad (11)$$

Design Calculations. For machined rings made of hardened and tempered special steel and infrequent maximum stress, the permissible stress $\sigma_{z\,perm} = 1000\ \text{N/mm}^2$ can be assumed and permissible compressive stress $\sigma_{d\,perm}$ about 20% higher ($E = 2.1 \cdot 10^5\ \text{N/mm}^2$).

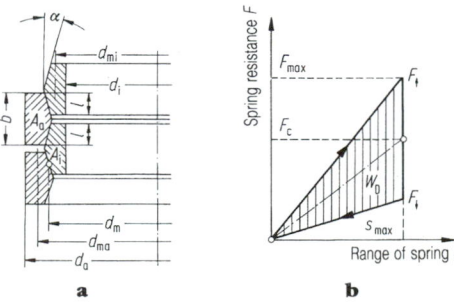

Figure 3. Ring spring: **a** cross-section, **b** characteristic curve before locking.

Microstructure. Depends on lubrication (including lifetime lubrication). For mass-produced products see manufacturer details in **Table 1**.

2.2.2 Leaf Springs and Laminated Leaf Springs

Leaf Springs

Uses. As spring elements in slides, anchors and catches in locks, as contact springs in switches and as guide springs.

Basic Shapes (**Table 2**). As a rectangular spring (**Table 2a**) with a rectangular cross-section of thickness *t* and breadth *b*, constant over its length, or as a triangular (**Table 2b**) or trapezoid spring (**Table 2d**) with constant thickness *t* and linearly variable breadth $b(x)$ or as a parabolic spring (**Table 2c**) with constant breadth *b* and parabolic progression of height $b(x)$ or as a rectangular parallel spring (**Table 2e**).

Design Calculations. Formulae for permissible transverse force F_{perm}, deformation *s* or permissible deformation s_{perm} independent of the transverse force *F* or the permissible bending nominal stress $\sigma_{b\,perm}$, spring rate *c*, cushioning effect *W* and volume energy storage efficiency factor η_a: **Table 2**. If the breadth *b* is very great as compared with the thickness *t*, then the modulus of elasticity can be substituted in the formulae by $E/(1 - v^2)$, with Poisson's ratio of transversal contraction $v \approx 0.3$ (see B3). The triangular spring and the parabolic spring are beams of uniform marginal bending stress (see B2.4.5). If the rectangular parallel spring is used for the vertical support of a vibrating table weighing $G = mg$, then the astatic pendulum effect must be allowed for when calculating the natural angular frequency $\omega_e = \sqrt{c/m - g/l_{red}}$, where l_{red} is the curvature of the radius of the leg of trajectory of the mass guided in parallel by the supporting leaf springs, $l_{red} \approx 0.82l$; in the case of suspension from vertical leaf springs the minus sign after the root symbol must be changed to a plus sign.

Microstructure. In order to keep stress concentration to a minimum the fixed ends must be rounded off and packing, such as paper, plastic, brass and copper or copperplating (or zinc-plating) must be provided in the fixed area. Fixing holes must be located at a distance of at least 3*t* from the fixed end of the spring leaf. Cover plates should be at least 3*t* thick. Stress concentration can be avoided by adjusting the thickness or breadth (**Fig. 4**).

Laminated Leaf Springs

Uses. As spring and steering systems in agricultural, rail and road vehicles.

Structural Shapes. Preformed elliptical leaf springs with rectangular cross-section and longitudinal ribs as described in DIN 11 747 for single or twin axle agricultural trailers; performed trapezoid and parabolic springs as described in DIN 2094 for road vehicles as described in DIN 70 010; preformed parabolic springs as described in DIN 5544 Parts 1 and 2 and shown in **Fig. 5** for rail vehicles.

Design Calculations. In accordance with **Table 2**, these should take into consideration the spring combination required by the load (A4.1). As an initial approximation, combined springs of equal plate thickness and range of spring can be regarded as contiguous (with the same neutral fibre). Friction, which cannot be calculated and is to a great extent dependent on the lubrication and surface finish of the leaves, has the (limited) advantage that it provides damping and the disadvantage compared with other damping systems that impact noise is transmitted

Table 2. Basic shapes and design formulas for the macrostructure of leaf springs

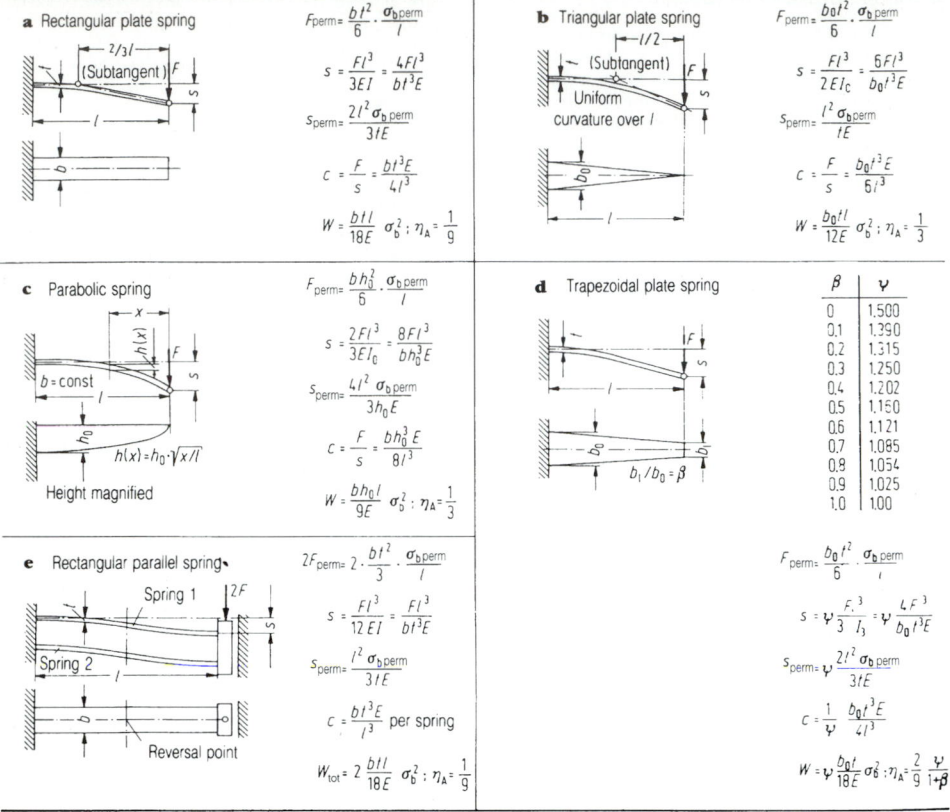

undamped. Reference values for permissible stress are provided in GDR Standard TGL 39249.

Microstructure. For design information on spring ends and initial stress points see standards, **Table 1** as well as [10, 46]. **Figure 5** shows a multiple-leaf parabolic spring for goods wagons, which has been designed to cope with stress [13]. At low loads only the main spring is bearing, but once a certain range of spring has been exceeded the auxiliary spring comes into action, which results in an (axially compressed) progressive characteristic curve. The characteristic curve follows the shape shown in **Fig. 2a**. In order to increase fatigue strength the oil-hardened special steel 50 CrV4 spring leaves are so designed that the laminated spring leaves do not come into contact with each other in high-stress parabolic regions; the spring leaves are hardened to $R_m = 1450$ to 1600 N/mm², provided they have been shot-peened on the tension side and corrosion-proofed on all sides using zinc dust paint. For experimentally obtained fatigue strength values see [13].

2.2.3 Spiral Springs and Helical Torsion Springs

Spiral Springs

Uses. As mainsprings in clocks and watches and as readjusting springs in electrical measuring devices as in DIN 43 810.

Structural Shapes (**Table 3a**). Archimedean spiral as shown in **Table 3a** with rectangular cross-section and both ends of spring fixed or DIN 8287 spring strip coiled in a spiral round a barrel arbor (shaft).

Design Calculations (**Table 3a**). The increase in stress within the cross-section of the spring due to the curvature is not allowed for in the equations, because in general the spring index $w =$ radius of curvature/(half strip thickness) is large enough and if stress is present in the direction of winding a compressive stress with a higher permissible load is acting there. Reference values for permissible load as for helically coiled spiral springs.

Microstructure of springs and fastening ends for mainsprings: see DIN 8287.

Helically Coiled Spiral Springs

Uses. For returning or exerting pressure on levers, covers and similar ("mousetrap spring").

Structural Shapes. As in DIN 2088 with firmly fixed spring legs or static leg located on a spindle as shown in **Fig. 3b**. Spring index $w = D_m/d = 4$ to 15.

Design Calculations (**Table 3b**). Owing to the clamping characteristics, the bending stress is almost uniform in the helix area. If, exceptionally, no dynamic load is acting in the helix direction, the factor k_b, which takes into account

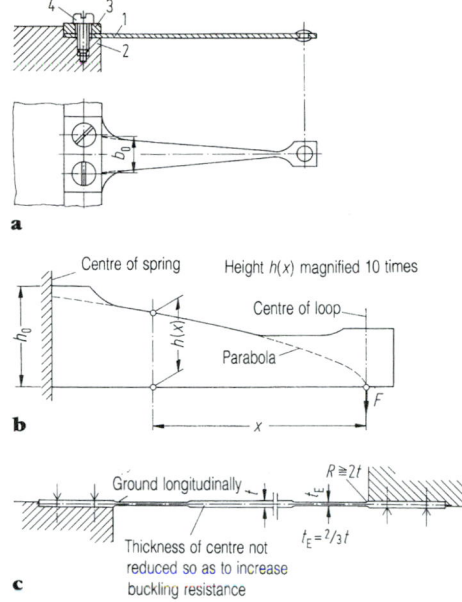

a

Centre of spring   Height $h(x)$ magnified 10 times

Centre of loop

Parabola

b

Ground longitudinally $R \geq 2t$

Thickness of centre not reduced so as to increase buckling resistance

c

$t_E = \frac{2}{3}t$

Figure 4a–c. Microstructure of leaf springs under cyclic stress: **a** *1* triangular plate spring (with tension width extended to $2b_0$), *2* clamping surface with shoulder, *3* cover plate, *4* screw (lacquered tight); **b** variation in thickness in a Brüninghaus parabolic spring; **c** leaf spring (guide spring) fixed at both ends, stress concentration allowed for by reducing thickness at both ends to $2/3t$.

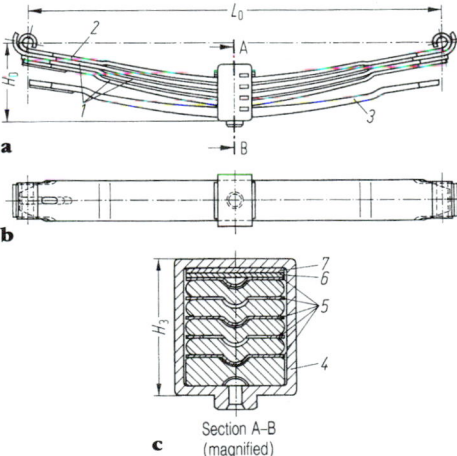

a

b

c

Section A–B (magnified)

Figure 5a–c. Two-stage parabolic spring for goods wagon: **a** side view, **b** top view, **c** cross-section at centre; *1* leaf spring, *2* main leaf spring (shot-peened on train side), *3* auxiliary spring, *4* spring band, *5* lining (galvanised), *6* gib-headed key, *7* taper key.

the increase in stress at the inside edge must be included in the calculations for circular springs $k_b = 1 + 0.87d/D_m + 0.642(d/D_m)^2$. Permissible loads are as in DIN 2088 or simply use values for torsionally stressed helical compression springs increased by factor of 1.42. The stresses present where the wire is bent at the legs must also be checked.

Table 3. Basic shapes and design formulas for the macrostructure of spiral springs and torsion springs under uniform bending stress

a Spiral spring with rectangular cross-section, fixed at both ends	**b** Torsion spring with circular cross-section located on spindle, fixed at both ends
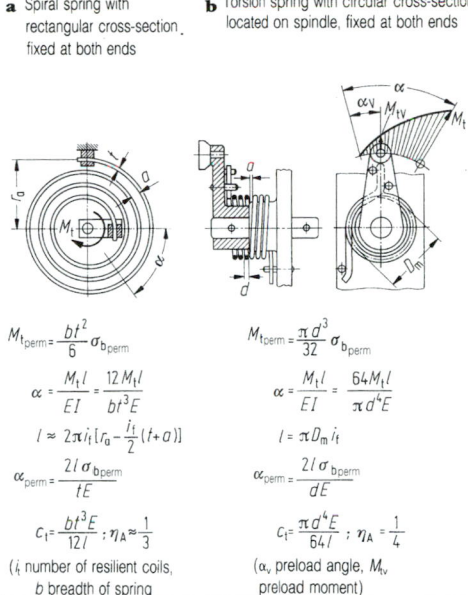	
$M_{t\,perm} = \dfrac{bt^2}{6}\sigma_{b\,perm}$	$M_{t\,perm} = \dfrac{\pi d^3}{32}\sigma_{b\,perm}$
$\alpha = \dfrac{M_t l}{EI} = \dfrac{12 M_t l}{bt^3 E}$	$\alpha = \dfrac{M_t l}{EI} = \dfrac{64 M_t l}{\pi d^4 E}$
$l \approx 2\pi i_t [r_a - \dfrac{i_t}{2}(t+a)]$	$l = \pi D_m i_t$
$\alpha_{perm} = \dfrac{2l\,\sigma_{b\,perm}}{tE}$	$\alpha_{perm} = \dfrac{2l\,\sigma_{b\,perm}}{dE}$
$c_t = \dfrac{bt^3 E}{12\,l}$; $\eta_A \approx \dfrac{1}{3}$	$c_t = \dfrac{\pi d^4 E}{64\,l}$; $\eta_A = \dfrac{1}{4}$
(i_t number of resilient coils, b breadth of spring	(α, preload angle, M_{tv} preload moment)

Microstructure. If the spring is located on a spindle as in **Table 3b**, then clearance is required between the spring and the guide (diameter of mandrel ≈ 0.8 to $0.9D_i$). For more precise details, also for spring rates, with worked examples, see DIN 2088.

2.2.4 Conical Disc (Belleville) Springs

Uses. Because little space is required (these are usually racked into columns), and/or great forces are acting at small deflection, they are used as clamping elements in equipment and tools to operate valves, for buffer and shock absorber springs, to support machines and bed plates, to make longitudinal and tolerance adjustments, and in similar applications.

Types. Typical DIN 2093 Belleville springs are designed as conical discs which can be axially loaded. They are manufactured with and without supporting surfaces (**Fig. 6**).

Macrostructure. $D_e/D_i \approx 2$; Series A is expressed as $D_e/t \approx 18$, $h_0/t \approx 0.4$; Series B is expressed as $D_e/t \approx 28$,

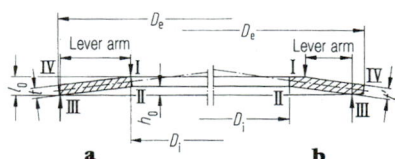

a   b

Figure 6. Single Belleville spring and points of intersection of the stresses to be calculated according to Almen-László (as in DIN 2092): **a** without supporting surfaces, group 1 ($t < 1$ mm) and group 2 (1 mm $\leq t \leq 6$ mm); **b** with supporting surfaces, group 3 (4 mm $< t \leq 14$ mm). Designation of a Belleville spring of series A with external diameter $D_e = 40$ mm, group 2: Belleville spring to DIN 2093–A40 GR2.

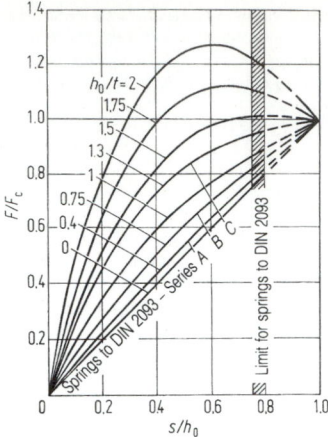

Figure 7. Path of the spring characteristics calculated according to Almen-László at different ratios h_o/t (DIN 2092). Calculated elastic force for $s = h_o$: $F_c = 4Et^3h_o/[(1 - v^2)K_1D_e^2]$.

$h_o/t \approx 0.75$; D_e and D_i are given tolerances of h12 and H12 respectively; load capacity is standardised in the region of $D_e = 8$ mm to 250 mm with $F_{max} \approx 120$ N to 120 kN at a range of spring $s \approx 0.75 h_o$.

Design Calculations. When force is transmitted over circular curves I and III as in **Fig. 6**, the approximate formulae for $h_o/t \leq 0.4$ (Series A) are

$$F \approx \frac{4E}{(1 - v^2)} \frac{(t^3s)}{K_1D_e^2} \quad \text{or} \quad c \approx \frac{4E}{(1 - v^2)} \frac{t^3}{K_1D_e^2} \,, \quad (12)$$

$$\sigma_{I, II} \approx \mp F K_3/t^2 \quad \text{and} \quad \sigma_{III, IV} \approx - (D_i/D_e) \, \sigma_{I, II}, \quad (13)$$

where F is the spring force, c is the spring rate, σ is the edge stress with special steels being expressed as $4E/(1 - v^2) = 905.5$ kN/mm² as in DIN 2092. The dimensionless correction values for $D_e/D_i = 2$ governed by the diameter ratio are: $K_1 = 0.69$, $K_3 = 1.38$.

The non-linearities of the springs cannot be ignored after $h_o/t > 0.4$; the formulae devised by Almen and László [1] as contained in DIN 2092 are sufficiently accurate for this in the case of Belleville springs without supporting surfaces. The spring characteristics in **Fig. 7** are the result when these equations are evaluated. The typical distribution of stresses [48] represented in **Fig. 8** shows that the maximum calculated tensile stresses occur at the base of the Belleville springs at points II and III, irrespective

of the location of the stress pole, whereas the maximum compressive stress can be expected at point I.

In DIN 2093 Belleville springs which are only loaded *statically* without load variation or with occasional load variations at wide intervals of time and at less than 10^4 stress cycles, the calculated compressive stress may be σ_I at $s = 0.75h_o$ up to $\sigma_I = 2000$ to 2400 N/mm² without fear of any significant permanent set.

Where *cyclic* loading is present between the spring range limits s_o and s_u, the associated maximum and minimum stresses σ_{IIo} (σ_{IIIo}) and σ_{IIu} (σ_{IIIu}) must be checked for compliance with, for example, the stress cycle limits of endurance and fatigue strength diagrams reproduced in **Fig. 9**. For worked examples of static or rarely variable loading and of cyclic loading see DIN 2092.

Microstructure. Measured characteristics deviate from the calculated characteristics to a greater or lesser degree owing to contact conditions in the points of support or supporting surfaces (rolling down, slipping). By means of layers of individual disc springs in parallel (spring assemblies), or alternate individual disc springs in series or spring assemblies (spring columns), the characteristics can be varied as well as made to be progressive if a locking facility is provided by spacers or steps on the guide pin to prevent, e.g. deformations over $s \approx 0.75h_o$. In spring assemblies in particular it is not possible after this point to ignore the friction that is dependent on surface finish and lubrication. For details of the various possibilities for characteristic progression see the literature contained in DIN 2092 and catalogues of the disc spring manufacturers. Guides and supports for disc springs should where possible be case-hardened (ideal depth ≈ 0.8 mm) and have a minimum hardness of 55 HRC. Surfaces of the guides should be smooth and where possible ground, clearance for the guides being standardised at about 1 to 2% of the guide diameter. Where cyclic loading is present, springs must be prestressed by at least $s_u = (0.15$ to $0.20) \, h_o$ in order to prevent incipient cracks resulting from tensile residual stresses caused by permanent set at point I.

2.2.5 Torsion Bar Springs

Uses. For the elastic coupling of driving elements, measuring torsional force, in torque wrenches or as antiroll bars in vehicles.

Types. Basic shapes shown in **Table 4**; either with circular cross-section as in DIN 2091 or with rectangular cross-section, both also bundled.

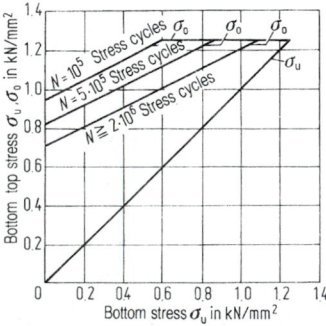

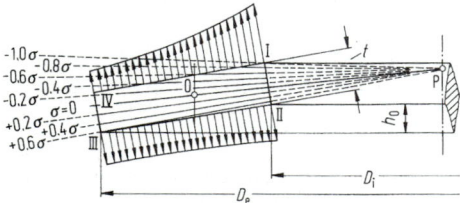

Figure 8. Stress distribution along the edges of the cross-section and lines of equal normal stress in a Belleville spring cross-section after Lutz [43]; P load-dependent pole of stress on axis of Belleville spring.

Figure 9. Endurance and fatigue strength diagram for Belleville springs to DIN 2093 with 1 mm $\leq t \leq 6$ mm in spring sets with a maximum of six alternately combined single Belleville springs (99% probability of survival, room temperature).

Table 4. Basic shapes and design formulas for the macrostructure of torsion bar springs

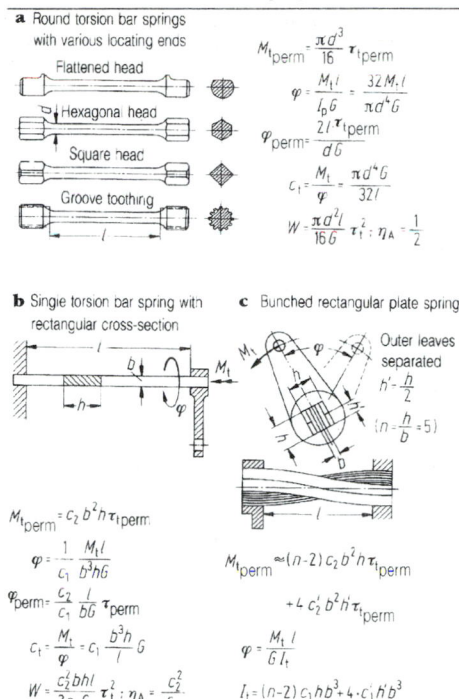

a Round torsion bar springs with various locating ends

Flattened head

Hexagonal head

Square head

Groove toothing

$$M_{t\,perm} = \frac{\pi d^3}{16}\,\tau_{t\,perm}$$

$$\varphi = \frac{M_t l}{I_p G} = \frac{32\,M_t l}{\pi d^4 G}$$

$$\varphi_{perm} = \frac{2\,l\,\tau_{t\,perm}}{d\,G}$$

$$c_t = \frac{M_t}{\varphi} = \frac{\pi d^4 G}{32\,l}$$

$$W = \frac{\pi d^2 l}{16\,G}\,\tau_t^2 ; \eta_A = \frac{1}{2}$$

b Single torsion bar spring with rectangular cross-section

$$M_{t\,perm} = c_2\,b^2 h\,\tau_{t\,perm}$$

$$\varphi = \frac{1}{c_1}\frac{M_t l}{b^3 h G}$$

$$\varphi_{perm} = \frac{c_2}{c_1}\frac{l}{b G}\,\tau_{perm}$$

$$c_t = \frac{M_t}{\varphi} = c_1\frac{b^3 h}{l}\,G$$

$$W = \frac{c_2^2\,b h l}{2 c_1\,G}\,\tau_t^2 ; \eta_A = \frac{c_2^2}{c_1}$$

c Bunched rectangular plate springs

Outer leaves separated

$$h' = \frac{h}{2}$$

$$(n = \frac{h}{b} = 5)$$

$$M_{t\,perm} \approx (n-2)\,c_2\,b^2 h\,\tau_{perm}$$

$$+ 4\,c_1'\,b^2 h'\tau_{perm}$$

$$\varphi = \frac{M_t l}{G I_t}$$

$$I_t = (n-2)\,c_1 h b^3 + 4\cdot c_1'\,h' b^3$$

Values for c_1 (c_1^*) and c_2 (c_2^*) see **B2, Table 7**

Design Calculations. As in **Table 4** or in greater detail for circular cross-section in DIN 2091. Expressions are given below for steels to DIN 17 221 where hardening strength is $R_m = 1600$ to 1800 N/mm² and modulus of shear $G = 78\,500$ N/mm² under static load for non-preset bars $\tau_{t\,perm} = 700$ N/mm² and for preset bars $\tau_{t\,perm} = 1020$ N/mm²; fatigue strength under fluctuating stresses ($N = 2 \cdot 10^6$) for preset bars with ground and shot-peened surfaces can be 740 N/mm² for ∅20 mm and as much as 550 N/mm² for ∅60. For endurance and fatigue strength values governed by the mean stress and standard values for relaxation or creep see DIN 2091.

Microstructure. Designs for torsion bar heads using a square or hexagonal profile or gear toothing are standardised in DIN 2091 (gear toothing in DIN 5481 or SAE J 498 b). These are mainly used for bars which are only stressed in one direction of rotation. The minimum diameter of head d_F should be at least 1.25 to 1.30 times the diameter of the bar d. Due to high notch sensitivity in the high-tensile spring material, freedom from notches and grooves and compressive residual stress to be aimed for through, for example, shot-peening. In springs subject to pulsating stress a favourable, not merely near-surface, residual stress condition can be set up by means of presetting, i.e. deformation beyond the yield point towards the subsequent operating stress [11]. The active spring length l_t is calculated taking into account the circular shoulders before the head as in DIN 2091. Permanent corrosion proofing can be easily applied to torsion springs (**Table 4a**), because when suitably anchored these operate with-

out wear or friction. Stabilisers can also be designed with grommets at the crimped ends unless the ends are simply bent to a leg shape [12].

In order to provide accurate calculations for torsion bar springs with a rectangular cross-section, the additional tensile and compressive stresses set up by torsion resulting from bulging force must also be considered (see B2.5.5). No pure torsion is present in compound springs, e.g. four round bars in parallel or even rectangular springs (**Table 4c**), with the result that relative motion occurs, especially in compound rectangular springs; they cannot therefore be permanently protected against wear and corrosion.

2.2.6 Helical Compression Springs, Helical Extension Springs

Uses. As contact, release and restoring springs in clutches, brakes, valves, switches, brush holders and similar, or as suspension springs in vehicles and bearing spring for machine beds.

Types. Compression or extension springs as shown in **Table 5** are detailed in DIN 2089 Parts 1 and 2 respectively. For sizes of loops or hooks for extension springs see DIN 2097.

Design Calculations. Formulas for round wire cross-section in **Table 5** corresponding to B2.5, in which shear stresses resulting from transverse action and normal stresses are ignored when calculating deformation of the spring. The nominal shear stress τ is determined using the mean coil diameter D and the wire diameter d: $\tau = (FD/2)\,(\pi d^3/16)$; the edge stress $\tau_K = k\tau$ increased by the curvature of the wire at the edge of the cross-section inside the spring is determined using the stress factor k, which is a function of the spring index $w = D/d$. Operating stress is determined under static and semi-static loading without taking into consideration factor k, but with factor k when under dynamic loading. The expression for the usual spring indices $w = 4$ to 20 is $k \approx 1.4$ to 1.07. In addition to the nomograms in DIN 2089 Part 2, the straight-line diagram **Fig. 10** has proved its worth in providing a brief overview of the mutual dependencies of the various types of spring, in particular for variant calculations. Taking values of D, d, the helical force F and the range of spring per coil (s/n) are read off in the preferred number layout independently of the nominal shear stress τ. For rough calculations it is recommended that in the first instance the expression $\tau = 500$ N/mm² should be used for steel springs. The values for a spring with spring index $w = 20$ have been entered in the straight-line diagram as an example.

Note. In order to avoid any misapprehension concerning the manufacture or finishing of helical springs, the printed form in DIN 2099 Part 1 should be consulted to obtain information on compression springs and the printed form in DIN 2099 Part 2 to obtain information on extension springs.

Helical Compression Springs

Microstructure. As a rule, helical compression springs are coiled to the right and in spring sets they are coiled alternately to the right and left, whereby the outside spring is usually coiled to the right. In order to obtain uniformity of all coils when compressing the spring to solid, the total number of coils should if possible end on ½, especially where the number of coils is small (DIN 2096 Part 1). The spring ends are attached and in order to transmit force have their surface either ground or left unmachined, which in the case of larger wire diameters requires a suit-

Table 5. Basic shapes and design formulas for cylindrical helical compressive and extension springs made of round wires

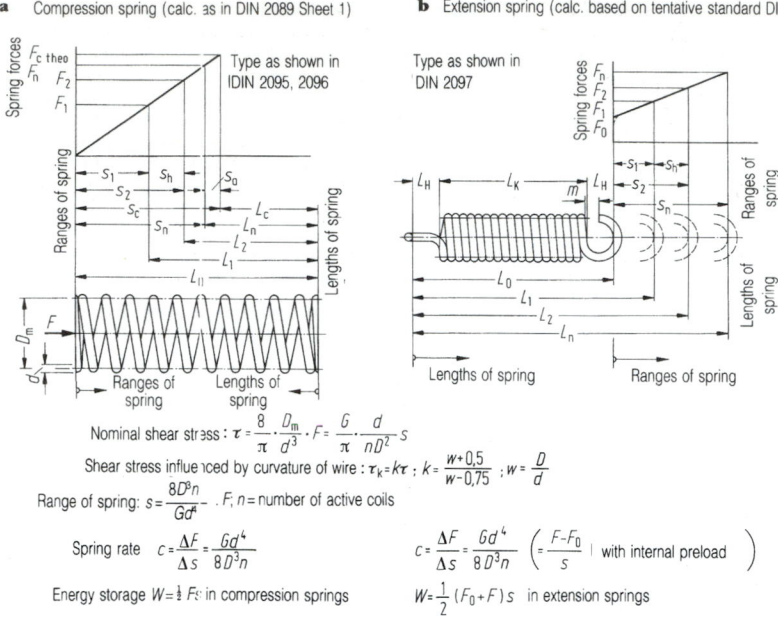

a   Compression spring (calc. as in DIN 2089 Sheet 1)   **b**   Extension spring (calc. based on tentative standard DIN 2089 Sheet 2)

Nominal shear stress: $\tau = \dfrac{8}{\pi}\cdot\dfrac{D_m}{d^3}\cdot F$, $F = \dfrac{G}{\pi}\cdot\dfrac{d}{nD^2}\cdot s$

Shear stress influenced by curvature of wire: $\tau_k = k\tau$; $k = \dfrac{w+0,5}{w-0,75}$; $w = \dfrac{D}{d}$

Range of spring: $s = \dfrac{8D^3n}{Gd^4}\cdot F$, $n = $ number of active coils

Spring rate $c = \dfrac{\Delta F}{\Delta s} = \dfrac{Gd^4}{8D^3n}$ $c = \dfrac{\Delta F}{\Delta s} = \dfrac{Gd^4}{8D^3n}\left(=\dfrac{F-F_0}{s}\right)$ with internal preload

Energy storage $W = \tfrac{1}{2}\,Fs$ in compression springs $W = \dfrac{1}{2}(F_0 + F)\,s$ in extension springs

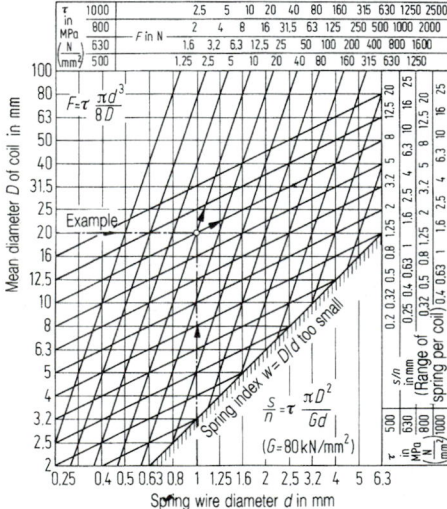

Figure 10. Straight-line diagram of mutual dependencies of the various helical spring data, after H. R. Thomsen. Example: $d = 1$ mm, $D = 20$ mm, $\tau = 500$ N/mm^2; $F = 10$ N, $s/n = 8$ mm.

able ring plate. The number of non-springing end turns required mainly depends on the manufacturing process. The total number of coils n_t where there are n springing coils in cold-formed springs as in DIN 2095 is: $n_t = n + 2$ and in hot-formed springs as in DIN 2096: $n_t = n + 1.5$. The minimum distance between the active coils S_a/n under maximum operating load depends on the type of

load as well as the manufacturing process. *Reference values*: $S_a/n \approx 0.02(D + d)$ under static load or $\approx 0.04(D + d)$ under dynamic load; for further details see DIN 2089 Part 1. For manufacturing reasons it must be possible to compress all springs to solid length, expression for solid length L_c.

Additional details for calculations in **Table 5** for cold- and hot-formed steel compression springs with quality specifications as in DIN 2095 and DIN 2096 Parts 1 and 2 are also summarised in DIN 2089 Part 1. The following limits for production and operating loads must be noted for cold-formed, patented and cold-drawn spring wire belonging to Class C and D in DIN 17 223 Part 1: The permissible nominal shear stress for solid length is $\tau_{c\,perm} = 0.56R_m$, where R_m is the minimum tensile strength dependent on the diameter of the wire. The permissible nominal shear stress under static or semi-static operating load is limited by the *relaxation*, i.e. the loss of force at constant clamping length, warranted by the application. For the results of relaxation tests see DIN 2089 Part 1; with larger diameters of wire (6 mm) and higher temperatures (80°C) considerable percentage losses of force (15%) were detected after 48 hours in cold-formed compression springs, even under long-term acceptable maximum stresses of $\tau = 800$ N/mm^2. In order to assess dynamic loads in the fatigue strength range (number of stress cycles $N = 10^4$ to 10^7) and in the endurance strength range (number of stress cycles $N \geq 10^7$) Goodman diagrams are used in which the permissible marginal maximum stress τ_{kO} is placed over the marginal minimum stress τ_{kU} and from which the acceptable stress cycle τ_{kU} can be read. **Figure 11** shows a Goodman diagram for springs which have not been shot-peened. The permissible stress cycle of these springs can be increased by 20% by shot-peening. Compression springs with a wire

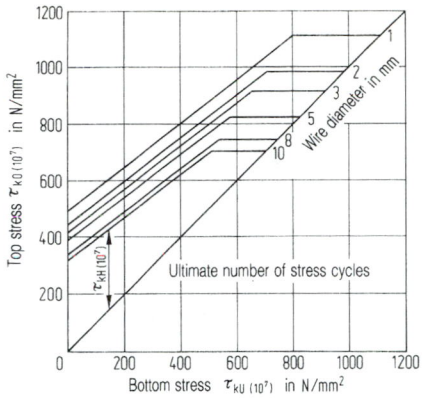

Figure 11. Goodman diagram as in DIN 2089 Part 1 for cold-formed helical compression springs made of patented and cold-drawn spring steel of Class C and D as in DIN 17 223 Part 1, not shot-peened.

diameter over 17 mm are no longer cold-formed but manufactured only by hot forming using for example hot-rolled hardenable steels to DIN 17 221. According to requirements, steel with a rolled or machined-down (i.e. turned, shelled or ground) surface is used as a primary material. Shot-peening is used to increase acceptable cyclic stress under dynamic load. DIN 2089 also contains formulae for calculating *transverse resilience*, *axial compression*, *natural frequency* and *shock load* [20, 54, 5].

Progressive Helical Compression Springs, as occasionally required in motor vehicle construction, may be made from bars with a conical taper at both ends with variable coil pitch or constant wire diameter with variable coil diameter, but not cylindrical. During spring deflection some of the coils are continuously being pressed into a solid, thereby being prematurely released as a spring element [11, 19, 46, 65].

Helical compression springs are occasionally used in the form of *spring nests* with two (or three) concentric springs coiled alternately to the right and left in order to make optimum use of available space. Careful centring of individual spring ends and sufficient radial clearance between the springs must be provided [63]. The springs in parallel should be so arranged that at maximum range of spring they are subject to the same amount of stress and have approximately the same solid length. The spring index $w = D/d$ must then be equal for the individual springs. The force and energy borne by individual springs are in the ratio of the square of their diameter d. The small advantage derived from the third concentric spring makes it hardly ever worth the expense, apart from the fact that the inside spring cannot often be designed to be non-buckling.

Cylindrical Helical Extension Springs

Microstructure. Loop and hook shapes for cold-formed extension springs given in DIN 2097. In springs with loops the total number of coils is governed by the position of the loops; where endpieces have been screwed or rolled on, the total number of coils exceeds the number of resilient coils by the number of coils blocked by the rolling or screwing on of endpieces. If extension springs have been preloaded the coils are squared, but this is not necessarily so if they are not preloaded.

For further additional calculations as in **Table 5** for cold- and hot-formed steel extension springs see DIN 2089 Part 2. In addition to the given rebound space, it is first and foremost the cushioning action to be achieved and the maximum spring resistance F_n which are the deciding factors for calculation and design. For cold-formed extension springs made from patented and cold-drawn spring steel wire of Classes A to D as specified in DIN 17 223 Part 1, the nominal shear stress is $\tau_{n\,perm} = 0.45R_m$, where R_m is the minimum tensile strength dependent on the diameter, when static or semi-static loading is present. As a space-saving measure, cold-formed extension springs are usually coiled with an internal preload F_o so that their theoretical stress–strain curve runs as shown in **Table 5b** (theoretical spring nominal shear stress $\tau_0 \leq 0.1R_m$).

Worked Examples in DIN 2089 Part 2. If possible extension springs are to be avoided under cyclic loading. This is because the stress peaks in the hook ends can only be calculated with a degree of uncertainty, the reason being that their surface cannot be strengthened by shot-peening, owing to the coils that usually lie close together in the unloaded condition and because, unlike in helical compression springs, a fatigue fracture may result directly in sequential failure. If extension springs have to be used under cyclic loading then they should only be cold-formed extension springs, suitably provided with screwed cover plates as in DIN 2097 [4, 55].

2.3 Rubber Springs and Anti-vibration Mountings

Rubber springs [80–98] are machine elements, whose high resilience is determined by the elasticity of the elastomers (rubber) used as well as their shape and connection to metal parts.

2.3.1 Rubber and Its Properties

For essentials of elastomers see D4.8. Information on outstanding characteristics for types of natural and synthetic rubber which can be used for spring elements is summarised in **Table 6**.

Deformation of a rubber spring consists of elastic distortion and *creep* dependent on stress amplitude and time. To creep under static load can be added a *permanent set* under dynamic load during the first $5 \cdot 10^5$ stress cycles. After load alleviation and a reflux due to residual stresses a noticeable, material-related *set* may remain (DIN 53 517, DIN 53 518). The appearance of creep (flow) and permanent set is considerably more marked in synthetic rubber compounds than in highly elastic natural rubber compounds; they are governed by temperature in the same way as the damping which is the result of the same physical relation. Even highly elastic rubber compounds begin to creep considerably at 80 °C. "Rubber" can be readily described in its range of applications using rheological models [88]. In general, different rheological models are required for the *modulus of shear G* and the *modulus of compression K*.

The *modulus of compression K* indicates the relative change in volume under all-round compression. Linearly elastic materials are expressed as $K = E/(3 - 6v)$ and $E = 2G(1 + v)$, where v is the transversal contraction factor. Elastomers with small deformations and low speeds of load application are expressed as $v \approx 0.5$ and $E \approx 3G$; the modulus of compression K may, for example, amount to 1280 N/mm² with a modulus of shear G of 18 N/mm²

Table 6. Survey of the elastomers used in rubber springs and their main characteristics

Elastomers with letter symbols as used in DIN ISO 1629 and example of trade name	Styrene-butadiene rubber	Natural rubber (polyisoprene)	Butyl rubber (bromine, chlorine rubber)	Ethylene propylene diene rubber	Chloroprene rubber	Chloro-sulphonyl polyethylene rubber	Nitril butadiene rubber	Polyester urethane rubber	Methyl vinyl silicon rubber	Polyacrilate rubber (PA)	Fluorocaoutchouc
	SBR	NR	BIIR CIIR	EPDM	CR	CSM	NBR	AU EU	MVQ	ACM	FPM
Characteristics	Buna	Rubber	Butyl	Buna AP	Neoprene	Hypalon	Perbunan	Vulkollan	Silopren	Cyanacryl	Viton
Shore A hardness, shA (DIN 53505)	30 to 100	20 to 100	40 to 85	40 to 85	20 to 90	50 to 5	40 to 100	65 to 95	40 to 80	55 to 90	65 to 90
Elongation at tear (DIN 53504)	100 to 800	100 to 800	400 to 500	150 to 500	100 to 800	200 to 50	100 to 700	300 to 700	100 to 400	100 to 350	100 to 300
Working temperature range in °C	−50 100	−55 90	−40 120	−50 130	−40 100	−20 120	−40 100	−25 80	−60 200	−20 150	−20 200
Resistance to hydrocarbons	low	low	low	moderate	moderate	moderate to good	good		good	very good	excellent
Creep resistance	very good	excellent	medium	good	good	medium	very good	good	good	good	good
Damping	good	moderate	very good	good	good	very good	very good	good	good	very good	highly dependent on temperature
Adhesion to metal	good	excellent	moderate	moderate	good	moderate	very good	very good	medium	medium	good
Special characteristics	–	combustible	good resistance to acid	excellent resistance to ozone	–	good resistance to acid	–	susceptible to water at 40 °C	non-flammable	combustible (in bright conditions)	resistance to silicone fluid
Price level	100	85	125	120	250	270	170	400	800	350	1000

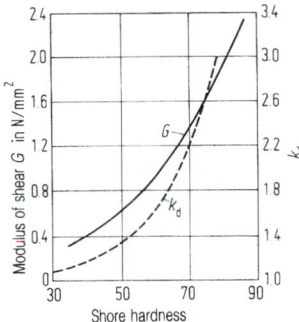

Figure 12. Modulus of shear G and dynamic factor k_d of rubber (natural rubber) in relation to Shore A hardness [84].

($v = 0.493$), at which point rubber has a practically incompressible reaction. This must be taken into account when designing and fitting.

As a result of the relatively high shear strains that can be withstood, the spring characteristics can be used into the non-linear range. Hooke's law therefore applies only approximately over the whole range of applications, even at low rates of load application. In practice, in order to distinguish qualities of rubber Shore A hardness as in DIN 53 505 – shA for short – is used. This can be measured with a consistent margin of error of at most ± 2 shA. The Shore A hardness can be given an estimated value for a modulus of shear G as shown in **Fig. 12**. **Figure 13** shows a characteristic curve for a rubber element under cyclic loading. New rubber springs are generally harder than those that have already been subjected to dynamic loading.

The damping forces acting in the rubber can only be considered as proportional to speed within very narrowly defined frequency ranges. Even in highly elastic qualities with low Shore hardness excesses of up to as much as 20% over the modulus of elasticity and the modulus of shear measured under continuous loading in the normal vibration frequency range of 25 to 50 Hz; in greater Shore

hardnesses in the range 54 to 72 shA the excess may amount to 40 to 60%. For this reason a distinction must be made between static spring rate c and dynamic spring rate c_{dyn} when dealing with rubber elements. To simplify matters, an interrelationship $c_{dyn} = k_d c$ exists between the two characteristic values. The standard value is expressed by the fact that the factor k_d which increases little with frequency lies between 1.1 and 1.6 in a typical hardness range of 35 to 60 shA but may be considerably higher at Shore hardnesses over 60 (**Fig. 12**). To determine more accurate characteristic values for the elastic-viscous properties which are governed by frequency, amplitude of deformation, mean deformation and temperature, see DIN 53 513 or [88].

2.3.2 Basic Types of Rubber Spring

Uses. In increasing volumes for *vibration isolation* in motor construction and as elastic fasteners and joints in machine construction as they can be ideally adapted to meet design requirements [84, 91].

Types. Rubber springs may be used as free-formed, compact elements such as simple cylindrical rubber blocks with $d = h$ as vibration isolation or as elements in combination. In the case of combined springs sufficient compression must be used in the active surfaces to ensure that stresses are transmitted to the rubber as uniformly as possible without preventing deformation. Rubber springs are usually made as so-called steel-rubber springs, as for example shown in **Table 7**, in which metal surfaces intimately bonded to the rubber during vulcanisation provide perfect force transmission. Such springs are mass-produced and listed in manufacturers' catalogues with their relatively reliable stiffness and strength values [94-98]. They should not be specified for new applications without a thorough consultation with the manufacturer. Steel-rubber springs in shear are preferably used under medium loads as soon as greater spring ranges or lower natural vibration frequencies are specified. Rubber springs under compressive load are used for large loads whenever a high degree of stiffness in the direction of the load is allowed or desired. Rubber springs under tensile load are used if very small masses are to be suspended and vibration-isolated, because they have the advantage of a particularly favourable sound insulation. For more types see VDI Guideline 2062.

Design Calculations (**Table 7**). Reference values for permissible stresses are given in **Table 8**. In general, static shear strain must not exceed tan $\gamma = 0.2$ to 0.4; compression strain should be less than $\varepsilon = 0.1$.

Microstructure. In steel-rubber springs in shear (**Table 7a, b**) the thickness/length ratio $t/l \ll 0.25$ (DIN 53 513) should remain constant so that additional normal stresses acting on the metal surfaces which transmit shear can be kept to a minimum; the characteristic curve is then also to a great extent linear and service life is increased. Compressive preloading of the springs also serves to prevent tensile stresses at the surface periphery (**Fig. 14**; [87]). In springs under torsional shear stress (**Table 7c, d**) there is no increase in stress at the periphery of the load-transmitting surfaces, which means they become deformed due to shear to a greater extent than steel-rubber springs limited in the direction of shear. If possible they should be designed as bodies having the same shear stress (**Fig. 15**; [92]).

The stiffness of rubber springs under compressive stress may be increased if thin metal plates are inserted parallel

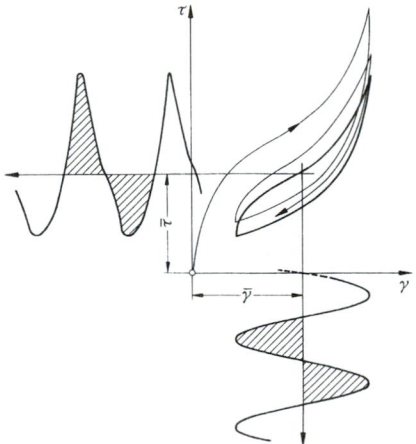

Figure 13. Hysteresis loops of a sinusoidally deformed rubber spring [ISO 2856-1975(E)] for initial load cycles and final load cycle with stress and extension mean value lines.

Table 7. Types of steel–rubber springs with design fundamentals

Spring type	Shape of spring, load	Equations, remarks	Range of application, remarks
Disc spring in parallel shear **a**	Shear stress distribution	$s \approx \dfrac{F\,t}{l\,b\,G}$ $\tau_n \approx F/(l\,b)$ $\approx G\,s/t$ $s_{perm} \approx t\,\gamma_{perm}$ $\eta_a \approx 1$ if $l \gg t$ $F_{perm} \approx b\,l\,G\,\gamma_{perm}$	In the range $s/t \approx \gamma \le 20°$ ($s \le 0.35\,t$) the characteristic curve is practically straight. At the edges at I to IV, $\tau = 0$. From then on τ first increases beyond τ_n. At I and III tensile stress is superimposed. at II and IV compressive stress.
Cylindrical spring in axial shear **b**		$s \approx \dfrac{F\,\ln\,(d_a/d_i)}{2\pi\,l\,G}$ $\tau_{ni} = \dfrac{F}{\pi\,d_i\,l}\ ;\quad \tau_{na} = \dfrac{F}{\pi\,d_a\,l}$ $s_{perm} = \dfrac{d_i}{2}\ln\dfrac{d_a}{d_i}\,\gamma_{perm}$ $F_{perm} = \pi\,d_i\,l\,G\,\gamma_{perm}$	Linearity up to $\gamma_{ni} = \dfrac{\tau_{ni}}{G} \le 20°$ If height of rubber l reduces with the reciprocal of the diameter, then $l\,d_i = l_a\,d_a$, $\tau_{ni} = \tau_{na}$ and $s \approx \dfrac{F\,(d_a - d_i)}{2\pi\,d_i\,l_i\,G}\ ;\quad \eta_A = 1$ (Springs under the same stress)
Disc spring in torsional shear **c**	 $(l_a/l_i = d_a/d_i)$	$\varphi \approx \dfrac{24\,M_t\,l_a}{\pi\,G\,(d_a^4 - d_i^3\,d_a)}$ $\tau = \varphi\,\dfrac{d_a}{2\,l_a}\,G = \varphi\,\dfrac{d_i}{2\,l_i}\,G$ $\varphi_{perm} = \dfrac{2\,l_a}{d_a}\,\gamma_{perm}\ ,\ \eta_a = 1$ $M_{t\,perm} = \dfrac{\pi\,G\,(d_a^3 - d_i^3)}{12}\,\gamma_{perm}$	Valid for $\varphi \le 20°\cdot\dfrac{2\,l_a}{d_a}$ If $t_i = t_a = t$ then $\varphi = \dfrac{32\,M_t\,t}{\pi\,(d_a^4 - d_i^4)\,G}$ Where t_a and φ_{perm} are the same $M_{t\,perm}$ for $d_a/d_i = 2$ is 0.8 times the spring as drawn.
Cylindrical spring in torsional shear **d**		$\varphi \approx \dfrac{M_t}{\pi\,l\,G}\left(\dfrac{1}{d_i^2} - \dfrac{1}{d_a^2}\right)$ $\tau_i = \dfrac{2\,M_t}{\pi\,d_i^2\,l}\ ;\quad \tau_a = \dfrac{2\,M_t}{\pi\,d_a^2\,l}$ $\varphi_{perm} = \dfrac{(d_a^2 - d_i^2)}{2\,d_a^2}\,\gamma_{perm}$ $M_{t\,perm} = \dfrac{\pi\,G\,d_i^2\,l}{2}\,\gamma_{perm}$	If width of rubber l reduces with the reciprocal square of the diameter $l_i\,d_i^2 = l_a\,d_a^2$, then $\tau_i = \tau_a$, $\eta_A = 1$ and $\varphi = \dfrac{2\,M_t}{\pi\,l_i\,G\,d_i^2}\ln\dfrac{d_a}{d_i}$ $\varphi_{perm} = \ln\dfrac{d_a}{d_i}\,\gamma_{perm}$ (Linearity up to $\gamma \approx 40°$)
Rubber buffer under compressive load **e**		$s \approx \dfrac{4\,F\,h}{E_{calc}\,\pi\,d^2}$ $F_{perm} = \dfrac{\pi\,d^2}{4}\,\sigma_{perm}$ Under steady load $s_{perm} = 0.1\,h$, otherwise creep Stress concentration factor $= \dfrac{\pi\,d^2/4}{\pi\,d\,h} = \dfrac{d}{4\,h}$	

to the compression area by vulcanisation or pressure (**Fig. 16**; [87]), thereby preventing still further the transverse strain of the rubber as it occurs through the external compression areas. The transverse obstruction caused by the non-slipping compression areas is determined by the *stress concentration factor k*, the ratio of loaded area of rubber to free rubber surface (**Table 7e**). Thin metal plates can also be used to carry off heat and consequently lower temperature in elastomer springs under cyclic loading. The damping capacity of elastomers causes high temperatures internally (heat cavities), which can be predicted using modern methods of calculation such as the finite-element method [88].

Note. Further design characteristics may be considered after first discussing the matter with manufacturers and

tapping their varied experience. Manufacturers must also be asked to advise the permissible load capacity of rubber springs for each individual case if this is not given in the manufacturer's catalogue.

2.4 Fibre Composite Springs

Fibre composite springs [100–106] are said to combine the advantages of both metal springs (high load capacity, small clearance, low relaxation) and rubber springs

Table 8. Reference values for the rough calculation of permissible loads and deformations of rubber springs (k = stress concentration factor as in **Table 7e**; permissible alternating stresses approximately one-third to one-half of the permissible static loads) as in [16]

Shore hardness	Density (t/m³)	Modulus of elasticity E_{st} under compression (N/mm²)		Modulus of shear G_{st} (N/mm²)	Dynamic factor k_d	Permissible static deformation (% under permanent static load)		Permissible stress (N/mm²) under permanent static load		
		$k=\frac{1}{4}$	$k=1$			Compression	Shear, tension	Compression $k=\frac{1}{4}$	Compression $k=1$	Shear, tension
30	0.99	1.1	4.5	0.3	1.1	10 to 15	50 to 75	0.18	0.7	0.2
40	1.04	1.6	6.5	0.4	1.2		45 to 70	0.25	1.0	0.28
50	1.1	2.2	9.0	0.55	1.3		40 to 60	0.36	1.4	0.33
60	1.18	3.3	13.0	0.8	1.6		30 to 45	0.5	2.0	0.36
70	1.27	5.2	20.0	1.3	2.3		20 to 30	0.8	3.2	0.38

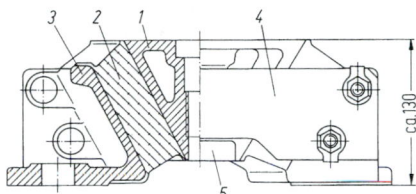

Figure 14. Engine bearer for locomotive and marine diesel engines in cross-section and top view in [87]: *1* internal component (casting) with screw thread and transverse force transmission by means of adjusting ring, *2* rubber component under shear and compressive stress, *3* angle brackets (castings), *4* tension webs, *5* back stop on internal component *1*.

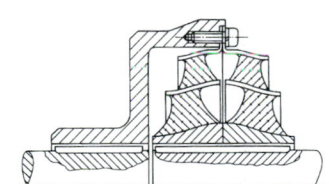

Figure 15. Torsionally elastic shaft coupling as in [92].

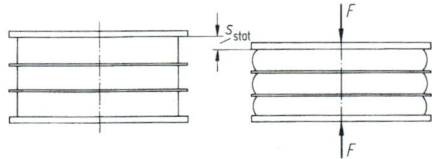

Figure 16. Steel–rubber spring under compressive load with vulcanised-in spacer plates, which to a great extent prevent transverse strain as in [87]. (Resultant modulus of elasticity: $E_R = KG$, where $K = 19.5$ for stress concentration factor $k = d/4b = 1.5$.)

(lighter weight, damping capacity). The load capacity and stiffness are governed by the fibres (usually glass fibres, but also aramid fibres and carbon fibres) and the matrix (usually polyester or epoxy resin). The properties of the composite material vary, depending on the proportion of

the fibre volume (30 to 60%) and can be adjusted accordingly. Chemical and mechanical compatibility of the components must be ensured under the ambient conditions prevailing during manufacture, storage and operation, e.g. monitoring of dampness and temperature.

Uses. For leaf springs in vehicle construction, for high-performance sports equipment, for electrically insulating supports in electric machines, for components in aircraft construction.

Types. Springs under tensile and bending stress with basic shapes as described in F2.2.1 and F2.2.2 and metal-reinforced points of force transmission.

Design Calculations. Owing to anisotropy of mechanical properties it is the composite matrix which generally determines strength. The simple design calculations which apply to metal springs, which in general do not take into account an assessment of shear stresses, can be used at best as an initial basis for comparison when component tests are being carried out on composite springs. Moreover, it must always be ascertained whether the matrix provides sufficient resistance to buckling. For more detailed literature see [100–106].

Microstructure. Force is transmitted, when using tension bars, by means of two metal spools round which the fibre is practically endlessly coiled; the distance between the spools is maintained by a compressively stiff structural element. A similar design was tested for mass-reducing connecting rods (carbon fibre/aluminium) [102]. In leaf springs too, loops and hooks of formed steel strips are used [105]. In general, care must be taken that the glass fibres embedded longitudinally are not cut through. Screw fittings are also to be avoided, because owing to the relaxation in the composite the preload required for the screw cannot be maintained without special measures.

2.5 Gas Springs

The principle of gas springs (air springs) [110–114] is based on the compressibility of a volume of gas (air) trapped in a container.

Uses. In motor vehicle manufacture to give non-linear characteristics and for level control, in pneumatic clutches [111].

Types. Piston air spring similar to air pump with constant cross-secton A and variable air column height h. The compression of the air column by deflection s causes increase in external compression p_0 (internal compression at $s = 0$) to ultimate compression p. The gasket required for the piston results in a frictional force being set up, thereby producing energy losses. Friction is absent in cushion-type pneumatic springs. These can also be combined with a fluid friction damper.

Basic Factors. Equation of state for gases $pv_n = \text{const}$, where p is absolute compression, v is specific volume and

n is a polytropic exponent as shown in C7. A non-linear spring characteristic

$$F = p_0A \left(-1 + 1/(1 - s/b)^n\right)$$

is obtained for piston air springs if friction is not taken into account.

For further details see VDI Guideline 2062, p. 2, and [110-114].

3 Couplings, Clutches and Brakes

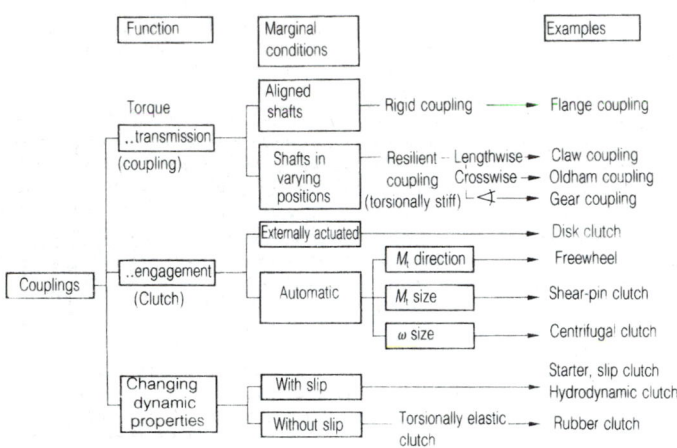

Figure 1. Functions of shaft couplings.

K. Ehrlenspiel, Munich

3.1 Survey, Functions

Couplings are used to *transmit* torques in aligned and non-aligned shafts and often, in addition, to improve the *dynamic characteristics* of a drive mechanism and to *engage* the torque. In **Fig. 1** and **Fig. 2** a survey is provided of the various functions and methods of transmitting torque [1, 2]. Unlike gears of torque converters, slip couplings or clutches have the same (fixed) amplitude of torque M_t at input and output. As the difference in power $M_t \Delta\omega$ can only be converted into heat, problems with

heat storage and cooling are very much to the fore, together with the effects of wear.

Principles of Selection [3]. Transmittable nominal and peak moment, elasticity, damping, puncture resistance, maximum speed, reversing characteristics, play, resonant frequency, fatigue strength of the coupling; types and second moment of area of the machines to be coupled, vibrations, progress of moments in time; possibility of stalling the machine. Shaft position: permissible radial, axial and angular displacements (**Fig. 4f**), permissible axial and radial forces; axial fixing, dimensions, bore tolerances, weight, position of centre of gravity, ease of balancing; influence of critical speed. For *clutches*: fre-

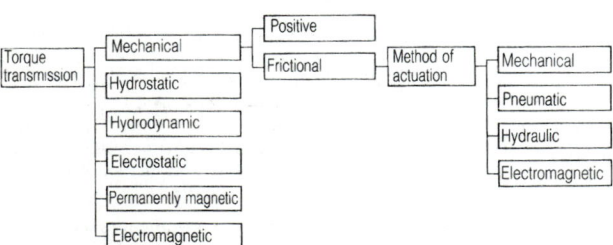

Figure 2. Methods of transmitting torque in couplings.

quency of engagement, heating, cooling, engagement period, automatic disengaging and re-engaging, engagement conditions, engagement forces, methods of engagement, residual moment after disengaging, susceptibility to chattering, permanent slip, "wet or dry" coupling, ambient temperature. *Operating characteristics*: possibility of axial expansion, adjustability of alignment, accident safety features, wear control, useful life, replacement of worn parts, insulation from structure-borne noise and electric current, sensitivity to dust, damp, oil, chemicals; maintainability, inspection interval, delivery time for spare parts.

Selection Parameters. Survey of manufacturers in [4].

Non-engaging Couplings. In these, as shown in **Appendix F3**, **Fig. 1a**, the external diameter is a function of the torque, according to type: $D_a = \text{const} M_{KN}^{0.3}$. The external length L_a and the weight G share this characteristic of the external diameter (**Appendix F3**, **Fig. 1b**). In spite of dispersions which extend into the next type it is clear that the *size of the coupling* grows with increasing torsional elasticity. An opposite tendency can be noted at permissible *speeds*, although with still greater dispersions: $n = \text{const} M_{KN}^{-0.27}$. Couplings can be classified as higher-speed type (index a) and intermediate-speed type (index b) [5].

Manufacturing Details for Clutches are given in **Appendix F3**, **Fig. 2** [6].

3.2 Permanent Torsionally Stiff Couplings

3.2.1 Rigid Couplings

These are generally cheap and small-scale couplings which are used to solve alignment and bending vibration problems in very simple drive mechanisms at maximum torques and revolutions (forged flanges).

Types (Fig. 3a–c). (**a**) The *flange coupling* transmits torque with frictional locking provided by preloaded setscrews ($M_{max} = 10^6$ Nm, $n_{max} = 8000$ min^{-1}). It is possible to effect a radial disconnection where there is a two-piece intermediate faceplate. Special types are also produced with a conical seat and hydraulic pressure fit. (**b**) The *split muff coupling* makes it possible simply to expand the frictionally engaged transmitting shells where

the radial dimensions are small ($M_{max} = 0.3 \cdot 10^6$ Nm, $n_{max} = 1700$ min^{-1}). Setscrews are used for safety reasons. It is not suitable for alternating or intermittent loads. (**c**) *Radial tooth coupling.* This is the smallest self-centring coupling with radial teeth, which does, however, require considerable axial prestressing. It is free from play and suitable for alternating torques. (In **Fig. 3c**: bevel wheel connected to a tubular shaft.)

3.2.2 Torsionally Stiff Self-Aligning Couplings

Torsionally stiff self-aligning couplings can compensate for axial, radial or angular shaft displacement and are used if an isogonal transformation is required and the torsional vibration characteristics are not to be altered. (Causes of shaft displacement include inaccurate alignment/assembly, thermal expansion, elastic deformation and structural misalignment.) Torsionally stiff self-aligning couplings are smaller than torsionally elastic ones but, unlike these, they have to be lubricated (exception: diaphragm couplings).

Types (Fig. 4a–f). (**a**) The *positive contact coupling* with axial engaging dogs compensates for axial misalignment only but can be made in the form of a clutch. The *parallel crank coupling* [4] is very short for large radial displacements of parallel shafts and makes possible isogonal transformation. The short *Oldham coupling* [7] provides isogonal transformation, but owing to problems with wear transmits only small torques ($\Delta K_r = 1$ to 5 mm, $\Delta K_w = 1$ to 3°). (**b**) The *universal joint* [8, 9] permits deflection angles up to 40° but transforms a uniform angular velocity ω_1 into an angular velocity ω_2 pulsating at $2\omega_1$: $\omega_2 = \omega_1 \cos\beta/(1 - \sin^2\beta \sin^2\alpha_1)$ (where β is the deflection angle $\triangleq \Delta K_w$ as in DIN 740 and α_1 is the rotation angle of shaft 1). The maximum/minimum values are $\omega_{2max} = \omega_1/\cos\beta$; $\omega_{2min} = \omega_1 \cos\beta$ and the degree of angular irregularity

$$U = (\omega_{2max} - \omega_{2min})/\omega_1 = \tan\beta \sin\beta.$$

(**c**) In the *Cardan shaft* [9–12] this pulsation is counteracted by means of a second universal joint. There are three requirements for this: the forks of the connecting shaft must be in one plane, the deflection angle should be $\beta_1 = \beta_2$, and the driving and the driven shafts should be in one plane. Because of dynamic loads a large deflection angle β reduces the transmittable power. (**d**) *Homokinetic joints* [10, 13] are uniform and are very short at deviation angles up to 48°. (**e**) *Double tooth couplings* [14–19] transmit the torque ($M_{max} = 5 \cdot 10$ Nm, $n_{max} = 10$ min^{-1}) by means of a straight external/internal row of teeth which can also be crowned, thereby making possible angle deviations $\Delta K_w < 1°$ per tooth. The permissible parallel displacement of the shafts ($L \tan \Delta K_w$) is proportional to the space between the rows of teeth L. The maintenance time spent on lubrication to ensure operational safety is considerable and is the main disadvantage apart from the uncertainty of axial and radial responses on the bearings. Slowly turning double-tooth couplings are grease-lubricated, whereas for increased revolutions once-through oil lubrication is used. The advantages are small size, non-susceptibility to overloads and suitability for a high number of revolutions [20]. The permissible unit pressure in the engaging tooth surface is 15 N/mm^2 [9] for unhardened, tempered steel. The *spring shackle coupling* [21] compensates for angular, axial and, in the case of two joints, radial displacement by means of shackle assemblies under tensile stress screwed to the half coupling on alternate sides. Like the diaphragm coupling it is lubrication and maintenance-free and there-

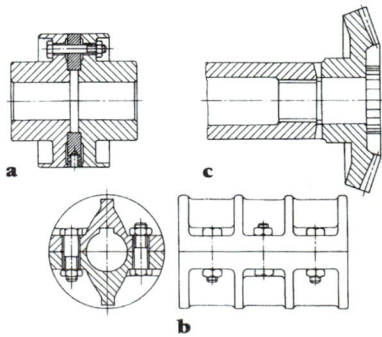

Figure 3. a–c. Permanent torsionally stiff couplings [4] (see text).

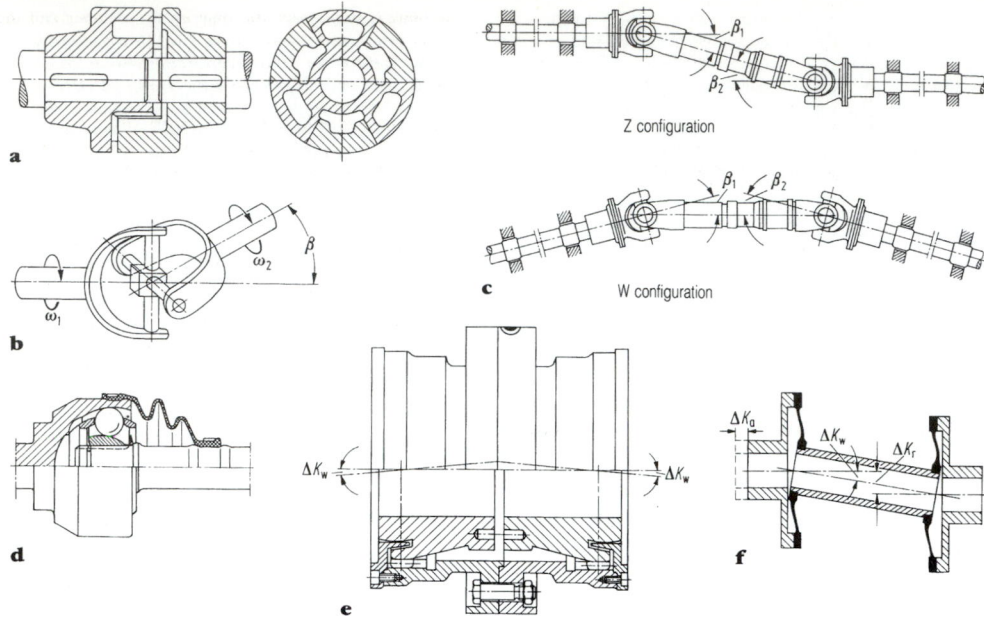

Figure 4. a–f. Torsionally stiff self-aligning couplings (see text).

fore suitable for higher temperatures. (**f**) The *diaphragm coupling* compensates for the shaft displacements mentioned with regard to the spring shackle coupling by means of elastic deformation of metal rings ($\Delta K_w = 0.5$ to $1°$ per half, $\Delta K_a = 1$ to 5 mm). The responses of force can be calculated. Its susceptibility to overload puts it at a disadvantage to tooth couplings [22].

3.3 Permanent Elastic Couplings

Elastic couplings transmit rotary motion without slip, and are primarily used for reducing torque fluctuations or vibrations, thereby preventing damage to machinery. They also make it possible to compensate for any misalignments that may for example result from thermal expansion, structural displacements or inaccurate assembly. They are mainly used in machinery with wide torque fluctuations such as piston engines, rolling mills, winding engines, etc.

3.3.1 Elastic and Damping Characteristics

The elastic and damping characteristics of an elastic coupling alter the *dynamic properties* of a drive system. *Torsional vibrations* are reduced by the elastic accumulating effect of the transmitting devices. A large twist angle reduces peak torque M_t or torsional vibration at the start of a given piece of work $\Delta W = \Delta(\frac{1}{2} M_t \varphi)$. *Torsional vibrations* are damped by means of "internal" (material) damping as in elastomer couplings or by "external" (frictional) damping as in many resilient metal couplings. The *torsional frequencies of resonance* are distorted by the elastic coupling so that they are placed outside the operating speed range.

Elasticity. This is produced by metal springs or elastomers (rubber, plastics). The relative parameters are tor-

sional rigidity of the spring $C_T = dM_t/d\varphi$ (tangents on the characteristic line of the spring; **Fig. 5**) [23], the axial and radial rigidity of the spring C_a and C_b respectively and the angular rigidity of the spring C_w (see DIN 740). Elastomer couplings in particular become more rigid with increasing torque (progressive characteristic line), whereas resilient metal ones require special design features. Unlike resilient metal couplings the torsional rigidity of elastomer couplings increases with the frequency and the preload $C_{Tdyn} \approx 1.2$ to $1.4 C_{Tstat}$ (at ≤ 50 Hz) (**Fig. 6**) and decreases with rising temperature, increasing amplitude and ageing. The increase in temperature may be brought about by the internal damping operation, assisted by poor thermal conductivity. Despite these drawbacks it is mainly elastomer couplings that are used, as, unlike most resilient metal couplings, they do not have to be lubricated and are therefore practically maintenance-free. *Elastomers* include in particular natural and synthetic rubber (buna, buna-N, neoprene), polyurethane (Vulkollan), polyamide, fluorine-based elastomers (Viton). They should be stressed in shear or compression but not in tension. The maximum ambient temperature for elastomer

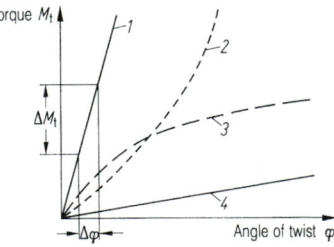

Figure 5. Typical spring characteristics of elastic couplings [2]: *1* linearly stiff, *2* progressive, *3* decreasing, *4* linearly resilient.

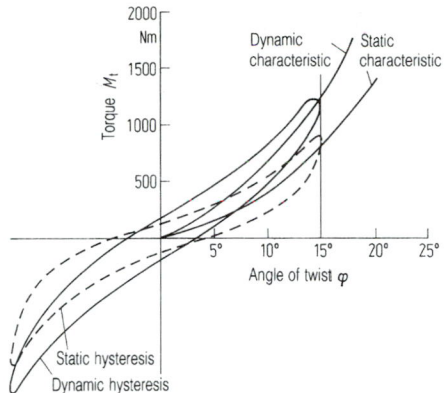

Figure 6. Static and dynamic hysteresis loop of a reinforced disc clutch where $f > 1$ Hz [2].

couplings is, at < 80 to 100 °C, considerably lower than in resilient metal couplings at < 120 to 150 °C.

Damping. Damping of the coupling for the most part relies on the material damping of the elastomers used and on the friction coefficients in the contact surfaces [2]. The "*relative damping*" $\psi = A_D/A_{cl}$ (**Fig. 7**) is specified as a damping parameter in DIN 740 Part 2. It is dependent on material, temperature, stress amplitude, variable stress component, stress frequency and action time and in rubber couplings is in the region of $\psi = 0.8$ to 2. Considerable damping values can also be obtained in resilient metal couplings through frictional and viscous forces.

3.3.2 Layout Design Principles, Vibration Characteristics [2, 24-26]

An elastic coupling must be designed in such a way that the loads and temperatures that occur do not exceed the permissible values in any operating state. DIN 740 Part 2 lays down three methods for designing the coupling:

(a)   Rough calculation using experimental values obtained from manufacturer.
(b)   Rough calculation based on a linear dual mass oscillator.
(c)   More advanced methods of calculation [27-30].

The second method of calculation can be used if the

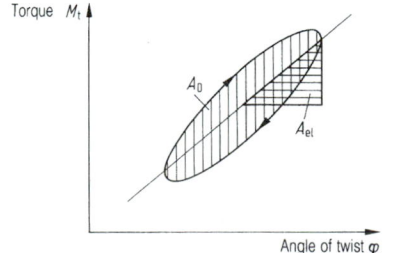

Figure 7. Relative damping: A_D damping force during a vibration cycle, A_{el} elastic strain energy.

coupling is in practice the only elastic member and the layout can be reduced to a dual mass system in respect of torsional vibration. In this case the following method of calculation applies, which is partly based on DIN 740:

(a) The permissible *nominal torque* M_{KN} of the coupling must be at least as great as the nominal moment M_{AN} on the drive side or M_{LN} on the load surface:

$$M_{AN}S_\vartheta \le M_{KN} \ge M_{LN}S_\vartheta.$$

The operating temperature is taken into account by temperature factor $S_\vartheta = 1$ to 1.8 (at -20 °C to $+80$ °C, depending on material).

(b) The permissible *maximum torque* M_{Kmax} of the coupling must be at least as great as the peak torques M_{AS} or M_{LS} that occur in operation as a result of torsional vibrations on the drive side and load surface, taking account of the inertia of masses J_A or J_L, the impact coefficients S_A or $S_L = 1.6$ to 2.0, the starting coefficient $S_z = 1$ to 2 and the temperature coefficient S_ϑ:

$$M_{AS} \frac{J_L}{J_A + J_L} S_A S_Z S_\vartheta \le M_{Kmax} \ge M_{LS} \frac{J_A}{J_A + J_L} S_L S_Z S_\vartheta.$$

(c) When the *resonance* passes through quickly with the peak torques M_{Ai} or M_{Li} being stimulated at the drive side and load surface, M_{Kmax} must not be exceeded:

$$M_{Ai} \frac{J_L}{J_A + J_L} V_R S_Z S_\vartheta \le M_{Kmax} \ge M_{Li} \frac{J_A}{J_A + J_L} V_R S_Z S_\vartheta.$$

Resonance coefficient $V_R \approx 2\pi/\psi$; index i: stimulation of the order of i.

(d) If loading is due to a *fatigue stress moment* with amplitudes of M_{Ai} or M_{Li} the permissible *alternating torque* M_{KW} must not be exceeded:

$$M_{Ai} \frac{J_L}{J_A + J_L} V S_\vartheta S_f \le M_{KW} \ge M_{Li} \frac{J_A}{J_A + J_L} V S_\vartheta S_f.$$

Frequency coefficient S_f: for frequency

$$f \le 10 \text{ Hz}: S_f = 1$$

$$f > 10 \text{ Hz}: S_f = \sqrt{f/10}.$$

The amplification ratio V for an excited dual mass oscillator indicates the amplification of the torque acting with the exciting frequency f_i:

$$V = \sqrt{\frac{1 + \left(\frac{\psi}{2\pi}\right)^2}{\left(1 - \frac{f_i^2}{f_e^2}\right)^2 + \left(\frac{\psi}{2\pi}\right)^2}}$$

Natural frequency f_e is calculated using the second moments of area J_A and J_L of the drive side and load surface respectively and torsional stiffness C_{Tdyn} at

$$f_e = \frac{1}{2\pi} \sqrt{C_{Tdyn}\left(\frac{1}{J_A} + \frac{1}{J_L}\right)}.$$

It should not coincide with torsion-raising frequencies f_i such as the operating frequency or multiples of it (distance, e.g. $\pm 20\%$). It should be noted that asynchronous engines irrespective of their nominal RPM excite at the standard frequency (50 Hz) when starting [5, 31, 32]. Many couplings (Cardan, double-toothed) can excite at twice the operating frequency. If $f_i < \sqrt{2f_e}$, then the elastically coupled machine runs more smoothly than the excited one. When the resonance passes through, the moment being set up becomes smaller, the greater the damping ψ.

(e) The displacement characteristics of the coupling (ΔK_a, ΔK_r, ΔK_w: **Fig. 4f**) axially, radially and angularly must be greater than the shaft displacement which occurs in practice (ΔW_a, ΔW_r, ΔW_w). At coupling stiffnesses C_a, C_r and C_w displacement causes restoring forces and moments to act on adjacent components, which must be checked for permissibility [33, 34]. Good alignment, especially with continuous running and high RPM, is the key to extending the service life of the coupling.

3.3.3 Types

Resilient Metal Couplings. The various types can be distinguished mainly by their use of different types of spring (torsion angle $\varphi = 2$ to $25°$) at different levels of damping: **Fig. 8a–d**. Further, the linear spring characteristics can be changed into mainly progressive ones by structural means, e.g. in the coil spring coupling by means of axially tapering "teeth". In this way, when the torque increases the free length of spring is shortened, ΔK_r a few mm, $\Delta K_w = 1$ to $2°$, ΔK_a up to 20 mm.

Elastomer Couplings with Medium Elasticity (Fig. 9a–b). These have twisting angles $< 5°$ and are either (a) *pin couplings* which have cylindrical, spherical or grooved elastomer bushings or (b) *jaw couplings* the components of which are stressed in bending or in compression. Further properties: ΔK_r a few mm, $\Delta K_w = 1$ to $2°$, ΔK_a up to 20 mm.

Elastomer Couplings with High Elasticity (Fig. 9c–e). These are couplings with torsion angles of $\varphi = 5°$ to $30°$ at nominal torque ($\Delta K_r = 6$ to 10 mm, $\Delta K_w = 8°$, ΔK_a 10 to 15 mm). These couplings are usually distinguished by the large amount of rubber they contain. Types include *bead couplings* (**c**) with a transversely slit, tyre-like bead, which can degenerate into a disc in flange couplings (flywheel attached). The spring characteristics are usually linear as also in *disc couplings* (**d**) which have

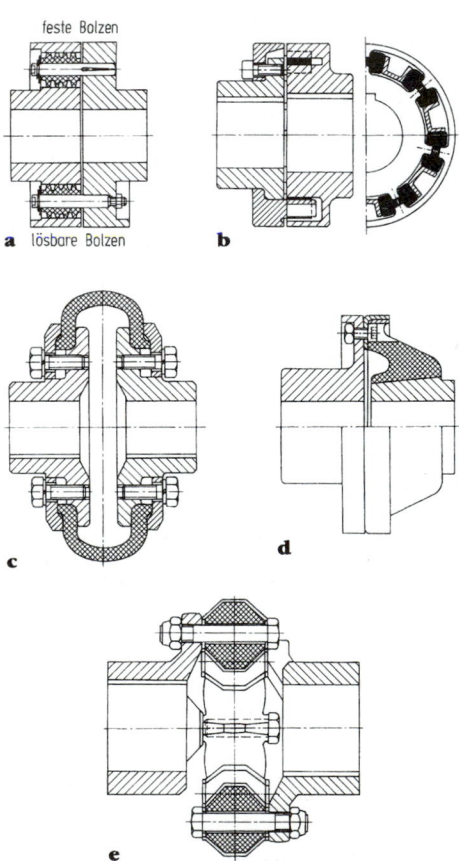

a lösbare Bolzen   **b**

c

d

e

Figure 9. Elastomer couplings. **a** Pin coupling (Elco coupling, Renk); progressive characteristic ($\varphi = 2°$ to $3°$) through shaped, preloaded rubber sleeves [33]. **b** Claw coupling with compressed elastic elements (N-Eupex coupling, Flender); puncture-proof, spring characteristic progressive (φ up to 2.5°). **c** Highly elastic reinforced coupling (Periflex, Stromag) with annular rubber tyre cut open at right angles to the periphery; spring characteristic approximately linear ($\varphi = 5°$ to $12°$). **d** Highly elastic disc coupling (Kegelflex coupling, Kauermann) with moulded on rubber disc; linear spring characteristic can be changed by varying type of rubber (φ up to 10°). **e** Highly elastic intermediate ring coupling (Ortiflex coupling, Ortlinghaus) with radially preloaded six-cornered rubber ring with steel sleeves moulded in; spring characteristic progressive (φ up to 10°).

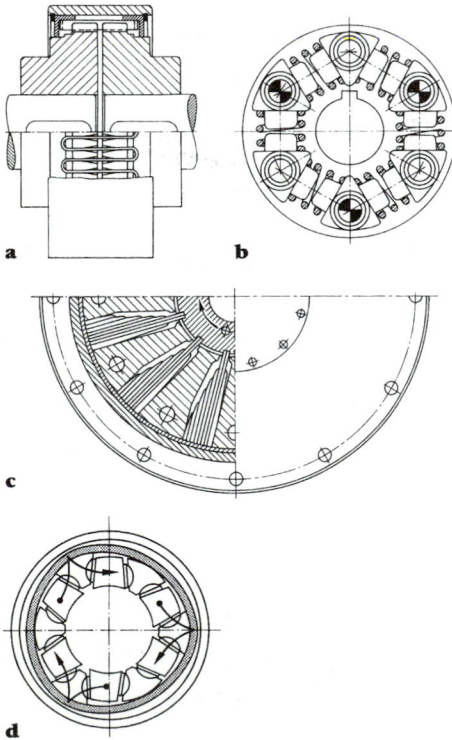

a   **b**

c

d

Figure 8. Resilient metal couplings: **a** coil spring coupling (Malmedie–Bibby) with structurally constrained progressive characteristic curve ($\varphi = 1.2°$), **b** coil coupling (Cardeflex) with tangential, preloaded coil springs (φ up to 5°), **c** Geislinger coupling with leaf spring assembly arranged radially; frictional and adjustable oil damping by means of oil displacement from spring chambers (φ up to 9°), **d** Keilflex coupling with three coupling claws at both the driving and driven ends. Torque is transmitted by means of sliding keys fastened to an external elastically deformable split ring ($\varphi = 2°$ to 3°).

a vulcanised rubber disc. *Spider couplings* (**e**), on the other hand, usually have progressive characteristics. A six- or eight-cornered rubber ring, reinforced at the corners by metal sleeves, is screwed alternately to each half-body in the periphery. These couplings are used in a similar way to *inside disc couplings* ("Hardy" discs) in motor vehicle construction as maintenance-free "rubber universal joint shafts".

3.3.4 Type Selection

Simple constant-speed drive mechanisms (including electric motors, centrifugal pumps and fans) are coupled to *elastomer couplings* with medium elasticity ($\varphi < 5°$) in order to compensate for starting impulse and faulty shaft positioning as they are both cheap and maintenance-free. *Highly variable-speed drive mechanisms* (piston engines, crushing mills, presses, rolling mills) or the *transfer of resonance speed* require highly elastic couplings ($\varphi = 5°$ to $30°$). Highly elastic couplings (elastomer or metal) are particularly suitable for large *shaft displacements. Large axial displacements* in particular can be readily controlled using pin and jaw couplings. *Disruptive strength*, i.e. the capacity to transmit torque even when the elastic elements have been destroyed (a requirement for lift drive mechanisms) is a particular feature of pin and claw couplings. *Permissible speeds* are generally lower with torsionally elastic couplings than with torsionally stiff ones (e.g. toothed and diaphragm couplings).

3.4 Clutches

Externally actuated (mechanical, hydraulic and inductive) clutches may be classified according to the following criteria [1, 4, 35 to 39].

Operating Principle. *Closing* clutches transmit torque in the operating condition, whereas *opening* clutches interrupt torque flow when actuated. In electromagnetically actuated clutches *load current actuated* clutches are referred to as closing and *static current actuated* clutches as opening (definition according to VDI Guideline 2241).

Type of Actuation. *Electromagnetically* or *centrifugally* actuated clutches (hydraulic or pneumatic) permit remote operation and facilitate automation, unlike *mechanically* actuated clutches.

In mechanical clutches the main method of classification is to distinguish between *positive and frictional* torque transmission (cf. **Fig. 2**).

3.4.1 Positive (Interlocking) Clutches (Dog Clutches)

Positive clutches can be engaged only when the shafts are at a standstill or running synchronously. They can however be disengaged when running even under torque, providing the disconnecting forces do not become too great. They are very small and are therefore often cheap. In most cases they permit axial shaft displacement under rather frequently occurring displacement forces (frictional forces).

Types. The *positive-contact clutch* (**Fig. 4a**) is mostly used in general mechanical engineering, whereas the *gear*

clutch [40] is mainly used in gearbox construction, as it results in the smallest diameters. The *double-gear clutch* and still more so the *elastic pin and positive-contact clutches* (**Fig. 9a, b**) make it possible to compensate for shaft positional errors. The *magnetic tooth clutch* (**Fig. 10**) which has radially running teeth is frequently used and can be engaged by magnetic force in precisely aligned shafts (one seated inside the other), sometimes even at low relative speeds.

3.4.2 Friction Clutches

Externally actuated friction clutches are used in the *transmission of torque* from a driving to a driven shaft using external energy to produce a magnetic flux. The connection is provided by mechanical *frictional engagement* (definition according to VDI Guideline 2241).

Types. They may be classified according to *number* and *arrangement of frictional surfaces* in *single surface clutches* (**Fig. 11a**), *dual-surface (single-disc) clutches* (**Fig. 11b**), *multiple surface (multiple-disc) clutches* (**Fig. 11c**), *cylinder* and *cone clutches* (**Fig. 11d**). These clutches may either be made *dry* or *wet* (oil lubricated) [41].

Multiple-disc clutches (**Fig. 11c**) are small and low-cost and are therefore the most used. They are however susceptible to overheating (heat transferred from the set of discs) and have a rather long reaction time, owing to the action paths accumulating with the number of friction discs [42]. They also have a relatively high idling moment, because the discs separate only imperfectly [43]. If this should be small and the heat transfer large, clutches with clearly defined lines of separation and large heat-emitting surfaces are used. These are *single-surface, dual-surface, cone* and *cylinder clutches* (**Fig. 11a, b, d**).

Clutch-brake combinations (**Fig. 12**) constitute a combination of clutch and brake in a structural unit. They are particularly suitable for *frequent engaging and releasing* and *rapid engaging*. In order to attain the shortest intervals between actuations, overlaps can be selected when actuating (separately) the clutch and the brake.

The *magnetic particle clutch* (**Fig. 13**) is an *electromagnetically actuated friction clutch*, in which magnetisable particles transmit power by means of *friction* in a space between the driving and driven member (in an electromagnetic field) [44]. The degree of *slip* in the coupling is governed by the intensity of the magnetic field. The

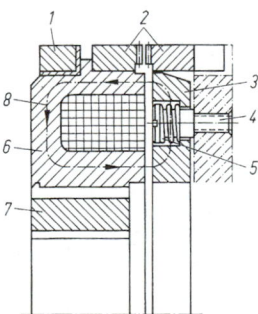

Figure 10. Electromagnetically actuated gear clutch with slip ring (Zahnradfabrik Friedrichshafen AG) [4]: *1* slip ring, *2* gear rim, *3* anchor plate, *4* spring bolt, *5* spring, *6* magnet, *7* tapered bushing, *8* course of magnetic flux.

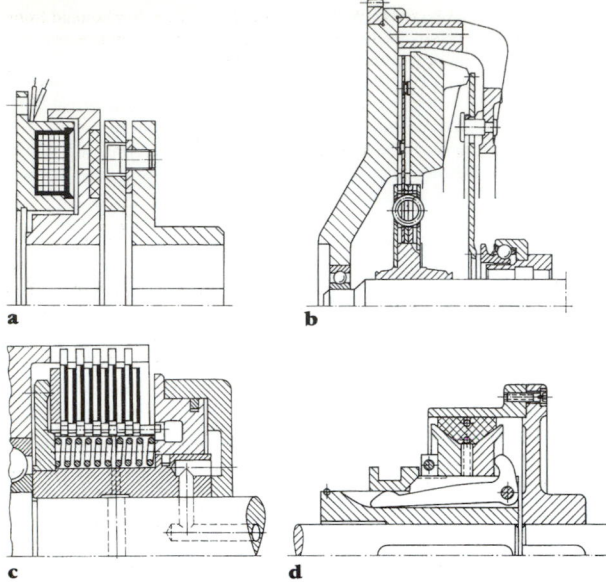

a   b

c   d

Figure 11. Types of friction clutches: **a** electromagnetically actuated single-surface clutch without slip ring (Ortlinghaus), **b** mechanically actuated single-disc clutch (diaphragm spring clutch) for commercial vehicles (Sachs), **c** hydraulically actuated multiple-disc clutch (Ortlinghaus), **d** mechanical (Conax, Desch).

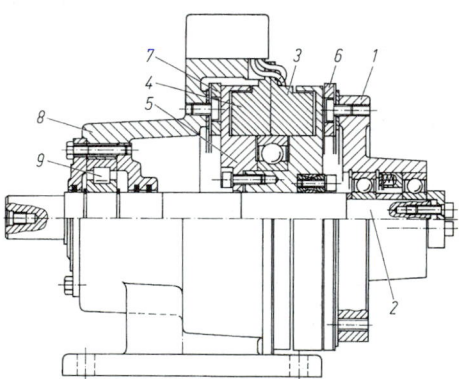

Figure 12. Electromagnetic clutch brake combination (Zahnradfabrik Friedrichshafen AG) [41]: *1* drive, *2* drive/shaft, *3* magnet (clutch), *4* magnet (brake), *5* rotor assembly, *6* anchor (clutch), *7* anchor (brake), *8* casing, *9* bearing.

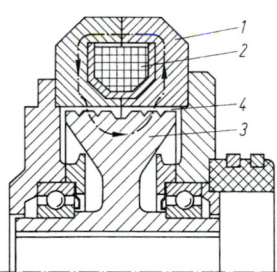

Figure 13. Magnetic particle clutch (AEG-EMG) with magnetic flux indicated [37]: *1* iron bodywork with *2* magnetic toroidal winding, *3* armature, *4* air gap containing magnetic particles.

clutch makes smooth starting possible and with a suitable control mechanism it can be used as an overload clutch. Achieving the permissible amount of slip is a matter of dissipating the heat.

Types of Actuation. *Hydraulic* actuation results in particularly small, wet-running clutches whereas the *pneumatic* actuation of dry clutches results in short engagement times but requires a free shaft end to allow application of pressure. *Mechanical* engagement produces long clutches. *Electromagnetic* actuation provides a very good control facility, good remote operation and simple energy supply.

Methods of Operation. *Dry* clutches are simple and economic and require little maintenance. They have a low idling moment (e.g. 0.05% of the nominal moment), show little tendency to chatter when engaged and have short

reaction times. *Wet* clutches are used if the environment cannot be made oil-free (gearboxes) or if minimal wear combined with good heat dissipation (flow-through lubrication!) have been specified. The *disadvantages* of wet running friction systems are low sliding friction coefficients and a relatively high idling moment.

Friction materials (**Appendix F3, Table 1**) should produce as little waste as possible between μ_0 and μ, because in this way the slip-stick effect (chattering) can be avoided. This is especially important in the dry condition.

3.4.3 Transient Slip in Friction Clutches During Engagement [35, 37, 41, 45, 46]

Some of the principles of clutch calculations are said to be shown in the schematic model of a load to be accelerated from ω_{20} to ω_{11} (e.g. impact crusher, stamping machine drive, winding drum) (**Fig. 14**). The motor producing the driving torque M_A possesses mass moment of inertia J_A and rotates at angular velocity ω_{10}. The load (load moment M_L, mass moment of inertia J_L, angular velocity ω_{20}) can be connected to the drive mechanism by

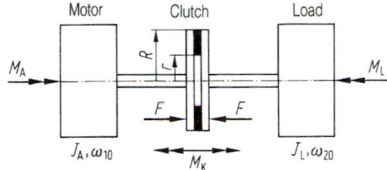

Figure 14. Schematic model of a drive system.

means of the clutch (characteristic moment M_K, outside radius of frictional surfaces R, inside radius r, contact pressure F). **Figure 15** shows the basic clutch movement in schematic form.

Before the clutch is actuated the *idling moment* M_r is present in the drive system (incomplete separation). After actuation and the *reaction delay* t_{11}, torque transfer is built up during the *build-up time* t_{12}. The *slip moment* M_s acting on the clutch band after the buildup time is made up of the *load moment* M_L and a moment M_a required to overcome the inertia of masses in movement. The moment M_s must therefore be greater than M_L by M_a in order to enable the speed of the load to be increased (cf. **Fig. 15**). M_s is not constant and depends, among other things, on *running speed*, *temperature of frictional surfaces* and *marginal structural conditions*. At a differential speed of zero the *synchronous moment* M_{syn} is soon formed. When the driving and driven member are synchronised the (in this case constant) load moment M_L is present.

For the *experimental evaluation* of existing clutches the measured progression of the torque as shown schematically in **Fig. 15** is idealised and approximated using a *linear buildup* (in the time t_{12}) with, subsequently, constant moment M_K.

The *characteristic moment* M_K can either be read from the line of ordinates in **Fig. 15** or be determined in a simplified way in Eq. (1) [41]:

$$M_K = C \pm \sqrt{C^2 - B}$$

$$\text{when} \quad C = \frac{M_L t_3 + J_L (\omega_{10} - \omega_{20})}{2t_3 - t_{12}} \tag{1}$$

$$\text{and} \quad B = \frac{t_{12} M_L^2}{2t_3 - t_{12}}.$$

Equation (1) applies to $M_L = \text{const}$ and $\omega_{10} = \omega_{11} = \text{const}$, i.e. the speed of the motor does not drop when the clutch is engaged. The length of the buildup time t_{12} is specific to the type of clutch and actuation, whereas the slip period t_3 depends on the load, among other things.

$$t_3 = \frac{J_L(\omega_{10} - \omega_{20})}{M_K - M_L} + \frac{t_{12}}{2}\left(1 + \frac{M_L}{M_K}\right). \tag{2}$$

In Eq. (2) the slip period t_3 increases as the load (J_L, M_L) gets bigger and the buildup time (t_{12}) lengthens, whereas a large characteristic moment M_K reduces the contact time.

The *engagement force* Q, which is converted to heat during engagement, is produced when characteristic moment M_K is present and there is a difference between angular velocities at $\Delta Q = \Delta \omega\, M_K\, \Delta t$. The mean over the entire slip period t_3 is $\Delta \omega \approx (\omega_{10} - \omega_{20})/2$. Using the simplifications in Eq. (1), the engagement force then becomes

$$Q = \frac{(\omega_{10} - \omega_{20})^2}{2} \cdot \frac{J_L}{1 - \dfrac{M_L}{M_K}} + \frac{(\omega_{10} - \omega_{20})}{2} t_{12} M_L. \tag{3}$$

The engagement force is made up of the *static engagement force* Q_{stat} resulting from the load moment and the *dynamic engagement force* Q_{dyn} to overcome the inertia of masses J_L (**Fig. 15** above). Thus, if there is an acceleration from $\omega_2 = 0$ to $\omega_2 = \omega_{10} = \omega_{11}$ then half of the energy applied during engagement is converted to heat. As the speed of the motor usually falls and the synchronisation of the clutch discs is achieved sooner a lower frictional force is produced. If, after engagement, M_s first rises slowly frictional force is increased, as until $M_s = M_L$ is reached there is no increase in speed. t_3 likewise increases. In the case of *step-by-step* engagement (as against a single engaging action) in gearboxes (motor vehicles) Q is reduced.

If the frictional force becomes low, $1 - (M_L/M_K)$ must become large, i.e. $M_K \gg M_L$. At a given M_L there is a requirement for a *"hard"* clutch to minimise thermal stress. This reduces the slip period t_3, but the clutch may in certain circumstances produce *sharp shock loads*. The opposite extreme, of having too *soft* a clutch $M_K \to M_L$, results in gentle clutch engagement but also a high degree of heating. With a combination of long slip period t_3 and frequent engagement, thermal stress can lead to the destruction of the clutch. The heat resulting from a single clutch engagement is mainly dependent on *angular velocity difference* and *frictional unit pressure*. In the case of *repeated* engagement the temperature of the frictional surfaces increases with the frequency of engagement.

Values for the maximum permissible thermal stress Q_E for a *single* engagement and Q_{perm} for *repeated* engaging are specified by the clutch manufacturer. Maximum values Q_E are dependent on the *frictional surface material* and

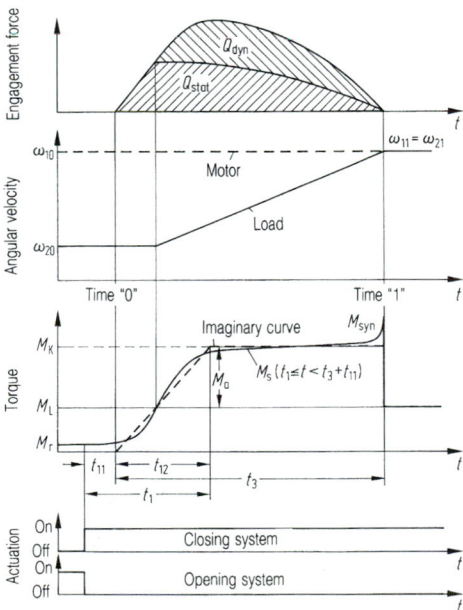

Figure 15. Engagement of externally actuated friction clutches [as in 35, 41].

the *thermal capacity* of the clutch, whereas Q_{perm} are mainly dependent on *cooling* and *heat emission*.

Values acquired empirically or by means of costly mathematical processing for Q_E and Q_{perm} can be shown as *characteristic curves* for specific clutches (**Fig. 16**). This shows the permissible engagement force Q_{perm} (per engaging action) as a function of the *engagement frequency* S_h. The *transitional engagement frequency* $S_{h\ddot{u}}$ produces a *characteristic value* on the curve and is specified by the clutch manufacturer. Using the parameters Q_E and $S_{h\ddot{u}}$, the permissible thermal stress Q_{perm} can be specified as a function of the engagement frequency S_h [35, 41]

$$Q_{perm} = Q_E \left[1 - \exp(-S_{h\ddot{u}}/S_h) \right] \qquad (4)$$

3.4.4 Layout Design of Friction Clutches [37, 41, 47, 48]

A clutch is basically designed according to the *maximum moment* to be transmitted and the *engagement force* that produces it, and so the thermal stress is usually the deciding factor for selecting the appropriate size [49, 50].

The *moment to be transmitted* is governed by the nominal moment of the drive motor and operating machine where allowance must be made for cyclic variations (e.g. piston engines) or the moment of tilt (2 to 3 M_N) in squirrel-cage motors. The *engaging and disengaging torque* of a clutch $M_s > M_L$ is generally smaller than the *transmittable moment* $M_{\ddot{u}}$, which is set up in frictional surfaces which are relatively static in relation to each other. In particular, for wet clutches the *sliding friction coefficient* μ is smaller than the *static friction coefficient* μ_0 (cf. **Appendix F3, Table 1**). For the practical design of a friction clutch the specified moment M_K in Eq. (5) is used:

$$M_K = F\mu z r_m. \qquad (5)$$

This makes it possible to specify iteratively the required *contact pressure* F, the *number of frictional surfaces* z and the required *mean radius* of the frictional surfaces $r_m = (R + r)/2$ (cf. **Fig. 14**). If, for example, the diameter of the clutch should be small then the number of frictional surfaces or the contact pressure (for maximum permissible contact pressure cf. **Appendix F3, Table 1**) can be increased. In this way a variety of clutch types can be designed.

In **Fig. 15** the reaction delay t_{11} must be noted at engagement time $t_{tot} = t_{11} + t_3$. To simplify layout design buildup time, t_{12} can be disregarded when calculating slip period t_3 (cf. Eq. (2))

$$t_3 = \frac{J_L (\omega_{10} - \omega_{20})}{M_K - M_L}. \qquad (6)$$

Engagement force Q can be determined as follows:

$$Q = \frac{(\omega_{10} - \omega_{20})^2}{2} \cdot \frac{J_L}{\left(1 - \dfrac{M_L}{M_K} \right)}. \qquad (7)$$

The *surface engagement force at a single engagement* q_A is then determined as follows:

$$q_A = \frac{Q}{A_{Rg}}, \qquad (8)$$

where A_{Rg} denotes the entire frictional surface of the clutch

$$A_{Rg} = A_R z = \pi(R^2 - r^2)z. \qquad (9)$$

The surface engagement force q_A can be compared with the *permissible surface engagement force at a single engagement* q_{AE}: $q_A < q_{AE}$ (cf. **Appendix F3, Table 1**). It is also possible to make a comparison between the *actual* and the *permissible surface frictional force* ($\dot{q}_A$ and $\dot{q}_{AO}$) (cf. **Appendix F3, Table 1**):

$$\dot{q}_A = \frac{q_A}{t_3} = p_R v_t \mu < \dot{q}_{AO}, \qquad$$

where p_R denotes the frictional surface pressure, v_t the running speed and μ the sliding friction coefficient.

3.4.5 Size Selection of Friction Clutches

If a clutch to be purchased is being designed for a specific application the exact requirements must first be specified. Reference [36] provides a questionnaire to assist in the selection of clutches.

The required characteristic moment M_K of the clutch can be estimated on the basis of the load moment M_L, the (reduced) mass moment of inertia J_L, the angular velocity difference $\Delta\omega$, the approximate slip period t_3 and buildup time t_{12} required (cf. F3.4.3, Eq. (1)).

Using this value a specific clutch is sought from a selected manufacturer. The *buildup time* t_{12} should be apparent from the catalogue. It is then possible, by using Eq. (2) and (3), to accurately determine the *slip period* t_3 and the *engagement force* Q. If a drop in the speed of the drive motor and the mass moment of inertia (with gears) is to be allowed for when the clutch is engaged, then more detailed literature [37, 41] must be obtained. The calculated engagement force Q can be compared with the permissible values Q_E (catalogue) for the clutch selected. A larger clutch must be selected when $Q > Q_E$. In the case of frequent engaging and disengaging the permissible engagement force must be determined using Eq. (4) and compared with the actual engagement force. This is only possible if $S_{h\ddot{u}}$ is known for a specific clutch. If only characteristic curves are provided as in **Fig. 16**, they also can be used to specify the clutch required (cf. [35]).

3.4.6 Selection Criteria [6, 51]

A selection of the types of clutch commercially available is shown in **Fig. 17**. A characteristic moment $M_K = 500$ Nm and a speed of $n = 1500$ min^{-1} were used as a basis for comparison. The chart therefore enables a comparison to be made between the diameter and length of the clutches and between the permissible engagement force Q_E at a single engagement and the limiting value $Q_E S_{h\ddot{u}}$ at very high engagement frequency. With the exception of the commercial vehicle clutch (opening clutch)

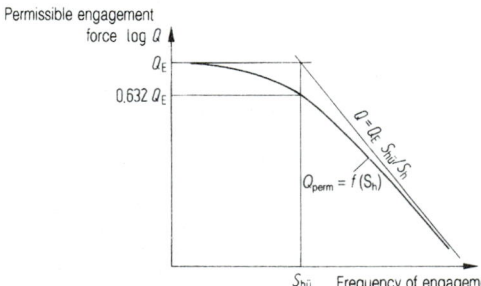

Permissible engagement force log Q

Q_E

$0.632\ Q_E$

$Q = Q_E \cdot S_{h\ddot{u}}/S_h$

$Q_{perm} = f(S_h)$

$S_{h\ddot{u}}$ Frequency of engagement

Figure 16. Permissible engagement force as in Eq. (4) as a function of the frequency of engagement [35].

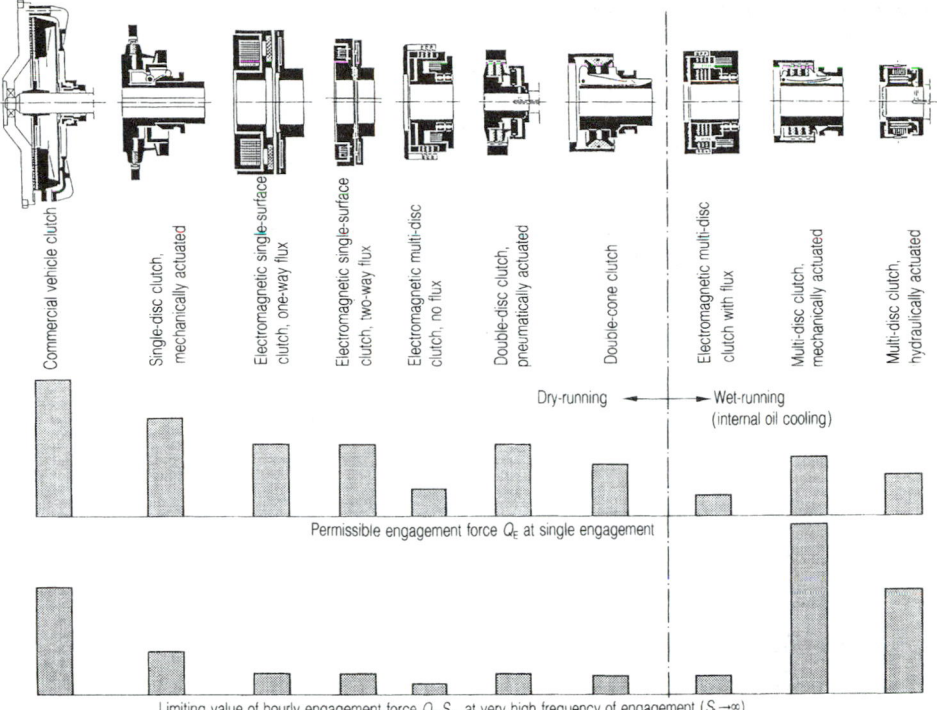

Figure 17. Comparison of types of friction clutches for a characteristic moment $M_K = 500$ Nm and a speed $n = 1500$ min⁻¹ [36].

they are closing clutches. Also shown in **Appendix F3**, **Fig. 2**, is the engaging moment region of various externally actuated friction clutches.

Methods of Operation and Actuating Systems, Properties

Single-Surface Clutches. Dry-running friction pairings are preferred so as not to obtain too great a diameter at a given torque. A closed axial magnetic flux within the clutch is only possible through electromagnetic actuation; rapid reaction with short lifts; low idling moment.

Single-, Double-Disc Clutches. Again dry running for larger torques; all methods of actuation can be used, but hydraulic actuation is usually avoided owing to the danger of leakage (friction linings are oil lubricated); efficient cooling (cooling ribs), rapid reaction, low idling moment, relatively chatter-free (materials with decreasing μ/v_r characteristics).

Multiple-Disc Clutches [42, 43]. Small size even with large torques, engagement under load, effective cooling but only possible with oil circulation, i.e. wet running; all methods of actuation possible. In the case of magnetic flux controlled discs (electromagnetic actuation) only specific friction pairings can be selected. Rapid reaction can be achieved in wet running by means of thin oil, oil mist or grooves in the discs; low idling moment achieved by corrugated discs. Longer useful life, i.e. little wear in wet running.

Cone Clutch. Suitable for high torques and engagement forces in dry running, actuation usually mechanical or pneumatic.

3.4.7 Brakes

Brakes are clutches with an idle driven member and 100% slip. Like the clutches there are mechanical, hydraulic, pneumatic and electric brakes (cf. **Fig. 2**). According to their function they can be classified as locking brakes, stop and regulating brakes as well as dynamometric brakes [37]. Various types of brake are shown in **Fig. 18**. The *calculations* for mechanical, engaging and disengaging friction brakes are carried out much like clutch calculations: characteristic moment M_K is replaced by braking moment and accelerating torque M_a by decelerating torque. Principles of calculation for drum and disc brakes are contained in Part 1 of DIN 15 434.

Types. In principle all types of clutches are also available as brakes (cf. **Figs 17, 13** and **11**). *Shoe brakes* can be divided into internal (**Fig. 18c**) and external shoe brakes (**Fig. 18b**) (vehicles, hoists). *Band brakes* (**Fig. 18a**) require only low operating forces on account of the self-energising effect of contact friction. In addition to the type shown in **Fig. 18a** there are also band brakes with multiple wrapping as well as internally acting band brakes. *Disc brakes* (**Fig. 18d**) have favourable cooling characteristics, especially if they are manufactured with internal air supply. *Hydraulic* (water, oil) and *electric* brakes (generators) which easily dissipate the resulting energy are non-wearing *dynamometric brakes. Induction brakes* (**Fig. 18e**), by means of a static live coil in the rotor, induce current which causes a force to act counter to the direction of rotation. In these eddy current brakes the braking moment is to a large extent governed by rotational speed [52].

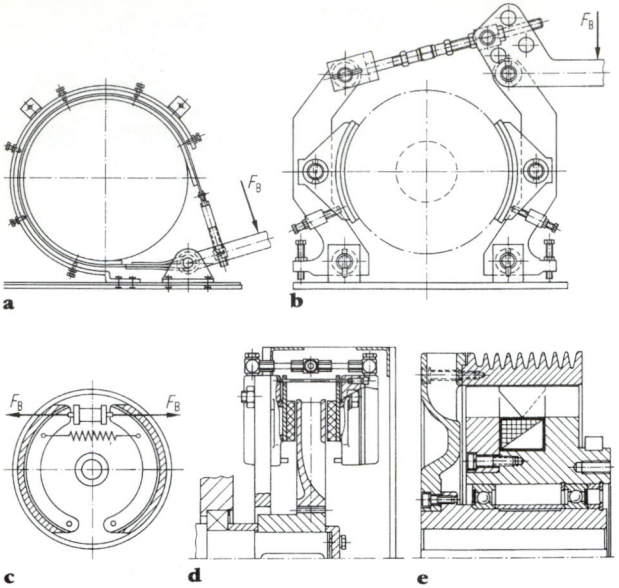

Figure 18. Types of brake (operating force F_B partially shown): **a** band brake [37], **b** external shoe brake (double) [37], **c** internal shoe brake (drum brake, simplex), **d** pneumatically operated disc brakes (Ortinghaus), **e** induction brakes with fan wheel (Stromag).

3.5 Automatic Clutches

3.5.1 Torque-Sensitive Clutches (Slip Clutches)

These are safety clutches which protect machinery from damage as they do not exceed a predetermined torque. In this way the unnecessary oversizing of machinery to meet peak moments can also be avoided [53].

Types. In principle, all friction clutches with fixed clutch force can be used as *slip clutches*. It is important that the normal force should not alter significantly with the wear pattern (even spring characteristic) and the clutches should be monitored for permanent slip. *Block clutches* usually have spring-loaded tapers or pins which disengage when the torque limit is reached. With overload release snap pin and snap ring clutches [54] it must be borne in mind that the ultimate moment may be greatly diffused unless special measures are adopted.

Both slip and block clutches may be used to disconnect electromechanical or electric switches in order to switch off the drive motor.

3.5.2 Speed-Sensitive Clutches (Centrifugal Clutches)

These are clutches which allow smooth starting so that electric motors or combustion engines accelerate first of all and only then drive the machine. *Starting clutches* make it possible to design for a reduction in motor size or even electricity supply for machines with a high second moment of inertia or load moment.

Types (Fig. 19). *Centrifugal clutches* [55] with *segments* (**Fig. 19a**) [9] if provided with retaining spring will transmit a moment only after a specific speed has been reached. *Sprag clutches* (**Fig. 19b**) [9] throw powder, balls or rollers against the casing of the driven member by means of a star-shaped rotor so that the moment to be transmitted increases quadratically with the speed of the drive motor. At nominal speed these clutches, unlike *hydrodynamic* clutches [56], are free of slip and leakage. *Starting* in an asynchronous motor (characteristic curve

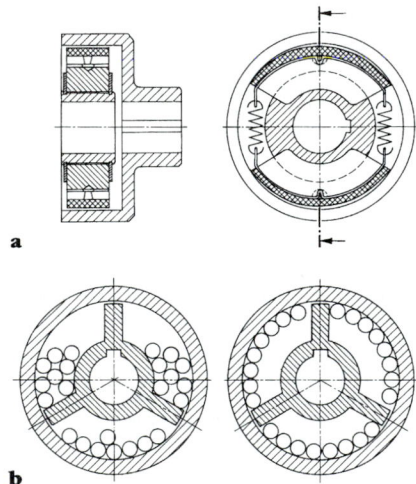

b

Figure 19. Speed-sensitive couplings (see text).

M_M in **Fig. 20**) when using a sprag clutch is practically load-free (only M_K) and the machine is at rest up to the intersection *1* of the clutch characteristic M_K and the load characteristic M_L. The motor remains at point *2* and accel-

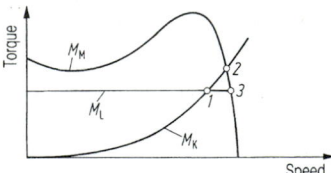

Figure 20. Characteristic curves of an asynchronous motor M_M and a centrifugal clutch M_K; load moment M_L.

erates the machine up to synchronous running. Subsequently all the parts reach the operating speed at point *3*. A disadvantage of these clutches compared with spring-loaded slip clutches is that in practice they only operate on quick-running shafts.

3.5.3 Directional (One-Way) Clutches, Overrun Clutches

The engaging procedure depends on the direction of the relative rotary motion between the driving and driven member: it is prevented in *one* direction of the relative rotation (locked condition) but not in the *other* direction (freewheeling condition). Freewheel clutches have the following *functions* (they cannot usually be distinguished by structural shape) [9, 57, 58]: *return stop* (for conveyor belts, pumps, automatic gearboxes for motor vehicles, fans); *overrun clutch* (for multi-motor drives, starting motor drives, bicycle hubs); *step-by-step freewheel* (for shaping machines, feed mechanisms, ratchet mechanisms).

Types. For simple tasks: *Ratchet freewheels* (ratchet wheels, ratchet drills) engage the drive positively in one direction of rotation. There are also friction-driven, noiseless ratchets. *Grip freewheels* [4, 59], on the other hand, grip noiselessly at each location at greater clutch velocities and with smaller dimensions. *Grip roller freewheels* (**Fig. 21**) are often provided with internal spiders with which

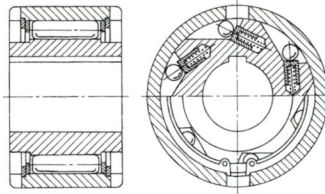

Figure 21. Grip roller freewheel with internal spider and individual cushioning.

individual cushioned rollers are pressed into the wedge-shaped pockets. *Clamp freewheels* [9, 60, 61], which are the same size, transmit more torque but are less robust. They have out-of-centre clamping jaws between circular cylindrical slideways. Wear-reducing additives in the lubricant [63] have the greatest influence on useful life [62] and accuracy of engagement. Wear can be reduced in *return stops* by centrifugal lift [9]. It is essential to have a perfect radial and axial bearing (there are assemblies with rolling bearings) [4, 9]. Clutches can also be actuated externally: disengaging (full freewheeling condition), re-engaging, full locking, engaging only during a revolution (one-stop clutch [64]). *Friction freewheels* are friction clutches (discs, cones) which are forced in one direction along a sharp thread. If helical gear clutches are used to transmit torque these are *toothed freewheels*.

4 Rolling Bearings

H. Peeken, Aachen

Rolling bearings are ready-to-fit machine parts. They consist of rolling elements running on inner and outer race rings and a cage, which keeps the rolling elements apart.

4.1 Fundamentals

4.1.1 Material Stresses and Fatigue in Rolling Contact

In rolling contact under load, the effect of "flattening" results in a contact surface the size and stress of which can be calculated according to the Hertzian equations. Hertzian theory applies to solid and isotropic bodies with material showing elastic behaviour. The bearing surface arising in rolling contact is assumed to be plane and small relative to the dimensions of the elements (for a detailed description of the Hertzian equations, see B4).

To calculate material stress on the basis of Hertzian compression, hypotheses are used for main shear stress, deformation energy and alternating (orthogonal) shear stress. In **Figs 1** and **2** the comparative stresses σ_v in the three hypotheses are shown, relative to the maximum Hertzian compression p_0, for line contact.

Consequently the maximum material stress value occurs under the tangential plane. Minimal differences arise in its absolute magnitude and depth. Structural changes in the rolling bearing material, e.g. plastic deformation (shearing strains) or the so-called butterflies, which occur below an angle of approximately 45°, indicate that the *process of fatigue* (formation of incipient crack points, cracking,

crack growth, crumbling of material particles (paring and pitting when lubricated)) is initiated at points of material inhomogeneity owing to the shear stress.

In the case of line contact (**Fig. 2**) the greatest shear stress $\tau_{max} = 0.304p_0$ occurs at a distance of $0.78b$ from the surface at the point $x = 0$ (where b is half the width of the rectangular bearing-surface area); point contact $\tau_{max} = 0.31p_0$, distance $0.47b$. With surfaces which are permanently subject to rolling action the pulsating shear stress may be regarded as the dynamic load limit. Since the maximum shear stress is proportional to the Hertzian compression, its calculation is sufficient for the evaluation of the state of stress. With mixed friction, as well as with liquid friction, tangential stresses arise from bearing friction in addition to normal load in the contact zone. This results in an increase in the stress maximum migrating to the surface (**Fig. 3**).

4.1.2 Load Distribution

The load distribution in a loaded rolling bearing is dependent upon the elastic deformations at the contact points of the individual rollers. The calculation of this load distribution and the maximum rolling contact load exert a decisive influence on the determination of the bearing's capacity rating C. **Figure 4** shows as an example a loaded single-row inclined ball bearing. The rolling contact forces act in the direction of the pressure angle α, while the load component F_r forms angle β with F. If β does not exceed a particular magnitude, only part of the race is under load. The load per rolling contact is determined by the elastic deformations at the contact points. According to the Hertzian equations (see B4) for point contact $Q_\psi/Q_{max} = (\delta_\psi/\delta_{max})^{3/2}$; Q_ψ is the rolling contact load at the point ψ, Q_{max} the maximum rolling contact load, δ_ψ

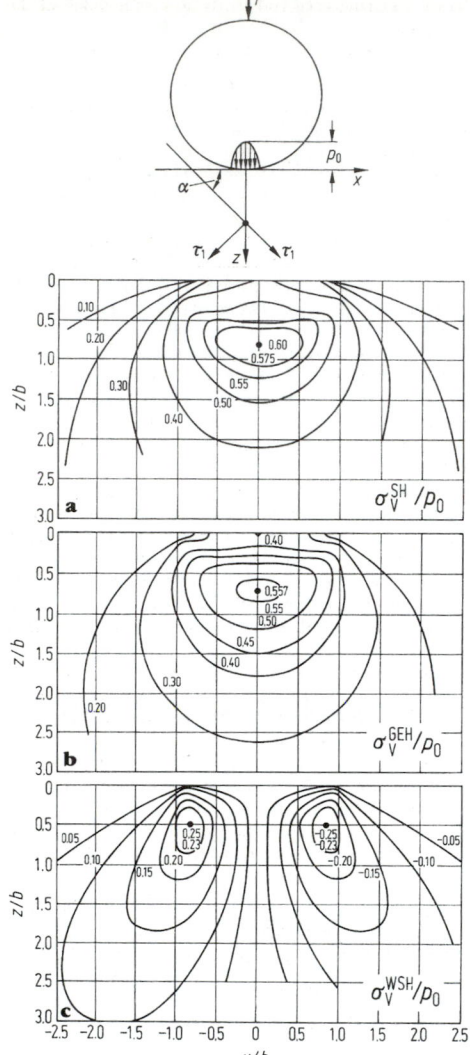

Figure 1. Undimensioned comparative stresses σ_v/p_0 [1]: **a** main shear stress hypothesis, **b** deformation energy hypothesis, **c** alternating shear stress hypothesis.

the displacement of the bodies (rolling elements) at point ψ, and δ_{max} the maximum displacement.

From the balance between the rolling contact forces and the external load follows the relationship between Q_{max} and the radial force F_r and axial force F_a. If, for example, in the inclined ball bearing being considered half the bearing circumference is under load ($\varepsilon = 0.5$ in **Fig. 4**), then for point contact $Q_{max} = 4.37 F_r/(z \cos a)$; z is the number of rolling elements. In the case of line contact (e.g. a single-row tapered roller bearing) the load distribution follows $Q_r/Q_{max} = (\delta_r/\delta_{max})^{1.08}$. For $\varepsilon = 0.5$ the maximum rolling element load $Q_{max} = 4.06 F_r/(z \cos \alpha)$. With $\alpha = 0°$ these equations can also be used for single-row ball and roller bearings that have no play. Together with the material characteristics, information can be derived from them about the load capacities (cf. F4.3.1 and F4.3.2).

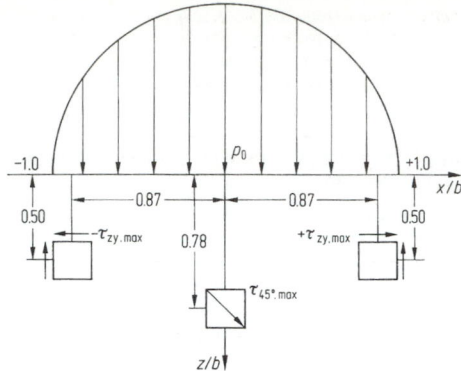

Figure 2. Shear stresses under the surface according to the main shear stress and alternating shear stress hypotheses for line contact with Hertzian compression [1].

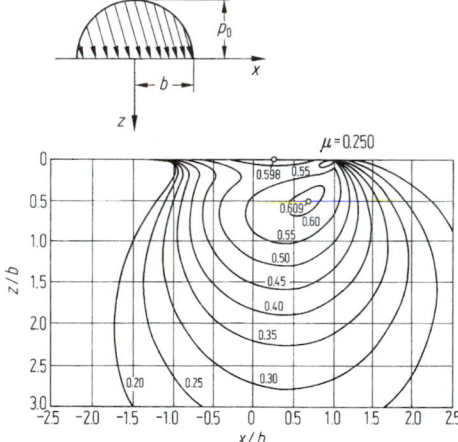

Figure 3. Material stress with line contact, normal and tangential load [2].

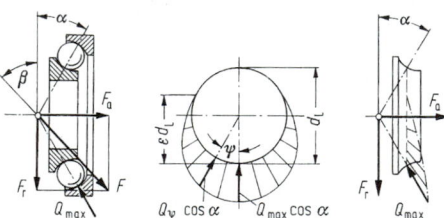

Figure 4. Load distribution in single-row inclined ball bearing: α pressure angle, d_L race diameter, F_a axial force, F_r radial force, β angle of direction of bearing load F, Q_ψ rolling element load, ψ position angle of rolling element, Q_{max} maximum rolling element load, εd_L extent of rolling race load.

4.1.3 Designations of (Standard) Rolling Bearings

Rolling bearings are designated in accordance with DIN 623 Part 1 by symbols consisting of *prefixes*, *base symbols* and *suffixes*. Parts of complete rolling bearings are designated by *prefixes*, e.g. K cage with rolling con-

Table 1. Base symbol for rolling bearings

Bearing series

	Size series	Symbol for bearing bore
		See DIN 623
Type of bearing See DIN 623	Width or height series	Diameter series
	See DIN 616	

tacts, L free race, R race with rolling elements, S stainless steel.

The *base symbol* designates the type and size of the bearing. It consists of two symbols or groups of symbols (**Table 1**).

The dimensions (bore d, external diameter D, width B, edge clearances r_{1min}, r_{2min}) of rolling bearings are arranged in such a way that several widths and external diameters are allocated to each bearing bore in order to cover a large load range (DIN 616; ISO 104). For radial bearings the grading is by width series (7, 8, 9, 0, 1, 2, 3, 4, 5, 6) and diameter series (7, 8, 9, 0, 1, 2, 3, 4, 5). By combining the two figures (B before D!) the size series is formed (**Fig. 5**). In addition, size plans are applicable to tapered roller bearings and axial bearings (height series 7, 9, 1, 2; diameter series 0, 1, 2, 3, 4, 5). For bore diameters of 20 to 480 mm the bore number is given. Except for bearing sizes up to $d = 17$ mm, bore d in mm is obtained by multiplying the bore number by 5.

For example the base symbol 6204 means: grooved ball bearing, single row (bearing series 62), size series 02 (width series 0, diameter series 2), bore $d = 5.04 = 20$ mm from the width series $0B = 14$ mm and from the diameter series $2D = 47$ mm.

In the case of bore diameters below 20 and above 480 mm the bore code number is replaced by the millimetre figure (in part separated by a slash). For tapered roller bearings DIN ISO 355 lays down a new identification. The base identification begins with T for tapered radial roller bearing; then follows for contact angle α the angle series (2, 3, 4, 5, 7), the diameter series (B, C, D, E, F, G), the width series (B, C, D, E) and the three-digit bore diameter in mm.

The suffixes are used to designate the internal construction, external form, cage design, accuracy, bearing clearance and heat treatment. For more details see DIN 623 Part 1.

4.1.4 Fit and Bearing Clearance

The following aspects are of importance in the selection of fit:

Support of the bearing rings at their periphery to maintain the full load capacity of the bearing.

Radial and tangential and partly also axial fixing of the bearing.
Ease of installation and removal.

The first two features require an interference fit. Particularly with greater loads which cause an extension of the rings, and also with shock loads, tight fits are required. The temperature drop between the bearing rings, which occurs in virtually all operating conditions, is also relevant. Tolerances for normal bearing clearance are given in **Table 2**.

As the bearing rings are not very thick, rigid bearing seating and limited shape and running tolerances (straightness, roundness, parallelism and planarity of the shoulders), toleranced tighter than the diameter, are stipulated.

The tolerances of rolling bearings are standardised in DIN 620; ISO 492, 199, 5753, 582. Apart from tolerance class P0 (normal tolerance) the standard provides for tolerance classes P6, P6X, P5, P4 and P2. Bearings with these tightened tolerances are intended for very accurate shaft guides and very high rotating speeds.

A major application of bearings with tightened tolerances is in the operating spindles of machine tools. Bearings for this purpose are manufactured in the tolerance classes SP (special precision), UP (ultra-precision) and HG (high accuracy), in addition to in the standard tolerance classes. Tapered roller bearings dimensioned in inches are found in normal tolerance and in tolerance class Q3.

The term radial (axial) bearing clearance refers to the distance through which the bearing rings can be moved in a radial (axial) direction from one end position to the other. The bearing clearance should be chosen in such a way that there is no distortion of the bearing rings and surrounding parts. The radial bearing clearance is reduced as a result of the fit, particularly in the case of tighter fits, but also because of the temperature drop. This reduction must be taken into account when choosing the bearing clearance.

Bearing Clearance Groups (according to DIN 620 Sheet 4). C1 radial bearing clearance smaller than C2, C2 radial bearing clearance smaller than normal (C0), C0 normal radial bearing clearance, C3 radial bearing clearance greater than normal (C0), C4 radial bearing clearance greater than C3, C5 radial bearing clearance greater than C4.

4.2 Types of Rolling Bearings

4.2.1 Ball Bearings

Types of Construction. For predominantly radial loads: **Fig. 6a–g**.

(*a*) *Single-row grooved ball bearings* (DIN 625; ISO 15) can take radial and axial forces and are suitable for high speeds. Their angular adjustability is small. Bearing

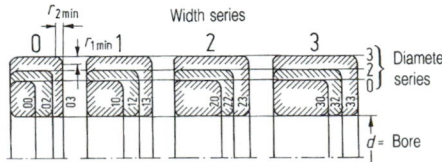

Figure 5. Structure of size plans for radial bearings.

Table 2. Tolerances for shaft and housing for normal bearing clearance [3]

	Shaft	Housing
Ball bearings	j5 to k5	J6
Roller and needle bearings	k5 to m5	K5

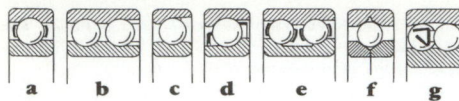

Figure 6. a–g. Ball bearing types for predominantly radial loads (see text).

positions that are not in alignment result in additional stresses which reduce the service life of the bearing. Grooved ball bearings are also manufactured with cover or sealing washers.

(*b*) *Double-row grooved ball bearings* (DIN 625; ISO 15) are manufactured with or without filling grooves. Bearings with filling grooves can therefore convey only slight axial forces. They are not suitable where there are angular errors.

(*c*) *Detachable groove ball bearings* (DIN 615; ISO 15) are only standardised up to a bore diameter of 30 mm. They have only one shoulder on the outer ring and can therefore be dismantled. Inner and outer ring are fitted separately. A transmission of axial forces is possible in one direction only.

(*d*) *Single-row inclined ball bearings* (DIN 628; ISO 15) take axial forces in one direction only. They are therefore located against another bearing in an O or X arrangement. Single-row inclined ball bearings cannot be dismantled.

(*e*) *Double-row inclined ball bearings* (DIN 628; ISO 15) take radial and axial loads in both directions as well as instantaneous loads. Their structure is equivalent to a pair of single-row inclined ball bearings in O arrangement. The bearings are delivered with very small amounts of play and so the fits used should not be too tight.

(*f*) *Four-point contact bearings* (DIN 628; ISO 15) are single-row inclined ball bearings which take axial forces in both directions. In axial section the contour of the raceways of the inner and outer ring consist of arcs which form ogives. The inner ring of the four-point contact bearing is divided so that a large number of balls can be accommodated.

(*g*) *Self-aligning ball bearings* (DIN 630; ISO 15) are double-row bearings with hollow spherical outer raceway which compensate for alignment errors and shaft deflections of up to 4°. Owing to the unfavourable osculation between balls and outer ring the axial load-bearing capacity is less than that of a grooved ball bearing.

Types of Construction. For predominantly axial loads; see **Fig. 7**.

Axial Grooved Ball Bearings (DIN 711, DIN 715; ISO 104) of the unidirectional type (axial force in one direction only) or the two-directional type take high axial forces. They are not suitable for radial loads. To achieve kinematically efficient rolling even at high rotational speeds a minimum axial load is required.

4.2.2 Roller Bearings

Types of Construction (Fig. 8a–g)

(*a*) *Cylindrical roller bearings* (DIN 5412; ISO 15) can transmit high radial forces but no or only slight axial forces. They can be dismantled; the inner and outer ring can thus be installed separately. The various types are distinguished by the arrangement of the flanges. Types NU and N are used as loose bearings. Type Nj has two flanges on the outer ring and one flange on the inner ring, so that small axial forces in one direction can be absorbed. To take small axial forces in both directions type NUP is used, with its two flanges on the outer ring, a fixed flange and a loose flange disc on the inner ring. The angular adjustability of cylindrical roller bearings is small. Between cylindrical shell surfaces and edge radiusing there is a spherical transition zone. This cylindrical–spherical profile prevents the occurrence of edge stresses and produces a modified line contact with comparatively reduced stress distribution. An almost constant compressive load application without stress peaks is achieved by the so-called logarithmic profile, which does not show any discontinuity in the profile curve.

(*b*) *Tapered roller bearings* (DIN 720; ISO 355) have a high load-carrying capacity and can take combined loads. They can be dismantled so that the inner and outer ring can be installed separately. Since they can take axial forces in one direction only, it is necessary to fit a second bearing in symmetrical opposition as a countersupport. The bearing clearance is adjusted on installation. Angular adjustability is slight and therefore attention must be paid to good alignment.

(*c*) *Self-aligning or barrel-shaped roller bearings* (DIN 635; ISO 15) are single-row roller bearings with angular adjustability (up to 4°), which are suitable for high radial loads. Axial loading capacity is small.

(*d*) *Self-aligning radial roller bearings* (DIN 635; ISO 15) are suitable for the heaviest loads. In this bearing two rows of barrel-shaped rollers run on the hollow spherical track of the outer ring. Alignment errors and shaft deflections are compensated for. The rollers are guided on fixed flanges so that axial forces also can be taken.

(*e*) *Axial cylindrical roller bearings* (DIN 722; ISO 104) take high axial forces in one direction. An axial minimum load is required for kinematically efficient rolling.

(*f*) *Axial self-aligning roller bearings* (DIN 728; ISO 104) for high axial forces and relatively high rotational speeds. Owing to their raceways inclined to the bearing axis they can also take radial loads which however must

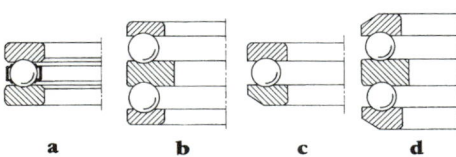

Figure 7. a–d. Axial grooved ball bearings: **a** unidirectional, **b** two-directional, **c** unidirectional with spherical housing disc (compensating for angular errors), **d** two-directional with spherical housing disc.

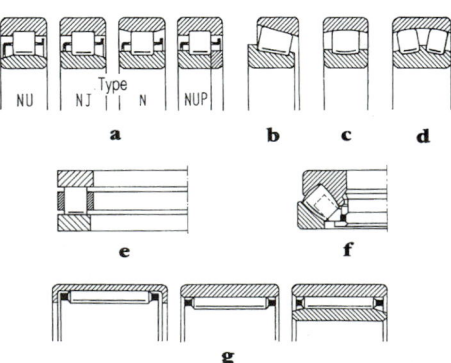

Figure 8. a–g. Types of roller bearing (see text).

not exceed 55% of the axial force. Because of the hollow spherical raceway the bearings can be adjusted by up to 2°. To ensure kinematically efficient rolling a minimum axial load is given.

(*g*) *Needle bearings* (DIN 617, DIN 618; ISO 1206, 3245) require small radial dimensions owing to limited space. They are particularly suited to shock loads and pivoting motions. Axial forces cannot be taken. They have a higher coefficient of friction than other types of rolling bearing. The needles are kept parallel by the cage.

4.2.3 Linear Rolling Bearings and Ball Splines

Types of Construction (Fig. 9a–c)

(*a*) *Ball guides* consist of external bush, cage with balls, and internal bush or shaft. As the cage only performs the half stroke, the axial movement of stroke is limited.

(*b*) *Ball splines* contain three or more ball grooves with return. Consequently the stroke is unlimited. Coefficient of friction $\mu = 0.002$ to 0.004. They are suitable only for rectilinear shaft guides.

(*c*) *Roller guides* as ladder-shaped flat cages or in the form of roller shoes are suitable as flat guides.

4.2.4 Materials

Rolling bearing steels; see D3.1.4. The cages are mainly pressed from sheet steel. Brass, light metal (aluminium alloys) and steel are used for the manufacture of *solid cages*. Solid cages are increasingly being made now of plastic (glass-fibre-reinforced polyamide PA66).

4.3 Load Capacity, Fatigue Life, Service Life

The bearing size required for any particular bearing is determined on the basis of the load capacity of the bearing in relation to the loads occurring and to the fatigue life and operating safety requirements. As a measure of load capacity the bearing computation uses the static load coefficient C_0 and the dynamic load coefficient C, which can be computed in accordance with DIN ISO 76 and DIN ISO 281 Part 1 or taken from the rolling bearing manufacturers' catalogues.

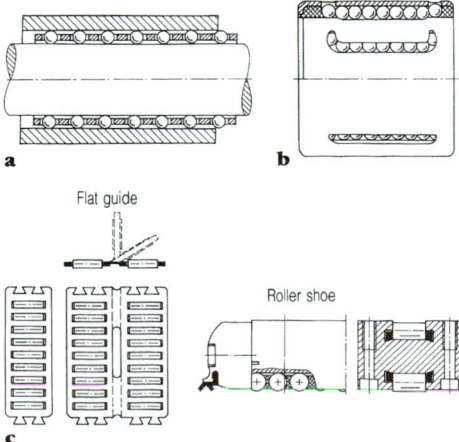

Figure 9. a–c. Linear rolling bearings and ball splines (see text).

If a bearing is stationary, swivels or rotates slowly, it is regarded as a statically stressed bearing, for which the static load capacity must be given. This also applies to dynamically stressed bearings which are subject to short sharp shocks. The dynamic load coefficient C is used with rotating bearings. The terms static and dynamic do not refer to changes in the external load.

4.3.1 Static Load Capacity

The static load coefficients C_{0r} for radial loads and C_{0a} for axial loads are static forces which are based on calculated stresses at the contact point in the centre of the most severely loaded contact point between rolling elements and raceway rated at 4600 MPa for self-aligning ball bearings, 4200 MPa for all other radial bearings, 4000 MPa for all radial roller bearings with radial load and 4200 MPa for axial ball bearings, 4000 MPa for all axial roller bearings with axial load.

Under these loads a permanent deformation of approximately 0.0001 times the diameter of the rolling element occurs at the contact points between rolling element and raceway.

To demonstrate the adequacy of a bearing's load capacity, the static factor $f_s = C_0/P_0$ is used. For bearings which are to run particularly smoothly and easily, a large factor f_s is required. The following figures are used:

> When the requirements regarding smoothness and friction characteristics are high, $f_s = 2$ to 2.5; with marked shock loads $f_s = 1.5$ to 2; with normal smoothness demands $f_s = 0.8$ to 1.2; with low smoothness demands and with vibration-free operation $f_s = 0.5$ to 0.8; for axial self-aligning roller bearings f_s should be ≥ 2, as the flange of the shaft disc is severely stressed.

Loads which are composed of a radial and an axial load must be converted into the equivalent static bearing load P_0. By this is meant in the case of radial bearings the radial load, and in the case of axial bearings the axial load, that would have caused the same permanent deformations in the bearing as the load actually applied. The equivalent static bearing load is obtained from the following two general formulae

$$P_0 = X_0 F_r + Y_0 F_a, \quad P_0 = F_r.$$

The greater of the two values is to be used. F_r is the radial component of the greatest static load, F_a the axial component of the greatest static load, X_0 the radial factor of the bearing and Y_0 the axial factor of the bearing, which can all be taken from **Tables 2** and **3** of DIN ISO 76 or from the rolling bearing catalogues. They differ for the various types of bearing.

4.3.2 Fatigue Life Under Steady Load and Speed

The dimensioning of a dynamically loaded rolling bearing is done on the basis of the *fatigue life* (DIN ISO 281). For an individual bearing it gives the number of revolutions that are executed by a bearing ring or disc in relation to the other bearing ring or disc, before the first sign of material fatigue (pitting) is visible on one of the two rings or discs or on the rolling element. Fatigue life must be distinguished from service life, which means the actually possible operating time of a bearing.

It is not possible to predict exactly the fatigue life of the individual rolling bearing, even with an accurate knowledge of the loading and operating conditions, since fatigue running times vary widely. An assessment can therefore only be made on the basis of statistics from a relatively large number of tests with the same bearings under the same test conditions.

Consequently the concept of *nominal life* L_{10} is used. It corresponds to the fatigue life in millions of revolutions reached or exceeded by 90% of a relatively large number of clearly the same bearings. Hence 10% of the bearings can fail earlier. The nominal life is calculated, using the life equation (DIN ISO 281), as

$$L_{10} = \left(\frac{C_r}{P_r}\right)^p \quad \text{for radial bearings,}$$

$$L_{10} = \left(\frac{C_a}{P_a}\right)^p \quad \text{for axial bearings,} \tag{1}$$

$$L_{10} \text{ in } 10^6 \text{ revolutions.}$$

The exponent p has a value of 3 for ball bearings, and a value of 10/3 for roller and needle bearings.

The dynamic radial (axial) load coefficient C_r (C_a) indicates for a rolling bearing the radial (axial) external load, of constant magnitude and direction, which the bearing can take theoretically for a nominal life of 10^6 revolutions. P_r (P_a) is the dynamically equivalent radial (axial) load whose magnitude and radial (axial) direction is constant and under the effect of which a rolling bearing would achieve the same nominal life as under the conditions actually prevailing. For radial bearings

$$P_r = XF_r + YF_a$$

and for axial bearings

$$P_a = FX_r + YF_a.$$

F_r is the radial component of the load and F_a the axial component. The radial factor X and the axial factor Y are established by DIN ISO 281 or the manufacturer's data. If the bearing turns at constant speed, the life in hours may be expressed by

$$L_{h10} = \frac{L_{10}}{n}. \tag{2}$$

This procedure for determining the L_{10} life is a comparative method whose certainty is the greater, the better the preconditions such as employment of a conventional rolling bearing steel and in practice normal operating conditions (a high degree of separation of the surfaces by the lubricant, no contamination in the lubrication clearance) are satisfied.

Recommendations in DIN ISO 281 enable improvements in rolling bearing steels and in production methods, and the effect of operating conditions, particularly more precise knowledge of the influence of lubrication on the process of fatigue, to be incorporated in the life calculation. Accordingly the attainable fatigue running time L_{na} by the modified life equation is

$$L_{na} = a_1 a_2 a_3 L_{10} \tag{3}$$

or, expressed in hours,

$$L_{hna} = a_1 a_2 a_3 L_{h10}. \tag{4}$$

Life Coefficient a_1 for Probability of Failure. For certain applications it may be desirable to calculate life for failure probabilities other than 10%. For this purpose the factor a_1 was introduced (**Table 3**).

Life Coefficient a_2 for the Material. The characteristics of the material have an influence on the life of a rolling bearing. This influence is covered by the coefficient a_2. Currently, however, under DIN ISO 281 the coefficient cannot be selected on the basis of quantifiable characteristics.

Table 3. Failure probability factor a_1

Probability of failure (%)	10	5	4	3	2	1
Fatigue running time	L_{10}	L_5	L_4	L_3	L_2	L_1
Factor a_1	1	0.62	0.53	0.44	0.33	0.22

Life Coefficient a_3 for Operating Conditions. Coefficient a_3 is used to take into account the suitability of the lubrication, and the conditions that cause changes in the material characteristics. Here too there are no quantitative estimates for a_3 in DIN ISO 281. Assuming that no greater probability of survival than the generally accepted figure of 90% is to be applicable, that the bearings are made from materials that were assumed for the specified dynamic load ratings, and that normal operating conditions apply, then $a_1 = a_2 = a_3 = 1$; in that case Eqs (1) and (3) are identical.

Beyond DIN ISO 281, the rolling bearing manufacturers offer expanded life calculations in which the coefficients are quantified. Correspondingly, the life coefficients a_2 for the material and a_3 for the operating conditions are combined to form a common factor a_{23} because of their mutual influence. As a function of $\kappa = \nu/\nu_1$ (ν is the operating viscosity of the lubricating oil, ν_1 the reference viscosity as a function of bearing size and speed from **Fig. 10**), it can be taken from **Figs 11** or **12**. The influence of the operating temperature on the material is taken into account by the temperature factor f_t as per **Table 4**. The life coefficient a_1 is taken unchanged from DIN ISO 281. The expanded life equation then reads

$$L_{na} = a_1 a_{23} f_t L_{10}, \tag{5}$$

$$L_{hna} = a_1 a_{23} f_t L_{h10}. \tag{6}$$

By including the fatigue strength of rolling bearings, the rolling fatigue theory according to Lundberg and Palmgren, from which the classical ISO-standardised equation for calculating L_{10} life originates, has been expanded according to [4] to become life L_{naa}, such that a fatigue

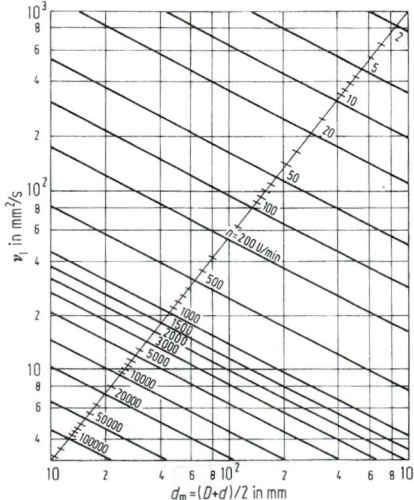

Figure 10. Kinematic reference viscosity ν_1 as a function of mean bearing diameter d_m and speed n.

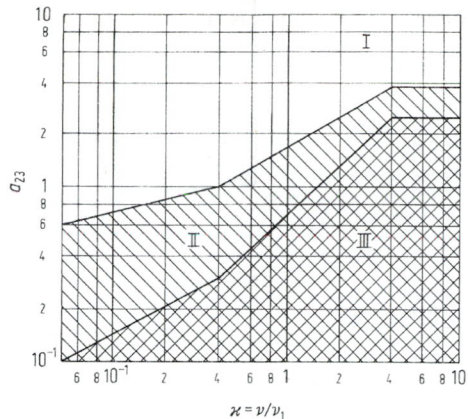

Figure 11. a_{23} graph in accordance with [3]: ν operating viscosity of lubricant, ν_1 reference viscosity.

Zone

  I: Transition to fatigue limit.
Condition: utmost cleanliness in lubrication clearance and not too high load if fatigue limit is aimed for.

 II: Good cleanliness in lubrication clearance.
Suitable additives in lubricant.

III: Unfavourable operating conditions, contaminants in lubricant, unsuitable lubricants.

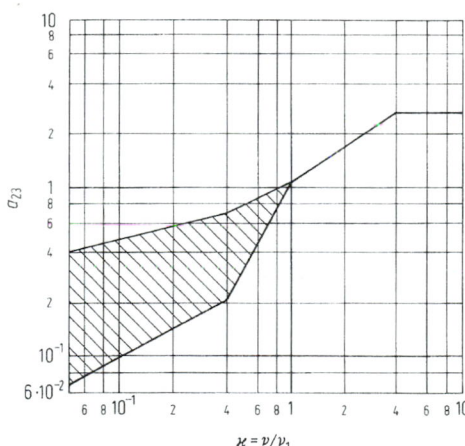

Figure 12. a_{23} graph in accordance with [4]. Grid surfaces when EP additives are used.

Table 4. Temperature factor f_t [3]

Operating temperature (°C)	Temperature factor f_t
150	1
200	0.73
250	0.42
300	0.22

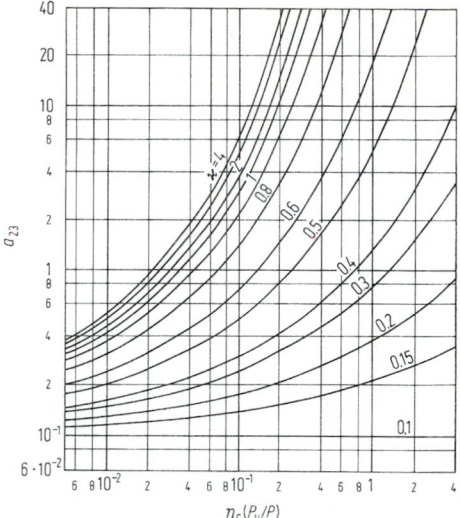

Figure 13. Coefficient a_{23} for radial ball bearings [4]. For $\kappa > 4$ the curve $\kappa = 4$ is to be used. As $\eta_c(P_u/P)$ tends towards 0, a_{23} tends towards 0.1 for all κ values.

limit load P_u is introduced, meaning the limit load up to which no fatigue occurs in the bearing. Values for P_u are given in the bearing tables [4]. Accordingly L_{naa} is governed by

$$L_{naa} = a_1 a_{23} f_t L_{10}. \qquad (7)$$

The values for a_{23} to be inserted in Eq. (7) may be taken from **Figs 13** to **16** for the different types of bearing as functions of $\eta_c(P_u/P)$, with κ as parameter. η_c registers various degrees of contamination (**Table 5**).

The figures are based on a general safety factor which depends on the type of bearing and is comparable with

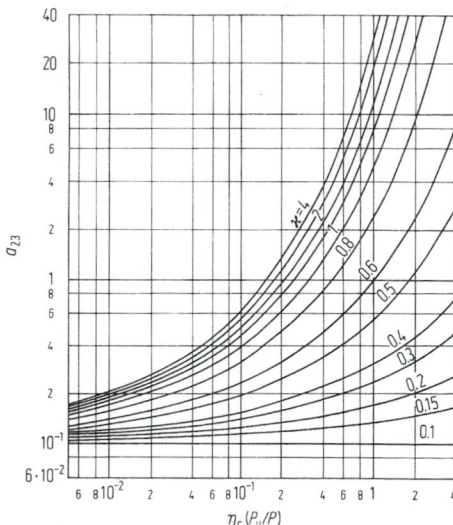

Figure 14. Coefficient a_{23} for radial roller bearings [4]. For $\kappa > 4$ the curve $\kappa = 4$ is to be used. As $\eta_c(P_u/P)$ tends towards 0, a_{23} tends towards 0.1 for all κ values.

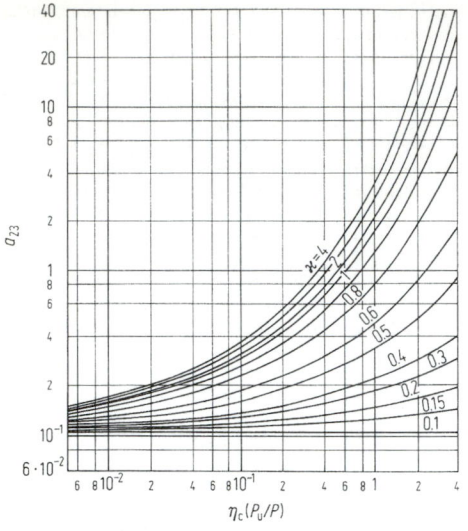

Figure 15. Coefficient a_{23} for axial ball bearings [4]. For $\kappa > 4$ the curve $\kappa = 4$ is to be used. As $\eta_c(P_u/P)$ tends towards 0, a_{23} tends towards 0.1 for all κ values.

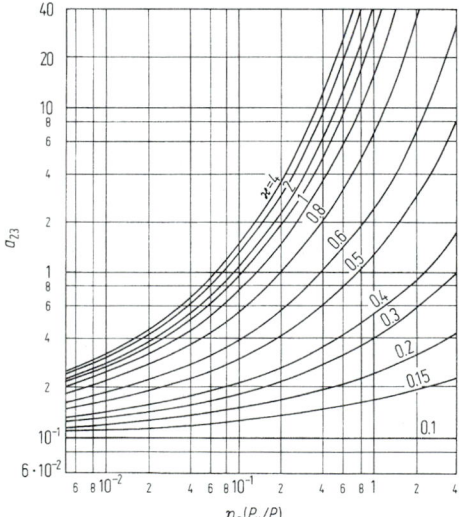

Figure 16. Coefficient a_{23} for axial roller bearings [4]. For $\kappa > 4$ the curve $\kappa = 4$ is to be used. As $\eta_c(P_u/P)$ tends towards 0, a_{23} tends towards 0.1 for all κ values.

the safety factors normal in fatigue strength analyses. The graphs are drawn for typical values of this safety factor and are applicable to lubricants without (extreme pressure) EP additives.

If lubricants with such additives are used a greater life may if applicable be achieved in the range $\kappa < 1$. The maximum possible life may be estimated by multiplying the coefficient a_{23} (without EP additives) by the factor from $(4 - 3\kappa)$ and inserting this higher factor a_{23} (with EP additives) in the formula for L_{naa}. It is however questionable whether, when there is contamination, a longer life can be achieved by EP additives anyway. If $\eta < 0.5$

Table 5. Coefficient η_c (guide values) for various grades of contamination [4]

Operating conditions	Coefficient η_c[a]
Highest cleanliness (contamination particle size of the same order of magnitude as thickness of lubricating film)	1
High cleanliness (corresponds to the conditions typical for grease-filled bearings with sealing discs on both sides)	0.8
Normal cleanliness (corresponds to the conditions typical for grease-filled bearings with cover discs on both sides)	0.5
Contamination (corresponds to the conditions typical for bearings without cover or sealing discs; coarse filtering of the lubricant and/or of contamination penetrating from outside)	0.5 to 0.1
Severe contamination[b]	0

[a] The quoted η_c values are only valid for typically solid contamination; life-reducing effects arising from the penetration of water and other fluids into the bearing are not taken into account here.
[b] With extremely severe contamination the wear predominates; in this case life is well below the calculated value for L_{naa}.

the use of the factor $(4 - 3\kappa)$ is therefore not recommended. If $a_{23}(4 - 3\kappa)$ becomes greater than the value of a_{23} for $\kappa = 1$ from the graph, this graphical value for $\kappa = 1$ is to be used.

4.3.3 Dynamic Load Capacity Under Varying Load and Speed

If a rolling bearing runs at varying speeds and varying load P, its fatigue life can be determined from Eq. (1), with the mean speed $n_m = q_1 n_1 + q_2 n_2 + \cdots + q_n n_n$ and the mean dynamic equivalent load P_m. For this purpose the total time period T under consideration is to be divided into individual periods t_i during which the constant loads P_i are effective. $q_i = t_i/T$ is the proportion of the fractional times t_i relative to the overall time T. The relationship for P_m can be derived from the assumption that the "fatigue resistance" $1/L$ used per unit of time is the same as the sum of the individual resistances a_n/L_n used per revolution share a_n:

$$1/L = a_1/L_1 + a_2/L_2 + \cdots + a_n/L_n$$

where $L_1, L_2 \cdots L_n$ are the fatigue life values which would have resulted in the respective operating conditions, e.g. loads $P_1, P_2 \cdots P_n$ (the Palmgren–Miner rule). It follows that

$$P_m = \left(P_1^p \cdot \sum_{i,i'} n_i q_{i'}/n_m + P_2^p \cdot \sum_{j,j'} n_j q_{j'}/n_m + \cdots \right.$$
$$\left. + P_n^p \cdot \sum_{k,k'} n_k q_{k'}/n_m \right)^{1/p}.$$

If speed and bearing load in the period T are clearly defined time functions $n(t)$ and $P(t)$, this gives for n_m and P_m

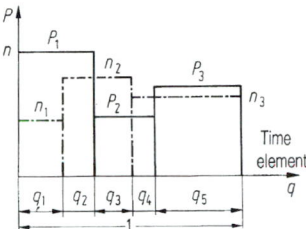

Figure 17. Example of dynamic equivalent bearing load and speed varying in increments with time.

$$n_m = \frac{1}{T} \int_0^T n(t)\, dt$$

and

$$P_m = \left(\frac{\int_0^T n(t)P^p(t)\, dt}{\int_0^T n(t)\, dt} \right)^{1/p}.$$

As an example, this gives for the speed and force graph shown in **Fig. 17**

$$P_m = \left(\frac{1}{n_m} \left[P_1^p (n_1 q_1 + n_2 q_2) + \right. \right.$$

$$\left. \left. P_2^p (n_2 q_3 + n_3 q_4) + P_3^p\, n_3 q_5 \right] \right)^{1/p}.$$

4.3.4 Service Life and Wear

Service life is the operating time actually possible during which the bearing fully performs the required function. When external influences are unfavourable, service life can be less than the calculated fatigue life. Thus errors in alignment between shaft and housing, dirt in the bearing, corrosion, excessive operating temperature or unsuitable lubricants may lead to the premature failure of the bearing through wear or fatigue. Given the multitude of installation and operating conditions it is not possible to determine service life accurately in advance. The most reliable

way to estimate service life is by comparison with similar installation situations. Eschmann [5] was able to develop a method of estimating wear time from extensive investigations on bearings. Bearing wear V, measured in μm, which has the effect of increasing play, may be recorded as wear factor f_v relative to a constant e_0 dependent on the bearing bore d, measured in μm. $f_v = V/e_0$ (**Fig. 18**). The wear time is directly dependent on the conditions at the rolling and sliding surfaces and on the running accuracy requirements of the bearing. The conditions in the contact surfaces (e.g. lubrication, dirt) are identified in **Fig. 18** by the fields a to k. The empirically permissible wear factors f_v and the fields identifying the operating conditions for various installation situations are compiled in **Appendix F4, Table 1**.

4.3.5 Choice of Required Fatigue Life

If the required life is known from the conditions of machine operation, Eq. (2) to (7) can be used to select the correct bearing size by determining the required load coefficient. If data on the required fatigue life are not available, however, guide values may be taken from **Appendix F4, Table 2**.

4.3.6 Limiting Speeds

The rotational speed limit of a rolling bearing for a particular application can only be determined in advance approximately. It depends on type and size of bearing, type of cage, bearing play, accuracy of bearing parts, bearing load and lubrication. Running tests show that the product of limiting speed n_s and mean bearing diameter $d_m = (D + d)/2$ is roughly constant for radial bearings up to $d_m \le 75$ mm. For larger bearings correction factors as in **Fig. 19** are introduced to cover the influences from bearing size f_1 and bearing load f_2. Thus the resulting formulae for estimating limiting speeds are (evaluation **Figs 20** and **21**): for radial bearings $n_s d_m = f_1 f_2 A$, for axial bearings $n_s \sqrt{DH} = f_1 f_2 A$. ($A$ is the type-dependent coefficient as per **Figs 20** and **21**, H the height of the axial bearing.) If the load is greater, i.e. lower life values arise, the values read off are multiplied by f_2. In **Figs 20** and **21** two values for A are given for each type of bearing, the "normal" speed limit, which can be achieved without special action using grease lubrication, and the

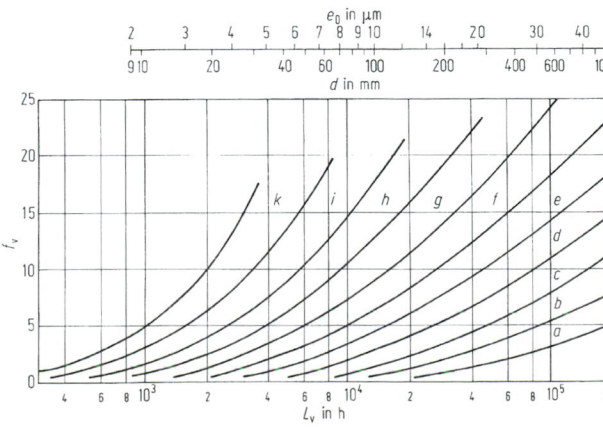

Figure 18. Nomogram for e_0, wear time L_v as a function of wear factor f_v and operating conditions a to k.

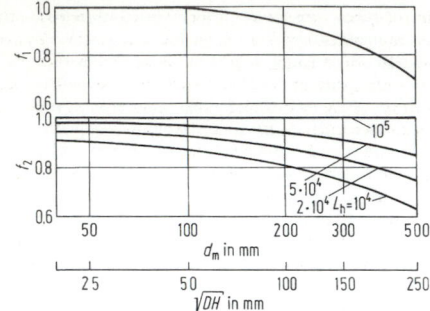

Figure 19. Correction factors f_1 and f_2 for calculation of limiting speeds. The curves for $\sqrt{DH}$ are valid for bearings of series 511, 2344(00) and 2347(00). For the latter two the value $H/2$ should be inserted instead of H.

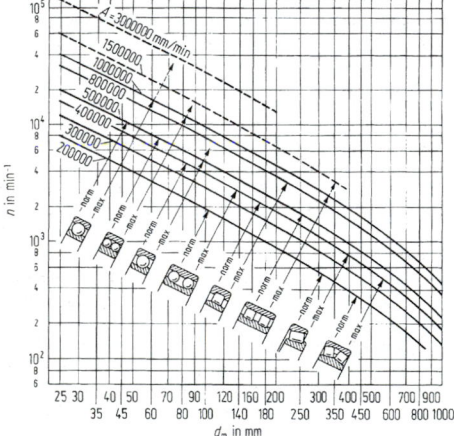

Figure 20. Approximate speed limits for radial bearings under a load corresponding to a nominal life of 100 000 hours.

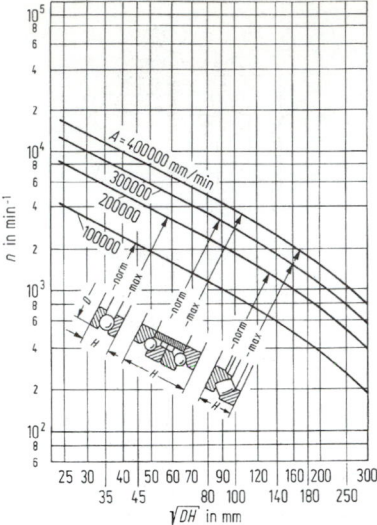

Figure 21. Approximate speed limits for axial bearings under a load corresponding to a nominal life of 100 000 hours.

"maximum" speed limit. The maximum speed limit can only be achieved with improved cage design, increased bearing clearance and by appropriate lubrication, e.g. oil spray lubrication, and favourable load and cooling conditions.

4.4 Lubrication of Rolling Bearings

Lubricants have the job of reducing friction and wear in rolling bearings by carrying oil into the contact areas in order to achieve a complete separation of the surface by a load bearing lubricating film (*liquid friction*). The state of lubrication is estimated by means of the equations of the EHD theory (elastohydrodynamic lubrication, **Fig. 22**) or by determining the viscosity ratio $k = v/v_1$ (cf. Section F4.3.2). Furthermore, lubricants protect rolling bearings from corrosion and the ingress of foreign matter.

4.4.1 Choice of Method of Lubrication

Since the type of lubrication influences the development of bearing and seal, the method of lubrication must be decided before the start of the design stage. Rolling bear-

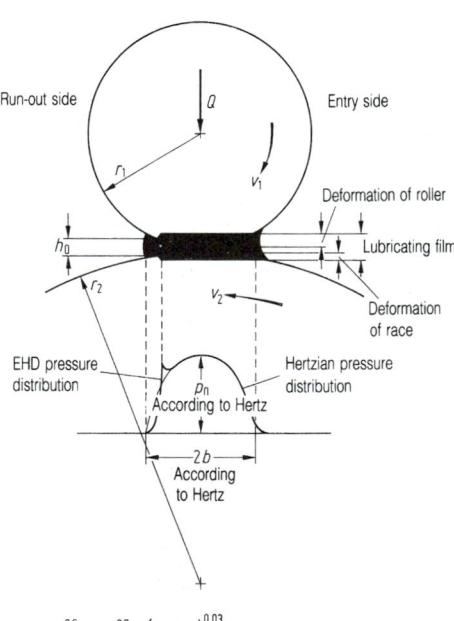

$$h_0 = \frac{0.1 \cdot \alpha^{0.6} \cdot (\eta \cdot v)^{0.7}}{\left(\frac{1}{r_1}+\frac{1}{r_2}\right)^{0.43} \cdot \left(\frac{Q}{l}\right)^{0.13}} \cdot \left(\frac{E}{1-\frac{1}{m^2}}\right)^{0.03} \text{ in } \mu m$$

Figure 22. Elastohydrodynamic lubricating film, example roller/inner ring [6]. b_0 (measured in μm) minimum lubricating film thickness in rolling contact, a (in mm²/N) pressure viscosity coefficient, η (in mPa s) dynamic viscosity, v (in m/s) = $(v_1 + v_2)/2$ velocities, r_1 (in mm) radius of inner ring race, Q (in N) roller load, l (in mm) roller length, E (in N/mm²) modulus of elasticity = 2.08×10^5 for steel, $1/m$ Poisson's constant = 0.3 for steel.

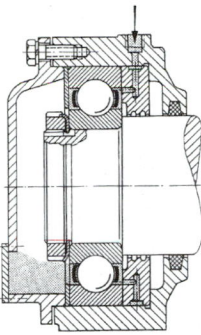

Figure 23. Greasing by means of disc with grease holes.

ings are mostly lubricated with grease (approximately 90% of rolling bearings). Expenditure for seals is low; design is simple [7]. Oil lubrication is used if other machine parts used in the design already require oil lubrication, or if oil is necessary for heat transmission. Solid lubrication is reserved for special applications. The choice of lubrication method depends on the operating conditions and the environmental influences. **Appendix F4 Table 3** gives common lubrication methods and systems as a function of the speed characteristic value nd_m in mm per min.

The reliability of the lubricating method also depends on the undisturbed supply of lubricant. The circulation of oil must be monitored and with oil sump lubrication oil level checks are necessary. Grease lubrication is reliable if the regreasing schedules are kept to.

Greasing. The filling amount is determined by the speed. The bearing cavities should be permanently smeared with grease so that all functioning surfaces receive lubricant. The housing area on both sides of the bearing should however remain free of grease at high speeds so that the grease expelled by the rolling elements can be contained there. In **Fig. 23** the passage of grease is directed by a disc with grease holes. In bearings with grease volume regulators (**Fig. 24**) greater quantities can safely be fed in. The grease volume regulator consists in principle of a disc that revolves with the shaft and spins off excess grease into side housing areas.

Oil Lubrication. There are two ways to obtain a low bearing temperature: a sparing or a very copious supply of oil. If cooling at high speeds is necessary, lubrication using small quantities of oil should always be preferred (*drip feed oil lubrication, oil mist lubrication* or *oil air lubrication*) because the friction losses in the bearing are then smaller.

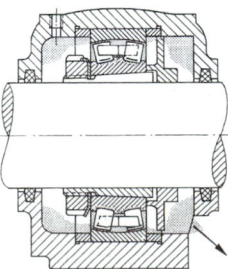

Figure 24. Rolling bearing with grease volume regulator.

In *oil spray lubrication* – for very high speeds – the most important function of the lubricant is to dissipate heat. The oil is sprayed into the bearing between cage and bearing ring by nozzles. Experience has shown that minimum oil jet speeds at 15 m/s are required. Drain passages are necessary to prevent oil stagnation.

Oil bath or oil immersion lubrication is suitable for low speeds. When not in motion the oil level should normally only reach to the middle of the lowest rolling element. A higher oil level may result in frothing and therefore deficient lubrication. The installation of a centrifugal disc intensifies the delivery of oil in the bearing so that higher speeds can be attained.

In *oil circulation lubrication (pressure lubrication)* the oil is fed in most cases along a direct pipe through the bearing. The advantages of circulation lubrication are heat dissipation and the flushing out of wear particles and also low construction cost when exploiting the delivery effect of bearings with asymmetrical cross-sections (roller bearings).

Cleanliness in the lubrication clearance, which is necessary for the endurance of the rolling bearing, can be achieved by filtering the circulating oil.

4.4.2 Choice of Oil

For oil lubrication mineral oils subject to minimum requirements under DIN 51 501 are suitable. Lubricating oils with improved resistance to ageing according to DIN 51 517 are preferred. Synthetic oils are reserved for special applications. The behaviour of the oils *vis-à-vis* sealing materials and plastics is to be checked. Speed, mean bearing diameter, load and temperature determine the choice of oil. With $P/C < 0.1$ and speeds $n < 0.66 n_{s\,oil}$ a lubricating oil with a kinematic operating viscosity of $\nu = 12$ mm^2/s is sufficient.

Figure 25 can be used to establish more precisely the operating viscosity ν as a function of bearing type and limiting speed ratio for oil lubrication $n_{s\,oil}/n$; it also gives the nominal viscosity of the oil at 40 °C. For bearings subject to very high loads $P/C > 0.2$ the next higher nominal viscosity grade is chosen. If there are no empirical values, the probable operating temperature is to be estimated. In normal conditions, unblended but preferably inhibited oils (DIN 51 502, code letter L) can be used. High loads $P/C > 0.1$, given a viscosity ratio $\nu/\nu_1 < 1$ and/or high sliding friction elements, demand oils with wear-reducing additives (DIN 51 502, code letter P or EP additives). For oil mist lubrication the atomization capability and oxidation resistance of the oil must be assured. Synthetic oils are used at extremely high or low temperatures. Silicon oils can only be employed with low loads ($P/C < 0.025$). Guide values for the oil quantity $\dot{V}$ as a function of rolling bearing external diameter D can be taken from **Fig. 26**. **Appendix F4, Table 4** gives characteristic values of various oils.

4.4.3 Choice of Grease

For the lubrication of rolling bearings greases of consistency classes 1, 2 and 3 (NGLI values) are mainly used. The choice of consistency is determined by bearing type, speed, operating temperature, position, starting torque, sealing effect and ease of delivery. The data in **Appendix F4, Table 5** may be used as guide values.

There is only an indirect relationship between the consistency of the greases and the speed characteristic nd_m. The decision factor is the basic oil viscosity so that the permissible speed characteristic values with greases of one consistency class may fluctuate within broad limits.

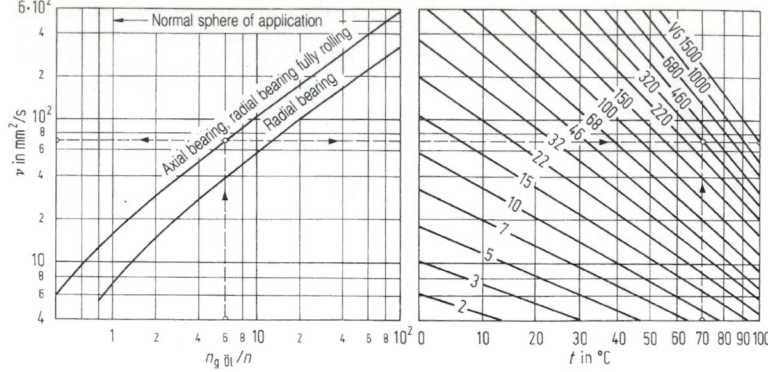

Figure 25. Determination of oil viscosity for rolling bearings. Example: radial bearing fully rolling: $n_{gOil}/n = 6$, $t = 70\,°C$ gives $v = 70\,mm^2/s$ and viscosity class VG 320.

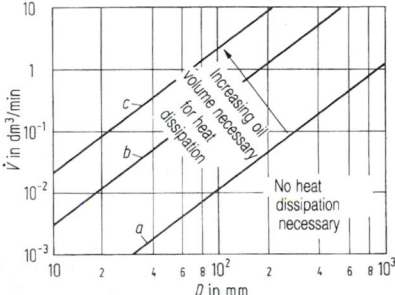

Figure 26. a–c. Volume of oil for circulation lubrication: a volume of oil sufficient for lubrication, b upper limit for bearings of symmetrical form, c upper limit for bearings of asymmetrical form.

Appendix F4, Table 6 gives an overview of the composition and characteristics of the most important types of grease. In fully filled rolling bearings the necessary quantity of grease is adjusted automatically according to speed. It must be possible for the excess grease to be taken up in spaces in the housing to the side of the bearing. For bearings with cover and sealing discs a filling quantity of around 30% has proved favourable considering service life and friction. The required quantity of grease may be calculated approximately from the following numerical equation:

$$G = f \cdot B d_m / 1000 \text{ in cm}^3,$$

where B = bearing width in mm and $d_m = (D + d)/2$ in mm.

$d \leq 40$ mm	40 to 100	100 to 130	130 to 160	160 to 200	> 200
$f = 1.5$	1.0	1.5	2.0	3.0	4.0

In the case of axial bearings the bearing height H is to be inserted instead of B.

Lubricating greases lose their characteristics and must be supplemented and/or renewed. For supplementing, the required period is termed the regrease time (t_{tm}) and for renewal the grease change time (t_{tw}). In regreasing, the used grease should as far as possible be removed from

the rolling bearing and replaced by fresh grease. It must be ensured that the used grease can leave the site of the bearing. It is an advantage to regrease when the rolling bearing is warm and turning.

Guide values for the regrease and grease change times of lithium grease when P/C is ≤ 0.1 are given in **Fig. 27**, with the following values to be used for the required coefficients K_L:

Type of bearing	Coefficient K_L
Radial grooved ball bearing	1.8
Inclined ball bearing	1.4
Axial grooved ball bearing	1.2
Cylindrical roller bearing	1.0
Axial cylindrical roller bearing	1.0 (with cooling)
Axial cylindrical roller bearing	0.3
Needle bearing	1.0
Tapered roller bearing	0.8
Self-aligning radial roller bearing	0.6
Cylindrical roller bearing fully rolling	0.5

The values are valid for atmospheric environmental conditions up to temperatures of $+70\,°C$, measured at the outer race ring. At higher temperatures the stress on the

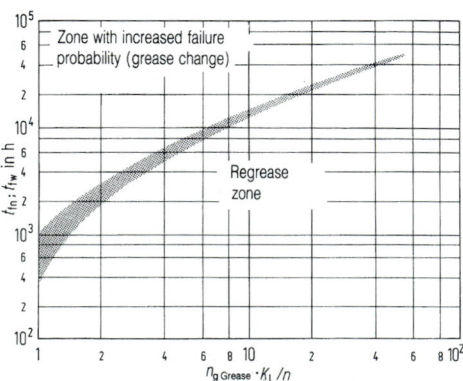

Figure 27. Regrease time and grease change time for lithium soap greases valid for $P/C \leq 0.1$.

greases increases considerably. Shorter greasing times must be expected; for roughly every 15 °C increase in temperature (from 70 °C) the greasing time drops to half the initial value.

4.5 Friction and Heating

In rolling bearings the frictional work is composed of the following elements: friction between rolling elements and raceways including losses by material damping, friction between rolling elements and cage, and between cage and guide surfaces, friction between rolling element end faces and flanges in the case of roller bearings, filling resistance of the lubricant, ventilation [windage] losses and resistance from foreign bodies. These influences generate the friction torque on the rolling bearing, which is dependent not only upon bearing load but also on bearing type, lubrication, load direction and rotational speed.

With a load ratio $P/C = 0.1$, good lubrication and normal operating conditions, the friction torque may be calculated roughly by

$$M = \tfrac{1}{2}\mu_i Fd.$$

Figure 28 gives coefficients of friction μ_i for rolling bearings with normal bearing clearance and sparing lubrication, as a function of purely radial bearing load [8].

A more accurate calculation is allowed by

$$M = M_0 + M_1,$$

where M_0 represents the load-independent friction torque and M_1 the load-dependent friction torque.

$M_0 = 10^{-7} f_0 \, (vn)^{2/3} d_m^3$ for $vn \geq 2000$ mm²/s · min⁻¹

$M_0 = 160 \cdot 10^{-7} f_0 d_m^3$ for $vn < 2000$ mm²/s · min⁻¹

where M_0 = the load-independent friction torque in N mm, f_0 = the coefficient, dependent on lubrication and bearing type ($f_0 = 0.75$ to 20), n = the speed in min⁻¹, v = the operating viscosity in mm²/s, $d_m = (d + D)/2$ = the mean diameter.

$$M_1 = f_1 P_1 d_m,$$

where M_1 = the load-dependent friction torque in min⁻¹, f_1 = the coefficient, dependent on bearing type and load ($f_1 \leq 0.002$), and P_1 = the load governing the friction torque (see [4]).

Bearing friction, friction of seals, external heating and

the dissipation of heat to the surroundings and/or to the lubricant influence the operating temperature of the bearings. The bearing temperature follows from

$$P_R = \phi_U + \phi_0,$$

with

$$\phi_U = \alpha A (T_L - T_U)$$

and

$$\phi_0 = Q c_\rho (T_A - T_E).$$

Here, ϕ_U, ϕ_0 = the flow of heat dissipated to the surroundings and to the oil respectively, α = the heat transfer coefficient; A = the size of surface transmitting heat, Q = the volume flow of oil, c = the specific heat capacity (1.7 to 2.4 kJ/(kg K)), ρ = the oil density, T_L, T_U, T_A, T_E = the bearing temperature, ambient temperature, oil outlet temperature and oil inlet temperature respectively. For rolling bearings the heat transfer coefficient α is some 50% greater than for plain bearings.

4.6 Design of Rolling Bearing Assemblies

4.6.1 Mounting and Arrangement of Bearings

Bearings must be mounted, so that there is sufficient bearing clearance between rolling elements and race rings. Assembly is aided by pressure oil (force-feed union) and tapered seating, especially with large bearings. When rolling bearings are being assembled, care must be taken to ensure that the assembly forces are not transmitted through the rolling elements. Bearing arrangement (**Fig. 29**) for a shaft bearing is preferably as *fixed bearing* and *loose bearing*. The fixed bearing undertakes the axial guiding by fixing the inner and outer ring, while the loose bearing is freely adjustable. Axial adjustability is by bearing type (e.g. roller bearing type NU) or by movability of the

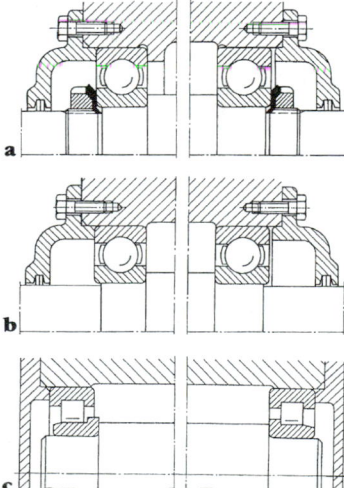

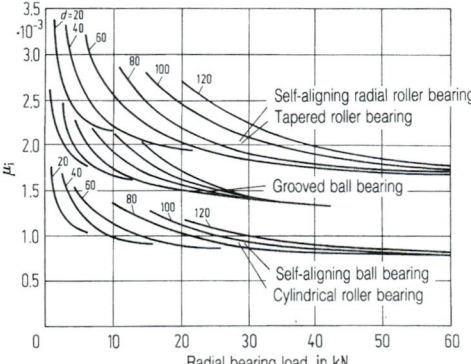

Figure 28. Friction coefficient μ_i for rolling bearings (medium series) with normal bearing clearance and sparing lubrication in accordance with [8].

Figure 29. a–c. Bearing arrangement: **a** fixed and loose bearing with peripheral load for inner ring and point load for outer ring (with turning load for outer ring, loose bearing has sliding fit on inner ring); **b** support bearing (with peripheral load for outer rings, the latter are fixed axially on both sides and the inner rings move in one direction); **c** floating bearing.

outer ring (point load outer ring) or inner ring (point load inner ring).

In a *support bearing* each bearing position takes on the axial force in one direction. The advantages of the support bearing lie in the simple construction of the bearing seats and accurate guiding in the axial direction, but there is the risk of axial distortion owing to thermal expansion in operation. With a floating bearing axial guiding is deliberately dispensed with. End-float is limited to approximately 0.5 to 1 mm.

Adjusted bearings (usually constructed with inclined bearings) offer the capability of a precise adjustment of a clearance or initial stress. A distinction is made between *O-arrangement* (small tilt play) and *X-arrangement* (small angular mobility). Allowing for the distance between the two rolling bearings, the arrangements are equivalent to the pairings shown in **Fig. 30**.

Bearings Prestressed and Installed in Pairs give little play and little resilience as individual bearings. Installation and load distribution possibilities: **Fig. 30**. Further configuration guidelines [8] and examples of installation from the manufacturers.

4.6.2 Selection of Fits

The fit is to guarantee secure attachment and uniform support of the bearing rings, as only then can the load capacity be fully utilised. The greater the bearing load, the greater the fit interference that should be chosen.

If the fit is too loose there is the danger of the ring shifting when there is radial load (bearing ring revolving relative to the direction of load). Guide values for selecting the fit at the bearing seat positions, as a function of type of load, type of bearing, shaft diameter, load and displaceability, together with data on deviations of shape, may be found in the manufacturers' catalogues. The reduction in the radial clearance owing to an interference fit can be allowed for in the choice of clearance group for the bearing. In principle, both rings should be rigidly fitted. In loose bearings the fit of a ring must allow some displacement. The ring with point load should always have the loose seating. With point load the load is permanently directed on to the same point of the ring.

4.6.3 Seals

Rolling bearings can function without trouble and achieve high service life only if they are protected by effective seals throughout their entire operating time, so that penetration of dirt and loss of lubricant is prevented.

Non-contact Seals Functioning Without Wear. Fig. 31a–c gives examples of seals in decreasing order of effectiveness from the range of *non-contact* seals functioning *without wear*.

(a) *Labyrinth seals*, ideally with automatic regreasing and grease nipple regreasing, suitable even at high speeds for grease or oil mist lubrication.

(b) *Spring sealing washers* work without wear after running in. They are best suited to rigid, non-dismantlable bearings.

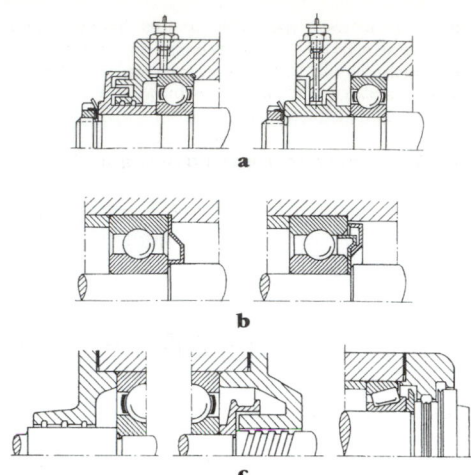

Figure 31. a–c. Non-contact seals for rolling bearings (see text).

(c) *Gap seal with grease grooves* for grease lubrication for high speeds. The sealing effect increases with the length of the gap. The use of the gap seal for circulation lubrication requires spray edges, spray rings or also oil delivery thread.

Grinding Seals Subject to Wear (Fig. 32a–d)

(a) *Slide ring seal* made of synthetic carbon or metal for oil lubrication with automatic adjustment by spring parts suitable for peripheral speeds of up to ≈ 15 m/s.

(b) *Radial sealing rings* are suitable for peripheral speeds of 8 to 12 m/s with oil or grease lubrication. The seal is provided by sleeves with sealing lips made of plastic, which are pressed on to the shaft by loop springs. The sealing lip is subject to wear and its life is therefore limited. If the seal is to prevent foreign matter penetrating the bearing, the sealing lip must point to the outside.

(c) *Felt rings* are used for peripheral speeds of up to 4 m/s. Their performance deteriorates over time, owing to decreasing ring elasticity and splitting. They are therefore only used where dirt accumulation is low.

(d) For designs with limited installation space *rolling bearings with cover discs* (suffix Z) and *rolling bearings*

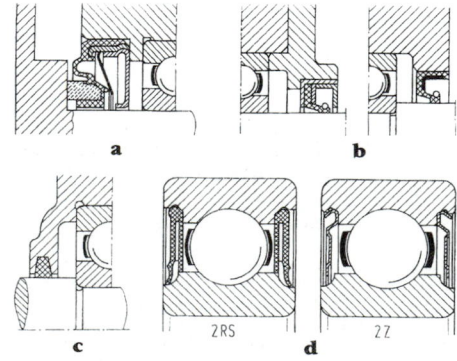

Figure 32. a–d. Grinding seals for rolling bearings: **a** slide ring seal; **b** radial sealing rings of various types; **c** felt sealing ring; **d** bearing with cover discs, bearing with sealing discs.

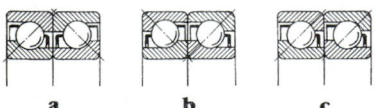

Figure 30. a–c. Paired inclined ball bearings: **a** O-arrangement, **b** Z-arrangement, **c** tandem arrangement.

with sealing discs (suffix RS) are used. Both seals lie at the bearing periphery. Cover discs may be designated non-grinding seals and sealing discs grinding seals. Bearings with discs and grease filling on both sides are considered maintenance-free bearings.

4.6.4 Influence of Bearing Housing Design on Fatigue Life

The distribution of forces on the rolling elements is dependent to a large degree upon the design of the bearing housing. The load-dependent elastic deformations of housing, bearing components and shaft produce variations in the theoretical load distribution on which the calculation is based. With diminishing housing wall thickness the maximum load on the rolling elements increases and fatigue life decreases [9, 10]. In the housing design a favourable load distribution can be achieved by variation of force introduction and stiffenings.

5 Plain Bearings

H. Peeken, Aachen

5.1 Fundamentals of Plain Bearing Design

5.1.1 Hydrodynamic Lubrication

The sound functioning of machines demands wear-resistant design and construction of bearings in order that bearing forces are transmitted reliably and at temperatures that are still permissible. Wear-resistance exists if the sliding surfaces are separated from each other by a load-bearing film. In the plain journal bearing, for example, there is a load-bearing film when the shaft is positioned eccentrically. The pump action of the rotating shaft delivers the lubricant into the bearing clearance and with convergent bearing clearance brings about the build-up of oil pressures (**Fig. 1**). The eccentricity of the shaft adjusts itself in operation in such a way that the integral of the oil pressures of the external bearing load F holds the equilibrium. An interruption in the bearing surface by lubrication grooves in the support zone reduces the load capacity. When the cylindrical plain bearing is loaded, the angular position β of the relative eccentricity ε as a function of width ratio B/D follows a semicircular function.

The oil feed occurs conveniently in the unloaded zone. Ideally, this is the region behind the smallest film thickness h_{min}. The oil is quickly sucked into the bearing by the

underpressure here and in this way prevents admission of air and consequent foaming.

5.1.2 Friction Regimes in Plain Bearings

The possible friction regimes in plain bearings may be explained using the *Stribeck curve* (friction coefficient f plotted against the angular speed ω for constant bearing temperature ϑ) (**Fig. 2**). The friction coefficient f is defined as $f = 2M_r/(FD)$. For $\omega = 0$ (point A) there is contact between shaft and bearing shell. Here the law of solid-body friction applies in approximate terms, so that the friction coefficient $f_0 = \mu = F_r/F$ is determined by the shaft-shell material pair. The friction coefficient drops with increasing rotational speed and reaches the friction minimum at $\omega = \omega_{tr}$ (point B). When $\omega > \omega_{tr}$, f rises again. The point C divides the area of *mixed friction* (which is associated with wear) in which in addition to liquid friction there is still solid contact, from the area of *liquid friction*. Only in the liquid friction zone is operation without wear possible, so that the operation point D must always lie to the right of C. The illustrated Stribeck curve in **Fig. 2**, curve a, is valid for constant bearing temperature and therefore constant viscosity. In practice, regimes with $\vartheta \neq$ const often occur. The increasing temperature has the effect, by virtue of the decreasing lubricant viscosity η, of compensating for the friction value rising with speed (**Fig. 2**, curve b) so that there can thus be a roughly constant friction value.

5.2 Calculation of Plain Journal Bearings Under Steady Radial Load

5.2.1 Wear Safety, Conditions for Full Film Lubrication

The hydrodynamic pressure distribution $p = f(\varphi, z)$

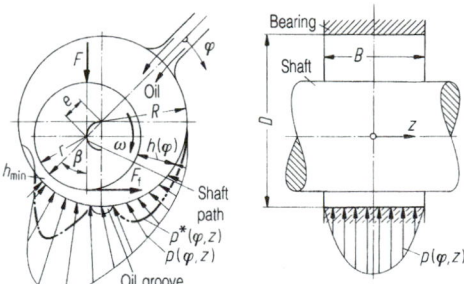

Figure 1. Plain bearing schematic with pressure distribution. Designations: F bearing load, R bearing shell radius, r shaft radius, D bearing diameter, B bearing width, p oil pressures in sliding space, p^* oil pressures with an oil groove in the load-bearing zone, φ and z coordinates, e eccentricity, h lubrication clearance height, h_{min} minimum lubrication clearance height, ω angular speed of shaft, β angle of direction of shaft displacement, $C = 2(R - r)$ operating bearing play, $2e/C = \varepsilon$ relative eccentricity, $\psi = C/D$ relative operating bearing play, F_r friction force.

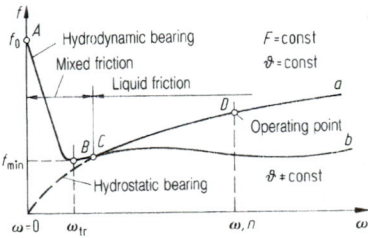

Figure 2. Stribeck curve (schematic), f coefficient of friction, M_i friction torque of bearing, f_0 friction coefficient of solid body friction, ω_{tr} angular speed at transition to mixed friction, ϑ bearing temperature. Curve a for $\vartheta =$ const, curve b for $\vartheta \neq$ const.

follows from the solution of the Reynolds differential equation

$$\frac{\partial}{\partial\varphi}\left(\frac{h^3}{\eta}\frac{\partial p}{\partial\varphi}\right) + r^2\frac{\partial}{\partial z}\left(\frac{h^3}{\eta}\frac{\partial p}{\partial z}\right) - 6Ur\frac{\partial h}{\partial\varphi} = 0. \quad (1)$$

Notation is as per **Fig. 1**, with U as initial speed at the shaft $h = C(1 + \varepsilon\,\cos\varphi)/2$ as the idealised clearance height without taking into account deformation roughness). The integration of the pressure distribution in non-dimensional representation gives the hydrodynamic load capacity in the form of the *Sommerfeld coefficient*

$$So = \overline{p}\psi^2/(\eta\omega) \quad (2)$$

as a function of the relative eccentricity ε and the width ratio B/D. (Here $\overline{p} = F/(B\cdot D)$ = mean surface pressure, $\psi = C/2R = C/D$ = relative bearing play.)

Figure 3 shows the load capacity according to Eq. (1) in the form of the Sommerfeld coefficient So with variable ε and B/D under conditions of incompressible Newtonian lubricant, laminar flow, absolutely rigid and smooth sliding surfaces, and axial parallel clearance. In order to ensure sound running, the operating point D of the plain bearing must lie within the non-wear zone of liquid friction, without a permissible maximum temperature being exceeded. Freedom from wear accordingly exists if the operational angular velocity ω shows an adequate safety margin relative to ω_{tr}: $\omega > \omega_{tr}$. It is assumed with generally adequate accuracy that ω_{tr} defines the transition to mixed friction. The recommended size of the safety margin between ω and ω_{tr} is given by

$$\frac{\omega}{\omega_{tr}} = \left|\sqrt{9 - 3\{U\} + \{U\}^2}\right|, \quad (3)$$

where $\{U\}$ is the numerical value for initial speed in m/s. The angular speed at the transition to mixed friction follows according to [1] for $0.5 < B/D < \infty$:

$$\omega_{tr} = F/(C_{tr}\eta V), \quad (4)$$

with $V = \pi D^2 B/4$ as the bearing volume and η the viscosity of the lubricant at operating temperature. C_{tr} is a constant described by the clearance geometry $C_{tr} = 2/(\pi\psi h_{lim})$, in which the smallest lubricant film thickness h_{lim} and the relative bearing play ψ appear. If we insert for example $\psi = 2\cdot10^{-3}$ and $h_{lim} = 10/3\;\mu m = 10/3\cdot10^{-6}$ m, it follows that $C_{tr} \approx 1\cdot10^8$ 1/m. With this value for C_{tr} the calculation remains mainly within the safe margin. With optimal bearing design, low roughness levels and run-in bearings, higher C_{tr} values can be achieved. With heavily loaded bearings, whose sliding space is influenced by deformations of shaft and shell, C_{tr} is dependent on load [2].

Wear safety may also be derived via the ratio F_{tr}/F. F_{tr} is the overload of the bearing at the boundary with mixed friction. For $0.5 < B/D < \infty$:

$$F_{tr} = C_{tr}\eta V\omega. \quad (5)$$

With plain bearings under dynamic loads, wear safety is given by the ratio of the smallest clearance to the smallest lubricant film thickness which is still permissible h_{min}/h_{lim}. If, with bearings under static load, deformation influences are disregarded, the minimum clearance h_{min} is generated by $h_{min} = C(1 = \varepsilon)/2 > h_{lim}$. The relative eccentricity ε may be determined from **Fig. 3** when the Sommerfeld coefficient is known. In order that no contact between the sliding surfaces should occur in the operating point, the condition $h_{min} > h_{lim}$ is to be complied with. h_{lim} takes into account the sum of the roughness of bearing and shaft, any shape defects of bearing and shaft, shaft skewing in the event of alignment errors and – particularly with rigid bearing construction – shaft deflections. Guide values for h_{lim} may be taken from **Fig. 4** [3]. **Figure 4** shows that the production roughnesses both of shaft R_{zJ} and of bearing R_{zB} are dependent on the bearing diameter D.

The roughnesses of the harder component (generally the shaft) which take the peripheral load are a determining factor in calculating the minimum permissible lubricant film thickness at the transition to mixed friction h_{lim}. The roughness of the bearing changes more than that of the shaft as it is run in. For this reason the following approximation is also valid:

$$h_{lim} = 1.5R_{zJ} + 0.5R_{zB} + \text{deviations of shape.} \quad (6)$$

In establishing ω_{tr} in accordance with Eq. (3), no allowance was made for the mean surface pressure $\overline{p}_{St}$ with which the mixed friction zone is traversed. It does have relevance for the life of a plain bearing whether the mixed friction zone is traversed with a high or low surface pressure. Therefore the limiting value $\overline{p}_{St}U_{tr} = 25\cdot10^5$ W/m^2 is inserted as the permissible friction work in the mixed friction zone. $U_{tr} = \omega_{tr}D/2$ is the peripheral speed at the transition to mixed friction. The permissible limiting value

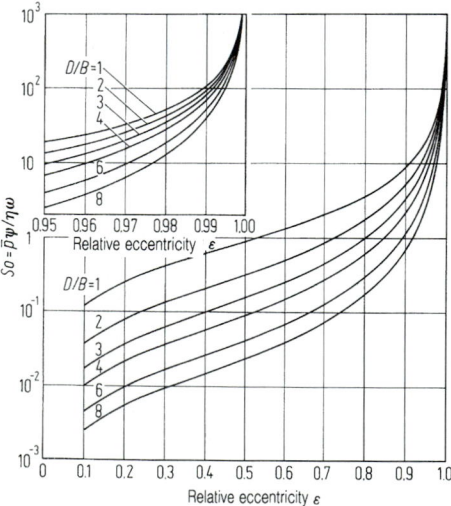

Figure 3. Sommerfeld coefficient for fully enclosed journal bearings as a function of B/D and ε in accordance with [20].

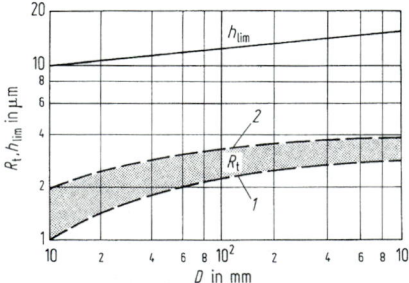

Figure 4. Lowest lubricating film thickness h_{lim} and peak-to-valley height R_t as a function of bearing diameter D [3]; *1* shaft, *2* bearing.

$\bar{p}_{\text{St lim}}$ is determined by the bearing material used (guide value $\bar{p}_{\text{St lim}} = 50 \cdot 10^5$ N/m^2).

5.2.2 Calculation of Bearing Temperature

The calculation of wear safety in accordance with Eqs (4) and (5) requires the operating viscosity and thereby also the operating temperature to be known.

The oil temperature in the bearing is derived from the balance between the friction work and the heat quantities per unit of time which are carried off by convection from the free surfaces A of bearing and shaft: $\alpha A(\vartheta - \vartheta_{\text{amb}})$ or by the lubricant: $Q_{\text{cl}}C_{\text{p}}(\vartheta_{\text{ex}} - \vartheta_{\text{c}})$.

$$P_{\text{f}} = M_{\text{f}}\omega = fFU = \alpha A(\vartheta - \vartheta_{\text{amb}}) + Q_{\text{cl}}C_{\text{p}}(\vartheta_{\text{ex}} - \vartheta_{\text{c}}). \tag{7}$$

The relative friction coefficient f/ψ of a 360 ° bearing may, when $B/D \approx 1$, be determined approximately by the following equations:

$$So \le 1: f/\psi = K/So; \quad So \ge 1: f/\psi = K/\sqrt{So}. \tag{8}$$

With pressure oil lubrication and a loose oil ring (**Fig. 6**), $K = 3$ is to be inserted; with a tight oil ring (DIN 118) $K = 4$ is to be inserted. More precise values for f/ψ are obtained from integration of the Reynolds equation as a function of ε and B/D in accordance with [1].

(a) Heat Dissipation by Convection. If heat is dissipated only by convection, e.g. in bearings with oil sump and good internal wetting of the bearing housing, the relationships for bearing temperature follow from Eq. (7). For $So > 1$ (heavy load range):

$$\left. \begin{array}{l} \vartheta - \vartheta_{\text{amb}} = [4.25U/(\alpha A)] \sqrt{FUB} \cdot \sqrt{\eta} = W \cdot \sqrt{\eta}; \\[2mm] W = [4.25U/(\alpha A)] \sqrt{FUB}. \end{array} \right\} \tag{9}$$

For $So \le 1$ (rapid action range):

$$\vartheta - \vartheta_{\text{amb}} = \frac{6BU^2}{\alpha A \psi} \cdot \eta = W^* \cdot \eta; \quad W^* = \frac{6BU^2}{\alpha A \psi}. \tag{10}$$

For the heat-emitting surface of the bearing A which is determined from the bearing design, in the machine-mounted situation we can put approximately $A = 15$ to $20\ BD$. The coefficient of heat transfer α with moving air and adjustable speed w is derived in m/s for bearing housings of any size from the equation

$$\alpha = 7 + 12\sqrt{w}$$

in W/(m^2 °C). As there is always air movement in machine rooms $w = 1.2$ m/s at least should be inserted; the resulting α value is at least $\alpha = 20$ (W/m^2 °C).

Since, in addition to temperature, the viscosity is often also unknown, to calculate the two unknowns η and ϑ for the operating regime the temperature dependence of the viscosity of the oil used must be known. For the viscosity–temperature behaviour of lubricating oils we can put

$$\eta = a \exp [b/(\vartheta + 95)]. \tag{11}$$

According to Rodermund [4], the constants are derived from:

$$a = \eta_X \exp (-b/887); \quad \eta_X = 1.8 \cdot 10^{-4} \text{ Pa s}$$

$$b = 159.56 \ln (\eta_{40}/\eta_X); \quad \eta_{40} \text{ in Pa s}$$

(η_{40} = the nominal viscosity at $\vartheta = 40$ °C). The dependence of viscosity on pressure is disregarded for plain bearings.

An explicit determination of temperature ϑ from Eqs (9), (10) and (11) is not possible. It can be done graphically with the coordinate charts for $So > 1$ and for $So < 1$ (see **Appendix F5, Table 1** and **Table 2**). These charts illustrate the viscosities (in Pa s) of oils of various viscosity classes = viscosity grades VG as per DIN 51 519 and the straight lines $W = $ const and/or $W^* = $ const dependent on temperature for the ambient temperature $\vartheta_{\text{amb}} = 20$ °C. The viscosity grade of an oil is derived from the kinematic viscosity $\nu = \eta/\rho$ (ρ = mass density of oil) at 40 °C reference temperature. Every point of intersection of the temperature rise straight lines designated by W and W^* with one of the plotted standard oils is a solution to the set of equations and gives, as intersect coordinates, the desired bearing temperature ϑ_{20} (for $\vartheta_{\text{amb}} = 20$ °C) and the associated operating viscosity η. For determining the bearing temperature the coordinate chart for $So > 1$ can be initially used. If the subsequent calculation produces $So > 1$, the bearing temperature has already been found, if it gives $So < 1$ the calculation is to be repeated with the coordinate chart $So < 1$.

At environmental temperatures of $\vartheta_{\text{amb}} \ne 20$ °C the relevant bearing temperature as a function of the bearing temperature ϑ_{20} determined for $\vartheta_{\text{amb}} = 20$ °C may be found in **Fig. 5** separately for $So > 1$ and $So < 1$. The bearing temperature should not exceed a limiting value of $\eta_{\text{limit}} \approx 60$ to 70 °C, as otherwise the oil degrades more rapidly. (High temperatures, e.g. in thermal engines, require special additives in the oil.) The calculated bearing temperature is taken as a mean housing temperature at which the heat arising in the bearing can be dissipated by convection. The temperature which varies over the lubrication clearance differs only slightly from this temperature when there is good equalisation of temperature.

Load capacity diagrams [9, 10] give a brief overview of the complete operating range.

Example Bearing load $F = 10$ kN, $n = 1500$ rev min^{-1}, $\vartheta_{\text{amb}} = 20$ °C, $D = 100$ mm, $B/D = 0.8$, ISO VG22 DIN 51 519, $\alpha = 20$ W/(m^2 °C), $A = 20BD = 0.16$ m^2, $U = 7.85$ m/s, $\psi = \sqrt[4]{U/2.5} \cdot 10^{-3} = 1.3 \cdot 10^{-3}$, $\omega = 157$ s^{-1}, $C_{\text{tr}} = 1 \cdot 10^8$ l/m.
In accordance with Eq. (9): $W = 826$ °C/$\sqrt{\text{Pa s}}$: **Appendix F5, Table 1**: $\vartheta_{20} = 82$ °C and $\eta_{\text{operating}} = 0.0052$ Pa s. Check for So as per Eq. (2): $So = 2.6 > 1$, therefore assumption correct; no check necessary for $So < 1$. Check for transition zone (wear) as per Eq. (4): $\omega_{\text{tr}} = 30.6$ s^{-2}, $\omega/\omega_{\text{tr}} = 5 < \sqrt{9 - 3\{U\} + \{U^2\}} = 6.86$, no wear safety.

(b) Heat Dissipation by the Lubricant. If heat is dissipated from the bearing mainly by the lubricant, the Sommerfeld coefficient is determined approximately with the viscosity η effective in the clearance, and this is derived from the temperature ϑ average from between entry temperature ϑ_{c} and exit temperature ϑ_{ex}.

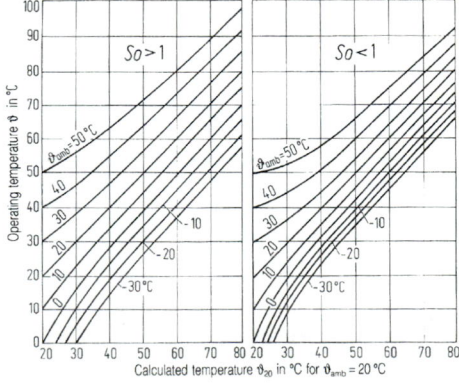

Figure 5. Bearing temperature at ambient temperatures $\vartheta_{\text{amb}} \ne 20$ °C.

With higher eccentricities the development of pressure shifts more to the smallest clearance, so that the mean temperature ϑ to be inserted moves more towards the oil exit temperature. If we disregard the heat dissipated by convection in Eq. (7) (adiabatic case), the oil exit temperature is given by $\vartheta_{ex} = fFU/\dot{Q}_{cl}C_p) + \vartheta_e$ and the mean bearing temperature is $\vartheta = (\vartheta_{ex} + \vartheta_e)/2$.

($\dot{Q}_{cl}$ cooling oil quantity, C_p specific heat, for oil $C_p = 1.8 \cdot 10^6$ N/m/(m³ °C).) With closed-circuit cooling of the oil in an oil radiator, temperature differences of $\vartheta_{ex} - \vartheta_e = 10$ to 20 °C can be achieved in normal designs of radiator.

5.2.3 Required Oil Flow Rate

To build up a load-bearing film an oil throughput $\dot{Q}$ is required, which is given by the difference between the quantities flowing in the widest clearance and in the narrowest clearance

$$\dot{Q} = [1 - 0.223(B/D)^2] \cdot BUC \cdot \varepsilon/2. \quad (12)$$

If the Sommerfeld coefficient is known, the relative eccentricity ε is given by **Fig. 3**. In the case of pressure oil lubrication in addition to the oil supply resulting from the turning of the shaft (Eq. (12)) there is the amount delivered through the sliding space by oil pressure p_z.

If the oil is fed through a bore with diameter d at the point where the clearance is widest, the following applies approximately in accordance with [7]

$$\dot{Q}_B = \frac{\pi D^3 \psi^3 p_z}{48\eta \ln (B/d)} (1 + \varepsilon)^3.$$

In the case of pressure oil lubrication and oil feed through a bore the overall amount of cooling oil is governed by $\dot{Q}_{cl} = \dot{Q} + \dot{Q}_B$. An increase in $\dot{Q}_{cl}$ over $3\dot{Q}$ does not bring greater heat dissipation.

5.2.4 Relative Journal Bearing Clearance

The relative bearing clearance $\psi = C/D$ necessary in operation is chosen as a function of the sliding speed U. As a guide the following numerical equation is used: $\psi = \sqrt[4]{U/2.5} \cdot 10^{-3}$; U in m/s. These values may be deviated from by ± 25%. The following information is given with respect to the upper and lower limits for ψ:

Operating conditions	Lower ψ range for	Upper ψ range for
Bearing material	Soft, low modulus of elasticity, white metal	Hard, high modulus of elasticity, bronzes
Area load	Relatively high	Relatively low
Bearing width	$B/D \leq 0.8$	$B/D \geq 0.8$
Mounting	Self-adjusting	Rigid
Load transmission	Rotating (peripheral load for bearing shell)	Static (point load for bearing shell)
Treatment/machining	Very good	Good
Hardness difference between journal and bearing material	≥ 100 HB	≤ 100 HB

With increasing bearing temperature the thermal expansion of shaft and bearing, which reduces bearing play, is to be compensated for by an additional clearance (installation > operating clearance). For bearings with free expansion ability and only small temperature

difference $\vartheta - \vartheta_{amb} = \Delta\vartheta < 20$ °C the operating clearance roughly corresponds to the installation clearance. For bearings with free expansion ability and $\Delta\vartheta > 20$ °C owing to friction heat or heat flux it is accepted approximately that the bore does not change and the heating only has the effect of expanding the shaft.

5.3 Calculation of Plain Journal Bearings Under Variable Radial Load

Here bearing load (according to magnitude and direction) and angular speeds of bearing and shaft are functions of time (e.g. bearings in internal combustion engines). It follows that shaft position, friction, lubricant throughput and load safety of the bearing are dependent on time. If the functions of bearing load and angular speeds are periodic, closed shaft centre paths result. The calculation of these paths starts from Eq. (1), which is extended by the term $- 12r^2 \, \partial h/\partial t$ and is gradually solved.

The practical calculation of the displacement paths may be facilitated by the application of approximation functions [11] for the Sommerfeld coefficients So_D and So_V attainable by turning and displacement. With periodically loaded bearings the iteration is to be carried out until closed path curves result.

5.4 Turbulent Film Flow

At high peripheral speeds the laminar film flow in plain bearings changes into a turbulent film flow after a critical Reynolds number Re_{crit} is exceeded. Turbulence commences if

$$Re = \rho UC/2\eta \geq Re_{crit} = 41.3/\sqrt{\psi}, \quad (13)$$

with ρ the mass density of the lubricant.

With turbulent flow too the plain bearing characteristic values may be determined theoretically by means of a modified Reynolds equation [12]. The equation contains position-dependent correction factors for viscosity, so that in tangential and in lateral direction higher viscosities apparently take effect. Depending on the type of bearing and the eccentricity ε the turbulence zone begins above $Re = 300$ to 1000. With turbulence bearing temperature rises owing to the increased lubricating film friction, but so do the lubricating film pressure and, accordingly, bearing load capacity.

5.5 Calculation of Plain Thrust Bearings

In the case of thrust bearings the convergent clearances needed for hydrodynamic pressure development are produced by machining, tiltable or elastic segments of a rectangular or circular shape (**Fig. 6**). Sufficiently broad spaces are to be left between the segments so that the exiting warm oil can be replaced by fresh oil (use scraper if necessary).

The choice of bearing type is determined by the operating conditions. If there are high surface pressures, and frequent speed changes under load are expected, bearings with tiltable pads are to be preferred as the optimum wedge inclination with correct support adjusts itself automatically and the wear occurring when the bearing is

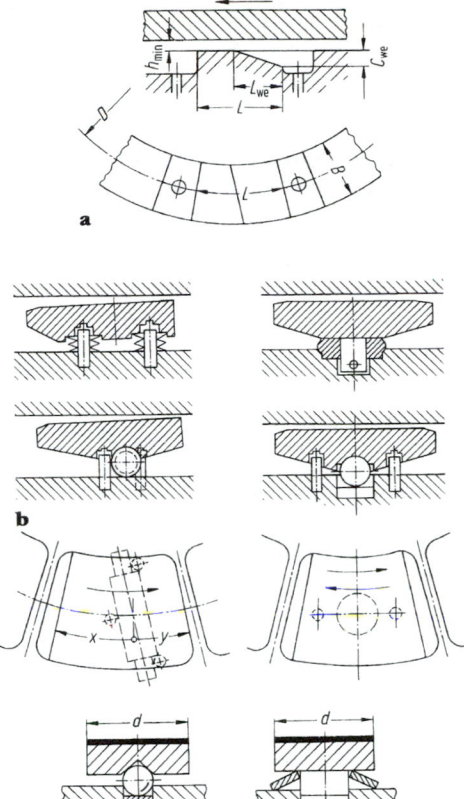

Figure 6. Design variants for plain thrust bearings. **a** Solidly work-ed-in wedge surfaces; L_{we} wedge length, D_m mean diameter of thrust bearing and C_{we} depth of worked-in wedge surfaces. **b** Rigid and elastic support of pads for constant and changing direction of rotation. **c** Rigid and elastic support of circular pads for constant and changing direction of rotation; d diameter of circular pad.

starting up or slowing down does not produce any change in clearance geometry.

When dimensioning a thrust bearing, the main dimensions (Z load-carrying surfaces and/or pads of breadth B and length L or diameter d) are to be distributed in such a way that with $\bar{p} = F/(ZLB)$ or $\bar{p} = 4F/(Z\pi d^2)$, $U = \omega D/2$, an average effective viscosity η and $h_{min} > h_{lim}$ the load factor $\bar{p}h_{min}^2\,(\eta UB)$ assumes the appropriate value.

In **Fig. 7a** the load factors $(\bar{p}h_{min}^2/(\eta UB))$ and friction factors $f\sqrt{\bar{p}B}/(\eta U)$ are plotted as functions of h_{min}/C_{we} for various breadth ratios L/B valid for the lubricating wedge without rest surface. $h_{min}/C_{we} = 0.5$ to 1.2 and $L/B = 0.7$ to 1.5 give broadly optimal conditions. The smaller values for L/B are more favourable for heat dissipation. With tilt-able pads, by setting the axis of rotation at $a = 0.42L$ from the run-out edge the optimum ratio $h_{min}/C_{we} \approx 0.8$ is obtained, at which the greatest load factor is achieved independent of operating regime. On the other hand, with solidly worked-in wedged surfaces of depth C_{we}, variations in operating regime (different h_{min}) also bring about changes in the load factor.

Generally, the most frequently occurring operating regime is chosen as the design point for the bearing. A check is to be made as to whether the wedge angle

resulting from the load factor via the ratio h_{min}/C_{we} can still be produced. In the case of tilting segment bearings for both directions of rotation the support is concentric. Owing to the occurring thermal and elastic deformations approximately 80% of the load capacities as per **Fig. 7a** are achieved [12].

Figure 7b gives load and friction factors for the lubricating wedge with optimum rest surface element ($W_{we}/W = 0.8$). In [13] the load capacities for various other clearance shapes are compared. Here the orifice plate bearing with parallel step clearance gives very good results. Bearings with tilting circular blocks are dealt with in [14], and with centrally supported circular pads in [12]. For accurate calculations the elastic and thermal curvature of the pads is to be taken into account [12].

Figure 8 shows a combined thrust and journal tilting segment plain bearing for the mounting of a turbo compressor.

Particularly at high peripheral speeds ($U > 25$ m/s) the temperature difference between entry and exit rises with increasing pad length L, so that the mean viscosity in the clearance drops and therefore values smaller than 1 are chosen for L/B. In this case it is also recommended that there be a reduction in the surface utilisation $\Phi = ZLB/(\pi DB) < 0.8$, and direct injection of the lubricant against the alignment disc, because this improves the mixture between the fresh oil and the warm oil leaving the preceding pad. For bearings with solidly worked-in wedge surfaces improved cooling is achieved by increasing the wedge pitch to $h_{min}/C_{we} \leq 0.25$.

In the lower speed range, friction heat can still be dissipated to the surroundings by convection via the bearing housing that contains the oil filler. For determining the mean bearing temperature the same considerations apply as for the plain journal bearing. For the friction value of thrust bearings, broadly for all operating regimes (cf. **Fig. 7**), $f \approx 3\sqrt{\eta U/(pB)}$ applies in the case of rectangular segments and $f \approx 3.3\sqrt{\eta U/(pD)}$ in the case of circular segments.

For heat dissipation by convection at the bearing housing the mean bearing temperature ϑ can also be determined graphically with the coordinate chart at **Appendix F5, Table 1**. The value to be used for W here is $W_{ax} = 3U\sqrt{ZFUL/(\alpha A)}$ for rectangular segments and $W_{ax} = 2.92U\sqrt{ZFUd/(\alpha A)}$ for circular segments.

At sliding speeds of $U > 25$ m/s friction heat can no longer be dissipated to the surroundings by convection alone. In that case return cooling of the oil heated in the bearing is necessary. Heat conduction by the alignment disc and pad is generally disregarded here. The temperature at entry into the load-bearing clearance is then derived from the mixture between the heated oil coming from the previous pad and the oil fed between the pads and coming from the cooler. Owing to the mixing of oil flows the temperature at which the oil enters the lubricating clearance (ϑ_1) is always higher than the temperature at which the fresh oil is delivered to the bearing. Stationary temperature conditions are produced in the bearing if the friction heat generated in the bearing per unit of time is removed from the bearing either by convection or by return cooling of the lubricating oil. Therefore with ϑ_s as the oil sump temperature, ϑ_e the temperature of the return-cooled oil, ϑ_{amb} the ambient temperature and P_f the friction work, we have

$$\text{convection:}\quad P_f = fFU = \alpha A(\vartheta_s - \vartheta_{amb}),$$
$$\text{return cooling:}\quad P_f = fFU = \dot{Q}_{cl}C_p\,(\vartheta_s - \vartheta_e),\quad (14)$$

with $\dot{Q}_{cl}$ as the quantity of cooling oil.

To cover the mix between the hot oil leaving the pad

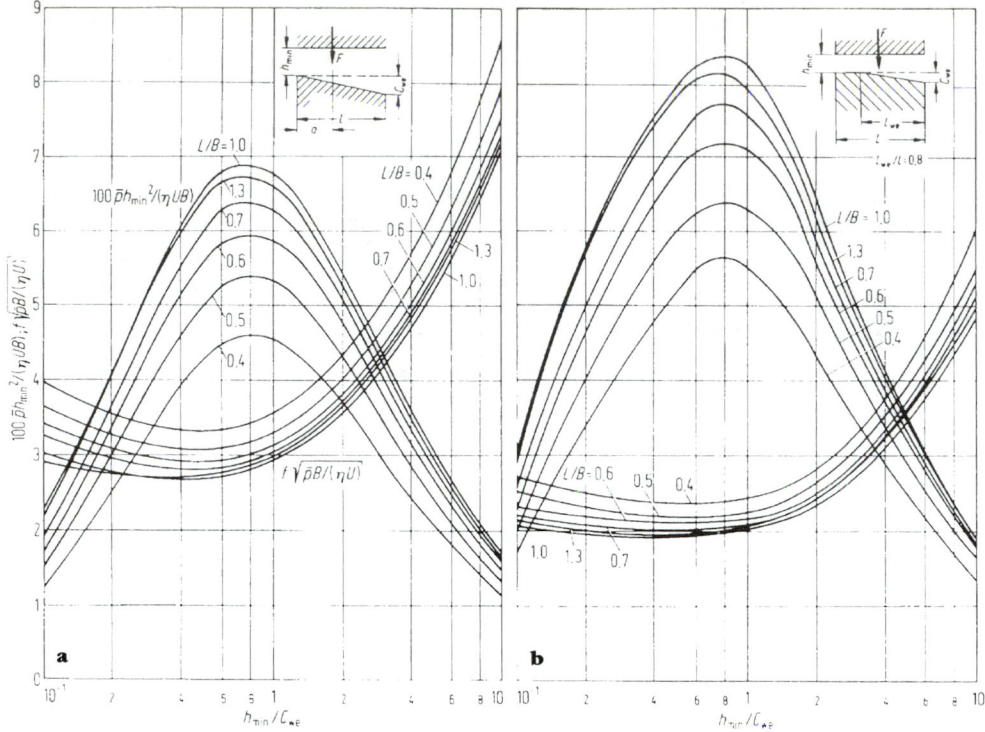

Figure 7. Load and friction factors $100\bar{p}h_{min}^2/(\eta UB)$ and $f\sqrt{\bar{p}B/(\eta U)}$ for the lubrication wedge: **a** without rest surface for various width ratios, **b** with optimal rest surface ($L_{we}/L = 0.8$) for various width ratios.

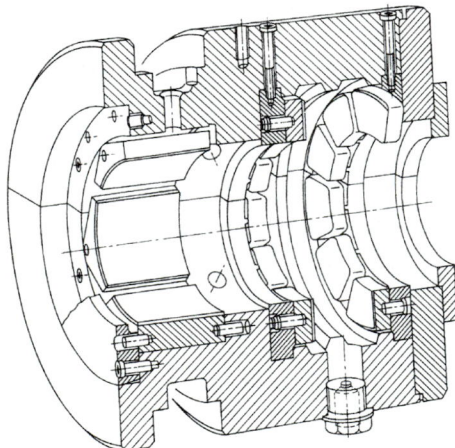

Figure 8. Combined axial and radial tilting segment bearing for a turbo-compressor (Demag-Verdichtertechnik, Duisburg), shaft removed.

and the cooled oil supplied the mixing factor m is introduced, for which $m = 0.4$ to 0.6 is recommended. $m = 0$ means that the lubricant leaving the lubricating clearance fully enters the input clearance of the following pad. Only the lateral flow $\dot{Q}_3$ is replaced by fresh oil. $m = 1$ means that no mixing takes place. The entire oil supply to the segments is with fresh oil.

Taking into account the mixing processes the maximum temperature in the lubricating film is

$$\vartheta_2 - \vartheta_s = \frac{P_f}{ZC_p\left(\dot{Q}_1 - \dfrac{\dot{Q}_3}{2}\right)\left[1 - m\left(1 - \dfrac{\dot{Q}_3}{\dot{Q}_1}\right)\right]}. \quad (15)$$

With the temperature at clearance entry ϑ_1 from the relationship

$$\vartheta_1 - \vartheta_s = m\left(1 - \frac{\dot{Q}_3}{\dot{Q}_1}\right)(\vartheta_2 - \vartheta_s), \quad (16)$$

the mean lubricating film temperature/bearing temperature is given by

$$\vartheta = (\vartheta_1 + \vartheta_2)/2. \quad (17)$$

The entry temperature of the return-cooled oil is given by

$$\vartheta_e = \vartheta_s - (10 \text{ to } 15) \,°C. \quad (18)$$

The derivation of these equations presupposes that the oil quantity $\dot{Q}_3$, leaving the lubricating wedge at the side only, warms up on average to $(\vartheta_2 + \vartheta_1)/2$.

The oil quantity entering the lubricating wedge, $\dot{Q}_1$, and the oil quantity leaving as a lateral flow, $\dot{Q}_3$, may be determined from **Fig. 9**. As an example, for pads with optimum support without rest surface ($a = 0.42L$ measured from the run-out edge) $b_{min}/C_{we} \cong 0.8$.

The minimum oil quantity required by the bearing for hydrodynamic load transmission, namely the load-bearing oil quantity $\dot{Q}_T$, is derived from $\dot{Q}_1$

$$\dot{Q}_T = Z\dot{Q}_1. \quad (19)$$

Wear safety of the thrust bearing is provided as long as

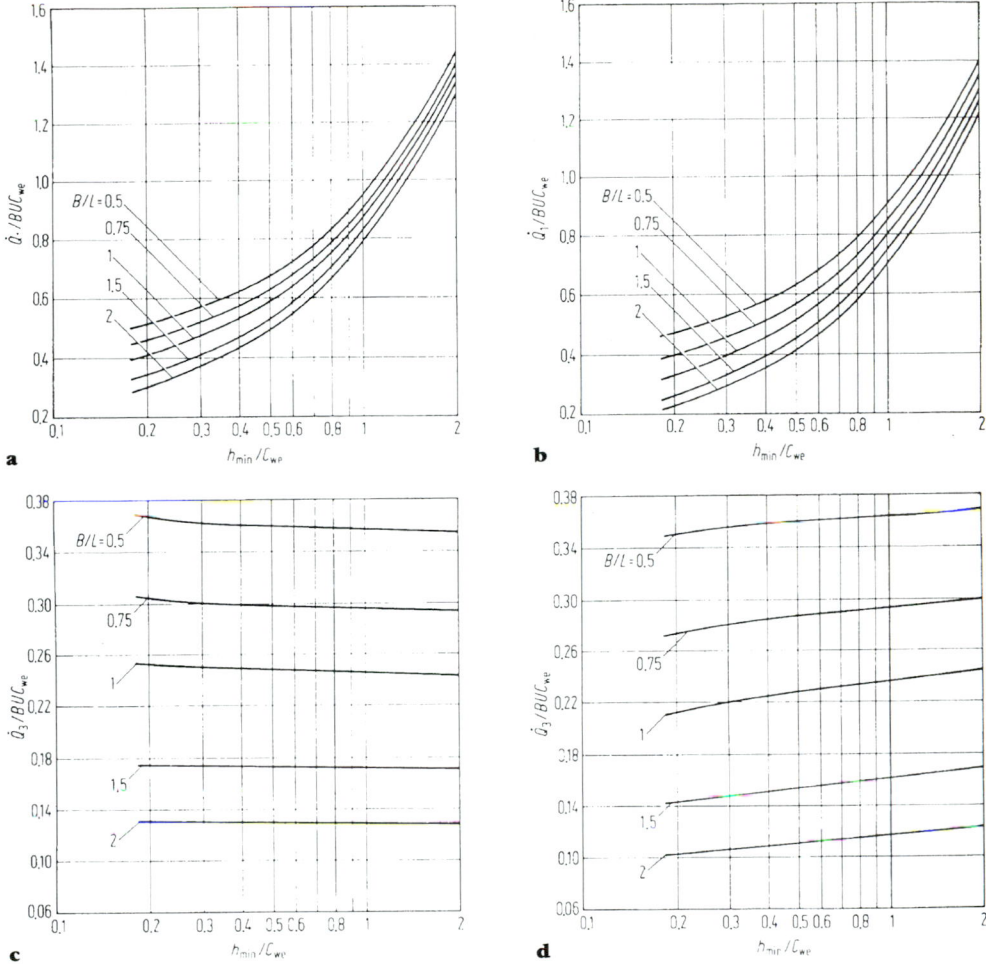

Figure 9. a–d. a Flow factors $\dot{Q}_1/BUC_{we}$ (entry to wedge clearance) for the lubrication wedge, without rest surface for various width ratios B/L. **b** Flow factors $\dot{Q}_1/BUC_{we}$ (entry to wedge clearance) for the lubrication wedge, with optimum rest surface ($L_{we}/L = 0.8$) for various width ratios. **c** Flow factors $\dot{Q}_3/BUC_{we}$ (lateral flow) for the lubrication wedge, without rest surface for various width ratios B/L. **d** Flow factors $\dot{Q}_3/BUC_{we}$ (lateral flow) for the lubrication wedge, with optimum rest surface ($L_{we}/L = 0.8$) for various width ratios.

the minimum film thickness in operation h_{min} does not exceed the minimum film thickness at the transition to mixed friction $h_{min,tr}$. The ratio $h_{min}/h_{min,tr}$ is a measure of the safety margin considered necessary from the wear limit. $h_{min,tr}$ is determined, in addition to the roughnesses, primarily by production tolerances, assembly errors and elastic and thermal deformations. As running-in operations are only possible to a limited degree, $h_{min,tr}$ is generally greater with thrust bearings than with plain journal bearings. Guide values for the selection of $h_{min,tr}$ are given by

$$h_{min,tr} = \sqrt{DR_{zj}/3000}$$

for closed bearings with solidly worked-in wedge surfaces, and

$$h_{min,tr} = \sqrt{DR_{zj}/12\,000}$$

for tilting segment bearings

$\qquad\qquad$ (20)

(R_{zj} is the averaged peak-to-valley height for the alignment disc).

The minimum film thickness in operation h_{min} may be calculated for the lubricating wedge with and without rest surface approximately by

$$h_{min} = \sqrt{0.06\eta UB/p}. \qquad (21)$$

Guide values for the smallest permissible minimum lubricating film thickness for thrust bearings are given by

$$h_{min,lim} = \sqrt{5D} \cdot 10^{-5} \text{ m}. \qquad (22)$$

For the mixed friction zone to be traversed smoothly when the bearing runs out/slows down it is crucial that the peripheral speed on reaching the mixed friction zone does not exceed the limit $U_{tr} = 1.5$ to 2 m/s.

In addition to a constant dead load, thrust bearings in many cases also have to bear loads which are dependent on rotational speed (e.g. in turbomachinery). If the thrust bearing only has a constant dead load to bear, the mixed friction zone is only crossed when the bearing is starting up and slowing down. The operating speed of the bearing must therefore be greater than the transition speed by a corresponding safety margin. If on the other hand the load on the bearing is only attributable to kinetic forces, e.g. fans with horizontal shaft, or in ships' thrust blocks, the load on the bearing decreases more quickly than the bearing's load capacity, so that there is no lower mixed friction limit. Here however a mixed friction limit is reached in the upper speed range, so the operating speed must keep a sufficiently large safety margin relative to this limit. If the bearing load consists of both dead weight and kinetic forces (e.g. pumps with vertical shaft), mixed friction zones occur in the upper and lower speed range, and sufficient safety margins must be observed relative to them.

5.6 Form Design of Plain Bearings

The bearing design creates the preconditions for the hydrodynamic lubrication on which the bearing calculation is based.

5.6.1 Influence of Design on the Configuration of Sliding Surfaces

In the support of shafts the bending line generally shows an inclined position which can be increased further still by alignment errors, and by offset and angle error of the bearing blocks. The *edge loading* thereby caused produces a reduction in the bearing's load capacity and requires design measures to adjust the bearing to the distortion of the shaft. This can be done by a one-off adjustment on assembly by means of spherical mounting of the bearing body or by continuous adjustment to shaft inclination by a tilting mounting. With bearings fitted in engines, self-adjustment can be obtained by eccentric support of the bearing body (**Fig. 10a**) or by the use of an elastic membrane (**Fig. 10b**). As with inclination, shaft bending in the bearing also results in a reduction of bearing load capacity. The load capacity reductions, which are particularly visible where $B/D > 0.3$, require design measures so that there are as far as possible parallel axial clearances. With bearings of greater breadths satisfactory results are given by resilient design of the bearing support brackets so that optimum adjustment to shaft deformation is achieved. Incorporating the elastic deformations of support brackets and shaft in the bearing calculation [16] it is possible to give, for example for concentrically arranged bearing support brackets, optimal wall thicknesses s_L/D as a function of support width b/B (**Fig. 11**). However, the studies [17] show that with cylindrical bracket shapes only a relative optimum in respect of load capacity can be achieved. Further increases in load capacity can only be achieved in operation by running-in procedures. A more precise adjustment of shaft deformation to bearing shell set is produced by a taper cone, but here the bearing

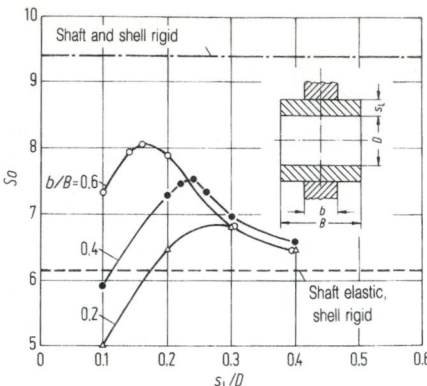

Figure 11. Load capacities *So* of cylindrical plain bearings with centre bar of various widths b/B as a function of the bearing wall thickness s_L/D.

edges must display sufficient rigidity. With thrust bearings the inclined positions of the footstep may be balanced by elastic design of the bearing ring support or by seating the segments on elastic elements (**Fig. 6b**).

5.6.2 Lubricant Supply

Furthermore, the bearing design has to fulfil the requirement for an adequate lubricant supply. With free-standing pedestal bearings, the oil supply is effected by loose or fixed lubricating rings (**Fig. 12**) or by circulation lubrication (**Fig. 13**).

The operating limits of *loose* lubricating rings are approximately $U = 20$ m/s and of *fixed* lubricating rings $U = 10$ m/s. The lubricant is delivered most effectively in the zero pressure zone or in the underpressure zone (divergent clearance) in order to prevent the tendency to froth. Outside these areas the resistances opposing bearing oil flow must be taken into account. The distribution of the lubricant within the sliding space is through well-rounded axial oil grooves or oil recesses in the zero pressure zone of 0.7 of the bearing width. Individual oil feed holes are generally not adequate. Annular grooves in the bearing centre give a good oil supply, but divide the bearing into two halves of lesser load capacity. Oil grooves in the load-bearing zone (**Fig. 1**) disturb the pressure buildup and are to be avoided in rotating, as opposed to oscillating, bearings. In combined thrust-journal bearings the lubricant supply to the thrust bearing is often provided by the oil flowing off from the side of the journal bearing. In the case of bearings under variable load best results have come from annular grooves for the oil supply and axial grooves in the unloaded zone. Oscillating bearings (e.g. piston pin bearings) have for the supply of oil

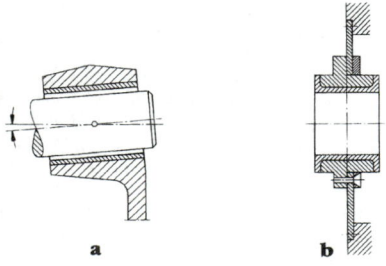

a   **b**

Figure 10. Adjustment capability of plain bearings by **a** elastic deformation with eccentric support, **b** elastic deformation of a membrane.

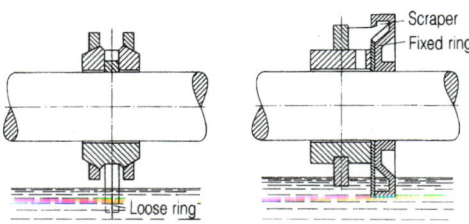

Figure 12. Lubrication variants.

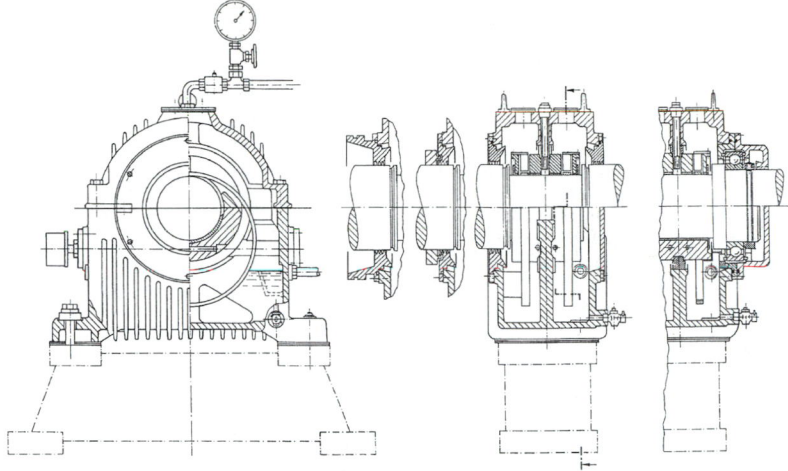

Figure 13. Plain bearing with oil circulation and ring lubrication (AEG-Telefunken).

several grooves running straight or obliquely in the pin and terminated at the bearing ends, these grooves being supplied with lubricant via a hole. If oil foaming in the bearing cannot be avoided then the frothed-up oil which no longer meets the incompressibility requirement must be removed from the bearing by scrapers. At the same time the scrapers prevent the hot oil that is draining out from re-entering the sliding space [20].

5.6.3 Bearing Cooling

The cooling effect of the lubricant can be improved by increasing oil throughput, e.g. by partly hollowing out the upper bearing shell. Heat dissipation can also be improved by suitable guiding of the hot oil draining out in the housing of a pedestal bearing. Additional bearing cooling is possible by the installation of cooling channels or pipes carrying oil or water [18].

5.6.4 Bearing Materials

The bearing material [21-24] must in conjunction with lubricant and shaft material display good sliding and dry-running properties, and adequate behaviour in respect of wear, running in and bedding in.

As regards sliding properties the *wetting power* of the lubricant on the sliding surfaces, i.e. the ability to penetrate narrow gaps, is of major importance. This applies particularly if in the mixed friction range (starting up and slowing down) the generated film pressures are not yet sufficient to fill the clearance. *Running-in behaviour* identifies the ability of plain bearing materials to match surface condition and configuration of running surfaces to each other by abrasion, in such a way that design and production errors, which manifest themselves as alignment errors, deformations and other deviations from specified shape (undulation or roughness) are compensated for. The matching of the bearing surfaces in operation is assisted by the elastic resilience of the bearing materials. With steel as the shaft material, sliding characteristics and running-in behaviour decrease in the following order: white metal on lead base, white metal on tin base, lead bronze, gun metal, tin bronze, special brass. *Dry-running property* is the capacity of a bearing metal to keep the bearing operable for a short time without great damage even if the lubrication fails. In dry running

residual oil and any solid lubricants present (graphite, molybdenum sulphide) help. Mainly, however, the dry-running properties are determined by the material characteristics of the metals. Low-melting-point metals with little hardness which melt with local heating and thus reduce friction have the best dry running properties. *Bedding-in behaviour* identifies the ability to embed dirt or wear particles into the surface. The risk of damage to the sliding surfaces can thereby be mitigated. Materials that bed in well (e.g. white metal) in no way obviate the requirement to protect the bearing from the entry of dirt and to keep the lubricant clean by filtering. Wear arises in plain bearings in the area of mixed friction (e.g. during the starting-up and slowing-down phases). Quantitative wear values for bearing metals depending on the operating conditions are only available in limited measure. Wear resistance declines from the bronzes via brass, AlPb bronzes, gun metal, AlZn alloys and cadmium alloys down to the white metals. Plain bearing materials must transmit the forces acting on the bearing to the surrounding construction for an adequate lifetime (load capacity).

The pressure distribution causes within the bearing material a three-dimensional stress state which can lead to fatigue when there is variable loading. In addition, the temperature gradients produce thermal stresses on the plain bearing material. In this respect it should be noted that strength reductions occur above all in bearing materials with low melting point. In the case of statically stressed bearings the material design is undertaken using permissible mean surface pressure. **Appendix F5, Tables 3** and **4** give chemical composition, strengths, spheres of application and permissible mean surface pressures for a selection of bearing metals and plastics.

With bearings under variable load the functional reliability of the material is conditional upon the fatigue strength not being exceeded. Calculation statements regarding the endurance design of plain bearing material determine from the available three-dimensional stress state, using a suitable strength hypothesis, a comparative stress which is compared with the fatigue strength characteristic value determined experimentally [25].

5.6.5 Bearing Shells (Bearing Liners)

Plain bearing shells are furnished with bearing metal by the lining or centrifugal methods or are manufactured in a multi-component version. In particular cases the bond between bearing metal and supporting construction is also made by solder or adhesive. For bearings under high dynamic stresses, such as engine bearings, *multi-component bearings* – generally manufactured by the band lining or roll cladding methods – are used. With a *three-component bearing* the bearing metal layer (e.g. lead bronze or aluminium bronze) bonded to the steel backing is also provided with an electroplated white metal layer to improve its running behaviour. Since strength increases with declining layer thickness, bearings under high loads require very thin layers. Three-component bearings therefore possess extremely thin white metal layers, approximately 0.02 mm thick. Since lead bronze or CuAl alloy as a carrier material possesses adequate fatigue strength, the layer thickness is not of such overriding importance. Owing to the low ductility relative to white metal the layer thickness must be between 0.4 and 1 mm depending on the bearing dimensions. The thinner the bearing metal layer, the less the bearing is able to compensate for example for alignment errors owing to plastic deformations in the bearing metal, and the more important is the correct design of support shell and/or bearing block.

5.6.6 Special Materials for Plain Bearings

Sintered metals made of Fe powder or bronze (Bz) powder pressed and sintered in part with additions have a volume of voids of 0 to 3% or 10 to 45%. Bearings with volumes of voids > 10%, which possess an oil supply through oil saturation, can run without maintenance. Operation is limited to low sliding speed of 0.3 to 0.5 m/s at $\bar{p}$ up to 600 N/cm². The stress capacity of the bearings is limited by heat dissipation, so that at higher speeds the load capacity drops.

Graphited Bearing Metals. Fe or Bz sintered metals impregnated with graphite (up to 10%) are used as oil-free bearings in the food and textile industries. In the chemical and electrical industry bearings made of pressed colloidal graphite are also used, often also impregnated with metal powder. Graphite bearings have good sliding properties and are therefore also used as air bearings [1].

Plastics and Rubber. In contrast with metallic materials, here the tendency to scuffing/seizing is totally absent. The materials possess low thermal conductivity, low modulus of elasticity and are susceptible to high temperatures. Rubber is often used in water lubrication. With chemically aggressive media ceramics or carbon are used as bearing materials.

5.7 Lobed and Multi-pad Plain Bearings

With small Sommerfeld coefficients, bearings show increasingly unstable running behaviour (e.g. high speed steam and gas turbines, turbo-compressors), such that the required quiet running necessitates the use of multiple sliding surface bearings with three or more faces at which the shaft is held by several pressure peaks, whose load capacities add up geometrically (**Fig. 14**). Multiple sliding surface bearings also include bearings with lobed clearance. Multiple sliding surface bearings are also used for precise and largely load-independent radial shaft mounting. The calculation of load capacity, friction, spring and damping constants is ascribed to that of a part bearing with arbitrary eccentricity and angular position [28, 29]. The shaft or rotor movement around the position of equilibrium may be determined with the four

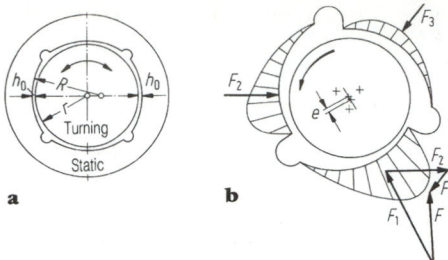

Figure 14. Multiple-sliding-surface bearings: **a** for both directions of rotation (Gleitlagergesellschaft, Göttingen), **b** with three faces for one direction of rotation with pressure distribution and load capacities for $\varepsilon = 0.6$ (Caro Metallwerke, Vienna).

spring constants and the four damping constants, which depend on the static operating state of the bearing, by the solution of the movement equations of the rotor.

The calculation of the spring and damping constants is generally based on a linear disturbance formula for the oil pressure components. It should be noted that with multiple sliding surface bearings, as a result of variable load, bearing capacity fluctuations occur which for their part can lead to induced oscillation [30]. In order that the bearing temperature does not become unacceptably high in the case of fully enclosed bearings, relatively large degrees of play are required which however promote the change from laminar to turbulent clearance flow (with resulting increased bearing friction). High friction losses may be reduced with radial tilting segment bearings, in which the running surfaces only partly enclose the shaft. With approximately point support of the segments the bearings are relatively insensitive to shaft skew. With bearings subject to high loads the segments may be produced in differing lengths. **Figure 15** shows the construction of a three-segment bearing for a large steam turbine. For the calculation of this bearing taking into account the dependence of viscosity on temperature and pressure, see [31].

For certain applications *spiral groove bearings* are used as axial and radial guide elements [32, 33].

5.8 Bearing Seals

The sound functioning of plain bearings requires adequate sealing of the bearing interior, in order to prevent the

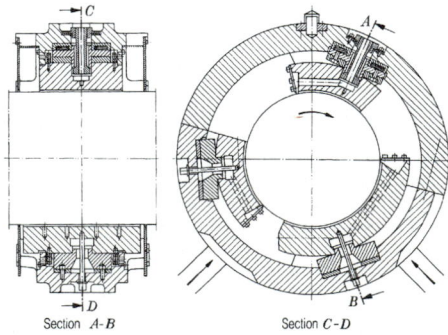

Figure 15. Tilting segment journal bearing of a large steam turbine (BBC).

ingress of foreign matter. For common types of seal, see F4.6.3.

5.9 Dry Bearings

Oil-free plain bearings are employed as maintenance-free bearings for moderate loads and speeds where for economic reasons expense on lubricants is not possible (household, agricultural and office machines) or where lubricants are undesirable for reasons of cleanliness (food and textile machines). With low loads *air bearings* are possible [1]. Sintered metal materials, plastics such as polyamides, polythenes and phenolic resins, and carbon are used as bearing materials. The most favourable running characteristics come together with hardened steel shafts of low roughness. By incorporating metal powder into the plastics their thermal conductivity can be improved [34–39].

The calculation of dry bearings covers bearing temperature, mechanical load capacity, and thereby life. As the sound operation of these bearings is crucially determined by heat dissipation at still permissible bearing temperature, the operating limits are given by $(\bar{p}U)_{\text{perm}}$ values.

For bearings with oscillating swivel movements, in large areas of machine construction ball and socket joints, consisting of outer and inner ring with spherical sliding surfaces, are used [40].

5.10 Bearing with Hydrostatic Jacking Systems

Hydrostatic starting aids facilitate the transition through the mixed friction zone when large machines are started up. By means of a pump the high-pressure oil is delivered through a non-return valve into the recess or groove provided for the purpose. Experience indicates that a pressure of 5 to 6 times $\bar{p}$ is required for the suspension of the shaft. It is only possible to prove suspension by means of a precise calculation [41]. The oil connection in the bearing is to be designed in such a way that penetration of the pressure oil between bearing metal and support shell is prevented [42].

5.11 Hydrostatic Bearings

In hydrostatic bearings pressure oil is produced outside the bearing with a pump and fed to pressure chambers. The lubricant flows off through narrow gaps whose size b is determined by the load. Hydrostatic bearings also function at zero speed, so that the branch of the Stribeck curve lying to the left of the transition point C (**Fig. 2**) in the area of mixed friction (the dotted line) is missing. However, the ability to operate without wear at any speed has to be achieved at the expense of increased effort for the oil supply by comparison with bearings working hydrodynamically. There are several possibilities for generating pressure in the pressure chambers: (a) a displacement pump for each pressure chamber, (b) common pressurisation for all the pressure chambers, with a throttle connected in front of each pressure chamber to stabilise the bearing. Capillary tubes or diaphragms are used as throttle elements. The diaphragms have the disadvantage of a higher risk of clogging. Oil supply systems that deliver constant amounts of oil per chamber make throttles for stabilisation superfluous.

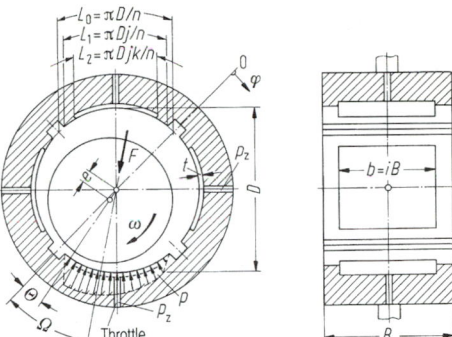

Figure 16. Hydrostatic journal bearing (schematic) with oil drain grooves and four recesses.

5.11.1 Journal Bearings

In hydrostatic journal bearings [43–45] bearings with (**Fig. 16**) and without oil drain grooves (**Fig. 17**) are used. The hydrostatic journal bearing shown in **Fig. 17** is supplied with pressure oil by a common annular groove from all the recesses via built-in capillary throttles.

The approximate calculation of the bearings is based on the statement of flow equilibrium at each bearing recess. If there are no oil return grooves the equalising flows in tangential direction between the recesses are to be taken into account. For bearings with oil return grooves a simplified calculation method is given which presupposes a large recess depth $t \approx 5C$ and requires a relatively small effort (C bearing clearance). The illustration uses the following factors: $u = \delta^4/(c^3\lambda)$ throttle factor, $w = \eta\omega/(p_{\text{amb}}\psi^2)$ speed factor and $v = p_z/p_{\text{amb}}$ pressure ratio, where δ = the diameter of capillary tubes, λ = the length of capillary tubes, η = the operating viscosity, p_{amb} = the ambient pressure, p_z = the feed pressure, c = the radial bearing clearance, and other designations are as in **Fig. 1**.

The recess geometry of the bearing with Z recesses is described by the reduction factors $i = b/B$, $j = L_1/L_0$ and $k = L_2/L_1$ as in **Fig. 16**. With laminar flow in the gaps we have for the mean surface pressure of the bearing and $z \geq 4$:

$$\bar{p} = \frac{F}{BD} = \frac{\pi j}{16}(1 + k)(1 + i)p_z\left\{\frac{1}{H_1} - \frac{1}{H_2}\right\}\Lambda,$$

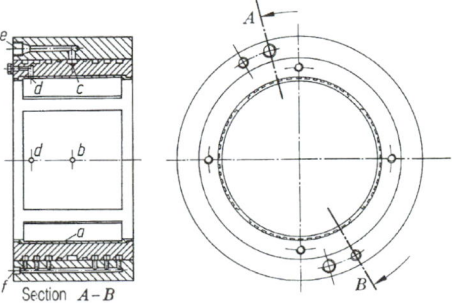

Figure 17. Hydrostatic journal bearing without oil drain grooves with four recesses at the circumference (Konings, Swalmen, Netherlands): *a* bearing lining made of plastic, *b* pressure oil feed, *c* capillary tubes, *d* pressure measurement connection, *e* pressure oil supply, *f* connection for additional cooling circuit.

with

$$H_1 = 1 + \frac{64}{3\pi u} G(1 - \varepsilon)^3, \quad H_2 = 1 + \frac{64}{3\pi u} G(1 + \varepsilon)^3,$$

$$G = \frac{Z(1 + i)(B/D)}{\pi j(1 - k)} + \frac{\pi j(1 + k)}{Z(1 - i)(B/D)}.$$

The factor Λ which it contains covers the curvature of the recess:

$$\Lambda = \frac{2Z}{\pi j(1 + k)} \sin\left[\frac{\pi j}{2Z}(1 + k)\right].$$

For $Z = 3$ recesses, $\bar{p}_{3T} \approx (2/3)\bar{p}$ applies. In rough terms, $\bar{p} \approx (0.2$ to $0.3)p_z$ can be expected for normal $\varepsilon = 0.5$ to 0.6. **Figure 18** shows the mean surface pressure $\bar{p}$, relative to p_z, plotted against the eccentricity ε in comparison with measurements. The simultaneous insertion of exact calculations shows that the influence of the bearing installation angle Ω (the angle between bearing reference line and load) is of subordinate importance. The illustration in **Fig. 18** is based on the optimum throttle characteristic value

$$u_{\text{opt}} \, \pi/G = \left(\frac{64}{3}\right)(1 - \varepsilon^3)^{3/2}.$$

which guarantees the maximum possible load capacity. **Figure 19** shows the influence of the recess number Z on the load capacity of the bearing. At higher speeds the bearing's capacity increases due to hydrodynamic influences. Oil consumption is given by

$$\dot{Q}_{\text{tot}} = p_z \frac{\pi u c^3}{128\eta}\left[4 - \frac{1}{H_1} - \frac{1}{H_2} - \frac{2}{1 + (64/(3\pi))(G/u)}\right].$$

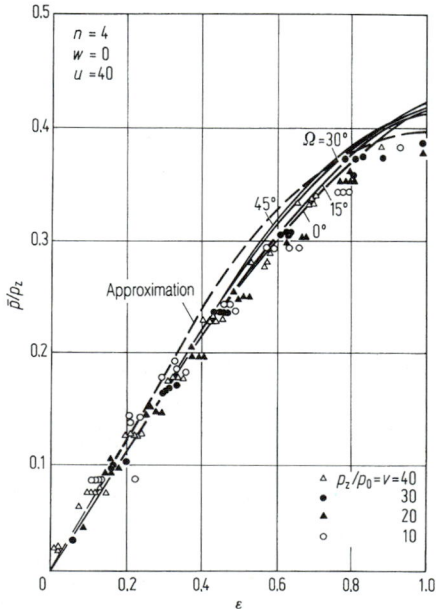

Figure 18. Surface pressure $\bar{p}$, relative to feed pressure p_z, as a function of ε with the installation angle Ω as parameter. The graph is valid for $n = 4$, $w = 0$, $u = 40$, $i = 0.75$, $j = 0.85$, $k = 0.64$, $v = 10, 20, 30, 40$.

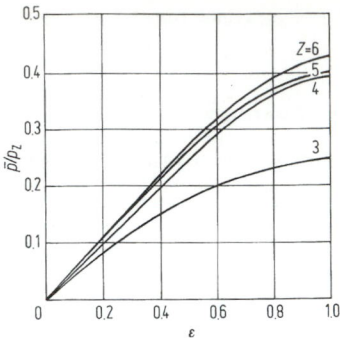

Figure 19. Relative load capacity $\bar{p}/p_z$ of hydrostatic journal bearings as in **Fig. 16** as a function of ε with various recess numbers Z.

The relative friction coefficient is, up to $\varepsilon = 0.6$, rendered very well by

$$\frac{f}{\psi} = \frac{\pi}{So} \cdot \frac{j(1 - k)}{Z} \cdot \left(1 + k \cdot \frac{1 - i}{1 - k}\right) \cdot$$

$$\left[\frac{1}{1 - \varepsilon} + \frac{1}{1 + \varepsilon} + 2\right].$$

Thus the friction work of the bearing $P_f = fFU$ and pump output $P_p = p_z\dot{Q}_{\text{tot}}/\xi$ may be determined (ξ = the efficiency of the pump). Studies on bearing optimisation show that the lowest energy dissipation from friction and pump output is reached at $So = 1$ and $\varepsilon = 0.5$ to 0.6. The friction work declines as web widths decrease; however, web widths should not fall below a particular size, so that bearing force when stationary can be transmitted without damage to the webs ($i = k = 0.7$ to 0.8, $j = 0.85$). Bearing temperature and thereby operating viscosity is determined on the basis of the heat balance as per Eq. (4).

5.11.2 Thrust Bearings

Figure 20 shows a single-acting circular hydrostatic thrust bearing with capillary tubes as throttles. With laminar flow in the bearing the bearing's capacity is given by

$$F = \frac{\pi}{2} \frac{1 - \rho^2}{\ln(1/\rho)} r_a^2 \frac{p_z}{H};$$

with

$$H = \frac{p_z}{p_T} = 1 + \frac{64}{3u \ln(1/\rho)};$$

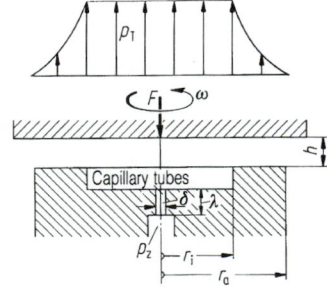

Figure 20. Single-acting hydrostatic thrust bearing with capillary tubes as throttles. r_i recess radius, r_a external radius.

$$u = \frac{\delta^4}{b^3\lambda}; \quad \rho = r_i/r_a.$$

The resulting oil throughput of the bearing is

$$\dot{Q} = \frac{\pi}{6} \cdot \frac{p_T b^3}{\eta \ln(1/\rho)}.$$

The total energy dissipation P is given by the sum of friction work P_f and pump output P_p: $P = P_f + P_p$, where

$$P_f = (\pi/2)(1 - \rho^4)\eta\omega^2\, r_a^4/b,$$

$$P_p = p_z \cdot \dot{Q}/\xi = \frac{2\ln(1/\rho)}{3\pi(1-\rho^2)^2} \cdot \frac{F^2 b^3}{\eta r_a^4} \cdot \frac{H}{\xi}.$$

The minimum possible energy dissipation is achieved if

the undimensioned factor Ψ approximately assumes the numerical value 1.

$$\Psi = \frac{Fb^2}{\eta\omega r_a^4} = \sqrt{\frac{H}{\zeta}}$$

$$= \sqrt{\frac{3\pi^2\rho^3(1 - \rho^2)^3}{4\rho\ln(1/\rho) - (1/\rho)(1 - \rho^2)}}.$$

For the clearance b the permissible minimum value b_{minlim} is to be inserted. The pressure ratio should be no greater than $H = 3$. Only for high rigidity requirements may greater pressure ratios H be chosen. In any case, two-thirds of maximum rigidity is already obtained with $H = 3$. The energy dissipation increases with $\sqrt{H}$. If F and ω are fixed $\Psi = 1$ then can be achieved by variation of η and r_a. Care should be taken to ensure that the viscosity does not fall below a minimum value of 15 mPa s, so as not to endanger the oil pump. For the calculation of double-acting hydrostatic thrust bearings, see [45].

---

6 Belt and Chain Drives

H. Mertens, Berlin

6.1 Types, Uses

Belt and chain drives are used to convert rotary speeds and torques between two or more non-coaxial shafts, even when there are considerable distances between shafts, at low construction costs. Traction is provided by endless flat belts, V-belts, synchronous belts or chains which are wound round the pulleys or sprockets on driving or driven shafts, thereby transmitting peripheral speeds and forces [1].

Frictionally Engaged Belt and Chain Drives. These require a minimum initial tension in order to maintain frictional engagement. The rotary speed is converted when the layout is correct, having a limited, load-dependent slip rate (slip due to elongation) and a near constant (**Fig. 1**) or infinitely variable (e.g. **Fig. 8c**) speed ratio.

Positive Belt and Chain Drives. These too require a type-dependent minimum initial tension to achieve optimum running characteristics combined with a long service life and/or to prevent faults in transmission (skipping teeth) (**Fig. 2**). They then produce a constant speed ratio, leaving aside the usually slight mismatching of the torque transmission with the frequency of the advancing teeth or chain links (polygon effect).

As they can be easily twisted, flat belts, V-belts and synchronous belts make it possible to construct spatial drives with non-parallel shafts (**Fig. 3d, e**). Steel chains

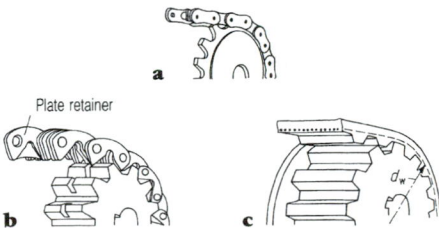

Plate retainer

Figure 2. Positive belts and chains: **a** roller on bushing chain on sprocket, **b** inverted tooth chain on sprocket, **c** synchronous belt on synchronous pulley.

are only suitable for drives between parallel shafts. The centrifugal forces increasing as the peripheral speed v of the belt or chain increases reduce the peripheral forces which can be transmitted. Maximum power is therefore transmitted at an optimum peripheral speed v_{opt} of the belt or chain, which however is usually dependent on the minimum size of pulley.

6.2 Flat Belt Drives

6.2.1 Forces in Flat Belt Transmissions

Peripheral force is transmitted between belt and pulley by means of shear stresses. Eytelwein's expression for the limiting case for slip throughout the arc of belt contact (shear slip, see A1.11.2) applies: $F_1'/F_2' = e^{\mu\beta}$, where F_1' and F_2' are the belt surface forces without centrifugal force and the arc of contact $\beta[\text{rad}] = (\pi/180)\,\beta[\text{degrees}]$: see **Fig. 4**. During normal operation the belt on each pulley first passes through a static arc β_r in which the belt does not slip on the pulley and then an active arc $\beta_w = \beta - \beta_r$. Shear stresses are transmitted by means of static friction in the static arc and by sliding friction in the active arc [2]. If one leaves aside the shear stress transmission in the static arc, then Grashof's expression for belt surface force ratio $F_1'/F_2' = e^{\mu\beta_w}$ applies. In design calculations the

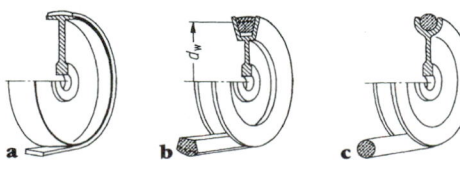

Figure 1. Friction belts: **a** flat belt, **b** V-belt, **c** round belt. Each with pulley.

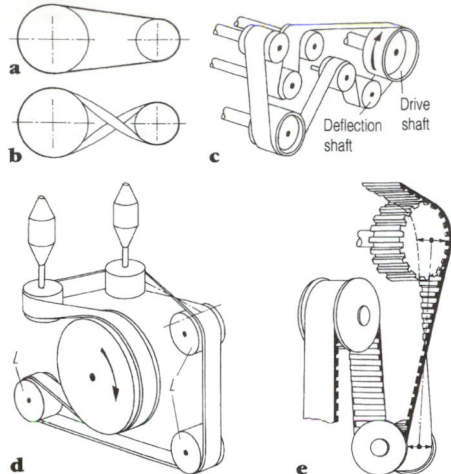

Figure 3. Planar (**a** to **c**) and spatial (**d** and **e**) drives: **a** open-belt drive, **b** crossed-belt drive, **c** multiple-shaft drive with flat belt, **d** spatial flat-belt drive with three guiding idler pulleys L, **e** spatial synchronous belt drive.

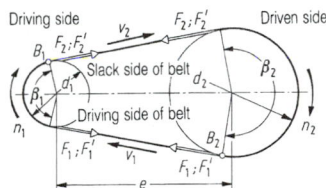

Figure 4. Designations on open-belt drive with index 1 for the smaller pulley.

full arc of contact β of the smaller pulley is assigned to the *design load*

$$F_1'/F_2' = m = e^{\mu\beta}. \tag{1}$$

The centrifugal forces acting in the arcs of belt contact to reduce the load pressure are supported by the free side of the belt, thereby acting as a uniform centrifugal force $F_f = \rho v^2 A = q v^2$ throughout the belt (ρ = the average thickness, A = the cross-section of the belt, q = the mass of a belt per unit length). Effective belt surface forces $F_1' = F_1 - F_f = mF_2'$; $F_2' = F_2 - F_f = F_1'/m$; peripheral force (effective load)

$$F_u = F_1 - F_2 = F_1' - F_2' = F_1'\,(1 - 1/m),$$

maximum belt surface force

$$F_{max} = F_1 = F_1' + F_f = F_2' + F_u + F_f.$$

The *elastic force of the shaft* F_w which generally does not point towards the bisecting line of the angle of β, but which does govern the bearing load, as shown in **Fig. 5**, is expressed as

$$F_w = \sqrt{F_1'^2 + F_2'^2 - 2F_1'\,F_2'\cos\beta_m}. \tag{2}$$

The *pull factor* Φ denotes the minimum elastic force required by the shaft to produce the peripheral force as a function of the coefficient of friction μ and arc of contact β:

$$\Phi = F_u/F_w = (m - 1)/\sqrt{m^2 + 1 - 2m\cos\beta}. \tag{3}$$

The *yield k* denotes the peripheral force F_u obtainable

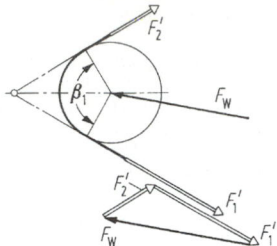

Figure 5. Forces acting on a pulley.

using the permissible belt surface force F_1' as a function of μ and β

$$k = F_u/F_1' = 1 - (1/m). \tag{4}$$

The reduction of yield which occurs as the arc of contact decreases is expressed by the arc factor c_β, which is related to $\beta = \pi$ or 180°. Arc factor $c_\beta = k_\beta/k_\pi$, where μ = const; where $\beta < \pi$: $c_\beta \ge \beta/\pi = \beta[\text{degrees}]/180$.

6.2.2 Stresses

Single-ply Flat Belts. The stresses for single-ply belts are obtained from the forces and the belt cross-section $A = bs$. For multiple belts these stresses can only be considered as hypothetical, calculated average values:

Belt tensioning $\sigma_1 = F_1/A,\ \sigma_2 = F_2/A.$

Effective stress $\sigma_n = F_u/A = \sigma_1 - \sigma_2.$

Centrifugal stress $\sigma_f = F_f/A = v^2.$

The bending stress is obtained from the bending elongation in the arc of contact of the smaller pulley. Bending stress $\sigma_b = E_b\varepsilon_b = E_b s/d_{w1}$ (E_b = modulus of elasticity on bending, ε_b = belt elongation on bending, s = belt thickness).

Maximum stress:

$$\sigma_{max} = \sigma_1 + \sigma_b = \sigma_2 + \sigma_n + \sigma_b. \tag{5}$$

With quarter-turn (crossed in alternate directions) and crossed-belt drives the belt undergoes an additional crosswise tensioning σ_s at its edges, so that $\sigma_{max,s} = \sigma_1 + \sigma_b + \sigma_s$.

Multiple Flat Belts. In multiple flat belts (**Fig. 10**), which comprise a high-strength, stress-bearing ply Z, a friction ply R for transmitting frictional force internally and frequently also an outer ply D or a further running ply (for multiple drives) on the outside of the belt, widely varying stresses are produced in the individual plies when elongation occurs. When bent the position of the neutral bending surface in the belt depends on the thickness and modulus of elasticity of the individual plies. **Figure 6**

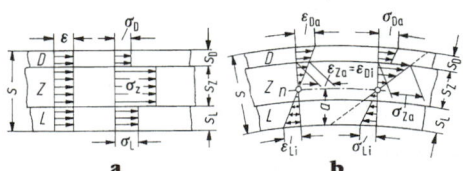

Figure 6. Stresses and strains in multiple belts. **a** Under tensile load. **b** Under bending load (*n* neutral surface).

shows the qualitative stress distribution under tensile and bending stress.

To simplify matters, in practice only the peripheral force per belt width F_u^* that is permissible for the particular type of belt is used as a basis for the design of multiple belts as well. This also takes into account the alternating bending stress that can be withstood in respect of minimum permissible pulley diameter d_{min} and the maximum permissible bending frequency f_B assigned to it. During bending the neutral fibre is assumed to be at the centre of the thickness of the belt at $s/2$; the elongation ε under tensile stress is calculated using an average modulus of tension (EA^*): $\varepsilon = F^*/(EA^*)$.

6.2.3 Geometrical Relations

The effective running diameter d_w of a belt is determined by the position of its neutral bending surface in the arc of contact. To simplify matters when making approximate calculations the pulley diameter d can be used instead of d_w. The expression for single-ply belts is: $d_{w1} = d_1 + s$; $d_{w2} = d_2 + s$; this is much the same for multiple belts.

Open Belt Drives (Fig. 4)

Arc of contact:

$$\beta_1 = 2 \arccos [(d_2 - d_1)/2e]; \quad \beta_2 = 360° - \beta_1';$$

Belt length (length of neutral fibre when stretched):

$$L_w = 2e \sin (\beta_1/2) + (d_{w1}\beta_1 + d_{w2}\beta_2) (\pi/360°).$$

Approximate formula for *shaft centre distance e* at a given belt length

$$e \approx \left(p + \sqrt{p^2 - q}\right)$$

where

$$p = 0.25L_w - \pi(d_{w1} + d_{w2})/8$$

and $q = (d_{w2} + d_{w1})^2/8$. The increase Δe of the shaft centre distances to prestretch the belt by $\varepsilon_o = \Delta L/L$ is obtained from a calculation for L_w and $(1 + \varepsilon_o)L_w$ or $\Delta e \approx (\varepsilon_o L_w/2)/\sin(\beta_1/2)$.

Crossed Belt Drives (Fig. 3b with designations as in **Fig. 4)**

Arc of contact:

$$\beta_1 = \beta_2 = \beta_{kr} = 360° - \beta_R, \quad \text{where}$$

$$\beta_R = 2 \arccos [(d_{w1} + d_{w2})/(2e)].$$

Length of the crossed belt (centreline):

$$L_{kr} = 2e \sin (\beta_R/2) + (d_{w1} + d_{w2}) \beta_{kr} (\pi/360°).$$

The expression $e \geq 20b$ is recommended because of twisting stresses σ_s. Owing to reverse bending service life shorter than with open belt drive.

Quarter-turn Belt Drive (Fig. 7). Angle of crossing $\delta \neq 0°$. Length of the centreline of the quarter-turn belt where $\delta = 90°$:

$$L_{90} \approx 2e + d_{w1}(\pi + \hat{\gamma})/2 + d_{w2}(\pi + \hat{\varphi})/2,$$

where $\tan (\gamma/2) = d_{w1}(2e)$ and $\tan (\varphi/2) = d_{w2}/(2e)$. Note design dimensions e_1 and e_2 $(\leq b/2)$, so that the belt winds on in the correct plane of the pulley! The side of the belt running off must be at an angle (below 25°) to the plane of the pulley, the direction of motion is irreversible. The expressions $e \geq 20b$ and $e \geq 2(d_w)_{max}$ are recommended because of twisting stress σ_s.

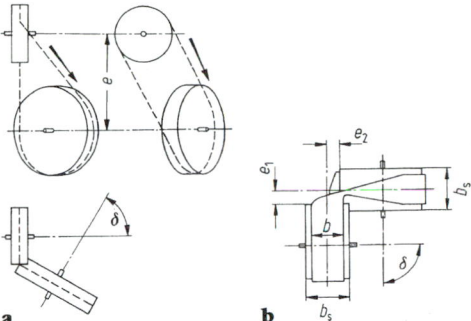

Figure 7. Belt geometry in crossed-belt drive: **a** crossed at an obtuse angle, **b** crossed at right angles.

6.2.4 Kinematics, Power, Efficiency

Belt Speeds

$$v_1 = \pi n_{an} d_{w,an}; \quad v_2 = \pi n_{ab} d_{w,ab}. \tag{6}$$

As a result of the greater elongation the speed v_1 of the driving side of the belt must be somewhat greater than the speed v_2 of the slack side of the belt in order to maintain steady operation. In practice, the equilibrium between the elongations of the driving and slack sides of the belt is obtained by slip due to elongation in the effective arcs of the driving and driven pulley. *Slip due to elongation* is produced when $\psi = \varepsilon_1 - \varepsilon_2 = (\sigma_1 - \sigma_2)/E = \sigma_n/E \approx (v_1 - v_2)/v_1$. The speed ratio i is therefore slightly load-dependent in normal operation:

$$i = n_{an}/n_{ab} = d_{w,ab}v_1/(d_{w,an}v_2)$$
$$\approx d_{w,ab}/[d_{w,an} (1 - \sigma_n/E)]. \tag{7}$$

In idle running the expression $i \approx d_{ab}/d_{an}$ applies.

Bending frequency (number of alternate bends per s),

$$f_B = z_s v/L_w = (z_s \pi d_{w1} n_1)/L_w, \tag{8}$$

where z_s = number of pulleys.

Torques follow from the forces on the side of belt:

$$M_1 = F_u d_{w1}/2; \quad M_2 = F_u d_{w2}/2.$$

Power:

$$P_{an} = 2\pi M_{an} n_{an}; \quad P_{ab} = 2\pi M_{ab} n_{ab}. \tag{9}$$

Rated power $c_B P_{an}$ where c_B is the service factor as in **Table 1** for initial design without calculations for vibration (based on DIN 2218).

Efficiency $\eta = P_{ab}/P_{an} = M_{ab}/(M_{an}i) \approx (1 - \sigma_n/E) = 1 - \psi$. In practice, leaving aside bearing friction and windage, efficiency is solely dependent on the slip as the peripheral force of each side of the belt on both pulleys is assumed to be equal. Efficiencies at best point $\eta = 0.96$ (chrome leather) and 0.98 (elastomer running layer).

6.2.5 Coning Action of Flat Belts, Tensioning

Step-Cone Pulleys in Variable-Speed Drives. The edge of the belt on a step-cone pulley reaches a higher speed on the larger diameter than the smaller one, with the result that the next section of belt is canted towards the larger diameter, thereby starting to run on to a larger running diameter d_L (**Fig. 8a**). Any section of belt that does not slip in the arc of contact must compensate for the differing speeds by means of elongations. It must assume the shape of a truncated conical surface and be

Table 1. Service factor c_B for obtaining an approximation of the dynamic characteristics of drive and machine and the daily operating time for open belt drives without idlers

Service factor c_B

Mode of operation of the driving machine	Mode of operation of the driven machine			
	Uniform	Near-uniform	Moderate shock	Heavy shock
Uniform	$1 + 0.04q + r$	$1 + 0.24q + r$	$1 + 0.44q + r$	$1 + 0.64q + r$
Moderate shock	$1 + 0.14q + r$	$1 + 0.38q + r$	$1 + 0.62q + r$	$1 + 0.86q + 1.2r$
Heavy shock	$1 + 0.24q + r$	$1 + 0.52q + r$	$1 + 0.78q + 1.2r$	$1 + 1.06q + 1.5r$

where $q = 1.1$ and $r = 0$ for positive chain drives,
$q = 1.0$ for synchronous belts
$q = 0.5$ for flat belts and V-belts, } and
$r = 0$ for up to 10 h operation per day,
$r = 0.1$ for between 10 and 16 h operation per day,
$r = 0.2$ for over 16 h operation per day.

The low q values of c_B for flat belts and V-belts are based on the assumption that infrequent short-term overloads are partially compensated for by slip action. In the case of positive belt and chain drives it must be made certain that the rated power covers maximum peak loads, including moments of inertia and shock!

Examples of mode of operation of the *driving machine*

Mode of operation	Driving machine
Uniform	Electric motors with low starting torque (up to 1.5 × nominal torque), water and steam turbines, internal combustion engines with eight cylinders or more.
Moderate shock	Electric motors with medium starting torque (1.5 to 2.5 × nominal torque), internal combustion engines with between four and six cylinders.
Heavy shock	Electric motors with high starting and braking torque (over 2.5 × nominal torque), hydraulic engines, internal combustion engines with up to four cylinders.

Examples of mode of operation of the *driven machine*

Mode of operation	Driven machine
Uniform	Small masses to be accelerated. Typewriters, light-duty conveyor belts, domestic appliances.
Near-uniform	Moderate masses to be accelerated. Light-duty ventilation fans, light to medium-duty wookworking machines, belt conveyors for ore, coal and sand, agitators (liquid, semi-liquid), turning, drilling and grinding machines, textile machinery, printing machinery, centrifugal pumps, washing machines.
Moderate shock	Moderate masses to be accelerated. Heavy-duty conveyor systems, screw conveyors, mixing machines, industrial ventilation fans, generators and exciters, centrifuges, rubber processing machines, hammer mills.
Heavy shock	Large masses to be accelerated. Reciprocating pumps and compressors with velocity fluctuation $< 1 : 80$; ball rollers and gravel mills, edge mills, shearing machines, stamping machines, rolling mills for non-ferrous metals, stone crushers.

virtually bent edgewise (**Fig. 8b**). Equilibrium is achieved if the bending moment at A caused by this bending strain is compensated by oblique tension acting on the side of the belt (**Fig. 8c**). Axial misalignment is approximately 0.6 of the belt width, the exact misalignment being obtained after a short running-in period.

Flat Belt Drives with Constant Speed Ratios. The pulleys of conventional open and crossed flat belt drives are made with slightly arched running surfaces to DIN 111 (ISO/R 100) (**Table 2**) in order to guide axially the belt that always tends towards the largest diameter of the pulley. In the case of open belt drives with horizontal shafts, the smaller pulley can be made cylindrical where the speed ratio $i > 3$. The preconditions for satisfactory running of the belt are: axial parallelism of both shafts, concentrically running pulleys, alignment of the largest diameters of convex pulleys in one plane, edges of the belt within the breadth of the pulley $b_s > b$, smooth pulley running surfaces to DIN 111. "Non-slip", porous or rippled surfaces or adhesive anti-stripping agents prevent natural slip due to elongation in the effective arc, increase wear and may excite longitudinal vibrations by means of stick-slip effects.

Three-dimensional Belt Drives (**Fig. 3d, e**) accept cylindrical pulleys. To provide a secure belt guide in quarter-turn belt drives ($\delta = 90°$) the following is recommended:

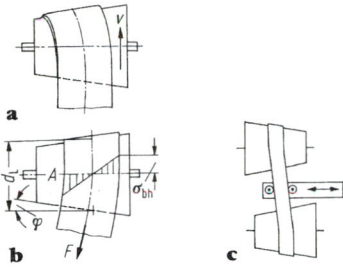

Figure 8. **a** Belt running axially on to the larger diameter. **b** Equilibrium when belt runs tangentially on to cone pulley. **c** Drive with two cone pulleys for continuously variable transmission.

Table 2. Recommended curve heights b as in DIN 111

d_1 in mm	b in mm for $b_s \leq 250$ mm	b in mm for $b_s > 250$ mm
up to 112	0.3	0.3
up to 140	0.4	0.4
up to 180	0.5	0.5
up to 224	0.6	0.6
up to 355	0.8	0.8
up to 500	1.0	1.0
up to 710	1.2	1.2
up to 1000	1.2	1.5
up to 1400	1.5	2.0
up to 2000	1.8	2.5

breadth of pulley $b_s = 2b$, centre distance of the central plane of the pulley from its respective counter-rotating wheel e_1, $e_2 = (0.2$ to $0.5)b$ (**Fig. 7b**), $d_2/d_1 = 1$ to 2.5, $e \geq 20b$.

Production of Tensioning. The minimum required shaft load F_w to ensure frictional engagement can be produced using the processes shown in **Fig. 9a–d** by:

(**a**) *Belt Stretch with Fixed Distance Between Centres.* The belt length is dimensioned so that the belt is tensioned by elastic elongation when it is wound on to

the pulleys. When the distance between centres is adjustable (e.g. drive motor to slide rails) tensioning can also be produced after winding on by increasing the distance between centres. If the distance between centres is fixed, the belt length remains constant in all operating conditions. The belt surface forces F' and the shaft elastic forces F_w are therefore reduced by centrifugal force. The degree of stretching must therefore be specified as larger in terms of σ_f in order to ensure the required frictional engagement for the operating speed. The shaft load rises slightly with increased torque and is determined by the exact distribution of the elongation [2]. As the stretch is to be maintained for long periods of operation, this tensioning process is particularly suitable for belts with high dimensional stability, e.g. multiple belts with polyamide or polyester fabric plies; it is the tensioning process that is mainly used for this purpose.

(**b**) *Tensioning Shaft.* The shaft load F_w is applied to the traversing shaft by means of weights or (flexible) springs, suitable for belts with time-dependent elongation under load. Note, however, that in addition to the greater expense involved there is also the danger of vibration.

(**c**) *Tensioning Idler at the Slack Side of the Belt.* The movable spring-loaded or weighted tensioning idler produces constant belt surface force F_2 in all operating conditions. If the tensioning idler is used on the outer side of the belt, the arc of contact β is increased and at the same time the arc factor c_β is improved. The additional tensioning idler does, however, increase bending frequency, thereby lowering the permissible useful stress at higher belt speeds. Its diameter should be greater than $d_{1,\min}$ to take the service life of the belt into consideration and its running surface should always be cylindrical. As this tensioning process results in rather low loads on the side of the belt and the shaft with small torques, it is suitable for drives that operate mainly underdriving and belts with time-dependent elongation. Here too, the risk of vibrations must be noted. If a *fixed* (adjustable) tensioning idler is used on the slack side of the belt to adjust the stretch and also to increase β, the same operational characteristics arise as in the tensioning process shown in **Fig. 9a**.

(**d**) *Self-tensioning with Double Tensioning Idler* [3]. The tensioning idlers in the driving and slack side of the belt have a fixed (adjustable) distance between

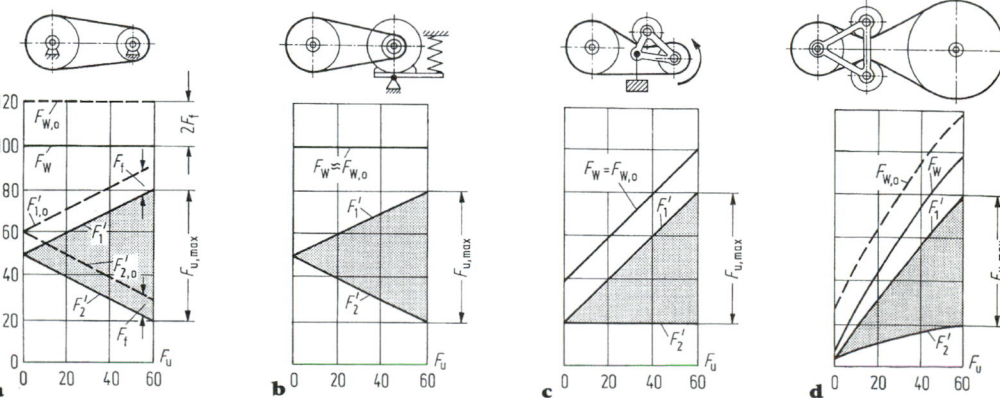

Figure 9. Dependence of tangential forces and shaft load F_w on the peripheral force F_u at constant speed with different tensioning methods **a** to **d** (for $\beta_1 = \beta_2 = 180°$). Index 0: Forces when stationary.

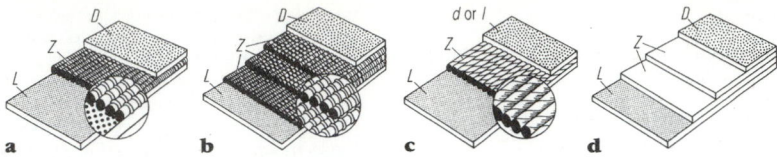

Figure 10. Construction of multiple belts: **a** single-ply fabric belt, **b** multiple-ply fabric belt, **c** rayon cord belt, **d** band belt with wide tension bands, most commonly used type; D outer layer, Z tension member, L running layer.

Table 3. Construction and use of belts as shown in **Fig. 10** (for standard values, see manufacturers' data)

Belts	a	b	c	d
Tension member[a]	PA, B	B, PA, E	E	PA
Running layer[a]	PU	G or balata	G or CH	G or CH
Manufacture	Endless, in lengths	Cut to size from roll, endless, vulcanised at joint	Endless, in lengths	Cut to size from roll, endless, cemented at joint
Use	High speeds, grinding spindles	Robust, for low power ratings	Multiple pulley drives, maximum speed up to 1000 kW	Robust, most common type, up to 6000 kW for twin- and multiple-shaft drives
v_{max} (m/s)	70	20 to 50	100	70
$d_{1,min}$ (mm)	15	150	20	63
$f_{B,max}$ at d_{min} (l/s)	10 to 20(50)[b]	10 to 20	30(100)[b]	30(80)[b]
$F^*_{u,max}$ (N/mm)	10	30	48	48(110)[b]
Maximum elongation ε in operation (%)	3	2 to 4	1.8	3
Ambient temperature range (°C)	−20 to +70	−20 to +70	−40 to +80	−20 to +80

[a]PA polyamide, E polyester, B cotton, CH chrome leather, PU polyurethane, G elastomer (rubber).
[b]Values in brackets only after consultation with manufacturer.

centres, they are guided, with little friction, on a circuit round a pulley bearing and must therefore each have the same axial loads in each operating condition. This however requires differing angles of contact for the tensioning idlers and results in self-tensioning. In the case of a part-load low belt surface forces F'_1 and F'_2 are set up, and with them low shaft loads. The belt surface forces and shaft loads rise with increasing peripheral force F_u. Careful matching of the tensioning idler geometry with distance between centres, pulley diameter and belt elasticity is required in the case of this tensioning process which is also effective for alternating drive direction (brakes). The relatively simple design is suitable for drives up to very high power with predominantly part-load operation and belts without significant time-dependent elongation,

although here too there is a greater tendency for transverse vibrations than in designs with fixed tensioners.

6.2.6 Materials

Owing to their limited strength, short service life and considerable elongation during operation, leather belts, which at one time were in common use, have been replaced by synthetic multiple belts (composite belts). These belts are either manufactured in endless form to a suitable length or heat-bonded at the point of application at their ends, which have been cut diagonally and bevelled to make them endless. **Figure 10** and **Table 3** show the construction and materials of common types of belt and **Table 4** the characteristic values of the materials used in flat belt plies.

Table 4. Material characteristics of flat-belt tension members

Material	R_m (N/mm²)	$E_{tension}$ (N/mm²)	ρ (kg/m³)	Ultimate tensile strength (%)	Coefficient of friction on grey cast iron and steel
Rayon cord	900	700 000	1400	15	
Polyamide band	500	150 000	1140	20 to 25	
Leather, high-grade	30 to 50	300 to 500	900	30	0.3 to 0.7
Leather, normal	20 to 30	100 to 300	1000	30	0.3 to 0.7

6.2.7 Calculation

The limits on permissible stress for belts are not set by their tensile strength but by disintegration (failure due to fatigue) and, in the case of inadequate tensioning, by wear. In flat belts the tensile strength R_m therefore amounts to 10 to 20 times the permissible operating tension σ_n. Damage to belts is accelerated by higher temperatures and greater flexing work, i.e. by greater bending frequencies and smaller bending radii. The permissible operating load is determined by experiment. The rough design of an open flat belt drive of the most frequent type (as shown in **Fig. 10d**) is based on the permissible (index*) nominal peripheral force F_{uN}^* relative to 1 mm belt width, with an allocated minimum permissible pulley diameter $d_{1,min}$ of the smaller pulley, as shown in **Appendix F6, Table 1**. The belt speed v_{max} and the bending frequency $f_{B,max}$ as in **Table 3** should not be exceeded.

Where d_1 is the diameter of the smallest pulley, β_1 the arc of contact, c_β the arc factor, b the belt width and n_{an} the drive speed, the following applies to belts as shown in **Fig. 10d** on the basis of manufacturers' data [4]:

Permissible relative peripheral force
$$F_{u,perm}^* \approx c_\beta F_{uN}^* \cdot (2 - d_{1,min}/d_1);$$
Rating $c_w P_{an} \leq F_{u,perm}^* \, b d_{w,an} \pi n_{an}$;
Belt width $b \geq c_\beta P_{an}/(F_{u,perm}^* \, d_{w,an} \pi n_{an})$.

Refinements can be expected in the calculation in Eq. (11) in respect of V-belts. If a belt drive is specified with fixed distance between centres as in **Fig. 9a**, the belt must be mounted using elastic seating stretch. If when operating with $F_{u,perm}^*$ the sum $(F_1' + F_2') = k_v F_{u,perm}^* b$ is selected and allowance is made for the centrifugal force during operation as in **Fig. 9a**, the stretch ε_a can be calculated as follows:

$$\varepsilon_a = \Delta L/L = \varepsilon_o + \varepsilon_f = [(k_v/2) F_{u,perm}^* + F_f^*]/(EA^*),$$

where $F_f^* = \rho' v^2$; (EA^*) and ρ' are as in **Appendix F6, Table 1**. Reference values for $k_v = (m + 1)/(m - 1)$ with m as in Eq. (1), e.g. for $\beta_1 = \pi$ and $\mu = 0.51$: $K_v = (5 + 1)/(5 - 1) = 1.5$ or $\mu = 0.4$: $k_v = 1.8$. Belt length slack, i.e. smaller by the amount of stretch:

$$L = L_w/(1 + \varepsilon_a).$$

Shaft load produced by tensioning when stationary with additional load F_f^* and

$$\left.\begin{array}{l} F_1 = F_2 = [(k_v/2)F_{u,perm}^* + F_f^*]b = \varepsilon_a(EA^*)b \\[6pt] F_{wo} = F_1 \sqrt{2(1 - \cos \beta_1)} = 2F_1 \sin (\beta_1/2). \end{array}\right\} (10)$$

Compare the bending frequency f_B with the permissible bending frequency $f_{B,max}$ for minimum pulley diameter $d_{1,min}$ given in manufacturer's data.

Example Open-belt drive for rotary piston blower, where $d_1 = 315$ mm, $d_2 = 800$ mm, $e = 870$ mm, $n_1 = 1450$ min^{-1}, $P_1 = P_{an} = 90$ kW. If $d_w \approx d$, then according to Eq. (6) $v = 23.9$ m/s, $i = d_{ab}/d_{an} = d_2/d_1 = 2.54$; $\beta_1 = 147.6°$; $\beta_2 = 214.4°$; $L_w = 3559.5$ mm. Belt type for $d_{1,min} < d_1 = 315$ mm as in **Appendix F6, Table 1**: Type 40, where $d_{1,min} = 280$ mm and $F_{uN}^* = 35$ N/mm. Arc factor for rough dimensioning $c_\beta \approx \beta_1/180° = 0.82$. Then $F_{u,perm}^* = 0.82 \cdot 35 \cdot (2 - 280/315) = 31.89$ N/mm and if $c_\beta \approx 1.23$ as in **Table 1**, then $b = 1.23 \cdot 90 \cdot 10^6/(31.89 \cdot 315 \cdot \pi \cdot 1450/60)$ mm $= 145.1$ mm ≈ 140 mm. Stretch (for $k_v = 1.76$) $\varepsilon_a = [(1.76/2) \cdot 31.89 + 4.5 \cdot 23.9^2 \cdot 10^{-3}]/2000 = 0.0153$, thus $L = 3559.5/1.0153 = 3505.8$ mm ≈ 3505 mm. Shaft load as in Eq. (10) $F_{wo} = 2 \cdot 0.0153 \cdot 2000 \cdot 140 \cdot 0.9165 = 7860$ N; if $k_v = 2.11$, then $F_{wo} = 9.293$ N. Bending frequency $f_B = 2 \cdot 24\,000/3559.5 = 13.5$ s$^{-1} < f_{B,max}$; reference value $f_{B,max} = 30$ s^{-1} for $d_{1,min}$ for belt calculation. The task may also be accomplished by belt type 28

where $d_{1,min} = 200$ mm and larger belt widths. With this smaller type of belt the pulley diameters d_1 and d_2 can be reduced and higher shaft loads withstood.

Vibration characteristics of the belt drive must also be taken into consideration when making a final decision using the calculations provided in DIN 740 for *resilient shaft couplings* and *string vibration*. Belt manufacturers should always be consulted in respect of particular cases; see manufacturers' literature [11].

6.3 V-Belts

6.3.1 Uses and Characteristics

V-belts (**Fig. 1b**) make use of the frictional transmission of motion and power over medium shaft centre distances [5]. They are guided firmly in sheaves in all working positions, as well as over short stretches and in angular drives. Almost all types are also suitable for coupling (tensioning the V-belt while the drive wheel is running by means of a radially moving shaft or tensioning idler). Dimensions for the basic types have been standardised internationally; see **Appendix F6, Table 2**. For further types for special purposes, see **Fig. 12**.

The frictional transmission of peripheral force takes place solely through the tapered sides of the belt cross-section. Direct contact with the bottom of the groove results in reduction of transmittable peripheral force, shear slip and damage due to overheating. The shaft centre distance must be adjustable by x amounts as in ISO 155 or manufacturers' data; as a rough guide $x \geq +0.03L_w$ is sufficient for tensioning and retensioning the belt and $|x| \geq 0.015L_w$ for installing the belt over the rim of the sheave without using force. The effective diameters d_w (**Fig. 1b**) and corresponding effective widths b_w (**Fig. 12a and Appendix F6, Table 2**) of the belt and the sheave denote the position of the non-flexing tensile ply in the V-belt cross-section. They should as far as possible agree with the corresponding reference diameter d_r and the reference width b_r of the sheave (this does not apply to V-ribbed belts to DIN 7867). The wrap angle α is specified as a function of d_r, owing to the lateral strain in the belt. Frequent (f_B) and large $(1/d_w)$ bending strains increase the internal heating of the belt and reduce its power rating over the same service life (**Fig. 11a**). Preconditions for a long service life are: continuous maintenance (inspection) of the correct tension, accurate alignment of the sheaves and smooth surfaces of the sheaves, $d_{w,min}$ and shaft centre distance e no smaller than necessary. Avoid reverse bending (back bend idler). Where they are unavoidable, idlers should be designed as sheaves, where $d_w > d_{w,min}$.

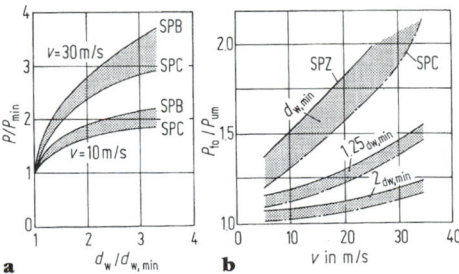

Figure 11. Power rating of narrow V-belts to DIN 7753 with the same service life [5, 8]. **a** Encased V-belt. **b** Relation to power rating P_{fo} of raw-edged narrow V-belt to power rating P_{um} of encased narrow V-belt. $d_{w,min}$ as in **Appendix F6, Table 2**.

Operating Limits. Ambient temperatures $= -30$ to $80\,°C$ (-55 to $70\,°C$); $i_{max} \approx 10$; $e \approx (0.7$ to $2)$ $(d_{w1} + d_{w2})$; $F_w = (1.5$ to $2.5)F_u$; power ratings up to $P_{max} > 1000\,kW$ (up to 35 parallel belts), $\eta_{max} = 0.97$ for single belts; η_{max} up to 0.95 for V-ribbed belts.

6.3.2 Types and Sizes

The various types are distinguished by the geometrical shapes of the belt cross-section and the sizes by their internal construction. **Figure 12a-i** shows the most common types of V-belts:

(**a**) *Endless V-Belts* to DIN 2215 (also classic V-belt). $b_0/h \approx 1.5$ to 1.6; cross-sections designated according to width b_0; for V-belted sheave dimensions and materials see DIN 2211 and DIN 2217. Examples of use: types 5 and 6 for laboratory equipment, precision mechanics; 8 and 10 for domestic appliances; 13 to 22 for machine construction (medium rev/min) and farm machinery; 25 to 40 for heavy-duty drive mechanisms with large shaft centre distances, large sheave diameters, low speeds and severe operating conditions.

(**b**) *Open-Ended V-Belts* to DIN 2216. Cut lengths, strong fabric plies, with holes pre-punched for belt connector, for medium peripheral speeds. P_{max} up to 15% lower, $d_{w,min}$ up to 15% greater than with endless belts to DIN 2215 having the same cross-section. Greater residual elongation so frequent retensioning or shortening required. Belt ends are joined by belt connector after seating, simple installation on awkward drives where endless belts cannot be installed. Can be easily stored.

(**c**) *Endless Narrow V-Belts* to DIN 7753, $b_0/h \approx 1.2$ to 1.4 with narrow sheaves to DIN 2211 (dimensions and materials). These transmit greater power than V-belts of the same effective width as described in DIN 2215. They constitute the most commonly used type of belt.

(**d**) *Endless Broad V-Belts* for industrial variable-speed drives to DIN 7719. $b_0/h = 2.8$ to 3.25. Groove angle $\alpha = 24$ to 30 °. Smaller angles produce a greater regulating range as well as a risk of self-locking (the V-belt wedging in the sheave groove). Approximately 20% less power is transmitted than with V-belts of the same height in cross-section as described in DIN 2215. DIN 7719 does not apply to variable-speed drives in motor vehicles or farm machinery. Regulating range $i_{max}/i_{min} = 4$ to 12 possible with two adjustable sheaves.

(**e**) *Cogged V-Belts.* V-belts as shown in **a** and **d**, with transverse grooves cut in the inner face of the cross-section to increase flexibility, make it possible to have smaller sheave diameters and take up less space, with power being only slightly reduced. The grooves do however cause periodical running-on shocks and noise, unless the transverse grooves are unequally spaced.

(**f**) *Endless Hexagonal V-Belts* for farm machinery (*twin V-belts*) to DIN 7722. $b_{max}/h \approx 1.3$. For planar multiple-shaft drives with counter-rotating sheaves. Power transmitted as in V-belts to DIN 2215 with equal maximum cross-sectional width. Used in medium-duty drives (combine harvesters) down to light equipment (gardening equipment, street-sweeping machines).

(**g**) *Raw-edged V-Belts.* Cross-sections as in DIN 2215 and DIN 7753 Part 1. They have only one outer covering ply of fabric, but – unlike the other types of belt – no fabric shell on the bearing edges or the "cogged" inner face. The belt substructure, a polychloroprene–rubber mixture, is very flexible and elastic and is reinforced for increased expansion forces (preliminary tension) with supporting planes arranged at right angles to the running direction. They transmit greater power using small sheave sizes and high speeds (**Fig. 11b**), cope with smaller sheave diameters (approximately 0.7 to 0.8 $d_{w,min}$ as in **Appendix F6, Table 2**) than encased V-belts (therefore also require less working space for the same power) and are less susceptible to oil, heat, slippage and abrasion.

(**h**) *Multiple Narrow V-Belts* (power belts). These comprise up to five narrow or classic V-belts of equal length (within the set), which are banded tightly together by shrouding. Shrouding prevents twisting or significant vibration of individual belts within the set. Grooved belts to DIN ISO 5290.

(**i**) *Ribbed V-belts* to DIN 7867. Further development of multiple V-belts towards flat belts. Five cross-sections with space between ribs in mm: PH 1.60; PJ 2.34; PK 3.56; PL 4.70; PM 9.40. PK preferred for motor vehicle construction, PJ, PL and PM preferred for industrial belt drives and PH for special applications. Up to 60 polychloroprene ribs wide without casing which completely fill the grooves in the appropriate pulleys. Tension cord made of low-stretch rayon cord. Refer to manufacturers' data for power rating with additional speed ratio per rib. Peripheral speeds up to $v \approx 60\,m/s$, depending on the cross-section. The smaller sheave diameters and higher speed ratios per step than in V-belts reduce working space required, whilst there is quieter running and greater uniformity of motion; deflection possible.

6.3.3 Calculation

In order to calculate the rated power P_N of open V-belt drives as a function of service life, a numerical equation obtained from test results, as given in ISO 5292, is being increasingly used. When characteristic quantities are introduced this equation may be formulated more clearly:

$$P_N = c_\beta P_0 \cdot \frac{v}{v_0} \cdot \left[1 + K_2 \left(1 - \frac{d_{w,min}}{d_{w1}} \cdot \frac{1}{K_i} \right) \right.$$
$$\left. + K_3 \left[1 - \left(\frac{v}{v_0} \right)^2 \right] + K_4 \ln \left(\frac{v_0}{v} \cdot \frac{L_w}{L_0} \right) \right], \quad (11)$$

where the arc factor $c_\beta = 1.25 \cdot (1 - 5^{-\beta_1/180})$; arc of contact β_1 of the smaller pulley; rated power P_0 at peripheral speed v_0 for minimum pulley diameter $d_{w,min}$ at speed ratio $i = 1$ ($\beta_1 = 180°$) and belt length L_0; rated power

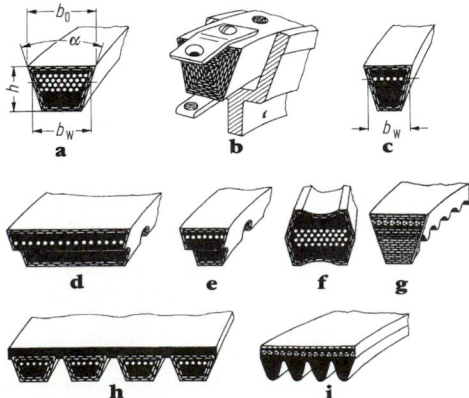

Figure 12. Types of V-belt. For **a** to **i**, see text.

P_N at peripheral speed v for effective diameter of the smaller pulley d_{w1} at speed ratio $i \neq 0$ ($\beta_1 \neq 180°$) and belt length L_w; $K_i \approx 1.124 - 0.124 \exp[-3(i-1)]$ and $i \geq 0$. An evaluation of details provided in a manufacturer's catalogue is given in **Appendix F6, Table 2**, as an initial guide. As only deviations from the manufacturers' information arise in a particular case, the information provided by the V-belt manufacturers must be used when checking calculations. Standards DIN 2218 and DIN 7753 may also be used as a guide. The correct dimensioning of a belt drive is dependent on a series of factors and environmental conditions. It is therefore recommended, especially where there are difficult problems with drives, that the experience of the manufacturers in this sector, i.e. V-belt and drive manufacturers, be called upon; see manufacturers' literature [11].

The rated power $c_B P_{an} \leq z P_N$ for z belts running in parallel is determined according to **Table 1** using estimated values for c_B so that the required number of belts is $z \geq c_B P_{an}/P_N$. Calculation of all other operating dimensions as per flat belts.

6.4 Synchronous Belts

6.4.1 Design, Characteristics and Uses

Synchronous belts (**Fig. 13**) have teeth on one or both sides, with which they transmit peripheral force positively without slip (**Fig. 2c**). The body of the belt consists of neoprene or polyurethane with the pulling strands made of high-tensile glass fibres or steel, kevlar or rayon cord which are wound helically in the case of belts which are manufactured in standard lengths, normally in endless form. The pulling strand determines the neutral bending plane and its length is also the effective length L_w of the belt, it runs on the effective diameters $d_{w1,2} = z_{1,2} p_b/\pi$ round the synchronous pulleys (crowned rims) where the tooth numbers are z_1, z_2 and the tooth pitch is p_b. Synchronous belts (toothed belts) are maintenance-free if correctly set, no lubrication required. At greater speeds, power ratings, initial tensions and belt widths meshing noises occur, basic frequency $f_0 = n_1 z_1$. Owing to their positive transmission of motion synchronous belts are suitable for drives which maintain the speed ratio (e.g. timing gears), and in the case of toothing on both sides, for multiple-shaft drives with counter-rotating pulleys, and in the case of greater axial distances, for spatial drives (**Fig. 3e**).

Standards. DIN 7721 and DIN/ISO 5296 regarding dimensioning and measurement of effective length.

6.4.2 Design Hints

In planar drives the synchronous belts must be guided axially by lateral rims on both sides of at least one crown pulley or alternate sides of two crown pulleys. To permit installation and initial tensioning there should be a shaft

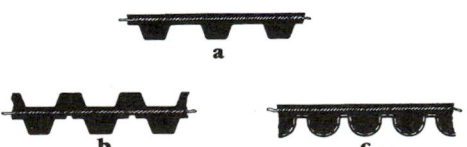

Figure 13. Cross-sectional shapes to toothed belts: **a, b** single and double toothed to DIN 7721 with metric increments and DIN/ISO 5296 with inch increments, **c** HTD (high torque drive) cross-section.

or belt tightener which can be moved radially. Where the shaft centre distance is fixed the crown pulleys are mounted together with the installed belt. As far as possible tensioning idlers should be designed as crown pulleys ($d_w > d_{w1}$) and arranged internally so as to avoid deflection at the slack side of the belt, but not spring-mounted as there should be no stretching of the belt if it is laid out correctly. Recommended limiting values: $e \approx (0.5$ to $2)$ $(d_{w1} + d_{w2})$, $d_1/b \geq 1$. In the case of spatial belt drives the straight line between the winding-on and running-off point must at the same time be the line of intersection of both wheel planes so that the belt is only twisted and not pulled off to the side (see **Fig. 3e**); lateral rims may be dispensed with; shaft centre distance per 90° of twist $e_{90} \geq 12b$.

Operating Limits. Ambient temperature $= -40$ to 90 °C; $p_{max} = 400$ kW; $v_{max} = 40$ (Type T20) to 80 (T5) m/s. $f_{B,max} \approx 100$ s^{-1}; $i_{max} \approx 12$; $\eta_{max} \approx 0.98$.

6.4.3 Calculations

Calculation of L_w (approximate), e and v as for flat-belt drive; exact: $L_w = p_b z_b$, where $z_b =$ number of belt teeth; number of engaging teeth $z_{e1} = z_1 \beta_1/360°$ (rounded off to whole number); speed ratio $i = z_2/z_1$; selection of belt according to rated power and number of teeth $z_1 > z_{1,min}$ with power data for reference width b_{s0} as in **Appendix F6, Table 3** and width factor $k_w = (b_s/b_{s0})^{1.14}$ as in ISO 5295 and initial load factor $k_z = 1$ for $z_{e1} \geq 6$ or $k_z = 1 - 0.2(6 - z_{e1})$ for $z_{e1} < 6$.

With the power rating

$$c_B P_{an} \leq k_z P_0 \frac{v}{v_0} \frac{b_s}{b_{s0}} \left\{ 1.5 \left(\frac{b_s}{b_{s0}}\right)^{0.14} - 0.5 \left(\frac{v}{v_0}\right)^2 \right\}$$

and $v = n_1 z_1 p_b = n_2 z_2 p_b$, the minimum required belt width b_s is obtained. Maximum belt widths $b_{s,max} \approx (4$ to $10)p_b$. Recommended shaft preload $F_{wo} \approx F_u$. The service factor c_B must be increased in accordance with manufacturers' data rather than **Table 1** for step-up transmissions for $1/i \approx 1.24$. Higher power ratings can be transmitted using HTD (high torque drive) belts [6] and RPP belts (belts with a parabolic cross-section) [9] as a further development of trapezoidal toothed belts and AT belts [10] as strengthened T types. An additional deciding factor for the selection of belts, particularly in automobile construction, is that the development of noise should be kept to a minimum. This is what is aimed for by modifying the shape of trapezoidal teeth. See manufacturers' literature [11].

6.5 Chain Drives

6.5.1 Types, Characteristics and Uses

Chain drives (**Fig. 2a, b**) transmit power loads up to 200 kW per single chain positively and without slip using low peripheral speeds between parallel shafts and where there are more than two shafts also by counter-rotation. Power loads of more than 500 kW are possible using multiple-strand chains (made with up to 12 strands but mostly with up to 3 strands). Where the smaller sprocket has a small number of teeth, the rotary transmission becomes uneven owing to the rhythmically variable run-off or nip point of the chain, or the *polygon effect* as it is known. Resulting from this are periodically variable rim speeds, excitation of vibration and noise at higher chain speeds. This is moderated by increasing the number of teeth and

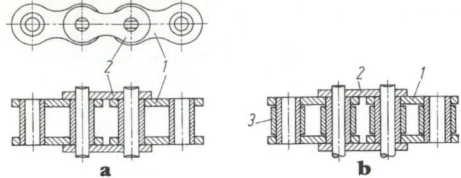

Figure 14. Transmission chains: **a** single bushing chain, **b** single roller chain; *1* roller link plate with bushings held in place, *2* pin-link plate with pins, *3* movable roller.

reducing the spacing. On the other hand, by reason of its longitudinal elasticity the chain moderates starting shocks. The service life of a chain is limited by the maximum wear-induced elongation it can withstand and is reduced by inadequate lubrication, dirt and stress due to shock and vibration. The most common types are the *bushing chain* (**Fig. 14a**) to DIN 8154 (inside a closed transmission casing with very good lubrication), the *roller chain* (**Fig. 14b**) to DIN 8187 and DIN 8188 (the most frequently used type; the lubricated roller reduces wear and noise) and the *inverted tooth chain* (**Fig. 2b**) (to DIN 8190) (silent running at high peripheral speeds). For other *steel link chains* see DIN 8194 which covers structural shapes and nomenclature (German, English, French).

6.5.2 Design Hints

Where possible, measure shaft centre distances to an exact number of chain links (pitch p) in order to avoid bent links. The centre distance should be such that the arc of contact is at least $120°$ to pinion, normal: $e = 30$ to $50p$. The sag on the slack side of the chain should be approximately 1% of the centre distance. The maximum allowable wear-induced elongation of the chain Δl should generally not exceed 3% of the original chain length l, in sprockets with more than 67 teeth only $\Delta l/l \leq 200/z_2$ as a percentage. However, with fixed shaft centre distance without a tensioning device, only $\Delta l/l \leq (0.6 \text{ to } 1/5)\%$ compensation for wear is allowable on the chain using transversely movable shafts. For a fixed shaft centre distance, a cylindrical tensioning idler (up to $v = 1$ m/s) or a tensioning wheel is used, both on the slack side, lightly loaded by springs or weight. Owing to the polygon effect only sprockets with at least 17 teeth should be selected. For medium to high speeds or maximum permissible load, the pinion should have hardened teeth and if possible 21 teeth. Sprockets should normally have a maximum of 150 teeth. The preferred number of teeth is 17, 19, 21, 23, 25, 38, 57, 76, 95 or 114. If the chain drive is positioned at an inclination towards the horizontal greater than $60°$,

then the required chain tension is provided by tensioning idlers, tensioning wheels or other suitable devices. In tensioning and deflection wheels at least three teeth should be engaged at any one time. A speed ratio i of 3 to 7 times is considered good, but ratios above 10 times are possible. The amount of lubrication required depends on the type of chain and the chain speed; *for notes on roller chains* see DIN 8195: for a DIN 8188-08A-1 chain with a pitch p of 12.7 mm, for example, the following applies: oil applied by oil can or brush, up to $v \approx 0.7$ m/s (uncertain, at least once daily); drip-feed lubrication, up to $v \approx 3.9$ m/s (drip-fed oil for each series of link plates with 2 to 6 drops each per minute); oil bath (maximum oil level up to lowest roller centre) or centrifugal lubrication up to $v \approx 8.4$ m/s; pressure circulating lubrication, if necessary with filter and oil cooler, up to $v_{max} \approx 19$ m/s (with constant oil flow on the inner face of the slack side as well as the tight side of the chain; also for cooling the chain). After a single lubrication, efficiency drops off quickly with increasing operating time; $\eta_{max} < 0.97$. The maximum power transmission at $v_{opt} = n_0 z_1 p$. For reference values for n_0 where $z_1 = 19$ and 15 000 operating hours with speed ratio $i = 3$ with 100 chain links as in DIN 8195 see **Appendix F6, Table 4**.

6.5.3 Calculations

Chain speed $v = n_1 z_1 p = n_2 z_2 p$, geometrical diameter (roller centres) $d_{w1,2} = p/\sin(180°/z_{1,2})$, chain length $l = Xp$, where X is the number of links (whole, exact number). $X \geq X_0$ where $X_0 = 2e/p + (z_1 + z_2)/2 + p(z_2 - z_1)^2/(4e\pi^2)$, centre distance

$$e \approx \frac{p}{4}\left[\left(X - \frac{z_1 + z_2}{2}\right)\right.$$
$$\left. + \sqrt{\left(X - \frac{z_1 + z_2}{2}\right)^2 - 2\left(\frac{z_2 - z_1}{\pi}\right)^2}\right].$$

The pitch p of the roller chains as in DIN 8187 (European type, Code B) and DIN 8188 (American type, Code A) is standardised in inch increments; see **Appendix F6, Table 4**.

The power rating P_0 corresponds to the speed n_0; in accordance with DIN 8195 the following applies to $n_1 \leq n_0$, $i \leq 7$:

$$P_N \approx P_0 \left(\frac{n_1}{n_0}\right)^{0.9} N^{0.97} \left(\frac{z_1}{19}\right)^{1.073} \left(\frac{i}{3}\right)^{0.18} \left(\frac{e}{40p}\right)^{0.26},$$

with rated power $c_B P_{an} \leq P_N$, by which means the service factor can be estimated from **Table 1** or even DIN 8195; $N = 1$ for single-strand chain, $N = 2$ for double-strand chain, $N = 3$ for triple-strand chain. See manufacturers' literature [11].

---

7 Friction Drives

H. Peeken, Aachen

7.1 Mode of Operation, Definitions

Friction drives, also known as *ratio traction drives*, are uniform-motion, frictionally engaged transmissions [1] in

which, in contrast to belt drives, contact does not take place over a large area but more or less in the form of points or lines. The area of the contact surfaces caused by flattening and the distribution of pressure can be determined by applying Hertz's law (see B4). The theory of *Stribeck's contact pressure* applies in the case of non-rigid, non-metallic materials. The transmission of moments

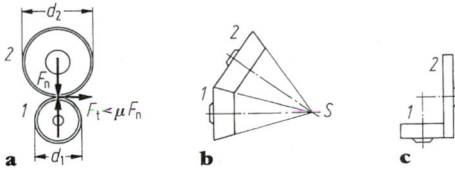

Figure 1. Forces and transmission in friction wheels: **a** with parallel axes, **b** with intersecting axes, without sliding friction, **c** with intersecting axes, with slip in the contact line.

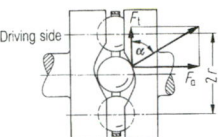

Figure 3. Device for producing a torque-dependent axial force $F_a = F_t \tan \alpha = (M/r) \tan \alpha$.

is brought about by means of peripheral forces F_t, which act between the rotationally symmetrical wheels under contact pressure F_n (**Fig. 1a**). An *effective coefficient of friction* $\mu_N = F_t/F_n$ (**Table 2**) is defined which is always smaller than the actual coefficient of friction μ. The tangential energy storage efficiency factor is therefore $v_t = \mu_N/\mu$.

The axes of rotation are usually situated in one plane in order to avoid the skewing that occurs with inclined axes. In the case of variable-speed drives, however, a certain sliding motion (see F7.3.1) must be taken into account. A purely rolling motion is only possible if the tips of both pitch cones coincide (**Fig. 1b**). The transmission is defined by the speed ratio of driving shaft (index 1) and driven shaft (index 2):

$$i = n_1/n_2 = d_2/d_1.$$

In the literature the reciprocal $i = n_2/n_1$, signed if necessary, is also to be found for the speed ratio, especially that of variable-speed drives. The speed of the driving motor, n_1 which in practice is often constant, serves as the reference quantity with the result that $(n_2 = 0)$ does not become $i = \infty$ when the driven shaft is idle.

7.2 Types, Examples

In their simplest form, friction drives consist of two rotating bodies directly attached to the driving and driven shafts. In order to reduce the high contact pressure, which in this case must be fully absorbed by the bearings, pairings with *higher coefficients of friction* (**Fig. 2**) are preferred. Special characteristics can be obtained by means of designs using connecting links, which, although having the disadvantage of the serial mounting of two contact points in the power flow, does allow the parallel mounting of a number of connecting links. In this way

the power can be increased and the bearing load reduced (e.g. planetary arrangement to reduce radial forces). In variable drives, the driving and driven shaft can then be fixed spatially, and the sliding motion can be minimised over the whole range of adjustment.

Contact pressure F_n is either produced by elastic force, in which case it is generally constant and slippage is made possible in the event of overload, or it grows as the load increases. Conditional on the principle used, the force is load-dependent (**Fig. 5b, d**) or else it is strongly influenced by torque-dependent clamping devices as shown in **Fig. 3**. In this way, the speed ratio changes only slightly as the load varies, the drive being *"torsionally stiff"*.

7.2.1 Friction Drives with Fixed Ratio

In all applications that do not require synchronous running, friction drives with fixed ratio find themselves in direct competition with positive drive types such as toothed gear drives. They are distinguished by their simple construction, which allows economic designs and may at the same time perform the task of an overload clutch. They also perform a dual function in the bearing arrangement and driving of large tubular vessels.

As the geometry of the contact area does not alter with time, there is no fear of periodical excitement of vibrations (contact shock, fluctuations due to teeth sticking), in contrast to toothed gear drives. It is therefore possible to produce drives with very low noise-levels (**Fig. 4**) and very high speeds (e.g. up to $16\,000 \text{ s}^{-1}$ in texturing machines) are also attainable at the speed increasing ratio.

7.2.2 Continuously Variable Traction Drives

The lack of positive locking in traction drives makes possible a continuous variation of their speed ratio within the limits i_{min} and i_{max}. This property is characterised by the *adjustment ratio* $\varphi = i_{max}/i_{min}$. By combining this with a planetary drive to make an adjustable coupling drive (see F8.9) the adjustment ratio can be enlarged or narrowed down as desired, so that for example it is possible to reverse the direction of rotation with every type.

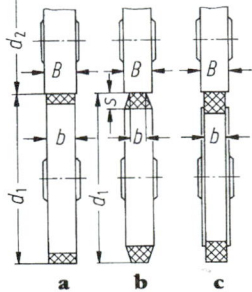

Figure 2. Friction wheels with friction coatings, where $B > b$: **a** hard organic friction coating, **b** vulcanised rubber friction ring, **c** tensioned rubber friction ring.

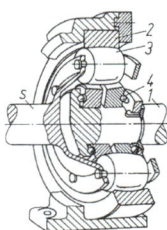

Figure 4. Planetary friction drive as in [11]: *1* drive shaft for split sun wheel, *2* fixed outer ring, *3* spherical planet wheels, *4* device for torsionally adapting the two axially movable and rotatable halves of the planet wheel to shaft *1* (cf. **Fig. 5**), *s* planet carrier as main drive pinion.

Variable-speed drives are often supplied as complete drive units with mounted asynchronous motors by which the range can be extended via pole-changing control. In most cases, drive-side step-down gear systems can be mounted which are used to obtain any desired speed range. **Figure 5** shows a selection of common operating principles. (Drives in **Fig. 5a** are dry-running with plastic friction ring, all others with lubricated steel rolling elements.) The great variety is due to the various demands that are made of friction drives, such as economic efficiency (price, operating efficiency, service life), idle adjustment, adjustment up to $n_2 = 0$, etc.

The selection of a suitable variable-speed drive for a specific application is made on the assumption that the drive must meet the torque requirement of the machine over the entire speed range. The course of the driven moment, designated as the driven centreline, over the speed n_2 is therefore an important characteristic of the variable-speed drive. At constant drive speed n_1, the characteristics of the types shown in **Fig. 5** through various ranges (**Table 1**) of the schematic driven centreline as shown in **Fig. 6** can be demonstrated. The steady torque which in many types is constant over a certain

range of adjustment *II* often may not be transmitted at extreme speed ratios (ranges *II* and *III*), as for example the permissible pressures according to Hertz's law are exceeded by smaller radii of curvature or the drilling motion results in increased wear. Moreover, the often hyperbolic torque decrease over range *III* is caused by the limited driving power.

7.3 Principles of Calculation

7.3.1 Sliding Motion

In order to calculate the relative motion in the contact area the friction wheels involved are replaced by cones tangential to the contact surface, which is assumed to be flat. In general, the tips of these pitch cones do not meet at a point in the contact plane, as shown in **Fig. 7**. The peripheral speeds are only identical at point *P*, the difference between them increasing along the surface line. This motion, superimposed on the purely rolling motion, can be described by a relative rotation at angular velocity ω_b, normal to the contact plane. Generally speaking, the rela-

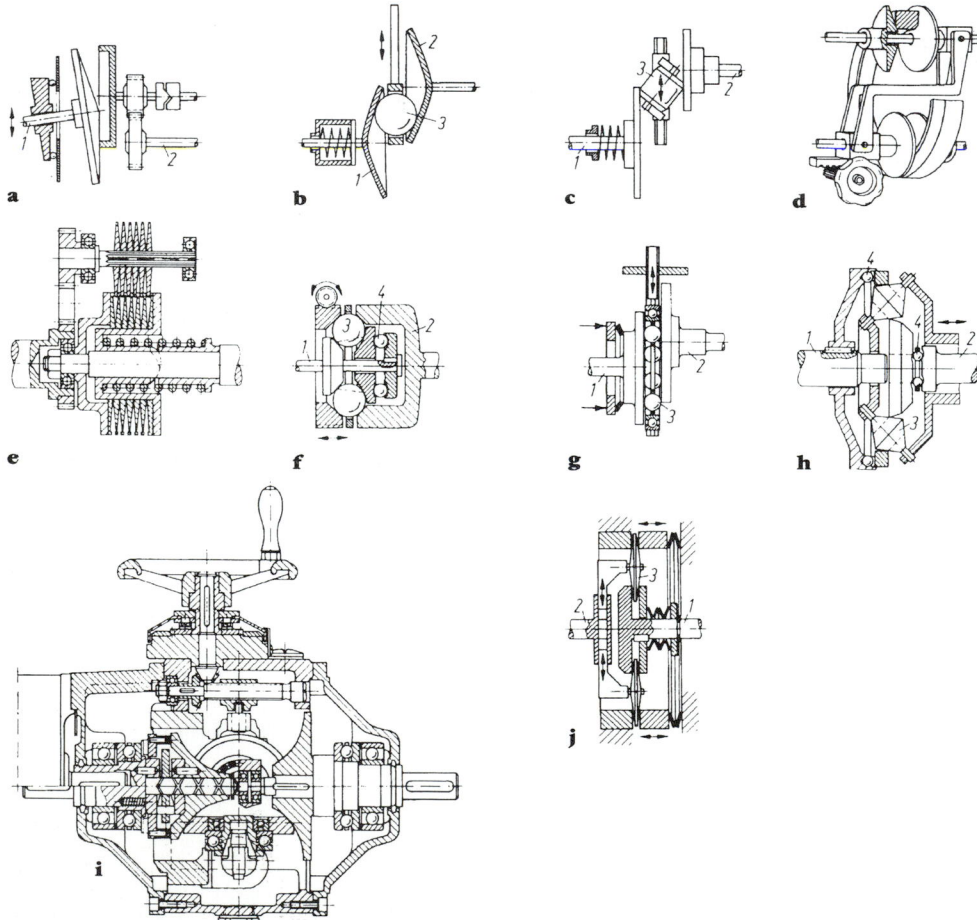

Figure 5. Schematic drawing of some traction drives (cf. Table 1): *1* drive mechanism, *2* main drive pinion, *3* connecting member, *4* device for torsionally adapting the rollers.

Table 1. Ratings of traction drives (**Fig. 5**) in accordance with manufacturers' catalogues (as at 1989). Values for largest and smallest of each type with flange-mounted drive motor, $n_1 = 24$ l/s

Fig. no.	Description (manufacturer)	$P_{2\,max}$ kW	$M_{2\,max}$ Nm	$\varphi = \dfrac{(n_2/n_1)_{max}}{(n_2/n_1)_{min}}$	$\eta_{max} = \dfrac{P_2}{P_{el}}$	Characteristic ranges
5a	Cone/friction-ring drive (SEW, Stöber, Flender-Himmelwerke)	10 0.08	75 2.4	1.25/0.25 = 5 1.1/0.22 = 5	0.9 0.7	II, III
5b	Hollow cone-ball drive (Heynau)	0.15^b 0.05	0.6 0.36	2/0.22 = 6 3/0.33 = 9	0.61 0.55	II, III
5c	Cone-disc drive (Unicum)	103 0.15	1407 3.8	0.86/0.43 = 2 2.40/0.2 = 12	0.92 0.92	II, III
5d	"H-drive" ring-sheave drive (Heynau)	3.2 0.2	43 3.0	3/0.33 = 9 3/0.33 = 9	0.79 0.79	II, III
5e	"Beier drive" cone pulley-ring drive (Sumitomo)	120^a 0.2	3440 3.2	1.3/0.33 = 4 0.8/0.2 = 4	0.8 0.8	III
5f	Ball-ring drive (Planetroll, Neuweg)	5.76 0.02	150 1.2	0/0.39 = ∞ 0/0.39 = ∞	0.77 0.7	I, II, III
5g	Ball-disc drive (PIV, Riemers)	2.36^b 0.086^c	13.4 2.0	1.2/0 = ∞ 1.2/0 = ∞	0.79 0.72	III
5h	Double cone-ring drive (Kopp)	68^d 0.8^a	1200 18	1.2/0.2 = 6 1.2/0.12 = 10	0.9 0.9	I, III
5i	Toroidal drive (Arter)	10.4 0.14	120 2	2.21/0.29 = 7.75 2.14/0.21 = 10	0.95 0.8	III
5j	"Disco" planetary sheave/ring drive (Lenze)	18.6 0.12	300 2	0.67/0.13 = 5 0.67/0.11 = 6	0.86 0.85	III

$^a n_1 = 12.5$ l/s.
$^b n_1 = 47$ l/s.
^{c}with transmission.
^{d}without drive motor.

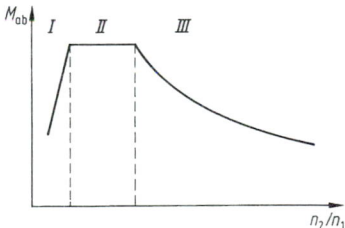

Figure 6. Schematic driven curve for traction drives as shown in Fig. 5. The existing ranges for the individual types are given in Table 1.

tive motion between rolling elements 2 and 1 is obtained using the vector equation $\omega_{rel} = \omega_2 - \omega_1$. By sectioning vertically parallel to the contact surface the sliding and rolling speeds being sought can be determined:

$$\omega_b + \omega_w = \omega_2 - \omega_1$$

with the values

$$\omega_b = |\omega_2 \sin \alpha_2 \pm \omega_1 \sin \alpha_1|$$

A plus sign is used when P is between S_1 and S_2,

$$\omega_w = |\omega_2 \cos \alpha_2 \pm \omega_1 \cos \alpha_1|$$

and a minus sign when a pitch cone is a hollow cone.

The *sliding/rolling ratio* ω_b/ω_w denotes the extent of the sliding motion and its associated losses. It is determined by the type and varies over the range of adjustment (e.g. 0 to 15 (**Fig. 5a**) and 0 to 0.5 (**Fig. 5i**)).

7.3.2 Slip Rate

The size and shape, i.e. the semiaxes a and b of Hertz's contact ellipse, are determined, among other things, by the main radii of curvature of the rolling elements at the contact point. In the planes fixed by the axes of rotation these are the radii ρ_1 and ρ_2. The plane perpendicular to them and to the contact surface produces conic sections with radii of curvature ρ_1' and ρ_2' at the contact point.

Where sliding motion is present, the peripheral speeds of the rolling elements are identical only at one point, the centre of rotation P. Consequently, its position determines the relative speed ratio. During idle running, P is at the centre M of the contact ellipse (**Fig. 7a**), by means of which the speed ratio $\omega_{02}/\omega_{01} = r_{01}/r_{02}$ is fixed. Frictional forces which, although producing a moment about P, fail to produce any resulting peripheral force for reasons of symmetry, are set up in the direction of the slipping speeds.

When moments are being transmitted and the position of the contact surface is invariable, the centre of rotation must as a consequence lie outside the centre M [2]. The integral action of the frictional forces $\mu p\,dA$ towards the circumference then produces the desired tangential force F_t. Further, a sliding moment M_b is set up about P. These sectional reactions can be combined into a resultant force F_t whose line of application passes through the hypothetical point of application of force K. This is expressed as $M_b = F_t l_N$. In order to minimise the sliding moment the contact surface should be as small as possible. Point contact is therefore preferred where sliding motion is present. The peripheral speeds of both rolling elements again coincide at P to produce the speed ratio under load

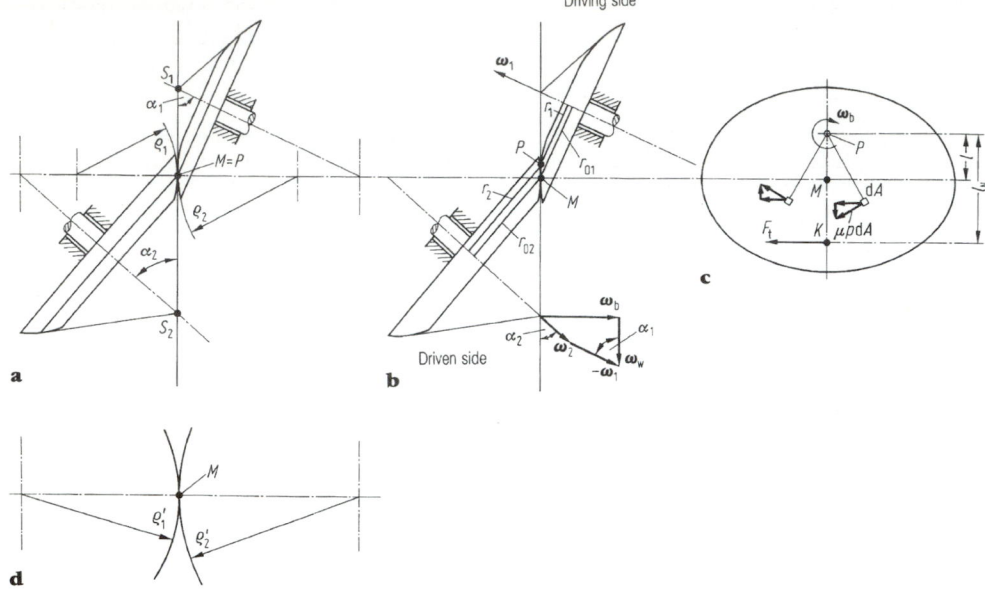

Figure 7. Roller contact with sliding motion: **a** idle running, **b** under load, **c** magnified contact ellipse with frictional forces in the direction of the running speed, displacement of the centre of rotation P by l on the occurrence of a peripheral force F_t, **d** Upturned sectional representation of **a** with main radii of curvature ρ_1' and ρ_2'.

$$\omega_2/\omega_1 = r_1/r_2.$$

The relative change in speed ratio compared with idle running is described as rolling slip s_w

$$s_w = \frac{\omega_{02}/\omega_{01} - \omega_2/\omega_1}{\omega_{02}/\omega_{01}} = 1 - \frac{r_1/r_2}{r_{01}/r_{02}}$$

$$= 1 - \frac{(r_{01} - l\sin\alpha_1)/(r_{02} + l\sin\alpha_2)}{r_{01}/r_{02}},$$

$$s_w = 1 - \frac{(r_{01} - l\sin\alpha_1)/r_{01}}{(r_{02} + l\sin\alpha_2)/r_{02}}.$$

In the event of constant contact pressure F_n and invariable coefficient of friction μ the slippage becomes greater with increasing load, i.e. increasing excursion from the centre l. Large wheel diameters and small cone angles α have a beneficial effect on slip characteristics.

Calculation methods for determining the peripheral forces to be transmitted and the length l which determines the kinematics usually assume a geometry unaffected by tangential forces and a pressure distribution in the Hertz contact surface. For the simplest case of a constant coefficient of friction, phase diagrams [2, 3] are available which show graphically the mutual dependency of the influencing variables l, l_N, a, b and v_t.

In the case of traction drives without sliding motion (e.g. **Fig. 1a, b**) the basic theory cannot be applied. A displacement of the centre of rotation continuously increasing with increasing load cannot be signalled, because it has not been defined in idle running on account of the peripheral speeds that are constant over the whole contact surface. These cases can only be worked out if a coefficient of friction is assumed which varies with running speed. Accordingly, the running speed not only determines the direction of frictional forces but also their variable quantity.

More recent theories [4] take this influence into con-

sideration, especially for the most common case of lubricated Hertz contact surfaces. The simultaneous calculation of elastic deformations and hydrodynamic processes characterises these EHD (elastohydrodynamic) contacts. The course of the pressure in the contact area resembles the Hertz pressure distribution with maximum values of some 1000 N/mm². This radically alters the properties of the lubricant in the gap. In particular, special friction wheel oils known as traction fluids [5] solidify, thereby enabling the surfaces to separate (gap width < 1 μm [6]) when the maximum permissible shear stress in the order of $\tau = 100$ N/mm² is attained at the same time. **Figure 8** shows measured friction curves for a conventional mineral oil with a suitable high naphthene content and a synthetic traction fluid at various sliding/rolling ratios.

Independently of the rolling slippage under examination here, an alteration occurs to the speed ratio caused by an alteration of the friction radii resulting from load-dependent elastic deformations. It is possible to conceive of designs in which rolling slip may even be fully compensated for in this way.

Exceptionally, the slip values s_w of finished variable-speed drives are between 1.5 and 5% above this when under nominal load.

7.3.3 Power Rating and Efficiency

Power ratings of the types of drive shown in **Fig. 5** in accordance with manufacturers' catalogues for the largest and smallest of each type are shown in **Table 1**. The *power* shown is the mechanical power P_2 available at the driven shaft and the total efficiency derived from it is calculated on the basis of absorbed electric power P_{e1}.

Besides the Hertzian stress as delimited by material strength and frictional wear, temperatures which rise as a result of inefficient heat elimination with larger size determine the power limit of traction drives.

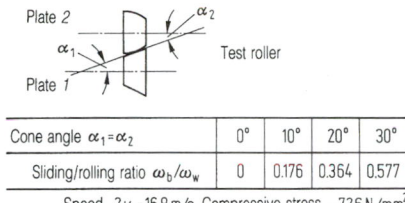

Cone angle $\alpha_1 = \alpha_2$	0°	10°	20°	30°
Sliding/rolling ratio ω_b/ω_w	0	0.176	0.364	0.577

Speed $2v_1 = 16{,}8$ m/s Compressive stress $= 726$ N/mm²

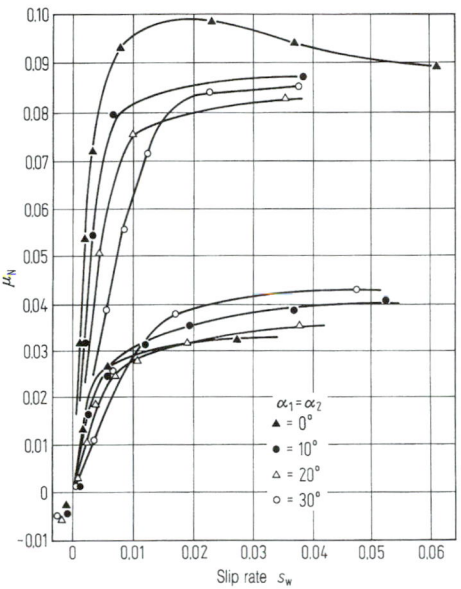

Figure 8. Friction curves as in [7] of a naphtha-based mineral oil and a synthetic friction wheel oil (higher μ_N values) at various sliding-rolling ratios.

When they are of equal weight, and thus shaft and bearing capacity are approximately equal, the power rating of traction drives is smaller by an order of magnitude than that of toothed wheel drives (**Fig. 9**) as these are capable of bearing the full normal force F_n when their contact surfaces are under an equal amount of stress, whereas friction drives can only bear μF_n as a peripheral force.

Power losses occur especially in the bearings and in frictional contact itself. Only in roller pairings without sliding motion can the frictional power be given directly. The difference between the peripheral speeds is approxi-

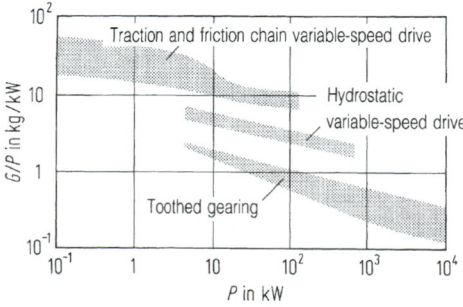

Figure 9. Power–weight ratio comparison of traction drives.

mately equal over the contact surface and has at the contact point in idle running the value

$$\Delta v = \omega_1 r_{01} - \omega_2 r_{02} = \omega_1 r_{01} (1 - \omega_2 r_{02}/\omega_1 r_{01}) = \omega_1 r_{01} s_w.$$

The frictional power is therefore $P_v = \Delta v \mu_N F_n = \omega_1 r_{01} s_w \mu F_n.$

The associated coefficients of friction and slip values μ_N and s_w are obtained, for example, from available friction curves or roughly calculated using the data provided in **Table 2**. If sliding motion is present, the frictional power can be estimated as in [8] in the following manner. First, the slip rate associated with the existing power ratio $\mu_N = F_1/F_n$ from the friction curve for sliding motion and placed in the above equation. The effective coefficient of friction for this slip rate is, however, selected from the slip curve without sliding motion. Only part of this high coefficient of friction is utilised for the transmission of peripheral force when sliding motion is present, the rest being attributed to sliding friction losses. A more accurate method of calculation is to be found, for example, in [4].

7.3.4 Combinations of Materials in Use

Table 2 shows a selection of friction wheel materials in use together with standard values for calculations. In the case of metallic materials the permissible Hertzian stress $p_{H perm}$ is given, otherwise it is the Stribeck contact pressure $k_{perm} = F_n/(bd_1)$, cf. **Fig. 2b** or $k_{perm}^* = F_n/(d_0 b)$ where $d_0 = d_1 d_2/(d_1 + d_2)$ (**Fig. 2a**). The effective coefficients of friction μ_N indicated contain a certain, common margin of safety. Details are in accordance with [8], otherwise sources are as indicated.

The demands made on friction pairings in respect of high resistance to rolling and wear with a simultaneously high coefficient of friction cannot be met simultaneously to best advantage. Because of the favourable point contact in variable-speed drives, it is almost exclusively all-steel drives that are to be found there. Friction drives with a fixed speed ratio, on the other hand, usually have line contact and can be cheaply designed with elastomer friction wheels, because any shaft and bearing loads that occur are light. Lubricants and dirt must be kept away from the running surfaces so that a high coefficient of friction can be guaranteed.

7.4 Hints on Use and Operation

Friction drives with *fixed speed ratios* are often used in precision drives to transmit low power ratings. If the wheels are removed they have the effect of a clutch (tape recorders). If provided with a flexible rubber friction coating they are particularly quiet. With hardened, precision-ground and lubricated steel friction wheels they are silent, but with fast-running, dry metallic friction pairings they are noisy.

Variable-speed friction drives are used to drive machinery and equipment whose driving speed is intended to be continuously variable (agitators, smooth-running belt conveyors) as well as to maintain a constant speed by manual adjustment or automatic control of the speed ratio. The range of adjustment should be specified as small as possible so that full use can be made of it. In this way localised wear, i.e. the formation of tracks during prolonged running, is avoided by means of a uniform speed ratio. An exception to this is the drive shown in **Fig. 5f**, as the ball tracks also vary with each rotation when the speed ratio is uniform [9]. With slow-running drives it is usually better to have a small size with an additional over-

Table 2. Characteristics of some material pairings

Pairing	Lubricant	P_{Hperm}, k^*_{perm}, k_{perm} (N/mm^2)	Effective coefficient of friction μ_N	Associated slip rate s_w (%)
Hardened steel		Point contact		
Hardened steel for sliding/rolling ratio				
$\omega_b/\omega_w = 0$	Naphtha-based friction wheel oil	$P_{\text{Hperm}} = 2500$ to 3000	0.03 to 0.05	0.5 to 2
$= 1$		$P_{\text{Hperm}} = 2000$ to 2500	0.025 to 0.045	1 to 2
$= 10$		$P_{\text{Hperm}} = 300$ to 800	0.015 to 0.03	4 to 7
$\omega_b/\omega_w = 0$	Synthetic friction wheel lubricant	$P_{\text{Hperm}} = 2500$ to 3000	0.05 to 0.08	0 to 1
$= 1$		$P_{\text{Hperm}} = 2000$ to 2500	0.04 to 0.07	1 to 3
$= 10$		$P_{\text{Hperm}} = 300$ to 800	0.02 to 0.04	3 to 5
Grey cast iron–steel GG 26–St 70	Paraffin-based friction wheel oil	Line contact $P_{\text{Hperm}} = 450$	0.02 to 0.04	1 to 3
Grey cast iron–steel GG 21–St 70 GG 18–St 50 (Crane wheels, DIN 15070)	Dry	Line contact $P_{\text{Hperm}} = 320$ to 390	0.1 to 0.15	0.5 to 1.5
Rubber friction wheels to DIN 8220	Dry	Line contact $v < 1$ m/s: $k^*_{\text{perm}} = 0.48$	0.6 to 0.8	6 to 8
Coating vulcanised onto St [12]		$v = 1$ to 30 m/s: $k^*_{\text{perm}} = 0.48/v^{0.75}$		
Coating pressed on		$v < 0.6$ m/s: $k^*_{\text{perm}} = 0.48$ $v = 0.6$ to 30 m/s: $k^*_{\text{perm}} = 0.33/v^{0.75}$	0.6 to 0.8	6 to 8
Organic friction material	Dry	Line contact $k_{\text{perm}} = 0.8$ to 1.4	0.3 to 0.6	2 to 5

drive transmission at the front and underdrive transmission at the rear than a heavy construction without an auxiliary drive, because the efficiency of friction drives increases in proportion to the speed range [10]. If only a small adjustment ratio is required for precision control, a planetary adjustable coupler drive should be used (see F8.9.6) in which the variable-speed drive has to transmit only a part of the total power and can therefore be specified small.

In most manufactured drives, contact pressure increases, either as determined by the type of construction or as a result of torque-dependent pressing devices, with increasing load. Over the range of partial loads the rolling elements can be released, thereby avoiding considerable wear due to slippage when overloaded. In order to reduce the risk of failure in the event of a heavy overload, many manufacturers supply their drives with auxiliary slip clutches.

8 Gearing

H. Winter, Munich
(Section 8.9 by **H. W. Müller**, Darmstadt)

Advantages. Transmission of motion (precision equipment) and power (up to 85 000 kW in one pairing) without backlash. Relatively compact systems. High efficiency (restrictions for worm gears and spiral gears should be respected).

Disadvantages. Rigid power transfer (flexible coupling provided if applicable), vibrations due to meshing, e.g. chatter marks in machining processes. Counter-measures: finer tooth quality, helical gearing, belt-driven stage, etc.

Gear Pairs (Fig. 1). Parallel shafts: *spur gears*, simplest to manufacture, most reliably controlled, up to highest power levels and speeds (internal gear dearer, limited manufacturing possibilities, in some circumstances "flying pinion", mainly for epicyclic drive trains). Intersecting shafts (usually less than 90°): *bevel gears*. Small pinion offset: *hypoid gears*, extreme-pressure lubricants required because of longitudinal sliding in point contact [1]. Large pinion offset (centre distance): crossed helical gears, for small forces (point contact) other than for small intersection angles. *Worm gears* for high bearing capacity (line contact) with lower transmission levels; self-locking in certain circumstances when power flow is reversed.

Noise Behaviour (see F3). *High sliding fractions are advantageous:* worm gears (up to 10 dB lower noise level attainable than for spur gear systems), hypoid gears. Noise levels for fine-quality highly-stressed spur gear systems can be decisively reduced only by going over from spur toothing to helical gearing (total contact ratio > 2.5). For low-stressed gears (precision equipment), the influence

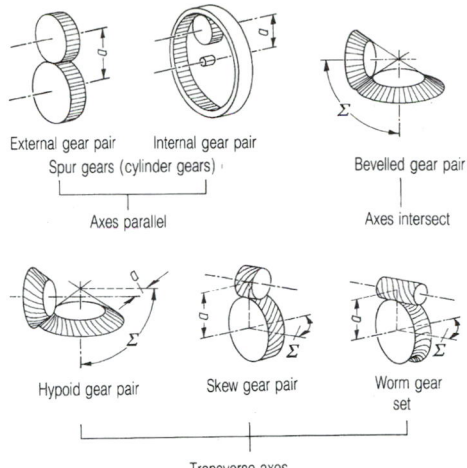

External gear pair · · · Internal gear pair

Spur gears (cylinder gears)

Axes parallel

Bevelled gear pair

Axes intersect

Hypoid gear pair · · Skew gear pair · · Worm gear set

Transverse axes

Figure 1. Gear pairs.

of tooth accuracy is predominant. For low power levels, plastic gears (metal pinions), noise reduction up to 6 dB; plastic–plastic pair, up to 12 dB as against steel–steel [20].

Efficiency, η. Under full load, including Planck's losses, storage losses and seal losses for oil lubrication: single-stage spur gear system with roller bearings approximately 98% (1% loss each shaft), for best quality (turbo gears) up to 99%, slow-running, grease-lubricated spur gear stage, cast $\eta = 93\%$, milled 95%; bevel gearing 97%; hypoid gears 85 to 96%, worm gear 30 to 96% (see F8.8.3). Coefficient of friction for oil-lubricated tooth flanks $\mu_M = 0.03$ to 0.07.

Overall efficiency $\eta = \eta_1, \eta_2$ to where η_1 is the efficiency of the first stage, etc. Efficiency considerably lower for partial load and start-up (lower temperature).

8.1 Spur and Helical Gears – Gear Tooth Geometry

A pair of gears must transmit rotary motion *uniformly* from shaft a to shaft b: $\omega_a/\omega_b = $ const. This happens when two imaginary *pitch cylinders* roll over each other (**Fig. 2**). The tooth profiles must be produced in such a way that this condition is adhered to.

8.1.1 Rule of the Common Normal

Figure 3 applies to flat gears (see F9.3.2): the circumferential speeds of the two pitch circles must be identical at

the contact point – *pitch point C*. Instead of rotating around O_1 and O_2, gear 2 (pitch circle 2) is allowed to roll over the *stationary* gear 1 (pitch circle 1). Thus every point on gear 2 – even the instantaneous contact point, Y_2 – executes a rotary movement around the corresponding *instantaneous centre of rotation* – pitch point C. To ensure that flank 2 neither rises above flank 1 nor cuts into it while this is happening, common tangents TT in $\underline{Y}$ must also be tangents to the circle with a radius $\overline{CY}$ around C; i.e., TT must be vertical to YC – for every pitch position:

The contact normal must always pass through the pitch point.

Spatial Gearing. The motion is accordingly transferred in the same way, even if the rule of the common normal is adhered to for only *one* transverse engagement position and the contact point moves around over the *width* during the rotary movement. For helical gearing with overlap ratio, Eq. (13) $\varepsilon_\beta m > 1$. Wildhaber–Novikov gearing (see F8.1.8).

8.1.2 Transmission Ratio, Gear Ratio, Torque Ratio

Transmission

$$i = \omega_a/\omega_b = n_a/n_b = r_b/r_a \quad (\textbf{Fig. 2}). \qquad (1)$$

Entire reduction ratio $i = i_1, i_2...$, where i_1 is the transmission in the first stage, etc.

Gear Ratio (for spur gears = radius ratio)

$$u = z_2/z_1 = r_2/r_1 = \omega_1/\omega_2 \quad \text{always} > 1. \qquad (2)$$

u required for calculation of replacement radii of curvature (see F8.1.7).

Transmission into low (gear 1 driving): $i = u$.

Transmission into high (gear 2 driving): $i = 1/u$.

Pitch point C therefore divides centre distance a in inverse ratio to the angular velocities; see Eq. (6). For gears with transmission that is *not constant* (e.g. elliptical gear wheels), C must change its position on the centreline O_1–O_2 as per Eq. (1).

Moment Ratio

$$i_M = M_b/M_a. \qquad (3)$$

In practice, for high-efficiency power gears $i_M = i$, but not for many clock gears (see F8.1.8).

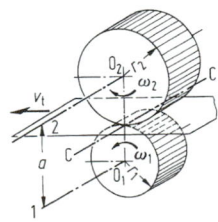

Figure 2. Pitch cylinders with joint pitch plane: *1* axis of small gear (pinion); *2* axis of large gear (wheel); pinion driving: $\omega_1 = \omega_a$, $\omega_2 = \omega_b$; wheel driving: $\omega_2 = \omega_a$, $\omega_1 = \omega_b$; CC instantaneous axis = axis of instantaneous rotation.

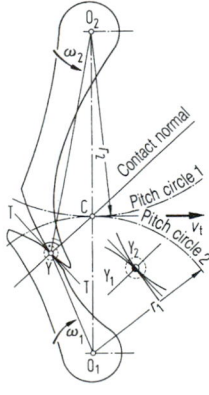

Figure 3. For basic requirement of gear tooth system.

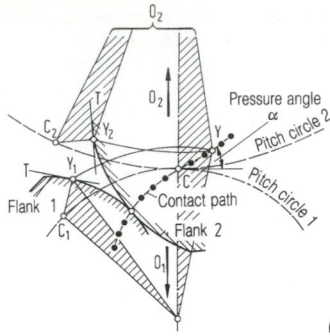

Figure 4. Point-for-point determination of path of contact and conjugate tooth profile.

8.1.3 Geometric Construction for Path of Contact and Conjugate Tooth Profile

Flank 1 and pitch circle given (**Fig. 4**). Normal at point Y_1 intersects pitch circle 1 at C_1. If gear 1 is rotated through the triangle $Y_1C_1O_1$ until C_1 coincides with C, then Y is a point on the path of contact (geometrical location of all contact points), since YC is the flank normal. Rotating the triangle YCO_2 backwards around the segment $CC_2 = CC_1$ leads Y to the point on the conjugate tooth profile, Y_2, allocated to Y_1.

8.1.4 Tooth Traces and Tooth Profiles

Tooth Traces (Fig. 5). *Spur toothing* for low circumferential speeds; advantage: no axial forces, simple manufacture, suitable for sliding gears; disadvantage: runs less quietly. *Helical toothing* for higher bearing capacity and circumferential speed due to more uniform transmission under load, runs quietly; disadvantage: axial forces. *Double helical toothing* makes it possible to neutralise the axial forces. Disadvantage: gap for tool to run out, load distribution not always reliable, axial vibrations in certain circumstances. *Note*: transverse rolling motion and sliding motion take place even with helical toothing.

Single Toothing. Simple tooth profile of one gear preset. Profile of mating gear to be designed as per F8.1.3, or given profile to be reproduced by tool in hobbing operation [1].

Toothing Pair. Generation of toothing by hobbing a common *crown toothing reference profile*: for spur toothing, this is for a flat plate – i.e. a rack (e.g. **Fig. 10**); for bevel toothing for a flat gear – a crown gear. Reference profile and counter-profile are *not* identical, two tools are required [1].

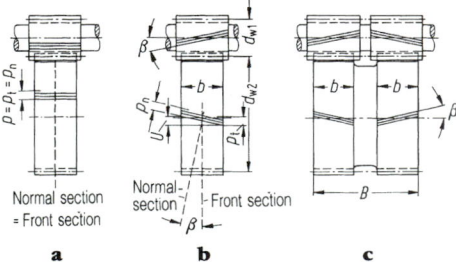

Figure 5. Spur gears: **a** spur gear, **b** helical gear, **c** double helical gear.

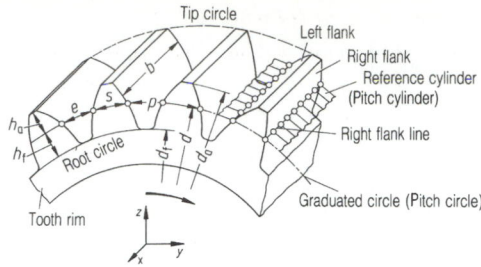

Figure 6. Description and dimensions of spur gearing.

Toothing for Change Gears. Profile and counter-profile (rack tool for gear and mating gear) of crown toothing are *identical* here, so that only *one* tool is needed to manufacture all the gears, which can also all mesh with one another, if the profile centreline = the rolling curve during manufacture. Involute change gears [4].

8.1.5 General Relationships for All Tooth Profiles

Figures 6 and **7**. The equations also apply to helical gears (henceforth referred to as *//Hel. . .//*). Transverse values (**Fig. 5**) are denoted by the index t and normal values by n. For spur toothing, the indices t and n can be dispensed with. For specifications for internal gears, see F8.1.7.

Pitch, p. Distance between two adjacent flanks on pitch circle. If p is determined by standardised module $m = p/\pi$, the relevant circle will be described as a *graduated circle*. (For involute gear teeth, graduated circle, if applicable, $\neq$ pitch circle.)

$$p = \pi d/z = \pi m,$$
$$//Hel.\colon p_n = p_t \cos\beta = \pi m_n; \; p_t = \pi m_t//. \tag{4}$$

Pitches of pinion and gear must coincide.

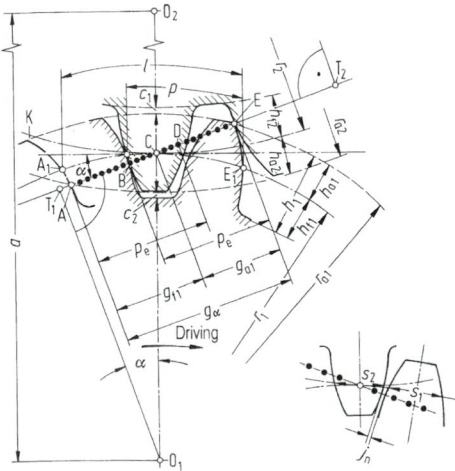

Figure 7. Gearing dimensions of spur gear pair (involute gear): B single internal contact points, the leading pair of teeth just out of contact (point E); D single external contact point: subsequent pair of teeth just in contact; B the single external contact point for gear 2.

Graduated Circle Diameter

$$d_1 = 2r_1 = z_1 p/\pi = z_1 m, \; d_2 = 2r_2 = z_2 p/\pi = z_2 m,$$
$$//Hel.\!: d_1 = z_1 p_t/\pi = z_1 m_t, \; d_2 = z_2 p_t/\pi = z_2 m_t //. \tag{5}$$

Centre Distance (Fig. 2)

$$a = r_1 + r_2 = m(z_1 + z_2)/2 = mz_1\,(1 + u)/2$$
$$//Hel.\!: \text{with } m = m_t //. \tag{6}$$

For involute gear teeth, see Eq. (30, 33).

For internal gears z_2, d_2, a is negative (see F8.1.7).

Modulus, m. Important standard values. Addenda and dedenda are usually selected in relation to m. To limit the number of tools, m_n should be selected from standard range. **Table 1** $//Hel.\!: m_t = m_n/\cos\beta//$. (In the UK and USA, diametral pitch is usually $P_d = z/d$. With d in inches: m in mm $= 25.4/P_d$.)

Depths of Tooth. Addendum h_a (normal $= m$), dedendum, h_f

$$(\text{normal} = 1.1\,m \text{ to } 1.3m). \; //Hel.\!: \text{with } m = m_n //. \tag{7}$$

Depth of tooth $h = h_a + h_f$.

Working depth of teeth $h_w = h_{a1} + h_{a2}$.

Tip Diameter

$$d_a = d + 2h_a = 2d - d_{f\,\text{meeting wheel}} - 2c. \tag{8}$$

Root Diameter

$$d_f = d - 2h_f. \tag{9}$$

Tip Clearance, c. Distance between addendum circle and dedendum circle of mating gear (normal $= 0.1m$ to $0.3m$). $//Hel.\!:$ where $m = m_n //$,

$$c_1 = h_1 - h_w = a - (d_{a1} + d_{f2})/2,$$
$$c_2 = h_2 - h_w = a - (d_{a2} + d_{f1})/2. \tag{10}$$

Tooth Thickness in graduated circle

$$s = p - e \quad \text{for space width } e \tag{11}$$

s_1 and s_2 are made smaller than the nominal dimension by the *tooth thickness margin*, A_s. This gives rotary backlash

$$j_c = p - s_1 - s_2, \quad normal \; backlash \tag{12}$$

$j_n = j_t \cos\alpha$; shortest distance between non-working flanks; required in order to avoid jamming from heating, from swelling (plastics!), or as a result of manufacturing tolerances. $//Hel.\!: j_n = j_t \cos\alpha_n \cdot \cos\beta//$. Reference values for A_s as per **Table 4**.

Transverse Path of Contact, g_α. For contact of utilised part of line of action. Normally bounded by addendum circle, or earlier with undercuts (**Figs 7, 11**).

Angle of Action, I. Travel from start to end of meshing, A_1 to E_1, on pitch circle (**Fig. 7**).

Transverse Contact Ratio, ε_α. Ratio of angle of action to pitch. For uniform transmission of motion with spur toothing, $\varepsilon_\alpha = l/p > 1$ required; usually 1.1 to 1.25 needed (even for helical gearing). For $\varepsilon\alpha$ in involute gears, see F8.1.7.

Pressure Angle, α. Angle between tangents to pitch circle at C and corresponding contact normals YC (**Figs 4** and **7**); for α in involute gears, see F8.1.7. $//Hel.\!:$ tan $\alpha_t = \tan\alpha_n/\cos\beta//$.

Contact profile, active profile (**Fig. 7**). The part of tooth profile AK used for contact.

Additional Variables for Helical Gears

Spread (for Helical Gears), U. Distance between end points of a tooth brace over the width, measured on the graduated circle arc. $U = b \tan\beta$ (**Fig. 8**).

Tooth Alignment. Right-hand: β positive; left-hand: β negative. For external toothing, tooth alignments of pinion and gear must be *opposed*; for internal gears they must be *the same*.

Overlap Ratio $\varepsilon_\beta = U/p_t = b \sin\beta/(m_n\pi)$. $\tag{13}$

Uniform transmission of motion possible, even for small depths of tooth (limiting case zero), if $\varepsilon_\beta > 1$.

Total Contact Ratio $\varepsilon_\gamma = \varepsilon_\alpha + \varepsilon_\beta$. $\tag{14}$

8.1.6 Sliding and Rolling Motion

According to law of motion (see A21.2), absolute velocity in direction of contact tangents, TT (**Fig. 9**)

$$w_a = \omega_a \rho_a = (v_t/r_a)\,(r_a \sin\alpha \mp g_y)$$
$$= v_t\,(\sin\alpha \mp g_y/r_a),$$
$$w_b = \omega_b \rho_b = (v_t/r_b)\,(r_b \sin\alpha \pm g_y)$$
$$= v_t\,(\sin\alpha \pm g_y/r_b). \tag{15}$$

Upper sign for contact point on dedendum flank $\overline{a}$ or tip $\overline{b}$, lower sign on addendum flank $\overline{a}$ or root $\overline{b}$ of the tooth.

$+$ at tip ($\overline{a}$ or $\overline{b}$); $-$ at root ($\overline{a}$ or $\overline{b}$).

Cumulative Speed. This is important for lubrication pressure (see F8.3):

$$v_\Sigma = w_a + w_b = v_t\,[2\sin\alpha \mp g_y\,(1/r_a + 1/r_b)] \tag{16}$$
$$= v_t\,[2\sin\alpha \mp g_y\,(1 + 1/i)/r_a]$$

Minus sign at root $\overline{a}$ or tip $\overline{b}$; plus sign at root $\overline{b}$ or tip $\overline{a}$.

Cumulative Factor

$$K_\Sigma = v_\Sigma/v_t = [2\sin\alpha \mp g_y\,(1 + 1/i)/r_a]. \tag{16a}$$

Table 1. Module series (DIN 780 and ISO standard 54-1977). Without signs: distortion range I, with signs ⟩⟨: range II

Module m (mm)

1	⟩1.75⟨	⟩3.5⟨	⟩7⟨	⟩14⟨	25	⟩45⟨
⟩1.125⟨	2	4	8	16	⟩28⟨	50
1.25	⟩2.25⟨	⟩4.5⟨	⟩9⟨	⟩18⟨	32	
⟩1.375⟨	2.5	5	10	20	⟩36⟨	
1.5	⟩2.75⟨	⟩5.5⟨	⟩11⟨	⟩22⟨	40	
	3	6	12			

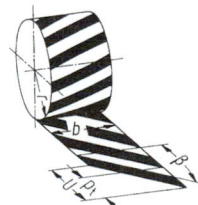

Figure 8. Overlap length, U, and helix angle, β, on a helical gear (DIN 3960).

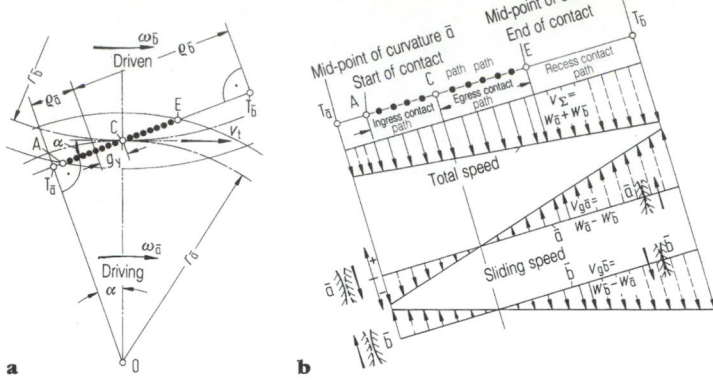

Figure 9. Speeds on tooth flanks. **a** Dimensions for calculation: index a : driving; b: **b** speeds of flank contact points during engagement.

Sliding Speed. This is important for heating, seizing stress (see F8.5.1):

$$v_{ga} = w_a - w_b, \quad v_{gb} = w_b - w_a = -v_{ga},$$
$$v_g = \mp v_t g_y (1/r_a + 1/r_b). \tag{17}$$

Sliding Factor, K_g

$$K_g = v_g/v_t = \mp g_y (1/r_a + 1/r_b) = \mp g_y (1 + 1/i)/r_a. \tag{18}$$

Minus sign at root $\bar{a}$ or $\bar{b}$, plus sign at tip $\bar{a}$ or $\bar{b}$. The sign indicates the direction of the friction force (**Fig. 9b**).

8.1.7 Involute Teeth

Used almost exclusively in machine-building. Simple precise manufacture is possible in small cut processes (straight-flanked reference profile; **Fig. 10**). Change gear characteristics; uniform transmission of motion is possible with the same tool even with centre distance variations, different tooth forms and centre distances, by means of profile offset. The direction and size of the tooth normal

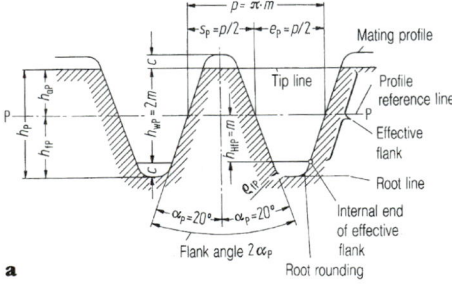

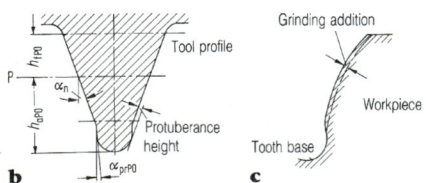

Figure 10. Basic profile of involute gear: **a** basic rack as per DIN 867; **b** protuberance tool as per [49], $\alpha_{prPO} \approx (0.3$ to $0.6)\alpha_n$ (addendum, b_{aPO} of basic tool profile corresponds to dedendum, b_{fP} of basic gear profile): **c** tooth flank generated by **b**.

force (bearing force) is constant during contact (see K5.2).

Characteristics of Involute Gears. The path of contact is a straight line with an angle of pressure α; the actual profiles of tooth flanks are circular involutes, the tooth flanks of crown gears (racks) being straight, while those of external geared wheels are convex and those of internal geared wheels concave.

Circular involutes are described by the points on a straight line, the "generatrix", which rolls over a circle, the "base circle".

The straight-flanked *reference profile* is standardised for machine manufacture in DIN 867 (**Fig. 10a**); for corresponding tool profiles I and II for finish machining, together with III and IV for the pre-machining of gears, see DIN 3972. Suitable and properly proportioned gears are available for most applications in this area. For a reference profile for light engineering, see DIN 58 400.

Special Cases. The protuberance profile (**Fig. 10b**) cuts the tooth root cleanly, in order to avoid notches due to gear grinding. A greater depth of tooth ($b_w \approx 2.5m$ instead of $2m$) is used for particularly quiet-running gears (with high gearing, beware risk of seizing!). Angle of pressure $15°$ with adjustable centre distances (higher transverse contact ratio).

ISO and abroad. Standards: ISO 53; AGMA 201.02, 207.06; BS 436.

Involute Function. For the calculation of numerous variables of involute gears, e.g. the tooth thickness at a particular point, it is advantageous to use the involute function, "inv α" (referred to in speech as "involute α"), which can be obtained from tables as a function of α.

$$\text{inv } \alpha = \tan \alpha - \hat{\alpha}. \tag{19}$$

Gear Variables for Involute Gears. The general relationships in F8.1.5 apply. For further dimensions, see **Fig. 7**.

Base Circle:

$$r_{b1} = r_1 \cos \alpha, \ r_{b2} = ur_{b1},$$
$$//\text{Hel.:} \ r_b = r \cos \alpha_t //. \tag{20}$$

Base Pitch: $p_a = p \cos \alpha = p_b$.

Base Circle Pitch: //Hel.: transverse base pitch, $p_{at} = p_t \cos \alpha_t$.

Normal Pressure Distribution:

$$p_{en} = p_n \cos \alpha_n //. \tag{21}$$

Radii of Curvature. //*Hel.*: transverse// as per **Figs 7** and **9a**:

$$\left.\begin{aligned}
\rho_{C1} &= \overline{T_1C} = 0.5d_{b1}\tan\alpha_w = 0.5d_1\sin\alpha_2, \\
\rho_{C2} &= \overline{CT_2} = u\rho_{C1},\ \rho_{A2} = \overline{AT_2} = 0.5(d_{a2}^2 - d_{b2}^2)^{\frac{1}{2}}, \\
\rho_{E1} &= 0.5(d_{a1}^2 - d_{b2}^2)^{\frac{1}{2}},\ \rho_{B1} = \overline{T_1B} = \rho_{E1} - p_{et}, \\
\rho_{B2} &= \overline{BT_2} = \alpha\sin\alpha_w - \rho_{B1}
\end{aligned}\right\} \quad (22)$$

where $d_b = 2r_b, d_a$ (**Fig. 6**), α_w = the effective pressure angle, //*Hel.*: $\alpha_w = \alpha_{wt}$//. (For *internal* gears, ρ with index 2 is *negative*!)

Contact Path:

$$g_\alpha = g_f + g_a$$

with root contact path 1:

$$g_f = \overline{AC} = \rho_{A2} - \rho_{C2}$$

and tip contact path 1:

$$\left.\begin{aligned}
g_a &= \text{CE} = \rho_{E1} - \rho_{C1}, \\
g_\alpha &= 0.5d_{b1}\,[(d_{a1}/d_{b1})^2 - 1]^{\frac{1}{2}} + \\
&\quad u[(d_{a2}/d_{b2})^2 - 1]^{\frac{1}{2}} - \tan\alpha_w\,[u + 1]),
\end{aligned}\right\} \quad (23)$$

//*Hel.*: $\alpha_w = \alpha_{wt}$//.

Transverse Contact Ratio:

$$\varepsilon_\alpha = g_\alpha/p_e,\ //Hel.: \varepsilon_\alpha = g_\alpha/p_{et}//. \quad (24)$$

Tooth Thickness at radius r_y (transverse value)

$$\left.\begin{aligned}
s_y &= 2r_y\,(s/2r + \text{inv}\,\alpha - \text{inv}\,\alpha_y) \\[4pt]
&\text{with } \alpha_y \text{ from } \cos\alpha_y = r_b/r_y = r\cos\alpha/r_y
\end{aligned}\right\} \quad (25)$$

given s and α at radius r.
At tip:

$$s_{an} > 0.2m_n \quad \textbf{(Figs 13 and 14)}.$$

Shaft Pitch a_y from tooth thickness with zero-play contact (transverse value):

$$\left.\begin{aligned}
a_y &= a\cos\alpha/\cos\alpha_y \\[4pt]
&\text{with } a \text{ as in Eq. (6) and } \alpha \text{ from} \\[4pt]
\text{inv}\,\alpha_y &= \text{inv}\,\alpha + [z_1(s_1 + s_2) - 2\pi r_1]/[2r_1\,[z_1 + z_2)]
\end{aligned}\right\} \quad (26)$$

with s_1 at radius r_1, s_2 and r_2 (Eq. (27)). α with r_1 and r_2.

Undercut (**Fig. 11**). For small numbers of teeth, the addendum flank of the rack undercuts the tooth root of the gear when the point of intersection H lies below T_1.

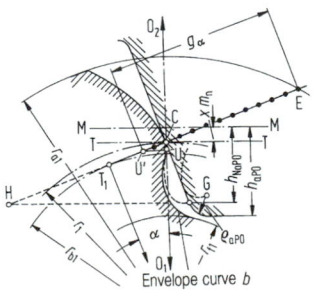

Figure 11. Undercut: contact cannot begin before U'; for the remaining transverse path of contact g_α is as per [1].

The path of the rounded rack tip (relative tooth crest track) cuts U-shaped involutes while rolling; corresponding point on path of contact: U'.

Undercutting can reduce the contact ratio (**Fig. 11**; "harmful" undercut) and weakens the tooth root. The maximum number of teeth follows from the condition that H coincides with T_1.

$$z_G = 2\cos\beta\,(b_{NaPO} - xm_n)/(m_n\sin^2\alpha_t),$$

where

$$b_{NaPO} = b_{aPO} - \rho_{aPO}(1 - \sin\alpha_n);\ \text{see } \textbf{Fig. 11}.$$

Moving the tool off (positive profile offset x), reducing b_{NaPO} or using a helical gear can therefore avoid undercutting, i.e. reduce the maximum number of teeth (**Figs 13** and **14**).

Gearing with Modified Profile (normal case of involute gears). During manufacture, tool reference profile backed off from graduated circle (radius r) by an amount xm (profile offset = $+ xm$) or moved forward ($- xm$) and rolled at this dimension. Base circle radii $r_b = r\cos\alpha$ remain unaltered. Thus undercutting can be avoided, larger radii of curvature, thicker tooth roots, and the maintenance of specific centre distances are possible with standardised modulus.

Contact ratio usually lower, radial force greater as a result of larger effective pressure angle. Only small change in tooth form with large numbers of teeth.

Dimensions of Gears with Modified Profiles

Tooth Thicknesses at geometrical radius, r:

$$s = m'\pi'2 + 2x\tan\alpha) + A_s$$

with (negative) tooth thickness margin, A_s; reference values for A_s, **Table 4** (see F8.2);

$$//Hel.: s_n = s_t\cos\beta = m_n(\pi/2 + 2x\tan\alpha_n) + A_{sn}//. \quad (27)$$

Root Diameter:

$$\left.\begin{aligned}
d_f &= d + 2xm - 2b_{fp}, \\
&//Hel.: \text{with } m = m_n//.
\end{aligned}\right\} \quad (28)$$

Tip Diameter:

$$\begin{aligned}
d_a &= 2a - d_{fmeeting\ wheel} - 2c \\
&= d + 2xm + 2b_{aP} + 2km, \quad (29)
\end{aligned}$$

//*Hel.*: with $m = m_n$//, b_{fP}, b_{aP}, c (**Fig. 10a**).

km, addendum height alteration (= telescoping, **Fig. 12**), Eq. (32), to maintain the negative values of the tip clearance for external gear pairs (positive for internal gear pairs, usually set to zero).

Centre Distance:

$$\begin{aligned}
a &= 0.5m(z_1 + z_2)\cos\alpha/\cos\alpha_w \\
&= a_d\cos\alpha/\cos\alpha_w, \quad (30)
\end{aligned}$$

//*Hel.*: with $m = m_t = m_n/\cos\beta$; $\alpha = \alpha_t$; $\alpha_w = \alpha_{wt}$//, a_d

Centre Distance of Zero Gear. Manufacturing tolerance ($\pm$ centre distance variation, $A_m = A_{m1} + A_{m2}$) increases or reduces backlash. For reference values for A_{m1}, A_{m2}, see **Table 4** (see F8.2).

Effective Pressure Angle α_w. This is made up of

$$\text{inv}\,\alpha_w = \text{inv}\,\alpha + 2\tan\alpha(x_1 + x_2)/(z_1 + z_2), \quad (31)$$

//*Hel.*: inv $\alpha_{wt} = $ inv $\alpha_t + 2\tan\alpha_n(x_1 + x_2)/(z_1 + z_2)$//.

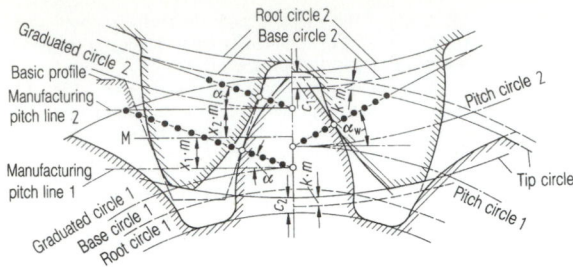

Figure 12. Gearing with modified profile (staggered teeth). *Left*: gearing for gear and mating gear with joint basic profile (note: *no flank contact!*). *Right*: operating position of gearing after pushing together and alteration of addendum, $k \cdot m$ (*note*: no joint basic generating profile).

Addendum Height Alteration:

$$km_n = a - a_d - m_n(x_1 + x_2), \qquad (32)$$

where a_d (centre distance of zero gear) is as per Eq. (33). For reference profile as per DIN 687: $\alpha = 20°$, $\cos \alpha = 0.940$, $\tan \alpha = 0.364$, $\text{inv } \alpha = 0.0149$.

Zero Gear:

$$x_1 = x_2 = 0, \alpha_w = \alpha,$$
$$a = a_d = 0.5m(z_1 + z_2), \qquad (33)$$
$$//\text{Hel.}: \alpha_{wt} = \alpha_t//.$$

V Zero Gear:

$$x_1 = -x_2, \alpha_w = \alpha, a = a_d.$$

For elimination of undercutting and strengthening of pinion at expense of gear when $u \neq 1$.

V-Gears:

$$x_1 + x_2 \neq 0.$$

Many suitable profile offset systems [4, 5].

Recommendations

Flexible rule, which leads to balanced gear systems, in DIN 3992. Selection from $(x_1 + x_2)$ as per **Fig. 13a**; a as per Eqs (31) and (30), round up if necessary and determine accompanying $(x_1 + x_2)$ as per Eqs (30) and (31) or from given a; distribute $(x_1 + x_2)$ as per pairing line (**Fig. 13b** or c).
  Simple rule: 05 gear as per DIN 3994 and 3995, $x_1 = x_2 = + 0.5$; change gears; $a = F(z_1 + z_2)m$; number F established through $(z_1 + z_2)$ and β; F and important gear data can be obtained from DIN 3995.

Additional Specifications for Involute Helical Gears. Here too, the lines of contact are straight lines, but run obliquely over the tooth flanks and wander over the tooth width during contact. The *profile offset* is given in multiples of the *normal plane pitch*; for selection of profile offset, see F8.1.7, corresponding virtual number of teeth z_{nx} as per Eq. (34).
  The normal tooth form of an involute spur gear with a *virtual number of teeth*, z_{nx}, is similar:

$$z_{nx} = z/(\cos^2 \beta_b \cos \beta) \approx z/\cos^3 \beta, \qquad (34)$$

and is used in the selection of profile offsets, for determining the geometrical limits (e.g. tip thickness) and for calculating the strength.

Back Helix Angle β_b

$$\left. \begin{array}{l} \text{from } \tan \beta_b = \tan \beta \cos \alpha_t, \\ \text{or } \sin \beta_b = \sin \beta \cos \alpha_n. \end{array} \right\} \qquad (35)$$

For special gears with pinion tooth numbers 1 to 4, see [6].

Additional Specifications for Involute Internal Gears. All the gear geometry equations can be used unaltered if the tooth number of the internal geared wheel, z_2, is made *negative*. All calculated values of the diameter thus become negative, as do the gear ratio and the centre distance of an internal gear pair. (But the absolute values are to be shown in the drawings!) Profile offset towards the tip - so with internal gears *inwards* is described as positive. Only the diameter of the root circle can be obtained from the generating tool:

$$d_{f2} = 2a_o - d_{mo},$$

where a_o = the centre distance during the cutting of the teeth and d_{ao} = the gear shaper cutter tip diameter.

Selection of Profile Offset. Favourable: gear pair with reference centre distance, $x = \pm 0.5$ to 0.65. For $z_2 < -40$ (extreme -26), $z_1 \geq 14$ (extreme 12) and $z_1 + z_2 \leq -10$, respect conditions for manufacture and mounting (radial assembly). For other gear pairs with reference centre distance, see DIN 3993. Staggered teeth do not essentially increase bearing capacity, but allow for greater freedom in configuration, and of course make it necessary to check meshing interference, tip thicknesses and space widths (**Fig. 14**). For planet gears, select planet tooth number, z_P, 0.5 to 1.5 smaller than results for zero gear from z_z (sun gear) and z_H (internal geared wheel). x_z and x_P are determined using Eqs (30) and (31) and distributed, for example as per **Fig. 13**; the aim should be to obtain $x_P + x_H \leq 0$. For direction of pitch with helical gear, see F8.1.5. Full account of geometrical relationships: DIN 3993 [8–10].

8.1.8 Other Tooth Profiles (Besides Involute) and Gears for Non-uniform Transmission

Cycloid Gears. Flank forms originate from rolling of two pitch circles on the circles of contact. Scarcely used now except for vane-type pumps, since they are difficult to manufacture accurately (individual hob for each number of teeth), sensitive to centre distance variations and not true to moment (see F8.1.2).

Circarc Gears. Higher standards in precision equipment manufacture (high degree of transmission, constant moment ratio - Eq. (3), small bearing forces, hardly any lubrication, large amount of backlash, i.e. no jamming due to dirt; though not conformal transmission, high contact ratio), can be attained by using circarc gears and involute gears (DIN 58405) than by using cycloids. Since not conformal, transverse contact ratio always 1.
  Teeth thickness on pinion (generally steel) usually smaller than on wheel (generally brass or plastic). Bending stress at root of gear generally decisive for bearing capacity. Manufacture from metals usually by means of hobbing, sometimes copy-milling; from plastics by means of injection moulding [11].

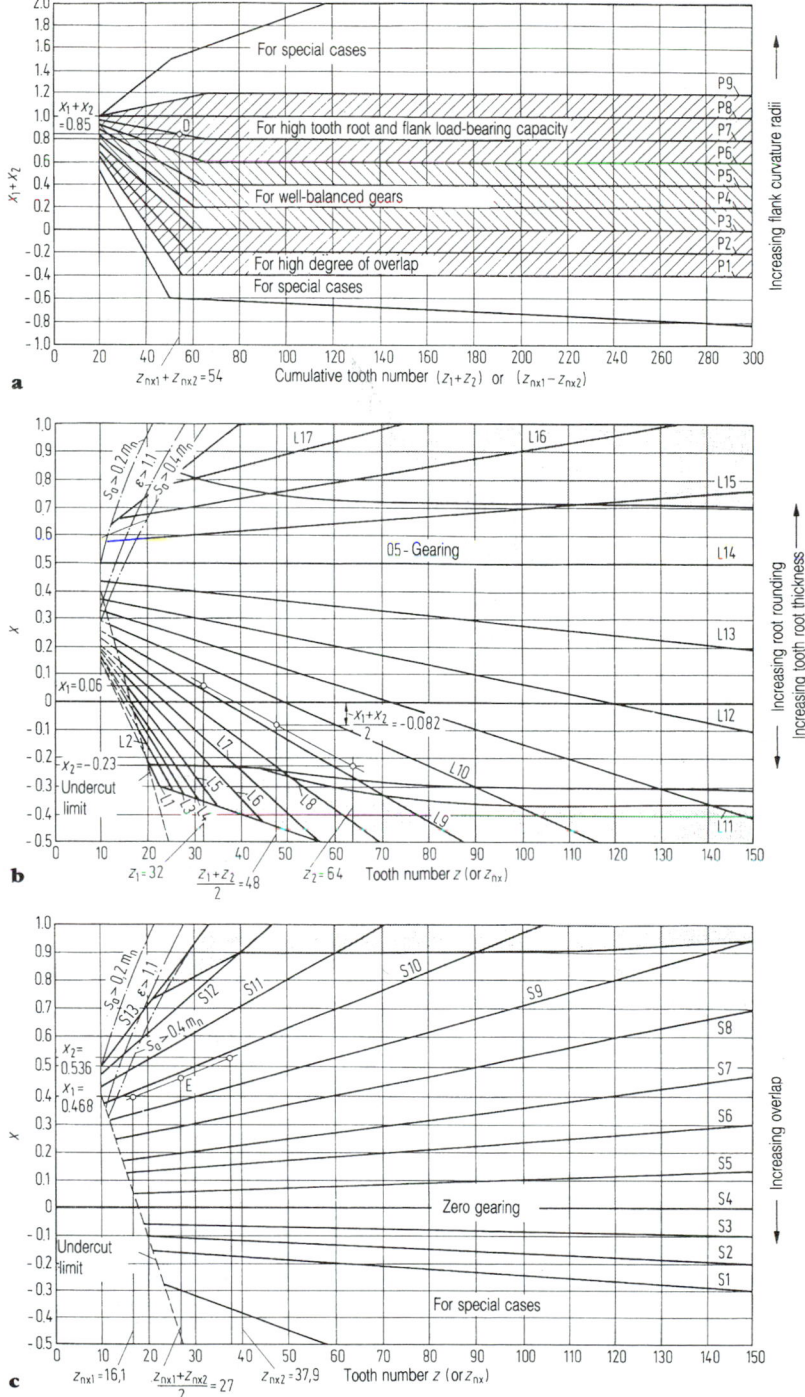

Figure 13. Profile offset selection (DIN 3992). **a** Recommendations for profile offset factors sum. **b** and **c** Recommendations for total profile offset distribution: **b** transmission into slow, **c** transmission into rapid. For example, see text. Grey area: danger of contact interference. For distribution of $x_1 + x_2$ with $z_2 > 150$, $z_2 = 150$ can be used.

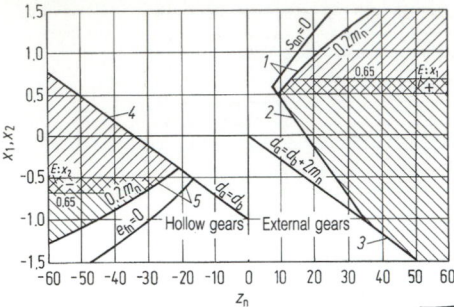

Figure 14. Practicable profile offsets for internal gear pairs with basic profile (DIN 867). E: x_1, E: x_2: recommended range for zero gear pairs with reference centre distances. Limits: 1 through minimum pinion tooth tip thickness (see Fig. 13), 2 through undercut, 3 and 4 through minimum tip diameter, 5 through minimum tooth root space width of hollow gear (DIN 3960).

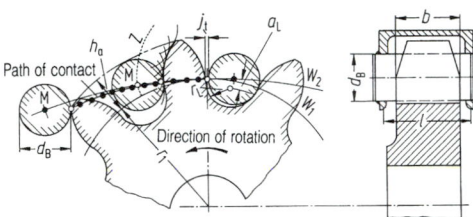

Figure 15. Mangle gear. Design for path of contact and tooth flank, dimensions.

Mangle Gears. Used for heavy-duty turntables with large diameters, rack-and-pinion jacks (**Fig. 15**). For rolling of W_2 on W_1, M describes curve Z; equidistant with bolt radius gives pinion flank.

Reference Values. Lowest pinion tooth number min. $z_1 \approx 8$ to 12 for circumferential speed $v_t = 0.2$ to 1.0 m/s; pin diameter $d_s \approx 1.7m$; addendum, $b_m \approx m(1 + 0.03z_1)$; tooth width $b \approx 3.3m$, average bearing length of pin $l \approx b + m + 5$ mm; space radius, $r_1 \approx 0.5d_B$; distance, $a_L \approx 0.15m$; backlash, $j_t \approx 0.04m$. *Bearing capacity* according to practical experience: **Table 2**.

Wildhaber–Novikov (W-N) Gears

Tooth Forms. In their basic form, pinion flank consists of convex arc and gear-tooth flank of concave arc, with radius $p_1 = p_2$ around pitch point C, **Fig. 16**. Contact on entire arc only in this meshing position, i.e. no transverse contact ratio present. Uniform transmission of motion possible only using helical gear with intermittent contact ratio $\varepsilon_\beta > 1$. In order to avoid bearing edges at tip or root due to centre distance variations, p_2 is made rather larger

Table 2. Guide data for mangle gear of crane slewing gears with pinions made of St 70 and bolts made of St 60 for heavy duty operation [51]

Peripheral force	(kN)	20	30	40
Pinion tooth number z_1	—	9	9	9
Module m	(mm)	21	25	30
Tooth width b	(mm)	80	90	110
Bolt diameter d_B	(mm)	35	45	50

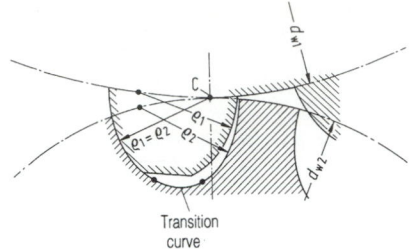

Figure 16. W-N gearing. Pinion flank convex, gear flank concave (left: basic form; right: practical execution, $p_2 > p_1$.

than p_1 – point contact. In rotary transmission the contact point wanders over the tooth width.

Individual tools (for each modulus and helix angle) required for pinion and gear for gearing with convex tip profile and concave root profile [1, 12, 13].

Bearing Capacity. Hertzian ellipticity surface is spherical surface. Corresponding expansion is greater than that in lateral sense, owing to close fitting. Pressure areas wandering over tooth width favourable for formation of lubrication pressure; friction effect small. Transverse sliding speed same for all flank contact points. Wear consequently uniform (favourable for running-in lapping).

Flank Bearing Capacity (from comparison of Hertzian compression values), torque approximately two to three times as high as for involute gears.

Tooth Root Bearing Capacity about the same as for involute gears. Point application of force entails risk of corner breakages at $\varepsilon_\beta \approx 1$ and breaks in the tooth centre (single tooth contact) at $\varepsilon_\beta > 1.2$.

Operating Performance. Favourable noise and oscillation behaviour with precise, rigid construction. Pitch and flank line variations lead to jerking when meshing begins. Under certain circumstances, centre distance variations (to which deformation can also contribute) bring about considerable displacement of the meshing at the tip or root, i.e. an increase in the flank and root stress, together with greater running noise.

Eccentric Gears. [38–42].

Non-circular Gears [53–57].

8.2 Tooth Errors and Tolerances, Backlash

Tooth accuracy to be specified in terms of grade as per DIN 3961 to 3967! Grade 1: highest precision, grade 12 lowest. Examples: master gears Q2 to 4; marine gears and turbo-gears Q4 to 6; heavy machine construction, Q6 to 7; smaller industrial gears, crane control gears and belt gears, Q6 to 8; slow, open gears, Q10 to 12; slewing rings Q9 (cast > Q12). For large tooth widths, additional specification of a contact pattern required (no interchangeable manufacture!). Flank line corrections or profile corrections if necessary, i.e. known variations for balancing deformations effective [1].

Tolerances for *individual errors* (profile, pitch, concentricity, flank lines): DIN 3962; for *composite errors* (tangential composite error and radial composite error): DIN 3963. Checking tangential composite error or radial composite error is frequently sufficient for acceptance of

Table 3. Estimating flank lines angular deviation, $f_{H\beta}$: for precise values, see DIN 3961: $f_{H\beta} = H_\varphi \cdot 4.16 b^{0.14}$; tables: DIN 3962

Quality grade	3	4	5	6	7	8	9	10	11	12
Factor H_φ		0.57	0.76	1		1.32	1.85	2.59	4.01	6.22 9.63 14.9

gears. For tolerances for *centre distances*: DIN 3964; for tooth thicknesses: DIN 3967. For $f_{H\beta}$, see **Table 3**.

For precision levels attainable using various manufacturing and heat treatment processes, and a comparison of the DIN grades with the ISO and AGMA grades, see **Fig. 17**. Recommendations for selection of *tooth thickness margins*, A_{sne}, *tooth thickness tolerances*, T_{sn} and *centre distance margins*, A_a: **Table 4**.

Thus *theoretical backlash*:

$$j_n = -(A_{sn1} + A_{sn2}) \pm A_a \tan \alpha_n; \qquad (36)$$

max. j_n where $A_{sn} = A_{sne} - T_{sn}$ and $+ A_a$,

min. j_n where $A_{sn} = A_{sne}$ and $- A_m$.

Theoretical *rotary* backlash, $j_t = j_n / \cos \beta$.

Acceptance backlash usually smaller due to manufacturing variations.

Operational backlash, e.g. during startup, due to gears warming up more rapidly than housing, under certain circumstances significantly smaller than $j_{n,t}$; see **Table 4**, footnote a.

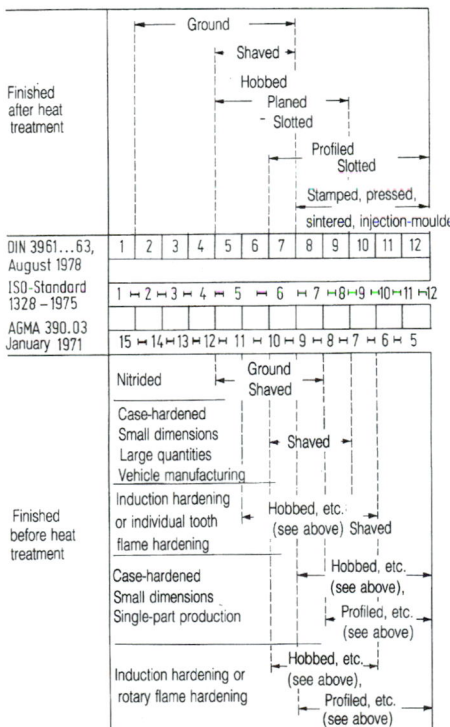

Figure 17. Gearing quality and manufacturing processes (approximate allocation of DIN, ISO and AGMA grades as per adjacent pitch error, $m = 6$, $d = 75$ to 150 mm). For manufacturing processes, see S5.2.

Table 4. Recommendations[a] for upper tooth thickness deviations A_{sne}, upper tooth thickness tolerances T_{sn} (DIN 3967) and centre distance deviations A_a (DIN 3964)

	$A_{sne}-$ series[b]	$T_{sn}-$ series[c]	Centre distance deviations A_a
Cast turntables DIN > 12	$2 \cdot a$	29(30)	10
Turntables (normal play)	a	28	9
Turntables, converters (little play)	bc	26	9(8)
Plastic machines	c to cd	25	7
Locomotive drives	cd	25	7
General machine construction, heavy-duty machine construction, non-reversing	b	26	7
Vehicles	d	26	7
General machine construction, heavy-duty machine construction, reversing, shearing, running gear	c to e	24 to 25	6 to 7
Machine tools	f	24 to 25	6
Agricultural tractors, combine-harvesters	e	27 to 28	8

[a] In case of lack of experience, check as per DIN 3967, Appendix A, in particular temperature influence. As per proposals by A. Seifried, Friedrichshafen.
[b] Condition: for each gear $|A_{sne}| \geq |A_a|$.
[c] Condition: $T_{sn} > 2$ tooth thickness variation R_s as per DIN 3962; check!

8.3 Lubrication and Cooling

Lubricating Film Thickness. The minimum lubricating film thickness at the pitch point is a suitable indicator of the lubrication condition, as per EHD theory. For steel gears, as per Oster on the basis of [14], with the gear ratio u, the numerical value equation

$$h_C = 0.003 \, [(au)/(u+1)^2]^{0.3} \cdot (v_0 v_1)^{0.7} \cdot (p_C/840)^{-0.26}$$
$$\text{in } \mu\text{m} \quad (37)$$

$$\left(p_C = Z_H Z_E \sqrt{\frac{F_t}{d_1 b} \cdot \frac{u+1}{u}} \text{ for Eq. (48)} \right)$$

is valid as an approximation. The lubricating film material viscosity, v_0, in mm²/s, is obtained from the bulk temperature

$$\vartheta_0 = \vartheta_L + 7400 [P_{vz}/(ab)]^{0.72}$$
$$\approx \vartheta_L + 2.2 \cdot 10^{-4} \, (\varepsilon_\alpha m/a)^{0.72} \, v_t^{0.576} \, P_C \text{ in } °\text{C}.$$

Here a = the centre distance and b = the width in mm, v_t = the circumferential speed in m/s, $\vartheta_L \approx$ the oil temperature in °C, P_{vz} = the tooth power loss, from Eq. (38), in kW, P_c = the Hertzian compression in N/mm² (see Eq. (37)) and e_α = the transverse contact ratio. The specific lubricating film thickness,

$$\lambda = \frac{h_C}{(R_{a1} + R_{a2})/2},$$

can be used for qualitative evaluation.

$\lambda > 2$: mainly hydrodynamic lubrication, hardly any wear;

$\lambda < 0.7$: applications in many sectors of industry, marginal lubrication predominates. Check risk of micro-pitting!

Lubricant and Lubrication Method

Instructions for selection: see **Table 5**.

Lubricant Viscosity (DIN 51502) or worked penetration (DIN 51804), dependent on temperature: *manual application*; NLGI class 1 to 3 adhesive lubricant (NLGI = National Lubricating Grease Institute). Central lubrication system: NLGI 1 to 2 lubricating grease (transportable); *spray coating*: NGLI 00-0 liquid grease (sprayable); *splash lubrication*: NLGI 000-0 liquid grease (free-flowing); *lubricating oil viscosity*: reference values as per **Fig. 18**. (Influence of roughness, temperature, type of operation [1]. EP additives where danger of seizing exists; synthetic oils (low coefficient of friction, high viscosity index, expensive) under extreme operating conditions.

For Lubricating Devices, housing connections: see F8.10.4.

Thermal Economy. Power loss P_v should not exceed cooling capacity P_K. For small to medium-sized gears, air cooling through housing walls (cooling area, A, in m^2) and temperature difference between housing and ambient air of $\vartheta_G - \vartheta_a$ in K, normally sufficient. Remove excess power loss by water cooling.

$$P_V = P_{VZ} + P_{VL} + P_{VD}. \qquad (38)$$

Roughly Gear losses $P_{VZ} = 0.5$ to 1% of nominal power for each stage (with $v > 20$ ms, gear losses independent of load are also to be taken into account [1]). Storage losses P_{VL} (see F5.5 and F6.1.2). P_{VD} other losses, seals (see F5.6.3 and F6.6.7).

Cooling capacity (heat emission) of housing:

$$P_{KG} = \alpha A(\vartheta_G - \vartheta_\infty) \quad \text{with} \quad \alpha = 15 \text{ to } 25 \text{ W/(m}^2\text{ K)} \quad (39)$$

with static air and unimpaired convection (lower limits: higher levels of dirt and dust, low speeds, large gears). For fans on rapid-running shafts, α is increased by a factor of f_K: spur gear with one fan, $f_K \approx 1.4$; two fans, $f_K \approx 2.5$; bevel gear with one fan, $f_K \approx 2.0$. Influence of wind speed and insolation considerable.

8.4 Materials and Heat Treatment – Gear Manufacture

(For worm gears, see F8.8.) For bearing capacity of materials and corresponding quality requirements, see **Table 14**. In addition, *costs* for materials and heat treatment, *machinability* and/or *workability*, *noise behaviour*, *number of units* (manufacturing process) are decisive (in many cases the only important factor) in the selection.

Typical Examples from Various Applications

Gears for Tackle, Instruments, Domestic Appliances, etc. (i.e. for the transmission of motion or small forces). Alloys of Zn, Mg, Al. Thermoplastics (injection moulding); automatic steels, structural steels; malleable alloys of Al, Zn, Cu, laminated plastic, thermoplastics (extruding machines, cold drawing, presses or punches, or millers); sintered metals (final sintering).

Vehicle Gears. Alloyed carburising steels – milled or shaped, shaved – case-hardened – (if necessary, ground instead of shaved); low-alloy heat-treatable steels – milled or spliced, shaved – carbonitrided.

Turbo-gears, Marine Gears. Alloyed heat-treatable steels – milled, ground if necessary; Al-free nitriding steels – milled, shaved (or ground) – gas-nitrided (ground if necessary); alloyed carburising steels – milled – case-hardened – ground.

Table 5. Selection of lubricants and type of lubrication

Circumferential speed (m/s)	Lubricant	Type of lubrication	Gear format	Special features
Up to 2.5	Adhesion lubricants	Apply by hand	Open[a]	Provide for covering hood wherever possible[b]
Up to 4 (if necessary 6)	Fluid grease	Spray lubrication		
Up to 8 (if necessary 10)				Note c
				Note d
Up to 15	Oil	Splash lubrication[c]	Closed	With perforated sheet walls, splash lubrication possible, cooling fins
Up to 25 (if necessary 30)				
Over 25 (if necessary 30)		Injection lubrication		Note f
Up to 40		Mist lubrication		For low-stress intermittent service

[a]For example, cement mills, rotary kilns.
[b]Spraying amounts, spraying times [1]; lubricate bearings separately.
[c]Especially for intermittent service, partially for lifetime filling (no oil tightness required!).
[d]Immersion depths in *operating condition*, 3 to max. $6m$ at $v > 12$ m/s, absolutely necessary to check c for low values.
[e]Spraying oil adequate for above gears and bearings if v_1^2 in (m/s)^2/d in m > 5, otherwise wiper needed. Oil quantities approx. 5 to 10 l/loss-kW; for oil level check ventilation, see F8.10.4. Large-scale operations, friction bearing operations, vertical operations – usually injection lubrication here too.
[f]Injection lubrication: if oil pressure varies upwards or downwards, up to $v = 25$ m/s *in front of* contact, above 25 m/s *in front of and behind* engagement. Injection quantities as per heat balance, roughly: $Q_e = 0.8$ to 1.0 l/min per cm tooth width. Total amount of oil, $Q = Q_e$ (0.5 to 2.5 min) with external tank, $Q = Q_e$ (4 to 30 min). Oil pressure in front of nozzle, approx. 0.8 to 1.0 bar, higher for turbo-gears; for fittings, see F8.10.4; for coolers and filters, see [1].

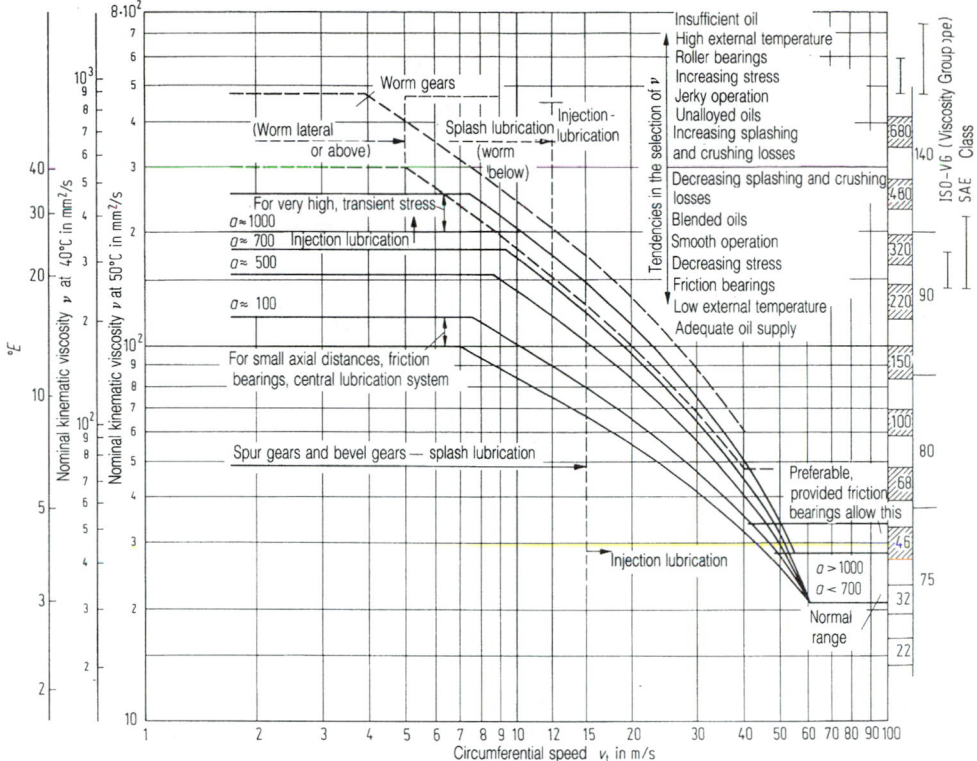

Figure 18. Selection of lubrication oil viscosity for spur gears, bevel gears and worm gears. Shown is the approximate allocation of ISO and SAE viscosity classes; distortion classes appear hatched. Splash lubrication is also possible at higher v_t if centrifuged oil is fed to the tooth entrance through fins or oil guide plates.

Large-Toothed Gears, Turntables. Alloyed cast steel (attention should be paid to scrap risk due to cavities), alloyed carburising steels (rolled) – milled – induction or flame-hardened if necessary.

Gearing for General and Construction Industry. Alloyed and unalloyed cast steel – hobbed, shaped, or planed. Alloy carburising steel, hobbed or similar – case-hardened – ground (if necessary, finish milled using carbide metal hobs, honed if necessary). Al-free nitriding steels – hobbed or similar (if necessary, shaved or ground, lapped if necessary) – gas-nitrided. Non-alloy and alloyed heat-treatable steels – hobbed or similar, shaved – bath-nitrided. Non-alloy and alloyed heat-treatable steels – hobbed or similar – induction-hardened – or rotary flame-hardened. Non-alloy and alloyed heat-treatable steels – hobbed or similar – induction-hardened or individual teeth flame-hardened – (ground if necessary).

Materials and Heat Treatment

Principles of Selection. GG grey iron, GGG spheroidal graphite iron, GS cast steel – for instructions see **Table 14**. Special cast iron with suitable heat treatment equivalent to heat-treatable steels (attention should be paid to machinability!) [15].

Heat-treatable Steels – Unhardened. Gears – and drive units – larger, heavier and more expensive than with gear-cutting processes involving hardening. However: heat treatment (before gear cutting) free from risk, no change in dimensions *after* gear cutting, usually no gear grinding necessary; relatively soft material compensates for design and manufacture faults earlier during running in; manual

dressing of tooth flanks possible; safety against fracture usually more than sufficient.

Carburising Steels – Case-Hardened. Expensive, but root and flank strength levels controllable for small to medium-sized gears up to highest hardness range (HRC = 58 to 62). Distortion due to hardening requires gear grinding in single-part production (normal: $d \le 900$ mm; $m \le 25$ mm; extreme d up to 3000 mm; m up to 36 mm). For coarser grades unground (see **Fig. 17**) (usually $d \le 250$ mm, $m \le 6$ mm; with limitation $d \le 500$ mm, $m \le 10$ mm).

Quenching cracks certainly avoidable – reliable hardening process.

Heat-Treatable Steels – Rotary Hardening (flame or induction hardening). Cost-effective for small to medium-sized gears (normal: $d \le 200$ mm, $m \le 6$ mm: extreme, d up to 1500 mm, m up to 18 mm), certainly controllable in medium hardness range (HRC = 45 to 56), higher risk of cracking thereafter. Uniform gear quality only with constant material values and heat treatment kept constant [15].

Heat-Treatable Steels – Individual Tooth Hardening – Twin-Flank Hardening (flame or induction hardening). Cost-effective for large gears (d up to about 3000 mm, $m > 8$ mm); controllable in medium hardness range (HRC = 45 to 56). Careful preparation (hardness sampling), constant, i.e. continuously monitored hardness regulating data required. Little distortion, gear grinding usually not necessary. Tooth base unhardened, reduced root strength [16].

Refining Steels – Individual Tooth Hardening – Space Hardening (flame or inducation hardening). Tooth base also hardened. Cost-effective for large gears in medium hardness range (as for twin-flank hardening, but flame only at $m > 16$ mm) (HRC = 45 to 52, possibly 56). Low hardness risk (quenching cracks) only with appropriate preparation and monitoring, long years of experience, suitable materials and optimum hardness conditions (hardness sampling). Little distortion, but frequent pitch errors at start of hardening; gear grinding often required [15].

Al-free Nitriding Steels, Heat-Treatable Steels, Carburising Steels – nitrided (Long-Duration Gas-Nitrided). Low distortion, difficult process. Normal: nitriding depth, nhd, ≈ 0.3 mm, $d < 300$ mm, $m \le 6$ mm; more difficult: nhd ≈ 0.6 mm, $d < 600$ mm, $m < 10$ mm. Lower strength values should be specified for larger d and m with nitriding steels! Here, and with thin-walled gears, gear grinding usually needed after nitriding due to distortion. High strength attainable with certainty only with special grade of material, long years of experience, optimum manufacturing and inspection equipment. Otherwise, strength can vary widely. Nitriding steels especially sensitive to impacts and bearing edges. $< 15 \ \mu$m white layer should be aimed for.

Heat-Treatable Steels – Nitro-carburised (Short-Duration Gas-Nitrided). New low-distortion process, avoids many short-duration bath nitriding problems [17], has largely replaced it. Low overload capacity.

Heat-treatable Steels – Nitro-carburised (Short-duration Bath-nitrided). Low-distortion process. Normal: $d < 300$ mm, $m < 6$ mm; more difficult: d up to 600 mm, m up to 10 mm. Practically no diffusion zone, i.e. reduced load capacity, if white layer ($< 30 \ \mu$m thick) worn.

Heat-treatable Steels – Carbo-nitrided. Hardness penetration depths (nitrogenous martensite layers) 0.2 to 0.6 mm. Highest core strength possible to support thin hardened case. Suitable for small gears made in large quantities.

8.5 Load Capacity of Spur and Helical Gears

8.5.1 Types of Tooth Damage and Remedies

For definitions and origins, see DIN 3979; cf. **Fig. 19**.

Forced Rupture. Usually from accidents, jamming or similar; difficult to estimate forces involved. *Remedy*: overload protection, breaking pieces.

Endurance Failure. Fatigue fracture after comparatively long running periods above endurance limit, usually from notches, quenching cracks, faulty material or faulty hardness treatment in tooth root. *Remedies*: higher moduli, effective pressure angle (profile offset), root rounding (avoid grinding notches), surface hardening (especially case-hardening), shot-blasting, precise gear-cutting, tooth end-relief or crowning to relieve ends of teeth.

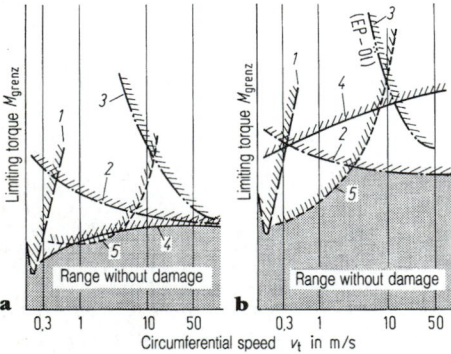

Figure 19. Main load-bearing capacity limits of gears: **a** heat-treatable steel, **b** carburising steel; *1* wear limit, *2* tooth break limit, *3* seizing limit (hot seizing), *4* pitting limit, *5* micro-pitting limit.

Pitting. Crumbling, especially between dedendum and pitch circles, from excess flank pressure. Initial pitting ends heat-treatable steel local overloads and stops – and so is harmless. Progressive pitting destroys tooth flanks. *Remedy*: Large radii of curvature (profile offset), surface hardening (especially case-hardening) (**Fig. 19**), more viscous oils, precise gear-cutting, low flank roughness.

Micropitting. Numerous microscopically small incipient cracks and breaks, optical impression of a grey fleck. Remedy through improved lubrication conditions (also influence of additive) [57].

Hot Seizing. Grooves and seizing marks in high sliding speed range as a result of limiting temperature conditioned by material and lubricant. Remedy through lower moduli, tip and root relief, nitriding, low flank roughness (running-in), especially effective: extreme-pressure oils (oils with chemically active additives).

Cold Seizing. Groove wear with removal of large amounts of material at low circumferential speeds. Remedy through better gear precision, smoother tooth flanks, more viscous lubricants, tip relief.

Abrasion Wear. Laminar removal of material, especially at tip and root, often decisive at low circumferential speeds ($v < 0.5$ m/s) as a result of insufficient lubrication pressure being generated. Can be remedied through high lubricant viscosity, certain synthetic lubricants, many extreme pressure additives, MoS suspension, surface hardening or nitriding. Important: Equal flank hardness on pinion and gear.

8.5.2 Checklist

Before design work begins, all requirements for and influences on the functioning of the gear are to be listed. Often decisive for success or failure. Instructions: **Table 6**.

8.5.3 Guide Data for Gear Rating

Gear data (transmission, modulus, centre distance, diameter, contact ratio (see F8.1.2, F8.1.4, F8.1.5, F8.1.7).

Pinion Diameter d_1. From the simplified characteristic value for the rolling pressure, $K^* = [F_t (u + 1)/(bd_1u)]$, it follows that

$$d_1 \ge \sqrt[3]{\frac{2M_1}{K^* (b/d_1)} \frac{u + 1}{u}}. \tag{40}$$

In contrast to DIN 3990 and other gear standards, the torque is described as M instead of T, in order to maintain uniformity for all areas of application.

Experimental values for K^* as per gears manufactured; examples are given in **Table 7**. For the selection of material and heat treatment see F8.4. For heat-treatable steels, the hardness of the pinion material selected should be approximately HB = 40 above that of the gear material.

Tooth Width b (as per guide data for b/d_1, **Table 8**). For larger widths, flank line corrections required to balance out deformations. Overlap ratio: attention should be paid to Eq. (13).

Number of Teeth and Modulus. Standard pinion tooth numbers – **Table 9**, with modulus determined using Eq. (5). Minimum moduli conditioned by risk of tooth corner break, **Table 10**. Standardised modulus range: **Table 1**.

When the modulus has been determined, check

Table 6. Checklist for gear drives (plus sketch with connection dimensions)

Effect on: sealing A, application factor B, manufacture F, gear model G, housing H, construction, cooling/heating K, bearing L, lubrication S, gearing V, permissible stress Z

1. *Main functions* required for draft calculation

o Drive/power takeoff speeds (transmission constant, switch step tolerance); direction of rotation constant/alternating................ Z

o Type of operating machine, type of drive machine... B

o Customer specifications for main functions: type of gear (spur gears, bevel gears, etc.), type of fitting (stationary, slip-on, flange gear, etc.), other (application factor, multi-motor drive, flywheels, drive/power takeoff left-hand/right-hand/optional) (see also 2.4) .... K

o Position of working machine to drive machine (position of drive shaft to power take-off shaft of gear, variable position, limits), type of gear, if applicable centre distance.. K

o Power, permanent operational moment, nominal moment of working/drive machine, maximum moment, starting moment or similar...................................... Z

2. *Other functions* required for draft, calculation and design

2.1 *Operating data*

o Number of machine startups.................... B

o Consequences of a damage incident (endangering human life, loss of production) .... Z

o Overturning, starting and switching-off moments of machine, level, quantity and duration of shocks in operation, peak moment, catastrophe moment.................. B

o Running time per day (% operating time) Z

o Overload safety, switching-off moment..... B

o Reversal of direction of force (reverse operation)... Z

2.3 *Forces on gear*

o Axial forces on drive and power takeoff shafts (e.g. toothed-gear type coupling)...... ....H, L, V

o Forces on housing..................................H, L

o Radial forces on drive and power takeoff shafts (e.g. chain sprocket, belt pulley)...... ....H, L

o Return barrier..S

2.5 *Lubrication*

o Heating (for startup)

o Cooling (fresh water, salt water, brackish water or air, temperature); central cooling system or individual cooling

o Lubricant freely selectable/specifications

o Provision through central lubrication system (lubricant, viscosity, pressure) or individual gear lubrication

2.2 *Manufacturing data*

o Restrictions on materials selection (machinability, delivery time) .... Z

o Dimensions and weight restrictions due to machine tools, furnace dimensions, hardening devices...................................F, Z

o Tools available ....F, V

2.4 *Customer requirements: specifications, acceptance conditions*

o Type of couplings on drive and power takeoff...L, V

o Calculation specifications (e.g. classification societies, factory specifications) .... Z

o Form of shaft journals on drive and power takeoff (flange forged on – hole circle, adjustment spring or similar corrected for oil pressure unit)

o Noise, efficiency, guarantee (type of test run)..V, H, F

o Design (forged, welded, shrunken tooth rims; shaft–hub connection; cast, welded housing)..K, H

o Accident prevention specifications ...... K, Z

2.6 *Environment, location*

o Location (hall, covered, in open air)........... ....A, S, K

o Restrictions for assembly, installation, space, weight, transportation, dirt, dust, foreign bodies, spray water, water vapour.. ....A, H, F

o Foundation (e.g. steel frame, rigid concrete): separate (joint with drive and power takeoff) ....H

o Temperature (max., min.), insulation...K, S

---

whether sufficient rim thickness is available under the tooth root with the pinion mounted (adjusting spring or similar) (see **Fig. 49**), and whether the remaining shaft section is adequate with a geared shaft.

Spur Gearing – Helical Gearing. For properties, see F8.1.4. If low-noise running is required, and for jerky operation, it is preferable to go over to helical gears and finer quality. For average conditions:

Spur: Up to $v_t = 1$ m/s Q10-12, up to 5 m/s Q8-9, up to 20 m/s Q6-7.

Helical or Double Helical: Required for Q8 hardened gears or more precise, otherwise increased danger of tooth corner breaks and no

advantages are obtained through helical gear. For coarser grades and material for casting also used with unhardened steels (can withstand running-in, surplus safety against fracture).

Up to $v_t = 2$ m/s Q10-12 unhardened, Q7-8 hardened,

Up to $v_t = 5$ m/s, Q8-9 unhardened, Q7 hardened,

Up to $v_t = 20$ m/s, Q6-7, above $v_t = 40$ m/s with Q4-5.

Helix Angle. *Single helical gear* $\beta = 6$ to 15° (limitation of axial force). Check overlap ratio, Eq. (13): up to $v_t = 20$ m/s; $\varepsilon_\beta \gg 1.0(0.9)$; $\varepsilon_\gamma \geq 2.2$; above 40 m/s: $\varepsilon_\beta \geq 1.2$, $\varepsilon_\gamma \geq 2.5$. *Double helical gear* only if single helical gear too wide or axial forces too great: $\beta = 20$ to 30° *Note*: Only fix one shaft axially and check whether axial

Table 7. K^* factors of manufactured spur gears (for nominal power, unless otherwise stated) as per company specifications and [1, 2, 47, 48]. Material: steel (unless otherwise stated). Heat treatment: v, quenched and subsequently drawn; eh, case-hardened; n, nitrided. Machining: f, milled, planbed, slotted; s, shaved; g, ground

Application Drive/power takeoff	v (m/s)	Pinion Material Heat treament Machining	Hardness	Gear Material Heat treament Machining	Hardness	K^* factor (N/mm²)	Comments
Turbine/Generator	>20	v, f	225 HB	v, f	180 HB	0.80	
	>20	n, s	>60 HRC	n, s	>60 HRC	2.0	$K_A \approx 1.1$[a]
	>20	eh, g	>58 HRC	eh, g	>58 HRC	2.8	
E-Motor/ Industrial operation 24-hour operation	5	v, f	210 HB	v, f	180 HB	1.2	
		v, f	350 HB	v, f	300 HB	2.0	
		eh, g	>58 HRC	eh, g	>58 HRC	4.4	
							$K_A \approx 1.3$[a]
	10	v, f	210 HB	v, f	180 HB	1.0	
		v, f	350 HB	v, f	300 HB	1.8	
		eh, g	>58 HRC	eh, g	>58 HRC	4.0	
E-Motor/Large-scale operation (lifts, rotary furnaces, mills)	<5	v, f	225 HB	v, f	180 HB	0.6	
		v, f	260 HB	v, f	210 HB	1.0	
							$K_A \approx 1.6$
	7.5	eh, g	>58 HRC	v, f	320 HB	1.5	
Converter (for *maximum* moment)	0.3	v, f	260 HB	GS, f	180 HB	1.3	*Not* catastrophe moment
E-motor/ Machine tools Hobbing machines	22 0.3	eh, g eh, g	>58 HRC >58 HRC	eh, g eh, g	>58 HRC >58 HRC	3.0 9.0	For peak moment *seldom arising*
Milling machines (spindle head)	22	eh, g	>58 HRC	Cast polyamide 12 g, f	75 Shore hardness (SH)D	0.70	
E-motor/Crane hoisting unit (for *max.* weight- lifting capacity and continuous operation)	10 to 14 4 to 8 2 to 4 0.5 to 2	v, f v, f v, f v, f	230/280 HB 230/280 HB 230/280 HB 230/280 HB	v, f v, f v, f v, f	190/230 HB 190/230 HB 190/230 HB 190/230 HB	1.1 1.3 1.6 1.8	Stage 1 Stage 2 Stage 3 Stage 4
E-motor/Gripper hoisting unit (for *max.* gripper closing moment)	12 6 3	eh, g eh, g eh, g	>58 HRC >58 HRC >58 HRC	eh, g eh, g eh, g	>58 HRC >58 HRC >58 HRC	7.0 11.0 15.0	Stage 1 Stage 2 Stage 3
E-Motor/Small industrial operation	<5	v, f v, f	350 HB 350 HB	Laminated plastic Polyamide		0.53 0.35	
E-Motor/Small units	<5 <3 <3	v, f v, f Brass, aluminium	200 HB 200 HB	Zinc die casting Brass, aluminium Brass, aluminium		0.20 0.20 0.10	

[a]Application factor for calculation.

forces are being introduced from outside (causing non-uniform force distribution!). In general, arrowhead should run behind. Attention should be paid to limits of manufacture (e.g. hob exit) (see F8.10.3).

Basic Profile   See **Fig. 10**.

Profile Offset   See F8.1.7.

Bearing Forces (Fig. 20). Tooth normal force, $F_t/\cos \alpha_{wt}$ acts as lateral force, axial force, $F_x = F_t \tan \beta$ on lever arm r on shaft. Radial and thrust bearing forces at A and B should be determined from this in accordance

Table 8. Maximum values for b/d_1 of fixed spur gears with rigid foundation[a]

Straight and helical gears; twin-sided, symmetrical mounting:	
Normalised (HB $\leq$ 180)	$b/d_1 \leq 1.6$
Quenched and subsequently drawn (HB $\geq$ 200)	$b/d_1 \leq 1.4$
Case-hardened or surface-layer hardened	$b/d_1 \leq 1.1$
Nitrided	$b/d_1 \leq 0.8$
Double helical gear	$B/d_1 \leq 1.8$ times b/d_1 values above; for B see **Fig. 5**
Twin-sided, asymmetrical bearings	80% of values above
Same size pinion and gears (broad-faced steel gears and $i = 1$)	120% of values above
Floating bearings	50% of values above

[a]For lighter models on steel frame approx. 60% of values.

Table 9. Standard pinion tooth numbers, z_1. Lower range for speeds, $n < 1000$ min^{-1}, upper range for $n > 3000$ min^{-1}

Transmission	1	2	4	8
Quenched and subsequently drawn to 230 HB	32 to 60	29 to 55	25 to 50	22 to 45
Above 300 HB (and hard/quenched and subsequently drawn)	30 to 50	27 to 45	23 to 40	20 to 35
Cast iron	26 to 45	23 to 40	21 to 35	18 to 30
Nitrided	24 to 40	21 to 35	19 to 31	16 to 26
Case-hardened (or surface-hardened)	21 to 32	19 to 29	16 to 25	14 to 22

$z = 12$	Lowest practical tooth number for power gear (mating tooth number $\geq$ 23)
$z = 7$	Lowest tooth number for transmission of motion for reference profile as per DIN 867, straight gear
$z = 5$	Lowest tooth number for transmission of motion for reference profile as per DIN 58 400 (light engineering), straight gear
$z = 1$ to 4	For transmission of motion, possible with staggered gears or helical gears, $\varepsilon_\alpha < 1$ [7]

with bearing distances. Attention should be paid to overturning moment of axial forces in calculating radial forces!

8.5.4 Evaluation of Load Capacity

A check should be carried out to see whether the gear has adequate theoretical margins of safety against all failure limits, in so far as a rough estimate as per Eq. (40) and **Table 7** is not sufficient.

Basic Concepts. Calculation is based on the nominal circumferential force exerted on the tooth of a faultless

Table 10. Minimum values for m_n

DIN gear quality	Bearings	min m_n or m_t
11 to 12	Steel construction, light housing	$b/10$ to $b/15$
8 to 9	Steel construction or floating pinion	$b/15$ to $b/25$
6 to 7	Well supported in housing	$b/20$ to $b/30$
6 to 7	Exactly parallel, rigid housing	$b/25$ to $b/35$
5 to 6	$b/d_1 \leq 1$, exactly parallel, rigid mounting	$b/40$ to $b/60$

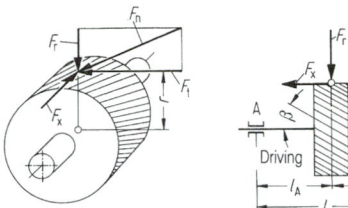

Figure 20. Tooth force components for calculating bearing forces.

rigid gear, average lubrication conditions, and strength values which have been determined using standard reference test gears under standard test conditions.

Different conditions are present in reality: external additional forces due to jerky starting and load variations; internal additional forces due to tooth forming errors and deformation effects; model influence, lubrication (circumferential speed; viscosity, roughness); root rounding, etc. The effect of these deviations is determined through influence factors.

Input Quantities. See diagram with example (p. F136).

$$\text{Peripheral force} \quad F_t = 2M/d = 2P/(d\omega); \quad (41)$$

$$\text{Peripheral velocity} \quad v_t = 0.5d\omega = \pi dn. \quad (42)$$

Application for simplified calculation for industrial gears: DIN 867 basic profile: $\alpha_0 = 20°$, $h_{a0}/m = 1.25 \pm 0.05$, $\rho_{a0}/m = 0.25 \pm 0.05$. Number of teeth in pinion: $20 \leq z_1 \leq 50$. Average to high load: $K_A F_t/b \geq 200$ N/mm tooth width. Operation in subcritical range: see **Fig. 21**. Transverse contact ratio: $1.2 < \varepsilon_\alpha < 1.9$. $v_t > 1$ m/s. Roughness in root rounding $R_z < 16$ µm. Lubricant as per **Table 5** and **Fig. 18**. Continuously operating gears. For helical gears, $\varepsilon_\beta \geq 1$.

For *divergent* conditions, calculations as per DIN 3990 [1].

Force Factors

(These determine the controlling force per mm of tooth width, valid for all limits of stress.) The factors depend on the controlling circumferential force; they are calculated approximately as follows: K_v with grade of gear cutting and $K_{H\beta}$ or $K_{F\beta}$ with circumferential force $F_t K_A K_v/b$. Many force factors become 1 with small errors and high external circumferential forces.

Application Factor K_A. This takes into consideration the additional forces introduced by the drive or the power takeoff. For guide data, see **Table 11**. If the maximum moment is being used for the calculation (see **Table 11c**), then $K_A = 1$ should be used.

Dynamic Factor. This takes into account internal dynamic additional forces: **Fig. 21**.

Width Factor $K_{H\beta}$, (flank). $\approx K_{F\beta}$ (root) takes into account influence of manufacturing tolerances, f_{ma}, and cumulative deformation, f_{shg}, on the force distribution over the tooth width:

$$F_{\beta y} = x_\beta F_{\beta x} = x_\beta \, (f_{ma} + f_{shg}) \quad (43)$$

is determined, and $K_{H\beta}$ ($\approx K_{F\beta}$) is derived from **Fig. 22**. For x_β, see **Table 12**: $f_{shg} \approx 1.33 f_{sh}$.

Set $f_{ma} \approx f_{H\beta}$ of a gear as per **Table 3** or as per special specification. f_{shg} as per approved gears, **Table 13**; the construction should be made suitably rigid. In case of any doubt, deformation should be checked – especially of pinion shaft. Check can be carried out under load, in accordance with bearing pattern, using bearing pattern lacquer insoluble in oil (DIN 3990).

Face Factors $K_{H\alpha}$ (Flank) and $K_{F\alpha}$ (Root) take into account non-uniform distribution of circumferential force on engaged pairs of teeth as a result of pitch and form deviations.

For rough estimates or coarse gears under low stress:

Table 11. Application factors for gear drives
a For industrial gears ($n < 3600$ min^{-1}, $(z_1 v_t / 100) \cdot [u^2/(1 + u^2)]^{\frac{1}{2}} < 10$ with v_t in m/s)

Method of operation of drive machine (for examples, see Table 11b)	Method of operation of driven machine			
	Uniform	Moderate shocks	Average shocks	Strong shocks[a]
Uniform	1.00	1.25	1.50	1.75
Light shocks	1.10	1.35	1.60	1.85
Moderate shocks	1.25	1.50	1.75	2.0
Strong shocks	1.50	1.75	2.0	2.25 or above

[a]Nitrided gears not suitable in general

b Examples of method of operation of drive machines

Method of operation	Drive machine
Uniform	Electric motor, steam turbine, gas turbine in uniform operation (low start-up moments seldom arising)
Light shocks	Steam turbines, gas turbines, hydraulic motors, electric motors (larger startup moments arising more frequently)
Moderate shocks	Multi-cylinder combustion motor
Strong shocks	Single-cylinder combustion motor

Table 11. Continued

c Examples of method of operation of driven machines

Method of operation	Driven machines
Uniform	Current generators, uniformly loaded belt conveyors or plate conveyors, worm conveyors, light lifts, machine tool feed drives, ventilators, turbo-compressors, agitators and mixers for substances of uniform density, punches, when laid out in accordance with maximum cutting moment
Moderate shocks	Non-uniformly loaded belt conveyors or plate conveyors, machine tool main drives, heavy lifts, crane slewing gears, heavy duty centrifuges, agitators and mixers for substances of non-uniform density, metering pumps, piston pumps with several cylinders
Average shocks	Mixers for interrupted operation with rubber and plastics, light ball mills, woodworking, single-cylinder piston pumps
Strong shocks	Bucket chain drives, sieve drives, dipper dredgers, heavy-duty ball mills, rubber kneaders, smelting equipment, heavy-duty metering pumps, rotary boring equipment, pan grinders

o The table values apply only to the nominal moment of the working machine. Thus, the nominal moment of the drive motor can be used in its place, provided this corresponds to the moment requirement of the working machine.
o The values apply only for gears that are not operating in the resonance range, and only with uniform power consumption levels. In applications involving unusually severe stresses, motors with high starting torque, intermittent service or operation under extreme repeated sudden loads, gears must be checked for static strength and time strength.
o If special application factors are required for specific gears then they are to be used.
o Torque values resulting from inertia moments are to be respected in a brake. Moreover, these are decisive for the maximum gear stress.
o For a hydraulic coupling between motor and gear, the K_A values for moderate, average and strong shocks are reduced if the characteristic of the coupling permits this.

$$\left. \begin{array}{l} K_{H\alpha} = 1/Z_\varepsilon^2 \geq 1.2; \\ K_{H\alpha} = \varepsilon_{\alpha n} \geq 1.4. \end{array} \right\} (44)$$

$$\left. \begin{array}{l} K_{F\alpha} = 1/Y_\varepsilon \geq 1.2; \\ K_{F\alpha} = \varepsilon_{\alpha n} \geq 1.4. \end{array} \right\} (45)$$

We are calculating Z_ε (see Eq. (50)) and Y_ε (see Eq. (54)) on the safe side here.

For gears under normal load (fatigue failure margin of safety, $S_F \leq 2$, pitting margin of safety, $S_G \leq 1.3$), DIN grade 8 or finer with spur gears, or 7 or finer with helical gears:

$$K_{H\alpha} = K_{F\alpha} \approx 1. \quad (46)$$

Safety Against Pitting

The flank compression (Hertzian compression; see B4.2) at the pitch point must be smaller than the permissible compression, with the condition:

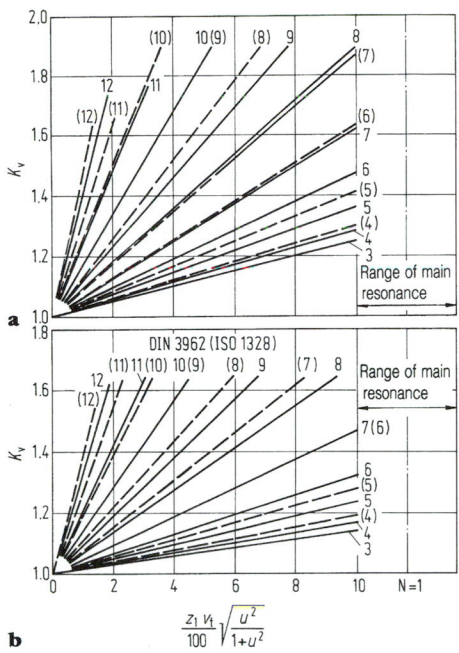

Figure 21. Dynamic factor K_v (DIN 3990/ISO/DIS 6336): **a** spur gears, **b** helical gears with $\varepsilon_\beta \geq 1$ (for $\varepsilon_\beta < 1$, see DIN 3990, [1]).

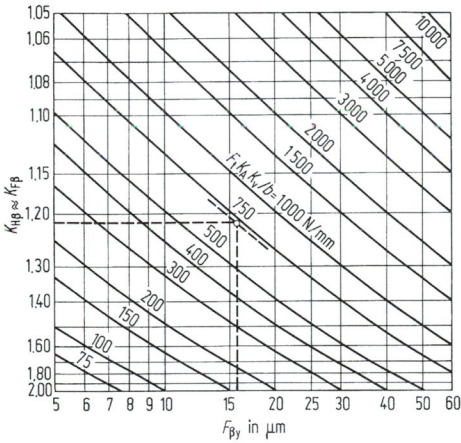

Figure 22. Width factor, $K_{K\gamma}$ ($\simeq K_{F\beta}$) (DIN 3990/ISO).

$$S_H = \sigma_{H\lim} Z_X/(\sigma_{HO} \sqrt{K_A K_v K_{H\beta} K_{H\alpha}}) \geq S_{H\min} \quad (47)$$

where $\sigma_{H\lim}$ is the continuous rolling strength as per test bench experiments and experiences with gears manufactured (**Table 14**).

σ_{HO} is the *nominal value of flank compression*:

$$\underbrace{\sigma_{HO} = Z_H Z_E \sqrt{\frac{F_t}{d_1 b} \frac{u+1}{u}} Z_\varepsilon Z_\beta = Z_H Z_E \sqrt{K^*} Z_\varepsilon Z_\beta}_{p_c}. \quad (48)$$

Table 12. Run-in characteristics for Eq. (43)

Material	$\sigma_{H\lim}$ N/mm²	$x_\beta{}^a$
Cast iron		0.45[b]
Heat-treatable steel	400	0.20[b]
Heat-treatable steel	800	0.60[b]
Heat-treatable steel	1200	0.73[b]
Case-hardened or nitrided		0.85

[a]Applies to any $F_{\beta x} = (f_{ma} + f_{shg})$ at $v \leq 5$ m/s, for $F_{\beta x} < 80$ μm at 5 m/s $\leq V_t < 10$ m/s, for $F_{\beta x} < 40$ μm at $v_t \geq 10$ m/s. At larger $F_{\beta x}$ see DIN 3990 [1]
[b]If necessary, carry out linear interpolation, and also where rack and pinion are made of different materials

Table 13. Guide data for permissible flank line variations due to cumulative deformation, f_{shg}, in μm (for gear pair in gear unit)

Tooth width (mm)	up to 20	Above 20 up to 40	Above 40 up to 100	Above 100 up to 200	Above 200 up to 315	Above 315 up to 560	Above 560
Very rigid gear (e.g. stationary turbo-gear)	5	6.5	7	8	10	12	16
Average rigidity (usually industrial gears)	6	7	8	11	14	18	24
Flexible gears	10	13	18	25	30	38	50

For pliant, flexible gears (e.g. welded single-web gears and small helix angles, with small hub diameters and small hub widths), use f_{shg} from column 2 for calculation.

p_c is the Hertzian compression at pitch point;
Z_X is the *amplitude factor* for pitting strength (**Fig. 23**).
Z_H, the *zone factor*, determines the curvature at the pitch point:

$$Z_H = \sqrt{\frac{2 \cos \beta_b \cos \alpha_{wt}}{\cos^2 \alpha_t \sin \alpha_{wt}}}. \quad (49)$$

Z_E, the *elasticity factor*:
St/St: $Z_E \approx 190 \sqrt{N/mm^2}$, St/GG: $Z_E \approx 165 \sqrt{N/mm^2}$, GG/GG: $Z_E \approx 145 \sqrt{N/mm^2}$.
Z_ε, the contact ratio factor; Z_β, the helix factor:

$$Z_\varepsilon = \sqrt{(4 - \varepsilon_\alpha)/3} \quad \text{for spur gearing,}$$

$$Z_\varepsilon = \sqrt{1/\varepsilon_\alpha} \quad \text{for helical gearing } (\varepsilon_\beta \geq 1), \quad (50)$$

$$Z_\beta = \sqrt{\cos \beta}.$$

u, the gear ratio, z_2/z_1, is negative for internal gear pairs.

For lubricants and viscosity values other than those as per

Table 14. Standard gear materials, application, strength

No.	Type, treatment		δ^a	Application, properties	HB Flanks	σ_{Hlim} (N/mm²)^g	σ_{FE} (N/mm²)^g
1	Grey iron	GG 20		For complicated gear forms, cost-	180	300^h	80^h
	DIN 1691			effective, easily machinable,			
2		GG 25		silencing – susceptible to shock	220	340^h	110^h
3	Black malleable	GTS 35	12%	For small dimensions, properties	150	350^h	280^h
4	iron	GTS 65	3%	between GG and GS	220	440^h	310^h
	DIN 1692						
5	Spheroidal	GGG 40	12%	For large dimensions also;	180	390 to 470^h	280 to 370^h
6	graphite iron	GGG 60	4%	properties between GG and GS,	250	490 to 570^h	330 to 430^h
7	DIN 1693	GGG 100^k	4%	flame and induction hardening also possible	350	700	520
	Unalloyed cast steel			More cost-effective for large dimensions than rolled or forged			
8	DIN 1681	GS 52.1	18%	gears – difficult to cast (cavities,	160	320^h	250^h
9		GD 60.1	15%	casting stresses)	180	340^h	270^h
10	Standard	St 37	25%		120	320^h	250^h
11	structural steels	St 50	20%	Easily welded, no defined join	160	370^h	280^h
12	DIN 17 100	St 60	15%		190	430^h	300^h

	DIN 17 200 heat-treatable steels (also as cast steel^m)		\multicolumn{6}{l}{R_m in N/mm² for heat treatment cross-section^b as per}

No.		material	20∅	50∅	100∅	250∅	500∅	1000∅	HB Flanks	σ_{Hlim}	σ_{FE}
13		Ck 45 N^c	720	680	650				190	430 to 530^{h,m}	320 to 400^{h,m}
14		34 CrMo 4 V^d	980	880	800	700			270	530 to 710^{h,m}	430 to 580^{h,m}
15		42 CrMo 4 V^e	1080	960	870	740			300	580 to 770^{h,m}	450 to 620^{h,m}
16		34 CrNiMo 6 V	1190	1050	940	790			310	590 to 780^{h,m}	460 to 620^{h,m}
16A		30 CrNiMo 8 V		1160	1050	800;1200^f	1000^f		320	600 to 790^{h,m}	470 to 640^{h,m}
16B		34 NiCrMo 12.8 V				1300^f	1200^f	1100^f	350	650 to 840^{h,m}	490 to 650^{h,m}

No.	Type, treatment		Application, properties	HB Flanks	σ_{Hlim}	σ_{FE}
17	Heat-treatable steels, flame-	Ck 45	*Rotary* hardening, small dimensions, $b < 20$			Root also hardened
18	hardened or induction-hardened	34 CrMo 4	Rotary or individual tooth hardening			
19		42 CrMo 4	*Rotary* hardening (individual tooth	50 to 55 HRC	1000 to 1230	500 to 750
20		34 CrNiMo 6	hardening), *individual* tooth hardening, *not* sensitive to cracks, for high core strength at unhardened tooth root			Root not hardened 300 to 450
21	Heat-treatable steels and	42 CrMo 4 V	Nht < 0.6; R_m > 800; $m < 16$; Rather suitable for running-in, less	48 to 57	780 to 1000	520 to 740 HRC
22	carburising steels, nitrided	16 MnCr 5 V	edge-sensitive than 31 CrMo V 9 Nht < 0.6; R_m > 700; $m < 10$			
23	Nitriding steels, nitrided	31 CrMo V 9 V	Standard steel Nht < 0.6; R_m > 900; $m < 16$; Edge-sensitive for		1120 ($m < 16$)	
24		14 CrMo V 6.9 V	Nht > 0.6; R_m > 900; $m < 16$	60 to 63 HRC	1250 ($m < 10$)	560 to 840
26	Heat-treatable steels and	C 45 N	Low distortion, more favourable	42 to 45 HRC	650 to 760	460 to 600
27	carburising steels,	16 MnCr 5N	price: $d < 300$; $m < 6$			
28	nitro-carburised	42 CrMo 4 V	Higher core strength and surface hardness: $d < 600$; $m < 10$	52 to 55 HRC	650 to 800	460 to 640
29	Carbo-nitrided	34 Cr 4 V	Core strength up to 45 HRC, vehicle gears	55 to 60 HRC	1100 to 1350	600 to 900

Table 14. Continued

No.	Type, treatment		Application, properties	HB Flanks	$\sigma_{H\lim}$ (N/mm²)^g	σ_{FE} (N/mm²)^g
30	Carburising steels	16 MnCr 5	Standard steel: normal up to $m = 20$	58 to 62 HRC	1300 to 1500	620 to 1000
31	DIN 17 210	15 CrNi 6	For large dimensions above $m = 16$;			
32	Case-hardened	17 CrNiMo 6	Under sudden load above $m = 5$			

[a]Breaking elongation as dimension for viscosity.
[b]For lower third or scatter range.
[c]Budget-priced, easily machinable; for advantageous smoothable black-white structures, $\sigma_{H\lim}$ to 700.
[d]Easily welded.
[e]Standard steel for medium-sized and large gears.
[f]Obtainable.
[g]*Upper* limiting values for $\sigma_{H\lim}$ and σ_{FE} for quality industrial gears (steel production monitoring, high degree of purity, ground tooth flanks, acceptance as per factory certificate, many years of experience with carefully supervised heat treatment, comprehensive inspection of surface hardness, progression of the hardening process, structure, etc.)
Lower limiting values and values given without scatter range safely obtainable. They apply to materials taken from stock and involve limited checks on the main material data and heat treatment data.
[h]Carry out linear interpolation in case of hardness variations in groups nos. 1/2, 3/4, 5 to 7, 8/9, 10 to 12, 13 to 16 B.
[k]Austempered.
[m]$\sigma_{H\lim}$ and σ_{FE} approx. 80 N/mm² lower for GS.

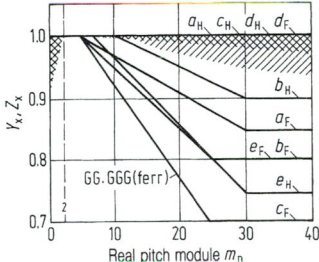

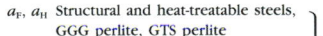

a_F, a_H Structural and heat-treatable steels,
    GGG perlite, GTS perlite
b_F, b_H Edge-hardened steels ⎫ continuous
c_F, c_H Grey iron, GGG ferr. ⎬ strength
d_F, d_H All materials under static stress ⎭
e_F, e_H Nitrided steels

Figure 23. Size factor for tooth root strength (index F). Size factor for pitting resistance (index H) as per DIN 3990, ISO/DIS 6336.

Table 5 and **Fig. 18**, take influence of viscosity and speed as per DIN 3990 into account. Use 85% $\sigma_{H\lim}$ for milled tooth profiles (roughness influence).

For hardened ground mating gears, $\sigma_{H\lim}$ of heat-treated gears can be increased by the mating of material factor, Z_w:

$$Z_w = 1.2 - (HB - 130)/1700 \qquad (51)$$

using the HB of the heat-treated gear.

Equation (48) applies to helical gears where $\varepsilon_\beta \geq 1$; it can also be used for helical gears where $\varepsilon_\beta = < 1$ and spur gears, provided profile offset values are selected as per **Fig. 13**. Otherwise, see DIN 3990. Convert to specific internal point of application, B (see **Fig. 7**) (DIN 3390) [1], at $z_{n1} < 20$: σ_{HD}.

Minimum margin of safety $S_{H\min}$: for guide data, see **Table 15**.

Micro-pitting (see [58]) approximately: $\lambda_{crit} \approx 0.7$. At $\lambda > \lambda_{crit}$, according to experience so far, micro-pitting should not be expected. (λ: see F8.3.)

Margin of Safety Against Fatigue Failure

The local stress arising at the tooth root (if the notch effect is taken into account) must be less than the permissible stress. So the condition arises:

Table 15. Guide data for safety factors

Damage limits	Continuous strength		
Load acceptance	Maximum moment[b]	Nominal moment × Application factor	
(a) – (b) – (c)	(a)	(b)	(c)
Pitting safety $S_{H\min}$	0.5 to 0.7	1.0 to 1.2	1.3 to 1.6
Tooth break safety $S_{F\min}$	0.7 to 1.0	1.4 to 1.5[a]	1.6 to 3.0[a]

(a) For calculations with *maximum* moment against *continuous* strength (e.g. shearing, pressing, converters, hoisting units); values apply to heat-treated or case-hardened gears (avoid nitriding).
(b) Normal case (usually industrial gears); plant gears to fulfil higher specifications; values in upper range.
(c) High reliability, critical cases (very high endurance, high risk of damage, high consequent costs, no spare parts, no overload safety devices – e.g. large-scale gears, turbo-gears, marine gears, aircraft gears).
[a]Adequate safety (approx. 1.5) against maximum moment provided (e.g. start-up shocks).
[b]Theoretical working moment.

$$S_F = \sigma_{FE} \, Y_X/(\sigma_{FO} \, K_A K_v K_{F\beta} K_{F\alpha}) \geq S_{F\min}. \qquad (52)$$

Here $\sigma_{FE} = \sigma_{F\lim} \cdot 2.0$; $\sigma_{F\lim}$ is the nominal flexural fatigue strength of the standard reference test gear with stress correction factor ($\approx$ notch configuration number) = 2.0; guide data for σ_{FE} as per test bench experiments (see **Table 14**).

Y_X is the *size factor* for tooth root strength; see **Fig. 23**.

σ_{FO} is the nominal value of the basic stress:

$$\sigma_{FO} = \frac{F_t}{bm_n} \, Y_{FS} Y_\varepsilon Y_\beta. \qquad (53)$$

Y_{FS}, the *tip factor*, determines tooth form, including notch form, when force is applied to the tip. The basic profile is as per DIN 867; see **Fig. 24**.
Y_ε, the contact ratio factor, determines the conversion

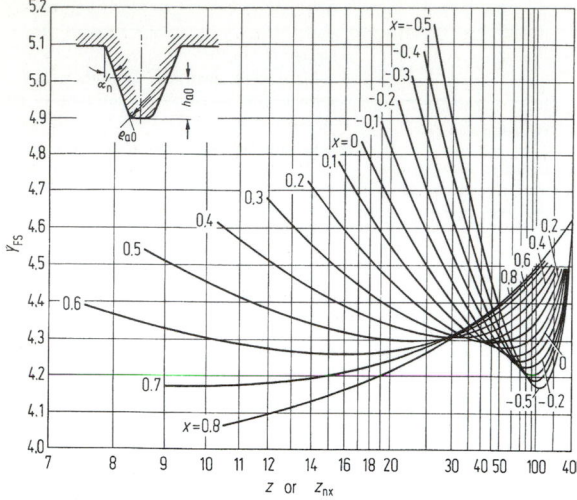

Figure 24. Tip factor (DIN 3990, ISO/DIS 6336). Y_{FS} ($= Y_{Fa} \cdot Y_{Sa}$) for basic profile: $\alpha_n = 20°$, $b_a/m_n = 1$, $b_{a0}/m_n = 1.25$, $p_{a0}/m_n = 0.25$; for rack, $Y_{FS} = 4.62$; for internal spur gears with $p_F = p_{a0}/2$: $Y_{FS} = 5.79$.

when a force is applied at the specific external point of application (for helical gears, Eq. (34) applies for the normal replacement gear).

Y_β is the helix factor.

$$Y_\varepsilon = 0.25 + 0.75/\varepsilon_{\alpha m}, \left.\begin{array}{c}\\\\\end{array}\right\} \quad (54)$$
$$Y_\beta = 1 - \beta°/120 \geq 0.75.$$

For large root rounding-off values, the stress concentration index must be taken into account (DIN 3990) [1]. Note the influence of greater roughness, grinding notches, shotblasting, grinding out notches [18].

Margin of Safety Against Hot Seizing and Cold Seizing

Subsequent remedial measures are often possible (see F8.5.1) [1]. For calculation, see [1] and DIN 3990, ISO/DIS 6336.

Margin of Safety Against Sliding Abrasion

This is necessary for speeds below 0.5 m/s. According to [19], increased wear should be expected if theoretical minimum lubricating film thickness (Eq. (37)) is less than 0.1 μm (maximum wear occurs at approximately 0.01 to 0.02 μm). For remedial measures, see F8.5.1. On calculation, see [1].

For the calculation of *timing gears*, gears with *load peaks* that rarely occur on or with *unified loads*, see [1, 19].

Example

Calculation of load-bearing capacity of first spur gear stage of a stirring machine. Drive: electric motor. □ = value on diagram.

Given. Motor speed $n_1 = 1000$ min^{-1}; power $P_1 = 51$ kW; quieter running required (see also **Table 3**). Centre distance a preset
□ Gear grade 6 as per DIN 3962 (see also **Table 3**); $F_{H\beta} = 10$ μm.
□ Basic profile as per DIN 867, $\alpha_n = 20°$ (**Fig. 10**).
□ Gear material: pinion 16 MnCr 5 (**Table 14**, no. 30), gear 42 CrMo 4 V (**Table 14**, no. 15)
□ Hardness: pinion 60 HRC, gear 300 HB
□ Flank machining (roughness): ground, $R_a = 0.5$ μm (corresponding to $R_z \approx 3$ μm)
□ Roughness at tooth root: $R_a \leq 2$ μm (corresponding to $R_z \approx 12$ μm.).

Geometry:	Gear 1	Gear 2	Unit
□ Normal pressure angle, α_n		20	°
□ Real pitch module, m_n		3.5	mm
□ Centre distance, a		180	mm
□ Tooth width, b		53	mm
□ Number of teeth, z	36	63	—
— Gear ratio, u		1.75	—
□ Helix angle, β		12	°
□ Profile offset factor, x	0.5	0.3686	—
□ Reference diameter, d, Eq. (5)	128.815	225.426	mm
□ Root number, d_f, Eq. (28)	123.5	219.2	mm
□ Tip diameter, d_a, Eq. (29)	139.0	234.7	mm
where $b_{fP} = 1.25$ m;			
$c = 0.25$ m;			
□ Real pressure angle, α_t, (see F8.1.5, pressure angle)		20.4103	°
□ Base circle, d_b, Eq. (20)	120.728	211.274	mm
□ Contact pitch, p_{et}, Eq. (21)		10.535	mm
□ Effective pressure angle α_{wt}, Eq. (31)		22.7462	°
□ Transverse path of contact, g_α, Eq. (23)		15.9	mm
□ Transverse contact ratio, ε_α, Eq. (24)		1.51	—
□ Overlap ratio, ε_β, Eq. (13)		1.00	—
□ Total contact ratio, ε_γ, Eq. (14)		2.51	—

Calculation of Load-Bearing Capacity

Circumferential force, Eq. (41), $F_t = 7.561$ N.

K* factor, Eq. (40) $= 1.74$ as per **Table 7**, adequately dimensioned.

Circumferential speed, Eq. (42): $v_t = 6.7$ m/s.

Lubricating oil viscosity at 40 °C, **Fig. 18**: $v_{40} \approx 1.3 \cdot 10^2$ mm²/s, ISO-VG 220.

Force Factors

Application factor: $K_A = 1.3$ used (see also **Table 11**).

Dynamic factor: $K_v \approx 1.08$ as per **Fig. 21b**, where $(v_t \cdot z_1/100) \cdot [u^2/(1 + u^2)]^{1/2} = 2.1$.

Width factor, $K_{H\beta}$ ($\approx K_{F\beta}$):

Initial characteristic value as per **Table 12** for $\sigma_{H \lim} = 750$ N/mm²/inclusive, going to $x_\beta = 0.55/0.85$, $f_{ma} \approx f_{H\beta} = 10$ μm (gear grade 6, see above), flank line deviation through cumulative deformation: $f_{shg} = 8$ μm as per **Table 13**. With Eq. (43): $F_{\beta y} = 12.6$ μm.

Figure 22, with $F_t K_A K_v / b = 200$ N/mm: $K_{H\beta}$ ($\approx K_{F\beta}$) ≈ 1.6.

Face factor, $K_{H\alpha}$ and $K_{F\alpha}$: helical gear, DIN grade ≤ 7, Eq. (46): $K_{H\alpha} = K_{F\alpha} = 1$.

Margin of Safety Against Pitting

Zone factor, Eq. (49) with β_b as per Eq. (35), α_t, α_{wt}: $Z_H \approx 2.3$.

Elasticity factor, for St/St: $Z_E \approx 190 \sqrt{\text{N/mm}^2}$.

Contact and helix factors (Eq. (50)): $Z_\varepsilon Z_\beta \approx 0.8$.

Nominal value of flank compression (Eq. (48)): $\sigma_{HO} = 466$ N/mm².

Size factor (**Fig. 23**): $Z_X = 1$

Pitting fatigue strength (**Table 14**), fixed, for pinion $\sigma_{H \lim} = 1500$ N/mm², for 300 HB gear $\sigma_{H \lim} = 750$ N/mm².

Mating of material factor (gear) (Eq. (51)): $Z_W = 1.1$.

Safety factor for pitting (Eq. (47)): pinion $S_H = 2.1$, gear $S_H = 1.2$. As per **Table 15** sufficient.

Margin of Safety Against Fatigue Fracture

Tip factor (**Fig. 24**): $Y_{FS1} \approx 4.32$, $Y_{FS2} \approx 4.35$ (with Eq. (34)): $z_{n1} = 38.3$, $z_{n2} = 67$).

Contact and helix factors (Eq. (54)): $Y_\varepsilon Y_\beta \approx 0.67$.

Nominal value of basic stress (Eq. (53)): $\sigma_{FO1} = 157$ N/mm², $\sigma_{FO2} = 158$ N/mm².

Basic fatigue strength, fixed as per **Table 14**, for pinion $\sigma_{FE} = 900$ N/mm², for gear $\sigma_{FE} = 600$ N/mm².

Size factor (**Fig. 23**): $Y_X = 1$.

Safety factor for fatigue failure (Eq. (52)): pinion $S_{F1} = 3.4$, Gear $S_{F2} = 2.2$. As per **Table 15** sufficient.

8.6 Bevel Gears

Greater efficiency as against worm gears, and more cost-effective at higher power levels (as straight bevel gear pair). More complicated as against spur gears (high offset, shaft angle deviations, axial position of gear and pinion, deflection with flying pinion). *Counter-measures:* limitation of tooth width, crown gearing, lapping and pairing pinion and gear together, axial setting of pinion, roller bearings (little bearing play), rigid housing (see F8.6.5).

8.6.1 Straight Bevel Gears

Normal up to $v = 6$ m/s, ground at up to 50 m/s (aircraft construction).

Tooth Form. Straight-sided basic crown gear, manufactured using planing tool with straight cutting – leads to

cutting teeth of *octoid* form [3], and therefore to profile offset only for *gear pairs with reference centre distances* (see F8.1.7). Moreover, reinforcement of the pinion at the cost of the gear by means of tooth thickness alteration is possible (lateral profile offset), as well as differing flank angles on the front and rear flanks or tooth height alteration with identical tools (separate planing tools for the two flanks!). Tooth height generally diminishing towards apex of cone [50].

Gear Dimensions (Fig. 25). Dimensions on *external* reference cone (back cone): index e. The tooth form (on the back cone, RB) is approximately the same as that of a spur gear with the radii r_{v1} and r_{v2} on the generatrices of the back cone.

$$\text{Axial angle } \Sigma = \delta_1 + \delta_2 \text{ usually } \Sigma = 90°. \quad (55)$$

Reference cone angle, δ, from
$$\tan \delta_1 = \sin \Sigma/(u + \cos \Sigma), \quad (56)$$

$$\text{For } \Sigma = 90°: \tan \delta_1 = 1/u, \quad \tan \delta_2 = u. \quad (57)$$

$$\text{External cone distance } R_e = 0.5 \, d_e/\sin \delta; \quad (58)$$

$$\text{for } \Sigma = 90°: R_e = (d_{e1}/2) \sqrt{u^2 + 1}. \quad (59)$$

External reference diameter
$$d_{e1} = z m_{e1}, \quad d_{e2} = z m_{e2} \quad (60)$$

with modulus on back cone, m_e.

Gear ratio
$$u = z_2/z_1 = d_{e2}/d_{e1} = \sin \delta_2/\sin \delta_1, \quad (61)$$

for $\Sigma = 90°$; see Eq. (57).

Tip diameter
$$d_{ae1} = d_{e1} + 2_{ae1} \cos \delta_1, \quad (62)$$
$$d_{ae2} = d_{e2} + 2_{ae2} \cos \delta_2, \quad (63)$$

normally:
$$h_{ae1} = m_e \, (1 + x_h); \, h_{ae2} = m \, (1 - x_h). \quad (64)$$

Dimensions on internal reference cone: index i instead of e.

Alternative Spur Gears, based on mean tooth width – decisive for load-bearing capacity calculation (**Fig. 25**).

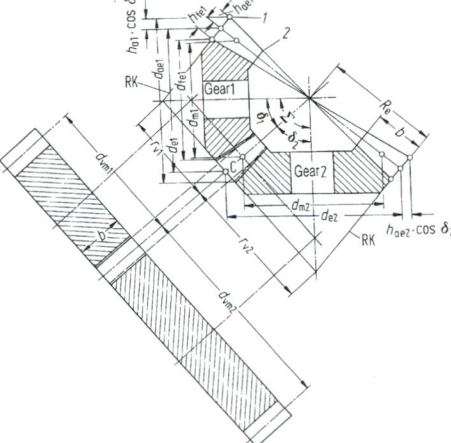

Figure 25. Bevel gear pair and alternative spur gears for calculating loadbearing capability. *1* heel, *2* toe.

Table 16. Guide data for selection of pinion tooth number[a], tooth width and profile offset factor[b] in bevel gears

u	1	1.12	1.25	1.6	2	2.5	3	4	5	6
z_1	18 to 40	18 to 38	17 to 36	16 to 34	15 to 30	13 to 26	12 to 23	10 to 18	8 to 14	7 to 11
b/d_1	0.212	0.226	0.240	0.284	0.336	0.404	0.474	0.615	0.75	0.75
x_h	0	0.03	0.06	0.12	0.18	0.24	0.28	0.36	0.42	0.45

Limiting values: $b/R_c \leq 0.3$; $b/d_1 \leq 0.75$; $b/m \leq 10$. For helical and spiral-toothed bevel gears $\varepsilon_\beta \geq 1.5$.

[a]For spiral-toothed, hardened bevel gears, select z_1 more on lower limit, for spur-toothed, unhardened bevel gears more on upper limit.
[b]For spur-toothed bevel bears with staggered O-toothing ($x_{h1} = -x_{h2}$) and normal tooth height ($b_{gP} = b_{fP} = m$, Fig. 10). Profile offset for helical or spiral-toothed gears about 85% of these values.

$$d_{m1} = d_{e1} - b\sin\delta_1, \quad \delta_{m2} = u d_{m1}, \qquad (65)$$

for

$$\Sigma = 90°: d_{m1} = d_{e1} - (b/\sqrt{u^2 + 1}). \qquad (66)$$

$$d_{vm1} = d_{m1}/\cos\delta_1, \; d_{vm2} = d_{m2}/\cos\delta_2, \qquad (67)$$

for

$$\Sigma = 90°: d_{vm1} = d_{m1}\sqrt{(u^2 + 1)/u^2},$$

$$d_{vm2} = d_{vm1} \cdot u^2. \qquad (68)$$

$$m_m = d_{m1}/z_1 = d_{m2}/z_2 = m_{vm} = d_{vm1}/z_{v1}$$

$$= d_{vm2}/z_{v2}. \qquad (69)$$

$$z_{v1} = z_1\sqrt{(u^2 + 1)/u^2}, z_{v2} = z_{v1} \cdot u^2. \qquad (70)$$

Recommendations for selection of number of teeth, modulus, tooth width, profile offset; see **Table 16**; backlash: **Table 17**. Basic profile: see **Fig. 10**, ISO 677.

Load-Bearing Capacity for alternative spur gears to be determined as per Eqs (65) to (70), using $F_t = 2M_1/d_{m1}$. Guide data for $K_{\beta\alpha} = (K_{H\beta}K_{F\alpha})$ as per Eqs (47) and (52), owing to greater uncertainties for bevel gears and the restricted contact pattern (crown gear):

$K_{\beta\alpha} = 2.0$ when pinion and gear supported on both sides;

$K_{\beta\alpha} = 2.2$ with floating pinion and ring gear supported on both sides;

$K_{\beta\alpha} = 2.5$ with floating supported pinion and ring gear.

Check: in no operating condition should contact pattern lie at end of tooth (see F8.6.5).

Bearing Forces. See F8.6.4.

8.6.2 Helical and Spiral Bevel Gears

Low-noise running; milled or planed and lapped up to $v = 40$ m/s; ground at up to 80 m/s (maximum 130 m/s); pay attention to axial forces!

The relationships in (F8.6.1) apply to the gear geometry for the transverse values of the bevel gears and alternative spur gears, i.e. $m = m_t = m_n/\cos\beta$.

Table 17. Normal flank play values for bevel gears and worm gears

Module m	Up to 1.6	Above 1.6 Up to 5	Above 5 Up to 16	Above 16
Flank play	(0.08 to 0.04)m	(0.05 to 0.03)m	(0.04 to 0.03)m	(0.03 to 0.02)m

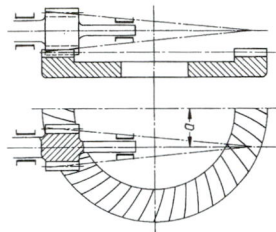

Figure 26. Crown wheel gear with axial offset, α, as per Dubbel, Taschenbuch für Maschinenbau, 13th edition.

Helical Gears. Tooth height dimensions such as b_{ac}, b_{fc}, are given dependent on the *real pitch* module, but the profile offset is frequently given in multiples of the transverse module. For selection of number of teeth, profile offset factor, x_h and remaining gear variables, see **Table 16**.

Spiral Bevel Gears. Helix angle variabie over width, flank line course, tooth heights, helix angle largely conditioned by manufacturing process (see K5.2). So layout and calculation as per machine manufacturer's specifications.

8.6.3 Special Gears

Crown Gears (Fig. 26). Pinion is spur wheel or helical gear, crown gear is generated by shaping, using pinion-type cutter; pinion offset also possible. Load-bearing capacity low [2].

Bevelled Spur Gears (Fig. 27). Spur wheels or helical gears with profile offset variable over width. As per **Fig. 27a**, suitable for setting for contact free from back-

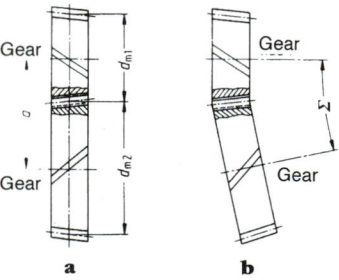

Figure 27. Tapered spur gears: **a** as spur gearing (parallel axes); **b** as bevel gear pair (axial angle Σ).

Table 18. Calculation of tooth force components on bevel gear

Spiral direction and direction of rotation[b] of driving gear	Axial force	Radial force
Right-hand spiral, rotating to right	Driving gear $$F_x = \frac{F_t}{\cos \beta}(\tan \alpha_n \sin \delta + \sin \beta \cos \delta)$$	Driving gear $$F_r = \frac{F_t}{\cos \beta}(\tan \alpha_n \cos \delta - \sin \beta \sin \delta)$$
or Left-hand spiral rotating to left	Driven gear $$F_x = \frac{F_t}{\cos \beta}(\tan \alpha_n \sin \delta - \sin \beta \cos \delta)$$	Driven gear $$F_r = \frac{F_t}{\cos \beta}(\tan \alpha_n \cos \delta + \sin \beta \sin \delta)$$
Right-hand spiral, rotating to left	Driving gear $$F_x = \frac{F_t}{\cos \beta}(\tan \alpha_n \sin \delta - \sin \beta \cos \delta)$$	Driving gear $$F_r = \frac{F_t}{\cos \beta}(\tan \alpha_n \cos \delta + \sin \beta \sin \delta)$$
or Left-hand spiral rotating to right	Driven gear $$F_x = \frac{F_t}{\cos \beta}(\tan \alpha_n \sin \delta + \sin \beta \cos \delta)$$	Driven gear $$F_r = \frac{F_t}{\cos \beta}(\tan \alpha_n \cos \delta - \sin \beta \sin \delta)$$

[a] Use values of angles β, α and σ for which the stresses are specified.
[b] Spiral direction and direction of rotation seen from bevel tip outwards.

lash, as per **Fig. 27b**, for small pitch angles that cannot be set on bevel gear cutting machines [21–23].

Hypoid Bevel Gears. Bevel gears with intersecting axes (**Fig. 1**). Execution with spiral toothing throughout, as per machine manufacturer's specifications [25–28].

8.6.4 Bearing Loads

Calculation of force components as per **Table 18** and **Fig. 28**. Attention should be paid to overturning moments of axial forces in calculating radial bearing loads.

8.6.5 Design Hints for Bevel Gears

For pinions mounted on shaft: gear rim thickness under toe ≥ 2 m (take groove into account if applicable). Bearing interval as per **Fig. 28**: $l_1 = (1.2$ to $2)d_1$, where $u = 1$ to 2; $l_1 = (2$ to $2.5)d_1$ where $u = 3$ to 6; one bearing as close as possible to pinion tip; $l_2 > 0.7d_2$. Contact pattern under full load approximately $0.85b$ (tooth ends free) with high-precision teeth and housing and rigid construction, otherwise smaller (approximately $0.7b$). Select direction of skew in such a way that axial force pushes pinion away from contact (guarding against face backlash). Mounting must permit axial shifting of pinion and gear (setting of contact pattern and face backlash).

Tooth widths of pinion and gear as similar as possible (approach edges!). Ensure oil feed to rear pinion bearing.

Hubs and gear-cutting tolerances DIN 3965, gear specifications in drawings DIN 3966.

8.7 Crossed Helical Gears

Characteristics (see F8, introduction), application: tacho drives, small units, textile machines, centrifuges and similar [1, 43–49].

8.8 Worm Gears

Characteristics (see F8, introduction): standard transmission in a stage 5 to 70 into low, 5 to 15 into high.

Automatic locking with driving gear (i.e. $\eta' \leq 0$) conditions efficiency, $\eta < 50\%$ with driving worm! Any alteration in worm requires alterations in tool (for gear pairing, see F8.1.4).

Main application (economic efficiency) up to centre distance of $a \approx 160$ mm, n up to 3000 min^{-1}, practicable up to $a = 2$ m and 1000 kW power. Low-play duplex worms for dividing gear [29].

Types of Pairing (**Fig. 29**). On most customary cylindrical worm gear (**Fig. 29a**). For double enveloping worm gear pair, see [29]; cone drive worm gear see [30].

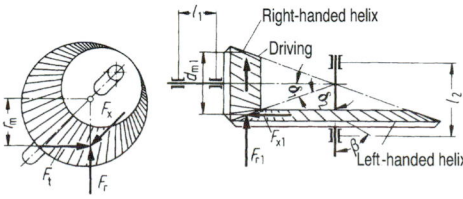

Figure 28. Tooth force components for calculating bearing forces.

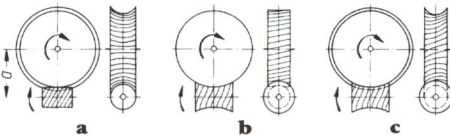

Figure 29. Types of worm gear pairs: **a** cylindrical worm gear (cylinder worm – globoid worm wheel), **b** contrate worm gear (enveloping worm – spur gear); **c** double enveloping worm gear pair (enveloping worm – globoid worm wheel).

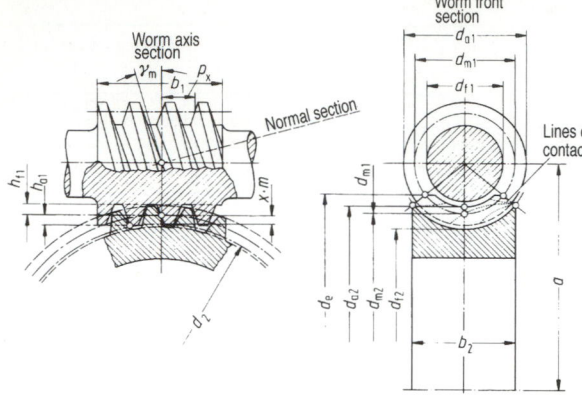

Figure 30. Defining quantities of a cylindrical worm gear.

Flank Form follows from manufacture (see K5.2). ZA, ZN, ZK and ZI worms differ only slightly in efficiency and flank load-bearing capacity. ZC (hollow flank) worms are therefore rather more favourable, but more sensitive to load variations (worm deflection).

8.8.1 Cylindrical Worm Gear Geometry

For axial angle $\Sigma = 90°$: initial sizes are average worm diameter d_{m1} and tooth profile in axial section (**Fig. 30**). With other axial angles, the analogous relationships for cylindrical helical gears apply (see F8.7).

Equations follow from the relationships between *rack profile* of worm (in axial section) and worm gear (sign: Z) or from considering worm as *helical gear* (sign: S) or as *threaded spindle* (sign: G).

Main Dimensions and Gear Data

$$\text{Transmission: } i = n_a/n_b \qquad (71)$$

S (with driving worm $= n_1/n_2$).

$$\text{Gear ratio: } u = z_2/z_1 \qquad (72)$$

(with driving worm $= i$).

Centre distance:

$$a = (d_{m1} + d_{m2})/2$$
$$= (d_{m1} + d_2 + 2xm)/2. \qquad (73)$$

Z Profile offset, x: Since a rack (= axial section of worm) Z is not altered by profile offset, only the worm gear can have a profile offset, $x = x_2$, as a result of which the pitch lines of the rack are displaced, while the pitch circle (= graduated circle) of the gear remains unaltered. For selection of profile offset, see F8.8.4.

Axial pitch modulus:

G

$$m = m_{x1} = m_{t2} = p_x/\pi$$
$$= p_{z1}/(\pi z_1) = d_{m1} \tan \gamma_m/z_1. \qquad (74)$$

Diameters:

$$d_{m1} = 2a - d_{m2}, \qquad (75)$$

$$d_{a1} = d_{m1} + 2m, \qquad (76)$$

Z $$d_{a2} = d_{m2} + 2m\,(1 + x), \qquad (77)$$

For normal worm profile, $2m$ is the usual common tooth height.

$$d_2 = z_2 m = d_{m2} - 2xm,$$
$$d_e = d_{a2} + m, \qquad (78)$$
$$d_{f1} = d_{m1} - 2(m + c_1), \qquad (79)$$
$$d_{f2} = d_{m2} - 2(m + c_2). \qquad (80)$$

(graduated circle = pitch circle)

See note on Eqs (76) and (77). (81)

Tip clearance usually $c_1 = c_2 \approx 0.2m$.

Pitch angle:

S

$$\tan \gamma_m = mz_1/d_{m1} = d_2/(ud_{m1}), \qquad (82)$$

Z

$$\tan \gamma_m = [(2a/d_{m1}) - 1]z_1/(z_2 + 2x). \qquad (83)$$

$$\text{Sliding speed: } v_g = \pi d_{m1} n_1/\cos \gamma_m. \qquad (84)$$

For ZI worms, the relationships for involute helical gears also apply (see F8.1.7) with $\beta_m = 90° - \gamma_m$.

Lines of Contact (C-lines)

Contact points and tooth form of gear can be calculated or constructed from given axial section profile, A, of worm for given pitch circle (= graduated circle) of gear in accordance with basic requirement of gear tooth system (see F8.1.1).

The same applies for every section, P, parallel to the worm axial section. *C-lines* are thus obtained; example, see **Fig. 30**. Since the tooth profile of the worm in section P differs from that in the axial section, here too there is another counter-profile.

For construction, see [1], for calculation [32, 33].

8.8.2 Tooth Loads, Bearing Loads

Calculation of circumferential force F_t from torque M and power P. For relationships see **Figs 30** and **31**. Tooth

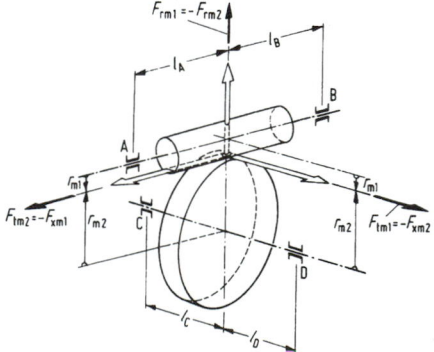

Figure 31. Tooth forces on a worm gear.

loads of gears with offset profiles are also specified for d_m [1].

$$F_{tm1} = F_{tm2} \tan(\gamma_m + \rho_z) = -F_{xm2}. \qquad (85)$$

To calculate the bearing loads, it is sufficient to insert ρ_z as always being positive, with a value of between 3 and 5° (see F8.8.3).

$$F_{tm2} = -F_{xm1}, \qquad (86)$$

$$F_{rm1} = F_{rm2} = F_{tm2} \tan \alpha_x. \qquad (87)$$

Bearing loads can be derived from these load components, radii and distance between bearings (**Fig. 31**). In this context, attention should be paid to overturning moments:

$$M_{K1} = F_{tm2}d_{m1}/2, \; M_{K2} = F_{tm1}d_{m2}/2. \qquad (88)$$

Similarly, any external transverse forces on input or output shafts should be taken into account.

8.8.3 Efficiency

For guide data, see **Table 19**. Preferences based on scatter ranges specified there:

gear material Cu-Sn-bronze
    more favourable than grey, iron, Al-bronze, brass;
hardened ground worm
    more favourable than hardened, milled worms;
ZC worms more favourable than other tooth forms;
high viscosity, suitable synthetic oils
    more favourable than low viscosity, mineral oils
    (attention should be paid to running-in characteristics);
large pitch (multiple and thin worms)
    (attention should be paid to bending)
more favourable than small pitch
    (single thick worms).

8.8.4 Rating and Evaluation of Load Capacity

To begin with, all requirements and influences with regard to stresses and functions should be carefully clarified. Compare check list for spur gears (**Table 6**).
    Dimensions are determined and the margins of safety, S_H and S_F and the worm deflection, δ, are checked, together, if applicable, at high speeds, with the margins of safety for temperature, S_T, and wear, S_W, as per [1], and the values obtained are corrected if necessary.

Centre Distance a, Transmission i and Power P_1 Specified

Select *number of teeth*, z_1, according to experience [BS 721] (a in mm). Numerical value equation:

Table 19. Cumulative efficiency η in % of cylindrical worm gears (average value), roller bearings, standard mineral oil (Italic figures: self-locking or possible self-locking)

Worm speed min⁻¹	Transmission				
	5	10	20	40	70
15	77 to 82	68 to 73	60 to 65	*47 to 52*	*30 to 35*
150	84 to 89	76 to 81	72 to 77	58 to 63	*43 to 48*
1500	91 to 96	88 to 92	81 to 87	75 to 80	64 to 69

$$z \approx (7 + 2.4a^{\frac{1}{2}})/u, \qquad (89)$$

Round number of teeth, z_1, up or down to next whole number; then z_2 as per Eq. (72).

Note. The z_2/z_1 ratio, which is not a whole number, makes it easier to manufacture the gear with a fly cutter and reduces the harmful effect of pitch fluctuations. The running noise decreases as the number of gear teeth increases; $z_2 \geq 30$ as far as possible with $\alpha_x = 20°$ and normal tooth height.

Selection of Diameter/Centre Distance Ratio, d_{m1}/a (**Fig. 32**). Pay attention to tendencies of S_H, δ and η. Thus, with a view to as high a degree of efficiency as possible, a low value for d_{m1}/a is aimed at, but attention should be paid to deflection, owing to the risk of worm shaft breakage.
    Then $d_{m1} = a \; (d_{m1}/a)$ and $\tan \gamma_m$ as per Eq. (82). Finally, a check should be made to determine whether available tools (especially hobs) can be used. This usually also determines the tooth form.

Recommendation for Profile Offset Factor, x.

ZI worms: $-0.5 \leq x \leq +0.5$, preferably: $x \approx 0$;
ZC worms: $0 \leq x \leq 1.0$, preferably: $x \approx 0.5$.

Additional Variables. m as per Eq. (74), d_2 in accordance with Eq. (78), d_{m1} as per Eq. (76), d_{m2} as per Eq. (77), d_{f1} as per Eq. (80), d_{f2} as per Eq. (81), d_{m2} in accordance with Eq. (75).
    Guide data for additional dimensions (see **Fig. 30**):

$$d_e \approx d_{a2} + m, \; b_1 \approx 2m(z_2 + 1)^{\frac{1}{2}}, \qquad (90)$$
$$b_2 \approx 2m[1 + (d_{m1}/m + 1)^{\frac{1}{2}}].$$

Worm (d_{m1}, z_1, m) and Transmission i Given

Interesting if hobs are available to cut the gear. It should also be noted that *one* worm (i.e. *one* hob as well) can be used for various transmissions and provides for different centre distances to this end.
    First determine z_2 as per Eq. (72) and select x_2, d_{m2} as per Eq. (78) and a as per Eq. (73). Then continue as described above.

Gear Moment M_2, Speed n_2, Transmission i, Given

Calculate centre distance, a, from Eq. (91) and the variables given there. Round up a to next highest value of range as per DIN 3976. Then continue as described above.

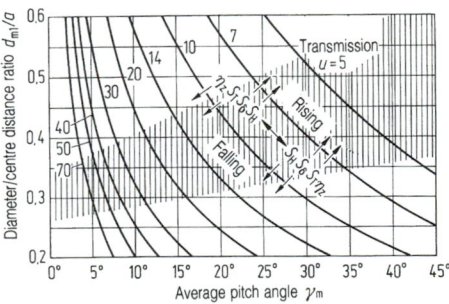

Figure 32. d_{m1}/a ratio. Solid curves as per Eq. (83) for $x = 0$; dotted lines define field of industrially made worm gears; ↑ increases, ↓ decreases.

Table 20. Material characteristic values for worm gears

Standard	Worm gear material		$R_{p0.2min}$ (N/mm²)	R_m (N/mm²)	HB	δ_s (%)	Elastic modulus (N/mm²)	$\left(\dfrac{Z_E^c}{\sqrt{N/mm^2}}\right)$	$\sigma_{Hlim}{}^a$ (N/mm²)	$U_{lim}{}^b$ (N/mm²)
DIN 1705	G-CuSn 12		140	260	80	12	88 300	147	265	115
	GZ-CuSn 12	SK 12	150	280	95	5	88 300	147	425	190
	G-CuSn 12 Ni		160	280	90	14	98 100	152	310	140
	GZ-CuSn 12 Ni	SK 12 Ni	180	300	100	8	98 100	152	520	225
	G-CuSn 10 Zn		130	260	75	15	98 100	152	350	165
	GZ-CuSn 10 Zn		150	270	85	7	98 100	152	450	190
—	GZ-CuSn 14		200	300	115	4	92 700	150	370	180
DIN 1709	G-CuZn 25 Al 5		450	750	180	8	107 900	157	500	565
	GZ-CuZn 25 Al 5	SoMs	480	750	190	5	107 900	157	550	605
DIN 1714	G-CuAl 11 Ni^{d,e}		320	680	170	5	122 600	164	250	402
	GZ-CuAl 11 Ni^{d,e}		400	750	185	5	122 600	164	265	502
	GZ-CuAl 10 Ni	WIA	300	700	160	13	122 600	164	660	377
DIN 1691	GG-25^{e,f}		120	300	250		98 100	152	350	150
DIN 1693	GG-70^{e,f}		500	790	260	5.5	175 000	182	490	628

[a] Applies to case-hardened worms (ground, HRC 60 ± 2); for heat-treated, unground worms: values for σ_{Hlim} multiplied by 0.75; for grey iron worms, multiply values for σ_{Hlim} by 0.5.
[b] Applies to $\alpha_m = 20°$, for $\alpha_n = 25°$ multiply values by 1.2, for alternating stress multiply values by 0.7.
[c] For steel worms. For grey iron worms: $Z_E = [2.86\,(E_1 + E)\,(E_1 E)]^{-\frac{1}{4}}$ with E_1 for grey iron, E as per table.
[d] Driven only with mineral oil.
[e] For low sliding speeds (manual operation).
[f] Pearlitic.

Calculation of Margin of Safety Against Pitting, S_H

Numerical value equation:

$$S_H = \sigma_{Hlim}\, Z_h Z_n / (Z_E Z_p\, \sqrt{1000 M_2 K_A / a^3}) \ge S_{Hlim} \quad (91)$$

(for units, see tables and M_2 in N m).

For σ_{Hlim}, rolling strength, and Z_E, elasticity factor, see **Table 20**.
Service life factor, $Z_h = (25\,000/L_h)^{1/6} \le 1.6$ with L_h in h.
Load cycle factor, $Z_n = [1/(n_2/8 + 1)]^{1/8}$; contact factor, Z_p: for guide data, see **Fig. 33**. Application factor, K_A: see **Table 11**.

S_{Hmin}, depending on reliability of specifications and consequences in case of damage, = 1 to 1.3.

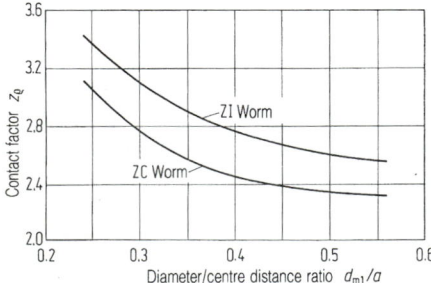

Figure 33. Contact factor Z_p for ZI worms ($\alpha_0 = 20°$, $-1 \le x \le 0.5$) (also approximately, for ZK, ZA and ZN worms) and ZC worms ($\alpha = 24°$, $0.3 \le x \le 1.2$) [1, 33].

Calculation with variable load and speed, short-time operation [1]. Wear load-bearing capacity [1, 34].

Calculation of Margin of Safety Against Tooth Breakage, S_F

$$S_F = U_{lim}\, mb_2 / (F_{t2} K_A) \ge S_{Fmin}; \quad (92)$$

for U_{lim}, limiting value of U factor, see **Table 20**; for K_A, application factor, see **Table 11**; $S_{Fmin} \approx S_{Hmin}$.

Checking Worm Deflection

The deflection, δ, of the worm must be limited in order to avoid problems with contact (for failure to meet basic requirement of gear tooth system, see F8.1.1) and increased contact pattern dislocation.

Assumptions. Shafts with diameter d_{m1} on two supports with intervals as per **Fig. 31** under load from net force from F_{tm1} as per Eq. (85), F_{rm1} as per Eq. (87) and M_{K1} as per Eq. (88).
Limiting value, $\delta_{lim} = m/100$ to $m/250$, depending on individual initial capability of material pairing and requirement for efficiency.

8.8.5 Embodiment Design, Materials, Bearings, Accuracy, Lubrication, Assembly

Embodiment Design for housings, shafts, seals: see F8.9. For example, see **Fig. 34**.
Position of worm for splash lubrication, as low as possible, with $v_1 < 10$ m/s, also at side, for $v_2 < 5$ m/s also above; for injection lubrication, any position.

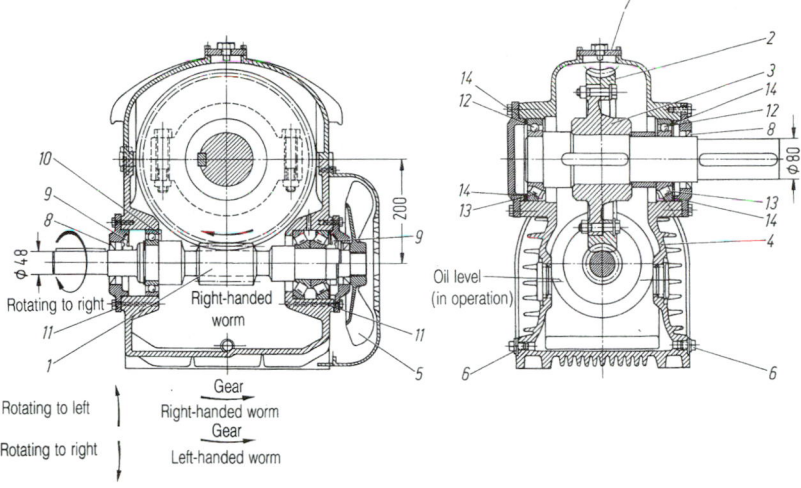

Figure 34. Worm gears (Flender, Bocholt). Rated power, 24.5 kW, $n_1 = 1500$ min^{-1}, $i = 20$. *1* ZC worm, 16 MnCr 5, case-hardened, ground; *2* gear rim, GZ – CuSn 12 Ni; *3* hub St 37; *4* housing GG 20 with horizontal fins; *5* ventilator; *6* oil drain; *7* inspection cover with ventilation; *8* radial sealing ring (sealing inwards); variable sealing of worm shaft shown; *9* additional sealing rings; *10* centrifugal disc; *11* oil return (shown offset); *12* separable ball bearing (for light duty); *13* tapered roller bearing (for heavy duty); *14* shim rings for axial adjustment of gear. Dimensions in mm.

Optimum Worm made of carburising steel (58 to 62 HRC) or heat-treatable alloy steel, carburised surface hardened (HRC < 56), at $v_0 < 3$ m/s also unhardened.

For power gears usually as right-hand solid worm, **Fig. 34**. For cost-effective gear under low load, also slip-on hollow worm. Bearing interval as small as possible (deflection!):

$$l = (1.3 \text{ to } 1.5)a \qquad (93)$$

Worm Gear Rim for power gears, bronze moulded by centrifugal action (GZ-CuSN 12 or GZ-CuSn 12 Ni) most suitable, having initial capability and with little tendency to be corroded. Al-bronze, special brass for low gliding speeds only (danger of seizing, increased sliding abrasion). Grey iron for low speeds only (manual operation), relatively favourable for pairings with grey iron worms.

Gear Rim usually screwed to hub using dowel screws; for press-on or cast gear rims, see [1].

Bearing Interval of gear shaft not too small (tipping risk!): $l_2 = l_c + l_D = (0.5 \text{ to } 0.7)d_2$ (see **Fig. 31**).

Mounting to be on roller bearings throughout, sliding bearings for low running noise only (e.g. for lifts).

Worm Shaft. For small and medium-sized dimensions, mountings with separable ball bearings or tapered roller bearings, range 313. For large dimensions, fast and loose mountings, for example with double-row angular ball bearings.

Gear Shaft. Grooved ball bearings, range 63, or tapered roller bearings, range 302, 322.

Accuracy, Flank Backlash. As grades not standardised to date, guidelines as per DIN 3961 to 3964, *individual and cumulative deviations*: grade 4 to 5 for precise dividing gears, sighting mechanisms and similar; grade 5 to 6 for lifts and smooth-running gears with $v_t < 5$ m/s; grade 8 to 9 for normal industrial sectors; grade 10 to 12 for auxiliary drives, manual drives and similar with $v_t < 3$ m/s.

Set contact pattern on outgoing side (lubrication wedge!).

Running-in (with nominal moment, low speed, thin oil) increases efficiency and flank load-bearing capacity, but is economically possible only in special cases.

Flank Backlash is approximately as per **Table 17**. For low-play gears see [28].

Lubrication. For a guide to the selection of oil viscosity and type of lubrication, see **Fig. 18**. Grease lubrication only at $v_t < 0.8$ m/s or for intermittent service (heat removal), abrasion in grease, attention should be paid to difficult grease changes! Mineral oils with mild extreme pressure additives simplify running-in. Synthetic oils make low coefficients of friction possible, i.e. high efficiency and high limiting heat output; running-in behaviour usually less favourable. Oil change after running-in, then after approximately 3000 hours, then approximately once a year [35].

8.9 Epicyclic Gear Arrangements

H. W. Müller, Darmstadt

8.9.1 Kinematic Fundamentals, Terminology

Epicyclic gear arrangements differ essentially in only one aspect from the usual simple transmission gearing. While with transmission gearing the housing, together with the gears housed within, is rigidly connected to a base, with epicyclic gear arrangements it is mounted on the base so that it is able to rotate, and is provided with an additional shaft (**Fig. 35**). So, from positive transmission gearing with a degree of operation $F = 1$, there emerges a *differential or superimposed gear arrangement* with a degree of operation $F = 2$. (The degree of operation of a gear lays down how many movements of any type can and should be assumed for it in order to determine its state of motion clearly.) Here the original housing is shrunk onto

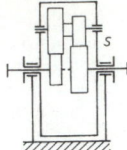

Figure 35. Development of an epicyclic gear from a transmission gear.

a support that carries only the gear mountings. Protection and oil tightness are ensured by a new housing. Kinematically, however, this is a part of the base. The torque of the new support shaft, s, is identical with the supporting moment of the original gear housing.

In this way, epicyclic gear arrangements can be developed from toothed-wheel and friction gears (*epicyclic gears*), hydrostatic gears, traction mechanism gears, linkages having only turning and sliding pairs and other gears [58]. Epicyclic gears (for the most frequent formats, see **Fig. 36**) are also described as "*planet gears*" and their gearwheels, with circulating axes, as "*planet gearwheels*" or "*planets*".

If the new support shaft, s, is momentarily or permanently secured or designed to be stationary, then the epicyclic gear arrangement reverts to "*stationary transmission*" with the "*stationary gear ratio*", i_{12}, using any of its "*stationary transmission shafts*", numbered 1 and 2 and the "*stationary efficiency*", η in both possible directions of the power flow (pfl) during the exchange from drive to power takeoff:

Stationary gear ratio, $i_{12} = \left(\dfrac{n_1}{n_2}\right)_{(n_s = 0)}$.

Stationary efficiency η_{12} for drive on shaft 1, power takeoff on 2.

Stationary efficiency η_{21} for drive on shaft 2, power takeoff on 1.

The indices 1, 2 and s each correspond to the shafts so described with their gears or the support. For speed ratios or speed transmissions, the sequence of indices indicates: first index numerator, second index denominator: for efficiency values; first index drive shaft, second index power takeoff shaft. Planet gears are described using p and the index of the gear that they mesh with in each case - **Fig. 36**.

This individual indexing of the epicyclic gear shafts using 1, 2 and s simplifies calculation and makes it easier, and makes it possible, for example, to analyse the operating behaviour of all models of epicyclic gear arrangements using a single simple computer program [59, 60].

In a stationary transmission gear, the power is transferred through the mesh exclusively as "*rolling power*", P_w, during the meshing of the gears with their "*rolling speeds*", n_{w1} and n_{w2}. While this is happening, the tooth friction power loss P_{vz} disappears as lost heat.

Should only the support be set in motion, at the speed n_m, with the transmission gear *stationary*, the entire gear arrangement rotates, including the two stationary transmission shafts, at this speed, like a coupling, with no internal relative motion. This can involve the transmission of "*coupling power*", p_k, without any loss, at the "*coupling speed*", n_s. Should a coupling speed, n_s, be superimposed on an *operating* transmission gear, then the typical operating condition arises of a planet gear with three running shafts with speeds of n_s, $n_1 = n_{w1} + n_s$ and

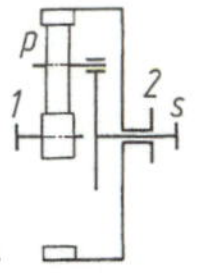

a	A:	$i_{12} = -1.2 \ldots -11.3$	
	B:[a]	$i_{12} = z_2/z_1$	
	C:	$\eta_{12} \approx \eta_{21} \approx 0.985$	
	D:	$(|z_2| + |z_1|)/q = g$	
b	A:	$i_{12} = -0.54 \ldots -1 \ldots -53$	
	B:[a]	$i_{12} = (z_2/z_{p2})(z_{p1}/z_1)$	
	C:	$\eta_{12} \approx \eta_{21} \approx 0.985$	
	D:	$(|z_{p1} z_2| + |z_1 z_{p2}|)/qt = g$	
c	A:	$i_{12} = -1$	
	B:	$i_{12} = -z_2/z_1$	
	C:	$\eta_{12} \approx \eta_{21} \approx 0.98$	
	D:	$(|z_2| + |z_1|)/q = g$	
d	A:	$i_{12} = 1 \ldots 41$	
	B:	$i_{12} = (z_2/z_{p2})(z_{p1}/z_1)$	
	C:	$\eta_{12} \approx \eta_{21} \approx 0.98$	
	D:	$(|z_{p1} z_2| - |z_1 z_{p2}|)/qt = g$	
e	A:	$i_{12} = 1 \ldots 2.7$	
	B:[a]	$i_{12} = (z_2/z_{p2})(z_{p1}/z_1)$	
	C:	$\eta_{12} \approx \eta_{21} \approx 0.99$	
	D:	$(|z_{p1} z_2| - |z_1 z_{p2}|)/qt = g$	
f	A:	$i_{12} = 1.2 \ldots 17.6$	
	B:[a]	$i_{12} = -z_2/z_1$	
	C:	$\eta_{12} \approx \eta_{21} \approx 0.975$	
	D:	$(|z_2| - |z_1|)/q = g$	
g	A:	$i_{12} = -0.2 \ldots -17.6$	
	B:	$i_{12} = -z_2/z_1$	
	C:	$\eta_{12} \approx \eta_{21} \approx 0.99$	
	D:	$z_1 . z_2$ variable	

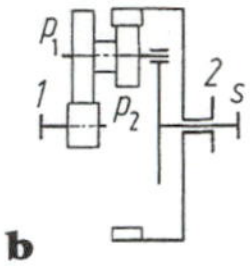

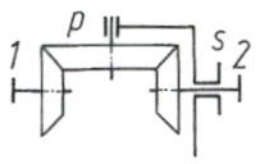

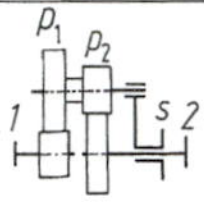

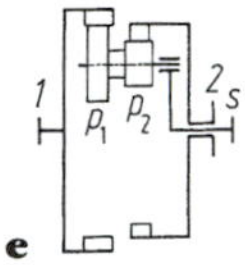

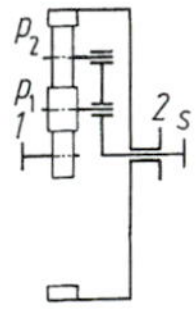

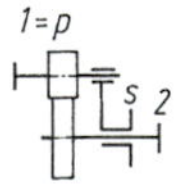

[a] tooth numbers of hollow gears are negative, see DIN 3960

Figure 36. Most frequent types of planet gear: **a–c** minus gears, **d–f** plus gears, **g** open planet gear. z = tooth numbers. A = possible range of stationary gear ratio for $q = 3$ planets/planet sets on periphery, approximately equal tooth root stress for all gears, $z_{min} = 17$, $z_{max} = 300$. B = stationary gear ratio. $C = \eta_{12} = \eta_{21}$ with $\eta_{wa} = 0.99$ for a spur gear stage, $\eta_{wi} = 0.995$ for a hollow gear stage. D = tooth number conditions for uniform arrangement of q planets/planet sets on periphery with $\pm g$ = whole number, t = largest common divisor of z_{p1} and z_{p1} of a stage planet.

$n_2 = n_{w2} + n_s$. Here the rolling power, P_w, which can be transferred only between gear shafts 1 and 2, and the coupling power, P_k, transferred without loss between all three shafts, also overlap at the same time.

Vice versa, the rolling speeds of a gear operating with three shafts can be derived as $n_{w1} = n_1 - n_s$ and $n_{w2} = n_2 - n_s$, as also can the stationary gear ratio

$$i_{12} = \frac{n_1 - n_s}{n_2 - n_s}. \tag{94}$$

Rearranged, this *basic speed equation*, valid for all types of epicyclic gear, can be simplified to give

$$n_1 - n_2 i_{12} - n_s(1 - i_{12}) = 0. \qquad (95)$$

While the *gear ratio*, i_{12}, of a positive transmission gear is unchangingly fixed by its geometrical data, e.g. the gear diameter, *any* two speeds can be preset for a three-shaft epicyclic gear and can determine its state of motion. The speed ratios arising as a result of such speeds can no longer be described in terms of format-dependent "gear ratio", i, but are called *"free speed ratios"*, k. Particular attention should be paid to this difference, because both variables can appear in *one* equation. Thus, for example, for any preset free speeds, n_1 and n_2, Eq. (95) gives:

$$k_{1s} = n_1/n_s = (1 - i_{12})/(1 - i_{12}/k_{12}).$$

However, should one of the three shafts be fixed, e.g. $n_2 = 0$, or $n_1 = 0$, then the gear arrangement becomes positive again, and Eq. (95) gives the *"epicyclic gear ratios"*

$$i_{1s} = 1 - i_{12}, \quad i_{2s} = 1 - 1/i_{12} \qquad (96)$$

as well as their reciprocal values, which describe the rotation of the support and the planet gears. The shaft not referred to in the index of a specific gear ratio, i, is not moving.

8.9.2 Generalisation of Calculations

The two stationary transmission shafts and the support shaft of an epicyclic gear are kinematically at the same level. Thus, Eq. (95) can also be written in a general form [58]:

$$n_a - n_b i_{ab} - n_c(1 - i_{ab}) = 0, \qquad (97)$$

where a, b and c can be replaced by 1, 2 or s in any arrangement. From this follows **Table 21**, showing the direct calculation of any free speed ratio, k, or $1/k$, assuming that any stationary gear ratio or epicyclic gear ratio, i, and any free speed ratio, k, or $1/k$, of the gear arrangement are known. But it also follows, as a wider consequence of this, that the equations for *all* operating data, and thus also for torque values, power values and efficiency values, remain valid if the indices of the shafts are exchanged in any way, provided that it is the same change for all equations.

Thus, the equations given below for simple epicyclic gear arrangements for the calculation of torque values,

power values and efficiency values are also valid for any compound gear trains, provided these have a basic ratio of $F = 2$, with three external connecting shafts, a, b and c, and provided their speeds and torque values are not reciprocally dependent on one another, as sometimes happens with hydrodynamic converters. It is of no importance here which three members or shafts, out of the many elements present in the gear arrangement, are selected as external connecting shafts.

For gear arrangements with non-uniform transmission, e.g. linkages having only turning and sliding pairs, the equations are valid, in each case, for only *one* relative arrangement of the elements in their *positive* kinematic chain [61] and the associated instantaneous gear ratio between two of the three connecting shafts.

The complete interchangeability of the indices is the basis for a principle that is useful in the building up of gears:

Should any stationary or epicyclic gear ratio of an epicyclic gear coincide with any stationary or epicyclic gear ratio of another epicyclic gear, then the two gears are kinematically at the same level, i.e. both have the same six gear ratios, but as a rule they differ in efficiency.

For an example of kinematically equivalent gears, see **Fig. 37**.

8.9.3 Sign Conventions

The following sign conventions apply in the analysis and building up of epicyclic gears:

Speeds. All speeds of parallel shafts with the same direction of rotation have the same signs. The positive direction of rotation ($n > 0$) can be attributed to any shaft. Speeds with the opposite direction of rotation are then negative. It follows from this that

gear ratios i and free speed ratios k are positive if shafts are running in the same direction (i, k > 0), and negative if shafts are running in opposite directions (i, k < 0).

The direction of rotation of required speeds is then determined in accordance with the same rule from their signs, obtained as per Eq. (95), Eq. (97) or **Table 21**.

Torque Values. A torque value is positive ($M > 0$) if the torque is acting on the gear in the direction of rotation defined as positive; if it is acting in the opposite direction, it is negative ($M < 0$).

Power Values. It follows from the above definitions that driving power fed to a gear is always positive

Table 21. Generally valid conversion of free speed ratios, k or $1/k$, of gear with a known stationary or epicyclic gear ratio, i_{ab}. For a and b, use indices of known gear ratio, for c index of remaining shaft

Desired	In relation to free speed ratio		
	k_{ab} or k_{ba}	k_{bc} or k_{cb}	k_{ca} or k_{ac}
$k_{ab} =$	$1/k_{ba}$	$k_{cb}(1 - i_{ab}) + i_{ab}$	$\dfrac{k_{ac} \cdot i_{ab}}{k_{ac} - 1 + i_{ab}}$
$k_{bc} =$	$\dfrac{1 - i_{ab}}{k_{ab} - i_{ab}}$	$1/k_{cb}$	$\dfrac{k_{ac} + i_{ab} - 1}{i_{ab}}$
$k_{ca} =$	$\dfrac{1 - i_{ab}k_{ba}}{1 - i_{ab}}$	$\dfrac{1}{1 - i_{ab}(1 - k_{bc})}$	$1/k_{ac}$

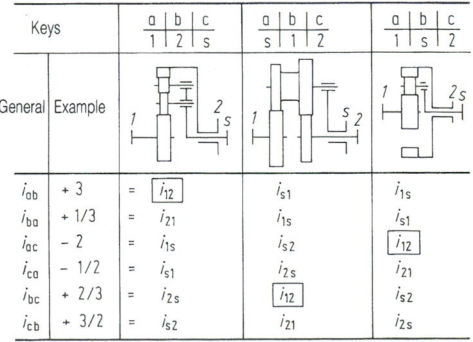

Keys	a | b | c : 1 | 2 | s	a | b | c : s | 1 | 2	a | b | c : 1 | s | 2
General | Example			
i_{ab}	$+ 3$	$= i_{12}$	i_{s1}	i_{1s}
i_{ba}	$+ 1/3$	$= i_{21}$	i_{1s}	i_{s1}
i_{ac}	$- 2$	$= i_{1s}$	i_{s2}	i_{12}
i_{ca}	$- 1/2$	$= i_{s1}$	i_{2s}	i_{21}
i_{bc}	$+ 2/3$	$= i_{2s}$	i_{12}	i_{s2}
i_{cb}	$+ 3/2$	$= i_{s2}$	i_{21}	i_{2s}

Figure 37. Example of three kinematically equivalent planet gears.

$(P_{an} = 2\pi M_{dr} n_{an} > 0)$, because a drive shaft always assumes the direction of rotation in the sense of rotation of the driving torque. Power takeoff values, in contrast, are negative $(P_{ab} < 0)$, because the external power takeoff moment acting as a brake on the gear is operating in the opposite direction to the power takeoff sense of rotation. Power losses are negative, as they constitute a power takeoff $(P_v < 0)$.

8.9.4 Torques, Powers, Efficiencies

Torques. The *ratio* of the torques is determined through the stationary gear ratio i_{12} and the stationary efficiency, η_{12} and η_{21}. It does not alter if any coupling speeds n_s are superimposed (without loss) on an operating transmission gear.

From the equilibrium conditions there follows the moment equilibrium

$$M_1 + M_2 + M_s = 0. \tag{98}$$

For the transmission gear, it follows from the power balance that, in the two power flow directions:

Drive on 1: $M_2 n_2 = - M_1 n_1 \eta_{12}$,

Drive on 2: $M_2 n_2 = - M_1 n_1 / \eta_{21}$.

If the two efficiencies are combined in the expression η_0^{w1}, the torque ratios can be formulated, independently of the power flow:

$$\frac{M_2}{M_1} = - \frac{n_1}{n_2} \eta_0^{w1} = - i_{12} \eta_0^{w1}. \tag{99}$$

Equations (98) and (99) give

$$\frac{M_s}{M_1} = i_{12} \eta_0^{w1} - 1, \tag{100}$$

$$\frac{M_s}{M_2} = \frac{1}{i_{12} \eta_0^{w1}} - 1. \tag{101}$$

Since the power balance of the transmission gear is identical to the balance of the rolling power if the support is circulating, these equations are also valid for epicyclic gears. Here the exponent w1 follows from the sign of the rolling power P_{w1} of the shaft 1: If $P_{w1} > 0$, the rolling power flows from shaft 1 to shaft 2, if $P_{w1} < 0$, from 2 to 1. From that follows the definition of η_0^{w1} for the calculation:

$$M_1^* (n_1 - n_s) \cdot 2\pi \begin{cases} >0: \text{w1} = +1 \rightarrow \eta_0^{w1} = \eta_{12} \\ <0: \text{w1} = -1 \rightarrow \eta_0^{w1} = 1/\eta_{21} \end{cases} \tag{102}$$

where M_1^* is the preset torque, or is calculated, if M_2 or M_s is preset, from Eq. (99) or (100), using $\eta_0^{w1} = 1$. Equations (99) and (100) indicate that the ratios of the three shaft moments to one another are determined only by the stationary gear ratio, i_{12}, and the stationary efficiencies, η_0^{w1}, and are thus constant for either of the two rolling power flows.

$$M_1 : M_2 : M_s = f(i_{12}, \eta_0^{w1}) = \text{const.} \tag{103}$$

This equation, characteristic for differential gears, is valid irrespective of the individual speeds involved, even when one shaft is stationary. If power is transferred between three machines through the three shafts of an epicyclic gear, then Eq. (95) must be fulfilled for the speeds, and Eqs (98) and (103) for the torque values. This arrangement leads to an operating condition in which the reciprocal coordination of speeds and torques, though

still free, results from the characteristics $M = f(n)$ of the three machines [58]. Should this fail to bring about a stable condition, the equipment begins to race or stops. Should one of the torques M equal zero (e.g. the machine be disconnected), then the remaining moments also equal zero, as per Eq. (103), the gear runs idle, and power transmission is not possible. For a list of torque equations see **Table 22**.

In accordance with Eq. (98), one of the three shaft moments must have the opposite sign to the other two and must be equal to the sum of the other two. This shaft is called the *cumulative shaft*, the other two being the *differential shafts*. For epicyclic gears with negative stationary gear ratios (*minus gears*), the support shaft is always the cumulative shaft, and with positive stationary gear ratios (*plus gears*) it is the slower-running stationary transmission shaft. If the cumulative shaft is stopped, negative transmission always occurs on the two differential shafts running, owing to their torques being in the same direction, and positive transmission occurs if a differential shaft is stopped. Thus any simple epicyclic gear can generate two reciprocal negative transmissions and four reciprocal positive transmissions in pairs.

Powers. With M in N m (kN m) and n in s^{-1}, the shaft outputs and the power loss, P_v, become:

$$P_1 = M_1 n_1 2\pi \text{ W (kW)}, \tag{104}$$

$$P_2 = M_2 n_2 2\pi \text{ W (kW)}, \tag{105}$$

$$P_s = M_s n_s 2\pi \text{ W (kW)}, \tag{106}$$

$$P_v = M_1 (n_1 - n_s) 2\pi (1 - \eta_0^{w1}) \text{ W (kW)}. \tag{107}$$

A characteristic feature of the epicyclic gear is that the shaft outputs, P_1 and P_2, are the sums (superpositions) of the rolling and coupling powers. With $\omega = 2\pi n$, the following equations are obtained:

Shaft output = rolling power + coupling power

$$P_1 = P_{w1} + P_{k1} = M_1 (\omega_1 - \omega_s) + M_1 \omega_s,$$

$$P_2 = P_{w2} + P_{k2} = M_2 (\omega_2 - \omega_s) + M_2 \omega_s,$$

$$P_s = P_{ks} = M_s \omega_s.$$

Depending on the speeds selected, the rolling and coupling powers can have the same sign or opposing signs, i.e. have power flows which move in the same direction or in opposite directions. The shaft outputs, P_1 and P_2, can thus constitute the sum of these two partial outputs or the difference between them. In the first case, the dissipative rolling power remains smaller than the shaft output, and the total efficiency becomes greater than the stationary efficiency. But with partial power flows going in opposite directions, the rolling power can be as much greater than the shaft power as is desired; the total efficiency then becomes correspondingly lower than the stationary efficiency. It can even become negative, and thus lead to the automatic locking of the gear – see F8.9.5. This con-

Table 22. Formulae for torque values
With w1 = + 1: $\eta_0^{w1} = \eta_{12}$ or w1 = − 1: $\eta_0^{w1} = 1/\eta_{21}$, w1 from **Table 23** for transmission gears, from **Table 24** for superposition gears, or from Eq. (102)

$M_1 + M_2 + M_s = 0$ $M_1 : M_2 : M_s = f(i_{12}, \eta_0^{w1}) = \text{const}$
$M_2/M_1 = -i_{12} \eta_0^{w1}$ $M_s/M_1 = i_{12} \eta_0^{w1} - 1$ $M_s/M_2 = 1/(i_{12} \eta_0^{w1}) - 1$

sideration of the partial outputs gives an insight into the operating behaviour of a simple planet gear, but it is not required for the calculation of the operating data. By the superposition of *any* rolling and coupling powers in *any* planet gear, *any* of the six possible power flows can be generated: three each with either *1*, *2* or *s* as the only drive shaft and with two power takeoff shafts (power division), or with *1*, *2* or *s* as the only power takeoff shaft and with two drive shafts (*power totalisation*). Which of these is to be the only drive or power takeoff shaft (*total output shaft*, TOS) is decided by the stationary gear ratio, i_{12}, and a completely free choice of speed ratio, k (**Table 24**).

If the TOS of a superposition gear is preset by a layout having a single drive shaft (motor) and two power takeoff shafts (machines), then the speed ratios, k_{12}, k_{1s} and k_{2s}, must lie within the range given for them in **Table 24**. If the speeds are preset for a superposition gear with two motors and one machine connected, and if in addition the power takeoff shaft is also the TOS, then this is a power totalisation system. However, if one of the two motors is connected to the TOS, then it alone is driving the gear, while the other motor, together with the machine, must form a power takeoff and must be operated as a hypersynchronous brake – cf. F8.9.7.

Efficiency. With the general definitions,

$$\text{Efficiency } \eta = -(P_{ab}/P_{an}) = 1 - \zeta, \qquad (108)$$

$$\text{Loss } \zeta = -(P_v/P_{an}) = 1 - \eta, \qquad (109)$$

the total efficiency of a planet gear with two or three shafts running

$$\eta_{tot} = 1 + \frac{P_v}{\Sigma P_{an}} = 1 + \frac{M_1(n_1 - n_s)2\pi(1 - \eta_0^{w1})}{\Sigma P_{an}} \qquad (110)$$

is displayed, with P_v as per Eq. (107), and one or two shaft outputs as per Eq. (104), which are shown as drive outputs by their *positive sign*. With a self-locking gear, however, a power takeoff shaft, the torque of which takes on a positive sign only as a consequence of self-locking (see F8.9.5), may *not* be taken into consideration. The minus signs in the definition equations (Eqs (108) and (109)) are needed so that η and ζ should take on a positive value, as usual, in spite of the negative quotients (P_v, $P_{ab} < 0$).

The efficiency of transmission gears can be expressed using the stationary gear ratio and the stationary efficiency alone. For superposition gears, a free speed ratio must be added, e.g. k_{12}, which characterises the power flow (**Tables 23**, **24**) [58], with the relevant equation depending on the associated power flow in each case.

Simple planet gears, representing *stationary transmission*, are practically loss-symmetrical, like standard gear train gears, i.e. $\eta_{12} = \eta_{21}$. For *epicyclic gears*, however, especially for plus gears, the efficiency values in the two power flow directions can be very different, owing to the superposition of the rolling and coupling powers. For minus gears, the epicyclic efficiency values are always higher than the stationary efficiency values.

In Eq. (110) and **Tables 23** and **24** – as also in the other literature – it is assumed that, with a circulating support, the load-controlled tooth friction losses and planet bearing losses in the transmission of the rolling power P_w are as large as for stationary transmission. These losses alone form the basis of the calculation. Additional churning losses and ventilation losses arising when the support is also circulating, losses due to sealing ring friction and

influences due to lubricating oil guidance can, if necessary, be taken into account *after* the calculation of η_{tot}.

In the determination of the stationary efficiency, only the *load-controlled* losses referred to may be brought into play. Should more precise specifications not be available, then for practical calculations it is sufficient to assume a rolling efficiency $\eta_{wa} \approx 0.99$ for a pair of external helical gears and $\eta_{wi} = 0.995$ for a hollow gear stage with internal gearing. The stationary efficiency, η_{12}, is determined from this as the product of the rolling efficiency of the individual gear stages – cf. **Fig. 36**; for the more precise determination of the degree of effect, see [69].

8.9.5 Self Locking and Partial Locking

Self locking (sl) means that a gear cannot be moved, whatever the size of the drive moments, but instead remains stationary, internally blocked, because its friction power loss, P_v, would be larger than the driving power in the state of motion. However, it does operate if the additional power required to overcome the friction, or the "*release moment*" needed to release the jamming is fed into it through the power takeoff shaft in the power takeoff direction of rotation. A self-locking gear made to run in this way must be driven on all running shafts and produces only friction power loss.

Example Self-locking hoist mechanisms must be driven on what are actually the power takeoff shafts to reduce a (driving) load.

As a rule, self locking occurs only in one area of the possible power flows of a self-locking gear. For power flows in self locking systems, the torque M_j of a locked

Table 23. Efficiency levels for epicyclic transmission gears (For simple toothed wheel planet gears, $\eta_{12} \approx \eta_{21}$, for coupled planet gears, determine $\eta_{1\,\text{II}}$ and $\eta_{\text{III}\,1}$ separately; first index drive shaft, second power takeoff shaft.)

i_{12}	<0	0 to 1	>1
η_{1s}	$\dfrac{i_{12}\eta_{12} - 1}{i_{12} - 1}$	$\dfrac{i_{12}/\eta_{21} - 1}{i_{12} - 1}$	$\dfrac{i_{12}\eta_{12} - 1}{i_{12} - 1}$
w1	+1	−1	+1
η_{s1}	$\dfrac{i_{12} - 1}{i_{12}/\eta_{21} - 1}$	$\dfrac{i_{12} - 1}{i_{12}\eta_{12} - 1}$	$\dfrac{i_{12} - 1}{i_{12}/\eta_{21} - 1}$
w1	−1	+1	−1
η_{2s}	$\dfrac{i_{12} - \eta_{21}}{i_{12} - 1}$	$\dfrac{i_{12} - \eta_{21}}{i_{12} - 1}$	$\dfrac{i_{12} - 1/\eta_{12}}{i_{12} - 1}$
w1	−1	−1	+1
η_{s2}	$\dfrac{i_{12} - 1}{i_{12} - 1/\eta_{12}}$	$\dfrac{i_{12} - 1}{i_{12} - 1/\eta_{12}}$	$\dfrac{i_{12} - 1}{i_{12} - \eta_{21}}$
w1	+1	+1	−1

Table 24. Efficiency levels for superposition gears and allocation of ranges of k_{12}, k_{1s} and k_{2s} to positions of totalisation output shaft, TOS [58]; PF power flow

i_{12}	k_{12}	k_{1s}	k_{2s}	TOS	PF	Efficiency level η_{tot}	w1	PF	Efficiency level η_{tot}	w1
<0	$<i_{12}$	$>i_{1s}$	<0	1	$1<\frac{2}{s}$	$\dfrac{k_{12} - i_{12} + i_{12}\eta_{12}(1 - k_{12})}{k_{12}(1 - i_{12})}$	$+1$	$\frac{2}{s}>1$	$\dfrac{k_{12}\eta_{21}(1 - i_{12})}{\eta_{21}(k_{12} - i_{12}) + i_{12}(1 - k_{12})}$	-1
	i_{12} to 0	<0	$>i_{2s}$	2	$2<\frac{1}{s}$	$\dfrac{k_{12} - i_{12} + \eta_{21}(1 - k_{12})}{1 - i_{12}}$	-1	$\frac{1}{s}>2$	$\dfrac{\eta_{12}(1 - i_{12})}{\eta_{12}(k_{12} - i_{12}) + 1 - k_{12}}$	$+1$
	0 to 1	0 to 1	1 to i_{2s}	s	$s<\frac{1}{2}$	$\dfrac{(k_{12} - i_{12}\eta_{12})(1 - i_{12})}{(k_{12} - i_{12})(1 - i_{12}\eta_{12})}$	$+1$	$\frac{1}{2}>s$	$\dfrac{(k_{12} - i_{12})(\eta_{21} - i_{12})}{(k_{12}\eta_{21} - i_{12})(1 - i_{12})}$	-1
	>1	1 to i_{1s}	0 to 1	s	$s<\frac{1}{2}$	$\dfrac{(k_{12}\eta_{21} - i_{12})(1 - i_{12})}{(k_{12} - i_{12})(\eta_{21} - i_{12})}$	-1	$\frac{1}{2}>s$	$\dfrac{(k_{12} - i_{12})(1 - i_{12}\eta_{12})}{(k_{12} - i_{12}\eta_{12})(1 - i_{12})}$	$+1$
0 to 1	<0	0 to i_{1s}	i_{2s} to 0	s	$s<\frac{1}{2}$	$\dfrac{(k_{12} - i_{12}\eta_{12})(1 - i_{12})}{(k_{12} - i_{12})(1 - i_{12}\eta_{12})}$	$+1$	$\frac{1}{2}>s$	$\dfrac{(k_{12} - i_{12})(\eta_{21} - i_{12})}{(k_{12}\eta_{21} - i_{12})(1 - i_{12})}$	-1
	0 to i_{12}	<0	$<i_{2s}$	2	$2<\frac{1}{s}$	$\dfrac{k_{12} - i_{12} + \eta_{21}(1 - k_{12})}{1 - i_{12}}$	-1	$\frac{1}{s}>2$	$\dfrac{\eta_{12}(1 - i_{12})}{\eta_{12}(k_{12} - i_{12}) + 1 - k_{12}}$	$+1$
	i_{12} to 1	>1	>1	1	$1<\frac{2}{s}$	$\dfrac{k_{12} - i_{12} + i_{12}\eta_{12}(1 - k_{12})}{k_{12}(1 - i_{12})}$	$+1$	$\frac{2}{s}>1$	$\dfrac{k_{12}\eta_{21}(1 - i_{12})}{\eta_{21}(k_{12} - i_{12}) + i_{12}(1 - k_{12})}$	-1
	>1	i_{1s} to 1	0 to 1	1	$1<\frac{2}{s}$	$\dfrac{\eta_{21}(k_{12} - i_{12}) + i_{12}(1 - k_{12})}{k_{12}\eta_{21}(1 - i_{12})}$	-1	$\frac{2}{s}>1$	$\dfrac{k_{12}(1 - \eta_{12})}{k_{12} - i_{12} + i_{12}\eta_{12}(1 - k_{12})}$	$+1$
>1	<0	i_{1s} to 0	0 to i_{2s}	s	$s<\frac{1}{2}$	$\dfrac{(k_{12}\eta_{21} - i_{12})(1 - i_{12})}{(k_{12} - i_{12})(\eta_{21} - i_{12})}$	-1	$\frac{1}{2}>s$	$\dfrac{(k_{12} - i_{12})(1 - i_{12}\eta_{12})}{(k_{12} - i_{12}\eta_{12})(1 - i_{12})}$	$+1$
	0 to 1	0 to 1	i_{2s} to 1	2	$2<\frac{1}{s}$	$\dfrac{\eta_{12}(k_{12} - i_{12}) + 1 - k_{12}}{\eta_{12}(1 - i_{12})}$	$+1$	$\frac{1}{s}>2$	$\dfrac{1 - i_{12}}{k_{12} - i_{12} + \eta_{21}(1 - k_{12})}$	-1
	1 to i_{12}	>1	>1	2	$2<\frac{1}{s}$	$\dfrac{k_{12} - i_{12} + \eta_{21}(1 - k_{12})}{1 - i_{12}}$	-1	$\frac{1}{s}>2$	$\dfrac{\eta_{12}(1 - i_{12})}{\eta_{12}(k_{12} - i_{12}) + 1 - k_{12}}$	$+1$
	$>i_{12}$	$<i_{1s}$	<0	1	$1<\frac{2}{s}$	$\dfrac{k_{12} - i_{12} + i_{12}\eta_{12}(1 - k_{12})}{k_{12}(1 - i_{12})}$	$+1$	$\frac{2}{s}>1$	$\dfrac{k_{12}\eta_{21}(1 - i_{12})}{\eta_{21}(k_{12} - i_{12}) + i_{12}(1 - k_{12})}$	-1

power takeoff shaft j, and thus the takeoff output P_j, reverse their signs, by comparison with a frictionless drive ($\eta_{12} = \eta_{21} = 1$). But the power takeoff output, P_j, which thus becomes positive, does not become a "real" drive output. Thus, for example, the bearing flanks remain the same as if j were a power takeoff shaft. They do not change over to the other side, which is the bearing side for a "real" drive. The "power takeoff output" which has become positive may therefore not be used as P_{an}, but only as P_{ab} in Eqs (108) to (110)!

A *negative efficiency value* for the *running condition* in the locked power flow direction is thus established as the criterion for self locking.

Self locking occurs as reciprocal blocking of all three connecting shafts if the full release moment required for

the locked *power takeoff shaft* to run is not supplied from outside. If the locked shaft here is the *single power takeoff shaft*, with two or three shafts connected, then the gear is completely blocked and stationary (self locking, $\eta < 0$). If it is one of two power takeoff shafts, then the gear rotates, internally blocked, and transfers power, like a coupling, through the unlocked power takeoff shaft, with $i = 1$, $\eta = 1$ (partial locking).

Partial locking is possible only in a gear capable of self locking, with three connecting shafts, and characterised by $\eta > 0$, as is described for further operational examples in [58, 60].

Epicyclic gears are capable of self locking only if either

1. Their stationary transmission is self locking in one or

both power flow directions (e.g. in a vehicle differential with self locking worm gears [60]), or

2. Their stationary gear ratio lies in the range $\eta_{12} < i_{12} < 1/\eta_{21}$.

In case (1), the shafts 1 and/or 2 are power takeoff shafts capable of self locking. Self locking or partial locking occurs if the *rolling power flow* of the gear coincides with the locked power flow of the stationary transmission. In case (2), the support shaft is the power takeoff shaft capable of self locking. Self locking or partial locking occurs only if it forms a power takeoff shaft for the gear. The change in the sign of its torque as a result of friction also occurs for the stationary transmission if the support shaft remains stationary, with only the direction of effect of its support moment being reversed.

8.9.6 Hints for Design

Planet gears display some special design features by comparison with simple transmission gears [62]. By means of torque division through q, on the periphery of mounted planet gears or planet gear sets, the transferable power of planet gears or similarly constructed stationary transmission gears, known as branching gears or star gears, can be increased by the factor q if uniform support is ensured for all gears by such a statically over-determined arrangement, for example by making the elastic flexibility in the gear area greater than the actual dimensional differences here. When $q = 3$ (sets of) planets, the gear is *statically determinate* at the periphery, if one of the three gear elements, 1, 2 or s, as often happens, is centred under load without support in the gear housing, through meshing alone. Nevertheless, additional *dynamic* stresses are present – see [63]. None of the above calculations is influenced by the number, q, of planets/planet sets involved. Uniform distribution of several planets on the periphery is geometrically possible only if the tooth number conditions as per **Fig. 36** are integrally fulfilled (for other gear shapes, see [58]). For *"stage planets"* (**Fig. 36b, d, e**) precise reciprocal positional coordination of the two planet gear rims and marking of the pair of teeth meshing in the mounting position are required. Gears with simple planets are thus easier to manufacture. When the life of the planet bearings is being calculated, the centrifugal forces of the planets are to be taken into account, and their *relative speeds*, as against the support, are to be used as basic data [64]. For gears as per **Fig. 36**, these are $(n_{p1} - n_s) = (n_1 - n_s)z_1/z_{p1} = (n_{p2} - n_s) = (n_2 - n_s)z_2/z_{p2}$. For gears as in **Fig. 36a, c, f**, insert $z_{p1} = z_{p2} = z_p$ and $n_{p1} = n_{p2} = n_p$.

8.9.7 Design of Simple Epicyclic Drive Trains

Transmission Gearing

Example $i_{\text{theoretical}} = +3$, smallest number of teeth, $z_n = 19$, $q = 3$ planets on periphery. There are three possible stationary or epicyclic gear ratios as per Eq. (96), each with suitable formats as per **Fig. 36**:

$i_{\text{theoretical}} = i_{12} + 3$, formats **d, f**

$i_{\text{theoretical}} = i_{1s}; i_{12} = 1 - 1i_s = 1 - 3 = -2$, formats **a, b**

$i_{\text{theoretical}} = i_{s1}; i_{12} = 1 - 1/i_{s1} = 1 - 1/3 = 2/3$, formats **d, e**

$i_{\text{theoretical}} = i_{21}, i_{s2}, i_{2s}$ give the same gears, with the designations 1 and 2 exchanged. Suitable format: gears as per **Fig. 36a**, with $i_{12} = -2$, lead to simplest construction (see **Fig. 37**). Determination of tooth numbers: equations B and D as per **Fig. 36a** must initially be fulfilled, and $a_{1p} = a_{2p}$ must apply for the centre distances. For a zero gear ($x_1 = x_2 = 0$), it follows that:

$z_2 = i_{12}z_1 = (-2)34 = -68, \quad a_{1p} = a_{2p} = (z_1 + z_p)/2 = (|z_2| - z_p)/2;$

thus

$z_p = (|z_2| - z_1)/2 = 17. (z_1 + |z_2|)/ = (34 + 68)/3 = 34$

integrally. The installation condition is fulfilled; if it is not fulfilled, $z_{\min}$ and approximate centre distances should be varied using profile offset – see F8.1.7. Finally, the calculation of the modulus as per F8.5 and the draft design should be carried out, with the centrifugal forces acting on the planet gear bearing being taken into account.

Superposition Gears

For each superposition gear, in addition to its stationary gear ratio, i_{12}, and two speeds, n, or a free speed ratio, k, the total output shaft is determined and moreover the power flow (PF) and the total efficiency, η_{tot}, are established using a torque. Thus a desired PF can be coordinated with preset speeds only in limited ranges of free speed ratios, k - see **Table 24**. The range limits are characterised in each case by one shaft remaining stationary during stationary or epicyclic transmission, or by the "coupling point" ($n_1 = n_2 = n_s$).

Speeds Constant. Should three constant *speeds*, n_a, n_b and n_c, be preset, then the stationary gear ratio required for this, $i_{12} = i_{\text{theoretical}}$, is obtained from Eq. (94). If n_a, n_b and n_c are used in the six possible combinations as n_1, n_2 and n_s, then the same three pairs of reciprocal stationary gear ratios are obtained, i.e. the same three gears as they are obtained from each of these individual stationary gear ratios by the variation of $i_{\text{theoretical}}$, as in transmission gears and as per **Fig. 37**. It follows from the kinematic equivalence of these three gears that in each of them the shaft with the same speed, n_a, n_b or n_c, is the total output shaft. Thus, if three speeds are preset, the power distribution between the associated shafts is fixed, and is indeed independent of where and how these shafts are arranged in the gear format finally selected - see **Table 24**.

Example $n_a, n_b, n_c = 18, 9, 12 \, s^{-1}$. With, for example, $n_1 = 9$, $n_2 = 12$, $n_s = 18$, using Eq. (94) it follows that: $i_{12} = 1.5$, $k_{12} = 9/12$, and in **Table 24** under $i_{12} > 1$ and $k_{12} = 0$ to $1 \rightarrow$ TOS is shaft 2, i.e. the shaft with $n = 12 \, s^{-1}$.

If two constant *speed ratios* are preset, e.g. k_{ab}, k_{cb}, then $i_{\text{theoretical}} = i_{ab}$ is calculated from **Table 21**, and three stationary gear ratios, together with suitable formats, as for transmission gears.

Speeds Infinitely Adjustable. For a superposition gear with infinitely variable speeds, calculations are carried out for each of its two speed setting *limits*, one identified with ° and the other with * (the identifications are interchangeable), as with constant speeds. In a layout as per **Fig. 38**, the following speeds are permanently associated with the gear shafts a, b and c:

n_a, variable power takeoff speed, adjustment ratio

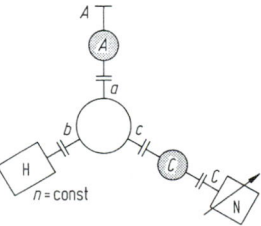

Figure 38. Symbol of a superposition gear with infinitely variable output speed. H constant-speed main motor; N auxiliary motor with infinitely adjustable speed; A, C possible supplementary gear positions.

$$\varphi_a = n_a^\circ/n_a^* = k_{ab}^\circ/k_{ab}^* \qquad (111)$$

n_b = preset constant (main motor, H)

n_c, adjustably preset (auxiliary motor, N)

$$\varphi_c = n_c^\circ/n_c^* = k_{cb}^\circ/k_{cb}^*. \qquad (112)$$

For a speed reversal within an adjustment range, $\varphi < 0$. Either of the minimum and maximum ratios k can be classified as $^\circ$ or *, as desired; there are thus four possible combinations: a φ_a, selected as desired, with two reciprocal φ_c's selected as desired (gives two solutions) and the reciprocal φ_a with the same φ_c (gives the same two solutions). From each, by using Eq. (97) on the adjustment limits, $^\circ$ and *, two equations can be derived for the determination of the stationary or epicyclic gear ratio, i_{ba} (with $n_c = 0$) of a suitable planet gear for either a given k_{ab}^* or a given k_{cb}^*:

$$i_{ba,a} = \frac{1 - \varphi_c}{k_{ab}^*(\varphi_a - \varphi_c)} \qquad (113a)$$

or

$$i_{ba,c} = 1 + \frac{1 - \varphi_a}{k_{cb}^*(\varphi_a - \varphi_c)}. \qquad (114a)$$

The speed ratio, k, not preset in either case, is obtained for both limits, $^\circ$ and * (as well as for any intermediate speeds selected), from the adjustment function valid for the entire speed range:

$$k_{cb,a} = \frac{1 - i_{ba}k_{ab}}{1 - i_{ba}} \qquad (113b)$$

or

$$k_{ab,c} = \frac{1 - k_{cb}(1 - i_{ba})}{i_{ba}}. \qquad (114b)$$

The limiting speed ratios, $k_{cb,a}^\circ$, $k_{ab,a}^*$, or $k_{ab,c}^\circ$, $k_{ab,c}^*$, calculated in this way certainly give the preset adjustment ratio φ_c or φ_a, but do not usually give the desired speed ratios $k_{theoretical}$. An adjustment to shaft a using *one* additional *transmission gear*, A, is thus required, if Eq. (113) has been used for calculation, or to c through C, using Eq. (114) - see **Fig. 38**. The gear ratio of such a *supplementary gear*, depending on position, is

$$i_{Aa} = k_{ab,theoretical}/k_{ab,c} \quad \text{or} \quad i_{Cc} = k_{cb,theoretical}/k_{cb,a}.$$

The A or C position of such a supplementary gear influences the absolute speed in the gear, so that, after the detailed calculation of all possibilities, it is placed at the point that leads to the most favourable speeds and torques in the entire installation. The most suitable planet gear is selected from the four solutions for $i_{ba} = i_{theoretical}$, with three possible solutions in each case, as in F8.9.7. The distribution of the driving power to the main motor H and the auxiliary motor N can be calculated for the solutions found as per F8.9.4. It is the same for several solutions, right down to the influence of the efficiency, for, according to [65], if operation is considered as being free of losses ($\eta = 1$), it is dependent only on the adjustment ratios φ_a and φ_c: with the power of the auxiliary motor N with reference to the total drive power being

$$\varepsilon_0 = P_c/(P_b + P_c) = -P_c/P_a,$$

$$\varepsilon_0^* = \frac{1 - \varphi_a}{1 - \varphi_c} \quad \text{and} \quad \varepsilon_0^\circ = \frac{\varphi_c}{\varphi_a}\varepsilon_0^* \qquad (115)\,(116)$$

apply at the adjustment limits.

The favourable combinations of φ_a and φ_c can be estimated using $\varepsilon_0 = -0.5 \ldots +0.5$ before work begins on

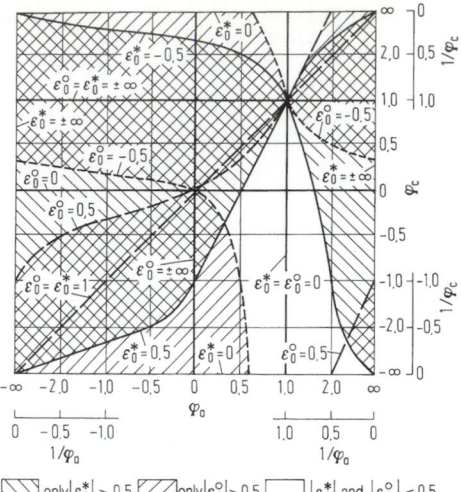

only $|\varepsilon_0^*| > 0.5$ only $|\varepsilon_0^\circ| > 0.5$ $|\varepsilon_0^*|$ and $|\varepsilon_0^\circ| < 0.5$

Figure 39. Power ratio, ε_0, plotted against combination of adjustment ratio, φ_a and φ_c of a superposition gear as per **Fig. 38**.

the layout (**Fig. 39**). For $\varepsilon_0 < 0$, the auxiliary motor, N, runs as a generator, $P_c < 0$.

Example with solutions *1* and *2*: required $n_a = 66$ to 40 s^{-1}, $n_b = 25$ s^{-1}, $n_{c,1} = 33$ to -50 s^{-1}, $n_{c,2} = 50$ to 33 s^{-1}. Whence $\varphi_a = n_a^\circ/n_a^* = 66/40 = 1.65$, $\varphi_{c,1} = n_c^\circ/n_c^* = 33/-50 = -2.0$. Using Eq. (113a): $i_{ba,1} = (1 + 0.66)/[1.60(1.65 + 0.66)] = 0.45$, $i_{ba,2} = (1 + 1.52)/[1.6(1.65 + 1.52)] = 0.50$. $k_{cb,a1}^\circ = (1 - i_{ba,1}k_{ab})/(1 - i_{ba}) = (1 - 0.45 \cdot 1.6)/(1 - 0.45) = 0.51 \neq K_{cb,theoretical}^*$. Supplementary gear, C: $i_{Cc} = \pm 2.0/0.51 = \pm 3.92$. $\varepsilon_{0,1} = (1 - 1.65)/(1 + 0.66) = -0.39$ (auxiliary motor runs as generator), $\varepsilon_{0,1}^\circ = -0.66(-0.39)/1.65 = 0.16$ (auxiliary motor runs on low drive power, as desired). $\varepsilon_{0,2}^* = (1 - 1.65)/(1 + 1.52) = -0.258$, $\varepsilon_{0,2}^\circ = -1.52(-0.258)/1.65 = 0.24$.

8.9.8 Compound Planetary Trains

Gear Symbols and Shaft Denotation

Gear symbols as per **Fig. 40**, though they contain only the information required for the calculation (position of shafts and their couplings) make it easier to keep track and simplify the analysis and synthesis of compound planetary trains considerably. The shafts of all sections of a compound planetary train are henceforth referred to as *1*, *2* and *s*, with the addition of one dash for the shafts of the second gear (*1'*, *2'*, *s'*) and two dashes (*1"*, *2"*, *s"*)

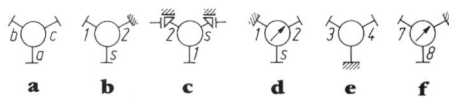

a   b   c   d   e   f

Figure 40. Symbols for epicyclic gears: **a** With support shaft in any or unknown position, **b** Shaft 2 designed to be stationary, **c** shafts 2 and s can be connected or disconnected or fixed by braking, **d** epicyclic adjustment gears with infinitely variable stationary gear ratio, e.g. hydrostatic epicyclic adjustment gears, **e** simple adjustment gears with stationary housing and two connecting shafts, specified by figures > 2, **f** simple adjustment gears with infinitely variable gear ratio, stationary housing and shaft specifications > 2, e.g. V-belt adjustment gears.

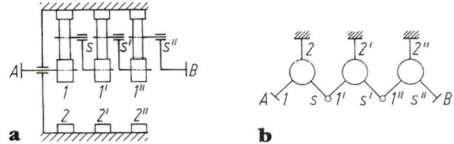

Figure 41. Example of three-stage series planet gear: **a** Diagram, **b** symbol with shaft specifications transferred from **a**, here $i_{AB} = i_{Is}$ $i_{1's'}$, $i_{1''s''}$; $\eta_{AB} = \eta_{1s}$, $\eta_{1's'}$, $\eta_{1''s''}$, $\eta_{BA} = \eta_{s''1''}$, $\eta_{s'1'}$, η_{s1}.

for the shafts of any third planet gear present, etc. (**Figs 41**, **42**). Thus all equations given previously, including those for **Tables 21** to **24**, or an existing computer program [60] can be directly used for every section. The dashes added to identify the sections are ignored during calculation in each case and then restored.

Formats of Compound Planetary Trains

Series Planetary Trains. (**Fig. 41**) are planet transmission gears connected in series, each with *one* fixed shaft in order to bring about higher gear ratios with high efficiency. A reduction in the space requirement and the best total efficiency are obtained with minus gears as per **Fig. 36a, b**. Calculation of entire reduction ratio and total efficiency analogous to simple multi-stage transmission gears (see F8, Introduction and F8.1.2).

Coupled Planet Gears. (**Fig. 42**) consist of two planet gears, which are connected to one another by *two shafts each*. As transmission or superposition gears, such trains provide for particularly low unit weights and unit volumes for gear ratios up to $i > |50|$ [59, 66]. If provided with external connecting shafts, *I*, *II* and *S*, in accordance with **Fig. 42b** to **d**, coupled planet gears have three connecting shafts with the degree of freedom $F = 2$, like a simple planet gear. Thus, as a set of gears it also has the same operating behaviour, and can be calculated precisely in the same way, using the same equations and **Tables 21** to **24**, if the indices 1, 2 and s are used instead of the

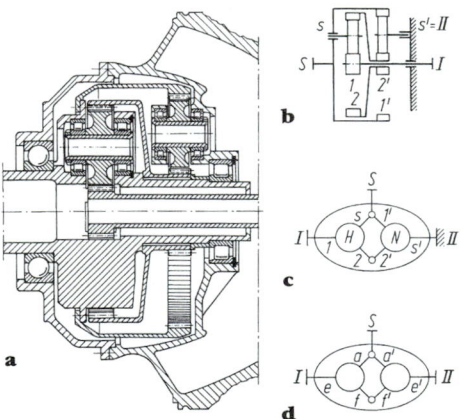

Figure 42. Example of coupled planet gear as turbo-prop. Reduction gear [68]. **a** Sectional drawing. **b** Diagram with shaft specifications. **c** Gear symbol with correctly positioned shaft specifications transferred from **b**. **d** Symbol of a coupled planet gear with function-oriented specifications for its shaft as per its position: *a*, *a'* connected coupling shaft; *f, f'* free coupling shaft; *e, e'* individual shafts; *I, II, S* external connecting shafts specified as for simple epicyclic gear.

analogous shaft denotations *I, II* and *S* [58]. If the connected coupling shaft *S* is fixed, then the set of gears acts as a series train like a stationary transmission, and its "series gear ratio" (analogous to a stationary gear ratio), $i_{I\,II}$, together with its series efficiency (analogous to stationary efficiency) $\eta_{I\,II}$ and $\eta_{II\,I}$ can be determined as for series trains (**Fig. 41** - see example).

If coupled planet gears are operating as a superposition gear, then their two sections are equivalent in their functions. If one of its individual shafts, e.g. shaft II, **Fig. 42b**, **c**, is fixed, then the associated section acts as a transmission gear and can be formed by a planet gear with a stationary shaft or by a simple transmission gear with a stationary housing. Here, as the "auxiliary gear", N, its only task is to preset the speed ratio, $k_{2s} = i_{2'1'}$ of the "main gear" H, which is connected to the external connecting shafts. The external gear ratio of the coupled planet gears, $i_{Is} = k_{ea}$, can be calculated using **Table 21**. If the function-oriented denotations as per **Fig. 42d** are replaced by the general denotations (see F8.9.2), e.g. $e \to a, a \to b, f \to c$, then, in **Table 21**, first column, the speed ratio desired becomes $k_{ea} = k_{ab} = k_{cb}$ $(1 - i_{ab}) + i_{ab}$, and converts back to the original denotations as per **Fig. 42d**:

$$i_{IS} = k_{ea} = k_{fa}(1 - i_{ea}) + i_{ea}, \tag{117}$$

where i_{ea} is the gear ratio of the main gear when the shaft *f* is intended to be stationary.

Example For gears as per **Fig. 42**: $i_{12} = -4.3, i_{1'2'} = -0.36$. So in the previous equation $k_{fa} = k_{2s} = i_{2'1'} = 1/-0.36 = -2.778$ and $i_{ea} = i_{Is} = 1 - i_{12} = 1 + 4.3 = 5.3$, and using Eq. (117) $i_{rs} = k_{ea} = -2.778(1 - 5.3) + 5.3 = 17.24$. The same result is obtained, together with the efficiency, if the coupled planet gears are generated analogously to a simple planetary train: as per **Fig. 42d** and **c** and Eq. (96),

$$i_{I\,II} = i_{ef} \cdot i_{f'e'} = i_{12} \cdot i_{2's'}$$

$$= i_{12} \cdot (1 - 1/i_{1'2'}) = -4.3 \cdot (1 - 1/-0.36) = -16.24.$$

Whence, using Eq. (96),

$$i_{IS} = 1 - i_{12} = 1 - (-16.24) = 17.24.$$

Series efficiency:

$$\eta_{I\,II} = \eta_{ef} \cdot \eta_{f'e'} = \eta_{12} \cdot \eta_{2's6} = 0.985 \cdot 0.989 = 0.974,$$

with

$$\eta_{2's'} = (i_{1'2'} - \eta_{2'1'})/(i_{1'2'} - 1)$$

as per **Table 23**, and with

$$\eta_{12} = \eta_{21} = \eta_{1'2'} = \eta_{2'1'} = 0.985.$$

Whence, using **Table 23**, under $i_{12} < 0$:

$$\eta_{IS} = (i_{I\,II} \eta_{I\,II} - 1)/(i_{I\,II} - 1)$$

$$= (-16.24 \cdot 0.974 - 1)/(-16.24 - 1) = 0.976.$$

If friction is neglected, the power flowing through the auxiliary gear is dependent only on the gear ratios, and can easily be estimated using denotations as per **Fig. 42d**: if the power ratio is defined as

$$\varepsilon_0 = \frac{\text{driving power of auxiliary gear}}{\text{driving power of coupling gear}}$$

$$= \frac{P_f}{P_1} = \frac{P_{a'}}{P_S} = [58] \text{ applies} \tag{118}$$

$$\varepsilon_0 = 1 - i_{ea}/k_{ea} = 1 - i_{ea}/i_{IS} \quad \text{or else}$$

$$\varepsilon_0 = (1 - 1/i_{IS})/(1 - 1/i_{I'a'}).$$

For the example in relation to **Fig. 42**, $\varepsilon_o = 0.693$ using these equations.

Adjustable Coupled Gears (**Fig. 43**) are coupled planet gears which, as auxiliary gears, contain an adjustable gear with an infinitely adjustable gear ratio, $i_{r'a'}$, and thus also offer an infinitely adjustable overall gear ratio, t_{IS}. Their adjustment ratio φ (adjustment range) is infinitely selectable for a specific adjustment ratio φ' of the auxiliary gear N with a suitable layout for the main gear H. As a rule, a commercially available adjustable gear is used as the auxiliary gear, the housing of which, being a fixed "support shaft" corresponds to the individual shaft e' of the auxiliary gear. As with coupled planet gears with constant gear ratios, a separate calculation is carried out for the two limiting gear ratios of the adjustment range. Here all variables associated with one another at a selected limiting gear ratio are identified using °, while the corresponding variables at the other end are identified using *. Thus the adjustment ratios φ for the coupling gear and φ' for the auxiliary gear, are defined as follows:

$$\varphi = i^{\circ}_{1S}/t^*_{1S}, \ \varphi' = i^{\circ}_{r'a'}/i^*_{r'a'}. \tag{119}$$

If the speed is reversed within an adjustment range, φ and/or φ' is negative. The load on the auxiliary gear (which it is desired to keep low) can already be estimated from the adjustment ratios (**Fig. 44**) for operation conceived as being *friction-free* (index °): at the adjustment limits, with ε_o as per Eq. (118):

$$\varepsilon^*_0 = (\varphi - 1)/(\varphi' - 1), \ \varepsilon^{\circ}_0 = \varepsilon^*_0 = \varepsilon^*_0 \varphi'/\varphi.$$

To obtain the preset adjustment ratios φ and φ', a planet gear should be designed with the gear ratio i_{ea} between the shafts e and a, with the shaft f designed to remain stationary. Depending on whether the limiting gear ratio is assumed to be i^*_{1S} or $i^*_{r'a'}$, the result obtained is either

$$i_{ea} = i^*_{1S(\varphi - \varphi')/(1 - \varphi')} \tag{120}$$

or

$$1/i_{ea} = 1 + (1 - \varphi)/[i^*_{r'a'}(\varphi - \varphi')]. \tag{121}$$

Whichever limiting gear ratio was not selected, $i^*_{r'a'}$ or i^*_{1S}, it is then obtained from $k_{fa} = i_{r'a'}$, using Eq. (117). As

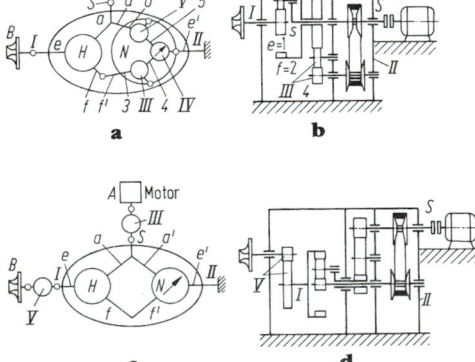

Figure 43. Coupled adjustment gear with infinitely adjustable V-belt gear [70]. **a** Symbol of format with supplementary gears, *III* and *V*, belonging to auxiliary gear. **b** Gear diagram of coupled adjustment gear as per **a** with supplementary gear, *III*. **c** Symbolic representation with external supplementary gears, *III* and *V*. **d** Gear diagram of gear as per **c** with supplementary gear, *V*, and additional two-stage gear with $i = 1$ for centre distance bridging.

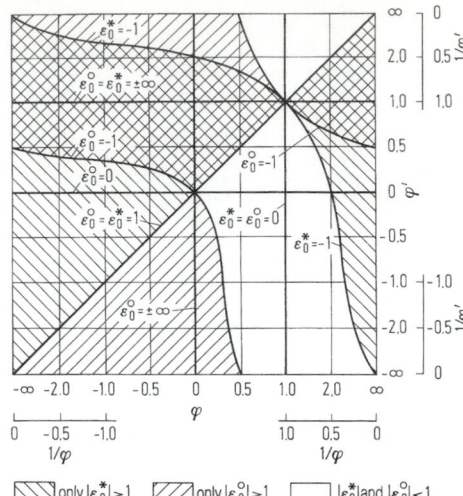

only $|\varepsilon^*_0| > 1$ **only** $|\varepsilon^{\circ}_0| > 1$ $|\varepsilon^*_0|$ and $|\varepsilon^{\circ}_0| < 1$

Figure 44. Power ratio ε_0 plotted against combination of adjustment ratios φ of coupled adjustment gear and φ' of its infinitely adjustable auxiliary gear.

a rule, it differs from the desired theoretical gear ratio $i_{theoretical}$, so that supplementary gears *III* and/or *V* are required, as per **Fig. 43a** and **b** in a layout using Eq. (120) or as per **Fig. 43c** and **d** using Eq. (121). The correspondence between t_{IS} and $i_{r'a'}$ is closer for any operating point within the adjustment range using Eq. (117) – see [70].

Reduced Coupled Planet Gears are coupled planet gears in which the supports form the free coupling shafts, ff', of the two gear sections (**Fig. 42d**) and can thus be combined into one element. Moreover, the gears of the two gear sections mounted on the connected coupling shaft and the planet gears meshing with them are of the same size; they can thus be reduced to a single gear pair [58, 67] – **Fig. 45**. However, a given reduced coupling gear can be expanded into three different coupled planet gears, depending on whether shaft A, B or C is considered as the coupling shaft, S, connected to the gear. All three have the same speed behaviour in relation to the shafts A, B and C, and are thus kinematically equivalent. However, their efficiency values can differ considerably from one another. The only simple coupling gear which "has the same effect" as the reduced coupling gear is the one for which one of the individual shafts, *I* and *II*, is a differential shaft for their gear section, while the other is a totalisation shaft (F8.9.4) [58, 60]. At the same time, this has the highest efficiency. It can be determined using a simple formulation [60]: if a stationary gear ratio, $i_{xy} > 1$, then y is a totalisation shaft. Otherwise, i.e. even with a negative gear ratio, y is a differential shaft. Let the shaft S in **Fig. 45b** to **d** be identified as x and the part gears *I*, *II* and *III*, each connected to *I* and *II*, be identified, each in succession, as y. Then i_{12} or i_{21} becomes i_{xy}. In **Fig. 45**, the totalisation shafts are marked by double dashes in the symbols. The combination in **Fig. 45c** proves to be the coupled planet gear with the same effect, which is now analysed, taking the place of the reduced coupling gear, as was described in connection with **Fig. 42**.

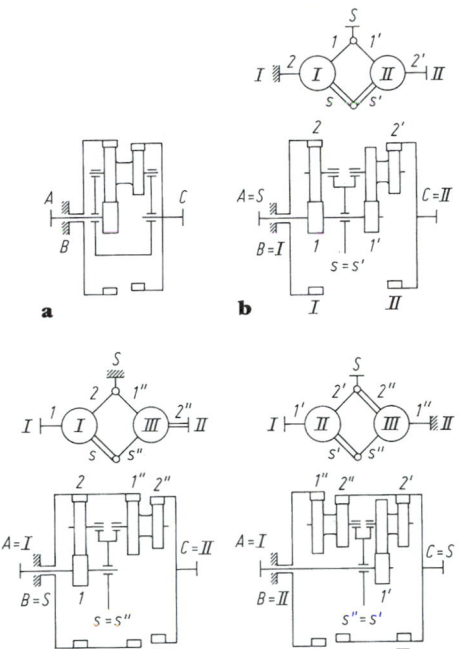

Figure 45. Reduced coupled planet gear. **a** Diagram of reduced coupling gear. **b-d** Schematic representation and symbols (with double dash for totalisation shaft) of three kinematically equivalent simple coupled planet gears derivable from it with **c** as the one with the same effect.

8.10 Design of Geared Transmissions

The rules and guide data given here are the bases for many constructions used in the machine building industry for average ratios. It is useful to round up the dimensions so obtained. Other dimensions are known from experience to be within specific ranges, or are useful or necessary for individual pieces of research. If possible, strength and rigidity should be calculated.

8.10.1 Types

Spur Gears

Normal Format (as in **Fig. 46a** and **b**). Simple, reliable in operation, with easy access. For larger, multi-stage gears, symmetrical format as per **Fig. 46c** – larger total tooth width, compact.

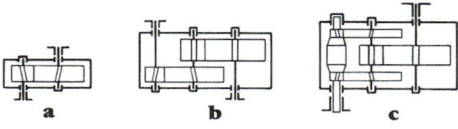

Figure 46. Gear with laterally offset drive and power take-off. **a** Single-stage for $i < 6(8)$. **b** Two-stage for $6 < i < 25(35)$, pinion of first stage mounted so that torsion and bending work against each other. **c** Heavy-duty gear.

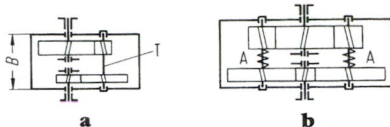

Figure 47. Gear with coaxial drive and power takeoff: **a** without, **b** with torque division; B large overall length, T long intermediate shaft, A torsionally elastic shafts.

Coaxial Drive and Power Takeoff. First stage badly used in format as per **Fig. 47a**. Smaller and lighter gears obtained using torque division, for example as per **Fig. 47b** (but more complicated, load balancing elements required); planet gears (see F8.9).

Distribution of total gear ratio for condition: minimum total gear volume, free selection of b/d or b/a (check as per **Table 8**); index I, first stage, etc. For σ_{Hlim} values, see **Table 14**.

Two-stage gears:

$$u_1 \approx 0.8 \, (u\sigma_{\text{Hlim I}}/\sigma_{\text{Hlim II}})^{2/3}. \tag{122}$$

Three-stage gears:

$$u_1 \approx 0.6 u^{4/7} \, (\sigma_{\text{Hlim I}}/\sigma_{\text{Hlim II}})^{2/7} \, (\sigma_{\text{II lim I}}/\sigma_{\text{Hlim II}})^{4/7} \tag{123}$$

$$u_{\text{II}} \approx 1.1 u^{2/7} \, (\sigma_{\text{Hlim II}}/\sigma_{\text{Hlim I}})^{4/7} \, (\sigma_{\text{Hlim II}}/\sigma_{\text{Hlim III}})^{2/7}. \tag{124}$$

Total

$$u = u_1 u_{\text{II}} \cdots . \tag{125}$$

Straight Bevel Gear Pairs

More rigid and more cost-effective than bevel gear pairs for $i > (3 \text{ to } 5)$ as per **Fig. 48** (large ring gears, thin pinion shafts). Usually bevel gears in first stage (for larger moments spur gears are more cost-effective and less sensitive in second and third stages); exception – rapid-running gears with high noise requirements [1] or modular gears [35].

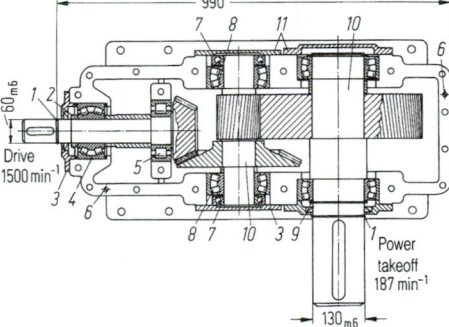

Figure 48. Straight bevel gear pair (Lohmann & Stolterfohlt, Witten). Rated power $P = 280$ kW, splash lubrication, 35 l oil, weight without oil 495 kg, oil level check with dipstick, spur gears case-hardened and ground, bevel gears case-hardened and lapped; *1* spring rings as stop for coupling hubs, *2* shaft nut, locking ring and washer, *3* depth of fit for cover adjusted at fitting, *4* oil feed from catch pocket, *5* NUP bearing in stator bore H7, *6* adjusting pin, *7* axial bearing, *8* radial bearing, *9* spring ring with angular shim, *10* H6/u6 shrink fit, *11* cover sealed with sealing paste.

Worm Spur Gears

Economical from $i > 12$, depending on size. Worm gears in first stage as far as possible (efficiency, noise, size); exception – if spur pinion mounted directly on motor shaft, e.g. in gear motors (no coupling, no separate pinion mounting needed).

8.10.2 Connection to Driving and Driven Machine

Electric motor often flanged directly on gear for *gear motors* up to 50 kW (usually 0.4 to 4 kW) (no coupling, no separate setting up, no alignment).

For *higher power levels*, usually separate setting up, connection to motor and machine through compensating couplings (see F4). Considerable forces can be introduced through transverse misalignment and offset angles or protruding couplings, axial movements of the motor armature and the power takeoff – in spite of compensating couplings – (pay attention to two half-arches in dimensioning bearings, housing, shafts and force distribution!). This does not apply to plug (slip-on) gears for the power take-off shaft, and with flanged motors it does not apply to the drive side either. The gear power takeoff shaft is permanently connected to the shaft of the machine and the gear rides on it. Gear weight and transverse forces from the support moment must be absorbed by this shaft and a torque support.

8.10.3 Detail Design and Measures of Gears

Finished cast gears – including gear-cutting – (also injection cast gears) for small dimensions, light loads and large quantities of components, if necessary with cast-on cams, dogs, etc., for heavy loads also finish-forged (e.g. differential bevel gears). In machine construction for small and medium-sized dimensions, usually *fully-turned* or *contour-turned* disc gears; for larger dimensions *welded* gears have largely forced out cast, shrinkage and dust constructions (even for alloy steels up to 300 HB or possibly 340 HB) (see F8.4).

Gear Types

For $d < 500$ mm and series – drop-forged, for individual manufacture – solid discs or web gears (light construction) made of forged round stock; for $500 < d < 1200$ mm, disc gears or web gears, free-form forged and/or contour-turned if necessary, even for larger dimensions with severe safety requirements; for $d > 700$ mm usually welded ($b/d < 0.15$ to 0.20: single-disc gears, above that twin-disc gears, $b > 1000$ to 1500 mm: triple-disc gears). Transition at lower values under heavy load, thick bandage, vertical shaft, if high axial rigidity required (large β), for finer gear quality (rigidity during gear-cutting!).

General Embodiment Design Rules (Fig. 49). If b_R undershoots the limiting value given here, the gear must be cut into the shaft. With shrunk-on, thin tooth rims, pay attention to shrinkage stress and tooth root stress [37]. Always check whether clamping is possible for gear cutting and gear grinding. For solid gears and disc gears see **Fig. 50**; for welded gears see **Fig. 51**.

For *specifications* for gears and gear body dimensions in drawings, see DIN 3966 and DIN 7184.

8.10.4 Embodiment Design of Gear Cases

Usually overall housing as bearing construction – for examples see **Figs 48** and **34**.

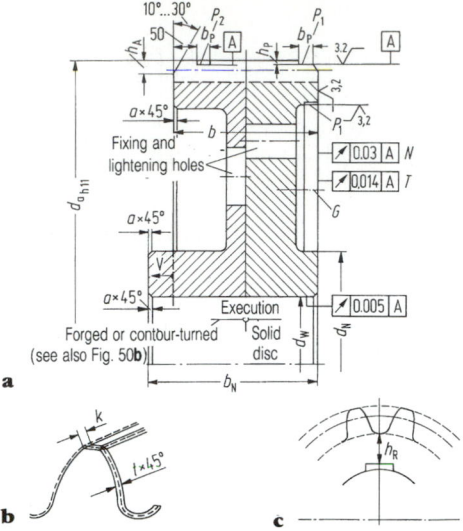

Figure 49. Gear body dimensions – general.

a for *relief* of tooth ends:

at $b > 10m$: $b_A \approx m$, at $b < 10m$: $b_A \approx 1 + 0.1m$.

P_1 *locating faces* (internal or external) for gears which cannot be generated on shafts or tensioning spindles, from diameter approximately 700: $b_P \approx 0.1$ mm, $b_P \approx 10$ mm. *2*, locating face, P_2, at $b > 500$ mm.

Axial runout: N at $v_t \leq 25$ ms, T at $v_t > 25$ m/s.

Sprocket holes, clamping bores and lightening holes:

Quantity n: $d_a < 300$: – (clamping through bore),

$300 < d_a < 500$: $n = 4$,
$500 < d_a < 1500$: $n = 5$,
$1500 < d_a < 3000$: $n = 6$,
$d_a > 3000$: $n = 8$;

(Check clamping facilities of work site) – no bores in high-speed gears; threaded blind holes, G, for transportation of solid disc gears heavier than 15 kg.

Hub diameter, $d_N = (1.2$ to $1.6)d_w$ (depending on material, shrinkage: small values for large d_w); hub width, $b_N \geq d_w$ and $b_N \geq d_a/6$ (for helical gears, check tipping through neutralisation of play or shrink fit slipping). Avoid V-shaped projection of hub (cf. Fig. 50).

b For *protection* against *transportation damage*:

Edge break, $a \approx 0.5 + 0.01d_w$.

Tip edge break, $k \approx 0.2 + 0.045m$.

Face edge break, $t \approx 3k$.

Edge *rounding* with radius $\approx k$ or t for strictest requirements (e.g. aircraft gears) and nitrided gears (see also F8.4).

c *Basic hub thickness*:

Unhardened or nitrided, $b_R > 2.5m$.

Case hardening, flame hardening, induction hardening, flank hardening or space hardening, $b_R > 3.5m$.

Rotary flame hardening or rotary induction hardening, $b_R > 6m$ (pay attention to position of adjusting spring and to shrinkage strain).

For *surface hardening*, specify which areas must remain soft, e.g. threaded holes, possibly bores).

For larger gears, from time to time rigid lower cases with mounted bearing upper sections. Upper sections then have only protective function, and must be easily inspected [1].

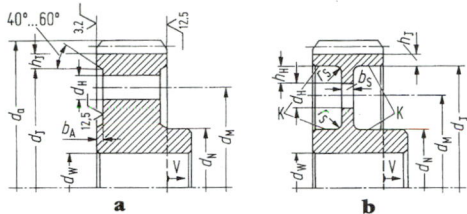

Figure 50. Dimensions of turned or forged and turned gear bodies. Avoid V-shaped projection of hub (machining of plane surface, stacked mounting).

a *Normal form* (provided there is no limitation on weight) made from forged round steel; lateral boring because of machining costs, and to ensure support from chuck only at $(d_J - d_N)/2 > 25$ mm.

Cost-effective for unhardened and hardened gears (low machining volume, little distortion due to hardening)

$b_J \geq 3m$; $b_A = 0.5 + 0.1m$, max. 2 mm.

Transverse bores (quantity: **Fig. 49**): $d_M \approx 0.55(d_N + d_J)$, $d_H \approx d_a/20 \geq 30$ mm, edge distance between bores $\geq 0.8d_H$, d_N; see **Fig. 49**.

b *Light construction* format (e.g. aircraft and spacecraft, small working load, as per prototype test): $b_H > 2r_S$; $d_H = (0.1$ to $0.2)d_a$. Number of bores: see **Fig. 49**; $b_J = b_R \geq 1m$ (as per **Fig. 49c**); $r_s \approx t$ (as per **Fig. 49b**); d_N, see **Fig. 49**; d_M, see **Fig. 50a**.

Unhardened, nitrided, case-hardened (very light): $b_s = 1.5m$.

Flame-hardened or induction-hardened, case-hardened (less light): $b_J \approx b_R$ as per **Fig. 49c**; $b_s = 2m + 0.15b$.

Drop-forged or free-form forged: generatrices, K: 5 to 10° tapered.

General Embodiment Design Rules

Cast Housings with more than three sections preferably made of GG 20, large gears GG 18 (easily castable, little shrinkage or distortion, easily machinable), GGG 40, GS 38.1 (weldable!) (higher strength, more difficult to process). Pay attention to greater heat expansion and lower rigidity for light metals.

Welded Housings make it possible to save weight (rigidity through fins or profile); suitable for individual manufacture and impact stress. Material usually St 37-1 or 2 (high-stress: St 52-3).

Undivided Housings preferred for small gears; installation through lateral apertures. Incidentally, *horizontal joints* in shaft plane favourable for sealing, installation, inspection.

Bearing Bolts should be designed in accordance with static tooth root load-bearing capacity. Tighten to 70 to 80% R_e. Provide for at least two *alignment pins* ($d \approx 0.8$ flange screw diameter), for larger gears, others near the bearings. Secure *screws* in *inside of gear* with wire. Provide for at least two opposing threads for *forcing screws*.

Foot screws should be calculated from support moment of gear. For steel frame foundations *alignment pins* and **adjusting screws** (with fine threads) useful in gear root.

Distance between gears and housing walls large enough to avoid jamming of fragments and high pumping of oil. Distance between gears, and between gears and housing walls, laterally and on diameter as per numerical value equation,

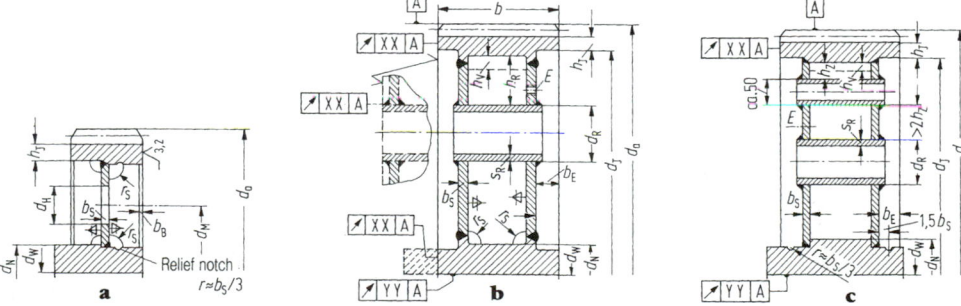

Figure 51. Dimensions of welded gear bodies. Web plates and fins are usually made from St 37.1 or 37.2, hubs from St 52.3. Welded joints are formed as per stress and manufacturing possibilities. $b_J = b_R$ as per **Fig. 49c**; d_H, d_M, b_H as per **Fig. 50b**; quantity of bores or pipes, see **Fig. 49**.

a Single-disc gear: $b_s \approx 0.12d_a + (5$ to 10 mm), depending on difficulty of operation, thicker if need be, if precise manufacture (clamping) complicated.

Without lateral fins, if $\beta < 10°$.

With lateral fins, in general, if $\beta > 10°$; thickness of lateral fins = $0.6b_s$; $b_s \approx 1.5b_s$; $r_s = 1.5b_s$ (at least 10 mm).

Quantity of fins:

at $10° < \beta < 20°$ = quantity of bores,

at $\beta > 20°$ = double quantity of bores.

b, **c** Twin-disc gears: $b_s \approx 0.08d_a + (5$ to 10 mm), depending on difficulty of operation; $b_E \approx b/7$; $i_R > 40$ mm; $d_R = (0.12$ to $0.20)$ $(d_J - d_N)$, at least 50 mm: $s_R = (0.3$ to $0.5)b_s$ for small to large pipe diameters. Stiffening fins between pipes approx. $0.8b_s$ thick; $b_v \approx 2b_s$; r_s and quantity of fins as in **a**. For other dimensions and quantity of stiffening fins see **Fig. 49**.

E vent hole, approx. Ø6 mm; after stress-free annealing weld up or close with screw.

b Format for $d_a < 2000$ mm. $b_R > 40$ mm. Form b1 for hub jutting forward or back (dashed). Then support to cut teeth on gear rim and pipe. Safe (costly) welded connections for high dynamic stresses. (Also useful for types in **Fig. 51a** and **c**.)

c Format for $d_a > 2000$ mm. Smaller pipe near tooth rim ($h_z = 40$ mm; as small as possible) to let clamping screw through; larger pipe to let clamping mushroom head through. Other dimensions as **b**.

Table 25. Guide data for dimensions of gear housings (L = longest housing length in mm)

Component		Designation	Cast construction[b]	Welded construction
Wall thickness for bottom box		$w_w{}^a$		
(a) Unhardened gearing	GG		$0.007L + 6^c$	$0.004L + 4$
	GGG, GS		$0.005L + 4$	
(b) Hardened gearing	GG		$0.010L + 6^c$	$0.005L + 4$
	GGG, GS		$0.007L + 4$	
	minimal		GG, GGG: 8; GS: 12	4
	maximal		50	25
Stressed top box, bearing cover		$w_o{}^d$	$0.8w_w$	$0.8w_w$
Unstressed hood		w_H	$0.5w_w$	$0.5w_w$
Reinforcement and cooling fins		w_R	$0.7 \times$ Thickness of walls to be reinforced	
Flange thickness		$w_F{}^c$	$1.5w_w$	$2w_w$
Flange width (projecting part)		b_F	$3w_w + 10$ mm	$4w_w + 10$ mm
Continuous base strip with recess		w_L	$3w_w$ (Wall thickness w_w)	
Continuous base strip without recess		w_L	$1.8w_w$	$3.5w_w$
Continuous transverse base strip		w_Q	$1.5w_z'$	$1.5w_L$
Base strip width (projecting part)		b_L	$3.5w_w + 15$ mm	$4.5w_w + 15$ mm
External diameter of bearing housing		D_G	$1.2 \times$ External bearing diameter	
Bearing bolt diameter[f]		d_s	$2w_w$	$3w_w$
Flange bolt diameter[g]		d_F	$1.2w_w$	$1.5w_w$
Distance between flange bolts		L_F	$(6 \text{ to } 10)d_F{}^h$	$(6 \text{ to } 10)d_F{}^h$
Foundation bolts[i]		d_U	$1.6w_w$	$2w_w$
Inspection cover bolt		d_D	$0.8w_w$	$6w_w$

[a]For gears from approx. $L = 3000$ mm bottom box often double-walled with approx. 70% of above wall thickness.
[b]Lifting taper approx. 3°.
[c]For turbo-gears: + approx. 10 mm (vibration and noise damping).
[d]Thicker if necessary, in accordance with noise level required.
[e]For bolts and nuts.
[f]As close as possible to bearing.
[g]Lifting screws of same thickness.
[h]Depending on density requirements.
[i]Quantity $\approx 2 \times$ quantity of bearing screws.

$$s_A \approx 2 + 3m + B \text{ with}$$

$$B = 0.65 \, (v_t - 25) \geq 0 \, (v_t \text{ in m/s})$$

to ground about $2s_A$, provided oil supply adequate. For injection lubrication large drain aperture important: diameter approximately (3 to 4)s_A.

For splash lubrication, *oil drain plug* (with magnetic spark plug if necessary, see below) at lowest point. Gradient of gear base toward drain aperture 5 to 10%.

Locating Faces for larger gears on narrow sides of bottom flange, jutting out approximately 120×40 mm, for large gears at external bearings also. Contact pattern can be repeatedly set using spirit level.

Machining of Flange Surfaces, $R_z = 25$ μm, bearing points and bearing faces, $R_z = 16$ μm, inspection cover, root surfaces, $R_z = 100$ μm.

Inspection Cover should permit inspection of entire meshing over entire tooth width, together with lubricating oil supply. If danger of loosening exists, provide hinged cover and hinged bolt (e.g. on crane control gears).

Through-holes should be avoided to inside of housing (oil tightness).

Lifting Lugs, Eye-bolts or similar should be provided to allow the removal of the top box and to lift the gear (on the lower box).

Ventilation for pressure balance with filter (against dirt and moisture) at highest point (pay attention to direction of splashing!). For splash lubrication, *sight glass* or dipstick required. The dipstick can be provided with a *magnetic sparking plug* (wear check). For injection lubrication, connections for monitoring *oil pressure, rate of flow and temperature* [1].

Housing Dimensions are determined by inherent stability (not strength). For guide data, see **Table 25.**

8.10.5 Bearings

Roller bearings are preferred throughout. *Friction bearings* are only used for high-speed gears (about $v_t > 30$ m/s), very large dimensions or specially quiet running.

Bearings need to be as *tight* as possible to gears (for minimum distance see F8.10.4), but the minimum distance between bearings should be $0.7d_2$ (effect of centre distance variations, bearing rigidity, overturning moment due to axial force).

Avoid *overhanging*. If necessary, select distance between bearings approximately two to three times overhang, shaft diameter > overhang.

For *double helical gears*, fix only one shaft axially, in general the gear shaft (with larger dimensions; larger axial forces often introduced from outside through them).

For *small* gears, grooved ball bearings, fast and loose bearings usually economic, for *medium* sizes grooved ball bearings as fixed bearings, cylindrical roller bearings as movable bearings or tapered roller bearings in O arrangement (provided distance between bearings not too large).

For *spur* or helical gears with $F_a/F_t \leq 0.3$, cylindrical roller bearings possible. Take up *high axial forces* using separate thrust bearings:

Four-point contact bearings (also when axial force is reversed).

Self-aligning roller bearings up to $F_a/F_t = 0.55$. *Note*: At $F_a/F_t > 0.1$ to 0.25 centre bearing, but not below this;

if applicable, pay attention to angular deviation when axial force reversed and axial play relatively large. *Double-row tapered roller bearings* suitable for high axial forces and changes of direction, **Fig. 34**.

Adjustable bearings, e.g. with eccentric cases, used to set contact pattern for large and high-speed gears.

Bearing lubrication for series gears using oil spray or oil collection pockets, out of which oil or bores ($d \approx 0.01 \times$ external bearing diameter, minimum 3 mm) run behind bearing. For large and high-speed gears, usually injection lubrication (oil nozzle diameter ≥ 2.5 mm owing to danger of stopping up, corresponding to approximately 3 l/min); ensure oil return from cavity behind bearing through bore ($d \approx 0.03 \times$ external bearing diameter, at least 10 mm or several bores) (at height of lower roll body, thus providing oil supply for startup).

9 Kinematics

H. Kerle, Brunswick

9.1 Systematics of Mechanisms

9.1.1 Fundamentals

Definition of Mechanisms. Mechanisms are systems for converting or transmitting movements and forces (torque). They consist of at least three links, one of which must be defined as the *fixed link* [1]. For the sake of completeness, a distinction is drawn between *kinematic chain*, *mechanism* and *motor mechanism*. The chain becomes a mechanism when one of the links of the former is selected as the fixed link. The mechanism becomes a motor mechanism when one or more links of the former is driven.

Mechanism Structure. Structural studies of the type, number and configuration of links and the joints which connect them usually begin with the kinematic chain. There are *open* and *closed* kinematic chains, with or without *multiple power-transmission paths* (**Fig. 1**).

Points on links of *planar* mechanisms describe paths in mutually parallel planes; points on links of (generally) *spatial* mechanisms describe spatial curves or paths in mutually non-parallel planes; *spherical* mechanisms are special spatial mechanisms with point paths on concentric spheres (**Fig. 2**).

A *kinematic pair* comprising two contiguous links (or sections thereof) determines the *joint*. Planar mechanisms require planar joints with up to two *degrees of freedom* (translation and rotation), while spatial mechanisms on the other hand very often also require spatial joints with up to five degrees of freedom in addition to planar joints (**Fig. 3**). For example, turning and turning-and-sliding pairs are characterised by a shaft and bore, sliding pairs by hollow and solid prisms, screw joints by a nut and bolt, ball-and-socket joints by the ball and socket. *Lower* kinematic pairs or sliding joints are in surface contact (e.g. shaft and bore), *higher* pairs in linear (e.g. cam plate and

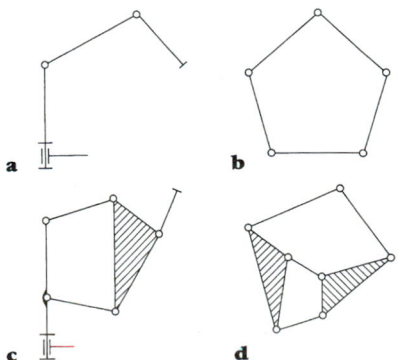

Figure 1. Kinematic chains: **a** open, **b** closed, **c** open with multiple-power transmission paths, **d** closed with multiple-power transmission paths.

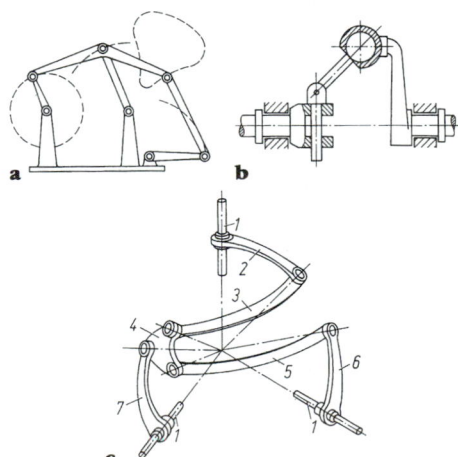

Figure 2. Sample mechanisms: **a** planar, **b** general spatial (shaft coupling); **c** spherical; *1* fixed link.

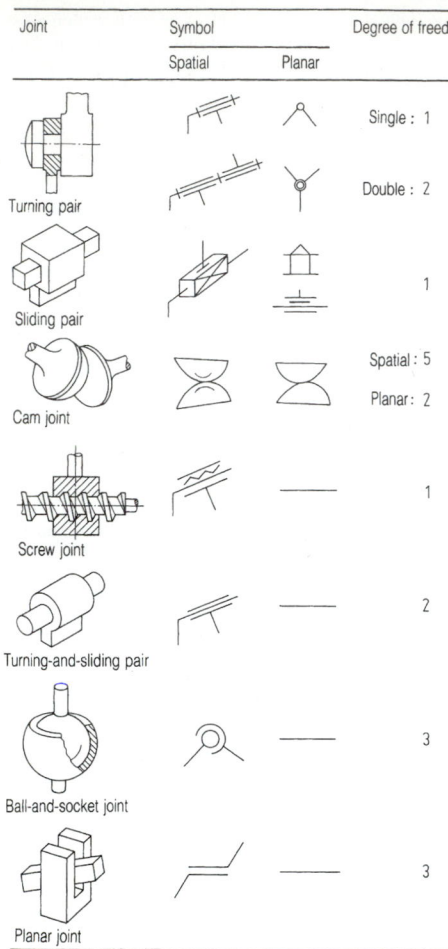

Figure 3. Joints and joint symbols.

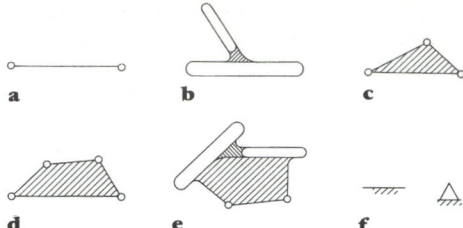

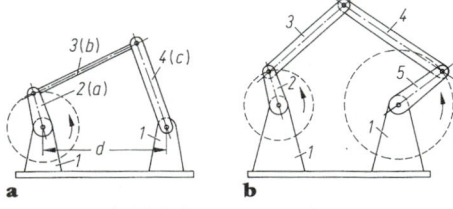

Figure 4. Link symbols for planar mechanisms: **a** binary (n_2) link with two turning elements, **b** binary (n_2) link with two sliding elements; **c** ternary (n_3) link with three turning elements, **d** quaternary (n_4) link with four turning elements, **e** quaternary (n_4) link with two turning and two sliding elements, **f** fixed link.

roller) or point contact (e.g. ball on a plate). *Form-closed* joints ensure contact between the elements by means of an appropriate form, while *force-closed* joints require one or more additional forces to maintain permanent contact.

In the case of *planar* mechanisms which usually have turning-and-sliding pairs it is expedient to divide the mechanism links according to the number of element sections into binary (n_2-), ternary (n_3-) and quaternary (n_4-) links (**Fig. 4**), especially as in addition a planar cam joint can be replaced kinematically by a binary link (cf. F9.1.2).

Degree of Mechanism Freedom. The degree of freedom F of a mechanism is a function of the number n of links (including the fixed link), the number g of joints with the respective degree of joint freedom f and the *freedom of motion* b:

$$F = b(n - 1) - \sum_{i=1}^{g} (b - f_i). \tag{1}$$

For mechanisms which are generally spatial $b = 6$ is inserted, and for spherical and planar mechanisms $b = 3$. If, in addition, individual links can be moved without hav-

ing to move the entire mechanism (e.g. a roller with a rotary bearing on a cam plate), F is reduced by these *identical degrees of freedom. Grübler's equation*

$$F = 3(n - 1) - 2g \tag{2}$$

applies to planar mechanisms that only have turning-and-sliding joints with $f = 1$. $F = 1$ means constrained motion, e.g. for the four-bar linkage (**Fig. 5a**) where $n = 4$ and $g = 4$. For a five-bar linkage (**Fig. 5b**) where $n = 5$ and $g = 5$, $F = 2$ applies. The degree of freedom of a mechanism indicates the minimum number of drives or input impulses a mechanism must receive in order to fulfil a function which is calculable in advance. Where $F = 2$, motion must be induced at two independent locations (e.g. main and regulating drives) or two independent forces or moments act as input impulses (differential gear mechanism or self-adjusting mechanism). Correspondingly higher minimum requirements apply in the case of $F > 2$.

9.1.2 Types of Planar Mechanism

Four-Bar Turning-Pair Linkages. A four-bar linkage is capable of rotation if *Grashof's criterion* is met: the sum of the lengths of the shortest and longest links must be less than the sum of the lengths of the other two links. There can only be one "shortest" link (l_{min}) but up to three "longest" (i.e. identical lengths). Depending on which of the four lengths a, b, c, d (**Fig. 5a**) is l_{min}, the resulting mechanism is either a crank-and-rocker ($l_{min} = a$, c), a double crank ($l_{min} = b$) or a double rocker with rotating coupler ($l_{min} = d$). Four-bar linkages that cannot rotate are designated non-rotatable double rockers. All relative rocking motions occur symmetrically to the adjacent link. There are internal and external rockers. Non-rotatable four-bar linkages can only contain one "longest" element but up to three "shortest" [2]. The third group comprises *folding* mechanisms where each of two link pairs are identical in length, e.g. parallel-crank mechanisms [3].

Figure 5. Planar turning-pair linkages: **a** four-bar linkage ($F = 1$), **b** five-bar linkage ($F = 2$).

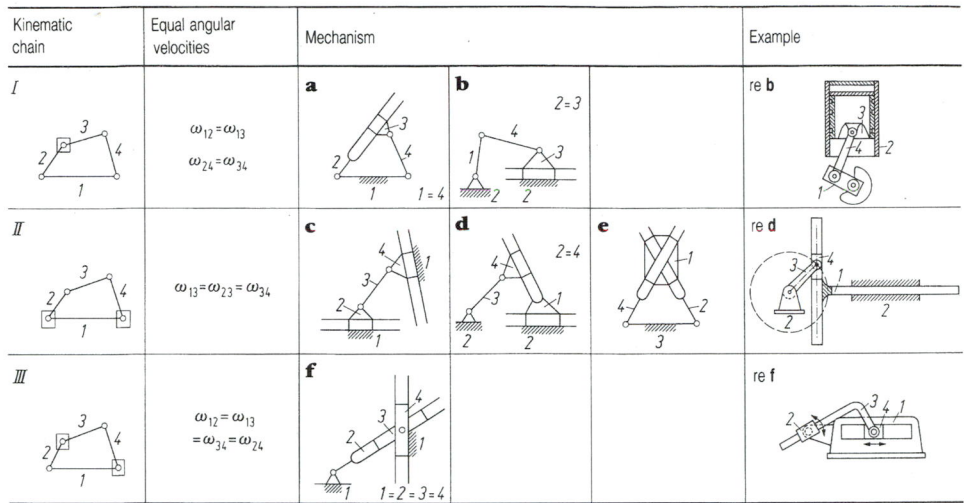

Figure 6. Four-bar sliding-pair linkages: **a** inverted slider crank, **b** slider crank, **c** ellipsograph, **d** scotch yoke, **e** rotating scotch yoke (Oldham coupling), **f** sliding-pair loop.

Four-Bar Sliding-Pair Linkages. If turning pairs are replaced by sliding pairs this results in sliding-pair chains and mechanisms. *Loop* motions occur if the sliding joint connects two moving elements. Three chains (**Fig. 6**) are derived from the four-bar linkage (kinematic chain of each four-link mechanism): chain *I* with one sliding joint, chain *II* with two adjacent sliding joints and chain *III* with two diagonal sliding joints. As a result of *kinematic inversion* (link inversion and change of fixed link), the three chains yield six four-bar sliding-pair linkages. Each sliding pair causes equal angular velocities, regardless of the mechanism dimensions, e.g. in the case of chain *I*, $\omega_{12} = \omega_{13}$ and $\omega_{24} = \omega_{24}$. The following is generally applicable: $\omega_{ij} = \omega_{ji}$ is the angular velocity of the link *i* relative to the link *j*. Sliding-pair mechanisms are therefore sometimes transmissions which transform speed uniformly (constant transmission ratios).

Multi-link Mechanisms. For each group of kinematic chains with the same number of links and the same degree of freedom there is a clearly definable number of different chains and mechanisms. **Figure 7** shows six-link con-

strained chains ($F = 1$) based on *Watt's* and *Stephenson's chains* (variants as a result of changing the fixed link) with a number of practical examples. If *double joints* are used a further five different chains can be added. The equations of synthesis (**Fig. 8**) yield eight-link constrained chains with two quaternary and six binary links, with one quaternary, two ternary and five binary links and with four ternary and four binary links. If multi-joint links are also considered there are according to Hain 60 different eight-link constrained chains, yielding a total of 330 mechanisms as a result of kinematic inversion.

Cam Mechanisms. The standard cam mechanisms are three-link cam mechanisms consisting of the cam, follower (rocker) and frame. The cam and follower touch at the *cam joint* (point of contact *K*) – in many cases an additional tracer element, e.g. a follower-mounted roller with an identical degree of freedom, improves the operational properties without changing the kinematics; the frame links the cam and follower [4]. Generally, the frame is the fixed link *1*, the cam is the input member *2*, and the follower is the output member *3*.

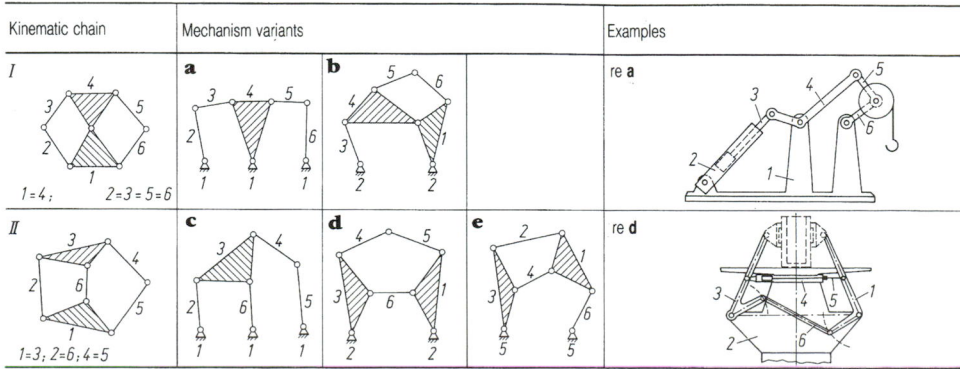

Figure 7. Six-link constrained kinematic chains and sample mechanisms (*I* Watt chain, *II* Stephenson chain).

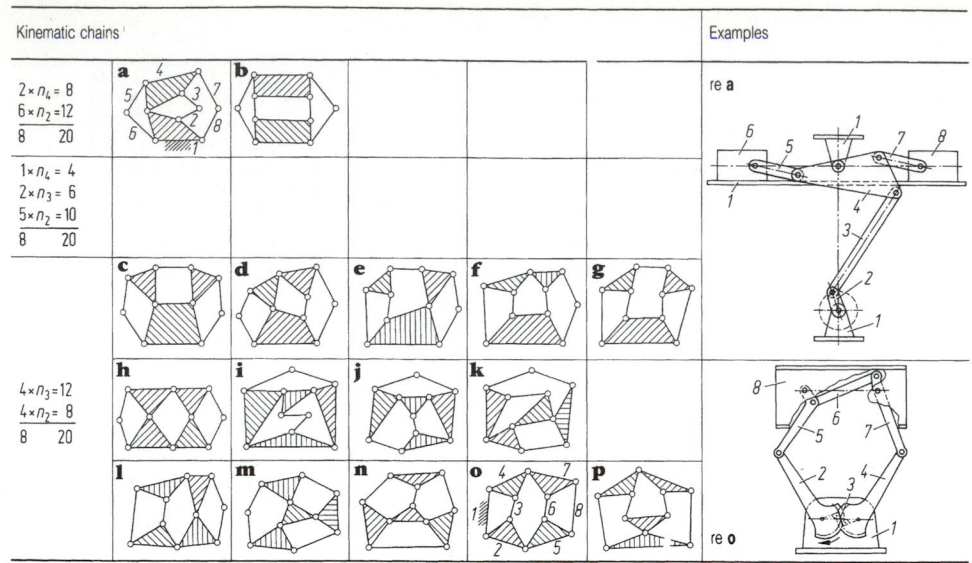

Figure 8. Six-link constrained kinematic chains and sample mechanisms.

All three-link cam mechanisms can be derived by means of a *fixed-link change* from the three-link cam-pair chain with turning-and-sliding pairs, which in turn originates from a corresponding four-link chain (*equivalent chain*) (**Fig. 9**) [5]. In this equivalent chain a binary link connects the centres of curvature of the cam and follower or tracer element which are currently correlated at the point of contact K. The *"three-pole theorem"* that is well known in kinematics states that the relative motions of three links *i, j, k* (random link numbers) are determined by the three *instantaneous centres of velocity ij, ik* and *jk* (double and multiple joints represent degenerated pole line sections in one point) that lie on a line (pole line). This theorem is of particular significance for cam mechanisms, both for systematic classification (equivalent mechanisms, sliding and rolling cam mechanisms), for analysis (velocity

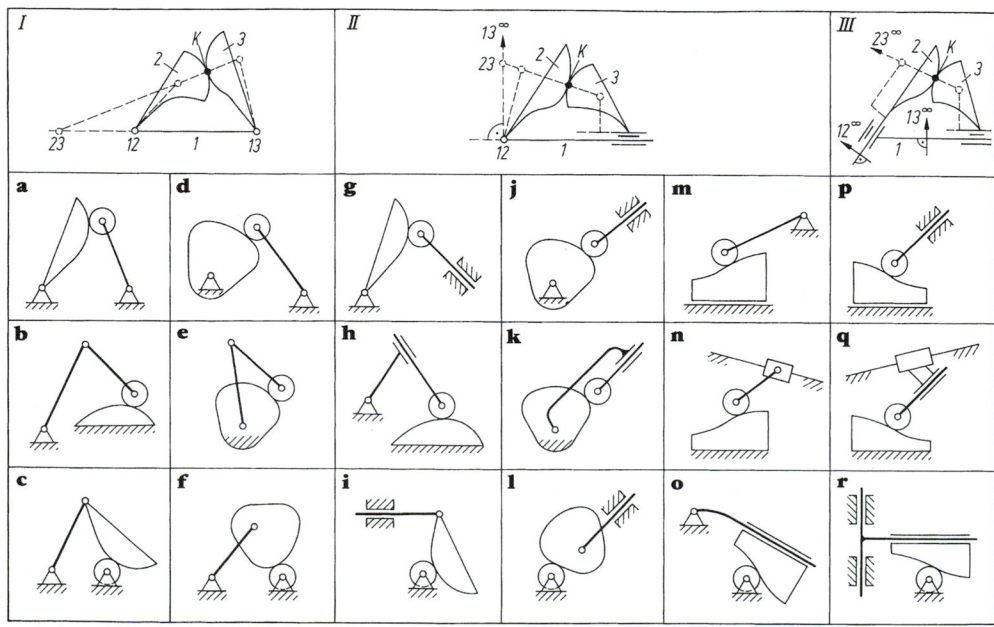

Figure 9. Systematics of three-link cam mechanisms with turning-and-sliding pairs.

determination) and for synthesis (determination of principal dimensions).

In general, each chain with turning pairs and at least four links produces cam-pair chains if a binary link is replaced by a cam joint. If the connecting joint between this binary link and the adjacent link is a rotary joint [6, 7] the concomitant cam is a closed cam about which full rotation takes place; if an oscillating link is present, only a partially tracked cam (slotted link) can be provided which permits ahead and reverse rotation of the follower in the slotted link. The interchangeability between chains and mechanisms with turning and cam pairs (equivalent mechanism theory) is applicable as far as the acceleration stage of kinematic calculation methods (see F9.2).

In general, there is *sliding* and *rolling* of the contiguous links in accordance with the two degrees of freedom with a (planar) cam joint; most cam mechanisms are therefore *sliding cam mechanisms*. In the special case of *roller cam mechanisms* pure rolling occurs in the cam joint because the instantaneous centre *23* in a three-link cam mechanism (**Fig. 9**) is coincident with the point of contact *K*. *Tooth gear mechanisms* with two meshing cam flanks are easily classified as sliding cam mechanisms.

9.2 Analysis of Mechanisms

9.2.1 Kinematic Analysis of Planar Mechanisms

Graphic-Computational Method

Transfer Functions of Four-Bar Linkages

Correlation of Locations. With linkages in general and four-bar linkages in particular, it is important to specify certain relative locations for two mechanism links. This coordination is known as a *"zero-order transfer function"*. In the case of a slider crank with kinematic offset *e*, the instantaneous location of the sliding block *c* as the output link is correlated with the location of the crank *a* as the input link as a function of the cam angle φ (**Fig. 10a**):

$$s = a \cos \varphi + \sqrt{b^2 - (a \sin \varphi - e)^2}. \qquad (3)$$

For the inverted slider crank (**Fig. 10b**), the location ψ of the sliding bar *c* characterises the relationship to the location of the crank *a*

$$\psi = \psi^* + \arccos(e/m^*). \qquad (4)$$

The following applies in the case of a four-bar turning-pair linkage as shown in **Fig. 10c**:

$$\psi = \psi^* - \arccos\left(\frac{m^{*2} + c^2 - b^2}{2m^*c}\right). \qquad (5)$$

The following apply for Eqs (4) and (5):

$$\psi^* = 180° - \arccos\left(\frac{d - a \cos \varphi}{m^*}\right)$$

and

$$m^* = \sqrt{a^2 + d^2 - 2ad \cos \varphi}.$$

Velocity State as First-Order Transfer Function. In the case of the slider crank (**Fig. 10a**) the sign-oriented (directional) *"radius of translation"* m represents the velocity v_8 of the sliding block relative to the crank's angular velocity ω_a

$$m = \text{ÜF1} = v_B/\omega_a = ds_B/d\varphi. \qquad (6)$$

The radius of translation as the first-order transfer function (ÜF1) of the sliding block can be measured perpendicular to the sliding direction as the distance between the relative pole *Q* and the crank pivot A_0.

For the inverted slider crank (**Fig. 10b**) and the four-bar turning-pair linkage (**Fig. 10c**) ÜF1 of the link *c* is expressed by the angular velocity ratio ω_c/ω_a or reciprocal transmission ratio $1/i$ with the pole pitches q_a and q_b:

$$\text{ÜF1} = \omega_c/\omega_a = d\psi/d\varphi - 1/i = q_a/q_b. \qquad (7)$$

The pole *Q* corresponds to the pitch point of two meshed gearwheels and can lie both inside (external gearing) and outside (internal gearing) the path A_0B_0.

Acceleration State as Second-Order Transfer Function. The second-order transfer function (ÜF2) can be determined with the aid of the collineation angle λ and ÜF1. Kinematic derivation is based on the law that the velocity of the relative pole *Q* on the linear frame extension A_0B_0 is a measure of the acceleration of the output link *c*. The following equation applies for the sliding block of the slider crank (**Fig. 10a**) with λ as the angle between the coupler *b* (on the inverted slider crank between the normal to the sliding direction) and the collineation axis *k* as the link between the two instantaneous centres *P* and *Q*:

$$\text{ÜF2} = d^2s/d\varphi^2 = \text{ÜF1}/\tan \lambda. \qquad (8)$$

In the case of an inverted slider crank and a four-bar turning-pair linkage, the following applies for the ÜF2 of the link *c* (**Fig. 10b** and **c**):

$$\text{ÜF2} = d^2\psi/d\varphi^2 = \text{ÜF1}(1 - \text{ÜF1})/\tan \lambda. \qquad (9)$$

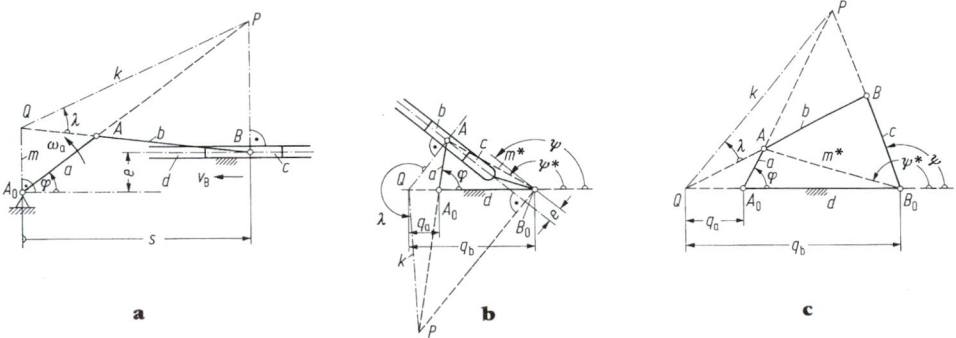

Figure 10. Geometric principles of transfer functions: **a** slider–crank; **b** inverted slider–crank, **c** four-bar turning-pair linkage.

With the aid of the transfer functions it is possible in turn to determine the acceleration a_B of the sliding block or the angular acceleration α_c of the link c of an inverted slider crank and four-bar turning-pair linkage

$$a_B, \alpha_c = \ddot{U}F2\omega_a^2 + \ddot{U}F1\alpha_a. \qquad (10)$$

The angular acceleration of the crank a is designated α_a.

The rotating inverted slider crank and the rotating four-bar turning-pair linkage can be used for two different principal motions, i.e. to generate oscillating and rotating drive motions. Oscillating $(d > a + e)$ and rotating $(d < a + e)$ inverted slider cranks and four-bar turning-pair linkages are available as crank-and-rocker and double-crank mechanisms. Oscillating inverted slider cranks and crank-and-rocker mechanisms are used for reciprocating motion, and rotating inverted slider cranks and double cranks for non-uniform rotation, e.g. as *compound mechanisms* [3, 8].

Loop Iteration Method

The structure of the mechanism to be examined is inserted in the complex number plane (**Fig. 11**). The complex number

$$z = x + iy = r \exp(i\varphi), i = \sqrt{-1}, \qquad (11)$$

then describes the connecting line between two pivot points. Initially a specified original position for the driving member(s) $-r = r_{an}$ for a reciprocating cam as drive element and $\varphi - \varphi_{an}$ for a crank as drive element – is postulated together with appropriately estimated parameters of location (paths r_j and/or angles φ_j (expressed in radians) for the other links from

$$r_j^* = r_j + \Delta r_j, \; \varphi_j^* = \varphi_j + \Delta\varphi_j. \qquad (12)$$

The deviations Δr_j and/or $\Delta\varphi_j$ of these estimated values from the exact values are calculated iteratively as unknowns in a linear equation system until they no longer exceed a specified positive value. Then r_{an} or φ_{an} is increased by an increment, where the previously iterated location of the mechanism serves as the new estimated location etc. [9]. The iteration is based on the "*closed conditions*" of the polygons or loops replacing the mechanism, from the complex number z_j:

$$\varepsilon_k = \sum_{j=1}^{m} (z_j) = \sum_{j=1}^{m} [r_j \exp(i\varphi_j)] = 0; k = 1(1)p \; (13)$$

(summation of m joint spacings). Equation (13) must be repeated p times. The number p of independent loops is derived independently of the degree of freedom F of a mechanism with n links and g joints with a degree of freedom $f = 1$ to be [10]

$$p = g - (n - 1). \qquad (14)$$

For the mechanism in **Fig. 11** this becomes $p = 7 - (6 - 1) = 2$ and consequently

$$\varphi_{an} = \varphi_2 = \varphi_2^* \; \text{(drive equation)},$$

$$r_2 \exp(i\varphi_2) + r_3 \exp(i\varphi_3) - r_8 \exp(i\varphi_8) - ir_1 - r_6 = 0,$$

$$r_7 \exp(i\varphi_7) + r_5 \exp(i\varphi_5) - r_4 \exp(i\varphi_4) - ir_1 - r_6 = 0.$$

With the angles β_2 and β_4 constant, $\varphi_7 = \varphi_2 + \beta_2$ and $\varphi_8 = \varphi_4 + \beta_4$. The lengths r_j are also constant apart from r_6 and are specified like φ_{an}.

The closed conditions mean that $2p$ (real and imaginary parts) transcendental equations are available to determine the same number of parameters of location. A Taylor's series expansion for

$$z_j^* = z_j + \Delta z_j, \qquad (15)$$

which only takes account of the first-order series terms, results in the following iterative specification after insertion in Eq. (13):

$$\Delta r_{an} \text{ or } \Delta\varphi_{an} = 0 \; \text{(drive equation)} \qquad (16a)$$

$$\sum_{j=1}^{m} [\exp(i\varphi_j) \Delta r_j + ir_j \exp(i\varphi_j) \Delta\varphi_j] = -\varepsilon_k;$$
$$k = 1(1)p. \qquad (16b)$$

A linear equation system is derived in this way from the real and imaginary parts of Eq. (16a) and (16b).

$$\mathbf{K} \, \Delta\mathbf{e} = \mathbf{b}_L \qquad (17)$$

with a $(2p + 1) * (2p + 1)$ matrix of coefficients $\mathbf{K}$ for the components of the correction vector $\Delta\mathbf{e}$ which contains the deviations Δr_j and/or $\Delta\varphi_j$, where $j = 1(1)m$. After each iteration step there is an improvement in the initial approximation $\mathbf{b}$ – comprising the real and imaginary parts of the complex sums ε_k in Eq. (13) – in accordance with Eq. (12). The sums ε_k (control option and termination criterion) disappear for the accurately calculated mechanism location. The value of the determinant of the matrix of coefficients $\mathbf{K}$ must be continuously monitored. If Eq. (17) cannot be resolved, either a closed condition has been violated or a special position of the mechanism with poor transmission properties in respect of motions and forces has been reached. A change of determinant sign indicates a change in the installation location.

To determine the velocities and accelerations the closed conditions – Eq. (13) – are derived once or twice as a function of time. This generates two further linear equation systems with the known matrix of coefficients $\mathbf{K}$ which now only have to be resolved once

$$\mathbf{K} \, \dot{\mathbf{e}} = \mathbf{b}_v \qquad (18)$$

and

$$\mathbf{K} \, \ddot{\mathbf{e}} = \mathbf{b}_A. \qquad (19)$$

The vectors $\dot{\mathbf{e}}$ and $\ddot{\mathbf{e}}$ contain the velocities $\dot{r}_j$ and/or $\dot{\varphi}_j$, and the accelerations $\ddot{r}_j$ and/or $\ddot{\varphi}_j$, where $j = 1(1)m$; the vector $\mathbf{b}$ contains only zeros apart from the drive velocity $\dot{r}_{an}$ or $\dot{\varphi}_{an}$; in the main, the vector $\mathbf{b}_a$ contains normal and Coriolis acceleration terms.

Module Method

This method has shown itself to be particularly user-friendly for mechanisms comprising "*double hinges*" (two

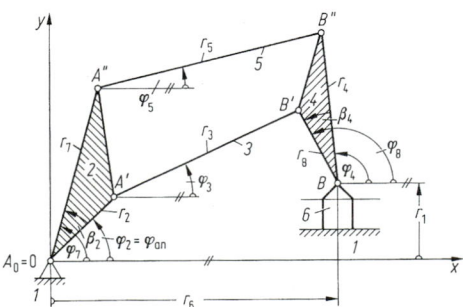

Figure 11. Six-bar linkage with multiple power transmission paths $(F = 1)$: 2 input crank, 6 output reciprocating cam.

jointed binary links) with turning-and-sliding pairs. A further condition for this is that the drive variables (travel or rotational angle, usually relative to the fixed link) are available as functions of time. The structure of a constrained eight-bar linkage (double press) shown schematically in **Fig. 12** contains the relatively simple kinematic modules "*input crank*" A_0A', "*double hinge with three turning pairs*" $A'C'C_0$, $C_0C'C''$, $A'A_0A''$ and "*double hinge with output reciprocating cam*" $C''D$, $A''B$ [11-13]. The output variables A (coordinates x, y of a link point P and angle w of a link with temporal derivatives) of a module are either variable inputs EV for the subsequent module or final results. Constant input variables EK represent, for example, link point pitches l, crank radii r, static offsets v and location parameters K. A ternary link with three turning pairs (links 2 and 6 in **Fig. 12**) can be seen as being derived from a double hinge with three turning pairs.

"Input Crank" Module. Calculation of the coordinates x, y in m with temporal derivatives $\dot{x}$, $\dot{y}$ in m/s and $\ddot{x}$, $\ddot{y}$ in m/s^2 of the link point P when the following are given: coordinates x_0, y_0 in m of the fixed link point P_0, the crank radius r in m, the angular location w in degrees or rad, the angular velocity $\dot{w}$ in rad/s and the angular acceleration $\ddot{w}$ in rad/s^2; see **Fig. 13a**.

Input $r, P_0 [x_0, y_0]; W[w, \dot{w}, \ddot{w}]$

Output $P[x, y, \dot{x}, \dot{y}, \ddot{x}, \ddot{y}]$

Calculation method:

$$x = x_0 + r\cos(w), y = y_0 + r\sin(w), \quad (20)$$

$$\dot{x} = (y_0 - y)\dot{w}, \dot{y} = (x - x_0)\dot{w}, \quad (21)$$

$$\ddot{x} = (y_0 - y)\ddot{w} + (x_0 - x)\dot{w}^2, \quad (22a)$$

$$\ddot{y} = (x - x_0)\ddot{w} + (y_0 - y)\dot{w}^2. \quad (22b)$$

"Double Hinge with Three Turning Pairs" Module. Calculation of the coordinates x, y in m with temporal derivatives x, y in m/s and x, y in m/s^2 of the link point P when the following are given: coordinates x_1, y_1, x_2, y_2 in m with temporal derivatives $\dot{x}_1$, $\dot{y}_1$, $\dot{x}_2$, $\dot{y}_2$ in m/s and $\ddot{x}_1$, $\ddot{y}_1$, $\ddot{x}_2$, $\ddot{y}_2$ in m/s^2, the pitches l_1, l_2 in m of the fixed link points P_1, P_2 and the location parameter K ($K = +1$) if the order of the points P_1P_2P is oriented such that it is mathematically positive, otherwise $K = -1$); see **Fig. 13b**.

Input: $l_1, l_2, K; P_1 [x_1, y_1, \dot{x}_1, \dot{y}_1, \ddot{x}_1, \ddot{y}_1],$

                  $P_2 [x_2, y_2, \dot{x}_2, \dot{y}_2, \ddot{x}_2, \ddot{y}_2]$

Output: $P [x, y, \dot{x}, \dot{y}, \ddot{x}, \ddot{y}]$

Calculation method:

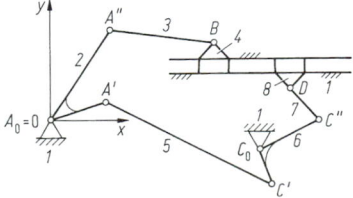

Figure 12. Eight-bar linkage ($F = 1$) assembled from simple modules. 2 input crank, 4 and 8 output reciprocating cams.

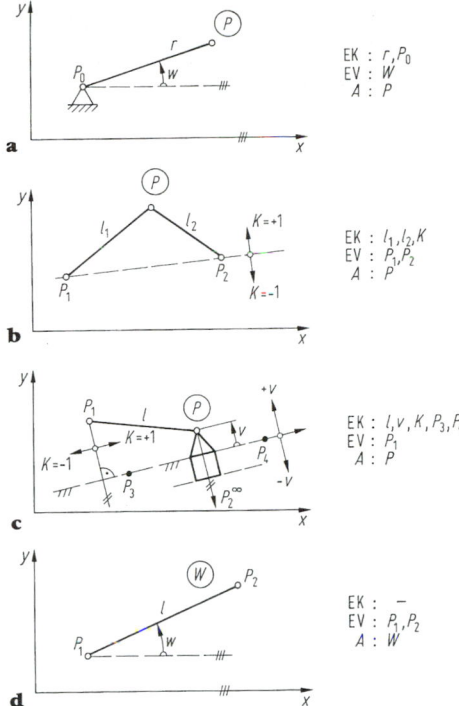

Figure 13. Planar kinematic modules: **a** input crank, **b** double hinge with three turning pairs, **c** double hinge with output reciprocating cam, **d** link with two turning elements.

$$H_1 = x_2 - x_1, H_2 = y_2 - y_1, H_3 = H_1^2 + H_2^2,$$

$$H_4 = l_1^2 - l_2^2 + H_3, H_5 = K\sqrt{4H_3 l_1^2 - H_4^2},$$

$$H_6 = (H_1H_4 - H_2H_5)/(2H_3),$$

$$H_7 = (H_2H_4 + H_1H_5)/(2H_3),$$

$$x = x_1 + H_6, y = y_1 + H_7, \quad (23)$$

$$H_8 = \dot{x}_2 - \dot{x}_1, H_9 = \dot{y}_2 - \dot{y}_1,$$

$$H_{10} = H_6 - H_1, H_{11} = H_7 - H_2,$$

$$H_{12} = H_8H_{10} + H_9H_{11},$$

$$H_{13} = H_7H_{10} - H_6H_{11},$$

$$H_{14} = H_7H_{12}/H_{13}, H_{15} = -H_6H_{12}/H_{13},$$

$$\dot{x} = \dot{x}_1 + H_{14}, \dot{y} = \dot{y}_1 + H_{15}, \quad (24)$$

$$H_{16} = \ddot{x}_2 - \ddot{x}_1, H_{17} = \ddot{y}_2 - \ddot{y}_1,$$

$$H_{18} = (H_{14} - H_8)H_7 + H_{10}H_{15}$$

$$\quad - (H_{15} - H_9)H_6 - H_{11}H_{14},$$

$$H_{19} = (H_{14} - H_8)H_8 + H_{10}H_{16} + (H_{15} - H_9)H_9$$

$$\quad + H_{11}H_{17},$$

$$\ddot{x} = \ddot{x}_1 + (H_{12}H_{15} + H_7H_{19} - H_{14}H_{18})/H_{13}, \quad (25a)$$

$$\ddot{y} = \ddot{y}_1 - (H_{12}H_{14} + H_6H_{19} + H_{15}H_{18})/H_{13}. \quad (25b)$$

"Double Hinge with Output Reciprocating Cam" Module. Calculation of the coordinates x, y in m with temporal derivatives $\dot{x}$, $\dot{y}$ in m/s and $\ddot{x}$, $\ddot{y}$ in m/s^2 of the link point P when the following are given: coordinates x_1,

y_1 in m with temporal derivatives $\dot{x}_1$, $\dot{y}_1$ in m/s and $\ddot{x}_1$, $\ddot{y}_1$ in m/s², the sliding lines and direction with the coordinates x_3, y_3 and x_4, y_4 in m of two points P_3, P_4, the pitch l in m of the link points P_1, P, the static offset v in m ($v > 0$ if the order of the points P_3P_4P is oriented such that it is mathematically positive, otherwise $v < 0$) and the location parameter K ($K = +1$ if the order of the points P_1P is oriented such that it is mathematically positive, otherwise $K = -1$); see **Fig. 13c**.

Input: l, v, K, P_3 [x_3, y_3], P_4 [x_4, y_4];

         P_1 [x_1, y_1, $\dot{x}_1$, $\dot{y}_1$, $\ddot{x}_1$, $\ddot{y}_1$]

Output: P [x, y, $\dot{x}$, $\dot{y}$, $\ddot{x}$, $\ddot{y}$]

Calculation method:

$$H_1 = x_4 - x_3, \; H_2 = y_4 - y_3,$$

$$H_3 = H_1^2 + H_2^2, \; H_4 = H_1/\sqrt{H_3},$$

$$H_5 = H_2/\sqrt{H_3},$$

$$H_6 = x_3 - x_1, \; H_7 = y_3 - y_1,$$

$$H_8 = H_4H_7 - H_5H_6 + v,$$

$$H_9 = H_6^2 + H_7^2 - v^2 - l^2 + (2vH_8),$$

$$H_{10} = H_4H_6 + H_5H_7, \; H_{11} = K\sqrt{H_{10}^2 - H_9} - H_{10},$$

$$x = x_3 + H_4H_{11} - vH_5, \; y = y_3 + H_5H_{11} + vH_4, (26)$$

$$H_{12} = (x - x_1)/l, \; H_{13} = (y - y_1)/l,$$

$$H_{14} = H_4H_{12} + H_5H_{13},$$

$$H_{15} = (\dot{x}_1H_{12} + \dot{y}_1H_{13})/H_{14},$$

$$\dot{x} = H_4H_{15}, \; \dot{y} = H_5H_{15}, \qquad (27)$$

$$H_{16} = (\dot{x}_1H_5 - \dot{y}_1H_4)/H_{14},$$

$$H_{17} = (H_{16}^2/l - \ddot{x}_1H_{12} - \ddot{y}_1H_{13})/H_{14},$$

$$\ddot{x} = -H_4H_{17}, \; \ddot{y} = -H_5H_{17}. \qquad (28)$$

"Link with Two Rotary Elements" Auxiliary Module. Calculation of the angular location w in degrees or rad, the angular velocity $\dot{w}$ in rad/s and angular acceleration $\ddot{w}$ in rad/s² of the mechanism link when the following are given: coordinates x_1, y_1, x_2, y_2 in m with temporal derivatives $\dot{x}_1$, ($\dot{y}_1$), $\dot{x}_2$, ($\dot{y}_2$) in m/s and $\ddot{x}_1$, ($\ddot{y}_1$), $\ddot{x}_2$, ($\ddot{y}_2$) in m/s² of the link points P_1, P_2 (value in ( ) as alternatives); see **Fig. 13d**.

Input: P_1 [x_1, y_1, $\dot{x}_1$, ($\dot{y}_1$), $\ddot{x}_1$, ($\ddot{y}_1$)],

           P_2 [x_2, y_2, $\dot{x}_2$, ($\dot{y}_2$), $\ddot{x}_2$, ($\ddot{y}_2$)]

Output: W [w, $\dot{w}$, $\ddot{w}$]

Calculation method:

$$l = \sqrt{(x_2 - x_1)^2 + (y_2 - y_1)^2},$$

$$w = \arccos\left[(x_2 - x_1)/l\right] \, \text{sign} \, (y_2 - y_1),$$

$$-180° \le w \le +180°, \qquad (29)$$

$$\dot{w} = (\dot{x}_2 - \dot{x}_1)/(y_1 - y_2) = (\dot{y}_1 - \dot{y}_2)/(x_1 - x_2), (30)$$

$$\ddot{w} = [\ddot{x}_1 - \ddot{x}_2 + \dot{w}^2(x_1 - x_2)]/(y_2 - y_1)$$

$$= [\ddot{y}_1 - \ddot{y}_2 + \dot{w}^2 (y_1 - y_2)]/(x_1 - x_2). \qquad (31)$$

9.2.2 Kinetostatic Analyses of Planar Mechanisms

No account is taken initially of friction in calculating the forces transmitted in the joints between the mechanism

links, i.e. in a sliding pair the joint force acts perpendicular to the sliding direction, in a cam pair in the direction of the normal at the point of contact. It is further postulated that the input member moves at a constant velocity or angular velocity Ω. The input force or input torque required for this can be determined.

The joint forces in the joint jk between two link elements j and k are always revealed in pairs by means of a section through the joint jk. If G_{jk} represents the joint force from the link j to the link k, $G_{jk} = -G_{kj}$ applies both to the direction of the joint force as vector and to the components X_{jk} and Y_{jk} in the x and y directions; see **Fig. 14**. In accordance with the three conditions of planar statics, the joint forces on one link k are in equilibrium with the other forces and moments acting on link k. These also include the inertia force - in components $-m_k\ddot{x}_k$ and $-m_k\ddot{y}_k$ - at the centre of gravity S_k (mass m_k in kg), which moves in the x and y directions respectively with the accelerations $\ddot{x}_k$ and $\ddot{y}_k$, and the moment of inertia $-J_k\varphi_k$ (mass moment of inertia J_k in kg m² relative to the centre of gravity) of the link rotating in the x-y plane with the instantaneous angular acceleration φ_k.

For a ternary input link labelled link number 2, mounted such that it can rotate in fixed link 1 and connected to links l and m by means of rotating pairs and on which act, in addition to the inertia effects (centrifugal force only in this case), the input moment M_{an}, a supplementary moment M_2 and an external force F_2 at point P_2, the conditions of equilibrium for $\varphi_2 = \varphi_{an} = \Omega t$ (time t) as per **Fig. 15a** are:

$$X_{12} + X_{12} + X_{m2} = -m_2r_2\Omega^2 \cos(\varphi_2 + \gamma_2)$$

$$-F_2 \cos(\tau_2), \qquad (32)$$

$$Y_{12} + Y_{12} + Y_{m2} = -m_2r_2\Omega^2 \sin(\varphi_2 + \gamma_2)$$

$$-F_2 \sin(\tau_2), \qquad (33)$$

$$-X_{12}l_{21} \sin(\varphi_2) - X_{m2}l_{2m} \sin(\varphi_2 + \beta_2)$$

$$+Y_{12}l_{21} \cos(\varphi_2) + Y_{m2}l_{2m} \cos(\varphi_2 + \beta_2) + M_{an}$$

$$= -F_2p_2 \sin(\tau_2 - \varphi_2 - \varepsilon_2) - M_2. \qquad (34)$$

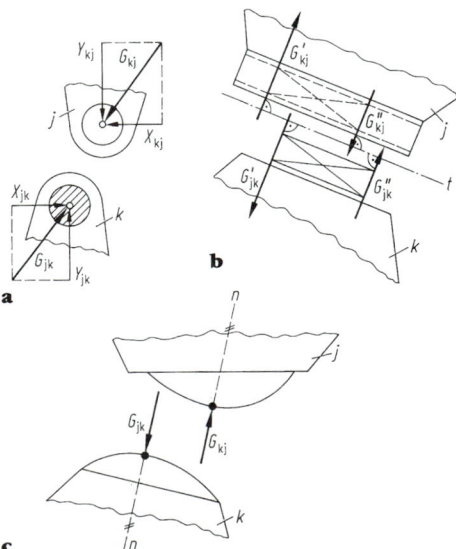

Figure 14. Forces in a frictionless joint. **a** Turning pair, **b** sliding pair (sliding direction t), **c** cam joint (normal direction n).

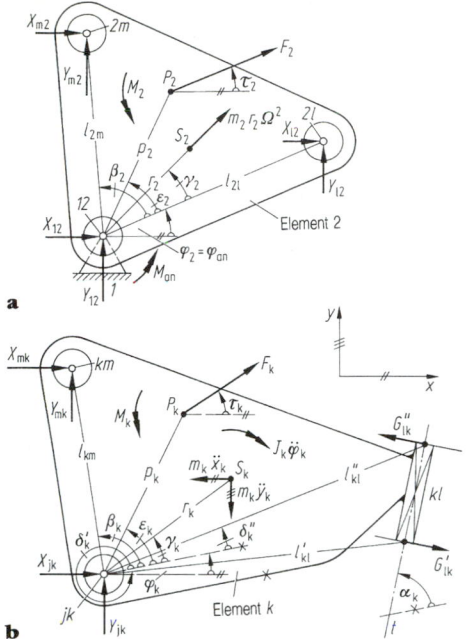

Figure 15. Forces and moments on ternary links with turning and sliding elements: **a** input link, **b** link in general motion.

For a ternary link k subjected to general motion and connected to links j and m by means of turning pairs, and to which link l is connected by a sliding pair, the following equilibrium applies (**Fig. 15b**):

$$X_{jk} + (G'_{lk} - G''_{lk}) \sin (\varphi_k + \alpha_k) + X_{mk}$$
$$= m_k \ddot{x}_k - F_k \cos (\tau_k), \tag{35}$$

$$Y_{jk} - (G'_{lk} - G''_{lk}) \cos (\varphi_k + \alpha_k) + Y_{mk}$$
$$= m_k \ddot{y}_k - F_k \sin (\tau_k), \tag{36}$$

$$- G'_{lk} l'_{kl} \cos (\alpha_k - \delta'_k) + G''_{lk} l''_{kl} \cos (\alpha_k - \delta''_k)$$
$$- X_{mk} l_{km} \sin (\varphi_k + \beta_k) + Y_{mk} l_{km} \cos (\varphi_k + \beta_k)$$
$$= m_k r_k [\ddot{y}_k \cos (\varphi_k + \gamma_k) - \ddot{x}_k \sin (\varphi_k + \gamma_k)]$$
$$- F_k p_k \sin (\tau_k - \varphi_k - \varepsilon_k) - M_k + J_k \ddot{\varphi}_k. \tag{37}$$

In general, the angles and lengths cited are constant, with the exception of φ_2, φ_k, τ_2, τ_k. This can be applied to binary links by setting the appropriate link pitches and concomitant link forces and link force components at zero.

For the actuated $n - 1$ links of a n-bar linkage with a degree of freedom F, g_1 turning-and-sliding pairs and g_2 cam pairs, $3(n - 1)$ linear equations must be developed for F input variables (force or torque), $2g_1$ and g_2 joint forces or components

$$3(n - 1) = 2g_1 + g_2 + F. \tag{38}$$

Allowing for $G_{kj} = - G_{jk}$, $X_{kj} = - X_{jk}$ and $Y_{kj} = - Y_{jk}$, the linear equation system

$$Ax = r \tag{39}$$

is derived for each mechanism position with the unknown vector x, which contains the joint forces or their compo-

nents and the input variables, with the matrix of coefficients A, which can be reduced to a "core matrix" by deleting those columns that only contain one element other than zero and the related lines, and with the vector r, which largely comprises the known (given) forces and moments.

9.2.3 Kinematic Analysis of Spatial Mechanisms

A closed analytical representation of the kinematics of spatial mechanisms is only possible in individual cases [14–17]. An iterative method – cf. F9.2.1 – based on spherical coordinates (spatial polar coordinates r_j, a_j, b_j) for each mechanism link j [10, 18, 19] in the vector form

$$r_j = r_j e_j \tag{40a}$$

with the length r_j and the unit vector

$$e_j = \begin{bmatrix} \cos (\alpha_j) & \cos (\beta_j) \\ \cos (\alpha_j) & \sin (\beta_j) \\ \sin (\alpha_j) & \end{bmatrix}, \tag{40b}$$

is therefore recommended (**Fig. 16a**). The description of the spatial mechanism structure (example in **Fig. 16c**) is based on the "*vectorial equivalent system*", **Fig. 16d**. The constant coordinates are the sizes, while the variable coordinates are the mechanism's location- and time-dependent kinetic quantities with temporal derivatives (velocities and accelerations) which have to be calculated; the time-dependent input values r_{an} or α_{an} or β_{an}, which have to be given in accordance with the degree of freedom F (Eq. (1)), are also variable. The closed condition

$$\sum_j (r_j) = 0 \tag{41}$$

must be evaluated p times (p as per Eq. (14)). The location of the kinetic axes (e.g. rotary, sliding and helical axes) relative to each other, which must be constantly maintained during motion, can be expressed either by scalar products

$$e_j \cdot e_l = \cos (\lambda_{jl}), \tag{42}$$

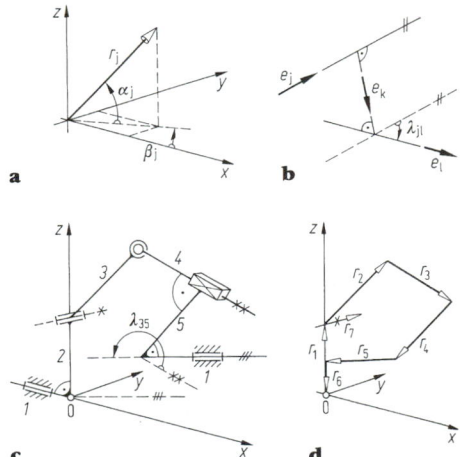

Figure 16. Kinematic analysis of spatial mechanisms: **a** spherical coordinates, **b** basic vectors of crossing and intersecting axes of motion, **c** shaft coupling as example, **d** vectorial equivalent system for **c**.

or by vector products

$$\boldsymbol{e}_j \times \boldsymbol{e}_l = \boldsymbol{e}_k \sin (\lambda_{jl}), \qquad (43)$$

where the shaft angle λ_{jl} = const. (**Fig. 16b**). To this end, either the $\boldsymbol{r}$ vectors already defined in Eq. (14) are used or new ones are introduced, e.g. $\boldsymbol{r}_7$, in **Fig. 16d**.

The evaluation of Eqs (41) to (43) is carried out iteratively with the aid of the Taylor's series expansions terminated after the first-order terms

$$\left. \begin{array}{l} \boldsymbol{e}_j^* = \boldsymbol{e}_j + \boldsymbol{e}_{j,\alpha}\Delta\alpha_j + \boldsymbol{e}_{j,\beta}\Delta\beta_j, \\[4pt] \boldsymbol{e}_{j,\alpha} = \partial\boldsymbol{e}_j/\partial\alpha_j, \; \boldsymbol{e}_{j,\beta} = \partial\boldsymbol{e}_j/\partial\beta_j, \end{array} \right\} \qquad (44a)$$

$$\boldsymbol{r}_j^* = r_j\boldsymbol{e}_j^* + \boldsymbol{e}_j\Delta r_j. \qquad (44b)$$

If the exact values and are inserted in Eqs (41) to (43), a linear equation system can be constructed for the corrections Δr_j, $\Delta\alpha_j$ and $\Delta\beta_j$ of the estimated values $\boldsymbol{e}_j$ and $\boldsymbol{r}_j$. The process is begun with an initial position for the input link and relevant estimates for the mechanism's kinetic quantities in accordance with the drawing or approximate calculation; the sufficiently precise iterated location provides the estimated values for the next location after incrementation of the input variable, etc. The values of the velocity and acceleration stage can be determined from the temporal derivatives of Eqs (41) to (43) which have been carried out once or twice.

9.2.4 Running Quality of Mechanisms

The running quality of the mechanisms depends on the geometric and kinematic variables, design and material properties of the links and joints and on the interplay of forces and power flow in the mechanism [20, 21]. For planar mechanisms at least, the transmission angle and the dynamic properties are important parameters in respect of power flow.

Transmission Angle

The divergence from the optimum 90° transmission angle is an indicator of the quality of motion transmission from drive to drive. The transmission angle is defined as the angle μ between the tangent t_a to the absolute path of the pivot point of the *joint guide link* [22] (output links mounted with bearings in the fixed link are always joint guide links) and the tangent t_r to the relative path of the (transmission) link driving the joint guide link with respect to the input link. In the case of four-bar turning-pair linkages this is also the angle μ_{34} between the links *3* and *4* (**Fig. 17a**) when link *2* is the input link; in the case of a slider crank the direction *1434* is replaced by the normal to the sliding direction. Excessively small μ values indicate that there is a danger of jamming.

Account may have to be taken of a number of transmission angles, which can only be determined with knowledge of the instantaneous centre configuration, in the case of multi-link mechanisms with multiple power-transmission paths. With the six-bar linkage ($F = 1$) represented schematically in **Fig. 7b** $\mu_{56}^{(2)}$ applies for motion transmission from the input link *2* to the output link *6*; in the reverse direction, however, with link *6* as the input link and *2* as the output link the angles $\mu_{23}^{(6)}$, $\mu_{35}^{(6)}$ and $\mu_{45}^{(6)}$ apply.

Dynamic Properties

The power flow during a motion period in a mechanism under load can change its direction continuously; with

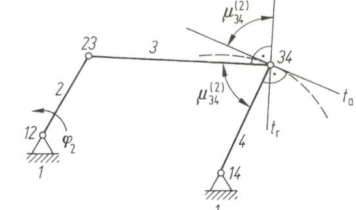

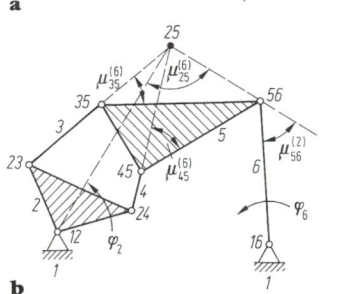

Figure 17. Transmission angle: **a** four-bar linkage, **b** six-bar linkage with multiple power-transmission paths.

linkages, therefore, the transmission angle only has a limited significance in respect of the term "*opposite fixed angle*". High-speed linkages should be evaluated on the basis of their dynamic properties, taking account of the influence of both inertia and external loading [23, 24].

9.3 Synthesis of Mechanisms

The synthesis of mechanisms (dimensional synthesis) is used to identify suitable mechanisms for the given transmission and guidance tasks of points and link locations [25]. This process uses an analysis of the systematic and of the geometric–kinematic properties of the mechanisms. Appropriate analytical programs are also employed alongside certain synthesis processes to determine suitable mechanisms using the "synthesis by means of iterative analysis" method.

9.3.1 Four-Bar Linkages

Favourable Oscillating Motions for Transmission and Acceleration

A four-bar linkage (**Fig. 18**) in the form of a crank-and-rocker mechanism converts rotary motion to oscillating

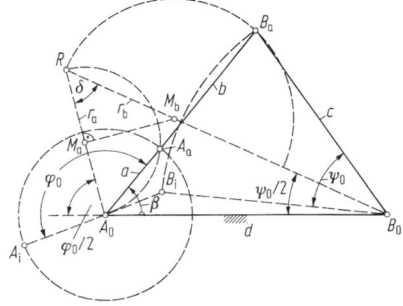

Figure 18. Geometric principles of Alt's dead-centre positions for crank-and-rocker mechanisms.

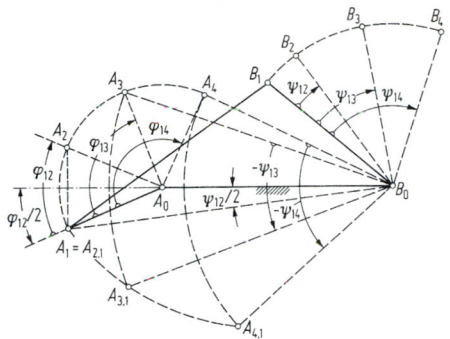

Figure 19. Synthesis of a four-bar turning-pair linkage for given angular positions.

or five positions if two such curves intersect. Simpler procedures are available if the special positions are used [30]. Programmable calculators permit the calculation of the dimensionally synthesised curves without application of Burmester's theorem by means of automatic iterations [31, 32]. Further options are offered by *point location reductions* [33].

Example The three angles φ_{12}, φ_{13} and φ_{14}, are to be correlated with the angles ψ_{12}, ψ_{13} and ψ_{14} (**Fig. 19**). The angles $\varphi_{12}/2$ and $\psi_{12}/2$, for example, are applied to A_0B_0 at A_0 and B_0 respectively and their free sides intersect at A_1. With a crank length of $\overline{A_0A_1}$ the crank positions A_0A_2, A_0A_3, A_0A_4 are defined with the concomitant φ angles. The points A_2, A_3, A_4 are rotated about B_0 in the opposite direction to the given ψ angles, i.e. $-\psi_{12}$, $-\psi_{13}$ and $-\psi_{14}$, resulting in the points $A_{2.1}$, $A_{3.1}$, $A_{4.1}$, of which $A_{2.1}$ is coincident with A_1. The circle drawn through the three points $A_1 = A_{2.1}$, $A_{3.1}$, $A_{4.1}$ yields as its centre point the pivot position B_1 and thus all the dimensions of the mechanism sought in its position *1*. In the initial design phase

motion. The crank angle φ_0 is correlated with the oscillating angle ψ_0. An infinite number of crank-and-rocker mechanisms is available for φ_0 and ψ_0 [26]. $\psi_0/2$ and $\varphi_0/2$ applied to $d(A_0B_0 = d)$ at B_0 and A_0 respectively give the intersection point R. The perpendicular bisector of A_0R at M_a intercepts B_0R at M_b. Circles with $M_aR = r_a$ and $M_bR = r_b$ are geometric loci for crank positions A_a and rocker positions B_a of a crank-and-rocker mechanism at the *bottom dead centre position* $A_0A_aB_aB_0$ with an arbitrary angle β. If d is assumed, the dimensions are

$$a = 2r_a \cos (180° - \beta - \varphi_0/2), \qquad (45)$$

$$b = 2r_b \cos (180° - \delta - \beta - \varphi_0/2) \qquad (46)$$

and

$$c = \sqrt{d^2 + (a + b)^2 - 2d(a + b) \cos \beta}. \qquad (47)$$

β enables the determination of the optimum crank-and-rocker mechanism for transmission [27], the optimum adjustment means for transmission with variable φ_0 and ψ_0, the optimum six-bar series linkage for transmission and the optimum crank-and-rocker mechanism for acceleration with the smallest maximum acceleration for outward or return travel.

There are a similar design and appropriate results for the slider crank in respect of the optimum dimensions for transmission and acceleration [27–29].

Correlation of Angles

The *Burmester circle point and centre point curves* allow four (homologous) positions in a plane to be governed

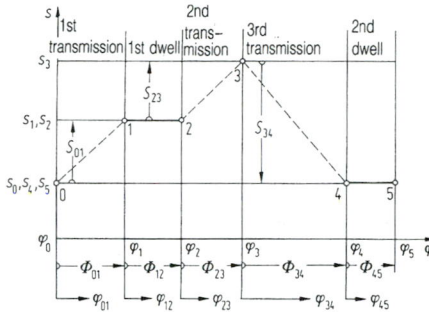

a

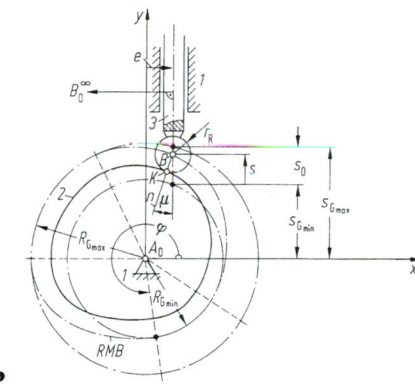

b

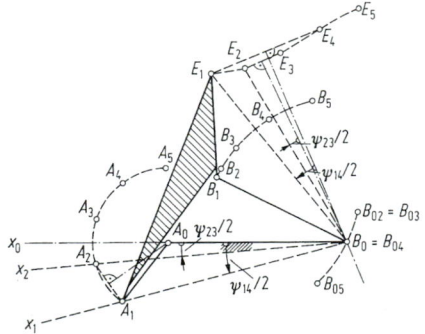

Figure 20. Synthesis of a four-bar turning-pair linkage for given coupler point positions.

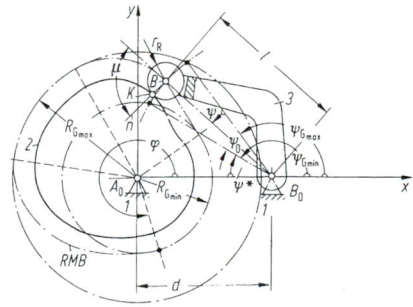

c

Figure 21. Notation in planar three-link mechanisms: **a** motion plan, **b** mechanism with follower roller, **c** mechanism with lever roller.

a pivot point B_1, i.e. a link length B_0B_1, can be selected instead of A_1, together with other correlated initial angle pairs. With six-bar linkages it is possible to define six or, under certain conditions, even eight correlated angle pairs with appropriately extended point position reduction.

Generation of Given Planar Curves

In theory a given planar curve can be generated in nine points exactly by the coupler curve of a four-bar turning-pair linkage. To date, practical procedures for general locations are only known, as in the example below, for seven points.

Example If five points E_1 to E_5 are given on a curve (**Fig. 20**) the perpendicular bisectors of the paths E_1E_4 and E_2E_3 intersect at B_0, from which an arbitrary ray x_0 emanates. The rays x_1, x_2 are marked on this such that with x_0 they include the angles $\psi_{14}/2$ and $\psi_{23}/2$, which are formed by the perpendicular bisectors and B_0E_1 and B_0E_2. $E_1A_1 = E_2A_2$ can be marked off with A_1 on x_1 and A_2 on x_2 at an arbitrary equal length. The perpendicular bisector of A_1A_2 intersects x_0 at A_0, and the circle about A_0 can be drawn through A_1 and A_2 on which A_3, A_4, A_5 are located as intersections of the circles about E_3, E_4, E_5 with E_1A_1 as the radius. The points B_{02} and B_{05} are found with $\Delta E_1A_1B_{02} = \Delta E_2A_2B_0$, $\Delta E_1A_1B_{05} = \Delta E_5B_5B_0$. The equivalent would occur with the points A_3 and A_4 to give $B_{03} = B_{02}$ and $B_{04} = B_0$ as point position reductions. The circle through the three points $B_0 = B_{04}$, $B_{02} = B_{03}$ and B_{05} yields its centre point as the point position B_1 and thus the mechanism sought in its position 1. Other E points can also be paired initially, thereby achieving a different intersection B_0. Since the ray x_0 and the lengths E_1A_1 can be arbitrarily postulated the coupler curve can be made to pass through seven E points also.

9.3.2 Cam Mechanisms

Three-link cam mechanisms with the frame as the fixed link [4] are generally used to generate periodic motions with dwells (output link standstills) and transitions with favourable acceleration properties. The task assigned within the overall machine system determines the *"motion plan"* (**Fig. 21a**) of a cam mechanism with motion broken down into individual steps *ik*. Thus the drive motion s for a follower roller (**Fig. 21b**) or ψ for a lever roller is a function of the drive motion φ (rotary angle of the cam plate). A lever's rotary angle ψ expressed in radians can be converted to a follower stroke with the aid of the relation $s = l\psi$ (lever length $l = B_0B$). With the exception of the dwells, a "motion law" is assigned to each motion step in a standardised wording, i.e. relative to

the partial stroke $S_{ik} = s_k - s_i$ or $\Psi_{ik} = \psi_k - \psi_i$ and partial rotational angle $\Phi_{ik} = \varphi_k - \varphi_i$ [34, 35]

$$(s - s_i)/S_{ik} = f_{ik} [(\varphi - \varphi_i)/\Phi_{ik}] = f(z). \qquad (48)$$

The functions $f(z)$ are mainly exponential functions $f(z) = A_0 = A_1z + A_2z^2 + \cdots + A_xz^2$ or trigonometric functions $f(z) = A \cos (vz) + B \sin (vz)$ or combinations of the two. The boundary values of the derivatives according to the rotational angle φ or the first-order or second-order transfer functions at the locations i and k determine the type of motion task, and it is essential that they be assimilated shock-free (no jump from s' or ψ') and jerk-free (no jump from s'' or ψ''). Further quality criteria arise from the maximum values of the following derivatives of the standardised laws in accordance with z:

Velocity coefficient	$C_v = \max (|f'|)$
Acceleration coefficient	$C_a = \max (|f''|)$
Jerk coefficient	$C_j = \max (|f'''|)$
Static torque coefficient	$C_{H\,stat} = C_v$
Dynamic torque coefficient	$C_{H\,dyn} = \max (|f'f''|)$.

The smallest values of the selected function $f(z)$ are the optimum in each case.

There is an infinitely large number of cam profiles for a specified motion task, of which the optimum for transmission purposes (minimum deviation of transmission angle m from $90°$) can be identified. The "principal dimensions" for this are, in the case of a cam mechanism with roller follower, the offset e and the radius of the "basic circle" $R_{G\,min}$ of the roller centre-point path or the "basic stroke" $S_{G\,min}$ or, in the case of a cam mechanism with a roller lever, the lever length l and the "basic angle" $\psi_{G\,min}$ or ψ^* between the lever and the fixed link.

9.4 Special Mechanisms

For details of special mechanisms to achieve particular motion tasks, sometimes with unusual design boundary conditions, please refer to the specialist literature and the relevant VDI Codes: spatial linkages and cardan shafts [36–40], spatial cam mechanisms [41, 42], indexing mechanisms [43–46, 49] and wheel crank mechanisms which are combinations of linkages and at least two wheels for rotary-dwell motion and pilgrim-step motion [47–49].

10 Crank Mechanisms

K.-H. Küttner, Berlin

A crank mechanism (**Fig. 1**), referred to in kinematics as a slider crank (see F9, **Fig. 6a**), converts the oscillating motion of a piston 1 with a gudgeon pin B via the connecting rod 2 into rotary motion of the crank 3 with the webs, crankpin and crankshaft journal K and M, or vice versa. Its task is power transmission – when it is usually called a drive mechanism – and in control applications which also require special designs. Such drive mechanisms are used in piston engines, in presses and in hydraulic and pneumatic drives.

Designs

The crosshead drive mechanism (**Fig. 1a**), rated power ≤ 1800 kW and speed ≤ 1000 min^{-1}, comprises the cross-

head 4 for particular guidance with journals, piston rod 5 and piston 1. The trunk piston system (**Fig. 1c**) is used for rated powers of ≤ 420 kW and speeds of ≤ 10000 min^{-1}. Special piston forms illustrated are the eccentric (**Fig. 1e**) for control systems, which was developed from an extension of the crankpin because of the small eccentricity, and the equivalent cam mechanism (**Fig. 1d**) used for refrigerator compressors.

10.1 Kinematics

The dimensions of the crank mechanism (**Fig. 2**) are determined by the crank radius r and the connecting rod length l or by the stroke $s = 2r$ and the connecting rod ratio $\lambda = r/l$. The stroke is the piston displacement

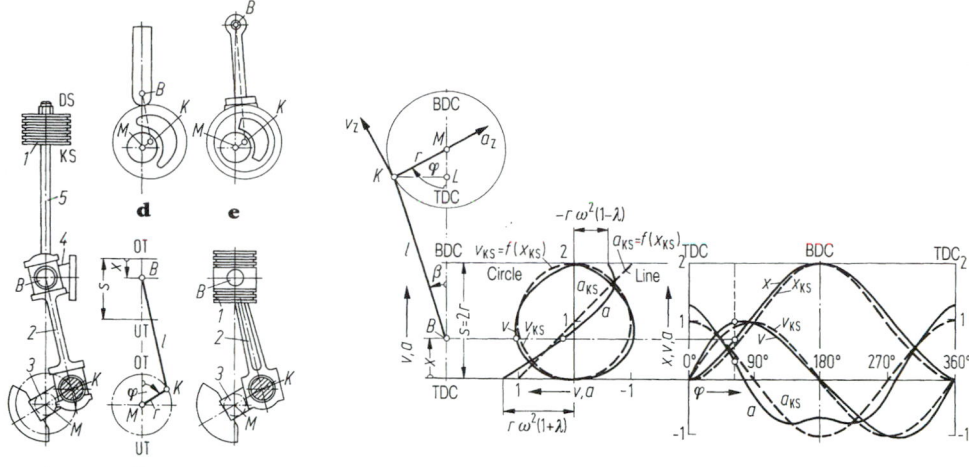

Figure 1. a–e. Crank mechanism forms (broken line marks positions in operation).

Figure 2. Kinematics of crank mechanism: $\lambda = 1/3$ withdrawn, $\lambda \to \infty$ broken line; scales for $r = 1$ m and $\omega = 1\ \mathrm{s}^{-1}$.

between the top (TDC) and bottom (BDC) dead centres. The motions proceed from the piston displacement x and the crank and connecting rod angles φ and β. They are measured from TDC downwards in which B, K and M are on the centreline. The following apply for rotation in the time T, i.e. the angle $\varphi = 2\pi$ at a constant angular velocity $\omega = 2\pi/T$, with rotational speed n, mean piston velocity c_m, crankpin velocity v_z and acceleration a_z:

$$n = 1/T = \omega/(2\pi), \qquad c_m = 2s/T = 2sn, \qquad (1), (2)$$

$$v_z = \qquad \omega r = \pi ns, \qquad a_z = r\omega^2 = v_z\omega. \qquad (3), (4)$$

The crank displacement, velocity and acceleration have the period $T = 1/n$, act on the centreline and are positive from B to M.

10.1.1 Piston Displacement

As shown in **Fig. 2**, the piston displacement is

$$x = r(1 - \cos\varphi) + l(1 - \cos\beta)$$

$$= r[1 - \cos\varphi + (1 - \sqrt{1 - \lambda^2 \sin^2\varphi})/\lambda]. \quad (5)$$

where $\lambda = r/l$, $\sin\beta = \lambda \sin\varphi$ or $\cos\beta = \sqrt{1 - \lambda^2 \sin^2\varphi}$ from the triangles BKL and LKM. If the expression under the radical is expanded in accordance with Taylor's series, then

$$x = r[1 - \cos\varphi + (\lambda/2)\sin^2\varphi + (\lambda^3/8)\sin^4\varphi$$

$$+ (\lambda^5/16)\sin^6\varphi + \ldots]. \qquad (6)$$

Approximate values are yielded if only the first three or two terms of Eq. (6) are used.

$$x_K = r[1 - \cos\varphi + (\lambda/2)\sin^2\varphi], \quad x_{KS} = r(1 - \cos\varphi).$$
$$(7), (8)$$

Equation (8) applies exactly for the crosshead crank (see F9, **Fig. 6d**) or for $\lambda = 0$ or without the term $l(1 - \cos\beta)$ in Eq. (5). The errors $x - x_K \approx r/200$ and $x - x_{KS} \approx r/6$ for a large $\lambda = 1/3$ decrease as λ decreases.

10.1.2 Piston Velocity

The piston velocity, assuming $\varphi = \omega/t$ and Eq. (5), is

$$v = \frac{dx}{dt} = \omega\frac{dx}{d\varphi} = r\omega\frac{\sin(\varphi + \beta)}{\cos\beta}$$

$$= r\omega\left(\sin\varphi + \frac{\lambda}{2}\frac{\sin 2\varphi}{\sqrt{1 - \lambda^2 \sin^2\varphi}}\right). \qquad (9)$$

It follows from Eq. (6), taking account of the goniometric equations, that

$$v = r\omega\left[\sin\varphi + \left(\frac{\lambda}{2} + \frac{\lambda^3}{8} + \frac{15\lambda^5}{26}\right)\sin 2\varphi\right.$$

$$\left. -\left(\frac{\lambda^3}{16} + \frac{3\lambda^5}{64}\right)\sin 4\varphi + \frac{3\lambda^5}{256}\sin 6\varphi \mp \ldots\right]. \quad (10)$$

Approximate values are yielded by Eqs (7) and (8)

$$v_K = r\omega\left[\sin\varphi + (\lambda/2)\sin 2\varphi\right], \quad v_{KS} = r\omega\sin\varphi. \quad (11), (12)$$

The errors $v - v_K = r\omega/207$ and $v - v_{KS} = r\omega/6$ for $\lambda = 1/3$ decrease as λ decreases. The velocity changes its sign at the dead centres and is a function of the crankpin velocity $v_z = r\omega$. Its extreme values $v_{max} \approx r\omega\sqrt{1 + \lambda^2}$ are approximately $\beta \approx 56.5°\cdot\lambda$.

10.1.3 Piston Acceleration

The piston acceleration, assuming $\varphi = \omega t$ and Eq. (9), is

$$a = \omega\frac{dv}{d\varphi} = r\omega^2\left[\frac{\cos(\varphi + \beta)}{\cos\beta} + \frac{\sin\beta\cos^2\varphi}{\sin\varphi\cos^3\beta}\right]$$

$$= r\omega^2\left[\cos\varphi + \lambda\frac{\cos 2\varphi + \lambda^2\sin^4\varphi}{\sqrt{(1 - \lambda^2\sin^2\varphi)^3}}\right]. \quad (13)$$

From Eq. (10), harmonic analysis of the acceleration yields

$$a = r\omega^2 \left[\cos\varphi + \left(\lambda + \frac{\lambda^3}{4} + \frac{15\lambda^5}{128} \right) \cos 2\varphi \right.$$

$$\left. - \left(\frac{\lambda^3}{4} + \frac{3\lambda^5}{16} \right) \cos 4\varphi + \frac{9\lambda^5}{128} \cos 6\varphi \pm \dots \right]. \quad (14)$$

Approximate values are yielded by Eqs (11) and (12).

$$a_K = r\omega^2 (\cos\varphi + \lambda\cos 2\varphi), \quad a_{KS} = r\omega^2 \cos\varphi. (15), (16)$$

The maximum errors are $a - a_K = r\omega^2/50$ and $a - a_{KS} = r\omega^2/2.83$ for $\lambda = 1/3$.

Curve (Fig. 2). Accelerations achieve the exact values according to Eq. (13) at the TDC (OT) ($\varphi = 0°$, $\beta = 0°$) and BDC (UT) ($\varphi = 180°$, $\beta = 180°$).

$$a_{K,OT} = r\omega^2 (1 + \lambda), \quad a_{K,UT} = -r\omega^2 (1 - \lambda), \quad (17)$$

where $a_{K,OT}$ is always the maximum, $a_{K,UT}$ is only the minimum, however, for $\lambda \leq 1/4$. Where $\lambda > 1/4$ there are two minima $|a_{K,min}| > a_{K,UT}$ symmetrically to BDC. Acceleration is a function of the crankpin acceleration in accordance with Eq. (4). Its zero point is at the maximum velocity.

10.2 Dynamics

Periodically fluctuating fluid pressure, inertia, weight and frictional forces occur in crank mechanisms. The fluid pressure or primary forces depend on the working process involved, with whose period $T_a = a_T/n$ they change, and on the piston area. The number of cycles is $a_t = 2$ for four-stroke motors, otherwise $a_T = 1$. These forces are transferred through the machine frame and drive mechanism and transmitted as a rotary force at the crankpin K and as a torque to the crank MK (**Fig. 1a**). The inertia forces of the drive mechanism or the secondary forces are proportional to the square of the angular velocity, have the period $T = 1/n$ and are transmitted from the drive mechanism to the base and the environs, where they act as vibration exciters. They are therefore of great importance, although their mean value per revolution is zero. The weight forces are negligible at high speeds. The frictional forces depend on many influences, e.g. working pressures, drive mechanism masses and condition, finish and lubrication of bearings. Experiments must therefore be conducted to identify them.

10.2.1 Fluid Pressure Forces

With trunk pistons or simple disc pistons (**Fig. 1b**) of area A_K, the fluid enclosed by the cylinder, case and piston exerting pressure p and the constant air pressure p_a on the rear of the piston generate the fluid pressure force

$$F_S = (p - p_a)A_K. \quad (18)$$

This acts on the piston and drive mechanism. An opposite force of equal magnitude is exerted via the case on the cylinder and frame. With double-action machines (**Fig. 1a**) the forces

$$F_{DS} = p_{DS}A_{DS} \quad \text{and} \quad F_{KS} = p_{KS}A_{KS} = p_{KS} (A_{DS} - A_{St}),$$

thus in total

$$F_S = (p_{DS} - p_{KS}) A_{DS} + (p_{KS} - p_a) A_{St} \quad (19)$$

act in addition to the piston rod force $F_{St} = p_a A_{St}$ on the case or crank side DS and KS respectively.

10.2.2 Inertia Forces

Forces. Where there is oscillating and rotary motion in the drive mechanism with masses m_o and m_r the following forces occur:

$$F_o = m_o a \quad \text{and} \quad F_r = m_r r\omega^2. \quad (20)$$

Oscillating Forces (Fig. 3). Substituting the acceleration a from Eq. (14), the forces of the first, second, fourth, sixth etc. orders are given for the multiples of the crank angle φ.

$$F_o = m_o\, r\omega^2 \sum_{k=1}^{n} f(\lambda) \cos k\varphi = \sum_{k=1}^{n} F_k,$$

$$k = 1, 2, 4, 6, 8, \dots, n; \quad F_I = m_o r\omega^2 \cos\varphi;$$

$$F_{II} = m_o r\omega^2 \left(\lambda + \frac{\lambda^3}{4} + \frac{15\lambda^5}{128} \right) \cos 2\varphi; \quad (21)$$

$$F_{IV} = -m_o r\omega^2 \left(\frac{\lambda^3}{4} + \frac{3\lambda^5}{16} \right) \cos 4\varphi;$$

$$F_{VI} = \frac{9\lambda^5}{128} m_o r\omega^2 \cos 6\varphi.$$

These periodic forces counteract the piston acceleration in the centreline. They are positive from the crankshaft to the piston [2]. Sufficiently precise values are yielded by a_K as per Eq. (15).

$$\left. \begin{array}{l} F_o = F_I + F_{II}; \quad F_I = m_o r\omega^2 \cos\varphi; \\ F_{II} = \lambda m_o r\omega^2 \cos 2\varphi. \end{array} \right\} \quad (22)$$

Although the deviation of the approximation is only 0.46% for $\lambda = 1/3$ and $\varphi = 0$, the more precise values are significant in the case of resonance of poorly damped oscillations [3]. The amplitudes or the magnitudes of the vectors rotating at angles φ or 2φ are

$$P_I = m_o r\omega^2 \quad \text{and} \quad P_{II} = \lambda m_o r\omega^2 = \lambda P_I.$$

Their projections to the centreline yield the forces $F_I = P_I \cos\varphi$ and $F_{II} = P_{II} \cos 2\varphi$. The extreme values of F_I are $\pm P_I$ with $\varphi = 0°$ or $180°$, i.e. at the top or bottom dead centres, while those of F_{II} are $\pm P_{II}$ with $\varphi = 0°$, $90°$, $180°$, $270°$ and $360°$. F_I forces are zero at $90°$ and $270°$, and F_{II} at $45°$, $135°$, $225°$ and $315°$. The F_o forces reach the maximum $P_I + P_{II}$ at TDC and the value $- P_I + P_{II}$ at BDC which is at a minimum at $\lambda < 1/3.8$.

Rotary Forces. $F_r = m_r r\omega^2$ are centrifugal forces at the rotating crank of constant magnitude but changing angle φ. Their locus curve is a circle, and their components are $F_r \cos\varphi \sim F_I$ in the centreline and $F_r \sin\varphi$ perpendicular to this. Equilibrium (**Fig. 4b**) is achieved by means of balance weights on the crank webs opposite the crankpin with the centre of gravity S_G, the radius r_G and the mass $m_G = m_r r/r_G$ [1, 4] (on multi-cylinder machines, see F1.1).

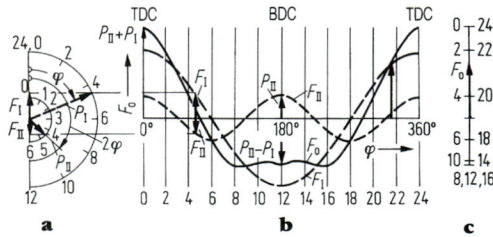

Figure 3. Inertia forces (oscillating): **a** vectors, **b** time curve, **c** locus curve.

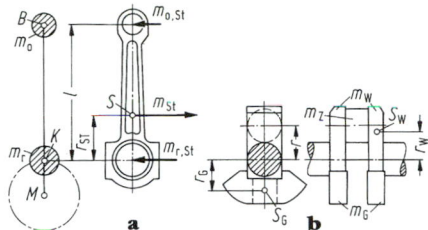

Figure 4. a and **b**. Masses and balance weights.

Masses

Connecting Rod (Fig. 4a). This moves with oscillating motion at the gudgeon pin B and with rotary motion at the crankpin K. With the rod length l and the distance r_{St} between the point K and the centre of gravity S through which its mass m_{St} passes, then

$$m_{o,St} = m_{St}r_{St}/l, \quad m_{r,St} = m_{St}\,(l - r_{St})/l. \quad (23)$$

follows approximately as in the calculation of the bearing forces. For rods of the same design $\lambda = r/l$ and r_{St}/l are constant. Under the usual conditions of $\lambda = 1/4$ and $r_{St}/l = 1/3$ then $m_{o,St} \approx m_{St}/3$ and $m_{r,St} \approx 2m_{St}/3$ apply.

Crankshaft (Fig. 4b). To simplify the centrifugal force calculation, the crank webs with the mass m_w and the distance r_w of their centre of gravity S_w from the rotational axis M are reduced to the crank radius r. Since the centrifugal force cannot change during this, then

$$m_{red,w} = m_w r_w/r. \quad (24)$$

Oscillating Mass. Assuming the masses m_K for the piston, m_{KS} for the piston rod, m_{Kr} for the crosshead and $m_{o,St}$ for the connecting rod as per Eq. (23), then

$$m_o = m_K + m_{KS} + m_{Kr} + m_{o,St}, \ m_o = m_K + m_{o,St} \quad (25)$$

for the crosshead or trunk piston system.

Rotating Mass. With the crankpin mass m_Z, masses m_{red} and $m_{r,St}$ as per Eqs (23) and (24), then

$$m_r = m_Z + m_w(r/r_w) + m_{r,St}. \quad (26)$$

10.2.3 Resultant Forces

The resultant of the fluid pressure and inertia forces F_s and F_o has the period a_T/n. It passes through the drive mechanism and housing, i.e. through the case, cylinder and frame; see **Fig. 5a, b**.

Piston. The piston force acts at the centreline (**Fig. 5a**)

$$F_K = F_s - F_o. \quad (27)$$

The direction from B to M is positive. For simplicity's sake, the piston absorbs the entire inertia force F_o; frictional and weight forces are disregarded. Its curve, especially the extreme values, is a function of the working process, the speed and the $F_{o,max}/F_{S,max}$ ratio. Zero positions are at $F_s = F_o$.

Gudgeon Pin. The piston force F_K is divided here into the connecting rod force F_{St} at its centreline and the normal force F_K perpendicular to the cylinder centreline.

$$F_{St} = F_K/\cos \beta, \quad F_N = F_K \tan \beta \quad (28), (29)$$

Zero positions (**Fig. 6a**): for $F_K = 0$, $F_{St} = F_K = 0$; furthermore, $F_K = 0$ at $\beta = 0°$ or $\varphi = 0°$, $180°$ and $360°$.

Crankpin. The connecting rod force is divided here into the radial force F_K in the crank direction and the tangential force F_T perpendicular to this. In accordance with Eqs (28) and (29), then

$$F_T = F_K \sin (\varphi + \beta)/\cos \beta, \quad (30)$$

$$F_R = F_K \cos (\varphi + \beta)/\cos \beta. \quad (31)$$

The positive tangential force acts in the direction of rotation.

Torque. The changes in $M_d = F_T r$ cause major speed fluctuations which are balanced by flywheels (see J1.1). There are zero positions at $F_K = 0$ and at $\varphi = 0°$, $180°$ and $360°$ etc., thus at $\beta = 0°$. These are the TDC and BDC in which an engine does not start up automatically. The points where $F_o = 0$ and $F_K = F_s$ (in **Fig. 6b** at $\varphi = 76.4°$ and $283.55°$ for $\lambda = 1/3.8$) and the mean torque are independent of rotational speed since the inertia forces, thus also their mean tangential force, are zero (on multi-cylinder machines see J1.1).

Housing (Fig. 5b). This absorbs the drive mechanism forces. On the case, F_S acts on the cylinder and F_K on the guideway. In the frame the bearing absorbs F_{St} as per Eqs (27) and (29). The moment $M_d = F_N a = F_T r$ is absorbed by the frame anchor bolts or used for torque measurements in motors with self-aligning bearings. At the cylinder centreline the force F_S on the case is countered by the force F_K. To ensure equilibrium, the base must absorb the rest, i.e. inertia forces F_o. This also applies to F_r, which is not contained in F_K.

10.2.4 Forces in Drive Mechanism Components

Although the inertia forces relieve the drive mechanism components at TDC at high rotational speeds, they are

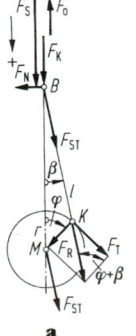

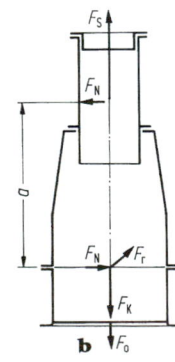

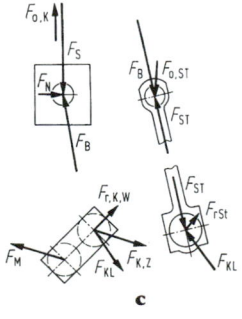

Figure 5. Forces in piston engine: **a** drive mechanism, **b** housing, **c** individual components.

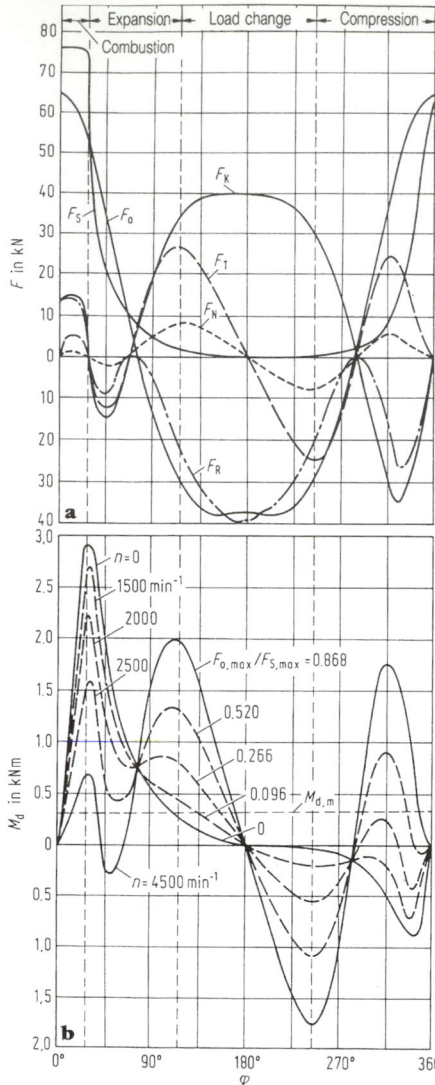

$$F_{BL} = -F_{St} = -\sqrt{F_K^2 + F_N^2}. \tag{33}$$

At the dead centres $F_{BL} = F_K$, since $F_N = 0$. At TDC, $F_{BL} = m_o r\omega^2 (1 + \lambda) - Fs$, and BDC $F_{BL} = m_o r\omega^2 (1 - \lambda)$. In addition to F_{St} or FT and F_R, the big end also absorbs $F_{r,St} = m_{r,St} r\omega^2$. Thus

$$F_{KL} = \sqrt{F_T^2 + (F_R - F_{r,St})^2}. \tag{34}$$

At the dead centres here $F_T = 0$ and $F_R = F_K$ as per Eqs (30) and (31). Thus at TDC

$$F_{KL} = F_R - F_{rSt} = F_S - m_o r\omega^2 (1 + \lambda) - m_{r,St}\, r\omega^2$$

and at BDC, when $F_S = 0$

$$F_{KL} = m_o r\omega^2 (1 - \lambda) - m_{r,St}\, r\omega^2.$$

Crankshaft. The forces $-F_{KL}$ and $F_{r,K,w} = m_{K,w}\, r\omega^2$ act on the crankpin. Where $F_r = F_{r,w} + F_{r,St}$, it follows that

$$F_{K,z} = \sqrt{F_T^2 + (F_R - F_r)^2}. \tag{35}$$

The crankshaft journal absorbs the opposed force which is

$$F_M = -F_S + m_o r\omega^2 (1 + \lambda) + m_r r\omega^2 \text{ at TDC} \quad \text{and}$$

$$F_M = -m_o r\omega^2 (1 - \lambda) - m_r r\omega^2 \text{ at BDC}.$$

10.3 Components of Crank Mechanism

There is a very wide range of embodiments of the individual components. Their designs are determined by the machine size, working process, medium, rotational speed, bearings, loads, materials and production processes [1, 6–8].

10.3.1 Crankshafts

The crankshaft (**Fig. 7**) consists of the crank throws with the journals *1* running in the main bearings, the crankpins *2* for the connecting rod and the webs *3* to connect the pins and hold the balance weights *4*. The coupling *5* transmits the torque while the shaft extension *6* drives the auxiliary systems. The mean distance between two throws is called the cylinder pitch $a = (1.2 \text{ to } 1.6)D$ where D is the piston diameter.

Designs. With in-line diesel engines (**Fig. 7a**), which large forces bear, there is a journal between each pair of crankpins, thus with z cylinders there are $z + 1$ main bearings. With an even number of cylinders and small forces (**Fig. 7b**) oblique webs connect two crankpins each, resulting in $1 + z/2$ main bearings. With fan-type machines (**Fig. 7d**), up to five connecting rods are positioned adjacent to each other on a crankpin, while with V-engines (**Fig. 7c**) there is also one crank throw per connecting rod. Throws with attached pins (**Fig. 7e**), i.e. for two strokes, are also found with mechanical compressors. Overhung cranks (**Fig. 7f**) – used with very small high-speed machines – require very strong webs and two bearings because of their free design. Crankshafts (**Fig. 7a**) made of case-hardening or tempering steel, usually forged in several drop forges, often have hardened journals. With cast shafts (**Fig. 7b**) of malleable cast iron or cast steel the low material strength is balanced out by the design because of the ease of deformability during casting. Crankshafts which are too large even for hammer forging are fabricated, i.e. pins and webs are joined by shrink fitting. Plain bearings are easily split, as a result of

Figure 6. Force and torque curves of a two-stroke diesel engine. **a** Forces at $n = 4500\ \text{min}^{-1}$, **b** torque with changing rotational speed.

strongly present at BDC where the fluid pressure forces are usually negligible ($F_{SK} \to 0$). Account is also taken, for this reason, of the masses attributed to the individual components (**Fig. 5c**) [5].

Piston. The resultant of the forces F_S, F_N acting on the gudgeon pin as per Eq. (29) and $F_{o,K} = m_K F_o/m_o$, where m_K is the piston mass, is

$$F_B = \sqrt{(F_S - F_{o,K})^2 + F_N^2}. \tag{32}$$

At the dead centres, $\varphi = \beta = 0$ or $\varphi = 180°$ and $\beta = 0°$, thus $F_N = 0$ as per Eq. (29). At TDC $F_B = F_S - m_K r\omega^2 (1 + \lambda)$, and at BDC, if $F_S = 0$, then $F_B = m_K r\omega^2 (1 - \lambda)$.

Connecting Rod. The forces $+ F_B$ and $F_{o,St} = m_{o,St} F_o/m_o$ act on the little end. When $F_o = F_{o,K} + F_{o,St}$, then

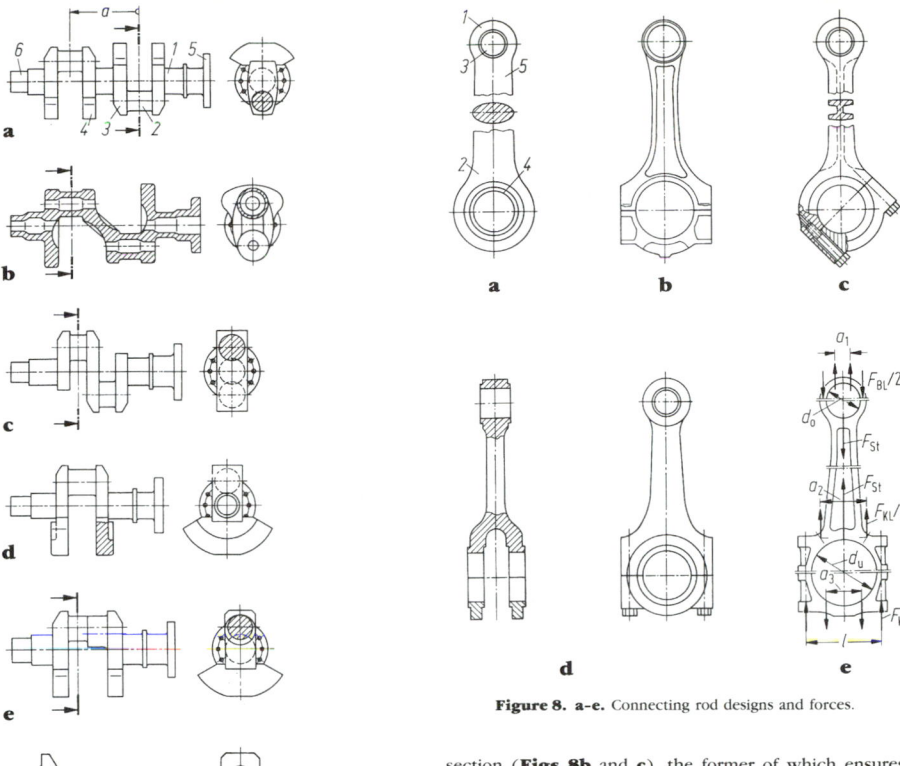

Figure 7. a-f. Crankshaft designs for two-cylinder machines.

Figure 8. a-e. Connecting rod designs and forces.

which the crankshaft (**Fig. 7a**) can comprise a single component. Roller bearings are usually not split because of heavy wear at the die line. For this reason, the journals are extended beyond the crankpins or the shafts are fabricated.

10.3.2 Connecting Rods

Connecting rods (**Fig. 8**) which form the connection between the crosshead or piston and the crankshaft, consist of the little and big ends 1 and 2 containing the bearings 3 and 4 and the shaft 5. They transmit the rod force and are made of case-hardened or cast steel or light alloy. Because of their low tensile strength, grey iron rods are only admissible in single-action two-stroke engines. The outer points of the rod describe a violin-shaped curve when in motion. This curve defines the space which must be left for the engine unit in the frame and cylinder.

Designs. The simplest form with undivided ends (**Fig. 8a**) requires overhung cranks or fabricated crankshafts. Split big ends enable one-piece crankshafts to be used. Oblique splits (**Fig. 8c**) are found with the strong big ends needed in diesel engines because of their high ignition forces to enable the connecting rod to be withdrawn through the cylinder bore. The big or little ends can be branched (**Fig. 8d**) with crosshead or V-engines. If the inertia forces are of major proportion, i.e. with high-speed units, the shaft is designed with an H or J cross-

section (**Figs 8b** and **c**), the former of which ensures better merging into the big and little ends. Otherwise a rectangular, circular or oval cross-section, which is also easier to manufacture, is sufficient, although the necessary moment of inertia requires a relatively large mass.

Bearings. These are usually plain bearings comprising a support shell approximately 2 to 5 mm thick with a coating 0.25 to 0.5 mm thick. They are made of gunmetal for the gudgeon pin and white metal or high-lead bronze for the crankpins. Roller bearings for the gudgeon pin (usually needle bearings, with cylindrical roller bearings for crankpins) require undivided connecting rods with large ends, even if the rolling elements run directly on the hardened pins or journals or in the connecting rod ends. The bearings are lubricated through bores in the rod or crankshaft or, in the case of gudgeon pin bearings, oil is also injected.

10.3.3 Pistons

Pistons are made of grey iron, cast steel or light alloy. Their crown absorbs the fluid pressure forces. The skirt ensures straight movement of the piston and provides a seal for the medium and lubricant. With gaseous fluids the piston is heated significantly, i.e. it is subjected to severe mechanical and thermal stresses. The heat is dissipated by means of the piston rings. Because the piston is heated more than the cylinder, the piston skirt must be ground to the appropriate shape or contain expansion-regulating elements to ensure that there is play in operation. Contact between the piston and the case is prevented by the axial play.

Designs

Trunk Piston (**Fig. 9a**). With motors, the crown 1 often accommodates valve cavities or a combustion chamber or,

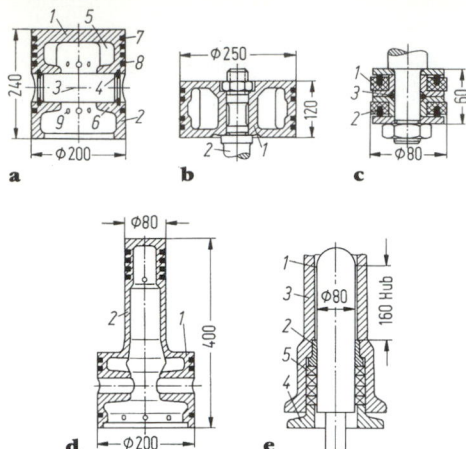

Figure 9. a–e. Piston designs.

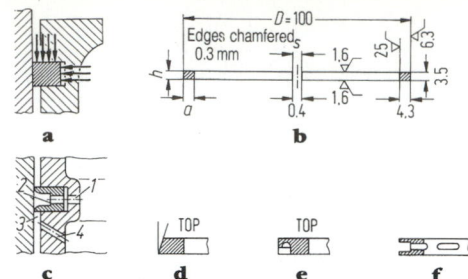

Figure 10. a–f. Piston rings.

with large motors, ducts for the oil cooling system. The skirt 2 is ground to an appropriate shape or contains the expansion-regulating elements to ensure that there is play in operation. It houses the floating gudgeon pin 3 held by circlips 4 in its bosses 6 reinforced with ribs 5. The piston rings 7 and the scraper rings 8 with their drainage bores 9 are located around the skirt circumference. The material used is, for example, aluminium with silicon, manganese and nickel admixtures of low density (≈ 2.85 g/cm³), good thermal conductivity (≈ 1.25 W/(cm K)) and high alternate bending strength (≈ 80 N/mm²). The high coefficient of thermal expansion (≈ 2.5 per °C), however, requires particular design measures.

Disc Piston (**Fig. 9b**). Its boss 1 accommodates the piston rod 2. It is designed for double-action machines and has two sets of piston rings. Several discs are screwed together to fabricate pistons in order to utilise one-piece sealing elements such as carbon rings or grooved sleeves. In hydraulic pistons (**Fig. 9c**), grooved sleeves 1 with back-up rings 2 seal the cylinder while the rubber ring 3 seals the rod.

Differential Piston (**Fig. 9d**). This has up to three stages for compressors. With the exception of one stage, the piston surfaces are of annular design. The trunk-piston embodiment shown here is for a two-stage compressor with the first and second stages 1 and 2 [4].

Plunger (**Fig. 9e**). The skirt 1 slides past the guide bush 2 of the cylinder 3. The packing 5 which is adjustable from the outside by means of the gland 4 seals the plunger. The size of the seal area makes this design long and heavy, and it can be used only with self-lubricating and slow-running machines such as hydraulic pumps [4].

Piston Rings. The special grey iron, rectangular cross-section rings (**Fig. 10b**) are slotted and sit in the piston grooves. Their contact face, either parallel or oblique, determines the leakage losses but must accommodate the expansion of the entire ring without seizing. Their axial play is negligible so that the grooves do not deflect. The rings are subjected to the greatest bending stress (≈ 400 N/mm²) when running over the cylinder surface. Their hardness is less than that of the cylinder walls in order that, as the easier to replace, they wear more rapidly. DIN 24 909 specifies compression rings for sealing purposes, with tapered rings (**Fig. 10d**) used when running the motor in and scraper rings (**Fig. 10e** and **f**) to remove excess oil. Their dimensions are standardised. The legend "TOP" on the ring must be positioned towards the piston crown.

Action. The compression ring (**Fig. 10a**) is pressed via its internal face against the cylinder wall and via its flank against the opposite side of the groove by the medium, thereby ensuring a seal. The oil scraper-ring (**Fig. 10c**) distributes the injected oil to the wall from the inside of the piston via bores 1 and slots 2, while the edge 3 scrapes off the used oil which drains away via the bore 4.

10.3.4 Strength Calculation

Calculations are carried out using the simple strength-of-materials equations for certain cross-sections during the design phase or using finite elements (FEM) for the whole assembly in a concluding calculation (see B8).

Approximate Calculation. This refers to the locations of greatest stress and potentially dangerous cross-sections and is helpful in the initial design stage.

Crankshafts (**Fig. 11**). The crankpins and journals 1 and 3 are subjected to torsional stresses, while the webs 2 must also withstand tensile and compressive loads. The greatest stress comes from alternate bending in the fillet between the web and the crankpin towards the rotational axis. The load acting here is $F = F_R + F_r$, where the radial and rotary forces can be calculated with the aid of Eqs (20) and (31). The pitch line for bending caused by the tangential force is also located here. The substantial fatigue notch effects can be reduced if the following conditions are observed for the radius of curvature p, pin/journal diameter d, web width b and height h: $p/d \geq 30.05$, $b/d = 1.2$ to 1.8 and $h/d = 0.3$ to 0.5. Oil bores should be sited well away from the areas of curvature. Further improvements can be ensured by relieving notches, tapers, enlargements and filleted bores (see D1.5). At a load $F = F_R + F_r$, bearing pitch l and pitch e of the dangerous cross-section and the stress concentration factor α (see D1.5), the bending stress is

$$M_b = Fe/2, \quad \sigma_b = \alpha M_b/W. \tag{36}$$

Connecting Rods (**Fig. 8e**). The ends are subjected to bending stresses like a beam mounted on two supports. In accordance with Eqs (33) and (34), the maximum load at TDC in trunk piston machines is

$$F_{BL} = m_o r \omega^2 (1 + \lambda) - F_S \quad \text{and}$$

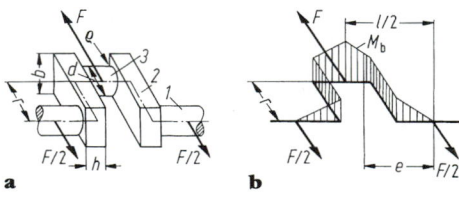

Figure 11. Crank throw: **a** construction, **b** bending moments.

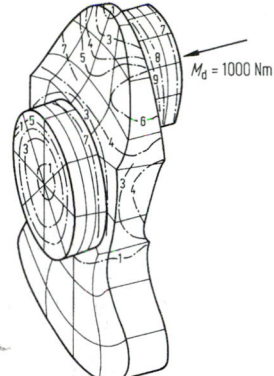

$M_d = 1000$ Nm

Figure 12. Crankshaft half-throw of a car with division into elements and lines of constant relative torsional stress (Daimler Benz AG).

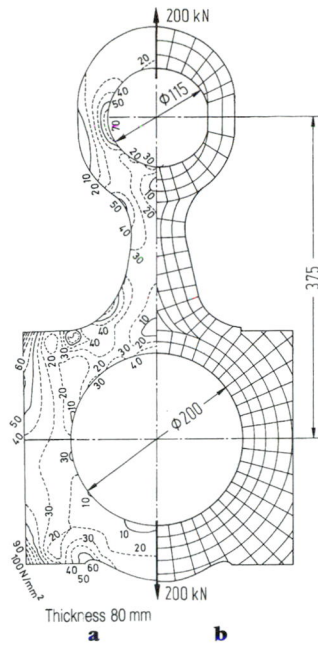

200 kN

Ø115

Ø200

200 kN

Thickness 80 mm

a   b

Figure 13. Connecting rod of a crosshead compressor (Borsig AG, Deutsche Babcock): **a** curves of equal stress, **b** division into elements.

$$F_{KL} = - F_{BL} - m_{r,St}\, r\omega^2.$$

These forces act on the ends in the direction of the shaft when $m_o r\omega^2 (1 + \lambda) < F_s$. In *double-action machines* the maximum for the crank side KS (see **Fig. 1a**) is at BDC. Here, however, the masses and fluid pressure forces expressed in F_{BL} and F_{KL} have the same sign. They are thus cumulative and are directed away from the shaft. This results in larger ends and cases. Their lever arms are $a_1 = d_o/2$ and $a_2 = d_u/2$ where d_o and d_u are the internal diameters of the little end and big end respectively. The case tension bolts are tightened to $\approx 2F$ and $\approx 70\%$ of the yield point where their shaft diameter is $\approx 80\%$ of the thread root diameter. In the case of roller bearings, the cubic mean of F_{KL} applies [5].

Pistons. The crown represents a plate subjected to bending forces with a uniformly distributed load from the pressure. This plate is clamped at the edges in the case of a trunk piston and in the centre in the case of a disc piston. The skirt is subjected to internal pressure stress like a pipe. The initial design of pistons for thermal machines is based on empirical values because the stresses involved are difficult to quantify. Trials are then conducted to improve the performance of the finished pistons.

Finite Elements (FEM). Finite element analysis is used to calculate the tensile, compressive, bending and torsional stresses on drive mechanism components and their resulting deformation, and to determine their vibration properties. It also records the interaction of

these stresses with the crankcase via the bearings. Several thousand volume elements with nodes are frequently necessary for each component. However, this enables curves of equal stress and deformation to be derived for the entire component. Because the method is time-consuming and expensive, specialised computer programs such as "Nastrans" are available, and a combination of FEM and CAD is advantageous for visualising the result. Simpler methods, such as the integral equation method, are also in use. The elements and curves in **Figs 12** to **14** are greatly simplified.

Crankshafts. The torsional stresses in the half throw (**Fig. 12**) were calculated using boundary element modelling. The results are then transferred to the entire crankshaft by a special program. The relative torsional stress curves shown indicate a marked increase in the crankpin fillet. The interaction between the shaft, bearings and

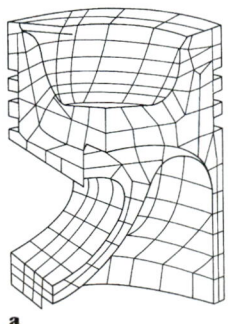

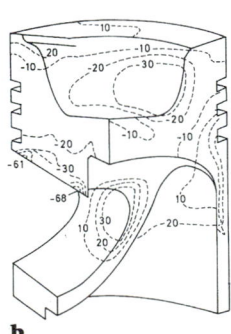

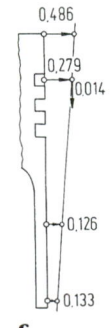

0,486

0,279

0,014

0,126

0,133

a   b   c

Figure 14. Solid-shaft piston, diameter, 115 mm, for a diesel engine: **a** division into spatial elements, **b** mean stress s_a in N/mm² with warm pistons, **c** wall expansion in mm (Mahle GmbH).

crankcase with non-linear lubricating film properties is the subject of further study.

Connecting Rods. The effects of fluid pressure and inertia forces are calculated. Peak stresses of 100 N/mm² were found on the tension bolt seats and 70 N/mm² at the upper bore of a large compressor connecting rod cut from sheet metal (**Fig. 13**).

Pistons. The stresses and deformations caused by fluid pressure and inertia forces as per Eqs (18) and (20) and by temperature are at their greatest on diesel engine trunk pistons. With peak pressures of up to 150 bar, mean piston speeds of 9 m/s and power outputs of 35 kW per cylinder, mean temperatures of ≈ 400 °C occur in the centre

of the piston crown and ≈ 150 °C on the underside of the skirt. The solid-shaft piston (**Fig. 14**) is divided into about 700 spatial elements with 4200 nodes for finite-element analysis. The gudgeon pin boss is the most severely stressed part during operation. The mean load at the lower ring groove is $\sigma_m = -61$ N/mm² with deflection $\sigma_a = -58$ N/mm²; for its inner edge $\sigma_m = -68$ N/mm² and $\sigma_a = -55$ N/mm². In the piston hollow, $\sigma_m = -30$ N/mm² and $\sigma_a = -20$ N/mm². The upper and lower edges expand by 0.486 mm and 0.133 mm, respectively. FEM enables very complicated piston calculations to be carried out, but is not suitable in the event of plastic deformation. Further aids are the strain gauge technique, photoelasticity and holography, which only record mechanical loads.

11 Appendix F: Diagrams and Tables

Appendix F1

Appendix F1 Table 1. Basic, additional, and explanatory symbols for the description of welds according to DIN 1912

a Basic symbols for welds

No	Designation	Illustration[1]	Symbol	No	Designation	Illustration[1]	Symbol
1	Flange weld						
2	I-weld			13	Linear Seam[2]		
3	V-weld						
4	Semi V-weld			14	Steep V-weld		
5	Y-weld			15	Semi-steep V-weld		
6	Semi Y-weld			16	Face weld		
7	U-weld						
8	Semi U-weld			17	Surface weld		
9	Butt weld						
10	Fillet weld			18	Oblique weld		
11	Slot weld			19	Fold weld		
12	Spot weld						

[1] The illustration is only to explain the form of the weld
[2] Also rolled seam weld where appropriate

b Additional symbols

Surface form	Additional symbol
Concave	
Plane	
Convex	

c Explanatory symbols

Run and type of weld	Explanatory symbol
Welds all around (e.g. fillet weld)	
Mounting welds	

d Basic and additional symbols (examples)

Weld type	Illustration	Symbol
V-weld with plane surface		
Double-V weld (X-weld) with convex surface		
Fillet weld with concave surface		
Butt V-weld with plane surface		
I-weld with double-sided curved surfaces (e.g. bead weld)		

Appendix F1 Table 2. Rules and standards for welds

Area of application	General calculation base	Weld calculation
1. Engineering	Rules for the classification and building of machinery and seagoing ships[a]	Section: "Welded connections and welding techniques"[a] Also recommended: DV 952: see under vehicle manufacture DIN 15 018: see under materials handling
2. Vehicle manufacture	Loading and safety of railway vehicles[c]	DV 952: rules for welding in private factories (welded vehicles, machines and equipment[b])
3. Shipbuilding	Rules for the classification and building of steel seagoing ships	
	1980 edition, vol. 1, Classification rules for seagoing ships, building rules for hulls[a] Classification rules for steel seagoing ships, refrigeration equipment, steel inland water ships, 1980 Edition	1980 edition, vol. III, Materials – welding: chapter 7, section 1, Shipbuilding – welding rules[a]
4. Tank construction	DIN 4119: Surface cylindrical sheet tank building from metallic materials	DIN 18 800 T1: Steel construction
5. Pressure vessel construction	DIN 3396: Surface high-pressure gas containers: recommendations for construction, equipment, installation, testing, commissioning, and operation	DIN 18 800 T1: Steel construction
	TRD (Technische Regeln Druckgase – Technical rules for high-pressure gases)[d] Instruction leaflet of the Pressure Vessel Research Group (AD-notes)[d]	
	Rules for the classification and building of machinery for seagoing ships, section, containers and apparatus under pressure[a]	Section: "Welded connections and processing of welds"[a]
6. Boiler and boiler-tube construction	Technical rules for steam boilers[d] Rules for the classification and building of steel seagoing ships, 1980 edition: volume II, Machinery, electrical, refrigeration[a] volume III, Materials – welding: chapter 7, section 2, Engineering – Welding Rules[a]	
7. Containers for inflammable liquids	DIN 6608: Horizontal steel containers for underground storage of liquid mineral oil products DIN 6616: Horizontal steel containers for surface storage of liquid mineral oil products DIN 6618: Vertical steel containers for surface storage of liquid mineral oil products DIN 6625: Site-assembled steel containers for surface storage of heating and diesel oils	AD notes
	Rules for inflammable liquids (VbF): see National Law 1970 p. 689. Technical rules for inflammable liquids (TRbF)	
8. Gas Mains	DIN 2470 Part 1: Gas mains from steel tube with operating pressure up to 16 bar Part 2: Gas mains from steel tube with operating pressure over 16 bar	DIN 2413: Steel tubes, calculation of wall thickness in terms of internal pressure
9. Building Construction	DIN 1050: Structural steel: calculations and structural design DIN 4112: Portable buildings DIN 4114: Steelwork, stability (cracking, tipping, buckling)	DIN 18 800 T1: Steelwork DIN 18 801: Steel structures
	DIN 18 808: Steel construction, frameworks of hollow sections under direct static loading	DIN 18 800 T1: Steelwork DIN 18 808: Steel construction, frameworks of hollow sections under direct static loading
	DIN 4113: Aluminium structures under direct static loading	Recommendations of the Institute for Building Construction. Notices from the Institute for Building Construction, 1972; 3(5): 1-6

Appendix F1 Table 2. Continued

Area of application	General calculation base	Weld calculation
10. Bridge construction	DIN 1073 (DIN 18809): Steel road bridges: basics for calculation DS 804: Rules for railway bridges and other engineering structures (VEI)[b]	DIN 4101: Welded steel road bridges (additional DIN 4100 and DS 804) DS 804: Rules for steel railway bridges and other engineering structures (VEI)[b]
11. Materials handling	DIN 4118: Materials handling equipment for mining	DIN 18800 T1: Steel structures
	DIN 4132: Craneways, steel frames: basics of calculation, structural design and equipment DIN 15018: Cranes, basics of steel frames	
12. Scaffolding	DIN 4420: Working and safety scaffolding (in connection with DIN 18800 T1, DIN 18808)	Effective: DIN 18800 T1: Steel structures, and DIN 18808: Steel structures, frameworks of hollow sections under direct static loading

[a]Germanischer Lloyd.
[b]German State Railways.
[c]Light gauge railway design 1970; 14: additional vol. 2.
[d]Vereinigung der Technischen Überwachungvereine (Association of Supervisors).

Appendix F1 Table 3. Weld seams of various quality levels: DIN 15018

Weld type	Weld quality	Welding method	Symbolic example	Testing and error-free finish	
				Test method	Abbreviation
Butt weld	Extra quality	Root cleaned out, top butt-welded, sheet processed in stress direction, no end-crater		Distortion-free weld test on 100% of weld length, e.g. X-ray	P 100
	Normal quality	Root cleaned out, top butt-welded, no end-crater		As for extra quality but with only pull-test with maximum $\sigma_z \geq 0.8 \times$ perm σ_{zD} in range for pulsating tensile stress with maximum $\sigma_z \geq 0.8 \times$ perm σ_{zD} in alternating range with maximum $\sigma_z \geq 0.8 \times$ perm σ_{zD} or with maximum $\sigma_d \geq 0.8 \times$ perm σ_{dD}	P 100
				Distortion-free test of most important usual welds in samples of at least 10% of the length of each weld, e.g. X-ray	P
Double semi-V-weld with double fillet	Extra quality	Root cleaned out, through-welded, kerf-free weld-run, processed as necessary		Distortion-free test for lamination and grain distortion in the weld region of the appropriate sheet transverse to tension plane, e.g. X-ray	D
	Normal quality	Width of remaining joint in the root to 3 mm or to 0.2 times thickness of welded component. Smaller value to apply.			
Fillet weld	Extra quality	Weld-run kerf-free, processed as necessary			
	Normal quality				

Appendix F1 Table 4. Metric ISO threads, coarse and fine series (as DIN 13 Part 12, Part 12 Additional Sheets, and Part 28)

Nominal diameter d	Coarse thread			Fine thread			Fine thread (extra fine)		
	Pitch P	Core section A_3	Stressed section A_s	Pitch P	Core section A_3	Stressed section A_s	Pitch P	Core section A_3	Stressed section A_s
(mm)	(mm)	(mm²)	(mm²)	(mm)	(mm²)	(mm²)	(mm)	(mm²)	(mm²)
4	0.7	7.75	8.78	(0.5)	9.01	9.79	(0.35)	10.02	10.6
5	0.8	12.69	14.2	(0.75)	13.07	14.5	(0.5)	15.12	16.1
6	1	17.89	20.1	(0.75)	20.27	22.0	(0.5)	22.79	24.0
8	1.25	32.84	36.6	1	36.03	39.2	(0.75)	39.37	41.8
10	1.5	52.30	58.0	1.25	56.29	61.2	0.75	64.75	67.9
12	1.75	76.25	84.3	1.25	86.03	92.1	1	91.15	96.1
(14)	2	104.7	115	1.5	116.1	125	1	128.1	134
16	2	144.1	157	1.5	157.5	167	1	171.4	178
(18)	2.5	175.1	193	1.5	205.1	216	1	221.0	229
20	2.5	225.2	245	1.5	259.0	272	1	276.8	285
(22)	2.5	281.5	303	1.5	319.2	333	1	338.9	348
24	3	324.3	353	2	364.6	384	1.5	385.7	401
(27)	3	427.1	459	2	473.2	496	1.5	497.2	514
30	3.5	519	561	2	596.0	621	1.5	622.8	642
(33)	3.5	647.2	694	2	732.8	761	1.5	762.6	784
36	4	759.3	817	3	820.4	865	1.5	916.5	940
(39)	4	913	976	3	979.7	1 028	1.5	1 085	1 110
42	4.5	1 045	1 121	3	1 153	1 206	1.5	1 267	1 294
(45)	4.5	1 224	1 306	3	1 341	1 398	1.5	1 463	1 492
48	5	1 377	1 473	3	1 543	1 604	1.5	1 674	1 705
(52)	5	1 652	1 758	3	1 834	1 900	2	1 928	1 973
56	5.5	1 905	2 030	4	2 050	2 144	2	2 252	2 301
(60)	5.5	2 227	2 362	4	2 384	2 485	2	2 601	2 653
64	6	2 520	2 676	4	2 743	2 851	2	2 975	3 031
(68)	6	2 888	3 055	4	3 127	3 242	2	3 374	3 434

Nominal diameter d	Fine thread (fine 1)			Fine thread (fine 2)			Fine thread (extra fine)		
	Pitch P	Core section A_3	Stressed section A_s	Pitch P	Core section A_3	Stressed section A_s	Pitch P	Core section A_3	Stressed section A_s
(mm)	(mm)	(mm²)	(mm²)	(mm)	(mm²)	(mm²)	(mm)	(mm²)	(mm²)
72	6	3 287	3 463	4	3 536	3 568	2	3 799	3 862
(76)	6	3 700	3 889	4	3 970	4 100	2	4 248	4 315
80	6	4 144	4 344	4	4 429	4 566	2	4 723	4 794
(85)	6	4 734	4 945	4	5 038	5 190	2	5 352	4 530
90	6	5 364	5 590	4	5 687	5 840	2	6 020	6 100
(95)	6	6 032	6 270	4	6 375	6 540	2	6 727	6 810
100	6	6 740	7 000	4	7 102	7 280	2	7 473	7 560
(105)	6	7 488	7 760	4	7 869	8 050	2	8 259	8 350
110	6	8 273	8 560	4	8 674	8 870	2	9 084	9 180
(115)	6	9 100	9 400	4	9 519	9 720	2	9 948	10 100
(120)	6	9 965	10 300	4	10 404	10 600	2	10 852	11 000
125	6	10 869	11 200	4	11 327	11 500	2	11 795	11 900
(130)	6	11 813	12 100	4	12 290	12 500	2	12 777	12 900
140	6	13 818	14 200	4	14 334	14 600	2	14 859	15 000

Appendix F1 Table 5. Nominal values for metric ISO trapezoidal threads (selection) as DIN 103 Part 4 (pitch as preferred series DIN 103 Part 2)

Thread nominal diameter d (mm)	Pitch P (mm)	Flank diameter $d_2 = D_2$ (mm)	Nut external diameter D_4 (mm)	Bolt root diameter d_3 (mm)	Nut root diameter D_1 (mm)	Bolt core section $\pi d_3^2/4$ (mm²)
10	2	9.0	10.5	7.5	8.0	44
12	3	10.5	12.5	8.5	9.0	57
16	4	14.0	16.5	11.5	12.0	104
20	4	18.0	20.5	15.5	16.0	189
24	5	21.5	24.5	18.5	19.0	269
28	5	25.5	28.5	22.5	23.0	398
(30)	6	27.0	31.0	23.0	24.0	415
32	6	29.0	33.0	25.0	26.0	491
36	6	33.0	37.0	29.0	30.0	661
40	7	36.5	41.0	32.0	33.0	804
44	7	40.5	45.0	36.0	37.0	1018
48	8	44.0	49.0	39.0	40.0	1195
(50)	8	46.0	51.0	41.0	42.0	1320
52	8	48.0	53.0	43.0	44.0	1452
(55)	9	50.5	56.0	45.0	46.0	1590
60	9	55.5	61.0	50.0	51.0	1964
(65)	10	60.0	66.0	54.0	55.0	2290
70	10	65.0	71.0	59.0	60.0	2734
(75)	10	70.0	76.0	64.0	65.0	3217
80	10	75.0	81.0	69.0	70.0	3739
90	12	84.0	91.0	77.0	78.0	4657
100	12	94.0	101.0	87.0	88.0	5945

Appendix F1 Table 6. Minimum thread depths in blind-hole threads [78]

	Recommended single thread depth for class				
	8.8	8.8	10.9	10.9	12.9
Thread fineness d/P	< 9	≥ 9	< 9	≥ 9	< 9
Nut material hard Al alloy, AlCuMgl	1.1d	1.4d		—	
Grey iron GG 25	1.0d	1.25d		1.4d	
Steel St 37, C15M	1.0d	1.25d		1.4d	
Steel St 50, C35M	0.9d	1.0d		1.2d	
Steel cast with	0.8d	0.9d		1.0d	

$$R_{\mathrm{m}} > 800 \left[\frac{N}{mm^2}\right]$$

Appendix F1 Table 7. Tensile force F_{Sp} and tightening torque M_{Sp} for bolts and screws with metric ISO coarse threads as DIN 13, Sheet 13 and heads as DIN 912 or 931, for friction coefficient $\mu_G = \mu_K = 0.12$ over 90% of length (according to VDI Guideline 2230)

Dimension	F_{Sp} (N)			M_{Sp} (N m)		
	8.8[a]	10.9[a]	12.9[a]	8.8[a]	10.9[a]	12.9[a]
Bolt						
M 4	4050	6000	7000	2.8	4.1	4.8
M 5	6600	9700	11400	5.5	8.1	9.5
M 6	9400	13700	16100	9.5	14.0	16.5
(M 7)	13700	20100	23500	15.5	23.0	27.0
M 8	17200	25000	29500	23.0	34.0	40.0
M 10	27500	40000	47000	46.0	68.0	79.0
M 12	40000	59000	69000	79.0	117.0	135.0
M 14	55000	80000	94000	125.0	185.0	215.0
M 16	75000	111000	130000	195.0	280.0	330.0
M 18	94000	135000	157000	280.0	390.0	460.0
M 20	121000	173000	202000	390.0	560.0	650.0
M 22	152000	216000	250000	530.0	750.0	880.0
M 24	175000	249000	290000	670.0	960.0	1120.0
M 27	230000	330000	385000	1000.0	1400.0	1650.0
M 30	280000	400000	465000	1350.0	1900.0	2250.0
Screw ($d_T = 0.9 \cdot d_s$)						
M 5	4500	6600	7800	3.8	5.5	6.5
M 6	6300	9300	10900	6.5	9.5	11.1
(M 7)	9500	14000	16400	10.9	16.0	18.5
M 8	11800	17300	20200	16.0	23.0	27.1
M 10	18900	27500	32500	32.0	47.0	55.0
M 12	27500	40500	47500	55.0	81.0	95.0
M 14	38000	56000	65000	88.0	130.0	150.0
M 16	53000	79000	92000	135.0	200.0	235.0
M 18	66000	94000	110000	195.0	280.0	320.0
M 20	86000	123000	144000	280.0	400.0	460.0
M 22	109000	155000	182000	380.0	540.0	630.0
M 24	124000	177000	207000	480.0	680.0	800.0
M 27	166000	236000	275000	720.0	1020.0	1190.0
M 30	200000	285000	335000	970.0	1400.0	1600.0

[a]Classes according to DIN ISO 898 T1.

Appendix F1 Table 8. Surface pressure p_G (N/mm²) for compressed components of various materials (according to VDI Richtlinie 2230)

Material	Pull stress R_m (N/mm²)	Surface pressure p_G (N/mm²)
St 37	370	260
St 50	500	420
C 45	800	700
42 CrMo 4	1 000	850
30 CrNiMo 8	1 200	750
X 5 CrNiMo 18 10[b]	500 to 700	210
X10 CrNiMo 18 9[b]	500 to 750	220
Stainless dispersion hardened materials	1 200 to 1 500	1 000 to 1 250
Titanium, unalloyed	390 to 540	300
Ti-6Al-4V	1 100	1 000
GG 15	150	600
GG 25	250	800
GG 35	350	900
GG 40	400	1 100
GGG 35.3	350	480
GDMgAl 9	300 (200)	220 (140)
GKMgAl 9	200 (300)	140 (220)
GKAlSi6Cu 4	—	200
AlZnMgCu 0.5	450	370
Al 99	160	140
Glass-fibre-reinforced materials	—	120
Glass-fibre-reinforced materials	—	140

[a]With mechanised tightening surface pressure can be up to 25% less.
[b]With cold fastened materials surface pressure is considerably higher.

Appendix F3

Appendix F3 Table 1. Notes on common friction pairings [36]

	Wet			Dry			
Friction pairs	Sinter bronze–steel	Sinter iron–steel	Paper–steel	Hardened steel–hardened steel	Sinter bronze–steel	Organic coatings–grey iron	Nitrided steel–nitrided steel
Coefficient of friction							
Sliding μ	0.05 to 0.10	0.07 to 0.10	0.10 to 0.12	0.05 to 0.08	0.15 to 0.30	0.3 to 0.4	0.3 to 0.4
Static μ_0	0.12 to 0.14	0.10 to 0.14	0.08 to 0.10	0.08 to 0.12	0.2 to 0.4	0.3 to 0.5	0.4 to 0.6
Ratio μ_0/μ	1.4 to 2	1.2 to 1.5	0.8 to 1	1.4 to 1.6	1.25 to 1.6	1.0 to 1.3	1.2 to 1.5
Technical data (standard values)							
Max. sliding speed v_R (m/s)	40	20	30	20	25	40	25
Max. friction surface pressure p_R (N/mm²)	4	4	2	0.5	2	1	0.5
Permissible surface work for individual q_{AE} (J/mm²)	1 to 2	0.5 to 1	0.8 to 1.5	0.3 to 0.5	1 to 1.5	2 to 4	0.5 to 1
Permissible surface friction power $\dot{q}_{AO}$ (W/mm²) (see VDI 2241, Sheet 1, Section 3.2.2)	1.5 to 2.5	0.7 to 1.2	1 to 2	0.4 to 0.8	1.5 to 2.0	3 to 6	1 to 2

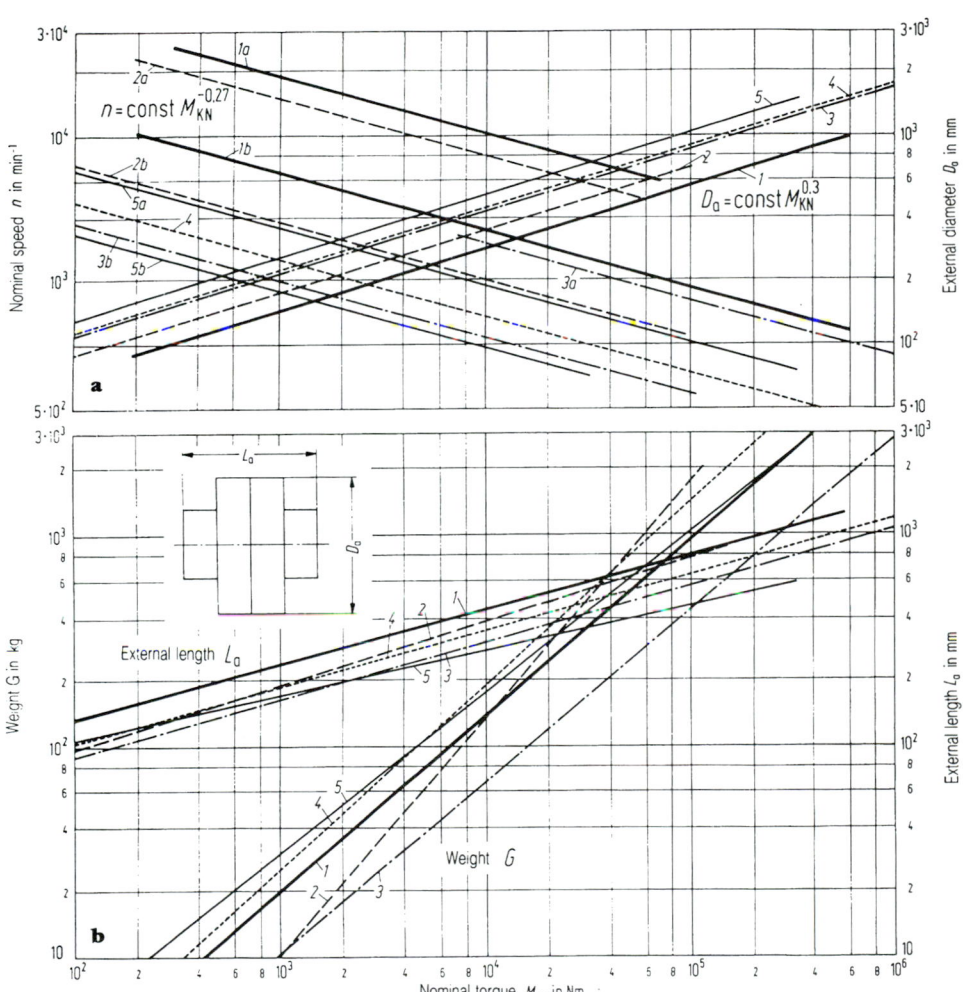

Figure 1. Coefficients for non-switchable couplings [5]: **a** speed n or external diameter D_a; **b** weight g or length L_a as in catalogue; *1* double-tooth couplings, *2* membrane and spring shackle couplings, *3* metal–elastic (rotary elastic) couplings, *4* medium elasticity elastomeric couplings, *5* high elasticity elastomeric couplings, *a* high-speed, *b* medium-speed types.

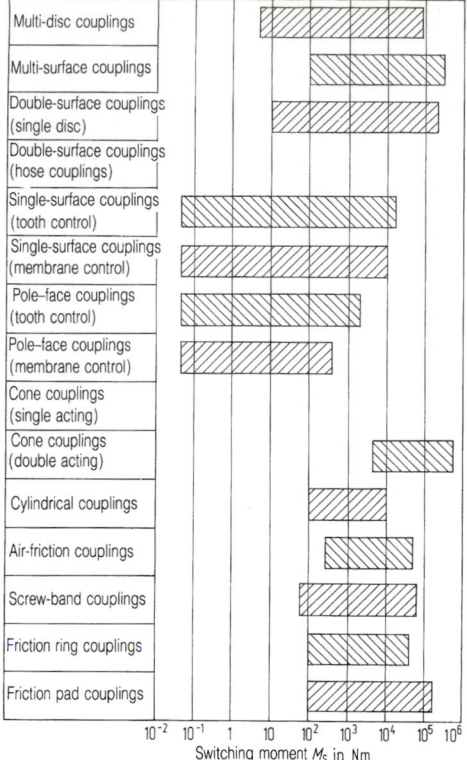

Figure 2. Switching moment ranges for types of remote control friction-switched couplings [6].

Appendix F4

Appendix F4 Table 1. Operating ratios *a* to *k* and operating factors f_v for various assemblies

Assembly	Operating ratio fields as in **Fig. 18**	Wear factor f_v
Vehicle clutches	*g–k*	5 to 8
Vehicle wheel bearings	*h–i*	4 to 6
Trucks	*f–h*	12 to 15
Railway freight cars	*c–d*	8 to 12
Railway rolling stock	*c–d*	3 to 6
E-series motors	*c–d*	3 to 5
Large motors	*b–d*	3 to 5
Turning and milling spindles	*a–b*	0.5 to 1.5
Machine tools	*c–d*	3 to 8
Drive motors	*d–e*	3 to 8
Large-scale machinery	*c–d*	6 to 10
Blowers	*f–h*	5 to 8
Circumferential pumps	*d–f*	3 to 5
Compressors	*d–f*	3 to 5
Jaw crushers	*f–g*	8 to 12
Papermaking machinery	*b–c*	7 to 10

Appendix F4 Table 2. Practical values for necessary service-life

	h
Road vehicles (full load)	
Passenger cars	900 to 1 600
Trucks and buses	1 700 to 9 000
Rail vehicles	
Cargo truck axle bearings	10 000 to 34 000
Tramway vehicles	30 000 to 50 000
Railway carriages	20 000 to 34 000
Locomotives	30 000 to 100 000
Gearing for railway rolling stock	15 000 to 70 000
Agricultural machinery	2 000 to 5 000
Constructional machinery	1 000 to 5 000
Electric motors	
for household equipment	1 500 to 4 000
Series motors	20 000 to 40 000
Large motors	50 000 to 100 000
Machine tools	15 000 to 80 000
Gearing for general engineering	4 000 to 20 000
Large gears	20 000 to 80 000
Ventilators, blowers	12 000 to 80 000
Gear pumps	500 to 8 000
Crushers, mills etc.	12 000 to 50 000
Paper and print machinery	50 000 to 200 000
Textile machinery	10 000 to 50 000

Appendix F4 Table 3. Choice of lubricants [6]

Lubricant	Lubrication	Lubrication equipment	Constructional details	Attainable speed coefficient $n \cdot d_m$ (min^{-1} · mm)[a]	Type of bearing used
Solid lubricant	For-life lubrication	—	—	≈ 1500	Mainly deep-groove ball bearings
	Regreasing	—	—		
Grease	For-life lubrication	—	—	≈ 0.5 · 10⁶ ≈ 1 · 10⁶ for certain special greases, period as **Fig. 27**	All bearing types, except axial spherical roller bearings, depending on speed and grease type
	Regreasing	Hand pressure, grease guns	Input via grease nipples or grease controllers, exit space for used grease		
	Spray lubrication	Commercial lubrication equipment[b]	Input via tubes or bores, exit space for used grease		Lower friction and better noise properties with special greases
Oil (large amounts)	Oil sump lubrication	Dipstick or standpipe bevel control	High-volume housing. Overflow pipes, fitting for control equipment	0.5 · 10⁶	All bearing types, sound damping depending on oil viscosity, higher bearing friction due to oil splash losses, good cooling. Removal of wear particles by circumferential and spray lubrication
	Oil circulation lubrication through self-lubrication of the bearing or bearing elements		Oil inlet bores, high-volume bearing housing. Supply components appropriate to circumferential velocity and oil viscosity. Check supply action of bearing	Must be determined	
	Oil circulation lubrication	Circulation lubrication equipment[b]	Very large bores for oil inlet and outlet	≈ 1 · 10⁶	
	Oil injection lubrication	Circulation lubrication equipment with spray nozzles	Oil inlet through directed nozzles, exit through sufficiently large bores	Tested to 4 · 10⁶	
Oil (minimal amount)	Oil impulse lubrication, oil drop lubrication	Commercial lubrication equipment,[b] drop generator, oil spray lubrication equipment	Outlet bores	≈ 1.5 · 10⁶ depending on type of bearing, oil viscosity, oil quantity or type of construction	All bearing types. Noise damping depending on oil viscosity, friction depending on oil quantity and oil viscosity
	Oil-mist lubrication	Oil-mist generator,[c] or oil collector	Extraction equipment if necessary		
	Oil-air lubrication	Oil-air lubrication equipment[d]	Extraction equipment if necessary		

[a]Depending on type of bearing and constructional ratios.
[b]Central lubrication equipment consisting of pumps, container, filter, piping, valves, throttles. Circulation equipment with oil return, possibly with cooler. Commercial equipment with time controlled dose valves for small quantities (5 to 10 mm³/hub).
[c]Oil-mist equipment consisting of container, mist generator, leads, back-pressure jets, control, compressed air equipment.
[d]Oil-air lubrication equipment consisting of pumps, containers, leads, volumetric oil-air dose distributor, nozzles, control, compressed air equipment.

Appendix F4 Table 4. Coefficients for various oils [6]

	Mineral oil	Polyalphaolefin	Polyglycol (water insoluble)	Ester	Silicone oil	Alkoxyfluoro-oil
Viscosity at 40 ° (mm²/s)	2 to 4500	15 to 1200	20 to 2000	7 to 4000	4 to 100 000	20 to 650
Use for oil sump temperature (°C) to	100	150	100 to 150	150	150 to 200	150 to 220
Use for oil exit temperature (°C) to	150	200	150 to 200	200	250	240
Pourpoint in °C	−20[b]	−40[b]	−40	−60[b]	−60[b]	−30[b]
Ignition point (°C)	220	230 to 260[b]	200 to 260	220 to 260	300[b]	—
Evaporation loss	Moderate	Low	Moderate to high	Low	Low	Moderate to bad[b]
Water resistance	Good	Good	Good,[b] difficult to separate because of similar density	Moderate to good[b]	Good	Good
Tar content	Moderate	Moderate to good	Good	Good	Very good	Moderate to good
Pressure–viscosity coefficient (m²/N)[c]	1.1 to 3.5 · 10⁸	1.5 to 2.2 · 10⁸	1.2 to 3.2 · 10⁸	1.5 to 4.5 · 10⁸	1.0 to 3.0 · 10⁸	2.5 to 4.4 · 10⁸
High-temperature suitability (≈ 150 °C)	Moderate	Good	Moderate to good[b]	Good[b]	Very good	Very good
High-load suitability	Very good[a]	Very good[a]	Very good[a]	Good	Bad[b]	Good
Compatibility with elastomers	Good	Good[b]	Moderate, tested on coatings	Moderate to bad	Very good	Good
Price ratios	1	6	4 to 10	4 to 10	40 to 100	200 to 800

[a]With EP-additives.
[b]Dependent on oil type.
[c]Measured to 200 bar. Height dependent on oil type and viscosity.

Appendix F4 Table 5. Criteria for selection of greases [6]

Criteria for selection of greases	Properties of selected greases
Operating conditions Speed coefficient $n \cdot d_m$ Load coefficient P/C	Grease selection according to Appendix F4 Table 6
Running requirements Low friction, even at start	Grease of consistency classes 1 to 2 with synthetic low-viscosity oil base.
Low and constant friction under continuous running, but higher starting friction permissible	Grease of consistent classes 3 to 4, grease amount less than 30% of free-bearing space, or grease of consistency classes 2 to 3, grease amount less than 20% of free-bearing space
Low running noise	Filtered grease (high purity) of consistency class 2, with especially high provision for low noise, very well filtered grease of consistency classes 1 to 2 with high-viscosity oil base
Constructional ratios Position of bearing axis transverse or vertical	Adhesive grease of consistency classes 2 to 3
Outer ring rotating, inner ring stationary or operated by bearing centrifugal force	Grease of consistency classes 3 to 4 with high thickening agent content
Evaluation Frequent regreasing	Thin grease of consistency classes 1 to 2
Occasional regreasing, for-life lubrication	Squeeze-stable grease of consistency classes 2 to 3, service temperature considerably higher than operating temperature
Environmental ratios High temperature, for-life lubrication	Temperature-stable grease with synthetic base oil and with temperature-stable (or synthetic) thickening agent
High temperature, regreasing	Grease that does not build up sediment at high temperature
Low temperature	Grease with thin synthetic oil base and suitable thickening agent, consistency classes 1 to 2
Dusty conditions	Firm grease of consistency class 3
Condenser water	Emulsified grease, e.g. sodium or lithium soap
Water spray	Water shedding grease, e.g. calcium soap
Aggressive media (acids, alkalis etc.)	Special grease recommended by roller bearing or lubricant supplier
Ionising radiation	Up to energy levels $2 \cdot 10^4$ J/kg, roller bearing grease to DIN 51825 Up to energy levels $2 \cdot 10^7$ J/kg, as recommended by bearing manufacturer
Vibratory stresses	Lithium EP grease of consistency class 2, frequent regreasing Under moderate vibration conditions, barium-complex soap of consistency class 2 with firm lubrication properties, or lithium soap of consistency class 3.
Vacuum	To 10^{-5} mbar, roller bearing grease to DIN 51825; to higher vacuums as recommended by manufacturer.

Appendix F4 Table 6. Rolling bearing greases and their properties

No.	Thickener	Base oil	Operating temperature °C[a]	Water resistance	Remarks
1	Sodium soap	Mineral oil	− 20 to + 100	Not resistant	Emulsified with water otherwise possibly fluid
2	Lithium soap[b]	Mineral oil	− 20 to + 130	Resistant to 90 °C	Emulsified with small amount of water, thinner with larger amounts, multi-use grease
3	Lithium-complex soap	Mineral oil	− 30 to + 150	Resistant	Multi-use grease with high temperature resistance
4	Calcium soap[b]	Mineral oil	− 20 to + 50	Very resistant	Good sealing against water, penetrating water not taken up
5	Aluminium soap	Mineral oil	− 20 to + 70	Resistant	Good sealing against water
6	Sodium-complex soap	Mineral oil	− 20 to + 130	Resistant to approx. 80 °C	Suitable for high temperatures and loads
7	Calcium-complex soap[b]	Mineral oil	− 20 to + 130	Very resistant	Multi-use grease suitable for high temperatures and loads
8	Barium-complex soap[b]	Mineral oil	− 20 to + 150	Resistant	Suitable for high temperatures and loads and also speeds (depending on base oil viscosity): resistant to evaporation
9	Polycarbamide[b]	Mineral oil	− 20 to + 150	Resistant	Suitable for high temperatures, loads and speeds
10	Aluminium-complex soap[b]	Mineral oil	− 20 to + 150	Resistant	Suitable for high temperatures and loads as well as speeds (depending on base oil viscosity)
11	Bentonite	Mineral oil and/or Ester oil	− 20 to + 150	Resistant	Gel grease, suitable for high temperatures at low speeds
12	Lithium soap[b]	Ester oil	− 60 to + 130	Resistant	Suitable for low temperatures and high speeds
13	Lithium-complex soap	Ester oil	− 50 to + 220	Resistant	Multi-range lubricant for wide temperature application
14	Barium-complex soap	Ester oil	− 60 to + 130	Resistant	Suitable for high speeds and low temperatures, resistant to evaporation
15	Lithium soap	Silicone oil	− 40 to + 170	Very resistant	Suitable for high and low temperatures at low loads up to medium speeds

[a]Depending on type of bearing and lubricant. Through selection of appropriate mineral oils the cold operating condition of the grease can be improved 1 to 10 (e.g. − 30 °C in special cases down to − 55 °C).
[b]Also with EP additives.

Appendix F5

Appendix F5 Table 1. Graph for determination of bearing temperature β_{20} to $\beta_{aab} = 22\,°C$ for $So < 1$

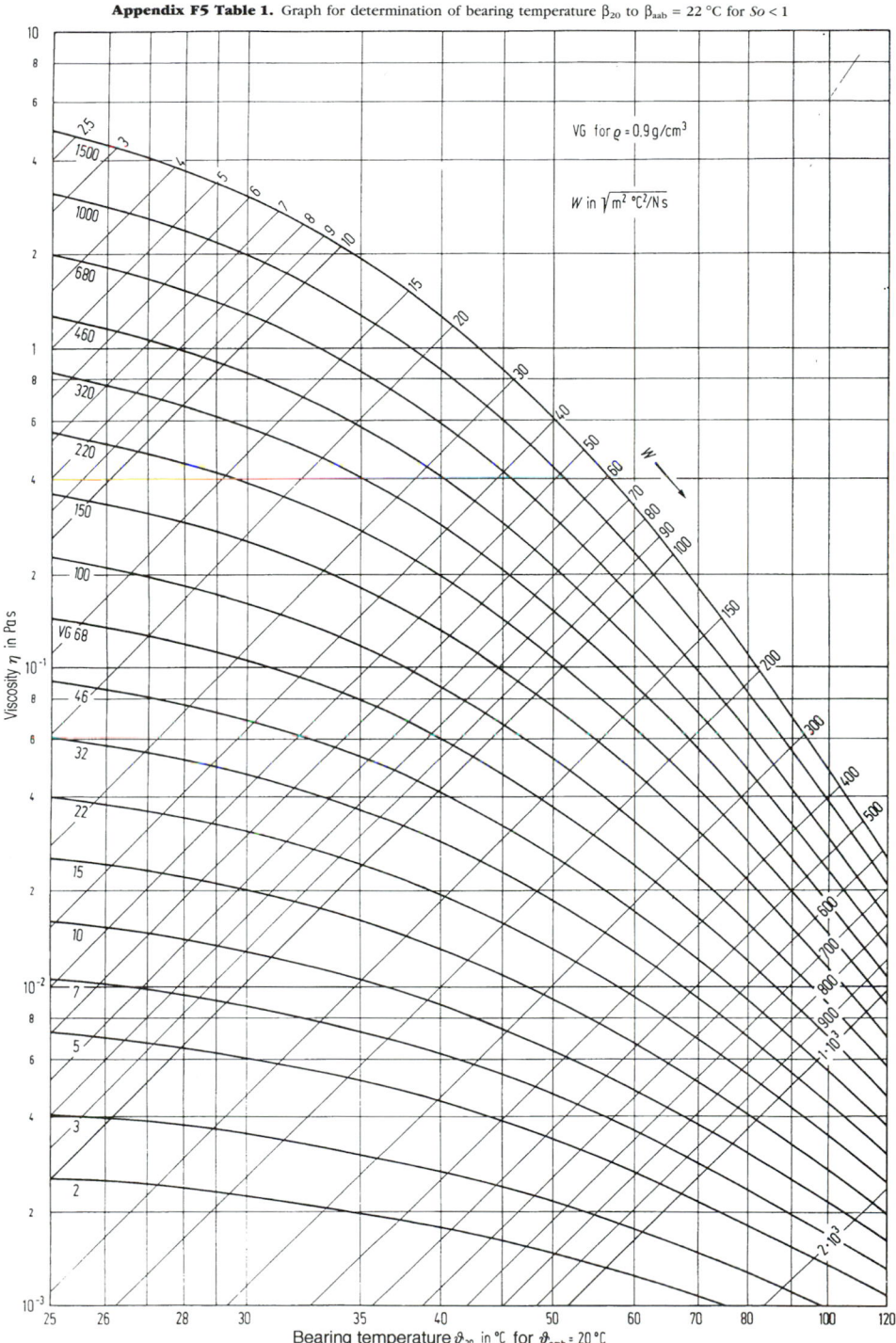

Appendix F5 Table 2. Graph for determination of bearing temperature β_{20} to $\beta_{aab} = 22\,°C$ for $So < 1$

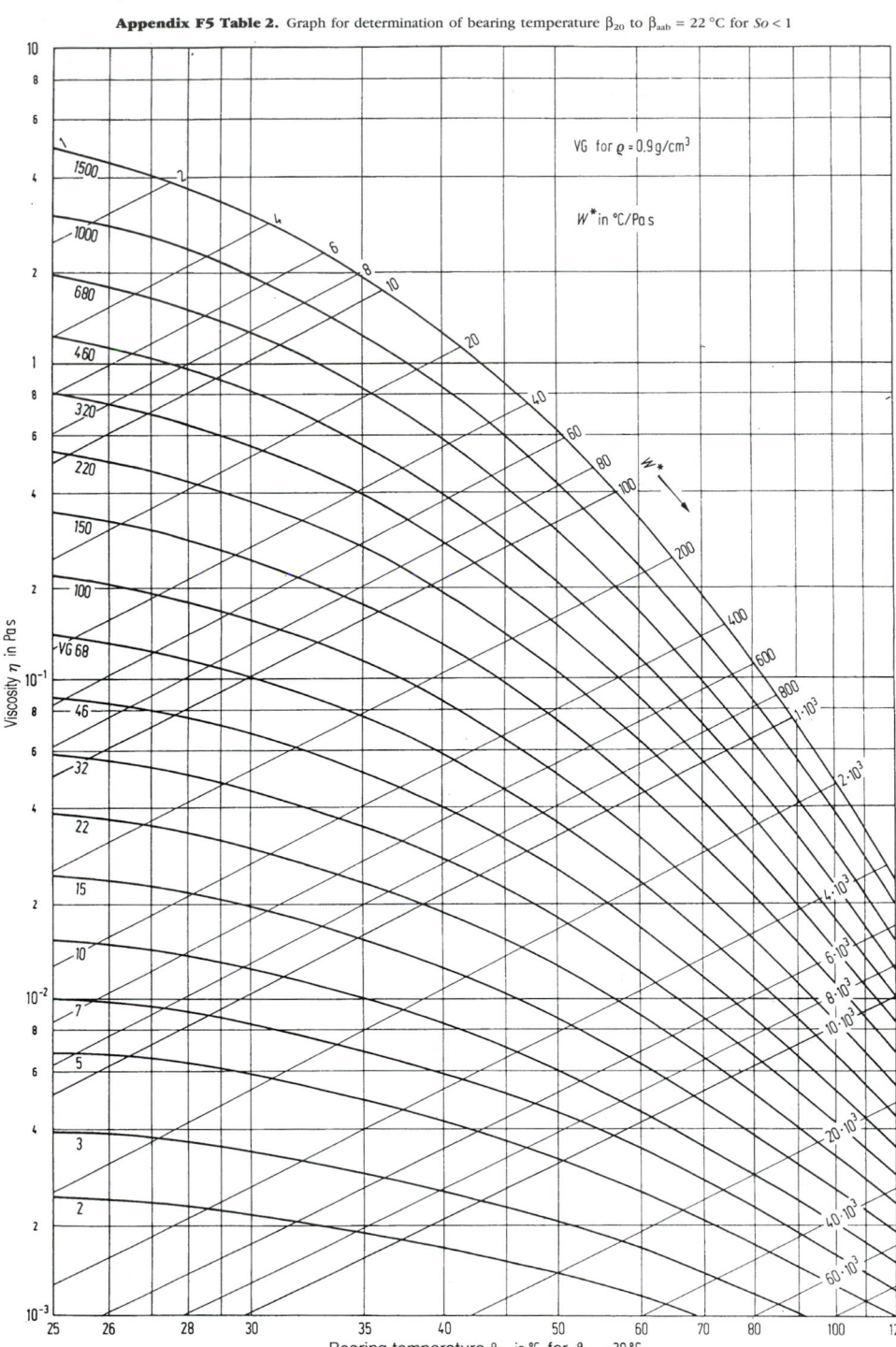

VG for $\varrho = 0.9\,g/cm^3$

W^* in $°C/Pa\,s$

Viscosity η in Pas

Bearing temperature ϑ_{20} in $°C$ for $\vartheta_{amb} = 20\,°C$

Appendix F5 Table 3. Maximum permissible loadings of friction materials $\bar{p}$ bar (N/mm^2) (lower limit for unhardened shaft and high sliding speed, upper limit for hardened shaft and low sliding speed)

Bearing material	Brinell hardness HB (N/mm^2)	$\bar{p}$ (N/mm^2)
Lg Pb Sn 10	230	5 to 25
Zinc alloy	700	5 to 28
Aluminium alloy	750	5 to 50
Tin bronze, gunmetal	700	8 to 55
G Pb Bz 25	300	6 to 80
Special alloys	950	7 to 50

Appendix F5 Table 4. Chemical and technological properties of bearing materials

Material		LgPbSb 14	LgPbSn 6 (WM 5)	LgSn 80	LgSn 80 F	GCdNi	SnPbBz 15	SnPbBz 13	SnPbBz 11
Chemical composition (%)		83.8 Pb; 14 Sb; 1 Sn;1 As; 0.2 Cu	75.8 Pb;15 Sb; 6 Sn; 0.5 As; 1.2 Cu;1 Cd; 0.5 Ni	2 Pb; 12 Sb; 80 Sn; 6 Cu	0.5 Pb; 11 Sb; 80 Sn; 9 Cu	98.4 Cd; 1.6 Ni	15 Pb; 2.5 Sn; 79.5 Cu; 3 Ni	13 Pb; 5 Sn; 79 Cu; 3 Ni	11 Pb; 8 Sn; 77.5 Cu; 3.5 Ni
Hardness and hot hardness HB (N/mm^2)	20 °C	180	256	274	278	340	513	675	863
	50 °C	160	210	232	237	289	491	658	803
	100 °C	130	142	133	154	197	466	649	786
	150 °C	80	81	73	76	115	445	626	769
Apparent yield point $R_{p0.2}$ (N/mm^2)		27	29	61.8	66.7	78.5	84.4	120	163
Tensile strength R_m (N/mm^2)		70	58	89.3	76.5	129	136	192	209
Modulus of elasticity E (N/mm^2)		29 500	29 900	55 700	57 900	54 200	81 500	84 000	85 100
Crushing yield point, $\sigma_{d0.2}$ (N/mm^2)	20 °C	34	47	61.8	67.7	69.7	76.5	109	138
	100 °C	21	27	37.3	42.2	50	64.8	95.2	116
Crushing resistance σ_{dB} (N/mm^2)	20 °C	141	137	189	157	285	515	661	701
	150 °C	95	85	121	102	226	420	524	666
Flexural resistance, σ_{bB} (N/mm^2)		60	115	167	132	241	277	351	348
Bonding resistance according to Chalmers (N/mm^2) (steel C10)		—	56.9	39.2	45.1	104	155	196	216
Fatigue strength under reverse bending stress, σ_{bv} (N/mm^2)		30	28	27.5	30.4	32.4	60.8	76.5	87.3
Thermal conductivity, 20 to 100 °C (kJ/(m h °C))		—	76.9	137.2	141.6	250.2	185.4	177.9	188.5
Average coefficient of expansion, 20 to 100 °C ($10^{-6}/°C$)		25	24	22	22	19	18	18	18

Appendix F6

Appendix F6 Table 1. Flat belts (Siegling, Hannover) Extremultur 80/85G (elastomer) or L (chrome-leather). Tensile modulus ($EA*$), belt thickness s, mass per unit length ρ', * refers to belt thickness approximately 1 mm

Type no.			10	14	20	28	40	54	80
$d_{1,min}$	(mm)		63	100	140	200	280	385	540
F^*_{uN}	(N/mm)		8	12.5	17.5	25	35	48.5	67.5
$EA* =$	(N/mm)		500	700	1000	1400	2000	2700	4000
F^*/ε									
s	G	(mm)	1.5	1.7	2.5	2.9	3.5	4.3	5.7
	L	(mm)	2.2	2.7	3.0	3.7	4.5	5.7	7.5
ρ'	G	(kg/m²)	1.5	11.7	2.7	3.1	3.8	4.7	6.1
	L	(kg/m²)	2.1	2.4	3.1	3.6	4.5	6.1	7.4

Appendix F6 Table 2. V-belt dimensions (selection) and belt coefficients for assessment of permissible nominal power P_N from Eq. (11) and manufacturers' data [6, 7], valid for speeds of smaller disc $n \le n_{1,max}$ and $v \le v_{max}$. Profile designation according to DIN 7753 T1 (corresponding ISO) or DIN 2215 (number) or ISO (chapter)

Profile designation		Working width	Draw length	Power	Speed	Speed	Rotational speed	Diameter	Power factors		
DIN	ISO	b_w (mm)	L_0 (mm)	P_0 (kW)	v_0 (m/s)	v_{max}[d] (m/s)	n_{max}[d] (min⁻¹)	$d_{w,min}$ (mm)	K_2	K_3	K_4
										$=(1-K_4)/2$	
SPZ[c]	SPZ	8.5	1600	1.90	19.50	44.0	8000	63	4.610	0.250	0.500
SPA[c]	SPA	11	2500	3.57	20.73	44.0	6000	90	4.268	0.270	0.460
SPB[c]	SPB	14	3550	9.43	25.95	41.9	5000	140	2.832	0.330	0.340
SPC[c]	SPC	19	5600	21.54	28.15	44.5	3500	224	2.339	0.353	0.294
6[ab]	Y	5.3	315	0.21	12.22	35.2	12000	20	4.730	0.200	0.600
10[ac]	Z	8.5	822	0.53	13.02	32.7	6000	45	4.725	0.250	0.500
13[ac]	A	11	1730	1.04	11.52	33.5	6000	71	5.950	0.160	0.680
17[ac]	B	14	2283	2.99	16.42	33.4	4000	112	4.113	0.240	0.520
22[ac]	C	19	3802	8.28	21.02	33.4	2850	180	2.725	0.300	0.400
32[ac]	D	27	6375	21.45	24.54	34.2	1450	315	1.994	0.330	0.340
40[ac]	E	32	7182	30.18	24.62	33.5	1200	450	1.713	0.350	0.300

[a]Agrees in general with maximum width b_0 according to **Fig. 12a**.
[b]Rough cut.
[c]With woven cover.
[d]Upper limit of manufacturers' data.

Appendix F6 Table 3. Values for commonly used synchronous belts for estimating according to manufacturers' data with glass-fibre strands Gf [6] and steel strands St [10]

Type	Fibre	Pitch p_b (mm)	P_0 (kW)	v_0 (m/s)	b_{s0} (mm)	$z_{1,min}$ for n_1 (min^{-1})	v_{max} (m/s)	n_{max} (min^{-1})
XL	Gf	5.080	3.61	29.85	25.4	$10\left(\dfrac{n_1}{950}\right)^{0.20}$	25.4	10 000
L	Gf	9.525	4.72	28.96	25.4	$12\left(\dfrac{n_1}{950}\right)^{0.30}$	46	6 000
H	Gf	12.700	16.33	38.94	25.4	$16\left(\dfrac{n_1}{950}\right)^{0.24}$	61	6 000
XH	Gf	22.225	16.94	29.85	25.4	$20\left(\dfrac{n_1}{950}\right)^{0.17}$	50	4 400
XXH	Gf	31.750	20.31	29.21	25.4	$22\left(\dfrac{n_1}{950}\right)^{0.15}$	50	3 000
T 2.5	St	2.5	0.95[b]	25.00	25.4	10/18[a]	(25)	15 000
T 5	St	5.0	9.38[b]	86.16	25.4	10/15[a]	80	15 000
T 10	St	10	16.32[b]	65.98	25.4	12/20[a]	60	15 000
T 20	St	20	22.91[b]	50.18	25.4	15/25[a]	45	6 000

[a]Higher minimum number of teeth with contra-bending.
[b]Calculation for 6 load-bearing teeth (manufacturers' data $z_{c\,max} = 15$).

Appendix F6 Table 4. Standard roller chains (selection)

DIN 8187	DIN 8195		DIN 8188 DIN 8195			
Chain no.	P_0 kW $\approx$	n_0 min^{-1} $\approx$	Chain no.	P_0 kW $\approx$	n_0 min^{-1} $\approx$	p mm
06 B	3.5	1700				9.525
08 B	7.5	1400	08 A	8.5	1950	12.7
10 B	11.0	1200	10 A	14.8	1550	15.875
12 B	14.7	1050	12 A	19.0	1300	19.05
16 B	32.0	680	16 A	34.2	980	25.4
20 B	47.5	500	20 A	54.0	720	31.75
24 B	68.0	350	24 A	70.0	550	38.1
28 B	78.0	300	28 A	85.0	440	44.45
32 B	92.0	250	32 A	105	320	50.8
40 B	120	180	40 A	120	205	63.5
48 B	140	125	48 A	100	100	76.2
56 B	160	80				88.9
64 B	160	54				101.6
72 B	124	30				114.3

12 References

F1 Connections. [1] DIN 7190: Pressverbände, Berechnungsgrundlagen und Gestaltungsregeln. Berlin: Beuth 1988. – [2] DIN 4768, T I mit Beiblatt 1: Ermittlung der Rauheitsmessgrössen R_a, R_z, R_{max} mit elektrischen Tastschnittgeräten; Grundlagen und Umrechnung der Messgrössen R_a in R_z und umgekehrt. – [3] DIN 7154 Pt 1 and Pt 2: ISO-Passungen für Einheitsbohrung; Toleranzfelder, Abmasse und Passtoleranzen, Spiele und Übermasse. – [4] DIN 18 800 Pt 1: Stahlbauten, Bemessung und Konstruktion, März 1981. – [5] DIN 254: Kegel, Begriffe und Vorzugswerte. – [6] DIN 1488: Kegelige Wellenenden mit Aussengewinde, Abmessungen. – [7] DIN 1449: Kegelige Wellenenden mit Innengewinde, Abmessungen. – [8] DDR-Standard TGL 19 340 Pt 4: Maschinenbauteile; Dauerschwingfestigkeit; Kerbwirkungszahlen β_k für Achsen und Wellen. – [9] VDI-Richtlinie 2029: Presspassungen in der Feinwerktechnik, October 1958. – [10] Kollmann F G. Welle-Nabe-Verbindungen. Konstruktionsbücher, vol. 32. Springer, Berlin; 1984. – [11] Leidich F. Beanspruchung von Pressverbindungen im elastischen Bereich und Auslegung gegen Dauerbruch. Diss., TH. Darmstadt, 1983. – [12] Seefluth K. Dauerfestigkeit an Wellen-Naben-Verbindungen. Diss., TU, Berlin, 1970. – [13] Galle G. Tragfähigkeit von Querpressverbänden. Schriftenreihe Konstruktionstechnik (ed. Beitz W), no. 4, TU, Berlin, 1981. – [14] Kreitner L. Die Ausbreitung von Reibkorrosion und von Reibdauerbeanspruchung auf die Dauerhaltbarkeit zusammengesetzter Maschinenenteile. Diss., TH Darmstadt, 1976. – [15] Häusler N. Zum Mechanismus der Biegemomentübertragung in Schrumpfverbindungen. Diss., TH Darmstadt, 1974. – [16] Müller H W. Drehmoment-Übertragung in Pressverbindungen. Konstruktion 1962; 14: 47-52, 112-15. – [17] Lundberg G. Die Festigkeit von Presssitzen. Kugellager 1944; 19: 1/2, 1-11. – [18] SKG-Zeitschrift: Der Druckölverband. Schweinfurt, 1977. – [19] Kollmann F G. Die Auslegung elastisch-plastisch beanspruchter Querpressverbände. Forsch Ingenieurwes 1978; 28: 1-11. – [20] Gamer U, Kollmann F G. A theory of rotating elastoplastic shrink fits. Ing Arch 1986; 56: 254-64. – [21] Schmid E A. Theoretische und experimentelle Untersuchung des Mechanismus der Drehmomentübertragung von Kegelpressverbindungen. VDI-Fortschr Ber series 1, no. 16, 1969. – [22] Michl: Statisch überbestimmte Flanschverbindungen mit gleichzeitigem Reib- und Formschluss. Diss., TU Berlin (1988). – [23] Roland G. Beistrag zur Konstruktion von Fördertrommeln mit durchgeführter Welle in Spannsatzausführung. VDI-Fortschr Ber series 13, no. 24, 1984. – [24] BIKON-Technik: Welle-Nabe-Verbindungen. Grevenbroich, 1989. – [25] BIKON-Technik: Grundwissen, Hinweise Lieferprogramm. Grevenbroich, n.d. – [26] Deutsche Star: Toleranzringe. Schweinfurt, 1988. – [27] Fenner. Taper-Lock-Spannbuchsen. Nettetal-Breyell, 1988. – [28] Hochereuter & Baum. DOKO Spannelemente. Ansbach (ohne Jahr). – [29] Ringfeder. Spannsätze, Krefeld, 1988. – [30] Ringfeder. Spannelement. Krefeld, 1988. – [31] Ringfeder. Schrumpfscheiben. Krefeld, 1988. – [32] Ringspann. TOLLOK Konus-Spannelemente. Bad Homburg, 1989. – [33] Ringspann. Sternscheiben und Spannscheiben für Welle-Nabe-Verbindungen. Bad Homburg, 1989. – [34] SKF Kugellagerfabriken. Druckölverband. Schweinfurt, 1977. – [35] Spieth-Maschinenelemente. Druckhülsen. Esslingen, n.d. – [36] Spieth-Maschinenelemente. Fabrikations-Programm. Esslingen, n.d. – [37] Stüwe. Schrumpfscheiben-Verbindung. Hat-

tingen, 1989. – [38] Lenze, Südtechnik. ETP-Spannbuchsen für Wellen-Nabenverbindungen. Waiblingen, n.d. – [39] Handbuch Antriebstechnik. Tabellenwerte-über Lieferanten und Produktdaten. Krausskopf: erscheint jährlich. – [41] DIN-Taschenbuch 43: Mechanische Verbindungselemente 2, Bolzen, Stifte, Niete, Keile, Stellringe, Sicherungsringe. Beuth, Berlin, 1988. – [42] DIN-Taschenbuch 69: Stahlhochbau (Normen, Richtlinien). Beuth, Berlin; 1986. – [43] DIN 1073: Stählerne Strassenbrücken. Berechnungsgrundlagen. – [44] DIN 4113 T1: Aluminiumkonstruktionen unter vorwiegend ruhender Belastung; Berechnung und bauliche Durchbildung 1980. – [45] DIN 15 018 Pt 1: Krane; Grundsätze für Stahltragwerke; Berechnung. DIN 15018 Pt 2: Krane; Stahltragwerke; Grundsätze für die bauliche Durchbildung und Ausführung. DIN 15 018 Pt 3: Krane, Grundsätze für Stahltragwerke; Berechnung von Fahrzeugkranen. – [46] DIN 18 800 Pt 1: Stahlbauten, Bemessung und Konstruktion. – [47] LN 29 730: Nietrechnungswerte bei statischer Beanspruchung für Universal-Nietverbindungen. – [48] LN 29 731: Nietrechnungswerte bei statischer Beanspruchung für Senknietverbindungen. – [49] DASt Bau-Richtlinien für Verbindungen mit Schliessringbolzen im Anwendungsbereich des Stahlhochbaus mit vorwiegend ruhender Belastung. Deutscher Ausschuss für Stahlbau, Cologne, 1970. – [50] Vorläufige Richtlinien für Berechnung, Ausführung und bauliche Durchbildung von gleitfesten Schraubenverbindungen (HV-Verbindungen). Stahlbau-Verlag, Cologne. – [51] Dampfkesselbestimmungen. III. Techn. Vorschriften, Pt 3, TUV, Essen, 3rd edn, Cologne. – [52] Stahl im Hochbau. Verein Deutscher Eisenhüttenleute, Düsseldorf. – [53] Aluminium-Taschenbuch, 14th edn (Aluminium-Zentrale Düsseldorf), Aluminium-Verlag, Düsseldorf, 1988. – [54] Niemann G. Maschinenelemente, vol. 1, 2nd edn, Springer, Berlin, 1981. – [55] Kollmann F G. Welle-Nabe-Verbindungen, Springer, Berlin, 1984. – [56] Heide W. Untersuchungen an Kerbstiften und Kerbstiftverbindungen. Diss., TU Hannover, 1969. – [57] Hoffmann G. Technologische Probleme der Nietung und ihre Auswirkung auf die Dauerfestigkeit. Luftfahrttechnik 1962; 8: 90-8. – [58] Hummel O H. Nieten. KEM 1977; 14: 90-2. – [59] Militzer O. Rechenmodell für die Auslegung von Wellen-Naben-Passfederverbindungen. Diss., TU, Berlin, 1975. – [60] Munz D. Bruchmechanikkonzepte für Zeitfestigkeitsberechnungen. In: VDI Ber 661 – Dauerfestigkeit und Zeitfestigkeit, VDI-Verlag, Düsseldorf, 1988. – [61] Netz H. Dampfkessel, 7th edn, Stuttgart, 1972/3. – [62] Neuber H. Über die Berücksichtigung der Spannungskonzentration bei Festigkeitsberechnung. Konstruktion 1968; 20: 245-51. – [63] Sollmann H. Ein Beitrag zur Elastizität der Bolzen-Laschen-Verbindung. Wiss Z TU Dresden 1965; 14: 1417-24. – [64] Steinhardt O, Valtinat G. Hochfeste, vorgespannte Schliessringbolzen in Stahlbau. Maschinenmarkt 1970; 76: H. 44. – [65] Valtinat G. Untersuchung zur Festlegung zulässiger Spannungen und Kräfte bei Niet, Bolzen- und HV-Verbindungen aus Aluminium-Legierungen. Aluminium 1971; 47: 735-40. – [66] Willms V. Auslegung von Bolzenverbindungen mit minimalem Bolzengewicht. Konstruktion 1982; 34: 63-70. – [67] Firmendruckschriften zu Stiftverbindungen: W. Hedtmann KG, Hagen-Kabel (Spannhülsen) 5800 Hagen 1. – Kerb-Konus Ges., Dr C. Eibes & Co., 8454 Schnaittach/Opf. (Kerbstifte). – W. Prym, 5190 Stolberg (Spiralspannstifte). – C. Vogelsang GmbH, 5800 Hagen 5 (Spannhülsen). – [68] Firmendruckschrift zu Polygon-Verbindungen: Fortuna-Werke Maschinenfabrik: Fortuna-Polygon-Verbindungen, Stuttgart-Bad Cannstatt (ohne Jahr). - [69] Firmendruckschrift zu Axiale Sicherungselemente: Seeger-Orbis GmbH, 6240 Königstein. - [70] Pahl G, Heinrich J. Berechnung von Sicherungsringverbindungen – Formzahlen, Dauerfestigkeit, Ringverhalten. Konstruktion 1987; 39: 1-6. - [71] Beitz W, Pfeiffer B. Einfluss von Sicherungsringverbindungen auf die Dauerfestigkeit dynamisch belasteten Wellen. Konstruktion 1987; 39: 7-13. - [72] Firmendruckschriften zu Nietverbindungen: Gebr. Titgemeier, Gesellschaft für Befestigungstechnik, 4500 Osnabrück (leaflets on HUCK bolts with locking ring; HUCK blind rivets. POP and POP-blind cup-rivets, blind riveting nuts and bolts; GETO expanding rivets in plastics). - Gesipa - Blindniettechnik GmbH, 6082 Mörfelden-Walldorf (leaflets on blind rivets in steel, copper and aluminium alloys). - Honsel, Alfred: Nieten- und Metallwarenfabrik, 5758 Fröndenberg/Ruhr (leaflets on tubular and semi-tubular rivets, brake and clutch-lining rivets). - [73] DIN-Taschenbuch 10: Mechanische Verbindungselemente (Schrauben, Massnormen), Beuth, Berlin, 1985. - [74] DIN-Taschenbuch 45: Gewindenormen. Beuth, Berlin, 1985. - [75] DIN-Taschenbuch 55: Mechanische Verbindungselemente 3 (Technische Lieferbedingungen für Schrauben und Muttern), Beuth, Berlin; 1985. - [76] VDI-Richtlinie 2230 Paper 1: Systematische Berechnung hochbeanspruchter Schraubenverbindungen – Zylindrische Einschraubenverbindungen. VDI-EKV-Ausschuss Schraubenverbindungen, Beuth, Berlin, 1986. - [77] DIN 18800 Pt 1: Stahlbauten, Bemessung und Konstruktion (auch gleitfester Verbindungen mit hochfesten (HV-) Schrauben). Beuth, Berlin, March 1981. - [78] Blume D, Jllgner K H. Schrauben-Vademecum, 7th edn. Bauer & Schaurte Karcher GmbH Neuss/Rhein, 1988. - [79] Wiegand H. et al. Schraubenverbindungen, 4th edn, Konstruktionsbücher, vol. 5. Springer, Berlin, 1988. - [80] Kübler K H, Mages W. Handbuch der hochfesten Schrauben, Girardet, Essen, 1986. - [81] Kober A. Schäden an grossen Schraubverbindungen - Spannungsanalyse - Bruchmechanik - Abhilfemassnahmen. Maschinenschaden 1986; 59: 1-9. - [82] Agatonovic P. Verhalten von Schraubenverbindungen bei zusammengesetzter Betriebsbeanspruchung. Diss., TU Berlin, 1973. - [83] Bauer C D. Ungenügende Dauerhaltbarkeit mitverspannter federnder Elemente. Konstruktion 1986; 38: 59-62. - [84] Birger J A. Die Stauchung zusammengeschraubter Platten oder Flansche (russ.). Russ. Eng. J. 1961; no. 5: pp. 35-8. Extract in: Konstruktion 1963; 15: 160. - [85] Esser J. Verriegelungsrippen an Sicherungsschrauben und Muttern. Ingenieurdienst no. 34, Bauer & Schaurte Karcher GmbH, Neuss/Rhein, 1986. - [86] Galwelat M. Rechnerunterstütze Gestaltung von Schraubenverbindungen. Diss., TU Berlin, 1979. - [87] Galwelat M. Programmsystem zum Auslegen von Schraubenverbindungen. Konstruktion 1979; 31: 275-82. - [88] Jende S, Knackstedt R. Warum Dehnschrauben? VDI-Z 1986; 128: 111-19. - [89] Kober A. Zum betriebsfesten Dimensionieren grosser Schraubenverbindungen unter schwingender Beanspruchung mit besonderem Bezug auf den Abmessungsbereich M 220 DIN 13. Maschinenschaden 1987; 60: 1-8. - [90] Koenigsmann W, Vogt G. Dauerfestigkeit von Schraubenverbindungen grosser Nenndurchmesser. Konstruktion 1981; 33: 219-31. - [91] Michligk T. Statisch überbestimmte Flanschverbindungen mit gleichzeitigem Reib- und Formschluss. Diss., TU Berlin, 1988. - [92] Neuendorf K. Ein Balkenmodell für die Berechnung des elastostatischen Verhaltens hochbean-

spruchter Schraubenverbindungen. Diss., TU Berlin, 1975. - [93] Spiess D. Das Steifigkeits- und Reibungsverhalten unterschiedlich gestalteter Kugelschraubtriebe mit vorgespannten und nicht vorgespannten Muttersystemen. Diss., TU Berlin, 1970. - [95] Thomala W. Beitrag zur Dauerhaltbarkeit von Schraubenverbindungen. Diss., TH Darmstadt, 1978. - [95] Yakushev A J. Effect of manufacturing technology and basic thread parameters on the strength of threaded connections, Pergamon Press, New York, 1964. - [96] VDI-Richtlinie 2232: Methodische Auswahl fester Verbindungen. Systematik, Konstruktionskataloge, Arbeitshilfen. VDI-EKV-Ausschuss Verbindungstechnik, Beuth, Berlin, 1990.

F2 Elastic Connections (Springs) [1] Almen J O, László A. The uniform-section disk spring. Trans ASME 1936; 58: 305-14. - [2] Baumgartl E et al. Zur Dauerfestigkeit vernickelter Schraubendruckfedern. Draht 1967; 18: 582-91. - [3] Betz A. Federgelenke. VDI-Z 1966; 108: 51-4, 93-6. - [4] Brunn H. Vergleich der Beanspruchung von Öse und Wickelkörper bei der Schraubenzugfeder mit angebogenen Ösen. Draht 1969; 20: 661-3. - [5] Busse L. Schwingungen zylindrischer Schraubenfedern. Konstruktion 1974; 26: 171-6. - [6] Bussien, Automobiltechnisches Handbuch, Cram, Berlin, 1965. - [7] Buttler K. Reibungsfedern Bauart Ringfeder im Maschinenbau. Konstruktion 1970; 22: 149-53. - [8] Bühl P. Zur Berechnung von Tellerfedern mit Auflageflächen. Draht 1966; 17: 753-7. - [9] Bühl P. Zur Spannungsberechnung von Tellerfedern. Draht 1971; 22: 760-3. - [10] v. Estorff H-E. Einheitsparabelfedern für Kraftfahrzeug-Anhänger. Brüninghaus-Information no. 2, 1973. - [11] v. Estorff H-E. Technische Daten Fahrzeugfedern. Pt 1, Drehfedern. Stahlwerke Brüninghaus Werdohl, 1973. - [12] v. Estorff H-E. Technische Daten Fahrzeugfedern. Pt 3. Stabilisatoren. Stahlwerke Brüninghaus Werdohl, 1969. - [13] v. Estorff H-E. Parabelfedern für Güterwagen. Techn Mitt Krupp 1979; 37: 109-15. - [14] Federn K. Beherrschung und Ausnutzung von Schwingungen als Konstruktionsaufgabe. VDI-Z 1958; 100: 1220-32. - [15] Federn K. Dämpfung elastischer Kupplungen (Wesen, Frequenz- und Temperaturabhängigkeit. Ermittlung). VDI-Ber 1977; 299: 47-61. - [16] Federn K. Federnde Verbindungen (Federn). In: Dubbel: 16th edn, Springer, Berlin, 1987. - [17] Fischer F, Vondracek H. Warmgeformte Federn, Konstruktion und Fertigung. Hoesch AG, Hohenlimburg, 1987. - [18] Friedrichs J. Die Uerdinger Ringfeder (R). Draht 1964; 15: 539-42. - [19] Go G D. Problematik der Auslegung von Schraubendruckfedern unter Berücksichtigung des Abwälzverfahrens. Automobil Ind 1982; 3: 359-67. - [20] Gross S. Berechnung und Gestaltung von Metallfedern. Springer, Berlin, 1960. - [21] Gross S. Zylindrische Schraubenfedern mit unveränderlichem Verhältnis von Einheitskraft zu Belastung. Draht 1963; 14: 483-7. - [22] Gross S. Drehschwingungen zylindrischer Schraubenfedern. Draht 1964; 15: 530-4. - [23] Hegemann F. Über die dynamischen Festigkeitseigenschaften von Blattfedern für Nutzfahrzeuge. Diss., TH Aachen 1970. - [24] Hempel M. Untersuchungen an Eindraht-Schraubenfedern. (Pt I: Werkstoffe und Fertigung. Prüfung und Federkraft, Bruchhäufigkeit und Federverformung. Pt II: Dauerfestigkeitsprüfungen an Federdrähten und Federn). Konstruktion 1953; 5: 335-43 und 1954; 6: 60-9. - [25] Hertzer K H. Über die Dauerfestigkeit und das Setzen von Tellerfedern. Diss., TH Brunswick, 1959. - [26] Heuer P-J. Entwicklungstendenzen bei Fahrzeugfedern. ATZ 1966; 68: 241. - [27] Hoesch, Stahlfeder-Handbuch. - [28] Hoesch-Federn, Technische Blätter der Hoesch Werke Hohenlimburg-Schwerte. - [29] Hoff H. Federstähle. Werkstoff-Handbuch Stahl u. Eisen, Q 11. Verlag Stahleisen, Düsseldorf. – [30] Huhnen J. Entwicklungen auf dem Federngebiet. Draht 1966; 17: 669–81 and 1967; 18: 592–612. – [31] Huhnen J. Unmögliche "Schraubenzugfedern" jetzt verwirklicht. Draht 1975; 26: 595–9. – [32] Huhnen J. Schraubenzugfedern, Vorschläge zur Weiterentwicklung der DIN 2089 Paper 2. Draht 1987; 38: 218–22. – [33] Hübner W. Deformationen und Spannungen bei Tellerfedern. Konstruktion 1982; 34: 387–92. – [34] Kaiser B. Einfluss des Kugelstrahlens auf die Schwingfestigkeit von Federelementen. Draht 1987; 38: 116–20. – [35] Kaiser B. Dauerfestigkeitsuntersuchungen an biegebeanspruchten Federn aus Federbandstahl. Draht 1987; 38: 675–80. – [36] Keding H. Möglichkeiten zur Erhöhung der Schwingfestigkeit von Fahrzeugblattfedern aus Federstahl 55SiMn7. Kraftfahrzeugtechnik 1970; 104. – [37] Keil E. et al. Versuche mit Federstäben bei statischer und schwingender Verdrehbeanspruchung. Konstruktion 1969; 21: 61–8. – [38] Keitel H. Die Rollfeder – ein federndes Maschinenelement mit horizontaler Kennlinie. Draht 1964; 15: 534–8. – [39] Kranz A. Beitrag zur Beschreibung der Eigenschaften geschichteter Trapez- und Parabelfedern, Diss., Uni Hannover 1983. – [40] Kreutzer A. Warmsetzen von Schraubendruckfedern. Draht 1984; 35: 386–9. – [41] Kuhn P R. Über den Einfluss der Endwindungen auf das Verhalten schraubenförmiger Druckfedern. Draht 1969; 20: 206–12. – [42] Loeper B. Nicht-zylindrische Schraubenfedern im Automobilbau und deren Berechnung. Autom.-techn. Z 1974; 76: 385–90. – [43] Lutz O. Zur Berechnung der Tellerfeder. Konstruktion 1960; 12: 57–9. – [44] Lutz D. Auswahl, Konstruktion und Erprobung von Federungssystemen für die Niveauhaltung von Kraftfahrzeugen. Diss., TU, Munich, 1973. – [45] Manual on design and manufacture of coned disk springs or Belleville springs. Special Publications Dept, SAE SP-63. – [46] Merkblatt Stahl no. 394: Fahrgestellfedern (Tragfedern für Strassenfahrzeuge und ihre Berechnung). Beratungsstelle für Stahlverwertung, Düsseldorf, 1974. – [47] Muhr und Bender. Mubea Tellerfedern-Handbuch. Attendorn, 1987. – [48] Muhr K-H, Niepage P. Zur Berechnung von Tellerfedern mit rechteckigem Querschnitt und Auflageflächen. Konstruktion 1966; 18: 24–7. – [49] Muhr K-H, Niepage P. Eine Methode zur schnellen und einfachen Berechnung von Tellerfedern mit Auflageflächen. Konstruktion 1967; 19: 109–11. – [50] Muhr K-H. Einfluss von Eigenspannungen auf das Verhalten von Federn aus Stahl unter schwingender Beanspruchung. Stahl u Eisen 1968; 88: 1449–55 and 1970; 90: 631–6. – [51] Muhr K-H, Niepage P. Über die Reduzierung der Reibung in Tellerfedernsäulen. Konstruktion 1968; 20: 414–17. – [52] Muhr K-H et al. Warmvorsetzen vermindert die Relaxation von Tellerfedern. Konstruktion 1975; 27: 468–71. – [53] Murasch P-J. Kreuzfederelemente als Gelenke für kleine Schwenkbewegungen, ihr kinematisches Verhalten und ihre Belastbarkeit. Diss., TU Berlin, 1972. – [54] Niepage P. Beitrag zur Frage des Ausknickens axial belasteter Schraubendruckfedern. Konstruktion 1971; 23: 19–24. – [55] Niepage P. Zur rechnerischen Abschätzung der Lastspannungen in angebogenen Schraubenzugfeder-Ösen. Draht 1977; 28: 9–14, 101–8. – [56] Niepage P, Muhr K-H. Nutzwerte der Tellerfeder im Vergleich mit den Nutzwerten anderer Federarten. Konstruktion 1967; 19: 126–33. – [57] Ottl D. Schwingungen mechanischer Systeme mit Strukturdämpfung. VDI-Forschungsheft no. 603. VDI-Verlag, Düsseldorf, 1981. – [58] Palm J, Thomas K. Berechnung gekrümmter Biegefedern. VDI-Z 1959; 101: 301–08. – [59] Palm J. Gekrümmte Trapez-Biegefedern. VDI-Z 1960; 102: 653–59. – [60] Schremmer G. Über die dynamische Festig-

keit von Tellerfedern. Diss., TU Brunswick, 1965. – [61] Schremmer G. Dynamische Festigkeit von Tellerfedern. Konstruktion 1965; 17: 473–79. – [62] Schremmer G. Die geschlitzte Tellerfeder. Konstruktion 1972; 24: 226–29. – [63] Svoboda Z. Zur Berechnung von Federsätzen. Draht 1969; 20: 592–98. – [64] Tautenhahn W, Otto P. Parabelfeder mit progressiver Krümmung für Nutzfahrzeuge. ATZ 1983; 85: H. 7/8. – [65] Ulbricht J. Progressive Schraubendruckfeder mit veränderlichem Drahtdurchmesser für den Fahrzeugbau. ATZ 1969; 71: H. 6. – [66] Ulbricht J. Volumennutzungsgrad von warmgeformten Federn. Estelberichte 1974; 9: H. 3; see also Ind Anz 1974; 96: 1663–8. – [67] Wahl AM. Mechanische Federn. Triltsch, Düsseldorf, 1966. – [68] Walz K. Werkstoffprobleme bei Federn. Draht 1968; 19: 604–12, 783–97. – [69] Walz K. Tellerfedern aus Kunststoff. KEM 1969; 6: 58, 61–3. – [70] Walz K. Korrosionsprobleme bei Federn. Technica 1974; no. 25. – [71] Walz K. Stahlfedern und Werkstoffabbau. Draht 1976; 27: 91–8. – [72] Walz K. Gestaltung von Tellerfedersäulen für den Werkzeug-, Vorrichtungs- und Maschinenbau. Bleche-Rohre-Profile 1976; 23: 134–9. – [73] Wanke K. Korrosionsschutz von Federn. Draht 1964; 15: 103–13. – [74] Wanke K. Beitrag zum Vorsetzen (Voreinrichten) von Schraubenfedern bei Raumtemperatur bzw. erhöhten Temperaturen (Warmsetzen). Draht 1964; 15: 309–17. – [75] Wernitz W. Die Tellerfeder. Konstruktion 1954; 6: 361–76. – [76] Wernitz W. Neuere Erkenntnisse über Eigenschaften von Tellerfedern. TH Brunswick: JMF-Bericht 1960; 41. – [77] Wiesecker-Krieg J. Federstähle, Röchling-Burbach Techn Mitt 1972; 31. – [80] Battermann W, Köhler R. Elastomere Federung – Elastische Lagerungen. Ernst & Sohn, Berlin, 1982. – [81] Becker G. W. et al. Elastische und viskose Eigenschaften von Werkstoffen. Deutscher Verband für Materialprüfung (VDM). Beuth, Berlin, 1963. – [82] Benz W. Elastische Lagerung auf geneigt angeordneten Gummipuffern. MTZ 1967; 28: 28–34. – [83] Gamer U. Genaue Berechnung der Gummi-Torsionsfeder. Forsch Ing Wes 1973; 39: 13–16. – [84] Göbel E F. Gummifedern, Berechnung und Gestaltung, 3rd edn. Konstruktions-Bücher, vol. 7. Springer, Berlin, 1969. – [85] Göbel E F. Gummifedern als moderne Konstruktionselemente. Konstruktion 1970; 22: 402–6. – [86] Joos R. Übersicht über verschiedene Dämpfungsmechanismen. Feinwerktechnik/Messtechnik 1976; 84: 219–28. – [87] Jörn R. Lang G. Gummi-Metall-Elemente zur elastischen Lagerung von Motoren. MTZ 1968; 29: 252–8. – [88] Kümmlee H. Ein Verfahren zur Vorhersage des nichtlinearen Steifigkeits- und Dämpfungsverhaltens sowie der Erwärmung drehelastischer Gummikupplungen bei stationärem Betrieb. Diss., TU Berlin, 1985 und VDI-Fortschrittsber 1/136, VDI-Verlag, Düsseldorf, 1986. – [89] Lipinski J. Fundamente und Tragkonstruktion für Maschinen. Bauverlag, Wiesbaden, 1972. – [90] Malter G, Jentzsch J. Zur Abhängigkeit des E- bzw. des G-Moduls von der Beanspruchung. Plaste Kautschuk 1975; 22: 30–2. – [91] Malter G, Jentzsch J. Gummifedern als Konstruktionselement. Maschinenbautechnik 1976; 25: 109–12, 121, 225–8. – [92] Pinnekamp W, Jörn R. Neue Drehfederelemente aus Gummi für elastische Kupplungen. MTZ 1964; 25: 130–5. – [93] Walz K. Tellerfedern aus Kuntstoff. KEM 1969; 6: 58, 61–3. – [94] Continental-Schwingmetall: Katalog der Continental Gummi-Werke AG, D-3000 Hannover. – [95] Gimetall: Katalog der Firma Metzeler AG, D-8990 Lindau. – [96] Phoenix-Metallgummi: Katalog der Phoenix Gummiwerke AG, D-2100 Hamburg 90. – [97] Simrit. Catalogue no. 400, January 1976. (Werkstoffbeschreibungen) and Katalog Megulastik der Firma Freudenberg, D-6940 Weinheim. – [98] Werkstoff-

Gummi. Veröffentlichung der Dätwyler AG, Schweizerische Kabel, Gummi- und Kunststoffwerke, CH-6460 Altdorf-Uri. – [100] Hansen J. Faserverbundwerkstoffe, vol. 3. Dokumentation des BMFT. Springer, Berlin, 1986. – [101] Mallik PK. Static mechanical performance of composite elliptic springs. Trans ASME J Eng Mater Technol 1987; 109: 22-6. – [102] Ophey L. Faser-Kunststoff-Verbundwerkstoffe. VDI-Z 1986; 128: 817-24. – [103] Schütz D. et al. Werkstoffmechanik (Faserverbundwerkstoffe). In: LBF-Bericht no. TB-108, 1988. – [104] Publication of GKN Vandervell Ltd, London SW1A 1DB, 1987. – [105] Publication on Composite Springs, Hoesch-Iscar Faserverstärkte Federn GmbH, Eisenstadt/Österreich, 1987. – [106] Kunststoff-Federn (GFK), Krupp Brüninghaus GmbH, Werdohl, 1987. – [110] Behles F. Zur Beurteilung der Gasfederung. ATZ 1961; 63: 311-14. – [111] Die Gasfeder. Technical information from the Stabilus GmbH, Koblenz, 1983. – [112] Hamaekers A. Entkoppelte Hydrolager als Lösung des Zielkonflikts bei der Auslegung von Motorlagern. Automobil Ind 1985; 5: 553-60. – [113] Reimpell J. C. Fahrwerktechnik, vol. 2. Vogel, Würzburg, 1975, p. 207. – [114] Spurk J H, Andrä R. Theorie des Hydrolagers. Automobil Ind 1985; 5: 553-60.

Standards and Guidelines. DIN-Taschenbuch 29: Federn, Normen. Beuth, Berlin, 1985. – DDR-Standard TGL 39 249: Ermüdungsfestigkeit; Schraubenfedern; Blattfedern; Kennwerte und Diagramme. – DIN-VDE-Taschenbuch 47: Kautschuk und Elastomere. Physikalische Prüfverfahren, 5th edn, Beuth, Berlin, 1988. – DIN 740 Pt 2: Nachgiebige Wellenkupplungen: Begriffe und Berechnungsunterlagen, August 1986. Beuth, Berlin. – DIN 53 440: Prüfung von Kunststoffen und von schwingungsgedämpften geschichteten Systemen – Biegeschwingungsversuch. Pt 1: Allgemeine Grundlagen zur Bestimmung der dynamisch-elastischen Eigenschaften stab- und streifenförmiger Probekörper; Pt 2: Bestimmung des komplexen Elastizitätsmodul; Pt 3: Bestimmung von Kenngrössen schwingungsgedämpfter Mehrschichtsysteme. January 1984. Beuth, Berlin. – DIN 53 445: Prüfung von polymeren Werkstoffen: Torsionsschwingungsversuch. August 1986. Beuth, Berlin. – DIN 53 505: Prüfung von Kautschuk, Elastomeren und Kunststoffen; Häteprüfung nach Shore A und D. June 1987. Beuth, Berlin. – DIN 53 513: Prüfung von Kautschuk und Elastomeren. Bestimmung von visko-elastischen Eigenschaften von Elastomeren bei erzwungenen Schwingungen ausserhalb der Resonanz. January 1983. Beuth, Berlin. – DIN 53 531 Paper 1: Prüfung von Elastomeren; Trennversuch an Elastomer-Metall-Verbindungen; Prüfung an einer Metallplatte. December 1972; Pt 2: Prüfung von Kautschuk und Elastomeren; Trennversuch an Elastomer-Metall-Bindungen. June 1981. Beuth, Berlin. – DIN 53 533: Prüfung von Elastomeren; Prüfung der Wärmebildung und des Zermürbungswiderstandes im Dauerschwingversuch (Flexometerprüfung). Pt 1: Grundlagen; Pt 2: Rotationsflexometer; Pt 3: Kompressions-Flexometer. August 1975, Beuth, Berlin. – ISO 2856-1975 (E): Elastomers – General requirements for dynamic testing. International Organization for Standardization, Geneva, 1975. – ISO/TC 108-DB 5405: Nomenclature for Specifying Damping Properties of Materials. ISO/TC 108 (Secr. 108) 185. August 1975. – ISO/TC 61/WG 2 – Draft Proposal: Plastics – Terminology for Characterizing the Damping Properties of Solid Polymers. American National Standards Institute (ANSI), September 1975. – VDI-Richtlinie 2062: Schwingungsisolierung; Paper 1: Begriffe und Methoden; Paper 2: Isolierelemente. January 1976. Beuth, Berlin.

F3 Couplings, Clutches and Brakes. [1] VDI-Richtlinie 2240: Wellenkupplungen. VDI-Verlag, Düsseldorf, 1971. – [2] Peeken H, Troeder C. Elastische Kupplungen. Springer, Berlin, 1986. – [3] Hinz R. Verbindungselemente: Achsen, Wellen, Lager Kupplungen. VEB Fachbuchverlag, Leipzig, 1984. – [4] Schalitz A. Kupplungsatlas, 4th edn, Ludwigsburg: AGT-Verlag 1975. – [5] Ehrlenspiel K, Henkel G. Membrankupplungen als drehstarre, biegenachgiebige Ganzmetallkupplungen. VDI-Berichte 299, 1977. – [6] Buschhaus D. Rechnergestützte Auswahl von schaltbaren Wellenkupplungen. Diss., TU Berlin, 1976. – [7] Tochtermann W, Bodenstein F. Konstruktionselemente des Maschinenbaus, Pt 2. Springer, Berlin, 1979. – [8] VDI-Richtlinie 2722: Homokinetische Kreuzgelenkgetriebe einschliesslich Gelenkwellen. VDI-Verlag, Düsseldorf, 1982. – [9] Dittrich O, Schumann R. Anwendungen der Antriebstechnik, vol. II: Kupplungen. Krausskopf, Mainz, 1974. – [10] Schmelz F, Graf v. Seherr-Thoss H-C, Aucktor E. Gelenke und Gelenkwellen. Springer, Berlin, 1988. – [11] Hartz H. Antriebe mit Kreuzgelenkwellen, Pt 1: Kinematische und dynamische Zusammenhänge. Antriebstechnik 1985; 24: 72-5. – [12] Hartz H. Antriebe mit Kreuzgelenkwellen. Pt 2: Probleme und ihre Lösungen. Antriebstechnik 1985; 24: 61-9. – [13] Schütz K H. Gleichlauf-Kugelgelenke für Kraftfahrzeugantriebe. Antriebstechnik 1971; 10: 437-40. – [14] Benkler H. Zur Auslegung bogenverzahnter Zahnkupplungen. Konstruktion 1972; 24: 326-33. – [15] Fleiss R, Pahl G. Radial- und Axialkräfte beim Betrieb von Zahnkupplungen. VDI-Berichte 1977; 299. – [16] Heinz R. Untersuchung der Zahnkraftund Reibungsverhältnisse in Zahnkupplungen. Konstruktion 1978; 30: 483-92. – [17] Pahl G, Strauss E, Bauer H P. Fresslastgrenze nichtgehärteter Zahnkupplungen. Konstruktion 1985; 37: 109-16. – [18] Pahl G, Müller N. Temperaturverhalten ölgefüllter Zahnkupplungen. VDI-Berichte 1987; 649: 157-77. – [19] Stotko H. Moderne Entwicklungen bei Bogenzahn-Kupplungen. Konstruktion 1984; 36: 433-7. – [20] Basedow G. Zahnkupplungen für hohe Drehzahlen. Antriebstechnik 1984; 23: 18-21. – [21] Jarchow F, Sturmath R. Tragfähigkeit und Federsteifen von Wellenkupplungen mit federnden Laschengelenken. Konstruktion 1979; 31: 33-40. – [22] Henkel G. Membrankupplungen – Theoretische und experimentelle Untersuchung ebener und konzentrisch gewellter Kreisringmembranen. Diss., Univ. Hannover 1980. – [23] Röper R, Japs D. Bestimmung der statischen Momentkennlinie elastischer Wellenkupplungen. Antriebstechnik 1980; 19: 403-7. – [24] Beitz W. Untersuchungen der elastischen und dämpfenden Eigenschaften drehelastischer Kupplungen und ihre Dauerfestigkeit. Diss., TU Berlin, 1961. – [25] Klingenberg R. Experimentelle und analytische Untersuchungen des dynamischen Verhaltens drehnachgiebiger Kupplungen. Diss., TU Berlin, 1977. – [26] Gnilke W. Zur Grössenauswahl drehnachgiebiger Kupplungen. Maschinenbautechnik 1982; 31: 537-40. – [27] Benner J. Experimentelle Untersuchungen des mechanischen Verhaltens drehnachgiebiger Wellenkupplungen und Entwicklung eines Ersatzmodells. Diss. RWTH Aachen, 1984. – [28] Kümmlee H. Ein Verfahren zur Vorhersage des nichtlinearen Steifigkeits- und Dämpfungsverhaltens sowie der Erwärmung drehelastischer Gummikupplungen bei stationärem Betrieb. Fortschritt-Berichte VDI series 1, no. 136. VDI-Verlag, Düsseldorf, 1986. – [29] Peeken H et al. Angenäherte Bestimmung des Temperaturfeldes in elastischen Reifenkupplungen. Konstruktion 1986; 38: 485-9. – [30] Troeder C et al. Berechnungsverfahren von Antriebssystemen mit drehelastischer Kupplung. VDI-Berichte 1987;

649: 41–68. – [31] Hartz H. Anwendungskriterien für hochdrehelastische Kupplungen. Pt 1: Antriebsarten und deren Besonderheiten. Antriebstechnik 1986; 25: 47–52. – [32] Peeken H et al. Beanspruchung elastischer Kupplungen in Antriebssystemen mit Asynchron-Motoren. Antriebstechnik 1979; 18: 484–9. – [33] Peeken H, Troeder C. Auswirkungen des Wellenversatzes bei elastischen Kupplungen. VDI-Berichte 1977; 299. – [34] Heyer R, Möllers W. Rückstellkräfte und -momente nachgiebiger Kupplungen bei Wellenverlagerungen. Antriebstechnik 1987; 26: 43–50. – [35] VDI-Richtlinie 2241 Paper 1: Schaltbare fremdbetätigte Reibkupplungen und -bremsen. VDI-Verlag, Düsseldorf, 1982. – [36] VDI-Richtlinie 2241 Paper 2: Schaltbare fremdbetätigte Reibkupplungen und -bremsen. VDI-Verlag, Düsseldorf, 1984. – [37] Niemann G, Winter H. Maschinenelemente, vol. III, 2nd edn. Springer, Berlin, 1983. – [38] Winkelmann S. Klassenmerkmale, Anwendungsfelder und Trends bei schaltbaren, mechanischen Kupplungen. VDI-Berichte 1987; 649: 273–87. – [39] Orthwein W. Clutches and brakes. Dekker, New York, 1986. – [40] Appelhoff H. Elektromagnet-Zahnkupplungen und Periflex-Wellenkupplungen für die Hütten- und Schwermaschinenindustrie. Antriebstechnik 1986; 25: 31–4. – [41] Winkelmann S, Harmuth H. Schaltbare Reibkupplungen. Springer, Berlin, 1985. – [42] Federn K, Beisel W. Betriebsverhalten nasslaufender Lamellenkupplungen. Antriebstechnik 1986; 25: 47–52. – [43] Korte W. Betriebs- und Leerlaufverhalten von nasslaufenden Lamellenkupplungen. VDI-Berichte 1987; 649: 335–58. – [44] Korte W, Rüggen W. Magnetpulverkupplungen. asr-digest für angewandte Antriebstechnik 1979; 3: 47–9. – [45] Hasselgruber H. Der Schaltvorgang einer Trockenreibungskupplung bei kleinster Erwärmung. Konstruktion 1963; 15: 41–5. – [46] Duminy J. Beurteilung des Betriebsverhaltens schaltbarer Reibkupplungen. Diss., TU, Berlin, 1979. – [47] Steinhilper W. Der zeitliche Temperaturverlauf in schnell geschalteten Reibungskupplungen und -bremsen. Diss., TH Karlsruhe, 1963. – [48] Steinhilper W. Der Kraftfluss in unter Last geschalteten Lamellen-Kupplungen und das übertragbare Drehmoment. Konstruktion 1967; 7: 262–7. – [49] Pahl G, Zhang Z. Dynamische und thermische Ähnlichkeit in Baureihen von Schaltkupplungen. Konstruktion 1984; 36: 421–6. – [50] Pahl G, Oedekoven A. Kennzahlen zum Temperaturverhalten von trockenlaufenden Reibungskupplungen bei Einzelschaltung. VDI-Berichte 1987; 649: 289–306. – [51] Ernst L, Rüggen W. Richtige Auswahl von Kupplungen und Bremsen. Antriebstechnik 1982; 21: 616–19. – [52] Schneider R. Elektromagnetische Hysteresekupplung. VDI-Berichte 1987; 649: 435–47. – [53] Hoppe F. Das Abschalt- und Betriebsverhalten von mechanischen Sicherheitskupplungen. Diss., TU Münich, 1986. – [54] Rettig H, Hoppe F. Sicherheitskupplung mit Brechringen für Schwermaschinenantriebe. Antriebstechnik 1986; 25: 48–53. – [55] Fleissig M. Untersuchungen zum Drehmomentverhalten von Fliehkraftkupplungen. VDI-Z 1984; 126: 869–72. – [56] Körber E. Hydrodynamische Anlauf- und Rutschkupplung mit konstanter Füllung. VDI-Berichte 1979; 299: 171–7. – [57] Stölzle K, Hart S. Freilaufkupplungen. Springer, Berlin, 1960. – [58] Timtner K. Freilaufkupplungen für zukunftsorientierte Anwendungen. Antriebstechnik 1986; 25: 31–5. – [59] Jorden W. Gebrauchsdauer von Klemmfreilaufkupplungen. Konstruktion 1972; 24: 485–91. – [60] Timtner K. Die Berechnung der Drehfederkennlinien und zulässigen Drehmomente bei Freilaufkupplungen mit Klemmkörpern. Diss., TH, Darmstadt, 1974. – [61] Peeken H, Hinzen H. Funktionsfähigkeit und Gebrauchsdauer von Klemmkörperfreiläufen im Schaltbetrieb. Antriebstechnik 1986; 25: 35–40. – [62] Schlattmann J. Lebensdauerermittlung von Klemmrollenfreiläufen aufgrund von Werkstoff-Verformung, -Ermüdung und Wälzverschleiss. Fortschritt-Berichte VDI Reihe 5 no. 200. VDI-Verlag, Düsseldorf, 1986. – [63] Tönsmann A. Verschleiss und Funktion – Der Einfluss des Schaltverschleisses auf die Schaltgenauigkeit von Klemmrollenfreiläufen. Diss., Univ. Paderborn, 1989. – [64] Bollmann E. Die Eintouren-Rollenkupplung – ein vielseitiges Schaltelement. Antriebstechnik 1973; 12: 101–06.

Standards and Guidelines. DIN 115: Schalenkupplungen. – DIN 116: Scheibenkupplungen. – DIN 740: Nachgiebige Wellenkupplungen. – DIN 15 431–15 437: Trommel- und Scheibenbremsen. – DIN 42 955: Toleranzen für Befestigungsflansche für elektrische Maschinen, zulässige Lageabweichungen. – DIN 71 751–71 752: Gabelgelenke. – DIN 71 802; 71 803; 71 805: Winkelgelenke. AGMA Standards: 510.03: Nomenclature of flexible couplings. – 516.01: Metric dimensions for gear coupling flanges. – 9000-C90: Flexible couplings – potential unbalance classification. – 9001-A86: Lubrication of flexible couplings. – 9002-A86: Bores and keyways for flexible couplings (inch series). – 9003-A91: Flexible couplings – keyless fits.

F4 Rolling Bearings. [1] Schlicht H et al. Ermüdung bei Wälzlagern und deren Beeinflussung durch Werkstoffeigenschaften. FAG-Wälzlagertechnik; 1987-1. – [2] Stöcklein W. Aussagekräftige Berechnungsmethode zur Dimensionierung von Wälzlagern. Wälzlagertechnik. Pt 2: Berechnung von Lagerungen und Gehäusen in der Antriebstechnik. Kontakt und Studium; B. 248. Expert-Verlag, Grafenau, 1988. – [3] FAG Standardprogramm. Katalog WL 41 510/2 DB. 1987. – [4] SKF Hauptkatalog. Katalog 4000 T. 1989. – [5] Eschmann P. Das Leistungsvermögen der Wälzlager. Springer, Berlin, 1964. – [6] FAG Kugelfischer George Schäfer: Schmierung von Wälzlagern. Publ. no. WL 81 115 DA; 1985 edition. – [7] Druckschrift SKF (Werkzeugmaschinenlager). – [8] Jürgensmeyer: Gestaltung von Wälzlagerungen. Springer, Berlin, 1953. – [9] Münnich H. Auswirkungen elastischer Verformungen auf die Krafteinleitung in Wälzlager. Kugellager-Z. no. 155, pp. 3–12. – [10] Sommerfeld H, Schimion W. Leichtbau von Lagergehäusen durch günstige Krafteinleitung. Z. Leichtbau der Verkehrsfahrzeuge 1969; H. 3: 3–7.

Company documents: FAG, Schweinfurt. – INA, Herzogenaurach. – NSK, Ratingen. – NTN, Erkrath-Unterfeldhaus. – SKF, Schweinfurt. – SNR, Stuttgart.

Standards and Guidelines: DIN-Taschenbuch no. 24: Wälzlager, 5th edn, Beuth, Berlin, 1985. – DIN 611: Übersicht über das Gebiet der Wälzlager. – DIN 615: Schulterkugellager; ISO 15. – DIN 616: Masspläne; ISO 104. – DIN 617: Nadellager mit Käfig; ISO 15. – DIN 618: Nadelülsen-Nadelbuchsen; ISO 3245. – DIN 620: Toleranzen; ISO 15. – DIN 622: Tragfähigkeit von Wälzlagern. – DIN 623: Bezeichnungen. – DIN 625: Rillenkugellager. – DIN 628: Schrägkugellager. – DIN 630: Pendelkugellager; ISO 15. – DIN 635: Tonnenlager-Pendelrollenlager; ISO 15. – DIN 711: Axial-Rillenkugellager; ISO 104. – DIN 715: zweiseitige Axial-Rillenkugellager. – DIN 720: Kegelrollenlager; ISO 355. – DIN 722: Axial-Zylinderrollenlager; ISO 104. – DIN 728: Axial-Pendelrollenlager; ISO 104. – DIN 736–739: Stehlagergehäuse für Wälzlager; ISO 113-2. – DIN 981: Nutmuttern; ISO 2982. – DIN 4515: Spannhülsen; ISO 1012. – DIN 5401: Kugeln; ISO 3290. – DIN 5402: Zylinderrollen-Walzen-Nadeln; ISO 3096. – DIN 5404: Axial-Nadelkränze. – DIN 5405: Radial-Nadelkränze; ISO 3030, 3031. – DIN 5406: Sicherungsbleche; ISO 2982. – DIN 5407: Walzenkränze. – DIN 5412: Zylinderrollenlager; ISO 15. – DIN 5416: Abziehhülsen; ISO 113-1. – DIN 5417: Sprengringe; ISO 464. – DIN 5418: Anschlussmasse. – DIN 5419: Filzringe-Ringnuten für Wälzlagergehäuse. – DIN 5425: Passungen für den Einbau. – DIN 51 825: Wälzlagerfette. – DIN-ISO 76: Statische Tragzahlen. – DIN-ISO 281: Dynamische Tragzahlen. – DIN ISO 355: Metrische Kegelrollenlager.

F5 Plain Bearings. [1] Vogelpohl G. Betriebssichere Gleitlager. Springer, Berlin, 1967. – [2] Spiegel K. Über den Einfluss elastischer Deformationen auf die Tragfähigkeit von Radialgleitlagern. Schmiertechnik und Tribologie 1973; 20: 3–9. – [3] VDI-Richtlinien 2204 E Papers 1–4: Düsseldorf, 1990. – [4] Rodermund H. Berechnung der Temperaturabhängigkeit der Viskosität von Mineralölen aus dem Viskositatsgrad. Schmiertechnik und Tribologie 1978; 25: 56–7. – [5] Hakansson B. The journal bearing considering variable viscosity. Report no. 25. Inst. of Machine Elements, Chalmers Univ. of Technology, Götesborg, Sweden, 1964. – [6] BMFT Projektleitung Material- und Rohstofforschung (ed.): Tribologie: Reibung – Verschleiss – Schmierung. B. 1 to 12. Springer, Berlin, 1988. – [7] Roemer E. Öldurchsatz, Öltemperatur und Lagerspiel von Gleitlagern mit Druckschmierung. VDI-Z 1961; 103: 743-7 and 790-4. – [8] Roemer E. Der Einfluss der Temperatur auf das Lagerspiel eines Gleitlagers. Konstruktion 1961; 13: 262-7. – [9] Noack G. Berechnung hydrodynamisch geschmierter Gleitlager dargestellt am Beispiel der Radiallager. Gleitlagertechnik 1. Tribotechnik vol. 49. Expert-Verlag, Grafenau, 1981. – [10] Peeken H. Zustandsschaubild für Gleitlager. Konstruktion 1968; 20: 169-76. – [11] Butenschon H J. Das hydrodynamische, zylindrische Gleitlager endlicher Breite unter instationärer Belastung. Diss., Univ. Karlsruhe, 1976. – [12] Glienecke J, Han D C. Gleitlager-Turbulenz. Forschungsber. der Forschungsvereinigung Verbrennungskraftmaschinen (FVV). H. 265. Frankfurt, 1983. – [13] Fricke J. Das Axiallager mit kippbeweglichen Kreisgleitschuhen. VDI-Forschungsheft 567. Düsseldorf, 1975. – [14] Rost U. Die Berechnung des ebenen Kreisgleitschuhs. Ing.-Arch. 1969; 38: 1-14. – [15] Fricke J. Berechnung und Auslegung von hydrodynamischen Axialgleitlagern. Gleitlagertechnik 2, Tribotechnik, vol. 163. Expert-Verlag, Grafenau, 1986. – [16] Peeken H, Knoll G. Zylindrische Gleitlager unter elastohydrodynamischen Bedingungen. Konstruktion 1975; 27: 176-81. – [17] Peeken H. Rechnerunterstützte Konstruktion von Maschinengehäusen zur Optimierung von Steifigkeit, Festigkeit und Betriebssicherheit hydrodynamischer Gleitlager. Konstruktion 1982; 34: 229-38. – [18] Droste K. Schmierungsgerechte Konstruktion. VDI-Ber 1966; 111: 15-19. – [19] VDI Richtlinie 2201: Düsseldorf. – [20] Lang O R, Steinhilper W. Gleitlager. Springer, Berlin, 1978. – [21] Hilgers W. Abhängigkeit der Lagerwerkstoffeigenschaften in Verbundlagern von der Stützkörperkonstruktion. VDI-Ber 1975; 248: 149-58. – [22] Roemer E. Lagerschalen aus Bandmaterial für die Motorenindustrie. Auto-Industrie 1961; 3-7. – [23] Hilgers W. Erkennung der Ursache von Schäden an dickwandigen Verbundlagern. Goldschmidt informiert 1978; 3: H. 45, 70-89. – [24] Hilgers W. Lagermetalle. Goldschmidt informiert 1970; 2: H. 11, 2-24. – [25] Peeken H, Salm T. Drittschicht-Dauerfestigkeit. Forschungsber. der Forschungsvereinigung Verbrennungskraftmaschinen (FVV). H. 403. Frankfurt, 1987. – [26] VDI-Richtlinie 2203: Düsseldorf, 1970. – [27] Hilgers W. Lagerwerkstoffe für höhere Forderungen. VDI-Nachrichten 1975; no. 38. – [28] Pollmann E. Das Mehrgleitflächenlager unter Berück-sichtigung der veränderlichen Ölviskosität. Konstruktion 1969; 21: 85-97. – [29] Frössel W. Berechnung von Gleitlagern mit radialen Gleitflächen. Konstruktion 1962; 14: 169-80. – [30] Gersdorfer O. Tragkraft und Anwendungsbereich von Mehrflächenlagern. Konstruktion 1962; 14: 181-8. – [31] Ott H H. Kippsegment-Radiallager mit Schmiermittel von veränderlicher Viskosität. VDI-Ber 1975; 248. – [32] Muyderman E A. Constructions with spiral-groove bearings. Wear 1966; 9: 18-141. – [33] Hübner W, Hallstedt G. Berechnung und Anwendung von Spiralrillen-Kalottenlager. Konstruktion 1972; 24: 393-7. – [34] Hentschel G. Hochbelastbare Trockengleitlager. Antriebstechnik 1976; 15: 522-8. – [35] Erhard G, Strickle E. Gleitelemente aus thermoplastischen Kunststoffen. Z. Kunststoffe 1973; 63: 4-5. – [36] Hachmann H, Strickle E. Polyamide als Gleitlagerwerkstoffe. Konstruktion 1964; 16: 4. – [37] Detter H, Holocek K. Der Reibwiderstand und die Beanspruchung von feinmechanischen Lagern im Trockenlauf bei kleinen Gleitgeschwindigkeiten. Feinwerktechnik 1970; 74: H. 11. – [38] BASF: Kunststoffe in der Prüfung. Werkstoffblatt 3110, 1 October 1975. – [39] Kayser H D. Hinweise zur Dimensionierung von Gleitlagern aus Kunststoff oder Kunstpressstoffen im Trockenlauf. VDI-Ber 1975; 248. – [40] Hentschel G. Wartungsfreie Gelenk- und Schwenklager (Oszillationslager). VDI-Ber 1975; 248: 137-42. – [41] Dietz R, Herfeld J. Die Berechnung hydrostatischer Radiallager für die Drehzahl n = 0 unter Berücksichtigung veränderlicher Spalthöhe. Maschinenmarkt 1964; H. 3, 18-24. – [42] Rasmus W. Hydrostatische Anfahrhilfe. Goldschmidt-Mitt. 1974; H. 30, 46/47. – [43] Peeken H, Benner J. Berechnung von hydrostatischen Radial- und Axiallagern. Goldschmidt informiert, Gleitlagertechnik 1984; 2: H. 61, 42-148. – [44] Peeken H, Heil M. Das optimale hydrostatische Axiallager. Konstruktion 1972; 24: 381-6. – [45] Heil M. Die Auslegung optimaler hydrostatischer Axiallager für hohe Tragiasten. Konstruktion 1974; 26: 227-31.

Standards and Guidelines: DIN 38: Lagermetallausguss in Gleitlagern. – DIN 118: Stehglleitlager mit Ringschmierung. – DIN 149: gerollte Buchsen für Gleitlager. – DIN 322: Schmierringe. – DIN 502/3: Flanschlager. – DIN 504: Augenlager. – DIN 505/6: Deckellager. – DIN 648: Gelenklager; ISO 7124. – DIN 1591: Schmierlöcher – Schmiernuten – Schmiertaschen; ISO 12 128. – DIN/ISO 4384: Härteprüfung an Lagermetallen. – DIN/ISO 4386: Prüfung der Bindung metallischer Verbundgleitlager. – DIN/ISO 6279, 4281/2/3: Lagerwerkstoffe. – DIN 7473/74: Gleitlager ungeteilt/geteilt mit Lagermetallausguss. – DIN 7477: dazu Schmiertaschen. – DIN 8221: Buchsen für Gleitlager nach DIN 502/3/4. – DIN 31 651: Gleitlagerkurzzeichen und Benennungen; ISO 7904-1. – DIN 31 652: Berechnung von hydrodynamischen Radial-Gleitlagern; ISO 7902-1. – DIN 31 654: Hydrodynamische Axial-Gleitlager im stationären Betrieb. – DIN 31 661: Schäden. – DIN 31 670: Qualitätssicherung von Gleitlagern. – DIN 31 690: Gehäusegleitlager. – DIN 31 692: Schmierung. – DIN 31 696: Segmentaxiallager. – DIN 31 697: Ring-Axiallager. – DIN 31 698: Gleitlager-Passungen; ISO 12129-1. – DIN 50 282: Gleitverhalten von Werkstoffen. – DIN 71 420/24: Zentralschmierung.

F6 Belt and Chain Drives. [1] Dittrich O et al. Anwendungen der Antriebstechnik, vol. III. Krausskopf, Mainz, 1974. – [2] Halbmann W. Zum Schlupf kraftschlüssiger Umschlingungsgetriebe. VDI-Fortschrittsber series 1, no. 145. VDI-Verlag, Düsseldorf, 1986. – [3] Neu K. Untersuchungen zum Betriebsverhalten offener und selbstspannender Flachriemengetriebe. Diss., Univ. Stuttgart, 1979. –

[4] Siegling: 3000 Hannover 1 (publications on high-strength flat belts, transfer and process drives, spindle drives, corrugated and conveyor belts). – [5] Müller H W. Anwendungsbereiche der Keilriemen in der Antriebstechnik. In: Arntz-Optibelt-Gruppe Höxter: Keilriemen. Heyer, Essen, 1972. – [6] Continental: 3000 Hannover 1 (publications on V-belts, V-ribbed belts, toothed belts and HTD belts). – [7] Optibelt: 3740 Höxter 1 (publications on transmission components, ribbed belts). – [8] Müller H W. Zugmittelgetriebe. In: Dubbel: 16th edn. Springer, Berlin, 1987. – [9] Pirelli: 8752 Kleinostheim (publications on toothed belts). – [10] Mulco: 3000 Hannover 1 (publications on toothed belts). – [11] Handbuch Antriebstechnik: Table of manufacturers and product data. Krausskopf: appears annually.

F7 Friction Drives. [1] VDI-Richtlinie 2155: Gleichförmig übersetzende Reibschlussgetriebe, Bauarten und Kennzeichen. VDI-Verlag, Düsseldorf, 1977. See also "Note on Standards" at the end of this section. – [2] Lutz O. Grundsätzliches über stufenlos verstellbare Wälzgetriebe. Konstruktion 1955; 7: 330-5, 1957; 9: 169-71, 1958; 10: 425-7. – [3] Overlach H, Severin D. Berechnung von Wälzgetriebepaarungen mit ellipsenförmigen Berührungsflächen und ihr Verhalten unter hydrodynamischer Schmierung. Konstruktion 1966; 18: 357-67. – [4] Gaggermeier H. Untersuchungen zur Reibkraftübertragung in Regel-Reibradgetrieben im Bereich elasto-hydrodynamischer Schmierung. Diss., TU Munich, 1977. – [5] Matzat N. Einsatz und Entwicklung von Traktionsflüssigkeiten. Synthetische Schmierstoffe und Arbeitsflüssigkeiten. 4th Int. Koll., Technische Akademie Esslingen, January 1984, 16.1-16.26, paper no. 16. – [6] Johnson K. L., Tevaarwerk J. L. Proc Roy Soc A (1977). – [7] Winter H., Gaggermeier H. Versuche zur Kraftübertragung in Verstell-Reibradgetrieben im Bereich elasto-hydrodynamischer Schmierung. Konstruktion 1979; 31: 2-6, 55-62. – [8] Niemann G., Winter H. Maschinenelemente, vol. III, 2nd edn, Springer, Berlin, 1983. – [9] Basedow G. Stufenlose Nullgetriebe schützen vor Überlast und Anfahrstössen. Antriebstechnik 1986; 25: 20-5. – [10] Schroebler W. Praktische Erfahrungen mit speziellen Reibradgetrieben. Tech Mitt 1968; 61: 411-14. – [11] Hewko L. O. Roller traction drive unit for extremely quiet power transmission. J Hydronautics 1968; 2: 160-7. – [12] Bauerfeind E. Zur Kraftübertragung mit Gummiwälzrädern. Antriebstechnik 1966; 5: 383-91.

F8 Gearing. [1] Niemann G., Winter H. Maschinenelemente, vols II and III, 2nd edn. Springer, Berlin, 1989/6. – [2] Dudley D. W., Winter H. Zahnräder. Springer, Berlin, 1961. – [3] Keck K. F. Die Zahnradpraxis, parts 1 and 2. Oldenbourg, Munich, 1956 and 1958. – [4] Winter H. Die tragfähigste Evolventen-Geradverzahnung. Vieweg, Brunswick, 1954. – [5] Dudley D. W. Gear handbook. McGraw-Hill, New York, 1962. – [6] Richter W. Auslegung profilverschobener Aussenverzahnungen. Konstruktion 1962; 14: 189-96. – [7] Roth K. Evolventenverzahnungen für parallele Achsen mit Ritzelzähnezahlen von 1 bis 7. VDI-Z 1965; 107: 275-84. – [8] Piepka E. Eingriffsstörungen bei Evolventen-Innenverzahnung. VDI-Z 1970; 112: 215-22. – [9] Clarenbach J et al. Geometrische Auslegung von zylindrischen Innenradpaaren – Erläuterung zum Normentwurf DIN 3993. Antriebstechnik 1975; 651-8. – [10] Erney G. Auslegung von Evolventen-Innenverzahnungen. Antriebstechnik 1975; 14: 625-9. – [11] Naville R. Die Theorie der Verzahnung der Uhrwerktechnik. Microtechnic 1967; XXI: 506-9, 587-90. – [12] Niemann G. Novikov-Verzahnung und andere Sonderverzahnungen für hohe Tragfähigkeit. VDI-Ber 1961; 47: 5-12. – [13] Shotter B A. Experiences with Conformal/WN-

gearing. World Congress on Gearing, Paris, 1977, vol. I, p. 527. – [14] Dowson D, Higginson G R. Elasto-hydrodynamic lubrication. Pergamon Press, Oxford, 1966. – [15] Johansson M et al. Austinitisches-bainitisches Gusseisen als Konstruktionswerkstoff im Getriebebau. Antriebstechnik 1976; 15: 593-600. – [16] Winter H, Weiss T. Tragfähigkeitsuntersuchungen an induktions- und flammgehärteten Zahnrädern, parts I and II. Antriebstechnik 1988; 27: 45-50, 57-62. – [17] Walzel H. Kann das Nikotrierverfahren das Badnitriren ersetzen? TZ für prakt Metallbearb 1976; 70: 291-4. – [18] Winter H, Wirth X. Einfluss von Schleifkerben auf die Zahnfussdauertragfähigkeit oberflächengehärteter Zahnräder. Antriebstechnik 1978; 17: 37-41. – [19] Rhenius K T. Betriebsfestigkeitsrechnungen von Maschinenelementen in Ackerschleppern mit Hilfe von Lastkollektiven. Konstruktion 1977; 29: 85-93. – [20] Rettig H, Plewe H-J. Lebensdauer und Verschleissverhalten langsam laufender Zahnräder. Antriebstechnik 1977; 16: 357-61. – [21] Krause W. Untersuchungen zur Geräuschverhalten evolventenverzahnter Geradstirnräder der Feinwerktechnik. VDI-Ber 1967; 105. – [22] Gavrilenko V A., Bezrukov V I. The geometrical design of gear transmissions comprising involute bevel gears. Russ Eng J 1976; 56: 34-8. – [23] Beam A S. Beveloid gearing. Mach Design 1954; 220-38. – [24] Hiersig H M. Zylinderräder mit Rechts- und Linksflanken von ungleicher Steigung. Konstruktion 1979; 31: 7-11. – [25] Keck K F. Die Bestimmung der Verzahnungsabmessung bei kegeligen Schraubgetrieben mit 90° Achswinkel. ATZ 1953; 55: 302-8. – [26] Coleman W. Hypoidgetriebe mit beliebigen Achswinkeln. Automotive Ind. June 1974. – [27] Richter M. Der Verzahnungswirkungsgrad und die Fresstragfähigkeit von Hypoid- und Schraubenradgetrieben. Diss., TU Munich, 1976. – [28] Winter H, Richter M. Verzahnungswirkungsgrad und Fresstragfähigkeit von Hypoid- und Schraubenradgetrieben. Antriebstechnik 1976; 15: 211-18. – [29] Heyer E. Spielfreie Verzahnungen besonders bei Schneckengetrieben. Industriebl 1954; 54: 509-12. – [30] Macabrey C. Globoid-Schneckengetriebe "Cone-Drive". TZ prakt Metallbearb., pt I, 1964; 58: 669-72; pt II, 1965; 59: 711-14. – [31] Jarchow F. Stirnad-Globoid-Schneckengetriebe. TZ prakt Metallbearb 1966; 60: 717-22. – [32] Wilkesmann H. Berechnung von Schneckengetrieben mit unterschiedlichen Zahnprofilen. Diss., TU Munich, 1974. – [33] Holler R. Rechnersimulation der Kinematik und 3 D-Messung der Flankengeometrie von Schneckengetrieben und Kegelrädern. Diss., RWTH Aachen, 1976. – [34] Mathiak D. Untersuchungen über Flankentragfähigkeit, Zahnfusstragfähigkeit und Wirkungsgrad von Zylinderschneckengetrieben. Diss., TU, Munich, 1984. – [35] Hecking L. Schneckengetriebe im Kranbau, dima 1967; 3: 39-41. – [36] Hofmann E. Neuartige Kegelradgetriebemotoren und Kegelradgetriebe. Antriebstechnik 1978; 17: 271-5. – [37] Lechner G. Zahnfussfestigkeit von Zahnradbandagen. Konstruktion 1967; 19: 41-7. – [38] Grodzinski P. Eccentric gear mechanisms. Mach Design 1953; 25: 141-50. – [39] Miano S V. Twin eccentric gears. Prod Engng 1962; 33: 47-51; see also [52]. – [40] Benford R L. Customized motions. Mach design 1968; 40: 151-4. – [41] Federn K et al. Drehschwingprüfmaschine für umlaufende Maschinenelemente. Konstruktion 1974; 26: 340-9. – [42] Mitome K., Ishida K. Eccentric gearing. Trans ASME J Engng Ind 1974; 94-100. – [43] Naruse C. Verschleiss, Tragfähigkeit und Verlustleistung von Schraubenradgetrieben. Diss., TH Munich, 1964. – [44] Wetzel R. Graphische Bestimmung des Schrägungswinkels für das treibende Rad bei Schraubentrieben mit gegebenem Wellenabstand. Werkst Betr 1955; 88: 718-

9. - [45] Jacobsen U A. I. Crossed helical gears for high speed automotive applications. Inst Mech Engng Proc. Autom Div 1961/2; 359–84. - [46] Rohonyi C. Berechnung profilverschobener, zylindrischer Schraubenräder. Konstruktion 1963; 15: 453–5. - [47] Henriot G. Engrenages. Dunod, Paris, 1980. - [48] Seifried A. Über die Auslegung von Stirnradgetrieben. VDI-Z 1967; 109: 236–41. - [49] Siefried A, Bürkle R. Die Berührung der Zahnflanken von Evolventenschraubenrädern, Werkst Betr 1968; 101: 183–7. - [50] Maag-Taschenbuch. MAAG AG, Zürich, 1985. - [51] Pohl F. Betriebshütte, vol. I, Abschn Kegelradbearbeitung und Maschinen für Kegelradbearbeitung. Ernst & Sohn, Berlin, 1957. - [52] Ernst H. Die Hebezeuge, vol. I. Vieweg, Brunswick, 1973. - [53] Chironis N. P. Gear design and application. McGraw-Hill, New York, 1967; includes papers from: Bloomfield B. Noncircular gears, pp. 158–63; Rappaport S. Elliptical gears of cyclic speed variations, pp. 166–68; Miano S. V. Twin eccentric gears, pp. 169–73. - [54] Cunningham F, Cunningham D. Rediscovering the noncircular gear. Mach Design 1973; 45: 80–5. - [55] Ludwig F. Verwendung eines Koppelgetriebes zum Herstellen wälzverzahnter Ellipsenräder. VDI-Ber 1956; 12: 139–44. - [56] Ferguson R. J. et al. The design of a stepless transmission using noncircular gears. Mech Mach Theory 1975; 10: 467–78. - [57] Yokoyama Y. et al. Dynamic characteristic of the noncircular planetary gear mechanisms with nonuniform motion. Bull ISME 1974; 17: 149–56. - [58] Winter H, Schönnenbeck G. Graufleckigkeit an einsatzgehärteten Zahnrädern: Ermüdung der Werkstoffrandschicht mit möglicherweise schweren Folgeschäden. Antriebstechnik 1985; 24: 53–61. - [59] Müller H W. Die Umlaufgetriebe, Berechnung, Anwendung, Auslegung. Springer, Berlin, 1971. - [60] Müller H W. Einheitliche Berechnung von Planetengetrieben. Antriebstechnik 1976; 15: 11–17, 85–9, 145–9. - [61] Müller H W. Programmierte Analyse von Planetengetrieben. Antriebstechnik 1989; 28: [62] Müller H W. Ungleichmässig übersetzende Umlaufgetriebe. VDI Fortschrittsber. Series 1, 1988; 159: 49–64. - [63] Jarchow F. Entwicklungsstand bei Planetengetrieben. VDI-Ber 1988; 672: 15–44. - [64] Winkelmann L. Lastverteilung in Planetengetrieben. VDI-Ber 1988; 672: 45–74. - [65] Potthoff H. Anwendungsgrenzen vollrolliger Planetenrad-Wälzlager. VDI-Ber 1988; 672: 245–64. - [66] Müller H W. Überlagerungssysteme. VDI-Ber 1986; 618: 59–78. - [67] Dreher K. Rechnergestütze Optimierung von Planeten-Koppelgetrieben. Diss., Darmstadt, 1983. - [68] Schnetz K. Reduzierte Planeten-Koppelgetriebe. Diss., Darmstadt, 1976. - [69] Brass E A. Two stage planetary arrangements for the 15 : 1 turboprop reduction gear. ASME Paper 60-SA-1 (1960). - [70] Schoo A. Verzahnungsverlustleistungen in Planetenradgetrieben. VDI-Ber 1988; 627: 121–40. - [71] Müller H W. Anpassung stufenloser Getriebe an die Kennlinie einer Maschine. Also: Optimierung der Grundanordnung stufenloser Stellgetriebe. Maschinenmarkt 1981; 90: 1968–71, 2183–5.

ISO Standards: ISO 53: Bezugsprofil für Stirnräder für den allgemeinen Maschinenbau und den Schwermaschinenbau. - ISO 677: Bezugsprofil für geradverzahnte Kegelräder für den allgemeinen Maschinenbau und den Schwermaschinenbau. - ISO 701: Internationale Verzahnungsterm- inologie: Symbole für geometrische Grössen. - ISO/R 1122: Vokabular für Zahnräder; Geometrische Begriffe. - ISO/R 1122, Add. 2: Vokabular für Zahnräder; Geometrische Begriffe, Schneckengetriebe. - ISO 1328: Stirnräder mit Evolventenverzahnung - ISO Genauigkeitssystem. - ISO 1340: Stirnräder; Angaben für die Bestellung. - ISO 1341: Geradverzahnte Kegelräder,

Angaben für die Bestellung. - ISO 2203: Zeichnungen; Darstellung von Zahnrädern.

DIN Standards: DIN Taschenbuch 106. Antriebstechnik 1. Normen über die Verzahnungsterminologie. Beuth, Berlin and Köln: 1981. - DIN 37: Zeichnungen; Darstellung von Zahnrädern. - DIN 780: Modulreihe für Zahnräder; Moduln für Stirnräder und Zylinderschneckengetriebe. - DIN 783: Wellenenden für Zahnradgetriebe mit Wälzlagern. - DIN 867: Bezugsprofil für Stirnräder (Zylinderräder) mit Evolventenverzahnung für den allgemeinen Maschinenbau und den Schwermaschinenbau. - DIN 868: Allgemeine Begriffe und Bestimmungsgrössen für Zahnräder, Zahnradpaare und Zahnradgetriebe. - DIN 3960: Begriffe und Bestimmungsgrössen für Stirnräder (Zylinderräder) und Stirnradpaare (Zylinderradpaare) mit Evolventenverzahnung. - DIN 3961: Toleranzen für Stirnradverzahnungen; Grundlagen. - DIN 3962: Toleranzen für Stirnradverzahnungen; Zulässige Abweichungen einzelner Bestimmungsgrössen. - DIN 3963: Toleranzen für Stirnradverzahnungen; Zulässige Wälzabweichungen. - DIN 3964: Toleranzen für Stirnradverzahnungen; Gehäuse-Toleranzen. - DIN 3966: Angaben für Verzahnungen in Zeichnungen; Angaben für Stirnrad-(Zylinderrad-) Evolventenverzahnungen und Geradzahn-Kegelradverzahnungen. - DIN 3967: Getriebe-Passsystem; Flankenspiel, Zahndickenabmasse und Zahndickentoleranzen. - DIN 3970; Lehrzahnräder zum Prüfen von Stirnrädern. - DIN 3971: Verzahnungen; Bestimmungsgrössen und Fehler an Kegelrädern. - DIN 3972: Bezugsprofile von Verzahnwerkzeugen für Evolventenverzahnungen nach DIN 867. - DIN 3975: Begriffe und Bestimmungsgrössen für Zylinderschneckengetriebe mit Achsenwinkel 90°. - DIN 3976: Zylinderschnecken; Abmessungen, Zuordnung von Achsabständen und Übersetzungen in Schneckengetriebe. - DIN 3978: Schrägungswinkel für Stirnradverzahnungen. - DIN 3979: Zahnschäden an Zahnradgetrieben; Bezeichnung, Merkmale, Ursachen. - DIN 3990: Tragfähigkeitsberechnung von Stirnrädern. - DIN 3991: Tragfähigkeitsberechnung von Kegelrädern. - DIN 3992: Profilverschiebung bei Stirnrädern mit Aussenverzahnung. - DIN 3993: Geometrische Auslegung von zylindrischen Innenradpaaren. - DIN 3994: Profilverschiebung bei geradverzahnten Stirnrädern mit 05-Verzahnung, Einführung. - DIN 3995: Geradverzahnte Aussen-Stirnräder mit 05-Verzahnung. - DIN 3998: Benennungen an Zahnrädern und Zahnradpaaren. - DIN 3999: Kurzzeichen für Verzahnungen. - DIN 58 400: Bezugsprofil für Stirnräder mit Evolventenverzahnung für die Feinwerktechnik. - DIN 58 405: Stirnradgetriebe der Feinwerktechnik. - DIN 58 420: Lehrzahnräder zum Prüfen von Stirnrädern der Feinwerktechnik. - DIN 58 425: Kreisbogenverzahnungen für die Feinwerktechnik. - DIN 45 635 T 23: Geräuschmessung an Maschinengetrieben.

VDI Guidelines: VDI-Richtlinie 2060: Beurteilungsmassstäbe für den Auswuchtzustand rotierender starrer Körper. - VDI-Richtlinie 2159: Getriebegeräusche; Messverfahren - Beurteilung - Messen und Auswerten, Zahlenbeispiele. - VDI-Richtlinie 2546: Zahnräder aus thermoplastischen Kunststoffen. - VDI-Richtlinie 3720: Lärmarm konstruieren.

F9 Mechanism Engineering, Kinematics. [1] VDI-Richtlinie 2127 (Entwurf): Getriebetechnische Grundlagen: Begriffsbestimmungen der Getriebe (1988). - [2] Braune R. Die Bedeutung des kürzesten und des längsten Gliedes für die systematische Betrachtung ebener viergliedriger kinematischer Ketten. Ind-Anz 1971; 93: 2258–60. - [3]

VDI-Richtlinie 2145: Ebene viergliedrige Getriebe mit Dreh- und Schubgelenken: Begriffserklärungen und Systematik (1980). – [4] VDI-Richtlinie 2147: Ebene Kurvengetriebe: Begriffserklärungen (1962). – [5] Volmer J. (ed.). Getriebetechnik; Leitfaden, 3rd edn. Vieweg, Brunswick, 1989. – [6] Hain K. Ermittlung der Umlauf- und Schwingbewegungen in durchlauffähigen sechsgliedrigen Getrieben. Grundl Landtechn 1966; 16: 129–39. – [7] Hain K. Systematik und Umlauffähigkeit drei- und mehrgliedriger Kurvengetriebe. Konstruktion 1967; 19: 379–88. – [8] Hain K. Rechenprogramme für beschleunigungsgleiche Getriebe mit unterschiedlichen Hauptbewegungen. Werkstatt Betrieb 1976; 109: 73–80. – [9] Shigley J. E., Uicker J. J. jr. Theory of machines and mechanisms. McGraw-Hill, Tokyo, 1980. – [10] Lohe R. Beeinflussung der Laufeigenschaften durch Massenverteilung bei ebenen und räumlichen Getrieben. VDI-Ber 1980; 374: 135–45. – [11] Kerle H et al. Berechnung und Optimierung schnellaufender Gelenk- und Kurvengetriebe. Expert Verlag, Grafenau, 1981. – [12] Braune R. Entwurf Richtlinie VDI 2729: Modulare kinematische Analyse ebener Gelenkgetriebe mit Dreh- und Schubgelenken (1989). – [13] Lütgert A, Braune R. KAMOS – ein interaktives Entwicklungswerkzeug zur Analyse, komplexer Koppelgetriebe. VDI-Ber 1989; 736: 119–50. – [14] VDI-Richtlinie 2138: Räumliche Kurbelgetriebe: Umformung von Drehbewegung in Schwingschubbewegung (1959). – [15] VDI-Richtlinie 2139: Räumliche Kurbelgetriebe: Umformung von Drehbewegung in umlaufende Drehschubbewegung (1959). – [16] VDI-Richtlinie 2723: Vektorielle Methode zur Berechnung der Kinematik räumlicher Getriebe (1982). – [17] VDI-Richtlinie 2724: Berechnung der Kinematik viergliedriger Getriebe: Ein Rechenprogramm (1986). – [18] Lohe R. Berechnung und Ausgleich von Kräften in räumlichen Mechanismen. Fortschr-Ber VDI-Z, series 1, no. 103 (1983). – [19] Ahlers W. Zur Bestimmung der Lageführungsgrössen von Manipulatoren am Beispiel einer Operationsleuchte. Konstruktion 1986; 38: 81–6. – [20] Matthaei H. Über den Leistungsfluss in Kurbelgetrieben. Konstruktion 1966; 18: 45–9. – [21] Marx U. Ein Beitrag zur kinetischen Analyse ebener viergliedriger Gelenkgetriebe unter dem Aspekt Bewegungsgüte. Fortschr-Ber VDI-Z series 1, no. 144 (1986). – [22] Müller H W. Beurteilung periodischer Getriebe mit Hilfe des "Übertragungswinkels". Konstruktion 1985; 37: 431–6. – [23] Stündel D. Das dynamische Laufkriterium bei Gerätemechanismen. Feingerätetech 1974; 23: 507–9. – [24] Kerle H. Dynamische Maschinenanalyse mit Hilfe programmierbarer Tischrechner. Forsch Ing-Wes 1980; 46: 149–53. – [25] Dittrich G. Systematik der Bewegungsaufgaben und drundsätzliche Lösungsmöglichkeiten. VDI-Ber 1985; 576: 1–20. – [26] Alt H. Das Konstruieren von Gelenkvierecken unter Benutzung einer Kurventafel. VDI-Z 1941; 85: 69–72. – [27] VDI-Richtlinie 2130: Getriebe für Hub- und Schwingbewegungen: Konstruktion und Berechnung viergliedriger ebener Gelenkgetriebe für gegebene Totlagen (1984). – [28] VDI-Richtlinie 2125: Ebene Gelenkgetriebe: Übertragungsgünstigste Umwandlung einer Schubschwing- in eine Drehschwingbewegung (1987). – [29] VDI-Richtlinie 2126 (Entwurf): Ebene Gelenkgetriebe: Übertragungsgünstigste Umwandlung einer Drehschwing- in eine Schubschwingbewegung (1986). – [30] Kracke J. Massbestimmung ebener viergliedriger Kurbelgetriebe für die Sonderfälle von vier Übereinstimmungen. Diss., TU Brunswick, 1972. – [31] Braune R. Ein Beitrag zur Masssynthese ebener viergliedriger Kurbelgetriebe. Diss., RWTH Aachen, 1980. – [32] Hain K. Konstruktionsdaten-Auswahl für das Gelenkviereck durch Computer Dialog. technica 1976; 25: 791–8. – [33] Hain K. Punktlagenreduktion als getriebesynthetisches Hilfsmittel. Maschinenbau/Betrieb, Beil Getriebetechn 1943; 11: 29–31. – [34] VDI-Richtlinie 2143, paper 1: Bewegungsgesetze für Kurvengetriebe: Theoretische Grundlagen (1980). – [35] VDI-Richtlinie 2143, paper 2: Bewegungsgesetze für Kurvengetriebe: Praktische Anwendung (1987). – [36] Duditza F. Kardangelenkgetriebe und ihre Anwendungen. VDI-Verlag, Düsseldorf, 1973. – [37] Hiller M. Analytisch-numerische Verfahren zur Behandlung räumlicher Übertragungsmechanismen. Fortschr-Ber VDI-Z series 1, no. 76 (1981). – [38] VDI-Richtlinie 2156: Einfache räumliche Kurbelgetriebe: Systematik und Begriffsbestimmungen (1975). – [39] VDI-Richtlinie 2154: Sphärische viergliedrige Kurbelgetriebe: Begriffserklärungen und Systematik (1971). – [40] VDI-Richtlinie 2722: Homokinematische Kreuzgelenkgetriebe einschliesslich Gelenkwellen (1982). – [41] Hain K. Entwurf übertragungsgünstigster Zylinderkurven- und Kegelkurvengetriebe. Werkstatt Betrieb 1978; 111: 93–8. – [42] Zakel H. Geometrie, Kinematik und Kinetostatik des Kurvengelenks räumlicher Kurvengetriebe. Diss., RWTH Aachen, 1983. – [43] Lichtwitz O. Getriebe für aussetzende Bewegungen. Springer, Berlin, 1953. – [44] Eckerle R. Optimale Auslegung von Malteser-Schaltwerken. Feinwerktechn 1969; 73: 482–7. – [45] Hain K. Erzeugung von Schrittbewegungen durch Planeten-Kurven-Getriebe. Antriebstech 1973; 12: 315–22. – [46] VDI-Richtlinie 2721: Schrittgetriebe: Begriffsbestimmungen, Systematik, Bauarten (1980). – [47] Hain K. Die Erzeugung gegebener Kurven mit Hilfe von Räderkurbelgetrieben. Feinwerktech 1949; 53: 81–9. – [48] Volmer J. Räderkurbelgetriebe. VDI-Forsch 1957; 461: 52–5. – [49] Neumann R., Watzlawik P. Synthese von Räderkoppelschrittgetrieben mit Hilfe von Kurventafeln, Maschinenbautech 1974; 23: 52–9.

F10 Crank Mechanism. [1] Bensinger W D, Meier A. Kolben, Pleuel und Kurbelwelle bei schnellaufenden Verbrennungsmotoren, 2nd edn. Springer, Berlin, 1961. – [2] Biezeno C. B., Grammel R. Technische Dynamik, vol. 2. Springer, Berlin, 1971. – [3] Haffner K E, Mass H. Torsionsschwingungen in der Verbrennungskraftmaschine. Springer, Vienna, 1985. – [4] Küttner K H. Kolbenmaschinen, 5th edn. Teubner, Stuttgart, 1984. – [5] Köhler G, Rögnitz H. Maschinenteile, vol. 2, 7th edn. Teubner, Stuttgart, 1986. – [6] Lang O R. Triebwerke schnellaufender Verbrennungsmotoren. Springer, Berlin, 1966. – [7] Mayr F. Ortsfeste Dieselmotoren und Schiffsdieselmotoren, 3rd edn. Springer, Vienna, 1960. – [8] Sass F. Bau und Betrieb von Dieselmaschinen, vol. 1, 2nd edn. Springer, Berlin, 1948. – [9] Maas H, Klier H. Kräfte Momente und deren Ausgleich in der Verbrennungskraftmaschine. Springer, Vienna, 1981.

Standards and Guidelines: Kolbenbolzen für Kraftfahrzeugbau DIN 73 124 für Dieselmotoren, DIN 73 125 für Ottomotoren. Kolbenringe (the first numbers given apply to mechanical engineering, the second-named numbers apply to automotive engineering) DIN 34 109 and DIN 70 909 Übersicht, Allgemeines; DIN 34 110 and DIN ISO 6620 Rechteckringe; DIN 34 111 and DIN ISO 6620 Minutenringe; DIN 6662 Trapezringe; DIN 34 130 and DIN ISO 6624 Nasenringe; DIN 34 146 and DIN ISO 6625 Ölschlitzringe; DIN 34 147 and DIN ISO 6625 Dachfasenringe; DIN 34 148 and DIN 70 948 Gleichfasenringe.

Note on Standards

F1.1.2, page F8: see also the following related (R) or equivalent (E) publications. For DINs 1913, 8554, 8555, 8556: ISOs 544 (R), 636 (R), 2560 (R), 3580 (R), 3581 (R); for DIN 1913: ISO/DIS 2560 (R), prEN 20 544 (E); for DINs 8554, 8555, 8556: prEN 20 544 (R); for DIN 8556: ISO/DIS 11 837 (R); for DIN 8559: ISOs 636 (R), 864 (R), prEN 440 (E).

F1.1.2, page F9: see also the following related publications. For DINs 1732, 1733, 1736: ISOs 544, 636, 2560, 3580, 3581; for DIN 1732: also prEN 20 544; for DIN 8573 T1: ISO 1071.

F1.1.4, page F11: see also the following related (R) or equivalent (E) publications pertaining to DIN 1912. T1: ISO 2553 (R); T2: ISO 6947 (R); T4: no equivalent; T5: ISO 2553 DAD 1 (R), ISO 2553 (R). Publications pertaining to DIN 8563. T1: ISO 6213 (E); T2: ISO 6213 (R), ISI/DIS 3834 (R); T3: ISO 5817, DIN EN 25 817; T30: ISO 10042 (R), prEN 630 (E); T104: prEN 719 (E); T105: prEN 288-5 (E), ISO/DIS 9956-5 (E); T106: prEN 288-6 (E), ISO/DIS 9956-6 (E); T107: prEN 288-7 (E), ISO/DIS 9956-7 (E); T108: prEN 288-8 (E), ISO/DIS 9956-8 (E); T110: prEN 729-1 (E), ISO/DIS 3834-1 (E); T111: prEN 729-2 (E),

ISO/DIS 3834-2 (E); T112: prEN 729-3 (E), ISO/DIS 3834-3 (E); T113: prEN 729-4 (E), ISO/DIS 3834-4 (E); T120: prEN 1011 (E).

F1.1.5, page F12: see also the following related publications. For DIN 1745: ISO 6361-2, ISO/DIS 6361-2 DAD 1; for DIN 1746 Part 1 and DIN 1747: ISO 6362-2 and ISO/DIS 6362-2; for DIN 1746 Part 2 and DIN 1748 Part 1: ISOs 6362-1 and 6363-1; for DIN 1748 Part 4: ISO 6362-4.

F1.1.5, page F13: see also the equivalent DIN V ENV 1993 (Eurocode 3). DIN 15 018-related publications are: Part 1: ISO 4301-4, ISO 8686-1, ISO/DIS 8686-1, ISO 4301-5, ISO 8686-5, ISO/DIS 8686-3; Part 2: no equivalent; Part 3: ISO 8686-1, ISO/DIS 8686-1, ISO 8686-5, ISO/DIS 8686-3.

F7, page 200: see the following related British Standards: BS 3092: 1973 (1988): Specification for main friction clutches, main power take-off assemblies and associated attachments for internal combustion engines. BS 3170: 1972 (1991): Specification for flexible couplings for power transmission. BS 6613: 1985 (1981): Methods for specifying characteristics of resilient shaft couplings. BS AU 203a: 1988: Specification for dimensions of couplings between power take-offs and ancillary driven units on commercial road vehicles.

G

Hydraulic and Pneumatic Power Transmission

R. Röper, Dortmund

1 Fundamentals of Fluid Power Transmission Systems

1.1 The Flow Process

The specific energy of a moving fluid (liquid or gas) is described by the Bernoulli equation:

$$Y_f = E/\dot{m} = h + \frac{u^2}{2} + gz + \int \frac{\partial u}{\partial t} \cdot ds.$$

The continuity equation applies for the steady-flow condition:

$$\dot{m} = \text{const.} = \rho A u.$$

In the special case of incompressible fluids, $\dot{V} = Au$. Fluids can transfer signals by means of an energy state (e.g. a given pressure), or flow strength and energy by transfer of specific energy as mass flow. The conversion from mechanical to fluid energy takes place in a steady flow process, as in **Fig. 1** (throughput direction is positive):

$$Y_m = P_m/\dot{m} = Y_{f2} - Y_{f1}$$
$$= h_2 - h_1 + (u_2^2 - u_1^2)/2 + g(z_2 - z_1).$$

With $h_2 - h_1 = \Delta h_{12} = (\Delta h_s)_{12} + P_{v.12}/m$ (index s = isentropic), the irreversible component of the specific work is

$$(\Delta h_s)_{12} = (\Delta u^2)_{12}/2 + g \cdot \Delta z_{12},$$

giving the irreversible loss as $P_{v.12}/m$.

For energy transfer in a fluid drive, the specific enthalpy h of the remaining components becomes negligible. The specific work is then simplified as

$$Y_m = P_m/\dot{m} \approx h_2 - h_1 = (\Delta h_s)_{12} + P_{v.12}/\dot{m}, \quad \text{or}$$
$$Y_m = (\Delta h_s)_{12} \, \eta_t^{\pm 1} \quad (+1 \text{ motor}, -1 \text{ generator}).$$

i.e. the fluid work is the differential of work done by output and input shear forces. The power transfer then follows for constant flow strength $\dot{m}$ or $\dot{V}$

$$P = dE/dt = \dot{V} \cdot \Delta p_{12} \quad \text{(uniform flow)}.$$

Mean working velocities up to 5 m/s are found.

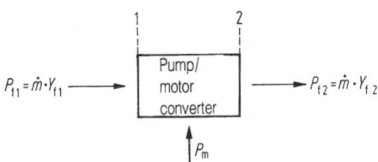

$P_{f1} = \dot{m} \cdot Y_{f1} \longrightarrow$ Pump/ motor converter $\longrightarrow P_{f2} = \dot{m} \cdot Y_{f2}$

$\uparrow P_m$

Figure 1. Schematic of open-flow process.

Designation	Pressure range	Application
Low pressure	30 to 50 bar	Machine tools (feed mechanisms)
Medium pressure	Up to 170 bar	Transport equipment, building machinery, traversing gear
High pressure	200 to 450 bar	Presses, tensioners, aircraft hydraulics

Under unsteady conditions, the compressibility of the working fluid must be taken into account. Mean values for the compressibility of hydraulic oils are:

$$\beta = - \, dV/Vdp = (7 \text{ to } 4.5) \cdot 10^{-5} \, \text{bar}^{-1}$$

at 250 to 20 bar oil pressure and 80 to 20 °C oil temperature.

For unsteady conditions, calculations are more complicated, since masses, resistances and spring characteristics are distributed, not concentrated. Differential equations are non-linear. Linearisation gives approximations to within ± 20% of the critical frequency, or accurate results with matrix methods [1]. By change of state, it can also be expressed as

$$\Delta h_{12} = \int_1^2 v \, dp = \left(\int_1^2 v \, dp \right)_s + P_{v.12}/\dot{m}$$

(see **G2 Fig. 2**).

Nomenclature: A flow cross-sectional area; E fluid energy; g acceleration due to gravity; h specific enthalpy; m mass; $\dot{m}$ mass flow; p pressure; P_m mechanical power; P_v power loss; s length of flow path; u flow velocity; v specific volume; $\dot{V}$ volumetric flow; Y_f specific hydraulic energy (work); Y_m specific mechanical work; z geodetic altitude; ρ density of fluid; η_t overall efficiency.

By using high energy density (pressure), large forces or moments in linear or rotary motion may be obtained with small dimensions. This gives low specific weight (< 1 kg/kW for the complete apparatus) and moment of inertia (rapid reversing). The lower volumetric flow resulting from the use of higher pressures allows the use of smaller dimensions in control units, and produces low transmission losses (loss ~ $\dot{V}$ or $\dot{V}^2$). Losses in the flow process also occur as pressure losses. The energy loss causes the temperature of the fluid to rise.

1.1.1 Hydraulic Power Transmission

The various oils and other fluids used in hydraulic equipment are virtually incompressible. The equation of state is therefore identical with the isochore

$$(\Delta b_s)_{12} = v \int_1^2 dp = \Delta p_{12}/\rho.$$

1.1.2 Pneumatic Power Transmission

Gases are extremely compressible, and working velocities are therefore uneven due to variations in expansion. Gases are mainly used for power transmission only for subordinate purposes (small machine tools), partly at polytropic conditions pv^n = const., where n = 1.3 to 1.35, and frequently without expansion at full pressure operation. Accurate positioning can only be achieved via mechanical equipment, and the main applications are in tensioning and press tools. Working velocities are very high. Single-stage compression limits the available pressure range, particularly as compression to higher pressures causes a large temperature rise.

Designation	Pressure range	Application
Low pressure	Up to 1 bar	Control equipment
High pressure	6 to 10 bar	Presses, tensioners, transport and machine tools

For precise velocity, input to pneumatic equipment on machine and hand tools, parallel or post hydraulic control (pneumo-hydraulic), or electronically operated throttle controls may be used.

1.2 Hydraulic Fluids

Mineral oils, water-based fluids, or water-free synthetic fluids are used as working fluids in hydraulic systems. The DIN 51 524 or ISO TC 131 HL and HLP fluids are mineral oils defined in DIN 51 519 or ISO 3448 for viscosity classes v = 7, 10, 15, 22, 32, 46 and 68 · 10^{-6} m/s at 40 °C. Mineral oil HL contains additives to improve resistance to high thermal loadings and to increase corrosion protection. The HPL oils have extra additives to improve properties in regions of mixed friction, while the HLPD oils have other additives to counter the effect of entrained water.

Also, HD, S1, and S3 engine oils are sometimes used as hydraulic fluids.

For viscosity and temperature characteristics of hydraulic oils see **Appendix G, Fig. 1**. The variation of viscosity with pressure is negligible at pressures below p = 200 bar. The selection of oil is made according to the operational viscosity needed for the components (manufacturer's recommendation) and the mean operating temperature, with the temperature of the oil being likely to exceed that of the surroundings by 30 to 50 K. An additional selection criterion is a good starting capability after an extended period at low temperature. High viscosity can lead to suction difficulties. Normal operating viscosities range between 20 and 60 · 10^{-6} m²/s. New developments in oils with polymer additives (high-VI oils) show less dependence of viscosity on temperature, allowing cold-start temperatures to be lower by about 10 °C.

In situations where oil leaks may be ignited (forges, foundries, coal-mines), non-flammable fluids have to be used. Fluids containing water can prevent sudden ignition by the buildup of a layer of steam. These include the following: HFA – oil–water emulsions (see G6); HFB – water–oil emulsions with water contents up to 60%; HFC: aqueous solutions of polymers, preferably polyalkylene–glycol–water solutions with up to 60% water. When water-containing fluids are used, loadings must be reduced because of lower wear resistance, especially in the case of rolling friction, and special filters must also be used. Category HFD comprises water-free fluids, non-flammable because of their chemical composition, e.g. only phosphate esters. They exhibit good wear resistance, but before use, toxicity and effect on sealing materials should be tested.

1.3 Systematology

1.3.1 Structure and Operation of Fluid Transmissions

In fluid drives, generators (pumps, compressors), motors, and control units are connected in a circuit in which the working fluid circulates (**Fig. 2**). Because of the high pressures involved, only positive-displacement machines are suitable. Because of this, and the incompressibility of the hydraulic fluid, there is in hydrostatic drives a volumetric link between the drive and the drive components, i.e. the transmission of hydraulic drive is virtually independent of load (parallel characteristic). In contrast, the compressibility of air under pressure is very much greater. Pneumatic drives exhibit series characteristics.

The pressure medium is transferred by pipes, which feature allows considerable latitude in the location of drive, motor and control unit. Distances can be up to 30 m with hydraulic drive, and as much as 150 m with pneumatic drive.

The control system regulates the transmission of work and limits system loadings. It can work directly on the fluid flow by switching flow channels, by changing or splitting the direction of flow, or indirectly by changing the geometry of pumps and motors. The function may be either engaged (pressure control, position control), or disengaged. Control systems also work by cylinder cutoff. They can operate either directly or indirectly.

There are thus good possibilities for remote control and automation by a combination of electrical and electronic control media.

1.3.2 Classification of Fluid Transmissions

Energy transmission using compressible gas or liquid flow gives an almost unlimited variation of energy flowing through the system to transmit force or torque and linear or angular velocity. A classification can be made according to the following external constraints.

Transmission

Power Transmissions transfer input power over the widest possible range to produce required force moments

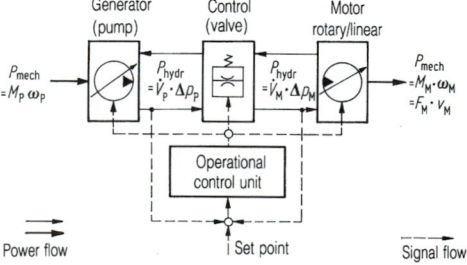

Figure 2. Block diagram of fluid drive.

and the desired linear or angular velocity at the work location. Because of their high power, high efficiencies are possible.

Position Transmissions transmit accurately input signals and errors to the work location for control and regulation. Accuracy of information is most important; efficiency can be neglected.

Operation

Depending on function, piping components can have varying significance:

Power Mechanisms transfer power from generator to work location. It is important to have good efficiency over long ranges (e.g. traversing gear).

Force Mechanisms deliver large forces or moments to the work location; efficiency is less important (e.g. presses, shears, clamps).

Feed Mechanisms usually act against only small forces to transmit advance motion with high accuracy in both position and velocity. Efficiency is mostly unimportant (e.g. machine tool feeds, copying controls).

Type of Output Motion

In hydraulic drives, similar and different types of machine may be arbitrarily combined. Systems can therefore be classified as follows, according to the more important output motion designed for individual cases:

Rotary. Where the angle of rotation of the output shaft is infinite.

Reciprocating. Where the angle of rotation of the output shaft is limited to a given value.

Linear. Where output motion is in a straight line.

Function

Self-Controlled Systems. These amplify or distribute forces, or transmit force to remote locations. Input force is the muscular force of the operator (e.g. hydraulic brakes).

Remote-Controlled Systems. These are the more usual type of hydraulic and pneumatic drive. Mechanical energy is input from some external source, transmitted as fluid pressure, and released at the work location. Only control actions such as switching involve operator commands.

Servo Systems. Here input force (simple mechanical force, or forces from measurement) is amplified analogically by the addition of input energy (e.g. turbine governors, power steering and braking systems).

1.3.3 Transmission Layouts

The internal construction of a hydraulic drive is determined by application, operating conditions, and consequent requirements of pumps and motors. Positive-displacement machines consist of a main stator and rotor, the function of which then determines the necessity for the other components (internal/external rotor).

With interconnected drives, pumps and motors may be separated, whereas in compact drives they will be located within the same casing (only hydrostatic drives).

Remote transmissions are, as seen in **Fig. 3**, possible

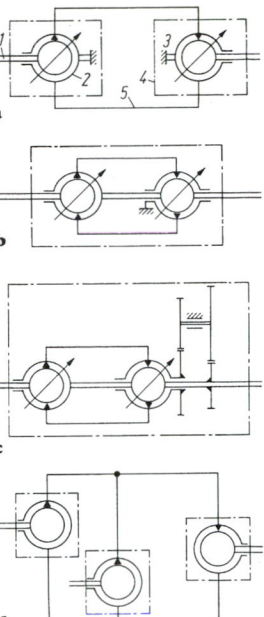

Figure 3. Classification of fluid drives: **a** remote-control stationary transmission (*1* internal rotor, *2* housing, *3* fixed point (control), *4* subassembly, *5* fluid piping), **b** compact hydrostatic circulating drive, **c** combined drive with hydrostatic main and mechanical auxiliary drive, **d** remote-controlled drive assembly.

only as fixed transmissions, i.e. one component of the machine is fixed relative to the foundation (frame). Epicyclic transmissions occur by mechanical connection of machine components in compact transmissions. Coupled transmissions are the combination of hydrostatic transmissions with mechanical (geared) transmissions, and are usually compact. Their construction results from a three-shaft (epicyclic) transmission and a twin-shaft parallel transmission, with which two of the shafts of the epicyclic transmission are connected. Depending on requirements, the main or parallel transmission may be hydrostatic (see VDI Guideline 2151).

Because of the connection between the machines, power can be transmitted between the required components only via fluid flow: *collected transmissions* are supplied by several pumps connected in parallel; in *distributed transmissions* power is transmitted from several motors in series or parallel (differential).

1.3.4 Symbols

In drawings of hydraulic and pneumatic transmissions and control systems, components are represented by graphical symbols. These ignore actual shape and represent only function. The representation in the circuit is at the rest condition of the mechanism, or, if this is not possible, in the home position of the control system.

Shape, meaning and use of symbols are given in ISO 1219. Further symbols, especially for piping (DIN 2429, ISO 4067–1), can also be used. For the most important symbols in use, see **Appendix G, Fig. 2**.

2 Components of Hydrostatic Transmissions

2.1 Pumps

2.1.1 Synopsis

Hydraulic pumps are either rotary pumps (rotary piston) or lift pumps (linear piston) with fixed or adjustable displacement volume (**Fig. 1** and **Table 1**). In practice, application determines the type of pump used. Permissible operating pressure (in economic use and over full service life) is determined by pump type and consequent component loading, etc. The second important criterion is the form of chamber, i.e. the magnitude of the lift volume relative to the size of the machine. With the generally adopted rectangular cell section of rotary pumps, tolerances are more difficult to determine. The pressure-dependent internal leakage losses limit the range of application of low and medium pressure machines. Since cylindrical fits are simpler to manufacture, linear piston machines are

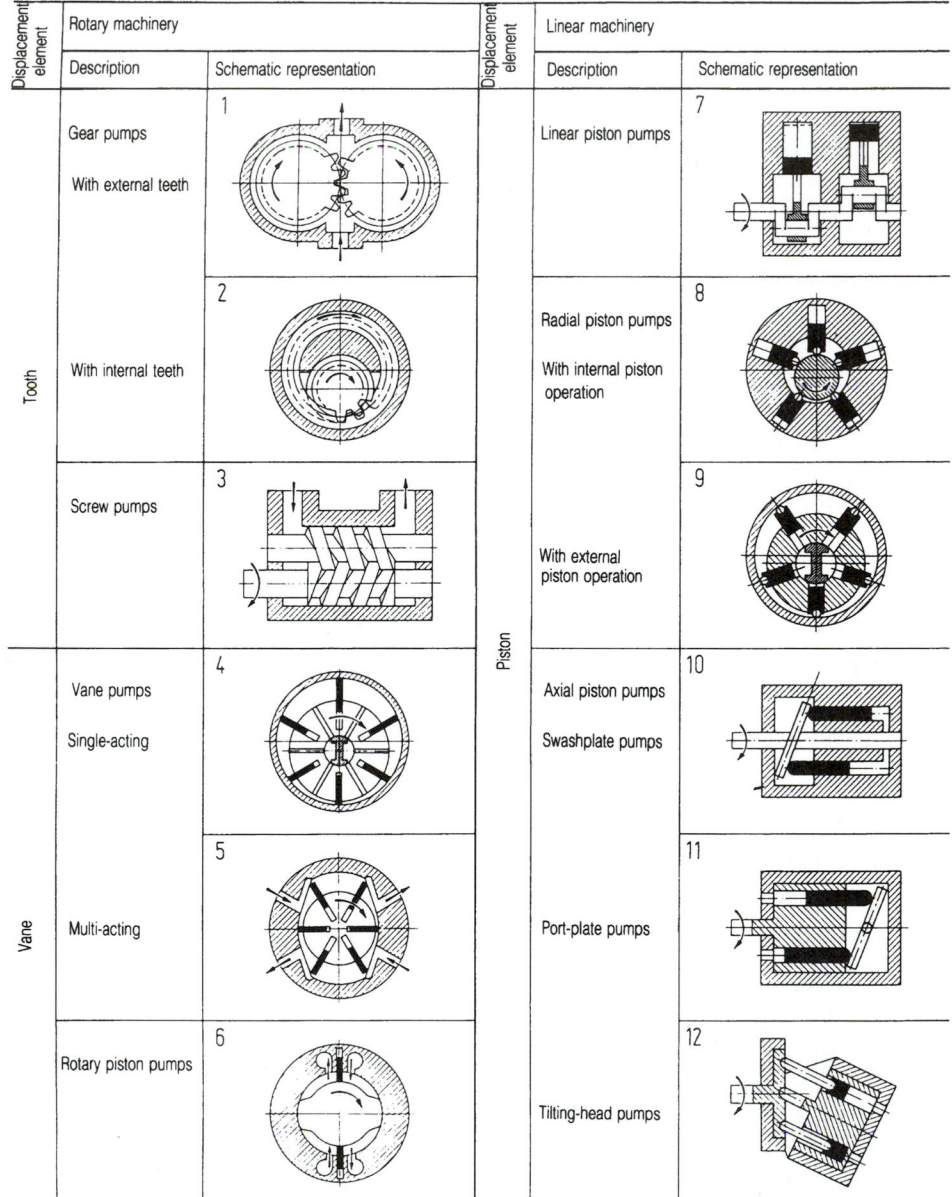

Figure 1. Overview of commercial hydraulic pumps.

Table 1. Common operating values for hydraulic pumps (system numbers as **Fig. 1**)

No.	Displacement volume cm³/rev	Pressure range bar	Speed rev/min	Preferred oil viscosity 10^{-6} m²/s
1, 2	0.4 to 1200	Up to 200 Internal gear pump to 350	1500 to 3000 (Up to 3500)	40 to 80
3	2 to 800	Up to 200	1000 to 5000	80 to 200
4	30 to 800	Up to 100	500 to 1500	30 to 50
5	3 to 500	Up to 160 (200)	500 to 3000	30 to 50
6	8 to 1000	Up to 160	500 to 1500	30 to 50
7	800	Up to 400	1000 to 2000	20 to 50
8, 9	0.4 to 15000	Up to 630	1000 to 2000	20 to 50
10, 11, 12	1.5 to 3600	Up to 400	500 to 3000	30 to 50

more suitable for the higher pressure ranges. In contrast, application has little effect on rotational speed. The general rule is that permissible speed is determined by physical size.

Rotary Pumps

These deliver the working fluid by uniform rotation in a chamber, the volume of which is cyclically altered through the form of the confining walls or the movement force of a vane. The processes of suction and delivery occur concurrently rather than consecutively as would be the case in a piston pump. Adjustable displacement volume is only possible with single-acting vane pumps.

Lift Pumps

These are characterised by the separation of the drive shaft from the pumping cylinder, whose volume is cyclically altered by the action of a linear piston. The delivery volume can be adjusted by the driving gear geometry, or in the control system. Because of the internal flow reversal of the fluid, valves are necessary between the cylinder and the intake and delivery pipes.

2.1.2 Characteristics and Power Rating

The displacement volume, which is equal to the lift volume, V_l, is determined by the geometric properties of the machine, and is usually expressed in cm³/rev. Assuming complete filling of the lift volume at suction, we have:

Theoretical Supply $\dot{V}_{th} = nV_l = \omega V_0$

(n = rotational speed; $\omega = 2\pi n$; V_0 = basic volume = $V_l/2\pi$). For pressure drop between inlet (I) and outlet (O) $\Delta p = p_I - p_O$, supply gives a theoretical pump moment

$$M_{th} = \Delta p V_l/2\pi = \Delta p V_0.$$

The actual relationships between loss-related power transmission are shown in **Fig. 2**. Mechanical drive power

$P_m = M\omega$, because of friction in the mechanism and between compressor elements is reduced by the friction power loss $P_{v,r} = M_r\omega$ to

Displacement Power $P_u = (M - M_r)\omega$.

This is transmitted by the displaced volumetric flow and may be divided into the displacement power P_{th} for Δp and the hydraulic power loss

$$P_{v,h} = \dot{V}_{th}\Delta p_h = M_h\omega.$$

This brings together the flow losses and the (very small) compression work. Thus we have

$$P_m = P_{th} + P_{v,r} + P_{v,h} \quad \text{or} \quad M = M_{th} + M_r + M_h.$$

Both types of loss arise in the mechanism and, since they are dimensionally similar, are jointly expressed in the

Hydro-mechanical Efficiency

$$\eta_{hm} = P_{th}/P_m = 1 - (P_{v,r} + P_{v,h})/P_m$$
$$= 1/[1 + (P_{v,r} + P_{v,h})/P_{th}].$$

The pressure drop Δp causes a leakage flow $\dot{V}_v$ through the gap, so that the displacement volumetric flow is reduced to the

Actual Supply Flow $\dot{V} = \dot{V}_{th} - \dot{V}_v$

and thus affects the power loss $P_{v,v} = \dot{V}_v\Delta p = P_{th} - P_h$.

Volumetric Efficiency. This is given by the equation

$$\eta_v = P_h/P_{th} = 1 - P_{v,v}/P_{th} = 1 - \dot{V}_v/\dot{V}_{th}.$$

The balance of the transfer of mechanical drive power in the hydraulic pump power $P_h = \dot{V}\Delta p$ is brought together in

Total Efficiency

$$\eta_t = P_h/P_m = 1 - \Sigma P_v/P_m$$
$$= (1 - P_{v,rh}/P_m)[1 - P_{v,v}/(P_m - P_{v,rh})] = \eta_{hm}\eta_v.$$

Magnitudes depend on operating conditions, and they are usually expressed as characteristic curves as in **Fig. 3**.

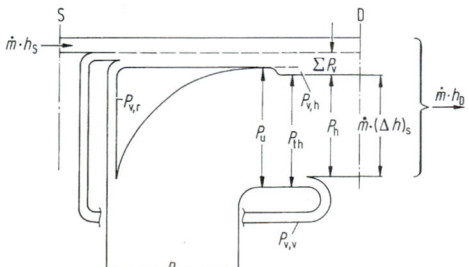

Figure 2. Power flow diagram for a hydraulic pump (see Section G1.1 for an explanation of the nomenclature).

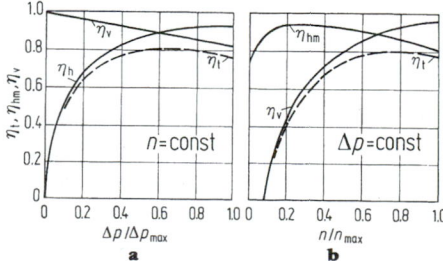

Figure 3. Typical curves of efficiency for constant-flow pump, in relation to **a** operating pressure, **b** speed.

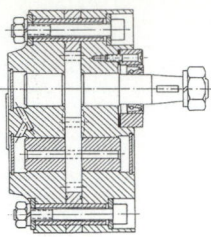

Figure 4. Simple gear pump with plate construction (Robert Bosch, Stuttgart).

2.1.3 Gear-Type Pumps

Gear-type pumps (rotary displacement machines) consist of at least two intermeshing rotors which cause displacement via tooth meshing. Two subsidiary types are gear pumps and screw pumps.

Gear Pumps

These may be further subdivided into external gear pumps, in which at least two externally toothed wheels mesh, and internal gear pumps in which one internally toothed gear wheel meshes with at least one externally toothed gear wheel (**Fig. 1**). There is a further division into simple pumps with one pair of wheels, and multiple units in series (one central drive gear, several secondary wheels) or parallel (several sets of wheels on the same shaft). Multiple units are used to supply several separate circuits (multiflow pumps), or as cut-off pumps.

The *displacement volume* is

$$V_1 = \frac{\pi b}{4} \, [d_{a_1}^2 + d_{a_2}^2 \cdot z_1/z_2 - d_{w_1}^2 \, (1 + z_2/z_1)$$
$$- (1 + z_1/z_2)\pi^2 m^2 \cos^2 \alpha_p/3]$$

(d_a external diameter, d_w pitch circle diameter, z number of teeth, m modulus, b width, α_p flank angle of the basic tooth profile).

The supply flow will have a pulsation rate dependent on the number of teeth. The *irregularity* $\delta = (\dot{V}_{max} - \dot{V}_{min})/\dot{V}$ depends mainly on the number of teeth. Flow pulsation causes pressure changes in the pressure strokes and is the main source of noise.

External Gear Pumps. These usually have involute teeth, with a contact ratio greater than unity. The volume pressurised by the meshing teeth is discharged via grooves in the compression space. There are normally two equal-sized wheels. The irregularity $\delta = 25$ to 10% for numbers of teeth z between 9 and 20.

Figure 4 illustrates plate construction. Usually simple journal bearings are used, but roller bearings may be used in larger pumps. Operating pressure (normally less than 100 bar) can be increased to more than 200 bar by press-

ure compensation using pressure fields made up of movable bearing elements. Efficiency in continuous use is $\eta_t = 0.85$ to 0.75 (normal performance) and up to $\eta_t > 0.9$ with pressure charging.

Internal Gear Pumps. These have better mesh ratios and thus an irregularity in delivery flow of only $\delta \approx 3$ to 5%. Extending the inlet and outlet zones over a greater range of angles ensures good filling and delivery ratios and low noise. Ring pumps consist of two wheels, where the number of teeth on the internal gear wheel is 1 greater than on the pinion. Operating pressure is below 100 bar. High-pressure pumps have a sickle-shaped seal between the rotors with operating pressure < 350 bar.

Screw Pumps. These have pulsation-free and low-noise characteristics (normal construction: see **Fig. 1**), and run at higher speeds and pressures. They are used in lift machinery and fine-work machine tools. Problems are manufacturing cost, relatively lower volumetric efficiency (< 0.8%), and delivery of high-viscosity fluids.

2.1.4 Vane-Type Pumps

With vane pumps (typical construction: see **Fig. 1**), the compression stroke is divided by a number of vanes that can retract into the rotor (vane pump) or the stator (rotary piston pump). The compression stroke is formed by the relative motion between rotor and stator. Their advantages over gear pumps are: small pulsation, low noise, low specific weight (0.4 to 0.6 kg/kW), and higher operating speed.

Vane Pumps

These operate on the principle of a rotor eccentrically mounted within a casing and provided with radial slots in which vanes slide (**Fig. 5**). These vanes are compressed against the outer wall by centrifugal force, spring pressure, or the working pressure, to form cells that expand and contract with the movement of the rotor.

It follows that the compression volume is

$$V_1 = 4\pi r_m e b.$$

Adjustable pumps are constructed so that the eccentricity e can be altered while in operation, and the supply for a given pump speed and direction varied or even reversed.

Multi-vane, single-stage pumps work up to a maximum operating pressure of 160 (250) bar. By arranging two opposing inlet and outlet chambers, radial force on the rotor can be cancelled out. Multiple-vane pumps reach efficiencies of 0.85 to 0.9 or more, while single-vane pumps are around 0.6 to 0.85.

Rotary Piston Pumps

These work on the principle of stationary vanes and a rotating cam. The rotor has two curved, polished cams set at 90°, which rotate in two opposed chamber rings,

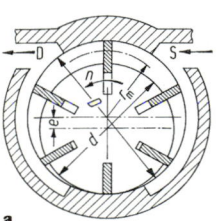
a

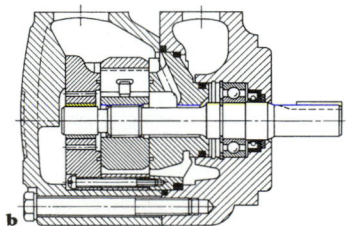
b

Figure 5. Vane pumps: **a** flow diagram, **b** example of construction.

separated by a partition. Operating pressure is 175 bar, and efficiency > 0.9.

2.1.5 Piston Pumps

Piston pumps (positive-displacement machines) have advantages over rotary machines in their lower leakage losses owing to good cylinder sealing, and the possibilities for use as variable-displacement pumps with high operating pressures by altering the drive geometry. Because of the small displacement, the pistons are usually driven by cams (eccentrics). The drive can be applied to either the central shaft or the cylinder block. Control is by slide or conventional valves. Slide valves are commonly used in rotary block designs; otherwise axial and cylindrical designs have rotary slide valves or eccentrically driven longitudinal slide valves. Disadvantages are the leakage losses and the higher noise levels due to compressive shock with the necessary connection of the cylinders with the pressure channels. If on the other hand the self-controlled valve opens when the pressures are equal, the pumps run more quietly, especially at higher pressures and have a higher volumetric efficiency because of the positive valve seating. However, it is not possible to reverse the flow, and at higher speeds valve overlap leads to poorer volumetric efficiency. The flow from a single-cylinder pump is basically sinusoidal. The equal division of supply from a multi-cylinder pump

$$\dot{V}_{th} = nV_1 \quad (n = \text{rotational speed})$$

is overlaid by a flow pulsation, the irregularity of which (δ) depends on the number of cylinders:

i	3	4	5	6	7	8	9
δ%	14	32.5	5	14	2.5	7.8	1.5

The *displaced volume* is $V_1 = iAH$ (i number of pistons, A piston area, H piston lift).

Pumps are therefore designed with odd numbers of cylinders. The most important types are radial and axial pumps. Series piston pumps are not widespread; they are however found as variable-displacement pumps with swashplate control (Bosch Presspump, cf. diesel injection equipment) in test machinery.

Radial Piston Pumps

In this type the pistons move along radii of a circle, the centre of which is a rotating shaft. Pumps with internal eccentrics, driven by an eccentric shaft, and with a fixed piston block, have valve or slide valve control. This is the preferred structure for constant-displacement pumps, but some very large pumps are still controlled by adjustment of the eccentric drive (Exzentra, Stuttgart). The use of twin cylinder blocks side-by-side driven by 180° opposed eccentrics facilitates drive-shaft balance. Drives with external eccentrics are used exclusively with rotating cylinder blocks and internal control (port control).

Constant-Displacement Radial Piston Pumps

With these the eccentric is enclosed in a roller bearing, the outer ring of which acts on the pistons by friction during the pressure stroke. The kinematically limited sliding occurs only during the suction stroke. Depending on the form of the load ring, there can be up to eight equal-sized cylinder units in one plane, connected together via a peripheral channel or in groups (multi-flow pump) leading to the output. Control is by spring-loaded needle valves on the suction side, and ball valves on the pressure side. As the casing also defines the suction area, a smaller supply is necessary. Operating pressures of up to 600 bar are employed.

Variable-Displacement Pumps. Figure 6 shows the main type of construction, in which the cylinder block rotates about a fixed central shaft. This shaft has twin channels bored through it for the oil supply, and forms a control slide valve in the working plane. The pistons are transversely loaded by a crosshead drive, and are held by gudgeon pins and slides against the outer race, which is driven by the roller bearing and rotates with it. The external eccentric can move relative to the casing pins, to give stepless adjustment of the eccentricity from $+ e$ to $- e$, and with it the magnitude and direction of the supply for any given rotational speed.

The rotational speed is limited because of the considerable inertial forces, which are additional to the pressure forces acting on the race. Working pressures range up to 450 bar, with a typical mid-range efficiency of 90%.

Axial Piston Pumps

The pistons of the axial piston pump are arranged in a circle parallel to the axis of the cylinder drum and are driven by the motion of an inclined plate (adjustable in the case of variable-displacement machinery) relative to the cylinder (see **Fig. 1**). The following layouts are possible:

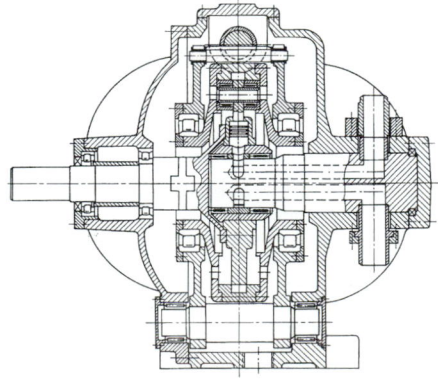 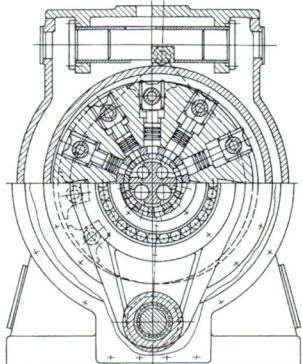

Figure 6. Adjustable radial piston pump with external piston operation and internal control (Wepuko Hydraulik, Metzingen).

Swashplate Pumps. Driveshaft and cylinders are coaxial; the cylinders are fixed and the shaft transmits motion via a swashplate (wobbleplate). These pumps are usually valve-controlled, in which case the flow direction is irreversible, but the shaft can turn in either direction. Mainly used as constant-displacement pump.

Port-Plate Pumps. The driveshaft and the cylinder are coaxial, with the cylinder driven by the driveshaft. The inclined port-plate is fixed. Valve control is by the cylinder; flow direction is reversible without reversing direction of rotation.

Tilting-Head Pumps. The driveshaft is tilted relative to the cylinder, with the cylinder and bearing plate driven. Valve control is by cylinder, with flow direction reversible (see **Fig. 7a, b**). Operating pressures vary from 180 to 220 bar, with peak pressure over 400 bar, and drive speed up to 3500 rev/min, depending on size.

Typical performance and efficiency curves are shown in **Fig. 7c**. The displacement volume is given by

$$V_l = iAH = i\pi(d^2/4)D \sin \alpha$$

(d piston diameter, D diameter of piston assembly, i number of pistons, α tilt angle).

2.2 Hydraulic Motors

Depending on output motion, hydraulic motors can be classified as rotary reciprocating (limited angle of rotation) or linear (cylinders). Rotary motors follow all the constructional principles of rotary compressors and valve-controlled piston pumps detailed in Section G2.1. They usually have constant-displacement volume, and are only used as variable-displacement machines in exceptional cases. The power balance for a hydraulic motor (G2.1.2) is shown thus:

The *hydraulic power* $P_H = \dot{V}\Delta p$ is reduced by the *leakage losses* $P_{v,v} = \dot{V}_v \Delta p$ from the *theoretical power* $P_{th} = \dot{V}_{th}\Delta p = (\omega V_0 M_{th})/V_0$. Therefore:

Volumetric efficiency: $\eta_v = P_{th}/P_H = 1 - (\dot{V}_v/\dot{V})$.

The *hydraulic power loss* $P_{v,h} = \dot{V}_{th}\Delta p_h = \omega M_h$, and the *mechanical power loss* $P_{v,r} = M_r\omega$ can be combined as $P_{v,rh} = P_{vr} + P_{v,h}$.

Mechanical motor power $P_H = P_{th} - P_{v,rh} = P_{th}\eta_{hm} = M\omega$.

Overall efficiency $\eta_t = P_m/P_h = \eta_v\eta_{hm}$.

For the distribution of losses and actuating variables, see the notes in Section G2.1.2.

Gear Motors

These have poor starting characteristics under load and their range of application is limited to higher speeds. For slow-running applications, gear motors with flanged-gearwheel reduction gears ($i_G = 6$ to 18) are acceptable. Internally toothed gears without separators are better. The inner rotor has one tooth fewer than those on the outer ring. When both rotate, control is by fixed, sickle-shaped grooves (Gerotor). With a fixed outer ring, the inner wheel performs additional peripheral motion, controlled by rotary slide valves (Orbit).

Vane Motors

Vane-type motors can be used as high-speed units, possible with reduction gearing, or as slow-running units with multiple admission.

Piston Motors

All types of slide-valve controlled axial and radial piston pumps are equally good as hydraulic motors. They can effectively be classified in speed ranges:

Slow-speed	$n = 1$ to 150 rev/min
Medium-speed	$n = 10$ to 750 rev/min
High-speed	$n = 300$ to 3000 (6000) rev/min

For high speeds, axial piston motors are preferable, with reduction gears for lower output speeds. Slow-speed motors are usually radial. For a given drive torque they have a lower moment of inertia and therefore have better dynamic properties than geared motors. Some irregularity is noted at lower speeds. Pressures and efficiencies are as for pumps.

Reciprocating Motors

These work over limited angles (maximum $720°$), and the reciprocating motion is either produced directly (vane motors, with a moving vane in a divided annular cylinder, working angle $< 300°$), or indirectly from a linear piston via gearing (for rack and pinion drive see **Fig. 8**).

Linear Motors

These can be either single- or double-acting, depending on cylinder design. Single-acting cylinders, as their name implies, can exert force only in one direction. Piston rod

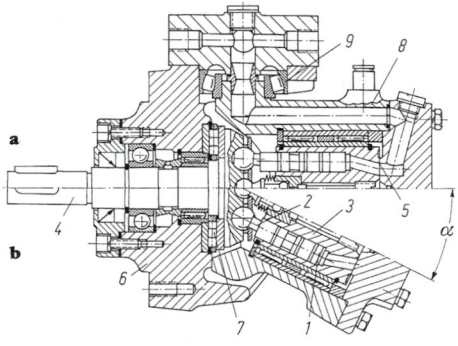

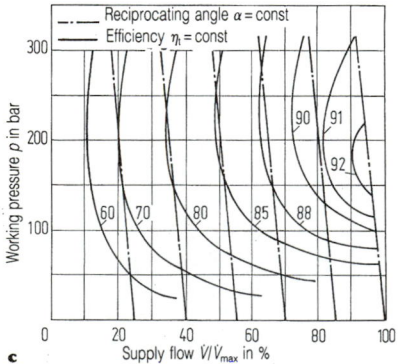

Figure 7. Adjustable tilting-drum axial-piston pump (Mannesmann-Rexroth GmbH, Horb): **a** zero cutting, **b** flow conditions at tilt angle α_0 set at $90°$ (*1* piston, *2* piston rod, *3* cylinder block, *4* shaft with drive flange, *5* control surface, *6* bearing flange, *7* axial cylinder roller-bearing, *8* cylinder housing, *9* reciprocating bearing), **c** curves of flow and efficiency.

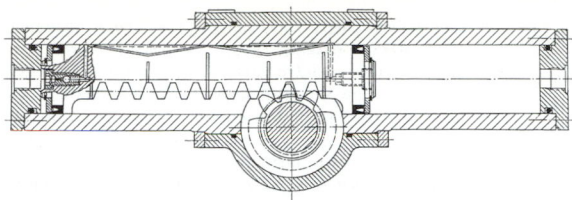

Figure 8. Reciprocating motor with straight-line working piston motion and rack-and-pinion operation.

and piston are one unit and sealed in the piston rod guides. Stroke is about 2.5 × piston rod diameter, and return force has to be provided by external means, or by a spring. Double-acting cylinders can be pressurised on either side to provide motion in both directions. Piston area on the piston-rod side is smaller by the area of the piston rod A_{St} than the piston area A_k, so that different forces are produced in either direction, or different speeds are obtained for the same pressure and suction flow. The area ratio is $\varphi = A_k/(A_k - A_{St})$.

Calculations: *Piston force*: stroke $F_D = \eta_D p A_k$, return $F_z = \eta_z p A_k/\varphi$.

Speed: $v_D = \dot{V}/A_k$ $v_z = \dot{V}/(A_k - A_{St}) = v_D \varphi$.

For rapid traverse (bottom and piston-rod side strokes equally pressurised, effective area therefore equal to that of the piston rod):

$$v_E = \dot{V}/A_{St} = v_D \varphi/(\varphi - 1).$$

Losses due to pressure-dependent sealing friction F_r and inflow pressure losses Δp_h are incorporated into the cylinder efficiency (A = operational area):

$$\eta = [(p - \Delta p_h)A - F_r]/pA.$$

In double-acting cylinders, for outward stroke $(A = A_k)\eta_D = 0.9$ to 0.95, for return stroke $(A = A_k/\varphi)\eta_z = 0.85$ to 0.9, and for rapid traverse 0.2 to 0.4. For stroke terminal velocities greater than 0.1 m/s, thrust buffers are necessary.

Main cylinder dimensions are defined by piston diameter $d_k = (12$ to $400)$ mm as well as φ values of 1.25; 1.6; 2; 2.5; and 5.

Structural Guidelines. Cylinders must not be load-bearing, and must experience no bending moments or transverse forces. The load should be supported by the shortest path compatible with function so as to avoid deflection. Examples of correct mounting are shown in **Fig. 9**.

2.3 Valves

These are switching devices in the flow of hydraulic power between pumps and motors with either on–off or adjustable action.

Classification by Function: *directional control valves* (directing oil flow); *shut-off valves* (stopping flow); *pressure control valves* (regulating pressure); *flow control valves* (regulating flow).

Classification by Principle of Operation: *seat valve* (sealing element can be a ball, cone, or plate; leak-proof sealing) or *slide valve* (rotary or longitudinal slide; versatile as shutoff valves).

Classification by Type As *single valves* in piping systems, as *block valves* with equal size and continuous main channels running through them forming a single block, as *valve blocks* with several valves in the same casing, and as *plate valves*. Plate valves are widely used, as their simple construction and interconnection permit their use in circuit blocks (see **Fig. 11**).

2.3.1 Directional Control Valves

Into this category come all valves that transfer external positional movements between the connections and thus determine the course and direction of the oil flow. In the majority of cases they have pure switching functions (on–off). It is possible to control the flowrate through throttling (static position function), but because of the associated losses this is only suitable for low-power equipment.

Valves can be designated according to the number of switched connections and the number of switch positions (e.g. 4 connections with 3 switch positions gives a "4/3 valve").

Valve connections are designated as follows: P pressure connection; L leakage connection; A, B, work connections; R, T discharge connections; Z, Y, X, control connections.

Seat Valves. These are insensitive to type of fluid used and to contamination, very reliable, and suitable for high pressures. Their disadvantages are high operating forces, and the necessity for individual operation of the closing components so as to ensure proper closing. With directly controlled valves inlet, nominal diameter is limited to < 4 and switching function is simple (2/2 and 3/2 directional control valves). Large sections (often up to nominal diameter 100) are possible with indirect operation. Modular construction (cartridges) is possible with 2/2 directional function (**Fig. 10**), allowing extensive circuits within a

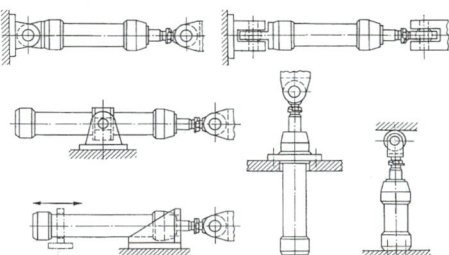

Figure 9. Examples of cylinder layouts.

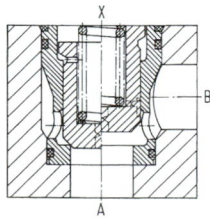

Figure 10. Modular 2/2 distribution valves. A, B working connections; X control connection.

single, suitably bored-out block. Operation is by an external control valve with X-connections.

Slide Valves. These are very common because the pistons can switch several ways simultaneously and can therefore permit different switching arrangements depending on design. Their construction principle is illustrated in **Fig. 11**. Piping connections are made by means of drilled or cast channels in the annular slot. For each position, the slotted slide piston makes a connection between different pipe fittings. Static pressures are equalised by equal areas in the slide chamber, flow forces by slot design. Opening characteristics are influenced by indents on the piston shoulder, while alteration of the switching characteristic is possible by chamfering the piston (e.g. continuous two-way connection). Opposed location of the control edges of the slide and casing slots (overlap) affects switching characteristics. With negative overlap, more strokes are briefly interconnected, i.e. there is a danger of unintentional activation of the motor, but there is also better sensitivity to flow control and building of pressure peaks in switching moving masses. Positive overlap gives better protection against leakage losses.

Operating pressures of slide valves range up to 350 bar. Leakage losses at high pressures are noticeable, so that, e.g. running motors are safer with shuttle valves. Some exhibit greater flow resistance (nominal 3 to 8 bar, according to manufacturers' data).

Operating Methods for Directional Control Valves

Valves are provided with and without preferred switching positions; the so-called impulse valves remain in the switched condition after removal of the control command (memory function), otherwise they move back to the rest position under spring pressure, or, in large installations, hydraulic pressure. Valves are switched manually or mechanically, by hydraulic or pneumatic pressure, or electro-magnetically. Direct electromagnetic operation is limited to about 3 kW hydraulic power because of the relatively small magnetic switching force. Larger valves are operated by oil pressure via small flanged pilot valves taking oil either from the working circuit (self-controlled) or from a distant source (remote-controlled). Operating pressures are up to ≈ 4 bar. Magnetic switches can be either dry (operating against oil pressure), or wet (submerged in oil), for either AC or DC. Voltages are usually 24, 48, 180, or 220 V. Maximum switching power is about 100 W.

2.3.2 Shuttle Valves

Shuttle valves allow flow in only one direction. Construction is on the seat-valve principle, the simplest form being a spring-loaded ball valve. Because of their leakproof closure, shuttle valves are often used as stop valves for cylinders under load. In such cases, pressure relief is via auxiliary pistons, controlled by auxiliary cones for large cross-sectional areas (see **Fig. 12**). Opening pressure against spring can vary between 0.5 and 3 bar. In feeder valves, which require very low opening pressures, operation occurs by the weight of the cone itself, and they can therefore only work in the vertical position.

2.3.3 Pressure Control Valves

The essential characteristic of these valves is a special function whereby switching occurs when a predetermined pressure is reached, usually against spring resistance. The connection can be continuous (via change of throttle cross-section) or discontinuous (switching).

Pressure-Limiting Valve. With this type of valve, oil is allowed to flow freely into the tank up to a predetermined pressure, after which any slight increase in pressure rapidly opens the throttle and so limits system pressure. With directly controlled pressure-limiting valves, oil pressure lifts the cone off its seat against a spring force. Pressure increases sharply with flow, giving an operational limit above the "saturation" point, depending on the preset pressure. Alternating static and dynamic forces acting on the cone tend to produce flutter, but this is overcome by damping.

Preset Valves. For large flows, these are constructed as shown in **Fig. 13**. The main cone is held in the "closed" position by a weak spring and backpressure, until a small, directly controlled pressure-limiting valve opens and pressurises the rear face (with throttle between admission and rear face). When the rear face is pressurised via an additional 2/2 directional control valve, the preset valve can function as a rotary valve.

Remote control can be effected via connection of further preset and loading valves at X.

Pressure Control Valves. These maintain constant downstream pressure independent of higher upstream

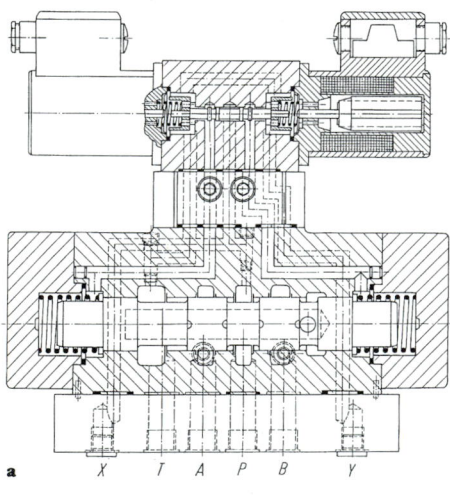

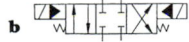

Figure 11. Preset 4/3 distribution valve with electromagnetic operation (Mannesmann-Rexroth, Lohr): **a** design, **b** symbol.

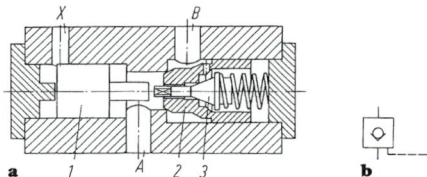

Figure 12. Resettable non-return valve (Mannesmann-Rexroth, Lohr): **a** design; *A*, *B* working connections, *X* control connection; *1* release piston, *2* main cone, *3* preset cone; **b** symbol.

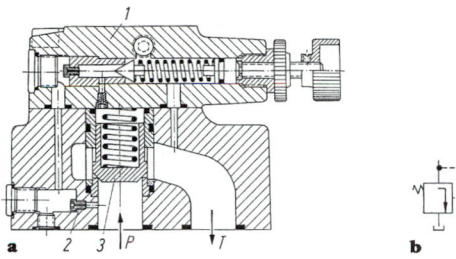

a 2 3 P T **b**

Figure 13. Preset pressure limiting valve for plate construction (Mannesmann-Rexroth, Lohr): **a** design (*1* preset valve, *2* throttle, *3* main cone), **b** symbol.

pressure by throttling, if necessary by additional release of outflow (3-way pressure control valve).

Pressure Switching Valves. With these valves, new paths for the working fluid are opened when the predetermined pressure is reached. The self-controlled type continue to switch in a subordinate circuit while maintaining pressure in the main circuit (emergency valves, servo valves). Remote-controlled valves, depending on pressure, switch to another working circuit, or allow depressurised free circulation in it (cutoff valves, load-storage valves).

2.3.4 Flow Control Valves

Flow control valves are the simplest way of controlling the rate of motion in a hydraulic drive. In principle, they are continuously operating throttle valves with adjustable throttle area. Depending on this area, and on the square of the flow velocity, a pressure difference arises in the valve that is part of the total pressure drop in the system.

Single Throttle Valves. Control is from outside and is therefore unrestricted (see **Fig. 14**). With fixed pressure level and given motor load the residual pressure drop over the throttle remains at a level equivalent to a fixed pressure flow. Changes in circuit pressure or motor load cause changes in flow. Throttling should occur over the shortest possible section (aperture), otherwise there is a strong oil viscosity (temperature) effect. When it is important to maintain constant working speed, flow regulators should be used. Here, with flow as the measured variable, there is a limited adjustment of the throttle section, and hence a readjustment of the valve pressure drop for constant flow.

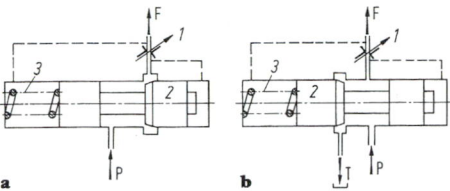

a P **b** T P

Figure 15. Schematic of flow regulators: **a** 2-way design, **b** 3-way design; P pressure connector, F working connector, T outlet connector, *1* measurement throttle, *2* throttle piston, *3* spring.

Two-Way (Main) Flow Regulators. Construction is as shown in **Fig. 15**. The flow-measuring orifice *1*, adjustable for flow strength, generates a pressure drop Δp_m (approx. 3 bar), balanced by throttle piston *2* against spring *3*. This is adjusted for a pressure drop $\Delta p_k = (p_p - \Delta p_m) - p_F$. Motor load fluctuations first produce small changes of $\dot{V}$. The measurement orifice pressure drop altered by $\Delta p_m'$ thus moves *2* to a new throttle position $\Delta p_k'$, which then makes Δp_m = const., i.e. $\dot{V}$ = const. The pump supply excess then flows back into the tank through the pressure-limiting valve.

Only unidirectional flow is possible in a flow regulator; if flow control in both directions is required, non-return valves must be used. With large motor fluctuations, the controller should be on the discharge side.

Three-Way Flow Controllers. These provide constant motor inlet flow by diverting excess flow from the pump (see **Fig. 15**). Construction is mainly as described above, except that the throttle piston *2* opens an additional outlet. It is only possible to use these on the motor inlet side. Accuracy is 2 to 5%.

Flow Distributors are constructed on similar principles. The pressure drops in two parallel measurement orifices are balanced by a single throttle piston, which then throttles both or if necessary the differentially loaded motor branches. A very good flow distribution for both flow directions is possible by parallel circuits of two-gear motors with mechanically coupled shafts.

2.3.5 Proportional Valves

Magnetically operated distribution valves act only with a discrete switching function (i.e. digitally). With proportional valves, analogue conversion of electrical signals into hydraulic actions is possible. The magnetic force determined by the feed current is balanced against the effect of pressure or a pressure drop, and thus a proportionality between the electrical input parameter and the output pressure or flow strength is established. There is no feedback between mechanical components (cost-

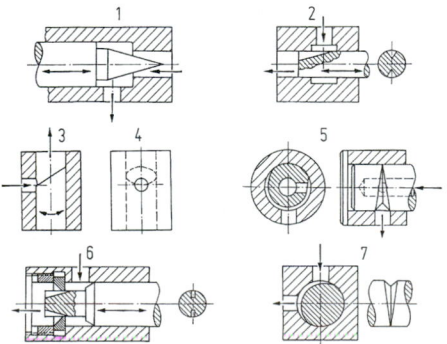

Figure 14. Common throttle designs.

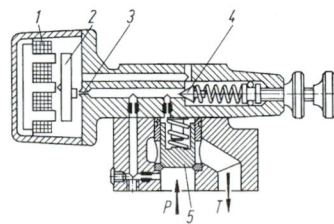

P 5 T

Figure 16. Proportional DBS pressure valve (Mannesmann-Rexroth, Lohr): *1* magnetic system, *2* baffle plate, *3* nozzle, *4* preset valve, *5* main piston.

effective compared with servotechnology). Reproducible accuracy is ± 2%. Construction and function are as shown in **Fig. 16**. The magnetic force operates on the nozzle–deflector-plate system and produces a control flow proportional to chamber pressure (i.e. system pressure). Function is affected by flow strength, thereby limiting application, e.g. as a precontrol unit for a pressure-limiting system.

2.4 Hydraulic Equipment

For connecting pumps, motors, and valves (hydraulic circuits), tubing of steel (DIN 2391, ISO 3304, DIN 2413, ISO/DIS 10 400), or synthetic rubber with fibre reinforcement, pipe couplings (DIN 2367), oil containers ($V_T \sim (3$ to 5) capacity per min), filters (fineness 10 to 30 to 60 μm), and hydraulic accumulators are necessary [3].

3 Structure and Function of Hydraulic Transmissions

3.1 Hydraulic Circuits

The circulation of pressurised fluid in a hydrostatic drive is referred to as a circuit, which may be open or closed, and with or without a feed pump. A circuit should be protected against overload by at least one pressure-limiting valve, while closed motor circuits with small permissible loadings should have their own pressure-limiting valves downstream of the distribution valve.

3.1.1 Open Circuits (Fig. 1a)

Open circuits incorporate an oil reservoir. The pumps always feed in the same direction, and the outlet flow from the motor back to the tank is unpressurised. Changes in the working direction of the motor are brought about by diverting the flow through a four-way distribution valve. Hydraulic circuits with fixed-displacement pumps may or may not have unpressurised circulation under idle conditions, but variable-displacement pumps usually revert to zero supply. Advantages of open circuits are the way in which excess heat is carried away by the flow, as well as cooling and cleaning of the oil in the tank. The

main disadvantage is the unidirectional flow. The motor braking power, which is developed when for example load falls off, can only be offset by throttling at outlet (exit flow throttle valve, or, with high flow rates, a special brake valve).

3.1.2 Closed Circuits (Fig. 1b)

In a closed circuit, low-pressure oil from the motor outlet is led back directly to the pump suction side. Oil can flow in either direction; that is to say, not only is energy supplied to the motor by the pump, but the motor can also act as a brake on the pump and so feed back to the prime mover. Thermal loading is considerably lower than in open circuits for this reason.

Change of motor direction is either by distribution valve (unidirectional circuits) or by changing the direction of pump supply with a reversing mechanism (circuits with variable flow direction). Closed circuits have a feed pump that pressurises the appropriate low-pressure circuits to 3 to 8 bar. This is to avoid cavitation on the suction side of the main pump, to reduce leakage losses in the main circuit, and to exchange feed-pump excess flow with that in the main circuit for cooling and cleaning (without a high-pressure cooler and filter, a flush valve is needed; 5 in **Fig. 1b**). Feed flow is about 10% of main flow. Depending on flow direction, the circuit is protected by one or two cross-connected pressure-limiting valves. For idle flow, the main pump circulation valve is used (with variable machines for about 4% of $\dot{V}_{max}$).

3.1.3 Semi-closed Circuits

With closed circuits incorporating hydraulic cylinders, the different bottom and shaft-side volumes have to be taken into account. Depending on working direction and area ratio φ, large differential flows may have to be introduced or removed from the circuit. Flush valve and feed pump are calculated accordingly, and if necessary a back-pressure valve 2 is provided for the main pump (**Fig. 1b**).

3.2 Operation of Hydraulic Transmissions

3.2.1 Starting Process

The pressure in the working circuit is controlled primarily by motor load: load pressure is Δp_F. To this must be added the pressure associated with overcoming mechanical friction Δp_r, flow losses in pipes and valves Δp_h, and, where flow velocity is not constant, with accelerating masses Δp_a. The pump pressure $\Delta p_p' = \Delta p_F + \Delta p_r + \Delta p_h + \Delta p_a$ is limited to Δp_{PLV} by the pressure-limiting valve. A greatly simplified representation of the starting conditions is

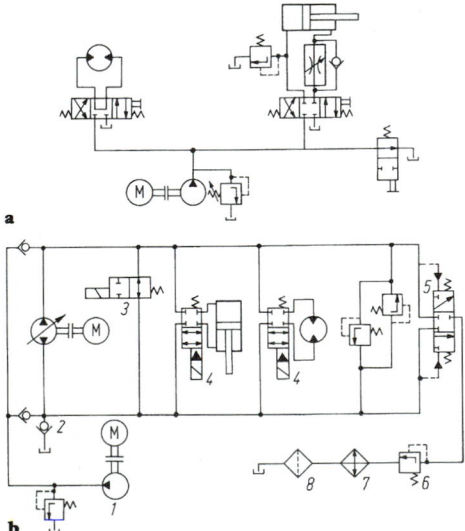

Figure 1. a Open circuit with rotary motor and flow control hydraulic cylinder connected in parallel; unpressured circulation in idle condition. **b** Closed circuit with variable flow direction with rotary motor and hydraulic cylinder; *1* feed pump, *2* return valve, *3* circulating valve, *4* motor distribution valve, *5* flushing valve, *6* preset valve, *7* cooler, *8* filter.

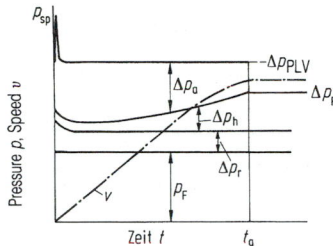

Figure 2. Start-up of hydraulic drive. t_a acceleration time; p_{PLV} pressure-limiting valve adjustment; p_F load pressure; p_r friction pressure, p_h hydraulic losses, p_a acceleration pressure; p_P, p_P' pressures at the pump; v velocity.

given in **Fig. 2**. As long as the motor has not reached the speed corresponding to the pump supply flow, the excess flow is diverted to the pressure-limiting valve, and the pressure in the circuit is Δp_{PLV}. The flow losses Δp_h increase with supply flow $\Delta \sim \dot{V}$ or $\sim \dot{V}^2$; motor acceleration arises from the pressure Δp_a available after subtracting the usable loading pressure Δp_F and the pressure used in overcoming friction Δp_r, up to the limiting pressure Δp_{PLV}.

A more accurate consideration takes account of the energy involved in oil compression and in elastic deformation of circuit components, especially in circuits with large oil contents or with elastic components (reservoirs, hoses), which lead to transient effects and increased acceleration times. Once the motor has reached its final speed, pump pressure falls back to Δp_p. When the motor is suddenly started from idle, a pressure peak p_{sp} occurs before the pressure-limiting valve can respond because of the mass effect of the oil flow, the pump and the drive motor. This peak can be reduced by elastic components (hoses, reservoirs), and rapidly reacting pressure-limiting valves, but is most safely counteracted by the rise-delay time associated with the opening characteristics of the circulating valve.

3.2.2 Formal Description

Under steady conditions, the motor generates *mechanical power* $P_{ab} = P_{mM} = M_M \omega_M$, or $F_M v_M$ to overcome working and friction conditions, and therefore a *hydraulic power* $P_{hM} = \dot{V}_M \Delta p_M = P_{mM}/\eta_{tM}$. The *operating power* $P_{zu} = P_{mP}$ is converted to *hydraulic power* $P_{hP} = \dot{V}_p \, \Delta p_p = P_{mP} \eta_{tP}$, and covers the *transmission losses* (pressure Δp_{hL} and flow losses $\dot{V}_v$) in the circuit apart from the *motor power*:

$$\dot{V}_M = \dot{V}_P - \dot{V}_v = \dot{V}_P(1 - \dot{V}_v/\dot{V}_P) = \dot{V}_P \eta_{v\ddot{u}},$$

$$\Delta p_M = \Delta p_P - \Delta p_{hL} = \Delta p_P \eta_{h\ddot{u}}.$$

Overall Efficiency

$$\eta_t = P_{ab}/P_{zu} = \dot{V}_M \cdot \Delta p_M \eta_{tM} \eta_{tP}/\dot{V}_P \cdot \Delta p_P = \eta_{v\ddot{u}} \eta_{h\ddot{u}} \eta_{tP} \eta_{tM}.$$

The heat generated by flow friction losses $\dot{Q}_\vartheta = P_\zeta$ $(1 - \eta_t)$ must be removed by convection in the pipes or the tank, or if necessary by an additional cooler. The allowable excess temperature of the oil over ambient is around 30 to 50 K. The definitions

Speed Ratio

$$\nu = \omega_M/\omega_P = (V_{OP}/V_{OM}) \, (1 - \dot{V}_v/\dot{V}_P) \, \eta_{vP} \eta_{vM},$$

Torque Ratio

$$\mu = M_M/M_P = (V_{OM}/V_{OP}(1 - \Delta p_{hL}/\Delta p_P) \, \eta_{hmP} \eta_{hmM}$$

show that the gear ratio is affected (also during operation) by two factors:

(a) Changing V_{OP}/V_{OM} = variable-speed drive,
(b) Changing $\dot{V}_v/\dot{V}_P$ = throttle-controlled drive.

3.3 Control

3.3.1 Variable-Speed Drive Units

Depending on type of control, these may be classified as:

Primary Control. Pump adjustable from zero to maximum feeds a constant-displacement motor.

Secondary Control. Pump delivers a constant supply to a variable-speed motor.

Compound Control. Both components are controllable; control can be series or parallel.

Primary-control drives are very widespread, though the use of primary–secondary drives is increasing. The working diagram is shown in **Fig. 3**. During the primary phase, motor torque M_M and circuit pressure Δp remain constant. From startup, the pump, running at fixed speed, delivers a flow $\dot{V}_P$, which increases by adjustment, and the drive speed n_M and power P increase proportionally. At n_1, the pump is at maximum, and further increase of output speed is possible only by secondary control, i.e. by reducing the motor volume V_{OM}. When maximum pressure Δp_{max}, and hence constant power P, is reached, this adjustment produces a hyperbolic decrease in motor torque M_M. If the drive power is less than the "gear-angle power" $P_E = \Delta P_{max} \dot{V}_{max}/\eta_{tP}$, then there is a power limit on primary control, which only happens with a corresponding pressure reversal.

Secondary control takes place via speed-controlled or positionally controlled hydraulic motors, fed from pressure mains (as comparable electric motors are fed from electric mains). Pressure-controlled pumps and if necessary hydraulic accumulators keep circuit pressure constant. Load torque adjustment at constant speed is achieved by adjustment of motor swept volume, i.e. the power change produces different values of motor displacement. Advantages over the flow controlled machinery described above are: better time response, especially over greater distances; parallel connection of several units possible without mutual interaction; braking energy fed back into circuit.

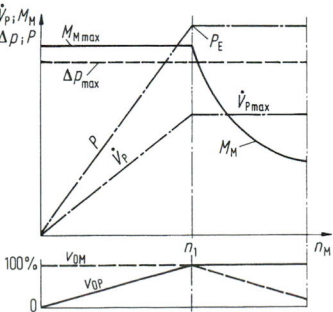

Figure 3. Characteristics of a drive with primary–secondary control.

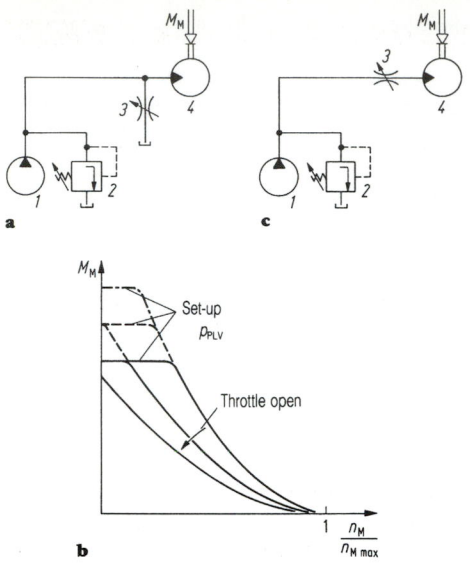

Figure 4. Throttle-controlled drive, circuit and drive characteristics: **a** bypass throttle drive; **b** drive characteristic for **a**; **c** main flow throttle drive. M_M load moment, *1* pump, *2* pressure-limiting valve, *3* throttle, *4* motor.

3.3.2 Throttle-Controlled Drives

For low powers (< 5 kW), hydraulic drives with low-cost fixed-displacement pumps and motors are more economic, though of course their working speeds have to be controlled by diverting flow from the main circuit. With bypass throttled circuits (**Fig. 4a**), the diverted flow is led from the throttle into the tank. As motor loading M_M increases, the circuit working pressure also increases and the diverted flow, or in other words the output speed n_M, is reduced. Flexibility is affected by throttle setting, and maximum torque by the setting of the pressure-limiting valve.

With primary-flow throttling as in **Fig. 4c**, the fraction $\dot{V}_M$ of the pump supply flow reaches motor *4* via throttle *3*, and the outlet flow $\dot{V}_v$ returns to the tank via pressure-limiting valve *2*. Pump pressure Δp_P, kept constant by adjusting *2*, is reduced in the throttle to about ΔP_{Dr} of the pressure Δp_M required by the motor loading. Motor volumetric flow $\dot{V}_M$ and hence speed n_M are therefore controllable by adjusting throttle area A_{Dr}, since $\dot{V}_M \sim A_{Dr} \sqrt{\Delta p_{Dr}}$.

As load moment M_M increases, working speed n_M falls as a consequence of shift of pressure drop $\Delta p_{Dr} \rightarrow \Delta p'_{Dr} = \Delta p_P - \Delta p'_M$. Maximum load moments are preset by adjustment of the pressure-limiting valve.

Both types of drive are only usable when there is little requirement for speed constancy (saws, woodworking machinery). Their main advantage is overload protection because of their great flexibility under load [4].

It is also possible to have parallel connection of several units on one load-sensing pressure main (modified throttle control).

3.3.3 Automatic Control of Variable-Displacement Pumps

Since adjustment of flow volume depends mainly on loading (pressure), the term "controller" for the adjustment device is not really accurate.

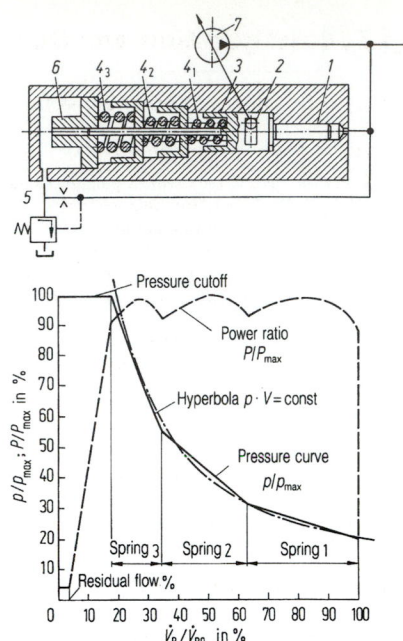

Figure 5. Schematic and mode of operation of a power-regulating valve with pressure cutoff: *1* control piston, *2* control pin, *3* spring guide, 4_1 to 4_3 spring series, *5* outlet valve, *6* pressure piston, *7* control pump.

Zero-Flow Control. This is the simplest type for the situation where a given pressure has to be attained, but no more flow is available (presses etc.). The pumps are maintained at maximum via spring pressure. Once this pressure is exceeded, the pressurised control piston of the pump moves back to lower supply. To ensure cooling and lubrication under maximum pressure, minimum flow is maintained at 4% of main flow. Pressure increases linearly and the supply is reduced, i.e. the power curve is parabolic. With the *preset* type, the pressure characteristic is horizontal, and power falls off linearly (constant pressure control).

Constant-Power Control. By setting up a number of springs so that the force of the maximum total pressure is equal to the spring force of the next set, or where extra springs are stepped in after a predetermined interval, a hyperbolic adjustment $P = \Delta p \dot{V} = $ const. (precise load limitation) is obtained via a series of tangents (see **Fig. 5**). Preset types provide better dynamics, and allow additional functions, e.g. controlled starting, to be fulfilled.

Pressure Cutoff. If no more flow is available after reaching a given high pressure on the hyperbolic power curve, the controller can be combined with a pressure cutoff device (**Fig. 5**). The spring packs 4_1 to 4_3 of the power controller act against an oil-pressure loaded piston *6*. When the maximum pressure is reached, cutoff valve *5* opens and releases the load on the piston. The pump then reverts to idle condition because of the pressure on control piston *1*.

4 Configuration and Design of Hydraulic Transmissions

4.1 Hydraulic Circuit Arrangements

4.1.1 Remote Drive Transmissions

If feed motion only is required with a plunger cylinder, then circuits with 3/2 distribution valves will suffice (**Fig. 1a**). Flow is led from the pump into the cylinder, but, at idle, pump and cylinder are switched into the return flow circuit. If the cylinder has to stop at each intermediate position, a 3/3 distribution valve with a closed central position is required (**Fig. 1b**).

Normal switching of the motor results with a 4/3 valve as shown in **G3 Fig. 1**. Valve arrangements with cross-switching either A–B–T or A–B–P (for rotary motors only A–B) at the middle position permit external motor adjustment, e.g. to control units (floating installations).

Rapid feed of machine tools is most often provided by cutoff pumps (**Fig. 2a**). The increasing pressure at the beginning of the working stroke switches in the fast-action pump with its large supply to the low-pressure circuit. With cylinders, the different values of the bottom and shaft-side areas of the piston face are used. The fast-action valve 4 in **Fig. 2b** initially opens both cylinder connections, and the piston rapidly moves forward; since the pump flow is only acting on the shaft side, $v_E = \dot{V}/A_{st}$. The pressure increase then switches the fast-action valve out, so that only the bottom side of the piston is exposed to the action of the supply flow (step function corresponding to cylinder area ratio in valve needed).

With *multi-motor drives* in parallel, it must be noted that if several machines are switched on simultaneously only the motor with the smallest loading will advance, and that non-return valves must be provided to prevent the other motors reversing. Simultaneous feed of all motors in such cases can be achieved by including flow regulators. The pump circulating valve is so arranged that it closes under either hydraulic or mechanical control upon operation of a given distribution valve.

Series Circuits are possible when the distribution valve can be loaded on the back-pressure side with the full working pressure (connection T); the circulating valve is not needed. Simultaneous

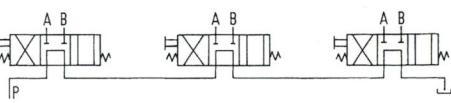

Figure 3. Circuit series switching; multiple motors with 4/3 distribution valves.

switching of several motors is not permissible, otherwise opposing pressure and flow conditions would prevail (see **Fig. 3**). Use of 6/3 valves (**Fig. 4a**) prevents possible error switching, since the operation of a valve shuts off the downstream motor from the flow. Arranging these valves in blocks permits freely interchangeable series–parallel circuits (**Fig. 4b**).

Because of the problems associated with it, reference should be made to the literature [5] for applications of *synchronised switching*.

4.1.2 Variable-Speed Drive Units

In compact drives, pumps and one or more motors are built into a single housing. Closed circuits are generally used. The housing contains all necessary auxiliaries, such as feed pumps, pressure-limiting and flushing valves, and positional and other controls; it also acts as an oil reservoir.

4.2 Design of Hydraulic Circuits

The design of hydraulic circuits is carried out as follows:

(a) Record and set in chronological order mechanism function; establish cycle time and working speed.
(b) Evaluate working principle (linear, reciprocating, or rotary motion).
(c) Record forces (torques) as a function of time; establish pressure range. These are selected so that the permissible working ranges of available hydraulic series components are used so far as possible (cost minimum), while allowing a safety factor for overload (around 10 to 15%).
(d) Select motors as (a), (b) and (c).
(e) Calculate flow strengths by drawing a volume–time diagram using data from (a) and (d);
(f) Select pumps for pressure range and supply (dimensions, number, fixed or variable), decisions on storage units.
(g) Select control type (manual, partly or fully automatic), or controllers.
(h) With decisions (e) and (g), select valves (dimensions, type of circuit, operation), establish

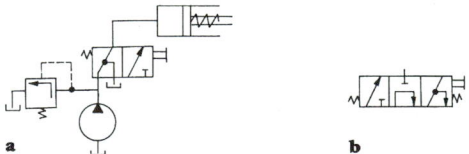

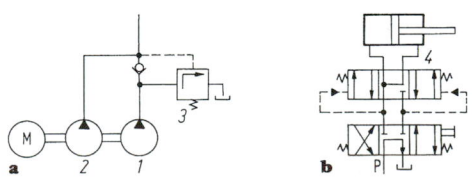

Figure 1. a Simple plunger cylinder switching with circulation, end-stop motion. **b** Valve design with intermediate position.

Figure 2. Rapid-feed switching: **a** with rapid-feed switching pump, **b** using piston area ratios with rapid-action valve; *1* rapid-action pump, *2* working feed pump, *3* cut-off valve, *4* rapid-feed valve.

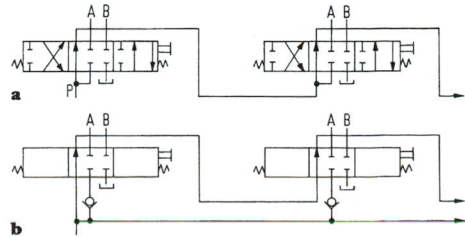

Figure 4. a Safety circuit series switching with 6/3 distribution valves. **b** Extension for series–parallel switching.

pipe diameters (permissible oil velocities up to 1.5 m/s in suction piping, 3 to 6 m/s in pressure piping in pressure range 100 to 400 bar).

(i) Calculate output power and losses: drive power; establish loss balance including heat removal.

After this, considerations of economy, ease of assembly and maintenance, operational safety, and other factors (climatic conditions, service and maintenance staff) must be taken into account.

The most important interpretative aid is the volume–time diagram for the cycle time t_T (t_T with recurring working strokes: time for one working cycle), in which the displacement flow of the motor is additive (as below) (see **Fig. 5**).

Time t_1 to t_2: pressurising with two cylinders 1, simultaneous loading of pressure reservoir 2. Displacement volume $2V_1 + V_2$ (V_2 from storage diagram).

Time t_3 to t_4: rapid feed of working cylinder 3 from $\frac{1}{3}$ displacement, volumes $V_3/3$.

Time t_4 to t_5: working feed cylinder 3 in idle stroke, volumes $2V_3/3$.

Time t_6 to t_7: rapid-return cylinder 3 for full stroke, volumes V_3/φ.

Time t_8 to t_9: depressurisation, volumes = 0, pressure cylinder return via spring force.

Time t_9 to t_T: change workpiece, volumes = 0.

$V_M/(t_{i+1} - t_i) = \dot{V}$ = flow strength = required pump flow, since motor displacement flow + opposed pump supply flow = 0. The diagram is especially suitable for determining the storage circuit flow (cycle time t_T large compared with motor working time). Pump supply is then $\dot{V}_P = \Sigma V_M/(0.9t_T)$, i.e. the pump can be made as

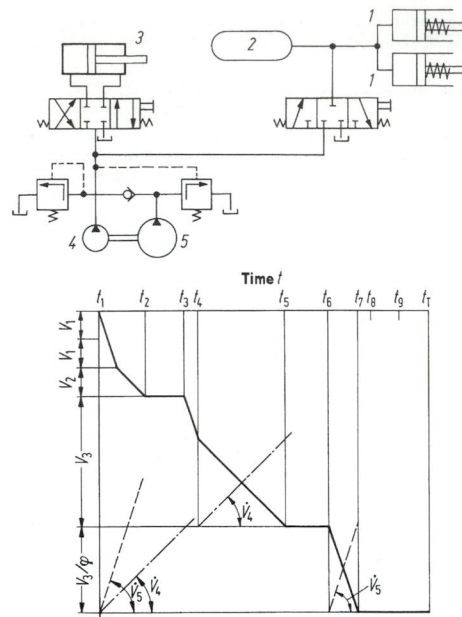

Figure 5. Switching diagram and displacement diagram for press circuits: 1 pressure cylinder, 2 pressure reservoir, 3 working cylinder, 4 working pump, 5 rapid-action pump.

small as possible and storage volume = difference (displacement line − supply line) from the zero line.

5 Pneumatic Installations

Properties of pneumatic drives are as follows:

Advantages

Rapid operation because of high flow speed (up to 40 m/s in piping) and low mass of working medium. High switching frequency (hammers, etc.).

Highly compressible working medium (air), with virtually constant pressure forces even with change of position (pressure cylinders, shock absorbers).

Insensitivity to temperature changes, although there is the danger of condensed water freezing in control valves in open-air plant.

Low piping costs, because air is simply blown off after giving up its energy to the control valves. Also small leakages are unimportant, and there is no danger of contaminating sensitive materials (foodstuffs, etc.).

Usually cheap to install, since in many cases there is already a compressed-air main which can be used.

Disadvantages

Because of the compressibility of the working medium, use is restricted to mechanisms with mechanically or otherwise limited end stops.

Pneumatic drives are only suitable for relatively low powers because of low working pressure.

5.1 Pneumatic Components

Compressors. Pneumatic equipment is almost exclusively fed from compressed air mains with a centralised compressor unit.

Motors. Rotary motors are usually of vane or gear type. As they do not use the expansion work of the compressed air (full-pressure machinery), efficiency is low. They are also usually very noisy, making outlet silencers necessary. Oscillating motors are predominantly of the toothed-rack type. Linear motors (cylinders) are in principle identical in structure to the hydraulic cylinder, except that they are rather lighter owing to the lower pressures involved. For small strokes, membrane cylinders are suitable, in which the piston is replaced by a rubber or plastics membrane loaded and supported by an intermediate piston rod and cylinder jacket. For larger strokes a roller membrane is used. For cutting, press, and die work, in which the working volume is filled for only short periods, percussion cylinders (which use expansion energy of the compressed air) are more economical than full-pressure cylinders. Here, the compressed air is fed into a pre-chamber and on starting led through a large opening in the cylinder bottom. During this expansion, it encounters the large-area piston face and imparts high kinetic energy to the system.

Valves. In construction, function and operation, pneumatic valves are generally similar to those used in hydraulic units. However, because of the lower pressures and higher flow velocities involved, they can have smaller dimensions and incorporate lighter materials such as aluminium and plastics. There is increasing use of seat valves, because of greater operational safety and little requirement for lubrication.

For complex circuits, disc-type distribution valves are used. For small units, piston slide valves set in control bores, or running in elastomeric sealing elements, are used. Larger units usually have O-ring seals on the piston or in the housing, because lapping-in of the mating parts would be prohibitively expensive.

Series Installations. Compressed air for pneumatic units should be filtered free from particles of dust and scale, should be dry, and should carry in mist form the lubricant necessary for the equipment drive. Furthermore, its pressure should remain constant at the correct level independent of the mains pressure. Installations are therefore provided with what is usually called a "service unit", consisting of a combined filter, pressure regulator and oil mist generator.

Filter. This consists mainly of a combination of a vortex chamber (to centrifuge out large particles and droplets), with a metal-mesh, textile, or sinter filter downstream. Dirt and condensed water are collected in a transparent plastics container, which enables the level of contamination to be checked.

Pressure Regulator. This adjusts the pressure downstream of the throttle area by means of a membrane held against spring resistance. Increasing secondary pressure reduces the throughput cross-section; additional equalisation pistons to compensate for primary pressure give even more accurate control.

Oil Mist Generator. This works on the carburettor principle: pressure drop across a nozzle draws oil from a transparent storage tank and injects a fine mist into the airflow. Adjustment of the oil–air ratio is made via throttles in the air nozzle and the oil pipe. Some units use a so-called "micro-oiler", by which oil droplets that are too large are removed from the airflow by diverting it inside the oiler.

5.2 Circuits

Automated units with sequence control are cheaper to produce than those with individual initiation of the work-

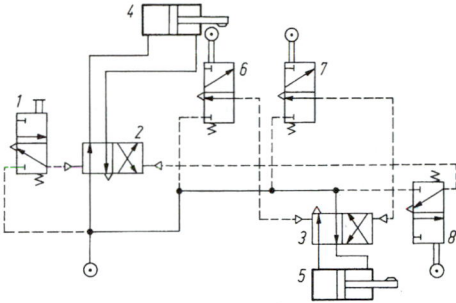

Figure 1. Simple sequence switching of two pneumatic cylinders with impulse valves: *1* start-valve; *2, 3* 4/2 distribution impulse valves; *4, 5* cylinders; *6, 7, 8* push-button valves, roller-operated.

ing cycle; they are also safer in function, because the initiation of the next stage is connected to the completion of the previous one (sequential circuit). This type of equipment can either be electrically controlled, or can be fully pneumatic, with push-button valves to trigger impulse distribution valves. The latter type has the advantage that the whole unit takes compressed air from a single source and so is less subject to disturbances (**Fig. 1**). The starter valve *1* is pressed and this puts the impulse valve *2* into the state that allows the piston in cylinder *4* to advance. At the end of its stroke, the piston reaches contact *7*, which switches impulse valve *3* to operate cylinder *5*. This in turn reaches contact *8* at the end of its working stroke, which sets cylinder *4* into reverse via valve *2*, and the piston, when it reaches contact *6*, gives the feed command to cylinder *5* via valve *3*. The unit thus returns to its initial position, from which another operation of starting valve *1* will initiate a new working cycle.

Storage units in pneumatic circuits act like capacitors, allowing time-dependent operations to be included.

The disadvantage of unequal feeds in pneumatic drives with varying loads, as well as the sometimes excessive feed velocities, can be countered through a combination with hydraulic control equipment. Here the compressed air provides the feed force, but the hydraulics provide the feed velocity (hydropneumatic feed units). Constant feed and if necessary position control are also possible using electronically controlled proportional throttle valves.

6 Water Hydraulic Systems

Water has certain advantages as a working fluid that favour its use in hydraulic systems. Applications include units in which for safety, or because of environmental considerations, the use of mineral oil is illegal (e.g. mining), or where the inherent safety qualities of water (fire protection, waste management), or its economics, favour its use.

There are two distinct areas of application, depending on basic design. What may be termed *water hydraulic systems* are those specially developed to use water; they tend to be large and expensive, such as in slow-running plunger pumps, large presses, lifting equipment, or mining construction. The other category, *industrial water hydraulics*, contains simply oil-hydraulic equipment adapted to use water.

The working medium is HFA fluid with a water content

> 95%. Oil–water emulsions need to be carefully checked for, e.g., micro-organisms, pH value, or separation. Single-phase solutions of synthetic concentrates do not give problems. Additives are required for corrosion protection, especially in steam condensation areas. Properties are essentially determined by the water content: low contaminant absorption (high demands on filter); operating temperature limited to + 2 to 50 °C (ice formation, cavitation); viscosity only slightly (50%) greater than water; extremely poor lubrication properties.

Use in commercial applications imposes considerable restrictions. Working pressure and speeds must be reduced by about 40%, and the service life of rolling bearings can fall by a factor of 20. Where sliding bearings are used, there is little hydrostatic support. Leakage losses are

3 to 10 times higher than with oil, requiring larger pumps. Distribution valves above nominal size 10 may only be seat valves (cartridge-V; see **G2, Fig. 10**), but commercial pressure-limiting, pressure-regulating, and throttle valves can readily be used. Units with commercial components are relatively problem-free up to about 70 bar with HFA fluid. There is no difficulty with sealing, but the use of zinc and cadmium, or paper (filters) are not permissible. Newly developed pumps for higher operating pressures have encapsulated, self-lubricated bearings or pressurised hydrostatic bearings. Price in relation to standard pumps

is 1.5 to 2, with much the same ratios for unit and energy costs.

Acquisition costs for HFA are about 25% of those of oil, but the additional costs of maintaining the working fluid need to be taken into account. No general recommendations can therefore be made, except that individual decisions must be taken on a cost basis in which the primary definitive costings for acquisition and operation are offset against secondary advantages (e.g. reduced environmental damage).

7 Appendix G: Diagrams and Tables

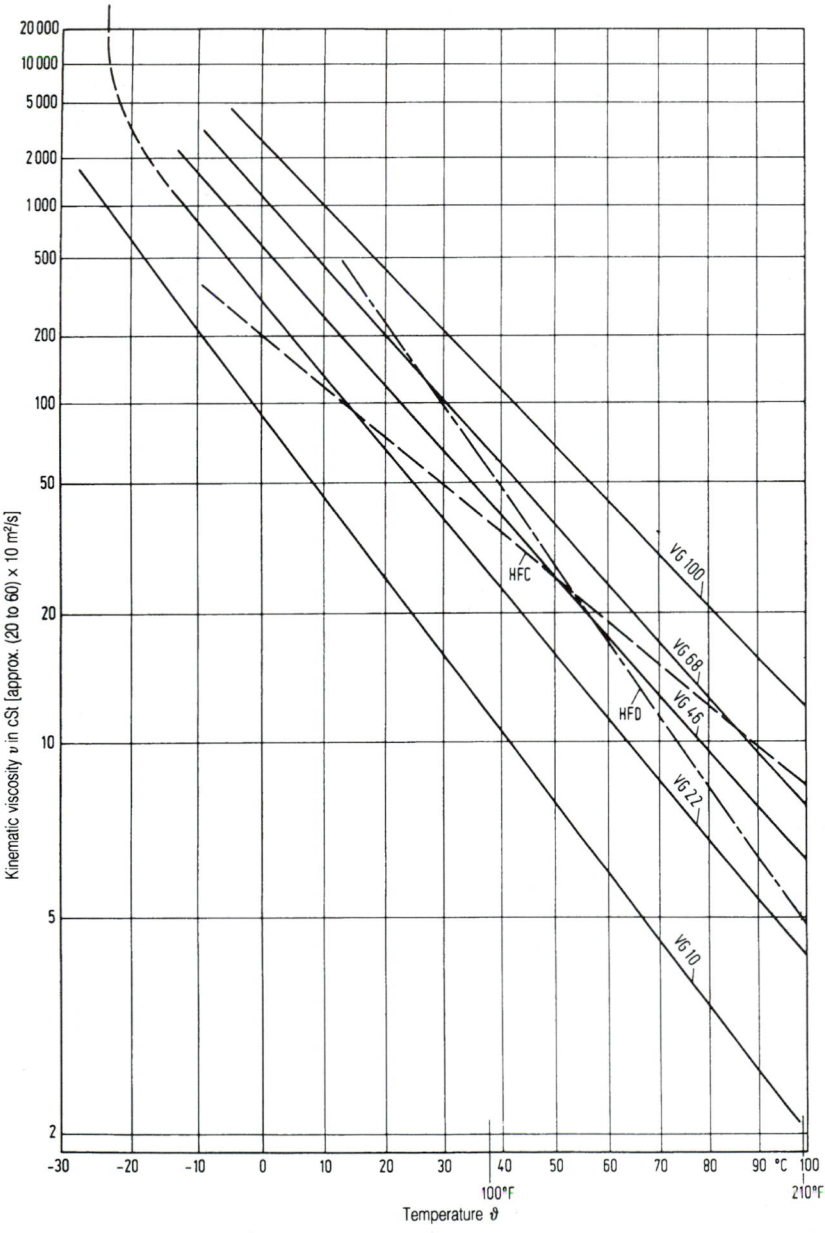

Figure 1. Viscosity–temperature diagram for hydraulic fluids.

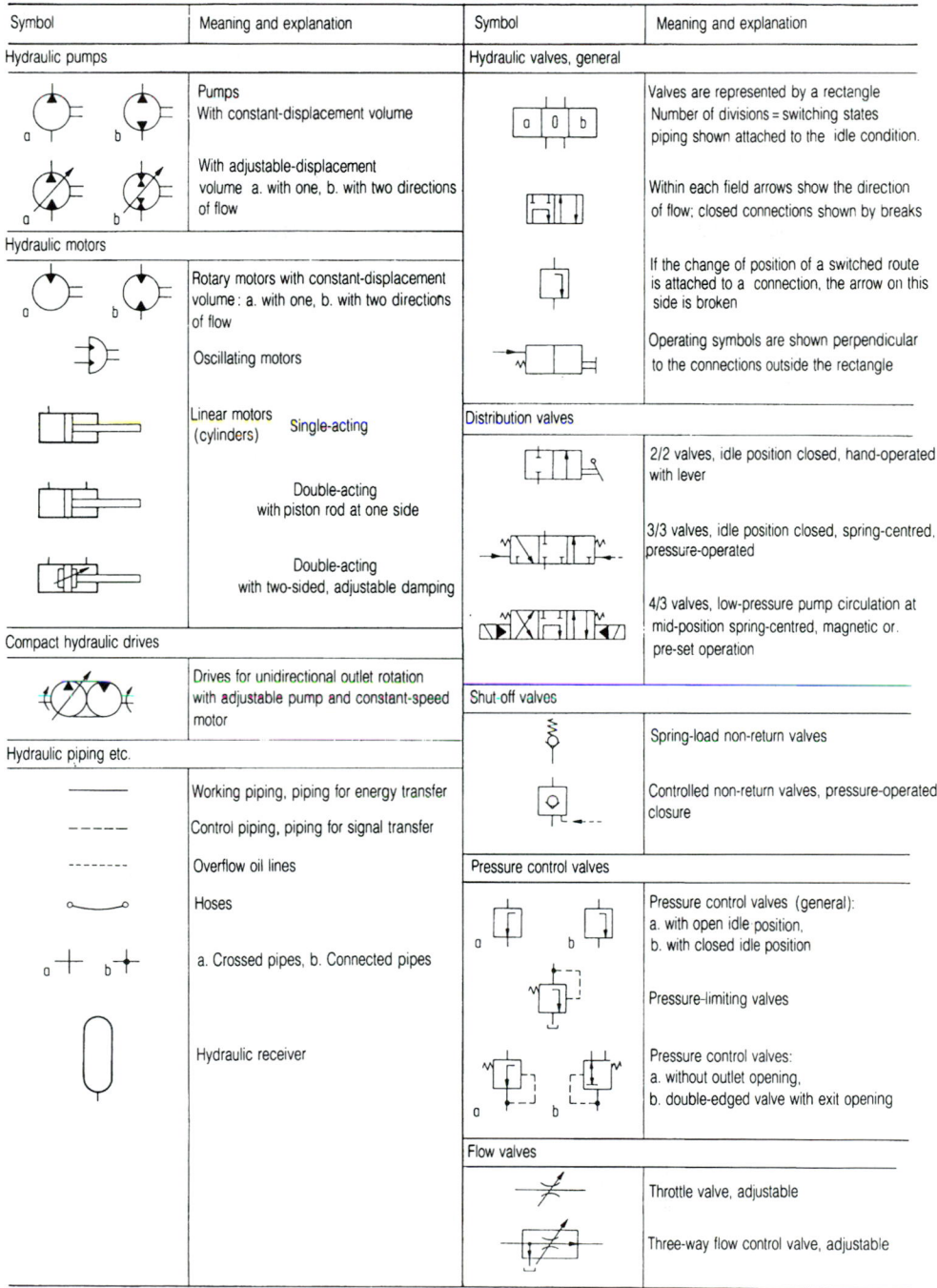

Symbol	Meaning and explanation	Symbol	Meaning and explanation
Hydraulic pumps		**Hydraulic valves, general**	
	Pumps With constant-displacement volume With adjustable-displacement volume a. with one, b. with two directions of flow		Valves are represented by a rectangle Number of divisions = switching states piping shown attached to the idle condition. Within each field arrows show the direction of flow; closed connections shown by breaks
Hydraulic motors			If the change of position of a switched route is attached to a connection, the arrow on this side is broken Operating symbols are shown perpendicular to the connections outside the rectangle
	Rotary motors with constant-displacement volume : a. with one, b. with two directions of flow Oscillating motors		
		Distribution valves	
	Linear motors (cylinders) Single-acting Double-acting with piston rod at one side Double-acting with two-sided, adjustable damping		2/2 valves, idle position closed, hand-operated with lever 3/3 valves, idle position closed, spring-centred, pressure-operated 4/3 valves, low-pressure pump circulation at mid-position spring-centred, magnetic or. pre-set operation
Compact hydraulic drives		**Shut-off valves**	
	Drives for unidirectional outlet rotation with adjustable pump and constant-speed motor		Spring-load non-return valves Controlled non-return valves, pressure-operated closure
Hydraulic piping etc.			
	Working piping, piping for energy transfer Control piping, piping for signal transfer Overflow oil lines Hoses a. Crossed pipes, b. Connected pipes Hydraulic receiver	**Pressure control valves**	
			Pressure control valves (general): a. with open idle position, b. with closed idle position Pressure-limiting valves Pressure control valves: a. without outlet opening, b. double-edged valve with exit opening
		Flow valves	
			Throttle valve, adjustable Three-way flow control valve, adjustable

Figure 2. Symbols for hydraulics and pneumatics according to DIN ISO 1219 (selection).

8 References

General. Pippenger and Koff. Fluid-Power Controls. McGraw-Hill, 1959. – Streeter and Wylie. Fluid Mechanics. McGraw-Hill, 1985. – White FM. Fluid Mechanics. McGraw-Hill, 1986. – Boxer. Fluid Mechanics, Work Out Series - Macmillan, London, 1988.

G1 to G6. [1] Feldman DG. Untersuchung des dynamischen Verhaltens hydrostatischer Antriebe. Konstruktion 1971; 23: 420–8. – [2] Schlösser WMJ. Über den Gesamtwirkungsgrad von Verdrängerpumpen. o + p 1968; 12: 415–20. – [3] Röper P. Die Dynamik des HydroSpeicherkreislaufes. Konstruktion 1968; 20: 341–9. – [4] Röper P. Hydrogetriebe mit Stromteilung Sammelschrift. V. Konferenz über Hydraulische Antriebe. Ceskoslovenská Vedecko-Technická Spolecnost, Prague, 1981. – [5] Zoebl H. Schaltpläne der Ölhydraulik. Krausskopf, Mainz, 1973. – [6] Rechten AW. Fluidik. Springer, Berlin, 1976.

Standards and Specifications. ISO 1219: Schaltzeichen (replacement for DIN 24300). – DIN 24312: Druck, genormte Druckwerte, Begriffe. – DIN 20021/22: Schläuche mit Geflecht-Einlage. – DIN 24334: Hydrozylinder, Hauptmasse. – DIN 24335: Pneumatikzylinder. – DIN 24340: Hydroventile, Lochbilder. – DIN 24346: Fluidtechnik, Ausführungsgrundlagen. – VDMA 23417: Schwerentflammbare Druckflüssigkeiten, Richtlinien. – VDMA 24320 (DIN-E 24320): Schwerentflammbare Druckflüssigkeiten, HFA. – VDI 2152: Hydrostatische Getriebe.

H | Components of Thermal Apparatus

H. Gelbe, Berlin

1 Fundamentals

1.1 Heat Exchanger Characteristics

Heat exchangers are devices that transfer heat between two or more fluid flows in the direction of the temperature gradient. Their aim is to change the state of these fluids (cooling, heating, changing the state of aggregation and/or other physical properties), and to assist in making processes economic (waste heat utilisation). Their distinguishing characteristics are:

Mode of Operation. A distinction is made between continuous-flow (recuperative) and discontinuous-flow (regenerative) heat exchangers.

Heat Transfer. This can take place directly ('without wall', or heat-transfer contact) or indirectly (transfer by thermal conduction through separating walls). Examples of direct heat transfer include injection condensers, separating stages for the thermal separation of mixtures of materials, and solar-powered distillation plant. Boilers, piping systems or vessels are heated directly by flames or flue gases, occasionally with the aid of a heat-transfer medium (organic heat-transfer medium, molten salts or metals).

State of Aggregation of Fluids. A distinction is made between pieces of apparatus with flow and without phase change (preheaters, air coolers, flue-gas-driven superheaters and others) and those with phase change (condensers, evaporator equipment, evaporator coolers and others). Calculations become more difficult if phase changes have to be anticipated on both sides (as e.g. in an evaporator–condenser).

Temperature and Pressure. Depending on their use, heat exchangers may be classified as either low-temperature (down to −100 °C), normal (50 to 500 °C) or high-temperature (up to ≈ 1400 °C, e.g. waste-heat boilers in petrochemistry), as well as vacuum, low-pressure (a few bar), high-pressure (100 to 500 bar), or ultra-high-pressure (some 10^3 bar).

Mode of Construction. Equipment featuring tube bundles (plain tubes, hairpin tubes, field tubes and tube registers) is among the most widely used. Further subdivisions are according to the method of attachment (tube plates, headers) and the guiding of the bundles (spiral tube, wrapped bundle). Apart from these, plate, spiral, double-shell and lamellar heat exchangers are also employed.

Size. Compact heat exchangers are those with heat-transfer areas exceeding 700 m² per m³ of volumetric structure (space vehicles, aircraft).

1.2 Thermodynamic and Fluid Dynamic Design

The aim is to obtain a large thermal power transfer $\dot{Q}/A$ under conditions of optimum or maximum permissible pressure loss (see H1.2.3) and where the sum of the costs for the equipment, the required building and the energy including generation and transfer (pumps, pipelines) is to be minimised.

1.2.1 Thermodynamic Design of Recuperators

The thermodynamic design is based on the transfer equation (see C10.2) $\dot{Q} = k \cdot A \cdot \Delta t_M$, where $\dot{Q}$ is calculated from the balancing equations $\dot{Q} = \dot{m}_1 \, c_{p1} (t_1 - t_2) = \dot{m}_2 \, c_{p2} (t'_1 - t'_2)$ (see **Fig. 1**).

Heat Transfer Coefficient. Because $1/k = 1/\alpha_1 + \delta/\lambda + 1/\alpha_2$ (see C10.2), k is always smaller than the lowest value of α. Hence this smallest value must be improved by influencing the flow (crossflow, generation of turbulence) by increasing the velocity (increasing the number of tubular or shell ducts or deflection plates increases the pressure loss) or by insertion of fins or lamellae (especially in the case of gases with low α-values). Determination of the heat-transfer coefficient α as a function of the phase condition and the mode of flow of the fluids as well as of the surface geometry (plate, interior or exterior of the pipe, smooth, undulating, finned) and the position of the apparatus (horizontal, vertical) is carried out with the help of power series of non-dimensional characteristic numbers (see C10.4 and [1]). Leakage and bypass flow as well as non-uniform flow in tube bundles can be taken into account by means of correction factors [1]. Estimated k-values are given in

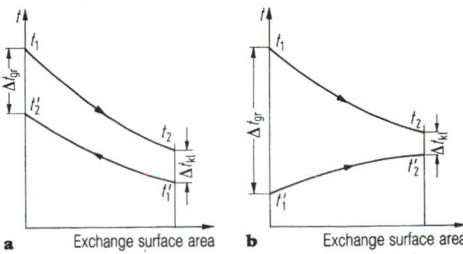

Figure 1. Temperature dependence in both media. **a** Counterflow. **b** Parallel flow.

Table 1. Attention must be paid to the degradation of heat transfer due to accumulations of dirt.

Temperature Calculation. The distribution of the fluid temperature and the mean integral temperature difference Δt_M depend on the mode of flow (parallel flow, counterflow, crossflow, crossmixing, combinations of parallel flow and counterflow in multi-pass and coupled apparatus) as well as on the intensity of the heat transfer. (Transfer units $N = k \cdot A / \dot{m} c_p$.) If Δt_{gr} and Δt_{kl} are the large and the small temperature difference for parallel flow and counterflow (**Fig. 1**) then the following formula applies:

$$\Delta t_M = \frac{\Delta t_{gr} - \Delta t_{kl}}{\ln(\Delta t_{gr}/\Delta t_{kl})} \tag{1}$$

To check the calculations for a given heat exchanger ($k \cdot A$ known), two temperatures are sufficient to determine all the others with the help of **Fig. 2** (required temperatures circled; given temperatures not). The quantities A and B for parallel flow and counterflow respectively (subscripts Gl and ct respectively) are given by

$$
\left.
\begin{aligned}
A_{Gl} &= \frac{t_1 - t_2}{t_1 - t_1'} = \frac{w}{w + W}\left[1 - \exp\left[-\left(\frac{1}{w} + \frac{1}{W}\right)kA\right]\right], \\
B_{Gl} &= \frac{t_1 - t_2'}{t_1 - t_1'} = \frac{W}{w + W}\left[\frac{w}{W} + \exp\left[-\left(\frac{1}{w} + \frac{1}{W}\right)kA\right]\right],
\end{aligned}
\right\} \tag{2}
$$

$$
\left.
\begin{aligned}
A_{ct} &= \frac{t_1 - t_2}{t_1 - t_1'} = \frac{1 - \exp\left[\left(\frac{1}{w} - \frac{1}{W}\right)kA\right]}{1 - \frac{W}{w}\exp\left[\left(\frac{1}{w} - \frac{1}{W}\right)kA\right]}, \\
B_{ct} &= \frac{t_1 - t_2'}{t_1 - t_1'} = \frac{1 - \frac{W}{w}}{1 - \frac{W}{w}\exp\left[\left(\frac{1}{w} - \frac{1}{W}\right)kA\right]},
\end{aligned}
\right\} \tag{3}
$$

where W and w are the thermal capacities of the fluid flows with the higher and the lower temperature respectively. Here w or $W = \dot{m} c_p$, where $\dot{m}$ is the mass flow and c_p the thermal capacity. Further, t_1, t_2 and t_1', t_2' are the inlet and outlet temperatures of the hotter and the colder medium respectively. Flows that differ from parallel flow or counterflow result in smaller values for Δt_M

Table 1. Estimated k-values in W/(m² K) for heat exchangers of tube-bundle (shell and tube) type (VDI-Wärmeatlas, paragraph C1-4)

Gas ($\approx$ 1 bar) against gas ($\approx$ 1 bar)	5 to 35
Gas, high pressure (200 to 300 bar) surrounding the tubes	150 to 500
Gas, high pressure (200 to 300 bar) in the tubes	
Fluid against gas ($\approx$ 1 bar)	15 to 70
Fluid against fluid	150 to 1200
Heating vapour surrounding the tubes, fluid in the tubes	300 to 1200
Evaporator: heating steam surrounding the tubes	
(a) Natural circulation, depending on viscosity	300 to 1700
(b) Forced circulation	900 to 3000
Condensers:	
Cooling water through tubes	
Organic vapours and NH₃ surrounding tubes	300 to 1200
Steam turbine condenser (pure H₂O steam, brass tubes)	1500 to 4000

Counterflow	Crossflow	Parallel flow	
$t_1 \longrightarrow \boxed{t_2}$	$\overset{\boxed{t_2''}}{\underset{t_1''}{t_1 \rightarrow}}$	$t_1 \longrightarrow \boxed{t_2}$	$t_2 = t_1 - A\,(t_1 - t_1')$
$\boxed{t_2'} \longleftarrow t_1'$	t_1''	$t_1' \longrightarrow \boxed{t_2'}$	$t_2' = t_1 - B\,(t_1 - t_1')$
$\boxed{t_1} \longrightarrow t_2$	$\overset{t_2''}{\underset{\boxed{t_1}}{\rightarrow t_2}}$	$\boxed{t_1} \longrightarrow t_2$	$t_1 = t_2 + \dfrac{A}{A - B}\,(t_2' - t_2)$
$t_2' \longleftarrow \boxed{t_1'}$	$\boxed{t_1''}$	$\boxed{t_1'} \longrightarrow t_2'$	$t_1' = t_2 - \dfrac{1 - A}{A - B}\,(t_2' - t_2)$
$\boxed{t_1} \longrightarrow t_2$	$\overset{\boxed{t_2''}}{\underset{t_1}{\rightarrow t_2}}$	$\boxed{t_1} \longrightarrow t_2$	$t_1 = t_1' + \dfrac{1}{1 - A}\,(t_2 - t_1')$
$\boxed{t_2'} \longleftarrow t_1'$	t_1''	$t_1' \longrightarrow \boxed{t_2'}$	$t_2' = t_1' + \dfrac{1 - B}{1 - A}\,(t_2 - t_1')$
$t_1 \longrightarrow \boxed{t_2}$	$\overset{t_2''}{\underset{t_1}{\rightarrow t_2}}$	$t_1 \longrightarrow \boxed{t_2}$	$t_2 = t_1 - \dfrac{A}{B}\,(t_1 - t_2')$
$t_2' \longleftarrow \boxed{t_1'}$	$\boxed{t_1''}$	$\boxed{t_1'} \longrightarrow t_2'$	$t_1' = t_1 - \dfrac{1}{B}\,(t_1 - t_2')$
$\boxed{t_1} \longrightarrow \boxed{t_2}$	$\overset{t_2''}{\underset{\boxed{t_1}}{\rightarrow \boxed{t_2}}}$	$\boxed{t_1} \longrightarrow \boxed{t_2}$	$t_1 = t_1' + \dfrac{1}{1 - B}\,(t_2' - t_1')$
$t_2' \longleftarrow t_1'$	t_1''	$t_1' \longrightarrow t_2'$	$t_2 = t_1' + \dfrac{1 - A}{1 - B}\,(t_2' - t_1')$
$t_1 \longrightarrow t_2$	$\overset{\boxed{t_2''}}{\underset{t_1}{\rightarrow t_2}}$	$t_1 \longrightarrow t_2$	$t_1' = t_1 - \dfrac{1}{A}\,(t_1 - t_2)$
$\boxed{t_2'} \longleftarrow \boxed{t_1'}$	$\boxed{t_1''}$	$\boxed{t_1'} \longrightarrow \boxed{t_2'}$	$t_2' = t_1 - \dfrac{B}{A}\,(t_1 - t_2)$

Figure 2. Temperature distribution for counterflow, crossflow and parallel flow (Plank).

$$\Delta t_M = \varepsilon \, \Delta t_{M,f} \tag{4}$$

The correction factor ε is normalised to the mean temperature $\Delta t_{M,f}$ for counterflow, which has been calculated for the same starting and final temperatures. In **Fig. 3** ε is plotted as a function of the operating characteristic $S = (t_2' - t_1')/(t_1 - t_1')$ for cross-counterflow, with $R = w/W = (t_1 - t_2)/(t_2' - t_1')$ as a parameter. Here, fluid Fl

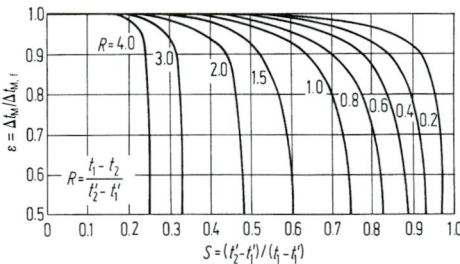

Figure 3. Mean temperature difference for the special case of cross-counterflow (Plank).

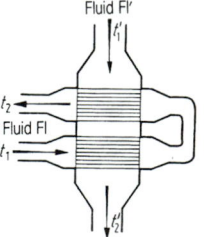

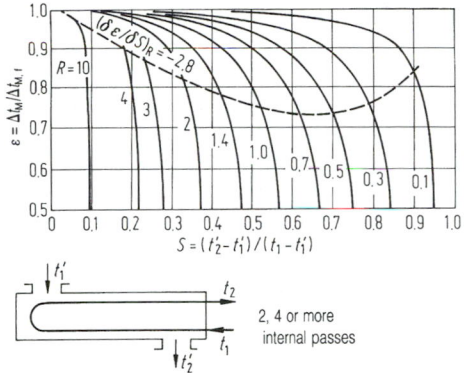

Figure 4. Correction factors ε for a 1,2-heat exchanger and the technically acceptable region [2].

is not thoroughly mixed, while fluid Fl' is. **Figure 4** shows an analogous diagram for a 1,2 heat exchanger ($p = 1$ pass through shell, $r = 2$ passes through tubes). For $R = 1$, efficiency = operating characteristic S cannot exceed 0.57. The dashed line connects points of constant slope and separates the region where the curves have steep slopes (high sensitivity to perturbation of the operating conditions) from the region of technically acceptable values of ε [2] (see H1.3).

1.2.2 Thermodynamic Design of Regenerators

Heat transfer takes place during two periods (the heating and cooling periods). The changeover sequence is shown in **Fig. 5**.

Mode of Construction with static or moving storage mass (Ljungström-type) [1, 3]. Further distinguishing characteristics are type and structure of the storage mass as well as changeover time. The storage mass temperature is subject to periodic fluctuations. The calculation of heat transferred during a *full* period in *one* regenerator is carried out [1, 3] in accordance with relation

$$Q_{per} = k\,\Delta t_M (T_1 + T_2), \tag{5}$$

with $k/k_0 = 1$ for the fundamental oscillation (approximation) and $k/k_0 < 1$ when taking harmonics into consideration as in [1, 4] and with

$$1/k_0 = (T_1 + T_2)\left[\frac{1}{\alpha_{1m}T_1} + \frac{1}{\alpha_{2m}T_2} + \left(\frac{1}{T_1} + \frac{1}{T_2}\right)\frac{\delta}{\lambda_B}\,\Phi\right];$$

where, with the usual units shown in brackets:

k_0 (W/m² K) = heat-transfer coefficient for fundamental oscillation;

A (m²) = total heat transfer area of a regenerator;

T_1, T_2 (s) = duration of hot or cold period respectively;

Δt_M (K) = logarithmic temperature difference using the inlet temperatures t_h', t_c' or outlet temperatures t_h'', t_c'' respectively;

α_{1m}, α_{2m} (W/m² K) = heat-transfer coefficients referred to mean temperature of storage mass;

δ (m) = thickness of a storage mass element (i.e. diameter if shape is cylindrical or spherical);

λ_B (W/m K) = thermal conductivity of storage mass;

Φ = auxiliary function as in **Fig. 6**;

a (m²/s) = thermal diffusivity of storage mass.

1.2.3 Calculation of Pressure Drop

The size of a heat exchanger is decisively determined by the pressure drop. For this reason, calculation of the pressure drop is one of the first steps in design because it determines the geometry of the tube bundle (diameter and length of the bundle and the tubes, flow cross-sections) required for the thermodynamic design. The frictional pressure loss in fully formed tubular flow contains contributions ζ_v from the inlet and the outlet pressure losses and from losses due to internal structures and deflections (see A6.2). The total pressure drop in the tubes for one tubular pass is given by

$$\Delta p_{tot} = \frac{\rho}{2}w^2\left(\lambda\frac{L}{d} + \Sigma\zeta_v\right). \tag{6}$$

For values of $\lambda(Re, d/k)$ see **A6 Fig. 8**; for reference values for ζ_v see A6.2.4. If the number of tubular passes is n_T, then for constant volume flow and if $w \sim n_T$ the pressure drop in terms of that occurring in one tubular pass is given by

$$\Delta p_{tot\,n_T} = n_T^3 \cdot \Delta p_{tot\,1}. \tag{7}$$

To calculate the pressure drop for flow through multiple-row tube bundles in cross-flow (external space of

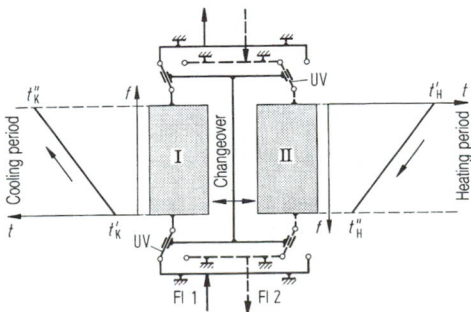

Figure 5. Circuit diagram for regenerators (shown for gas). Indices: ' inlet; " outlet; H, period of heating; K, period of cooling. Fl 1 and Fl 2, cold and hot flow of medium. UV, Changeover valve.

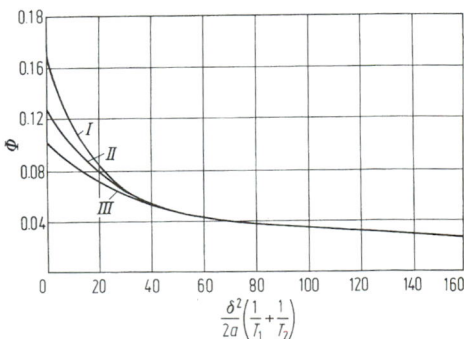

Figure 6. Auxiliary functon Φ for the calculation of the heat-transfer coefficient (Hausen): *I*, plate; *II*, cylinder; *III*, sphere; δ, thickness of plate or diameter; a, thermal diffusivity.

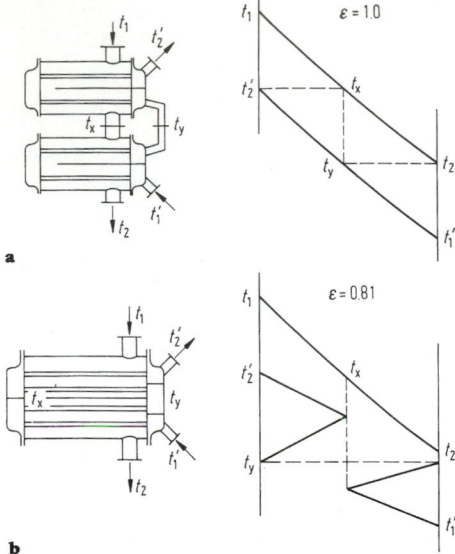

a

b

Figure 7. Series arrangement of **a** two reverse-current (2,2) and **b** two 1,2 apparatuses to form one 2,4 heat exchanger.

heat exchanger with deflection baffles) reference should be made to the literature [1]. In Eqs (6) and (7) λ is the frictional coefficient of the tube, ζ_v the drag coefficient, w the fluid velocity, ρ the fluid density, L and d the length and inside diameter, respectively, of the flow channel. The influence of frictional pressure drop on the heat-transfer coefficient α of a smooth tube or tube bundle carrying flow along its length is given approximately by Grassmann [5] for Re numbers > 6000 as

$$\alpha = (K/d_h^{0.127})(\dot{V}\,\Delta p/A)^{0.291} \qquad (8)$$

where K is the medium constant, d_h the hydraulic diameter of the flow channel, $\dot{V}$ the volume and A the area of heat transfer. From the transfer equation $\dot{Q} = kA\,\Delta t_M$ and from Eq. (8) and using $k^* = k/\alpha \approx$ const. (k^* depends on the resistances of both fluids and on the roughness) Eq. (9) is obtained:

$$A = \left(\frac{\dot{Q}}{k^*K\Delta t_M}\right)^{1.41} \frac{d_h^{0.179}}{(\dot{V}\,\Delta p)^{0.41}} \qquad (9)$$

This allows an estimation of the area A if the pressure drop is given.

1.3 Heat Exchanger Flow Arrangements and Operating Characteristics

If the correction factors of Eq. (4) become too small (see **Fig. 4**), a series arrangement of 1,2, 1,4 or 2,4 heat exchangers may be considered. This makes possible an increase of the effective temperature gradient up to values in close proximity to the optimum. For an economic assessment, the costs of this solution (compact method of construction, improved α-values, smaller ε-values) must be compared with the cost of a counterflow apparatus.

Such a comparison is shown in **Fig. 7**. The change in temperature $S = 0.67$, which permits only 1,1 apparatus in counterflow, is distributed over two 1,2 apparatuses, each with $S_1 = S_2 = 0.5$: $\varepsilon = 0.81$ for $R = 1$. Apparatuses with more than $p = 2$ shell passes are barely capable of economic manufacture and operation.

If the transfer equation and the balance equations (see H1.2.1) are divided by the greatest temperature difference $t_1 - t_1'$ and by $\dot{m}_1 \cdot c_{p1}$, or, where appropriate, $\dot{m}_2 \cdot c_{p2}$, six non-dimensional characteristic quantities, S_1, S_2, N_1, N_2, R and Θ, are obtained. For a given flow arrangement, two of these characteristics determine the other four. The following applies to the operating characteristic:

$$S_1 = N_1 \cdot \Theta = S_2 \cdot N_1/N_2 = S_2 \cdot R, \qquad (10)$$

where (similarly for S_2, N_2)

$$S_1 = \frac{t_1 - t_2}{t_1 - t_1'}; \quad N_1 = \frac{k \cdot A}{\dot{m}_1 \cdot c_{p1}}; \quad \Theta = \frac{\Delta t_M}{t_1 - t_1'} \qquad (11)$$

The effect of the flow configuration on operating characteristics is shown in **Fig. 8**. A detailed description of the effect of different flow configurations and changeover alternatives on power transfer and calculation of temperature behaviour is given by Martin [6]; tables in [1].

1.4 Efficiency, Exergy Losses

1.4.1 Efficiencies

The degree of reversibility is a measure of the thermodynamic perfection of an apparatus or process [5].

$$\eta_R = \dot{E}_\omega/\dot{E}_\alpha \qquad (12)$$

is the ratio of the exergy flow $\dot{E}_\omega$ leaving the balance region for the environment (useful power) to the consumed exergy flows $\dot{E}_\alpha$. In order to assess the effect of the heat exchanger on the process, the overall process quality index

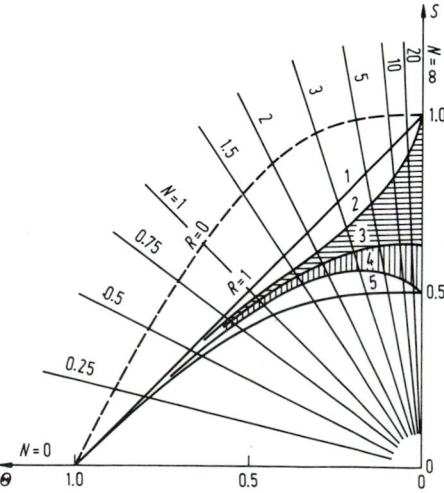

Figure 8. Effect of the flow configuration on power transfer for flows of equal capacity $|R| = 1$ (continuous line) and $R = 0$ (dashed line): 1, counterflow; 2, ideal; 3, single-sided; 4, crossflow, mixing both sides; 5, parallel flow [6].

$$\eta_G = \Sigma \dot{E}_\alpha / \dot{E}_a, \qquad (13)$$

given by Glaser [7] can be used, for which the sum of the consumed exergy flows of all equipment with loss-free heat exchangers has to be determined.

1.4.2 Exergy Losses

Exergy losses are mainly caused by the following processes: finite temperature differences, thermal conduction or inverse mixing, pressure drops, and heat exchange with the surroundings (insulation losses). Detailed examples for heat exchangers can be found in [8].

Losses Due to Finite Temperature Differences

If a quantity of heat $\dot{Q}$ flows from the absolute temperature T to the temperature T', the corresponding exergy loss is given by

$$\dot{E}_v/\dot{Q} = (T_u \cdot \Delta T)/[T(T - \Delta T)] \qquad (14)$$

where $\Delta T = T - T'$. For heat exchangers, approximate calculations using the mean logarithmic values (isobar, constant values of c_p) can be carried out:

$$T = (T_1 - T_2)/\ln(T_1/T_2). \qquad (15)$$

Similarly for T'. Subscripts 1 and 2 relate to the inlet and outlet temperatures respectively. Equation (14) is illustrated in **Fig. 9**. The losses increase steeply with decreasing temperature (low-temperature technology). In order to limit these, small temperature differences are required.

Thermal Conduction or Inverse Mixing

Owing to the usually high flow velocities, losses due to molecular or turbulent axial transport processes are small. For thermal inverse mixing as a result of designs featuring multiple passes, see H1.3.

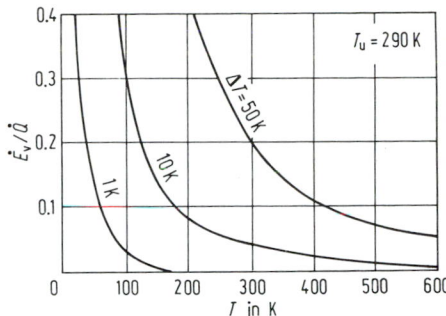

Figure 9. Exergy losses due to finite temperature differences.

Losses Due to Friction. The specific exergy loss is

$$e_v = -T_u \int_1^2 (v/T) \, dp. \qquad (16)$$

The losses increase with the specific volume v of the fluid and with decreasing temperature. For determination of the losses [5], the pump power can be used as an approximation. An accurate analysis is necessary [8], because under certain circumstances a part of the lost energy can be regained as heat.

Insulation Losses. If $T' = T_u$ and $\dot{Q} = (\lambda/\delta)A(T - T_u)$ are substituted in Eq. (14) (where λ is the thermal conductivity and δ the thickness of the insulation), Eq. (17) is obtained [5]:

$$\frac{\dot{E}_v}{(\lambda/\delta) \cdot A} = \frac{(T - T_u)^2}{T}. \qquad (17)$$

At low temperatures, good insulation becomes cost-effective.

---

2 Apparatus and Piping Components

2.1 Basis for Design Calculations

Permissible Operating Pressure p_B. This is from 10 to 20% higher than the maximum operating pressure that can occur under the least favourable operating conditions. This determines the acceptance conditions in accordance with the Pressure Vessel Regulations, the assignment to test groups, the calculating pressure, the test pressure and the threshold pressure of the safety valve.

Calculating Pressure p. In general, this is the permissible operating pressure. Static pressures due to the feed media are taken into account only if they increase the stress on the components by more than 5%. In general, calculations should not use the pressure difference between internal and external pressures, but both pressures separately (the exception is in cases of partial vacuum).

Calculating Temperature. This is the highest wall temperature to be reached (at least 20 °C) under normal conditions as indicated in **Table 1**.

Characteristic Strength Value. This is the lower of the following two values (at the calculating temperature): tensile yield point R_m or $0.2 \times$ limit $R_{p0.2/\vartheta}$ and long-term creep strength $R_{p/100\,000\,\vartheta}$ for 100 000 h (see D2.2).

Safety Factor. This takes care mainly of uncertainties in the calculation assumptions and guarantees that during water pressure tests at the normal pressure of $1.3p_B$, a sufficient margin is maintained relative to the characteristic strength value at 20 °C. This amounts to $S = 1.1$ at the test pressure for rolled and forged steels and to $S = 1.5$ at the calculating temperature and pressure.

Table 1. Calculating temperature [1]

Type of heating	Calculating temperature
None	Highest temperature of the feed media
By gases, vapours or liquids	Highest temperature of the heating medium
Fire, flue gas or electrical heating	Highest temperature of the feed medium +20 °C for screened wall +50 °C for direct contact with wall, however at least 250 °C

Allowances. c_1 for shortfall in the wall thickness resulting from the manufacturing process; to be determined in each individual case; c_2, wear and corrosion allowance (at least 1 mm – not required if adequate protection against the effects of the feed medium exists, when $s_c \geq 30$ mm or for heat exchanger tubes).

2.2 Cylindrical Shells and Tubes Under Internal Pressure

$D_a/D_i < 1.2$. The required wall thickness is given by

$$s = \frac{D_a p}{2v(K/S) + p} + c_1 + c_2 , \qquad (1a)$$

where D_a and D_i are outer and inner diameters, K is the permitted material stress, p the calculating pressure, S the safety factor and c_1 and c_2 the thickness allowances for the wall. The reduction correction factor is $v = 1.0$ for seamless welded shells and 0.8 to 1.0 for welded joints in the shell, depending on the rating of the welded seam in accordance with the AD Code of Practice HPO.

$1.2 < D_a/D_i \leq 1.5$. The wall thickness is given by

$$s = \frac{D_a p}{2.3(K/S) - p} + c_1 + c_2 . \qquad (1b)$$

Tubes $D_a/D_i < 1.7$. The tube wall thickness is determined by internal pressure, ease of handling during transport and installation, sag between supports, possibilities of external damage (mechanical, corrosion), type of tube joint, working load and restrained thermal expansion.

For $D_a < 200$ mm, wall thickness under internal and external pressure is calculated in accordance with the AD Code of Practice B1 and Eq. (1a).

The calculation for steel tubes with regard to internal pressure distinguishes three regions as DIN 2413: I, mainly static stress up to 120 °C; II, mainly static stress above 120 °C; and III, dynamic stress. For region I, the required wall thickness is given by

$$s = \frac{D_a p}{2v \, (K/S)} + c_1 + c_2 . \qquad (2a)$$

The safety factor S depends on the elongation at rupture. For $\delta_5 = 20\%$, $S = 1.6$ with an acceptance certificate in accordance with DIN 50 049, and $S = 1.75$ without. In the case of seamless tubes with special-quality specifications, the safety rating of the welding seam v lies between 0.5 and 1.0, depending on quality. For region II, the calculation is carried out in accordance with

$$s = \frac{D_a p}{(2K/S - p)v + 2p} + c_1 + c_2 . \qquad (2b)$$

The safety factor S with respect to the thermal yield point (tensile) = 1.5 (in accordance with DIN 50 049), otherwise $S = 1.7$.

Thermal Stresses $D_a/D_i < 1.7$. The change in length Δl caused by the temperature differences $\vartheta - \vartheta_0$ between operating and installation temperature is given by

$$\Delta l = \alpha \cdot l_0(\vartheta - \vartheta_0) . \qquad (3a)$$

If the change in length is restrained, an axial stress

$$\sigma_\vartheta = E \cdot \alpha(\vartheta - \vartheta_0) \qquad (3b)$$

will occur; l_0 is the installation length, α the coefficient of thermal expansion (see C3.1.2) and E the modulus of elasticity. If pressure forces exist, attention must be given to buckling of the tube.

If temperature differences occur in the wall owing to heating or cooling, tangential and axial stresses will occur at the inner and outer axes (subscripts i and a respectively), which are of equal magnitude in both cases, positive at lower temperatures and negative at higher ones:

$$\sigma_{\vartheta_i} = \frac{\alpha}{2} \frac{E}{1-v} (\vartheta_a - \vartheta_i) \frac{3D_a + D_i}{2(D_a + D_i)} ,$$

$$\sigma_{\vartheta_a} = \frac{\alpha}{2} \frac{E}{1-v} (\vartheta_a - \vartheta_i) \frac{D_a + 3D_i}{2(D_a + D_i)} \qquad (4)$$

From these, the maximum internal and external stationary stresses can be calculated approximately:

$$\sigma_{v,i} = \frac{p(D_a + s_c)}{2.3 \cdot s_c} + \sigma_{\vartheta_i} ,$$

$$\sigma_{v,a} = \frac{p(D_a - 3s_c)}{2.3 \cdot s_c} + \sigma_{\vartheta_i} . \qquad (5)$$

Here, s_c is the actual wall thickness, v the lateral contraction coefficient, and ϑ the temperatures. These approximate formulas are sufficiently accurate in practice as long as the greater of the two comparative stresses $\sigma_{v,i}$ and $\sigma_{v,a}$ is always considered, i.e. as long as the following holds true:

$$\sigma_{\vartheta_i} \geq \frac{p(D_a + s_c)}{4 \cdot s_c} ,$$

$$\sigma_{\vartheta_o} \geq -\frac{p(D_a - 3s_c)}{4 \cdot s_c} . \qquad (6)$$

All equations apply to unclamped cylinders without additional external or bearing forces. Under any shortfall of the conditions expressed by Eq. (6), or in the presence of additional axial stresses, the comparative stresses must be calculated from the previously summed main stresses. Stationary thermal stresses may exceed K/S, provided they occur alone ($\sigma_{v,\,max} \leq 2K/S$).

According to Eq. (5), superimposed stresses arising from pressure and temperature differences will lead to large peak stresses at the internal axis if the gradients are opposed ($p_i > p_a$, $\vartheta_i < \vartheta_a$) (disadvantageous!), but to more uniform stress distributions if the gradients have the same sense (test whether under certain conditions $\sigma_{v,a} > \sigma_{v,i}$). According to Eq. (4), thermal stresses increase with increasing wall thickness at constant temperature difference $\vartheta_a - \vartheta_i$. At a given quantity of heat $\dot{Q}$ and length of tube l_0, the temperature difference must also increase with increasing wall thickness, owing to the increasing resistance of thermal conduction:

$$\vartheta_a - \vartheta_i = \frac{\dot{Q}}{2\pi l_0 \lambda} \ln \frac{D_a}{D_i} . \qquad (7)$$

The thermal stresses increase in logarithmic form while the pressure stresses decrease. The summed comparative stresses form pronounced maxima, which are displaced in the direction of lower wall thicknesses as the thermal stresses increase.

2.3 Cylindrical Shells Under External Pressure

For cylindrical shells under external pressure with $D_a \geq 200$ mm, calculations have to be carried out with respect to denting and plastic deformation if $D_a/D_i \leq 1.2$.

Table 2. Safety correction factor against elastic buckling in the presence of external pressure [1]

$(s_e - c_1 - c_2)/R$	0.1	0.01	0.005	0.003	0.001
S_k	3.0	3.5	3.7	4.0	5.5

Elastic Indentation. If the length of the shell is L then the maximum permissible operating pressure (notation: see Eq. 1a) is given by

$$p = \frac{E}{S}\left[\frac{2.0}{(n^2-1)A^2}\frac{s_e - c_1 - c_2}{D_a}\right.$$
$$\left. + 0.733\left(n^2 - 1 + \frac{n^2 - 1.3}{A - 2}\right)\left(\frac{s_e - c_1 - c_2}{D_a}\right)^3\right], \quad (8)$$

where $S = 3$ and $A = 1 + (nD_a / 2L)^2$. It is required that $n \geq 2$ and $n \geq D_a/2L$, where n is the number of indentation waves that can occur along the circumference on failure (**Table 2**).

Plastic Deformation. If the length of the shell is L then the highest permissible operating pressure for $D_a/L \leq 5$ (notation: see Eq. 1a) is given by

$$p = 2.0\frac{K}{S}\frac{s_e - c_1 - c_2}{D_a} \bigg/$$
$$\left[1 + 0.015u\left(1 - 0.2\frac{D_a}{L}\right)\frac{D_a}{s_e - c_1 - c_2}\right]. \quad (9)$$

The usual value u of the circularity error is 1.5. For $D_a/L > 5$ the maximum permissible operating pressure is the lesser of the two following values:

$$p = 2.0\frac{K}{S}\frac{s_e - c_1 - c_2}{D_a} \quad \text{and} \quad p = 3.0\frac{K}{S}\left(\frac{s_e - c_1 - c_2}{L}\right)^2, \quad (10)$$

with the safety factor $S = 1.6$ for rolled and forged steels.

2.4 Flat End Closures and Tube Plates

Flat plates are always used if the pressures or pressure differences are small, or when it is necessary to have a flat separating face. This is the case in the majority of tube bundle apparatuses and for the covers and closures of high-pressure vessels. Where no flatness requirement exists, it should be determined whether the separating or closure function can be fulfilled by domed components. These permit a more economic use of material.

Tube plates are used in the range of sizes from 100 to 4500 mm. As flat and hole-free closures of very large vessels or apparatus, they are also used up to 8000 mm. The thickness of tube plates varies from a lower limit (membrane closure) of a few millimetres to an upper limit of 650 mm for steam generation in nuclear power stations. Tube divisions lie between $1.2d$ and $1.5d$ and the number of holes between the limits of 10 to 10^4 (the latter in steam generators of nuclear power stations). Besides the circular plate (most frequently used), rectangular or elliptic plates, annular plates or flat closures with flanged rims are also employed.

Flat-walled components can, in principle, be implemented unreinforced or reinforced by profiling or by tie-rods. The plate thickness may vary in the radial direction.

In fixed-tube apparatus (see H3), the most frequent form of application, the tube plates are mutually reinforced by tubes that have been welded or rolled in.

Wall Thickness. For flat end closures and tube plates this is calculated from

$$s = CD\sqrt{\frac{ps}{Kv}} + c_2 \quad (11)$$

(notation: see Eq. 1a).

If t is a tube division the reduction correction for plates with re-entrant tubes (U-tubes) is given by

$$v = (t - d_a)/t \,,$$

and for plates with fully supporting tubes (fixed plate, floating head) by

$$v = (t - d_i)/t \qquad \text{for} \quad d_a/d_i \leq 1.2 \quad \text{and}$$
$$v = (t - 0.833d_a)/t \qquad \text{for} \quad d_a/d_i > 1.2,$$

where d_a is the external diameter of the hole or tube and d_i the internal diameter.

Calculation Correction and Diameter. The values of these, C and D, depend on the type of end closure, its connection to the casing, and the arrangement of the tubes. If no additional edge moment is present, then for full flat plates and uniformly perforated plates with re-entrant tubes (hairpin tubes) C depends on the plate support. C lies between 0.32 and 0.35 for plates with fixed support and between 0.40 and 0.45 for plates with loose support. For welded-in plates see [1]. If, in the case of fixed-plate apparatus, the supporting action of the tubes is to be taken into account, the permissible buckling force for the tubes must be observed.

If an edge moment acting in the same sense is present, some of the correction values will increase considerably. In the case of rolled-in or welded-in tubes, the tensile force F exerted on an individual tube must be transmitted to the tube plate. The formula $10F/(\sigma_w d_a) \geq s_w \geq F/(\sigma_w (d_a - d_i))$ applies to the standard rolled length s_w, where 12 mm $\leq s_w \leq$ 40 mm. The permissible stress σ_w of the rolled joint is 150 N/mm² if the rolled joint is smooth, 300 N/mm² if it is grooved, and 400 N mm² if it is flanged. In the case of welded-in tubes, the thickness of the welding seam g at the shear cross-section must be at least $g = 0.4F/(d_a K/S)$.

2.5 Domed End Closures

The shapes of domed end closures lie between the limiting cases of flat and hemispherical closures. In Central Europe, torispheric closures predominate, consisting of a spherical cap (radius R) and a rim (radius r) (**Fig. 1**). Well-known constructions are the dished end ($R = D_a$, $r = 0.1D_a$) and the three-centre arch end closure ($R = 0.8D_a$, $r = 0.154D_a$). In general, $R \leq D_a$, $r = \geq 0.1D_a$ or $r \geq 30s$ applies, where s is the required wall thickness of the domed closure. The height of the rim h_1 should be not less than $3.5s$ for dished and $3.0s$ for three-centre arch

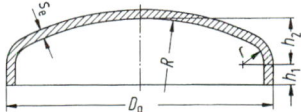

Figure 1. Domed end closure.

closures. For closures that consist of a flange part and a spherical cap part that have been welded together, a minimum distance x must be maintained between the welding seam and the flange. For dished end closure, $x = 3.5s$ and for three-centre arch closures $x = 3.0s$, but must be at least 100 mm.

In the Anglophone countries, the ellipsoidal shape predominates, usually with a 2 : 1 ratio of the axes. In all cases, dome end closures allow better utilisation of material than flat closures. In comparison with hemispherical closures, they offer the advantage of smaller height and, frequently, better access. The dimensions lie between the limits of 50 and 12 000 mm.

Connection between end closures and neighbouring components should, if possible, be implemented in the form of butt joints. Transitions in cross-section must be implemented in conical form.

The rules for calculation apply to domed end closures with a cap radius $R \leq D_a$, a rim knuckle radius $r \geq 0.1D_a$ and a wall thickness $s_e \geq 0.001D_a$ ($s_e \geq 2$ mm). In the presence of external pressure, the safety factor must be increased by 20%.

Required Wall Thickness. With the calculation correction factor β and other factors, as in Eq. (1a), the required wall thickness is given by

$$s = \frac{D_a p \beta}{4 v K/S} + c_1 + c_2. \tag{12}$$

Calculation Correction Factors. With $x = (s_e - c_1 - c_2)/D_a$ and $y = d_i/D_a$, the following applies to domed end closures:

Dished. $\beta = \max \ (1.9 + 0.0325/x^{0.7} + x; \ 1.9 + 0.933y/\sqrt{x})$, *Three-Centre Arch* $\beta = \max \ (1.55 + 0.0255/x^{0.625}; \ 1.55 + 0.866y/\sqrt{x})$, where $0.001 \leq x \leq 0.1$ and $0 \leq y \leq 0.6$. The formulas apply to cutouts with diameter d_i in the region of the flange and outside the crown region of $0.6D_o$ of the cap.

Buckling. In the presence of internal pressure, the end closures are dimensioned adequately against buckling in the flange region if the following apply:

Dished end closures, $p \leq 41.6E[(s_e - c_1 - c_2)/D_a]^{2.24}$
Three-centre arch, $\quad p \leq 33.3E[s_e - c_1 - c_2)/D_a]^{2.34}$ $\tag{13}$

where E is the modulus of elasticity (see B1.2). In the presence of external pressure, additional safety against elastic indentation must be provided:

$$p \leq 0.366 \frac{E}{S_k} \left(\frac{s_e - c_1 - c_2}{R} \right)^2. \tag{14}$$

The safety correction factor S_k can be obtained from **Table 2**.

2.6 Cutouts

The equations apply to vessels with internal pressure and for $D_o D_i \leq 1.2$, with a distinction between disc-shaped and tube-shaped reinforcements. Within limits, these are mutually interchangeable, or they may be used simultaneously.

Reduction Correction Factor. For a wall thickness s_S of the connection tube of diameter d_i, required wall thickness s_A at the edge of the cutout and with

$x = d_i / \sqrt{(D_i + s_A - c_1 - c_2) (s_A - c_1 - c_2)}$ and
$y = (s_S - c_1 - c_2)/(s_A - c_1 - c_2)$

the reduction correction factor is given as follows:

Cylinders $\quad v_A = 2/(2 + x) + 2.52y^{1.66}/(10 + x);$
Spherical shells
$\quad\quad v_A = 2/(2 + x) + 1.28y^{1.75}/(4.5 + x);$ $\tag{15}$

with $0 \leq x \leq 8$ and $0 \leq y \leq 2$.

Width and Thickness. The width of a disc-shaped reinforcement must be at least

$$b \geq \sqrt{(D_i + s_A - c_1 - c_2) (s_A - c_1 - c_2)}$$

and $\quad\quad\quad\quad b \geq 3s_A.$

The thickness of the reinforcement must not exceed the actual thickness s_e of the end closure. The length l_s of a tube-shaped reinforcement must be at least

$$l_6 = 1.25 \sqrt{D_i + s_S - c_1 - c_2} \ (s_S - c_1 - c_2).$$

The ratio of the wall thicknesses should be

$$(s_S - c_1 - c_2)/(s_A - c_1 - c_2) \geq 2.$$

Mutual influence of two cutouts can be neglected if the distance between their closest points $l \geq 2 \sqrt{(D_i + s_A - c_1 - c_2) (s_A - c_1 - c_2)}.$

2.7 Flange Joints

2.7.1 Bolts

At operating temperatures above 300 °C or operating pressures exceeding 40 bar, tension bolts must be used. Here, only those bolts will be rated as tension bolts whose shank diameter $d_s \leq 0.9d_K$ or whose dimensions comply with DIN 2510. Bolts with a full-length thread are rated as stud bolts. If possible, bolts smaller than M10 should not be used. The number of bolts should be as large as possible (ratio of interval between bolts to bolt hole diameter, $t/d_L \leq 5$).

Stress Conditions. As seen in **Fig. 2**, the following forces are applied to a flange: F_R, longitudinal force; F_p, force on annular cross-section due to internal pressure; F_D, jointing force; F_S force arising from bolting. The flange must withstand these forces. Forces that may occur due to a bending moment in connected piping systems are, as a rule, not taken into account. In compliance with AD Code of Practice B7 and also if test pressures exceed $1.3p_B$, the forces at the bolts must be evaluated in accordance with the operating and installation conditions before pressure is applied.

At the permissible material stress K and the safety cor-

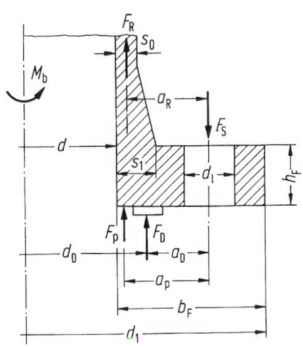

Figure 2. Forces applied to a fixed flange.

Table 3. Safety correction factor S for bolted connections [1]

		Operating condition	Installation and test condition
Materials of known UTS and safety in proportion to UTS or $\sigma_{B/100\,000}$	For tension bolts e.g. to DIN 2510	1.5	1.1
	For stud bolts e.g. to DIN 2509 or DIN 931	1.8	1.3
Materials of unknown UTS with safety in proportion to UTS		5.0	3.0

rection factor S (**Table 3**), the thread diameter d_k of a stud bolt and the shank diameter d_s of a tension bolt in a connection employing n bolts are both given by:

$$d_s \quad \text{and} \quad d_k = \sqrt{\frac{4SF_S}{\pi \varphi Kn}} + c \qquad (16)$$

For a support surface created by a metal-removing process or an equivalent surface, the quality rating factor φ can be put at 1.0, otherwise at 0.75. The design allowance c, for the operating condition, must be $c = 3$ mm up to M24 and $c = 1$ mm from M52 or corresponding diameter onwards. Linear interpolation must be used in the intermediate region, where $c = 0$ mm for tension bolts. The temperature used for calculation is around 30 °C below the highest temperature of the working medium for connections of loose flange to loose flange, around 25 °C for fixed flange to loose flange and around 15 °C for fixed flange to fixed flange.

Jointing Forces. These depend on the shape, the jointing material and the operating conditions (pressure and temperature). If k_0 and k_1 are the effective width of the gasket during installation and operation respectively, d_D the mean diameter of the gasket, p the pressure to be sealed off and $K_{D,\vartheta}$ the deformation stress of the gasket (see **Table 4**), then the required sealing force is $F_D = \pi d_D k_1 p S_D$ with $S_D \geq 1.2$ for operating conditions and $1 \leq S_D \leq 1.2$ for test conditions. The required initial compression force is $F'_{Do} = \pi d_D k_0 K_{D,20}$. For gaskets of soft materials with $pD \leq 10$ bar · m,

$$F''_{Do} = 0.2F'_{Do} + 0.8\sqrt{(0.25\pi d^2 p + F_D)F'_{Do}} ,$$

applies if $F''_{Do} < F'_{Do}$. In the case of metal gaskets $F_{max} = \pi d_D k_0 k_{D,\vartheta}$.

Width of Gasket. Under operating conditions for liquids, this is $0.5b_D \leq k_1 \leq 1.1b_D$ if the gaskets are of soft material or soft metal material and $0.5b_D \leq k_1 \leq 1.8b_D$ for gases and vapours, depending on shape and material. For

Table 4. Deformation stress $K_{D\vartheta}$ of metallic sealing materials [1]

Sealing material	$K_{D,\vartheta}$ in N/mm² at					
	20	100	200	300	400	500 °C
Aluminium, soft	100	40	20	(5)		
Copper	200	180	130	100	(40)	
Soft iron	350	310	260	210	170	(80)
Steel St 35	400	380	330	260	190	(120)
13 CrMo 44	450	450	420	390	330	280
Austenitic steel	500	480	450	420	390	350

flat metallic gaskets, $k_1 = b_D + 5$ mm and for other types of metallic gasket 5 mm $\leq k_1 \leq 6$ mm. For the effective gasket width under installation conditions, $0.8b_D \leq k_0 \leq b_D$ applies to flat metallic gaskets, and $0.16 \leq k_0/k_1 \leq 0.33$ to other metallic gaskets depending on shape (**see Appendix H2, Table 1**).

2.7.2 Flanges

The bolting force to be taken up by the flange (see **Fig. 2**) is $F_S \geq F'_{Do}$ during initial compression and $F_S \geq F_D + F_R + F_p$ during pressure testing and operation.

Weakening of the flange by the bolt hole is accounted for in the calculation by a calculating diameter d'_L ('reduced bolt-hole diameter') (**Fig. 4**).

Loose Flange (Fig. 3a). The resistance moment with $b = d_a - d_i - 2d'_L$ and with the reduced bolt-hole diameter d'_L (**Fig. 4**) must be at least

$$W = 0.7874 b_F^2 b \geq F_S \frac{S}{K}(d_t - d_4)/2 . \qquad (17)$$

Welded Flange (Fig. 3b). Here,

$$W = 0.7042\,[b_F^2 b + (d_i + s_1)s_1^2] \geq F_S \frac{S}{K}a \qquad (18)$$

applies, with $b = d_a - d_2 - 2d'_L$; $a = 0.5(d_t - d_i - s_1)$ for test and operating conditions and $a = 0.5(d_t - d)$ for installation conditions.

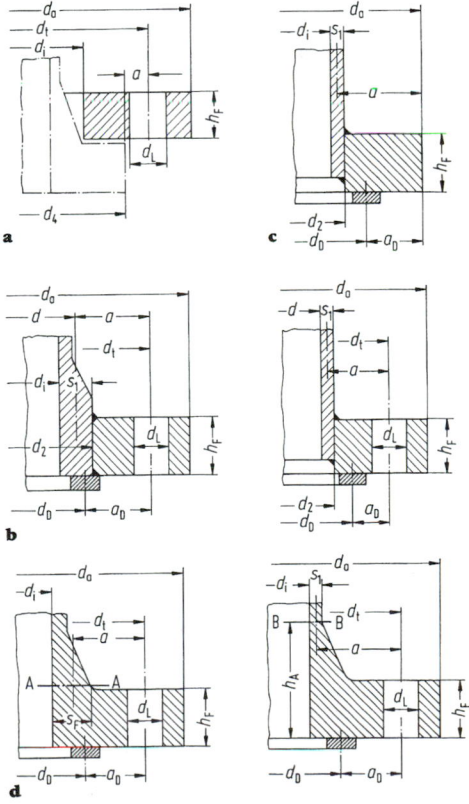

Figure 3. Flange types. **a** Loose backing flange. **b** Welded flange. **c** Welded collar. **d** Weldneck flange.

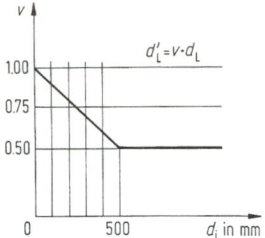

Figure 4. Reduced diameter of bolt hole.

Welded Collar (Fig. 3c). Equation (18) applies where d_o is used in place of d_t and $d_L = 0$ is entered.

Weldneck Flange (Fig. 3d). Here the section moduli W for cross-section A–A and B–B must be verified by calculation.

$$D_i \leq 1000 \text{ mm}$$

Section A–A: $W = 0.7874(b_F^2 b + (d_i + s_F)s_F^2 \geq F_S(S/K)a,$

Section B–B: $W = 0.7874(b_F^2 b/B^2 + 0.75(d_i + s_1)s_i^2)$
$$\geq F_S\,(S/K)a,$$

where $B = \dfrac{b + (s_F - s_1)B_1}{b + (s_F + s_1)B_1(B_1 + 2)}$ with $B_1 = (b_A - b_F)/b_F,$

$b = d_o - d_i - 2d'_L$ and $a = 0.5(d_t - d_i - s_1)$ under pressure testing and operating conditions and $a = 0.5(d_t - d_D)$ under installation conditions.

$$1000 \text{ mm} \leq D_i \leq 3600 \text{ mm}$$

Section A–A: $W = 0.943(b_F^2 b + (d_i + s_F)s_F(0.8s_F + 0.1b_F)$
$$\geq F_S(s/K)a,$$

Section B–B
$W = 0.943(b_F^2 b/B^2 + 1.5(d_i + s_1)s_i^2) \geq F_S(S/K)a$

with B, b and a as for $D_i \leq 1000$ mm. However, $b_A - b_F$ must be $\geq 0.6\,b_F$ and $s_F - s_1$ must be $\geq 0.25b_F$.

2.8 Piping

2.8.1 Pipe Diameter

The internal pipe diameter d is calculated from the continuity equation with volume flow $\dot{V}$ and pipe cross-section corresponding to the chosen flow velocity v as $d = \sqrt{4\dot{V}/(\pi v)}$. If $\dot{V}$ is given, v must be chosen so that the operating and piping system costs are low and d corresponds to standard values. Large v implies a small pipe diameter, small fittings, low costs of insulation and painting, but with, on the other hand, high pressure-drops (greater costs of pumps, higher operating costs) and higher noise levels.

The economic diameter is that leading to the lowest sum of plant and operating costs, taking into account the efficiency of utilisation (= operating time/(operating time + down time)).

Standard velocity values are to be found in [6, 7] and in **Appendix H2, Table 2**.

2.8.2 Flow Losses

Incompressible fluids cause pressure drops; compressible fluids (gases) cause pressure drops, volume expansion and accelerations. Heat exchange with the surroundings depends on insulation.

Pressure drops are made up from losses in straight lengths of pipe and losses occurring in bends and fittings (individual resistances); see A6.2 and [1]. For pressure-drops in steel pipe see **Appendix H2, Fig. 1**, and for fittings see H2.9.1 and **Appendix H2, Fig. 2**.

2.8.3 Types, Standards, Materials

General

Important standards and regulations for piping systems:

DIN 2400 Piping systems; summary of standards for planning, design and materials.

DIN 2401 T1 (Part 1), Components subject to internal and external pressure stress; pressure and temperature data, concepts, standard pressure ranges. T2 (Part 2), Piping systems; pressure ranges, permissible operating pressures for piping components of ferrous materials.

DIN 2402 Piping systems; nominal sizes; grading.

DIN 2406 Piping systems; letter symbols; pipe categories.

DIN 2408 T1 (Part 1), Piping systems for process plant; planning and design data.

DIN 2413 Steel pipes; calculation of wall thickness with respect to internal pressure.

DIN 4279 T1 (Part 1), Tests for internal pressure of pressure piping systems for water; general data. T2 to T10 (Parts 2 to 10), Tests for internal pressure of pressure piping systems for water, various materials.

ISO 4200 Seamless and welded pipe; summary of dimensions.

Pressure Vessels V, Regulations relating to pressure vessels, May 1989.

VdTüV Code of practice sheets relating to various test procedures for piping systems. Maxmilian-Verlag, Herford.

DVGW Worksheets for the construction of gas and water pipeline systems. ZfGW-Verlag Frankfurt a.M.

Nominal Pressure PN. This is the pressure in standard piping components based on a given starting material mentioned in the relevant standards and designed for a temperature of 20 °C. It corresponds to the permissible operating pressure at the lowest operating temperature of a material. Staging of the nominal pressures PN in bars (units not given):

1	1.6	2.5	4	6
10 12.5	16 20	25 32	40 50	60 80
100 125	160 200	250 320	400 500	600 800
1000	1600	2500	4000	6000

Nominal Size DN. This is the reference parameter (characteristic indicator) for parts fitting each other, e.g. pipes with formed parts, or fittings. The nominal size is stated without units; it corresponds approximately to the internal size in mm.

Summary of Types of Pipe (DIN 2410). General data relating to *welded pipe* of unalloyed or low-alloy steels DIN 1626: Commercial grade: For general requirements of piping systems and vessels as well as of equipment manufacture. Up to 120 °C: For liquids up to 25 bar, for air and non-hazardous gases up to an operating pressure of 10 bar; up to 180 °C: For saturated steam up to 10 bar. Materials: St 33, St 37, St 42. With Regulations for commercial grades: For more demanding applications, suitable for bending, flanging and similar; up to 120 °C: up to 64 bar, over 120 and up to 300 °C, also up to an operating pressure of 64 bar if wall temperature in °C multiplied by pressure in bar ≤ 7200; with special acceptance without specified limit. Specially tested pipe with quality specifications:

For especially demanding requirements; up to 300 °C without specified limitation of the operating pressure.

General data regarding *seamless tube* of non-alloyed or low-alloy steel DIN 1629: Applications and materials similar to DIN 1626.

Precision Steel Tube. Seamless (DIN 2391, for all pressures, 4 to 120 mm external diameter), welded (DIN 2393, for all pressures, 4 to 120 mm external diameter), welded and single cold-drawn (DIN 2394, up to PN 100, 6 to 120 mm external diameter) for high-precision applications, especially surface conditions, low wall thicknesses. Designation and material: tube 30 × 2 DIN 2391 St 35 (or St 45 or St 55) bright drawn, soft, hard, soft annealed, etc.

Threaded Tube. This is seamless or welded, medium-heavy (DIN 2440) and heavy (DIN 2441) of St 33-1 or St 33-2.

Seamless Steel Tube (DIN 2445, 2448, 2449, 2450, 2451, 2456 and 2457). This is made of different steels St 00 to St 52 (corresponds to DIN 1629) with 10.2 to 558.5 mm external diameter. At the same external diameter lower wall thicknesses than DIN 2240, e.g. for outside diameter = 60.3 mm, according to DIN 2448, wall thickness = 2.9 mm standard (however, a large choice is possible), as compared with wall thickness = 3.65 mm according to DIN 2440. It is available up to PN 100, hence suitable for the most diverse purposes in mechanical engineering and equipment manufacture.

Welded Steel Tube (DIN 2458). This is made of St 33 to St 52-3 steels for all nominal pressures with 10.2 to 1016 mm outside diameter and even smaller wall thicknesses than DIN 2448, e.g. at $d_o = 60.3$ mm, $s = 2.3$ mm standard (however, an equally large choice as with DIN 2448, hence extended field of application).

Steel Tube for Gas and Water Piping Systems. This is seamless (DIN 2460) and made of various steels: St 00 for gas up to PN 1 and water up to PN 25, St 35 for gas up to PN 100 and water up to PN 64; 88.9 up to 508 mm external diameter. Welded tube (DIN 2461) is of St 33 for gas up to PN 1 and water up to PN 20, St 37-2 for gas up to PN 80 and water up to PN 64; 88.9 to 2020 mm external diameter. Surface protection: external protection is by bituminous materials with glass-fibre matting and whitewashed; internal protection is by bitumen coating, linseed oil, cement mortar or other materials forming protective coatings. Applications are gas or water piping systems external to buildings in or above the ground.

Steel Tube for Long-Distance Pipelines. This is employed for flammable liquids and gases (DIN 17 172) for all pressures from 100 mm external diameter upwards.

Cast-Iron Pipe

Pressure Pipe. For water up to PN 16 threaded sockets DIN 28 511 (DN 40 to DN 600), stuffing box sleeves DIN 28 512 (DN 500 to DN 1200), lead-joint sockets DIN 28 513 (DN 40 to DN 1200), flanges DIN 28 514 and

28 516 (DN 40 to DN 1200) and TYTON sleeves DIN 28 516 (DN 50 to DN 600).

Pressure Pipe of Ductile Iron (DIN 28 610) with threaded sockets (water up to PN 40, DN 80 to DN 600), stuffing box sleeves (water up to PN 25, DN 500 to DN 1200), or TYTON sleeves (water up to PN 40, DN 80 to DN 600), for gas up to PN 1.

Other Pipe Materials

Copper. DIN 1754 for external diameter 3 mm (wall thickness 1 mm max.) up to 419 mm (wall thickness 4 mm max.); material: copper DIN 17 671 with strength specification F20 ($\sigma_B = 200$ to 250 N/mm², $\delta_5 = 40\%$) up to F37 ($\sigma_B = 360$ N/mm², $\delta_5 = 3\%$), usual F30 ($\sigma_B = 290$ to 360 N/mm², $\delta_5 = 6\%$).

Aluminium. DIN 1795, preferred dimensions for piping systems of high purity Al, pure Al, and Al wrought alloy with external diameter 3 mm (wall thickness 1 mm max.) to 273 mm (wall thickness 5 mm max.).

Polyvinylchloride (PVC) Hard (unplasticised PVC (uPVC)) for drainage installations, vent ducting, water and gas piping systems. For general quality specifications see DIN 8061; for dimensions see DIN 8062: External diameter 5 mm (wall thickness 1 mm max.) up to 1000 mm (wall thickness 29.2 mm max.). For standards relating to chemical stability see DIN 16 929.

Other Plastics [8]. DIN 8072 soft polyethylene tube. DIN 8074 high-density polyethylene tube. DIN 8077 polypropylene tube. DIN 16 868 and DIN 16 869 T1 (Part 1), glass-fibre-reinforced polyester resin tube. DIN 16 870 and DIN 16 871 T1 (Part 1), glass-fibre-reinforced epoxy resin tube.

2.8.4 Pipe Fittings

For Steel Pipe

Flanged Connections (**Fig. 3** and **Fig. 5**). These are preferred for high-pressure lines and because they are easily dismantled. DIN 2500 gives a summary for steel and cast iron. For mating dimensions, see DIN 2501.

Standards for Flange Shapes

Fig. 5a and **b**:

DIN 2558 (PN 6; DN 6 to 100),
DIN 2561 (10, 16; 6 to 40),
DIN 2566 (10, 16; 6 to 100).

Fig. 5c:
DIN 2573 (6; 10 to 500),
DIN 2576 (10; 10 to 500).

Fig. 5d:
GG: DIN 2530 (1; 10 to 4000),
DIN 2531 (6; 10 to 3600),

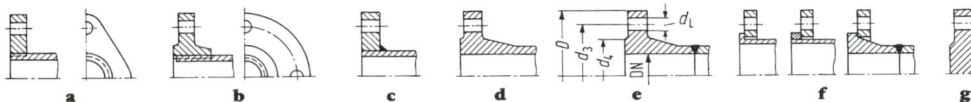

Figure 5. Flange forms. **a** Screw-on flange, oval, smooth. **b** Screw-on flange with boss, round. **c** Smooth-face flange for soldering or welding. **d** Flange of GGL, GS or GGG. **e** Weldneck flange. **f** Loose backing flange. **g** Blind flange.

DIN 2532 (10; 10 to 3000),
DIN 2533 (16; 10 to 1000),
DIN 2534 (25; 10 to 500),
DIN 2535 (40; 10 to 400); *GGG*: DIN 28 604 (10; 40 to 1200),
DIN 28 605 (16; 40 to 1200),
DIN 28 606 (25; 40 to 600),
DIN 28 607 (40; 40 to 400); *GS*: DIN 2543 (16; 10 to 2200),
DIN 2544 (25; 10 to 2000),
DIN 2545 (40; 10 to 1600),
DIN 2546 (64; 10 to 1200),
DIN 2547 (100; 10 to 700),
DIN 2548 (160; 10 to 300),
DIN 2549 (250; 10 to 300),
DIN 2550 (320; 10 to 250),
DIN 2551 (400; 10 to 200).

Fig. 5e:

DIN 2630 (1; 10 to 4000),
DIN 2631 (6; 10 to 3600),
DIN 2632 (10; 10 to 3000),
DIN 2633 (16; 10 to 2000),
DIN 2634 (25; 10 to 1000),
DIN 2635 (40; 10 to 500),
DIN 2636 (64; 10 to 400),
DIN 2637 (100; 10 to 350),
DIN 2638 (160; 10 to 300),
DIN 2628 (250; 10 to 300),
DIN 2629 (320; 10 to 250),
DIN 2627 (400; 10 to 200).

Fig. 5f:

DIN 2641 (6; 10 to 1200),
DIN 2642 (10; 10 to 800),
DIN 2655 (16; 10 to 500),
DIN 2656 (25; 10 to 400).

For Welded-on or Formed Collar.

DIN 2673 (10; 10 to 1200),
DIN 2674 (16; 10 to 500),
DIN 2675 (25; 10 to 500),
DIN 2676 (40; 10 to 500),
DIN 2667 (160; 100 to 250),
DIN 2668 (250; 100 to 250),
DIN 2669 (320; 100 to 250).

Fig. 5g:

DIN 2527 (6 to 100; 10 to 500).

Screwed Connections. Steel fittings for the chemical and shipbuilding industries: see DIN 2980 and 2982. Detachable connections for connecting to apparatus in frequent need of repair or possible modifications using flat gaskets (Klingerite gaskets) or conical sealing surfaces metal-to-metal contact, **Fig. 6a**. See also DIN 2353 and DIN 3930: Solderless screw connections with olives; **Fig. 6b**. Advantage of this screw connection: high-pressure loading (up to DN 630, easy installation, small space requirement, suitable for various tube qualities.

Welded Connections. Welded tubes have the advantages of unchanged fluid-tightness (hence the testing of the welded seam for fluid-tightness by X-rays or ultrasound in the case of important long-distance pipelines) and – in contrast to flanged connections – of low heat loss. Branch components and those providing change of direction or

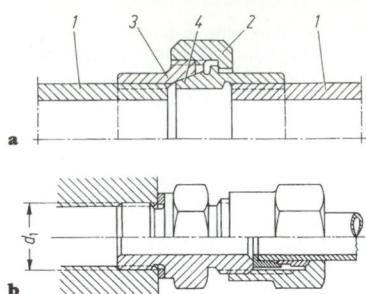

Figure 6. Screwed connectors for pipes. **a** *1* Steel pipe. *2* Union nut. *3* Female seating cone. *4* Male seating cone. **b** DIN 3930 mechanical compression fitting.

cross-section of every type are also made from tubular material. Modern piping systems mostly feature only flanged or screwed connections for their fittings. If welding is not carried out carefully in pipes of small nominal sizes (approximately below DN 50), reduction of cross-section could occur and hence increase in flow resistance.

Processes. Gas welding (for unalloyed or low-alloy steels with wall thicknesses of up to about 3 mm); arc welding (for wall thicknesses above 3 mm); inert-gas welding or submerged-arc welding (for automatic welding of large piping systems); see DIN 8564: welding of piping systems; steel piping systems, fabrication, testing of welded seams.

Additional standards, guidelines and Regulations must be observed [8].
DIN 2559 T1 (Part 1), weld preparation for the seam, guidelines for end preparation forms.
DIN 8558 T1 (Part 1), guidelines for welded connections to steam boilers, vessels and piping systems; examples of implementation.
DIN 8560 certification of welders.
DIN 8563 T1 (Part 1), quality control of welding work; T3 (Part 3) fusion-welded connections for steel.

For Cast-Iron Pipe

Socket Connections (**Fig. 7**). These are preferred for cast iron (GG) and cast iron globular (GGG). Direction of flow from the socket end to the plain end of the pipe. Advantage: quick installation; disadvantages: exact pipe installation length required, sensitive to longitudinal forces.

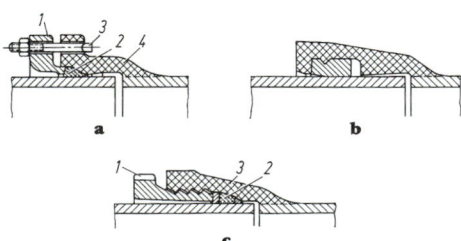

Figure 7. Socket connections. **a** Stuffing box. *1* Stuffing box ring follower. *2* Sealing ring. *3* T-bolt with nut. *4* Stuffing box socket. **b** Plug socket. **c** Screwed socket. *1* Screwed ring follower. *2* Sealing ring. *3* Screwed socket.

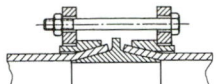

Figure 8. Connections for plastics pipe.

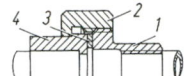

Figure 9. Screwed connections for PVC pipe. *1* Screwed sleeve. *2* Union nut of uPVC or annealed cast iron (GTW) or Cu-Zn alloy. *3* Flat ring seal. *4* Nipple sleeve, glued on.

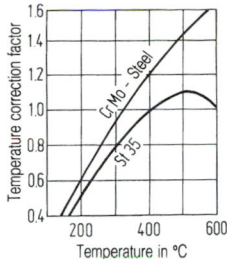

Figure 11. Correction factor for temperature conversion.

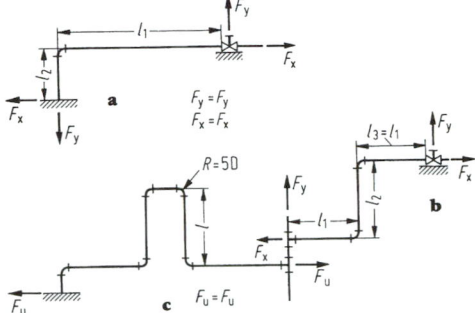

Figure 10. Simple expansion compensators. **a** Pipe leg. **b** Z-bend. **c** U-bend.

For Copper Pipe

Flanged and Screwed Connections. These are similar to those for steel pipe, but with different pressure ranges (strength).

Welded Connections. These are very extensively used in equipment manufacture.

For PVC and Other Plastics Pipe

Flanged Connections. See DIN 8063 for large diameters with loose backing flanges (mostly metal; **Fig. 8**).

Screwed Connections (**Fig. 9**). See DIN 8063.

Welded and Adhesive Connections. For processes, see DIN 19 533. PVC is mostly hot-air welded using filler rod; polyethylene is welded by fusion jointing. PVC can be glued using preformed or glued-on adhesive sleeves (similar to soldering sleeves). Adhesives are usually of the soluble type (tetrahydrofuran). Polyethylene cannot be glued.

2.8.5 Expansion Compensators

Expansion compensators take up thermally induced changes in length (see Eq. (3) and **Appendix H2,**

Fig. 3), between two fixed points. From the constructional point of view, they may be classified as follows:

Expansion Compensation by Piping Layout (without additional fittings; **Fig. 10**). Anchors should be at fittings where possible. If large temperature differences are present, piping must be installed prestressed to counteract thermal expansion (e.g. for compressive stresses when piping heats up, install under tension – cold-set). Customary prestress equals 50% of expected stress [9].

Length *l* of piping leg for steel with pipe outside diameter *D* and change of length Δl is $l = 0.0065 \sqrt{D \, \Delta l}$; for copper $l = 0.0032 \sqrt{D \, \Delta l}$; for calculation see [11].

Approximate Calculation of the Anchor Forces. This is carried out with numerical equations for St 35 at a temperature of 400 °C with 50% prestressing and a radius of curvature $R = 5d$. For conversion to other temperatures and materials, see **Fig. 11**.

Return Bends. $F_u = 10I\Delta l/(l^3 C)$ in N, where Δl is the total expansion between the anchors in cm, *I* is the axial geometrical moment of inertia of the pipe in cm⁴ and *C* is the correction factor as in **Fig. 12a**.

Piping Legs. $F_x = b_1 I/l^2$, $F_y = b_2 I/l^2$ in N.

Z-Bends. $F_x = b_3 I/l^2$, $F_y = b_4 I/l^2$ in N. *I* in cm⁴ applies to both, $l = l_x + l_y$ in m is the total length of the piping legs. Correction values b_1 to b_4 are as in **Fig. 12b**.

Expansion Compensation Through Special Components [10].

Expansion Loops (**Fig. 13a**) are, like return bends, very safe in operation, requiring no maintenance. They do, however, take up a great deal of space. They are suitable for outside use. Implementation may be effected as plain, corrugated or folded pipes. They should be arranged, if possible, so that the apex of the loop itself does not move, but is restrained by a sliding support. The anchor force is as for return bends.

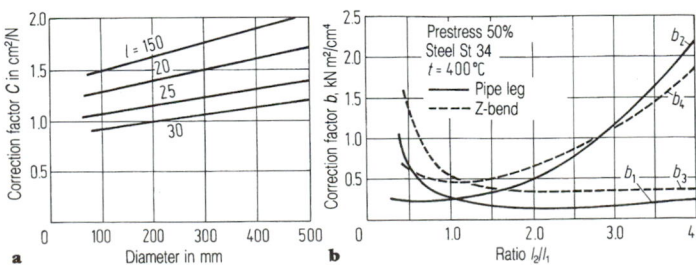

Figure 12. Correction factor for the calculation of axial pipe forces. **a** U-bend. **b** Z-bend and pipe leg.

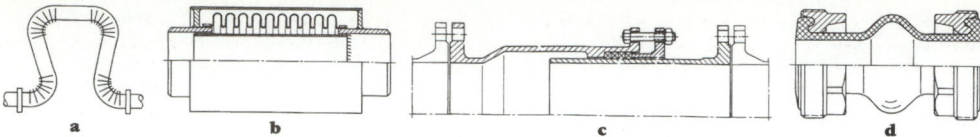

Figure 13. Expansion compensators. **a** Expansion loop. **b** Axial compensator with internal pipe (bellows compensator). **c** Telescopic compensator. **d** Rubber compensator.

Bellows Compensators are expansion compensators, require no maintenance and take up the least space. Lenticular compensators with few but high corrugations are usual for very large diameters (about DN 5000). Single- or multi-layer bellows (**Fig. 13b**) with many low corrugations, made up of single or multi-layer cold-worked sheet steel with large expansion capabilities, are used for high pressures (DN 600: PN 100; DN 250: PN 250).

Rubber Compensators (**Fig. 13d**) of various kinds are used for DN 40 to DN 400 and for temperatures of 100 °C at PN 10.

Telescopic Compensators (**Fig. 13c**) are prefabricated. The inner tube is smoothed and sometimes hard-chromed so as to reduce frictional resistance. *Packing Materials*: Permanently elastic Buna seals require no maintenance and can be used for almost all media. Plastic seals (adipose hemp for water, lead lamellae–asbestos for gas) must be after-sealed.

Articulatory Compensators, apart from axial expansion, also take up lateral deformation. Attention must be paid to axial forces during installation!

2.8.6 Pipe Supports

With reference to the pipe and surroundings (e.g. buildings), the purpose of pipe supports is to fix free piping systems to make them operationally secure.

Hangers are intended to carry the pipes. The drop must be adjusted accurately and a certain movement allowed. Designs go as far as 'constant hangers' where the forces caused by expansion are kept constant by means of preset springs and articulated levers.

Supports fulfil the same function as hangers, with the difference that the forces are restrained from below (**Fig. 14**).

Anchors serve to fix the direction of expansion unequivocally; they are capable of resisting forces and

moments. The force acting on a fixed point is usually the resultant of forces acting in different directions.

Guides fulfilling the function of guide points supplement the anchors and allow axial movement and in part also torsional movement; see **Fig. 14** [10].

2.9 Shutoff and Control Valves

2.9.1 General

Function

Valves (pipe closers) in piping systems serve as follows:

Shutoff elements prevent the flow of a fluid. They must close tightly against the fluid flow and, in order to avoid percussive stresses, in such a manner that the velocity is not reduced to zero suddenly (exception: quick-action gate valves).

Control Elements (globe valves) are intended to affect the volume flow as a function of parameter to be controlled.

Safety Elements open a cross-section for pressure relief if pressure becomes excessive.

Types of Construction (Summary)

Fittings (DIN 3211) used are as follows:

Valves. A shutoff element (plate, cone, piston, ball) is lifted in the direction of flow to open a cylindrical annular cross-section as a free-flow cross-section (**Fig. 15a**). Valve-like shutoff elements, where, because of particularly favourable flow conditions or particular corrosiveness of

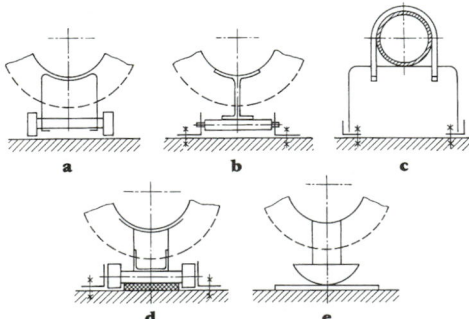

Figure 14. Pipe supports. **a** Pipe trolley. **b** Roller support. **c** U-bolt. **d** Cylinder bearer. **e** Dome head.

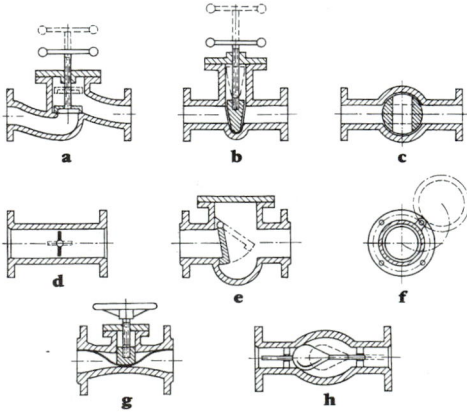

Figure 15. Basic types of shut-off element. **a** Valve. **b** Slide valve. **c** Cock (or ballvalve). **d** Rotary disc in pipe. **e** Flap valve. **f** Blank swing disc. **g** Diaphragm valve. **h** Teardrop clack valve.

the fluid, a membrane is depressed, are the diaphragm valve (**Fig. 15g**) and the cylindrical piston valve (**Fig. 15h**) in which the flow configuration has circular symmetry.

Slide Valves. The shutoff element (a circular plate with faces arranged either in parallel or to form a wedge) is moved at right-angles to the direction of flow to open up a crescent-shaped or circular free-flow cross-section (**Fig. 15b**).

Cocks or Rotary Slide Valves. The shutoff element (a ground truncated cone or a sphere with lateral bore) is rotated about its axis at right-angles to the direction of flow and opens up a lenticular or circular cross-section (**Fig. 15c**).

Flaps or Butterfly Valves. A disc, initially normal to the direction of flow, is turned about a hinge or about its own axis, which positions it parallel to the pipe axis in the pipeway; or a disc swings out of the pipeway on one flange bolt, thereby opening up the entire pipe cross-section (**Figs 15d-f**).

Slide valves and cocks that open up the full pipe cross-section are suitable when pull-through elements (pigs) are used to separate fluids conveyed in different forms, or for cleaning.

See **Table 5** for a comparison of the advantages/disadvantages of these construction forms.

Materials

Body materials are selected on the basis of flow medium requirements (erosion, corrosion), operating temperature (heat resistance), and operating pressure (strength, possibly resistance against bulging). For the selection of metallic materials see DIN 3339. About 80% of all bodies are cast, mainly from grey cast iron but also from cast steel and non-ferrous castable materials (brass and gunmetal in installation engineering). In the chemical and water-treatment industries there has been a sharp increase in castings made of plastics (usually injection-moulded). Some parts of valves made from steel are produced by drop-forging (at high pressure).

Grey Cast Iron

For water, steam, oil and gas, lined with rubber or enamel for aggressive media (GGL denotes cast iron alloy; GGG cast iron globular); GGL-20 to PN 16 at 120 °C, GGL-25 to PN 16 (25) at 300 °C; GGG-45 to 70 for feed water and live steam up to PN 40 at 450 °C.

Cast Steel (GS). GS-C25 for steam, water and hot oil up to PN 320 at 450 °C, easy to weld; GS-20 MoV 84 for steam and hot oil up to PN 400 at 550 °C, weldable; GS-X 12 CrNiTi 18.9 for acid-proof and hot valves.

Steel. C 20 for drop-forged bodies, yokes and flap screws, weldable; 50 CrV 4 for flanges, spindles, bolts and nuts up to 520 °C, conditionally weldable; X 20 Cr 13 for parts of valves subjected to high mechanical stresses, barely weldable; X 10 CrNiTi 18.9 with very good chemical stability (organic and mineral acids), weldable; X 10 CrNiMoTi 18.10 in the presence of highly aggressive acids and for higher temperatures, also for cold valves down to −200 °C, weldable.

Non-ferrous Metals. G-Cu64Zn, G-CuSn10, G-CuSn5 Zn7, G-AlMg3 and others, potable-water valves, physiologically acceptable, Al alloys, seawater proof (shipbuilding), also for the chemical industry.

Plastics and Others. Hard PVC (unplasticised PVC – uPVC), polyamide, PTFE, and silicones as well as ceramics for the chemical and hygiene industries.

Hydraulic Properties

In the case of sharp changes of direction, fittings (valves) cause large pressure drops, a desirable property where they are used as control elements. The flow-resistance coefficient ζ_R and velocity v are referred to the cross-sectional area of the connection A_R. The volume $\dot{V}$ is given by $\dot{V} = A_R \sqrt{2\Delta p/\rho\zeta_R}$, where $\Delta p = \zeta_R \rho v^2/2$ is the flow pressure drop. At large Reynolds numbers ($Re > 10^5$), ζ_R changes only slightly (for ζ_R values see **Appendix H2, Fig. 2**). For completely open shutoff elements, ζ_R of 0.2 to 0.3 can be assumed [12].

The value of k_v as defined in VDI/VDE Guidelines 2173 for control valves and in VDI/VDE Guidelines 2176 for control flaps, is important in control engineering as are the basic forms of control-valve characteristics [13]. Here the valve characteristics at constant Δp across the valve must be distinguished from the operating characteristics, which are affected by the ratio of the pressure drop across the valve to the total pressure drop in the pipe as a function of the flow [14].

2.9.2 Valves

Regardless of their function, valves are manufactured as straight-seat, slanted-seat or angle valves.

Straight-seat Valves (**Fig. 16**). These provide the most advantageous arrangement in piping systems, with easy operation and maintenance and uniform stressing of the valve elements but entail a high pressure-drop.

Slanted-seat Valves (**Fig. 17**). These possess a low resistance coefficient ζ_R.

Angle Valves. These can be advantageous if the additional function of an elbow is required, but mean higher pressure-drops. For dimensions see DIN 3202.

Table 5. Advantages and disadvantages of the various forms of construction

Property	Valves	Sliders	Cocks	Flaps
Flow resistance	moderate	low	low	moderate
Opening/closing time	medium	long	short	medium
Wear-rate of seat	good	moderate	poor	moderate
Suitability for changing direction of flow	moderate	good	good	poor
Installed length	large	small	medium	small
Installed height	medium	large	small	small
Range of use up to	medium DN maximum PN	maximum DN medium PN	medium DN medium PN	maximum DN only small PN
Suitability for throttling	very good	poor	moderately good	good

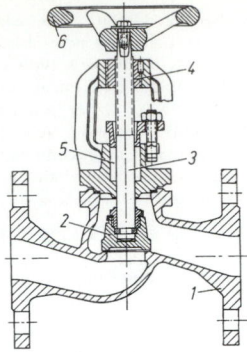

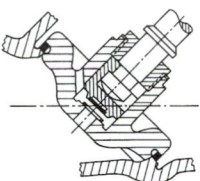

Figure 16. Straight-seat valve (J. Erhard).

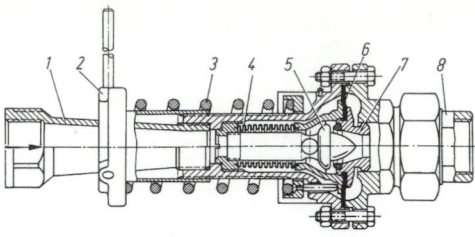

Figure 19. Axial pressure-reduction valve (Samson). *1* Coupling nipple. *2* Set-point adjuster. *3* Spring. *4* Metal sealing bellows. *5* Cone. *6* Working diaphragm. *7* Seat. *8* Connection nipple.

in one piece; the stem is located in a self-sealing cover, the shape of the body is advantageous for flow, and the stem nut is rotated (handwheel height is constant).

Valve Types for Various Purposes

Changeover Valve. This is employed for fluid flow that is to be directed through alternative piping systems.

Non-return Valve (prevents backflow). Fluid flow possible only against the force of a spring or a weight.

Pressure-Reducing Valve. Inlet pressure is reduced to an adjustable outlet pressure (lower pressure) which is kept constant with great accuracy independent of the inlet pressure and the flow rate. For example (**Fig. 19**), if outlet pressure falls owing to increased flow or falling inlet pressure, the diaphragm (*6*) with seat (*7*) moves to the right, thereby opening a larger orifice.

Float Valve. Here a hinged float raises or lowers the valve stem or the valve plate.

Steam Trap (**Fig. 20**). This drains off the liquid phase (e.g. condensate from saturated-steam equipment), float drainers, thermal drainers, thermodynamic drainers.

Safety Valve. This prevents an increase in the operating pressure to values above the permissible pressure (see AD Code of practice A2). The threshold pressure equals the permissible operating pressure. The valve may be weight-loaded (very accurate) or spring-loaded (for a compression spring, the valve force increases as the valve lifts).

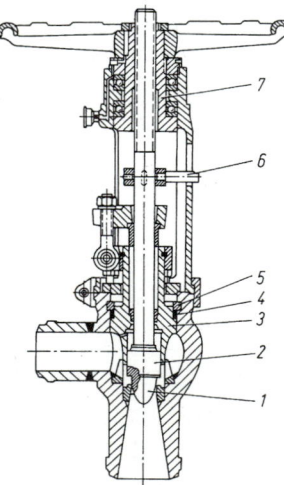

Figure 17. Canted-seat block and bleed valve.

Valve Components. **Figure 16** shows: *1*, valve body (cast, forged, welded or moulded manufacture); *2*, valve disc with seat rings (plate shape, conical or parabolic); seat rings of rubber, cast iron (GG), copper alloys, high alloy steels, stellite or nitride steel depending on fluid, pressure and temperature; *3*, valve stem, and *4*, nut; *5*, stuffing box for sealing off the valve stem; *6*, valve or stem drive (handwheel; electromotive, hydraulic, pneumatic or electromagnetic drive with remote control).

For large-seat cross-sections, a prelift valve (block-and-bleed valve) which reduces the opening force may be useful (**Fig. 17**). **Figure 18** shows a high-pressure control valve. This is a forged valve, with control cone and stem

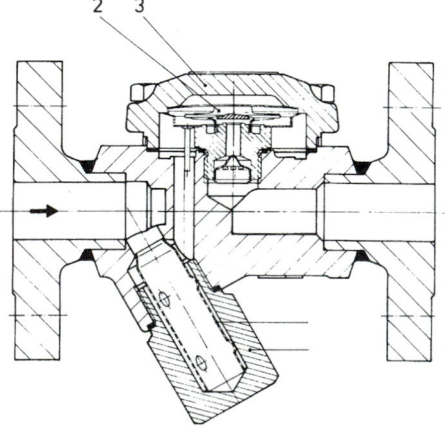

Figure 18. High-pressure forged control valve (Sempell). *1* Control cone. *2* Stem guide. *3* Cover, self-sealing. *4* Uhde–Bredtschneider seal with: *5* Divided ring. *6* Valve stem register to prevent stem turning. *7* Rotatable stem nut.

Figure 20. Thermal action steam trap with regulating diaphragm (GESTRA AG). *1* Body. *2* Regulating diaphragm. *3* Bonnet. *4* Backstroke cone. *5* Dirt collector. *6* Sieve holder/support.

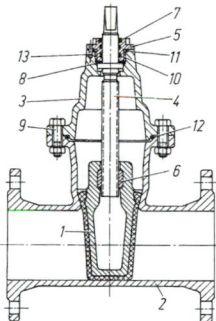

Figure 21. Shut-off gate valve. *1* Sealing wedge. *2* Body. *3* Bonnet. *4* Stem. *5* Lock nut. *6* Stem nut. *7* Dust ring. *8* Slip ring. *9* Hexagonal bolt. *10* to *12* O-rings. *13* Grooved cylinder dowel.

Quick-Action Valve. This shuts off the system in the case of pipe fracture or similar damage. Direct shutoff movement is by means of a spring, a weight or a pneumatic force (principle of the closed-circuit current).

2.9.3 Gate Valves

Field of Application. Large nominal sizes, high flow-velocities, low to medium nominal pressures, small installed lengths (see DIN 3202). A survey is given in DIN 3200.

Components. With the exception of seats and seals these are similar to valve components (see **Fig. 16**). **Figure 21** shows a simple shutoff gate valve with an internal stem nut (which is liable to seize up owing to dirt and high temperatures), and O-ring seals in place of a stuffing box.

Forms of Construction (Fig. 22). Depending on the shape of the bonnet flange, gate valves may be defined as *round-body* (large installed length, high pressure-strength of the bonnet connector), *oval* (shorter construction, low pressure-strength or increased wall thicknesses) or *flat valves* (further reduction in length, frequent strengthening of the cover connector by means of ribbing, particularly for large nominal sizes).

For a summary of materials and limiting conditions of application of gate valves, see DIN 3352 and [12].

In contrast to other valves, gate valves are always suit-

able for both directions of flow; however, they can be used only as shutoff elements. Generally they are implemented as straight-through devices (there are no angle types). The type of sealing is most important, since the stem force does not act directly on the sealing faces.

Figure 22a: Construction is simple; in the closed position a plate is seated by the line pressure. Sealing action is low during opening movement, and wear may occur owing to sliding friction. Used in long-distance gas pipelines.

Figure 22b: Raising the spectacle plate exposes the orifice. Where necessary, sealing is by spring-loading. Used in gas and oil systems (and also where dust contamination is liable to occur).

Figure 22c: In this common construction, shutoff takes place by pushing a rigid, wedge-shaped valve disc between the seats of the valve body. The stem force enhances the sealing action. Used frequently in the low and medium pressure ranges.

Figure 22d: Two parallel moving sealing discs are forced onto their seats at the end of the closing movement by the action of toggle levers or wedges. This results in a greatly reduced sliding action and hence reduced wear.

Figure 22e: In this improved form of the wedge gate valve, two wedge-shaped sealing discs, capable of moving relative to each other, are pushed by means of a hemispherically shaped pressure piece onto the seat faces with great force. This is a robust form of construction with high sealing capability and low wear – up to PN 400.

The gates are actuated by hand or else via transmission gearing, on electric motor with a gearbox or hydraulic or pneumatic actuators.

Standards

DIN 3204 Flat wedge gate valves (PN 4; DN 40 to 300).
DIN 3216 Flat wedge gate valves in cast iron (GG) (1.6 to 10; 350 to 1600).
DIN 3226 Round wedge gate valves in cast iron (GG) (PN 16).
DIN 3228 Flat wedge gate valves in cast steel (GS) (PN 10).
DIN 3229 Oval wedge gate valves in cast steel (GS) (PN 16).

2.9.4 Cocks (Rotary Gate Valves)

Advantages are simple and robust construction, low space requirement, possibility of rapid shutoff and changeover, low flow losses, possibility of conversion to multi-way cock with several connections. Disadvantages are large sealing faces which slide on each other, causing wear. The frictional forces, depending on the preloading of the plug, manufacturing quality of the sealing faces, lubricants, and on the type and temperature of the fluid, are relatively high.

Further included in the group of taper-plug cocks are the *gland cock* used, in particular, in the chemical industry for poisonous media (body closed off below, plug sealed and held by packing and follower), the *lubricated cock* for aggressive viscous and contaminated media in coking plant and in the petrochemical industry (in this case the plug is lubricated via a groove and lubricating chamber), the *easy-turn* cock for viscous media like latex (in this case the plug is slightly raised before turning and pressed back into its seat after turning), and the *multi-way cock*, e.g. a three-way or four-way cock, for switching the flow to various directions.

An important technical onward development is the *ballvalve* (**Fig. 23**). The sealing component in this case is a ball with a cylindrical bore for straight-through flow.

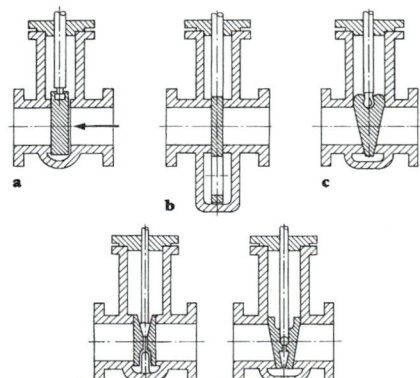

Figure 22. Forms of gate sealing. **a** Plate gate. **b** Spectacle shutoff gate. **c** Wedge gate. **d** Parallel double-plate gate. **e** Double-plate wedge gate.

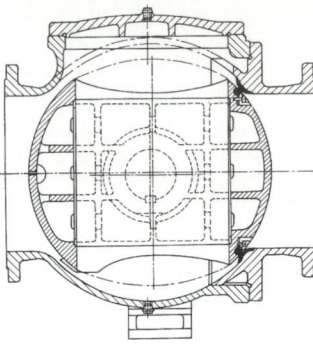

Figure 23. Ballvalve for large diameter pipelines.

This is practically free of flow resistance (resistance coefficient $\zeta_R = 0.03$ when the ballvalve is fully open and is approximately equal to the resistance of a similar length of pipe). Such ballvalves are produced in sizes from DN 80 to DN 1400 and for PN 10 to PN 64.

2.9.5 Flap Valves (Butterfly and Clack Valves

Flap (butterfly) valves, with a construction similar to that of **Fig. 24**, are used as *shutoff valves*, *throttle* (*control*) *valves* and, more rarely, as *safety valves* in the water supply industry (pumping stations and filtration plant), in power stations (cooling circuits), in the chemical industry (service water, also acidic and alkaline media) and in sewage treatment (purification plant, pumping stations). They are used in increasing measure in potable water dis-

tribution and in long-distance gas pipelines where they take the place of oval gate valves. Like gate valves, flap (butterfly) valves close off tightly against liquids. Flap (butterfly) valves are produced in the larger nominal sizes (DN 5300), generally for PN 4 to DN 2400 and for PN 16 up to DN 1200. Their space requirement is not much larger than the pipe cross-section. The flap (disc) can be actuated by hand, by an electric motor through a spur-wheel segment or worm drive, or by means of a hydraulic piston and, where necessary, a dead weight to amplify the action or to balance the flow forces. In general, the disc is so arranged that the half of the disc pointing upstream moves downward on closing (so that the closing force is amplified by hydraulic action). *Non-return flap* (*clack*) *valves* serve as safety elements; the flap (clack) is held open by the flow. Under no-flow conditions or on reversal of the pressure, the flap (clack) closes helped by its dead weight and, if necessary, slowed down by an oil brake (dashpot).

2.10 Seals

Seals are intended to stop leakage of fluids through the gaps of parts connected to each other (normally flanges; see H2.7.1). They must be easily deformable in order to compensate for the roughness of the sealing faces and must be sufficiently strong to withstand initial compression and internal pressures. Temperature and chemical stability must be considered, as well as the prevention of electrochemical decomposition of metallic seals or attack on the contact faces through electrochemical anodic action.

A summary on seals, their function and designation is given in DIN 3750.

2.10.1 Static Contact Seals

Figure 25 gives an overview of the most important types of seal. They are distinguished by whether they are (a)

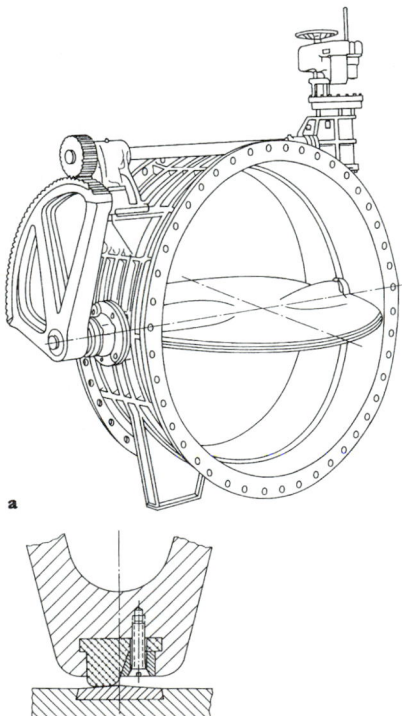

Figure 24. a Butterfly valve (Bopp & Reuther). **b** Lens-shaped plate with rubber sealing rings, body seal insert of stainless steel.

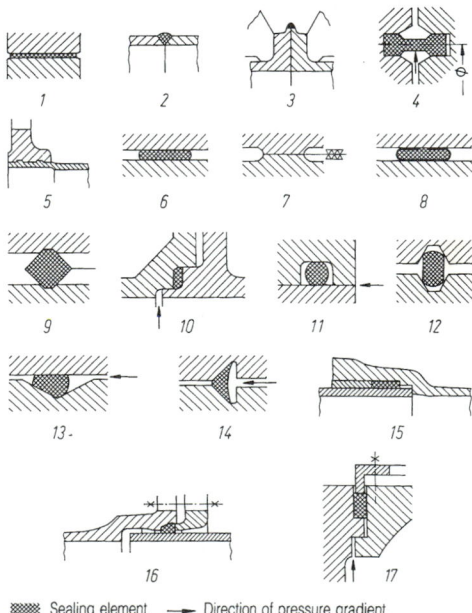

Figure 25. Static contact seals [15].

non-removable or conditionally removable (c.r.) and (b) removable. Interposed between them are (*1*) material contact joints with sealing materials or adhesives. Group (a) comprises (*2*) welded joint, (*3*) welded lip seal (c.r.), (*4*) push-fit (c.r.), (*5*) rolled joint. Group (b) encompasses (*6*) flat seal (soft or hard), (*7*) face-to-face joint (without jointing material), (*8*) compound material flat seal, (*9*) edge seal (plastic deformation), (*10*) fluid seal, (*11*) circular seal (O-ring of soft material or metal, resilient deformation), (*12*) seals of hard material (ring joint, resilient), (*13*) self-acting seals of soft material (compressed by internal pressure), (*14*) self-acting seals of hard material (delta ring), (*15*) to (*17*) seals of stuffing box type. Forms of seal implementation with sealing characteristics are as specified in DIN 2505, see **Appendix H2, Table 1**.

Flat Seals. These are discs, rings or gaskets that adapt to the jointing faces across the whole of their width. They consist either of a uniform material like asbestos board or paper (see DIN 3752; 0.1 to 10 mm thick; application up to 500 °C) or It-plates (asbestos with inorganic fillers and an elastomer binder) in accordance with DIN 3754 for cold and hot water as well as oil, steam, saline solutions, etc., 0.5 to 4 mm thick, capable of loading up to 300 bar at 300 °C, made of several materials such as laminated metal foil (Al, Cu), or faced with sheet metal, or entirely of metal (see H2.7.1). For flat seals as flange seals see **Fig. 26**.

Profiled Seals (**Fig. 25**, *9* and *10*). These are discs or rings which, because of the form of their cross-section, do not make contact across the whole of their width, thereby achieving a higher bearing pressure. They consist of elastomeric materials, soft metals or material combinations and are – depending on the material – suitable for high pressures (PN 400) and high temperatures (about 500 °C) (single use only).

Toroidal Sealing Rings (O-Rings). These are rings of circular cross-section made of elastic materials or metals which, with low prestressing during installation, are able to form seals, aided by the operating pressure (**Fig. 25**, *11* and *13*). *Dimensions*: see DIN 3770 (d_1 = 2 to 800 mm; d_2 = 1.6 to 10 mm). *Applications*: oils; water; air; glycol mixtures from −50 to +200 °C and medium pressure (suitable for repeated use).

High-Pressure Seals. (a) Small DN (pipe) (see **Appendix H2, Table 1**): grooved seal, ring-joint seal (frequent

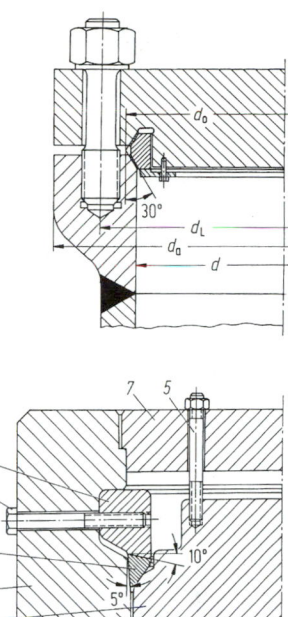

a

b

Figure 27. a Double conical seal. **b** Uhde-Bredtschneider seal. *1* Cover. *2* Wedge-section sealing ring. *3* Head of container. *4* Divided ring. *5* Fixing studs. *6* Pre-tensioning screws. *7* Retaining ring.

opening), lenticular seal (lens ring); (b) large DN (apparatus flange) (see **Fig. 25**): delta ring (*14*), gap seal (*17*), or, as in **Fig. 27a**, double-cone seal self-acting with an intermediate layer of 0.3 to 1 mm aluminium foil, and Uhde-Bredtschneider seal (**Fig. 27b**) – pressure-aided, requires no bolts or expensive flanges.

2.10.2 Dynamic Contact Seals

Stuffing Box Seals (Packing)

Packings are sealing elements which seal off cylindrical faces in relative motion against fluids and gases. The stuffing box seal (**Fig. 28**) consists of (*1*) the fixed part of the casing with the stuffing box space, (*2*) the sealing

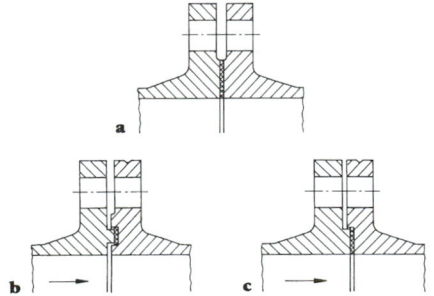

a

b   **c**

Figure 26. Flat seals (gaskets) and flange sealing faces [5]. **a** Smooth raised-face flange and gasket to DIN 2690 (PN 1 to 6, 10, 16, 25, 40). **b** Tongue and groove flanges to DIN 2512 and gasket to DIN 2691 (PN 10, 16, 25, 40, 64, 100). **c** Male and female spigot flanges to DIN 2513 and gasket to DIN 2692 (PN 10, 16, 25, 40, 64, 100).

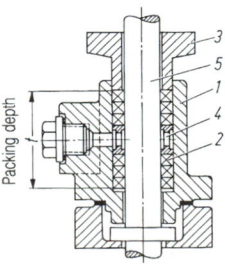

Figure 28. Stuffing box seal (Goetze).

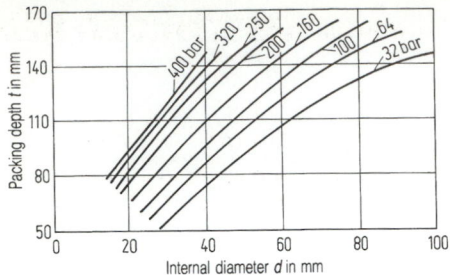

Figure 29. Depth of packing space for laminated packing rings (Goetze).

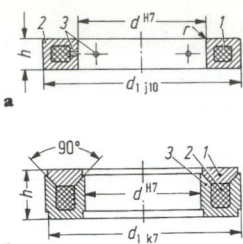

Figure 30. Packing rings (Goetze). **a** Hollow ring. *1* Lead or copper. *2* Graphite lubricant. *3* Radial holes. **b** Bevelled lip ring. *1* Bevelled washer. *2* Soft material infill. *3* Bevelled lip U-ring.

material (packing), (*3*) the collar plate (follower) screwed into the casing (flange or thread; providing adjustment), (*4*) the intermediate lantern ring (to distribute lubricating oil if necessary) as well as (*5*) the shaft or stem capable of rotary and axial movement. Packings can be used at relatively low sliding velocities (up to about 0.3 m/s), high temperatures (up to about 520 °C), high pressures (up to about 300 bar) and shaft diameters of 10 to 200 mm; the external diameter of the packing may be from 18 to 245 mm (up to 800 mm for expansion compensators in gas lines). Sealing works on the principle that screwing down in the axial direction causes lateral deformation and pressure against the cylindrical sealing faces. The width of packings of soft material is equal to $\sqrt{d}$ for small diameters d of the stem and = $2\sqrt{d}$ for large ones.

Laminated Packing Rings (**Fig. 29**) These are made from corrugated metal intercalations such as soft lead, copper, nickel, or chrome-steel, embedded in layers in asbestos or cotton. The rings have an inclined split and can therefore be bent open and placed around the shaft. If several rings are fitted, the splits must be staggered. In the case of gases, the seal should be improved by lubricating oil so that friction is reduced.

Hollow Ring of Lead or Copper (**Fig. 30a**). These may be undivided or divided into two. The lead or copper casing is filled with graphite lubricant, which penetrates through small radial holes for self-lubrication; ground bearing-faces are required. A typical application is in hydraulic press pumps.

Foil Packing Ring. Cotton wrapped with aluminium foil.

Bevelled Lip Packing Ring (**Fig. 30b**). Axial stresses are transmitted to the running surface as a result of the wedge

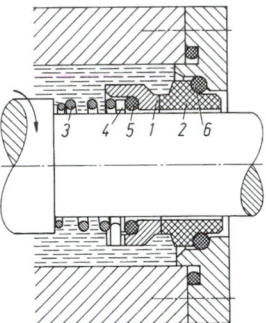

Figure 31. Axial face seal (Burgmann). *1* Rotating mechanical seal. *2* Stationary counter ring. *3* Compression spring. *4* Washer. *5* Sealing ring. *6* Bearing ring.

form. Perfectly functioning external lubrication is required. Such seals are suitable for very high pressures (in excess of 400 bar) in autoclaves, press and ultra-high-pressure pumps.

Rotating Mechanical Seals. Axial and radial rotating mechanical seals have increasingly displaced stuffing box packings for rotating shafts. **Figure 31** shows the principles of construction of an axial rotating seal. This can be applied to shaft diameters of 5 to 500 mm, pressures of 10^{-5} mbar to 450 bar, circumferential velocities of more than 100 m/s and temperatures from −200 to +450 °C. For various configurations, leakage losses, axial face seal closure, friction losses and operational integrity, see [15, 16].

3 Types of Heat Exchanger

3.1 Tube-Bundle (Shell-and-Tube) Heat Exchangers

The shell-and-tube heat exchanger is used in many branches of industry because of its versatile applicability to gaseous and liquid media within wide ranges of temperature and pressure. **Figure 1** shows a DIN 28 183 heat exchanger featuring fixed tube ends, one tube-side and one shell-side pass each, a shell compensator, and dished ends in the form of vessel domes. If the tubes and shell are of the same material and the temperature differences are not too great, thin tube ends can be used owing to the support action of the tubes. Easy mechanical cleaning of the tube space (inner space of tubes and dome) is possible.

An exchanger with hairpin tubes (U-tube exchanger, **Fig. 2a**) requires thicker tube plates. This is constructed with two tube passes and, as counterflow types, with two shell passes (partitions must be watertight) and flat head closure. The floating-head version (**Fig. 2b**), in contrast to the fixed-tube version, can deal with greater temperature differences between tubes and shell. There is easy cleaning of the shell and tube space. Construction is by means of four tube passes and one shell pass. The inner dome is sealed by means of a divided counter-flange. The pullthrough exchanger (**Fig. 2c**), in contrast to the exchanger shown in **Fig. 2b**, allows removal of the tube bundle without dismantling the floating head.

Disadvantages. These are large clearances between bundle and shell, bypass flow, and installation of internal slide rails for removal of the tube bundle (see **Fig. 3**).

Important Standards

DIN 28 008 Tolerances.
DIN 28 080 Saddles.
DIN 28 180 Steel tubes for heat exchangers.
DIN 28 182 Tube sections and tube connections.
DIN 28 191 Flanged floating head.

Tube bundle exchangers are also employed for phase changes: concentrators, evaporators with forced and natural circulation, condensers (see H4), waste-head boilers with steam generation. A falling film evaporator for econ-

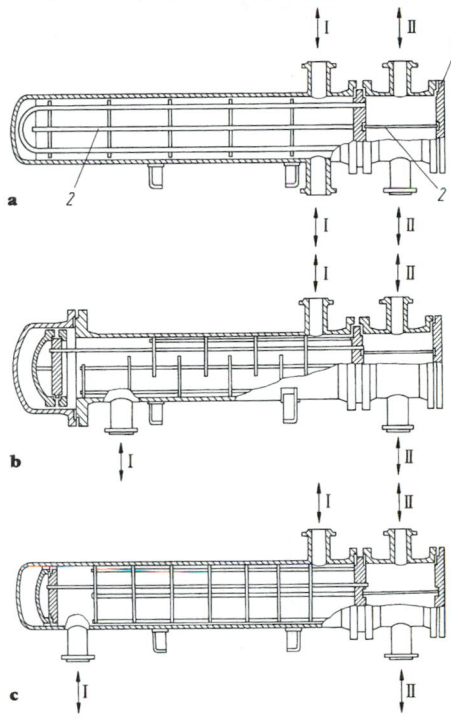

Figure 2. Various designs of tube-bundle (shell and tube) exchanger (Dupont). I, Shell-side medium. II, Tube-side medium. *1* Flat end (head closure). *2* Partition.

omic concentration is shown in **Fig. 4**. After entry of the fresh solution, a thin liquid film is produced on the inner faces of the tubes by a suitable device. This film flows downwards under gravity, together with the vapour produced. In this way a given concentration of the solution is achieved during a single pass through the heat exchanger, if the exchanger has been suitably designed.

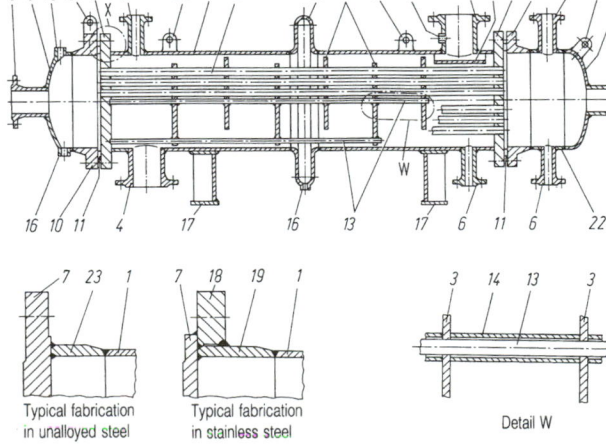

Typical fabrication in unalloyed steel   Typical fabrication in stainless steel

Detail X

Detail W

Figure 1. Tube bundle (shell and tube) heat exchanger with two fixed tube plates [1], as in DIN 28 183. *1* Shell. *2* Tubes. *3* Segment baffle. *4* Shell mount. *5* Vent connection. *6* Drainage connection. *7* Tube end, tube plate. *8* Dome mount. *9* Dome end. *10* Dome flange. *11* Seal. *12* Compensator. *13* Tie-rod. *14* Spacer. *15* Vent boss. *16* Drainage boss. *17* Support. *18* Shell flange. *19* Flange stub. *20* Impingement plate. *21* Lifting eye. *22* Dome shell. *23* Shell/tube-plate stub.

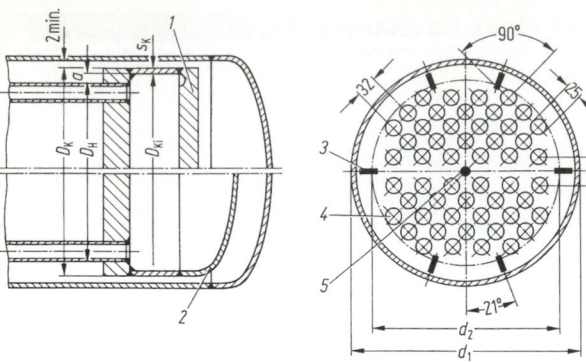

Figure 3. Welded floating head [1] to DIN 28 190. Two tube passes (passages), nominal diameter 350 mm, enveloping ring diameter 288 mm. *1* Made with flat plate. *2* Fabricated with domed end. *3* Slide rail 30 × 10 flat. *4* Internal tube. *5* Tie-rod, 12 mm diameter.

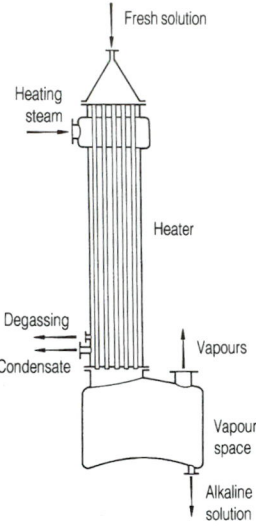

Figure 4. Falling-film evaporator (Wiegand).

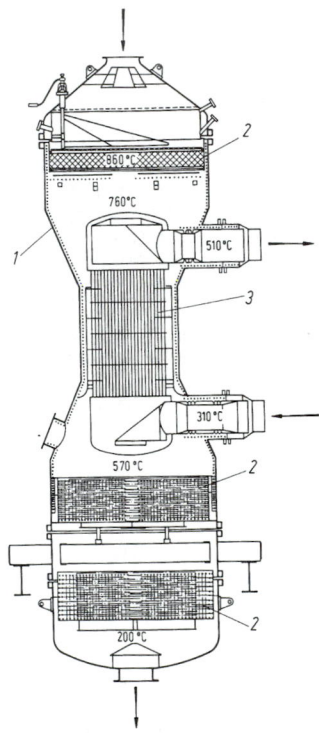

Figure 5. Heat exchanger in nitric acid installations with built-in residual gas heater (Steinmüller). *1* Water-carrying spiral tube as wall protection. *2* Heat-exchanger packs (spiral tube for steam generation). *3* Residual gas heater.

In **Fig. 5** the walls of the heat exchanger are protected from unacceptably high temperatures by built-in tube spirals (steam generation). Cooling of the gases takes place in the built-in fixed-tube exchanger by heat transfer to another process gas.

Because of the high thermal stresses in thick-walled tube ends at high temperature differences, special designs are required [2]. A cooled tube end supported by a cradle is shown in **Fig. 6**. By suitable flow guidance, uniform cooling is achieved and deposits from the water (peak temperatures) are avoided. At very high temperatures (1000 to 1500 °C), additional sleeve tubes must be provided to protect the tubes and plates, or the tubes must be cooled individually at the gas entry end [2, 3].

3.2 Other Types

These include heat exchangers with finned surfaces [4], which reduce thermal resistance in particular in the presence of gas flow (air coolers). Finned surfaces are also employed in evaporators (low fins) if large heat-transfer coefficients occur on the hot side, e.g. in the case of steam condensation.

Plate-type heat exchangers [4], which can be used in versatile ways to vary the flow of the media and can be cleaned easily (making them useful in the food industry), also need to be mentioned. They consist of a pack of profiled plates, separated by soft seals held together by means of a clamp. Large transfer areas can be accommodated in a small volume.

Spiral heat exchangers (**Fig. 7**) represent a special form. They are produced by the winding of two or four plates, furnished with spacing bolts, around a stable core of up to 2 m diameter. The front faces are closed off by endplates and soft seals.

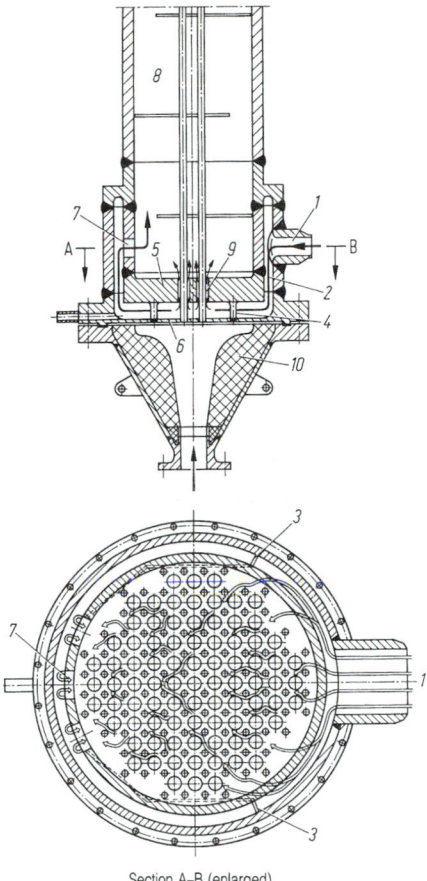

Section A–B (enlarged)

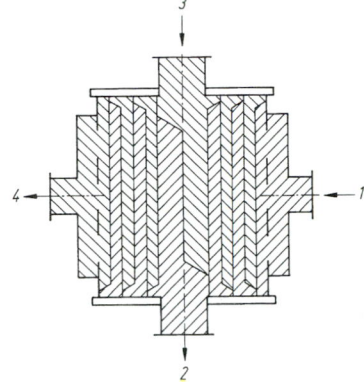

Figure 6 *(left)*. Waste-heat boiler with water-cooled and cradle-supported double bottom (Borsig). *1* Water inlet. *2* Distributor gap. *3* Baffle plates. *4* Support. *5* Load-bearing cooled bottom. *6* Supported membrane bottom. *7* Water inlets. *8* Evaporation chamber. *9* Tube gap. *10* Brick lining.

Figure 7. Spiral heat exchanger in counterflow configuration (Kapp). *1*, *2* Inlet and outlet of cold medium. *3*, *4* Inlet and outlet of warm medium.

Advantages. These are high flow-velocities, high heat-transfer coefficients (1500 to 2300 W/(m² K)), compact construction (20 to 70 m²/m³), dirt resistance and easy cleaning.

Disadvantages. These are low pressures and temperatures (up to 20 bar and 400 °C).

4 Condensers and Reflux Coolers

4.1 Principles of Condensation

When a vapour cools below its saturation temperature, or dew point, it changes to a liquid.

Areas of Application. In the case of condensers, this means the production of the highest possible vacuum (steam engines), the recovery of the condensate as a valuable liquid (distillation plant), and the precipitation of exhaust vapours damaging to the environment (vapours containing corrosive materials), as well as heating (water vapour as a heat-carrying medium).

Cooling Agents. Water, air, refrigerating brine, and other substances which can take up heat, act as cooling agents.

Types. These, and their working principles, are as follows:

Surface Condensers. Vapours are condensed by indirect contact with a cooling agent, usually via cooling surfaces consisting of tubes (a form of construction known as 'closed').

Injection (Mixed) Condensers. Vapours are brought into direct contact with injected cooling water and precipitated.

Direct Air Cooling. In air-cooled condensers of open construction, vapours are turned into liquid by heat transfer to the ambient air.

Indirect Air Cooling. Water is used as the cooling medium in surface or injection condensers and then transfers the heat to the air via cooling towers or watercourses.

Surface and air-cooled condensers allow the recovery of pure condensates and the production of a higher vacuum than do mixed condensers (air is dissolved in the injected water!); they are particularly useful for the precipitation of vapours of no commercial value. For heating and evap-

oration, the closed mode of construction of surface evaporators is required.

Non-Condensable Gases. These build up at the points
of lowest pressure (lowest temperature) and there form
a layer of increasing thermal resistance. Since the vapours
must diffuse through these to gain the cooling surface, the
vacuum becomes weaker. At constant total pressure, the
partial pressure and the process-promoting temperature
gradient between the vapour temperature and the coolant
temperature are reduced. Condensers must therefore be
vented at pressures above atmospheric and, when
operating under vacuum, must be kept free from inert gas
by pumping off.

4.2 Surface Condensers

4.2.1 Thermodynamic Design

Heat To Be Removed

$$\dot{Q} = \dot{m}_D(h_D - h_K) = \dot{m}_w c_w(t_2 - t_1). \tag{1}$$

Cooling Surface of a Condenser

$$A = \dot{m}_D(h_D - h_K)/k\Delta t_M . \tag{2}$$

In Eqs (1) and (2), $\dot{m}_D$, $\dot{m}_w$ are vapour and coolant mass
flow respectively, h_D and h_K are the specific enthalpy of
the vapour and condensate respectively, c_w is the thermal
capacity of the coolant, t_1 and t_2 are the inlet and outlet
temperatures of the coolant respectively, k is the heat-
transfer coefficient and Δt_M is the mean temperature difference (see H1.2.1).

Heat-Transfer Coefficient k (see C10.2). This is usually determined from the heat transfer on the coolant side,
because the heat-transfer coefficients on the condensation
side are large – particularly in the case of water vapour;
k increases with coolant velocity and reducing tube diameter. For steam condensation with cooling water flow in
the tubes of between 1.5 to 2.5 m/s, $k \approx 3000$ to
4000 W/(m² K) (see H1.2.1). The cooling surface area A
obtained from Eq. (2) is divided up into design units and
k is recalculated on the basis of the geometrical data so
obtained [1, 2]. In this process, separate account must
be taken of the effects both of layers of dirt and of inert
gases [3].

Superheated Steam. In this case a film of condensate
forms on the wall when the wall temperature is equal to
or lower than the saturation temperature of the steam;
the heat-transfer coefficient for condensation itself (see
C10.4.2) changes only slightly during the process. The
ranges for vapour cooling (dry wall) and condensate cooling require separate calculations.

4.2.2 Condensers in Steam Power Plant

In steam power installations, the aim is the generation of
the largest possible pressure and heat gradients. Owing to
the large specific volume of steam under vacuum, large
inlet cross-sections are required so that the pressure drops
do not exceed the gain in gradient. Economically attainable end pressures p_1 are 0.1 bar for piston engines, 0.025
bar for turbines (assuming low temperatures t_1 of the
cooling water, which vary with locality and season).
Values of t_1 and p_1 applying to central Europe: ground
water 10 to 15 °C and 0.03 bar, river water 0 to 25 °C and
0.04 bar, recooled water 15 to 30 °C and 0.06 bar. The
pressure p_1 is from 0.005 to 0.01 bar higher than the satu-

ration vapour pressure corresponding to the outlet temperature of the cooling water. The cooling water mass
flow $\dot{m}_w \approx 70\dot{m}_D$ for steam turbines, $\dot{m}_w \approx 40\dot{m}_D$ for
piston engines. If t_D is the temperature of the saturated
vapour at the cooling waste outlet, then the relation
$t_D - t_2 = 3$ to 5 K applies. Supercooling of the condensate
$t_0 - t_K < 3$ K, since otherwise inert gases will be dissolved
and returned to the circuit. The pumping off of the inert
gases must be effected at the coldest point (lowest total
pressure), which must be screened against steam ingress
(see H4.2.4).

4.2.3 Condensers in the Chemical Industry

Surface condensers for the recovery of valuable condensates downstream of columns and reactors are cooled
either by water or by air (see H4.4). Products that need to
be reheated or evaporated are also employed in increasing
measure as cooling agents in order to save energy. Water,
as the cooling agent, flows in the tubes (providing a better
means of cleaning), pure condensate media on the shell
side of bundles (giving a greater cross-section and lower
pressure drops). This demands particular consideration in
the case of vacuum operation, which is employed for tem-
perature-sensitive substances.

Heat-Transfer Coefficients. For condensing organic
media, these are lower than those for steam. If it is a matter of reaching high values, condensation on horizontal
tube bundles is more advantageous than on vertical ones.
This applies above all at low boiling temperatures and
short tube lengths. Heat transfer is improved if the flow is
lateral to the bundle. On the side of the cooling medium,
conditions for good heat transfer must be created through
high water velocities and the avoidance of dirt layers. This
is best accomplished by automatic cleaning devices of
brush or sphere type. If water conditions are unfavourable, coating of the tubes, to provide smooth surfaces,
may help.

Heating and Evaporation of Media. This is often
carried out in condensers where steam from a power
station is used as a source of heat. Since, under these circumstances, condensate heat-transfer coefficients are
high, the conveying of phases and the geometrical layout
on the evaporation side are determined by the smaller
values that occur in most cases. For this reason, vertical
bundle arrangements with condensation at the columns
on the shell side of circulation condensers are commonly
encountered, but horizontal bundle arrangements may
also be found in reboilers of 'kettle' type, where condensation occurs on the tube side. For exact calculations concerning such evaporator condensers, the distribution of
the heat-transfer coefficient for condensation and evaporation must be determined section by section.

Multi-Component Mixtures. The calculations require
particular effort if such mixtures condense (partial condensers, reflux condensers), possibly with the aid of inert
gases, and multi-component products are preheated on
the side of the cooling media (part evaporators, flash
evaporators). In this case, boiling or precipitation temperatures, which vary along the exchanger, must first be
determined from equilibrium calculations and then plotted in an enthalpy–temperature diagram. In exothermically operating reactors, such heat exchangers serve for
the heating up of reaction materials by the reactive product. Owing to the high temperatures (thermal stresses)
and the danger of contamination, floating-head
exchangers are preferred (see H3).

4.2.4 Design Considerations

Low-Pressure Saturated Vapour Condensers

These are constructed mainly as horizontal tube-bundle configurations.

Low Pressure-Drops. For large steam volumes, a further steam chamber is located below the large entry connection tube and wedge-shaped steam passages in the upper tube bundle (**Fig. 1**). This leads to the danger of partial pressure minima occurring in the part bundles at which inert gases collect, which impede heat transfer if the gas is not pumped away. It would be more advantageous to narrow the condenser in the downward direction or to provide it with tube sections that taper downwards [2]. However, this has not gained favour for reasons of design and cost.

Inert Gas Removal. This is done exclusively from the coolest point (pressure minimum), with a minimum steam contribution. In accordance with [4], the best solution is to arrange for the pumping off to take place at the centres of the part-bundles through tubes running the length of the bundles with many suction orifices. Baffles screen against steam ingress; dead corners must be avoided.

Prevention of Supercooling of the Condensate (Fig. 1). This is accomplished by deflectors (*2*), which keep the condensate away from the cooling tubes. Vapour traps or suction pumps continually drain the condensate.

Design. Shells: > 500 mm dia. welded from sheet steel, length ≈ 2 × dia. Tube ends: steel (or brass if water contains acid or salt) 20 to 30 mm thick. Tube dia.: 15 to 25 mm. Tube sections: (1.4 to 1.5) × external dia., tapered downwards. If there is condensation, no deflectors are required on the shell side. In order to avoid vibrations, supporting sheet-metal plates must be fitted at intervals of tube dia. × (50 to 70). For design in relation to vibrations see [1, 5]. Thermal expansion must be taken up by expansion compensators or by preformed S-shaped tubes (the

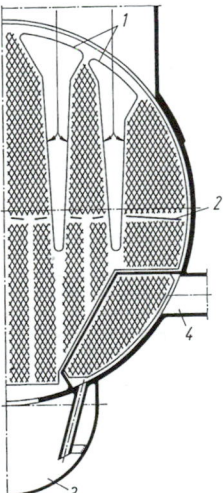

Figure 1. Single-flow Balcke–Dürr condenser. Diameter 2.5 m, length 12 m. Tubes: number 4960, dimensions 19.05 mm × 0.89 mm, spacing 26 mm. Vapour 44.2 kg/s, water 2.1 m³/s, power 97 MW. *1* Vapour slots in tube-support wall. *2* Condensate deflector slats. *3* Condensate outlet. *4* Air suction tube connection.

S-bends should be located at the supporting plates). In double-flow versions, one half can be cleaned without shutdown of the installation. A safety exhaust valve must be provided at the vapour inlet.

Condensers in the Chemical Industry

Vapour passages are not normally required (higher pressures, higher pressure-drops). Length: ≈ dia. × (3 to 4), depending on the pressure drops and the layout of flow passes. An advantageous design consists of central vapour entry with division of the flow by a separating partition in the longitudinal direction of the bundle (i.e. split flow) [6]. Tubes dia.: 18 to 25 mm. Tube section: outside dia. × 1.3 to 1.5, to DIN 28 182. Floating-head condensers may be manufactured in the simplest pullthrough versions (see H3) using a strip gasket, because space is required for the distribution of the vapour and the collection of the condensate. If two tube plates are sufficient, the U-tube type is also suitable (tube passes cannot be cleaned). If condensation takes place in the tubes, the tube bundle is inclined for easier drainage of condensate. Apart from tube bundles, there are also spiral tube, double tube and surface irrigation coolers employed as condensers.

4.3 Injection (Direct-Contact) Condensers

By injecting finely atomised cooling water into the vapour, greater heat-transfer coefficients can be obtained in comparison with surface condensers. These are measured in [7] for freely falling films, jets and drops as well as for pressure atomisation. In the latter case, values of $k = 100\,000$ W/(m² K) were measured for droplets of 0.6 mm diameter and velocity of 15 m/s at a thermal flux density of 230 000 W m². The values reduce considerably with decreasing drop velocity or increasing dwell time, as well as with decreasing condenser pressure and increasing inert gas content (50% reduction at 1% gas content by mass). Since the area of the phase boundary per unit volume is also large, the dimensions of the direct contact condensers are smaller than those of surface condensers. Internal structures to increase the area of contact and the dwell time are relatively inexpensive.

The specific cooling water requirement $\dot{m}_W/\dot{m}_D$ is calculated from Eq. (1). Since $t_2 = t_K$, $\dot{m}_W/\dot{m}_D$ at 15 to 30 kg/kg is smaller than in the case of surface condensers. For large power capacities and low pressures, the counterflow configuration (dry pumping-off of the inert gases at the head) is more economic than the parallel flow configuration (wet pumping-off). Discharging of the condensate and cooling water is usually carried out by means of a liquid receiving tank or via a water-jet pump, also, in the case of parallel flow configurations, via a jet condenser.

Owing to the mixing of cooling water with the condensate, this efficient process can be applied only to vapours without commercial value. An exception is the Heller process [8], where vapour in an injection condenser is precipitated only by its own condensate, which has previously been cooled by air in a dry cooling tower. This indirect air cooling process (see H4.6) is applied only when there is a shortage of water. The injection condenser is only a third the size of a surface condenser of the same power. On the other hand, the investment costs of condensate cooling are considerable. By means of a threefold flow rate of air, the same vacuum is obtained as with a wet cooling tower [9].

4.4 Air-Cooled Condensers

In the case of water shortage, apart from indirect methods, direct air cooling, requiring smaller surfaces, is applied in increasing measure. In most cases cooling takes place by blowing air against the finned external faces using fans or, more rarely, by natural air flow. Owing to legal requirements, slow-running, low-noise fans with wide fan blades are in increasing use. Investment costs are higher than those of surface condensers. However, if air cooling is compared with surface condensers and the recooling plant is taken into consideration, the investment costs are approximately equal, but the running costs for air cooling are lower as long as the product temperature lies above 60 °C.

Installations for Power Stations. These are built with power capacities up to about 1100 t/h of condensate (400 MW). The tube bundles can be arranged vertically, horizontally or inclined (in the form of an A-frame or a V-frame) and, in order to save space, can be mounted above pipe bridges or on the tops of buildings. The A-frame arrangement (**Fig. 2**), where steam is supplied from above, is in widespread use, producing parallel flows of steam and condensate. The decreasing condensing power of the tube rows that are situated in the already warm airflow is compensated for by the narrower fin spaces (*1* in **Fig. 2**).

Under frost conditions and during vacuum operation, liquid in the lower ends of the tubes is liable to freeze owing to the formation of dead zones through reflux streaming in the tubes with complete condensation and the occlusion and concentration of inert gas. In such cases, steam supplied from below (counterflow), giving poorer heat transfer, may be a solution. Alternatively, a combination of both configurations may be employed, which ensures that in the parallel-flow condenser, situated upstream, partial condensation occurs in all tubes, thereby preventing supercooling of the condensate. If operating conditions vary, it is safer to provide each tube row with a separate collector.

Refineries and Chemical Works. Here too, air-cooled condensers find increasing acceptance. In the form of head condensers for distillation columns, they are now being built for cooling rates of up to 40 GJ/h.

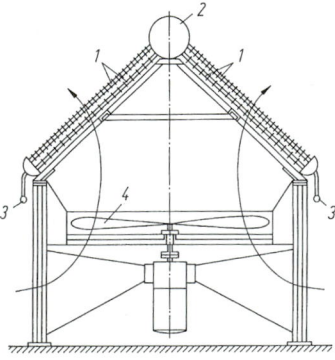

Figure 2. Air-cooled condenser in A-frame configuration. *1* Finned tubes with varying spacing between fins. *2* Vapour supply. *3* Condensate drain. *4* Fan.

4.5 Auxiliary Equipment

The condensate and the air which enters with the vapour, the cooling water and through leakage (piping systems, stuffing boxes of machines) must be continually pumped away from the condensers. Wet-air pumps are used almost exclusively for direct-contact condensers in the counterflow configuration; they convey condensate and air simultaneously at atmospheric pressure. In larger installations, dry-air pumps are employed with separate pumps for the condensate.

4.5.1 Air Ejectors

As far as turbines or steam piston engines are concerned, if no empirical values are available for design, average air quantities of about 0.1 to 0.25 of the maximum condensate mass percent may be assumed. Even with good cooling of the air to t_L, the vapour flow $\dot{m}_D$ pumped away with the air is greater than the air flow $\dot{m}_L$:

$t_D - t_L$ (K)	0	1	2	4	6
$\dot{m}_D / \dot{m}_L$ (kg/kg)	∞	12 to 13	5 to 6	2 to 3	1.2 to 1.5

Low values occur for total pressures of about 0.02 bar, high values for 0.1 bar. Cooling must be so designed that $t_D - t_L > 4$ K, where t_D is the saturated vapour temperature at the condenser inlet.

The most widely used pumps are water-jet and steam-jet pumps. Apart from these, water-ring air pumps are employed, if the pressures are not too low, which can also be operated using low-vapour-pressure sealing liquids if necessary. Oil-filled rotary piston pumps are suitable for high-vacuum applications. The advantage of the jet pump over mechanical pumps is that it can be manufactured from special materials for dealing with corrosive media.

Water-Jet Air Pumps. These are built only as single-stage types. The air is compressed isothermally because the heat of compression is conveyed to the water. Theoretical vacuum (depending on the inlet temperature of the water) is achieved to 98%. Efficiency is low owing to the high water consumption (20 to 40 m³ per kg of air), but greater than in the case of vapour jet ejectors. To obtain greater power, parallel operation should be employed. Water-jet pressure should be above 2 bar.

Advantages. The large pumping capacity means rapid attainment of the operational state of the condenser, while the simple mode of construction, with no moving parts, bestows high operational integrity, the more so because there is no sensitivity to contaminated water.

Disadvantages. Such devices are liable to power losses and to loss of condensate.

Vapour-Jet Air Ejectors. In these, a pressurised steam jet is brought to supersonic velocity by expansion in a Laval nozzle. Air is drawn in by subatmospheric pressure. The maximum single-stage compression ratio is 1 : 7, corresponding to a partial vacuum of 0.15 bar in conveying to atmosphere. In the case of low pressure, operation involves several stages where condensable components are precipitated in intermediate condensers in order to save energy and to reduce the size of the following stages. Turbine condensate (feedwater preheating) can be used for cooling. **Figure 3** shows a two-stage version. Direct condensers as well as surface condensers can be employed. In many cases a water-jet air pump or a water-

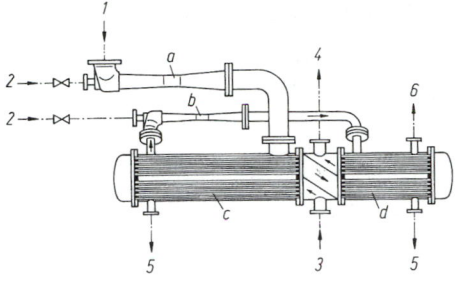

Figure 3. Vapour-jet air ejector (type: Körting), two-stage with intermediate condensation. *a, b* Vapour-jet ejector, first and second stage respectively. *c, d* Surface double condenser, first and second stage respectively. *1* Suction connection. *2* Operating steam. *3* Cooling water inlet. *4* Cooling water outlet. *5* Condensate drain. *6* Air vent.

ring air pump as the last stage (pressure range 0.2 to 1.0 bar) offers greater economy.

Advantages. These are the same as for water-jet air pumps. In addition, suction pressures are lower and recovery of the condensate is possible.

Disadvantages. An additional condenser is required, and efficiency is lower.

4.5.2 Cooling-Water Pumps and Condensate Pumps

Cooling-Water Pumps. For greater flow rates (up to 15 000 m³/h) and low conveying pressures (0.8 to 2 bar), these are usually of centrifugal type. In power stations, duplicate drives are by steam turbine and electric motor for reasons of safety. Allocation is usually to several parallel sets of pumps; power matching occurs without change of speed by switching in or switching out.

Condensate Pumps. These are also designed as rotary pumps. They are over-sized in case of leakage in the cooling water pipes. A positive suction head is required. Such a pump is usually implemented in a multi-stage version, and is often arranged on a common shaft with the cooling-water pumps. The power requirement of the condensation plant is 0.4 to 0.5% of the normal power of a prime mover.

4.6 Indirect Air Cooling and Cooling Towers

If no fresh water is available or inlet flow temperatures have been specified, indirect heat transfer to the air must be accomplished via an intermediate cooling agent (almost exclusively water). Depending on water flow, the towers are termed 'dry' or 'wet'. In the case of dry cooling towers, the heat is transferred by convection over cooling surfaces (i.e., a finned tube bundle) which have a higher thermal resistance and a smaller enthalpy gradient than those obtained in the wet process. In the latter case, the heat is transferred mainly by evaporation of water. About 1 to 2% of the flow to be cooled is lost and must be replaced.

The environmentally friendly dry process requires greater ventilation power and higher investment costs. It comes increasingly into its own where the disadvantages of the wet process – damp mist, spray, formation of ice, extra water, encrustation and corrosion have a deleterious

effect and where closed circuits are unavoidable, e.g. in nuclear power stations or in the Heller process using direct condensers (see H4.3). The planned combination of both processes [10, 11] allows change from the preferred dry operation in winter (where the characteristic depends on air temperature) to partial wet operation in summer (where the characteristic depends on wet-bulb temperature).

4.6.1 Types

Open cooling ponds (0.5 m³/h of water to be cooled per m² of ground area over a cooling interval of 10 K) or graduation works with 1 to 2 m³/(h m²) have by now become a rarity. Nowadays the choice for large power capacities is mainly a closed cooling tower built of concrete (5 to 10 m³/(h m²)); for small units, small coolers are made from plastics (up to 4000 m³/h), while for medium power capacities, cellular cooling towers are built of prefabricated reinforced concrete sections.

Types of Cooling Tower. A distinction is made between towers with forced and natural draught. At base load, natural-draught cooling towers are more economic for large power operation than are those with fans in spite of higher investment costs. *Dimensions*: These structures measure approximately 110 m in diameter and 150 m in height with concrete walls 14 cm thick (power 1000 to 1200 MW for wet cooling towers), while greater dimensions, which would be required for dry towers of the same power (200 to 300 m diameter and height), can be realised cost-effectively by means of the catenoid form of construction (**Fig. 4**). The paraboloid form with constriction is intended to prevent ingress of cold air from above [12, 13].

Forced-Draught Cooling Towers. These are economical only as discharge cooling towers (cooling of the discharge water flow in summer) for peak-load operation of large power installations (up to 100 000 t/h). Normal application is for small to medium generator power capacities. The diameter of the round cooling towers is up to 70 m, that of the fans up to 26 m; operating power is 0.55 kW/m² of droplet-receiving ground area. A pressure-generating fan is used for the sake of low noise (with sliding sound screens at the air inlet); the diameter is 7 to 8 m. Tower height is under 50 m. The shell is easy to manufacture. Costs of construction are 10 to 15% less, power requirement 10 to 15% higher, than in the case of an induction fan.

Wet Cooling Towers. The water to be cooled is pumped to distributing troughs in the lower third of the tower and then flows atomised by nozzles; or, distributed by overflow gutters to internal fittings, it flows in thin layers towards the air current.The water drop is 6 to 8 m. Air enters from below. Configurations with counterflow using drip plates are usual in Europe, with crossflow the norm in the USA.

4.6.2 Design Calculations

Heat transfer in dry cooling towers is calculated as in finned-tube coolers. In towers with natural draught, the air flow rate is calculated from the balance between buoyancy and pressure drop [1]. For wet cooling towers, the main equation [14] derived by Merkel applies approximately.

Heat Flow. The heat extracted from the water by the air through heat transfer and evaporation is given by

$$d\dot{Q} = \dot{m}_w c_w \, dt_w = \beta_x (h'_L - h_L) dA \tag{3}$$

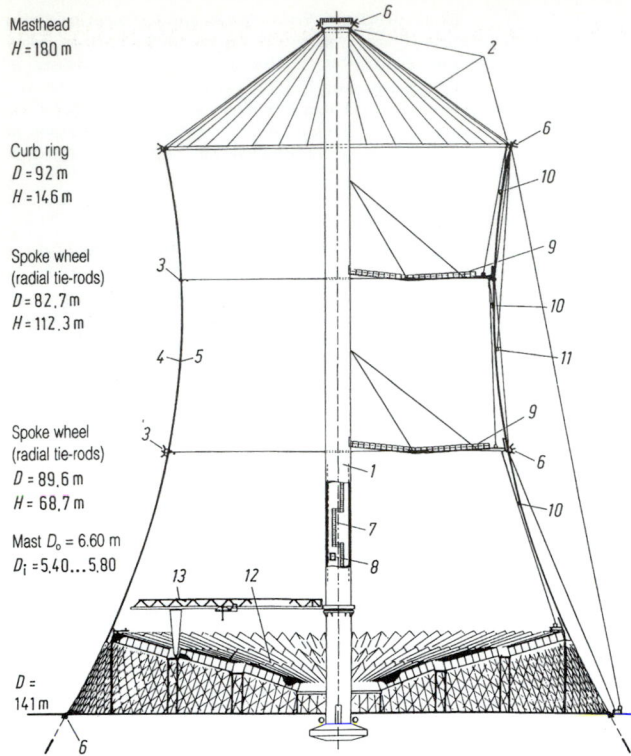

Masthead
$H = 180$ m

Curb ring
$D = 92$ m
$H = 146$ m

Spoke wheel
(radial tie-rods)
$D = 82.7$ m
$H = 112.3$ m

Spoke wheel
(radial tie-rods)
$D = 89.6$ m
$H = 68.7$ m

Mast $D_o = 6.60$ m
$D_i = 5.40...5.80$

$D =$
141 m

Figure 4. Catenoid dry-type cooling tower – Schmehausen [12]. *1* Mast. *2* Spoke wheel (radial stays). *3* Spoke wheels (radial tie-rods). *4* Shell. *5* Lining. *6* Aircraft warning light. *7* Access ladder. *8* Lift. *9* Catwalks. *10, 11* Internal and external access for inspection. *12* Heat exchanger, A-frame configuration. *13* Crane.

and is proportional to the boundary area of the phase dA between water and air and the difference of the enthalphy b_L'' of saturated air at the temperature t_w of the water and the enthalpy b_L of the air. β_x is the mass transition coefficient (mass flow per unit area).

Lewis Number. If α is the heat-transfer coefficient and c_{Lm} the mean thermal capacity of the moist air, then

$$Le = \alpha / (\beta_x c_{Lm}) \ . \tag{4}$$

applies. $Le \approx 1$ for evaporating water.

Merkel Number. Neglecting changes in the quantity of water,

$$Me = \frac{\beta_x A}{\dot{m}_w} = \int \frac{c_w - dt_w}{b_L'' - b_L} \ . \tag{5}$$

follows from Eq. (3). The Merkel number is derived by an averaging process for b_L'' and b_L [1] or graphically (Sherwood).

Number of Transfer Units. With mass flows $\dot{m}_L$ for air and $\dot{m}_w$ for water, this is given by

$$NTU = \beta_x A / \dot{m}_L = \dot{m}_w \cdot Me / \dot{m}_L \ . \tag{6}$$

The loss of water is made up from $\beta_x A$ plus the entrained droplets $(0.3 \cdot \beta_x A)$ and must be replaced by additional treated water. The cooling effect increases with increasing water temperature. Below the limiting cooling temperature (wet-bulb temperature), which is lower for unsaturated air than the dry air temperature, water cannot be cooled. For the usual mean values of 15 °C and 70% relative humidity measured for air, the limiting cooling temperature is about 12 °C. The usual interval of the cooling region $t_{w2} - t_{w1} = 10$ K. For the calculation and plotting of the changes of state in a Mollier b, x diagram, see C6.2.2. Values for $\beta_x A$ or NTU may be found in the literature or may be determined empirically [1].

5 Appendix H: Diagrams and Tables

Appendix H2 Table 1. Sealing data for gases and vapours in accordance with DIN 2505 [4]

Type of seal	Form of seal	Designaiton	Material	Sealing data		
				Initial deformation		Operating condition
				k_0 mm	$k_0 \cdot K_D$ N/mm	k_1 mm
Seals of soft materials		Flat seals as DIN 2690 to DIN 2692	Rubber PTFE It (asbestos with inorganic fillers and an elstometer binder	– –	$2\,b_D$ $25\,b_D$ $b_D \dfrac{200}{\sqrt{b_D\,h_D}}$	$0,5\,b_D$ $1,1\,b_D$ $b_D\left(0,5+\dfrac{5}{\sqrt{b_D\,h_D}}\right)$
Seals of metal and soft materials		Spiral asbestos seal	Asbestos/ steel		$50\,b_D$	$1,3\,b_D$
		Corrugated ring seal to DIN 2698	Al Cu, Ms Soft steel	–	$30\,b_D$ $35\,b_D$ $45\,b_D$	$0,6\,b_D$ $0,7\,b_D$ $1\,b_D$
		Sheet metal encased seal	Al Cu, Ms Soft steel	– –	$50\,b_D$ $60\,b_D$ $70\,b_D$	$1,4\,b_D$ $1,6\,b_D$ $1,8\,b_D$
Metal seals		Flat metal seal		$1 \cdot b_D$	–	$b_D + 5$
		Diamond-edge metal seal	–	1	–	5
		Oval-profile metal seal		2		6
		Round-section metal seal		1,5		6
		Ring-joint seal	–	2	–	6
		Lenticular seal to DIN 2696		2		6
		Crest profile seal to DIN 2697		$0,5\sqrt{Z}$	–	$9 + 0,2 \cdot Z$

Z = number of crests

Appendix H2 Table 2. Standard velocity values in m/s [6]

Hot vapour ($v = 0.025$ m³/kg)	35	to 45
Hot vapour ($v = 0.2$ m³/kg)	50	to 60
Saturated vapour, also pipes in piston engines	15	to 25
Gas (long-distance pipelines)	5	to 10 to 20
Gas (domestic piping)	1	
Air (STP)	10	to 40
Compressed air	2	to 10
Oil (long-distance pipelines[a])	1	to 2
Fuel lines in internal-combustion engines	approx. 20	
Lubricating oil lines[a] in internal-combustion engines	0.5	to 1
Water: Suction side of pumps[b]	0.5	to 1 to 2
Delivery side of pumps	1.5	to 2 to 4
Domestic piping	1.5	to 2.5
Long-distance pipelines	1.5	to 3.5
For water turbines	2	to 4 to 8

[a] Attention to viscosity.
[b] Danger of cavitation.

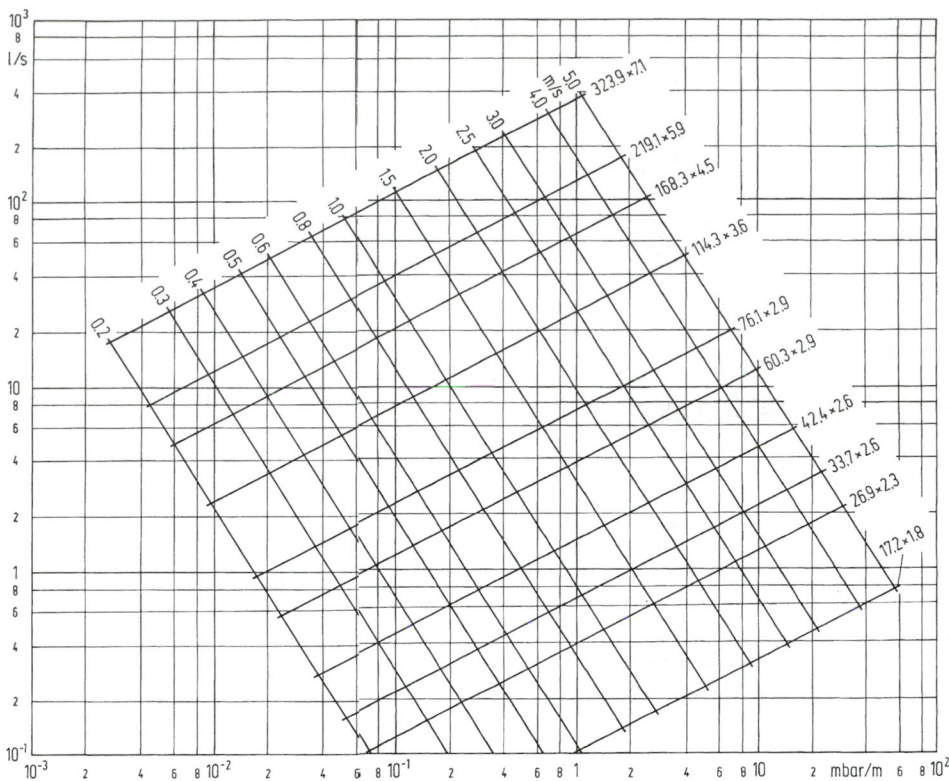

Figure 1. Pressure drops in steel tubes DIN 2448 for cold water (+10 °C).

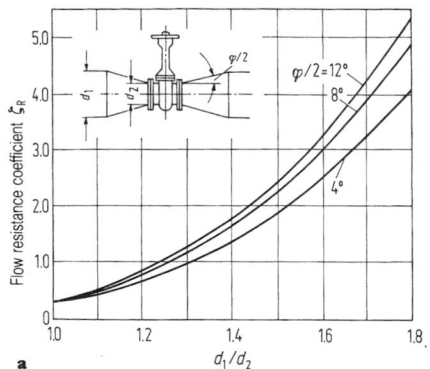

	Nominal size in mm									
	25	32	40	50	65	80	100	125	150	200
Straight-through valves										
Free flow	1.5	1.4	1.3	1.0	1.0	1.0	1.3	1.3	1.3	1.6
Type Boa	2.1	2.2	2.3	2.3	2.4	2.5	2.4	2.3	2.1	2.0
DIN	4.0	4.2	4.4	4.5	4.7	4.8	4.8	4.5	4.1	3.6
Angle valves										
Type Boa	1.6	1.6	1.7	1.9	2.0	2.0	1.9	1.7	1.5	1.3
DIN	2.8	3.0	3.3	3.5	3.7	3.9	3.8	3.3	2.7	2.0
Non-return flap (clack) valve	1.9	1.6	1.5	1.4	1.4	1.3	1.2	1.0	0.9	0.8

Figure 2. Flow resistance coefficient ζ: **a** of shut-off slide valves with reducers, **b** of valves and flaps [17].

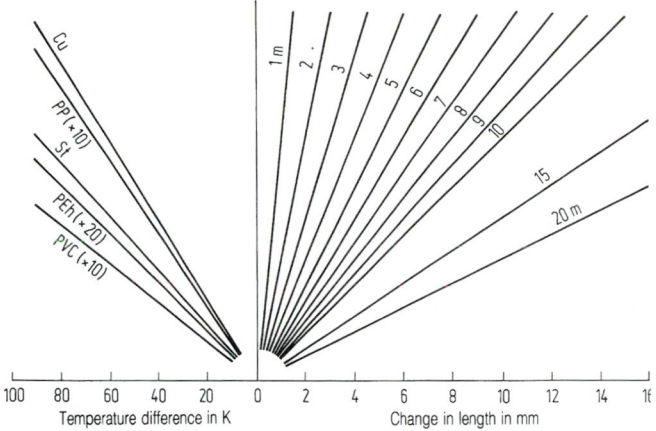

Figure 3. Change in length of various materials as a function of temperature. PEh = polyethylene hard.

6 References

H1 Fundamentals. [1] VDI-Wärmeatlas, 6th edn. VDI, Düsseldorf, 1991. – [2] Ahmad S, Linnhoff B, Smith R. Design of multipass heat exchangers: an alternative approach. Trans ASME/J Heat Transfers 1988; 110: 304–9. – [3] Rummel K. Die Berechnung der Wärmespeicher auf Grund der Wärmedurchgangszahl. Stahl und Eisen 1928; 48: 1712-25. – [4] Hausen H. Wärmeübertragung im Gegenstrom, Gleichstrom und Kreuzstrom, 2nd edn. Springer, Berlin, 1976. – [5] Grassmann P. Physikalische Grundlagen der Verfahrenstechnik, 3rd edn. Salle, Frankfurt, 1982. – [6] Martin H. Wärmeübertrager. Thieme, Stuttgart, 1988. – [7] Glaser H. Der thermodynamische Wert und die verfahrenstechnische Wirkung von Wärmeaustauschverlusten, Chem Ing Techn 1952; 24: 135–41. – [8] Gregorig, R. Wärmeaustausch und Wärmeaustauscher, 2nd edn. Sauerländer, Aarau, 1973.

H2 Apparatus and Piping Components. [1] AD-Merkblätter: Richtlinien für Werkstoff, Herstellung, Berechnung und Ausrüstung von Druckbehältern. Loseblatt-Sammlung. Heymann, Cologne. – [2] Klapp E. Festigkeit im Apparate- und Anlagenbau. Werner, Düsseldorf, 1970. – [3] Titze H. Elemente des Apparatebaues, 2nd edn. Springer, Berlin, 1967. – [4] Schwaigerer S. Festigkeitsberechnung im Dampfkessel, Behälter- und Rohrleitungsbau, 4th edn. Springer, Berlin, 1983. – [5] Tochtermann W, Bodenstein F. Konstruktionselemente des Maschinenbaues, pt 1, 9th edn. Springer, Berlin, 1979. – [6] Richter, H. Rohrhydraulik, 5th edn. Springer, Berlin, 1971. – [7] Zoebl H, Kruschik J. Strömung durch Rohre und Ventile. Springer, Vienna, 1978. – [8] Grassmuck J, Houben KW, Zollinger RM. DIN-Normen in der Verfahrenstechnik. Teubner, Stuttgart, 1989. – [9] Richarts F. Berechnung von Festpunktbelastungen bei Fernwärmeleitungen. Heiz Luft Haustech 1955; 6: 220. – [10] Merkblatt 333. Halterungen und Dehnungsausgleicher für Rohrleitungen. Beratungsstelle für Stahlverwertung, Düsseldorf. – [11] Wagner W. Rohrleitungstechnik, 2nd edn. Vogel, Würzburg, 1983. – [12] Armaturen-Handbuch der Fa. KSB, Frankenthal. – [13] Früh KF. Berechnung des Durchflusses in Regelventilen mit Hilfe des k_v-Koeffizienten. Regelungstechnik 1957; 5: 307. – [14] Ullmanns Encyklopädie der techn. Chemie, vol. 4, 4th edn. Verlag Chemie, Weinheim, 1974, pp 258-67. – [15] Trutnovsky

K. Berührungsdichtungen, 2nd edn. Springer, Berlin, 1975. – [16] Mayer E. Axiale Gleitringdichtung, 7th edn. VDI, Düsseldorf, 1982.

H3 Types of Heat Exchanger. [1] Grassmuck J, Houben KW, Zollinger RM. DIN-Normen in der Verfahrenstechnik. Teubner, Stuttgart, 1989. – [2] Klapp E. Apparate- und Anlagentechnik. Springer, Berlin, 1980. – [3] Becker J. Ausführungsbeispiele für Wärmeaustauscher in Chemieanlagen. Verfahrenstechnik 1969; 3: 335-40. – [4] Shah RK. Classification of heat exchangers. In Heat Exchangers, Advanced Study Institute book. Hemisphere, Washington, 1981, pp 9-46.

H4 Condensers and Aftercoolers. [1] VDI-Wärmeatlas. Berechnungsblätter für den Wärmeübergang, 5th edn. VDI, Düsseldorf, 1988. – [2] Dornieden, M. Zur Berechnung ein- mehrgängiger Rohr-bündel-Kondensatoren. Chem Ing Techn 1972; 44: 618-22. – [3] Schrader H. Einfluss von Inertgasen auf den Wärmeübergang bei der Kondensation von Dämpfen. Chem Ing Techn 1966; 38: 1091-4. – [4] Grant IDR. Condenser performance – the effect of different arrangements for venting noncondensing gases. Brit Chem Eng 1969; 14: 1709-11. – [5] Chen SS. Flow induced vibration of circular cylindrical structures. Hemisphere, Washington, 1987. – [6] TEMA-Standards of Tubular Exchanger Manufacturers Association, 6th edn. New York, 1978. – [7] Kopp JH. Über den Wärme- und Stoffaustausch bei Mischkondensation. Diss. ETH. Juris-Verlag, Zürich, 1965. – [8] Forgo L. Probleme der Mischkondensatorkonstruktion bei Luftkondensationsanlagen System Heller. Energietechn 1967; 17: 302-5. – [9] Schröder K. Das neue Dampfkraftwerk. Brennst-Wärme-Kraft 1963; 15: 140-2. – [10] Berliner P. Kühltürme. Springer, Berlin, 1975. – [11] Vodicka V, Henning H. Überlegungen zur optimalen Gestaltung eines Nass-/Trockenkühlturms unter dem Gesichtspunkt der Minimierung des sichtbaren Schwadens. Brennst.-Wärme-Kraft 1976; 28: 387-92. – [12] Schlaich J, Mayr G, Weber P, Jasch E. Der Seilnetzkühlturm Schmehausen. Bauing 1976; 51: 401-12. – [13] Über den Windeinfluss bei natürlich belüfteten Kühltürmen. Balcke-Dürr: Die aktuelle Information no. 10-5/1976. – [14] Merkel F. Verdunstungskühlung. VDI-Forschungsh. no. 275. VDI, Düsseldorf, 1925.

Machine Dynamics

D. Föller, Frankfurt-on-Main; K. H. Küttner, Berlin; R. Nordmann, Kaiserslautern

1 Crank Operation, Forces and Moments of Inertia, Flywheel Calculations

K. H. Küttner, Berlin

Forces and moments acting on the piston and other power unit components are used in calculations involving the power unit (see F10.3), its smooth operation, crankshaft torsional vibration [1] (see J2), environmental mass effects, and resonance phenomena [2].

1.1 Graph of Torque Fluctuations in Multi-Cylinder Reciprocating Machines

This is affected by type of construction, offset of cranks, the oscillating power-unit masses and cylinder pressure, as well as the firing sequence [3] in the case of engines.

Pressure Graph. This is obtained [4] as $p = f(\varphi)$ from a cathode ray oscillogram or as $p = f(x)$. The non-dimensional

$$\xi = \frac{x}{r} = 1 - \cos \varphi + \frac{\lambda}{2} \sin^2 \varphi + \frac{\lambda^3}{8} \sin^4 \varphi + \dots \quad (1)$$

quantity expresses the distance x travelled by the piston in terms of the crank angle φ, where in most cases the first three terms will be sufficient. Its steps $\Delta\varphi$ numbered $k = 0$ to i at the period $\varphi_A = 360° a_T$ for the working cycle are then given by $\Delta\varphi = \varphi_A/k$. The number of strokes $a_T = 2$ in the case of a four-stroke engine, otherwise $a_T = 1$. For a compressor with $\varphi_A = 360°$, $\Delta\varphi = 6°$ for $k = 60$. Its TDC (top dead centre) lies at $\xi = 0$, i.e. at $k = 0$ or 60, the BDC (bottom dead centre) at $\xi = 2$ and $k = 30$. For new designs without diagrams, the latter must be derived from the ideal process. In the case of compressors, this can be accomplished with the help of the simplified Seiliger process [5] by replacing compression and expansion by polytropes, and in the case of pumps by the suction and pressure characteristics. If the media are gases, the volumes are given by the p, V diagram [5]:

$$\begin{aligned} \text{Engine:} \quad & V = [1/(\varepsilon - 1) + \xi/2] \cdot V_{\text{h}}, \\ \text{Compressor:} \quad & V = [\varepsilon_0 + \xi/2] \cdot V_{\text{h}}. \end{aligned} \quad (2)$$

In the case of double-acting cylinders and symmetric diagrams, F10 Eq. (19)

$$p_{\text{KS}} = (2 - \xi) p_{\text{DS}} \quad \text{applies.}$$

Torque. With the piston force $F_K = F_s - F_0$ and the difference between the gas and the oscillatory inertia forces (as F10.2.1 to F10.2.3), where in most cases two terms are sufficient, the torque of a power unit is given by

$$M_{\text{d}} = F_T r = F_K(\varphi) \cdot \left(\sin \varphi + \frac{\lambda}{2} \frac{\sin 2\varphi}{\sqrt{1 - \lambda^2 \sin^2 \varphi}} \right), \quad (3)$$

with period $\varphi_A = 360° \, a_T$ and the forces equal to zero in F10.2.3. At increasing revolutions, the inertial forces at first reduce the gas forces only to exceed them later, a situation which also affects the torque fluctuations (see **F10 Fig. 8** and **Fig. 1c**).

Total Torque. This applies to machines with z cylinders and equal piston and crankshaft offsets as in the case of engines [6], where the angle φ is measured with respect to the angular position of the crank 1. For an in-line machine (**Fig. 1a**) this is

$$M_{\text{d tot}} = \sum M_{\text{d}}[\varphi + (k - 1)\varphi_P]. \quad (4)$$

The total moment repeats with the periodicity $\varphi_P = \varphi_A/z$, i.e. with the angle between two cranks. Here, the torque fluctuations decrease with an increasing number of cylinders. For fan-type and radial machines and angles $\gamma_F = 180°/z$ and $\varphi_s = 360°/z$ respectively between the neighbouring cylinder centrelines, the following applies:

$$M_{\text{d tot}} = \sum M_{\text{d}} [\varphi + (k - 1)\gamma]. \quad (5)$$

Equal firing intervals are only possible for radial engines with an odd number of cylinders. For the W compressor with $\gamma_F = 60°$ (**Fig. 1a**), the following applies:

$$M_{\text{d tot}} = M_{\text{d}}(\varphi) + M_{\text{d}}(\varphi - 60°) + M_{\text{d}}(\varphi - 120°).$$

In the case of multi-stage compressors, the torque function of the individual stages differ. Owing to the coupling between the power unit and the driven unit, the torques of both must be taken into account (**Fig. 1**). In order to investigate the vibrations, the harmonic analysis of the diagram of **Fig. 1d** must be applied.

Mean Torque. The following equation applies:

$$M_{\text{dm}} = \frac{1}{\varphi_P} \int_0^{\varphi_P} M_{\text{d tot}} \, d\varphi; \quad (6)$$

it is determined by means of numerical integration. In the steady state it is equal to the mean value of the coupled machine and independent of the inertial forces.

Flywheel. This compensates for the maximum of individual energy fluctuations (**Fig. 2**):

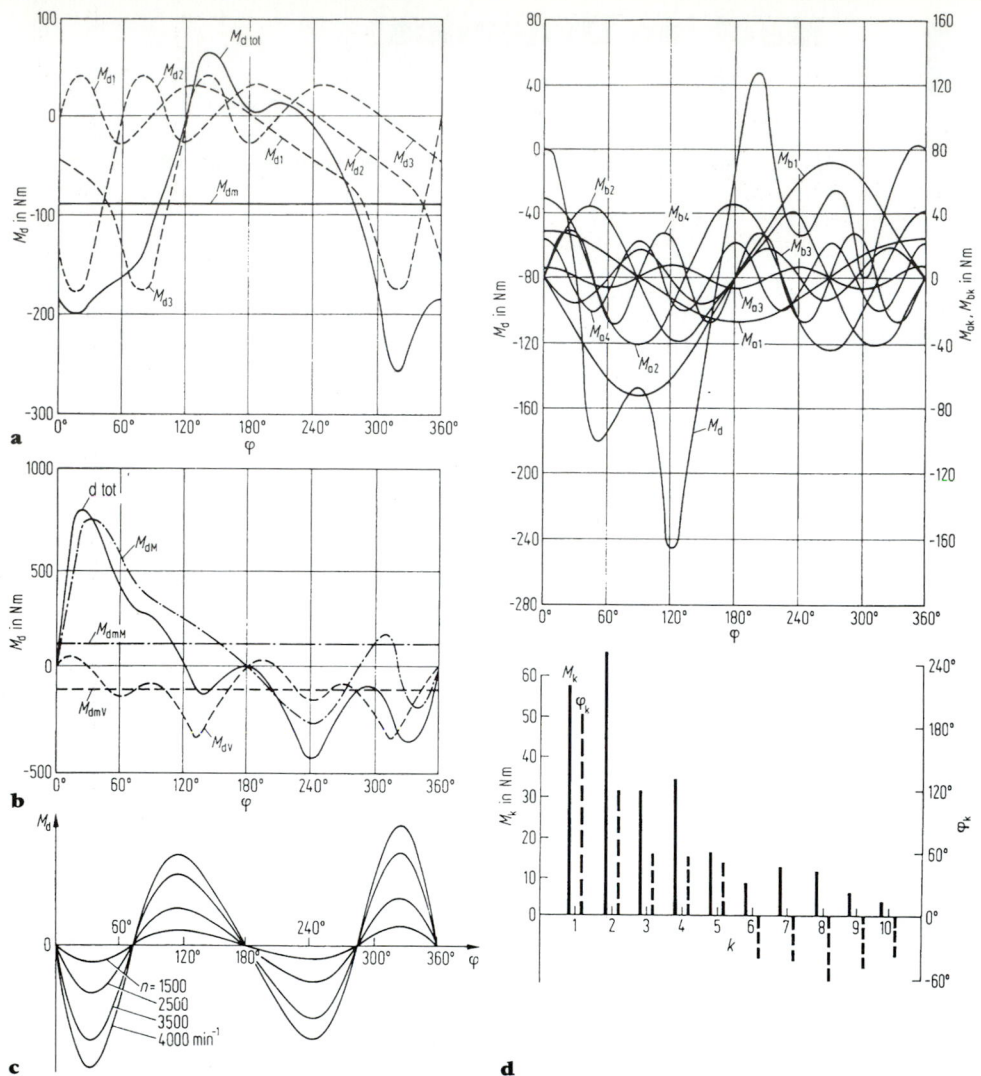

Figure 1. Torque diagrams. **a** Single-stage W compressor, $\gamma_F = 60°$. **b** Four-stroke engine with single-stage V compressor with two banks of two cylinders each; $M_{dmM} = -M_{dmV}$, $\varphi_{PM} = \varphi_{PV}$. **c** Two-stroke engine idling. **d** Harmonic analysis of the torque of a two-stage compressor with spectrum of the instantaneous torque amplitudes and their phase angles $M_k = \sqrt{M_{ka}^2 + M_{kb}^2}$, or tan $\varphi_k = M_{ak}/M_{bk}$ respectively.

$$W_s = \int_{\varphi_k}^{\varphi_{k+1}} (M_d - M_{dm})\,d\varphi . \tag{7}$$

Here the φ_k or φ_{k+1} occurs at the point where $M_d = M_{dm}$.

Moment of Inertia. It follows from the energy conservation law, with $W_{smax} = J\,(\omega_{max}^2 - \omega_{min}^2)/2$, the mean value $\omega_m = (\omega_{max} + \omega_{min})/2$ and the irregularity factor $\delta = (\omega_{max} - \omega_{min})/\omega_m$ that, in accordance with **Table 1**,

$$J = \frac{W_s}{\delta\omega_m^2} = \frac{W_s}{4\pi^2\,\delta n^2} . \tag{8}$$

This also takes into account the coupled machine and power unit, and is supplied by the flywheel, the function

of which also includes control [5]. Reference values for four-stroke engines [6] are obtained from the indicated power P_i and the constant k from **Table 2**, using

$$J = k\,\frac{P_i}{\delta(n/100)^3} . \tag{9}$$

Thus at constant power, the moment of inertia decreases with the third power of the rate of revolution, the number of cylinders and the irregularity factor.

Design. The flywheel (**Fig. 3**) consists of k discs of width b_k and outer and inner diameters D_k and d_k respectively, and has a density ρ. Its mass and moment of inertia are therefore given respectively by

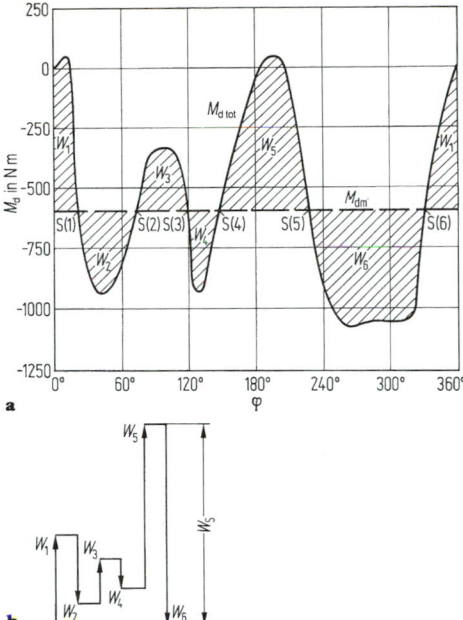

Figure 2. Determination of available power: **a** torque, **b** energy curve.

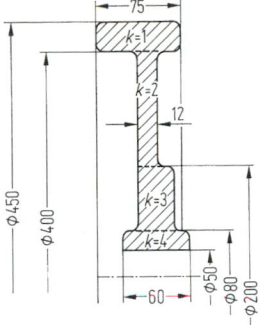

Figure 3. Disc-type flywheel.

Table 1. Reference values of the irregularity factor

Ship's propeller	1/30
Pumps and blowers	1/30 to 1/50
Machine tools	1/50
Piston compressors	1/50 to 1/100
Vehicle engines	1/150 to 1/300
Generators:	
– AC	1/125 to 1/300
– DC	1/100 to 1/200
Aircraft engines	1.1000

Table 2. Constant k in kg m^2/(kW min^3) for four-stroke engines

Number of cylinders	1	2	3	4	5	6	7	8
Diesel engine	17.5	7.2	4.3	0.92	1.63	0.54	0.73	0.49
Spark-ignition engine	6.0	2.5	1.3	0.5	0.24	0.12	—	—

$$m_s = \frac{\pi}{4} \rho \sum (D_k^2 - d_k^2) b_k,$$

$$(10)$$

$$J = \frac{1}{8} \sum m_k (D_k^2 + d_k^2) = \frac{\pi}{32} \rho \sum (D_k^4 - d_k^4) b_k,$$

where $D_{k+1} = d_k$.

Accordingly, the outer ring has the greatest effect and accounts for about 90% of the moment of inertia for disc-type flywheels and 95% for those of spoke type [7, 8]. In order to utilise material as efficiently as possible, the outer diameter should be as large as the stresses due to centrifugal force allow. The limits lie at peripheral speeds of $u = 50$ m/s for grey cast-iron flywheels and $u = 75$ m/s for cast-steel flywheels.

Example. Torque diagram of a single-stage compressor with trunk piston. This is based on the simplified calculation of the pressure curve from the p, V diagram (**Fig. 4**).

Back Expansion 3-4. Using Eqs (1) and (2) it follows from the polytropic equation

$$pV^n = p_2 V_3^n \quad \text{with} \quad V_3 = V_s = \varepsilon_0 V_h$$

that

$$p = p_2 \left(\frac{\varepsilon_0}{\varepsilon_0 + \xi/2} \right)^n.$$

It occurs between p_2 and $p_1 = p_2/\psi$, or between $\xi = 0$ and $\xi_{back} = 2\varepsilon_0 (\psi^{1/n} - 1)$, respectively.

Induction 4-1. This takes place along the isobar p_1 between $\xi = \xi_{back}$ and $\xi = 2$.

Compression 1-2. Using the polytropic equation $pV^n = p_1 V_1^n$, where $V_1 = V_s + V_h = (1 + \varepsilon_0) V_h$, the following is obtained:

$$p = p_1 \left(\frac{1 + \varepsilon_0}{\varepsilon_0 + \xi/2} \right)^n$$

This covers the region from p_1 to $p_2 = \psi p_1$, i.e. from

$$\xi = 2 \text{ to } \xi_{com} = 2[(1 + \varepsilon_0)\psi^{-1/n} - \varepsilon_0].$$

Exhaust 2-3. This occurs between $\xi = \xi_{com}$ and $\xi = 0$, at constant pressure p_2.

Moment. The force $F_s = (p - p_a)A_k$ exerted by the medium is obtained from F10 Eq. (18) with the pressure $p = f(\varphi)$ of the

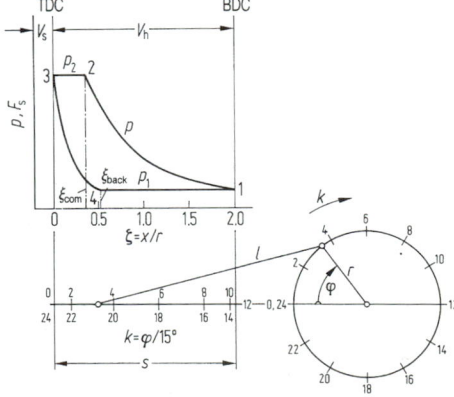

Figure 4. Determination of the pressure curve of a piston compressor.

individual parts of the operating cycle. With the inertial forces as in F10 Eq. (22), $F_o = m_o r \omega^2 (\cos \varphi + \lambda \cos 2\varphi)$, i.e. with $F_k = F_s - F_o$ as in F10 Eq. (27), the moment is obtained from Eq. (3) (see M_{d1} in **Fig. 1**).

1.2 Forces and Moments of Inertia

In multi-cylinder machinery, resultants are found by vector addition of the forces of each power unit and the moments produced by them. These are the rotating forces $F_r = m_r r \omega^2$, or the first- and second-order forces $F_I = m_o r \omega^2 \cos \varphi = P_I \cos \varphi$ and $F_{II} = \lambda P_I \cos 2\varphi$, respectively, obtained from F10 Eqs (21) and (22), as well as the moments M_r, M_I and M_{II}. Their addition is carried out graphically [9] or analytically in accordance with the position of the cranks and the position of the centrelines. In the case of engines, the masses m_r and m_o of the power units are constant according to F10 Eqs (25) and (26), as are the distances a between the cylinders and the difference $\Delta \alpha_k$ of the crank offset. Their centroidal axis SS lies in the middle of the crankshaft (**Fig. 5**). Multi-stage compressors have different pistons [10] and hence masses m_o. These forces and moments cause vibrations in the power unit and the machine [11]; in particular, torsional vibrations in the crankshaft [12].

1.2.1 Analytical Methods

These are particularly suitable for the programming of computers [5].

In-line Machines. Using the stroke number a_T, the distance b_k and the offset α_k of a k crankshaft of z cylinders (**Fig. 5a**) is given by

$$\alpha_k = (n-1)360° \, a_T/z,$$
$$b_k = [0.5(z+1) - k]a = \nu_k a. \tag{11}$$

The index $k = 1$ to z designates the power units along the crankshaft starting at the coupling and the index $n = 1$ to z determines the angle α_k and is counted in the direction of rotation.

Rotating Moments. Their vertical and horizontal components are given respectively by

$$M_{rx} = F_r \, \Sigma b_k \sin(\varphi + \alpha_k) \quad \text{and}$$
$$M_{ry} = F_r \, \Sigma b_k \cos(\varphi + \alpha_k),$$

Using the non-dimensional constants

$$c_{r1} = \Sigma \nu_k \cos \alpha k \quad \text{and} \quad c_{r2} = \Sigma \nu_k \sin \alpha_k, \tag{12}$$

the resultant and its angular position are calculated from

$$M_{rres} = \sqrt{M_{rx}^2 + M_{ry}^2}, \quad \tan \chi = M_{rx}/M_{ry} = c_{r2}/c_{r1}; \tag{13}$$

the rotating moments using $c_r = \sqrt{c_{r1}^2 + c_{r2}^2}$ are

$$M_{rres} = F_r a c_r \quad \text{and} \quad \alpha_L = 90° + \varphi + \chi. \tag{14}$$

Moments of the mth Order. With the force amplitudes $P_{mk} = F_{mk}/\cos(m\varphi)$, as in F10 Eq. (21), which act along the centreline of the cylinder,

$$M_{mres} = \Sigma P_{mk} \, b_k \cos m(\varphi + \alpha_k) \quad \text{applies.}$$

The maximum is obtained from $dM_{mres}/d\varphi = 0$

$$\tan m\varphi = - \Sigma P_m b_k \sin m\alpha_k / \Sigma P_m b_k \cos m\alpha_k, \tag{15}$$

depending on the angle φ for its calculation and direction. If the pistons, i.e. the forces P_{mk}, are identical, then using the constants

$$c_{m1} = \Sigma \nu_k \cos m\alpha_k \quad \text{and} \quad c_{m2} = \Sigma \nu_k \sin m\alpha_k \tag{16}$$

the instantaneous values and their maximum respectively are given by

$$\left. \begin{array}{l} c_m = \sqrt{c_{m1}^2 + c_{m2}^2} \\ M_{mres} = P_m \, a(c_{m1} \cos m\varphi - c_{m2} \sin m\varphi) \\ M_{mmax} = P_m a c_m. \end{array} \right\} \tag{17}$$

They occur at the crankshaft angle

$$\varphi = \tan^{-1}(-c_{m2}/c_{m1})/m, \tag{18}$$

where $c_{11} = c_{r1}$ and $c_{12} = c_{r2}$.

Forces. For these $b_k = a\nu_k = 1$ applies in Eqs (11) and (16). This means that the constants are

$$k_{m1} = \Sigma \cos m\alpha_k \quad \text{and} \quad k_{m2} = \Sigma \sin m\alpha_k. \tag{19}$$

Most Advantageous Crank Sequence. The forces vanish if the mth order crank angular position diagrams with the

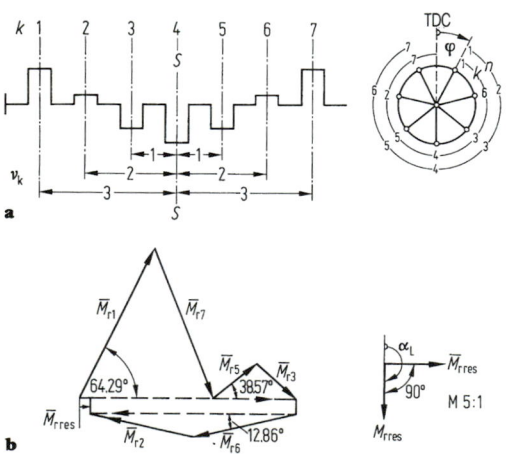

Figure 5. Seven-cylinder in-line engine: **a** crank schematic with first- and second-order stars, **b** vectorial determination of the resulting rotating moments.

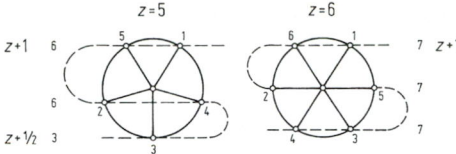

Figure 6. Most favourable crank sequence for two-stroke engines with even and odd numbers of cylinders.

Table 3. Extreme values of the inertial forces of V machines $P_1 = m_r r \omega^2$ and $P_{II} = \lambda P_1$

γ in °	F_{Ia}/P_I	F_{Ib}/P_I	F_{IIa}/P_{II}	F_{IIb}/P_{II}
30	1.867	0.134	1.673	0.259
45	1.707	0.293	1.307	0.541
60	1.50	0.50	0.866	0.866
75	1.259	0.741	0.411	1.176
90	1.0	1.0	0	1.414
120	0.5	1.50	0.5	1.50
180	0.0	2.0	0	0

angles $m\alpha_k$ are symmetrical (**Fig. 5a**). Two-stroke engines (**Fig. 6**) have their smallest moments if their first-order crank angular position diagram passes through the sequence 1, z, 2, $z - 1$, n, $n(z - n + 1)$ [9, 13]. In four-stroke engines the moments cancel each other if the angle α_k and the length of their lever arms b_k are the same for any two cranks.

Engines. In the case of two-cylinder engines, the centre-lines of the power units A and B displaced by the width of a pushrod form the angle $\gamma = \varphi_A + \varphi_B$ (**Fig. 7**). The vertical and horizontal components of the first-order force are then

$$F_{1x} = (F_{1A} - F_{1B}) \sin (\gamma/2) \quad \text{and}$$

$$F_{1y} = (F_{1A} + F_{1B}) \cos (\gamma/2)$$

respectively, because $\varphi_A = \gamma/2 - \varphi_k$ and $\varphi_B = \gamma/2 + \varphi_k$, with $F_{1A} = P_{1A} \cos \varphi$ and $F_{1B} = P_{1B} \cos \varphi_B$.
This makes the resultant and its angular position

$$F_1 = \sqrt{F_{1x}^2 + F_{1y}^2} \quad \text{and} \quad \tan \alpha_1 = F_{1x}/F_{1y}, \quad (20)$$

respectively.

If the piston masses are equal then

$$F_{1x} = 2P_1 \sin^2(\gamma/2) \sin \varphi_k \quad \text{and}$$
$$F_{1y} = 2P_1 \cos^2(\gamma/2) \cos \varphi_k. \quad (21)$$

At $\gamma = 90°$, Eqs (20) and (21) show that $F_1 = P_1$ and $\alpha_L = \varphi$. The first-order forces can be balanced by counter-weights at the countermass extensions. Their extreme values occur when $\cos \varphi = 1$ and 0 respectively. They rep-

resent the semi-axes of the ellipses in accordance with Eq. (21) and are therefore given by

$$F_{1a} = 2P_1 \cos^2(\gamma/2) \quad \text{and} \quad F_{1b} = 2P_1 \sin^2(\gamma/2). \quad (22)$$

Their position is vertical or horizontal respectively and for $\gamma < 90°$ F_{1a} is the maximum and F_{1b} the minimum (see **Table 3**). For second-order forces with the component $F_{IIA} = P_{IIA} \cos 2\varphi_A$ and $F_{IIB} = P_{IIB} \cos 2\varphi_B$, the following apply:

$$F_{IIx} = (F_{IIA} - F_{IIB}) \sin (\gamma/2) \quad \text{and}$$

$$F_{IIy} = (F_{IIA} + F_{IIB}) \cos (\gamma/2).$$

The resultant and its position angle are given by $F_{II} = \sqrt{F_{IIx}^2 + F_{IIy}^2}$ and $\tan \alpha_{II} = F_{IIx}/F_{IIy}$ respectively. For pistons of equal mass,

$$F_{IIx} = 2P_{II} \sin(\gamma/2) \sin \gamma \sin 2\varphi_k,$$
$$F_{IIy} = 2P_{II} \cos(\gamma/2) \cos \gamma \cos 2\varphi_k. \quad (23)$$

Their extreme values, which occur at $\cos 2\varphi_k = 1$ and 0 respectively, are

$$F_{IIa} = 2P_{II} \cos (\gamma/2) \cos \gamma \quad \text{and}$$
$$F_{IIb} = 2P_{II} \sin (\gamma/2) \sin \varphi_k. \quad (24)$$

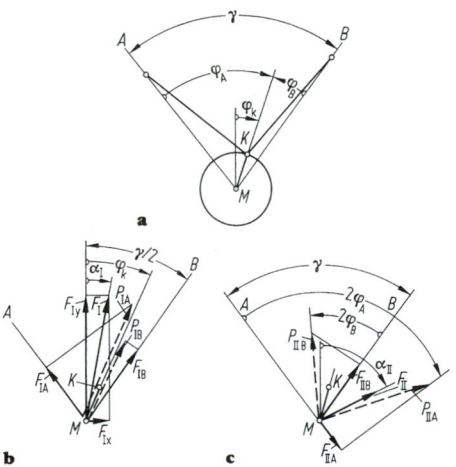

Figure 7. V machine: **a** arrangement of the power units, **b** determination of the first-order force from its components, **c** vectorial determination of the second-order force.

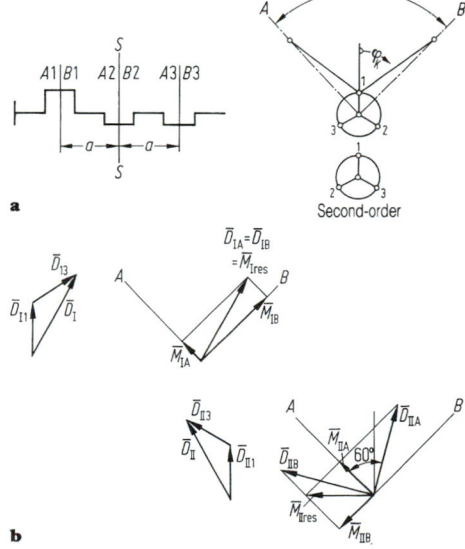

Figure 8. V in-line machines: **a** schematic construction and first-order moments, **b** second-order crank angular position diagram with moments.

Table 4. Free forces and moments of inertia of various cylinder arrangements (compiled from [3, 6, 11, 12, 16])

Designation	2 cylinders in line [1,2,3]	2 cylinders in line [1,2,3]	2 cylinders opposed [1,2,3]	2 cylinders, 45° V [1,2,3,4]	2 cylinders, 60° V [1,2,3,4]	2 cylinders, 90° V [1,2,3,4]
First-order crank angular, position diagram / Schematic of crankshaft						
Crankshaft construction	2 cranks	2 cranks	2 cranks	1 crank	1 crank	1 crank
Firing interval	180° – 540°	360° – 360°	360° – 360°	405° – 315°	420° – 300°	450° – 270°
Free forces (no compensation) First-order	0	$2P_I$	0	v.$1.707P_I$; h.$0.293P_I$	v.$1.5P_I$; h.$0.293P_I$	v. and h. $1.0P_I$
Second-order	$2P_{II}$	$2P_{II}$	0	v.$1.31P_{II}$; h.$0.34P_{II}$	v. and h. $0.865P_{II}$	v.$0P_{II}$; h.$1.414P_{II}$
Free moments (no compensation) First-order	$a \cdot P_I$	0	$b \cdot P_I$	$b \cdot F_I$	$b \cdot F_I$	$b \cdot F_I$
Second-order	0	0	$b \cdot P_{II}$	$b \cdot F_{II}$	$b \cdot F_{II}$	$b \cdot F_{II}$
Higher order free forces	$2(P_{IV}+P_{VI}+...)$	$2(P_{IV}+P_{VI}+...)$	0	$0.765P_{IV}$; $0.765P_{VI}$	$\sqrt{3}P_{IV}$	$\sqrt{2}P_{IV}$; $\sqrt{2}P_{VI}$
Higher order free moments	0	0	$b \cdot (P_{IV}+P_{VI}+...)$	$0.765 \cdot b \cdot P_{IV}$; $0.765 \cdot b \cdot P_{VI}$	$b \cdot P_{IV}$; $b \cdot 1/2\sqrt{3}P_{VI}$	$b \cdot 1/2\sqrt{2}P_{IV}$; $b \cdot 1/2\sqrt{2}P_{VI}$
Counterweights: usual number	2	2	2	2	2	2
magnitude	$<(F_r+0.5P_I)$	$F_r+0.5P_I$	$<(F_r+0.5P_I)$	$1/2(F_r+...P_I)$	$1/2(F_r+...P_I)$	$1/2(F_r+...P_I)$
effort	large	large	large	large	large	large
Torsional vibrations, critical / Torsional vibration behaviour	0.5; 1.5; 2; 2.5;... / good	1; 2; 3;... / good	1; 2; 3;... / good	see [10,13]	see [10,13]	see [10,13]
General dynamic behaviour	serviceable	serviceable	serviceable	moderate	moderate	serviceable
Assessment	serviceable	serviceable	serviceable	moderate	moderate	serviceable

Designation	3 cylinders in line [1,2,3,5]	4 cylinders in line [1,2,3]	4 cylinders in line [1,2]	4 cylinders, 2×180° v [1,2,3,4]	4 cylinders opposed [1,2,3]
First-order crank angular, position diagram / Schematic of crankshaft					
Crankshaft construction	3×120° cranks	4 cranks	2·2 cranks 90° offset	2 cranks	4 cranks
Firing interval	240° – 240°	180° – 180° – 180° – 180°	Z.T. 90° – 90° – 90° – 90°	180° – 180° – 180° – 180°	180° – 180° – 180° – 180°
Free forces (no compensation) First-order	0	0	0	0	0
Second-order	0	$4P_{II}$	0	0	0
Free moments (no compensation) First-order	$\sqrt{3} \cdot a \cdot P_I$	0	$\sqrt{2} \cdot a \cdot P_I$	$a \cdot F_I$	0
Second-order	$\sqrt{3} \cdot a \cdot P_{II}$	0	$4 \cdot a \cdot P_{II}$	$2b \cdot F_{II}$	$2b \cdot P_{II}$
Higher order free forces	$3P_{VI}$	$4(P_{IV}+P_{VI}+...)$	$4P_{IV}$	0	0
Higher order free moments	$\sqrt{3} \cdot a \cdot P_{VI}$	0	$4 \cdot a \cdot P_{VI}$	$2b \cdot F_{IV}$; $2b \cdot F_{VI}$	$2b \cdot P_{IV}$; $2b \cdot P_{VI}$
Counterweights: usual number	4	4	4	4	4
magnitude	$<(F_r+1/2P_I)$	$\ll(F_r+1/2P_I)$	$F_r+0.5P_I$	$1/2F_r+...1/2P_I$	$\ll(F_r+0.5P_I)$
effort	small	moderate	large	moderate	small
Torsional vibrations, critical / Torsional vibration behaviour	1.5; 3; 4.5;... / good	2; 4; 6;... / moderate	4; 6; 8;... / good	2; 4; 6;... [10,13] / moderate	2; 4; 6;... / good
General dynamic behaviour	medium	good	moderate	poor	good
Assessment	medium	medium	moderate	poor	good

Designation	4 cylinders 2×90° V [1,4,5]	4 cylinders 2×90° V [1,3,4,6]	4 cylinders 2×90° V [1,3,4,6]	4 cylinders 60° V [1,3,7]	5 cylinders in line [1,2,3]
First-order crank angular, position diagram / Schematic of crankshaft					
Crankshaft construction	2 cranks	2 cranks	2 cranks 90° offset	2×120° cranks 60° offset	5×72° cranks
Firing interval	90° – 180° – 270° – 180°	90° – 270° – 90° – 270°	180° – 90° – 270° – 180°	180° – 180° – 180° – 180°	5×144°
Free forces (no compensation) First-order	0	$2F_I$	$\sqrt{2}F_I$	0	0
Second-order	v. $0P_{II}$; h. $2\sqrt{2}P_{II}$	$2F_{II}$	0	$2\sqrt{3}P_{II}$	0
Free moments (no compensation) First-order	$a \cdot F_I$	$b \cdot F_I$	$a/2\sqrt{2}F_I$; $b/2\sqrt{2}F_I$	$a \cdot P_I$	$0.449 \cdot a \cdot P_I$
Second-order	$2b \cdot F_{II}$	0	$2a \cdot F_{II}$	$b \cdot P_{II}$	$4.98 \cdot a \cdot P_{II}$
Higher order free forces	0	$2F_{IV}$; $2F_{VI}$	$2F_{IV}$; $2F_{VI}$	$2\sqrt{3}(P_{IV}+P_{VI})$	0
Higher order free moments	$2b \cdot F_{IV}$; $2b \cdot F_{VI}$	$b \cdot \sqrt{F_{IV}}$; $b \cdot \sqrt{F_{VI}}$	$b \cdot \sqrt{F_{IV}}$; $b \cdot \sqrt{F_{VI}}$	$b(P_{IV}+P_{VI})$	$0.449 \cdot a \cdot P_{IV}$; $0.449 \cdot a \cdot P_{VI}$
Counterweights: usual number	4	4	4	4	5
magnitude	$1/2(F_r \cdot P_I)$	$1/2F_r+1/2P_I$	$1/2F_r+1/2P_I$	$F_r+1/2P_I$	$F_r+1/2P_I$
effort	moderate	small	small	small	medium
Torsional vibrations, critical / Torsional vibration behaviour	0.5; 1.5; 2.5;... [10,13] / good	1; 3; 4; 5;... [10,13] / good	0.5; 1; 1.5; 2.5;... [10,13] / good	2; 4; 6;... / moderate	1; 1.5; 2.5; 3.5; 4;... / moderate
General dynamic behaviour	moderate	moderate	moderate	moderate	moderate
Assessment	moderate	moderate	moderate	moderate	moderate

Continued

Table 4. Continued

Designation	6 cylinders in line [1,2,3]	6 cylinders in line [1,2,3,5]	6 cylinders 60° V [1*,2,3]	6 cylinders 60° V [1*,2,3]	6 cylinders opposed [1,2,3]
First-order crank angular, position diagram Schematic of crankshaft					
Crankshaft construction	6×60° cranks	6×120° cranks	6×60° cranks	3×180° cranks 120° offset	6×180° cranks 120° offset
Firing interval	120°-120°-180°-120°-120°-60°	6×120°	6×120°	6×120°	6×120°
Free forces (no compensation) First-order Second-order	0 0	0 0	0 0	0 0	0 0
Free moments (no compensation) First-order Second-order	0 $2\sqrt{3}\cdot a\cdot F_{II}$	0 0	0 $3/2\cdot a\cdot F_{II}$	0 $3/2\cdot a\cdot F_{II}$	0 0
Higher order free forces Higher order free moments	$6\,P_{VI}$ 0	$6\,P_{VI}$ 0	$3\sqrt{3}\,F_{VI}$ $3/2\cdot a\cdot F_{IV}$; $3/2\cdot b\cdot F_{VI}$	$3\sqrt{3}\,F_{VI}$ $3/2\cdot b\cdot F_{VI}$	0 $3\cdot b\cdot P_{VI}$
Counterweights: usual number magnitude effort	6 $F_r+1/2\,P_I$ medium	6 $\ll(F_r+1/2\,P_I)$ small	6 $F_r+1/2\,P_I$ large	6 $1/2\,F_r+1/2\,P_I$ medium	6 $<(1/2\,F_r+1/2\,P_I)$ small
Torsional vibrations, critical Torsional vibration behaviour	1.5; 3; 4.5; 6;... moderate	3; 6; 9;... good	3; 6; 9;... moderate	3; 6; 9;... serviceable	3; 6; 9;... good
General dynamic behaviour	moderate	good	moderate	serviceable	good
Assessment	moderate	good	moderate	serviceable	good

Designation	6 cylinders 3×90° V [1*,2,3,6]	6 cylinders 3×120° V [1*,2,3]	6 cylinders 3×180° V [1*,3,6]	7 cylinders in line [1,2,3]	8 cylinders in line [1,2,3]
First-order crank angular, position diagram Schematic of crankshaft					
Crankshaft construction	3 cranks 120° offset	3 cranks 120° offset	3 cranks 120° offset	7×51,43° cranks	8×90° cranks 1×45° offset
Firing interval	150°-90°-150°-90°-150°-90°	6×120°	120°-120°-60°-120°-120°-180°	7×102,86°	90°-90°-90°-90°-90°- 90°-135°,two-stroke 8×45°
Free forces (no compensation) First-order Second-order	0 0	0 0	0 0	0 0	0 0
Free moments (no compensation) First-order Second-order	$\sqrt{3}\cdot a\cdot F_I$ $\sqrt{6}\cdot a\cdot F_{II}$	$1.5\sqrt{3}\cdot a\cdot F_I$ $1.5\sqrt{3}\cdot a\cdot F_{II}$	$2\sqrt{3}\cdot a\cdot F_I$ 0	$0.267\cdot a\cdot P_I$ $1.006\cdot a\cdot P_{II}$	$0.448\cdot a\cdot P_I$ 0
Higher order free forces Higher order free moments	$3\sqrt{2}\,F_{VI}$ $\sqrt{6}\cdot a\cdot F_{IV}$; $3/2\sqrt{2}\cdot b\cdot F_{VI}$	$3\,F_{VI}$ $3/2\cdot a\cdot F_{IV}$; $\sqrt{3}\cdot b\cdot F_{VI}$	0 $3\cdot b\cdot F_{VI}$	0 $9,845\cdot a\cdot P_{IV}$; $0,263\cdot a\cdot P_{VI}$	0 $16\cdot a\cdot P_{IV}$
Counterweights: usual number magnitude effort	6 $1/2\,F_r+1/2\,P_I$ medium	6 $1/2\,F_r+1/2\,P_I$ medium	6 $<(1/2\,F_r+1/2\,P_I)$ good	7 $F_r+1/2\,P_I$ large	8 $(F_r+1/2\,P_I)$ large
Torsional vibrations, critical Torsional vibration behaviour	1.5; 3; 4.5;... good	3; 6; 9;... good	0.5;1.5; 2.5; 3.5; 4.5;... moderate	1; 2.5; 3.5; 4.5; 6; 7; 8; moderate	2; 2.5; 3.5; 4; 4.5;... moderate
General dynamic behaviour	good	serviceable	serviceable	serviceable	serviceable
Assessment	moderate	poor	serviceable	serviceable	serviceable

Designation	8 cylinders in line [1,2,3]	8 cylinders 4×90° V [1*,3,4,6]	8 cylinders 4×180° V [1*,3,4]	8 cylinders opposed [1,2,3]	8 cylinders 60° V [1,2,3]
First-order crank angular, position diagram Schematic of crankshaft					
Crankshaft construction	4×180° cranks, 2×90° offset	4 cranks, 90° offset	4 cranks, 180° offset	4×180° cranks, 90° offset	4×30° cranks, 90° offset
Firing interval	8×90°	8×90°	4×180° two stroke	8×90°	8×90°
Free forces (no compensation) First-order Second-order	0 0	0 0	0 0	0 0	0 0
Free moments (no compensation) First-order Second-order	0 0	$\sqrt{10}\cdot a\cdot F_I$ 0	0 $4\cdot b\cdot F_{II}$	0 0	$(3.054\pm0.818)\,a\cdot P_I$ 0
Higher order free forces Higher order free moments	$8\,P_{IV}$ 0	$4\sqrt{2}\,F_{IV}$ $2\sqrt{2}\cdot b\cdot F_{IV}$	0 $4\cdot b\cdot F_{IV}$; $4\cdot b\cdot F_{VI}$	0 $4\cdot b\cdot P_{IV}$; $4\cdot b\cdot P_{VIII}$	$4\sqrt{3}\,P_{IV}$ $2\cdot b\cdot P_{IV}$
Counterweights: usual number magnitude effort	8 $(F_r+1/2\,P_I)$ large	8 $F_r+1/2\,P_I$ medium	4 $<(F_r+1/2\,P_I)$ small	4 $F_r+1/2\,P_I$ moderate	8 $F_r+1/2\,P_I$ medium
Torsional vibrations, critical Torsional vibration behaviour	4; 8; 12; ... moderate	4; 8; 12;... medium	2; 4; 6;... medium	4; 8; 12;... serviceable	4; 8; 12;... serviceable
General dynamic behaviour	serviceable	good	moderate	good	serviceable
Assessment	serviceable	good	serviceable	good	serviceable

Here, F_{IIa} is the maximum and F_{IIb} the minimum, when $\gamma < 60°$ (see **Table 3**).

The rotating forces are given by F10 Eq. (26) as

$$F_r = m_{rV} r^2 \omega^2 \quad \text{with} \quad m_{rV} = m_{rKW} + 2m_{rSt}. \tag{25}$$

V In-Line Engines (Fig. 8). If the piston masses are equal the components of the first-order moments are, from Eqs (17) and (22),

$$M_{Ix} = 2P_I a \sin^2 (\gamma/2)(c_{r1} \sin \varphi + c_{r2} \cos \varphi),$$
$$M_{Iy} = 2P_I a \cos^2 (\gamma/2)(c_{r1} \cos \varphi - c_{r2} \sin \varphi), \tag{26}$$

with $c_{I1} = c_{r1}$ and $c_{I2} = c_{r2}$.

Using Eq. (17) with $m = II$, the following apply for the second-order moments:

$$M_{IIx} = 2P_{II} a \sin \gamma \sin (\gamma/2)(c_{II1} \sin 2\varphi + c_{II2} \cos 2\varphi), \tag{27}$$
$$M_{IIy} = 2P_{II} a \cos \varphi \cos (\gamma/2)(c_{II2} \cos 2\varphi - c_{II2} \sin 2\varphi),$$

The resultant and its angular position follow from Eq. (13).

The extreme values of the first-order moments are given by

$$M_{Ia} = 2P_I ac_r \cos^2 (\gamma/2)$$
$$\text{and} \qquad M_{Ib} = 2P_I ac_r \sin^2 (\gamma/2), \tag{28}$$

with $c_r = \sqrt{c_{r1}^2 + c_{r2}^2}$.

For the second-order moments using c_{II1} and c_{II2} as given by Eq. (19) and with $c_{II} = \sqrt{c_{II1}^2 + c_{II2}^2}$ the following apply:

$$M_{IIa} = 2P_{II} ac_{II} \cos \gamma \cos \gamma/2$$
$$\text{and} \qquad M_{IIb} = 2P_{II} ac_{II} \sin \gamma \sin (\gamma/2). \tag{29}$$

The rotating moments are calculated using Eq. (14) and the mass m_{rV}, as in the case of the in-line machine.

Table 4 (see pp. J6 and J7) shows the forces and moments of inertia of the most important engine types.

Example. Forces and moments of inertia for an engine with crank sequence 1, 6, 3, 4, 5, 2, 7 of simple in-line and V in-line construction with 60° and 90° V angles.

In-Line Engine. In accordance with Eq. (11) and **Fig. 5**, the crank offset and lever arm for $z = 7$ cylinders are

$$\alpha_k = (n - 1)51.43° \quad \text{and} \quad v_k = b_k \, a = 4 - k.$$

The crank angle $\varphi = 51.43°/2 = 25.72°$. From **Table 5**, calculated by using these values and employing Eqs (12) and (19), it follows that $c_{r1} = 0.1160$ and $c_{r2} = 0.2407$ or $c_r = 0.2672$ and $k_{r1} = k_{r2} = 0$

respectively. This leads to the resultant and/or the maximum first-order moment

$$M_{rres}/(F_r a) = M_{1res}/(P_I a) = 0.2672.$$

From Eq. (14), the vector of the rotating moment with $\tan^{-1}$ $(0.2407/0.116) = 64.28°$ has the angular position

$$\alpha_L = 90° + 25.72° + 64.28° = 180°.$$

The maximum first-order moment occurs at the crank angles $\varphi = -64.26°$ and $115.75°$, i.e. after rotation of crank 1 by $90°$. The moment is zero at $\varphi = 64.28°$ and $154.28°$. For the second-order moment, **Table 5** is recalculated for $2\alpha_k$. From Eq. (16) this yields $c_{II1} = 0.7862$ and $c_{II2} = 0.6270$, i.e. $c_{II} = 1.006$ and $k_{II1} = k_{II2} = 0$.

The maximum second-order moment is $M_{IIresmax}/(\lambda P_I a) = 1.006$. It occurs at $\tan^{-1} (-c_{II2}/c_{II1}) = 38.57°$ when $\varphi = (90 - 38.57)°$ $= 25.71°$, i.e. in the position shown in the drawing. In the graphical solution (**Fig. 5b**) as under J1.2.2, it follows from the moment diagram that

$$M_r = 2F_r a (3 \cos 64.28° + \cos 38.57° - 2 \cos 12.86°)$$
$$= 0.2672 \, F_r a.$$

Here the vector $\overline{M}_{res}$ must still be rotated clockwise by $90°$. No forces occur, because $k_{r1} = k_{r2} = k_{II1} = k_{II2} = 0$, i.e. the crank stars are symmetrical.

V In-Line Engines. If the vee-angle $\gamma = 60°$, the extreme values of the first-order moments are given by Eq. (28) as

$$M_{Ia}/(P_I a) = 2 \cdot 0.2672 \cos^2 30° = 0.4008 \quad \text{and}$$
$$M_{Ib}/(P_I a) = 2 \cdot 0.2672 \sin^2 30° = 0.1336,$$

and the second-order moments in accordance with Eq. (29) as

$$M_{IIa}/(\lambda P_I a) = M_{IIb}/(\lambda P_I a) = 2 \cdot 1.006 \cos 30° \cos 60°$$
$$= 0.8712.$$

For the vee-angle $\alpha = 90°$, the corresponding calculations yield

$$M_{Ia}/(P_I a) = M_{Ib}/(P_I a) = 0.2672;$$
$$M_{IIa}/(\lambda P_I a) = 0 \quad M_{IIb}/(\lambda P_I a) = \sqrt{2}$$

Fan-Type and Radial Engines. Here, the power units (**Fig. 9**) are distributed over a half or a complete circumference. The angle between two centrelines is $\gamma_F = 180°/z$ and $\gamma_s = 360°/z$ respectively where they are offset by the

Table 5. Calculation of forces and moments of inertia of an in-line engine (see example)

n	k	α_k in °	$\cos \alpha_k$	$\sin \alpha_k$	v_k	$v_k \cos \alpha_k$	$v_k \sin \alpha_k$
1	1	0.0	1.0	0	3	3.0	0.0
2	6	51.43	0.6235	0.7818	−2	−1.2470	−1.5636
3	3	102.86	−0.2225	0.9750	1	−0.2225	0.9750
4	4	154.29	−0.9010	0.4339	0	0.0	0.0
5	5	205.72	−0.9010	−0.4339	−1	0.9010	0.4339
6	2	257.15	−0.2225	−0.9750	2	−0.4450	−1.9500
7	1	308.58	0.6235	−0.7818	−3	−1.8705	2.3454
			0	0		0.1166	0.2407
			$= k_{\gamma 1}$	$= k_{\gamma 2}$		$= c_{\gamma 1}$	$= c_{\gamma 2}$

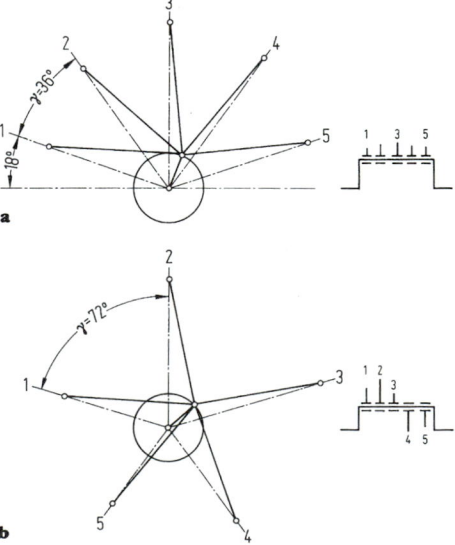

Figure 9. a Fan-type machine. **b** Radial machine.

Table 6. Fan-type and radial machines: constant first-order inertial forces rotating with the crank

z	Fan-type γ in °	Radial γ in °	F_1/P_1
5	36	72	2.5
4	45	90	2.0
3	60	120	1.5
2	90	—	1.0
2	—	180	2.0

Table 7. Moments of inertia for opposed-cylinder machines

z	Crank angular position diagram		$\frac{M_r}{F_r e}$	$\frac{M_{I max}}{P_I e}$	$\frac{M_{II max}}{P_{II} e}$
2			1	1	1
4			$\sqrt{2}$	$\sqrt{2}$	0
4			0	0	2
6			0	0	0

width of a pushrod. Here, up to five are located on one crankpin. They feature lightweight casings and the resultant of the first-order forces (see **Table 6**) rotates with the crank and can therefore be counterbalanced easily by weights at its countermass extensions. In the case of larger cylinder numbers, it is usual to employ a master connecting rod with pivoted-link connecting rods [14].

Opposed-Cylinder (Flat) Engines. Here, the power units 1 and 2, 3 and 4 (**Fig. 10a**) are allocated to the double cranks A, B, etc. If the masses are equal, the double cranks are free of forces, their moments are equal (**Fig. 10b**) and independent of the inter-cylinder distances. Equations (14), (17) and (18) then apply to the double cranks. This mode of construction with small moments due to mass (**Table 7**) is now mainly used for horizontal compressors.

1.2.2 Graphical Methods

These show plainly the magnitude of the forces and moments [15, 16] and are particularly suitable for non-symmetric machines. The polygons resulting from this method can also be evaluated by computer using vector programs.

General Methods. The rotating forces are plotted in the direction of the cranks with which they rotate. In the case of oscillating forces, the auxiliary vectors $P_I = m_o r \omega^2$ and $P_{II} = \lambda P_I$ take the directions of their cranks or twice the latters' angle and are then projected onto the centreline of their power unit (**Fig. 11**).

In-Line Engines. Here, the first- and second-order crank angle position diagrams with their crank angles and their double angles respectively serve to simplify the plot-

ting of the vectors (**Fig. 11b**). In the case of the moments, these are rotated in a counter-clockwise direction into the cranks to simplify matters (**Fig. 11c**), and are marked with a cross-stroke. For negative lever arms as in Eq. (11), or if the cranks are located to the right of the centroidal plane, the vectors must be plotted in the opposite direction from that of the cranks. The rotated auxiliary vectors are then projected and added to the first- or second-order moments just as in the case of the forces.

1.2.3 Compensation of Forces and Moments

Forces and moments can cause dangerous resonance phenomena. They must therefore be compensated within the machine or be avoided by tuning the foundations [17, 18].

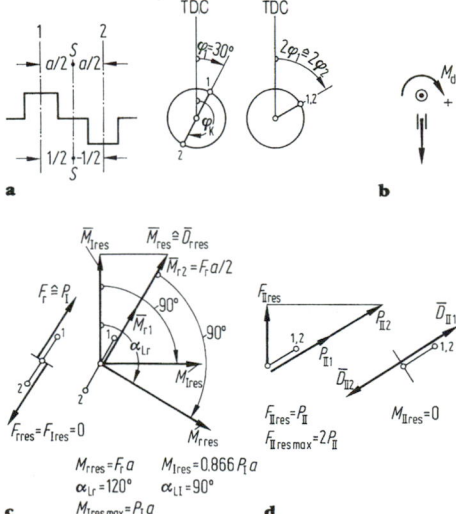

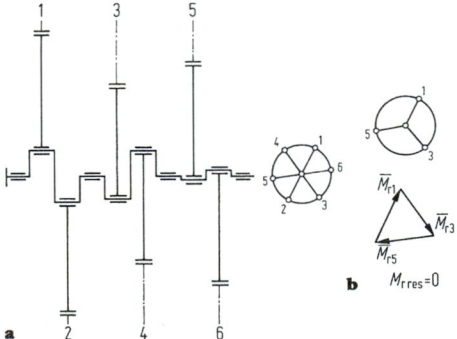

Figure 10. Opposed 6-cylinder machines: **a** power unit arrangement with crank angular position diagram, **b** moment star with polygon.

Figure 11. Graphical methods (two-cylinder in-line engine): **a** schematic of crankshaft and stars, **b** definition of moments; forces and moments, **c** and **d** first- and second-order.

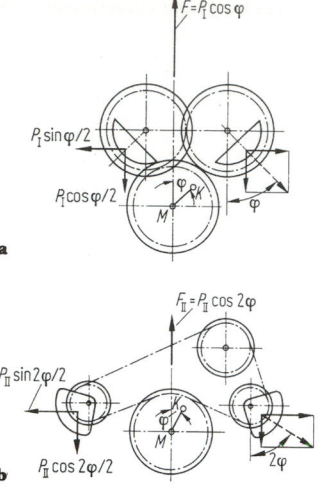

a

b

Figure 12. Compensation of oscillating forces: **a** counter-rotating gears for first-order forces, **b** Lancaster drive for second-order forces.

Rotating Masses. Their forces and moments are compensated by counterweights at one or all cranks (see **F10 Fig. 5**). If the forces vanish, it is sufficient to place counterweights at the crank countermass extensions to compensate for the moments; however, this leaves moments internal to the shaft in existence [6].

Oscillating Masses. These are compensated by weights moving in the opposite direction or rotating at twice the rate of revolution (**Fig. 12a**). Their components, which are normal to each other, compensate for the masses and the free centrifugal forces. They are driven by the crankshaft and lie in the centroidal plane below it, so that no additional moments are generated. In order to compensate for the moments, these weights are situated in front or behind the crankshaft. They are driven by a pinion from the shaft extension using an auxiliary shaft (**Fig. 14**). In the case of a Lancaster drive (**Fig. 12b**), this is accomplished by means of a toothed vee-belt.

The oscillating forces of all orders of crosshead power units are compensated by the weight $m = m_K + m_{KS}$

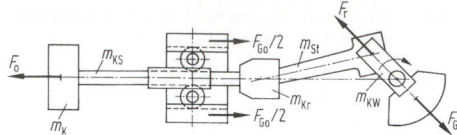

Figure 13. Crosshead power unit with complete mass compensation.

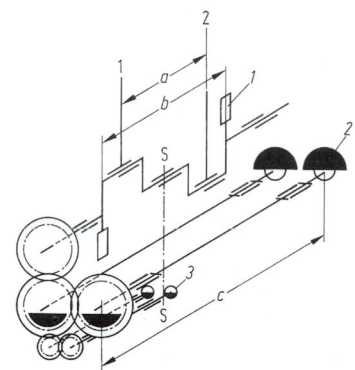

Figure 14. Compensation of mass effects by counterweights: *1* at the crank countermass extensions for rotating moments, *2* at the shaft ends for first-order moments, *3* in the centroidal plane for second-order forces.

$+ m_{Kr} + m_{ost}$ moving in a direction opposite from that of the piston (**Fig. 13**). Its two tooth racks are driven from the piston rod via two pinions. This arrangement is suitable only for lower rates of revolution and requires long power units.

Example. Compensation of the forces and moments of a two-stroke engine with two cylinders (**Fig. 14**) whose masses m_g of the counterweights are to be determined when the distances b and c are known. For the rotating moments, the relation $m_g = m_o(r/r_g) \cdot (a/c)$ applies to each counterweight 1, since $m_g r_g \omega^2 b = m_r r \omega^2 a$.

For the first-order moments of the two rotating weights 2, the relation $m_g = m_o(r/r_g) \cdot (a/c)$ applies. For the second-order forces, the relation for weights 3 is $m_g = 0.5 \lambda m_o(r/r_g)$, since $m_g r_g (2\omega)^2 = 2P_{II} = 2\lambda m_o r \omega^2$.

2 Vibrations

R. Nordmann, Kaiserslautern

2.1 The Problem of Vibrations in Machines

Machine dynamics in general deals with the interaction between *forces* and *motions* in machines. Here, apart from the desired dynamics that are necessary for the machine's function, there are also undesirable dynamic effects. These arise because machines and components possess elastic properties and mass and thus form systems capable of oscillations. If forces occur that change with time or movements of the base, these result in *machine vibrations*. Although compared with the required move-

ments, such movements are in general small, they can become dangerous under certain circumstances. So-called *resonance phenomena* are particularly hazardous and occur when an exciting frequency corresponds to the natural frequency of the machine structure thereby amplifying the vibration amplitudes. *Self-excited vibrations*, which are maintained by the presence of a source of energy, also represent continued dynamic loading.

Machine vibrations always constitute a problem when excessive stressing of the materials takes place. If the permissible tensile values of materials are exceeded, they may be damaged. In order to guarantee the functional integrity of machines, deformation limits must also be observed. Thus *rotor vibrations* in turbines and electric motors must not become so large that they exceed the clearance between rotor and housing. Vibrations are also inimical to the environment. This applies not only to oscillatory movements often perceived as unpleasant but above all to the noise caused by vibrations (e.g. *body drumming*). Finally, vibrations have a deleterious effect on the quality of manufactured articles in production processes. In the case of machine tools, therefore, attempts are made to keep the relative movement between tool and workpiece as small as possible.

A typical example of machine vibrations is found in vehicle engines, where the dynamics of the machine functions are of interest because of the crank drive. The problem involves the motions of the pistons and the crankshaft under the effect of the forces applied by the gas pressure (see F10).

The crankshaft itself is a system capable of oscillations that occur as flexural and torsional vibrations excited by the gas pressure and inertial forces transmitted via the connecting rod (see J1). This can be accompanied by resonance effects if the frequency of excitation is now the natural frequency of the crankshaft. In order to avoid dangerous vibration conditions, it is therefore important to know the amplitude and frequencies of the exciting forces as well as the dynamic properties of the crankshaft (natural frequencies, damping, eigenvectors).

Engineers must be concerned with the problem of machine vibrations during the development and design of machines, as well as during the testing and later operation (machine supervision and diagnosis). Modern mathematical and experimental aids are of special importance in this respect.

2.2 Some Fundamental Concepts

By way of introduction, some important concepts in the field of machine vibrations will be explained.

2.2.1 Mechanical Equivalent System

It is advisable to base all investigations of the vibrating engine on certain model concepts. For this reason the real system is assigned a mechanical *equivalent system* (vibration model), which permits the neglect or idealisation of certain features (see J2.6) and which is made up from simple mechanical elements (e.g. masses, dampers, springs, rods, beams etc.). It possesses a certain number N of degrees of mechanical freedom (displacements, angles).

2.2.2 Equations of Motion, System Matrices

If the fundamental mechanical equations (Newton, d'Alembert's principle of virtual work; see A3) are applied to the mechanical equivalent system, then the equations of motion that express the interdependence of a time-dependent input parameter $F(t)$ and the output parameters $x(t)$ are obtained. These equations may be linear or non-linear. In the case of many practical tasks, in particular, when dealing with small oscillations about an equilibrium point, linear models will suffice.

We confine ourselves here to the description of linear time-invariant vibration systems with deterministic input parameters $F(t)$. For a treatment of *non-linear* systems see A4.3 and [1, 2]. Under the above conditions it is always possible to obtain a system of linear, time-invariant, second-order equations of motion independent of the number of degrees of freedom prevailing in each case (time-invariant means that M, D and K are not time-dependent):

$$M\ddot{x}(t) + D\dot{x}(t) + Kx(t) = F(t). \qquad (1)$$

Here:

M = the Quadratic $N \times N$ mass matrix. M contains the inertial coefficients of the system and is symmetric.

D = the Quadratic $N \times N$ damping matrix. D contains the damping coefficients of the system. D may also be asymmetric (gyroscopic effects, sliding bearings and seal gap forces).

K = the Quadratic $N \times N$ stiffness matrix. K contains the stiffness coefficients of the system. K may also be non-symmetric (rotating forces, sliding bearings and seal gap forces).

$F(t)$ = the $N \times 1$ vector of the time-dependent exciting forces. Displacement or acceleration excitations can always be transformed into force excitation.

$x(t)$ = the $N \times 1$ vector of time-dependent displacements or angles. $\dot{x}$, $\ddot{x}$ are the associated velocities and accelerations respectively.

The equations of motion, Eq. (1), express the balance of forces and moments and take inertial forces into account. They are generally valid under the above conditions (linearity, time-invariant matrices), and can be applied to different types of machine as well as to different types of vibration (flexural vibrations, torsional vibrations etc.).

It is natural to use a graphical representation of the vibratory system. This can be done with a block diagram which relates input to output parameters (**Fig. 1**). Certain input parameters $F(t)$ enter the system as exciting forces (e.g. out-of-balance forces, process forces, impact forces etc.) or as base excitations (floor disturbances). The system transforms the inputs in accordance with its transfer characteristics and responds with the output parameters $x(t)$ or the velocities $\dot{x}(t)$ and the accelerations $\ddot{x}(t)$ derived from these. The transfer characteristics are determined by the structure of the system, i.e. by the appropriate physical laws, and by the system parameters M, D and K. If M, D and K as well as the excitation vector $F(t)$ are known (for instance periodic or pulse excitation of certain degrees of freedom), then first the natural oscillation parameters and then the response parameters $x(t)$ can be determined by calculation (see J2.2 and J2.7).

2.2.3 Model Parameters: Natural Frequencies, Modal Damping, Eigenvectors

Natural Frequencies. Every linear oscillatory system has a particular natural frequency behaviour, which is determined by its natural frequencies, damping factors and eigenvectors.

For instance, if a short disturbance in the form of a pulse $F_k(t)$ is applied to the ventilator rotor shown in **Fig. 2**, the oscillatory system will subsequently undergo natural oscillations which are composed of several component vibrations ($n = 1, 2, \ldots, N$):

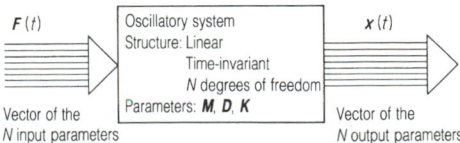

Figure 1. Block diagram of an oscillatory system with physical parameters.

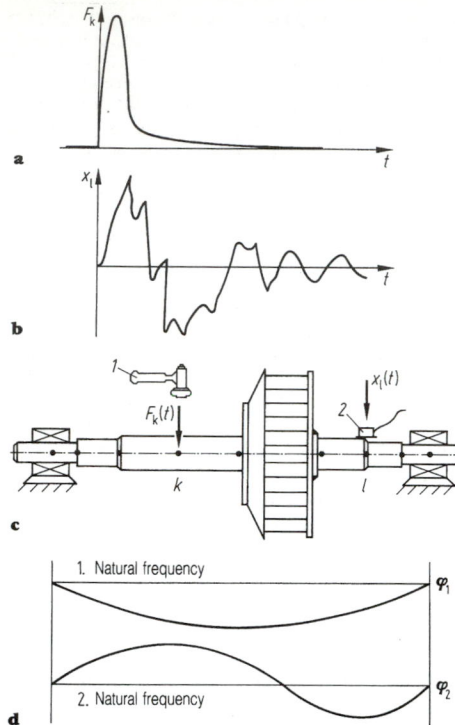

Figure 2. Natural oscillation parameters of a ventilator rotor: **a** time function of the force, **b** time function of the decrement of the amplitude of oscillation, **c** construction principles of the ventilator rotor (*1* impact force generator, *2* vibration sensor), **d** time function of the eigenvectors.

$$x(t) = \sum_{n=1}^{N} A_n\, e^{\alpha_n t} \left\{ \varphi_n^{Re} \cos(\omega_n t + \psi_n) \right.$$
$$\left. - \varphi_n^{Im} \sin(\omega_n t + \psi_n) \right\}. \tag{2}$$

Each component vibration consists of an exponential function which describes the decrement or, in the case of unstable systems, the buildup of the component; and trigonometric functions that describe the vibration behaviour.

The nth partial solution encompasses: ω_n, the natural frequency (s^{-1}); $-\alpha_n$, the decrement factor (s^{-1}); φ_N^{Re}, φ_N^{Im} the real and imaginary part of the *eigenvectors* φ; and A_n, ψ_n, constants; these are determined by the boundary conditions (impact).

By measurement of the pulse response $x_l(t)$ or the acceleration $\ddot{x}_l(t)$ of degree of freedom l, the natural frequency parameters ω_n, α_n can be determined by signal analysis and on processing further signals, the eigenvector components φ_n^{Re}, φ_n^{Im} as well. These are described as modal parameters.

A knowledge of these quantities is extremely important, because they characterise the dynamic properties of a system capable of oscillations. With their help, it is possible to determine, *inter alia*, at what frequency resonance effects can be expected and the magnitude of the resonance amplitudes (damped). The eigenvector indicates what deformation occurs when the system oscillates at its appropriate natural frequency. This deformation occurs

approximately if the system is harmonically excited with a frequency that corresponds to one of the natural frequencies.

Eigenvalue Analysis. The modal parameters may be obtained mathematically by making $F(t) = 0$ on the right-hand side of Eq. (1) (homogeneous equations), and using

$$x(t) = \varphi e^{\lambda t},$$
$$\dot{x}(t) = \lambda \varphi e^{\lambda t}, \tag{3}$$
$$\ddot{x}(t) = \lambda^2 \varphi e^{\lambda t}$$

to define the eigenvalue problem

$$(\lambda^2 M + \lambda D + K)\varphi = 0. \tag{4}$$

If behaving in a purely oscillatory manner this has the solution

$$\lambda_n = \alpha_n + i\omega_n;\ \lambda_n^* = \alpha_n + i\omega_n \text{ eigenvalues}, \tag{5}$$
$$\varphi_n = \varphi_n^{Re} + i\varphi_n^{Im};\ \varphi_n^* = \varphi_n^{Re} - i\phi_n^{Im} \text{ eigenvectors}. \tag{6}$$

In many practical cases, it is difficult to construct a *damping matrix*. In the case of weakly damped structures, which often occur in machine design (rotors in roller bearings with elastic, torsion and flexural properties, turbine blades, steel foundations) the assumption of *"modal damping"* suffices. The eigenvalue problem is initially solved for the undamped system ($D = 0$) in the purely real form

$$(K - \omega^2 \cdot M)\varphi = 0 \tag{7}$$

and the natural frequencies ω_n and the corresponding real eigenvectors φ_n are then obtained. Any damping that does not enter this calculation is estimated or determined experimentally. A decrement factor $-\alpha_n$ or a modal damping value (degree of damping) $D_n = -\alpha_n/\omega_n$ is then assigned to each natural angular frequency ω_n.

In practice, the following parameters are most frequently used:

$$f_n = \omega_n/2\pi \qquad \text{natural frequency (Hz)}, \tag{8}$$
$$D_n = -\alpha_n/\omega_n \qquad \text{modal damping (-)}, \tag{9}$$
$$\varphi_n \qquad \text{real eigenvector.} \tag{10}$$

Some numerical values for the modal damping D in %:

Material/Components	D in %
Steel	0.1
Cast iron	1.8 to 2.0
Rubber (natural caoutchouc)	1 to 8
Steel structures	0.2 to 1.5
Reinforced concrete structures	4
Steel foundations for turbines without floor damping	0.5 to 1.5
Steel foundations for turbines with floor damping	1.5 to 3.0

Knowledge of modal damping is particularly important in determining amplitudes of resonance oscillations caused by the excitation force $F(t)$.

Figure 2 shows the two first eigenvectors φ_1 and φ_2 with their associated natural frequencies f_1 and f_2 for the ventilator rotor with roller bearings. The first form of

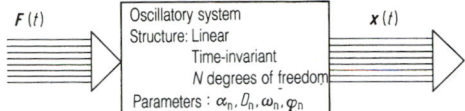

Figure 3. Block diagram of a vibration system with modal parameters.

natural oscillation is visually similar to the static bending mode; the second mode of oscillation with a single node is described as the *S-mode*. In contrast to complex eigenvectors which occur when damping is taken into account, the ratio of the eigenvector components always indicates a constant deformation figure in the case of real eigenvectors.

The indicated simple procedure is not permissible in dealing with an oscillatory system capable of self-excitation, as for instance in the case of rotating machinery featuring sliding bearings and seal gaps (pumps, turbines, compressors). In these cases it is necessary to solve the eigenvalue problem Eq. (4) and to assess the stability behaviour using the eigenvalues so obtained (see J2.7.4).

2.2.4 Modal Analysis

In a way analogous to **Fig. 1**, the relation between input parameters $F(t)$ and output parameters $x(t)$ can also be expressed by modal parameters (**Fig. 3**). If all natural frequencies ω_n, eigenvectors φ_n and damping factors $(-\alpha_n)$ or their modal damping factors D_n are known, they can be used to calculate the vibrations. In the case of a system capable of self-excitation, the set of eigenvectors on the left is required as well [1, 2]. This calculation procedure is also described as "modal analysis" because the modes (eigenvectors) enter into the calculations. An advantage of this method is that the equations of motion Eq. (1), which were originally coupled, can be decoupled by making use of certain orthogonal properties of the eig-

envectors. This simplifies the calculation, and a physical interpretation of the dynamic processes is made easier.

The concept of modal analysis is now also employed in the determination of modal parameters from measurements. The basis of the process is the representation of the system responses as a function of the modal parameters and the exciting frequency (**Fig. 4**). In matching analytical *system responses* (frequency response of the model) to the measured system responses (measured frequency response), the modal parameters are changed until good matching between model and measurement is obtained. The result yields the required modal parameters.

In a measurement process in general, test forces (pulse, sinusoidal, noise) are input into the system and the oscillatory response is registered at the individual measuring points. The measured frequency responses are calculated from the time signals after they have been converted to the frequency domain by Fourier analysis (see J2.4.2), and these are then used in the matching process that is required for the calculation of the modal parameters [13].

2.2.5 Frequency Response Functions of Mechanical Systems, Amplitude and Phase Characteristics

Definition. If a linear vibration system which is described by the equations of motion Eq. (1) has its degree of freedom k excited by a harmonic exciting force

$$F_k = \hat{F}_k \sin \Omega t, \tag{11}$$

where $\hat{F}_k$ is the constant force amplitude and Ω is the exciting frequency (all other forces being absent), then, after its transient state has passed, the system responds with movements that are also harmonic (**Fig. 4**). All response parameters can be expressed by means of the vector $x(t)$:

$$x(t) = \begin{bmatrix} x_1(t) \\ x_2(t) \\ \vdots \\ x_l(t) \\ \vdots \\ x_N(t) \end{bmatrix} = \begin{bmatrix} \hat{x}_1 \\ \hat{x}_2 \\ \vdots \\ \hat{x}_l \\ \vdots \\ \hat{x}_N \end{bmatrix} \begin{bmatrix} \sin(\Omega t + \varepsilon_{1k}) \\ \sin(\Omega t + \varepsilon_{2k}) \\ \vdots \\ \sin(\Omega t + \varepsilon_{lk}) \\ \vdots \\ \sin(\Omega t + \varepsilon_{Nk}) \end{bmatrix} \tag{12}$$

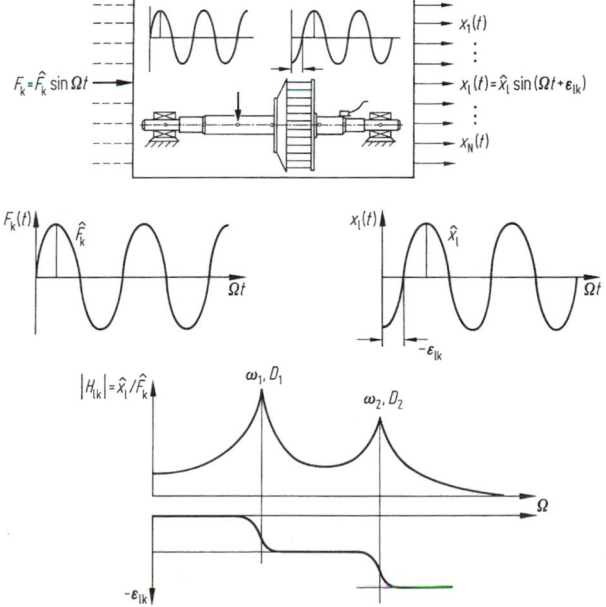

Figure 4. Harmonic excitation of a linear vibration system.

The response for each degree of freedom is characterised by an amplitude and a phase angle in relation to the excitation force. For example, for the degree of freedom l,

$$x_1(t) = \hat{x}_1 \sin{(\Omega t + \varepsilon_{1k})}. \qquad (13)$$

$\hat{x}_1$ as well as ε_{1k} (ε_{1k} is negative) depends on the frequency of excitation. Therefore, the expression

$$\hat{x}_1 (\Omega)/\hat{F}_k \qquad (14)$$

is called the amplitude–frequency response (between l and k), and

$$\varepsilon_{1k} \qquad (15)$$

is the phase–frequency response (between l and k).

In practice, both functions can often be combined in the complex frequency response

$$\bar{H}_{1k} = (\hat{x}_1/\hat{F}_k)\ e^{i\varepsilon_{1k}} = |\bar{H}_{1k}|\ e^{i\varepsilon_{1k}}. \qquad (16)$$

Since the quotient of the quantities $\hat{x}_1/\hat{F}_k$ represents a compliance parameter (displacement/force), $\bar{H}_{1k}(\Omega)$ is often described as a complex compliance frequency response. **Figure 4** shows qualitatively the dependence of the amplitude $|\bar{H}_{1k}| = \hat{x}_1/\hat{F}_k$ (amplitude response) and the phase ε_{1k} (phase response) on the exciting frequency Ω. The significance of the frequency response functions becomes particularly clear if the plot of the amplitude response is followed. If the *exciting frequency* Ω lies close to a *natural frequency* (ω_1, ω_2 ... ω_N) (resonance case) the response amplitude $\hat{x}_1$ reaches a maximum whose magnitude in each case depends on the associated damping (α_1, α_2, ... α_N; D_1, D_2 ... D_N respectively) (large damping, low amplitude increment). In the region of the resonance frequency the phase angle ε_{1k} (here defined as negative) changes relatively steeply.

Calculation of Frequency Responses and Harmonic and Periodic Systems Responses. If the equations of motion Eq. (1) are known, together with the matrices **M**, **D**, **K**, then the complex transfer function $\bar{H}_{1k}/(\Omega)$ can be calculated by introducing the harmonic excitation function $F_k/(t)$ into the complex force function:

$$F_k(t) = \hat{F}_k\ e^{i\Omega t} = \hat{F}_k\ (\cos{\Omega t} + i \sin{\Omega t}), \qquad (17)$$

where in the case of excitation at a single point, only the kth component in the force vector has finite magnitude:

$$\boldsymbol{F}(t) = \hat{\boldsymbol{F}}\ e^{i\Omega t}; \quad \hat{\boldsymbol{F}} = \{0,0, \ ... \ \hat{F}_k, 0, \ ... \ 0\}. \qquad (18)$$

Putting Eq. (18) into Eq. (1) yields

$$\boldsymbol{M}\ddot{\boldsymbol{x}} + \boldsymbol{D}\dot{\boldsymbol{x}} + \boldsymbol{K}\boldsymbol{x} = \hat{\boldsymbol{F}}\ e^{i\Omega t}. \qquad (19)$$

Using the complex formulation and its time derivatives,

$$\boldsymbol{x} = \hat{\boldsymbol{x}}\ e^{i\Omega t},$$
$$\dot{\boldsymbol{x}} = i\Omega\hat{\boldsymbol{x}}\ e^{i\Omega t}, \qquad (20)$$
$$\ddot{\boldsymbol{x}} = -\Omega^2\ \hat{\boldsymbol{x}}\ e^{i\Omega t},$$

the complex system of equations follows:

$$(\boldsymbol{K} - \Omega^2\ \boldsymbol{M} + i\boldsymbol{D})\hat{\boldsymbol{x}} = \hat{\boldsymbol{F}}; \qquad (21)$$

from this it is possible to determine the vector $\hat{\boldsymbol{x}}$ of the complex system response associated with any given excitation frequency Ω by solving the system of complex linear equations Eq. (21), provided that the **M**, **D**, **K** matrices and the force vector $\hat{\boldsymbol{F}}$ are known. The components of $\hat{\boldsymbol{x}}$ have the form

$$\hat{x}_1 = \hat{x}_1\ e^{i\varepsilon_{1k}} \qquad (22)$$

and contain both the amplitude $\hat{x}_1$ and the phase ε_{1k}. If this calculation is repeated for other frequencies Ω, further values of the frequency response function $\bar{H}_{1k}(\Omega)$ are obtained.

For a system with N mechanical degrees of freedom (displacement and angles), there are $N \times N$ frequency responses (N degrees of freedom and N responses).

The total matrix $\bar{\boldsymbol{H}}(\Omega)$ of all frequency response functions $\bar{H}_{1k}(\Omega)$ ($1 = 1 ... N$; $k = 1 ... N$) is obtained from the inversion of the complex (dynamic) stiffness matrix $\bar{\boldsymbol{K}}(\Omega) = \boldsymbol{K} - \Omega^2\boldsymbol{M} + i\Omega\boldsymbol{D}$:

$$\bar{\boldsymbol{H}}(\Omega) = (\boldsymbol{K} - \Omega^2\boldsymbol{M} + i\Omega\boldsymbol{D})^{-1} = \qquad (23)$$

$$\begin{bmatrix} \bar{H}_{11} & \bar{H}_{12}... & \bar{H}_{1k}... & \bar{H}_{1N} \\ \bar{H}_{21} & \bar{H}_{22}... & \bar{H}_{2k}... & \bar{H}_{2N} \\ \bar{H}_{N1}...... & & \bar{H}_{Nk}... & \bar{H}_{NN} \end{bmatrix}.$$

The problem of harmonic excitation and therefore of harmonic oscillations plays an important part in machine dynamics. Knowing the frequency response functions of a system, excitation frequencies at which particularly large response amplitudes occur can be assessed.

An important application in rotating machinery is where harmonic excitation forces rotate with the angular frequency Ω (frequency of rotation) caused by *out-of-balance* masses. If the out-of-balance force vector (out-of-balance forces are proportional to Ω^2; see J2.5) is substituted in Eq. (1), and the effect of the rate of revolution in the system matrices is included, special frequency response functions are obtained which describe the response amplitudes of flexural oscillations for the rotating shaft as a function of the exciting frequency. Since exciting frequency is equal to rotational speed, the term "critical angular frequency" is used when the angular frequency corresponds to a natural frequency of the system.

If the exciting forces of a system contain several simultaneous excitation frequencies, as with periodic functions, then the amplitudes obtained from the frequency responses of the various exciting frequencies can be superposed with their correct phases to yield the total response. Periodic excitation forces or moments are found for instance in the case of gas pressure and inertial forces in the operation of the crankshaft of a vehicle engine (see J1).

2.3 Basic Problems in Machine Dynamics

In the treatment of vibration problems in machines many questions arise. In the following survey it will be shown briefly that the problems occurring in various types of machine can be reduced to a few fundamental problems. In the explanations, an oscillatory system block diagram (**Fig. 1**) and the associated equations of motion, Eq. (1), are used.

2.3.1 Direct Problem

The direct problem is the task most frequently posed in practice and usually associated with the design phase of a new development. Here, the system to be investigated is usually given in the form of a design drawing (**Fig. 5a**). The basic task to be solved consists of calculating the time dependence of the system responses $\boldsymbol{x}(t)$ from the known critical time-dependence of the forces $\boldsymbol{F}(t)$ and the system

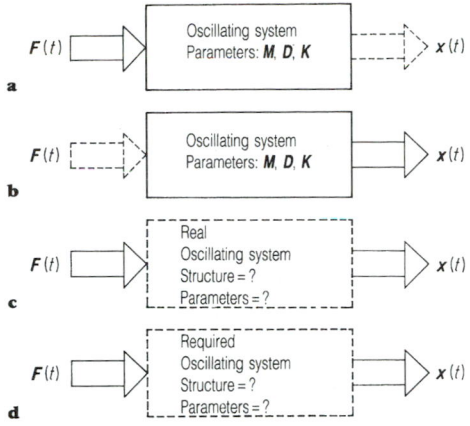

Figure 5. Fundamental problems of machine dynamics: **a** direct problem, **b** input problem, **c** identification problem, **d** optimisation problem.

properties in the form of matrices M, D and K which are also known. The natural frequencies and modes are usually determined in a preliminary step. In accordance with [2] the following procedure is recommended for this important machine dynamic analysis:

1. **Listing of All Load Cases (Excitation Forces).** Load case of normal operation; loads arising from perturbations.
2. **Idealisation of the Structure.** Providing a mechanical equivalent system that represents the dynamic behaviour for the various load cases with sufficient accuracy. Decision regarding the type of modelling (multi-body systems, finite elements). Number of mechanical degrees of freedom.
3. **Derivation of Equations of Motion.** In the case of discrete systems (multi-body systems, finite-elements) with linear system properties, the linear system of differential equations already given in Eq. (1) is obtained:

$$M\ddot{x} + D\dot{x} + Kx = F(t).$$

4. **Solution of Equations of Motion.** The homogeneous solution of the linear equations of motion defines the natural frequency parameters and the stability of the system. Then the particular solutions for the various load cases that describe the forced vibrations must be calculated. Where necessary, several load cases must be superposed.
5. **Graphical Presentation of Results.** In order to present clearly the often enormous mass of data, the time dependence of displacements, the accelerations or shear loads and the frequency dependence of the amplitudes (frequency response) are shown by the computer in graphical form.
6. **Evaluation and Interpretation of Results.** Several questions must be answered by the results, e.g.: Is the structure capable of standing up to the stresses occurring in all load cases? Is the system stable? Are resonance points closely approached? Where necessary, flaws must be removed, components changed or other material employed.

2.3.2 Input Problem

Here, the question that arises is the reverse of that of the direct problem, in that now the time dependence of the

system response $x(t)$ is given, and if the system properties M, D, K are also known then the question concerns the time dependence of the input parameters $F(t)$ (**Fig. 5b**).

A widely known application example of this problem is the *balancing* of rotors. Here the out-of-balance forces of a rotating shaft must be offset by mass compensation at the rotor, so that the bearings are not subject to forces at rotational frequencies. The vibration signals $x(t)$ are measured before the balancing process.

2.3.3 Identifying the System Parameters

In the identification problem, the question is the determination of the equations (structure), including the system parameters, which describe the behaviour of the system from the measured input and output signals (**Fig. 5c**). Since criteria relating to the structure of the equations are frequently known (e.g. linearity, time-invariance, number of degrees of freedom) or assumptions regarding these are made, the task is reduced to one of so-called parameter identification.

Here, test forces $F(t)$ (pulse forces, step changes of force, harmonic or random excitation forces) are impressed on the vibration system in question and the resulting system response $x(t)$ is registered. With the aid of the measured input quantities $F(t)$ and output quantities $x(t)$ the required system parameters can be determined by a process of estimation, taking into account the known input–output relations (structure). Here, procedures in the time as well as in the frequency domain are applied.

In more extended vibration systems it is particularly problematical to determine the system matrices M, D, K completely by identification of parameters. Since in general the parameters for simple mechanical elements (rods, beams, plates) can be obtained fairly well by calculation, experimental parameter determination is limited to certain system components with complex force displacement laws that in most cases possess only a few degrees of freedom. In the design of machines, such components are for instance sliding bearings, gap seals, couplings, etc., which often have a major influence on the vibration behaviour of the complete system and for which spring and damping coefficients are therefore required.

Modal analysis has become very significant in machine dynamics. In this identification procedure, the modal parameters of mechanical systems can be obtained from measured frequency responses.

2.3.4 Design Problem

The design problem is to construct a system so that certain desired output parameters $x(t)$ are obtained from given excitation parameters $F(t)$ (**Fig. 5d**). The task is to create an optimum dynamic system. Frequently in these cases, the structure is given as well, so that parameter optimisation has to be carried out only within certain limits.

2.3.5 Improving Machine Vibration

This task often occurs in the practical operation of machines. Here some of the partial tasks already described have to be carried out.

Machine vibrations are undesired phenomena that must not exceed certain limiting values. If the displacements $x(t)$ are too large, the dynamic condition of the machine has to be improved and this can be accomplished in four steps. First the output signals $x(t)$ are measured and analysed in the time and frequency domain. Excessive vibrations could be caused by excessive excitation $F(t)$ or poor system properties (ω_n, α_n, φ_n). Therefore, as a second step, dynamic properties of the system are systematically investigated. Then, system properties can be identified (identification problem) by suitable test signals $F(t)$, and the corresponding measured output signals $x(t)$. A calculation model that represents the dynamic properties of the machine investigated with sufficient accuracy, can be matched to these results. The last step is now to find by simulation calculations those system modifications that are most effective in leading to a reduction of the vibrations. Here optimisation algorithms are applied that take account of the prevailing boundary conditions in each case (design problem).

2.4 Representation of Vibrations in the Time and Frequency Domains

2.4.1 Representation of Vibrations in the Time Domain

Machine vibrations manifest themselves by the time-varying displacements of individual points of the machine, which either repeat themselves or decrease or build up in one single process (*natural oscillations* of limited duration), or else proceed in an irregular (*stochastic*) manner.

The time dependence of vibration processes is part of the study of kinematics (see A2). This deals particularly with the time dependence of individual components of $x(t)$. Since, however, the excitation forces $F(t)$ are also time-dependent we include them in the considerations. Thus, in accordance with the block diagram of **Fig. 1**, the study deals with the analysis of the signals that enter and leave the vibration system.

Classification. Figure 6 shows a classification of important vibration signals, where the "oscillating" quantity is here generally designated $x(t)$ (scalar). These can be roughly divided into deterministic and stochastic signals, where the deterministic signals are here treated as the more important. They are subdivided once more into periodic and non-periodic oscillations. Harmonic sine and cosine functions are elementary signals belonging to the periodic category. General periodic signals are built up from sine and cosine components whose frequencies are multiples of a fundamental frequency Ω_0. The non-periodic signals are represented for instance by damped harmonic oscillations (natural oscillations), the pulse function and the step function.

All the signals in **Fig. 6** have in common that they are represented as time-dependent. While all the deterministic signals can be described by mathematical functions, random signals are not uniquely determined. It has turned out to be useful to characterise the various ways by which signals depend on time by mean values [1].

Mean Values. The temporal linear mean value of $x(t)$ is called the equivalent value,

$$\bar{x}(t) = \frac{1}{T} \int_0^T x(t)^2 \, dt, \qquad (24)$$

where T is the period of observation, the duration of the period in periodic signals. The mean-square value is

$$\overline{x^2(t)} = \frac{1}{T} \int_0^T x(t)^2 \, dt, \qquad (25)$$

from which the root-mean-square (r.m.s.) value is derived:

$$x_{\text{eff}} = \sqrt{\overline{x^2(t)}} = \sqrt{\frac{1}{T} \int_0^T x^2(t) \, dt}. \qquad (26)$$

For harmonic signals frequently encountered in practice, the mean value $\bar{x}(t) = 0$ and the root-mean-square value is about 70% of the peak value: $x_{\text{eff}} = \sqrt{2}/2\hat{x}$.

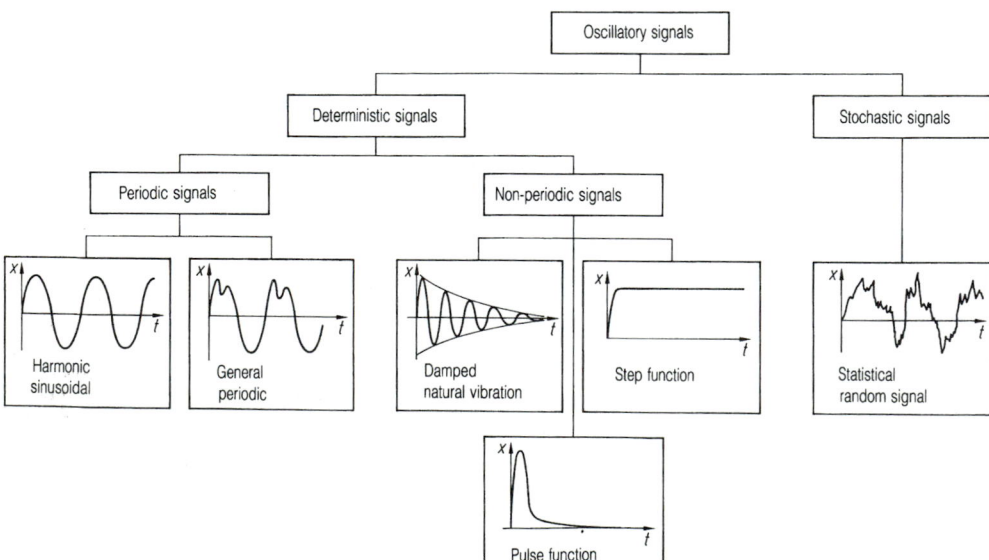

Figure 6. Classification of oscillatory signals.

2.4.2 Representation of Vibrations in the Frequency Domain

In order to interpret input quantity $F(t)$ and output quantity $x(t)$ of an oscillating system more accurately, they can also be represented in the frequency domain as $x(\Omega)$ and $F(\Omega)$. Here $\Omega = 2\pi f$ is an angular frequency in s^{-1} and f is the frequency in Hz. Representation in the frequency domain is often more informative because the frequency contributions of a vibration and their relationship to the dynamic properties of a system are clearer.

By using *Fourier analysis*, it is possible to transform from the time domain to the frequency domain. The representation in both domains becomes clear by the simple example of a harmonic sinusoidal oscillation, as shown in **Fig. 7**. The sinusoidal oscillation

$$x(t) = \hat{x} \sin (\Omega t + \varepsilon) \qquad (27)$$

is determined by the amplitude $\hat{x}$, the angular frequency Ω and the phase angle ε relative to zero. In the frequency domain the value of $\hat{x}$ is plotted as a function of angular frequency Ω in an amplitude diagram $\hat{x}(\Omega)$, and the value of ε as a function of Ω in a phase diagram.

Fourier Analysis of Periodic Oscillations. According to Fourier's theorem, every periodic function $x(t)$ of period $T = 2\pi/\Omega_0$ can, under certain conditions, be expressed unequivocally as a sum of sine and cosine functions with the angular frequencies $\Omega_0, 2\Omega_0, 3\Omega_0 \ldots$:

$$x(t) = x_0 + \sum_{n=1}^{\infty} \{s_n \sin n\Omega_0 t + c_n \cos n\Omega_0 t\}$$

$$= x_0 + \sum_{n=1}^{\infty} \{\hat{x}_n \sin (n\Omega_0 t + \varepsilon_n)\}, \qquad (28)$$

where

$$x_0 = \frac{1}{T} \int_0^T x(t) \, dt \text{ is the arithmetic mean,}$$

$$\left. \begin{array}{l} s_n = \dfrac{2}{T} \displaystyle\int_0^T x(t) \sin n\Omega_0 t \, dt \\[3mm] c_n = \dfrac{2}{T} \displaystyle\int_0^T x(t) \cos n\Omega_0 t \, dt, \end{array} \right\}$$

are the Fourier coefficients ($n = 1, 2, \ldots, \infty$),

$$\Omega_0 = 2\pi/T$$

is the fundamental frequency (angular frequency),

$$\hat{x}_n = \sqrt{s_n^2 + c_n^2}$$

are the values of the Fourier amplitude spectrum, and

$$\varepsilon_n = \tan (c_n/s_n)$$

are the values of the Fourier phase spectrum.

Example. **Figure 8** shows, as an example, a simple periodic function with two sinusoidal components in the time and frequency domains. This type of oscillatory signal can occur in rotating machines where the fundamental frequency Ω_0 coincides with the angular frequency (out-of-balance oscillations) and with twice the angular frequency $2\Omega_0$ caused, e.g., by the departure from perfect circularity of a shaft cross-section (rotor of a generator, shaft with a crack). Numerical values: $x_0 = 0$; $\hat{x}_1 = s_1 = 20 \ \mu m$; $\hat{x}_2 = s_2 = 10 \ \mu m$; $c_1 = c_2 = 0$.

Fourier Analysis of Non-periodic Processes. A transition from periodic to non-periodic processes is obtained when limiting values for infinitely long periods T are considered. In such a case the fundamental frequency has infinitely small values $(\Omega_0 \to d\Omega)$ and the higher harmonics are closely spaced. This leads to a continuous spectrum. The time function can now be expressed by means of the Fourier integral

$$x(t) = \int_{-\infty}^{\infty} x(\Omega) \, e^{i\Omega t} \, d\Omega. \qquad (29)$$

Here the complex spectral function $x(\Omega)$ is the Fourier transform of the time signal $x(t)$:

$$x(\Omega) = \int_0^{\infty} x(t) \, e^{-i\Omega t} \, dt. \qquad (30)$$

Example. **Figure 9** shows qualitatively the contribution to the Fourier transform $|x(\Omega)|$ by three non-periodic signals. The first two are frequently used as test signals for the artificial excitation of oscillatory systems. The values of the spectral function $|x(\Omega)|$ of the pulse function (**Fig. 9a**) remain almost constant over an extended region. The position of the zero transit $|x\Omega| = 0$ depends on the pulse length (hard or soft pulse). In the case of the *step function* (**Fig. 9b**) the greater part of the energy is distributed over the lower frequencies. Hence systems with low natural frequencies are strongly excited. A very interesting result is found in the case of the third signal (**Fig. 9c**). This is the *pulse response function* (weighting function) of an oscillator, i.e. the natural frequency of the system after a short pulse. If this function is transformed into the frequency domain, the corresponding frequency distribution

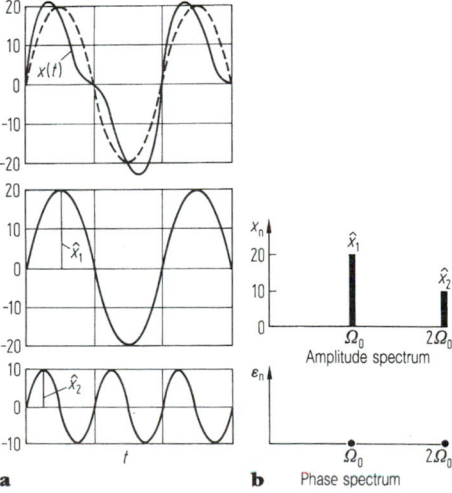

a   **b**   Phase spectrum

Figure 8. Periodic function with two sinusoidal functions ($\hat{x}_1 = 20 \ \mu m$; $\hat{x}_2 = 10 \ \mu m$; $\varepsilon_1 = 0$; $\varepsilon_2 = 0$): **a** time domain, **b** frequency domain.

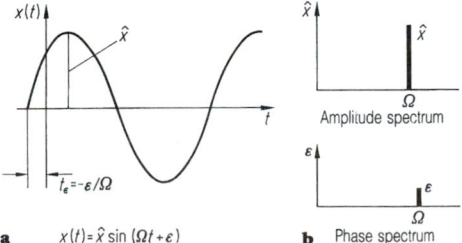

a $\quad x(t) = \hat{x} \sin (\Omega t + \varepsilon)$ **b** Phase spectrum

Figure 7. Representation of a sinusoidal oscillation in the time and frequency domains: **a** time domain, **b** frequency domain.

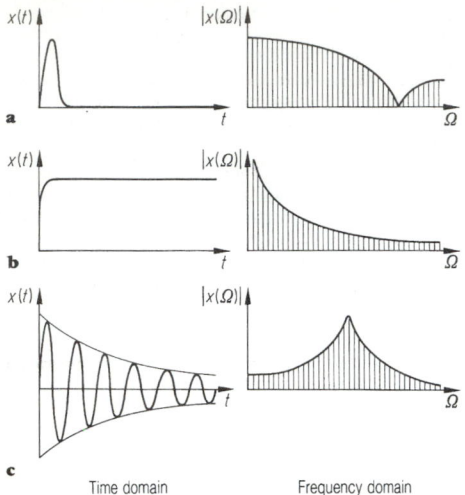

a

b

c

Time domain   Frequency domain

Figure 9. Spectral function $|x(\Omega)|$ for three non-periodic functions: **a** pulse function, **b** step function, **c** pulse response function.

functions already defined in J2.2.5 are obtained. **Figure 9** shows the frequency distribution for an oscillator with one degree of freedom.

2.5 Origin of Machine Vibrations, Excitation Forces F(t)

Vibrations in machines can have very different origins. In [5], a subdivision according to the originating mechanism is presented which differentiates between free, self-excited, parameter-excited, and forced vibrations. The various cases are best explained by the equations of motion Eq. (1) using the block diagram (**Fig. 1**). These presuppose linear time-invariant oscillatory systems. The various origins for oscillatory movements $x(t)$ are shown clearly in **Fig. 10**.

2.5.1 Free Vibrations

Free vibrations occur only when a system is left to itself after receiving a pulse and is not exposed to any further external influences (see A4.1). Moreover, no further energy is supplied to the system. In the equations of motion, the right-hand sides representing the excitations are equal to zero ($\boldsymbol{F}(t) = 0$, homogeneous system of

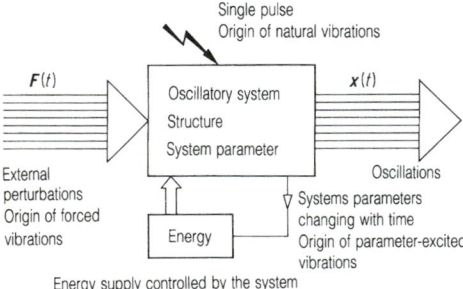

Single pulse
Origin of natural vibrations

$F(t)$ | Oscillatory system Structure System parameter | $x(t)$

External perturbations Origin of forced vibrations

Energy

Oscillations

Systems parameters changing with time Origin of parameter-excited vibrations

Energy supply controlled by the system Origin of self-excited vibrations

Figure 10. Origin of machine vibrations.

equations). The frequencies of oscillations are determined by the system properties ($\boldsymbol{M}$, $\boldsymbol{D}$, $\boldsymbol{K}$), distinguishing between damped and undamped oscillatory systems. In the ideal case, free from damping, an exchange between kinetic and potential energy takes place (permanent oscillations). In the real case, the vibrations always decrease in the presence of genuine damping (see **Figs 2, 6** and **9**).

2.5.2 Self-excited Vibrations

These are a special case of natural vibration. As with free vibrations, no external excitation is present in the equation of motion ($\boldsymbol{F}(t) = 0$). However, the oscillating system obtains energy from an energy source at the same frequency as the natural frequency. The energy input can lead to a buildup of (self-excited) oscillations if this is not prevented by countervailing damping forces. Strictly speaking, the self-excited vibrations have a non-linear character. However, linearised equations may be used in order to assess the stability at equilibrium. The tendency to self-excitation of an oscillatory system is recognised from the skew-symmetric contributions of the *stiffness matrix* $\boldsymbol{K}$ (circulatory forces) which is opposed by the *damping factors* ($\boldsymbol{D}$ matrix). In machine construction, examples of self-excited oscillations are found, *inter alia*, in rotating shafts with sliding bearings and sealing gaps. Here the energy input comes from the rotation of the shaft.

2.5.3 Parameter-excited Vibrations

Parametrically excited oscillations are characterised in that the oscillating system has time-dependent, mostly periodic, parameters. The presupposition of time-invariant equations of motion is then no longer satisfied, and the matrices are in general time-dependent: $\boldsymbol{M}(t), \boldsymbol{D}(t), \boldsymbol{K}(t)$. As a consequence, damped and undamped as well as stimulated oscillations can occur.

For instance, rotors of electrical machines often have cross-sectional forms with strongly differing bending moments in two directions normal to each other (e.g. two-pole rotors of synchronous machines). When the shaft turns, the vertical stiffness of the shaft in a fixed coordinate system changes periodically with time. The stiffness matrix $\boldsymbol{K}$ of the rotor is therefore time-variant. Rotational speeds that produce a buildup of parameter-excited vibrations must therefore be avoided.

2.5.4 Forced Vibrations

Forced vibrations (see A4.1), which probably occur most frequently in practice, are caused by external perturbations, and their time behaviour is determined by them. These perturbations are represented as exciting forces (moments) by the vector $\boldsymbol{F}(t)$ on the right-hand side of the equations of motion. They depend only on the time t and not on the displacement $x(t)$ of the oscillatory system itself. With regard to the exciting functions, the periodic functions and, as a special case, the harmonic functions are of particular interest in vibration practice. Next to these, pulse functions (perturbation through impact), step functions (switch-on processes) and random functions are of great importance.

Perturbations enter the system either as forces (moments) or as displacements or accelerations of the base. For instance, considerable exciting forces can occur in machines in the form of inertial forces through the translational or rotational movement of their masses. Other important excitations are caused by the coupling of mechanical systems with surrounding work media (gas, vapour) or with electrical systems (motors, generators) where strict coupling may be approximated by purely time-dependent perturbation functions. Perturbations in the environment of machines (ceilings of buildings, foundations of buildings) manifest themselves as excitations of the base of the oscillatory system. In regions subject to earth tremors, for instance, important machines and machine aggregates (e.g. cooling pumps in nuclear power stations) must retain their functional integrity even under strong external influences. Some important cases of excitations are presented and discussed in the following paragraphs.

Excitation Through Harmonic Out-of-balance Forces. In the construction of turbomachinery, flexural oscillations of rotating shafts are, in most cases, caused by out-of-balance forces. A clear explanation of excitation by out-of-balance forces is given by the example of a turbine rotor that has been idealised to the form of a disc (**Fig. 11**). Owing to manufacturing inaccuracies and non-uniform blade distribution, the centre of gravity S of the disc and the point at which the shaft passes through the disc W do not coincide. These two points have a fixed distance e from each other, described as the mass eccentricity, which represents a relatively small quantity compared with the turbine rotor diameter. During the operation of a machine, the mass eccentricity can increase because of deposits, by erosion, or through fracture of a blade. The product of the mass of the rotor m and the mass eccentricity e is known as the out-of-balance moment $U = me$.

Rotation of the shaft generates the centrifugal force

$$F = me\Omega^2 \tag{31}$$

which, as a result of the rotation of S about the centre of the shaft W, acts in the direction of the line connecting W and S (centrifugal acceleration) and which rotates at the angular frequency Ω. The magnitude of the force increases with the square of Ω. An observer in a fixed coordinate system sees the two components of the centrifugal force as periodic or, more accurately, as harmonic functions:

$$F_{\text{hor}} = me\Omega^2 \cos \Omega t,$$
$$F_{\text{vert}} = me\Omega^2 \sin \Omega t. \tag{32}$$

In the case of complex rotors, the out-of-balance force is distributed continuously along the shaft axis so that, apart from the force amplitudes, the relative angular positions have to be taken into account as well. Since the real distribution of the out-of-balance forces is never known accurately, certain sample distributions are assumed in the vibration calculations (e.g. distribution in accordance with eigenforms). Discrete out-of-balance force contributions must be attributed to the various degrees of freedom used in the calculations.

The out-of-balance forces excite the shaft as well as the bearing blocks, the foundations and the housing, to harmonic oscillations with the angular frequency Ω of the shaft.

In practice the intention will always be to keep the out-of-balance exciting forces as small as possible. This is achieved by the process of balancing in which suitable balancing weights are attached to the rotor. In balancing, it must be ascertained whether the rotor to be balanced is to be regarded as rigid or as elastic. Further details regarding the practice of balancing and the quality of balancing can be found in [6, 7].

Excitation Through Inertial and Gas-generated Forces in Piston Machinery. In the power units of piston machines (four-stroke engines, two-stroke engines, piston compressors), inertial forces (see J1.3) caused by parts in longitudinal motion (pistons, parts of the connecting rod etc.), and by gas forces at the piston, occur in addition to the out-of-balance forces caused by rotating parts (crankshaft). This can lead to the excitation of considerable vibrations of individual components or of the entire engine [8, 9] (see F10). In most cases, these forces vary periodically with engine speed (fundamental frequency Ω_0 = angular frequency), but the gas-generated forces in four-stroke engines have a period of two revolutions because a combustion stroke occurs in the cylinder of a four-stroke engine only once every other revolution.

Of the various vibration phenomena in piston engines, crankshaft vibration needs special attention, so that its stresses do not lead to a fracture. For crankshaft vibration calculations, the time-varying exciting forces acting on the crankshaft which result from the above-mentioned inertial and gas forces must be known. The following details apply to the stationary condition (constant revolutions). The most important relations are best explained by means of a single-cylinder power unit (four-stroke engine). They can easily be applied to multi-cylinder engines.

The resulting force $F_K(t)$ acting on a piston is composed of the force $F_G(t)$ due to gas pressure and the inertial force $F_M(t)$ (**Fig. 12**) (see F10.2.3):

$$F_K(t) = F_G(t) + F_M(t). \tag{33}$$

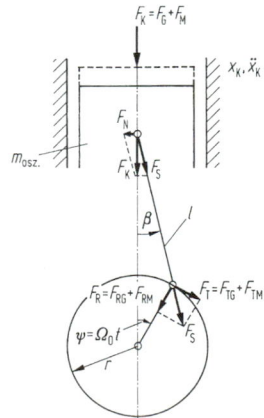

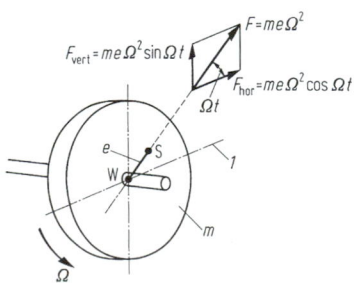

Figure 11. Out-of-balance forces of a rotating disc: e mass eccentricity, Ω angular frequency, m mass, 1 zero axis for the angle Ωt.

Figure 12. Forces occurring during crankshaft operation: ψ crank angle, r crank radius, β angle of excursion, l length of connecting rod.

The piston force can be resolved geometrically into the normal force $F_N(t)$ and the connecting-rod force $F_S(t)$, where the connecting-rod force at the crankpin can once more be subdivided into the tangential component $F_T(t)$ and the radial component $F_R(t)$ (see F10.2). These are the exciting forces for the crankshaft which lead to torsional and flexural oscillations. These may again be subdivided into components due to gas-pressure forces and those due to inertial forces:

$$F_T(t) = F_{TG}(t) + F_{TM}(t),$$
$$F_R(t) = F_{RG}(t) + F_{RM}(t). \tag{34}$$

For each calculation, the force ratios F_T/F_K and F_R/F_K applying to the two types of force (gas forces, inertial forces) are needed. These are periodic functions which represent the geometry of the crank operation

$$\frac{F_T}{F_K} = \frac{\sin(\psi + \beta)}{\cos \beta} =$$
$$B_1 \sin \psi + B_2 \sin 2\psi + B_4 \sin 4\psi + \dots \tag{35}$$

with

$$B_1 = 1,$$
$$B_2 = \lambda/2 + \lambda^3/8 + \dots,$$
$$B_4 = -\lambda^3/16 - 3\lambda^5/64 - \dots .$$

$$\frac{F_R}{F_K} = \frac{\cos(\psi + \beta)}{\cos \beta}$$
$$= A_0 + A_1 \cos \psi + A_2 \cos 2\psi + A_4 \cos 4\psi + \dots , \tag{36}$$

with

$$A_0 = -\lambda/2 - 3\lambda^3/16 - \dots,$$
$$A_1 = 1,$$
$$A_2 = \lambda/2 + \lambda^3/4 + \dots,$$
$$A_4 = -\lambda^3/16 - \dots$$

($\psi = \Omega_0 t$ crank angle, Ω_0 angular frequency of the crankshaft, β angle of excursion, $\lambda = r/l$ connecting-rod ratio). The four individual parts of Eq. (34) can now be stated as follows:

$$\left.\begin{array}{l} F_{TG}(t) = F_G(t) \cdot (F_T/F_K), \\ F_{TM}(t) = F_M(t) \cdot (F_T/F_K) , \end{array}\right\} \tag{37}$$

$$\left.\begin{array}{l} F_{RG}(t) = F_G(t) \cdot (F_R/F_K) , \\ F_{RM}(t) = F_M(t) \cdot (F_R/F_K) . \end{array}\right\} \tag{38}$$

However, the inertial force $F_M(t)$ as well as the gas-pressure force $F_G(t)$ are also periodic functions under stationary operation.

The inertial force $F_M(t)$ is given, for instance, by the product of the oscillating mass m_{osc} (mass of piston, mass contribution of the connecting rod), and the piston acceleration $\ddot{x}_k(t)$ and can be expressed by the following Fourier series:

$$F_M(t) = -m_{osc}\ddot{x}_k = -m_{osc}r\Omega_0^2 (C_1 \cos \psi + C_2 \cos 2\psi$$
$$+ C_4 \cos 4\psi + C_6 \cos 6\psi + \dots), \tag{39}$$

with

$$C_1 = 1,$$
$$C_2 = \lambda + \lambda^3/4 + 15\lambda^5/128,$$
$$C_4 = -\lambda^3/4 - 3\lambda^5/16 - \dots,$$
$$C_6 = 9\lambda^5/128 + \dots .$$

From Eqs (37) and (38) and making use of (35), (36) and (39), the inertial tangential force and the inertial radial force can be calculated as

$$F_{TM} = m_{osc} r\Omega_0^2 \sum_{k=1}^{\infty} T_k \sin k\psi,$$

with

$$\left.\begin{array}{l} T_1 = \lambda/4 + \lambda^3/16 + 15\lambda^5/512 + \dots , \\ T_2 = -1/2 - \lambda^4/32 - \lambda^6/32 \dots , \\ T_3 = -3\lambda/4 - 9\lambda^3/32 - 81\lambda^5/512 - \dots , \\ T_4 = -\lambda^2/4 - \lambda^4/8 - \lambda^6/16 - \dots , \\ T_5 = 5\lambda^3/32 + 75\lambda^5/512 + \dots . \\ F_{RM} = m_{osc} r\Omega_0^2 \left(R_0 + \sum_{k=1}^{\infty} R_k \cos k\psi \right). \end{array}\right\} \tag{40}$$

with

$$\left.\begin{array}{l} R_0 = -1/2 - \lambda^2/4 - 3\lambda^4/16 - 5\lambda^6/32 - \dots , \\ R_1 = -\lambda/4 - \lambda^3/16 - 15\lambda^5/512 - \dots , \\ R_2 = -1/2 + \lambda^2/2 + 13\lambda^4/32 + 11\lambda^6/32 + , \\ R_3 = -3\lambda/4 - 3\lambda^3/32 - 9\lambda^5/512 - \dots , \\ R_4 = -\lambda^2/4 - 5\lambda^4/16 - 5\lambda^6/16 - \dots . \end{array}\right\} \tag{41}$$

Similarly, the forces F_{TG} and F_{RG} resulting from gas-pressure forces on the piston may be obtained. If, for instance, discrete values of the force $F_G(t)$ are available over a period, then these are multiplied as in Eqs (37), (38) and thereafter a harmonic analysis of the force components F_{TG} and F_{RG}, so determined, is carried out. Here the different fundamental frequencies generated by the two-stroke engine (Ω_0) and the four-stroke engine ($\Omega_0/2$) must be taken into account.

Figure 13 shows the results of the harmonic analyses of the radial force $F_{RG}(t)$ and the tangential force $F_{TG}(t)$ of a four-stroke engine. In all cases the values shown refer to the piston area A_k.

For multi-cylinder power units, it is generally assumed that all cylinders are equal and function in the same way, and that therefore the forces are the same for all cylinders. The forces for the various cylinders, however, suffer a time phase difference since the ignition points do not coincide. This phase difference yields different harmonic coefficients for the exciting forces for the different cylinders [8, 9], which can be derived from the stated values for the single-cylinder engine.

Excitation by Electrical Perturbation Moments. In electrical machines (motors, generators) considerable electrical perturbation moments may occur which can excite torsional oscillations along the entire length of the shaft. These will be introduced here in the form of perturbations in a power generator turbine set. In the stationary state, the torques exerted by the driving turbines and the braking generator are in equilibrium. This

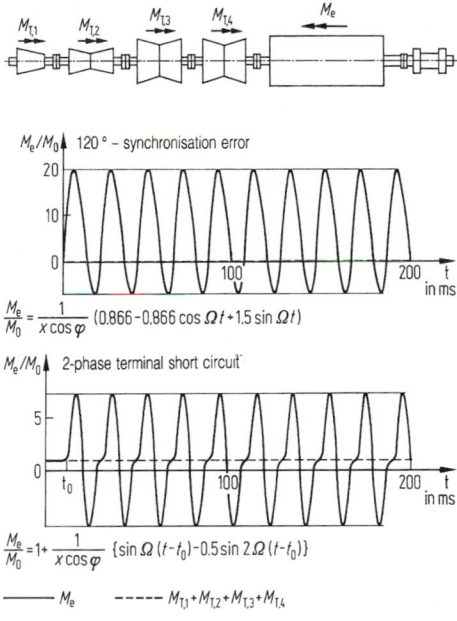

$$\frac{M_e}{M_0} = \frac{1}{x \cos \varphi} (0.866 - 0.866 \cos \Omega t + 1.5 \sin \Omega t)$$

$$\frac{M_e}{M_0} = 1 + \frac{1}{x \cos \varphi} \{\sin \Omega (t - t_0) - 0.5 \sin 2\Omega (t - t_0)\}$$

——— M_e - - - - - $M_{T,1} + M_{T,2} + M_{T,3} + M_{T,4}$

Figure 14. Air-gap moment $M_e(t)$ in a generator.

120° synchronisation error:

$$M_e(t) = \frac{M_0}{\cos \varphi} \cdot \frac{1}{x_d'' + x_{TR} + x_N}$$

$$\cdot \{0.866 - 0.866 \cos \Omega t + 1.5 \sin \Omega t\}.$$

where x_d'' = the subtransient reactance of the generator, x_{TR} = the transformer reactance, x_N = the mains reactance, in each case normalised to the generator impedance, $\cos \varphi$ = the power factor, M_0 = the nominal moment and Ω = the mains frequency.

With nominal torque M_0 and the alternating contributions varying with frequency of rotation or twice that frequency respectively, the time-invariant contribution is clear. The given exciting moments must be inserted at the appropriate place into the excitation vector $F(t)$ of the shaft equations of motion.

The torques of drives with electrical speed control are increasingly important. Here, exciting torques that pulsate can occur as a consequence of energy supply via converters (voltage changes) because these cause harmonics in current and voltage. In [11], exciting frequencies as a function of the rate of revolution are given for two of the most widely employed types of drive (slip-ring motor with static converter cascade, current converter with synchronous motor drive).

2.6 Mechanical Equivalent Systems, Equations of Motion

In order to obtain calculated solutions or to interpret the results of measurements, mechanical equivalent systems are needed that reproduce the true dynamic behaviour with sufficient accuracy. The way to proceed in formulating the model is shown in **Fig. 15**. The starting point is a consideration of the real system (design drawing) where it must be stated, *inter alia*, where the system limits are to be drawn. After formulating the task and its limits, the equivalent mechanical system can be

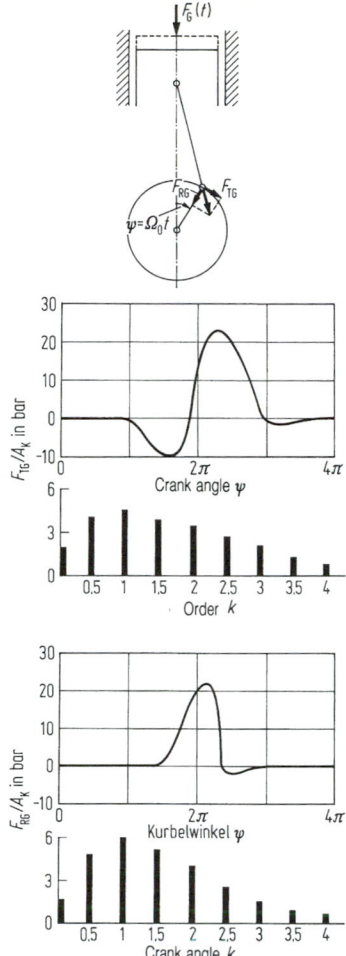

Figure 13. Harmonic analyses of the tangential force F_{TG} and the radial force F_{RG} for a cylinder of a four-stroke engine.

equilibrium can be severely disrupted by electrical faults in the mains or at the generator, or by switching and synchronisation processes. The moment exerted by the generator then contains additional constant and oscillating components.

Studies show that the greatest shaft loading occurs if the terminals are short-circuited or during faulty synchronisation with an angular error of 120°. For this reason, only these cases are used in most of the appropriate design standards and regulations. **Figure 14** shows the time-dependence of the air gap moment $M_e(t)$ of a generator in relation to the nominal moment M_0 for an undamped two-pole terminal short circuit and for a 120° synchronisation error. The time dependence can be calculated from the following equations [10]:

Two-phase short circuit at the terminals:

$$M_e(t) = M_0 + \frac{M_0}{\cos \varphi} \cdot \frac{1}{x_d'' + x_{TR}}$$

$$\cdot \{\sin \Omega(t - t_0) - 0.5 \cdot \sin 2\Omega(t - t_0)\}. \quad (42)$$

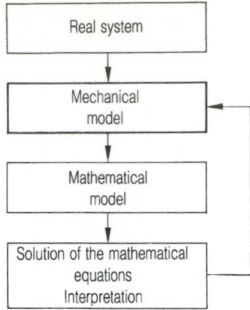

Figure 15. Procedure for the formulation of a model.

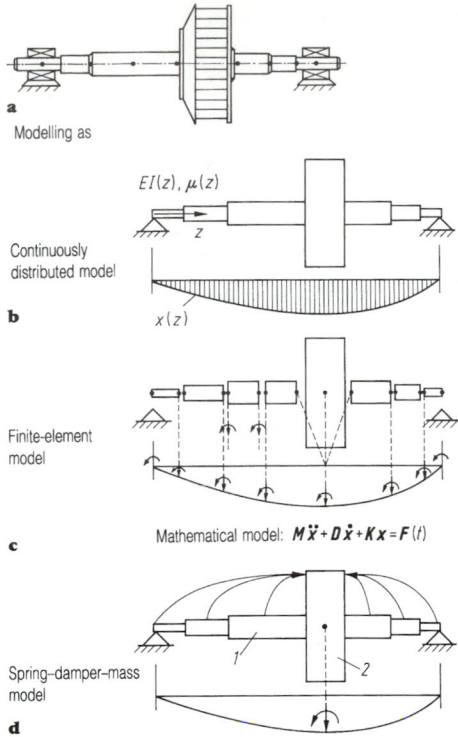

Figure 16. Possibilities of model formulation using the example of a shaft and turbine rotor: **a** real system; modelled as: **b** continuously distributed model; **c** finite-element model; **d** spring–damper–mass model; *1* flexural spring, *2* mass.

set up. Here, simplifications based on assumptions and experiences or observations derived from similar systems are made. The mechanical model should be as simple as possible, but should contain all essential features. Model and reality should conform closely in as much as the required information is concerned. The aim is a minimum model that gives relevant data regarding the dynamic behaviour of the system while making use of the smallest possible number of degrees of freedom.

In addition to the *equivalent mechanical system*, it is also desirable to establish the corresponding mathematical model, which in the case of oscillatory systems often leads to a system of differential equations with constant coefficients (see Eq. (1)). These equations can be solved subsequently, and the results interpreted and discussed. If necessary, and where doubts regarding the solutions remain or where strong discrepancies from the real behaviour exist, the mechanical model must be revised.

In formulating a mechanical equivalent system, the system structure is defined, and then the appropriate system parameters may be determined (**Fig. 1**).

2.6.1 Structure Definition

Several questions arise in the definition of a structure. It is first necessary to choose between a continuous system with distributed masses and stiffness, and a discrete system. This leads to partial differential equations in the first case, and to ordinary differential equations in the latter case. It is also important to decide whether or not linear of non-linear relations apply. Further, there is the question of how many degrees of freedom are required, what elements (springs, masses, dampers, rods, beams, plates, etc.) the system is to consist of, and what boundary conditions apply.

Figure 16 shows various possibilities of modelling using the example of a shaft and a turbine wheel (see **Fig. 2**). The continuous system, with its infinite number of degrees of freedom, is a structure that closely approximates to reality, because the continuous distribution of masses and stiffnesses is taken into account. However, in the case of complex systems the solution of the accompanying partial differential equations is possible only with a very considerable effort and is therefore not feasible for practical purposes.

Good approximate solutions can be found using discrete systems. In the discrete process, which has already become classical engineering procedure, the continuously distributed masses are replaced by point masses or rigid bodies, and these are countered by massless springs and dampers (spring–mass–damper system, lumped mass models). As is shown in the examples, certain lumped

masses (rotors) suggest themselves for representation by point masses or rigid bodies to which neighbouring smaller masses (mass of the shaft) can be added in proportion. Springs, massless torsion rods and bending beams, among others, are employed as elastic connecting elements.

Recently, finite-element analysis has become extremely significant (see B8). The FE method is very versatile and allows the treatment of any arbitrary one-, two- or three-dimensional oscillatory systems. Any boundary conditions and any distribution of mass, stiffness and damping are also permitted. Each element is treated on its own, and the dynamic behaviour is described in the form of a force-displacement relation using forces and moments with displacements and torsions, respectively, at the nodal points (**Fig. 16**). This is accomplished by so-called initial functions in which the mechanical degrees of freedom at the nodal points appear as free parameters. The properties of the elements are gathered together in mass, damping and stiffness matrices. This expresses clearly that the properties of inertia, damping and stiffness are all taken into consideration simultaneously in a finite element. Finally, the elements are connected at the nodal points taking into account all boundary and transition conditions and formed into the total structure.

The appropriate mathematical model takes the same form as the discrete engineering model, and in general leads to a set of ordinary differential equations already presented in Eq. (1), if the system behaves linearly.

2.6.2 Parameter Definition

If the structure of the oscillatory system and hence the form of the mathematical equations are given, the next step is to determine the values of the system parameters, or the elements of the matrices M, D and K, respectively. In determining the parameters, important information is taken from the design drawings (dimensions, material parameters, masses) and the laws of mechanics (moments of inertia, flexural rigidity, torsional rigidity) are applied. However, for several machine elements or mechanisms (sliding bearings, seals, couplings) properly developed theoretical models describing their dynamic behaviour do not yet exist. In such cases, empirical procedures are often unavoidable and attempts are made to determine the unknown parameters of individual system components by a (parameter) identification process [12, 13].

2.6.3 Examples of Mechanical Equivalent Systems. Spring-Mass-Damper Models

Unrestrained Torsional Oscillators with Two Rotating Masses. The torsional oscillatory behaviour of machine equipment can often be described in good approximation by a linear mechanical equivalent system with two rotating masses with a torsional spring and a torsional damper between the two masses (**Fig. 17**). Θ_1 and Θ_2 are the moments of inertia of the two machines (e.g. electric motor–compressor) about the axis of rotation and k_1 and d_1 represent the torsional stiffness of the spring and the torsional damping constant respectively of the connecting shaft or a coupling with torsional elasticity between them. The moments of inertia of a given body about a fixed axis is given by $\Theta = \int r^2 \, dm$ and the torsional stiffness of a cylindrical rod by $k = G I_\mathrm{T}/l$ ($G = $ the shear modulus, $I_\mathrm{T} = $ the polar moment of inertia, $l = $ the length of the rod). In general, details regarding stiffness and damping properties of the coupling can be obtained from the manufacturer (non-linearities in the couplings must be taken into account).

If x_1 and x_2 designate the two degrees of freedom of rotation and $M_1(t)$, $M_2(t)$ the exciting moments applied to the rotating masses, then the equations of motion shown in **Fig. 17** can be derived. They are used, for

example, to calculate the torque maxima in the drive shaft (coupling) that result during startup with an asynchronous electric motor [14].

Model of a Shaft with Rotor (Ventilator). The oscillatory system introduced in **Fig. 2** can be used to represent a simple model for the calculation of low-frequency flexural oscillations. To this end, the mass is considered to be concentrated in the rotor and the elasticity of the shaft and the bearings are taken as a single entity. In the case of the rotor, torsion as well as excursion must be included in order to take account of the effects of rotary inertia (**Fig. 18**).

In writing down the equation of motion for this model, the gyroscopic terms for the shaft as well as the inertia and stiffness terms resulting from the law of angular momentum must be taken into account. The complete equation of motion is therefore

$$
\underbrace{\begin{bmatrix} m & 0 & 0 & 0 \\ 0 & \Theta_\mathrm{ä} & 0 & 0 \\ 0 & 0 & m & 0 \\ 0 & 0 & 0 & \Theta_\mathrm{ä} \end{bmatrix}}_{M} \cdot \underbrace{\begin{bmatrix} \ddot{x}_1 \\ \ddot{x}_2 \\ \ddot{x}_3 \\ \ddot{x}_4 \end{bmatrix}}_{\ddot{x}} +
\underbrace{\begin{bmatrix} 0 & 0 & 0 & 0 \\ 0 & 0 & 0 & -\Omega\Theta_\mathrm{p} \\ 0 & 0 & 0 & 0 \\ 0 & \Omega\Theta_\mathrm{p} & 0 & 0 \end{bmatrix}}_{D} \cdot \underbrace{\begin{bmatrix} \dot{x}_1 \\ \dot{x}_2 \\ \dot{x}_3 \\ \dot{x}_4 \end{bmatrix}}_{\dot{x}}
$$

$$
+ \underbrace{\begin{bmatrix} k_{11} & k_{12} & 0 & 0 \\ k_{21} & k_{22} & 0 & 0 \\ 0 & 0 & k_{11} & -k_{12} \\ 0 & 0 & -k_{21} & k_{22} \end{bmatrix}}_{K} \cdot \underbrace{\begin{bmatrix} x_1 \\ x_2 \\ x_3 \\ x_4 \end{bmatrix}}_{x} = \underbrace{\begin{bmatrix} F_1 \\ F_2 \\ F_3 \\ F_4 \end{bmatrix}}_{F(t).} \quad (43)
$$

The elements of the stiffness matrix can be calculated from the data on unit deformations and the determination of the corresponding forces. The matrix D contains gyroscopic effects that are proportional to angular frequency Ω and polar moment of inertia Θ_p. The inertial matrix contains the masses m and the equatorial moments of inertia $\Theta_\mathrm{ä}$ in its diagonal. More detailed directions for the derivation of the equations of motion are to be found, *inter alia*, in [15].

2.6.4 Examples of Mechanical Equivalent Systems: Finite-Element Models

Finite-Element Models of a Turbogenerator. Turbine sets for electrical power generation of up to 1200 MW are no longer a rarity. The shaft is about 35 m long, weighs approximately 220 t and rotates at 50 revolutions per second in order to generate electricity at mains frequency. The worst torque stresses in the rotor are

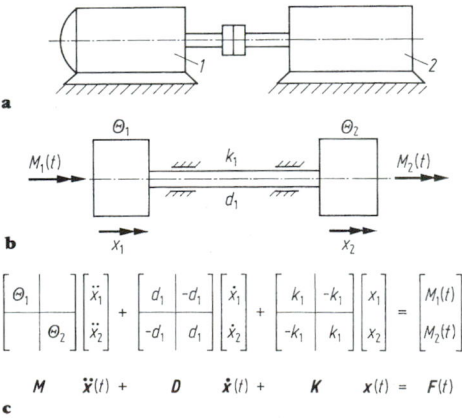

Figure 17. Unrestrained rotary oscillations with two rotating masses: **a** machine set (*1* electric motor, *2* compressor), **b** equivalent system, **c** equation of motion.

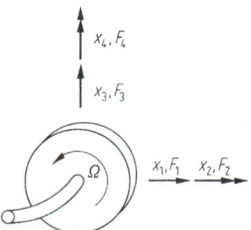

Figure 18. Excursion of a singly equipped rotor disc.

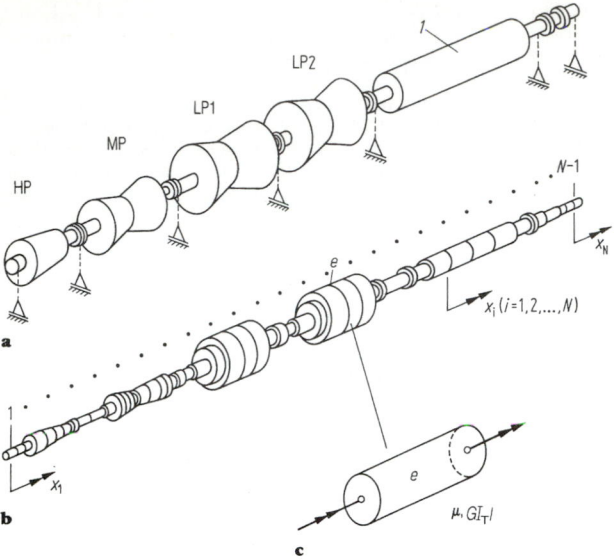

Figure 19. Representation of the real-system turbogenerator as a finite-element model: **a** arrangement (construction) (*1* generator, HP high pressure, MP medium pressure, LP low pressure), **b** mechanical model, **c** torsion element.

caused by torsional oscillations brought about by electrical perturbations in the generator (see J2.5.4) or in the grid. During the design of the machine, the designer must calculate as accurately as possible the stresses at the shaft cross-sections resulting from these cases. Since the rotor system of a turbine generator unit represents a mechanical system with several shafts, a finely detailed model is required for accurate mechanical predictions. Since the shaft has to be subdivided into 200 to 300 elements for this purpose, the finite-element model suggests itself as a mechanical equivalent system [2, 10].

Figure 19 shows, apart from the real system of a turbogenerator with the turbines and the generator, also the appropriate FE model with $N-1$ cylindrical torsional elements. The following constant entities belong to any given "finite" element e of constant cross-section: μ^e = the rotating mass assigned to the element, GI_T^e = the torsional stiffness, and l^e = the length of the element.

With the help of local initial functions which are substituted into the work integrals (principle of virtual work), there can be constructed for each element an *element stiffness matrix*

$$K^e = \frac{GI_T}{l^e} \begin{bmatrix} 1 & -1 \\ -1 & 1 \end{bmatrix} \tag{44}$$

and an *element mass matrix*

$$M^e = \mu^e \, l^e \begin{bmatrix} 1/3 & 1/6 \\ 1/6 & 1/3 \end{bmatrix} \tag{45}$$

which are of the second order owing to the two local degrees of freedom (one angle of rotation for each element node). The torsional oscillations of the total system are globally described by the angles of rotation x_i

which, in each case, are introduced at the nodal points (interfaces between two elements). A system of $(N-1)$ elements has N global degrees of freedom gathered together in the vector x.

The total matrices M and K are built up from the superposition of the elementary matrices. In the present chain-like structure, simple overlapping results, leading to a form of band matrix that is advantageous as regards storage requirements and calculation time. Damping constants in these systems are generally defined as "modal" (see J2.2.3). In branched systems with sets of gears the structure is not quite so simple, but does not present a serious problem when the well-tried FE method is employed.

Finite-Element Model of a Multi-Stage Centrifugal Pump. The trend of development in centrifugal pumps, similar to that of other machines, tends to higher rates of revolution, lighter construction and greater efficiency. For that reason, dynamic behaviour is of ever greater importance, principally in terms of impeller flexural oscillation. Finite-element analysis is used in the majority of cases where, apart from the inertial and stiffness properties of the beam elements (shaft), other factors must be taken into account, namely the fluid forces acting on the rotor in sliding bearings, on seal gaps and on balancing pistons, as well as the interaction between the impeller and the guide vane ring (**Fig. 20**).

In the case of the beam elements, four degrees of freedom per node are employed in order to take account of torsions as well as excursions. Further, shear deformation, gyroscopic effects and damping associated with the materials can also be taken into consideration. Pump impellers are, as a rule, considered to be rigid discs.

Sealing gaps in centrifugal pumps make spaces with

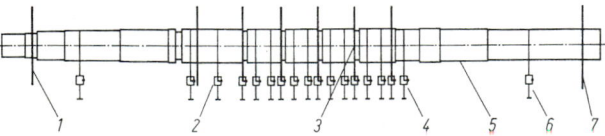

Figure 20. Representation of a multi-stage centrifugal pump as a finite-element model: *1* axial bearing (journal), *2* balancing piston, *3* impeller, *4* seal gap (interactions: impeller-ring of guide vanes), *5* shaft, *6* sliding bearing, *7* coupling.

unequal pressures fluid-tight. Here, leakage loss through the gap, which is approximately 200 to 300 μm wide, is tolerated because the advantages of low friction and low wear are more important. The sealing gap has, however, a considerable effect on the vibration behaviour. The surrounding fluid exerts forces on the moving rotor with its excursions (radial displacements x_1, x_2 and associated speeds $\dot{x}_1$, $\dot{x}_2$ and accelerations $\ddot{x}_1$, $\ddot{x}_2$ respectively) which contribute in large measure to the out-of-balance oscillations and the stability behaviour of the machine. These forces can be described by inertial, damping and stiffness coefficients in the form of a linear force-displacement relation:

$$\begin{bmatrix} m_{11} & m_{12} \\ m_{21} & m_{22} \end{bmatrix}\begin{bmatrix} \ddot{x}_1 \\ \ddot{x}_2 \end{bmatrix}+\begin{bmatrix} d_{11} & d_{12} \\ d_{21} & d_{22} \end{bmatrix}\begin{bmatrix} \dot{x}_1 \\ \dot{x}_2 \end{bmatrix}+$$
$$\begin{bmatrix} k_{11} & k_{12} \\ k_{21} & k_{22} \end{bmatrix}\begin{bmatrix} x_1 \\ x_2 \end{bmatrix}=\begin{bmatrix} F_1 \\ F_2 \end{bmatrix}. \quad (46)$$

To determine the dynamic coefficients, any of a number of theories may be used that attempt to describe the flow in the gap by means of various initial formulations [16, 17]. All theories have in common that they describe a displacement outward from the central position. The matrices have a skew-symmetric structure which is confirmed by measurement. In the case of a hydrodynamic sliding bearing (see F5), the shaft is supported by a pressure field set up by the rotation of the shaft. This results in a strongly non-linear relation between the forces and the movement of the shaft relative to the housing. Linearisation is possible for small movements:

$$\begin{bmatrix} d_{11} & d_{12} \\ d_{21} & d_{22} \end{bmatrix}\begin{bmatrix} \dot{x}_1 \\ \dot{x}_2 \end{bmatrix}+\begin{bmatrix} k_{11} & k_{12} \\ k_{21} & k_{22} \end{bmatrix}\begin{bmatrix} x_1 \\ x_2 \end{bmatrix}=\begin{bmatrix} F_1 \\ F_2 \end{bmatrix}. \quad (47)$$

The dynamic coefficients d_{ij} and k_{ij} result from the solution of the Reynolds differential equation or from empirical studies. They are usually given in non-dimensional form as a function of the Sommerfeld number So [18] (see F5.2). Since the static bearing load F_L is a function of the bearing compliances, the weight forces and the hydraulic forces acting on the pump impeller, the problem of determining the stationary position of the shaft for each condition of operation is non-linear. The interactions between the impeller and the ring of guide vanes can be described in a similar way.

Taking into account all the above effects, an equation of motion for the centrifugal pump can be derived by superposition of the equations for the elements

$$M\ddot{x}+D\dot{x}+Kx=F. \quad (48)$$

The matrices M, D and K have a band structure and are, in general, asymmetric. In addition, some of the matrix elements depend on speed.

2.7 Application Examples for Machine Vibrations

The solutions of the equations of motion (natural vibrations, forced vibrations) can be discussed with the help of a few examples. This will reveal effects that occur frequently in machine dynamics.

2.7.1 Torsional Oscillator with Two Rotating Masses

Natural Vibrations and Modal Entities. The equation of motion for the undamped torsion model with two rotating masses (**Fig. 17**) was given in matrix form:

$$\begin{bmatrix} \Theta_1 & 0 \\ 0 & \Theta_2 \end{bmatrix}\cdot\begin{bmatrix} \ddot{x}_1 \\ \ddot{x}_2 \end{bmatrix}+\begin{bmatrix} k & -k \\ -k & k \end{bmatrix}\begin{bmatrix} x_1 \\ x_2 \end{bmatrix}=\begin{bmatrix} M_1 \\ M_2 \end{bmatrix}$$
$$M \quad\cdot\quad \ddot{x} \quad+\quad K \quad\cdot\quad x \quad=\quad F. \quad (49)$$

If no external excitations are present, the vibrations of the system are described by the homogeneous equation of motion

$$M\cdot\ddot{x}+K\cdot x=0. \quad (50)$$

The solution is obtained using the relation $x=\varphi\cdot e^{i\omega t}$. It consists of the natural frequencies ω_n and eigenvectors φ_n which result from the eigenvalue problem

$$\begin{bmatrix} k-\omega^2\Theta_1 & -k \\ -k & k-\omega^2\Theta_2 \end{bmatrix}\cdot\begin{bmatrix} \varphi_1 \\ \varphi_2 \end{bmatrix}=0$$
$$(K-\omega^2 M)\cdot\varphi=0. \quad (51)$$

The characteristic equation is obtained in the form $\det\{K-\omega^2 M\}=0$:

$$\omega^2(-k(\Theta_1+\Theta_2)+\omega^2\Theta_1\Theta_2)=0. \quad (52)$$

From this the natural frequencies are calculated as

$$\omega_{1,2}=0,$$

$$\omega_{3,4}=\pm\sqrt{\frac{k(\Theta_1+\Theta_2)}{\Theta_1\Theta_2}}=\pm\sqrt{\frac{k}{\Theta_1}+\frac{k}{\Theta_2}}. \quad (53)$$

If these results are resubstituted into the eigenvalue problem, the corresponding eigenvectors are obtained:

$$\varphi_{1,2}=\begin{pmatrix} 1 \\ 1 \end{pmatrix};\ \varphi_{3,4}=\begin{pmatrix} 1 \\ -\Theta_1/\Theta_2 \end{pmatrix}. \quad (54)$$

Discussion of the results shows some interesting aspects. Since the system is not bound to any number of degrees of freedom, natural frequencies with value zero ($\omega_{1,2}=0$) result. They are rigid-body displacements as shown by the corresponding eigenvectors. In this modal displacement the torsional composite shaft does not deform, and no internal strains appear. The other two solutions represent elastic modal movements. Their natural frequencies and modal forms depend on the two inertial rotary masses Θ_1 and Θ_2 and the stiffness k.

Figure 21 shows the modes of oscillation. In the special case of $\Theta_1=\Theta_2$ the system is symmetrical and the natural frequency corresponds to that of a single mass oscillation with spring stiffness $2k$. If one mass becomes

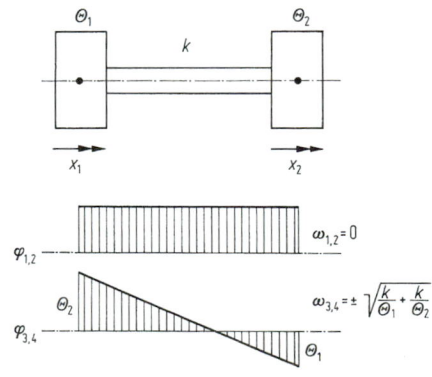

Figure 21. Modes of oscillation for rotary oscillators with two degrees of freedom.

very large in comparison with the other, this remains at rest and the natural frequency corresponds to that due to rigid clamping at that point.

Forced Vibrations. If external forces (moments) act, the torsional oscillation with two masses is described by an inhomogeneous differential equation. In order to simplify matters, damping is not taken into account:

$$\boldsymbol{M} \cdot \ddot{\boldsymbol{x}} + \boldsymbol{K} \cdot \boldsymbol{x} = \boldsymbol{F}. \tag{55}$$

If the system is excited by a force $\boldsymbol{F}(t)$ the solution consists of an homogeneous contribution $\boldsymbol{x}_{\text{hom}}$ and a particular contribution $\boldsymbol{x}_{\text{part}}$. The forced contribution to the solution $\boldsymbol{x}_{\text{part}}$ results from the solution of the inhomogeneous equation of motion by an initial formulation in the manner of the right-hand side of the equation. In the case of a sinusoidal dependence of the force $\boldsymbol{F}(t) = \hat{\boldsymbol{F}} \sin(\Omega t)$, $\boldsymbol{x}_{\text{part}} = \hat{\boldsymbol{x}} \sin(\Omega t)$ is obtained, with

$$\hat{\boldsymbol{x}} = \begin{bmatrix} k - \Omega^2 \Theta_1 & -k \\ -k & k - \Omega^2 \Theta_2 \end{bmatrix}^{-1} \hat{\boldsymbol{F}}$$

$$\hat{\boldsymbol{x}} = \frac{1}{\Omega^2(\Omega^2 \Theta_1 \Theta_2 - k(\Theta_1 + \Theta_2))}$$

$$\cdot \begin{bmatrix} k - \Omega^2 \Theta_2 & k \\ k & k - \Omega^2 \Theta_1 \end{bmatrix} \hat{\boldsymbol{F}}. \tag{56}$$

For certain exciting frequencies, the excursions increase steeply. Such is the case at $\Omega = \omega_{1,2} = 0$, the reason for which is the absence of damping of the oscillations and resonance, if the exciting frequency Ω agrees with the next natural frequencies $\Omega = \omega_{3,4}$. Further, the excursions at the point of excitation can vanish if, for instance, the mass Θ_1 is excited with the frequency $\Omega^2 = k/\Theta_2$, or vice versa. **Figure 22** shows the plot of the torsional vibration amplitudes $\hat{x}_1$, $\hat{x}_2$ as a function of the exciting frequency Ω. Since any periodic function $\boldsymbol{F}(t)$ can be represented by a sum of harmonic functions, the oscillations due to such excitations can be given as the sum of several contributions of the above force.

For non-periodic excitation, closed solutions are frequently found. Complex force functions can be solved by numerical methods, or calculated using polygons that are closed piecewise.

2.7.2 Torsional Vibrations of a Turbosystem

The case of the shaft assembly train of a turbosystem represents a rather more complex example and is of central importance in any power station. Apart from flexural vibrations, torsional vibrations in particular form a decisive criterion for plant reliability. Once again, calculations are carried out using finite-element analysis. Here the tur-

bine set is frequently subdivided into several hundred elements in order to reproduce the vibration behaviour with sufficient accuracy (**Fig. 19**).

Natural Vibrations and Modal Entities. The natural vibrations of the system are described by the equation of motion without external exciting forces. The D matrix is omitted in the analysis of the natural vibrations because of the weak damping:

$$\boldsymbol{M}\ddot{\boldsymbol{x}} + \boldsymbol{K}\boldsymbol{x} = 0. \tag{57}$$

The mass matrix $\boldsymbol{M}$ is always definite positive, since all the matrix elements used represent masses. The stiffness matrix $\boldsymbol{K}$ is semi-definite positive, since a rigid body displacement of the "torsional train" is possible. In accordance with the $(N \times N)$ order of the matrix, N natural frequencies and modal forms from the solution of the eigenvalue problem are obtained:

$$(\boldsymbol{K} - \omega^2 \boldsymbol{M}) \, \varphi = 0. \tag{58}$$

The solution itself can only be carried out economically with the help of numerical algorithms, distinguishing between direct and iterative procedures. The eigenvalues and eigenvectors are determined by iterative processes from initial values that are changed by iteration, until a predetermined termination condition is satisfied.

Example. The modal entities of a 600-MW turbosystem are being considered, with the torsion train subdivided into 250 torsion elements. Since torsional vibrations are frequently only very weakly damped, it is sufficient to consider the undamped system. **Figure 23** shows the five lowest natural frequencies ($f_n = \omega_n/2\pi$) and the normalised eigenvectors of the turbosystem. The rigid body form of the zero natural frequency is not shown. In the first natural mode, the HP (high-pressure), MP (medium-pressure) and LP1 (low-pressure) turbine oscillate at 18.19 Hz with respect to the LP2 turbine and the generator. The natural mode has a zero transit in the region of the coupling (vibration node). A further node is added in each further natural mode. The low natural modes are distributed over the entire shaft train, while in the case of the higher frequencies only individual part rotors are in oscillation.

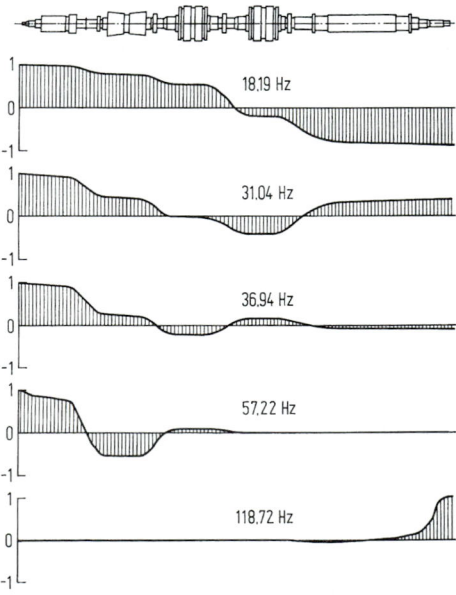

Figure 23. Natural frequencies and natural vibration modes for the turbosystem.

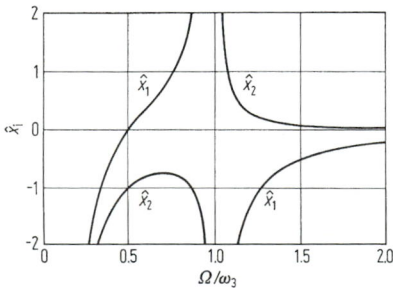

Figure 22. Normalised amplitudes of torsional vibrations of a torsional oscillator with two degrees of freedom as a function of the normalised excitation frequency.

Forced Vibrations. Because of the many degrees of freedom, the solution of the equation of motion for forced vibrations not caused by harmonic excitations is very time-consuming and frequently numerically inaccurate. It is possible to decouple the equations by coordinate transformation when, as a rule, the number of equations is also considerably reduced (modal analysis; see J2.2.4). Once the decoupled equations have been solved, they can be retransformed, thus obtaining the required results. Decoupling is accomplished with the aid of the so-called modal matrix Φ, which is made up from the calculated eigenvectors. This leads to generalised equations for simple inertial oscillators, which can be solved very effectively. It is also evident from the right-hand side of a modal

equation how intensively the vibration of this natural mode is excited. In the modal calculation of the forced vibrations, damping in modal form is also taken into consideration.

Example. The short-circuit response of the above 600-MW turbo-system is being considered. The angle of torsion of each degree of freedom is composed of the superposition of the part solutions of the modal single mass oscillators. With the aid of the element matrices it is also possible to determine the shear moments from the calculated torsions that are critical for the design of the shaft train. **Figure 24** shows the contribution to these moments from the various natural vibration modes. Their sum results in a maximum loading of the coupling of the generator amounting to four times the nominal moment.

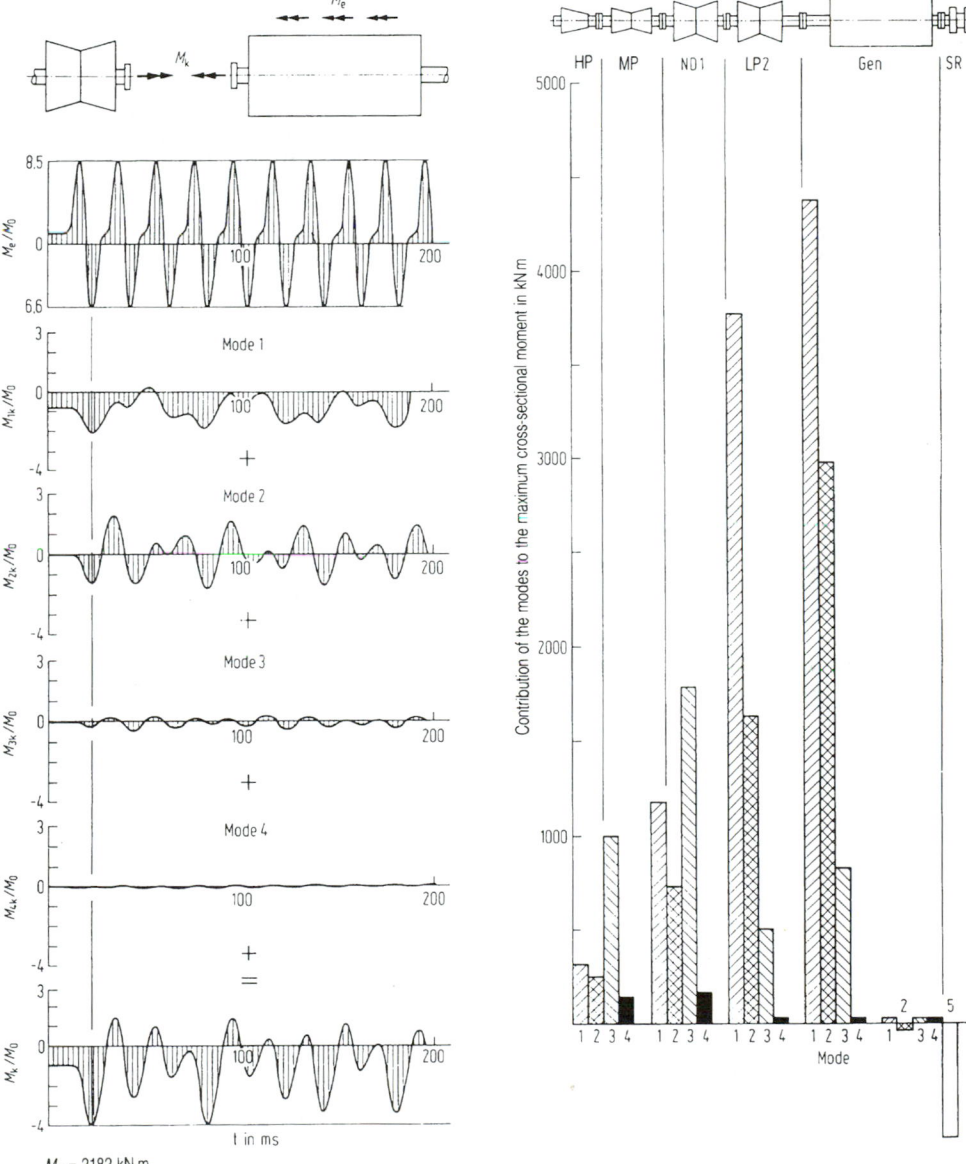

$M_0 = 2182 \ kN \ m$

Figure 24. Cross-sectional moments in the shaft of a turbosystem following a short circuit.

2.7.3 Flexural Vibrations of a Shaft with Rotor

In calculating the flexural vibrations of rotors, the gyroscopic effect often plays an important part. In the model for the simpler rotor introduced above, the equation of motion was characterised by a skew-symmetrical gyroscopic matrix D (see J2.6.3).

Natural Vibrations and Modal Entities. A harmonic initial formulation of the homogeneous equation of motion leads to a characteristic equation, from which four natural angular frequencies ω can be calculated:

$$m\Theta_{\ddot{a}}\omega^4 - m\Theta_p\Omega\omega^3 - (k_{22}m + k_{11}\theta_{\ddot{a}})\omega^2$$

$$+ k_{11}\theta_p\Omega\omega + (k_{11}k_{22} - k_{12}^2) = 0. \qquad (59)$$

In each case the corresponding modal forms represent circular movements of the point at which the shaft passes through the rotor, where the torsions may be interpreted in a similar manner. Derivation of an explicit presentation of the results for eigenvalues ω requires considerable effort. A qualitatively striking feature is that ω does not depend only on the physical system parameters but also on the angular frequency Ω. This dependence is the more

pronounced the more strongly the rotor tilts during a natural vibration and the greater the moment of inertia [15].

Forced Vibrations. Since the natural modes in this model lie along circular paths, it is important in considering forced vibrations to observe the sense of rotation along such circular paths. The out-of-balance excitation which, in the case of rotors, is frequently the dominant one, also moves along circular paths about the axis of rotation. This means that only those natural modes are excited that rotate in the same sense as the shaft. These are termed ganged natural frequencies, and, in the other case, counter-rotational natural frequencies [15].

2.7.4 Flexural Vibrations of a Multi-Stage Centrifugal Pump

The flexural vibrations of a complex structure are considered via another example. The model of a multi-stage centrifugal pump is characterised by asymmetrical systems matrices M, D and K (see J2.6.4).

Natural Frequencies and Modal Entities. The solution of the homogeneous differential equation yields

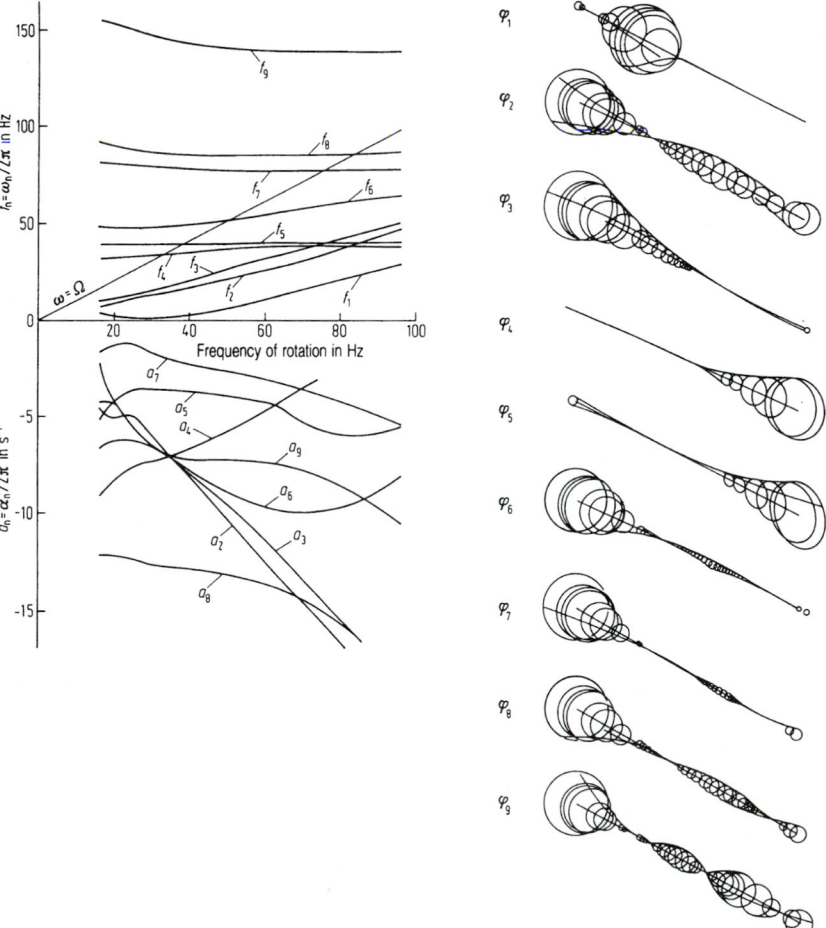

Figure 25. Eigenvalues as a function of the rate of revolution and modes for a multi-stage centrifugal pump.

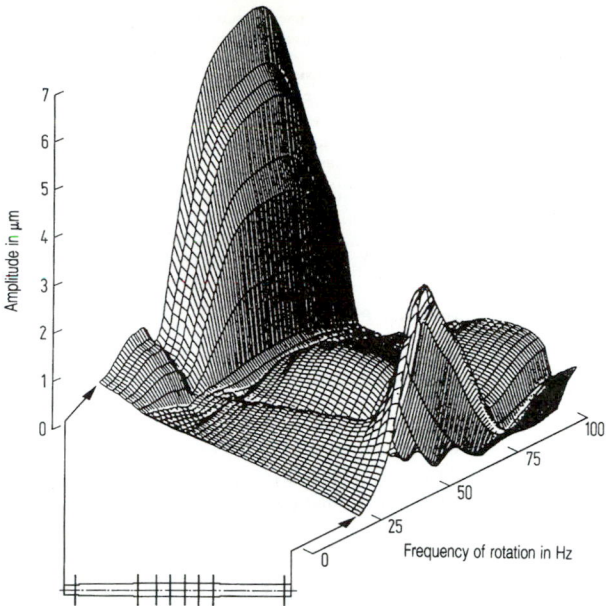

Figure 26. Amplitudes of the out-of-balance vibrations of a centrifugal pump.

complex eigenvalues λ_n and natural modes φ_n, which always appear as complex conjugate quantities:

$$\lambda_n = \alpha_n + i\omega_n; \quad \varphi_n = \varphi_n^{Re} + i\varphi_n^{Im}. \qquad (60)$$

Modern numerical methods can exploit the special band structure of the matrices and can also, if necessary, calculate some or all eigenvalues and eigenvectors. The imaginary part ω_n of an eigenvalue gives the natural angular frequency of the system. The corresponding real part α_n is a measure for the damping experienced by a free vibration.

If $\alpha_n > 0$, the oscillatory movements $x(t)$ increase, i.e. the pump is unstable. The eigenvectors describe the natural modal vibrations. Since complex excursions cannot be visualised, the two occurring complex conjugate eigenvectors are taken together and transformed into a real representation. The result is elliptical paths at the various nodes, so that the mode of oscillation can change during a period.

A typical representation of eigenvalues as a function of the rate of revolution and of the complex eigenvectors is shown in **Fig. 25**. Possible resonance points occur at the

point of intersection of the plots of the natural frequencies and the start-up line ($\omega = \Omega$), and the intensity of resonance amplification can be judged by the damping factor α_n at the appropriate point.

Forced Vibrations. The most important forced vibrations in such machines arise from out-of-balance excitations. In this case the vector $F(t)$ is a harmonic function depending on the out-of-balance moment me and the rate of revolution (see J2.5.4). The steady-state condition $x(t)$ results from the solution of a complex system of equations and describes elliptic paths for the various nodal points of the model similar to the natural modes

$$(K - \Omega^2 M + i\Omega D)\,\hat{x} = \hat{F}. \qquad (61)$$

As a rule, the main interest is directed to the maximum amplitudes that can occur at the various points. These are shown in **Fig. 26** as a function of the rate of revolution. The strong excursions of the end of the shaft at the resonance points that were also noticeable in **Fig. 25** can be noted.

3 Acoustics in Mechanical Engineering

D. Föller, Frankfurt-on-Main

Machine acoustics deals with the occurrence and reduction of the noise caused by machines. It is a branch of the science of acoustics applied to the requirements of mechanical engineering.

3.1 Basic Concepts

An extensive collection of concepts from the entire field of acoustics can be found in [1]. Here, those that are most important for the acoustics of machines will be explained.

Noise, Acoustic Pressure, Acoustic Pressure Level

Mechanical oscillations with frequency components in the audible range (16 to 16 000 Hz) are described as *noise*. Oscillations in air and gases are called airborne noise, those in liquids (water, oil) fluid-borne noise, and those in solids (i.e. also in mechanical structures) structure-borne noise.

Noise in air and other gases as well as in liquids only propagates as compression waves. The alternating pressure $p(t)$ that is superimposed on the static pressure is called the *acoustic pressure* and is the most important

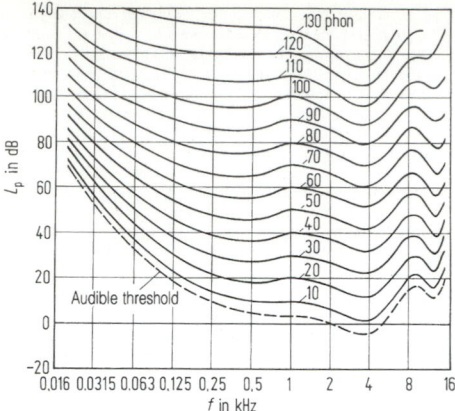

Figure 1. Normal curves of equal acoustic intensity (DIN 46630, Sheet 2).

measurable quantity for airborne and fluid-borne noise. Measuring probes are microphones or pressure sensors respectively. The acoustic pressure is given as an effective value $\tilde{p}$ averaged over a certain time and within a certain frequency interval. The range for $\tilde{p}$ within which the human ear can perceive sound lies between $\tilde{p}_{HS} = 2 \cdot 10^{-4}$ μbar $= 2 \cdot 10^{-5}$ N/m² (audible threshold at 1000 Hz) and $P_{SG} = 2 \cdot 10^2$ μbar $= 2 \cdot 10^1$ N/m² (pain threshold). Since $\tilde{p}$ can range over six orders of magnitude, and in order to arrive at smaller numerical values, acoustic pressure is not stated in absolute microbars or N/m², but in a relative measure of power known as the *acoustic pressure level* L_p in decibels (dB). This is defined as

$$L_p = 10 \lg (\tilde{p}^2/\tilde{p}_0^2) \text{ dB} = 20 \lg (\tilde{p}/\tilde{p}_0) \text{ dB}, \quad (1)$$

where $\tilde{p}_0 = 2 \cdot 10^{-4}$ μbar $= 2 \cdot 10^{-5}$ N/m², the internationally agreed reference value for the effective value (r.m.s. value) of the acoustic pressure. The dynamic range of the ear (at 1000 Hz) is then $\Delta L_{ear} = L_{p,SG} - L_{p,HS} = 120$ dB. At the frequency of 1000 Hz the acoustic pressure level in dB is the same as the *acoustic intensity level* in phon. The dependence of the acoustic intensity on frequency and acoustic pressure level is represented by curves of equal intensity (**Fig. 1**).

The weighting curves for acoustic pressure level sensors (**Fig. 2**) are the mirror-image-like approximations for selected curves of equal intensity. With their help it is

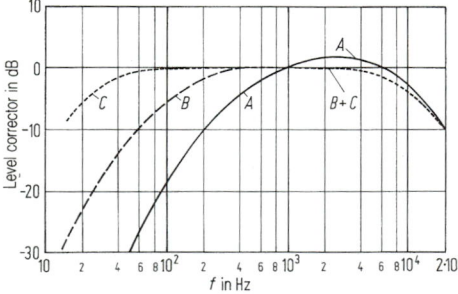

Figure 2. Weighting curves for acoustic level meters. The curves, with their difference relative to zero, state to what degree the acoustic pressure level at a given frequency is evaluated at a lower or higher value than would have been the case without weighting.

possible to match the sensitivity of a sound-level sensor to the natural sound sensitivity of the human ear. In practice it is usual to carry out the weighting in accordance with curve A, the acoustic pressure level so weighted being given in dB(A).

The acoustic pressure level measured in the vicinity of a machine is not an entity specific to the machine but depends on the distance of the measuring point from the machine. In the case of machines with pronounced directional properties, it also depends on the location of the measuring point relative to the machine as well as on the size and nature of the room (wall reflection).

Acoustic Power Level

In contrast to acoustic pressure level, acoustic power level L_p is an entity specific to the machine. It is given by

$$L_p = 10 \lg (P/P_0) \text{ dB}, \quad (2)$$

where it is usual to put $P_0 = 10^{-12}$ W as the value to which the acoustic power P is referred. The determination of the acoustic power level L_p of a machine is always based on measurements of the acoustic pressure level L_p, because $P \sim \tilde{p}^2$ (DIN 45635).

Structure-Borne Noise, Velocity, Velocity Level

In solids, in contrast to gases and liquids, compressive stresses occur in addition to tensile stresses. Therefore, in the case of *structure-borne noise*, transverse waves moving parallel to the shear strain appear in addition to longitudinal or compression waves. Various other types of wave are composed of these two, of which flexural waves are the most important. These cause the largest displacement normal to the surface of a body, e.g. a plate, and in general make the strongest contribution to machine acoustic radiation, because only these normal oscillations can be transferred to the surrounding medium. They are excited not only by the transverse forces but also by bending moments. The most important measurable variable for structure-borne noise is the velocity of oscillation or the structure-borne noise *velocity* $v(t)$ normal to the radiating surface of a noise generator. Calculation of the velocity $v(t)$ from the acceleration $a(t)$, obtained from measurements using piezoelectric sensors, is carried out using the relation

$$a(t) = \frac{d}{dt} v(t), \quad (3)$$

so that

$$\tilde{v}(f) = \tilde{a}(f)/(2\pi f) \quad (4)$$

applies for the r.m.s. values in a frequency band that is not too wide (e.g. third octave band) and whose centre frequency is f. The velocity too is usually stated as a *velocity level* L_v, given by

$$L_v = 10 \lg (\tilde{v}^2/\tilde{v}_0^2) \text{ dB} = 20 \lg (\tilde{v}/\tilde{v}_0) \text{ dB}. \quad (5)$$

The reference value $\tilde{v}_0$ has been chosen as $5 \cdot 10^{-8}$ m/s.

Radiation Coefficient

This relates the radiated power P to the mean-square velocity $\overline{v^2}$ averaged over the total radiating surface S and is defined as a non-dimensional factor [2],

$$\sigma(f) = \frac{P(f)}{\rho_L c_L S \overline{v^2}(f)}, \quad (6)$$

where ρ_L is the density of air and c_L the velocity of sound in air. The mean-square velocity $\overline{v^2}$ is determined from

$$\overline{v^2}(f) = (1/S) \sum_{i=1}^{N} S_i \bar{v}_i^2(f) \approx (1/N) \sum_{i=1}^{N} \bar{v}_i^2(f), \quad (7)$$

where $\bar{v}_i$ is the velocity of the ith of a total of N (equally large) area elements S_i, or the velocity at the ith measuring point on $S = \Sigma \, S_i$.

Equation (6) is at the same time the definition of and the rule for measurement of the radiation coefficient.

Receptance, Impedance, Structure-Borne Noise Factor

The (structure-borne noise) *receptance* is the ratio of generated velocity v to exciting force F and is, for instance, termed input admittance $b_E = v_e(x_0)/F$ when v_e is the velocity at the point of excitation x_0, or transmission admittance $b_Ü = v(x)/F$, when v is measured at any other required point x. The inverse of the receptance b is called impedance $Z = 1/b = F/v$. The *mean* square of the transmission receptance $b_{Ü,m}^2$ is important in radiation and is determined from the mean-square velocity $\overline{v^2}$ (Eq. (7)) of the entire surface from

$$b_{Ü,m}^2(f) = \overline{v^2}(f)/\bar{F}^2(f,x_0). \quad (8)$$

The above expression multiplied by the radiating area S, $Sb_{Ü,m}^2$ is called the *structural noise factor* [8]. Investigation of structure-borne noise also requires the measurement of forces. In view of the range of acoustic frequencies, piezoelectric transducers are preferred for this purpose. The *force level* L_F is defined as

$$L_F = 10 \lg \, (\bar{F}^2/\bar{F}_0^2) \, \text{dB} = 20 \lg \, (\bar{F}/\bar{F}_0) \, \text{dB}, \quad (9)$$

where $\bar{F}_0 = 1$ N.

Equation (8) may then be rewritten in a way that is simpler to use from a practical viewpoint:

$$10 \lg \, (b_{Ü,m}^2/b_{Ü0}^2) \, \text{dB} = \overline{L}_v - L_F.$$

Spectra

The radiation coefficient and receptance are frequency-dependent properties of a machine structure. Their determination as shown by Eqs (6) and (8) also requires that the force, velocity and acoustic power be ascertained as a function of frequency. This *frequency analysis* is carried out by harmonic analysers which measure the r.m.s. value of the time-variant signals within a fixed or a variable frequency interval. The most frequently used types are third octave analysers and narrow-band sinusoidal tone or FFT analysers. They yield the amplitudes of third-octave or narrow-band spectra. The phase information lost in the case of analysers using filters normally plays no part in noise measurements.

Sound generation processes in machines run periodically with speed, so that velocity and acoustic pressure are also periodic. Periodic processes possess a *line spectrum*. This means that apart from the fundamental frequency f_0 = frequency of rotation in s^{-1}, whole number multiples (harmonics) nf_0 also appear. The amplitudes of these spectral lines for $f = nf_0$ are given by the (continuous) *pulse spectrum*, which is determined by the time dependence (pulse form) of the corresponding measured quantity over one period. For the mathematical relation between these two spectra and the resulting measurement technique see [3].

It is frequently the case that several working processes with mutual phase displacement take place during one revolution of the shaft, e.g. in gear trains (meshing of teeth) or in hydraulic pumps. The fundamental frequency f_0 is then given by frequency of rotation multiplied by number of teeth, or frequency of rotation by number of cylinders, and the line spectrum of such machines contains the corresponding harmonics nf_0. It happens only too frequently that this main spectral line is found in measured spectra surrounded by a series of subsidiary lines, whose interval in each case is $\Delta f = f_0$. These occur when the time dependencies of the sound-generating processes per working cycle are not exactly equal (amplitude modulation) and/or the frequency of rotation fluctuates (frequency modulation). Apart from this, measured spectra often have a background of a continuous, more or less wide-band *noise spectrum*. This is caused by stochastic processes, e.g. frictional forces.

Shielding, Damping

Damping is the transformation of acoustic energy into heat. Shielding of airborne noise is achieved by means of porous or fibrous absorption materials with a high coefficient of absorption α. Damping of structure-borne noise is achieved by means of sound deadening layers or composite metal sheets with a high loss factor η.

Shielding is the prevention of sound propagation. In the case of airborne noise, this is achieved by sound insulating walls (e.g. casings), in the case of structure-borne noise, by step changes in impedance (e.g. decoupling by rubber or spring elements). The *transmission loss* ΔL_D is defined as the difference in level between locations in front of and behind a "shielding barrier", i.e. by

$$\Delta L_D = L_{before} - L_{behind}.$$

The *insertion loss* ΔL_E by contrast describes the effect of the sound insulating method and is defined as the difference in level evaluated at a position on the far side of a barrier in relation to the point of sound generation with and without the sound insulating measure,

$$\Delta L_E = L_{without} - L_{with},$$

e.g. the level difference of a machine with and without casing. It is always true that $0 \leq \Delta L_E < \Delta L_D$. In order to make $\Delta L_E > 0$, energy must be destroyed by damping on the sound generation side of the sound barrier. This is required by the law of conservation of energy: no barrier effect without damping.

3.2 The Generation of Machine Noise

Airborne Noise

This is caused by machine processes, which in turn cause fluctuations in the air pressure and hence airborne noise. Such noise is generated [4] by the following:

Aeropulsive Sources. These displace the surrounding air in a pulsating manner (examples are induction and exhaust noise).

Aerodynamic Sources. These occur where sound is generated as a consequence of interrupted air flow. This manifests itself in a locally varying velocity profile (e.g. zones downstream of cascades, jet zones behind nozzles), which causes pressure fluctuations near objects moving at right angles to the flow, and these are propagated as airborne noise (e.g. fans, perforated disc sirens) or in an eddy zone which forms in the wake of an obstruction in the flow and directly generates sound (e.g. flow noise).

Thermodynamic Sources. Here the energy supply resulting from exothermic processes causes the expansion of numerous small-volume elements of gas with a corresponding buildup of pressure (e.g. noise from explosions, or electric discharges).

Structural Noise Generation

In most machines, the greatest contribution is made by the initial excitation of noise in the elastic structure of the machine, which is then transferred from the external surfaces to the surrounding air and radiated as airborne noise. Structure-borne noise can be excited in two ways [4], as follows.

Force-generated noise is caused by changing motive forces which elastically deform the stressed parts, thereby exciting mechanical oscillations in the machine structure. This forced excitation does not occur exclusively at the point of action of the force, e.g. at the pressure face of a cylinder, but also encompasses other parts that are included in the *transmission path of the force*.

Velocity-generated noise, on the other hand, occurs in those parts of the machine structure that are not themselves stressed by the transmission of changing motive forces but are fixed to vibrating parts excited by these forces. They undergo base excitation at the points or edge of attachment where the vibration displacement or the vibration velocity is imposed upon them. This, as a rule, is the case with oil pans, protective panels and other mounted and relatively light components.

Noise-Generating Forces

These occur in numerous forms in mechanical engineering. The most important forms are: alternating pressure forces as in i.c. engines, compressors, hydraulic machines and piping systems; inertial forces such as out-of-balance forces in rotating machine parts, or inertial forces in reciprocating components (coupling links in cam drives, slides in working machines); inertial and spring forces in cam gears; alternating forces caused by discontinuous force transfer such as the tooth-transmitted forces in gearing or the support forces in roller bearings; impulsive and impact forces on impact of machine parts as, for instance, when clearances are exceeded during roller bounce on cams, or of valves bouncing on valve seats; magnetic forces in electric motors, generators and transformers.

Furthermore, all those forces give rise to noise which occurs during the processing of workpieces: separating forces in metal-removing production processes (milling, turning, drilling, grinding, sawing), as well as those occurring during cutting and stamping; deformation forces in material-forming manufacturing processes (forging, riveting, hammering, pressing).

These working forces $F(t)$ can be unequivocally resolved into an amplitude and a phase spectrum by Fourier analysis. The amplitude spectrum shows the exciting amplitudes as a function of the frequency f, and is called the *excitation spectrum* $F(f)$. For the assessment of changing forces see [23].

The Behaviour of Body Noise in Structures

Structural noise amplitudes are proportional to the excitation amplitudes, but their ratio is not the same for all frequencies. For instance, amplified structure-borne vibrations occur in frequency ranges where the machine structure has natural frequencies. A measure of the *frequency behaviour body noise* of a structure when excited by forces is the structural noise factor $Sh_{0,m}^2(f)$ of Eq. (8).

Closed numerical solutions for the structural noise factor are obtainable only in those cases where it is possible

to develop the solutions of the (inhomogeneous) differential equations of vibration (e.g. the differential equation for flexural waves; see A4.2.4) in terms of (orthogonal) eigenfunctions of the oscillatory system (e.g. a plate) (theorem of series expansion). Complex structures, including three-dimensional structures, can be dealt with by the finite-element method (FEM), but the calculation of entire frequency plots requires a great deal of computer time. A first formulation for an estimate of the structural noise factor of plates, based on the theory of structural noise and treated in detail in [2], can be found in [3]. An improved estimate with regard to frequency dependence is investigated in [5]. According to this, the structural noise factor below the first natural frequency of a plate structure $(f < f_1,$ quasi-static region) increases at $+ 20$ dB/decade, whereas in the neighbourhood of the natural frequency $(f > f_1)$ it decreases on average at $- 10$ dB per decade (**Fig. 3**). The expected value of the structural noise factor (independent of the point of excitation x_0) averaged over frequency, for $f < f_1$, is given according to [8] by

$$\overline{\langle Sh_{0,q}^2 \rangle} = \frac{1}{16\pi} \frac{1}{1 + \eta^2 \frac{f^2}{f_1^2}} \frac{1}{m'' \sqrt{m''B''}} \sim 1/B''^2 \sim 1/h^6,$$

and for $f > f_1$ by

$$\overline{\langle Sh_{0,e}^2 \rangle} = \frac{1}{16\pi f \, \eta m'' \sqrt{m''B''}} \sim 1/(m'' \sqrt{m''B''}) \sim 1/h^3,$$

where $m'' = \rho h$ is the mass, $B'' \approx Eh^3/12$ is the stiffness per unit area, ρ is the density, E is the modulus of elasticity, η is the loss factor and h is the plate thickness. The application to ribbed plates and box-shaped machine casings is treated in [6].

Radiation of Airborne Noise

The structural vibrations normal to a radiating surface are not transformed into airborne vibrations equally efficiently at all frequencies. A measure for the *radiation behaviour* of a radiator is the radiation coefficient $\sigma(f)$ of Eq. (6).

Even radiators whose surfaces vibrate entirely in phase, as for instance the zeroth-order spherical radiator, or the piston membrane, possess a frequency-dependent radiation coefficient. For the above-mentioned spherical radiator this is given [7] by

$$\sigma_0 = \frac{f^2/f_0^2}{1 + f^2/f_0^2} \approx \begin{cases} f^2/f_0^2, & \text{for } f < f_0, \\ 1, & \text{for } f > f_0, \end{cases} \tag{10}$$

where $f_0 = c_L/(2\pi R)$ is the cutoff frequency of the spherical radiator and R the radius of the sphere. The radiation coefficient σ_{KM} of the piston membrane is described quite well by the two asymptotic approximations of Eq. (10) if R and f_0 of the equivalent spherical

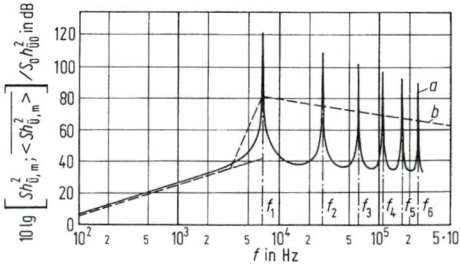

Figure 3. Structural noise factor of a circular plate excited at its centre ($R = 60$ mm, $h = 10$ mm, $\eta = 10^{-4}$ [5]): *a* exact plot of $Sh_{0,m}^2(f)$, *b* smoothed plot (expected value) $\overline{\langle Sh_{0,m}^2 \rangle}$.

radiator (equal volume flow) are determined. If S_{KM} is the area of the piston membrane then $R = (S_{KM}/2\pi)^{1/2}$ applies.

According to [8], the radiation coefficient σ_P of a *plate* which has been excited to flexural oscillation attains at most the radiation coefficient σ_{KM} of its equivalent piston membrane below the limiting frequency f_g but may be reduced below σ_{KM} between the transmission frequency $f_0 \approx f_0^2/f_g$ and the limiting frequency f_g (short-circuit region) as a result of the interaction with regions of the plate vibrating out of phase (**Fig. 4**).

The reason for this is that the propagation velocity c_B of flexural vibrations is frequency dependent (dispersion): whereas c_L = constant and hence $\lambda = c_L/f \sim 1/f$, $c_B \sim \sqrt{f}$ and thus $\lambda_B = c_B/f \sim 1/\sqrt{f}$ (**Fig. 5**). The frequency at which $\lambda_L = \lambda_B$ is called the (flexural wave) *limiting frequency*, f_g. As soon as $\lambda_L > \lambda_B$, i.e. $f < f_g$, acoustic coupling occurs between the wave peaks and the wave troughs and this brings about a more or less pronounced hydrodynamic short circuit and correspondingly reduced radiation. For this case, the actual radiation coefficient for the plate is given according to [8] by

$$\sigma_P = \min (\sigma_{KM}, \sigma'_P)$$

$$\sigma'_P \approx f_0 f^{1/2}/f_g^{3/2} \quad \text{for } f < f_g/2 \qquad (11)$$

$$f_g = \frac{c_L^2}{2\pi} \sqrt{\frac{m''}{B''}} \sim \sqrt{\frac{\rho b}{E b^3}} \cdot \frac{1}{b}. \qquad (11a)$$

σ_{KM} only depends on the size of the plate via f_0, whereas the radiation coefficient σ_P in the short-circuit region also depends on the plate thickness, because

$$\sigma'_P \sim 1/f_g^{3/2} \sim (B''/m'')^{3/4} \sim b^{3/2}.$$

The radiation coefficient of (box-shaped) *machine casings* can be calculated from the radiation coefficients of the plates which form the *parts* of the casing [8]. It is given by the mean radiation coefficient $\bar{\sigma}_P$ of the component surfaces, which are to be regarded as independent with respect to radiation as long as λ_L is less than half the perimeter of the box (when the component plates are decoupled). When λ_L exceeds the perimeter of the housing, the component surfaces are coupled. The radiation coefficient of a box can then be estimated in accordance with [8] as

$$\sigma'_0 = 1.13 \frac{f^{5/2}}{c_L} \overline{U/f_g^{3/2}},$$

where U is the perimeter and f_g the limiting frequency of the individual component plates from which the mean value $\overline{U/f_g^{3/2}}$ of all the

plates has been calculated. The radiation coefficient of a box-shaped casing is then given by

$$\sigma_K = \min (\sigma'_0, \overline{\sigma}_P, 1).$$

Basic Equations for Machine Acoustics

The basic equation of machine acoustics can be derived using the definitions given in J3.1. The radiated acoustic power in a given frequency range with centre frequency f, when excited by a force, is

$$P(f) = \rho_L c_L \, \sigma(f) S b_{0,m}^2 (f) \, \tilde{F}^2(f). \qquad (12)$$

Here ρ_L is the density and c_L the velocity of noise in air, $\sigma(f)$ the coefficient of radiation, $Sb_{0,m}(f)$ the structural noise factor and $\tilde{F}(f)$ the r.m.s. value of the exciting force, which is known from the excitation spectrum. Written in level form, and taking account of weighting curve A which describes the frequency-dependent sensitivity of the ear, Eq. (12) becomes

$$L_{PA}(f) = L_F(f) + [10 \lg (Sb_{0,m}^2(f)/S_0 \, b_{0,o}^2)$$
$$+ 10 \lg \sigma(f)] \text{ dB} + \Delta L_A(f), \qquad (13)$$

where $\tilde{F}_o = 1$ N, $b_{0o} = 5 \cdot 10^{-8}$ m/s N, $S_0 = 1$ m^2 and $\Delta L_A(f)$ is the level correction for the A weighting as in DIN 45633.

The structural noise factor and the coefficient of radiation are thus "*weighting functions*" and, through their filtering action, the airborne noise spectra of the machine noise are derived from the excitation spectra of the exciting forces. Finally, the total level L_{PA} of the radiated acoustic power is obtained if the levels $L_{PA}(f_i)$ occurring at all the exciting frequencies f_i as calculated by Eq. (13) are added logarithmically, i.e. with respect to power, to yield

$$L_{PA} = 10 \lg \left(\sum_i 10^{L_{PA}(f_i)/10 \text{dB}} \right) \text{ dB}. \qquad (14)$$

The total level of several individual noise sources (independent of and decoupled from each other) is also calculated by this law of logarithmic level addition. In order to do so, the acoustics levels $L_{PA,i}$ of the i individual sources are put into the exponent of Eq. (14).

3.3 Methods for Reducing Machine Noise

Survey

Methods for reducing machine noise are subdivided as follows.

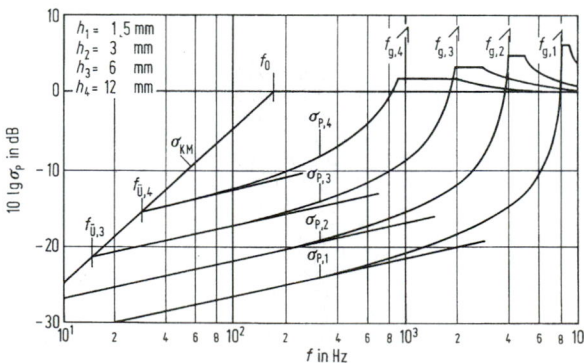

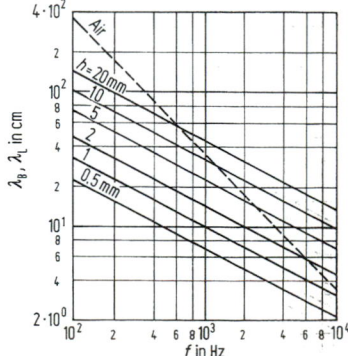

Figure 4. Radiation coefficient σ_P of plates of size 1000 × 700 mm^2 as a function of thickness b [8], with piston radiation region $f < f_0$, short-circuit region $f_0 < f < f_g$, region of full radiation $f > f_g$.

Figure 5. Flexural wavelength λ_B and wavelength in air λ_L as a function of frequency f and plate thickness b, valid for steel and approximately valid for aluminium [2].

Primary Measures. These act directly on the generation of noise, that is to say, are applied to the noise source itself. In the case of airborne noise, this means an intervention in the generation of fluctuating pressures; in the case of structure-borne noise it means a reduction of the exciting forces (exciting spectrum) as well as alteration of the structural noise behaviour and the radiation (admittance and coefficient of radiation).

Secondary Measures. These are intended to reduce the airborne noise after generation and thus act either on the emission of noise from the machine (silencers, casing) or reduce the noise emission, e.g. at the operating point, by space-acoustic means. Secondary measures are meant to block the acoustic energy (barrier action) and then to destroy as large a part of it as possible by absorption (damping).

Rules for low-noise design can be found in [11, 12], practical examples in [4, 13, 14].

Measures Applied to Sources of Airborne Noise

Aeropulsive noise generation due to pressure equalisation processes is reduced, when pressure equalisation, e.g. via a valve, proceeds not in sudden bursts but slowly and steadily.

Aerodynamic noise can be reduced by avoiding interruptions to the air flow. At the inlet of flow machines, care must be taken to see that a constant velocity profile is maintained (no struts, gratings, guide vanes in front of fan rotors, inlet flow kept turbulence-free). If shadow zones cannot be avoided, then parts that move at right angles to the flow (e.g. a rotor), must not be located directly behind the obstruction, because at this point local variations in the velocity profile are at their greatest. For instance, the optimum distance from fan rotors according to [1] is calculated as $S_{ax} = 0.03(u/\text{m s}^{-1})^2$ mm, where u is the peripheral velocity of the rotor.

Silencers

Dissipative silencers, like absorption and relaxation silencers, convert acoustic energy directly into heat. Both are relatively broad-band, generating only a little back pressure. At low frequencies, however, they require relatively large cross-sectional dimensions of the acoustically effective lining and a rather extensive length if large insertion losses are to be achieved.

Impedance silencers, which include all types of resonance and interference silencers, primarily bring about a screening of the sound in that they introduce points with step changes in impedance into the flow channel (step changes in cross-section, coupled resonators, bypass lines) at which the sound waves are reflected. These reflections are associated with a correspondingly high back pressure; they occur, however, only at certain frequencies. Impedance silencers have to be tuned and are effective over relatively small bandwidths.

Further information on silencer function can be found in [15] and on design in [4].

Changing Excitation Spectra

The excitation spectrum above the frequency $f \approx 2/\tau$ due to noise generating motive forces, is determined by the first and second and occasionally also the higher *differential coefficients* of the force–time functions (τ is the duration of the application of the individual force during one

period) [3]. Hence, if the slope dF/dt and the curvature d^2F/dt^2 of the time dependence of the exciting force can be reduced, noise levels from as low a frequency as a few hundred hertz can be decreased without affecting the working principle of the machine. A simple estimating procedure for the excitation spectrum that is developed in [3] and shown graphically in [16–18] makes it possible to ascertain which properties of a force–time function must be altered in order to achieve a reduction in noise.

Noise excitation by *pulse forces* is the smaller [3] the smaller the transmitted pulse $A = mv$ (i.e. the smaller the moving mass m and/or the smaller the relative velocity v between machine parts that impact each other). Reduction of the pulse m_1v_1 to m_2v_2 lowers impulsive excitation by

$$\Delta L = [20 \lg (m_2/m_1) + 20 \lg (v_2/v_1)] \text{ db}, \quad (15)$$

i.e. by -6 dB when halving the mass m or the velocity v, by -12 dB, when both together. In extending the duration of the pulse from $\tau_{s,1}$ to $\tau_{s,2}$ a reduction in the pulse excitation may also be achieved. From $f_{s,1} = 1/\tau_{s,1}$ onwards, this amounts to

$$\Delta L = 20 \lg (\tau_{s,1}/\tau_{s,2}) \text{ dB}, \quad (16)$$

i.e. to -6 dB on doubling τ_s. Longer pulse durations are achieved by means of elastic intermediate layers or by increased compliance at the point of impact. The pulse is then damped by spring action and hence becomes "softer".

It is generally true [3] that the longer the pulse duration τ of an individual force application, the sooner the excitation spectrum flattens off. For this reason, simultaneous excitation (small τ) is avoided by applying the wedge principle (spiral cutter, oblique shearing, clipping punch with ridge cut, helical gearing).

Measures Applied to the Machine Structure

The aim of these measures must be to keep the amplitude of structural noise excitation force as low as possible at radiating surfaces. Power flow due to the motive forces must therefore be confined to a narrow, massive and rigid region and not allowed to reach radiating outer surfaces. The principle of *functional separation* [11] should be applied. Forces must be absorbed in the interior of the machine, and walls with protective and sealing functions must be attached to force-transmitting structures by means that insulate them from structural noise.

At the points where the forces are applied, *input impedance* should be increased by means of mass concentration [13, 19]. At the points of base excitation (velocity-excited structural noise) on the other hand, decoupling by means of rubber or spring elements (*structure-borne sound insulation*) must be applied.

Increasing mass m'' (density of materials, additional masses), stiffness B'' (E modulus, ribbing) as well as the thickness h of the walls of the casing affected by *flexural force* (flexure $B'' \sim Eh^3$) reduces the structural noise factor $S b_{0,m}^2$. At the same time, the radiation coefficient σ_P in the short-circuit range is also reduced by m'' but is increased by B'' and h. The effect on noise generation can therefore be assessed only by considering $S b_{0,m}$ and σ simultaneously (**Fig. 6**). Since the effect evidently differs in magnitude depending on frequency range, the excitation spectrum $F(f)$ must further be weighted by $S b_{0,m}^2$ and σ if possible noise reductions are to be ascertained.

If the exciting force leads primarily to tensile loading, the structural noise behaviour is mainly determined by the tensile stiffness ($\sim E b^1$). In comparing different materials

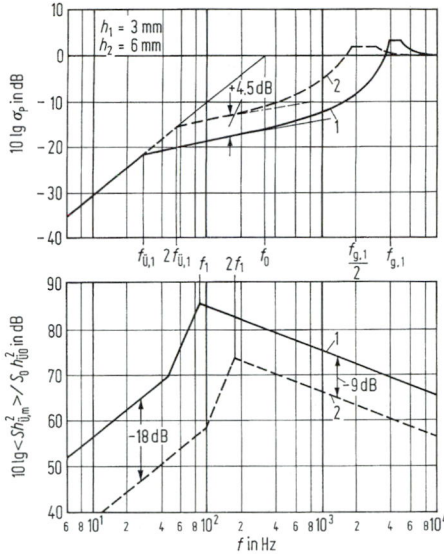

Figure 6. The effect of doubling of the thickness b on the coefficient of radiation σ_p and the structural noise factor $\overline{\langle Sb_{\bar{u},m}^2 \rangle}$ of a steel plate of size 500×350 mm^2, $\eta = 0.1$.

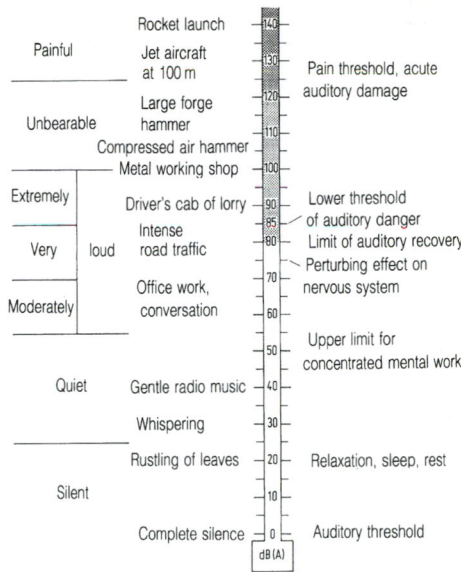

Figure 7. Noise conditions, perception and effect.

$(m'' \sim \rho, B'' \sim E)$, the effect of the dimensioning design rules that apply to the thickness b in order to achieve rigidity must also be considered. Relevant examples can be found in [4, 17, 18, 24, 25], and examples applying to product lines in [24, 26].

Damping of structural noise by sound-insulating layers and composite metal sheets reduces the structural noise factor in the natural frequency range of the structure in accordance with the expression

$$\overline{(Sb_{\bar{u}e}^2)} \sim 1/\eta, \tag{17}$$

where η is the loss factor and acts as a measure of the damping [2]. The reduction in noise through increasing the loss factor η_1 to η_2 is thus on average

$$\Delta L = 10 \lg (\eta_1/\eta_2) \text{ dB},$$

i.e. -3 dB if the damping is doubled. In this case, however, attention must be paid to the fact that the output damping of a structure is, as a rule, not determined by damping in the material used but by frictional damping in mounting and contact regions (screw assemblies, tolerances, bearings, etc.), which is often even greater than the damping in cast iron, and, further, that in relation to noise radiation an optimum loss factor exists in the short-circuit region [4, 17, 18].

Enclosure of Machines

High-powered machinery or numbers of smaller machines often generate noise at dangerous levels (**Fig. 7**). Noise-induced hearing damage is therefore one of the main industrial injuries nowadays [22]. In view of legal requirements, the designer is therefore faced with the task of designing low-noise machines and manufacturing processes. However, primary noise reduction, in the sense of the design possibilities already discussed, is not always sufficient to meet the required limits. In this context, enclosure, the function of which is based on shielding and damping, offers itself as an effective secondary measure.

Shielding. The input attenuation ΔL_E as a measure of the effectiveness of an enclosure (see J3.1) depends on the (mean) sound attenuation factor R of the wall (transmission attenuation, property of the wall), on the (mean) absorption α as a measure of the (necessary) losses in the enclosure (absorbent lining, internal losses of the walls excited to structure-borne sound, absorption at high frequencies by the enclosed volume of air) and on the aperture ratio $q = S_O/(S_O + S_W)$, which is able to decrease considerably the possible effectiveness of the enclosure (keyhole effect). S_O is the entire unavoidable aperture area (shaft penetration, piping systems etc., transportation openings) whose transmission attenuation factor D can be matched to the wall attenuation factor R by means of silencers placed upstream. S_W is the remaining wall area of the enclosure. [20] gives ΔL_E as

$$\Delta L_E = \{-10 \lg [(1-q) \cdot 10^{-R/10 \text{ dB}}$$
$$+ q \cdot 10^{-D/10 \text{ dB}}] + 10 \lg [\alpha \cdot (1-q) + q]\} \text{ dB}. \tag{18}$$

Since the wall attenuation factor R and the coefficient of absorption α (and in the case of silencers also D) are frequency-dependent, the input attenuation is frequency-dependent: $\Delta L_E = \Delta \tilde{L}_E(f)$. For this reason, the machine structural power spectrum must always be weighted using the "filter curve" $\Delta L_E(f)$ of the enclosure in order to determine overall level reduction. The assessment of a (frequency-)averaged attenuation value for R and also for ΔL_E is not a measure of the attainable attenuation noise reduction and is often too high.

Below its limiting frequency $f_g \approx 12\,000$ Hz/(b/mm) the wall attenuation factor R for enclosure walls made from sheet steel of thickness b and shortest edge length l lies [4, 20] between $\min(R_{min}, R_{max})$ and R_{max}, where

$$R_{min} = [15 \lg (f/\text{Hz}) + 5 \lg (b/\text{mm})$$
$$+ 10 \lg (l/\text{m}) - 15] \text{ dB}, \tag{19}$$

$$R_{max} = [20 \lg (f/\text{Hz}) + 20 \lg (b/\text{mm})$$
$$- 27.5] \text{ dB}. \tag{20}$$

With regard to α, it should be noted that sheet metal enclosures without absorbent lining possess absorption coefficients $\alpha \approx 10^{-2}$ that are practically independent of frequency and therefore, for $b = 1$ mm, have a considerable insertion loss $\Delta L_E(f) > 10$ dB (see [20]). In the case of absorbent lining, the following applies [20]:

$$\alpha \approx \begin{cases} 0.85 \text{ to } 0.95 & \text{for } f > f_d \\ \sqrt{f/f_d} & \text{for } f < f_d/2, \end{cases} \tag{21}$$

where f_d is given by $f_d = 5460$ Hz cm/d. If the largest possible value for α above a certain frequency f' is required, and with it the smallest possible loss of shielding action, then $f' = f_d$ should be substituted and the required thickness d of the absorption layer, and hence the specific flow impedance, calculated. This is given by

$$\Xi_{opt} = (80 \text{ to } 240 \text{ g s}^{-1} \text{ cm}^{-2})/d. \tag{22}$$

A suitable absorption material can then be chosen.

Soundproof Boxes. These are of the usual form of machine enclosure (**Fig. 8**).

Lining. This is provided by sheet metal, 1 to 1.5 mm thick, with an absorption layer 10 to 15 mm thick. Its sound attenuation is 6 to 12 dB from 125 to 8000 Hz. Material data are: panel weight 0.5 to 0.7 N/m^2, temperature range -25 to $+80$ °C and thermal conductivity 0.4 to 0.6 W/(m K). The surfaces must be resistant to aggressive media like petroleum, mineral oils, hydrochloric acid, caustic soda, and solvents.

Construction. The boxes should be as light as possible and as simple as possible to assemble. They should be watertight outside and inside and satisfy local fire regulations. They must be equipped with observation windows and be easily accessible at important points. To dissipate any heat generated by the machine, it is frequently necessary to provide an acoustically damped air-conditioning plant. Further, attention is required at the entry points of the supply lines for the operating media and control elements. They must be implemented in flexible form for machines with intense heat radiation or those mounted on resilient supports. They must be acoustically tight, since the acoustic level increases sharply in the interior of the enclosure.

Directions for the construction of enclosures and attenuating feedthroughs, as well as working diagrams for their design, can be found in [4, 21].

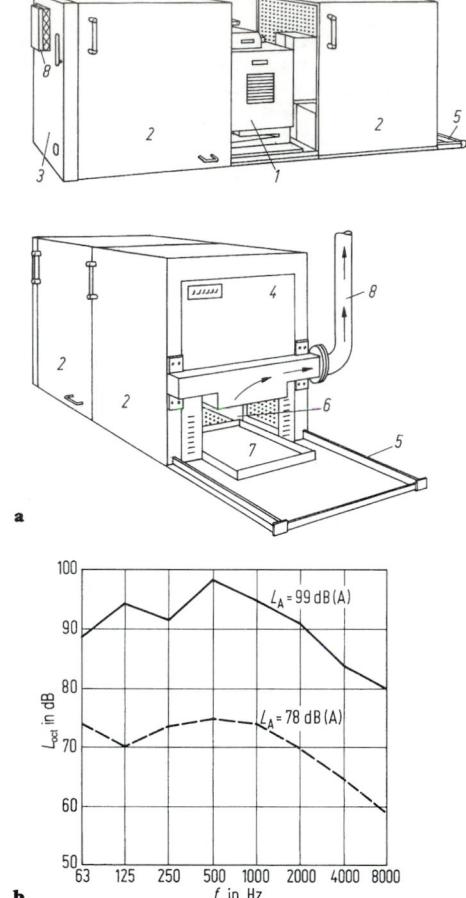

Figure 8. Single-casing movable enclosure for a high-speed wire bolt press: **a** construction (*1* high-speed wire bolt press, *2* movable part of enclosure in tunnel form, *3* fixed rear wall, *4* fixed front wall, *5* guide frame, *6* bolt delivery opening (implemented as a sound-damping channel), *7* bolt-catching receptacle, *8* outlet or exhaust air duct for heat dissipation and oil disposal with forced ventilation), **b** acoustic octave level L_{oct} as a function of frequency f: full line, without enclosure; dashed line, with enclosure.

4 References

J1 Crank Operation, Forces and Moments of Inertia [1] Haug K. Die Drehschwingungen in Kolbenmaschinen. Springer, Berlin, 1952. – [2] Krämer E. Maschinendynamik. Springer, Berlin, 1984. – [3] Maas H. Gestaltung und Hauptabmessungen der Verbrennungskraftmaschine. In List H. Die Verbrennungskraftmaschine, vol 1. Springer, Vienna, 1979. – [4] Woschni G. Thermodynamische Auswertung von Indikatordiagrammen electronisch gerechnet, MTZ 25/7, 1964; 284–9. – [5] Küttner KH. Kolbenmaschinen, 5th edn. Teubner, Stuttgart, 1984. – [6] Maas H, Klier H. Kräfte, Momente und deren Ausgleich in der Verbrennungskraftmaschine. In List H. Die Verbrennungskraftmaschine, vol 2. Springer, Vienna, 1981. – [7] Hasselgruber H.

Massnahmen zur Verbesserung der Laufruhe von Verbrennungskraftmaschinen insbesondere von Schleppermotoren. Landtechnik 1965; 15: no. 1. – [8] Schmidt F. Schwungräder für Grossdieselmotoren. VDI-Z 1930; 74: 230. – [9] Sass F. Bau und Betrieb von Dieselmaschinen, vol 2. Springer, Berlin, 1957. – [10] Fröhlich F. Kolbenverdichter. Springer, Berlin, 1961. – [11] Haffner KE, Mass H. Theorie der Triebwerkschwingungen in der Verbrennungskraftmaschine, vol 3. Springer, Vienna, 1984. – [12] Haffner KE, Mass H. Torsionsschwingungen in der Verbrennungskraftmaschine. In List, H. Die Verbrennungskraftmaschine, vol 4. Springer, Vienna, 1985. – [13] Krämer O, Jungbluth G. Bau und Berechnung von Verbrennungsmotoren, 5th edn. Springer, Berlin, 1983. –

[14] Mickel E, Sommer P, Wiegand H. Berechnung und Gestaltung der Triebwerke schnellaufender Kolbenkraftmaschinen. Konstruksionsbücher, vol 6. Springer, Berlin, 1942. - [15] Köhler G, Rögnitz H. Maschinenteile, pt 2, 7th edn. Teubner, Stuttgart, 1986. - [16] Schrön H. Die Dynamik der Verbrennungskraftmaschine, 2nd edn. Springer, Vienna, 1947. - [17] Waas H. Federnde Lagerung von Kolbenmaschinen. VDI-Z 1937; 26: June. - [18] Lang G. Zur elastischen Lagerung von Maschinen durch Gummifederelemente. MTZ 24/17. 1963.

J2 Vibrations. [1] Krämer E. Maschinendynamik. Springer, Berlin, 1984. - [2] Gasch R, Knothe K. Strukturdynamik, vol 1. Springer, Berlin, 1987. - [3] Holzweissig F, Dresig H. Lehrbuch der Maschinendynamik. Springer, Vienna, 1979. - [4] Schiehlen W. Technische Dynamik. Teubner, Stuttgart, 1986. - [5] Magnus K. Schwingungen. Teubner, Stuttgart, 1976. - [6] Kellenberger W. Elastisches Wuchten. Springer, Berlin, 1987. - [7] Federn K. Auswuchttechnik. Springer, Berlin, 1977. - [8] Maas H, Klier H. Die Verbrennungskraftmaschine. vol 2, Kräfte, Momente und deren Ausgleich in der Verbrennungskraftmaschine. Springer, Vienna, 1981. - [9] Kuhlmann P. Schwingungen in Kolbenmaschinen. VDI-Bildungswerk, Schwingungen beim Betrieb von Maschinen BW 32.11.07, VDI-Gesellschaft Konstruktion und Entwicklung, 1980. - [10] Schwibinger P. Torsionsschwingungen von Turbogruppen und ihre Kopplung mit den Biegeschwingungen bei Getriebemaschinen. Fortschrittber. VDI, Düsseldorf, 1987. - [11] Grgic A. Torsionsschwingungsberechnungen für Antriebe mit elektrisch drehzahlgeregelten Wechselstrom-Motoren. VDI-Ber 603, 1986. - [12] Natke HG. Einführung in die Theorie und Praxis der Zeitreihen- und Modalanalyse. Vieweg, Brunswick, 1983. - [13] Ewins DJ. Modal testing: theory and practice. Research Studies Press, Taunton, 1984. - [14] Peeken H, Troeder C, Diekhans G. Beanspruchung elastischer Kupplungen in Antriebssystemen mit Asynchronmotoren. Antriebstechnik 1979; 18. - [15] Gasch R, Pfützner H. Rotordynamik. Springer, Berlin, 1975. - [16] Diewald W. Das Biegeschwingungsverhalten von Kreiselpumpen unter Berücksichtigung der Koppelwirkungen mit dem Fluid. Fortschrittber. VDI, Düsseldorf, 1989. - [17] Dietzen FJ. Bestimmung der dynamischen Koeffizienten von Dichtspalten mit Finite-Differenzen-Verfahren. Fortschrittber. VDI, Düsseldorf, 1988. - [18] Glienicke J. Feder- und Dämpfungskonstanten von Gleitlagern für Turbomaschinen und deren Einfluss auf das Schwingungsverhalten eines einfachen Rotors. Diss., University of Karlsruhe, 1966.

J3 Machine Acoustics. [1] Schmidt H. Schalltechnisches Taschenbuch, 2nd edn. VDI, Düsseldorf, 1976. - [2] Cremer L, Heckl M. Körperschall - Physikalische Grundlagen und Technische Anwendungen. Springer, Berlin, 1967. - [3] Föller D. Untersuchung der Anregung von Körperschall in Maschinen und der Möglichkeiten für eine primäre Lärmbekämpfung. Diss., TH Darmstadt, 1972 or Forschungshefte Forschungskuratorium Maschinenbau eV, part 15. Maschinenbau, Frankfurt-on-Main, 1972. - [4] Föller D. et al. Geräuscharme Maschinenteile - Die Entstehung von Maschinengeräuschen und konstruktive Massnahmen zu ihrer Verminderung. Forschungshefte Forschungskuratorium Maschinenbau eV, part 26. Maschinenbau, Frankfurt-on-Main, 1974. - [5] Kassing W. Untersu-

chungen zum Schwingungs- und Körperschallverhalten rotationssymmetrischer Maschinenstrukturen und übertragung der Ergebnisse auf die Geräuschentwicklung von Axialkolbeneinheiten. Diss., TH Darmstadt, 1975 or Forschungshefte Forschungskuratorium Maschinenbau eV, part 42. Maschinenbau, Frankfurt-on-Main, 1976. - [6] Welp EG. Untersuchung des Körperschallverhaltens von Platten- und Kastenstrukturen mit der Methode der Finiten Elemente. Diss., TH Darmstadt, 1977 or Forschungshefte Forschungskuratorium Maschinenbau eV, part 70. Maschinenbau, Frankfurt-on-Main, 1978. - [7] Skudrzyk E. Die Grundlagen der Akustik. Springer, Vienna, 1954. - [8] Föller D. Die Geräuschabstrahlung von Platten und kastenförmigen Maschinengehäusen. Forschungshefte Forschungskuratorium Maschinenbau eV, part 78. Maschinenbau, Frankfurt-on-Main, 1979. - [9] Maidanik G. Response of ribbed panels to reverberant acoustic fields. J Acoust Soc Amer 1962; 34: 809-26. - [10] Sennheiser J. Ein Modell zur Bestimmung der Schallabstrahlung von Platten unterhalb der Grenzfrequenz. Acustica 1975; 32: 244-54. - [11] Müller HW, Föller D. Regeln für lärmarme Konstruktionen. Konstruktion 1976; 28: 333-9. - [12] VDI 3720. Sheet 1: Lärmarm Konstruieren. Allgemeine Grundlagen, 1980; Sheet 2: Beispielsammlung, 1982. - [13] Schmidt K-P. Lärmarm Konstruieren - Beispiele für die Praxis. Forschungsbericht no. 129 der Bundesanstalt für Arbeitsschutz und Unfallforschung. Wirtschaftsverlag Nordwest, Wilhelmshaven, 1974. - [14] Heckl M. Lärmarm Konstruieren - Bestandsaufnahme bekannter Massnahmen. Forschungsbericht no. 135 der Bundesanstalt für Arbeitsschutz und Unfallforschung. Wirtschaftsverlag Nordwest, Wilhelmshaven, 1975. - [15] VDI 2567: Schallschutz durch Schalldämpfer, 1971. - [16] Föller D. Ein Verfahren zur quantitativen Beurteilung der Geräuschanregung in Maschinen. Tagungsbericht Akustik und Schwingungstechnik, Stuttgart, 1972, pp 418-21. VDE, Berlin, 1972. - [17] Föller D. Maschinenakustische Probleme in neuerer Sicht. Tagungsbericht DAGA '73, Aachen, pp 57-75. VDI, Düsseldorf, 1973. - [18] Föller D. Maschinenakustische Berechnungsgrundlagen für den Konstrukteur. VDI-Ber 1975; 239: 55-65. - [19] Schroeder P-J. Konstruktive Lärmminderungsmassnahmen an einer Doppelständer-Exzenterschmiededepresse. VDI-Ber 1977; 278: 135-45. - [20] Fecher F. Abschätzung der Lärmminderung mittels raumakustischer Massnahmen und Kapseln. Konstruktion 1976; 28: 341-6. - [21] VDI 2711: Schallschutz durch Kapselung, 1978. - [22] Connert W. Lärmminderung - eine aktuelle Gemeinschaftsaufgabe. VDI-Ber 1977; 278. - [23] VDI 3720, Sheet 7 (Developed): Lärmarm Konstruieren - Beurteilung von Wechselkräften bei der Schallentstehung, 1989. - [24] Storm R. Untersuchung der Einflussgrössen auf das Akustische übertragungsverhalten von Maschinenstrukturen. Diss., TH Darmstadt, 1980 or Forschungshefte Forschungskuratorium Maschinenbau eV, part 84. Maschinenbau, Frankfurt-on-Main, 1980. - [25] Storm R. Möglichkeiten zum geräuscharmen Konstruieren bei krafterregten Maschinenstrukturen in Leichtbauweise. Tagungsbericht DAGA '81, Berlin, pp 337-40. VDE, Berlin, 1981. - [26] Storm R. Zur Abschätzung des akustischen übertragungsmasses von krafterregten Maschinenstrukturen in Baureihen mit geometrischer Ähnlichkeit. Tagungsbericht DAGA '81, Berlin, pp 333-6. VDE, Berlin, 1981.

K Manufacturing Processes

K. Herfurth, Düsseldorf; L. Kiesewetter, Cottbus; J. Ladwig, Stuttgart; G. Mauer, Aachen; W. Reuter, Aachen; G. Seliger, Berlin; K. Siegert, Stuttgart; H. K. Tönshoff, Hanover; G. Spur, Berlin; H.-J. Warnecke, Stuttgart; M. Weck, Aachen

1 Survey of Manufacturing Processes

H.K. Tönshoff, Hanover

1.1 Definition and Criteria

Manufacturing is the production of workpieces of a geometrically defined shape (Kienzle).

Unlike the other production technologies, i.e. process technology (chemical, thermal or mechanical process technology) or energy technology, manufacturing technology produces products distinguished by *material* and *geometric* characteristics.

The selection of a manufacturing process is determined by four basic criteria:

Main Technology. This means the sizes and shapes that can be produced and the materials that can be worked by means of a manufacturing process.

Error Technology. This means the errors in dimension, shape, position and surface that are caused by manufacturing (error geometry). Besides the microgeometric form of a technical surface with its deviations from the mathematically geometric desired shape, manufacturing processes produce physical and chemical changes in peripheral zones [1]. Quality of manufacturing means manufacturing within predetermined error tolerances.

Economic Efficiency. The quantities to be manufactured per unit of time (output), the cost of preparation (preparation costs) and of repeating the order (order repetition costs), the unit costs (directly assignable to an individual item) and the secondary costs (including storage costs) determine typical areas of use of competing manufacturing processes. In this context, the *flexibility* of a manufacturing process (flexibility of volume and flexibility of adaptation) is growing in importance, so that not only are the productivity and capacity utilisation of a manufacturing plant satisfactory but so too are the requirements with regard to the flow time of a product through the plant, the capital tied up in stock, and delivery reliability [2].

Adaptation of Work to People. Manufacturing processes and means should be designed so that people and the environment suffer the least possible nuisance or harm. Pollution limits (noise, vibrations, noxious substances) and safety standards should be complied with.

Each of the four fundamental criteria should be observed to an equal extent.

Production engineering products, modules and components are manufactured in sequences of operations (*manufacturing steps*). Rationalisation to improve economic efficiency and quality therefore not only should be applied to individual operations or manufacturing steps but also should aim at an overall optimum. To achieve this, opportunities for *adaptation, substitution* and/or *integration* (the A-S-I method) may be sought (**Fig. 1**) [3]. Adaptation is the favourable coordination of consecutive processes, e.g. production of unmachined parts by forging, followed by machining. The development of tools and machine tools may cause the substitution of one manufacturing process for another, e.g. replacement of grinding by hard turning. Integration of manufacturing steps shortens the sequence of operations and is often linked to direct cost savings, and certainly with shortened flow times and reduced control expense (indirect costs). Complete machining of components on multi-axis lathes or machining centres are current examples.

1.2 Classification

According to Kienzle [4], today's and tomorrow's many and varied manufacturing processes can be classified into six main groups (**Fig. 2**) according to the criteria "*changing of material cohesion*" (creation, preservation, increase and reduction) and "*changing of material properties*". These main groups are: primary shaping, forming, cutting, joining, coating and changing of material properties. The main groups are subdivided into groups, e.g. *cutting* into severing, machining with geometrically defined cutting edges, machining with geometrically undefined cutting edges, metal removal, dismantling and cleaning. Within the groups, the manufacturing processes themselves are distinguished by subgroups. Index num-

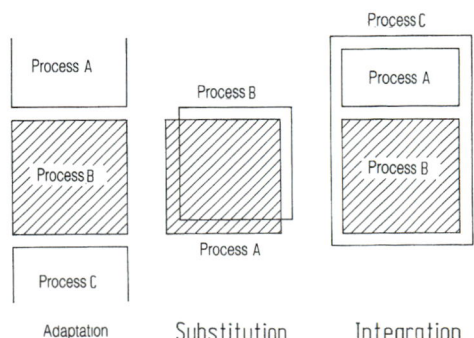

Figure 1. ASI method of rationalisation.

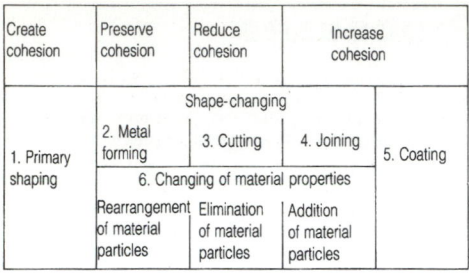

Create cohesion	Preserve cohesion	Reduce cohesion	Increase cohesion	
1. Primary shaping	Shape-changing			5. Coating
	2. Metal forming	3. Cutting	4. Joining	
	6. Changing of material properties			
	Rearrangement of material particles	Elimination of material particles	Addition of material particles	

Figure 2. Classification of manufacturing processes (DIN 8580).

bers are assigned to this system according to the rules of decimal classification.

Standards and Codes DIN 3960: Definitions, parameters and equations for involute cylindrical gears and gear pairs. DIN 3971: Definitions and parameters for bevel gears and bevel gear pairs.
DIN 3975: Terms and definitions for cylindrical worm gears with shaft angle 90°.
VDI code 3333: Hobbing of spur gears with involute profile. VDI-Verlag, Düsseldorf, 1977.
DIN 8580: Classification of manufacturing methods.
DIN 8593: Manufacturing process joining.
Draft VDI code 2861 B1 (9.80): Assembling and handling; characteristics of industrial robots; designation of coordinates.
Draft VDI code 2861 B2 (5.82): Assembling and handling; characteristics of industrial robots; application-related characteristics.

2 Primary Shaping

K. Herfurth, Düsseldorf

2.1 General

According to DIN 8580, primary shaping is the manufacturing of a solid body from an amorphous material by creating cohesion. Thus primary shaping serves to give a component made from a material in *amorphous condition* an initial *form*. Amorphous materials are gases, liquids, powders, fibres, chips, granules, solutions, melts and the like. Primary shaping may be divided into two groups with regard to the form of the products and their further processing:

1. Products produced by primary shaping which will be further processed by forming, severing, cutting and joining. The final product no longer resembles the original product of primary shaping in form and dimensions, i.e. a further material change in form and dimensions is accomplished by means of other main groups of manufacturing processes.
2. Products produced by primary shaping which essentially have the form and dimensions of finished components (e.g. machine parts) or end-products, i.e. their form essentially corresponds to the purpose of the product. The attainment of the desired final form and dimensions usually requires only operations that fall into the main process group "cutting" (machining).

The production of cast parts from metallic materials in the foundry industry (*castings*), from metallic materials in powder metallurgy (*sintered parts*) and from high-polymer materials in the plastics processing industry has major advantages for economic efficiency:

The production of cast parts is the shortest route from the raw material to the finished part. It bypasses the process of forming and all the associated expense. The final form of a finished component with a mass ranging from a few grams to several hundred tonnes is practically achieved in one direct operation.
The production of cast parts by primary shaping from the liquid state allows the greatest freedom of design. This cannot be achieved by any other manufacturing process.

Primary shaping also enables processing of materials which cannot be achieved by means of other manufacturing methods. The direct route from the raw material to the pre-form or the end-product results in a favourable material and energy balance.
The continual further development of primary shaping processes increasingly permits the production of components and end-products with enhanced practical characteristics, i.e. cast parts with lower wall thicknesses, lower machining allowances, narrower dimensional tolerances and improved surface quality.

In the following, primary shaping of metallic materials from the liquid state in foundry technology, of metallic materials from the solid state in powder metallurgy, and of high-polymer materials (plastics) from the plasticised state or from solutions is discussed on a common basis with regard to the fundamental technological principles, the discussion being restricted to subjects relevant to mechanical engineering.

For a better appreciation of the relevant principle of action, many detailed technological operations are omitted that, while they are vital to the specific manufacturing technology, are of minor importance. Furthermore, in discussing the specific primary shaping processes, only products with a simple form are referred to, as the diversity of the possible geometric forms cannot be described here.

Only the most important primary shaping processes are selected, as the large number of technological processes and process variants means that it is impossible to provide anything like a complete description. The processes are selected first according to their technical importance and second according to the principle of action.

Materials technology problems will only be mentioned briefly, although they are vital in order to understand the technological processes, their applicability and efficiency, and the changes in material properties brought about by the technological processes.

Process Principle in Primary Shaping

In the processes of primary shaping, the technological manufacturing process essentially comprises the following steps:

Supply or production of the raw material as an amorphous substance.

Preparation of a material state ready for primary shaping.

Filling of a primary shaping tool with the material in a state ready for primary shaping.

Solidification of the material in the primary shaping tool.

Removal of the product of primary shaping from the primary shaping tool.

These individual steps are discussed in detail below.

Material State Ready for Primary Shaping. In primary shaping of *metallic materials* from the *liquid state*, the raw materials (pig-iron, scrap, ferroalloys and the like) are melted in a metallurgical melting furnace by means of thermal energy. The melting furnaces are usually physically separated from the primary shaping tool. The molten metal is carried by means of transfer vessels (ladles) to the primary shaping tools, termed *moulds* in the foundry industry, and cast there.

In primary shaping of *high-polymer materials* from the *plasticised state*, bulk raw materials (granules, powder) are fed after proportioning into a preparation device which is usually integral with the primary shaping tool. There, thorough mixing, homogenising and plasticising of the material to be processed are accomplished under the action of heat and pressure. When solutions are used, these are produced in a mixing unit and then poured into the primary shaping tool. In primary shaping of metallic and also high-polymer materials from the *solid state*, the bulk raw materials (metal powder, plastic powder or plastic granules) are poured straight into the primary shaping tool, where they sinter, or first become plastic and then solidify under the action of pressure and thermal energy.

Primary Shaping Tools. The primary shaping tool contains a *hollow space* which, with the allowance for contraction, usually corresponds to the form of the product (unmachined part) to be manufactured, but may be smaller or larger than the resulting unmachined part. Furthermore, primary shaping tools often contain systems of channels (runners) for feeding the material in the state ready for primary shaping. The allowance for contraction corresponds to the dimensional changes which occur in the material to be processed from the moment of solidification to its cooling to room temperature.

In the production of cast parts, a distinction is made between primary shaping tools for once-only and for repeated use. Primary shaping tools for *once-only use* are only used for primary shaping of metallic materials from the liquid state in foundry technology; they are termed *expendable or "dead" moulds.* Only one product (casting) can be manufactured, as the mould is subsequently destroyed. However, primary shaping tools for *repeated use (permanent moulds)* are also used in foundry technology. A larger quantity of cast parts can be produced. The primary shaping technologies for processing of high-polymer materials and powder metallurgy use only primary shaping tools for repeated use. Primary shaping tools for repeated use are usually made of metallic, and more rarely of non-metallic, materials. Primary shaping tools for once-only use (dead moulds) are made with the aid of patterns.

Filling the Primary Shaping Tools. Filling of the primary shaping tools with the material ready for primary shaping may be accomplished by means of the following principles of action: under the influence of *gravity, elevated pressure* or *centrifugal force* and by *displacement.* The material to be processed can be put into the primary shaping tools in solid, pourable form (e.g. powder), as molten metal in the case of metallic materials, or in plasti-

cised condition, as a solution or as a paste in the case of high-polymer materials.

Change of State Ready for Primary Shaping into the Solid State of Aggregation. *Liquid metallic materials* (molten metals) change by *crystallisation* to the solid state of aggregation on cooling owing to the removal of heat.

Thermoplastics are cooled in the primary shaping tool after forming. As a result of temperature reduction, which is accomplished either by heat removal in cooled tools or in downstream equipment (cooling baths), the plastic mass passes through the following states: plastic–rubberlike–elastic-solid. In setting by cooling, secondary valency bonds are restored. This process is repeatable; therefore thermoplastics can be restored to the plastic state by reheating.

Thermosetting plastics or *thermosets* (cross-linkable plastics) are cured after forming by a hardening process. Primary valency bonds form, and the plasticised mass solidifies directly under the effect of heat and/or pressure. The curing is an irreversible chemical process: thermosets disintegrate on reheating without needing to pass through a plastic state. Fundamental chemical reactions during solidification are polymerisation, polycondensation and polyaddition.

In primary shaping of *high-polymer* materials, if solutions are used then the transformation to the solid state may be accomplished by the physical process of solvent evaporation.

In primary shaping by *sintering*, a process of shrinkage of the internal and external surface area of a body formed from powder by pressure takes place. Powder particles that are in contact are joined by the formation or reinforcement of bonds (material bridges) and/or by reducing the pore volume; at least one of the material constituents involved remains solid throughout the process. The bonding of the porous pressed body of powder takes place mainly through *diffusion mechanisms*.

In connection with the description of the technological aspects of primary shaping, further details of the processes that occur as a material changes from the state ready for primary shaping to the solid, dimensionally stable state are dispensed with (see D3.1.1 and D3.1.2).

2.2 Shaping of Metals by Casting

2.2.1 Manufacturing of Semi-finished Products

This group of primary shaping processes involves the production of initial and intermediate products which are further processed by, for instance, metal forming (plastic deformation).

Ingot Casting Processes

Here, ingots, slabs, wirebars, etc. are produced in permanent moulds made of metal (usually cast iron). These products are converted by metal forming (rolling, forging, pressing, wire drawing, etc.) into a semi-finished product (sheet, plate, section, wire) that no longer resembles the original ingot in form and dimensions. In ingot casting a distinction is made between *top pouring* (downhill casting, **Fig. 1a**), where the mould is filled by directly pouring the molten metal in from above, and *bottom pouring* (uphill casting, **Fig. 1b**), where one mould or several moulds simultaneously (group casting) are filled from below by means of a distribution system (pouring gate and runner bricks).

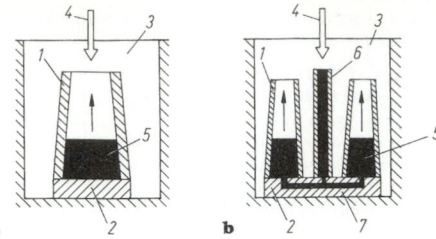

Figure 1. Ingot casting methods: **a** downhill casting, **b** uphill casting; _1_ ingot mould, _2_ baseplate, _3_ casting pit, _4_ molten metal supply, _5_ molten metal, _6_ pouring gate, _7_ runner bricks.

Procedure. The prepared moulds are set up in the casting pit as illustrated. They are filled with the liquid metal, which solidifies in them. The moulds are stripped from the ingots, which are taken away.

Continuous Casting Processes

In these processes, which are used to produce either intermediate products for metal forming or semi-finished products, the primary shaping tool (continuous mould, casting roller, casting belt, casting wheel) is always smaller than the product of primary shaping.

With Continuous Mould. In this casting process, a bath of the molten metal is fed into a stationary continuous mould, where the solidification begins.

Depending on the design, a distinction is made between _batch_ or _continuous vertical_ (**Fig. 2a**) and _horizontal_ continuous casting systems (**Fig. 2b**). On leaving the continuous mould the resulting continuous casting (solid or hollow section) is cooled until it solidifies completely. The continuous casting is usually cut into defined lengths at intervals; like ingots from ingot casting, these are further processed by metal forming.

Travelling Primary Shaping Tools. In this continuous casting process, metal-forming equipment for rolling or drawing is installed directly following the casting plant, thus dispensing with the manufacturing stages of metal forming. In this case there is usually no cutting of the continuous castings into sections.

Strip and Wire Rod Casting Plants

In _vertical uphill casting_ between two casting rollers (**Fig. 3a**) the molten metal is fed from below between two casting rollers. Solidification takes place between these two rollers, and the finished continuous casting (a strip) emerges vertically upward from these rollers.

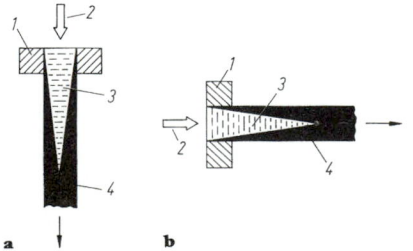

Figure 2. Continuous casting machine [4]: **a** vertical, **b** horizontal; _1_ open-ended mould, _2_ molten metal supply, _3_ molten metal, _4_ solidified billet.

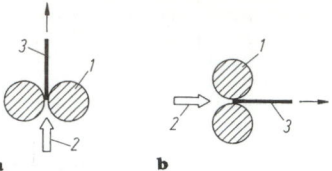

Figure 3. Strip casting machine [4]: **a** vertically uphill, **b** horizontal: 1 pair of casting rollers, 2 molten metal supply, 3 solidified strip.

In _horizontal casting_ (**Fig. 3b**), both the feeding of the molten metal and the discharge of the solidified continuous casting (strip) take place horizontally. In casting between a _casting roller_ or a _casting wheel_ having the profile of the desired strip or rod and an endless casting belt (**Figs 4a** and **4c**), the molten metal solidifies between the casting roller/wheel and the casting belt and emerges into the open air. In casting in _belt moulds_ (two endless, rotating casting belts), solidification is accomplished with the aid of further rotating equipment to restrict the product laterally between these casting belts (**Fig. 4b**); the solidified continuous casting then emerges into the open air as a strip.

2.2.2 Manufacturing of Cast Parts

Manufacturing of cast parts is accomplished with primary shaping processes by means of which a practically finished component, e.g. a machine part or an end-product, is produced without metal forming. The product's shape and dimensions do not undergo any further significant change; however, primary shaping is followed by other manufacturing processes, e.g. cutting (turning, planing, milling, drilling), to obtain a component ready for fitting. The intention is to perfect and further develop the primary shaping techniques in order to, for instance, reduce the amount of machining work to a minimum. **Table 1** gives a survey of the moulding and casting processes.

Use of Expendable Primary Shaping Tools (Moulds)

This technique, which is only used in primary shaping of metallic materials from the liquid state in foundry technology, uses a _pattern_ to produce the expendable primary shaping tool. Depending on the type of pattern used, a

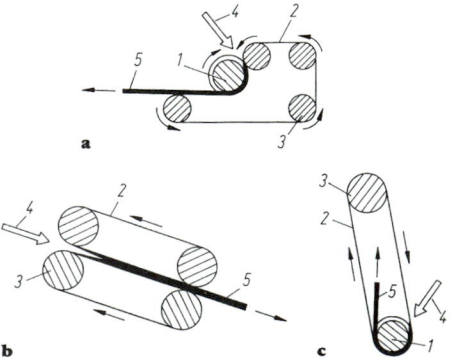

Figure 4. Casting machines [4]: **a** strip casting machine (rotary process), **b** strip casting machine (Hazelett process), **c** wire rod casting machine; _1_ casting wheel, _2_ casting belt, _3_ guide rollers, _4_ molten metal supply, _5_ solidified strip or wire rod.

Table 1. Survey of moulding and casting processes [2]

Type of mould	Expendable moulds						Permanent moulds			
Type of pattern	Permanent patterns				Expendable patterns		No pattern			
Process	Hand moulding	Mechanical moulding	Shell moulding	Ceramic moulding	Precision casting (lost-wax process)	Full mould casting	Pressure die casting	Chill casting	Centrifugal casting	Continuous casting
Materials	All metals	All metals	All metals	All metals	All metals	All metals	Al-, Mg-, Zn-, Cu-, Sn- or Pb-based die casting alloys (iron-based materials under development)	Light metals, special copper alloys, high-grade zinc, lamellar and nodular graphite cast iron	Lamellar and nodular graphite cast iron, cast steel, light metals, copper alloys	Lamellar and nodular graphite cast iron, cast steel, copper and copper alloys, aluminium and aluminium alloys
Weight range (approx. values)	No limit, available transport facilities and melting capacity determine maximum weight	Up to several tonnes, restricted by size of machines	Up to 150 kg	Up to 1000 kg	1 g up to several kg (up to 100 kg in special cases)	No limit (maximum transportable weight); particularly suitable for heavy components	Al alloys: up to 45 kg Zn alloys: up to 20 kg Mg alloys: up to 15 kg Cu alloys: up to 5 kg (limited by size of pressure die casting machine)	Up to 100 kg (more in special cases)	Up to 5000 kg	Up to several tonnes
Quantity range (approx. values)	Single items, small production runs	Small to large production runs	Medium and large runs	Single items, small to medium runs	Single items, small runs. Series production of suitable components	Series production. Life of mould: Zn ≈ 500000 castings Mg ≈ 100000 castings Al ≈ 80000 castings Cu ≈ 10000 castings	Series production. Life of mould: Al ≈ 100000 castings	Series production. Life of mould: 5000 to 100000 castings depending on size of workpiece, casting material and type of mould	Length of billet depends on machine	
Tolerance range[a]	2.5 to 5%	1.5 to 3%	1 to 2%	0.3 to 0.8%	0.3 to 0.7%	3 to 5%	0.1 to 0.4%	0.3 to 0.6%	1%	0.8%

[a] For 500 mm nominal size (approx. values), dependent on degree of accuracy, material, size of workpiece, shape. For material-specific tolerances see DIN 1680 and DIN 1683 to DIN 1688.

distinction is made between processes using a *permanent pattern* and those using an *expendable pattern*. A permanent pattern can be used to make many expendable moulds, but an expendable pattern can only be used for one expendable mould. Expendable patterns are also made in an appropriate primary shaping tool.

The patterns are similar in shape to the case part to be manufactured, but are larger by the *allowance for contraction* of the material to be cast. They also incorporate the *machining allowances*, which will subsequently be eliminated by machining of the casting with the aim of achieving accuracy in dimensions, shape and position, as well as *tapers* to enable the pattern to be removed from the mould. Most patterns have a *pattern joint*, i.e. they consist of at least two parts (pattern halves); in addition, for castings with hollow spaces the pattern has *core marks* for insertion of the cores in the mould.

In the case of permanent patterns for making an expendable mould for casting, these patterns or their sections made from metals, high polymers or wood are used to make the moulds by the sand moulding, template moulding or shell moulding process.

Hand Moulding (Fig. 5). *Mould.* This is expendable (i.e. used only once). The medium used may be natural or artificial sand, with or without a synthetic resin binder, CO_2 sand, cement sand, or a moulding compound. It is worked by hand.

Pattern. Patterns for repeated use, and patterns and core boxes, are made to DIN 2522, those in quality classes H1a and H1 being mainly made of hardwood plywood, those in quality classes H2 and H3 of sawn timber and those in quality classes S1, S2 and S3 of foamed plastic.

Process Characteristics. Hand moulding denotes the production of a sand mould without using a moulding machine. The mould consists of the external parts for the external profile and the internal parts for the internal profile. Hollow spaces in the casting are formed by cores placed in the mould.

The principle of moulding is illustrated in **Fig. 5**. First of all the bottom half of the two-part pattern is moulded. After turning the moulding box over, the top half of the pattern and the pouring gate and risers are placed in position and the top mould is made. The top box is lifted off, the pattern halves are removed from the mould and the core is inserted. The halves of the mould are joined and the casting is made.

Casting Materials. All metals and alloys that are castable with the current technology.

Weight of Castings. The maximum transportable weight and the melting capacity determine the maximum weight.

Number of Castings. Single items, small production runs.

Tolerances. From about 2.5 to 5%.

Mechanical Moulding (Fig. 6). *Mould.* Expendable (used only once). Natural sand, artificial sand, sand with synthetic resin binders, CO_2 sand. Preparation on moulding and core moulding machines. Used in semi-automatic and fully automatic production lines.

Pattern. Patterns and core boxes are made to DIN 1511, in quality class H1 mainly of hardwood plywood, in quality classes M1 and M of metal, in quality classes K1 and K2 of plastic.

Process Characteristics. Mechanical moulding is characterised by a semi-automatic or fully automatic manufacturing operation for efficient production of ready-to-cast sand moulds. The casting process is often incorporated into the production line. The main stages are: moulding station, core insertion section, casting section and cooling section. The emptying station releases the cast moulds. The moulding station may consist of one automatic moulding machine for complete moulds or of two or more for making separate top and bottom boxes. There are also boxless moulding units, where the moulds are made using only a frame, which is withdrawn after compacting the sand.

Casting Materials. All metals and alloys that are castable with the current technology.

Weight of Castings. Limited by the size of the moulding machines: up to approximately 5000 kg.

Number of Castings. Owing to the mechanical preparation, mechanical moulding is suitable for series and mass production of quantities of 1000 and multiples thereof.

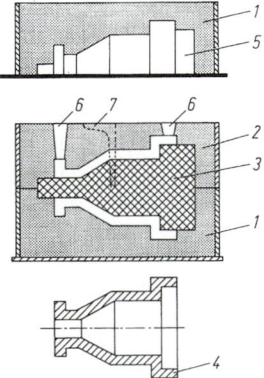

Figure 5. Hand moulding [3]; *1* bottom half of box (drag), *2* top half of box (cope), *3* core, *4* casting, *5* plate with wooden pattern half, *6* risers, *7* pouring gate.

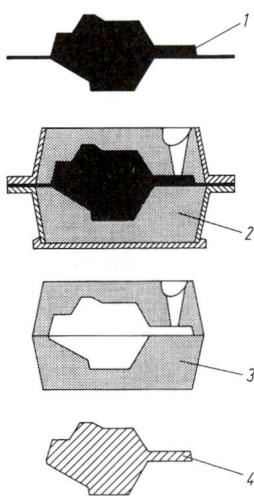

Figure 6. Mechanical moulding [3]; *1* plate with metal pattern, *2* compaction of the moulding sand in a frame, *3* boxless mould ready for casting, *4* casting.

Tolerances. From about 1.5 to 3%.

Suction Moulding (Fig. 7) *Mould.* Expendable (used once only), made from wet artificial casting sand.

Pattern. Wood, plastic, metal.

Process Characteristics. The process is characterised by the formation of a vacuum by withdrawing air from the mould space and the incoming moulding sand. This accelerates the sand, which spreads over the wall of the pattern. The sand can be subsequently pressed against the pattern. Advantages of the process are optimum mould compaction around the pattern, no shadow effect with plane surfaces, decreasing hardness of compacted sand from the inside to the outside, high surface quality, dimensionally stable castings, reduced cleaning. This process should not be confused with vacuum moulding.

Casting Materials. Iron, steel and aluminium.

Weights of Castings. From about 0.1 to 120 kg.

Number of Castings. Small, medium and large production runs.

Tolerances. Conventional to DIN 1683; maximum offset of compacted sand 0.3 mm.

Shell Moulding (Fig. 8). *Mould.* Expendable (used only once). Resin-coated sands or sand/resin mixtures.

Pattern. Patterns for repeated use, heatable metal patterns and metal core boxes.

Process Characteristics. These moulds are shell moulds with walls only a few millimetres thick. The mould material is poured onto the heated metal pattern. This cures the synthetic resins in the mould material, solidifying the mould. The result is a self-supporting, stable shell mould. Shell moulds are often moulded in one piece and then divided. After putting in the cores, the two halves of the mould are glued together. The shell moulding process is used in various stages of mechanisation and automation. This process is used not only for making moulds for shell casting, but also for producing hollow

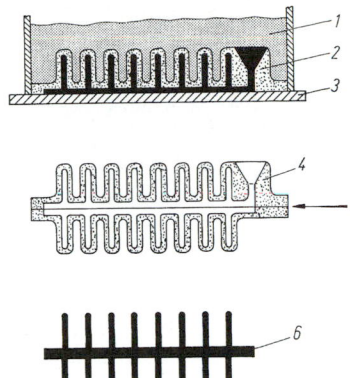

Figure 8. Shell moulding [3]; *1* loose synthetic-resin-coated sand, *2* hardened synthetic-resin-coated sand, *3* heated metal pattern, *4* shell mould, *5* glued seam, *6* casting.

shell cores for sand and chill casting. These cores are produced on special core moulding machines. Shell casting offers high dimensional accuracy with excellent surface quality.

Casting Materials. All metals and alloys that are castable with the current technology.

Weight of Castings. Up to ≈ 150 kg.

Number of Castings. Medium to large production runs.

Tolerances. From about 1 to 2%.

Ceramic Moulding (Fig. 9). *Mould.* Expendable (used only once), made of highly refractory ceramic similar in kind to mould materials for precision casting.

Patterns. Reusable, made of metal, plastic or specially varnished wood.

Process Characteristics. A slip consisting of highly refractory substances is poured around the pattern; these

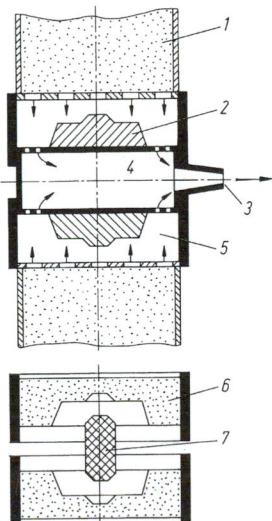

Figure 7. Suction moulding [3]; *1* moulding sand, *2* pattern, *3* air connection, *4* vacuum, *5* mould space, *6* sand mould, *7* core.

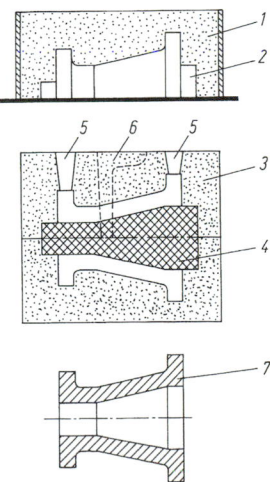

Figure 9. Ceramic moulding [3]; *1* pasty ceramic moulding compound, *2* plate with pattern half, *3* hardened split ceramic mould, *4* inserted core, *5* risers, *6* pouring gate, *7* casting.

substances then harden by chemical reaction. Often only one layer is poured, which is then back-filled with "normal" moulding sand. After removing the pattern, the ceramic is fired or skin-dried (Shaw process). To keep ceramic moulding, which is relatively expensive, to a minimum, it is usually only the parts of the mould that will be cast in finished or near-finished form that are made from special ceramic. In the case of components for fluid flow machines, these are the three-dimensionally curved parts; in the case of tools, these are the contours that require no finishing by machining or which, after hardening, will be finished only by electrical discharge machining or grinding. Castings from ceramic moulds have no casting skin in the conventional sense and are among the precision casting processes that, as the technology develops, are becoming more and more widely used owing to their efficiency.

Casting Materials. All metals and alloys that are castable with the current technology, especially iron-based materials.

Weight of Castings. From about 0.1 to 2500 kg, depending on the production equipment.

Number of Castings. Single items, small and medium runs, also larger runs in the case of fluid flow machines.

Tolerances. Up to about 100 mm ± 0.2%, over 100 mm ± 0.3 to 0.8% of nominal dimensions.

Vacuum Moulding (Fig. 10). *Mould.* Expendable (used only once); plastic sheet vacuum-moulded to the contours of the pattern, back-filled with fine-grained, binder-free quartz sand and sealed with a covering sheet. Dimensional stability is preserved by means of a vacuum of 0.3 to 0.6 bar.

Pattern. Permanent patterns, not subject to significant wear. Quality classes 1 and 2, mainly made of wood. Core boxes according to the core manufacturing method.

Process Characteristics. The process is characterised by the use of a vacuum for both deep drawing the pattern sheet over a pattern with nozzle holes and maintaining the stability of the mould. A moulding box equipped with suction systems is connected by a pipe to the vacuum grid. The fine, binder-free sand with which the mould box is filled is compacted by vibration. After applying a cover sheet, the air is evacuated from the sand and the mould thus becomes rigid. The mould is constantly connected to the vacuum grid before, during and after casting. To empty the mould, the vacuum is switched off and the sand and cast parts drop out of the mould box without additional force. The advantages of the process are: high, reproducible dimensional accuracy with outstanding surface quality; the mould seam at the joins and core marks is very small; tapers can be entirely dispensed with in certain areas of the casting.

Casting Materials. All metals and alloys that are castable with the current technology.

Weight of Castings. Restricted by the equipment available, not by the process.

Tolerances. 0.3 to 0.6%.

Casting Under (High) Vacuum. *Mould.* Expendable (used only once), shell moulds (for investment casting) and precision casting moulds made of special mould materials.

Pattern. Wax for investment casting, also of metal, plastic or the like depending on the type of mould.

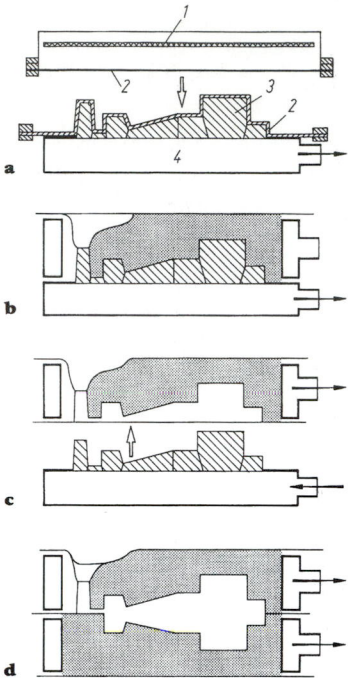

Figure 10. Vacuum moulding [3]; *1* heating, *2* plastic sheeting, *3* pattern, *4* vacuum box. **a** The plastic sheeting is softened by means of a sheet-type heating element and drawn tightly against the pattern by vacuum through holes. **b** The mould box is placed on top, filled with sand, pre-compacted, and the top of the box is covered with plastic sheeting. **c** A vacuum is applied to the mould box, compacting the sand. By switching off the vacuum, the mould box can be easily lifted off the pattern. **d** The top and bottom halves of the box are joined. The vacuum is maintained during casting.

Process Characteristics. Titanium and zirconium are among the reactive metals that have high affinities with oxygen, nitrogen and hydrogen in the molten state. This is even the case when present as alloying constituents in appropriate percentages in, e.g., molten nickel. All these alloys must therefore be produced and cast under defined conditions, normally under high vacuum.

The new mould ceramics, e.g. those made of yttrium and zirconium oxides, resist attack by reactive metals and melts. However, these special ceramics are not (yet) required for nickel-base alloys that are only alloyed with titanium, aluminium, etc.

To optimise quality and structure, the castings are usually isostatically pressed at high temperature by means of the HIP process.

Casting Materials. Alloys based on (in order of importance) nickel, titanium, cobalt, iron and zirconium.

Weight of Castings. Approx. 0.01 to 100 kg and more, depending on the manufacturing equipment.

Number of Castings. Small series to fairly large production runs.

Tolerances. ± 0.3 to ± 0.8% of nominal size, depending on the moulding process.

Precision Casting (Fig. 11). *Mould.* Expendable (used only once), made of highly refractory ceramic; sin-

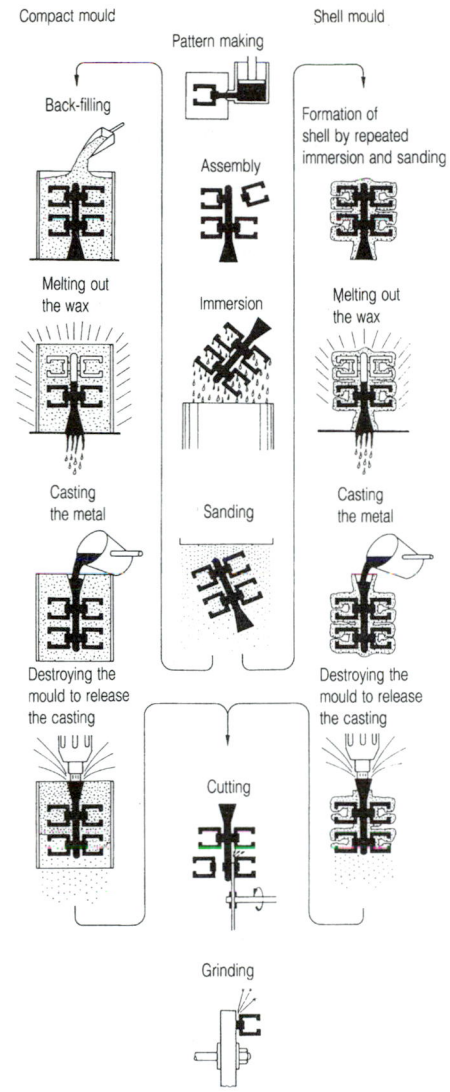

Figure 11. Schematic diagram of sequence of operations in precision casting [5].

gle or group pattern with runners, combined to form casting units ("clusters" or "trees").

Pattern. Made by injection moulding from special waxes or the like, thermoplastics or mixtures thereof.

Process Characteristics. The distinguishing features are the expendable patterns, the one-piece moulds and the casting in hot moulds (≈ 900 °C for steel). A casting skin in the conventional sense does not form. The patterns are injection-moulded in single or multiple tools made of aluminium, steel or soft metal, for which an original pattern is required. The most suitable injection moulding tool in each particular case is chosen according to the planned total quantity, the form of the casting and the nature of the pattern material. The formation of certain undercut contours may require the use of water-soluble or ceramic cores, for which a supplementary tool is used. The pat-

terns are assembled into clusters by means of casting systems, usually again employing injection moulding. The method of this assembly is crucial for the quality of the castings and for efficiency. Viscous ceramic coatings which cure by chemical reaction are then applied to these clusters. For aluminium, special plasters are also used. After melting out (lost wax process) or dissolving away the pattern material, the resulting one-piece moulds are fired. Casting takes place in moulds which are usually still hot from firing, so that narrow cross-sections and fine profiles "turn out" cleanly. Precision casting, with its tight tolerances and high surface qualitty, is the casting technology that offers the greatest freedom of design coupled with high quality.

Casting Materials (in order of importance). Steels and alloys based on iron, aluminium, nickel, cobalt, titanium, copper, magnesium or zirconium, including aerospace materials, produced at atmospheric pressure or under vacuum.

Weight of Castings. 0.001 to 50 kg, also up to 150 kg and over depending on manufacturing equipment.

Number of Castings. Small series to large production runs, depending on the complexity and/or machinability of the workpiece concerned.

Tolerances. ± 0.4 top $\pm 0.7\%$ of nominal dimensions.

Full Mould Casting (Fig. 12). *Mould.* Expendable (used only once), mould material usually self-curing.

Pattern. Expendable, foam material.

Process Characteristics. One-piece pattern made of foam material (polystyrene). Shape and dimensions match the part to be cast (taking into account allowance for contraction). The pattern need not be removed from the mould after mould-making. The heat of the molten metal flowing into the full mould vaporises the pattern, which is continuously replaced by cast metal. Mould joints and cores are usually unnecessary. Bolts, sleeves, lubrication

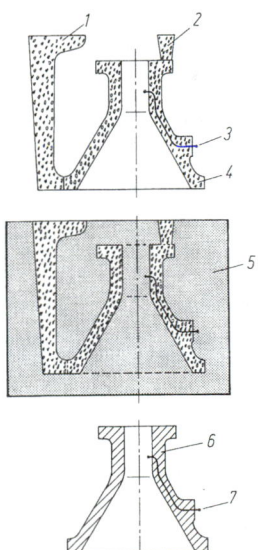

Figure 12. Full mould casting [3]; *1* pouring gate, *2* riser, *3* pipe to be integrally cast, *4* one-piece polystyrene foam pattern, *5* mould, *6* casting, *7* integrally cast pipe.

lines, etc. can be integrally cast in. The absence of mould tapers reduces the weight of the casting. The time and cost of manufacture are only a fraction of those encountered with a wooden pattern.

Materials for Casting. All metals and alloys that are castable with the current technology, especially those with high casting temperatures.

Weight of Castings. From approximately 50 kg up to maximum transportable weight; especially suitable for large components.

Number of Castings. Single pieces, small production runs.

Tolerances. From about 3 to 5%.

Magnetic Moulding. *Mould.* Expendable (used only once), iron granules.

Pattern. Expendable, foam material.

Process Characteristics. Magnetic moulding is a type of full mould casting. The casting units prefabricated from foam material (patterns with pouring gates and runner) are coated with a refractory ceramic (similar to the shell moulds for precision casting). They are then back-filled with pourable iron granules in a mould box. By applying (or switching on) a d.c. magnetic field, the iron powder becomes rigid and thus supports the casting unit. After casting and solidification of the metal, the magnetic field is switched off, causing the iron granules to become pourable again. Then the casting is removed; the iron granules can be reused.

Casting Materials. All metals and alloys that are castable with the current technology. As the thermal conductivity of the magnetisable mould material is higher than that of quartz sand, the cooling rate of the castings is higher and leads to a finer metallographic structure. The properties in use are especially improved in the case of steel castings.

Number of Castings. Single items, small production runs.

Tolerances. From about less than 3 to 5%.

Use of Permanent Moulds

Chill Casting (Fig. 13). *Mould.* Permanent mould, cast iron or steel, cores made of steel.

Pattern. None required.

Process Characteristics. Casting takes place by gravity in permanent metal moulds. These moulds are made in two or more parts for removal of the finished casting. The higher thermal conductivity of the metal mould compared with moulding sand brings about faster cooling of the solidifying molten metal. The result is a relatively fine-grained, dense structure with better mechanical properties than parts made by sand casting. High dimensional accuracy, excellent surface quality and good reproduction of contours characterise the chill casting. This process fully meets the specifications for gas- and liquid-tight valves, owing to the production of a dense structure. A rapid, efficient casting sequence, with machining generally being unnecessary or requiring only small machining allowances, are further features of this process.

Casting Material. Chill casting alloys, DIN 1709 copper–zinc alloys, DIN 1714 copper–aluminium alloys, DIN 1725 aluminium alloys, DIN 1729 magnesium alloys, DIN 1743 high-grade zinc alloys, also copper, copper–chromium alloys, super-eutectic aluminium–silicon alloys, lamellar and nodular graphite cast iron. The standard chill casting alloys are denoted by the symbol GK.

Weight of Castings. Non-ferrous metals and cast iron up to ≈ 100 kg, more depending on equipment. Cast iron for certain purposes up to ≈ 20 t (= 20 000 kg).

Number of Castings. From about 1000 and multiples thereof, depending on the material being cast (e.g. Al ≈ 100 000 castings).

Tolerances. From about 0.3 to 0.6%.

Low-Pressure   Chill   Casting   (Fig.   14). *Mould.* Permanent mould, cast iron or steel.

Pattern. No pattern required.

Process Characteristics. Casting is carried out under pressure (usually with compressed air) in permanent metal moulds. These moulds are made in 2 or more parts for removal of the finished casting. The higher thermal conductivity of the metal mould compared with moulding sand brings about faster cooling of the solidifying molten metal. The result is a relatively fine-grained, dense structure with better mechanical properties than parts made by sand casting. The distinguishing feature is the application of pressure, which dispenses with the need for risers on the casting. High dimensional accuracy, excellent surface quality and good contour reproduction together with a rapid, efficient casting sequence and considerable machining economies are further features of this process. Gas- and liquid-tight valves can be efficiently manufactured owing to the dense structure of the casting.

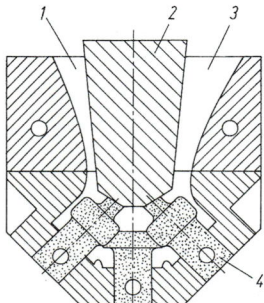

Figure 13. Chill casting (composite mould with metal and sand cores, the latter for undercuts) [3]; *1* riser, *2* metal core, *3* pouring gate, *4* sand core.

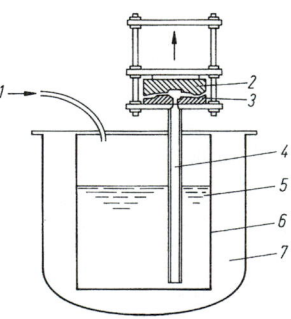

Figure 14. Low-pressure chill casting [3]; *1* air or gas, *2* moving block, *3* fixed block, *4* ascending pipe for molten metal, *5* molten metal, *6* crucible, *7* heating.

Casting Materials. Light metal, especially aluminium alloys.

Weight of Castings. Up to ≈ 70 kg.

Number of Castings. From about 1000 and multiples thereof.

Tolerances. From about 0.3 to 0.6%.

Pressure Die Casting (Fig. 15). *Mould.* Permanent mould, usually high-tensile hot forming tool steel or special metals.

Pattern. No pattern required.

Process Characteristics. The distinguishing feature of this process is that the molten metal is forced into the two-part permanent mould at high pressure and relatively high speed in pressure die casting machines. Two types of process are distinguished, namely:

1. *Hot-chamber process.* Here the die casting machine and the holding furnace for the molten metal form a unit. The casting assembly is immersed in the molten metal. In each casting operation, a precisely predetermined volume of molten metal is forced into the mould. The hot-chamber die casting process is especially suitable for lead, magnesium, zinc and tin. The output of components manufactured by this process is considerable, but varies according to the size of the component and the casting material.
2. *Cold-chamber process.* In this process the die casting machine and the holding furnace for the molten metal are separate. After being taken from the furnace, the molten metal is poured into the cold pressure chamber and forced into the mould. The pressure chamber is mounted directly on the runner-side mould block. This process is chiefly suitable for aluminium- and copper-based alloys, as these would attack the steel casting assembly when molten if the hot-chamber process were employed. Cold-chamber die casting machines do not achieve the rates of output of hot-chamber machines, owing to the nature of the process. Pressure die casting is today one of the most efficient casting processes around. The machines are mostly semi-automatic or fully automatic. Pressure die-cast parts have smooth, clean surfaces and edges. They are extremely dimensionally accurate. Therefore only the fitting and bearing surfaces, at most, require machining. Very low machining allowances result in short machining times.

Casting Materials. Materials suitable for pressure die casting are: DIN 1709 copper–zinc alloys, DIN 1714 copper–aluminium alloys, DIN 1725 aluminium alloys, DIN 1729 magensium alloys, DIN 1741 lead alloys, DIN 1742 tin alloys, DIN 1743 high-grade zinc alloys. The standardised pressure die casting alloys are denoted by the symbol GD. For the *hot-chamber process*: lead, magnesium, zinc and tin alloys. For the *cold-chamber process*: particularly aluminium- and copper-based materials.

Weight of Casting. Up to 45 kg for light alloys, up to 20 kg for other materials, depending on the material being cast and the working dimensions of the die casting machines.

Number of Castings. Varies widely depending on the material being cast. Example: Zn alloys ≈ 500 000 castings.

Tolerances. From about 0.1 to 0.4%.

Centrifugal Casting (Fig. 16). *Mould.* Permanent, water-cooled cast iron or steel mould.

Pattern. None required.

Process Characteristics. The centrifugal casting process is used to manufacture hollow products having a rotationally symmetrical hollow space and an axis coinciding with the axis of rotation of the centrifugal casting machine. The external form of the casting is determined by the shape of the mould. The internal form is determined by the effect of the centrifugal force of the rotating mould. The wall thickness of the casting depends on the quantity of molten metal supplied. A variant of the process is centrifugal mould casting, which produces finished hollow or even massive castings using rotating moulds. Composite centrifugal casting is also possible, as is centrifugal casting with a flange. The condition on delivery of centrifugal-cast Fe, Ni and Co-based alloys is normally (at least) pre-turned.

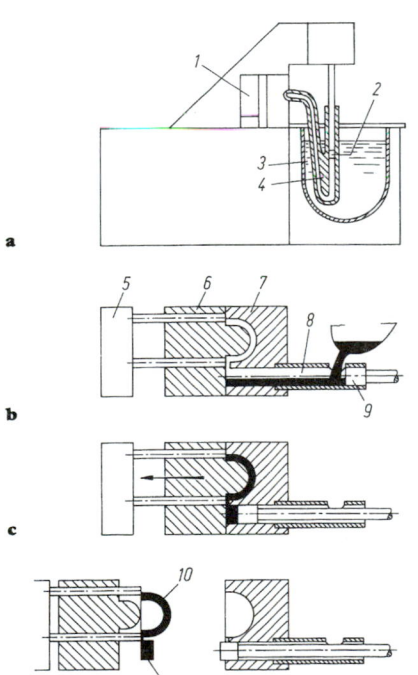

Figure 15. Pressure die casting [3]: **a** hot-chamber process, **b** to **d** cold-chamber process, **b** filling of casting chamber, **c** plunger forces molten metal into die, **d** ejection of casting; *1* die, *2* plunger, *3* crucible with molten metal, *4* casting vessel, *5* ejection, *6* moving block, *7* fixed block, *8* pressure chamber, *9* plunger, *10* casting, *11* slug.

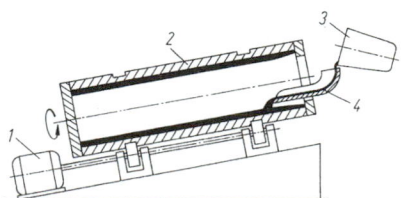

Figure 16. Centrifugal casting [3]; *1* drive, *2* mould, *3* pouring ladle, *4* pouring basin.

Casting Materials. Especially cast iron, cast steel, heavy and light metals.

Weight of Castings. Up to about 5000 kg.

Number of Castings. From about 5000 to over 100 000, depending on the mould and the casting material. In special cases, e.g. castings of stainless steel and the like, also single pieces and small production runs (from ≈ 100 mm internal diameter upwards).

Tolerances. About 1%.

Composite (or Integral) Casting (Fig. 17). *Mould.* Metal mould, e.g. for centrifugal composite casting.

Pattern. None.

Process Characteristics. These types of process are used: to cast structural parts from two or more different metallic materials which are firmly joined together. At least one material is poured in the molten state into a mould, which may also be part of a product to be manufactured; for composite casting of various metals and/or alloys in molten or semi-solid condition, e.g. in centrifugal casting; and for casting in, casting round and lining solid components, which may be made not only of metal but also of ceramic. The bond may be formed by shrinkage, positive force or both.

Casting Materials. All metals and alloys that are castable with the current technology.

Weight of Castings. Up to about 50 kg and over, depending on the manufacturing equipment.

Number of Castings. Medium and large production runs.

Tolerances. From about 0.1 to 0.6%, depending on the process.

2.2.3 Guidelines for Design

Forming by casting enables design ideas to be turned into reality to a particularly high degree, owing to the *extensive freedom of design* that it offers. A design appropriate for manufacturing, which contributes decisively to the efficient production of a casting, can generally be achieved only by close collaboration between the design engineer and the founder. Forming by casting differs from other forming processes in that the material only receives its shape, material structure and quality after cooling, with shrinkage – which may sometimes be considerable – in the liquid state and during solidification, and appreciable contraction in the solid state (**Fig. 18, Table 2**). The *contraction in the solid state* should be accounted for by means of a suitable allowance (*allowance for contraction*). The alloying specifications often suffer considerable deviations, owing to obstruction of contraction

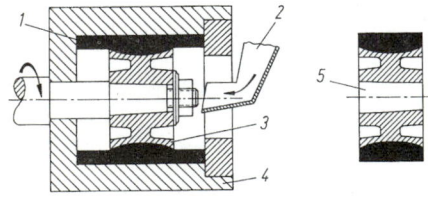

Figure 17. Composite (or integral) casting [3]; *1* composite material for casting, with expendable heads which are removed by cutting, *2* pouring basin, *3* clamped-on hub, *4* mould, *5* composite casting (two different materials).

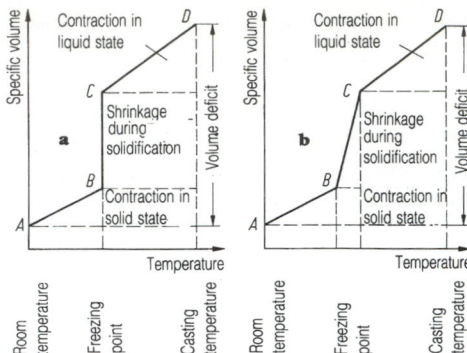

Figure 18. Schematic diagram of contraction of metallic materials during cooling from the molten state: **a** for pure metals and eutectic alloys, **b** for non-eutectic alloys [6].

Table 2. Contraction of various casting materials (approximate values)

Material	Liquid (max. %)	Solid (max. %)
Lamellar graphite cast iron	3	1
Nodular graphite cast iron	5	2
Cast steel	6	3
Malleable cast iron	5.5	2
Copper alloys	4	2
Aluminium alloys	5	1.25

by ribs, projections, more or less flexible cores and mould parts (**Table 3**). Provided that they fall within the acceptable dimensional variations or are compensated for by machining allowances, as is usually the case with small castings, they do not present a problem. With large castings, though, empirical values for the deviations due to contraction have to be taken into account when making the pattern. If there is a one-sided obstruction of contraction, e.g. due to the mould or even the shape, especially in the case of longer castings (different cross-sections along their length and consequently different cooling rates and thermal stresses), the castings would distort unless the pattern is curved in the opposite direction. Large wheel centres, for instance, are not infrequently split, e.g. to prevent unacceptable out-of-roundness. Thermal stresses that are not reduced by plastic deformation may, besides distortion, also result in undesirable "relief by cracking". Therefore if insufficient attention is paid to contraction of the material being cast at the design stage, taking into account the possibilities afforded by gating of moulds and risering, pipes (shrinkage cavities), shrinkage voids, sinks, hot (pipe) cracks, distortion and stress cracks may form.

Manufacture-Orientated Design

Correct design of changes in wall thickness with a view to shrinkage during solidification and contraction in the solid state:

Wall Thickness Graduations should permit directional solidification.

Junctions formed by the meeting of two or more walls form concentrations of material, i.e. hot zones. Therefore

Table 3. Guide values for linear contraction and possible deviations [7]

Casting material	Guide value %	Possible deviation %
Lamellar graphite cast iron	1.0	0.5 to 1.3
Nodular graphite cast iron, unannealed	1.2	0.8 to 2.0
Nodular graphite cast iron, annealed	0.5	0.0 to 0.8
Cast steel	2.0	1.5 to 2.5
Austenitic manganese steel	2.3	2.3 to 2.8
White malleable cast iron	1.6	1.0 to 2.0
Black malleable cast iron	0.5	0.0 to 1.5
Aluminium casting alloys	1.2	0.8 to 1.5
Magnesium casting alloys	1.2	1.0 to 1.5
Casting copper (electrolytic)	1.9	1.5 to 2.1
CuSn casting alloys (cast bronzes)	1.5	0.8 to 2.0
CuSn–Zn casting alloys (gunmetal)	1.3	0.8 to 1.6
CuZn casting alloys (cast brass)	1.2	0.8 to 1.8
CuZn (Mn, Fe, Al) casting alloys (special cast brasses)	2.0	1.8 to 2.3
CuAl (Ni, Fe, Mn) casting alloys (cast aluminium bronzes and cast multicomponent aluminium bronzes)	2.1	1.9 to 2.3
Zinc casting alloys	1.3	1.1 to 1.5
Babbitt (Pb, Sn)	0.5	0.4 to 0.6

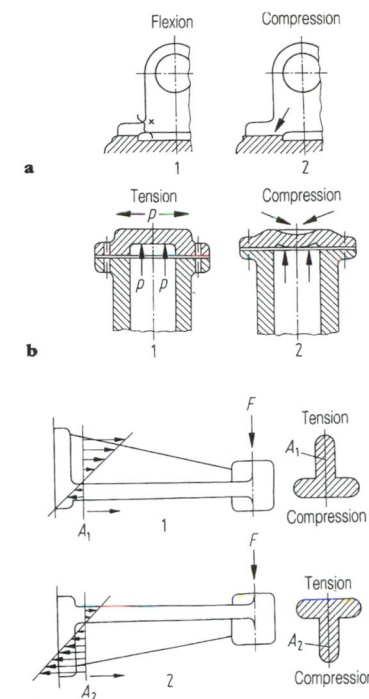

Figure 19. Examples of stress-orientated casting design for a material having higher compression strength than tensile strength according to [6]. **a** Pedestal; *1* flexurally stressed – inadequate bearing surface, *2* compressively stressed – bearing surface widened. **b** Cylinder cover; *1* tensionally stressed – poor design, *2* compressively stressed – good design. **c** Wall bracket arm; *1* poor cross-sectional arrangement, *2* stress-absorbing cross-sectional arrangement.

they should be separated as far as possible, or designed for efficient casting by narrowing the cross-section. Concentrations of material, especially at locations that are inaccessible to feeding, lead to piping.

Sudden Changes in Wall Thickness should be avoided, as they produce high thermal stresses due to different cooling rates. In addition, there is often increased obstruction of contraction by the mould. The risk of formation of hot cracks ("pipe cracks" between liquidus and solidus temperatures) and stress cracks (during further cooling in the solid state) is therefore high. Locations prone to cracking can be protected by ribs.

Sharp Corners additionally cause a heat buildup (sand-edge effect) and, accordingly, not only hot cracks but also porosity due to contraction as well as drawholes.

For a summary of design recommendations see **Fig. 20**.

Stress-orientated Design

When designing castings, the main stresses occurring during manufacture have to be taken as a basis. Here, the freedom of design offers excellent adaptation to the technical requirements. Forming by casting permits the efficient manufacture of parts of the most complicated kind with *high strength in relation to shape*. The stress condition of the design can often be made more favourable by suitable ribbing or slight modification (**Figs 19 and 20**).

It is important to know the load-bearing capacity of the materials to be cast. For approximate values for lamellar graphite cast iron see **Fig. 21**.

Examples. A casting made from GG-15 grey cast iron (top horizontal grey bar) with a wall thickness of 10 mm or a test bar diameter of 20 mm (vertical line) has a tensile strength of approx. 22 dN/mm², a hardness of approximately 220 HB and an initial

modulus of elasticity of almost 10 000 dN/mm². For a wall thickness of 45 mm, on the other hand, the tensile strength is approximately 10 dN/mm², the hardness is approximately 130 HB and the modulus of elasticity almost 8000 dN/mm².

If, however, the tensile strength of this 45-mm-thick wall is 22 dN/mm², a strength of approximately 180 HB and an E_0 of approximately 11 500 dN/mm² should be expected. The material grade GG-30 should be selected. For a wall thickness of 10 mm, this cast iron has a tensile strength of approximately 35 dN/mm², a hardness of approximately 260 HB and an E_0 of approximately 13 000 dN/mm².

2.2.4 Preparatory and Finishing Operations

Melting of Materials for Casting. For transforming the metal to be cast and the additives into the molten state, a wide variety of melting equipment – e.g. shaft (cupola), crucible and hearth-type furnaces – is available. These furnaces are heated with coke, gas, oil or electricity. The most important types of melting equipment are: for *cast iron, including malleable cast iron*: cupola (shaft) furnaces, induction furnaces, rotary kilns (oil-fired); *cast steel*: electric arc furnaces, induction furnaces; *non-ferrous metal castings*: induction furnaces, electrically, gas- or oil-heated crucible furnaces.

Cleaning of Castings. The moulds are emptied by means of emptying jiggers. The sand adhering to the casting is generally removed by means of abrasive blasting equipment employing, without exception, steel shot or steel grit made from wire.

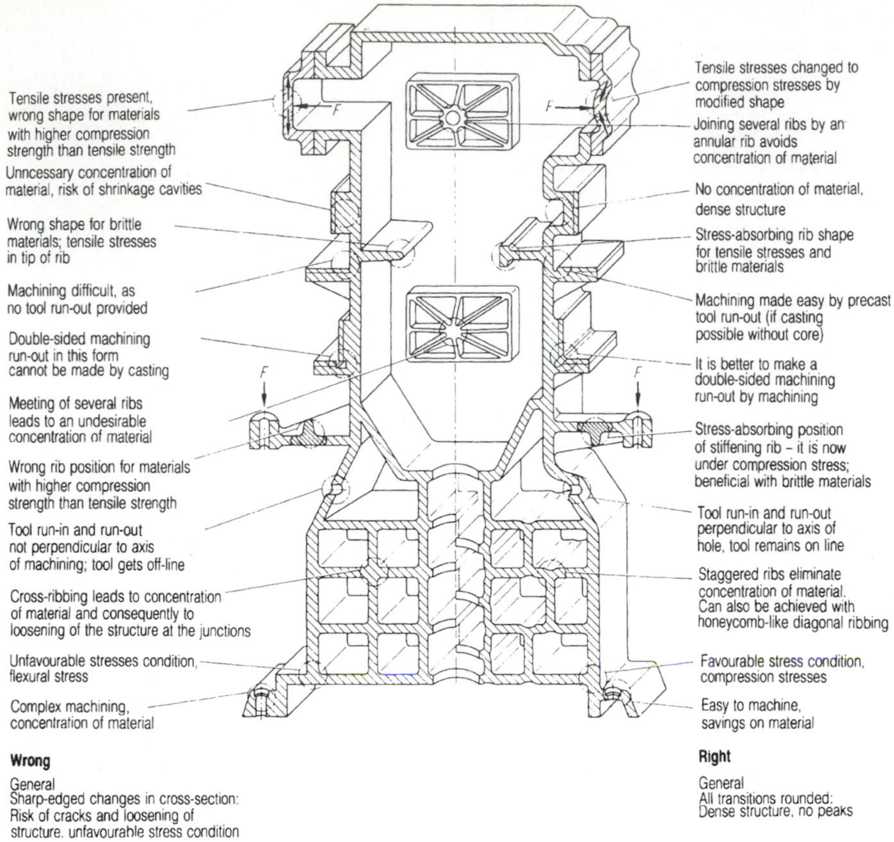

Tensile stresses present, wrong shape for materials with higher compression strength than tensile strength
Unncessary concentration of material, risk of shrinkage cavities
Wrong shape for brittle materials; tensile stresses in tip of rib
Machining difficult, as no tool run-out provided
Double-sided machining run-out in this form cannot be made by casting
Meeting of several ribs leads to an undesirable concentration of material
Wrong rib position for materials with higher compression strength than tensile strength
Tool run-in and run-out not perpendicular to axis of machining; tool gets off-line
Cross-ribbing leads to concentration of material and consequently to loosening of the structure at the junctions
Unfavourable stresses condition, flexural stress
Complex machining, concentration of material
Wrong
General
Sharp-edged changes in cross-section: Risk of cracks and loosening of structure, unfavourable stress condition
Tensile stresses changed to compression stresses by modified shape
Joining several ribs by an annular rib avoids concentration of material
No concentration of material, dense structure
Stress-absorbing rib shape for tensile stresses and brittle materials
Machining made easy by precast tool run-out (if casting possible without core)
It is better to make a double-sided machining run-out by machining
Stress-absorbing position of stiffening rib – it is now under compression stress; beneficial with brittle materials
Tool run-in and run-out perpendicular to axis of hole, tool remains on line
Staggered ribs eliminate concentration of material. Can also be achieved with honeycomb-like diagonal ribbing
Favourable stress condition, compression stresses
Easy to machine, savings on material
Right
General
All transitions rounded: Dense structure, no peaks
F

Figure 20. Illustration of important design guidelines [11].

Approx. plate thickness in mm
5
7.5
10
15
30
45
Test bar diameter in mm
10
15
20
30
60
90
80 kN/mm²
150 N/mm²
90
200
150
GG-15
100
1
2
250
GG-20
110
175
300
GG-25
120
350
GG-30
3
4
200
130 kN/mm²
400 N/mm²
225
GG-35
250
275
Tensile strength
Modulus of elasticity
Hardness HB
1.2
1.1
1.0
0.9
0.8
0.7
0.6
Degree of saturation S_c
4.0
3.6
3.2
2.8
2.4
2.0
1.6
1.2
0.8
0.4
0
Surface-to-volume ratio in cm⁻¹

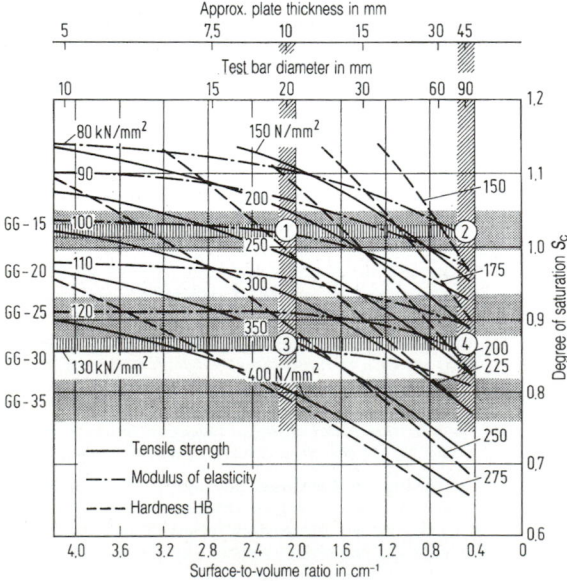

Figure 21. Chart illustrating mechanical properties of lamellar graphite cast iron. Relationship between chemical composition, rate of cooling and mechanical properties (tensile strength, hardness, modulus of elasticity) in the casting (wall thickness) and the separately cast test bar [9, 10]. Each point on the diagram signifies a specific combination of mechanical properties for a specific material. It also determines the material grade to be selected.

Heat Treatment. Many materials only obtain the physical and technological characteristics required in use from heat treatment. This treatment requires the use of electrically heated or gas- or oil-fired furnaces in continuous or batch operation. Their size is matched to the size and quantity of the castings and their mode of operation to the wide variety of heat treatment processes (see D3).

Inspection and Testing Methods. The diverse demands made on the casting, which become greater with advances in technology, and the trend towards light-weight construction and thus more efficient use of materials inevitably lead to stringent requirements with regard to casting quality, with particular emphasis on consistency. Inspections of the process and the castings begin with checking of the metallic and non-metallic feedstocks and end with the final inspection of the castings. Materials and workpieces are mainly tested by means of the non-destructive testing methods such as radiographic, ultrasonic, magnetic powder and liquid penetrant testing. Destructive tests, e.g. tensile, notched-bar impact and bending, are usually carried out with specimens cast either separately or as an appendage to the casting; in exceptional cases, specimens taken from the casting itself may be used.

2.3 Forming of Plastics

For the materials properties of plastics see D4.

Thermoplastics (in the form of injection moulding compounds) account for the greater part by far of production of moulded parts and semi-finished products compared with *thermosetting plastics* (compression moulding compounds and casting resins). Primary shaping may be accomplished by *gravity* (static) *casting* or *centrifugal casting*, but is more frequently done by compression moulding and, most of all, *injection moulding* and *extrusion*. The moulding technology is described in DIN 16700. Important factors for discontinuous (fixed-cycle) production of moulded parts (compression moulding, injection moulding) and continuous production of sections, films, sheet, etc. (extrusion) from moulding compound (powder, granules, chips, etc.) are the "processing parameters", e.g. melting range, viscosity, melting index, flow behaviour, disintegration temperature range, stated by the manufacturers (cf. D4).

2.3.1 Casting of Plastic Sheet

In casting of plastic sheet, the material to be processed flows by gravity from a storage tank with a controllable slot at the base onto a slowly rotating drum underneath the storage tank (drum casting, **Fig. 22b**) or an endless

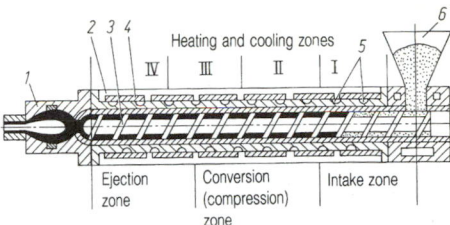

Figure 23. Extruder [4]; *1* primary shaping tool (die), *2* cylinder, *3* screw, *4* heating elements, *5* cooling channels, *6* feeder head.

belt made of copper (belt casting, **Fig. 22a**). The sheet thus formed passes through a drying zone where the solvent evaporates, causing the sheet to solidify, and is removed from the rotating drum or belt with a stripper device.

2.3.2 Extrusion

The characteristic feature of extrusion is that the material to be processed is continuously forced, in the form of moulding compound, in plasticised condition from a pressure chamber through a suitably shaped extrusion tool (die), emerging through a nozzle into the open air. The result is strip, rigid or flexible tubing, solid sections, fibres or sheet in a continuous stream.

The purpose of the extruder (**Fig. 23**) is to receive, compress and preheat the moulding compound (granules, powder) in the intake zone; to plasticise the moulding compound in the conversion zone; and to homogenise and compress the moulding compound and expel it from the extruder at the correct temperature in the expulsion zone. To transform the processed material into its final solid state, the extruded material has to be cooled with air or water after leaving the extrusion tool.

Besides the continuously operating screw extruders, there are also discontinuously operating ram extruders, which deliver similar products in individual sections. Extruders are also used as plasticising equipment for injection moulding, calendering and blow moulding. The last-named process, however, begins with a defined form which simply undergoes a further change with the material in rubber-like elastic condition; it is therefore regarded as a forming process.

2.3.3 Calendering

Calendering is the term given to primary shaping of sheet or film from preheated, pre-plasticised moulding compound between rotating rollers. The pre-plasticised material coming from the preparation unit (e.g. an extruder, **Fig. 23**) is fed by means of conveying equipment between the heated rollers of the calender (**Fig. 24**), where it undergoes final homogenising and plasticis-

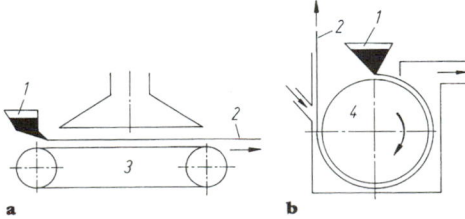

Figure 22. Plastic sheet casting machine [4]: **a** belt process, **b** drum process; *1* material (plastic compound) supply, *2* sheet, *3* casting belt, *4* casting drum.

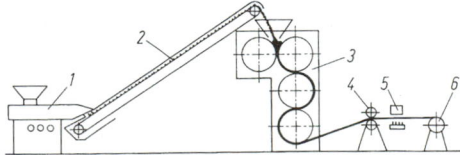

Figure 24. Calendering unit [4]; *1* extruder, *2* belt conveyor, *3* four-roller calender, *4* cooling rollers, *5* thickness meter, *6* reeling up.

ing and receives the desired thickness. After leaving the last calendering roller, the sheet or film passes over cooling rollers to harden it.

2.3.4 Laminating

In laminating, webs of supporting materials (paper or fabric) are soaked with high-polymer materials (resin carriers and resins; thermoplastic materials may also be used) and converted to laminates by pressing between heated plates as the primary shaping tool (mould). Depending on the desired thickness, several resin-soaked webs are placed on top of each other, covered with pressing plates on both sides and pressed into semi-finished products in multiplaten presses. Between the heated pressing plates, the resin is plasticised, completely impregnates the webs and solidifies. In the production of tubes on this principle, the resin-coated webs are wound onto a mandrel and hardened by the action of heat and, usually, pressure as well. The most important laminates are resin-bonded paper and fabric as well as vulcanised fibre, which are supplied as boards, coiled, compression-moulded or non-compression-moulded round tubes, solid bars and strip.

2.3.5 Injection Moulding

The characteristic feature of this process is that the plasticised material (injection moulding compound) is injected into either a cooled primary shaping tool (injection moulding tool) in the case of *thermoplastics* where the material solidifies due to cooling or into a heated tool in the case of *thermosets* at high pressure and solidifies there under the action of pressure. In **Fig. 25** the material to be processed is fed to the heating cylinder of the extruder as pourable *granules* or *powder*. It is plasticised in the heating cylinder and injected in this state through a nozzle into a closed primary shaping tool at a pressure of 80 to 180 N/mm², where the material solidifies. The screw or piston is withdrawn, the injection moulding tool is opened and the injection-moulded part is removed.

2.3.6 Compression Moulding

The characteristic feature of compression moulding is that the material to be processed (compression moulding compound) is softened in the primary shaping tool (compression moulding tool) under the action of heat and pressure, fills the hollow space with the tool closed and then solidifies. The heated primary shaping tool is filled with a quantity of compression moulding compound (powder, pellets, granules) corresponding to the mass of the moulded part, usually in preheated condition. At a pressure of 8 to 80 N/mm² the compound fills the hollow space and begins to solidify. When sufficient cross-linkage of thermosets or sufficient cooling of thermoplastics has taken place, the mould is opened and the moulded part ejected. Compression moulding can be used to produce plastic components both with and without fillers.

2.3.7 Transfer Moulding

Transfer moulding is a primary shaping process in which the material to be processed (transfer moulding compound) is plasticised in a pressurised cylinder (loading chamber) and then transferred to a closed primary shaping tool (transfer moulding tool), where it solidifies. The loading chamber is filled with a measured, suitably pelletised quantity of the preheated transfer moulding compound, which has to correspond to the mass of the moulded part, the inlet and the distributor.

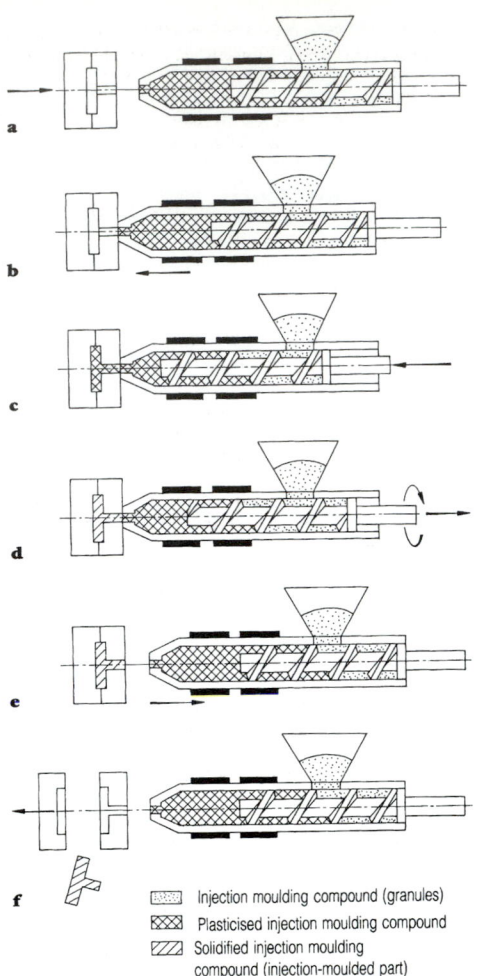

	Injection moulding compound (granules)
	Plasticised injection moulding compound
	Solidified injection moulding compound (injection-moulded part)

Figure 25. Injection moulding [4]. **a** Injection moulding tool is closed; **b** Nozzle is placed in position. **c** Moulding compound is injected and compressed. **d** Injection-moulded part solidifies. Moulding compound is metered in and plasticised. **e** Nozzle is withdrawn. **f** Injection moulding tool is opened and injection-moulded part is ejected.

2.3.8 Expanding

Production of parts from cellular materials plays a role in the case of high-polymer materials. The resulting parts contain only a fractional amount of the actual material, while a high proportion of their volume consists of hollow spaces (bubbles, voids). In the expanding of high-polymer materials, a distinction is made between three methods of working:

1. First, a carrier foam is formed by stirring air into a foaming agent (e.g. soap solution). Into this foam is poured the solution of a hardenable plastic, which spreads over the lamellae of the carrier foam and solidifies there (*churning process*).

2. Two substances are mixed together that react with each other either immediately or only under the action of heat with liberation of gas, foam the material to be processed and then solidify (*mixing process*).

3. To the plastic to be processed is added a special foaming agent, which is mixed with the molten material at atmospheric or elevated pressure. Cooling produces an expandable mixture. On reheating the foaming agent expands or decomposes, resulting in an expanded material whose structure is fixed by cooling.

2.4 Forming of Metals and Ceramics by Powder Metallurgy

2.4.1 General

For terminology see DIN 30 900. Powder metallurgy comprises the production of powders from metals, metal alloys and metal compounds (e.g. carbides, borides, silicides, nitrides, oxides and metals) and their conversion into semi-finished and finished components. In this primary shaping process, powder with a particle size – depending on the manufacturing process – of less than 0.5 mm ($\approx$ 0.1 to 500 μm) is usually mechanically compacted (pressed) in moulds and generally converted into rigid finished parts by *sintering* under shielding gas at high temperature. Pressing takes place at room temperature or, in some cases, at elevated temperature (hot pressing) in moulds made of wear-resistant or high-temperature steel. The sintering temperature (to obtain cohesion of the particles by diffusion) is of the order of $\frac{2}{3}$ to $\frac{4}{5}$ of the absolute melting point of the metal in the case of unary systems, but is often above the melting point of the lowest-melting component in the case of polynary systems. If the powder mixture has a heterogeneous structure, it is entirely possible for a small quantity of the liquid phase to be present. It is essential to avoid extensive melting. Bronzes and the like, for instance, are sintered at 600 to 800 °C, ferroalloys at 1000 to 1300 °C, hard metals at 1400 to 1600 °C and the high-melting metals like molybdenum, tungsten and tantalum at $\approx$ 2000 to 2900 °C. As the compression pressure ($\approx$ 1 to 10 kbar), the sintering time and the sintering temperature increase and the particle size decreases, the density increases up to that of practically non-porous material. Consequently, technologically desirable porosities can be deliberately established (**Fig. 26**).

2.4.2 Uses

Powder metallurgy is only economic for large production runs, because of the expensive pressing tools, and then chiefly for smaller components (less than one to a few thousand grams) of the simplest possible shape, owing to the poor mould-filling ability compared with casting, the limitations with regard to sufficient and above all uniform

Proportion of pore volume		Examples of uses
Pore volume	Density	
Up to 60 %		Filters
Up to 30 %		Oil-impregnated journal bearings
Up to 20 %		Components
Up to 15 %		Fairly high-strength components
Up to 5 %		High-strength components

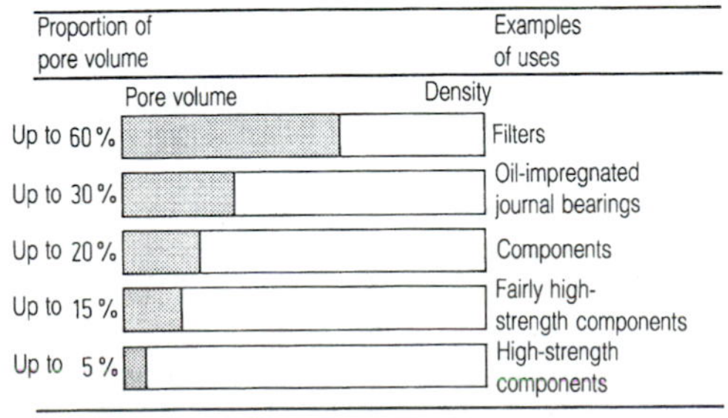

Figure 26. Specifically variable porosities of sintered parts with regard to their use.

compaction, and the low strength in unsintered condition. Disadvantages are the high capital expenditure requirements for presses, tools and kilns, the complex volume change conditions during pressing and sintering (for solid products, up to 20% linear contraction during sintering), the relatively limited design possibilities and the generally lower strength and toughness compared with cast parts. Advantages are the low manpower requirements, the high yield, the high dimensional accuracy (after calibration) and surface quality and, in particular, the materials technology possibilities that are only feasible with powder metallurgy.

Important *technical applications* that can only (or more easily) be achieved by sintering occur in the following fields:

1. *High-melting metals* like molybdenum, tungsten, tantalum, niobium. With fusion metallurgy, in addition to the high temperature further problems are caused by (in some cases undesirable) strong reactions between the molten metals and the refractory crucible or the kiln lining, and by high gas solubility.

2. *Hard metals as cutting materials.* Production of a composite-metal-like structure from brittle hard metals like tungsten, molybdenum and tantalum carbides and a tough bonding metal like cobalt, which is liquid at the sintering temperature.

3. *Composite products* made from unalloyable or difficult-to-alloy constituents, e.g. metal-bearing carbon bushes made from copper and graphite with the good conductivity of copper and the excellent sliding properties of graphite; materials for contacts with the high hardness of high-melting tungsten and molybdenum and the good conductivity of low-melting copper and silver; "diamond metals" made by homogeneous sintering of fine-grained hard materials such as diamond particles or corundum into a tough metallic matrix.

4. *Filters* and porous (oil-impregnated, self-lubricating and in some cases sealed-for-life) bearings with evenly distributed, interconnected pores; pore size and volume can be deliberately adjusted over a wide range.

5. *Alloys* of a metal with a high melting point and a metal whose boiling point is exceeded at this temperature, i.e. one with a high vapour pressure (e.g. iron, cobalt, nickel on the one hand and zinc, cadmium, lead, etc. on the other).

6. Where *very brittle materials* that are difficult or impossible to machine are to be converted (e.g. iron–aluminium–nickel–cobalt–copper-based permanent magnets or iron–aluminium–chromium-based brittle high-alloy steels), and machining would be comparatively time-consuming and expensive with other forming processes (e.g. mass-produced small components made of ferrous and non-ferrous metals), or a very high purity and a homogeneous composition are required (which cannot always be ensured with melting and the inevitable batchwise operation).

2.4.3 Technology

The sequence of manufacturing operations for sintered components can be divided into four: powder production, forming, sintering and after-treatment.

Powder Production. This is carried out with particle sizes of $\approx$ 1 μm to 0.5 mm by mechanical methods (crushing, grinding, granulating, atomising), physical methods (condensation) and chemical methods (reduction, electrochemical and electrolytic processes, decomposition).

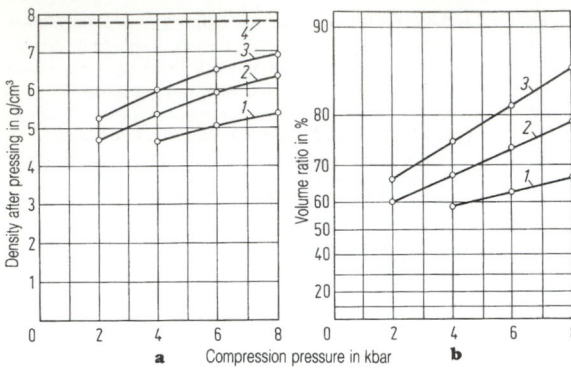

Figure 27. Compression behaviour of metal powders [12]. **a** Density after pressing versus compression pressure. **b** Volume ratio versus compression pressure. Hametag iron powder < 0.3 mm: *1* unannealed without additives, *2* annealed without additives, *3* annealed with 0.5% lithium stearate, *4* theoretical density (non-porous material, 7.86 g/cm³).

Forming. Forming into semi-finished and finished components is mainly carried out by cold pressing in wear-resistant moulds, but also, for better compressibility, by hot pressing and pressure sintering, as well as by explosive compaction and, lastly, also by extrusion and powder rolling. Besides these, forming is also practised without compaction by simple gravity flow of powder or with suspensions of powder in liquids (slip casting), the result being a sintered component with high porosity even after sintering (e.g. metal filters).

As the compression pressure is increased, the compaction ratio and consequently the density and space filling increase too (**Fig. 27**). As the rate of pressure transmission in powder mixtures is not uniform as with liquids, in order to achieve optimum uniformity of compaction, and thus a homogeneous material condition, the height of the compact or its height/diameter ratio is limited to ≈ 2 : 1 or, in favourable special cases, to 3 : 1. Homogeneous compaction of complex components inevitably requires expensive tools with plungers at varying heights.

Cold Pressing Method (**Fig. 28**). Care should be taken when designing components to ensure that they can be manufactured at all by pressing and are as simple as possible to make.

Guidelines for design (see [8]): adherence to dimensional limits and conditions: height/width < 2.5, wall thickness > 2 mm, holes > 2 mm. Insufficient tolerances should be avoided: holes ≥ IT7, width > IT6, height ≥ IT12. Sharp edges, acute angles and tangential transitions should be avoided.

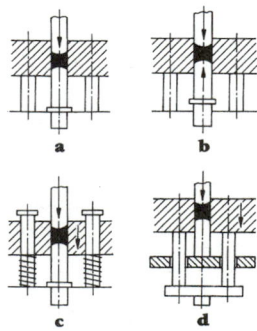

Figure 28. Cold-pressing processes [13]: **a** uni-directional pressing, **b** bi-directional pressing, **c** pressing with sprung sheath, **d** pull-off method.

Hot Pressing. This is chiefly used with powders made from brittle materials for better compaction. *Pressure sintering*, which is often used for composite materials, produces high compaction at a relatively low compression pressure. Better compaction also results from *isostatic pressing*, in which a specimen in a closed plastic or rubber envelope is subjected to very high pressure from all sides in a compression liquid by applying pressure with a plunger. Even higher pressures, required with difficult-to-compress powders, can be achieved by explosive compaction.

Powder Rolling. In this continuous compaction process for producing strip, pressures of several kilobars, as in conventional cold pressing, are applied in the roll gap. Lead bronzes and other composite materials that cannot be manufactured by fusion metallurgy are already being produced by this method on an industrial scale. Extrusion enables bulk or pre-compacted powders, especially those made from low-melting metals (e.g. aluminium), to be converted to various sections and to tube with a practically non-porous material.

Forming Without Compaction. This enables sintered products with high porosity and of simple shape to be manufactured, depending on the pouring method. In slip casting, fine powders are generally mixed with the smallest possible quantity of water or other liquid to form a castable slurry and poured into porous moulds which absorb the liquid, resulting in a practically dry porous moulded product which is then sintered. Among the materials suitable for this process are powders made from nickel, copper, bronze or ferroalloys. Another major advantage is that it can be used to convert difficult-to-compact compounds like oxides, nitrides and silicides, from which it may only be possible to manufacture components, including complex ones, by this method.

Sintering. The particles of powder are permanently bonded by diffusion in muffle, hood-type and often through-type furnaces under shielding gas (hydrogen, NH₃ reforming gas and (to prevent carburisation) partly burnt methane, city gas or producer gas), and, occasionally, under vacuum. Heating takes place electrically or, with very high-melting metals, by direct current passage.

Stages in this solidification process are the adhesion of the particles (even at room temperature, promoted by high compaction), bridge formation between the particles, and compaction of the moulded product, sometimes into a material with practically closed pores.

The operations *pressing* (pouring), *sintering* and *calibration* (after-pressing to increase dimensional accuracy)

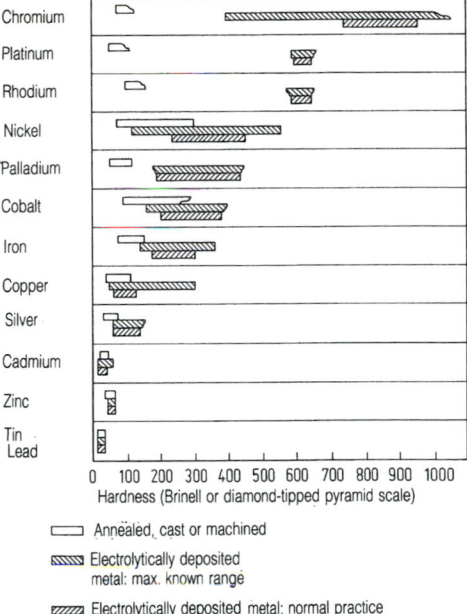

Annealed, cast or machined

Electrolytically deposited
metal: max. known range

Electrolytically deposited metal: normal practice

Figure 29. Hardness of various electrolytically deposited metals.

may be performed in varying combinations to a certain extent, e.g. in:

Single pressing. Pressing–demoulding–sintering (also calibration where necessary).
Double pressing (high density, improved mechanical properties). Pressing–demoulding–sintering–pressing–sintering (also calibration where necessary).
Pressure sintering. Sintering in the mould (for difficult-to-press powders; high density with comparatively low compression pressure)–demoulding–(calibration where necessary).

Aftertreatment. The sintered mouldings may be given their final, ready-for-use condition by a variety of aftertreatment methods, depending on the application, e.g. by forming without cutting (calibration or pressing, rolling, drawing), machining, surface treatment to protect against corrosion or to increase wear resistance (chromising, car-

burising, nitriding, etc.), heat treatment and impregnation (increasing strength by filling the pores with low-melting metal, or with oil in the case of self-lubricating bearings).

2.5 Other Methods of Primary Shaping

2.5.1 Electroforming

In electroforming, a metal deposit is produced on a moulded cathode by electrodeposition. This deposit is either removed as a solid product (electro) or, occasionally, remains bonded to the mould (core electro). Electroforming is suitable for both single items and for series production. The mould, a negative of the part to be formed, is produced by machining, casting or pressing. A positive original mould, on which the negative moulds are made by electroforming, is used when it is not possible to make a negative mould. A positive mould is also the starting point if a sizeable quantity of negatives has to be manufactured for series production.

Examples of materials deposited from aqueous and organic solutions and from salt melts for making shells are: iron, chromium, tungsten, niobium, copper, nickel, cobalt, tin, aluminium, silver, gold, and alloys of nickel–cobalt, nickel–chromium, nickel–manganese, cobalt–tungsten, cobalt–tungsten–nickel–iron. Copper, nickel and, increasingly, nickel–cobalt alloys have in practice attained prime importance, in the last-named case not least because of the favourable combination of mechanical, wear and corrosion properties (**Fig. 29**). Advantages are the extremely high copying accuracy and, with the appropriate mould quality, the high dimensional accuracy. Peak-to-valley heights as small as 0.5 μm can still be copied exactly. The parts manufactured by electrodeposition are on the whole able to satisfy the specifications as to, e.g., high surface quality, uniform layer thickness and required properties in use (structural parts, stencils for screen printing, spinnerets, household articles like pots, shakers, sieves and filters, including micro-filters).

2.5.2 Autocatalytic Plating

Whereas in electroforming the metal ions are reduced by absorption of electrons at the cathode (external current source) and are deposited there as metal or metal alloy, in autocatalytic plating the reduction is accomplished by electron exchange with reactants which, in this non-electrical deposition process, inevitably change to a higher level of oxidation. This requires the use of catalysts.

3 Metal Forming

K. Siegert, Stuttgart

3.1 Classification and Introduction

Metal forming, according to DIN 8580, is the deliberate *alteration* of the *shape*, *surface* and *material properties* of a *workpiece* while *preserving* its mass and cohesion.

The workpiece is generally made of pure metal, a metal alloy manufactured by fusion or powder metallurgy, or a composite material.

Classification of Metal Forming Processes. There are various possibilities, as follows.

One possibility is to classify them according to the predominantly *active stresses* (loads). Thus they are classified as follows:

Forming under compressive conditions (DIN 8583)
Forming under a combination of tensile and compressive conditions (DIN 8584)
Forming under tensile conditions (DIN 8585)
Forming under bending conditions (DIN 8586)
Forming under shearing conditions (DIN 8587)

Another possibility is classification into *sheet/plate forming* processes and *massive forming* processes.

A very important question is whether or not forming

produces a *change in strength*. Therefore a distinction is made between processes where forming produces *no change* in strength, those where a *temporary* change in strength occurs during forming, and processes where forming leads to a *permanent* change in strength.

Depending on whether the workpiece is heated before forming, the term *cold forming* or *hot forming* is used (DIN 8582). In cold forming processes, where the workpiece is introduced into the forming process at room temperature, with metallic materials whose recrystallisation temperature is significantly greater than room temperature, there is generally an increase in yield strength and ultimate tensile strength accompanied by a decrease in elongation after fracture as deformation increases. This is termed *work-hardening*.

Furthermore, a distinction can be made according to the *method of applying force*, viz. forming processes with *direct application of force* and those with *indirect application of force*. For instance, wire drawing, where the drawing force is applied to the forming zone by means of the already-drawn wire, is a process of indirect application of force. Forging, where the force is directly applied to the forming zone by means of the tool, is consequently a process of direct application of force.

The Forming Process. This is determined by several factors: the *workpiece*, the *tool*, the *lubricant*, the *surrounding medium* and the *machine* (including process control). Furthermore, mechanised or automatic transfer of the workpiece into, out of and between the tools should be taken into account.

The tribological systems of the forming process are determined by the workpiece, the tool, the lubricant and the surrounding medium (**Fig. 1**; cf. D5).

In *describing the workpiece* (e.g. metallographic structure, temperature, geometry, surface as well as technological data such as yield point, tensile strength, elongation after fracture and stress/strain curve), the following conditions should be referred to:

On delivery
Immediately before forming
During forming
Immediately after forming
After ageing at room temperature or after heat treatment

For the initial parameters of the forming process, the condition of the workpiece immediately before forming is relevant.

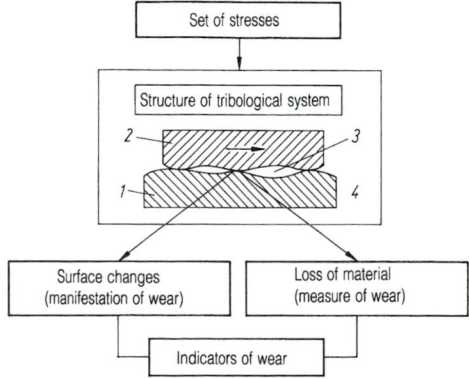

| Set of stresses |
| Structure of tribological system |
| Surface changes (manifestation of wear) | Loss of material (measure of wear) |
| Indicators of wear |

Figure 1. Tribological system according to DIN 50 320; *1* primary object, *2* opposing subject, *3* intervening substance, *4* surrounding medium.

It is important to view the forming process as a link in the *"manufacturing chain"* of a component. Thus the manufacture of the blank to be formed influences the forming process significantly. For example, forging of a cast blank can be optimised only if the alloying constituents, the structure resulting from casting and the heat treatment prior to forging are known. However, the method of further processing or treatment after forming, such as heat treatment, subsequent forming operations, machining, surface treatment, etc. should also be known in order to optimise the forming process, as the entire manufacturing chain determines the characteristics of a component. Optimisation of the forming process should therefore be carried out in the knowledge of and in coordination with the preceding and subsequent manufacturing processes.

3.2 Fundamentals of Metal Forming

3.2.1 Flow Stress

Flow of a material takes place when *permanent deformation* is caused by a specific stress condition. The *flow stress* k_f (also known as the yield strength) is, in a uniaxial tensile test, the *tensile force F* in relation to the respective instantaneous *cross-sectional area A* at which the material flows, i.e. undergoes permanent deformation:

$$k_f = F/A. \qquad (1)$$

(*Note*: with $\sigma = F/A_0$, the force F is expressed in relation to the original cross-sectional area A_0.)

3.2.2 Characteristics of Material Flow

The *logarithmic deformation (degree of deformation)* describes the *magnitude* of the deformation. In the Cartesian system of coordinates, the following applies:

$$\varphi_l = \ln \frac{l_1}{l_0}; \quad \varphi_b = \ln \frac{b_1}{b_0}; \quad \varphi_h = \ln \frac{h_1}{h_0}. \qquad (2)$$

In the polar system of coordinates, the following is obtained:

$$\varphi_l = \ln \frac{l_1}{l_0}; \quad \varphi_r = \ln \frac{r_1}{r_0} = \varphi_t = \ln \frac{r_1}{r_0}. \qquad (3)$$

If a body with the dimensions l_0, b_0, h_0 is transformed by metal forming into a body with the dimensions l_1, b_1, h_1, at constant volume

$$l_1 b_1 h_1 = l_0 b_0 h_0$$

or

$$\frac{l_1}{l_0} \cdot \frac{b_1}{b_0} \cdot \frac{h_1}{h_0} = 1. \qquad (4)$$

Taking logarithms gives the following:

$$\ln \frac{l_1}{l_0} + \ln \frac{b_1}{b_0} + \ln \frac{h_1}{h_0} = 0. \qquad (5)$$

Using Eq. (2), Eq. (5) can be written as

$$\varphi_l + \varphi_b + \varphi_h = 0. \qquad (6)$$

The total of the logarithmic deformations thus equals zero: $\Sigma \varphi = 0$.

The *rate of deformation* is the derivative of the logarithmic deformation in reaction to time:

$$\dot{\varphi} = d\varphi/dt. \qquad (7)$$

The *acceleration of deformation* is the derivative of the rate of deformation in relation to time:

$$\ddot{\varphi} = \mathrm{d}\dot{\varphi}/\mathrm{d}t. \qquad (8)$$

3.2.3 Flow Criteria

The change from *elastic* deformation to *permanent plastic* deformation is described by *flow criteria*. In elementary metal forming theory, Tresca's shear stress hypothesis is generally applied (cf. B9.2.2). This says that flow occurs when the maximum possible shear stress $\tau_{\max}$ reaches the flow stress in shear k of a material:

$$\tau_{\max} = k. \qquad (9)$$

It can be seen from Mohr's circle (cf. B1.1) that

$$\tau_{\max} = \tfrac{1}{2}\,(\sigma_{\max} - \sigma_{\min}) \qquad (10)$$

where $\sigma_{\max}$ is the most positive and $\sigma_{\min}$ the most negative principal stress. For the uniaxial stress condition ($\sigma_1 \neq 0$, $\sigma_2 = \sigma_3 = 0$),

$$\sigma_{\max} = \sigma_1 = F/A = k_\mathrm{f};$$

$$k_\mathrm{f} = 2\tau_{\max} = (\sigma_{\max} - \sigma_{\min}). \qquad (11)$$

This relationship is referred to as *"Tresca's shear stress hypothesis"*. According to this hypothesis, the *principal deformation* φ_g is the numerically largest logarithmic deformation

$$\varphi_\mathrm{g} = \{|\varphi_1|;|\varphi_2|;|\varphi_3|\}_{\max}. \qquad (12)$$

A further hypothesis frequently used in metal forming is the *maximum distortion energy theory* formulated by von Mises and Henky (cf. B1.3.3). This says that flow occurs when the elastic shape modification energy reaches a critical value. With the principal stresses σ_1, σ_2, σ_3 the following applies:

$$k_\mathrm{f} = \sqrt{\tfrac{1}{2}\,[(\sigma_1-\sigma_2)^2 + (\sigma_2-\sigma_3)^2 + (\sigma_3-\sigma_1)^2]}. \qquad (13)$$

If the mean stress is

$$\sigma_\mathrm{m} = \tfrac{1}{3}\,(\sigma_1 + \sigma_2 + \sigma_3), \qquad (14)$$

Eq. (13) gives

$$k_\mathrm{f} = \sqrt{(3/2)\,[(\sigma_1-\sigma_\mathrm{m})^2 + (\sigma_2-\sigma_\mathrm{m})^2 + (\sigma_3-\sigma_\mathrm{m})^2]}. \qquad (15)$$

In pure shear stress,

$$k_\mathrm{f} = \sqrt{3}\,\tau_{\max}. \qquad (16)$$

According to the maximum distortion energy theory, the principal deformation φ_g is

$$\varphi_\mathrm{g} = \sqrt{(2/3)(\varphi_1^2 + \varphi_2^2 + \varphi_3^2)}. \qquad (17)$$

The principal deformation φ_g calculated according to the maximum distortion energy theory is also termed *comparative deformation* φ_v.

Law of Flow. For isotropic materials, according to [1] the following relationship applies between the principal stresses σ_1, σ_2, σ_3 and the associated logarithmic deformations, taking Eq. (14) into account:

$$\varphi_1/\varphi_2/\varphi_3 = (\sigma_1 - \sigma_\mathrm{m})/(\sigma_2 - \sigma_\mathrm{m})/(\sigma_3 - \sigma_\mathrm{m}). \qquad (18)$$

Therefore, if a principal stress equals the mean stress σ_m, the associated deformation is zero.

3.2.4 Flow Curve

The flow stress k_f of a material that is required to achieve and sustain flow depends on the *principal deformation* φ_g, the *rate of principal deformation* $\dot{\varphi}_\mathrm{g}$ and the *temperature* ϑ of the material to be formed:

$$k_\mathrm{f} = f(\varphi_\mathrm{g}, \dot{\varphi}_\mathrm{g}, \vartheta). \qquad (19)$$

In high-speed forming, k_f also depends on the *acceleration of principal deformation* $\ddot{\varphi}_\mathrm{g}$.

In cold forming of metallic materials at forming temperatures well below the recrystallisation temperature,

$$\vartheta \ll \vartheta_{\text{recryst}}, \qquad (20)$$

For most materials (e.g. low-alloy steels, copper, brass, aluminium) the flow stress k_f depends only on the principal deformation φ_g:

$$k_\mathrm{f} = f(\varphi_\mathrm{g}). \qquad (21)$$

It should be noted, however, that with large deformations and rates of deformation, even in cold forming (original temperature of material being formed = room temperature) such high temperatures may occur in the area of the forming zone (e.g. cold extrusion of aluminium) so that the conditions of Eq. (20) no longer apply. If Eq. (20) applies, for most metallic materials the flow curve can be described by the approximation

$$k_\mathrm{f} = a\varphi^n, \qquad (22)$$

where $k_\mathrm{f} \geq R_{\mathrm{p}0.2}$ or R_{eH} (see D2.1).

The exponent n is termed the *consolidation index*, as it determines the rise of the flow curve. A high index n indicates that the material's hardness increases very rapidly with increasing deformation.

As Eq. (22) is only an approximation, indication of the range $\varphi_{\mathrm{g}1} \leq \varphi_\mathrm{g} \leq \varphi_{\mathrm{g}2}$ to which an index n applies is recommended. If the flow curve is depicted on a double logarithmic scale, a straight line with the gradient n results for Eq. (22) (**Fig. 2**).

For hot forming, it is generally the case that the flow stress falls as the temperature increases and rises as the rate of principal deformation φ_g increases. The influence of the principal deformation φ decreases with higher degrees of deformation at elevated temperatures (**Fig. 3**).

A *plot of the flow curve* is generally made at room temperature for a uniaxial tensile test in the zone of uniform elongation [3] and for a uniaxial compression test [4]. At elevated temperatures and high degrees of deformation, flow curves are generally determined using a compression test and a torsion test. Other methods are given in [5] and [6].

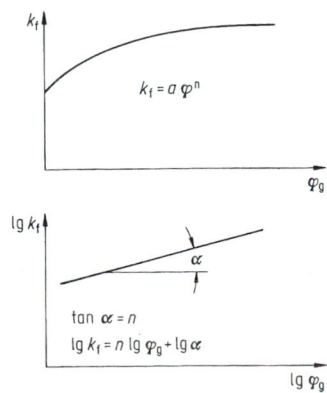

Figure 2. Typical flow curve for $\vartheta \ll \vartheta_{\text{recr}}$.

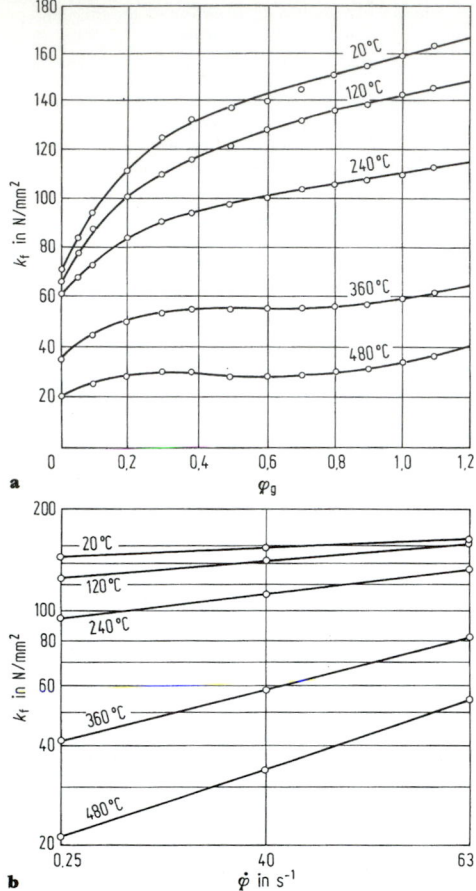

a

b

Figure 3. Flow curves of Al99.5 [2]: **a** flow stress in relation to principal deformation φ_g at $\dot{\varphi} = 4$ s^{-1}, **b** flow stress in relation to the rate of principal deformation $\dot{\varphi}_g$ at $\varphi_g = 1$.

3.2.5 Anisotropy

Anisotropy exists if a material exhibits *direction-specific* properties. In sheet metal forming, *vertical anisotropy* r is defined as the ratio of the logarithmic deformation of the width to the logarithmic deformation of the thickness in a uniaxial tensile test:

$$r = \varphi_b/\varphi_s. \qquad (23)$$

If $r > 1$, longitudinal flow of the material occurs more from the width of the sheet than the thickness. If $r < 1$, the material flows more from the thickness of the sheet. In sheet metal forming the highest possible values for r are aimed at. It should be noted, however, that r generally depends on the position of the sample in relation to the direction of rolling.

In general, r_0 for a sample at 0° to the direction of rolling, r_{45} for a sample at 45° to the direction of rolling and r_{90} for a sample at 90° to the direction of rolling are determined. If $r_0 \neq r_{45} \neq r_{90}$, *earing* (**Fig. 4**) occurs in *deep-drawing* of rotationally symmetrical pots, i.e. the height of the pot is not constant over its circumference,

$$Z = \frac{b_g - b_T}{(b_B + b_T)/2} \cdot 100\%. \qquad (24)$$

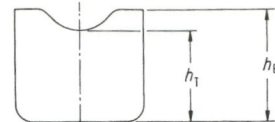

Figure 4. Earing as a result of planar anisotropy.

The vertical anisotropy is often denoted by the *mean vertical anisotropy*

$$r_m = (r_0 + 2r_{45} + r_{90})/4. \qquad (25)$$

For denoting the suitability of a sheet metal for deep-drawing, however, this indication seems appropriate only to a limited extent. A better way is to denote the *vertical* anisotropy by the r_{min} value. The suitability of a sheet metal for drawing of rotationally symmetrical pots with as little trimming waste as possible is denoted by the *planar* anisotropy:

$$\Delta r = r_{max} - r_{min}. \qquad (26)$$

3.2.6 Formability

Formability means the plastic deformation that a specific material is able to withstand in the forming zone up to the point of fracture at a specific stress condition, temperature and rate of deformation. Other parameters such as acceleration of deformation may also be relevant at extremely high rates of deformation. Formability, measured for instance as the deformation on fracture, depends greatly on the stress condition. The more negative the mean stress according to Eq. (14) or, in other words, the greater the mean compressive stress, the greater is the formability [7]. Here, however, the principal stress σ_2 is also relevant if $\sigma_1 > \sigma_2 > \sigma_3$ applies. At the same mean stress σ_m the formability is greatest when σ_2 becomes equal to σ_3. It decreases as σ_2 becomes larger and is smallest when $\sigma_2 = \sigma_1$ [8].

Figure 5 illustrates the deformation on fracture as a measure of formability by means of the mean stress in relation to k_f.

Warning: In processes with indirect application of

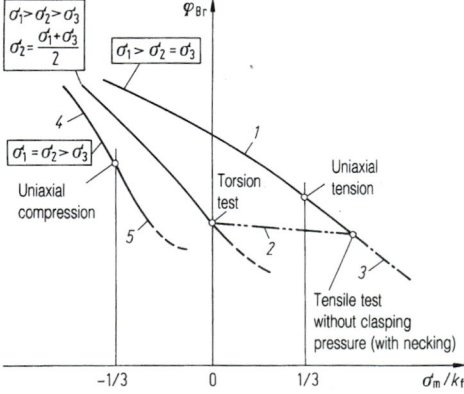

Figure 5. Deformation on fracture as a measure of formability over the mean stress σ_m in relation to k_f [8]; *1* tensile tests with clasping pressure, *2* torsion tests with axial tensile stress, *3* notched bar tensile tests, *4* compression tests with clasping pressure, *5* compression tests with radial tensile stress.

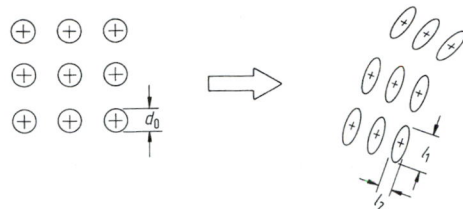

Figure 6. Deformation analysis in sheet metal forming by grid measurement.

force, the failure case "fracture" generally occurs *outside* the forming zone. This is a question of the limits of deformation, which are generally process-specific.

3.2.7 Forming Limit Curve (FLC)

In sheet metal forming, the deformations are often analysed by marking the sheet with a grid of circles (diameter of circles e.g. 4.5 mm) before forming and measuring the resulting ellipses after forming [9] (**Fig. 6**).

The forming limit curve is obtained by plotting against one another the deformations φ_1 and φ_2 in the plane of the sheet at which necking and fracture occur. The larger deformation is plotted above the smaller deformation (**Fig. 7**).

This curve applies only if the path of deformation, until failure due to necking or fracture, occurs at a constant ratio of φ_1 to φ_2. It should be noted that deformation of the thickness φ_s is calculated from Eq. (6) as

$$\varphi_3 = \varphi_s = -[\varphi_1 + \varphi_2]. \qquad (27)$$

Figure 8 shows a forming limit curve for φ_1, φ_2 and $\varphi_3 = \varphi_s$ for the start of necking [10].

3.3 Theoretical Models

The elementary theory of plasticity (cf. B9) originates from work by Siebel, Karmann, Sachs and Pomp [11–14]. This elementary theory was revised, generalised and expanded by Lippmann and Mahrenholtz [15, 16] (cf. also [17, 18]).

Three fundamental models are taken as the basis (**Fig. 9**). The following assumptions are made for the discussion which follows:

Homogeneous forming (pure elongations/shearing strains) in the individual strips, discs and tubes. The principal axes correspond to the axes of the bodies. The strips and discs remain flat during forming, the tubes remain

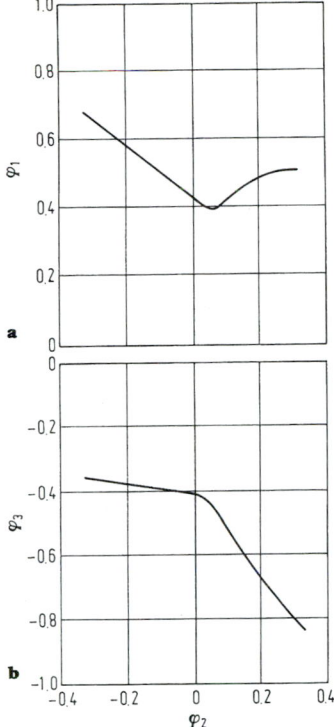

Figure 8. a Forming limit curve. **b** Deformation of thickness according to Eq. (6).

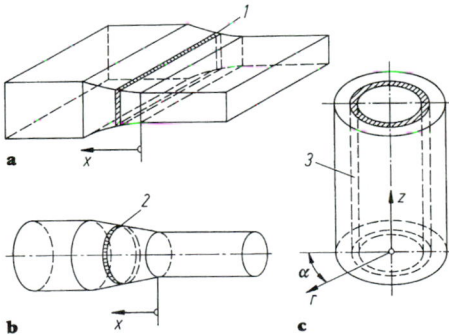

Figure 9. Basic models of elementary forming theory: **a** strip model, **b** disc model, **c** tube model; *1* strip, *2* disc, *3* tube.

cylindrical. In this way of looking at the situation, the real conditions are deliberately ignored.

Homogeneous, Isotropic Material

Friction According to Coulomb's Law of Friction. Friction is constant over the area of contact between tool and workpiece

$$\tau = \mu\, p_n \quad (\mu = \text{const}). \qquad (28)$$

Although forces of gravity and inertia can be allowed for in the model, they are usually negligible. They are not allowed for in this description.

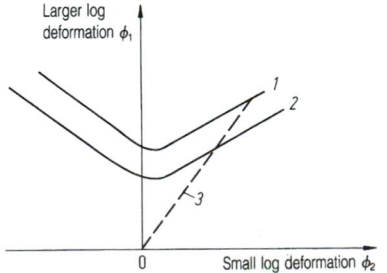

Figure 7. Forming limit curve (FLC); *1* fracture, *2* necking, *3* strain path.

The flow stress k_f is constant over the strip, disc or tube.

Strip Model

This model was developed by Siebel and von Karman. According to **Fig. 9**, a strip of the material being formed is considered, which is formed from two parallel boundary surfaces a differentially small distance apart which are bounded at top and bottom by the forming tool (**Fig. 10**).

As this strip width dx is required to be differentially small, the top and bottom boundaries can be described by straight lines to be regarded as tangents to the tool contour. The angle of these tangents to the horizontal is a function of x, as is the height of the strip:

$$\alpha_1 = f(x); \quad \alpha_2 = f(x); \quad h = f(x).$$

In the strip theory model the strip is assumed to be formed in such a way that the two cross-sectional surfaces bounding the strip remain flat and parallel to each other. Every time this theoretical model is applied to a specific metal forming process, these assumptions should be checked as to what extent they adequately describe what actually happens.

If the forces of inertia acting on the material being formed are ignored, only stresses that act directly on the cross-sectional surfaces and those that act directly on the boundary surfaces have to be taken into account. If there is friction between the material being formed and the tool, *edge shear stresses* occur at the boundary surfaces.

Shear stresses may also occur at the cross-sectional surfaces if the tool contour forces the strip to undergo sudden deformation. Thus **Fig. 10** shows that as the strip enters the forming zone, α jumps from 0 to α_1 or α_2 and jumps back to $\alpha = 0$ as the strip leaves the forming zone. These shear stresses occurring at the cross-sectional surfaces are not considered for the time being. They are taken into account later as *"losses due to shear strain"* (or, in the more recent literature [15], *"tangential adjustments"*).

According to **Fig. 11**, the stresses acting are therefore the compressive stresses on the cross-sectional surfaces p_x and $(p_x + dp_x)$, the compressive stresses on the boundary surfaces p_{n1} and p_{n2}, and the edge shear stresses τ_1 and τ_2.

According to Coulomb's law of friction,

$$\tau_1 = p_{n1}\mu_1, \quad \tau_2 = p_{n2}\mu_2. \tag{29}$$

From the stresses, the forces acting directly and tangentially on these surfaces are obtained by multiplying by the respective surface areas (**Fig. 12**). By resolving the forces into horizontal and vertical forces, the following apply at the boundary surfaces (**Fig. 13**):

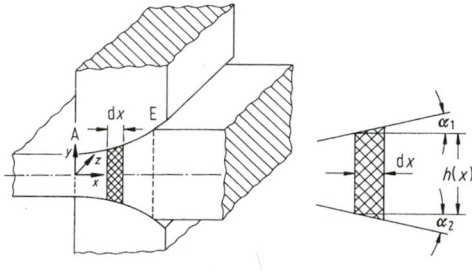

Figure 10. Strip model.

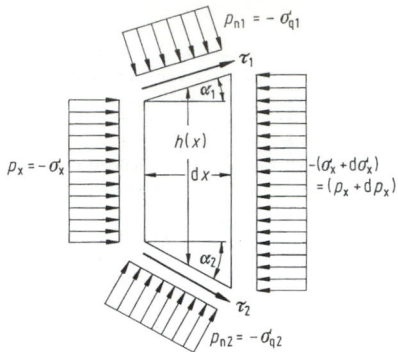

Figure 11. Stresses acting on strip element.

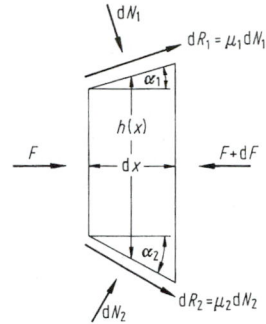

Figure 12. Forces acting on strip element.

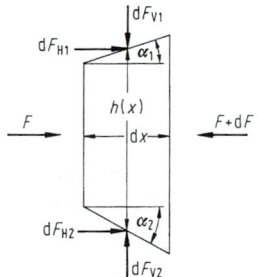

Figure 13. Resolution of forces acting on strip element.

$$dF_{H_1} = p_{n_1} \, dx \, dz(\tan\alpha_1 + \mu_1);$$
$$dF_{H_2} = p_{n_2} \, dx \, dz(\tan\alpha_2 + \mu_2). \tag{30}$$

$$dF_{V_1} = p_{n_1} \, dx \, dz(1 - \mu_1 \cdot \tan\alpha_1);$$
$$dF_{V_2} = p_{n_2} \, dx \, dz(1 - \mu_2 \cdot \tan\alpha_2). \tag{31}$$

If $p_y = dF_V/(dx \, dz)$ is defined as the vertical compressive stress, with $\tan\rho = \mu$ one obtains:

$$dF_{H_1} = p_{y1} \, dx \, dz \tan(\alpha_1 + \rho_1);$$
$$dF_{H_2} = p_{y2} \, dx \, dz \tan(\alpha_2 + \rho_2). \tag{32}$$

Thus the horizontal and vertical forces at the boundary surfaces of the strip elements that are plotted in **Fig. 13** can be described by Eqs (30), (31) and (32).

Disc Model

The strip model is based on a strip of depth dz. For metal forming processes with an axially symmetrical forming zone (e.g. extrusion, wire drawing), it is advisable to imagine that in the forming zone the material being formed consists of "discs" (**Fig. 9**). These discs have a differentially small thickness and a defined outside diameter. A defined inside diameter is also used where applicable.

For the disc model it is assumed that the discs are formed in such a way that the cross-sectional surfaces remain plane and parallel to each other. For solid cross-sections the contour can be described by

$$D = f(x). \tag{33}$$

To avoid misunderstandings in differentiation and integration, the diameters are designated by D_0, D_1 and $D = f(x)$ (**Fig. 14**).

The boundary surface of the disc is formed by the tangential surface at the contour of the forming zone at point x. The gradient of these tangential surfaces is obtained from

$$d(D(x))/dx. \tag{34}$$

If x is the angle of slope of the tangent and also a function of x according to Eq. (34), then $\alpha = f(x)$. For *conical* forming zones, α = const. Thus the following applies to the boundary surface d$A(x)$ of the disc element:

$$dA(x) = D(x)\pi \, dx/\cos \alpha. \tag{35}$$

As in the strip model the boundary surface d$A(x) =$ dz d$x/\cos \alpha$, Eq. (35) gives the following by analogy with Eq. (30):

$$dF_H = p_n D(x)\pi \, dx(\tan \alpha + \mu), \tag{36}$$

and by analogy with Eq. (31),

$$dF_r = p_n D(x)\pi \, dx(1 - \mu \tan \alpha), \tag{37}$$

and gives the following with the radial compressive stress p_r by analogy with Eq. (32):

$$dF_H = p_r \, dx \, D(x)\pi \tan(\alpha + \rho). \tag{38}$$

3.4 Stresses and Forces in Selected Metal Forming Processes

3.4.1 Upsetting of Cylindrical Parts

Use is made of the tube model, Fig. 15. The equilibrium of forces gives, with $\sigma_r = \sigma_t$, d$r \cdot$ d$x = 0$, $\sin(d\alpha/2) \cong dx/2$, Tresca's hypothesis and assumption of Coulomb friction,

$$\frac{d\sigma_r}{dr} + \frac{\mu}{b}\sigma r - \frac{2\mu}{b}k_f = 0. \tag{39}$$

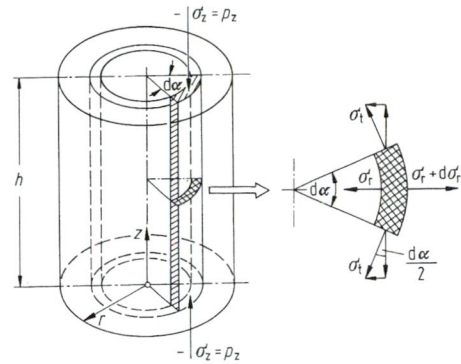

Figure 15. Stress conditions in tube model.

Solving this first-order differential equation, taking into account that $\sigma_r = 0$ at $r = d/2$ and $p_z = -\sigma_r$, gives:

$$p_z = k_f \cdot \exp\left[\frac{2\mu}{b} ((d/2) - r)\right]. \tag{40}$$

By series expansion and truncation after the first term, this gives

$$p_z = k_f \left[1 + \frac{2\mu}{b}((d/2) - r)\right]. \tag{41}$$

Under frictionless conditions ($\mu = 0$),

$$p_z = k_f. \tag{42}$$

The upsetting force F_z is given by integration of Eq. (41) over the compressed area

$$A_D = A_0 b_0/b \tag{43}$$

as

$$F_z = k_f A_D \left[1 + \frac{1}{3}\frac{\mu d}{b}\right]. \tag{44}$$

At the end of the operation, i.e. when d_1 and b_1 are attained,

$$F_{z_{max}} = F_{z_1} = k_{f_1} A_0 \frac{b_0}{b_1}\left[1 + \frac{1}{3}\frac{\mu d_1}{b_1}\right]. \tag{45}$$

If the resistance to deformation is designated as

$$k_{W_1} = k_{f_1}\left[1 + \frac{1}{3}\frac{\mu d_1}{b_1}\right]. \tag{46}$$

Eq. (45) can also be written as:

$$F_{z_{max}} = F_{z_1} = k_{W_1} A_{D_1}. \tag{47}$$

3.4.2 Upsetting of Square Parts

Using the strip model, the following is obtained by analogy with Eq. (45):

$$F_{z_{max}} = F_{z_1} = k_{f_1} A_0 b_0/b_1\left[1 + \frac{1}{2}\frac{\mu b_1}{b_1}\right]. \tag{48}$$

3.4.3 Wire Drawing

In drawing the wire, the original diameter of the wire $D_0 = D_E$ is reduced to the diameter $D_1 = D_A$. The *die*, also known as the *drawing hole*, is the forming tool here. The drawing force acts on the emerging wire and is transmitted by this means to the forming zone. This process

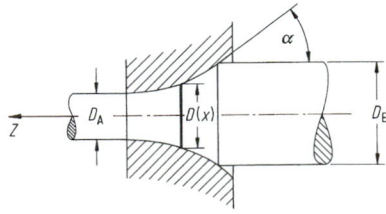

Figure 14. Geometrical conditions applying to disc model (example: wire drawing).

is therefore one with *indirect* action of force (**Fig. 16**). Characteristic geometrical values are

$$A_E = \pi D_E^2/4; \quad A_A = \pi D_A^2/4; \quad A(x) = \pi D(x)^2/4. \quad (49)$$

$$l_u(D_E - D_A)/2 \tan \alpha; \quad (50)$$

$$D(x) = D_A + 2x \tan \alpha. \quad (51)$$

If the shear strain losses are ignored and frictionless conditions are assumed ($\mu = 0$), the so-called ideal stress σ_{id} is obtained. It is given by Eq. (36) and the shear stress hypothesis as:

$$\sigma_{id}(x) = \sigma_x(x) = 2k_{fm} \cdot \ln D_E/D(x). \quad (52)$$

Here, k_{fm} is the arithmetical mean of the flow stress on entering the forming zone k_{fE} and the flow stress in the cross-section under consideration $k_f(x)$:

$$k_{fm} = (k_{fE} + k_f(x))/2. \quad (53)$$

At the exit from the die, the following applies:

$$\sigma_{id_{max}} = \sigma_{id}(x = 0) = 2\bar{k}_f \ln (D_E/D_A), \quad (54)$$

where

$$\bar{k}_f = (k_{fE} + k_{fA})/2. \quad (55)$$

k_{fA} is the flow stress in the exit plane of the forming zone.

Multiplied by the relevant cross-sectional area, the ideal drawing force is given with the aid of Eq. (54) as:

$$F_{id} = \sigma_{id_{max}} A_A = \bar{k}_f \, \varphi_{g_{tot}} A_A. \quad (56)$$

Here the logarithmic total principal deformation is

$$\varphi_{g_{tot}} = 2 \ln (D_E/D_A). \quad (57)$$

Under conditions of friction ($\mu \neq 9$), the following applies to conical dies on the basis of Eq. (36):

$$\sigma_x(x = 0) = \bar{k}_f \left(1 + \frac{\tan \alpha}{\mu}\right)\left[1 - \left(\frac{D_A}{D_E}\right)^{\frac{2\mu}{\tan\alpha}}\right]. \quad (58)$$

If the shear strain losses, i.e. the angular distortions of the "discs" on entering and leaving the forming zone, are taken into account, an axial stress portion σ_{disc} is necessary for this. According to [19, 20] this is

$$\sigma_{disc} = \frac{1}{3} \tan \alpha (k_{fE} + k_{fA}), \quad (59)$$

or, using Eq. (55),

$$\sigma_{disc} = \frac{2}{3} \bar{k}_f \tan \alpha. \quad (60)$$

If this portion is added to Eq. (58), the result is [19]

$$\sigma_{x_{tot}} = \bar{k}_f\left\{\left[1 + \frac{\tan \alpha}{\mu}\right]\left[1 - \left(\frac{D_A}{D_E}\right)^{\frac{2\mu}{\tan\alpha}}\right] + \frac{2}{3}\tan\alpha\right\}. \quad (61)$$

By series expansion and interruption after the first element, for the small angles α that occur in wire drawing the following relationship established by E. Siebel is obtained:

$$\sigma_{x_{tot}} = \sigma_x(x = 0) = \bar{k}_f\varphi_{g_{tot}}\left[1 + \frac{\mu}{\hat{\alpha}} + \frac{2}{3}\frac{\hat{\alpha}}{\varphi_{g_{tot}}}\right]. \quad (62)$$

The drawing force is given as

$$F_{tot} = \frac{\pi \cdot D_A^2}{4} \bar{k}_f \, \varphi_{g_{tot}}\left[1 + \frac{\mu}{\hat{\alpha}} + \frac{2}{3}\frac{\hat{\alpha}}{\varphi_{g_{tot}}}\right]. \quad (63)$$

where $|\hat{\alpha}|$ is the angle α as a radian measure. The optimum angle $|\hat{\alpha}|_{opt}$ is given by Eq. (63) and $dF_{tot}/d\alpha = 0$ as

$$\hat{\alpha}_{opt} = \sqrt{1.5\mu \, \varphi_{g_{tot}}}. \quad (64)$$

3.4.4 Extrusion

In extrusion, the blank is pushed through a forming tool (*die*). According to DIN 8583, the extrusion processes consist of the forming processes *tapering*, *extrusion* and *extrusion moulding*.

Stresses in the Forming Zone

The following discussion of the stresses in the forming zone applies to all three processes, assuming that a *conical die* is used (**Fig. 17**). If the elementary theory is taken as a basis, the geometrical and kinematic conditions in the forming zone are identical. The forming zone is bounded by the wall of the die, the entry plane ($x = l_u$) and the exit plane ($x = 0$). The difference compared with wire drawing lies in the action of the force. In similar fashion to wire drawing, for *tapering* the axial compressive stress in the inlet plane is

$$P_{x_E} = P_x(x = l_u)$$

$$= \bar{k}_f\left\{\left[1 + \frac{\tan \alpha}{\mu}\right]\left[\left(\frac{D_E}{D_A}\right)^{\frac{2\mu}{\tan\alpha}} - 1\right] + \frac{2}{3}\tan\alpha\right\}. \quad (65)$$

For small angles α, $\tan \alpha \approx |\hat{\alpha}|$.

By series expansion and interruption after the first element, Eq. (65) gives

$$P_{x_E} = \bar{k}_f \, \varphi_{g_{tot}}\left[1 + (\mu/\hat{\alpha}) + \frac{2}{3}(\hat{\alpha}/\varphi_{g_{tot}})\right]. \quad (66)$$

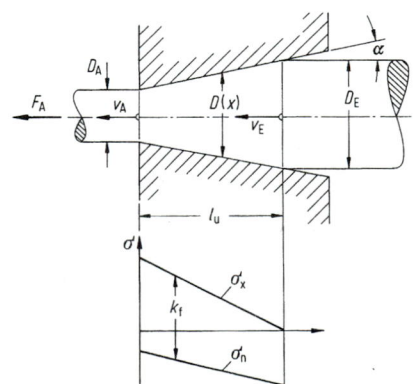

Figure 16. Geometrical conditions and basic pattern of stresses in wire drawing with a conical die.

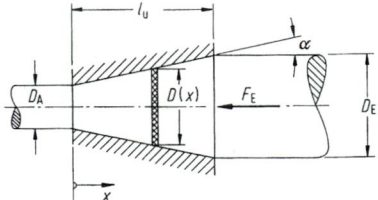

Figure 17. Geometrical conditions in extrusion – conical die.

where $\varphi_{g_{tot}} = 2 \ln (D_E/D_A)$ and $\bar{k}_f = (k_{fE} = k_{fA})/2$.

In wire drawing and tapering, the die angles α divided by 2 are relatively small; thus, for

$$p_n = p_r[1/(1 - \mu \tan \alpha)]$$

it can be assumed that $p_n \approx p_r$.

This is unacceptable for *extrusion*. As $\alpha \gg 0$, $p_n \neq p_r$ and $p_r = k_f + p_x$.

This gives, taking into account Eqs (55) and (22),

$$p_{x_E} = p_x (x = l_u) = \bar{k}_f \frac{C_1}{C_1 - 1} \left[\left(\frac{A_E}{A_A} \right)^{C_1 - 1} - 1 \right], \quad (67)$$

where $C_1 = (1 + \mu \cdot c \tan \alpha)/(1 - \mu \tan \alpha)$.

If Eq. (67) is expanded to include the axial stress portion σ_{disc} (Eq. (60)), the result using Eq. (55) is

$$p_{x_E} = \bar{k}_f \left\{ \frac{C_1}{C_1 - 1} \left[\left(\frac{A_E}{A_A} \right)^{C_1 - 1} - 1 \right] + \frac{2}{3} \tan \alpha \right\}. \quad (68)$$

This is the axial compressive stress in the die entry plane.

Stresses in the Material Being Formed Outside the Forming Zone

Consideration is given to the stress condition in the cylindrical, upset part of the ingot preceding the forming zone $x_E \leq x \leq x_E + l_0 - s$ according to **Fig. 18**: for the edge shear stress acting on the surface of the ingot in this area, with Coulomb friction the following applies:

$$\tau_R = \mu_0 p_r \quad (69)$$

An upper limit applies if the material being formed is sheared off within a boundary layer:

$$\tau_R = \tau_{crit} = k_{f0}/2. \quad (70)$$

The radial compressive stress p_r is calculated from Tresca's hypothesis as

$$p_r = p_x - k_f. \quad (71)$$

To determine the axial compressive stress P_x in the range $0 \leq x \leq l_0 - s$, the equilibrium of forces at a cross-sectional disc is examined:

$$dp_{p_x} = 4(\tau_R/D_E) \, d\bar{x}. \quad (72)$$

For shearing off, the result is as follows, using Eq. (70) and taking Eq. (68) into account:

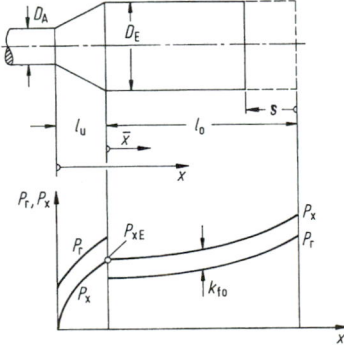

Figure 18. Curves of axial compressive stress P_x and radial compressive stress p_r over x, assuming Coulomb friction.

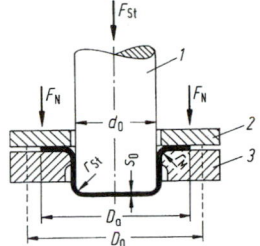

Figure 19. Schematic diagram of deep drawing with punch applied: *1* punch, *2* blank holder, *3* drawing die, D_0 outside diameter of blank, d_0 punch diameter, s_0 thickness of blank, F_{st} force of punch, F_N force of blank holder, r_{st} punch edge radius, r_M die radius.

$$p_x(\bar{x}) = 2(k_{f_0}/D_E)\bar{x} + p_{x_E}. \quad (73)$$

The compressive stress at the end of the ingot is obtained from this:

$$p_x(\bar{x} = l_0 - s) = 2(k_{f0}/D_E) (l_0 - s) + p_{x_E}. \quad (74)$$

If Coulomb friction applies, the following is obtained from Eq. (72), using Eq. (69) and taking Eq. (68) into account:

$$p_x = p_{x_E} \exp [4(\mu_0/D_E) \bar{x} + k_{f0} \{1 - \exp [4(\mu_0/D_E) \bar{x}]\} \quad (75)$$

(cf. [23-25]).

The resulting compressive stress at the end of the ingot is:

$$p_x(\bar{x} = l_0 - s) = p_{x_E} \exp[4(\mu_0/D_E) (l_0 - s)]$$
$$+ k_{f0} \{1 - \exp[4(\mu_0/D_E) (l_0 - s)] \} \quad (76)$$

This relationship was determined by Eisbein [23] and Sachs [24] (cf. [25]).

The ram force F_{St} is given by multiplying the compressive stress at the end of the ingot by the cross-sectional area of the ingot:

$$F_{St} = p_x (\bar{x} = l_0 - s) \cdot (\pi D_E^2/4). \quad (77)$$

Equation (74) or (76) should be inserted in this equation, depending on the friction conditions.

3.4.5 Deep Drawing

Deep drawing is a method of forming a flat sheet-metal blank into a hollow part. **Figure 19** illustrates the tool arrangement and the terminology for deep drawing of rotationally symmetrical parts. The *forming zone* is the area of sheet underneath the *blank holder* up to the exit from the curvature of the drawing die. **Figure 20** shows

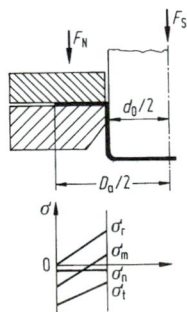

Figure 20. Schematic diagram of stress pattern for deep drawing with punch applied.

the basic stress pattern for this process. It can be seen that the normal stress σ_n intersects the mean stress σ_m. According to the law of flow as in Eq. (18), therefore, at this point no deformation takes place in the direction of thickness. It can also be deduced that an increase in sheet thickness towards the flange edge and a decrease in sheet thickness towards the radius of the drawing die entry take place. On average, however, the surface remains approximately constant in deep drawing:

According to [26], the total drawing force $F_{tot} = F_{St}$ is

$$F_{St} = F_{id} + F_{RN} + F_{RZ} + F_{rb}. \tag{78}$$

F_{id} is the ideal force required for no-loss forming:

$$F_{id} = \sigma_{id}\pi\, d_0 s_0 \tag{79}$$

with $\sigma_{id} = |\sigma_r| = k_{fm} \ln (D_a/d_0)$ and $k_{fm} = (k_{fi} + k_{fa})/2$. Here, k_{fi} is the flow stress at the drawing die run-out for $r = d_0/2$ and k_{fa} is the flow stress at the outside flange diameter for $r = D_a/2$.

These are determined from

$$\varphi_{g_i} = \ln \sqrt{(D_0^2 + d_0^2 - D_a^2)/d_0^2}, \quad \varphi_{g_a} = \ln D_0/D_a. \tag{80}$$

F_{RN} is the frictional force arising between the blank and the drawing die and between the blank and the blank holder. According to Paknin [26],

$$F_{RN} = 2\mu F_N\,(d_0/D_a). \tag{81}$$

Here, $F_N = \pi/4(D_0^2 - d_0^2)p_n$ for the pressure of the blank holder p_n, which is set as constant throughout the drawing process. According to Siebel,

$$p_n = (0.002 \text{ to } 0.0025)\,[(\beta_0 - 1)^2 + 0.5(d_0/100s_0)]\, R_m,$$

where $\beta_0 = D_0/d_0$ drawing ratio.

F_{rb} is the frictional force arising between the workpiece and the curvature of the drawing die:

$$F_{rb} = (e^{\mu\pi/2} - 1)(F_{id} + F_{RN}). \tag{82}$$

F_{rb} is the reverse bending force that is necessary after leaving the drawing die:

$$F_{rb} = \pi d_0 s_0 k_{fi}\,(s/4r_m). \tag{83}$$

According to **Fig. 21**, the drawing force F_{St} has a maximum which, according to [26], is $h/h_{max} = 0.4$ for most metallic materials.

3.5 Technology

3.5.1 Stretch Forming

Stretch forming is used to produce large sheet-metal parts (VDI 3140). A distinction is made between:

Simple Stretch Forming. The sheet-metal blank is clamped at two opposing sides. Forming is accomplished by the movement of the former (**Fig. 22a**). Owing to the friction between the former and the blank, uniform distribution of the elongations over the component is prevented. Failure occurs between the clamping jaws and the areas not in contact with the formers.

Tangential Stretch Forming (**Fig. 22b**). This enables uniform distribution of the elongations over the workpiece and a higher degree of forming in the central area. The sheet-metal blank is clamped in vertically and horizontally movable jaws and prestressed with them until the blank has undergone plastic deformation of 2 to 4%. If the formed component is to be embosssed, the stretch forming device can be incorporated in a single-acting press with countermould (**Fig. 23**).

The Cyril-Bath process, in which the clamping jaws can be moved horizontally and vertically by CNC, enables the central area of large, drawn sheet-metal components to undergo a higher degree of forming and thus enables a higher degree of work-hardening to be obtained [27, 28]. Sheet metal for stretch forming should have the highest possible consolidation index n, so that the deformations are distributed as uniformly as possible over the component and a premature local tear is avoided (cf. K3.2.4). The friction between the sheet and the stretch forming tool should be as low as possible ($\mu \rightarrow 0$). On materials see D3.1.4.

3.5.2 Deep Drawing

A distinction is made between:

First-operation drawing (**Fig. 19**)

Second-operation drawing (DIN 8584; **Fig. 24**)

Whereas in stretch forming the shape of the formed part is obtained by enlargement of the surface area at the expense of the thickness, because the sheet is clamped at the sides and cannot continue flowing, in deep drawing the sheet thickness over the drawn component is approximately constant in first-operation drawing, which means that the surface area of the blank equals the surface area of the drawn product. A minimum holding-down force F_N is required in order to suppress puckering underneath the blank holder (cf. K3.4.5). If possible, the sheet metal should have a Δr of zero so as to avoid earing. r_{min} should be as large as possible, as should the index n (cf. K3.2.5). To reduce friction at the interfaces of blank holder, sheet and drawing die and also in the area of the curvature of

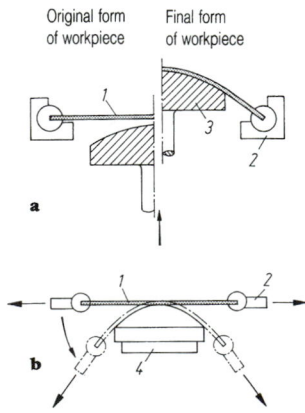

Original form   Final form
of workpiece   of workpiece

a

b

Figure 22. Stretch forming: **a** simple stretch forming, **b** tangential stretch forming; *1* workpiece, *2* clamping jaw, *3* former, *4* tool.

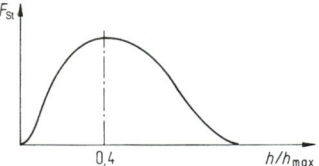

F_{St}

0,4 h/h_{max}

Figure 21. Curve of punch force F_{St} over the specific drawn product height.

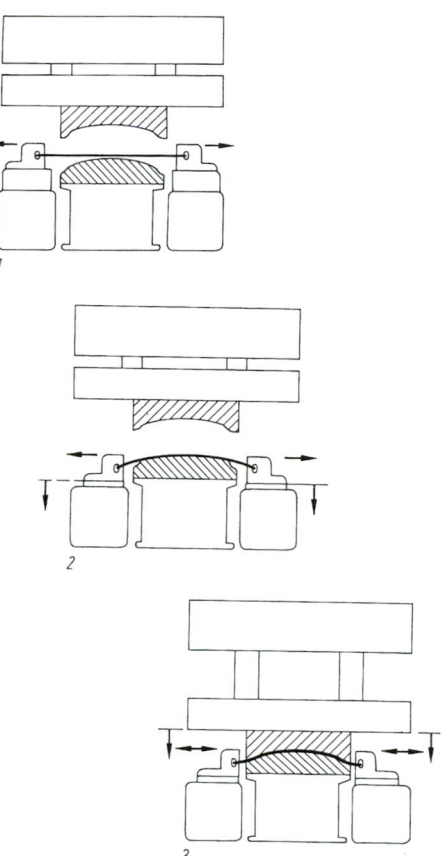

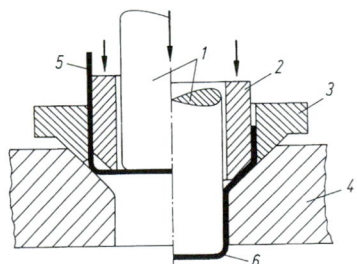

Figure 23. Cyril-Bath process (Cyril-Bath Company). *1* to *3*: sequence of operations.

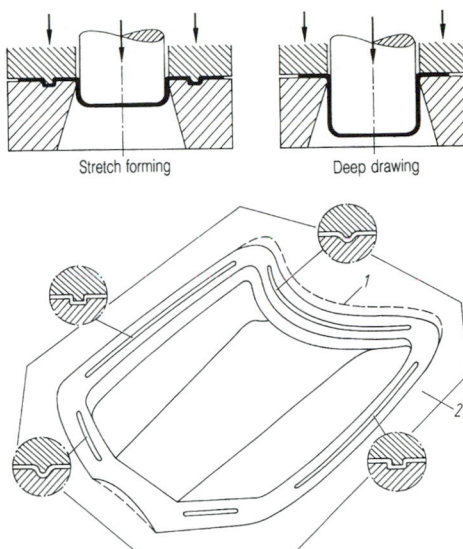

Figure 25. Drawing of large non-symmetrical parts by a combination of stretch forming and deep drawing; *1* shape of blank, *2* drawing frame.

drawing (**Fig. 25**). The flow of material underneath the blank holder is influenced by drawing channels, the shape of the blank, zones of higher and lower surface pressure and a specifically directed lubricant supply.

In drawing with a *single-acting* press, a drawing device is placed in the bed of the press. This can be designed as a pneumatic or hydraulic drawing device [29, 30]. **Figure 26** shows a tool arrangement for drawing with a *double-acting* press.

3.5.3 Bending

Bending is one of the most commonly used methods of sheet metal forming. Its uses range from mass production

Figure 24. Second-operation drawing [5]; *1* punch, *2* blank holder, *3* support ring, *4* drawing die, *5* cup after first drawing, *6* cup being redrawn.

the drawing die, the coefficient of friction should be as small as possible ($\mu \to 0$). If the coefficient of friction between the sheet and the drawing die is relatively large ($\mu \to 1$), the drawing force applied can be increased by means of frictional forces and a larger limit drawing ratio obtained. On materials see D3.1.4.

Drawing of non-symmetrical parts (e.g. car body panels) is by a combination of stretch forming and deep

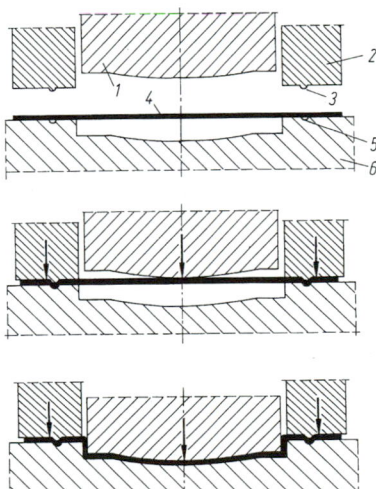

Figure 26. Schematic diagram of drawing of large, irregular, sheet-metal panels with a double-acting press; *1* punch, *2* blank holder, *3* drawing tongue, *4* sheet, *5* drawing channel, *6* bed.

of small parts to fabrication of single items in shipbuilding and industrial plant construction [31]. Apart from sheets, bending is performed above all on tubes, wire and bar of widely differing cross-sections. Cold forming is used in most cases; only in exceptional cases, i.e. for large cross-sections or very small bending radii, is the material heated to reduce the forces required for forming or to enable higher degrees of deformation to be achieved with a given material. Elementary bending theory [32] is discussed in B2.4.

A phenomenon typical of bending is elastic *spring-back*. After removal of the load, the bending angle is smaller and the radius of the bent part larger than under load. In bending without radial stress, the spring-back on removal of the load is given by the spring-back ratio (**Fig. 27**):

$$K = \frac{\alpha_R}{\alpha} = \frac{\left(r_i + \dfrac{s_0}{2}\right)}{\left(r_{iR} + \dfrac{s_0}{2}\right)}.$$

The degree of spring-back depends on the material (modulus of elasticity, yield point, index n), the stress condition under which forming is carried out, and the initial deformation of the part to be bent. Depending on whether plastic deformation is successfully produced over the full cross-section of the workpiece to be bent, or whether elastic deformations predominate in the neutral fibre zone, the degree of spring-back is large or small. Spring-back can be reduced, prevented or compensated for by the following methods:

Limiting the tolerances for the sheet parameters $R_{p0.2}$, index n and sheet thickness as a prerequisite for reproducible conditions.

Over-bending.

After-pressing in a forging die.

Subsequent forming under tensile stress to obtain the final geometry, i.e. further forming under such high-tensile stresses that the whole cross-section undergoes plastic deformation.

Superposition of tensile stresses during bending.

In determining the blanks for bent products, empirical formulas have to be relied on at present, as precise determination of the geometry of the bent component is not yet possible [33]. If the bending radius is less than a material-specific minimum radius $r_{i\,min}$, cracks occur in the outer fibre. Data on minimum bending radii for steel sheet are given in DIN 6935 in relation to the material, sheet thickness, and bending axis position in relation to the rolling direction of the sheet.

Bending Methods [32]

Free Bending and Folding. Technologically important methods are free bending with a three-point support or free bending of a sheet clamped at one side with a beam applied to the projecting part. If this beam turns through an angle, the method is called *folding*.

Bending in a Vee-die (V-bending). Here, two sub-operations occur in succession [35]. First, free bending takes place until the sides of the product being bent are in contact with the walls of the die ($\alpha = \alpha_G$, **Fig. 28a**) or until $r_i < r_{St}$ (r_i = inside radius of component being bent, r_{St} = radius of die). This is directly followed by further pressing in a die. The shape of the bent product is essentially altered to the shape of the tool. If r_{St} is small, the workpiece is overbent until the bending legs lie against the die (**Fig. 28c**). If the load is removed in this position, the bending angle α may still be larger than the die angle α_G. The inside radius steadily becomes smaller during further pressing. The same phenomena occur with regard to the bending angle applicable to further pressing both if the die radius r_{St} (or r_{St}/s_0) is large and if it is small. The precision of the bent components can be improved during further pressing, but this requires large forces (**Fig. 29**).

Example. Folding to produce a shape (**Fig. 30**).

Channel or U-Bending. This denotes the simultaneous bending of two sides joined by a web, generally through 90°, in a die to produce a U-shaped component (**Fig. 31**). A distinction is made between U-bending *with* and *without* a pressure pad. In U-bending *without a pressure pad*, the bellying of the web can be largely cured by further pressing. This leads to an increase in force at the end of the operation [34]. Influencing factors are the curvature of the die, its depth and the tool gap. In bending *with a pressure pad* (where the force of the pressure pad is approximately one-third of the bending force), the web remains flat during forming.

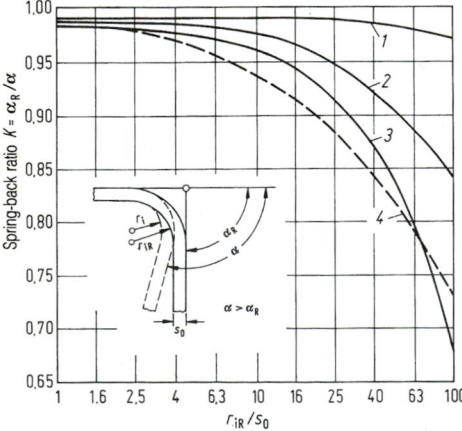

Figure 27. Spring-back ratio K of various materials in relation to the bending radius [5]; *1* Al99.5w, *2* St 1404, *3* St 1203, *4* CuZn33w.

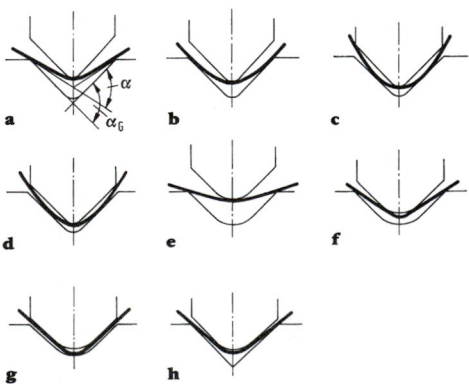

Figure 28. Bending in a 90° vee-die [55]. **a** to **d** Small punch radius: **a** free bending, **b** end of free bending, **c** end of overbending, **d** bending back. **e** to **h** Large punch radius: **e** free bending, **f** continued bending with two-point contact at punch radius, **g** start of further pressing, **h** further pressing in a semi-open die.

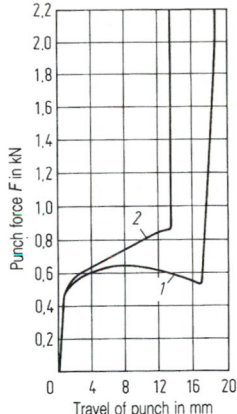

Figure 29. Force–travel curve for bending in a 90° vee-die for a small and a large punch radius r_{St} [5]; *1* r_{St} = 2 mm, *2* r_{St} = 15 mm (die width w = 42 mm, s_0 = 2 mm, material St 1404).

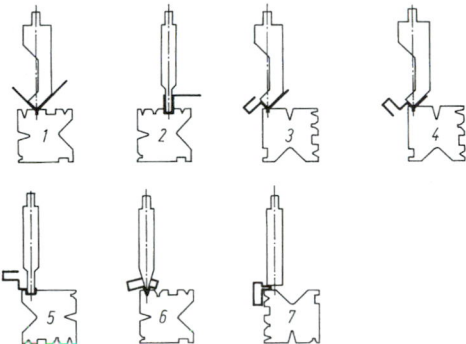

Figure 30. Steps in making a sheet-metal shape by folding.

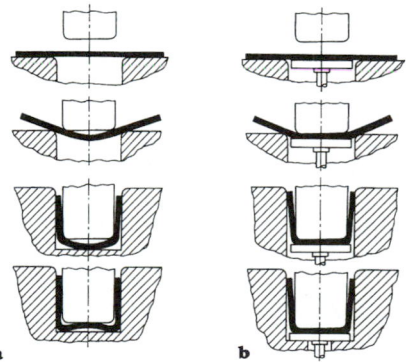

Figure 31. U-Bending [33]: **a** without a pressure pad, **b** with a pressure pad.

Folding. Here, one side of the component being bent is clamped tightly and the other side is bent with a folding beam (**Fig. 32**). If the smallest bending radius is larger than the radius of curvature of the clamping beam, *free bending* takes place. The pattern of the bending force exhibits two clearly defined zones, depending on the

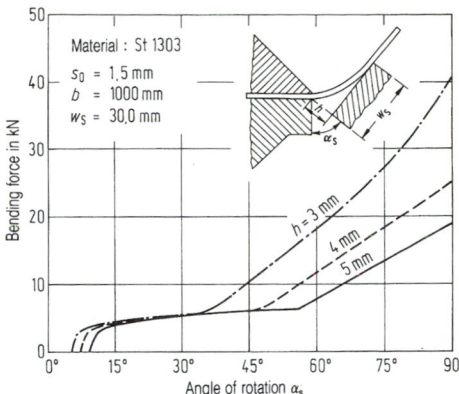

Figure 32. Bending force as a function of the angle of rotation α_s for different positions of the folding beam [36].

angle of rotation α_s. At small angles of rotation, the force required is low and increases only slowly, owing to a large effective lever arm. This increase is a consequence of the hardening of the material.

3.5.4 Superplastic Forming of Sheet

Special metallic materials (chiefly eutectics and eutectoids) with an extremely fine-grained structure can withstand extremely large degrees of deformation (by over 1000%) under the following conditions: $T_u > 0.5 T_s$ (T_s = melting point, T_u = forming temperature in K) and low rates of deformation $\dot\varphi$ (usually < 10^{-2} 1/s).

The material properties required for technically useful superplastic materials are: an extremely fine-grained structure (grain size < 10 μm); high resistance to void formation by avoiding coarse inclusions, especially at the grain boundaries; low flow stress values k_f at high rates of deformation $\dot\varphi$ (cf. K3.2.3).

Superplastic forming is chiefly accomplished by means of blowing. The tool may take the form of either a positive mould (*male mould method*, **Fig. 33**) or a negative mould (*female mould method*, **Fig. 34**). The ratio of the drawing depth to the smallest plane dimension is limited to $h/b < 0.4$ for the female mould method and $h/b < 0.6$ for the male mould method [37]. The normal sheet thicknesses range from 0.5 to 3 mm. In the superplastic state, the flow stresses for e.g. aluminium and titanium alloys range from 4 to 20 N/mm², i.e. forming pressures of 0.5 to 200 bar are required. The formed workpieces are practically free from intrinsic stresses and are thus free from spring-back. Approximate values for Al parts are given in [37], and for materials and parameters to be met in [38 to 42].

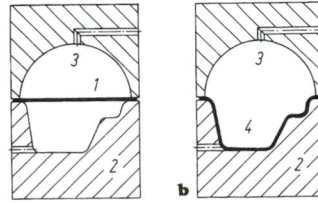

Figure 33. Male mould process. **a** Sheet metal blank; *1* in place. **b** Workpiece; *4* moulded; *2* moulding tool, *3* pressure space.

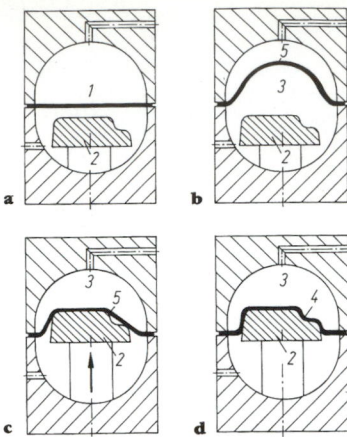

Figure 34. Female mould process. **a** Blank *1* in place. **b** Forming the blank into a bubble by gas pressure *3*. **c** Drawing over the mould *2*. Showing *5* intermediate stages of forming. **d** Final shape of workpiece *4*.

3.5.5 Upsetting

Upsetting is a basic method of forging and cold massive forming, e.g. of fasteners (DIN 8583). It is important in theoretical studies as a model process.

The limits to upsetting are:

The *degree of upsetting* (logarithmic deformation) as the limit of formability: $\varphi = \ln(l_1/l_0)$ **(Fig. 35)**. In cold upsetting of steel, $\varphi_{max} = 1.6$ should not be exceeded, regardless of the number of upsetting stages.

The *upsetting ratio* $s = l_0/d_0$ as the limit to prevent buck-

ling. For free upsetting (cold), this is $s \leq 2.3$. Larger values require several stages. **Figure 36** shows the manufacture of a cap bolt in several stages by tapering, initial upsetting and final upsetting.

The force–displacement curve for upsetting rises steeply towards the end of the operation, which has a particularly large effect with smaller cap heights. Good filling of the mould and low burr formation are requirements for hot upsetting in a die. A special form of cold upsetting is *embossing*, which is performed as *smooth embossing* (surface quality) or as *sizing* (thickness tolerance).

Recourse is made to *hot upsetting* only with difficult components, to keep the forming forces small. Owing to the given force–displacement curve, presses with a fixed way of travel are particularly suitable for upsetting.

3.5.6 Forging

The basic forging processes belong to the manufacturing processes cutting, metal forming, and joining. There are processes for changing cross-sections (*straightening, broadening, solid or cup extrusion, upsetting, heading*), for changing direction (*bending, folding, twisting*), for producing hollow spaces (*opening out, perforating, opening out with a hollow mandrel, indirect impact extrusion of hollow items*), for cutting (*parting, deflashing, piercing, pressing-knife cutting, chiselling, slitting*) and for joining (*shrinking, welding*) if elements of complex forgings have to be united into the complete workpiece. Drop forging processes are *heading, closed die*

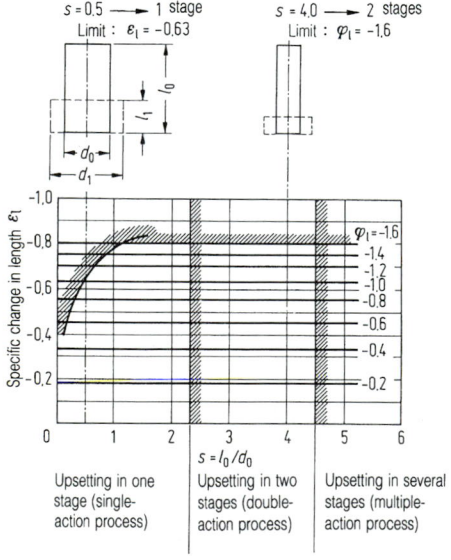

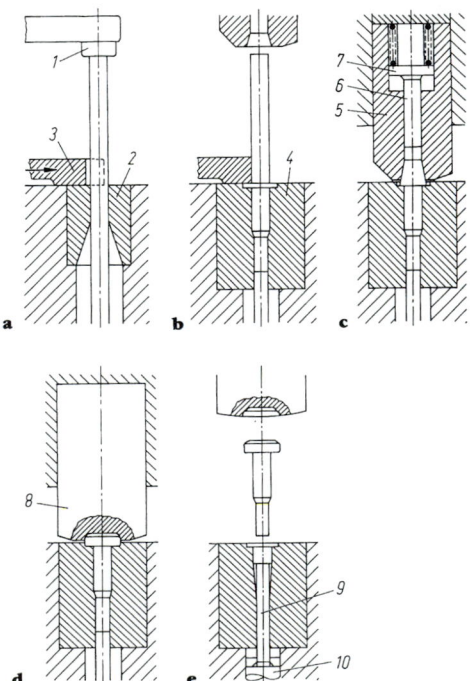

Figure 35. Process limits for cold upsetting of steel Cq35.

$\varphi_l = -1.6$	Limit of formability of material
$\varepsilon_l = f(s)$	Limit to elevated tool loading in minor upsetting operations
$s = 2.3$	Buckling point in single-action process
$s = 4.5$	Buckling point in double-action process

Figure 36. Operations in making a cap bolt (transverse feed press): **a** feeding of material and shearing off with a shearing blade in separate stages, **b** feeding of blank up to die, **c** insertion into die and initial upsetting, **d** final upsetting, **e** ejection; *1* stop, *2* shearing die, *3* shearing blade, *4* die (reducing die), *5* initial upsetting tool, *6* ejection pin (ram), *7* ejector (ram), *8* final upsetting ram, *9* ejection pin (die), *10* ejector (die) (VDI – Guideline 3171).

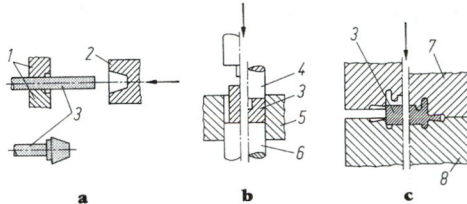

Figure 37. Process of drop forging in the narrower sense [5]: **a** heading, **b** closed die forging, **c** open forging; *1* clamping jaw, *2* heading die, *3* workpiece, *4* ram, *5* container, *6* ejector, *7* upper block, *8* lower block.

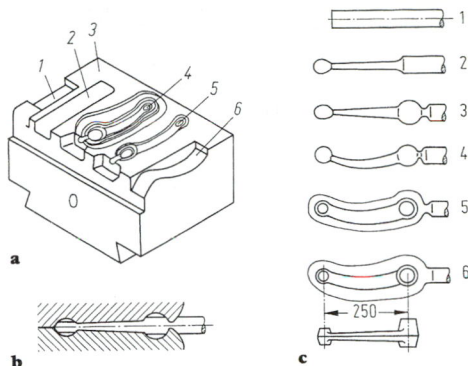

Figure 39. Sequence of operations for forging in a multiple cavity die [44]. **a** die; *1* stretching impression, *2* rolling impression, *3* impact surface, *4* final impression, *5* preliminary forging impression, *6* bending impression. **b** Section through rolling impression. **c** Sequence of operations; *1* initial form, *2* stretched workpiece, *3* rolled workpiece, *4* bent workpiece, *5* workpiece after preliminary forging, *6* finished drop forging.

forging and *open forging* (**Fig. 37**). In *hot forging*, the blanks are heated to a temperature above the recrystallisation temperature (850 to 1250 °C for steel), so that no permanent increase in the hardness of the material occurs. The production of blanks for forging comprises selection of the mill products, cutting the mill products into sections by shearing, breaking, sawing or parting, followed where necessary by setting (closed die forging to produce flat parallel ends) and heating the blank to forging temperature.

Hammer Forging. This process is generally used for one-off and small-scale production of parts with a mass ranging from 1 kg to 350 t. Typical operations are illustrated in **Fig. 38**. Workpieces produced by hammer forging usually have to be finished by machining.

Drop Forging. Here, the blank is formed into the finished workpiece via several intermediate shapes [43]. The sequence of operations consists of mass distribution, preshaping of the cross-section (often by hammer forging) and closed die forging (**Fig. 39**), which consists of the basic operations upsetting, spreading and rising (**Fig. 40**). The blank and the intermediate shapes should be matched to the finished part in such a way that the best possible fibre pattern is achieved (**Fig. 41**). In preforming, the flash, which severely affects the process, is removed in the last stage of work by deflashing. *Close-tolerance forging* enables the production of forgings of high dimensional accuracy (IT9 to IT11, compared with IT12 to IT16) and better surface quality by performing at least one operation in a closed die and/or by means of forming at *warm temperatures* (600 to 900 °C for steel). *Precision forging* (e.g. under shielding gas with precise temperature control) produces ready-to-fit workpieces of even higher accuracy for selected machine parts (e.g. turbine blades, bevel wheels).

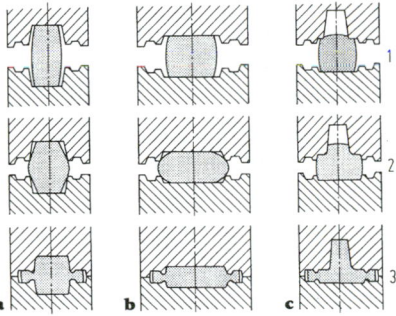

Figure 40. Basic types of operations for filling die impressions [5]: **a** upsetting, **b** widening, **c** rising; *1* upsetting, *2* applying pressure, *3* filling.

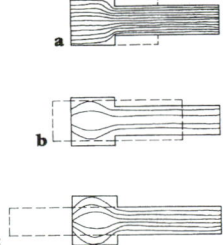

Figure 41. Selection of blank, process and pattern of fibres for forged parts [5]: **a** stretching, **b** stretching and upsetting, **c** upsetting.

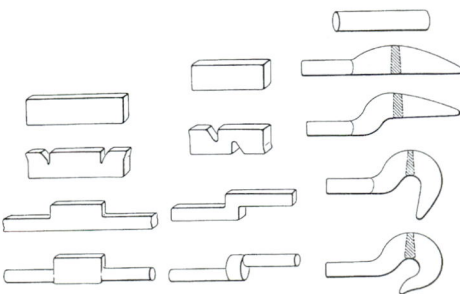

Figure 38. Application of basic processes for hammer forging of steel in sequence [5].

Important types of forging die are, in the case of *open dies*, the *full die* as a single- and multiple-impression die, the *insertion die* as a single die and a die with several identical impressions, and the *multiple-cavity die*. *Closed dies* are dies without a flash gap with one or – in the case of multiple cavity dies – several parting lines. The high thermal and mechanical stresses (heating to 700 °C, stresses up to 1000 N/mm²) mean that tool life is limited. Conventional steels for drop forgings are low-alloy hot forming tool steels, e.g. 55NiCrMoV6, 56NiCrMoV7, 57NiCrMoV77, for full dies and high-alloy hot forming tool steels, e.g. X38CrMoV51, X37CrMoW51, X32CrMoV33, for die inserts. (cf. D3.1.4).

Cold Forging. This essentially comprises the processes of extrusion, also upsetting, embossing (q.v.) and – for smaller components made from steel and non-ferrous metals – also open forging and closed die forging. Here, forming takes place at room temperature without preheating.

3.5.7 Extrusion

A survey of the extrusion processes is given in **Fig. 42**. A distinction is made between cold extrusion and hot extrusion. *Cold extrusion* denotes the pressing of ingots which are placed in the press without preheating. *Hot extrusion*, generally (as this is the norm) referred to simply as *extrusion*, denotes the pressing of ingots which are preheated before entering the press.

Direct Extrusion

The ingot is first upset in the container until it assumes the diameter of the container bore. It is then forced through the die by the ram. Relative motion arises between the ingot and the container, so work has to be done to overcome friction.

With Lubricant (Fig. 42a). Friction is reduced by a film of lubricant between the ingot and the container, which is facilitated by conical dies. Uses are in hot extrusion of steel and cold extrusion of aluminium alloys.

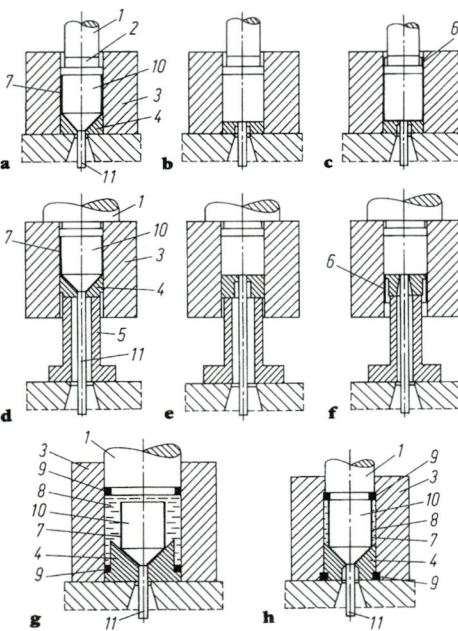

Figure 42. Schematic diagram of extrusion processes. **a** to **c** Direct extrusion: **a** with lubricant, **b** without lubricant or skull, **c** without lubricant, with skull. **d** to **f** Indirect extrusion: **d** with lubricant, **e** without lubricant or skull, **f** without lubricant, with skull, **g** hydrostatic pressing, **h** hydrafilm process. *1* Ram, *2* pressure disc, *3* container, *4* die, *5* hollow ram, *6* skull, *7* lubricant, *8* hydrostatic medium, *9* gasket, *10* ingot, *11* extrusion. Processes **a**, **d**, **g** and **h** are mainly for cold extrusion; processes **b**, **c**, **e** and **f** are mainly for hot extrusion.

Without Lubricant or Skull (Fig. 42b). This requires higher compression owing to greater friction between the ingot and the container and between the die and the ingot. Therefore this process is generally used for hot extrusion. The clearance between the dummy block and the container is such that no "skull" (the outer zone of ingot sheared off during extrusion) can form. Owing to the friction between the ingot and the wall of the container, depending on the amount of friction and the thermal conditions the outer zones of the ingot are obstructed to such an extent during movement of the ingot in the container that the core of the ingot flows ahead to a greater or lesser extent. This makes it possible to prevent the outer zones of the ingot from flowing into the forming zone preceding the die. This effect is exploited in extrusion of light metals. For instance, ingots with a continuous-cast surface are extruded up to certain ingot lengths in such a way, depending on the cross-sectional shape and the ratio of the container cross-section to the product cross-section, that the outer zones of the ingot do not become part of the extruded product but remain in the discard.

Without Lubrication and with Skull (Fig. 42c). If it is desired to ensure that no contaminated or oxidised outer zones of the ingot become incorporated in the extruded product, extrusion with a skull is carried out. In this method, the clearance between the container bore and the pressure disc is such that the outer zones of the ingot adhere to the container as a skull of metal. Thus only the inner part of the ingot is extruded. A disadvantage is the need to clear out the skull.

Indirect Extrusion

In indirect extrusion, too, the ingot is first upset in the container [45]. Here, a short closing ram closes the container on one side and the die, which is supported against a fixed hollow ram, enters the container from the other side. The ingot and the container move together during extrusion, so no relative motion and thus no friction occurs between the ingot and the container. A disadvantage is the hollow ram, the internal bore of which limits the circumscribed circle of the extruded product.

With Lubricant (Fig. 42d). Friction between the die and the material being formed and between the die and the container is reduced, provided that the ingot is lubricated before processing and conical dies are used [46, 47].

Without Lubricant or Skull (Fig. 42e). In this process there is no friction between the ingot and the container in indirect extrusion. It is therefore suitable as a substitute for direct extrusion in cases where, in direct extrusion, one or both of the following occurrences would take place:

1. The force required to overcome the friction between ingot and container causes the total extrusion force to rise so rapidly that the extrusion force or the total extrusion force, in terms of the cross-sectional area of the container, restricts the use of this extrusion method too severely.
2. The amount of heat resulting from the work done due to friction greatly lowers the rate of extrusion and/or the product quality.

In the skull-less process the clearance between the die and the container is set in such a way that the forces required to overcome the friction between die and container can be kept negligible compared with the forming force, and no skull (a thin film of extruded material covering the walls of the container) can form between the die and the container. Because in this process the outer zones of the ingot become part of the extruded product, since they are not retained by friction on the ingot surface as in direct extrusion, either turned ingots should be used, or the ingots should have continuous-cast surfaces of adequate quality.

Without Lubricant and with Skull (Fig. 42f). The advantage of this process is that ingots with contaminated and oxidised outer zones can also be extruded, as the outer zones remain behind in the skull. The clearance between the die and the container is made so large that the outer zones of the ingot adhere to the container walls as a skull [48]. A disadvantage, again, is that the skull has to be cleared out.

Hydrostatic Extrusion

In this process, the ingot is surrounded in the container by a compression medium (hydrostatic medium) **(Fig. 42g).** As the ram advances and the hydrostatic medium is compressed, the ram does not touch the ingot. Therefore the speed at which the ingot moves towards the die during extrusion is not equal to the speed of the ram, but is proportional to the displaced volume of hydrostatic medium [49]. Other features of the process are [49-52]: low liquid friction at the surface of the ingot, sealing

between the ingot and the conical die by extrusion pressure, separate lubrication of the ingot is unnecessary if the hydrostatic medium has lubricating qualities, otherwise the ingot must be lubricated before use [53]. This process is used chiefly for cold extrusion. In hot extrusion, there is the problem of high thermal stress on all the components, and temperature control is also necessary.

The Hydrafilm Process

This process is also known as the "thick-film" process. The quantity of hydrostatic medium is kept so low that the ram can touch the ingot during extrusion [54, 55]. The die abuts against the container, as ingot and container are separated only by a film of liquid [52] **(Fig. 42h).** Furthermore, ingot speed and ram speed are practically the same, so extrusion can be interrupted at any time by stopping the motion of the ram. In this process, too, the ingot may be coated with a separate lubricant [56]. The process is mainly used for cold extrusion, but may also be used to a limited extent for hot extrusion, as additional equipment for the hydrostatic medium is unnecessary if use is made of hydrostatic media which can be applied to the ingot at normal pressure in the solid condition before placing the ingot in the extruder. The hydrostatic medium becomes viscous under extrusion pressure [56].

4 Cutting

H.K. Tönshoff, Hanover

4.1 General

Cutting is manufacturing by changing the shape of a solid object. The cohesion of the material is locally destroyed. The final shape is contained in the original shape. Disassembly of assembled (joined) objects is considered to be part of cutting (according to DIN 8580).

The *main group "cutting"* can be divided into seven groups:

Severing (DIN 8588)
Machining with geometrically well-defined tool edges (DIN 8589 Part 0)
Machining with geometrically undefined tool edges (DIN 8589 Part 0)
Chipless machining (DIN 8590)
Disassembly (DIN 8591)
Cleaning and
Evacuation (DIN 8592)

Cutting by *severing* and *machining* is accomplished with the *mechanical* action of a tool on a workpiece. In *chipless* cutting, material particles are removed from a solid object by non-mechanical means. In cutting by *cleaning*, undesirable substances or particles are removed from the surface of a workpiece. Cutting by *evacuation* means the removal of gases from enclosed spaces; it is generally employed in connection with another manufacturing process such as electron beam welding or coating by ion plating.

4.2 Machining with Geometrically Well-defined Tool Edges

4.2.1 Fundamentals

Machining is manufacturing by *cutting*. Particles of material are mechanically removed as chips from a blank or workpiece by the cutting action of a tool. In machining with geometrically well-defined tool edges, the number of cutting edges, the shape of the cutting tips and the position of the cutting edges in relation to the workpiece are known and describable (in contrast to machining with geometrically undefined tool edges, e.g. grinding). **Figure 1** illustrates important processes in this group. The processes are distinguished according to the *cutting motion* (cutting speed v_c), the *feed motion* (rate of feed v_f) and the resulting *effective motion* (effective speed v_e).

The feed and cutting direction vectors fix the working plane. The angle between the two vectors is termed the *feed direction angle* φ, while the angle between the effective direction and the cutting direction is termed the *effective direction angle* η. The relationship

$$\tan\eta = \frac{\sin\varphi}{(v_c/v_f) + \cos\varphi}$$

applies to all processes.

The mechanical operation of cutting particles off the workpiece, i.e. chip formation, can best be described with reference to the orthogonal process (two-dimensional deformation). The *wedge* is described by the *rake angle* γ, the *clearance angle* α and the edge radius r_B. The penetration of the wedge causes the material to undergo plastic deformation. **Figure 2** shows the zones of plastic deformation by way of example during flowing chip formation. Five zones can be distinguished:

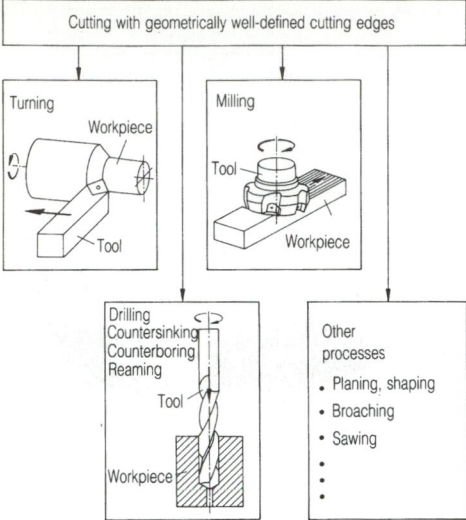

Figure 1. Processes of machining with geometrically well-defined tool edges.

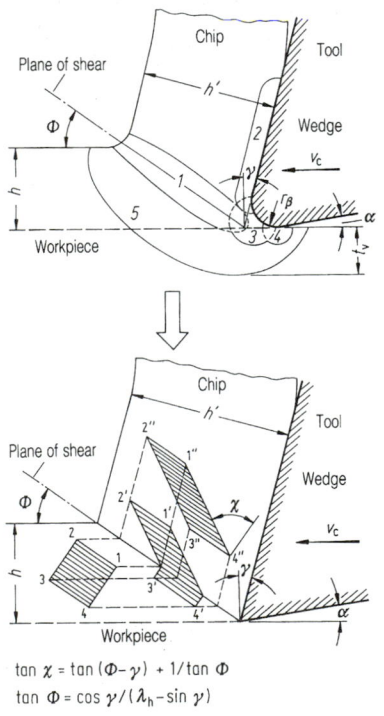

$$\tan \chi = \tan(\Phi - \gamma) + 1/\tan \Phi$$

$$\tan \Phi = \cos \gamma / (\lambda_h - \sin \gamma)$$

$$\lambda_h = h'/h$$

Figure 2. Effective zones in chip formation and derived model of deformation in the shear plane; 1 primary shearing zone, 2 secondary shearing zone at the face, 3 secondary shearing zone at the pressure and cutting zone, 4 secondary shearing zone at the flank, 5 deformation advance zone.

The primary shearing zone comprises the actual area of chip formation by shearing.

In the secondary shearing zones ahead of the face and at the flank, forces of friction act between the tool and the workpiece, causing plastic deformation of these layers of material.

In the deformation advance zone, chip formation causes stresses to act which lead to plastic and elastic deformation of this zone.

In the pressure and cutting zone, the material is deformed and cut under high compressive stresses.

Because of these mechanisms, the stress thickness h in the undeformed condition changes to the *chip thickness* h', resulting in the *chip compression* $\lambda_h = h'/h$. The plane of shear encloses the *angle of shear* φ with the cutting speed vector. The angle of deformation χ denotes the shearing of a particle which has passed through the plane of shear. Besides flowing chip formation, other kinds of chip may form:

With *flowing chips* there is a steady flow of material from the workpiece. The deformation of the material is continuous. Periodic changes may occur in the intensity of the deformation – usually at higher cutting speeds. Lamellae form in the chip, which may be pronounced up to cutting of the material and formation of chips [1].
Shearing chip formation occurs when the formability of the material is exceeded in the shearing zone and locally concentrated shearing takes place without complete cutting of the material. Chip formation is uneven.
Tearing chip formation occurs in materials with low formability, e.g. lamellar graphite cast iron. The interface between the chip and the workpiece is irregular.

Built-up edges may form when machining ductile, work-hardening materials at low cutting speeds and sufficiently steady chip formation (flowing chip formation). They are parts of the material that have been severely deformed and work-hardened in the area of the compression zone and have become welded to the curvature of the cutting edge and the face under high pressure, thus becoming part of the cutting tip [2].
In the chip formation zone the *cutting energy* supplied is completely converted. It is calculated as

$$E_c = F_c l_c$$

(F_c = cutting force, l_c = travel in direction of cut).

The cutting energy is made up of forming and shearing energy E, friction energy at the rake face E_γ, friction energy at the flank E_α, surface energy to form new surfaces E_T, and kinetic energy due to chip deflection E_M.
The energy converted in machining one unit of volume is

$$e_c = E_c/V_w$$

(e_c = specific energy, V_w = volume of metal machined). Like E_c, the individual components of E_c can be expressed in relation to V_w.

Example. A numerical estimate shows that most of the cutting energy is converted into forming and friction energy. For $e_c = 2760$ N mm^{-2}, for a sample calculation where chip thickness $h = 0.1$ mm, cutting speed $v_c = 60$ m min^{-1}, rake angle $\gamma = +10°$, chip compression $\lambda_h = 3.9$, the energy components are as follows: specific forming and shearing energy $e_\varphi = 2010$ N mm^{-2}, specific friction energy at the face $e_\gamma = 745$ N mm^{-2}, specific surface energy $e_T = 2 \cdot 10^{-2}$ N mm^{-2}, specific kinetic energy $e_M = 1 \cdot 10^{-2}$ N mm^{-2}. In this example the specific friction energy at the flank e_α is included in e_γ, as the two cannot be determined separately.

From the specific energy e_c introduced, the specific cutting force k_c can be derived as an index for calculating the cutting force:

$$k_c = F_c/A = F_c/(bb),$$

where A = the undeformed chip cross-section A, b = the undeformed chip width and b = the undeformed chip thickness;

$$e_c = E_c/V_w = P_c/Q_w = (F_c v_c)/(A v_c) = k_c,$$

where P_c = the cutting performance, Q_w = the chip volume over time, and F_c = the cutting force.

Thus the specific cutting force k_c can be seen as an energy-related variable. (The application and determination of k_c are discussed in detail in K4.2.2.) The energy introduced into the chip formation zone is almost entirely converted into heat, with a small remnant being turned into intrinsic stress in the chip and the workpiece (spring energy). This produces high temperatures in the cutting tip, which is thus subjected to mechanical and thermal stress. The *surface forces* and the *temperature distribution* are shown in **Fig. 3** [3]. From this, stresses can be calculated. In **Fig. 4** only the direct tensile stresses which are especially critical for high-temperature ceramic cutting materials are plotted. Mechanical and thermal stress, assisted by chemical reactions, cause wear.

The *stress* on the cutting tip depends on various influences. Besides control parameters like cutting speed, feed and depth of cut as well as environmental influences such as the cooling lubricant, the workpiece in particular influences tool wear.

Kinds of wear [4], **Fig. 5** (cf. D5.4).

Fractures and cracks; these occur in the area of the cutting edge due to excessive mechanical or thermal stress.
Mechanical abrasion, mainly caused by hard inclusions such as carbides and oxides in the material.

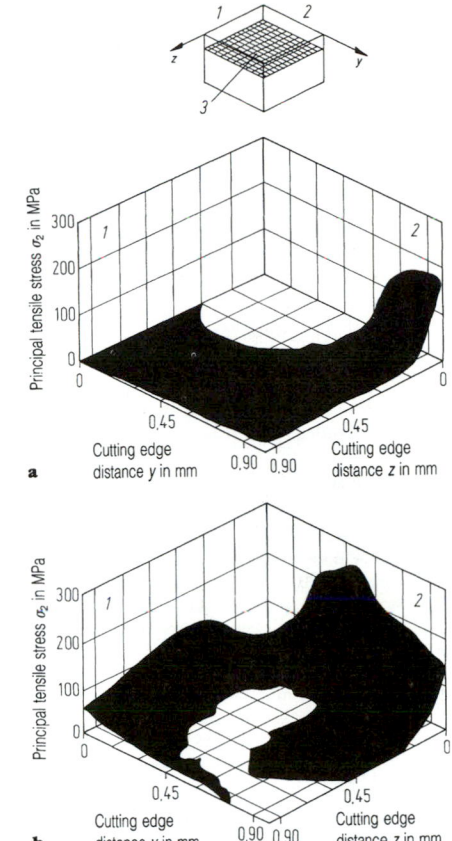

Figure 4. Calculated principal tensile stress distributions in Al_2O_3 ceramic indexable inserts, under **a** mechanical and **b** thermal stress [3]; *1* flank, *2* face, *3* calculating plane.

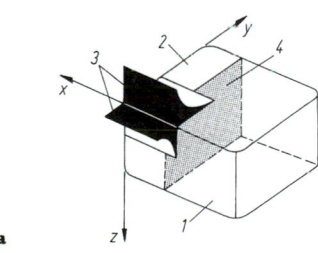

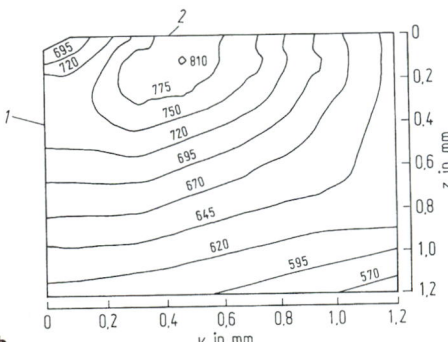

Figure 3. a Assumed force and **b** temperature distribution in a cutting tip; *1* primary flank, *2* face, *3* surface loads, *4* calculating plane.

Stress with		Types of wear
Long-term effect	Short-term effect	
Constant mechanical stress	Alternating mechanical stress	Abrasion
		Adhesion
		Fracture
Constant thermal stress	Alternating thermal stress	Flaking
		Cracking
Internal chemical attack	Surface chemical attack	Diffusion
		Oxidation

Figure 5. Typical kinds of wear and their causes (cf. D6).

Plastic deformation occurs if the cutting material has inadequate resistance to deformation but sufficient toughness.
Adhesion is the shearing off of pressure-welded points between the material and the chip, the point of shearing being located in the cutting material.
Diffusion occurs at high cutting speeds and mutual solu-

bility of the cutting and workpiece materials. The tool material is weakened by chemical reaction, becomes detached and is removed.

Oxidation also occurs only at high cutting speeds. The cutting material oxidises in contact with atmospheric oxygen and the structure is weakened.

The *machinability* of a workpiece is determined by the composition of the material, its structure in the machined area, the preceding metal forming/primary shaping operation, and the heat treatment it has undergone.

Machinability is evaluated according to the following criteria:

Tool wear
Surface quality of the workpiece
Machining forces
Chip form

The nature of the machining job should be taken into account in weighting the criteria.

4.2.2 Turning

According to DIN 8589E, turning is defined as machining with a continuous (usually circular) cutting motion and any desired feed motion in a plane at right angles to the cutting direction. The axis of rotation of the cutting motion maintains its position relative to the workpiece regardless of the feed motion. **Figure 6** illustrates some important turning processes.

In the following, *longitudinal cylindrical turning* is taken as an example of a turning process. Terminology, names and designations for describing the geometry at the cutting tip are laid down in DIN 6580 and ISO 3002/1. **Figure 7** shows the surfaces and cutting edges defined for the cutting tip.

The angles shown in **Fig. 8** serve to determine the position and shape of the tool in three dimensions: the *entry*

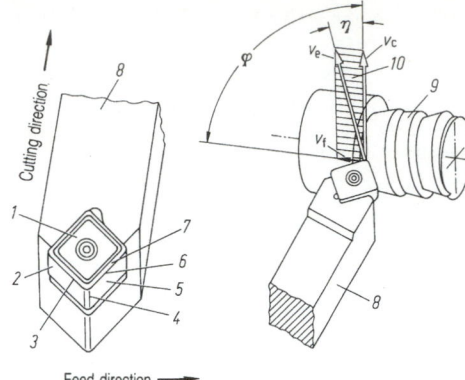

Figure 7. Terminology at cutting tip and directions of motion of tool (DIN 6580, ISO 3002/1); *1* indexable insert, *2* secondary flank, *3* secondary cutting edge, *4* nose, *5* primary flank, *6* primary cutting edge, *7* face, *8* tool holder, *9* workpiece, *10* working plane, v_c cutting speed, v_e effective speed, v_f rate of feed, φ feed direction angle, η effective direction angle.

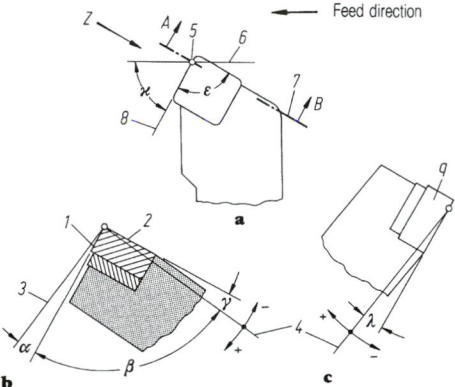

Figure 8. Turning tool angles (DIN 6581): **a** main view, **b** cross-section *A–B* (orthogonal plane of tool), **c** view *Z* (onto cutting edge plane); *1* flank, *2* face, *3* tool cutting edge plane, *4* tool reference plane, *5* point on cutting edge under consideration, *6* assumed working plane, *7* tool wedge measuring plane, *8* cutting edge plane of primary cutting edge, *9* cutting tip.

angle κ is the angle between the primary cutting edge and the working plane. The *edge angle ε* is the angle between the primary and secondary cutting edges and is predetermined by the cutting edge geometry. The *angle of inclination λ* is the angle between the cutting edge and the reference plane and is apparent when looking down onto the primary cutting edge. The *clearance angle α*, *wedge angle β* and *rake angle γ* are the angles measured in the wedge measuring plane and total 90°. The values of the relevant tool angles are determined from approximate value tables in relation to the workpiece and cutting materials and the machining process. **Table 1** shows some values for machining of steel. The entry angle *κ* influences the shape of the undeformed chip cross-section to be removed and thus the power required for the machining process (**Fig. 9**).

The chips flowing over the face of the tool have a different *bulk volume* depending on chip type and form. The

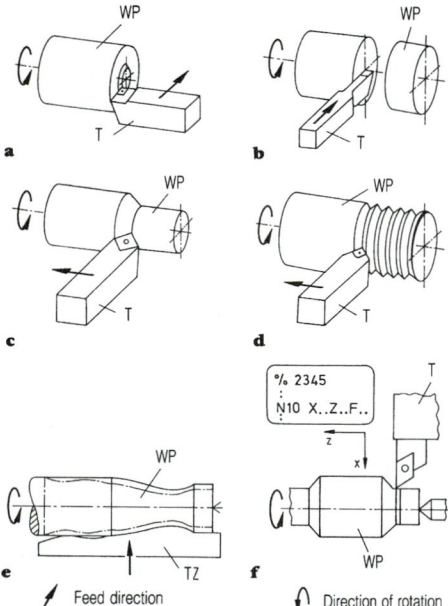

Figure 6. Turning processes (DIN 8589 Part 1): **a** facing, **b** parting, **c** turning, **d** thread turning, **e** profile turning (workpiece contour is duplicated in tool), **f** form turning; WP workpiece, T tool.

Table 1. Normal values for tool angles in machining of steel

Cutting material	Cutting tip geometry					
	Rake angle γ	Clearance angle α	Angle of inclination λ	Entering angle κ	Edge angle ε	Edge radius r_ε
High-speed steel	$-6°$ to $+20°$	$6°$ to $8°$				
Cemented carbide	$-6°$ to $+15°$	$6°$ to $8°$	$-6°$ to $+6°$	$10°$ to $100°$	$60°$ to $120°$	0.4 to 2 mm
Ceramic insert	$-6°$ to $0°$	$6°$ to $8°$	$-6°$ to $0°$	$45°$ to $100°$		0.8 to 2 mm

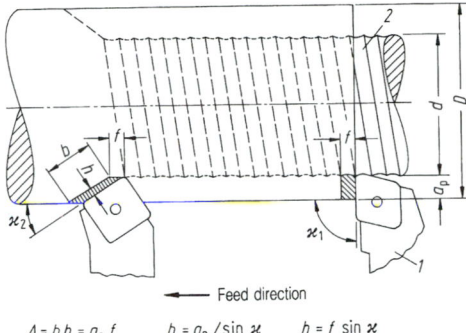

← Feed direction

$$A = bh = a_p f \qquad b = a_p /\sin \varkappa \qquad h = f \sin \varkappa$$

Figure 9. Cut and chip variables in turning; *1* tool, *2* workpiece.

parameter is the *chip space coefficient RZ*, the ratio of the time-related chip volume Q_w to the bulk volume Q'. Here,

$$RZ = Q'/Q_w,$$

$$Q_w = a_p f v_c = a_p f D\pi n.$$

The chip space coefficient indicates the *"bulkiness"* of the chips. It is used to dimension machine tool working spaces, chip conveying equipment, and chip spaces of tools. The chip space coefficient *RZ* may have widely differing values depending on the shape of the chip (**Fig. 10**). The more brittle the material, the lower is this value. Brittleness can be influenced via the composition of the material. For steel, higher sulphur contents (more than 0.04%, free-cutting steel with 0.2% S) have a beneficial effect. However, this may impair the toughness of the material in the transverse direction, depending on the form of the dispersed sulphides [5]. Chip-breaking steps sintered onto the face of the tool or attached chip breakers produce additional chip deformation, i.e. additional stress on the chip material, and deflect the chip against an obstacle in the direction of flow (**Fig. 11**). The chip is bent by contact with the cut surface of the workpiece or the flank of the tool and breaks (secondary chip breaking in contrast to tearing chip formation or lamellar chip formation with material cutting (cf. K4.2.1), where the chips leave the chip formation zone as small fragments). Favourable chip forms can also be achieved by selecting the appropriate machine setting data such as rate of feed and depth of cut (**Fig. 12**).

Every material resists the penetration of the tool during chip removal. This has to be overcome by means of a force, the machining force F. This force is analysed by resolving it into its three components (**Fig. 13**):

$$F_c = k_c a_p f = k_c bh.$$

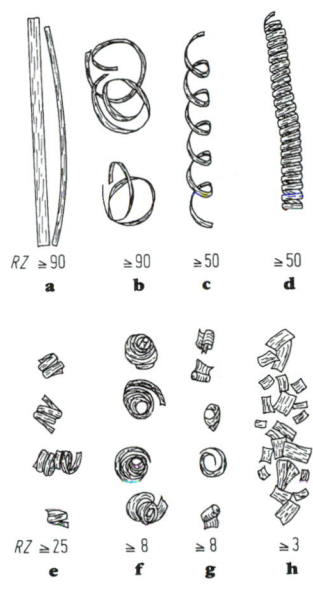

$RZ \geq 90$
a

≥ 90
b

≥ 50
c

≥ 50
d

$RZ \geq 25$
e

≥ 8
f

≥ 8
g

≥ 3
h

Figure 10. Chip shapes (iron and steel testing sheet 1178-69): **a** ribbon chips; **b** snarl chips; **c** flat helical chips; **d** long, cylindrical helical chips; **e** helical chip fragments; **f** spiral chips; **g** spiral chip fragments; **h** discontinuous chips.

← Cutting speed

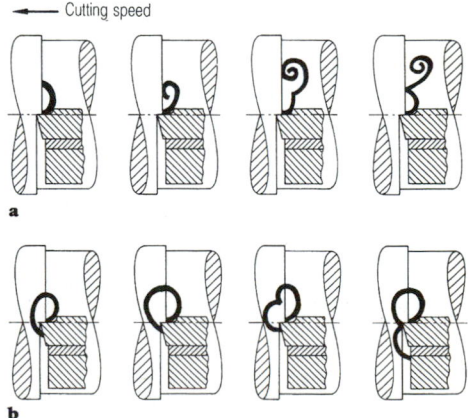

a

b

Figure 11. Effect of chip shape stages: **a** contact with cut surface, **b** contact with flank.

Material   C35N
Cutting speed $v_c = 100$ m/min
Cutting material   HM P25

α	γ	λ	ε	$\varkappa$	r_ε
6°	−6°	−6°	90°	70°	0.8 mm

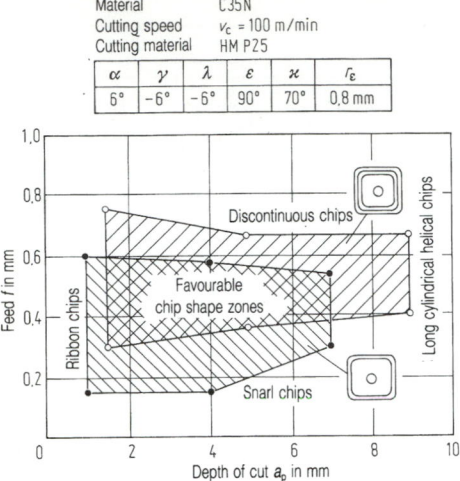

Figure 12. Areas of favourable chip shape with tools having chip-breaking grooves (according to König).

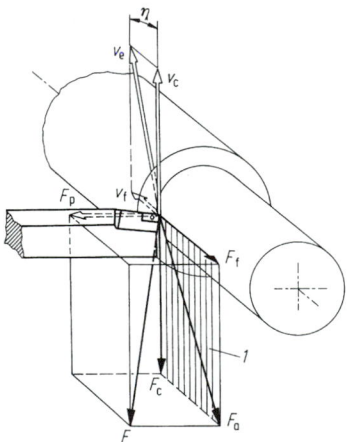

Figure 13. Components of machining force (DIN 6584); *1* working plane.

It is known from tests that the specific cutting force k_c is also a function of the undeformed chip thickness h. It can be seen from the log-log representation (**Fig. 14**) that [6]

$$k_c = k_{c\,1.1}\,h^{-m_c}.$$

Here, $k_{c\,1.1}$ is the "primary value of the specific cutting force", i.e. k_c at $h = 1$ mm (indices 1.1 due to $k_{c\,1.1} = F_c/1.1$ at $b = 1$ mm and $h = 1$ mm). The exponent m_c indicates the increase and is the "incremental value of the specific cutting force". Kienzle's cutting force formula can also be written

$$F_c = k_{c\,1.1}\,bh^{1-m_c}.$$

$k_{c\,1.1}$ and $1 - m_c$ are listed for various ferrous materials in **Appendix K4, Table 1**. A direct comparison of the $k_{c\,1.1}$ values to indicate the machinability or the energy required for machining is unacceptable, as the incremen-

Material   20 MnCr 5BG
Cutting speed $v_c = 100$ m/min
Depth of cut $a_p = 3$ mm
Cutting material   HM P10

α	γ	λ	ε	$\varkappa$	r_ε
5°	6°	0°	90°	70°	0.8 mm

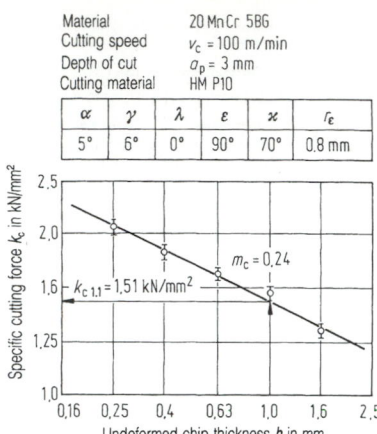

Figure 14. Specific cutting force as a function of undeformed chip thickness.

tal values m_c may vary widely. It follows from $m_c < 1$ that for a given machining cross-section, the cutting force and power requirements increase at smaller undeformed chip thicknesses. The physical reason lies in the higher proportion of friction applicable to smaller undeformed chip thicknesses (cf. K4.2.1).

k_c depends on other variables apart from the material and the undeformed chip thickness. Additional influencing factors are therefore applied. The influencing factors for the cutting speed K_v, the rake angle K_γ, the cutting material K_{ws}, the cutting edge sharpness K_{wv}, the cooling lubricant K_{ks} and the workpiece shape K_f are also shown in **Appendix K4, Table 1**.

The *passive force* F_p as a further component of the machining force (**Fig. 13**) produces no work, as it is perpendicular to the working plane, and motion occurs only in this plane. However, it is important for the dimensional accuracy and accuracy of shape of the machine–workpiece–tool system. The third component is the *feed force* F_f.

The passive force F_p and the feed force F_f can be combined into the *resultant cutting force* F_D. For thin, undeformed chip cross-sections ($b \gg h$), the resultant cutting force is perpendicular to the primary cutting edge. From this it follows that

$$F_f/F_p = \tan \kappa.$$

For normal values of b and h it can roughly be assumed that

$$F_D \approx (0.65 - 0.75)\,F_c,$$

by means of which F_f and F_p are to be determined. More exact determination is achieved by means of exponential functions corresponding to the cutting force formula. The exponents and principal values are shown in **Appendix K4, Table 1**.

The surface finish is determined by the profile of the cutting edge which produces the workpiece surface and by the feed. From the shaping of the cutting edge corner radius r_ε, the theoretical surface roughness $R_{t,\,th}$ can be geometrically determined as $R_{t,\,th} = f^2/(8r_\varepsilon)$.

This value should be regarded as a minimum, which increases due to vibrations, especially at higher rotational and cutting speeds, on the formation of built-up edges (cf. K4.2.1) and with the progressive wear of the cutting edge.

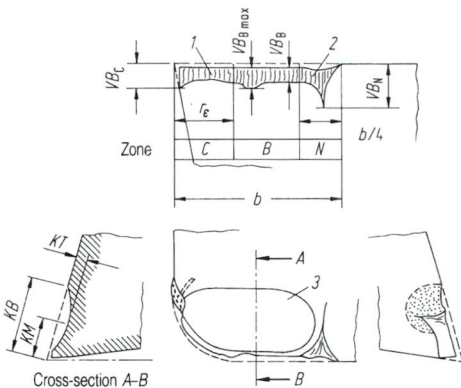

Figure 15. Forms of wear in turning (ISO 3685): *1* flank wear, *2* notching, *3* cratering; *C, B, N* zones, *KB* crater width, *KM* distance from crater centre to point of cutting tip, *KT* crater depth.

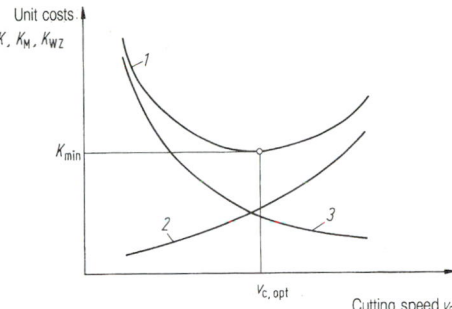

Figure 16. Manufacturing costs as a function of cutting speed v_c; *1* unit costs *K*, 2 machine-specific unit costs K_M, 3 tool-specific unit costs K_{WZ}.

The tool is subject to mechanical stress due to the machining force, thermal stress due to heating and chemical attack due to the interaction of the cutting material, the workpiece material and the surrounding medium. This results in wear on the cutting tool (cf. K4.2.1). Typical forms of wear are illustrated in **Fig. 15**. In addition, cutting edge wear, rounding of the cutting edge and scoring may occur on the secondary cutting edge. The type of wear that determines the end of tool life (tool life criterion) is determined by the respective use. Weakening of the wedge due to crater wear, or an increase in the proportion of the machining force accounted for by friction due to flank wear, are critical in roughing. Cutting edge wear leads to changes in workpiece dimensions, and flank wear or scoring impair surface quality and determine the end of tool life in finishing. The end of tool life is often set at a width of wear land of 0.4 mm or a crater depth of 0.1 mm. The flank is divided into three zones for more precise identification of the wear.

For a specific cutting-material–workpiece-material combination and a given tool life criterion, *tool life* depends chiefly on the cutting speed according to an exponential function (Taylor's equation shown on a straight line on a log–log graph) [7]:

$$\frac{T}{T_0} = \frac{v_c^k}{C}.$$

Here, T_0 and v_c are reference values, T_0 normally being set at $T_0 = 1$ min. *C* is the cutting speed for a life of $T_0 = 1$ min.

The Taylor straight line is plotted on the basis of a wear/tool life turning test to ISO 3685. This lays down appropriate set values for high-speed steel, cemented carbides of all machining categories (cf. K4.2.6) and ceramic inserts. It is usually sufficient to determine the width of the wear land and/or the crater depth as well as the distance from the crater centre to the point of the cutting tip. **Table 2** shows, for various materials, normal values of the gradient exponent *k* and the cutting speed *C* for a tool life of $T = 1$ min and a width of wear land of 0.4 mm.

For metal cutting machines the optimum cutting speed has to be established according to commercial criteria (**Fig. 16**). The optimum cutting speed in relation to time is

$$v_{c\ opt} = C(-k-1)\ t_{WZ}^{1/k}.$$

Optimisation of the cutting speed to minimise unit costs takes into account not only the tool changing time t_{WZ} but also the tool costs per cutting edge K_{WZ} and the hourly machine rate K_M:

$$v_{c,\ opt} = C(-k-1)\ (t_{WZ} + (K_{WZ}/K_M))^{1/k}.$$

4.2.3 Drilling

Drilling is a metal cutting process with a rotary cutting motion (primary motion). The tool, the *drill*, performs a

Table 2. Coefficients for determining Taylor straight lines

Taylor function $v_c = C \cdot T^{1/k}$	Uncoated cemented carbide		Coated cemented carbide		Oxide ceramic (steel) Nitride ceramic (cast metal)	
	C (m/min)	k	C (m/min)	k	C (m/min)	k
St 50-2	299	−3.85	385	−4.55	1210	−2.27
St 70-2	226	−4.55	306	−5.26	1040	−2.27
Ck 45N	299	−3.85	385	−4.55	1210	−2.27
16MnCrS 5 BG	478	−3.13	588	−3.57	1780	−2.13
20MnCr 5 BG	478	−3.13	588	−3.57	1780	−2.13
42CrMoS 4 V	177	−5.26	234	−6.25	830	−2.44
X155CrVMo 12 1G	110	−7.69	163	−8.33	570	−2.63
X40CrMo V 5 1G	177	−5.26	234	−6.25	830	−2.44
GG-30	97	−6.25	184	−6.25	2120	−2.50
GG-40	53	−10.0	102	−10.0	1275	−2.78

feed motion in the direction of the axis of rotation; **Fig. 17** shows common drilling processes. In drilling into solid metal, either through holes or blind holes may be produced. The tool used is usually a *twist drill*. Enlarging a drilled hole is done with twist drills or with *countersinks or counterbores* having two or more cutting edges. *Step drills* produce stepped holes. They usually have multiple cutting edges; for manufacturing reasons, not every cutting edge has to support all parts of the contour (e.g. one cutting edge may break the edge of a step, while the adjacent one produces a flat surface). *Centre drills* are special-profile drills with a thinner spigot and a short, stiff drill section in order to develop a good centring effect. *Combination centre drills and countersinks* cut a ring into the metal, a cylindrical centre hole being drilled at the same time. *Taps* are used to cut threads. *Reaming* is a hole-enlarging process with a small undeformed chip thickness, for producing holes of precise size and shape with a high-quality surface.

For drilling holes with diameters of 1 to 20 mm, with drilling depths up to five times the diameter, the twist drill is the tool most commonly used (**Fig. 18**). The twist drill consists of the shank and the body. The shank is used to clamp the drill in the machine tool. It is straight or tapered. If high driving torques are to be transmitted, tangential flat surfaces transmit the force. The body has a complex geometry, which can be modified to adapt the drill to the respective machining duty. Essential parameters are the *profile* and the *core thickness*, the *flute geometry* and the *helix angle*, i.e. the pitch of the flutes, the *land* and the *point angle*. Of these, the land and the point angle can be influenced by the user. The profile of the twist drill is designed in such a way that the flutes provide the maximum possible space for chip removal while ensuring that the drill is able adequately to withstand torsional stress. These two main requirements may be accompanied by others, such as production of favourable chip shapes, which have led to a variety of special profiles and which enable the drilling process to be adapted to particular boundary conditions. Material also has to be removed ahead of the centre of the twist drill.

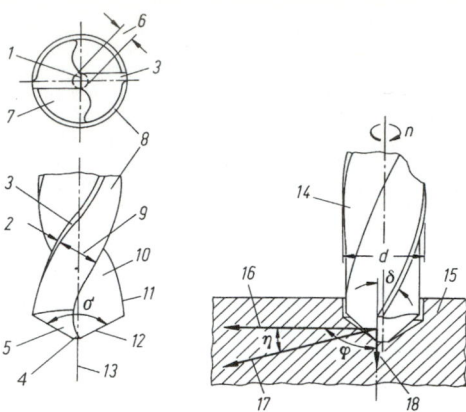

Figure 18. Terminology and mode of operation of twist drill (DIN 6580, 6581, 1412): n rotational speed, δ helix angle, d diameter, σ cone angle, φ feed direction angle, η angle of effective direction of cut; *1* transverse cutting edge (angled part of primary cutting edge), *2* margin width b, *3* margin of secondary flank, *4* nose, *5* primary flank, *6* centre thickness K, *7* flute, *8* secondary flank, *9* land, *10* face, *11* secondary cutting edge, *12* primary cutting edge, *13* tool axis, *14* tool, *15* workpiece, *16* cutting motion, *17* effective motion, *18* feed motion.

This is achieved by the transverse cutting edge, which links the two primary cutting edges.

Along the primary and secondary cutting edges, the rake angle γ, which is the most important variable influencing the drilling process, is not constant but decreases from the outside to the inside, even before the primary cutting edge is reached. In **Fig. 19**, the rake angles at three points on the cutting edge are shown by plotting the

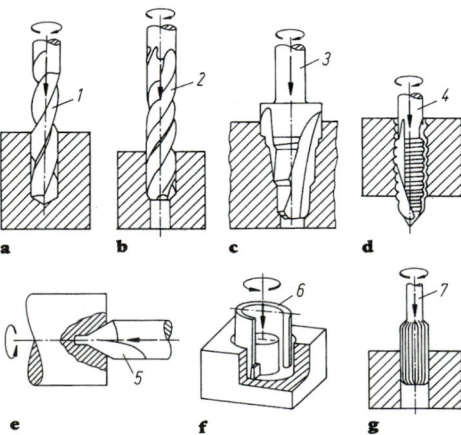

Figure 17. Drilling processes (DIN 8589). **a** Drilling into solid metal; *1* twist drill. **b** Hole enlarging; *2* twist-type counterbore with three cutting edges. **c** Counterboring with multiple diameters; *3* step drill. **d** Centre drilling; *4* centre drill. **e** Centre drilling and countersinking; *5* combined centre drill and countersink. **f** Tapping; *6* tap. **g** Reaming; *7* machine reamer.

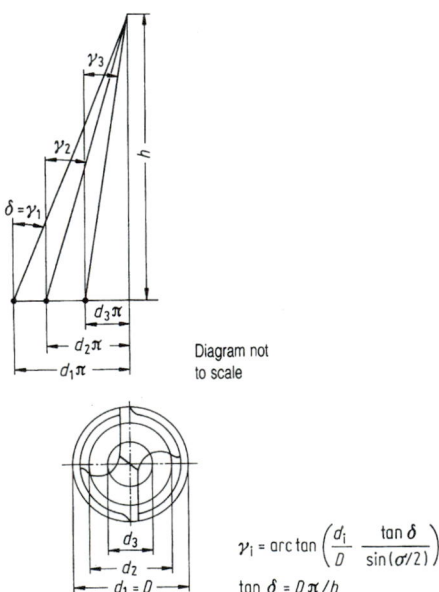

Figure 19. Rake angle at primary cutting edge of twist drill: b flute pitch, σ cone angle, δ helix angle, D drill diameter, d_i diameter at relevant point on cutting edge i, γ_i rake angle at relevant point on cutting edge i.

$$\gamma_i = \arctan\left(\frac{d_i}{D}\cdot\frac{\tan\delta}{\sin(\sigma/2)}\right)$$

$$\tan\delta = D\pi/h$$

pitch b of the flute over the development of the circles associated with the diameters [8]. At the outside diameter it is identical to the helix angle δ and decreases in direct proportion to the diameter. Negative rake angles may occur even before the primary cutting edge is reached. Preceding the transverse cutting edge, the rake angles are sharply negative. Here the workpiece material has to be displaced radially. Negative rake angles and the material displacement effect generate high pressures in the area of the transverse cutting edges. To alleviate this effect, twist drills are pointed. The centre of the drill is tapered by profile grinding in the direction of the flute and towards the drill tip on a conical or similar surface. In this way, the rake angle at the transverse cutting edge is increased and/or the transverse cutting edge is shortened.

The most important type of wear on the twist drill is flank wear at the nose. This wear, which is mainly caused by abrasion, produces an increase in the torsional stress on the drill, as higher machining forces are present in the nose area. This torsional stress may cause the drill to break. Worn twist drills are therefore reground until the damaged area of the secondary cutting edge margin is removed.

Machining Forces. The *forces* and *moments* present during drilling are calculated on the basis of Kienzle's approach [7, 9]. **Figure 20** illustrates the metal cutting geometry and the forces during drilling. The forces occurring at each cutting edge, which are assumed to act in the middle of the cutting edge, are resolved into their components F_c, F_p and F_f. The cutting forces F_{c1} and F_{c2} generate, via the lever arm $D/4$, the cutting moment

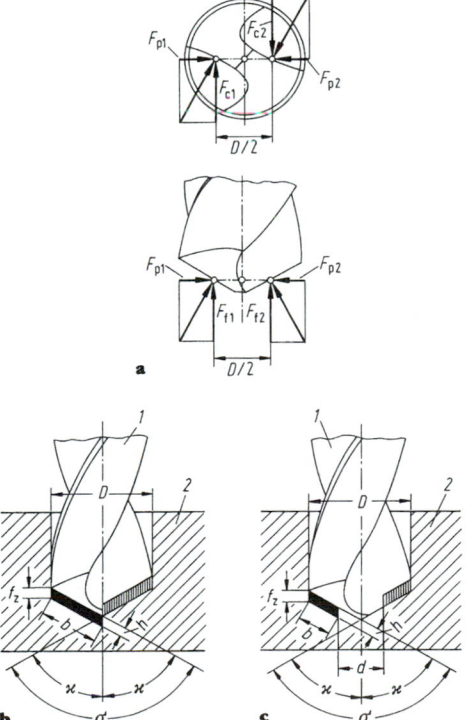

a

b

c

Figure 20. Metal cutting geometry and machining forces in drilling: **a** forces, **b** drilling into solid metal, **c** hole enlarging; *1* tool, *2* workpiece.

$$M_c = (F_{c1} + F_{c2})D/4, \quad F_{c1} = F_{c2} = F_{cz},$$
$$M_c = F_{cz}D/2.$$

The feed forces F_{f1} and F_{f2} are added together to give F_f

$$F_f = F_{f1} + F_{f2}, \quad F_{f1} = F_{f2} = F_{fz}, \quad F_f = 2F_{fz}.$$

The passive forces F_{p1} and F_{p2} cancel each other out in the ideal case, i.e. with a symmetrical drill. If there are errors of symmetry, F_{p1} and F_{p2} generate interference forces which impair the quality of the hole. The cutting force per cutting edge works out as

$$F_{cz} = bh^{(1-m_c)} k_{c1\cdot1}, \quad h = f_z \sin \kappa, \quad b = D/(2 \sin \kappa).$$

By analogy, the feed force is

$$F_f = Dh^{(1-m_f)} k_{f1\cdot1}.$$

Values are given in **Appendix K4, Table 2**. The feed forces are heavily dependent on the form of the transverse cutting edge. They can be greatly reduced by grinding the point. Wear causes them to rise to twice the original value or more.

Surface quality in drilling with twist drills corresponds to roughing with $R_z = 10$ to 20 μm. The roughness can be reduced by reaming. Another possibility is to use drills made entirely of cemented carbide. When drilling solid metal, surface qualities, dimensional accuracy and accuracy of shape like those obtained with reaming are achieved.

Short-hole Drilling

Short-hole drilling, with drilling depths of $L < 2 \cdot D$, covers a large proportion of bolt hole drilling, through hole drilling and tapping. For this, short-hole drills with indexable inserts may be used for diameters from 16 to over 120 mm. Their advantage compared with twist drills is the absence of a transverse cutting edge and the increase in cutting speed and feed rate achieved with indexable cemented carbide or ceramic inserts. Owing to the asymmetrical machining forces, the use of short-hole drills requires rigid tool spindles such as are common on machining centres and milling machines. The higher rigidity of the tool enables pilot drilling of inclined or curved surfaces. Accuracies of IT7 are achieved without further work [11].

4.2.4 Milling

Classification of Milling Processes

In milling, the necessary relative motion between the tool and the workpiece is achieved by means of a circular cutting motion of the tool and a feed motion perpendicular to or at an angle to the axis of rotation of the tool. The cutting edge is not continually in engagement. It is therefore subject to alternating thermal and mechanical stresses. The complete machine-tool–workpiece system is *dynamically* stressed by the interrupted cutting action.

Milling processes are classified according to DIN 8589 on the basis of the following:

The nature of the resulting workpiece surface.
The kinematics of the metal cutting operation.
The profile of the milling cutter.

Milling can be used to produce a practically infinite variety of workpiece surfaces. A distinguishing feature of a process is the cutting edge (primary or secondary) that

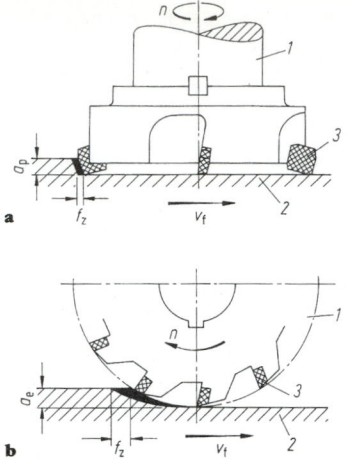

Figure 21. Comparison of face milling and peripheral milling: **a** face milling – workpiece surface produced by secondary cutting edge, **b** peripheral milling – workpiece surface produced by primary cutting edge; *1* tool, *2* workpiece, *3* cutting edge.

produces the workpiece surface (**Fig. 21**): in *face milling* it is the *secondary cutting edge* located on the face of the milling cutter, while in *peripheral milling* it is the *primary cutting edge* located on the circumference of the milling cutter.

A distinction can be made on the basis of the feed direction angle φ, **Fig. 22**: in *downcut milling*, the feed direction angle φ is > 90°, thus the cutting edge of the milling

cutter enters the workpiece at the maximum undeformed chip thickness, while in *upcut milling* the feed direction angle φ is < 90°, thus the cutting edge enters at the theoretical undeformed chip thickness $h = 0$. This initially results in pinching and rubbing.

A milling operation may include both upcut milling and downcut milling. The principal milling processes are summarised in **Fig. 23**.

Plain Face Milling with End Milling Cutter

The kinematics of metal cutting and the relationship of the metal cutting forces during milling will be discussed with reference to plain face milling with an end milling cutter. Further milling processes are described in [12].

Kinematics of Metal Cutting. To describe the process, it is necessary to distinguish between the engagement variables and the metal cutting variables. The *engagement variables*, which are expressed in relation to the working plane, describe the interaction of the cutting edge and the workpiece. The working plane is described by the cutting speed vector v_c and the rate of feed vector v_f. In milling the engagement variables are (**Fig. 24**): depth of cut a_p, measured at right angles to the working plane; cutting engagement a_c, measured in the working plane at right angles to the feed direction; and feed of the cutting edge f_z, measured in the feed direction.

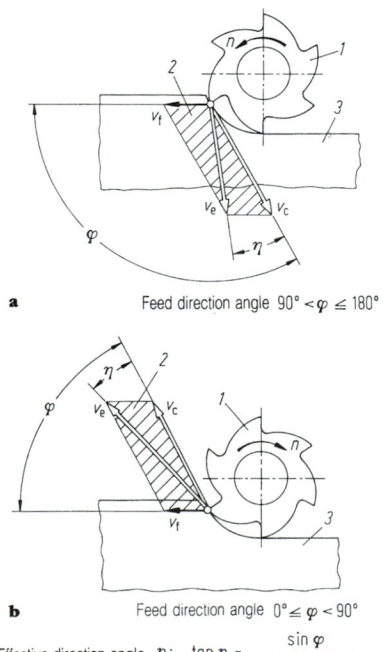

a   Feed direction angle 90° < φ ≤ 180°

b   Feed direction angle 0° ≤ φ < 90°

Effective direction angle η: $\tan \eta = \dfrac{\sin \varphi}{v_c / v_f + \cos \varphi}$

Figure 22. Comparison of **a** downcut milling and **b** upcut milling (DIN 6580 E); *1* milling cutter, *2* working plane, *3* workpiece.

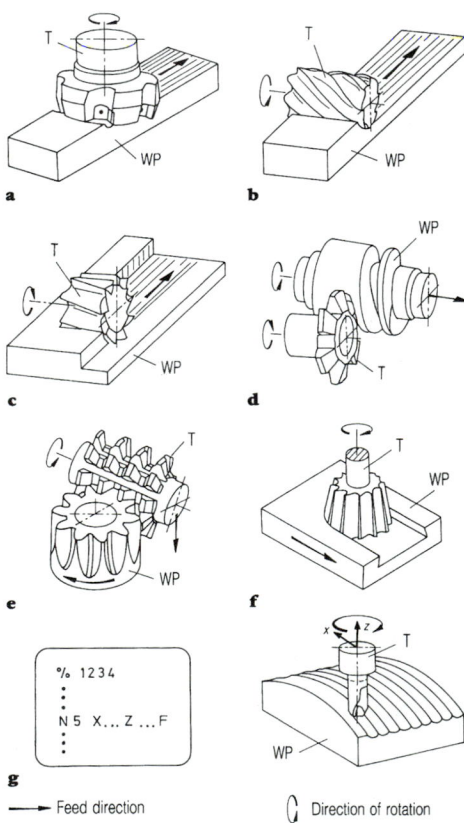

— Feed direction   ◖ Direction of rotation

Figure 23. Milling processes (DIN 8589). Plain milling: **a** face milling, **b** peripheral milling, **c** combined face and peripheral milling, **d** thread milling, **e** hobbing, **f** profile milling, **g** form milling, WP workpiece, T tool.

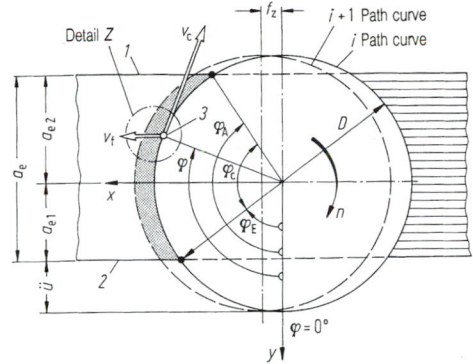

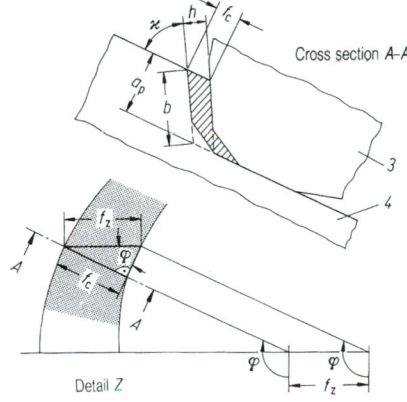

Figure 24. Engagement variables in plain face milling with an end milling cutter; *1* exit plane, *2* entry plane, *3* tool cutting edge, *4* workpiece.

For a full description of the kinematics of metal cutting, the following data are required: milling cutter diameter D, number of teeth z, tool projection $\ddot{u}$ and the cutting edge geometry (side rake angle γ_f, back rake angle γ_p, side clearance angle α_f, back clearance angle α_p, entry angle κ_r, angle of inclination λ_s, cutting edge radius r, and land).

Owing to the interrupted cut, and the entry and exit conditions of the cutting edge, and the types of contact, are especially important for the milling process. The types of contact describe the nature of the first and last contacts of the cutting edge with the workpiece. They can be determined from the entry and exit angles and the tool geometry. It is especially bad if the cutting edge tip is the first point of contact.

From the engagement variables can be derived the *metal cutting variables*, which indicate the dimensions of the layer of metal to be removed from the workpiece. The metal cutting variables are not identical to the chip variables, which describe the dimensions of the resulting chips. The cutting edges describe cycloids in relation to the workpiece. As the cutting speed is significantly higher than the rate of feed, they can be approximated by circular paths. In this way of looking at the subject, the undeformed chip thickness is (**Fig. 24**):

$$b(\varphi) - f_z \sin \kappa \cdot \sin \varphi.$$

With the undeformed chip width $b = a_p/\sin \kappa$, the undeformed chip cross-section is

$$A(\varphi) = bb(\varphi) = a_p f_z \sin \varphi.$$

The time-related chip volume is $Q = a_e a_p v_f$.

The undeformed chip thickness is a function of the entry angle φ and is thus not constant as in turning. The evaluation of the milling process is based on the mean undeformed chip thickness

$$b_m = (1/\hat{\varphi}_c) \int_{\varphi_E}^{\varphi_A} b(\varphi) \, d\varphi$$

$$= (1/\hat{\varphi}_c) f_z \sin \kappa (\cos \varphi_E - \cos \varphi_A).$$

Machining Force Components. The machining force required for chip formation has to be absorbed by the cutting edge and the workpiece. According to DIN 6584, the machining force F can be resolved into an active force F_a, which lies in the working plane, and a passive force F_p, which is at right angles to the working plane. The direction of the active force F_a changes with the entry angle φ. The components of the active force can be expressed in relation to the following directions (**Fig. 25**):

Direction of cutting speed v_c: the components cutting force F_c and perpendicular cutting force F_{cN} relate to a co-rotating system of coordinates (tool-specific components of the active force).

Direction of rate of feed v_f: the components feed force F_f and perpendicular feed force F_{fN} relate to a fixed system of coordinates (workpiece-specific components of the

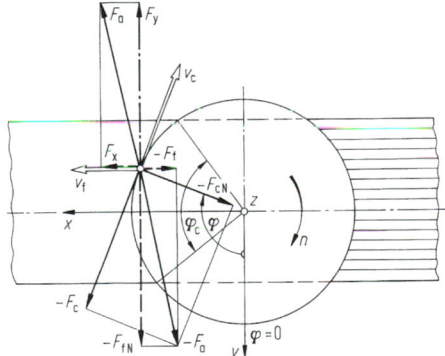

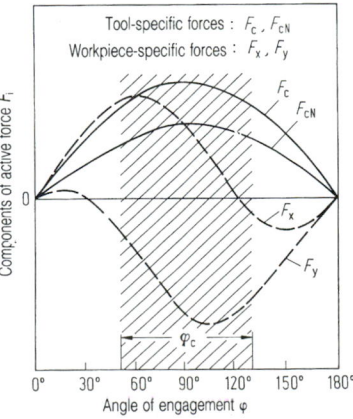

Figure 25. Components of machining force in plain face milling with an end milling cutter.

active force). For converting the active force from the fixed system of coordinates into a co-rotating system, the following apply:

$$F_c(\varphi) = F_f(\varphi) \cos \varphi + F_{fN}(\varphi) \sin \varphi,$$

$$F_{cN}(\varphi) = F_f(\varphi) \sin \varphi - F_{fN}(\varphi) \cos \varphi,$$

$$F_x(\varphi) = F_f(\varphi),$$

$$F_y(\varphi) = F_{fN}(\varphi).$$

This transformation is important if, for instance, the cutting force F_c is to be measured with a three-component force-measuring platform to which the workpiece is fixed. **Figure 25** shows the pattern of the components of the active force in the tool- and workpiece-specific systems of coordinates for axial milling with an end milling cutter.

Relationship of Machining Forces. Kienzle's machining force equation [7] can also be applied to milling. For the machining force components cutting force F_c, perpendicular cutting force F_{cN} and passive force F_p,

$$F_i = Ak_i, \quad \text{where } i = c, cN, p,$$

where A = the undeformed chip cross-section and k_i = the specific machining force. Owing to the wide range of undeformed chip thicknesses that is covered by milling (the undeformed chip thickness depends on φ), Kienzle's relationship applies only to certain areas. The undeformed chip thickness range of 0.001 mm $< b <$ 1.0 mm is divided into three sections (**Fig. 26**) [13, 14]. For each section, a straight line can be determined, which is established by the parameters: the main value of the specific machining force and the incremental value. For the specific machining force the following apply:

$$= k_{i\,1.0,\,0.01} \cdot b^{-m_{i\,0.01}} \quad \text{for 0.001 mm} < b < 0.01 \text{ mm}$$

$$k_i = k_{i\,1.0,\,0.1} \cdot b^{-m_{i\,0.1}} \quad \text{for 0.01 mm} < b < 0.1 \text{ mm}$$

$$= k_{i\,1.1} \cdot b^{-m_i} \quad \text{for 0.1 mm} < b < 1.0 \text{ mm}$$

where i = c, cN, p.

Thus the machining force for milling with an end milling cutter is

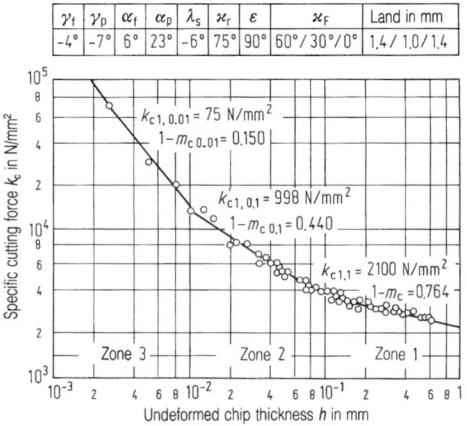

Workpiece material	Ck 45 N						
Cutting material	HM P25						
Cutting conditions	v_c = 190 m/min						

γ_f	γ_p	α_f	α_p	λ_s	$\varkappa_r$	ε	$\varkappa_f$	Land in mm
-4°	-7°	6°	23°	-6°	75°	90°	60°/30°/0°	1.4 / 1.0 / 1.4

$k_{c1,\,0.01}$ = 75 N/mm^2
$1 - m_{c\,0.01}$ = 0.150

$k_{c1,\,0.1}$ = 998 N/mm^2
$1 - m_{c\,0.1}$ = 0.440

$k_{c1,1}$ = 2100 N/mm^2
$1 - m_c$ = 0.764

Zone 3   Zone 2   Zone 1

Specific cutting force k_c in N/mm^2

Undeformed chip thickness h in mm

Figure 26. Specific cutting force in plain face milling with an end milling cutter [13].

$$F_i = bk_{i\,1.1} \cdot b^{1-m_i}, \quad \text{where } i = c, cN, p.$$

The corresponding component of the machining force can be calculated for milling if the main value of the specific machining force component and the incremental value for the workpiece/cutting material pair and the cutting conditions are available. The machining indices for axial plane face milling with a milling head are given in **Appendix K4, Table 3** for a number of workpiece materials and cutting conditions [7, 13, 15]. Often, though, to estimate the machining force during milling, machining indices obtained from turning will have to be used.

Milling machine capacity is designed on the basis of the average machining force

$$F_{im} = bk_{i\,1.1} \cdot b_m^{1-m_i} K_{ver}K_\gamma K_v K_{ws}K_{wv},$$

where i = c, cN, p.

In this equation, b_m is the mean undeformed chip thickness, K_{ver} = 1.2 to 1.4 is the correction factor for the manufacturing process (the factor takes into account the fact that the machining indices were obtained from turning tests), K_γ is the correction factor for the rake angle (cf. turning), K_v is the correction factor for the cutting speed (cf. turning), K_{wv} is the correction factor for tool wear (cf. turning) and K_{ws} is the correction factor for the cutting material (cf. turning).

Research into plain face milling shows that the influence of wear on the machining force components cannot be ignored [14].

Vibrations. Depending upon the elasticity frequency response of the complete milling machine–milling cutter–workpiece system, the metal cutting forces generate vibrations which may affect surface quality and tool life. According to the method of their generation, these vibrations are divided into separately excited and self-excited vibrations [cf. J2].

Separately Excited Vibrations. In the case of separate excitation, the complete system vibrates at the frequency of the exciting forces. The intermittent cutting action of milling means that the cutting edges are not constantly in engagement. With a multiple-edged milling cutter, the number of cutting edges in engagement at any one time should be taken into account. Depending on the ratio a_c/D, z_{iE} cutting edges are in engagement, the following relationship being valid:

$$z_{iE} = (\hat{\varphi}_c/2\pi)z, \quad \text{where } \varphi_c/2 = a_c/D.$$

The mean cutting force acting on the milling cutter and thus on the spindle of the milling machine is

$$F_{cm} = z_{iE} F_{cmz},$$

where F_{cmz} is the mean cutting force of a cutting edge.

Superimposed on the mean cutting force is a dynamic force element. The larger the value of z_{iE}, the smaller is the force amplitude; if z_{iE} is an integer, the cutting force amplitude is at a minimum. The dynamic force element leads to separately excited vibrations between the workpiece and the milling cutter.

Self-excited Vibrations. With self-excitation, the complete system vibrates at one or more eigenfrequencies, without an external interference force affecting the system.

Special importance attaches to self-excited vibrations which arise because of the regenerative effect and are also referred to as "regenerative chatter". The chatter is caused by variations in cutting force due to changes in undeformed chip thickness [16].

Chatter can be influenced by varying the cutting speed, depth of cut, feed rate and cutting edge geometry.

Wear. Owing to the intermittent cutting action in milling, the cutting material is subjected to alternating thermal and mechanical stresses. As a result, not only face and flank wear but cracking in the cutting tip may determine tool life. **Figure 27** depicts the flank wear on the primary cutting edge and the crater depth for plain face milling with an end milling cutter. Approximate data for selecting the setpoint values are given in **Appendix K, Table 4**. With the development of cubic boron nitride, the finish-machining of hardened materials by milling has been further developed [17–19]. Depending on the cutting conditions, surface roughnesses comparable to those obtained in grinding are achieved. In grinding, accuracy of shape is achieved by spark-out. As there has to be a minimum undeformed chip thickness in milling, shape defects occur which can be attributed to the following influencing variables [20]: environment, operating behaviour of the milling machine, inhomogeneities in material hardness, heating of the workpiece due to metal cutting, and a change in intrinsic stress in the edge zone of the workpiece.

Form Milling

Hollow-mould tools such as deep-drawing tools are manufactured by both chip-forming and chipless machining processes. Milling plays a central role as a controlled shaping process. The essential characteristic in form milling is the number of actively controlled axes, a distinction accordingly being made between *three-axis milling* and *five-axis milling* (**Fig. 28**). In five-axis milling, not only the tip but the axial direction of milling cutter are continuously and simultaneously controlled relative to the workpiece coordinate system. As a rule, three-axis milling is performed with a convex milling cutter, and five-axis milling is performed with an end milling cutter. The profile of the milled grooves determines the productivity and quality of the process (little finishing being required with a small profile depth). It is formed by machining a curved surface in parallel lines and depends on the milling cutter geometry, the workpiece geometry and the method of working. For a given groove depth t_R, five-axis milling with an end milling cutter produces significantly larger

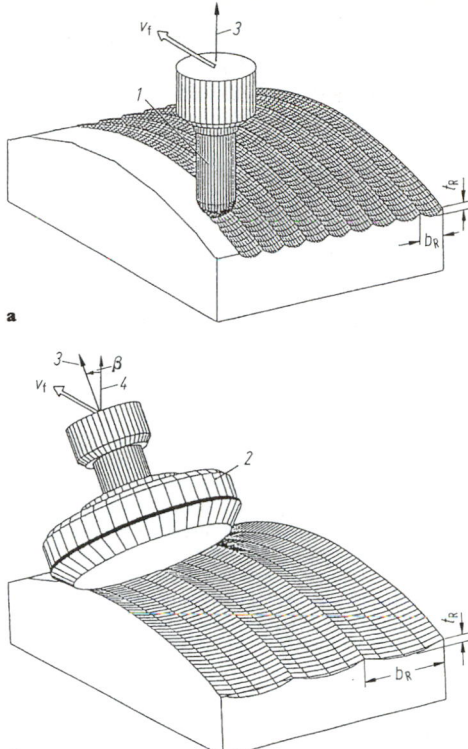

a

b

Figure 28. Form milling by **a** three-axis milling and **b** five-axis milling; *1* convex milling cutter, *2* end milling cutter, *3* tool axis, *4* surface normal.

groove widths b_R than three-axis milling with a convex milling cutter [21].

4.2.5 Other Processes: Planing and Shaping, Broaching, Sawing

Planing and Shaping

DIN 8589 Part 4 distinguishes between planing and shaping. Chip removal is accomplished during the working stroke with a single-point cutting tool. The following return stroke restores the tool to its original position. The feed is intermittent, usually at the end of a return stroke.

In *planing*, the workpiece performs the cutting and return motion. Feed and engagement are accomplished by the tool (**Fig. 29**). In *shaping*, the tool performs the cutting and return motion, while feed and engagement are accomplished by the workpiece or the tool.

The reciprocating motion of the workpiece (in planing) or the tool (in shaping) produces high inertia forces and limits the cutting speed. As a guideline for the cutting speed, for machining steels the ranges v_c = 60–80 m/min (for roughing) and v_c = 70–100 m/min (for finishing) are well established for cemented carbide tools.

Frequently used special methods are hobbing by planing and shaping for manufacturing involute gears (cf. K5.2.1).

Broaching

In broaching (DIN 8589 Part 5), material is removed using a multiple-pointed tool, the teeth of which are one behind

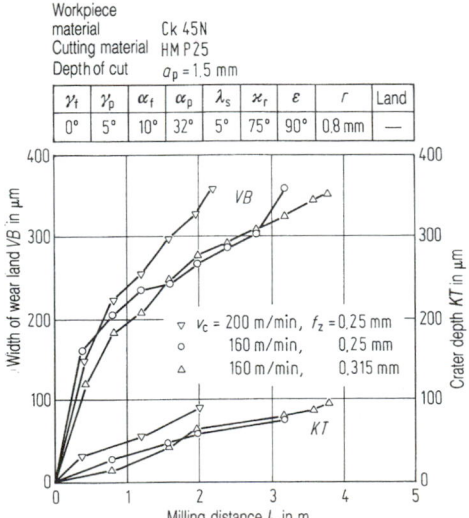

Workpiece material	Ck 45N							
Cutting material	HM P25							
Depth of cut	a_p = 1.5 mm							

γ_f	γ_p	α_f	α_p	λ_s	$\varkappa_r$	ε	r	Land
0°	5°	10°	32°	5°	75°	90°	0.8 mm	—

▽ v_c = 200 m/min, f_z = 0.25 mm
○ 160 m/min, 0.25 mm
△ 160 m/min, 0.315 mm

Figure 27. Progress of wear in plain face milling with an end milling cutter.

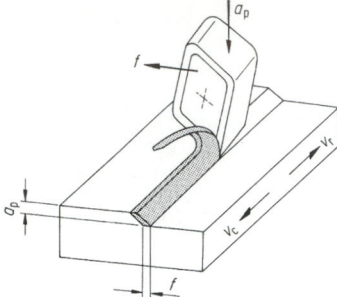

Figure 29. Planing: a_p depth of cut, f feed, v_c cutting speed, v_T return speed.

the other and are successively stepped by one layer of the metal to be removed. Thus no feed motion is required, as it is "built into" the tool, as it were. The cutting motion is translatory or, in certain circumstances, helical or circular.

The advantages of the process are the high machining capacity and the possibility of finishing workpieces with a single tool. Furthermore, high surface qualities and dimensional accuracies with tolerances up to IT7 can be achieved. Owing to the high cost of tools, the main areas of use are series and mass production; a new tool is required for each changed workpiece shape.

A basic distinction is made between *internal broaching* and *external broaching* (**Fig. 30**). In internal broaching, the broaching tool (broach) is pushed or pulled through a hole, while in external broaching the tool is moved across the external surface.

Broaches are divided into roughing, finishing and sizing teeth sections. Normal undeformed chip thicknesses in flat broaching of steels are $b_z = 0.01$ to 0.15 mm for roughing and $b_z = 0.003$ to 0.023 mm for finishing. Broaching of cast materials is carried out to a thickness of $b_z = 0.02$ to 0.2 mm in the roughing section and $b_z = 0.01$ to 0.04 mm in the finishing section [22].

Cutting speeds are restricted by the hardness of the chosen cutting material at high temperature and by the efficiency of the machine. The cutting material most commonly used, high-speed steel (HSS), permits only low cutting speeds owing to the decrease in high-temperature hardness at approximately 600 °C; the capacity of the process can be increased by using TiN-coated HSS or cemented carbide. Cutting speeds of $v_c = 1$ to 30 m/min are used, with speeds of up to 60 m/min in exceptional cases. High cutting speeds require high drive power outputs to accelerate and brake the tool and the broach slide, causing a disproportionate increase in equipment costs [23]. Vibration problems also become greater, especially with thin internal broaches.

In broaching, mainly mineral oils are used for lubricating and cooling in the contact zone area, but above all to prevent the formation of built-up edges and to carry away the chips. They usually contain EP (extreme pressure) additives, which are nowadays mostly chlorine-free [22].

Sawing

Sawing is metal cutting with a multiple-pointed tool having a small width of cut for severing or slitting workpieces. The rotational or translatory principal motion is performed by the tool (DIN 8589 Part 6). The teeth of the tool are offset in alternate directions. By this means the kerf is widened in relation to the saw blade, reducing friction between tool and workpiece.

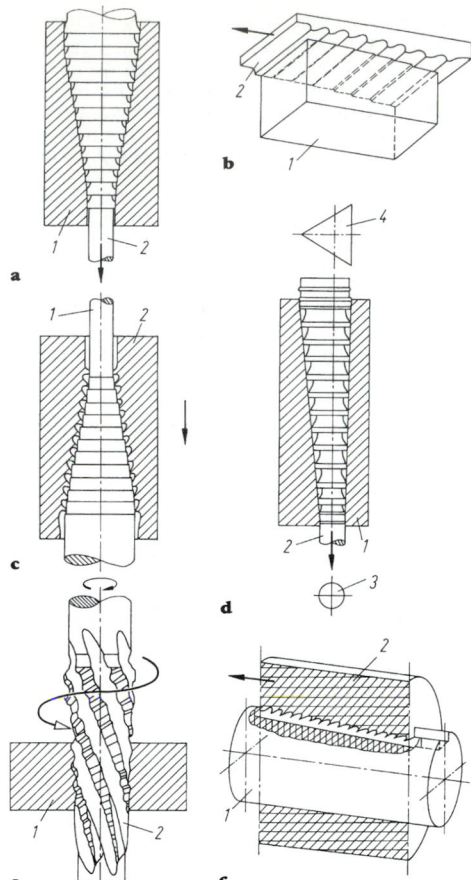

Figure 30. Broaching: a internal cylindrical broaching, **b** external surface broaching, **c** external cylindrical broaching, **d** internal profile broaching, **e** thread broaching, **f** external groove broaching; *1* workpiece, *2* tool, *3* original cross-section, *4* final cross-section.

Band sawing is sawing with a continuous, usually straight cutting motion of a rotating endless band. The motions and cutting parameters are shown in **Fig. 31**.

The normal cutting speeds with high-speed steels lie in the range $v_c = 6$ to 45 m/min with feed rates per tooth of $f_z = 0.1$ to 0.4 mm. If bands with inserted cemented-carbide teeth are used, the cutting speed can be increased to 200 m/min for steels and up to 2000 m/min for light metals.

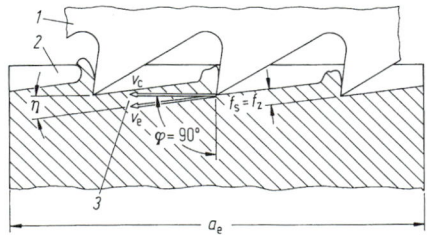

Figure 31. Cutting variables in band sawing: *1* band saw, *2* workpiece, *3* working plane, v_c cutting speeed, f_z tooth feed, v_e effective speed, a_e engagement distance, f_s cutting feed.

In *reciprocating sawing* (hacksawing), a tool of finite length clamped in a holding frame is used. The feed motion is carried out intermittently only as the tool advances or at a constant perpendicular force.

Circular sawing is sawing with a continuous cutting motion using a circular saw blade. In terms of kinematics and metal cutting technique, circular sawing resembles peripheral milling.

4.2.6 Cutting Materials

Tools for machining with geometrically well-defined tool edges consist of the cutting tip, the holder and the shank (**Fig. 32**). Holders and shanks are designed according to constructional and organisational requirements, such as mating dimensions of the machine, the nature and extent of tool storage and tool changing, or the geometry of the workpiece. The cutting tip is responsible for chip removal. It is subject to mechanical and thermal stresses and chemical attack. The tip consequently wears (for types of wear cf. K4.2.2).

A basic dualism applies to all cutting materials. They are either hard and wear-resistant – but are then less tough and stable under periodically changing loads as well as under unstable cutting conditions and in intermittent cutting. Or they are better able to bear variable mechanical or thermal stresses, but are then less resistant to wear. To overcome this restricting dualism, various cutting materials are manufactured as *composite materials*. Coating with hard-wearing carbides or oxides produces a *separation of functions*: the physically (PVD – physical vapour deposition) or chemically (CVD – chemical vapour deposition) vapour-deposited layers provide wear protection, while the underlying, tougher substrate performs the supporting function, even under dynamic load conditions.

Cutting materials comprise: plain and alloy steels (still important for manually guided tools), high-speed steels, cemented carbides, ceramics and superhard cutting materials (diamond and boron nitride) (cf. D3.1.4).

High-Speed Steels. High-speed steels are used for drilling, milling, broaching, sawing and turning tools. Their high temperature hardness (up to approximately 600 °C) is far superior to that of tool steels (**Fig. 33**). Their hardness results from their basic martensitic structure and from interspersed carbides: tungsten carbides, tungsten-molybdenum carbides, chromium carbides, vanadium carbides. Accordingly, high-speed steels can be divided into four groups (high-speed steels are designated by S followed by W%–Mo%–V%–Co%):

Steels with 18% W,	e.g. S 18-1-2-10
Steels with 12% W,	e.g. S 12-1-4-5
Steels with 6% W, 5% Mo	e.g. S 6-5-2
Steels with 2% W, 9% Mo	e.g. S 2-9-2-8

High-speed steels are standardised in "Stahl-Eisen-Werkstoffblatt" (iron and steel data sheet) 320. The through hardenability of tools with large cross-sections is increased

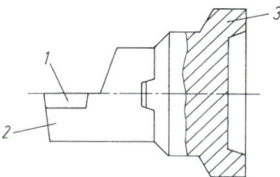

Figure 32. Parts of a metal cutting tool: *1* cutting tip, *2* holder, *3* shank.

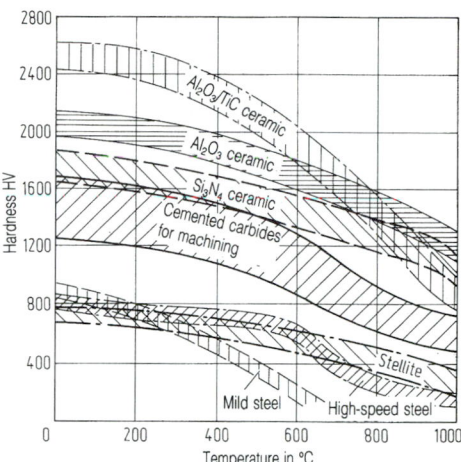

Figure 33. High-temperature hardness of cutting materials.

with molybdenum and/or by alloying with chromium. Tungsten increases the wear resistance and tempered strength, vanadium the wear resistance (but is difficult to grind when hard) and cobalt the high-temperature hardness. High-speed steels are manufactured by casting. This determines their structure and segregations. These disadvantages can be overcome by means of powder metallurgy (sintered high-speed steels). PM steels have enhanced edge strength and cutting edge durability. They are used for thread-cutting and reaming tools. At high vanadium carbide contents they are easier to grind than cast high-speed steels. They have not yet gained wide acceptance because of their higher cost.

High-speed steels are usually coated by PVD (reactive ion plating at low temperatures so as to remain below the tempering temperature). Simple shapes such as indexable inserts can be treated by CVD followed by rehardening. The coating material is titanium nitride (TiN, gold-coloured). The life of coated tools (drills, taps, hobs, form turning tools) is increased by two to eight times.

Cemented Carbides. Cemented carbides are two-phase or multi-phase alloys manufactured by powder metallurgy with a metallic binder. The materials used are tungsten carbide (WC: α-phase), titanium carbide and tantalum carbide (TiC, TaC: γ-phase). The binder is cobalt (Co: β-phase) with a content of 5 to 15%. Nickel and molybdenum binders (Ni, Mo) are also used in so-called "cermets" (also cemented carbides). Higher β-phase contents increase toughness, higher α-phase contents increase wear resistance and higher γ-phase contents enhance wear resistance at high temperature. Cermets have high edge strength and cutting edge durability. They are suitable for finishing under stable cutting conditions. The manufacture of cemented carbides by powder metallurgy permits considerable freedom in the choice of constituents (in contrast to casting).

Cemented carbides retain their hardness up to over 1000 °C (**Fig. 33**). They can therefore be used at higher speeds (three or more times as high) than high-speed steels. According to DIN 4990/ISO 513, cemented carbides are classified into the metal cutting application groups P (for long-chipping, ductile ferrous materials), K (for short-chipping ferrous metals and for non-ferrous metals) and M as a universal group (for ductile cast iron and for ferritic and austenitic steels). Each group is subdivided into toughness and wear resistant grades by adding a number; for example, P02 stands for very hard-wearing cemented carbide, P40 for tough cemented carbide. The metal cutting application groups do not indicate the material's composition. The classification is done by the manufacturer.

Cemented carbides are coated with titanium carbide (TiC), titanium nitride (TiN), aluminium oxide (Al_2O_3) or chemical or physical combinations of these substances. The coatings are usually applied by CVD. They are used to achieve longer tool lives or higher cutting speeds. They broaden the range of use of a grade. Coated cemented carbides should not be used for non-ferrous metals, high-nickel ferrous materials or – owing to the edge rounding caused by manufacturing – in precision or ultra-precision machining (cermets are better for this purpose). Intermittent cutting and milling requires coatings of especially high bonding strength, which can be influenced by process control during coating.

Ceramic Inserts. Ceramic inserts are single-phase or multi-phase sintered hard materials based on metal oxides, carbides or nitrides. They are distinguished from cemented carbides by the absence of metallic binders and exhibit high hardness even at temperatures above 1200 °C. Ceramic inserts are therefore generally suitable for machining at high cutting speeds, usually exceeding 500 m/min.

The use of *aluminium oxide ceramic* is restricted by its lower bending strength and fracture toughness compared with cemented carbide. In intermittent cutting and alternating mechanical and thermal stresses, microcracking, crack growth with peeling or total fracture can occur. This effect greatly depends on the nature and composition of the ceramic. The change from single-phase materials (Al_2O_3) to multiphase materials has improved toughness considerably: Al_2O_3 containing 10 to 15% ZrO_2 (transformation reinforcement [24]) or Al_2O_3 with TiC (dispersion ceramic). The main uses are turning of lamellar graphite cast iron, under stable conditions, at cutting speed > 500 m/min. Turning of steel is feasible. Additions of up to 40% TiC to the Al_2O_3 ceramic (black mixed ceramic) increase toughness and edge strength. These are used for hard machining and the finishing of cast iron.

Silicon nitride (Si_3N_4) exhibits ideal cutting material qualities (high strength, hardness, oxidation resistance, thermal conductivity and resistance to thermal shock, owing to strong covalent bonding of the elements). Here there is no limitation, because of a lack of fracture toughness. Si_3N_4 is used as a cutting material in three versions: sintered Si_3N_4 ($\rho = 3.1$ g/cm^3, $R_m = 650$ MPa), hot-pressed Si_3N_4 ($\rho = 3.2$ g/cm^3, $R_m = 700$ MPa), and as the material system Y-Si-Al-O-N. The manufacture and use of Si_3N_4 is limited by the sintering auxiliaries (e.g. magnesium oxide, yttrium oxide) that are at present necessary. They determine the glassy phases in the cutting material. When machining steel or ductile cast iron, failure occurs due to severe wear. In contrast, Si_3N_4 is suitable for turning and milling of grey cast iron, and also for highly intermittent cutting actions and for the turning of high-nickel materials.

Superhard Cutting Materials. These are polycrystalline diamond (PCD) and boron nitride (PBN). The materials are synthesised at high pressure and temperature. PCD is supplied as an approximately 0.5-mm layer on cemented carbide. It is used to machine aluminium and aluminium alloys, especially easily-earing AlSi alloys, fibre-reinforced plastics, graphite, and non-ferrous metals; it cannot be used for steel, owing to the high rate of chemical wear. PBN, on the other hand, is chemically stable towards iron. It is used for hardened iron and steel and is supplied as a solid product or as an approximately 0.5-mm layer on cemented carbide. Monocrystalline (natural) diamond is used for precision and ultra-precision machining (turning, milling) of Al and Cu alloys with extremely sharp cutting edges ($r_\beta < 1$ μm).

4.3 Machining with Geometrically Non-defined Tool Edges

4.3.1 Fundamentals

Machining with geometrically non-defined tool edges is cutting by the mechanical action of cutting edges on the material (DIN 8580, third group of the main group "cutting"). The cutting edges are formed by *grains of hard material*. They are shaped and arranged irregularly. The cutting edge geometry is therefore not described with reference to a single grain. The individual cutting edge is geometrically non-defined. The processes are broken down into the following subgroups:

Grinding with a rotating tool
Belt grinding
Reciprocating grinding
Honing
Lapping
Barrel polishing
Machining by abrasive blasting (DIN 8200)

The common factor in these processes is that the grains of hard material generally form several cutting edges. The important cutting edge angles for chip formation, the *clearance angle* α, the *rake angle* γ and the *wedge angle* β, are only indicated by means of statistical parameters such as means or distributions. On average, sharply negative face angles and large contact and friction zones are formed between the grain and the workpiece. The cutting edges penetrate only a few micrometres into the material. The undeformed chip thickness distribution depends on the position of the cutting edges in the mixture of grains (microtopography of the cutting edge zone) and the geometry of the machined workpiece surface. Not only chip removal but also elastic and plastic deformations without chip removal take place.

High normal forces result between tool and workpiece at the predominantly negative rake angles of the cutting edges. They lead to elastic deformations in the machine (stretching of the frame and deflection of the spindle), the tool and the workpiece. The deformations may markedly exceed the normal small feed motions. Therefore a distinction should be made between the theoretical and the actual feed motion (**Fig. 34**).

Machining processes with geometrically non-defined tool edges are frequently used as final machining processes for workpieces subject to exacting quality requirements. **Figure 35** shows a comparison of various precision machining processes with regard to operational results and efficiency. It can be seen that the grinding processes achieve high rates of metal removal, while honing and lapping are able to produce the best surface qualities.

The cutting edges are formed by the contours of grains of the hard material. The materials used are hard–brittle materials such as zirconium–corundum (ZrO_2 with Al_2O_3), corundum (Al_2O_3), silicon carbide (SiC), boron carbide (B_4C), boron nitride (BN) and diamond (C); their hardnesses are shown in **Fig. 36**. However, diamond is unsuitable for machining steel, as there is a high chemical

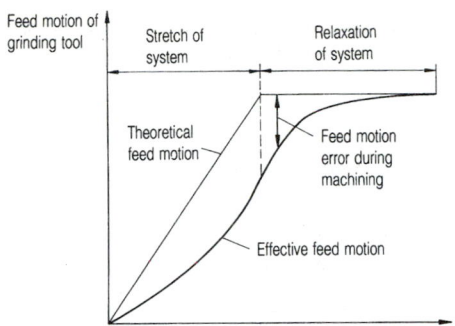

Figure 34. Feed motion errors in precision machining due to elastic deformations in the machine–tool–workpiece system.

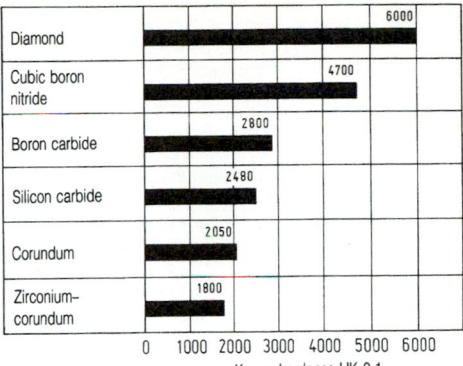

Figure 36. Knoop hardness of various hard materials.

affinity between diamond and iron which leads to rapid tool wear.

Sorting of the grains according to size is accomplished by screening (DIN 69 100). The basis of all standards (cf. **Appendix K4, Table 5**) is the mesh of the screens through which the abrasive grains pass. The average grain size is determined by the shape (angularity) of the individual grains. Below a certain grain size, sorting can be carried out by settlement out of a slurried suspension of water and grains. The grains are either bonded together into a tool (grinding, honing) or are used loose (lapping, abrasive blasting).

The bonding material (bond) is chosen according to the requirements of the machining process and of the grain material. Inorganic (ceramic, silicate, magnesite), organic (rubber, synthetic resin, glue) and metallic bonds (bronze, steel, cemented carbide) are used; the most popular bonds are those made from ceramic or synthetic resin. In manufacturing a tool, its structure can be influenced to a certain extent by varying the proportions of grains, bond and pore volume.

The *chip formation mechanism* when using geometrically non-defined tool edges differs from that applicable to machining with geometrically well-defined tool edges (**Fig. 37**). A characteristic feature of this process is the often sharply negative rake angle of the individual grains. In Phase 1, this causes elastic deformation of the material.

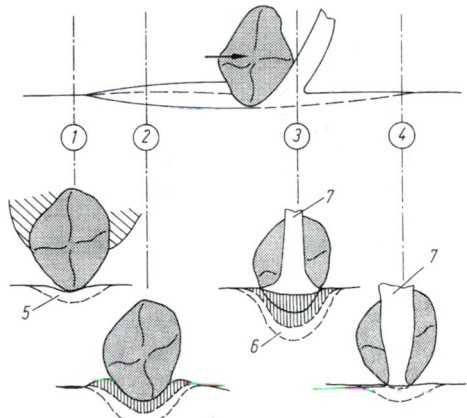

Figure 37. Phases of chip formation: *1* elastic deformation, *2* elastic and plastic deformation ("ploughing"), *3* elastic and plastic deformation ("ploughing") and deformation of chip by shear (cutting deformation), *4* elastic and shear deformation of chip, *5* elastic deformation zone, *6* plastic deformation zone, *7* chip.

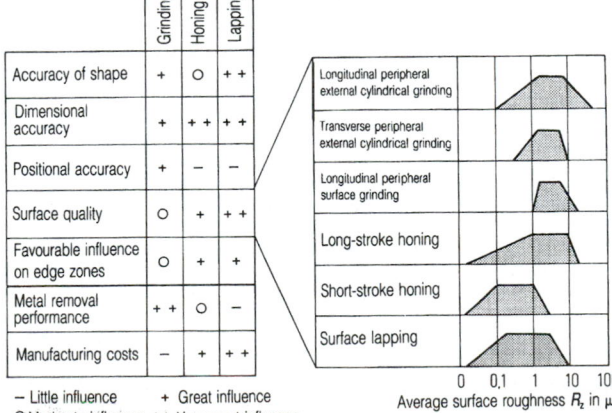

	Grinding	Honing	Lapping
Accuracy of shape	+	O	+ +
Dimensional accuracy	+	+ +	+ +
Positional accuracy	+	−	−
Surface quality	O	+	+ +
Favourable influence on edge zones	O	+	+
Metal removal performance	+ +	O	−
Manufacturing costs	−	+	+ +

− Little influence   + Great influence
O Moderate influence ++ Very great influence

Longitudinal peripheral external cylindrical grinding

Transverse peripheral external cylindrical grinding

Longitudinal peripheral surface grinding

Long-stroke honing

Short-stroke honing

Surface lapping

0   0,1   1   10   100
Average surface roughness R_z in μm

Figure 35. Economic and technological comparison of various precision machining processes.

In Phase 2 plastic deformation occurs, while in Phase 3 the actual chip removal takes place. A large amount of friction occurs between the individual grains and the workpiece.

The mechanical energy supplied is almost exclusively converted into heat. **Figure 38** shows a qualitative distribution of the heat flows around an individual grain. Most of the heat generated flows into the workpiece, while a small proportion finds its way into the grain, the bond and the surroundings (cooling lubricant, atmosphere). Temperature increases in the workpiece may harm its edge zone. This is manifested in thermally induced inherent stresses, structure changes or cracks, which influence its subsequent behaviour in service. The use of readily heat-conducting granular materials (CBN, diamond) and bonds reduces the proportion of the heat that passes into the workpiece [25].

In machining with geometrically non-defined tool edges, the use of *cooling lubricant* is important for the end result. The cooling and lubricating effect can reduce tool wear. Furthermore, the temperature of the workpiece is reduced and the danger of damage to the edge zones is decreased. The lubricants used are non-water-miscible (oils) and water-miscible (emulsions, solutions) cooling lubricants (DIN 51 385), the effect of which can be further improved with additives (polar and EP additives to improve the lubricating effect, antifoaming agents, biocides and rust inhibitors). The cooling effect depends on physical parameters: specific heat capacity c in kJ/kg K, heat transmission coefficient α in W/m^2 K, thermal conductivity λ in W/m K, heat of evaporation l_d in kJ/kg and surface tension σ in N/m.

The lubricating effect is described by the tribological parameters of the cooling lubricant.

4.3.2 Grinding with Rotating Tool

Processes. Grinding is divided in DIN 8589 Part 11 into six processes according to the shape of the surfaces pro-

duced. **Figure 39a** shows the classification and **Figs 39b** to **j** show examples of various motion classifications and tool shapes.

Chip Formation. Material is removed by the penetration of abrasive grains into the material along a flat path. Owing to the generally unfavourable shape of the cutting edge, the actual chip formation is accompanied by friction and displacement processes. The process is evaluated by calculating statistical averages. **Figure 40** shows in simplified terms how a comma-shaped chip is formed by the successive action of two cutting edges. While grain *1* has travelled the path AB, the centre of the grinding wheel has moved from 0 to 01. The next grain *2* will travel the path CD. In the process, the thickness of an average chip increases from 0 up to b_{max}. A simple relationship for the average undeformed chip thickness b is obtained by applying the continuity relationship $v_{ft}a_ca_p$

$$= v_cCV_{sp}a_p:$$

$$\bar{b} = \frac{v_{ft}}{v_c} \cdot \frac{1}{bC} \sqrt{\frac{a_c}{d_{eq}}},$$

with $\bar{l} = \sqrt{a_cd_{eq}}$, $V_{sp} = \bar{l}\bar{b}\bar{b}$ and $d_{eq} = \dfrac{d_wd_s}{d_w \pm d_s}$ (+ external cylindrical grinding, − internal cylindrical grinding) or

$$\bar{b} = \sqrt{\frac{v_{ft}}{v_c} \cdot \frac{1}{rC} \sqrt{\frac{a_c}{d_{eq}}}} \quad \text{with } r = \frac{\bar{b}}{\bar{b}}.$$

Here, $\bar{b}$ = average (undeformed) chip thickness, $\bar{l}$ = average (undeformed) chip length, $\bar{b}$ = average (undeformed) chip width, v_{ft} = rate of feed of workpiece, v_c = cutting speed, a_c = depth of cut, infeed, a_p = width of engagement (width of grinding), d_{eq} = equivalent grinding wheel diameter, d_s = grinding wheel diameter, d_w = workpiece diameter ($\rightarrow \infty$ in surface grinding), C = number of active cutting edges per unit of surface area of grinding wheel, r = ratio of average chip thickness to average chip width. The maximum chip thickness b_{max} is twice the average chip thickness $\bar{b}$ thus determined.

Owing to technical difficulties in measuring the number and distribution of grains, the equivalent chip thickness b_{eq} is often used as a parameter for evaluating the grinding process [26, 27].

$$b_{eq} = a_cv_{ft}/v_c.$$

Composition of Grinding Wheels. A grinding wheel consists of *grains, bonding material (bond)* and *pores*. The specification of a grinding wheel is standardised to DIN 69 100. Grinding wheels made of diamond or cubic boron nitride (CBN) are not covered by this standard. They consist of a backing material to which the grinding layer is applied. Usual layer thicknesses are 2 to 5 mm.

Grinding wheel wear may take place in the grains and the bond. Various kinds of wear and means of sharpening are shown in **Fig. 41**.

Limits to the Process. Restrictions on the process arise if the original data, e.g. *dimensional accuracy, accuracy of shape, surface quality* or *condition of workpiece edge zones*, do not lie within the required limits. The interaction of the various influencing variables, such as workpiece, machine setting data, tool, cooling lubricant, etc., may be extremely diverse.

Mechanical or thermal overstressing of the material in the grinding process may adversely affect the characteristics of a ground component [28]. Typical grinding

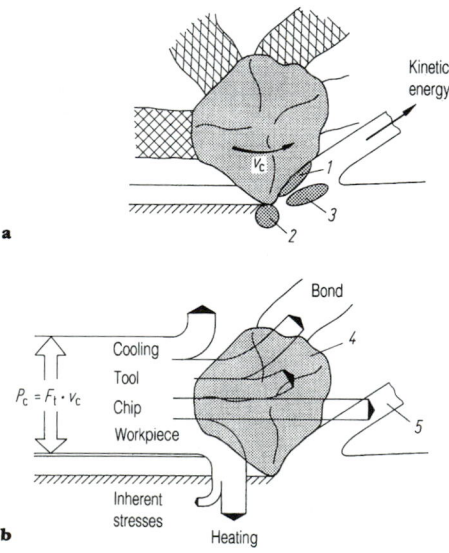

Figure 38. Energy conversion: **a** effects of energy conversion, **b** energy flows; *1* friction, *2* cutting, *3* shearing, *4* grain, *5* chip.

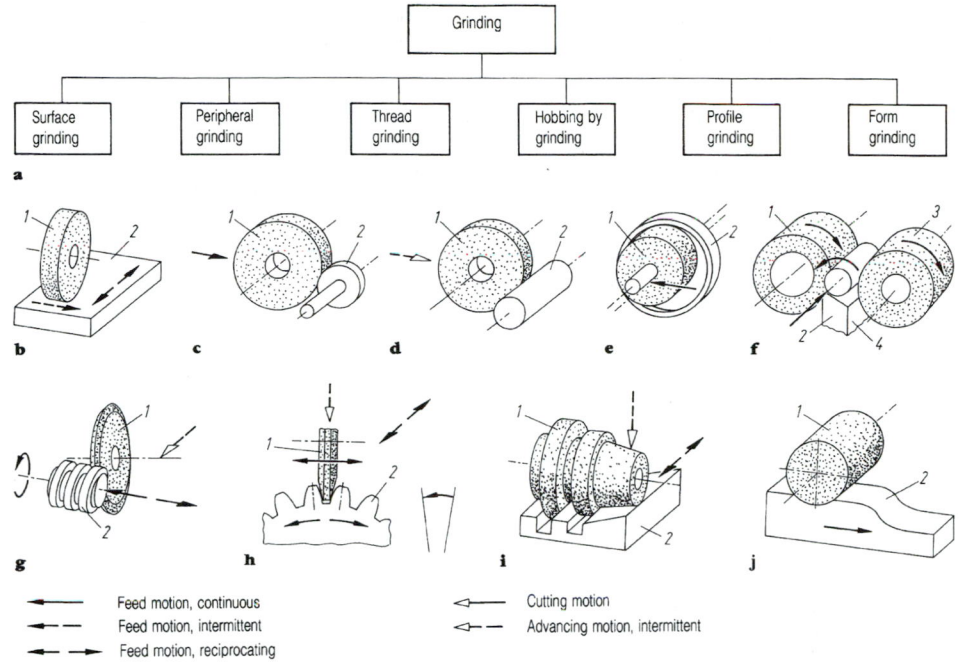

Figure 39. Grinding processes, schematic (DIN 8589): **a** classification, **b** longitudinal peripheral surface grinding, **c** transverse peripheral external cylindrical grinding, **d** longitudinal peripheral external cylindrical grinding, **e** transverse peripheral internal cylindrical grinding, **f** centreless throughfeed grinding, **g** longitudinal external thread grinding, **h** discontinuous external hobbing by grinding, **i** longitudinal external profile grinding, **j** form regrinding; *1* grinding wheel, *2* workpiece, *3* regulating wheel, *4* work rest.

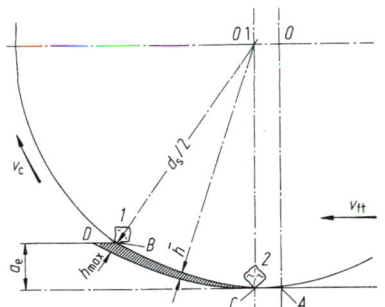

Figure 40. Engagement relationships in surface grinding (explanations in text).

defects attributable to poor process control are chatter marks, inherent tensile stresses [29], grinding burn and cracks in the workpiece.

Dressing. The purpose of dressing is to give the grinding wheel the required *profile* and *concentricity* (profiling) and to produce the necessary *grinding wheel topography* with sharp grains (sharpening). Both operations are generally performed at the same time by passing a dressing tool over the face of the grinding wheel. A survey of common *stationary* and *rotating* dressing tools is contained in [30]. The main components of dressing tools are elements coated with diamond particles; however, there are also diamond-free steel and ceramic elements or surfaces. Electrochemically bonded grinding tools coated

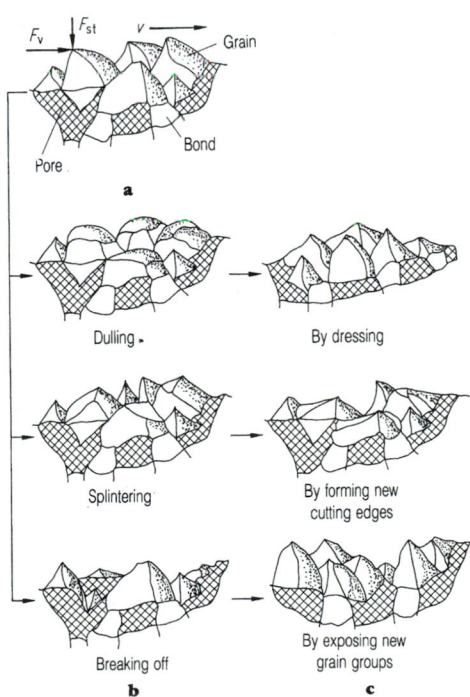

Figure 41. Types of wear and means of sharpening: **a** sharp grinding wheel, **b** types of wear, **c** means of sharpening.

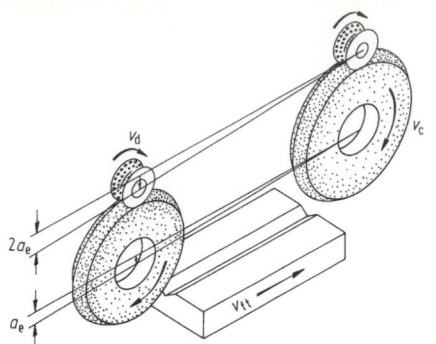

Figure 42. Principle of grinding with continuous dressing (CD grinding).

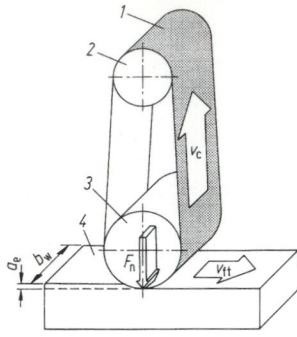

Figure 44. Schematic representation of process variants in surface belt grinding: v_c, cutting speed, v_{ft} workpiece speed, a_e working engagement, b_w grinding width, F_n perpendicular force; 1 grinding belt, 2 deflection roller, 3 contact roller, 4 workpiece.

with only one abrasive layer cannot be dressed. They have reached the end of their life when this abrasive layer is worn down.

A special position is occupied by grinding with continuous dressing (CD grinding) (**Fig. 42**). Here the dressing tool, generally a diamond dressing roll, is in engagement during grinding and is continuously advanced radially to the grinding wheel. As a constant grinding wheel profile and a uniform grinding wheel topography with sharp cutting edges are permanently ensured, the time-related chip volume can be increased considerably [31]. With the aid of the machine control system, the dressing tool and the grinding wheel have to be advanced in relation to the workpiece in such a way that the decrease in the diameter of the grinding wheel is compensated for.

Development Trends. Grinding has developed from a traditional precision machining process to improve dimensions, shape and surface quality into a very versatile and efficient manufacturing process.

New grinding processes such as creep feed grinding, high-speed grinding and grinding with continuous dressing (CD grinding), the growing use of the superhard abrasives diamond and cubic boron nitride (CBN), together with CNC and sensor technology, have equally contributed towards increasing the performance of this manufacturing process.

4.3.3 Belt Grinding

Belt grinding is grinding with tools on a bed (DIN 8589). Belt grinding can be broken down as shown in **Fig. 43** according to the surfaces that can be produced.

Surface belt grinding predominates in industrial applications (**Fig. 44**). Belt grinding is normally performed at a constant perpendicular force F_n (pressing force). In this way, consistent surface qualities can be produced. The time-related chip volume is determined by the sharpness of the belt (of the active cutting edges). In belt grinding with constant working engagement a_c, a constant time-related chip volume is removed. Surface quality depends on the condition of the cutting edges. This process is especially suitable for removing large volumes of material (time-related chip volumes) [32].

The movement of the process variables during belt grinding stems from the changes to the cutting edges during working. In contrast to grinding wheels, grinding belts generally consist of a single layer of abrasive. Therefore the cutting edge zone changes over the period of use, owing to progressive abrasive and bond wear.

Over the *life* of the grinding belt, three phases can be distinguished (**Fig. 45**). In the calibration phase, a rapidly changing grinding-in process takes place. The small number of grains that project furthest out of the cutting

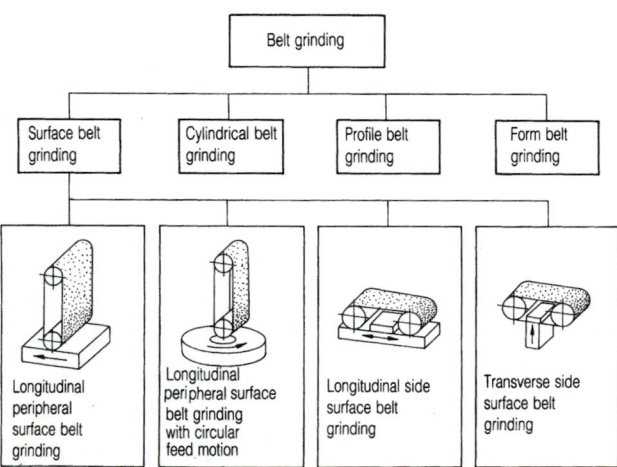

Figure 43. Survey of belt grinding processes with detailed classification of surface belt grinding (DIN 8589 Part 12).

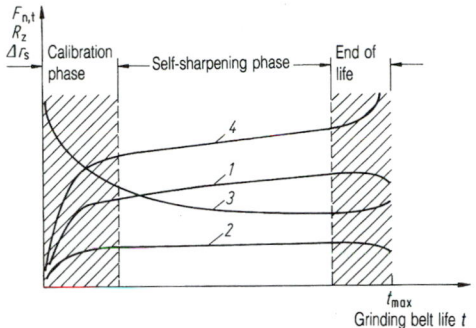

Figure 45. Process behaviour of a grinding belt during its service life: *1* perpendicular force F_n, *2* tangential force F_t, *3* average surface roughness R_z, *4* radial wear Δr_s.

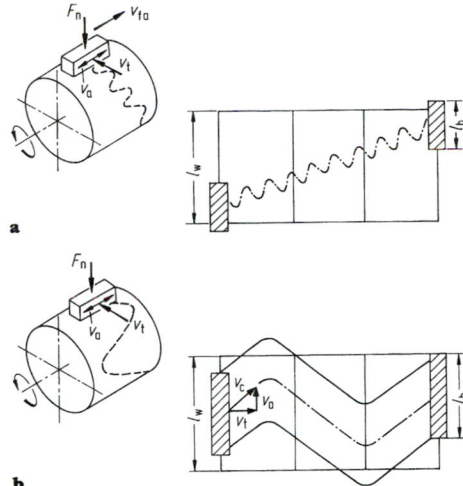

a

b

Figure 46. Geometry and kinematics in **a** short-stroke and **b** long-stroke external cylindrical honing [34]. v_{fa} axial rate of feed, l_w workpiece length, l_h length of honing stone.

edge zone owing to the belt manufacturing process very quickly break off. As a result, radial wear initially rises rapidly and deeper lying grains engage in cutting. The increase in the number of active cutting edges causes a rise in the grinding forces and a decrease in the average surface roughness R_z.

The self-sharpening phase is characterised by continuous wear combined with formation of new cutting edges. The increase in the number of grains actively participating in the cutting process leads to a regular increase in the grinding forces and a decrease in the depth of roughness. The end of belt life is marked by rapidly advancing grain and bond wear, ultimately leading to failure of the grinding belt.

Grinding belts consist of four elements: backing (paper, fabric), first-coat and top-coat bonding materials (phenolic resins), and abrasive grains (corundum, zirconium-corundum, silicon carbide). Scattering of the grains on the first coat of bonding material is carried out in an electrostatic field. This ensures that the abrasive grains are aligned perpendicularly. An even grain distribution and thus high reproducibility in belt manufacture compared with conventional gravity scattering can be achieved [33].

The grinding belt is supported in the working zone by contact elements. A contact disc is used in peripheral grinding, while a contact shoe or beam is used in side grinding. Hard contact rolls made of aluminium or steel are particularly suitable for roughing with coarse grinding belts (grade 36; 50) to transmit the relatively high grinding forces. Soft, rubber-sheathed contact rolls are used in finishing with fine grinding belts. They absorb the shocks caused by grinding belt wear [32].

Conventional applications of belt grinding are grinding of individual sheets and sheet coils, deflashing and deburring, and grinding down of excess metal in welded joints. A recent advance is the successful use of heavy-duty belt grinding in the motor industry and elsewhere as a substitute for turning and milling of components made of grey cast iron or aluminium alloys [32, 33]. Some principal grinding data are given in **Appendix K4, Table 6**.

4.3.4 Honing

Honing is performed with a multiple-point cutting tool consisting of bonded abrasive grains using a cutting motion consisting of two components of which at least one is oscillating. The principal honing processes are external cylindrical honing, internal cylindrical honing, and surface honing. According to the oscillation amplitude, a further distinction can be made between two main

groups: long-stroke honing and short-stroke honing (**Fig. 46**) [34].

Long-stroke honing employs a large oscillation amplitude at a low frequency; in *short-stroke honing* the oscillating motion is performed at low amplitude and a correspondingly high frequency. The path curves in **Fig. 45** depict the motion of a honing strip over a developed workpiece surface.

Owing to the superimposed motion during honing, the workpiece surface exhibits intersecting tracks of the cutting grains, the two tracks enclosing an angle α (**Fig. 47**). The magnitude of the overlap angle α is determined by the selection of the ratio of the axial (v_a) and tangential (v_t) cutting speed components. For workpieces without longitudinal and transverse grooves, the angle α is generally 45°. The cutting speed v_c can be calculated by means of the aforementioned speed components

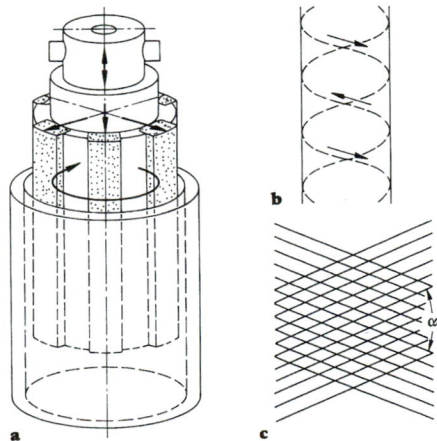

a

b

c

Figure 47. Working operation in long-stroke honing: **a** working principle, **b** honing movement of tool, **c** surface structure (α overlap angle).

according to $v_c = (v_a^2 + v_t^2)^{1/2}$, The cutting speed normally does not exceed $v_c = 1.5$ m/s [35, 36].

During the cutting motion, the honing stones are pressed against the workpiece face to be machined at a perpendicular honing force F_n, which may be generated by means of various feed systems (**Fig. 48**). In force-dependent feeding, a defined hydraulic pressure p_{oil} is set on the machine. The resulting advancing force F_z is transmitted to the honing stones via an advancing pin and cones. In path-dependent feeding, defined feed paths are generated, e.g. with a stepping motor [37], from which the perpendicular force F_n at the honing stones results.

Important variables influencing the result of honing are type of abrasive, grain size, type of bond, hardness and impregnation of the honing stones. The types of abrasive can be broken down into the conventional abrasive materials corundum and silicon carbide and the superhard abrasive materials diamond and cubic crystalline boron nitride (CBN). Grain size influences time-dependent chip volume and surface quality. The achievable surface roughnesses are $R_z = 1$ μm for long-stroke honing and $R_z = 0.1$ μm for short-stroke honing. Dimensional accuracies and accuracies of shape of the machined workpieces of 1 to 3 μm are achieved. Unlike in grinding, the grains bonded in the honing stone are stressed on more than one axis owing to the oscillating motion. Honing tools are therefore self-sharpening.

Cooling lubricants are used in honing as well as in grinding. Owing to the low cutting speed, however, heating is minimal, so the cooling effect plays a minor role. The surface contact between the honing stone and the workpiece requires instead a friction-reducing lubricating effect. Therefore pure oil, with additives if required, is generally used.

The applications of honing can also be categorised according to whether long-stroke or short-stroke honing is used. Long-stroke honing is generally used for internal-cylinder workpieces, e.g. piston contact surfaces in internal combustion engines. Short-stroke honing is mainly used for machining small cylindrical components, e.g. running surfaces of inner and outer rings and rollers of rolling-contact bearings [34].

4.3.5 Other Processes: Lapping, Inside Diameter Cut-off Grinding

Lapping

Lapping is defined in DIN 8589 as metal cutting with loose grains of abrasive distributed in a paste or a liquid, the lapping mixture, which is carried on a generally shape-imparting counterpart (lap), the individual grains following cutting paths which are random. Lapping processes are broken down into surface lapping, cylindrical lapping, hole lapping and ultrasonic machining (**Fig. 49**).

Surface lapping or *plane parallel lapping* is carried out on single- or twin-disc lapping machines. The lapping discs act as the holder for the lapping abrasive. They are chiefly made of perlitic cast materials or hardened steel alloys.

The abrasive consists of the lapping powder and the fluid in the ratio 1 : 2 to 1 : 6. Particles of silicon carbide, corundum, boron carbide or diamond are used as lapping powder. The type of grain for a particular application is determined by the material to be machined. Grain sizes ranging from 5 to 40 μm are generally used. Besides high-viscosity oils or similar fluids, water with suitable additives has seen growing use as a vehicle for the abrasive in recent years. The purpose of lapping fluids, among other things, is to cool the workpiece and ensure chip removal from the effective zone.

Lapping is a precision or ultra-precision machining process for producing functional surfaces of optimum surface quality. Surface roughnesses up to $R_t = 0.03$ μm, flatnesses < 0.3 μm/m and plane parallelisms of up to 0.2 μm are achieved. Typical applications of lapping are the machining of precision cemented carbide tools, calliper gauges or hydraulic rams. A special category of lapping is ultrasonic machining, which is particularly suitable for machining hard-brittle materials, e.g. fully sintered ceramic components [38, 39].

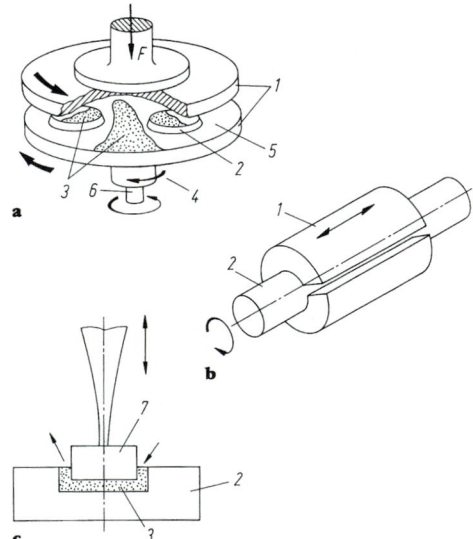

a

b

c

Figure 49. Lapping processes (DIN 8589 Part 15): **a** plane–parallel lapping, **b** lapping of external cylinders, **c** ultrasonic machining; *1* lap, *2* workpiece, *3* abrasive, *4* lapping disc drive, *5* lapping cage, eccentric, mounted in bearings, *6* cage drive, *7* vibrating tool.

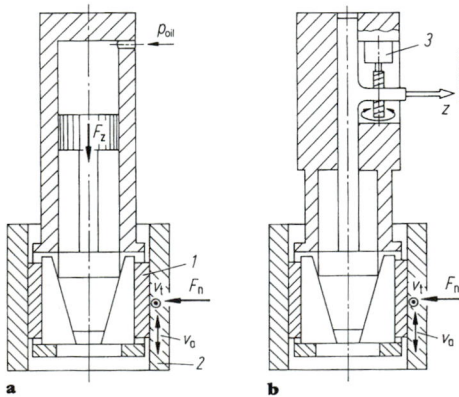

a   b

Figure 48. Force- and path-dependent feed devices for honing [37]: **a** force-dependent, **b** path-dependent; *1* honing stone, *2* workpiece, *3* step motor.

Inside-Diameter Cutoff Grinding

Inside-diameter (ID) cutoff grinding is a high-precision fin-ish-machining process for hard–brittle materials. It is used to cut rod-shaped materials into thin slices (**Fig. 50**).

Besides its applications to optical materials (glasses, glass ceramics), magnetic materials (samarium–cobalt, neodymium–iron–boron), ceramics and crystals for solid-state-type lasers, this process is used above all for semicon-ductor materials. Single-crystal silicon rods are cut into thin slices termed "wafers" (see K5.3).

Compared with conventional abrasive cutting pro-cesses, the loss of material in the cutting gap can be reduced by approximately 80% by means of a narrow width of cut. This is a decisive advantage, especially for expensive, high-grade materials.

To achieve the narrow widths of cut that are typical of this process, a comparatively unconventional tool is used for ID cutoff grinding. The basic body of the tool consists of a cold-rolled circle of high-tensile stainless steel with a thickness of 100 to 170 µm.

The inside edge of the cutting blade is coated by elec-trodeposition with a diamond layer in a nickel-base bond; this layer forms the droplet-shaped cutting edge. Grain sizes commonly range from 45 to 130 µm. Accordingly, the width of cut extends from 0.29 to 0.7 mm. Natural diamond is generally used as the cutting material. CBN may also be used for special applications. Workpieces with a diameter of up to 200 mm can be cut.

To achieve the stiffness at the cutting edge that is neces-sary for cutting, the cutting blade, which can be compared to an eardrum, is clamped at its outside edge with a spe-cial clamping device. This widens the ID cutting blade radially until the tangential stresses at the inside edge reach around 1800 N/mm². In cutoff grinding, the work-piece undergoes a radial feed motion relative to the rotat-ing tool [40, 41].

4.4 Chipless Machining

4.4.1 Survey

Chip-producing machining processes operate by the mechanical effect of cutting edges on the workpiece.

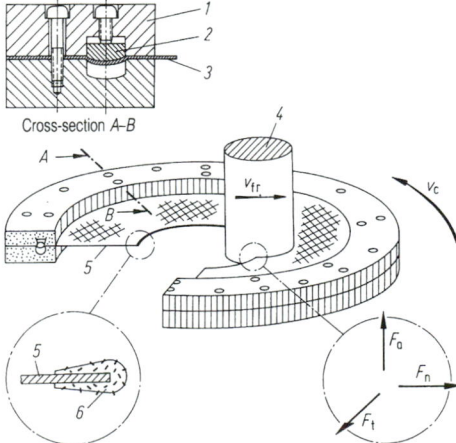

Figure 50. Principle of ID cut-off grinding; *1* clamping ring, *2* holding ring, *3* ID cutting blade, *4* Si crystal, *5* blade centre, *6* diamond-coated cutting edge, v_c cutting speed, v_{fr} radial feed rate, F_n, F_t, F_a process forces.

They therefore depend on workpiece properties such as strength, hardness, resistance to wear or toughness. Chip-less machining processes utilise *thermal, chemical* or *elec-trochemical processes* for forming. They do not depend on the mechanical properties of the materials. They are well established for machining of materials that are diffi-cult or impossible to machine mechanically (highly tem-pered and hardened tool steels, nickel-base alloys or superhard materials such as diamond or cubic boron nitride). They are also used to machine complex, hard-to-reach or very small (microtechnology) surfaces and con-tours.

According to DIN 8590, chipless machining is manufac-turing by removal of material particles from a solid object without mechanical action (cf. **K5, Fig. 43**).

Thermal machining is characterised by the removal of material particles in solid, liquid or gaseous condition by the action of heat. The particles removed are carried away by mechanical and/or electromagnetic forces. This sub-group is further broken down according to the energy source that supplies the external heat required for cutting.

The principle of *chemical milling* is based on chemical reaction of the material with an active medium to form a compound that is volatile or easily removable. The conver-sion of material takes place by a direct chemical reaction.

Electrochemical machining is accomplished by the reaction of metallic materials with a dissociated, electro-conductive active medium by the action of an electric cur-rent to form a compound that is soluble in the active medium or precipitates out. The current flow is initiated by an external power source.

4.4.2 Electro-discharge Machining (EDM)

Electro-discharge machining is used for machining of elec-troconductive materials in a dielectric. This is achieved by creating repeated discharges between an electrode and the workpiece in rapid succession [42] (**Fig. 51**), as fol-lows. In the first phase, the dielectric is ionised at the point where the gap is narrowest (highest field strength) (t_1). A discharge channel is formed cumulatively (ionisation by collision). The discharge current builds up, and the voltage falls to the physically dependent gap volt-age of approximately 25 V (t_2). In the third phase, the plasma is heated in the widening discharge channel. Tem-peratures of approximately $10 \cdot 10^5$ K arise, owing to con-striction of the discharge. Small quantities of material are melted at the ends of the arc (electrode and workpiece) (t_3). At the end of the pulse, the superheated molten metal evaporates explosively (t_4). The energy per pulse determines the crater size and the effect on the workpiece edge zone [43].

The *dielectric* has the following functions: insulation of workpiece and electrode, establishment of favourable ion-isation characteristics, constriction of the discharge chan-nel, carrying away of the particles removed, and cooling of electrode and workpiece. Hydrocarbons are used as dielectrics.

The spark energy is produced by a generator (static pulse generators are used exclusively nowadays). The pulse is controlled with an electric switch. The current is restricted by the impedance Z; the pulse length is adjust-able from 1 to 2000 µs. The pulse-to-space ratio $T = t_i/t_p$ is variable from 0.1 to 0.5, the no-load voltage U_i can be varied between 60 and 300 V, and the pulse current I_e is adjustable within the range 1 to 300 A.

Electro-discharge machining may be carried out in vari-ous alternative forms (**Fig. 52**). In *cavity sinking by EDM*, the tool is an electrode with the negative shape of

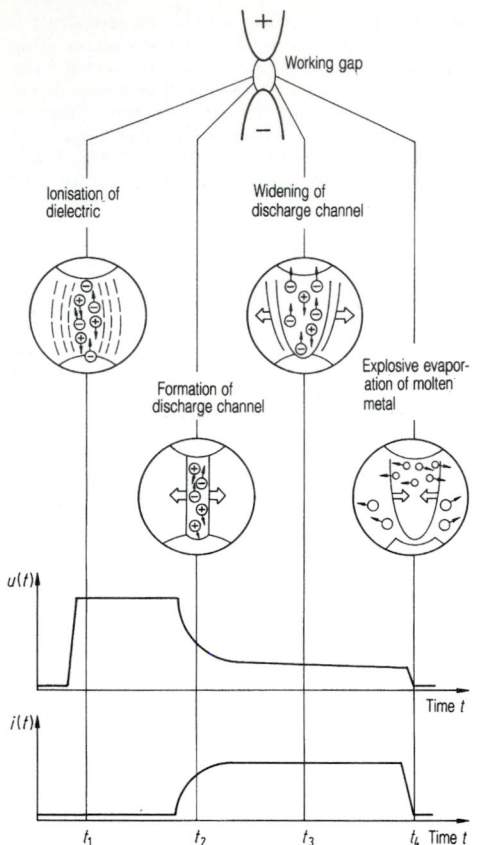

Figure 51. Phases of spark discharge [44].

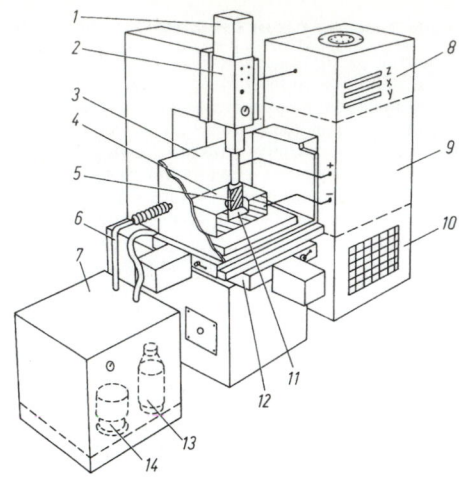

Figure 53. Parts of an EDM cavity sinking machine [45]: *1* feed drive, *2* working head, *3* work container, *4* workpiece, *5* electrode, *6* return pipe, *7* dielectric supply unit, *8* axial drive control system, *9* generator, *10* power supply, *11* spark gap, *12* compound table, servomotor-driven, *13* filter, *14* pump.

the cavity to be produced. An EDM cavity sinking machine consists of the machine tool, generator, control unit for the axial drives and the dielectric unit (**Fig. 53**). Drives in three spatial directions perform the positioning and advancing of the electrode. By monitoring the electrical data at the spark gap, its width is adjusted to the setpoint value (≈ 10 to $80\ \mu$m) in a highly dynamic manner. The rate of feed is determined by the progress of the machining process and cannot be preset [44].

Productivity and end result are determined by the electrode material, the dielectric and the electrical settings (current, pulse duration, pulse-to-space ratio and polarity). The machining process is divided into several

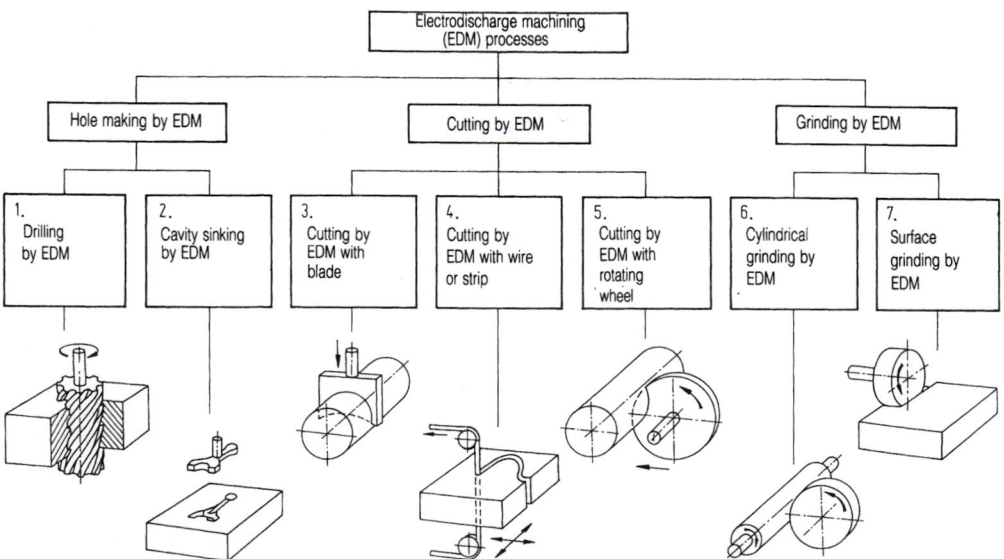

Figure 52. Classification of electro-discharge machining (EDM) processes (according to VDI standard 3400).

roughing and finishing processes. In roughing, removal rates of $Q_w = 600$ mm³/min at $I_c = 60$ A and low relative wear (2 to 5%) are achieved. Finishing is carried out at low currents and with brief discharges. Surface qualities of $R_a = 0.3$ μm and dimensional deviations of less than 10 μm can be achieved. The thermal machining process affects the workpiece edge zone to a depth of 5 to 50 μm. An amorphous structure may form there. Intrinsic tensile stresses occur in the layer close to the surface, reducing the dynamic strength of the material. Electrodes are made from materials with high melting points and/or high thermal conductivity. Common materials are copper and graphite; sintered tungsten–copper materials are used in special cases [45] (**Appendix K4, Table 7**).

Cavity sinking by EDM is used for making hollow moulds for primary shaping and metal forming tools. To the original EDM cavity sinking process with only one vertical feed motion have been added planetary EDM and path-controlled EDM (**Fig. 54**). In planetary EDM, a rotary motion of the electrode is superimposed on the downward vertical motion. This achieves better flushing, even distribution of electrode wear and a uniform undersize of the roughing and finishing electrodes. Wider possibilities are offered by path-controlled EDM; electrodes of simple shape can be controlled to produce complex shapes.

In *cutting by EDM*, a running wire electrode is moved along a curved path relative to the workpiece. The cutting gap is produced by electrical discharge machining. Cutouts of any desired contour are produced in flat components (**Fig. 55**). For diagonal prismatic cutouts the wire guides may be offset against one another. An EDM cutting machine comprises the actual machine tool with the wire feed, the generator, the control unit for the axial drives and the dielectric preparation unit. The end result depends essentially on the cutting wire. The usual wire diameter is 0.25 mm. The wire running speed is up to 300 mm/s, the generator current is between 15 and 100 A, removal rates are up to 350 mm³/min, dimensional accuracies and accuracies of shape are better than 0.01 mm, surface roughnesses are $R_a = 0.3$ μm. Cutting by EDM is used in tool manufacture, e.g. for making punching, injection moulding and extrusion tools.

4.4.3 Laser Cutting

In laser cutting, light energy is generated in an optical resonator and transmitted to the material by absorption in

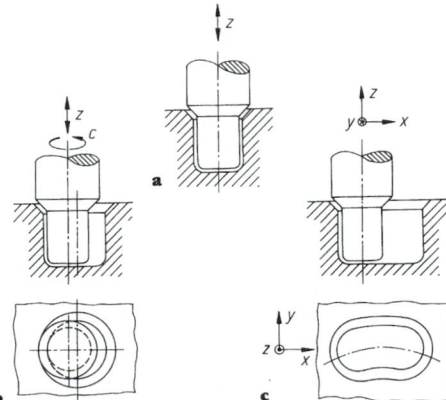

Figure 54. a Cavity sinking by EDM, **b** planetary EDM, **c** path-controlled EDM; x, y, z, c: relative electrode motion.

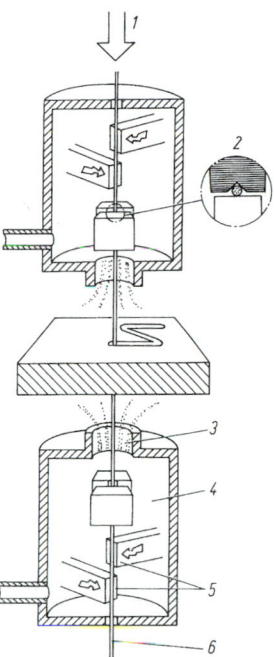

Figure 55. Principle of cutting by EDM [44]; *1* wire feed, *2* inverted-V guidance principle, *3* flushing nozzle, *4* flushing chamber, *5* power supply, *6* cutting wire.

the form of heat. (LASER = Light Amplification by Stimulated Emission of Radiation.)

For cutting tools, only CO_2 lasers, Nd : YAG lasers and, recently, excimer high-capacity lasers are employed, owing to the extremely high beam powers required [46–48]. The beam characteristics of these lasers that are important for material machining are summarised in **K5, Table 2**.

Laser cutting of metallic materials requires intensities of $> 10^6$ W/cm², which are achieved by focusing the laser beam with the aid of lenses or mirrors [49]. The thermal material removal process, which is directed into the depths of the material, produces a kerf in the material when feed motion is applied. The principle of laser cutting is illustrated in **Fig. 56**.

The material that is melted (*laser fusion cutting*), burnt (*laser flame cutting*) or vaporised (*laser sublimation cutting*) at the focal point of the laser beam, depending on the intensity and the length of interaction, is expelled from the kerf by a stream of gas emitted from a nozzle coaxially to the optical axis. The cutting gas also serves to protect the sensitive focusing optics from spattered material.

In laser flame cutting, oxygen or oxygen-rich gas is used as the cutting gas; at higher cutting speeds, however, this leads to oxidation of the cut surfaces owing to the introduction of additional exothermic energy. In contrast, in the other laser cutting processes mentioned above inert gases (e.g. argon, nitrogen) are used as cutting gases, which results in a slower cutting speed; however, they produce an oxide-free cut [50].

The relative motion between the laser beam and the workpiece that is required to produce a continuous kerf is achieved in practice in various ways. For laser cutting of small, easily handled components, the latter are generally

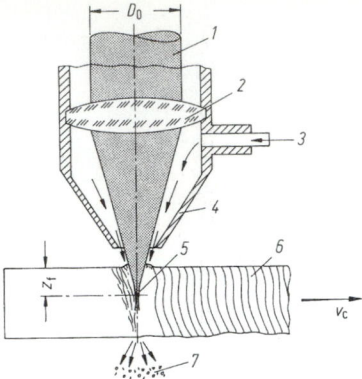

Figure 56. Principle of laser cutting: v_c cutting speed, z_f focus position, *1* laser beam (wavelength λ, laser power P_L, mode, pulse frequency f_p, pulse duration t_i), *2* focusing lens (focal length f), *3* cutting gas (gas pressure p_g, type of gas), *4* cutting tip (shape, diameter), *5* diameter of focal spot d_f, *6* workpiece, *7* expelled material, D_0 diameter of the unfocused light beam.

moved underneath the stationary laser beam, e.g. with the aid of an X, Y coordinate table. For laser machining of larger workpieces, the laser unit including the cutting tip is either moved across the stationary workpiece, or a movable system of mirrors is guided together with the cutting tip ("flying optics") between the fixed laser unit and workpiece. For Nd : YAG lasers alone, flexible optical fibres may be used for guiding the beam [51, 52].

The machining process is influenced by many different process parameters, the most important of which are given in **Fig. 57** together with their definitions. The maximum achievable cutting speed in relation to laser power and material thickness is shown in **Fig. 58** by way of example for structural steel using a CO_2 laser. These are average values calculated from data supplied by various users. In addition, the achievable cutting speeds of other metallic and non-metallic materials for a (CO_2) laser power of $P_L = 500$ W are summarised in **Table 3**.

There is no current standard for the definitions for determining the quality of laser cuts and their measuring instructions; however, this is often done on the basis of a guideline issued by the CIRP-STC-"E" expert group [53]. This guideline is based on the definitions shown in **Fig. 57**.

High-powered lasers of the type mentioned above generally fall into laser (protection) class 4, which has the highest hazard level (except for laser machining systems with closed working chambers, which are equipped with additional safety facilities such as interlock systems and radiation-absorbing protective windows). Allocation to this safety class means that even diffusely reflected laser radiation is a hazard to the skin and the human eye. Comprehensive instructions on radiation safety of lasers are laid down in DIN VDE 0837 and accident prevention regulation 46.0 (VBG 93).

4.4.4 Electrochemical Machining (ECM)

The basic principle of electrochemical machining corresponds to an electrolytic cell. Electrolyte solution flows at high speed between the workpiece (anode) and the tool (cathode); the gap between the electrodes is 0.05 to 1 mm. Hydrogen ions are discharged at the cathode. Metal ions react at the anode with OH ions from the water, for-

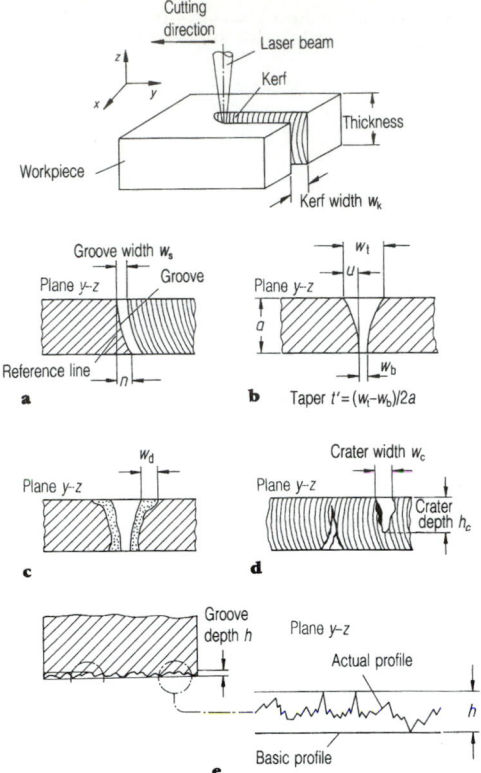

Figure 57. Definitions for evaluating quality of cut [53]: **a** groove deflection n, **b** unevenness u, **c** width of affected zone w_d (in Germany: including matrix restoration and heat-affected zone), **d** cratering, **e** groove depth $h(z)$.

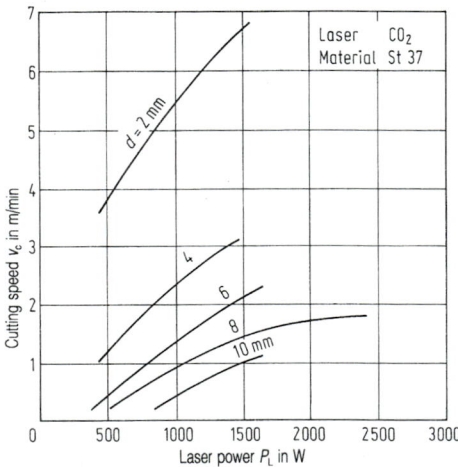

Figure 58. Cutting speed versus laser power for various material thicknesses.

ming metal hydroxide compounds which settle as a sludge. A widespread process is mould deflashing (**Fig. 59**). The tool electrode has to be matched to the workpiece. The flash is preferentially removed owing to the maximum current density being present at that point.

Table 3. Machining parameters for laser cutting of various materials. Laser CO_2/500 W, lens focal length $f = 5$ in.

Material	Thickness (mm)	Cutting gas/ pressure (MPa)	Cutting speed (m/min)
PMMA (Plexi)	4	Air / 0.06	3.5
Rubber	3	N_2 / 0.3	1.8
Asbestos	4	Air / 0.3	1.6
Deal	3	N_2 / 0.15	5.5
Eternit	4	Air / 0.3	0.8
AlTi ceramic	8	N_2 / 0.5	0.07
Aluminium	1.5	O_2 / 0.2	0.4
Titanium	3	Air / 0.5	2
CrNi steel	2	O_2 / 0.45	1.9
Magnetic steel sheet	0.35	O_2 / 0.6	7
Grey cast iron	3	N_2 / 1	0.9

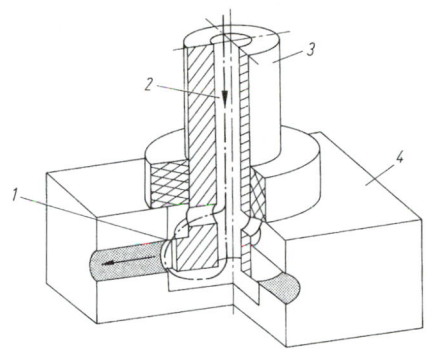

Figure 59. Electrochemical mould deflashing (according to DIN 8590); *1* flash, *2* flow of electrolyte, *3* tool electrode (cathode), *4* workpiece (anode).

4.4.5 Chemical Machining

Chemical machining is accomplished by a chemical reaction of the material with a liquid or gaseous medium. The product of reaction is gaseous or easily removable. An example of chemical machining is thermal deburring. This process consists of a thermal component (heating up the material) and a chemical one (burning of the material). In thermal deburring, metallic or non-metallic workpieces are pressed with a closing plate under a bell-shaped deburring chamber (**Fig. 60**). Oxygen and fuel gas (natural gas, methane or hydrogen) are fed into the chamber at a controlled rate. The gas pressure and the mixing ratio determine the amount of material removed.

While the mixture is being burnt, the temperature briefly reaches 2500 to 3000 °C. Parts of the workpiece with a large surface area and a small volume (low heat capacity) are burnt (oxidised). The burrs must be thinner than the thinnest parts of the workpiece. After deburring, the temperature of the workpieces is 100 to 160 °C.

4.5 Shearing and Blanking

K. Siegert and J. Ladwig, Stuttgart

4.5.1 Classification

According to DIN 8588, the processes of severing – mechanical cutting of workpieces without formation of amorphous material – are divided into *blanking*, *wedge-action cutting* (single-blade cutting, cutting with two approaching blades, cleaving), *tearing* and *breaking* (**Fig. 61**). Particularly in sheet metal working, blanking predominates, often as a preparatory, finishing or inter-

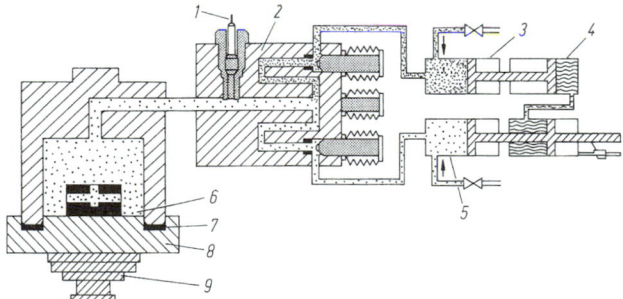

Figure 60. Parts of a thermal deburring unit (according to Thilow); *1* spark plug, *2* mixing block, *3* fuel gas feed cylinder, *4* gas injection cylinder, *5* oxygen feed cylinder, *6* deburring chamber, *7* gasket, *8* workpiece holder, *9* closing plate.

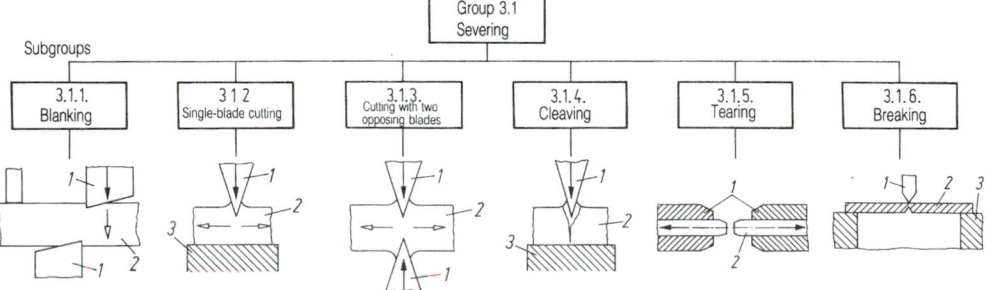

Figure 61. Severing processes (DIN 8588); *1* tool, *2* workpiece, *3* rest.

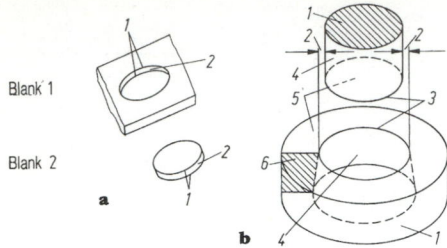

Figure 62. Blanking: workpiece and tool terminology. **a** work-piece [55]; *1* cut edges, *2* cut surface. **b** Tool; *1* tool, *2* cutting gap, *3* cutting edge, *4* flank, *5* pressure surface, *6* wedge.

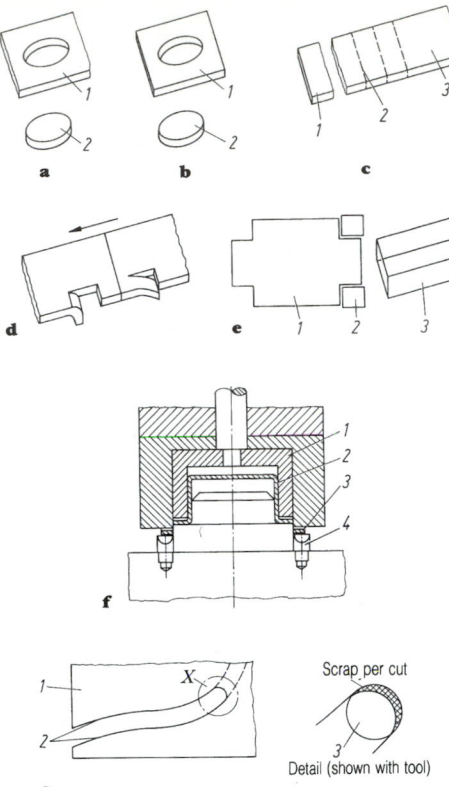

Figure 64. a Cutting out; *1* scrap, *2* blank. **b** Piercing; *1* work-piece, *2* scrap. **c** Cutting off; *1* workpiece, *2* line of cut, *3* metal strip. **d** Slitting. **e** Notching; *1* workpiece, *2* scrap, *3* finished part. **f** Trimming; *1* ejector, *2* finished part, *3* scrap, *4* edge cutter. **g** Nibbling; *1* workpiece, *2* cut edges, *3* tool [55].

mediate operation in metal forming. There is a certain affinity with metal forming processes in that the cutting processes involve plastic deformation. The previously common designation *punching* is no longer contained in the standard [54].

In principle, the tool-related terms contain the word *cutting* (cutting edge, cutting surface), while the work-piece-related terms have the word *cut* (cut edge, cut surface), **Fig. 62**.

Blanking processes are divided according to the nature of the line of cut into processes with a *closed line of cut* and those with an *open line of cut* (**Fig. 63**). Whereas closed cutting is performed on presses using punches and dies, open lines of cut are produced not only with these tools but also with longitudinal and circular blades on special machines (cf. K3). The processes with a closed line of cut include cutting out and piercing (**Figs 64a, b**). *Cutting out* produces the complete external shape in one operation. *Piercing* produces an internal shape in the workpiece.

The processes with an open line of cut encompass cutting off, notching, slitting and trimming (**Figs 64c to f**).

Cutting off is the severing of a part from the mill product (sheet or strip) or the semi-finished product.

Notching is cutting out parts of the surface at internal and external edges.

Slitting is partial severing of the workpiece without removal of material. It is generally used as preparation for metal forming.

Trimming is used to sever material attached to the work-piece that is not required on the finished product.

A special position is occupied by *nibbling* (**Fig. 64g**). In this process, the workpiece is gradually severed along a line of cut of any desired shape by means of a simple punch.

4.5.2 Technology

Application of Force. On the impact of the punch, the vertical punch force F_S and, with increasing depth of cut, the horizontal force F_H are produced (**Fig. 65**). The resolution of the punch force F_S and the reaction force F_S' leads to the punch-side forces F_V and F_H on the one hand and F_V' and F_H' on the other, which act on the blanking

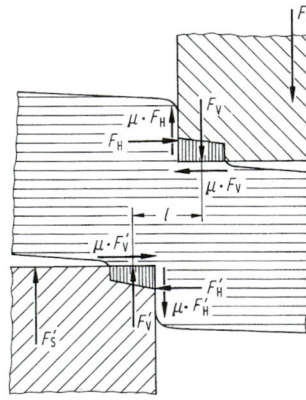

Figure 65. Action of forces in blanking (punching), explanations in text [54].

die. Owing to the distance *l* between the points of application of the forces, a moment is generated which causes the workpiece to bend.

Sequence of Blanking (Punching) and Formation of Cut Surfaces. These depend on the tool geometry –

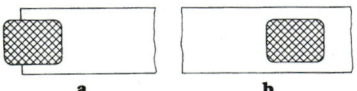

Figure 63. Blanking: **a** open, **b** closed.

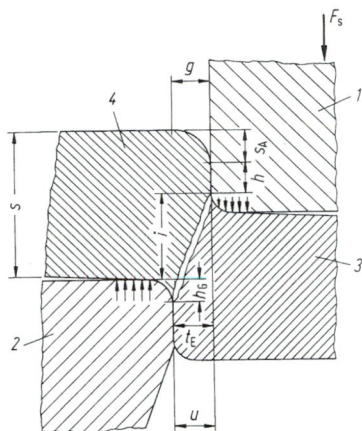

Figure 66. Cutting process in blanking [54]; *1* punch, *2* die, *3* blank, *4* metal sheet, *u* cutting gap, s_A height of edge indentation, *g* width of edge indentation, *h* height of zone of cut, *i* height of zone of fracture, b_G height of burr, t_E tear depth, *s* sheet thickness, F_s cutting force.

cutting gap *u* (**Fig. 66**), rounding or dulling of the cutting edge – and on the material and the characteristics of the mill product – sheet thickness *s*, mechanical properties, chemical composition and metallographic structure.

The sequence of blanking is characterised by the following phases: **Fig. 67** [55].

Owing to the influence of the vertical force, elastic deformation occurs first of all. The sheet bulges under the punch and partly lifts off the face of the die. Then the sheet undergoes local plastic deformation, producing permanent bulging of the sheet. The edge indentation is formed in the upper side of the sheet and in the blank. In the next cutting phase the material is sheared off, producing the smooth-cut part of the cut surface. In the residual cross-section the tensile stresses increase, leading to the formation of the first incipient cracks starting from the cutting edge of the die. Further incipient cracks then form in the sheet at the edge of the punch. At the moment of

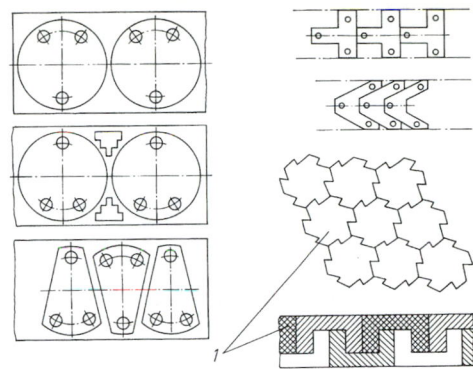

Figure 68. Full use of material in blanking [55]; *1* scrap-free shapes.

cracking, the maximum actual shear stress reaches the shear fracture point, leading to cracking [55].

When punching out parts from sheet metal, the aim is to use as much of the metal strip as possible (**Fig. 68**). The beginnings and ends of metal strips generally produce additional scrap; therefore the aim is to punch out the blanks directly from the coil. A range of CAD systems are available which enable computer-aided optimisation of the cutting of the blank (nesting drawings).

4.5.3 Forces and Energies

Among the most important parameters for the design or selection of presses is the maximum actual cutting force. The maximum cutting force is influenced by sheet thickness, punch geometry, tensile strength of the sheet metal, tool wear and the cutting gap *u* (**Fig. 69**). It should be noted that the burr height b_G depends on the cutting gap.

The maximum cutting force is determined according to an empirical equation in which $k_s \approx 0.8 R_m$ and $A_s = l_s s$:

$$F_{s\,max} = A_s k_s.$$

Here, l_s is the length of cut, *s* the sheet thickness and R_m the tensile strength of the material.

Details of the factors influencing the maximum cutting force are shown in **Table 4**. The maximum cutting force can be reduced if the effective line of cut l_s is shortened. The engagement of the punches may also be staggered over time (**Fig. 70**). As a result of the horizontal forces between the sheet and the punch, withdrawal forces,

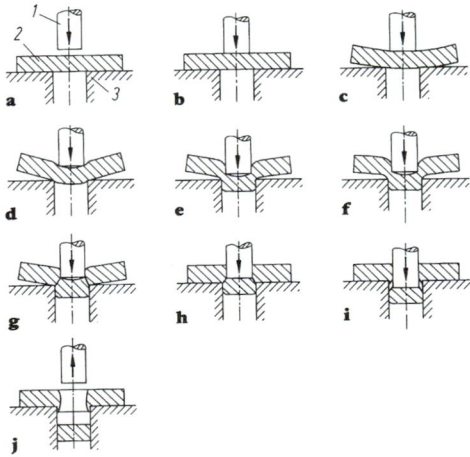

Figure 67. Sequence of events in blanking (punching) [55]; *1* punch, *2* sheet, *3* die.

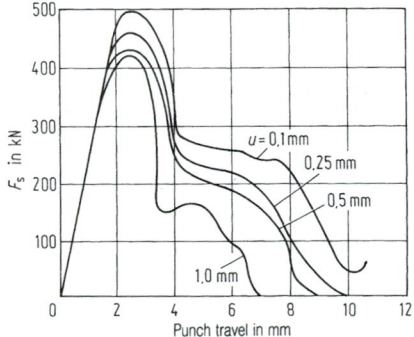

Figure 69. Cutting force F_s versus punch travel for various cutting gaps *u* [55].

Table 4. Variables influencing cutting force

Influencing variable increases	Max. cutting force F_s or specific cutting force k_s
Cutting gap u	k_s falls
Sheet thickness s	k_s falls
Punch diameter d_{St}	k_s falls
Tensile strength R_m	Rule of thumb: $k_s = 0.8 \cdot R_m$
Tool wear	F_s rises to 1.6 times original value

$F_s = l_s s k_s$
l_s length of line of cut

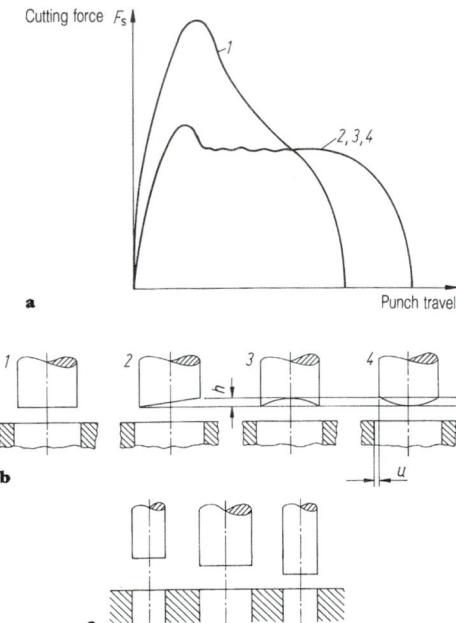

a

b

c

Figure 70. a Cutting-force–travel curve in relation to shape of cutting edge (**b**). **c** Staggering of punch action over time (progressive blanking) [54].

which are influenced by the cutting gap, the punch dimensions, and the thickness and mechanical properties of the sheet metal, arise when the punch is withdrawn (**Table 5**).

The cutting energy is influenced by the tool geometry and the workpiece properties to a far greater extent than the maximum cutting force. It decreases as the width of the cutting gap increases and increases as the sheet thickness increases.

Table 5. Relationships of forces in blanking

Side force/cutting force	0.02 to 0.2
Withdrawal force/cutting force	0.01 to 0.4
Ejector force/cutting force	0.005

4.5.4 Workpiece Properties

The blanks may exhibit a range of defects (**Fig. 71**): the *shape defects* edge indentation, tear depth, and burr height. With parts having small outside dimensions compared with the sheet thickness, deviations from flatness may occur. Edge indentation is constant only in the case of circular parts; in projections with small radii, it may be as much as 30% of the sheet thickness [55]. The tear depth depends on the cutting gap and the material. Burr formation on blanks is a consequence of cutting edge wear and the resulting change in the course of the crack.

Dimensional defects arise in the event of dimensional inaccuracies of tools and/or, with multi-stage tools, as a consequence of feeding errors. The *positional defects*, which are usually deviation from parallel alignment, are caused by incorrect relative positioning of the tool elements. This may arise from inaccuracies in tool manufacture, tilting and shifting of the press slide, or feeding errors in the case of multi-stage tools. *Angle defects* of the cut surfaces result from angular spring, which is particularly severe in the case of C-frame presses.

Owing to the plastic deformation at the beginning of the process, an increase in hardness occurs directly at the cut edges. The depth of the increase in hardness and the area over which this occurs depend on the material. Various investigations show that, with steel sheets, an increase in hardness of 2.0 to 2.2 times the original hardness may result over a distance of 30 to 50% of the sheet thickness from the cut surface. On materials, see D3.1.4.

4.5.5 Tools

Types. Blanking tools are described according to the type of guidance of the cutting elements relative to each other as open, plate-guided and pillar-guided tools (**Fig. 72**). They are suitable in the order stated for small, medium and large quantities (**Table 6**). However, the guiding precision of the press greatly influences the quality of the cut.

Depending on the requirements, the blank is cut out of a strip of sheet metal in one or more stages. Accordingly, a distinction is made between *single-stage* and *multi-stage* blanking tools. Tools for combinations of blanking and forming operations are termed *compound multi-stage tools*. In a single-stage tool, all the cut surfaces are produced in one operation. This is generally possible with simple blanks. A finished blank is thus produced at each stroke of the press. The precision of the blank is determined by the precision of the tool.

For difficult parts with narrow land areas, the workpiece is generally fabricated in several stages in a *multi-stage tool*. The blank remains connected to the metal strip during its passage through the stages and is cut out only in the final stage. With multi-stage tools, the precision of the blank is determined not only by the precision of the

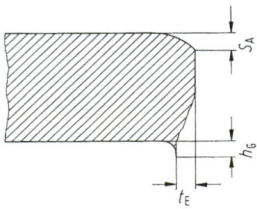

Figure 71. Shape defects on cut blanks. Height of edge indentation s_A, burr height b_G, tear depth t_E [55].

Table 6. Types of punches

	Free punch	Plate-guided punch	Pillar-guided punch
Guide	Press ram	Guide plate Press ram	Pillar guide frames (DIN 9812, 9814, 9816, 9819, 9822)
Advantage	Cheap due to simplicity of design	No positional defects; little risk of buckling of thin punches; high tool life	Very accurate workpieces, low wear; no positional defects; simple and cheap to set up
Disadvantage	Difficult to set up in press; high wear if installation inaccurate	Blanking tool element (punch) is used as guide element	Displacement forces and tilting moments arise from off-centre clamping, expensive tools required
Use	Small quantities	Fairly large quantities	Large quantities

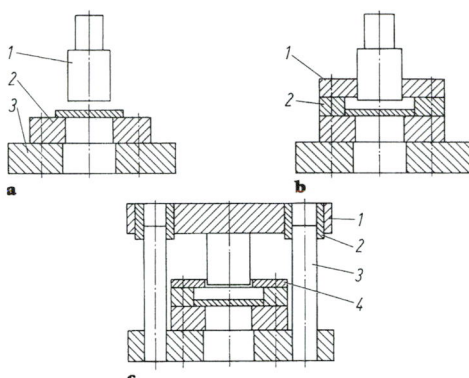

Figure 72. Types of blanking tools. **a** Free punch; *1* punch, *2* die, *3* bolster plate. **b** Plate-guided punch; *1* punch guide plate (stripper), *2* guide strip. **c** Post-guided punch; *1* top, *2* guide bushing, *3* guide post, *4* stripper.

tool but also by the accuracy of the strip feed. This is ensured by means of pilot punches or pins [56].

Tool Position in the Press. Wherever possible, the tools should be positioned in such a way that the resultant of the individual forces passes through the middle of the press. In this way, moments caused by eccentric loading and resulting defects in the precision of the workpiece as well as increased tool wear are avoided. In press design it is assumed that the resultant force acts at the centre of gravity of the lines of cut. The cutting gap, which influences the formation of the cut surfaces and the path of the cutting force, is established according to the requirements applicable to the cut surface – appearance, precision, further processing, and function [57]. For approximate values see **Appendix K4, Table 8**.

Blanking Elements. The punches are designed both to exert pressure and to prevent buckling (in piercing). Some punch designs are shown in **Fig. 73**. The inside

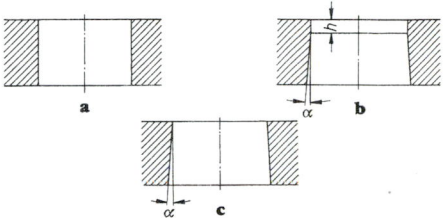

Figure 74. Internal die face shapes. See text for explanations [55].

faces (**Fig. 74**) of dies should be at less than 90° to the bearing surface if the blanked part has to be ejected in the direction opposite to the direction of cutting. Otherwise, clearance angles in the range $1° \leq \alpha \leq 45°$ are usual, depending on sheet thickness. The height of the 90° inside face (**Fig. 74b**) is between 2 and 15 mm. At the design stage, a means of regrinding the blanking elements should be provided. For materials for blanking tools, see **Appendix K4, Table 9**.

4.5.6 Special Blanking Processes

If flat parts with smooth, crack-free cut surfaces and high dimensional accuracy are required, the blanks have to be either finished or cut out by means of special processes.

Reblanking. This process forces the workpieces to be reblanked through a die whose inside dimensions are smaller than the workpiece by approximately twice the thickness of the layer of material to be peeled off (**Fig. 75**).

Precision Blanking. The distinguishing feature of this process is that immediately before the workpiece is blanked, a knife-edged ring is pressed into the sheet a short distance away from the line of cut from one or both sides depending on the sheet thickness. During blanking, an ejector serving as a pressure pad prevents the bulging

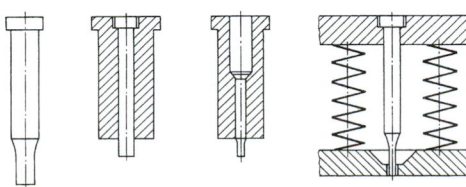

Figure 73. Designs of piercing punches and punch guides.

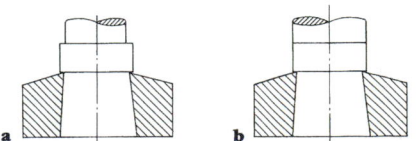

Figure 75. Reblanking of cut parts according to [58]. **a** Punch smaller than die. **b** Punch larger than die.

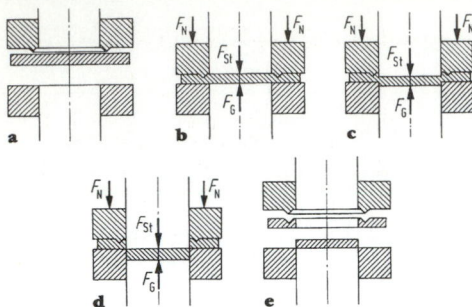

Figure 76. Sequence of events in precision blanking [58]. **a** Original position. **b** Pressing of knife-edged ring. **c** Blanking with opposing force exerted by pressure pad. **d** End of blanking. **e** Sheet stripped and blank ejected. F_N Knife-edged ring force, F_G pressure pad force, F_{St} punch force.

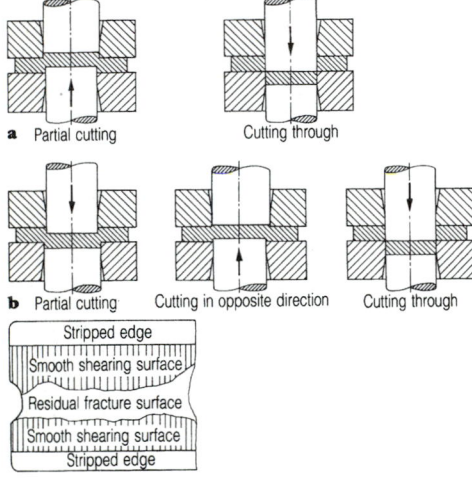

Figure 77. Counterblanking [56]: **a** two-stage, **b** three-stage, **c** cut surface.

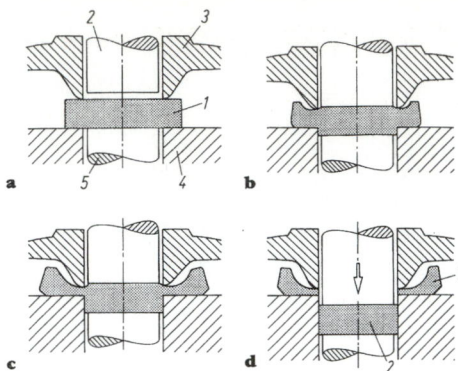

Figure 78. Upset blanking [55]. **a** Contact of punches; *1* sheet, *2* blanking punch, *3* upsetting punch, *4* die, *5* ejector. **b** Upset blanking. **c** End of upset blanking. **d** Cutting out; *1* scrap, *2* blank (workpiece).

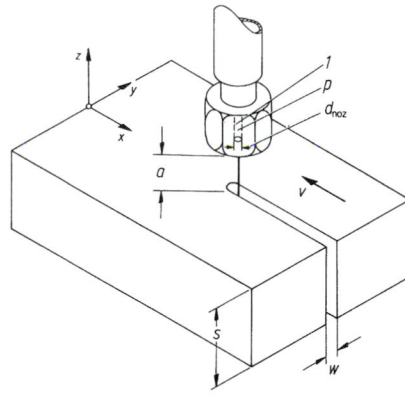

Figure 79. High-pressure water jet cutting; *1* cutting nozzle, *p* pump pressure, d_{noz} nozzle diameter, *d* depth of cut, *a* gap between nozzle and workpiece, *s* workpiece thickness, *w* kerf width, *v* rate of feed [55].

of the blanks that usually occurs in normal blanking (**Fig. 76**) [59].

The effect of the knife-edged rings is important: they generate compressive stresses in the shearing zone at right angles to the direction of cut. In this way the proportion of the cut surface that undergoes plastic deformation by shearing is increased, improving the precision of the cut.

A triple-acting press is required for precision blanking, owing to the knife-edged ring force and the pressure pad force that have to be exerted in addition to the cutting force.

Materials generally suitable for precision blanking are aluminium and its alloys, copper, brass with a copper content ≥ 63%, plain steels containing ≤ 1% C, case-hardened steels, low-alloy hardened and tempered steels, and ferritic and austenitic stainless steels.

Counterblanking. Here, two or three blanking stages acting in opposing directions are employed (**Fig. 77**). In the first stage, a partial cut is made to the point just before cracking begins. The second stage, performed in the opposite direction, also puts a stripped edge on the other

side of the workpiece. Sometimes the workpiece is again only partly cut in this second stage and not cut through until the third stage. The advantage of this process is that burr-free cut surfaces are formed on both the outer and the inner blank and both can be put to use. For this, either a single-stage or a multi-stage tool is generally required.

Upset Blanking. Upset blanking enables smooth, burr-free cut surfaces to be produced. First, the sheet is sheared off with a hollow outside punch as in **Fig. 78**. The remaining sheet thickness is then cut with the actual blanking punch and lastly the cut workpiece is expelled with the ejector. The process is also suitable for cutting phenolic or epoxy resin laminates and glass-fibre-reinforced plastics.

Jet or Beam Cutting. In these cutting processes, the material is removed along a designated workpiece contour by the effect of an active medium concentrated into a jet or an active energy source supplied in the form of a beam. The processes employing an *active medium* use a beam having mass as the "tool" (*plasma cutting* and *water jet cutting*, **Fig. 79**), whereas the processes employing *active energy* use a quasi-massless beam (*laser cutting*).

5 Special Technologies

5.1 Thread Production

G. Spur, Berlin

5.1.1 Single-Point Thread Turning

Single-point thread turning is a spiral turning operation to produce a thread using a single-point cutting tool. Both external and internal threads can be produced. When using a universal lathe, the feed drive is accomplished via a lead screw,

$$n_w P_w = n_L P_L,$$

where n_w = the rotational speed of workpiece, P_w = the thread pitch of workpiece, n_L = the rotational speed of lead screw and P_L = the pitch of lead screw.

Conventionally controlled automatic lathes are equipped with guide devices (feed curves, guide collets) to start the feed motion. On numerically controlled lathes the feed drive is separated from the main drive. The feed motion is computer-controlled and is accomplished with the aid of a recirculating ball screw. As the thread is produced in several stages, the tool has to be advanced repeatedly at the same point on the circumference of the workpiece. For this purpose the angular position of the main spindle is recorded with mechanical devices on conventionally controlled machine tools and by means of digital transducers on numerically controlled machines.

The *thread turning tools* correspond to the *thread profile*. There are *shank-type form-cutting tools, round form-cutting tools* and *relieved round form-cutting tools* (**Fig. 1**). The side rake angle of the tool is generally $\gamma_f = 0°$; the face is adjusted to the centre of the workpiece. To obtain the necessary side clearance angle of $\alpha_f = 6$ to $8°$ for a round form-cutting tool, the centre of the tool lies a defined distance h above the centre of the workpiece (**Fig. 1b**). At this point the round form-cutting tool has to have the desired thread profile. In the case of the relieved round form-cutting tool, the required side clearance angle is produced by the relief (**Fig. 1c**). Regrinding is carried out on the face, which is positioned radially relative to the tool axis.

The effective side clearance angle is $\alpha_{fe} = 3$ to $5°$ (**Fig. 2**). It depends on the thread pitch. A symmetrical ground surface is adequate for smaller pitches (**Fig. 2a**). For larger pitches the shank-type form-cutting tool should be ground at differing angles (**Fig. 2b**). This produces different effective side wedge angles $\beta_{fe\,1}$ and $\beta_{fe\,2}$, which may lead to unfavourable cutting conditions. To avoid this, the shank-type form-cutting tool is positioned at an angle (**Fig. 2b**). Profile distortions arise, however, which have to be compensated for by contouring the shank-type form-cutting tool accordingly. Precision threads are produced with two cutting tools which each machine one flank.

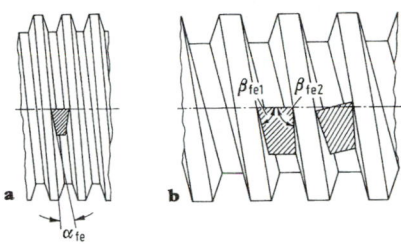

Figure 2. Shape of shank-type form-cutting tool for different thread pitches: **a** for small thread pitches; **b** for large thread pitches.

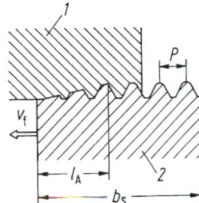

Figure 3. Thread chaser: l_A bevelled length, v_f feed rate, b_s chaser width, P pitch, *1* workpiece, *2* chaser.

5.1.2 Thread Chasing

Thread chasing is turning to produce a thread with a tool having multiple cutting edges in the feed direction. The *chaser* is usually bevelled at the leading side (**Fig. 3**). It may be engaged either radially or tangentially. Chasing is frequently employed on turret and automatic lathes with the use of a guide device. The chaser is guided by a guide collet or a chasing curve. Single- and multiple-start internal and external threads can be chased. Several thread pitches may be turned with the same chasing curve by means of a change gear train [1].

5.1.3 Thread Cutting with Dies

This is spiral turning to produce a thread using a tool with *multiple* chasers. *Solid (non-opening)* and *opening dies* are generally used for thread cutting by hand and for threads with low precision requirements. Solid dies (**Fig. 4**) may

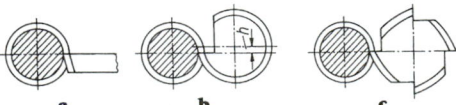

Figure 1. Single-point thread-turning tools: **a** shank-type form-cutting tool; **b** round form-cutting tool; **c** relieved round form-cutting tool.

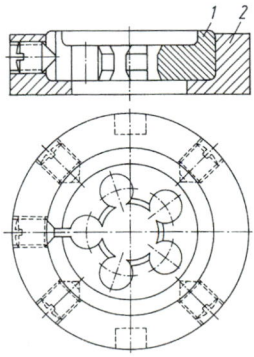

Figure 4. Threading die in a stock: *1* threading die, *2* stock.

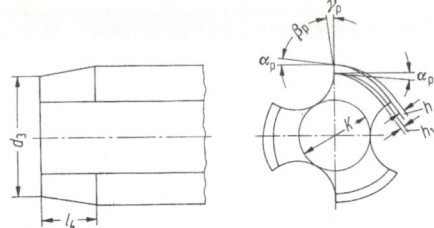

Figure 5. Shape of cutting edges of a tap: α_p back clearance angle (flank relief angle), α_{p1} back clearance angle on tapered end, β_p back wedge angle, γ_p back rake angle, h relief, h_1 relief on tapered end, d_3 diameter of tapered end, l_4 length of tapered end, K centre diameter.

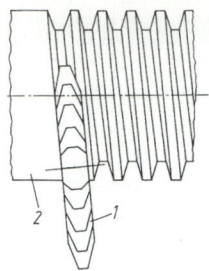

Figure 7. Long-thread milling: *1* tool, *2* workpiece.

be slotted or closed. Opening dies usually have four radially or tangentially positioned chasers. These are adjustable and interchangeable for different thread diameters and pitches and can be opened after cutting, so that the unscrewing necessary with solid dies is dispensed with.

For series production, self-opening dies with radially or tangentially adjustable chasers are used. If the die hits a stop which marks the end of the thread length, the chasers open automatically.

5.1.4 Tapping

Tapping is spiral drilling with a tap to produce an internal thread. The geometrical cutting edge shape of a tap is shown in **Fig. 5**. Taps are relief-ground to reduce the work done by friction. The back clearance angle (relief angle) is $\alpha_p = 1$ to $5°$; the back rake angle γ_p is 0 to $3°$ for grey cast iron, 3 to $15°$ for steel and 12 to $25°$ for aluminium alloys.

The chamfered end of the tap performs most of the cutting, while the rest of the tool is mainly for guidance and is slightly tapered (1 : 1000).

Hand and machine taps are used. Hand taps consist of a set of several taps, selected according to the material to be cut. Sets of three taps are normally used. The cutting work is split between the individual taps as follows: taper tap 50%, plug tap 30% and bottoming tap 20%. Machine taps are normally used as single-cut taps. Poor chip removal and the resulting risk of tool breakage necessitate low cutting speeds in machine tapping.

Figure 6 shows common designs of machine taps. Short-chipping materials are cut mainly with straight-flute taps (**Fig. 6a**). Curling taps (**Fig. 6b**) achieve better chip removal when machining through holes. Tapping in sheet metal is carried out with taps having short flutes (**Fig. 6c**). Taps with sharply helical flutes (**Fig. 6d**) promote good chip removal in bottoming operations with small runouts [2].

5.1.5 Thread Milling

Long Thread Milling. Here, the length of thread that can be produced is independent of the tool (**Fig. 7**). Disc-shaped, relieved *profile milling cutters* are used, the profile of which has to be adjusted in the case of large pitches. The axis of the milling cutter is angled in relation to the workpiece axis according to the thread pitch. If the partial thread angle is less than $10°$, profile distortions arise from the lateral free-cutting of the milling cutter. Either upcut or downcut milling may be used. Long thread milling is used with longer threaded spindles. The cutting speeds of relieved profile milling cutters made of high-speed steel are $v_c = 4$ to 20 m/min for steel, depending on its tensile strength [2].

Thread whirling is another method of long thread milling. The principle of thread whirling – also known as *thread peeling* or fly milling – is illustrated in **Fig. 8**. One to four cutting tools with their tips pointing towards the centre of rotation of the holder rotate around the workpiece on an eccentric orbit. The whirling head is inclined in relation to the workpiece at the pitch angle. The tools may be positioned radially or tangentially. Such thread whirling devices can also be used on centre lathes. In thread whirling with cemented carbide, the cutting speed for steel, depending on its tensile strength, is $v_c = 100$ to 125 m/min [3]. The peripheral speed of the workpiece is between 0.5 and 4 m/min. Thread whirling may be performed in either the upcut or the downcut direction. For internal threads the required orbit should be 2 to 5 mm smaller than the minor diameter. Internal threads and external threads with small thread depths are cut using only one tool.

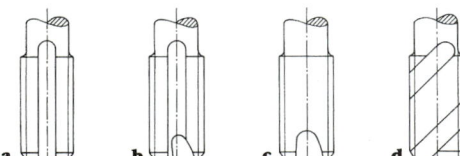

Figure 6. Tap shapes (see text for explanations).

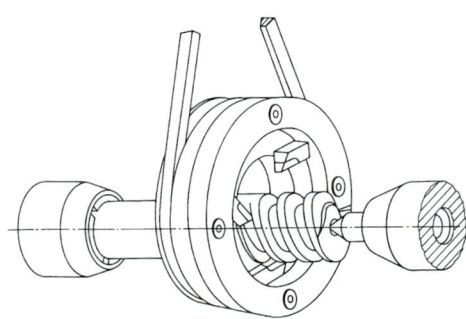

Figure 8. Principle of thread whirling.

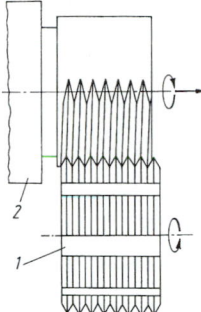

Figure 9. Short-thread milling. *1* tool, *2* workpiece.

Short Thread Milling. This is done with roll-shaped thread milling cutters (**Fig. 9**). These have adjacent thread profiles with no pitch, the spacing of which corresponds to the thread pitch. One tool can only produce threads of the same pitch, but to various diameters. During about 1/6 of a workpiece *revolution*, the milling cutter is radially advanced to the required thread depth and displaced axially during a further workpiece revolution.

Large-profile multiple start threads, e.g. worms, may be economically manufactured by hobbing. The tool moves on rolling contact along a line parallel to the axis of rotation at the periphery of the workpiece.

5.1.6 Thread Grinding

The three most important thread grinding processes are illustrated in **Table 1**.

Longitudinal Spiral Grinding. The thread starts are ground in succession using a *single-profile* grinding wheel. The grinding wheel is inclined in relation to the workpiece axis according to the thread pitch. Every possible pitch can be ground. The small machining force components promote the high degree of precision achievable, but the grinding time is relatively long.

In *multiple-profile* longitudinal spiral grinding, there are

several thread profiles next to one another on the grinding wheel according to the thread pitch. These profiles are stepped on the chamfer of the grinding wheel. The feed distance depends on the length of the thread and the width of the grinding wheel. Threads with a collar cannot be produced. The pitch range is $P = 0.8$ to 4 mm.

Transverse Spiral Grinding. In multiple-profile transverse spiral grinding, the grinding wheel is fed inward to the full thread depth during a one-quarter revolution of the workpiece. The thread is finished during a further revolution with simultaneous axial displacement by the amount of the pitch. The machining force components are relatively large. Only thread lengths up to about 40 mm can be ground. The pitch range is $P = 0.8$ to 4 mm. Multiple-profile transverse spiral grinding results in the shortest grinding times.

Centreless thread grinding may be performed by the throughfeed method or the transverse grinding method. The profiled grinding wheel is pivoted according to the helix angle of the pitch diameter.

The grinding wheels have to be dressed to produce the desired thread profile at the circumference. Single-profile grinding wheels are dressed with diamond dressing tools with a single layer of grains, while multiple-profile grinding wheels are dressed with shaped rollers made of hardened tool steel or with rotating shaped diamond rollers.

5.1.7 Electro-discharge Machining of Threads

Electro-discharge machining of threads is used for difficult-to-cut materials, usually for producing internal threads. The tool electrode, made of brass, copper or steel, bears the thread profile and screws itself into the workpiece, which has usually already been centre-drilled.

5.1.8 Thread Rolling

In thread rolling with flat dies (**Fig. 10a**), the pair of dies carries the opposite profile to the thread with the helix angle of the thread. One die is fixed and the other is movable. The workpiece is rolled between the two dies under the effect of friction forces, forming the thread over its entire circumference. The rolling dies have a chamfered lead-in and lead-out and a straight grooving section.

Table 1. Working methods and accuracies in thread grinding [4]

	Longitudinal spiral grinding with single-profile grinding wheel	Longitudinal spiral grinding with multiple-profile grinding wheel	Transverse spiral grinding with multiple-profile grinding wheel
Feed distance l_f	$l_f > L$	$l_f > L + b_S$	$l_f > P$
Workpiece revolutions i_w	$i_w = l_f/P$	$i_w = l_f/P$	$i_w > 1$
Grinding wheel width b_S	—	—	$b_S > L + P$
Accuracies: Pitch diameter	± 2 μm	Finishing ± 4 to 5 μm Rough-working ± 10 to 15 μm	± 10 to 20 μm
Partial thread angle	± 5′	± 5′ to 10′	± 10′
Pitch over a length of 25 mm	± 2 to 3 μm	± 5 μm	± 5 μm (25 mm wheel width)
Pitch over a length of 300 mm	± 5 μm	± 10 μm	

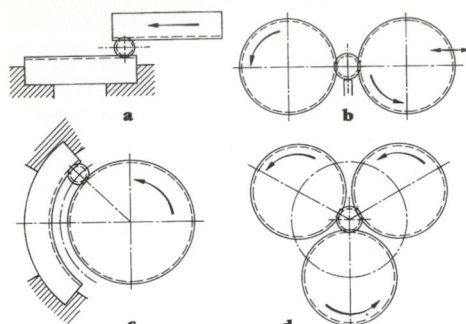

Figure 10. Thread rolling processes [5] (see text for explanations).

Thread rolling with roller-type dies (**Figs 10b–d**) may be carried out with infeed or through-feed of the blank. Either short or long threads may be rolled, even right up to the collar. For infeed rolling, the rollers have the required thread profile with the same helix angle but the opposite direction of twist. The workpiece is either held with a guide (centreless) or clamped between centres. The workpiece remains axially at rest during rolling, apart from slight compensating movements. As the roller-diameter–workpiece-diameter ratio and the number of turns have to be precisely matched, a pair of rollers can be used only for a specific thread. For through-feed rolling, the rollers have adjacent pitchless thread profiles. They are pivoted about their horizontal longitudinal axis through the required helix angle. In this way the axial feeding of the workpiece by the depth of pitch is accomplished in one revolution. The rollers may also be used to a limited extent for various workpiece diameters. This method, though, results in a lower pitch accuracy than does infeed rolling.

Self-opening thread rolling heads usually have three rollers. They are advanced axially towards the workpiece and are automatically pulled onto the workpiece by the inclined position of the rollers. When the required thread length is reached, the rolling head opens automatically and can be withdrawn.

5.1.9 Thread Forming

Thread forming is the indentation of a thread into a workpiece using a tool with a helical effective surface. The process (**Fig. 11**) resembles tapping in terms of kinematics, however the tool has no flutes and its cross-section is shaped like a rounded polygon with three or more forming webs. The torques to be exerted are considerably higher than in tapping and depend heavily on, among other things, the diameter of the forming hole and the

cooling lubricant used [6]. Lubricating channels may be cut in the tools to improve cooling lubricant delivery.

5.1.10 Thread Pressing

Thread pressing (**Fig. 12**) is mostly used to produce cylindrical threads in fairly thin sheets. The thread is pressed into the workpiece with two shaped rollers.

5.2 Gear Cutting

M. Weck, G. Mauer and W. Reuter, Aachen

5.2.1 Cutting of Cylindrical Gears

Fundamentals

Figure 13 gives a survey of gear manufacturing processes. Initial gear cutting is mainly carried out by hobbing, generating by shaping, and broaching. Generating by planing may be used to cut large gears. Finishing before heat treatment is performed by shaving. Finishing processes after heat treatment are generating by grinding or form grinding. The processes are divided into form-cutting processes and generating processes (**Fig. 14**).

Form-Cutting Processes. The tool (*disc or end milling cutter, shaping tool, broach, grinding wheel*) has the profile of the tooth space. The spaces are produced individually. To machine the next tooth space, the gear blank is turned through the circular pitch (individual indexing process). For each gear blank design (number of teeth, modulus, meshing angle, helix angle, addendum modification, and tooth adjustment) an appropriate tool profile is required (cf. F8.1). Tool profiling for helical gear cutting is complicated, as the line of contact between the gear blank and the tool is a three-dimensional curve that cannot be derived from the transverse profile of the gear blank in a simple manner (it also depends on the tool diameter). The grinding wheel profile has to be calculated with a computer.

Generating Processes. A rolling motion is produced between the gear blank and the tool during machining by kinematic linkage (self-contained gear train, electronic control loop). The flank shape (involute) is formed as the envelope of the straight-sided flank tool edge (**Fig. 15**; cf. F8.1.7). The involute shape is generated by rolling coupling of linear motion (translatory component of generating) with the rotation of the workpiece (rotational component of generating). The tools used are: *hobs, rack-shaped cutters, disc and taper grinding wheels, grinding worms*. With the involute profile, the tools have a more universal range of uses than with the form-cutting process (independent of the number of teeth, the helix angle or the addendum modification). Continuous generating is possible with a worm-shaped tool (*hob, grinding worm*) or a gearwheel-shaped tool (*cutting wheel, peeling wheel, shaving wheel*). *Rack-shaped cutters and disc, flat or taper grinding wheels* are used to machine one or more tooth spaces (generating with indexing). When the mesh-

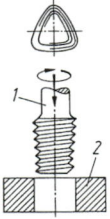

Figure 11. Thread forming: *1* thread forming tool, *2* workpiece.

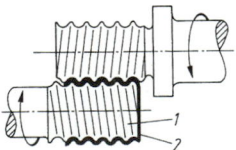

Figure 12. Thread pressing: *1* pressing roller, *2* workpiece.

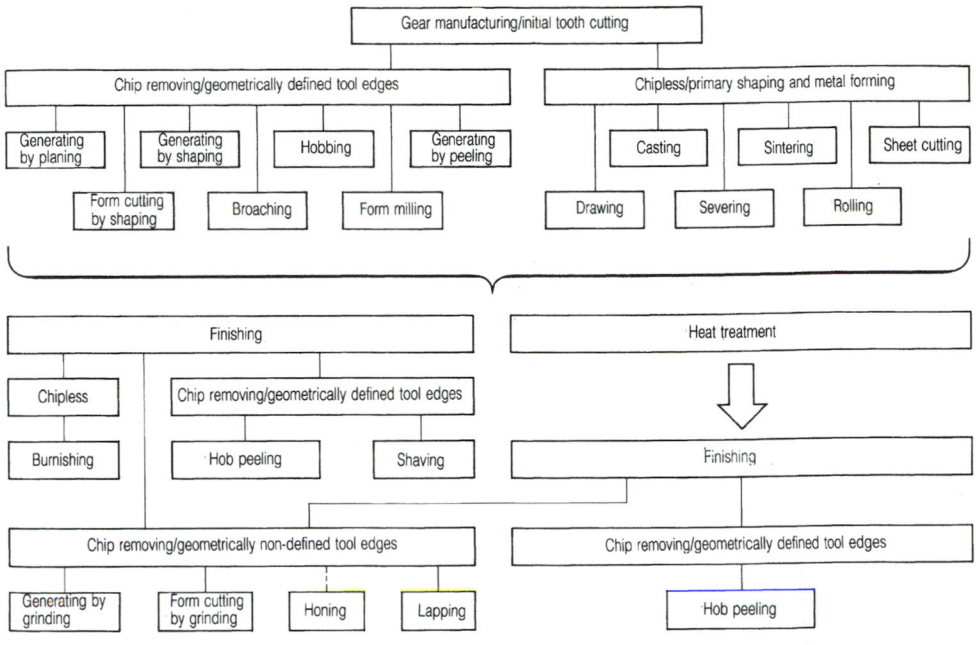

Figure 13. Gear manufacturing processes.

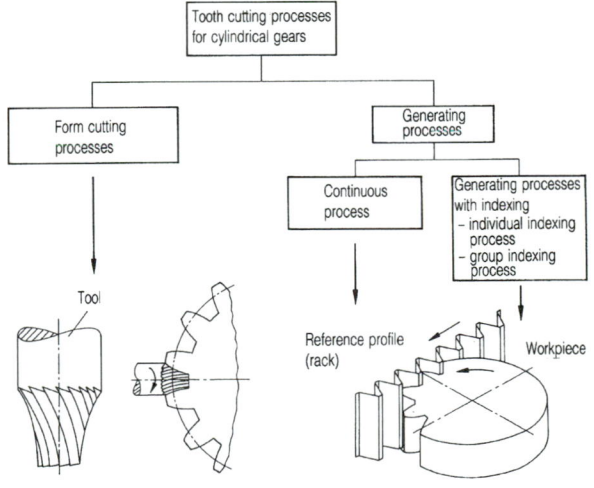

Figure 14. Cylindrical gear manufacturing processes.

ing zone has been machined, the workpiece is turned through one or more circular pitches and the generating operation is repeated (reversing).

Form Milling

Uses. Large-pitch or large-diameter gear blanks, gear blanks with non-generatable profiles, gear blanks with large gear cutting tolerances and for initial machining; also in the manufacture of single units (spur teeth can be produced on conventional universal milling machines with an indexing attachment).

Machine. The tool motor directly drives the form milling cutter. Precise indexing equipment is required. For helical teeth, the rotary motion of the gear blank is derived from the feed motion of the tool (generation of spiral motion in the gear blank-specific coordinate system). The spiral motion depends on the helix angle of the gear blank.

Tool. *End* or *side-and-face milling cutters* (also fitted with cemented carbide cutting tips). In profiling the tool, the grinding wheel is guided by a template, a cam mechanism or numerical control. The cutting capacity is high, because the tool cuts along the entire length of the profile.

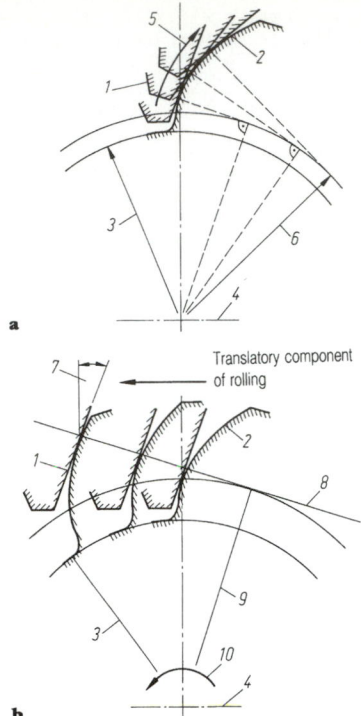

a

b

Figure 15. Production of the involute-shaped tooth flank. **a** Theoretical production principle. **b** Production principle in machine; *1* tool cutting edge, *2* tooth flank, *3* dedendum circle radius, *4* gear blank axis, *5* rolling feed, *6* base circle radius, *7* pressure angle, *8* line of action, *9* pitch circle radius, *10* rotational component of rolling.

Generating by Planing

Principles. An oscillating rack-shaped cutter rolls with the gear blank (**Fig. 16**). The cutting motion is accomplished by the reciprocating motion of the rack. The feed motion is accomplished by radial plunging or tangential feeding of the rack and the rolling motion of the blank. During the return stroke, the rack is lifted off the blank. The rolling feed takes place cyclically at the top of the return stroke of the tool. Only one group of teeth can be machined at any one time. The rack is next drawn back and the gear blank then indexed by the corresponding number of teeth (group generating process with indexing).

Uses. Initial cutting and finishing of large-size spur and helical gears of high-tensile material and large tooth width.

Machine. The rack-shaped cutter is attached to a slide which can be pivoted in the helix direction of the gear teeth. The rolling motion is accomplished by means of a cradle and a turntable. The rolling feed is transmitted to the cradle and the turntable by the ram stroke motor via the rolling feed crank and the rolling gears. With smaller machines, the tool travel is accomplished by means of a crank mechanism.

Tool. Involute gears are cut with a relieved, straight-sided flank involute rack; the rake angle is adjusted by hollow grinding of the face. The tool is cheap and is easy to manufacture with a surface grinding machine. The wear rate is low and tool changing is simple.

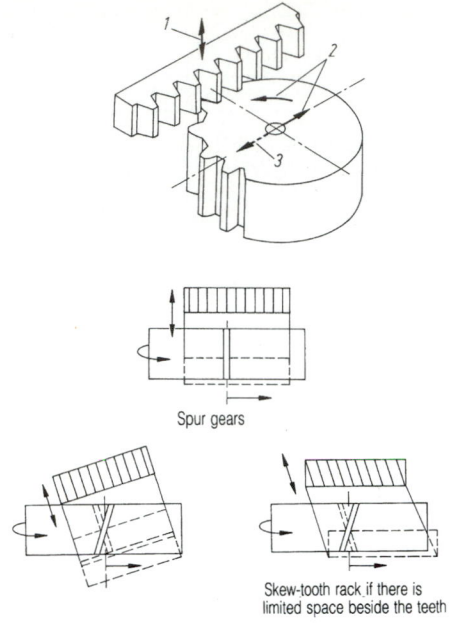

Spur gears

Skew-tooth rack if there is limited space beside the teeth

Helical gears

Figure 16. Principle of generating by planing; *1* cutting motion, *2* feed motion, *3* reversing motion.

Gear Shaping

Principles. The rotation of the gear-shaped tool (cutting wheel) is matched to the rotation of the gear blank by a kinematic linkage so that both elements revolve like a pair of cylindrical gears (**Fig. 17**). The cutting motion is accomplished by the stroke of the cutter. The feed motion is accomplished by radial feeding up to the depth of cut and by a roling motion (rolling speed in relation to stroke frequency).

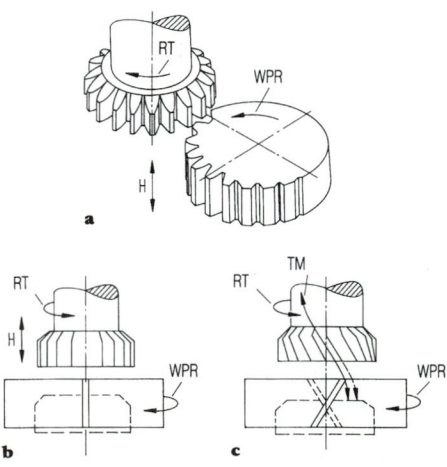

a

b   **c**

Figure 17. Principle of gear shaping. Cutting wheel in engagement with straight-toothed gear blank, H tool travel, RT tool rotation, WPR workpiece rotation, TM helical tool motion in cutting of helical teeth. **a** Principle. **b** Spur gears. **c** Helical gears.

On the return stroke of the cutting wheel, a lift-off motion has to take place to prevent collision between the continually revolving cutting wheel and the gear blank (the flank of the gear blank is not yet fully profiled and therefore cannot roll, continuous radial feeding takes place during the stroke). The direction of lift may be determined with reference to the gear blank flank by tangential offsetting of the cutter and gear blank axes. For cutting helical gears, a spiral motion corresponding to the helix angle has to be superimposed on the stroke.

Uses. These are the initial cutting and finishing of internal gear teeth, teeth with too small an axial idle travel distance of the tool for hobbing (collar following teeth, step wheels), and short teeth.

Machine (Fig. 18).

Mechanical Drive. The rolling drive drives the cutting wheel and the gear blank. The kinematic coordination of the rolling feed of the cutting wheel with that of the blank is accomplished by the table change gears. A separate reciprocating drive generates an oscillating ram motion. The lift-off motion is controlled via the reciprocating drive and the lifting cam.

NC Drive. Each axis has a separate drive. The rolling linkage is accomplished electronically in the control system. The lifting cam is also linked to the stroke. The spiral motion is produced mechanically via an inclined guidance sleeve.

Tool. The cutting wheel is made of high-speed steel (HSS) or coated TIN. The helix angle of the cutting wheel depends on the gear blank helix angle. For cutting helical teeth, the face is often ground in such a way that it is perpendicular to the helix direction (step grinding). The flanks of the cutting wheel teeth are relieved in such a way that the cutter can generate the desired gear profile in every transverse position (i.e. after every regrinding of the face).

Gear Hobbing

Principles. The rotation of the gear blank is matched by kinematic linkage to the rotation of the worm-shaped tool (hob) so that both elements rotate like a worm and wheel. Through the additional superimposition of a feed motion (axially, radially, radially–axially, tangentially or axially–tangentially to the gear blank cylinder) the hob cuts material from the tooth spaces. **Figure 19** illustrates the engagement of the hob with the blank. A rolling motion is produced by superimposition of an imaginary tangential motion of the hob teeth on the rotational motion of the blank (during a revolution of the hob, tangentially offset cutter teeth of the worm start to engage in succession). The gear blank profile is composed, in polygonal fashion, of enveloping cuts.

Axial Method. The hob feed direction is axial to the gear blank cylinder. This is the most common method of cutting cylindrical gear wheels on a hobbing machine.

Diagonal Method. The hob feed direction is both axial and tangential to the gear blank cylinder. It is used to cut cylindrical wheels. The tangential motion of the cutter must be balanced by additional rotation of the gear by an equal amount (analogous to the rack rolling with the gear blank).

Tangential Method. The hob feed direction is tangential to the gear blank cylinder. It is used for cutting worm wheels. The kinematics are as in the diagonal method (no axial feed).

Radial Method. The hob feed direction is radial to the gear blank cylinder. It is used for cutting the teeth of worm wheels and very thin wheels.

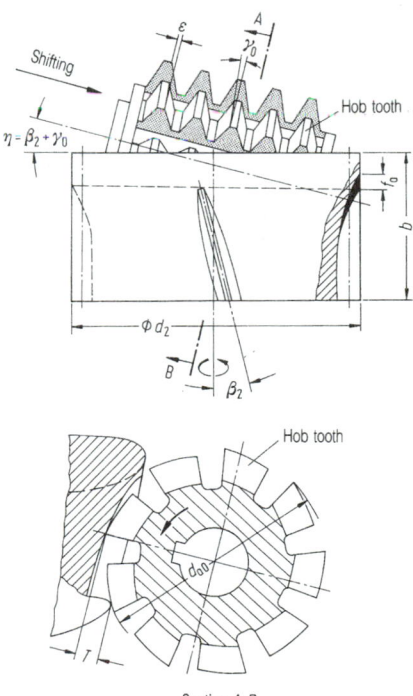

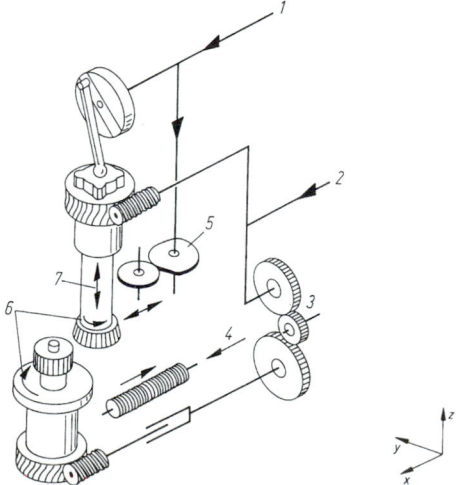

Figure 18. Diagram of mechanism of a gear shaping machine: *1* reciprocating drive, *2* rolling drive, *3* turntable change gears, *4* radial feed, *5* lift-off cam wheel, *6* rolling feed, *7* reciprocating motion.

Figure 19. Terminology for hob–workpiece pair. Gear: d_2 gear diameter, z_2 number of teeth, β_2 helix angle, b gear width. Hob: d_{a0} hob diameter, z_0 number of starts, γ_0 pitch angle, ε axial pitch, i number of teeth. Machining: η pivoting angle ($\eta = \beta_2 \pm \gamma_0$), f_a axial feed, T depth of cut.

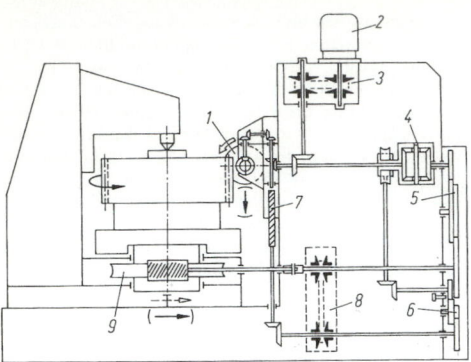

Figure 20. Gear train of a hobbing machine; *1* hob, *2* main motor, *3* gearbox, *4* differential, *5* indexing change gears, *6* differential change gears, *7* feed spindle, *8* feed gearbox, *9* indexing worm gear.

Uses. Initial cutting of automotive gears in series production, initial cutting and finishing of soft, quenched and tempered and hardened (with a cemented carbide peeling hob) large gear teeth up to approximately 4000 mm outside diameter, also initial cutting and finishing of worm wheels and special teeth (axial compressor rotors, serrated or splined shaft teeth, chain wheels).

Machine (Figs 20 and 21)

Conventional Drive. All the motions are derived from the main motor. The rotational speed of the hob (cutting speed) may be varied by means of a steplessly adjustable drive. The rotation of the hob and the gear blank are linked by means of the indexing change gear train. The relationship of the gear blank rotation to the cutter rotation is the same as the relationship of the number of starts of the hob (number of teeth) to the number of teeth of the workpiece. This adjustment is made by means of the indexing change gears. The axial feed is taken from the worm shaft of the table drive mechanism. This feed is steplessly adjustable.

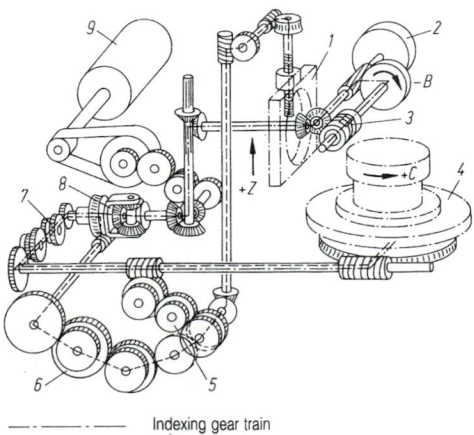

—·——·—— Indexing gear train

————— Axial differential gear train

Figure 21. Generating module of a conventional hobbing machine; *1* axial cradle, *2* flywheel, *3* hob, *4* workpiece table, *5* axial feed change gears, *6* differential change gears, *7* indexing change gears, *8* differential, *9* main motor.

To produce helical gear teeth, the axial feed must be linked to the rotation of the workpiece and the hob. To achieve this, the feed motion is returned to the table gear train via the differential gear and the differential. The outer cage of the differential, which is stationary during cutting of spur teeth, moves in line with the translations in the differential gear train and superimposes an additional motion on the table gear train. Depending on the pitch direction of the gear blank teeth, this results in an increased or reduced table speed to produce helical teeth on the gear blank.

NC Drive. **Figure 22** shows a diagram of the drive system of an NC gear hobbing machine. All the axes are driven by separate motors. The kinematic linkage is accomplished by means of the control computer. From the information supplied by the hob angle measuring system and, for helical gear teeth, by the axial drive measuring system, the setpoint value for the table rotation is calculated in the computer. Comparison with the information on the workpiece motion gives a difference which is compensated for via the table drive.

Tool. For cutting involute gear teeth, the enveloping surface of the hob is an involute screw with straight-sided flanks which is broken by flutes at right angles to the involutions. The teeth are relieved so as to form flanks at the tips and sides of the teeth which permit regrinding of the face at a constant rake angle and tooth profile (radial regrinding). Tooth profiles are standardised as reference profiles (i.e. the normal section of the rack) in DIN 3972.

Solid Hobs. These are made in one piece from high-speed steel (HSS) with or without TIN coating. For tilting-tooth hobs, the cutting teeth are ground in ancillary devices as involute screws and then tilted into the working position in the tool body, producing clearance angles at the tip and sides. The teeth are made of HSS or cemented carbide, while the body is made of tool steel.

Cutter Plate Hobs. The teeth are relieved in the body as with solid hobs. The hob tooth length is extensively regrindable owing to back supports. The teeth are made of HSS or cemented carbide, while the body is made of tool steel.

The stress on the cutting edge varies along the zone of engagement between cutter and gear blank, so there is no uniform distribution of wear in the longitudinal direction of the hob. This is remedied by gradual tangential shifting of the working range of the hob when it is out of engagement after the permitted maximum width of wear land has been reached, or by continuous tangential shifting by the diagonal method.

Gear Finishing

Gear finishing is carried out by *shaving* if the teeth are in a soft condition (before heat treatment) and by *grinding, hob peeling* or *peeling gear shaping*. The main purpose of fine machining is to remove the geometrical deviations on the gear blanks such as envelope cutting and feed markings (**Fig. 23**). The making of tooth flank corrections is growing in importance. As **Fig. 24** shows, relieving the tip zone or making corrections in the direction of tooth width can improve the running and stress behaviour of the gears. Topological corrections can be used to make specific adjustments in the dynamic running characteristics.

Shaving with a Shaving Cutter

Principles. The initially cut gear blank rolls with a gear-shaped tool (shaving cutter) with intersecting axes and

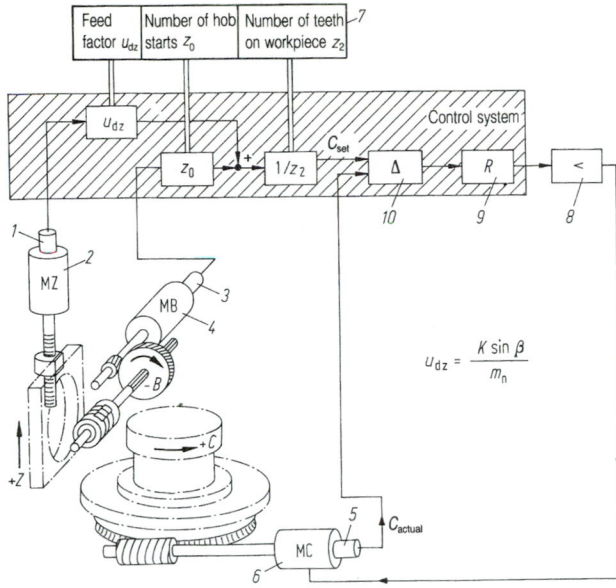

Figure 22. Electronic generating module of an NC hobbing machine: *1* angle measuring system Z, *2* axial drive, *3* angle measuring system B, *4* hob drive, *5* angle measuring system C, *6* table drive, *7* input data, *8* booster, *9* regulator, *10* subtraction, *K* machine constant, β helix angle, m_n normal modulus.

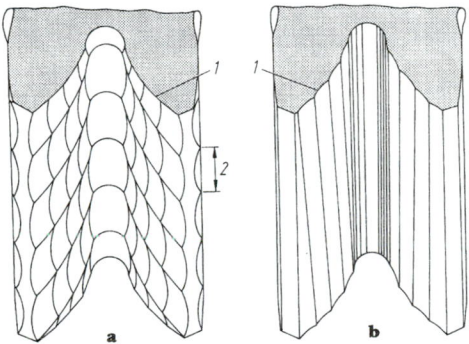

Figure 23. Deviations from the enveloping cut in initial gear cutting: **a** hobbed tooth flanks, **b** tooth flanks generated by shaping; *1* deviations from the enveloping cut, *2* axial feed.

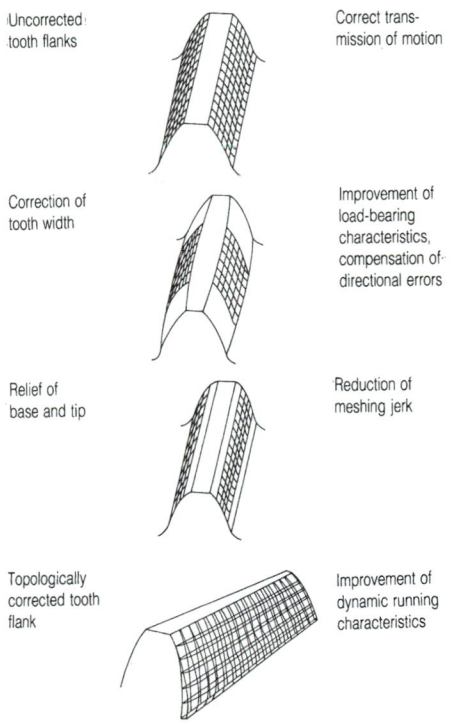

Figure 24. Correction of tooth flank geometry.

without kinematic linkage. The shaving cutter drives the gear blank. Intersection of the axes (conditions corresponding to a spiral toothed gear train) gives rise to sliding of the shaving wheel tooth flank on the gear blank tooth flank in vertical and longitudinal directions of the tooth.

On the flank of the shaving cutter are cutting teeth. If the flank of the cutting wheel runs over the flank of the gear blank with application of force, chip removal takes place. The feed motion is axial (parallel shaving), tangential (transverse shaving), diagonal (diagonal shaving) or radial (plunge shaving) relative to the gear blank cylinder. **Figure 25** illustrates the shaving cutter engaging with the gear blank.

Uses. Fine machining of soft spur and helical gear teeth after initial cutting, series manufacture of automotive gears, improvement of surface roughness and tooth-cutting errors, and reduction of distortion due to hardening by initial correction of the gear blank flanks.

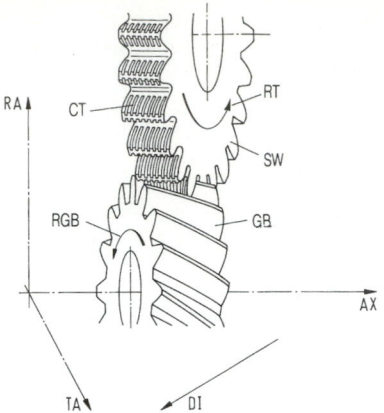

Figure 25. Shaving wheel SW meshing with helical-tooth gear blank GB (after Buschhoff K. Thesis, RWTH Aachen, 1975). RGB rotary motion of gear blank, RT rotary motion of tool, AX axial direction relative to gear blank, TA tangential direction relative to gear blank, RA radial direction relative to gear blank, DI diagonal direction relative to gear blank, CT cutting tooth.

Gear Tooth Grinding

Principles. Finishing of mainly hardened or quenched and tempered gear teeth (improvement of surface roughness and tooth cutting errors, elimination of distortion due to hardening, and corrective grinding). By analogy with the initial gear cutting methods, the grinding methods are classified into form-cutting methods, continuous methods and generating methods with indexing (**Fig. 26**). The motions correspond in principle to those of the initial cutting processes. Owing to the lower cutting forces and higher cutting speeds, machine designs have to be adapted to these circumstances. Corundum and CBN grinding wheels are used as tools.

Form Grinding. The conditions are similar to those of form milling. The grinding wheel is dressed to the gear blank profile according to the desired line of contact. This is easy with spur gear teeth but complicated with helical gear teeth. **Figure 27** illustrates the possible grinding wheel positions. The advantages of the process are a high metal removal capacity due to line contact over the full width of the flank and minimal profile errors (no kinematic rolling errors during machining). It is suitable for internal gear teeth.

Generating by Grinding with Indexing. The grinding wheel flank embodies the flank of an imaginary rack which rolls with the gear blank flank to be generated (conditions analogous to generating by planing). **Figure 28** shows a machine for generating by grinding with indexing with a double-conical grinding wheel. From the kinematic point of view, this machine may be compared with a machine for generating by planing. The rolling feed takes place continually during grinding. It is a general-purpose machine, used mainly in small- and medium-scale series production and for grinding of large gear teeth.

Besides the already mentioned generation of the rolling motion by means of a self-contained rolling gear train, in gear tooth grinding the rolling motion may be generated by means of pitch blocks, rolling belts and cradles.

Continuous Generating by Grinding. The conditions are similar to those of hobbing, but the hob is replaced by a grinding wheel of larger diameter, the outer

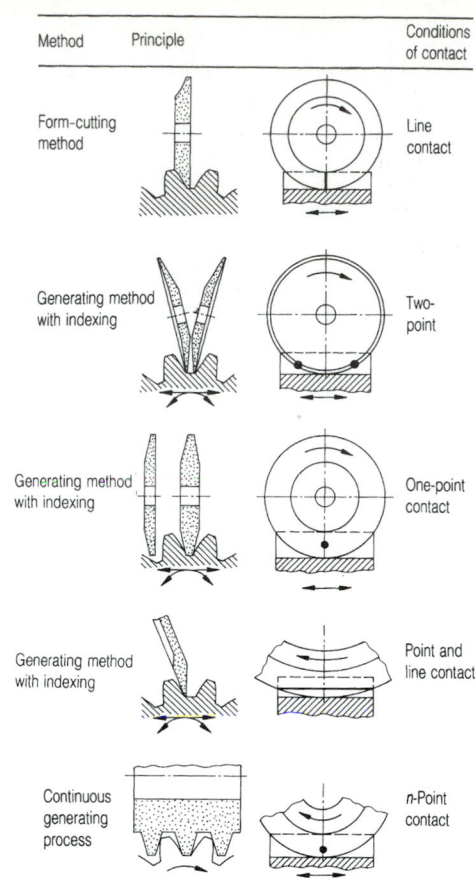

Figure 26. Types of gear tooth grinding.

Method	Principle		Conditions of contact
Form-cutting method			Line contact
Generating method with indexing			Two-point
Generating method with indexing			One-point contact
Generating method with indexing			Point and line contact
Continuous generating process			n-Point contact

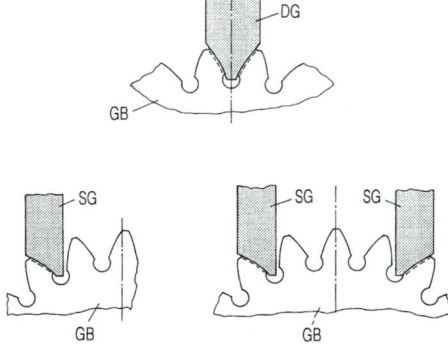

Figure 27. Position of form grinding wheel relative to gear teeth (after Dudley DW, Winter H. Zahnräder. Springer, Berlin, 1961): DG double-flank grinding wheel, SG single-flank grinding wheel, GB gear blank.

shell of which is dressed in the form of an involute screw. **Figure 29** shows the machine construction with six NC axes. The tools are corundum or CBN grinding wheels with a flange-mounted polishing worm. By means of the diagonal method (simultaneous axial and tangential

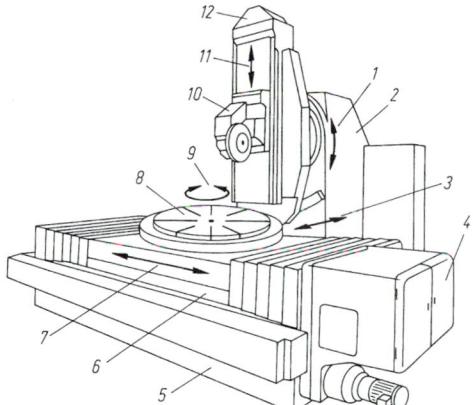

Figure 28. Machine for generating by grinding with indexing, with double-conical grinding wheel: *1* helix angle adjustment, *2* stand, *3* infeed, *4* change gear box, *5* machine bed, *6* workpiece slide, *7* table feed, *8* workpiece table, *9* table rotation, *10* grinding slide, *11* tool travel, *12* grinding slide support.

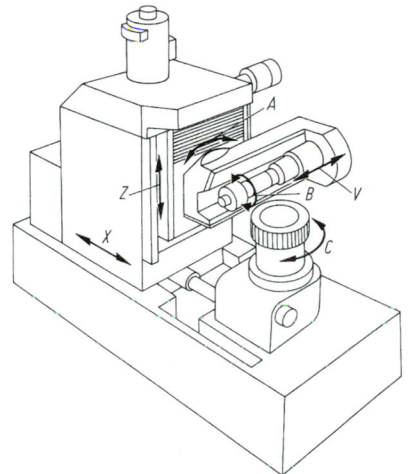

Figure 29. Axes of motion of a continuous-type machine for generating by grinding: *X* radial feed, *Z* axial feed, *V* tangential feed, *A* pivoting angle of tool, *B* tool rotation, *C* workpiece rotation.

feeding), tooth flank corrections can be made in the direction of tooth height and width with the aid of a profile which is variable over the length of the involution. Continuous generating by grinding is used in large-scale series manufacturing.

5.2.2 Cutting of Worms

Principles

Four blank shapes of cylindrical worms are standardised according to DIN 3975 (**Fig. 30**):

Flank Shape A. Trapezoidal tool cutting edges lie in the axial plane. The axial profile of the worm is trapezoidal and has straight-sided flanks.

Flank Shape N. Trapezoidal tool cutting edges lie in the normal plane. The normal profile of the worm is trapezoidal and has straight-sided flanks.

Flank Shape K. The axial profile of the disc-shaped, tapered rotary tool lies in the normal plane. Owing to the three-dimensional line of contact between the tool and worm flanks, the axial profile of the tool is not reproduced in the normal plane of the worm; in the case of a tool profile with straight-sided flanks, therefore, the normal profile of the worm is convex.

Flank Shape I. This corresponds to a helical-toothed cylindrical wheel, with involute profile in the transverse plane. The flanks of the worm are produced by form cutting or generating.

Form Milling and Form Grinding with Disc-shaped Tool

The conditions are the same as in the cutting of helical-toothed cylindrical wheels. If a tapered tool profile with straight-sided flanks is used, only flank shape K is possible (**Fig. 31**). The other flank shapes are possible if the tool profile takes into account the conditions of contact with the worm flank. With a flat tool, flank shape I is possible if the tool axis is pivoted in the normal plane of the worm and tilted through the generating angle.

Form Turning

To produce flank shape A or N, the trapezoidal turning tool cutting tips are guided in the axial or normal plane of the worm with an axial motion linked to the rotation of the worm (generation of a spiral motion in the coordinate system of the workpiece) (**Fig. 32**). Flank shape I is produced if trapezoidal turning tool cutting tips lie in a plane which is tangential to the base cylinder of the involute worm. Flank shape N can also be approximated with a small-diameter end milling cutter or side-and-face milling cutter with straight-sided flanks.

Hobbing and Hob Peeling

The conditions are the same as in hobbing of helical-toothed cylindrical gears. The flank shape is I. The worm blank, instead of the hob, and the tool (peeling wheel), instead of the gear blank, are clamped into the generating machine. The cylindrical worm performs a tangential motion. Worm rotation, tangential motion and peeling wheel rotation are kinematically linked.

To produce an enveloping worm, the peeling wheel is fed in the radial direction of the hobbing machine. Flank shapes A or I are possible. The enveloping worm is matched to the curvature of the circumference of the worm gear. During manufacture, the tool cutting edge must rotate about the centre of the worm gear in kinematic linkage with the rotary motion of the worm gear. **Figure 33** illustrates the motion relationships.

5.2.3 Cutting of Worm Gears

Principles

The flank of the worm gear is a helical surface; the basic body is *globoid* in shape. The flanks are produced by generating the enveloping body of the tool corresponding to the worm with which the worm gear is to be paired.

Radial Method

A radially fed cylindrical hob plunges into the worm gear until the centre-to-centre distance between the worm and the worm gear is reached. The effective hob length must cover the profile formation zone of the worm gear. This

Designation	Illustration	Tools and tool incidence for worm cutting
Flank shape A (ZA-type worm)	Axial section A–A — Straight-sided flanks	Turning tool with trapezoidal profile Axial incidence parallel to axis
Flank shape N (ZN-type worm)	Normal section N–N — Straight-sided flanks	Turning tool with trapezoidal profile Normal incidence at right angles to direction of tooth space
Flank shape I (ZI-type worm)	Normal section N–N — Convex	Hob or grinding wheel with straight-sided flank profile

Incidence as for manufacturing of involute-toothed helical gears |
| Flank shape K (ZK-type worm) | Normal section N–N — Convex | Disc cutter or grinding wheels with trapezoidal profile

Normal incidence at right angles to direction of tooth space |

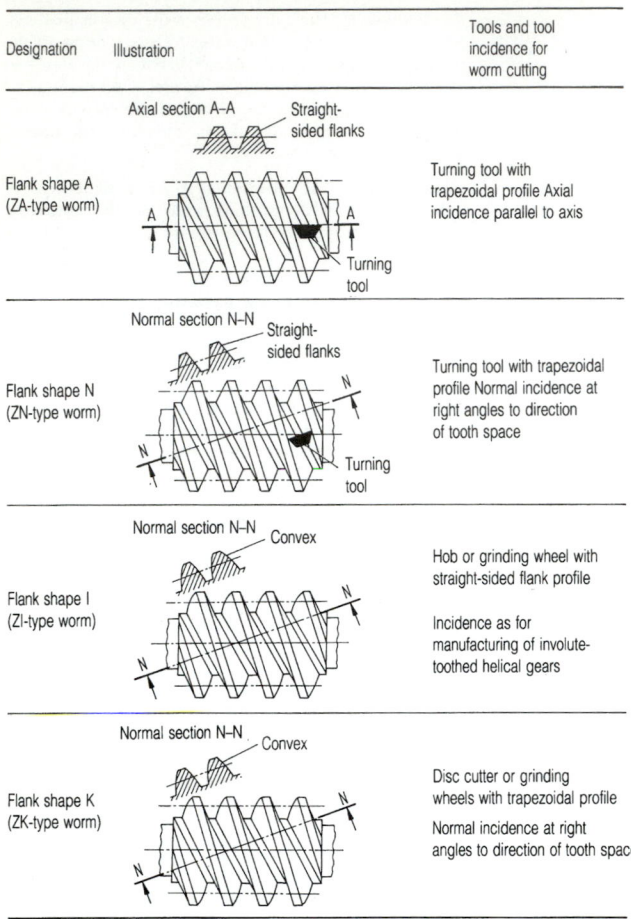

Figure 30. Flank shapes of cylindrical worms, standardised according to DIN 3975.

method is suitable only for worm gears with an 8° lead angle. At larger lead angles, the hob cuts away parts of the flank which belong to the helical surface at the full depth of cut before the final centre-to-centre distance is reached (**Fig. 34a**).

Tangential Method

A hob with a chamfer (i.e. a tapered part of the hob; its purpose is to spread the load on the cutting edge by increasing the addendum of the hob, and to shorten the tangential approach distance) is rolled tangentially past the dedendum circle cylinder of the worm gear with the same centre-to-centre distance as that of the worm and the worm gear. Only one hob tooth need have the full profile. It is therefore possible to manufacture worm gears by the tangential method using a fly cutter (**Figs 34b, c**).

Radial–Tangential Method

This method combines the advantages of the radial method (short feeding distance) and the tangential method (precise flank formation). It comprises radial plunge cutting until the centre-to-centre distance between the worm and the worm gear is reached, followed by tangential feeding. Straight hobs shorter than the worm gear profile formation zone are possible (**Fig. 34d**).

5.2.4 Bevel Gear Cutting

Principles

Bevel gears are used to transmit motion between axes which intersect or cross. The basic bodies are *cones* in the case of gear pairs without axial displacement and *hyperboloids* in the case of axially displaced gears. Any shaft angle is possible; in practice, however, the shaft angle is usually 90°. During manufacture (by generating), both wheels of a bevel gear pair roll with the imaginary generating wheel (crown gear); the tool embodies one tooth flank, one tooth or several teeth of the generating wheel. The teeth are described by the profile in the vertical tooth direction and by flank lines in the longitudinal tooth direction. The gear profile depends on the tool profile and the relative motion between the tool and the gear to be manufactured. Flank lines result from the kinematics of the generating process (straight, arc-shaped, epicycloid-shaped or involute-shaped teeth). The cutting motion is performed in the longitudinal tooth direction.

Shaping Methods

The tool (milling head, planing tool, side-and-face milling cutter, end milling cutter, or grinding wheel) has the profile of the tooth space. The tooth spaces are produced

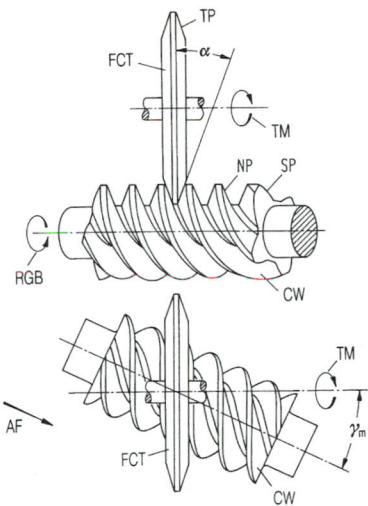

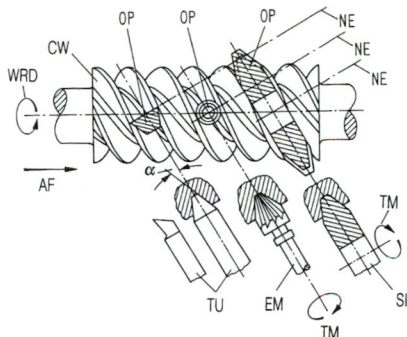

Figure 32. Trapezoidal, straight-sided-flank form-cutting tools (TU turning tool, EM end milling cutter, SI side-and-face milling cutter) in engagement with cylindrical worm CW (DIN 3975): α generating angle, NE normal plane of worm, OP cross-sectional tool profile, TM cutting motion of tool, RGB rotary motion of gear blank, AF axial feed motion of worm.

Figure 31. Disc-shaped, tapered form-cutting tool FCT in engagement with cylindrical worm CW (DIN 3975): α generating angle; γ_m centre pitch angle of worm; TP straight-sided-flank, trapezoidal tool profile, NP slightly concave normal profile of worm, TM cutting motion of tool, RGB rotary motion of gear blank, AF axial feed motion of worm.

individually or continuously. During manufacture, an imaginary tooth of the generating wheel machines the material of the tooth space by means of a straight or curved cutting motion *4* (**Fig. 35**).

Generating Methods

The gear profile is formed as the envelope of the tool cutting edge. The motion *2* of the generating wheel is matched to the rotary motion *1* of the gear blank by a kinematic linkage as if a gear and a mating gear were revolving in a bevel gear train. A tool profile with straight-sided flanks is usually used. Continuous generating is possible with a milling head or a bevel hob (tapered worm). Generating with indexing is carried out using planing tools, side-and-face milling cutters, and disc or cup grinding wheels.

Principle of an NC Bevel Gear Generating Machine

The cutting motion of the tool *D* (**Fig. 36**) and the rotation of the cradle *A* supply speed sensor pulse

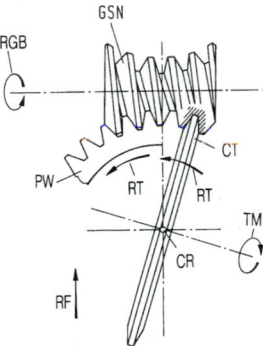

Figure 33. Disc-shaped, tapered form-cutting tool CT and peeling wheel PW in engagement with enveloping worm GSN (after Thomas AK. Zahnradherstellung. Hanser, Munich, 1965). RGB rotary motion of gear blank, RT rotary motion of tool, CR centre of rotation = centre of worm, TM cutting motion of tool, RF radial feed motion.

sequences for the continuous control of the gear blank rotation *B*, to achieve the kinematic linkage predetermined by the indexing and rolling transmissions. During the infeed phase there is no rolling, only plunge feeding *X*. The machine is set up by means of electronic positioning axes: workpiece positioning *Y*, workpiece pivot-

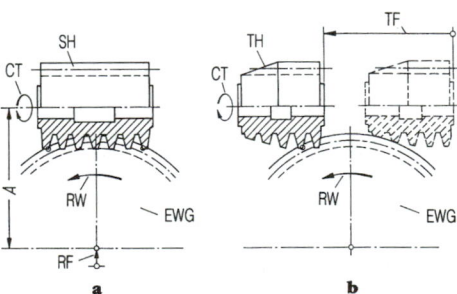

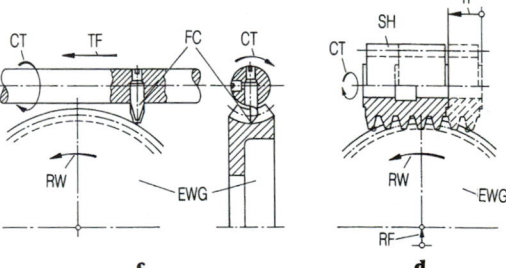

Figure 34. Worm gear cutting methods: **a** radial method, **b, c** tangential methods, **d** radial–tangential method; RW rotary motion of worm gear, CT cutting motion of tool, *A* centre-to-centre distance between worm and worm gear, SH straight hob, TH tapered hob, FC fly cutter, TF tangential feed motion, RF radial feed motion, EWG enveloping worm gear.

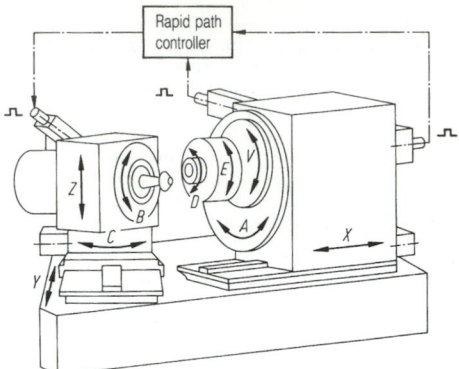

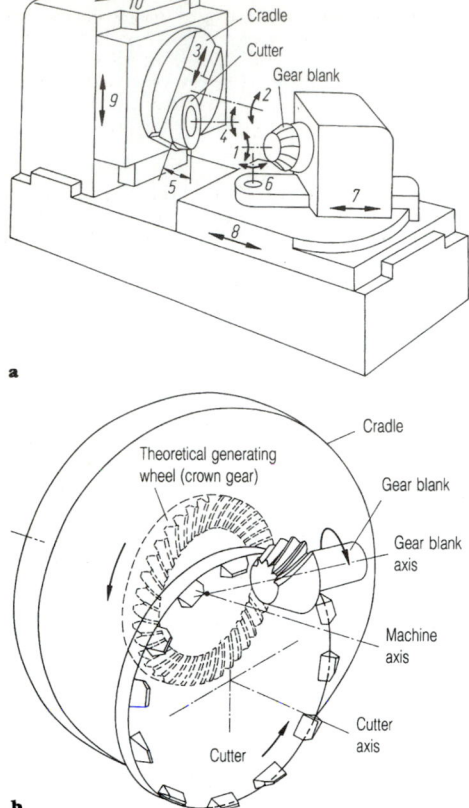

Figure 36. Design of an NC bevel gear hobbing machine (according to Messrs Klingelnberg, Hückeswagen): *B* continuous NC gear blank rotation, *D* cutter rotation, *A* cradle rotation (alternating with feed motion *X*). NC machine adjustment axes: *Y* workpiece positioning axis, *C* workpiece pivoting axis, *V* cutter eccentricity adjustment, *E* cutter positioning axis, *Z* manual axial displacement adjustment.

Figure 35. Basic system configuration for profile milling and hobbing of bevel gears: **a** degrees of freedom of machine adjustment and movement, **b** gear-blank-generating wheel-tool configuration; *1* rotary motion of gear blank, *2* rotary motion of generating wheel (cradle), *3* adjustment of eccentricity of multitooth cutter, *4* cutting motion of cutter, *5* adjustment of angle of inclination of cutter, *6* adjustment of machine axis angle, *7* axial adjustment of gear blank, *8* feed motion, *9* adjustment of axial displacement.

edges. The rolling motion and indexing are the same as in generating by planing (generating method with indexing).

Cyclo-Palloid Process (Klingelnberg). This process is used for manufacturing spiral-tooth bevel gears. The flank lines of the generating wheel are epicycloids (**Fig. 37**). The ratio of the cutting motion *4* of the cutter to the continuous indexing motion of the gear blank (**Fig. 35**) is the same as the ratio of the number of teeth to be cut on the gear blank to the number of starts of the cutter. The rotary motion of the cradle *2* and the additional rotation of the gear blank are performed in the same ratio as the number of teeth to be cut in the gear blank and the number of teeth on the generating wheel. The crowning is generated by an indexed multitooth cutter with various orbit radii.

Spiroflex Process (Oerlikon). This process is used for manufacturing helical-toothed bevel gears. The flank lines and the process of generation are analogous to the cyclo-palloid process, but with no indexed multitooth cutter; instead, the crowning is generated by the inclination of the cutter spindle *5* (**Fig. 35**).

Generating with a Bevel Helical Milling Cutter

Palloid Process (Klingelnberg). This process is used for manufacturing helical-toothed bevel gears. The bevel heli-

ing axis *C*, cutter distance adjustment *V* and cutter positioning axis *E*. There is a manual axial displacement adjustment *Z*.

Bevel Gear Cutting Processes

Generating by Planing. This process is used for manufacturing straight-tooth or helical-tooth bevel gears. One or two planing tools (a straight-sided flank cutting edge may be used) perform a reciprocating cutting motion. The planing slide is mounted in the cradle in place of the cutter (**Fig. 35**). The gear train for linking the cutting motion of the tool with the motion of the table is dispensed with. The rolling motion is produced by rotation of the cradle and balancing rotation of the gear blank. When a tooth space has been completed, indexing is carried out and the cradle returns to its initial position (generating method with indexing).

Generating with a Multitooth Rotating Cutter

Cutting of Curved Teeth (Gleason). This process is used for manufacturing curved-tooth bevel gears. The multitooth cutter has straight-sided flank or spherical cutting

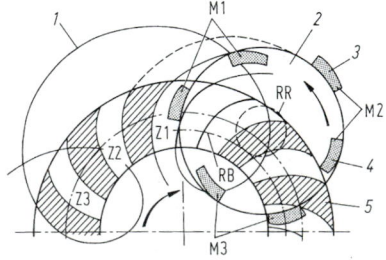

Figure 37. Formation of epicycloid flank line on generating wheel (continuous generating process) from the cutting tip paths by rolling of the rolling circle RR on the base circle RB; cutting tip groups M1, M2, M3, consisting of outer and inner cutting tips, cut successive tooth spaces Z1, Z2, Z3; *1* epicycloid flank line of generating wheel, *2* triple-start multitooth cutter, *3* outer cutting tip, *4* inner cutting tip, *5* generating crown gear.

cal milling cutter replaces the multitooth cutter on the cradle. The flank lines of the generating wheel are involutes (**Fig. 38**) which arise from the kinematic linkage between the cutting motion of the cutter and the rotation of the gear blank. The indexing motion of the gear blank is continuous. The generating feed is accomplished in such a way that starting from position *1*, the cutter plunges into the gear blank as far as position *2* and clears to position *3*.

Grinding of Bevel Gears

This is done to improve the surface finish and to remove the distortion due to hardening and the tooth-cutting errors. Principle: in grinding of straight- and helical-toothed bevel gears, the profile of the grinding wheel (disc wheel) embodies the profile of the imaginary generating tooth. For bevel gears with curved flank lines, the body of the grinding wheel (cup wheel) corresponds to the envelope of the multitooth cutter. The motion relationships are essentially the same as in the bevel gear cutting processes.

5.3 Manufacturing in Precision Engineering and Microtechnology

L. Kiesewetter, Cottbus

5.3.1 Introduction

As miniaturisation progresses, precision engineering has to solve special problems with regard to both design and manufacturing technology. Precision engineering is not a kind of "mechanical engineering in miniature", but has a technical character of its own which derives from the

smallness of the components, the high absolute precision, the signal-orientated mode of operation and the mass production that is typically encountered here. These characteristics necessitate the use of specific functional elements, production processes and highly refined materials. Accordingly, there is often a close interrelationship with other disciplines, especially physics, optics and electronics (**Fig. 39**). Design and manufacturing technology in these fields is concerned with small objects such as instruments, measuring and control system components, data processing devices, clocks and watches, balances, small drives, and even toys. The trend within precision engineering is in the direction of *microtechnology*, meaning components and systems that are manufactured by the production methods of semiconductor technology but which functionally take far greater account of structuring in the direction of the third dimension.

The product range of precision engineering therefore ranges from geometric bodies with very tight tolerances and high-quality surface finishes to mass-produced appliance technology articles and further to microtechnology, which offers maximum precision but works with extremely small dimensions. The associated manufacturing technology has to encompass the areas that lead on the one hand to extremely accurate one-off pieces and on the other to high-precision mass-produced articles. The former culminates in all the processes of *fine machining* [10, 11].

By means of high-precision machines and tools, conventional machining techniques lead to extremely precise surfaces and tight tolerances. In addition, small-size designs have long been used as machines for precision engineers. For the manufacturing process of *turning*, table-top machines and bench lathes without integral stands are available. For large quantities, automatic lathes are used which process bar stock and use coils of wire for extremely large quantities. An analysis of the motions shows that machines of a completely reversed design are feasible (**Fig. 40**), where the workpiece performs translatory motion analogous to feeding and the tools perform the radial and rotational motions in a rigid tool plane. Starting material in coil form is up to 30% cheaper than bar stock.

Modern manufacturing processes of precision engineering, however, are often based on the application of new

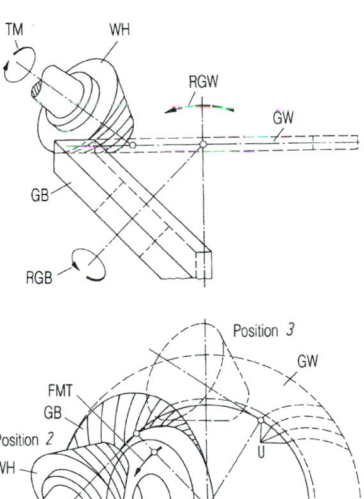
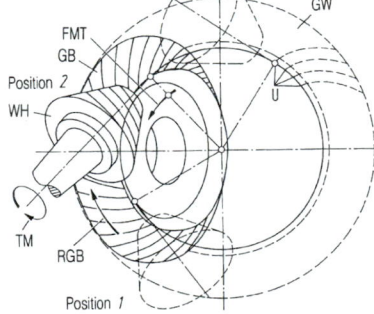

Figure 38. Bevel worm hob WH in engagement with hypoid bevel gear blank GB (palloid process, according to Messrs Klingelnberg): GW generating wheel, TM cutting motion of tool, RGB rotary motion of gear blank, RGW imaginary rotary motion of generating wheel, FMT rolling feed motion of tool, U point of origin of involute.

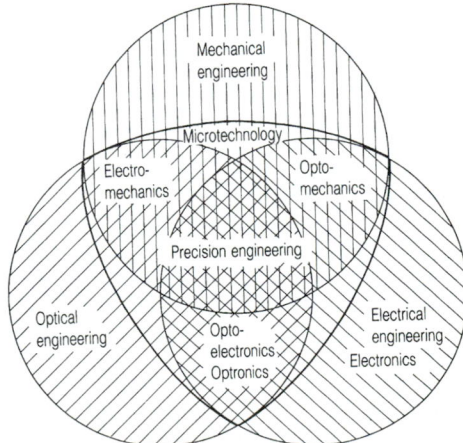

Figure 39. Schematic diagram of the sectors of application of important disciplines within precision engineering.

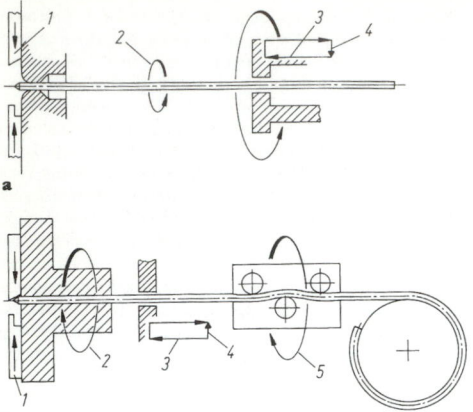

a

b

Figure 40. a Longitudinal turning and **b** turning from the coil: *1* tool, *2* angular velocity of cutting, *3* feed step = component length, *4* tensioning step, *5* angular velocity for straightening.

physical effects [12]. To understand these manufacturing processes, the physical effects on which they are based must be known.

One of the primary tasks in mass production by precision engineering is to optimally adapt the products to economic production.

As up to 70% of the manufacturing costs of a precision engineering product are accounted for by *assembly* and the associated *quality assurance*, in the field of product design special emphasis must be placed on design for *ease of handling*. Comparisons of the cost of alternative solutions must not end at the production of the components, but extend up to the condition in which the component is operational in its final position. For this purpose the *positional probabilities* of the components have to be calculated, the *supply functions* and *supply equipment* selected [13] and the *level of automation* precisely matched to the manufacturing task.

Technological progress is particularly expected in the field of *microminiaturisation*. It is necessary to create a synthesis of innovations in design science and materials technology in conjunction with newly developed and not yet developed unconventional manufacturing methods [14].

5.3.2 Laser Beam Processing

Physical Principles

In 1960 T. H. Maiman (USA) succeeded in achieving an inversion of the population numbers of discrete energy levels with lingering periods of a few milliseconds. The first LASER (acronym for light amplification by stimulated emission of radiation) had been invented.

In contrast to thermal radiation sources, a laser emits amplified and correspondingly intensive, highly monochromatic light of high coherence in space and time. The almost parallel beam of light has the properties of sharp directionality, high coherence length, good focusability down to almost one wavelength and extremely high power densities. These densities are achievable only up to values of 10^{15} W/cm² if the stored energy of the laser is withdrawn in pulses and focused on small focal spots. 10^6 to 10^7 W/cm² is the power density at which most materials evaporate. Thus manufacturing technology appears as the preferred field of application of the laser [12, 15–17].

A laser is a device for producing inversion of population numbers in different energy levels with lingering periods of a few milliseconds in the metastable band. Such systems are pumped with continuous or pulsed light or with a d.c. or a.c. voltage to produce a gas discharge. To emit laser light, precisely defined absorption and emission bands are passed through, which on broad excitation leads to the poor efficiency encountered here and on emission leads to very narrow-frequency spectra and thus to the defined wavelength of the light beam.

Applications. The types of laser most suitable for manufacturing technology applications are *solid-state lasers* and *gas lasers* (**Table 2**) [18]. Of the solids, the most important are ruby (Al_2O_3 as host material doped with Cr^{3+} ions, $\lambda = 0.69\ \mu m$), glass and garnet ($Y_3Al_5O_2$, YAG for short, as host material doped with active Nd ions, $\lambda = 1.06\ \mu m$), while the CO_2 laser occupies the primary position among the gas lasers (CO_2 mixed with N pumping atoms, $\lambda = 10.6\ \mu m$). Both may be operated continuously or in pulses. Recently, so-called *excimer lasers* have come to be used more and more in manufacturing. Excimers are diatomic excited molecules in high-pressure gas which consist of a noble gas and a halogen atom. On disintegration they emit light with a particularly short wavelength of 193 to 248 nm, i.e. in the u.v. range. Therefore

Table 2. Types of laser and their applications in manufacturing technology

Type of laser	Power range in W	Mode of operation	Application
Excimer laser (0.193 μm/ 0.248 μm)	$5 \cdot 10^6$ to $3 \cdot 10^7$	pulsed (15 to 30 ns)	Chipless machining, notching, photochemistry, spectroscopy
He–Ne laser (0.632 μm)	$< 10^0$	continuous	Measuring systems
Ruby laser (0.693 μm)	$1 \cdot 10^4$ to $4 \cdot 10^4$ $1 \cdot 10^2$ to $2 \cdot 10^2$	pulsed (1 to 10 ms) continuous	Chipless machining
Nd–YAG (1.06 μm)	10^6 $1.5 \cdot 10^3$	pulsed (1 to 10 ns) continuous	Chipless machining Joining
CO_2 laser (10.6 μm)	$< 5 \cdot 10^6$ 2 to $2.5 \cdot 10^4$	pulsed (1 to $1 \cdot 10^5$ μs) continuous	Cutting, joining, chipless machining, surface treatment

they are ideally suited to the machining of even more precise dimensions.

For manufacturing technology, the laser beam is often focused with the wavelengths of adapted lens systems and guided to the point of action by means of beam deflection systems or optical fibres.

Being a "tool" that is not subject to wear, lasers are suitable for welding, notching, engraving, sheet cutting, drilling and for altering the properties of various materials such as metal, glass, silicon, diamond, ceramic, plastic, paper and textiles.

At the often low energy levels, high power densities are achievable only with small working ranges. Alongside this, the trend is towards CO_2 lasers with 25-kW beam capacities for applications in machine manufacturing. These have cooled mirrors and aerodynamic output windows to prevent heat losses in the area partitioned off against the low pressure.

Welding. The work is performed in air or under shielding gas, mainly with Nd or CO_2 lasers. The welding equipment is perfected, especially for microwelding; seam and spot welding are performed in the millisecond range [19]. It is especially important to coordinate the geometry, choice of material and manufacturing technology in order to minimise the high reflection losses in the surfaces at the start of machining. In addition, the laser beam must be prevented from reflecting back into itself. Laser beams can also be used through transparent walls, e.g. behind glass. *Microsoldering* is regarded as a process requiring particularly sensitive execution in making microcontacts with high packing density in microtechnology. In this fairly complex process, the increased absorption of laser light on the melting of the solder is sensed from the measurement of the thermal radiation by means of infrared sensors and the laser beam is switched in fractions of 0.1 s [20].

Drilling. Practically all materials can be drilled, even quite hard materials like glass, corundum and diamond, at power densities of 10^7 to 10^8 W/cm^2 and machining times of 10^{-4} to 10^{-6} s [21]. Owing to the beam acoustics, cylindrical holes can be drilled only under limited aspect conditions. The material must vaporise, but the formation of plasma must not screen the laser beam. Because of the small photon mass, the beam penetrates the surface only to a depth of a fraction of a micrometre; thus the material is removed layer by layer.

Sheet Cutting. The CO_2 laser is ideally suited to sheet cutting. Most industrial materials can be cut, often under gases such as inert gas or oxygen due to the high continuous wave power; sheets up to 5 mm thick can be cut at power densities of 10^8 W/cm^2 and at speeds of 6 m/min. Not only steels and metal alloys but also organic materials and ceramics can be cut. The two last-named materials in particular can be effectively notched with the laser by introducing stresses into the workpieces by cutting rows of holes in such a way that on subsequent bending the folds run in the desired directions.

Chipless Machining. The laser beam is used as an accurate material removing tool for trimming, e.g. for tuning of tuning forks and quartz crystals or for adjusting hybrid-connected resistors and capacitors to their setpoints. For resistors, initial tolerances of −10% can be trimmed to 1%, whereas for quartz tuning forks, accuracies of 10^6 are achieved by evaporation of extremely thin gold layer zones. Extremely precise chipless machining of plastics is accomplished with, e.g., pulsed excimer lasers up to 250 W. Polymer laminates are photochemically removed without imposing thermal stresses on the base materials, even in narrow edge zones. Chipless machining in the sense of evaporation of four-component sintered compacts rotating in a vacuum chamber may also be carried out with excimer lasers having a frequency of 30 Hz, a pulse duration of 40 ns and λ = 248 nm. With a 5-min cycle time for evacuating, heating to high temperature, vapour deposition and dismantling, substrates of Sr, Ti, O_3 for computer chips can be coated with superconducting polycrystalline films consisting of yttrium–barium–copper oxide ($YBa_2Cu_3O_7$) which withstand current densities up to 1.5 kA/cm^2.

Coating. Ceramic substrates can be given a structured metal coating by means of the laser chemical vapour deposition (LCVD) technique [22]. This is achieved in a chamber at low pressure with u.v. lasers which release the metal atoms from gaseous metallo-organic compounds by pyrolysis or photolysis in the surface zone by direct programmed inscribing or by means of templates. Metals such as Au, Rn, Pd or Os are used. A further area of emphasis for lasers is *remelt coating*. Here the surfaces of workpieces are completely or locally treated, chiefly with CO_2 lasers at 500 W, in such a way that coating materials previously formed by thermal coating or application of powders, pastes or solids are alloyed with the surface or diffused into it. Stereolithography, perhaps better termed multilayer laser polymerisation in manufacturing technology, appears to be becoming an efficient method of manufacturing three-dimensional prototype shaped parts from polymer materials. Using HeCd lasers, only the top 0.05 to 0.15-mm thick liquid monomer layers are inscribed with laser light and approximately 70% cross-linked. Then the component is lowered in the bath of monomer by the thickness of the layer and the next layer is structured with defined geometrical data. The three-dimensional component is ultimately obtained by post-curing in an oven under u.v. light.

5.3.3 Electron Beam Processing

Physical Principles

In machining of workpieces with electron beam devices, a concentrated beam of highly accelerated electrons is directed at the point of action in a vacuum. It is perhaps surprising that the effect often resembles that of a laser beam, although the rest mass of electrons alone is $3 \cdot 10^5$ times greater than that of a photon and the same energy level is achieved at acceleration voltages of only approximately 2 V. With acceleration voltages of 200 kV in electron beam generators, the *deep welding effect* and the dependence of the manufacturing process on the density of the material to be machined can thus be explained. The beam generator consists of a hot cathode, an anode and the control electrode (Wehnelt cylinder) [23, 24]. The latter focuses and switches the beam intensity up to power densities of 10^9 W/cm^2 in the focal spot. The beam may be shaped on the way to the point of action and guided and deflected without inertia by electrostatic or electromagnetic deflection equipment. The smallest focal spot diameters are less than 1 μm.

Applications. Three types of electron beam machines are distinguished: *high-vacuum*, *semi-vacuum* and *atmospheric machines*. Chamber-type machines, phased machines with turntables and batch-operated through-type machines are used for series production. The controllability of capacity, focusing area and beam direction in conjunction with automatic finding of the point of action by measuring the intensity of revertively controlled electrons permits a broad spectrum of applications, particularly in precision engineering. Extremely precise *spot* and *seam welding* can be carried out under vacuum, as can the *removal* of material by *cutting, drilling, piercing, engraving* and *fusion notching* [25, 26]. The materials used are metals, alloys, ceramic, precious stones and the like. Only a small amount of thermal stress occurs around the point of action and the processing in vacuum maintains a high material purity. The duration of beam positioning and the effective duration of e.g. spot joining or metal removal is a few milliseconds.

One of the main applications of electron beams is vapour deposition in a vacuum (**Fig. 41**) [12]. *Thin films* for opto-electronic components, in semiconductor technology, for film capacitors, large-area coating of window glass and the like are produced by this means. Vapour deposition by means of electron beams shows in this case the advantages of minimal crucible contamination and the possibility of varying the properties of the film by using different deposition materials in several crucibles which have to be individually controlled by the electron beam with regard to intensity and time. Evaporation rates ranging from 1 g/h to 100 kg/h permit high working speeds, even when producing thicker layers of over 10 μm [27, 28].

A non-thermal application of electron beam processing is found in electron beam lithography, where the beam is used to inscribe mask structures in photosensitive layers. This is the most important technique for producing the master masks in IC technology and micromechanics.

5.3.4 Ultrasonic Processing

Physical Principles

Ultrasound is an elasto-mechanical vibration above the hearing threshold. It ranges from 20 kHz to beyond the megahertz range and is generated from electrical energy by means of *piezoelectric* or *magnetostrictive transducers*.

Applications. In manufacturing technology, ultrasound is used for *cleaning* and for *joining processes* such as *welding*, *riveting* and *embedding*, in instrument technology and in medicine. There is no other process that

can achieve reproducibility, 100% cleaning efficiency and machining times ranging from seconds to minutes. This is accomplished using austenitic stainless steel 18/8 troughs in the 20 to 40 kHz range to which are attached transducers leading to a sound field of maximum possible homogeneity. The latter's main effect is cavitation, which at 20 W/l chiefly occurs at contaminated material and induces pressures of over 1000 bar [29-31].

Machining in the defined effective range of the workpieces is accomplished by means of the ultrasonic equipment illustrated in **Fig. 42**. An acoustic head mounted at the vibration node transmits its sympathetic vibration to a booster and, to increase the amplitude, into the effective tool range of the machine by means of a sonotrode. Here amplitudes of 5 to 35 μm are produced at 20 to 40 kHz.

In *welding* of metals, the shearing effect of the surfaces to be joined is exploited, while in *joining* of plastics the compression and tension phases within the thermoplastic materials are made use of. The preferred metal is aluminium, the oxide skin of which performs considerable frictional work until it is fully destroyed. Thus ICs with 27-μm thick aluminium wires are contacted by "wedge-bonding" on a large scale [32].

The phenomenon of ultrasound offers the following advantages: short welding and post-clamping times of

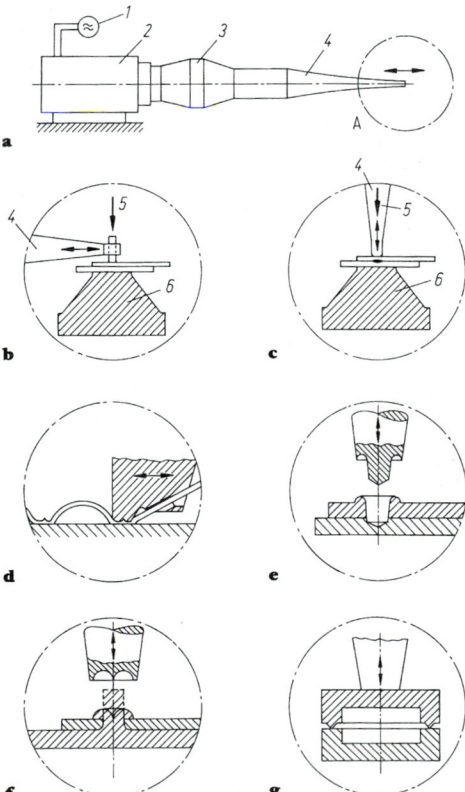

b   **c**

d   **e**

f   **g**

Figure 42. **a** Schematic construction of an ultrasonic device and the most important applications: *1* of the HF generator, *2* transducer, *3* booster, *4* sonotrode, *5* force, *6* anvil. **b** Principle of ultrasonic metal welding. **c** Principle of ultrasonic plastic welding. **d** Principle of ultrasonic wedge bonding. **e** Principle of ultrasonic spot welding. **f** Principle of ultrasonic riveting. **g** Principle of ultrasonic far-field welding.

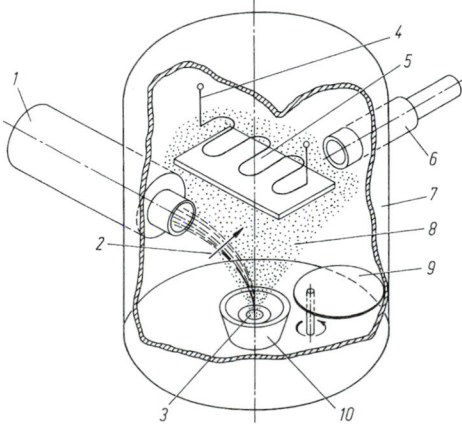

Figure 41. Principle of electron beam vapour deposition: *1* electron gun, *2* magnetic field **B**, *3* evaporated material, *4* substrate heating, *5* substrate, *6* vacuum pump, *7* recipient, *8* vapour stream, *9* evaporator diaphragm, *10* crucible (cooled).

about 1 s, the possibility of joining parts with widely differing wall thicknesses, high strengths, no pretreatment, and no structural changes in the material. With plastics, joining is performed in the near-field range of 6 mm and in the far-field range [33].

Ultrasonic cavity sinking, also termed *ultrasonic drilling*, is based on the machining of hard, brittle materials with an abrasive suspension acting in the effective zone between the workpiece and the end of the sonotrode, which is the tool. Machining can be carried out with a relative tool wear of 1%, and material removal rates of 1200 mm³/min, chiefly in hard, non-conductive materials unsuited to electro-discharge machining, which is normally the alternative manufacturing process [12]. Glass, diamond and materials used in the gem-cutting and semiconductor industries are machined. The feed force in the direction of oscillation must permit the tool to lift off during the decompression phase to create space for the material removed to be washed away and fresh abrasive suspension, such as oxides and carbides, to be supplied.

5.3.5 Electro-discharge Machining, Electrochemical Machining, Metal Etching

In the survey of manufacturing processes according to DIN 8580 (cf. K1), the chipless machining processes (DIN 8590) can be classified according to **Fig. 43**. The most interesting processes in precision engineering are marked with an asterisk; ultrasonic processing is a purely mechanical process, while the effective mechanism of the beam-based processes often lies in thermal effects. For shaping small components, the processes of *electro-discharge machining*, *electrochemical machining* and *metal etching* which are listed in VDI Codes 3400 and 3401 are often also used. What they have in common is that electric current, sometimes with "localised voltaic cell formation", is responsible for their effect. In contrast

to dry processes such as ion-beam etching, these processes take place in liquid active media [34–37].

Electro-discharge Machining. As electric arc machining produces inexact copies, under the heading of *electrical machining*, *electro-discharge machining* is often used, which takes the form of removal of material or migration of material between electrically conductive contacts. According to the definition in VDI Code 3402: "Electrical machining comprises the removal of electroconductive materials caused by electrical discharges between electrodes under a working medium for machining purposes." The electrodes are the shaping tool and the workpiece to be machined. Electrical machining therefore represents the electrical alternative to ultrasonic processing. The polarities of the workpiece and the tool have to be borne in mind deliberately to achieve low relative tool wear. The sparks across a gap are temporary local discharges, the effect of which on the workpiece surface is characterised by the *pinch* and *skin effects*. The machines, which are operated with pulse or relaxation generators, can carry out the processes of *cavity sinking*, *wire EDM*, *grinding* and *sawing* [38].

Electrochemical Machining. This is an electrochemical process in which metal atoms of the anode pass into solution under the influence of a d.c. voltage of about 20 V in aqueous solutions of salts or acids as electrolytes. It is the reverse of electroplating, in which a material migration of one gram-equivalent is caused by 96 487 C. To determine the geometry, the electrolyte is fed through an insulated nozzle at a velocity of up to 30 m/s and achieves very high removal rates at current densities of 250 A/cm². With regard to specific machines and applications, the processes can be divided into *electrochemical etching*, *surface removal* at up to 40 cm³/min and, by analogy and in conjunction with chip-forming processes,

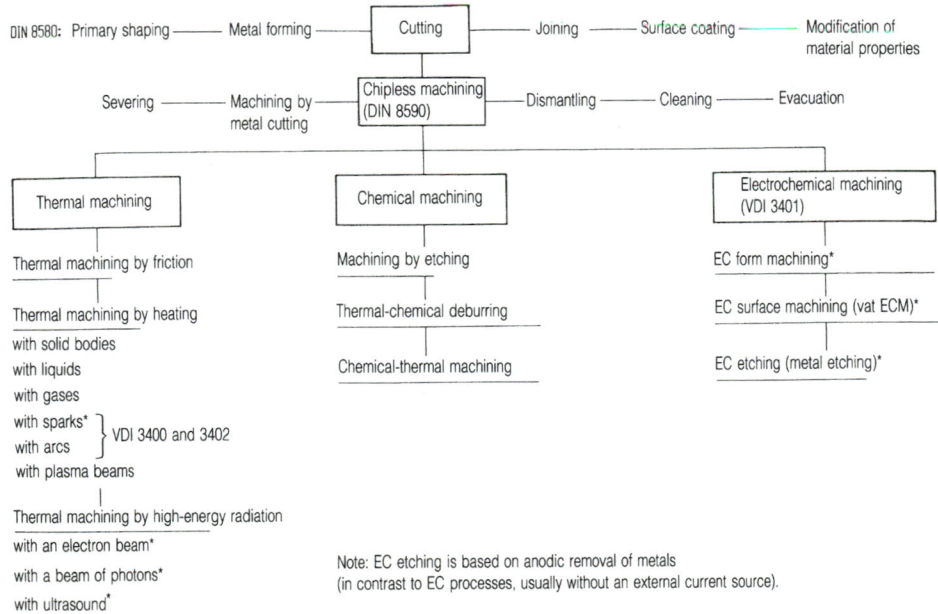

Figure 43. Classification of chipless manufacturing processes according to DIN 8580 and 8590, VDI Codes 3400, 3401 and 3402.

shaping by electrochemical machining (e.g. EC grinding). Thus they are at the same time processes for achieving component geometries and surfaces with roughnesses down to $R_t = 0.5\ \mu m$ with no burr [39].

Metal Etching. This is carried out with an external current source applied to the component, which serves as the anode. However, voltaic cell formation also takes place locally in the electrolyte, e.g. in the HCl or $FeCl_3$ bath in the case of copper components. Other etching solutions include ammonium persulphate, sulphuric acid, nitric acid, hydrofluoric acid, copper chloride and caustic soda. The machining is then carried out by direct reaction of the etching medium with the component material, often with liberation of hydrogen or an oxygen reaction. This process is often used to manufacture complex shapes in the form of foils or sheet metal components or for structuring the strip conductors of printed circuits (foil etching, shape etching). Etching is carried out by *immersion* or with *centrifugal* or *spray etching devices*; etching rates of up to 50 μm are achieved. Defined structures can be produced if a masking layer (etching resist) is applied to the surfaces that are not to be etched. Isotropically acting etchants cause undercutting of the resist, which leads to narrower components or lead geometries. More accurate sheet metal components are therefore obtained with double-sided coating and etching; attention needs to be paid to the overlay. This process is especially efficient with layer thicknesses up to 0.2 mm and with exacting requirements with regard to geometry and freedom from burr or in small production runs.

5.3.6 Coating Processes

Coating is normally used for *decoration* and *protection* of surfaces [12, 40–42]. In precision engineering, however, the layer – particularly if it is structured – often becomes the performer of the *function*, while the coated material becomes the *substrate*. Depending on the requirements, electroconductive, semiconducting, insulating, superconducting, soft and hard magnetic, hard-wearing and self-lubricating surfaces are required. According to DIN 8580, coating is the application of an adherent layer of an amorphous substance to a workpiece. In view of the possibilities offered by manufacturing technology particularly for making coatings, the implantation or burial of coating material must be included here.

With regard to function and manufacture, a distinction is made between *thin* coatings ranging from 0.01 nm to 1 μm and *thick* ones of greater depth. The coating material may be present in any state of aggregation: in the gas phase, the liquid phase (electroplating [43]) and the solid particulate material [44]. An interesting variant of wide-area coating with a monomolecular film thickness is the *Langmuir–Blodgett process*, in which the finely distributed coating material is floating on a liquid and fully and directionally wets the substrate when lifted out of the liquid [45].

All processes, especially for coating of thin-walled substrates, depend on minor internal stresses, which may result from disorder and/or the bringing in of foreign atoms and different coefficients of expansion. Moreover, an important evaluation criterion is the bonding strength of the coating, which is produced by the bonding forces between the coating material and the substrate. In the case of glass or ceramic substrates, the bonding forces to the desired metal coating can be deliberately increased by means of reactive intermediate layers of metal in the nature of layers of bonding agent consisting of Ti or Cr.

Thin coatings are produced by means of the PVD and

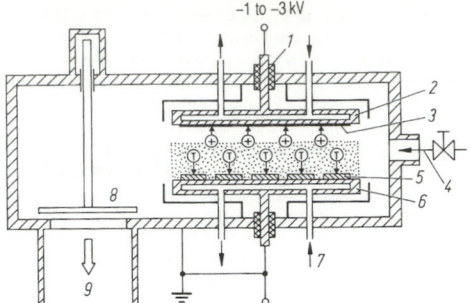

Figure 44. Schematic diagram of a diode sputtering unit: *1* insulation, *2* cathode with magnetic field, *3* target, *4* argon, *5* substrates, *6* holding device, *7* cooling, *8* valve, *9* pump system.

CVD processes. Physical deposition from the *gas phase (PVD)* comprises vacuum metallising, sputtering and ion implantation, together with their reactive variants.

In *vacuum metallising*, coating material is deposited in straight-line propagation from the vaporiser to the substrate in a vacuum chamber according to a cosine distribution law.

Sputtering is a purely mechanical process in which gas ions strike the coating material – the target – in the vacuum chamber and "lever out" the atoms (with maximum effect at an angle of incidence of 45° to 60°), which are accelerated towards the substrate in an a.c. or d.c. field. **Figure 44** illustrates a magnetron system in which free electrons are kept away from the substrate by means of directional magnetic fields. Thus sputtering is accomplished with the substrate at a fairly low temperature.

In *ion implantation*, ions in an electric field are accelerated to such a degree that they penetrate deep into the surface of the substrate and thus alter the properties of the material; it is clear that if the technical conditions are changed slightly, one can very quickly jump between the main categories of DIN 8580 within the processes.

Chemical vapour deposition (CVD) denotes the deposition of coating material from the vapour phase by means of the activation energies in the nature of thermal CVD, plasma CVD, photon CVD and laser-induced CVD. For example, silicon layers may be produced according to the reaction equation $SiH_4 \rightarrow Si + 2H_2$ at a rate of 0.5 μm/min.

Often it is required to coat metals or insulating substrates with plastics. **Table 3** gives a summary of the most

Table 3. Processes for coating metals with plastic

Starting material		
Plastic paint	Plastic powder	Plastic film
Painting	Powder coating	Roller coating
Brushing	Fluidised bed sintering	Calendering
Spraying	Electrostatic fluidised bed sintering	Roller melting process
Dipping	Flame spraying	Extrusion coating
Curtain coating	Electrostatic powder	Film laminating
Centrifuging	coating (flock coating)	

common manufacturing methods. Of these, an interesting process is "resist centrifuging", where, for instance, for the lithography technique, an excess quantity of photosensitive resist is placed with a dispenser in the centre of the substrate to be coated and then centrifuged to produce defined dry film thicknesses in a few seconds, depending on speed and duration. The fields of application are IC technology, microtechnology and liquid crystal technology [46].

5.3.7 Production of Plane Surface Structures

For the products of precision engineering, the feasibility of producing plane surface structures is a determining factor for the high *packing density* to be achieved [12, 47]. A characteristic feature is always the job of structuring a surface or its coating into surface elements or paths in such a way that its properties are fundamentally and clearly different from those of the surroundings. This binary statement may relate to any chemical and physical properties; a simple example is the printed circuit board [48, 49]. It should always be assumed, however, that plane surface structures exist on or within a substrate and the lateral extents of the overall layout are considerably larger than its measurements in the third dimension, height. Regardless of this, there are practical examples where, within a cross-sectional area, the depth dimensions are greater than those of the widths (vertical structure). If DIN 8580 is followed, plane surface structures are produced by coating, chipless machining and changing substance properties. The oldest examples are found in the production of written and printed matter by means of *letterpress*, *gravure*, *flatbed* and *screen printing*, while new methods are encountered in all *lithography* and *moulding processes* for producing video discs, printed circuit boards, thick-film and thin-film circuitry, solid-state circuitry and especially for producing masks for these manufacturing methods.

In *screen printing*, a squeegee is drawn over a gauze stencil, forcing ink or electroconductive paste through the parts of the surface where the mesh is open [50, 51]. By this means, bonding frames 2 to 3 µm thick and 200 µm wide in LC manufacture can be made just as accurately as the printed gold strip conductors on ceramic substrates for manufacturing multilayer circuits on ceramic substrates with alternating strip conductor levels and ceramic insulating layers [52]. In accurately fitting multilayer techniques, special attention must be paid to the problems of the overlay. In *impact printing processes*, the contrast-producing ink is applied to the paper by mechanical pressure. *Thermal printing processes* (so-called *non-impact processes*) apply the structuring material by the action of heat at low mechanical pressure. Complete freedom in the design of the structure, with dependence on the software alone, is achieved in printing with *laser printers*, which produce electrostatic charge images and thus come under the heading of flatbed printing. With the use of light, however, one comes closer to entirely new techniques and the feasibility of producing considerably finer structures still.

The techniques of *photolithography* use extremely precise masks with the scale copy of the structural data in optical beam paths in lithographic coatings of the substrate in the contact method or with an extremely small mask gap (proximity). The principle of photolithography (photoresist technique), as used to produce printed circuit boards up to production of structures in the submicrometre range of silicon technology, is shown in **Fig. 45**. Method **b** can also be carried out as a reciprocal variant, in which the metallised photosensitive coatings are lifted off.

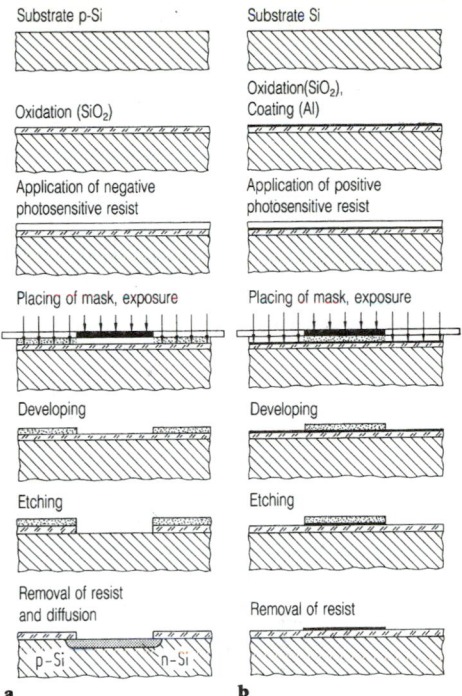

Substrate p-Si   Substrate Si

Oxidation (SiO$_2$)   Oxidation(SiO$_2$), Coating (Al)

Application of negative photosensitive resist   Application of positive photosensitive resist

Placing of mask, exposure   Placing of mask, exposure

Developing   Developing

Etching   Etching

Removal of resist and diffusion   Removal of resist

p-Si   n-Si

a   **b**

Figure 45. Principle of photolithography: **a** selective etching and diffusion with negative resist, **b** selective metal coating with positive resist.

The limits to the photolithographic processes lie in the achievable edge sharpness and the desired fineness of the structures; both these are limited or only achievable at all [53] by the wavelength of the light, any diffraction that occurs, and interference [54]. It appears possible to increase the gradation by means of special photosensitive dyes which are applied to the layer of photosensitive resist to a thickness of 0.3 µm. The satisfaction of the desire for ever finer structures (1-megabit chips) with strip widths in fractions of a micrometre is a challenge and a stimulus to manufacturing technology, for the "packing density" on a substrate is inversely proportional to the square of the structural width. Technical circumstances such as copying errors, thermal expansion and intrinsic stresses due to "retention" of the substrates permit sufficiently precise matching of the mask and the substrate only within limited areas of the surface. Large complete fields can consequently be achieved only by "tacking together" small subfields by the step-and-repeat method. Laser systems can achieve positioning accuracies of 0.1 µm.

X-ray lithography uses a "wavelength window" of 0.2 to 4 nm for exposing special photosensitive resists to produce structure widths of less than 0.5 µm. This process is described in K5.3.8 as a specimen application. In the picture of the wave–particle dualism, the electron beam has an even shorter wavelength ($\lambda < 0.1$ nm) [55].

Electron beam methods are of fundamental importance to lithography for another reason. Whereas in electron beam cathode projection on an image scale of 1 : 1 the electrons are liberated directly from the mask by u.v. light, in electron beam projection the image scale is reduced by means of further optoelectronic systems.

The most interesting variant of this application is the *electron beam inscriber*. It offers the possibility of inscribing directly with the electron beams, with control by software and with an adjustable focal spot size, to produce the extremely precise masks required elsewhere and for "direct inscribing" of structures and substrates for producing prototypes or small runs for product trials.

5.3.8 Manufacturing of Microstructures

In [56] a process for manufacturing quartz tuning forks for watches is described, which can be chemically etched out of an SiO_2 wafer 125 μm thick at 85° in the nature of "wafer batch processing" [57]. The bath of hydrofluoric acid and ammonium fluoride etches the quartz in a highly anisotropic manner, depending on the directions of the crystal axes. This means that with an etching rate of approximately 4 μm/min in the z-direction, an extremely small amount of undercutting can be achieved in the x- and y-directions. Thus, by "leading" in defined axial directions, not only the lateral dimensions of etched components but also the consequences in the third dimension can be predetermined and used for the geometric designs of the microstructures. Microelements and modules composed of them are products of microtechnology, which are themselves made up of elements of *microelectronics*, *micromechanics* and *microoptics*. Microtechnology products therefore include *semiconductor circuits*, *integrated optical and optoelectronic systems*, *sensors* made of *silicon*, *microjets*, *microactors*, *subminiaturised mechanical*, *electrical and optical connections*, and *switches*. All the components are distinguished by extremely small dimensions in the sub-micrometre range, an integrated structure and a system concept which is vitally necessary here. Electronic system components have been produced as integrated circuits for many years, so it was logical, in the spirit of monolithic or hybrid integration, to use the materials and technologies for micromechanical parts as well [14, 58].

In production technology, microtechnology comprises special *coating*, *lithography* and *etching techniques*.

Silicon Technology

A material of supreme importance is the monocrystalline material *silicon*, which is extremely attractive because of its mechanical properties such as minimal attenuation, absence of fatigue, maximum crystal purity, an electrical conductivity that can be determined by doping, etchability and coatability. The most important thing about this material, however, is the possibility of *spatially* exploiting the third dimension, that of *depth*, micromechanically with special *selective* and *anisotropic etchants* in relation to the crystal orientation. This anisotropic and isotropic etching, for the production of extremely fine structures and masks not only with lasers but also with electron beams and X-rays and the erosive methods, and electrocoating processes, justifies the great efforts made in the field of microtechnology [59]. The starting material for silicon processing is a *wafer*, from which many identical elements are produced by batch processing. The designing and structuring of the wafer is accomplished by means of a wide variety of manufacturing processes and equipment which, while mostly familiar from IC technology, have to be specially adapted to the thicknesses in question [60]. The more finely the lateral structures are to be resolved, the more exacting are the demands made on the lithography processes; here, use is made of every type of radiation, such as light, X-rays and particle radiation.

Besides processes employing a focused beam such as

an electron beam, even finer resolutions in the sub-micrometre range are achieved with X-rays, which are formed tangentially to the acceleration sections of electron synchrotrons and are conducted into vacuum tubes. They are emitted as a wide beam from a window covered with plastic sheet with Gaussian intensity distribution at a distance of approximately 10 mm.

Thus lithographic exposures can be achieved through beryllium masks with absorber zones by the proximity method, either directly or, at a greater distance, oscillating together in the vertical direction.

Structuring in the direction of the coating thickness, i.e. to determine the geometry of the components within the silicon wafer, is achieved by means of *additive* and *subtractive techniques*. The former permit the building up of insulating layers, e.g. SiO_2 or Si_3N_4, or doped semiconductor layers, as well as coating with metals such as Al, Al/Si, Al/Si/Cu or organic material and glasses. The technologies for this are epitaxial processes for monocrystalline silicon, chemical deposition from the gaseous phase and condensation of the products of decomposition (CVD), thermal oxidation, or vacuum deposition and sputtering.

Specific *material removal* from silicon and coating structures is achieved by means of the *wet chemical etching processes*, which are isotropic and thus produce identical etching rates in all spatial directions, as well as extensive undercutting. Important influences on the etching arise from the nature and freedom from defects of the etched material, the maskings and the orientation in relation to the crystal axes, the etchants with regard to temperature and age, and the external influences such as cleanliness. The advantages of the dry etching processes are the often higher material removal rate, the excellent structure resolution and an often-observed anisotropy or directional dependence.

Material removal can also be accomplished by *ion bombardment* in low-pressure chambers. One of the main representatives of these equipment categories is a *sputter-etching device*, which is familiar in the inverse mode of operation as a coating device. Here a negative potential is applied to the component to be etched in a low-pressure plasma consisting of a chemically inert gas such as argon. The positive argon ions dislodge molecules and atoms from the substrate, which is partially protected in other places with photosensitive resist. In ion beam milling, a beam of argon ions is formed in a chamber and accelerated to 0.5 to 1 keV, striking the substrate under conditions of high vacuum. By adding a chemically reactive constituent, this process becomes *reactive ion beam etching* with initialising by the ion bombardment. Plastics can chiefly be structured by adding O_2, e.g. webs 1.5 μm wide and 30 μm thick. So far, Si, SiO and Al can be easily processed with etching rates of 0.1 to 1 μm/min and adjustable profiles or gradients of the sidewalls and aspect ratios of 10 : 1. The aspect ratios are always understood as the depth in the direction of processing in relation to the channel or web width of the structure.

Anisotropic Silicon Etching. Silicon has a lattice structure like that of diamond, as shown in **Fig. 46** [14]. With Miller indexing of the orientated single crystals, there are preferred planes which are removed at widely varying rates with anisotropically acting etching solutions such as the alkalies KOH and NaOH or ethylenediamine with catechol and water or hydrazine and water. This anisotropic behaviour is caused by the lattice structure of the crystal and the associated varying bonding forces. Because the energy required to detach a silicon atom is greatest in the direction of the 111 plane, this direction is preferably

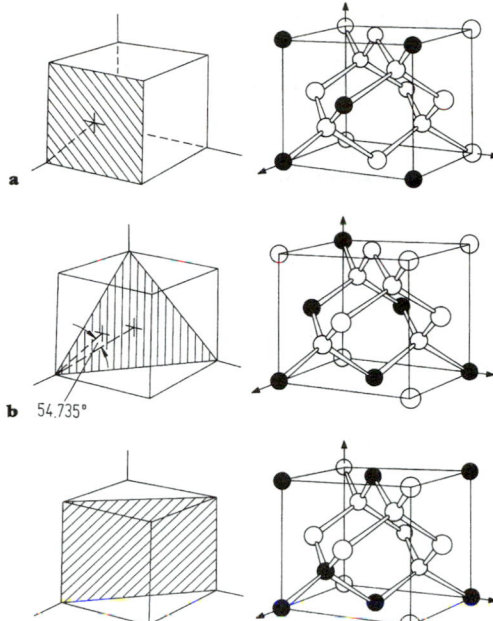

Figure 46. Lattice structure of silicon. **a**, 100 plane; **b**, 111 plane; **c**, 110 plane.

maintained. Here the material removal rates are over a hundred times lower than in the other crystal directions 100 and 110. The production engineer is now faced with the task of placing the wafer surface in relation to the crystal axes in such a way that the etching produces the required geometries. A characteristic angle of 54.735° will very often determine the geometry of the components. This is the angle between the 111 and 100 planes with their widely differing etching rates. For if the surface of the wafer forms the 100 plane, four areas are formed in an etching window which, depending on etching depth and wafer thickness, may run out to a point. Double-sided etching with a precise overlay then leads to double-coni-cal penetrations (**Fig. 47**). At etching rates of 5 to 150 μm/h in the desired etching direction, the processes take a long time. Therefore the masking layers are made of SiO_2 or Si_3N_4. Boron-doped silicon with 10^{20} boron atoms per cubic centimetre is also resistant to such etching solutions. If, however, self-supporting tongues and the like are to be produced, it must be borne in mind that the lattice constant and atomic radius of boron differ from those of silicon, and this doping process leads to internal

stresses – compressive stresses in the case of boron. How-ever, methods of compensating for these internal stresses by doping with, e.g., Ge atoms have become known [61]. As in general the "crystal determines" what is feasible with regard to manufacturing technology, it is extremely important that even within the wafer plane, the mask structures are aligned with the crystal direction. One sol-ution so far has been to determine the direction of the crystal axes precisely on the "flat" of the wafer in aniso-tropic etching tests. For clarification, it should be imagined in this context that with prolonged etching, a window opening of any shape in the wafer always leads to an etching geometry whose shape lies through the tan-gents to the window opening in the direction of the crystal axes. Internal corners form sharp edges, whereas convex structures are undercut (**Fig. 48**). Etched walls pointing vertically into the thickness of the wafer are achieved if they are formed by the 111 plane of the crystal.

Silicon microelements have to fit into a geometrical con-cept for the component periphery. To join multilayer sys-tems made of, e.g., silicon and glass, the anodic bonding method has emerged, in which Pyrex glass is chemically joined to silicon at approximately 300 °C at low pressure and voltage through becoming conductive in the joint gap. By anisotropic etching, three-dimensional micro-elements can be made from a solid silicon wafer 0.5 to 0.8 mm thick by chipless machining.

Other Materials

Production of extremely accurate metal and plastic components of approximately identical thickness by means of a *"buildup technique"* is accomplished with the *"LIGA" process*, derived from the manufacturing steps lith-ography, electroforming and moulding [62].

Figure 49 illustrates the process sequence, in which a resist structure which can be easily modified by radiation physical means is first irradiated with high-intensity paral-lel X-rays for several hours via a mask at approximately 40 μm distance. Depending on the resist material, the irradiated and unirradiated areas are selectively removed by developing, leaving structures with extremely fine

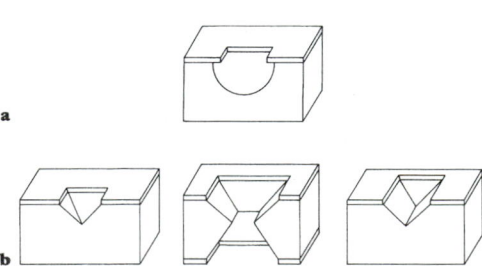

Figure 47. Isotropic and anisotropic etching of silicon.

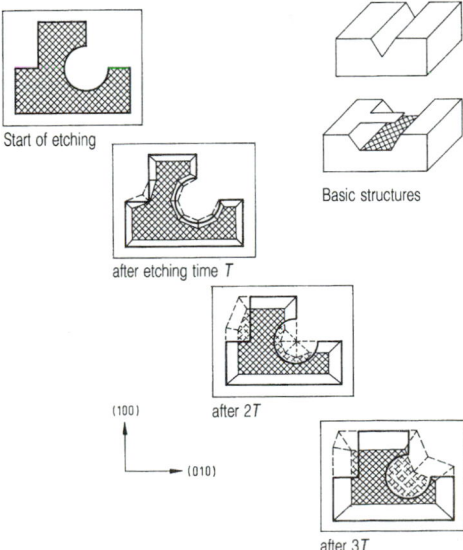

Figure 48. Undercutting at convex corners (Si technology).

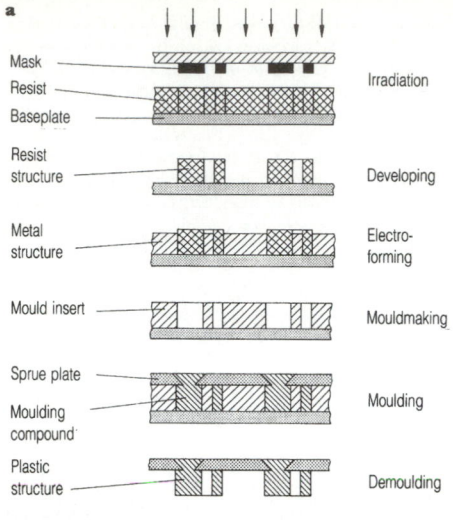

a

Mask

Resist — Irradiation

Baseplate

Resist structure — Developing

Metal structure — Electro-forming

Mould insert — Mouldmaking

Sprue plate —

Moulding compound — Moulding

Plastic structure — Demoulding

b

500 μm

Figure 49. LIGA process: **a** manufacturing steps, **b** plastic honeycomb structure. The wall thickness is 4 μm, the height of the structure 350 μm (Photo: KfK).

resolution owing to the short wavelength of the X-rays. At lateral dimensions of a few micrometres film thicknesses of several hundred micrometres can be produced. The gaps or clearances, now precisely represented, can be filled with metals such as nickel. Identical height is achieved by mechanical finishing. After removing the resist, the result is a metal mould which can be used as a spray-coating mould for an infinite quantity of plastic components. The plastic elements can of course be used in their turn as tools for further electroforming.

While silicon technology can be mainly recognised as a direct method for manufacturing workpieces by wafer technology, the LIGA process is more likely to develop into a method of producing extremely accurate moulding tools.

5.4 Surface Coating

H. K. Tönshoff, Hannover

Surface coating is the application of an *adherent layer* of *amorphous material* to a workpiece (DIN 8580).

The coating and the substrate (base) form a *sandwich* of different materials. This enables a separation of functions: the *coating* assumes contact functions like protection against chemical or corrosive attack of tribological stress, influences the friction behaviour or serves optical or decorative purposes; the *substrate* performs support functions, for which its properties can be adapted to the specific stress regardless of the contact behaviour. This degree of freedom, which is achieved by combining the properties of the coating and the substrate, is the reason for the growing interest in surface coating technology.

Multilayer coatings can be used to achieve further property-specific advantages, e.g. reduction of the coefficient of friction with the top contact layer, followed by diffusion-blocking layers and layers to increase the bonding strength with the substrate.

In principle, three zones are distinguished: the coating zone, the bonding zone joining the coating to the base and the substrate as the shape-giving, supporting body. Coatings are applied to metals, ceramics, single crystals, glasses and plastics. Depending on the material composition and the coating process, the coating and bonding zones take an almost infinite variety of forms (**Table 4**).

According to the state of aggregation of the amorphous substance to be applied, a distinction is made between coating from the *gaseous* or *vapour* state, that from the *liquid* or *powder* (or solid) state, and that from the *ionised* state with film thicknesses from less than 1 μm to more than 100 μm. Coating from the gaseous or vapour state may be achieved by *physical* processes (physical vapour deposition, PVD) or *chemical* processes (chemical vapour deposition, CVD).

PVD Processes. These consist of three phases [63]: evaporation of the coating material, transfer from the source to the substrate, and condensation on the substrate. The gaseous state is achieved by heating – *evaporation* – (emission energy of particles low, < 0.5 eV; transfer vacuum high, 10^{-4} Pa) or by particle bombardment – *atomis-*

Table 4. Examples of coatings

Process	Coating			Application
	Material	Thickness (μm)	Hardness (HV)	
PVD, ion plating	TiN	3 to 8	2300	Drills, milling cutters Cutting tools Metal forming tools
CVD	TiC	7	3500	Indexable inserts
CVD	TiC	4	3500	Antifriction bearings/nuclear engineering
Plasma spraying	Cemented carbide	50 to 300	1600	Nuclear components
Currentless deposition	Ni dispersion	10 to 100	550	Cylinder liners
Electroplating	Cr	10 to 50	900	Piston rods

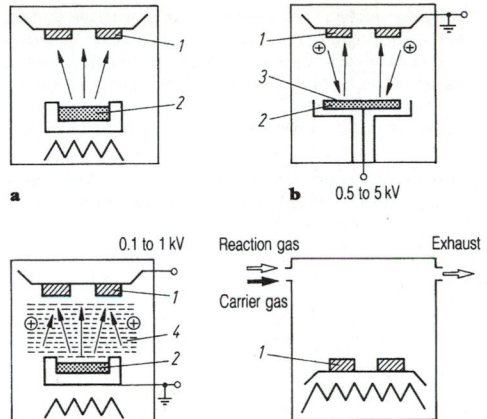

Figure 50. Coating from the vapour phase: **a** evaporation (PVD), **b** sputtering (PVD), **c** ion plating (PVD), **d** chemical vapour deposition (CVD); *1* substrate, *2* coating material, *3* cathode, *4* plasma.

ation (sputtering) – (emission energy high, < 40 eV, transfer vacuum lower, 1 to 10^{-3} Pa) (**Fig. 50**). With vapour deposition, condensation takes place without a major change in the temperature of the substrate, whereas in sputtering a large change in temperature occurs owing to the high kinetic energy of the particles. *Ion plating* combines advantages of vapour deposition and sputtering (**Fig. 50c**). The substrate carries a negative potential; the plasma is formed by glow discharge at a vacuum of 1 to 10^{-1} Pa and a particle energy of 10 to 100 eV. The high impact energy removes foreign coating matter at the same time. In all PVD processes, the process temperature is < 500 °C, with a trend towards lower process temperatures so as not to affect the base material.

CVD Processes (**Fig. 50d**). These are based on chemical reactions of gases. The process temperatures range from above 700 to 1500 °C. Here too the trend is towards lower temperatures. The reaction takes place between the metal joining gas (e.g. $TiCl_4$) and the reactive gas (e.g. CH_4); the substrate (e.g. cemented carbide) may act as a catalyst. A third, inert or reducing, gas performs the transfer of the reaction gases. (In the example, TiC is precipitated [64].) In CVD coating, the energy is supplied by heating of the substrate (by radiation) and, lately, also by plasma discharge or by means of lasers. A controlled laser beam is able to produce coating patterns, permitting local variations in properties.

Coating from the liquid state includes the application of organic coatings by brushing or spray painting, cold enamelling, buildup welding and laser coating. Explosion cladding, roll-bonding and powder spray coating are examples of coating from the solid state. *Powder coating* is used for corrosion protection or optical surface treatment. Thermosetting plastics (based on epoxy polyester and acrylic resin) are applied to workpieces in an electrostatic field and the powder is baked at 150 to 220 °C. In *fluidised bed sintering*, heated workpieces are immersed in fluidised powder (based on polyamide, polyvinyl chloride and polyethylene). The powder melts to form a protective coating, the thickness of which is determined by the immersion time.

In *electroplating*, coating takes place from the ionised state. The coating materials are Cr, Ni, Sn, Zn, Cd, etc. Pure metals or alloys are electrolytically deposited from an aqueous solution (an exception is aluminium, which is deposited from a non-aqueous solution). Metal ions are discharged and deposited at the cathode; at the anode they pass into solution (where the anode is soluble). Deposition takes place according to Faraday's law, $m = kIt$, where m = the deposited mass, I = the current, t = time, k = a material constant. The rate of deposition is 0.2 to 1 µm/min.

6 Assembly

G. Seliger, Berlin

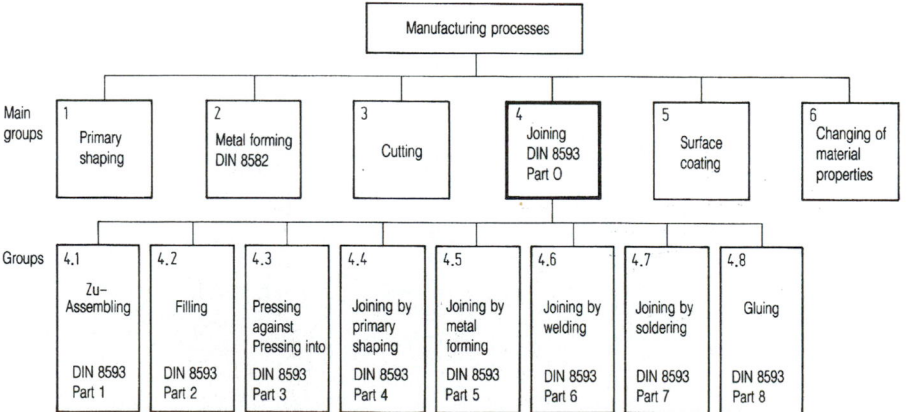

Figure 1. Classification and subdivision of the manufacturing process "joining" according to DIN 8593.

6.1 Definitions

Assembling. Generic term for all operations directed towards the joining together of geometrically defined objects. Amorphous material may be additionally used for this purpose [1–3]. The manufacturing process *joining*, which accomplishes the actual process of creating a connection between two or more parts, should be regarded as the main function of assembly.

Joining. Joining should not be equated with assembling. Although assembling is always carried out by means of joining processes, it includes the subsidiary functions of *materials handling*, *adjusting*, *inspection* and *special operations*. As main group 4 in the overall system of manufacturing processes according to DIN 8580, joining is classified into nine groups (**Fig. 1**) [4].

Materials Handling. According to VDI Code 2860, Sheet 1 (draft), materials handling is the *establishment*,

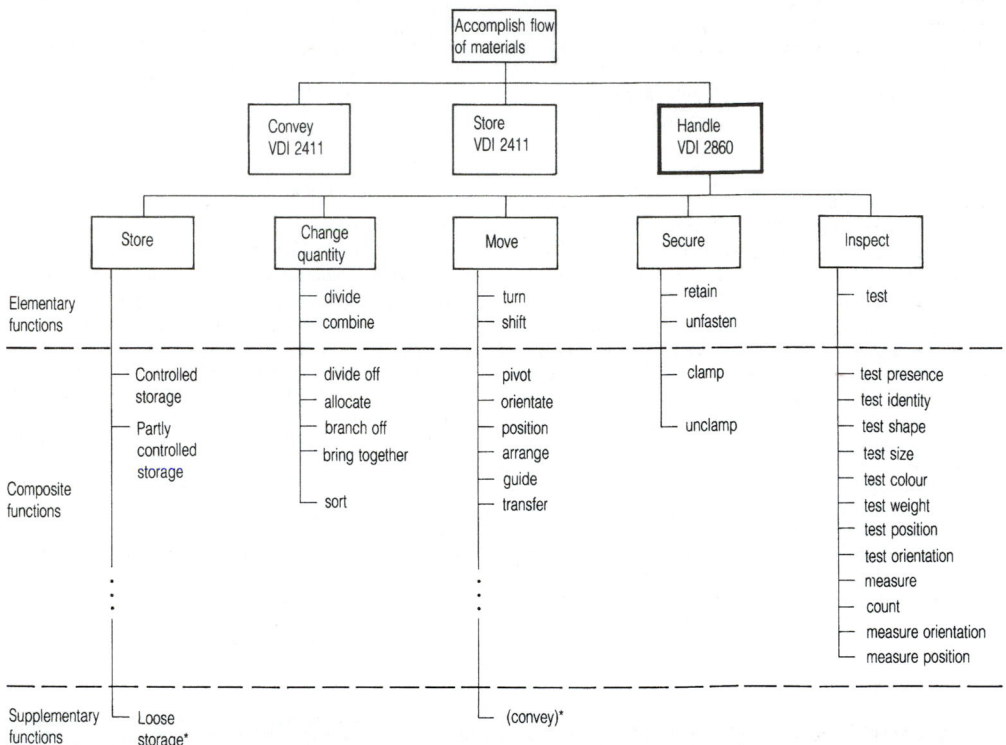

Figure 2. Classification of materials handling according to VDI Code 2860, Sheet 1 (draft): **a** subfunctions; **b** breakdown of handling equipment into groups according to main functions.

defined *changing* or temporary *preservation* of a pre-determined physical location of geometrically defined objects in a system of reference coordinates. The physical location of an object in the system of reference coordinates is defined by its *orientation* and *position*. The orientation of an object is the angular relationship between the axes of the object's own system of coordinates. The position of an object is the place that a specific point on the object occupies in the system of reference coordinates [5]. Materials handling is divided into the following functions (**Fig. 2**):

Storage (keeping of quantities).
Changing of quantities.
Moving (establishing and changing a defined physical location).
Securing (preserving a defined physical location).
Inspection (measuring and examining completed handling operations) [5].

Adjusting. Generic term for all activities that are routinely necessary during or after product assembly in order to *compensate for* unavoidable *deviations* for manufacturing technology reasons with the aim of achieving specified functions, functional accuracies or product characteristics within preset limits [1].

Inspection. Inspection is divided into *measuring* and *testing*. Testing is ascertaining whether specific characteristics or conditions are met. The result is binary in nature, e.g. of the type good/bad or yes/no. Measuring is the term used when characteristics or conditions are ascertained by means of a predetermined reference value. Inspection occurs as a subfunction in all manufacturing sequences and stages [5].

Special Operations. These comprise activities that cannot be assigned directly to one of the above-named functions, but are nevertheless considered essential constituents of assembly. Examples are application of fluxes or securing of nuts with varnish [1, 3].

6.2 Tasks of Assembly

At the interface with development and marketing, assembly as the final stage of the manufacturing process becomes a *logistical* orientation point of works management. In assembly, a technology- and procedure-related coordination of the productive factors takes place. *Technologically*, the ability of the products to function is demonstrated in assembly. *Organisationally*, the elasticity of

production in the face of demand fluctuations in the market is demonstrated in assembly. Product design and production resource planning for ease of assembly harbours a great potential for rationalisation. **Figure 3** illustrates the siting of assembly between market, development, design and manufacturing [6].

Assembly in production is necessary for various reasons, e.g.:

Achievement of function-related mobility.
Combination of various material properties.
Simplification of manufacturing.
Replaceability of wearing parts.
Reduction of manufacturing costs.
Testability.
Increasing the variety of different models.
Weight reduction [7].

6.3 Realisation of Assembly

The Assembly Process

This process is accomplished by the interaction of product-specific, production-resource-specific and cycle-specific variables. The product is described by parts lists and by the geometrical and technological characteristics of the components and modules to be assembled. The cycle is technologically determined by the individual assembly operations and their interrelationships. These can be illustrated graphically with the aid of the priority graph. A priority graph is a critical-path-like depiction of sub-operations of assembly and their sequential relationships (**Fig. 4**). Organisationally, the cycle structure is determined by the *production programme* and the *control of assembly*. Control of assembly refers to the coordination and control of the cycle in order to complete the end-products on schedule and in the specified quantity and quality. The production resources comprise all the function-performers in their interaction in performing the tasks of assembly.

Assembly Planning

The goal of systematic assembly planning is to assist the planner in the individual planning phases from analysis through drafting and design up to the introduction of assembly systems. Information technology tools may be used to model assembly processes in order to increase reliability of planning and productivity.

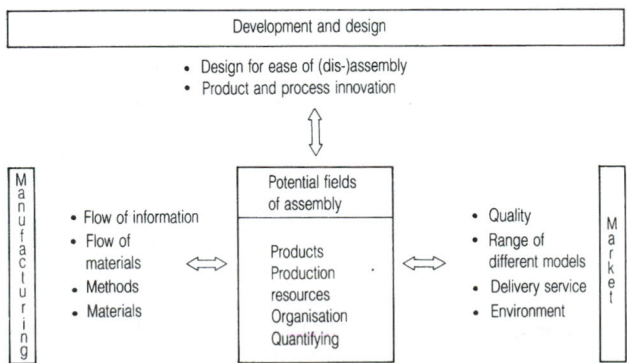

Figure 3. Siting of assembly between the market, development, design and manufacturing.

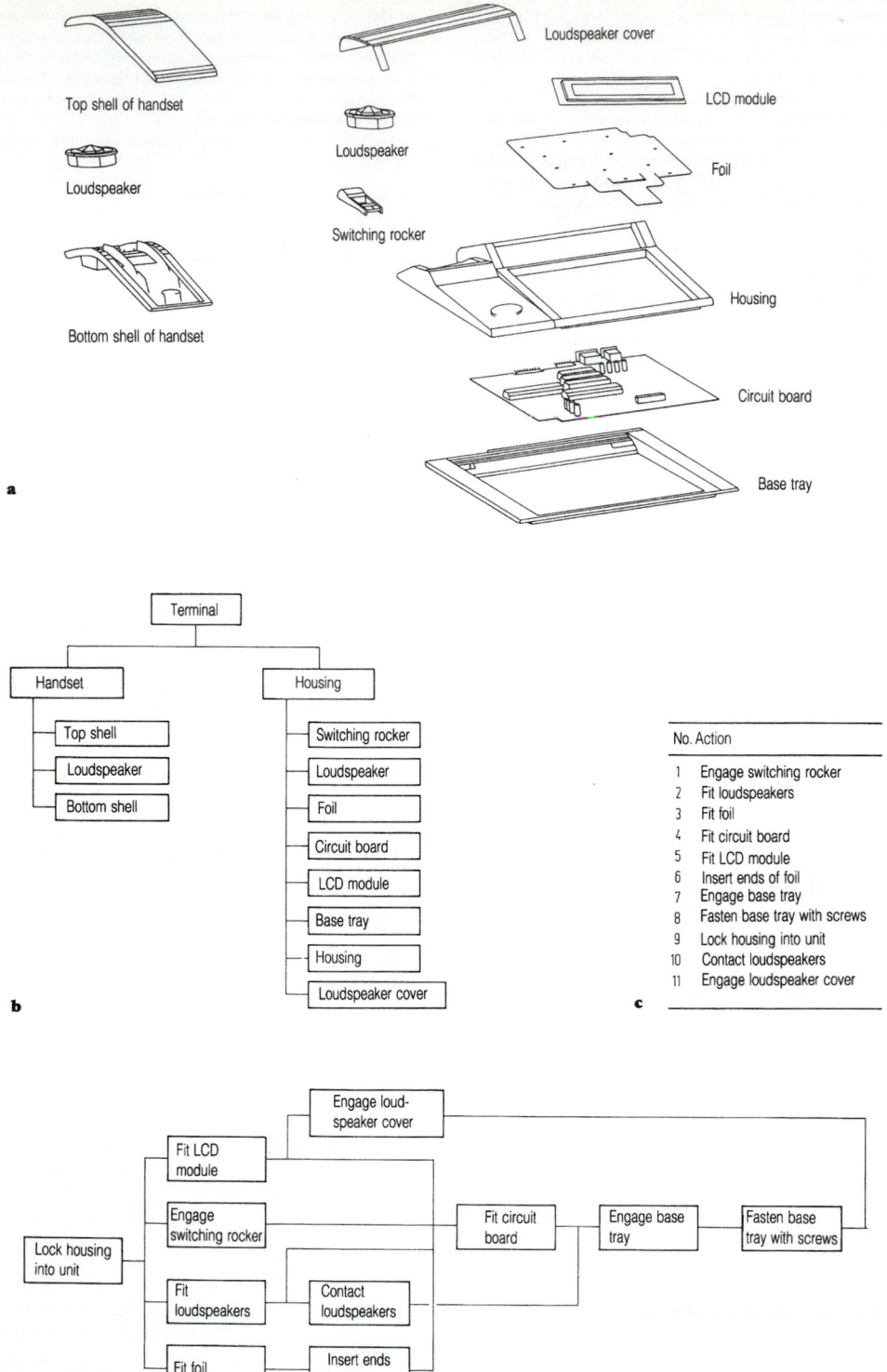

Figure 4. Sample case of an assembly task – communication terminal (telephone): **a** exploded view, **b** parts list of structure, **c** operations for assembling the housing, **d** priority graph for housing assembly.

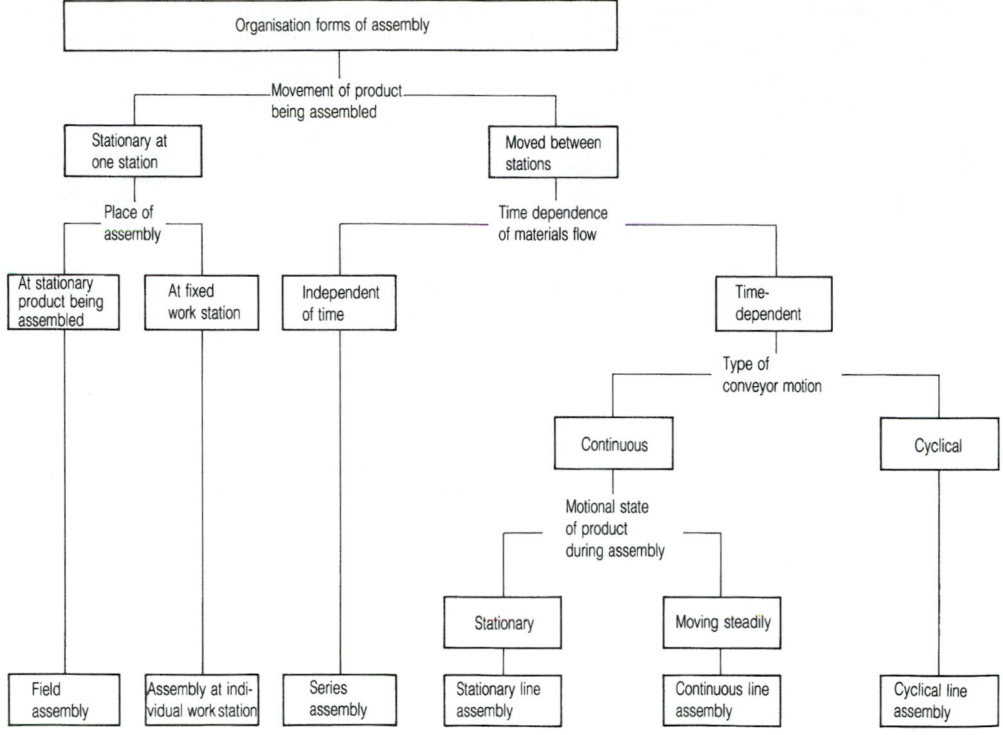

Figure 5. Organisation forms of assembly [1].

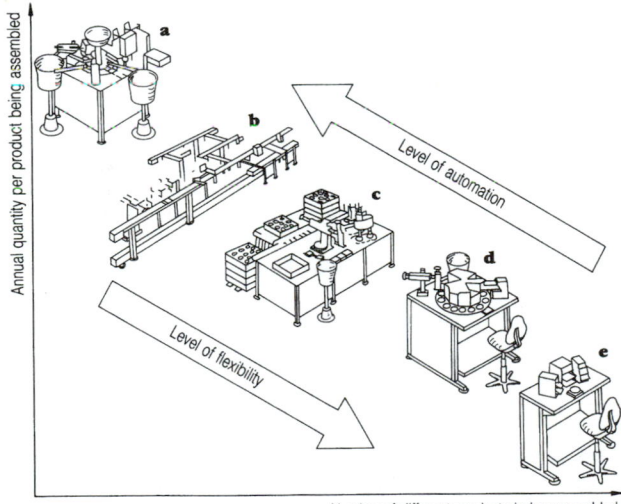

Annual quantity per product being assembled

Level of automation

Level of flexibility

Number of different products being assembled

Figure 6. Fields of application of different assembly equipment [9]: **a** automatic assembly machine, **b** flexibly automated assembly line, **c** flexibly automated assembly station, **d** mechanised individual workstation, **e** manual individual workstation.

Organisational Forms of Assembly

Assembly systems may be classified according to the *movement* of the product being assembled into locally concentrated systems and systems distributed over several stations (**Fig. 5**) [1]. A distinction is made between *quantity-based* and *type-based division*. Quantity-based division is the parallel performance of the same assembly operations, while type-based division is the sequential performance of different assembly operations at the respective capacity points.

Assembly Systems

The variety of the components, their joining behaviour and different tasks of assembly lead to a diverse range of assembly systems [7].

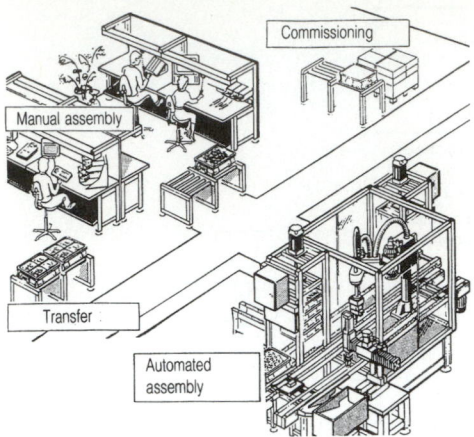

Figure 7. Integrated mechanical and automated assembly (Bosch GmbH).

Depending on the *quantity* to be produced and the *design* of the product, the entire task of assembly is broken down on a quantity or type basis. *Flexibility* and the *level of automation* need to be adjusted according to economic criteria (**Fig. 6**). For the assembly of different products on an assembly system, a low flexibility requirement is desirable. Through product design for ease of

assembly, joining behaviour, joining and handling kinematics, methods of supply, components, modules and joining sequences can be largely standardised for the product range to be assembled.

Automated Assembly

Automation of assembly is intended to increase *efficiency* and *productivity*. Reduction of employee stress and increasing of product quality are also important. Automatic assembly devices are technical equipment items by means of which assembly operations can be automated, either fully or with manual assistance [3]. Automated assembly systems consist of assembly stations, the links between them and the peripherals [2]. The characteristics of automated assembly systems are:

The nature of their structure.
The flexibility achieved with the assembly system.
The extent of the automated areas [1].

For efficient integration of manual and automatic assembly stations, *standardisation of the materials flow* is essential. In the case of physically separated manual and automatic assembly, standard transfer containers are necessary for direct transfer of components without intermediate handling (**Fig. 7**). Scanning elements for position recognition and the use of coding with mobile data storage media or *bar codes* permit automated transfer. The use of standard transfer systems can achieve an integrated flow of materials. Product-specific devices facilitate automated positioning and orientation of the workpieces [8].

7 Production and Works Management

H.-J. Warnecke, Stuttgart

7.1 Job Planning

Job planning comprises the sum total of all measures, including the preparation of all the necessary documents and the provision of production resources, which minimise the cost of manufacturing products by means of planning, control and supervision (definition according to Germany's AWF – *Ausschuss für wirtschaftliche Fertigung*: Committee for Efficient Manufacturing). Job planning is

subdivided into *production planning* and *production control*.

7.1.1 Production Planning

Production planning comprises all the one-off measures. These measures relate to the design of the product, the production engineering, the planning and the provision of the production resources and finish with clearance for production (definition according to the AWF).

Tasks of Production Planning (Table 1)

The main task is the preparation of the *job schedule*. Alongside the drawing and the parts list, the job schedule

Table 1. Scope of production planning

	Advice on manufacturing technology	Methods planning	Materials planning	Operations and time scheduling	Planning of production resources	Cost scheduling
Activities	Advising of design dept. on designing workpieces for ease of manufacture and assembly Checking of drawings	Planning of new methods and procedures Preparation of planning documents Experimental procedure comparisons	Specification of material (unmachined dimensions and shape) Optimising of waste Material storage planning	Preparation of job schedule Product classification Work instructions Design of: working methods workplaces Allowed time system	Production facility Machines Jigs and fixtures Tools Measuring and testing equipment Conveying equipment and storage facilities	Cost of materials Cost of working equipment Labour costs Initial cost accounting Follow-up cost accounting

Quantity ordered	Unit pcs.	Batch size	Date of issue		Processing time		Job schedule no.		
2500	pcs.	500	15.9.64		4 weeks		10 29 13		
Product name					Drawing no.		Type.		Sheet no.
Lid					630 - 310 32		630		1
Material quantity		Unit	Material Lid blank GG-22		Size/model no.		Cost type account		
500		pcs.	Lid blank GG-22		630 - 310 31				
Cost loc.	Work station	Opera-tion no.	Operation		Tool Jig/fixture	Wage group	t_r	t_e	
255	Turret lathe	1	Bore holes dia. 121.5$^{+0.5}$ and dia. 120 Face end Surface		Four-jaw chuck	6	48	7.84	
256	Raboma	2	Drill, debarr and counter-sink four holes dia. 13		Drilling jig dia. 13 drill Countersink dia. 25	5	42	6.0	
258	Horizontal milling machine	3	Mill two surfaces		Milling jig Shell end Milling cutter dia. 50	5	72	3.0	

Figure 1. Job schedule for a manufacturing operation with chip removal, according to Sonnenberg.

is a further basic document in the technical organisation of the firm. The *job schedule data* comprise drawing, parts list and contract data. The information which a job schedule should contain is determined by the tasks to be accomplished in the various divisions of the firm.

Job Schedule. This generally contains the following information (**Fig. 1**).

Title Block Data. Part name, part number, material of construction or raw material, dimensions, batch size range, name of person in charge, date, clearance indication or validity.

Operation-Describing Data. Operation number, cost location, designation of operation, machine number, machine name, necessary tools, fixtures and testing facili-

ties, wage group, preparation time, time per unit, and notes where necessary.

The time data are usually based on the breakdown of the allowed time according to REFA (**Fig. 2**). Allowed times are target times for operations performed by people and production resources. In the case of mainly manual operations, *systems of predetermined times* are often used. These are methods by which times can be determined, with the aid of time tables, for the performance of those operational elements which can be fully influenced by people (e.g. manual assembly). The best-known methods are *MTM* (methods time measurement) and *work factor*.

Application. The job schedule is intended first and foremost as working instructions for manufacturing. However, the job schedule data are also the basis for the following:

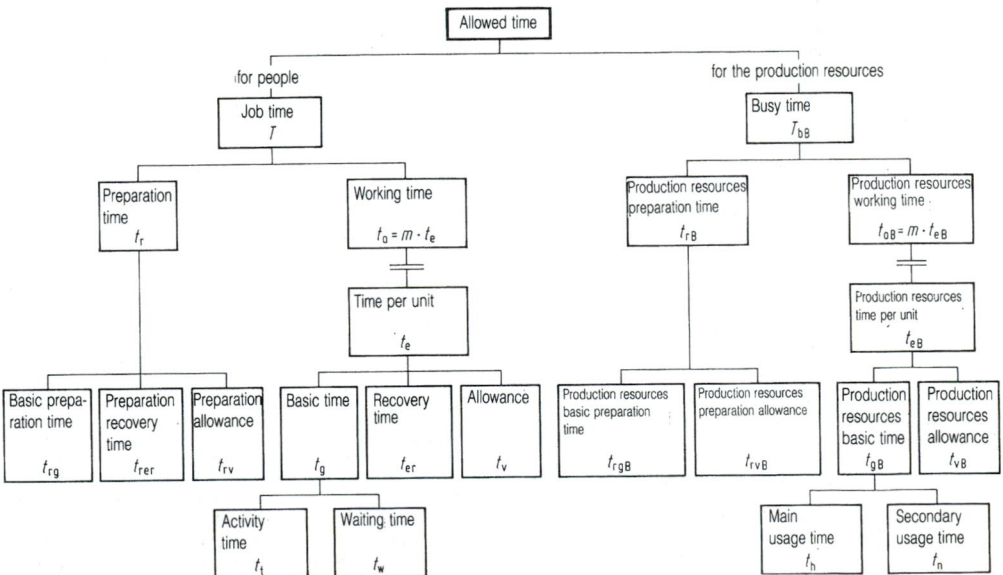

Figure 2. Breakdown of allowed time according to REFA (REFA-Verband für Arbeitsstudien und Betriebsorganisation e.V., Darmstadt); m = number of units.

Time scheduling of operations, determining the capacity requirements of machinery and personnel, materials planning, planning of production resources and procurement.

Preparation of contract documents, routing cards, wage slips, production resources supply lists, materials supply lists.

Initial, interim and follow-up cost accounting, valuation of repeat work and rejects.

Long-term planning tasks, organisation of data management if using EDP.

Computer-Aided Production Planning

With regard to the use of computers, a distinction has to be made between *job schedule management* and *job schedule issuing*.

Job Schedule Management. This expression is used when the job schedule data are ascertained by the planner in the conventional way and recorded on a form. The job schedule can then be input into the computer and stored in a job schedule master file. The stored job schedules may be output at any time, with the addition of up-to-date customer contract data if necessary. The advantage of computerised job schedule management lies in the fact that the stored job schedule data are available as input data for further computer programs, e.g. for time scheduling and expediting as well as materials and time management.

Job Schedule Issuing. In this case, the computer performs some of the activities of the job scheduler (**Fig. 3**). Based on a description of the production task, the job schedule data are calculated by machine and the job schedules prepared by means of a programmed planning logic and the corresponding files. The program systems so far developed for computer-aided job schedule issuing can be traced back to two principal planning methods: the "variant principle" and the "replanning principle" [1, 2].

Variant Principle. Here, for workpieces of the same kind a standard solution is developed in the form of a basic type with an associated job schedule, which embodies the respective individual solutions by means of variations of the basic type within preset limits. An up-to-date job schedule can be prepared by varying geometrical and/or

technological input data by which the working cycle is influenced.

Replanning Principle. This is based on a universally valid analysis of the manufacturing process. Based on a description of the blanks and the finished parts, the job schedule data are calculated by means of an inter-plant planning logic. Alternative solutions may be optimised according to predetermined target criteria (minimum cost, minimum time, or optimum alternative strategy in the event of capacity bottlenecks).

7.1.2 Production Control

Production control comprises the measures required to perform a contract in the sense of production planning (definition according to the AWF). It plans and supervises the flow of the contracts, particularly in the area of production. Its special responsibility lies in efficient capacity utilisation, fixing of key dates and contract fulfilment.

Production control, with its two functions *materials management* and *time management*, is an integral part of the operational or techno-organisational information systems [3]. **Figure 4** illustrates the main functions of such a system, which is particularly suitable for contract flow control. The production process and the associated planning, control and supervisory system form a unit in the form of a control loop.

The large quantities of data to be processed and the necessary high-speed transmission of the control information are increasingly leading to the use of EDP systems in production control. Furthermore, such computerised systems enable complex planning models and methods to be applied which may significantly improve the profitability of the operational processes and thus the firm's trading result [4].

Materials Management

The task of materials management is to *plan*, *control* and *supervise* the materials in the form of modules, individual parts, raw materials and consumables. It follows from this that the most important goal of materials management is to ensure high availability of the materials required for component production and for assembly. Under the aspect of cost minimisation, further objectives are low capital tie-up by virtue of low warehouse and short-term stocks, low planning and procurement costs, and high machine capacity utilisation due to coordinated provision of materials.

The stated objectives are partly conflicting. The desired direction must be established as part of the stock-keeping policy of the firm. **Figure 5** shows the various subtasks of materials management, on which the following comments may be made [5].

Stock-Keeping Policy. Establishment of guidelines for the level of availability of the material and in-process stocks, the maximum level of capital tie-up and frequency of ordering; planning of minimum and safety stocks for every stock item; and coordination of procurement and production.

Demand Calculation

Gross Demand Calculation. Calculation of the gross demand for modules, components, raw materials (secondary demand) and consumables (tertiary demand) on a quantity and due date basis by derivation from the customer orders received and/or from the current production programme (primary demand), or by extrapolating or estimating the demand trend based on past demand.

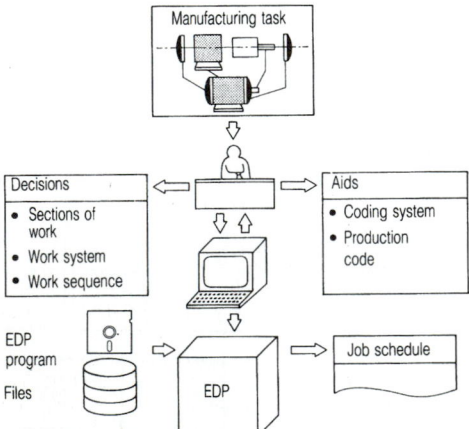

Figure 3. Sequence of computer-assisted job schedule issuing.

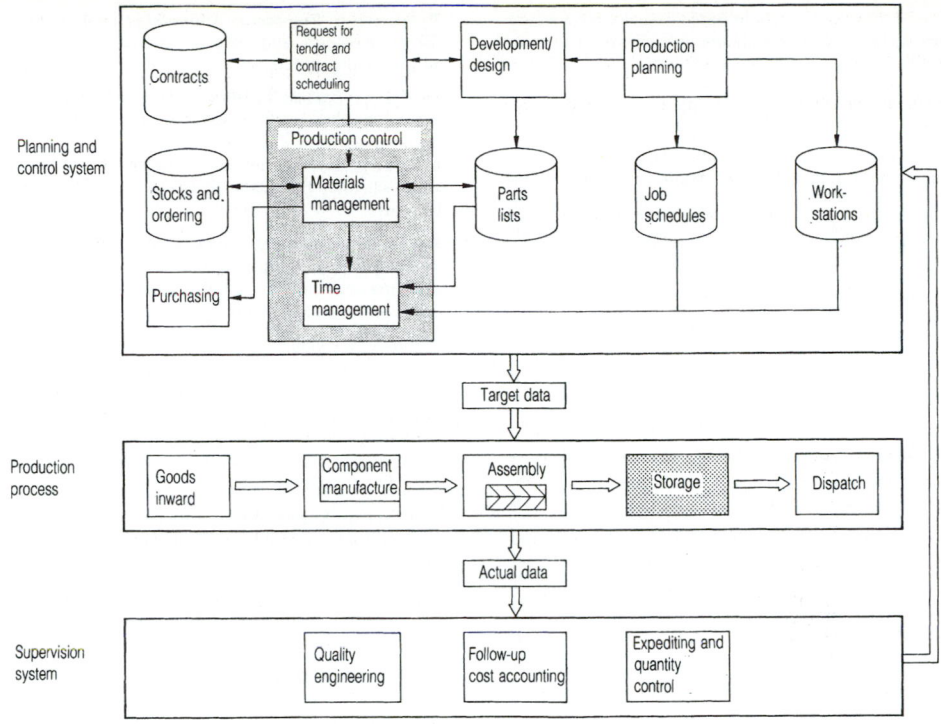

Figure 4. Structure of operational information systems [5].

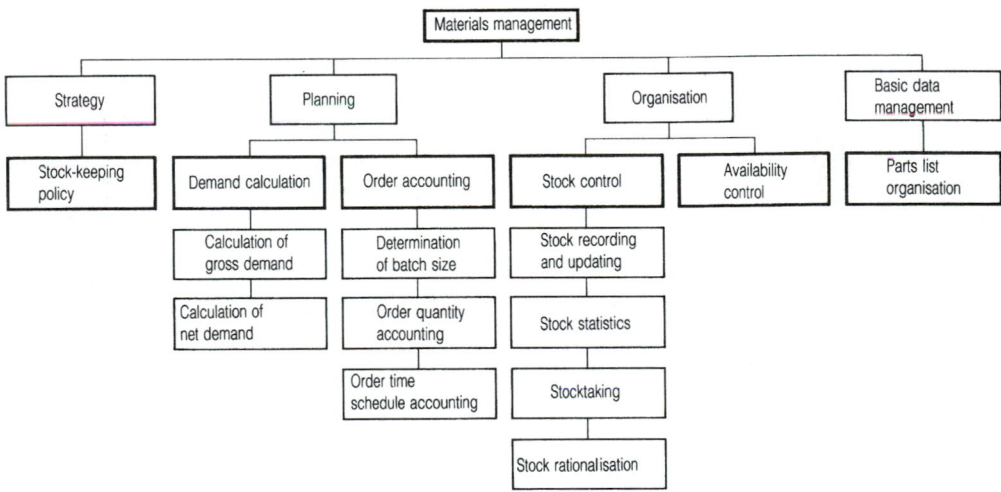

Figure 5. Scope of materials management.

Net Demand Calculation. Calculation of the net demand by offsetting the gross demand against the corresponding stock level, and possibly against workshop, order and reserved stocks.

Order Calculation

Batch Size Determination. Calculation of economic production run quantities (optimum order and production batch sizes), taking into account preparation and storage costs, the production rate and similar factors.

Order Quantity Calculation. Establishment of the quantities to be produced on the basis of the calculated optimum production run quantity and the forecast future demand situation.

Calculation of Order Dates. Calculation of the order dates based on the dates when the respective goods are required, taking into account processing and replacement times and the calculated optimum production run quantities.

Stock Management

Stock Recording and Updating. Recording and accounting of additions to and withdrawals from stock in terms of quantity and value, separately for different types of stock.

Stock Statistics. Records and statistical analyses of stocks and consumption, e.g. according to types of material and components as well as products.

Stocktaking. Recording of the actual stock level and comparison with the stock accounts, making corrections if necessary.

Stock Rationalisation. Checking of stock levels in respect of items with an above-average storage time ("slow movers"), possibly initiating their scrapping.

Availability Control. Checking whether the available stock levels satisfy the requirements for the planned production and assembly jobs.

Organisation of Parts Lists. Management of the product structure data and preparation of parts lists and component use records of various kinds.

Time Management

The field of time management encompasses the planning, control and supervision of all the firm's manufacturing operations. It consists essentially of the chronological assignment of production orders to machines or workstations. This task is characterised by the goal of on-schedule completion of products at the lowest possible cost.

From this the following subgoals of time management, some of which coincide with those of materials management, can be derived: high utilisation of the available capacity (production resources and manpower), short processing times for production orders and low capital tie-up in the firm's current assets.

The goals of "high capacity utilisation" and "short processing time" conflict ("operations planning dilemma"). **Figure 6** shows the subtasks of time management [6].

Job Scheduling Strategy. Establishment of priority rules according to the goals to be achieved as a matter of priority.

Scheduling of Processing. Calculation of the starting, interim and completion dates based on the operation and transfer times, without taking the available capacity into account.

Production Sequence Planning. Establishment of the chronological and physical production sequence for all production orders.

Survey of Capacity Utilisation. Periodic totalling of the utilisation values per production capacity (capacity group or individual capacity), which are derived from the scheduling of processing; graphic representation of the utilisation situation per planning period and/or production capacity.

Job Distribution. Allocation of contracts and job documents to the individual capacities; establishment of the definitive starting date for the work and the definitive order of the individual operations.

Capacity Balancing. Coordination of the orders with processing schedules (capacity demand) with the capacity actually available (capacity supply) by deferring entire orders or individual operations (physical capacity balancing allocates alternative capacity to the order or operation, while in chronological balancing new production deadlines are established).

Production Supervision. Supervision of the measures taken by means of place-specific and time-specific expediting.

Sequence Planning. Establishment of the sequence in which orders waiting for production capacity are to be processed (assignment of job and operation priorities).

Job Schedule Organisation. Management of the job schedules by amendment, deletion and addition of job schedule data.

7.2 Manufacturing Systems

7.2.1 The System "Manufacturing"

The performance of a manufacturing task requires various coordinated operations which are carried out by subsystems of manufacturing. **Figure 7** shows the functional structure of manufacturing and the linking of the dynamic subsystems by the flow of materials, energy and information. The subsystems, which are installed singly or in multiple, perform the following subfunctions (by analogy with [7]):

Work System. Changing of geometrical and/or material characteristics of the workpieces in line with the manufacturing task (e.g. with the aid of a machine tool).

Control System. Processing, transmission and storage of technical and/or organisational information.

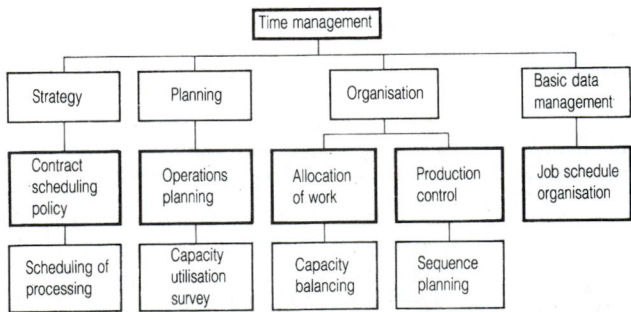

Figure 6. Scope of time management.

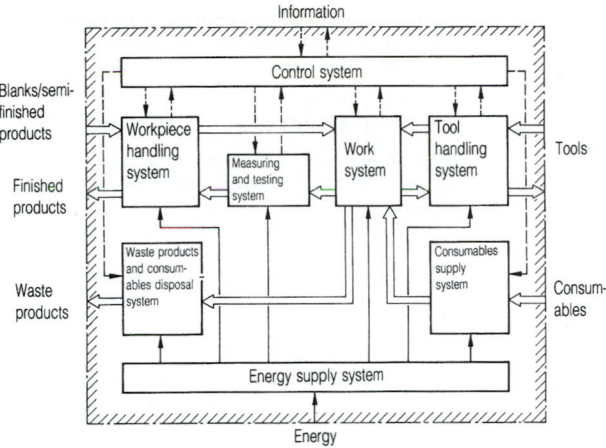

Figure 7. Functional structure of the system "manufacturing".

Energy Supply System. Conversion, transmission and storage of the energy required in all subsystems.

Workpiece Handling System. Storage, supply, positioning, clamping and transfer of the workpieces.

Tool Handling System. Storage, supply, clamping and changing of the tools.

Measuring and Testing System. Comparison of the actual values with predetermined target values.

Consumables Supply System. Supply of the consumables required for the manufacturing process in the work system (e.g. coolant).

Waste and Consumables Disposal System. Removal of the unconsumed consumables an the waste generated during the manufacturing process (e.g. chips).

Because of technological progress, the demands made on people in the system "manufacturing" have changed as follows:

Relief of physical stress on human beings through mechanisation of the energy supply system and the work system, partial relief from control of the work system, and mechanisation of transport tasks, and full relief from manual activity and control function in mass production through the use of production resources such as transfer lines; the direct link between people and the work system has been abandoned.

In series manufacturing, this state still has to be achieved (e.g. by using "flexible manufacturing systems").

The trend is in the direction of the "automatic factory", which enables maximum productivity and quality while minimising the direct ties between people and the manufacturing process.

7.2.2 Automation of Materials Handling Functions

For an analytical description of handling processes, these processes are broken down into individual handling functions (in accordance with VDI Code 3239: Supply Functions). Each function can be represented by a symbol and an associated code number. The functions that are important for the use of handling equipment are described by the characteristic functions, the performance of which may require several installed functions (cf. E4.3.10).

The automation of materials handling processes means

that account needs to be taken of the handling characteristics of the workpieces (objects to be handled), the conditions of the respective manufacturing equipment and the technical possibilities of handling equipment, as well as their interdependence [8]. Owing to the wide range of these influences, handling devices are usually tailor-made individual solutions. The associated high cost of development often means that automation is possibly only for frequently recurring handling tasks (large-scale series production or mass production) if standardised handling equipment cannot be used, the further development of which is favoured by the progress achieved in control technology and information processing.

Feeding, delivery, transfer and similar handling functions are performed with insertion devices, programmable handling devices ("industrial robots") and telemanipulators.

Teleoperators. These are remote-controlled manipulators [9] without program control. Control is performed by a human controller, who takes the necessary decisions and initiates the motions. Teleoperators are strength, performance and range boosters for the human handling characteristics. If an appropriate communication system is available, the teleoperator may be set up and work at any desired distance from the human controller. In industry, heavy load manipulators are used where it is desired to relieve people of heavy physical work, but control of the motion sequences must remain in the hands of human operators (e.g. in nuclear engineering, marine engineering, space engineering).

Insertion Devices. These are mechanical handling devices, usually equipped with grabs, which perform predetermined motion sequences according to a fixed program [10]. They work at presses ("iron hands"), on assembly lines, in the packaging industry, etc., i.e. wherever the same handling task is to be performed over a long period.

Industrial Robots. In contrast, these are automatic handling devices equipped with grabs or tools which are designed for industrial use and which are programmable in several axes of motion (**Fig. 8**) [11]. The difference between them and insertion devices lies in their programmability and in their usually more sophisticated kinematics.

Materials handling tasks can normally be automated only with the aid of robots if some of these tasks

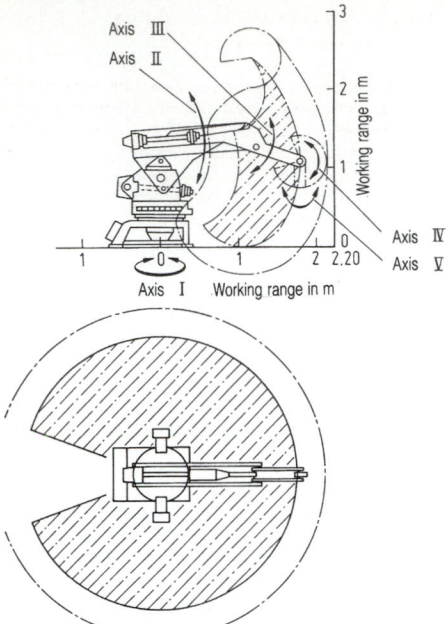

Figure 8. Programmable handling unit – industrial robot (Volkswagenwerk AG). The working space (hatched areas) results from the rotary motion of axis I and the pivoting motion of axes II and III; the working space can be extended by means of the hand axes IV (dot-and-dashed line) and V (rotary motion).

(especially arranging in component manufacture, arranging and positioning in assembly) are taken over by other equipment.

For *arranging*, there are two possibilities.

The required arrangement state is produced by placing every workpiece in a predetermined location and position.

The arrangement is recognised, and the location and position of every workpiece are determined. For this, sensors record specific features of the workpieces, a simple control system processes these data with the aid of a predetermined "internal model", i.e. a program, and derives from this signals for controlling the handling device. In the process, the volume of data is reduced to the degree acceptable for solving the problem at hand, thus achieving simple, rapid processing.

The tasks that are performed today by industrial robots can be divided into workpiece handling and tool handling [12].

Tool Handling

Coating. Painting, enamelling and spray application of adhesive.

Spot welding. Welding of car bodies.

Continuous welding. Guiding of a welding torch (in gas-shielded welding, often with the aid of sensors which find the beginning of the seam and guide the torch in the middle of the welding groove during welding).

Workpiece and Tool Handling

Machining. Deburring, grinding, polishing, water jet cutting, cleaning of castings, laser or plasma cutting.

Assembly. Joining, gluing, screwing, pressing in, joining by diffusion, riveting, soldering, mounting electronic components on printed circuit boards.

Workpiece Handling

Handling at presses, forging machines, pressure die casting and injection moulding machines. Loading and unloading of machine tools or test systems. Palletising, commissioning, interlinking of two or more machines.

7.2.3 Transfer Lines and Automated Production Lines

In automated production, the Taylor principle of division of labour is particularly marked in transfer lines. A transfer line is a production line in which workpieces are passed from machining station to machining station in a cycle (automatic workpiece processing). Transfer lines are designed as special machines for machining workpieces, usually in large quantities, which are very similar in their manufacture. Their typical field of application is the motor vehicle industry. Although transfer lines are single-purpose systems, they are largely assembled from packaged units on the modular construction principle (cf. L5.12). These packaged units are self-contained modules, each of which embodies one or more subfunctions of a machine tool. A distinction is made between basic units, main units and supplementary units [13]. To ensure that units from different manufacturers are interchangeable, the principal main and mating dimensions of the various equipment sizes are standardised in DIN 69 512 *et seq.*

Rigid Interlinking. The stations of a transfer line (**Fig. 9**) are rigidly interlinked. The characteristics of this are: common control of the interlinking device and the machining stations; processing of workpieces on an even cycle prescribed by the slowest work cycle; a fault in one machining station causing the entire production system to shut down.

Loose Interlinking. In comparison, the characteristics of *loose interlinking* are: independent work cycles of the machining stations (no common cycle); greater freedom in installing and positioning the machining stations; a workpiece reservoir as a breakdown buffer with physical and chronological unlocking of the workpieces between the machining stations; the ability of the preceding and following machining stations to remain in operation in the event of a breakdown; and consecutive downtimes not being cumulative provided the capacity of the breakdown buffers is sufficient.

Series machines are often connected by loose interlinking; it also enables manual and automatic machining stations to be uncoupled in production lines. Combinations of loose and rigid interlinking are found especially in automatic production lines with many stations, machining stations with the same technological operations and productivity being rigidly interlinked. In this way the advantages of the direct, short workpiece passage on the one hand and the section-by-section bridging of downtimes by means of buffers are partially combined.

7.2.4 Flexible Manufacturing Systems

A flexible manufacturing system consists of several individual machines that are generally numerically controlled and loosely interlinked and is able, owing to the linkage with regard to material flow and information technology,

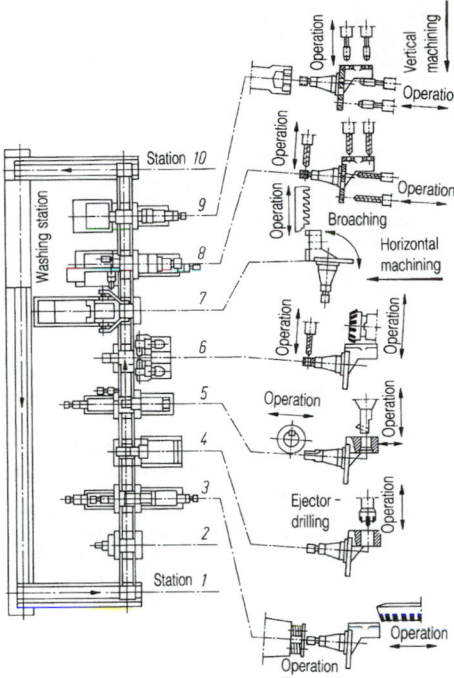

Figure 9. Transfer line for machining steering knuckles; outline and machining sequence (Mauser Schaerer GmbH). The jig carriages are returned to the loading and unloading station (station *1*) by means of inclined hoists; cycle time 0.96 min.

following *functions* or equipment items usually form part of a flexible manufacturing system:

Workpiece changing (by means of handling devices in the case of rotationally symmetrical workpieces, by means of workpiece carriers (pallets) and pallet-changing devices in the case of prismatic workpieces).

Tool changing (changing of individual tools, multiple-spindle drill heads or tool magazines).

Consumable supply and disposal, chip removal system, automatic control of subsystems (NC, CNC), adaptive control.

NC data distribution with a DNC computer (direct numerical control data distribution computer).

Buffering and storage equipment (depending on the workpiece spectrum, machine set-up and type of interlinking), automatic washing and cleaning of workpieces, automatic measurement of workpieces.

Computer control of the entire system including transfer control.

Computer-aided capacity and scheduling calculation (organisational control) by a higher-ranking production control computer.

Online operational data logging.

Automatic fault diagnosis (monitoring).

Figure 10 illustrates the fundamental structure of a flexible manufacturing system. The machinery and equipment control systems perform the processing of NC programs and transfer data and the logging of the operational data. A higher-ranking process control computer performs the tasks of distributing and managing the NC programs (DNC cf. L2), control of the flow of materials and logging of further operational data. The process control computer is linked to a central in-house computer which carries out the tasks of production planning, production control and the management information system (MIS).

to machine workpieces automatically in medium-sized and small batches down to a minimum batch size of one. Various workpieces undergo machining at the same time, passing through the system on various paths. The machining stations within the system may substitute for or complement each other with regard to the installed production functions. Substituting stations have the advantage of high utilisation of the system as regards time, optional workpiece passage and high flexibility of the overall system, as they can each alternately perform the same machining tasks and have the same technological functions and working geometry. Complementary stations only perform machining steps that cannot be performed on other stations because of differing technical functions and working space geometry. Systems with complementary stations have a high technical utilisation, realise the line principle and have high productivity.

The *interlinking* of the machining stations in flexible manufacturing systems may take the following form [14]:

Interlinkage by means of a single transfer vehicle (e.g. a high-bay storage and retrieval unit, or a mobile industrial robot).

Interlinkage by means of several transfer vehicles (e.g. an inductively controlled industrial truck).

Interlinkage by means of a fixed conveyor line (e.g. a friction-driven roller conveyor).

Depending on the level of development of a system, the

7.3 Quality Engineering

7.3.1 Scope of Quality Assurance

Quality engineering is responsible for the lion's share of the quality assurance tasks. These tasks comprise all the organisational and technical activities for ensuring high product quality while taking profitability into account. They can be divided into the following sub-activities in accordance with DIN 55 350 Part 11 and [15]:

Quality Management. Performs the overall management task to establish and implement the quality policy.

Quality Planning. Selects quality features, classifies and weights them, and puts all the individual requirements as to the condition of a product into concrete form, taking into account the feasibility of their realisation.

Quality Testing. Ascertains the extent to which a unit meets the quality requirement. The quality requirement corresponds to specified and assumed requirements imposed on a unit. It may be documented in bills of quantities, specifications, drawings and the like. Quality testing particularly involves the planning and performance of tests.

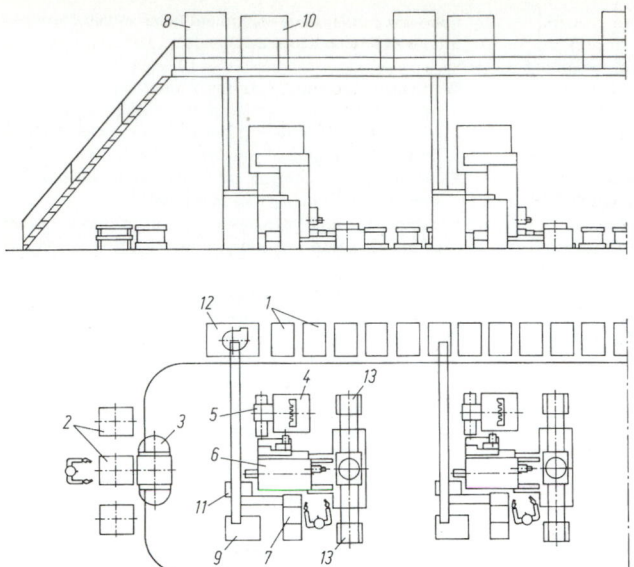

Figure 10. Flexible manufacturing system for non-rotating parts (Burkhardt und Weber Group): *1* pallet store, *2* loading and unloading station, *3* pallet transfer vehicle, *4* high-capacity tool store, *5* tool transfer device, *6* horizontal machining centre (working distances 1250 × 1000 × 800 mm), *7* CNC control, *8* IC adaptor, *9* hydraulic unit, *10* thyristor unit with power pack, *11* cooling unit, *12* washing station, *13* pallet transfer station.

Quality Control. Performs preventive, supervisory and corrective tasks in order to meet the quality requirement. It generally analyses the quality testing and corrects processes.

7.3.2 Quality Systems

Quality systems establish the structural and procedural organisation for performing the quality assurance tasks. It describes the responsibilities, methods, processes and means for implementing the quality management. The individual contributions of activities or processes to quality and planning, implementation and utilisation phases may be represented as a quality circle (**Fig. 11**). Quality systems are normally documented in quality assurance manuals. The requirements for proof applying to such systems are standardised in DIN ISO 9001 to 9003.

7.3.3 Methods and Procedures

To perform the quality assurance tasks, numerous methods have been developed and applied in industry in recent years. Some of the most important are:

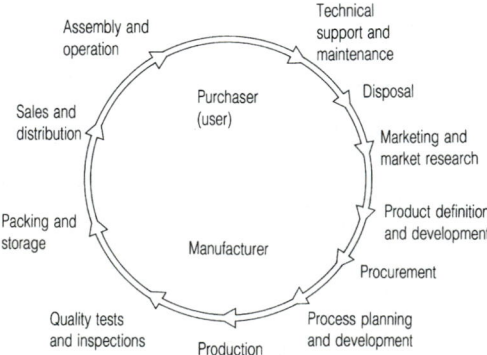

Figure 11. Quality circle.

FMEA (Failure Mode and Effect Analysis). FMEA [16] is a method of estimating, at the planning stage, the risk of failure of a product component, a process stage or a system. Failure modes and their causes and effects are mostly determined, systematically recorded and valued in teamwork. If the valuation index, or the risk priority number, exceeds an acceptable maximum, then measures are drawn up to prevent or detect the potential failure or its cause.

Test Planning [15]. For planning the quality tests, testing characteristics are selected and the testing systems, the frequency, and the method and location for their supervision are established. Planning of the testing frequency is carried out by statistical methods. As part of this, random sampling systems are used to evaluate a batch (cf. DIN 40 080).

CAQ (Computer-Aided Quality Control). A high proportion of the quality assurance tasks for planning and performing tests can be assisted with CAQ systems. Standard functions of such systems are the preparation of testing schedules, testing order management, test data logging and analysis of, e.g. measurements, failures, quality costs. A CAQ system, as a subsystem in the entire spectrum of operational systems, should at least be combined with CAD (drawing data) and PPS (contract data) systems.

SPC (Statistical Process Control) [17]. Mathematical statistical methods are an essential element of quality assurance in manufacturing technology (cf. *test planning*). Direct control of the quality of the manufacturing process is achieved by means of quality control cards for variable and attributive characteristics. Observation of statistical variables such as mean, span or standard deviation enables the capabilities of machines and processes to be assessed. The computer systems used for this purpose also permit automatic data logging and control of measuring operations.

7.3.4 Testing Systems

In mechanical engineering, it is mainly testing systems for geometrical length testing that are used. A distinction has to be made between representations of measurements (gauge blocks, gauges, yardsticks) and measuring means (measuring tools, measuring instruments and measuring devices). As the level of automation in manufacturing technology increases, computerised measuring instruments are being used. Flexible-use CNC-controlled coordinate measuring instruments [18] with mechanical or optical data recorders are especially widespread. The use of a computer enables the control programs to be written away from the machine. This increases the main utilisation time of the capital-intensive appliances. The facility to generate the control program straight from CAD data rationalises programming and avoids errors in transferring data caused by the manual input of data taken from the drawing.

7.4 Operational Costing

7.4.1 Fundamentals of Operational Costing

The operational accounting organisation has the task of recording and monitoring all the operations of procurement, production, sales and financing in terms of quantity and value. It is institutionally broken down into financial accounting, costing, statistics and budgeting.

Costs are the consumption in value terms of goods and services for the production and sale of operational outputs and for maintaining the necessary operational readiness [19]. The purpose of costing is to keep a check on the profitability of the output production process by recording, distributing and allocating the costs incurred in performing the firm's objectives.

In detail, costing is the basis for [20] *cost accounting* (bid price, price limit), *manufacturing control* (comparison of costs and profits, comparison of budgeted and actual costs), *operational planning* and *operational policy*.

Costing as a whole is divided into three areas (**Fig. 12**) [19]:

Accounting by Cost Type. This is used to record the costs in full detail by type.

Cost Location Accounting. This is carried out with the aid of a manufacturing cost sheet (MCS); it distributes the costs that are not directly attributable to the product (overheads) among the cost locations.

Cost Unit Accounting. In the form of time-based cost unit accounting (operating statement), this calculates the profit as the profit per period, while item-based cost unit accounting (cost accounting) determines the cost per product.

The cost type account and the cost location account are accounts for specific periods. The time-based cost unit account also relates to a specific period. The item-based cost unit account, on the other hand, is an account for a specific item.

Costing is essentially performed in the following stages [20]: (1) *recording* of the costs by type of cost, (2) *allocation* of the costs to cost locations or cost units, and (3) *use* of the costs to measure operational activity for monitoring operational behaviour and/or for planning purposes.

Costs can generally be broken down according to two aspects: (1) cost breakdown into direct costs and overheads according to their attributability to a cost unit; (2) cost breakdown into fixed and variable costs according to their response to changes in employment.

7.4.2 Types of Cost

Accounting by cost type covers all the costs incurred in a firm in the procurement, storage, production and sale of operational outputs during a work period. In addition, it delimits the costs *vis-à-vis* the expenses of the firm as a whole. The importance of accounting by cost type lies in its subdivision of the overall costs into individual types of cost and the resulting possibility of attributing the individual costs to cost locations and cost units according to their origin (cf. E2.5.3).

According to the most important operational functions, a distinction is made between *procurement* costs, *storage* costs, *production* costs, *administrative* costs and *marketing* costs.

According to their source, five types of cost can be dis-

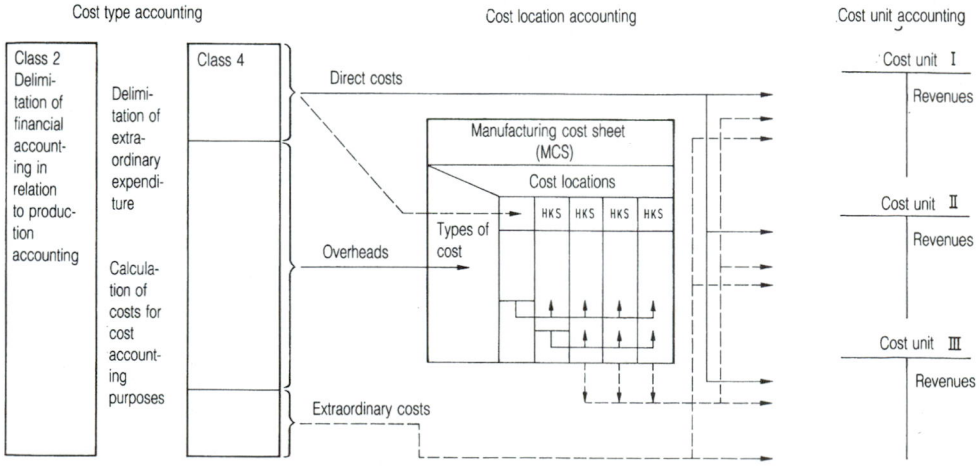

Figure 12. Costing system according to Schönfeld.

tinguished [20]: (1) *labour costs* (wages, salaries, subsidiary wage costs, entrepreneurial income), (2) *material costs* (cost of raw materials and supplies), (3) *cost of capital* (interest, depreciation, capital risks), (4) *cost of outwork* (cost of repairs, transport services) and (5) the *costs of human society* (taxes in the nature of costs, fees, contributions).

7.4.3 Cost Location Accounting and the Manufacturing Cost Sheet

Cost location accounting lies between *accounting by cost type* and *cost unit accounting* (cost accounting).

In firms with a diverse production programme, it enables the *overheads* to be allocated to the cost units in a way which best corresponds to their origin. While the direct product costs can be attributed directly to the cost unit even without cost location accounting, in the case of the overheads the lack of a cost location account would make for a very imprecise distribution of costs. By the establishment of cost locations (accounting sectors) within a firm, the overheads can be recorded location by location and allocated to the products with the aid of a special distribution code according to the demands made on the location by the product (**Fig. 12**). As individual cost locations (e.g. energy generation) pass on in-house outputs to other cost locations (e.g. manufacturing), a compensation of the in-house outputs must be performed as part of cost location accounting. On a formal basis, cost location accounting is carried out with the aid of a manufacturing cost sheet (MCS), which lists types of cost and cost locations in tabular form as rows and columns. **Table 2** shows an example of the design of a manufacturing cost sheet, in which the compensation of the in-house services has been dispensed with in order to simplify matters.

The tasks of the MCS are:

Distribution of the primary overheads among the cost locations according to origin.

Allocation of the costs of the general cost locations to subsidiary cost locations.

Allocation of the costs of the ancillary cost locations to the main cost locations.

Calculation of the overhead surcharge rates for every cost location by comparing direct costs and overheads.

Rechecking of the costs charged, i.e. calculation of the difference between the envisaged costs charged and the actual costs incurred.

Control of the efficiency of the cost centres by calculating indices [22].

7.4.4 Calculation of Machine-Hour Rate

The calculation of the machine-hour rate represents the furthest-reaching breakdown of the cost locations in cost location accounting. Here, the cost locations are individual machines. The total costs of a machine are termed *machine costs*. The purpose of such in-depth cost location accounting in the form of the machine-hour rate is to achieve increased precision in charging the overheads.

The machine-hour rate is calculated according to VDI Code 3258 – costing with machine-hour rates – by applying the calculated machine costs to the planned or average customary annual period of use T_N in h/yr:

$$K_{MH} = \frac{K_A + K_Z + K_R + K_E + K_I}{T_N}.$$

Here, K_A is the depreciation for costing purposes in money/yr. It is calculated from the replacement value (including installation and startup costs) according to

Table 2. Example of the structure of a simple manufacturing cost sheet according to [21]

Types of cost	Figures in accounts	Cost locations					
		Production division			Materials division	Administration division	Marketing division
		Production cost location I	Production cost location II	Production cost location III			
Salaries	2 600	300	400		200	1 200	500
Ancillary pay	1 800	800	200	200	300	100	200
Social welfare expenditure	900	300	150	150	50	180	70
Consumables	500	100	100				300
Office supplies	400				100	200	100
Outside repairs	400			300			100
Energy consumption	350	50	100	50	20	80	50
Depreciation	250	40	60	50	20	40	40
Taxes	100					100	
Postage	150					50	100
Advertising expensees	350						350
Misc. expenses	200	10	40	50	10	30	60
Total overheads	8 000	1 600	1 050	800	700	1 980	1 870
Direct labour costs	4 500	2 000	1 000	1 500			
Direct material	5 300				5 350		
Production costs						14 000	14 000
Overhead surcharge rates		80.0%	105.0%	53.3%	13.1%	14.1%	13.4%

Cost type accounting   Cost location accounting

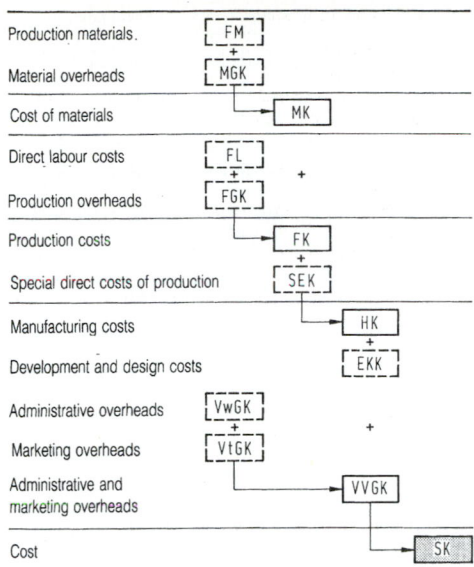

Production materials. ⌐ FM ⌐
                     +
Material overheads   ⌐ MGK ⌐

Cost of materials ──────── MK

Direct labour costs ⌐ FL ⌐
                     +   +
Production overheads ⌐ FGK ⌐

Production costs ──────── FK
                          +
Special direct costs of production ⌐ SEK ⌐

Manufacturing costs ──────── HK
                             +
Development and design costs ⌐ EKK ⌐

Administrative overheads ⌐ VwGK ⌐
                         +   +
Marketing overheads   ⌐ VtGK ⌐

Administrative and ──────── VVGK
marketing overheads

Cost ──────── SK

Figure 13. Procedure for surcharge cost accounting.

business principles and applied to the expected useful life of the machine. K_Z is the interest for costing purposes in money/yr. It is applied at the normal interest rates for long-term debt. To simplify the calculation and for ease of comparison, the interest is calculated on half the replacement value. K_R is the space costs in money/yr. They are usually applied to the floor area taken up by the machine including the necessary secondary areas. They include depreciation and interest on buildings and plant, building maintenance expenses, and the cost of light, heating, insurance and cleaning. K_E is the energy costs in money/yr. They are calculated for electricity, gas, water, etc. on the basis of the actual average annual figures. K_I is the maintenance costs in money/yr. They are to be calculated for regular servicing and for uncapitalised repairs as average annual values over extended periods. The dif-

fering repair-proneness of various types of machine should be taken into account.

7.4.5 Cost Accounting

The purpose of cost accounting is to allocate the costs incurred in producing the operational outputs and in selling these outputs commercially and in-house. Cost accounting may form the basis for: *price calculation* (initial cost accounting), *price review* (follow-up cost accounting), *profit calculation*, performance of *comparative calculations* and *output valuation*.

Wherever several products with various material and manufacturing labour costs are produced by various manufacturing processes, surcharge cost accounting is used. This cost accounting method is based on separate allocation of the direct costs and overheads to the cost units. The direct costs are directly charged to the cost units with individual vouchers (e.g. material requisition slip), while the overheads are charged indirectly by means of overhead surcharges (cf. MCS). The procedure for calculating the cost price with the aid of surcharge cost accounting can be demonstrated by means of the diagram in **Fig. 13**.

7.5 Basic Ergonomics

The subject of ergonomics is human work. *Work* in this sense is regular human activity directed at the creation of a permanent result, using one's physical, mental and psychological powers.

Accordingly, ergonomics is concerned with the expressions of the characteristics of human work (*loads*) and their physical, mental and psychological effects on people (*stresses*). The results of ergonomic studies are used to create or change working conditions (workplaces, working procedures, environmental influences) in such a way that they can be termed *humane* in the broadest sense of the word [23]. The adaptation of the working environment to people by methods engineering contrasts with the adaptation of people to the requirements of the work. This process can be assisted by instruction and training measures (cf. E4.3.7).

The starting data for designing workplaces are the dimensions of the human body. For this purpose, mean

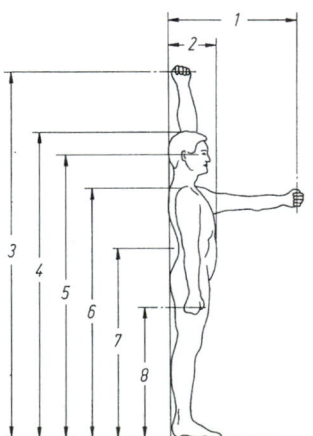

	Percentiles (dimensions in cm)					
	Male			Female		
	5 %	50 %	95 %	5 %	50 %	95 %
1 Forward reach	66.2	72.2	78.7	61.6	69.0	76.2
2 Depth of body	23.3	27.6	31.8	23.8	28.5	35.7
3 Upward reach (both arms)	191.0	205.1	221.0	174.8	187.0	200.0
4 Body height	162.9	173.3	184.1	151.0	161.9	172.5
5 Height of eyes	150.9	161.3	172.1	140.2	150.2	159.6
6 Height of shoulders	134.9	144.5	154.2	123.4	133.9	143.6
7 Height of elbow above standing surface	102.1	109.6	117.9	95.7	103.0	110.0
8 Height of hand above standing surface	72.8	76.7	82.8	66.4	73.8	80.3
9 Width of hips, standing	31.0	34.4	36.8	31.4	35.8	40.5
10 Width of shoulders	36.7	39.8	42.8	32.3	35.5	38.8

Figure 14. Body measurements of German adults – standing – according to DIN 33402.

values and distributions of body measurements have been calculated in serial studies with representative samples (cf. DIN 33 403). Some of these body measurements are shown in **Fig. 14** for standing adults and **Fig. 15** for seated adults. In physical work, it is usually the transmission of forces from the worker to the workpiece by means of tools and equipments that is to the fore. These tools and equipments must be designed so as to permit the largest possible forces to be transmitted with little load on the worker. On machines, levers, handwheels, pushbuttons, etc. should be positioned in such a way that their operation approximates to a natural movement [24].

Informational work can be broken down into *information receiving*, *information processing* and *information issuing*. Information is received via the sensory organs, mostly by sight and hearing and to a lesser extent by touch, smell or taste. The receiving of information via the eye can take place only if the associated signals are supplied in the field of vision. The field of vision is a circle whose diameter d_G (in metres) increases linearly with the distance from the eye a (in metres) according to the numerical value equation $d_G = 1.64a$.

The working area requires *illumination*, the strength of which depends on the type of work to be performed (**Table 3**). Further information can be found in DIN 5034 and DIN 5035.

For *visual* means of information (indicating instruments) there are design possibilities of varying worth. For instance, round instruments are preferable to rectangular displays. **Figure 16** compares common types of display [25].

Issuing of information to technical systems is generally accomplished by means of operating elements. Speech input is a possible alternative.

Human efficiency is also affected by the environmental conditions (climate, noise, dust) in which the work is done. The climate in working premises is described by the air temperature, the relative humidity and the air motion velocity (**Table 4**). For an air space of 15 m³ per person, the supply of fresh air should be at least 30 m³ per person an hour, even in the case of very light work [26].

Table 3. Illumination required for specific visual tasks

Level	Nominal illumination (lx)	Classification of visual tasks
1	15	
2	30	Orientation; temporary stay only
3	60	
4	120	Easy visual tasks; large details with high contrast
5	250	
6	500	Normal visual tasks; medium-size details with medium contrast
7	750	
8	1000	Difficult visual tasks; small details with low contrast
9	1500	
10	2000	Very difficult visual tasks; very small details with very low contrast
11	3000	
12	5000	Special cases, e.g. illumination of operating field

Almost all work operations generate noise in some form. The noise is measured with instruments standardised to DIN IEC 651 by methods laid down in DIN 45 635. The statutory order on health and safety at work (Arbeitsstättenverordnung) specifies 55 dB(A) as the maximum in premises where the work is mainly intellectual in nature, 70 dB(A) for simple office work and 85 dB(A) for industrial workplaces.

Gases, dust and vapours are subject to MAC (maximum allowable concentration) or TLV (threshold limit value) values. They indicate the concentrations of noxious substances which, at an effect duration of 8 hours a day, are not harmful to health even over a prolonged period [27].

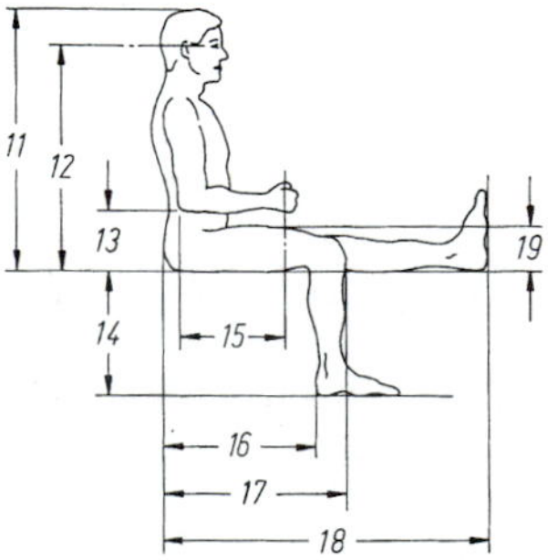

		Percentiles (dimensions in cm)					
		Male			Female		
		5 %	50 %	95 %	5 %	50 %	95 %
11	Body height, seated	84.9	90.7	96.2	80.5	85.7	91.4
12	Height of eyes, seated	73.9	79.0	84.4	68.0	73.5	78.5
13	Height of elbow above seat	19.3	23.0	28.0	19.1	23.3	27.8
14	Length of lower leg with foot (seat height)	39.9	44.2	48.0	35.1	39.5	43.4
15	Distance from elbow to gripping axis	32.7	36.2	38.9	29.2	32.2	36.4
16	Sitting depth	45.2	50.0	55.2	42.6	48.4	53.2
17	Length from bottom to knee	55.4	59.9	64.5	53.0	58.7	63.1
18	Length from bottom to sole of foot	96.4	103.5	112.5	95.5	104.4	112.6
19	Height of upper leg	11.7	13.6	15.7	11.8	14.4	17.3
20	Width across elbows	39.9	45.1	51.2	37.0	45.6	54.4
21	Width of hips, seated	32.5	36.2	39.1	34.0	38.7	45.1

Figure 15. Body measurements of German adults – seated – according to DIN 33 402.

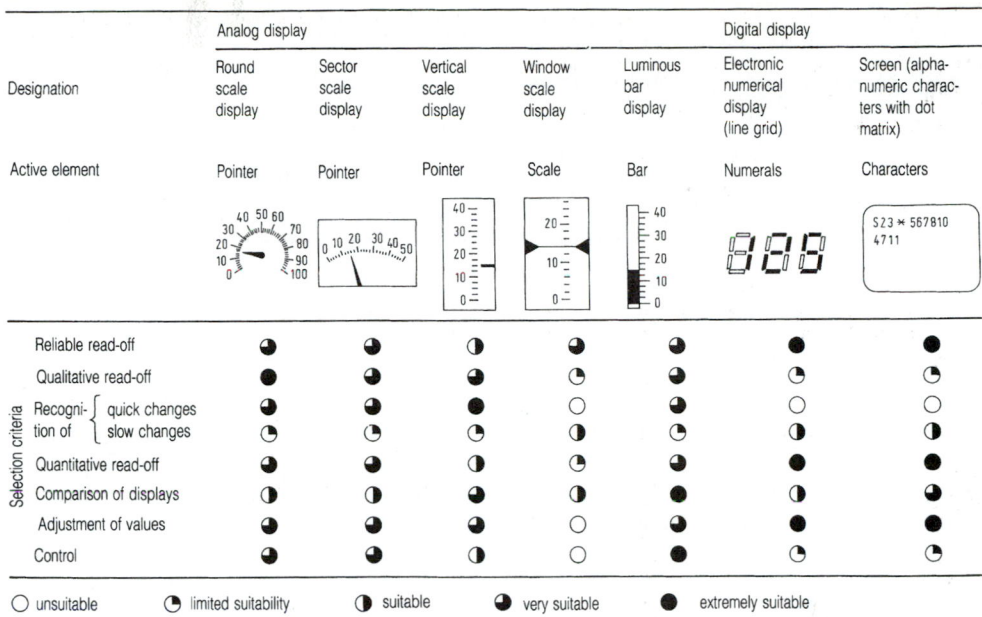

Designation	Analog display					Digital display	
	Round scale display	Sector scale display	Vertical scale display	Window scale display	Luminous bar display	Electronic numerical display (line grid)	Screen (alpha-numeric characters with dot matrix)
Active element	Pointer	Pointer	Pointer	Scale	Bar	Numerals	Characters

Selection criteria	Round scale	Sector scale	Vertical scale	Window scale	Luminous bar	Electronic numerical	Screen
Reliable read-off	◑	◕	◔	◕	◕	●	●
Qualitative read-off	●	◕	◕	◔	◕	◔	◔
Recognition of — quick changes	◕	◕	●	○	◕	○	○
Recognition of — slow changes	◔	◔	◔	◑	◔	◑	◑
Quantitative read-off	◕	◕	◑	◔	◕	●	●
Comparison of displays	◑	◑	◕	◑	●	◑	◕
Adjustment of values	◕	◕	◕	○	◕	●	●
Control	◕	◕	◑	○	●	◔	◔

○ unsuitable   ◔ limited suitability   ◑ suitable   ◕ very suitable   ● extremely suitable

Figure 16. Comparison of common analog and digital displays according to [25].

Table 4. Climatic data for specific activities

Activity	Air temperature (°C)			Rel. humidity (%)			Air motion velocity (m/s)
	min.	opt.	max.	min.	opt.	max.	max.
Office work	18	21	24	30	50	70	0.1
Light manual work, seated	18	20	24	30	50	70	0.1
Light manual work, standing	17	18	22	30	50	70	0.2
Heavy work	15	17	21	30	50	70	0.4
Very heavy work	14	16	20	30	50	70	0.5

8 Appendix K: Diagrams and Tables

Correction values:

Cutting speed correction factor

$$K_v = \frac{2.023}{v_c^{0.153}} \text{ for } v_c < 100 \text{ m/min}$$

for v_c = 20 to 600

$$K_v = \frac{1.380}{v_c^{0.07}} \text{ for } v_c > 100 \text{ m/min}$$

Rake angle correction factor

$$K_\gamma = 1.09 \text{ to } 0.015 \cdot \measuredangle \,° \text{ (steel)}$$

$$K_\gamma = 1.03 \text{ to } 0.015 \cdot \measuredangle \,° \text{ (castings)}$$

Cutting material correction factor

K_{ws} = 1.05 (HSS)

K_{ws} = 1.0 (cemented carbide)

K_{ws} = 0.9 to 0.95 (cutting ceramic)

Tool wear correction factor

K_{wv} = 1.3 to 1.5

K_{wv} = 1.0 (work-sharp cutting edge)

Cooling lubricant correction factor

K_{ks} = 1 (dry)

K_{ks} = 0.85 (non-water-miscible cooling lubricant)

K_{ks} = 0.9 (cooling lubricant emulsion)

Workpiece shape correction factor

K_f = 1 (external turning)

K_f = 1.2 (internal turning)

Appendix K4 Table 1. $k_{c\,1.1}$ and $1 - m_c$ values for ferrous materials

Cutting conditions	Cutting speed Depth of cut Cutting materials				$v = 100\ \text{n min}^{-1}$ $a_p = 3\ \text{mm}$ Cemented carbide P10		
		α	γ	λ	ε	κ	Γ_ε
	Steel	5°	6°	0°	90°	70°	0.8 mm
	Castings	5°	2°	0°	90°	70°	0.8 mm

Material	R_m (N/mm²)	Specific machining forces $k_{i\,1.1}$ (N/mm²)					
		$k_{c\,1.1}$	$1-m_c$	$k_{f\,1.1}$	$1-m_f$	$k_{p\,1.1}$	$1-m_p$
St 50-2	559	1499	0.71	351	0.30	274	0.51
St 70-2	824	1595	0.68	228	-0.07	152	0.10
Ck 45 N	657	1659	0.79	521	0.51	309	0.60
Ck 45 V	765	1584	0.74	364	0.27	282	0.57
Ck 60 N	775	1686	0.78	285	0.28	259	0.59
Ck 60 V	873	1662	0.77	337	0.29	249	0.53
40 Mn 4V	755	1691	0.78	350	0.31	244	0.55
37 MnSi 5V	892	1656	0.79	239	0.31	249	0.67
18 CrNi8BG	618	1511	0.80	318	0.27	242	0.46
30 CrNiMo8V	971	1704	0.82	337	0.46	371	0.88
34 CrNiMo6V	1010	1686	0.82	291	0.37	284	0.72
34 Cr 4 V	902	1536	0.78	327	0.36	222	0.59
41 Cr 4 V	961	1596	0.77	291	0.27	215	0.52
16 MnCr 5N	500	1411	0.70	406	0.37	312	0.50
16 MnCr 5BG	500	1575	0.81	391	0.30	324	0.54
20 MnCr 5N	588	1464	0.74	356	0.24	300	0.58
20 MnCr 5BG	588	1523	0.76	356	0.33	271	0.52
34 CrMo 4V	1000	1632	0.80	276	0.34	172	0.48
42 CrMo 4V	1138	1773	0.83	354	0.43	252	0.49
50 CrV 4V	1050	1698	0.78	295	0.28	195	0.44
Ck 35 V	622	1527	0.72	344	0.25	291	0.46
Ck 55 N	661	1396	0.65	316	0.16	255	0.42
55 NiCrMoV6V	1141	1595	0.71	269	0.21	198	0.34
100 Cr 6	624	1726	0.72	318	0.14	362	0.47
GG 30	HB= 206	899	0.59	170	0.09	164	0.30

Appendix K4 Table 2. Machining forces for drilling [9, 10]

Material	R_m (N · mm⁻²)	$1-m_c$	$k_{c\,1.1}$ (N · mm⁻²)	$1-m_f$	$k_{f\,1.1}$ (N · mm⁻²)
18 CrNi 8	600	0.82 ± 0.04	2690 ± 230	0.55 ± 0.06	1240 ± 160
42 CrMo 4	1080	0.86 ± 0.06	2720 ± 420	0.71 ± 0.04	2370 ± 230
100 Cr	710	0.76 ± 0.03	2780 ± 220	0.56 ± 0.07	1630 ± 300
46 MnSi 4	650	0.85 ± 0.04	2390 ± 250	0.62 ± 0.02	1360 ± 100
Ck 60	850	0.87 ± 0.03	2200 ± 200	0.57 ± 0.03	1170 ± 100
St 50	560	0.82 ± 0.03	1960 ± 160	0.71 ± 0.02	1250 ± 70
16 MnCr 5	560	0.83 ± 0.03	2020 ± 200	0.64 ± 0.03	1220 ± 120
34 CrMo 4	610	0.80 ± 0.03	1840 ± 150	0.64 ± 0.03	1460 ± 140
Grey cast iron					
Up to G-22	–	0.51	504	0.56	356
Over G-22	–	0.48	535	0.53	381

Appendix K4 Table 3. Main and incremental values for axial plain face milling

Material	Cutting material	Cutting speed v_c (m min^{-1})	Cutting edge geometry	Main and incremental values for spec. machining force in axial face milling					
				$k_{c\,1.1}$ (N mm^{-2})	m_c	$k_{c\,N\,1.1}$ (N mm^{-2})	m_{cN}	$k_{p\,1.1}$ (N mm^{-2})	m_p
St 52-3N	HM P25	120	negative	1831	0.29	809	0.54	705	0.41
			positive	1469	0.25	447	0.57	174	0.56
Ck 45N	HM P25	190	negative	1506	0.45	708	0.62	653	0.52
X22CrMoV121	HM P40	120	positive	1533	0.29	497	0.70	164	0.77

Cutting edge geometry	γ_f	γ_p	α_f	α_p	λ_s	K_r	ε	K_F	Land (mm)
negative	$-4°$	$-7°$	$6°$	$23°$	$-6°$	$75°$	$90°$	$60°/30°/0°$	1.4/0.8/1.4
positive	$0°$	$8°$	$9°$	$29°$	$8°$	$75°$	$90°$	$45°/0°$	0.8/1.4

Appendix K4 Table 4. Guide values for milling of ferrous materials. The guide values apply to pre-machined workpieces and stable machining conditions. In the case of unstable conditions or in the presence of mill scale, forging scale or casting skin, the stated cutting conditions should be reduced accordingly. f, feed/cutting edge; v, cutting speed

Appendix K4 Table 4a. Negative wedge geometry; $\gamma = -6°$

Cutting material	f_z mm/U, v_c m/min	Unalloyed steel and steel castings HB 110 to 200	HB 200 to 265	HB 265 to 450	Alloyed steel and steel castings HB 125 to 200	HB 200 to 265	HB 280 to 345	Stainless steel and steel castings HB 110 to 265	HB 265 to 340	Grey cast iron HB 130 to 200	≤B 200 to 280	Nodular cast iron HB 125 to 230	White heart malleable cast iron HB 150 to 180	Black heart malleable cast iron HB 130 to 180
P10	f_z / v_c	0.1 to 0.4 / 80 to 250		0.1 to 0.4 / 90 to 165	0.1 to 0.4 / 115 to 190	0.1 to 0.2 / 135 to 170	0.1 to 0.2 / 95 to 120	0.1 to 0.2 / 90 to 180						
P25	f_z / c_c	0.1 to 0.8 / 70 to 205	0.1 to 0.8 / 70 to 180	0.1 to 0.8 / 55 to 140	0.1 to 0.8 / 70 to 165	0.1 to 0.8 / 65 to 150	0.1 to 0.8 / 40 to 105	0.1 to 0.8 / 75 to 210	0.1 to 0.8 / 85 to 155					
P40	f_z / v_c	0.1 to 0.8 / 35 to 130		0.1 to 0.8 / 35 to 90	0.1 to 0.8 / 45 to 105	0.1 to 0.8 / 45 to 100	0.1 to 0.8 / 25 to 70	0.1 to 0.8 / 55 to 130						
M15	f_z / v_c									0.1 to 0.5 / 90 to 200	0.1 to 0.4 / 80 to 150		0.1 to 0.3 / 70 to 130	0.1 to 0.5 / 90 to 200
K01	f_z / v_c									0.1 to 0.2 / 120 to 180	0.1 to 0.2 / 100 to 160		0.1 to 0.15 / 100 to 150	0.1 to 0.2 / 120 to 220
K10	f_z / v_c									0.1 to 0.2 / 100 to 170	0.1 to 0.2 / 100 to 170	0.1 to 0.2 / 170 to 195	0.1 to 0.3 / 60 to 100	0.1 to 0.5 / 70 to 170
K20	f_z / v_c									0.1 to 0.8 / 50 to 120	0.1 to 0.8 / 30 to 90	0.1 to 0.8 / 50 to 95		

Appendix K4 Table 4b. Positive wedge geometry; $\gamma = +6°$

Cutting material	f_z mm/U v_c m/min	Unalloyed steel and steel castings			Alloyed steel and steel castings		Stainless steel and steel castings		Grey cast iron		White heart malleable cast iron	Black heart malleable cast iron
		HB 110 to 200	HB 200 to 265	HB 265 to 450	HB 200 to 265	HB 280 to 345	HB 110 to 265	HB 265 to 340	HB 130 to 200	HB 200 to 280	HB 150 to 180	HB 130 to 180
P10	f_z	0.1 to 0.2		0.1 to 0.2	0.1 to 0.2							
	v_c	80 to 250		50 to 90	80 to 120							
P25	f_z	0.1 to 1.2	0.1 to 0.4	0.1 to 0.4	0.1 to 0.6	0.1 to 0.4	0.1 to 0.4	0.1 to 0.4				
	v_c	120 to 200	80 to 160	30 to 90	70 to 130	65 to 110	40 to 90	95 to 130				
P40	f_z	0.2 to 1.5		0.2 to 0.6	0.2 to 0.8		0.1 to 0.6					
	v_c	35 to 135		30 to 70	40 to 80		40 to 80					
M15	f_z								0.1 to 1.2	0.1 to 1.0	0.1 to 0.6	0.1 to 1.2
	v_c								90 to 200	80 to 150	70 to 130	90 to 200
K01	f_z								0.1 to 0.2	0.1 to 0.2	0.1 to 0.2	0.1 to 0.2
	v_c								120 to 180	100 to 160	100 to 150	120 to 220
K10	f_z								0.1 to 0.4	0.1 to 0.4		
	v_c								80 to 135	60 to 110		
K20	f_z								0.1 to 0.4	0.1 to 0.4		
	v_c								80 to 150	55 to 110		

Appendix K4 Table 5. Standards for grain size of abrasives (according to DIN 69 100, VDI Code 3394)

Average spacing of two screening machines (μm)	US Standard ASTM E11 — Range Narrow	Broad	FEPA grain size designation — Range Narrow	Broad	Designation according to DIN 848 — Range Narrow	Broad	Average grain diameter (corundum SiC) (μm)
420							300
			D426				
	40/50			D427		D350	
		40/60					
			D356				
							150
297							
					D280		
	50/60		D301			D250	
							125
250							
	60/70		D251		D220		
		60/80		D252			
210							105
	70/80		D213		D180		
							90
177							
	80/100		D181			D150	
149					D140		75
	100/120		D151				
125							62
	120/140		D126		D110	D100	
105							45
	140/170		D107		D90		
							37
88							
	170/200		D91			D70	
							31
74					D65		
	200/230		D76				
							27
63							
	230/270		D64		D55		22
53							
	270/325		D54			D50	
44					D45		18
	325/400		D46				
37					D35		
32	300/500						

Appendix K4 Table 6. Principal data for belt grinding

Material	Roughing			Finishing		
	Manipulated variables		Result	Manipulated variables		Result
	Specific rate of metal removal Q_w' (mm³/mm · s)	Cutting speed v_c (m/s)	Measured surface roughness R_z (μm)	Specific rate of metal removal Q_w' (mm³/mm · s)	Cutting speed v_c (m/s)	Measured surface roughness R_z (μm)
Steel for anti-friction bearings (62 HRC)	Up to 150	40 to 60	>40	2 to 8	30 to 40	5 to 8
Grey cast iron	Up to 200	30 to 50	>50	5 to 10	20 to 30	8 to 10
AlSi alloys	Up to 100	20 to 40	>60	1 to 5	30 to 40	10 to 12

Appendix K4 Table 7. Material pairs in electro-discharge machining

Electrode material	Workpiece material	Machining operation	Electrode polarity	Quality of machining	Electrode wear	Notes
Copper-tungsten alloy	Copper-tungsten alloy	Roughing Finishing	negative negative	good good	reasonable reasonable	Low metal removal capacity
Copper-tungsten	Steel	Roughing Finishing	positive positive	good good	low low	Used for small, high-precision press tools
Copper-tungsten	Tungsten carbide	Roughing Finishing	negative negative	good good	moderate moderate	Used for small, high-precision press tools
Copper alloy	Steel	Roughing Finishing	positive positive	good good	low low	High cost of electrodes, ideal for small steel workpieces
Copper alloy	Tungsten carbide	Roughing Finishing	negative negative	good good	moderate moderate	Ideal for tungsten carbide workpieces
Graphite	Cast iron	Roughing Finishing	positive negative	good good	low moderate	Higher metal removal capacity with negative polarity, but electrode wear is higher
Graphite	Copper	Roughing Finishing	negative negative	moderate moderate	reasonable reasonable	Low metal removal capacity
Graphite	Nimonic	Roughing Finishing	positive negative	good good	low low	Good metal removal rate Reasonable metal removal rate
Graphite	High-speed steel	Roughing Finishing	negative negative	moderate moderate	reasonable reasonable	Moderate metal removal rate Reasonable metal removal rate
Graphite	Stainless steel	Roughing Finishing	positive negative	moderate moderate	moderate moderate	Good metal removal rate
Graphite	Steel	Roughing Finishing	positive negative	good good	low low	Good metal removal rate High metal removal rate
Graphite	Stellite	Roughing Finishing	negative negative	moderate moderate	reasonable reasonable	Moderate metal removal rate
Graphite	Tungsten carbide	Roughing Finishing	negative negative	moderate moderate	high high	Moderate metal removal rate; sparking is a problem if t_i is too long
Steel	Steel	Roughing Finishing	positive positive	poor poor	reasonable reasonable	Low metal removal rate Use for special purposes only

(continued)

Appendix K4 Table 7. Continued

Electrode material	Workpiece material	Machining operation	Electrode polarity	Quality of machining	Electrode wear	Notes
Tungsten carbide	Steel	Roughing	positive	moderate	low	Low metal removal rate
		Finishing	positive	good	low	
Tungsten carbide	Nimonic	Roughing	positive	moderate	low	Reasonable metal removal rate
		Finishing	positive	good	low	For machining small openings
Aluminium	Steel	Roughing	positive	good	low	Stability doubtful for some grades of
		Finishing	positive	poor	high	Al
Aluminium	Tungsten carbide	Roughing	negative	poor	high	Not universally recommended
		Finishing	negative	very poor	very high	
Brass	Copper	Roughing	negative	good	reasonable	Use for special purposes only
		Finishing	negative	good	reasonable	
Brass	Steel	Roughing	negative	good	high	For narrow openings
		Finishing	negative	good	high	Reasonable metal removal rate
Brass	Stellite	Roughing	negative	poor	high	Unstable
		Finishing	negative	moderate	high	Reasonable metal removal rate
Brass	Titanium	Roughing	negative	good	high	Reasonable metal removal rate
		Finishing	negative	good	high	
Brass	Tungsten carbide	Roughing	negative	moderate	high	Not universally recommended
		Finishing	negative	moderate	high	
Copper	Aluminium	Roughing	positive	good	low	Low metal removal rate
		Finishing	positive	good	low	
Copper	Brass	Roughing	positive	good	reasonable	Use for special purposes only
		Finishing	positive	good	reasonable	
Copper	Cast iron	Roughing	positive	good	low	Reasonable metal removal rate
		Finishing	positive	good	low	
Copper	Copper	Roughing	positive	poor		Not recommended
		Finishing	positive	poor		
Copper	Graphite	Roughing	positive	moderate	reasonable	Use for special purposes only
		Finishing	positive	moderate	reasonable	
Copper	Nimonic	Roughing	positive	good	low	Reasonable metal removal rate
		Finishing	positive	good	reasonable	Good metal removal rate
Copper	Stainless steel	Roughing	positive	moderate	reasonable	Stability doubtful for some grades of
		Finishing	positive	moderate	reasonable	steel
Copper	Steel	Roughing	positive	good	low	Good metal removal rate. Never use
		Finishing	positive	good	low	negative polarity
Copper	Stellite	Roughing	positive	good	reasonable	Reasonable metal removal rate
		Finishing	positive	good	reasonable	
Copper	Tungsten carbide	Roughing	negative	good	high	Reasonable metal removal rate
		Finishing	negative	good	high	Stability becomes problematic if t_i is too long

Appendix K4 Table 8. Guide values for the relationship die clearance/sheet or plate thickness

Sheet/plate thickness (mm)	Tensile strength of material (N/mm²)			
	<250	250 to 400	400 to 600	>600
Independent of sheet/plate thickness	0.03	0.04	0.05	0.06
<1	0.025	0.025	0.03	0.035
1 to 2	0.03	0.03	0.035	0.04
2 to 3	0.035	0.035	0.04	0.045
3 to 5	0.04	0.04	0.045	0.05
5 to 7	0.045	0.045	0.05	0.055
7 to 10	0.05	0.05	0.055	0.06

Appendix K4 Table 9. Conventional materials for cutting tools and field of application

Tool material	Approx. service hardness HRC, HV	Sheet/plate thickness (mm)	Characteristics
1. Cold work steel X 155CrVMo12 1 X 165CrMoV12 X 210CrW12 X 210Cr12 X 210CrCoW12 S 6-5-2	62 up to 65 HRC	up to 4 mm	Materials with low toughness and high wear resistance for shearing hard, thin sheets
90MnV8 105WCr6	60 up to 64 HRC	4 up to 6 mm	Low-distortion materials of average toughness and wear resistance
45WCrV7 60WCrV7 X 45NiCrMo4 X 50CrMoW9 11 X 63CrMoV5 1	56 up to 63 HRC	over 6 mm	Tough materials for absorbing high stress peaks in shearing of thick plates; lower wear resistance to abrasive wear mechanisms
2. Cemented carbides GT 15[a] GT 20[a] GT 30[a] GT 40[a] THR-F[a]	1450HV 1300HV 1200HV 1050HV 1500HV	up to 1 mm	Brittle materials for shearing thin plates; extremely high wear resistance to predominantly abrasive wear mechanisms
3. Hard material alloys Ferro-Titanit-C-Special[b] Ferro-Titanit-WFN[b] S 6.5.3 (ASP 23)[c] CPM 10V[d] CPM Rex M 4[d]	68 to 71 HRC 68 to 71 HRC 61 to 65 HRC 61 to 64 HRC 61 to 65 HRC	up to 8 mm	Hard-wearing materials with high ductility due to a homogeneous structure.

9 References

K1 Survey of Manufacturing Processes. [1] Tönshoff HK. Randzonenbeeinflussung durch Spanen und Abtragen. Ann CIRP 1974; 23: 187-8. – [2] Wiendahl H-P. Belastungsorientierte Fertigungssteuerung. Hanser, Munich, 1987. – [3] Tönshoff HK. Processing alternatives for cost reduction. Ann CIRP 1987; 36: 445-7. – [4] Kienzle O. Begriffe und Benennungen der Fertigungsverfahren. Werkstatttechnik 1966; 56: 169-73.

K2 Primary Shaping. [1] Hilgenfeldt W, Herfurth K. Tabellenbuch Gusswerkstoffe. VEB Deutscher Verlag für Grundstoffindustrie, Leipzig, 1983. – [2] ZGV. Giessen heute. Ed. Zentrale für Gussverwendung, Düsseldorf, 1974. – [3] Guss Produkte '89. Hoppenstedt, Darmstadt. – [4] Herfurth K. Einführung in die Fertigungstechnik. Kapitel Urformen. VEB Verlag Technik, Berlin, 1975. – [5] Feinguss für alle Industriebereiche. Ed. Zentrale für Gussverwendung, Düsseldorf, 1984. – [6] Leitfaden für Gusskonstruktionen. Ed. Zentrale für Gussverwendung. Giesserei-Verlag, Düsseldorf, 1966. – [7] Verein Deutscher Giessereifachleute (VDG). Giesserei-Kalender 1977. Giesserei-Verlag, Düsseldorf, 1976. – [8] Pahl G, Beitz W. Konstruktionslehre – Handbuch für Studium und Praxis, 2nd edn. Springer, Berlin, 1986. – [9] Patterson W, Döpp R. Betriebsnomogramm für Grauguss. Giesserei 1960; 47: 175-80. – [10] Colland A. Giesserei, techn.-wiss. Beih. 1954; 14: 709-26, and 1955; 15: 767-99. – [11] ZGV-Mitteilungen. Düsseldorf, 1976. – [12] Eisenkolb F. Einführung in die Werkstoffkunde, vol. V: Pulvermetallurgie, 2nd edn. VEB Verlag Technik, Berlin, 1967. – [13] Technikum für berufliche Bildung des Ministeriums für Erzbergbau, Metallurgie und Kali (Technical Centre for Vocational Training of the Ministry of Ore Mining, Metallurgy and Potash). Lehrbuch Metallurgie. VEB Deutscher Verlag für Grundstoffindustrie, Leipzig, 1971.

Standards and Codes. DIN 1680 Part 1: Rough castings; general tolerances and machining allowances; general. Part 2: Rough castings; general tolerance system.
DIN 1683 Part 1: Steel raw castings; general tolerances; machining allowances.
DIN 1684 Part 1: Malleable iron raw castings; general tolerances, machining allowances.
DIN 1685 Part 1: Raw castings made from nodular graphite cast iron; general tolerances, machining allowances.
DIN 1686 Part 1: Rough castings of grey iron with flake graphite; general tolerances, machining allowances.
DIN 1687 Part 1: Heavy metal alloy raw castings, sand castings; general tolerances, machining allowances. Part 3: Rough castings of heavy metal alloys, gravity die castings; general tolerances, machining allowances. Part 4: Heavy metal alloy raw castings, pressure die castings; general tolerances.
DIN 1688 Part 1: Light metal alloy raw castings, sand castings; general tolerances, machining allowances. Part 3: Light metal alloy raw castings, gravity die castings; general tolerances, machining allowances. Part 4: Light metal alloy raw castings, pressure die castings; general tolerances.
DIN 1690 Part 1: Technical delivery conditions for castings made from metallic materials; general conditions.
Ferrous cast materials:
DIN 1691: Grey iron with flake graphite.
DIN 1693: Cast iron with nodular graphite.
DIN 1694: Austenitic cast iron.
DIN 1695: Abrasion-resisting alloy cast iron.
DIN 1692: Malleable cast iron; concepts, properties, final inspection.

DIN 1681: Cast steels for general engineering purposes.
DIN 17 245: Ferritic steel castings creep-resistant at elevated temperatures.
DIN 17 445: Stainless steel castings.
DIN 17 465: Heat-resisting steel castings.
SEW 410: Stainless steel castings (Iron and Steel Material Data Sheets).
SEW 685: Steel castings tough at sub-zero temperatures.
SEW 510: Heat-treatable steel castings with wall thicknesses up to 100 mm.
SEW 515: Heat-treatable steel castings with wall thicknesses over 100 mm.
SEW 595: Cast steel for crude oil and natural gas plants.
SEW 471: Heat-resisting cast steel.
SEW 390: Non-magnetisable cast steel.
SEW 835: Cast steel for flame and induction hardening.

Light Metal Casting Materials. DIN 1725 Sheet 2: Aluminium casting alloys.
DIN 1729 Sheet 2: Magnesium casting alloys. Heavy metal casting materials.
DIN 1705: Copper-tin and copper-tin-zinc casting alloys (cast tin bronze and gunmetal), castings.
DIN 1709: Copper-zinc alloys castings (brass and special brass castings).
DIN 1714: Copper-aluminium casting alloys (cast aluminium bronze), castings.
DIN 1716: Copper-lead-tin casting alloys (cast tin-lead bronze), castings.
DIN 17 655: Unalloyed and low-alloy copper materials for casting; castings.
DIN 17 658: Copper-nickel casting alloys.
DIN 1743: High purity zinc casting alloys (Part 1: Ingot metals; Part 2: Castings).
DIN 1741: Lead alloys for pressure die castings.
DIN 1742: Tin alloys for pressure die castings.
DIN 17 730: Nickel and nickel-copper casting alloys.

K3 Metal Forming. General reference: Lange K. Handbook of Metals Forming. McGraw-Hill, New York, 1985. [1] Henky H. Z angew Math Mech 1924; 4: 323-34. – [2] Bühler H, Höpfner HG, Löwen J. Die Formänderungsfestigkeit von Aluminium und einigen Aluminiumlegierungen. BBR 1970; 11: 645-9. – [3] Krause K. Formänderungsfestigkeit der Werkstoffe beim Kaltumformen. In: Grundlagen der bildsamen Formgebung. VDEh, Düsseldorf, 99-145. – [4] Kienzle O, Bühler H. Das Plastometer, eine Prüfmaschine für Staucheigenschaften von Metallen. Z Metallkd 1964; 55: 668-73. – [5] Lange K. Lehrbuch der Umformtechnik, vol. 1, 2nd edn, 1984; vol. 2, 1988; vol. 3, 1990; vol. 4, 1993. Springer, Berlin. – [6] Müller G. Formänderungsfestigkeit beim Umformen in der Wärme. In: Grundlagen der bildsamen Formgebung. VDEh, Düsseldorf, pp. 146-61. – [7] Siebel E. Grenzen der Verformbarkeit. Mitt für die Mitglieder der Forschungsgesellschaft. Blechverarbeitung 1952; 16: 177-84. – [8] Stenger H. Über die Abhängigkeit des Formänderungsvermögens metallischer Werkstoffe vom Spannungszustand. Thesis, RWTH Aachen, 1965. – [9] Hasek V. Untersuchung und theoretische Beschreibung wichtiger Einflussgrössen auf das Grenzformänderungsdiagramm. Blech-Rohr-Profile 1978; 25: 213-20, 285-92, 493-9, 620-7. – [10] Siegert K. Grenzen des Ziehens von Karosserieteilen. Werkst Betrieb 1985; 118: 709-13. – [11] Siebel E. Kräfte und Materialfluss bei der bildsamen Formänderung. Stahl Eisen 1925; 45: 139-41. – [12] Siebel E. Die Formgebung im bildsamen Zustand. Stahleisen, Düsseldorf, 1932. – [13]

Sachs G. Zur Theorie des Ziehvorgangs. Z angew Math 1927; 235-6. - [14] Siebel E., Pomp A. Zur Weiterentwicklung des Druckversuches. Mitt K-Wilh-Inst Eisenf 1928; 10: 55-62. - [15] Lippmann H, Mahrenholtz O. Plastomechanik der Umformung metallischer Werkstoffe, vol. 1. Springer, Berlin, 1967. - [16] Ismar H, Mahrenholtz O. Technische Plastomechanik. Vieweg, Brunswick, 1979. - [17] Lippmann H. Die elementare Plastizitätstheorie der Umformtechnik. Bänder Bleche Rohre 1962; 374-83. - [18] Spur G, Stöferle T. Handbuch der Fertigungstechnik, vol. 2. Hanser, Munich, 1983. - [19] Körper F, Eichinger A. Die Grundlagen der bildsamen Formgebung. Mitt K-Wilh-Inst Eisenf 1940; 22: 57-80. - [20] Pawelski O. Grundlagen des Ziehens und Einstossens. In: Grundlagen der bildsamen Formgebung. VDEh, Düsseldorf, pp. 384-433. - [21] Sachs G. Zur Theorie des Ziehvorgangs. Z angew Math Mech 1927; 7: 235-36. - [22] Lippmann H. Theorie der Einstoss- und Strangpressvorgänge. Bänder Bleche Rohre 1963; 223-5. - [23] Eisbein W. Kraftbedarf und Fliessvorgänge beim Strangpressen. Thesis, TH Berlin, 1931. - [24] Sachs G. Spanlose Formgebung der Metalle. In: Handbuch der Metallphysik, vol. 3, Lief. 1, 1937. - [25] Rathjen C. Untersuchungen über die Grösse der Stempelkraft und des Innendruckes im Aufnehmer beim Strangpressen von Metallen. Thesis, RWTH Aachen, 1966. - [26] Panknin W. Die Grundlagen des Tiefziehens im Anschlag unter besonderer Berücksichtigung der Tiefziehprüfung. Bänder Bleche Rohre 1961; 133-43, 201-11, 264-71. - [27] Siegert K. Ziehen von flachen Karosserieteilen, Verfahren-Maschinen-Werkzeuge. VDI-Z 1989; 131: no. 4. - [28] Cyril Bath Company. Streckziehen von Karosserieteilen. Werkstatt und Betrieb 1965; issue 3. - [29] Neuere Entwicklungen in der Blechumformung, ed. K. Siegert. DGM-Informationsgesellschaft mbH, Oberursel, 1990. - [30] Siegert K. Zieheinrichtungen im Pressentisch einfach wirkender Pressen. In [29]. - [31] Zünkler B. Biegeumformen. In: Spur G, Stöferle T. Handbuch der Fertigungstechnik, vols 2 and 3. Hanser, Munich, 1985. - [32] Ludwik P. Technologische Studie über Biegung. Techn Blätter 1903; pp. 133-59. - [33] Oehler G. Biegen. Hanser, Munich, 1963. - [34] Zünkler B. Rechnerische Erfassung der Vorgänge beim Biegen im V-Gesenk. Ind Anz 1966; 88: 1601-5. - [35] Kienzle O. Untersuchungen über das Biegen. Mitt DFBO 1952; 57-65. - [36] Fait J. Grundlagenuntersuchungen zur Ermittlung von Kenngrössen für das CNC-Schwenkbiegen. Ind Anz 1987; 109: 45-6. - [37] Eichner AJ. Superplastisches Fertigen komplexer Formstücke. Werkstatt und Betrieb 1981; 114: 715-18. - [38] Winkler P-J, Keinath W. Superplastische Umformung, ein werkstoffsparendes und kostengünstiges Fertigungsverfahren für die Luft- und Raumfahrt. Metall 1980; 34: 519-25. - [39] Pischel H. Superplastisches Blechumformen. Werkstatt und Betrieb 1989; 122: 165-9. - [40] Bunk W, Kellerer H. Neue Fertigungsverfahren zur Verbesserung der Wirtschaftlichkeit. Aluminium 1985; 61: 247-51. - [41] Richards JH. Einsatz superplastisch umgeformter Blechbauteile im Bauwesen. Aluminium 1987; 63: 360-7. - [42] Hojas M, Külein W, Siegert K, Werle T. Herstellung von superplastischen Aluminiumblechen und deren Verarbeitung mit numerisch gesteuerten Pressen. In [29]. - [43] Lange K, Meyer-Nolkemper H. Gesenkschmieden, 2nd edn. Springer, Berlin, 1977. - [44] Bruchanow AW, Rebelski AV. Gesenkschmieden und Warmpressen. Verlag Technik, Berlin, 1955. - [45] Rathjen C. Die historische Entwicklung des Strangpressverfahrens. Ind Anz 89. 1967; 47: 17/2. - [46] Ziegler W, Siegert K. Spezielle Anwendungsmöglichkeiten der indirekten Strangpressmethode. Metall 1977; 31: 845-51. - [47] Ruppin D, Müller K. Kalt-Strangpressen von Aluminium-Werkstoffen mit Druckfilmschmierung. Aluminium 1980; 56: 263-8, 329-31, 403-6. - [48] Ziegler W, Siegert K. Indirektes Strangpressen von Leichtmetall. Metallkunde 1973; 64: 224-9. - [49] Pugh H Li D. The mechanical behaviour of materials under pressure. Applied Science Publishers, London, 1971. - [50] Hornmark N, Ermel D. Kupferumhülltes Aluminium, ein neuer Werkstoff für die industrielle Fertigung von Kompoundleitern. Draht-Welt 1970; 56: 424-6. - [51] Fiorentino RJ, Richardson BD, Sabrow AM, Boulger FW. New Developments in Hydrostatic Extrusion. Proc Int Conf Manuf Techn 1967; 25/28: 941-54. - [52] Fiorentino RJ, Sabrow AM, Boulger FW. Advances in hydrostatic extrusion. Tool Mfng Engr 1973; 00: 77-83. - [53] Pugh H Li D, Donaldson GHH. Hydrostatic extrusion - a review. Ann CIRP 1972; 21/2: 000-0. - [54] Fiorentino RJ, Meyer GE, Byrer TG. Some practical considerations for hydrostatic extrusion. Metallurgia and Metal Forming 1974; 00: 210-13, 296-9. - [55] Fiorentino RJ, Meyer GE, Byrer TG. Technical and economic potential of hydrostatic extrusion over conventional extrusion. Preliminary reports for the symposium "Neue Verfahren für die Halbzeugherstellung". Deutsche Gesellschaft f Metallkunde, 1973. - [56] Fiorentino RJ, Meyer GE, Byrer TG. The thick-film hydrostatic extrusion process. Metallurgia and Metal Forming 1972; 00: 200-3.

K4 Cutting. [1] Patzke M. Einfluss der Randzone auf die Zerspanbarkeit von Schmiedeteilen. Thesis, University of Hannover, 1987. - [2] Warnecke G. Spanbildung bei metallischen Werkstoffen. Fertigungstechnische Ber, vol. 2. Resch, Gräfelfing, 1974. - [3] Bartsch S. Verschleissverhalten von Aluminiumoxidschneidstoffen unter stationärer Belastung. Thesis, University of Hannover, 1988. - [4] Tönshoff HK. Schneidstoffe für die spanende Fertigung. wt-Z Ind Fert 1982; 72: 201-8. - [5] Knorr W. Bedeutung des Schwefels für die Zerspanbarkeit der Stähle unter Berücksichtigung ihrer Gebrauchseigenschaften. Stahl und Eisen 1977; 97: 414-23. - [6] Kienzle O, Victor H. Die Bestimmung von Kräften und Leistungen an spanenden Werkzeugmaschinen. VDI-Z 1952; 94: 299-305. - [7] Taylor FW. On the art of cutting metals. Trans Am Soc Mech Engrs 1907; 28: 30-351. - [8] Gawehn H. Das Spanwinkelproblem des Spiralbohrers. Maschinenbau und Betrieb 1931; 00: 440-6. - [9] Spur G. Beitrag zur Schnittkraftmessung beim Bohren mit Spiralbohrern unter Berücksichtigung der Radialkräfte. Thesis, TU Brunswick, 1961. - [10] Hütte, Taschenbuch für Betriebsingenieure, vol. 1: Fertigung, 5th edn. Ernst, Berlin, 1957. - [11] Tuffentsammer K. Kurzlochbohren mit unterschiedlichsten Werkzeugen möglich. Ind Anz 1980; 102: 100, 38-41. - [12] Victor H, Müller M, Opferkuch R. Zerspantechnik, vols I-III. Springer, Berlin, 1985. - [13] Kamm H. Beitrag zur Optimierung des Messerkopffräsens. Thesis, University of Karlsruhe, 1977. - [14] Victor HR. Zerspankennwerte. Ind Anz 1976; 98: 1825-30. - [15] Müller M. Zerspankraft, Werkzeugbeanspruchung und Verschleiss beim Fräsen mit Hartmetall. Thesis, University of Karlsruhe, 1982. - [16] Roese H. Untersuchung der dynamischen Stabilität beim Fräsen. Thesis, RWTH Aachen, 1967. - [17] Borys WE. Vergleichsuntersuchungen zum Einsatz hochharter polykristalliner Schneidstoffe beim Fräsen. Thesis, University of Hannover, 1984. - [18] Chryssolouris G. Einsatz hochharter polykristalliner Schneidstoffe beim Drehen und Fräsen. Thesis, University of Hannover, 1984. - [19] Töllner K. Fräsen von hochharten Eisenstoffen. wt-Z Ind Fert 1982; 72: 493-6. - [20] Tönshoff HK, Bussmann W. Formfehler bei der Hartbearbeitung: Fräsen gehärteter Führungsflächen. Ind Anz 1988; 110: 29,

35–6. – [21] Tönshoff HK, Hernándes-Camacho J. HFF-Ber 10. Hannover, March 1987, pp. 1–20. – [22] Laufer HJ. Einsatz von Prozessmodellen zur rechnerunterstützten Auslegung von Räumwerkzeugen. Thesis, University of Karlsruhe, 1988. – [23] Opferkuch R. Die Werkzeugbeanspruchung beim Räumen. Thesis, University of Karlsruhe, 1981. – [24] Dworak U. Mechanical strengthening of alumina and zirkonia ceramics through the introduction of secondary phases. Sci Ceramics 1987; 9: 543. – [25] Choi H-Z. Beitrag zur Ursachenanalyse der Randzonenbeeinflussung beim Schleifen. Thesis, University of Hannover, 1986. – [26] Kurrein M. Die Messung der Schleifkraft. Werkstattstechnik 1927; 20: 585–94. – [27] Snoeys R. The mean undeformed chip thickness as a basic parameter in grinding. Ann CIRP 1971; 20: 183–6. – [28] Tönshoff HK, Brinksmeier E, Choi HZ. Messung und Berechnung mechanischer und thermischer Werkstoffbeanspruchungen beim Schleifen. Jahrbuch Schleifen, Honen, Läppen und Polieren, 53rd edn. Vulkan, Essen, 1985, pp. 31–47. – [29] Brinksmeier E. Randzonenanalyse geschliffener Werkstücke. Thesis, University of Hannover, 1982. – [30] Saljé E. Abrichtverfahren mit unbewegten und rotierenden Abrichtwerkzeugen. Jahrbuch Schleifen, Honen, Läppen und Polieren, 50th edn. Vulkan, Essen, 1981, pp. 284–98. – [31] Saljé E. Abrichten während des Schleifens – Grundlagen, Leistungssteigerungen, Wirtschaftlichkeit. Jahrbuch Schleifen, Honen, Läppen und Polieren, 53rd edn. Vulkan, Essen, 1985, pp. 1–30. – [32] Tönshoff HK, Dennis P. Hochleistungsbandschleifen – ein massgebendes Verfahren. Werkstattstechnik 1988; 78: 665–9. – [33] König W, Tönshoff HK, Fromlowitz J, Dennis P. Belt grinding. Ann CIRP 1986; 35: 487–94. – [34] Mushardt H. Modellbetrachtungen und Grundlagen zum Innenrundhonen. Thesis, TU Brunswick, 1986. – [35] Tönshoff T. Formgenauigkeit, Oberflächenrauheit und Werkstoffabtrag beim Langhubhonen. Thesis, University of Karlsruhe, 1970. – [36] Saljé E, Möhlen H, See vM. Vergleichende Betrachtungen zum Schleifen und Honen. VDI-Z 1987; 129, (1): 66–69. – [37] See vM. Prozessoptimierung beim Honen. Jahrbuch Schleifen, Honen, Läppen und Polieren, 55th edn. Vulkan, Essen, 1988, pp. 401–14. – [38] Spur G, Simpfendörfer D. Neue Erkenntnisse und Entwicklungstendenzen beim Planläppen. Jahrbuch Schleifen, Honen, Läppen und Polieren, 55th edn. Vulkan, Essen, 1988, pp. 469–80. – [39] Nölke H-H. Spanende Bearbeitung von Siliziumnitrid-Werkstoffen durch Ultraschall-Schwingläppen. Thesis, University of Hannover, 1980. – [40] Tönshoff HK, Brinksmeier E, Schmieden VM. Grundlagen und Theorie des Innenlochtrennens. Jahrbuch Schleifen, Honen, Läppen und Polieren, 55th edn. Vulkan, Essen, 1988, pp. 481–93. – [41] Brinksmeier E, Schmieden vW. Werkzeugaufspannung und Prozessverlauf beim ID-Trennschleifen. Ind Diamanten Rundsch 1988; 00: 214–19. – [42] Weckerle D. Prozessstörungen bei der funkenerosiven Metallbearbeitung. Tech. Mitt. F Deckel AG, Munich, 1985. – [43] Schmohl H-P. Ermittlung funkenerosiver Bearbeitungseigenspannungen in Werkzeugstählen. Thesis, TU Hannover, 1973. – [44] Wijers JLC. Numerically controlled diesinking. EDM-Digest 1984; 00: 9/10. – [45] Schumacher B, Weckerle D. Funkenerosion – Richtig verstehen und anwenden. Technischer Fachverlag, Velbert, 1988. – [46] Tönshoff HK, Seemrau H. Laser beam machining in new fields of application. Proceedings ASME Symposium, Chicago, December 1988. – [47] Tönshoff HK, Bütje R. Excimer laser in material processing. Ann CIRP 1988; 37: 000–0. – [48] Dickmann K, Emmelmann C, Hohensee V, Schmatjko KJ. Excimer-Hochleistungslaser in der Materialbearbeitung. Laser Magazin, pt 1, 1987; issue

3: 26–9, and pt 2, 1987; issue 4: 34–44. – [49] Bimberg D. Laser in Industrie und Technik, vol. 13. Expert, Grafenau, 1985. – [50] Semrau H. Erzeugen von oxidfreien Schnittflächen durch Laserstrahlschneiden. Thesis, University of Hannover, 1989. – [51] Beske EU, Meyer C. Schweissen mit kW-Festkörperlasern. Laser Magazin 1989; (3): 42–6. – [52] Beske EU. Handhabung einer Lichtleitfaser zum Führen eines Nd-Yag-Laserstrahls. Laser und Optoelektronik 1989; 21(3): 60–1. – [53] König W, Schmitz-Justen C, Trasser Fr-J, Willerscheid H. Provisional list of terminology for laser beam cutting. Ann CIRP 1988; 37: 675–80. – [54] Spur G. Handbuch der Fertigungstechnik, vol. 2/3: Stöferle, T. Hanser, Munich, 1985. – [55] Lange K. Umformtechnik: Handbuch für Industrie und Wissenschaft, vol. 3. Springer, Berlin, 1990. – [56] VDI Code 3368. Die clearance, punch and die plate size for blanking punches. VDI-Verlag, Düsseldorf, 1965. – [57] Tschätsch H. Taschenbuch Umformtechnik: Verfahren, Maschinen, Werkzeuge. Hanser, Munich, 1977. – [58] Guidi A. Nachschneiden und Feinschneiden. Hanser, Munich, 1965. – [59] VDI Code 3345. Fine blanking (with further literature). VDI-Verlag, Düsseldorf, 1980.

Standards and Codes. DIN 2310: Thermal cutting.
DIN 4990: Groups of application of carbides for machining by chip removal.
DIN 6580: Movements and geometry of the chip removing process.
DIN 6581: Reference systems and angles on the cutting part of the tool.
DIN 8200: Blasting techniques.
DIN 8580: Manufacturing methods.
DIN 8589: Manufacturing processes cutting.
DIN 8590: Manufacturing processes removal operations.
DIN 69 100: Bonded abrasive products.
DIN 51 384: Cooling lubricants.

ISO (International Organisation for Standardisation).
ISO 513: Application of carbides for machining by chip removal.
ISO 3002: Basic quantities in cutting and grinding. Part 1: Geometry of the active part of cutting tools. Part 3: Geometric and kinematic quantities cutting.
ISO 3685: Tool life testing with single point turning tools.

VDI Codes. VDI Code 3332: Chip breakers at carbide-tipped lathe tools.
VDI Code 3335: Groups of application in the field of chip removal and operating angles for turning on a lathe with carbide metal tools.
VDI Code 3400: Electrical discharge machining – concepts, methods, application.
VDI Code 3401, Sheet 2: Electrochemical machining – bath electrolytic machining.
Stahl-Eisen-Prüfblatt 1160: Chip removing tests, general basic concepts.

K5.1 Thread Production. [1] Spur G. Mehrspindel-Drehautomaten. Hanser, Munich, 1970. – [2] Stock-Taschenbuch. R. Stock AG, Berlin, 1979. – [3] Stender W. Schälen von Gewindespindeln. Brochure from Messrs Waldrich, Coburg. – [4] Druminski R. Analytische und experimentelle Untersuchungen des Gewindeschleifprozesses beim Längs- und Einstechschleifen. Thesis, TU Berlin, 1977. – [5] Lickteig E. Schraubenherstellung. Verlag Stahleisen, Düsseldorf, 1966. – [6] Siebert H. Werkstattblatt 501: Gewindefurchen. Hanser, Munich, 1970.

K5.3 Manufacturing in Precision Engineering and Microtechnology. [10] Degner W, Böttger HC. Handbuch Feinbearbeitung. Hanser, Munich, 1979. – [11] Grünwald.

Fertigungsverfahren in der Gerätetechnik. VEB Verlag Technik, Berlin, 1980. - [12] Schweizer W, Kiesewetter L. Moderne Fertigungsverfahren der Feinwerktechnik, ein Überblick. Springer, Berlin, 1981. - [13] Lotter B. Wirtschaftliche Montage. Handbuch für Elektrogerätebau und Feinwerktechnik. VDI-Verlag, Düsseldorf, 1986. - [14] Heuberger A. Mikromechanik. Mikrofertigung mit Methoden der Halbleitertechnologie. Springer, Berlin, 1989. - [15] Weber H, Herziger G. Laser. Grundlagen und Anwendungen. Physik-Verlag, Weinheim, 1972. - [16] Bimberg D. Laser in Industrie und Technik, 2nd edn. Expert, Sindelfingen, 1985. - [17] Kiesewetter L. Laser – ein Werkzeug der Feinwerktechnik. Wissenschaftsmagazin TU-Berlin, issue 9, 1986, 33-7. - [18] Herziger G. Werkstoffbearbeitung mit Laserstrahlung. Feinwerktechnik + Messtechnik 1983; 91: 156-63. - [19] Seiler P. Festkörper-Impulslaser zum Fügen und Abtragen im Mikrobereich. Company publication of Carl Haas, Schramberg. - [20] Moller W. Laser-Mikrolöten mit Temperatur- und Zeitsteuerung. Optoelektronik Mag 1988; 4: 684-9. - [21] Benninghoff H. Werkstoffbearbeitung mit dem Laser. Tech Rundsch 1989; 6: 26-31. - [22] Brochure issued by Kammerer GmbH, Pforzheim-Huckenfeld, 1989. - [23] Schiller S, Heisig U, Panzer S. Elektronenstrahltechnologie. Stuttgart, Wissenschaftliche Verlagsanstalt, 1977. - [24] Dobeneck vD. Die Elektronenstrahltechnik – ein vielseitiges Fertigungsverfahren. Feinwerktechnik + micronic 1973; 77: 98-106. - [25] Behnisch H. Einsatz des Elektronen- und Laserstrahls in der Schweiss- und Schneidtechnik. Technica (CH) 1976; 25: 1341-7. - [26] Schulz H. Schweissen von Sondermetallen. Deutscher Verlag für Schweisstechnik, Düsseldorf, 1971. - [27] Schiller S, Panzer S. Thermische Oberflächenmodifikation metallischer Bauteile mit Elektronenstrahlen. Metall 1985; 39: 227-32. - [28] Schiller S, Heiseg U, Frach P. Elektronenstrahlbedampfen. In: Sudarshan TS, Surfacing technologies handbook. Marcel Dekker, New York, 1987. - [29] Lehfeldt W. Ultraschall. Vogel, Würzburg, 1973. - [30] Matauschek J. Einführung in die Ultraschalltechnik. VEB Verlag Technik, Berlin, 1962. - [31] Millner R. Ultraschalltechnik. Grundlagen und Anwendungen. Physik-Verlag, Weinheim, 1987. - [32] Dorn L. Schweissen in der Elektro- und Feinwerktechnik. Expert, Grafenau/Württ., 1984. - [33] Abel F. Ultraschall in der Kunststoff-Fügetechnik. Herfürth GmbH, Hamburg, 1979. - [34] Berger A. Elektrisch abtragende Fertigungsverfahren. VDI-Verlag, Düsseldorf, 1977. - [35] Degner W, Böttger HC. Handbuch Feinbearbeitung. Hanser, Munich, 1979. - [36] Grünwald F. Fertigungsverfahren in der Gerätetechnik. VEB Verlag Technik, Berlin, 1980. - [37] Degner W. Elektrochemische Metallbearbeitung. VEB Verlag Technik, Berlin, 1984. - [38] Janicke J. Anwendungstechniken der Funkenerosion. Tech Rundsch 1975; 31: 10-11. - [39] Schadach P. Elektroerosive und elektrochemische Metallbearbeitungsverfahren. VDI-Z 1975; 117: PT32-PT37. - [40] Haefer RA. Oberflächen- und Dünnschicht-Technologie. Pt 1: Beschichtungen von Oberflächen. Springer, Berlin, 1987. - [41] Simon H, Thoma M. Angewandte Oberflächentechnik für metallische Werkstoffe. Hanser, Munich, 1985. - [42] Czichos H. Konstruktionselement Oberfläche. Konstruktion 1985; 37: 219-27. - [43] Paatsch W. Technologische Eigenschaften galvanisch abgeschiedener Schichten. Galvanotechnik 1985; 75: 1234-41. - [44] Frey H. Dünnschichttechnologie. VDI-Verlag, Düsseldorf, 1987. - [45] Ikeno H. Electrooptic bistability of a ferroelectric liquid crystal device prepared using polyimide Langmuir–Blodgett orientation films. Jap J Appl Phys 1988; 27: L475. - [46] Kiesewetter L, Gleske G. Bauform und

Fertigungsverfahren für Flüssigkristall-Anzeigen. Berlin-Tronics 10. Verlag für technische Publikationen, Berlin, 1988, 4-8. - [47] Hanke H-J, Fabian H. Technologie elektronischer Baugruppen. VEB Verlag Technik, Berlin, 1977. - [48] Joachim F-W. Kupferplattiertes Invar als Metallkern in Leiterplatten mit einstellbarem Wärmeausdehnungskoeffizienten. Feinwerktechnik und Messtechnik 1986; 94: 507-9. - [49] Huber B. Leiterplatten- und Hybridtechnologien im Vergleich. Feinwerktechnik und Messtechnik 1986; 94: 215-20. - [50] Duppen vJ. Handbuch für den Siebdruck. Verlag der Siebdruck, Lübeck, 1981. - [51] Scheer HG. Siebdruck und Elektronik-Druckformherstellung in der Elektronik. IS + L 1983/4 (August). - [52] Steinberg JJ, Horowitz SJ, Bacher RJ. Herstellen von Mehrlagenschaltungen mit niedrig sinterden grünen Keramikfolien. EPP Hybridtechnik, October 1986, pp. 43-7. - [53] Lehmann HW, Gale T. Submikrongitter. Tech Rundsch 1989; 00: 46-53. - [54] 0.4 µm-Strukturen mit normaler Optik. Elektronik 1984; 17: 22. - [55] Jagt JC, Whipps PW. Elektronenempfindliche Negativlacke für VLSI. Philips Tech Rundsch 1981; 39: 368-75. - [56] Staudte HJ. Proceedings 27th Annual Symposium on Frequency Control, 1973, 50-4. - [57] Zwingg W. Miniaturquerschwinger und -Quarzsensoren. Jahrbuch der Deutschen Gesellschaft für Chronometrie e.V., vol. 36. Stuttgart, 1985. - [58] Johansson S. Micromechanical properties of silicon. Acta Universitatis Upsaliensis, Faculty of Science, Uppsala, 1988. - [59] Petersen KE. Silicon as a mechanical material. Proc IEEE 70: 1982; 420-57. - [60] Hohm D. Mikromechanik eröffnet neue Wege zu elektroakustischen Wandlern. Spektrum der Wissenschaft 1988; 00: 38-50. - [61] Herzog H-J, Csepregi L. X-ray investigation of boron- and germanium-doped silicon epitaxial layers. I. Electrochem Soc 1984; 131: 000-0. - [62] Becker EW, Ehrfeld W. Das LIGA-Verfahren. Phys B1 1988; 44: 166-70.

K5.4 Surface Coating. [63] Pulker HK. Verschleissschutzschichten unter Anwendung der CVD/PVD-Verfahren. Expert, Sindelfingen, 1985. - [64] Günther KC. Advanced coating by vapour phase processes. Ann CIRP 1989; 38: 645-55.

K6 Assembly. [1] Spur G, Stöferle T (ed.). Handbuch der Fertigungstechnik, vol. 5: Fügen, Handhaben, Montieren. Hanser, Munich, 1986. - [2] Wranecke H-J, Schraft RD (ed.). Handbuch – Handhabungs-, Montage- und Industrierobotertechnik, vol. 3: Montagetechnik. Verlag Moderne Industrie, Munich, 1984. - [3] Lotter B. Wirtschaftliche Montage. Ein Handbuch für Elektrogerätebau und Feinwerktechnik. VDI-Verlag, Düsseldorf, 1986. - [4] DIN 8593: Manufacturing production processes joining. Classification, subdivision, concepts. Beuth Verlag, Berlin, 1985. - [5] VDI Code 2860, Sheet 1, Draft: Assembly and handling units. Handling functions, handling units, terminology, definitions and symbols. VDI-Verlag, Düsseldorf, 1982. - [6] Seliger G (ed.). Montagetechnik. gfmt, Munich, 1989. - [7] Andreasen, Kähler, Lund. Montagegerechtes Konstruieren. Springer, Berlin, 1985. - [8] Deutschländer A. Integrierte rechnerunterstützte Montageplanung. Series: Produktionstechnik Berlin, vol. 72. Hanser, Munich, 1989. - [9] Severin F. Flexibel automatisierte Montageeinrichtungen – Innovationspotential der achtziger Jahre (pt 1). ZwF 1982; 77: 529-40.

K7 Production and Works Management. [1] Olbrich W. Arbeitsplanerstellung unter Einsatz elektronischer Datenverarbeitungsanlagen. Thesis, RWTH Aachen, 1970. - [2] Warnecke HJ, Hirschbach O, Metzger H. Rechnerunterstützte Montagearbeitsplanerstellung. wt-Z ind Fertig

1975; 65: 147–52. – [3] Warnecke HJ, Graf H, Kunerth W. Stand und Entwicklungstendenzen technisch organisatorischer Informationssysteme. In: Hansen HR. Informationssysteme im Produktionsbereich. Oldenbourg, Munich, 1975. – [4] Hahn R, Kunerth W, Roschmann K. Fertigungssteuerung mit elektronischer Datenverarbeitung. Beuth, Berlin, 1973. – [5] Graf H. Methodenauswahl für die Materialbewirtschaftung in Maschinenbau-Betrieben. Thesis, University of Stuttgart, 1977. – [6] Rabus G, Nakonzer K. Analyse der Fertigungssteuerungsaufgaben im Hinblick auf den EDV-Einsatz. Unpublished research report of the Institut für Produktionstechnik und Automatisierung (IPA), 1976. – [7] Scharf P. Strukturen flexibler Fertigungssysteme. Gestaltung und Bewertung. Krausskopf, Mainz, 1976. – [8] Frank E. Handhabungseinrichtungen. Krausskopf, Mainz, 1975. – [9] Dröge KH. Telemanipulatoren – Stand der Technik. Paper for the 5th workshop of the Institut für Produktionstechnik und Automatisierung (IPA): "Erfahrungsaustausch Industrieroboter". Stuttgart, 1975. – [10] Warnecke HJ, Schraft R-D. Einlegegeräte zur automatischen Werkstückhandhabung. Krausskopf, Mainz, 1973. – [11] Warnecke HJ, Schraft R-D. Industrieroboter. Krausskopf, Mainz, 1989. – [12] Schweizer M. Robotertechnik. Bibliothek der Technik, vol. 1. Verlag moderne industrie, 1987. – [13] Gerlach B. Spanende Sonderwerkzeugmaschinen. Techn. Verlag Grossmann, Stuttgart, 1977. – [14] Warnecke HJ, Gericke E, Vettin G. Auslegung der Verkettungseinrichtungen flexibler Fertigungssysteme mit Hilfe der Simulation. Proceedings of the CIRP Seminars on Manufacturing Systems 1976; 5: 155–64. – [15] Masing W. Handbuch der Qualitätssicherung, 2nd edn. Hanser, Munich/Vienna, 1988. – [16] VDA: Qualitätskontrolle in der Automobilindustrie – Sicherung der Qualität vor Serieneinsatz. Verband der Automobilindustrie e.V., Frankfurt, 1977. – [17] Ford Motor Company publication: Statistische Prozessregelung. Leitfaden Eu880b, April 1986. – [18] Warnecke HJ, Melchior KW, Ahlers R-J, Kring J. Handbuch Qualitätstechnik: Methoden und Geräte zur effizienten Qualitätssicherung. moderne industrie, Landsberg/Lech, 1987. – [19] Warnecke HJ, Bullinger H-J, Hichert R. Kostenrechnung für Ingenieure. Hanser, Munich, 1979. – [20] Mellerowicz K. Kosten und Kostenrechnung, vol. 1, 5th edn. De Gruyter, Berlin, 1973. – [21] Bussmann KF. Industrielles Rechnungswesen. Poeschel, Stuttgart, 1963. – [22] Warnecke HJ, Bullinger H-J, Hichert R. Wirschaftlichkeitsrechnung für Ingenieure. Hanser, Munich, 1980. – [23] Institut für angewandte Arbeitswissenschaft e.V. (ed.). Arbeitsgestaltung in Produktion und Verwaltung: Taschenbuch für den Praktiker. Bachem, Cologne, 1989. – [24] Bullinger H-J, Solf JJ. Ergonomische Arbeitsmittelgestaltung I: Systematik/Forschungsbericht no. 196, Bundesanstalt für Arbeitsschutz, Dortmund. Bremerhaven, Wirtschaftsverlag NW, 1979. – [25] Neudörfer A. Anzeiger und Bedienteile: Gesetzmässigkeiten und systematische Lösungssammlungen. VDI Verlag, Düsseldorf, 1981. – [26] Lange W. Kleine Ergonomische Datensammlung, 4th edn, ed. Bundesanstalt für Arbeitsschutz TÜV Rheinland, Cologne, 1985. – [27] Schmidtke H (ed.). Lehrbuch der Ergonomie, 2nd edn. Hanser, Munich, 1981.

Standards and Codes. DIN 5034: Daylighting of interiors (principles).

DIN 5035: Artificial lighting of interiors.

DIN 5036: Radiometric and photometric properties of materials.

DIN 33 402: Body dimensions of adults.

DIN 40 080: Sampling procedures and tables for inspection by attributes.

DIN 45 635: Measurement of airborne noise emitted by machines.

DIN 69 513–69 643: Machine tools (various subheadings).

DIN IEC 651: Second level meters.

Manufacturing Systems

B. Behr, Aachen; E. Dannenmann, Stuttgart; I. Dorn, Berlin; G. Pritshow, Stuttgart; K. Siegert, Stuttgart; G. Spur, Berlin; M. Weck, Aachen; T. Werle, Stuttgart.

1 Machine Tool Components

M. Weck and **B. Behr**, Aachen

1.1 Fundamentals

1.1.1 Function Structure

System Structure

Manufacturing systems are classified according to DIN 8590. Subsystems are machine tools which are defined according to DIN 69 651 as "mechanised and partially or fully automated manufacturing systems which can generate a prescribed form or change in the workpiece as a result of the relative motion between the *tool* and *workpiece*". Individual and multiple machine tool systems consist of one or more basic machine tool systems and other operational and auxiliary systems.

The assemblies (i.e. drives, frame components, tool carriers and workpiece carriers) required for the execution of the *basic operation* form the *basic machine tool system*. The designs of the various tool and workpiece carriers range from rigid tables to a number of combinations of linear and rotary guides and/or bearings depending on the machine tool structure. Tools and workpieces are held or clamped onto the respective carriers. The embodiment design of the mechanical interfaces between the equipment components and the machine is determined by the interchangeability and flexibility of the machine tools in a variety of different machining tasks. The complete machine tool system consists of various components of machine tool and machine tool flow systems, depending on the degree of automation, and there are a number of similarities between the components for the execution of the *handling*, *transport* and *storage* operations and the components of the basic machine tool system. Handling systems are linked to the basic system at the relevant clamping points (**Fig. 1**).

Action Pair, Effective Motion

A workpiece with a specific basic form is transformed into a given form as a result of the relative motion between tool, workpiece and process-related power transmission, i.e. separating and re-forming. The technical quality of a workpiece is determined by the dimensional accuracy and surface finish. Technological advances in machine tool components are constantly producing improvements in machining accuracy (**Fig. 2**).

Effective motion is composed of *cutting motion*, *infeed motion* and *feed motion*. These take the form of *linear*, *rotary*, *continuous*, or *intermittent* motion. This prod-uces a *three-dimensional working space* in proportion to the size of the feed or infeed axis and the work length, where applicable, i.e. in planing, shaping and slotting, and metal-forming machines. This space is cylindrical in turning machines and circular grinding machines; in milling, drilling, shaping and slotting machines it is normally cuboid. The working space in metal-forming machines is determined by the maximum travel and maximum tool surface area at right angles to the direction of travel.

Rotating motion normally occurs in the form of cutting motion in cutting machine tools, e.g. turning, drilling and milling. The required speed range is limited by the maximum and minimum cutting speed required and the largest and smallest workpiece and/or tool diameter. For each machining task an optimum rotational speed may be specified at which the most economic cutting speed is achieved. Higher cutting speeds are being achieved all the time with the improved efficiency of cutting materials. These high speeds make great demands on the constructions of spindle–bearing systems (**Fig. 3**). For example, a rotational speed of $n = 12\,500\ \text{min}^{-1}$ is required for a cutting speed of $2\,000\ \text{m/min}$ at a milling rate of $d = 50\ \text{mm}$, which represents a critical load for a roller bearing of diameter 100 mm or over.

To allocate effective motion purely in terms of the form of the workpiece would be misleading, however. The execution of the required motion with the workpiece carrier and tool carrier may be effected in a wide variety of ways with kinematic reversal, where the components of effective motion may be interchanged. This results in various types of machines which make an extremely wide range of demands on the components of linear and rotary motion, namely the guides. Rational arrangements may be derived from the individual machining task, as may the specific requirements for automatic workpiece and tool change facilities. These types range from machines where motion is confined to the tool carrier, through the various intermediate stages with their respective combinations, to those where motion is effected by the workpiece carriers.

Motion is normally actuated by separate *main* and *feed drives* and, less commonly, the feedrate is derived from the main spindle. *Gears* alter rotational speeds and speed–torque characteristics. *Transmission components*, such as threaded spindles and toothed belts, transmit the motion to the tool and/or workpiece carrier, normally consisting of slides with linear motion.

The forces generated by the production process at the active point, including friction force and force due to weight, are taken up by *guides and bearings* and conducted to assemblies such as slides, spindle drive boxes and tailstocks. The flow of force is completed through the

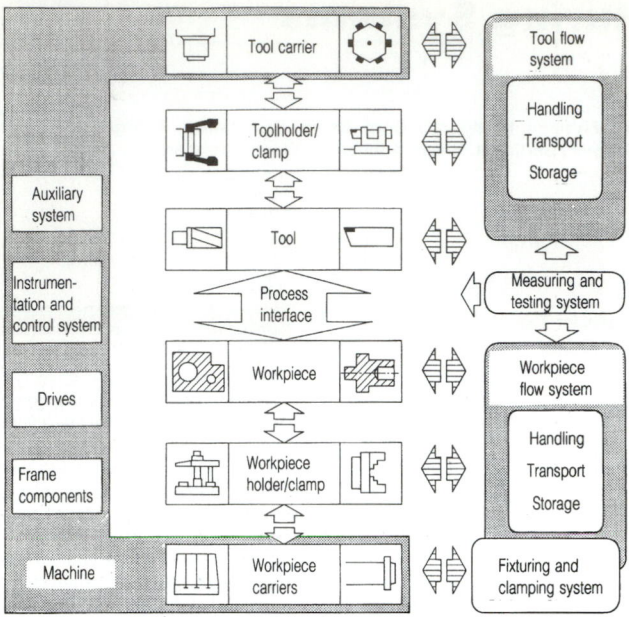

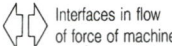
Interfaces in flow of force of machine

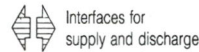
Interfaces for supply and discharge

Figure 1. System setup for machine tools and equipment accessories.

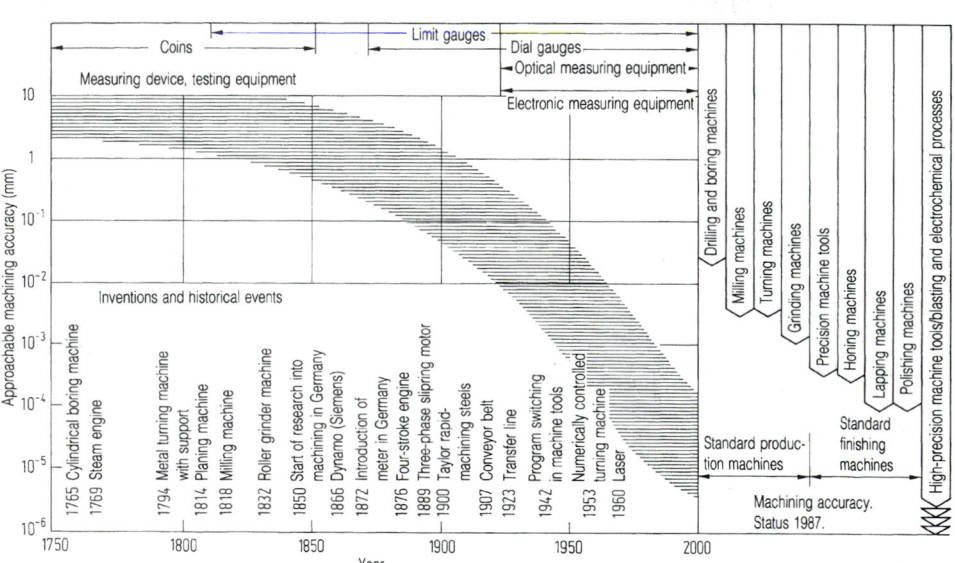

Figure 2. Chart showing historical progression of machining accuracy achieved in machine tools.

frame components, such as the columns and bed, which also provide the connection to the foundations. Static, dynamic and thermal loads produce elastic deformation in individual components which may result in surface imperfections on the workpiece or may affect economic efficiency.

1.1.2 Mechanical Characteristics

The static, dynamic and thermoelastic characteristics of a machine tool, regarding either the whole assembly or an individual part, may have a substantial effect on the process capability and quality of manufacture which is achievable with that tool.

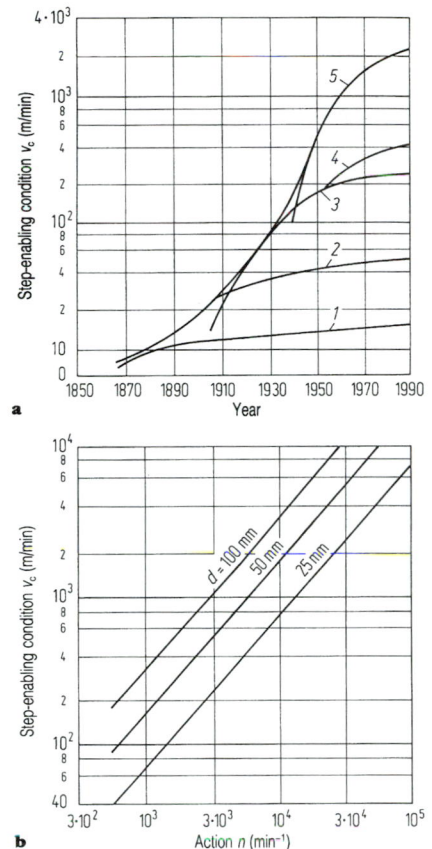

a

b

Figure 3. Trends **a** in cutting speeds and **b** in rotational speeds in machine tools in steel machining.

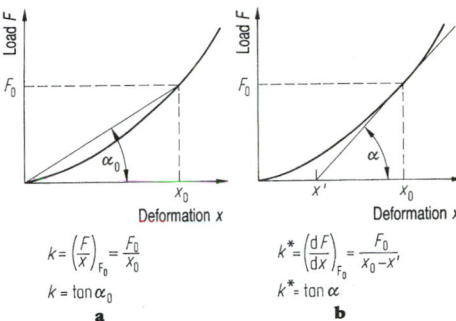

$$k = \left(\frac{F}{x}\right)_{F_0} = \frac{F_0}{x_0}$$

$$k = \tan \alpha_0$$

a

$$k^* = \left(\frac{dF}{dx}\right)_{F_0} = \frac{F_0}{x_0 - x'}$$

$$k^* = \tan \alpha$$

b

Figure 4. Definition of rigidity rate, **a** with secants, **b** with tangents.

rigidity rates k_i of the components involved, calculated from the sum of their respective compliance rates $1/k_{tot}$, as reciprocal values of their rigidity rates, thus: $1/k_{tot} = \Sigma \ 1/k_i$. The total machine is, thus, always "less stiff" than its most compliant part within the flow of force. The rigidity normally occurring at the cutting point in cutting machine tools is between 20 and 500 N/μm and between 10^4 and 10^5 N/μm in metal-forming machines measured between the ram and machine tool table.

Criteria for Dynamic Load

The dynamic characteristics of a machine tool are determined primarily by the static rigidity, spatial distribution and size of the component masses, as well as by system damping. Specific spatial *natural oscillation forms* are produced for the structure of each machine and/or part at certain individual frequencies, depending on these characteristics. It is essential for the description of the dynamic characteristics of complex machine tool structures of these types that the natural oscillation forms are known. This enables the individual components which predominantly cause the natural oscillations to be determined (weak point analysis). **Figure 5** shows the natural oscillation form of a bed-type milling machine for an individual supply frequency of 105 Hz. A bending oscillation can be seen in the vertical section of the column, and

Criteria for Static Load

The static characteristics of a machine tool are represented by elastic deformation, which occurs under a continuously rated load over time, i.e. process forces and forces due to weight. This consequently produces the *static rigidity k* as the most important factor. It is a measure for the resistance to changes in form and is shown as the ratio of load F to displacement x of the part in the direction of the application of force: $k = dF/dx$. The dependence of the deformation x on the force brought under load F is represented in the form of characteristic curves (**Fig. 4**; see A4.1 and F2.1). The theoretical relation is linear: $k = F/x$ (spring rate).

In practice, however, a progressive relation occurs as a result of a number of contact surfaces between the parts. There are two definitions for rigidity at a tool centre point. The secant is taken from the origin to the point under observation F_0, x_0 in the first diagram (**Fig. 4a**) and in the second diagram (**Fig. 4b**) the slope of the tangent to the characteristic curve at the point under observation F_0, x_0 is introduced. Depending on the type of load, this is known as the tensile, compressive, bending or torsional rigidity; the last (k_t) is shown as the ratio of torque M to the angle of rotation φ, $k_t = dM/d\varphi$.

The resulting rigidity k_{total} at the point of application of force is always obtained by carrying over the individual

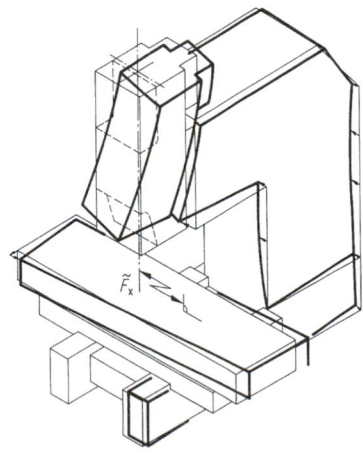

Figure 5. Natural oscillation forms of a bed-type milling machine.

slight torsion occurring in the horizontal section of the column.

In order to visualise the dynamic characteristics, one should imagine the machine tool divided up into mass elements elastically coupled to one another. Equilibrium conditions for *exciter forces F(t)*, displacement-related *spring energy*, rate-controlling *damping forces* and proportional gain *inertial forces* may be described by a system of differential equations. Dynamic characteristics of existing machines and frames may be obtained experimentally by exciting with different frequencies f [60]. The *compliance frequency response* $1/k_{dyn} = x_{dyn}/F_{dyn}$ is obtained from the ratio of the dynamic displacement x_{dyn} and exciter force F_{dyn} at the point of application of force and the phase displacement φ between the force sensor and displacement transducer. It may be shown separately as amplitude response and phase response or as a vector diagram (circle diagram). **Figure 6** shows measured frequency response curves and circle diagrams for an assembly with two resonant frequencies. Static compliance may be determined at $f = 0$. Resonant dynamic compliance is approximately 2 to 5 times higher than static compliance depending on system damping. Natural frequencies of a minimum of a factor of approximately 1.2 to 1.4 should be located outside the exciter frequency range generated by cutting forces or feed drives for example, to prevent resonant oscillation via external exciters. There is a danger of regenerative *chatter* [61] and destruction of tools and workpieces in dynamically weak machines. High individual frequencies are achieved by setting high static rigidity rates and minimising mass. These

are to be distributed in such a way that high-mass structures such as gears and motors are located at rigid points, i.e. the bed or the column underbody, where possible. Damping is to be as high as possible. The design of the joints and guides is critical, with particular attention to aspects such as oil film.

Damping may also be affected by the selection of materials, e.g. polymer concrete has a higher material damping effect than cast iron and this, in turn, has a higher value than steel. Both sand infill, where the casting core is not removed, and concrete infill are useful for increasing damping. In welding constructions the joints within the welded joint act as suppressors.

Criteria for Thermal Loads

The thermal characteristics of machine tools may be described in terms of the thermoelastic relative displacement at the active point between workpiece and tool as a result of the effects of heat sources. These displacements are determined by all parts in the thermal chain of effect and their thermal deformation properties. Variable temperature distributions over time (*isotherm lines*) and, consequently, time-dependent deformations occur in the parts as a result of *internal heat sources* present inside machine tools, such as bearings, motors, gears, process heat, etc., and *external heat sources* acting on machine tools, e.g. temperature of surrounding components, sun's rays, day–night temperature fluctuations, etc. **Figure 7** shows the potential heat sources and attributive effects of these thermal influences, using a knee-type milling machine as an example.

The typical temperature distributions occurring in the parts due to the heat sources are determined by the specific thermal material properties, such as heat capacity and heat conductivity, and the conditions for heat exchange from the surroundings or adjacent parts. The connections of the individual parts in relation to the machining position, the relative position of the individual parts and the interactive effect of the deformation of the parts may be cumulative, or may also have a balancing effect; nevertheless, they all have an effect on deformation resulting from the temperature distribution at the cutting point in conjunction with the heat expansion coefficient. Interactive compensation of thermal-related displacement in relation to the machining point may be exploited intentionally by way of a targeted design in terms of the heat sources (*thermosymmetrical* construction).

1.2 Drives

Drives are required mainly in machine tools for cutting and feed motion [1–9]. These are normally employed as separate drives for each individual motion, particularly with numerically controlled machines and less frequently as a common drive with power takeoff gears. Stepless drives are becoming more popular because of their greater compatibility with machining specifications. A distinction is made between electrical, hydraulic and pneumatic drives, depending on the type of drive circuitry and power supply (DIN 24 300) and hybrids such as electrohydraulic drives.

The term *drive* includes assemblies such as motors, power transformers, gears and transmission components.

1.2.1 Motors

Electrical Three-Phase Slipring Motors

These are traditionally employed in machine tools as *asynchronous motors* in conjunction with gear cone trans-

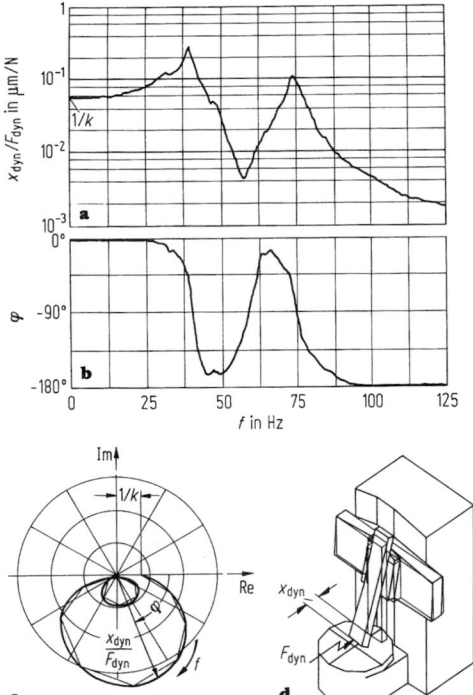

Figure 6. Compliance frequency response of a revolving turning machine with two resonant frequencies, measured on excitation of the plunger by F_{dyn}. **a** Amplitude response. **b** Phase response. **c** Circle diagram. **d** Oscillation form.

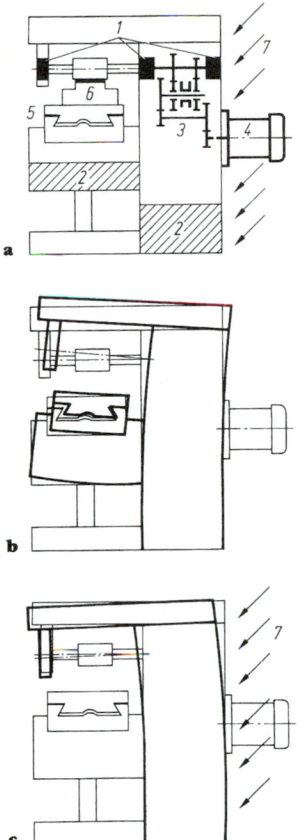

a

b

c

Figure 7. Examples of thermal-related deformation in a milling machine. **a** Principal heat sources: *1* bearings, *2* gear and hydraulic oil, *3* gears, clutches, *4* pumps, motors, *5* guides, *6* cutting point and swarf, *7* external heat transfer. **b** Deformation due to internal heat sources. **c** Deformation due to external heat transfer.

mission units (see L1.2.2). The controlled asynchronous motor is being selected increasingly for use as the main spindle drive of the machine, whereas the *synchronous motor*, even in its controlled form, is now being employed in specialised feed tasks. Both types of motor have a broad speed range (10^3 to 10^4), so that changeover gears are not normally required [2].

Squirrel-Cage Motors

These are the most common type of construction, as they require little maintenance and offer stable operation under rated conditions. However, they have a high starting current with a low starting torque. Thanks to the variety of squirrel-cage rotor designs, the motor may be adapted to suit the characteristics of the machine tool. *High-torque cage motors* (deep-bar squirrel-cage motors) have an extremely high breakaway torque with a low starting current and are therefore suited to direct-on-line starting. *High-resistance squirrel-cage motors* have the highest specified breakaway torque (as high as the breakdown torque) and a low starting current.

The rotational speed n may be changed in three-phase slipring motors by changing the number of poles p, as the speed is dependent on the number of poles p and the

frequency f, according to $n = 2f/p$. Pole-changing motors may be designed for all speeds with constant torque or with constant power.

Modern asynchronous motor applications cater for speed-controlled operations. Such drives are known as *servodrives*. Speed control is determined by instantaneous position and magnetic field strength. The magnetic field and torque are generated to the desired level independently of one another by the controlled input of stator currents. The combined effect of the two factors produced on the stator results in a constant rotor speed. The regulated asynchronous motor is based on so-called field-vector control (**Fig. 8**); see also L2 [10–12].

Example. **Figure 8** shows the relation between the field and stator winding coordinates. The field-vector control takes the established relationships as the basis, according to which the torque is controlled via the torque-producing current components and the speed is controlled by the flow-producing current components. The effect of the temperature on the rotor time constant and the magnetic saturation on the motor parameters represent the limits of the concept. Control of these influencing factors may further improve the quality of the regulated asynchronous motor [13].

Figure 9 shows an asynchronous motor of the squirrel-cage rotor type designed as a servomotor. To compensate for the relatively complex control procedure associated with the servo amplifier, there are advantages, such as the fact that it requires no maintenance and has a broad speed range under field control. The latter characteristic enables the speed to be adjusted over a broad range with a constant power output, see **Fig. 10**. This is why the regulated asynchronous motor is becoming increasingly popular as a main spindle motor. Up to 80 kW of power and speeds of up to 8000 min⁻¹ are adequate for main spindle drives. The speed may reach 14 000 min⁻¹ for feed ranges.

Slipring Motors. These are installed in machine tools with larger drive outputs and those with flywheel drives, because the energy stored in the flywheel may be released during the working stroke when the motor speed is reduced.

Synchronous Motors. These have developed from the permanently excited d.c. motor, where the role of the stator and rotor are interchanged. In synchronous motors the electrically generated exciter field in the stator rotates in relation to speed. Permanent magnets are installed in the rotor, and the primary windings are attached to the stator to generate the field of rotation. The separation of the current supplying the stator windings is effected in relation to the rotor position angle which has to be measured for this to be achieved. Sensors for rotor position and speed measurement are typically non-contact transmitters, so that electrical torque transmission is not required via collectors or brushes from the stator to the rotor or vice versa. The attributes of the design of synchronous motors, also known as brushless d.c. motors, therefore include lack of any need for maintenance and low heat buildup. At the same time, a more expensive electronic control system is needed than with d.c. motors, but the performance characteristics are broadly similar.

Example. **Figure 11** shows the principle of a six-pole, permanently-excited synchronous motor (the supply frequency is three times as high as the torque frequency of the motor). In rotor position *1* the current flows into winding phase *a* and out of winding phase *c*, whereas in position *2* the current flows into winding phase *b* and out of winding phase *c*. The polarities set with a time switch in this way in the stator windings produce a torque in the same direction on the magnetised rotor, which sets the rotor in motion in a clockwise direction. Current supply of the winding phases is carried out alternately synchronised at 60° to the rotor position. The current supply of each phase winding occurs periodically with

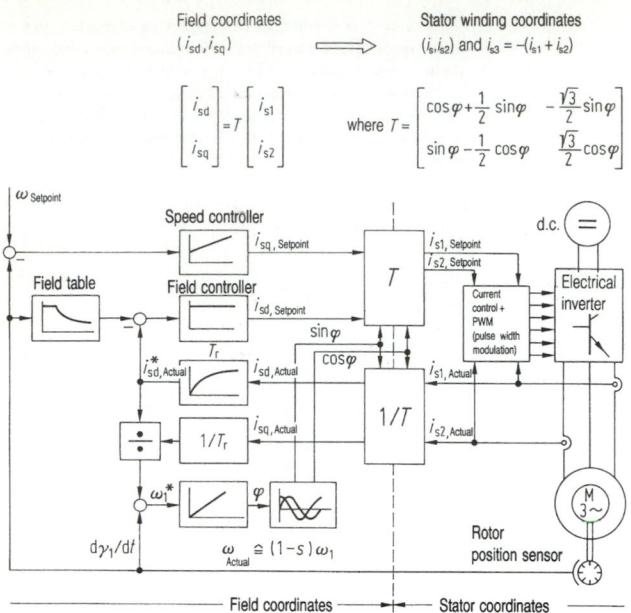

Field coordinates
(i_{sd}, i_{sq})

Stator winding coordinates
(i_{s1}, i_{s2}) and $i_{s3} = -(i_{s1} + i_{s2})$

$$\begin{bmatrix} i_{sd} \\ i_{sq} \end{bmatrix} = T \begin{bmatrix} i_{s1} \\ i_{s2} \end{bmatrix}$$

where $T = \begin{bmatrix} \cos\varphi + \frac{1}{2}\sin\varphi & -\frac{\sqrt{3}}{2}\sin\varphi \\ \sin\varphi - \frac{1}{2}\cos\varphi & \frac{\sqrt{3}}{2}\cos\varphi \end{bmatrix}$

Figure 8. Asynchronous motor control based on the principle of field orientation. ω rotational speed, i current, i_{sd} flow-producing current components, i_{sq} torque-producing current components, φ field coordinate angle, γ rotor position angle, s slip, T_r rotor time constant, T transformation matrix.

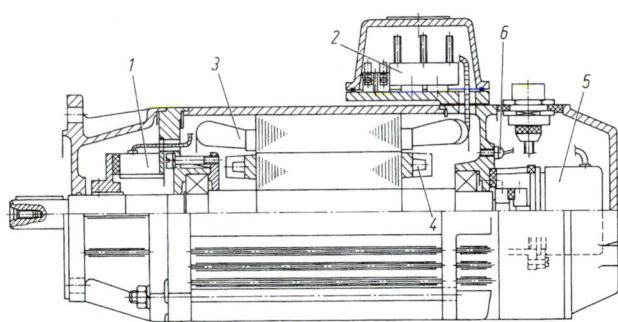

Figure 9. Structure of an asynchronous motor of the squirrel-cage rotor type (ABB): *1* holding brake, *2* motor and brake connections, *3* stator winding, *4* rotor winding, *5* measuring system, *6* thermistor.

a period of 40° supply and 20° rest, where these figures refer to the rotor revolution. In electrical terms, this produces angle deviations of 120° and 60°.

The responses of the induced voltage in the individual phase windings and at the terminals are also shown in **Fig. 11**. The response of the induced voltage within the current-free rest phase is, in principle, insignificant for constant power output. The trapezoidal response occurs as a result of skewing in the magnet plates on the rotor which is carried out to avoid problems with slotting [14, 15].

A basic distinction is made in synchronous motors between supply with sinusoidal and square-wave currents. The advantage of square-wave current supply lies in the simplified signal forming and the use of a simple sensor for detecting the position of the rotor. Two different rotor position sensors may be employed for sinusoidal current supply, depending on the degree of accuracy required. The sensor with only three sensor elements detects the initial points for the U, V and W supply in each case, whereas the cyclical sensor operating with absolute values also provides additional information on the absolute rotor position in addition to the three sensor elements for detecting the initial points for more accurate sinusoidal supply. In general, supply with sinusoidal currents may cause attenuation in the harmonic waves and thereby result in high synchronisation in the drive [15, 16].

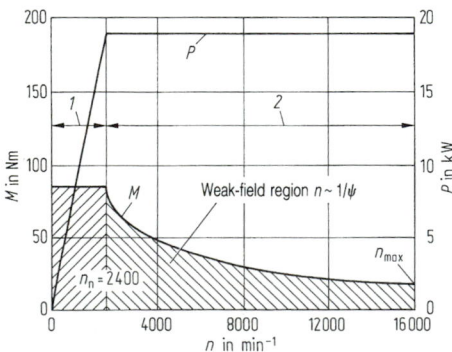

Figure 10. Typical characteristic curve of an asynchronous motor (AMK). *P* output, *M* torque, n_n rated speed, n_{max} maximum speed, ψ magnetic field strength, *1* constant torque, *2* constant power.

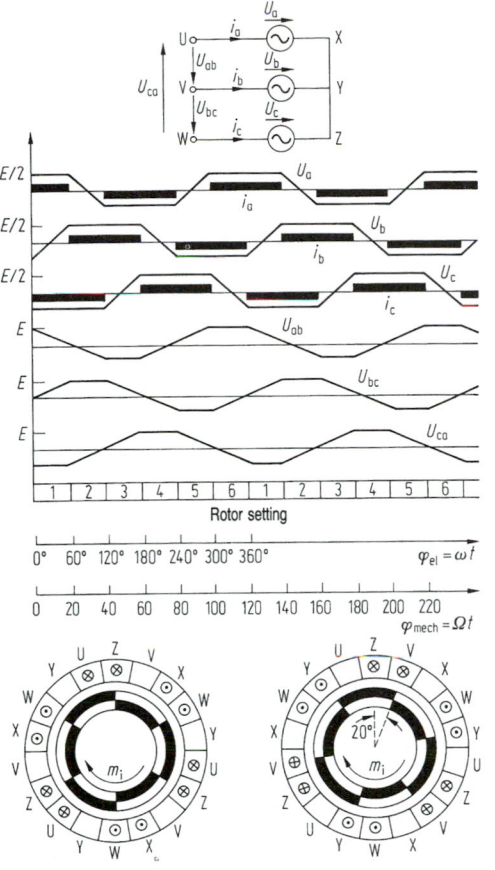

Figure 11. Function diagram of a permanently-excited synchronous motor (Bosch). $i_{a,b,c}$ winding phase current, $U_{a,b,c}$ winding phase voltage, $U_{ab,bc,ca}$ terminal voltage, U, V, W, X, Y, Z motor terminals, m_i internal torque, E intermediate circuit voltage.

Electronic circuitry may be produced with the synchronous motor control as shown in **Fig. 11** (see also **Fig. 12**). In this control concept, current is supplied in quasi-sinusoidal or (more precisely) trapezoidal form. The resolution required in the rotor position sensor for genuine sinusoidal supply would have to be far higher than that produced by three sensor elements offset to 60°. This is the only way constant-phase displacement or phase coincidence can be guaranteed between the induced voltage and phase winding current.

There is no commutation limit for synchronous motors which is comparable to that in d.c. motors. The power tends to be restricted by the servoamplifier instead. A typical characteristic curve field of a synchronous motor is shown in **Fig. 13** for the purpose of comparison with the d.c. motor. The standard speed range is up to 3000 min⁻¹, with a maximum output of up to 10 kW.

Asynchronous Linear Motors. These have low power at low speeds with less power efficiency and generate high levels of heat [17]. They are suitable for use in forging machines as pile hammer drives and in flexible manufacturing processes as transport drives for workpiece pallets.

New developments in asynchronous linear motors are

characterised by the fact that the servo characteristics have been much improved. The boundary circle frequency extends from 500 to 100 s⁻¹, with speeds of up to 3 m/s and maximum acceleration even up to 10 g [18]. It is already clear that the use of servodriven linear motors of the asynchronous type is set to increase dramatically in the future.

Electrical d.c. Motors

Electrical Shunt Motors. These are characterised by their high speed constant under load, and are employed where possible in main and feed drives for flywheel drives, with additional series field windings due to their stepless speed control characteristics. The high starting current in the armature circuit is limited by preliminary resistors and/or by thyristor power units. The direction of rotation may be changed by interchanging the armature or field connections.

The speed may be increased by increasing the armature voltage with constant torque and/or field control with constant power and reduced torque. In the Ward–Leonard system the rotational speed which may be achieved amounts to up to $B \approx 40$, with a thyristor power unit and speed control in the main drives. Armature control range of $B_A > 50$ is standard, and in feed drives with servomotors the range is even wider. The heat dissipation is lower at low speeds and external ventilation is therefore required.

Permanently Excited d.c. Motors. These are employed exclusively in feed drives with speed control [19]. They may be employed as shunt-characteristic motors with permanent field excitation. The speed may be adjusted via the armature voltage, while the power supply is by way of thyristor or transistor electrical inverter units from the three-phase supply with smoothing reactors in series. Speed control is effected via a reverse tachometer (coupled directly to the main shaft), which enables the speed to be reduced to virtually nil while maintaining high rotational regularity, thereby producing a broader control range of $B \approx 10^3$ to 10^4, suitable for continuous-path control operations. Special-purpose constructions (**Fig. 14**) reveal improvements in dynamic characteristics with the minimal mass moment of inertia compared with conventional constructions. Rotor windings have high specific loading and high phase spacing. High rates of increase of current due to low armature inductance, highest possible starting torques (3 to 10 times rated torques, depending on the current limiter of the inverter and switching frequency), with short-term high current overload capacity, enable extremely high startup times of 5 to 50 ms. **Figure 15** shows the structure of a d.c. motor control system with a transistor amplifier.

One special problem associated with d.c. motors is limiting the transmittable current. The reason for this is the type of current transmission which takes place via *brushes* and *collectors*. This provides a natural boundary for the maximum transmittable current, which may still be transmitted without damaging the contact elements. This characteristic is demonstrated in **Fig. 16**. The commutation limit is dependent on speed and decreases rapidly at increasing speeds. A speed-dependent current limiter is usually installed in the servoamplifier to compensate for this. The ratio between the maximum available torque and the rated torque is thereby effectively reduced. The characteristics for the motor govern its design, as the motor is normally designed for rated operation.

One long-term handicap with d.c. motor applications

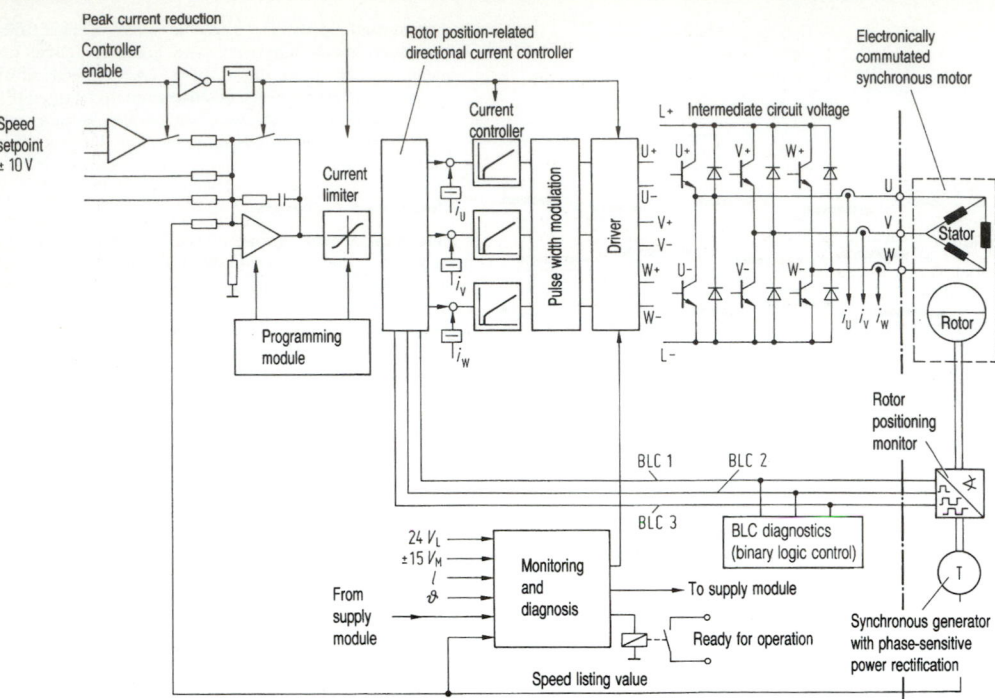

Figure 12. Structure of a synchronised motor control system (Indramat).

has been the relatively short lifetime of the commutator, which incurs high maintenance costs owing to the frequent replacement of worn parts. This has led to a preference in favour of the three-phase slipring motor over the past decades. Commutator technology has advanced in recent years, however, and lifetimes of over 30 000 hours are now being achieved. Direct-current technology is therefore rising in popularity again [20].

Bar-Type Rotors. These have a thin, unslotted rotor with a uniform winding and high winding density, rotational speeds of up to 3000 min^{-1} (in certain con-

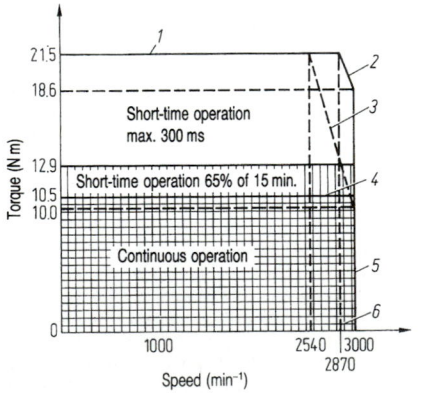

Figure 13. Characteristic curve field of a synchronous motor (Indramat). *1* maximum torque, *2* voltage limiter at rated current, *3* voltage limiter at 15% supply undervoltage, *4* continuous torque, *5* maximum speed, *6* buckling speed.

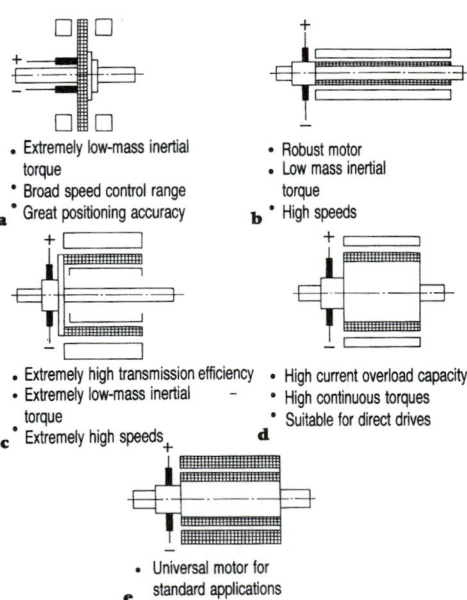

a
• Extremely low-mass inertial torque
• Broad speed control range
• Great positioning accuracy

b
• Robust motor
• Low mass inertial torque
• High speeds

c
• Extremely high transmission efficiency
• Extremely low-mass inertial torque
• Extremely high speeds

d
• High current overload capacity
• High continuous torques
• Suitable for direct drives

e
• Universal motor for standard applications

Figure 14. Constructions of d.c. motors: **a** disc-type rotors, **b** bar-type rotors, **c** hollow-type rotors, **d** slow-type rotors, **e** conventional construction.

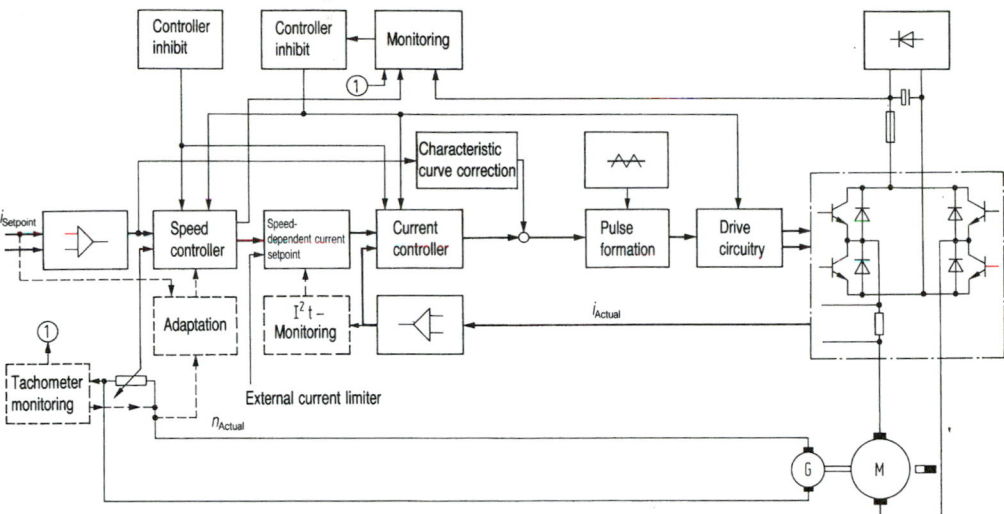

Figure 15. Structure of d.c. motor drive circuitry with transistor amplifier (Siemens). *n* speed of rotation, *i* current, G tacho-generator, *M* motor.

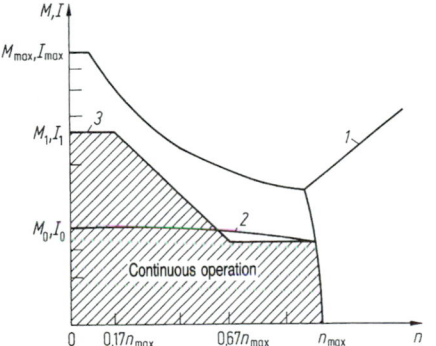

Figure 16. Progression of speed-dependent current limiter (ABB): *1* commutation limit, *2* characteristic curve of motor, *3* current limiter, e.g. $I_0 = 10$ A, $I_1 = 30$ A, $I_{max} = 100$ A, $n_{max} = 2500$ min^{-1}, $M_0 = 8$ N m, $M_1 = 21$ N m, $M_{max} = 69$ N m.

ditions up to 14 000 min^{-1}) and require backlash-free gearing connected in series (rapid-type rotor).

Disc-Type Rotors. These have a lightweight, fibreglass-reinforced, non-ferrous, synthetic resin disc-type rotor with bonded-on current conductors running between permanent magnets. Rated speeds are 2100 to 4500 min^{-1}, maximum speeds up to 6000 min^{-1}.

Hollow-Type Rotors. These have a bell-shaped winding assembly which is surrounded both inside and outside by the field system.

Slow-Type Rotors. These have large numbers of poles (torque motors), and typically an annular, slotted, large-diameter rotor. The speed range of $n < 1$ min^{-1} to 1200 min^{-1} enables the provision of a direct link to feed spindles without intermediate gearing (freedom from backlash) for high torques of up to 4000 N m.

Electrical Stepper Motors. These have three, five or more stators and may execute angular and position increments with the appropriate field control and, as

such, may act both as motors and as measuring devices. Highest speeds are approximately 3600 min^{-1}. The start-stop frequency is limited in relation to external inertial torque, with no loss of increment. Owing to their low torques, they find application only as feed motors in smaller machine tools and as controlled auxiliary drives. Hydraulic torque amplification is required for larger machine tools.

Linear stepper motors were developed for sequence switching in five-stator rotating stepper motors. They are used in small, double-axis, point-controlled machine tools (e.g. boring and drilling machines fitted with Fujitsu motors and can achieve increments of 0.1 mm).

Hydromotors

Rotational Hydromotors. These (G2.2) are mainly found in machine tool applications in feed drives, acting as direct main drives only on special-purpose machines. The most common constructions (including pump operation) are gear-type machines, vane-type and radial machines, and axial and rotating piston machines. Typical applications are in pump motor systems with stepless speed adjustment or as electrohydraulic motors.

Figure 17 shows the structure of an electrohydraulic feed drive based on the positive displacement principle, and a system under constant pressure. Index 1 shows the supply side under constant system pressure. On the delivery side, the adjustable hydromotor supplied direct from the mains supply produces linear motion in the machine slide with the aid of a threaded spindle. The hydromotor setting is actuated via the adjusting piston, which is controlled in turn by the output of the positional controller, the feedback on the regulating distance y and the spindle speed n_2. The oil flow $\dot{V}_Q$ through the valve regulates a double-sided acting cylinder piston, which alters the intake volume of the hydraulic motor according to the rotational speed and/or setpoint position of the slide to be controlled. In practice, the displacement controller is energy-saving, as the adjustable pump which is driven by an electrical control signal generates hydraulic power only as and when the drive (delivery) requires it. One disadvantage, however, is that the timing is somewhat slow, as

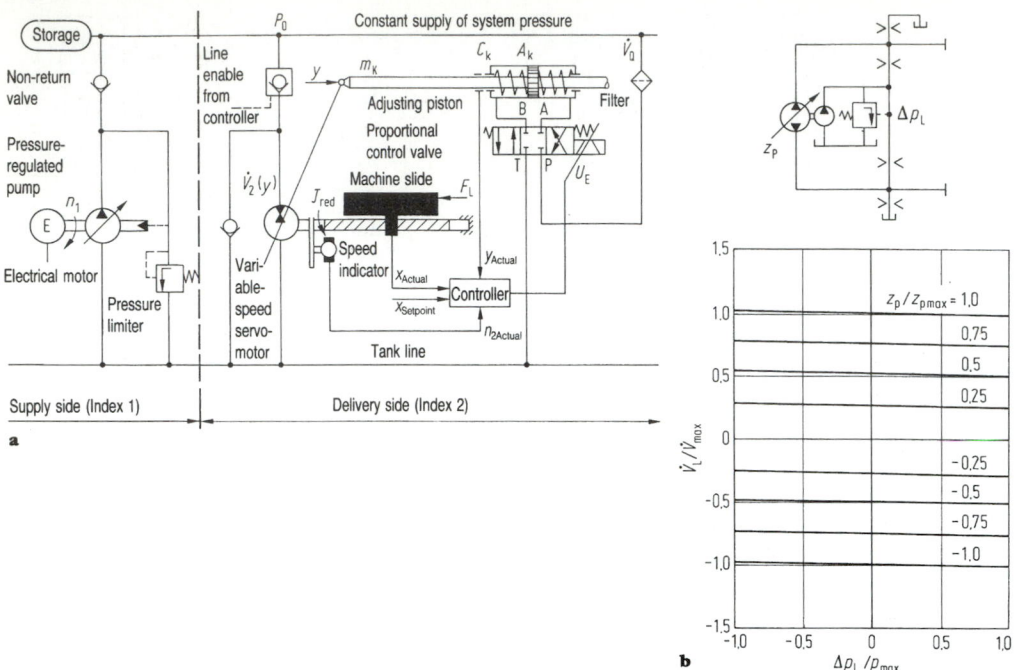

Figure 17. Electro-hydraulic feed drive based on displacement control principle [21, 22]. **a** Circuit arrangement: p pressure, $\dot{V}$ volumetric flow, n speed, x path, U_E control voltage, J_{red} reduced mass moment of inertia, F_L force under load, T, P, A, B valve connections, y regulating distance. **b** Characteristics of servo-control pump: Δp_L pressure change under load, Z_p pump setting, V_L volumetric flow under load.

fairly large masses have to be moved over long distances (e.g. 10 to 100 kg mass over a distance of some 10 to 100 mm). This is why this control principle is mainly of interest in high-power applications [21–24].

Hydraulic Linear Motors (Hydrocylinders). These are employed in machine tools, as the main drives of planing, shaping and slotting and broaching machines, in presses, in the feed drives of grinding machines, in crossover turning machines and machining centres, and also for auxiliary drives, e.g. for automatic tool change in machining centres or on workpiece transport systems in transfer lines.

Electrohydraulic Motors. The electrohydraulic servomotor consists of a hydromotor or hydrocylinder which is driven via a flanged electrohydraulic servovalve. It is suitable for the continuous-path direct drive of feed spindles thanks to its excellent dynamic characteristics, which consist of low mass inertial torque, high torque, short startup times and speed ranges of $B \approx 10^3$ to 10^4 (with tachometer feedback).

Figure 18 shows the structure of an electrohydraulic feed drive based on the principle of rheostatic control under constant system pressure. The proportional control valve and the servomotor form the drive which moves the slide via a threaded spindle. The slide position x_{actual} and the motor speed n_{actual} are obtained and fed back to the positional controller and/or speed controller. The system deviation regulates the volumetric flow V_L to the motor via the valve and the speed is adjusted accordingly.

Rheostatic control is characterised by its extremely good dynamic behaviour, but also by its low level of efficiency due to high energy loss as a result of the inductance effect. This high dynamic response may be attributed

to the movement of low mass over very short distances (e.g. 0.1 kg mass over a distance of approximately 0.1 to 1 mm) in the valves. Rheostatic control is normally employed in applications at ratings ranging up to 10 kW. Displacement control may be substituted for even larger power ratings [21–24].

Example. Figure 19 shows the combined roller vane motor with a two-stage servovalve. The input quantity of the valve is the low control current i, while the output quantity is the proportional oil flow q_A and/or q_B which may be translated into proportional speed in the motor. Power amplification is between 10^3 and 10^5. Control current i causes displacement of the flapper (nozzle–baffle-plate system, Stage I) via control coils and armature. As a result, different pressures are produced on the left and right side of the piston valve, causing this to be displaced (Stage II). Depending on the setting of the piston, the oil under pressure flows to A or B. Constant regulators and nozzles (adjustable-gap inductors) form the basic bridge connections. Extremely fine-grade filtering of the oil is required for the sensitive inductor system. The effect of the inductor causes pressure loss and extreme heat generation in the valve, and coolers are therefore required in most cases.

The roller vane motor converts the direct tangential force of the internal supply of oil under pressure into continuous torque with a positively driven rotor and rotary slide valves. Two diagonally arranged rotary slide disc valves each revolve in relation to the internal diameter of the rotor to seal the rotary pressure chambers. The rotary slide disc valves are thereby driven by the motor shaft via the planetary gears at a ratio of 2 : 1, and lens-shaped depressions are provided for the passage of the rotor. This arrangement is compact and fully symmetrical, thereby providing uniform operation even at low speeds.

Electrohydraulic Stepping Motors. These are employed where the torque of electrical stepping motors is not adequate for feed drives. Hydraulic torque amplifiers are then coupled to the stepping motor shaft

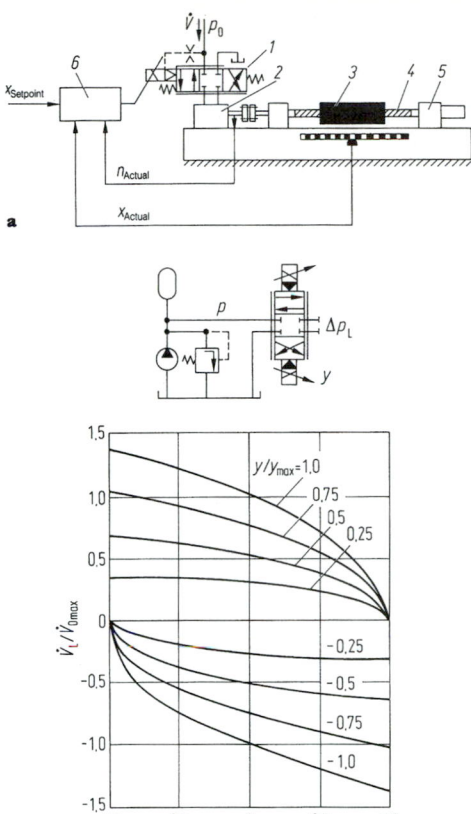

Figure 18. Electrohydraulic feed drive based on the principle of rheostatic control [21, 22]. **a** Structure: *1* proportional control valve, *2* servomotor, *3* slide, *4* spindle, *5* bearing, *6* controller, p_0 constant system pressure, $\dot{V}$ volumetric flow, n_{Actual} spindle speed, *x* slide position. **b** Characteristics: *y* regulating distance, p_0 system pressure, Δp_L pressure change under load, V_L volumetric flow under load.

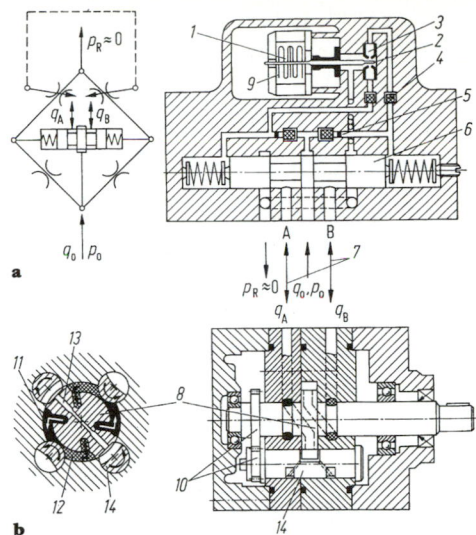

Figure 19. Electrohydraulic servomotor. **a** Electrohydraulic servovalve (Moog) with equivalent cross-section. **b** Roller vane motor (Hartmann) in longitudinal section and cross-section (actual size ratio ≈ 1 : 5). *1* control coils, *2* flapper, *3* hydraulic amplifier, *4* filter, *5* constant inductor, *6* piston valve, *7* consumer's terminals, *8* rotor, *9* rotary disc valve, *10* planetary gears, *11* oil under pressure, *12* oil return, *13* leakage oil, *14* roller vanes; p_0 (≈ 140 bar) pressure of supply unit, q_0 available oil volume, p_R (≈ 0) oil return.

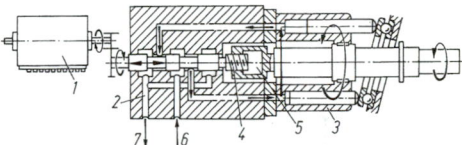

Figure 20. Electrohydraulic stepper motor (Fujitsu). *1* electrical stepper motor, *2* servovalve, *3* hydromotor, *4* bolt, *5* nut, *6* connection for oil under pressure, *7* oil return line.

(**Fig. 20**), these consisting of a servovalve and hydrometer. These two form the internal hydromechanical servo-control circuit. The piston valve of the servovalve and the hydromotor shaft are mechanically coupled via a bolt and nut connection. The system deviation is produced at this point. The rotary action of the piston valve produced by the stepping motor or angle deviation of the hydromotor shaft causes axial displacement of the piston valve, the oil flow released driving the motor impeller until the piston valve is screwed back to the midpoint setting.

1.2.2 Transmission Units

Mechanical Gears

In machine tool construction, these gears primarily reduce the typically high speeds of the motors to achieve the operating speeds and to generate defined feed motion in the tool supports [28]. Distinctions are made between gears for uniform and non-uniform transmission (see F9).

Gear Cone Transmission Units [25]. The smallest functional group in the gear train consists of a single pair of gears, where the gear and mating gear are located on

different shafts. Transmission ratio *i* is given by $i = n_{driving}/n_{driven}$ = driving speed/driven speed. The gear ratio *u* is given by $u = 1/i = z_1/z_2$ = number of driving gears/number of driven gears (see F8.1).

Types. There are a number of gear-switching mechanisms, e.g. pickoff gears, change gears [26, 27], sliding gears, and mechanically and electrically operated clutches. The smallest switching mechanism is the two-speed basic gearbox with two driven speeds, while the next largest unit is the three-speed gearbox with three driven speeds (**Fig. 21**).

The arrangement of the sliding gears for *sliding-gear transmission units* is possible on both the drive and the output shafts. The lighter gears are displaced in such a way that less mass is to be moved and shorter gearshift forks are required owing to the smaller diameter.

Switching takes place via clutches in *power shift gears*. It is therefore possible to carry out switching under loads and in a rotating condition.

Gears with a number of driven speeds are produced by switching the basic gearboxes in sequence. Two two-speed basic gears (**Fig. 22**) switched in sequence produce a four-speed gear with four driven speeds, for example.

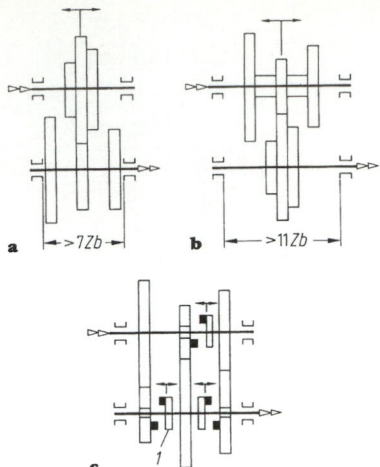

Figure 21. Three-speed basic gearbox. Sliding-gear transmission: **a** standard arrangement; **b** expanded arrangement, Zb width of tooth; **c** power shift gear, I clutch.

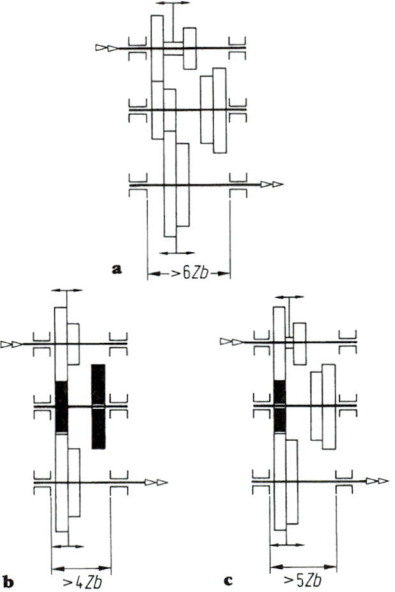

Figure 22. Four-speed three shaft gear: **a** basic gearbox; **b** single mesh gearing; **c** double mesh gearing, Zb width of tooth.

Meshed gears are employed to produce smaller overall lengths and save on gearwheels. These are composed of one or more wheels of different driving gears. The meshed gearwheels are marked by shading areas in **Fig. 22**. As the meshed gearwheels are engaged with two gears, the same module has to be employed for all three wheels. The size of the module is determined by the driving gear with the largest torque, as a result of which larger shaft–centre distances may be produced under certain conditions. A shorter overall length in axial terms therefore means an increase in the gear in radial terms.

One construction commonly employed is transmission gearing (**Fig. 23**), which consists of three shafts and is

always switched via a clutch. The flow of force either passes directly from shaft I to shaft III or first to shaft II and from there to shaft III. In the first case, the gears are engaged without effect, so that driving and driven speeds are equal. In the second case, a large overall transmission ratio is achieved by switching two pairs of gears in sequence. A small overall volume is produced as a result of the design structure (feedback of flow of force to the coaxial shaft III). Transmission gearing is normally attached to the driven shaft to provide operation with the gearing for as long as possible at high speeds, i.e. at small torques.

Design. There are graphics aids for the *design* of speed rates for stepped and continuously variable transmission units which simplify the task considerably (**Fig. 24**) [28]. The *modular network* (**Fig. 24a**) forms the basis of the preliminary gearing design. It shows the various possible different divisions for the progressive ratios within the transmission unit and the individual switching blocks. The modular network is always drawn symmetrically. Each connecting line corresponds to a transmission speed. The size of the individual transmissions is not determined at this point.

In addition to the modular network, there is also the *speed flow diagram* (**Fig. 24b**) showing the speed of each shaft and size of the transmission. The output speeds are shown in the speed flow diagram in geometric progressions (φ = constant) at equal intervals using a logarithmic scale. This interval may be viewed both as the ratio between two consecutive speeds and the power of φ. The transmission is characterised by the gradients of the connecting lines of the speeds of two consecutive shafts. In **Fig. 24b** these are, for example

$$i_1 = \varphi^0 = 1,$$

$$i_2 = \varphi^1 = n_4/n_3, \quad i_3 = \varphi^0 = 1,$$

$$i_4 = \varphi^2 = n_4/n_2 = n_3/n_1.$$

Other aids employed in gearing design are the *gearwheel arrangement* (**Fig. 24c**) and the *flow of force diagram* (**Fig. 24d**). The gearwheel arrangement shows

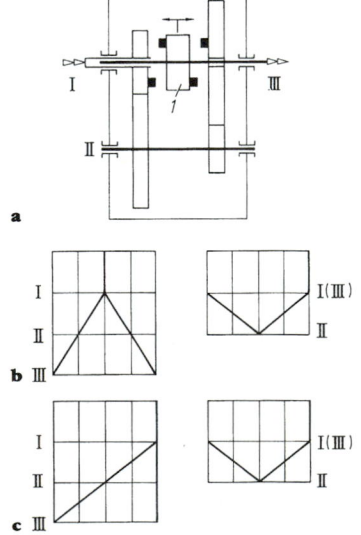

Figure 23. Transmission gearing: **a** gearwheel arrangement; **b** modular network, I clutch; **c** speed flow diagram.

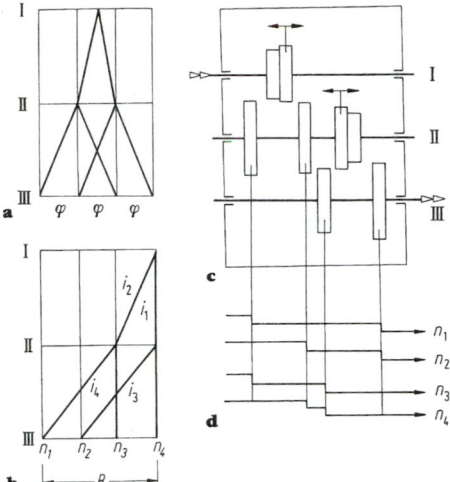

Figure 24. Basic requirements for gearing design. **a** Modular network. **b** Speed flow diagram, B speed range. **c** Gearwheel arrangement. **d** Flow of force diagram.

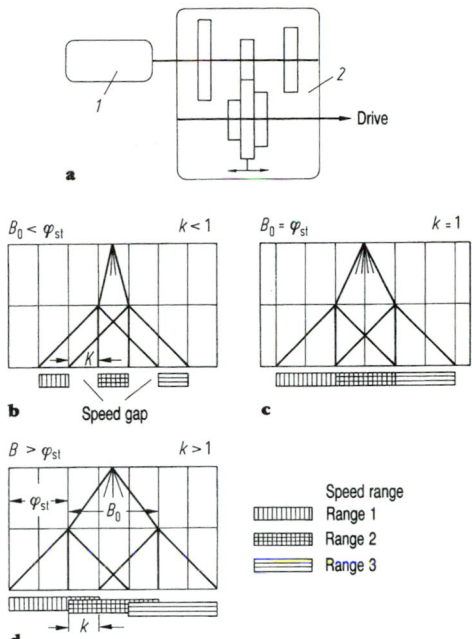

Figure 25. Combination of controllable electrical motor with range transmission. **a** Block diagram: *1* controllable d.c. motor, *2* range transmission. **b** Negative overlap. **c** No overlap. **d** Positive overlap.

the arrangement and number of shafts, gearwheels and any clutches which may be employed. Symbols are employed to illustrate the construction. The flow of force diagram shows which wheels transmit the flow of force in the individual switching positions. The flow of force diagram may also be employed to establish how the individual blocks are to be switched to produce a specific output speed.

Further developments in the control of electronic drives have enabled an increasing variety of combinations of stepless electrical drives to be employed with stepped transmissions as the main drives for machine tools (**Fig. 25**). The operating speed B_0 of the stepless drive may be extended by a stepped transmission connected in series. A speed coverage of $k > 1$ is desirable, so that all speeds within the operating speed range B_0 may be achieved. Thus:

$$B_{tot} = B_0 B_{St},$$
$$k = B / \varphi_{St}$$

where B_0 = the operating speed range of the d.c. motor, B_{St} = the operating speed range of the stepped transmission, φ_{St} = the progressive ratio of stepped transmission, k = the speed overlap.

Example. Figure 26 shows the structural design of a twelve-speed clutch-controlled gear in the spindle box of a numerically controlled turning machine with the respective speed flow diagram. The drive consists of a pole-changing three-phase slipring motor with a V-belt and V-belt pulley at shaft I. The speed switching is carried out via slipring-free electromagnetic multiple-disc clutches K_1 to K_5. The figures in the speed flow diagram and at the gearwheel speeds show the transmission ratios. They should amount to between 0.5 (speed-increasing) and 4 (speed reduction) depending on the performance and the overall size. The progressive ratio amounts to $\varphi = 1.6$ in the example selected; smaller progressive ratios are calculated in the main operating range of $\varphi^{0.5} = 1.25$.

Flexible Drive and Friction Drive

Belt Drives (see F6). These are suitable for the transmission of rotary motion between the motor and gears or directly to the main spindle in machine tool construction.

Their advantages include shock suppression and protection against over-ranging.

Rapid traverse speed in spindles commonly uses smooth-running belt transmission. The tension force is taken up by separate bearings at slow speeds via toothed wheels. Flat belt transmission is employed for the highest speeds and lowest torques, e.g. for grinding spindles, otherwise V-belt drives are the norm, with toothed belts and nylon toothed belts less common, the latter also usable in conjunction with oil. High-torques applications may feature a number of belts switched in series.

Stepped speed pulleys are installed in high-speed spindles and with lower power ratings, e.g. in small-scale drilling and boring machines and small, rapid-turning machines. These have significant downtime for belt adjustment. Even belt tension is achieved at all speeds, where the shaft–centre distance is given by $a \geq 10(d_{max} - d_{min})$, where d_{max} is the maximum belt pulley diameter and d_{min} is the minimum. The total belt pulley diameter must remain constant by comparison. An idler pulley must be provided for smaller shaft–centre intervals.

Chain Transmission (see F6). This is normally used only in conjunction with roller chains in machine tool construction for auxiliary and transport motion. Low-noise, inverted-tooth chains are also employed in feed and main spindle drives of small-scale automatically controlled machines.

Stepless chain-driven gears are employed primarily in main drives of up to 40 kW. With positive, multiple-disc chain drive the speed range B is up to 6, with roller chain construction up to 10. Non-positively driven roller chain is also suitable for higher speeds. Wheel gears are frequently connected in series to extend the speed range (see the

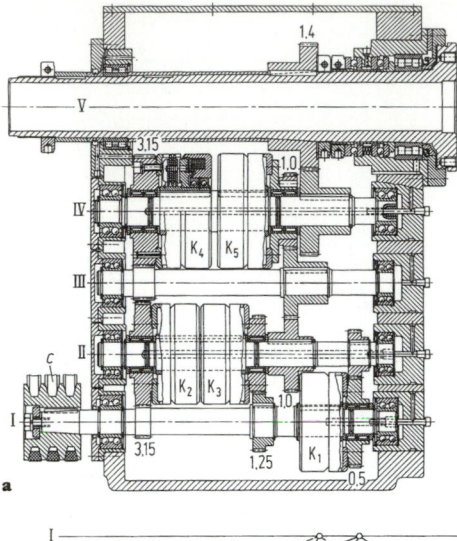

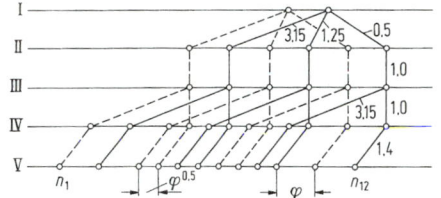

Figure 26. a Twelve-speed clutch-controlled gear of a numerically controlled turning machine (Gildemeister, Bielefeld). **b** Speed flow diagram, K_1 to K_5, slipring-free electromagnetic multiple-disc clutches. c Drive via pole-changing three-phase slipring motor and V-belt. Figures are transmission ratios.

planetary gearing in **Fig. 27**). Specifically compact construction is available via torque division.

Frictional Gears (see F7). These are infinitely variable, and are employed in main and feed drives in medium to small-scale boring, drilling and turning machines, where a limited speed range of $B < 5$ is adequate at high speeds.

Crank Mechanisms (see F9 and F10). These are installed in machine tools for straight-line back-and-forth motion, where non-uniform speeds are admissible or even desirable [29].

Slider–Crank Mechanisms. These have equal back-and-forth times, i.e. 50% non-productive time. Consequently they are rarely employed in metal-*cutting* machine tools, although they are frequently in use in metal-*forming* machine tools. In this case the adjustable crank journal is extended to form an eccentric cam. The connecting rod may be strained to buckling and is consequently designed to be short and compact. The rotating joint (connecting rod journal) consists of ball bearings in a ball cup. A flywheel is provided to ensure even operation of the machine tool.

Crank–Rocker Mechanisms. These are employed in shaping and slotting machines. In **Fig. 28** the shaping and slotting spindle is driven by a rocker arm with a toothed wheel section. The plunger stroke may be adjusted on the crankwheel (eccentric cam).

Slider–Crank Mechanisms. These are used in horizontal shaping and slotting machines as oscillating or rotating sliders to achieve more rapid return times (see F9). It is normally necessary to calculate the dynamic forces only for reversal points; static forces may be determined from diagrams.

Slider–crank mechanisms are employed in infeed areas of shaping and slotting machines for step-by-step infeed motion via stops or a locking wheel. The crank or wrist pin may be adjusted at this point.

Electric Transmission

The Ward–Leonard speed control system traditionally employed in many d.c. motor control applications has become outdated and is being superseded by power electronics, i.e. thyristor drives. Stepless electronic drives may now be employed in high-precision feed drives, e.g. in gear-cutting machines for toothed wheel production with control electronics (see K5, Fig. 22).

Hydraulic Gears

These use pressurised liquid, normally oil (see G3), for power transmission. The hydraulic gears employed in machine tools are virtually all hydrostatic ones. These remain virtually unaffected by the kinetic energy of the liquid flow, in contrast to hydrodynamic gears. The liquid serves purely for the transmission of the compressive forces. With hydraulic transmission stepless adjustment of

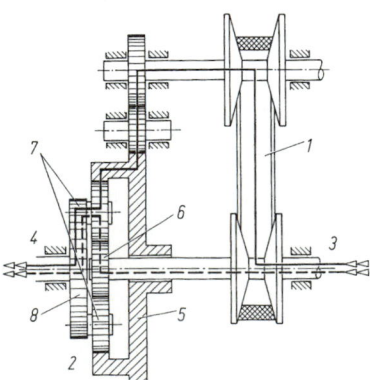

Figure 27. Stepless (PIV = peak inverse variable) chain transmission *1* with planetary gearing connected in series *2* (arrows mark the flow of force at the torque division), *3* drive, *4* output, *5* internal gear, *6* sun wheel, *7* planetary gear wheels, *8* web.

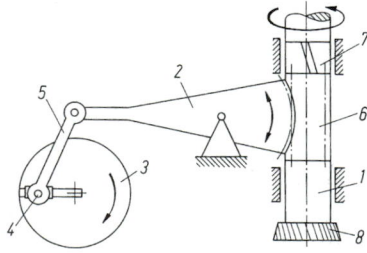

Figure 28. Drive of the shaping and slotting spindle *1* of a gear-shaping machine (Lorenz, Ettlingen) with crank–rocker mechanism *2*; *3* eccentric disc (drive), *4* adjustable crank journal, *5* connecting rod, *6* cylindrical toothed rack, *7* helical guide bushing, *8* cutting wheel.

the output speed may be carried out within broad limits. Constant operating speeds are achieved, also bumpless changeovers, and the oil pressure may be exploited for clamping and control motion and for braking.

The hydropumps and motors employed are circulating gear-type or vane-type pumps with constant or variable delivery volumes or straight-stroke piston pumps. Pumps and motors may be of similar or different constructions. Rotary drive and output motion or rotary drive with linear reciprocating output motion may be obtained, depending on the composition [30–32, 34].

Hydraulic Gears with Rotary Drive and Output. These are used in broaching, planing and surface grinding machine tool applications. Both pump and motor are located in a single housing and may normally be adjusted independently of one another.

The power characteristics of a hydraulic gear are similar in principle to those in electrical transmission. The selection and design of the oil circuit are important (see G3, **Fig. 1**).

The pump removes the entire delivery stream from the tank in an *open circuit* (**Fig. 29a**), whereas the return oil from the motor, minus the leakage oil, is fed back to the pump in a closed circuit.

In a *closed circuit* (**Fig. 29b**) the motor is "hydraulically clamped", so its torsional rigidity is higher than in an open circuit. The closed circuit is therefore suitable for braking, rapid change in the direction of rotation and feed drives where the machine tool table has a tendency to stick or slip. Precautions have to be taken to ensure that the heated oil in the circuit is continuously replaced by oil from the tank or cooled with additional aggregates to achieve the required heat dissipation.

Hydraulic Gears with Rotary Input and Linear Output. These may be used for the main motion in planing, slotting, broaching and flat grinding machines and presses, for feed drives in constructional units and automatically controlled machinery, and for auxiliary and tension motion in devices. The oil supply to the cylinder is via a constant pump in the reactance circuit or a servo-pump with a rapid-traverse switch.

Reactance Circuit (**Fig. 30**). This is equipped with a constant delivery pump (see G3.3.2). Fine tuning of the oil flow is via an inductor, with free flow on the return of piston provided via a directional valve. Where possible the inductor should not be installed in forward motion, since back pressure then builds up via forces within the table and there is a danger of chatter occurring with fluctuating forces. Where the inductor is located in the return motion, back pressure builds up which stabilises the piston (back pressure control). This back pressure is even more effectively controlled where a variable-displacement

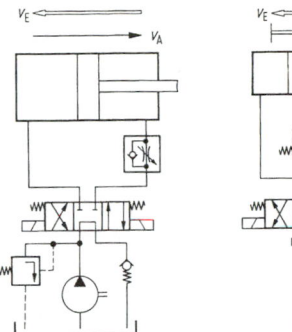
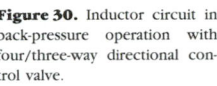
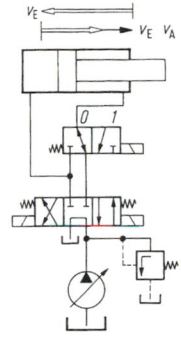

Figure 30. Inductor circuit in back-pressure operation with four/three-way directional control valve.

Figure 31.   Rapid-traverse switch with four/three-way directional control valve and additional three/two-way relay valve.

pump is installed as a gauge box in place of the inductor. Although the reactance circuit is cheaper and simple, and extremely reliable in operation, it is less effective, as the difference between the continuous flow of oil and that consumed in the cylinder is discharged via the pressure-reducing valve causing heat dissipation.

Rapid-Traverse Gears. In cylinders with single-sided piston rods (**Fig. 30**), larger piston surface areas are normally employed to produce greater power and slower operating speeds v_A, with higher rapid-traverse speeds v_E in smaller ring surface areas with less power. Where the rapid traverse also applies to the direction of operation, an additional relay valve is employed (**Fig. 31**). The cylinder chambers are interconnected in position *1*, where the oil is exchanged and the total volume of oil conveyed by the pump responds to the smaller differential surface area (corresponding to the piston rod cross-section). A smaller volume of oil is required as a result, so fixed-displacement pumps are frequently employed instead of variable-displacement ones, with a considerable saving in costs.

Pneumatic Gears

These are normally employed in machine tools as cylinders for automatic clamping, auxiliary and transport motion (see G5 and [35]). Advantages are simple installation, extremely reliable operation and high operating speeds of up to 3 m/s. Disadvantages include low rigidity of air cylinder, non-uniform motion with fluctuations in forces under load and frictional forces (though these may be overcome with hydraulic regulator), with intermediate positions difficult to contol. With larger air cylinders consumption costs are high. Noise is generated by air being discharged, but this can be avoided with the aid of a silencer.

Standard supply pressure p is 4 to 6 bar, with a maximum of 10 bar. Piston force F is given by the formula $F = \eta p A_w$, where A_w is the effective cross-section. Efficiency η is 0.8 to 0.95, depending on the pressure and the size of the cylinder.

Single-acting Cylinder. This is employed for clamping, lifting, extracting, etc. There is a standard production size with a stroke of up to 100 mm. Return action is brought about by a spring mechanism or dead weight.

Double-acting Cylinder. This type may also be fitted with a penetrating piston rod. Where uniform operating feed is required, pneumatic action must be combined with

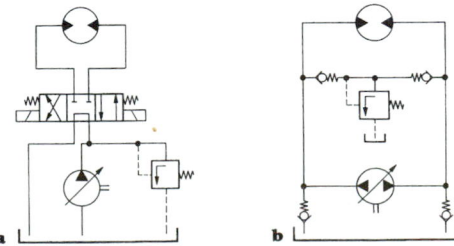

Figure 29. Oil circuits. **a** Open circuit with servocontrol pump and four/three-way directional control valve. **b** Closed circuit without directional control valve with reversing servopump.

hydraulic action, either in separate cylinders (**Fig. 32a**) or in a common cylinder (**Fig. 32b**). The latter construction is more compact [33].

1.2.3 Mechanical Feed Drive Components

All parts and machine components located in the flow of force between the motor and tool or workpiece constitute mechanical feed drive components. The following feed drive components are significant: gearing for translating rotary motion into linear motion; gearing for speed/torque alignment; couplings; bearings; joints. The design of these mechanical feed drive components contributes considerably to the performance and precision of a numerically controlled machine tool.

The main criteria for their design are as follows:

- High geometric and kinematic precision
- High rigidity and freedom from backlash
- High initial resonant frequency
- Low mass moment of intertia and mass of moving mechanical parts.

There are also requirements relating to adequate damping and low friction and linear transmission characteristics of the components.

Feedscrew and Nut Drive

This is the machine tool component most commonly employed for translating rotary motion into linear motion in feed drives of machine tools. *Acme thread spindles* (see F1.6.3) with bronze nuts are used for simple systems, and *ballscrew mechanisms* are employed in modern, high-precision, numerically controlled machine tools (**Fig. 33**).

The ballscrew mechanism is ideally suited to the requirements of the transmission characteristics of feed drive components. It embodies the following principal advantages:

- Extremely high mechanical efficiency ($\eta = 0.95$ to 0.99) owing to low roller friction ($\mu = 0.01$ to 0.02)
- No stick/slip effect (slide back)
- Low wear and hence long service life
- Low heat buildup
- High positioning accuracy and repeatability as a result of adequate rigidity and the elimination of backlash
- High traversing speed

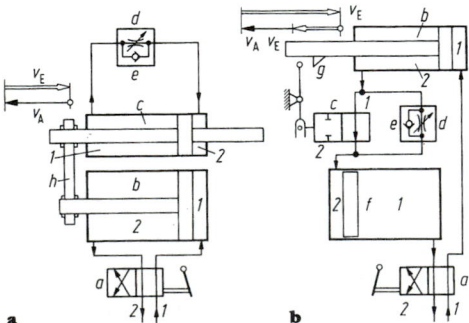

a

b

Figure 32. Pneumatic-hydraulic feed units. **a** Separate cylinder: *b* air cylinder, *c* oil brake cylinder. **b** common cylinder: *b* operating cylinder, *f* air-oil actuator. Working cycles with **a**: slow feed: air *a1-b1-web b*, oil *c1-d-e2*; rapid return: air *a2-b2-b*, oil *c2-e-c1*. Working cycles with **b**: rapid forward travel: air *a1-b1*, oil *b2-c1-f2*; feed rate: air *a1-b1*, oil *b2-d-f2*; return: air *a2-f1*, oil *f2-e-b2*.

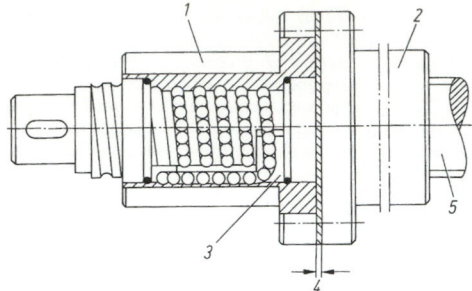

Figure 33. Ballscrew with backlash elimination. *1* first ball nut, *2* second ball nut, *3* ball return, *4* preload spacers, *5* ballscrew.

The low system damping is a disadvantage, however.

As the ball bearings roll along the retaining grooves of the spindle and the nut, they move in a tangential and axial direction. The ball bearings, therefore, require a recirculating mechanism (see F4).

The ballscrew and nut assembly cannot, however, be manufactured to provide the full elimination of backlash. To achieve this, i.e. minimum backlash on reversal and high overall rigidity, the ballscrew and nut assembly must be preloaded. Double or single nuts are used for this. Where double nuts are used, the preload is achieved by pushing the two halves of the two nuts together or apart using preload spacers (**Fig. 33**). Single nuts are preloaded in an axial arrangement where the relevant ball bearing turns are offset by a distance Δl. The rigidity of the system is directly dependent on the preload force generated and on the number of load bearing ways. A minimum feed must be maintained in addition to any external loading in order to produce the required rigidity in the system.

The *spindle bearing* should also be mentioned as another important component of the ballscrew mechanism. Its function is to guide the spindle in a radial direction and take up the feed force in the axial direction simultaneously, while the spindle deformation and dislocation must be kept within admissible levels at the same time. This is why the main requirements of ballscrew bearings include high axial load-bearing capacity, high rigidity, low axial backlash, low bearing friction, high rotational speeds and a high degree of freedom from vibration. These various individual criteria take on a greater or lesser significance depending on the individual application. Whereas the rigidity of the bearing plays a major role in large-scale milling machines with high cutting forces, friction in preloaded bearings forms the main criterion for low-load grinding machines. A low-friction bearing is also essential at high speeds in this type of application.

Thrust angular ball bearings or *roller and needle bearings* are normally employed in threaded spindle bearing mechanisms (see F4). Thrust angular ball bearings have a wide angle of compression (60°) and are therefore able to withstand high axial loads. They are placed against a second bearing, which acts as the return guide because of their single-sided action. Thrust angular ball bearings should preferably be employed in pairs or in groups in X, O or tandem arrangements (**Figs 34** and **38b**). In order to prevent misalignment or compensate for this more easily, it is recommended that matched bearings are installed in the X arrangement because of the small area of support. Roller and needle bearings are employed as complete needle-thrust cylindrical roller bearing units. The special

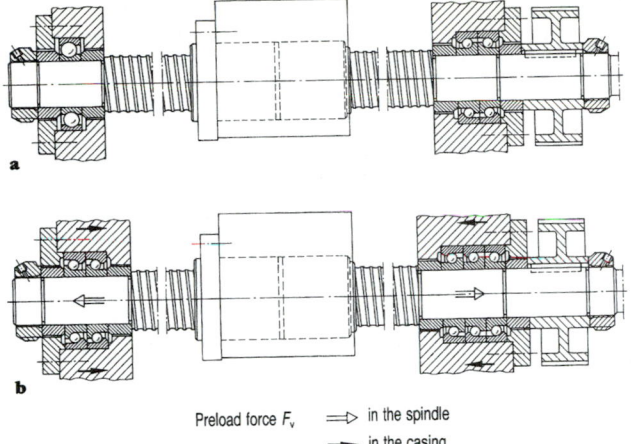

Preload force F_v ⟹ in the spindle
→ in the casing

Figure 34. Examples of bearings for ball screws (SKF, Schweinfurt). **a** Feed spindle bearing for low loads, single-sided clamping. **b** With high rigidity, double-sided clamping.

collar in the thrust bearing simultaneously acts as the outer ring of the needle bearing. The width of the inner ring is thus adjusted to fit the outer ring with the respective axial cylinder roller collars, in such a way that a specific axial preload may be achieved by tightening the groove nut (**Figs 38** and **40**). Both types of bearing require grease or oil lubrication.

Bearings are constructed in different ways, depending on the loading requirements. A threaded spindle permanently fixed on one side in an axial direction with one free end is the standard solution for small loads (**Fig. 34a**). A rigid spindle guide is required in feed drives with high rigidity requirements, and double-sided clamping is preferable (**Fig. 34b**). Thrust angular ball bearings are used to achieve high rigidity at both ends as back-to-back bearing seats in a tandem arrangement. The spindle is distended as a result of this and the preload of the ball screw mechanism is consequently increased. Spindle feed with rigidly clamped spindles should be designed so that it is not offset by the operating forces or the spindle expansion occurring as a result of frictional heat.

These various different types of bearing require axial rigidity characteristics for the ball screw drive which are dependent on the traverse path of the feed slide. The rigidity of the arrangement in traditional types of bearings with a fixed bearing and a moving bearing is reduced hyperbolically according to the distance of the slide from the fixed bearing. Where the second bearing seat is also designed as a fixed bearing, a rigidity curve which is laterally reversed may be superimposed, so that a symmetrical curve is produced (**Fig. 35**). The overall rigidity is thereby considerably greater where there are two thrust roller bearings and this is virtually constant over a large area at the centre of the spindle.

The following constructional rules should be observed in the design of rigid spindle bearings in general:

- Needle bearings and roller bearings should be used in preference to ball bearings where possible owing to their line contact and hence greater rigidity.
- Thrust bearings should always be preloaded.
- Rigid connections should link separable surfaces (rigid jointed bolts).
- Bearing and spacer rings are to be avoided where possible, so as to obtain the smallest possible number of contact surfaces which reduce rigidity.

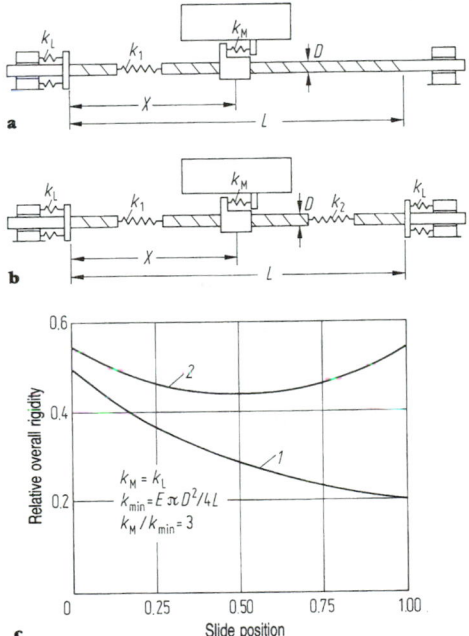

Figure 35. Rigidity rate characteristics **c** of a spindle drive with **a** single-sided fixed thrust bearing and **b** double-sided fixed thrust bearing; *1* single thrust bearing, *2* two thrust bearings.

- Fitting surfaces and spacer surfaces should be ground to provide a high contact area ratio and high rigidity.

The structural design of the ballscrew mechanism should be carried out in conjunction with the preset parameters relating to loading, traverse path, traversing speed and positioning accuracy according to the criteria of rigidity, tensile strength, bending, critical operating speed, mass moment of inertia and service life. This normally concerns the spindle diameter where there ultimately has to be a compromise between the rigidity requirements and the mass moment of inertia.

Example. The required torque of the spindle $M_{sp} = F_{ax}h/(2\pi\eta)$ with thread pitch h and efficiency η with a given axial force F_{ax}. For ballscrew spindles $\eta = 0.8$ to 0.95, for Acme thread spindles $\eta = 0.2$ to 0.55, corresponding to a pitch angle of between 2 and 16°. The relation between linear speed v and rotational speed n_{sp} of the spindle is given by $n_{sp} = v/h$.

Rack-and-Pinion Drive

Long feed spindles would be severely deformed by the axial load and dead weight where there are long traverse paths, e.g. in long turning automatic lathes, longitudinal milling machines and apron-type horizontal boring machines. They have a tendency to buckle. There is also the danger that the spindle rotational frequency may fall within the natural bending frequency range of the spindle. It is therefore recommended that *rack-and-pinion drives* are employed where the traverse path is greater than 4 m. Any length of feed path may be achieved with jointed toothed rack sections. The overall rigidity of the rack-and-pinion drives is always independent of the traverse path in such a case, and is primarily determined by the proportions of torsional rigidity from the pinion shaft and the pinion–toothed-rack couple.

The power transmission at the pinion is characterised by the extremely low rotational speed and high torques, and additional gear speeds are therefore required. The construction of the entire drivetrain should be torsionally rigid and free from backlash. Backlash elimination is achieved by preloading two equal trains of gears A and B (**Fig. 36**), where their pinions are engaged in a single toothed rack at the high speed. Backlash is eliminated in the last three gear speeds by reciprocal preloading of the divisions. The last three speeds of the two trains of gears are preloaded, i.e. free from backlash, as a result of the axial shifting of a helical toothed pinion shaft with the two gearings via spring washers or hydraulic pistons facing in the opposite direction of skew. Gearing faults may thereby be offset. The slide is driven via the train of gears A, depending on the direction of travel of the slide, whereby the train of gears B is tightened free from backlash, or vice versa.

Worm–Rack Gearing

Worm–rack gearings are frequently employed instead of the rack-and-pinion drive where multiple step gearing is to be avoided with long traverse paths. Work–rack systems are constructed with a hydrostatic lubrication system to minimise friction (**Fig. 37**). The worm is equipped with oil-pressure pads which are only fed with oil under pressure in the area affected on the tooth surfaces in the rack internally via a steady distributor (control plate). The rack tooth surfaces are plastic-faced. The high-precision construction is achieved by way of a second casting process from an impression of a master worm before the plastic applied is left to harden. The oil-pressure pads on the worm flanks are produced in the milling process. In more modern constructions these pads may be positioned directly in the rack tooth surfaces, so that these may be manufactured economically directly during the casting process by way of wax-coated material bonded to the rack tooth surfaces. The supply of oil under pressure is still provided via the worm.

Feedgear Mechanisms

Additional *feedgear mechanisms* are installed in feed drives between the motor and ballscrew or rack-and-pinion shaft to cut the high motor speeds to match suitable spindle or pinion speeds with higher torques and to further reduce the slide-side mass moment of inertia in terms of the engine shaft. The gear mechanisms should be torsionally rigid, with low inertia and no rotational backlash.

Toothed gears should therefore have wheels of small diameter, because this increases to the fourth power in terms of the mass moment of inertia. One way to eliminate backlash in practice is with tangential preloading of the meshing wheels. A toothed wheel is divided for this purpose. The two halves are twisted against one another until the desired backlash elimination is set with the counter wheel from the width of the two toothed wheels (**Fig. 38a**). Otherwise, it is possible to position the toothed wheels or wheel shafts in adjustable eccentric bushes. The axle distance may be varied by turning the eccentric bush until the tooth surface backlash is eliminated.

Nowadays, *synchronous belt drives* are being used in many applications instead of toothed-wheel drives, in conjunction with threaded spindle-and-nut systems (see F6), where additional speed is essential on construction grounds (**Fig. 38b**). The synchronous belt drive satisfies the requirements for feedgear mechanisms employed in numerically controlled machine tools as far as rigidity, power transmission and precision are concerned, thereby providing a particularly economical solution. Tensioners made of fibreglass or stranded steel provide a high level of tensile strength, bending strength and minimal strain.

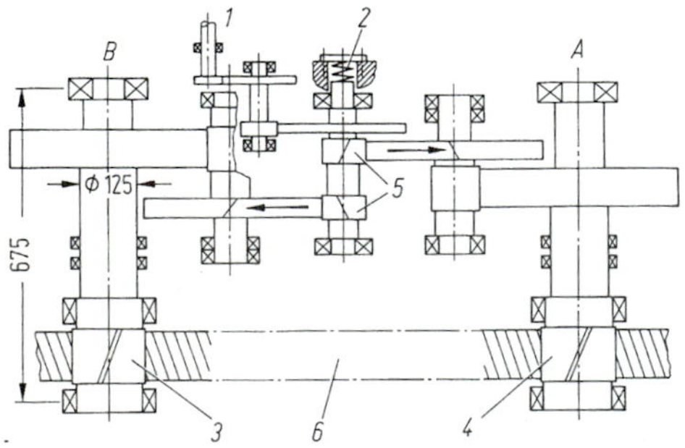

Figure 36. Backlash-free feed drive with rack-and-pinion system. A, B gear divisions, *1* drive, *2* bracing mechanism, *3, 4* with rack, *6* braced pinion (output), *5* positive and negative bevel gearing.

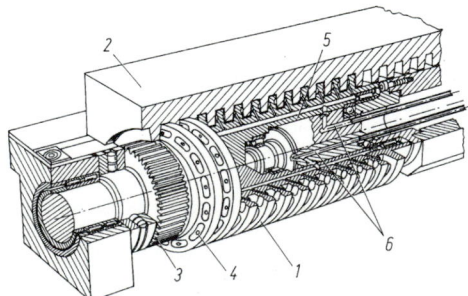

Figure 37. Hydrostatic worm rack gearing (Waldrich, Coburg). *1* worm, *2* worm rack, *3* drive wheel, *4* oil pressure pockets, *5* oil distributor, *6* supply of oil under pressure for front and/or rear tooth surface.

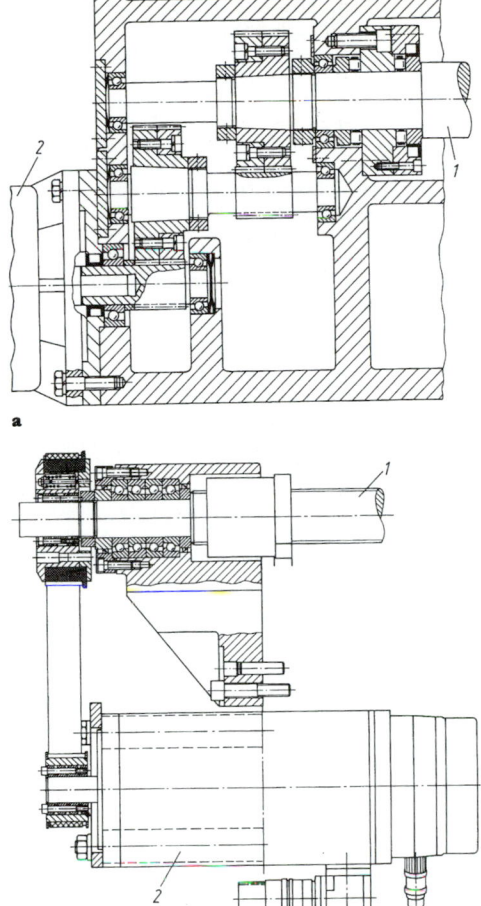

a

b

Figure 38. a Backlash-free transmission gears for feed spindles (Scharmann, Mönchengladbach): *1* feed spindle shaft, *2* direct current actuator motor. **b** Feed drive of bed-type milling machine with integrated toothed-belt gear mechanism (Maho AG).

The toothed belt is preloaded to increase rigidity and prevent backlash. The toothed-belt pulleys are made of aluminium to improve their dynamic response. The high material damping of the toothed belt material ensures low-oscillation transmission to the motor actuator motion. Moreover, the toothed-belt drive also provides for substantially more favourable structural design options, owing to the larger shaft–centre distance. This has led to feed drive concepts with smaller footprints and thereby smaller machine designs as a whole. The toothed-belt drive is ultimately also the most favourable manufacturing solution in terms of cost owing to the small number of parts involved. The spindle bearing should be designed to provide exceptionally high axial rigidity in this drive concept, and the same applies for the radial and tilting rigidity.

The special-purpose *harmonic drive* and *Cyclo* feedgear mechanisms satisfy the requirement of providing the highest possible speed transmission with a compact construction, high rigidity and coaxial input and output drive. The *harmonic drive* gear mechanism (**Fig. 39a**) may be constructed as a can-type construction (**a1**) or a

flat-type construction (**a2**). Both consist in principle of a rigid cylindrical ring with internal gear teeth (circular spline *1*) which is permanently jointed to the housing. Inside this ring there is an elastic steel bushing with external gear teeth (flexspline *2*) which are pressed into the internal gear teeth of the circular splines and caused to rotate, thereby displacing an elliptical cam connected with the drive via a tightened piston bearing (wave generator *3*), generating rotary motion between two opposite points in the large elliptical axis. A relative turn is produced between the circular spline and the flexspline owing to a difference of two in the number of gear teeth between these two, which is directly transmitted to the output in the can-type construction via the flexspline and in the flat construction via the flexspline and the dynamic spline (*4*). The drive and output move in opposite directions. The transmission ratio i is obtained from the number of gear teeth z in the circular spline and the flexspline,

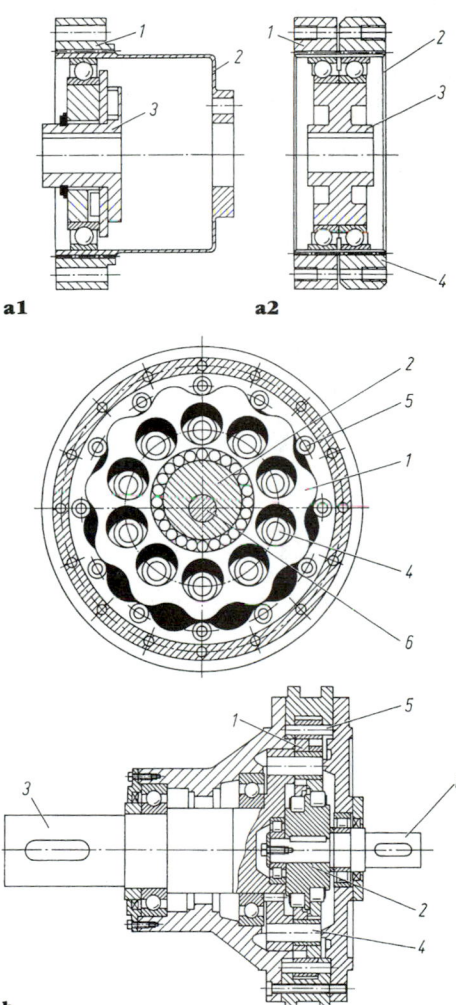

a1   **a2**

b

Figure 39. Types of high-transmission gear mechanisms. **a** Harmonic drive (Harmonic Drive System GmbH, Limburg); **a1** cam-type construction; **a2** flat-type construction; **b** Cyclo gear mechanism (Cyclo Getriebebau Lorenz Braren GmbH, Markt Indersdorf); explanations in text.

thus $i = z_{Fl}/(z_{Fl} - z_{Ci})$. Transmission ratios of $i = 50$ to 320 and output torques of $M = 1.5$ to 4000 N m may be achieved. The gear mechanism possesses extreme torsional rigidity and also freedom from backlash, owing to the large meshing area which amounts to 15% of the total number of gear teeth.

In the cyclo gear mechanism (**Fig. 39b**), a cam plate *1* is driven via an eccentric cam *2* (drive shaft *6*) and revolves around a rigidly mounted ring. Each point on the plate thereby describes a cycloid curve. Rotary motion is generated at a far lower speed in the opposite direction on the plate and is dependent on the ring–plate diameter ratio.

In order to prevent any sliding when rolling, the plate in the cyclo gear mechanism is fitted with an enclosed cycloidal train, thereby providing an external form, and the ring is replaced by bolts arranged in a circle. Each cam plate thereby has one cam curve segment fewer than there are bolts in the ring of bolts. The curve tracks on the cam plate engage positively in the rollers *5* of the rigid outer ring and move on rolling contact with this. The reduced rotary motion of the cam plate is transmitted to the output shaft *3* via the bolts *4* which engage in drill holes in the same plate. The transmission ratio is determined by the number of cam curve segments in the cam plate.

Rollers are mounted onto the bolts of the ring of bolts *5* and output shaft *3* which generate power transmission via their pure rolling action between the cam plate and ring of bolts *5* and cam plate and retaining bolts *4* of the output shaft. Friction losses, noise generation and wear are thereby reduced to a minimum. The gear mechanism is fitted with two cam plates offset by 180° driven via a double eccentric cam, thereby providing a mechanical balance.

These high-transmission compact gear mechanisms are employed as intermediate gearing in feed drives or for driving rotating tables, tool magazines and tool changers. Compact gearing mechanisms also constitute one of the main components of articulated joint drives in robot applications. The criteria of high transmission within the smallest available construction, concentricity, high dynamic response, low backlash, high torsional rigidity and high excess load-bearing capacity provide highly dynamic drives with extremely low backlash and high positioning accuracy and repeatability.

Couplings

Special-purpose flexible couplings may be employed to connect two shaft ends, in particular those of the motor shaft and ballscrew in feed drives, which still have a high level of rigidity in a perpendicular direction. The rotary motion in the perpendicular direction is thereby transmitted with great accuracy. Radial and axial displacement of the shaft ends and angular displacement are tolerated to a limited degree in this coupling (see F3).

Friction-locked clutches (e.g. bellows and diaphragm couplings) are normally employed in high-precision feed drives. They satisfy most aptly the high requirements of torsional rigidity, freedom from backlash and low mass moment of interia. Their structural design is determined by the torque to be transmitted, the shaft diameter and the torsional rigidity.

Slip couplings are employed as effective *protection* for numerically controlled machine tools against damage from excess loads and impacts following tool breakages and programming and operating errors, where the effective torque in a drive train is limited to a maximum limit value.

Where this value is exceeded, the flow of force is interrupted to ensure the reliable protection of the endangered parts from any damage.

Slip couplings are constructed as spring-loaded conical friction clutches and dog clutches. They are frequently integrated into the spindle-side toothed-belt pulley in feedscrew-and-nut drives within an established synchronous belt drive. The coupling is mounted on the shaft journal of the ballscrew and connected to this via a conical clamp *12* which is frictionally engaged with this and free from backlash (**Fig. 40**).

In normal operation the torque transmission in the dog clutch is carried from the toothed belt plate *1* and the flanged ring bolted to it *2* via the ball bearings *3* to the driving collar *4*. The ball bearings are thereby fed to the narrow-tolerance penetration drill holes in the driving collar and pressed into the conical collars of the flanged ring head-on by the spring washer *5* and index ring *6*. The spring washers *5* are preloaded with the aid of the setting screw. The torque to be transmitted may thereby be adapted to suit the respective operating conditions.

The driving collar twists towards the flanged ring where there is excess loading, whereby the ball bearings are pressed out of the conical collars of the flanged ring against the force of the spring washer. The four-point contact bearing *8* takes on the bearing function between the moving belt pulley and rigid driving collar. The flow of force is interrupted as a result. The clutch continues to slip until the torque falls back below the prescribed critical limit value. It then re-engages automatically. The excess loading is detected immediately via the proximity switch *9* so that the drive may be switched off.

Collision Force Calculations and Estimates. The arrangement of the slip coupling in the flow of force depends on the one hand on the position of the parts to

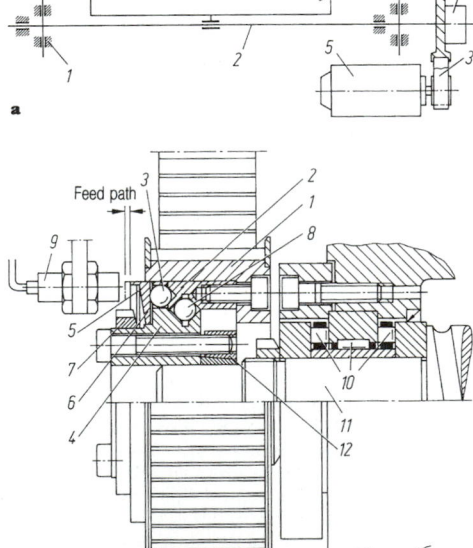

a

b

Figure 40. Integrated slip coupling for feed drives with synchronous belt drive (Jakob, Kleinwallstadt). **a** General arrangement: *1* spindle bearing, *2* ballscrew, *3* synchronous belt drive, *4* slip coupling, *5* servomotor, *6* machine tool table. **b** Slip coupling: *10* spindle bearing, *11* spindle (further details in text).

be protected and on the other on the position of the machine components that cause the high collision forces. These forces are brought about primarily by two mechanisms. First, mass forces are released in a case of collision with the sudden delay of impact on a machine axis; they are determined by the kinetic energy of the moving machine parts, e.g. slide, workpiece, spindle, motor, etc. Second, in a case of collision, motor torque increases rapidly by three to ten times the rated torque depending on the type of motor. The mass forces and peak torque of a motor are added together to provide a total force on impact which may cause elastic deformation, and in the worst case even permanent deformation or breakdown in the machine components located in the flow of force. An estimate of the total force on impact may be derived on the basis of a simulated model.

1.3 Frames

1.3.1 Requirements, Types

Frames and frame components are *load-bearing* and *supporting foundations* of machine tools [36]. They carry and guide the parts required for *relative* motion between the workpiece and tool, e.g. supports, gears, motors and control systems. The basic shape and dimensions of these parts are determined by the *working space*, the level of *process forces* and the required *degree of accuracy* (rigidity). The necessary access to the machine for maintenance, service and assembly purposes must also be avoided.

Because of production and assembly considerations the frames themselves are in many cases manufactured from a number of individual parts which are then bolted together at the joints and even cemented together in certain cases. Frames consist of *bed-type, column-type, table-type, knee-type* and *crossbar* designs. (See L4.2.2 and L5 for examples of machine tool frames.)

The *structure* of the bed of the machine and *position* of the operating spindle are important structural considerations for turning machines (**Fig. 41**).

Flat-bed construction *1* is used mainly in large-scale turning machines (rolling and turning machines). The angled-bed construction *2* permits the hot swarf and the coolant lubrication to be discharged or removed from the working space, so reducing any danger of clogging or thermal overloading of the machine bed. Front-bed turning machines *3* are particularly suitable for chucking part processing with automatic tool change. Vertical turning machines based on the column construction *4* have the advantage of being particularly suitable for accommodating and processing large parts without any bending stress being sustained by the spindle. The spindle (workpiece) may be positioned both parallel *5* and vertical *6* to the base.

Figure 42 shows the major horizontal and vertical drilling and milling machine constructions, structured according to their frame design (knee-type, bed-type and H-frame type) and the number of axes in the tool carrier or workpiece carrier. The *knee-type design* is employed only in small-scale machine tools owing to the mass to be moved in a perpendicular direction. *Bed-type milling machines* are employed for processing heavier workpieces. In contrast to the knee-type, in this design the table is mounted on a rigid machine bed. A distinction is made in bed-type designs between *cross-table* and *cross-bed* constructions. In cross-table constructions the workpiece carrier, i.e. the table, has two directions of motion

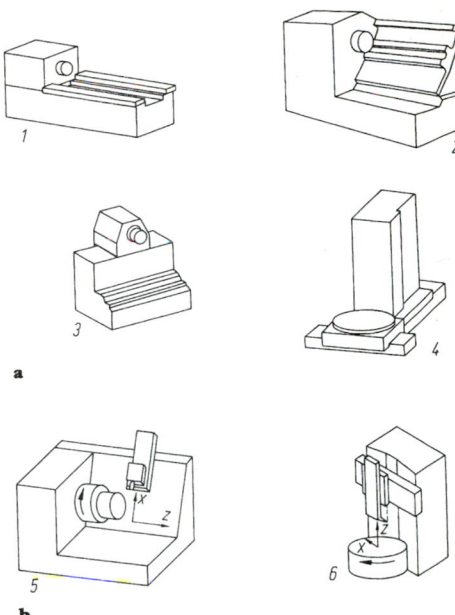

a

b

Figure 41. Classification of turning machines based on their frame constructions. **a** Bed-type designs. **b** Relative position between tool and workpiece (explanations in text).

which are vertical in relation to one another, and the positioning of the cross-table on the wide guideways of the bed produces high static and dynamic rigidity. "Cross-bed" construction is where two vertical directions of feed are provided on the bed, with one assigned to the tool-carrying group (normally column-type frames) and the other to the workpiece-carrying group (normally table-types).

The *H-frame* construction (also known as the double column frame with crossbar) provides a particularly stable and suitable construction for higher cutting power with bulky workpiece items. Of the two designs available, the *longitudinal table-type design* (**Fig. 42b**) is fitted with a table that may be moved in one direction only and the bed is twice as long as the table. All coordinated motion vertical to the feed movements of the bench is carried out by the machine tool. In contrast, the *gantry* design employs a permanently fixed clamping plate and a moving gantry, with the advantage that the machine has only to be as long as the longest workpiece to be machined and/or the clamping plate, whereas machines with moving tables require twice the length. Both the constructions just described have to be capable of high accuracy in final machining and possess a high cutting capacity for premachining requirements. A distinction is also made between crank presses and eccentric presses in terms of the frame construction (see L3 and L4). There are also open, discharging *C-frames* and closed *O-frames* in two-column designs.

The C-frames have the drawback that they open up as a result of the forming force (**Fig. 43a**), which may result in misalignment in the two halves of the tool, but the advantage that access is provided to the working space from three sides on the machine. Closed-frame constructions (**Fig. 43b**) are employed primarily for medium-scale and larger constructions and where the machine tool requires a particularly rigid and accurate

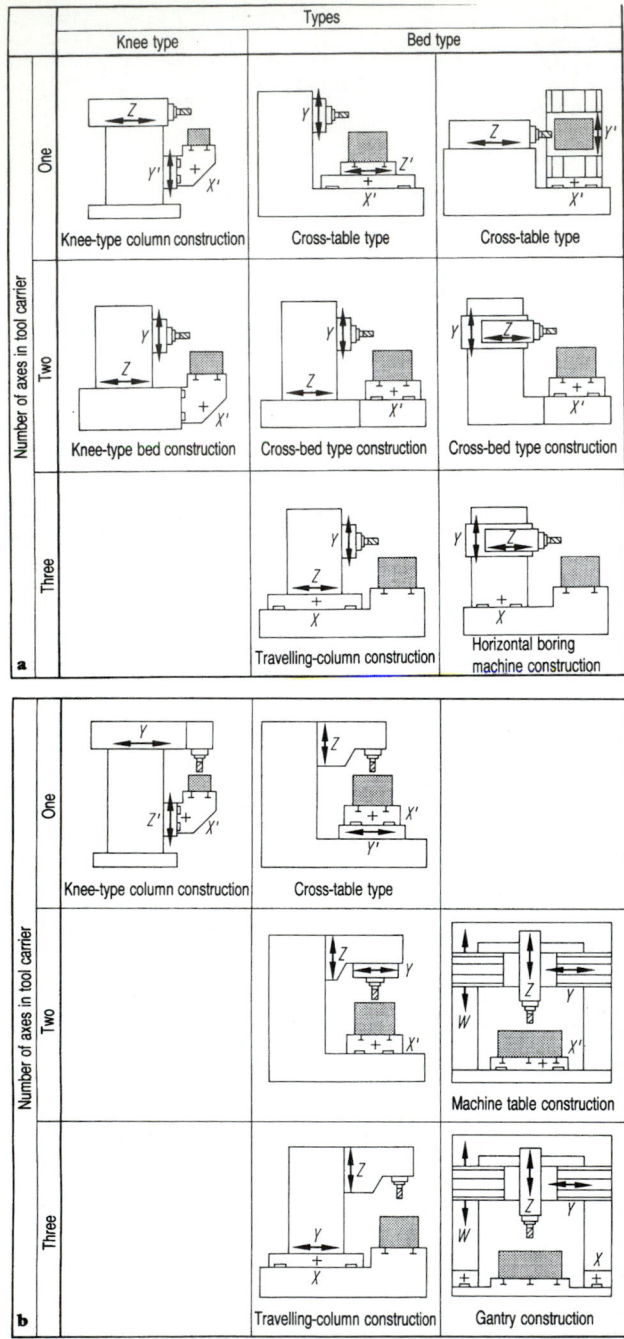

Figure 42. Designs of drilling and milling machines: **a** horizontal, **b** vertical design.

guideway to contain the forces generated during the forming process.

1.3.2 Materials for Frames

Steel, cast steel and *cast iron* may all be employed as materials for frames and frame parts. More recently, *polymer concrete* has been proving increasingly popular in smaller machine frames. **Table 1** shows the most important physical characteristics of the above materials for frames.

Advantages of Steel Construction. The coefficient of elasticity is about twice that of cast iron, resulting in material savings and lower weights. There are no modelling costs, making steel especially suitable for products made to

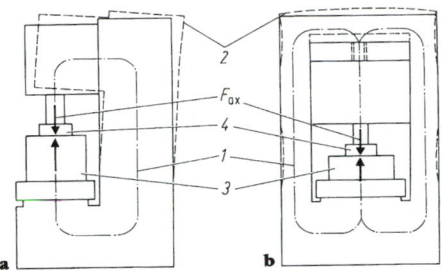

Figure 43. **a** Single column and **b** double column frames: *1* flow of force and *2* displacement characteristics, *3* workpiece, *4* tool, F_{ax} axial components of machining force.

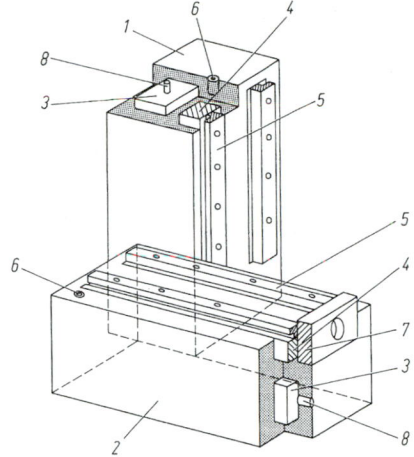

Figure 44. Polymer concrete machine frame: *1* polymer concrete machine columns, *2* polymer concrete machine bed, *3* polyurethane foam core (suds tank), *4* bolted trim (cemented in subsequently), *5* guideways (bolted on), *6* threaded bush, *7* bearing flange (cemented in subsequently), *8* conduit for retaining core (suds).

order. Steel constructions may be of either *plate type* or *cell type*. The plane method relates to cast constructions: plates or moulded units made of thick sheet metal are welded together with ribs to frames. This method is frequently found in presses, cutters and similar machines where cast iron is not rigid enough. Frames made on the cell-principle consist of a wide range of individual cells constructed from thin welded sheets, which gives a considerable saving in weight while maintaining a high level of rigidity. The heat-generating capacity is correspondingly smaller, owing to the minimum of material employed, which results in a greater danger of thermoelastic deformation (see L1.1.2).

Advantages of Cast Iron. These are its high damping capacity, good sliding characteristics in guideways, good machining properties and good dimensional stability. These are further enhanced by employing special formulas to achieve good casting properties and different wall thicknesses and a high degree of stability ($R_m = 400$ N/mm² and over), using nodular cast iron with a high coefficient of elasticity.

Advantages of Polymer Concrete. This material is frequently employed in the manufacture of small and medium-sized (< 5 m) frame components, specifically machine beds.

It possesses an even higher damping capacity than cast iron, and therefore higher dynamic stability. It has a lower heat conductance and higher heat-generating capacity than other materials, and so is insensitive to short-term fluctuations in temperature. There is a wide variety of applications. Embedded parts, e.g. clamping surfaces for bolting on cover plates, motors, spindle boxes, etc. may be positioned in such a way that no subsequent reworking is required. Conduits, cable and ducting channels for the power supply may be incorporated directly into the casting. Modular systems provide a simple range of options, while simple and rapid modification of forms is possible

and hence rapid production of a second cast following constructional modifications. Parts to be positioned accurately (e.g. guideways) are subsequently cemented in place in premachined grooves or furrows using mortar [37]. An example is shown in **Fig. 44**.

1.3.3 Embodiment Design of Structural Components (Frames)

The embodiment design of structural components (frames) is determined on the basis of the requirement for *static* and *dynamic rigidity* and the *lowest possible material consumption*. These are, therefore, constructed to be rigid and lightweight, by increasing the moment of inertia with the appropriate design of the cross-section. Open cross-sections and penetrations should be avoided because these reduce rigidity considerably. In addition, the construction should be as sturdy as possible, as the width between supports and projection will have a substantial effect on all bending stress. The bending and torsional rigidity of frame parts may be increased with the appropriate ribbing. **Figure 45** shows the ribbing types most frequently employed in column-type parts.

In cases *A* to *D*, longitudinal ribbing is provided. The columns shown under *E* to *H* are fitted with cross-rails. The relative bending and torsional rigidity of the different column ribbing shown in **Fig. 46** are calculated using the finite-element method (see B8).

Table 1. Physical characteristics for frame materials in machine tools [37]

Material	Coefficient of elasticity	Specific weight	Heat expansion coefficient	Specific heat-generating capacity	Heat conductance	Rigidity area
	E (N/mm²)	γ (N/dm³)	α (1/K)	C (J/(g K))	λ (W/(m K))	σ (N/mm²)
Steel	$2.1 \cdot 10^5$	78.5	$11.1 \cdot 10^{-6}$	0.45	14 to 52	400 to 1300
Nodular cast iron	$1.7 \cdot 10^5$	74.0	$9.5 \cdot 10^{-6}$	0.63	29	400 to 700
Cast iron	0.5 to $1.1 \cdot 10^5$	72.0	$9.0 \cdot 10^{-6}$	0.54	54	100 to 300
Polymer concrete	$0.4 \cdot 10^5$	23.0	10 to 20.10^{-6}	0.9 to 1.1	0.9 to 1.1	10 to 15

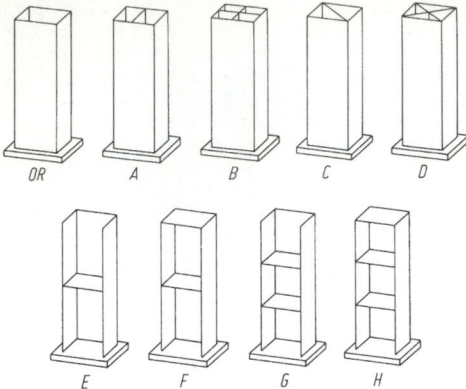

Figure 45. Types of ribbing in columns.

The *longitudinal ribs* improve the bending rigidity of the part by increasing the axial geometric moment of inertia. Vertical ribs running parallel to the outer walls do not provide any notable improvement in torsional rigidity. The torsional stress occurring in frame components normally results from a couple acting on the guideways, which can cause extreme cross-sectional distortion. Diagonal ribs in the longitudinal rib design are particularly suitable for preventing this cross-sectional distortion.

Horizontal ribs (cross-rails and end-plates) also effectively help to prevent cross-sectional distortion with torsional loading via a couple. Horizontal ribs have virtually no effect on bending rigidity. Nevertheless, they may pro-

vide considerable stiffening in the walls as protection against local denting and warping and thereby contribute towards inhibiting local deformation at the points of application of force.

Example. **Figure 47** shows a right-angled cross-section of a column-type frame, where the walls are ribbed in accordance with the cell-type design. As the interior of the column remains free to take a counterweight, discontinuous cross-rails are employed to increase the torsional rigidity. Longitudinal ribs increase the bending rigidity. The guideways are not positioned in the vicinity of the side walls,

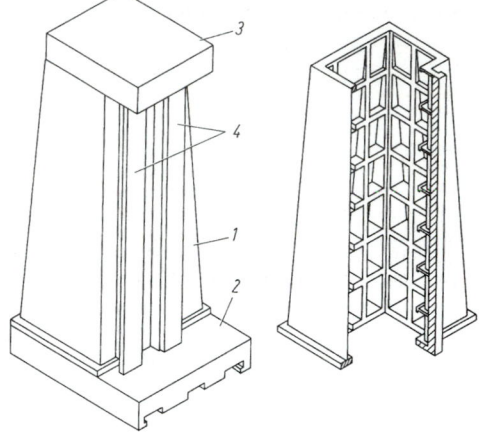

Figure 47. Frame of a horizontal drilling and milling machine with ribbing: 1 column, 2 bed slide, 3 end plate or hood, 4 guideways for spindle box.

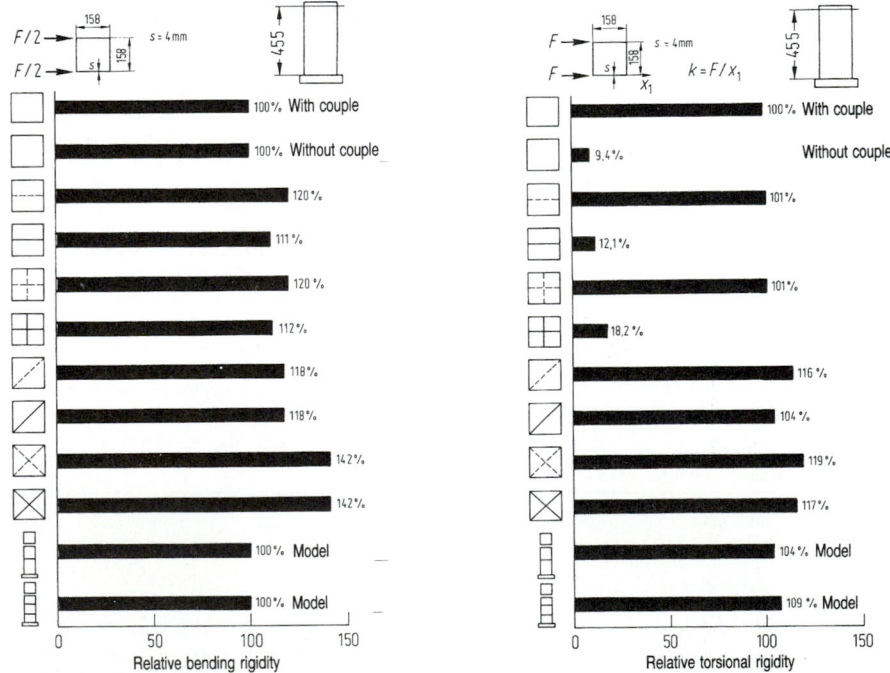

Figure 46. Relating bending and torsional rigidity in various types of ribbing (finite element method calculations).

but are supported by the cross-rails. Circular cross-sections are particularly torsionally stiff and do not cause distortion under load conditions. The manufacture of these forms is complex, however.

Example. Figure 48 shows the cast-iron construction of the travelling column of a roller milling machine. The transverse ribbing and the diagonal longitudinal ribs provide sufficient bending and torsional rigidity for the column. The connection of the diagonal longitudinal ribs to the front wall in the vicinity of the guideways provides an even load distribution from the guideways over the entire column and prevents excessive local deformation at the points of application of force.

1.3.4 Calculation and Optimisation

Computer technology is extremely useful for making effective predictions about the characteristics of a machine tool in the construction and development phase. A high-performance user software package is a basic requirement for this type of computer application. The software normally employed for calculating the mechanical characteristics of frame components, such as static, dynamic, thermal and stress analysis, consists of programs which are based on the finite-element method (FEM) (see B8). Static and dynamic displacement and/or compliance and rigidity, natural frequencies and temperature distri-

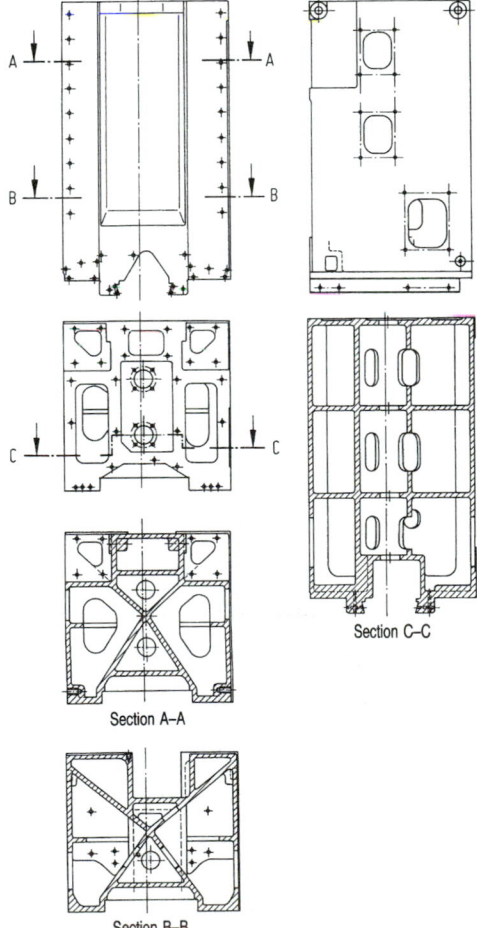

Section C–C

Section A–A

Section B–B

Figure 48. Machine columns with diagonal ribbing and cross-rails.

butions may be calculated for given heat sources and thermo-elastic deformation. The execution of a structural analysis with FEM involves the *data acquisition* phase (preprocessing), *calculation and evaluation of results* (postprocessing). Optimisation strategies based on the FEM method are becoming increasingly popular in structural analysis. Such systems programs serve to optimise the weight and rigidity of mechanical structures and to minimise peak loads at the margins of the curve. They are capable of automatically varying the geometric component parameters, such as wall thickness within specific limits, so that the optimisation target (optimum) is achieved. A complete FEM analysis must be performed at each iteration stage. One of the chief targets with cutting machine tools is to minimise deformation occurring at the cutting point during machining. This means optimising frame components in cutting machine tools with respect to maximum rigidity for a given total weight.

Example. Figure 49 shows the travelling columns of a drilling and milling machine. The target of this optimisation was to minimise deformation at the structural point *P* in the vicinity of the right-hand guideway. The loads shown affect the columns depending on the operating forces, causing slight bending and extreme torsion in the column. The external and ribbed wall thicknesses of the column are determined as the optimisation parameters. Since the columns are designed as a welded structure, the optimisation parameters may only assume eight discrete wall thicknesses of between 8 and 40 mm with a number of restrictions. The wall thickness distribution is shown in **Fig. 49c** before and after the optimisation calculation, with a given material consumption (weight) in the form of a bar chart. The effect of the redistribution of the volume of material on the deformation of the guideways in the column is shown in **Fig. 49d**. Where a double optimisation algorithm was employed (University of Lüttich, Fleury, Braibant), reductions in deformations of up to 17% were established compared with the original structure.

In the case of metal-forming machine tools, e.g. presses, where the forces within the component are vital, as is a sufficient degree of rigidity, local excess loads which tend to occur as a result of fatigue in discontinuous cross-sectional junctions (drill holes) should be noted specifically. These may result in the breakdown of an entire machine in a substantial number of cases.

Example. Figure 50 shows the optimisation of the curve of a C-frame press to minimise peak loads occurring at the margins of the curve. It is clear that the peak loads at 235° may be reduced by around 30%.

1.4 Linear and Rotary Guides and Bearings

Linear and rotary guides and bearings in machine tools have the function of providing a precise, linear trajectory for the execution of the cutting and feed motion of specific components, such as slides, spindle boxes, rams of presses, spindle sleeves, etc. The weights of the guided components and workpieces also have to be carried and process forces have to be absorbed with the minimum of deformation [38].

Important requirements for the linear and rotary guides and bearings of machine tools are high machining accuracy and high performance in the long term, with low manufacturing and operating costs [39]. In order to satisfy these requirements, the linear and rotary guides and bearings must have the following characteristics:

– Low friction and freedom from stick/slip as a requirement for accurate positioning at low feed power.
– Low wear and protection against seizure, so that long-term accuracy may be maintained.

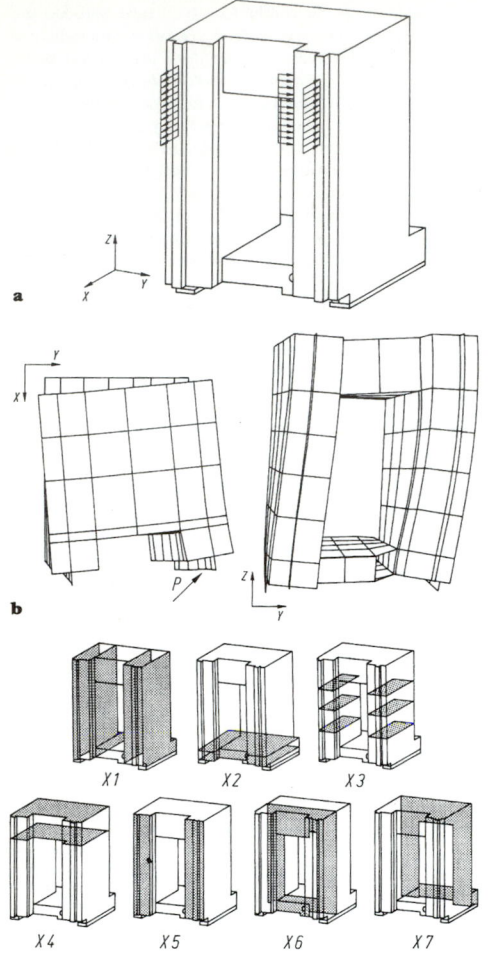

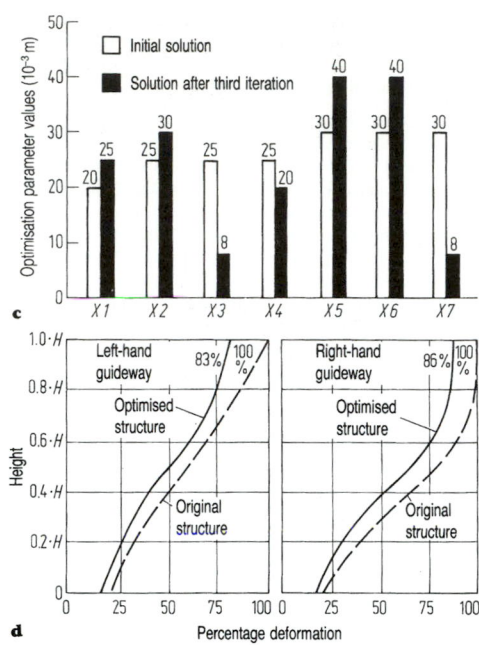

Figure 49. Minimisation of deformation in machine tool columns due to variations in wall thickness with a constant total weight. Optimisation: minimum deformation at structural point P. Restrictions: same weight, use of sheet thicknesses of 8, 10, 12, 15, 20, 25, 30 and 40 mm. **a** Schematic diagram, **b** deformed structures, **c** optimisation parameters $X1$ to $X7$, **d** deformation of guideways (in space).

– High rigidity and low clearance of linear and rotary guides and bearings and/or freedom from backlash to minimise changes in position of guided components.
– Good damping in load-bearing direction and direction of motion to prevent excessive oscillation in feed drives and tendency of machine tools to chatter.

Other criteria also affect the machining accuracy and the performance of the machine tools and must therefore be taken into consideration; these include power loss and thermal characteristics depending on heat conductance, protection against swarf, contaminant or coolant penetration, and obstruction in the linear and rotary guides and bearings.

Manufacturing and operating costs are determined primarily by the choice of the guide and bearing principle. **Figure 51** shows linear and rotary guides and bearings classified according to the physical principle on which they are based and/or the type of lubrication and lubrication system together with the friction characteristics.

The manufacturing costs (**Fig. 52**) may be reduced by rationalising the manufacturing process, using preprocessed and/or standard components and selecting suitable materials, such as plastic side linings which may be preformed. Another important consideration is the ease of

assembly of a complete linear and rotary guide and bearing system [39].

Operating safety and fault tolerance with the capacity to withstand excess loads also affect the operating costs of linear and rotary guides and bearings. The maintenance requirement and resistance to contamination of the various basic guide and bearing principles and methods are also criteria which affect operating costs and must therefore be taken into consideration in their selection.

1.4.1 Linear Guides

Flat Slideways. These are the most common design employed in machine tools irrespective of the design principle. They are designed to transmit high weight, mass and cutting forces mainly vertical to the slideway (**Fig. 53**). Holding strips are provided to prevent the slide from lifting or skewing. Backlash elimination is provided for horizontal slideways by adjustable gib-strips. Aspect ratios vary between 1 : 40 and 1 : 100. See **Fig. 53e, f** for adjustment controls.

Dovetail Slideways (**Fig. 53b**). These prevent the slide from lifting by setting the side surface area at an angle of 55°. They may be adjusted via gib-strips arranged

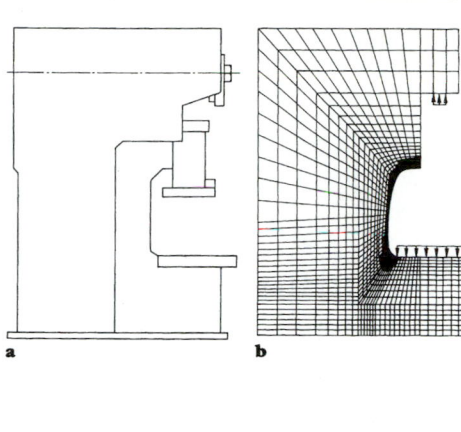

a

b

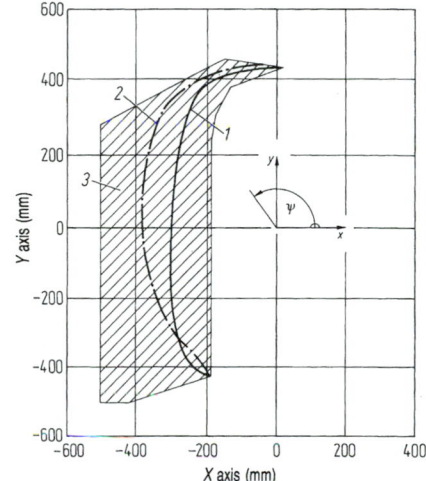

c

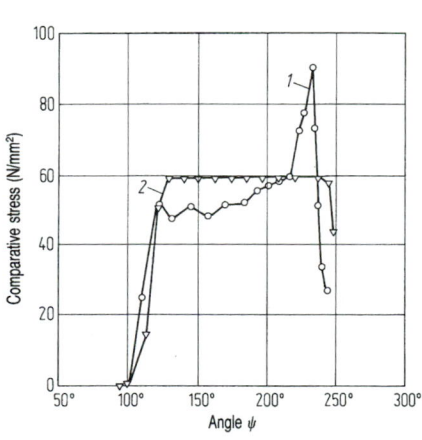

d

Figure 50. Optimisation of curve in C-frame press. **a** Block diagram. **b** Finite element network (line loads each 800 kN respectively). **c** Curves: *1* prior to optimisation, *2* following optimisation, *3* within the admissible variation range. **d** Stress characteristics at the margins of the curve: *1* prior to optimisation, *2* following optimisation, where the boundaries along the angle ψ are to be plotted as evoluted (anticlockwise).

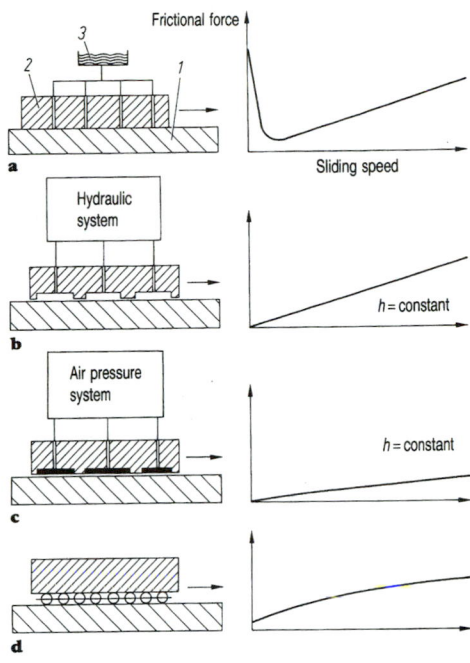

a

b

c

d

Figure 51. Principles of guides and friction characteristics. **a** Hydrodynamic slide: *1* bed, *2* slide, *3* oil sump. **b** Hydrostatic slide. **c** Aerostatic slide. **d** Roller slideway.

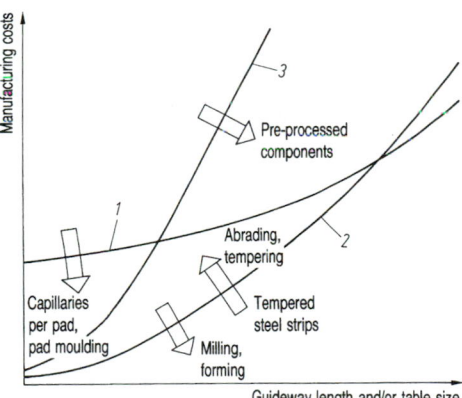

Figure 52. Trends of manufacturing costs of various guide principles. *1* Hydrostatics, one pump per pad, cast iron/cast iron, milling/milling. *2* Slideway cast iron/cast iron, grinding/grinding. *3* Roller slideway, fittings for recirculating rollers/tempered steel holding strips.

at an angle. Advantages compared with the flat guides are low height and good damping qualities. Designs may also be employed with a slope on one side and flat guides on the other (**Fig. 53c**). Use dovetail guides in short planing, slotting and shaping and small milling machines, otherwise only on slides for auxiliary and feed motion. They are mainly employed as slideways.

Vee and Flat Slideways (Fig. 53d) (V- and flat shape). These take up force in two directions. A flat form is used as a slideway for the main support in small and

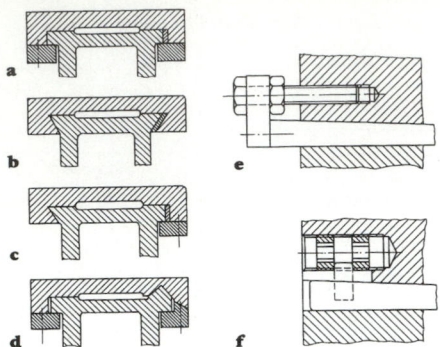

Figure 53. Types of linear and rotary guides and bearings. **a** Flat slideway with holding strip and adjustable gib-strip. **b** Dovetail slideway with gib-strip. **c** Flat slideway and dovetail combination. **d** Vee and flat (cambering) slideway combination with adjustable holding strip. Gib-strips may be adjusted: **e** externally, **f** internally via hexagonal socket screw.

medium-sized turning machines, and also in combination with a flat slideway. Protection against lifting is via holding strips, which may be adjusted in pitch for clearance.

Cylindrical Ways. These are installed as directional guides (e.g. drilling spindle sleeves) or slideways with shaft sleeves (Spieth sleeves) with adjustable clearance, or as roller slideways (see F4). They have the advantages of being simple to manufacture and possessing high guide accuracy, but are complex in assembly (shaft–centre distance) and are only suitable for limited lateral loads.

Hydrodynamic Lubrication Slides. These are most commonly employed in the machine tool construction sector. The grounds for this are the high damping factor and a high degree of accuracy and rigidity in conjunction with relatively low construction and manufacturing costs [38]. The relatively high frictional forces may adversely affect the feed drives.

Material Mix. **Figure 54** shows the material mixes employed in slideways and combined roller slideways/slideways [40]. In this case, 30% cast-iron–cast-iron material mixes and 28% cast-iron–plastic material mixes are employed respectively, whereas the remaining mixes are far less widely employed. Cast-iron and epoxy resin and Teflon-based plastics (PTFE) are normally used for the moving part of the guideway (slide). The fixed part (bed) is normally of cast iron, or in a few cases steel (Ck 45, 16MnCr5 or 90MnV8).

Manufacturing and Processing. The manufacture of plastic-coated guideways is carried out by bonding on plastic sheeting or using moulding technology. In the moulding process, the plastic compound is applied to the roughly prepared sliding surface and then pressed in before it hardens with the opposing component which has been pre-finished and sprayed with a parting compound (application technique). In order to achieve the correct shape for the guideway and an even layer of compound, positioning and spacing strips between the two parts are adjusted before embedding. The excess compound is squeezed out of the gaps by the force of the weight and additional loading. In the injection moulding technique, the lining is formed by pressing the plastic compound into the space between preset and adjustable components (**Fig. 55**). Good adhesion is achieved between the plastic

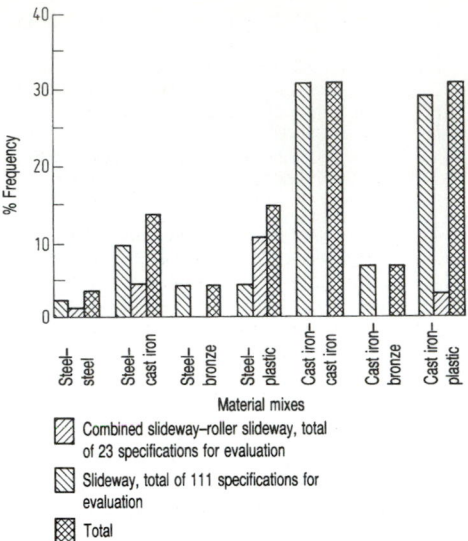

Figure 54. Material mixes employed in slideways and combined slideways/roller slideways.

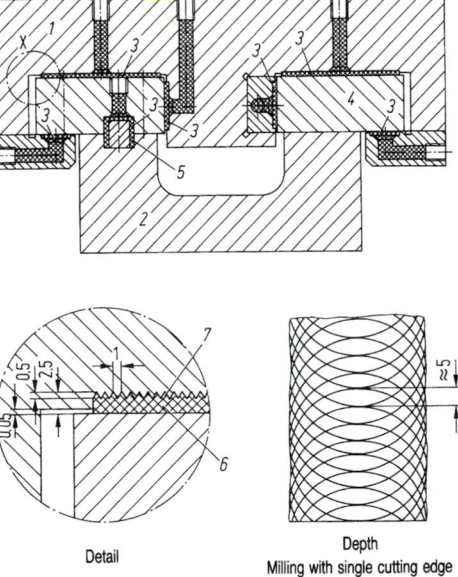

Figure 55. Injection moulding techniques for plastic-coated slideways (SKC-Gleitbelagtechnik): *1* slide, *2* bed, *3* plastic sliding surface, *4* press-fit drill holes, *5* bolted guide rail and spline.

and slide by planing with a diamond-point tool or milling with a single cutting edge.

The majority (some 60%) of the slideways which are primed or sprayed with compound are abraded following hardening to form the oil pads. A smaller proportion (some 25%) are put to use without any further processing [40]. Four finishing processes, i.e. abrading, circumferential grinding, end milling and fine milling, are undertaken in the most popular guideway material, cast iron, whereas steel is normally processed only with circumferential grinding and end milling.

Load-bearing guideways should be tempered to counteract seizure or wear. Cast-iron tempering is performed with flame hardening or induction hardening or casting permanent mould tempering (Brinell hardness, HB = 4.5 to 6 kN/mm^2). Surface tempering steel guides (Rockwell Hardness C, HRC = 58 to 63) are available in cylinders, block strips, plate or spring band steel.

Tribological Characteristics. In the tribological survey (see D6) of friction and wear, the collective load must be taken into account in all cases [41]. The collective load includes the type of motion (sliding, rolling, etc.), the time-related sequence of motion (continuous, oscillating, etc.), and the loading parameters (normal force F_N, speed v, temperature and length of load t_B). The characteristics of the parent and offspring substances with their materials and surface structures and the type, viscosity and volume of the precursor are particularly important.

The *frictional characteristics* of the different guide principles and of slideways with their various materials and surface structures are shown in **Fig. 56** [39]. The hydrostatic slide has the lowest friction coefficient. The friction coefficients in hydrodynamic slides are significantly higher than those in hydrostatic slides and roller slideways. The surface structures have a substantial effect on the frictional characteristic curve (Stribeck curve) in this type of guide. The use of the machining characteristic for circumferential grinding on the fixed under-surface section (bed) and the moving surface section (slide) reveals a sharp fall in the friction coefficient at increasing speeds (characteristic curve *1*). This encourages the undesirable stick/slip characteristic (slideback) at low feed rates. A part of the slideway, preferably the slide, should have tool marks crossing the slideway to alleviate this steep fall [42]. This may also be achieved with end grinding or preferably end milling (characteristic curves *2*, *3* and *4*). In this case, the overall level of the friction coefficient in the bottom section of the sliding speed range is significantly lower, and the stick/slip tendency is alleviated as a result. A favourable frictional characteristic curve, and an appropriately lower stick/slip tendency, is achieved with resin and PTFE (Teflon) infills using bronze (characteristic curves *5* and *6*). Teflon also permits dry operation, but has low compressive strength (edge tear resistance).

Table 2 shows the test results relating to the wear characteristics of different slideways [43]. The wear of lubricated, untempered cast-iron slideways is around 1 to 3 μm per sliding couple for a 60-km slide path with a load of 50 N/cm^2, which corresponds to an operating life of around five years in single-shift operation. Tempering metal guides do not cause any substantial reduction in wear in lubricated sliding load operation. The moulded plastics employed nowadays adversely affect the gap height change (i.e. the gap becomes smaller) by around 3 μm because of their swelling properties. Because coolant emulsions may also inadvertently find their way onto the guideway during the manufacturing process in addition to the slideway oil required, higher swelling values may be anticipated in general in plastics.

Extremely soft guide materials, such as pure PTFE, suffer excessively high wear under the normal loads in machine tool construction of 50 N/cm^2. Lower wear values may be achieved, while favourable frictional characteristics are retained, by employing suitable additives, e.g. pulverised bronze.

Lubrication of hydrodynamic slideways is an important aspect with respect to wear. Most machine tools (up to 80%) are fitted with pulsed lubrication systems. Continuous dropper oil lubrication and manual lubrication are

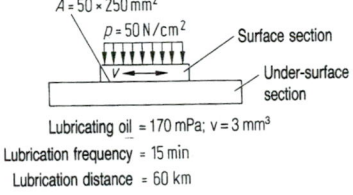

$A = 50 \times 250 \text{ mm}^2$
$p = 50 \text{ N/cm}^2$

Surface section

Under-surface section

Lubricating oil = 170 mPa; $v = 3 \text{ mm}^3$
Lubrication frequency = 15 min
Lubrication distance = 60 km

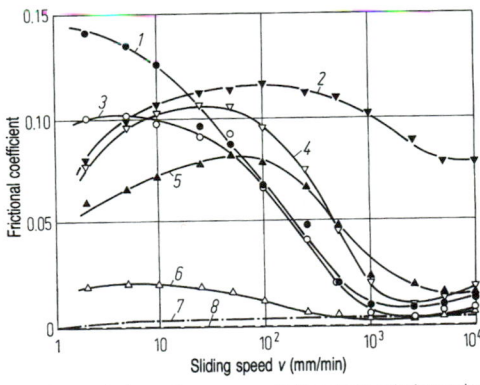

Surface section/processing	Under-surface section/processing
1 Cast iron 25/ circumferential grinding	Cast iron 25/circumferential grinding
2 Cast iron 25/end milling HM hard metal	Cast iron 25/circumferential milling
3 Cast iron 25/end grinding	Cast iron 25/circumferential grinding
4 Cast iron 25/end milling with cutting ceramics	Cast iron 24/circumferential grinding
5 Moulded resin/moulding	Cast iron 25/circumferential grinding
6 PTFE with bronze/ circumferential grinding	Cast iron 25/end milling HM hard metal

Figure 56. Frictional characteristics of different guides. *1–6* slideways, *7* roller slideways, *8* Hydrostatic slides.

Table 2. Linear wear rate in μm for slide path of 60 km

Operating process[a] Surface section/ under-surface section	Material mix in slideways (surface section/under-surface section)			
	Cast iron/ cast iron	Cast iron/ cast iron[b]	Resin[e]/ cast iron	PTFE compound/ cast iron[c]
1	2.7/1.1 (2.0/0.6)[d]			
2	1.8/1.7			
3	2.3/1.7			
4	2.5/0.6	1.5/0.6		
5			−2.8/0.8 (−1.2/0.6)[d] (−1.7/0.5)[f]	3.5/0.3

[a] Test setup and machining process in accordance with **Fig. 5b**.
[b] Tempered. [c] With bronze. [d] With slide path of 20 km. [e] Filled.
[f] With slide path of 5 km.

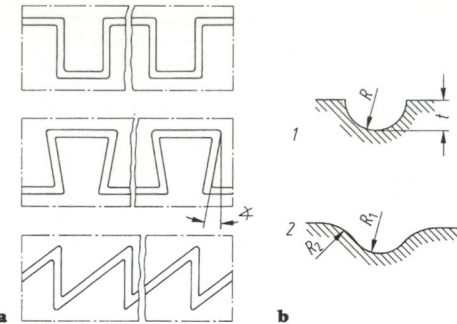

Figure 57. Lubrication groove forms. **a** Forms. **b** Lubrication groove cross-sections; cross-section 2 occurs more frequently than 1.

employed only in a very few cases. Slideway oils with viscosities of $n_{50} = 30 \times 10^{-3}$ to 80×10^{-3} N s/m² are used in lubrication. **Figure 57** shows lubrication groove forms and cross-sections of slideways. To prevent dirt from penetrating them, slideways have wipers (normally plastic constructions) mounted on them.

Hydrostatic Lubrication Slides. The slides of the machine components fed in this guideway construction are separated from one another by an oil film which is under pressure and is maintained via an external oil supply system [38]. The compressed oil is fed over the supply drill holes onto hydrostatic pads and flows away into the parallel gap between the sliding surfaces, thereby losing pressure. The oil supply is provided via either a separate pump for each pad (**Fig. 58a**) or a common pump at constant pressure with hydraulic cutoff to the pads via preliminary regulators, and normally takes the form of capillary tubes (**Fig. 58b**). The first case provides higher

rigidity and excess load capacity (**Fig. 59**, curve 1). At low power loss, as in the second case, the manufacturing costs are lower at half the operating rigidity (**Fig. 59**, curve 2).

Advantages of Hydrostatic Slides. These are: wear-free operation, hence no startup friction and only slight friction without backslip (stick/slip effect) in the range of feed rates; very good damping characteristics due to an even oil film across the guideway; high rigidity in small footprints, achievable within generous limits.

Slides with Aerostatic Lubrication. *Gas-lubricated* bearings operate on the same functional principle as those with fluid lubrication. The differences between them mainly concern the characteristics of their lubricants.

The advantages include very low friction, low heat generation, very good repeatability and low construction costs due to the omission of seals and a lubrication return system. The disadvantages are larger construction sizes, lower damping, inadequate emergency running properties and increased costs involved in manufacture and air treatment. Self-excited pneumatic instabilities ("air hammer") may occur owing to the compressibility of the lubricant, but may be corrected via structural measures, and by restricting the supply pressure to 4 to 10×10^5 Pa. The extremely narrow bearing gap of around 10 μm requires very high machining accuracy and low static, dynamic and particularly thermal-related displacement. The calculations for aerostatic bearings, based on the assumption of viscous gap flows, are carried out with the aid of the Navier–Stokes equations.

Figure 60 shows an aerostatic slide bearing with a rotational table. In order to reduce the manufacturing costs, the slide is preloaded with spring-loaded support rollers which have an extremely low level of rigidity compared with the aerostatic bearings.

Roller Slideways. Linear guides in roller slideways are used for a wide range of applications, as are standard slideways. They have the advantages over slideways of smooth operation due to roller friction, low startup resistance, absence of stick/slip and freedom from maintenance. Their disadvantages compared with hydrostatic and hydrodynamic guides is their low damping normal to the direction of motion [38]. Combined roller–standard slideways are frequently employed for this reason.

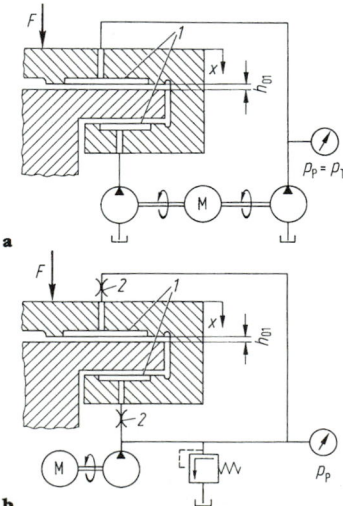

Figure 58. Oil supply to hydrostatic compressed oil pads 1 over multiple pumps **a**, common pumps **b** and capillary regulators 2.

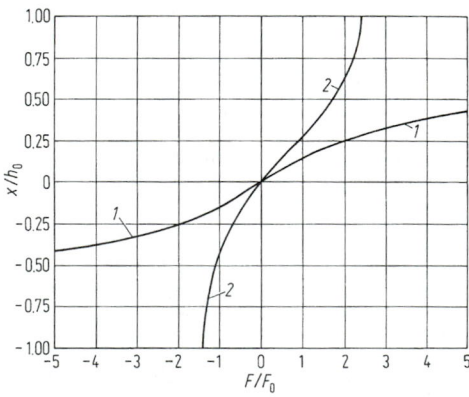

Figure 59. Displacement–load characteristic curves for hydrostatic slides as shown in **Fig. 5**: h_0 gap height in initial condition, F_0 internal preload force of holding strip. Technical specifications: surface area ratio $\varphi = 0.6$, initial gap height ratio $\lambda = 1.0$, regulator ratios $\xi_1 = 1.67$ and $\xi_2 = 0.6$.

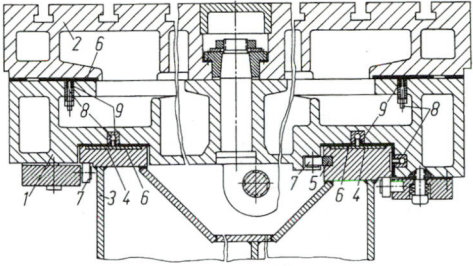

Figure 60. Aerostatic slide *1* in table base with rotating table *2* (Wotan, Düsseldorf), *3* bed, *4* cemented in tempered steel plates, *5* cemented in steel strip, *6* cemented in plastic plates, *7* spring-loaded support rollers, *8* air supply, *9* inlet stream aperture with jets as regulator.

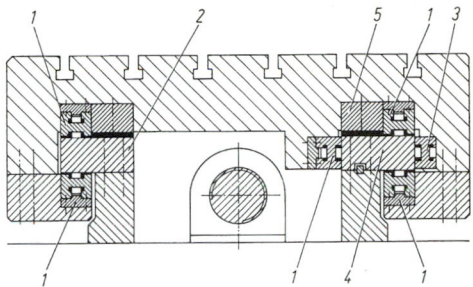

Figure 63. Slide conveyor of a milling machine: *1, 2* fittings for recirculating rollers, *3, 4* guideway, *5* damping strip (INA Lineartechnik).

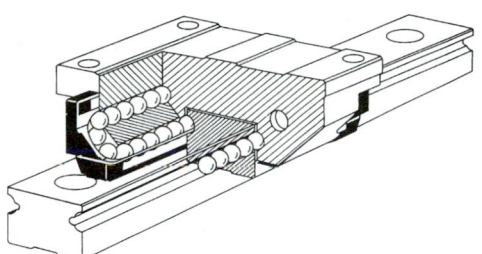

Figure 61. Recirculating ball bearing assembly.

Roller slideway components are available in a variety of sizes and precision classes. Ball bearings, cylindrical rollers and needles may be used as rollers (see F4). Rotating components containing a return for the rollers with one, two or four rows which are especially suited for long traverse paths may be incorporated to achieve greater rigidity where preloading is employed as well (**Fig. 61**). The ball-bearing-shaped rollers with two or four-point contact (pointed arc) may also be fitted between the fitting for the recirculating rollers and the guide rail (**Fig. 62**).

Example. **Figure 63** shows four roller slideway components (fittings for recirculating rollers) positioned on top of one another in the slide guide which take up the main load of the horizontal slide. The damping strips are located adjacent to these components. The remaining oil in the capillary gap between the strip and the guideway acts as an oscillation damper.

1.4.2 Rotary Guides, Bearings

The suitability of various types of bearings for standard application criteria in machine tool construction is shown in **Table 3** [51]. **Table 4** provides a summary of the resultant areas of application of the various bearing systems.

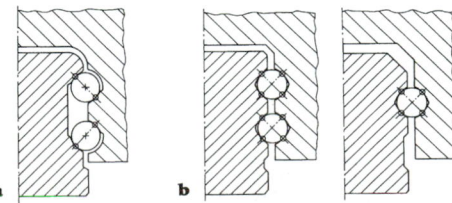

Figure 62. Ball bearing guides: **a** with 2-point and **b** with 4-point contact of ball bearings (Deutsche Star, Schweinfurt).

Table 3. Comparison of types of bearings

	Roller bearing	Hydrodynamic bearing	Hydrostatic bearing	Aerostatic bearing	Magnetic bearing
Speed limit	◕ [a]	◕	●	●	●
Service life	◕	◕	● [b]	● [b]	● [b]
Freedom from vibration	◕	◕	●	●	●
Damping	○	◕	●	◐	◕
Rigidity	◕	◕	●	◐	◕
Lubrication (cost)	○ [c]	◕	●	○	✕
Power loss	◐	●	●	○	○
Price (acquisition, maintenance)	○ [c]	◐	◕	◐	●

[a] Dependent on lubricating system and type of roller bearing
[b] Unlimited in fault-free operation
[c] Medium with oil lubrication

● Extremely high　◕ High　◐ Average　○ Low

Main Bearings. These serve to guide and take up the force of the rotating components which generate part of the cutting or shaping motion. The highest requirement is freedom from vibration for spindle bearings employed as boring, milling, turning and grinding spindles. This is why the dimensions of the various components in the spindle bearing system, i.e. the bearing, the spindle, the casing, etc., are given such narrow tolerances [53]. In addition to the rotational speed limit, the required speed range and characteristics also affect the selection of the bearing. Where the roller bearing is employed, the lubrication principle to be employed (i.e. minimum oil and grease volumes or oil coolant lubrication) is to be selected to correspond to the speed of the application, system loading and admissible power loss [54-56].

Roller and bottom bracket bearings normally have to transmit the most power in small constructions. They are therefore frequently constructed as sliding bearings [52].

Feed Spindle Bearings. These make great demands on the thrust bearings at extremely high operating speeds in terms of accuracy and loading, and therefore always take the form of rolling bearings which are preloaded.

Gear Bearings. Shafts, gear hubs, etc. operate in these as components of wheel gears. They normally transmit

Table 4. Areas of application of various bearing systems in machine tools

		Roller bearing	Hydrodynamic bearing	Hydrostatic bearing	Aerostatic bearing	Magnetic bearing
Boring, milling, grinding and turning spindles	Standard milling machines	●	◐	◐	○	○
	High-speed milling [c]	● [b]	◐	◐	●	●
	Internal cylindrical grinding	● [b]	◐	◐	●	●
	External cylindrical grinding	●	● [a]	● [a]	◐ [a]	◐ [a]
	Turning	●	◐ [a]	● [a]	◐ [a]	◐ [a]
	Boring	●	●	●	○	○
Tables on boring and milling machines		●	●	●	○	○
Roller and ball bearings		●	●	●	○	○
Feed spindles		●	○	◐	○	○
Gear shafts		●	◐	◐	○	○

[a] Where surface roughness of less than 0.2 μm is required
[b] Limited suitability with greasing
[c] $nd_m > 10^6$ mm/min

○ Unsuitable   ◐ Limited suitability   ● Extremely suitable

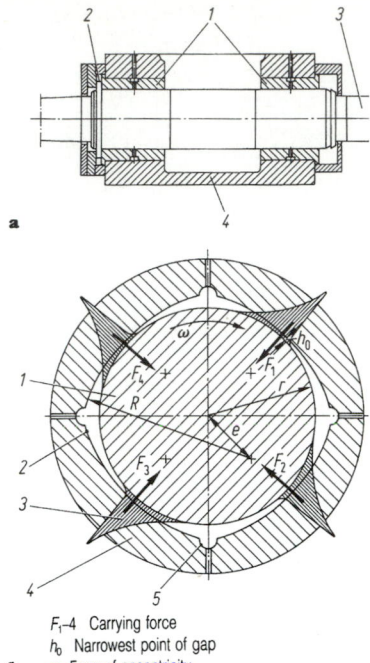

F_1–4 Carrying force
h_0 Narrowest point of gap
b e Form of eccentricity

Figure 64. a Grinding spindle bearing with hydrodynamic multiple sliding valves: *1* bearing bushes, *2* thrust bearing, *3* spindle, *4* spindle box. **b** Cross-section of multiple sliding valve: *1* spindle, *2* sliding surface, *3* pressure envelope, *4* bearing shell, *5* oil groove.

relatively high speeds over a small to medium speed range. Applications are found in standard rolling bearings with small relative speeds and in small constructions; sliding bearings made of bronze or cast iron are also used.

Form

Sliding Bearings with Hydrodynamic Lubrication
[38, 45–47, 49] (see F5). These are employed in machine tools as main bearings where high accuracy and good damping characteristics are required at high, virtually constant operating speeds, i.e. wear-free operation in viscous friction area, or where high power is to be transmitted in small constructions to mixed friction areas in some cases.

Circular sectional bearings are used in heavy machine tool constructions, such as in rolling machines, large-scale turning machines and eccentric presses.

Multiple sliding valves (see F5.6) are used as spindle bearings for low-power, high-speed grinding, fine boring and finish cutting machines. These bearings should be avoided in multi-directional or frequent startup applications with fluctuating power.

Sliding bearing bushes are normally designed with a conical form for controllability and are press-fitted or in some cases cemented in. To prevent strain on the edges, these must be installed accurately and have minimal shaft bending deflection. Slide materials [48] include tin or lead plating or bronze. The surface of the shaft must be tempered, ground and superfinished. Surface roughness and circularity errors are between 1 to 2 μm. Diameter tolerance is h7 to h8, bearing clearance is 0.4 to 3‰ of diameter and length/diameter ratio is 0.5 to 1.

Example (see **Fig. 64**). The grinding spindle bearing [50] shown is centred by two multiple sliding valves with fixed sliding surfaces with the clamping effect of four pressure envelopes.

Sliding Bearings with Hydrostatic Lubrications
[38, 44] (see F5.10). These are employed as the main bearings in grinding, finish-turning, and boring and milling machines where high loads have to be taken up and high rotational speeds are required. However, almost any operating characteristics may be achieved at will with the

appropriate selection of structural parameters. These advantages are countered by the high costs of an oil supply system and safety precautions required in case of breakdowns. Low oil viscosity should be used to maintain lower levels of friction loss and heat buildup at high sliding speeds (15 m/s and above) and with a small oil gap (around 30 μm). Careful installation and consideration of the shaft bending resistance are required, because dry friction may occur as a result of tilting.

Example (See **Fig. 65**).

Roller Bearings (see F4). These have a broad area of application in machine tool constructions, owing to their adaptability to extremely high requirements, such as high accuracy in continuous operation, high load-bearing capacity and rigidity, and high operating range at high speeds with low heat buildup. These requirements are satisfied by a combination of the advantageous roller, retainer and bearing surface design and arrangement, bearing clearance and/or preload, lubrication and choice of quality classification. Roller bearings are standardised and therefore less costly, and are of a simple design.

Roller bearings are employed in spindle bearings up to the highest precision classification defined in DIN 620.

A small to medium preload should be applied to the bearing to provide the highest possible rigidity and concentricity levels and the lowest possible wear [57, 58]. Lubrication of roller bearings is essential because otherwise they would break down after a short period of operation, and anyway the bearing temperature would be too high. Grease is used to lubricate small to medium-sized bearings. Where the bearings are employed in high-speed applications, with operating speed characteristic values

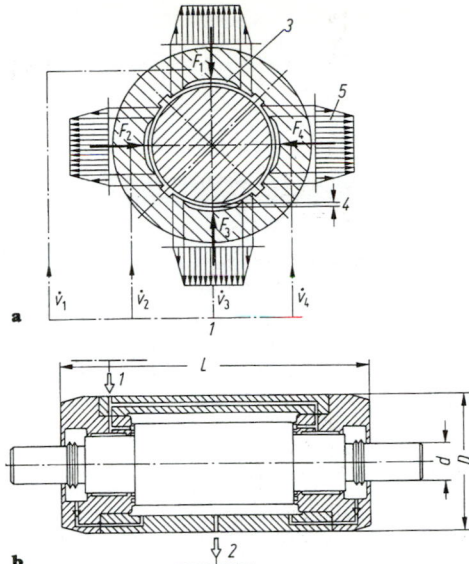

a

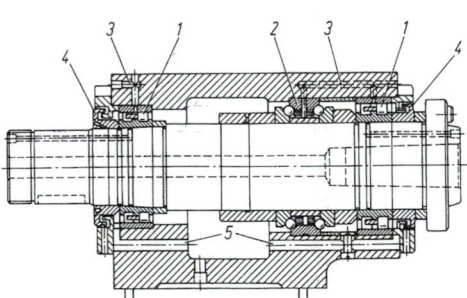

b

Figure 65. Hydrostatic spindle in bearing (FAG, Schweinfurt): *1* oil supply, *2* oil drain, *3* hydrostatic pad, *4* gap, *5* pressure envelope, *V* oil volume. **a** Cross-section of pressure envelope (resulting force of pressure). **b** Longitudinal section.

nd_m of 0.5 to 1×10^6 mm/min (d_m = mean bearing diameter), oil injection lubrication or minimum oil lubrication is recommended in most cases owing to the increased levels of operational safety these offer [54, 55]. In injection lubrication, a substantial volume of oil is guided around a cooled oil circuit and therefore also serves to cool the bearing. Where precision angular-contact ball bearings are employed, as is normally the case in high-speed spindle bearings, this greasing method may be employed with operating speed characteristic values of up to $nd_m = 0.8 \times 10^6$ mm/min. Special-purpose synthetic greasing is then required, with accurately regulated metering according to the roller bearing. In addition, an accurate input process is essential in this case, together with gradual increases in speed and intermittent operation [54, 55].

Cylindrical roller bearings are frequently employed as radial bearings (**Fig. 66**), the rollers producing high rigid-

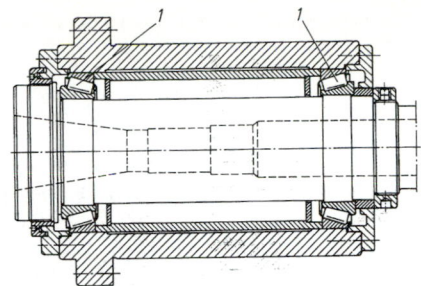

Figure 67. Milling machine spindle (SKF, Schweinfurt): *1* taper roller bearing.

ity and damping, particularly in two-row design. Clearance adjustment is via a taper roller seat on a spindle.

Taper roller bearings provide a means for adjustment by way of the axial advance of a bearing ring (**Fig. 67**). They possess good damping characteristics, but rotational speed has an upper limit imposed by the rim friction of the rollers. The O-shaped arrangement of the taper roller bearings helps to compensate for the temperature expansion.

Thrust angular ball bearings cater for operating speed characteristics of up to $nd_m = 5 \times 10^5$ mm/min with low preloading (**Fig. 66**).

Thrust cylindrical roller bearings are used where there are substantial axial forces and the rotational speeds are not too high ($nd_m \leq 0.4 \times 10^5$ mm/min), e.g. for table bearings in large-scale turning machines or feed spindle bearings. In the case of the latter, the spindle should be supported axially at both ends to improve its overall rigidity.

Axial grooved ball bearings are used to transmit axial power, and are therefore little used as spindle bearings. In order to be able to take up high axial spindle loads, thrust angular ball bearings are employed in preference.

Angular-contact ball bearings operate at high speeds. The low rigidity of these bearings, particularly in an axial direction, may be increased by a series of bearings (up to four) arranged back-to-back in a tandem arrangement

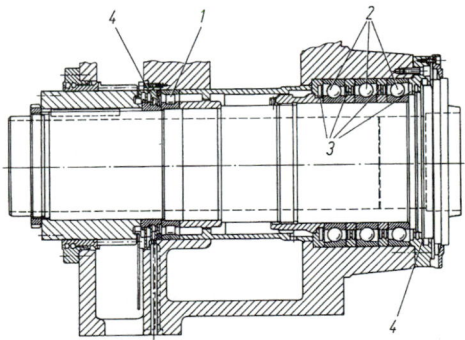

Figure 66. Milling spindle (SKF, Schweinfurt): *1* cylindrical roller bearing, *2* thrust angular ball bearing, *3* oil supply, *4* labyrinth seal, *5* oil drain.

Figure 68. Main spindle of turning machine (FAG, Schweinfurt): *1* cylindrical roller bearing, *2* angular-contact ball bearing, *3* grease chambers, *4* labyrinth seal [59].

which are preloaded with up to two single thrust bearings (**Fig. 68**). Angular-contact ball bearings are frequently employed in combination with one or more rows of cylindrical roller bearings. Where the spindles to be supported are to be operated at extremely high speeds (operating speed characteristics $nd_m = 1$ to 2×10^6 mm/min), angular-contact ball bearings should be used in preference to all other types.

2 Control Systems

G. Pritschow, Stuttgart

2.1 Fundamentals of Control

2.1.1 Definitions of Control

DIN 19 226 and IEC 50/chapter 351 define *control* as a process within a system whereby one or more variables as inputs affect outputs, depending on the individual system environment. The term *control* is also used to define equipment where a combination of mechanical and electronic engineering and information technology, in jargon *technotronics*, nowadays forms the basis of automation in machine tool construction. The control system is an essential part of a machine tool, allowing an operating process to be carried out independently according to a specified program. DIN 19 237 and IEC 117-15 part 15 provide a specification for the term [21], defining it in terms of types of information layout and signal processing.

2.1.2 Information Layout

A distinction is made in information layout between *analog*, e.g. cam-throttle control, cam-type control, copying control, and *digital*, e.g. NC, systems. The latter operate with digital (quantised) signals which are normally represented in *binary* form (bivalent). A further category of control classification in the information layout is the distinction between digital and binary control systems. Binary control systems operate primarily with binary signals, which do not normally form part of data displayed in numerical form.

2.1.3 Program Control and Function Control

Machine functions such as motion and switching functions which may be accessed manually are known as *manual control* systems; where these are accessed through the various steps of a program stored in memory they are known as *program control* [1]. Digital program control systems have a processor which interprets the user program step by step.

Program control is used to process *source statements* into individual function requests and to coordinate the function process automatically. Where the control status is determined by a time factor, for example with the guide of a tool cutter over a cam plate, where the angle of rotation is a function of time, then the term *time control* (e.g. cam-throttle control) applies. All other program control systems are *process-control*-based, i.e. the conditions enabling advance to the next step of the program are dependent on specific process variables being satisfied, such as path, temperature, force, etc. Positioning control systems are frequently employed in the control of machine tools, the commonest alternative being *numerical control* [3, 4].

Conversion of machine functions requested manually or via a program is carried out via *function control* (**Fig. 1**). This analyses the functions requested according to a predetermined series of machining sequences and initiates their execution. Function control may also include program control, depending on the complexity of the tasks set. This term also covers a wider, more general context, as program control is ultimately used to convert logical functions (**Fig. 2**). In this case, function control consists of measuring elements and actuators. Actuators are elements which have a direct influence on the system or process as the output of the control or monitoring device. Elements which are actuated include e.g. hydro and electrical motors, hydraulic and pneumatic adjusting rotors, couples and transmission units. Where the operating program is based on process control, measuring elements are attached to the machine, e.g. position monitoring systems, which signal the status of the process to the control system, i.e. the position of the tool. This enables machining sequences to be initiated or halted in relation to the path travelled or to specific positions.

2.1.4 Input and Output of Signals

A signal at the input of a functional component is described as an input or incoming signal; in the same way, signals at the output are called output or outgoing signals. Signals are normally controlled by input and output elements before and after processing. The following functions apply for these elements:

- *Input element*: resetting, re-forming, directing, isolating potential, conditioning, converting (analog/digital, digital/analog).
- *Output element*: amplifying, converting, storing, decoupling.

Input and output elements may be omitted where the processor technology is modified to suit the signal

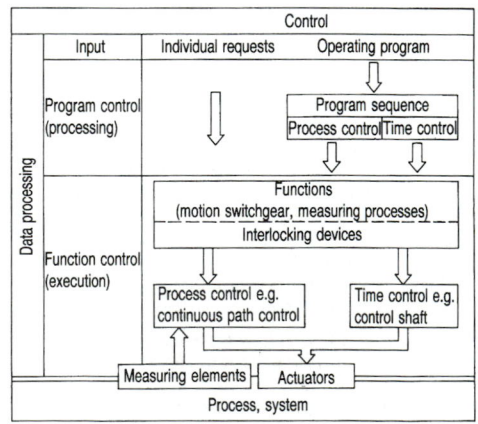

Figure 1. Control structure.

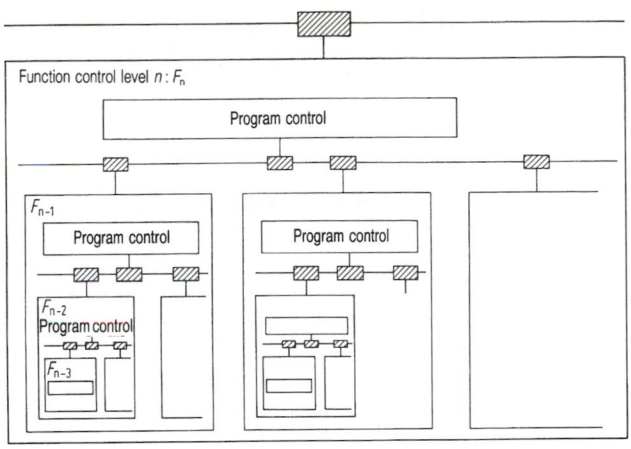

Figure 2. Function and program control.

environment within the controller (system-compatible signals).

2.1.5 Signal Forming

Input and output signals in a control system are signals from a signal-forming generator. A distinction is made between the following, depending on the signal:

- *reporting* – signal to inform operators about status or process (acoustic or visual signalling as specified in DIN 19 235 and IEC 73);
- *acknowledgement* – signal which has a direct effect on a command.

2.1.6 Signal Processing

Each control function may be classified as either *signal input*, *signal processing* or *signal output*, irrespective of the scope and level of control. Signal processing takes the form of either logic control or sequence control.

Logic Control. This means that output signals are assigned to specific input signals in the context of logic operation. Signal processing is carried out via *basic logic elements*. The following are examples of basic logic elements:

- logic elements: AND, OR, NOT;
- monoflops for pulse contracting, signal delay or signal stretching;
- flip-flops, such as RS flip-flops (R = reset, S = set), D bistable flip-flops and JK flip-flops.

Sequence Control. Control systems with positive step-by-step sequences are known as *sequence control systems*. A distinction is made here between control with time control or process control step-enabling conditions. The control problem may be described in the form of a sequence cascade (**Fig. 3**).

The following are important attributes of sequence control systems based on process control:

- Only one sequential control element is set.
- The step enabling condition is only dependent on conditions following the current step.
- The safety interlocking function operates independently of the sequence cascade.
- Extensive control tasks frequently require a number of

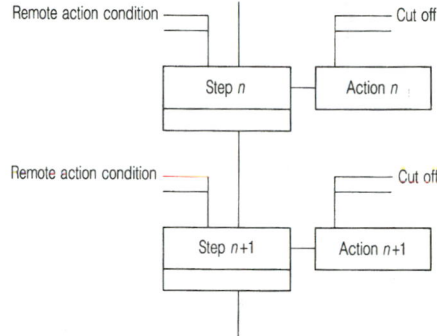

Figure 3. Description of sequence cascade.

sequence cascades, which may be derived from the structure as shown in **Fig. 4**.

There is another distinction based on time control of signal processing:

Clocked Control. With this, signal processing occurs only in the individual control elements at specific times which are synchronised by a clock pulse. This process is particularly useful where different signal propagation times in various parts of the control system and their spread would mean that any control result which may be produced could be ambiguous. It is employed chiefly in electronic control.

Asynchronous Control. Asynchronous signal processing is based on demand and operating time, and not linked to a fixed time. The type of control ensures that no operating time-dependent errors occur in signals which may influence one another. This normally takes the form of a prescribed series of signals, where data processing is only enabled after specific signals for release have been given and a subsequent operation may only be initiated in conjunction with an "operation completed" message relating to the previous operation.

2.1.7 Control Programs

Characteristics. Control programs include all the commands and directives required for signal processing, as a result of which a system (process) is influenced in accord-

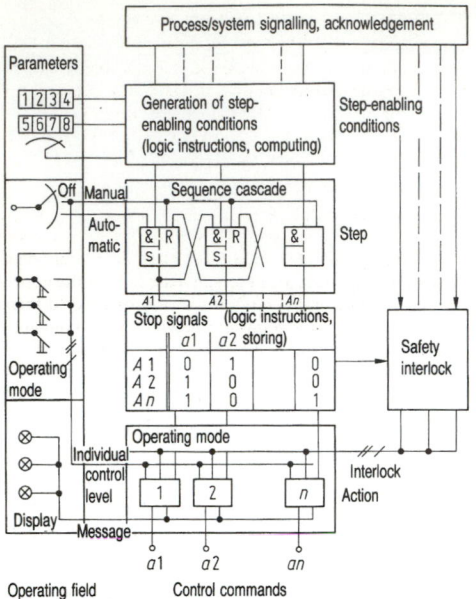

Figure 4. Structure of process control system.

ance with the task-related conditions. This may occur in various forms. Rigid systems operate with fixed programs, where a selection may be made between a number of programs. Where programs are changed frequently, it is more practical to employ interchangeable, freely programmable memory. This may apply in the case of e.g. cam plates, cams, stops or notched bars in mechanical control systems, while in electrical control systems applications may include program cylinders, crossbar distribution panels, punched tape or electronic data carriers.

Exchangeable programs produced by the users of the processes to be controlled are known as *user programs*. Electronic control systems require additional internal *system programs* to enable them to interpret and process these user programs.

Program-Dependent Hardware. Classifications are provided in DIN 19 237 [21] and IEC 117-15 part 15 for *hard-wired programmed controllers* and *programmable logic controllers (PLC)* (**Fig. 5**). Hard-wired programmed controllers may either be *permanently programmed*, i.e. invariable, e.g. owing to their hard-wired

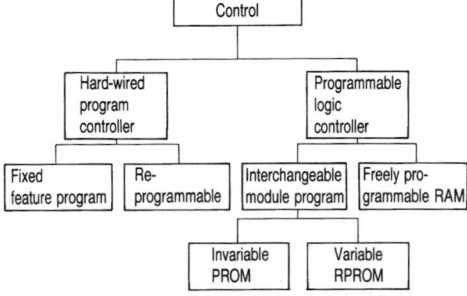

Figure 5. Control classification in accordance with DIN 19 237 (functional as for program implementation).

connections or pcb connections, or *programmable*, i.e. variable, by way of plug-in line connections, diode matrices, plug boards, variable crossbar distributor panels or interchangeable modules. In the case of PLCs, a distinction is always made between freely programmable control systems and programmable controllers with interchangeable memory. The programmable memory in *freely programmable electronic control systems* is a read/write memory or RAM (random-access memory), where the entire contents may be changed at will without any mechanical intervention and to any degree however small.

Programmable controllers with interchangeable memory on the other hand have a ROM (read-only memory), where the contents, once programmed, may be only changed by mechanical intervention in the control system after they have been programmed. A distinction is made here between controllers with read-only memory that are programmed following manufacture at the factory and may be changed subsequently (RPROM = reprogrammable ROM) and those that may be programmed only once during or following manufacture and are subsequently invariable (PROM = programmable ROM).

2.1.8 Organisation of Control

Great importance is attached to hierarchical process control system organisation in industrial applications. The various control systems assigned to the different hierarchical levels are as follows:

Individual Control. A process within a control system is normally influenced via actuator intervention at the level of the individual controller. This is the smallest control element which monitors the drive components and may be operated either manually or from a superior level.

The appropriate operating mode is determined via specific input in each individual case. Where operation is selected with a volatile command, the individual controller is assigned a control signal memory to determine the time of issue of the command. The operation (command input) is only activated where the safety interlock conditions are satisfied and a release is triggered, where this is applicable. The individual controller normally includes a monitoring and reporting sector for reporting the operating status of an actuator (steady state, transition, not ready, fault).

All the individual control systems (the drive controllers) together make up the individual control level (drive control level).

Cell Control. The functional unit required to control a subprocess is known as the "cell control". It is superior to the individual control system forming part of the subprocess (the drive controller). A number of cell controllers may be arranged hierarchically in order of importance where this is required for the process enabling conditions. The cell controller group constitutes the cell controller level.

Primary Control. Primary control is superior to cell control and is used to control the complete process.

This division into individual, cell and primary control is structured according to functional units, where the nearest superior level is the control level of the inferior level in each case.

Control Levels in Manufacturing Systems

Classification of control tasks in terms of levels leads to decentralisation of the data processing tasks and thereby to less complex system parts with their own separate data management and configurable standardised interfaces and modular software. The advantages of the autonomous system parts are wider accessibility over the entire system and simplified conditions for commissioning or modifications.

The controller tasks are hierarchically structured in production engineering in terms of primary, cell and machine controller levels. This classification also represents a structure based on functional units (**Fig. 6**).

The above designation may not be adopted for all applications. The cell control level may be linked to the primary control level, depending on the size of the production system, and may also undertake machine control tasks where the system hardware requirements are appropriate.

The control tasks in a chained production system may not, therefore, necessarily be permanently assigned to the above-mentioned levels; nevertheless, they normally follow a structure which is similar to the one illustrated in the following example:

Primary Control Level

– Control data generation for workpiece and tool flow (internal planning).
– NC program management.
– System mapping.
– Production data acquisition (PDA)/machine data acquisition (MDA) for composing and editing for display, documentation and controlling.

Cell Control Level

– Tool data management.
– NC program organisation.
– Acquisition and evaluation of PDA/MDA data.
– Equipment synchronisation at machine controller level.
– Evaluation of measured data and control where applicable.

Machine Control Level

– Manual operation/initialisation.
– Program debugging.
– Processing tool offsets.
– Generation of axis motion.
– Processing logic functions.
– Monitoring and diagnostic functions.
– Measuring processes.
– PDA/MDA data acquisition.

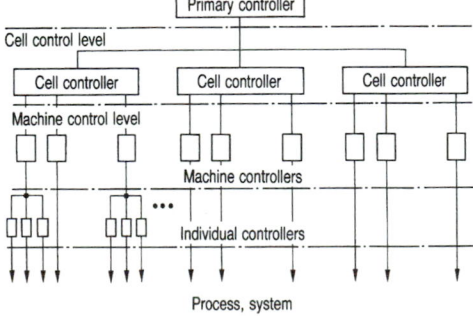

Figure 6. Controller hierarchy.

2.1.9 Databases and Link Structures in Manufacturing

The requirements for manufacture are as follows: to be able to satisfy market demand flexibly, and to frequently supply up to lot size *1*. This means that the response time between order and delivery must be minimised via the deceptively simple formula:

turnaround time/processing time → *1*

with the maximum degree of flexible automation of the means of production. Only if planning and implementation at all production levels, including the relevant databases and data sinks, are computer-based and if the appropriate databases and computer systems assigned to them are integrated into a computer network via a communications system may this be achieved. The short response times which may be achieved between the areas involved, plus data integrity and the power of the decentralised yet interlinked process engineering closed-loop control circuits are of considerable help in achieving this ambitious target. **Figure 7** shows which areas are principally affected by this in terms of the areas in a plant that generate data that may be combined within a CIM system (CIM = computer-integrated manufacturing). The data provision requirement for integrated data processing can only be attained with the aid of individual database systems which are networked and coordinated. In addition to the actual transmission technique employed, there are also the questions of data management and data content which have yet to be addressed.

The following *basic functions* are required for a CIM solution:

– EDP equipment linking via networks with standardised network protocols;
– uniform data formats for communication between sectors;
– coordination of flow of information.

2.1.10 Safety Standards

No fault or error occurring in automation systems may result in any danger to human life, and those who design, install and operate such systems have a responsibility to provide safety functions in addition to executing control and monitoring tasks. Current engineering practice is to be observed for industrial control technology in the form of:

– The law relating to the safety of operating equipment (Gerätesicherheitsgesetz GSG, Bundesgesetzblatt (German Civil Code) Pt 1, p. 717.
– Accident prevention regulations in accordance with the Commercial Trade Association (VBG) Register (VBG regulations).
– Safety regulations: Rules and principles relating to ZH-I Register (Workers' cooperative guidelines on work safety).
– Technical standards: DIN 31 000/VDE 1000 General principles for safety structures for technical products; DIN 57 113/VDE 0113, IEC 68, IEC 529, IEC 536, IEC 742 and IEC 801: Regulations for electrical equipment in machining and finishing machines with rated currents of up to 1000 V; VDI/VDE 3541, Sheets 1–3: Control systems with interlocking functions.

2.2 Means of Control

2.2.1 Mechanical Memories and Control Systems

Cam-Throttle Control. Cam gears are frequently employed for generating path and feedrate variation, i.e.

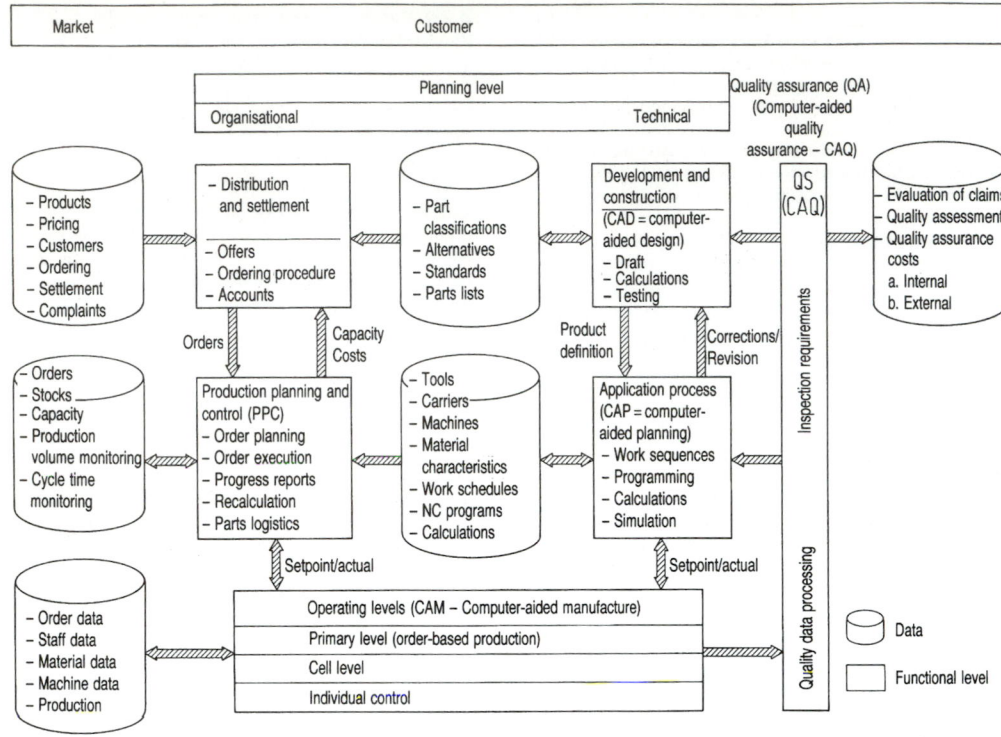

Figure 7. Databases in works organisation.

curves represent mechanical reference systems for path and feedrate variation. The required motion displacement is transmitted via the sensor head of the transmission element to the part to be moved, e.g. to the machine tool slide during one revolution, depending on the path and feedrate. These curves may be either three-dimensional (drum-type cams) or two-dimensional (plate cams) (**Fig. 8**).

One important area of application for cam-throttle control is in automatically controlled turning machines or presses. Process control is carried out automatically via plate cams and cams which are positioned on control shafts and normally rotate at constant speeds (time-program control). The curves form program references for the paths and feedrates, where the relations are as follows:

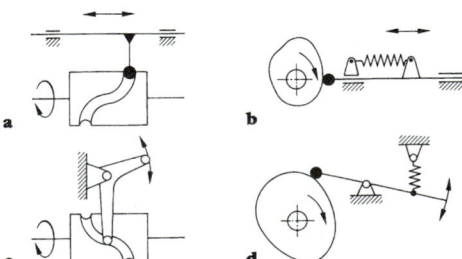

a   **b**   **c**   **d**

Figure 8. Principles of cam-throttle control: **a** pushing – drum-type cam (form-fit), **b** pushing – plate cam (adhesion), **c** turning – drum-type cam (form-fit), **d** turning – plate cam (adhesion).

Path reference. Stroke $\Delta s = f(\alpha)$; $\Delta s_{max} = r_{max} - r_{min}$.

Feedrate reference. $v = \omega(ds/d\alpha)$; $\omega = 2\pi/T$,
$$\alpha = \omega t.$$
Here α = the angular position of the curve,
  r = the cam plate radius,
  ω = the angular velocity,
  T = the period of rotation.

These transmit the feedrate required to the actuator and the moments and/or power required for acceleration. The transmission mechanism consists of mechanical components such as rollers, levers, ball bearings, guideways and springs.

Cam Control. Cams move a ram by travelling over it, thereby triggering a mechanical, hydraulic or pneumatic logic function. Electrical spring-operated limit switches trip logic functions accurately. Examples of logic functions include main spindle speed changes, coolant supply circuitry, switches for feed motion, etc.

The cams are normally attached to cam strips or cam rollers which generally have several cam tracks and may be clamped at any point at will (**Fig. 9**). The cam strip attached to the slide or tool bed serves as a reference position for the cam program controller, whereas the cam roller rotating with the control shaft represents time-program control. Cam tracks are constructed in the form of a rectangular, T-shaped or dovetailed groove.

Tracer Control. Tracing (copying) denotes a machining process where the tool motion is controlled by a control curve or surface area (model, template), so that the profile of the sample is transferred to the workpiece. Trac-

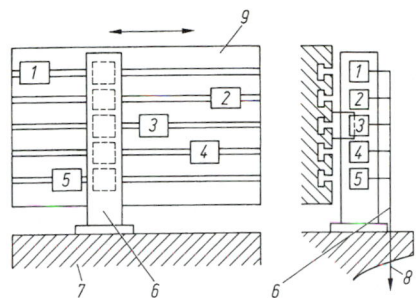

Figure 9. Continuous path cam-type control: *1* to *5* adjustable cams, *6* cam limit switch, *7* bed, *8* message, *9* slide.

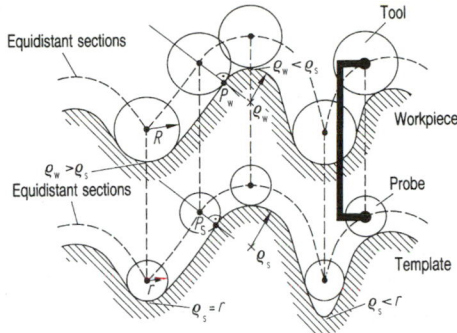

Figure 11. Touch trigger probe transmission from sample curve to workpiece via equidistant sections with different probe and tool radius. R tool radius, r probe radius, P point, ρ profile radius, suffix S template, suffix W workpiece.

ing is employed for manufacturing workpieces which are complex to create (e.g. shaped tools), specifically in small-scale series production, but is being superseded increasingly by NC technology.

A distinction is made in tracing between single-, double- and triple-axis kinds. In *single-axis tracing* the feed motion of the tracing slide is only controlled in one axis, whereas it passes through the machine in the other direction of axis with a constant sliding feed. Similarly, two or three axes of motion are controlled in *double- or triple-axis tracing*, where both processes permit three-dimensional tracing. The slide motion in double-axis control then runs along the third axis with the sliding feed (= 2½-axis transport), e.g. for traverse milling and multi-layered rotational milling (**Fig. 10**).

Transmission from flat curves in samples to the workpiece is effected via equidistances (parallel curves). The curved radii of the sample curve, the probe radius r and the tool radius R have to be viewed as critical limit values and in relation to profile distortion. Thus, $r \neq R$ leads to distortions in transmission. This means for the case $R < r$ (**Fig. 11**) that the profile radius of the workpiece for concave profiles is greater (smaller in the case of convex profiles) than that of the template. A probe radius of the template of $p_s < r$ will result in coarser workpiece profiles.

2.2.2 Fluidics

Fluidics (see G) operate with *compressed air* or *hydraulic oil* [5–7]. These are applied where fluid drives are used because their special characteristics and the control tasks are simple. This saves on conversion from one energy form to another. The introduction and end of

motion is normally signalled via directional control valves (**Fig. 12**), where the speed of the motion is set via flow valves. The directional control valves are activated either mechanically, directly from the system, or electromag-

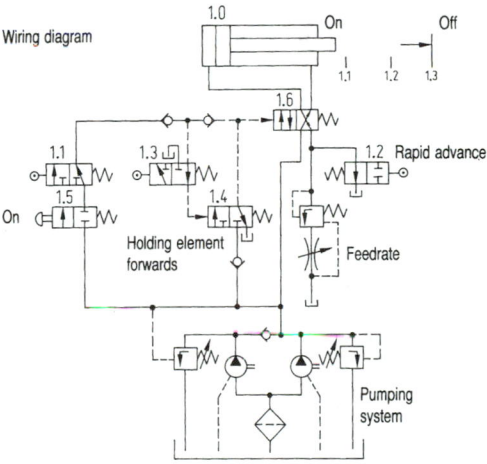

Control system flowchart (path–time diagram)

No.	Designation	Position	Work process 1 2 3 4 5 6 7 8 9 10
1.0	Cylinder	Off / On	
1.1	Three/two-way directional control valve	1 / 0	
1.2	Three/two-way directional control valve	1 / 0	
1.3	Three/two-way directional control valve	1 / 0	
1.4	Two/two-way directional control valve	1 / 0	
1.5	Two/two-way directional control valve	1 / 0	Manual initiation
1.6	Four/two-way directional control valve	1 / 0	

Figure 12. Switching examples for hydraulic drill feed equipment.

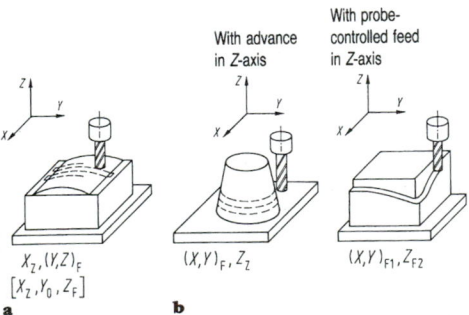

Figure 10. Profile milling methods: **a** traverse milling; **b** contour milling; F sensor, z line feed.

netically, on the basis of electrical signals. Occasionally, pressure-related switching is also used. The combination of electrical signal processing and hydraulic oil mechanical advantage is known as *electrohydraulics*. Simple logic operation and ease of handling of electrical signals are combined in this case with the high mechanical advantage and good time response of hydraulic oil drives.

The combined effect of directional control valves and flow valves may be produced with servovalves. These consist of an electric motor with a single- or multiple-step piston cylinder system, enabling the restrictors to be adjusted continuously depending on the excitation of the motor. Continuously variable speed drives may be set up with servovalves, where their characteristics may be improved specifically with the employment of a speed control mechanism (servohydraulics). With *servohydraulics*, speed adjustment takes place via an analog adjustment to the restrictors which affects the oil flow. The same aim, i.e. an adjustment to the oil flow, may be achieved with digital *hydraulics* by opening and closing the oil flow via a rapid switching valve at high frequencies and the time intervals after which this may be opened or closed are variable.

Fluid drives are available as rotary or linear motors. Transmission units are not always required, in contrast to electric motors, on account of the high pressure of the fluid media and the associated high torques.

The ratio of the moment of intertia to the torque which corresponds to the startup time of the motor, is lower in *rotary hydraulic motors* than in the majority of rotary electromagnetic motors. As the favourable ratio is also due to the low moment of inertia, the additional moment of inertia of the machinery to be driven also needs to be taken into account in the final calculation.

Linear hydraulic motors may be manufactured extremely competitively on the piston-and-cylinder principle. They also save on the mechanical transmission of rotary motion with their linear motion, which is more frequently required.

Hydraulic oil linear drives provide for high loads at average speeds (up to 1 m/s) with adequate rigidity. Their rigidity is directly proportional to the total enclosed oil volume, and may be improved by increasing the operating pressure. Higher operating pressures at the same output produce lower manufacturing tolerances (due to the associated losses), however, and therefore more expensive components. Thus, the dependence of the operating characteristics on the temperature is increased.

Pneumatic linear drives are cheap and extremely fast. They are designed for positioning rates of between 1 and 10 m/s in borderline cases. The forces to be generated are restricted by the dimensions and limited operating pressures, normally $<6 \times 10^5$ Pa (<6 bar). Their rigidity is low, and the same applies to the potential speed adjustment. They are therefore used almost entirely for positioning motion.

The use of *hydraulic oil drives* in a machine requires a *hydraulic generator* which converts electrical energy into hydraulic energy. For maintenance purposes it is advisable to install this generator with the control valves in a hydraulic control cabinet, where possible. The control circuits may therefore be composed of individual elements or a number of valves may be connected to form a common control block (*hydraulic block*). Where there are high requirements as regards time response, the switching elements should be arranged directly adjacent to the drives to reduce the volume of oil to be switched.

Hydraulic oil systems are described in hydraulic draw-

ings and parts lists. The symbols to be employed for this are listed and explained in DIN 24 300 and/or ISO 1219.

2.2.3 Electrical Control

Electrical control systems are designed as either contact-controlled or electronically controlled.

Contact Control. Large outputs may be switched easily and rapidly via contacts. These are also suitable for binary circuits (DIN 19 237), where a change in the system status is carried out via a change in a binary-coded signal. The combination of contacts and an electromagnetic drive is known as a *contactor* or *relay*.

As both the switching of three-phase motors and the operation of actuators are frequently effected via contactors, linking may also be carried out with additional contacts to these elements. The load factor and logic level are amalgamated within the system. Where the control systems are not too elaborate in the function control sector, contact controls are a suitable solution. However, it should be noted that the number of switches for contactors is limited mechanically to around 10^6 to 10^7 switches and the electrical switching capacity of the contact itself has to be considered. Furthermore, contactors require switching times of 10 to 200 ms which have to be taken into account in high-speed processes, e.g. with cutoff at high speeds. The switching times are dependent on the type and output capacity of the equipment and are affected by the magnetic flux.

Contact controls are described in circuit diagrams and parts lists and are structured with programmed connections (**Fig. 13**). The opportunities for rationalising the manufacturing control system are therefore limited. Symbol representation and the standards to be applied in connection with their use are contained in DIN 40 703, DIN 40 713 and IEC 117-3. The VDE regulations are also to be observed in association with their practical design, as these define state-of-the-art technology. Specifically, these include VDE 0100ff and VDE 0113, which cover systems for processing and finishing machines. The switching devices to be used are to correspond to specific switching classifications.

The control voltage in contact controls should amount to 220 V. It is not normally advisable to switch voltages of under 24 V using contacts owing to switching jitter. Direct currents should only be switched with the contacts specifically designed for the purpose, and protective equipment should be used to prevent arcing.

Electronic Control. Where the data processing extends beyond simple logic tasks, electronically operated control is normally employed. This is used both for binary (bit) and digital (word) signal processing. The processing of simple functions such as an AND or OR logic with simple functions within a semiconductor component is carried out in the same way as the operation of a counter or a digital/analog converter. Electronic control systems are unlimited in terms of the number of switches or their useful service life, they provide extremely rapid switching (nanoseconds or microseconds) and this is achieved with a low load factor. PLCs (*programmable logic controllers*) are employed as the hardware solution for bit and word processing of process and logic-orientated control problems. In addition, digital control is used for special applications using computers (and microcomputers). In both cases, the control algorithm is achieved via a program. In certain cases, *hard-wired program controllers* are still employed. These are only economical in large volumes with the same control systems due to the development, pre-production and inspection costs associated with pcb-

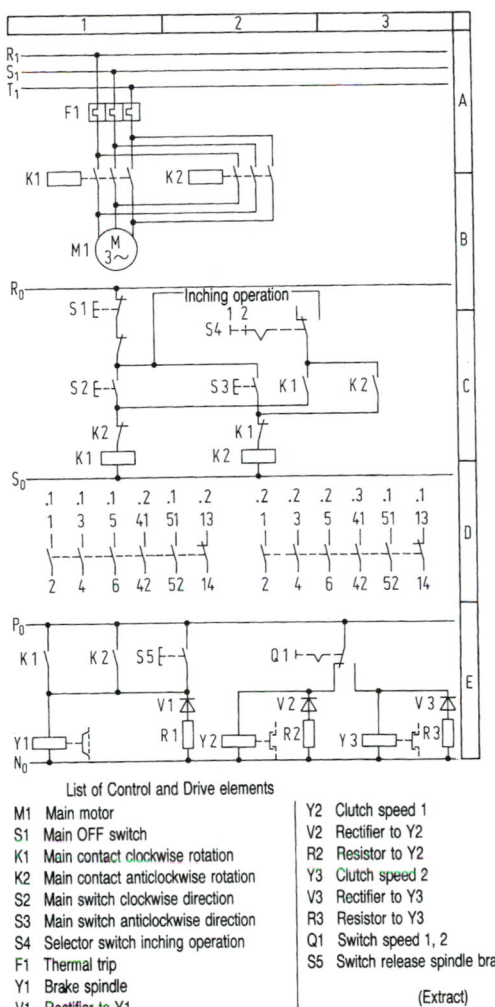

List of Control and Drive elements

M1	Main motor	Y2	Clutch speed 1
S1	Main OFF switch	V2	Rectifier to Y2
K1	Main contact clockwise rotation	R2	Resistor to Y2
K2	Main contact anticlockwise rotation	Y3	Clutch speed 2
S2	Main switch clockwise direction	V3	Rectifier to Y3
S3	Main switch anticlockwise direction	R3	Resistor to Y3
S4	Selector switch inching operation	Q1	Switch speed 1, 2
F1	Thermal trip	S5	Switch release spindle brake
Y1	Brake spindle		
V1	Rectifier to Y1		(Extract)
R1	Resistor to Y1		

Figure 13. Example of contact control: spindle drive of turning machine.

type data carriers. Electronic control systems may be subject to mechanical or electromagnetic interference from peak currents owing to their rapid switching and the status conditions that may be stored. Measures to counteract this include careful proof testing of the mains supply equipment, sufficient printed conductors and screening of the equipment itself. Inputs and outputs should be kept free of interference via a low-pass filter and be decoupled mechanically if necessary. Care should also be taken to ensure the reference potential is well-defined with adequate earthing.

As the control equipment is only heated slightly, owing to the low power factor in signal processing electronics, it is easy to provide protection for the electronic components against dust and humidity. The output signals may be set to the level required for the actuators by switching amplifiers or infinitely variable power amplifiers. Actuators are thereby frequently driven via interconnected contact controllers. Continuously variable power amplifiers are required for the relevant variable-speed d.c. drives. These may also be constructed as thyristors or transistor regulators.

Programmable logic controllers (PLC) and numerical control (NC) are dealt with in greater detail below in view of their significance in the control of manufacturing systems.

2.3 Programmable Logic Controller (PLC)

According to VDE Guideline 2880, the term "programmable logic controller (PLC)" is defined as follows: a programmable-memory automation system with user-orientated programming language which is employed primarily for control purposes [14].

This type of controller normally consists of a bit or word based processor with RAM, ROM and PROM (**Fig. 14**), where specific software enables control problems to be described in a user-orientated programming language (circuit diagram, logic diagram, Boolean algebra, status graphs, sequence cascade).

2.3.1 Structure

Simple PLC controllers are almost all programmable as far as *logic* operations are concerned. As such operations in the control sector are based on bit data, the processors preferably operate with word lengths of a single bit.

Each logic operation to be executed by the controller is arranged in a fixed order in the memory and may be

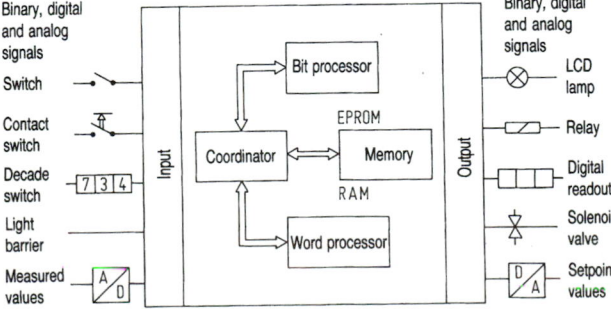

Figure 14. Structure of programmable controller with bit and word processor.

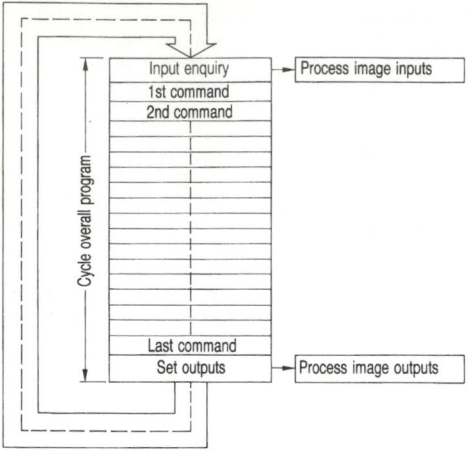

Figure 15. Program organisation with process I/O image.

accessed cyclically and processed accordingly (**Fig. 15**). The overall PLC program consists of a *system program* and *user programs*. The system program encompasses all the commands and directives of internal device restart functions and forms a permanent part (EPROM) of the PLC. This program may not be modified by the user. The user program is processed cyclically by the PLC. When the program has run through to the end once, the cycle begins again.

Inconsistencies, which may be generated when an input is accessed which changes state during the course of a cycle, may also be avoided by way of a program organisation with a process I/O image. All outputs are updated simultaneously at the end of a cycle and their values are stored as a process image in an integral memory corresponding to the simultaneous access of all inputs and storage of these values in a process image.

2.3.2 Programming

PLC programming is similar to the standard descriptions of those of conventional controllers which are more or less standardised. Programs are classified as follows:

1. Ladder programming in accordance with DIN IEC 65A (SEC) 67 [23] and IEC 1131 (circuit diagram, graphics).
2. Mnemonic designation in the form of a command listing in accordance with DIN 19 239 (alphanumeric).

3. Logic diagram in accordance with DIN 19 239 Symbols [22] (graphics, logic diagram).
4. Boolean equations in accordance with DIN 19 239 (mathematical description).
5. Flow chart programming in accordance with DIN 40 719, Sheet 6, IEC 113, IEC 750, DIN IEC 65A (SEC) 67 and IEC 1131.
6. Flow diagram in accordance with DIN 66 001, ISO 1028-1, ISO 2636-1 and ISO 5807-1.
7. Use of higher programming language in accordance with DIN IEC 65A (SEC) 67 and IEC 1131.

Whereas programming types 1 to 4 form part of *logic control*, types 5 and 6 are classified as *process control*. Program type 7 is employed for complex control tasks. The types described under 1 to 4 produce the same type of processing within the PLC. Only their representation is different (**Fig. 16**).

The latest PLC programming systems provide direct graphics programming of *sequence cascade representation* (**Fig. 17**). The French standard guideline known as GRAFCET has been adopted by many manufacturers and much of the new standard guidelines forming DIN IEC 65 A (Draft) and IEC 1131 are based on this.

The most important elements in the language are the *steps* which specify actions or conditions in accordance with DIN 40 719, Sheet 6 IEC 113, IEC 750 and *step-enabling conditions* (transitions T) which designate the point of deactivation of a step and the transfer to the next step (activation).

In addition to a straight sequence in a chain, it is also possible to represent branches and synchronisation of branched chains. Branches may run both forwards (parallel sequences, jumps) and backwards (loops).

Two time variables are allocated to each step: the *wait time*, which describes the minimum lapse in the step irrespective of the action time of the step, and the *check time*, which specifies the maximum length of time a step may take. At the end of the check time the program reverts to hold mode and generates an error message.

This type of representation provides a description of all machine construction control problems and has been introduced by several PLC manufacturers.

2.4 Numerical Control (NC)

2.4.1 Definition

The concept of numerical control was developed at the Massachusetts Institute of Technology (MIT) in 1951 for process engineering tasks. "Numerical" means that the input of the control data takes the form of numbers. These are represented in binary code and may be processed

Ladder programming	List of instructions	Logic diagram	Boolean equation
Programming with graphics symbols as in circuit diagram Corresponding to DIN 19 239	Programming with mnemonic abbreviations for function designations. Corresponding to DIN 19 239	Programming with graphics symbols Corresponding to IEC 117–15 DIN 40 700 DIN 40 719 DIn 19 239	Programming with mathematical description Corresponding to DIN 19 239

Ladder programming
E1 E2 E3 A1
┤├┤/├ ┤ ├┤( )├
E4
┤/├
E5
┤ ├
	List of instructions
	U E 1
	UN E 2
	U E 3
	ON E 4
	O E 5
	= A 1
		Logic diagram
		E1—
		E2—○ &
		E3—
		  E4 —○ >=1
		  E5 — —A1
			! E 1
			&N E 2
			& E 3
			/N E 4
			/ E 5
			= A 1

Figure 16. Types of coding for PLC programming.

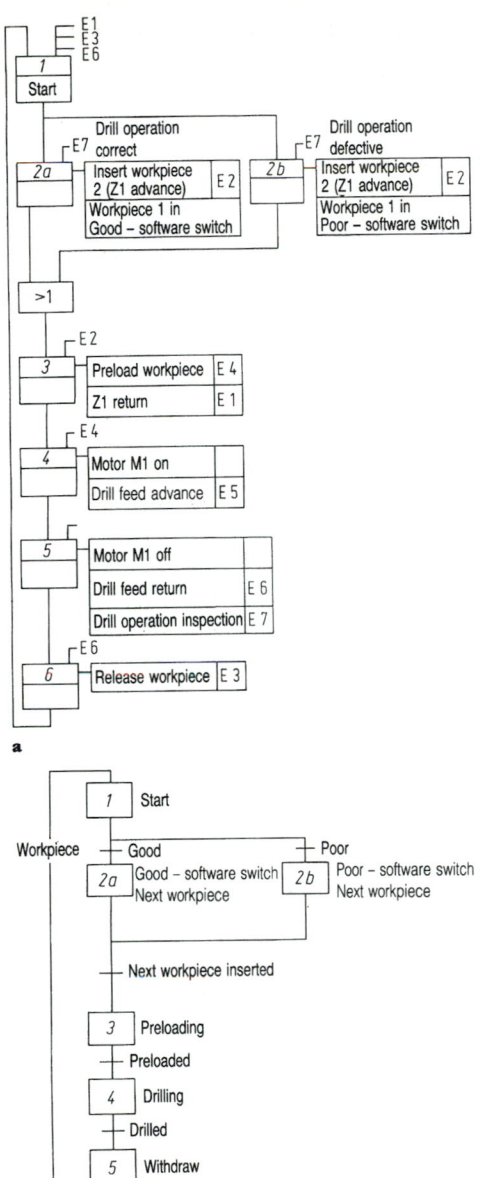

Figure 17. a Flow diagram in accordance with DIN 40 719 Part 6 for automatic drilling station. **b** Flow diagram from GRAFCET designed to address the same problem.

directly from the controller. Numbers have to be input to describe the workpiece geometry (path data) and technological specifications relating to the tools and operating speeds (alter statements), in numerical form in each case. This figure designation is characterised by an address letter prefix (DIN 66 025, ISO DP 6983/3 and ISO 646). Any control in which the path data are entered in numerical

form is considered to be numerical control, irrespective of the input device or data storage system [8–10].

2.4.2 Programming

Workpiece programming describes the generation of workpiece-related control data for numerical control (**Fig. 18**). Path data and alter statements are to be input to a data storage device in a predetermined order. One data storage device which used to be frequently employed was punched-hole tape, because it is robust and the sequence of operations is invariable; a large volume of data could be stored cheaply and read stage by stage as the process was being carried out. The use of inexpensive electronic memory has enabled large volumes of data to be stored far more simply nowadays, so that the input of entire programs is normally carried out either manually, e.g. via magnetic data storage media, or via data links from a main host computer.

NC programs may be either generated *on-line*, i.e. via the operator directly on the machine (shop-floor programming), or *off-line*, as part of the operational planning stage. Workpiece descriptions in the form of construction drawings (**Fig. 19**) or CAD data serve as output data for program generation where one of the more complex process engineering programming languages such as EXAPT is employed. The source program is translated into CLDATA (cutter location data) code with the aid of a processor (conversion program). The processor thereby processes the geometric data and supplements the technological processing requirements using materials and workpiece data files. A post processor converts the computer-independent CLDATA code for use with a specific NC machine.

2.4.3 Data Interfaces

As can be seen from **Fig. 20**, a further data interface is relevant to numerical control systems in addition to the NC program interface as given in DIN 66 025, ISO DP

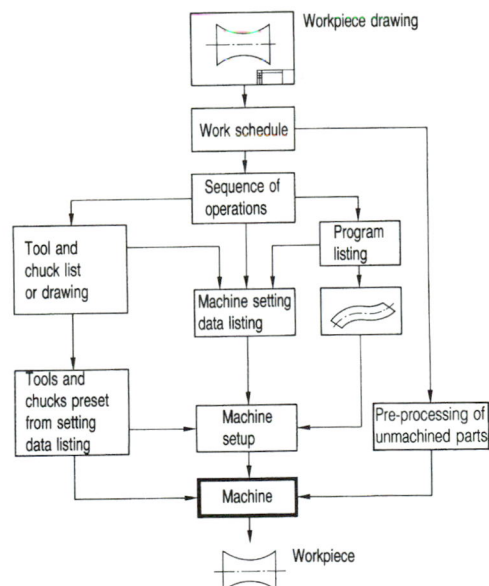

Figure 18. Flow of data from drawing to machining (conventional method).

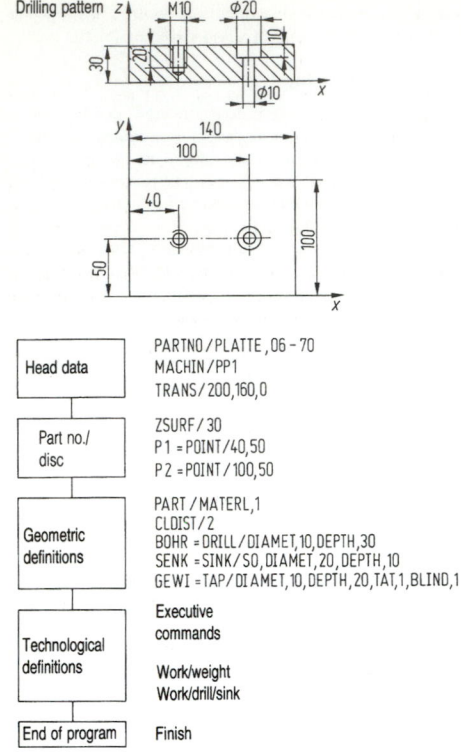

Head data	PARTNO / PLATTE , 06 – 70 MACHIN / PP1 TRANS / 200,160,0
Part no./ disc	ZSURF / 30 P1 = POINT / 40,50 P2 = POINT / 100,50
Geometric definitions	PART / MATERL, 1 CLDIST / 2 BOHR = DRILL / DIAMET, 10, DEPTH, 30 SENK = SINK / SO, DIAMET, 20, DEPTH, 10 GEWI = TAP / DIAMET, 10, DEPTH, 20, TAT, 1, BLIND, 1
Technological definitions	Executive commands Work/weight Work/drill/sink
End of program	Finish

Figure 19. EXAPT parts program.

6983/3 and ISO 646 – namely, the CLDATA (cutter location data) interface for computer-independent programming of machine tools, e.g. with the complex programming language EXAPT (cf. L2.4.2). CLDATA may only be run on a control system when it has been assigned control-related parameters in contrast to DIN 66 025, ISO DP 6983/3 and ISO 646. This either takes place via the post processor referred to above, which compiles the CLDATA in a form corresponding to the specifications in DIN 66 025, ISO DP 6983/3 and ISO 646 (**Fig. 21**), or the NC controller has an interpreter which executes the translation and modification line by line as part of the program control.

2.4.4 Control Data Processing

The control data programmed and stored in memory is processed in the numerical controller to produce *position setpoints* for the individual axes or output in the form of *switching commands*. Continuous motion in a number of axes takes the form of continuous computerised stage-by-stage output of separate position setpoints which are synchronised with the process. The position setpoints of each axis are compared with the respective *actual positions*. A positioning control speed is obtained from the *position tolerances* by multiplying them by a factor which is the same in all axes (vector amplification K_V (s^{-1})). Different position setpoints in the individual axes lead to different position tolerances, known as *following errors*, and consequently to different speeds which are necessary for motion over different track angles (**Fig. 22**). Impulses are generated for stepper motors from position setpoints.

The calculation of the position setpoints from the programmed control data is carried out according to fixed methods of calculation and is known as *interpolation*. The aim of interpolation is to reduce the volume of control data to a level which is sufficient for generating a combination of workpiece contours at will from simple straight lines, circles or segments of a parabola. Reducing the volume of control data frees control data storage media and input devices. Linear and rotary interpolation suffices for the majority of applications, i.e. the control data input is then used for the interpolation points,

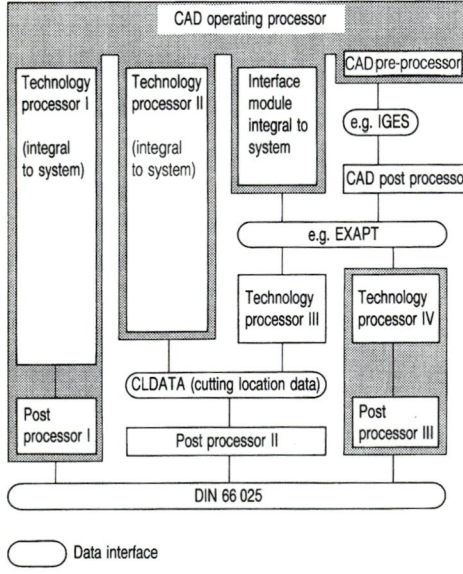

Figure 20. Data interfaces in control technology.

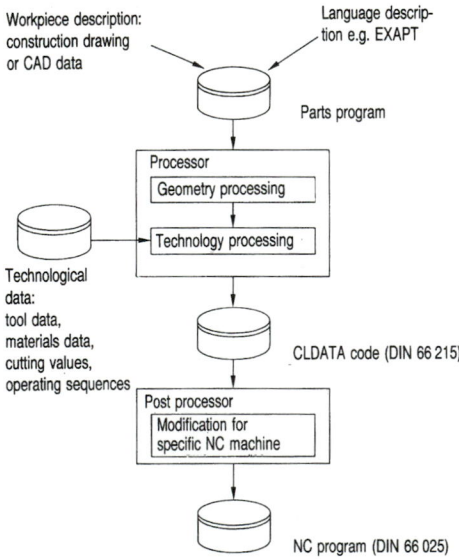

Figure 21. Computer-based NC programming.

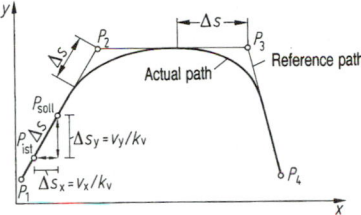

Figure 22. Continuous path NC control for linear interpolation and directional change in path: Δs following error, k_v acceleration, v_x and v_y speeds.

between which intermediate values are calculated on these curves as position setpoints, so that a position setpoint is output every 5 to 10 ms to each axis. In the less common method used for three-dimensional curves, i.e. parabola interpolation, a compensating parabola is set using three points in each case.

Tolerance calculations are normally to be carried out prior to interpolation (i.e. coordinate transformation, tool length offsets and radius compensation, etc.). Position adjustment may be effected via a separate external module, or in some cases it is carried out from the NC control within the control software. In the latter case, scanner systems provide computerised solutions, where discrete position control is executed via computer programs and the position control circuit within the control computer is closed.

Alter statements are normally only stored in numerical control and output in sequence via a defined interface (VDI 3422) to the subordinate function control units. Where programmable memory control systems are employed in function controllers, the interface may be omitted as defined in VDE 3422, where the PLC is integrated directly within the NC.

2.4.5 Numerical Basic Functions

Where the functions of an NC system are divided up into function-related parts, there are four basic tasks as shown in **Fig. 23** which represent the minimum scope of functions within the system. These include operating and control data input/output, NC data management, composing

and editing, and dissemination, technological data processing and geometrical data processing (GEO).

These basic functions are described briefly below.

Operation and Control Data Input/Output. Man/machine communication is becoming more and more essential in the numerical control sector. New developments are taking place in terms of the operator interface with menu-driven technology, graphics screens, windows functions, etc. Operation and programming facilities are becoming more and more complex. This part of the system software already accounts for over half of the overall system in modern NC systems. The following main functions all form part of the operator environment of an NC system nowadays:

- NC program memory storage (including any associated management operations).
- Operation and operator guidance.
- Plausibility checks for input data.
- Editing functions.

NC Data Management and Data Composing and Editing. The main tasks forming part of this functional unit include:

- Preparation of NC records for decoding and display.
- Decoding NC records (ASCII character conversion into internal control image).
- Resolution of operating cycles and subprograms, parameter calculation.
- Execution of tolerance calculations (tool length offset, tool radius compensation).

Technological Data Processing. Technological data processing is responsible for the execution of the alter statements (= technological instructions) which activate the switching of main spindle speeds, feedrates, tool change systems, coolant supply, etc., for example via the individual control level.

Geometrical Data Processing (GEO). Geometrical data processing encompasses all the basic functions required for path generation. A path is generated via the primary control motion of the individual axes. The functional units controlling setpoint generation

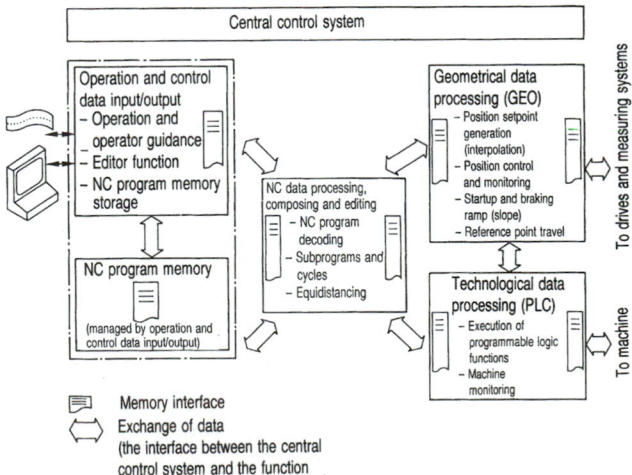

| Memory interface |
| Exchange of data (the interface between the central control system and the function blocks is not shown in the drawing) |

Figure 23. Examples of function blocks in numerical control.

(interpolation), setpoint adjustment (slope) and position control are required when adjusting the position of an axis. The next section (L2.4.6) goes into these functions in greater detail.

2.4.6 Position Adjustment

Position Setpoint Generation

Position setpoints are obtained for the individual axes of the controlled system from geometric input data in numerical control. Distinctions are made in relation to the kinematic sequences between three types of control system, namely point-to-point control, line motion control and continuous path control.

In *point-to-point control* the point defined by the setpoint may be approached via any path at will, as the tool is not operative during startup (**Fig. 24a**). The shortest path is normally selected to save time; the geometric shape of a tool only affects the traverse path in exceptional cases. This type of control constitutes the simplest form of numerical control and is normally employed in applications with drilling and boring machines, spot welding machines and tooling machines for electronic components.

Line motion control is related to point-to-point control, the difference being that the tool may be operative during installation. The sequence of motions is thereby carried out parallel to the axes of motion of the machine and the operating speed may be predetermined (**Fig. 24b**). In special cases, parallel operation of two or three axes at the same speed is possible (motion of less than 45°).

Where processing of arbitrary two- or three-dimensional curves is required, as is normally the case in milling machines, turning machines and flame-cutting machines, for example, *continuous path control* is employed. Its characteristics are comparable to those of tracer control. Where a tool traverses from point A to point B, for example (**Fig. 24c**), the tool follows the function shown, where $y = f(x)$, and is operative. The relative motion between workpiece and tool is thus continuously variable in terms of size and direction. The slide motion is therefore to be controlled at a minimum of two coordinates during processing.

Position setpoint generation (interpolation) for continuous-path control is explained in greater detail below.

The input data required for the desired traverse path are already present in digital form in NC machine tools with continuous path control, as has already been mentioned. The position setpoints are generated from the geometrical and motion data as position reference variables in the form of a finely stepped function of path/time by

the interpolator, which is a computer device. This function is converted by the position control system, which adjusts the individual machine slides in accordance with the position reference variables (**Fig. 25**).

The path produced by the position reference variables is primarily dependent on the interpolation process (single-stage, two-stage), the interpolation increment and the interpolation calculation process.

These three influencing factors are detailed briefly below.

Interpolation Process. In *single-stage interpolation* the interpolation points are calculated directly as reference variables for the position control. There is a drawback in this as regards the calculation input required for non-linear interpolation, e.g. circular interpolation. This may be reduced significantly by *two-stage interpolation*. A rough interpolation calculation is carried out initially when using this method, where interpolation points are generated at larger intervals. This is subsequently followed by a detailed interpolation calculation in the form of a simple linear interpolation where the intervals are cut by half (**Fig. 26**). A polygon that is preselected in this way is smoothed by the low-pass characteristics of the drives and does not, therefore, produce a continuous curve.

Interpolation Increment. Interpolation may be carried out in the form of:

- *fixed time increments*: the path to be travelled is predetermined for each interpolation cycle;
- *fixed path increments*: the interpolator outputs individual path elements in the form of the smallest unit of travel in accordance with the prescribed speed.

Interpolation Calculation Process. The following processes may be employed: step search process, digital differential analyser (DDA), direct function calculation and recursive function calculation.

Both the step search and DDA processes are interpolation processes with fixed-path increments and require standard measuring systems, normally consisting of special hardware interpolators for increment lengths in the resolution area, and are now outdated.

Direct or recursive function calculations are both interpolation processes with fixed time increments and may be implemented relatively simply using microprocessors and are, therefore, in popular use. The latter process provides for relatively simple calculations, although

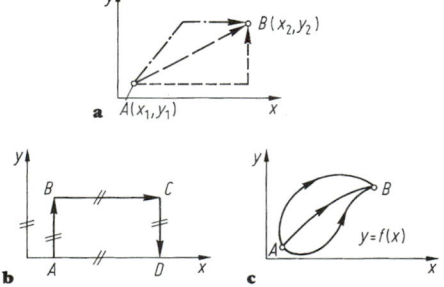

Figure 24. Types of NC control: **a** point-to-point control, **b** line motion control, **c** continuous path control.

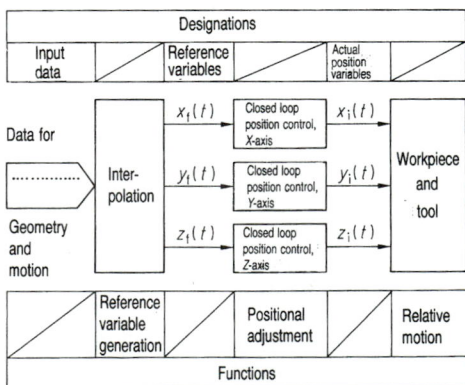

Figure 25. Signal flow diagram for generating relative motion between workpiece and tool.

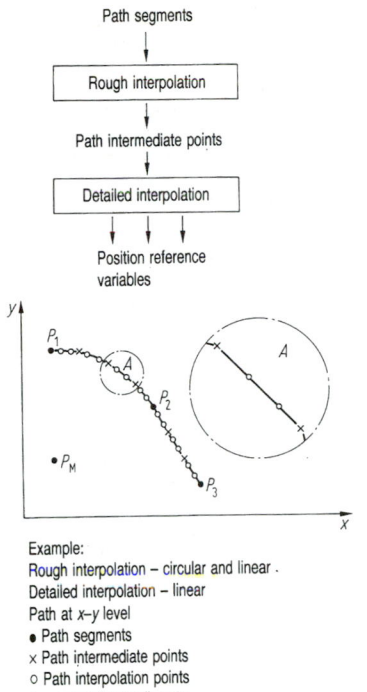

Example:
Rough interpolation – circular and linear .
Detailed interpolation – linear
Path at x–y level
● Path segments
x Path intermediate points
o Path interpolation points
(corresponding to discrete
time position reference variables)

Figure 26. Principles of two-stage interpolation.

the calculation has to be carried out with a greater degree of accuracy because of error propagation.

The calculation of interpolation intermediate points using the two-function calculation processes is illustrated in **Fig. 27**, based on the example of two-dimensional linear interpolation (also known as straight-line interpolation). If a parameter τ is increased in the interpolation cycle T by one increment $\Delta\tau$ in each case, this produces a constant path speed of v_B in a Cartesian-based working space. The size of the increment $\Delta\tau$ is proportional to the programmed path speed v_B and inversely proportional to the spatial traverse path s:

$$s = \sqrt{a_1^2 + b_1^2}, \quad \Delta\tau/1 = \Delta t/T, \quad v_B = s/T.$$

Linear interpolation

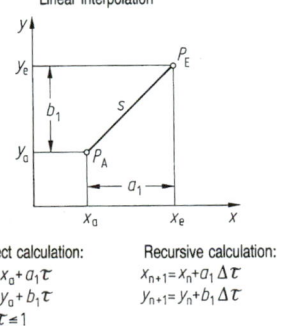

Direct calculation:
$x = x_a + a_1\tau$
$y = y_a + b_1\tau$
$0 \le \tau \le 1$

Recursive calculation:
$x_{n+1} = x_n + a_1\Delta\tau$
$y_{n+1} = y_n + b_1\Delta\tau$

Figure 27. Function calculation of linear interpolation.

Thus, the parameter $\Delta\tau = (v_B/s)\Delta T$ applies, where ΔT is the interpolation cycle time and T is the overall travel time.

Conversion of Spatial Coordinates into Axis Coordinates

Geometrical programming is normally carried out in the x, y, z coordinates of the Cartesian system. Where the axis coordinates are not identical to the main coordinates of the Cartesian system, the relevant conversion from a spatial to an axis system of coordinates is carried out by the control computer. This normally takes the form of a matrix operation. The conversion should be carried out within the interpolation cycle and may involve substantial calculation for multiple-axis machines, such as five-axis milling or six-axis robot guidance. This is why the principle of rough interpolation is frequently employed with the three-dimensional system of coordinates and detailed interpolation is used in axis coordinate systems. The flow of data from interpolation in spatial coordinates (Cartesian coordinates) to the reference variable of the individual axes may be seen in **Fig. 28**.

Where conversion is omitted altogether owing to the amount of calculation involved, and interpolation is only carried out in the axis coordinates, substantial discrep-

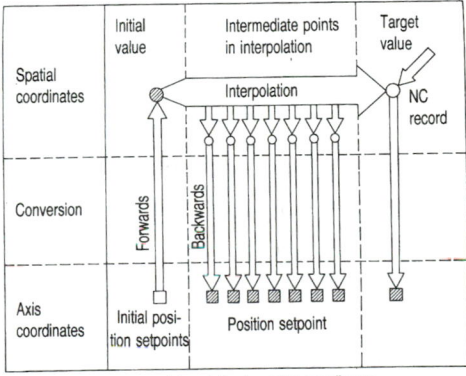

Interpolation in spatial coordinates

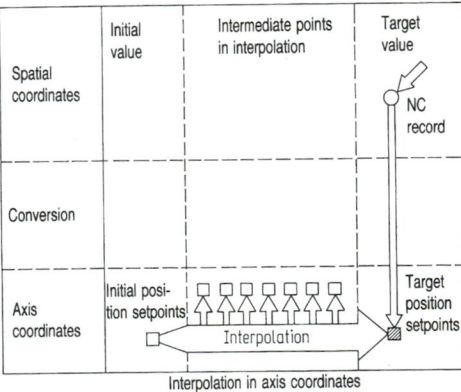

Interpolation in axis coordinates

▨ ⊘ Converted coordinates
□ ○ Given coordinates

Figure 28. Flow of data on interpolation to spatial coordinates and to axis coordinates.

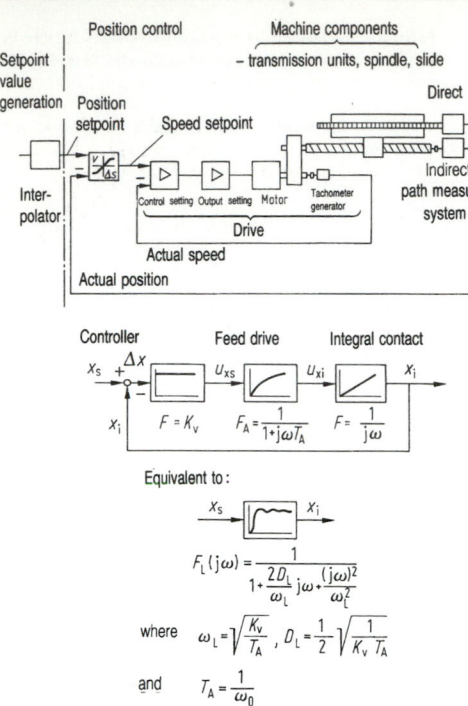

Figure 29. Simplified model of closed-loop position control axis.

ancies are likely to occur compared with the desired three-dimensional path (linear, circular).

Position Control

The relative motion between the tool or measuring instrument and workpiece is effected by the primary control motion of a minimum of two axes in continuous-path-controlled NC machine tools [11].

Figure 29 shows the structure of a closed-loop position control axis using a simplified control structure chart, where the drive is shown as a system of the first order and the position controller is typically designed as a P-controller with the acceleration K_V.

In order to prevent excessive oscillation, damping of $D_L = 0.7$ is recommended for the position control circuit described, i.e. the drive time constant T_A dictates the

potential acceleration K_V. The K_V factor in turn determines the following error, e.g. via the relation $\Delta s_x = \dot{x}_i K_V^{-1}$ for the x-axis with Δs_x in relation to the velocity $\dot{x}_i$ = constant. Where this following error Δs_x is compensated by suitable circuitry, the additional following error for constant acceleration $\ddot{x}_i$ = constant, where $\Delta s_x = \ddot{x}_i K_V^{-1} T_A$, may be calculated. As can be seen, the drive constant only has a direct effect on the following error in acceleration procedures. These relations enable the effects of a following error to be derived under a variety of influencing factors for linear travel in a Cartesian configuration. Contour distortions in linear travel may then be avoided, if both the acceleration factor K_V and the drive time constant T_A are equal in both axes.

2.5 Equipment for Position Measurement at NC Machines

Position-measuring systems in NC machines register a given line motion as an analog geometrical figure and produce this in the form of a digital position value. These are essential components of the position control circuit and their accuracy also helps determine the manufacturing quality of a machine tool. Their structure consists of a material measure, e.g. in the form of a scale, readout and analysing unit (cf. terms employed in instrumentation [24, 25]).

The following terms are considered to be particularly significant: resolution capacity, accuracy, sensitivity towards external influences and expansion capacity.

2.5.1 Types of Position Data Registration

These measuring systems vary depending on their particular characteristics (**Fig. 30**).

2.5.2 Measuring Spot and Data Sensing

The distinction between direct and indirect measurement depends on the position of the measuring spot [18].

The measuring system is located directly on the machine slide in *direct* measurement.

In *indirect* measurement, intermediate contacts transmit the change in position or line motion to a measuring system which is normally of a rotational design. Intermediate contacts may constitute the main spindle with a measuring gear and rack-and-pinion gearing, where necessary. Indirect measurement is often carried out on constructional or cost-effective grounds. Poor machining accuracy in the intermediate contacts and changes in size due to

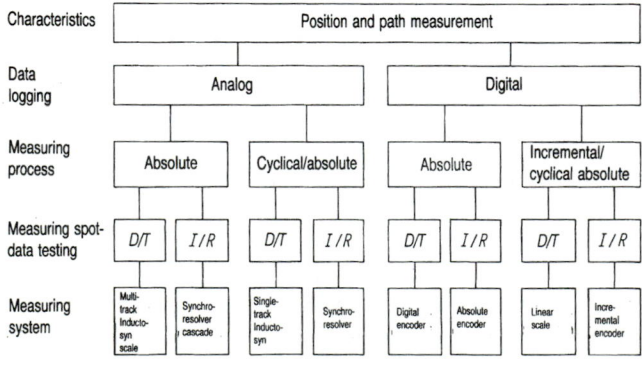

Characteristics	Position and path measurement							
Data logging	Analog				Digital			
Measuring process	Absolute		Cyclical/absolute		Absolute		Incremental/ cyclical absolute	
Measuring spot-data testing	D/T	I/R	D/T	I/R	D/T	I/R	D/T	I/R
Measuring system	Multi-track Inducto-syn scale	Synchro-resolver cascade	Single-track Inducto-syn	Synchro-resolver	Digital encoder	Absolute encoder	Linear scale	Incre-mental encoder

Figure 30. Types of position and path measurement: *D* direct, *I* indirect, *R* rotary, *T* linear.

temperature influencing factors have a direct effect on results of measurements, however. This is why direct measurement is more accurate than indirect measurement. Options for installation in-line motion or position measuring systems and error-influencing factors are shown in **Fig. 31**.

Error-influencing factors are basically always smaller, the better the measuring apparatus satisfies the conditions of the Abbe coefficient (after Abbe, 1890). According to this, the sample and comparative line motion should be arranged in alignment with the direction of measurement. The alignment angle error is then only of the second order in the measuring result.

2.5.3 Digital Data Logging

In digital data logging, the dimension to be measured is in a quantitative form. A distinction should be made between two different function principles, namely incremental and absolute measuring systems. The functional principles of incremental measuring systems are discussed first below.

Digital Incremental Measuring Systems

The digital incremental measuring system is based on the subdivision of the path into parts of equal size (increments) and the principle of incremental dimensioning. The measurement of a line (of an angle) is carried out by adding the increments on an incremental grating or incremental plate detected by a scanner device in relation to direction. To distinguish between two adjacent increments, the individual increments are alternately assigned different physical characteristics. The scanner device then provides the so-called count pulse from the relative motion in terms of the increment which is added up by the counter. The majority of incremental measuring systems use photoelectric pulse transmitters. In addition, pulse transmitters with magnetic sensors are also employed (e.g. gear wheel and magnetic pulse transmitter) where the requirements for resolution are not high.

There are a number of different constructions of photoelectric pulse transmitters, namely transmitted-light operation, front-lighting operation, polygon reflecting operation, and multi-prismatic operation.

The main functional units of these measuring systems include incremental scale, grating, lighting unit, counter device and direction discriminator.

The *incremental scale* is characterised by a regular series of markings. Depending on the design, these relate to light-permeable and non-light-permeable, reflecting and non-reflecting fringes. The pitch (grating constant) may be extremely finely stepped (up to 1 μm).

The *grating* forms part of the scanner and determines the form of the signal to be evaluated. There are three types of grating (**Fig. 32**). In the first form, the grating pitch is the same as the scale pitch and is parallel to this. When in motion the entire grating area alternates between light and dark. In the second form, where the grating area is positioned at an angle, Moiré fringes are generated. In the third form, the grating pitch is not the same. The light and dark zones occurring run lengthwise to the direction of scale.

The same aim is achieved with all these embodiments,

Direct data logging	Indirect data logging	
	Drive element and transformer separate	Drive element and transformer identical
Measuring system operates in linear direction and is directly coupled with the longitudinal travel of machine slide.	Conversion from longitudinal travel to rotary motion via gear rack and pinion	Longitudinal travel logging via rotary motion of operating spindle
Error-influencing factors Temperature Pitch errors Distance and angle errors Errors at intersection points of scales	Error-influencing factors Pitch errors of gear rack and pinion Eccentricity of pinion Errors in any transmission units employed Temperature Pickup error Sudden load change in gear rack	Error-influencing factors Elastic deformation of spindle Angle errors Play Obstruction of the spindle Pickup error Temperature

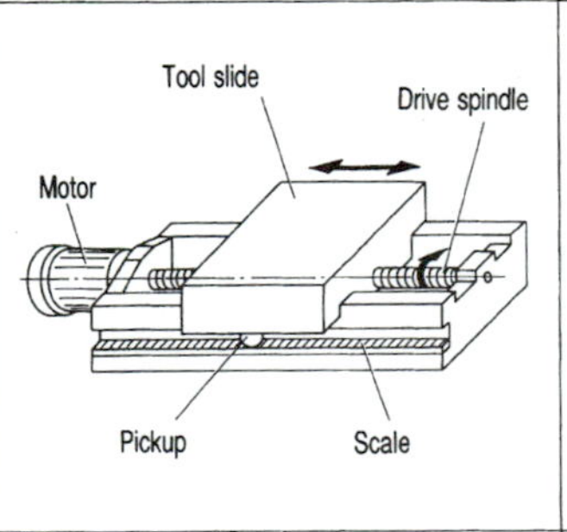

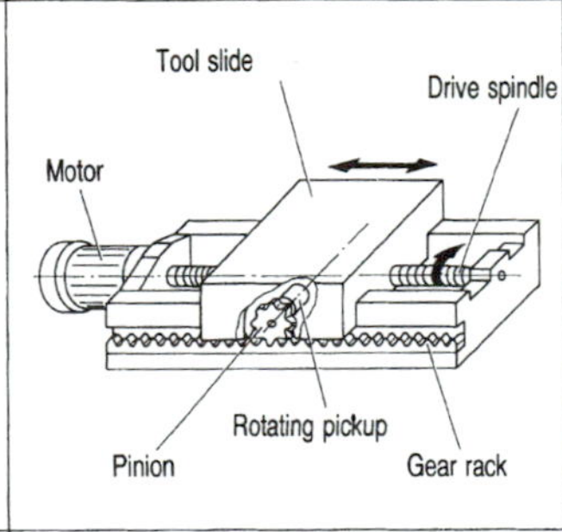

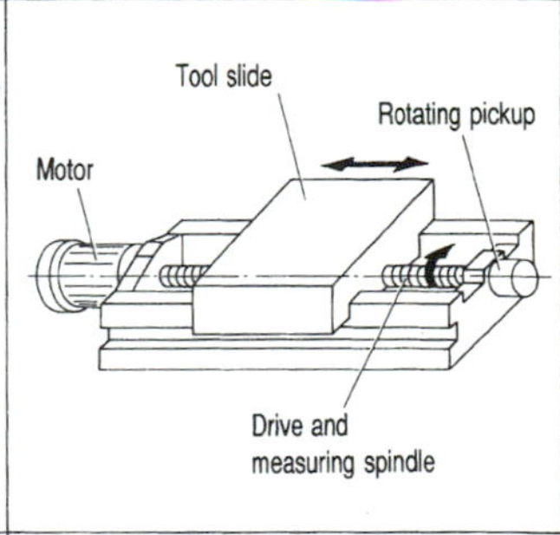

Figure 31. Error-influencing factors in direct and indirect data logging [2].

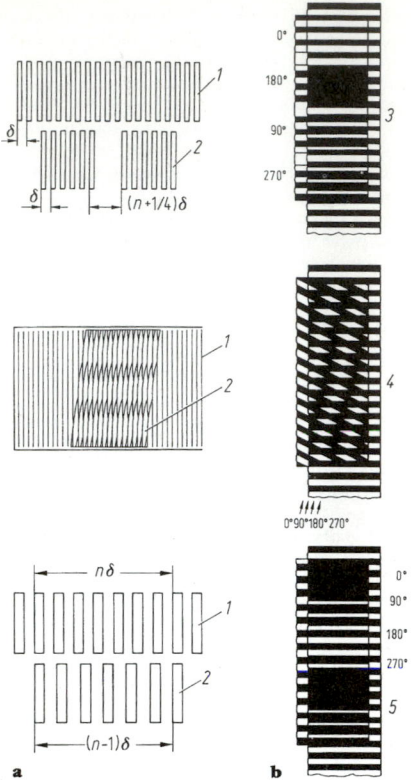

Figure 32. Designs of gratings in incremental measuring systems: **a** construction; **b** output signal; *1* scale, *2* gratings, *3* light/dark, *4* Moiré fringe, *5* vernier fringe.

namely that the finely graded light/dark line structure of the scale is converted into large areas of light/dark zones. This is needed for scanning with the light-sensitive elements.

Transmitted-Light Optical Diffraction Systems

The method of evaluation is explained in detail here and applies for all transmitted-light systems. The light travels from the light source via an optical element which generates parallel light through the grating and the grating plate on the photoelectric elements (**Fig. 33**). The photoelectric elements generate virtually sinusoidal signals period-

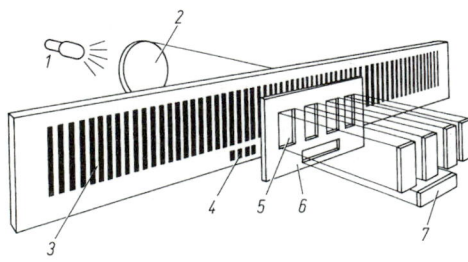

Figure 33. Operating principle of transmitted-light optical diffraction process: *1* light source, *2* condenser, *3* scale grating, *4* reference mark, *5* scanner grating, *6* scanner disc, *7* silicon photoelectric diodes.

ically with any motion in the scale in relation to the scanning unit. These signals are displaced by 90° (= $\frac{1}{4}$ pitch period or grating constant) in relation to one another, in order to be able to carry out directional recognition. After preamplification, the two analog measuring signals (between 1 : 4 and 1 : 64) may also be subdivided via a voltage-multiplier connection. Two right-angled pulse sequences displaced electrically by 90° are generated from the analog signals in a transducer switched downstream. Directional recognition and pulse counting are carried out in a counter circuit with directional logic for display or processing purposes. The counter level provides a measure for the path travelled in relation to the point of reference after completion of the reference point travel. The maximum traversing speed is limited by the maximum counting frequency.

In addition to the incremental measuring system described for linear motion, there is one for rotary motion which operates on the same operating principle. For standard applications the revolution counters have between 1000 and 3000 lines per revolution and for precision applications up to 3600 lines are possible, providing an angular resolution of up to 0.5″.

Digital Absolute Measuring Systems

In digital absolute measuring systems, a single measured value based on a fixed datum is assigned to each line motion component. These data are marked from the datum onwards by a uniquely recognisable code word. The datum is set mechanically. A datum offset may be carried out mechanically or by adding a sum to the position measured value. A *linear encoder* serves as the material measure in linear position measuring systems and a *rotary encoder* is used in rotational measuring systems. The physical possibilities of measuring the codes and the scanning process correspond to those employed to generate increments in incremental systems. Photoelectric scanning is used in preference in this area as well.

However, there are problems with ambiguity in scanning the increment in this instance, owing to the simultaneous change of state in a number of tracks on transition of data from one measured value to another.

There are three ways of preventing this:

1. introduction of an additional clock track;
2. use of single-step code;
3. use of double scanning or V-scanning.

Regarding 1, the additional pulsed track ensures that the measurement reading is only enabled when the transition has definitely been completed in all tracks.

Regarding 2, the Gray code is the best single-step code. It has the advantage that only one signal change takes place on transition from one increment to the next one in line.

Regarding 3, double scanning is employed in preference in dual or binary–decimal encoded increments, in order to render pitch errors in the increments and the scanning arrangement ineffective.

Two scanners are installed in each code track, apart from the finest-stepped, one of which is employed as the reference track. The advantage of the dual system is exploited for this purpose, where on generation of the signal *L* in a track the right datum line is always critical in the subsequent track. Where 0 is generated in a track, the left datum line is always critical in the subsequent track. Two scanners are, therefore, always arranged in the subsequent track and are displaced by half the pitch width of the preceding track (**Fig. 34**). This generates a field of tolerance which is large enough to allow for the transition from one value to another in the tracks. The scanner used in each case is selected from the preceding track.

This V-scanner is more costly to employ than the Gray code, yet the tolerances of the tracks may be greater where the values are higher. This advantage is extremely important for rotational measuring systems with rotary encoders switched in sequence owing to the precision of the intermediate gearing.

2.5.4 Analog Data Logging

Analog data logging is characterised in that a measuring signal value may be assigned to each measured value of the data volume continuously.

In the simplest case, the change in resistance is employed in relation to the length of the electrical conductor to generate an electrical measuring signal. In practice, potentiometers are switched as voltage dividers.

Measuring devices of these types (linear or rotational) are only employed in special cases, e.g. for rough positioning as they are not wear-free and their resolution capacity is limited. Their linearity is not normally higher than 1% either.

Inductively operating measuring systems have become more popular for analog position and line motion measurement, i.e. in the form of *synchro-resolvers* (or *rotary transducers*) and *Inductosyn* scales.

Synchro-resolvers These are rotational measuring systems where angles are registered inductively without contact being made. They normally consist of a rotating transformer with a rotor (impeller) and stator (column) (**Fig. 35**).

The arrangement shown in **Fig. 35a** has a single-phase winding stator and rotor in each case. This has no practical significance, although the input and output voltage proportions are the most clearly represented in this case.

Where the alternating voltage is $u_1 = U_1 \sin \omega t$ at the stator winding, the magnetic flux generated in the rotor winding induces an amplitude-modulated voltage of the same frequency, where $u_2 = (U_1 \sin \omega t) \cos \alpha = U_2^* \sin \omega t$.

As **Fig. 35a** shows, the time-related amplitude modification is modulated by the angle-dependent change in $\cos \alpha$. The envelope reflects the respective angular position and the modulated voltage causes a phase shift of $180°_{el}$ at the voltage zero in this envelope. This produces a clear relation between amplitude and angle.

Figure 35b shows a synchro-resolver with a single-phase rotor winding and two stator windings physically displaced by 90° as an example of the many possible winding arrangements and types of circuitry available in practice.

The voltage induced in the rotor winding is as follows,

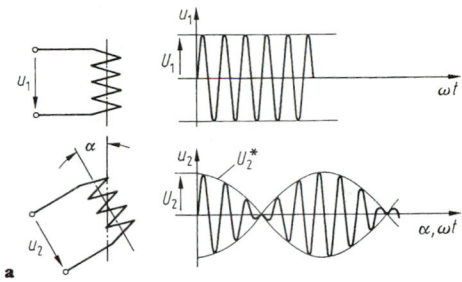

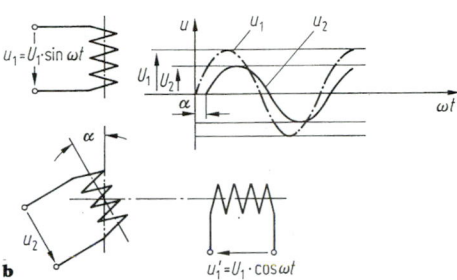

b $u_1' = U_1 \cdot \cos \omega t$

Figure 35. Synchro-resolver (principle): **a** with stator coil, **b** with two stator coils.

where a cos and a sin voltage are applied to the stator windings:

$$u_2 = [U_1 \cos \alpha] \sin \omega t + \left[U_1 \cos \left(\alpha + \frac{\pi}{2}\right)\right] \cos \omega t$$

$$= [U_1 \cos \alpha] \sin \omega t - [U_1 \sin \alpha] \cos \omega t$$

$$= U_1 \sin (\omega t - \alpha)$$

The voltage at the secondary winding changes constantly with the spatial angle in the phase position, compared with the voltage at one of the primary windings. A phase discriminator produces a signal proportional to α.

Inductosyn Scale. This is an inductively operating scale based on the synchro-resolver principle. Square-wave divisions are used on the scale on the printed circuit board with Inductosyn systems (**Fig. 36**).

A contactless slide with two square-wave windings spatially displaced by 90° back-to-back moves above this. This device operates in exactly the same way as a two-phase

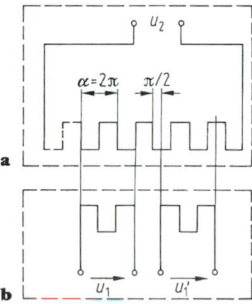

Figure 36. Principle of Inductosyn scale system. **a** Linear scale ≙ rotor in synchro-resolver. **b** Cursor ≙ stator in synchro-resolver.

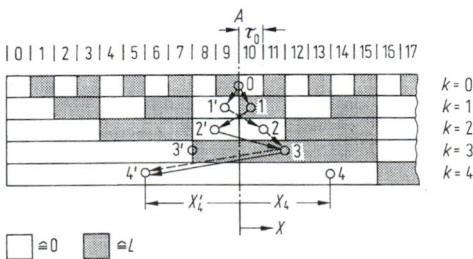

Figure 34. Double scanning for binary-coded absolute encoder.

synchro-resolver, as the two coils which are both displaced by 90° within the pitch also generate a specific field vector. Inductosyn scales all have pitches of $\frac{1}{10}''$ or 2 mm. Resolution is thereby provided in the lower micrometre range. Absolute measurements can only be carried out from within the pitch.

2.5.5 Laser Interferometer

The laser interferometer is an extremely high-precision measuring system for scanning and ranging machine tools, i.e. scanning, ranging and calibrating scales, scanning and ranging the accuracy of roller ball spindles and rack gears, etc. It is also employed in high-precision and large-scale machine tools and as a position measuring system [20].

In a Michaelson interferometer a light source emits a coherent monochromatic beam of light of the same frequency and phase length which is split into two parts by a semitransparent mirror. The two beams are reflected onto two mirrors with total reflection, where one mirror is fixed and serves as a reference line and the other is movable and is located on the device under test. Where the reflected parts of the beam meet they interfere with one another. This causes a reciprocal attenuation or amplification according to the respective mutual phase position in the photoelectric receiver. Pulses are formed from the signal received, where their number is proportional to the path. The precision of the laser interference process depends considerably on the stability of the light source length, which itself is affected by ambient conditions, such as air pressure, temperature, humidity, CO_2 content in the air and the operating condition of the laser. There are various ways of increasing the stability of the waves, which are particularly well suited to the dual-frequency laser process.

A measuring error of approximately 1 μm/m is achieved with such stable laser interferometers, and linear speeds of 18 m/min may be recorded. A resolution of 5×10^{-3} μm may be achieved via frequency multiplication.

3 Shearing and Blanking Machines

K. Siegert and **T. Werle**, Stuttgart

3.1 Shearing Machines

Plate Shears. These are used for shearing strips or making linear blanking cuts in sheet metal. A blankholder with appropriately shaped cutting blades top and bottom is used to drive one or both cutting edges to generate cutting surfaces which are free of flash where possible and which run at right angles to the sheet metal (**Fig. 1**). Where the top cutter runs parallel to the bottom one, the cut edge produced is slightly angled. Angled or oscillating cutters improve the right-angled shape of the cut surface. Crank mechanisms or toggle drives and variations on these constructions are employed as gearing in these machine tools. Hydraulic drives are also used. Nowadays, CNC-controlled machine tools are available for all types of drives where the cutting angle, blade clearance and maximum cutting power may be pre-programmed.

Strip Shears. These cut strips in continuous operation (**Fig. 2**). They may be used in applications with sheet

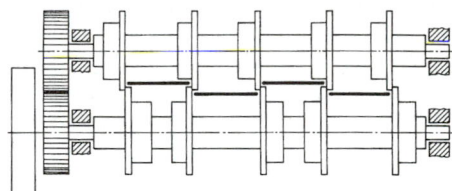

Figure 2. Functional diagram showing principle of strip shears.

thicknesses of up to a maximum of 6.5 mm and minimum widths of 40 mm per strip.

Circular Shears and Curved Shears. These are designed for cutting along curved lines. The diameter D of the cutting tools (**Fig. 3**) must not exceed a specific threshold limit value ($D < 120s_0$), which represents the sheet thickness, because of the flexibility required for extreme curves.

Billet Shears and Billet Parters. These are used for the production of unmachined parts, e.g. for *drop forging*. In this case, the emphasis is placed on the production of a constant volume of cut, unmachined parts. A complex measuring and control device is required for setting the feed length stop owing to the variations in the thickness of the billet.

Shears for Unmachined Parts. Used specifically for processing unmachined parts in cold forging, these oper-

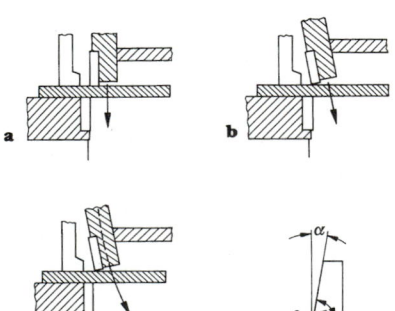

Figure 1. Plate shears. **a** Parallel top cutting edge. **b** Angled top cutting edge. **c** Oscillating top cutting edge. **d** Angle on shearing blade.

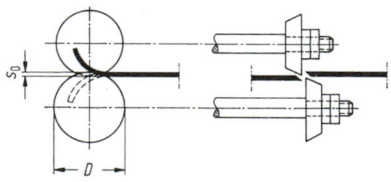

Figure 3. Functional diagram showing principle of curved shears.

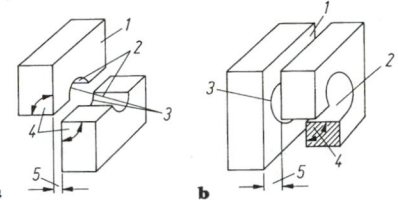

Figure 4. Shears for unmachined parts: **a** open cutting edge, **b** closed cutting edge; *1* flank, *2* support, *3* cutter, *4* wedge angle, *5* blade clearance.

ate at a relatively high number of strokes per minute. They shear off either the unmachined parts using rolled wire, or rolled bars with an open cutting edge (**Fig. 4**), or cropping bars with a closed cutting edge. Crank mechanisms are employed as the standard drive.

3.2 Blanking Machines

The typical force/path curve (**Fig. 5**) illustrates the requirement for machines with high-rated cutting forces even for relatively small machining capacities, although the energy capability has to be greater for blanking sheet metal with higher ductile yields, owing to the longer cutting distances involved, than with sheet metal with lower ductile yields and the same stability. High-speed crank presses are employed for blanking and even small-stroke, high-speed, toggle-joint presses and hydraulic presses with stroke arrestors are used in special-purpose applications. Cutting impact damping is carried out as an additional process in the interests of noise abatement in blanking machines, where the fall-through of the slide is prevented at the end of the process. Stroke rates of up to 800 min^{-1} are normal for mechanical automated blanking machines.

Automatic slotting machines with toggle mechanisms which oscillate around their line position achieve stroke rates of 1300 min^{-1}. These therefore require correspondingly high-performance and accurate feed mechanisms (feedrate accuracy ± 0.01 mm, average continuous speed of up to 120 m/min) in conveyor operation or for precision part mechanisms for cutting slots in sheet metal stators or rotors of electric motors. Modern precision automatic punching machines can cater for stroke rates of up to 1800 min^{-1} with a rated cutting force of 200 kN thanks to the use of CNC controllers. Such stroke rates are only possible with optimised slideways (where the slide is guided at conveyor level to prevent tipping - **Fig. 6**), mechanical balance (compensation for dynamic forces in high-speed presses - **Fig. 7**), and complex sensory analysis (monitoring all essential influencing parameters) and

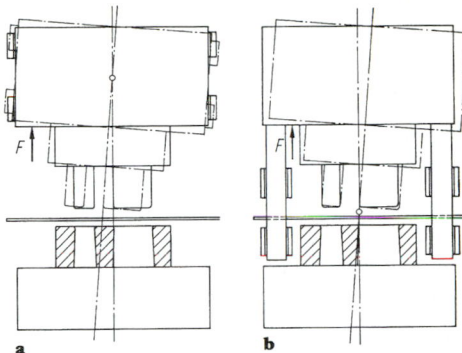

Figure 6. Alternative slideways in high-speed presses. **a** Schematic diagram of slideway above conveyor level. **b** Schematic diagram of slideway at conveyor level.

adjustment of the insertion depth of the cut. Insertion depth of cuts, automatic slide stops at the upper dead centre and feed lengths are monitored by computer, whereas band thickness, bandwidths and tool parameters such as the band feed height, feedrate accuracy of the roller feed mechanism and cutting power are monitored by the machine tools (see **Fig. 8**) [1]. The latter also provides information on tool wear by comparing actual and setpoint forces. Presses should have as rigid a construction as possible in order to minimise wear, and the slideway guide should be optimised. High-speed presses are particularly suitable for use in flexible punching centres. **Table 1** shows industrial punching techniques classified according to characteristics such as accuracy, areas of application and stroke rates [2].

3.3 Nibbling Machines

These are available in all sizes, to incorporate all the various intermediate stages of automation, from manual nibblers to numerically controlled sheet metal working centres. In the case of stationary machines the workpiece

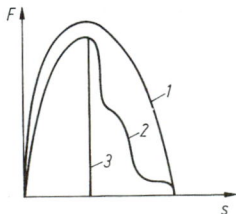

Figure 5. Force–path curve during blanking: *1* precision blanking, *2* sheet metal blanking with high ductile yields, *3* sheet metal blanking with low ductile yields.

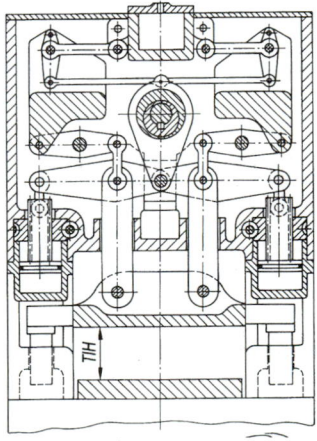

Figure 7. Function of mechanically balanced automatic punching machine (Bruderer), with four-point support of slide and punching force reduction on eccentric cam. *TIH* = tool installation height.

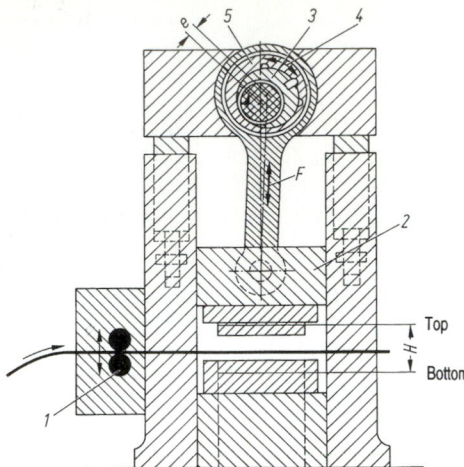

Figure 8. Diagram of eccentric press with feed mechanism; setting adjustment is via numerically controlled actuators. *F* compression force, *H* stroke, *e* eccentricity, *1* roller feed, *2* slide, *3* eccentric bush, *4* slide strike adjustment, *5* eccentric cam.

is moved in relation to the tool [3]. Sheet metal working centres have a tool data store from which a variety of standardised tools may be selected at will for automatic operation. The numerical control system generates control commands for moving and positioning the workpiece via the positioning table, switching commands for initiating and repositioning strokes (individual continuous strokes) and commands for tool changes. The body of the machine consists of a protruding throat-type frame, or occasionally a straight-sided press (see L1.3). The slide is normally driven by a crank mechanism.

3.4 Beam Cutting Machines

These generate individually programmed geometrical blanks or unmachined part cuts (see K4.5.6) using a beam

Table 1. Classification of industrial punching techniques [2]

Punching techniques	Characteristics
High-performance punching techniques	Punched parts from strips of 0.1 to 3 mm thick, high accuracy, high number of strokes (up to 1800 min^{-1}), extremely high piece numbers (e.g. plug-in contacts, electronic system data carriers, laminated rotors and stators)
Conventional punching techniques	Punched parts from strips of between 1 and 6 mm thick, average accuracy (e.g. machine components, fixtures and fittings), average number of strokes (up to 400 min^{-1})
Precision blanking & punching techniques	Punched parts with smooth cutting surface (R_a = 0.8 to 2.4 μm) of between 1 and 15 mm thick, high degree of accuracy and wide variety of shapes, including three-dimensional (e.g. coupling halves, trip levers, gearwheel sections)
Nibbling and laser blanking techniques	Punched parts from 0.5 to 3 mm (and thicker) sheet metal or blanks for lot sizes of "one" or more, predominantly large-scale, bulky parts. Table motion CNC-controlled (e.g. blanked sheet metal casings)
Large-scale punching techniques	Punched parts of 0.3- to 2-mm-thick sheet metal blanks in large sizes (e.g. coach parts)

of bundled active media or active energy. They frequently form elements of a manufacturing station, and are specifically suited to small and medium-scale blanking operations owing to their high degree of flexibility.

Water torches can produce tolerances of ≥ 0.1 mm, while *laser cutting tool equipment* may have tolerances of up to ≥ 0.005 mm.

4 Presses and Hammers for Metal Forging

J. Siegert and **E. Dannenmann**, Stuttgart

These are designed to activate forming tools, carry out reciprocal guiding of tool parts and provide the deformation forces, moments and energy input requirements for the process. **Figure 1** shows the forming machine classifications. Presses are the most frequently employed machine tools for unit loads and are here dealt with first of all.

4.1 Characteristics of Presses and Hammers

The characteristics describe the attributes of a forming machine. Compared with the requirements of the forming process, these enable the most suitable machine tool to be selected for the relevant process. Three groups of characteristics are relevant for presses: *energy and force*

characteristics, time characteristics and *precision characteristics* (**Table 1**).

In addition to these characteristics and their numerical values (characteristic values), there are also mechanical data which are essential for press applications, such as the height of lift of the slide or pile hammer, the dimensions and design of the installation area inside the machine tool, the connection ratings, the space requirement and the weight. Construction sizes have been standardised for a wide range of presses (DIN 55 170, DIN 55 181, ISO/DIS 9188-1989, DIN 55 184, ISO 6898/1984, DIN 55 185, ISO/DIS 9189-1989, DIN 55 222).

The main *force and energy characteristics* associated with forming machines are the slide force F_{St} and energy capability E_M. These values always have to correspond as a minimum requirement to the forming force F and forming action W required for the process, in order that the process may be carried out on the machine. In addition to

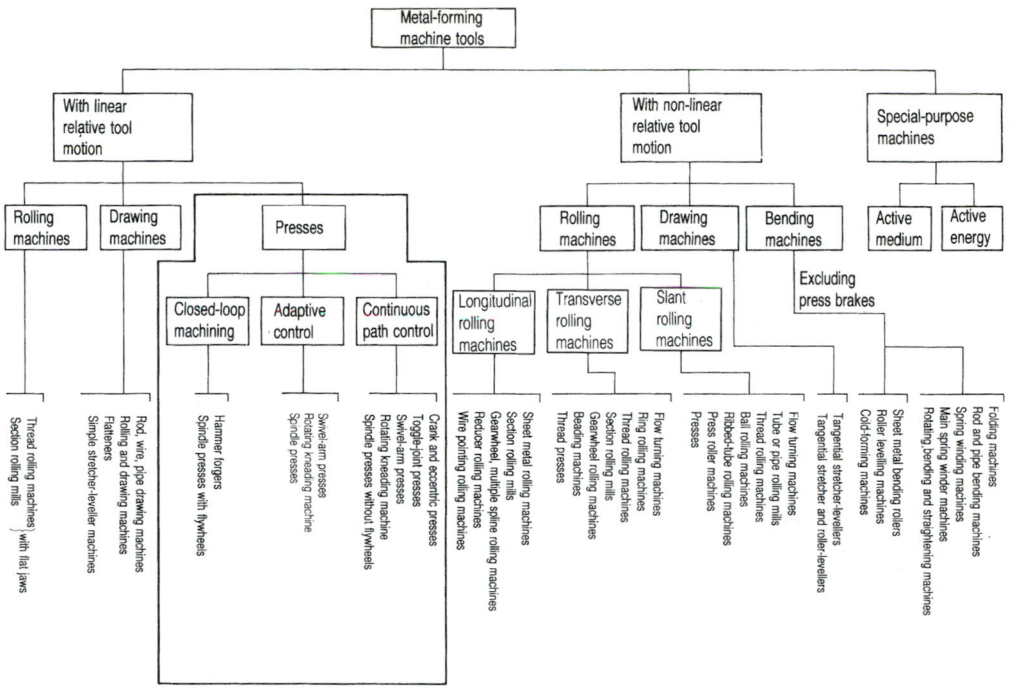

Metal-forming machine tools

- With linear relative tool motion
  - Rolling machines
  - Drawing machines
  - Presses
    - Closed-loop machining
    - Adaptive control
    - Continuous path control
- With non-linear relative tool motion
  - Rolling machines
    - Longitudinal rolling machines
    - Transverse rolling machines
    - Slant rolling machines
  - Drawing machines
  - Bending machines
    - Excluding press brakes
- Special-purpose machines
  - Active medium
  - Active energy

Bottom labels: Thread rolling machines) with flat jaws / Section rolling mills) with flat jaws; Simple stretcher-leveller machines; Flatteners; Rolling and drawing machines; Rod, wire, pipe drawing machines; Spindle presses with flywheels; Hammer forgers; Spindle presses; Rotating kneading machine; Swivel-arm presses; Spindle presses without flywheels; Rotating kneading machine; Swivel-arm presses; Wire pointing rolling machines; Reducer rolling machines; Gearwheel, multiple spline rolling machines; Section rolling mills; Sheet metal rolling machines; Thread presses; Beading machines; Gearwheel rolling machines; Flow turning machines; Ring rolling machines; Thread rolling machines; Section rolling mills; Ball rolling machines; Thread rolling machines; Tube or pipe rolling mills; Flow turning machines; Presses; Press roller machines; Ribbed-tube rolling machines; Tangential stretcher and roller-levellers; Tangential stretcher-levellers; Cold-forming machines; Roller levelling machines; Sheet metal bending rollers; Rotating, bending and straightening machines; Main spring winder machines; Spring winding machines; Rod and pipe bending machines; Folding machines

Figure 1. Classifications of forming machines (*D* continuous process, *E* plunge process).

forming forces and action, additional force and action may also be required for the operation of additional aggregates, such as drawing machines, blankholders, ejectors, etc. Spring action stored within the machine–tool system should also be taken into account, depending on the process.

Presses are distinguished according to the way in which the machine generates the characteristic deformation forces and energy: continuous-path control, adaptive control and closed-loop machining (**Fig. 2**).

Time characteristics describe machine-dependent process times and speeds, such as impact and stroke sequence time, pressure contact time and tool speed.

Precision characteristics provide a reference for the workpiece speeds which may be achieved with a forming machine. There are characteristics for no-load machines (accuracy of manufacture) and machines under load.

These guidelines for accuracy of manufacture relate to the geometric dimensions of the installation area inside the machine tool and accuracy of motion of the slide and are specified in DIN 8650, ISO 6899-1984 and DIN 8651 for continuous-path control presses according to the type and size of the relevant machine tool construction. Accuracy characteristics of machines under load are defined in DIN 55 189 for mechanical (continuous-path control) and hydraulic presses, and describe the offsets in the tool-carrying surface areas under load compared with no-load conditions.

An offset is only generated in the direction of operation (v_{totZ}) under average load in presses with symmetrically structured frame constructions (O-frames) and driving mechanisms (**Fig. 3**).

This is composed of the initial offset v_{aZ}, generated by clearance and load-related elastic deformation v_{elZ} in the individual press components (**Fig. 4**). Rigidity c_Z is one of the accuracy characteristics and is obtained from the linear section of the offset curve, where $c_Z = \Delta F_Z/\Delta v_{elZ}$.

Eccentric loading (**Fig. 5**) leads to tilting between the table and slide vertical to the direction of operation and a shift in the centres of the table and slide (offset), irrespective of the design of the frame. The overall tilt k_{tot} is composed of k_a (mechanical balance of guide clearance) and elastic tilt k_{el} (frame, slide and driving mechanism deformation) (**Fig. 6**). Breakdown rigidities c_{kA} and c_{kB} about the x- or y-axis are defined as accuracy characteristics from the linear section of the tilt curve at $c_{kA} = \Delta F_Z \cdot \Delta lY/\Delta k_{elA}$, or $c_{kB} = \Delta F_Z \cdot \Delta l_X/\Delta k_{elB}$ (Δl_X, Δl_Y eccentricity of load application in X, Y direction, set at 10% of the productive slide depth or width). The characteristic for the offset vertical to the direction of operation (shift) is the distance of the central vertical line of the slide in relation to the central vertical line of the table, measured as half the distance between the table clamping area and the slide area (**Fig. 5**). This is obtained as the overall offset $v_{totX} = v_{aX} + v_{elX}$, or $v_{totY} = v_{aY} = v_{elY}$ at a force under load (F_Z) of 50% of the rated force (F_N) and an eccentricity $(\Delta l_{X(Y)})$ of 10% of the productive slide width or depth (**Fig. 7**).

4.2 Mechanical Presses

In continuous-path control presses (**Fig. 2a**) the machine slide runs through a path prescribed by the kinematics of the main gearing. The latter is driven by an electric motor via a flywheel and clutch. Intermediate gears may be arranged between the flywheel and main transmission unit. The force emitted by the slide F_{St} is dependent on the slide setting h. The dominant characteristics thereby

Table 1. Characteristics of presses

Characteristics:		
Force and energy F_s		Slide force, energy produced by machine at any point in process
	F_N	Rated force, force determining design of machine
	F_{imp}	Force on impact, compression force at highest speed on impact excluding effective energy losses (closed-loop machining presses)
	$F_{max\ perm}$	Maximum admissible force (in continuous operation) (spindle presses)
	E_M	Energy capability
	E_N	Rated energy capability, maximum amount of energy available for operating cycle
	W_F	Spring action, potential energy stored in machine and tool in operating process
Time	t_H	Time to impact, stroke sequence time, duration of pile hammer or slide stroke before machine is ready for next stroke
	n_H	No. of impacts, no. of strokes, reciprocal value of stroke sequence time, in presses with crank mechanisms equal to crankshaft speed n_K
	v_{wz}	Tool speed at any point of process (normally equal to slide speed v_{St})
Accuracy	No load machines	Flatness and parallelism of tool clamping surfaces, right-angled slide motion in relation to table surface
	Machines under load	c_z Spring rate in operating direction
		c_{kA}, c_{kB} Stable rigidity
		v_{tot}, v_{totV} Offset vertical to direction of operation (shift)

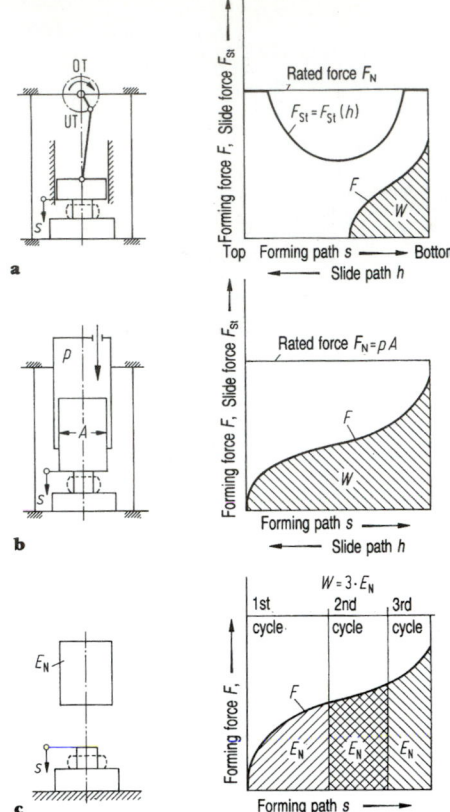

Figure 2. Principles of presses: **a** continuous-path control, **b** adaptive control, **c** closed-loop machining.

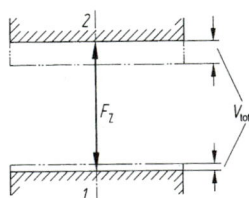

Figure 3. Offsets in symmetrically structured press frames and centre loading (DIN 55 189): *1* table, *2* slide, F_z force under load, v_{totz} overall shift between table and slide.

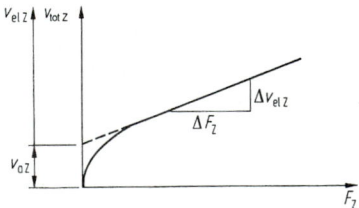

Figure 4. Offset v_z as a function of the force under load F_z (offset curve, DIN 55 189).

consist of the curve of the slide force in relation to the slide path, i.e. $F_{St} = F_{St}(b)$, and the relevant maximum admissible value, the rated force F_N, which the parts affected by the flow of force are designed for. The energy requirement of an operating cycle is almost exclusively covered by the energy released by the flywheel. The energy capability is a further significant characteristic and is determined by the design of the flywheel and operating mode, i.e. the maximum available rated energy capability E_N in continuous-stroke operation. In single-stroke operation a higher energy capability $E_E = 2E_N$ may be employed owing to the relatively brief machining cycle and, consequently, the lower thermal loading of the drive motor.

4.2.1 Types

Presses are known as crank or cam mechanisms (**Fig. 8**) depending on the type and structure of the main gearing. *Cam mechanisms* are limited to a small rated force F_N; nevertheless, they provide virtually unlimited possibilities for the sequence of movements. *Crank mechanisms* are

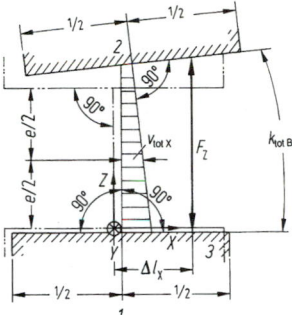

Figure 5. Tilt and offset vertical to the direction of operation in eccentric loading (DIN 55 189): *1* operational side, *2* slide, *3* table.

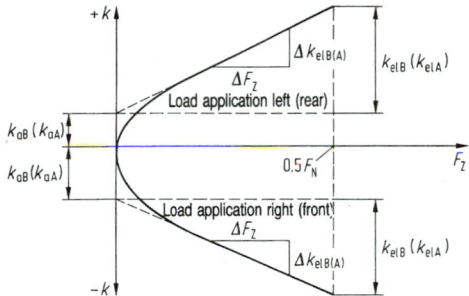

Figure 6. Tilt k as a function of the force under load F_z with a given eccentricity $\Delta l_{x(Y)}$ (tilt curves, DIN 55 189); k_{aA}, k_{aB} = initial tilt.

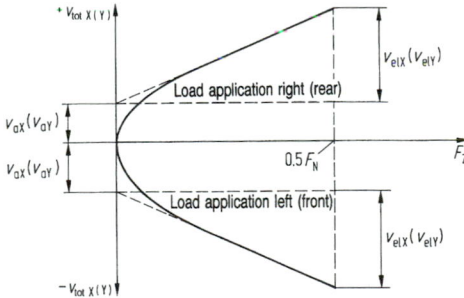

Figure 7. Offset $v_{totX(Y)}$ vertical to direction of operation as a function of the force under load F_z with a given eccentricity $\Delta l_{x(Y)}$ (DIN 55 189); v_{aX}, v_{aY} = initial offset.

Type of gearing		Structure
Crank mechanism	Simple — Slider–crank mechanism	
	Complex — Slide–crack toggle mechanism	
	Swivel mechanism	
Cam mechanism		

Figure 8. Structures of main gearing.

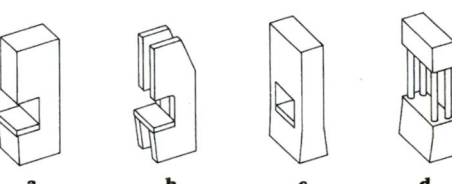

Figure 9. Shapes and constructions of frames for continuous-path control presses. **a** C-type frame, single-column construction. **b** C-type frame, double-column construction. **c** O-type frame, two-column construction. **d** O-type frame, pillar-type construction.

divided into the simple and the complex. The popular such presses are those with *slider–crank mechanisms*, either with a fixed overall stroke (crank presses) or a variable overall stroke (eccentric presses). Complex crank mechanisms are employed where a large F_{St} is required with a small stroke (toggle mechanisms) or where reduced machining speeds are desirable in the working space (swivel mechanisms).

4.2.2 Assemblies

Frames (Fig. 9).

C-frames come in single- and double-column design, and may be vertical, inclinable, horizontal or in some cases have tension rods. These are normally used for presses with small-to-medium-scale rated forces.

O-frames may be of two-, three- or four-column design with penetrations in the side columns for tool changes and workpiece supply and discharge, and are occasionally found with pillar design; these are medium-sized presses, normally of a unitary construction. For large-scale presses, multi-part two-column frames are employed, with a table, side columns and cross-beams connected via tensioning rods. Frames may be of cast-iron, cast-steel or, increasingly nowadays, of welded steel plate construction.

Drive. The flywheel is normally driven via three-phase asynchronous sliping motors. The stroke rate adjustment for presses with small to average rated forces is via gears between the drive motor and flywheel, while in presses with high rated forces it is via variable-speed drive motors.

Clutch, Brake. Nowadays, clutches are mainly of the friction-locked type (friction clutches), and of single- or multiple-disc construction (F3). The forces brought under load are normally applied via a pressure medium (air, or occasionally oil). Brakes are basically similarly in design to clutches. The forces brought under load are generated from spring mechanisms, for reasons of safety. Positive clutches (rotating wedge-and-pin couplings) are today becoming less popular because of current safety requirements.

4.2.3 Dynamics and Kinematics

Slide path h and slide speed v_{St} are dependent on the crank angle α in *slider crank mechanisms* (**Fig. 8**; see F10.1). This can be simplified, so that the following applies for the slide force F_{St}:

$$F_{St} = M_K/(r \sin \alpha)$$

where M_K = the crank torque,
    r =the crank diameter and
    α =the crank angle.

F_{St} has a minimum value (F_{Stmin}) where $\alpha = 90°$ ($h \approx H/2$, $H = 2r$ overall stroke) and at the end position ($\alpha = 0°$, $\alpha = 180°$) tends towards infinity. For the components in the flow of force, the slide force has to be limited to a finite value, the rated force F_N, either over a specific crank angle (rated angle of force α_N), or over a specific slide path (rated force path h_N), before bottom dead centre. The force is limited via excess load protection (shear plate, hydraulic die, force sensor element reacting in response to machine control). The magnitude of rated force angle α_N or rated force path h_N depends on the structure and the area of application.

According to DIN 55 170 and the standards valid up to 1985 for eccentric presses with throat-type frames, the drive is to be constructed such that the rated force F_N at $\alpha_N = 30°$ (corresponding to $h_N = 0.073\ H_{max}$) is available for the maximum stroke H_{max} (standard design). Further standard designs depend on the area of application include drop forging $\alpha_N = 10°$, blanking $\alpha_N = 20°$, extruding $\alpha_N = 45°$, deep-drawing up to $\alpha_N = 75°$. Critical limits for slide forces for $\alpha > \alpha_N$ or $h > h_N$ are provided by $F_{St} = F_N\ (\sin \alpha_N/\sin \alpha)$ (**Fig. 10a**).

The design specification in more recent standards relating to construction sizes of continuous-path control presses is based on the rated force path h_N. For presses with throat-type frames (DIN 55 184), rated force paths are specified in the region of 2 mm $< h_N <$ 9 mm. For presses with straight-sided frames (DIN 55 181 and ISO/DIS 9188-1989), requirements for rated force paths of h_N = 3, 5, 7, 12.5 and 25 mm are specified. In the latter case, critical limits for slide forces for the various designs are shown in **Fig. 10b**.

The rated energy capability in continuous stroke operation (E_N) is normally $E_N = F_N \times h_N$.

The overall stroke in eccentric presses is normally variable within the range of H_{max}/H_{min} = 10. The direction and magnitude of F_{St} changes as a result of the stroke adjustment (**Fig. 11**), as does the slide speed v_{St}, although the energy capability remains unaffected.

In *slider–crank toggle mechanisms* (**Fig. 8**) with tension or pressure applied to the connecting rod, the slide motion is delayed near bottom dead centre. As a result, the slide forces in the range $h > h_N$ are lower than with slider–crank mechanisms of the same design as regards the rated force path h_N and overall stroke H (**Fig. 12**).

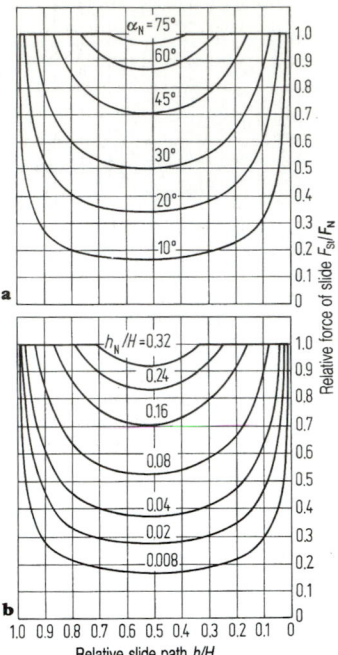

Figure 10. Critical values of slide forces in relation to slide path for various designs of crank press. **a** Design in accordance with rated angle of force α_N. **b** Design in accordance with rated force path h_N ($\lambda = 0.1$).

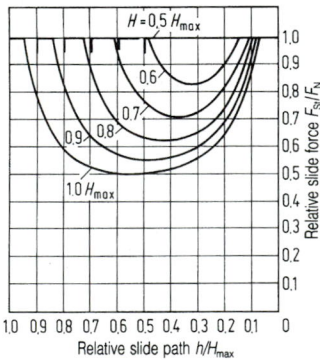

Figure 11. Critical values of slide forces in relation to slide path with stroke adjustment (design: $\alpha_N = 30°$ for $H = H_{max}$).

Swivel Mechanisms (**Figs 8** and **18a**). These provide low and virtually constant slide speed v_{St} in the working space. No-load paths are traversed rapidly (**Fig. 13**). Higher slide forces are available for $h > h_N$ compared with slider–crank mechanisms with a rated force path h_N of the same size (**Fig. 14**).

4.2.4 Applications

Continuous-path control presses constitute the majority of forming machines employed in unit load manufacturing, with a variety of construction requirements designed to suit the relevant individual application.

The high rigidity required for *massive forming* (drop forging, extruding), in view of the machining accuracy

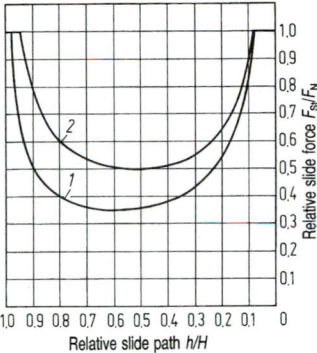

Figure 12. Critical values of slide forces in relation to slide path for slider–crank toggle mechanisms. Tension is applied to connecting rod (*1*) and simple slider–crank mechanism (*2*) with the same rated force path b_N and the same overall stroke H (b_N/H = 0.073).

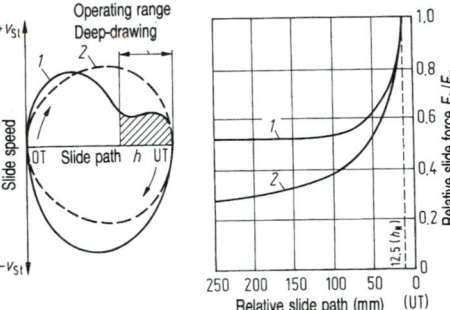

Figure 13. Slide speed in relation to slide path (Schuler): *1* swivel mechanism, *2* slider-crank mechanism.

Figure 14. Critical values of slide forces in relation to slide path (Schuler): *1* swivel mechanism, *2* slider-crank mechanism. Rated force path b_N = 12.5 mm.

and pressure contact time, is achieved by way of the shape of the frame (i.e. an O-frame with small column width) in conjunction with the design of the driving mechanism. The main drive shaft in drop forging presses is in the form of a rigid eccentric shaft with a short, wide connecting rod. A wedge press with a slider–crank toggle mechanism is used in extrusion machines. The slide drive in the wedge press (**Fig. 15**) is moved by the connecting rod via an interposed wedge (wedge angle 30°). Thus, only approximately half the slide force is applied to the connecting rod. The wedge prevents the slide from tilting as a result of the guide clearance. Wedge presses designed with rated forces of up to 125 MN are employed in the manufacture of long, precision-forged items. In addition to the most popular construction with vertical feed motion, there are also designs with horizontal feed motion for processing workpieces, with long shafts (horizontal upsetting presses) or die-shaped parts (e.g. machines for tube extrusion) providing advantages in workpiece handling. Presses with throat-type and O-type frames are employed in *sheet metal forming*. In throat-type frame presses (**Fig. 16**) the working space is freely accessible from three sides. This design is normally constructed with stroke adjustment, as it may be adapted easily to suit a variety of applications as a result (universal presses). Presses with O-frames are normally used for sheet metal forming owing to their relatively large column widths with multi-point slide drive; nowadays the majority are used in cross-shaft drives (**Fig. 17**). Deep drawing with a blankholder requires a device on the press for activating the blankholder in the tool. The tool blankholder is activated in two-way-action presses via blankholder slides (**Fig. 18b**) separated from the die slide with step-motion linkage (see **Fig. 19** for the sequence of motion of die and blankholder slides), in simple-action presses normally via pneumatic-powered drawing mechanisms. More recently, hydraulically powered drawing mechanisms

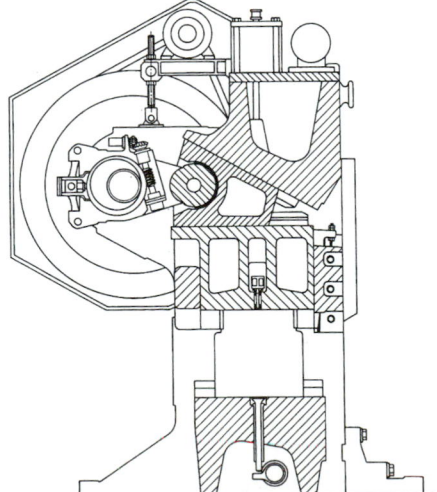

Figure 15. Wedge press (EUMUCO).

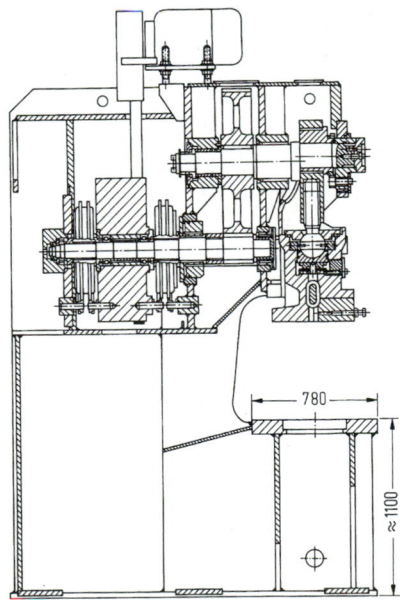

Figure 16. Eccentric press with C-type frame (Müller-Weingarten); F_N = 1600 kN, H_{max} = 160 mm, H_{min} = 20 mm, n_K = 50 min⁻¹.

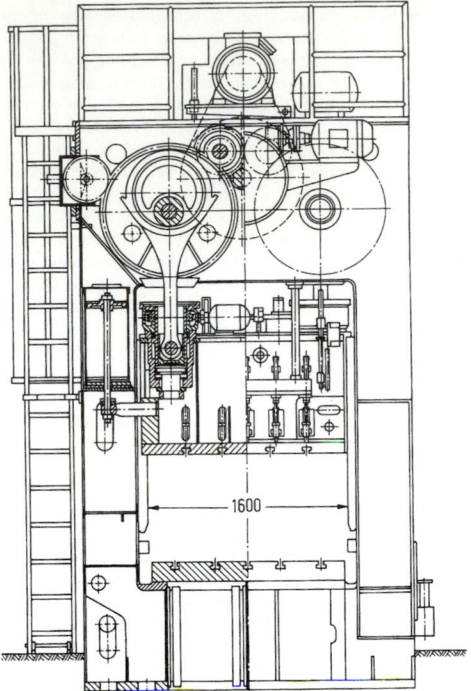

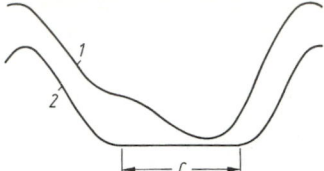

Figure 19. Sequence of motion of die slide (*1*) and blankholder slide (*2*) of press as shown in **Fig. 18** [2]. *C*, standstill phase of blankholder slide.

common of the two types, and operate on the principle of hydrostatics (see G1.1). The high-pressure energy of the medium under pressure (oil, water) is converted into mechanical action in the cylinders. The pressure p and the volumetric flow $\dot{V}$ are the most important characteristics of hydraulic drives. (See G2 for the design.)

The slide force F_{St} is determined by the pressure p and the piston surface area A; thus, $F_{St} = pA$. It is not, therefore, dependent on the slide setting (**Fig. 2b**). The maximum value of F_{St}, i.e. rated force F_N, may not be exceeded. F_N constitutes the most important force characteristic; the energy capability plays a secondary role in direct pump drives, as the energy required for the cycle is provided in sufficient quantities by the drive motor. In stored energy mechanisms the energy capability E_N is specified by the quantity of energy stored and thereby constitutes another important characteristic.

Hydraulic presses are easily adapted to the requirements of the process as regards force and machining requirements, speed and forming path. They are normally employed in processes with large force and/or machining requirements and long action paths with small to medium stroke rates, depending on the stroke size.

Figure 17. Two-column press with two-point transverse shaft drive (Schuler); $F_N = 1600$ kN, $H = 300$ mm, $n_K = 32$ min^{-1}.

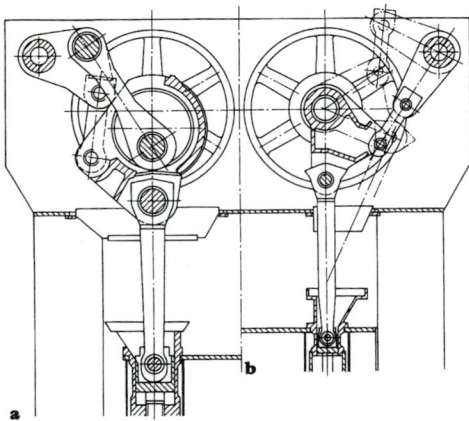

Figure 18. Basic structure of gearing in two-way action press [2]. **a** Swivel mechanism for die slide. **b** Stop-motion linkage for blank holder slide.

have been employed more frequently because the blankholder force may be more accurately reproduced and there is the option of making targeted adjustments in relation to the die path [3].

4.3 Hydraulic Presses

Adaptive control presses (**Fig. 2b**) consist of *hydraulic* and *pneumatic presses*. Hydraulic presses are the most

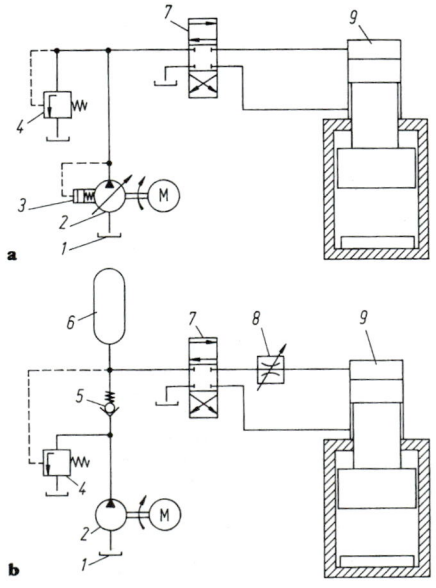

Figure 20. Basic layout of hydraulic circuit of press: **a** with volumetric flow source (direct pump drive), **b** with pressure source (stored energy drive); *1* vessel, *2* pump and motor, *3* speed governor, *4* pressure-reducing valve, *5* non-return valve, *6* hydraulic accumulator, *7* four/three-way valve, *8* throttle valve, *9* hydraulic cylinder of press.

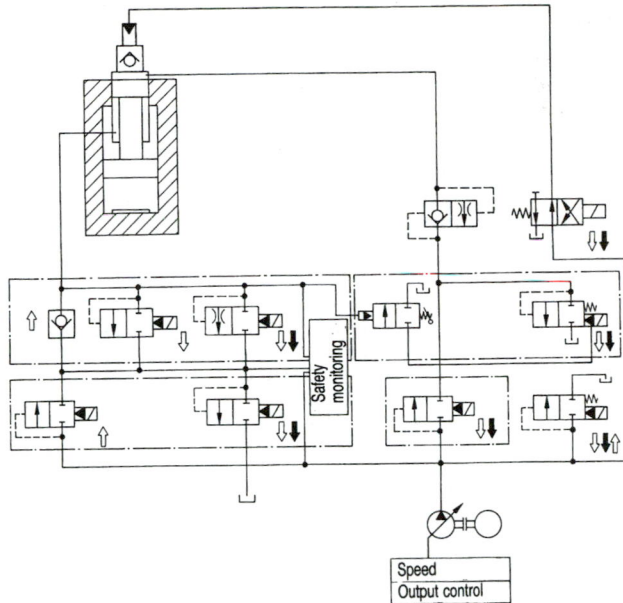

Figure 21. Hydraulic circuit of press with direct pump drive and safety monitoring devices to prevent the slide from sinking and accidental pressure buildup (SMG).

4.3.1 Types

There are various different types of drives:

Hydraulic Presses with Conveyor System (Direct Pump Drive). See **Fig. 20a** for the layout of the hydraulic circuit. For the basic design with a safety monitoring device (to prevent the slide from sinking, and unforeseeable pressure loss) see **Fig. 21**. Characteristics: the pump and the drive motor are designed for the maximum instantaneous output requirement of the pump. Oil is used as the medium under pressure. The slide speed is normally infinitely variable by means of adjustment to the delivery of the high-pressure pump.

Hydraulic Presses with Pressure System (Stored Energy Drives) (**Fig. 20b**). These are characterised by pumps and drive motors designed for average output. Oil or water is used as the medium under pressure.

There is an increased tendency towards direct pump drives owing to reduced stroke sequence times as a result of mechanisation and automatic tool handling.

4.3.2 Assemblies

In hydraulic presses, in addition to one- and two-column frames, pillar-type frames with 2 and 4 pillars are also frequently employed. The latter are normally used in presses with high-rated forces for hammer forging and extrusion.

Multiple-piston pumps (axial, radial and serial piston pumps) with a small stroke and piston diameter are used as high-pressure oil pumps. The various constructions include those achieving both constant and infinitely variable delivery rates. Flow and pressure may be adapted to suit the operating cycle via control devices (output rating, pressure and zero strike regulators). In addition, gear pumps may also be employed for constant delivery rates. Standard pressure levels p are between 2×10^7 and 3.15×10^7 Pa (200 and 315 bar), in exceptional cases even more.

Hydraulic accumulators directly pressurised with compressed air are used in stored energy drives, nitrogen batch storage, piston energy storage, or with water as the medium under pressure.

4.3.3 Applications

Hydraulic drives are frequently employed in machines for mass working and sheet-metal forming owing to the high level of controllability of the slide force and speed. Hydraulic presses for serial manufacture of *sheet metal parts* (with precision blanking and drawing (see **Fig. 22**) and press baking) and for *cold forging* (extruding, hobbing, embossing) almost exclusively with direct pump drive. *Forging presses* (hammer forging, drop forging of lightweight metals) with rated forces of up to approx. 30 MN and slide speeds of less than 80 mm/s also employ direct pump drives. Stored energy drives are preferred at higher rated forces with high slide speeds of up to 250 mm/s. *Hammer forging presses* often use pillar-type columns which provide easy access to the working space. There are special advantages in this respect for under-floor drives.

Extruders (**Figs 23** and **24**) are almost exclusively reserved for horizontal constructions with pillar-type frames.

Pneumatic drives in small-scale presses are limited to drawing, blanking, bending and riveting applications.

4.4 Hammers and Screw Presses

Closed-loop machining presses (**Fig. 2c**) consist of *hammers* and *flywheel spindle presses*. The main characteristics include energy capability E which is transmitted in full for each operating cycle, except in clutch spindle presses. Otherwise, the rated force F_N, the maximum (continuous) permissible force $F_{\text{max perm}}$ and the impact force F_{imp} are also important considerations in spindle presses.

4.4.1 Hammers

These are the cheapest forming machines for generating substantial forces and transmitting high-energy capability.

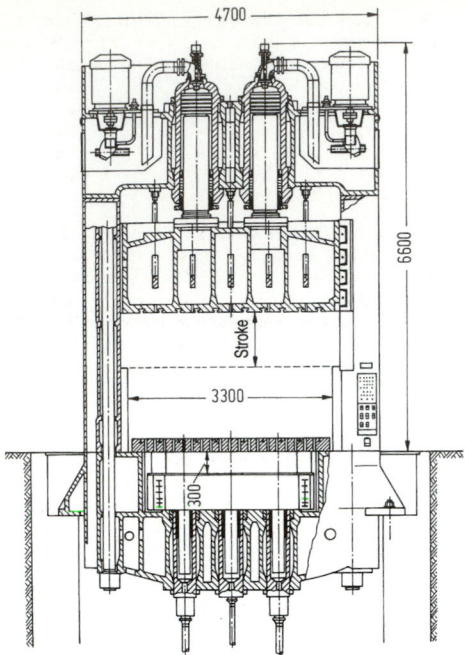

Figure 22. Hydraulic oil press with die cushion (Müller-Weingarten); $F_N = 6000$ kN.

Of simple construction, they are impossible to overload, as hammer frame and drive are not located within the flow of force in the operating cycle.

The forming process in hammers follows the law of impact. The potential energy E is translated into productive energy W_N and no-load operation W_V (pile hammer return and anvil bed losses). One characteristic of energy transmission is the efficiency of impact $\eta_S = W_N/E$. In theory the following therefore applies for the anvil hammer:

$$\eta_S = (1 - k^2)/(1 + m_B/m_S).$$

Here k is the number of impacts: in an upsetting operation $k = 0.1$ to 0.3, in drop forging $k = 0.6$ to 0.8. The ratio between the anvil mass m_S and pile hammer m_B affects the loads applied to the foundations and the return acceleration of the anvil (the jump of the item being forged). Minimum values are $m_S/m_B = 10$ to 20 with a permanently fixed anvil, $m_S/m_B = 3$ to 5 with a moving anvil.

Structures

There are *anvil hammers*, which include *drop hammers*, *double-acting hammers* and *counterblow hammers* (**Fig. 25**). Anvil hammers (**Fig. 26a**) have a permanently fixed anvil, counterblow hammers (**Fig. 26b**) have two pile hammers moving in opposition to one another.

Applications

The main areas of application are in hammer forging and drop forging and in special-purpose applications in embossing, hot extrusion and sheet metal forming. For the areas of application of the various constructions see **Fig. 27**.

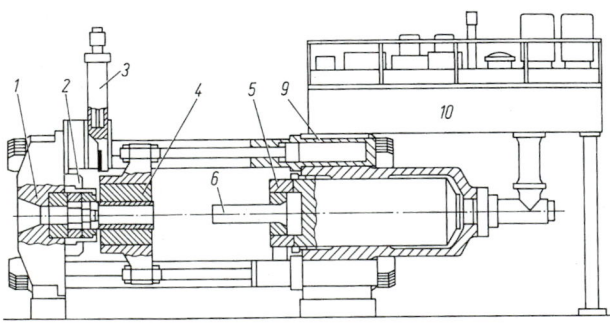

Figure 23. Direct press extruder (SMS Hasenclever): *1* counter bar, *2* tool slide or tool turret, *3* shears, *4* block pickup, *5* running bar, *6* stamp, *9* cylinder bar, *10* oil container with drive and controls.

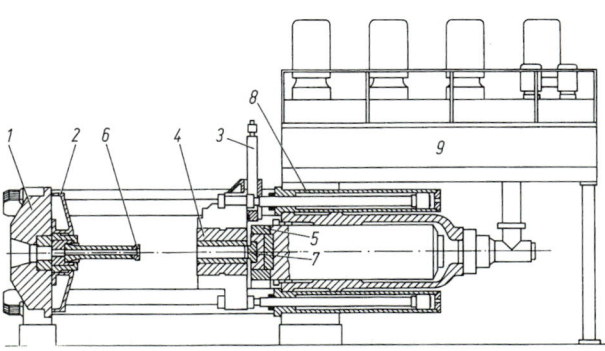

Figure 24. Indirect press extruder (SMS Hasenclever): *1* counter bar, *2* tool slide, *3* shears, *4* block pickup, *5* running bar, *6* matrix stamp, *7* seal, *8* cylinder bar, *9* oil container with drive and controls.

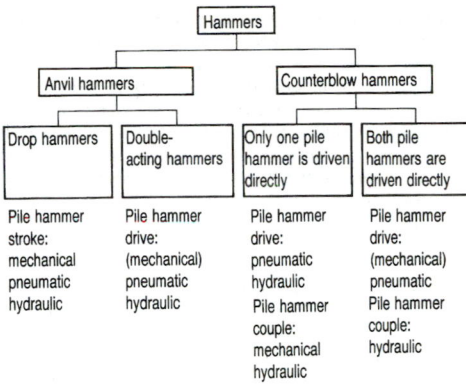

Figure 25. Classification of hammers.

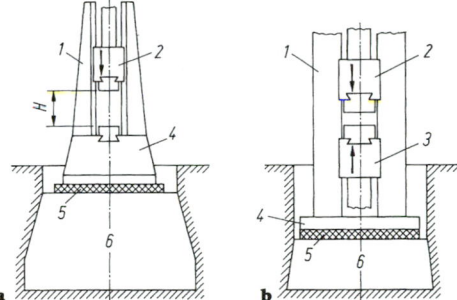

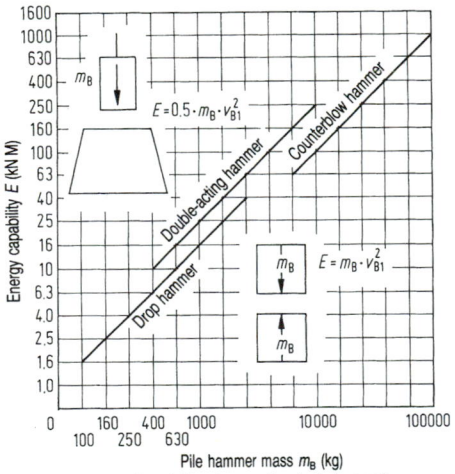

Figure 26. Basic principles of hammers: **a** anvil hammer, **b** counterblow hammer; *1* frame, *2* (top) hammer, *3* (bottom) hammer, *4* anvil or baseplate, *5* backing plate, *6* base.

Figure 27. Areas of application for anvil hammers (drop and double-acting hammers) and counterblow hammers for drop forging.

Drop Hammers. The energy capability (excluding friction and windage losses) is given by

$$E_N = m_B gH.$$

Here g = the acceleration due to gravity, H = the drop height. The stroke is limited to $H = 1$ to 1.6 m, in order to provide numbers of impacts n_H of 50 to 60 min^{-1}. Pile hammer impact speed is between 4 and 6.5 m/s. Technology has developed from strap-lift drop hammers and gravity drop hammers to pneumatic or hydraulic drop hammers, which have the advantages of low levels of wear on the stroke components and simple control and energy metering.

Double-Acting Hammers. These possess extra energy storage capacity in addition to the pile hammer in the form of compressed air, steam (0.6 to 0.7 MPa – 6 to 7 bar) or hydraulic oil (2 to 20 MPa or 20 to 200 bar). Hydraulically driven double-acting hammers (**Fig. 28**) are employed increasingly today owing to their favourable energy consumption figures. Their energy capability E_N is given by

$$E_N = (m_B g + p_{mi}A)H,$$

where p_{mi} = the average index operating pressure and A = the piston area. Double-acting hammers provide shorter strokes of $H = 0.4$ to 0.7 m at the same pile hammer impact speeds as drop hammers, and thus substantially higher numbers of impacts ($n_H = 55$ to 250 (450) min^{-1}, depending on the type of construction and type of drive).

Counterblow Hammers. These have only a third of the mass of double-acting hammers, with the same energy capability. Correspondingly smaller foundations are therefore possible. Design is the same in both vertical (more common) and horizontal feed motion. Both pile hammers are mechanically (conveyor) or hydraulically coupled in their movement (**Fig. 29**). In addition to traditional constructions where the mass of the top and bottom pile hammer are more or less equal, there are also more modern developments where the mass of the bottom pile hammer is significantly greater than that of the top pile hammer (**Fig. 30**). As a result, the stroke of the bottom pile hammer is much smaller than that of the top pile hammer, producing advantages in loading. The number of impacts depending on the type of drive may be between 30 and 120 min^{-1}.

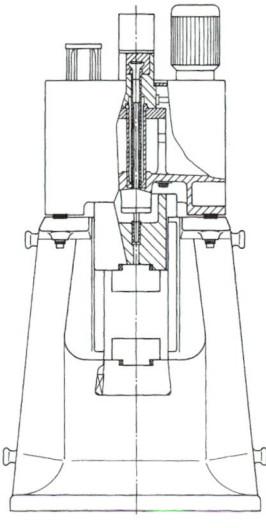

Figure 28. Hydraulically driven double-acting hammer (Lasco).

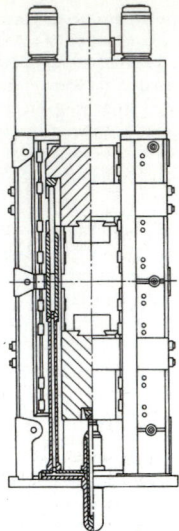

Figure 29. Counterblow hammer, hydraulic drive and pile hammer couple (Bêché & Grohs).

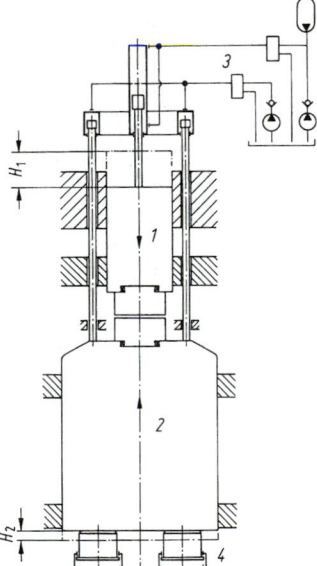

Figure 30. Drive system (schematic diagram) of counterblow hammer with unequal pile hammer mass (Lasco): *1* top pile hammer with mass m_1, *2* bottom pile hammer with mass m_2, *3* oil, *4* air, H_1 top pile hammer stroke, H_2 bottom pile hammer stroke, v_1 speed of top pile hammer, v_2 speed of bottom pile hammer, $m_1/m_2 = 1/4$, $H_1/H_2 = 4/1$, $v_1/v_2 = 4/1$.

4.4.2 Screw Presses

In the traditional construction of the (flywheel) screw press, the spindle and flywheel are either continuously positively connected or else friction-locked. Rotary motion of the flywheel and spindle over a treble or quadruple re-entrant winding of the coarse pitch (angle of pitch 12 to 17°) is translated into linear slide motion.

Where the tool makes an impact on the workpiece, kinetic energy is translated in full from the flywheel, spindle and slide into productive energy and no-load operation (lengthwise and torsional spring losses in spindle and frame and frictional losses on guide and spindle).

Energy transmission is characterised by the degree of effect of impact η_S.

One significant characteristic is energy capability E, given by

$$E = (J\omega_0^2/2) + (m_B v_{St}^2/2),$$

where J = the moment of inertia of flywheel and spindle,

ω_0 = the angle speed of flywheel or spindle on impact with workpiece,

m_B = the slide mass,

v_{St} = the speed on impact of slide on workpiece (normally between 0.5 and 1 m/s).

Otherwise, the rated force F_N, maximum admissible force (in continuous operation) $F_{max\,perm}$ and impact force F_{imp} are essential factors, as these compressive forces are applied to the spindle and the frame (depending on the type of construction). The impact force F_{imp} occurs where the total energy capability is translated into spring energy without deducting productive energy. F_{imp} may be estimated from the rated energy capability E_N and rigidity of the press in the direction of operation c_Z, where $F_{imp} \approx \sqrt{2c_Z E_N}$. The following normally applies for F_N, F_{imp} and $F_{max\,perm}$:

$$F_{imp} = 2F_N, \qquad F_{max\,perm} = 1.6F_N = 0.8F_{imp}.$$

Screw presses with a high energy capability (for hot metal forming) cannot be designed to be free of impact

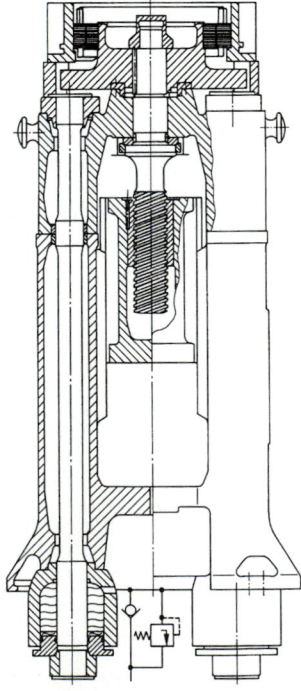

Figure 31. Direct drive screw press with hydraulic excess load protection (Müller–Weingarten).

effect on economic grounds. The forces occurring in the machine may be limited using hydraulic cushions between the tension rod nuts and the frame (**Fig. 31**), or a friction clutch between the flywheel and spindle, or via energy metering. Stroke rates of between 12 and approximately 65 min^{-1} may be achieved in screw presses, depending on the type of construction and size.

Types of Construction

The friction disc drive is the classic form of drive, with two or three continuously rotating flat lateral discs with a spindle with longitudinal motion, or two tapered lateral discs with a permanently fixed spindle (Vincent friction screw press). However, the high load application and the associated high wear to the friction linings are disadvantageous of such drives, and recent use has tended instead towards reversible electrical motors which drive the flywheel via friction rollers or pinion gears, or indeed are directly coupled (**Fig. 31**). Large-scale screw presses (the largest with energy capabilities of 4.5 MN m and impact forces of 315 MN) are driven via a number of electrical reversible motors or hydromotors (**Fig. 32**) positioned along the flywheel perimeter via pinion gears on the toothed flywheel rim. More modern developments have included the clutch screw press (**Fig. 33**) with a continuously rotating flywheel that is connected to the spindle via a switchable friction clutch to initiate the working stroke. When a preselected force is reached, the clutch disengages the flywheel from the spindle, and the return stroke of the slide is then carried out via a pull-back cylinder. The acceleration times of the slide are

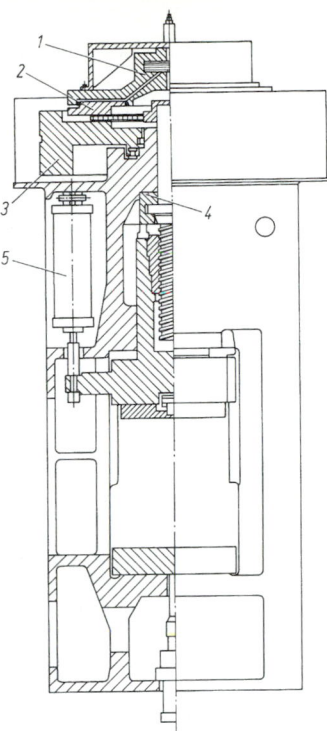

Figure 33. Screw press with continuously rotating flywheel and clutch (Siempelkamp): *1* clutch cylinder, *2* clutch piston, *3* flywheel, *4* thrust bearing, *5* pull-back cylinder.

consequently shorter, because the masses to be accelerated during the working stroke are small. The compressive forces may be limited via the frictional torque of the frictional clutch.

Application

Screw presses are found both in forging operations (drop forging of non-ferrous metals, production of precision forging parts) and in cold forging (cutlery manufacture, mintage and sizing, and coining) and sheet-metal forming (production of flat die parts from thick sheet metal).

4.5 Safety

Work on presses is subject to safety regulations. These are designed to prevent *accessibility* to the hazardous area (tool installation area) while the tools are making contact (e.g. via hand guards), to eliminate any accidental contact movement by the tools (press safety devices) and to limit noise emissions.

The rules and regulations and standards summarised in **Fig. 34**, from the relevant standard engineering practice according to §3 of the German law on working materials, apply for work safety on presses in general. Accordingly, hand guards may take the form of secured tools, permanently fixed screening of hazard areas, two-handed switching and no-contact protective equipment (photoelectric barriers). Press protective equipment includes safety clutches, safety brakes, safe control, follow-up monitoring, removable screening and devices for automatic pass dis-

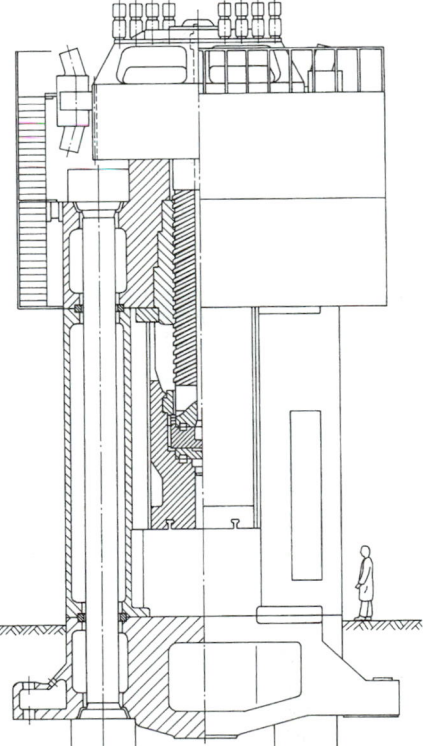

Figure 32. Large-scale screw press with hydromotor drive (SMS Hasenclever).

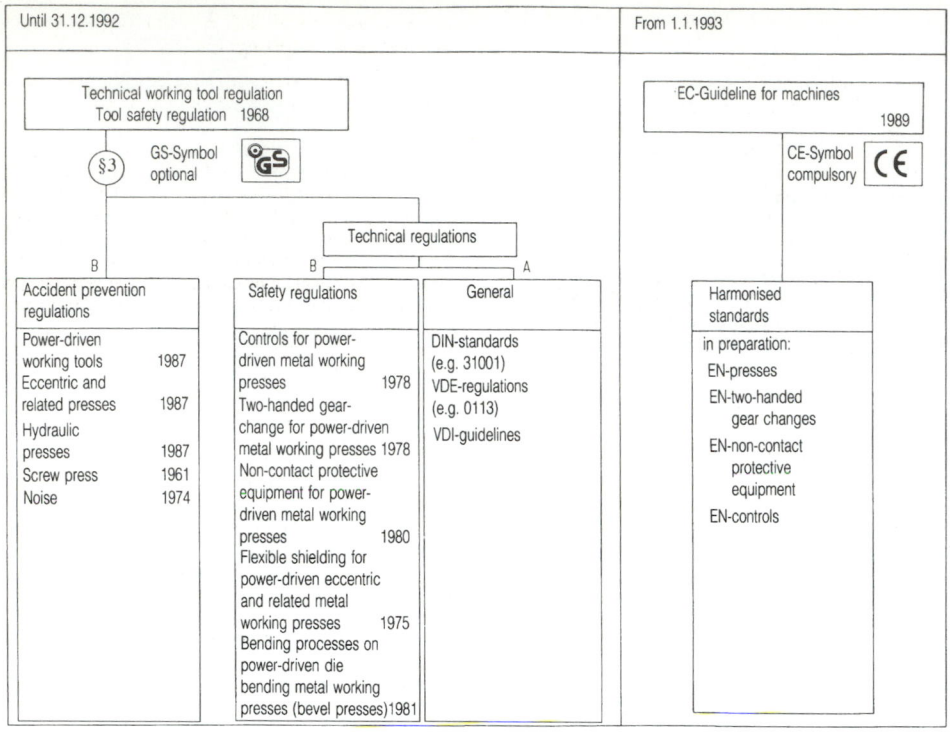

Figure 34. Safety regulations for working on presses.

abling. **Figure 16** shows the structural design of the latter protective measure. The operating brake is separated into two equal units operating independently of one another.

In the German accident prevention regulation on noise (VBG 121: German accident prevention regulations of the Employers Liability Insurance Association), the acceptable noise level specified is 90 dB (A), with levels in excess classified as "areas of excessive noise". In many areas of operation presses have as yet been unable to adhere to this acceptable noise level. In these cases, secondary noise reduction measures (such as partial or full encapsulation in enclosures are required), along with acoustic protection for each individual employee.

5 Metal-Cutting Machine Tools

G. Spur, Berlin

5.1 Lathes

5.1.1 General

Parts generated by rotation are manufactured on lathes. The workpiece performs a circular motion about an axis of rotation, while the tool carries out the feed motion in a plane perpendicular to the direction of cut. In special-purpose constructions, the tool may also be rotated. Power-driven tools are also used for light metal drilling and milling operations as well and are thus also suitable for end-face machining, slotting and eccentric milling, or drilling diagonal to the axis of the workpiece on the lathe.

Classification. Classification in terms of *universal lathes, automatic lathes, front-operated lathes, vertical boring and turning mills* and *special-purpose lathes* has come from the lathe construction industry itself. A system-atic classification may also be made according to the position of the main axis in vertical and horizontal machines, the number of spindles in single or multiple spindle machines and the type of control, i.e. manually operated machines, automatic mechanical PLC machines and numerically controlled machines.

Construction. The lathe processing system may be divided into subsystems, including workpiece systems, tool systems, kinematics systems, energy systems, information systems and auxiliary systems.

Workpiece Systems. These include the workpiece and workpiece clamping and support mechanisms. The most popular form of clamping is the *three-jaw chuck* (**Fig. 1**). It is employed in conventional machines as a manual chuck and in NC machines as a power-operated chuck in hydraulic, electrical or pneumatic operation. Lathe chucks with two, four or six jaws are also used in special cases. Other clamping devices consist of *collets, faceplates* and *lathe mandrels*. Long workpieces are held

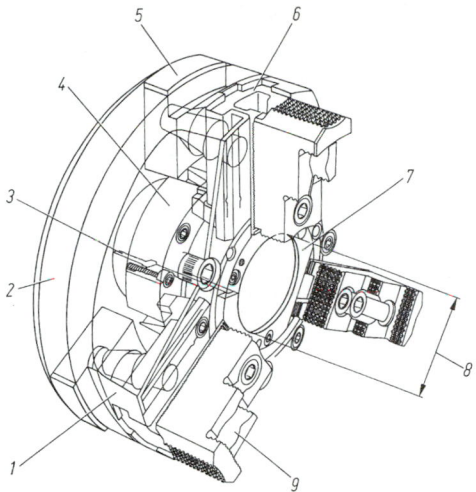

Figure 1. Power-operated wedge-type collet chuck with mechanical centrifugal force balance (Forkardt GmbH, Düsseldorf): *1* chuck body, *2* chuck adapter for universal spindle assembly, *3* threaded ring for connection to air line, *4* faceplate jaw, *5* centrifugal force, *6* base jaw, *7* protective bushing, *8* bar capacity, *9* standard jaw attachment.

between *points*, driven by *drivers* and supported by a *steady rest* (backrest) where applicable.

Tool Systems. These include the tool, tool clamp and tool carrier. In addition to rapid-change tool holders, *turrets* are in common use in program-controlled machines as tool carriers for between 4 and 16 tools each. The tool holders are carried in standard straight-shank carriers, T-bolts or V-guides. However, more and more tool change systems are being adopted where either only the cutting head or the entire tool including shaft and holder fixture are changed. The latter are also specifically designed for changing power-driven tools.

Energy Systems and Kinematics Systems. The kinematics system is divided into one system for generating the cutting motion and another for generating the feed motion that is transmitted directly from the energy system. Single-stage or reversible-pole three-phase motors are combined with multiple-stage mechanical switching mechanisms to cater for a broad speed range in *main drives*. D.C. motors with electrical speed adjustment with two- to four-stage electrically switched transmission units are normally coupled in numerically controlled machines. The *main spindle* is usually seated in a roller bearing. High requirements relating to thermal characteristics and high speeds may require special lubrication systems.

Electrical motors and occasionally hydraulic motors are employed in *feed drives* not derived from the main drive. The slideways normally run in *V-guides*, although enclosed flat-bed guides, circular guides and combinations thereof are also employed.

Information Systems. These are employed for the control of the interaction of functions between the subsystems. They incorporate external information such as machining programs or parts sequences into the manufacturing system and submit status reports to the primary control system. Control cams, stroke rates and trip cams are employed as mechanical data stores in automatic lathes for mass production. The greater flexibility offered by

numerical control has made this form more popular in small and medium-sized serial production.

Auxiliary Systems. These include functional systems such as coolant lubrication, swarf-clearing installations, and central lubrication.

The *frame* forms a definitive part of the basic shape of the machine as the carrier of the other assemblies, carrying the guides of the tool slideways in the form of the *bed*, and the main spindle and main drive in the form of the *headstock*. Cast, welded, and concrete constructions are employed. The *horizontal bed-type* machine, the standard manually operated machine, is being superseded predominantly by the *inclined-bed* construction in numerically-controlled machine tools because of the improved chip clearance. Lathes with a vertical main axis are designed as column-type machines.

5.1.2 Universal Lathes

Single spindle lathes are dominant in small and medium-sized serial production and are available in a variety of sizes. They are classified in terms of drive ratings and operating ranges. The operating range is determined by the maximum turning diameter and the longest distance between centres. The most common sizes have a maximum pitch of between 100 and 500 mm and a maximum distance between centres of between 250 and 1250 mm. Then come the small lathes and above this the large-scale lathes with pitches of up to approximately 2500 mm and distances between centres of up to approximately 10 m.

The manually operated *sliding and screw-cutting lathe* has the basic form of a universal lathe (**Fig. 2**). The main spindle is driven by a multiple-stage, change-speed gear drive, in order to be able to operate over a broad speed range at a constant output. The feed drive is taken from the main drive via the feedgear mechanism, feed rod and bed slide drive. The main spindle serves to provide the kinematic connection between the main drive and longitudinal feed in screw cutting operations.

With the use of turrets as tool carriers for the tools required for processing the workpiece, *turret lathes* were developed. A distinction is made between the *drum turret, turnstile turret, disc-type turret* and *flat-bed turret*, depending on the orientation of the tool axis and the operating axis (**Fig. 3**).

Special-purpose constructions include the block turret, cross-type turret and crown-type turret (**Fig. 3**). The manually operated turret lathe has been superseded to a large extent by numerically controlled machines. Mechanical program-controlled models are used in mass production environments as a single spindle lathe.

Copying Lathes. These use mechanical, electrical or hydraulic systems to scan a two- or three-dimensional standard shape which is subsequently retrieved to control the feed motion of the lathe cutting tool. Devices for out-of-true copying by lathe are also employed in addition to longitudinal copying lathes for shaft processing. Mechanical systems operate with direct power transmission and a lead cam or swivel guide bar (taper turning). Power-assisted systems operate with sensitive scanning devices and electrically or hydraulically controlled reproductive motion by the tool. Because NC continuous-path control systems carry out the same tasks, and the description of the workpiece contour in the NC program is less complex than the manufacture of a lead cam, lathe copying has also been mainly superseded by the use of NC machines.

In *numerically controlled universal lathes* the main

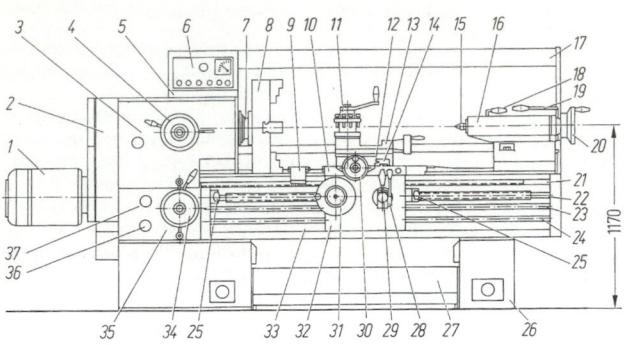

Figure 2. Engine lathe and bar lathe (Gebr. Boehringer, Göppingen): *1* flange motor drive, *2* feed drive, *3* headstock with main gearing, *4* speed switch, *5* switch cabinet, *6* operator's panel, *7* main spindle, *8* faceplate, *9* longitudinal feed stop, *10* saddle (longitudinal slide), *11* toolholder, *12* cross-slide, *13* top support, *14* transverse stop, *15* dead centre, *16* tailstock, *17* swarf guard, *18* lever for tailstock clamp, *19* tailstock clamping lever, *20* handwheel, *21* gear rack, *22* guide screw, *23* feed rod, *24* operating shaft, *25* remote-control lever, *26* base of bed, *27* swage box, *28* intermediate switch lever for direction of feed, *29* split nut, *30* handwheel for cross-slide, *31* handwheel for longitudinal slide, *32* lock box, *33* bed, *34* feed and thread-rolling drum, *35* feed box, *36* switch-over mm–inches, *37* engine-lathe–bar-lathe changeover lever.

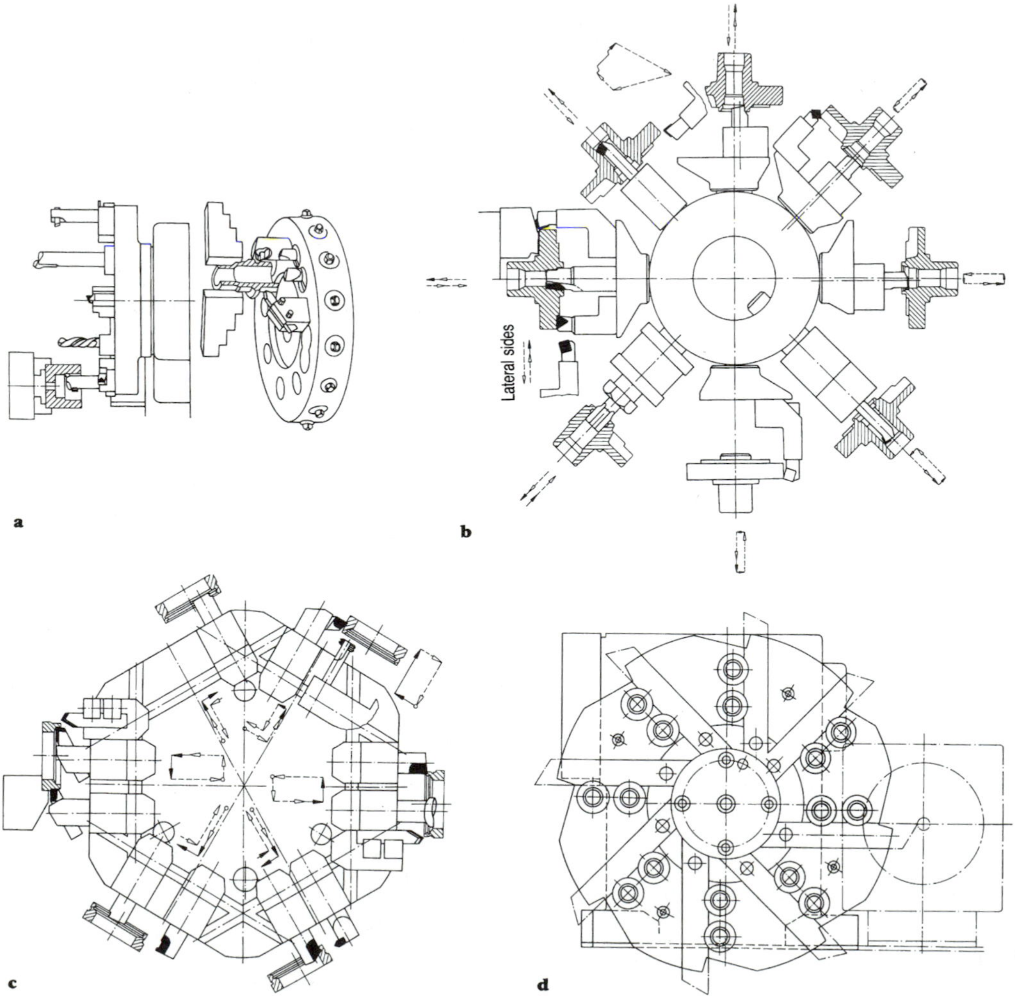

Figure 3. Constructions of turret heads. **a** Drum turret (Pittler, Langen). **b** Turnstile turret (Pittler, Langen). **c** Flatbed turret (Pittler, Langen). **d** Disc-type turret (= turnstile turret) (Gebr. Boehringer GmbH, Göppingen).

spindle is either driven directly by a speed-controlled main motor via a belt drive or with an intermediate gear mechanism (**Fig. 4**). The feed slides are positioned by separate path-controlled precision ball screw drives. For longer overall lengths a permanently fixed spindle with a notched drive (drive motor on saddle), or a pinion drive with a gear rack located on the saddle, is normally employed. The position of the angle of the spindle is verified using a rotational measuring system and coupled via the controller to the feed to synchronise the spindle rotation with the longitudinal feed in screw production.

CNC continuous-path control systems provide high productivity owing to their ease of operation and programming, such as the provision of subprograms for machining cycles, thread-cutting programs, automatic cut sectionalisation and simulated process graphics on the controller monitor.

5.1.3 Front-Turning Machines

Front-loading automatic chuck lathes are used to machine disc-shaped workpieces with pitches of up to 800 mm (**Fig. 5**). These machines are available as one- and two-spindle models, permitting multi-point machining, and they have electrohydraulic or numerical program control systems. Capstans, facing slides and cross-slides are employed. These machines are also constructed in the column-type design with a block turret and lateral slides for chucking operations and as *facing lathes* with a main bed guide which is normally positioned at an angle to the horizontal spindle axis for processing larger and bulkier workpieces on faceplates.

High output volumes may be achieved with automatic workpiece handling by way of two-arm systems for gripping, loading, turning and unloading in conjunction with workpiece stores.

5.1.4 Automatic Lathes

These provide automatic workpiece machining from rod materials or pre-formed chuck parts. One characteristic common to the various designs is *multi-point machining* which is supplemented in multiple spindle machines by *sectionalised machining*. The criteria distinguishing these include the number of spindles, the horizontal or vertical position of the main machine axis, the frame construction, the type of tool carrier, the coordination of the cutting and feed motion and the number of potential feed movements. In automatic machines employed in mass production with mechanical control, control cams and operating cams are used for the following functions: workpiece feed or workpiece advance, movement of longitudinal or cross-slide and tailstock and turret head, speed adjustment and change in direction of rotation, operation of clamping devices, movement of auxiliary or special purpose equipment, and moving headstock or spindle sleeve where required. The *main control shaft*, which completes one revolution for each completed workpiece, determines the sequence of the productive and idle time motion. It is supplemented by an *auxiliary control shaft* to generate independent idle time motion. Mechanical cam control may be supplemented or substituted by electrical, hydraulic or numerical control. Automation of the machine tool handling process, by way of system-compatible loading and magazine devices for rod or chuck parts, also permits a number of machines to be chained.

Single-Spindle Lathes. In mechanically controlled automatic lathes, all feed and switching motion is controlled by control cams and operating cams whose motion is derived from the main drive via complex gearing units (**Fig. 6**). As modifying the cams to suit the machining task to control tool paths and speeds is a complex business, mechanically controlled single-spindle drives are increasingly being superseded by numerically controlled machines.

Swiss bush-type automatic lathes are employed for manufacturing long, slender workpieces (**Fig. 7**) where the workpiece is fed close to the cutting point. Automatic cross-feed motion is carried out via tool carriers arranged in a turnstile formation in relation to the rotating axis and attached to a permanently fixed rocker column. The longitudinal feed is carried out by a moving headstock in the *Swiss system* and a moving back rest in the *Offenbach system*. Guide bushes are employed in the Swiss system to increase the degree of accuracy. Both numerically controlled systems and mechanical cam curve systems are in common use.

Multiple-Spindle Automatic Lathes. Multiple-spindle bar lathes or chuck lathes are used for automatic turned part mass production (**Fig. 8**).

Classification. There are constructions with or without tool carrier control with rotating workpieces or tools, a horizontal or vertical main machine axis and a variety of types of control. The size may be specified in terms of the spindle capacity diameter or chuck diameter and the number of spindles.

Construction. The frame normally takes the form of a single column or double-column construction with high rigidity depending on the working space, tool and workpiece motion, chip clearance and coolant lubrication system; it is also available in vertical construction for larger pitches. The main spindles are driven centrally; only bulky or unbalanced workpieces are processed from one or more sides on machines with rotating tools and a bed-type or column construction, where clamping takes the form of indexing plates or drums. Longitudinal slides and cross-slides and all auxiliary motion are controlled mechanically, such as indexing and latching spindle drums, bar feeding, bar stops, collet gripping or chucking and the operation of special-purpose equipment (**Fig. 9**). Path and switching data are stored by way of curves and cams on the central or branched control shaft which runs at the feed-rate for productive time motion and in rapid gear for idle time motion. Longitudinal slides and cross-slides are controlled in groups or individually. An increase in the working space can be attained with the help of additional equipment for shutting down spindles, variable speed drives, duplex switching with two clamping points, and rod and chuck gripping.

Numerical control and independent feed drive applications are made more complex in multiple-spindle lathes owing to the limited space involved. Numerically controlled slides are frequently only provided in critical machining centres for this reason. The *two-spindle lathe* (**Fig. 10**) provides the advantage of sectionalised machining and there is sufficient room available for it to be designed as a NC machine. The use of compact hydraulic feed drives permits the conventional structure to be retained and exploits the inherent flexibility of numerical control (**Fig. 11**).

5.1.5 Heavy-Duty Lathes

Large-scale, longitudinally arranged rotating parts are manufactured on special horizontal lathes. The structural

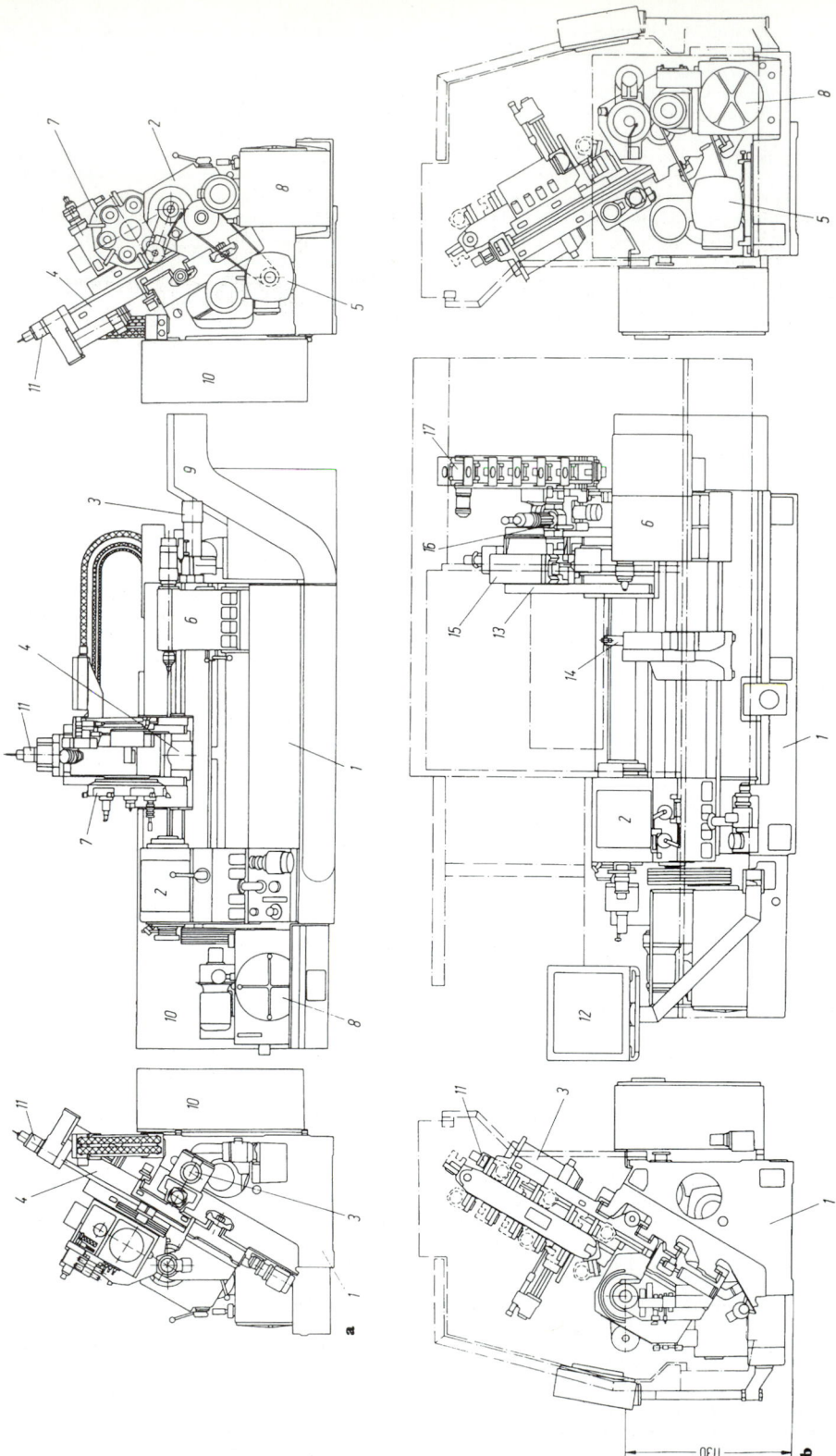

Figure 4. Numerically controlled lathe (Gildemeister AG – Max Müller, Hanover): **a** turret design; **b** tool magazine design: *1* machine bed, *2* headstock, *3* feed drive, *4* cross-slide, *5* d.c. main drive, *6* tailstock, *7* disc-type turret, *8* hydraulic accumulator, *9* chip conveyor, *10* switch cabinet, *11* path measuring elements, *12* operator's panel, *13* longitudinal slide, *14* back rest, *15* tool carrier, *16* tool grip, *17* tool magazine.

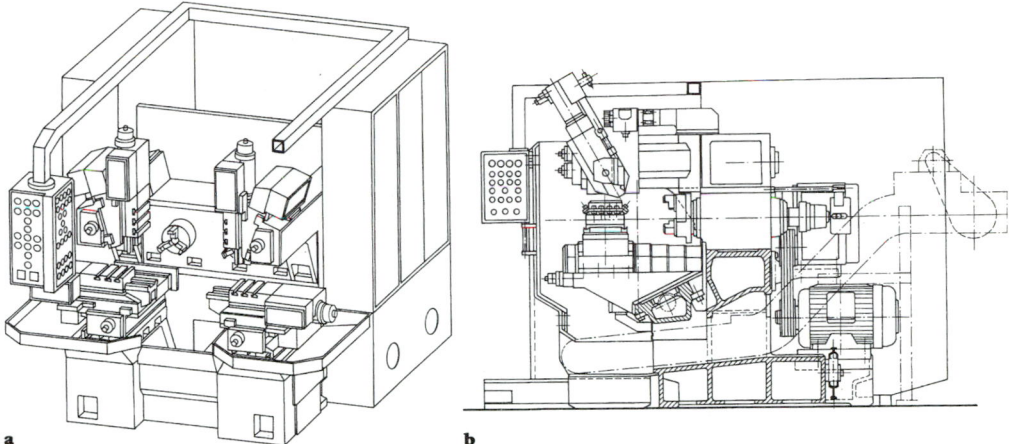

a   b

Figure 5. Double-spindle front-turning machine (J.G. Weisser & Söhne, St. Georgen): **a** overall layout, **b** section of front turning machine.

design of these follows in principle that of single-spindle universal lathes. As the sag of the workpiece has the least effect on the machining diameter when the direction of the tool feed is perpendicular to the direction of the force due to weight, the construction is always of flat-bed type. Additional milling attachments offer overall machining of large-scale parts and produce high rates of metal removal in *turning and milling* (**Fig. 12**).

Vertical Boring and Turning Mills. These are employed for machining heavy and bulky workpieces with a small length/turning-diameter ratio. The *faceplate* rotating around a vertical axis is characteristic as a carrier for the workpiece which is to be machined in the subsequent installation position and clamping device where possible. There are *single- and two-column machines* with stationary or moving frame components and a number of tool slides on the columns and crossbar (**Fig. 13**). The two-column machine in a double-column construction, or the single-column machine with a projecting cross-beam, is employed where there is a turning diameter of over 3 m. Double columns, single columns and faceplate undercarriages may be designed as moving parts to increase the turning diameter or to improve accessibility for servicing by crane. The faceplate diameter amounts to between 800 and 5000 mm under normal operating conditions, with a maximum of up to 18 m, while the maximum turning diameter is between 1400 and 5200 mm, or approximately 25 m. The loading capacity of the faceplate determines the permissible weight of the workpiece. Core and ring faceplates may be combined in large-scale machines. The feed is effected increasingly via individual numerical control drives. D.C. motors with drive output ratings of up to 200 kW are normally used as the main drive; positioning control and NC continuous-path control of the faceplate drive are available. Turret or ram slides supply the tool system. Operation is carried out via a suspended panel or in the case of large-scale turning machines via a travelling operating stand.

Additional attachments may be employed for special-purpose operating processes (**Fig. 14**).

The workpiece weight and dead weight of frame components causes tool shifts which should be taken into account in design planning.

5.1.6 Special-Purpose Lathes

These are used for workpieces where processing on standard machines is either impossible or uneconomical. Their construction depends on the size and type of the workpiece. The structure has to be functional and to conform with standard-sized components based on the mechanical assembly technique with a variety of sizes, where matching or reconstruction will be carried out as required. Turning machines for roll pins, crankshafts, turbine discs, cylindrical bushes, axles, cams, wheel sets, wheel discs, pipes, sleeves and out-of-true lathes are available depending on the specified manufacturing task.

5.1.7 Flexible Turning Centres

Numerically controlled machines may be adapted for machining the reverse side for overall machining of a comprehensive range of workpieces by expanding the kinematics system and with the use of power-driven tools and additional attachments.

The use of a second slide unit produces a *four-axis lathe*, which permits an increase in the output volume owing to sectionalised machining. A *C-axis* is frequently used for the angle positioning of the spindle in combination with power-driven tools for boring and milling work, while occasionally a *Y-axis* is also used as the third slide axis. **Figure 15** shows examples of structural forms which may be manufactured with a controlled C-axis in conjunction with tool drives. When the workpiece has been finished on the front, it may be taken up by a clamping device and the reverse side may be finished at a separate machining centre, for example in *reverse side machining*.

Figure 16 shows a two-spindle machine with one stationary and one moving spindle-and-turret system in each case. The CNC machine shown in **Fig. 17** with the rotating spindle head permits tool changes to take place outside the working space, thereby reducing non-productive time.

Where the turning machine is designed for semi-automatic operation with memory storage and handling systems for workpieces and tools, this is known as a *turning centre* (**Fig. 18**). Drum, chain and ring magazines are

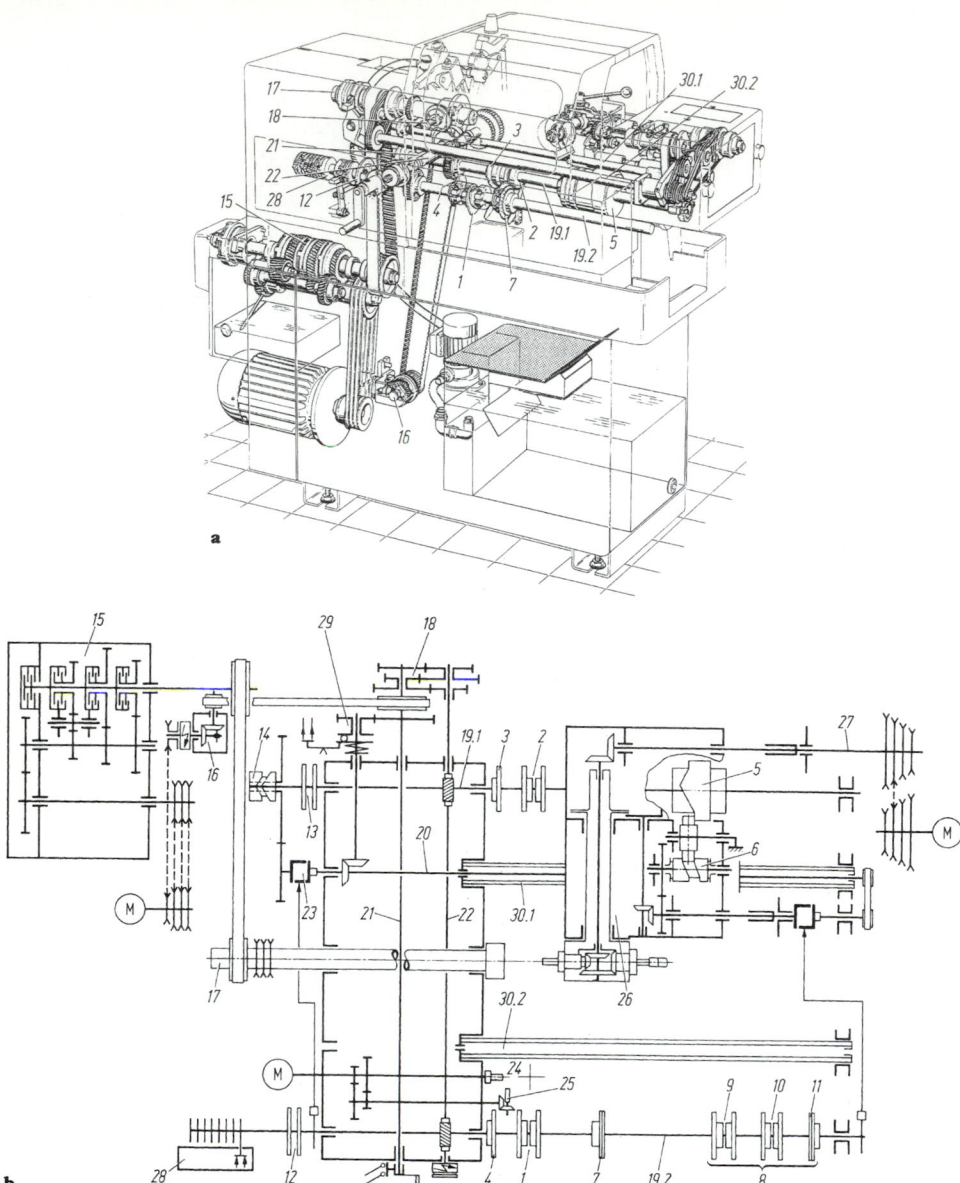

Figure 6. Mechanically controlled automatic single spindle short-stroke lathe with auxiliary control shaft and separate main control shaft (H. Traub AG, Reichenbach): **a** gearwheel layout, **b** gearwheel arrangement; *1* cam for tool slide S1 (front traverse support), *2* cam for tool slide S2 (rear traverse support), *3* cam for tool slide S3 (rear vertical support), *4* cam for tool slide S4 (front vertical support), *5* drum cam for drive stroke of turnstile turret, *6* cam for rapid return of turnstile turret, *7* cam for sorting device, *8* cam for gripper device, *9* cam for longitudinal stroke of gripper lever, *10* cam for rotating motion of gripper lever, *11* cam for clamping in gripper lever, *12* cam for front longitudinal lathe attachment, *13* cam for rear longitudinal lathe attachment, *14* cam of clamping drum for tool clamping or rapid clamping, *15* spindle controlling mechanism, *16* mitre gear, *17* main spindle, *18* feed drive, *19.1* rear control shaft, *19.2* front control shaft, *20* auxiliary control shaft, *21* operating shaft for feed drive, *22* worm gear shaft, *23* jaw clutch for workpiece rapid clamping, *24* back-drilling attachment and slotting device, *25* cross-drilling attachment, *26* turnstile turret with rapid return, *27* drive attachment for turnstile turret tools, *28* multi-position switch (multiple control switch), *29* excess load protection for auxiliary control shaft, *30.1* rear guide shaft, *30.2* front guide shaft.

used for external tool storage, and handling is effected via freely programmable systems which are frequently designed in double-column style in the same way as tool handling systems. There are a variety of automatic tool-change systems for changing the cutting head as a result of wear (**Fig. 19**). Tool data storage often takes place in pallets or conveyor stores.

Further measures for low operation systems and high

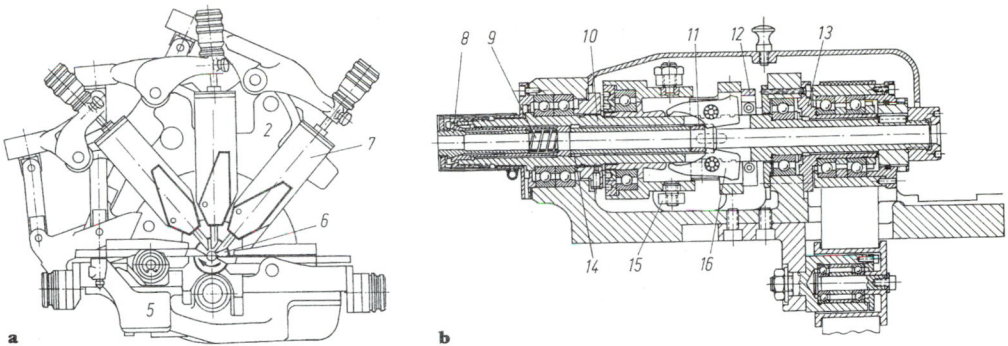

Figure 7. Tool column and headstock of longitudinal automatic lathe (H. Traub AG, Reichenbach): **a** tool or rocker column, **b** section of headstock design with longitudinal motion; *1* rocker, *2* guide bush, *3* tool slide, *4* collet, *5* relief spring, *6* feed spindle, *7* clamping bush, *8* tension ring, *9* spindle bearing free of shearing load, *10* spacer tube, *11* fork, *12* clamping lever.

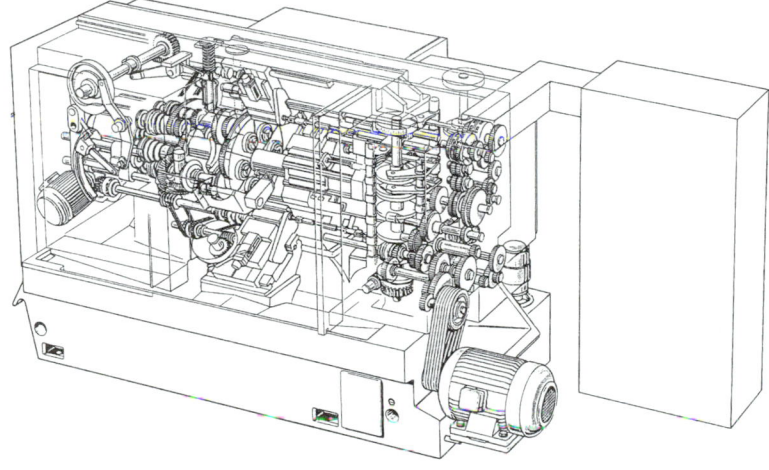

Figure 8. Six-spindle bar lathe with cross-slide (Gildemeister AG, Bielefeld).

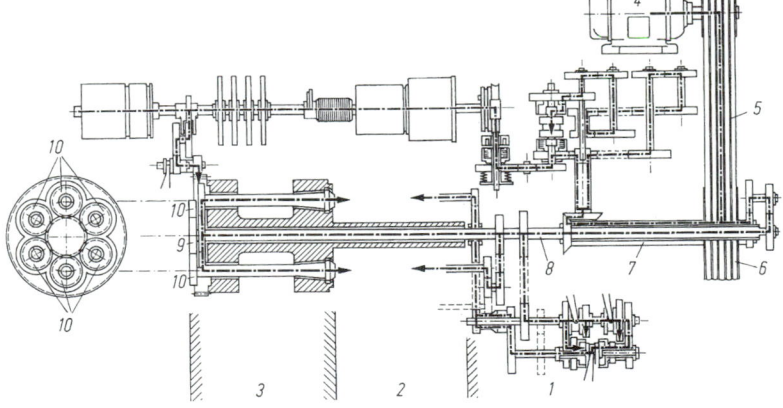

Figure 9. Gear wheel arrangement of multiple-spindle lathe: *1* drive column, *2* tool space, *3* spindle column, *4* motor, *5* V-belt, *6* drive pulley, *7* tubular shaft, *8* central fixed shaft, *9* central gear, *10* spindle gears.

flexibility include the use of systems for automatic change of clamping jaws or the entire clamping device and systems for tool monitoring and workpiece measurement.

5.2 Drilling and Boring Machines

5.2.1 General

Cutting motion and feed motion are coordinated according to the process, the tool or the workpiece in drilling and boring machines. Constructions are classified depending on the position of the drilling spindle and type of frame structure (**Fig. 20**). Other criteria for classification and selection include the clamping and operating range, rated drilling capacity, speed and feed range, machining accuracy, degree of automation and position and number of drill holes. Twist drills, counterbores, reaming bits, tap drills, drill heads, spade drills and special-purpose drills and boring bars in drilling and boring machines are

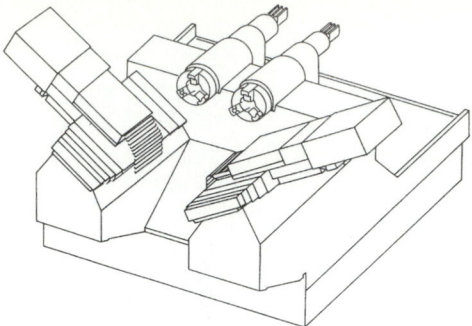

Figure 10. Two-spindle lathe.

used. The tools have cylindrical or tapered ends and fit into cylindrical or tapered sockets on the machine. In special-purpose machines arrangements are made for adjusting the length of the drilling.

5.2.2 Bench Drilling Machines

These are suitable for small-scale drilling operations and workpieces. The torque is transmitted from the motor shaft to the drilling spindle via belt transmission. Speed changes may be made by adjusting the belt or replacing the belt pulley; the feed motion is carried out manually via a start-handle shaft, feed pinion and gear rack drive.

5.2.3 Free-Standing Pillar Machines

The frame of these is designed as a free-standing hollow pillar, to which the drive casing and drilling spindle are attached either permanently or as part of a height-adjustable boring slide. The bottom section of the column carries the drill table which is height-adjustable and may be rotated and is not normally supported. Wheel gears or friction gears provide speed adjustment (**Fig. 21**). Feed is manual, or via a feed cam, or by way of worm gearing attached to the spindle sleeve unit.

5.2.4 Column-Type Drilling Machines

These are suitable for small and medium-sized workpieces. There is a moving boring slide located at the top of a

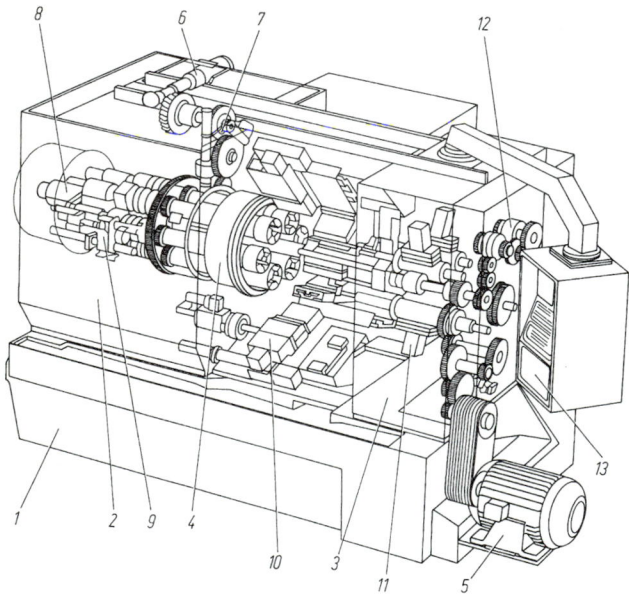

Figure 11. Multiple spindle lathe with hydraulic feed drives (Gildemeister AG, Bielefeld): *1* machine bed, *2* spindle column, *3* drive column, *4* spindle drum, *5* main drive motor, *6* hydromotor for spindle drum control, *7* spindle drum latching, *8* hydraulic cocking cylinder, *9* shutdown gate valve, *10* cross-slide with continuous path control, *11* feed drive sliding spool control valve, *12* screw cutting drive, *13* control panel with VDU.

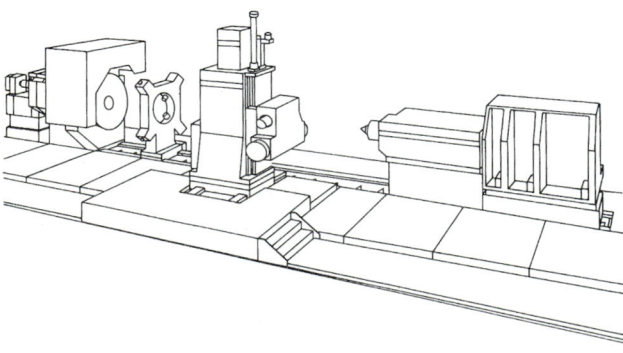

Figure 12. Large-scale lathe with milling attachment (Wohlenberg KG, Hanover).

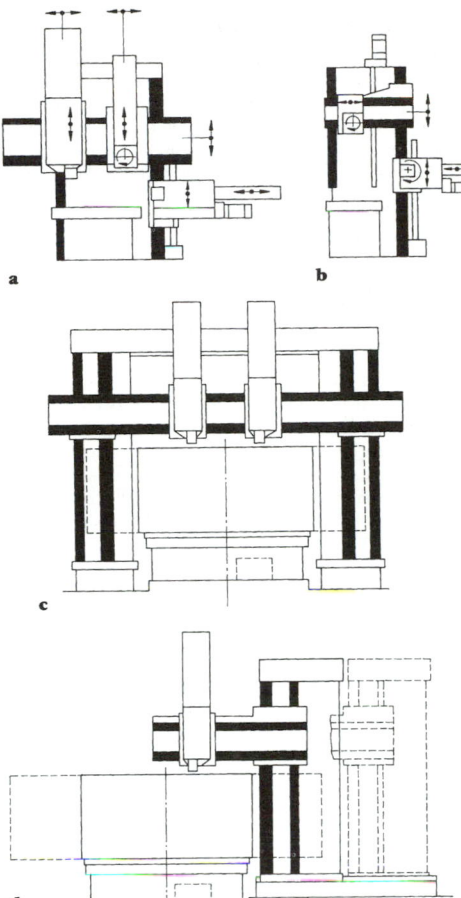

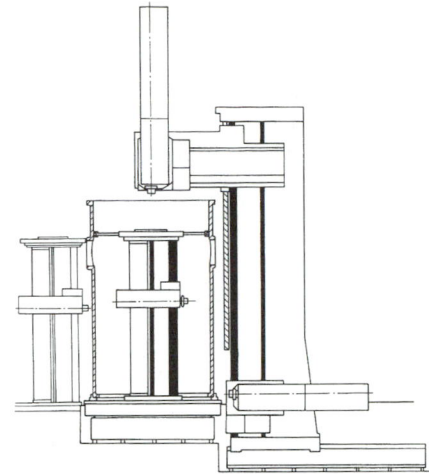

Figure 14. Single-column vertical boring and turning mill with moving column for external machining of reactor pressure vessels. Convertible auxiliary columns for internal machining of vessels on stationary faceplate centre.

Figure 13. Vertical boring and turning mill. **a** Single-column vertical boring and turning mill with left-hand ram slide, right-hand square turret slide and right-hand lateral slide. **b** Single-column vertical boring and turning mill with cross-slide and lateral slide and two square turrets. **c** Two-column machine with moving double column. **d** Single-column machine with moving column.

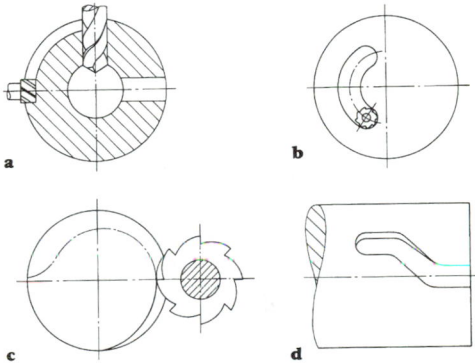

Figure 15. Boring and milling with C-axis control: **a** positioning control and line motion control, **b** line motion control, **c** C/X interpolation, **d** C/Z interpolation.

rectangular column and the bottom section supports the drill table, which is designed to cater for heavy workpieces. Tailstock or boring slides execute feed motion, while spindle speed is adjustable via V-belt gearing, charge gears and changeover gears or infinitely variable drives.

5.2.5 Multi-Spindle Drilling Machines

These are suitable for use with high volumes or frequent repetition of drilling jigs. The frame structure is often identical to that in heavy-duty column-type drilling machines. A multi-spindle *cutter* is used in place of a single-spindle drilling shaft and is driven via one or two shafts from the main spindles. The spindles are driven via gearwheels in machines with permanently fixed multi-spindle drilling heads, and via drive shafts in adjustable, *articulated spindle drilling machines*. Spindle guides in permanently fixed bearing plates (**Fig. 22**) have a higher rigidity, but the adjustable supporting beam guide adversely affects the reproducibility. A number of drilling jigs may be in use simultaneously for the spindle guide.

Separate drive trains with reversible-pole motors and

main and change gears permit speeds to be transmitted independently by the distributor gear mechanisms, which consist of the central drive shaft, planetary and intermediate gearing. The latter may be omitted, depending on the direction of rotation of the spindle and size of the gearing. Rapid advance motion may be generated by a three-phase motor, and feed motion may be generated via infinitely adjustable d.c. motors owing to the improved speed synchronisation. Multi-spindle drilling heads are also used in special-purpose machines with horizontal and angled directions of feed.

5.2.6 Radial Drilling Machines

Radial drilling or cantilever boring machines are employed for drilling large and bulky workpieces from scratch. The chief components are the baseplate, pillars, cantilever arm, drill slide and drill table (**Fig. 23**). The baseplate is designed as a rectangular section or in special-purpose designs as an angle plate, cross-baseplate, or double, turn-

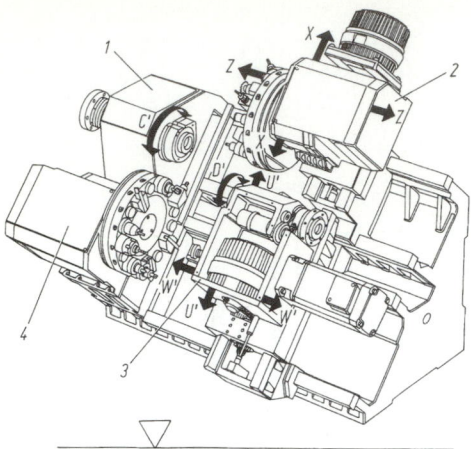

Figure 16. Two-spindle turning machine (Boley GmbH, Esslingen): *1* headstock I, *2* turret slide I, *3* moving spindle unit II, *4* stationary turret carrier II.

stile or circular baseplate. The machines are employed in small- and medium-scale serial production as universal machines. *Wall-slewing drilling machines* have a cantilever arm attached permanently to the wall or to the height-adjustable track.

5.2.7 Jig Boring Machines

Jig boring machines may be employed to produce bores, depressions and shallow notches to a high degree of accuracy without tracings or templates. These are employed in single-part production, e.g. in gauge engineering, tool construction and the construction of jigs and fixtures. Single-column machines are employed for small workpieces, larger double-column machines are available in column or free-standing pillar designs.

Single-column machines (**Fig. 24**) are designed with a cross-table and double-column machines have a longitudinal table. The boring slide is positioned horizontally on a crossbar between the two pillars or columns in backlash-free, anti-friction slideways. The vertical feed is introduced via a pinion on the spindle sleeve. A horizontal shaft drives two worm-type gears synchronously for raising and lowering the crossbar, which is arrested via the self-locking hydraulic flow constriction mechanism.

5.2.8 Turret Drilling Machines

The frame structure is similar to the column-type drilling machine. The drive and gearing are arranged either inside moving boring slides or separately. The spindle is driven via drive shafts and a disengaging clutch. Feed motion is normally executed via slides. The turret circuit is designed in the form of a cross-wheel with indexing pins or via cross-toothing (**Fig. 25**). Turnstile turrets are normally used as tool carriers, while tool changers may also be used. Machines with cross-tables are also designed for light milling operations. They may be expanded in conjunction with numerical control to providing machining centres.

5.2.9 Precision Drilling Machines

Characteristics include high static and dynamic rigidity, high damping factor, high degree of uniformity in motion and minimum change in temperature. As a result, diameter and position tolerances of IT6 to IT5 may be achieved and IT4 with the relevant input. The torque is transmitted via a specially positioned belt pulley and clutch onto the vertical or horizontal main spindle to prevent torsional forces and oscillation. Both three-phase motors and infinitely variable d.c. motors may be used. The feed motion is normally generated hydraulically and is either executed by the headstock or the carriage (workpiece).

5.2.10 Deep-Hole Drilling Machines

These are employed for the manufacture of drilled deep holes with a diameter/length ratio of between 1 : 3 and 1 : 200 and are structured in a similar way to lathes. A distinction is made between short and long bed-type machines, machines with rotating workpieces, rotating tools or counter-rotation of tool and workpiece with a horizontal spindle position and machines with rotating tools and a vertical spindle. The feed is executed via the ballscrew spindle or rack; the tool is guided into the drill bushing or pivot hole via a three-point support on support rails. Continuous supply of coolant and chip removal are required. Machine *assemblies* include workpiece spindle box, workpiece steady rest, guide slide with coolant lubrication supply, tool clamping bearing and tool spindle box.

5.2.11 Special-Purpose Drilling Machines

The simplest type of special-purpose drilling machine is a serial arrangement of the table-type, free-standing pillar-type and column-type drilling machines. The construction is modified depending on the manufacturing task, where

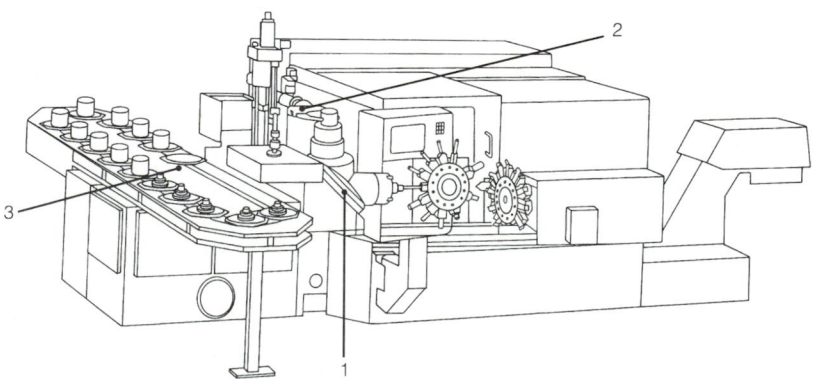

Figure 17. CNC chuck lathe (Index-Werke KG, Esslingen): *1* swivelling headstock, *2* hinged-arm loader, *3* oval-shaped conveyor.

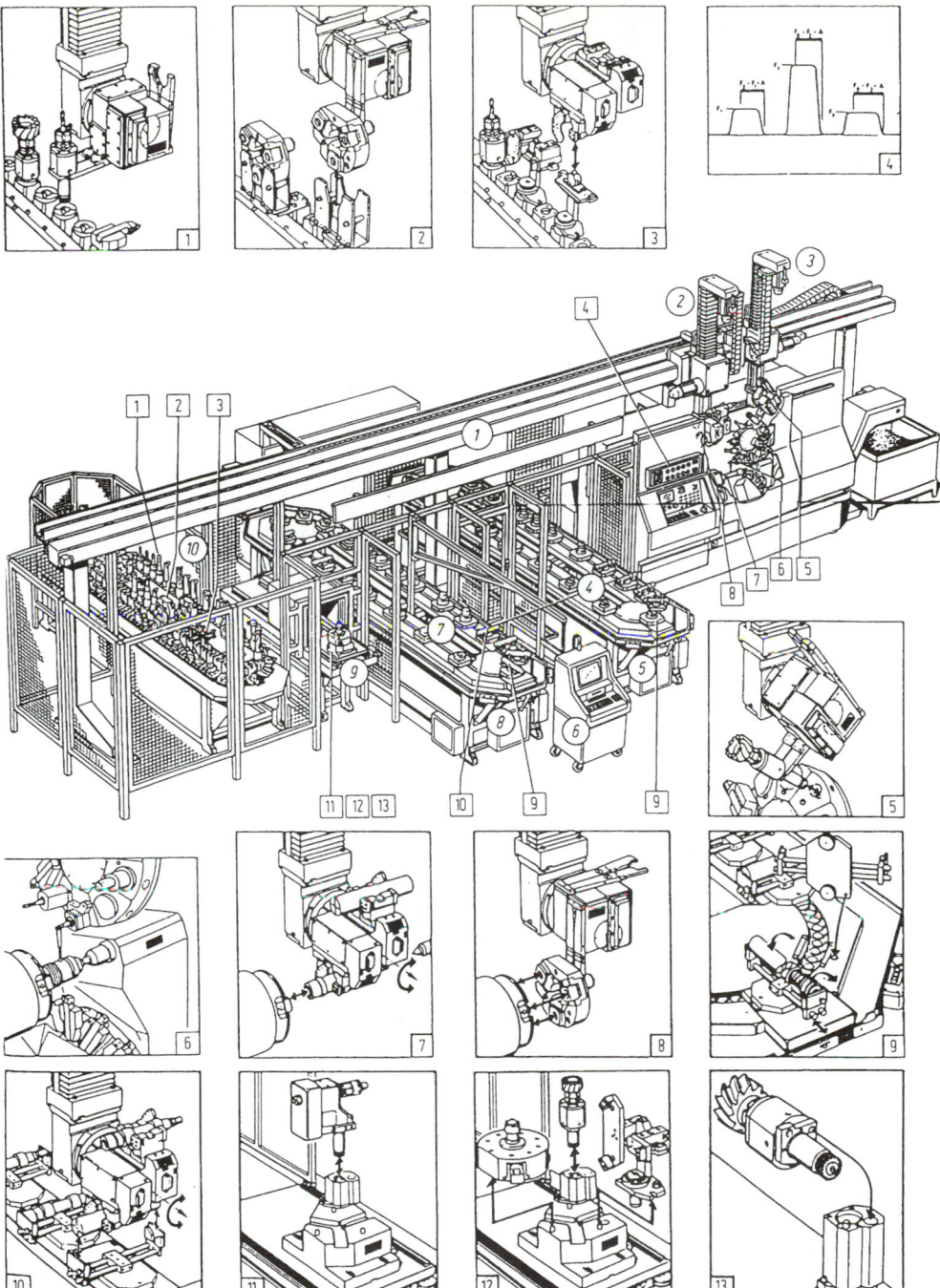

Figure 18. Flexible turning centre (Traub AG, Reichenbach). *Components*: *1* line double column, *2* workpiece gripper, *3* tool gripper, *4* workpiece store 1, *5* transfer tube for workpiece store 1, *6* tool-setting terminal, *7* workpiece store 2, *8* transfer tube for workpiece store 2, *9* transfer tube for tool store, *10* tool store for tools, clamping devices and grippers. *Operations*: 1: Remove tool, place tool in magazine. 2: Remove chuck jaw, place chuck jaw in magazine. 3: Change gripper chuck, place gripper chuck in magazine. 4: Tool monitoring. 5: Tool change. 6: Check workpiece dimensions, set tool offset. 7: Remove finished part, feed in unmachined part. 8: Change chuck jaw. 9: Outward transfer of finished part, inward transfer of unmachined part. Bar code marker on workpiece pallets, bar code reader for pallets. 10: Remove unmachined part, sort finished part. 11: Outward transfer/inward transfer of tool holder for cross-machining. 12: Outward transfer/inward transfer of tools. 13: Gripper plate on tool holder in accordance with DIN 69 880 with bar code marker, bar code reader for tools.

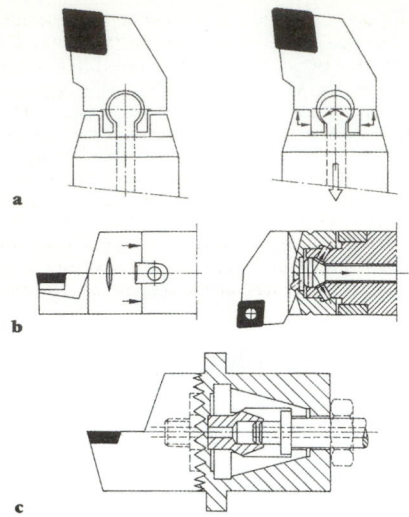

Figure 19. Changeover systems for cutting heads (Traub AG, Reichenbach). **a** Sandvik system, **b** Widea system, **c** Hertel system.

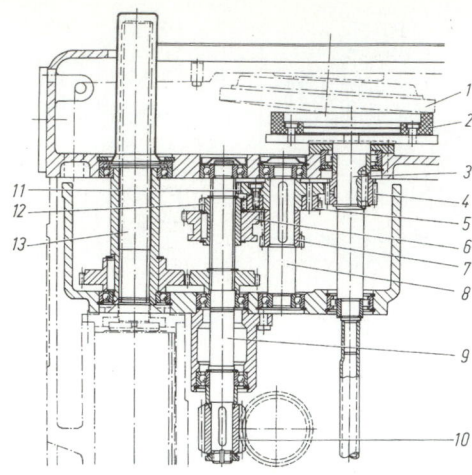

Figure 21. Infinitely variable speed control mechanism in free-standing pillar-type drill (WEBO-Hofheinz, Maschinenfabrik, Düsseldorf). The control range of the two-step gear is 10.4 (with reversible-pole motor, 20.8). The drive shaft *2* drives the intermediate shaft *8* with gearwheels *4* and *11* via the upper friction wheel *1* and the friction ring *2*. Torque is transmitted from the transmission gearing shaft *9* to the drilling spindle *13* via pair of gears *5* and *12* or *7* and *6*. The feed motion is derived from the pinion *10*.

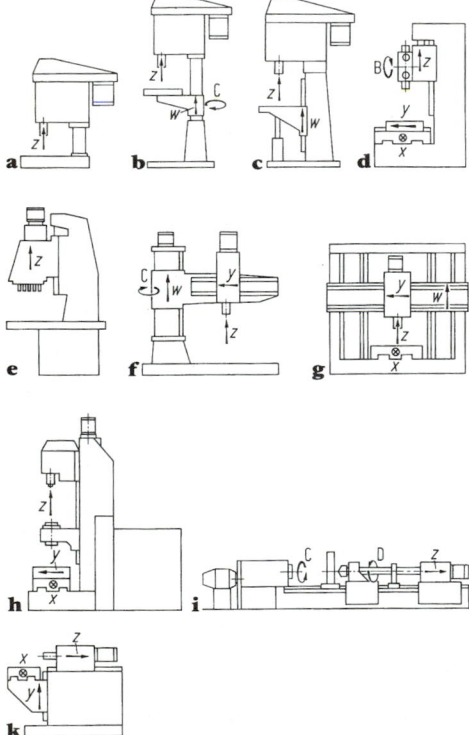

Figure 20. Schematic diagram of different drill constructions (designation of coordinates in accordance with DIN 66 217). **a** Table-type drill. **b** Free-standing pillar-type drill. **c** Column-type drill. **d** Turret drill. **e** Multi-spindle drill. **f** Radial drill. **g** Jig boring machine. **h** Vertical deep-hole drill. **i** Horizontal deep-hole drill. **k** Precision drill.

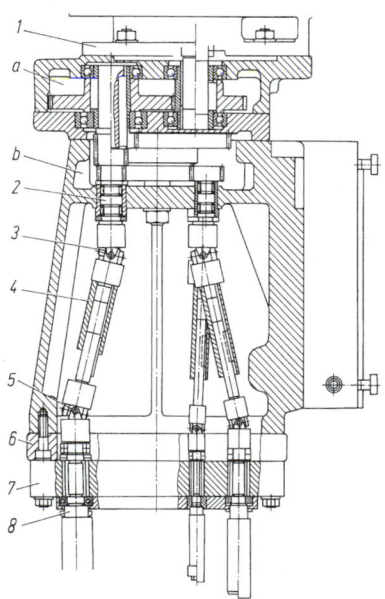

Figure 22. Spindle cutter of a multi-spindle drill with a stationary bearing plate (Bernhard Steinel, Werkzeugmaschinenfabrik GmbH & Co., Schwenningen): *1* flange for motor drive, *a* main gearing, *b* power takeoff gears (articulated spindle drive), *2* driving spindle, *3*, *5* joints, *4* driving bush for length offset, *6* spacer, *7* spindle plate, *8* drilling spindle.

a particular machining sequence is specified with the relevant direction of feed, specifications and optimisation for cutting quality and cycle time. There are single-path and multiple-path machines, rotary cycle machines with transit tables, rotating tables, ring benches or drums. Switched

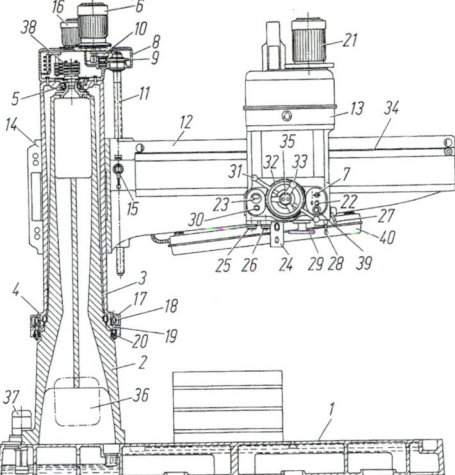

Figure 23. Structure of a radial drill (Raboma type radial boring machine, Hermann Kolb Maschinenfabrik, Cologne). The clamping plate *1* supports an internal free-standing pillar *2*, on which the pipe sleeve *3*, which may be rotated by 360°, is seated on bearings at *4* and *5*. The hoisting motor *6* is switched via *7* to lifting gear *8*, drives the lead screw *11* via a gear *9* with a safety clutch *10* and adjusts the height of the cantilever arm *12* and boring slide *13*. The cantilever lifting stroke automatically cuts out at the end position. The slotted cantilever arm clamp *14* opens and closes automatically before and after the height adjustment of the cantilever arm *12* (lifting stroke control). The lubrication of the lead screw *11* and the nuts, cantilever arm clamp *14* and clamp drive is via a central oil pump *15*. To clamp the cantilever arm *12*, the pipe sleeve *3* is held via a hydraulic clamping device (not shown) to the internal free-standing pillar *2*. The clamping motor *16* drives a gearwheel pump with an integral pressure relief valve. Piston motion is transformed into the rotary motion of the clamping ring *17*, causing eccentrically positioned cam levers *18* to lift the clamping bolts *19*, and the double bevel ring *20* to be activated – axial motion prevents any skew in the cantilever arm during clamping. The drilling motor *21*, switched via *22* and controlled with *23*, drives the drilling spindle *24* via variable gearing, from which the feed drive is derived. The hydraulically switched changeover and preselection of all drilling spindle speeds and feeds are adjusted with knobs *25* and *26*. The hand lever *27* controls the hydraulically switched multiple-disc clutch with an automatic spindle brake to the home position. The clutch operates with automatic delay when switched on for tooth-by-tooth gearwheel control. The starting handle *28* operates the feed clutch with overload protection and disengages the feed wheel *29* when switched on. There is automatic disconnection of the feed drive at both ends of the stroke of the drilling spindle *24*. The pressure switch *30* releases the drilling spindle *24* from the gearing for tool change. Butterfly-type levers *31* disengage the feed drive on the control head and permit rapid manual adjustment of drilling spindle *24*. The graduated dial *32* gives accurate adjustment of the drilling depth with trip range over the entire stroke of drilling spindle (impact boring). The drilling spindle weight is compensated via a counterweight. The boring slide *13* is adjustable via the handwheel *33* on the cantilever arm. Clamping is on a track held via a hardened steel band *34*, and this may be tightened with the clamping shoe eccentrically positioned on the clamping shaft, which is driven by the hydraulic clamping unit contained in the boring slide, with an automatic reset device. The clamping device for the rotating motion of the cantilever arm *12* and the slide *13*, activated by a pushbutton *35*, are interconnected via positively actuated electrical sequence switching. The power supply from the switch cabinet *36* is to the coolant submersible pump *37* and via sliprings *38* to an electrical control unit (not shown) in the rear of the cantilever arm, and thence to operating units and motors. The mushroom-type switch *39* is the "Main OFF" switch. Illumination of the operating field is by the direct-axis lamp *40*.

indexing tables are designed for automatic manufacturing processes. With the modular system, special-purpose drilling machines may be built up from the various drilling spindle and feed units. Multi-spindle drilling heads are employed for the manufacture of specific drilling patterns.

Routing drilling machines in bridge constructions are designed for processing large workpieces. The bridge runs on rigidly mounted tracks of the desired length. Multi-spindle drilling units may be employed with moving drilling spindles. One special-purpose construction is the numerically controlled *pcb (printed circuit board) drilling machine* where performance depends on the number of drilling cycles per time unit which may be achieved.

5.3 Milling Machines

5.3.1 General

Milling machines are characterised by three or more *axes of motion* which are assigned to the tool or workpiece. The position of the axes of motion determines the type of machine. Other criteria determining their classification include kinematic considerations and the frame construction. Technological advantages of specific milling processes and the frequency of their use have led to tried and tested designs (**Fig. 26**), where their characteristics include the main spindle diameter, table surface area, main spindle location and type of control. A distinction is made between *horizontal and vertical milling machines* depending on the position of the main spindle. Small workpieces with complex machining operations are carried out on machines with several types of table motion. In the case of large, bulky workpieces, it is preferable for the feed motion to be executed by the tool. A distinction is made accordingly between *knee-type* and *bed-type milling machines*. The tools are clamped directly or via a cutter arbor. Special-purpose machining tasks may require modifications using rotary tables, dividing heads, angle cutters, precision measuring devices and digital displays.

5.3.2 Knee-Type Milling Machines

These are designed as *horizontal, vertical or universal milling machines* (**Fig. 27**). These are particularly suitable for a variety of machining applications in individual and small-scale serial production because of the simple positioning of the workpiece in all machining directions and favourable accessibility. The machine frame consists of a baseplate and column which accommodates the main drive, main spindle and guide of the knee. Horizontal milling machines have an axially adjustable counter clamp on the column. The knee which carries the cross-slide and milling table is clamped during the milling process. The cross-feed may also be effected via a moving milling spindle box or motion of the spindle sleeve. The individual drive is increasingly tending to supersede the central drive, where the feed is derived from the main gearing and transmission is executed from a telescopic drive shaft to the knee. Heavy-duty knee-type milling machines have permanently fixed main spindles, while universal milling machines have a clamping table which may be rotated about the vertical axis. A vertical or universal milling head may often replace a counter clamp where the basic construction is the same.

Control Systems. A distinction is made between manual control by way of selector switches or pushbutton switches, program control with pre-programmed or freely programmable machining processes and numerical con-

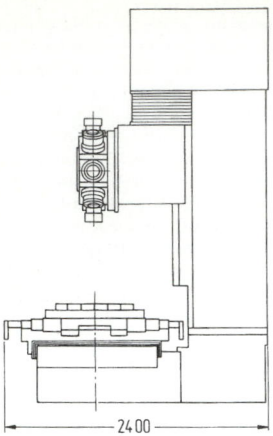

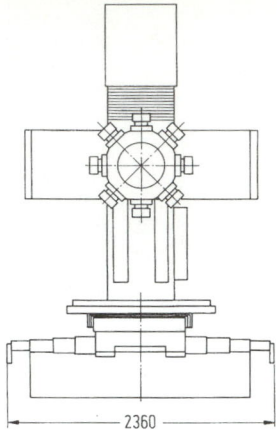

|← 2400 →|

|← 2360 →|

Figure 24. Single-column jig drill with an eight-spindle turret head (Hermann Kolb Maschinenfabrik, Cologne).

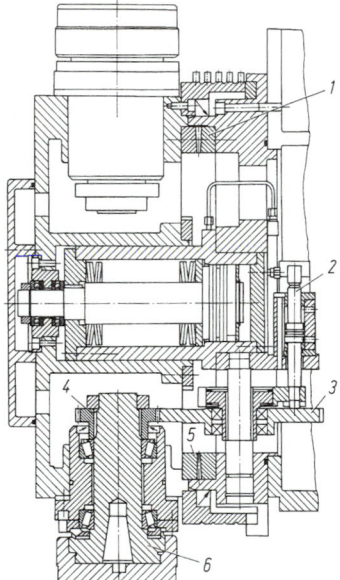

Figure 25. Turnstile turret (Hermann Kolb Maschinenfabrik, Cologne). The drilling spindle drive *6* is via the gearwheels *3* and *4*. The cross-toothed rings *1*, *5* are for indexing. The hydraulically powered motion is via the gearwheel *3* and the piston *2*.

trol with feed motion in all axes effected via d.c. motors and ballscrews.

5.3.3 Bed-Type Milling Machines

These are constructed in a number of different arrangements as *single-column and two-column machines* (**Fig. 26c, d**). The table is supported by way of a rigid machine bed, the milling slides are height-adjustable, while the torsionally stiff construction provides for good absorption of the cutting forces and workpiece weights. Load capacity is high, as is the accuracy of table guide.

Single-Column Bed-Type Milling Machine. With its milling unit guide vertical to the column, this is capable of applications similar to those of the knee-type milling

machine (**Fig. 28**). The main spindle is normally positioned vertically, or occasionally horizontally in the milling unit. The machine bed accommodates the cross-table in two inverted guideways or three-way flat guideways. Modified forms may have the moving machine column in the cross direction and the work table moving in the longitudinal direction. The milling slide is guided on the column and is tilt-free, partially supported and fixed with a clamping device.

Horizontal Plane Milling Machines. These are intended for the economic manufacture of large and long workpieces and are built up from a number of modules graded according to size. The central module is the long machine bed where the work table moves longways only. One or two columns are arranged along the side of the bed, depending on the level of expansion. Up to four milling machines may be employed for simultaneous machining of a number of surfaces, guided at the columns, with an additional arm or on the crossbar. The individual modules may be combined to produce a variety of different constructions. Modification may be based on standard classes of table widths, clamping lengths and clearance heights and equipment with different milling units. The double-column construction is characterised by its high rigidity and flexibility. The two columns which guide the crossbar in the rigid or moving double-column design are connected to the bed and traverse within a torsionally rigid frame. This either runs along guides attached on two sides to the bed or is laid completely separately in the base in machines with moving double columns (gantry style), whereas the workpiece clamping table is stationary. Two synchronised feed drives are required for the positioning motion of the crossbar and double column.

Other design characteristics of horizontal plane machines include replaceable milling units, weight compensation of the crossbar for sensitive vertical advance, infinitely variable speeds and feeds, motorised tool clamping, automatic clamping devices, central lubrication, automatic tool raising and lowering and a central control desk for all functions. Additional attachments provide for overall machining of workpieces in a single tool setting. Milling units with drive motors and switchgear are available in a number of designs, e.g. as slide milling units, tailstock milling units and rotating milling units.

Feed motion is executed via separate infinitely variable d.c. motors and ballscrews, while table or double-column

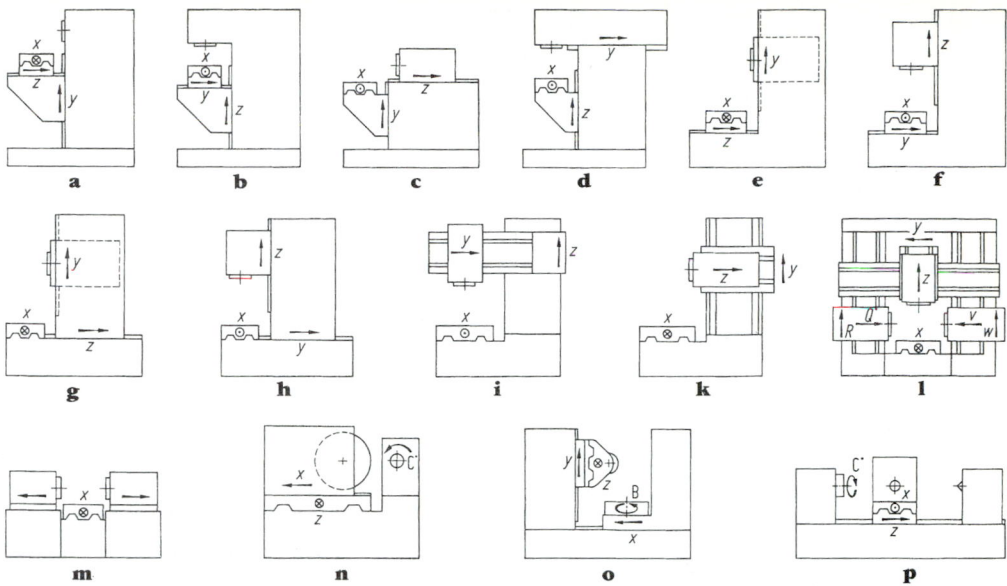

Figure 26. Types of milling machine (coordinate designation in accordance with DIN 66 217). Knee-type milling machines in horizontal and vertical design: **a, b** three axes in the knee or two axes in the knee and one axis in the spindle box; **c, d** bed-type milling machine in horizontal and vertical design; **e, f** two axes in the cross-table and one axis in the spindle box, or **g, h** one axis each in table, column and spindle box. Horizontal-plane milling machines: one axis in the table and two axes each in the horizontal or vertical milling units: **i** single-column horizontal-plane milling machine with cantilever arm; **k** single-column horizontal-plane milling machine; **l** double-column milling machine; **m** cross-milling machine; **n** cylindrical milling machine, with one rotary axis for workpiece feed. Special-purpose milling machines: **o** roller milling machine, with two axes and one rotating motion in milling unit, and two axes (one rotary axis) in table slide; **p** thread-cutting milling machine with one rotary axis for workpiece feed and two axes in the milling slide.

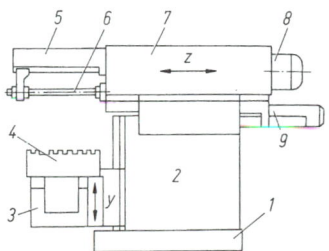

Figure 27. Horizontal knee-type milling machine, now also available with three-axis numerical control (DIAG (now trading as Werner & Kolb) Werk Fritz Werner, Berlin). The table clamping area is 1500×400 mm^2. The power of the main drive P is 12 (or 14) kW. There are 22 speeds of rotation, with $\varphi = 1.25$ from 22.4 to 2800 min^{-1}. Three separate drives give feedrates of from 6.3 to 3150 mm/min. The speed of rapid advance is 8000 mm/min, that in creep gear 2 mm/min. *1* baseplate, *2* column, *3* knee, *4* work table, *5* counter clamp, *6* cutter arbor, *7* spindle box, *8* main drive, *9* feed drive (Z-axis).

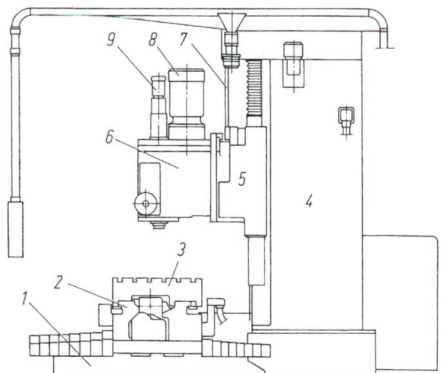

motion is normally effected via a helix or worm cutting gear rack.

Horizontal Milling Machines. These are designed as simplified horizontal-plane milling machines. Columns and projecting beds are normally employed which are bolted to a box bed on one or two sides and accommodate horizontal milling units. The main application is for horizontal-plane machining with cutting heads at high cutting rates in serial production. Feed motion is frequently limited to longitudinal table direction. Program control is standard for automatic tool advance and raising and lowering in

Figure 28. Single-column vertical milling machine (Droop & Rein, Bielefeld). The main spindle diameter is 130 mm. The working space is given by an x-dimension of 2300 mm, a y-dimension of 950 mm and a z-dimension of 800 mm. The table clamping area is 3000×1000 mm^2. The main drive output rating is 30 kW. The 18 speeds vary from 31.5 to 1600 min^{-1}. The feed is infinitely variable, in three ranges, from 3 to 3000 mm. *Frame*: *1* machine bed, *2* carriage, *3* table, *4* column, *5* saddle, *6* hydraulic-powered rotating milling unit, *7* chain for weight compensation. *Drive*: *8* d.c. current main motor, *9* tool clamping motor. *Control*: alternatives include manual input, programmable control, copy control and numerical control.

combination with automatic tailstock or slide clamping. Horizontal milling machines based on the modular system may be set up easily to accommodate specific user requirements when using simple main and feed drives.

5.3.4 Copy Milling Machines

These are suitable for the manufacture of complex three-dimensional shapes which are scanned by sensors from a master or model. The copy control systems differ, depending on the principle of operation or scanning method. Electrical, hydraulic and electrohydraulic systems have been gaining in popularity in recent years. The tool is guided along a curve determined by a model via continuous changes in two or three feed motions arranged vertically in relation to one another. The copy control amplifier generates the feed signals for the actuators from a signal in proportion to the tracer deflection. An automatic line switch with two scan feeds and one sliding feed permits three-dimensional machining with two-dimensional control as well. The model is scanned in lines. After each cycle, it is advanced by the preselected interval between lines. Three-dimensional control is suitable for three scan feeds with either two separate sensors or one special sensor which reacts simultaneously to both radial and axial deflection. Standard bed-type milling machines are frequently fitted with an arm for the sensor device. Special-purpose machines carry out specific copy machining.

5.3.5 Machines for Circular Milling

One of the characteristics of a machine for circular milling is the unit for the circular feed motion for producing cylindrical surfaces. The circular feed may be superimposed by a plunge-cut feed or longitudinal feed. For example, crankshaft circular milling.

Circular milling operations may also be carried out on standard vertical milling machines, a rotary table being required for the circulating feed motion instead of the machine table.

The main spindles of the milling unit in machines for circular milling and the workpiece spindles of the vice are arranged parallel to one another. The workpiece is caught in the centred scrolling jaws. Synchronous drive of the two workpiece spindles is carried out by a worm gear mechanism, vice and adjustable transverse milling unit positioned on a common machine bed. An additional slide is provided for lengthwise adjustment of the milling units and back rest.

5.3.6 Universal Milling Machines

These have a broad range of applications in machine tool construction and the construction of jigs and fixtures. They have a wide, finely-stepped speed and feed range, a high degree of machining accuracy and a variable modular program of additional attachments. It is therefore possible to carry out milling, drilling, turning, broaching and grinding within a single chucking operation. A variety of different clamping and parts devices is used to manufacture complex shapes.

Main Characteristic. The simple basic machine is of knee-type construction (**Fig. 29**). Height adjustment and lengthwise feed are via the knee, cross-feed via the milling headstock. A moving machine column may be added for machining heavy-duty workpieces, also a moving cross-feed milling headstock. All feed motion is assigned to the tool.

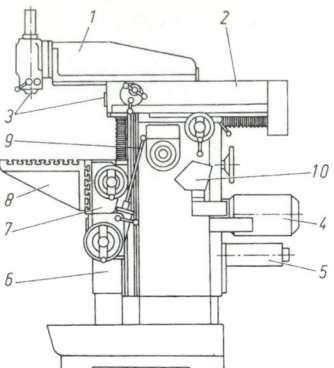

Figure 29. Universal milling and drilling machine tool (Friedrich Deckel AG, Munich). An extended version of the basic model with automatic gear switching and stepper motors for the feed motion, and equipped with numerical control (freely programmable memory via keyboard control or continuous-path control) designed for three or four axes. *1* moving vertical milling head, *2* spindle socket, *3* main spindle with extending spindle sleeve, *4* main drive motor (brake motor), *5* infinitely variable feed motor (d.c.), *6* knee slide with work table guides, *7* table slide, *8* angle table, *9* feed control, reciprocal interlocking with clamps, *10* control desk and high-resolution digital display.

Additional Attachments. These include a counter holder, a vertical milling head, an angular milling head, a rapid advance vertical milling head, a broaching device, a drilling head, a grinding attachment, an angle table (fixed, rotating), a transit table, a rotary table, a rotary table with optical adjustment, a part head, a swaging device, a helical milling attachment, and precision measuring equipment.

5.3.7 Special-purpose Milling Machines

These are intended for a single purpose, or for machining a limited number of different workpieces.

Thread-Cutting Milling Machines. The structure and kinematics of these are similar to those in machines for circular milling. Longitudinal thread-cutting milling machines have the same structure as screw-cutting lathes with an extra milling support. Short thread-cutting milling machines operate with multiple thread section cutters using the plunge-cut process.

Roller Milling Machines. These are used to cut spur gears, worm gears and special toothed gears. Their special characteristic is a roller train of gears and change gear box generating positive motion between workpiece and tool rotation.

Engraving Machines. These operate on the basis of a pantograph which may be moved on a plane or three-dimensionally. The transmission ratio between the sensor tip and milling head may be set by a slide and transmitted to the workpiece by scanning the guide template.

Slotting and Grooving Machines. These produce nuts, keyways, splined shafts and grooves from pre-programmed motions which are normally generated automatically.

Rotary Slotting Machines. These represent a special design for the manufacture of winding slots and air flutes in generator rotors and differ from the above-mentioned machines in view of their size.

Cam Forming and Profiling Machines. These serve to manufacture control cams with end-milling cutters by superimposing workpiece rotation over the milling table longitudinal motion. Control is effected via template scanning.

Cross-cut Milling Machines. These are machines composed of two milling units for two-sided machining of unmachined parts to a given length. Simultaneous centring is also an option.

5.4 Horizontal Boring and Milling Machines

The construction of these with a number of options for motion and adjustment permits the machining of bore holes in true alignment and of rotating bearing front surfaces and the machining of large, bulky items. A broad field of applications has produced a variety of designs and sizes (**Fig. 30**).

Construction. The drives and main spindles are designed accordingly for drilling and milling machining in the main. High speed and feed ranges are characteristic, with four or more axes of motion and a comprehensive range of additional attachments. Simultaneous execution of machining operations independently of one another is possible with a separate spindle bearing and faceplate bearing. Lathes with a high degree of part accuracy and backlash-free table drive systems permit bore holes to be machined in true alignment in the transition process. The structural design of the main spindle system consists of a main spindle bearing with a faceplate permanently attached to spindle bushing, or else the separate location of the main spindle and the faceplate which are then able to rotate at varying speeds. The machine bed in table drilling machines and milling machines accommodates a fixed column, the cross-slide and a steady rest. The spindle box with the retracting main spindle runs laterally along the side column or centrally along the arc-shaped column, and the height is adjustable. The cross-slide normally supports a rotating table. The drill rod is guided inside the adjustable steady rest for heavy-duty and precision boring operations with substantial excess lengths. The main drive is designed for a broad speed range and rapid changes in the direction of rotation, the main spindle for rapid braking. These drives normally take the form of d.c. motors with infinitely variable speed settings and mechanical transmission units switched in sequence. There are individual drives for all axes of motion, each with high rapid advance speed, short acceleration and braking times, a broad speed range and freedom from backlash. The individual machine slides are fitted with clamping devices. The use of this flexible type of machine is mainly restricted by the size and weight of the workpieces.

Routing drilling and milling machines are suitable for all heavy-duty workpieces. All possible settings and feed motion combinations are available in a column and a spindle box. The workpiece is clamped to a permanently fixed plate adjacent to the machine bed. There are two different types of column motion. In simple designs, the column is pushed along the bed at right angles to the main spindle axis. In order to guide the tool to the workpiece, the main spindle may be adjusted axially. An extended area of applications is achieved by way of the cross-motion of the column. In bearing sleeve drilling and milling machines, the main spindle is supported within a rigid extending sleeve bearing. As a result, greater absorption of cutting forces is possible with a larger spindle drilling surface area. Angular milling heads may be attached to the bearing sleeve. A wide variety of special-purpose constructions with rotating spindle boxes or turning columns have emerged for special operations.

More recent developments in horizontal drilling and milling machines have combined the advantages of these two types. The development of the machining centre is currently underway whereby adaptation to the machining task is carried out by both the tool and the workpiece.

5.5 Machining Centres

These are designed for the numerically controlled machining of complex prismatic workpieces in one tool setting. In addition to the various machining operations to be carried out, there are a number of manufacturing processes which are available, i.e. drilling and milling. Machining centres have a high level of flexibility in addition to a high level of automation. Typical is automatic operation encompassing both the machining sequence with all the path and switching functions and the rapid repeatability of individual machining processes and tool change from the corresponding size of magazine. The latest developments in machining centres include pallet-change devices, drilling head changers and tool magazine changers.

Machining centres have at least three numerically controlled linear axes, which may be supplemented by two rotary axes (**Fig. 31**). The type of assignment of the axes of motion to the workpiece carriers (table) or tool carriers (spindle) determines the type of machining centre. A horizontal or vertical position of the main spindle is available. The constructions of horizontal machining centres have emerged as a result of the different assignment of the axes of motion. **Figure 32** shows the most common types.

The main characteristics of the machining centre include the type and number of controlled axes, cutting power, speed and feed range, length of operating paths, table area and table load application, resolution, positioning tolerance and number of tools in magazines.

Tool Storage. Turrets or magazines are available. Up to 40 tools may be stored at full capacity in the disc-shaped revolving magazines. Larger magazines are normally designed as chaining magazines. In addition to the basic

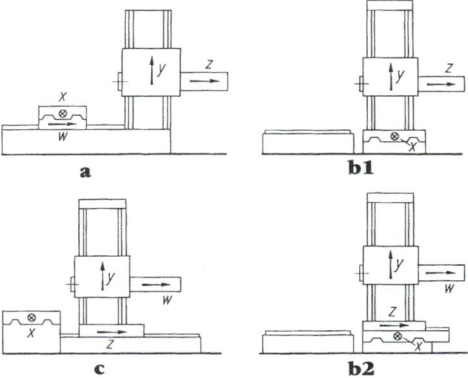

Figure 30. Schematic diagram of basic constructions of horizontal drilling and milling machines (coordinate designation in accordance with DIN 66 217): **a**, table-type drilling and milling machine; **b** routing drilling and milling machine, **1** standard design, **2** with column moving crosswise and endwise; **c** two-way design (Planer type).

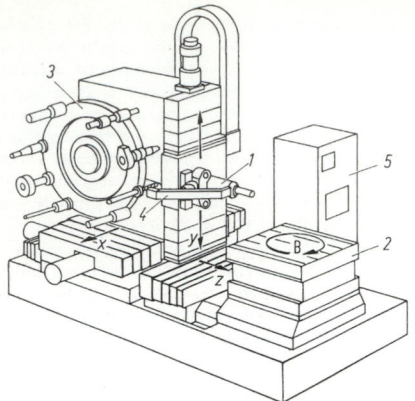

Figure 31. Definition of the direction of motion on a four-axis machining centre (DIAG (now trading as Werner & Kolb) Werk Fritz Werner, Berlin): *1* main spindle, *2* rotating table, *3* tool magazine, *4* tool changer, *5* numerical control.

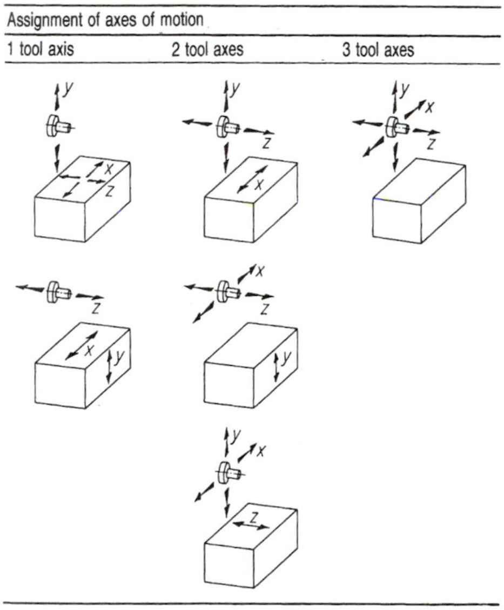

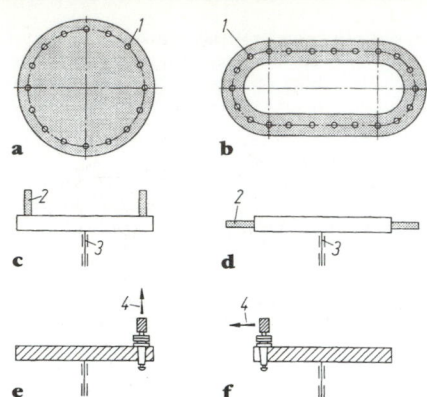

Figure 33. Basic types of tool magazines. *Constructions*: **a** revolving magazine, **b** chain magazine. *Position of tools*: **c** drum-shaped, **d** turnstile-shaped. *Direction of assembly*: **e** parallel, **f** vertical. *1* Tool location. *2* Tool. *3* Magazine axis. *4* Direction of assembly.

Assignment of axes of motion		
1 tool axis	2 tool axes	3 tool axes

Figure 32. Construction of horizontal machining centres.

shape of the magazine, the location of the tool and the direction of motion for assembly in relation to the magazine axis is also significant. A basic distinction may be made between turnstile and drum-shaped tool arrangements and parallel and vertical motion for assembly (**Fig. 33**).

The construction and arrangement of the magazine determine the kinematics of the tool change. Tool recognition is carried out via tool coding (the individual tools are coded), location coding (the magazine locations must be labelled with the tools in the correct order), or, since the introduction of CNC control, via electronic accounting (combination of both types of coding).

Dead time is affected by tool change time. This is dependent on the number of motions required (degree of freedom) and the size of the paths or angles which have

to be covered during a tool change. Simple changing units handle one tool each and are, therefore, only economic to employ in combination with a tool turret or as auxiliary grippers, as they carry out the entire tool change process, i.e. including the selection motion of the turret without any intermediate storage capacity in dead time. Double tool changers remove one tool from the main spindle and the magazine or intermediate store and reverse the two positions simultaneously. This enables shorter tool change times to be achieved.

Figure 34 shows a design where the arms of the tool changer are positioned at an angle of 90° to one another. This enables tools to be changed in a short time from a revolving magazine either attached at one side or on the column. In **Fig. 35** a tool change system is shown which is known as a universal unit and employs the principle of the single tool changer in combination with a double one. The tool change may also be carried out directly from the magazine. The correct motion of the magazine in the

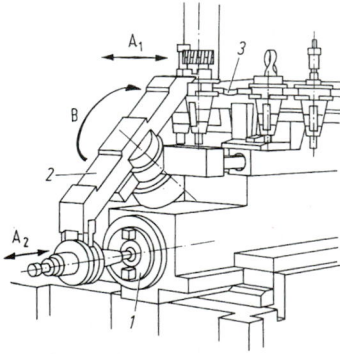

Figure 34. Double gripper for automatic tool change (Friedrich Deckel AG, Munich). The function sequence of the tool changer *2* begins with the gripper taking one tool each from the magazine *3* and the main spindle *1* respectively. The tools are removed from the main spindle and magazine simultaneously via motion in the direction of the Z-axis. It then swings through 180° to synchronise assembly motion in the opposite Z-direction with the release of the gripper finger. A Removal and motion for assembly in the direction of the Z-axis, A_1 on the magazine, A_2 on the main spindle. B Rotating motion.

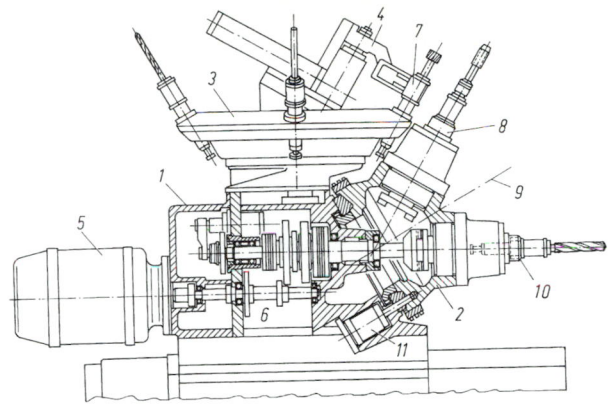

Figure 35. Universal machining centre (Burkhardt & Weber GmbH & Co. KG, Reutlingen). The housing *1* supports the double rotating head *2*, the revolving magazine *3* with space for 30 tools, the tool changer *4*, the main drive *5*, where $P = 11$ kW output, and the stepper gear mechanism *6*. To change tools, the tools are gripped by the gripper *7* and changed at point *8* between the double rotating head and the magazine. Where a new tool is needed, the double rotating head carries out a rotation about 180° around the axis *9*, so that the positions of points *8* and *10* are reversed. Indexing of the rotating head is carried out via piston *11*.

direction of the main spindle axis is a basic requirement for this.

Dead times for loading and unloading and setting up and clamping the workpieces may be run parallel to productive time where tool change devices are in use. Where pallet change systems are used the workpieces clamped on the pallet are brought to the working space via linear or rotary motion of the pallet in turn and subsequently taken out of this space. Pallet storage systems enable a number of workpieces to be prepared and thereby to provide for automatic tool changes within longer production periods (**Fig. 36**).

5.6 Planing, Shaping and Slotting Machines

5.6.1 Planing Machines

Single-Column Planing Machines. These provide for machining of bulky items (**Fig. 37**). Constituent modules are: bed, column and height-adjustable arm, with, where required, a removable auxiliary column for support. Workpieces protruding at the side may be held by a roller support bed arranged at the open side. The main drive of the table is by d.c. motor with clutch gearing and gear rack, or else hydraulic. The tool slides are arranged on the arm and column.

Two-Column Planing Machines. These have a closed double-column frame with a height-adjustable crossbar (**Fig. 38**). The tool slides are recommended for use on the crossbar, although they are also provided on the columns in large-scale machines.

Additional Attachments. Milling or grinding attachments (**Fig. 39**) may be placed on the crossbar with their own separate main and feed drives.

5.6.2 Shaping and Slotting Machines

Horizontal Shaping and Slotting Machines. These are employed primarily for machining surfaces on small- to medium-scale workpieces (**Fig. 40**). The cutting motion of the tool is carried out by the ram, guided along the top side of the baseframe. The workpiece is supported by a table arranged at the top end of the baseframe *1*. The ram *2* is driven either hydraulically or mechanically via a sliding block. The tool is taken up by the rotating bit-retaining head *3* which places the tool in a vertical position. The bit vent lifts the tool from the workpiece on

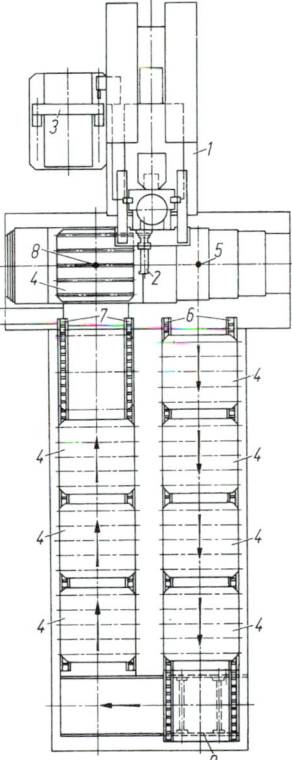

Figure 36. Pallet storage system (Hüller Hille GmbH, Ludwigsburg). Machining centre *1* is equipped with main spindle and tool *2* and tool store *3*. The pallets *4* are taken up and transported internally by the work table in the machine. During pallet changeovers, the pallet moves with the machined workpiece to point *5* and is shunted onto roller conveyor *6*. The next pallet leaves the roller conveyor *7* and is attached to the work table at point *8*. The pallets run through the pallet store in one direction only, transported via power-driven rollers. The bulkhead trolley *9* is used to transfer the pallets from roller conveyor *6* to roller conveyor *7*.

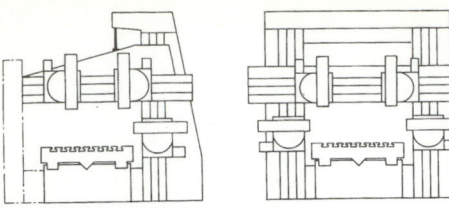

Figure 37. Single-column planing machine.

Figure 38. Double-column planing machine.

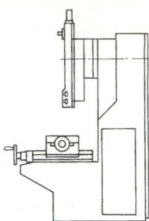

Figure 41. Vertical shaping and slotting machine with single-column and bed.

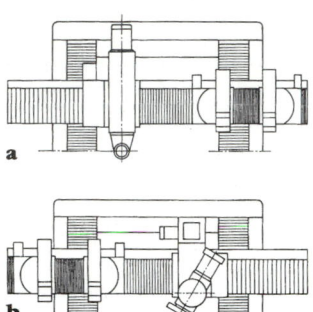

Figure 39. a Planing machine with milling attachment. **b** Planing machine with grinding attachment.

the return of the ram. The position of the ram stroke in relation to the table may be adjusted via a moving spindle. The sliding block *4* is connected at the top with the ram and at the bottom with a fixed point in the baseframe *1*. The sliding block wheel *5* drives the sliding block via a sliding block counterweight *6*. The rotating motion of the drive motor is transmitted to the sliding block via a cascade gear mechanism. The workpiece to be machined is clamped to the table *7* fitted with T-bolts which may be rotated around a horizontal axis in relation to the receiving table *8*. The feed motion is carried out by the receiving table on the support *9* in a cross direction. The support is height-adjustable.

Vertical Planing Machines. In these, the ram carries out either vertical motion or motion in two directions. In

smaller vertical planing machines (**Fig. 41**) columns and beds consist of one element and in larger machines these are composed of two parts. The workpiece is carried by a cross table or additional rotary table. In smaller machines with stroke lengths of up to 630 mm, the mechanical drive is predominantly used in conjunction with gear indexing mechanisms and crank rotary belts. Hydraulic drives with larger stroke lengths have advantages over mechanical drives with continuous infinitely variable speeds and impact-free control.

5.7 Broaching Machines

These are classified as *external* or *internal*, depending on the broaching process, and *vertical* or *horizontal*, depending on the position of the main axis. *Chained* and *special-purpose* machines are considered to be special constructions (e.g. for use in transfer lines). Classification is carried out in terms of size, cutting motion and standardised according to main dimensions and supply ratings. Advantages of vertical construction include low space requirement, no torsional stress on broaching tool as a result of dead weight, better coolant lubrication effect and good conditions for integration in transfer lines. Advantages of horizontal construction are low set-up height, potentially larger stroke lengths, simpler infeed of heavy-duty workpieces, and absence of any requirement for excavated foundations or an operating platform. The machining accuracy of broaching machines depends primarily on slideway of tool or workpiece. **Figure 42**

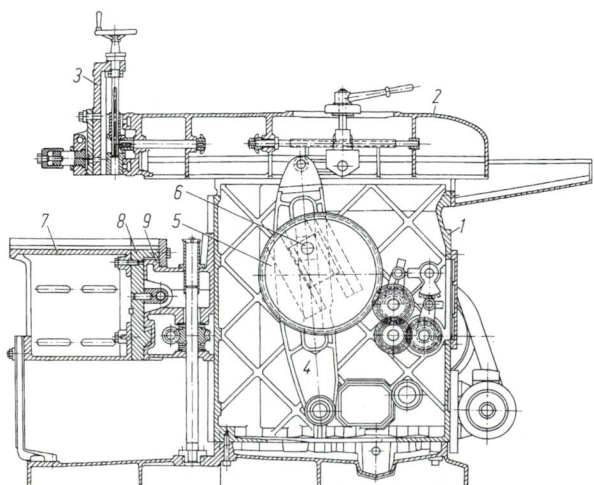

Figure 40. Horizontal shaping and slotting machine (Schlenker & Cie GmbH, Werkzeugmaschinenfabrik, Hornberg): *1* baseframe, *2* ram, *3* bit retaining head, *4* sliding block, *5* sliding block wheel, *6* sliding block counterweight, *7* table, *8* receiving table, *9* support.

shows the structural design of a vertical external broaching machine.

Broaching machines are employed in mass production. Chaining of broaching machines and transfer lines is possible using automation and feed systems. The frames of broaching machines have to be designed in such a way that they have high static and dynamic rigidity. The requirements are suitable ribbing, cell-type design and welded construction with damping surfaces.

In internal broaching machines dual-cylinder construction has the advantage over single-cylinder design in that the propelling force and the guides are on one level, so that the machine is subjected to less bending stress. In external broaching machines, bending oscillation is generated where the column oscillates vertically in terms of the direction of broaching. It is possible to obviate this problem with an extremely rigid joint between the column and the broaching machine.

Cutting motion in broaching machines is normally mechanical or hydraulic. Mechanical drives are employed in all chained broaching machines and in some cases in horizontal external broaching machines.

A variety of types of table are installed in broaching machines depending on the machining task.

Clamping devices are required only in exceptional cases for internal broaching; normally simple receivers or receiving lugs are enough to secure the workpiece. A workpiece receiver *3* is shown for internal broaching in **Fig. 43**. The workpiece *2* is pre-centred using three pins *1*, while positioning is carried out by inserting the broaching needle.

The clamping devices for external broaching are of a more complex structure. The main clamping system has

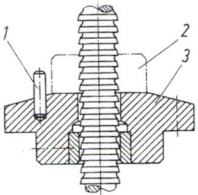

Figure 43. Workpiece receiver for internal broaching (Kurt Hoffmann, Maschinenfabrik, Pforzheim): *1* centring pin, *2* workpiece, *3* workpiece receiver.

to be self-locking, so that the clamping force is not reduced where there is a loss of operating power. A self-locking support system is shown in **Fig. 44**. The bolt *1* is pressed against the workpiece initially and secured by wedge *2* and bolt *3*.

Important potentially hazardous areas on broaching machines include the open, moving workpiece, the motion of the slide and parts table, and the motion of the loading and clamping equipment. The tool and cam strip are concealed beneath the working space; an enclosed superstructure is not normally required and may cause a hindrance during tool changes. Two-handed operation is common. Accidents due to the uncontrolled motion of tables are prevented by setpoint position control of those machine parts which are moved. Noise abatement is effected via the correct selection of the drive units and active insulation on installation of the broaching machine.

5.8 Sawing and Filing Machines

5.8.1 General

These separate and generate cuts and cutouts with flat or uniaxially curved surfaces in workpieces made of metal, wood, glass, ceramics, stone and plastics. Multiple cutting tools are made of tool steel, rapid-machining steel or hard metals. The rotary or linear, continuous or oscillating cutting motion is carried out by the tool, as is the feed motion. The area of application of sawing machines covers the entire range of single, serial and mass pro-

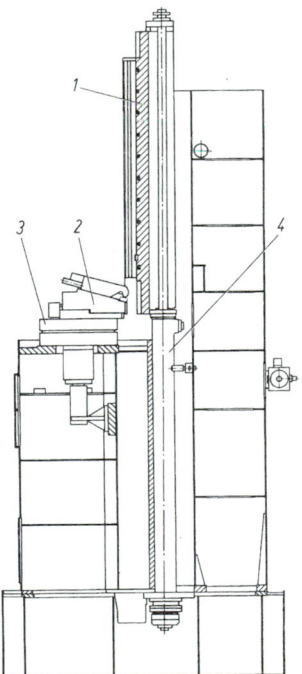

Figure 42. Overall layout using vertical external broaching machine as an example (Kurt Hoffmann, Maschinenfabrik, Pforzheim): *1* tool slide, *2* workpiece table, *3* rotating table, *4* hydraulic cylinder.

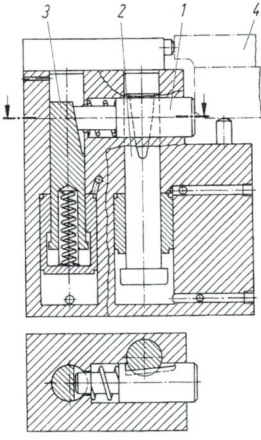

Figure 44. Self-locking workpiece support system (Kurt Hoffmann, Maschinenfabrik, Pforzheim): *1* bolt, *2* wedge, *3* securing bolt, *4* workpiece.

duction. Filing machines are used predominantly in tool, jigs and fixtures and equipment construction.

Friction cutting and blanking machines and electrical discharge machining, e.g. hot-wire cutting machines, are only related to sawing machines from the kinematic point of view; however, in terms of productions systems they are viewed as metal removal machines.

Classification. Sawing machines are classified in terms of cold and hot sawing machines. Circular, bandsawing and filing and bandsawing and power hacksawing machines for filing are classified in terms of kinematics. Other characteristics determining their classification include designation of the tool used, type of drive and control and level of automation.

5.8.2 Circular Sawing Machines

A variety of constructions (**Fig. 45**) are categorised in terms of the direction and type of feed motion which may be horizontal or vertical and linear or curved and may be carried out by a *sawing slide*. This consists of the sawing blade shaft for carrying the tool, gearing, drive motor, housing and guides, and is normally driven hydraulically via a cylinder and piston or hydraulic motor and ballscrew drive. Automatically activated feed adjustment prevents excess loading of the tool and machine specifically when sawing variable cross-sections, such as T-beams for example, by limiting the pressure in the hydraulic system and thereby the feedrate where the feed force is increased.

The self-activating workpiece loading which is required for automatic sawing machines is normally generated hydraulically, where the feed may only be switched on at the end of the clamping process (**Fig. 46**). In order to be able to cut badly deformed workpieces such as warm sheared billets smoothly, the vertical clamping device is designed as a float device in some machines. Hoisting mechanisms at the sectional clamping point prevent the sawing blade from becoming stuck after completing the cut. The clamping chucks are designed as parallel, horizontal or V-type constructions. Multiple clamping is more economical.

The material supply (workpiece feed) takes place on rollers or roller blocks which are activated as part of the automatic sequence of motion of the machine, normally against a rotating stop. The largest workpiece dimensions for circular and square-shaped materials and sections are dependent on the diameter of the cutting blade. The range of cutting speeds for steel and cast metal materials is between 5 and 40 m/min with HSS blades and between 60 and 200 m/min for hard metals. When sawing non-ferrous metals, 500 to 1600 m/min is standard. The maximum feedrate for steel and cast iron materials amounts to approx. 1250 mm/min and 2300 mm/min and more for non-ferrous metal machining. The required drive rating for large-scale sawing blade diameters of $d = 1120$ mm amounts to approximately 55 kW.

5.8.3 Bandsawing and Filing Machines

These are designed in horizontal and vertical constructions. The most important attributes include roller guides for the saw or file bands and their arrangement inside the machine frame or saw frame (**Fig. 47**). There are usually two rollers (three for wide clearances), with a coating (e.g. rubber) to preserve the tooth offset, to guide the continuous welded saw or file band or file chain. One of the rollers may be rotated to provide smooth band operation within narrow tolerances. The saw or file band normally runs between tempered or hard metal guides directly in front of and behind the cut. The band speed is mainly infinitely adjustable via belt drives and achieves rates of between 10 and 1200 m/min and more in universal machine applications. Drive ratings of under 1 kW are required for small cutting speeds at over 1000 m/min and 4 kW and over particularly for frictional cutting. Mechanical or hydraulic and sometimes numerically controlled feed units are employed depending on the machining task.

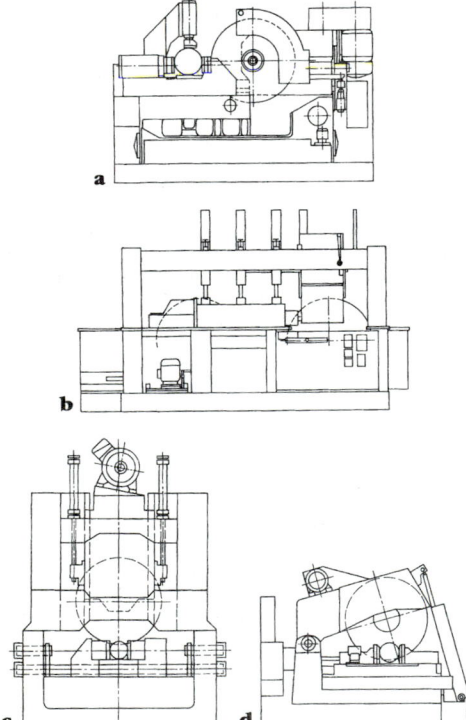

Figure 45. Designs of cold circular sawing machines. **a** Circular saw cutter with horizontal feed motion (Heller, Nürtingen). **b** Longitudinal cross-section circular sawing machine with horizontal feed motion (Treenjaeger, Euskirchen). **c** Circular sawing machine with vertical feed motion (Ohler, Remscheid). **d** Circular sawing machine with arc-shaped feed motion (Kaltenbach, Lörrach).

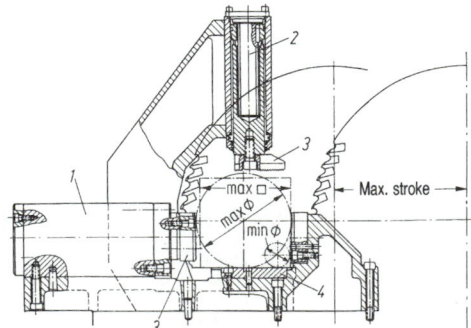

Figure 46. Hydraulically operated workpiece clamping device for cold circular saw (Heller, Nürtingen): *1* hydraulic horizontal clamping cylinder, *2* hydraulic vertical clamping cylinder, *3* case-hardened, removable clamping chucks, *4* workpiece support.

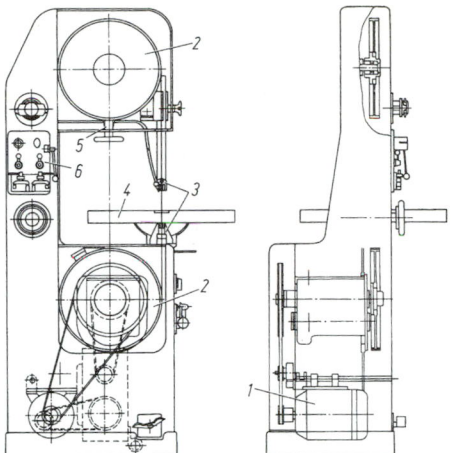

Figure 47. Belt saw and filing machine with two belt rollers and vertical belt guide (Mössner, Mutlangen): *1* drive motor and belt pulley, *2* belt rollers, *3* belt guide (removable and adjustable), *4* tilting table, *5* adjustment control for belt tension, *6* belt weld attachment.

5.8.4 Machines for Power Hacksawing and Filing

There are power hacksaws with horizontal cutting motion and machines for power hacksawing and filing with vertical cutting motion.

Power hacksaws are fitted with a ballscrew drive to generate oscillating stroke motion (**Fig. 48**). The feed motion is normally generated by force due to weight, or by hydraulics in automatic machine tools. The saw blade is raised during the rapid return with the relevant equipment. The double stroke frequency is between 30 and 150 min^{-1}, with an installed drive rating of up to 6 kW.

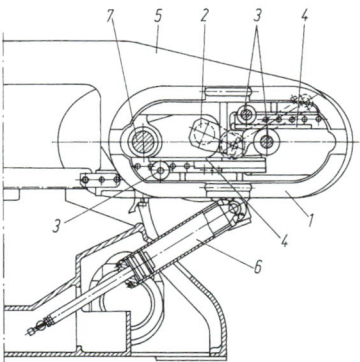

Figure 48. Drive of power hacksaw machine for curved shaping and cutting motion (Stolzer, Achern): *1* saw frame, *2* operating crank, *3* guide rollers, *4* guide rail, *5* saw bow, *6* hydraulic cylinder, *7* point of rotation of saw frame. *Kinematic generation*: The stroke motion of the saw frame *1* is generated with the operating crank *2*. The frame is guided with the rollers *3* along the guide rails *4*. The pressure in the operating stroke and lift in the return stroke of the saw bow *5* are achieved via the operation of the hydraulic cylinder *6*, where the rotation takes place about point *7*.

5.9 Grinding Machines

5.9.1 General

Classification. The main criterion is the type of surface produced on the workpiece. On *horizontal plane grinding machines* (surface grinding machines), flat surfaces are produced, while circular cylindrical surfaces are processed on *cylindrical grinding machines*. Other constructions include *screw thread grinding machines*, *gear grinding machines* and *form grinders* for machining specific forms specified by forming tools at will, and *copy grinding machines* for generating any form of surface via mechanical control of the feed motion.

Other criteria for classification include the position of the area to be machined, so that a distinction is made between *external* and *internal cylindrical grinders*, the most effective surface of the grinding wheel, which gives *cylindrical* and *surface grinders*, and the type of feed motion, which includes *face* and *cross grinders* (plunge-cut grinding machines). A further distinction is made, according to the type of tool carrier, between *centred grinding machines*, *clamping grinding machines* and *centreless cylindrical grinding machines*; depending on the area of application, these are known as *abrasive cutting machines*, *grinding machines*, *finishing machines* and *roughing machines*, where the latter are normally distinguished depending on the machining accuracy which may be achieved and the maximum metal removal rates over time.

5.9.2 Surface Grinding Machines

These are designed with a horizontal or vertical main spindle and a longitudinal or rotating table. Sectional wheels are normally used with rotary tables. The oscillation grinding method is frequently employed. Backlash-free prestressed roller bearings and electromechanical drives with infinitely variable d.c. motors and ballscrew drives are required for full-contour deep grinding operations.

Figure 49 shows a *longitudinal surface grinding machine* with a horizontal main spindle. The machine bed carries the sliding and roller bearing cross-slide for the longitudinal and cross motion. The feed motion is carried out electromechanically via controllable three-phase or d.c. drives which are either stepped or infinitely variable. The longitudinal table is driven hydraulically or electromechanically and is infinitely variable. The rapid-advance feed motion is generated via three-phase asynchronous motors or via solenoids with synchronous or stepper motors for rough or finish feed. The speed of the grinding wheel is frequently infinitely variable via statically or dynamically operating frequency transformeres.

Figure 50 shows a *rotary table surface grinding machine* with a horizontal main spindle. Alternatively, this may be constructed using a vertical spindle and sectional grinding wheels.

A number of design principles for surface grinding machines with horizontal main spindles and alternative motion assignment are shown in **Fig. 51**.

5.9.3 Cylindrical Grinding Machines

The *Norton process* is employed for short and medium-length workpieces with a vertical grinding headstock and workpiece moving lengthwise along the grinding wheel. Long workpieces are ground on machines based on the *Landis process*. The headstock with the wheel is thereby moved along the stationary workpiece, which reduces the length of bed required.

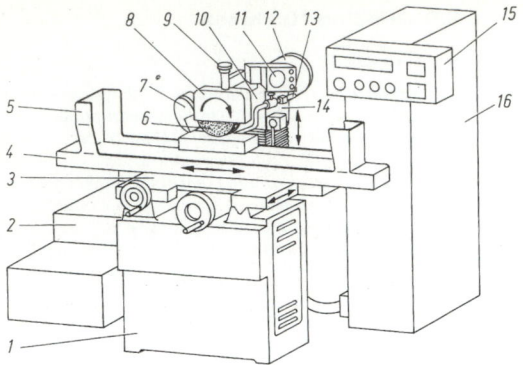

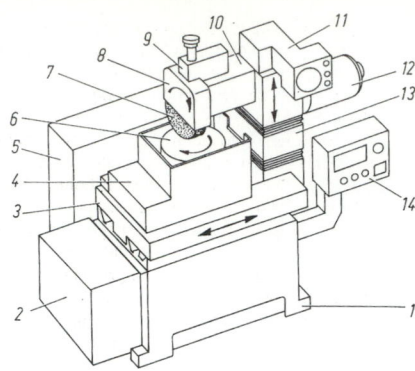

Figure 49. Longitudinal surface grinder with horizontal main spindle: *1* machine bed, *2* coolant lubrication pot, *3* cross-slide, *4* grinding table, *5* spray protection, *6* grinding wheel, *7* exhaust device, *8* protective cover, *9* trimming device, *10* main headstock, *11* feeder arm, *12* main spindle drive motor, *13* coolant lubrication supply, *14* free-standing pillar, *15* control panel, *16* control cabinet.

Figure 50. Rotating table surface grinding machine with horizontal main spindle: *1* machine bed, *2* hydraulic unit, *3* cross-slide, *4* spray protection, *5* control cabinet, *6* rotating table, *7* grinding wheel, *8* protective cover, *9* trimming device, *10* main headstock, *11* feeder arm, *12* main spindle drive, *13* free-standing pillar, *14* control panel.

The machine bed is frequently designed as a welded construction on small and medium-scale machines and the main spindles are positioned in hydrodynamic multi-surface friction bearings designed for circumferential speeds of the grinding wheel of up to 60 m/s. The machines are also frequently designed for plunge-cut as well as longitudinal grinding; *angled plunge-cut grinding* is especially common in mass production, where a lengthwise orientation of the workpiece cannot be avoided. A measuring control system is part of the standard equipment in medium-scale serial and mass production. There are a

number of longitudinal measuring heads for both automatic and manual positioning.

The main headstock or table in *universal cylindrical grinding machines* may be rotated for tapered grinding and a rotating internal grinding spindle may be accessed for machining drill holes.

The external diameter, rough shoulders or internal diameter of a workpiece may be machined on *NC cylindrical grinding machines*, and a variety of radii and forms may be ground using continuous path control. **Figure 52** shows the controlled axes of an NC external cylindrical grinding machine. The machining axes X and Z are assigned to the main headstock, the control of the diameter measuring head is carried out in the V-axis and the tool centre point of the grinding space is set via the W-axis. The length measuring head is guided to the shoulder to be measured via the U-axis.

Centreless Cylindrical Grinding Machines. These machines are employed in highly automated environments in mass production. The operating range is between 0.1

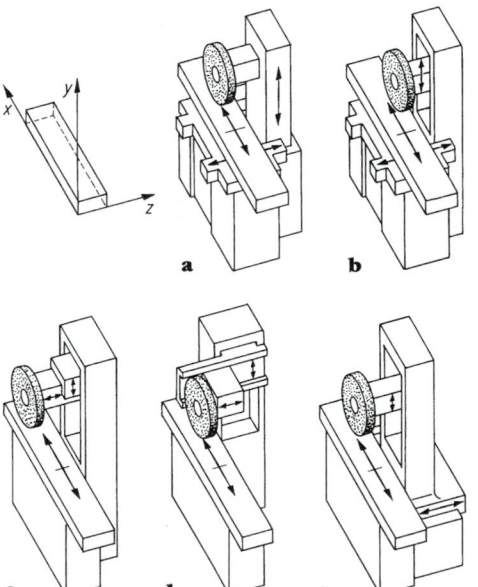

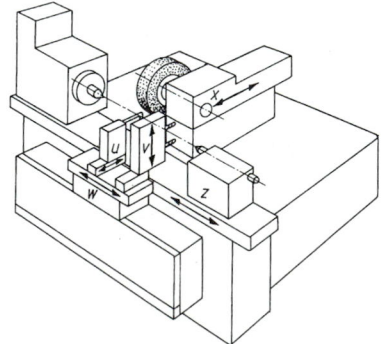

Figure 51. Design principles of surface grinding machines with a horizontal main spindle. **a** Support design I, plunging pillar, cross-slide. **b** Support design II, stationary pillar, cross-slide. **c** Traversing design I, stationary pillar, internal arm. **d** Traversing design II, stationary pillar, external arm. **e** Pillar slide design, table base with integrated pillar.

Figure 52. Controlled axes of NC external cylindrical grinding machine (Schaudt GmbH, Stuttgart-Hedelfingen): *X*-axis (main headstock), *Z*-axis (workpiece slide), *U*-axis (cross-positioning length measuring head), *V*-axis (control of diameter measuring head), *W*-axis (longitudinal positioning of diameter measuring head).

and 400 mm. The workpieces lie lengthwise on a receiver between the *grinding wheel* and *regulating wheel*. As the machining forces are absorbed efficiently by the regulating wheel adjacent to the workpiece, which is normally the same size as the grinding wheel, long or thin workpieces may also be ground without bending or torsional stress occurring at high rates of metal removal (**Fig. 53**). The *main headstock* is bolted rigidly to the machine bed, while the control headstock and workpiece receiver clamp are arranged on a slide which carries out the infeed motion. In designs with a main and control headstock slide arrangement, the grinding slide carries out the infeed motion and the workpiece receiver is stationary inside the machine bed and is only height-adjustable. This principle is applied above all for heavy-duty workpieces, so that the loading equipment does not require adjustment. The regulating wheel and grinding wheel are either floating or reversible. The regulating wheel is rotated by a small angle to generate a longitudinal feed for the workpiece with centreless grinding in continuous operation. Most of the machines are also capable of centreless plunge-cut grinding.

The diameters and widths of the grinding wheels go up to 650 mm, the regulating wheel always having a smaller diameter than the grinding wheel. Hydrodynamic multi-surface friction bearings are normally employed for both spindles. The spindle of the regulating wheel is frequently driven by an infinitely variable d.c. motor via a worm and a worm gear.

Automatic feed, loading and unloading devices are also integral to the machine tool system and chaining is also available. Other functions are also automatic, such as trimming the grinding wheel, offsetting the grinding wheel wear on the trimming attachment, and compensating for the trimming factor with the precision feed and measuring control system, in addition to workpiece transportation.

Internal Cylindrical Grinding Machines. The most important modules of an internal cylindrical grinding machine with an additional surface grinding attachment are shown in **Fig. 54**. The cross-slide *19* for the infeed motion with the drive motor *10* of the main spindles *11* is located on a crossbar *20* rigidly bolted to the machine bed. The lengthwise and feed motion is carried out via the table *1* on friction bearings in the machine bed onto which the workpiece headstock *3* is rigidly mounted. The spindle box may be rotated on the backing plate *16* to grind taper bores. The workpiece spindle is driven via an asynchronous motor *24* and sliding gear transmission. The surface grinding attachment *8* may be rotated into pos-

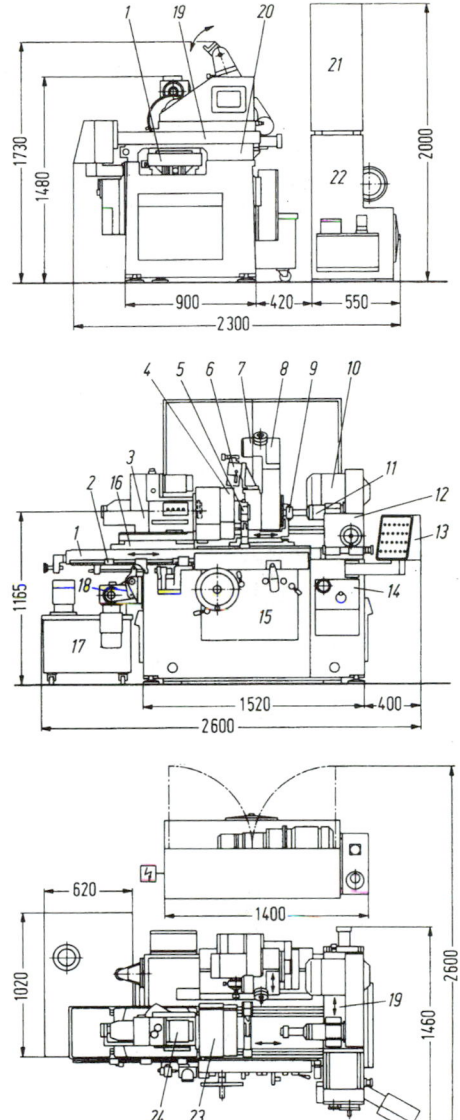

Figure 54. Internal cylindrical grinding machine assemblies (formerly Jung Schleifmaschinen, H. Gaub, Berlin): *1* longitudinal slide, *2* table stops, *3* workpiece spindle box, *4* clamping chucks, *5* trimming device for bore grinding, *6* trimming device for crown grinding wheel (surface grinding attachment), *7* crown grinding wheel for surface grinding attachment, *8* swivel surface grinding attachment, *9* grinding wheel for internal cylindrical grinding, *10* main spindle drive motor, *11* main spindle, *12* grinding support, *13* control desk, *14* feed system for cross-slide, *15* switch cabinet, *16* backing plate, *17* coolant lubrication reservoir, *18* short-stroke attachment, *19* cross-slide, *20* crossbar for cross-slide, *21* electrical switch cabinet, *22* hydraulics cabinet, *23* protective hood, *24* drive motor for workpiece spindle.

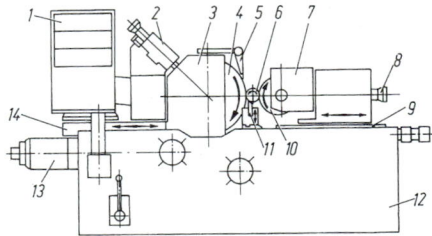

Figure 53. Centreless cylindrical grinding machine: *1* control desk, *2* trimming device for grinding wheel, *3* main headstock, *4* grinding wheel, *5* coolant lubrication supply, *6* workpiece, *7* regulating wheel headstock, *8* trimming device for regulating wheel, *9* regulating wheel headstock slide, *10* regulating wheel, *11* supporting arm, *12* machine bed, *13* feed device, *14* grinding headstock slide.

ition for machining, may be moved in a lengthwise direction on a bed baseplate and may be rotated by up to 10° to grind surface areas on workpieces with taper bores. A short stroke attachment *18* is employed to generate a low

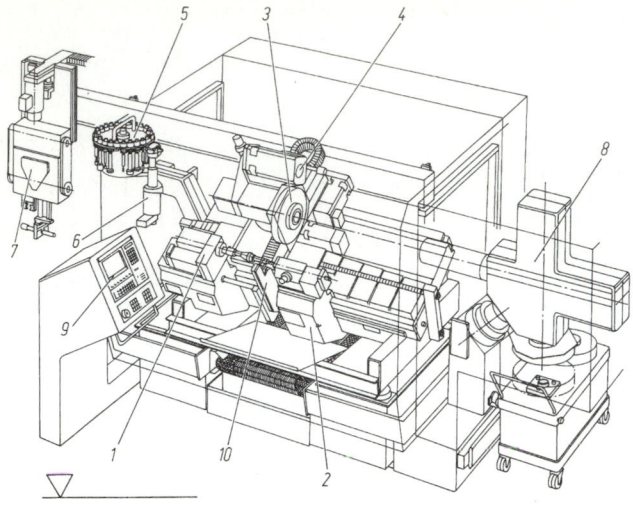

Figure 55. Flexible cell-type grinding centre for external, internal and form grinding in single chucking (Schaudt Maschinenbau GmbH, Stuttgart): *1* workpiece headstock, *2* tailstock, *3* external grinding spindle, *4* internal grinding spindle, *5* magazine for internal grinding points, *6* grinding point changer, *7* double-column loader for workpiece change, *8* automatic grinding wheel changer, *9* CNC control for up to five axes, *10* measuring system for continuous diameter and length measurement during grinding process.

level of oscillation in the table of up to 6 mm, which, in turn, is operated with an eccentric press driven by a brake motor via a worm gear.

5.9.4 Screw Thread Grinding Machines

Machines are classified according to single or multi-sectional lengthwise grinding and multi-sectional plunge-cut grinding of screw threads.

5.9.5 Gear Grinding Machines

Machines are classified according to the processes to be carried out by them. Part and continuous rolling processes are called after their inventors, i.e. Maag, Niles, Kolb and Reishauer.

5.9.6 Development Trends

The introduction to *high-speed grinding* and the latest technology requires higher static and dynamic rigidity in machines, high-powered drives, improved spindle bearings, coolant lubrication systems and safety devices, such as equipment for rapid balancing at operating speeds. There is a clear trend in favour of plunge-cut grinding with boron nitride and diamond wheels in the grinding machine range.

Other development trends include higher, infinitely variable peripheral speeds for workpiece and grinding, improved fine-feed systems, roller and hydrostatic guides, increasing employment of diamond trimming rollers, infeed, loading and unloading equipment designed for the relevant workpiece and improved control systems (measuring controllers, NC controllers, AC systems). Machines for external and internal grinding have emerged as part of the overall machining process, as is the case with the machine shown in **Fig. 55**, where both headstocks are arranged parallel to one another on a rotating table.

5.10 Honing Machines

5.10.1 Long-Stroke Honing Machines

Figure 56 shows the classifications of long-stroke honing machines. The horizontal design is normally employed for

manual honing machines. The rotating motion is executed by the honing tool and the typical stroke required for honing is carried out by hand. Production times are minimised by way of mechanically generated stroke motion. Honing of external cylindrical areas is performed by rotating the workpiece via the main spindle. The stroke motion with an external honing tool may be manual or mechanical. The horizontal arrangement is also employed for machining large-scale workpieces, especially for internal honing of long pipework. The rotating and stroke motion may be assigned to the workpiece or the tool, or both.

The required motion in vertical long-stroke honing machines is executed by the tool. The main spindle drive is powered from the three-phase motor by way of an infinitely variable frictional gear mechanism and V-belt. The stroke motion is carried out hydraulically via a cell pump with a variable volumetric flow. The stroke length, stroke position and stroke rate are infinitely variable. Standard peripheral speeds for a honing tool are between 15 and 40 m/min; axial speeds are between 12 and 25 m/min. The corresponding honing stone contact pressure is between 20 and 200 N/cm^2 for honing stones made of aluminous abrasive or silicon carbide, or between 200 and 350 N/cm^2 for cuboid crystalline boron nitride (CBN) and between 300 and 600 N/mm^2 for sintered diamond honing strips. Long-stroke honing machines are designed for

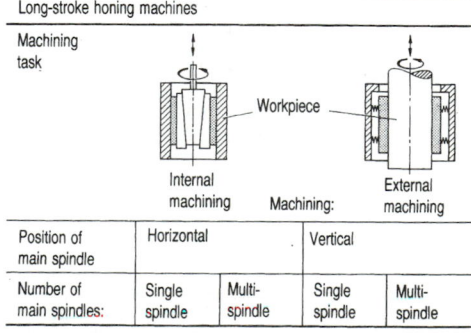

Long-stroke honing machines				
Machining task				
Position of main spindle	Horizontal		Vertical	
Number of main spindles:	Single spindle	Multi-spindle	Single spindle	Multi-spindle

Figure 56. Classification of long-stroke honing machines.

machining drill hole diameters of up to 1200 mm and lengths of up to 12 000 mm (24 000 mm with repositioning).

Mechanical, pneumatic, hydraulic or electrohydraulic motion may supply the infeed of the interlocking honing tool. Graduated feed initiates machining of the tightest points and areas of a bore first. Infeed time intervals are variable and may be adapted to the prevailing conditions. Compliance with specific machining diameters is ensured by using a programmable interval timer or mechanical or pneumatic measuring devices. These measuring devices have the advantage that the wear of the honing stone is taken into account.

Clamping of the honing tool and workpiece is carried out in such a way that there is offset motion for minimal axis displacement. This may also be achieved by suspending the honing stone in an oscillating position. This is in a "floating" position in small or medium-sized workpieces or where the workpiece can withstand slight pressure, and the honing stone is connected rigidly to the honing spindle. Infeed attachments for the workpieces simplify loading and chaining in long-stroke honing machines in transfer lines. The external honing tool is clamped to the operating table and the workpiece is connected to the main spindle for external honing with long-stroke honing machines.

5.10.2 Short-Stroke Honing Machines

Short-stroke honing machines (**Fig. 57**) are also known as *superfinishing machines, superhoning machines* or *orbital grinders*, and are used for machining internal and external surfaces.

Figure 58 shows the structure of a centreless short-stroke honing machine in continuous operation. An oscillating head *1* is used to generate the sinusoidal oscillating motion of the honing stones and is guided by a toolholder to two pillars connected with a crossbar. Height adjustment of the oscillating head is by means of a ballscrew. In continuous operation, machining is normally carried out with a number of honing stones of different grades and hardnesses positioned behind one another so that the surface quality is improved gradually. The honing stone is carried and guided in stone guides *2* which are installed in the required number on the oscillating head. Each stone guide consists of a cylinder via which a piston effects the

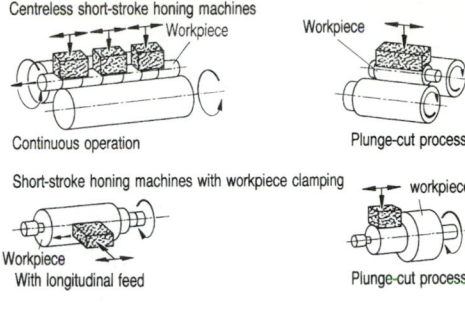

Centreless short-stroke honing machines

Continuous operation   Plunge-cut process

Short-stroke honing machines with workpiece clamping

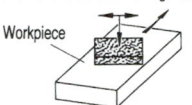

With longitudinal feed   Plunge-cut process

Short-stroke surface honing machines

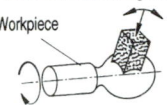

Short-stroke form honing machines

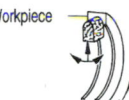

Internal short-stroke honing machines

Workpiece

Figure 57. Classification of short-stroke honing machines.

rise and fall of the honing stone. Control may be pneumatic or hydraulic. Contact pressure is individually infinitely variable, depending on the machining task. The workpieces are moved with a defined feed motion running parallel to the oscillating axis below the stone guides via transport rollers *3* moving in the same direction. Parallel continuous operation is made possible by the form and adjustment capacity of the transport rollers (angle of incline 0.5 to 2°). In special cases, e.g. short-stroke honing of tapered rollers, the transport rollers are not angled back

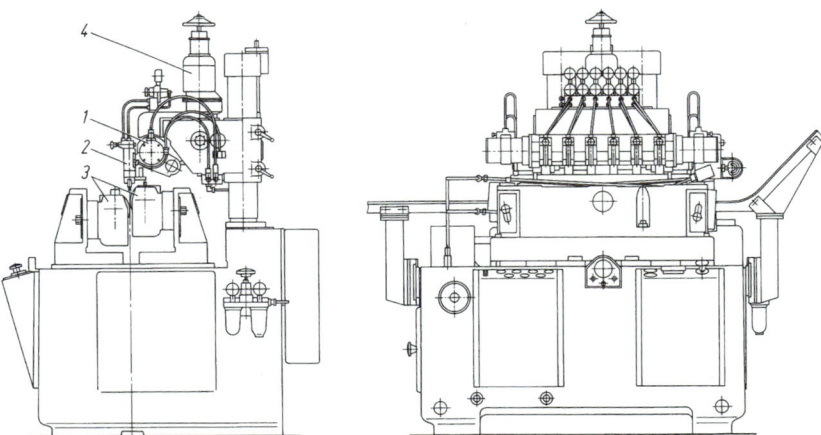

Figure 58. Structure of centreless continuous operation short-stroke honing machine (Supfina, Remscheid): *1* oscillating head, *2* slotted link, *3* transport rollers, *4* mechanical oscillation generation.

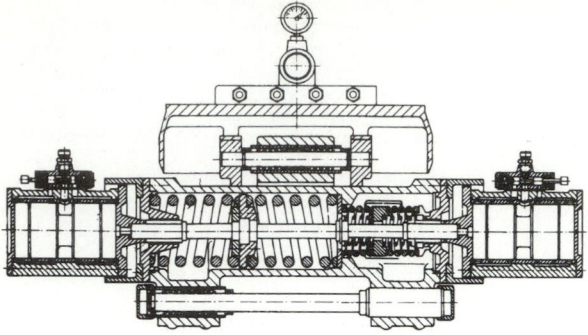

Figure 59. Oscillating head of pneumatic drive driven via two exciter pistons (Supfina, Remscheid).

to back and continuous operation is provided via a suitable roller form. Oscillation frequencies are between 4 and 45 Hz, and stone contact pressure is approx. 25 to 100 N/cm².

The oscillating motion may be generated mechanically by translating rotary motion into linear motion via an eccentric shaft. Another principle is shown in **Fig. 59**. The oscillating head thereby consists of a triple-mass oscillating system with a pneumatic drive driving two self-synchronising exciter pistons. Frequency and amplitude adjustment is carried out by way of a pressure relief valve with a vibration damper downstream. The stroke length is infinitely variable within a range of ± 15 mm.

Internal short-stroke honing machines are normally equipped with clamping devices which are modified for the workpiece and machining task. There are also special-purpose machines for short-stroke honing of flat and curved surfaces. There are short-stroke honing machines which may be designed for attachment to other machine tools, e.g. turning machines for precision machining in single-part manufacture and small-scale serial production. These additional attachments are equipped with one or more stone guides and clamped to the tool carrier instead of the tool. Longitudinal and plunge-cut machining is possible. The generation of the cutting motion is carried out on the basis of the principles described.

5.11 Lapping Machines

5.11.1 General

Mechanical lapping is classified as shown in **Fig. 60**.

5.11.2 Single-Wheel Lapping Machines

See **Fig. 61** for the basic structure. A baseframe accommodates the lapping table, consisting of an underbody fitted with a lapping wheel or lapping segments, depending on the size of the machine. Abrasive wheels supported by lateral guide arms run along the lapping table. These sharpen the lapping wheel continuously during the main operating process as a result of the friction conditions created by the turning motion (*friction coupling*). This ensures even wear of the lapping tool. The abrasive wheels also serve as carriers of the workpieces, which are laid in the machine without any clamps. Where necessary, the workpieces are hydraulically or pneumatically loaded with additional weight to maintain the desired pressure for the lapping process. The infeed of the lapping abrasive is carried out via one or more supply reservoirs. Turbomixers and agitators provide a homogeneous mix for the lapping abrasive (normally a suspension of water and

Type of machine	Arrangement of main spindle	Machining task	Characteristic machine size
Single-wheel lapping machine	Vertical	Workpiece  Surface lapping	Lapping wheel diameter
Twin-wheel lapping machine	Vertical	Workpiece	
Triple-wheel lapping machine	Vertical	With parallel faces Workpiece 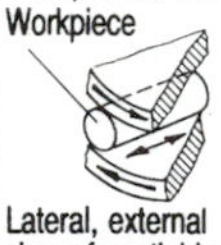 Lateral, external circumferential lapping with line contact	
Spherical lapping machine	Vertical Horizontal	Workpiece Lapping of ball bearings	
Internal and external cylindrical lapping machines	Vertical	Workpiece 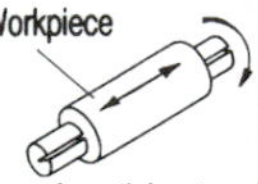 Circumferential external cylindrical lapping	Diameter and length of internal or external surface for lapping
	Horizontal	Workpiece  Circumferential external cylindrical lapping with surface contact	
Oscillating lapping machine	Vertical	Workpiece 	Travel paths
	Horizontal	Internal and external surfaces, as desired	

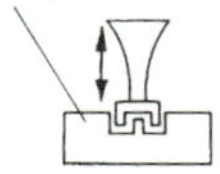

Figure 60. Classification of lapping machines according to DIN 8589 T15.

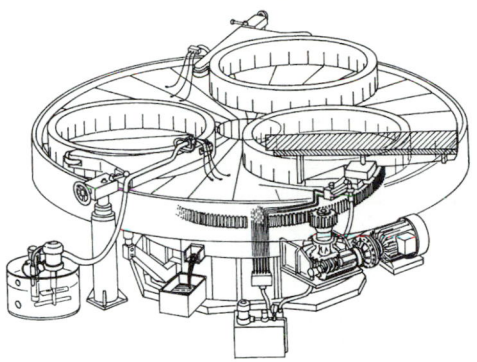

Figure 61. Single-wheel lapping machine (Waldrich Coburg, Coburg).

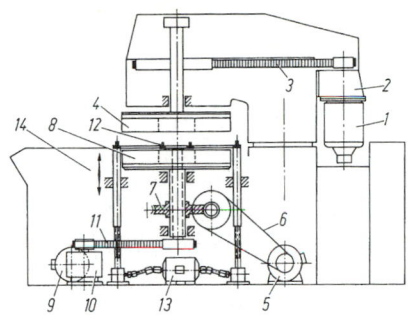

Figure 62. Structure of twin-wheel lapping machine (Peter Wolters Maschinenfabrik GmbH & Co., Rendsburg): *1* motor for top lapping wheel drive, *2* gear mechanism, *3* toothed belt drive, *4* top lapping wheel, *5* motor for bottom lapping wheel drive, *6* V-belt drive, *7* worm gearing, *8* bottom lapping wheel, *9* motor for workpiece drive, *10* worm drive, *11* toothed belt drive, *12* internal gear ring, *13* motor for lowering external gear ring, *14* external gear ring lowering device.

lapping powder, i.e. aluminous abrasive, silicon carbide or boron carbide).

Control devices and jets in the lapping abrasive circuit provide a metering system and may be adapted to suit the individual machining task. Single-wheel lapping machines are designed with lapping wheel diameters of between 350 and 5000 mm.

5.11.3 Twin-Wheel Lapping Machines

These are used for surface lapping, lapping with parallel faces and external cylindrical lapping (**Fig. 62**). Lapping wheel diameter is between 250 and 1000 mm. Up to four separate coupled or decoupled speed drives are available for all types of motion. Speeds may be stepped or infi-

nitely variable. The base contains the drives for the bottom lapping wheel *8*, for the internal gear ring *12* and for lowering the external gear ring *14*. The bottom lapping wheel is arranged on a thrust bearing which is prestressed via grooved ball bearings and cup springs and is driven by a three-phase motor *5* with a V-belt *6* and worm gearing *7*. The top lapping wheel is arranged to oscillate, and the top main spindles are arranged on a pivoting arm. Workpiece drive is for *positively driven operation* of workpieces, resulting in even wear of lapping wheels and a high degree of uniformity of workpieces *9*, *10*, *11*. The

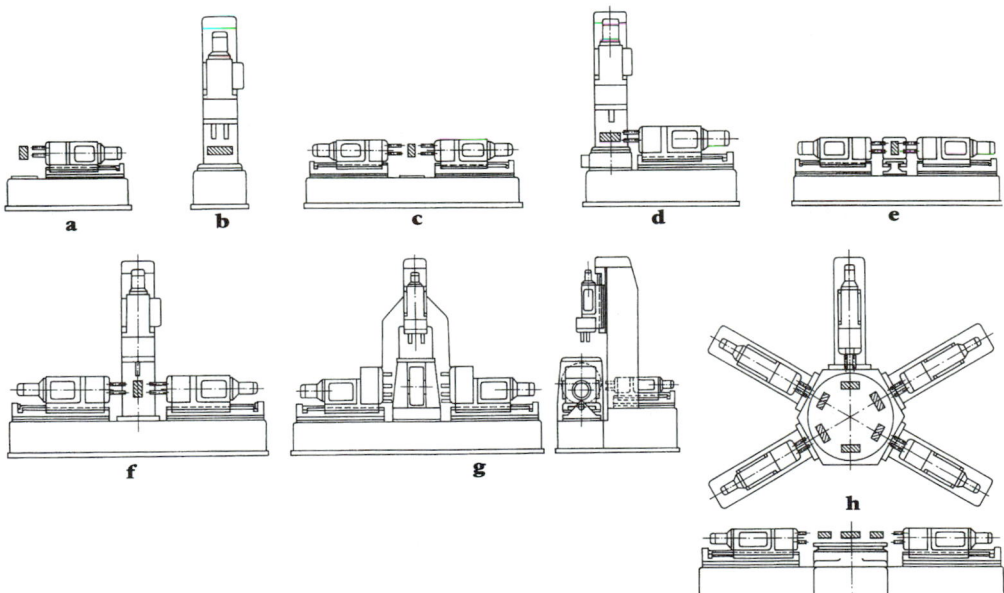

Figure 63. Arrangement of modules in special-purpose machines. **a** Single-path machine, horizontal, with permanently fixed table or with horizontal linear feed motion. **b** Single-path machine, vertical, with permanently fixed table or with horizontal linear feed motion. **c** Two-path machine, horizontal, with permanently fixed table or with horizontal linear feed motion. **d** Two-path machine, horizontal and vertical, with permanently fixed table. **e** Three-path machine, horizontal, with permanently fixed table. **f** Three-path machine, horizontal and vertical, with permanently fixed table. **g** Three-path machine, horizontal and vertical, with vertical rotary feed motion. **h** Five-path machine, horizontal and vertical, with horizontal rotary feed motion with five work stations and one loading/unloading station.

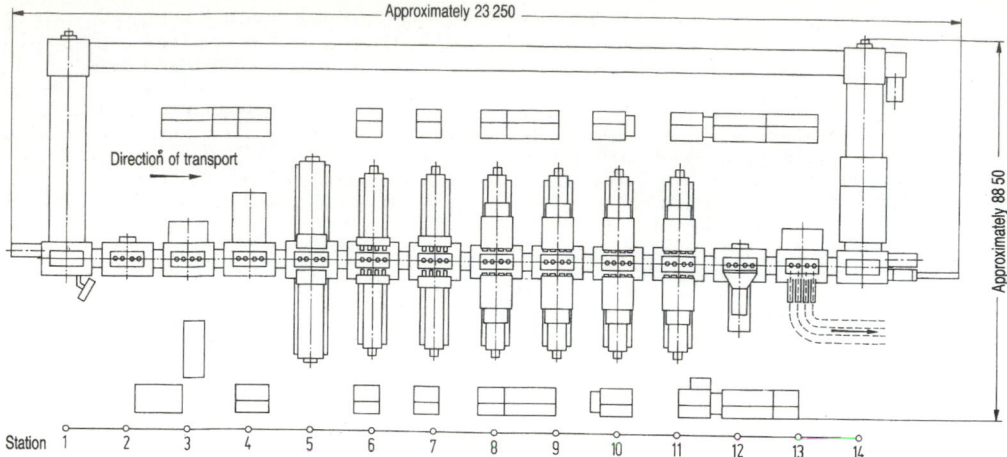

Figure 64. Fourteen-station transfer line for machining articulated shafts (Hüller Hille GmbH, Werkzeugmaschinen, Ludwigsburg).

workpiece carriers arranged between the lapping wheels are driven by the internal gear ring and roll around the external gear ring, which is normally stationary. The workpieces are thereby moved onto cycloidal paths between the lapping wheels. The drive shaft for the internal gear ring is guided by two angled ball bearings arranged around the Z-axis in conjunction with a needle-type bearing. The external gear ring is lowered to change the workpiece carrier. The lapping pressure is built up hydraulically or pneumatically. The pressure level may be varied via a PLC system.

Specific workpiece dimensions and tolerances are normally maintained using a programmable interval timer system. The lapping time required is set on the basis of his-

toric values. An additional further option is an indirect measuring control system. Here the distance between the two lapping wheels is measured and the machine is switched off having reached a set level. This type of control is often employed in association with greater admissible workpiece tolerances, because the lapping abrasive film thickness and the lapping wheel wear are included as the measurement.

5.11.4 Triple-Wheel Lapping Machines

Triple-wheel lapping machines have two bottom lapping wheels arranged adjacent to one another, with a diameter of between 380 and 810 mm. The top lapping wheel may be accessed and rotated in turn over both bottom wheels.

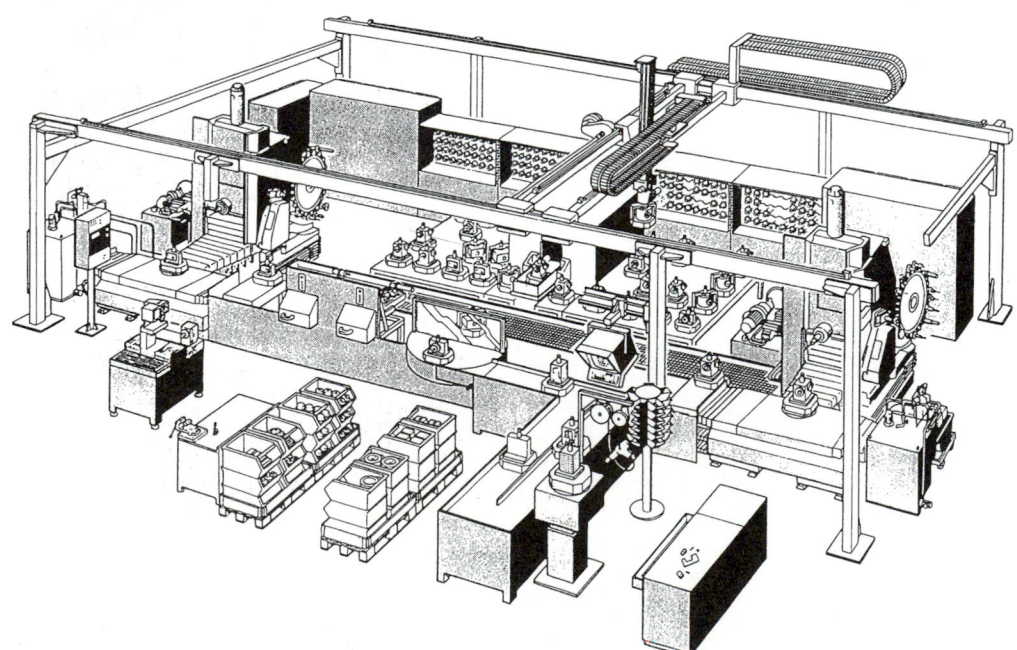

Figure 65. Flexible manufacturing system with two machining stations for prismatic workpieces (Fritz Werner, Berlin).

When a pair of wheels are in operation, the disengaged lapping wheel may be loaded and unloaded.

5.11.5 Spherical Lapping Machines

Spherical lapping machines are constructed in much the same way as twin wheel lapping machines. Only one lapping wheel is driven in this design, with a diameter of between 100 and 1200 mm; its arrangement may be vertical or horizontal. One of the two lapping wheels normally has concentric V-shaped grooves, whose dimensions match the dimensions of the ball bearings to be lapped and semi-circular wear is generated during lapping. As the ball bearings have to pass between the lapping wheels several times in a different groove each time, infeed attachments are required which also perform a mixing function at the same time. Preliminary machining operations employ bonded abrasive wheels, while precision machining uses cast lapping wheels and freely flowing lapping abrasive suspension.

5.12 Multi-machine Systems

Special-purpose machines and *transfer lines* are employed to obtain high productivity in mass production. The structure takes the form of modular components with standardised main and fixing dimensions, which may be combined to produce special-purpose machines of differ-ent designs in horizontal, vertical or any angular arrangement. **Figure 63** shows a number of forms of single and multi-path machines, machines with stationary tables and with linear or rotary workpiece infeed motion. A transfer line is shown in **Fig. 64** for machining articulated shaft parts. The tools pass through the machining, washing and dust extractor stations in continuous operation and transport from station to station is simultaneous and fully automatic. The cycle time is determined by the station with the longest machining time.

With the growing variety of parts and falling unit volumes in serial production, requirements for flexibility in manufacturing systems are increasing. Manufacturing machines with tool magazines and tool change systems are employed in stations in the *flexible transfer line*, as a result of which production of a wide range of parts may be carried out, instead of single-purpose machines with a fixed arrangement of tools.

Flexible cell-type manufacturing systems are employed in serial production of average lot sizes which contain workpiece and tool storage systems as well as handling, measuring and monitoring systems in addition to the numerically controlled machining centre for prismatic or rotary-based parts. *Flexible manufacturing systems* are created by networking these types of cell-type manufacturing system, where work stations substitute or supplement one another and provide non-cycled manufacture independently of lot sizes. **Figure 65** shows a flexible manufacturing system with two work stations for prismatic workpieces.

6 Welding and Soldering (Brazing) Machines

L. Dorn, Berlin

(For welding and soldering see F1.1 and F1.2.)

6.1 Arc Welding Machines

Requirements. A number of specific electrical requirements have to be satisfied relating to the welding power source for arc ignition and maintenance of the welding arc in welding operations:

- High no-load voltage compared with arc burning voltage (safe ignition).
- Rapid restoration of supply following drop short circuits (rapid reignition).
- Short-circuiting current slightly higher than welding current (low spatter welding).

Static Characteristic. This describes the change in the source voltage U in relation to the level of the welding current I (**Fig. 1**). The volt–ampere settings (operating point A) generated during welding correspond to the point of intersection between the static characteristic setting ($1, 4$) and the arc characteristic ($2, 3$) rising as the arc length increases.

Where a steeply falling curve is produced (4), changes in arc length (voltage) produce small changes in current. This assists in providing an even heat supply in manual arc welding. In internal submerged arc welding using thicker wire, the voltage change is exploited to maintain a constant arc length with an adjustable feed motor (known as "external control").

With a shallow volt–ampere output curve 1, small changes in the arc length (voltage) cause large changes in current. Where consumable electrodes with a higher electrode current density are employed, the melt rate changes according to the heat output of the arc, and consequently so does the arc length with a constant-speed wire feed. This is used in gas-shielded metal-arc welding and submerged arc welding with thinner wires to maintain a constant arc length with variations in the torch-to-work distasnce (known as internal control).

Dynamic Characteristic. This describes the power source characteristic with different short-term load characteristic changes, such as those generated on ignition and in drop short circuits (**Fig. 2**).

In power sources with excessive initial current surge I_{kst}, the electrode tends to adhere to the workpiece on ignition, and with drop short circuits large amounts of

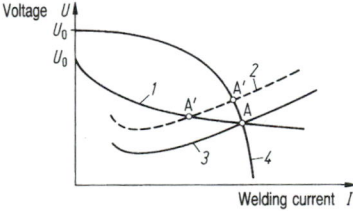

Figure 1. Change in operating point by increasing arc length with steep and flat static power source characteristic: *1* flat characteristic, *2* long arc, *3* short arc, *4* steep characteristic.

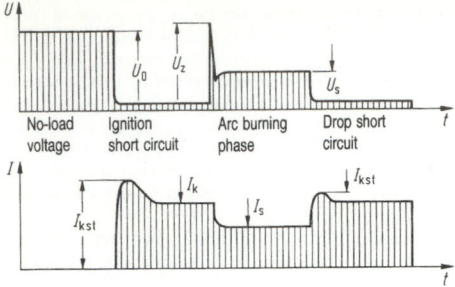

Figure 2. Time-based progression of current and voltage with short-term changes in loading as a result of contact ignition and drop short circuits.

spatter are generated; also, where the initial current surge is too low, the heat generation is not adequate to ensure reliable ignition. The voltage on restoration following dipping should return to the full no-load voltage U_0 as soon as possible in order for the power source to provide reliable reignition. In power supplies with high inductance levels in the welding current circuit, e.g. welding rectifiers with constant current booster, there is a rapid rise in voltage to above U_0, whereas in rotating welding generators the no-load voltage is only reached after a delay. However, it is also possible to generate a short-term voltage surge U_z directly on restoration following dipping with structural modifications, thereby guaranteeing rapid reignition.   In *a.c. welding* the arc is interrupted in rapid succession even without drop short circuits, and has to be reignited. The rapid voltage rise required for this may be achieved using *welding transformers* with open-joint restrictors or special magnet core combinations which produce a rectangular-shaped voltage curve in contrast to any of the usual sinusoidal forms. An extremely rapid voltage rise is achieved by superimposing high-frequency, high-voltage pulsing (3000 V) using arc ignition and stabilisation equipment.

Setting Range of Welding Current. This is obtained from the point of intersection between the arc characteristic and the static characteristics at the highest and lowest setting level.

The following approximations apply for the arc characteristic in the form of equations with numerical values:

Coated electrodes	$U = 20 + 0.04I$.
TIG welding	$U = 10 + 0.04I$.
MIG welding	$U = 14 + 0.05I$.
Submerged arc welding	$U = 20 + 0.04I$ or
	$U = 14 + 0.05I$.

Welding Current and Welding Duty Cycle [*ED*]. The power supply is not normally permanently under load during welding, instead the loading is discontinuous, corresponding to a specific relative setting time:

welding duty cycle

$$= \frac{\text{time under load}}{\text{time under load} + \text{break time}} \times 100\%$$

This is why the admissible welding current for the machine heating system is not specified just for a welding duty cycle of 100% (continuous operation), but also for a welding duty cycle of 35 or 60%, in accordance with standard practice in manual welding operations. Where

the welding duty cycle in an operation differs from this, the maximum current may be obtained as follows:

$$I_2 = I_1 \sqrt{ED_1/ED_2}.$$

Admissible No-Load Voltage. No-load voltage is limited on the grounds of safety, namely to 113 V peak value in d.c. sources and 113 V peak value in welding transformers, and 80 V actual value. No-load voltages of no more than 68 V peak value and 48 V actual value are admissible for welding in boilers and restricted spaces using welding transformers. In fully mechanised welding plants, 141 V peak value and 100 V a.c. actual value are permitted, by comparison.

Design

The characteristic attributes and examples of welding current source applications are compared in **Table 1**.

Welding Transformers. These are normally connected to two external conductors in three-phase current supply; for smaller outputs they may also be connected to one external conductor and the mid-point conductor. The falling characteristic is produced by changing the coupling between the coils (*stray field transformer*), generating a secondary magnetic connection by introducing a plunger core (*stray core transformer*), or by changing the inductance using a *control throttle* downstream. A stepless, remote current control is provided by way of adjustable throttles (*transducers*) using a pre-magnetised iron core.

Welding Rectifiers. The mains voltage is transformed via a three-phase transformer and set to the desired characteristic via a control throttle or transducer. A semiconductor rectifier set rectifies the welding current. Additional current damping throttles are frequently installed to improve the dynamic characteristics.

Motor-Generator Welding Set. The motor–generator welding set is designed for a rotating d.c. generator driven by either an electric motor or a combustion engine. Motors and generators (including fans) are arranged around a common shaft in so-called single-shaft converter sets. Generators are used primarily on construction sites due to their mains-free operation. Differentially wound generators, quadrature-axis generators and stray field generators all generate a sharply falling characteristic in different ways, where the main field (and thereby the induced voltage) is weakened by an opposing field increasing in line with the welding current. More advanced brushless generators require very low maintenance because sliding-action contacts and commutator segments are no longer used.

Multiple Spot Welding Units. Where a large number of welding units are each operated with a very low duty cycle, supply via a multiple spot welding unit may be economically advantageous. In contrast to the setting equipment, this employs generators or rectifiers with constant voltage characteristics and high electric output ratings. The individual setting of the desired falloff characteristic at the welding spot takes place via leakage or multistep inductors for a.c. current, and for d.c. current via multi-step, high-wattage resistors.

Electronic Welding Current Sources. Thyristor or transistor current sources are characterised by their accurate and rapid controllability and wide range of potential applications, e.g. pulse-welding.

Table 1. Mains supply ratings, operating characteristics, costs and areas of application of welding current sources

	Welding transformer	Welding rectifier	D.C. converter
Mains supply	Single-phase	Three-phase	Three-phase
Mains loading	Asymmetrical	Symmetrical	Symmetrical
Feedback to mains	Undamped	Undamped	Damped
Effect of mains fluctuations	Proportionate	Proportionate	Insignificant for partial loads
Degree of effectiveness	80 to 90%	65 to 85%	45 to 60%
No-load pickup	0.06 to 0.15 kW	0.1 to 0.25 kW	1 to 3 kW
Output rating, without compensation	$\cos \varphi$ 0.4 to 0.6	$\cos \varphi$ 0.5 to 0.75	$\cos \varphi$ 0.85 to 0.9
Ignition characteristics	Satisfactory	Good	Good
Weld characteristics	Satisfactory to good	Good to very good	Good to very good
Application potential	Limited	Universal	Universal
Arc blow effect	Insignificant	Extreme	Extreme
Power loss in welding line	High	Low	Low
Noise emission	Insignificant	Low	High
Comparison of acquisition costs	50%	80%	100%
Operating costs	Minimal	Low	High
Maintenance and repair costs	Minimal	Low	High
Electric welding	Occasionally for B and C types	All types of electrodes	All types of electrodes
Submerged arc welding	No high base flux	All flux types	All flux types
MIG/MAG welding	Not common	Reverse polarity welding	Reverse polarity welding
TIG (tungsten inert gas) welding	Light metals	Steel, non-ferrous heavy metals	Steel, non-ferrous heavy metals

6.2 Resistance Welding Machines

Resistance welding machines may be stationary machines (**Fig. 3**) and multi-point equipment including portable weld guns and gun welding heads. These are classified as *spot*, *projection*, *seam-welding* and *butt welding machines* depending on the process. The task of supplying electrode forces and welding current requires a combination of mechanical and electrical functions.

Mechanical Functions. Electrode forces are variable over a broad range, thereby providing the best solution

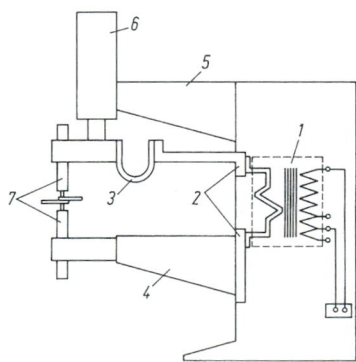

Figure 3. Schematic layout of spot welding machine: *1* transformer, *2* busbars, *3* live spring, *4* lower arm holder and lower arm, *5* upper arm, *6* compressed air cylinder and ram guide, *7* electrode holder with electrodes.

for the relevant welding task. The machine frame and electrode arms are to be designed with high rigidity. Bending under the electrode forces would cause a reciprocal shift in the electrodes during the expansion phase of the material. The rigidity is particularly important in projection welding machines in order to provide an even current distribution simultaneously over the projections to be welded.

Rapid closing motion of the electrodes is required to optimise production rates. The electrode should make full contact to minimise operating noise and electrode wear.

The adjustable electrode system is to be designed to be as lightweight as possible, in order that the low-inertia electrode can follow the elastic material in the expansion phase. Short-term release of interlocking between electrode and workpiece may have undesirable effects due to overheating.

Electrical Functions. The welding machine is to discharge the highest possible output current within a short period. The maximum short-circuit current measured where the electrodes collide is an extremely important characteristic. At the same time, low values are desirable for the output current to maintain the lowest possible installed load. This means that energy losses from the transformer, output circuit and welding current control should be minimised.

A high continuous load capacity of the machine, specifically of the welding current transformer, should be guaranteed in serial production. Continuous output, i.e. average output which can be taken up in the long term without excessive temperatures being recorded, should be taken into consideration when selecting seam and butt welding machines.

Welding current control is designed to generate brief welding periods of a few hundredths of a second and maintain these accurately. In addition, the full electrode force must be applied at the onset of the flow of current.

Three-Phase Welding Machines. Three-phase machines based on the principle of output side rectification or input side frequency transformation are becoming more popular than traditional single-phase types (**Fig. 3**), owing to the more reliable mains supplies and improved welding characteristics in some cases.

Welding Guns. These are a form of mobile welding equipment which is guided either manually (normally with weight redistribution) or by industrial robot.

6.3 Soldering and Brazing Equipment

Mechanised Soldering and Brazing Plants

The soldering and brazing process may be mechanised simply with suitable solder feed mechanisms, e.g. as *brazing alloy preform*, *solder powder*, *soldering paste* or as *braze-bonded cladding*. Normally, rotating tables or conveyor slides are employed for guiding the workpieces through the heating cycle. The heating is carried out by gas torches, induction coils, infrared radiation or by way of electrical resistors.

Rotating current transformers with mean frequencies of between 1 and 10 kHz are preferred for induction brazing of thicker parts. Core generators with frequency ranges

of > 100 kHz are more suitable for thin parts. Coil and panel inductors may be employed as inductors, depending on the shape of the workpiece.

The structure of resistor soldering and brazing machines corresponds to that of welding machines to a large extent. The soldering heat is generated in the workpiece itself with internal resistor heating. Copper electrodes are employed with short heating times. With carbon resistor heating, the heat is retained for longer and is preferably generated in graphite electrodes.

Oven Furnace Soldering

The soldering furnaces are either gas or oil-fired, or electrically-heated. The latter has the advantage of maintaining even temperatures accurately with a defined inert gas atmosphere in the furnace area. Furthermore, a distinction is to be made between discontinuous operation furnaces, such as retort-type furnaces, pit-type furnaces and bell furnaces, and continuous operation furnaces. Flux may be omitted in certain circumstances where inert gas reduction is carried out, e.g. H_2–CO/CO_2 mixtures. Either hot wall furnaces or cold wall furnaces are used with heated resistors or induction heating for flux-free vacuum brazing with pressures of between 10^{-1} and 10^{-6} mbar.

Dip Soldering and Wave Soldering

These are processes for the soft soldering of electrical component terminals onto the circuit-board conductors of printed circuit boards via immersion in a solder bath, or a solder shaft which simultaneously carries out the tasks of feeding and heating the solder.

7 Industrial Robots

G. Spur, Berlin

7.1 Systematics of Handling Systems

Handling systems consist of production machines which are designed to handle objects with suitably dedicated devices, such as grippers, tools, or baffle plates (see K6.2.2).

Computer-controlled universal handling systems with *programmable* sequences of operations are more popular in extended operations than other constructions. This group may be classified in terms of systems with *continuous path control*, such as cams, limit switches or fixed stops, and systems with *point-to-point control* via controlled setpoint input.

Continuous-path control devices only allow for actuation in two different positions in each axis of motion. On the other hand, with point-to-point control systems, as many positions as are required may be actuated for each axis of motion. The number of positions is limited only by the capacity of the setpoint memory in the controller. In computer-controlled handling systems flexibility is tending to increase with the improved programming facilities.

Flexible handling systems (industrial robots; **Fig. 1**) are production machines designed for automatic object handling with suitably designed tools and may be programmed in a number of axes of motion in terms of orientation, position and sequence of operations.

As the mechanical structure of handling systems may

be represented in the form of kinematic chains, the number of degrees of freedom within a handling system is equal to the number of links in the kinematic chain which may be moved independently, assuming that each articulated joint may only have one single degree of freedom.

Links, levers and drives are employed as elements in the creation of kinematic systems. Each individual combination of link, lever or drive is described as an axis of motion. Each axis of motion corresponds to a degree of freedom in the kinematic chain. Rotary or linear axes are generated as a result of the use of revolute or prismatic links (see F9.1).

Handling systems are used to change the orientation and position of three-dimensional objects. *Six independent movements* (degrees of freedom) are required to adjust the position of a three-dimensional object. If this is based on a closed Cartesian system of coordinates, three linear degrees of freedom are required to adjust the position of an object and three rotary degrees are required for the orientation of the object. Three-dimensional motion as is required in tool handling systems may be generated via *linear motion* and *rotary motion*. Three degrees of freedom are required to position an object within a specific three-dimensional Cartesian system of coordinates. The independent motion required for this may be achieved by way of a suitable arrangement of a minimum of three controllable axes.

The combination of linear and rotary axes and their arrangement determines the working space in the tool handling system and may be defined by way of the system

Axis combination	Coordinate designation		Working spaces
3 linear axes	Cartesian coordinates	Cuboid shaped	Axis designation
2 linear, 1 rotating axis	Cylindrical coordinates	Cylindrical	
1 linear, 2 rotating axis	Spherical coordinates	Spherical	
3 rotating axes	Revolute coordinates	Toroidal	
m linear, n rotating axes	e.g. Cartesian and revolute coordinates	Redundant kinematic system	

Figure 1. Working spaces and coordinates of motion in handling systems.

of coordinates of motion or the geometry of its areas of contact. **Figure 1** shows these relations in these basic three-axis systems. More advanced configurations with more than six controllable axes have been introduced very recently to add to these four basic types [1].

Examples of these are the *vertical column* or *track-mounted robots* and *robots on rotary bases*. These additional axes of motion extend the working space in the handling system and interference prevention and optimised motion sequences have also been developed, now that degrees of frequency have been largely superseded.

Robots may also be set up as systems with *distributed axis control*, with a combination of six-axis robots and rotating and tilting benches for seam welding applications, for example (**Fig. 2**), or multiple articulated joint handling with cooperating robots.

7.2 Components of Robots

The drive units of the active articulated joints must satisfy the high requirements which it is somewhat difficult to combine in view of the dynamic motion characteristics of the robot. These requirements include low inertia, low

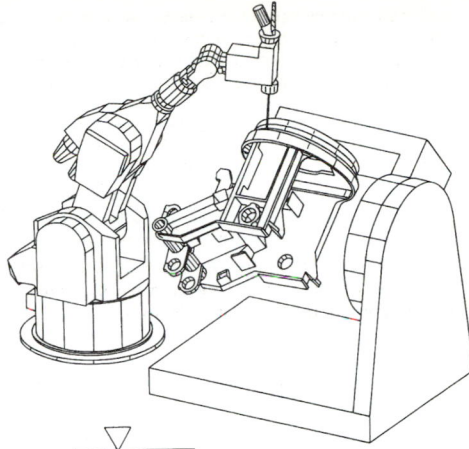

Figure 2. Robot systems with distributed axis control (KUKA, Augsburg).

output weight, high pulse rate, high short-term overload capacity, and high resolution both over the path setting range and speed range.

Electrical, hydraulic and pneumatic drives are commonly in use (see L1.2.1).

D.C.-powered machines are mainly employed in disc-type rotors and squirrel-cage induction rotor constructions. More recently, there has been a trend in favour of the use of brushless, wound d.c.-powered motors with rare-earth element solenoids, and asynchronous motors. These machines require very little maintenance and have a better power-to-weight ratio. The input required in terms of power electronics is normally compensated for by the advantages specified.

Electric motors are used in combination with high reduction gearing, such as harmonic drives, worm-type gearing, or planetary gearing (see L1.2.2).

Non-reduction direct drives are in limited use in a few specific applications.

Hydraulic cylinder drives cater for extremely high driving forces with relatively low dead weight. They are recommended for driving large-scale kinematic systems for this reason.

Pneumatic cylinders are also designed for extremely high speeds, but the air compressibility factor prevents them from following accurate paths. These drives are employed mainly in supply and loading systems with adjustable stops.

The internal robot sensors provide accurate and rapid data logging of the actual position of the articulated joint. The robot controller thereby obtains the feedback difference between the actual and setpoint setting, as a result of which this is translated by the position feedback control system into the appropriate manipulated variables.

Internal sensors may be classified in terms of test results as linear or rotary values, or in terms of data logging as digital and analog, and in terms of the measurement process as incremental and absolute values.

Synchro-resolvers and digital recorders are employed in practice in industrial robot applications.

Synchro-resolvers are measuring systems based on the induction principle with direct angle data logging. These may also be used as indirect path measuring sensors. Specific advantages of the resolver include their compact design, high resolution capacity and resistance to wear.

Digital locators record the measured value as a whole number multiple of the angle or path increment. A distinction is made between *incremental* (relative) and *encoded* (absolute) measuring systems, depending on whether the value is obtained as an increment or from a linear encoder readout. Absolute digital signals are relatively costly to construct if they are to provide long traverse paths with high resolution. Incremental digital signals are cheaper than absolute ones and have an unlimited measuring range in principle. The cumulative results of measuring errors and loss of the reference point during power loss are two of the main disadvantages of incremental measuring systems.

7.3 Kinematic and Dynamic Models

7.3.1 Kinematic Model

Six coordinates which are independent of one another as a function of time are required to describe the general three-dimensional motion of a rigid end effector. Three of these coordinates determine the original position of the effector coordinates. The remaining three determine the orientation of the effector coordinate system in relation to the fixed frame of reference. These coordinates are designated as *external X-coordinates*. The joint coordinates (inner coordinates) are selected as generalised coordinates in structures without branches or links in the kinematic chain of the robot members. The relationship between external and internal coordinates is defined by a non-linear vector-type pattern $X = f(q)$. The concrete form of the vector function f is dependent on the selection of the internal and external system of coordinates and the structure under observation. In practice, both coordinate systems are in common use. On the other hand, robot motion is normally planned in terms of external coordinates, control and monitoring being carried out in terms of joint coordinates.

Basic Kinematic Problems. There are basically two problems associated with the kinematic analysis of the robots, as follows.

Direct Problem. The corresponding position and orientation of the end effector has to be obtained in terms of external coordinates for given joint positions.

Inverse Problem. In order to generate the desired path in terms of external coordinates, the required internal coordinates have to be defined.

The Denavit–Hartenberg convention for describing and modelling the kinematic robot structure is employed to solve both of these problems. Each link in the kinematic chain is then assigned to an object-related system of coordinates, as follows.

The transformation of coordinates between two adjacent joints depends purely on the coordinate of the connecting member, the transformation $f(q)$ between the base and the effector joint of the unbranched kinematic chain of all member coordinates. The solution to the inverse problem, $q = f^{-1}(X)$, is normally ambiguous and may only be obtained in a closed form for specific robot structures. Explicit solutions in a closed form are possible, for example, for structures with "spherical manipulation", where the axes for articulated hand operation intersect at a point [1]. Numerical solutions should normally be applied.

7.3.2 Dynamic Model

The system of differential equations of motion for the robot may be produced in the form of the *Newton–Euler equations* for *holonomous* systems,

$$H(q)q + b(q)q + G(q) = P.$$

Here $H = (H_{ij})$ represents the *n*-dimensional inertia matrix, $b = (b_1 \ldots b_n)$ represents the vector of generalised constraining forces (centrifugal and compound centrifugal forces), $G = (G_1 \ldots G_n)$ represents the vector of generalised forces due to weight and P represents the vector of driving forces [2]. There are consequently two basic dynamic problems in connection with this model:

Direct Problem. The robot motion $q(t)$ has to be obtained for a given driving force $P(t)$.

Inverse Problem. The required driving forces $P(t)$ have to be determined before the desired robot motion $q(t)$ can be generated.

Manual calculation of the dynamic equations for industrial robots is extremely time-consuming and susceptible to errors. For this reason, a number of algorithms and correspondingly efficient numerical or mathematical program packages have been developed to create mathematical models for robots. Mathematical programs provide the equations of motion for the relevant task in hand in a form which minimises the input required for their arithmetic evaluation.

7.4 Characteristics, Accuracy

The specific attributes of the robot machine tool in terms of its design and sequence of motions require that specific characteristics and the process associated with their designation are defined in order to make robots comparable and to enable modifications to be estimated, e.g. as a result of wear. The following characteristics and the respective designation processes specify characteristics for industrial robots in the VDE Guideline 2861, as follows:

Working space, useful load, speed, acceleration, travel time, repeatability, reproducibility, contouring accuracy, programming accuracy.

ISO Standard DP 9283 represents a further development. An increase in the degree of accuracy of robots which is currently around a factor of 10 less accurate than machine tools will open up opportunities for numerous new applications in sectors with considerable potential for future expansion such as assembly work, for example. As any increase in the degree of accuracy by way of modifications on the part of the robot mechanics would lead to a considerable increase in costs, efforts have been made to increase the degree of accuracy over the past few years by identifying systematic errors and taking these into account in the mathematical and physical models on which the control is based or in compensation processes. This enables minor discrepancies in the geometric/kinematic data, elasticity values, or transmission-influencing factors to be accounted for numerically and be taken into consideration in the transformation calculations between external (effector position and orientation) and internal (link position) coordinates [15].

7.5 Industrial Robot Control Systems

The task of an industrial robot control system consists of controlling one or more handling systems in accordance with the handling and processing task required for the technical process. Motion sequences and actions are specified as part of a user program which is executed by the controller. It obtains process data via sensors and is therefore in a position to adapt the predefined processes, motion and actions to the changing or environmental conditions or those previously unknown to a certain extent. In addition, an industrial robot control system has to satisfy specific requirements as regards operating modes, user operation and programming, as well as monitoring and safety functions [1, 5].

Industrial robot control systems are based on microcomputers to a large extent nowadays, sometimes using multiprocessor technology (see L2). Interfaces to manufacturing communications systems, e.g. MAP (Manufacturing Automation Protocol), are available for connection to primary control and programming systems. More and more links are being established using serial bus systems, e.g. field bus and bit bus, to peripheral processes in the same way, e.g. in welding control and conveyor systems, also providing a means for external sensor analysis.

Software Components of a Robot Control System (Fig. 3). Data transfer is carried out via the *communications module* to other control systems (industrial robot controllers, cell-based computers, primary computers). Specifically, loading of user programs to the robot controller and the transfer of status data and reports to other control systems (DNC operation) are carried out in this way. A common communications standard was laid down in ISO 9506, known as the manufacturing message specification (MMS) for the various classes of equipment [3, 4].

Sequence Control. The user program of an industrial robot contains motion commands, effector commands, sensor commands, program sequence control commands, arithmetic commands and technological commands. The execution of the user program is organised in sequence control which is normally identical to the so-called interpreter. This means a program which reads the instructions of the user program or a relevant code, e.g.

IRDATA [6], or decodes the instructions generated by a compiler and accesses and coordinates the respective execution routines or segments.

The task of *motion control* is to generate the relevant reference variables for the servomechanism to drive the handling systems involved in the motion, i.e. robots, rotary table, kinematic elements and other auxiliary axes, using the given motion sequence provided by the user program and user data. *Point-to-point control* (PTP) enables a sequence of discrete three-dimensional points to be logged and actuated. The path of motion of the end-effector between these three-dimensional points is not explicitly specified. This enables extremely efficient time-controlled motion characteristics to be achieved. This method is used for tasks where the precise path progression is not important, e.g. for handling and spot welding tasks. *Continuous-path control* (CP = continuous path) offers the option of traversing mathematically defined paths of motion within the task range. The continuous-path control computer (interpolator) obtains a number of intermediate values on the three-dimensional curve programmed in terms of the points corresponding to a given path function (straight-line, circle or higher polynomial) and speed function and reports them to the servomechanism at a given rate. Continuous-path control is used, for example, in track-mounted robot welding and deburring applications.

The *servomechanism* has the task of traversing the robot axes in accordance with the current position setpoints. Axis-related intermediate points are recomputed within a narrower time grid within the axis controller (precision interpolation). The values of the axis angles or paths are translated into motor currents, voltages or increments and output to the actuators. Actuation of the axis positions is monitored and computed using the actual position feedback reported by the path and angle monitoring systems.

Sensor data processing involves the reception of signals and data, e.g. coordinate values of objects, from sensors inside the robot (path and angle monitoring systems, force/torque sensors) and outside it (proximity sensors, recognition system). These data are required and processed by different levels of the robot control systems, i.e. sequence control, motion control and axis control. The shortest response times to external conditions may be achieved where the relevant sensor data are reported at axis control level (e.g. force/torque sensor analysis, collision monitoring).

Action control executes the action commands of the user program which normally relate to objects to be handled or the control of peripherals. Depending on the action command, it executes a combined logic operation consisting of internal and external control process signals (i.e. motion status, limit switch, light barriers, responses from other controllers) and produces the drive signals for the binary actuators (such as the contactors, single drives and values) or generates commands for peripheral manufacturing devices (such as start, stop, and synchronisation).

The *operation* control segment supports functions such as selection of the operating mode, input of operating parameters, program start/stop, loading/storing programs. A robot controller can be operated in two basic operating modes which are normally divided up into further submodes. The user may access all operating elements in the controller which are required to operate the robot and to create or modify handling programs in "set-up" mode. Only a few simple operating functions may

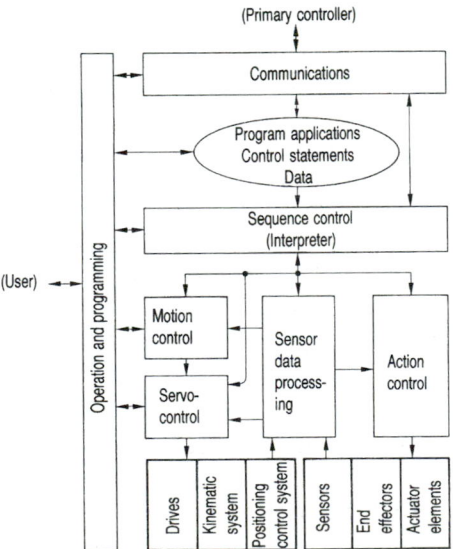

Figure 3. Software components of industrial robot controller.

be executed in "automatic" mode, e.g. program selection, start, stop, and resume. Information on the current handling program and operating instructions and error messages are displayed via the current handling program.

Programming components are used for user program generation, maintenance and management. The function modules required for program generation, such as the editor, debugger and compiler may form part of the control system or may be exported to another computer, e.g. a PC. An interactive programming component enables programs or selected positions for the robot or the effector in the teach-in process to be entered or programs of motion to be tested.

7.6 Programming

7.6.1 Programming Procedures

Programming procedures constitute the planning procedures for generating user programs. According to VDI Guideline 2863 (IRDATA), a user program constitutes a sequence of commands aimed at executing a given manufacturing task [7]. Programming procedures enable user programs to be developed and provide the relevant programming aids for this purpose (see L2).

Programming procedures may be divided up into *direct* programming (*on-line* systems), *indirect* programming (*off-line* systems) and *hybrid* programming [8].

Direct Programming. This is characterised by the fact that user program generation is carried out using the robot system. The consequence of this is that the manufacturing system is not available for production during the programming or testing period, resulting in high idle times during installation. Integration of operational, computer-based data systems is only possible on a limited scale. The quality of the user programs is dependent on the experience of the programmer to a large extent. There are subsidiary categories within these procedures, known as playback, teach-in and sensor-based processes:

In *playback programming*, programming of an operational sequence is carried out by manually guiding the robot along the desired three-dimensional curve. The actual positional values (axis points) within a defined time or path scale are thereby carried over into the user program. A special lightweight programming arm may be employed to program motion, which is extremely similar to human working methods in dynamic terms.

One typical application is programming paint spraying robots [9].

In *teach-in programming* the motion data are generated by tracking the desired three-dimensional coordinates with the aid of a hand-held programmer or operating field, and these points are subsequently programmed by activating a function key. Other motion commands may also be input via the keyboard, such as speed and acceleration data or type of control (point-to-point or continuous-path control).

Sensor-based programming, which is now becoming popular, may be categorised in terms of automatic sensor control and manual sensor control [10]. The former is based on rough motion data (such as start and target point) and the workpiece is scanned automatically by the robot using sensors. In the second system the robot is guided along the desired three-dimensional curve by the operator using a sensor or light pen. In contrast to playback programming, where the robot plays a passive role, in this case sensor signals are fed direct to the control circuits in the robot control systems. These initiate the

active sequence of the operating tasks. In sensor or manually guided programming, the path travelled is automatically stored in memory. This occurs by programming path interpolation points in accordance with given criteria, such as the desired degree of accuracy [11].

Indirect Programming. This is characterised by the fact that the user program generation is carried out separately from the robot system on independent computer control systems. It requires a computerised model of the robot system and the system environment. Programming and testing of the user program is reassigned to operational preliminary planning and therefore becomes a part of the production planning process. Integration of operational data systems and intelligent, computer-based aids are provided as support for the programmer.

A distinction is made between textual and CAD-based processes in indirect programming employed in *off-line programming procedures*.

Textual programming procedures request geometric data input via a keyboard, as is standard practice with computer and NC programming languages. A further development has taken place with textual input and graphics support. More advanced programming procedures provide direct CAD support for geometric definition and motion.

CAD-based programming procedures support geometric models of the components involved in the manufacturing process. Geometric model generation is carried out via CAD systems. Functions are provided on the graphics screen for defining positions to be covered and travel paths. Integrated simulated modules allow for visual display of the motion executed by the robot. CAD-based programming procedures are therefore characterised by their graphics representation.

A further distinction is made between motion-based (explicit) and task-based (implicit) programming procedures [12].

In *explicit* programming procedures, all the actions of the robot, especially the motion, including execution parameters (e.g. speed, acceleration), are specified by the programmer. Thus all travel paths and positions to be covered require designation and should take into account freedom from collisions.

In *implicit* programming procedures, programming is not carried out by designating the travel path, but by describing the handling task. The path data are derived from the programming procedure automatically using a

Table 1. Attributes of direct and indirect programming procedures

Direct programming	Indirect programming
– Real-time robot system and system environment required	– Computer model of robot system and system environment required
– Production system not available during programming	– Programming during operational planning as part of production planning process
– User program testing on real system	– Simulated program testing
– Limited integration with operational data systems	– Full integration with operational data systems
– Quality of user programs dependent on experience of programmer	– Programmer support via intelligent, computer-based help utilities

model of the robot cell (model environment). Programming procedures based on such systems are currently undergoing research.

Hybrid Programming. This represents a combination of direct and indirect programming procedures (**Table 1**).

The program sequence is subsequently determined by the indirect programming procedure. The motion control section of the program may be defined using the teach-in or playback system or by sensor-guiding.

7.6.2 Off-line Programming Procedures

In 1986 the US National Bureau of Standards published a study on the existing off-line robot programming procedures all over the world. Approximately 95 different robot programming languages were named, not all of which may be employed in off-line programming pro-

cedures according to modern standards. Robot programming languages have been classified based on an analysis of the language and system structures [13].

Local Control Off-line Programming Procedures

Program generation with systems of this type is carried out in the same programming language as is employed in the robot controller. Textual program generation is simplified with the use of user-friendly editing programs (e.g. menu-driven) and improved communications facilities.

Such off-line programming procedures enable a complete program to be developed which describes the sequence of the handling or machining task, communications with process peripherals and their synchronisation with the motion of the industrial robot. The definition of geometric data from the program of motion, i.e. the position and orientation of the end-effector system, is carried

Table 2. Requirements of robots systems in forming and shaping processes

Requirements	Areas of application for integrated robot manufacturing stations		
	Forming	Shaping	
	Preforming	Re-forming	Cutting
Process implementation			
Workpiece handling: - End-effector and workpiece location and takeup of process forces - Protection from environmental effects, process residues - Motion sufficiently accurate for machining purposes	Deleted	Limited applications due to high process forces	Grinding, brushing, laser beam
Tool handling: - Tool location and takeup of reactive forces - Energy signal guide to end of arm - Continuous-path control, teach-in programming and process data feedback normally required	Not known	Limited application due to high reactive forces, exception: jointing in re-forming process	Grinding spindle, brushing tools, drill unit, milling unit, laser device, fluidised jet head
Object handling			
- High process speeds - Average positioning accuracy - Large working space - Point-to-point control frequently adequate - Object presence check - Low number of axes - Flexible gripper system	Loading deleted Unloading Destacking workpieces to diecasting, gravity diecasting, injection moulding machines and sinter presses	Pallet transportation of workpieces to forging presses, bending machines Substantial geometrical modification	Pallet transportation of workpieces to machine tools Swage, cooling water, burring
Testers			
- Tactile and visual sensors - Program synchronisation with test results - Compliance with specified environmental conditions	Geometry Process material testing, e.g. blowholes, surface quality	Geometry Dimensions, e.g. wall thickness, transition radius	Geometry Dimensions, surface quality, process materials

out subsequently either in the form of numerical specifications or by way of subsequent teach-in programming on the actual robot.

All systems within this category carry out a syntactical examination of the user programs which are developed, so that only user programs with the correct syntax may be loaded into the robot controller. Although a number of systems check for compatibility of the specified end-effector positions with the robot kinematics (compliance with the axis travel ranges), no verification may be made to ensure proper function of the user programs which have been developed on the real robot system (e.g. relating to collision). This is why program simulations with the relevant graphics support are required, although these do not normally form an integral part of systems of this type.

Owing to the lack of facilities for geometrical definition and the limited simulation and test support in a large number of cases, the use of local control off-line programming

procedures is only foreseen within the framework of hybrid programming.

CAD-based Off-line Programming Procedures.

Simulated graphics support for programming and user program testing is, however, a characteristic for this category. Systems of this type are either based on existing CAD systems, expanded to encompass a robot-specific module, or on special developments with integrated graphics functions [14].

The functional capacity of these systems is not restricted to the actual programming of an industrial robot, but provides options for modelling, programming and simulation of the entire production centre. They therefore constitute a planning tool for robot-based systems. Such system applications are simplified thanks to the existence of libraries for robot and controller models. In addition, aids for the definition of new robot models or modifications

Table 3. Requirements of robot systems in treatment and assembly processes

Requirements	Areas of application for integrated robot manufacturing stations		
	Treatment		Assembly
	Changing material characteristics	Coating	Jointing
Process implementation			
Workpiece handling:	Magnetising	Object replacement	– Contact pressing and press-fitting object
– End-effectors and workpiece location and takeup of process forces	Laser heat treatment		– Jointing in re-forming process
– Protection from environmental effects, process residues			– Folding
– Motion sufficiently accurate for machining purposes			– Inserting
Tool handling:	Magnetising button	– Spray painting	– Welding with tongs, burners, laser
– Tool location and takeup of reactive forces	Laser device	– Fibreglass plastics	
– Energy signal guidance to end of arm	Nitration unit on robot	– Polyurethane coating	– Riveting, clamping
– Continuous-path control, teach-in programming and process data feedback normally required		– Cement application	– Bolting, bonding
		– Plasma spraying	– Gauge needle
			– Sensors
Object handling			
– High-speed system	Loading and unloading of tunnel-type furnace trucks and heat treatment equipment	Loading and unloading trays and drying shelves in galvanising and enamelling processes	– Pallet transportation of small assemblies
– Average positioning accuracy			– Loading and unloading automatic assembly machines
– Large working space			
– Point-to-point control frequently adequate	– Induction system		– Jointing machines:
– Object presence check	– Immersion baths		• Riveting machines
– Low number of axes	– Furnaces		• Automatic welding machines
– Flexible gripper system			• Bordering presses
Testers			
– Tactile and visual sensors	Material properties, e.g. hardness	Thickness of coat, surface quality	Force and torque adjustment
– Program synchronisation and test results			
– Compliance with specified environmental conditions			Inspection for compliance

to existing models are frequently provided. Program generation is generally effected either in system-specific advanced programming languages or in robot control-based languages. Post-processors carry out translations of programs for the various controllers. The geometric data for the user programs may be derived from the workpieces and other relevant components using CAD models. CAD-based systems provide the most flexible means of support for this.

Simulated testing of user programs generated off-line requires a computerised model (model generation) of the robot and its operational environment as regards all the relevant aspects, such as control, kinematics, form and communications models. The aim is to ensure that user programs of the type undergoing testing can be operated on the actual system with the least possible amount of modification. The operation sequences are presented in the form of a visual display via computerised graphics to assist the user.

The robot and associated system environment reference data are taken as the basis for programming and simulation. The two systems tend to be subject to tolerance errors and other errors, so that a user program generated off-line cannot normally be expected to be fully operable immediately. Robot and system offset measurement and error compensation are requirements for this [15].

7.7 Main Applications and Selection of Robots

Industrial robot applications are mainly centred around spot welding, track-mounted robot welding, coating, assembly and machine loading technology [16].

The process-related requirements for industrial areas of application are compared in **Tables 2** and **3** [17]. The most essential system attributes include adequate working space, number of additional axes, rated load, traverse rate and scope of functions to be executed. The latter encompasses the type of control (point-to-point or continuous-path), program length, and required number of signal inputs and outputs.

Requirements for characteristic system attributes for industrial robots in a variety of applications are shown in **Table 4**.

The sheer volume of influencing factors to be taken into account in terms of the manufacturing process and handling activities when planning an industrial robot applications means that a systematic, methodical process of elimination is essential [18]. Planning tasks are processed in stages with a certain number being executed simultaneously, however. The specification has to contain the basic planning data requirements in the form of fixed values for all those affected; in the final design of the lay-

Table 4. Requirements of characteristic system attributes for industrial robots in a variety of fields of application

Industrial robot application	Requirement							
	Load-bearing capacity (kg)	Number of axes	Max. feedrate (mm/s)	Repeatability (± mm)	Class of control	Sensor function		Required time for sensor data processing (ms)
						During movement	Intermittent	
Handling:								
Machine loading	10 to 40	4 to 5	—	0.5	PTP	—	B	—
Pallet transportation	10 to 40	3 to 5	—	1.5	PTP	—	B, IDP	—
Assembly								
Small part assembly	2 to 5	4 to 5	—	0.1	CP	B, (F)	B, IDP	—
Adhesive application	5 to 10	5 to 6	600	0.5	Circ.-CP	B	B	—
Spot welding	40 to 90	5 to 6	—	1.0	PTP	B	B	—
Track-mounted robot welding	5 to 10	6	20	0.5	Circ.-CP	S, F, T	B	< 70
Continuous-path treatment processing:								
Coating	5 to 10	5 to 7	1500	2.0	PTP (CP)	B	B	—
Shaping:								
Milling	10 to 60	6	40 to 100	0.2	Circ.-CP	S, F, T	B, geom.	< 30
Water torch cutting	10 to 60	6	20 to 250	0.1	Circ.-CP	(F)	B, (IDP)	(< 25)
Belt grinding, rotary grinding	10 to 60	6	60 to 120	1.0	Circ.-CP	S, F, T	B, geom.	< 20
Laser beam machining	10 to 60	6	100 to 300	0.05	Circ.-CP	S, F	B, (IDP)	< 15

B: Binary signal processing
Circ.-CP: Continuous-path control with circular interpolation
CP: Continuous-path control
F: Contour sequence functions
geom.: Continuous-path geometry adjustment
IDP: Image data processing
PTP: Point-to-point control
S: Search functions
T: Technological value adjustment
( ) Requirement in certain cases

out planning phases, equipment selection and peripheral configurations have to be run through individually until a satisfactory solution is reached. A preliminary decision should be made on the basis of "independent design requirements". The ultimate selection of the robot and detailed definition of all peripherals may only be carried out during the course of the layout planning. The individual optimum arrangement of all system components may then be obtained with the aid of selected industrial robots during this planning phase.

Safety, compact design, delivery, accessibility for main-tenance and repairs and modification and expansion costs should be taken into account in the selection of robot systems.

Computer-based planning systems have been developed to enable alternative solutions to be provided relating to operating materials, accessibility of three-dimensional points, collision hazards and display of execution time, sizing of memory and options for optimising performance in relation to the sequence of installation, parts and tool provision, and flow of materials based on CAD systems and simulated systems [19].

8 References

L1 *Machine Tool Components* [1] *Stute G.* Regelung an Werkzeugmaschinen. München: Hanser, Munich, 1981. - [2] *Weck M.* Werkzeugmaschinen, vol. 3, Automatisierung und Steuerungstechnik, 3rd edn. VDI-Verlag, Düsseldorf, 1989. - [3] *Weck M, Ye G.* Elektrische Stell- und Positions-antriebe - Systemaspekte und Anwendungen bei Werkzeugmaschinen. ETG-Fachber. 27. Berlin: VDE-Verlag 1989, pp 217-231. - [4] *Bederke HJ.* Elektrische Antriebe und Steuerungen. Teubner, Stuttgart, 1969. - [5] *Birett H.* Der elektrische Antrieb von Werkzeugmaschinen, 2nd edn, WB 54. Springer, Berlin, 1951. - [6] *Dutcher JL.* Maschinengestaltung und Regelantrieb für numerische Steuerungen. General Electric, Waynesboro, 1963. - [7] *Lehmann W, Geisweid R.* Elektrotechnik und elektrische Antriebe, 7th edn. Springer, Berlin, 1973. - [8] *Opitz H.* Auslegung von Vorschubantrieben für numerische gesteuerte Maschinen. RWTH Aachen, Werkzeugmaschinenlabor, 1969. - [9] *VEM-Handbuch.* Die Technik der elektrischen Antriebe. VEB Verlag Technik, Berlin, 1964. - [10] *Kennel.* Ein mikroprozessorgeregelter Hauptspindelantrieb mit Feldorientierung für Werkzeugmaschinen. Bosch Kolloquium Antriebstechnik, Solothurn, 1986. - [11] *Blaschke F.* Das Verfahren der Feldorientierung zur Regelung der Drehmaschine. Diss. TU, Brunswick, 1974. - [12] *Vogt G.* Digitale Regelung von Asynchronmotoren für numerisch gesteuerte Fertigungseinrichtungen. Springer, Berlin, 1985. - [13] *Pfaff G.* Neue Entwicklungen bei elektrischen Servoantrieben. tz für Metallverarbeitung (1984) H.8, 15-21. - [14] *Zimmermann P.* Bürstenlose Servoantriebe für Werkzeugmaschinen. wt-Z. ind. Fert. 73 (1983) 629-32. - [15] *Henneberger G.* Servoantriebe für Werkzeugmaschinen und Roboter. Stand der Technik, Entwicklungstendenzen. Conf. Proc. ICEM, Munich, Sept. 1986. - [16] *Polymotor.* Firmendruckschrift FASTTACT, Genoa, 1987. - [17] *Wolters P.* Lageregelung für asynchrone Linearmotoren. Ind. Anz. 99 (1977) 129ff. - [18] *Götz FR.* Hochdynamische Antriebe Umrichtergespeiste Drehstromservoantriebe und Linearmotoren. ETG-Fachber. 27, VDE-Verlag 1989, pp 159-167. - [19] *Stute G.* Untersuchungen über die Verwendbarkeit von Gleichstromnebenschlussmotoren als Vorschubantriebe für numerisch gesteuerte Werkzeugmaschinen. VDW-Ber. TU-Stuttgart, Inst. für Steuerungstechnik, 1971. - [20] *Wilhelmy L.* LongLife-Tachody-namos im Vergleich zu anderen modernen Drehzahlistwertaufnehmern für die Antriebs- und Regelungstechnik. ETG-Fachber. 27. VDE-Verlag, Berlin, 1989, pp 147-57. - [21] *Backé W.* Umdruck zur Vorlesung "Grundlagen der Ölhydraulik". Inst. für hydraulische und pneumatische Antriebe und Steuerungen, RWTH, Aachen, 1988. - [22] *Backé W.* Umdruck zur Vorlesung "Servohydraulik". Inst. für hydrau-lische und pneumatische Antriebe und Steuerungen, RWTH, Aachen, 1986. - [23] *Backé W.* Fluidtechnische Realisierung ungleichmässiger periodischer Bewegungen. Ölhydraulik und Pneumatik Mai (1987) 22-8. - [24] *Backé W.* Neue Möglichkeiten der Verdrängerregelung. Tagungsunterlagen zum 8. Aachener Fluidtechnischen Kolloquium, vol. 2, 1988, pp 5-59. - [25] *Rögnitz H.* Stufengetriebe an Werkzeugmaschinen, 4th edn., WB 55. Springer, Berlin, 1965. - [26] *Riegel F.* Rechnen an spanenden Werkzeugmaschinen, vol. 1, 5th edn. Springer, Berlin, 1964. - [27] *DIN 781:* Zähnezahlen für Wechselräder. Beuth, Berlin, 1973. - [28] *Schöpke H.* Grundlagen der Konstruktion von Werkzeugmaschinengetrieben. Westermann, Brunswick, 1960. - [29] *Rögnitz H.* Getriebe für Geradwege an Werkzeugmaschinen, 2nd edn, WB 101. Springer, Berlin, 1964. - [30] *Dürr A. Wachter O.* Hydraulik in Werkzeugmaschinen, 6th edn. Hanser, Munich, 1968. - [31] *Ebertshäuser H.* Bauelemente der Ölhydraulik, O+P TB 3. Krausskopf, Mainz, 1974. - [32] *Krug H.* Flüssigkeitsgetriebe bei Werkzeugmaschinen, 2nd edn. Springer, Berlin, 1959. - [33] *Wiessner H.* Über pneumatisch-hydraulische Vorschubeinheiten. wt 55 (1965) 163ff. - [34] *VDI-Richtlinie 3230.* Technische Ausführungsrichtlinien für Werkzeugmaschinen und andere Fertigungsmittel, H-Hydraulische Ausrüstung. VDI-Verlag, Düsseldorf, 1967. - [35] *VDI-Richtlinie 3229.* Technische Ausführungsrichtlinien für Werkzeugmaschinen und andere Fertigungsmittel, P-Pneumatische Ausrüstung. VDI-Verlag, Düsseldorf, 1967. - [36] *Weck M.* Werkzeugmaschinen, vol. 1. Maschinenarten, Bauformen und Anwendungsbereiche, 3rd edn. VDI-Verlag, Düsseldorf, 1988. - [37] *Sahm D.* Reaktionsharzbeton für Gestellbauteile spanender Werkzeugmaschinen. Diss. RWTH, Aachen, 1987. - [38] *Weck M.* Werkzeugmaschinen, vol. 2, Konstruktion und Berechnung. VDI-Verlag, Düsseldorf, 1990. - [39] *Rinker U.* Werkzeugmaschinen-Führungen, Ziele künftiger Entwicklungen, VDI-Z 130 (1988). - [40] *Weck M, Rinker U.* Einsatz von Geradführungen an Werkzeugmaschinen. Ind. Anz. 79 (1981). - [41] *DIN 50 320:* Verschleiss, Begriffe, Systemanalyse von Verschleissvorgängen, Gliederung des Verschleissgebietes. Beuth, Berlin, 1979. - [42] *Weck M, Rinker U.* Reibungsverhalten von Gleitführungen. Einfluss der Oberflächenbearbeitung. Ind. Anz. 28 (1986) - [43] Untersuchungen am Lehrstuhl für Werkzeugmaschinen. RWTH, Aachen, 1988. - [44] *Weck M, Miessen W.* Optimierung und/oder Berechnung hydrostatischer Radial- und Axiallagerungen. KfK-CAD 77. Kernforschungszentrum, Karlsruhe, 1979. - [45] *Peeken H, Benner J.* In: Goldschmidt informiert. Aus der Arbeit der Th. Goldschmidt AG, no. 52, 1980. - [46] *VDI-Richtlinie 2201:* Gestaltung von

Lagerungen, vols 1 and 2. VDI-Verlag, Düsseldorf, 1975. – [47] *VDI-Richtlinie 2202*; Schmierstoffe und Schmiereinrichtungen für Gleit- und Wälzlager. VDI-Verlag, Düsseldorf, 1970. – [48] *VDI-Richtlinie 2203*. Gestaltung von Lagerungen, Gleitwerkstoffe. VDI-Verlag, Düsseldorf, 1964. – [49] *VDI-Richtlinie 2204*. Gleitlagerberechnung hydrodynamischer Gleitlager für stationäre Belastung. VDI-Verlag, Düsseldorf, 1968. – [50] *Bräuning H*. Einsatz mehrflächiger hydrodynamischer Gleitlager in Schleifspindellagerungen. Ind. Anz. 59 (1973). – [51] *Weck M, Koch A*. Vergleich von Hauptspindel-Lager-Systemen. Vortrag am Lehrgang: Konstruktion von Spindel-Lager-Systemen für die Hochgeschwindigkeits-Materialbearbeitung an der TAE-Esslingen, 1988. – [52] *Korrenn H, Kleinhenz G, Voll H*. Spindellagerungen in Werkzeugmaschinen. Teil 1: Wälzlager; Teil 2: Gleitlager. Klepzig Fachberichte 12 (1972) and 3 (1973). – [53] *Brändlein J*. Eigenschaften von wälzgelagerten Werkzeugmaschinenspindeln. FAG-Publikation no. WL02113 DA. – [54] *Ophey L*. Entwicklung schnellaufender, wälzgelagerter Hauptspindeln für Werkzeugmaschinen. VDW Forschungsberichte 1986. – [55] *Weck M, Koch A*. Experimentelle Untersuchung von Hochgeschwindigkeits-Spindel-Lager-Systemen mit Wälzlagern. Vortrag am Lehrgang: Konstruktion von Spindel-Lager-Systemen für die Hochgeschwindigkeits-Materialbearbeitung an der TAE-Esslingen, 1988. – [56] *Weck M, Ophey L*. Wälzgelagerte Spindel-Lagersysteme für die Hochgeschwindigkeits-Hochleistungs-Bearbeitung. Ind. Anz. 37 (1987). – [57] *Weck M, Steinert T*. Konstruktive Auslegung der Wälzlagerung schnellaufender Werkzeugmaschinen-Spindeln. Vortrag am Lehrgang: Konstruktion von Spindel-Lager-Systemen für die Hochgeschwindigkeits-Materialbearbeitung an der TAE-Esslingen, 1989. – [58] *Giebner E*. Die Auslegung von Arbeitsspindellagerungen. SKF Publikation no. WTZ 83 06 20. – [59] *Fritz E, Haas W, Müller HK*. Abdichtung von Werkzeugmaschinenspindeln. Konstruktion 41 (1989). – [60] *Weck M*. Werkzeugmaschinen, vol. 4, Messtechnische Untersuchung und Beurteilung, 2nd edn. VDI-Verlag, Düsseldorf, 1985; – [61] *Weck M, Teipel K*. Dynamisches Verhalten spanender Werkzeugmaschinen. Springer, Berlin, 1977.

L2 *Control Systems* [1] *Berthold H*. Programmgesteuerte Werkzeugmaschinen. VEB-Verlag Technik, Berlin, 1975. – [2] *Weck M*. Automatisierung und Steuerungstechnik. Werkzeugmaschinen, vol. 3. VDI-Verlag, Düsseldorf, 1989. – [3] *Herold H-H, Massberg W, Stute G*. Die numerische Steuerung in der Fertigungstechnik. VDI-Verlag, Düsseldorf, 1971. – [4] *Simon W*. Die numerische Steuerung von Werkzeugmaschinen. Hanser, Munich, 1971. – [5] *Ammann J*. Grundlagen der Pneumatik und Hydraulik, 3rd edn. Halscheidt, Heidenheim, 1973. – [6] *Dürr A, Wachter O*. Hydraulik in Werkzeugmaschinen. Hanser, Munich, 1968. – [7] *Hemming W*. Steuern mit Pneumatik. Archimedes, Kreuzlingen, 1970. – [8] *Binder D*. Interpolation in numerischen Steuerungen. Springer, Berlin, 1979. – [9] *Schmid D*. Die numerische Bahnsteuerung. Springer, Berlin, 1979; – [10] *Walker B*. Konfigurierbarer Funktionsblock Geometriedatenverarbeitung in numerische Steuerungen. ISW Forschung und Praxis, vol. 68. Springer, Berlin, 1987. – [11] *Pritschow G*. (ed.). Die Lageregelung an numerisch gesteuerten Maschinen. Selbstverlag FISW-GmbH, Stuttgart, 1986. – [12] *Stute G*. Regelung an Werkzeugmaschinen. Hanser, Munich, 1981. – [13] *Egner R*. Hochdynamische Lageregelung mit elektrohydraulischen Antrieben. ISW Forschung und Praxis, vol. 74. Springer, Berlin, 1988. – [14] Speicherprogrammierbare Steuerungsgeräte. VDI-Ber. 481 (1983). – [15] *Storr A*. Planung und Steuerung flexibler Fertigungssysteme. Selbstverlag ISW, Stuttgart, 1984. – [16] *Spur G, Krause F-L*. CAD-Technik. Hanser, Munich, 1984. – [17] *Spur G, Stute G, Weck M* (eds). Fortschritte der Fertigung auf Werkzeugmaschinen, vol. 4 (6). Rechnergeführte Fertigung (Beiträge zur Weiterentwicklung der Automatisierungs-technik). Hanser, Munich, 1977 (1983). – [18] *Walcher H*. Digitale Lagemesstechnik. VDI-Verlag, Düsseldorf, 1974. – [19] *Philips AG*. Philips-Linear-Mess-system LM SIV. Company document, Eindhoven. – [20] *Hewlett-Packard*. Laser Transducer Systems Computer Interface Electronics. Company document, California, USA, 1976. – [21] *DIN 19 237*. Steuerungstechnik, Begriffe. – [22] *DIN 19 239*. Speicherprogrammierbare Steuerungen, Programmierung. – [23] *DIN/IEC 65A (SEC) 67*. Speicherprogrammierbare Steuerungen, T.3: Programmiersprachen. – [24] *VDI/VDE-Richtlinie 2600*; Messtechnik, paper 2. VDI-Verlag, Düsseldorf, 1973. – [25] *DIN 1319*; Grundbegriffe der Messtechnik.

L3 *Shearing and Blanking Machines* [1] *Hellwig W*. Automatisierung in der HochleistungsStanztechnik, VDI-Ber. 694 (1988) 251–73. – [2] *Hellwig W*. Entwicklungsfortschritte in der Stanzerei. Bänder Bleche Rohre 31 (1990) 1, 73–8. – [3] *Oehler/Kaiser*: Schnitt-, Stanz- und Ziehwerkzeuge, 5th edn. Springer, Berlin, 1966.

L4 *Presses and Hammers for Metal Forging* [1] *Lange K* (ed.). Umformtechnik – Handbuch für Industrie und Wissenschaft, vol. 1: Grundlagen, 2nd edn. Springer, Berlin, 1984. – [2] *Doege E. u.a.* Tiefziehen auf einfach- und doppeltwirkenden Karosseriepressen unter Berücksichtigung des Gelenkantriebs. Werkstatt u. Betrieb 104 (1971) 737–47. – [3] *Siegert K*. Einfachwirkende mechanische Karosseriepressen mit hydraulischer Zieheinrichtung im Pressentisch. ZwF CIM-Zeitschrift für wirtschaftliche Fertigung und Automatisierung 83 (1988) Sondernummer 24–6.

L7 *Industrial Robots* [1] *Spur G, Auer BH, Sinnig H*. Industrieroboter. Hanser, Munich, 1979. – [2] *Vukobratovic M, Kircanski M*. Scientific fundamentals of robotics 3: Kinematics and trajectory synthesis of manipulation robots. Springer, Berlin, 1986. – [3] ISO. Manufacturing message specification (MMS). ISO 9506, 1989. – [4] ISO: Robot companion standard to MMS. ISO/TC 184/SC 2/WG 6 N6&, 1988 (Draft). – [5] *Duelen G*. Robotersteuerungen. Automatisierungstechnische Praxis 30 (1988) 4–10. – [6] *VDI*: Industrial robot data (IRDATA). VDI 2863. 1987. – [7] *VDI-Richtlinie 2860*: Handhabungsfunktionen, Handhabungseinrichtungen, Begriffe, Definitionen, Symbole. VDI-Verlag, Düsseldorf, 1982. – [8] *Spur G*. Stand der Programmiertechnik für Industrieroboter. Vortrag am FTK '88, Stuttgart, 5–6 Oktober 1988, und Werkstatttechnik, Sonderheft FTK 17, 1988. – [9] *Prager K-P*. Kopplung externer Programmiersysteme für Industrieroboter. Reihe Produktionstechnik Berlin, vol. 33. Hanser, Munich 1983. – [10] *Pritschow G, Gruhler G*. Selbstprogrammierung von Industrierobotern durch Führung im geschlossenen Sensorregelkreis. VDI-Ber. 598. VDI-Verlag, Düsseldorf, 1986. – [11] *Balling G, Fuehrer D*. Einfache Programmerstellung für Roboter durch sensorgesteuerte Raumpunktgenerierung, Energie & Automation 9 (1987) 12–14. – [12] *Rembold U, Frommherz B, Hörmann K*. Programmiertechnik für Industrieroboter – Stand und Tendenzen. Techn. Rundsch. (1986) H. 25, 96–108. – [13] *Hocken R, Morris G*. An overview of offline robot programming systems. Ann. CIRP 35 (1986). – [14] *Spur G, Kirchhoff U, Bernhardt R, Held H*. Computer-aided application program synthesis for industrial robots. In CAD-based programming for sensor-based robots. Nato

Advanced Research Workshop, 4–6 July, Il Ciocco, Italy, 1988. – [15] *Duelen G, Held J, Kirchhoff U.* Approach for the estimation of kinematic parameters and joint stiffness of industrial robots. Robotics and flexible automatization. Proc. 5th Yugoslav Symp. Applied Robotics and Flexible Automatization, Bled, Yugoslavia, 1–4 June 1988. – [16] *Spur G et al.* Anforderungsprofile für die Weiterentwicklung der Robotertechnik. 4th Conference "Jurob 88", Ljubljana, Yugoslavia, 11–12 April 1988. – [17] *Severin F.* Planung der Flexibilität von roboterintegrierten Bearbeitungs- und Montagezellen. Hanser, Munich, 1987. – [18] *Furgac I.* Aufgabenbezogene Auslegung von Robotersystemen. Hanser, Munich, 1985. – [19] *Deutschländer A, Severin F.* Rechnerunterstützte Layout-Planung für Industrieroboteranwendungen. ZwF 81 (1986) 515–22.

Index

Abbe coefficient **L**49
Abrasives, grain size **K**114
Absorption number **C**34
Acceleration
  instantaneous centre of **A**25
  particle **A**19
    natural components **A**22
Accident prevention, *see* Safety
      standards
Accuracy, machining **L**1
Ackeret-Keller process **C**20
Acme thread **F**36
  spindles **L**16
Acoustics
  basic concepts **J**29-31
  power level **J**30
  see also Noise
Acrylate rubbers (ACM) **D**59
Acrylonitrile butadiene rubbers
    (NBR) **D**59
Acrylonitrile butadiene styrene
    (ABS) **D**56
Action control, robot **L**103
Adaptation of work to people,
      manufacturing processes
      K1
Adhesive bonding **F**21
Adiabatic systems **C**2
  changes of state **C**5, **C**14
  entropy changes **C**7
  exponent **C**11
Adjusting, assembly **K**93
Aero-emulsions **D**74
Aerodynamic drag **A**65
Aerofoils **A**66-68
Aesthetics, design for **E**19
Aggregation states **K**3
Air cooling, indirect **H**27
Air ejectors **H**26-27
Air entrainment, lubricating oils
      D74
Air pumps **H**26
Air resistance, of vehicles **A**65
Aircraft, steels for **D**40
Airy's stress function **B**37
Alitising **D**32
Aluminisation **D**32
Aluminium and aluminium alloys
      D45-46, **D**103-109
  as alloying element **D**33
  aluminium bronze **D**44, **D**102
  casting alloys **D**46, **D**109-110
  weldability **F**8
Amino plastics (MF, UF) **D**57
Amplitude-frequency response **J**14
Analog data logging **L**51-52

Analysis process, problem-solving
      E4
Anchors, piping **H**13
Anergy **C**8
Angular momentum
  conservation law **A**30, **A**32, **A**38
  equation **A**30, **A**32
Anisotropy **K**22
Annealing **D**30
  coarse-grain **D**30
  recrystallisation **D**30
  soft **D**30
  steel strength values **D**80
Annular discs, under in-plane loads
      B41
Anti-vibration mountings **F**61-62
Antoine's equation **C**10
  constants table **C**36
Anvil hammers **L**62
Arc welding machines, *see* Welding
      machines, arc
ASA (plastic) **D**56
A-S-I method (adaptation,
      substitution and/or
      integration) **K**1
Aspect ratio **A**68
Assembly process **K**91
  automated **K**96
  design for **E**19
  planning **K**93
  systems **K**95-96
Asynchronous control **L**35
Attenuation measurement **D**24
Austenite **D**26
Autocatalytic plating **K**19
Automation
  assembly **K**96
  materials handling **K**101-102
Avogadro's constant **C**9
Axial locking devices **F**32-33
Axis coordinates **L**47

Bach's correction factor **B**6
Backlash **L**18
Ball bearings, *see* Bearings, ball
Ballscrew mechanisms **L**16
Ballvalves **H**17
Bandsawing **K**48
  machines **L**88-89
Bar stress
  constant axial load **B**7
  notched bar **B**7
  temperature variation **B**7
  variable axial load **B**7
  variable cross section **B**7
Basic design **E**13

Bauschinger effect (fatigue
      strength) **B**56
Beam cutting (material removal), *see*
      Jet cutting
Beam elements, use of FEM **B**52
Beam welding **F**3-4
Beams
  arbitrary loads **B**11
  bending stresses **B**12-16
  continuous **B**33-34
  deflection **B**20-26
  equal bending loads **B**16
  forces and moments **B**8, **B**11-12
  highly curved **B**19
  lateral buckling **B**48
  line loads **B**10
  plane angled **B**11
  plane curved **B**11
  single loads **B**10
  single moments **B**10
  slightly curved **B**26
Bearing pressures, *see* Stress, contact
Bearing seals **F**98
Bearing shells/liners **F**97-98
Bearings
  ball **F**77-78, **L**16, **L**33-34
  dry **F**99
  feed drive **L**16-18
  hydrostatic jacking systems **F**99
  hydrostatic journal **F**99-100
  hydrostatic thrust **F**100-101
  lobed **F**98
  machine tool **L**1, **L**25-26, **L**31-
      34
  materials **F**97, **F**98, **F**191
  multi-pad plain **F**98
  oil flow rate **F**92
  plain
    design **F**89
    form design **F**96-98
    friction **F**89
    lubricant supply **F**96-97
  plain journal
    steady radial loading **F**89-92
    variable radial loading **F**92
  plain thrust **F**92-96
  reactions **A**34
  roller **F**78-79, **L**16, **L**32-33
  steels for **D**39
  rolling
    assembly design **F**87-89
    fatigue life **F**79-82, **F**83
    friction **F**87
    fundamentals **F**75
    heating **F**87
    limiting speeds **F**83
    load capacity **F**79, **F**82

Bearings (*cont.*)
    lubrication **F**84-86
    service life **F**83
    types **F**77-79
    sliding **L**32
    temperature calculations **F**91,
        F189-190
    types of **A**7
Beat (vibrations) **A**42
Belleville springs, *see* Conical disc
        springs
Bellows compensators **H**14
Belt drives **F**101
    flat **F**101-107
      calculations **F**107, **F**192
      constant speed ratios **F**104-
        105
      materials **F**106
      step-cone pulleys **F**103-104
      tensioning **F**105-106
      variable-speed drives **F**103-
        104
    machine tool **L**13
    synchronous **F**109, **F**193, **L**18
    toothed, *see* belt drives,
        synchronous
    V- **F**107-109
      calculations **F**108-109
      types and sizes **F**108, **F**192
Belt grinding **K**54-55
    data table **K**115
Beltram's equations **B**36
Bending arrow **B**45
Bending line, effect of shear strains
    B25-26
Bending sheet forming process
    K29-30
    methods **K**30-31
Bending strength test method **D**20
    plastics **D**60
Bending stresses
    double **B**14
    graphical determination of
        moment **B**11
    highly curved beams **B**19-20
    oblique loading **B**14
    simple loading **B**12-13
    straight beams **B**12-16
Benedict-Webb-Rubin equation (gas
      properties) **C**9-10
Bernoulli's criteria **B**12, **B**56
Bernoulli's equation **A**51-52, **A**53,
    A62, **A**66, **G**1
Bessel functions **A**47
Betti's Law (equilibrium conditions)
    B54, **B**55
Biaxial moment of area **B**12
Billet shears/parters **L**52
Bingham Medium (fluid flow) **A**59
Biot index **C**28
Black body **C**33
Blade rows, *see* Cascades
Blades, *see* Aerofoils
Blanking machines **L**53-54
Blanking processes **K**61, **K**62
    defects **K**64
    sequence **K**62-63
    special **K**65-66

    tools **K**64
Blasius formula (fluid flow) **A**53
Blister packing method **D**65
Blockboard **D**53
Blocks, modular **E**22-23
Blowing (of plastics) **D**63
Body measurements, representative
    K107-108
Boiler plate steels **D**36
Boiling point **C**2
Boiling process **C**32
Bolted connections, *see* Connections,
      bolted
Bolts and nuts **F**36-38
    large **F**45
    resilience **F**39
    steels for **D**39, **D**85-89
    stress in **B**8
    in thermal apparatus **H**8
Bonding
    adhesive, *see* Adhesive bonding
    of plastics **D**65
Boring machines **L**75-79
    horizontal **L**83-84
Boron (as alloying element) **D**33
Boron nitride (PBN), cutting
      material **K**50
Boronising **D**32
Boundary element method (BEM)
    B53
    beam **B**54
    discs, plates and dishes **B**54
Boundary layer hardening **D**31
    steels for **D**35
Boundary layer theory (fluid flow)
    A63, **A**64
Boundary zone, heat transfer **C**28
Boussinesq's formulae (stresses)
    B37, **B**38
Brainstorming method (for problem
      solving) **E**5
Brakes **F**73
    for presses **L**58
Brasses, *see under* Copper and
      copper alloys
Brazing, *see* Soldering
Brazing equipment **L**100
Bredt-Batho theory (thin-walled
      tubes) **B**28
Bredt's equations
    first **B**28
    second **B**30
Bricks **D**49-50
    fireproof **D**50
Brinell hardness test **D**20-21
Brittle materials, stress distribution **B**5
Broaching machines **L**86-87
Broaching process **K**47-48
Bronzes, *see under* Copper and
      copper alloys
Buckingham's *Π*-theorem **A**71-72
Buckling
    bars **B**45-48
    beams **B**48
    columns **B**47
    domed end closures **H**8
    energy method **B**46-47
    frames **B**47

    inelastic (plastic) **B**46, **B**50
    plates **B**48-49
    rings **B**47
    safety value **B**45
    shells **B**49-50
    torsional **B**47-48
Building board **D**53
Buildup technique, *see* LIGA process
Burmester circle/centre point
      curves **F**167
Burn characteristics, plastics **D**62
Butterfly valves, *see* Valves, flap
Butyl rubbers (IIR) **D**59

C-frames, press **L**57
Cables **A**12-13
    point load **A**13
    uniform load **A**13
Calendering, of plastics and rubbers
    D63, **K**15
Caloric state variables **C**9
    table of **C**49
Calorific value
    fuels **C**49, **C**50
    gross **C**17
    net **C**17
Calorisation **D**32
Cam control **L**38
Cam mechanisms
    kinematics **F**159, **F**168
    press **L**57
Cam-forming machines **L**83
Cam-throttle control **L**37-38
Capacity utilisation **K**100
Capillary tension **A**49
Carbides, cemented **K**49-50
Carbo-nitriding **D**31
Carbon (as alloying element) **D**33
Cardan shaft **F**65
Carnot cycle **C**19-20
Carnot factor **C**8
Cartesian coordinates **L**47
Cascades **A**68
Case hardening **D**31
    steels for **D**35, **D**80
Cast iron, *see* Iron, cast
Cast parts, manufacturing **K**4-12
Cast steel, *see* Steel, cast
Castigliano's theorem **B**26
Casting process
    aluminium alloys for **D**46, **D**109-
      111
    centrifugal **K**11-12
    cleaning operations **K**13
    continuous **K**4
    guidelines for design **K**12, **K**14
    ingot **K**3
    inspection and testing **K**15
    machines for **K**4
    manufacturing of parts **K**4-12
    melting of materials **K**13
    plastic sheet **K**15
    semi-finished products **K**3-4
    steel **D**28
    strip and wire rod **K**4
Catenary **A**12-13
Cauchy number **E**20

Cauchy-Riemann differential
    equations **A**62
Cauchy's Similarity Law **A**70
Cavity sinking, ultrasonic **K**85
Cell control **L**36
Cell-type manufacturing systems **L**97
Cellulose derivatives (CA, CP, CAB)
    D56-57
Celsius Scale **C**2
Cementite **D**26
Cements **D**51, **D**112
Centre of gravity **A**13-14
Centre of rotation, instantaneous
    A24-25
Centre-of-mass law **A**30, **A**36
Centrifugal casting **K**11-12
Centrifugal clutches **F**74
Centrifugal stresses, rotating
    components **B**43-44
Centroid gyro **A**38
Ceramic materials **D**49-50, **D**111
    inserts **K**50
    moulding **K**7-8
Cermets, see Carbides, cemented
Chain drives **F**101, **F**109-110, **F**193
Chain transmission, machine tool
    L13-14
Chains, see Cables
Changes of state
    adiabatic **C**5, **C**14
    gases and vapours
        at rest **C**14-15
        in motion **C**15-16
    irreversible **C**4
    isentropic **C**14
    for primary shaping **K**3
Channel bending, see U-bending
Chatter, machine tool **L**4
Chemical analysis methods **D**23
Chemical industry, use of
    condensers **H**24, **H**25
Chemical milling/machining **K**57,
    K61
Chemical vapour deposition (CVD)
    D48, **K**49, **K**86, **K**91
Chill casting **K**10
    low pressure **K**10
Chip formation
    grinding process **K**52
    machining **K**35-36
    non-defined tool edges **K**51
    turning **K**38-39
Chipboard **D**53
Chipless machining
    laser beam **K**83
    processes classification **K**85
Chromating **D**49
Chromising **D**32
Chromium (as alloying element)
    D33
Churchill & Chu equation (free
    convection) **C**32
Circuits, hydraulic **G**12, **G**15-16,
    L61
Circular sawing **K**49
    machines **L**88-89
Circular shears **L**52-53
Clack valves, see Valves, flap

Clamp joints **F**27-28
Clamping/support mechanisms,
    lathe **L**67
Classification systems, item **E**30-31
Clausius–Clapeyron equation
    (entropy of evaporation)
    C12
Clausius–Rankine process (reversible
    cyclic) **C**21
Clausius's inequality **C**7
Clevis joints **F**29
Clocked control **L**35
Closures
    domed **H**7-8
    flat end **H**7
Cloud point, lubricating oils **D**73
Clutches **F**65, **F**69-73
    automatic **F**74-5
    dog **F**69
    freewheel **F**75
    friction **F**69-70
        engagement slip **F**70-72
        layout design **F**72
        operating methods **F**73
        size selection **F**72
        type selection **F**72
    overrun **F**75
    press **L**58
    slip clutches **F**74
Co-generation, see Combined power
    and heat generation
Coating processes **K**86-87
    laser beam **K**83
    on metal **D**48-49
    non-metallic **D**48-49
    plastics on metals **K**86
    surface **K**90-91
Cobalt (as alloying element) **D**33
Cocks (piping) **H**17
Cold forging **K**34, **L**61
Cold forming processes **K**20
    effect on material properties **D**9
Colebrook–Nikuradse diagram **A**54
Colebrook's formula (pipe friction)
    A54
Collision forces, coupling **L**20
Collision time **A**39
Colour assessment, of plastics **D**62
Combined power and heat
    generation **C**23
Combined stresses
    axial load and torsion **B**32
    bending and axial load **B**31
    bending, axial load, shear and
        torsion **B**32
    bending and shear **B**32
    bending and torsion **B**32
    shear and torsion **B**32
Combustion processes **C**16-18
    temperature in **C**17-18
Commutators, d.c. motor **L**8
Compacting, hot-cold (heat
    treatment process) **D**32
Compensation effects, overload **E**14
Complex velocity potential **A**62
Compliance frequency response **L**4
Components, relative expansion
    E17

Composite design **E**19
Composite (integral) casting **K**12
Compressed air, pneumatic
    installations **G**17
Compressible fluids, see Fluids,
    compressible
Compression bars (spring) **F**52
Compression moulding **K**16
Compression refrigeration, see
    Refrigeration process,
    compression
Compression strength test method
    D20
    plastics **D**60
Compressors **G**16
Computer aided design (CAD)
    systems **L**105-106
Computer aided production
    planning **K**98
Computer aided quality control
    (CAQ) **K**104
Computer integrated manufacturing
    (CIM) system **L**37-38
Computer numerically controlled
    (CNC) systems **L**68
Conceptual design **E**11-12
Concrete **D**50
    blocks and slabs **D**51
    lightweight **D**51
    prestressed **D**51
    reinforced **D**50-51
Condensation process **C**10, **C**32
    principles **H**23-24
Condensers
    air-cooled **H**26
    in chemical industry **H**24, **H**25,
        H26
    direct-contact **H**25
    injection **H**25
    low-pressure saturated vapour
        H25
    for power stations **H**26
    in steam power plant **H**24
    surface **H**24
Conduction, thermal **C**26-27
Configuration guidelines **D**66
Confrontation, problem-solving **E**4
Conical disc springs **F**55-56
Connecting rods **F**173, **F**174, **F**176
Connections
    adhesive **F**21
    bolted **F**34
        loading **F**40-45
        strength **F**42-45
        tightening **F**38-40
    elastic, see Springs
    friction **F**23-28
    positive **F**28-34
    type selection **F**47-50
Consolidation index (material
    hardness) **K**21
Constant-power control **G**14
Constraints, task **E**4
Constructional interrelationships **E**4
Contact
    conformal **D**68
    contraformal **D**68
    stress **B**8

Contact controls **L**40–41
Contact corrosion **E**18
Contactor **L**40
Continuous casting processes **K**4
Continuous-path control **L**46, **L**58
   robot **L**103
Contraction coefficients **D**77
Control data processing **L**44–45
Control programs **L**35–36
Controls
   asynchronous **L**35
   cam operated **L**38
   clocked **L**35
   electrical **L**40–41
   electronic **L**40–41
   hydraulic transmissions **G**13–14
   levels of **L**37
   means of operating **L**37–38
   organisation of **L**36–37
   robot **L**102–104
   safety standards **L**38
   signal processing **L**35
   system structure **L**34–35
Convection heat transfer **C**26, **C**30–33
Coolants, refrigeration **C**22
Cooling lubricants, machining
   processes **K**52
Cooling ponds **H**27
Cooling towers **H**27
   design calculations **H**27–28
Coordinate conversion **L**47
Coplanar forces **A**2
   equilibrium conditions **A**6
Copper, weldability **F**8
Copper and copper alloys **D**33,
   D43, **D**93, **D**94
   –aluminium alloys (aluminium
      bronzes) **D**102
   –lead-tin alloys **D**103
   –tin alloys (tin bronzes) **D**101–102
   –zinc alloys (brasses) **D**43–44,
      D95–98, **D**99–100
   bronzes **D**44
Core zone remelting **D**28
Coriolis acceleration **A**25
Correction values table (cutting
   processes) **K**109
Corrosion
   design for avoidance **E**17–18
   types **D**5, **E**17–18
Corrosive media, effect on material
   properties **D**10
Cost accounting **K**107
Cost centre calculation **E**8
Cost location accounting **K**106
Cost type accounting **K**105
Costing **E**9–10
   operational **K**105–107
Costs, types of **K**105
Cottered joints **F**30
Couette flow **A**63–64
Couette viscosimeter **A**64
Coulomb's law of friction **A**42, **K**23
Counterblow hammers **L**62, **L**63
Couples **A**1
Couplings **L**20, **F**64–65

elastic **F**65–69
   resilient metal **F**68
   rigid **F**65
   self-aligning **F**65
   torsionally stiff **F**65
CR, *see* Polychloroprene rubbers
Crack ductility, test methods **D**22
Crack formation/propagation **D**3–4
   weld joints **F**6
Crank mechanisms **A**24
   components **F**172–176
   designs **F**168
   dynamics **F**170–172
   kinematics **F**168–170
   machine tool **L**14
   press **L**57
Crank-rocker mechanisms **L**14
Crankshafts **F**172, **F**174, **F**175
   oscillatory system **J**11, **J**19
Creep **B**56
   strength calculations **D**15
Cremona force diagram **A**11
Crevice corrosion **E**17
Critical loads, approximation
   methods **B**46–47
Culmann's auxiliary force **A**3
Curing process, thermo-setting
   plastics **K**3
Curved shears **L**52–53
Cutouts, pressure vessel **H**8
Cutter location data (CLDATA) **L**44
Cutting off process **K**62
Cutting processes **K**35
   speeds **L**1
   steels for **D**38
Cutting tools, materials for **K**49,
   K50, **K**117
Cyclic process **C**19
Cyclo gear mechanisms **L**19–20
Cylinders
   bound hollow **B**42
   elliptical **B**42
   pneumatic **L**15
   rotating **B**44
   thick-wall **B**43
Cylindrical shells
   under external pressure **H**6–7
   under internal pressure **H**6
Cylindrical ways **L**28
Cyril-Bath process (stretch forming)
   K28

D'Alembert's Principle **A**29, **A**36–37, **B**43
   constrained motion **A**31
   hydrostatic paradox **A**62
Dalton's Law **C**23
Damping
   machine tool **L**4
   modal **J**12
   vibration **A**42–45, **J**12
Damping couplings **F**67
Databases, manufacturing system
   L37–38
Debye temperatures **C**13
   table of **C**49
Decimal-geometric series **E**20
Deep drawing

plasticity theory **K**27–28
   technology **K**28–29
Deep-hole drilling machine **L**76
Deflection
   beams **B**20–26
   oblique bending **B**25
   slightly curved beams **B**26
Deformation **E**15, **E**16
   material flow **K**20–21
   plastic **H**7
   technology **B**56
   test methods for determining
      D19
Deformation energy theory **D**5
Degrees of freedom **A**23
Delphi method (for problem
   solving) **E**5
Demand calculation, material **K**98–99
Denavit-Hartenberg convention
   (robot structure) **L**102
Density **A**49
Design process **E**10–11
   basic **E**13
   conceptual, *see* Conceptual
      design
   detail, *see* Detail design
   embodiment, *see* Embodiment
      design
   redesign **E**13
   variants **E**13
Detail design **E**12
Deutsche Institut für Normung
   (DIN) Standards, *see* DIN
      Standards
Development and design, problem-
   solving **E**4
Diathermal walls **C**2
Diathermic body **C**33
Die, wire drawing **K**25
Diesel engine **C**18
Differential design **E**19
Diffusion coatings **D**48
Diffusion mechanisms, sintering **K**3
Diffusor flow **C**16
   drag coefficient **A**56
Digital absolute measuring systems
   L50–51, **L**102
Digital data logging **L**49–50
Digital incremental measuring
   systems **L**49–50, **L**102
Digital locators **L**102
Dilatant liquids **A**59
Dilatometer measurement **D**24
Dimensions, types of **E**25
DIN (Deutsche Institut für
   Normung) Standards **E**23–24
Dip soldering **L**100
Dipole current flow **A**62
Disc friction **A**65
Disc model, plasticity theory **K**25
Discs
   rotating **B**43–44
   rotation of two **A**26
   skewed angle rotation **A**34
   under in-plane loads **B**41

Discursive methods, problem-solving **E**5
Displacement control, motor **L**10
Displacement method, FEM **B**50
Displacement power, pump **G**5
Distortion **B**3
Distribution laws, sample evaluation **D**17-18
Dittus–Boelter–Kraussold equation (heat transfer) **C**31
DNC operation, robot control **L**103
Double-acting hammers **L**62, **L**63
Drag, solid bodies **A**65
Drag coefficient **A**55-57
Drag force **A**67-68
Drawing process **L**59
Drilling machines **L**73
  bench **L**74
  column-type **L**74-75
  deep-hole **L**76
  free-standing piller **L**74
  multi-spindle **L**75
  precision **L**76
  radial **L**75-76
  special purpose **L**76-79
  turret **L**76
Drilling process **K**41-43
  laser beam **K**83
  short-holes **K**43
Drives
  control of **L**7, **L**9-10, **L**13
  hydraulic **G**15-16, **L**61-62
    control **G**13-14
  machine tool **L**1, **L**4
  pneumatic, *see* Pneumatic drives
  press **L**57-58
Drop forging **K**33
Drop hammers **L**62-63
Drop short circuits, welding **L**97-98
Ductility **D**7
  at fracture **D**3
  crack **D**22
  processes affecting **D**8-10
Dulong–Petit rule **C**13
Dynamic loading, strength calculations **D**13
Dynamic similarity **A**70-71
Dynamic stress, weld joints **F**16
Dynamic viscosity **A**53

Economic efficiency, manufacturing processes **K**1
Effective motion **L**1
Effective principle **E**2, **E**4
Efficiency **A**28
  heat exchangers **H**4
Eigenvectors **J**12, **J**13
Elastic buckling, *see* Euler buckling
Elastic couplings, *see* Couplings, elastic
Elastic expansion, scale factor **A**69, **A**70
Elastic hardening material **B**56
Elastic limit **D**19
Elastic modulus, test methods for determining **D**19
Elasticity **B**36

Elasto-plastic material **B**56
Elastomer couplings **F**68-69
Elastomers **D**54, **D**58-59
  polyurethane (PUR) **D**59
  thermoplastically processable (TPE) **D**59
  for vibration reduction **F**66-67
Elastoviscous substances **A**59
Elbow pieces, drag coefficient **A**55
Electric slag remelting **D**28
Electrical discharge in gas cutting **F**18
Electrical machines, steels for **D**40, **D**41, **D**94
Electrical perturbation moments, excitation by **J**20
Electrochemical machining (ECM) **K**57, **K**60, **K**85
Electrodischarge machining (EDM) **K**57-59, **K**85
  cavity sinking **K**58, **K**59
  cutting by **K**59
  material pairs table **K**115-116
  thread production **K**69
Electroforming **K**19
Electrohydraulics **L**40
  drives **L**10-11
Electron beam processing applications **K**83-84
  inscriber **K**88
  lithography **K**84, **K**87
  micro-analysis method **D**23
  principles **K**83
  vapour deposition **K**84
Electroplating **K**91
Ellipsoid of inertia **A**34
Elongation **B**3, **B**4
  permanent limit **B**5
Embodiment design **E**12
  basic rule **E**13
  guidelines **E**16
  principles **E**14-16
Emission index **C**34
  table of **C**53
Empirical temperatures **C**2
EN Standards (European Standards) **E**23
Enamelling **D**49
Encoders, digital absolute measuring systems **L**50
Energy **C**3
  in closed systems **C**4
  dissipated **C**8
  forms of **C**3-4
  internal (stored) **C**4, **C**11, **C**12
  in open systems **C**4-5
Energy conservation law **A**36, **A**38
Energy efficiency, overall **C**18
Energy equation **A**29, **A**30
Energy method, buckling load **B**46-47
Energy transfer, fluid drives **G**1
Engesser–von Kármán's curve (buckling) **B**46

sizes and formats **E**27
types **E**25-27
Engineering surfaces **E**24-25
Engines
  diesel **C**18
  fan-type, force calculations **J**8-9
  flat, force calculations **J**9
  internal combustion **C**18
  machine vibrations **J**11
  pressure graph **J**1
  radial, force calculations **J**8-9
  two-cylinder, force calculations **J**5
  V-in-line, force calculations **J**8, **J**9
Engraving machines **L**82
Enthalpy **C**5, **C**11, **C**12
  humid air **C**24
Entropy **C**6, **C**12
  generation **C**6
Entropy flux **C**6
Environment, effect on material properties **D**10
Epicyclic gears **A**26
Epoxy resins (EP) **D**58
EPS, *see* Polystyrene, expanded
Equal embodiment strength principle **E**16
Equations of motion
  machine vibrations **J**11, **J**25-29
  see also Motion
Equilibrium conditions
  coplanar forces **A**6
  forces in space **A**5
  thermal **C**2
  types of **A**7
Ergonomics **E**18, **K**107-108
Ericson process **C**20
Error technology, manufacturing processes **K**1
ETFE (plastic) **D**57
Ethylene propylene rubbers (EPM, EPDM) **D**59
Euler buckling **B**45
Euler datum system (fluid mechanics) **C**1
Euler's equations of motion **A**23, **A**37, **A**51, **A**60
  for stream filament **A**51
Euler's rope friction formula **A**17-18
Euler's similarity parameters **A**71
Euler's velocity formula **A**23
Evaluation procedure, solution **E**6, **E**7, **E**8
Evaporation process **C**33
Excitation spectrum, noise **J**31, **J**34
Exergy **C**7
  closed system **C**7-8
  and heat **C**8
  losses **C**8, **H**5
  open system **C**8
Expanding process, plastics forming **K**16
Expansion
  thermal **C**12
    design for **E**17
    solids **C**12, **C**48
    table of coefficients **C**48

Engineering drawings
  conventions and scales **E**27
  dimensioning **E**27
  lines and lettering **E**27

Expansion compensators **H**13
  bellows drag coefficient **A**56
  loops **H**13
Exponential equations, use of **E**21–22
External forces **A**1
Extruders (machines) **L**61
Extrusion processes **K**15
  direct **K**34
  hydrafilm **K**35
  hydrostatic **K**35
  indirect **K**34–35
  plasticity theory **K**26–27
  plastics **D**63

Factors of safety, *see* Safety factors
Fahrenheit Scale **C**2
Fail-safe principle **E**16
Failure
  causes of **D**2
  stress failure criterion **B**6
  under alternating stress **D**3–4
  under complex stress conditions **D**4–5
  under continuous stress **D**2–3
Fatigue crack corrosion **E**18
Fatigue endurance limit **D**6, **D**13
Fatigue life, rolling bearings **F**79–82, **F**83
Fatigue notch factor **D**12
Fatigue strength **D**6
Fatigue test, plastics **D**61
Feed drive components **L**16
Feedgear mechanisms **L**18–20
Feedscrew mechanisms **L**16
FEP (plastic) **D**57
Fibre-reinforced components, manufacture **D**63
Field-vector control **L**5
Filing machines **L**88–89
Fine machining **K**81
Finishing, plastic components **D**67
Finite element analysis, drive mechanisms **F**175
Finite-element method (FEM) **B**50–53
  applications **B**52–53
  displacements **B**51
  elongations **B**51
  machine tool design **L**25
  oscillatory systems **J**22
  shears **B**51
  stresses **B**52
Fit systems **E**25
Fits and tolerances **E**21, **E**25
Fixed points, temperature scale **C**3, **C**34, **C**35
Flame hardening **D**31
Flammability, of plastics **D**62
Flange joints **F**42
  pressure vessel **H**8, **H**8–10
Flashpoint, oil **D**73
Flat-bed frames, machine tool **L**21
Flexibility matrix method, FEM **B**50
Flexible manufacturing systems **L**97, **K**102–103
Flexible turning centres **L**94
Floating bodies, stability **A**50

Flow
  criteria for material flow **K**21
  incompressible fluids **A**60
  laminar **A**52–53
  multi-dimensional
    inviscid fluids **A**60–63
    viscous fluids **A**63–68
  non-Newtonian fluids **A**59
  non-stationary **A**51
  pipe losses **H**10
  plane potential **A**62–63
  process theory **G**1
  stationary **A**51
  twisted bar analogy **B**31
Flow controllers **G**11
Flow curve, material **K**21
Flow distributors **G**11
Flow of force, *see* Force transmission
Flow law, isotropic materials **K**21
Flow regulators **G**11
Flow straighteners, drag coefficient **A**57
Flow stress **K**20
Fluid drives/transmissions **G**2–3, **L**40
Fluidics **L**39–40
Fluids
  compressible **A**53
  incompressible **A**60
Fluorine rubber (FKM) **D**59
Flywheel design **J**1–3
FMEA (Failure Mode and Effect Analysis) **K**104
Foam moulding, thermoplastic (TSG) **D**58
Foaming process **D**64
Fog range **C**24
Folding (metal bending) **K**30
Force compensation principle **E**15
Force of gravity, work **A**28
Force transmission **E**14
  by friction **F**23–28
  diagram **L**12, **L**13
Forces **A**1
  combinations of
    concurrent **A**2–3
    non-concurrent **A**3–5
  coplanar systems **A**2, **A**4
  excitation of out-of-balance **J**19
  method for statically indeterminate systems **B**33
  resolution of
    concurrent **A**3
    non-concurrent **A**4
  spatial systems **A**2, **A**5
Forging processes **K**32–34, **L**61, **L**63, **L**65–66
  aluminium alloys **D**45
  steels for **D**38
Form drag, *see* Pressure resistance
Form hardening, austenitic **D**32
Form milling **K**47
  gear cutting **K**71
Form numbers **D**78
Formability **K**22
Formed parts, design and tolerances **D**66–67

Forming
  cold, steels for **D**38
  hot, steels for **D**38
  powder metallurgy **K**18
  theory **K**23–25
Forming limit curve **K**23
Forming machines **L**54–57
Fourier analysis
  non-periodic processes **J**17
  periodic oscillations **J**17
Fourier index **C**28
Fourier's Law **A**71, **C**26, **C**27
Fracture mechanics **D**2, **D**3–4, **D**7
  test methods **D**22
Fracture strain, test methods for determining **D**19
Frames
  machine tool **L**2, **L**21, **L**57
    embodiment design **L**23–25
    materials **L**22
Framework, pin-jointed **A**10–12
Freedom of motion, mechanism **F**158
Freeing principle **A**7
Freezing point **C**2
Frequencies
  natural **J**11–13
  response functions **J**13–14
Friction
  sliding **A**15
  static **A**15
  systems analysis **D**71–72
  tube **A**53–55
  types of **D**67, **D**68
Friction corrosion **E**18
Friction coupling **L**94
Friction drives
  calculations **F**112–115
  definitions **F**110
  types **F**111–112
  use and operation **F**115–116
Friction losses, heat exchanger **H**5
Friction of motion, *see* Friction, sliding
Friction pairings, table **F**182
Frictional force, work **A**28
Frictional gears **L**14
Frictional resistance **A**65
Froude's Model Law **A**70
Froude's Similarity Law **A**70
Fuels
  calorific values table **C**49, **C**50
  composition **C**50
  liquid **C**50
Full mould casting **K**9
Function control **L**34
Functional interrelationships **E**2, **E**4

Gallery method (for problem solving) **E**5
Galvanic coatings **D**48
Gamma-ray test procedure **D**25
Gas constant, universal **C**9
Gas cutting processes **F**17–18
Gas fusion welding **F**2
Gas lubrication, bearings **L**30
Gas phase deposition coatings **D**48
Gas turbine system

Gas turbine system (*cont.*)
  closed **C**20
  open **C**18
Gas-shielded arc welding **F**2–3
Gases
  combustion characteristics **C**50
  ideal **C**9
    caloric properties **C**10–11
    flow **C**15–16
    mixtures of **C**23, **C**24
    physical properties **C**52
    radiation of **C**34
  real **C**9–10
    caloric properties **C**11
Gaskets **H**9
Gate valves **H**17
GE hypothesis (deformation) **K**21
Gear cone transmission units **L**11–12
Gear cutting
  by hobbing **K**73–74
  by planing **K**72
  by shaping **K**72–73
  form-cutting processes **K**70
  generating processes **K**70
  worms **K**77–78
Gear manufacture
  finishing **K**74–77
  grinding **K**76, **L**92
  materials and heat treatment
    F126–128
Gear pairs **F**116
Gear ratio **F**117
Gearbox, three-speed **L**11
Geared transmissions
  bearings **F**156–157
  cases/housings **F**154–156
  connections **F**154
  design **F**154
  types **F**153–154
Gears
  bevel **A**27
    cutting
      by generating methods **K**79
      by grinding **K**81
      by milling **K**80–81
      by shaping **K**78–79
      straight **F**137–138
  bevelled spur **F**138–139
  circarc **F**122
  compound planetary trains
    F150–152
  cooling **F**126
  crossed helical **F**139
  crown **F**138
  cycloid **F**122
  cylindrical **K**70–77
  efficiency **F**117
  epicyclic **F**143–152
  frictional **L**14
  grinding machine for **L**92
  helical **F**128–137
  hydraulic **L**14–15
  involute **F**120–122
  lubrication **F**125–126
  machine tool **L**1
  mangle **F**124
  mechanical **L**11
  noise behaviour **F**116

  planet, *see* Gears, epicyclic
  pneumatic **L**15
  spiral bevel **F**138
  spur **F**128–137
  superposition **F**149–150
  tooth geometry **F**117–124
  Wildhaber-Novikov (W-N) **F**124
  worm **F**139–143, **K**77–78
Gearwheel arrangement **L**12–13
Geometrical data processing (GEO)
  L45
Geometrical features **E**2
Geometrically similar series **E**21
German Standards, *see* DIN
    Standards; VDE regulations;
    VDI guidelines
Gibbs's fundamental equation **C**6
Glass **D**51–52
Gnielinski's equation **C**31
Göttingen-type profile (polar curve)
  A68
Graetz solution (infinite series) **C**31
GRAFCET (programming standard
    guideline) **L**42
Grain characteristics, for machining
    K50–51, **K**114
Graphite-bearing metals, bearings
    F98
Grashof number **C**30, **C**32
Grashof's criterion (linkages) **F**158
Gratings
  incremental measuring systems
    L49
  incremental scanner **L**49
Gravitational forces, scale factor
    A69, **A**70
Gray code, digital absolute
    measuring systems **L**51
Greases, *see* Lubricants, greases
Green's function (influence
    coefficient) **B**54
Grey cast iron, weldability **F**8
Grey radiators **C**33–34
Grids and screens, drag coefficient
    A57
Grinding machines **L**89
  centreless cylindrical **L**90–91
  cylindrical **L**89–90
  developments **L**92
  gear **L**92
  internal cylindrical **L**91
  screw thread **L**92
  surface **L**89
Grinding processes
  belt **K**54–55
  gear finishing **K**76
  inside-diameter cutoff **K**57
  rotating tool **K**52–54
  thread production **K**69
Grinding wheels **K**52
  dressing **K**53–54
Grübler's equation (degrees of
    freedom) **F**158
Guided gyro **A**38
Guides
  linear **L**25–31
  machine tool **L**1
  rotary, *see* Bearings

Gyroscopic motion **A**38
  zero-force **A**38

H-frames, machine tool **L**21
Hacksawing **K**49
Hacksawing machines **L**89
Hagen-Poiseuille formula (volume
    flow) **A**53
Haigh fatigue strength curve **D**6,
    D77
Hamilton's Principle (dynamics)
    A32–33
Hammer forging **K**33
Hammers **L**54, **L**61–63
Hand moulding **K**6
Handling systems **L**101
  computer controlled **L**101
Hard soldering **F**19
Hardening treatment **D**29
Hardness, of plastics **D**60
Hardness test methods **D**20
Harmonic drive gear mechanisms
    L19
Harmonic oscillations **J**14
Hausen's formula **C**31
Head loss **A**53
Heat **C**4
Heat conductivity **C**26
  table of values **A**52
Heat conductivity resistance, *see*
    Thermal resistance
Heat exchangers
  characteristics **H**1
  efficiency **H**4
  finned-surface **H**22
  flow arrangements **H**4
  operating characteristics **H**4
  plate-type **H**22
  recuperative, *see* Recuperators
  regenerative, *see* Regenerators
  shell-and-tube **H**21–22
  spiral **H**22
  tube-bundle **H**21–22
Heat flux **C**26
Heat penetration coefficient **C**29
Heat pump process **C**19
  compression **C**22–23
Heat transfer **C**26, **C**27–28, **H**1
  cooling tower **H**27
  with phase change **C**32–33
  without phase change **C**31–32
Heat transfer coefficient **C**27–28,
    H1, **H**24
Heat transmission, instationary **C**28–
    30
Heat treatment processes **D**30–32,
    K15
  effect on material properties **D**9,
    D29
Heating index **C**23
Hedström coefficient (fluid flow)
    A59
Heller process (condensation) **H**25
Helmholtz equation for vorticity
    A61
Helmholtz Laws **A**61
Hencky's Law (stress/strain) **B**57

Henneberg's Rod Transposition
    Process **A**11
Henry–Dalton law **D**74
Herpolhode **A**25
Hertzian contact stresses **B**38, **D**10,
    D69
    arbitrarily curved surface-plane
        contact **B**38-39
    arbitrarily curved surfaces
        contact **B**39
    cylinder-cylinder contact **B**38
    cylinder-plane contact **B**38
    sphere-plane contact **B**38
    sphere-sphere contact **B**38
Hertzian fundamental bending
    frequency **A**43
Hertzian theory (compression
    behaviour) **F**75
Hierarchical viewpoint **E**5
High vacuum casting **K**8
High-polymer materials, state for
    shaping **K**3
High-speed steels **K**49
Hobbing process **K**47
    gear cutting **K**73
Hole-based fit systems **E**25
Homogenizing **D**30
Homokinetic joint **F**65
Honing **K**55-56
Honing machines
    long-stroke **L**92-93
    short-stroke **L**93-94
Hooke's Laws **B**36
Hooke's Model Law (for similarity)
    A70
Hot forming processes **K**20
    temperature-controlled (heat
        treatment process) **D**32
Hot-dip coatings **D**48, **D**49
Humid air **C**24
    changes of state **C**25, **C**26
    table of characteristics **C**51
Hydrafilm process (extrusion) **K**35
Hydraulic circuits **G**12, **G**15-16,
    L61
Hydraulic fluids/oils **G**2
    characteristics **G**2, **G**18
    compressibility **G**1
Hydraulic gears, see Gears, hydraulic
Hydraulic linear motors, see
    Hydrocylinders
Hydraulic systems, water **G**17-18
Hydraulic transmissions, see
    Transmissions, hydraulic
Hydrocylinders **L**10
Hydrodynamic lubrication slides
    L28-30
Hydrodynamics **A**51
    flow **C**31
Hydromechanical efficiency, pump **G**5
Hydromotors **L**9-10
Hydrostatic lubrication slides **L**30
Hydrostatic transmissions, see
    Transmissions, hydraulic
Hydrostatics **A**49-50
Hysteresis loops **B**56, **D**6

ID cutoff grinding **K**57
Ideal mirror body **C**33
Identification (ID) number, item
    E30
IEC recommendations (International
    Electrotechnical
    Commission) **E**23
Π-theorem, Buckingham's **A**71-72
Impact **A**39
    eccentric **A**40
    normal **A**39
    oblique **A**40
    rotary **A**40
Impact coefficient **A**39
Impact tests **D**19
    on plastics **D**60
Impetus **A**29
In-line machines, force calculations
    J4
Inclusions, effect on material
    properties **D**8
Incompressible fluids, see Fluids,
    incompressible
Indentation, elastic **H**7
Induction heating **D**31
Inductosyn scale **L**51-52
Industrial robots, see Robots
Infiltration alloys **D**41
Influence coefficient, see Green's
    function
Influence lines **A**11
Informational work, human
    response **K**108
Ingot casting processes **K**3
Injection moulding **K**16
    plastics **D**62
Insertion devices, mechanical
    handling **K**101
Inspection and testing, casting
    process **K**15
Insulating materials, heat
    conductivity values table
    C52
Insulation losses, heat exchanger
    H5
Integral design **E**19
Interference fits **F**24-27
Internal combustion engines **C**18
Internal forces **A**1
Interpolation, see Position setpoints,
    generation
Interrelationships, engineering
    systems **E**2, **E**4
Intuitive methods, problem-solving
    E5
Inviscid fluids **A**60-63
Ion implantation **K**86
Iron
    aluminium **D**43
    cast **D**41-43, **D**92
        austenitic **D**43
        gray **D**42
        lamellar-graphite **D**42
        malleable **D**42
        pipes and fittings **H**11-12
        spheroidal-graphite (SG) **D**42
        white **D**42
    chromium **D**43

    silicon **D**43
    sintered **D**41
Iron base materials **D**26
Iron–carbon diagram **D**26
Irregulatory factor, flywheel inertia
    J2
Irreversibility principle **C**4, **C**6
Isentropic changes of state **C**14
ISO recommendations (International
    Organisation for
    Standardisation) **E**23
ISO Standard DP 9283 (industrial
    robots) **L**103
ISO thread
    metric **F**35-36
    force and torque table **F**181
    tables of sizes **F**179, **F**180
Isobaric (constant pressure) changes
    of state **C**14
Isochorous (constant volume)
    changes of state **C**14
Isothermal transformation (heat
    treatment process) **D**32
Item numbering systems **E**30-31

Jäger's load curve (bearing stress)
    B46
Jet cutting **F**18, **K**66
    machines **L**54
Jet flow **A**58-59, **C**16
    impact force **A**60
Jig boring machines **L**76
Job distribution **K**100
Job production control **K**98-100
Job production planning **K**96-98
Job schedule **K**96-98
    management **K**98, **K**100
    strategy **K**100
Joining processes **K**91-92
    see also Bonding; Connections
Jointing forces, flange **H**9
Joule process (reversible cyclic)
    C21
Journals, stress in **B**8

Kelvin Scale **C**2
Kepler's Second Law **A**30
Kienzle's approach
    drilling forces **K**43
    machining forces **K**46
Kinematic pair **F**157
Kinematic viscosity **A**53
Kirchoff boundary transverse forces
    B54
Kirchoff's Law **C**33-34
Knee-type frames, machine tool **L**21
Kutta–Joukowski equation (lift
    force) **A**63, **A**66

Labour costs **E**8
Ladder stability **A**16
Lagrange datum system (fluid
    mechanics) **C**1
Lagrange's equations (dynamics)
    A32

Lambert's cosine Law **C**33
Laminar flow **A**52-53
Laminating process **K**16
Lancaster drive **J**10
Landis process (grinding) **L**91
Langmuir–Blodgett coating process **K**86
Laplace potential equation **A**62
Lapping machines **L**94
 single-wheel **L**94-95
 spherical **L**97
 triple-wheel **L**96-97
 twin-wheel **L**95-96
Lapping process **K**56
Laser beam processing
 applications **K**82-83
 cutting processes **F**18, **K**59-60, **K**66
 principles **K**82
 surface hardening **D**31
Laser chemical vapour deposition (LCVD) **K**83
Laser cutting tool equipment **L**54
Laser interferometer **L**52
Lathes **L**66
 automatic **L**69
 C-axis **L**71
 classification **L**66
 construction **L**66
 copying **L**67
 facing **L**69
 four-axis **L**71
 front-turning **L**69
 heavy-duty **L**69, **L**71
 multi-spindle **L**69, **L**71
 numerically controlled **L**69, **L**73-75
 single-spindle **L**69
 sliding and screw-cutting **L**67
 special-purpose **L**71-72
 turret **L**67
 two-spindle **L**69
 universal **L**67
 Y-axis **L**71
Laval's compression ratio **C**15, **C**16
LD converter process (steelmaking) **D**27
Lead and lead alloys **D**47, **D**111
 as alloying element **D**33
 lead bronze **D**44
Ledeburite **D**26
Lever Law, *see* Moment-of-Momentum Law
Lewis number (heat transfer) **H**28
Lift
 aerofoils **A**66
 hydrostatic **A**50
LIGA process (build-up technique) **K**89
Limit cooling temperature **C**26
Limit slenderness **B**45
Limitation method (sample evaluation) **D**18
Line motion control **L**46
Liquid fuels, combustion characteristics **C**50
Liquids
 ideal **A**51

inviscid **A**60-63
Newtonian **A**51
non-ideal **A**51
physical properties table **C**52
pressure distribution **A**49
viscous **A**59
Lists, use of **E**5-6
Lithography processes **K**87
Ljungström-type regenerator **H**3
Load conditions
 fundamental **D**1
 pressure-thrust **D**2
 redistribution **E**14
Load-time functions **D**1
Loadbearing capabilities
 random loading **D**14
 single stage dynamic loading **D**13-14
 static loading **D**13
 under creep conditionss **D**15
Logic control **L**35, **L**42
Logical functions **E**2
Long-duration tests **D**25-26
Loss factors **A**54-57
Love's stress function **B**36
Lubricant friction **A**64
Lubricants
 additives **D**74
 choice of, table **F**185
 consistency classes **D**120
 film thickness **D**69-70, **F**126
 greases **D**75
  rolling bearings **F**85-87, **F**188
  table **F**187
 oils **D**72-75
  table **F**186
 SAE viscosity categories **D**120
 selection **E**18
 solid **D**75
Lubrication
 aerostatic **L**30
 elastohydrodynamic (EHD) **D**68-70
 gears **F**125-126
 hydrodynamic **D**69, **F**89
 rolling bearings **F**84-86

Machinability **D**24, **K**38
Machine dynamics **J**14-16
Machine enclosures, noise reduction **J**35
Machine equipment, mechanical equivalent system **J**23, **J**25
Machine noise
 generation of **J**31-35
 shielding **J**31, **J**35
Machine tools
 design of **L**25
 frames **L**21-25
 systems function structure **L**1-2
 systems mechanical characteristics **L**2-4
 thermal characteristics **L**4
Machine vibrations **J**14-16
 forced **J**16-21
 free **J**16
 paramter-excited **J**18
 self-excited **J**18

Machine-hour rate calculation **K**106-107
Machining forces
 drilling **K**110
 milling **K**45-46, **K**111
 table **K**110-111
Machining indices **K**46
Machining processes
 chipless **K**57
 with defined tool edges **K**35-50
 numerically controlled centres **L**83-85
 steels for **D**38
 without defined tool edges **K**50-57
Machining tools, materials for **K**49
Mach's Similarity Parameters **A**71
Macrostructure, investigation of **D**24
Magnesium alloys **D**46, **D**107, **D**110
Magnetic fracture testing **D**25
Magnetic moulding **K**10
Magnetic properties, steel sheets **D**89
Magnetic sensor pulse transmitters **L**49
Main technology, manufacturing processes **K**1
Maintainability, design for **E**19
Malleable cast iron, weldability **F**8
Manganese (as alloying element) **D**33
Manufacture-orientated design **E**19
 casting **K**12, **K**14
Manufacturing Automation Protocol (MAP) **L**103
Manufacturing communications systems **L**103
Manufacturing costs **E**8
 cost sheet (MCS) **K**106
Manufacturing message specification (MMS) **L**103
Manufacturing systems **K**100-103
 flexible **L**97, **K**102-103
 function structure **L**1-2
 mechanical characteristics **L**2-4
 multi-machine **L**97
 process classification **K**1
Martens thermal dimensional stability **D**61
Martensite hardening **D**32
Massive forming **L**58
Material service life, *see* Fatigue limit
Materials
 characteristics **D**78
 costs **E**8, **E**9
 design values **D**5-8
 features **E**2
 flow characteristics **K**20
 for machining tools **K**49
 selection of **D**53-54
 state for primary shaping **K**3
Materials handling **K**92-93
 automation of **K**101-102
Materials management, manufacturing processes **K**98-100
Materials testing
 fundamentals **D**17-18

Materials testing (*cont.*)
    methods **D**18–26
Mathieu's differential equation **A**49
Maurer diagram (cast iron
        structure) **D**41
Maximum distortion energy theory
        K21
Maxwell medium, *see* Elastoviscous
        substances
Maxwell's law **A**45
Measurement
    direct/indirect **L**48–49
    as part of inspection **K**93
Mechanical equivalent systems **J**11,
        J21–22
    examples **J**23–5
    parameter definition **J**23
    structure definition **J**22
Mechanical moulding **K**6
Mechanical work **C**3
Mechanisms
    crank-and-rocker **F**166–167
    dynamic properties **F**166
    kinematic analysis **F**161–163,
        F165–166
    kineostatic analysis **F**164
    running quality **F**166
    synthesis **F**166–168
    systematics **F**157–161
    see also Crank mechanisms
Melting curve **C**13
Membrane stress theory **B**41, **B**42
Membranes, vibrationnuss **A**47
Merkel number **H**28
Meshed gears **L**11
Metal cutting kinematics **K**44–45
Metal etching **K**86
Metal forming processes
    classification **K**19–20
    fundamentals **K**20–23
Metallic materials, state for shaping
        K3
Metallographic investigation
        methods **D**23–24
635 method (for problem-solving)
        E5
Methods time measurement (MTM)
        K97
MF (plastic) **D**58
Michaelson interferometer **L**52
Microminiaturisation **K**82
Microstructures
    investigation of **D**24
    manufacture of **K**88
Microtechnology (in manufacturing)
        K81–82, **K**88
Milling machines **L**71, **L**79
    bed-type **L**80–81
    circular operation **L**82
    copy system **L**82
    horizontal **L**81–82
    horizontal plane **L**80
    knee-type **L**79–80
    special-purpose **L**82–83
    universal **L**82
Milling processes **K**43–47
    guide values for ferrous materials
        K112–113

vibrations **K**46
Modal analysis **J**13
Modular network, gearing design
        L12
Modular parts lists **E**30
Modular systems **E**22–23
Modulus of elasticity **B**4, **D**5
    plastics **D**60
    typical values **D**77
Mohr's circle **B**3, **B**4
Mohr's failure criterion **B**6
Mohr's method, beam deflection
        B22, **B**25
Mol (unit symbol) **C**9
Molecular heat, gases **C**37
Molecular weight **C**9
Mollier curves **C**12
    humid air **C**24–25
Molybdenum (as alloying element)
        D33
Moment of area **B**12
Moment of displacement **A**1
Moment of impulse, *see* Angular
        momentum equation
Moment of inertia
    mass **A**36
    rotated axes **A**34–35
Moment of linear momentum, *see*
        Angular momentum
Moment-of-momentum Law **A**34,
        A36, **A**37
Moments **A**1
Momentum equation **A**28, **A**31
    incompressible fluids **A**60
Momentum Law **A**31
Motion
    of the centroid **A**30
    compound **A**23
    in cylindrical helix **A**22
    equations of **A**37
    non-uniformly accelerated **A**20
    particle **A**19–22
    plane **A**20–22, **A**24
        circular **A**22
        in polar coordinates **A**21
    relative **A**25–27
    rigid body **A**22–27
    rolling **A**36
    rotary **A**34
    spatial **A**22, **A**23, **A**37–38
    spherical **A**24
    uniform **A**19
    uniformly accelerated **A**19
Motion control
    machine tool **L**1
    robot **L**103
Motor control **L**5–8
Motor-generator welding set **L**98
Motors
    asynchronous **L**4, **L**7
    d.c. **L**7–9
    hydraulic **G**8–9
    machine tool **L**4–10
    pneumatic **G**16
    slipring **L**6
    squirrel-cage **L**5
    stepper **L**9
        electrohydraulic **L**10

synchronous **L**6
Moulding processes **K**4–6
    ceramic **K**7–8
    hand **K**6
    high vacuum **K**8
    magnetic **K**10
    mechanical **K**6
    plastics **D**63
    shell **K**7
    suction **K**7
    vacuum **K**8
Moulds, casting **K**3
Multi-component mixtures,
        condensation **H**24
Multi-degree-of-freedom systems,
        vibration **A**44–47
Multi-machine systems **L**97
Multi-point machining **L**69

Natural frequency **A**45–46
Navier–Stokes equations **A**63
Navier's equations (motion) **B**36
Needle bearings **L**16
Neutralisation capacity, lubricating
        oils **D**73
Neutralisation number (NZ) **D**73
Newton–Bertrand similarity law **A**70
Newton–Euler equations, robot
        movement **L**102
Newtonian fluids/lubricants **D**73
Newton's law of shearing stress **A**49
Newton's laws of motion **A**28–29,
        A30
Newton's similarity law **A**70
Nibbling machines **L**53–54
Nibbling process **K**62
Nickel, weldability **F**8
Nickel and nickel alloys **D**47
    as alloying element **D**33
Nikuradse's formula **A**54–55
Niobium (as alloying element) **D**33
Nitriding **D**31–32
    steels for **D**35, **D**80
Nitro-carburisation **D**31
No-load voltage, welding **L**98
Node intersection procedure **A**12
Nodular graphite cast iron,
        weldability **F**8
Noise **J**29
    airborne **J**31–33
    control regulations **L**66
    damping **J**31
    machine **J**31–36
    shielding **J**31
    spectra **J**31
    structure-borne **J**30–31
Nominal pressure (PN), piping **H**10
Nominal size (DN), piping **H**10
Non-destructive testing **D**24–25
Non-ferrous metals, physical
        characteristics table **D**115
    see also under specific names
Normalising **D**30
Norton process (grinding) **L**89–90
Notch concept (fatigue life) **D**6
Notch effect, constructional design
        D11, **D**12

Notch impact bending test **D**7, **D**21-22
  plastics **D**60
Notch impact ductility **D**21
Notch sensitivity factor **D**12
Notch stress conditions **D**11
Notching process **K**62
Nozzle, drag coefficient **A**56
NR (natural rubber) **D**59
Nuclear energy plants, steels for **D**40
Number of transfer units (NTU) **H**28
Numerical control (NC)
  basic functions **L**34, **L**45-46
  data processing **L**45-46
  definition **L**42-43
  of lathes, *see* Lathes, numerically-controlled
  workpiece programming **L**43
Nusselt number **C**30, **C**31, **C**32
Nusselt's similarity law **A**71
Nuts, *see* Bolts and nuts

O-frames, press **L**57
O-rings **H**19
Objectives, task **E**4
Octahedral stress **B**3
Offenbach system (lathe) **L**69
Offsets, forming machines **L**55
Oils
  hydraulic, *see* Hydraulic fluids/oils
  lubricating, *see* Lubricants, oil
Oldham coupling **F**65
Omega method **B**46
One-degree-of-freedom systems, vibration **A**40-44
Open arc welding **F**2
Open channel flow **A**58
Operability, design for **E**19
Operation control, robot **L**103-104
Operational costing **K**105-107
Optical diffraction systems **L**50
Orbital grinders, *see* Honing machines, short-stroke
Order calculation **K**99
Organisational plans **E**5
Oscillating extension test **D**26
Oscillation
  of machine tools **L**3
  pendulum **A**41
  rotary **A**41
Oscillator **A**32
Oscillatory signals **J**16
Oscillatory system block diagram **J**11, **J**14
Oseen formula (fluid resistance) **A**64
Oxidisation (as coating) **D**48-49
Oxygen cutting **F**17

Packing rings **H**20
Packing (stuffing box seals) **H**19-20
Pain threshold, noise **J**30
Painting **D**49
Pallet storage system **L**85
Parallel key assembly **F**30-31

Parallelogram of forces **A**5
Paris equation (crack propagation) **D**4
Particle dynamics **A**19-22, **A**28-30
  on helical curve **A**29-30
  on inclined plane **A**29, **A**30
  suspension velocity **A**65
Parts lists **E**27, **E**28
Pascal's law **A**49
Patenting (heat treatment process) **D**32
Patterns, use of in moulding **K**4, **K**6
Pearlite **D**26
Péclet number **C**30
Péclet's similarity law **A**71
Pendulum oscillation **A**31-32, **A**41
Permanent moulds **K**10-12
Petukhov's equation, modified (heat transfer) **C**31
PFA (plastic) **D**57
Phase-frequency response **J**14
Phenolic resins (PF) **D**57
Phosphating **D**49
Photoelectric pulse transmitters **L**49
Photolithography **K**87
Physical analysis methods **D**23
Physical effects **E**2
Physical operative principle **E**2
Physical phenomena
  investigation of **E**5
  table of properties **C**52-53
Physical vapour deposition (PVD) **D**48, **K**49, **K**86, **K**90
Pin-jointed frames
  planar **A**10
  spatial **A**12
Pinned joints **F**28
Pipe flow calculations
  drag coefficient **A**56
  inviscid liquids **A**60-63
  losses **H**10
  Newtonian liquids **A**52-59
  non-Newtonian liquids **A**59
Pipes/piping
  diameter **H**10
  expansion compensators **H**13
  fittings **H**11-13
  heat conduction through wall **C**27
  heat transfer **C**31-32
  materials **H**11
  standards **H**10-11
  steels for **D**40
  supports **H**14
Piston machinery, excitation forces **J**19
Pistons **F**173-174, **F**175, **F**176
  movements **F**169-170
Pitot tube **A**52
Plain face milling **K**44-47
Planck's radiation Law **C**3
Plane stresses **B**37
Plane surface structures, production of **K**87
Planing machines **L**85-86
Planing process **K**47
Planning systems, computer-based **L**107-108

Plastic hinge **B**56
Plasticity theory **B**55-58, **K**23-25
Plastics **D**54
  for bearings **F**98
  chemical characteristics **D**61
  elastomers **D**58-59
  electrical characteristics **D**61
  forming **D**62, **D**63-66, **K**3
    calendering **K**15
    characteristics **D**61-62
    compression moulding **K**16
    design **D**66-67
    expanding **K**16-17
    extrusion **K**15
    injection moulding **D**62, **D**63, **K**16
    laminating **K**16
    sheet casting **K**15
    transfer moulding **K**16
  mechanical characteristics **D**60-61, **D**116-120
  physical characteristics **D**115-118
  structure and properties **D**54-55
  testing **D**59-62
  thermal characteristics **D**62
  types
    cellular **D**58-59
    fluorinated **D**57
    foamed **D**58-59
    thermoplastic **D**55-57
    thermosetting **D**57-58
Plastics parts
  finishing **D**67
  testing **D**62
Plate elements, use of FEM **B**53
Plate shears **L**52
Plates **B**39
  buckling **B**48-49
  circular **B**40
  elliptical **B**40
  rectangular **B**39
  thermal stresses **B**41
  thickness/die clearance table **K**117
  triangular **B**40
  under in-plane loads **B**41
Plating techniques **D**48
Plywood **D**53
Pneumatic drives/transmissions **G**2, **G**16-17
  graphical symbols **G**3, **G**19
Pohlhausen formula (laminar flow) **C**31
Poinsot's law (gyro setting) **A**38
Point-to-point control **L**46
  robot **L**103
Poiseuille parabola **C**31
Poisson's ratio **B**4, **D**5
Polar coordinates, plane motion **A**21
Polar curve **A**68
Polhode **A**25
Polyacetal resins (POMS) **D**55
Polyacrylate (PMMA) **D**56
Polyalkylene terephthalates **D**55
Polyamide (PA) **D**55
Polycarbonate (PC) **D**55

Polychloroprene rubbers (CR) **D**59
Polycrystalline diamond (PCD),
          cutting material **K**50
Polyester resins, unsaturated (UP)
          D57
Polyesters, linear (PET PBT) **D**55
Polyethylene (PE) **D**56
Polygon of velocities, *see* Centre of
          rotation, instantaneous
Polygon-type connections **F**32
Polyimide (PI) **D**56
Polymer concrete, uses of **L**23
Polymers **D**54
Polyolefins **D**56
Polyphenyl ether, modified (PPE)
          D55
Polyphenyl sulphide (PPS) **D**56
Polypropylene (PP) **D**57
Polystyrene (PS) **D**56
     expanded (EPS) **D**58
Polysulphone (PSU/PES) **D**56
Polytetrafluoroethylene (PTFE) **D**57
     as lubricant **D**76
Polytropic changes of state **C**14
Polyvinyl chloride (PVC) **D**57
     plasticised (PVC-P) **D**57
     unplasticised (PVC-U) **D**57
Position control **L**48–49
     fluid transmissions **G**3
     measuring systems for NC
          machines **L**48–52
     setpoints generation **L**46–47
Potential flows **A**61–62
Potential vortex, *see* Vortex line flow
Pour point, lubricating oils **D**73
Powder metallurgy **D**28–29, **K**17–19
     coating **K**91
     for cutting steels **K**49
Power **A**28
Power shift gears **L**11
Power stations, condensers for **H**26
Power transfer/transmissions, fluid
          drives **G**1, **G**2–3
Prandtl number **C**30, **C**32, **C**52
Prandtl–Eyring formula **A**59
Prandtl–Reuss's law (stress/strain)
          B57
Prandtl's Similarity Law **A**71
Prandtl's soap film analogy **B**31
Precession cone **A**38
Precision engineering **K**81–82
     casting **K**8–9
     forming machines **L**56
Preferred numbers, *see* Series,
          preferred numbers
Presses **L**54, **L**55–57
     flywheel spindle **L**64–66
     hydraulic **L**60–62
     mechanical **L**55–60
     pneumatic **L**60–62
     safety regulations **L**65–66
     screw, *see* Presses, flywheel
          spindle
     wedge **L**59
Pressing
     plastics **D**63
     steels **D**38
Pressure **A**49

in container **A**49
     cutoff control **G**14
     loss of **A**53
     resistance **A**65
Pressure die casting **K**11
Pressure welding **F**4–5
Prestressed shaft-hub connections
          F32
Primary functions **E**2
Primary shaping **K**2–3
     tools for **K**3
Prime cost **E**8
Printing processes **K**87
Problem-solving **E**4
     solution evaluation **E**6–10
     solution principles **E**5–6
Process control **L**42
Process quality index **H**4–5
Production control **K**98–100
Production planning **K**96–98
     computer-aided **K**98
     estimation of costs **E**8
     sequence **K**100
Production structure **E**4
Production supervision **K**100
Production technological tests **D**24
Profiling machines **L**83
Program simulations **L**105–106
Programmable logic controller
          (PLC) **L**41–42
     programming **L**42
Programming procedures **L**104–106
     control systems **L**34, **L**36
     direct **L**104
     hybrid **L**105
     indirect **L**105
     off-line **L**105–106
Projection, drawing standard **E**30
Protective systems **E**16
Pulleys
     resistance at **A**18
     rope friction **A**17
Pulsation test, longtime **D**25
Pulse transmitters, photoelectric **L**50
Pumps
     centrifugal **J**24–25, **J**28–29
     condensate **H**27
     cooling water **H**27
     gear-type **G**6
     hydraulic **G**4–5
     lift **G**5
     piston **G**7
     port-plate **G**8
     power rating **G**5
     rotary **G**5
     rotary piston **G**6
     screw **G**6
     swashplate **G**8
     tilting-head **G**8
     vane-type **G**6
     variable displacement **G**7, **G**14
Punching **K**62–63, **K**65
     steels for **D**38
Punching machines, automatic **L**53
PUR, *see* Elastomers, polyurethane
PVDF (plastic) **D**57

Quality assurance **K**103
Quality engineering **K**103–105

Quality systems **K**103
Quantity lists **E**28

Rack-and-pinion drive **L**18
Radiation coefficient (heat) **C**33
Radiation coefficient (noise) **J**30–
          31, **J**32
Radiation exchange number **C**34
Radiation process **C**26, **C**33–34
     effect on material properties **D**10
     of gas **C**34
Radius of inertia **A**36, **B**13
Random loading, strength
          calculations **D**14
Rapid-traverse gears **L**15
Ratio, choice of, *see* Series, preferred
          numbers
Ratio traction drives, *see* friction
          drives
Rationalisation, manufacturing
          processes **K**1
Rayleigh number (heat transfer)
          C32
Rayleigh quotient (buckling) **B**46
Rayleigh quotient (vibrations) **A**46
Reactance circuit **L**15
Reaction equations, combustion **C**17
Reaction foams (RSG) **D**58–59
Reciprocating machines
     multi-cylinder
          force calculations **J**4
          moment of inertia **J**4
          torque fluctuations **J**1–3
Rectangular thread **A**17
Recuperators, thermodynamic
          design **H**1–3
Recycling, design for **E**19
Redesign process **E**13
Reduced mass **A**36
Redundancy principle **E**16
Reference surface/profile **E**24
Refrigeration process **C**19
     compression **C**22
Regenerators, thermodynamic
          design **H**3–4
Regression analysis, cost estimation
          using **E**9
Reinforcing overall effect **E**14
Relative costs **E**9
Relative humidity **C**24
Relative motion **A**39
Relaxation, creep **B**56
Relaxation test **D**25
Relay, *see* Contactor
Reliability principles **E**15–16
Remelt coating, laser beam **K**83
Requirements list, design **E**10–11
Resistance
     rolling **A**18
     tractive **A**18
Resistance fusion welding **F**4
Resistance pressure welding **F**4
Resistance welding machines, *see*
          Welding machines,
          resistance
Resonance **A**42–43, **E**17
     design for avoidance **E**17
     engine compensation **J**9–10

Response functions, frequency **J**13–14

Reverse side machining **L**71

Reversible changes of state **C**4

Reversible processes **C**6

Reynolds numbers **A**53, **A**63, **C**30

Reynolds similarity law **A**70, **A**71

Rheopexy **D**73

liquids **A**59

Rheostatic control, motor **L**10

Ribbing, rigidity requirement **L**23

Rigid body dynamics **A**33–38

acceleration **A**37

general plane motion **A**36

main axes **A**34

moment of inertia **A**34

plane motion **A**36–37

rotation about a fixed axis **A**33

straight-line motion **A**28–30

systems **A**10

Rings, rotating **B**43

Ritter's Method of Sections **A**11

Ritz formula (vibrations) **A**46

Ritz formulation (buckling) **B**49

Riveted joints **F**33–34

Robots

accuracy characteristics **L**102

applications and selection **K**102–103, **L**107–108

components **L**101–102

control systems **L**102–104

dynamic model **L**102

kinematic model **L**102

programming **L**104–106

Rocket drive equation **A**33

Rockwell hardness test **D**21

Rod transposition process **A**11

Roller bearings, *see* Bearings, roller

Roller milling machines **L**82

Roller slideways **L**30–31

Rolling motion

inclined plane **A**36

resistance **A**18

Rope friction **A**17–18

Rotary slotting machines **L**82

Rotating components

centrifugal stresses **B**43–44

slide crank **A**27

Rotating tube, motion in **A**26, **A**39

Rotation

law of motion **A**34

rigid-body **A**22

about a point **A**24

plane representation **A**23

vectorial representation **A**22

Rotors

balancing **J**15

motor **L**9

shaft mechanical equivalent system **J**23, **J**28

Rötscher cones **F**38

Roughness, tube **A**54

Roughness values, surface **E**24–25

Roulette **A**25

Rubber **D**58–59

bearings **F**98

properties **F**59, **F**61

springs, *see* Springs, rubber

SAE viscosity categories **D**120

Safe-life principle **E**16

Safety factors

strength calculations **D**16

thermal apparatus **H**5–6

Safety principles **E**15–16

buckling **B**45

design for **E**18

Safety regulations/standards **E**24

automation system **L**37

health **K**108

laser cutting **K**60

press working **L**65–66

Safety stress, *see* Stress, permissible

Safety valve **H**16

Saint-Venant's theory (warping) **B**30

Sampling methods, cast metals **D**17

Saturation, vapour **C**10

Sawing machines **L**87–89

Sawing processes **K**48–49

Scale factors **A**69

Screen printing **K**87

Screw motion **F**35

Screw thread grinding **L**92

Screwed connections, piping **H**12

Screws **A**17

Seals **H**18

bearing **F**98

dynamic contact **H**19–20

rolling bearings **F**88–89

rotating mechanical **H**20

static contact **H**18–19

Seamless tube steels **D**36–37

Second moment of area, beam bending **B**13–14

Secondary functions **E**2

Section tables, pipes and rolled sections **B**59–75

Sectionalised machining **L**69

Seiliger process **J**1

Selection procedure, solution **E**6

Self-help principle, system **E**14

Self-locking

epicyclic gears **F**147–149

screws **A**17

Semi-finished products, manufacturing **K**3

Semi-infinite bodies, heat transfer **C**28–29

Semi-similar series **E**21

Sensor data processing **L**103

Sequence control systems **L**35

cascade representation **L**42

robot **L**103

Series, preferred numbers **E**20–21, **E**24

Service life

experimental determination **D**15

prediction **D**15

rolling bearings **F**83

Servodrives **L**5

Servohydraulics **L**40

Servomotor **L**5

Severing processes

classification **K**61–62

forces and energies **K**63–64

technology **K**62–63

tools **K**64

workpiece properties **K**64

Shaft-based fit systems **E**25

Shafts

critical speed **A**43–44, **A**45

torsional buckling **B**47

Shaping machines **L**85–86

Shaping process **K**47

primary **K**2–3

Shaving (gear finishing) **K**74–75

Shear centre **B**18

Shear stability, lubricating oils **D**73

Shear stress **B**1

deformation **B**25–26

distribution **B**16–19

in fasteners **B**18

forces **B**8, **B**16

transverse **B**7

Shearing machines **L**52–53

Sheet cutting, laser beam **K**83

Sheet forming

metal **L**59, **L**61

superplastic **K**31

Shell elements, use of FEM **B**52, **B**53

Shell moulding **K**7

Shells **B**41–42

bending rigid **B**42

buckling **B**49–50

under internal pressure **B**42

Sherardisation **D**32

Short-circuits, shaft-loading by **J**21

Siemens–Martin process (steelmaking) **D**27

Silencers **J**34

Silicon (as alloying element) **D**33

Silicon technology **K**88–89

anisotropic etching **K**88

Silicone rubbers (VOM) **D**59

Siliconising **D**32

Similarity mechanics **A**69–72, **E**20

cost estimation using **E**9–10

thermal **A**71

Sintered materials **D**41, **D**91

aluminium alloys **D**46

metal bearings **F**98

Sintering process **K**3, **K**17, **K**18–19

fluidised bed **K**91

635 method (for problem-solving) **E**5

Size effect, constructional design **D**11, **D**12

Skin packing method **D**65

Skull (extrusion product) **K**34

Slider-crank mechanisms **L**14

press **L**57, **L**58

Slideways **L**26–31

Sliding pairs, oil-lubricated **D**69

Sliding-gear transmission units **L**11

Slip couplings **L**20

Slitting **K**62

Slotting/grooving machines **L**82, **L**85–86

automatic **L**53

Smelting process, steelmaking **D**26–28

Smith creep–strength diagram **D**6, **D**9, **D**76

Soldering **F**18–20
  equipment for **L**100
  soft **F**19
Solid body filling, flow through **A**57
Solid fuels, combustion
    characteristics **C**50
Solid lubricants, *see* Lubricants, solid
Solids
  caloric properties **C**12–13
  heat conductivity values table
      C52
  physical properties table **C**53
Solution heat treatment **D**30
Solutions, evaluation procedures
    E6–10
Sommerfeld coefficient
    (hydrodynamic load
    capacity) **F**90
Sound, velocity of **C**16
Sound emission analysis **D**25
Sound-deadening measures **J**34–36
Soundproof boxes **J**36
Space frames **A**12
Spacecraft, steels for **D**40
Spark ignition engine **C**18
Special-purpose machines, mass
    production **L**97
Specific heat capacity **C**11
  of air **C**36
  measuring **C**13
Spectral analysis method **D**23
Speed flow diagram, gearing design
    L12
Spheres, thick-wall **B**43
Spindle bearings **L**16–18
Splined joints **F**31–32
Spot welding units **L**99
Spray coatings **D**48
Spring characteristic, linear **A**40
Spring force, work **A**28
Spring rate **A**41
Spring steel, *see* Steel, spring
Spring-back, elastic **K**30
Spring–mass system **A**40–41
Springs **F**50–51
  fibre composite **F**62–63
  gas (air) **F**63–64
  helical **F**57–59
  leaf **F**53–54
  metal **F**51–59
  ring **F**53
  rubber **F**59, **F**61–62
  spiral **F**54–55
Sputtering **K**86
Square-threaded screw **A**17
Stability **A**7, **E**16–17
  design for **E**16
  floating bodies **A**50
Stainless steel, *see* Steel, stainless
Staircase method (sample
    evaluation) **D**18
Standardisation process **E**23–24
Standards
  basic **E**24–25
  European **E**23
  German, *see* DIN Standards
  industrial **E**24
  international **E**23

sources of **E**23–24
use of **E**24
Stanton number **C**30
State variables tables
  ammonia **C**43
  carbon dioxide **C**34
  difluorodichloromethane **C**46
  difluoromonochloromethane **C**47
  monofluorotrichloromethane **C**45
  water and steam **C**40–42
Static loading, strength calculations
    D13
Static rigidity, machine tool **L**3
Static similarity **A**69–70
Statically determinate systems **A**10
Statically indeterminate systems
    A10, **B**32–33
  annular beams **B**35
  annular frames **B**35
  closed rectangular frame **B**34
  constrained frame **B**34
  continuous beams **B**33–34
  double jointed frame **B**34
  frame with semi-circular curves
      B35–36
  projecting frame **B**34
Statistical process control (SPC)
    K104
Steady state processes **C**4–5
Steam
  superheated **H**24
    heat transfer **H**24
Steam power plant **C**21
  use of condensers **H**24
Steam table **C**38–39, **C**40–42
Steam trap **H**16
Steel
  alloying elements in **D**33
  boiler-plate **D**36
  case-hardening **D**80
  cast **D**28, **D**40–41, **D**90
    heat-resisting **D**40–41, **D**90
    wear-resisting **D**41
  cryogenic, *see* Steel, low-
      temperature
  designation/classification **D**32–33
  forged **D**33
    pearlitic **D**34
  free-cutting **D**35–36
  heat-resisting **D**37, **D**81, **D**82,
      D83
  high-alloy **D**33
  high-temperature **D**36–37
  low-alloy **D**33
  low-temperature **D**38, **D**81
  magnetic **D**90
  maraging **D**35
  nitriding **D**35, **D**80
  non-ageing **D**34
  pressurised-hydrogen-resisting
      D37
  rolled **D**33
  sintered **D**41
  spring **D**39, **D**87–89
  stainless **D**36, **D**81
  structural **D**34, **D**79
    special **D**34
    weatherproof **D**34

weldable **D**34
  tool **D**38, **D**39, **D**84
Steel pipes/tubes **H**10
  connections **H**11–12
Steel–Iron Test Sheet 1570-1 **D**8
Steelmaking processes **D**26–29
  effect on material properties **D**8
Stefan–Boltzmann Law (radiation)
    C33
Steiner's Principles (parallel axes)
    A35–36, **B**13
Stephan & Preusser formula
    (evaporation pressure) **C**33
Stephan formula (laminar flow) **C**31
Stepless drives **L**13, **L**14
Stepper motors, *see* Motors, stepper
Stereolithography **K**83
Stock management **K**100
Stock-keeping policy **K**98–99
Stokes resistance formula **A**64
Stoneware **D**50
Strain **B**3–4
Strain energy **B**4, **B**26
Strain–time functions **D**1
Stream filament **A**51
Stream tube **A**51
Streamline **A**51
Strength properties
  annealing steels **D**78
  dynamic **D**12
  evaluation of **D**17–18
  static **D**11
  typical values **D**79
Strength theories **D**4
Stress **B**1
  alternating conditions **D**3–4
  axisymmetric **B**36–37
  bending, *see* Bending stress
  combined, *see* Combined stresses
  contact **B**8
  continuous conditions **D**2–3
  distribution of **B**5
  internal conditions **D**2
  one-dimensional **B**1
  permissible **B**4–5
  plane **B**37
  residual **B**56
  static conditions **D**5
  three-dimensional (spatial) **B**2
  two-dimensional (plane) **B**1–2
  see also Bar stress
Stress crack corrosion **E**18
Stress dynamic conditions **D**5–6
Stress failure criterion
  maximum principle **B**6
  maximum shear strain energy **B**6
  maximum shear (Tresca) **B**6
  Mohr's **B**6
Stress relief, heat treatment **D**30
Stress-orientated design, casting
    K13, **K**14
Stress-rupture test **D**25
Stress–strain diagrams, closed, *see*
      Hysteresis loops
Stress–strain laws, plastic theory
    B58–59
Stress–time functions **D**1
Stretch forming **K**28

Stribeck curve (friction state) **D**69, **F**89
Strings, vibrations **A**47
Strip model, plasticity theory **K**24
Strip shears **L**52
Strip and wire rod casting process **K**4
Stripping strength, thread **F**44
Structural parts lists **E**28, **E**30
Strutt's map (vibrations) **A**49
Styrene acrylonitrile copolymer (SAN) **D**56
Styrene butadiene rubbers (SBR) **D**59
Styrene butadiene (SB) **D**56
Subfunctions **E**2
Sublimation curve **C**13
Subsystems **E**1
Suction moulding **K**7
Sulphur (as alloying element) **D**33
Superfinishing machine, *see* Honing machines, short-stroke
Superhoning machines, *see* Honing machines, short-stroke
Superplastic forming **K**31
Support, types of **A**7
Support reactions
    planar **A**7-9
    spatial **A**9
Surface coating processes **K**90-91
Surface conditions, effect on material properties **D**9-10
Surface equation **A**30
Surface pressure **B**38-39, **D**10, **D**69
    table **F**180
Surface roughness, *see* Roughness values, surface
Surface tension **A**49
Surfaces, engineering, *see* Engineering surfaces
Swiss system (lathe) **L**69
Swivel mechanisms, press **L**59
Synchro-resolvers **L**51, **L**101-102
Synchronisation faults, shaft-loading by **J**21
Synthesis process (for problem-solving) **E**4-5
System programs **L**36, **L**42
Systematic approach **E**4-10
Systems
    composition **E**1
    coupling **E**1
    elements **E**1
    state of **C**1

Tantalum (as alloying element) **D**33
Taper-pinned joints **F**28
Tapping (spiral drilling) **K**68
Tasks, division of **E**14
Taylor function (tool life) **K**41
Technical work **C**4
Techno-economic evaluation **E**7
Technotronics **L**34
Teleoperators **K**101
Temperature
    conductivity **C**28
    effect on material properties **D**10

equalisation **C**29-30
    mixing **C**13
Temperature profile **C**29-30
Temperature scales **C**2-3
    international practical **C**3, **C**34
Tempering treatment **D**30
    steels for **D**35
Ten-point height **E**25
Tensile test method **D**18-19
    hot **D**19
    for plastics **D**60
Tension bars (spring) **F**52
Testing
    materials, *see* Materials testing
    as part of inspection **K**93
    of plastic parts, *see* Plastic parts, testing
    quality system **K**104-105
Tetmajer's method (inelastic buckling) **B**46
Thermal apparatus, design of **H**1-4, **H**5
Thermal deburring **K**61
Thermal efficiency, engines **C**18-19
Thermal inverse mixing **H**5
Thermal machining **K**57
Thermal power plants **C**20-22
Thermal resistance **C**26, **C**27
Thermal state variables **C**9-10
Thermoanalysis **D**24
Thermochemical treatments **D**31
Thermodynamic temperature scale, *see* Kelvin Scale
Thermodynamics
    First Law of **C**3-5
    processes **C**1, **C**2
    Second Law of **C**6-7
Thermomechanical treatments **D**32, **K**57, **K**61
Thermometer **C**2
    fixed points **C**3
Thermoplastic foams (TSG) **D**58, **D**64
Thermoplastics **D**54, **D**55-57
    aggregation state **K**3
    hot forming **D**64-65
Thermosets, *see* Thermosetting plastics
Thermosetting plastics **D**54, **D**57-58
    aggregation state **K**3
Thermosymmetrical construction, machine tool **L**4
Thixotropy **D**73
    liquids **A**59
Thomas process (steelmaking) **D**27
Thread locking devices **F**46-47
Thread production
    chasing **K**67
    die cutting **K**67
    EDM **K**69
    forming **K**70
    grinding **K**69
    milling **K**68
    pressing **K**70
    rolling **K**69
    tapping **K**68
    turning **K**67

whirling **K**68
Thread run-out **F**45
Thread-cutting milling machines **L**82
Threads
    fatigue loading **F**45
    types **F**35-36
Three-jaw chuck **L**67
Three-jointed arch **A**10
Three-pole theorem (kinematics) **F**160
Threshold strength **D**6
Throttle devices, drag coefficient **A**56
Throttling, flow **C**5
Thyristor drives **L**14
Tightening procedures (bolted connections) **F**38-40
Timber, *see* Wood
Time characteristics, forming machines **L**56
Time management
    control **L**34
    job schedule **K**97
    manufacturing processes **K**100
Tin and tin alloys **D**47, **D**111
    tin bronze **D**44, **D**101-102
    tin-lead-bronze **D**103
Titanium alloys **D**33, **D**46-47, **D**110
Tolerances, *see* Fits and tolerances
Tool handling
    automated **K**102
    carrier **L**1
    change time **L**84
    magazines/changers **L**83-85
    systems **L**101
Tool life **K**41
Tool steel, *see* Steel, tool
Tooth accuracy, gears **F**124-125
Tooth profiles, gears **F**118-119
Torque, work **A**28
Torque fluctuations, reciprocating machines **J**1-3
Torsion
    arbitrary cross section bars **B**30
    circular bars **B**27-28
    thin-walled tubes **B**28, **B**30
    with warping constraints **B**31
Torsion bar springs **F**56-57
Torsion limit **D**5
Torsional oscillations, examples **J**25-29
Total base number (TBN), lubricating oils **D**73
Tough materials, stress distribution **B**5
TPE, *see* Elastomers, thermoplastically processable
Tracer control **L**38-39
Tractive resistance **A**18
Trajectory, particle **A**19, **A**20-21
Transfer functions **F**161
Transfer law **A**69
Transfer lines
    automated **K**102
    mass production **L**96
Transfer moulding **K**16
Transformation operations **E**1

Translation, rigid-body **A**22
Transmission angle, kinematics **F**166
Transmission components, machine tool **L**1
Transmission electron microscopy **D**24
Transmission ratio, gears **F**117
Transmission units **L**11–15
  gearing **L**12
Transmissions
  geared, *see* Geared transmissions
  hydraulic **G**1–2, **G**4–9, **L**30–40
    graphical symbols **G**3, **G**19
    operation **G**12–13
  pneumatic, *see* Pneumatic transmissions
Transmitted-light optical diffraction systems **L**50
Trapezoidal threads **A**17
Tresca failure criterion **B**6
Tresca flow conditions **B**56
Tresca's shear stress hypothesis **K**21
Triangular threads **A**17
Tribology **D**67, **E**18
  system characteristics **D**71–72
Trimming **K**62
Triple point **C**10, **C**13
TSG (plastic) **D**58
Tube plates **H**7
Tubes
  bends **B**42
    use of FEM **B**53
  friction in **A**53–55
  model plasticity theory **K**25
  steels for **D**40
  under internal pressure **H**6
Tungsten (as alloying element) **D**33
Turbo-generator, mechanical equivalent system **J**23–24, **J**26–28
Turbosystem, vibrations **J**26
Turbulent film flow, bearings **F**92
Turbulent flow **A**52–53
Turning centre **L**71
  flexible **L**71, **L**77
Turning mills **L**71
Turning process **K**38–41
Turret drilling machines **L**76
Turrets (tool carriers) **L**67, **L**68
TÜV (Technical Monitoring Associations) regulations **E**18, **E**24

U-bending (metal forming) **K**30
UF (plastic) **D**58
Ultrasonic processing
  applications **K**84–85
  principles **K**84
  testing **D**25
Universal joint **F**65
Unsteady state processes **C**5
UP (plastic) **D**58
Upsetting process **K**32
  cylindrical parts **K**25
  square parts **K**25
User programs **L**36, **L**42
  generation of **L**104–106

User safety, *see* Safety regulations/standards

V-belt drives, *see* Belt drives, V-
V-bending (metal forming) **K**30
V-scanner, digital absolute measuring systems **L**51
Vacuum forming, thermoplastics **D**64–65
Vacuum metallising **K**86
Vacuum moulding **K**8
Vacuum-pouring technique **D**28
Value analysis **E**7, **E**10
Valves
  control **H**14–18
  directional control **G**9, **G**10
  drag coefficient **A**56
  flap **H**18
  flow control **G**11
  gate **H**17
  hydraulic **G**9–11
  materials for **D**39
  pneumatic **G**16–17
  pressure control **G**10–11
  proportional **G**11
  rotary gate **H**17
  seat **G**9
  shutoff, *see* Valves, control
  shuttle **G**10
  slide **G**10
  throttle **G**11
  types **H**15–17
Vanadium (as alloying element) **D**33
Vapour deposition, electron beam **K**84
Vapour pressure curve **C**10, **C**12, **C**13
Vapour Tables **C**12
Vapour-jet air ejectors **H**26–27
Vapour–liquid critical properties **C**35
Vapours **C**10
  caloric properties **C**11
  mixture with gases **C**24
Variable mass systems **A**33
Variant design **E**13
  value concept **E**8
VDE (Association of German Electrical Engineers) regulations **E**24
VDI (Association of German Engineers) guidelines **E**24
Velocity of sound **C**16
Venturi tube **A**52
Vertical boring mills **L**71
Vessels, flow from **A**57–58
Vibration isolation, *see* Anti-vibration mountings
Vibration model, *see* Mechanical equivalent system
Vibration reduction **F**66–67
Vibrations
  extensional **A**46
  flexural **A**45–46, **J**28–29
  forced **J**26, **J**27, **J**29
  forced damped **A**43, **A**45
  forced undamped **A**42, **A**44

free damped **A**42, **A**44
free undamped **A**40–42, **A**44
machine **J**10–11
  fundamentals **J**11–14
  problems **J**14–16
membrane **A**47
milling process **K**46
non-linear **A**48
parametrically excited **A**48
plates **A**47
representation in frequency-domain **J**17
representation in time-domain **J**16
string **A**47
torsional **A**46
Vickers hardness test **D**21
Vincent friction screw press **L**65
Virial coefficients **C**9
Virtual work principle **A**6, **A**29, **B**26–27
Viscoelasticity theory **B**56
Viscosity
  hydraulic fluids/oils **G**2, **G**18
  lubricating oils **D**73–74, **D**119, **D**120
Viscosity index **D**73
Viscous flow **A**52–59
Volume strain **B**4
Volumetric work **C**3–4
VOM, *see* Silicone rubbers
von Mises's failure criterion **B**6
von Mises's flow conditions **B**57
Vortex lines **A**61, **A**62
Vortex tubes **A**61
Vulcanisation **D**59

Wall, conduction through **C**26–27
Ward-Leonard speed control system **L**8, **L**14
Warping, twisted bar **B**30
Washers **F**37
Waste heat utilization **H**1
Water content, humid air **C**24
Water hydraulic systems **G**17–18
Water torches **L**54
Water-jet air pumps **H**26
Wave soldering **L**100
WCTFE (plastic) **D**57
Wear process **D**5, **D**68, **D**70
  design to limit **E**18
  during milling **K**47
  resistance tests **D**24
  systems analysis **D**71–72
  types of **K**37
Wear safety, plain bearings **F**90
Weathering tests, of plastics **D**62
Weber's similarity law **A**71
Wedge friction **A**16–17
Wehnelt cylinder **K**83
Weld joints **F**9–11
  safe loads **F**13–16
  strength calculations **F**11–13
Weld positioning **F**6
Weldability **F**5–6
Welded connections, piping **H**12
Welding current **L**98

Welding current (*cont.*)
  electronic sources **L**98
Welding duty cycle **L**98
Welding guns **L**100
Welding machines
  arc **L**97–98
  butt **L**99
  projection **L**99
  resistance **L**99–100
  seam **L**99
  spot **L**98, **L**99
  three-phase **L**100
Welding processes **F**1–5
  graphical symbols **F**11, **F**176
  laser beam **K**83
  plastic materials **D**65
  reliability **F**6–8
  rules and standards table **F**177–
    178
Welding rectifiers **L**98
Welding transformers **L**98
Whitworth thread **F**36
Wildhaber-Novikov gears **F**124
Wind pressures, on buildings **A**65
Wire drawing, plasticity theory
    K25–26

Wöhler curves (fatigue endurance)
    D15, **D**25
Wood **D**52
  permitted loads **D**113
  protection of **D**52
  strength characteristics **D**113
Wood-based materials **D**53
Woodruff keys **F**30–31
Work **A**27, **C**3
  displacement **C**4
  of dissipation **C**4
  force of gravity **A**28
  frictional force **A**28
  spring force **A**28
  technical **C**4
  torque **A**28
  total **A**28
Work factor **K**97
Work Law **A**30, **A**34
Work-hardening **K**20
Working environment **K**108
Working space, machine tool **L**1
Workpiece embodiment design **E**19
Workpiece handling
  automated **K**102
  carrier **L**1

Workplace design **K**107–108
Worm gear cutting **K**77–78
Worm-rack gearing **L**18

X-ray fluorescence analysis
    method **D**23
X-ray lithography **K**87
X-ray test procedures **D**25

Yield strength, *see* Flow stress
Yield stress, test methods for
    determining **D**20

Zero-flow control **G**14
Zero-order transfer function **F**161
Zeroth main law of thermodynamics
    C2
Zinc and zinc alloys **D**47, **D**111
Zweifel's cascade results **A**68–69